George V Goodding

Jan. 30 1942.

with best wishes for your welfare + happiness

Edwin C. Miller

McGRAW-HILL PUBLICATIONS IN THE BOTANICAL SCIENCES
EDMUND W. SINNOTT, CONSULTING EDITOR

PLANT PHYSIOLOGY

Selected Titles From

McGRAW-HILL PUBLICATIONS IN THE BOTANICAL SCIENCES

Edmund W. Sinnott, *Consulting Editor*

Babcock and Clausen—Genetics
Belling—The Use of the Microscope
Boysen Jensen—Growth Hormones in Plants
Braun-Blanquet and Fuller and Conard—Plant Sociology
Curtis—The Translocation of Solutes in Plants
Eames—Morphology of Vascular Plants
Eames and MacDaniels—Plant Anatomy
Fitzpatrick—The Lower Fungi
Gäumann and Dodge—Comparative Morphology of Fungi
Haupt—An Introduction to Botany
Haupt—Laboratory Manual of Elementary Botany
Hill—Economic Botany
Hill, Overholts, and Popp—Botany
Loomis and Shull—Methods in Plant Physiology
Experiments in Plant Physiology
Lutman—Microbiology
Maximov—Plant Physiology
Miller—Plant Physiology
Pool—Flowers and Plants
Seifriz—Protoplasm
Sharp—Cytology
Sinnott—Botany
Sinnott and Dunn—Genetics
Smith—Cryptogamic Botany
Vol. I, Algae and Fungi
Vol. II, Bryophytes and Pteridophytes
Fresh-water Algae of the U. S.
Swingle—Systematic Botany
Weaver—Root Development of Field Crops
Weaver and Bruner—Root Development of Vegetable Crops
Weaver and Clements—Plant Ecology
Wodehouse—Pollen Grains

There are also the related series of McGraw-Hill Publications in the Zoological Sciences, of which A. Franklin Shull is Consulting Editor, and in the Agricultural Sciences, of which Leon J. Cole is Consulting Editor.

PLANT PHYSIOLOGY

With reference to the green plant

BY

EDWIN C. MILLER, PH. D.

Professor of Plant Physiology, Kansas State College of Agriculture and Applied Science, and Plant Physiologist, Kansas Agricultural Experiment Station

SECOND EDITION
SECOND IMPRESSION

McGRAW-HILL BOOK COMPANY, INC.

NEW YORK AND LONDON

1938

PRINTED IN THE UNITED STATES OF AMERICA

THE MAPLE PRESS COMPANY, YORK, PA.

To my wife
DELLA SLYE MILLER

PREFACE TO THE SECOND EDITION

In the preparation of this edition, various topics have been enlarged or revised to include the investigations and findings that have been made since the publication of the first edition. Certain topics that were overlooked in the preparation of that edition have been added. New material has been presented under appropriate headings.

Although no claim to completeness is made, it has been the effort of the author to present the more important citations relating to each subject. He has tried to present impartially the data that have been reported. He does not feel qualified to speak with authority on a majority of the topics, and his own ideas, if presented, thus would be of little consequence.

In order to conserve space, the questions that concluded each chapter in the first edition have been omitted.

The author is indebted in various ways to several of his associates at the Kansas State College of Agriculture and Applied Science, and he takes this occasion to thank them. He wishes to thank especially Hugh G. Gauch, for his invaluable aid in assembling and typing the bibliographies and other portions of the manuscript, and for aid in editing it; John C. Frazier, for aid in editing the manuscript, for helpful advice, and for unselfish service in many ways; Dr. Fritz Moore, head of the Department of Modern Languages, and Mrs. Irene Moore, for checking, respectively, the German and the French references; and Miss Esther Lewis, for clerical help in preparing the last half of the manuscript.

The author is especially indebted to Professor L. E. Melchers, head of the Department of Botany and Plant Pathology, Kansas State College of Agriculture and Applied Science, for his kindness in allotting clerical and stenographic help for the revision of this text.

EDWIN C. MILLER.

KANSAS STATE COLLEGE OF AGRICULTURE AND APPLIED SCIENCE,
October, 1938.

PLANT PHYSIOLOGY

CHAPTER I

THE PLANT CELL

The cell is not only the structural unit but also the functional unit of the plant and animal body. Since plant physiology is the study of the functions of plants, a general knowledge of the morphological parts of the plant cell is necessary for a clear understanding of the various life activities of the plant. In this chapter the intention is to discuss the parts of the plant cell only in a general way, since a consideration of the more minute details belongs to the province of plant cytology. The parts of a living plant cell are generally classified as follows:

- Parts of the plant cell
  - Cell wall
    - Middle lamella
    - Primary wall
    - Secondary wall
  - Protoplasm
    - Cytoplasm
    - Nucleus
    - Plastids
      - Leucoplasts
      - Chloroplasts
      - Chromoplasts
    - Chondriosomes or mitochondria
  - Vacuole or cell sap

I. THE CELL WALL

A. Origin

The protoplasm of plant cells is with but few exceptions enclosed by a more or less solid membrane which has been secreted by it. This membrane, which is called the "cell wall," is relatively very thin when it is first formed but increases in thickness through the deposition by the protoplasm of new particles upon those already present. The wall may increase in thickness but little over its original measurements, or it may become so much thickened that it almost fills the cell cavity. Thus, for example, the amount of thickening in the cell wall of the flax fiber may amount to as much as 90 per cent of the area of the cell in cross section (Anderson, 1927). (See Fig. 2.) The cell wall in many cases is thickened irregularly, leaving thin places in it which are termed "pits," which facilitate the interchange of materials from one cell to another (Fig. 2*p*). The manner in which the original cell wall becomes thickened is little understood and may be different in different types of plant cells. Aldaba (1927) observed in the case of the development of the cell walls of the bast fibers of *Boehmeria* and *Linum* that the cell wall is formed by a series of hyaline membranes that gradually become transformed into cell-wall lamellae. The modified hyaline membranes resemble plasma membranes and react similarly towards stains and during plasmolysis. The unmodified hyaline membranes are in direct contact with the protoplasm, but the thickening and differentiating portions are separated from it by one or more subsequently formed layers. The mode of formation is suggestive of a process of transformations. It cannot be satisfactorily accounted for by the theories of apposition and intussusception.

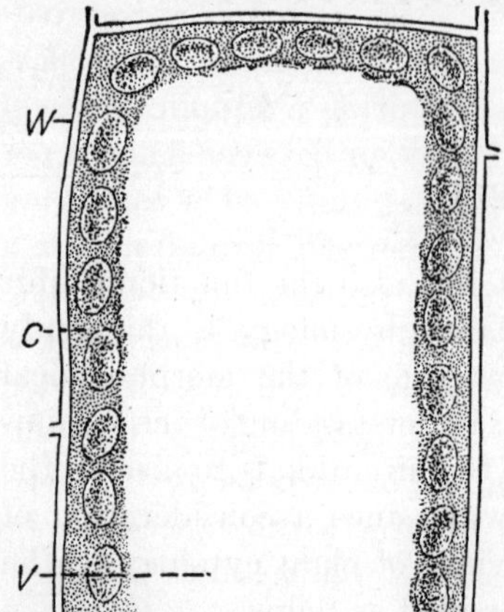

FIG. 1.—Diagram of a green parenchyma cell of a leaf showing the parts of a plant cell. *w*, cell wall. *c*, cytoplasm. *n*, nucleus. *p*, plastids. *v*, vacuole.

The formation of the cell wall of the cotton fiber has been studied intensely by Farr and Eckerson (1933, 1934) and Farr and Sisson (1934). They noted that, when a cotton fiber is only 5 to 6 days old, a thin layer of fibrils that give the cellulose reaction is closely affixed to the thin original wall. From the very beginning of the elongation of the fiber, there are present in the protoplasm ellipsoidal particles of cellulose covered with a thin layer of pectic substance. Toward the center of the cell these particles are separate or in short chains while in the extreme outer regions of the protoplasm next to the cell wall the chains are longer and in some instances have entered into the formation of fibrils. Regardless of the ultimate structure of these particles, they apparently represent standard units which build up the fibrils that compose the cell wall of the cotton fiber. This conclusion was reached because by various microscopic and submicroscopic procedures, including X-ray examinations, a relationship was definitely established between these more or less free cellulose particles in the cytoplasm and the compact cellulose membrane of the mature wall of the fiber. This behavior in cell-wall formation was found to be similar in various other plants and is considered to be of general occurrence.

Sponsler (1934) made observations of the dividing cells of *Rhizoclonium* in which the cross wall grows from the side walls toward the center of the cell. The growth rate of this cross cell wall is approximately 0.1μ per minute, and in the process monosaccharoses are apparently condensed into polysaccharose chains. During a 15-min. interval of growth there were deposited about 3,200 layers, or between 3 and 4 per second, each of the thickness of a monosaccharose unit. The bordering protoplasm evidently takes an active part in the wall formation, for a change in environment may practically stop the process.

Bonner (1935) believed that the young cell wall must be considered as living. According to him the growth of the cell wall cannot be regarded as a simple plastic stretching, because the alignment effect of the cell-wall micelles is partially or entirely lacking. The increase in the plasticity of the cell wall through the addition of growth substances can be traced back to a loosening up of the contacts of the cellulose micelles. He considered that growth may be a simultaneous action of stretching and stratification of cellulose micelles. Bonner and Heyn (1935) thought that the increase in cell-wall plasticity by growth substance might be due to the indirect effect of this substance upon the charges of the micelles of the cell wall. This might be brought about by certain ions entering the wall from the protoplasm thus increasing the micellar charges in the cell wall. These might cause a greater repulsion between the micelles and an increase of micellar hydration, thus increasing the plasticity of the wall. Investigations, however, indicated that growth substance had no effect upon the charges of the micelles.

That portion of the cell wall which is laid down during the early life of the cell is different in optical, chemical, and staining properties from that part of the wall which is later deposited. This central layer or plate is readily seen in those cell walls that have been greatly thickened, but it can be observed after proper treatment in nearly all cell-wall membranes, even when they are quite young. This central portion of the cell wall is called the "middle lamella" and is generally defined as the cell wall that was originally deposited by the protoplasm, although there is considerable uncertainty as to the length of the period during which this portion of the wall was formed. Allen (1901) studied the origin of the middle lamella in the cells of a considerable number of plants and concluded that the middle lamella consists of the layers first deposited by the protoplasm plus a certain amount of material, subsequently in contact with these layers, which is generally rich in pectic compounds as compared with the still later deposited strata.

The general terms "middle lamella," "primary wall" and "secondary wall" have been proposed to clarify the designation of the cell-wall parts. This terminology, however, is open to criticism (Kerr and Bailey, 1934).

Middle Lamella.—The amorphous isotropic material, largely, if not entirely, pectic compounds, first deposited by the cytoplasm.

Primary Wall or Cambial Wall.—The first anisotropic layer of the wall composed largely of cellulose and some pectic materials. Cells with walls of this type are capable of growth and extension. This wall has the capacity for undergoing reversible changes in thickness.

Secondary Wall.—The additional wall layers formed on the primary wall. Cells having secondary walls lose their potentiality for growth.

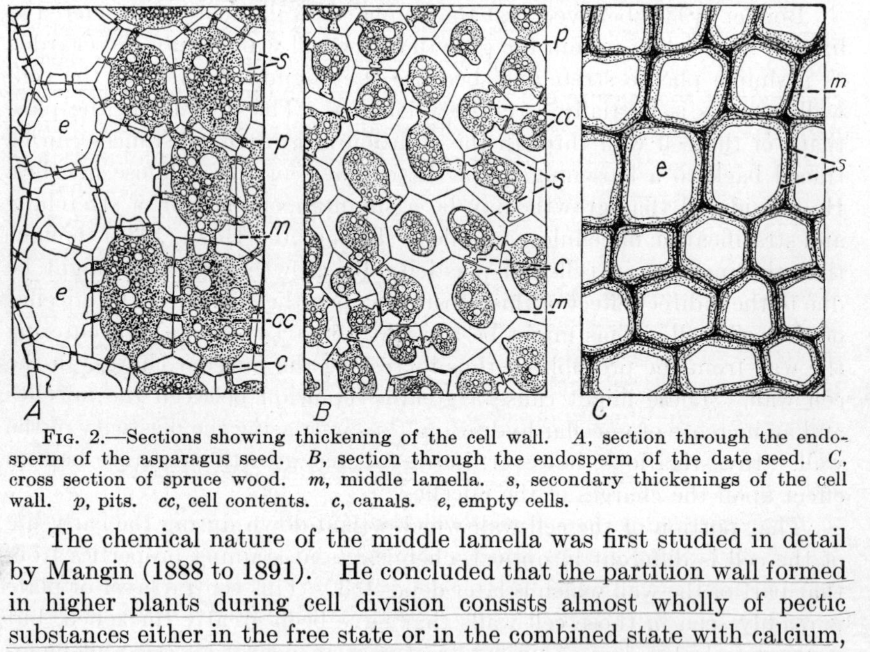

FIG. 2.—Sections showing thickening of the cell wall. *A*, section through the endosperm of the asparagus seed. *B*, section through the endosperm of the date seed. *C*, cross section of spruce wood. *m*, middle lamella. *s*, secondary thickenings of the cell wall. *p*, pits. *cc*, cell contents. *c*, canals. *e*, empty cells.

The chemical nature of the middle lamella was first studied in detail by Mangin (1888 to 1891). He concluded that the partition wall formed in higher plants during cell division consists almost wholly of pectic substances either in the free state or in the combined state with calcium, in which form it was considered the bulk of the middle lamella in mature cells.

Tupper-Carey and Priestley (1923) investigated the cell walls of the apical meristem of the stem and root and found the existence of cellulose in the walls of the meristem, although its presence is marked by association with other substances. Protein closely linked to it is considered to be the factor which prevents its detection with iodine and sulphuric acid. Pectin is present, but the middle lamella in the meristem is never of calcium pectate but is probably a mixture of pectin and protein. Fat is present in the cell walls of the meristem closely linked to cellulose and probably is responsible for the failure of chloriodide of zinc to give the cellulose reaction. Wood (1924, 1926), however, considered that there is not enough protein present in the cell wall to interfere with the cellulose-pectin reaction, since according to his tests not more than 0.001 per cent protein occurs in the cellulose cell walls of any of the plants examined.

Lüdtke (1931), in an examination of young and old winter wheat plants, found that a whole series of intermediate products is present in the younger cells from the simple saccharoses, as xylose and glucose, up to cellulose, thus indicating that the formation of this cell-wall constituent occurs in a series of transitional steps. Ritter (1925) considered that in woody tissue the middle lamella is not pectic in nature but is strongly lignified. Harlow (1927) observed in the young cortex unlignified pith and cambium region that the middle lamella is composed of pectin in one or more of its forms, while in woody tissue this layer is heavily lignified. There are two explanations for these observed changes in the older cell walls: (1) that the pectins are metamorphosed to form part of the lignin of the mature cell wall and (2) that the initial pectic structure may be completely covered with lignin substances which render it inert to the action of the usual pectic stains and solvents. A combination of these two processes might also occur. The evidence indicates that, at least in the softer tissues of plants, the middle lamella is composed, for the most part, of pectic substances that may become lignified in woody tissues. The middle lamella in most tissues dissolves readily under the action of potassium chlorate and nitric acid, which do not affect the later deposited portions of the cell wall. It resists the action of strong sulphuric and other mineral acids that cause the other portions of the cell wall to swell and finally dissolve.

B. General Physical and Chemical Properties

The chemical and physical structure of plant cell walls varies widely in different plant groups and in the same individuals. Thus the cell walls of root hairs, wood fibers, bast cells, cork cells, or epidermal cells have but few characteristics in common. All plant cell walls, however, have two general characteristics in common: (1) They are not chemically homogeneous but are composed of two or more chemically distinct substances. (2) They are not physically homogeneous but show stratification, striation, and other indications of structural variation.

1. Methods of Studying.—The present conception of the general structure of the cell wall is really a composite pattern built from the different insights that have been gained by the following methods (Scarth, Gibbs, and Spier, 1929; and Anderson, 1935):

a. Direct Observation of Untreated Walls.—This method reveals the presence of definite layers, surface markings, striations, radial lines, pits, and irregular thickenings.

b. Differential Staining of Walls.—The presence of the various compounds in the wall can be determined after this treatment. In addition some idea can be gained concerning the spatial distribution of the submicroscopic units in the wall (Czaja, 1930, 1934).

c. Differential Solubilities.—Extraction of the different compounds from the walls is accomplished by this method.

d. Ash Analysis.—Kind and amount of the various minerals in the walls are determined in this manner.

e. Use of the Spierer Lens.—Images produced by this lens have been supposed to indicate the presence of definite structural units of ultramicroscopic dimensions. It has been contended that the use of this lens provides further evidence for the micellar structure of the cellulose cell wall (Seifriz, 1931, 1934, and Thiessen, 1932). Seifriz (1936) considered that these microscopic units revealed in a Spierer picture are not the micelles the chemist has in mind, so he has termed these units "super micelles."

f. Measurement of Refractive Indexes.—The different lamellae of the wall have different refractive indexes so that by this method the prominence of certain portions of the wall is greatly emphasized.

g. Use of Fluorescent Light.—This detects the degree of lignification of the walls. They are irradiated with ultraviolet rays which on passing through lignified walls become visible due to fluorescence. The intensity of fluorescence varies directly with the degree of lignification.

h. Hydration.—By the use of suitable reagents the cell wall may be greatly swollen. In this swollen condition, structural relations appear that are invisible in the unswollen state. By this method Balls and Hancock (1919) and Balls (1923) were able to count the layers present in the cotton fiber.

i. X-ray Analysis.—This method has been used by numerous investigators and has probably contributed more to our knowledge of the ultimate structure of the cell wall than has any other single procedure. The actual size of the molecular aggregates in the wall, their distance apart, and their spatial relationships are shown.

j. Polarized Light.—This method is used to distinguish isotropic substances, those which show no double refraction, from anisotropic substances, which show this property. It also furnishes information regarding the spatial orientation of the submicroscopic units of the wall.

2. Theories of Cell-wall Structure.—In 1864 Nägeli suggested that the cell wall was composed of submicroscopic units, or micelles, arranged in definite layers. Each micelle was believed to be surrounded by a layer of water and thus separated from the adjacent micelles. The modern methods of the investigation have only served to emphasize its general accuracy and today it is probably more widely accepted than any other theory of cell-wall structure. The present theories have certain features in common and differ mainly in regard to the supposed size of the structural units concerned. These theories may be grouped under three headings:

a. The work of Frey (1926–1928), Frey-Wyssling (1930), and Astbury, Marwick, and Bernal (1932) with polarized light, and of Herzog (1925), Meyer and Mark (1928), Meyer (1929, 1931), and Clark (1930) with X-ray examinations indicates that the cell wall is composed of submicroscopic units or micelles of a crystalline nature. The crystalline micelles are separated from each other by a noncellulose hydrophilic colloidal material. The micelle is considered to be composed of a bundle of basic cellulose groups or molecules. The basic cellulose molecule appears to be a long chain of glucose residues (glucose minus a molecule of water). The length of the chain is variously estimated to be from 50 to 1,000 of these glucose residues (Meyer, 1929; Sponsler, 1929, 1930, 1933; Freudenberg, 1932; Seifriz, 1934; and Gibbs, 1935). These chainlike cellulose "molecules" are grouped in parallel rows into a bundle or micelle (Davidson and Richardson, 1936). The number of parallel chains present in a single micelle has been variously estimated to be from 40 to 60. Its length has been estimated to be from 15 to 100mμ and its diameter from 2.0 to 6.6mμ depending upon the source of the material. The size may vary in different plants, in different plant parts, or even in the same cell wall. The forces that hold the glucose residues together in long chains are the primary valence forces, while the forces holding the chains together in the micelle are those of molecular cohesion. In fiber cells the micelles are grouped into fibrils that collectively make up the cell wall. These fibrillar units may be parallel (Preston, 1934) or inclined to the long axis of the cell wall.

b. It was considered by Sponsler (1926–1933), and Sponsler and Dore (1936), that the evidence for the existence of micelles as definite entities in the cell wall is inadequate. They considered the wall to be composed of long parallel chains of glucose residues running lengthwise of the fiber. These chains are oriented in such a way as to form a spatial lattice of three dimensions. The cellulose wall is considered to be a complex system built up of very thin molecular layers each composed of long chains of glucose residues. Certain layers extend lengthwise of the wall, some extend across the wall, and other layers form definite angles with the lengthwise layers. A single fiber cell wall may be some 40,000 of these lattice units in thickness. The wall of a fiber cell like cotton consists then, according to Sponsler, of an enormous number of parallel molecular layers and not of a system of bricklike units as suggested by the micellar hypothesis.

c. According to Hess (1928), Lüdtke (1931), Farr and Clark (1932), Farr and Eckerson (1933, 1934), and Farr and Sisson (1934), the cell wall is composed of definite cellulose units microscopically visible, each unit being enclosed in a thin film of noncellulose material that serves to cement the units together. Farr and Eckerson (1933, 1934) considered this noncellulose constituent to be pectic material with its nature apparently varying in different walls. This theory of cell-wall structure finds support in the behavior of the walls during swelling, in dissolution by various reagents, and in the reactions to certain dyes. Certain cellulose-staining dyes frequently fail to impart color to cellulose walls except when these walls are subjected to mechanical pressure. It is presumed that this pressure ruptures the noncellulose constituent and permits the stain to reach the cellulose particles. Bailey and Kerr (1935), however, considered that there is no reliable evidence to indicate that the matrix of the cell wall is composed of entities of visible size bound together by a noncellulose material. They believed that such putative entities are actually heterogeneous fragments that are shredded or disrupted from an originally continuous and coherent matrix. If these are discontinuous in the structural pattern, they are confined to the submicroscopic field—to the realm of micelles and molecular chains.

C. Chemical Composition

The chemical nature of the cell wall is but little understood. The substances that compose it are complex, and their constitution is seldom known definitely. Furthermore, it is seldom known whether the cell wall is composed of a mixture of substances or whether these substances are chemically combined with each other. Any discussion of the chemical composition of the cell wall must, therefore, be rather meager and superficial. In general, however, it may be stated that the cell wall of the plant cell is composed of a mixture of celluloses, together with certain other compounds depending upon the nature of the cells in question. The following diagram shows the substances that may enter into the composition of the cell wall:

- Materials that may enter into the composition of the cell wall
  - Celluloses
    - Normal, typical, or true celluloses — α, β, γ types
    - Hydrocelluloses
    - Oxycelluloses
    - Compound celluloses—Lignocelluloses
    - Hemicelluloses or Pseudo-celluloses
      - Skeletal hemicelluloses — Pentosans (Xylan, Araban), Galactosans — In cell walls of wood and seed coats
      - Reserve hemicelluloses — Pentosans (Xylan, Araban), Mannosans, Galactosans — In cell walls of endosperm and young fibers of wood and bark
  - Suberin and cutin
  - Pectic substances
    - Pectic acid
    - Pectates
    - Pectose
    - Pectin
  - Other constituents — Resins, gums, tannins, minerals, coloring matters, proteins, fats, phospholipides, ethereal oils, and callose

The celluloses considered collectively may be defined as a group of relatively inert substances composed of polysaccharose units that constitute the major portion of the cell walls of plant cells.

1. Normal, Typical, or True Celluloses.—Normal cellulose is apparently the first cellulose formed by the protoplasm and is the one from which all the other celluloses are derived. It is a white hygroscopic substance that absorbs 6 to 12 per cent of water but loses it on being heated to 100°C. When heated with water under pressure to 260°C., it dissolves completely without disintegrating. Its general formula is $(C_6H_{10}O_5)_n$, so that it is chemically similar to but not identical with starch. Normal cellulose is insoluble in ordinary solvents but dissolves in ammoniacal cupric oxide (Schweitzer's reagent), or in zinc chloride in two times its weight of concentrated hydrochloric acid in the cold. It also goes into solution by heating 1 part of cellulose with 6 parts of zinc chloride in 10 parts of water. Concentrated sulphuric acid dissolves normal cellulose gradually, converting it into dextrin and then into glucose. If this solution as soon as made is diluted with water, a gelatinous precipitate is formed that gives a blue color with iodine. The same substance is formed by the action of chloriodide of zinc, which stains the compound at first a rose-red and later a violet color. When cellulose is acted upon with acetic anhydride and concentrated sulphuric acid, or is hydrolyzed under the influence of the enzyme cytase, which is secreted by bacteria and various fungi, it breaks down into cellobiose ($C_{12}H_{22}O_{11}$), an isomer of maltose. By further hydrolysis it is converted into glucose. Cellobiose stands in the same relation to cellulose as does maltose to starch. According to Haworth (1929) the sole constitutional difference between cellobiose and maltose is a difference in configuration, at the linkage between the glucose residues, similar to that which distinguishes the α- and β-stereoisomeric forms of methyl glucoside. Maltose is the alpha form, and cellobiose is the beta form.

Cotton fiber is the best example of almost pure normal cellulose occurring in nature. An analysis of American cotton fiber showed it to be composed of 91 per cent normal cellulose, 8 per cent water, and only 1 per cent of other materials, principally waxes, gums, minerals, and pectose derivatives. The cell walls of the younger plant cells are composed, for the most part, of normal cellulose or some of its derivatives. Wood (1924) studied the nature of the cellulose constituents of the cell wall of the cells of the young root tips of *Vicia faba* and *Helianthus annuus* as well as the parenchymatous tissues of the stems of these plants. He found that the cellulose constituents of the cell wall consist of normal cellulose, oxycellulose, and hydrocellulose, the proportion of each constituent varying with the age and kind of cell.

An illustration of the similarity of normal celluloses obtained from different sources was reported by Sanders and Cameron (1933). They found the cellulose of the cotton stalk and cusps to be the same type of cellulose found in cotton lint and spruce, pine, and poplar wood. However, the compounds termed "normal celluloses" are composed frequently of three different types or forms of cellulose, which have been designated as alpha, beta, and gamma celluloses. These differ from each other especially in regard to the action of alkalies and possess different degrees of dispersion, dehydration, and corresponding polymerization (Gibbs, 1935).

2. Hydrocelluloses and Oxycelluloses.—These celluloses are degradation products that are frequently found associated with normal celluloses in the cell wall. Hydrocellulose is apparently produced by partial hydrolysis and is a compound that possesses active carbonyl groups and shows a very strong absorptive power for different organic dyes. It may be prepared, according to Mehta (1925), by soaking cotton with as small a quantity as possible of cold 72 per cent sulphuric acid and treating almost immediately with 95 per cent alcohol. The hydrocellulose thus formed is washed free from acid and dried in the air. It has the formula $C_{12}H_{22}O_{11}$ but retains the fibrous structure of the original cellulose, can be rubbed into a powder, and stains blue with iodine. The formation of hydrocellulose is marked by an increase in reducing power, a fall in tensile strength and in viscosity in Schweitzer's reagent, and an increase in solubility in cold sodium hydroxide solution. The interrelations of these properties are always the same regardless of the acid treatments (Davidson and Richardson, 1936). The action of acid thus apparently is a hydrolysis of glucosidic linkages in the chain molecules, with the production of shorter chains and the liberation of reducing groups.

Oxycelluloses are mentioned by many authors to designate various compounds of an indefinite character which are produced by the action of various oxidizing agents on cellulose. The compounds of this type possess acidic properties, contain active carbonyl groups, and reduce Fehling's solution. The oxidation process always leads to a decrease of tensile strength, a fall in viscosity, and an increase in solubility showing that chain molecules are broken. Although in any given series of oxycelluloses these properties show definite relations to each other and to the chemical properties, the relations vary from one type of oxycellulose to another (Davidson and Richardson, 1936).

3. Compound Celluloses.—This is a general term that includes those compounds of the cell wall which consist of some form of cellulose combined or closely associated with a noncellulose constituent. The most common and widespread compound cellulose in the plant kingdom is lignocellulose. Lignocellulose is the chief constituent of wood, jute, and cereal grasses and straw. It is generally considered to consist of cellulose chemically combined with lignin, but some authorities are of the opinion that the lignin is simply an encrusting substance and consider its combination with the cellulose to be only a physical one.

The manner in which the lignin might be combined with the cellulose is a matter for speculation. Some have assumed that an esterlike union exists between an acidic group in the lignin and a hydroxyl of the carbohydrate, while others consider that an etherlike linkage exists between the lignin and the cellulose or other carbohydrates. It has also been shown that some of the lignin is combined in an acetal-like manner. It is very probable that all three of these combinations may occur, depending upon the kind of plant or the conditions under which the lignocellulose is formed (Phillips, 1934). It was also considered by Mehta (1925) that lignocellulose consists of a chemical combination of lignin with cellulose and related polysaccharoses as an aromatic glucoside. According to Candlin and Schryver (1928), and others, decarboxylation occurs when plant tissues lignify. There is also evidence that the formation of lignocellulose is an infiltration process. Fischer and Lieske (1928), Freudenberg and his workers (1931), and Harlow (1932) have noted that sections of wood still retain their structural features when treated with various reagents which remove either the cellulose or the lignin. According to Anderson (1935) the lignification of cellulose walls does not alter the X-ray patterns of cellulose. The process thus does not involve a modification of the cellulose units themselves but indicates that the lignin is deposited between the cellulose micelles, hence lignification involves a change in the intermicellar material and not in the micelles generally considered to compose the cellulose.

The lignocelluloses cannot be crystallized without decomposition and cannot be obtained in the form of true solutions, although colloidal solutions are possible. They are among the most inert plant compounds and probably have no role other than that of adding stiffness to the plant tissues. Lodged wheat stalks (Davidson and Phillips, 1930) and brash specimens of fir (Dadswell and Hawley, 1929) show a higher lignin content than the tougher ones. Lignification is not to be considered, however, as an indication of senescence, for Mehta (1925) found the lignin content of the buds of Oregon pine and spruce relatively high, and Norman (1933) noted that young cereal plants often contain appreciable amounts of this compound.

The location of lignocellulose in the cell wall is somewhat in dispute. Schellenberg (1896) seems to have been the first to note that the middle lamella can become lignified. Ritter (1925) considered that the greater percentage of lignocellulose in wood is located in the middle lamella of the cell wall. According to Harlow (1928) the secondary cell walls of soft wood are appreciably lignified, while those of the hardwoods with but few exceptions contain practically no lignin. The lignin of the secondary cell wall may be removed by mild chlorination without affecting to any great extent the lignin that comprises the middle lamella. In 1932 Harlow obtained additional evidence that most of the lignin in wood is located in the middle lamella, while Norman (1933) considered that at least 50 per cent of this compound is located therein. Kerr and Bailey (1933) considered that the middle lamella of lignified tissue is not composed entirely of lignin.

Lignocellulose is resolved completely into its constituents by a process of chlorination or by heating for 1 hr. with a 4 per cent sodium hydroxide solution at a pressure of 10 atmospheres. The lignin can then be precipitated by acidifying the alkaline liquid and isolated in the pure condition quantitatively by extraction with alcohol. Lignin as isolated from wood by the above method is a brown, amorphous, faintly acidic substance having an aromatic odor and melting at 170°C. It is insoluble in water at ordinary pressure but partially soluble under pressure and soluble in dilute alkalies and alcohol and is more readily attacked by oxidizing agents than cellulose. The lignins contain methyl, methoxyl, formyl, and acetyl groups and consequently have a higher carbon content than cellulose.

Lignin from different sources varies greatly in its methoxyl and acetyl contents. In barley (Phillips and Goss, 1935) the lignin from the young plants differs from that of mature plants in that the lignin from the former contains a much smaller percentage of methoxyl. As maturity is approached, the percentage of lignin increases and there is a rapid methylation of the hydroxyl groups. The general opinion is that lignin contains condensed, unsaturated aromatic nuclei perhaps with an aldehyde or ketone group and is substituted by hydroxyl, methyl, and possibly acetyl groups. Some consider lignin to be a polymerized and substituted ether. There is considerable discussion, however, whether lignin belongs to the aliphatic, aromatic, or heterocyclic series. It is a disputed question as to whether lignin contains a grouping related to coniferyl alcohol or aldehyde, although some investigators (Klason, 1910) support this hypothesis.

It should be kept in mind as Harlow (1927, 1928) has pointed out that the lignin of botanists may not be the same as the lignin of chemists. It is customary for the former to regard walls as "lignified" when they give color reactions with certain staining reagents, while the latter regard as lignin that material which is removable from the cell walls by certain chemical treatments. It has long been suspected that the color reactions are due to small amounts of substances of an aldehyde nature occurring with lignin but not necessarily a part of it.

The source of lignin is purely a matter of speculation. Lignified tissue contains lignin and hemicelluloses in relatively large amounts with only traces of pectin, while nonlignified tissues contain relatively large amounts of pectin, small amounts of hemicellulose and no lignin. It is thus considered by some that pectins are the substances from which lignins are formed. Phillips and Goss (1935) noted in the development of the barley plant that the percentage of the lignin, cellulose, and pentosans increased progressively with the age of the plant, but no indications were found that lignin increased at the expense of cellulose or pentosans. Smith (1935), on the other hand, considered that in the Kieffer pear the sugar that the leaves supply to the fruit is converted to pectins, to hemicelluloses, and finally to lignocellulose. There was a relative decrease in sugars during the lignification of this fruit. As shown by Crist and Batjer (1931), the first clear microscopic evidence of lignification appeared in 15 and 20 days after blooming in Bartlett and Kieffer pears, respectively. This lignification reached a maximum in from 3 to 5 weeks after its first appearance.

There is some evidence (Norman, 1933) of a relationship between lignins, certain resins, and the phenolic residues of tannins (Menon, 1935). The whole question of the formation of aromatic nuclei in nature, however, is a very perplexing one. The assumption is that they are not direct products of photosynthesis and must therefore be formed from pyran or furan rings of sugar.

There is at present no known commercial use of lignin, and it is estimated that 1,500,000 tons are wasted annually, mostly in the manufacture of paper pulp. In 1936 the production of a commercial compound known as "ralig" was announced. This compound is a 25 per cent solution of the lignin of hemlock, and it was stated that it was to be utilized in highway construction.

4. Hemicelluloses or Pseudocelluloses.—The term "hemicellulose" was first used by E. Schulze in 1892 to denote a group of substances in the cell wall which he considered very closely related to cellulose in chemical constitution and as intermediate substances in its development. The outstanding characteristics of this group of substances are their relative ease of hydrolysis by hot dilute mineral acids, under ordinary pressure, and their solubility in dilute alkali. They are commonly extracted from the cell wall by treatment with dilute alkali. This solution is then acidified and the hemicelluloses precipitated by the addition of ethyl alcohol. The properties of the hemicelluloses prepared after this manner are determined to a large extent by the concentration of the alkali used, by the conditions maintained during the extraction, and by the general technique used to precipitate and purify them (Burkhart, 1936). Schulze prepared these compounds from a large number of plant parts and obtained upon their hydrolysis the sugars, arabinose, xylose, galactose, and mannose. Most of his preparations contained more than one type of sugar unit. The term "hemicellulose" thus came to include a rather heterogeneous collection of substances in the cell wall, including galactosans, pentosans, and mannosans, which are the anhydrides of the abovementioned sugars. Waksman and Diehm (1931) suggested that hemicelluloses be defined as the noncellulose polysaccharose constituents of the cell wall.

From the work of numerous investigators including O'Dwyer (1923), Candlin and Schryver (1928), Norman (1929), Norris and Preece (1930), Preece (1930, 1931), and Anderson (1931), it has been found that many of the preparations of hemicellulose contain uronic acids in addition to the polysaccharoses that have been mentioned. The uronic acids that have thus been found are generally glucuronic and galacturonic, while mannuronic is less common. These acids possess a formaldehyde radical at

Investigators	Plant	Plant part	Compounds obtained by hydrolysis
Schulze (1893)	Corn	Seed coat	Xylose and galactose
	Lupine	Seed coat	Pentose and galactose
Grüss (1897)	Date	Seed	Mannose
Castoro (1906–1907)	*Ruscus aculeatus*	Seed	Mannose and arabinose
	Lupinus angustifolius	Seed coat	Galactose
	Lupinus albus	Seed coat	Galactose
	Cucurbita pepo	Seed coat	Xylose and galactose
Schulze and Pfenniger (1910)	*Pisum sativum*	Pod	Galactose and arabinose
	Phaseolus vulgaris	Pod	Galactose and arabinose
Tottingham, Roberts, and Lepkovsky (1921)	Apple	Twig	Xylose, glucose, and galactose
Cake and Bartlett (1922)	Asparagus	Seed	Mannose, glucose, fructose, and galactose
Schryver and Thomas (1923)	Various starches	Starch grains	Glucose
O'Dwyer (1923)	White oak	Wood	Xylose and arabinose
Link (1929)	Corn	Seedling	Xylose and glucose
Norris and Preece (1930)	Wheat	Wheat bran	Uronic acid, glucose, and pentose
Preece (1930)	Corn	Cob	Arabinose, xylose, methylpentose, and uronic acid
Anderson and Kinsman (1931)	Cotton	Seed hull	1-Xylose and 1-glycuronic acid
Widdowson (1932)	Apple	Developing fruit	Arabinose
Tottingham (1933)	Cherry, apple, peach, plum, pear	Wood	Glucose and xylose
Burkhart (1936)	Alfalfa	Xylem and phloem of roots	*d*-Xylose, *d*-glucose, and *d*-glucuronic acid

one end of the molecule and a carboxyl group at the other. The carboxyl group has been derived from the oxidation of the primary alcohol radical of the sugar molecule. The uronic acids associated with the hemicelluloses are apparently all in a combined form, being linked to anhydrous sugar units. Such compounds may rightfully be termed polyuronides. The hemicelluloses thus comprise a wide variety of compounds with pure polysaccharoses related to starch and inulin at the one extreme and pure uronic complexes at the other. Thus by the hydrolysis of these cell-wall constituents there may be obtained pure monosaccharoses, a mixture of monosaccharoses, or a mixture of monosaccharoses and uronic acids. A general list of the hydrolytic products obtained from various plant parts by different investigators is shown in the table on page 13.

Of all the various hemicelluloses, the pentosans have received the most consideration owing primarily to their greater abundance and to the ease of their determination by the furfuraldehyde method. The hemicelluloses may make up the entire cell wall or they may simply form incrustations on the cellulose framework of the wall. The former may occur in the cell walls of the endosperm or embryo in certain seeds, while the latter prevail in the walls of wood and bark fibers. Norman and Shrinkhande (1935) and Sands and Nutter (1935) noted in the cell walls of wheat and barley straw, hay, and woods of oak, fir, and mesquite that a large proportion of the hemicellulose is combined or closely associated with lignin.

The components of the cell wall designated as "hemicelluloses" rank next to cellulose in quantitative importance. They are found in the seed coats of beans, peas, and other legumes. They occur in the cell walls of the seeds of coffee, soybean, nasturtium, white lupine, date, onion, asparagus, vegetable-ivory palm, and many others (Mitchell, 1930). They are also found in pods of legumes, stony fruits, shells of nuts, hay, straw, and fodder; to some extent in the walls of the xylem elements of wood: and in certain fibers of the bark. In many of the cases mentioned the hemicelluloses undoubtedly function as reserve carbohydrates that are transformed into sugars when the need arises. Thus in the seed of the date, vegetable-ivory palm, and others, the hemicelluloses are the only reserve carbohydrates present and are transformed into sugars that are utilized by the young seedling. Tottingham, Roberts, and Lepkovsky (1921) and Murneek (1929) emphasized the importance of reserve hemicelluloses in the fruiting branches of apple twigs as reserve carbohydrates. The latter investigator found in the new growth of the fruit-bearing spur of the apple that starch constitutes only 1 to 4 per cent of the total dry matter, while the hemicelluloses represent 17 to 22 per cent. Similar proportions of these two groups of substances were found also in the leaves, flowers, and fruits. Murneek found that hemicelluloses were formed in relatively large quantities in developing fruits and considered that they were later hydrolyzed to form some of the sugar of the ripening fruit. Widdowson (1932), however, noted in developing apples a constant increase of hemicelluloses, but could find no time when they decreased in amount. He thus considered that they do not act as a reserve carbohydrate supply for the fruit.

The origin of the hemicelluloses is not definitely known. Malhotra (1930) claimed to have found them in young cell walls and in the protoplasm. Norman (1933) stated that there is apparently a close generic relationship between pectins and hemicelluloses and on purely chemical grounds it is possible to account for their formation by the protracted mild oxidation of linked hexose units. Norman and Norris (1930) considered that there is some evidence that hemicelluloses may be formed by the mild oxidation of pectin. Buston (1935) considered that by the oxidation of the free primary alcoholic groups of the hexoses, after the condensation of the sugars themselves to hexosans, there would be formed uronic anhydride units. By the decarboxylation of these units, pentosans would be formed, and stereochemical considerations show that by such a process arabans would arise from galactosans, while glucosans would pass through glucuronic anhydrides to xylans. It is evident that from a parent hexosan, even if based on a single sugar, a very wide variety of intermediate products could be formed, all of which might be classed as hemicelluloses. Sands and Gary (1933) showed there is no relationship between the hemicelluloses and gum of the mesquite that would indicate that one was the precursor of the other.

5. Suberin and Cutin.—These terms are used to designate those substances which render certain cell walls of the plant more or less impermeable to water. These two substances have frequently been called "adipo-" or "cutocelluloses" and were classified originally as compound celluloses, since it was assumed that they consisted of a cellulose base chemically combined or closely associated with compounds of a fatty or waxlike nature. The work of Gilson (1890), Van Wisselingh (1888 to 1892), von Schmidt (1903 to 1910), and Priestley and his coworkers (1921 to 1925), however, seems to show conclusively that suberin and cutin are not combined with a cellulose nucleus but are a collection of distinct compounds, although in some cases they may infiltrate a cellulose matrix.

The general properties of suberin and cutin may be stated as insolubility in and impermeability to water, considerable insolubility in fatty solvents, great resistance to concentrated sulphuric acid, but ready oxidizability by nitric or chromic acid and ready solubility in warm alkali. They are stained by fat stains such as Sudan III or scarlet red.

a. Suberin.—This is the substance that renders the cell walls of cork cells impervious to water, and on chemical grounds it may be regarded as an aggregate of variously modified forms (condensation products or anhydrides) of the suberogenic acids of which phellonic, suberic, and phloionic are representatives. Phellonic acid gives color reactions with iodine reagents, and this fact may account for the impression that cellulose is present in the suberin lamella. According to Priestley (1921) and Rhodes (1925), the distribution of suberin in the wall of the cork cell is as follows: Within each cell there is formed an inner layer which possesses fat-storing properties and which is called the "suberin lamella." Similarly, fatty substances impregnate the middle lamella of the cell wall. The continuous suberin lamella within the cell is laid down simultaneously in all parts of the cell and is undoubtedly formed from the contents of the cell itself which have accumulated at the surface of the protoplast. Between this suberin lamella and the protoplasm is located a layer of cellulose impregnated to some extent with fat but quite recognizable when all fat bodies are removed. The fatty substances in the walls of cork cells are thus not in a continuous sheet but are for the most part concentrated within a series of isolated lamellae lining individual cells and to some extent diffused throughout a carbohydrate basis of cellulose or of pectates. Rhodes (1925) found that neutral fat solvents remove only a small portion of the fatty substances of the cork cells of the potato and that of the fats so extracted only a small portion were glycerides. He removed the remainder of the fatty substances from the residual material by saponification with 3 per cent alcoholic sodium hydroxide. The fatty material thus removed is regarded as consisting largely of the salts of various organic acids, many of which have the characteristics of hydroxy-fatty acids. It seems evident that suberin owes its peculiar properties to the presence of normal and hydroxy-fatty acids in a state of combination and condensation and that the hydroxy acids are derived from the normal acids by some process of oxidation. Rhodes found that the chloroform-soluble material in the cork cells of the potato remained constant as the thickness of the cork cells increased, and he considered that this chloroform-soluble material is responsible for the staining of the cork tissue with Sudan III.

The formation of suberin in the healing of the cut or injured surfaces of the potato, sweet potatoes, bulbs, sugar beets, and various other plant parts, has received the attention of numerous investigators. In the white potato Priestley and Woffenden (1923), and Artschwager (1927) found that the first process in the healing of the cut tuber is the deposition of a fatty "suberin" layer formed by the oxidation and condensation of fatty substances that remain as a residue when the exuded sap dries at the cut surface. A supply of oxygen and a proper relative humidity of the air are necessary for the formation of this layer. Under suitable conditions this outer layer of suberin is formed within 24 to 48 hr. If the humidity is too low, the cut surface will dry and crack while the tuber will shrink from the loss of water. After a period of time varying with the prevailing temperature, relative humidity, solar radiation, and variety of tuber, a periderm is developed at varying depths beneath the cut surface. Peacock (1931) observed that skinned potatoes which were exposed during the early morning or late afternoon were not affected in their normal healing, while those exposed during the middle of the day were hindered in this process. When the injured portion of the tuber was placed on freshly stirred soil, a normal periderm developed regardless of the time of day. The relative humidity, light relations, and temperature under this condition are apparently optimum for the formation of periderm. Under optimum conditions of humidity Artschwager (1927) noted that the periderm cells appeared about the tenth day at 7°C., while at 21°C. they appeared after 2 days. Smith (1933) noted that in wounded potatoes suberization made its appearance on the second day in tubers stored at 68°F. but at 46°F. was not detected until the sixth day.

The sweet potato and the corms of gladiolus respond to wounding by a process of surface suberization and a subsequent formation of wound periderm in a manner similar to that of the potato tuber. Artschwager and Starrett (1931) observed in the sweet potato that temperatures of 22 to 35°C. and a relative humidity of 90 per cent or above were the best conditions for the formation of periderm. According to Weimer and Harter (1921) the production of a cork layer in a wound of the sweet potato is preceded by the formation of a layer of starch-free cells usually 3 to 10 cells deep, beneath the injured surface. Cross walls begin to appear in these cells from the second to the third day, and by the fourth to the sixth day a distinct layer of cork cells has formed a covering over the wound. No well-developed cork layer was produced over wounds under conditions existing in the storage house. Artschwager and Starrett (1933) noted that sugar beets respond to wounding by a suberization of the cut surface but show only a belated development of shallow wound periderm. Woodhead (1934) found that wounded plants of *Kleinia articulata* showed in their healing the same general sequence of events that occur in wounded potato tubers. Little suberization of the phelloderm layer occurs, however, and some of the phelloderm layers become collenchymatous. This formation of collenchyma is associated with the deposition of calcium oxalate crystals. Bloch (1935) noticed that the wound reaction in *Tradescantia fluminensis* varies according to the location of the injury so that the formation of suberized layers occurs only in certain regions of the plant.

b. Cutin.—Cutin is a substance that is more or less impervious to water and is found in a continuous external layer or lamella on the outer cellulose wall of the epidermal cells of leaves and stems. Lee and Priestley (1924) have shown that in some cases there may be a cellulose cutinized lamella impregnated with fatty deposits beneath this external layer of cutin or cuticle. The structure of the outer epidermal wall of the leaves of *Clivia nobilis* has been carefully studied by Anderson (1928, 1934). From the protoplasm outward the epidermal cell wall presents the following layers: (1) cellulose and pectic materials, (2) pectic substances, (3) cellulose and pectic material infiltrated with cutin, (4) cellulose infiltrated with cutin, and (5) an outer layer of pure cutin. This last layer is more resistant to saponification than other cutinized areas of the wall.

The thickness of the cuticle varies greatly for different plants and for different plant parts. Stevens (1932) noted in his study of 33 varieties of cranberries grown in Massachusetts during 1929 that the thickness of the cuticle on the fruits ranged from 9.9 to 13.7μ. This thickness was greater than that of the two following years and was found not to be correlated with the size or keeping quality of the fruit.

The cuticle of the leaves of many plants permits the loss of a considerable amount of water, and Crafts (1933) noted that acids and various plant poisons can readily penetrate it. The rate of penetration of these substances generally varies inversely as the thickness of the cuticle, although its composition may enter into the problem. Lee and Priestley (1924) considered that the cuticle is thickened by fatty substances migrating along the walls from interior tissues and that the mobility of the fatty substances in the walls may be modified by external conditions, which thus affect the thickness of the developing cuticle. Thoday (1933) believed that the formation of the cuticle depends primarily on the distinctive physiological activity of the epidermal cells and that it is not the relatively simple process of condensation of fatty materials. He fortified his opinion with the observation of Damm (1901) who noted that where the stem of *Viscum album* was cracked under the strain of growth a cuticle was developed by the outer cells.

Lee (1925) made a detailed macrochemical study of the cuticle of the petals of chrysanthemum and rose and of the petioles of rhubarb leaves. From her work it appears that cutin is a complex mixture of fatty substances, consisting of free fatty acids and condensation products of fatty acids, fatty acid combined with alcohol, soaps, unsaponifiable material, and resinous substances. A large amount of the fatty acids present are of the hydroxy-fatty type. These hydroxy-fatty acids are less soluble in chloroform than the normal fatty acids and are extracted from the cuticle only after a lengthy saponification. In the petals of chrysanthemum the chloroform extract of the cuticle contains 45 per cent of normal fatty acids and 16 per cent of hydroxy-fatty acids, while the saponified extract contains 5 per cent of the normal fatty acids and 80 per cent of hydroxy-fatty acids. Lee considered that the preponderance of the hydroxy-fatty acids is the result of oxidation processes taking place during the deposition of the cuticle. Since no glycerol has been identified, it seems that the fatty constituents of the cuticle have the characteristics of waxes rather than fats and have as their base monohydric alcohols rather than the trihydric alcohol glycerin. Legg and Wheeler (1925) extracted the cuticle of *Agave americana* and obtained four classes of compounds with the solvents used: (1) water-soluble material; (2) waxy material soluble in alcohol, benzene, or chloroform; (3) cellulose soluble in cuprammonia; and (4) the residue insoluble in the solvents used. This residue was considered as the characteristic constituent of the cuticle and was termed "cutin" by these investigators. More than 90 per cent of this cutin dissolved in alcoholic potash. The residue consisted for the most part of two semiliquid organic acids not hitherto reported as occurring in nature. One of these was termed "cutic acid" ($C_{26}H_{50}O_6$), and the other "cutinic acid" ($C_{13}H_{22}O_3$).

These investigators in 1929 reported that the cutin from *Agave rigida* on oxidation with dilute nitric acid gives suberic acid, a mixture of azelaic and sebacic acids, an acid ($C_{11}H_{20}O_3$) insoluble in water but soluble in cold alkali, and an acid ($C_{22}H_{12}O_3$) insoluble in both cold water and alkali. Chibnall (1931) and his workers found in the wax of apple peel that the amount of fatty acid obtained on saponification of the original crude wax was very small. The fatty acids found may be present as wax esters with long-chain alcohols or be derived from the residual amount of glycerides present in the crude wax. The major portion of the alcohols present is wax-free. The main general distinction between cutin and suberin that is now definitely known is the absence from the cutin of phellonic acid and glycerin.

6. Pectic Substances.—There are present in the cell wall—especially in the parenchymatous tissue of fruit, as pears, apples, currants, leaf stem of rhubarb, and fleshy roots, as carrots, beets, and turnips, as well as many others—a group of compounds that have the property of forming gels under certain conditions. These substances play an important part in the jam- and fruit-preserving processes, and it is to their peculiar properties that the "setting" of jams and jellies is due. It was formerly supposed that these substances consisted of a chemical combination of cellulose and certain mucilaginous compounds, so that they were classified as compound celluloses under the name of "pectocelluloses." It is now fairly well established, however, that these substances are not chemically combined with cellulose, although they are closely associated with it, so that they are classified under the general name of "pectic compounds" or "pectins."

These pectic substances are relatively unstable in contrast to the other constituents of the cell wall and their linkages are easily ruptured by both acids and alkalies. The formation of pectic substances in plants appears to occur when metabolism and growth are at their maxima. As growth slows and as maturity is reached the production of pectic materials decreases and those present in the wall are slowly converted into other substances. According to Anderson (1935) the pectic compounds at times apparently permeate the intermicellar spaces of the cellulose portions of the wall. Thus the removal of cellulose with Schweitzer's reagent leaves the cell wall in many cases unchanged in thickness indicating that the pectic compounds permeated the cellulose structure. Sometimes the removal of pectic materials brings about a striking lamellation of the wall, which suggests that definite lamellae composed mostly of pectic materials are present between the cellulose layers. All pectic compounds stain red with ruthenium red. With methylene blue they stain violet while the other cell-wall constituents stain blue or green. The membranes composed of pectic substances do not exhibit double refraction—they are isotropic—while cellulose membranes exhibit double refraction or are anisotropic. The pectic compounds are of variable composition but their principal components are galacturonic acid, galactose, arabinose, and methyl alcohol with the galacturonic acid forming in general from 65 to 85 per cent of the whole.

It has been considered that pectic substances, hemicelluloses, and lignin are interrelated and that the passage from pectic substances to lignin takes place through an intervening formation of certain hemicelluloses. No scheme, however, has been proposed concerning the method of this transformation. The common possession of a high methoxyl content by both soluble pectin and insoluble lignin has been regarded as evidence of their interrelation. There is apparently a close relationship between the pectic substances and the hemicelluloses. The pectins represent a type of highly oxidized hexosan with but little of the unchanged parent hexose and a small amount of the ultimate pentose constituent. The pectic substances are thus mainly distinguished from the hemicelluloses by containing approximately 70 per cent of uronic anhydride, while in the hemicelluloses the amount rarely exceeds 25 per cent. Support has been found for the suggestion that the pectins are the precursors of the hemicelluloses from the fact that the former predominate in young fleshy or unlignified tissues, while in the older lignified tissues the relative amount is reversed.

The literature concerning the pectic substances is voluminous and contradictory. Over no other constituent of the cell wall is there such lack of agreement among investigators in regard to terminology as over this aggregation of pectic compounds, so that it is impossible to summarize the literature on the pectic substances with any degree of satisfaction. The different terms applied to these compounds have been tabulated by Ahmann and Hooker (1925). The lack of agreement of the various investigators on this subject can probably be traced to two causes. In the first place, the term "pectins" includes, no doubt, a large aggregation of compounds rather than a few substances, so that the reactions obtained by one individual may be due to one substance and those by another individual to another substance. In the second place, the existence of many pectic compounds, which have been described by numerous investigators, must be regarded as uncertain. These pectic compounds are undoubtedly of a complex colloidal nature, and changes of physical properties and state of aggregation may occur without any definite change of chemical composition. On that account, solubility, the loss of solubility, or gelatinization—the tests that have so frequently been applied as determiners of pectic compounds—cannot be regarded as evidence in themselves of chemical change.

The real foundation of our knowledge of the pectic substances was laid by the researches of Fremy (1840 to 1848), and his general classification of these compounds is the one that has been in general use until the present time.

A study of the general properties of the pectic substances indicates that there are three distinct types of compounds as shown in the following outline:

Pectic substances	Pectic acid Salts of pectic acid or pectates	Soluble in water	Potassium pectate Sodium pectate
		Insoluble in water	Calcium pectate Magnesium pectate
	Esters of pectic acid	Insoluble in water	Pectose
		Soluble in water	Pectin

a. Pectic Acid and Pectates.—The key number of the group is pectic acid from which all the other pectic compounds are derived (Norman, 1933). Pectic acid from different sources contains galacturonic acid, galactose, and arabinose in different proportions. According to one formula, 4 molecules of galacturonic acid are linked to 1 molecule each of galactose and arabinose, the whole forming a closed, six-membered ring. In some material its composition seems to be 1 arabinose molecule to 4 of galacturonic acid. Pectic acid is only slightly soluble in water even in the presence of a considerable amount of electrolytes. It forms soluble pectates with alkalies and insoluble ones with metals of the alkaline earth, of which calcium pectate is the most widely distributed. The pectic materials in the middle lamella of the cell wall are for the most part calcium pectates. The relatively large amount of pectates in some tissues indicates that they may occur in other parts of the cell than the cell wall (Norman, 1929). Pectic acid is found especially in overripe and decaying fruits. It may be obtained by the saponification of pectin with dilute alkali, pectinic acids, with decreasing methyl alcohol content, occurring as intermediate or transition compounds in this reaction. Pectic acid and its salts possess the property of forming gels in the presence of calcium. Conrad (1926) has described a method for the simultaneous removal and quantitative separation and determination of protopectin and pectic acid and its salts. Protopectin or its hydrolytic product, pectin, was found in all the tissues studied, but pectic acid or pectates were found only in radishes. He considered that the many cases of pectates reported in plant tissue are probably the results of enzyme action during the preparation of the material for analysis.

b. Pectose.—The term "pectose" has been generally used since the time of Fremy to denote an insoluble pectic substance that occurs in the cell walls of green fruits and other immature plant parts and is considered to be the parent substance of pectin. The terms "protopectin" and "pectinogen" have been used by von Fellenberg (1918) and Clayson, Norris, and Schryver (1921), respectively, to denote pectose. It has been claimed by the majority of investigators that during the ripening of the fruit, pectose decomposes into a soluble compound generally known as "pectin," which consequently appears in increasing amounts in the expressed juice as the fruit matures.

Protopectin is considered by some as a glucosidelike pectin-cellulose or pectin-hemicellulose compound, although it has not been isolated, and there is no substantial evidence as to its existence. The extraction of protopectin-containing materials with boiling water or dilute acid yields pectin. Protopectin has been defined as the water-insoluble, acid-soluble, and acid-hydrolyzable parent substance of the pectins.

Tutin (1923) concluded from his investigations with the apple that pectose or protopectin does not exist and that pectin is present in the green fruits as well as the ripe ones, but it has not previously been extracted, since the tissues are extremely hard to reduce to a fine powder and also because certain insoluble compounds by their presence prevent the pectin from going through the filter. He claimed that he could with certain precautions obtain pectin from the green fruit. Tutin's work has been severely criticised by Carre (1925) who introduced evidence to show that his conclusions are in error.

c. Pectin.—The compound or aggregate of compounds termed "pectin" has the following characteristics: It is soluble in water and may be precipitated from this solution by alcohol. It cannot be crystallized, and its only known method of purification is treatment in aqueous solution with animal charcoal and reprecipitation with alcohol. The chemical composition of pectin has not been determined with any certainty. von Fellenberg (1918) established the fact that pectin is the methyl ester of pectic acid. Tutin (1921) considered pectin to be the dimethyl isoprophenyl ester of pectic acid. Schryver and Haynes (1916) considered from the analysis of pectin from turnip, apple, strawberry, and rhubarb stems that it has the formula $C_{17}H_{24}O_{16}$ with strong evidence that it contains one pentose group.

Pectin constitutes a great part of the pectic material in the cell wall and is located for the most part in the primary and secondary portions of the wall. According to Norman (1933) it may also occur in the cell sap. Pectin is the only one of the pectic substances that is of commercial value. The commercial uses of pectin are based on three of its characteristics—its jellying properties, its colloidal stabilizing properties, and its imbibing power for water. Under suitable conditions pectin has the property of giving a viscous solution when boiled with sugar and dilute acid. Upon cooling, this solution sets to a firm gel. The process of jelly or jam making involves boiling to concentrate the pectin and fruit acids present so that with added sugar a satisfactory jelly will be obtained. Prolonged or over-heating of the solution destroys the power of the pectin to gel. It is believed that the degree of esterification of the pectin is intimately connected with jelly formation (Norman, 1933). The general investigations concerning the practical application of the pectin-sugar gel formation are difficult of interpretation because the substances used have been neither ash-free nor well-defined salts.

Fremy also discovered the action of the enzyme pectase on pectin. This enzyme causes a solution of pectin to set to a stiff gel and this change occurs more readily in the presence of chalk. The production of the gel has been considered to be due to the formation of calcium pectate. Tutin (1921) found that the action of cold dilute alkali and of pectase on pectin are essentially identical. In both cases a salt of pectic acid is formed and methyl alcohol and acetone are eliminated. Kopaczewski (1925) found that when pectin from carrots and pectase from alfalfa leaves which had been purified by long dialysis were mixed, coagulation of the pectin no longer occurs. Coagulation, however, occurs upon the addition of salts of calcium, barium, strontium, magnesium, iron, or copper to the mixture. All these salts show power to coagulate pectin in the absence of pectase. According to this author, a neutral medium gives the best results from the action of pectase. He concluded that any purely chemical explanation of the coagulation of pectin is at present impossible.

Considerable work has been done on the transformation of the pectic substances in fruits during their ripening and storage. Haller (1929) stated that the softening of apples on the tree as maturity approaches is to some extent associated with a decrease in the percentage of protopectin and a corresponding decrease in total pectic substances. The amount of pectin, however, remains constant. The softening of apples in maturing cannot be accounted for entirely by the changes in the pectic constituents. In storage the softening of apples is apparently due to the conversion of pectose into pectin. In the ripening of grapes, Hopkins and Gourley (1930) found that the total amount of pectic materials falls off in a regular manner and that there is a conversion of the insoluble pectic materials to the soluble form. The pears stored at different temperatures the development of soluble pectin is the chief factor concerned in their softening (Emmett, 1929). The normal ripening processes in citrus fruits, according to Gaddum (1934), are accompanied by a gradual transition from the insoluble protopectins to the soluble pectins with a consequent decrease in cell-wall rigidity. This transition terminates finally in demethylated pectic acids so that overmature fruits contain no pectins.

In the ripening of bananas, peaches, strawberries, and dewberries, Conrad (1930) showed that pectin increases at the expense of pectose. Concurrently with this behavior there is liberated an unidentified furfural-yielding substance, soluble in 70 per cent alcohol, which increases progressively with the pectin. This substance is also liberated when material containing pectose is heated with dilute acid. Gerritz (1935) found that the insoluble pectic substances in June-dropped apples could be converted to pectin of good jellying qualities by the treatment of this immature fruit with 0.5 per cent hydrochloric, sulphuric, or tartaric acid.

7. Other Constituents of the Cell Wall.—The cell walls of the plant cell have infiltrations more or less of resins, gums, tannins, callose, fats, oils, coloring matters, ethereal oils, proteins, phospholipides, and the various inorganic salts of which potassium, sodium, silicon, and calcium are the more prominent. The kind and amount of these substances will depend upon the kind of plant and the conditions under which it was grown. The presence of phospholipides in the cell wall has been especially emphasized by Hansteen-Cranner (1914 to 1926). He considered that the plasma membrane bordering the cell wall is largely composed of these substances and that they are in direct and intimate connection with the phospholipides of the cell wall. The plasma membrane is thus considered not sharply delimited from the cell wall but is continuous with the phospholipides that saturate it. This, however, has been little considered by workers on the subject of plasmolysis, absorption, and general cell response. The occurrence of phospholipides in plants will be discussed in detail in Chap. X.

Uber and Goodspeed (1935) reported on preliminary methods for microincineration studies to determine the localization of the inorganic substances in the cell wall.

Callose is sometimes found in the cell walls of root hairs, in that portion of the wall bordering sieve pits, and in calcified cell layers as in cystoliths. The chemical nature of callose is not known, but its staining properties and its solubility distinguish it from other cell-wall substances. It dissolves in a solution of hydroxide, in calcium chloride, and in stannous chloride but will not dissolve in Schweitzer's reagent or in ammoniacal oxalate. It stains blue with a dilute solution of resorcin blue or aniline blue and takes a brilliant red stain in rosolic acid.

The gums are not true polysaccharoses but consist of a number of sugar molecules united to a central nucleus of a sugar-acid nature. The attachment seems to be of an ethereal or glucosidic type since the gums retain acidic properties and are capable of forming salts. They are usually found as salts of calcium, magnesium, and potassium. The classification of a substance as a gum is as yet mainly an arbitrary one and is not based on any constitutional knowledge but is decided by its origin and physical behavior together with the fact that it is of a carbohydrate nature. Judged from the products of hydrolysis, there is apparently little or no essential difference in structure between gums and hemicelluloses (Norman, 1929; Bailey and Norris, 1932).

8. Commercial Uses of Cell Walls.—The most important commercial component of the cell wall is cellulose, and it constitutes one of the most important raw materials of industry. The finished products of cellulose enter into the daily life and routine of every individual. The sources of cellulose may be classified according to the parts of the plant from which it is derived (Skinner, 1932).

a. Fibers of Seeds or Fruits.—The fibers are borne on seeds or on the inner walls of the fruit. Cotton is the principal and by far the most important of this class, which also includes the silk cottons.

b. Fibers from Stems Other than Trees.—In some cases these are from the inner bark while in certain other cases they are obtained from the sclerenchyma of the rind and that surrounding the bundles. In this group are found flax, hemp, jute, ramie, the cereal straws, esparto grass, and the stalks of corn and sugar cane.

c. Fibers from Leaves.—Examples of these are Manila hemp, sunn hemp, and sisal hemp.

d. The Wood of Trees.—From this source the useful elements vary greatly according to the species used.

The materials from these various sources are used in the manufacture of textiles, paper, cordage, and for the production of pure cellulose, from which are produced rayon, cellophane, and explosives.

II. THE PROTOPLASM

The protoplasm is the name given to that aggregate of matter in the cell which displays those series of phenomena that we term "life." Its perceptible characters show that it has a viscous slimy consistency, that it is neither a true solid nor a true liquid, and that it possesses considerable power of cohesion. It is heavier than water and shows no tendency to mix with the same when placed therein. Its viscosity ranges from a degree slightly more than that of water to the firmness of a fairly rigid gel (Seifriz, 1920). Although a certain degree of viscosity may characterize the protoplasm as a whole, it is always divided into regions more or less definite, *e.g.*, the plasma membrane, nucleus, or plastids which differ in viscosity from its general mass (Beams and King, 1935). Its general appearance seems to support the view that protoplasm is essentially a colloidal solution of the emulsion type. It consists of a water phase containing dissolved minerals, a fat phase, a protein phase, and numerous minor phases, so that protoplasm may be considered as a polyphase colloidal system. The colloidal nature of protoplasm is manifested in its power of adsorption, in its staining properties, and in the changes in its physical state under varied external conditions.

Protoplasm as viewed through the microscope has the appearance of an emulsion (Seifriz, 1935). It has also been considered an emulsion because in certain cases it behaves like an emulsion. Thus an electric current will cause an emulsion to become more permeable to certain ions, and electric currents have the same effect on living tissue (Dixon and Clark, 1932). This observation, however, may only mean that two rather diverse types of systems—an emulsion and a living jelly—show similar responses to the same environmental changes and not that the one is the same type of colloidal mixture as the other. Protoplasm is characteristically elastic, and pure emulsions are not. Protoplasm coagulates and emulsions do not. Scarth (1927) says that the impression of fluidity in protoplasm is illusory. According to Seifriz (1935) the properties of protoplasm that force us to recognize it to be essentially a jelly or a lyophilic colloidal system, whether firm or fluid, are elasticity, rigidity, extensibility, imbibition, water immiscibility, thixotropy, synaeresis, and coagulation. Many of these properties are greatly accelerated in protoplasm by the addition of alkalies and are reduced or often wholly eliminated by the addition of certain acids. This effect is produced regardless of whether these substances are added separately or whether they originate at the cathode or anode of an electric current (Seifriz, 1935). The fact that protoplasm appears to the naked eye or under the microscope as an emulsion is thus not proof that its ultimate arrangement is emulsive.

Concerning the chemical constitution of living protoplasm, but little is known. Strictly speaking, living protoplasms cannot be analyzed, since under any known method of chemical analysis the protoplasm is killed and dead matter only is being analyzed. The chemical analysis of protoplasm shows that it consists of from 80 to 90 per cent of water. The residue that remains after driving off the water consists of both inorganic and organic compounds. The inorganic substances make up from 5 to 7 per cent of the dry weight and consist chiefly of the chlorides, phosphates, sulphates, and carbonates of magnesium, potassium, sodium, calcium, and iron. The organic constituents are chiefly protein, carbohydrates, and fatty substances. The nitrogenous compounds make up from 40 to 50 per cent, and the carbohydrates and fats each from 12 to 14 per cent, of the dry weight of the protoplasm. The principal component part of protoplasm thus consists of proteins and of fats and phospholipides collectively termed "lipoids" (Lepeschkin, 1928). The carbohydrates and decomposition products of protein substances are considered only as admixtures because in some cases they are absent. The various salts, however, are always present, although in some cases in very small amounts. The compounds that enter into the composition of protoplasm are thus no different from those found elsewhere, which are always considered as nonliving material. Since protoplasm does not lose weight when it is killed, it is inferred that no elements or compounds escape from it at death and that it is the manner in which these various compounds are combined that gives the protoplasm the outstanding characteristics which distinguish it from dead matter. Death of living matter seems to be due to disorganization.

The most outstanding characteristic of protoplasm is its power of organization. Seifriz (1935) states that, although this characteristic cannot be physically interpreted, structure is the most fundamental requisite for its manifestation, and this structure must be both continuous and exceedingly labile. The protoplasm is able to carry on many different processes simultaneously without one interfering with another. To account for this it must be assumed that within the protoplasm there are delicate membranes consisting of nothing more than firm protoplasm traversing it in all directions. The fact that protoplasm does not mix with water implies structural continuity. On account of its colloidal nature, protoplasm presents a multitude of surfaces, and it is at these surfaces that the numerous chemical reactions occur.

It is of interest at this point to note the age that the living cells of plants may attain. MacDougal (1926), MacDougal and Long (1927), MacDougal and Smith (1928), MacDougal and Brown (1928), and Faul (1928) have made observations along this line. Living cells were observed in the medulla and cortex of the tree cactus (*Carnegiea*) and in the massive cactus *Ferocactus wislizenii* which had an age of at least 100 years. Ray cells of *Sequoia sempervirens* were found in the heartwood that was over 100 years of age, while living ray cells and tracheids 250 years old were observed in *Parkinsonia microphylla*. It was observed that the pentosans or mucilages which are abundant in the young cells of the cacti decrease with age. The glucose content increases, while the lipoid and protein content undergoes the least change.

Bristol (1916) claimed to have found dormant but living moss protonema in air-dry soils that had been stored in sealed bottles for nearly 50 years. Lipman (1934) believed that he observed unicellular blue-

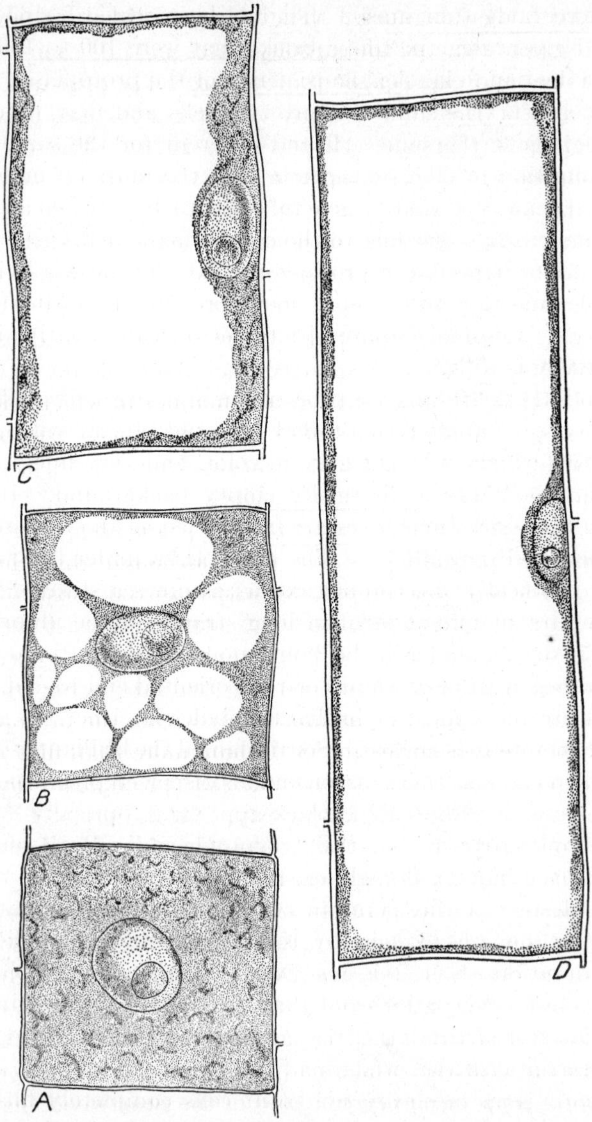

FIG. 3.—Stages in the development of the vacuole and protoplasmic membrane as shown by the cells of the growing region of a root of corn. *A*, an embryonic cell. *D*, a cell that has reached its full growth in length and in which the vacuole is fully formed, the protoplasm being in a thin membrane along the cell wall. *B* and *C*, intermediate stages in the enlargement of the cell, showing the formation of numerous vacuoles which finally coalesce to form a single large central one.

green and green algae in subsoil that had been sealed for 65 years, and single-celled green algae in adobe bricks that were 100 to 150 years old.

The physical and chemical consistency of the protoplasm is variable, and it may not be the same for any two cells and may be different in different regions of the same cell and different for the same cell under different conditions. The protoplasm is in the most intimate exchange with its environment so that its general constituency as well as its position in the cell may vary according to the conditions under which it is found.

The ultimate structure of protoplasm, like that of non-living matter, is not visible, and the more closely one approaches its ultimate structure the less easy it is to differentiate vitally between the relative importance of its constituents.

Seifriz (1931) in viewing the protoplasm of living onion cells under the oil immersion of a Spierer lens found he could observe two substances: one brightly illuminated, light gray in color, and very finely granular in texture, and the other an optically empty background. In quiescent protoplasm these two substances are intermixed as an emulsion and then present a mottled appearance. The protoplasm under tension, as when formed into a thread or when streaming, assumes a striated appearance due to the parallel arrangement of long strands of the illuminated substance. These strands may be continuous or discontinuous. In the latter case they may be made up of rods oriented end to end. Included particles occur and appear as brilliant globules in either the gray matter or the dark intervening surface. Seifriz named the brilliantly illuminated, gray-appearing, and at times discontinuous dispersed phase the "phaneroplasm" and the unilluminated, black-appearing, optically empty background or continuous phase, the "cryptoplasm." Both phaneroplasm and cryptoplasm flow, although apparently not at the same rate. Since the cryptoplasm is optically empty and thus cannot actually be seen, its streaming is made evident by the movement of included particles. In a consideration of the relative "vital" significance of phaneroplasm and cryptoplasm, Seifriz believed that the frequent discontinuity in the former and active streaming of the latter suggest that cryptoplasm is the more fundamental of the two.

The protoplasm in embryonic plant cells completely fills the entire cavity within the cell wall, but as the cells increase in size the protoplasmic content is altered by the appearance within the protoplasm of cavities filled with cell sap and called "vacuoles." These vacuoles are in most cases ultimately combined to form a single large sap cavity so that the protoplasm is spread out in a thin film along the cell wall (Fig. 3). The parts of the protoplasm of plant cells that are at this time definitely recognized under the ordinary microscope are the cytoplasm, nucleus, plastids, and chondriosomes.

A. The Cytoplasm

The cytoplasm may be considered as the less specialized portion of the protoplasm of the cell, since its functions are not so clearly defined as those of the other parts of the living substance. The cytoplasm makes up the general groundwork of the protoplasm and the greater portion of its bulk. Its physical and chemical characteristics are practically the same as those that have been mentioned in the discussion of the protoplasm. In general, the cytoplasm is the more fluid portion of the protoplasm in which the more highly differentiated morphological portions of the protoplasm are embedded and is thus that portion of the protoplasm that comes into the most intimate relationship with the environment of the cell. The surfaces of the cytoplasm that border on the cell wall and on the vacuole are generally modified in a greater or less degree in texture and consistency from its general mass. These modifications seem to consist of special layers or membranes of nongranular hyaline cytoplasm. Whether these membranes are the result of inherent specialization of the cytoplasm or simply the resultant of the physical forces of surface tension is not definitely known. The outside membrane next to the cell wall is generally called the "ectoplast" and the inside membrane bordering the vacuole the "tonoplast," while the more granular turbid bulk of the cytoplasm lying between these two membranes has been termed the "endoplasm." Any matter that enters the vacuole from the exterior of the cell or passes from the vacuole to the outside must pass through the cytoplasmic layer. The ectoplast and the tonoplast have been called by some authors the "plasma membranes" and have been considered by some authors to exercise a selective influence on the passage of materials into or out of the regions on which they border, but whether they are more specialized in this regard than the general mass of the cytoplasm has not been proved by any adequate experimental evidence. Concerning the physiological function of the cytoplasm but little is known. What little experimental work has been done seems to be in harmony with the hypothesis advanced almost 50 years ago by Bernard (1878) who maintained that the cytoplasm is the seat of destructive metabolism. If it is not thus involved, it would at least be safe to assume that it is a neutral region which translocates materials toward and away from the nucleus and plastids. It has been shown by Poirault (1893), Gardiner and Hill (1901), and Livingston (1933, 1935) that in many cases the cytoplasm of one cell is connected with that of the adjoining cell by delicate strands of cytoplasm that extend through the cell wall. These strands or fibrils are termed "plasmodesms" and lie in minute passageways through the cell wall. The plasmodesms are characteristic at least of those living cells which have cellulose or unlignified walls. These strands are the largest in the cells of mosses and the smallest in the cells of seed plants. They have been observed in the thick-walled cells of endosperms, in storage tissue, phloem, meristem, and in the medullary rays. The plasmodesms bear a close resemblance to the connecting strands of sieve tubes, and these strands are considered by some to be merely enlarged plasmodesms. Although the plasmodesms have been little recognized in plant physiology, it seems probable that they may play an important role in the transmission of stimuli and in translocation of materials from cell to cell.

B. The Nucleus

The nucleus is the most conspicuous organ of the protoplasm, and, owing largely to its behavior in the process of reproduction and cell division, cytologists have largely concentrated their attention upon it with the result that the nucleus is the best known body in the protoplasm in regard to its structure and morphology. A detailed discussion of the nucleus is beyond the scope of this work, and it is the intention to make only a few general statements concerning its structure and function. It is generally a rounded or ellipsoidal portion of the protoplasm whose structural basis is essentially similar to that of the cytoplasm. In its physical nature, the nucleus consists in part at least of a gel of higher viscosity than the cytoplasm. It is bounded by a distinct membrane within which is the nuclear sap or karyolymph. This substance is generally highly transparent and homogeneous and may occur as either a sol or a gel. In the karyolymph is embedded a network or reticulum that may be relatively uniform throughout or only fragmentary. In its chemical properties the nucleus is distinguished from the cytoplasm by the abundance of nucleoprotein that it contains. The nucleoproteins under proper procedure can be split into nucleic acid and a form of protein, the two existing in the nucleus in chemical combination. In its physiological functions the nucleus seems to be especially the center of the synthetic activities of the cell. The limited experimental evidence here again tallies with the opinion of Bernard (1878) that the nucleus is the organ of constructive metabolism.

C. The Plastids

The plastids are differentiated portions of the protoplasm of plant cells which perform certain physiological functions and are classed as protoplasmic organs on the same morphological and physiological plane as the nucleus of the cell. According to our present knowledge, the plastids or their primordia are present in every plant cell of the green plant, and the number of them formed in any given cell and the degree or form of differentiation that they assume will depend upon varied internal and external factors. Since the time of Schimper (1885) the plastids have been classified into three more or less distinct groups: the chloroplasts, the chromoplasts, and the leucoplasts.

1. The Chloroplasts.—The chloroplasts are the green plastids and owe their characteristic color to the presence of the pigment chlorophyll. They occur for the most part in those cells of the plant that are exposed to the sunlight. They vary much in shape and size in the algae, but for the most part, in the bryophytes, pteridophytes, and spermatophytes, they are small round or discoid bodies that more or less completely fill the peripheral cytoplasm of the plant cell. According to Eyster (1929), Mobius, in the examination of 215 species of plants belonging to many families of the bryophytes, pteridophytes, and spermatophytes, found 105 to have chloroplasts 5μ, 70 to have chloroplasts ranging from 3 to 5μ, 31 to have chloroplasts ranging from 5 to 7μ, and 9 to have chloroplasts from 7 to 10μ in their longest dimensions. In certain genetic strains of corn, Eyster (1929) found the chloroplasts in one strain to be only 3 to 4μ in length, while in other strains they ranged from 10 to 25μ in their longest dimensions, as compared with 6 to 8μ in most strains. He observed a direct correlation between the number and size of the chloroplasts in individual cells. The smaller the chloroplasts the greater the number in the cell. Our meager information concerning the internal structure of the chloroplast seems to indicate that it consists of a colorless protoplasmic groundwork in which are embedded the chlorophyll granules. Priestley and Irving (1907) considered that the chlorophyll is restricted to the peripheral ring of the chloroplast where it is held in the meshes of a network of protoplasm in two species of *Selaginella* examined by them. Zirkle (1926) examined the chloroplasts under normal conditions in a large number of higher plants in monochromatic light of a known wave length without any fixing or staining. He considered that the ground substance or stroma of the chloroplast is in the form of a hollow, flattened prolate spheroid surrounding a large central "vacuole." The stroma has a granular appearance that is due to numerous pores which connect the central vacuole with the cytoplasm surrounding the chloroplast. The pigments of the chloroplast are intimately mixed and evenly distributed throughout the protein ground substance. The starch inclusions of the chloroplasts are contained within the central vacuole. Zirkle also observed that in many leaves a constant differentiation of chloroplasts occurs. Some contain little starch and are mainly concerned in photosynthesis, while others contain much included starch and little green tissue and apparently function mainly as storage organs. It has not been adequately shown that the chloroplast or any of the other plastids have a specific membrane of their own like the plasma membrane, or that of the nucleus, although Zirkle (1926) determined that the chloroplast is surrounded by a more or less permanent sheath of nongranular cytoplasm. Harper (1919), however, considered that when the chlorophyll is dissolved from the chloroplast it has little to distinguish it from the adjacent cytoplasm. He considered that the plastid is to be regarded as a region of the protoplasmic complex rather than as a differentiated and definitely delineated body and that cytologically the function of the chloroplast is perhaps little more than an area of the cytoplasm impregnated or infiltered with chlorophyll. Weier (1933) defined the plastid of *Anthoceros* as a localized region of chlorophyll-impregnated cytoplasm.

The chloroplasts in the cells of the higher plants increase in number by the process of simple fission, which resembles the amitotic division of nuclei, but, so far as is known, the structure of the chloroplast undergoes no change in the process. The outstanding function of the chloroplasts is the formation of carbohydrates from carbon dioxide and water in the presence of sunlight. It is thus due primarily to the presence of the chloroplasts in certain of their cells that enable plants to manufacture their own food, while the members of the animal kingdom in whose cells the chloroplasts do not occur are totally unable to do so.

2. The Chromoplasts.—The chromoplasts are plastids that contain red, yellow, or orange pigments. They occur principally in the petals of flowers and in ripe fruits and may originate from chloroplasts or leucoplasts, as will subsequently be described. They have no special physiological function so far as is now known.

3. The Leucoplasts.—The leucoplasts are the colorless plastids and occur for the most part in those regions of the plant where light does not penetrate. They are of a denser consistency than the chloroplasts and are generally spherical in shape, although they may be greatly modified by the inclusions of starch and other compounds. Although the leucoplasts are colorless, they have the inherent capacity of forming coloring matter under certain conditions and thus being transformed into chloroplasts or chromoplasts. The term "leucoplasts" is generally used in a broad sense and is made to include not only the colorless plastids in the fully developed cells of the plant but also the colorless undifferentiated plastids of the embryonic cells which may eventually develop into chloroplasts or chromoplasts. The term also includes the colorless plastids of the epidermal cells and hairs which are apparently abortive chloroplasts. One of the chief functions performed by the leucoplasts, especially in the storage cells, is the formation of definite starch grains from the soluble carbohydrates that are present in the cell. The capacity for forming definite starch grains is thus possessed only by plant cells and does not occur at all in the animal kingdom, whose cells in no case possess leucoplasts. Mottier (1921) investigated the formation of the protein granules in the aleurone layer of the endosperm of corn and the formation of oil in the castor bean. He was of the opinion that leucoplasts may elaborate bodies other than starch grains and concluded from his cytological investigations that the protein and oil bodies found in the above storage cells owe their origin to plastids whose primordia are permanent organs of the cell.

The three different types of plastids seem to be homologous structures which in many cases display direct relationship. The primordia from which the chloroplasts and leucoplasts develop are identical, as far as our present knowledge goes, since they cannot be differentiated from each other by any staining or chemical reactions at present at our disposal. The more mature plastids under certain conditions change apparently from one into the other. The plastids of the tomato at a very early age may undoubtedly be classified as leucoplasts but soon develop into chloroplasts, which, upon the ripening of the fruit, become the chromoplasts that give the fruit its characteristic color.

4. The Origin of the Plastids.—There are two prevalent opinions in regard to the origin of the plastids in the cell. One group of investigators, including Gris (1857), Sachs (1875), Mikosch (1885), Belzung (1887 to 1891), and Stone (1932), holds the view that the plastids originate directly from the cytoplasm of the cell. According to them, the cells of the embryo of the mature seed contain no plastids. Their opinion is that, if any plastids were present in the young embryo, they lose their identity and disintegrate at the ripening of the seed and that at germination the cytoplasm in the cells gives rise to new plastids that function during the period of activity of the plant.

Stone (1932) studied the development of the chloroplasts in the young leaves of the potato plant. The young leaves observed contained green coloring matter that was diffused throughout the cell, and no definite bodies could be observed. The chlorophyll gathers into specific, delimited regions that eventually give rise to the chloroplasts. During their rounding off, these masses are often held together for a time by cytoplasmic connections that disappear as the plastids become fully formed. The process of chloroplast development is one of concentration and contraction rather than of expansion and growth since these original patches of chlorophyll may be larger than the resulting fully developed chloroplasts. The chloroplasts thus are considered to arise directly from cytoplasmic regions impregnated with chlorophyll.

The other group of investigators is inclined to the opinion first set forth by Meyers (1883), Schimper (1885), Bredow (1890), and Weier (1930 to 1933), that plastids never originate from the protoplasm in the cell but have an individual existence in the cell and multiply by division and are handed down from generation to generation. They consider that the fertilized egg contains plastids that have been derived from the parent plant. During the development of the egg into the embryo the plastids multiply and in this manner provide every cell of the embryo with plastids. The plastids go through the resting stage of the embryo intact and at germination increase in number and thus supply each cell of the new plant. In certain *Conjugatae*, *Chlorophyceae*, and *Archegoniatae*, the chloroplasts pass through the entire life cycle apparently unchanged. They are reproduced exclusively by the division of preexisting plastids and are apparently permanent cell organs. In the higher plants, however, the life history of the chloroplasts is more difficult and different, since in certain stages of the cycle no chloroplasts apparently occur but later reappear. They must then have originated from the cytoplasm or have passed through the preceding stage in an inconspicuous form. Although plastids in both the undifferentiated and differentiated state have been observed in the spores and fertilized eggs of several species of plants, the number of cases observed has been small owing to the difficulty encountered in proper differentiation by staining, so that, because of this lack of evidence in this link of the theory, it must remain more or less hypothetical. The work of later investigators has tended to strengthen this theory rather than to weaken it. Miller (1911) repeated the work of Famintzin (1893) on the origin of the chloroplasts in the cotyledons of the sunflower (*Helianthus annuus*) and showed that the plastids are present in their usual place in the cells of the seed but are very minute and that as the seed begins to germinate these plastids increase in size and divide by simple fission and develop into the chloroplasts of the green cotyledons of the seedling.

Mottier (1918) investigated the young tissues of *Pisum sativum*, *Zea mays*, and *Elodea canadensis* among others and concluded that the primordia of the leucoplasts and chloroplasts are permanent organs of the cell and that the egg and sperm have sufficient cytoplasm to carry these over into the new individual. Twiss (1919) traced in the root-tip cells of *Zea mays* from the embryonic cells backward an unbroken series of bodies varying in size from the most minute forms to plastids. She did not consider, however, that the observation of a series of gradations in size from mitochondria to the differentiated or mature chloroplasts is sufficient evidence that the mitochondria are the rudimentary plastids, since there is no way of showing that the rudimentary plastids are actually mitochondria and not young plastids.

Zirkle (1927, 1929) studied the origin of the plastids in the apical meristem of *Elodea canadensis*, *Lunularia vulgaris*, and *Zea mays*. He considered that the plastids develop from primordia in the meristematic tissue and that they are not reproduced *de novo* but reproduce by division, the division occurring in one plane only. The primordia are rod shaped, frequently found in strands end to end, and are often grouped around the nucleus. These enlarge, become green, and develop into chloroplasts. The immature plastids frequently contain starch inclusions. In the vascular bundles of corn many transitional stages can be observed in the course of development of mitochondria into plastids.

The origin of the plastids has been intensively and extensively studied in the leaves with various chlorophyll types including those that are entirely devoid of green pigment, variegated ones, and those in which the chlorophyll is evenly distributed throughout the leaf but lacks the intensity of greenness that is commonly observed.

Randolph (1922) made a cytological study of various chlorophyll types of maize in order to determine the status of the plastids in the cells of these strains. All the chlorophyll types examined were found to contain the same initial, minute primordia of plastids (protoplastids) of the same size and general appearance. In normal green plants the protoplastid first appears in the cell as a minute granule at the limit of visibility, gradually enlarging and developing chlorophyll until it becomes a mature chloroplast. In plants of the white virescent and other inheritance strains the unusual characters of the plants are due to the failure of the primordia initially present to develop into plastids with the normal size, color, or both. Randolph thought that the behavior of the primordia of the chloroplasts may in part at least be under the control of the nucleus of the cell. In the case of *Coleus,* which has variegated leaves, Stout (1915) observed that plastids are present in both green and yellow cells but that in yellow cells they are fewer in number, smaller in size, and somewhat distorted in shape and, further, that in extreme cases of yellow development nearly all the cells fail in the production of chlorophyll. Hein (1926) cut sections of the leaves of certain variegated plants so that they included both normal green and abnormal white and yellow areas and the marginal cells lying between. He found in the leaves of *Dieffenbachia seguine* (Schott) in going from the normal green cells through the critical zone to the pure white areas that the plastids decreased in number, size, and color until the cells showing no green were reached and that in these cells no evidence of chlorophyll bodies could be observed. In contrast to this, the nuclei of the various cells showed no modification in shape, size, or appearance. In the leaves of *Dracaena godseffiana* (Sanders) a more abrupt change from the normal green tissue to the colorless tissue was shown than in *Dieffenbachia,* but it is not absolutely sharp and definite, since the marginal cells show both normal and abnormal chloroplasts.

In the aberrant types of leaves that have been studied, it appears that the primordia of the plastids are normal but that these divergent forms are due (*a*) to a delayed development of the plastids, (*b*) to an arrestment of their development, (*c*) to an arrestment of development followed by degeneration, and (*d*) to a normal development of only a portion of the plastids (Küster, 1919; Zirkle, 1929; Eyster, 1933).

A study of chlorophyll inheritance has thrown considerable light on the development and subsequent behavior of the plastids. A summary of these observations has been made by Lubimenko (1926) and Priestley (1929). In some variegated plants the evidence indicates that the inheritance of the chloroplast mechanism can be interpreted in the terms of Mendelian segregation of certain pairs of unit characters. To the geneticist this indicates that at least part of the mechanism associated with the chloroplast inheritance is in the nucleus of the germ cell. Certain plants, however, afford evidence that the chloroplast mechanism is passed on by only one parent, and it is suggested that in these cases plastid gives rise to plastid through the germinal tract from this one parent to that of the opposite sex. In 15 experiments of crossing two pure lines of a species that differed markedly in chlorophyll content, there was found in 6 cases with the F_1 and succeeding generations a chlorophyll distribution that could be interpreted on Mendelian lines, while in the other 9 observations no such interpretation seemed possible. This suggests that both methods of chlorophyll inheritance may be involved. Chloroplasts may persist as units throughout the life cycle in the cytoplasm of the germ cell but factors segregating in the nucleus along Mendelian lines may determine their behavior during somatic development in succeeding generations.

D. The Mitochondria or Chondriosomes

In the cytoplasm of the cells of many plants representing all the great groups of the plant kingdom there have been described by numerous investigators small rod-shaped or granular bodies under the name of "chondriosomes" or "mitochondria." These minute bodies can seldom be observed in living fresh tissue and are, in general, not preserved by the usual methods of fixation, so that they are not detected by the ordinary methods of staining. It is due to these facts that their discovery was not made until comparatively recent times in botanical history. A complete and thorough review of the literature on chondriosomes is given by Cavers (1914). The literature on the chondriosomes is very confusing. Many of the investigators do not distinguish between the chondriosomes and the primordia of the plastids, and some (Guilliermond, 1911; Lewitsky, 1910; and others) consider that the bodies which they call chondriosomes give rise to the leucoplasts and chloroplasts. Rudolph (1912), Schmidt (1912), Mottier (1918), and others considered that the chondriosomes and the primordia of the plastids are entirely different structures and that in no case are chondriosomes changed into plastids. The evidence seems to indicate that there are present in the cytoplasm of plant cells certain definite minute bodies other than the primordia of the plastids. Guilliermond (1911 *et seq.*) and Mottier (1918) assigned to these bodies the same morphological rank as the nucleus of the cell. The latter considered that they are permanent organs of the cell and are transmitted from individual to individual through the cytoplasm of the sperm and egg. The mitochondria are considered by physiologists (Politis, 1921; Marston, 1923; Cowdry, 1926; and Horning and Petrie, 1927) to be the seat of both synthetic and hydrolytic degradation processes in plants. The phase boundary of the mitochondria and the surrounding cytoplasm is regarded as the seat of these processes. Marston (1923) from the reaction of the mitochondria to azine dyes concluded that they contain proteolytic enzymes and developed the theory that the mitochondria are the seat of enzymatic protein synthesis in the cell. He considered that the "water-poor" phases which exist at the surface of the lipoidal constituents of the mitochondria provide suitable conditions for synthetic activity. He considered that not only protein but also other foods and complex compounds may be synthesized in the same regions of the mitochondria. The function of the mitochondria in hydrolytic processes has been studied by Horning and Petrie (1927) in the germination of corn, wheat, and barley. During germination the mitochondria become more numerous apparently by simple fission in the scutellum and are secreted in large numbers from the epithelial cells into the adjacent starch-containing cells of the endosperm. These secreted mitochondria then aggregate around the starch grains prior to their corrosion. Their behavior appears to lend considerable support to the enzymatic conception of mitochondrial activity, that the starch-splitting enzyme is located within them or at their surfaces and is liberated therefrom when the bodies have reached the surface of the starch grains.

III. THE VACUOLE OR CELL SAP

A. Formation

As an embryonic plant cell begins to increase in size, the protoplasm soon develops cavities of varying sizes which become filled with water and materials in solution and suspension. These cavities with their contents are termed "vacuoles," and as the cell increases in size, these vacuoles usually coalesce to form a single large vacuole or cell cavity. The formation of the vacuole is apparently a much more complicated procedure than had formerly been considered. According to Guilliermond (1929), the vacuoles in the meristematic cells of flowering plants, as a result of the hydration and swelling of numerous and tiny mitochondria-like bodies, contain a thick colloidal solution. As the cell differentiates, these bodies swell into small, round vacuoles which fuse into one single large vacuole, making the bulk of the cell. The large liquid vacuoles originating from the mitochondria-like bodies by hydration of their contents ultimately contain a very dilute colloidal solution which stains faintly as a whole. The whole system of vacuoles in any given cell at any stage is termed the "vacuome." The colloidal substances in the vacuoles of flowering plants differ widely in different cases. According to Guilliermond, the vacuolar contents are most often composed of alcohol-soluble protein mixed with phenol compounds of the tannin group and closely related anthocyanin compounds, all of which absorb intra-vitam dyes.

The vacuome is considered to be always present in any cell, and it may appear in two altogether different aspects according to the age of the cell: (1) It may be distributed throughout the cytoplasm as numerous tiny thick bodies composed of a concentrated solution of colloids and showing as mitochondria-like grains or threads. (2) It may be in the form of one or a few large liquid vacuoles, where colloids are in weak solution, these vacuoles developing from the small, mitochondria-like bodies by hydration and coalescence. Either of these aspects may be converted into the other. The vacuoles may be made to give up water when the cell is dehydrated. They shrink and break down into thread-like thick drops, as is shown when a seed begins to lose water preparatory to maturing and as can be experimentally induced by plasmolyzing cells.

B. Composition

The vacuole or cell sap may contain the following materials:

1. Water.—As much as 98 per cent of the cell sap may consist of water.

2. Inorganic Salts.—Any of the inorganic salts that are absorbed by the plant from the soil may be found in the cell sap, but those which occur in the greatest abundance are the salts of sodium, potassium, magnesium and calcium.

3. Carbohydrates.—The carbohydrates that are generally found in the cell sap include cane sugar, maltose, glucose, fructose, dextrins, and inulin.

4. Nitrogenous Compounds.—The principal nitrogenous compounds are proteins, amino acids, amides, proteoses, and peptones.

5. Soluble Alkaloids and Glucosides.—These products, which appear to be waste products of the cell, include the tannins, amygdalin, anthocyanin, and the mustard-oil glucosides.

6. Gases.—Oxygen and carbon dioxide are the gases in the cell sap that are of the most physiological importance. The oxygen may come from the exterior, or if it is a green cell a portion of it may arise as a by-product in photosynthesis. Carbon dioxide occurs in all plant cells as an end product of respiration. It may also enter the cells from the outside, especially in green cells where it is used in photosynthesis.

7. Enzymes.—The principal enzymes that may be found in the cell sap are diastase, invertase, lipase, inulase, and proteolytic enzymes.

8. Organic Acids.—These compounds are present either as acids or in combination with mineral bases. The principal organic acids present in the vacuole are oxalic, malic, citric, tartaric, tannic, acetic, and formic. The last two seem to be very widespread in their distribution and are probably present in every living cell.

According to their origin, the materials found in the vacuole fall into three general groups: (*a*) the materials that have been absorbed by the cell from its exterior, (*b*) the materials that have been synthesized by the protoplasm, and (*c*) the waste products or excretions that arise in the protoplasm as a result of its life functions. It is evident, therefore, that the composition of the vacuole will be an extremely variable one, since the functions that give rise to the materials therein are dependent upon and influenced by a great many factors.

C. Concentration

The concentration of the cell sap is generally expressed in terms of osmotic pressure. By the osmotic pressure of the cell sap is meant the pressure that the water in the vacuole would be capable of exerting upon the protoplasm and the cell wall if the cell were placed in pure water. This definition of osmotic pressure is based on the assumption that the cell wall is absolutely rigid and that the protoplasm will allow water to enter freely and that neither will allow dissolved materials to escape from the vacuole. The osmotic pressure of the cell sap can also be defined by stating that it is the pull that the cell sap is capable of exerting upon pure water if it were separated from it by a perfectly semipermeable membrane. The magnitude of the osmotic pressure of the cell sap is commonly expressed in terms of atmospheric pressure. The osmotic pressure or pull of the cell sap varies directly with its concentration and is due to certain relationships of membranes and solutions that will subsequently be discussed in Chap. II.

1. Factors That Influence the Concentration of the Cell Sap.—The factors that influence the concentration of the cell sap are (*a*) the habitat of the plant, (*b*) the kind of plant, (*c*) the portion of the plant, (*d*) the age of the tissues, and (*e*) the time of sampling, *i.e.*, season, day or night.

a. The Habitat of the Plant.—It was shown by Stange (1892) with plants of wheat, beans, and peas, Hill (1908) with the root hairs of *Salicornia* and *Suaeda*, Roberts (1916) with the root hairs of numerous crop and vegetable plants, and Ohga (1926) with wheat and buckwheat seedlings that the osmotic pressure of the cell sap can be varied by changing the concentration of the medium in which the plants are growing. Roberts (1916) found that the concentration of the cell sap of the root hairs of numerous plants grown in moist air was equivalent to 5.1 to 7.7 atmospheres so that under this condition there was a minimum difference of four atmospheres in concentration between the inside and outside of the root-hair cell. By growing the roots of radish in a series of increasing concentrations of cane sugar from 0.02 M to 0.65 M, she showed that the cell sap of the root hairs increased in concentration directly with that of the external medium and maintained an osmotic pressure of from 4 to 6 atmospheres above it. Ohga (1916), studying wheat and buckwheat seedlings, observed that the maximum increase in osmotic pressure in the roots, due to placing them in sucrose solution, was about 0.8 M or 18.7 atmospheres.

McCool and Millar (1917) found that the osmotic pressure in the roots and tops of Canada pea and wheat and corn seedlings is indicative but not directly proportional to the density of the solution in which they are grown. The concentration of the soil solution was varied by adding solutions of $Ca(NO_3)_2$, $MgSO_4$, and K_3PO_4. The following figures show the variation of the osmotic pressure of the soil solution and the osmotic pressure of the corn roots growing therein:

Osmotic pressure of the soil solution, atmospheres	Osmotic pressure of sap of root cells, atmospheres
1.21	4.59
1.99	5.48
3.38	6.61
4.96	7.51
7.22	8.19

The application of the various mineral nutrients to the soil as fertilizers generally results in increased concentration of these elements in the cell sap (McCool and Weldon, 1928). It appears, however, that unless there is considerable contrast between two environments the osmotic value of the expressed sap cannot be taken as a safe criterion of the type of environment (Meyer, 1927).

Harris and Pascoe (1930) observed a general positive correlation between the salinity of the soil and the osmotic concentration of the cell sap in Prima Egyptian cotton and Lone Star upland cotton. This fact was indicated when measured in terms of freezing-point depression, chloride content, sulfate content, or electrical conductivity.

Magistad and Truog (1925) found that the application of fertilizer in the hill increases the osmotic pressure of the sap of corn plants which in turn lowers the freezing temperature of the plants from one to two degrees centigrade. This is often sufficient to prevent plants from being frozen by late spring frosts if they are not too severe. The greatest benefits in this regard would be derived from the application of fertilizers to peat, muck, or other soils of low-soluble salt content.

In a changing medium, plants differ widely in the extent to which they take up salts and increase the concentration. Thus Eaton (1927) found that the cell sap of wheat plants was 98 per cent more concentrated when growing on a soil with a higher salt content than on a nonsaline soil, while under the same conditions a difference of but 42 per cent was shown for the saltbush plant. Drabble and Drabble (1907) examined 50 species of plants growing under various conditions of water supply and found the osmotic pressure of the cells varied from 4 to 20 atmospheres. Their conclusions concerning the relation between environment and the osmotic pressure of the cell sap of plants have been verified by other investigators and can now be stated under the general law, that "In plants of the same species growing under the same conditions of water supply, the osmotic strength of the cell sap is generally the same, and in any area the osmotic strength of the cell sap varies indirectly with the physiological scarcity of water."

The osmotic pressure in the cell sap is more rapidly changed by the fluctuations in the moisture conditions of the site than by any other environmental factors (Korstian, 1924). The highest densities of cell sap occur on the sites with the lowest moisture supply regardless of whether this low supply of moisture is due to a direct lack of water or to a high concentration of solutes in the soil. Stoddart (1936) noted as the soil dried with the progress of the season that in all the prairie plants examined the osmotic pressure increased in their cell sap. This increase amounted to as much as 30 atmospheres or more in certain upland plants, while in plants growing in low moist habitats the increase in osmotic pressure was usually only 2 to 10 atmospheres. The water content of a plant is a fair indicator of osmotic pressure—plants with high pressures usually having a low water content and plants with low osmotic pressures usually having a high water content.

Beck (1929) observed that the cell sap of the assimilatory tissues manifests the greatest regularity, speed, and degree of response to drought of any of the plant tissues. The osmotic values of the epidermal cells show the least variations.

Stanfield (1932) found a correlation between the osmotic pressure and rate of evaporation in the leaves of *Pinus scopulorum*, the higher pressures occurring with the higher rates of evaporation. A difference of approximately 3 atmospheres was noted between the pressures occurring with the lowest and highest rates of evaporation.

Iljin (1929) noted that the osmotic pressure of the roots is closely related to the degree of moisture in the soil as shown by the following concentrations of the cell sap of plants from different habitats:

In swamps, 0.13 to 0.35 *M*
In meadow, 0.25 to 0.35 *M*
In steppe plants, 0.40 to 0.48 *M*
In exposed declivities, 0.55 to 0.60 *M*

Hibbard and Harrington (1916) found that the osmotic pressure of the cell sap of the root cells of corn was highest in those roots growing in the soil with the lowest moisture content (see also Whitfield, 1932; and McKay, 1935).

Fitting (1911) has given us the most complete record of the osmotic pressures in the cell sap of desert plants. He worked in the Sahara desert and examined 46 species of plants growing in oases, saline soils, rocky wastes, and sand dunes. His observations may be summarized briefly as follows:

Number of plants	Osmotic pressure	
	Salt equivalent	Atmospheres
10	3.0 N (KNO_3) or above	100 or above
16	1.5 N (KNO_3) or above	53 or above
24	1.0 N (KNO_3) or above	34 or above
5	0.3 to 0.6 N (KNO_3) or above	10 to 20 or above

The highest osmotic pressures were found in the plants growing in dry soil and no extremely high pressures were met with in moist nonsaline soil. The work of Fitting has been verified by Ilin and Nazarova (1915) and by Keller (1920) working in Russia.

Harris, Lawrence, and Gortner (1915 to 1921) and Harris (1934) studied the concentration of the cell sap of a large number of different plants growing under a wide range of conditions. Their investigations include not only plants growing in widely distributed geographical areas of North America but also those growing in different habitats of the same area. In an investigation of the spring flora of the mesophytic region of Cold Spring Harbor, N. Y., and that of the desert region of Tucson, Ariz., they reported the following: At Cold Spring Harbor 50 per cent of the pressures were equal to 10.5 atmospheres or lower, while at Tucson in the desert 50 per cent of the pressures equaled 15.7 atmospheres or above. At Cold Spring Harbor only 3 per cent of the plants exceeded a pressure of 20 atmospheres, while at Tucson 30 per cent of the pressures exceeded this figure. The maximum pressure observed at Cold Spring Harbor was 30 atmospheres, while the highest pressure at Tucson was 65 atmospheres. The average pressure of plants growing at Cold Spring Harbor was 14.4 atmospheres as compared to an average pressure of 38.8 atmospheres of those growing on the coastal slopes of Jamaica, B.W.I.

b. The Kind of Plant.—Fitting (1911) observed that the concentration of the cell sap is highest in trees and shrubs and lowest in the annuals. Harris and Lawrence (1916) in their work at Tucson, Ariz., reported the following average determinations of the cell-sap concentration:

Kind of plants	Osmotic pressure of cell sap, atmospheres
Trees and shrubs	28.10
Dwarfs and half shrubs	21.45
Perennial herbs	16.35
Winter annuals	14.73

The osmotic pressure of the sap of trees and shrubs growing at Tucson, Ariz., Jamaica, B.W.I., and Cold Spring Harbor, N. Y., was found by Harris and others (1921) to be 30.5 per cent higher than that of the herbaceous plants growing in the same region. Harris (1918) observed that the cell sap of epiphytic plants is only one-third to two-thirds as concentrated as that of the terrestrial plants. It was observed by him and is considered a general biological law that the conductivity of the sap of herbaceous plants is distinctly higher than in woody species. The concentration of ionized electrolytes is thus lower in woody forms than in herbaceous forms, although the reverse is true in regard to the total solutes. It was also observed by Korstian (1924) that the concentration of the cell sap of woody species is much higher than that of herbaceous species (see also Meyer, 1927).

The crowns, stalks, and roots of sorghums have a higher osmotic concentration than those of corn, while the leaf juices of sorghums have a lower freezing-point depression and a lower conductivity than corn.

Harris and Lawrence (1916) studied the osmotic pressure of the cell sap of the mistletoes growing on 20 different hosts. They found the average pressure for the parasites to be 14.43 atmospheres and that of the hosts to be 13.59 atmospheres. The osmotic pressure of the leafless species of mistletoe is distinctly lower than that of the leafy species.

The cell-sap concentration of the cells of *Phoradendron juniperinum*, parasitic on *Juniperus utahensis*, showed according to Harris, Pascoe, and Jones (1930) a pressure of 3.71 atmospheres in excess of the host plant. The chloride content of the cell sap of the parasite exceeds by 1.5 times that of the host. The almost universal absence of *Phoradendron californicum* as a parasite on *Covillea tridentata* is considered to be due to a higher osmotic pressure in the latter plant (Harris, Harrison, and Pascoe, 1930).

Harris and others (1925) found that the sulphate content of the leaves of the upland varieties of cotton is higher than that of the Egyptian variety, the differences ranging from 3 to 4 g. per liter. This is just the opposite from the chloride content, the Egyptian types taking up larger quantities of chlorides than the upland types. All of the Egyptian varieties considered have a higher osmotic pressure than the upland varieties.

c. The Portion of the Plant.—Hannig (1912) determined the relative concentration of the cell sap in leaves and roots of a large number of different plants. His observations are expressed as follows:

Percentage of plants examined	Ratio of the concentration of the sap of the leaves to that of the roots
14	1
51	1.25
12	1.5
23	2

Dixon and Atkins (1916) found that the osmotic strength of the cell sap of the leaves was greater than that of the roots. The concentration of the sap of the leaves of

Ilex aquifolium varied from 13 to 15 atmospheres, while that of the roots varied from 7 to 10 atmospheres. The leaves of *Iris* had a pressure of 13 atmospheres and the roots a pressure of 9 atmospheres. Lutman (1919) made an extended study of the osmotic pressures in the cells of the various organs of the potato at various stages of growth. He also observed the osmotic pressures in several plants closely related to the potato or closely resembling it in the manner of storing food. He found that the normal pressure in the seed tubers as they are taken from storage is between 7 and 10.3 atmospheres and that the sprouts which come from these tubers not in the soil exhibit a pressure slightly in excess of that of the tubers themselves. He found that the juice of the leaves of the young plant showed a higher osmotic pressure than that of the stalk and that the tubers as they develop maintain an almost constant pressure until maturity. The osmotic pressure of the sap of the growing tubers is always intermediate between that of the sap of the stalks and the roots whose osmotic pressure is always the least of any of the organs of the potato plant. Garden beets examined on Sept. 28 showed a pressure in the leaf sap of 8.83 atmospheres and a pressure in the root sap of 11.81 atmospheres. The sugar beets examined at the same time showed a pressure of 12.21 atmospheres for the sap of the leaves and of 17.18 atmospheres for the roots. According to Chandler (1914), the molecular concentration of the leaves of fruit trees is generally considerably greater than that of the fruit except in the case of some ripe fruits. The concentration of the cell sap of the leaves at different heights of insertion on the stem has been studied by Dixon (1914), Harris, Gortner, and Lawrence (1917), Korstian (1924), and Hurd-Karrer (1925). Dixon found that the leaves of wistaria at a height of 3 ft. above the ground had a concentration of 5.25 atmospheres, while the leaves at a height of 27 ft. had a sap concentration of 6.61 atmospheres. The leaves of the elm 1 ft. from the ground showed a sap concentration of 9.2 atmospheres, while those at a height of 10 ft. showed a pressure of 12.4 and 18.6 atmospheres in the shade and sun, respectively. Harris and others (1917) made determinations of the osmotic concentration of the leaf sap at different levels on 26 trees belonging to 12 species and concluded that the concentration of the leaves as measured by the freezing-point lowering method increases from lower to higher levels almost without exception. The following are the results that they obtained with *Betula lutea* and *Fagus grandifolia:*

Name of plant	Date	Height of leaves, feet	Osmotic pressure of cell sap, atmospheres
Betula lutea	July, 1914	11	12.6
Betula lutea	July, 1914	25	14.1
Betula lutea	July, 1914	39	15.1
Betula lutea	July, 1914	52	16.1
Betula lutea	July, 1914	66	15.6
Fagus grandifolia	August, 1913	19	17.33
Fagus grandifolia	August, 1913	64	21.92

Korstian also (1924) demonstrated an osmotic gradient in the cells of plants. In the case of the yellow pine the 2-year-old needles at the base of the leaves had a sap concentration equivalent to 17.3 atmospheres while at the top the osmotic pressure was equivalent to 18.5 atmospheres. Hurd-Karrer (1925) determined the concentration gradient in the juice of corn stalks as measured by the specific gravity of the expressed sap. The plants used were from 8 to 11 ft. high with large, but immature, ears. The sap from each internode showed a gradual gradation upward, the specific gravity of the sap of the first internode being 1.0347, while that of the thirteenth and fourteenth internode had a value of 1.0497. The electrical conductivity is always less at the higher levels than at the lower ones, which indicates that the greater concentration of the uppermost leaves is due to the increase of nonelectrolytes as sugar.

It was shown by Martin, Harris, and Jones (1931) that in the sorghums the highest osmotic concentration of the cell sap is in the terminal leaf and that there is a progressive decrease in the leaf juices from the top of the plant downward. The osmotic pressure of the leaves of the date palm in the Algerian desert is frequently 10 atmospheres higher than that of the roots (Killian and Faurel, 1934). It was stated by Iljin (1929) that the osmotic pressure of the roots is more closely correlated with the habitat than is that of the cell sap of the leaves.

d. The Age of the Tissues.—Dixon (1912) observed in the two evergreens, *Ilex aquifolium* and *Hedera helix*, that the osmotic pressure of the cell sap of the older leaves was from 2 to 4 atmospheres higher than that of the younger leaves.

Korstian (1924) in the case of the pine found that there was a difference of from 1 to 5 atmospheres in the osmotic pressure of the young and old leaves, the young having the lower concentration. Gail and Cone (1929) also observed this fact. In the mulberry it was observed by Lombardi (1932) that the older the leaves the higher the concentration of their cell sap. The conductivity of the sap of the older leaves was higher than that of the younger leaves and indicates that their greater concentration is due to the accumulation of electrolytes.

e. The Time of Sampling.—Since the sugars contribute to the solutes in the cell sap, the conditions that would affect their formation would modify its concentration. Shedd (1915) stated large differences in the composition of the sap of the wild grape were found when it was collected at the same point on the vine at different times during the same season. The mineral constituents were generally higher during the day and the sap had a more uniform condition during the night. Harris, Hoffman, and Lawrence (1925) noted an increase in the magnitude of the chloride concentration in Egyptian cotton plants as the season advanced. McCool and Millar (1917) found that the osmotic pressure of corn tops was lowest in the morning at sunrise and greatest at 1 P.M. The concentration of the cell sap of the leaves at sunrise was 6.06 atmospheres and at 1 P.M. it was 7.52. The osmotic pressure decreased after that period so that at 10 P.M. it amounted to only 6.72 atmospheres. The roots showed an osmotic pressure of 4.52 atmospheres at sunrise and a pressure of 6.05 atmospheres at 1 P.M. and then a decrease to 4.33 atmospheres at 10 P.M. These maximum pressures observed correspond in a general way to the maximum point of sugar production, and this fact apparently accounts for the increase in the osmotic pressures (Gail, 1926; and Greathouse, 1932). Korstian (1924) observed that the density of the cell sap changes during the day, increasing as the day advances and decreasing toward evening. The daily minimum occurs in the early morning, indicating a relationship between concentration and photosynthesis. The same fact apparently accounts for the differences observed by Dixon and Atkins (1912) on the concentration of the cells of leaves of a north exposure and those of a south exposure. In *Hedera helix* they found that the concentration of leaves of a north exposure averaged 8.7 atmospheres, while that of a south exposure averaged 9.3 atmospheres. They also observed that shaded or covered leaves always had a lower concentration than the ones exposed to the sunlight. This fact has also been observed by Lutman (1919) for the stem and leaves of potato, and by Halma and Haas (1928) for the leaves of Eureka lemons.

2. Cell-sap Concentration and Growth.—The relationship of cell-sap concentration and the rate of growth has received considerable attention.

It was observed by Chandler (1914) that the molecular concentration of the cortex sap is the lowest during the period of rapid growth and that the molecular concentration of the cortex sap is not so great during early summer with trees that have been forced into vigorous growth by heavy pruning as with trees not so pruned.

Reed (1921) investigated the relationship of growth and cell-sap concentration in the young trees of walnuts, apricot, and orange. The weekly increment in concentration, including the leaves, was determined. The more important facts observed by him are as follows: The rate of growth and sap concentration have a tendency to vary in opposite directions so that rapid growth is associated with a generally lower concentration of sap in the shoot while slower growth is accompanied by higher concentration of sap. There was a gradual increase in sap concentration as the season advanced. In the apricot tree the concentration of the sap continued to increase for some time after active growth had ceased. The average concentration of the cell sap of the apricot expressed in atmospheres by months was as follows: May, 11.84; June, 13.66; July, 14.34; August, 15.04; September, 15.18; and October, 16.48. Of several environmental factors measured, soil moisture was the only one having an obvious effect upon sap concentration. The addition of water to the soil usually diminished the concentration of the plant sap. A concentration gradient appears to exist in the shoot. The concentration of the sap in the apical portion of the stem was greater than that in the basal region. The sap concentration of shoots in trees heavily pruned was lower than that of shoots on trees not pruned because of the more rapid growth of the former. In the case of the lemon, however, Reed and Halma (1926) found that neither severe nor moderate pruning affects the sap concentration of leaves and shoots of lemon trees for more than a few weeks after the operation. In view of these results, these authors concluded that there is something about the lemon tree which operates to maintain osmotic pressure in the face of a growth rate undergoing extreme fluctuations.

D. The Acidity of the Cell Sap

1. General Discussion.—The cell sap as a rule has an acid reaction and only rarely is it alkaline and then only slightly so. A pH value on the alkaline side of 7 to 8 has been observed, while on the acid side the highest acid value so far observed has been 0.9 or a value equivalent to about 0.1 *N* sulphuric acid. The pH value of the cell sap of plants generally lies somewhere between 4.6 and 7.0 (Atkins, 1922; and Dunne, 1932). Oxalic, malic, and citric acids are generally the cause of this acid reaction. These acids are normally present for the most part in the form of their acid salts or of their neutral salts. The concentration of free acids in the cell sap is usually rather low.

The concentration of free malic acid in plant tissue rarely exceeds 0.05 per cent of the fresh weight and is usually much lower, although the total malate content may be as high as 2 per cent of the fresh weight. According to Jodidi (1933) citric acid in the Alaska pea equals 0.36 per cent of the oven-dried material. In the expressed sap from *Pelargonium*, Armstrong (1929) found oxalic acid 0.0573 *M*, malic acid 0.0337 *M*, citric acid 0.0066 *M*, and tartaric acid 0.003 *M*. Clague and Fellers (1934) reported that the benzoic acid content of 24 varieties of cranberries averaged 0.065 per cent.

In studying the acidity of the cell sap, four quantities must be considered (Bennet-Clark, 1933): (*a*) The hydrogen-ion concentration, which is very frequently termed the "actual acidity." This quantity refers to the number of free hydrogen ions in the cell sap. (*b*) The concentration of the free, undissociated acid. This term refers to the acids in the sap which are not ionized. (*c*) The "titratable acidity." This quantity includes the combined hydrogen ions of the acids and the acid salts in the cell sap as well as the free hydrogen ions therein. (*d*) "The total acidity." This is the total quantity of anions and molecules and is therefore best expressed as the number of mols or equivalents of the acid. This quantity is obtained by neutralizing and precipitating the acid as an insoluble salt. It is termed by some the "quantitative acidity." The difference between the total acidity and the titratable acidity may be termed the "nontitratable acidity." Ordinarily the quantity of acid formed or used during metabolism is obtained by determining the increase or decrease in the titratable acidity and under most circumstances this is equal to the change in quantity of total acid.

The oxalates and oxalic acid are probably present in almost all plants. The genera *Rumex*, *Oxyria*, *Geranium*, *Tropaeolum*, and *Oxalis* are especially rich in these two compounds. Most plant saps contain malic acid or malates (Bennet-Clark and Woodruff, 1935). The water extract of wheat, barley, rye, and corn plants at 2 weeks preceding heading contained aconitic, citric, malic, and malonic acids, with a trace of oxalic acid (Nelson and Hasselbring, 1931; and Nelson and Mattern, 1931).

The striking changes in the acidity of plant tissues that occur under natural conditions indicate that the organic acids play some important role in metabolism and that they are by no means waste products. It has been variously suggested that they may be converted into proteins, alkaloids, fats, or carbohydrates.

Plant	Actual acidity pH value estimate in normality	Total acidity by titration
Cranberry	0.004*N**	0.917*N*
Grapefruit	0.001*N*	0.192*N*
Apple	0.0004*N*	0.071*N*
Rhubarb (leaf stalk)	0.0006*N*	0.157*N*
Orange	0.00016*N*	0.094*N*
Pineapple	0.000035*N*	0.137*N*

* This means that a 0.004*N* acid if completely dissociated would give a pH reading equivalent to that observed.

The amount of titratable acidity and the value of the actual acidity of the cell sap of plants have been studied by a large number of investigators including Haas (1917); Atkins (1922); Smith and Quirk (1926); Martin, Rea, and Small (1926); and Dunne (1932).

Haas, who examined the more acid fruits, found the values of acidity for the cell sap as shown in the table on page 46.

Smith and Quirk (1926) determined the acidity of the cell sap of the shoots and leaves of *Begonia lucerna.* This plant is the most acid on record, the leaves being the most acid and the acidity decreasing from top to base. The juice of the leaves was twice as acid as lemon juice and practically as acid as 0.1 *N* hydrochloric acid. A considerable part of the acid is oxalic acid, although other acids are present. The hydrogen-ion concentration and total acidity of the sap of this plant ranged as follows: pH 0.9 to 1.36, 1.22 to 2.23, and 3.30 to 3.42 for the leaves, top, and base of shoots, respectively. The total acidity (undissociated salts) for the leaves, top, and base of shoots was estimated at +250 to 263, +99 to 140, and +41 to 56, respectively. The total acidity as here expressed in terms of Fuller's scale denotes the number of cubic centimeters of a *N* NaOH solution required per liter of juice to bring it from the given pH to pH 8.2 (phenolphthalein neutrality). Smith and Quirk also determined the acidity of the sap of other plants according to the following table:

Plant	Tissues	pH value	Total acidity	Plant	Tissues	pH value	Total acidity
Oxalis	Leaves	1.70	+224	Sugar cane	Soft top	4.88	+ 37
Rumex crispus	Leaves	3.53	+ 66	Bananas	Young petiole of midrib	4.74	+ 12
Wandering Jew	Leaves	5.47	+ 10				
Pineapple	Leaves	4.72	+ 40	White garlic	Bulbs	3.46	+165
Boston fern	Leaves	5.20	+ 38	Olives	Young shoots	5.22	+ 52
				Onions	Leaves	4.33	+ 30

The buffer systems in the cell sap have been studied by various workers. Small (1929) considered that inorganic phosphates, malates, oxalates, asparagine, and probably bicarbonates, act as buffers in the cell saps examined by him. During the growth of seedling wheat plants, Hurd-Karrer (1930) believed that asparagine was the substance in the sap of the etiolated seedlings that was responsible for the inflection of the titration curve near pH 8.0. The phosphates appeared to be the principal buffers between pH 6.0 and 7.5, while above pH 6.0 the buffering seemed largely due to the presence of an organic constituent. This was a solution of sodium malate alone or a combination of tartrate and citrate. Dunne (1932) considered that organic acids, amides, amino acids, phosphates, and sugars were, in the cases examined by him, the principal types of substances responsible for the buffering of cell saps.

2. Factors Influencing the Acidity of the Cell Sap.—Some of the factors that may influence the acidity of the cell sap are: (*a*) the reaction of the medium in which the plant grows, (*b*) the time of sampling, (*c*) previous treatment, (*d*) kind of tissue, and (*e*) age and vigor.

a. The Reaction of the Medium.—The effect of the changes in the soil solution upon the acidity of the cell sap is not yet definitely known, since the data which have been presented are very conflicting. Thus Truog and Meacham (1919) grew alfalfa, clover, lupine, sorghum, buckwheat, corn, and rape on unlimed acid soil and on limed acid soil. In 12 of the 16 cases the plant juices are more acid on the unlimed soils than on the limed ones and in the opposite cases the differences are quite small. Some of their data are shown in the following table:

Plant	pH value of juice from plants grown on	
	Unlimed soil	Limed soil
Alfalfa	5.78	6.03
Clover (*Trifolium pratense*)	5.69	5.83
Soybean	5.81	5.94
Corn	5.21	5.31
Buckwheat	4.02	4.05

Haas (1920) found that in most cases the effect of the limed soil on the sap of the tops of alfalfa, barley, garden beans, alsike clover, mustard, timothy, and wheat was to decrease the actual acidity. He considered that his data are in favor of the suggestion of Truog (1919) that the main specific harmful influence of soil acidity on certain plants is due to its influence in preventing the plant from securing rapidly enough the bases that are needed to neutralize and precipitate acids within the plant.

Bennet-Clark (1930) found that plants of *Crassula lactea* from cuttings of the same parent plant had different titratable acidities when grown on soils of different pH reactions. Thus the sap of plants which were grown in soils with a pH of 6.8 and 9.0, had titratable acidities respectively of 8.3 and 2.2 mg. equivalents per 100 grams fresh weight. Loehwing (1930) noted that the acidity of wheat plants grown on acid soils was much greater than that of plants grown on the same soils after the acidity was corrected by an application of calcium carbonate. Clevenger (1919), however, found that the acidity of the tops of oats, buckwheat, soybeans, and cowpeas grown in limed and unlimed acid sand was higher in those plants grown in the former medium, and his conclusions are the opposite of those obtained by Truog and Meacham (1919) and Haas (1920). Miyake and Masashi (1924) found that either phosphoric acid or potassium applied to the soil tends to raise the hydrogen-ion concentration of the cell sap of oats and flax, while nitrogen tends to lower it. More acid materials as determined by electrodialysis were found in tomato plants grown on acid soil and more basic materials in tomato plants grown in alkaline soils.

Hoagland and Davis (1925) found that the hydrogen-ion concentration of the cell sap of normal cells remained at a practically constant pH value of 5.2 when the cells were immersed in solutions of widely different reactions. They considered that the preservation of a definite reaction in the vacuolar sap may be essential to the normal functioning of the cell. Different types of cells may have different reactions so that the reactions of the cell sap expressed from a mass of tissue may not represent the reactions of any particular cell. Crozier (1919) noted that the cell sap of *Valonia* had a pH of 5.0 to 6.7, the average being about 6.0, while the pH value of sea water in which the plant grew ranged from 8.2 to 8.3. Masoni (1920) observed that juices of plants maintained an acid reaction several days after contact with calcareous soil or when in contact with an excess of pure calcium carbonate. Robertson and Smith (1931) reported that the acidity of the potato tuber is independent of environment.

b. The Time of Sampling.—According to Bennet-Clark (1933), the changes in the acidity of plant tissues are associated with diurnal and seasonal fluctuations in light intensity and temperature. The majority of succulent plants and some nonsucculent plants increase markedly in acidity during the night and decrease during the day. Kraus (1883) showed in *Bryophyllum calycinum* that this fluctuation in acidity was associated with a fluctuation of the carbohydrate content in the opposite direction. It appears that the nightly increase in acidity may be brought about by the formation of malic acid from sugar. Wolf (1931) showed, however, that the decrease in sugar content during the night was not parallel in most cases to the increase in the production of acid. Truog and Meacham (1919) found the juice of the alfalfa plant to be more acid in the morning than that of plants examined later in the day. In one case at 6 A.M. the pH value of the juice was 5.97 and at 2 P.M. 6.06. In another case at 8 A.M. the pH reading was 6.04 and at 2:30 P.M. it amounted to 6.11.

Clevenger (1919) followed the diurnal changes in the acidities of the leaves, stems, and roots of the cowpea during a 24-hr. period. The acidity of the leaves and stems was found to be highest in the morning and decreased during the day, while the acidity of the roots was found to be highest during the day.

Loehwing (1930) observed that the tissue fluids from entire wheat tops showed a diurnal acid periodicity in that the free acidity reached a minimum in the evening and a maximum in the early morning.

According to Reed and Halma (1926), the hydrogen-ion concentration of both lemon and orange trees shows a rather small amount of seasonal fluctuation. There is some increase in acidity, however, during the period of rapid vegetative growth when carbohydrates are presumably being extensively oxidized.

According to Rea and Small (1927) the pH of the pith and xylem of stem tissues tends to remain constant or to fluctuate only slightly throughout the year. The pH of the cortex, endodermis, and phloem, however, in most species tends to increase during the winter and decrease during the summer, the degree of variation differing with the species.

c. The Previous Treatment.—Gray and Ryan (1921) observed that the acidity of both navel and Valencia oranges was greatly reduced when a spray composed of soap, sodium carbonate, lead arsenate, and sulphur was applied to trees even for one season. This reduction in acidity amounted to more than 50 per cent in some cases and the arsenic compound seemed to be chiefly responsible for the effect, but the facts do not correlate the reduction of acidity with the local absorption of the spray by the fruit. The action of the spray is thought to be systemic, affecting the whole tree.

It was shown by Ruhland and Wetzel (1927) that a rich nitrogen manuring causes a considerable rise in the acidity of *Begonia* plants. Miller (1931) noted that the treatment of dormant potatoes with ethylene chlorohydrin invariably results in an increase in the pH value of the expressed juice. With the use of phosphate buffers, the addition of various amounts of ethylene chlorohydrin directly to the potato juice produces no change in the pH value until a concentration of 4 per cent is reached. Orange trees sprayed with lead arsenate produce a fruit juice which is less acid than that of the unsprayed fruit (Nelson and Mattern, 1932). According to Böning and Böning-Seubert (1932) the nutritional conditions have only a limited influence on the actual acidity of the sap of the tobacco leaf.

d. Kind of Tissue.—Gustafson (1924) showed that there was a hydrogen-ion concentration gradient in corn, squash, pole bean, pumpkin, and sunflower plants. This gradient is not always in the same direction in different species or in the leaves and stems of the same species. In corn, squash, and pole bean the older leaves had a higher hydrogen-ion concentration than the younger leaves, while in pumpkin and sunflower the reverse was the case. The bases of the stems of corn, sunflower, and pumpkin had a lower hydrogen-ion concentration than the tops.

Martin, Rea, and Small (1926) made a survey of 166 species of plants with especial regard to the young flowering stems and noted that the tissue reaction varied, in general, between pH 6.0 and 4.0. The tissue of the xylem and epidermis tended to be more acid than the others. Their work indicates that some tissues may remain constant in their reaction, while the reaction of others varies with the age of the plant.

Martin (1927) found in the sunflower that the cortex, root hairs, flower buds, and epidermis had pH values of 5.7, 4.2, 5.0, and 4.4, respectively. The epidermal hairs showed an alkaline reaction with a pH of 9.0 and 10.0. After examining almost 200 species of plants, Small (1929) concluded that in general the xylem tissue is more acid than the phloem tissue. Robertson and Smith (1931) noted that the acidity is not uniform throughout the potato tuber. In the early stages of development the underground stem and the basal end of the tuber are more acid than the other portions. When sprouting begins, the active buds have the most acid reaction.

e. Age and Vigor.—Hurd (1923) investigated the acidity of corn and its relation to vegetative vigor. She found that the concentration of titratable acid was always higher in the juice of the leaves than in that of the stalk regardless of the plant's vigor. The hydrogen-ion concentration was higher in the leaves than in the stalk in the vigorous plants only; in the stunted plants, it was greater in the stalks than in the leaves. The hydrogen-ion concentration of the tops of corn plants ranged from pH 5.0 to 5.6 and was inversely correlated with the degree of vegetative vigor induced by the environmental conditions affecting the different plots. The titratable acidity of these tops varied in the same manner, the values ranging from an average of 10 cc. of 0.05 *N* sodium hydroxide solution required to neutralize 10 cc. of juice from plants of the most stunted plot to 5 cc. required to neutralize the same quantity of juice from the most rapidly growing plants. The same author (1924) studied the course of the acidity changes in the cell sap of the wheat plant during different stages of growth. The acidity determinations were made on the sap at frequent intervals from the time the plants were 3 in. high until they were in the flowering stage at the age of about 6 months. The titratable acidity of the juice decreases progressively sometimes to one-half the initial concentration between the ages of 2 and 6 weeks. This period is followed by one of relatively low acidity extending up to the visible approach of maturity, when the acid concentration rises. The final acidity at maturity may be twice the highest seeding concentration and almost three times that of the least acid stage.

The hydrogen-ion concentration of the juice of the wheat plant does not decrease to any extent between the ages of 2 and 6 weeks. It is greatly increased during the preripening period and reaches a relatively high value at the flowering stage and later. This increasing acid concentration during the final stages of growth seems to be correlated primarily with the rate of drying rather than with head formation or kernel development.

Hurd (1924) observed that the electrometric titration curve of juice expressed from wheat plants changes progressively during the seedling stage and during the maturation period. Only minor changes, correlated with environmental factors, occur during the tillering stage and most of the shooting stage. The difference in the form of the titration curve for juice of seedlings in successive stages of development indicates differences in composition associated with increasing photosynthetic activity. The end of this period of change in the curve may be interpreted as the end of the seedling stage.

Gustafson (1924) showed that in *Zea mays*, *Cucurbita maxima*, *Helianthus sp.*, and *Bryophyllum calycinum* the total acid of the plant juice is not responsible for the hydrogen-ion concentration gradient found in plants and that there is no constant relation between total and actual acidity in these plants. The juice from young parts of plants requires more sodium hydroxide to neutralize it than does juice from older parts of the same plant, even when the former is nearer to the neutral point at the beginning than the latter.

Böning and Böning-Seubert (1932) stated that the actual acidity of the leaves of tobacco increases with age. Within a pH range of 3.0 to 6.0 the young leaves are more poorly buffered than are the fully developed older leaves.

Bibliography

AHMANN, C. F., and H. D. HOOKER, 1925: The estimation of pectin and a study of the constitution of pectin, *Mo. Sta. Res. Bull.* 77.

ALDABA, V. C., 1927: The structure and development of the cell wall in plants. I, Bast fibers of *Boehmeria* and *Linum*, *Am. J. Bot.*, **14**: 16–24.

ALLEN, C. E., 1901: On the origin and nature of the middle lamella, *Bot. Gaz.*, **32**: 1–34.

ALVARADO, S., 1923: Die Entstehung der Plastiden aus Chondriosomen in den Paraphysen von *Mnium cuspidatum*, *Ber. bot. Ges.*, **41**: 85–96.

ANDERSON, D. B., 1927: A microchemical study of the structure and development of flax fibers, *Am. J. Bot.*, **14**: 187–212.

———, 1927: Über die Struktur der Kollenchymzellwand auf Grund mikrochemischer Untersuchungen, *Sitzungsber. Akad. Wiss. Wien.*, Abt. I, **136**: 429–439.

———, 1928: Some recent work on the structure of the plant cell wall, *Ohio J. Sci.*, **28**: 299–314.

———, 1928: Struktur und Chemismus der Epidermus-Aussenwand von *Clivia nobilis*, *Jahrb. wiss. Bot.*, **69**: 501–515.

———, 1934: The distribution of cutin in the outer epidermal wall of *Clivia nobilis*, *Ohio J. Sci.*, **34**: 9–18.

———, 1935: The structure of the walls of the higher plants, *Bot. Rev.*, **1**: 52–75.

ANDERSON, E., 1931: Studies on the hemicelluloses. I, The evolution of carbon dioxide by plant materials and some hemicelluloses under the action of boiling 12 per cent hydrochloric acid, *J. Biol. Chem.*, **91**: 559–568.

——— and S. KINSMAN, 1931: Studies on the hemicelluloses. II, The composition of the hemicellulose from cotton seed hulls, *J. Biol. Chem.*, **94**: 39–49.

———, 1936: The isolation of pectic substances from wood, *J. Biol. Chem.*, **112**: 531–539.

Appel, O., 1906: Zur Kenntnis des Wundverschlusses bei den Kartoffeln, *Ber. bot. Ges.*, **24**: 118–122.

Armstrong, J. I., 1929: Hydrogen ion phenomena in plants. II, An investigation of the buffer complex of sap from the stems of *Pelargonium sp.*, *Protoplasma*, **8**: 313–343.

Artschwager, E., 1927: Wound-periderm formation in the potato as affected by temperature and humidity, *J. Agr. Res.*, **35**: 995–1000.

——— and R. C. Starrett, 1931: Suberization and wound-periderm formation in sweet potato and gladiolus as affected by temperature and relative humidity, *J. Agr. Res.*, **43**: 353–364.

——— and ———, 1933: Suberization and wound-cork formation in the sugar beet as affected by temperature and relative humidity, *J. Agr. Res.*, **47**: 669–674.

Astbury, W. T., T. C. Marwick, and J. D. Bernal, 1932: X-ray analysis of the structure of the wall of *Valonia*, *Proc. Roy. Soc.*, B, **109**: 443–450.

Astruc, A., 1903: Recherches sur l'acidité végétale, *Ann. Sci. Nat. Bot.*, **17**: 1–64.

Atkins, W. R. G., 1922: The hydrogen-ion concentration of plant cells, *Sci. Proc. Roy. Dublin Soc.*, **16**: 414–426.

———, 1922: The hydrogen-ion concentrations of some Indian soils and plant juices, *Agr. Res. Inst. Pusa Bull.* **136**: 1–12.

Bailey, I. W., and T. Kerr, 1935: The visible structure of the secondary wall and its significance in physical and chemical investigations of tracheary cells and fibers, *J. Arnold Arboretum*, **16**: 273–300.

Bailey, K., and F. W. Norris, 1932: The nature and composition of the mucilage of the seed of white mustard (*Brassica alba*), *Biochem. J.*, **26**: 1609–1623.

Baker, G. L., and R. Kneeland, 1935: Viscosity of apple pectin. Effect of commercial diastatic enzymes at 40°C., *Ind. and Eng. Chem.*, **27**: 92–94.

Balls, W. L., and H. A. Hancock, 1919: The existence of daily growth rings in the cell wall of cotton hairs, *Proc. Roy. Soc.*, B, **90**: 542–555.

———, 1923: Further observations on the cell wall structure as seen in cotton hairs, *Proc. Roy. Soc.*, B, **93**: 426–440.

Bamford, K. F., and W. G. Campbell, 1936: The determination of lignin in the analysis of woods, *Biochem. J.*, **30**: 419–428.

Beams, H. W., and R. L. King, 1935: The effect of ultracentrifuging on the cells of the root tip of the bean *Phaseolus vulgaris*, *Proc. Roy. Soc.*, B, **118**: 264–276.

Beck, W. A., 1929: The effect of drought on the osmotic value of plant tissue, *Protoplasma*, **8**: 70–126.

Beer, R., 1904: The present position of cell wall research, *New Phytol.*, **3**: 159–164.

Belzung, E., 1887: Recherches morphologiques et physiologiques sur l'amidon et les grains de chlorophylle, *Ann. Sci. Nat. Bot.*, VII, **5–6**: 179–311.

———, 1891: Nouvelles recherches sur l'origine des grains d'amidon et des grains chlorophylliens, *Ann. Sci. Nat. Bot.*, VII, **13**: 1–22.

Bennet-Clark, T. A., 1930: The metabolism of succulent plants, *thesis*, *Cambridge Univ.*

———, 1933: The role of organic acids in plant metabolism. I, *New Phytol.*, **32**: 37–71.

———, 1933: The role of organic acids in plant metabolism. III, *New Phytol.*, **32**: 197–230.

——— and W. M. Woodruff, 1935: Seasonal changes in acidity of the rhubarb (*Rheum hybridum*), *New Phytol.*, **34**: 77–91.

Bernard, C., 1878: "Leçons sur les phénomènes de la vie," 1st ed., 1878; 2d ed., 1885, Paris.

Bloch, R., 1935: Wound healing in *Tradescantia fluminensis*, *Ann. Bot.*, **49**: 651–670.

Böning, K., and E. Böning-Seubert, 1932: Wasserstoffionenkonzentration und Pufferung im Presssaft von Tobakblättern in ihrer Abhängigkeit von der Ernährung und Entwicklung der Pflanze. I, *Biochem. Z.*, **247**: 35–67.

Bonner, J., 1935: Zum Mechanismus der Zellstreckung auf Grund der Micellarlehre, *Jahrb. wiss. Bot.*, **82**: 377–412.

———, 1935: Some colloidal properties of the pectins, *Proc. Akad. Amsterdam*, **38**, No. 3: 1–11.

——— and A. N. J. Heyn, 1935: Cell elongation and the electrical properties of the cell wall, *Protoplasma*, **24**: 466–469.

———, 1936: The chemistry and physiology of the pectins, *Bot. Rev.*, **2**: 475–497.

Bowen, R. H., 1927: A preliminary report on the structural elements of cytoplasm in plant cells, *Biol. Bull.*, **53**: 179–195.

Brady, J., 1934: Some factors influencing lodging in cereals, *J. Agr. Sci.*, **24**: 209–232.

Bragg, W., 1930: Cellulose in the light of the X-ray, *Nature*, **125**: 315–322.

Brannon, M. A., 1913: Osmotic pressure in potatoes, *Bot. Gaz.*, **56**: 433–439.

Bredow, H., 1890: Beiträge zur Kenntniss der Chromatophoren, *Jahrb. wiss. Bot.*, **22**: 349–414.

Bristol, B. M., 1916: On the remarkable retention of vitality of moss protonema, *New Phytol.*, **15**: 137–143.

Brunstetter, B. C., and C. A. Magoon, 1930: A microelectrode for the rapid determination of the hydrion concentration of expressed juices from small amounts of plant tissue, *Plant Physiol.*, **5**: 249–256.

Burkhart, B. A., 1936: Hemicellulose constituents in alfalfa roots, *Plant Physiol.*, **11**: 421–428.

Buston, H. W., and V. H. Chambers, 1933: Some cell-wall constituents of *Cetraria islandica* ("Iceland moss"), *Biochem. J.*, **27**: 1691–1702.

———, 1935: Observations on the nature, distribution, and development of certain cell-wall constituents of plants, *Biochem. J.*, **29**: 196–218.

Cake, W. E., and H. H. Bartlett, 1922: The carbohydrate content of the seed of *Asparagus officinalis*, *J. Biol. Chem.*, **51**: 93–102.

Candlin, E. J., and S. B. Schryver, 1928: Investigations of the cell-wall substances of plants, with special reference to the chemical changes taking place during lignification, *Proc. Roy. Soc.*, B, **103**: 365–376.

Carre, M. H., and D. Haynes, 1922: The estimation of pectin as calcium pectate and the application of this method to the determination of the soluble pectin in apples, *Biochem. J.*, **16**: 60–70.

———, 1925: Chemical studies in the physiology of apples. IV, Investigations on the pectic constituents of apples, *Ann. Bot.*, **39**: 811–839.

———, 1925: The relation of pectose and pectin in apple tissue, *Biochem. J.*, **19**: 257–266.

Castoro, N., 1906: Beiträge zur Kenntnis der Hemicellulosen, *Z. physiol. Chem.*, **49**: 96–107.

———, 1907: Ueber die in den Samenschalen von *Cucurbita pepo* enthaltenen Hemicellulosen, *Z. physiol. Chem.*, **52**: 521–525.

Cavers, F., 1914: Chondriosomes (mitochondria) and their significance, *New Phytol.*, **13**: 96–106, 170–180.

Chandler, W. H., 1914: Sap studies with horticultural plants, *Mo. Sta. Res. Bull.* 14.

Chibnall, A. C., S. H. Piper, A. Pollard, J. A. B. Smith, and E. F. Williams, 1931: The wax constituents of the apple cuticle, *Biochem. J.*, **25**: 2095–2110.

CHRYSLER, H. L., 1934: Amounts of bound and free water in an organic colloid at different degrees of hydration, *Plant Physiol.*, **9**: 143–155.

CLAGUE, J. A., and C. R. FELLERS, 1934: Relation of benzoic acid content and other constituents of cranberries to keeping quality, *Plant Physiol.*, **9**: 631–636.

CLARK, G. L., 1930: Cellulose as it is completely revealed by X rays, *Ind. and Eng. Chem.*, **22**: 474–487.

CLAYSON, D. H. F., F. W. NORRIS, and S. B. SCHRYVER, 1921: The pectic substances of plants. Part II, A preliminary investigation of the chemistry of the cell walls of plants, *Biochem. J.*, **15**: 643–654.

———, and S. B. SCHRYVER, 1923: The hemicelluloses. I, The hemicellulose of wheat flour, *Biochem. J.*, **17**: 493–496.

CLEVENGER, C. B., 1919: Hydrogen-ion concentration of plant juices. I, The accurate determination of the hydrogen-ion concentration of plant juices by means of the hydrogen electrode, *Soil Sci.*, **8**: 217–226.

———, 1919: Hydrogen-ion concentration of plant juices. II, Factors affecting the acidity or hydrogen-ion concentration of plant juices, *Soil Sci.*, **8**: 227–242.

CONRAD, C. M., 1926: A biochemical study of the insoluble pectic substances in vegetables, *Am. J. Bot.*, **13**: 531–548.

———, 1930: A furfural-yielding substance as a splitting product of protopectin during the ripening of fruits, *Plant Physiol.*, **5**: 93–103.

———, 1931: Decarboxylation studies on pectins and calcium pectates, *J. Am. Chem. Soc.*, **53**: 1999–2003.

———, 1931: The decarboxylation of *d*-galacturonic acid with special reference to the hypothetical formation of *l*-arabinose, *J. Am. Chem. Soc.*, **53**: 2282–2287.

COWDRY, E. V., 1924: "General Cytology," University of Chicago Press.

———, 1926: Surface film theory of the function of mitochondria, *Am. Nat.*, **60**: 157–165.

CRAFTS, A. S., 1933: Sulfuric acid as a penetrating agent in arsenical sprays for weed control, *Hilgardia*, **8**: 125–147.

CRIST, J. W., and L. P. BATJER, 1931: The stone cells of pear fruits, especially the Kieffer pear, *Mich. Sta. Tech. Bull.* 113.

CROSS, C. F., and C. DORÉE, 1923: "Researches on Cellulose," Longmans, Green & Company, New York.

CROZIER, W. J., 1919: Intracellular acidity in *Valonia*, *J. Gen. Physiol.*, **1**: 581–585.

CULPEPPER, C. W., and H. H. MOON, 1933: Composition of rhubarb at different stages of maturity in relation to its use in cooking and canning, *J. Agr. Res.*, **46**: 387–402.

CZAJA, A. T., 1930, 1934: Untersuchungen über metachromatische Farbungen von Pflanzengeweben. I, *Planta*, **11**: 582*ff.*; II, **21**: 531–601.

DADSWELL, H. E., and L. F. HAWLEY, 1929: Chemical composition of wood in relation to physical characteristics. A preliminary study, *Ind. and Eng. Chem.*, **21**: 973–975.

DAMM, O., 1901: Ueber den Bau, die Entwicklungsgeschichte und die mechanischen Eigenschaften mehrjähriger Epidermen bei den Dicotyledonen, *Beih. bot. Cent.*, **11**: 219–260.

DAVIDSON, G. F., and W. A. RICHARDSON, 1936: The molecular structure of cellulose and starch, *Sci. Prog.*, **31**: 68–77.

DAVIDSON, J., and M. PHILLIPS, 1930: Lignin as a possible factor in lodging of cereals, *Sci.*, **72**: 401–402.

DIXON, H. H., and W. R. G. ATKINS, 1910: On osmotic pressure in plants, and on a thermo-electric method of determining freezing points, *Sci. Proc. Roy. Dublin Soc.*, n. s., **12**: 275–311.

——— and ———, 1912: Changes in the osmotic pressure of the sap of the developing leaves of *Syringa vulgaris*, *Notes Bot. School Trinity Col. Dublin*, **2**: 99–102.

——— and ———, 1912: Variations in the osmotic pressure of the leaves of *Hedera helix*, *Notes Bot. School Trinity Col. Dublin*, **2**: 103–110.

——— and ———, 1912: Variations in the osmotic pressure of the sap of *Ilex aquifolium*, *Notes Bot. School Trinity Col. Dublin*, **2**: 111–120.

——— and ———, 1913: Osmotic pressures in plants. I, Methods of extracting sap from plant organs, *Notes Bot. School Trinity Col. Dublin*, **2**: 154–165.

——— and ———, 1913: Osmotic pressures in plants. II, Cryoscopic and conductivity measurements on some vegetable saps, *Notes Bot. School Trinity Col. Dublin*, **2**: 166–172.

——— and ———, 1916: Osmotic pressure in plants. VI, On the composition of the sap in the conducting tracts of trees at different levels and at different seasons of the year, *Sci. Proc. Roy. Dublin Soc.*, **15**: 51–62.

———, and T. A. Bennet-Clark, 1932: Electrical properties of oil water emulsions, with special reference to the structure of the plasmatic membrane. II, *Sci. Proc. Roy. Dublin Soc.*, **20**: 211*ff*.

Dore, W. S., 1924: Recent progress in the chemistry of pectin and its industrial applications, *Ind. and Eng. Chem.*, **16**: 1042–1044.

Dorée, C. and E. C. Barton-Wright, 1927: Contributions to the study of lignin. I, Meta-lignin a new type of alkali lignin, *Biochem. J.*, **21**: 290–300.

Drabble, E., and H. Lake, 1905: The osmotic strength of cell sap in plants growing under different conditions, *New Phytol.*, **4**: 189–192.

———, and H. Drabble, 1907: The relation between the osmotic strength of cell sap in plants and their physical environment, *Biochem. J.*, **2**: 117–132.

Dunne, T. C., 1932: Plant buffer systems in relation to the absorption of bases by plants, *Hilgardia*, **7**: 207–234.

Eaton, F. M., 1927: The water requirement and cell-sap concentration of Australian salt brush and wheat as related to the salinity of the soil, *Am. J. Bot.*, **14**: 212–226.

Emmett, A. M., 1929: An investigation on the changes which take place in the chemical composition of pears stored at different temperatures, with special reference to the pectic changes, *Ann. Bot.*, **43**: 269–308.

Esselen, G. J., 1923: A few high spots in the chemistry of cellulose, *Ind. and Eng. Chem.*, **15**: 306–308.

Eyster, W. H., 1929: Variation in the size of plastids in genetic strains of *Zea mays*, *Sci.*, **69**: 48.

———, 1933: Plastid studies in genetic types of maize: argentia chlorophyll, *Plant Physiol.*, **8**: 105–121.

Faber, O., and B. Tollens, 1899: Untersuchungen über die Oxycellulose, *Ber. chem. Ges.*, **32**: 2589–2601.

Famintzin, A., 1893: Sur les grains de chlorophylle dans les graines et les plantes germeantes, *Bull. de l'Acad. imp. de Sciences de St. Petersbourg*, IV, **36**: 75–85.

Farr, W. K., and G. L. Clark, 1932: Cotton fibers. II, Structural features of the wall suggested by X-ray diffraction analyses and observations in ordinary and plane polarized light, *Contr. Boyce Thompson Inst.*, **4**: 273–296.

——— and S. H. Eckerson, 1933, 1934: The formation of cellulose membranes by microscopical particles of uniform size in linear arrangement, abst., *Am. J. Bot.*, **20**: 672; *Contr. Boyce Thompson Inst.*, **6**: 189–204.

——— and ———, 1934: Separation of cellulose particles in membranes of cotton fibers by treatment with hydrochloric acid, *Contr. Boyce Thompson Inst.*, **6**: 309–313.

——— and W. A. Sisson, 1934: X-ray diffraction patterns of cellulose particles and interpretation of cellulose diffraction data, *Contr. Boyce Thompson Inst.*, **6**: 315–321.

———, 1934: Cotton fibers. IV, Fiber abnormalities and density of the fiber mass within the boll, *Contr. Boyce Thompson Inst.*, **6**: 471–478.

Faul, J. H., 1928: Living cells in heartwood of trees, *Sci.*, **67**: 296.

Fink, B., 1925: Some considerations of protoplasm, *Ohio J. Sci.*, **25**: 99–113.

Fischer, F., and R. Lieske, 1928: Untersuchungen über das Verhalten des Lignins bei der natürlichen Zersetzung von Pflanzen, *Biochem. Z.*, **203**: 351–362.

Fitting, H., 1911: Die Wasserversorgung und die osmotischen Druckverhältnisse der Wüstenpflanzen, *Zeits. f. Bot.*, **3**: 209–275.

Franzen, H., and E. Keyssner, 1923: Kritisches über das Vorkommen der Äpfelsäure in der Pflanzen, *Biochem. Z.*, **135**: 183–216.

——— and F. Helwert, 1923: Kritisches über das Vorkommen der Citronensäure in der Pflanzen, *Biochem. Z.*, **135**: 384–415.

——— and ———, 1923: Kritisches über das Vorkommen der Weinsäure, *Biochem. Z.*, **136**: 291–305.

——— and R. Ostertag, 1923: Kritisches über das Vorkommen der Bernsteinsäure, *Biochem. Z.*, **136**: 327–335.

Fremy, M. E., 1840: Recherches sur la pectine et l'acide pectique, *J. de Pharm.*, **26**: 368–393.

———, 1859: Recherches chimiques sur la composition des cellules végétales, *Ann. d. Sci. Nat.*, IV, t. 12.

Freudenberg, K., F. Sohns, W. Durr, and C. Niemann, 1931: Ueber Lignin, Coniferylalcohol und Saligenin, *Cellulosechemie*, **12**: 263–276. Quoted by Harlow, 1932.

———, 1932: The relation of cellulose to lignin in wood, *J. Chem. Ed.*, **9**: 1171–1180.

Frey, A., 1926: Der heutige Stand der Micellartheorie, *Ber. bot. Ges.*, **44**: 564–570.

———, 1926: Die submikroskopische Struktur der Zellmembranen, *Jahrb. wiss. Bot.*, **65**: 195–223.

———, 1927: Der submikroskopische Feinbau der Zellmembran, *Die Natur. Wiss.*, **15**: 760–765.

———, 1928: La structure micellaire des membranes cellulaires, *Bull. Soc. Bot. France*, **75**: 950–956.

———, 1928: Über die intermicellar-Räume der Zellmembranen, *Ber. bot. Ges.*, **46**: 444–456.

Frey-Wyssling, A., 1930: Mikroskopische Technik der Micellaruntersuchung von Zellmembranen, *Z. wiss. Mikrosk.*, **47**: 1–46.

Gaddum, L. W., 1934: The pectic constituents of citrus fruits, *Fla. Sta. Bull.* 268.

Gail, F. W., 1926: Osmotic pressure of cell sap and its possible relation to winter killing and leaf fall, *Bot. Gaz.*, **81**: 434–445.

——— and W. H. Cone, 1929: Osmotic pressure and pH measurements on cell sap of *Pinus ponderosa*, *Bot. Gaz.*, **88**: 437–441.

Gardiner, W., and A. W. Hill, 1901: II, The histology of the cell wall with special reference to the mode of connection of cells, *Trans. Phil. Roy. Soc.*, B, **194**: 83–125.

Gerhardt, F., 1929: Effect of acid and alkaline hydrolysis on the estimation of hemicellulose and associated groups in young apple wood, *Plant Physiol.*, **4**: 373–383.

Gerritz, H. W., 1935: Extraction of pectin from apple thinnings, *Ind. and Eng. Chem.*, **27**: 1458–1459.

Gibbs, R. D., 1935: Studies of wood. I, The cell wall, *Can. J. Res.*, **12**: 715–726.

GILSON, E., 1890: La suberine et les cellules du liège, *La Cellule,* **6**: 63.

———, 1892: La cristallisation de la cellulose et la composition chimique de la membrane cellulaire végétale, *La Cellule,* **9**: 397–441.

GRAY, G. P., and H. J. RYAN, 1921: Reduced acidity in oranges caused by certain sprays, *Month. Bull. Dept. Agr. Calif.,* **10**: 11–33.

GREATHOUSE, G. A., 1932: Effects of the physical environment on the physico-chemical properties of plant saps and the relation of these properties to leaf temperature, *Plant Physiol.,* **7**: 349–390.

GRIS, A., 1857: Recherches microscopiques sur la chlorophylle, *Ann. Sci. Nat. Bot.,* **7**: 179–219.

GRÜSS, J., 1897: Studien über Reservecellulose, *Bot. Cent.,* **70**: 242–261.

GUILLIERMOND, A., 1911: Sur la formation des chloroleucites aux dépens des mitochondries, *C. R. Acad. Sci. Paris,* **153**: 290–292.

———, 1911: Sur les mitochondries des cellules végétales, *C. R. Acad. Sci. Paris,* **153**: 199–202.

———, 1911: Sur l'origine des leucoplastes et sur le processus cytologique de l'élaboration de l'amidon dans le tubercule de la pomme de terre, *C. R. Acad. Sci. Paris,* **153**: 1492–1494.

———, 1912: Sur les différentes modes de la formation des leucoplastes, *C. R. Soc. Biol. Paris,* **73**: 110–112.

———, 1914: Bemerkungen über die Mitochondrien der vegetativen Zellen und ihre Verwandlung in Plastiden. Eine Antwort auf einige Einwürfe, *Ber. bot. Ges.,* **32**: 282–301.

———, 1929: The recent development of our idea of the vacuome of plant cells, *Am. J. Bot.,* **16**: 1–23.

———, 1932: La structure de la cellule végétale: les inclusions du cytoplasme et en particulier les chondriosomes et les plastes, *Protoplasma,* **16**: 291–337.

———, 1932: La structure des cellules végétales à l'ultramicroscope, *Protoplasma,* **16**: 454–477.

GUSTAFSON, F. G., 1922: Studies of the hydrogen-ion concentration of plant juices, *Mich. Acad. Sci.,* **1922**: 49–53.

———, 1924: Hydrogen-ion concentration gradient in plants, *Am. J. Bot.,* **11**: 1–6.

———, 1924: Total acidity compared with actual acidity of plant juices, *Am. J. Bot.,* **11**: 365–370.

HAAS, A. R. C., 1916: The acidity of plant cells as shown by natural indicators, *J. Biol. Chem.,* **27**: 233–241.

———, 1917: The reaction of plant protoplasm, *Bot. Gaz.,* **63**: 232–235.

———, 1920: Studies on the reaction of plant juices, *Soil Sci.,* **9**: 341–370.

———, and F. F. HALMA, 1931: Sap concentration and inorganic constituents of mature citrus leaves, *Hilgardia,* **5**: 407–424.

HABER, E. S., 1928: The influence of soil reaction on the ionizable constituents of the tomato as determined by electro-dialysis, *J. Agr. Res.,* **37**: 101–114.

HABERLANDT, G., 1914: "Physiological Plant Anatomy," English Trans., 4th ed. Macmillan & Company, Ltd., London.

HALLER, M. H., 1929: Changes in the pectic constituents of apples in relation to softening, *J. Agr. Res.,* **39**: 739–747.

HALMA, F. F., and A. R. C. HAAS, 1928: Effect of sunlight on sap concentration of citrus leaves, *Bot. Gaz.,* **86**: 102–107.

HANSTEEN-CRANNER, B., 1914: Ueber das Verhalten der Kulturpflanzen zu den Bodensalzen. III, Beiträge zur Biochemie und Physiologie der Zellwand lebender Zellen, *Jahrb. wiss. Bot.,* **35**: 536–598.

———, 1919: Beiträge zur Biochemie und Physiologie der Zellwand und der plasmatischen Grenzschichten, *Ber. bot. Ges.*, **37**: 380–391.

———, 1922: "Zur Biochemie und Physiologie der Grenzschichten lebender Pflanzenzellen," Kristiania, 1922.

———, 1926: Untersuchungen über die Frage ob in der Zellwand lösliche Phosphatide vorkommen, *Planta Archiv. wiss. Bot.*, **2**: 438–445.

HARLOW, W. M., 1927: The chemical nature of the middle lamella, *N. Y. State Col. Forestry Syracuse Univ. Tech. Bull.* 21.

———, 1928: Contributions to the chemistry of the plant cell wall. II, Lignification in the secondary and tertiary layers of the cell wall of wood, *N. Y. State Col. Forestry Syracuse Univ. Tech. Bull.* 24.

———, 1932: Contributions to the chemistry of the plant cell wall. VI, Further studies on the location of lignin in the cell walls of wood, *Am. J. Bot.*, **19**: 729–739.

HARPER, R. A., 1919: The structure of protoplasm, *Am. J. Bot.*, **6**: 273–301.

HARRIS, J. A., J. V. LAWRENCE, and R. A. GORTNER, 1915: The osmotic pressure of the juices of desert plants, *Sci.*, **41**: 656–658.

——— and W. POPENOE, 1916: Freezing point lowering of the leaf sap of the horticultural types of *Persea americana*, *J. Agr. Res.*, **7**: 261–269.

——— and J. V. LAWRENCE, 1916: The cryoscopic constants of expressed vegetable saps as related to local environmental conditions in the Arizona deserts, *Physiol. Res.*, **2**: 1–49.

——— and ———, 1916: On the osmotic pressure of the tissue fluids of Jamaican Loranthaceae parasitic on various hosts, *Am. J. Bot.*, **3**: 438–455.

——— and ———, 1917: Cryoscopic determinations on tissue fluids of plants of Jamaican coastal deserts, *Bot. Gaz.*, **64**: 285–305.

——— and ———, 1917: The osmotic concentration of the sap of the leaves of mangrove trees, *Biol. Bull. Marine Biol. Lab. Woods Hole*, **32**: 202–211.

——— and ———, 1917: The osmotic concentration of the tissue fluids of Jamaican montane rain forest vegetation, *Am. J. Bot.*, **4**: 268–299.

——— and ———, 1917: Vegetable saps, *Carnegie Inst. Wash. Yearbook*, **16**: 131–132.

———, 1917: Osmotic concentration of tissue fluids in relation to geographical distribution, *Carnegie Inst. Wash. Yearbook*, **16**: 88–89.

———, R. A. GORTNER, and J. V. LAWRENCE, 1917: The relationship between the osmotic concentration of leaf sap and height of leaf insertion in trees, *Bull. Torrey Bot. Club*, **44**: 267–286.

———, 1918: On the osmotic concentration of the tissue fluids of phanerogamic epiphytes, *Am. J. Bot.*, **5**: 490–506.

———, R. A. GORTNER, and J. V. LAWRENCE, 1921: On the differentiation of the leaf-tissue fluids of ligneous and herbaceous plants with respect to osmotic concentration and electrical conductivity, *J. Gen. Physiol.*, **3**: 343–347.

———, J. V. LAWRENCE, and R. A. GORTNER, 1921: The osmotic concentration and electrical conductivity of the tissue fluids of ligneous and herbaceous plants, *J. Phys. Chem.*, **25**: 122–146.

———, R. A. GORTNER, W. F. HOFFMAN, J. V. LAWRENCE, and A. T. VALENTINE, 1924: The osmotic concentration, specific electrical conductivity, and chlorid content of the tissue fluids of the indicator plants of Tooele Valley, Utah, *J. Agr. Res.*, **27**: 893–924.

———, J. V. LAWRENCE, and Z. W. LAWRENCE, 1924: The chloride content of the leaf tissue fluids of Egyptian and upland cotton, *J. Agr. Res.*, **28**: 695–705.

———, 1925: The accumulation of chlorides in the leaf-tissue fluids of Egyptian cotton with the march of the season, *Proc. Soc. Exp. Biol. Med.*, **22**: 415–417.

———, C. T. Hoffman, and W. F. Hoffman, 1925: Sulfate content of the leaf-tissue fluids of Egyptian and upland cotton, *J. Agr. Res.*, **31**: 653–663.

———, W. F. Hoffman and others, 1925: The leaf-tissue fluids of Egyptian cotton, *J. Agr. Res.*, **31**: 1027–1035.

———, 1926: The relationship between the concentration of the soil solution and the physico-chemical properties of the leaf-tissue fluids of Egyptian and upland cotton, *J. Agr. Res.*, **32**: 605–649.

———, G. J. Harrison, and T. A. Pascoe, 1930: Osmotic concentration and water relations in the mistletoes, with special reference to the occurrence of *Phoradendron californicum* on *Covillea tridentata*, *Ecology*, **11**: 687–702.

——— and T. A. Pascoe, 1930: Further studies on the relationship between the concentration of the soil solution and the physico-chemical properties of the leaf-tissue fluids of cotton, *J. Agr. Res.*, **41**: 767–789.

———, ———, and I. D. Jones, 1930: Note on the tissue fluids of *Phoradendron juniperinum* parasitic on *Juniperus utahensis*, *Bull. Torrey Bot. Club*, **57**: 113–116.

———, 1934: "The Physico-chemical Properties of Plant Saps in Relation to Phytogeography. Data on Native Vegetation in its Natural Environment," 339 pp., University of Minnesota Press.

Hawley, L. F., and A. G. Norman, 1932: The differentiation of hemicelluloses, *Ind. and Eng. Chem.*, **24**: 1190–1194.

Haworth, W. N., 1929: "The Constitution of Sugars," London.

——— and E. L. Hirst, 1936: The chemistry of the carbohydrates and the glucosides, *Ann. Rev. Biochem.*, **5**: 81–100.

Haynes, D., 1914: The gelatinisation of pectin in solutions of the alkalies and the alkaline earths, *Biochem. J.*, **8**: 553–583.

Hein, I., 1926: Changes in plastids in variegated plants, *Bull. Torrey Bot. Club*, **53**: 411–419.

Herzog, R. O., 1925: Zur Erkenntnis der Cellulose Faser, *Ber. chem. Ges.*, **58**: 1254–1262.

Hess, K., 1928: Zur Frage des Aufbaues pflanzlicher Membrane, *Biochem. Z.*, **203**: 409–420.

Hibbard, R. P., and O. E. Harrington, 1916: The depression of the freezing point in triturated plant tissue and the magnitude of this depression as related to soil moisture, *Physiol. Res.*, **1**: 441–454.

Hill, T. G., 1908: Observations on the osmotic properties of the root hairs of certain salt marsh plants, *New Phytol.*, **7**: 133–142.

Hoagland, D. R., and A. R. Davis, 1925: Suggestions concerning the absorption of ions by plants, *New Phytol.*, **24**: 99–111.

Hooper, F. E., 1931: Studies on parenchymatous and vascular plant tissues, some analytical and specific gravity data, *Plant Physiol.*, **6**: 531–540.

Hopkins, E. F., and J. H. Gourley, 1930: Nitrogen fertilization and pectic materials in grapes, *Proc. Am. Soc. Hort. Sci.*, **27**: 164–169.

Horning, E. S., and A. H. K. Petrie, 1927: The enzymatic function of mitochondria in the germination of cereals, *Proc. Roy. Soc.*, B, **102**: 188–206.

Hurd, A. M., 1923: Acidity of corn and its relation to vegetative vigor, *J. Agr. Res.*, **25**: 457–469.

———, 1924: The course of acidity changes during the growth period of wheat with special reference to stem rust resistance, *J. Agr. Res.*, **27**: 725–735.

Hurd-Karrer, A. M., 1925: A concentration gradient in corn stalks, *J. Gen. Physiol.*, **9**: 341–345.

———, 1928: Changes in the buffer system of the wheat plant during its development, *Plant Physiol.*, **3**: 131–153.

———, 1929: Relation of leaf acidity to vigor in wheat grown at different temperatures, *J. Agr. Res.*, **39**: 341–350.

———, 1930: Titration curves of etiolated and of green wheat seedlings reproduced with buffer mixtures, *Plant Physiol.*, **5**: 307–328.

Ilin, V. S., and P. S. Nazarova, 1915: Osmotic pressure in roots and leaves of plants with respect to the moisture content of their habitats, *Bull. Acad. Imp. Sci. Petrograd.*, 6 ser., **9**: 749–768; *Exp. Sta. Rec.*, **36**: 733.

Iljin, W. S., 1929: Der Einfluss der Standortsfeuchtigkeit auf den osmotischen Wert bei Pflanzen, *Planta*, **7**: 45–58.

Jodidi, S. L., 1933: The presence of citric acid in the Alaska pea, *J. Am. Chem. Soc.*, **55**: 4663–4664.

Jungers, V., 1930: Recherches sur les plasmodesmes chez les végétaux, *La Cellule*, **40**: 7–81.

Kamerling, Z., 1897: Zur Biologie und Physiologie der Zellmembrane, *Bot. Cent.*, **72**: 49–54, 85–91.

Keller, B. A., 1920: The osmotic pressure of the cell sap of plants of different ecological types and different habitats, *Proc. Russ. Bot. Soc.*, **5**: 84–92.

Kerr, T., and I. W. Bailey, 1933: Composition of the so-called middle lamella in the cambium and its lignified derivatives, abst. *Am. J. Bot.*, **20**: 672.

——— and ———, 1934: Structure, optical properties, and chemical composition of the so-called middle lamella, *J. Arnold Arboretum*, **15**: 327–349.

Killian, C., and L. Faurel, 1934: Observations sur la pression osmotique des végétaux désertique et subdésertiques de l'Algérie, *Bull. Soc. Bot. France*, **80**: 775–778; *Biol. Abst.*, **9**: 1710.

Klason, P., 1911: "Beiträge zur Kenntniss der chemischen Zusammensetzung des Fichtenholzes," Berlin.

Kopaczewski, W., 1925: Sur la coagulation de la pectine, *Bull. Soc. Chim. Biol.*, **7**: 419–428.

Korstian, C. F., 1924: Density of cell sap in relation to environmental conditions in the Wasatch Mountains of Utah, *J. Agr. Res.*, **28**: 845–909.

Kozlowski, A., 1922: Critique de l'hypothèse des chondriosomes, *Rev. Gén. Bot.*, **34**: 641–658.

Kraus, G., 1883: Über die Wasserverteilung der Pflanzen. IV, Die Acidität des Zellsaftes, *Abh. naturf. Ges. zu Halle*, **16**: 141–205.

Küster, E., 1919: Über weissrandige Blätter und andere Formen der Buntblättrigkeit, *Biol. Cent.*, **39**: 212–251.

Lambrecht, E., 1926: Beitrag zur Kenntnis der osmotischen Zustandsgrössen einiger Pflanzen des Flachlandes, *Beitr. Biol. Pflanzen*, **17**: 87–136; *Biol. Abst.*, **5**: 394.

Lee, B., and J. H. Priestley, 1924: The plant cuticle. I, Its structure, distribution and function, *Ann. Bot.*, **38**: 525–545.

———, 1925: The plant cuticle. II, A macrochemical study, *Ann. Bot.*, **39**: 755–768.

Legg, V. H., and R. V. Wheeler, 1925: Plant cuticles. I, Modern plant cuticles, *J. Chem. Soc.*, **127**: 1412–1421.

——— and ———, 1929: Plant cuticles. Part I (continued), Modern plant cuticles, *J. Chem. Soc.*, **1929**: 2444–2449.

——— and V. W. Richard, 1929: Plant Cuticles. II, Fossil plant cuticles, *J. Chem. Soc.*, **1929**: 2449–2458.

Leontjew, H., 1927: Über das spezifische Gewicht des Protoplasmas. III, *Protoplasma*, **2**: 59–64.

LEPESCHKIN, W. W., 1928: The chemical and physical composition of protoplasm, *Sci.*, **68**: 45–48.

———, 1930: My opinion about protoplasm, *Protoplasma*, **9**: 269–297.

———, 1932: Influence of visible and ultraviolet rays on the stability of protoplasm, *Am. J. Bot.*, **19**: 547–558.

LEWITSKY, G., 1910: Über die Chondriosomen in pflanzlichen Zellen, *Ber. bot. Ges.*, **28**: 538–546.

———, 1911: Die Chloroplastenanlagen in lebenden und fixierten Zellen von *Elodea canadensis*, *Ber. bot. Ges.*, **29**: 697–703.

LINK, K. P., 1929: Chemical composition of corn (*Zea mays*) seedlings. I, The isolation of xylan and cellulose from the cell walls, *J. Am. Chem. Soc.*, **51**: 2506–2516.

———, 1929: The chemical composition of corn (*Zea mays*) seedlings. II, The isolation of a dextrin similar to the trihexosan obtained by the thermal depolymerization of potato starch, *J. Am. Chem. Soc.*, **51**: 2516–2522.

LIPMAN, C. B., 1934: Further evidence on the amazing longevity of bacteria, *Sci.*, **79**: 230–231.

LITVINOV, L. S., 1927: Chemical nature of the sap of plants, *Bull. Inst. Rech. Biol. and Sta. Biol. Univ. Perm.*, **5**: 311–329; *Biol. Abst.*, **4**: 45.

LIVINGSTON, L. G., 1933: The nature and distribution of plasmodesma in the tobacco plant, abst. *Am. J. Bot.*, **20**: 673.

———, 1935: The nature and distribution of plasmodesmata in the tobacco plant, *Am. J. Bot.*, **22**: 75–87.

LOEHWING, W. F., 1930: Effects of insolation and soil characteristics of tissue-fluid reaction in wheat, *Plant Physiol.*, **5**: 293–305.

LOMBARDI, L., 1932: Ricerche sulla pressione osmotica nelle pianta di *Morus alba* L., *Boll. Staz. Bachi-Gesilcolt. Ascoli-Piceno*, **10**: 11–28; *Biol. Abst.*, **8**: 872.

LUBIMENKO, V., 1926: Recherches sur les pigments des plastes sur la photosynthèse. I, La physiologie des plastes, *Rev. Gén. Bot.*, **38**: 307–328, 381–400.

———, 1927: Recherches sur les pigments des plastes sur la photosynthèse. II, Les pigments des plastes et leur transformation dans les tissues vivants de la plante, *Rev. Gén. Bot.*, **39**: 1–115.

———, 1928: Recherches sur les pigments des plastes sur la photosynthèse. III, La biologie de la photosynthèse, *Rev. Gén. Bot.*, **40**: 1–59.

LÜDTKE, M., 1931: Untersuchungen über Aufbau und Bildung der pflanzlichen Zellmembran und ihre stofflichen Komponenten, *Biochem. Z.*, **233**: 1–57.

———, 1931: Intermediate products of cellulose formation, *Cellulosechemie*, **12**: 307–310; *Chem. Abst.*, **26**: 4939.

———, 1935: Werden und Organisation der pflanzlichen Zellmembran, *Protoplasma*, **22**: 457–488.

LUTMAN, B. F., 1919: Osmotic pressures in the potato plant at various stages of growth, *Am. J. Bot.*, **6**: 181–202.

MA, R. M., 1930: Starch deposition in the sporogenous cells of certain mosses, *Bull. Torrey Bot. Club*, **57**: 525–533.

———, 1931: The chloroplasts of *Selaginella*, *Bull. Torrey Bot. Club*, **57**: 277–284.

MACDOUGAL, D. T., and H. A. SPOEHR, 1920: The components and colloidal behavior of plant protoplasm, *Proc. Am. Phil. Soc.*, **59**: 150–170.

———, 1926: Growth and permeability of century old cells, *Am. Nat.*, **60**: 393–415.

——— and F. L. LONG, 1927: Character of cells attaining a great age, *Am. Nat.*, **61**: 385–406.

——— and J. G. BROWN, 1928: Living cells two and one half centuries old, *Sci.*, **67**: 447–448.

——— and G. M. Smith, 1927: Long lived cells of the redwood, *Sci.*, **66**: 456–457.

Magistad, O. C. and E. Truog, 1925: The influence of fertilizers in protecting corn against freezing, *J. Am. Soc. Agron.*, **17**: 517–526.

Magness, J. R., 1920: Investigations in the ripening and storage of Bartlett pears, *J. Agr. Res.*, **19**: 473–500.

Mahood, S. A., and D. E. Cable, 1922: The chemistry of wood. IV, The analysis of the wood of *Eucalyptus globulus* and *Pinus monticola*, *J. Ind. Eng. Chem.*, **14**: 933–935.

——— and ———, 1922: Comparison of wood cellulose and cotton cellulose, *J. Ind. Eng. Chem.*, **14**: 727–731.

Malhotra, R. C., 1930: Microchemical study of hemicellulose in some plant cells with special reference to its distribution in the protoplasm, *J. Biochem.* (*Tokyo*), **12**: 341–349.

Mangin, M. L., 1888: Sur la constitution de la membrane de végétaux, *C. R. Acad. Sci. Paris*, **107**: 144–146.

———, 1889: Sur la présence des composés pectiques dans les végétaux, *C. R. Acad. Sci. Paris*, **109**: 579–582.

———, 1890: Sur la substance intercellulaire, *C. R. Acad. Sci. Paris*, **110**: 295–297.

———, 1890: Sur les réactifs colorants des substances fondamentales de la membrane, *C. R. Acad. Sci. Paris*, **111**: 120–123.

———, 1891: Observations sur la membrane cellulosique, *C. R. Acad. Sci. Paris*, **113**: 1069–1072.

———, 1893: Sur l'emploi du rouge de ruthénium en anatomie végétale, *C. R. Acad. Sci. Paris*, **116**: 653–656.

Marston, H. R., 1923: The azine and azonium compounds of the proteolytic enzymes, *Biochem. J.*, **17**: 851–859.

Martin, J. H., J. A. Harris, and I. D. Jones, 1931: Freezing-point depression and specific conductivity of sorghum tissue fluids, *J. Agr. Res.*, **42**: 57–70.

Martin, S. H., M. W. Rea, and J. Small, 1926: The reaction of plant tissues, *Brit. Assoc. Adv. Sci. Rept.*, **94**: 410.

———, 1927: The hydrion concentration of plant tissues. III, The tissues of *Helianthus annuus*, *Protoplasma*, **1**: 497–521.

Mason, T. G., 1920: On some factors affecting the concentration of electrolytes in the leaf sap of *Syringa vulgaris*, *Notes Bot. School Trinity Col. Dublin.*, **3**: 67–82.

Masoni, G., 1920: Acidity of plant roots and soil lime, *Staz. Sper. Agr. Ital.*, **53**: 421–437, *Exp. Sta. Rec.*, **49**: 121.

McBain, J. W., and D. A. Scott, 1936: Micellar structure as related to cellulose, *Ind. and Eng. Chem.*, **28**: 470–473.

McCool, M. M., and C. E. Millar, 1917: The water content of the soil and the composition and concentration of the soil solution as indicated by the freezing point lowerings of the roots and tops of plants, *Soil Sci.*, **3**: 113–138.

——— and M. D. Weldon, 1928: The effect of soil type and fertilization on the composition of the expressed sap of plants, *J. Am. Soc. Agron.*, **20**: 778–793.

McKay, E., 1935: Salt tolerance of *Ruppia maritima* in lakes of high magnesium sulfate content, *Plant Physiol.*, **10**: 425–446.

Mehta, M. M., 1925: Biochemical and histological studies of lignification. I, The nature of lignin; its physiological significance and its estimation in timbers, *Biochem. J.*, **19**: 958–978.

———, 1925: Biochemical and histological studies of lignification. II, Histological studies on the polysaccharides and aromatic constituents of the cell wall, *Biochem. J.*, **19**: 979–997.

MENKE, W., 1934: Chloroplasten-studien, *Protoplasma*, **21**: 279–298.
———, 1934: Chloroplasten-studien. 2, *Protoplasma*, **22**: 56–62.
MENON, S. R. K., 1935: The variation in the methoxyl and cellulose values of the fibre of the fruit of *Cocos nucifera* during its growth from infancy of the nut to its ultimate drying on the tree, *Biochem. J.*, **29**: 282–284.
MEYER, B. S., 1927: Studies on the physical properties of leaves and leaf saps, *Ohio J. Sci.*, **27**: 263–288.
———, 1929: Some critical comments on the methods employed in the expression of leaf saps, *Plant Physiol.*, **4**: 103–112.
MEYER, K. H., and H. MARK, 1928: Über den Bau der kristallinen Anteile der Zellulose, *Ber. chem. Ges.*, **61**: 593–614.
———, 1929: Die Chemie der Micelle und ihre Anwendung auf biochemische und biologische Probleme, *Biochem. Z.*, **208**: 1–31.
———, 1931: The molecular structure of the cell wall, *New Phytol.*, **30**: 1–11.
MEYERS, A., 1883: "Das Chlorophyllkorn in chemischer, morphologischer und biologischer Beziehung," Leipzig.
MIKOSCH, C., 1885: Über die Entstehung der Chlorophyllkörner, *Sitzungsber. Akad. Wiss. Wien.*, **92**: 168.
MILLER, E. C., 1911: The origin of the chloroplasts in the cotyledons of *Helianthus annuus*, *Bot. Gaz.*, **51**: 378–384.
MILLER, L. P., 1931: The effect of treatments with ethylene chlorhydrin on the pH of the expressed juice of potato tubers, *Contr. Boyce Thompson Inst.*, **3**: 321–335.
MITCHELL, E. M., 1930: A microchemical study of hemicelluloses of endosperms and cotyledons, *Am. J. Bot.*, **17**: 117–138.
MIYAKE, K., and A. MASASHI, 1924: On the influence of three manurial ingredients upon the hydrogen-ion concentration of the cell sap of oats and flax, *J. Biochem. (Tokyo)*, **4**: 317–321.
——— and Y. IMAI, 1935: Maternal transmission of mutated plastids in the Japanese morning glory, *Bot. Gaz.*, **96**: 571–574.
MOTTIER, D. M., 1918: Chondriosomes and the primordia of chloroplasts and leucoplasts, *Ann. Bot.*, **32**: 91–114.
———, 1921: On certain plastids with special reference to the protein bodies of *Zea, Ricinus* and *Conopholis*, *Ann. Bot.*, **35**: 349–365.
MURNEEK, A. E., 1929: Hemicellulose as a storage carbohydrate in woody plants with special reference to the apple, *Plant Physiol.*, **4**: 251–265.
NÄGELI, C., 1864: Über den inneren Bau der vegetabilischen Zellmembranen, *Sitzber. Bay. Akad. Wiss. München*, **1**: 282–323; **2**: 114–171.
NAYLOR, G. L., and B. RUSSELL-WELLS, 1934: On the presence of cellulose and its distribution in the cell walls of brown and red algae, *Ann. Bot.*, **48**: 635–641.
NELSON, E. K., and H. HASSELBRING, 1931: Some organic acids of wheat plants, *J. Am. Chem. Soc.*, **53**: 1040–1043.
———, and H. H. MATTERN, 1931: Some organic acids in barley, maize, and rye plants, *J. Am. Chem. Soc.*, **53**: 3046–3048.
NORMAN, A. G., 1929: The chemical constitution of the gums. I, The nature of gum arabic and the biochemical classification of the gums, *Biochem. J.*, **23**: 524–535.
———, 1929: The biochemistry of pectins, *Sci. Prog.*, **24**: 263–279.
———, and F. W. NORRIS, 1930: Studies on pectin. IV, The oxidation of pectin by Fenton's reagent and its bearing on the genesis of the hemicelluloses, *Biochem. J.*, **24**: 402–409.
———, 1933: Some aspects of the chemistry of the plant cell wall, *Sci. Prog.*, **28**: 229–245.

———, and S. H. Jenkins, 1933: A new method for the determination of cellulose based upon observations of the removal of lignin and other encrusting materials, *Biochem. J.*, **27**: 818–831.

———, 1934: The biological decomposition of pectin, *Ann. Bot.*, **43**: 233–243.

———, 1935: The hemicelluloses. I, Alcoholic sodium hydroxide as a pretreatment to extraction, *Biochem. J.*, **29**: 945–952.

——— and J. G. Shrinkhande, 1935: The hemicelluloses. II, The association of hemicellulose with lignin, *Biochem. J.*, **29**: 2259–2266.

———, 1936: The composition of some vegetable fibers with particular reference to jute, *Biochem. J.*, **30**: 831–838.

Norris, F. W., and S. B. Schryver, 1925: The pectic substances of plants. III, The nature of pectinogen and its relation to pectic acid, *Biochem. J.*, **19**: 676–694.

——— and I. A. Preece, 1930: Studies on hemicelluloses. I, The hemicelluloses of wheat bran, *Biochem. J.*, **24**: 59–66.

——— and C. E. Resch, 1935: The analysis of carbohydrates of the cell wall of plants. I, The relation between uronic anhydride content and furfuraldehyde yield, *Biochem. J.*, **29**: 1590–1596.

O'Dwyer, M. H., 1923: The hemicelluloses. III, The hemicellulose of American white oak, *Biochem. J.*, **17**: 501–509.

———, 1925: A note on the occurrence of a pectic substance in beech wood, *Biochem. J.*, **19**: 694–696.

———, 1928: Preliminary investigations on the constitution of the hemicelluloses of timber, *Biochem. J.*, **22**: 381–390.

Ohga, I., 1926: On changes of osmotic concentration in certain plants, *Bot. Mag. Tokyo*, **60**: 587–590.

Osterhout, W. J. V., 1906: The resistance of certain marine algae to changes in osmotic pressure and temperature, *Univ. Calif. Pub.* 2(8).

———, 1917: Tolerance of fresh water by marine plants and its relation to adaptation, *Bot. Gaz.*, **63**: 146–149.

Peacock, W. M., 1931: The healing of potatoes skinned during harvest as affected by temperature, humidity, and solar and sky radiation, *Proc. Am. Soc. Hort. Sci.*, **28**: 266–269.

Pfeiffer, M., 1933: Der osmotische Wert im Baum, *Planta*, **19**: 272–278; *Biol. Abst.*, **9**: 42.

Phillips, M., 1931: The isolation of normal propyl guaiacol as a degradation product of lignin, *Sci.*, **73**: 568–570.

———, 1934: Report on lignin, *J. Assoc. Off. Agr. Chem.*, **17**: 277–282.

———, 1934: The chemistry of lignin, *Chem. Rev.*, **14**: 103–170.

——— and M. J. Goss, 1935: The composition of the leaves and stalks of barley at successive stages of growth with special reference to the formation of lignin, *J. Agr. Res.*, **51**: 301–320.

Poirault, G., 1893: Recherches anatomiques sur les cryptogames vasculaires, *Ann. Sci. Nat. Bot.*, 7 ser., **18**: 113–256.

Politis, J., 1921: Du rôle du chondriome dans la formation des essences dans les plantes, *C. R. Acad. Sci. Paris*, **173**: 98–100.

Powell, D., 1925: The development and distribution of chlorophyll in roots of flowering plants grown in the light, *Ann. Bot.*, **39**: 503–515.

Preece, I. A., 1930: Studies on hemicelluloses. II, The hemicelluloses of maize cobs, *Biochem. J.*, **24**: 973–979.

———, 1931: Studies on hemicelluloses. IV, The proximate analysis of boxwood and the nature of its furfuraldehyde-yielding constituents, *Biochem. J.*, **25**: 1304–1319.

PRESTON, R. D., 1934: The organization of the cell wall of the conifer tracheid, *Phil. Trans. Roy. Soc. London*, **224**: 131–174.

PRIESTLEY, J. H., and A. IRVING, 1907: The structure of the chloroplast considered in relation to its function, *Ann. Bot.*, **21**: 407–413.

———, 1921: Suberin and cutin, *New Phytol.*, **20**: 17–29.

——— and L. M. WOFFENDEN, 1923: The healing of wounds in potato tubers and their propagation by cut sets, *Ann. App. Biol.*, **10**: 96–115.

——— and A. WORMALL, 1925: On the solutes exuded by root pressure from vines, *New Phytol.*, **24**: 24–38.

———, 1929: The biology of the living chloroplast. A critical abstract of Professor Lubimenko's review of recent Russian work, *New Phytol.*, **28**: 197–217.

PRINGSHEIM, H., and A. M. VON MERKALZ, 1919: Fermentversuche an Zelluloseabbauprodukten, *Z. physiol. Chem.*, **105**: 173–178.

——— and H. MAGNUS, 1919: Über den Acetylgehalt des Lignins, *Z. physiol. Chem.*, **105**: 179–186.

RANDOLPH, L. F., 1922: Cytology of chlorophyll types of maize, *Bot. Gaz.*, **73**: 337–376.

REA, M. W., and J. SMALL, 1927: The hydrion concentration of plant tissues. V, The tissues of *Vicia faba*, *Protoplasma*, **2**: 45–58.

——— and ———, 1927: The hydrion concentration of plant tissues. VI, Stem tissue reactions throughout the year, *Protoplasma*, **2**: 428–459.

REED, H. S., 1921: Growth and sap concentration, *J. Agr. Res.*, **21**: 81–98.

——— and F. F. HALMA, 1926: Further observations on the relations between growth and sap concentration in citrus trees, *J. Agr. Res.*, **32**: 1177–1195.

REINHARD, H., 1933: Ueber die Teilung der Chlorplasten, *Protoplasma*, **19**: 541–564.

RHODES, E., 1925: The chemical nature of the membrane of potato cork, *Biochem. J.*, **19**: 454–463.

RITTER, G. J., and L. C. FLECK, 1917: Chemistry of wood. V, Result of the analysis of some American woods, *J. Ind. Eng. Chem.*, **14**: 1050–1055.

———, 1925: The distribution of lignin in wood, *Ind. and Eng. Chem.*, **17**: 1194–1197.

ROBERTSON, I. M., and A. M. SMITH, 1931: A study of the hydrogen-ion concentration of the potato tuber, *Biochem. J.*, **25**: 763–769.

RUDOLPH, K., 1912: Chondriosomen und Chromatophoren, *Ber. bot. Ges.*, **30**: 605–627.

RUHLAND, W., and K. WETZEL, 1927, 1929: Zur Physiologie der organischen Säuren in grünen Pflanzen. III, V, *Planta*, **3**: 765–769; **7**: 503*ff*. Quoted by Bennet-Clark and Woodruff, 1935.

SANDERS, J. P., and F. K. CAMERON, 1933: Unit cell of cellulose in cotton stalks and cusps, *Ind. and Eng. Chem.*, **25**: 1371–1373.

SANDS, L., and W. Y. GARY, 1933: The hemicelluloses of mesquite wood, *J. Biol. Chem.*, **101**: 573–581.

——— and P. NUTTER, 1935: The hemicelluloses extracted from mesquite wood after chlorination, *J. Biol. Chem.*, **110**: 17–22.

SAYRE, J. D., and V. H. MORRIS, 1931: Use of expressed sap in physiologic studies of corn, *Plant Physiol.*, **6**: 137–148.

SCARTH, G. W., 1927: The structural organization of plant protoplasm, *Protoplasma*, **2**: 189–205.

———, R. D. GIBBS, and J. D. SPIER, 1929: Studies of the cell walls in wood. I, The structure of the cell wall and the local distribution of the chemical constituents, *Trans. Roy. Soc. Canada*, **23**: 269–279.

SCHAFER, E. R., and F. A. SIMMONDS, 1929: Physical and chemical characteristics of hemp stalks and of seed flax straw, *Ind. and Eng. Chem.*, **21**: 1241–1244.

Schellenberg, H. C., 1896: Beiträge zur Kenntniss der verholzten Zellmembran, *Jahrb. wiss. Bot.*, **29**: 237–266.

Schimper, A. F. W., 1885: Untersuchungen über die Chlorophyllkörper und die ihnen homologen Gebilde, *Jahrb. wiss. Bot.*, **16**: 1–246.

Schmidt, E. W., 1912: Pflanzliche Mitochondrien, *Progr. rei. Bot.*, **4**: 163–181.

Schorger, A. W., 1917: The chemistry of wood. I, Methods and results of analysis of some American species, *J. Ind. Eng. Chem.*, **9**: 556–561.

———, 1917: The chemistry of wood. II, Discussion of methods and results, *J. Ind. Eng. Chem.*, **9**: 561–566.

———, 1917: The chemistry of wood. III, Mannan content of the gymnosperms, *J. Ind. Eng. Chem.*, **9**: 748–750.

———, 1924: The constitution of cellulose, *Ind. and Eng. Chem.*, **16**: 1274–1275.

Schryver, S. B., and D. Haynes, 1916: The pectic substances of plants, *Biochem. J.*, **10**: 539–547.

——— and E. M. Thomas, 1923: The hemicelluloses. II, The hemicellulose content of starches, *Biochem. J.*, **17**: 497–500.

Schulze, E., 1892: Zur Chemie der pflanzlichen Zellmembranen. II, *Z. physiol. Chem.*, **16**: 387–438.

———, 1893: Zur Chemie der pflanzlichen Zellmembranen. III, *Z. physiol. Chem.*, **19**: 38–69.

——— and N. Castoro, 1902: Beiträge zur Kenntniss der Hemicellulosen. *Z. physiol. Chem.*, **37**: 40–53.

——— and ———, 1903: Beiträge zur Kenntniss der Hemicellulosen. II, *Z. physiol. Chem.*, **39**: 318–329.

——— and U. Pfenniger, 1910: Ueber das Vorkommen von Hemicellulosen in den Samenhülsen von *Pisum sativum* und von *Phaseolus vulgaris*, *Z. physiol. Chem.*, **68**: 93–108.

Schürhoff, P. N., 1924: "Die Plastiden," 225 pp., Figs. 57, Berlin.

Seifriz, W., 1920: Viscosity values as determined by micro-dissection, *Bot. Gaz.*, **70**: 360–387.

———, 1930: The alveolar structure of protoplasm, *Protoplasma*, **9**: 177–208.

———, 1931: The Spierer lens and what it reveals in cellulose and protoplasm, *J. Phys. Chem.*, **35**: 118–129.

———, 1931: The structure of protoplasm, *Sci.*, **73**: 648–649.

———, 1934: The origin, composition and structure of cellulose in the living plant, *Protoplasma*, **21**: 129–159.

———, 1935: The effects of acids and alkalies on the viscous and structural properties of protoplasm, *Physics*, **6**: 159–161.

———, 1935: The structure of protoplasm, *Bot. Rev.*, **1**: 18–36.

———, 1936: "Protoplasm," 584 pp., McGraw-Hill Book Company, Inc., New York.

———, 1936: Spierer lens and colloidal structure, *Ind. and Eng. Chem.*, **28**: 136–140.

Sharp, L. W., 1921: "Introduction to Cytology," 582 pp., McGraw-Hill Book Company, Inc., New York.

Shedd, O. M., 1915: Variations in mineral composition of sap, leaves, and stems of the wild grape vine and sugar-maple tree, *J. Agr. Res.*, **5**: 529–543.

Singh, L., 1922: A study of the relation of pectin and acidity in jelly making, *J. Ind. Eng. Chem.*, **14**: 710–712.

Skinner, H. J., 1932: Cellulose in industry, *Ind. and Eng. Chem.*, **24**: 694–704.

Small, J., 1929: "Hydrogen-ion Concentration in Plant Cell and Tissues," 421 pp., Protoplasma Monographien, II.

Smith, E. F. and A. J. Quirk, 1926: A begonia immune to crown gall with observations on immune or semi-immune plants, *Phytopath.*, **16**: 491–509.

Smith, O., 1933: The studies of potato storage, *N. Y. Sta.* (*Cornell*) *Bull.* 553.

Smith, W. W., 1935: The course of stone cell formation in pear fruits, *Plant Physiol.*, **10**: 589–612.

Sponsler, O. L., 1926: Molecular structure of plant fibers determined by X-rays, *J. Gen. Physiol.*, **9**: 677–695.

———, 1926: The structure of ramie cellulose as derived from X-ray data, *Colloid. Sym. Mono.*, **4**:174–202.

———, 1928: The molecular structure of the cell wall of fibers. A summary of X-ray investigations, *Am. J. Bot.*, **15**: 525–537.

———, 1929: Mechanism of cell wall formation, *Plant Physiol.*, **4**: 329–337.

———, 1930: Orientation of cellulose space lattice in the cell wall, *Protoplasma*, **12**: 241–254.

———, 1931: Orientation of cellulose space lattice in the cell wall. Additional X-ray data from *Valonia* cell wall, *Protoplasma*, **12**: 241–254.

———, 1933: The molecule in biological structures as determined by X-ray methods, *Quart. Rev. Biol.*, **8**: 1–30.

———, 1934: Rate of carbohydrate condensation on a cell wall, *J. Am. Chem. Soc.*, **56**: 1599–1601.

——— and W. H. Dore, 1936: X-ray studies on the structure of compounds of biochemical interest, *Ann. Rev. Biochem.*, **5**: 63–80.

Staehelin, M., 1919: Die Rolle der Oxalsäure in der Pflanze, *Biochem. Z.*, **96**: 1–49.

Stanfield, J. S., 1932: Osmotic pressure of leaves of *Pinus scopulorum* and certain environmental factors, *Bot. Gaz.*, **93**: 453–465.

Stange, B., 1892: Beziehungen zwischen Substratkonzentration, Turgor, und Wachsthum bei einigen phanerogamen Pflanzen, *Bot. Ztg.*, **50**: 253–259, 273–285.

Stern, A. L., 1904: The so-called hydro-cellulose, *J. Chem. Soc.* (London), **85**: 336–340.

Stevens, N. E., 1932: Thickness of cuticle in cranberry fruits, *Am. J. Bot.*, **19**: 432–435.

Stoddart, L. A., 1935, 1936: Osmotic pressure and water content of prairie plants, *Plant Physiol.*, **10**: 661–680; thesis, University of Nebraska, January, 1936.

Stone, W. E., 1932: The origin, development, and increase of chloroplasts in the potato, *J. Agr. Res.*, **45**: 421–435.

Stout, A. B., 1915: The establishment of varieties in Coleus by the selection of somatic variations, *Carnegie Inst. Wash. Pub.* **218**: 1–80.

Stratton, F. C., 1931: Changes in osmotic pressure of bananas during ripening, *Plant Physiol.*, **6**: 361–365.

Thatcher, R. W., 1922: The role of celluloses in plant life, *Sci.*, **55**: 80.

Thielman, M., and L. Berzin, 1927: Über den osmotischen Wert kultivierter Pflanzenzellen, *Arch. Exp. Zellforsch.*, **4**: 273–327; *Biol. Abst.*, **4**: 685.

Thiessen, R., 1932: Physical structure of coal, cellulose fiber, and wood as shown by Spierer lens, *Ind. and Eng. Chem.*, **24**: 1032–1041.

Thoday, D., 1933: Some physiological aspects of differentiation, *New Phytol.*, **32**: 274–287.

Tottingham, W. E., R. H. Roberts, and S. Lepkovsky, 1921: Hemicellulose of apple wood, *J. Biol. Chem.*, **45**: 407–414.

———, 1933: Character of hemicellulose in certain fruit trees, *Plant Physiol.*, **8**: 559–561.

TRUOG, E., and M. R. MEACHAM, 1919: Soil acidity. II, Its relation to the acidity of the plant juice, *Soil Sci.*, **7**: 469–475.

TUCKER, L. R., 1934: Soluble solids in the water melon, *Plant Physiol.*, **9**: 181–182.

TUPPER-CAREY, R. M., and J. H. PRIESTLEY, 1923: The composition of the cell wall of the apical meristem of stem and root, *Proc. Roy. Soc.*, B, **95**: 109–131.

——— and ———, 1924: The cell wall in the radicle of *Vicia faba* and the shape of the meristematic cells, *New Phytol.*, **23**: 156–159.

TUTIN, F., 1921: The behavior of pectin towards alkalis and pectase, *Biochem. J.*, **15**: 494–498.

———, 1923: A note on the hydrolysis of pectin, *Biochem. J.*, **17**: 83–84.

———, 1923: Pectin and its hypothetical precursor, protopectin, *Biochem. J.*, **17**: 510–514.

TWISS, W. C., 1919: A study of plastids and mitochondria in *Preissia* and corn, *Am. J. Bot.*, **6**: 217–235.

UBER, F. M., and T. H. GOODSPEED, 1935: Microincineration studies. I, Localization of inorganic elements in plant cell walls, *Proc. Nat. Acad. Sci.*, **21**: 428–433.

VAN WISSELINGH, C., 1888: Sur la pario des cellules subéreuses, *Arch. Néerl. des Sc. Exp. Nat.*, **22**: 253.

———, 1892: Sur la lamelle subéreuse et la subérine, *Arch. Néerl. des Sc. Exp. Nat.*, **26**: 304–353.

———, 1894: Sur la cuticularisation et la cutine, *Arch. Néerl. des Sc. Exp. Nat.*, **28**: 373–410.

VON FELLENBERG, T., 1918: Über die Konstitution der Pektinkörper, *Biochem. Z.*, **85**: 118–161.

VON SCHMIDT, M., 1903: Zur Kenntniss der Korksubstanz, *Sitzungsber. Akad. Wiss. Wien*, 112 abt., 11 b.; *Bot. Cent.*, **96**: 318.

———, 1910: Zur Kenntniss der Korksubstanz, *Monatshefte Chem.*, **31**: 347.

———, 1910: Zur Kenntniss der Korksubstanz. III, *Sitzungsber. Akad. Wiss. Wien*, **119**: 233–242.

WAKSMAN, S. A., and R. A. DIEHM, 1931: On the decomposition of hemicelluloses by microorganisms. I, Nature, occurrence, preparation, and decomposition of hemicelluloses, *Soil Sci.*, **32**: 73–95.

WARBURG, O., 1885: Ueber die Stellung der organischen Säuren im Stoffwechsel der Pflanzen, *Ber. bot. Ges.*, **3**: 280–289.

———, 1886: Über die Bedeutung der organischen Säuren für das Lebensprocess der Pflanzen, *Unters. bot. Inst. Tübingen*, **2**: 57–150.

WEBBER, H. A., 1929: Cellulose from corn stalks, *Ind. and Eng. Chem.*, **21**: 270–275.

WEIER, T. E., 1930: Notes on the plastid and other cytoplasmic bodies during sporogenesis and spermatogenesis in *Polytrichum commune*, *Proc. Nat. Acad. Sci.*, **16**: 536–543.

———, 1932: A comparison of the plastid with the Golgi zone, *Biol. Bull.*, **62**: 126–139.

———, 1933: A critique of the vacuome hypothesis, *Protoplasma*, **19**: 589–601.

———, 1933: On the structure of the *Anthoceros* plastid in reflected light, *Sci.*, **78**: 264–265.

WEIMER, J. L., and L. L. HARTER, 1921: Wound cork formation in the sweet potato, *J. Agr. Res.*, **21**: 637–647.

WELLS, S. D., 1929: Cellulose from cereal straws, *Ind. and Eng. Chem.*, **21**: 275–278.

WHITFIELD, C. J., 1932: Osmotic concentrations of chaparral, coastal sage brush and dune species of southern California, *Ecology*, **13**: 279–285.

WIDDOWSON, E. M., 1932: Chemical studies in the physiology of apples. XIII, The starch and hemicellulose content of developing apples, *Ann. Bot.*, **46**: 597–632.

Wiesner, J., 1877: Ueber das Vorkommen und die Entstehung von Etiolin und Chlorophyll in der Kartoffel, *Osterr. Bot. Z.*, **27**: 7–11.

Wood, F. M., 1924: Contributions to an investigation of the chemical nature of the cellulose membrane, *Ann. Bot.*, **38**: 273–298.

———, 1926: Further investigations of the chemical nature of the cell wall membrane, *Ann. Bot.*, **40**: 547–571.

———, 1931: Observations on the action of bromine on plant tissues, *Ann. Bot.*, **45**: 421–424.

Woodhead, N., 1934: Studies in growth and differentiation. V, Histological and metabolic changes during wound healing in *Kleinia articulata* Haw., *Ann. Bot.*, **48**: 467–480.

Wormall, A., 1924: The constituents of the sap of the vine, *Biochem. J.*, **18**: 1187–1202.

Wurdack, M. E., 1923: Chemical composition of the walls of certain algae, *Ohio J. Sci.*, **23**: 181–192.

Zirkle, C., 1926: The structure of the chloroplast in certain higher plants. I, II, *Am. J. Bot.*, **13**: 301–342.

———, 1927: The growth and development of plastids in *Lunularia vulgaris*, *Elodea canadensis* and *Zea mays*, *Am. J. Bot.*, **14**: 429–446.

———: 1929: Development of normal and divergent plastid types in *Zea mays*, *Bot. Gaz.*, **88**: 186–203.

———, 1937: The plant vacuole, *Bot. Rev.*, **3**: 1–30.

CHAPTER II

SOLUTIONS AND MEMBRANES IN RELATION TO THE PLANT CELL

In considering the relation of solutions and membranes to the physiological functions of the plant, the student should keep the two following salient facts clearly in mind as a background for the discussion: (1) Owing to the structural nature of the plant cell, all materials that enter or leave a plant or that move from one cell to another must pass through membranes. (2) Through the cell walls, protoplasm, and vacuoles of the living cells that compose a green plant, water is continuous from the exterior walls of the root hairs which come in contact with the soil particles to the exterior surface of the outer cells of the leaves which are exposed to the air. On this account and because membranes must be penetrated, all materials that enter a plant, whether from the soil or from the air, or that migrate from one part of the plant to another must do so dissolved in water. The entrance of water and dissolved substances into a plant and their translocation after they have once entered depend thus upon certain behaviors and relationships of membranes and solutions. In order to understand how materials enter a plant or how they migrate in it, one must have a considerable knowledge of the behavior of membranes and solutions, and for that reason it is the intention here to discuss in a brief and elementary manner some of the more important facts in regard to them.

I. SOLUTIONS

A. Nature

A solution may be defined as a homogeneous mixture of two or more substances having the same physical and chemical properties throughout, the proportion of whose components may be altered continuously within certain limits without producing any abrupt change in any property of the mixture. The term "true solution" is now frequently used to designate a mixture of the nature above described to distinguish it from a colloidal solution, which is a heterogeneous mixture occupying an intermediate position between a true solution on the one hand and the more pronounced heterogeneous mixtures called "suspensoids" on the other. The most important difference between a true solution and a colloidal solution is the difference in the degree of subdivision or dispersion of the dissolved substances. In a true solution the dissolved substance or substances are generally in the ionic or molecular condition and the magnitude of the particles is 1mμ[1] or less in diameter, while in a colloidal solution the degree of dispersion is not so great and the particles vary from 1mμ to 0.1μ in diameter. True solutions may be of three general types, gaseous, liquid, or solid; but in plant physiology we are concerned wholly with liquid solutions, and it is to this type that our attention will be devoted.

In a solution, that component which is present in the larger proportion is called the "solvent," while the component or components that are present in smaller amounts are termed the "solute" or "solutes." It is generally stated that the solute is dissolved in the solvent, but it should be remembered that this method of designation is purely arbitrary and does not necessarily imply that one component possesses greater solvent powers than the other. The case of a solution of two liquids that are miscible in all proportions, as alcohol and water, is illuminating in discussing this relation of solvent and solute. If we are considering a 1 per cent solution of alcohol in water, we say that the alcohol is dissolved in water, although it would be just as correct to say that the water is dissolved in the alcohol. If the concentration of alcohol is 51 per cent, it is spoken of as the solvent and the water is termed the solute. If there is a 50 per cent solution of alcohol in water, however, either liquid may be called the solvent and either the solute. Water is practically the only solvent that functions in plants, but the solutes concerned are numerous and include all the soluble inorganic salts that may be present in the soil, as well as oxygen, carbon dioxide, sugars, organic acids, and various other soluble organic compounds.

B. Concentration

There are numerous ways in which the concentration and composition of a solution may be expressed, but the four following are the ones most generally used.

1. In Terms of Percentage.—This method does not show the relative number of molecules of the solute that are present in the solution. Since many of the physical reactions of solutions are resultants of the relative number of molecules present in the solvent, the percentage method of stating the concentration of solutions is not a very desirable one.

2. In Terms of Gram-molecular Weight or Formula Weight per Liter of Solution.—A solution made up after this manner is termed a "molar," "formal," "molecular," "gram-molecular," or "molecular normal" solution. It is made by dissolving as many grams of a substance as there are units in its molecular weight so as to make 1 l. of solution.

[1] $1\mu = \frac{1}{1,000}$ mm. $1m\mu = \frac{1}{1,000,000}$ mm., or $\frac{1}{1,000}\mu$.

Thus, in the case of sulphuric acid (H_2SO_4), the molecular weight of which is 98.08, a molecular solution would contain 98.08 g. of sulphuric acid in 1 l. of solution. In the case of hydrochloric acid, the molecular weight of which is 36.46, a molecular solution would contain 36.46 g. of hydrochloric acid in 1 l. of solution.

3. In Terms of Equivalent Weight per Liter of Solution.—An equivalent solution is one that contains 1 g. atomic weight of hydrogen or its equivalent per liter. A solution made up after this manner is also termed an "equivalent normal solution." In the case of monovalent compounds, a molecular solution and equivalent solution are one and the same. In the case of bivalent compounds, however, an equivalent solution contains only one-half the molecular weight of the substance in grams dissolved in water so as to make 1 l. of solution. An equivalent solution of sulphuric acid would thus contain 49.04 g. of this acid in 1 l. of solution. All equivalent solutions, regardless of the substances that compose them, thus contain the same number of hydrogen ions or their equivalent.

4. In Terms of Osmotic Pressure.—In biological work the concentration of a solution is frequently stated in terms of the osmotic pressure that would be produced if the solution were separated from pure water by a perfectly semipermeable membrane. The value of this osmotic pressure is commonly expressed in units of atmospheric pressure.

C. Diffusion

One of the distinct characteristics of a solution is the property of diffusion or the ability of the molecules or ions of the solvent and solute to move more or less freely throughout it. For a clear understanding of the entrance and translocation of materials in plants, the following simple but important facts in regard to the behavior of solvent and solutes in a solution must be remembered: (1) The molecules or ions of the solutes and solvent move or diffuse constantly throughout the solution. (2) The molecules or ions of the solute and solvent move in the solution independently of each other. (3) When more than one solute is present in the solution, the molecules or ions of each solute move independently of the other solutes.[1] (4) More molecules or ions of a given solute move in a unit of time from the higher toward the lower concentration of the solute than in the opposite direction. (5) More molecules of the solvent move in a unit of time from the higher toward the lower concentration of the solvent than in the opposite direction. These properties of diffusion thus soon bring about a uniform distribution of the solute or solutes and solvent in any given volume of solution. These facts in regard to the

[1] This statement is not strictly correct. See Oswald (1891), Osborne and Jackson (1914), and Bowen (1931).

diffusion of the solute and the solvent can best be understood by considering a hypothetical case as diagramed in Fig. 4. Let us assume that we have two solutions *A* and *B* in which water is the solvent and sodium chloride and potassium nitrate the solutes. Solution *A* has a concentration 0.1 *N* sodium chloride and 0.3 *N* potassium nitrate, which makes a total concentration of four-tenths normal. Solution *B* has a concentration of 0.1 *N* potassium nitrate and 0.2 *N* sodium chloride, making a

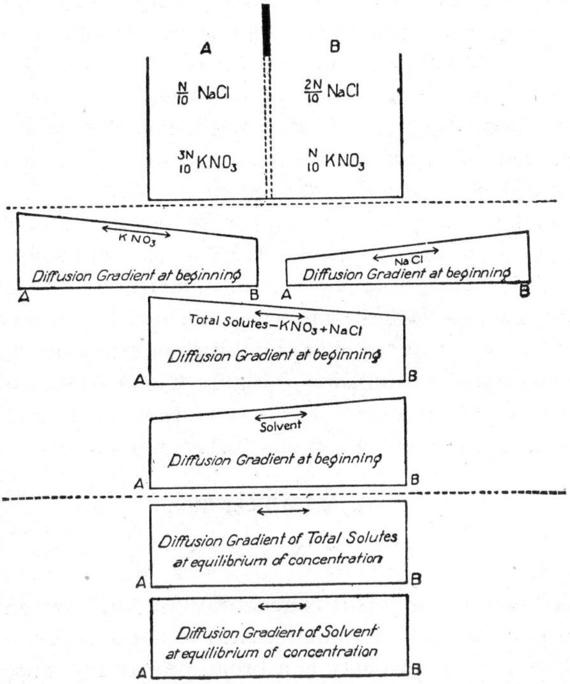

FIG. 4.—Diagrams showing the directions and relative rates of the diffusion of solvent and solutes. Description in the text.

total concentration of three-tenths normal. Now let solution *A* be placed in compartment *A* of a vessel and solution *B* in compartment *B* of the same vessel and the partition separating the two solutions be removed without disturbing them, thus allowing them to come in direct contact with each other. Let us now consider the behavior of the solutes and solvent when the two solutions thus come in contact.[1]

Let us consider first the behavior of the solutes. The number of molecules of sodium chloride moving from *A* to *B* in a unit of time is smaller than the number moving from *B* to *A*, since the concentration is greater in *B*. For the same reason the number of molecules of potassium

[1] For simplicity the two salts are considered here only in regard to their molecular concentration, and the fact of their ionization is not considered.

nitrate moving from *A* to *B* in a unit of time is greater than the number moving from *B* to *A*. The total number of molecules moving from *A* to *B* is greater in a unit of time than the number moving from *B* to *A*, since the total concentration of solutes in *A* (four-tenths normal) is higher than the total concentration of solutes in *B* (three-tenths normal). Let us now consider the behavior of the solvent. Since the total concentration of the solutes in solution *A* is equal to four-tenths normal and that in solution *B* to three-tenths normal, the concentration of the solvent in solution *A* is lower than the concentration of the solvent in solution *B*. The number of molecules of the solvent moving from *A* to *B* in a unit of time is therefore smaller than the number moving from *B* to *A*. These inequalities in the rates of diffusion of the solutes and solvent become less and less until the concentration of the two solutes becomes equally distributed throughout the solution. The independent rate of diffusion of the solutes and solvent of this solution can perhaps be better understood by considering the diagrams in Fig. 4, which represent the diffusion gradient or difference in concentration of each of the constituents. These diagrams show that the greater the difference in concentration of any component between two points or the steeper the diffusion gradient the more rapid the rate of movement or diffusion. Figure 4 also illustrates the rate of diffusion of the solutes and solvent between any two points in the solution after equilibrium of concentration has been reached.

II. MEMBRANES

A. Nature

The term "membrane" as it is generally used in relation to solutions means a thin sheet or layer of animal, plant, or artificial origin. The term as it is used in plant physiology is a broad one and includes on the one hand such definite tissues as the coats of seeds, scales of bulbs, or the epidermis of stems and leaves and, on the other hand, only the parts of the cell as the cell wall, the cytoplasmic layer, or even only the surface films of the latter bordering on the vacuole or on the cell wall. The consistency of these membranes thus varies from a semiliquid to a solid and ranges in thickness from a very delicate film that can be observed only with the high power of the microscope to a tissue whose thickness can easily be measured with the naked eye. In the performance of experiments to illustrate certain biological behaviors, artificial membranes of collodion, copper ferrocyanide, or parchment paper are frequently used, since it has been observed that artificial and natural membranes show in many cases striking similarities of behavior toward solutions.

Membranes differ greatly in their capacity for allowing substances to pass through them. A membrane is said to be permeable to a substance if it allows this substance to pass through and impermeable if it will not allow it to pass through. A membrane which will allow the solvent to pass through but not the solute is called a "semipermeable membrane." The capacity of a membrane for allowing substances to pass through it is termed its "permeability." Membranes differ not only in their permeability to a given substance but also in that their permeability is influenced by the general environmental factors, so that a membrane may have as many degrees of permeability toward a given substance as there are environmental conditions under which it may be placed. In speaking of the permeability of a membrane, therefore, it is necessary to define the system with which it is in contact.

None of the artificial membranes that have been made is perfectly semipermeable, although some of them prevent all but a trace of the solute from passing through. The nearest approach to a perfect semipermeable membrane is one of copper ferrocyanide in its relation to a solution of cane sugar in water. This membrane allows the water to diffuse through it readily but prevents almost completely the passage of the sugar. The cytoplasmic layer in plant cells behaves frequently as a more or less perfectly semipermeable membrane to various water solutions, although it is now considered that it is not nearly so semipermeable as formerly was supposed. There are some cases also where the cell walls behave as semipermeable membranes, as will be discussed later.

B. Imbibition of Water

The term "imbibition" as here used means the attraction of substances for water, a phenomenon that is especially characteristic of colloidal material in the gel condition. This process is best illustrated by the intake of water by dry wood, gelatin, starch grains, and dry seeds. Each particle of such substances has an attraction or affinity for water and will take it up from the surrounding medium against enormous pressures, which vary with the nature of the imbibing material and its water content (Shull and Shull, 1932). The water so attracted arranges itself around the particles in the form of films so that as the amount of water imbibed increases, the thickness of the watery films increases and thus the particles that compose the substance are crowded farther and farther apart. The limit to the distance that these particles may be separated from each other by the water thus taken up will depend upon the cohesive force of the particles of that substance. Thus in the case of wood the limit of separation is soon reached, while in the case of gelatin and similar substances the particles may go into colloidal solution. This phenomenon of imbibition of water is closely related to the behavior of membranes to watery solutions, since before there can be any passage of water, or of the solutes dissolved in it, through a membrane, the membrane must imbibe at least a certain amount of water. It is due to this characteristic imbibition of water by the protoplasm and the cell wall that water is continuous from the vacuole of one living cell to that of another. From the observed facts of the structure of the cell wall and the protoplasm and of the intake of water and of the migration of substances through them, we can crudely consider that these membranes are composed of a mixture

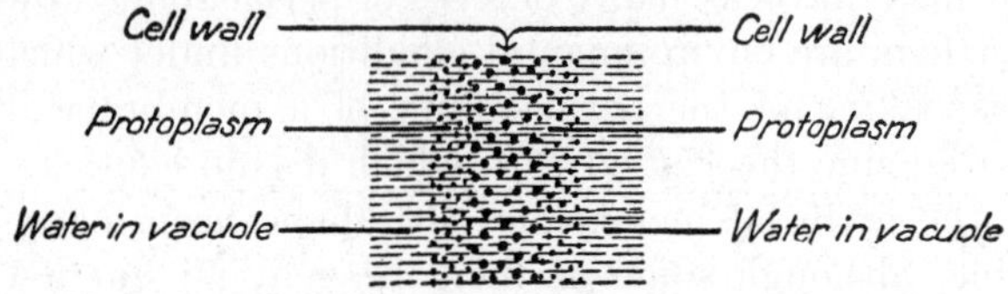

Fig. 5.—Diagram to illustrate that water is continuous from the vacuole of one cell through the protoplasm and cell wall to the vacuole of the adjoining cell. Description in the text.

of water and numerous and varied other substances. We can thus conceive that continuous waterways are provided along which the water-soluble substances may travel from cell to cell (Fig. 5).

C. Diffusion

When solutions are separated from each other or from the pure solvent by a membrane that is permeable to both the solvent and the solute or only to the solvent, certain phenomena occur which are not observable in the process of diffusion in solution when the membrane is not present. A discussion of this topic is necessary if we are to understand, in part at least, the intake, outgo, and translocation of materials in plants, since

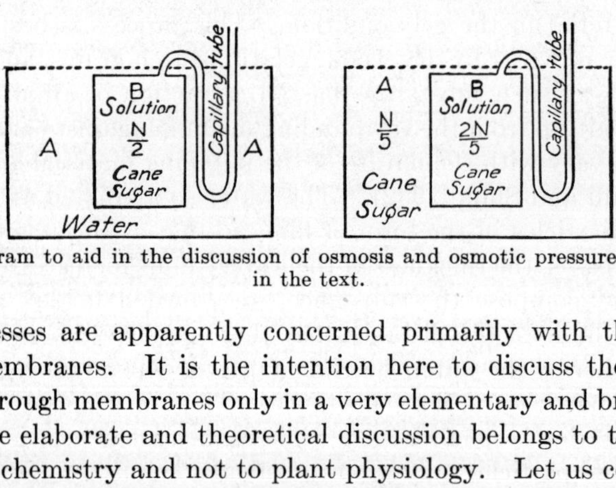

Fig. 6.—Diagram to aid in the discussion of osmosis and osmotic pressure. Description in the text.

these processes are apparently concerned primarily with the diffusion through membranes. It is the intention here to discuss the process of diffusion through membranes only in a very elementary and brief manner, since a more elaborate and theoretical discussion belongs to the province of physical chemistry and not to plant physiology. Let us consider first the behavior of the solvent and solute where the solution is separated from the pure solvent by a semipermeable membrane. In Fig. 6 let us consider that a half-normal solution of cane sugar completely fills a container *B*, the wall of which, or some portion of it, behaves as a semipermeable membrane to the sugar solution (in case only a portion of the wall is semipermeable, it is here assumed that the remaining portion is impermeable to both the solvent and the solute). Let us further suppose that a fine capillary tube be connected with the container in such a manner as to function as a manometer and that the vessel containing this sugar solution is immersed in distilled water in a larger container *A*. The solute will tend to diffuse very rapidly from *B* to *A* just as if no membrane intervened, for the diffusion gradient is steep, but the egress of the sugar from the vessel is prohibited, since the membrane is impermeable to it and equilibrium of concentration of the solute on both sides of the membrane cannot occur. Since the membrane is permeable to the solvent, water will readily move through it, diffusing more rapidly from *A* to *B* than it will from *B* to *A*, since the concentration of the solvent in *A* is higher than it is in *B*. The behavior of the solvent is thus similar to that which is attained when two solutions of unequal concentration are brought together without the intervention of a membrane, but with this difference, that the solute cannot diffuse into the pure solvent exterior to the membrane, so that the concentration of the solvent in *A* will always remain higher than it is in *B*.

Since water enters *B* more rapidly than it leaves, a hydrostatic pressure will be set up in *B* as indicated by the rise of the column of mercury in the manometer. Water will continue to move from *A* into *B* with a corresponding increase in the hydrostatic pressure until this pressure equals the force with which water tends to enter *B*. In the case of the half-normal solution of cane sugar, the hydrostatic pressure thus set up in *B* at 0°C. and 760 mm. of mercury will be approximately 11.2 atmospheres if the conditions are ideal and the membranes and walls of *B* are capable of sustaining such a pressure.

This discussion now brings us to the consideration of osmosis and osmotic pressure. The term "osmosis" has been used with such a variety of meanings by workers in biology that one is in doubt many times as to what it really signifies (Stark, 1928). The physicist and the physical chemist always consider an ideal condition for the illustration of this process similar to that just mentioned in the preceding discussion in which the solution is separated from the pure solvent by a perfectly semipermeable membrane. They would thus define osmosis as the passage of a solvent from one side of a semipermeable membrane to the other when the escaping tendency (or free energy) of the solvent on the two sides is unequal. The osmotic pressure would then be defined as the pressure difference which must be established upon the solution and pure solvent, respectively, in order to make the escaping tendency of the solvent the same for both of them. Osmotic pressure can be defined also as the equivalent of the hydrostatic pressure produced when solution and solvent are separated by a perfectly semipermeable membrane. It may likewise be expressed as the equivalent of the excess pressure that must be imposed on a solution to prevent the passage into it of the solvent through a perfectly semipermeable membrane (Findlay, 1913). Beck (1928) defines osmotic pressure as the difference of pressure on solution and solvent which produces a condition of equilibrium such that there is no tendency of the solvent to flow in either direction.

Let us now consider a case of diffusion through a membrane in which such ideal conditions do not prevail, since in plants a solution may be separated not only from the pure solvent but also from another solution not by a perfectly semipermeable membrane but by one that is permeable to both the solvent and solutes. This condition prevails in the cells of plants to as great extent as or greater than does semipermeability of membranes. For an illustration of the phenomena that occur under these conditions, let us assume (Fig. 6) that an apparatus *B* similar to that just described, but with a membrane permeable to both the solute and solvent, is filled with a two-fifths normal solution of cane sugar in water and immersed in a larger vessel *A* containing a fifth-normal solution of cane sugar in water. The solute under this condition will diffuse more rapidly from *B* to *A* than it will from *A* to *B* and thus tend to equalize the concentration of the solute on each side of the membrane just as it would act if the two solutions were in contact without the intervention of the membrane. The water diffuses more rapidly from *A* to *B* than it does from *B* to *A*, since the concentration of the solvent in *B* is lower than in *A*. The behavior of the solvent is thus similar to that which would occur if the two solutions were placed in contact without the intervening membranes, but with this difference that water as a general rule diffuses through a membrane at a more rapid rate than the solutes so that a hydrostatic pressure will be set up in *B* as indicated by the rise of the mercury in the manometer. This hydrostatic pressure, however, is only temporary, for as the concentration of the solute in *A* and *B* becomes more nearly equal, the solvent on both sides of the membrane more nearly reaches equilibrium and the hydrostatic pressure in *B* is diminished accordingly. When the concentration of the solute and solvent have become, respectively, the same in *A* and *B*, the rate of diffusion of the solvent from *A* to *B* is the same as it is from *B* to *A*, and as a result the hydrostatic pressure on the inside of *B* is reduced to zero.

It is evident from the above example that the phenomenon of osmosis and osmotic pressure may be exhibited when the membrane is not semipermeable. It would thus appear that osmosis should not be defined only in terms of semipermeable membranes but in terms of permeable membranes as well. Osmosis could thus be defined as the passage of the solvent from one side of a membrane to another where the escaping tendency of the solvent on the two sides is unequal. Because the diffusion of a salt solution through a membrane is similar to the diffusion of the water in which it is dissolved, there are those who believe that osmosis should be defined in terms of both the solvent and solute. Thus Stark (1928) proposed that osmosis be defined as diffusion through a membrane, the direction of the major movement being from a region of high concentration to a region of low concentration of the thing diffusing.

If the fifth-normal solution of cane sugar were placed in *B* and the two-fifths normal solution in *A*, the rate of water movement from *B* to *A* would be greater than from *A* to *B*, and as a result water would be withdrawn from *B*. Under such a condition the walls of *B* would contract or collapse unless they were strong enough to sustain the negative pressure developed therein. This phenomenon frequently prevails in plant cells under certain conditions, and their behavior under these conditions will be discussed in detail when the subject of plasmolysis is considered. The term "endosmosis" is used to designate the flow of either solute or solvent through the membranes of the plant cell into the vacuole, while the term "exosmosis" is used to designate the flow of solvent or solute outward from the cell vacuole. The connection of the phenomenon of osmosis with the water relations of the plant cell will be discussed in detail under the heading, Osmotic Relationship of the Plant Cell.

D. Membranes of the Plant Cell

With but few exceptions there are in each fully developed living plant cell two well-defined membranes through which materials must pass in moving from the vacuole or vacuoles of one cell to those of another or in entering the vacuoles of the external cells of the plant body from the medium in which the plant is growing. These two membranes are the cell wall and the cytoplasmic layer.

1. The Cell-wall or Nonliving Membrane.—With the exception of certain zoospores, sperms, unfertilized eggs, and in a few other cases, the protoplasm of plant cells is always enclosed by a definite compact solid membrane called the "cell wall." This membrane is frequently termed the "nonliving" membrane of the plant cell. The constituents that may enter into the composition of the cell wall have been named and described in considerable detail in Chap. I, so the intention here is to discuss only the general relationship of the cell wall to watery solutions.

a. Permeability to Water.—The cell wall is, for the most part, composed of an aggregation of colloidal particles in the gel condition, and like other gels it exhibits the power of imbibition and absorbs water when placed in contact with it. The amount of water thus absorbed before imbibitional saturation is reached varies greatly according to the constituents of the cell wall. Thus in the case of cutin and suberin the amount of water imbibed is very small, while in the case of the pectic substances the amount absorbed may be so great as practically to throw these substances into colloidal solution. Since all cell walls exhibit the power of imbibing water, it follows that none of them is absolutely impermeable to water. The suberized and cutinized cell walls are the least permeable to water, while the cell walls of embryonic tissue, root hairs, and parenchyma cells apparently offer the minimum retardation to water of any of the cell walls. The cell walls that are the most permeable to water, however, offer some resistance to water flow, and water never diffuses so readily through a cell-wall membrane as it would were there no membrane present. The principal factors that influence the rate of water movement through cell-wall membranes are the same as those that influence the behavior of practically all membranes. These main factors are the constitution of the membrane, the direction of flow, the concentration of the solutions in contact with it, and the temperature. The differences in the degree of permeability of the cell walls of different species of plants to water are shown by the work of Denny (1917) with seed coats and with the outer bulb scale of the onion. These membranes consist entirely of dead cells, so that with the possible exception of some nonliving material within the cavities of the cells, the behavior of the membranes toward the diffusion of water through them is due entirely to the nature of the walls of the cells composing them. The amount of water passing through equal areas (19.63 sq. mm.) of these membranes in a unit of time, when the external solution consisted of a saturated solution of sodium chloride with an osmotic pressure of approximately 375 atmospheres, is shown in the following table:

Membrane	Number of observations	Average amount of water passing per hour, milligram
Seed coat of grapefruit	3	0
Seed coat of *Cucurbita pepo*	4	4.0
Seed coat of *Cucurbita maxima*	5	9.3
Seed coat of cocklebur	4	18.2
Outer bulb scale of onion	6	24.0
Seed coat of almond	7	88.5
Seed coat of peanut	7	565.1

Denny investigated the nature of the constituents of these cell walls in order to determine if possible the cause of their differential permeability to water. His method consisted in determining the rate of flow of water through a given membrane and then after treating it with a given solvent to measure again the rate of water flow through it. The heating of the membranes in boiling water for 5 min. increased the permeability of the peanut and almond membranes but not that of the seed coats of the grapefruit or squash. The increase in permeability of the peanut coat averaged approximately 168 per cent, while the increase in permeability of the almond amounted to as much as 500 per cent. The treatment of these membranes by hot alcohol, hot acetone, and hot ether, which are solvents of fatty substances, increased greatly the permeability of the peanut, almond, squash, and cocklebur coats but not the grapefruit seed coats. The maximum increase in permeability of these membranes was 871, 242, 452, and 138 per cent, respectively, for the squash, peanut, almond, and cocklebur. After some of the membranes had been extracted with the fat solvents, they were immersed in the extract and exposed to evaporation so that they would again contain the materials that had been removed from them. Although the permeability of these membranes was reduced by this infiltration, it did not reach the low point exhibited by the membranes before their extraction. This fact indicates that the extracted materials which influenced the permeability of these membranes to water are distributed in an organized manner throughout the cell walls. This assumption is also strengthened by the fact that the permeability of some of these membranes is increased by simply heating them. Denny considered that the substances in the cell walls of the membranes mentioned above which are factors in determining their permeability to water are fatty substances and tannins. There was no evidence from his experiments that suberin plays any part in determining the permeability of these membranes. The squash seed coat, for example, is very resistant to the passage of water yet it has no distinctly suberized layers. Denny considered that the resistance of the seed coats of grapefruit to the passage of water is due to the presence in certain portions of the membranes of a large amount of fatty substances surrounded by thick pectinized walls. Denny made certain observations which showed that some of these membranes under consideration are more permeable to water in one direction than in another. Thus, for example, when the membrane of the almond was placed in an osmometer so that the water passed through the membrane, as it naturally would in entering the seed from the outside, the rate of flow was 48.5 mg. per hour. If the position of the membrane was reversed so that the water entered, as it would in passing from the seed to the external medium, the rate of flow was only 42.6 mg. per hour. In the case of the peanut membrane, the maximum percentage decrease of water flow from the "in" to the "out" amounted to 56 per cent. No difference in the rate of penetration of water in opposite directions could be observed in the case of the outer bulb scale of the onion and the seed coat of *Dioon edule*. This difference in the behavior of membranes is apparently correlated with their structure. The peanut and almond coats are composed of distinct layers and have surfaces of different physical and chemical natures, while the onion scale and *Cycad* membranes are apparently uniform in structure. A difference in the rate of water flow through membranes is characteristic of many living and nonliving animal membranes as well as many artificial ones.

According to Brauner (1930) the seed coat of *Aesculus hippocastanum* is 60 per cent more permeable when water passes from the outside to the inside than if in the opposite direction. The epidermis is the positive pole and a potential difference of several millivolts can be observed between the inner and the outer surface of the swollen membranes. The testa of this seed can thus be regarded as a "double membrane" containing electrolytes. This membrane consists of two layers differing in permeability and therefore the two sides are of different diffusion potentials. Because of this difference of potential, water flow across the membrane is promoted or hindered as the hydrodynamic potential varies with or against the water flow. The behavior of "double" membranes may have an important influence on the movement of water into and out of the plant cell and Denny asks the important question: "If this difference in the rate of diffusion is due to the presence of double membranes of different nature, or to the difference in surface on opposite sides, may not the plant cell itself show a difference in permeability in opposite directions, since such a system of double membranes is represented by the cell wall and ectoplast?"

Denny further made the very important observation in regard to the membranes with which he worked that when solutions of varying concentrations are placed on opposite sides of a membrane the relation between the rate of water movement and concentration difference is complex and that, in general, equal osmotic differences do not necessarily produce equal rates so that no mathematical relation can be noted between the concentration on opposite sides and the rate of water movement through the membranes.

b. Permeability to Solutes.—It has generally been supposed that if the cell wall were permeable to water, it would also be permeable to any solute dissolved in water. It has been shown, however, that there are numerous cell walls that are impermeable to various solutes, and this fact must be taken into consideration in studying the permeability of the plant cell. Brown (1907, 1909, 1915) studied especially the semipermeability of the covering of the barley grain. His first observation in this regard was that the blue pigment in the cells of the aleurone layer did not change in color even after the grains were soaked for several days in 1 per cent sulphuric acid, but, if the covering of the grain were punctured or otherwise injured, the pigment turned red at once, showing that the sulphuric acid had passed through the membrane. He further found that the seeds which had been soaked in the 1 per cent sulphuric acid absorbed water, swelled, and sprouted just as they did in distilled water, although as soon as the sprout ruptured the seed integument, the acid entered and turned the pigment red. It was further found that, when these seeds were placed in 4.9, 7.6, 9, 18, or even 36 per cent sulphuric acid, no acid entered after 44 hr., and these same seeds upon being washed and planted grew vigorously. The permeability of this membrane was studied in regard to copper sulphate, ferrous sulphate, potassium chromate, silver nitrate, and potassium ferrocyanide in a 5 per cent water solution. The covering was found to be semipermeable to all these, taking up water but as far as could be observed preventing the entrance of these solutes into the seed. The membrane was found to be permeable to mercuric chloride, acetic, formic, propionic, and butyric acids, ammonia, ethyl alcohol, and ethyl acetate among others. The degree of permeability to these substances, however, varied greatly. Ethyl alcohol, for example, entered at the same rate approximately as water, while acetic acid entered more rapidly, and ethyl acetate entered the most rapidly of the three. The seeds were subjected to boiling water for a 1-hr. period, but the results observed after this process were practically identical with those mentioned above, a fact which shows that this property of semipermeability lies in the cell walls of the covering of the grain and not in the protoplasm, since the latter was undoubtedly killed by this treatment. Brown also studied the rate of movement of various phenols through this seed membrane and concluded that when the osmotic pressures, vapor pressures, and viscosities of this series of solutions of soluble solutes are equal, their rate of diffusion through this membrane are inversely proportional to their surface tensions. The earlier work of Brown was followed by that of Schröder (1911), who found that the wheat grain possesses the same type of selective permeable coat as the barley grain, and by Tjebbes (1912), who observed that a part of the inner coat of the seed of the sugar beet shows the power of selective permeability. Becquerel (1907) showed that the dried seed coats of numerous plants are impermeable to various gases and to such penetrating substances as chloroform, absolute alcohol, and ether.

Shull (1913) extended the field of observation in regard to the permeability of seed coats and found that those of cocklebur, numerous grains, peach, apple, bean, and sunflower act as semipermeable membranes. His observations, however, were confined chiefly to the behavior of the seed coat of the cocklebur. This coat, he found, is composed of three rather definite layers, the outer of which does not function as a semipermeable membrane, the middle of which is several cells thick, and the inner of which is composed of a single row of cells. Both the middle and inner layers have osmotic properties, but neither functions so well separately as when they are together. The property of semipermeability in the case of the cocklebur coat is not dependent upon the protoplasmic membrane of the cells which compose it, since the same results are obtained after boiling the seeds or the membrane after it has been separated therefrom.

Shull tried the effect of soaking the dry cocklebur seeds in ether, chloroform, acetone, and absolute alcohol and found that no detectable amount of these substances penetrated through the seed coat and that the viability of the seed was not injured by this treatment. Thus, for example, seeds soaked in ether for 2 to 36 days and then dried and soaked in water and placed under germinative conditions after the testas were removed showed 87 per cent germination and the plants resulting were as vigorous as the controls. Shull investigated the permeability of this seed coat to approximately 40 different salts and found that sodium chloride, copper sulphate, potassium chromate, sodium thiosulphate, glycerol, hydrochloric acid, sugars, and tartaric acid were excluded while all the other compounds including silver nitrate, sodium nitrate, potassium nitrate, potassium chloride, ferrous sulphate, potassium hydroxide, sodium hydroxide, sulphuric, nitric, acetic, lactic, and citric acids were allowed to pass through, although the degree of permeability to these substances varied with the substance.

Experiments with a large number of seed types in 1.5 per cent solutions of potassium nitrate and manganese sulphate showed that when the seed coverings were intact any salts absorbed by the seeds were held superficially so that heavy losses in the amount of these salts resulted when the seeds were washed for 1 min. in water. When the seed coats were injured, a large amount of salts was absorbed but only a slight amount could be recovered by washing. Whitten (1914) found that the membranes of the corn seed are very impermeable to kerosene oils. Thus the grain may be immersed in kerosene for periods of 10 to 20 days without injury to the embryo.

The results obtained by Brown, Shull, Schröder, and others explain those reported by Giglioli (1895), who obtained a 20 per cent germination of alfalfa seed that had been in a saturated solution of mercuric chloride in absolute alcohol for 16 years.

Considerable work has been done to determine the structure, composition, and location of the portions of the seed coverings that prevent or retard the entrance of water and solutes into the seed. Collins (1918) considered that only a small part of the water absorbed by the grain of barley enters by the general surface of the grain, which he considered to be enveloped by cells with cuticularized cell walls. He believed that most of the water enters the micropyle or germinal region of the grain. It is here that the structure which is the seat of permeability must be sought, since he found that the solutes enter the grain as does the water through this restricted region. According to Brown (1931) the first intake of water by the seeds of *Lolium perenne* is through the micropyle. This diffuses upward and causes the endosperm to swell, which in turn induces greater permeability of the cuticular membranes of the seed coat. Although these membranes retard the absorption of water, their permeability is increased by the swelling and stretching of the seed.

In a study of the development of the testa of the wheat grain, Pugh, Johann, and Dickson (1932) noted the presence of suberin- or cutinlike compounds in the cell walls on the outer surface of one of the layers composing the testa. According to Hamly (1932) the impermeable region of the seed coat of *Melilotus alba* is formed by a layer of tightly appressed suberin layers.

The semipermeability of the grain of wheat is identified with the single layer of testa, and the cuticlelike membranes on its inner and outer surfaces (Brown, 1932). Microchemical tests showed that the testa is mucilaginous and that these cuticlelike membranes contain more true fat than does the cuticle of the normal foliage leaf.

In the seed the factors regulating the absorption of the solute would seem to be (1) electrical absorption in the cuticle-like membrane, (2) mechanical absorption in the testa and the tissues of the endosperm and embryo, (3) imbibitional pressure developed in the endosperm, and (4) the size of the intermolecular spaces of the semipermeable membranes. The solute will react to each of the above factors according to its own particular properties. It is thus extremely difficult to formulate general laws to cover the absorption of all chemical compounds.

Lavison (1910, 1911) studied the entrance of salts through the cell walls of the endodermis of the root. He found that the young cellulose periclinal walls of the endodermis are impermeable to the salts of iron, lead, copper, nickel, and mercury among others and that they behave in this regard like the protoplasm itself. We thus have a case in one of the most active translocation regions of the plant where the cell walls act in a semipermeable manner. Brooks (1917) found that the exterior cell wall of the epidermis of the inner surface of the bulb scale of the onion behaves as a semipermeable membrane toward a watery solution of calcium chloride, sodium chloride, aluminum chloride, sodium hydroxide, Bordeaux red, and eosin but that it allows a slight diffusion of hydrochloric acid through it. It is a striking fact that semipermeability to these substances is confined only to the exterior wall of the epidermal cell, while the other walls of the cell are permeable to them. The work that has thus far been done in studying the permeability of the cell-wall membrane makes it very evident that it is necessary to consider its influence in the interpretation of the permeability of plant tissues.

2. The Protoplasmic, Cytoplasmic, or Living Membrane.—One of the outstanding differences between the plant cell and the animal cell is the amount and distribution of the protoplasm within these units. An animal cell is composed almost entirely of protoplasm, and the growth of the cell simply means an increase in the amount of the living matter, which thus increases concurrently with the size of the cell. In the embryonic plant cell, the protoplasm occupies the whole space enclosed by the cell wall, but as the cell enlarges the protoplasm does not increase in amount proportionately but spreads out in a thin film around the inside of the cell wall and the outside of the vacuole which is developed under these conditions. A plant cell may increase in volume as much as 500 times or more and yet contain little more protoplasm than it did before it began to enlarge. This thin film of protoplasm thus forms a membrane through which materials entering or leaving the vacuole must pass and has been termed the "living membrane," the "protoplasmic membrane," or the "cytoplasmic membrane," the last term being used because the cytoplasm composes the greater portion of it.

a. Nature.—The protoplasmic membrane has more of the characteristics of a liquid than of a solid and evidently consists of a colloidal suspension and a true solution of numerous substances. It possesses considerable power of cohesion, is more or less elastic, and is stretched or held in its position by the pressure of the cell sap, as is evidenced by its rapid contraction inwardly as soon as the pressure within the vacuole is sufficiently reduced. Since the protoplasmic membrane is more or less liquid in its nature, it has been assumed that it exhibits certain surface phenomena which must here be discussed.

The surface of a liquid differs in physical properties from its general mass due to the fact that the molecules at the surface are not attracted equally on all sides. Any molecule throughout the general mass of the liquid is attracted equally in all directions by the molecules distributed around it and this mutual attraction results in keeping the pressure on any particular molecule uniform in every direction. A molecule at the surface of a liquid, however, is subjected to the attraction of the molecules in the interior and those adjoining it on the surface, but there is no attractive force at the exterior to balance this. As a result of this the surface layers of molecules are subjected to an inwardly directed attraction or pull at right angles to the surface. This phenomenon is called "surface tension" and because of it the surface of a liquid differs from its general mass in its behavior as a thin film or skin. Since the protoplasmic membrane is more or less of a liquid, it is generally assumed that its surfaces bordering on the cell wall and vacuole, respectively, exhibit characteristic surface-tension phenomena and also a difference in the degree of permeability as compared with the general mass of protoplasm.

The existence and nature of the surface layers of protoplasm have evinced much discussion, and probably no other topic in biophysics has been the subject of so much controversy. Some workers question its very existence, while others positively assert that an actual morphological and physiological definite surface layer can be clearly demonstrated. De Vries (1884, 1885) claimed that it has a morphological entity, Kite (1915) calls it a "hypothetical structure," while Fischer (1915) named it a "figment of the imagination," but Seifriz (1921) considered that the direct evidence from microdissection indicates a definite membrane on the surface of all protoplasm and considered that its thickness is about 0.1μ. The names used to designate this portion of the protoplasm are no less varied and numerous than the controversies concerning its existence. It has been called "surface layer," "surface film," "ectoplast," "phase boundary," "ectosarc," "hyaloplasm," and "plasma membrane." The term "plasma membrane" has been the one most commonly used. It was first used by Pfeffer (1877, 1890) to denote a surface layer or film of protoplasm that he considered to behave differently in regard to permeability from the general mass of the protoplasmic membrane. The term as it has been used by many authors and investigators is very loosely defined. In some cases it is employed only in the sense of a surface-tension film, in others it means a differentiated surface layer of the protoplasm of considerable depth, and by some it has been used to include the entire protoplasmic membrane. De Vries (1884) used the term "ectoplast" to denote the surface film of cytoplasm in contact with the cell wall and the term "tonoplast" to denote the surface bordering on the vacuole. Stiles (1924) suggested that the limiting surface of the protoplasm next to the cell wall be called the "external plasma membrane" and the surface bounding the vacuole the "internal plasma membrane."

Plowe (1931) in a study of plasmolyzed and unplasmolyzed cells concluded: (1) there is an external layer of the cytoplasm, the "plasma lemma" which, while fluid, is more elastic and more extensile than the general mass; (2) a similar but less elastic layer, the tonoplast, surrounds the vacuole; (3) the "mesoplasm" is much less extensile than the two membranes between which it lies; (4) the differential permeability of the protoplasm to water and dyes is a property of the surface layers rather than the mesoplasm; (5) the presence of differentially permeable membranes at the surfaces of the readily permeable mesoplasm makes the vacuolated plant cell a complex system, the behavior of which rests on osmotic as well as colloidal phenomena.

We must conclude therefore from Seifriz (1921), Stiles (1924), and Plowe (1931) that the surface of the protoplasmic membrane is not an invariable structure but will vary with the change in composition of the protoplasm and the external medium. The limiting surface of the protoplasmic membrane thus cannot be separated from the rest of the protoplasm, for it depends for its existence on and is closely related to the media that it separates, and it passes over into the phases on either side of it, sometimes abruptly and sometimes gradually, depending upon the environmental conditions and the composition of the mass of the cytoplasm and the surrounding medium.

The differences between the outer and inner surfaces of the protoplasmic membrane have been studied in two general ways:

1. *A Measurement of the Potential Difference between the Outer and Inner Layers.*—The works of Osterhout, Damon, and Jacques (1927), Osterhout and Harris (1928), Brooks and Gelfan (1928), Blinks (1929, 1932, 1935), Damon (1932), and Osterhout (1936) on the cells of *Valonia*, *Nitella*, and *Halicystis* show that when both surfaces of the protoplasmic membrane are in contact with the same solution there is a potential difference between them so that a current tends to flow through the electrometer from the positive to the negative surface. This potential difference ranges from a few to as many as 80 millivolts. The value of this potential difference, as well as the location of the negative and positive surfaces, is determined by the type of plant, the medium in which it is placed, and by the metabolism. Some of the substances or factors that may alter or even reverse the potential difference of the protoplasmic surfaces are distilled water, anesthetics, alcohol, other organic compounds, *e.g.*, guanidine and adrenaline, and temperature.

2. *A Comparison of the Effects of Chemicals upon the Two Surfaces.*—This comparison has been made by immersing plant tissues into chemicals, and by the injection of these same chemicals into the cells of another lot of tissue and noting any differences in effect upon the protoplasm. Thus Jacques and Osterhout (1928) found in *Nitella* and *Valonia* that the inner layer of the protoplasm was more sensitive to the action of chloroform than the outer layer. When manganese chloride was injected into the cells of *Valonia macrophysa*, they lived approximately one-half as long as when these cells were immersed in the same concentration of solution. This difference in longevity was not due to the injury by the needle used in injection since the immersed cells were stabbed in the same manner as the injected ones.

Skeen (1930) found that when certain compounds were injected into the root hairs of *Trianea bogotensis* they perished, but when the hairs were immersed in the same solution of these compounds they were uninjured for a considerable period of time. Osterhout and Harris (1928) found evidence in *Nitella* that the protoplasm in this case is made of layers differing considerably in their properties, each having a death curve of simple and regular form. These data indicate that the exterior portion or boundary of the protoplasm of these cells, under the conditions of this experiment, was of a different structure and composition from that portion bordering on the cell sap.

Whether or not the surface layers of the protoplasmic membrane differ in permeability from the greater portion of its bulk is of comparatively little moment. Many of the phenomena of diffusion that occur in plants, owing to the separation of solutions by the protoplasmic membrane, can be explained, partially at least, by assuming that as a unit it behaves as a permeable or impermeable membrane, as the case may be, without considering whether its qualities of permeability reside in its surface layers or in its general mass.

b. Permeability.—From our knowledge of the nature of protoplasm and its remarkable capacity to respond and adjust itself to its environment, we should suspect that as a membrane it would exhibit marked variations of permeability. It does, and so responsive is it to environmental conditions (only a few of which we at present comprehend) that the permeability of this membrane to a given substance may be strikingly different not only in adjoining cells but also in different parts of the same cell. Any statements, therefore, regarding the permeability of the protoplasmic membrane must be only relative to the given cell or portion of it, and on that account any discussion of its permeability to water or its solutes is inseparable from a discussion of the factors that influence permeability and the method of determining them.

The protoplasmic membrane imbibes a large amount of water (Seifriz, 1921), and as a general statement it can be said that it is readily permeable to water, but under certain conditions it is more so than under others. The greatest variations in permeability, however, prevail in the behavior of the protoplasmic membrane toward solutes, the gradations ranging from those cases where the protoplasm offers but little retardation to their passage on the one hand to practically absolute impermeability on the other. The degree of permeability depends primarily upon the environmental conditions, and these will be considered in detail when the factors influencing permeability are considered.

1. *Osmotic Relationships of the Plant Cell.*—In order to understand something of the factors that enter into the alteration of the permeability of the protoplasmic membrane as well as to comprehend the methods used in detecting these changes in permeability, it is necessary to have clearly in mind certain phenomena that occur in connection with the water relations of the plant cell. In a consideration of the osmotic relationships of a typically vacuolated plant cell, with an elastic cell wall, to the adjoining cells, or to the medium exterior to the plant, an understanding of the following factors, quantities, or forces is important:

(*a*) *Plasmolysis.*—In a mature living plant cell, the cell sap exerts a hydrostatic pressure against the protoplasm and the cell wall. As a consequence the protoplasm and cell wall are in close contact, and the cell is more or less expanded. If water is removed from the cell sap, the hydrostatic pressure against the protoplasm and the cell wall is released, and since they are elastic they contract. The cell wall soon reaches its limit of contraction, but the protoplasmic membrane which is more elastic will continue to contract as water is withdrawn and thus will move inwardly from its position along the cell wall.

The contraction of the protoplasmic membrane from the cell wall, owing to the release of the hydrostatic pressure of the cell sap against it, is called "evident plasmolysis." The term "plasmolysis" literally means "the loosening of the protoplasm." That condition in a plant cell when the cell wall has reached its limit of contraction but when the protoplasm has not yet receded from the wall is termed "incipient plasmolysis." The deficiency of water in the vacuole which causes plasmolysis in plant cells is generally due either to the difference in concentration of the solution external to the cell and that of the cell sap, or to the inability of the cell to replace the water that is being lost from it through evaporation.

(*b*) *Osmotic Value of the Cell Sap.*—By the osmotic value of the cell sap is meant the concentration of the osmotically active substances therein. This value may be expressed in terms of molar concentration, or in terms of the osmotic pressure, expressed in atmospheres, that the cell sap would be capable of producing if it were separated from pure water by a perfectly semipermeable membrane. It might also be appropriately named the "potential" or "latent" osmotic pressure of the contents of the vacuole, since it does not express the value of the osmotic pressure being exerted within the vacuole but rather that which such a solution would be capable of developing if it were placed under such ideal conditions.

It is evident from this definition that the osmotic value of the cell sap will be decreased as water enters and the cell expands, thus decreasing the concentration of the osmotically active substances. When water is withdrawn from the cell sap, the vacuole decreases in volume and the concentration of the osmotically active substances is increased, thus increasing the osmotic value of the cell sap. A plant cell is considered to function naturally as long as the protoplasm remains in contact with the cell wall. When evident plasmolysis occurs, the cell is under abnormal conditions which, if continued, will cause death. The smallest volume that the cell sap can attain and keep the protoplasm in contact with the cell wall is at the point of "incipient plasmolysis." The greatest volume that the cell sap can attain is when it is at equilibrium in distilled water. The osmotic value of the cell sap is thus the highest at incipient plasmolysis and the lowest when the cell is in equilibrium with distilled water.

A solution that has the same osmotic value as the cell sap is termed an "isotonic" solution, and if a plant cell is placed in such a solution it will not change in volume. A solution that has an osmotic value higher than that of the cell sap is called a "hypertonic" solution. A plant cell placed in such a solution will soon show evident plasmolysis. The term "hypotonic" is applied to a solution whose osmotic value is lower than that of the cell sap.

(*c*) *Osmotic Concentration of the External Solution.*—The term "external solution" as here used may refer to the cell sap of the cells adjoining the one under consideration or that which may obtain in the medium exterior to the plant. This external solution exerts an osmotic pressure that counterbalances to a greater or lesser degree that of the solution in the vacuole of the cell in question. If the osmotic concentration of the cell sap is higher than that of the external solution, water will be drawn into the cell with a force equal to the difference between the osmotic pressure of the cell sap and that of the external solution, provided that the cell wall and protoplasmic membrane exert no inward pressure upon the solution in the vacuole. If the osmotic value of the external solution is higher than that of the vacuole of the cell being considered, water will be withdrawn from the cell with a force equal to the difference between the osmotic pressure of the vacuole and the external medium, if the external medium is not subjected to any inwardly directed pressure.

This latter behavior can be observed in the laboratory by placing plant tissue in a solution, the concentration of which is higher than that of the cell sap of the cells which compose the tissue. Let us consider the phenomena that would prevail under these conditions if the protoplasmic membrane were (1) impermeable to the solutes of the exterior solution and those of the cell sap and (2) permeable to the solutes of the exterior solution but impermeable to those of the cell sap. It is assumed in both cases that the concentration of the solutes in the external solution is not of such a nature or of sufficient strength to be directly injurious to the protoplasm.

Let us now assume that a plant cell is placed in a solution whose osmotic value is higher than that of the cell sap and that the protoplasmic membrane is impermeable to the solutes of the exterior solution and of the cell sap (Fig. 7*A*). Under these conditions no diffusion of solutes through this membrane tending toward an equilibrium in their concentration could occur. Since the concentration of solutes in the exterior solution is higher than that in the vacuole of the cell, it follows that the concentration of solvent is lower in the former than it is in the latter and that as a result water will diffuse outwardly from the vacuole into the exterior solution more rapidly than it will move in the opposite direction. Hence water will be drawn from the vacuole of the cell and the hydrostatic pressure that is normally exerted against the protoplasm and the cell wall is lowered, and as a result these membranes contract inwardly. Since the cell wall is only slightly elastic, the limit of its contraction is soon reached, but this is not true of the protoplasmic membrane and it will continue further to contract inwardly and thus separate itself from the cell wall (Fig. 7*B*). This contraction of the protoplasm will continue until the osmotic concentration of the cell sap, owing to the unequal diffusion of water outwardly and also perhaps to the formation of new osmotically active materials by the protoplasm, has become equal to that in the external solution. When this state of equilibrium prevails, further contraction of the protoplasm will not occur, and, unless the protoplasm is capable of manufacturing osmotically active substances to increase further the osmotic concentration of the vacuole, the protoplasmic membrane will never regain its normal position, and death will shortly ensue.

Let us now consider the phenomena that occur if the protoplasm is permeable to the solutes of the medium in which the cell is placed but impermeable to the solutes within the vacuole. Since the solutes in the external solution are higher in concentration than those in the cell sap,

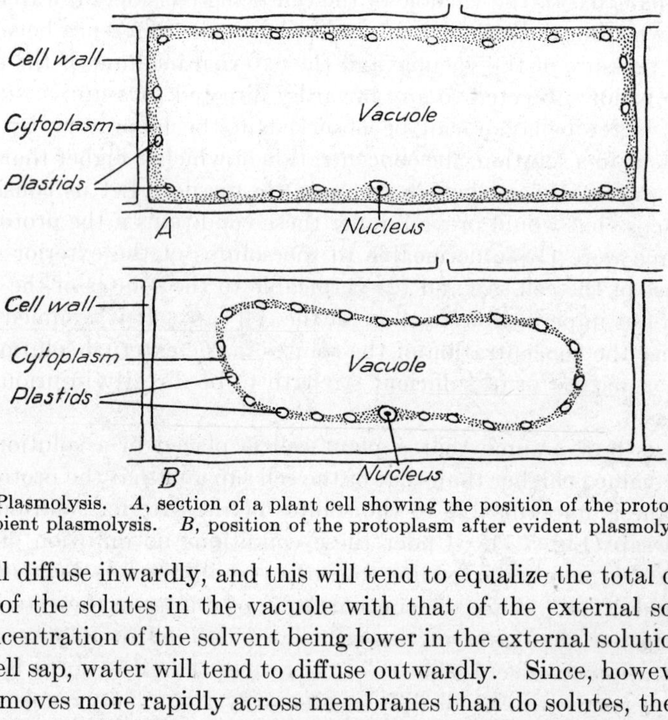

FIG. 7.—Plasmolysis. *A*, section of a plant cell showing the position of the protoplasm at incipient plasmolysis. *B*, position of the protoplasm after evident plasmolysis.

they will diffuse inwardly, and this will tend to equalize the total concentration of the solutes in the vacuole with that of the external solution. The concentration of the solvent being lower in the external solution than in the cell sap, water will tend to diffuse outwardly. Since, however, the solvent moves more rapidly across membranes than do solutes, the water of the cell sap will diffuse outwardly more rapidly than the solutes move inwardly. As a result, the hydrostatic pressure against the cell wall and the protoplasm will be reduced and these membranes will contract in the same manner as described in the previous example. The protoplasm will recede from the cell wall and continue to contract until, due to the diffusion outwardly of water and the diffusion inwardly of the solutes, the cell sap is in osmotic equilibrium with the external solution.

Since, however, the protoplasmic membrane is permeable to the solutes of the external solution, they will diffuse inwardly until the concentration of each solute is the same in the cell sap as it is in the external solution. As a result of this, the osmotic value of the cell sap will increase and water will move inwardly so that a hydrostatic pressure will be gradually set up which will force the contracted protoplasmic membrane against the wall into its normal position. Thus eventually the hydrostatic pressure against the protoplasmic membrane and the cell wall will be the same as that which prevailed before the cell was immersed in the solution.

According to King (1932) some cells die quickly when they are plasmolyzed, while others may remain alive for a few days to several weeks. The injury and recovery from plasmolysis will vary with the nature of the salt used, with the duration of plasmolysis, and with the kind of plant.

(*d*) *The Turgor Pressure and the Wall Pressure of the Cell.*—A vacuolated living plant cell under the ordinary conditions of plant life absorbs a sufficient amount of water to exert a hydrostatic pressure against the cell wall. Because the wall is elastic, it is stretched by the hydrostatic pressure, but its limit of extensibility is soon reached, and at that point the wall exerts a backward pressure against any further increase in the volume of the cell. When the limit of extensibility of the cell wall has been reached, its resilience or inward pressure is equal to the force with which water tends to enter the cell, and hence no more water will enter. When this process of expansion is under way or has reached its limit, the cell is rigid and is said to be "turgid." This condition or rigidity is spoken of as the "turgidity" of the cell.

The inwardly directed pressure of the cell wall against the protoplasm and the cell sap is termed the "wall pressure" of the cell. The hydrostatic pressure of the cell sap exerted against the cell wall is called the "turgor pressure" of the cell. This hydrostatic pressure is the equivalent of the osmotic pressure of the cell sap, but is designated as the turgor pressure only when it is considered in its relation to its pressure against the cell wall. When the cell has reached its limit of expansion, the wall pressure is at its highest possible value and is equal to the osmotic pressure, hydrostatic pressure, or turgor pressure at that time. At incipient plasmolysis the wall pressure and turgor pressure are zero. The cell-wall pressure is calculated at any point between incipient plasmolysis and the limit of cell expansion by assuming that any change in the pressure of the wall is proportional to the change in the volume of the cell. It appears, however, that such a relationship holds only approximately.

(*e*) *The Absorbing Power of the Cell.*—This term is used to designate the force with which water enters the plant cell under any given condition. In this relationship the terms suction tension, suction pressure, suction force, or suction of the cell have also been used. The use of these terms, however, has been adversely criticized by Shull (1927, 1932), and others, because water apparently moves into a plant cell and is in no sense sucked or pulled into it as these terms imply. The location and exact nature of the forces that cause water to enter the plant cell are not definitely known. These forces are in some manner the resultant of the reaction between the cell sap, the external solution, and the protoplasmic membrane separating these two solutions. In order to express the collective action of the forces involved in the intake of water by the cell, no term has been suggested or devised which is above criticism. The term "absorbing power" of the cell seems to be the least objectionable and is herein used.

The absorbing power of the plant cell, at any time, is equal to the osmotic value of the cell sap minus the osmotic value of the external solution minus the wall pressure.

Absorbing power = osmotic value of cell sap − (osmotic value of external solution + wall pressure)

When a plant is in equilibrium with pure water, the absorbing power of the cell is zero, while at the point of incipient plasmolysis it reaches its maximum value for any condition in which the protoplasm is in its normal position along the cell wall. The absorbing power of plant cells, under the conditions that generally prevail in the plant, is somewhere intermediate between the two extremes mentioned above.

The osmotic value of the cell sap at incipient plasmolysis has been wrongly termed by many investigators as the absorbing power of the cell under the conditions generally prevailing in the plant. These investigators considered the concentration of the cell sap at incipient plasmolysis to be the same as that which prevails in the cells of the plant under any condition. They thus did not take into consideration the shrinkage in the volume of the cell before incipient plasmolysis and the attending increase in the concentration of the cell sap. Ursprung (1916 to 1935), Ursprung and Blum (1916 to 1930), Molz (1926), Dumont (1926), Beck (1928 to 1935), Dixon (1930), Ernest (1931, 1934), Herrick (1933), and Lyon (1936) have studied the absorbing power of cells in different tissues under various conditions, and they have described and criticized the different methods that have been used in its determination.

(1) *Methods of Determining.*—Most of the information concerning the various osmotic forces of the plant cell has been obtained by using microscopic sections and noting the increase or decrease of the volume of the cell when solutions of various concentrations were applied. Ernest (1931) pointed out that the methods of determining the absorbing power of cells by means of sections are in error. In sections, the intact cells are in contact not only with the solution being used in the experiment but also with the cell sap of those cells which have been cut or ruptured in the sectioning of the material under observation. The absorbing power of the cells is thus not necessarily balanced by the solution that has been applied. The fact that the intact cells

expand abnormally because the adjoining cells are ruptured and their inwardly directed pressure is released also introduces another source of error in the use of sections.

The more simple method reported by Lyon (1936) seems to eliminate to a considerable extent the objections to the earlier methods. Blocks or strips of a tissue are immersed for some time in various concentrations of a sugar solution. The strips are measured microscopically for length before and after immersion. The concentration of the solution in which the strip neither lengthens nor shortens is a measure of the mean osmotic value of the cells, under the conditions prevailing in the plant. The first solution that gives the maximum shrinkage would be isotonic with the cell sap at incipient plasmolysis and would be a measure of the osmotic value and the absorbing power of the cell in that condition.

The solution that causes the first contraction of the tissue after it is in equilibrium with distilled water would indicate the osmotic value of the cell sap under that condition and would be the equivalent of the wall pressure of the cell at that time. The wall pressure is calculated at any point between these two extremes by assuming that the change in wall pressure is proportional to the change in volume of the cell.

(2) *Observations.*—As an example of the different quantities concerned in the osmotic relationship of the plant cell, those determined for the cells of the root of *Vicia faba* are given (Molz, 1926).

OSMOTIC VALUE, ABSORBING POWER, AND TURGOR PRESSURE AS DETERMINED IN THE ROOT CELLS OF *Vicia faba*

Tissue examined	Distance from growing point, millimeters	Osmotic[1] value of cell sap, atmospheres	Absorbing power of cells, atmospheres	Turgor pressure, atmospheres
Epidermis of cells of outer cortex......	1.5	14.3	7.4	3.9
	3.0	14.0	7.7	4.0
	5.0	13.3	10.0	2.7
	8.0	12.7	0.4	9.4

[1] At incipient plasmolysis.

These investigators found in the absorption zone of the root of *Vicia faba* an increase in the absorbing power of the cells from the epidermis toward the endodermis and then a marked decrease in the endodermis and wood parenchyma. It was also observed that the absorbing power of the cells of the upper portion of the stem and its leaves was higher than the cells of the same parts lower down. An increase in absorbing power from the base to the apex was found in the mid- and lateral veins of the ivy leaf. The observation of the absorbing power of a row of palisade cells that extended from the midvein into a portion of the leaf free from larger veins was especially noteworthy. From a minimum of 12 atmospheres near the midvein, the absorbing power rose regularly to a maximum of 32.6 atmospheres, which was attained in the two hundred and tenth palisade cell, from which point it decreased. In contrast to this regular rise in absorbing power, the osmotic value in the same row of palisade cells showed no regularity.

Buchinger (1927) observed that the absorbing power in 43 varieties of barley, varied from 23.4 to 35.2 atmospheres and that these differences were indicative of the geographical region to which the varieties are adapted. Herrick (1933) observed in *Ambrosia trifida* that if the competition for water was not severe neither the osmotic value nor the absorbing power of the cells reached a maximum in the upper part of the plant. Under these conditions there was no consistent gradient of these values throughout the plant. Increasing the internal competition for water, whether owing to seasonal or diurnal causes, resulted in an increase in the magnitude of the osmotic values throughout the plant and in the establishment of consistently increasing gradients of these values and absorbing power from the lower to the higher portions of the plant. Under conditions of severe competition for water, the osmotic value in any one leaf was essentially equal to the absorbing power of the same leaf. The osmotic value and absorbing power tended daily to reach maximum values between 1 and 3 P.M.

Ernest (1934) found that the absorbing power of the superficial mesophyll cells of the leaves of *Iris* maintain a remarkably constant value changing but little during a 24-hr. period. He observed a slight gradient of absorbing power from the water-conducting tissue to the surface tissue. He also observed that the mesophyll cells removed from a leaf that had been floating for some hours in water showed a considerable absorbing power which might amount to as much as 8 atmospheres. This absorbing power was due to the expansion of individual cells, which under natural conditions was resisted by the mutual pressure of the surrounding cells.

2. *Permeability and Absorption.*—The diffusion of any solute into a plant cell will depend not only upon the degree of permeability of the cell membranes to the substance in question but also on how much the equilibrium of the solute on the outside and inside of the cell is disturbed. On that account, therefore, the rate at which a substance enters a cell cannot be taken as a measure of the permeability of the cell membranes to that substance. Let us consider a case in which the protoplasmic membrane is only slightly permeable to a given solute, but as soon as this solute enters the cell it is chemically combined, precipitated, or adsorbed and thus rendered osmotically inactive. In that case, the concentration of this solute on the inside of the cell will always be lower than that on the outside and it will continue to enter, so that after a period of time a comparatively large amount of the substance has entered the cell even though the protoplasmic membrane is only slightly permeable to it. On the other hand, the protoplasmic membrane may be very permeable to a solute, but after its entrance into the cell, this solute may not be combined or adsorbed to any extent. In that case, equilibrium of concentration of this solute on the outside and inside of the cell will soon be reached and no more will enter. Thus after a period of time only a comparatively small amount of the solute has entered the cell, although the protoplasmic membrane is readily permeable to it. In discussing the methods used in determining the changes in the permeability of the protoplasmic membrane, the distinction between the degree of permeability of the membranes and the rate of absorption of a given solute must be kept clearly in mind. It should be remembered that these quantities have little to do with each other and that the rate of absorption is no criterion of permeability, although it has been so used by many investigators.

3. *Methods of Measuring the Permeability of the Protoplasmic Membrane to Solutes.*—The investigations of the changes in permeability of the protoplasmic membrane to solutes have shown marked differences in evidence secured by different methods and in the conclusions based thereon. These differences are in a large measure due to a lack of understanding of the limitations of the methods used and to assumptions which are unwarranted by the data obtained. A brief survey should thus be made of the methods that have been employed and their advantages and disadvantages mentioned so that the student may be able to judge better the value of the work that has been reported in this regard. The principal methods that have been used in the study of the permeability of the protoplasm are (*a*) chemical analysis of the tissue extracts or of the solutions external to the tissues; (*b*) observation of visible changes within the cells; (*c*) methods involving plasmolysis; and (*d*) electrical conductivity of tissues or masses of cells. These have been termed "indirect methods," since the cell contents are examined only in an indirect manner or not at all. A direct method has lately been devised that is applicable in certain cases, whereby the cell sap may be withdrawn directly from the cells and a quantity of material obtained sufficient to make a qualitative and in some cases a quantitative chemical determination.

(*a*) *Indirect Methods.* (1) *By Chemical Analysis.*—This method has been used primarily to determine what substances may enter or leave the plant cell through the protoplasmic membrane rather than a measure of the changes in its permeability. The method is more applicable in considering the amount of materials absorbed by a considerable portion of tissue or by the entire plant than to a consideration of a single cell or of a small amount of tissue. Two lines of procedure have been followed. One is to analyze the tissue or the tissue extract, and the other is to analyze the solution bathing the tissues. The principal objection to analysis of tissue extracts is that it is extremely difficult to obtain a sample that is representative of the cell sap by any of the present means of extraction. The analysis of the tissue or tissue extract also does not take into consideration the presence of salts in the intercellular spaces or in the cell wall where they may be held in solution or by adsorption. The analysis of the solution in which the plant is grown in order to determine the intake or outgo of solutes has many sources of error. The results in many cases can be observed only after several hours and, in many cases that have been reported, only after several days. Under such conditions some of the external cells are undoubtedly killed or are separated from the roots, especially the root tip, and thus give off their cell sap into the solution. Any increase in the diffusion outward of the solutes may be due not to increased permeability but to increased production within the cells under the conditions of the experiment. Brooks (1917) criticized the method, since it in no way distinguishes between the rate of absorption and the degree of permeability of the protoplasmic membrane. Steward (1928) concluded from a consideration of the effects of various treatments upon exosmosis that protracted leaching does not necessarily result in loss of semipermeability. It is evident thus that under certain conditions the chemical method may be an indicator at least of the degree of permeability of the protoplasmic membrane but that its limitations must be kept in mind.

(2) *By Visible Changes within the Cell.*—This method of studying permeability is of importance in the investigation of the diffusion of substances into the cell which will form therein precipitates, stains, crystals, or other compounds that may be observed. These substances include certain alkaloids that form intracellular precipitates; acids and alkalies that cause color changes of pigments within the cell; dyes that form precipitates; stains; salts that form crystals after their entrance; and sugar and glycerol whose penetration into the cells is made manifest by the formation of starch. This method of determining permeability is necessarily limited in its application, and it is also open to the objection that it is more a measure of the rate at which the substances are combined in the cell than a measure of the degree of permeability of the protoplasmic membrane to these substances.

(3) *By Plasmolytic Methods.*—Plasmolysis of a cell is caused by the inability of the solutes on the exterior of the cell to diffuse rapidly enough through the protoplasmio membrane to equalize the concentration of the cell sap with that of the exterior medium. This inequality of concentration may be due to the fact that the solutes are combined or rendered osmotically inactive after they have once entered the cell sap or that the protoplasmic membrane is not permeable to a sufficient degree to allow the solutes to enter rapidly enough. It is on the latter assumption that the plasmolytic methods for the studying of changes in permeability are based. Two methods have been the most generally used.

a) The Determination of Concentrations Necessary to Produce Plasmolysis.—If the protoplasmic membrane is only slightly permeable to the solute in the external solution, the attaining of equilibrium between the cell sap and a hypertonic solution takes place slowly, so that an exterior solution, the concentration of which is only slightly above that of the cell sap, will almost immediately cause plasmolysis. If, on the other hand, the protoplasmic membrane is readily permeable to the solute under consideration, the solute will enter the cell and equilibrium between it and the cell sap will be quickly restored and plasmolysis will not result. The external concentration of the solute must therefore be increased to such an extent that, even if the protoplasmic membrane is readily permeable to it, equilibrium will not be restored quickly enough to prevent plasmolysis. Thus if a cell or a portion of tissue under condition A is plasmolyzed, for example, by a 0.1 N sodium chloride solution and when placed under condition B it requires a fifth-normal solution of the same salt to produce plasmolysis, one may assume that the permeability of the protoplasmic membrane to sodium chloride is greater under condition B than it is under condition A.

b) The Determination of the Time Required to Recover from Plasmolysis.—If the formation of new osmotically active substances in the vacuole is not considered, the recovery of a cell from plasmolysis is due primarily to the diffusion of the solute of the exterior medium into the cell in sufficient amount to equalize the concentration of the cell sap and the exterior solution. Recovery from plasmolysis, therefore, will be more rapid if the protoplasmic membrane is readily permeable to the solute of the exterior solution than if it is only slightly permeable to it. Thus if a cell or the cells of a tissue are plasmolyzed by a hypertonic solution of sodium chloride and under condition A recovery from plasmolysis occurs in 30 min., while under condition B only 10 min. are required for recovery, it is considered that the permeability of the protoplasmic membrane to sodium chloride has increased under condition B as compared with what it was under condition A.

The main objection to the plasmolytic methods is that they do not take into consideration the possibility of the formation by the protoplasm of osmotically active substances in the cell sap, which if they were formed would have the same effect as solutes taken in from the external solution. The effect of ethyl alcohol on *Spirogyra* is of interest in this regard. Lepeschkin (1928) observed that the turgor pressure of this plant was decreased by a weak concentration of alcohol while a strong concentration of alcohol increased it to such an extent that the cells were torn to pieces. It is considered that the tannin in the cell sap is a colloidal solution in water, but in alcohol it becomes a molecular solution and thus increases the concentration of the cell sap. In a dilute solution of alcohol the permeability of the protoplasm is increased and the turgor pressure thus lessened.

(4) *By Electrical Conductivity of Tissues or Masses of Cells.*—"The conduction of an electrical current by a solution involves the passage through the solution of electrically charged ions of some substance. The rate at which the current will be conducted by the ions of a given salt will depend upon two factors, the potential gradient and the frictional or other resistance to the migration of the ions. If the potential gradient be kept constant we may follow fluctuations in the last factor by measurement of the current or by direct measurement of the electrical resistance. By the use of alternating currents of rather high frequencies, large effects due to the accumulation of ions at the surfaces impermeable to them are avoided" (Brooks, 1917). The electrical-conductivity method of measuring the permeability of the protoplasmic membrane is based on the assumption that the electrical conductivity of a tissue is to be regarded as a measure of the permeability of the protoplasmic membrane. If the current is forced through the living protoplasmic membrane in a solution, the resistance offered by this membrane to the passage of ions is considered to vary inversely as the permeability. This electrical method was first used by Bugarsky and Tangl (1897) and Stewart (1897) but has been modified, elaborated, and extensively used by Osterhout (1912, 1918) and his pupils. Much of the work of Osterhout in measuring changes in permeability by the electrical-conductivity method has been with the thallus of *Laminaria* after the following manner. One hundred to two hundred disks are cut from the fronds and packed together like a stack of coins. These form a cylinder 5 to 10 cm. high and about 1.3 cm. in diameter and are kept in position by means of glass rods. At each end of the cylinder and separated from it by a small amount of solution is a platinum electrode coated with platinum black and connected with an instrument for measuring the electrical conductivity. The general procedure is to immerse the stack of disks in a given solution and after drawing off the superflouus liquid to place the stack under the conditions mentioned above and measure the electrical conductivity. There is sufficient solution between each of the disks to make the circuit, and experiments have shown that they are not injured even by prolonged treatment.

The manner in which the electrical-conductivity method is used in determining changes in permeability can best be illustrated by a theoretical example. The resistance of a given column of disks after being immersed in a given solution *A* and then drained off, as has been previously mentioned, is 850 ohms. The same column after being immersed in solution *B*, which has the same conductivity as *A* but is of a different composition, and treated in the same manner has a resistance of 980 ohms. The assumption from the data is that the effect of solution *B* is to decrease the permeability of the protoplasmic membranes of the tissue under what it was in solution *A*. The question in dispute in regard to the reliability of this method is how far the electrical conductivity of tissue is a measure of the permeability of the protoplasmic membranes of the cells. Stiles and Jorgensen (1918) and Stiles (1924) have pointed out, however, that an increase in the conductivity of a tissue may be brought about in other ways than by an increase in the permeability of the protoplasmic membrane. They considered that changes in the content and nature of the electrolytes in the cell wall, protoplasm, and cell sap can have a marked influence on the alteration of the conductivity. Any factor such as diffusion, which would bring about an increase in the amount of free electrolytes in the cell parts or the breaking down of large complex molecules to smaller ones, which are more mobile and undergo more ionization, would increase the conductivity of the tissues. These changes might increase the permeability of the protoplasmic membrane, but the increase in conductivity could be due, in part at least, to other factors.

(*b*) *Direct Methods.*—The reliability of the various indirect methods that have been used in the study of permeability is open to question until they can be compared with results obtained by direct methods. In order to make direct determinations, it is necessary to remove the cell sap and then compare its composition with that of the solution exterior to the cell. The errors that are encountered in trying to analyze the extracts of cells and tissues have been enumerated above. The only known direct method is to use individual cells and to extract their cell sap without any contamination or change. Only a few cells, however, are large enough to give any satisfactory results by this method.

Wodehouse (1917) experimented with the marine alga *Valonia,* which has individual cells so large that the cell sap yields from 1 to 5 cc. of liquid. The cells were removed from the sea water, dried with filter paper, and then punctured with a needle. On pressing the cell wall, the cell sap exuded and was collected. Osterhout (1921) used the large multinucleate cells of a species of *Nitella* some of which reach a length of 6 in. and a diameter of $\frac{1}{32}$ in. Two methods were used for obtaining the cell sap. By one method the cell was placed on glue or filter paper and pierced with the point of a capillary tube which had been drawn out to a fine tip. The cell sap was by this means drawn up into the tube by capillary action and was not contaminated by the chloroplasts or protoplasm. The other method was to suspend the cell by the upper end, cut off the lower end, and bring it in contact with a glass slide and then to apply a slight pressure to the cell wall, beginning at the top. In this manner the cell sap flowed out on the slide, and by uniting the drops from several cells a sufficient quantity of material was obtained for examination. By this method Osterhout (1922, 1927) was able to compare direct observations of permeability behavior with that indicated by the electrical-conductivity method and concluded that the conductivity method gives reliable information concerning permeability. The direct method of studying permeability has also been used by Irwin (1922) and Hoagland and Davis (1923). The results obtained by these various investigators will be discussed in full in Chap. V when the intake of solutes by the plant will be considered.

4. *Methods of Measuring the Permeability of the Protoplasmic Membrane to Water.* The number of methods used in determining the permeability of the protoplasmic membrane to water are very few. Delf (1916) used the hollow leaves of the onion and the hollow scape of the dandelion for investigation. They lend themselves to this experiment because they present on their inside a natural surface of uncutinized thin-walled cells. An apparatus was employed by which a current of liquid could flow continuously through a cylinder 30 mm. in length of the tissue under investigation. It was so arranged that a solution of cane sugar could be immediately substituted for a water current, and vice versa, without altering the temperature of the tissue. The data on the passage of water through the protoplasmic membrane were obtained by observing the rate of shrinkage of the tissue under consideration when in contact with solutions that tended to plasmolyze them. An optical lever in connection with the tissue enabled alterations in length of 0.003 mm. or 1 part in 10,000 to be detected. An experiment of this nature cannot be carried out satisfactorily with strong plasmolyzing solutions because of the rapid shrinkage of the tissue. The solution must be sufficiently weak so that contraction will be comparatively slow and thus confine the experimentation to the strictly normal state of falling turgor without the actual separation of the protoplasmic membrane from the cell wall. The rate of shrinkage of the tissue under these conditions is the resultant of two factors. One of these is the

compressing force of the cell-wall action on the protoplasmic membrane, which causes exudation of water, and the other is the permeability of the protoplasmic membrane for water. The influence of the permeability factor can be obtained only by comparing the tissues under consideration when the compressing force is equal. In the experiments of Delf, the points of 30, 50, and 70 per cent of the total contraction were used, but the most acceptable series for the two kinds of tissue under consideration was the 50 per cent series.

The change-in-weight-and-volume method has been used by Stiles and Jorgensen (1917) to determine the rate of water intake by the tissue of the potato tuber and the carrot root. The intake of water was measured by immersing the tissue of known weight and volume in water and weighing the tissue and measuring the increase in volume at certain intervals. The change in weight under these conditions is attributed to the amount of water taken up, since the weight of solutes that might diffuse outward during the experiment is considered negligible in comparison to the weight of water involved. The tissue in these experiments was cut in cylinders 1.8 cm. in diameter by means of a cork borer and then into disks 0.2 cm. in thickness by means of a hand microtone. Each experiment was run in triplicate or quadruplicate with a set of 20 disks.

Huber and Höfler (1930) determined the permeability of the protoplasmic membrane to water by noting the rate of change in volume of the cells during plasmolysis and subsequent recovery in water. Baptiste (1935) immersed uniform disks of the potato tuber and carrot in water and followed its uptake by weighing these disks periodically at frequent intervals.

5. *Factors That May Influence the Permeability of the Protoplasmic Membrane.* (*a*) *Temperature.* (1) *Influence on the Permeability of the Protoplasmic Membrane to Water.*—Rysselberghe (1901) studied the changes in the permeability by the plasmolytic methods and also by the rate of shrinkage of tissue in a 0.73 *N* solution of cane sugar. He observed especially the pith of the cells of the elder, *Spirogyra,* and the epidermal cells of *Tradescantia.* He found that the effect of temperature is significant from 0 to 20°C. but that from 20 to 30°C. the permeability of the protoplasmic membrane to water remains almost constant, as shown in the following table:

Temperature, °C.	Rate	Temperature, °C.	Rate
0	1.0	20	7.1
6	1.9	25	7.5
12	4.5	30	8.0
16	6.3		

Lepeschkin (1906) found that the volume of liquid extruded by hairs on the leaves of *Phaseolus multiflorus* increased from 0 to 20°C. and decreased from 20 to 35°C.

Delf (1916) found that the permeability of the protoplasmic membrane to water as measured by the rate of tissue shrinkage of the onion leaf and dandelion scape in a dilute sugar solution is increased continuously by temperature up to 42°C.—the highest temperature investigated. If the rate of permeability at 25°C. is taken as unity, the following relative values were established:

Temperature, degrees	Permeability	
	Onion	Dandelion
5	0.36	
10	0.44	0.22
15	0.50	0.30
20	0.66	0.50
25	1.0	1.0
30	1.7	1.9
35	2.9	3.0
40	5.0	5.0

Delf criticized the results of Rysselberghe in regard to the permeability remaining constant above 20°C. She considered that the sugar solution used by Rysselberghe was too concentrated for the best results, since a solution of that strength causes such a rapid contraction of the protoplasmic membrane that the cell wall cannot follow it. From her observations Delf considered that there is an upper limit to the rate at which the mechanical tissue system of the cell walls can collapse and shrink. If this assumption is correct, then no matter how rapidly the protoplasm contracts under the influence of the external solution, the cell wall cannot follow at a rate above a certain limit. In Rysselberghe's experiments that limit seemed to have been reached at 20°C. with a 0.73 *M* cane-sugar solution. Stiles and Jorgensen (1917) found that the temperature coefficients for the rate of swelling of the tissues of the potato tuber and root of carrot in distilled water were as follows:

Temperature, degrees centigrade	Temperature coefficients	
	Potato	Carrot
10 to 20	3.0	1.3
20 to 30	2.7	1.6

The rate of swelling, however, is dependent not only on the permeability of the protoplasmic membrane to water but on the elasticity of the cell wall and the permeability of the protoplasm to the dissolved materials in the cell sap. The previous history of the tissue under consideration also has a great influence on the results obtained.

The experimental evidence seems to indicate that the permeability of the protoplasmic membrane to water increases continuously from the lower to the higher temperatures so that if the other factors are the same the intake or outgo of water from the cells increases with a rise in temperature.

(2) *Influence on the Permeability of the Protoplasmic Membrane to Solutes.*—Rysselberghe (1901) found that the permeability of the protoplasmic membrane of the epidermal cells of *Tradescantia discolor* to glycerin, potassium nitrate, and urea increases from 0 to 30°C. Lepeschkin (1905) observed that the protoplasm of the leaves of *Phaseolus multiflorus* increased in permeability to the solutes of the cell sap from 0 to 20°C. and decreased from 20 to 35°C. Eckerson (1914) investigated the influence of temperature on the permeability of the protoplasmic membranes of the cells of the roots of numerous plants. The results were obtained on the assumption that an increase in the concentration which is necessary to cause plasmolysis corresponds to an increase in the permeability of the protoplasm. Potassium nitrate was used, for the most part, as the plasmolyzing agent, and the tests for permeability were carried on under controlled temperatures. The permeability of the protoplasm of the cells of the radish root to potassium nitrate was found to increase from 10 to 18°C. but did not change from 18 to 24°C., increased from 24 to 40°C., and then decreased. The permeability of these cells to glucose was observed to increase gradually from 10 to 40°C. and then decrease. The permeability to sucrose remained constant until temperatures above 42°C. were reached. *Helianthus annuus* showed no increase in permeability with increase in temperature but a decrease from 20 to 40°C. The primary roots of *Phaseolus multiflorus* exhibited no increase in permeability from 13 to 20°C., while the secondary ones increased in permeability from 6 to 25°C. Both types of roots decreased in permeability from 25 to 35°C.

The table on page 104 is a summary of the work of Eckerson on the relation of temperature to the changes in permeability to potassium nitrate.

The work of Eckerson indicates that the changes in permeability due to changes in temperature vary with the species, with different temperatures, and with the plasmolyzing agent used.

Sen (1928) found that the permeability of the plasma membrane to ions increased with rising temperature from 20 to 30°C. in *Aloe* leaves and in *Bassela* stems. From 30 to 35°C. there was little or no increase, while beyond 35 or 40°C. and up to the lethal temperature, the permeability decreased but is reversible up to 40 or 45°C. At the lethal temperature, the plasma membrane becomes highly permeable, a change which is irreversible (see also Dixon, 1924).

PERMEABILITY VARIATIONS OF THE PROTOPLASMIC MEMBRANE TO POTASSIUM NITRATE AT DIFFERENT TEMPERATURES[1]

Cells of the root of	Degrees centigrade			
	+	0	+	−
Raphanus sativus	10 to 14	18 to 24	24 to 40	40 to 50
Pisum sativum	6 to 15	15 to 35	....	35 to 45
Sinapis alba	10 to 30	30 to 45		
Helianthus annuus	....	12 to 20	....	20 to 40
Phaseolus multiflorus:				
Primary roots	....	13 to 20	....	25 to 40
Secondary roots	6 to 25	....		25 to 40

[1] In the table a plus sign indicates increasing permeability; zero indicates no changes in permeability, while a minus sign indicates a decrease in permeability.

(*b*) *Light.*—Lepeschkin (1908) observed that light increases and darkness decreases the permeability of the protoplasm of the pulvinal cells of the legumes. The increase in the permeability of the protoplasmic membrane of these cells results in a decrease of their volume, while a decrease in permeability brings about an increased turgor pressure and an increase in volume. Tröndle (1909, 1910, 1918) found that the protoplasm of the cells of the leaves of *Tilia europaea*, *Buxus sempervirens*, *Salix babylonica*, and *Acer platanoides* are more permeable to sodium chloride in light than in darkness. The permeability of the cells of the leaves of *Buxus* and *Tilia* was found to be, respectively, 17.6 and 33 per cent greater in sunlight than in cloudy weather. If the leaves were held at the same temperature, increasing illumination increased the rate at which sodium chloride passed into the cell up to a certain maximum of light intensity that varies with the kind of plant. Lepeschkin (1930) studied the rate of accumulation of dye in the cell sap of the leaves of *Elodea* in light and in darkness. He concluded that the permeability of the protoplasm is increased in the light, and that not the mode of reception but the amount of light energy is the important factor. He noted that adsorption by the cell wall and protoplasm was not altered by light or darkness, and that the cells affected by light did not impart an increase in permeability to neighboring cells. The greater the change of the illumination, the greater was the change of permeability, but no proportionality between them was observed. The maximum change in permeability was observed when direct sunlight was diminished to 10 per cent of its full intensity. The rays most active in increasing permeability are those with a wave length of 320 to 420mμ, while the red rays are least active in this regard. Offord and d'Urbal (1931) found that sodium chlorate apparently penetrated *Nitella* more slowly when the light intensity was reduced. On the contrary, however, Ruhland (1911) found that light was without any measurable influence on the permeability of the protoplasmic membrane of the cells of the sugar beet to sucrose, glucose, and fructose.

According to Lepeschkin (1932), Zycha (1928) could detect no changes in permeability due to light; likewise, Jacques and Osterhout (1935) reported that the rate of entrance of potassium into *Nitella* cells was independent of illumination.

Hoagland and Davis (1923) observed that the alga *Nitella* absorbed more solutes from the surrounding medium in light than in darkness. They considered that light in this case is of primary importance as a source of energy in the process of absorption.

This assumption will be discussed in detail in Chap. V, when the absorption of solutes is considered.

(*c*) *The Nature of the Solutes.*—The influence of the various dissolved substances on the permeability of the protoplasm can best be understood by a summary of the results obtained by the various workers in that field. Osterhout (1911) found that a 0.2 *M* solution of calcium chloride just plasmolyzed the cells of *Spirogyra,* and a 0.38 *M* sodium chloride solution did the same. The cells, however, were not plasmolyzed by a 0.195 *M* calcium chloride solution or a 0.375 *M* sodium chloride solution. If, however, 10 cc. of a 0.195 *M* solution of calcium chloride is mixed with 1,000 cc. of the 0.375 *M* sodium chloride solution and Spirogyra cells placed therein, prompt plasmolysis occurs. The addition of the calcium chloride lowers the osmotic pressure of the solution, but in spite of that it plasmolyzes the cells. Thus by mixing two solutions, neither of which plasmolyzes, one is obtained that plasmolyzes very strongly. Osterhout (1915) used the electrical-conductivity method to determine the effects of various salts on permeability. Tissue of *Laminaria saccharina* in sea water had a net resistance of 770 ohms or a net conductance of 0.00130 reciprocal ohm. When the tissue was transferred to a 0.52 *M* solution of sodium chloride having the same conductivity as sea water, the resistance dropped to 580 ohms in 5 min. or to a net conductance of 0.001720 reciprocal ohm. The increase in permeability of the protoplasmic membrane thus amounted to 0.00042 reciprocal ohm or 32.3 per cent. This increase in conductance is not due to the increase in the number of ions of sodium chloride to which the protoplasmic membrane might be considered to be normally more permeable than those of ordinary sea water. For example, when the tissue was transferred from sea water to 0.52 *M* sodium chloride the conductance increased 25.2 per cent. The increase in sodium ions amounted to only 2 per cent, while the content of chlorine ions remained unchanged. Osterhout considered that this is evidence that there was a great increase in permeability of the protoplasm, which was induced by the sodium chloride. The effect of calcium chloride on the permeability of the protoplasm of *Laminaria* was found by the electrical-conductivity method as follows: The net resistance of a portion of tissue in sea water was 770 ohms. After transferring it to a 0.278 *M* solution of calcium chloride the net resistance rose within 10 min. to 1,260 ohms. This represents a decrease in permeability of 39.2 per cent. Osterhout found that potassium chloride, lithium chloride, ammonium chloride, sodium bromide, sodium iodide, sodium nitrate, potassium nitrate, and sodium sulphate all act, in general, like sodium chloride in increasing the permeability. All the bivalent cations investigated, *viz.*, magnesium, calcium, barium, strontium, manganese, cobalt, iron, zinc, cadmium, and tin, decreased permeability, and all the trivalent cations considered, *viz.*, lanthanum, iron, aluminum, cerium, and yttrium, did the same. Sodium hydroxide increased the permeability, while hydrochloric acid produced a rapid decrease followed at once by a rapid increase that continued until the death point was reached. Hydrogen is the only monovalent cation that was found able to decrease the permeability of the protoplasm.

These observations have been substantiated more recently by other investigators. Thus Ingold (1931) noted in the tissues of carrot, beet, and potato that the outward diffusion of solutes was increased by an immersion in a solution of potassium chloride, while treatments with the chlorides of calcium and lanthanum decreased the loss of solutes. Similar observations were made by Asprey (1935). Baptiste (1935) soaked disks of the potato and carrot in hypotonic solutions of the chlorides of potassium, sodium, ammonia, magnesium, and calcium, and then raised the absorbing power to a known, uniform level by controlled evaporation. The disks were then immersed in distilled water, and their rate of water intake determined. The effects of these cations on permeability to water were significant and followed the series,

$$K > NH_4 > Na > \text{Control} > Mg > Ca.$$

Kerr (1930, 1933) by microinjection into the root hairs of *Trianea bogotensis* and *Limnobium spongia* observed that sodium chloride and potassium chloride increased the fluidity of the protoplasm, while calcium chloride and magnesium chloride decreased it. Seifriz and Plowe (1930) noted that the chlorides of sodium, potassium, and lithium lowered while the chlorides of calcium and barium raised the elastic limit or extensibility of the protoplasm. DeHaan (1933) believed that the specific action of ions on the protoplasm is due to variations in its rate of swelling.

Raber (1920) investigated the effect of a series of anions on the permeability of *Laminaria agardhu* by means of the electrical-conductivity method. He used sodium sulphocyanide, iodide, nitrate, bromide, chloride, acetate, sulphate, tartrate, phosphate, and citrate in solutions having the same conductivity as a 0.51 *M* solution of sodium iodide. Measurements of the conductivity were taken, for the most part, at the end of 5, 25, 45, 85, and 105 min. and the effect of these anions upon the increase in permeability at the end of the experiment stood in the following order: $I > Br > SCN > Cl > NO_3 > \text{acetate} > SO_4 > \text{tartrate} > PO_4 > \text{citrate}$. From later work (1921) Raber considers that his results indicate that the effects upon permeability depend upon the valence of the anions regardless of whether the salts are organic or inorganic. He further found that bivalent and trivalent cations in combination with monovalent anions increase the electrical resistance of the protoplasm of *Laminaria*, but when combined with bivalent or trivalent anions the increase is less and may be entirely lacking. Chromium has a different initial effect upon the permeability of *Laminaria*, depending upon whether it occurs as the cation or the anion of the salt. If it is an anion, the first effect is a decrease in resistance, but if it is a cation, it increases the resistance. Osterhout (1913, 1916) observed that anesthetics, as ether and chloroform, increase or decrease permeability, depending upon the concentration used. Very low concentrations were found to produce little or no effect, but a point in concentration was soon reached in which a decrease in permeability could be maintained for several hours. This change is reversible and produces no injury. A higher concentration causes for a very short time a decrease and then an increase in permeability which is not reversible and continues until death ensues.

Stiles and Jorgensen (1917) studied the action of various organic substances on the permeability of plant cells by estimating the exosmosis of electrolytes from potato tissue especially immersed in watery solution of the substances. The exosmosis was estimated by the change in conductivity of the solution. They found that equimolecular solutions of different substances do not bring about the same exosmosis. In the homologous series of the monohydric alcohols, the more complex the molecule the greater the exosmosis produced by solutions of equimolecular strength. They were unable to find a critical concentration of solution below which exosmosis of electrolytes will not take place. True and Bartlett (1915) found that, when a plant like white lupine was grown in distilled water, leachings took place, and this exosmosis was increased in dilute solutions of sodium chloride and potassium chloride. In dilute solutions of calcium chloride, however, leaching stopped and absorption took place. The inference is that the permeability of the protoplasmic membrane of the roots of the plant was increased by the sodium chloride and potassium chloride and decreased by the calcium chloride.

Osterhout (1915) in experiments with *Laminaria saccharina* by the electrical-conductivity method showed that it is possible to bring about extreme changes in the permeability of the protoplasmic membrane without injury. Experiments were conducted over a period of 15 days in which the tissue each day was transferred from sea water to 0.52 *M* sodium chloride for 5 to 10 min. and then transferred again to sea water where it always regained its original permeability. The increase in permeability in the sodium chloride solution over that in sea water amounted to 20.3 per cent. The tissue that underwent this daily alteration in permeability retained its normal resistance in sea water until the tenth day, when it began to fall off. The control, however, showed the same results, so that the fall in resistance was due to other causes than the effects of the sodium chloride. The permeability was decreased 23 per cent by transference from normal sea water to sea water containing 0.01 *M* lanthanum nitrate for 5 min. After being again placed in normal sea water it regained its original permeability. The procedure was followed each day for 5 days, being repeated three or four times daily. At the end of the fifth day this tissue was from all appearances in as good condition as the control and had a slightly higher resistance. In another series of experiments covering a period of 3 days, the tissue was transferred from sea water to a 0.278 *M* solution of calcium chloride, which has the same conductivity as sea water, and then to a 0.52 *M* sodium chloride, and then again to sea water. The resistance of the tissue in sea water was 1,010 ohms; after 10 min. in the calcium chloride solution, it had risen to 1,500 ohms; and after being removed from the calcium chloride and transferred to the sodium chloride solution, it fell to 880 ohms. After 80 min. the tissue was again placed in sea water where it soon gained its normal resistance of 1,010 ohms. The permeability of the protoplasmic membrane in the calcium chloride solution decreased 39.2 per cent below the normal permeability in sea water, while in the sodium chloride solution it increased 20 per cent above normal. Regardless of these wide alterations of permeability, the tissue at the end of the experiments was to all appearances absolutely normal.

E. Theories of Membrane Action

Numerous theories have been put forward to explain the differential permeability of membranes but most of them can be grouped under one of the following four headings.

1. Sieve or Ultrafiltration Theory.—This theory assumes that the protoplasm acts as a sieve or ultra screen with pores that change in diameter according to surrounding conditions (Seifriz, 1936). The sieve thus is not a fixed one but is capable of constant change and adjustment. A change in the size of the pores can be brought about by aggregation or coagulation, by peptization or dispersion, by a change in sign and magnitude of electric charge, and by the reorientation of surface molecules. When small particles aggregate into large ones, the space between the particles will become larger, and the sieve will therefore become coarser. The reverse change takes place by peptization. A positively charged membrane would repel positive particles and thus keep them out, while a negatively charged membrane would permit them to enter. This theory was first put forward by Traube (1879) in order to explain the behavior of his precipitation membranes. It has received considerable support from the works of Küster (1911, 1918) and Ruhland (1912, 1913, 1914) on the intake of dyes by certain plant cells. These investigators studied the behavior of the cells of *Vicia faba, Allium cepa, Spirogyra, Helianthus annuus,* and *Coleus hybridus* with respect to over 80 different acidic and basic dyes.

Ruhland concluded that his work indicated that the entrance or nonentrance of a substance into the cell is related to the magnitude of its molecules or molecular aggregates and that the plasma membrane acts as an ultra filter. The behavior of the plasma membrane in the plant cells and of artificial parchment membranes toward dyes offers the strongest evidence for this theory, but there are numerous other cases where the observed facts are not in accord with the theory. Thus it has been shown that certain membranes are permeable to the large molecules of the alkaloids while they are impermeable to the relatively small molecules of the amino acids.

Thus in the case of glycerol and monoacetin, the molecules of which are very nearly equal in size, the latter enters the cell at a rate fifteen times more rapidly than the former.

Brooks (1930) considered that the behavior of the protoplasmic membrane of *Valonia macrophysa* to sea water indicated that the plasma membrane might consist of a mosaic of anion-permeable and cation-permeable areas that are of the nature of charged porous films. Such films exaggerate differences between the diffusion velocities of the ions to which they are permeable.

2. The Solution Theory.—This theory assumes that a membrane has definite solvent properties and that only those substances which are soluble in the membrane are capable of passing through it. The chief advocate of this theory in regard to the behavior of the protoplasmic membrane has been Overton (1895, 1896, 1899, 1900), who proposed the lipoid theory. This investigator from his observations on the permeability of the protoplasm to certain dyes and other organic substances concluded that the penetration of different substances into the protoplasm depends upon the solubility of these substances in the lipoids which with the proteins form the plasma membrane. This theory has only a very limited application, and the chief objection to it arises from the behavior of the numerous inorganic salts, which for the most part are insoluble in lipoid substances but which are known to enter the cells freely and in large amounts.

Apparently in many cases the question of the solubility of the substance overshadows other factors. Thus, as mentioned in the previous topic, the molecules of glycerin and monoacetin are very nearly of equal size, but the latter enters the cell at a rate fifteen times greater than does the former. Glycerin is very poorly soluble in lipoid material, while monoacetin is a good lipoid solvent. In this case solubility rather than size of molecule is apparently the determining factor in permeability. MacDougal (1928) in experiments with an artificial cell showed that the presence of a phospholipide such as lecithin alone lessens permeability and allows the attainment of a high degree of turgidity. Artificial cells, including both cholesterol and lecithin, showed certain behaviors in balanced solutions of sodium and calcium concurrent with those of living cells. It was found that ions penetrate a lipoidal cell more rapidly from solutions at pH 4.5 in contrast with the maximum for a lipoid-free cell at pH 5.4 of the solution. A lecithin-cholesterol suspension also stabilized the hydrogen-ion concentration of the cell content. This work of MacDougal indicates that certain lipoids in the protoplasm do influence the diffusion of organic and inorganic ions through it, but the exact manner in which they do so is not at present understood.

3. The Chemical Theory.—According to this theory the membrane is supposed to combine chemically with the substance to which it is permeable. This reaction is considered to be a reversible one, so that the diffusing substance is freed on the far side from its combination with the membrane. The chemical theory has received considerable support from numerous biological workers.

The temperature coefficient of permeability has been studied by numerous investigators in order to ascertain the nature of the absorption process through membranes. By the temperature coefficient of permeability is meant the increase in the permeability of the membrane for each 10°C. increase in temperature. The temperature coefficient for any reaction is commonly expressed by the symbol Q_{10}. That of chemical reactions is relatively high (2 to 3), and that of physical reactions is relatively low (1 to 1.5). Pfeffer (1877) measured the rate of water movement through the copper ferrocyanide membranes at different temperatures and found that from 7.1 to 17.6°C., $Q_{10} = 1.56$, and from 17.6 to 32.5°C., $Q_{10} = 1.27$. Krabbe (1896) observed that the temperature coefficient of permeability of the living pith cells of *Helianthus annuus* and of the roots of *Vicia faba*, from 0 to 20°C., was 2 to 2.5. He believed that this high coefficient indicated that purely physical forces were not operative but that the changes were due to the peculiar properties of the protoplasm. Rysselberghe (1901) studied the effect of temperature upon the permeability of the living protoplasm of the pith cells of *Sambucus nigra* and the epidermal cells of *Tradescantia* and found that from 0 to 30°C., Q_{10} averaged 2.0. Brown and Worley (1912) studied the rate of intake of water by the barley grains at different temperatures and obtained a temperature coefficient of 1.8 to 1.9. They considered that this indicated that chemical processes were involved in the penetration of water through the semipermeable seed coat of the barley grain. Shull (1920), however, found that the mean value of Q_{10} in *Xanthium* seeds was in one instance 1.55, in another 1.83, and in the case of split peas 1.6. He considered that his data do not support the theory of Brown and Worley but that the velocity of intake at any given moment in the seeds that were studied is approximately an inverse exponential function of the amount of water previously absorbed. He considered that his values do not indicate the absorption is conditioned by some single chemical change like simplifications of water to hydrone as the temperature rises but that absorption at different temperatures involves both physical and chemical changes.

Shull and Shull (1924), studying the absorption of water by seeds of corn, found that over the range from 5 to 55°C. the temperature coefficient averaged 1.54, somewhat above the coefficient for purely physical processes. The absorption at 55°C. was somewhat more than eight times as fast as at 5°C. whereas the chemical theory of absorption would call for a rate thirty-two times as great. Apparently many substances have a specific absorption behavior, and a rate law with wide applicability is not to be expected.

Osterhout (1935) found indications in *Valonia* that ammonia or ammonium hydroxide enters the cells by combining with a constituent of the protoplasm, and he considered that other electrolytes might also enter the cell by combining with one or more constituents of the protoplasm. His observations thus tend to substantiate the chemical nature of absorption.

Stiles and Jorgensen (1915) used 0.0011 *M* hydrochloric acid in which they placed plant tissue to observe the penetration of the hydrogen ions at various temperatures. They found that $Q_{10} = 2.22$ from 0 to 10°C. and 2.17 from 10 to 20°C. They believed that the high temperature coefficient thus obtained indicated that the process of absorption is a chemical and not a physical reaction. Osterhout (1917) criticized their conclusions on the ground that they, as many others, have not distinguished between permeability and absorption. The chemical process thus thought to have been observed might have been the union of the hydrogen ions with substances within the cell, and the permeability of the membranes thus might not have been considered. His experiments with the permeability of *Laminaria* gave a temperature coefficient of 1.3, which he considered suggested a physical reaction, although he was not positive. Denny (1917) measured the flow of water through the seed coat of *Arachis hypogaea* at temperatures ranging from 3.6 to 45°C. and found that Q_{10} ranged in value from 1.6 to 1.3, the coefficient being higher at the lower temperatures and lower at the higher temperatures, a fact that has been observed for a great many physical and chemical processes. The results obtained by Delf (1916), however, in measuring the temperature coefficient for increase of permeability for the onion leaf and the scape of the dandelion are in contrast to these. She found, for example, that the temperature coefficient between 10 and 20°C. was 1.5 and 2.3, respectively, for the onion leaf and dandelion scape, while from 20 to 30°C. it was 2.6 for the onion leaf and 3.8 for the dandelion scape. Denny (1917) makes the very important statement to the effect that the temperature effects on permeability thus observed may not be merely upon the membranes under consideration but also upon the cell contents. This latter effect may contribute to the total results so that the chemical reaction indicated by the relatively high temperature coefficient may have taken place in that part of the cell which is internal to the membrane under consideration. There seems thus to be no evidence that either physical or chemical processes are exclusively involved in the passage of materials through the membranes of the plant cell.

4. Colloidal Theories.—Since the protoplasmic membrane has been regarded as an emulsoid colloidal system, numerous hypotheses have been proposed to explain its changes in permeability upon the basis of changes in viscosity, phase inversion, or electrical reactions, all of which are common phenomena in colloidal suspensions or solutions.

The changes in viscosity, as suggested by Spaeth (1916), in the size of the colloidal particles, as offered by Lloyd (1915) and Free (1918), and the action of neutral salts on the colloidal membrane, as mentioned by Kaho (1921), without doubt have their effects upon the permeability of the protoplasm. These phenomena, however, need not be classified under a separate heading but may be considered simply as corollaries to the sieve theory.

The suggestion of Clowes (1916) that changes in permeability are brought about by the continuous and discontinuous phases changing places has received considerable attention. To support his theory he used data obtained in experiments on the effect of sodium hydroxide and calcium chloride in producing phase inversion in emulsion systems of olive oil and water.

However, there is no direct evidence whatsoever of phase reversal in the protoplasmic membrane. It is very unlikely that protoplasm could exist as a living substance if oil were the continuous phase, because so far as is known metabolic reactions take place only in an aqueous medium, and the amount of oil in the dispersed globules of an ultramicroscopic emulsion like the protoplasm is probably insufficient to enclose the aqueous medium.

No theory of membrane action has been proposed that will explain satisfactorily the behavior of the protoplasmic membrane. The passage of materials through it may occur to a certain extent after the manner proposed in the various theories that have herein been mentioned (Gurewitsch, 1929). When one considers, however, the highly complex nature of protoplasm, its rapid adjustment to its environmental surroundings, its highly specialized nature for different individuals, tissues, and even for the different parts of the same cell, it would seem most unlikely that any one theory could be proposed that would, under all conditions, cover its behavior toward the various substances that diffuse through it.

Bibliography

Asprey, G. F., 1935: On the relationship between exosmosis and salt absorption by potato tuber tissue previously treated with various salt solutions, *Protoplasma*, **23**: 497–504.

Bank, O., 1935: Zur Tonoplasten-frage. I, Die Natur der Vakuolenhaut. II, Die Reversibilität der Vakuolenform in Embryonalen Zellen, *Protoplasma*, **23**: 239–249.

Baptiste, E. C. D., 1935: The effect of some cations on the permeability of cells to water, *Ann. Bot.*, **49**: 345–366.

Beck, W. A., 1928: Osmotic pressure, osmotic value and suction tension, *Plant Physiol.*, **3**: 413–441.

———, 1929: Determining the osmotic value at incipient plasmolysis, *Trans. Am. Microsc.*, **48**: 204–208.

———, 1931: Variations in the O_g of plant tissues, *Plant Physiol.*, **6**: 315–323.

———, 1935: The effect of light on the O_g of plant tissues, I *Protoplasma*, **23**: 203–209.

Becquerel, P., 1907: Recherche sur la vie latente des graines, *Ann. Sci. Nat. Bot.*, IX, **5**: 193–320.

Blinks, L. R., 1929: Protoplasmic potentials in *Halicystis*, *J. Gen. Physiol.*, **13**: 223–229.

———, 1929: The direct-current resistance of *Valonia*, *J. Gen. Physiol.*, **13**: 361–378.

———, 1930: The variation of electrical resistance with applied potential. I, Intact *Valonia ventricosa*, *J. Gen. Physiol.*, **13**: 793–806.

———, 1932: Protoplasmic potentials in *Halicystis*. II, The effects of potassium on two species with different saps, *J. Gen. Physiol.*, **16**: 147–156.

———, 1935: Protoplasmic potentials in *Halicystis*. IV, Vacuolar perfusion with artificial sap and sea water, *J. Gen. Physiol.*, **18**: 409–420.

———, R. D. Rhodes, and G. A. McCallum, 1935: Protoplasmic potentials in *Halicystis*. V, The reversal of potential by unbalanced sodium chloride, *Proc. Nat. Acad. Sci.*, **21**: 123–125.

Blum, G., 1926: Untersuchungen über die Saugkraft einiger Alpenpflanzen, *Beih. bot. Cent.*, I., **43**: 1–100.

Bowen, J. L., 1931: Private communication.

Brauner, L., 1930: Über polare Permeabilität, *Ber. bot. Ges.*, **48**: 109–118.

Brooks, S. C., 1917: Methods of studying permeability of protoplasm to salts, *Bot. Gaz.*, **64**: 230–249.

———, 1917: A new method of studying permeability, *Bot. Gaz.*, **64**: 306–317.

———, 1917: Permeability of the cell walls of *Allium*, *Bot. Gaz.*, **64**: 509–512.

——— and S. Gelfan, 1928: Bioelectric potentials in *Nitella*, *Protoplasma*, **5**: 84–96.

———, 1930: The accumulation of ions in living cells—a non-equilibrium condition, *Protoplasma*, **8**: 389–412.

Brown, A. J., 1907: On the existence of a semipermeable membrane enclosing the seeds of the *Gramineae*, *Ann. Bot.*, **21**: 79–87.

———, 1909: The selective permeability of the coverings of the seeds of *Hordeum vulgare*, *Proc. Roy. Soc. London*, B, **81**: 82–93.

——— and F. P. Worley, 1912: The influence of temperature on the absorption of water by seeds of *Hordeum vulgare* in relation to the temperature coefficient of chemical change, *Proc. Roy. Soc. London*, B., **85**: 546–553.

——— and F. Tinker, 1916: The rate of absorption of various phenolic solutions by seeds of *Hordeum vulgare* and the factors governing the rate of diffusion of aqueous solutions across semipermeable membranes, *Proc. Roy. Soc. London*, B, **89**: 119–135.

——— and ———, 1916: Selective permeability: the absorption of phenol and other solutions by the seeds of *Hordeum vulgare*, *Proc. Roy. Soc. London*, B, **89**: 373–379.

Brown, R., 1931: The absorption of water by seeds of *Lolium perenne* L. and certain other *Gramineae*, *Ann. App. Biol.*, **18**: 559–573.

———, 1932: The absorption of the solute from aqueous solutions by the grain of wheat, *Ann. Bot.*, **46**: 571–581.

Buchinger, A., 1927: Saugkraftmessungen ("Osmotisches Verhalten") verschiedener Gerstensorten, *Fortschr. Landwirts.*, **2**: 344–350; *Biol. Abst.*, **3**: 468.

Bugarsky, S., and F. Tangl, 1897: Eine Methode zur Bestimmung des relativen Volums der Blutkörperchen und des Plasmas, *Zent. Physiol.*, **11**: 297.

Clowes, G. H. A., 1916: Protoplasmic equilibrium. I, Action of antagonistic electrolytes on emulsions and living cells, *J. Phys. Chem.*, **20**: 407–451.

Collins, E. J., 1918: The structure of the integumentary system of the barley grain in relation to localized water absorption and semipermeability, *Ann. Bot.*, **32**: 381–414.

Crozier, W. J., 1919: Intracellular acidity in *Valonia*, *J. Gen. Physiol.*, **1**: 581–583.

Damon, E. B., 1929: Dissimilarity of inner and outer protoplasmic surfaces in *Valonia*. II, *J. Gen. Physiol.*, **13**: 207–221.

——— and W. J. V. Osterhout, 1930: The concentration effect with *Valonia:* potential differences with concentrated and diluted sea water, *J. Gen. Physiol.*, **13**: 445–457.

———, 1932: Dissimilarity of inner and outer protoplasmic surfaces in *Valonia*. III, *J. Gen. Physiol.*, **15**: 525–535.

DeHaan, I., 1933: Protoplasmaquellung und Wasserpermeabilität, *Rec. trav. bot. Néerl.*, **30**: 236–329. Quoted by Baptiste, 1935.

Delf, E. M., 1916: Studies of protoplasmic permeability by measurement of rate of shrinkage of turgid tissues. I, Influence of temperature on the permeability of protoplasm to water, *Ann. Bot.*, **30**: 283–310.

Denny, F. E., 1917: Permeability of certain plant membranes to water, *Bot. Gaz.*, **63**: 373–397.

———, 1917: Permeability of membranes as related to their composition, *Bot. Gaz.*, **63**: 468–485.

De Vries, H., 1884: Eine Methode zur Analyse der Turgorkraft, *Jahrb. wiss. Bot.*, **14**: 427–601.

———, 1884: Zur plasmolytischen Methodik, *Bot. Ztg.*, **42**: 289–298.

———, 1885: Plasmolytische Studien über die Wand der Vacuolen, *Jahrb. wiss. Bot.*, **16**: 465–598.

Dixon, H., 1924: Variations in the permeability of leaf cells, *Sci. Proc. Roy. Dublin Soc.*, **17**: 349–356.

———, 1930: Über die Saugkraft, *Ber. bot. Ges.*, **48**: 428–432.

Dumont, A., 1926: Étude bibliographique et critique de la mesure de la concentration des liquides cellulaires végétaux, *Ann. Physiol. and Physico-chem. Biol.*, **2**: 215–256.

Eckerson, S., 1914: Thermotropism of roots, *Bot. Gaz.*, **58**: 254–263.

Eisenmenger, W. S., 1930: Toxicity of some aliphatic alcohols, *Plant Physiol.*, **5**: 131–156.

Ernest, E. C. M., 1931: Suction-pressure gradients and the measurement of suction pressure, *Ann. Bot.*, **45**: 717–731.

———, 1934: Studies in the suction pressure of plant cells. II, *Ann. Bot.*, **48**: 293–305.

———, 1934: The effect of intercellular pressure on the suction pressure of cells, *Ann. Bot.*, **48**: 915–918.

———, 1935: Factors rendering the plasmolytic method inapplicable in the estimation of osmotic values of plant cells, *Plant Physiol.*, **10**: 553–558.

Findlay, A., 1913: "Osmotic Pressure," 82 pp., Longmans, Green & Company, New York.

Fischer, M. H., 1915: "Oedema and Nephritis: A Critical, Experimental and Clinical Study of the Physiology and Pathology of Water Absorption in the Living Organism," 2d ed., John Wiley & Sons, Inc., New York.

Free, E. E., 1918: A colloidal hypothesis of protoplasmic permeability, *Plant World*, **21**: 141–150.

Fujita, A., 1925: Untersuchungen über elektrische Erscheinungen und Ionendurchlässigkeit von Membranen. I, Die Potentialdifferenz an der Apfelschale, *Biochem. Z.*, **158**: 11–27.

Giglioli, I., 1895: Latent vitality in seeds, *Nature*, **52**: 544–545.

Goldsmith, G. W., and J. H. C. Smith, 1926: Some physico-chemical properties of spruce sap and their seasonal and altitudinal variation, *Colorado Col. Pub. Sci. Ser.*, **13**: 13–71.

Gurewitsch, A., 1929: Untersuchungen über die Permeabilität der Hülle des Weizenkorns, *Jahrb. wiss. Bot.*, **70**: 657–706.

Hamly, D. H., 1932: Softening of the seeds of *Melilotus alba*, *Bot. Gaz.*, **93**: 345–375.

Hawthorne, W. C., 1930: Diffusion, osmosis, and osmotic pressure, *Sci. Month.*, **31**: 535–542.

Herrick, E. M., 1933: Seasonal and diurnal variations in the osmotic values and suction tension values in the aerial portions of *Ambrosia trifida*, *Am. J. Bot.*, **20**: 18–34.

Hoagland, D. R. and A. R. Davis, 1923: The composition of the cell sap of the plant in relation to the absorption of ions, *J. Gen. Physiol.*, **5**: 629–647.

——— and ———, 1923: Further experiments on the absorption of ions by plants. including observations on the effect of light, *J. Gen. Physiol.*, **6**: 47–62.

Hottes, C. F., and W. A. Huelsen, 1927: The determination of quality in sweet corn seed by means of the optical measurement of leached materials, *J. Agr. Res.*, **35**: 147–166.

HUBER, B., and K. HÖFLER, 1930: Die Wasserpermeabilität des Protoplasmas, *Jahr. wiss. Bot.*, **73**: 351–511.
INAMDAR, R. S., and K. V. VARADPANDE, 1929: The permeability of plant-cell membrane to sugar, *Nature*, **124**: 875–876.
INGOLD, C. T., 1931: On the effect of previous treatment with salt solutions on the subsequent outward diffusion of electrolytes from plant tissue, *Ann. Bot.*, **45**: 709–716.
IRWIN, M., 1922: Penetration of dyes into living cells, *Am. J. Physiol.*, **59**: *Proc.* 458.
———, 1923: The permeability of living cells to dyes as affected by hydrogen-ion concentration, *J. Gen. Physiol.*, **5**: 223–224.
———, 1923: The penetration of dyes as influenced by hydrogen-ion concentration, *J. Gen. Physiol.*, **5**: 727–740.
JACQUES, A. G. and W. J. V. OSTERHOUT, 1935: The kinetics of penetration. XI, Entrance of potassium into *Nitella*, *J. Gen. Physiol.*, **18**: 967–985.
KAHO, H., 1921: Ein Beitrag zur Permeabilität des Pflanzenplasmas für die Neutralsalze. IV, *Biochem. Z.*, **123**: 284–303; *Rev. Bot. Gaz.*, **81**: 350–353.
KERR, T., 1930: The effect of essential metals on injection into plant protoplasm and sap, *Am. J. Bot.*, **17**: 1043.
———, 1933: The injection of certain salts into the protoplasm and vacuoles of the root hairs of *Limnobium spongia*, *Protoplasma*, **18**: 420–440.
KING, R., 1932: Plasmolysis, *Proc. Ind. Acad. Sci.*, **41**: 135–136.
KITE, G. L., 1915: Studies on the permeability of the internal cytoplasm of animal and plant cells, *Am. J. Physiol.*, **37**: 282–299.
KOTOWSKI, F., 1927: Semipermeability of seed coverings and stimulation of seeds, *Plant Physiol.*, **2**: 177–186.
KRABBE, G., 1896: Über den Einfluss der Temperatur auf die osmotischen Processe lebender Zellen, *Jahrb. wiss. Bot.*, **29**: 441–498.
KÜSTER, E., 1911: Ueber die Aufnahme von Anilinfarben in lebende Pflanzenzellen, *Jahrb. wiss. Bot.*, **50**: 261–288.
———, 1918: Ueber Vitalfärbung der Pflanzenzellen, *Z. wiss. Mikros.* **35**: 95–100.
LAVISON, J., 1910: Du mode de la pénétration de quelques sels dans la plante vivante; rôle de l'endoderme, *Rev. Gén. Bot.*, **22**: 225–241.
———, 1911: Recherches sur la pénétration des sels dans le protoplasm et sur la nature de leur action toxique, *Ann. Sci. Nat. Bot.*, IX, **14**: 97–139.
———, 1911: Essai sur une théorie de la nutrition minérale des plantes vasculaires, *Rev. Gén. Bot.*, **23**: 177–211.
LEPESCHKIN, W. W., 1906: Zur Kenntnis des Mechanismus der aktiven Wasserausscheidung der Pflanzen, *Beih. bot. Cent.*, I, **19**: 409–452.
———, 1908: Über die osmotischen Eigenschaften und den Turgordruck der Blattgelenkzellen der Leguminosen, *Ber. bot. Ges.*, **26**: 231–237.
———, 1908: Zur Kenntnis des Mechanismus der Variationsbewegungen, *Ber. bot. Ges.*, **26**: 724–735.
———, 1909: Zur Kenntnis des Mechanismus des photonastischen Variationsbewegungen und der Einwirkung des Beleuchtungswechsels auf die Plasmamembran, *Beih. bot. Cent.*, I, **24**: 308–356.
———, 1928: The effect of ethyl alcohol on the turgor pressure of *Spirogyra*, *Am. J. Bot.*, **15**: 422–424.
———, 1929: The protective action of some substances on protoplasm, *Am. J. Bot.*, **16**: 207–218.
———, 1930: Light and the permeability of protoplasm, *Am. J. Bot.*, **17**: 953–971.
———, 1932: The influence of narcotics, mechanical agents, and light upon the permeability of protoplasm, *Am. J. Bot.*, **19**: 568–580.

Livingston, B. E., 1913: Osmotic pressure and related forces as environmental factors, *Plant World*, **16**: 165–176.
Lloyd, F. E., 1915: The behavior of protoplasm as a colloidal complex, *Carnegie Inst. Wash. Yearbook*, **14**: 66–69.
Loomis, W. E., 1931: A quantitative demonstration of osmotic equilibrium, *Plant Physiol.*, **6**: 365–367.
Lyon, C. J., 1936: Analysis of osmotic relations by extending the simplified method, *Plant Physiol.*, **11**: 167–172.
MacDougal, D. T., 1923: Permeability and the increase of volume contents of living and artificial cells, *Proc. Am. Phil. Soc.*, **62**: 1–26.
———, 1924: The arrangement and action of material in the plasmatic layers and cell walls of plants, *Proc. Am. Phil. Soc.*, **63**: 76–93.
——— and V. Moravek, 1927: The activities of a constructed colloidal cell, *Protoplasma*, **2**: 161–188.
———, 1928: Substances regulating the passage of material into and out of plant cells: the lipoids, *Proc. Am. Phil. Soc.*, **67**: 33–45.
Michaelis, L., and A. Fujita, 1925: Untersuchungen über elektrische Erscheinungen und Ionendurchlässigkeit von Membranen. II, Die Permeabilität von Apfelschale, *Biochem. Z.*, **158**: 28–37.
Molz, F. J., 1926: A study of suction force by the simplified method. I, Effect of external factors, *Am. J. Bot.*, **13**: 433–463.
———, 1926: A study of suction force by the simplified method. II, Periodic variations and the influence of habitat, *Am. J. Bot.*, **13**: 465–502.
Offord, H. R., and R. P. d'Urbal, 1931: Toxic action of aqueous sodium chlorate on *Nitella*, *J. Agr. Res.*, **43**: 791–811.
Oppenheimer, H. R., 1930: Kritische Betrachtungen zu den Saugkraftmessungen von Ursprung und Blum, *Ber. bot. Ges.*, **48**: 131–140.
Osborne, W. A., and L. C. Jackson, 1914: Counter diffusion in aqueous solution, *Biochem. J.*, **8**: 246–249.
Osterhout, W. J. V., 1911: The permeability of living cells to salts in pure and balanced solutions, *Sci.*, **34**: 187–189.
———, 1912: Permeability of protoplasm to ions and the theory of antagonism, *Sci.*, **35**: 112–115.
———, 1912: Reversible changes in permeability produced by electrolytes, *Sci.*, **36**: 350–352.
———, 1913: Some quantitative researches on the permeability of the plant cell, *Plant World*, **16**: 129–144.
———, 1913: The effect of anesthetics upon permeability, *Sci.*, **37**: 111–112.
———, 1913: The organization of the cell with respect to permeability, *Sci.*, **38**: 408–409.
———, 1914: The effect of alkali on permeability, *J. Biol. Chem.*, **19**: 335–343.
———, 1914: The effect of acid on permeability, *J. Biol. Chem.*, **19**: 493–501.
———, 1915: Extreme alteration of permeability without injury, *Bot. Gaz.*, **59**: 242–253.
———, 1915: On the decrease of permeability due to certain bivalent kations, *Bot. Gaz.*, **59**: 317–330.
———, 1915: The effect of some trivalent and tetravalent kations on permeability, *Bot. Gaz.*, **59**: 464–473.
———, 1915: Normal and abnormal permeability, *Am. J. Bot.*, **2**: 93–94.
———, 1916: The decrease of permeability produced by anesthetics, *Bot. Gaz.*, **61**: 148–158.

———, 1917: Similarity in the effects of potassium cyanide and of ether, *Bot. Gaz.*, **63**: 77–80.

———, 1917: Does the temperature coefficient of permeability indicate that it is chemical in nature? *Bot. Gaz.*, **63**: 317–320.

———, 1918: Decrease of permeability and antagonistic effect caused by bile salts, *J. Gen. Physiol.*, **1**: 405–409.

———, 1918: A method of measuring the electrical conductivity of living tissues, *J. Biol. Chem.*, **36**: 557–568.

———, 1919: A comparative study of permeability in plants, *J. Gen. Physiol.*, **1**: 299–305.

———, 1919: A comparison of permeability in plant and animal cells, *J. Gen. Physiol.*, **1**: 409–415.

———, 1921: Conductivity and permeability, *J. Gen. Physiol.*, **4**: 1–9.

———, 1921: Direct and indirect determinations of permeability, *J. Gen. Physiol.*, **4**: 275–283.

———, 1922: Some aspects of selective absorption, *J. Gen. Physiol.*, **5**: 225–231.

———, 1922: "Injury, Recovery, and Death in Relation to Conductivity and Permeability," 259 pp., J. B. Lippincott Company, Philadelphia and London.

———, 1927: Some aspects on bioelectrical phenomena, *J. Gen. Physiol.*, **11**: 83–99.

———, E. B. Damon, and A. G. Jacques, 1927: Dissimilarity of inner and outer protoplasmic surfaces in *Valonia*, *J. Gen. Physiol.*, **11**: 193–205.

——— and E. S. Harris, 1928: Protoplasmic asymmetry in *Nitella* as shown by bioelectric measurements, *J. Gen. Physiol.*, **11**: 391–406.

——— and ———, 1928: The death wave in *Nitella*. I, Application of like solutions, *J. Gen. Physiol.*, **12**: 167–186.

——— and ———, 1928: The death wave in *Nitella*. II, Application of unlike solutions, *J. Gen. Physiol.*, **12**: 355–361.

——— and ———, 1929: Note on the nature of the current of injury in tissues, *J. Gen. Physiol.*, **13**: 47–56.

——— and E. S. Hill, 1930: Negative variations in *Nitella* produced by chloroform and by potassium chloride, *J. Gen. Physiol.*, **13**: 459–467.

———, 1930: Calculation of bioelectric potentials. I, Effects of KCl and NaCl on *Nitella*, *J. Gen. Physiol.*, **13**: 715–732.

———, 1930: The accumulation of electrolytes, *J. Gen. Physiol.*, **14**: 285–300.

———, 1935: Mechanism of salt absorption by plant cells, *Nature*, **136**: 1034–1035.

———, 1935: How do electrolytes enter the cell? *Proc. Nat. Acad. Sci.*, **21**: 125–132.

———, 1936: Electrical phenomena in large plant cells, *Physiol. Rev.*, **16**: 216–237.

Oswald, W., 1891: "Solutions," viii + 316 pp., Transl. by H. M. P. Muir. Longmans, Green & Company, New York.

Overton, E., 1895: Über die osmotischen Eigenschaften der lebenden Pflanzen- und Tierzelle, *Vierteljahrschr. Nat. Ges. Zürich*, **40**: 159–201.

———, 1897: Über die osmotischen Eigenschaften der Zelle in ihrer Bedeutung für die Toxikologie und Pharmakologie, *Vierteljahrschr. Nat. Ges. Zürich*, **41**: 383–406.

———, 1899: Ueber die allgemeinen osmotischen Eigenschaften der Zelle, ihre vermuthlichen Ursachen und ihre Bedeutung für die Physiologie, *Vierteljahrschr. Nat. Ges. Zürich*, **44**: 83–135.

———, 1900: Studien über die Aufnahme der Anilinfarben durch die lebende Zelle, *Jahrb. wiss. Bot.*, **34**: 669–701.

Pfeffer, W., 1877: "Osmotische Untersuchungen: Studien zur Zell-mechanik," Leipzig.

———, 1890: Zur Kenntnis der Plasmahaut und der Vacuolen, *Abh. Sächs. Ges. Wiss.*, **16**: 187–343.

Plowe, J. Q., 1931: Membranes in the plant cell. I, Morphological membranes at protoplasmic surfaces, *Protoplasma*, **12**: 196–220.

———, 1931: Membranes in the plant cell. II, Localization of differential permeability in the plant protoplast, *Protoplasma*, **12**: 221–240.

Pugh, G. W., H. Johann, and J. G. Dickson, 1932: Relation of the semipermeable membranes of the wheat kernel to infection by *Gibberella saubinetii*, *J. Agr. Res.*, **45**: 609–626.

Pumphrey, R. J., 1932: Electrical properties of the inner epidermis of the onion scale, *Proc. Roy. Soc.*, B, **109**: 434–442.

Raber, O. L., 1920: A quantitative study of the effect of anions on the permeability of plant cells, *J. Gen. Physiol.*, **2**: 535–539.

———, 1920: The antagonistic action of anions, *J. Gen. Physiol.*, **2**: 541–544.

———, 1921: A quantitative study of the effect of anions on the permeability of plant cells. II, *Am. J. Bot.*, **8**: 366–369.

———, 1921: The effect of permeability on polyvalent cations in combination with polyvalent anions, *Am. J. Bot.*, **8**: 382–386.

———, 1921: The effect upon the permeability of (I) the same substance as a cation and anion and (II) changing the valency of the same ion, *Am. J. Bot.*, **8**: 464–470.

———, 1923: Permeability of the cell to electrolytes, *Bot. Gaz.*, **75**: 298–308.

Rubinstein, D. L., and V. Uspenskaja, 1934: Ueber den isoelektrischen Punkt der pflanzlichen Plasmahaut, *Protoplasma*, **21**: 191–225.

Ruhland, W., 1911: Untersuchungen über den Kohlenhydratstoffwechsel von *Beta vulgaris*, *Jahrb. wiss. Bot.*, **50**: 200–257.

———, 1912: Studien über die Aufnahme von Kolloiden durch die pflanzliche Plasmahaut, *Jahrb. wiss. Bot.*, **51**: 376–431.

———, 1912: Die Plasmahaut als Ultrafilter bei der Kolloidaufnahme, *Ber. bot. Ges.*, **34**: 139–141.

———, 1913: Zur Kritik der Lipoid- und der Ultrafiltertheorie der Plasmahaut nebst Beobachtungen über die Bedeutung der elektrischen Ladung der Kolloide für ihre Vitalaufnahme, *Biochem. Z.*, **54**: 59–77.

———, 1913: Zur Kenntnis der Rolle des elektrischen Ladungssinnes bei der Kolloidaufnahme durch die Plasmahaut, *Ber. bot. Ges.*, **31**: 304–310.

———, 1913: Zur chemischen Organisation der Zelle, *Biol. Cent.*, **33**: 337–351.

———, 1914: Weitere Beiträge zur Kolloidchemie und physikalischen Chemie der Zelle, *Jahrb. wiss. Bot.*, **54**: 391–447.

Rysselberghe, Fr. van, 1901: Influence de la température sur la perméabilité du protoplasme vivant pour l'eau et des substances dissoutes, *Bull. Acad. Roy. Belg. Cl. Sci.*, 173–221; *Rec. Inst. Bot. Univ. Bruxelles*, **5**: 209–248 (1902).

Schröder, H., 1911: Über die selektiv permeabele Hülle des Weizenkornes, *Flora*, **102**: 186–208.

Seifriz, W., 1921: Observations on some physical properties of protoplasm by aid of micro-dissection, *Ann. Bot.*, **35**: 269–296.

——— and J. Plowe, 1930: The effects of salts on the extensibility of protoplasm, *Am. J. Bot.*, **17**: 1046.

———, 1936: "Protoplasm," 584 pp., McGraw-Hill Book Company, Inc., New York.

Sen, B., 1928: The effect of temperature on the permeability of the protoplasmic membrane, *Proc. Roy. Soc.*, B, **103**: 272–288.

———, 1931: A method for measuring the change of permeability to ions of single cells under electric stimulation, *Ann. Bot.*, **45**: 527–531.

Showalter, A. M., 1929: The evolution of "plasmolysis," *Am. J. Bot.*, **16**: 523–524.

SHULL, C. A., 1913: Semipermeability of seed coats, *Bot. Gaz.*, **56**: 169–199.
———, 1920: Temperature and the rate of moisture intake in seeds, *Bot. Gaz.*, **69**: 361–390.
——— and S. P. SHULL, 1924: Temperature coefficient of absorption in seeds of corn, *Bot. Gaz.*, **77**: 262–279.
———, 1927: Suction forces in plant cells. A review, *Bot. Gaz.*, **83**: 213–214.
———, 1932: Textbook of plant physiology. A review, *Bot. Gaz.*, **93**: 108–109.
——— and S. P. SHULL, 1932: Irregularity in the rate of absorption by dry plant tissues, *Bot. Gaz.*, **93**: 376–399.
SKEEN, J. R., 1929: The tolerance limit of seedlings for aluminum and iron, and the antagonism of calcium, *Soil Sci.*, **27**: 69–80.
———, 1930: Experiments with *Trianea* on antagonism and absorption, *Plant Physiol.*, **5**: 105–118.
SPAETH, R. A., 1916 The vital equilibrium, *Sci.*, **43**: 502–509.
STARK, O. K., 1928: What is osmosis? *Sci.*, **67**: 556–557.
STEWARD, F. C., 1928: The maintenance of semipermeability in the plant cell during leaching experiments, *Proc. Leeds Phil. Soc.*, **1**: 258–270.
STEWART, G. N., 1897: Elektrische Leitfähigkeit thierischer Flüssigkeiten, *Zent. Physiol.*, **11**: 332–335.
STILES, W., and I. JORGENSEN, 1915: Studies in permeability. II, The effect of temperature on the permeability of plant cells to the hydrogen ion, *Ann. Bot.*, **29**: 611–618.
——— and ———, 1917: Studies in permeability. IV, The action of various organic substances on the permeability of the plant cell and its bearing on the Czapek's theory of plasma membrane, *Ann. Bot.*, **31**: 47–76.
——— and ———, 1917: Studies in permeability. V, The swelling of plant tissue in water and its relation to the temperature and various dissolved substances, *Ann. Bot.*, **31**: 415–434.
——— and ———, 1918: Quantitative measurements of permeability, *Bot. Gaz.*, **65**: 526–534.
———, 1922: The suction pressure of the plant cell. A note on nomenclature, *Biochem. J.*, **16**: 727–729.
———, 1924: "Permeability," 292 pp., Wheldon & Wesley, London.
TJEBBES, K., 1912: Keimproeven met suikerbietinzaad, *Inaug. Diss. Amsterdam.*
TRÖNDLE, A., 1909: Permeabilitätsänderung und osmotischer Druck in den assimilierenden Zellen des Laubblattes, *Ber. bot. Ges.*, **27**: 71–78.
———, 1910: Der Einfluss des Lichtes auf die Permeabilität der Plasmahaut, *Jahrb. wiss. Bot.*, **48**: 171–282.
———, 1918: Der Einfluss des Lichtes auf die Permeabilität der Plasmahaut und die Methode der Permeabilitäts-Koeffizienten, *Vierteljahrschr. Nat. Ges. Zürich*, **63**: 187–213.
URSPRUNG, A., and G. BLUM, 1916: Über die Verteilung des osmotischen Wertes in der Pflanze, *Ber. bot. Ges.*, **34**: 88–104.
——— and ———, 1916: Über die periodischen Schwankungen des osmotischen Wertes, *Ber. bot. Ges.*, **34**: 105–123.
——— and ———, 1916: Über den Einfluss der Aussenbedingungen auf den osmotischen Wert, *Ber. bot. Ges.*, **34**: 123–142.
——— and ———, 1916: Zur Methode der Saugkraftmessung, *Ber. bot. Ges.*, **34**: 525–539.
——— and ———, 1916, 1918: Zur Kenntnis der Saugkraft. I, II, *Ber. bot. Ges.*, **34**: 539–554; **36**: 577–599.
——— and ———, 1919: Zur Kenntnis der Saugkraft. III, *Ber. bot. Ges.*, **37**: 453–462.

——— and ———, 1920: Dürfen wir die Ausdrücke osmotischer Wert, osmotischer Druck, Turgordruck, Saugkraft synonym gebrauchen? *Biol. Zent.*, **40**: 193–216.

——— and ———, 1921: Zur Kenntnis der Saugkraft. IV, Die Absorptionszone der Wurzel. Der Endodermissprung, *Ber. bot. Ges.*, **39**: 70–79.

——— and ———, 1921: Zur Kenntnis der Saugkraft. V, Eine Methode zur Bestimmung des Widerstandes, den der Boden der Wasserabsorption durch die Wurzel entgegensetzt, *Ber. bot. Ges.*, **39**: 139–148.

———, 1923: Zur Kenntnis der Saugkraft. VII, Eine neue vereinfachte Methode zur Messung der Saugkraft, *Ber. bot. Ges.*, **41**: 338–343.

——— and G. Blum, 1924: Eine Methode zur Messung des Wand- und Turgordruckes der Zelle, nebst Anwendungen, *Jahrb. wiss. Bot.*, **63**: 1–110.

———, 1925: Einige Resultate der neuesten Saugkraftstudien, *Flora*, **118-119**: 566–599.

———, 1926: Über die gegenseitigen Beziehungen der osmotischen Zustandsgrössen, *Planta*, **2**: 640–660; *Biol. Abst.*, **3**: 1078.

——— and G. Blum, 1927: Eine Methode zur Messung der Saugkraft von Hartlaub, *Jahr. wiss. Bot.*, **67**: 334–348.

———, 1929: The osmotic quantities of the plant cell, *Proc. Int. Cong. Plant Sci.*, Ithaca, 1926, **2**: 1081–1094.

——— and G. Blum, 1930: Zwei neue Saugkraft-Messmethoden, *Jahrb. wiss. Bot.*, **72**: 254–334; *Biol. Abst.*, **6**: 2531.

———, 1935: Osmotic quantities of plant cells in given phases, *Plant Physiol.*, **10**: 115–133.

Walter, H., 1930: Saugkraft oder osmotischer Wert, *Z. Bot.*, **23**: 74–93.

Weber, F., 1929: Plasmolyse-Zeit-Method, *Protoplasma*, **5**: 622–624.

———, 1929: Plasmolysezeit und Lichtwirkung, *Protoplasma*, **7**: 256–258.

Whitten, J. H., 1914: The effect of kerosene and other petroleum oils on the viability of growth of *Zea mais*, *Bull. Ill. State Lab. Nat. Hist.*, **10**: 245–273.

Wodehouse, R. P., 1917: Direct determinations of permeability, *J. Biol. Chem.*, **29**: 453–458.

Zycha, H., 1928: Einfluss des Lichts auf die Permeabilität von Blattzellen für Salze, *Jahrb. wiss. Bot.*, **86**: 499–548.

CHAPTER III

THE ROOTS OF PLANTS

The root system of a plant includes all its roots considered collectively. The root systems are specific, *i.e.*, the root system of a given species, in its general appearance, can be distinguished from that of another type of plant. Weaver (1920) stated that the general characters of the root systems of species are often as marked and distinctive as are those of the aerial vegetative parts, and although these may be greatly modified when subjected to different environmental conditions, they still retain the characteristic impress of the species in its usual habitat. Kiesselbach and Weihing (1935) considered that root development in corn in subsequent hybrid generations exhibits heterosis as does stalk development. The root system of a given plant may, however, vary in its structure, extent, weight, number, and direction of its roots according to the conditions under which it is grown. The general behavior of roots in the soil is thus the resultant of the influence of many factors, the most important of which are moisture, nutrients, oxygen supply, temperature, physical texture of the soil, light, and gravity.

I. INFLUENCE OF ENVIRONMENT ON ROOTS

A. Influence of Soil Moisture on Root Development

The influence of varying amounts of moisture in the soil upon the roots has been studied along three general lines: (1) direction of root growth, (2) relative weight of roots and tops, and (3) lateral extent and depth of penetration into the soil. It is the purpose here to consider only the first two topics and to reserve the third for a later discussion under a separate heading.

1. Direction of Root Growth.—It is common knowledge that roots respond to the changes of moisture in the soil and that they are positive in their reaction as they turn from a lower to a higher water content. Some of the factors entering into this reaction were investigated by Hooker (1915) in working with the roots of *Lupinus albus*. He found that roots are positively hydrotropic in moist air and that they react only when the moisture of the air is between 80 and 100 per cent of saturation. He observed that the minimum, optimum, and maximum differences in moisture to which these roots reacted at 20°C. was a fall of 0.2, 0.4, and 0.5 per cent, respectively, per centimeter along a line at right angles to them. The best response of root reaction toward changes in moisture was thus obtained under conditions that would make a difference of only 0.04 per cent moisture between the opposite sides of a root 1 mm. thick. Hooker also found that under optimum conditions 6 hr. elapsed before the hydrotropic reaction was visible in the roots, and the sensitiveness of roots to moisture was exhibited chiefly at the root tip and to a lesser extent by the cells behind the root tip. From these facts it is evident that roots will turn toward or follow moisture in the soil only when they are in direct contact with it or in very close proximity to it.

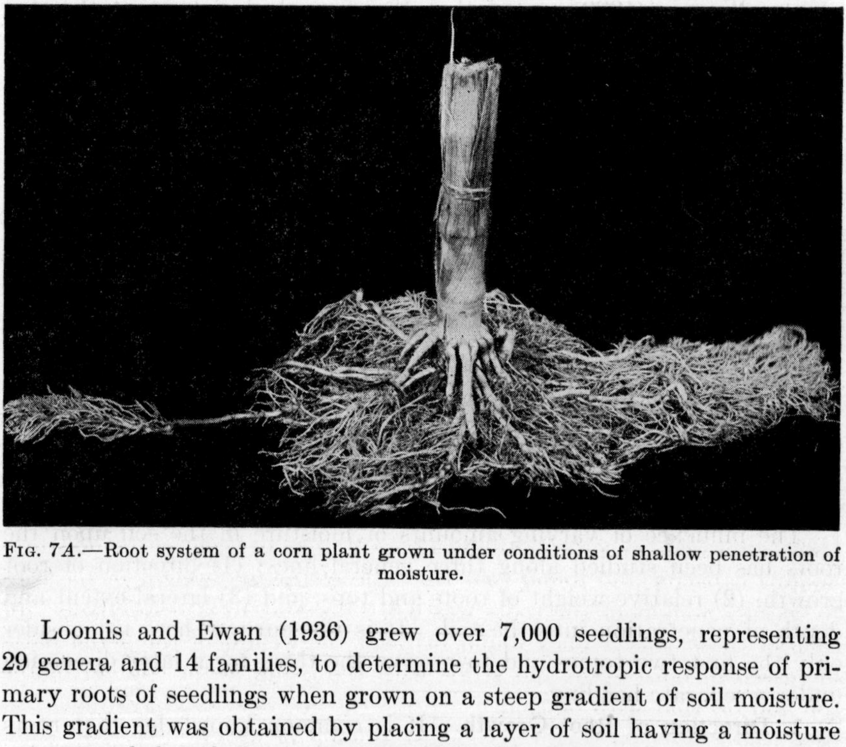

FIG. 7*A*.—Root system of a corn plant grown under conditions of shallow penetration of moisture.

Loomis and Ewan (1936) grew over 7,000 seedlings, representing 29 genera and 14 families, to determine the hydrotropic response of primary roots of seedlings when grown on a steep gradient of soil moisture. This gradient was obtained by placing a layer of soil having a moisture percentage below the hygroscopic coefficient in contact with a layer having a moisture percentage slightly below the moisture equivalent. The data obtained indicated that hydrotropism in roots is by no means comparable in occurrence and intensity of reaction with geotropism or phototropism. In some cases the results suggested that hydrotropic responses may depend upon a genetic factor present in some plants and absent from closely related species or varieties.

In this connection it should be mentioned that the roots of the vast majority of plants cannot penetrate a layer of dry soil in order to reach a supply of moisture that might be beneath, since, in a dry soil, water would be extracted from the roots, and they would perish as a result.

Considerable experimental work has been done to determine the ability of roots to penetrate dry soil. Magistad and Breazeale (1929) noted that in a soil at the wilting coefficient the roots of some plants were covered with a moist adhesive sheath. They believed that the moisture content of this sheath was probably maintained under such conditions by an exudation of water by the roots. This water, according to Breazeale (1930), is furnished in some cases by other roots that have penetrated moisture regions of soil. In experiments with wheat and barley, he obtained evidence which indicated that the roots of these plants could absorb water at lower depths and transmit it to other roots that were in soil with a moisture content at or about the wilting coefficient. It was assumed that when roots were supplied with water after this fashion they might penetrate soils from which at that time they could not obtain water. In this regard Shantz (1927) stated that there are certain trees of the African grasslands that have the ability to extend their roots into dry soil and in this manner prepare to absorb moisture rapidly when it comes.

Shantz, however, stated that the roots of ordinary farm crops do not possess this ability, and that they will penetrate only about the length of the root cap in a soil reduced to the wilting coefficient. Hendrickson and Veihmeyer (1931), working with sunflower, and Robertson, Kezer, Sjogren, and Koonce (1934), working with wheat, found that the roots of these plants do not penetrate into a soil that has been depleted of its moisture to or below the wilting coefficient (see Fig. 7*A*).

2. Relative Weight of Roots and Tops.—Tucker and Von Seelhorst (1898) grew oats in soil containing 42, 45, and 49 per cent of its moisture-holding capacity and found that the ratio of the dry weight of the tops to the dry weight of the roots was 6.6, 11.5, and 13.2, respectively, for each of these soils. They found (*a*) that the weight of the aerial parts increased in both grain and straw with the rising water content of the soil and (*b*) that with but one exception the greatest weight of roots was produced in the soil with the smallest amount of water.

Kiesselbach (1910) at the Nebraska Experiment Station grew corn plants in soil with a moisture content of 98, 80, 60, 40, and 20 per cent of saturation. The dry weight of the tops of these plants was, respectively, 8.6, 6.7, 7.2, 6.1, and 5.2 times the weight of the roots. He concluded that root development varies inversely as the soil-water content, and that plants which have their early growth in a relatively dry soil may be able to withstand drought better on account of having a greater surface exposed to the soil particles. The observations of the above investigators have further been confirmed by Polle (1910) working with wheat and barley, by Harris (1914) and Miller and Duley (1925) working with corn, and by Gordienko (1930).

Weaver (1925) grew Nebraska White Prize corn for 5 weeks in fertile loess soil at water contents of 9 and 19 per cent, respectively, above the hygroscopic coefficient. In the wet soil the area of the tops (including the stem and counting both surfaces of the leaves) was 82 per cent of that of the roots, but in the drier soil the tops had only 46 per cent as great an area as the roots. The absorbing area of the roots (exclusive of root hairs) was 1.2 times as great as the area of tops in the wet soil and 2.2 times as great as the area of the tops of plants grown in the drier soil.

Jones (1931), and Jones and Haskins (1935) found that the water content of soil in porous pots was lower than that in nonporous ones. A comparison of the root systems of plants grown in porous and nonporous pots invariably showed a larger root system in the porous pot. The ratio of tops to roots in the former case was 3.3, while in the latter it was 4.0.

B. Soil Nutrients and Root Development

The effect of the kind and amount of nutrients in the soil, or other medium in which the plant is grown, upon root development has been studied along the three following lines: the influence (1) on the direction taken by the roots, (2) on the number of roots, and (3) on the relative weight of roots and tops.

1. Direction of Root Growth.—The experimental evidence in regard to the response in direction of roots to nutrients is very limited. It is also difficult to interpret the data that have been obtained in this regard, since there is no way of knowing whether the observed curvatures of the roots are due to the attraction or repulsion of the nutrients in question or to the injury of portions of the root by them. The main work in this regard has been done by Newcombe and Rhodes (1904), Lilienfeldt (1905), and Porodko (1911 to 1914). Newcombe and Rhodes (1904) worked especially with the primary root of the seedlings of *Lupinus albus* and *Cucurbita pepo*. Blocks of gelatin, in which were dissolved the nutrients or chemicals to be investigated, were pressed gently, but closely, against the root on two sides, and after a given period observations were made to determine whether the reaction of the root to the dissolved substances had been positive, negative, or neutral. In one experiment with a block on one side containing only distilled water, and the block on the other containing a 1.5 per cent solution of disodium phosphate, all the 10 roots of *Lupinus albus* under observation after 24 hr. had grown into the block containing the solution of phosphate, where they quickly perished. In another case with distilled water on one side and a 0.01 per cent solution of copper acetate on the other, 13 of the 16 roots of *Lupinus albus* of the experiment had grown into the distilled water after a period of 24 hr. In the same manner the behavior of the roots was tested out when two unlike salts of equal osmotic pressure were brought at the same time against the opposite sides of the root. The chemicals used were calcium nitrate, potassium nitrate, and magnesium sulphate, and the following table shows the observed direction of growth for the roots of *Lupinus albus*:

	Neutral	
$KNO_3 \leftarrow 9$	2	$0 \rightarrow Ca(NO_3)_2$
$KNO_3 \leftarrow 10$	0	$1 \rightarrow MgSO_4$
$Ca(NO_3)_2 \leftarrow 0$	4	$10 \rightarrow MgSO_4$

The roots of *Cucurbita pepo* under the same conditions as those of *Lupinus albus* did not behave in the same manner, which indicates that the response of roots to a given chemical will depend to some extent upon the nature of the plant. Lilienfeldt (1905) repeated the work of Newcombe and Rhodes, using their method and one of his own after the following manner: Small containers 12 cm. high were filled with 3 per cent of gelatin in distilled water. As soon as it was sufficiently hardened a hole of about 20 cc. capacity was dug out of the center and filled with a solution of the chemical to be tested. The seedlings of *Lupinus albus* were carefully planted without injury by pushing the roots into the soft gelatin and so placing them that they were at distances varying from 5 to 50 mm. from the cavity. These were left in the dark for varying periods of time and the effect of the chemical diffusing through the gelatin on their direction of growth was observed. He obtained results identical with those of Newcombe and Rhodes when he used their method and with his own method the results are the same, with the exception of the salts of copper, zinc, mercury, and lead. By the method of Newcombe and Rhodes the roots reacted in a positive manner to these chemicals, but by the method of Lilienfeldt they responded negatively to them. The difference in results is evidently due to the amount of the salt that reached the root. In the method of Lilienfeldt, the amount of the poisonous salts that reach the roots by diffusing through the gelatin is much smaller than that which comes in contact with them when the salts are directly dissolved in the gelatin. A small amount of the poisonous salts stimulates growth on the contact side of the root and thus brings about a negative reaction in its behavior. A larger amount of the chemical would injure the cells of the root on the contact side and thus bring about a positive reaction in growth.

Kisser and Beer (1934) observed a similar behavior of seedling stems after the application of a salt solution to an incision on one side of the stem. High concentrations caused a positive curvature in all cases as a result of mechanical and osmotic effects. With low concentrations the negative curvatures were due to unequal growth following chemotropic response. The latter was most marked with lithium chloride, ammonium chloride, and manganese chloride, less with sodium chloride and potassium chloride, and did not occur with calcium chloride, strontium chloride, barium chloride, or magnesium chloride. Porodko (1911, 1914) worked with *Lupinus albus* and *Helianthus annuus* grown in the presence of various salts in agar as a diffusion medium and with solutions varying from one-thousandth normal to tenth normal. He found in the reaction to chemicals that the upper and lower portions of the root remained straight and that the intermediate portion was concerned in the curvature. He considered that the region in which chemotropic sensitivity of the root may be exhibited is limited to about 1 to 2 mm. of the root tip. He found that the intensity of the curvature depended upon the intensity of the diffusing forces or stimuli and that this, in turn, varies with the nature of the salt or diffusion substance used, with the concentration, with the thickness of the agar, and with the length of time the diffusion current acts. In general, the reaction as to the direction of the curvature manifested itself by a check in the growth and a positive curvature of the roots. This effect was observed in all the substances tested, whether they were electrolytes or nonelectrolytes.

Watenpaugh (1936) found that the development of alfalfa roots was definitely correlated with the pH reaction of the soil. Thus a pH of 4.8 definitely stopped the growth of roots, while they grew well in a medium with a pH of 5.0 or above.

Although the experimental evidence is rather contradictory in regard to the response in direction of growth to chemicals in the soil, two facts seem to be definitely established. In the first place, positive and negative reactions occur only within limited concentrations of the substances that have been used. In the second place, the roots must come in contact with the chemicals before they can in any way react to them. Roots cannot detect a substance needed for their growth and development and grow toward it, so that any nutrients that are applied to a soil must be so placed that they will come in contact with the roots in a solution of water if they are to be effective for the needs of the plant.

A practical application of this fact is observed in the application of fertilizer to corn. In a study of the development of the roots of corn seedlings, Millar (1930) noted that these roots tended to develop horizontally and to remain at comparatively shallow depths for a considerable time. Thus when the roots had reached a length of 9 in. they were only 5 in. from the surface in a sandy soil and about 3 in. from the surface in muck soil. Therefore, fertilizers placed on either side of the seed and slightly below it at the time of planting will be in the direct line of growth of the roots and should thus be used earlier by the plant than fertilizers placed directly above or below the seed.

Sayre (1934) considered that fertilizer should be placed in the soil at such a distance from the roots that they will escape for a few days the excessive soluble salt concentration in and immediately adjoining the fertilizer band. These salts diffuse rapidly from the fertilizer band into the soil solution and the excessive concentration of soluble salts therein is thus reduced so that the enriched region then becomes an excellent medium for root development.

2. Number of Roots.—One of the classic experiments used to demonstrate the effects of fertilizers upon the development of roots was that performed by Nobbe (1862) with corn plants growing in an infertile clay soil contained in glass cylinders. The experiment was so arranged that the nutrient salts ammonium sulphate, calcium nitrate, and dipotassium phosphate were added only to definite regions of the soil contained in each of six cylinders. After a growing period of 4 months, the plants were harvested and the root systems carefully isolated by a system of washing. The general form of the root systems remained normal and practically the same in all the experiments. The only difference that appeared was the local variation in the number of root branches. In the unfertilized portion of the soil the number of branches per unit of length of both primary and secondary roots remained small, while in the fertilized portions of the soil the number of branches of both primary and secondary roots was strikingly large. In the cylinder where the fertilizer was equally mixed throughout the soil mass the number of branches per unit length of both primary and secondary roots was the same in all parts of the vessel. Häveler (1892) grew corn and bean plants in vessels that contained alternate layers of sand and fertile soil rich in humus. He found that whenever the roots passed through the fertile soil a profuse branching of the several root orders occurred, but where the roots passed through the sand very little or no branching took place.

Frank (1893) grew corn and peas in such a manner that half of the roots of each plant grew in separate vessels. Each vessel contained the same amount of sand and was watered with the same nutrient solution, with the exception that to one vessel a portion of calcium nitrate was added, while the other received no nitrogen compounds. In the vessel containing the nitrogen-free medium, only a few small branch roots were formed, while the vessel containing the nitrogen compound was filled with a thrifty root growth. This same effect of nitrogen compounds on the secondary root formation of vetch, clover, corn, beans, and alfalfa was observed by Müller-Thurgau (1894).

Sayre (1934) found that roots of beans, cabbage, and tomatoes, coming in contact with fertilizers, developed an unusual number of fine fibrous roots and that this development was accompanied by stimulated plant growth. Rogers (1933) observed that the application of manure to gooseberries, currants, and pear trees increased both the root and top growth and the ratio of the weight of stem to roots.

Von Seelhorst (1902) carried on experiments to determine whether or not the use of fertilizers affected the number of roots of spring wheat, rye, peas, flax, beans, potatoes, beets, and barley. These plants were grown separately on two series of plots. One of the series had been fertilized annually for 25 years with fertilizers containing nitrogen, phosphorus, and potassium, although the amounts stated are not given, while the other series received no fertilizers during that period. The following table is a summary of the average number of roots found in an area of 314 sq. cm. at the various depths stated:

Soil	Number of roots at a depth in centimeters of				
	25	50	75	100	125
Fertilized	500	307	197	109	52
Unfertilized	459	206	158	92	43

Weaver, Jean, and Crist (1922) investigating the absorption of nitrates by roots found that where roots came in contact with a fertilized layer of soil they not only developed much more abundantly and branched more profusely, but also such a fertilized layer apparently retarded normal penetration into the soil below.

In a study of the effects of different fertilizer constituents on root development and anchorage of corn plants, Pettinger (1933) found: (*a*) the application of nitrogenous fertilizers to a soil deficient in nitrogen slightly increased the anchorage of corn plants when the supply of available potassium was ample but decreased this anchorage when the supply of potassium was limited; (*b*) the application of phosphorus or potassium to a soil deficient in these elements greatly improved the anchorage of the plants; (*c*) under the conditions of the experiments, both potassium and phosphorus increased root anchorage to a greater degree than did nitrogen, while potassium acted to a greater degree than did phosphorus; and (*d*) fertile soils were more productive of root growth than unfertile soils.

3. Relative Weight of Roots and Tops.—In general, it may be stated that the application of fertilizers increases the weight of the roots but that the weight of the tops generally increases more in proportion under such conditions than the roots so that the ratio of the dry weight of tops to the dry weight of the roots is generally higher in a fertile than in an infertile soil. Nitrogenous fertilizers especially increase the growth of tops as compared to the roots, while phosphorus as a rule increases the weight of the roots in a greater proportion than the tops. Harris (1914) found that the ratio of the weight of tops to the weight of roots of wheat seedlings was always greater with plants grown in concentrated than in dilute soil solutions. The concentrated soil solutions always produced more tops than the dilute, but the root growth was always greater in the dilute solution. The following table from the work of Harris (1914) shows that the proportion of roots and tops is affected more by a change in water content than by a change in the amount of nutrients:

Dry Matter of Wheat Tops and Roots at Various Stages When Grown in Soil with Different Amounts of Fertilizer and Moisture

Fertilizer	Per cent soil moisture	Boot stage, grams			Bloom stage, grams			Maturity, grams		
		Tops	Roots	*T/R*	Tops	Roots	*T/R*	Tops	Roots	*T/R*
None..........	30	35.3	15.8	2.2	50.5	10.7	4.7	74.9	9.3	8.1
None..........	15	15.2	10.0	1.5	19.3	8.8	2.2	29.2	9.5	3.1
Complete......	30	45.4	22.8	2.0	76.2	14.3	5.3	102.0	11.7	8.7
Complete......	15	27.8	20.7	1.3	39.9	12.4	3.2	52.3	13.2	4.0

Turner (1922) found that barley and corn showed significant increases in the ratio of tops to roots as the nitrate concentration of the solution was increased. The effect of the nitrates was shown to be independent of the concentration of the solution and the hydrogen-ion concentration. Flax plants did not show this increased ratio even after a period of 40 days or more. He concluded that the increased ratio of tops to roots which results from increasing the amount of nitrate in the solution may be explained on the basis of the increased use of carbohydrates in the tops, due to their combination with the nitrogen. This results in a decrease in the supply of the carbohydrates for the roots, which may bring about an absolute or relative reduction of root growth. It has been found that barley grown in the shade has a greater ratio of top weight to root weight than that grown in the unobstructed sunlight. Turner grew roots in pure cultures with a supply of sugar and found that nitrates did not check root growth directly but that they increased the growth of roots in all cases. Willard and McClure (1932) found with bluegrass that heavy fertilization with nitrogen greatly decreased the relative amount of underground parts.

It was found by Johnston and Miller (1934) that heavy leaf-rust infection on a susceptible variety of wheat resulted in a rapid and severe deterioration of the roots as indicated by discoloration, a decrease in the number of fibrous roots, and a marked loss in weight. Murphy (1935) found in oats that the ratio of roots to tops was greatly decreased by crown-rust infection, the decrease being in proportion to the duration and severity of infection.

Biswell (1935) showed that half shade favored the growth of seedlings of black walnut, buckeye, red oak, shellbark hickory, and hard maple but retarded the growth of honey locust, box elder, and sycamore. However, the root systems were deeper and more branched in all cases when the seedlings grew in full light.

Reid (1933) found that the ratio of the green weight of tops to the wet weight of roots in velvet bent grass under different degrees of sunlight was 1.1 for plants that were in the sun all day, 3.0 for plants that were in the sun all forenoon only, 3.3 for plants in the sun in the afternoon only, and 5.9 for those plants in the shade all day. For metropolitan bent grass the ratio was 3.3 in full sunlight and 8.3 in partial shade.

C. Oxygen Supply and Root Development

1. Aeration of the Soil. *a. Methods.*—Numerous benefits are generally derived by crop plants through the cultivation of the soil previous to their planting and at intervals during their growing season. The main benefits thus derived are (1) the conservation of moisture by increasing its reception and retention, (2) the eradication of weeds, (3) the correction of the physical texture of the soil, and (4) the adequate aeration of the soil. The aeration of the soil is beneficial to the plants in an indirect way, since an adequate supply of oxygen is necessary for the proper functioning of the bacteria that elaborate the various materials of the soil into a form in which they can be absorbed and utilized by the plants. In a direct way the aeration of the soil benefits the plant by furnishing a supply of oxygen, which is essential for the proper functioning of the protoplasm of the cells of the root. The general texture of the soil will determine to a great extent the benefits to be derived by increasing the oxygen supply by cultivation. Many soils are so light and porous that a deficiency of oxygen in them is not a limiting factor for growth, and it is a well-known fact that in many regions cultivation during the growing season is unnecessary for crop production provided the surface of the soil is kept free of weeds and is not too compact to absorb the rainfall. In very compact and in waterlogged soils, special methods in addition to cultivation must be applied in order to insure the proper aeration for normal plant development.

Howard (1915, 1916, 1917, 1918) has reported that special methods are necessary to bring about the proper aeration of the soil in the Bihar region of India. The soils in this region are composed of fine particles, become very compact upon wetting, and do not crack when the surface dries, thus preventing natural aeration. This condition is corrected by manuring with straw, by subsoiling, by the application of fragments of tile and brick, and by the proper sequence of crops. *Cajanus indicus*, the pigeon pea, confers a great benefit on a crop following it, especially tobacco. The root system of the pigeon pea grows to a great depth and is extensive laterally. White ants eat out the roots and leave air passages throughout the soil. In converting rough grassland into cultivated fields, sweet potatoes are planted as the first crop. They seem to thrive even if aeration is poor, and the swelling of the roots bursts open the soil and admits air, so that, when the roots are removed, the numerous cavities furnish aeration for other crops that would not otherwise grow.

Bergman (1920) found by growing plants in soil in pots that were submerged in water that *Impatiens*, *Pelargonium*, and *Coleus* were sensitive to a scarcity of oxygen. If, however, the water surrounding these pots was aerated the plants would grow normally. The following experiment is worthy of note: Two plants of *Impatiens* were potted in soil and the pots submerged in water. Several plants of *Philotria*, an aquatic plant, were placed in the water surrounding one of the pots but not in that surrounding the other. The evolution of oxygen from the photosynthetic activity of these aquatic plants aerated the surrounding water, and thus the water of the soil in the submerged pots, sufficiently to provide for the oxygen needs of the roots of the plants growing therein. As a result this plant was in good vegetative condition after two months, while the other plant whose roots were not thus aerated wilted after three days and shortly thereafter perished.

A sensitive and practicable method for comparing the oxygen-supplying powers of the soil atmosphere under different environmental conditions has been devised by Hutchins (1926). This method is based essentially on a colorimetric determination of the time required to produce a standard color change in an alkaline aqueous solution of pyrogallol by the absorption of oxygen. This oxygen is brought to the indicator in a continuous stream, having been absorbed from the environment in question by a very thin-walled, water-impregnated, porous porcelain conical-shaped absorber. The absorber has the capacity for absorbing oxygen and delivering it at least as rapidly as that element is supplied to the absorbing surface of the medium in which it is placed. The numerical results are termed "indexes" of the environmental oxygen-supplying power and are expressed as milligrams of oxygen supplied per hour through a square meter of absorbing area exposed to the environment under consideration.

By this method it was found that the oxygen-supplying power of the soil is higher the less its depth, the dryer, and the more loosely it is packed. With a very wet loamy garden soil the index value for a depth of 40 cm. is only about 0.7 per cent as great as that for a depth of 15 cm. The oxygen-supplying power varies also with the kind of soil and with the prevalence of aerobic organisms and readily oxidizable materials in the soil surrounding the absorber. Soil temperature and barometric pressure may also have an influence.

b. Effects on Plant Growth.—Two important observations have been made on the effect of soil aeration in the production of rice. As high as 20 per cent loss of the expected crop of rice has occurred due to a disease known as "straight head." In this trouble the flowers remain sterile and the panicles do not droop. Tisdale and Jenkins (1921) found that this condition of the plants was brought about mainly by the lack of proper soil aeration. Normal plants have few large roots and many branches and root hairs, while the plants affected with the disease show numerous large roots and very few branches and root hairs. For the control of this disease they recommended the irrigation of the plants ten days after they emerge or when they are 6 to 8 in. tall. The water should be allowed to drain off gradually, and if the soil is not free from water in 5 to 6 weeks after irrigation, it should be drained off and the land allowed to remain dry for 2 to 3 weeks. If the soil becomes dry enough to crack and the plants turn yellow, the results will be better. Water should then be turned on and allowed to remain until the end of the season. In the growing of swamp rice at Coimbatore, India, Harrison (1913) presented evidence to show that the surface film of algae and other organisms on these soils plays the leading part in the formation of the oxygen required by the roots of the rice. He believed that the aeration of these soils by the atmospheric oxygen is not so effective in promoting root aeration as is the aeration secured by the water that percolates through this surface film of algae. When certain rains fail to fall in the Bihar region, even if the soil is moist the rice crop is poor. This result is considered to be due to the failure of water to filter through the film algae and aerate the roots.

Albert and Armstrong (1931) found that a larger percentage of fruit buds were shed from cotton plants that were grown under conditions of high soil moisture than from plants grown under more nearly optimum conditions of soil moisture. The percentage of oxygen was lower and the percentage of carbon dioxide was higher in the soil with the higher content of moisture. They considered, therefore, that a high percentage of carbon dioxide and a low percentage of oxygen in the soil are associated with an increase in the percentage of shedding of the fruit buds of cotton.

c. Response of Roots.—Arker (1901) and Hall, Brenchley, and Underwood (1914) have found that the rate of root growth and top growth in both soil and water cultures may be increased by pulling a stream of air through the medium. The following table is a brief summary of the results obtained by Hall, Brenchley, and Underwood with barley and lupine plants grown in water cultures:

Plant	Treatment	Average dry weight, grams
Barley	Not aerated	1.31
	Aerated continually	2.12
Lupine	Not aerated	0.83
	Aerated continually	1.53

Bouygues (1928) noted that the roots of *Salix purpurea* stopped development after a short time when grown in an aqueous solution from which the air was excluded by an oily covering, but whenever the plants were removed to a solution in contact with air, the growth of roots was resumed.

Weaver and Himmel (1930) grew plants of the great bulrush (*Scirpus validus*), cattail (*Typha latifolia*), reed (*Phragmites communis*), and tall marsh grass (*Spartina michauxiana*) in dry soil, moist soil, alternately saturated and drained soil, and waterlogged soil. With the exception of *Scirpus*, the ratio of the dry weight of tops to roots was less in the drained than in the waterlogged soil. A response to poor aeration was shown in all experiments by the development of a shallow root system with fine, much-branched roots having a large surface in proportion to volume. Root growth increased with decreasing water content until moisture became a limiting factor. Similar observations were made by Dean (1933) on *Hibiscus*, *Acorus*, *Sagittaria*, and *Typha*, for in unaerated cultures the plants established shallow root systems, while in aerated cultures the roots ramified throughout the entire soil. Tincker (1935) noted with potted lilies that adequate drainage and aeration not only influenced the vigor of the leaves but also the size and color of the flowers.

Loehwing (1931) grew sunflower, flax, wheat, and soybeans in soil that was aerated for ½ to 5 hours daily. Two months after planting, the roots of the plants grown in aerated soil were more numerous, longer, and more fibrous than those grown in nonaerated soil. The total root surface of the aerated plants was at least twice that of the control plants. The root systems of aquatic plants, such as *Typha*, *Sagittaria*, and *Hibiscus*, also greatly increased in size in aerated media of sand, clay, or muck soils.

Heinicke (1932) considered that a low supply of oxygen is the primary cause of injury to apple trees when their roots are submerged.

Interference with the process of respiration is apparently the cause of the injury and eventual death of the roots due to the exclusion or depletion of the oxygen supply of the soil. The first visible effects of the lack of oxygen to the roots are the wilting of the plant and a lowering of the rate of transpiration. These effects are apparently due to the fact that the protoplasm does not function properly as an absorbing membrane. Hunter and Rich (1925) found that the rate of transpiration and the intensity of respiratory activity of the shoot of *Impatiens balsamina* are increased by the aeration of the soil about the root system.

It is suggested that these phenomena are due to the removal of carbon dioxide that has resulted from the respiration of roots and soil organisms. The concentration of this gas may become great enough to hinder or prevent absorption by the roots. The rate of elongation of the stem is governed by the degree of turgor pressure in the individual cells. It is thus conceivable that increased root activity in absorption, due to a lowering of the concentration of carbon dioxide by soil aeration, would tend to develop a higher degree of turgor pressure in the living cells of the shoot. This increased turgor would thus influence the rates of transpiration, growth, and respiration in that portion of the plant.

It has been shown by Cannon (1915 to 1925), Free and Livingston (1917), and Bergman (1920) that different species of plants vary widely in their response to the variations in the amount of oxygen in the soil, and these differences in responses are quite as significant as the responses to temperature and to available water. Cannon found that *Prosopis*—the mesquite—and *Opuntia*—a cactus—have roots that are unlike in their response to the oxygen supply in the soil. His experiments showed that if pure carbon dioxide filled the air space of the soil, growth ceased in the roots of both plants. The reduction of the oxygen to 12 to 15 per cent did not stop growth in the roots of *Prosopis* but did so in the roots of *Opuntia*. The roots of *Prosopis* continued to grow when only 2 per cent of oxygen was contained in the soil air. Free and Livingston (1917) found that the roots of the willow (*Salix nigra*) could endure successfully the almost complete exclusion of oxygen from the soil. Coleus and heliotrope were injured in a short time, so that within 12 to 24 hr. the absorption of water by the roots ceased.

2. Relation of Temperature to Oxygen Needs.—Cannon (1923 to 1925) has observed that the minimum oxygen supply is not a fixed amount but that it varies directly with temperature change. If the roots of *Potentilla anserina*, which inhabits swampy ground, are placed in a soil whose atmosphere consists of 1.2 per cent oxygen and the balance nitrogen gas, growth of the roots does not take place at 27 and 30°C., but at 18°C. it is about one-fourth normal for that temperature. In soils whose atmosphere contains 2 per cent oxygen, the growth rate is about one-fourth normal at 30°C., one-third normal at 27°C., and normal at 18°C. In the case of corn with the roots in soils containing 3 per cent oxygen, the rate of growth at 30°C. is about one-sixteenth normal, at 20°C. about one-fifth normal, and at 18°C. about one-third normal. When the percentage of oxygen is increased to 3.6, the rate of growth at 30°C. is about one-third the expected rate under normal conditions of aeration, while at 18°C., it is about two-thirds normal. These examples show that

> . . . there comes a point in the diminution of the oxygen content of the soil atmosphere when the growth of the root ceases because there is no longer a sufficient amount of this gas to supply the demands for energy correlated with physiological activities of higher temperatures.

On this account it can be inferred, for instance in the case of corn, that to attain a fair rate of growth at a time of high soil temperatures the aeration of the soil must be good, otherwise the rate of growth would be considerably cut down.

Collison (1935) noted that root elongation, and therefore absorption, assimilation, and respiration, took place in apple trees when the air temperature was below zero and apparently at a soil temperature not far above the freezing point.

3. The Sources of Oxygen.—The oxygen used by plants may have multiple origins. It may enter the plant directly from the atmosphere, it may enter the roots dissolved in water, or it may be liberated within the plant as a product of photosynthesis. The oxygen from the substratum may conceivably aerate the roots and perhaps the stem. That which is absorbed by the stem and leaves from the air may aerate the stem and perhaps the roots, while the oxygen that is liberated in photosynthesis may aerate both the stem and roots. The oxygen derived from photosynthesis has been designated "internal oxygen" by Cannon (1932). He noted in his experiments with sunflower and willow that in the majority of cases the roots showed a slower rate of oxygen consumption when the shoots were in sunlight than when they were in the dark. He considered that this indicated that the roots obtained a portion of their oxygen from the aerial parts exposed to sunlight. This indication was further substantiated by the work of Cannon, Demaree, and Purer (1933) who found in the willow that regardless of whether the temperature was high or low the expected drop in the rate of oxygen consumption by the roots occurred when the aerial portion of the plant was exposed to light.

4. General Observations.—Cannon (1917) has observed that an increased air supply to the roots of certain desert plants favored root branching. Miller (1916) and Weaver (1919) have noticed in isolating root systems in the field that a root upon passing through a cavity in the soil branches profusely so that in many cases it fills the entire cavity. This behavior of the roots is apparently due to the combined effects of moisture and the oxygen supply. The clogging of tile drains by roots is probably due to the same causes. The superiority of the roots of soybeans grown in aerated nutrient solutions over those in unaerated solutions was found by Allison and Shive (1923) to be due not to the higher dry weight of the substance produced but to the development of a more efficient absorbing system. The main roots in the aerated cultures were long and slender and thickly beset with well-developed laterals almost to the very tips.

The evidence in regard to the growth of roots toward an oxygen supply is mostly indirect. Many of the main roots of corn and sorghum plants growing in the field under ordinary agricultural conditions extend almost horizontally if there is sufficient moisture in the uppermost portion of the soil. The laterals arising from these horizontal roots have been observed by Miller (1916) to extend vertically upward to within ½ in. of the soil surface. This behavior is considered to be due to the influence of the supply of oxygen in the upper part of the soil. The same explanation is given to the behavior of roots that after being forced to grow downward to a considerable depth through a vertical tile have turned directly upward upon emerging from it. Ewart (1894) found that the roots of peas, hemp, and wheat grown over water deprived of oxygen either turned slightly upward or grew along the water surface, while if the water were aerated they grew directly downward into it.

Clements (1921) found in experiments with sunflowers growing in sealed pots which were glazed, with the exception of certain small, square surfaces, that in all cases the roots showed right-angled curvatures toward the unglazed squares.

Bennett (1904) worked with corn, squash, radish, peas, and lupine to determine whether any response could be obtained to oxygen and carbon dioxide from the roots of these plants. In general, the experiment consisted in allowing the gases to diffuse directly upon the roots growing in soil or water as a medium. Three hundred and sixty-two reactions were observed upon the roots of corn, radish, and squash growing in water upon which either carbon dioxide or oxygen was allowed to diffuse from one side only. Of this number 84 were positive, 72 were negative, and 206 were neutral. In an experiment in which carbon dioxide diffused against one side of the root and oxygen against the other, 166 observations were made of which 37 were positive in their reaction to oxygen, 22 were positive to carbon dioxide, and 107 were neutral to both gases. In the experiment in which plants growing in the soil were so arranged that oxygen could diffuse on one side of the primary root and carbon dioxide upon the other, the results showed no more than those mentioned above, since, of the 221 observations made, 45 reacted in a positive manner toward the oxygen, 48 in a negative way, and 128 in a neutral manner. The evidence in these three experiments refutes the idea that roots exhibit aerotropism.

D. Soil Temperatures and the Response of Roots

It is the intention here to discuss more particularly the response of roots as shown by curvatures induced by the changes in the temperature of the medium in which they are growing and to reserve for a subsequent chapter the effect of temperature changes upon the rate of root growth. Until 1914 our knowledge of the thermotropic curvature of roots was very confusing. The experimental results demonstrated that different plants exhibit little uniformity in behavior, since some roots were positive in their reaction at a given temperature, while others at the same temperature gave a negative or neutral reaction.

Eckerson (1914) investigated the response of roots when exposed to unequal temperatures on their two sides and arrived at results that cleared up the apparent discrepancies and contradictions in the results obtained by earlier investigators. In these experiments, Eckerson grew or placed the seedlings in sphagnum moss in a Ganong differential thermostat that furnished a gradation in temperatures from 5 to 58°C. If the roots curved toward a region of higher temperature, they were considered to have responded positively, while a curvature away from higher-temperature areas was listed as a negative response. After the roots of the radish had been in the thermostat 2 hr., they showed the following reactions in curvatures: 7 to 15° positive, 16 to 23° no curvatures, 24 to 36° positive, and 38 to 51° negative (see also Treitel, 1924). A study of the permeability of the cells of the roots of these plants at different temperatures by means of the plasmolytic method with potassium nitrate solutions showed that it increased from 10 to 18°, did not change from 18 to 24°, increased from 24 to 40°, and then decreased. A comparison of the changes in permeability and the response in curvatures at the different temperatures shows that in every case where the temperature induced an increased permeability there was a positive curvature, where no change in permeability occurred no curvatures resulted, while at temperatures causing a decrease in permeability, negative curvatures resulted. These observations show that with unequal temperatures on opposite sides of a root a curvature is produced only when the cells of the root are more permeable at one of the temperatures than the other. Those cells subjected to a temperature at which they are more permeable to dissolved substances are consequently less turgid, and this results in a shrinking of the tissue on that side of the root and a consequent mechanical curvature, and in all cases the more permeable side of the root becomes concave. Eckerson also studied the changes in the permeability of the cells of the roots of the various species of plants that had been investigated by Wortmann (1885) and by Klercker (1891) and found that the increase or decrease in the permeability at the various temperatures paralleled the positive or negative curvatures that had been observed by these investigators. She summed up her observations in the following statement: (1) Thermotropic curvatures of roots vary with the temperature and with the species. (2) The permeability of the cells of the roots to dissolved substances varies with the same factors, and wherever the thermotropic reaction of the root changes the permeability also changes. (3) Heat does not act as a stimulus but by affecting permeability is a direct factor in producing curvature, so that thermotropism is not to be considered a tropism but a turgor movement.

E. Soil Texture and Root Development

A hard, compact soil limits the extent of root systems to a marked degree, and Weaver (1919, 1920) observed that in a compact soil the roots are more or less contorted or kinked, while the branching is decidedly less than in a soil of loose texture.

Carlson (1925) in studying the effect of soil structure on the character of alfalfa root systems found that in compact soils all varieties and strains develop branch roots, while in open soil the taproots predominate. He also found that the different strains of alfalfa have inherent root characteristics which as a rule do not develop until after 3 or 4 months of normal growth. McCool (1928) observed the development of the roots of alfalfa, sweet clover, and red clover in three different types of soil. The results obtained showed that there were marked differences in the nature and extent of the roots of these plants. Thus the average length of alfalfa roots in one soil was 40 in. and in another soil only 8 in. Sweet-clover roots attained a length of 43 in. in one soil, 13 in. in another, and but 10 in. in a third soil.

Weaver and Crist (1922) made some observations on the relation of hardpan to root penetration. This soil layer, in general, delimits the depth of penetration of the roots of crop plants, but under exceptional moisture conditions they may penetrate the hardpan and extend 2 to 3 ft. beneath it. No modifications in root habits were apparent in the hardpan layer. Since roots will not grow into dry soil, the conclusion is that water penetrated the hardpan and the soil beneath.

Singh (1922) noted that the roots of wheat plants grew best in pure sand provided it was supplied with the proper amount of nutrients and water. At Rothamsted roots grew better in a mixture of sand and soil than in soil alone. Jensen (1931) showed that the type of soil has an important relation to the correct tillage methods of sugar cane.

It was noted by Taubenhaus, Ezekiel, and Rea (1931) that root strangulation of cotton occurs when the upper portions of the taproot and laterals of young seedlings are caught early in the season in a subsurface layer of hard, dry clay which prevents further development. Affected plants die when the constricted areas are killed by the hot dry soil or when the moisture supply that can be transported through the constricted area becomes inadequate for the needs of the plant.

In prairie soil in Nebraska, Weaver, Hougen, and Weldon (1935) studied the relation of root distribution to organic matter, in the case of *Andropogan scoparius* and *Andropogan furcatus*.

Except in the upper 6 in. of soil, there is an approximate linear relation between the amount of root material and the amount of organic matter in the various soil horizons.

F. The Response of Roots to Light

Some roots show a marked reaction in response to light. This is especially true of some members of the mustard family. The roots of these plants are those that are so frequently pictured in textbooks as examples of the negative action of roots to the stimulus of light. Many roots, however, show little or no response to light.

II. THE EXTENT OF ROOT SYSTEMS

This general subject can be conveniently discussed under two general headings: (*a*) the total number and combined length of the roots of a given plant, and (*b*) the extent of roots in the soil.

A. Total Number and Combined Length of Roots

Nobbe (1869) measured and counted the roots of winter wheat and rye and found that the roots of the first, second, third, and fourth order of 55-day-old plants growing in the soil numbered 16,000 for rye and 10,700 for wheat. The combined length of the roots of the rye plant measured 492 ft.; and that of the wheat plant, 283 ft. A mature wheat plant grown in the soil was found to possess 67,200 roots of the first four orders and to have a combined length of 1,730 ft.

Clark (1874) examined the roots of a squash plant growing in the greenhouse at the Massachusetts Agricultural College and estimated the total length of these roots at 15 miles. This estimate included not only the root system proper but also the roots with all their branches originating from the 70 nodes of the plant. In 1912 Gibson repeated this work with a closely related variety and found the combined length of both the main and nodal root systems to measure only 281 ft. at the time of full vegetative development. Hellriegel (1883) estimated the total length of the root systems of mature barley and oat plants grown in a fertile soil to be 134 and 168 ft., respectively. Weaver (1925) found that the root system of a single corn plant in the eighth leaf stage has from 8,000 to 10,000 laterals arising from 15 to 23 main roots. The area of the root system of winter wheat measured at various intervals from Sept. 30 to Nov. 29 exceeded the area of the aerial parts by 10 to 35 per cent, as shown by the following table:

Date	Total area of roots exclusive of root hairs, sq. cm.	Total area of tops, sq. cm.
Sept. 30	9.6	7.6
Oct. 10	31.4	22.5
Oct. 20	61.5	50.5
Oct. 30	105.1	89.0
Nov. 14	237.5	215.3
Nov. 29	309.7	280.0

Pavlychenko (1937) determined that the total extent of the roots of the first, second, and tertiary orders of mature wild oats, Marquis wheat, and prolific spring rye was 54, 44, and 49 miles, respectively, when the plants were spaced 10 feet from each other. However, when these plants were grown in 6-in. rows, with 18 to 20 plants per foot, the root systems were from 88 to 99 times smaller than in the former case.

Dittmer (1937) grew a plant of winter rye, *Secale cereale*, in soil, within a wooden container 12 in. square and 22 in. deep, for 4 weeks until the plant had reached a height of 20 in. The roots were then washed out and the number, area, and length of those of the first, second, third. and fourth orders determined. The number of root hairs, together with their area and extent, was also calculated. He found that the total length of the roots of the four orders was 387 miles with an area of 2,554 sq. ft. The total number of root hairs on the roots of this plant was calculated at about 14.5 billions, with a total length of 6,603 miles and an area of 4,321 sq. ft. The surface of the aerial parts of this plant totalled 51.5 sq. ft., so that the area of its subterranean parts was 130 times that of the tops.

B. Extent of Roots in the Soil

From 1880 to the present time, work has been done by Goff and Beckwith (1883 to 1888), Hays (1888 to 1894), King (1894 to 1893), Headden (1896), Georgeson and Payne (1897), Ten Eyck (1899, 1900, 1904), Cottrell (1902), Shepperd (1905), Schulze (1906), Rotmistrov (1907 to 1909), Miller (1916), Weaver (1915 to 1924), Collings and Warner (1927), Lee (1926), Arny and Johnson (1928), Gursky (1928 to 1929), Krassovsky (1929), Roxas and Villano (1930), Laird (1930), Farris (1934), West (1934), Sperry (1935), and Maximow and Kruzilin (1936) in studying the general distribution of the roots of cultivated plants in the soil and in perfecting methods for their isolation under field conditions. The general observations of these investigators and a brief summary of the conditions under which they worked are recorded in the table on page 139.

The roots of woody plants have been studied by Peren (1923), Allan (1926), Laitakari (1927), Gemmer (1928), Rogers and Vyvyan (1928 to

EXTENT OF ROOT SYSTEMS OF CULTIVATED PLANTS

Investigator, location, and date	Plants	Extent, inches		Remarks
		Vertical	Lateral	
Goff and Beckwith, New York (1883–1888)	18 garden plants	10 to 34	10 to 120	Clay loam 6 to 10 in. deep. Tenacious subsoil
Hays, Minnesota (1888)	Red clover	68	20	5-month-old plants. Soil fertile, 6 ft. deep
	Alfalfa	80	12	
	Mammoth clover	57	12	
King, Wisconsin (1892–1893)	Clover, oats, barley	48	....	Plants all mature. Clay-loam soil. Stiff subsoil
	Winter wheat, corn	48	24	
Headden, Colorado (1896)	Alfalfa	150	6	Plants 6 years old. Stiff clay soil
Georgeson and Payne, Kansas (1897)	White milo	....	72	In hills 8 ft. apart. Black loam soil 8 in. deep. Stiff clay subsoil
	White kafir	48	48 to 70	
	Corn	....	48	
Ten Eyck, North Dakota (1899–1900)	Spring wheat, oats	48	9	Soil a fertile loam 4 ft. deep. Oats and wheat in rows 18 in. apart
	Corn	42	42	
	Potatoes	18	24	
	Sugar beets	40	24	
Cottrell, Kansas (1902)	Alfalfa	120	....	Stiff subsoil. Did not find end of root
Ten Eyck, Kansas (1904)	Corn	48 to 60	....	Heavy clay-loam soil. Stiff clay subsoil
	Kafir	42	36 to 42	
	Winter wheat	48		
	Oats and barley	52		
Shepperd, North Dakota (1905)	Rye	36	....	Red clover 2 years old. Alfalfa 3 years old
	Spring wheat	48		
	Barley and oats	36		
	Red clover	48		
	Alfalfa	60		
Schulze, Germany (1906)	Rye	70		
	Wheat	74		
	Barley	70		
	Oats	90		
	Vetch	86		
	Potato	97		
Rotmistrov, Russia (1907–1909)	Corn	45	53	
	Potato	24	40	
	Yellow sorghum	42	44	
	Cucumber	42	33	
	Cotton	38	41	
	Rye	47	24	
	Wheat	41	37	
	Beet	58	44	
Miller, Kansas (1916)	Corn, kafir, milo	72	36 to 48	Grown on fall-irrigated, sandy loam 10 ft. deep
Weaver, Kansas, Colorado, South Dakota, Nebraska (1920–1921)	Rye	41	12	Short-grass plains
	Oats	41	9	
	Wheat	32	4 to 8	
	Rye	58	12	Mixed prairie
	Oats	50	9	
	Wheat	53	4 to 8	
	Rye	64	12	True prairie
	Oats	53	9	
	Wheat	65	4 to 8	
	Alfalfa	60	12 to 18	Grown on flood plain. Plants 116 days old. Eastern Nebraska
	Red clover	52	12	
	Sweet clover	64	12	
	Wheat, Durum	88	14	Upland silt-loam subsoil loose loess. Eastern Nebraska
	Wheat, Marquis	76	15	
	Barley	75	15	
	Corn	98	48	
	Potato	46	26	
Roxas and Villano, Philippines (1930)	Sugar cane	12 to 18	5	Aeration an important factor in determining depth
Laird, Florida (1930)	Bahia, Bermuda, St. Augustine, blue couch grasses	72 to 96		
Farris, New Jersey (1934)	Corn	24	24	Roots much less extensive than in semiarid regions
	Red clover (mature)	36		
	White potato	12		
West, Australia (1934)	Alfalfa	204	....	50 per cent of roots in upper 3.5 ft.
	Wheat	108	....	78 per cent of roots in upper 2.8 ft.
	Beans	36		
	Rice	36	....	82 per cent of roots in upper 4 in.
Sperry, Illinois (1935)	Big bluestem	50		
Maximov and Kruzilin, Russia (1936)	Spring wheat	....	....	80 per cent of roots in upper 40 cm. of soil, irrigated

EXTENT OF ROOT SYSTEMS OF TREES AND SHRUBS

Investigator, location, and date	Plants	Extent, ft. Vertical	Extent, ft. Lateral	Remarks
Peren, England (1923)	Cherry tree	....	30	15 years old
	Apple tree	....	24	15 years old
Allan, New Zealand (1926)............	*Podocarpus spicata*	....	65	Tree, 52 ft. high
Laitakari, Finland (1927)............	*Pinus sylvestris*	....	60 to 90	On sandy heaths
Gemmer, S.E. United States (1928).......	*Pinus palustris*	....	30	
Rogers and Vyvyan, England (1928)	Lane's Prince Albert apple	....	....	10 years old; 10 per cent of fibrous roots within 1 yd. from tree; 40 per cent more than 2.5 yd. from trunk
Traub and Stansel, Texas (1930)	Fig tree	....	50	5 to 6 years old
Holch, Nebraska (1931)............	Bur oak	5.7	Exceeded spread of tops	First-year seedlings
	Shellbark hickory	2.5		
	Red oak	2.4		
	Linden	1.2		
	Walnut	4.5		
Beckenbach and Gourley, Ohio (1932)	Apple tree	5	....	Roots under sod, continuous mulch, and cover crop of rye. First foot, 411 roots; second foot, 225 roots; third foot, 98 roots; fourth foot, 61 roots; fifth foot, 44 roots
Oskamp, New York (1932)	Cherry trees	6	....	20 years old; 38 to 73 per cent of roots in surface foot; 17 to 30 per cent in second foot; very few in fifth and sixth foot
Rogers, England (1932)	Bramley seedling apple	3	20	Wet clay soil, grassland, water table at 3.5 ft.; tree, 26 years old
Weaver and Kramer, Nebraska (1932)	Bur oak	14	20 to 60	Mature trees, 50 to 60 years old, 35 to 45 ft. tall, 1 to 1.5 ft. in diameter, and 40 ft. apart
Heyward, Florida (1933)	Long-leaved pine	14	75	90 per cent of all laterals occurred in surface foot
Nutman, East Africa (1933, 1934).........	*Coffea arabica*	6	....	Mature plants
Rogers, England (1933)............	Gooseberry	8	....	Water table at 11 ft.
	Black currant	7	....	
	5-year pear trees	11	....	
Harmon and Snyder, California (1934)	20 species of grapes	1 to 5	....	21 per cent of roots in first foot; 48 per cent in second foot; 25 per cent in third foot; and 6 per cent in fourth foot
Marth, Maryland (1934)	Stayman apple trees	....	12	70 per cent more roots in first 6 in. than in second 6 in.
Rogers and Vyvyan, England (1934)	Lane's Prince Albert apple trees	9.5	2 to 3 times extent of branches	Trees 10 to 11 years old; only 10 per cent of fibrous roots in square meter surrounding the trunk
Duncan, Indiana (1935)	*Rhus, Rosa, Rubus, Sassafras*, and others	1 to 3.5		
McQuilkin, New Jersey (1935)............	*Pinus rigida*	....	25 to 35	Trees, 12 to 30 years old
Rogers, British Columbia (1935)	McIntosh apple tree	5	28	Trees 19 ft. high, on irrigated soil
Yeager, North Dakota (1935)	31 species of trees and shrubs	4	0.4 to 2.1 times the height of the tree	When some water supplied, roots spread less and penetrate deeper. 97 per cent of roots in first 4 feet

1935), Traub and Stansel (1930), Holch (1931), Weaver and Kramer (1932), Oskamp (1932), Rogers (1932, 1933, 1935), Beckenbach and Gourley (1932), Nutman (1933, 1934), Heyward (1933), Marth (1934), Harmon and Snyder (1934), Duncan (1935), and others. A summary of their observations and of the conditions under which they were made is shown in the table on page 140.

The investigations of the distribution of the roots in the soil have been undertaken for two main purposes: (1) to determine the effects of cultivation upon roots and (2) to obtain information concerning the relations of the roots to the water supply of the soil. Most of the investigations have been made with the latter point in view, and a study of the tables shows that most of the work with cultivated plants has been conducted in more or less semiarid regions where rainfall is the limiting factor in crop production.

Gericke (1923) noted in wheat that plants with roots 50 to 60 cm. in length had 20 tillers. If these roots were cut back to 30 cm., only 17 tillers developed, and, if pruned to 4 in. (10 cm.), only four tillers developed. He thus considered that the extent of tillering is dependent upon the size of the root system.

Some of the most recent work relative to the effects of cultivation upon the roots is that of Kiesselbach, Anderson, and Lyness (1935). They found in corn that a considerable range in cultivation may be practiced without materially affecting the yield. Thus, shallow, medium, deep, and close cultivation in one direction yielded respectively 32.9, 34.0, 35.2, and 34.5 bu. per acre during an 11-year period. Shallow cultivation was 4 in. deep and 9 in. from the plant, deep cultivation was 6 in. in depth and 9 in. from the plant, and close cultivation was 5 in. deep and 7 in. from the plant. The grain yield was lowered 12 per cent by pruning the roots 6 in. deep and 7 in. from the plant, and pruning the roots at a distance of 4 in. from the plant and 6 in. deep lowered the yield 20 per cent.

Nedrow (1937) pruned the roots of Sudan grass at a depth of 5 or 10 in. every 10 days. The pruning of 5-in. depths reduced the weight of tops 56 per cent and the weight of roots 49 per cent. The three successive prunings of the roots at the 10-in. depth decreased the weight of tops and roots 34 and 37 per cent, respectively. Deuber (1936) pruned the roots of small oak trees that had been injured by gas but found that after 2 years the tops of the pruned trees showed no measurable improvement over those of the controls.

1. Vertical Penetration.—Cannon (1911) made a study of the root systems of over 50 annual and perennial plants growing in the desert region of Tucson, Ariz. He found that the root systems of the annuals rarely penetrated the soil more than 8 in., and that the greatest root development took place in the upper 2 to 3 in. of soil. He observed that the depth of root penetration depended upon the depth to which the rains moistened the soils. The cacti have two specialized root systems—a vertical one serving as an anchorage for the plant and a horizontal one that functions mainly in absorption. Most of the horizontal roots of cacti were found in the upper 4 in. of soil. Cannon (1913) also found a similar distribution of the roots of plants growing in the Algerian Sahara. Weaver (1919) found that the roots of *Yucca glauca* or Spanish bayonet, a plant very common in the semiarid regions of the Western plains of the United States, penetrated the soil to a depth of 18 in. only, while the horizontal roots extended to a distance of 30 ft. or more.

Weaver (1920) made an extensive study of the distribution in the soil of the roots of both cultivated and wild plants growing in the grassland formation that extends from the Missouri River to the foothills of the Rocky Mountains. Due primarily to variations in rainfall and evaporation, three more or less distinct plant associations are to be recognized in this area and have been classified by Clements (1920) as (*a*) the true prairie, (*b*) the short-grass plains, and (*c*) the mixed prairie. The true prairies occur in regions of sufficiently favorable soil- and air-moisture conditions to permit the growth of large numbers of dominant tall grasses with which are intermixed a great number of other herbs, the whole forming a luxuriant vegetation. The plants in this region grow under a mean annual precipitation of 28 in. or over, of which 20 in. comes during the growing season. The short-grass plains are characterized by a few short, sod-forming grasses, especially ***Bulbilis dactyloides*** (buffalo grass) and *Bouteloua gracilis* (grama grass). This short-grass association extends from southwestern Nebraska and the western half of Kansas through eastern Colorado into northwestern Texas, northern New Mexico, and Arizona. In these regions the evaporating power of the air is high and the rainfall is limited, averaging usually 20 in. or less. The mixed prairie occupies a region of greater precipitation and more favorable conditions for plant growth than the short-grass plains, and one of the distinctive features of this association is the intimate mixing of the tall with the shorter grasses.

From his 1,500 observations of the root systems of 180 species of plants growing under varied conditions of this grassland formation in 25 stations in Kansas, Colorado, South Dakota, and Nebraska, Weaver concluded that the longest and most deeply penetrating roots are found where there is considerable rainfall, where much of this rainfall penetrates into the soil, and where the water table is relatively low. He emphasizes the fact that the maximum depth to which a few roots will penetrate is not of so much importance in the water relations of the plant as is the working or average depth to which the majority of the roots penetrate. It is also not so much a question of the depth to which roots penetrate as what proportion of the root system is functioning in absorption. Weaver found that 65 per cent of the species in the true prairie reached depths greater than 5 ft.; 42 per cent in the mixed prairie and only 13 per cent of the species examined in the short-grass plains reached beyond that depth.

Most of the species of the plains area were characterized by a fine system of wide-spreading laterals. This type of growth is evidently a response to the moisture in the upper portions of the soil resulting from summer showers. Practically all the deep-rooted plants in the short-grass plains area are well adapted to absorb moisture from the upper portion of the soil. Of the 45 species examined by Weaver (1920) in this region the lateral extent of the roots ranged from ½ to 3½ ft. This

HEIGHT OF TOPS, WORKING DEPTH, AND MAXIMUM DEPTH OF ROOT PENETRATION OF RYE, OATS, AND WHEAT IN THE THREE REGIONS OF THE GRASSLAND FORMATION

Station and crop	Height of tops, feet	Working depth of roots, feet	Maximum depth of roots, feet	Remarks
Winter rye:				
Short-grass plains: 5 stations, Colorado and Kansas......	2.7	2.8	3.4	Silt-loam soil
Mixed prairie: 3 stations, Colorado and Kansas..........	3.3	3.7	4.8	Sandy loam and silt loam
True prairie: 3 stations, eastern Nebraska............	4.9	4.0	5.3	Mostly silt-loam soil
Oats:				
Short-grass plains: 3 stations, Colorado and Kansas........	2.4	2.7	3.4	Silt-loam soil
Mixed prairie: 4 stations, Colorado, Kansas, and S. Dakota	2.7	3.4	4.2	Silt and sandy loam
True prairie: 5 stations, eastern Nebraska............	2.8	3.4	4.4	Silt and clay loam
Wheat (turkey red):				
Short-grass plains: 7 stations, Colorado and Kansas....	2.1	2.3	2.7	Silt loam
Mixed prairie: 4 stations, Colorado, Kansas, and S. Dakota	2.8	3.6	4.4	Silt and sandy loam
True prairie: 6 stations, eastern Nebraska............	3.3	3.8	5.4	Silt loam

contrasts markedly with plants growing in the true prairie region where the lateral spread of roots ranges from 1 to 4 in. The working depth of the roots of the majority of plants in the short-grass plains region was found to be from 1 to 5 ft. Weaver concluded that the root habit of a species is usually altered greatly when plants are grown under distinctly different conditions and that, in general, the distribution of roots conforms strikingly with that of the soil moisture, although a few species are found that show little or no variation to a change in habitat. The effect of the moisture supply on the depth of root penetration can be illustrated by quoting the data obtained by Weaver (1920) in his examination of the root systems of oats, winter wheat, and winter rye growing in the three regions of the grassland formation (see table above).

The mean annual precipitation was 16.7 in. for the short-grass plains, 18.4 in. for the mixed prairie, and 28.8 in. for the true prairie.

The relative development of the roots and tops of plants growing in the short-grass plains and mixed prairie as compared to those growing in the true prairie region is shown in the following table:

Relative Development of Cereal Crop Plants and Relative Precipitation in the Grassland Area

Plant	Short-grass plain	Mixed prairie	True prairie
Rye:			
Height of tops	56	66	100
Working depth of roots	69	92	100
Maximum depth of roots	65	90	100
Oats:			
Height of tops	85	94	100
Working depth of roots	79	95	100
Maximum depth of roots	77	94	100
Wheat:			
Height of tops	64	85	100
Working depth of roots	61	93	100
Maximum depth of roots	51	80	100
Mean annual precipitation	58	64	100
Precipitation July, 1918 to July, 1919	73	77	100
Precipitation January, 1919 to July, 1919	50	53	100

If the proper conditions of soil texture, moisture, oxygen, and temperature prevail, there is apparently nothing to limit the growth of plant roots downward except the time limit in the length of the life of the plant. The roots of cultivated perennial plants as a rule extend more deeply than those of annuals. The maximum depth of penetration of the roots of corn, sorghums, and the small grains that have been investigated is between 5 and 8 ft. The roots of alfalfa have been observed to grow downward to depths of from 6 to 14 ft. depending upon the age of the plant and the nature of the soil in which they were grown. Cottrell (1902) stated that the roots of alfalfa in a good soil grow to a depth of 30 ft., but he gives no experimental evidence to support his statement. Kiesselbach, Russel, and Anderson (1929) found under the conditions existing in eastern Nebraska that 6-year-old alfalfa drew upon the moisture of the subsoil to a depth of 33 ft., indicating that some of the roots had penetrated to that depth. Duley (1929) at the Kansas Agricultural Experiment Station observed that alfalfa used the moisture of the subsoil to a depth of from 10 to 25 ft., depending upon the age of the plants. Coburn (1907) in his book on alfalfa mentions an instance in which the roots of alfalfa grew downward through the soil into a mining tunnel 129 ft. beneath. This great length was possible in this case, since the roots followed the crevices that were present in the soil and rock overlying the tunnel. The major portion of the root systems of the cultivated plants, however, is found in the upper 4 ft. of soil, and in many cases most of the roots are in the uppermost 2 to 3 ft. of soil. In the case of alfalfa, where the maximum depth of root penetration has been observed to be 16 ft. or more, the major portion of the root system was in the upper 6 ft. of soil. In observations on the roots of sugar cane growing in the Hawaiian Islands, Lee (1926) found that more than 58 per cent of the roots of the plants growing in furrows were above the 8-in. level of the soil and that more than 85 per cent of the roots were in the uppermost 2 ft. of soil. On the other hand, the few roots that extend downward into the maximum depth undoubtedly do absorb considerable moisture and in periods of severe drought may be a very important factor in furnishing sufficient moisture to prevent the death of the plant.

The vertical penetration of the roots of the woody plants that have been studied is relatively shallow. A survey of the table on page 140 shows that the maximum depth of penetration of bur oak, long-leaved pine, apple trees, cherry, grapes, gooseberry, and currant ranged from 1 to 14 ft. The plants, however, that have been studied were those growing for the most part in shallow soil or over a high water table, conditions that are conducive to a shallow development of roots.

2. Lateral Extent.—The lateral extent of the roots of plants will depend upon two main factors: (*a*) the amount of rainfall and (*b*) the thickness of the stand. The difference in the lateral extent of roots of plants growing in the more arid regions as compared with those growing where rainfall is more abundant has been discussed under the heading of the vertical depth of root penetration, but it should be repeated here that the lateral extent of roots is greater the less the rainfall (see Fig. 7*A*). The thickness of the stand is an important factor in determining the lateral extent of roots. A plant growing by itself in the open will extend its roots farther laterally than one that is closely surrounded by other plants. Miller (1916) found that the roots of corn and sorghum which were grown in rows 44 in. apart extended from one row to the other and then turned downward. The roots of a single plant of corn or sorghum under such conditions thus extend through a cylinder of soil 7 ft. in diameter and 6 ft. deep and are distributed through a volume of soil that totals 230 cu. ft. The longest roots observed in this experiment were 10 ft. in length. They were those that extended laterally until they reached the opposite row and then turned downward. From the table on the extent of roots it can be observed that the lateral extent of the roots of the smaller cereals is only 4 to 12 in., due to the fact that the plants are in very close rows. The long lateral extent of the roots of corn and sorghum that were investigated by Georgeson and Payne (1897) was due to the fact that these plants were grown in hills 8 ft. apart.

The lateral extent of the roots of woody plants that have been studied is shown in the table on page 140. These roots show a wide lateral extent which is apparently due to the fact that most of the plants were growing under conditions that prevented or hindered deep vertical penetration. The lateral roots of woody plants extend from two to seven times farther than the spread of their branches. The minimum lateral extent shown in the table is 12 ft., and the maximum is 90 ft.

3. The Depth from Which Roots Absorb Water and Nutrients.—The question concerning the depth from which the roots of plants absorb is a very important one from the standpoint of practical agriculture. The extensive investigations concerning the depth of root penetration have shown that a large proportion of the roots of the cultivated plants extend to 3 or 4 ft., a depth much greater than that from which absorption was supposed to occur. This had led certain investigators to doubt the accuracy of statements that are found in the literature on soils concerning the depth from which plants absorb. Russel (1917) stated that "only the upper 6 or 8 in. of the soil is suited to plant life and the subsoil plays only an indirect part in plant nutrition." There seems to be considerable indirect evidence to refute the above statement, especially from observations of plant life growing in the Great Plains area. In the first place, the number of roots found in any given area of soil is no indication of their activity. It is a common observation that the first roots formed in the upper portion of the soil become suberized or cutinized so that they are totally incapable of absorbing, even if that portion of the soil is moist. It has also been frequently observed that during periods of drought the upper foot of soil has been depleted of its moisture below the wilting coefficient so that little or no absorption occurs there, yet the plant remains turgid and functions normally if there is moisture at the lower depths.

Miller (1916) found, in growing corn and sorghums on fall-irrigated land under seasonal conditions in which the rainfall penetrated the surface mulch little or not at all, that the moisture content of the soil became depleted as the roots penetrated downward. When the corn was tasseling and the sorghums heading, the moisture content of the upper 2 ft. of soil was below the wilting coefficient, yet the plants remained perfectly normal and matured a crop of grain. Determinations at this

Soil-moisture Content and Depth of Root Penetration of Corn, Kafir, and Milo in 1914 at Garden City, Kan.

Date	Percentage of moisture at depth in feet of										Greatest depth of roots, feet		
	1	2	3	4	5	6	7	8	9	10	Corn	Kafir	Milo
June 5	22.9	22.5	22.1	22.7	23.5	21.2	16.1	17.0	16.4	14.5			
July 2	14.6	20.2	21.2	22.8	22.1	21.8	...	...	...	...	1	1.5	1.5
July 10	11.8	17.1	19.5	23.6	24.6	21.4							
July 21	10.6	13.3	14.5	19.4	22.8	21.7	...	...	...	...	3	2.75	3
July 29	8.7	13.1	13.5	16.8	21.4	21.0	...	...	...	...	4	4	4
Aug. 9	9.4	13.5	13.2	14.5	20.3	19.5							
Aug. 22	8.4	13.4	12.4	14.4	19.2	19.7	...	...	...	...	6	6	6
Sept. 6	7.7	12.2	11.4	12.9	15.6	19.1	...	...	...	...	6	6	6
Wilting coefficient	12.7	14.5	14.5	16.3	17.1	16.1	18.2	18.9	16.7	13.0			

time showed that the roots had drawn heavily on the moisture supply of the soil to a depth of 6 ft., as shown in the table on page 146.

Weaver, Jean, and Crist (1922) reported control experiments in which they determined the depth from which certain crop plants could absorb water and nitrates from the soil. They grew the plants in large containers. The soil was tamped firmly in layers separated from each other by wax seals 2 to 3 mm. in thickness, which were composed of a mixture of 85 per cent paraffin and 15 per cent petrolatum. These seals permitted the penetration of the roots as they pushed downward, but they prevented the upward ascent of water and nutrients from the lower portions of the soil to the depleted portions above. Each layer of soil could be analyzed for moisture and nutrients at the close of the experiment, and the depth from which absorption took place could thus be determined. These investigators found that oats absorbed water from a depth of 2.5 ft. and barley from a depth of 3.5 ft. before the blossoming stage. Corn absorbed large quantities of water from the third and fourth foot and some from the fifth foot. Potatoes absorbed water from a depth of 2.5 ft. Barley at the age of 54 days removed 168 to 145 p.p.m. of nitrates from 1- to 1.5- and 1.5- to 2-ft. levels, respectively. When in blossom at the age of 73 days, it had removed 168 p.p.m. from the 2- to 2.5-ft. level and at maturity 186 p.p.m. from the 2.5- to 3-ft. level. Corn removed 203, 140, and 118 p.p.m. at depths of 3, 4, and 5 ft., respectively.

Crist and Weaver (1924) found that barley plants under controlled conditions absorbed nutrients in large amounts at every level to 30 in. The absorption of nutrients at levels below the surface foot materially affects the quantity and quality of the yield, and the additive effect of a nutrient salt present in the lower levels is not lost, even when the surface foot is abundantly supplied with the same nutrients. The effects of the nutritive salts are most marked on the quantity and quality of yield at an early and late period in the development of the plant, *i.e.*, when absorption is effected largely from the first foot of soil as the crop is tillering and again when the younger portions of the longer roots are absorbing from the deeper levels at the time of heading. Millar (1925) studied the relative growth of oats and inoculated sweet clover on the different horizons of Fox sandy loam and Miami silt loam and found indications that different crops may have quite markedly different feeding powers on the various soil horizons.

He also concluded (1925, 1933) that in Michigan the corn plant draws only very sparingly on the soil horizons beneath the surface so that the subsoil contributes little nutrient to the growth of the plant. Aldrich, Work, and Lewis (1935) found that the pear tree extracted approximately 34 per cent of its moisture from the top foot of soil, 62 per cent from the upper 2 ft. and 84 per cent from the upper 3 ft. Approximately 35, 68, and 89 per cent of the total roots were in these respective regions. The most rapid extraction of moisture by the roots occurred between 2 and 8 ft. from the trunk where the greatest number of roots was located. According to Conrad and Veihmeyer (1929), in California, when the oil is not in contact with a free-water surface, the capillary movement of moisture from moist soil to drier soil is too limited in extent to be effective for use by the roots of plants. Unless the depleted moisture be returned to the dry soil by rain or irrigation, additional water can be obtained only by the elongation of the absorbing portion of roots into new moist soil.

This view is substantiated by the observations of Kiesselbach, Anderson, and Russel (1934), in Nebraska, and of Grandfield and Metzger (1936) in Kansas, on alfalfa. They found that within 2 years this plant depleted the soil moisture to a depth of 25 ft., and the restoration of this depleted water to the subsoil is practically impossible except under the most favorable conditions. After such depletion the alfalfa plants must depend on current rainfall for their water supply unless root penetration is sufficient to reach moisture at a lower depth.

4. Rate of Growth.—It was observed by Stevens (1931) in New Hampshire that the lateral roots of 4- to 6-year-old white pines showed no growth from Nov. 15 to Apr. 1 when grown in the open. When plants were grown in the greenhouse, however, the roots grew approximately as fast in winter as in summer. In sandy soil the growth of the roots of 6-year-old pine trees averages 18 to 20 in. annually, while in a clay soil the growth was only 5 to 10 in. There was no correlation between the amount of root growth and the amount of top growth. The average daily growth in the length of the roots from April to November varied from 0.0006 to 0.089 in.

Kinman (1932) found that the average maximum extension of the roots of the plum, peach, and apricot for a 2-week period was 3.2, 3.4, and 4.5 in., respectively. The greatest seasonal activity of roots preceded the critical period of need during fruit setting. It was noted by Heinicke (1935) that the amount of growth of roots of the apple is relatively small during the dormant period as compared to that made during the several weeks preceding leaf fall. He considered that root growth in the fall might considerably increase the absorption of nutrients for use the following spring. Such new roots produced late in the season might be killed by low temperatures during the winter or early spring.

III. THE WEIGHT OF ROOTS

The weight of the roots of some of the more common agricultural plants is here considered. It has been mentioned that the weight of the roots as compared to the weight of the tops will depend upon numerous factors, so that in discussing this subject any data given must be considered relative to the conditions under which the plants were grown. The data here presented were obtained by growing plants in large cans containing approximately 700 lb. of soil that was obtained from the surface foot of soil in the field. The soil was in good tilth and the water content was kept practically constant during the growing season. The quantity of soil was sufficient to produce plants comparable in every way to those growing in the field. The data were obtained at the Kansas Agricultural Experiment Station during the years 1923 to 1926.

Crop	Date	Number of plants per can	Number of observations	Dry weight of stem, leaves, and grain, grams	Dry weight of roots, grams	Weight of roots in comparison to tops, per cent
Corn, Kansas sunflower	1923	1	4	810.5	66.5	8.2
Kafir, Blackhull	1924	1	3	385.4	19.8	5.1
Oats, Kanota	1925	6	5	236.3	17.9	7.6
Wheat, Kanred	1926	6	5	358.5	12.9	3.5
Alfalfa, common (2 years old)	1926	3	3	242.0[1]	36.1	15.0

[1] Total for three cuttings during the season.

If the number of corn plants per acre is taken as 8,000, the dry matter left in the soil by the roots would approximate 1,125 lb. per acre. This together with the weight of the crowns from which the roots originate would make approximately ¾ ton of dry matter left in the soil.

Eaton (1931) found that root development of the cotton plant is largely influenced by the character of its growth and fruitfulness. Thus the weights of the roots of plants without bolls were approximately three times those of the control plants. Willard and McClure (1932) found that the dry weight of the aerial parts of bluegrass grown on unfertilized soil was 4,010 lb. per acre, while the underground parts weighed 2,060 lb. per acre. On fertilized soil the dry weight of the tops and roots was 8,470 and 2,120 lb., respectively, per acre.

Weihing (1935) noted that the secondary root system of corn tended to increase as the tops increased. The dry weight of the tops in the small, medium, and large varieties was respectively 6.4, 4.9, and 3.5 times that of the roots. It was estimated that 445, 1,080, and 1,420 lb. of dry roots per acre were left in the soil by the current crop of these respective plants.

From the experimental evidence at hand we must conclude that the deeper parts of the soil are not only suited to plant life but also play an exceedingly important part in it and deserve a careful consideration in the study of plant production.

IV. ROOT HAIRS

The root tissue that functions in the absorption of materials from the soil is limited to the single outer layer of epidermal cells. Absorption, however, is not carried on over the whole surface of the root system, since after varying periods of time the older portions of the roots become so modified by the deposition of cutin or the formation of cork cells that absorption through them can in no way take place. Usually, then, the absorption area of the root is confined to a definite zone that begins at or near the growing tip and extends backward for varying distances, depending upon the species of plant and the conditions under which it is growing. In most plants a varying number of the epidermal cells of the root undergo a modification that greatly increases their surface of absorption. This modification consists of a tubular outgrowth of the exterior wall of the epidermal cell and is termed a "root hair." The root hair is in no sense a root and it should not be confused with the term "hair roots," which is used to a considerable extent in agricultural literature to denote the small

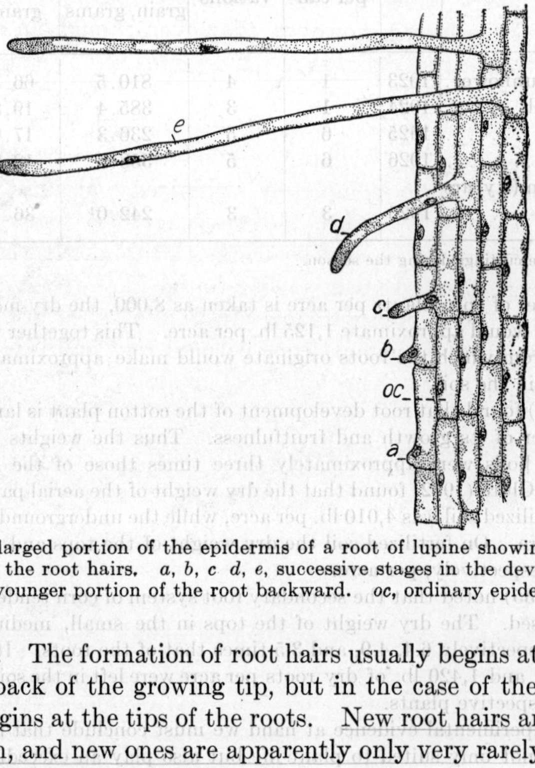

FIG. 8.—Enlarged portion of the epidermis of a root of lupine showing the origin and development of the root hairs. *a, b, c d, e,* successive stages in the development of root hairs from the younger portion of the root backward. *oc,* ordinary epidermal cell.

fibrous roots. The formation of root hairs usually begins at a distance of 1 to 3 mm. back of the growing tip, but in the case of the sedges, their formation begins at the tips of the roots. New root hairs arise in acropetal succession, and new ones are apparently only very rarely interpolated between the preexisting ones (Fig. 8).

Root hairs range in length from a fraction of a millimeter to 10 mm., but their length and general shape depend to a great extent upon the medium in which they are grown. In its passage through the soil the surface of the root hair retains the irregularity in the outlines of the objects to which it adheres. The changes in the shape of the root hair due to the various obstructive particles of the soil bring about a retardation of its general growth so that it often reaches only a small fraction of the length that it would attain if it could develop in moist air. The diameter of root hairs is about 10μ.

Root hairs increase the absorbing surface of the epidermal cells. Schwarz (1883) calculated that the absorbing surface of the roots was increased six times in corn and twelve times in peas by the root hairs that developed on these roots in moist air. He estimated the density of the root hairs on the roots of corn growing in moist air as 425 per sq. mm. while the number on the roots of the pea under the same conditions was 232 per sq. mm.

In most plants, each cell of the root epidermis is capable of forming a root hair, although under average conditions many do not so develop. It should be remembered, however, that even though an epidermal cell does not develop a root hair, it is still an effective organ for absorbing materials from the soil.

Popesco (1926) considered that the absorbing portion of the root lies between the part where elongation begins and the part where the walls of the endodermis are completely suberized. Hohn (1934) decided that the root hairs of *Vicia faba*, *Zea mays*, and *Tradescantia fluminensis* in water cultures are of no significance in the uptake of water under such conditions. Sierp and Brewig (1935) found in the roots of *Vicia faba* that the apical zone of approximately 5 mm. absorbed no water for transpiration and that the next zone without root hairs absorbed only an insignificant quantity. The zone of greatest water absorption was in the region 15 to 80 mm. from the apex. This is the region that ordinarily has the living root hairs.

A. Origin

Ordinarily it is not possible to predict which epidermal cells will develop root hairs. In some plants root hairs arise from a well-defined type of cell. It was first noted by Leavitt (1904) that certain higher cryptogams, monocotyledons, and water plants have short cells, interspersed among the longer epidermal cells of the root, which give rise to the root hairs. He named these short cells "trichoblasts." Bardell (1915) observed on the roots of corn, wheat, squash, beans, and peas, grown under various conditions, that the average length of the cells producing root hairs was less than the hairless epidermal cells. Bartoo (1929) also observed this peculiarity in the development of root hairs in *Schizaea rupestris*.

It was further observed by Cormack (1935) that the epidermis of the roots of certain species of *Brassica* was composed of long and short cells arranged in distinct rows, and that the root hairs were produced from the short cells. Wilson (1936) found that the root hairs of *Elodea canadensis*, *Potamogeton densus*, and *Alisma plantago* are produced from specialized cells. He considered that the two types of cells are differentiated early. The short cells are distinguished in their early stages by thicker walls, larger nuclei, and denser cytoplasm than the other epidermal cells. The differentiation is evident long before the root hairs are developed. The long cells are three or four times longer than the short ones. The proportion of these root-hair-forming cells laid down at the apex of the root is remarkably constant over a wide range of conditions. Cormack (1935) considered that the long cells have a metabolism that differs from that of the short cells. He based this assertion on the facts that their walls did not react with calcium, and that they had a considerably more acid reaction than the hair-forming cells.

B. Conditions Affecting the Formation of Root Hairs

The stimulus that initiates the production of root hairs is not definitely known. Some consider that the retardation of the longitudinal growth of the epidermal cells is necessary for the initiation of the root hairs. Others believe that their formation is an accompaniment of optimal conditions for growth. Both these theories may be correct in part, because the conditions that prevail when root-hair formation occurs may be opposite in different parts of the same cell.

The number of root hairs that will develop will thus depend primarily upon environmental factors. These factors include: the available water, temperature, obstructions, hydrogen-ion concentration, toxic substances, and certain elements.

The frequency of root hairs has generally been expressed as the number per unit area. Wilson (1936) considered that a better indication of the intensity of root-hair development would be an expression showing the proportion of the epidermal cells that produce hairs. Such an expression is analogous to the stomatal index, described in Chap. VII, and may be termed the "root-hair index." This may be expressed after the manner

$$\text{Root-hair index} = \frac{\text{Frequency of root-hair-forming cells}}{\text{Frequency of all epidermal cells}} \times 100$$

Snow (1904, 1905) found that a saturated soil tends to suppress root-hair production. Roots of corn and wheat grown in garden soil kept moist were found well covered with hairs after 7 days. Similar roots growing in a saturated soil for the same period showed few or no root hairs, but after this soil was allowed to dry, root hairs appeared as numerously as in the case of the moistened soil.

Spalding (1904) found better and more root-hair development in the creosote bush growing in a soil comparatively dry than when growing in a soil well supplied with water. Bardell (1915) grew corn, wheat, squash, beans, and peas in water and found that root hairs were produced under these conditions. She found that light conditions have no effect upon root-hair production.

In this connection it should be mentioned that the roots of *Lemna minor*, *Nymphaea alba*, *Hippurus vulgaris*, and others do not develop root hairs, while *Scirpus sylvaticus* and *Carex paludosa* growing under similar conditions produce them fairly abundantly. The roots of *Elodea canadensis*, *Acorus calamus*, *Nuphar luteum*, and others produce no root hairs so long as they are immersed in water but develop them abundantly upon entering the soil. Snow (1904, 1905) observed that high temperature with sufficient moisture tends to decrease the number of root hairs per unit of area by increasing the elongation of the epidermal cells. She also observed that the slower the growth the more root hairs were developed per unit area, and that not only did lower temperatures aid this, but also obstructions in the pathway of the root produced the same results.

Mer (1879, 1884) considered that kinking of the roots is due primarily to growth retardation that is accompanied by a good production of root hairs. Bardell (1915) noted that curves and swellings are favorable to root-hair formation, the hairs being most abundant on the concave side of the root.

The effect of bog water on the production of root hairs of the cuttings of *Tradescantia* was investigated by Rigg (1913). He found that practically no root hairs were produced normally by this plant when grown in bog water from Puget Sound. Of the specimens examined, 18 per cent produced no root hairs, 66 per cent produced hairs that were much stunted, and 16 per cent produced hairs only slightly stunted. He also found that effects on root-hair formation similar to those produced by bog water occurred in each of the following solutions: sea water diluted three times; carbolic acid, 0.001 per cent; formalin, 0.001 per cent; weak tea; and coffee. The root hairs developed normally in the filtered bog water and fairly well in the diluted bog water. Rigg believed that many plants are excluded from bogs on account of their inability to produce root hairs under those conditions, thus having a too limited absorbing surface. Addoms (1923) concluded that the abnormal root development and decreased growth of plants grown in nutrient solutions containing relatively large amounts of potassium dihydrogen phosphate are due to the coagulation of the protoplasm of the root hairs. This coagulation, which is accompanied with flocculation, was found to be induced by the hydrogen ions formed by dissociation of the phosphate. The hydrogen-ion concentration of the nutrient solutions used varied from pH 3.94 to 3.47.

Parker and Sampson (1930) found that when the plants of *Stipa pulchra* and *Bromus hordeaceus* were clipped frequently the roots were devoid of root hairs.

According to Cormack (1935) it has been shown by Magowan (1908), Hansteen-Cranner (1914), Kisser (1925), Farr (1925 to 1928), and Sorokin and Sommer (1929) that the presence of calcium is important for the development of root hairs. Mevius (1926), however, observed that in corn the development of root hairs is not related to the presence or absence of calcium. Pearsall and Wray (1927) found that the root hairs of *Eriophorum angustifolium* tend to be longer and more abundant in solutions of low calcium content.

Cormack (1935) found in studying the roots of Georgia collards that if the amount of calcium available for wall formation is reduced the root hairs frequently become inflated and branched, indicating that the cell walls were too soft to withstand the outward pressure. With further reduction in the amount of calcium there is no formation of root hairs, and the epidermal cells expand more or less symmetrically. The walls of the hairs in solutions minus calcium retain the pectic acid layer. Apparently the hydrogen-ion concentration plays a very important role in the development of root hairs, and this concentration is maintained at optimum by the presence of the proper amount of calcium. Pectic acid and calcium fail to unite if the medium is too acid, and the hydrogen-ion concentration must be such that the reaction between the calcium salt and the pectic acid of the cell takes place gradually. Cormack found that when the roots of Georgia collard were allowed to grow in a solution with sufficient calcium and with a pH of 7.2 to 8.2, the long cells produced typical hairs. When the medium was sufficiently alkaline, the walls hardened so rapidly that no hairs were produced. If the hardening of the walls of the epidermal cells is gradual, vertical elongation is arrested, and, owing to increasing pressure from within, the wall is pushed out at its softest point. In the case of Georgia collards this is at or near the distal end of the cell. The progressive hardening of the outer layer confines this soft spot to a narrow area at the growing tip of the hair where new wall is constantly being laid down. When the short cells begin to develop hairs they showed a pH value of 5.8 to 6.8, while the developing hairs varied from pH 6.6 to 7.2. The long cells in the same region showed a pH of 4.6 to 4.8.

C. Growth of Root Hairs

Haberlandt (1914) observed the manner of the longitudinal growth of the root hairs of *Cucurbita pepo, Pisum sativum,* and *Helianthus annuus* by dusting the hairs with rice powder. He found that the longitudinal growth is confined strictly to the dome-shaped tip of the hair. Haberlandt considered that this manner of growth greatly facilitates the power of the root hair to push itself between the solid particles of the soil.

Jeffs (1922) studied the rate of growth of root hairs in moist chambers in the dark at a temperature of approximately 22°C. He found that root hairs grow slowly at first, gradually increasing their rate of growth up to the time root elongation ceases, and that a definite relationship exists between the elongation of the root hair and the parent root. About the time root elongation ceases, the rate of growth of the root hairs in that region becomes stabilized, and there are only slight variations in their rate of growth until about 2 hr. before their growth ceases. The average rate of growth of root hairs of corn, each of which was on a different root, showed an initial rate of 6.5μ per 15-min. interval. This increased to 16.1μ per 15-min. interval at the end of the first hour and 33.6μ per 15-min. interval at the end of the third hour. After this there was no further root elongation in this region, and the root hairs grew at the average rate of 35μ per 15-min. period for the next 3 hr., after which they were studied no longer. It has been demonstrated that the rate of growth of a root hair diminishes very rapidly as it approaches maturity. The approximate length of five root hairs of the radish was 1,824μ at 2 hr. before growth ceased, when they were elongating at the rate of 44.5μ per 15-min. interval. At 1 hr. before growth ceased they were elongating at the rate of 36μ per interval, and at 15 min. before growth ceased their rate of elongation was only 15μ per ¼-hr. period. Growth ceased on an average of 2 hr. after the retardation in their growth rate could be observed.

Jeffs (1925) investigated the influence of light and temperature on the rate of elongation of root hairs and found that those of *Raphanus sativa* and *Sinapis alba* showed no noticeable growth reactions in the light intensities used. The root hairs of these two species grown under a constant temperature within the limits of 17 and 27°C. showed a decided reaction to temperature changes of from 1 to 6° produced within a period of from 5 to 15 min. A drop in temperature resulted in a decrease in the growth rate and sometimes even in a contraction of the root hair. A rise in temperature of 2.5 to 3° within the 5- to 15-min. period resulted in a temporarily increased growth rate, but this soon returned to that which characterized the root hair before the temperature was raised. Jeffs concluded that the effect of temperature upon the rate of growth of tissues is, in part at least, an effect upon the osmotic pressure and imbibition of the cell colloids, rather than simply an effect upon the rate of chemical reactions within the tissues.

Farr (1925) compared the rate of growth of the root hairs of *Tradescantia fluminensis* in tap water and in Knop's solution. The average growth per minute in tap water was 0.916μ, while that in Knop's solution was 0.437μ. The fastest growth recorded in tap water during any 10-min. period was 24μ as compared with 10.8μ in Knop's solution. The average growth per hour for any series in tap water was 91.8μ; while in Knop's solution it was 32.9μ. Farr's experiment shows that substances in solution may modify not only the form of root hairs but also their rate of elongation. Farr (1927 to 1928), in a series of the most extensive experiments yet undertaken on the growth of root hairs in solution, found that the root hairs of the Georgia collards (*Brassica oleracea*) do not grow at all during a period of 13 to 16 hr. after immersion in distilled water. When they do grow, however, they do so at a rate of 60μ per hour in tap water and about 50μ per hour in Knop's solution. They also do not grow at all during the period of 13 to 16 hr. after immersion in concentrations of calcium nitrate below 0.002 *M*. Their maximum growth rate of 61μ per hour in 0.012 *M* calcium nitrate is more or less closely approached when growing in concentrations between 0.007 and 0.012, but it drops abruptly at higher concentrations to 40μ at 0.013 *M*. Farr also observed that the graph representing the effects of varying concentrations of calcium chloride upon the growth rate of root hairs is almost identical with that for calcium nitrate in form and in the location of the optimum molar concentration. The importance of calcium in the root hair and in root growth is attributed largely to its being the only constituent of the cell wall that is of direct external origin, although it must be recognized that calcium also plays a part in antagonism, hydrogen-ion concentration, and permeability. In a study of the effect of the hydrogen-ion concentration in calcium solutions upon root-hair growth, it was found that the maximum elongation of the root hairs of the Georgia collards was 109.8μ per hour in 0.020 *M* calcium chloride at pH 6.9. The hydrogen-ion concentration range for root-hair elongation was from 3.4 to 11.9, varying with the concentration of the solution. It was also concluded that the entire root responds to changes in the hydrogen-ion concentration in much the same way as do the root hairs. Farr states that, in producing root hairs, roots exhibit the power of adaptation to a wider range of hydrogen-ion concentration than any other organ, organism, or physiological process that has thus far been studied, although enzymes have been found to tolerate a still wider range.

D. The Longevity of Root Hairs

The length of life of root hairs depends on the species of the plant and upon the moisture conditions of the soil. Root hairs of some species live for only a few days or weeks, while those of others exist for several years, although they do not apparently function in absorption for a period that long.

McDougall (1921) observed that the root hairs of *Gleditsia triacanthos* became thick walled and brown in color and that they persist as long as the root epidermis, which may be for several months. The older root hairs examined by him had a cell-wall thickness of 2μ, which is approximately four times the thickness of the cell wall of the root hairs of this plant when they are first formed. These hairs are only about 0.4 mm. in length and are stiff and rigid and do not shrivel upon exposure to the air even after several months. McDougall found that the root hairs of *Gymnocladus dioica* and *Cercis canadensis* sometimes become thick walled and brown, but that this is not nearly so characteristic of these as it is of *Gleditsia triacanthos*. Whitaker (1923) confirmed the work of McDougall and found that the root hairs in certain species of the *Compositae* persist for as long as 3 years. Weaver (1925) found in the case of wheat growing in the soil under field conditions that the root hairs did not begin to slough off from the oldest parts of the primary root until 7 weeks after planting and that at the end of 10 weeks more than 99 per cent of the root area still possessed functioning root hairs. In this case the soil was in excellent condition for their retention, but under conditions of a low water supply in the upper portion of the soil not only the older root hairs but also many of the more delicate branch roots would perish.

At 6-month intervals, Stoddart (1935) examined the roots of a group of prairie grasses 2 in. below the soil surface. During a 2-year period these roots were subjected to all variations of soil moisture from below the wilting coefficient to saturation, and to temperatures from 0° to 112°F. In all the plants observed the roots lived for at least a year and many in excess of 2 years, some new roots being produced each season.

E. The Cell Wall of Root Hairs

Schwarz (1883) found that the cell wall of the root hair of *Taxus baccata* or yew is of two parts—an inner layer that stains blue with chloriodide of zinc and an outer layer that stains yellow brown. He concluded that the contact between the root hairs and the soil particles is rendered more permanent and intimate by the gelatinization of the outermost layers of the root membrane. Hesse (1904) found that the thickness of the cell wall of the root hair of any plant varied with the medium in which it was grown.

The composition of the cell wall of the root hairs has been variously reported. Apparently it depends to a considerable degree upon the conditions prevailing at the time of their formation.

Ridgway (1913) observed the occurrence of callose in the cell wall of certain root hairs. Roberts (1916) found in a majority of the forms investigated that the walls of root hairs growing in moist air are made up of two parts—an inner membrane that gave the cellulose reaction with 75 per cent sulphuric acid and the potassium iodide iodine solution and an outer membrane that stained red with ruthenium red denoting pectic compounds. That this membrane is calcium pectate was shown by the fact that on the addition of ammonium oxalate the membrane broke down and calcium oxalate crystals were formed. Callose was observed at the tip of the root hairs in mustard, pea, and *Tradescantia* amongst others. The relative amount of the cellulose and pectin found in the cell walls of various root hairs is shown in the following table:

Plant	Nature of the cell wall of the root hair	
	Inner layer	Outer layer
Alfalfa	Thin cellulose	Thick pectin
Barley	Thin cellulose	Thick pectin
Corn	Thick cellulose	No pectin
Cabbage	Thin cellulose	Thin pectin
Pea	Thick cellulose	Thick pectin
Tobacco	Thin cellulose	Thin pectin
Mustard	Thin cellulose	Thin pectin

The calcium pectate membrane on the outer wall of the epidermal cell and the root hair is a continuation of the middle lamella of the other walls of the cell. There is thus a secretion of calcium pectate around the entire cell, but the layer on the external wall is thicker than that on the other walls of the cell. The layer of pectic material is thinner toward the tip than it is at the base of the root hair. The high efficiency of the root hair as an absorbing organ is thus due to the fact that the soil particles are held to it by this pectic substance.

2nd condition

Howe (1921) examined the root hairs of 20 economic plants growing in sand and in loam soil. Some of the plants examined were beans, cabbage, corn, cucumber, radish, peas, squash, tomatoes, and watermelons. She found no cellulose in the cell wall of the root hairs grown under these conditions. The wall of the root hair contained a layer of pectic material on the outside and on the inside a layer of callose that varied in thickness for different plants, but in all cases it was a little thicker at the tips than in the lateral walls. As the wall of the epidermal cell bulges to form the root hairs, the inner layer, which is composed of cellulose, apparently stretches to its capacity and breaks and no more cellulose is formed. The outer pectic layer of the wall, however, is not ruptured but is extended out to form the root hair. Howe found that the root hairs formed either in loam or in sand gave an acid reaction in all cases, but the root hairs produced in loam soil were usually somewhat the more acid. The pH values of the root hairs ranged from 6.0 to 6.8 in most cases. Whether the acidity of the root hair can be ascribed to the presence of pectic materials or to some other cause was not determined.

According to Cormack (1935) it was reported by Ziegenspek (1920) that the inner layer of cellulose in the cell wall of the root hair was replaced by amyloid at the tip, and he considered that this plastic amyloid made possible the growth of the hair. Hopmann (1930), however, reported the absence of amyloid from root hairs that had developed in water, but reported the presence of cellulose under all conditions over the whole inner surface of the wall of the hair. Cormack found in tomato, radish, endive, white mustard, corn, and Georgia collards that the cell wall of root hairs grown in moist air and in tap water had layers of cellulose and pectic materials continuous with corresponding layers in the epidermal cells. Tests were negative for all other wall constituents. On the dome-shaped tip of the root hair the pectic layer was composed either of pectic acid or of a softer layer of calcium pectate than the mature portion of the wall.

V. EXCRETIONS OF ROOTS

From the discussions in Chap. II concerning the factors affecting the permeability of the cell membranes, it can readily be conceived that numerous conditions, both internal and external, might arise under which the outward diffusion of the cell contents of the roots into the soil might occur. It is not the intention here to consider the possible loss of the inorganic salts from the roots but to discuss more particularly the excretion of those compounds that are waste products or synthetic substances of the protoplasm.

The two topics to be considered are (*a*) the excretion of acids, and (*b*) the excretion of enzymes.

A. Acids—Carbonic Acid

Liebig (1858) first observed the dissolving action of roots on limestone, and his observations were verified by Sachs in 1860. Sachs advanced two theories as to the cause of this dissolving action. He considered that it was brought about either by the liberation of carbon dioxide or by the formation of acid due to the decomposition of the cell walls of the cells of the root. Czapek (1896) studied the excretion of acids from roots by the use of plates made with mixtures of aluminum phosphate and plaster of Paris. Plates of this composition are soluble in hydrochloric acid, nitric acid, sulphuric acid, orthophosphoric acid, formic, oxalic, succinic, lactic, malic, citric, and tartaric acids but are insoluble in carbonic, acetic, propionic, and butyric acids. These plates showed no signs of etching when the roots of various plants came in contact with them, from which it was considered that none of these acids was excreted by the roots under the conditions of the experiment. Plants were grown in Congo red, which becomes brownish red when acted upon by carbon dioxide and bright blue through the action of acetic and propionic acids. The results showed that carbon dioxide alone was excreted by the roots. Kunze (1906) also observed that free mineral acids are not excreted from the roots of higher plants, and Haas (1916) showed that carbonic acid was the only acid excreted by the roots of sweet corn. Stocklasa and Ernest (1909) found that if the roots of plants have a normal oxygen supply no acids either organic or inorganic except carbonic acid are excreted. If, however, the oxygen supply is cut off or is limited, organic acids only are produced. They found that the roots of buckwheat and barley under limited supplies of oxygen produced formic and acetic acids, while the roots of oats and maize under like conditions excreted formic acid, and the roots of beets, oxalic acid. These organic acids arise from incomplete oxidation and under normal conditions of aeration would be oxidized to carbon dioxide and water.

1. Relative Amounts.—The evidence indicates that under normal conditions carbonic acid is the only acid liberated by the living cells of the roots. Coupin (1918) considered that this was excreted more particularly by the superficial cells of the rootlet between the root hairs and the root tip and by that portion from which the root hairs have disappeared.

Carbon dioxide is an end product of respiration, a process that is going on continuously in every living cell, so that the amount of carbon dioxide thus formed will depend, then, upon the environmental conditions and upon the nature of the plant. Since the intensity of respiration varies directly with the intensity of metabolic activity, it would also be expected that the amount of carbon dioxide excreted by the roots would vary during the growing season of that plant.

Kossowitch (1904) found that the roots of mustard in pots and water cultures produced increasing amounts of carbon dioxide up to the time of blooming, and Barakov (1910) showed that the carbon dioxide content of the soil was insignificant at the beginning of vegetative growth but increased rapidly and reached its maximum at the blooming period and declined rapidly to the minimum at the ripening period. He observed that the greater the plant growth the more carbon dioxide there was in the soil. Most of this came from the respiration within the roots and varied with different crops and different stages of growth. With lupines the maximum was reached at the blooming stage and with oats about 2 weeks before blooming. The maximum production of carbon dioxide by the roots of potatoes was reached after blooming, while its excretion by sugar beets reached its highest point at the period of maximum growth.

Lau (1906) observed that potatoes and legumes gave off more carbon dioxide than other crops, while Stocklasa and Ernest (1905) found that red clover, beet, and oat roots produced more carbon dioxide in the order named than the other plants considered.

Russell and Appleyard (1917) observed a considerable increase in the production of carbon dioxide in the soil during the period of rapid growth of the crop and another and larger increase at the time of ripening. Bizzell and Lyon (1918) found the greatest apparent production of carbon dioxide by oats to be at the time of blooming, after which a marked decrease occurred.

Stocklasa (1934) found that the evolution of carbon dioxide by the roots was highest with leguminous plants and lowest with wheat and barley. Turpin (1920) found that the oat crop increased the production of carbon dioxide in the soil. This increase became marked from the first month after seeding and increased to a maximum just previous to or after these plants had headed, following which decline took place. Millet produced about the same amount of carbon dioxide as oats. A cropped soil after harvest maintained a higher carbon dioxide content than bare soil, and it was considered that this was due to the decomposition of roots.

Smith and Brown (1931, 1932) considered that the production of carbon dioxide in the soil under crops is such a complex procedure that the amount present in the soil cannot serve as an index to the production of this gas by the roots.

Turpin considered that plants at their most active period of growth produce many times as much carbon dioxide as is produced by soil organisms, and that the excess of carbon dioxide in the soil of growing crops is due to the respiratory activity of the roots of the plants rather than to the decay of root particles from the crop. Newton (1924), however, considered that there is good evidence to prove that the quantity of carbon dioxide produced in soils by the decomposition of organic matter is greater than the quantity given off by the respiration of the roots of the growing crop.

2. Solvent Action.—The solvent power of the carbonic acid excreted by the roots upon the minerals of the soil is a matter of question. Hall (1906) considered that carbonic acid is the only acid excreted by the plant roots, and that it is capable of dissolving everything needed by the plant. Aberson (1910) considered that the mucilaginous covering of the root hairs contains a saturated solution of carbon dioxide sufficient to bring into solution the insoluble soil constituents with which it comes in contact, especially the phosphates. Pfeiffer and Blanck (1912), however, showed that in soils treated with phosphates the carbon dioxide given off by the roots of the plant does not form a solvent sufficiently powerful to account for all the mineral nutrients obtained by the plant from the soil. Fred and Haas (1919) studied the etching power of the roots of Canada field peas upon polished marble in sterile soil and in soil that had been inoculated with various bacterial cultures other than those of the nitrifying bacteria. It was found that in every case the marble slabs that had the greatest degree of etching were those in the soil that had been inoculated. They considered that the greater etching power of these roots in the presence of bacteria can be attributed to the normal carbon dioxide excretions of the living cells together with the carbon dioxide and acids set free from dead or dying root cells whose decomposition is accelerated by the presence of bacteria.

Parker (1924) is of the opinion that the carbon dioxide excreted by the roots is more effective in bringing plant food into solution within reach of the plant than a unit of carbon dioxide produced in the soil by the decomposition of organic matter. He, however, could find no relation between the carbon dioxide production of roots and the feeding power of the plants for calcium, magnesium, phosphorus, and potassium in a poor sandy soil. More carbon dioxide was excreted by the roots of cowpeas than by those of any other plant, while buckwheat roots gave off very little carbon dioxide but had the greatest feeding power of any of the plants used. For each gram of carbon dioxide excreted by the roots, buckwheat absorbed 41.5 mg. of calcium, while sorghum, cowpeas, and soybeans, respectively, absorbed 5.0, 12.7, and 21.2 mg. of this element for each gram of carbon dioxide excreted by the roots. It has long been recognized that species of plants differ widely in their ability to grow on infertile soils.

It is well known that beans, rye, and buckwheat can grow apparently to good advantage on poor land, while pennyroyal, sassafras, poverty grass, and others are indicators of impoverished soil upon which plants like corn, barley, wheat, blue grass, and others grow poorly or do not thrive at all. Dustman (1925) studied the rate of excretion of carbon dioxide from the roots of these two types of plants. He found in his experiments that the production of carbon dioxide whether estimated from the standpoint of dry matter of roots or unit area of root system was greater for buckwheat and rye than for corn and barley. He is of the opinion that the output of carbon dioxide in connection with the relative size and area of root systems is an important factor in aiding certain kinds of plants to secure the essential elements from the soil.

Stocklasa (1934) found that, when plants were grown in pulverized minerals, to which had been added a sufficient amount of nitrogen, the absorptions of phosphorus and potassium were in the order oats > rye > wheat > barley. The low absorption by barley was correlated with the smaller excretion of carbon dioxide by the roots of that plant. He believed that plants growing in radioactive soil can probably utilize insoluble minerals to a greater degree than similar plants in inactive soils.

Metzger (1928) also considered that the carbon dioxide excreted by roots exerts a measurable solvent action on soil nutrients. He used the concentration of bicarbonates as a means for determining the action of the carbon dioxide excreted by the roots. In 66 determinations with corn, kafir, wheat, and other plants, 53 showed higher concentration of bicarbonate in samples from around roots, 4 showed equal, and 9 lower concentrations than in samples taken at short distances from the root.

B. Enzymes

The question of the excretion by the roots of substances that are capable of oxidizing, reducing, or digesting materials in the soil has been investigated and is worthy of note. Molisch (1887) first demonstrated the oxidizing power of intact roots. He showed that there was considerable active secretion on the surface of grown roots and considered that this secretion had definite powers to effect changes in organic substances. Raciborski (1905) found that all the roots of the phanerogams examined by him had the property of extracellular oxidation. Schreiner and Reed (1909) considered that plant roots are able to carry on active extracellular oxidation chiefly by means of enzymes that they secrete. These investigators used wheat plants of the varieties Harvest Queen and Chul and found that their roots were able to oxidize organic substances such as chromogens at a fairly rapid rate. When the roots were grown in solutions containing α-naphthylamine, benzidine, vanillin, vanillic acid, or esculin, insoluble colored oxidation compounds were yielded which were deposited mainly upon the surface of the roots. Color changes that indicated oxidation were also observed when plants were grown in solutions containing phenolphthalein, aloin, and leuco-rosolic acid. The oxidizing power appeared to be most energetic in the region of the root where the root hairs are found and increasingly less active as that portion of the root becomes older. Schreiner and Reed found that the roots had greater oxidizing power in extracts of productive soil than in those of unproductive soil and that the presence of toxic organic substances in solution was extremely deleterious to the oxidizing power of the roots. Schreiner and Sullivan (1911) found that the roots carry on oxidation and reduction concurrently. The oxidation processes are weak in the young seedling but increase in strength as the seedling grows. The reduction processes are predominant in the early stages of the seedling growth but are less manifest at a later stage. Reduction is mostly intracellular, while oxidation, which seems to be the more prominent of these two properties of the roots, is mostly extracellular. The reducing power of the roots of grasses has also been observed by Sani (1919), and the oxidizing power of roots upon ammonium ferrosulphate has been observed by Borowski (1919). He also considered that this oxidizing ability is located in the region of the root hairs and at the growing tips. He found that the oxidizing ability of the roots of different plants varies. In his experiments the roots of *Sinapis* showed the weakest, while *Phaseolus* showed the greatest power of oxidation. Lyon and Wilson (1921) working with nutrient solutions in which corn, oats, peas, and vetch were growing could detect reducing substances by some tests but not by others. They were not able, however, to show to their satisfaction that oxidizing substances were present in these solutions. Neller (1922), on the other hand, found in all his experiments with soybeans, field peas, and buckwheat that these plants had a beneficial influence upon the oxidation activities in the substrata in which the plants were grown and suggested a symbiotic relationship between the soil oxidizing organisms and the roots of these growing plants.

The ability of the plant to absorb from the soil more or less complex organic compounds makes the excretion of digestive enzymes by the roots an interesting question. This subject has been investigated by Knudson and Smith (1919) and by Knudson (1920). They found that neither corn nor Canada field peas are capable of utilizing soluble starch directly or indirectly, nor is there any appreciable excretion of diastase by the roots of these plants under the conditions of the experiments. When these plants were grown in the presence of sucrose there was an increase in the amount of reducing sugars in the cultural solution, but incubation experiments gave negative results with respect to the presence of invertase. Knudson considered that the increase in reducing sugars in his experiments was due to their excretion from the roots into the solution rather than to the secretion of invertase with a consequent digestion of nonreducing sugars.

C. Organic Substances

The theory that plant roots excrete or eliminate into the soil certain waste products that are especially injurious to other plants is an old one. De Candolle (1832) advocated a theory of crop rotation on the basis of root excretions in which he claimed that plants excreted from their roots substances that are harmful to succeeding crops of closely related plants. His theory was based partly on his own observations and partly on the statement of earlier writers. Experiments carried out according to his directions tended rather to disprove than to support his hypothesis. The subject of root excretions was given very little attention until about 1900, when the matter was investigated by Pickering in England and by the

Bureau of Soils of the U. S. Department of Agriculture. The investigators in this department were at first inclined to favor the theory of root excretions to account for the deleterious substances that occur in soils. Later, however, it was considered that these detrimental substances might be caused by the decomposition of vegetable matter. In the case of growing plants, especially, it is considered that the decomposition of the contents of the disintegrating epidermal cells and the cells of the root cap, which are continually being lost into the soil, might produce the compounds that are toxic to other plants.

Fletcher (1912) reported that the roots of corn excreted a substance that was detrimental to the development of the plants of sesamum growing in adjoining rows. Shull (1932) grew plants of sesamum and sorghum together and noted that their roots intertwined without any injurious effects upon either plant. He concluded therefore that the roots of these plants did not secrete any substance that was detrimental to the growth of the other.

It is now rather well established that the roots of plants do not excrete substances that are beneficial or detrimental to either accompanying or succeeding crops. However, there are some cases in which the roots excrete substances that serve some definite purpose. Thus Saunders (1933) demonstrated that the host plants of the witchweed (*Striga lutea*) exude from their roots a substance that is necessary for the germination of the witchweed seeds. The precise nature of this substance is not known, but it is easily destroyed by high temperatures and loses its activating power on exposure to air. The liberation of this substance is not dependent upon the mineral nutrition of the host. Its origin is due apparently to the metabolic activity of the protoplasm of the host cells. Its action is upon the cell contents of the seed of the witchweed and not upon the seed coats.

VI. BENEFICIAL AND TOXIC EFFECTS OF ROOTS

It has been known from the earliest times that one species of plant may have a detrimental or beneficial influence upon other plants of the same or different species that are growing in close proximity to it or that follow it in succession. Let us consider first the beneficial influence of one plant upon another.

A. The Beneficial Influence of One Plant on Another

It has been known for several thousand years that grapes grow well in association with the elm tree and that certain grains produce better when grown in combination than when they are grown separately. Furthermore, mixed cropping apparently had a prominent place in primitive agriculture.

Dandeno (1908, 1909) grew squash and corn seedlings together and separately and found that the growth when grown together was greater than when grown separately. He also found that the growing of Canada thistle with oats, wheat, and barley was beneficial to them. Zavitz (1922) grew oats, barley, spring wheat, and peas separately and in various combinations for the production of grain for six years. Six mixtures having two classes of grain in each mixture, four having three classes of grain in each mixture, and one having all four classes of grain in combination were used each year and varieties were selected that matured uniformly. It was found with but one exception in the 11 mixtures that the grains grown in combination gave a greater yield per acre than the average production of the same grains when grown separately. The yield of straw was the highest in every instance in the case of the mixed grains. The different classes of grains exerted an influence on the productiveness of the mixtures in the following order: oats, barley, peas, and wheat—oats having the greatest effect and wheat least. It was found after 15 years of experimentation that the greatest yield of grain per acre in the case of barley and oats was produced by using a mixture of 48 lb. of barley and 34 lb. of oats per acre. The use of mixtures of different varieties of the same grain showed no beneficial results over the grains when grown by themselves.

The yields of grain and straw of the different mixtures and when grown separately are shown in the following table:

Varieties of grain grown in mixtures	Yield per acre			
	Straw, tons		Grain, pounds	
	Grown separately	Grown in mixture	Grown separately	Grown in mixture
Barley and oats	1.56	1.74	1,935	2,261
Barley, peas, and oats	1.47	1.67	1,489	2,101
Barley, wheat, and oats	1.47	1.72	1,683	2,067
Peas and oats	1.52	1.77	1,873	1,988
Barley, peas, wheat, and oats	1.43	1.71	1,682	1,955
Wheat and oats	1.52	1.68	1,624	1,921
Peas, wheat, and oats	1.44	1.73	1,642	1,860
Barley and peas	1.33	1.56	1,740	1,760
Barley, peas, and wheat	1.32	1.57	1,553	1,665
Wheat and barley	1.33	1.41	1,491	1,558
Peas and wheat	1.29	1.37	1,429	1,322

Lyon and Bizzell (1913) spaced plants of wheat, oats, and timothy so that they would not interfere with one another. After they had attained some size, seeds of different kinds of plants were placed between them. In the majority of the tests a larger yield of the primary crop resulted in the mixtures than when it was grown alone. These increases amounted to as little as two per cent and as much as 52 per cent. The plants that were particularly active in promoting better growth of the primary crop were mustard, buckwheat, maize, and red clover. In growing wheat and other plants in Knop's solution, Mukerji (1920) found that the solutions which had previously grown a set of plants supported better growth than the fresh solutions. It was noted by Peralta and Estioko (1923) that *Cyperus* and water lily (*Monochoria hastata*) are beneficial to the production of rice. Relevant to the possible production of beneficial substances by the roots of plants, it should be mentioned that Krogh and Lange (1930) showed that fresh-water algae lost into the surrounding water approximately 10 per cent of the organic matter synthesized, and that most of this organic matter came from dead and decomposed cells of these plants.

It was also noticed by Morrish (1934) that oats and barley produced a higher yield when they were grown together than they produced on an equal area, of which half was used for pure oats and the other half for pure barley.

Arny, Stoa, McKee, and Dillman (1929) under certain conditions and in some regions thought that mixtures of flax with wheat, oats, or barley, was beneficial in producing a greater yield but that under other conditions it was of no advantage.

It was reported by Thornton and Nicol (1934) that Italian rye grass grown in the presence of lucerne in sand for 18 weeks contained 2.25 times as much nitrogen as did the grass similarly grown except for the absence of lucerne. In another case, 0.33 gram of sodium nitrate was added per pot, and after a growth of 13 and 18 weeks the rye grass contained 2.5 and 5.5 times as much nitrogen, respectively, as was supplied by the added nitrate. They made no attempt to trace the course of the supposed transfer of nitrogen from this legume to the nonlegume. According to these authors, Virtanen (1933) found in pure sand, in which inoculated legumes were growing, the same percentage of combined nitrogen as occurred in the plants themselves.

According to Nicol (1935) it was observed by Kaserer (1911) that there was little or no interpenetration of the roots of plants of the same species when grown together, but that there was a marked interpenetration of roots when the plants of two different species were so grown. The maximum interpenetration was shown when a legume and nonlegume were grown together. Nicol found that the roots of alfalfa and of grass growing in sand were difficult to separate when these plants were grown together, but that no such difficulty was experienced when these plants were grown in single cultures.

B. The Deleterious Influence of One Plant on Another

It has long been a matter of common observation that the soil becomes more and more unproductive for a given crop that is grown continuously upon it, and also that the preceding crop may have a detrimental effect upon the succeeding one of a different species. And it has also been observed, especially in comparatively recent years, that plants growing in close relationships may have a detrimental influence on each other that cannot be wholly attributed to the competition of these plants for the raw materials of the soil and air or for the supply of sunlight.

Pickering (1903, 1917) and Bedford and Pickering (1914, 1919) reported their observations dating from 1898 upon the deleterious effects of one plant upon another. They considered that it is established with a reasonable amount of certainty that the detrimental effect of one growing plant upon another is of general occurrence. They developed a method for growing two plants in a single pot in order to study the influence of one upon the other. One of the plants was grown in a layer of soil contained in a tray that was superimposed on the soil in which the other plant was growing. The water was applied to the tray so that the lower soil received the drainage from the upper one. They found that pears, plums, apples, cherries, certain forest trees, mustard, tobacco, tomatoes, barley, clover, and some grasses are susceptible to this detrimental effect, which may be exerted by apple seedlings, mustard, tobacco, tomatoes, clover, and 16 varieties of grasses. This deleterious effect varies greatly but is especially manifest in the reduction of growth, which varies from 6 to 97 per cent. Fruit trees that are affected by the harmful products of certain grasses not only show reduced growth but also exhibit alterations in the coloring of the bark, leaves, and fruit that are unlike those attending other forms of ill-treatment. These investigators found the above results so manifest that expert fruit growers were unable to name correctly the varieties thus affected. Pickering (1917) considered that different individual plants of the same species when grown together exert a deleterious effect upon one another. For example, it has frequently been observed that when the seeds of a given plant are sown and then 3 or 4 days later more seeds of the same kind are sown among those already in the ground, the plants resulting from the seeds first sown outdistance those from the later-sown seeds in their general vigor and rate of growth out of all proportion to their respective ages. In the case of mustard, a difference of only 4 days in the date of planting showed 3 months later a difference of 60 to 70 per cent inferiority in growth of the plants from the seeds sown later. In one case, a difference in germination of 4 days of one-half of the mustard seeds in a given plot showed a reduction in yield by weight of 20 per cent.

Hedrick (1905) found that young peach trees were hindered in their development when grown in the same pot cultures with oats, potatoes, tomatoes, mustard, and rape even though the optimum conditions of moisture, aeration, and food supply were maintained. Dandeno (1908, 1909) observed that the growth of buckwheat was considerably deterred by Canada thistles when the two plants were grown together. He also observed that poplars and oaks were injurious to the growth of corn. The injurious effects extended to some distance from the trees, a fact which showed that shade and moisture were not the deciding factors. It has also been observed by Jones and Morse (1902), Cook (1921), Massey (1925), and Schneiderhan (1926) that walnut trees have a detrimental effect upon the growth and vigor of various plants. These trees prevent the growth of cinquefoil (*Potentilla fruticosa*) within a large area surrounding them and cause the wilting, stunting, and frequently the death of corn, tomatoes, potatoes, alfalfa, and apple trees when their roots come in contact or close proximity to those of the walnut.

It was found by Prianischnikov (1914) that when plants were grown in sand cultures there was a marked decrease in the amount of the harvest of the second and third crops. Perlata and Estioko (1923) found that zacate (*Leersia hexandria*), the most common forage plant in the Philippines, is detrimental to the rice that follows it. Rice is also detrimental to the rice crops that follow.

It is a common observation in the Great Plains area of the United States (Sewell, 1923) that wheat planted in the fall after the various types of sorghums (*Andropogon sorghum*) does not yield so well as wheat planted after corn. At the Kansas Agricultural Experiment Station, winter wheat grown after a kafir yielded three bushels less per acre per year over a period of six years than when grown after corn.

The probable causes for the detrimental influence of one plant or crop on another have been attributed to two general factors: (1) the depletion of the nutrients in the soil so that there is not a sufficient supply for the plants that follow, and (2) the production in the soil, by the decomposition of roots, stems, leaves, and the cells that are lost from the growing root, of compounds that are deleterious to the roots of the plants with which they come in contact. Let us now consider the evidence for the two general theories.

1. The Influence of the Depletion of the Nutrients in the Soil on Plant Growth.—It was noted by Livingston, Schreiner, Skinner, and Reed (1905 to 1912) that frequently there is no correlation between the mineral composition of soils and their crop-producing power, since the amount of nutrients removed from a given soil by a crop is not sufficient to account for the diminished yield of the following crop.

Hartwell and his workers (1918, 1927) have reported extensive work in regard to the influence of certain crops on those that follow. The general plan of their experiment was to grow 16 different crops on that number of parallel plots for two successive seasons and then every third year to grow one of these crops over the entire area. The plants that were grown were onion, potato, mangel beet, rutabaga, cabbage, buckwheat, corn, millet, oats, rye, carrot, redtop, timothy, alsike clover, and red clover. The crops that were grown uniformly over all the plots were onions in 1910, buckwheat in 1913, and alsike clover in 1916. The yield of onions following the various crops will serve as an example of the general facts observed. This was from 13 to 17 bu. per acre following cabbage, mangel beets, rutabaga, turnips, and buckwheat; 35 to 87 bu. following rye and potatoes; 131 to 178 bu. following corn, millet, onions, oats, and red clover; 240 to 314 bu. following squash, timothy, and alsike clover; and 406 to 412 bu. following mixed timothy and redtop and redtop alone. These field experiments were checked by pot experiments in the greenhouse, and the following conclusions drawn:

> The divergent effect of crops on those which follow seems not to be attributable at least principally to differences in the amount of nutrients removed by the crops grown previously. That is, the smallest yields may not occur after the crop which removed the largest amount of even the most needed nutrient.

They also found that the soil acidity was affected differently by the crops under consideration and that in the case of the onion, a plant that is sensitive to acid conditions, the best yields followed the crops giving rise to the least acidity. It was also found that the divergent effect of crops on those that follow was less if the soil acidity was corrected by liming. Garner (1925) and his workers from extensive studies on tobacco culture considered that the tobacco plant is particularly sensitive to the effects of preceding crops. It may be seriously retarded as a result of the effects of preceding crops of tobacco itself or of various other plants. Under intensive methods of production, when tobacco follows tobacco, reduced yields are not due, proportionately, to a reduction in the general productiveness of the soil.

Odland and Smith (1933) over a period of 23 years grew 10 uniform crops following each of 16 different crops. In some years very striking differences were obtained in the yields of the uniform crops following the various preceding ones, but at other times these differences were not significant. They considered that several factors were operative in causing these differences in growth. The variations in soil acidity caused by the growth of different crops showed sufficient correlation with the subsequent growth of other plants to prove that this is an important factor. In the case of corn, the relative removal of nitrogen by the preceding crops was correlated with yield but the results with other crops were not conclusive with respect to this factor.

It was observed by Wilkins and Hughes (1934) in Iowa that corn yielded 50.8, 49.0, and 47.8 bu. per acre after Sudan grass, soybeans, and oats, respectively. The Sudan grass produced over twice as much dry matter as soybeans and oats, removed over twice as much nitrogen as oats and considerably more than soybeans. Regardless of this, however, the Sudan grass plots after 13 years of cropping appeared to be just as productive as the plots of soybeans and oats. They believed that the same results would not have been obtained on infertile soils. Dodd and Pohlman (1935) in West Virginia found that the yield of buckwheat and potatoes was not affected differently by preceding crops of oats and soybean. Oats following oats, buckwheat, or corn were poorer than oats following wheat and potatoes. Oats, buckwheat, and potatoes were similar to wheat in their effects on the succeeding crop of soybeans. Russell and McDonald (1935) reported from England and Australia that continuous cropping of wheat is not detrimental to the soil nor is there any diminishing return, provided that fertilizer is supplied in sufficient quantities. However, on the incompletely fertilized plots there has been a falling-off in yield.

It is now generally accepted, at least frequently, that the detrimental effect of a crop on the succeeding one is not due to the exhaustion of the nutrients of the soil by the preceding crop.

2. The Influence of the Decomposition of Roots in the Soil on Plant Growth.—It is now universally accepted that the productivity of the soil may be decreased by reason of more or less complex organic substances that may be produced in it. Livingston, Britton, and Reid (1905) and Livingston (1909) found that the unproductive soils which they examined contained soluble organic substances that were toxic to wheat plants. They observed that these soils were greatly benefited without any alteration of the nutrient conditions by the use of practically insoluble bodies such as ferric hydrate and carbon black which apparently rendered the harmful substances inactive. Tannic acid and pyrogallol also benefited these soils. This beneficial action they considered due to the chemical combination of these compounds with the toxic substances, which so altered them that they became noninjurious to the plants. They considered that the beneficial effects derived from the use of stable manures on unproductive soils are due more to the addition of organic matter, which alters the toxic substances of the soil, than to the increased amount of salts added to the soil. Breazeale (1906) showed that the extracts of certain soils were injurious to wheat plants when grown in water cultures and that this toxicity could be overcome wholly or in part by the treatment of the extract with carbon black, calcium carbonate, ferric hydroxide, and other solids.

Schreiner, Skinner, Shorey, Reed, and their coworkers (1907 to 1912) in some of the outstanding work of the century in organic chemistry have studied the effects and nature of the organic compounds of the soil. They developed means of extraction and identification of many of these substances. The following are some of the compounds of the soil that have been isolated and identified: dihydroxystearic acid, picolinecarboxylic acid, agroceric acid, hydroxy-fatty acids, paraffinic acid, lignoceric acid, resin, glycerides of fatty acids, phytosterol, pentosan, pentose, histidine, arginine, pyrimidine, oxalic acid, succinic acid, purine. nucleic acid, xanthine, choline, adenine, creatine, tyrosine, vanillin, cumarin, and others. Some of these compounds were found to be noninjurious and even beneficial to plant growth, while others were found to be injurious. Among those which were injurious were cumarin, vanillin, dihydroxystearic acid, alanine, glycocoll, neurine, quinone, pyridine compounds, and tyrosine, the degree of injury depending upon the nature and concentration of the substances under observation. These investigators found that when absorbing agents or certain fertilizers were used, the toxicity of the above-mentioned compounds was greatly reduced. The various fertilizer salts acted differently in overcoming the harmful effects of certain compounds. To neutralize cumarin, phosphate fertilizers were the most effective. In the case of vanillin, nitrogenous fertilizers were the most efficient antidoting agents, while potassium salts were most effective in overcoming the toxic action of quinone. The action of fertilizers on nonproductive soils is not definitely known, but the results observed add weight to the evidence against the assumption that their effect is due altogether to the increase in plant nutrients that they add to the soil.

A glance at the list of the organic compounds that have been isolated from the soil shows that many of these substances are those which commonly result from the decomposition of animal and vegetable matter. The sources of the organic matter in the soil are the roots and stubbles of the plants which have previously grown in it and the cells which may be lost from the root tips of the plants growing therein. The kind of products that may arise from the decomposition of the organic matter in the soil will thus depend upon many and varied factors. Among these are the kind of plant that produced the organic matter, since each species of plant is made up of more or less specific compounds, and the age of the plants (Waksman and Tenney, 1928). The products produced will depend upon the amount of moisture in the soil, the thoroughness of aeration, the character of the microorganisms in the soil, and the mutual relation of the organic compounds and the mineral constituents. The factors that have just been mentioned are in turn influenced by cultural practices, fertilizer, drainage, irrigation, and inoculation. Thus the resulting compounds from plant material in the process of decomposition may vary greatly so that the substances which may arise from a given type of plant material under one set of conditions may be very toxic to plants, while those which may arise under another set of conditions may not injure the following plants at all. The aeration of the soil is a very important factor in determining or overcoming the toxic properties of many organic compounds. Pickering (1917) found that if the soil leachings that were toxic to plants were exposed to the air for 24 hr. their toxic effect disappeared entirely, and these leachings afterward became very beneficial. Schreiner and his coworkers (1906 to 1912) found that the detrimental influence of many of the organic compounds of the soil could be overcome by the addition of certain substances to oxidize more thoroughly the compounds in question and suggested that the beneficial effect of fertilizers upon plants might be due in part to their capacity as oxidizers.

The idea that the stubble and roots of certain plants may give rise to decomposition products that will be injurious to the crops which follow has been strengthened by the work of Sewell (1923), Breazeale (1924), Collison (1924), and Hawkins (1925). As has previously been mentioned, wheat does not yield so well when grown after a crop of sorghum as it does when grown after corn. In pot cultures where wheat received its moisture solely from water percolating through perforated trays (after the method of Pickering) in which corn and kafir were growing, the growth of the kafir plants inhibited the growth of the wheat. When kafir and corn were grown in these percolating trays, harvested, and then wheat planted so that it would receive the leaching from the trays, the kafir soil leachings depressed the growth of the wheat. From this Sewell concluded that there are decomposition products from the crop residue of kafir which have a retarding effect on the growth of the wheat that follows. Breazeale (1924) by growing wheat in water cultures containing the chopped-up pieces of the stubble of kafir found that a toxic property was developed during the decomposition of the stubble that was injurious to the wheat plants. He observed that this toxic body or bodies were shortly either volatilized or decomposed into nontoxic substances. The decomposition of the stubble is hastened by an increase in temperature, and a good nutrient solution can be produced from finely chopped sorghum stubble in 3 or 4 weeks. From a mixture of water and stubble that was toxic to wheat plants, Breazeale distilled off a poisonous compound that would kill wheat plants within a few hours. Hawkins (1925) found that the stalk residues of the sorghums were more detrimental to plants that follow than those from the roots. He also observed that certain crop plants such as field peas do not use the upper stratum of soil as do the sorghum plants and are not depressed in growth following a crop of sorghum so much as other crops such as vetch, which have root systems more like sorghums. The detrimental effects of sorghum disappear in a few months after the crop has been harvested. It was found by Conrad (1927) that sorghum roots may contain as much as fifteen times the amount of sugar as the roots of corn. When this is liberated into the soil after harvesting, it stimulates the growth of certain microorganisms which thus compete with the following plants for nitrogen and probably for other elements. This released sugar may also retard the process of nitrification. He considered, then, that the immediate cause of the deleterious effects of sorghums on crops that follow may be due to one or more elements becoming a limiting factor. He believed that the result can be overcome to a great extent by a system of cultivation that would lead to the rapid decay of the roots. The work of Wilson and Wilson (1928) suggests that the injurious aftereffects of sorghum may be associated with the comparative ease with which the roots are oxidized in the soil. This process, which is accompanied by an increase in the number of soil organisms and an increase in the assimilation of nitrate nitrogen, would tend to deplete the soil of available nitrogen.

McKinley (1931) found indications that sorghum residues, under favorable conditions and not in excess of 2 tons per acre, increase the yield of crops over those of untreated soils. The increased yield from these residues may or may not be less than that from corn or other plants. He considered that part of the decreased yield of crops following sorghums may be partially explained on the basis of the readily decomposable materials in the residues.

Watery extracts of cereal straws, timothy residues, and alfalfa were found by Collison (1924) to be very toxic to plants grown in them, and he considered that there is no doubt but that these and other plant residues are important contributors to the soil of compounds toxic to plant growth. Whether the amounts of material thus contributed to the soil would ever become toxic to a crop in the field would depend upon the nature of the soil and the treatment that it receives. Collison further suggests that if these toxic compounds are present in the fresh sap they might serve as inhibiting agents against parasitic fungi.

Davis (1928) in a search for the substances that might cause the toxic action of walnut trees upon plants growing near them succeeded in extracting from the hulls and roots of the walnut a substance that is apparently identical with juglone. This extract when purified and crystallized proved to be an exceedingly powerful toxin when injected into the stems of tomato and alfalfa plants.

The oxidation reduction system of juglone to hydrojuglone was studied by Davis (1931). The determination was made on the root-hair cells of *Trianea bogotensis*, and the data indicated that juglone would be reduced if injected into these cells. It is thought that an explanation of the resistance to walnut toxicity might be associated with the reduction of the toxic juglone to hydrojuglone, its nontoxic reductant. Whether this is the substance that exerts the toxic action by the roots in nature, however, is not known.

The problem of the toxic or beneficial effect of one plant on another is a complicated one and the causes are yet in doubt. A summary of this subject can best be stated in the words of Thatcher (1923):

> There is yet no positive proof of the nature of the causative agent or agencies for either the beneficial or injurious effect of one plant on another. It may vary widely in different cases and may be chemical or bacterial in character. Definite proof that observed injurious effects on a second crop are due to toxic chemical substances in the soil produced by or in association with the first crop has not been established.

Bibliography

Aberson, J. H., 1910: Ein Beitrag zur Kenntnis der Natur der Wurzelausscheidungen, *Jahrb. wiss. Bot.*, **47**: 41–56.

Addoms, R. M., 1923: The effect of the hydrogen ion on the protoplasm of the root hairs of wheat, *Am. J. Bot.*, **10**: 211–220.

———, 1927: Toxicity as evinced by changes in the protoplasmic structure of root hairs of wheat, *Am. J. Bot.*, **14**: 147–165.

Albert, W. B., and G. M. Armstrong, 1931: Effects of high soil moisture and lack of soil aeration upon fruiting behavior of young cotton plants, *Plant Physiol.*, **6**: 585–591.

Aldrich, W. W., R. A. Work, and M. R. Lewis, 1935: Pear root concentration in relation to soil moisture extraction in heavy clay soil, *J. Agr. Res.*, **50**: 975–988.

Allan, H. H., 1926: The surface roots of an individual matai, *N. Z. J. Sci. and Tech.*, **8**: 233–234. Quoted by Weaver, 1932.

Allison, R. V., and J. W. Shive, 1923: Studies on the relation of aeration and continuous renewal of nutrient solution on the growth of soybeans in artificial culture, *Am. J. Bot.*, **10**: 554–567.

Arker, J., 1901: Die Beeinflussung des Wachsthums der Wurzeln durch das umgebende Medium, *Bot. Cent.*, **87**: 433–434.

Arny, A. C. and I. J. Johnson, 1928: The roots of flax plants, *J. Am. Soc. Agron.*, **20**: 373–381.

———, T. E. Stoa, C. McKee, and A. C. Dillman, 1929: Flax cropping in mixture with wheat, oats and barley, *U. S. Dept. Agr. Tech. Bull.* 133.

Barakov, P., 1910: The carbon dioxide content of soils during different stages of growth of plants, *Zhur. Opuitn. Agron.* (*Russ. J. Expt. Landw.*), **11**: 321–343 (Russian); *Exp. Sta. Rec.*, **23**: 523.

Bardell, E. M., 1915: Production of root hairs in water, *Univ. Wash. Pub. Bot.*, **1**: 1–9.

Bartoo, D. R., 1929: Origin and development of tissues in root of *Schizaea rupestris*, *Bot. Gaz.*, **87**: 642–652.

Beckenbach, J., and J. H. Gourley, 1932: Some effects of different cultural methods upon root distribution of apple trees, *Proc. Am. Soc. Hort. Sci.*, **29**: 202–204.

Beckwith, M. H., 1885: Report of the assistant horticulturist, *N. Y. Sta. Rept.*, 261–267

Bennett, M. E., 1904: Are roots aerotropic? *Bot. Gaz.*, **37**: 241–259.

Bergman, H. F., 1920: The relation of aeration to the growth and activity of roots and its influence on the ecesis of plants in swamps, *Ann. Bot.*, **34**: 13–33.

Biswell, H. H., 1935: Effect of environment upon the root habits of certain deciduous forest trees, *Bot. Gaz.*, **96**: 676–708.

Bizzell, J. A., and T. L. Lyon, 1918: The effect of certain factors on the carbon dioxide content of soil air, *J. Am. Soc. Agron.*, **10**: 97–113.

Borowski, R., 1919: Beitrag zur Kenntnis des Oxydationsvermögens der Wurzeln der höheren Pflanzen, *Landw. Versuchs-Stat.*, **94**: 265–284.

Bouygues, H., 1928: À propos de la nécessité de l'oxygène de l'air sur l'accroissement des racines poussant dans l'eau, *Bull. Soc. Bot. France*, **75**: 469–472.

Breazeale, J. F., 1906: Effect of certain solids upon the growth of seedlings in water cultures, *Bot. Gaz.*, **41**: 54–63.

———, 1924: The injurious after effects of sorghum, *J. Am. Soc. Agron.*, **16**: 689–701.

———, 1930: Maintenance of moisture equilibrium and nutrition of plants at and below the wilting percentage, *Ariz. Sta. Tech. Bull.* 29.

Bryant, A. E., 1934: A demonstration of the connection of the protoplasts of the endodermal cells with the Casparian strips in the roots of barley, *New Phytol.*, **33**: 231.

Büsgen, M., and E. Münch, 1929: "The Structure and Life of Forest Trees," pp. 269–300. John Wiley & Sons, Inc., N. Y. Quoted by Weaver and Kramer, 1932.

Cannon, W. A., 1911: Root habits of desert plants, *Carnegie Inst. Wash. Pub.* 131.

———, 1912: Physical relations of roots to soil factors, *Carnegie Inst. Wash. Yearbook*, **11**: 61–62.

———, 1913: Botanical features of the Algerian Sahara, *Carnegie Inst. Wash. Pub.* 178.

———, 1915: On the relation of root growth and development to the temperature and aeration of the soil, *Am. J. Bot.*, **2**: 211–224.

———, 1916: Distribution of the cacti with especial reference to the role played by the root response to soil temperature and soil moisture, *Am. Nat.*, **50**: 435–442.

———, 1916: On the relation between the rate of root growth and the oxygen of the soil, *Carnegie Inst. Wash. Yearbook*, **15**: 74–75.

———, 1917: Root growth of *Prosopis velutina* and *Opuntia versicolor* under conditions of a small oxygen supply in the soil, *Carnegie Inst. Wash. Yearbook*, **16**: 82–83.

——— and E. E. Free, 1917: The ecological significance of soil aeration, *Sci.*, **45**: 178–180.

———, 1918: Root growth in desert plants and the oxygen supply of the soil, *Carnegie Inst. Wash. Yearbook*, **17**: 81–83.

———, 1923: The influence of the temperature of the soil on the relation of roots to oxygen, *Sci.*, **58**: 331–332.

———, 1924: A note on the relation of root growth in the soil to the oxygen supply: The growth ratio, *Ecology*, **5**: 319–322.

———, 1925: On the upper critical concentration of oxygen in root growth, *Sci.*, **61**: 118–120.

———, 1925: Physiological features of roots with especial reference to the relation of roots to the aeration of the soil, *Carnegie Inst. Wash. Pub.* 368.

———, 1932: Absorption of oxygen by roots when the shoot is in darkness or in light, *Plant Physiol.*, **7**: 673–684.

———, 1932: Experimental studies on the relation of roots to oxygen and on the aeration of plant tissues, *Carnegie Inst. Wash. Yearbook*, **31**: 294–296.

———, D. Demaree, and E. A. Purer, 1933: Evaporation, transpiration and oxygen consumption by roots, *Sci.*, **78**: 388–389.

Carlson, F. A., 1925: The effect of soil structure on the character of alfalfa root systems, *J. Am. Soc. Agron.*, **17**: 336–345.

Clark, W. S., 1874: Observations upon the phenomena of plant life, 22d *Ann. Rept. Mass. State Board Agr.*, 204–312.

Clements, F. E., 1920: Plant Indicators, *Carnegie Inst. Wash. Pub.* 290.

———, 1921: Aeration and air content. The role of oxygen in root activity, *Carnegie Inst. Wash. Pub.* 315.

Collings, G. H., and J. D. Warner, 1927: Root development of cotton on Cecil sandy loam during 1926, *J. Am. Soc. Agron.*, **19**: 839–842.

Collison, R. C., 1924: The presence of certain organic compounds in plants and their relation to the growth of other plants, *J. Am. Soc. Agron.*, **17**: 58–68.

———, 1935: Lysimeter investigations. IV, Water movement, soil temperatures, and root activity under apple trees, *N. Y. Sta. (Geneva) Tech. Bull.* 237.

Conrad, J. P., 1927: Some causes of the injurious after effects of sorghums and suggested remedies, *J. Am. Soc. Agron.*, **19**: 1091–1111.

——— and F. J. Veihmeyer, 1929: Root development and soil moisture, *Hilgardia*, **4**: 113–134.

Cook, M. T., 1921: Wilting caused by walnut trees, *Phytopath.*, **11**: 346.

Cormack, R. G. H., 1935: Investigations on the development of root hairs, *New Phytol.*, **34**: 30–54.

Cottrell, H. M., 1902: Growing alfalfa in Kansas, *Kans. Sta. Bull.*, **114**: 61–62.

Crider, F. J., 1933: Selective absorption of ions not confined to young rootlets, *Sci.*, **78**: 169.

Crist, J. W., and J. E. Weaver, 1924: Absorption of nutrients from the subsoil in relation to crop yield, *Bot. Gaz.*, **77**: 121–149.

——— and G. J. Stout, 1929: Relation between top and root size in herbaceous plants, *Plant Physiol.*, **4**: 63–85.

Curry, J., R. Platt, and S. F. Trelease, 1935: Does dilute heavy water influence biological processes? *Sci.*, **81**: 275–277.

Czapek, F., 1896. Zur Lehre von den Wurzelausscheidungen, *Jahrb. wiss. Bot.*, **29**: 321–390.

Dandeno, J. B., 1908, 1909: Mutual interaction of plant roots, *Rept. Mich. Acad. Sci.*, **10**: 32–36; **11**: 24–25.

Davis, E. F., 1928: The toxic principle of *Juglans nigra* as identified with synthetic juglone, and its toxic effects on tomato and alfalfa plants, *Am. J. Bot.*, **15**: 620.

———, 1931: Observations from the microinjection of acid-base and oxidation reduction indicators into the root-hair cells of *Trianea bogotensis*, *Am. J. Bot.*, **18**: 896.

Dean, B. E., 1933: Effect of soil type and aeration upon root systems of certain aquatic plants, *Plant Physiol.*, **8**: 203–222.

De Candolle, A. P., 1832: *Physiologie végétale*, **1**: 248–251; **3**: 1474–1475, 1493–1520, Paris.

Dittmer, H. J., 1937: A quantitative study of the roots and root hairs of a winter rye plant (*Secale cereale*), unpublished data, *Iowa State College*.

Dodd, D. R., and G. G. Pohlman, 1935: Some factors affecting the influence of soybeans, oats and other crops on the succeeding crop, *W. Va. Sta. Bull.* 265.

Duley, F. L., 1929: The effect of alfalfa on soil moisture, *J. Am. Soc. Agron.*, **21**: 224–231.

———, 1929: Unpublished data, *Kans. Sta.*

Duncan, W. H., 1935: Root systems of woody plants of old fields of Indiana, *Ecology*, **16**: 554–567.

Dustman, R. B., 1925: Inherent factors related to absorption of mineral elements by plants, *Bot. Gaz.*, **79**: 233–264.

Eaton, F. M., 1931: Root development as related to character of growth and fruitfulness of the cotton plant, *J. Agr. Res.*, **43**: 875–883.

Eckerson, S., 1914: Thermotropism of roots, *Bot. Gaz.*, **58**: 254–263.

Ewart, A. J., 1894: Oxytropic irritability of roots, *Trans. Liverpool Biol. Soc.*, **8**: 238; **10**: 190.

Farr, C. H., 1925: Root-hair elongation in Knop's solution and in tap water, *Am. J. Bot.*, **12**: 372–383.

———, 1925: The formation of root hairs in water, *Proc. Iowa Acad. Sci.*, **32**: 157–165.

———, 1927: Studies on the growth of root hairs in solutions. I, The Problem, previous work, and procedure, *Am. J. Bot.*, **14**: 446–457.

———, 1927: Studies on the growth of root hairs in solutions. II, The effects of the concentration of calcium nitrate, *Am. J. Bot.*, **14**: 497–515.

———, 1927: Studies on the growth of root hairs in solutions. III, The effect of concentrations of $CaCl_2$ and $Ca(OH)_2$, *Am. J. Bot.*, **14**: 553–565.

———, 1928: Studies on the growth of root hairs in solutions. IV, The pH-molar rate relations for collards in calcium chloride, *Am. J. Bot.*, **15**: 6–32.

———, 1928: Studies on the growth of root hairs in solutions. V, Root-hair elongation as an index of root development, *Am. J. Bot.*, **15**: 103–114.

———, 1928: Studies on the growth of root hairs in solutions. VI, Structural responses to toxic pH and molar concentrations of calcium chloride, *Am. J. Bot.*, **15**: 171–179.

———, 1928: Studies on the growth of root hairs in solutions. VII, Further investigations on collards in calcium hydroxide, *Bull. Torrey Bot. Club*, **55**: 223–247.

———, 1928: Studies on the growth of root hairs in solutions. VIII, Structural and intracellular features of collards in calcium nitrate, *Bull. Torrey Bot. Club*, **55**: 529–553.

Farris, N. F., 1934: Root habits of certain crop plants as observed in the humid soils of New Jersey, *Soil Sci.*, **38**: 87–112.

Fletcher, F., 1912: Toxic excreta of plants, *J. Agr. Sci.*, **4**: 245–247.

Frank, B., 1893: Die Assimilation des freien Stickstoffs durch die Pflanzenwelt, *Bot. Ztg.*, **51**: 139–156.

Fred, E. B., and A. R. C. Haas, 1919: The etching of marble by roots in the presence and absence of bacteria, *J. Gen. Physiol.*, **1**: 631–639.

Free, E. E., and B. E. Livingston, 1917: The effect of deficient soil oxygen on the roots of higher plants, *Johns Hopkins Univ. Circ.*, n. s., **3**: 182–185.

Garner, W. W., W. M. Lunn and D. E. Brown, 1925: Effects of crops on the yields of succeeding crops in the rotation with special reference to tobacco, *J. Agr. Res.*, **30**: 1095–1133.

Gautheret, R., 1933: Artificial cultures of corn root tip meristems, *C. R. Acad. Sci. Paris*, **197**: 85–87; *Exp. Sta. Rec.*, **73**: 23.

Gemmer, E. W., 1928: The root system of a long-leaf pine, *Sci. Month.*, **27**: 384.

Georgeson, C. C., and J. E. Payne, 1897: Investigations of the root development of some forage plants, *Kans. Sta. Bull.* **75**: 212–222.

Gericke, W. F., 1923: Further notes on effect of extent of root systems on tillering of wheat, *Bot. Gaz.*, **75**: 320–321.

Gibson, R. G. H., 1912: The extent of the root system of *Cucumis sativus*, *Ann. Bot.*, **26**: 951–952.

GOFF, E. S., 1883–1888: Report of the horticulturist, *N. Y. Sta. 2d Ann. Rept.*, 219; *3d Ann. Rept.*, 305–315 (1884); *5th Ann. Rept.*, 159–168 (1886); *6th Ann. Rept.*, 90–96 (1887); *7th Ann. Rept.*, 171–178 (1888).

GORDIENKO, M., 1930: Über die Beziehungen zwischen Bodenbeschaffenheit und Wurzelgestaltung bei jungen Pflanzen, *Landw. Jahrb.*, **72**: 125–139; *Biol. Abst.*, **6**: 2510.

GRANDFIELD, C. O., and W. H. METZGER, 1936: Relation of fallow to restoration of subsoil moisture in an old alfalfa field and subsequent depletion after reseeding, *J. Am. Soc. Agron.*, **28**: 115–123.

GURSKY, A. V., 1928–1929: The root systems of *Fraxinus excelsior*, *Fraxinus Pennsylvanica* and *Acer Negundo* on the black soils of Kuban, *Bull. App. Genetics and Plant Breeding*, **21**: 145–181.

HAAS, A. R. C., 1916: The excretion of acids by roots, *Sci.*, **44**: 685.

———, 1917: The excretion of acids by roots, *Proc. Nat. Acad. Sci.*, **2**: 561–566.

HABERLANDT, G., 1914: "Physiological Plant Anatomy," Eng. transl., 222 pp., Macmillan & Company, Ltd., London.

HALL, A. D., 1906: The solvent action of roots upon the soil particles, *Sci. Prog. Twentieth Cent.*, **1**: 51–57.

——— W. E. BRENCHLEY, and L. M. UNDERWOOD, 1914: The soil solution and the mineral constituents of the soil, *J. Agr. Sci.*, **6**: 278–301.

HANSTEEN-CRANNER, B., 1914: Über das Verhalten der Kulturpflanzen zu den Bodensalzen. III, *Jahrb. wiss. Bot.*, **53**: 536–598. Quoted by Cormack, 1935.

HARMON, F. N., and E. SNYDER, 1934: Grape root distribution studies, *Proc. Am. Soc. Hort. Sci.*, **32**: 370–373.

HARRIS, F. S., 1914: The effect of soil moisture, plant food and age on the ratio of tops to roots in plants, *J. Am. Soc. Agron.*, **6**: 65–75.

HARRISON, W. H., and P. A. S. AIYER, 1913: The gases of swamp rice soils; their composition and their relation to the crop, *Mem. Dept. Agr. in India, Chem. Ser.*, **3**: 65–106.

HARTWELL, B. L., and S. C. DAMON, 1918: The influence of crop plants on those which follow. I, *R. I. Sta. Bull.* 175.

———, F. R. PEMBER, and G. E. MERKLE, 1919: The influence of crop plants on those which follow. II, *R. I. Sta. Bull.* 176.

———, J. B. SMITH, and S. C. DAMON, 1927: The influence of crop plants on those which follow. III, *R. I. Sta. Bull.* 210.

HÄVELER, W., 1892: Ueber die Verwerthung des Humus bei der Ernährung der chlorophyllführenden Pflanzen, *Jahrb. wiss. Bot.*, **24**: 283–316.

HAWKINS, R. S., 1925: The deleterious effect of sorghum on the soil and on the succeeding crops, *J. Am. Soc. Agron.*, **17**: 91–92.

HAYS, W. M., 1888: Roots of clovers, *5th Bien. Rept., Univ. Minn., Suppl.* I, *Dept. Agr. Exp. Sta.*, 188–198.

———, 1894: The roots of crops, *N. Dak. Sta. Bull.* **10**: 47–50.

HEADDEN, W. P., 1896: Alfalfa, *Colo. Sta. Bull.* **35**: 47–59.

HEDRICK, U. P., 1905: The relationship of plants in the orchard, *Proc. West N. Y. Hort. Soc., 50th ann. meet.*, 17–28.

HEINICKE, A. J., 1932: The effect of submerging the roots of apple trees at different seasons of the year, *Proc. Am. Soc. Hort. Sci.*, **29**: 203–207.

———, 1935: Root growth in young apple trees made shortly before and after defoliation, *Proc. Am. Soc. Hort. Sci.*, **33**: 164–165.

HELLRIEGEL, H., 1883: Wurzel und Bodenvolumen, *Beiträge zu den naturwiss. Grundlagen des Ackerbaus*, 119–280.

Hendrickson, A. H., and F. J. Veihmeyer, 1931: Influence of dry soil on root extension, *Plant Physiol.*, **6**: 567–576.

Hesse, H., 1904: Beiträge zur Morphologie und Biologie der Wurzelhaare, *Inaug. Diss.*, Jena.

Heyward, F., 1933: The root system of long-leaf pine on the deep sands of western Florida, *Ecology*, **14**: 136–148.

Höhn, K., 1934: Die Bedeutung der Wurzelhaare für die Wasseraufnahme der Pflanzen, *Z. Bot.*, **27**: 529–564.

Holch, A. E., 1931: Development of roots and shoots of certain deciduous tree seedlings in different forest sites, *Ecology*, **12**: 259–298.

Hole, R. S., 1918: Recent investigations on soil aeration, *Indian Forester*, **1918**: 202.

Hooker, H. D., 1914: Thermotropism in roots, *Plant World*, **17**: 135–153.

———, 1915: Hydrotropism in roots of *Lupinus albus*, *Ann. Bot.*, **29**: 265–285.

Hopmann, O., 1930: Über das Vorkommen und die Mikrochemie des Amyloids in pflanzlichen Zellhauten, *Inaug. Diss.*, Münster. Quoted by Cormack, 1935.

Howard, A. and G. L. C. Howard, 1915: Soil ventilation, *Agr. Res. Inst. Pusa Bull.* **52**: 1–35.

———, 1916: Soil aeration in agriculture, *Agr. Res. Inst. Pusa Bull.* **61**: 1–22.

——— and G. L. C. Howard, 1917: The economic significance of the root development of agricultural crops, *Agr. J. India, Special Indian Sci. Congo Number.*

———, 1918: Recent investigations on soil aeration, *Indian Forester*, **1918**: 187.

Howe, C. G., 1921: Pectic materials in root hairs, *Bot. Gaz.*, **72**: 313–320.

Hunter, C., 1912: Some observations on the effect of soil aeration on plant growth, *Proc. Phil. Soc. Univ. Durham*, **4**: 183.

———, and E. M. Rich, 1925: The effect of artificial aeration of the soil on *Impatiens balsamina*, *New Phytol.*, **24**: 257–271.

Hutchins, L. M., and B. E. Livingston, 1923: Oxygen-supplying power of the soil as indicated by color changes in alkaline pyrogallol solution, *J. Agr. Res.*, **25**: 133–140.

———, 1926: Studies on the oxygen-supplying power of the soil together with quantitative observations on the oxygen-supplying power requisite for seed germination, *Plant Physiol.*, **1**: 95–150.

Jean, F. C., and J. E. Weaver, 1924: Root behavior and crop yield under irrigation, *Carnegie Inst. Wash. Pub.* 357.

Jeffs, R. E., 1922: The grand period of growth of root hairs, *Proc. Okla. Acad. Sci.*, (Okla. Univ.) **2**: 53–56.

———, 1925: The elongation of root hairs as affected by light and temperature, *Am. J. Bot.*, **12**: 577–606.

Jensen, J. H., 1931: Some studies on root habits of sugar cane in Cuba, *Sci. Contr. Tropical Plant Res. Found.* 21.

Johnston, C. O., and E. C. Miller, 1934: Relation of leaf rust infection to yield, growth, and water economy of two varieties of wheat, *J. Agr. Res.*, **49**: 955–981.

Jones, L. H., 1931: Flower pot composition and its effect on plant growth, *Mass. Sta. Bull.* **277**: 147–161.

——— and H. D. Haskins, 1935: Distribution of roots in porous and nonporous plant containers, *Plant Physiol.*, **10**: 511–519.

Jones, L. R., and W. J. Morse, 1902–1903: The shrubby cinquefoil as a weed, *16th Ann. Rept. Vt. Sta.*, 188–190.

Kaserer, H., 1911: Beobachtungen über die Bewurzelung der Kulturpflanzen bei Reinsaat und bei Mischsaat, *Z. landw. Versuchs. in Oesterreich*, **14**: 1022. Quoted by Nicol, 1935.

KERR, T., 1933: The injection of certain salts into the protoplasm and vacuoles of the root hairs of *Limnobium spongia*, *Protoplasma*, **18**: 420–440; *Biol. Abst.*, **9**: 750.

KIESSELBACH, T. A., 1910: Transpiration experiments with the corn plant, *Nebr. Sta. 23rd Ann. Rept.*, 125–139.

———, J. C. RUSSEL, and A. ANDERSON, 1929: The significance of subsoil moisture in alfalfa production, *J. Am. Soc. Agron.*, **21**: 241–268.

———, A. ANDERSON, and J. C. RUSSEL, 1934: Subsoil moisture and crop sequence in relation to alfalfa production, *J. Am. Soc. Agron.*, **26**: 422–442.

———, ———, and W. E. LYNESS, 1935: Cultural practices in corn production, *Neb. Sta. Bull.* 293.

——— and R. M. WEIHING, 1935: The comparative root development of selfed lines of corn and their F_1 and F_2 hybrids, *J. Am. Soc. Agron.*, **27**: 538–541.

KING, F. H., 1892–1893: Natural distribution of roots in field soils, *Wis. Sta. Rept.*, 112–121; 160–167 (1893).

KINMAN, C. F., 1932: A preliminary report on root growth studies with some orchard trees, *Proc. Am. Soc. Hort. Sci.*, **29**: 220–224.

KISSER, J., 1925: Über das Verhalten von Wurzeln in feuchter Luft, *Jahrb. wiss. Bot.*, **64**: 416–439. Quoted by Cormack, 1935.

——— and I. BEER, 1934: Chemotropic sensitivity of dicotyledonous seedlings, *Jahrb. wiss. Bot.*, **80**: 301–335; *Chem. Abst.*, **29**: 7384.

KLERCKER, J. A., 1891: Über calorotropische Erscheinungen bei einigen Keimwurzeln, *Oefvers. Vetensk. Akad. Förhandl. Stockholm*, **48**: 765–790.

KNIGHT, R. C., 1924: The response of plants in soil and in water culture to aeration of the roots, *Ann. Bot.*, **38**: 305–325.

KNUDSON, L., and R. S. SMITH, 1919: Secretion of amylase by plant roots, *Bot. Gaz.*, **68**: 460–466.

———, 1920: The secretion of invertase by plant roots, *Am. J. Bot.*, **7**: 371–379.

KOSSOWITCH, P., 1904: The quantitative determination of carbon dioxide produced by the roots of plants during the period of their development (Russian), *Zhur. Opuitn. Agron.*, **5**: 482–493 (quoted by Turpin, 1920).

KRASSOVSKY, I., 1929: The root system of different varieties of flax, *Bull. App. Bot. and Plant Breeding*, **22**: 43–100.

X KROGH, A., and E. LANGE, 1930: On the organic matter given off by algae, *Biochem. J.*, **24**: 1666–1671.

KUNZE, G., 1906: Ueber Säureausscheidung bei Wurzeln und Pilzhyphen und ihre Bedeutung, *Jahrb. wiss. Bot.*, **42**: 357–393.

LAIRD, A. S., 1930: A study of the root systems of some important sod-forming grasses, *Fla. Sta. Bull.* 211.

LAITAKARI, E., 1927: Männyn juuristo, morfologinen tutkimus, *Acta Forestalia Fennica*, **33**: 1–380. Quoted by Weaver and Kramer, 1932.

LAU, E., 1906: Beiträge zur Kenntnis der Zusammensetzung der in Ackerboden befindlichen Luft, *Inaug. Diss.*, Rostock.

LEAVITT, R. G., 1904: Trichomes of the root in vascular cryptogams on angiosperms, *Proc. Boston Soc. Nat. Hist.*, **31**: 273–313. Quoted by Cormack, 1935.

LEE, H. A., 1926: The distribution of the roots of sugar cane in the soil in the Hawaiian Islands, *Plant Physiol.*, **1**: 363–378.

LIEBIG, J., 1858: Ueber einige Eigenschaften der Ackerkrume, *Ann. Chem. Pharm.*, **105**: 109–144.

LILIENFELD, M., 1905: Ueber den Chemotropismus der Wurzel, *Ber. bot. Ges.* **23**: 91–98.

LIVINGSTON, B. E., J. C. BRITTON, and F. R. REID, 1905: Studies on the properties of an unproductive soil, *U. S. Bur. Soils Bull.* **28**.

———, C. A. Jensen, J. F. Breazeale, F. R. Pember, and J. J. Skinner, 1907: Further studies on the properties of unproductive soils, *U. S. Bur. Soils Bull.* 36.

———, 1909: An advance in root physiology, *Plant World*, **12**: 127–130.

———, 1923: Some physiological aspects of soil toxicity, *J. Am. Soc. Agron.*, **15**: 313–323.

Loehwing, W F., 1931: Effect of soil aeration on growth and root development, *Proc. Iowa Acad. Sci.*, **38**: 71–72.

Loomis, W. E., and L. W. Ewan, 1936: Hydrotropic responses of roots in soil, *Bot. Gaz.*, **97**: 728–743.

Lyon, T. L., and J. A. Bizzell, 1913: Is there a mutual stimulation of plants through root influence? *J. Am. Soc. Agron.*, **5**: 38–44.

——— and J. K. Wilson, 1921: Liberation of organic matter by roots of growing plants, *N. Y. Sta.* (*Cornell*) *Mem.* 40.

Magistad, O. C., and J. F. Breazeale, 1929: Plant and soil relations at and below the wilting percentage, *Ariz. Sta. Tech. Bull.* 25.

Magowan, F. N., 1908: The toxic effect of certain common salts in the soil on plants, *Bot. Gaz.*, **45**: 45–49. Quoted by Cormack, 1935.

Marth, P. C., 1934: A study of root distribution of Stayman apple trees in Maryland, *Proc. Am. Soc. Hort. Sci.*, **32**: 334–337.

Martin, T. L., 1927: Decomposition studies of alfalfa and sweet clover roots and straw, *Soil Sci.*, **24**: 309–316.

Massey, A. B., 1925: Antagonism of the walnuts (*Juglans nigra* L. and *J. cinerea* L.) in certain plant associations, *Phytopath.*, **15**: 773–785.

Maximow, N. A., and A. S. Kružilin, 1936: Influence of irrigation on the development of the root system of spring wheat plants, *Plant Physiol.*, **11**: 457–463.

McDougall, W. B., 1921: Thick-walled root hairs of *Gleditsia* and related genera, *Am. J. Bot.*, **8**: 171–176.

McKinley, A. D., 1931: Effects of sorghum residues on crop yields, *J. Am. Soc. Agron.*, **23**: 844–850.

McQuilkin, W. E., 1935: Root development of pitch pine with some comparative observations on shortleaf pine, *J. Agr. Res.*, **51**: 983–1016.

Mer, E., 1879: Recherches expérimentales sur les conditions de développement des poils radicaux, *C. R. Acad. Sci. Paris*, **88**: 665–668.

———, 1884: Nouvelles recherches sur les conditions du développement des poils radicaux, *C. R. Acad. Sci. Paris*, **98**: 583–586.

Metzger, W. H., 1928: The effect of growing plants on solubility of soil nutrients, *Soil Sci.*, **25**: 273–281.

Mevius, W., 1926: Kalzium-Ion und Wurzelwachstum, *Jahrb. wiss. Bot.*, **66**: 183–253. Quoted by Cormack, 1935.

Millar, C. E., 1925: The feeding power of plants in different soil horizons, *J. Am. Soc. Agron.*, **17**: 150–157.

———, 1925: Availability of nutrients in subsoils, *Soil Sci.*, **19**: 275–285.

———, 1930: Root systems of young corn plants in relation to fertilizer applications, *J. Am. Soc. Agron.*, **22**: 868–874.

———, 1933: Availability to corn of nutrients in the A_2 and B horizons of Hillsdale loam, *J. Am. Soc. Agron.*, **25**: 418–426.

Miller, E. C., 1916: Comparative study of the root systems and leaf areas of corn and the sorghums, *J. Agr. Res.*, **6**: 311–333.

———, 1916: The root systems of agricultural plants, *J. Am. Soc. Agron.*, **8**: 129–154.

Miller, M. F., and F. L. Duley, 1925: The effect of varying moisture supply upon the development and composition of the maize plant at different periods of growth, *Mo. Sta. Tech. Bull.* 76.

Minina, E. G., 1927: On the acidity of root secretions, *Bull. Inst. Rech. Biol. Univ. Perm.*, **5**: 233–258; *Biol. Abst.*, **4**: 62.

Mohammad, A., and R. B. Deshpande, 1929: Studies in Indian chillies. II, The root systems, *Agr. J. India*, **24**: 251–258. Quoted by Weihing, 1935.

Molisch, H., 1885: Ueber die Ablenkung der Wurzeln von ihrer normalen Wachsthums-richtung durch Gase (Aerotropismus), *Sitzungsber. Akad. Wiss. Wien.*, **90**: 111.

———, 1887: Über Wurzelausscheidungen und deren Einwirkung auf organische Substanzen, *Sitzungsber. Akad. Wiss. Wien.*, **96**: 84–109.

Molliard, M., 1913: Sur la sécrétion par les racines de substances toxiques pour la plante, *Soc. Bot. France Bull.* **60**: 442–446.

Morrish, R. H., 1934: Crop mixture trials in Michigan, *Mich. Sta. Spec. Bull.* 256.

Mukerji, J. N., 1920: The excretion of toxins from the roots of plants, *Agr. J. India*, **15**: 502–507.

Müller-Thurgau, H., 1894: Einfluss des Stickstoffs auf die Wurzel-bildung, *Jahrb. Versuchs-Stat. Wädensweil*, **4**: 48–52.

Murphy, H. C., 1935: Effect of crown rust infection on yield and water requirement of oats, *J. Agr. Res.*, **50**: 387–412.

Nedrow, W. W., 1937: Studies on the ecology of roots, *J. Ecology*, **18**: 27–51.

Neller, J. R., 1922: The influence of growing plants upon oxidation processes in the soil, *Soil Sci.*, **13**: 139–160.

Newcombe, F. C., and A. Rhodes, 1904: Chemotropism of roots, *Bot. Gaz.*, **37**: 23–35.

Newton, J. D., 1924: Measurement of carbon dioxide evolved from the roots of various crop plants, *Sci. Agr.*, **4**: 268–274.

Nicol, H., 1935: Mixed cropping in primitive agriculture, *Empire J. Exp. Agr.*, **3**: 189–195.

Nobbe, F., 1862: Über die feinere Verästelung der Pflanzenwurzeln, *Landw. Versuchs-Stat.*, **4**: 212–224.

———, 1869: Die Pflanzencultur in Wasser, und ihre Bedeutung für die Landwirthschaft, *Landw. Versuchs-Stat.*, **11**: 106–120.

Novez, A. E., 1930: On the distribution of tabular roots in *Ceiba* (*Bombocaceae*), *Proc. Nat. Acad. Sci.*, **16**: 339–344.

Nutman, F. J., 1933, 1934: The root system of *Coffea arabica*. I, II, III, *Empire J. Exp. Agr.*, **1**: 271–284; 285–296; **2**: 293–302.

Odland, T. E., and J. B. Smith, 1933: The effect of certain crops on succeeding crops, *J. Am. Soc. Agron.*, **25**: 612–618.

Oskamp, J., 1932: The rooting habit of deciduous fruits on different soils, *Proc. Am. Soc. Hort. Sci.*, **29**: 213–219.

Parker, F. W., 1924: Carbon dioxide production of plant roots as a factor in the feeding power of plants, *Soil Sci.*, **17**: 229–249.

———, 1925: The carbon dioxide content of the soil air as a factor in the absorption of inorganic elements by plants, *Soil Sci.*, **20**: 39–45.

Parker, K. W., and A. W. Sampson, 1930: Influence of leafage removal on anatomical structure of roots of *Stipa pulchra* and *Bromus hordeaceus*, *Plant Physiol.*, **5**: 545–553.

Pavlychenko, T. K., 1937: Quantitative study of the entire root systems of weeds and crop plants under field conditions, *J. Ecology*, **18**: 62–79.

Pearsall, W. H., and E. M. Wray, 1927: The physiology and ecology of the calcifuge habit in *Eriophorum angustifolium*, *Ecology*, **25**: 1.

Peralta, F. de, and R. P. Estioko, 1923: A tentative study of the effect of root excretion of common paddy weeds upon crop production of lowland rice, *Philippine Agr.*, **11**: 205–216.

Peren, G. S., 1923: Data on the lateral spread of the roots of fruit trees, *J. Pom. and Hort. Sci.*, **3**: 96–102.

Persecke, K., 1877: Über die Formveränderungen der Wurzel in Wasser und Erde, *Diss.*, Leipzig.

Pettinger, N. A., 1933: The effect of fertilizers, crop rotation, and weather conditions on the anchorage of corn plants, *Va. Polytech. Inst. Va. Sta. Tech. Bull.* 46.

Pfeiffer, F., and E. Blanck, 1912: Die Säureausscheidung der Wurzeln und die Löslichkeit der Bodennährstoffe in kohlensäurehaltigem Wasser, *Landw. Versuchs-Stat.*, **77**: 217–268.

Pickering, S. U., 1903: The effect of grass on apple trees, *J. Roy. Agr. Soc. England*, **64**: 365–376.

——— and Duke of Bedford, 1914: Effect of one crop upon another, *J. Agr. Sci.*, **6**: 136–151.

———, 1917: The effect of one plant on another, *Ann. Bot.*, **31**: 181–187.

——— and Duke of Bedford, 1919: "Science and Fruit Growing," Chap. 28, pp. 291–312.

Polle, R., 1910: Über den Einfluss verschiedener hohen Wassergehalts, verschiedener Düngung und Festigkeit des Bodens auf die Wurzelentwicklung des Weizens und der Gerste im ersten Vegetationsstadium, *J. Landw.*, **58**: 297–345.

Popesco, S., 1926: Recherches sur la région absorbante de la racine, *Buletinul Agriculturii (Rumania)*, IV, No. 10–12. 136 pp.

Porodko, T., 1911: Über den Chemotropismus der Pflanzenwurzeln, *Jahrb. wiss. Bot.*, **49**: 307–388.

———, 1912: Vergleichende Untersuchungen über die Tropismen. II, Mitt. Thermotropismus der Pflanzenwurzeln, *Ber. bot. Ges.*, **30**: 305–313.

———, 1914: Vergleichende Untersuchungen über die Tropismen. VI, Der relative chemotrope Wirkungswert von Alkali- und Erdalkalisalzen für Keimwurzeln von *Lupinus albus*, *Ber. bot. Ges.*, **32**: 25–35.

———, 1914: Vergleichende Untersuchungen über die Tropismen. VII, Der relative chemotrope Wirkungswert von Schwermetallsalzen für Keimwurzeln von *Lupinus albus*, *Ber. bot. Ges.*, **32**: 271–275.

———, 1914: Vergleichende Untersuchungen über die Tropismen. VIII. Bedingungen der chemotropen Krummungsreaktion bei den Pflanzenwurzeln, *Ber. bot. Ges.*, **32**: 275–282.

Pratt, I. W., 1930: Preliminary study of elongation of roots of Georgia collards as affected by sodium luminal, *Plant Physiol.*, **5**: 587–597.

Prianischnikov, D., 1914: Sur la question des excrétions nuisibles des racines. *Rev. Gén. Bot.*, **25**: 563–582.

Raciborski, M. M., 1905: Oxydierende und reduzierende Eigenschaften der lebenden Zelle, *Bull. Acad. Sci. Cracovie*, **1905**: 338, 668, 693.

Rastek, J. C., 1934: Evidence that some plant roots give off organic acids. *Proc. Am. Soc. Hort. Sci.*, **32**: 620–622.

Reid, M., 1930: The influence of nutritive conditions of seeds and cuttings upon the development of roots, *Rpt. and Proc. Ninth Int. Hort. Cong.*, Group A, 165–169.

———, 1933: Effects of shade on the growth of velvet bent and metropolitan creeping bent, *Bull. U. S. Golf Assoc., Green Sec.*, **13**: 131–135.

Ridgway, C. S., 1913: The occurrence of callose in root hairs, *Plant World*, **16**: 116–122.

Rigg, G. B., 1913: The effect of some Puget Sound bog waters on the root hairs of *Tradescantia*, *Bot. Gaz.*, **55**: 314–326.

——— and E. S. Harrar, 1931: The root systems of trees growing in sphagnum, *Am. J. Bot.*, **18**: 391–398.

Roberts, E. A., 1916: The epidermal cells of roots, *Bot. Gaz.*, **62**: 488–506.

Robertson, D. W., A. Kezer, J. Sjogren, and D. Koonce, 1934: Studies on the critical period for applying irrigation water to wheat, *Col. Sta. Tech. Bull.* 11.

Rogers, W. S., and M. C. Vyvyan, 1928: The root systems of some ten-year-old apple trees on two different root stocks and their relation to tree performance, *Ann. Rpt. East Malling Res. Sta.* **1926-1927**: 31–43.

———, 1932: Root studies. II, The root development of an apple tree in a wet clay soil, *J. Pom. and Hort. Sci.*, **10**: 219–227.

———, 1933: Root studies. III, Pear, gooseberry and black-currant root systems under different soil fertility conditions with some observations on root stock and scion effect in pears, *J. Pom. and Hort. Sci.*, **11**: 1–18.

——— and M. C. Vyvyan, 1934: Root studies. V, Root stock and soil effect on apple root systems, *J. Pom. and Hort. Sci.*, **12**: 110–150.

———, 1935: Root studies. VI, Apple roots under irrigated conditions, with notes on use of a soil-moisture meter, *J. Pom. and Hort. Sci.*, **13**: 190–201.

———, 1935: Soil factors in relation to root growth, *Trans. Third Int. Cong. Soil Sci.*, **1**: 249–253.

Rotmistrov, V., 1909: "Root Systems of Cultivated Plants of One Year's Growth," 57 pp., 22 pl., Odessa.

Roxas, M. L., and M. Villano, 1930: Studies on the development of the sugar-cane plant in the Philippines. Root and shoot development of M-1900, *Philippine J. Sci.*, **43**: 367–403.

Russell, E. J., 1917: "Soil Conditions and Plant Growth," 190 pp., Longmans, Green & Company, New York.

——— and A. Appleyard, 1917: The influence of soil conditions on the decomposition of organic matter in the soil, *J. Agr. Sci.*, **8**: 385–417.

Russell, J., and A. H. E. McDonald, 1935: Wheat after wheat. No inherent harm in continuous wheat cropping, *Agr. Gaz. N. S. Wales*, **45**: 183–184.

Sachs, J., 1860: Bericht über die physiologische Thätigkeit an der Versuchs station in Tharandt. II, Wurzel-Studien, *Landw. Versuchs-Stat.*, **2**: 1–31.

Saunders, A. R., 1933: Studies in phanerogamic parasitism with particular reference to *Striga lutea* Lour., *Dept. Agr. Union South Africa Sci. Bull.* 128.

Sayre, C. B., 1934: Root development of beans, cabbage, and tomatoes as affected by fertilizer placement, *Proc. Am. Soc. Hort. Sci.*, **32**: 564–571.

Schneiderhan, F. J., 1926: Apple disease studies in northern Virginia, *Va. Polytech. Inst. Va. Sta. Bull.* **245**: 29–30.

Schreiner, O., and H. S. Reed, 1907: Some factors influencing soil fertility, *U. S. Bur. Soils Bull.* 40.

———, ———, and J. J. Skinner, 1907: Certain organic constituents of soils in relation to soil fertility, *U. S. Bur. Soils Bull.* 47.

——— and ———, 1907: The production of deleterious excretions by roots, *Bull. Torrey Bot. Club*, **34**: 279–303.

——— and ———, 1908: The toxic action of certain organic plant constituents, *Bot. Gaz.*, **45**: 73–102.

——— and ———, 1909: The role of oxidation in soil fertility, *U. S. Bur. Soils Bull.* 56.

——— and ———, 1909: Studies on the oxidizing power of roots, *Bot. Gaz.*, **47**: 355–388.

——— and E. C. Shorey, 1909: The isolation of harmful organic substances from soils, *U. S. Bur. Soils Bull.* 53.

——— and ———, 1910: Chemical nature of soil organic matter, *U. S. Bur. Soils Bull.* 74.

——— and J. J. Skinner, 1910: Some effects of a harmful organic soil constituent, *U. S. Bur. Soils Bull.* 70.

——— and M. X. Sullivan, 1911: Reduction by roots, *Bot. Gaz.*, **51**: 121–130.

——— and ———, 1911: Concurrent oxidation and reduction by roots, *Bot. Gaz.*, **51**: 273–285.

——— and J. J. Skinner, 1912: The toxic action of organic compounds as modified by fertilizer salts, *Bot. Gaz.*, **54**: 31–48.

———, 1923: Toxic organic soil constituents and the influence of oxidation, *J. Am. Soc. Agron.*, **15**: 270–277.

Schulze, B., 1906: Studien über die Bewurzelung unserer Kulturpflanzen, *Festschrift*, 50 *Jubiläum Agr. Chem. Vers. und Kontroll Stat.*, Breslau, 67–95.

Schwarz, F., 1883: Die Wurzelnhaare der Pflanzen, *Untersuch. Bot. Inst. Tübingen*, **1**: 135–188.

Sewell, M. C., 1923: Effect of *Andropogon sorghum* on succeeding crops of *Triticum sativum vulgare*, *Bot. Gaz.*, **75**: 1–26.

Shantz, H. L., 1927: Drought resistance and soil moisture, *Ecology*, **8**: 145–157.

Shepperd, J. H., 1905: Root systems of field crops, *N. Dak. Sta. Bull.* **64**: 529–535.

Shull, C. A., 1932: Toxicity of root excretions, *Plant Physiol.*, **7**: 339–341.

Siemens, H. J., 1929: The development of secondary seminal roots in corn seedlings, *Sci. Agr.*, **9**: 747–759.

Sierp, H., and A. Brewig, 1935: Quantitative Untersuchungen über die Wasserabsorptionszone der Wurzeln, *Jahrb. wiss. Bot.*, **82**: 99–122.

Singh, K., 1922: Development of the root system of wheat in different kinds of soil and with different methods of watering, *Ann. Bot.*, **36**: 353–361.

Skinner, J. J., 1913: Illustration of the effect of previous vegetation on a following crop: cabbage after sesame, *Plant World*, **16**: 342–346.

Smith, F. B., and P. E. Brown, 1931: Soil respiration, *J. Am. Soc. Agron.*, **23**: 909–916.

——— and ———, 1932: Further studies on soil respiration, *J. Am. Soc. Agron.*, **24**: 577–583.

Snow, L. M., 1904: The effects of external agents on the production of root hairs, *Bot. Gaz.*, **37**: 143–145.

———, 1905: Development of root hairs, *Bot. Gaz.*, **40**: 12–48.

Sorokin, H., and A. L. Sommer, 1929: Changes in the cells and tissues of root tips induced by the absence of calcium, *Am. J. Bot.*, **16**: 23–39.

Spalding, V. M., 1904: Biological relations of certain desert shrubs. I, The creosote bush (*Covillea tridentata*) in its relation to water supply, *Bot. Gaz.*, **38**: 122–138.

Sperry, T. M., 1935: Root systems in Illinois prairie, *Ecology*, **16**: 178–202.

Stevens, C. L., 1931: Root growth of white pine, *Yale Forestry Bull.* 32.

Stewart, G. R., and J. C. Martin, 1921: Effect of various crops upon the water extract of a typical silty clay loam soil, *J. Agr. Res.*, **20**: 663–668.

Stieler, G., 1903: Ueber das Verhalten der Wurzelhärchen gegen Lösungen, *Diss.*, Kiel (abs. *Bot. Cent.*, **96**: 541 (1904)).

Stocklasa, J., and A. Ernest, 1905: Ueber den Ursprung, die Menge, und die Bedeutung der Kohlendioxyds im Boden, *Cent. Bakt.*, **2**: 14: 723–736.

——— and ———, 1909: Beiträge zur Lösung der Frage der chemischen Natur des Wurzelsekretes, *Jahrb. wiss. Bot.*, **46**: 55–102.

———, 1934: Decomposition of raw phosphate by the root secretions of individual culture plants, *Phosphorsäure*, **4**: 129–148; *Chem. Abst.*, **28**: 4100–4101.

Stoddart, L. A., 1935: How long do roots of grasses live? *Sci.*, **81**: 544.

Taubenhaus, J. J., W. N. Ezekiel, and H. E. Rea, 1931: Strangulation of cotton roots, *Plant Physiol.*, **6**: 161–166.

Ten Eyck, A. M., 1899: A study of the root systems of wheat, oats, flax, corn, potatoes, and sugar beets and of the soil in which they grew, *N. Dak. Sta. Bull.* **36**: 333–362.

———, 1900: A study of the root systems of cultivated plants grown as farm crops, *N. Dak. Sta. Bull.* **43**: 535–550.

———, 1904: The roots of plants, *Kans. Sta. Bull.* **127**: 199–252.

Thatcher, R. W., 1923: The effect of one crop on another, *J. Am. Soc. Agron.*, **15**: 331–338.

Thomas, W., 1930: The feeding power of plants *Plant Physiol.*, **5**: 443–491.

Thornton, H. G., and H. Nicol, 1934: Further evidence upon the nitrogen uptake of grass grown with lucerne. *J. Agr. Sci.* **24**: 540–543.

Tincker M. A. H., 1935: Experiments with lilies at Wisley. *R. H. S. Lily Year Book*, 68–75.

Tisdale, W. H., and J. M. Jenkins, 1921: Straighthead of rice and its control, *U. S. Farmers' Bull.* 1212.

Traub, H. P., and R. H. Stansel, 1930: The lateral root spread of the fig tree. *Proc. Am. Soc. Hort. Sci.*, **27**: 109–113.

Treitel, O., 1924: Thermotropismus bei Wurzeln, *Bot. Archiv.*, **7**: 375–388.

Tucker, M., and C. von Seelhorst, 1898: Der Einfluss, welchen der Wassergehalt und der Reichtum des Bodens auf die Ausbildung der Wurzeln und der oberirdischen Organe der Haferpflanze ausüben, *J. Landw.*, **46**: 52–63.

Turner, T. W., 1922: Studies of the mechanism of the physiological effects of certain minerals salts in altering the ratio of top growth to root growth in seed plants, *Am. J. Bot.*, **9**: 415–446.

Turpin, H. W., 1920: The carbon dioxide of the soil air, *N. Y. Sta (Cornell) Mem.* 32.

Venkatraman, R. S., and R. V. Vittal, 1928: Coimbatre seedling canes, *Agr. J India*, **23**: 37–38. Quoted by Weihing, 1935.

Virtanen, A. I., 1933: Nitrogen uptake of plants, *Nature*, **131**: 534 (1934); also, *Ann. Acad. Sci. Fenn. Ser A.* 1933, **36**, No. 12.

von Seelhorst, C., 1902: Beobachtungen über die Zahl und den Tiefgang der Wurzeln verschiedener Pflanzen bei verschiedener Düngung des Bodens, *J. Landw.*, **50**: 91–104.

Waksman, S. A., and F. G. Tenney, 1928: Composition of natural organic materials and their decomposition in the soil. III, The influence of nature of plant upon the rapidity of its decomposition. *Soil Sci.*, **26**: 155–171.

Weaver, J. E., 1915: A study of the root systems of prairie plants of southeastern Washington. I. *Plant World*, **18**: 227–248.

———, 1919: The ecological relations of roots, *Carnegie Inst. Wash. Pub.* 286.

———, 1920: Root development in the grassland formation. *Carnegie Inst. Wash. Pub.* 292.

——— and J. W. Crist, 1922: Relation of hardpan to root penetration in the Great Plains, *Ecology*, **3**: 237–250.

———, F. C. Jean, and J. W. Crist, 1922: Development and activities of roots of crop plant, *Carnegie Inst. Wash. Pub.* 316.

———, J. Kramer, and M. Read, 1924: Development of root and shoot of winter wheat under field environment, *Ecology*, **5**: 26–50.

———, 1925: Investigations on the root habits of plants, *Am. J. Bot.*, **12**: 502–510.

———, 1926: "Root Development of Field Crops," 290 pp., McGraw-Hill Book Company, Inc., New York.

——— and W. E. Bruner, 1927: "Root Development of Vegetable Crops," 351 pp. McGraw-Hill Book Company, Inc., New York.

——— and W. J. Himmel, 1930: Relation of increased water content and decreased aeration to root development in hydrophytes, *Plant Physiol.*, **5**: 69–92.

——— and J. Kramer, 1932: Root system of *Quercus macrocarpa* in relation to the invasion of prairie, *Bot. Gaz.*, **94**: 51–85.

——— and ———, 1935: Relative efficiency of roots and tops of plants in protecting the soil from erosion, *Sci.*, **82**: 354–355.

———, V. H. Hougen, and M. D. Weldon, 1935: Relation of root distribution to organic matter in prairie soil, *Bot. Gaz.*, **96**: 389–420.

Weihing, R. M., 1935: The comparative root development of regional types of corn, *J. Am. Soc. Agron.*, **27**: 526–537.

West, E. S., 1934: The root distribution of some agricultural plants, *J. Council Sci. and Ind. Res. Australia*, **7**: 87–93.

Whitaker, E. S., 1923: Root hairs and secondary thickening in the *Compositae*, *Bot. Gaz.*, **76**: 30–60.

Whitney, M., and F. K. Cameron, 1904: Investigations in soil fertility, *U. S. Bur. Soils Bull.* 23.

Wilkins, F. S., and H. D. Hughes, 1934: Effect of Sudan grass and of soybeans on yield of corn, *J. Am. Soc. Agron.*, **26**: 901–908.

Willard, C. J., and G. M. McClure, 1932: The quantitative development of tops and roots in bluegrass with an improved method of obtaining root yields, *J. Am. Soc. Agron.*, **24**: 509–514.

Wilson, B. D., and J. K. Wilson, 1928: Relation of sorghum roots to certain biological processes, *J. Am. Soc. Agron.*, **20**: 747–754.

Wilson, K., 1936: The production of root hairs in relation to the development of the piliferous layer, *Ann. Bot.*, **50**: 121–154.

Woodroof, J. G., and N. C. Woodroof, 1934: Pecan root growth and development, *J. Agr. Res.*, **49**: 511–530.

Work, R. A., and M. R. Lewis, 1936: The relation of soil moisture to pear tree wilting in heavy clay soil, *J. Am. Soc. Agron.*, **28**: 124–134.

Wortmann, J., 1885: Über den Thermotropismus der Wurzeln, *Bot. Ztg.*, **43**: 193–200, 209–216, 225–235.

Worzella, W. W., 1932: Root development in hardy and nonhardy winter wheat varieties, *J. Am. Soc. Agron.*, **24**: 626–636.

Yeager, A. F., 1935: Root systems of certain trees and shrubs grown on prairie soils, *J. Agr. Res.*, **51**: 1085–1092.

Zavitz, C. A., 1922: Grains grown in combination for grain production, *J. Am. Soc. Agron.*, **14**: 225–228.

Ziegenspek, H., 1920: Das amyloid jugendlicher Organe. Das Amyloid in den wachsenden Wurzelhaaren und seine Beziehungen zum Zellwachstum, *Ber. bot. Ges.*, **38**: 328–333.

CHAPTER IV

THE INTAKE OF WATER BY PLANTS

I. SOURCES OF THE PLANT'S WATER SUPPLY

It is an accepted fact that plants obtain through their roots from the soil the major portion of the water that is necessary for their needs. There is, however, considerable dispute in regard to the ability of plants to absorb water through their stem and leaves from the atmosphere, so the intention here is to review the experimental evidence.

Boussingault (1878) believed that the absorption of rain and dew by leaves is of common occurrence, while Burt (1893) and Dandeno (1901) demonstrated that certain leaves can absorb water when it is placed in direct contact with them. Ganong (1894) grew plants in soil in sealed containers and after allowing them to wilt badly placed them under condition favorable for the absorption of water by their leaves. In one experiment the wilted leaves were covered with moist filter paper, in another they were sprayed with water, in a third case the wilted plants were immersed entirely or partially in water and left for a considerable period of time, while in the fourth experiment they were placed in a saturated atmosphere. In all these experiments the wilted leaves showed no signs of regaining their turgidity. This indicated that the amounts of water absorbed by the leaves were of little significance in the welfare of the plant. Spalding (1906) immersed the cut branches of various species of desert plants in water for 3 hr. and found that some of these parts absorbed no water while others absorbed amounts ranging from 2 to 6 per cent of their original weight. The stems and leaves that did absorb water, however, lost it all in a few minutes when exposed to the air. The absorption of water by the stem and leaves was shown most strikingly by those plants which showed marked mesophytic tendencies. Spalding also investigated the ability of these plants to obtain moisture from the atmosphere. A review of the data obtained shows that of the 12 species of desert perennials subjected to the experiment, all exhibited some slight capacity for the absorption of water vapor from the atmosphere, but in general the amounts absorbed were inconsiderable in comparison with that given off in corresponding periods in dry air. Spalding considered that as yet there is no evidence to indicate that the small quantities of water absorbed are utilized in the body of the plant, since the promptness with which the water thus absorbed is returned to the air suggests a process purely physical in its nature and of no physiological significance. Spalding stated that the most important conclusion to be drawn from his experiments is the patent fact that the roots of desert plants constitute their only reliable active agent in providing a normal water supply no matter how far some of these plants have departed from the habits of the present-day mesophytes. In Australia all members of the genus *Atriplex* and to a less extent species of *Kochia baosia* and *Rhagodia* absorb an appreciable amount of water from saturated atmospheres (Wood, 1925). The much greater absorption in the species *Atriplex* appears to be correlated with the high salt content of the leaves of these plants, the percentage of salt ranging from 12 to 30 per cent as compared with 2 to 7 per cent for the other leaves in question. Henslow (1908) concluded that the stems and leaves of various horticultural plants immersed in water could absorb some water under these conditions, since the loss by transpiration subsequent to immersion was lower than it was previous to the immersion. His experiments on the absorption of dew by leaves is worthy of note. He used the leaves of 20 varieties of plants which were plucked in the evening between 4 and 5 P.M. and exposed to sunlight until nightfall, after which they were exposed to a heavy dew during the night. The specimens were dried with a cloth in the morning before the sunlight had had any effect and weighed. With but one exception these leaves showed a gain in weight of from 2 to 35 per cent over that of the previous evening. The conditions under which plants grow in the field would seem to eliminate, however, the absorption of any appreciable amount of moisture from the dew deposited on the surface of their leaves. Even when the moisture supply in the soil is extremely low and the plants are wilted badly during the day, the leaves rapidly regain their turgidity during the early evening so that loss of water from guttation frequently occurs during the night. It is thus difficult to conceive how the leaves could absorb any water of consequence through their surface under such conditions. In many semiarid regions where under extreme conditions the absorption of dew might be of some little benefit, the weather conditions are such that dew rarely if ever falls.

Various experiments have been conducted by Gates (1914), Grundmann (1923), Wetzel (1924), Nakajima (1928), Darrow and Sherwood (1931), Williams (1932), Kessler (1933), Brierley (1934), and Smyth (1934) to determine the amount of water absorbed by wilted leaves when submerged in water or when it is applied to their surfaces. Some of the plants that have been shown to absorb some water under such conditions include *Vitis, Lactuca, Solanum, Beta, Lycopersicum, Phaseolus, Trifolium,* and *Rubus.* In all, the leaves of more than 100 genera have been reported as capable of absorbing water. The uptake of water by leaves is influenced by (1) the permeability of the cuticle and epidermis, (2) the hairiness of the upper surface, (3) the ease of wetting the surface, and (4) the internal deficit of water in the parenchymatous cells bordering the epidermis.

There seems to be no doubt but that leaves in general are capable under certain conditions of absorbing moisture, but the conditions for such absorption seldom occur in nature. When such absorption of water does occur, the amount is so small compared to that lost by transpiration that it is generally of no physiological significance in the water economy of the plant.

The benefit to a plant of a light shower that does not penetrate to the roots or of a humid atmosphere is to reduce the evaporating power of the air so that the loss of water from the leaves is cut down to a rate that is lower than that of absorption by the roots, and as a result of this the cells of the leaves and stem become turgid and have the appearance of having absorbed water from the air. These statements refer more particularly to plants that are rooted in the soil. Epiphytic plants such as certain ferns, orchids, and hanging mosses, however, secure all their water from the atmosphere.

II. WILTING COEFFICIENT

A plant cannot remove all the moisture from the soil in which it grows, so that when a plant permanently wilts there will always be some moisture retained by the soil, regardless of how thoroughly roots may permeate it. A plant is said to be permanently wilted when its leaves have become so flaccid as a result of a deficiency in the soil moisture supply that they will not recover their normal turgidity when placed in an approximately saturated atmosphere without the addition of water to the soil. Briggs and Shantz (1912) first used the term "wilting coefficient" to denote the moisture content of the soil (expressed as percentage of the dry weight) at the time of the permanent wilting of the plants growing therein. These authors also found that the loss of moisture from the soil through the plant does not cease at permanent wilting but goes on to some extent even after the death of the plant—the plant during the drying stage acting simply as a medium for the transference of water from the soil to the air.

A. Methods of Determining the Wilting Coefficient

1. Direct Method. The Observed Wilting Coefficient.—The wilting coefficient is determined directly by growing plants in sealed containers until they are so wilted that they will not revive when placed in a saturated atmosphere. The soil is then sampled and the percentage of moisture on a dry basis determined. The wilting coefficient obtained after this manner is termed the "observed wilting coefficient." The direct method of determining the wilting coefficient has been used, for the most part, for seedling plants in small containers but has not been used successfully for plants at a more advanced vegetative state. In the case of more mature plants it is almost impossible to obtain a sharp end point for wilting, and in large containers the soil moisture is not evenly depleted throughout the soil mass.

Koketsu (1928) noted that the content of water in the leaves at the wilting point differed widely for different plants as did also the ratio of the critical water content of the leaves to the content at full turgidity. This ratio appeared to be a specific quality of a given plant and to indicate the power of resistance of the plant to wilting. He also noted that the water content of the leaves at various stages of wilting is affected by cultural methods. Breazeale (1930) stated that there is no direct method of determining the wilting coefficient which does not involve several errors. The wilting coefficient is largely a function of the layer of soil that lies in direct contact with the absorbing zone of the roots (Work and Lewis, 1936).

No accurate method of separating this layer of soil from the roots for a determination of its moisture content has yet been devised. On that account, if plants are grown in a certain volume of soil until they permanently wilt, and all the soil is then used in the moisture determination, the results obtained will represent the average percentage of soil moisture and not the percentage in the soil that has been in contact with the roots. As the amount of soil in contact with the roots may be relatively small in comparison with all the soil in the culture, the error may be highly significant.

Moinat (1932) considered that permanent wilting is not necessarily an accurate indicator of the wilting coefficient. Thus in the seedling of the bean a relatively large proportion of the water in the plant is to be found in the lower part of the hypocotyl. During wilting the transpiring leaves draw upon this reserve water in the stem when the supply from the roots is cut off so the critical point of soil moisture may be reached previous to permanent wilting.

2. Indirect Method. *a. The Calculated Wilting Coefficient.*—The determination of the wilting point, especially with plants growing in the field, is beset with many difficulties, among which are the direct evaporation from the soil, local variation in soil texture, lack of uniformity in root development, and the lack of conditions to produce wilting. Briggs and Shantz, therefore, upon the assumption that the wilting coefficient is practically the same for a given soil for any plant under any climatic condition, formulated several indirect methods of determining the moisture content of the soil at the wilting point, based upon the relationship of the wilting point to the moisture retentiveness of the soil as measured by certain physical methods. By these methods it is not necessary to work with the plants at all to determine the amount of moisture that would remain in the soil when they wilt permanently. By determining in the laboratory certain factors on the moisture retentiveness of a sample of soil in which the plants are growing the wilting coefficient can be calculated and is then termed the "calculated wilting coefficient." Of the methods thus formulated by Briggs and Shantz, three that have been used to a considerable extent in ecological and agronomic work are here considered.

1. *Relation to the Moisture Equivalent of the Soil.*—The moisture equivalent of the soil is the percentage of water that it can retain in opposition to a centrifugal force 1,000 times that of gravity. The ratio of the moisture equivalent to the wilting coefficient was found to be 1.84 ± 0.013, or, expressed in another way,

$$\text{Wilting coefficient} = \frac{\text{moisture equivalent}}{1.84 \pm 0.013}$$

Subsequent work by Veihmeyer and Hendrickson (1927, 1928), Veihmeyer, Osekowsky and Tester (1927), Hendrickson and Veihmeyer (1929), and Wadsworth and Das (1930) show that the wilting coefficient cannot be determined for all soils by using the factor 1.84. They concluded that, although there is a remarkable constancy of the residual moisture content for a given soil when permanent wilting is attained, a common factor to evaluate the amount of water that remains in all soils at permanent wilting cannot be used. These workers found the ratio of the moisture equivalent to the wilting coefficient to be as low as 1.3 for some soils and as high as 3.8 for others. Thus they concluded that the amount of water available for plant growth cannot be obtained from the moisture equivalent alone by using any single definite factor for all soils. The factor used must be determined for each soil under consideration because plants are apparently able to reduce the soil moisture content of different soils to different degrees of dryness, relative to the moisture equivalent before they become permanently wilted. Thus for accurate work the amount of moisture readily available to plants can be obtained only when the residual moisture at permanent wilting is known.

2. *Relation to the Hygroscopic Coefficient of the Soil.*—When a dry soil is placed in a saturated atmosphere it will absorb water vapor until a condition of approximate equilibrium is attained. The moisture content of a soil under such conditions is known as the "hygroscopic coefficient" of that soil. This term should not be confused with the term hygroscopic water content of the soil, which is used in some plant-physiological tests to signify the water content of "air-dry" soil. The mean ratio of the hygroscopic coefficient to the wilting coefficient was considered by Briggs

and Shantz to be 0.68 ± 0.012, or

$$\text{Wilting coefficient} = \frac{\text{hygroscopic coefficient}}{0.68 \pm 0.012}$$

3. *Relation to the Moisture-holding Capacity of the Soil.*—The moisture-holding capacity of a soil is the percentage of water that it can retain in opposition to the force of gravity when free drainage is provided. The experiments of Briggs and Shantz indicated that

$$\text{Wilting coefficient} = \frac{\text{moisture-holding capacity} - 21}{2.9 \pm 0.06}$$

b. The Index of Transpiring Power as an Indicator of Permanent Wilting in Plants.—It has been observed that the transpiration rate falls suddenly with wilting but rises thereafter to a considerable extent subsequent to a fall once more as desiccation occurs. This initial fall in the transpiration rate represents the increase in incipient drying within the leaves. During this initial decrease in transpiration the water columns in the conducting tissues remain intact. As the drying of the leaves continues, the strain on the columns of water increases so that the breaking point is eventually reached. When the water columns break, the tension is relieved and the transpiration rate is accelerated. When the water of the broken columns has been evaporated, the rate of transpiration will again fall. Bakke (1915) considered that the stage of permanent wilting represents the most intense wilting possible without the rupture of the water columns of the plant. He considered further (1918) that the transpiring power of plants as determined by standardized hygrometric paper gives an accurate knowledge of the internal water relations of the plant, and that the permanent wilting point of the plant corresponds to that point at which the transpiration rate indicates that the water columns of the stem have been ruptured.

B. Factors Influencing the Wilting Coefficient

The amount of water that remains in the soil at the time of permanent wilting will depend upon various factors, the chief of which are (1) the type of soil, and (2) the kind of plant.

1. Type of Soil.—Sachs (1859) was the first to observe that the type of soil determined the amount of water that remained when the plants wilted permanently. He grew tobacco plants in three different soils—a vegetable mold, a loam, and a sand—and found that they wilted permanently while 12.3, 8.0, and 1.5 per cent of water, respectively, remained in the soil. The most extensive work on the relation of the water remaining in the soil at the time of permanent wilting of the plant and the type of soil has been done by Briggs and Shantz (1912). These investigators determined the wilting coefficient for 20 different types of soil ranging from a coarse sand to a clay loam. The following table has been compiled by selecting a few examples at random from their 1,300 observations:

Plant	Wilting coefficient of				
	Coarse sand	Fine sand	Sandy loam	Loam	Clay loam
Moisture equivalent	1.55	5.5	12.0	18.9	27.4
Corn, Boone County, white	1.07	3.1	6.5	9.9	15.5
Sorghum, Dwarf milo	0.94	3.6	5.9	10.0	14.1
Wheat, Kubanka	0.88	3.3	6.3	10.3	14.5
Oats, Kherson	1.07	3.5	5.9	11.1	14.8
Pea, Canada field	1.02	3.3	6.9	12.4	16.6
Tomato, Golden Queen	1.11	3.3	6.9	11.7	15.3
Rice, Japan	0.96	2.7	5.6	10.1	13.0

These data show that the finer the soil particles the greater is the amount of water remaining in the soil when the plant dies from the lack of a moisture supply.

2. Kind of Plant.—It has generally been supposed that the various groups of plants differ in their ability to reduce the moisture content of the soil. It has also been assumed that the ability of certain plants to withstand drought is due to their power to obtain water by depleting the soil moisture to a lower point than could other plants which consequently perish from lack of moisture. These suppositions seem to be based, however, on little or no experimental evidence, and it is the intention here to consider the facts that have been observed in this regard. Briggs and Shantz (1912) found in their experiments that the plants with which they worked differ only slightly in their ability to deplete the moisture content of a given soil. If 100 is taken to represent the average wilting coefficient for all the plants considered, their results show that the value for corn would be 103, legumes 101, wheat 99, oats 99, sorghum 98, millet 97, barley 97, grasses 97, and rice 94. According to the figures just quoted, in a soil whose mean wilting coefficient is 10 per cent, a corn plant would wilt with a moisture content of 10.3 per cent, while the sorghums would wilt with a moisture content of 9.8.

Results quite comparable to these were obtained by Capalungan and Murphy (1930) for a relatively large number of crop plants in Oklahoma.

The advantage to be gained by this difference in depletion of soil moisture is very slight, and Briggs and Shantz state "that if the roots were distributed to a depth of 4 ft. this would correspond to an additional amount of water for the sorghum plants in the 4 ft. of soil equivalent to a rain of $^{3}/_{10}$ in., all of which penetrated to the root zone." They considered that this additional amount of moisture would be of slight benefit to a plant and would scarcely suffice to ameliorate conditions during a drought to any greater extent than a cool, cloudy day. Alway (1913) grew Red Fife and Kubanka wheat, milo, corn, Mexican beans, and the desert legumes, *Prosopis velutina, Acacia Greggi,* and *Acacia constricta* in water-tight galvanized-iron cylinders. These cylinders were 6 ft. long and 6 in. in diameter and contained known amounts of soil and water. The plants were allowed to grow without the further addition of water until they died. The cylinders were then opened and the moisture content determined at different depths. Alway found that, in their ability to exhaust the moisture of the soil before dying, wheat, corn, milo, and Mexican beans showed little difference, but that, in their ability to continue alive after first showing injury from drought, they exhibited marked differences. The interval between wilting and death in the case of the beans amounted to only a few days, while in the case of wheat and milo it amounted frequently to several weeks. He found that if no unfavorable conditions occurred before the death of the plants, the moisture content could be reduced by any of them almost to the hygroscopic coefficient. The desert legumes mentioned above remained alive after the water content of the soil had been depleted slightly below the hygroscopic coefficient after the above-mentioned annual crops had died. Although this does not mean that these plants can utilize for growth the small amount of this water that they obtain from the soil of so low a moisture content, it does indicate that this portion of water has a very high value for the maintenance of life.

The experimental evidence at hand thus indicates that, in the cases observed, one kind of plant cannot deplete the soil moisture in a given soil appreciably more than can another, before wilting permanently. The experimental work, however, has been done for the most part with seedling plants or with plants in a comparatively early stage of development (Crump 1913) and is open to criticism in that regard. It has been found difficult to determine the wilting point in the case of fully developed plants, since as the water content in the soil becomes low the lower leaves of the plant droop, dry up, and drop off while the upper ones continue to be turgid and function normally. Powers (1922) from observations of plants growing in the field under arid conditions concluded that the wilting point varies more than has commonly been supposed when judged in the light of crop appearance, soil moisture, and the yield of dry matter. It appears to vary more for different crops on a soil that is rather heavy in texture than on a soil of narrow moisture limits.

It has been shown that the cells of the roots, particularly of crop plants, show little difference in the tractile force that they exert upon the water of the soil. There are, however, growth and morphological differences that will enable certain plants to reduce the moisture of the soil more than others before permanent wilting occurs.

Wolfe (1926) showed that the back pull of the soil for water is approximately 25 atmospheres at the hygroscopic coefficient. Thus if the plant is obtaining any moisture from the soil under this condition, its roots are exerting an osmotic pull somewhat greater than 25 atmospheres. The distribution of the root system has a marked influence upon the degree of moisture depletion of the soil. Briggs and Shantz (1912) found that rice, a plant that grows for the most part in a saturated soil, depleted the soil moisture to a lower degree than did plants that grow in the more arid regions. They attributed this to the fact that the former plants possessed abundant fine fibrous roots that permeated all parts of the soil. This is no doubt true also in the case of the sorghums as compared to corn. Miller (1916) found that the sorghum plants have almost twice as many fibrous roots as have the corn plants, so that the soil is more thoroughly penetrated by roots, and the water in the soil thus has a much shorter distance to move before it enters the roots of the sorghums than it does in the case of the corn, where the roots are more sparsely distributed. Under mild atmospheric conditions where evaporation is low, these characteristics would be of little importance, but under conditions of exceedingly high evaporation, as encountered in the field in actual agricultural practice, they become very significant. While Briggs and Shantz found that the sorghum seedlings showed but little more ability to reduce the moisture content of the soil than did corn under the mild conditions of the greenhouse, the fact that the sorghum plants in the field survive under severe climatic conditions while corn wilts and dries up gives indirect evidence that these sorghum plants under these conditions can deplete the moisture content of the soil lower than can corn before permanent wilting occurs. These plants survive, not necessarily because their individual root hairs exert a greater tractile force on the water in the soil than does corn, but simply because the water in the soil has a shorter distance to move in order to come in contact with the roots of the sorghums and thus keeps the water lost from the leaves replaced and the turgidity of the leaves prolonged for a longer period than in the case of corn.

Apparently, the roots in the upper region of the soil can deplete the soil moisture below the wilting coefficient provided that there is a sufficient water supply in the lower levels and that enough roots penetrate therein to furnish the bulk of the water necessary to replace that lost from the leaves. The roots in the upper portion of the soil thus continue to absorb a small amount of water and pass it on to the stem, whereas, if this small amount of water in the upper portion of soil were all that was available, the plant would die from lack of moisture, and the water in the soil at that level would not be depleted to so low a degree.

The following table shows the calculated wilting coefficient and percentage of moisture at different depths under corn and milo at Garden

Depth of sample, feet	Per cent of moisture	Calculated wilting coefficient
1	7.7	12.7
2	12.2	14.5
3	11.4	14.5
4	12.9	16.3
5	15.6	17.1
6	19.1	16.1

City, Kan., at the close of the growing season, which extended from June 5 to Sept. 6, 1914.

These wilting coefficients were obtained by determining the moisture equivalent of the soil samples at these depths and using 1.84 as the factor, as given by Briggs and Shantz (1912). Consequently, the figures are subject to the previously mentioned criticisms of Veihmeyer and Hendrickson (1927, 1928), and Wadsworth and Das (1930).

Magistad and Breazeale (1929), and Breazeale (1930) considered that roots remain alive for a considerable period in a soil that has been depleted in moisture to the wilting coefficient because under these conditions the roots secrete around themselves a sheath with adhesive properties. They assumed that this sheath is kept at a moisture content near the wilting coefficient by an exudation of water from the root. They considered that this water is obtained from other roots which have penetrated into moist soil or from water stored in the plant.

A small leaf surface is no doubt an important factor in delaying the wilting of plants when the moisture supply of the soil is low. In general, the smaller the leaf surface the smaller the amount of water lost in a given time so that even the low rate of absorption that prevails when the lower limits of the water supply in the soil have been reached would suffice to replace the water lost by evaporation from a small leaf surface. Thus the soil moisture could be reduced to a lower level than would be possible if the amount of water lost was large owing to a large leaf surface. The case of the desert legumes mentioned by Alway (1913) illustrates this point. As the water supply in the soil became low their leaf surface was greatly decreased by the shedding of the lower leaves, while the upper ones continued to live until the water content had been reduced even below the hygroscopic coefficient.

C. Climatic Conditions and Wilting

The extensive observations of Briggs and Shantz (1912) on the wilting of plants were made for the most part upon plants growing in the greenhouse where the temperature was about 70°F. and where the rate of loss of water from the plants was relatively low. These investigators, however, considered that their observations warranted the conclusion that the wilting point for a given soil would be a constant regardless of the nature of the climatic conditions prevailing at the time of wilting.

Subsequent to the work of Briggs and Shantz, numerous investigators have studied the relation of the wilting point of plants to the climatic conditions prevailing at the time of wilting and the relationship of the observed wilting point to the calculated one obtained from the formulas mentioned above. Brown (1912) placed potted plants of *Martynia* (Devil's Claw) in positions having varying degrees of evaporation and observed the moisture left in the soil when these plants wilted from lack of water. The following table gives an example of his results:

Position	Evaporation from atmometer. Rate per hour previous to wilting, cubic centimeters	Soil moisture at wilting, per cent
Open	7.2	15.4
Lath shelter	4.1	11.3
Room	2.8	6.8
Moist chamber	0.9	5.2

He showed thus that the rate of evaporation previous to the time of wilting determined the percentage of moisture that the plant cannot extract from the soil. Brown transferred plants that had wilted under a high rate of evaporation to a position where a lower rate of evaporation occurred. After this transfer the plants revived quickly without the addition of water to the soil and removed therefrom a greater quantity of water. The case of three plants that were placed in the open, where evaporation was relatively high, and allowed to wilt badly, after which they were transferred to a moist chamber, where the evaporation was very low, will serve as an illustration of the facts observed (see page 199).

Caldwell (1913) repeated some of the work of Brown and reached the following conclusions concerning the values of the observed and calculated wilting points: When wilting of plants occurs in a partial shelter or in the open with relatively high evaporating rates, the ratio of the

	Hourly rate of evaporation from atmometer, cubic centimeters	Moisture remaining in the soil at the time of wilting, per cent
Plant 1:		
A[1]	4.2	11.3
B[2]	0.3	5.9
Plant 2:		
A	7.5	17.5
B	1.6	6.6
Plant 3:		
A	3.2	10.5
B	0.9	5.2

[1] *A* represents the open.
[2] *B* represents the moist chamber.

observed to the calculated wilting point is always greater than unity. When wilting occurs in a nearly saturated atmosphere, the ratio of the soil-moisture content at permanent wilting as found by actual observations to that obtained by calculations from physical constants is practically unity. Caldwell makes the significant statement that "for a series of plants grown in any one soil and wilted under a number of aerial conditions, as many different soil-moisture contents are obtained as there are sets of conditions." He observed that the moisture remaining in the soil at the time of wilting at the maximum evaporation rates at Tucson, Ariz., varied from 30 to 40 per cent in excess of that taking place when plants were wilted in damp chambers. Shive and Livingston (1914) grew potted plants of corn, common beans, and Spanish pepper under different evaporating conditions and determined the ratio of the actual water content at the time of wilting to the calculated moisture content. The following table shows the result they obtained with corn 4 weeks of age growing in a soil with a calculated wilting coefficient of 3.80:

Position	Evaporation for 2-hr. period previous to the wilting, cubic centimeters	Actual moisture content of soil at wilting, per cent	Ratio of actual to calculated moisture at wilting
Glass box	0.3	4.30	1.13
Lath shelter	2.4	5.78	1.52
Cheesecloth shelter	3.0	6.04	1.58
Open	3.7	6.16	1.68

The great bulk of the experimental evidence thus indicates that the amount of water remaining in a given soil when a plant dies from lack of moisture will depend upon the atmospheric conditions prevailing just previous to the time of wilting. The wilting of a plant is due to the excess of the water loss from the stem and leaves over the amount absorbed by the roots. The rate of water loss from a plant is influenced primarily by the evaporating power of the air. If it is high, the rate of water loss from the plant is relatively high; and, if it is low, the rate of water loss from the plant is relatively low. At the time that the water available to the plant has been depleted nearly to the critical stage, the rate of absorption by the roots becomes low. If the rate of water loss from the plant under such conditions is relatively high, the plant will soon die, even though the roots are still absorbing considerable moisture from the soil. If the rate of water loss, however, is slow, owing to mild atmospheric conditions, the roots can continue for a considerable length of time to absorb sufficient water from the soil to supply the loss from the leaves and as a consequence deplete the moisture in the soil to a point lower than they could if the evaporating rate were high.

D. Water-supplying Power of the Soil at the Time of Wilting

The rate at which water is delivered to the roots of plants by the soil has been termed the "water-supplying power" of the soil. The intention here is to consider the causes for the failure of a plant to obtain its needed water supply at the time it begins to wilt. The experimental work in this regard has been conducted along two main lines—(1) a study of the back pull of the water of the soil at the time of wilting, and (2) a study of the speed at which water moves from the soil to the roots at the lower limits of the water supply of the plant.

1. The Back Pull of the Soil for Water.—Shull (1916) investigated the back pull of the water of the soil at various percentages of moisture content by means of the amount of moisture that air-dry seeds can extract from it. Seeds will imbibe water under a high pressure or against an enormous back pull. Shull placed air-dry seeds of the cocklebur (*Xanthium pennsylvanicum*) in salt solutions of various concentrations and determined the percentage of moisture that they could imbibe under these different conditions. The first table on page 201 shows the intake of water by these seeds in the different concentrations after 48 hr.

Since the percentage of water absorbed by these seeds is an indicator of the force or back pull exerted by the external medium to its entrance, they were used by Shull after the following manner to measure the back pull exerted by the water of the soil at different percentages of water content. Sixty grams of soil of the desired moisture content were placed with seeds in 200-cc. bottles, sealed, and kept agitated for 15 days. At the end of that period the bottles were opened and the moisture contents of the seeds and of the soil were determined. The moisture content of the seeds and soil at the end of the experiment shows the equilibrium

Solution	Osmotic pressure or back pull of the water of the solution, atmospheres	Intake of water by the seeds, per cent
H_2O	0.0	51.6
0.1 *M* NaCl	3.8	46.3
0.2 *M* NaCl	7.6	45.5
0.3 *M* NaCl	11.4	42.0
0.4 *M* NaCl	15.2	40.3
0.5 *M* NaCl	19.0	39.0
0.6 *M* NaCl	22.8	35.2
0.7 *M* NaCl	26.6	32.8
0.8 *M* NaCl	30.4	31.1
1.0 *M* NaCl	38.0	26.7
2.0 *M* NaCl	72.0	18.5
4.0 *M* NaCl	130.0	11.8
Saturated NaCl	375.0	6.4
Saturated LiCl	965.0	− 0.3

relation of that soil-moisture content. The following abbreviated table shows the relation of the soil moisture in the subsoil of Oswego silt loam to the water intake of *Xanthium* seeds. This soil had a moisture equivalent of 35.2 and a calculated wilting coefficient of 19.1.

Moisture in soil, per cent air dry	Intake of moisture by seeds, per cent	Back pull of water of the soil, atmospheres
5.83	0.0 sat. LiCl	965
9.36	6.47 sat. NaCl	375
11.79	11.94 4 *M* NaCl	130
13.16	21.36 2 *M* NaCl	72
14.88	28.61 1 *M* NaCl	38
17.10	37.70 0.5 *M* NaCl	19
17.93	43.20 0.3 *M* NaCl	11.4
18.07	45.15 0.2 *M* NaCl	7.6
18.87	47.26 0.1 *M* NaCl	3.8
19.34	49.31	
20.04	50.50 sat.	0.0

An examination of this table shows that in the case of the subsoils of the Oswego silt loam the back pull of the water in the soil when it is air dry reaches the enormous force of 965 atmospheres. As the moisture content of the soil increases, this back pull becomes less and less, so that by the time the former has reached about 19 per cent, the wilting coefficient of that soil, the back pull of the soil water is somewhat less than that of a 0.1*M* sodium chloride solution, which is equivalent to between 3 and 4 atmospheres. Numerous types of soils were used ranging from sands to clay loams, and it was found that the average intake of water by the seeds from the seven different types of soil used was approximately 49 per cent at the percentage of soil moisture that represents their calculated wilting coefficient. Shull concluded that these results probably mean that the wilting coefficient represents a fairly definite water-holding power for the soil particle regardless of its size. Thus the force with which the soil particles of any type of soil withhold water from the root hairs of plants at the wilting coefficient is no greater than 3 to 4 atmospheres. Let us now consider how the soil forces at the wilting coefficient compare with the osmotic pull of the root hair. In Chap. I it was stated that their osmotic pressure varied with the conditions under which they were grown and that the work of Roberts (1916) showed that the root hairs adjust their osmotic pressure as the medium changes, keeping it approximately 4 atmospheres above that of the surrounding medium.

Stocker (1930) noted in this regard that the absorbing power of the cells of the roots of plants growing on the steppes of Hungary was generally only 2 to 7 atmospheres higher than the back pull of the soil for water. The root hairs of these plants die when the pull of the soil approaches 30 atmospheres. It thus seems reasonable to assume that as the soil dries out even below the wilting coefficient the root hairs maintain an osmotic pressure a few atmospheres in excess of the soil forces until these forces become relatively very high. Shull concluded that the wilting of the plant at the wilting coefficient of the soil cannot be due to the lack of moisture in the soil nor the lack of a gradient of forces tending to move the water toward the plant, but the wilting of the plants at this critical soil-moisture content must be due to the increasing slowness of the water movement from soil particle to soil particle and from these to the root hairs, the rate of movement falling below that necessary to replace that lost from the aerial parts even under conditions of low water loss from the plant.

2. The Rate of Water Movement in the Soil. *a. Methods of Measuring.*—Methods for measuring the water-supplying power of the soil have been studied and devised by Livingston and Hawkins (1915), Pulling and Livingston (1915), Livingston and Koketsu (1920), Mason (1922), Livingston and Ohga (1926), Hardy (1923), Baldwin (1928), and Wilson (1927). The method devised by Livingston and Koketsu (1920) has proved very satisfactory and gives promise of furnishing a means of predicting the need and time of application of water to plants. This method consists of pushing into the soil porcelain cones that have a known area of unglazed surface. The cones are weighed before insertion into the soil and after remaining there for a given period are brushed free from soil particles and weighed. In this manner the amount of water that they absorb is determined. The amount of water absorbed by these cones is taken as an index of the water-supplying power of a given soil under a given condition. Two points must be considered in the use of an absorbing instrument to measure the water-supplying power of the soil: (1) whether the absorbing surface is in such capillary contact with the soil as to allow the water to move from the soil to the instrument as rapidly as it arrives at the soil surface adjacent to it and (2) whether the absorbing surface itself is able to absorb water as rapidly as it is delivered to it by the soil. Numerous investigations by the authors mentioned on page 202 indicate that the soil points of Livingston and Koketsu (1920) are reliable in these respects.

Livingston and Koketsu (1920) studied *the critical water-supplying power of the soil* or the rate at which the water moves toward the plant when the moisture content has been depleted to or near the wilting coefficient. Their method of procedure was somewhat as follows: Plants of *Coleus blumei* and *Triticum sativum* were grown in 4-in. flowerpots and allowed to grow until they had reduced the soil moisture to or near the wilting point. At this point the dry unglazed porcelain cones were pushed into the soil and allowed to remain approximately 2 hr. These were then removed and their intake of water determined. The amount of water received by these cones, the area of which approximated 10 sq. cm., in a given time was taken as an index to the water-supplying power of the soil at that moisture content. The investigators worked with 12 types and mixtures of soils and found that the water-supplying power of the soil at the wilting coefficient does not vary with the physical make-up of the soil and is thus the same for sand, loam, humus, or any mixture of these.

b. Factors Influencing the Rate.—The two main environmental conditions that influence the relation of plants to water are (1) the capacity of the soil to deliver water to the plant roots, and (2) the capacity of the aerial surroundings to cause water loss by transpiration. The water-supplying power of the soil is dependent upon the temperature, upon the gradient of the moisture content, and upon the level of this gradient. The loss of water in transpiration is influenced not only by the climatic conditions but by the nature of the plant under consideration.

Kramer (1934) by means of soil points showed that an increase in the soil temperature increased the water-supplying power of the soil as shown in the table at the top of page 204.

Magistad and Breazeale (1929) mentioned that the rate of water movement in the soil is dependent not only upon the gradient of the moisture content but also upon the

Series A		Series B		Series C	
Temperature, Centigrade	Water absorbed, mg.	Temperature, Centigrade	Water absorbed, mg.	Temperature, Centigrade	Water absorbed, mg.
0.0	5.8	0.0	57.2	0.0	27.9
8.5	8.5	8.2	96.6	9.5	38.3
23.5	12.2	24.0	132.2	23.8	49.0
35.0	15.1	34.8	171.8	35.0	59.8
				43.0	66.5

amount of moisture present. Thus the rate of water movement in a soil from a region of 4 per cent moisture to one of 3 per cent is much lower than that from a water content of 20 per cent to one of 19 per cent although the gradient is the same in either case. It was first noted by Livingston and Koketsu (1920) that the average water-supplying power of the soils with which they worked was, at the wilting coefficient, 85 mg. during each 2-hr. period for the type of soil points used. They found that the movement of water when the soil was drain-saturated was 6,000 times more rapid than when the moisture content was reduced to the wilting coefficient.

The term "soil-moisture index" was used by Wilson and Livingston (1932) to express the water-supplying power of the soil under different plants. This term indicates the number of milligrams of water absorbed per square centimeter of soil point during 1 hr. An index value of 100 indicates that the soil can supply water to the absorbing surface at the rate of 8 mg. per square centimeter for the first hour. This value was taken as a basic point because, in a majority of the cases studied, drought injury became apparent at or near the time that the water-supplying power of the soil was reduced to this rate.

When death from lack of water was first noted, the soil-moisture indexes at a depth of 6 cm. varied considerably for different plants as shown below:

Plant	Soil-moisture index	Plant	Soil-moisture index
Tall oat grass	52	Perennial rye grass	38
Velvet bent grass	44	Timothy and red top	37
Meadow fescue	43	Bermuda grass	34
Brome grass	42	Alsike clover	30
Kentucky bluegrass	41	Red fescue	29

The influence of the plant on the water-supplying power of the soil has been studied also by Welton and Wilson (1931). They determined the water-supplying power of the soil under Kentucky bluegrass, narrow-leaved fescue, and Washington bent grasses at different moisture contents of the soil. In a majority of the 40 observations the water-supplying power of the soil under the narrow-leaved fescue was 60 per cent greater than that under Kentucky bluegrass. These results suggest that the former plant places a smaller demand on the available soil moisture and that this characteristic is probably a significant factor in enabling this species to survive drought periods better than bluegrass. Nutman (1933) found that the maximum rate of the entry of water into the roots of *Coffea arabica* in East Africa was 25 cc. per square meter per hour.

If the absorbing surfaces of the root are fixed in position and extent, the maximum rate of absorption is limited to any instant by the rate at which the soil system can supply water to these surfaces (Wilson and Livingston, 1932). Conditions in the plant, however, will control the rate of absorption of water as long as it is slower than the highest possible rate of the external supply. When an absorbing surface of a root initially comes in contact with a new system of soil films by the penetration of roots into untouched soil, absorption may be relatively rapid and the water-supplying power of the soil system may be greater for a time than the absorbing power of the root surface. This, however, does not long continue. Under conditions of a diminishing soil water supply the roots must advance into unpenetrated moist soil at an adequate rate if the aerial parts are to be supplied with sufficient moisture for their needs. The process of root extension thus becomes of great importance to plants in times of drought.

Veihmeyer (1927) and Veihmeyer and Hendrickson (1927, 1934) noticed that under comparable atmospheric conditions the rate of extraction of moisture by prune and peach trees was the same regardless of the moisture content of the soil above the wilting point. Apparently the roots of these trees were able to obtain water as readily when the soil-moisture content had been reduced almost to the wilting coefficient as when the soil was filled with water to its maximum field capacity. Lewis, Work, and Aldrich (1935), however, in Oregon found that the growth of the fruit of pear trees growing on heavy soil was reduced whenever the soil moisture in the major portion of the root zone was reduced below 70 per cent of the available capacity.

The work of the various investigators shows that the rate of movement of water in the soil from particle to particle toward the roots is one of the most important factors that influence the intake of water by the plant, especially when the water supply becomes low. This substantiates the statement of Livingston and Hawkins (1915) that, whatever may be the nature of the forces producing the migration of water from soil to plant, it seems that osmotic pressure does not play a prime and direct role in this movement, but the power of the soil to deliver water to root surfaces is the principal condition that determines the moisture supply to plants in nonsaline and nontoxic soils.

III. ENTRANCE OF WATER INTO THE ROOT

A. The Structure of the Root

In order to discuss the forces that are considered to act in the intake of water by the root, its structure must be clearly understood. From Fig. 9*A* it is seen that water must transverse three definite tissues in moving through the root from the soil to the stele or fibrovascular bundle and the cells enclosing it. The three tissues are the epidermis, the cortex, and the endodermis. The general nature of the epidermal cells of the root has been discussed in considerable detail in Chap. III and need not be considered here. The cortical cells are of the typical thin-walled vacuolated parenchyma type that is very common in plant tissue. The endodermis is the most outstandingly differentiated of the three tissues. It is a cylindrical layer of cells one layer thick which surrounds the fibrovascular bundle and tissues adjoining it (Eames and MacDaniels, 1925). The cells of this layer are living and are closely packed together and devoid of any intercellular spaces. The differentiations peculiar to the endodermis lie principally in the structure of the cell walls that compose it. In the young root these cells are thin walled, but the end and radial walls are thickened by a band or strip that is known as the "Casparian strip" (Fig. 9*B* and *C*). The development of the endodermis has recently been studied by Priestley, Rhodes, and North (1922, 1926)

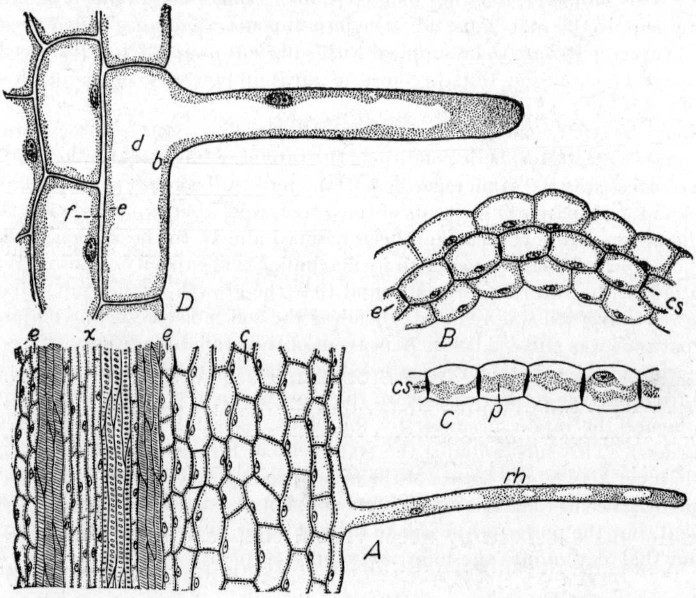

FIG. 9.—Drawings showing the cells of the root that are concerned in the absorption of water and its translocation to the central cylinder. *A*, longitudinal section of the absorbing portion of a root of the lupine; *rh*, root hair; *c*, cortex, *e*, endodermis; *x*, central cylinder. *B*, enlarged portion of the endodermis and adjoining cells; *e*, endodermis; *cs*, Casparian strip. *C*, the endodermis showing the Casparian strip and the attachment of the protoplasm to it at plasmolysis. *D*, diagram illustrating the manner in which water may be absorbed and translocated across the root. Description in the text.

especially in regard to the nature of the substances that compose the cell wall and the Casparian strip. They observed that the first noticeable differentiation of the endodermal cells from the apical meristem is the appearance of the Casparian strip, which is laid down even before cellulose can be detected in the walls of these cells. The wall of the cells of the endodermis with the exception of the Casparian strip develops into a thin cellulose wall that is soluble in sulphuric acid. The Casparian strip, however, does not so dissolve and remains intact when the other portions of the cell wall go into solution. It is more resistant to sulphuric acid and less resistant to oxidizing agents than the cellulose wall. The various reactions suggest the presence in the Casparian strip of basal substances that at a very early time in the development have been impregnated with complex substances containing derivatives of certain compounds that are apparently allied with suberogenic and cutinogenic acids and with substances giving reactions to lignins. The basal material of the Casparian strip is not definitely known. Priestley and Rhodes (1926) stated that upon treatment of this material with concentrated nitric acid it yielded a brown body, insoluble in water and dilute acids, and an ether-soluble body of a waxy appearance. Neither of these bodies, however, could be identified. The cell wall of the endodermal cells at a later stage of the development of the root becomes greatly thickened, lignified, and suberized, but from the standpoint of water absorption that later differentiation need not be considered.

The endodermal cylinder of the young root has thus in the Casparian strip an anatomical structure that renders difficult the radial passage of water through it by way of the radial walls. The passage of water through the endodermal cells of the young root must then take place through the protoplasm. The protoplasm of the endodermal cell is apparently closely attached to or interwoven with the Casparian strip, since it has been shown by De Rufz de Lavison that, when the cells are vigorously plasmolyzed, the protoplasm does not withdraw from the Casparian strip but remains suspended across the cell from strip to strip (Fig. 9*C*) after plasmolysis around the remainder of the cell wall has been complete (Bryant, 1934). The roots of certain plants have some cells in the endodermis that remain unsuberized and are termed "passage cells." It was observed by Scott (1928) that lead salts and dyes which cannot penetrate the suberized walls of the endodermis pass through these passage cells.

Priestley and Tupper-Carey (1922) obtained evidence to show that very little water can enter the root through the root apex. They stated that the meristematic tissue or cap closing the endodermal cylinder in the young root will prevent the passage of water under a pressure of 2 atmospheres. This tissue is also relatively impermeable to salts in solution. The stem apex, on the other hand, is very permeable to water and solutes. Its cell walls are composed of cellulose, while the cells that make up the apex of the root have walls of an unknown substance that apparently may be converted into cellulose by boiling with potassium hydroxide.

B. Modes of Water Entrance into the Roots

Two main forces are apparently concerned in the entrance of water into the root from the soil and its movement across it to the fibrovascular bundle. The forces are imbibitional forces and osmotic forces.

1. Imbibition.—The role of imbibition in the processes of plants has been shown by Shull (1924) and others to be of much import. The attraction or affinity of dry cell walls and protoplasm for water amounts in many cases to hundreds of and even 1,000 atmospheres (Shull, 1914 to 1916), and it has also been shown that very strong imbibitional forces may be developed within the plant even when the cell walls and protoplasm contain considerable water. If moisture is withdrawn from any region of an imbibing system, the surface tension of the watery films in that region is increased, and a corresponding force or pull is exerted upon the surrounding medium for the movement of water into the region of loss to restore the equilibrium of the surface tension of these films. This deficiency of water in any region of an imbibing system has been termed by Livingston and Brown (1912) "incipient drying" and by Renner (1912) "the saturation deficit."

All the external contacts of the plant that are concerned in the intake of water and also in water loss are made through the medium of imbibing surfaces. The cell wall of an epidermal cell of the young root is colloidal in nature and has a strong affinity for the water of the soil particles with which it is in close contact. The protoplasm that borders upon the cell wall likewise has a strong attraction for water. Let us now consider an epidermal cell of a young root in which the imbibitional and osmotic forces are in equilibrium with the soil and the surrounding cells. Let us now assume that water is withdrawn from the exterior surface of the cell wall (Fig. 9*Df*). A saturation deficit is set up at *f* and a pull of equal magnitude is exerted upon the water farther back in the cell wall. When the imbibing forces of the cell wall are thus set in action and cause a movement of water toward it, the supply of water must be secured from the protoplasm *e* with which it is in close contact. Holle (1915) found that during wilting the cell wall keeps in close contact with the protoplasm, thus permitting the continual withdrawal of water from it. The withdrawal of water from the protoplasm produces in it a saturation deficit that is satisfied by imbibing free water from the cell sap *d*. This action increases the absorbing power of the cell sap and causes the entrance of imbibed water from the protoplasm at *b*, which draws finally upon the exterior cell wall whose saturation deficit must be supplied by the water of the soil. The migration of water across the root to the endodermis could be explained, in part at least, under certain conditions after the same manner as has been described for the movement of water across the epidermal cell. Kunkel (1912) considered that imbibitional factors were the main ones in the intake of water by the root. It seems very evident, however, that imbibitional factors do not act independently but in connection with osmotic factors, since the imbibitional disturbances create osmotic disturbances and both together are equally important in the process of water intake and translocation across the root. Some water apparently may enter the cell wall of the epidermal cell and travel across the root through the cell walls of adjacent cells as far as the Casparian strip of the endodermis without the aid of the protoplasm or the cell sap. In such a case, imbibitional forces alone would be concerned. The intake of water by the young cells of the root where no vacuole is yet developed is apparently a purely imbibitional process. The amount of water, however, that may enter the plant under the direct influence of imbibitional forces alone is very small in amount compared to the intake through the vacuolated cells.

When transpiration is proceeding even at a low rate during the daylight hours the leaves tend to dry out on account of water being abstracted from them by evaporation, and a saturation deficit is produced. This results in the water of the vessels passing into a state of tension that is transmitted equally in all directions, so that under these conditions water is not being forced up through the fibrovascular bundles by the roots but is apparently being pulled up the stem under great stress by the forces operative in the leaves. Under these conditions when a stem is cut off water does not exude from the stump, and if the cut surface of the stump is immersed in water, water will be sucked back into it, showing that there is a negative pressure in the conducting vessels as far as the roots are concerned. Dixon (1910) and Renner (1912) considered that the water-absorbing power of the roots is directly related to the saturation deficit in the leaves and that the tensile stress developed in the cells of the leaves and conducting vessels on this account is transmitted to the root and has the ultimate effect of drying out the root surface. According to this theory of water intake in the absence of root pressure, the passage of water from the soil into the root is purely passive, and the root acts merely as a filter in the process. Of what importance the osmotic forces of the root are under the conditions of passive absorption is not known. Pulling and Livingston (1915) considered that the osmotic forces may influence the process of absorption in an indirect manner by maintaining the form of the roots and their contact with the water films of the surrounding soil. They also stated that it has been shown that roots do not take up water at rates proportional to the osmotic pressures of the cells, and that absorption may go on equally well with low or high turgidity of roots as long as the protoplasts of the cells involved are not deformed by plasmolysis.

2. Osmotic Forces.—The osmotic value of the cell sap of the epidermal cells of the roots is always higher than that of the soil solution, and apparently, as mentioned in Chap. I, this superiority is maintained as the concentration of the external medium is increased. The effective force, however, with which water will be drawn into the cell by osmotic action will be the difference between the osmotic value of the cell sap and the osmotic value of the external soil solution less the turgor pressure of the cell wall (Thoday, 1918). This absorbing power of the epidermal cell is equivalent to 1 atmosphere or more in the cases that have been observed. The conditions are fulfilled for the migration of water into the root, provided that there is an increasing gradient of absorbing power from cell to cell across the cortex to the fibrovascular bundle. The absorbing power of the cells of the root has been investigated by Ursprung and his workers, and they have determined that there is a gradient of absorbing power proceeding from the epidermal cell across the cortex to the endodermis and then a marked decrease in the endodermis and wood parenchyma. The following table is taken from Molz (1926):

ABSORBING POWER IN THE ABSORBING ZONE OF THE ROOT OF *Vicia faba*

Position of cells	Absorbing power, atmospheres	Position of cells	Absorbing power, atmospheres
Epidermis	0.7	Fifth row	2.8
Cortex cells		Sixth row	3.0
First row	1.4	Endodermis	1.7
Third row	1.5	Wood parenchyma	0.9
Fourth row	2.1		

In the intake of water by the root and in its movement across the same, osmotic forces always act in connection with imbibitional forces. There is in this connection as well as in the general movement of water through plant cells a tendency for these two forces to proceed toward equilibrium (Shull, 1924). Wherever imbibitional forces exceed osmotic forces, water will move into the imbibing matter, or if the osmotic forces are the larger, water will move from the imbibed material into the cell vacuole.

Under conditions of low transpiration or in the spring before the leaves are unfolded, water is forced into the conducting vessels of the root and up through the stem under pressures varying from a fraction of an atmosphere to several atmospheres. This pressure is evidenced by the bleeding of cut vines and branches of certain species of plants and is apparently connected with the exudation of water from the leaves of plants, which occurs under certain conditions. This pressure, which is set up in the fibrovascular bundles of the stem and root owing to the water that is being forced in them, is known as "root pressure." The forces that are involved in the transference of water across the endodermis into the fibrovascular bundle are not known. The effects of the deprivation of oxygen and of changes in temperature suggest that root pressure cannot be explained by means of osmotic and imbibitional forces alone and indicate that the vital relations of the cell involving energy transference are concerned (Blackman, 1921). The nature of the Casparian strip, which prevents the transference of water across the radial walls of the endodermal cells, would indicate that the water transference across the endodermis is under control of the protoplasm of these cells (Priestley, 1920). The manner in which the protoplasm brings about this migration of water into the bundles has not been determined experimentally, and any suggestions that have been made are purely theoretical. Most of the theories proposed assume that the processes involved are of a secretory or excretory nature. Since little or nothing is known, however, concerning the nature of the factors involved in the processes of secretion or excretion in either plants or animals, to state that water transference into the bundle from the endodermis is a process of secretion does not illuminate the matter at all. We must, therefore, in the words of Blackman (1921), have a much greater knowledge of cell dynamics than we possess at present before we can explain the exudation of water and root pressure with any degree of satisfaction.

Under conditions of little or no transpiration and an abundant water supply, the intake of water and its movement across the root can be explained by purely osmotic and imbibitional forces in the root, provided that a saturation deficit is created in the endodermis. At present, however, there is no satisfactory way of explaining how this saturation deficit is brought about by the endodermal cells.

Kramer (1932, 1933) noted that positive pressures were manifested at the cut surfaces of stems only when the roots were alive, and that the immersion of the living root systems into sucrose solutions stopped the exudation of water from cut stems. He further observed that transpiring woody and herbaceous plants would remain alive and unwilted for several days after their roots had been killed. Under these conditions, however, considerable quantities of water were absorbed from the soil through the dead root systems. When suction was applied to cut stems, much more water was taken up after the roots were killed than before. He believed that absorption by osmosis is too slow to replace the water lost by transpiration and that under conditions of high evaporation the role of living cells in absorption is apparently a passive one. They are important in absorbing surfaces in preventing the entrance of air into the vessels. It was thought by Reed and Bartholomew (1927) that the pectic materials of the cell walls of immature fruits absorb water by imbibition since the majority of such cells are some distance from the ultimate division of the bundle. Henderson (1934), however, found a correlation between the volume of water absorbed and the rate of respiration in the roots. This indicates that these organs expend energy in the process of absorption.

It is evident that the intake of water by the root and its movement across it is a very complex process and that numerous factors and forces are involved. Imbibition, osmotic forces, and passive absorption are evidently concerned in the process and under certain conditions act separately and under others jointly. That these forces can supply the plant with sufficient water to replace that lost during vigorous transpiration seems doubtful, and, in the words of Overton (1921), little is actually known about the forces that are operative in the passage of water from the soil through the root into the vessels in sufficient quantity to supply the transpiration needs.

Bibliography

Alway, F. J., 1913: Studies of the relation of the non-available water of the soil to the hygroscopic coefficient, *Nebr. Sta. Res. Bull.* **3**: 5–122.

Bakke, A. L., 1915: The index of foliar transpiring power as an indicator of permanent wilting in plants, *Bot. Gaz.*, **60**: 314–319.

———, 1918: Determination of wilting, *Bot. Gaz.*, **66**: 81–116.

Baldwin, H. I., 1928: A comparison of the available moisture in sod and open soil by the soil point method, *Bull. Torrey Bot. Club*, **55**: 251–255.

Blackman, V. H., 1921: Osmotic pressure, root pressure, and exudation, *New Phytol.*, **20**: 106–115.

Boussingault, J., 1878: Étude sur les fonctions physiques des feuilles: transpiration, absorption de la vapeur aqueuse, de l'eau des matières salines, *Ann. Chim. et Phys.*, Ser. 5, **13**: 289–394. Quoted by Williams, 1932.

Bowen, E. J., 1931: Water conduction in *Polytrichum commune*, *Ann. Bot.*, **45**: 175–200.

Breazeale, J. F., 1930: Maintenance of moisture equilibrium and nutrition of plants at and below the wilting percentage, *Ariz. Sta. Tech. Bull.* 29.

Brierley, W. G., 1934: Absorption of water by the foliage of some common fruit species, *Proc. Am. Soc. Hort. Sci.*, **32**: 277–283.

Briggs, L. J., and H. L. Shantz, 1911: Application of wilting-coefficient determinations in agronomic investigations, *J. Am. Soc. Agron.*, **3**: 250–260.

——— and ———, 1912: The wilting coefficient for different plants and its indirect determination, *U. S. Bur. Plant Ind. Bull.* 230.

——— and ———, 1912: The wilting coefficient and its indirect determination, *Bot. Gaz.*, **53**: 20–37.

——— and ———, 1912: The relative wilting coefficient for different plants, *Bot. Gaz.*, **53**: 229–235.

——— and ———, 1913: Die relativen Welkungskoeffizienten verschiedener Pflanzen, *Flora*, **105**: 224–240.

Brown, W. H., 1912: The relation of evaporation to the water content of the soil at the time of wilting, *Plant World*, **15**: 121–134.

Bryant, A. E., 1934: A demonstration of the connection of the protoplasts of the endodermal cells with the Casparian strips in the roots of barley, *New Phytol.* **33**: 321.

Burt, E. A., 1893: Do the leaves of our ordinary land plants absorb water? *Sci.*, **22**: 51–52.

Caldwell, J. S., 1913: The relation of environmental conditions to the phenomenon of permanent wilting in plants, *Physiol. Res.*, **1**: 1–56.

Capalungan, A. V., and H. F. Murphy, 1930: Wilting coefficient studies, *J. Am. Soc. Agron.*, **22**: 842–847.

Crump, W. B., 1913: Notes on the water content and the wilting point, *J. Ecol.*, **1**: 96–100.

Dandeno, J. B., 1901: An investigation into the effects of water and aqueous solutions of some of the common inorganic substances of foliage leaves, *Trans. Can. Inst.*, **7**: 238–350.

Darrow, G. M., and H. Sherwood, 1931: Transpiration studies on strawberries, *Proc. Am. Soc. Hort. Sci.*, **28**: 225–230.

Dixon, H. H., 1910: Transpiration and the ascent of sap, *Prog. rei Bot.*, **3**: 1–66.

Eames, A. J., and L. H. MacDaniels, 1925: "An Introduction to Plant Anatomy," pp. 100–103, McGraw-Hill Book Company, Inc., New York.

Ganong, W. F., 1894: On the absorption of water by the green parts of plants, *Bot. Gaz.*, **19**: 136–143.

Gates, F. C., 1914: Winter as a factor in the xerophily of certain evergreen ericads, *Bot. Gaz.*, **57**: 445–489.

Grundmann, W., 1923: Die Wasseraufnahme der Pflanzen durch die Blätter, *Z. Pflanzenernähr. Düng. u. Bodenk.*, **33**: 88–90.

Haines, F. M., 1932: A self-recording potometer with a note on rate of transpiration under pressure, *Ann. Bot.*, **46**: 1051–1060.

Halket, A. C., 1911: Some experiments on absorption by the aerial parts of certain salt plants, *New Phytol.*, **10**: 121–139.

Hardy, F., 1923: The soil-point method for directly estimating the water-supplying power of a soil in the field, *J. Agr. Sci.*, **13**: 355–360.

Henderson, L., 1934: Relation between root respiration and absorption, *Plant Physiol.*, **9**: 283–300.

Hendrickson, A. H., and F. J. Veihmeyer, 1929: Some facts concerning soil moisture of interest to horticulturists, *Proc. Am. Soc. Hort. Sci.*, **1929**: 105–108.

——— and ———, 1934: Irrigation experiments with prunes, *Calif. Sta. Bull.* 573.

Henslow, G., 1908: On the absorption of rain and dew by the green parts of plants, *J. Roy. Hort. Soc.*, **34**: 167–178.

Holle, H., 1915: Untersuchungen über Welken, Vertrocknen und Wiederstraffwerden, *Flora*, **108**: 73–126.

Kessler, O. W., 1933: Die Wasseraufnahme der Pflanzen durch die Blätter und ihre Bedeutung für die Feldberegnung, *Die Feldberegnung*, **38**: 111–118.

Koketsu, R., 1928: Variation of the water content of leaves as related to the wilting of plants, *J. Dept. Agr. Kyushu Imp. Univ.*, **2**: 93–116; *Exp. Sta. Rec.*, **61**: 62.

Kramer, P. J., 1932: The absorption of water by root systems of plants, *Am. J. Bot.*, **19**: 148–164.

———, 1933: The intake of water through dead root systems and its relation to the problem of absorption by transpiring plants, *Am. J. Bot.*, **20**: 481–492.

———, 1934: Effect of soil temperature on the absorption of water by plants, *Sci.*, **79**: 371–372.

Kunkel, L. O., 1912: A study of the problem of water absorption, *Mo. Bot. Gard. Ann. Rept.*, **23**: 26–40.

Lewis, M. R., R. A. Work, and W. W. Aldrich, 1935: Influence of different quantities of moisture in a heavy soil on rate of growth of pears, *Plant Physiol.*, **10**: 309–323.

Livingston, B. E., and W. H. Brown, 1912: Relation of the daily march of transpiration to variations in the water content of foliage leaves, *Bot. Gaz.*, **53**: 309–330.

——— and L. A. Hawkins, 1915: The water relation between plant and the soil, *Carnegie Inst. Wash. Pub.*, **204**: 1–48.

——— and R. Koketsu, 1920: The water supplying power of the soil as related to the wilting of plants, *Soil Sci.*, **9**: 469–485.

——— and I. Ohga, 1926: The summer march of soil-moisture conditions as determined by porcelain soil points, *Ecology*, **7**: 427–439.

———, T. Hemmi, and J. D. Wilson, 1926: Growth of young wheat plants in autoirrigated soils, as related to the water supplying power of the soil and to the adjustment of the autoirrigator, *Plant Physiol.*, **1**: 387–395.

MacDougal, D. T., 1925: Absorption and exudation pressures of sap in plants, *Am. Phil. Proc.*, **64**: 102–130.

Magistad, O. C., and J. F. Breazeale, 1929: Plant and soil relations at and below the wilting percentage, *Ariz. Sta. Tech. Bull.* 25.

Mason, T. G., 1922: The physiological humidity of the soil and its direct determination, *W. Ind. Bull.*, **19**: 137–154.

Miller, E. C., 1916: Comparative study of the root systems and leaf areas of corn and the sorghums, *J. Agr. Res.*, **6**: 311–331.

Moinat, A. D., 1932: Available water and the wilting of plants, *Plant Physiol.*, **7**: 35–46.

Nakajima, Y., 1928: Wirkung des der Luft entleerten Wassers auf die Wasseraufnahme verschiedener Körper, *Sci. Rept. Tôhoku Imp. Univ.*, Fourth Series, (Biol.), **3**: 279–298; *Biol. Abst.*, **4**: 686.

Nutman, F. J., 1933, 1934: The root system of *Coffea Arabica.* I, II, and III, *Empire J. Exp. Agr.*, **1**: 271–284, 285–296 (1933); **2**: 293–302 (1934).

Overton, J. B., 1921: The mechanism of root pressure and its relation to sap flow, *Am. J. Bot.*, **8**: 369–374.

Powers, W. L., 1922: Field moisture capacity and wilting point of soils, *Soil Sci.*, **14**: 159–167.

Priestley, J. H., 1920: The mechanism of root pressure, *New Phytol.*, **19**: 189–200.

———, 1922: Further observations upon the mechanism of root pressure, *New Phytol.*, **21**: 41–48.

——— and E. E. North, 1922: Physiological studies in plant anatomy. III, The structure of the endodermis in relation to its function, *New Phytol.*, **21**: 113–139.

——— and E. Rhodes, 1926: On the macro-chemistry of the endodermis, *Proc. Roy. Soc.*, B, **100**: 119–129.

Pulling, H. E., and B. E. Livingston, 1915: The water supplying power of the soil as indicated by osmometers, *Carnegie Inst. Wash. Pub.*, **204**: 49–84.

———, 1917: The rate of water movement in aerated soils, *Soil Sci.*, **4**: 239–268.

Reed, H. S., and E. T. Bartholomew, 1927: Water translocation in young fruits, *Sci.*, **66**: 382–385.

Renner, O., 1912: Versuche zur Mechanik der Wasserversorgung. I, Der Druck in den Leitungsbahnen von Freilandpflanzen. II, Über Wurzeltätigkeit, *Ber. bot. Ges.*, **30**: 576–580, 642–648.

Sachs, J., 1859: Ueber den Einfluss der chemischen und physikalischen Beschaffenheit des Bodens auf die Transpiration der Pflanzen, *Landw. Versuchs-Stat.*, **1**: 203–241.

Scott, L. I., and J. H. Priestley, 1928: The root as an absorbing organ. I, A reconsideration of the entry of water and salts into the absorbing region, *New Phytol.*, **27**: 125–141.

———, 1928: The root as an absorbing organ. II, The delimitation of the absorbing zone, *New Phytol.*, **27**: 141–174.

Shive, J. W., and B. E. Livingston, 1914: The relation of atmospheric evaporating power to the soil moisture content at permanent wilting in plants, *Plant World*, **17**: 81–121.

SHULL, C. A., 1913: Semipermeability of seed coats, *Bot. Gaz.*, **56**: 169–199.

———, 1914: Measurement of the internal forces of seeds, *Trans. Kans. Acad. Sci.*, **27**: 65–70.

———, 1916: Measurement of the surface forces in soils, *Bot. Gaz.*, **62**: 1–29; *Sci.*, **43**: 361.

———, 1924: Imbibition in relation to absorption and transportation of water in plants, *Ecology*, **5**: 230–241.

———, 1930: Absorption of water by plants and the forces involved, *J. Am. Soc. Agron.*, **22**: 459–471.

SMYTH, E. S., 1934: A contribution to the physiology and ecology of *Peltigera cania* and *P. polydactyla*, *Ann. Bot.*, **48**: 781–818.

SPALDING, V. M., 1906: Biological relation of desert shrubs. II, Absorption of water by leaves, *Bot. Gaz.*, **41**: 262–283.

STOCKER, O., 1930: Über die Messung von Bodensaugkräften und ihrem Verhältnis zu den Wurzelsaugkräften, *Z. Bot.*, **23**: 27–56; *Biol. Abst.*, **6**: 678.

THODAY, D., 1918: On turgescence and the absorption of water by the cells of plants, *New Phytol.*, **17**: 108–113.

VEIHMEYER, F. J., 1927: Some factors affecting the irrigation requirements of deciduous orchards, *Hilgardia*, **2**: 125–290.

——— and A. H. HENDRICKSON, 1928: Soil moisture at permanent wilting of plants, *Plant Physiol.*, **3**: 355–357.

WADSWORTH, H. A., and U. K. DAS, 1930: Some observations on the wilting coefficient of a selected Waipio soil, *Hawaiian Planters' Rec.*, **34**: 289–294; *Biol. Abst.*, **5**: 2659.

WELTON, F. A., and J. D. WILSON, 1931: Water-supplying power of the soil under different species of grass and with different rates of water application, *Plant Physiol.*, **6**: 485–494.

WETZEL, K., 1924: Die Wasseraufnahme der höheren Pflanzen gemässigter Klimate durch oberirdische Organe, *Flora*, **117**: 221–269.

WILLIAMS, H. F., 1932: Absorption of water by the leaves of common mesophytes, *J. Elisha Mitchell Sci. Soc.*, **48**: 83–100; *Biol. Abst.*, **8**: 1653.

WILSON, J. D., 1927: The measurement and interpretation of the water supplying power of the soil with special reference to lawn grasses and some other plants, *Plant Physiol.*, **2**: 385–440.

———, and B. E. LIVINGSTON, 1932: Wilting and withering of grasses in greenhouse cultures as related to water supplying power of the soil, *Plant Physiol.*, **7**: 1–35.

WOLFE, H. S., 1926: Surface forces of soils within the range of hygroscopic moisture, *Bot. Gaz.*, **82**: 195–207.

WOOD, J. G., 1925: The selective absorption of chlorine ions and the absorption of water by the leaves in the genus *Atriplex*, *Australian J. Exp. Biol. Med. Sci.*, **2**: 45–46; *Bot. Abst.*, **15**: 1164.

WORK, R. A., and M. R. LEWIS, 1936: The relation of soil moisture to pear tree wilting in heavy clay soil, *J. Am. Soc. Agron.*, **28**: 124–134.

CHAPTER V

THE INTAKE OF SOLUTES BY THE PLANT

Under this heading it is intended to discuss more particularly the entrance of the inorganic compounds into those cells of the plant which border upon the medium in which the plant is growing. The discussion will include not only the mode of entrance of the solutes but also the general influence of the environmental factors on the intake of materials by the plant as well as the nature and composition of the solutions that are commonly used in the study of plant nutrition.

I. MODE OF ENTRANCE OF SOLUTES

Numerous experiments have been performed to show that inorganic salts readily penetrate the protoplasmic membrane and enter the vacuole of the cell, but a simple experiment by Osterhout (1909) in this regard will suffice as an example. He noted that *Dianthus barbatus* could be grown with its root hairs entirely free from calcium oxalate if the seeds of this plant were simply placed on the surface of distilled water and allowed to grow. After the root hairs thus grown had reached the desired length they were transferred to a calcium solution and examined from time to time under the microscope for the appearance of calcium oxalate crystals. Controls in the distilled water were examined at the same time and treated in the same manner as the plants in the solution. In no case did the controls produce crystals, but calcium oxalate crystals were always produced in the root hairs when grown in the following solutions: 0.05*M* calcium sulphate, 0.005*M* calcium nitrate, 0.005*M* calcium chloride, and 0.001*M* calcium hydroxide, dilute sea water, and tap water. In each of these solutions the crystals of calcium oxalate made their appearance within the root hairs inside 4 hr. at 30°C. The shortest time required for the formation of crystals was 1½ hr. in the case of calcium sulphate solution. It was also found that the root hairs in contact with particles of precipitated calcium carbonate and tricalcium phosphate produced calcium oxalate crystals, but only after several days. It has generally been considered that inorganic salts penetrate the protoplasm in the dissociated or ionic form. Recently, however, work by Osterhout (1925), Irwin (1926), Cooper, Dorcas, and Osterhout (1929), and Osterhout, Kamerling, and Stanley (1934) indicates that, in some cases at least, certain compounds enter the plant cell through the protoplasm only in the undissociated form.

Brooks (1931) adversely criticized the molecular hypothesis of diffusion.

A. Passive or Physical Absorption

Passive or physical absorption may be defined as the absorption of solutes by cells according to the ordinary laws of diffusion.

It has generally been observed that the protoplasmic membrane of the cells bordering upon the soil or the medium in which the plants are growing is permeable to practically all the inorganic solutes present in the soil solution. It has thus been considered that the entrance of any given solute into the vacuole of these cells is a process of diffusion from the higher concentration of this solute in the soil to a lower concentration in the cell sap. If no disposition were made of the solute which entered, equilibrium in concentration between the cell sap and the soil solution in regard to this solute would be reached and no more would enter. If, however, some or all of this solute were removed from the cell, combined, precipitated, adsorbed, or otherwise rendered osmotically inactive, the concentration of the solute would continue to be lower in the cell sap than in the surrounding medium, so that it would continue to enter the cell. This principle can be illustrated by an experiment, as shown in Fig. 10. A solution of tannin is separated from a solution of ferric chloride by a membrane of collodion or parchment paper. The tannin will not diffuse through the membrane, but the ferric chloride will. As soon as the iron salt enters *A*, it is precipitated as iron tannate so that the concentration of ferric chloride in *A* never becomes equal to that in *B* and more will continue to enter. All the ferric chloride in *B* will pass through into *A*, provided there is a sufficient quantity of tannin in *A* to combine with the iron salt. Frequently the analysis of certain tissues has indicated a higher concentration of certain elements or ions in the cell sap than in the surrounding medium, and this fact has been explained until recently by assuming that a portion of the ions in question was adsorbed, combined, or precipitated so that the concentration was never greater than that of the surrounding medium.

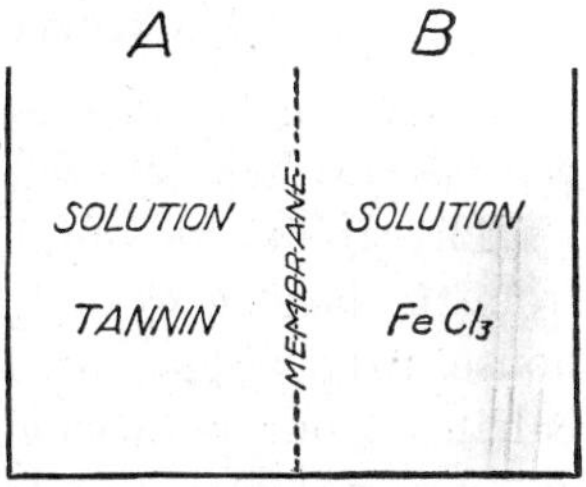

Fig. 10.—Diagram to illustrate the principle of physical absorption. Description in the text.

In a study of the relation between the composition of young orange trees and the concentration of nutrient solutions, Reed and Haas (1924) concluded, however, that there is apparently no evidence to support the assumption that the absorbed materials are converted into insoluble compounds as rapidly as they accumulate in the plant, or that absorption necessarily depends upon the precipitation of ions within the plant.

Gračanin (1932) observed in the plants of barley and corn that the amount of absorption of phosphorus ions by the roots was according to the ordinary physical laws of diffusion. Thus plants growing in solutions with concentrations of 0.001 and 0.05 per cent of phosphorus pentoxide, supplied as monocalcium phosphate, absorbed from equal concentrations of solutions the same quantity of phosphate ions regardless of the degree of illumination.

Examinations of the uncontaminated sap of individual cells show that, in some cases at least, the concentration of the osmotically active solutes is higher in the cell sap than in the surrounding medium. This indicates that the intake of solutes is not always so simple a process as has been mentioned in this topic.

B. Active or Physiological Absorption

Def. Active or physiological absorption is the absorption of solutes by cells after a manner contrary to the ordinary laws of diffusion.

The ability to obtain a sufficient quantity of uncontaminated sap from the large cells of *Valonia* and *Nitella* for chemical analyses (Wodehouse, 1917; Osterhout, 1922; and Hoagland and Davis, 1923) has greatly enhanced our knowledge concerning the concentration of the cell sap as compared to the external medium. Osterhout (1922) made an analysis of the cell sap of *Valonia* and found the concentration of certain elements in it as compared to the sea water in which it grew to be as follows:

Element	Concentration in sea-water, p.p.t.	Concentration in cell-sap, p.p.t.
Chlorine	19.6	21.2
Sodium	10.9	2.1
Potassium	0.5	20.1
Calcium	0.45	0.7
Magnesium	1.31	Trace
SO_4	3.3	0.005

The data show that some factor or factors prevented the sodium, magnesium, and sulphate ions from reaching as high a concentration in the cell sap as in the medium in which the cells were growing. The chlorine content was practically the same inside the cell as in the sea water. The most striking fact of the experiment was the high concentration of potassium in the cell sap as compared to the external concentration. Its concentration in the sap in this case was approximately forty times as high as in the sea water. Practically all the potassium in the cell sap was in the form of chlorides, so that this accumulation could not be attributed to its combination with other compounds.

Blinks and Jacques (1930) noted that the ratio of potassium to sodium in the cell sap of *Halicystis* was relatively low, amounting to only 0.0011, and Jacques and Osterhout (1932) found in *Valonia macrophysa* that potassium was the only cation that accumulated in the cell, and it became approximately 40 times more concentrated inside than outside. Cooper and Osterhout (1930) found that when 0.005M ammonium chloride was added to sea water containing cells of *Valonia macrophysa* ammonia soon appeared in the sap and reached as high a concentration as the potassium mentioned above.

The concentration of ions in the cell sap of *Nitella clavata* compared to that of the surrounding medium was investigated by Hoagland and Davis (1923). A summary of their work is given in the following table:

Items considered	K	Na	Ca	Mg	Cl	SO_4	PO_4	NO_3
	p.p.m.							
Pond water in which plants grew	Trace	5	31	41	32	31	0.4	34
Sap from cells	2,120	230	410	430	3,220	800	350	0
Ratio of cell sap concentration to external concentration	2,000	46	13	10	100	26	870	

They considered that the major portion of the inorganic elements in the cell sap was in the ionic state since its conductivity was slightly greater than that of a 0.1 N potassium chloride solution. By a comparison of the positive and negative ion equivalents it was possible to balance 80 to 90 per cent of the positive ions by negative ions. The remaining ions were ascribed to organic acids and other organic compounds.

The analysis of the sap showed that the concentration of each of the ions with the exception of NO_3 was higher than in the water in which the cells were growing. The accumulation of potassium was the highest and that of magnesium the lowest. The insoluble or combined elements found in the cell wall and protoplasm included calcium, magnesium, sulphur, silicon, iron, and aluminum, the calcium being the most abundant. No potassium was present in the insoluble form. Hoagland (1919) and Hoagland and Davis (1925) observed that the expressed sap of the leaves of barley growing in an ordinary cultural solution had a conductivity four to fifty times greater than that of the solution from which the plants obtained their inorganic nutrients. This also indicates that the concentration of the ions within the cell is higher than that of the surrounding medium.

In a study of the composition of the cell sap of the plants of corn, sorghum, and Sudan grass, Pierre and Pohlman (1933) found that the total content of electrolytes, as measured by conductivity, was higher in some cases and lower in others than the displaced soil solution. The amount of phosphate ranged between 150 and 400 p.p.m., silica averaged approximately 250 p.p.m., calcium 80 p.p.m., chlorides 90 p.p.m., while nitrates varied from a trace to 344 p.p.m. Phosphorus and silica were found in much higher concentrations in the sap than in the displaced soil solution, the concentration factors being 552 to 4,967, and 15.2 to 348, respectively. The concentrations of chlorine and calcium in the cell were considerably lower than those of the soil solution.

Stiles (1924) in a study of the absorption of salts by storage tissues concluded that the absorption of salt by them is not a simple process of diffusion of the salt through a cell membrane. The vital force that overcomes the physical equilibrium and enables the cell to accumulate salts in concentration greater than the outside is termed "epictesis" by Lapique (1925).

There are evidently many and varied forces concerned in physiological absorption, since the simplest living plant cell is a very complex system that changes continually in response to its environment. The soil in which a plant grows is a heterogeneous medium, and the plant responds to such a medium in a heterogeneous manner by effecting a sort of physiological integration of the various factors concerned (Hoagland, 1930).

Some of the various forces that may be concerned in the accumulation of solutes in the cell sap are: (1) the attraction of colloids for ions, (2) the amphoteric nature of protoplasm, (3) Donnan equilibrium, (4) ionic exchange, and (5) metabolic activity.

1. The Attraction of Colloids for Ions.—In the opinion of Breazeale (1923), the demand for the various elements originates in the tissues of the plants and is carried to the absorbing surface of the root by means of unsaturated colloidal compounds bearing a plus or minus electric charge. For example, let us suppose that the protoplasm in a leaf cell removes an atom of potassium from a colloidal compound and uses it in building up a permanent compound that is to be one of the final constituents of its tissues. This leaves the colloidal compound or molecule out of equilibrium and with a minus charge. This charge is transmitted by replacement and not by bodily movement down through the cells to the epidermal cells of the root and there appears as a minus charge. If an ion of potassium were present in the soil solution, it would be attracted to the negative charge. The potassium could in this manner be transported from the absorbing surface of the root to any part of the plant without the bodily movement of a potassium compound through the sap. Breazeale further considered that if an ion of potassium bearing a plus charge is removed from the soil solution by the root, this ion can be filled by replacement and the charge carried along through the soil, where the source of supply of potassium exists. If this is true, the plant will not be dependent upon the soil grains that come in contact with the epidermal cells of the root but may actually obtain its nutrients at a distance from the roots. The objections to this theory will be brought out in the following discussions.

2. Amphoteric Nature of the Protoplasm.—The proteins and protoplasm of a plant cell are ampholytes. They are thus substances that can behave as either cations or anions depending upon the hydrogen-ion concentration of the medium in which they are placed. The isoelectric point of an ampholyte is the pH value of the solution in which it is placed at which the ampholyte acts as neither an anion nor a cation. The isoelectric point is sometimes called the "neutral point" of the ampholyte in question. Thus an ampholyte on the acid side of its isoelectric point behaves as a cation and combines with anions, while on the alkaline side of the isoelectric point it behaves as an anion and combines with cations. Since the absorption of ions is intimately connected with certain characteristics of the protoplasm and since many of these characteristics are bound up apparently with the peculiar properties of its constituent proteins, it would seem possible that the amphoteric nature of the plant proteins would exert an influence on the absorption of ions (Robbins, 1923).

Lapique (1925) considered that the outer surface of the protoplasm of a plant cell may be in contact with a medium that is either neutral or alkaline to it, while at the vacuole it may be in contact with a medium that is on the acid side of its isoelectric point. As the protoplasm rotates or circulates in the cell, every part of it is alternately in contact with the exterior medium; when it is on the alkaline side of its isoelectric point, it combines with cations and releases anions. As circulation continues it comes in contact with the acid cell sap, releases its cations, and combines with anions. This theory accounts in part for the "heaping up" or accumulation of cations in the cell sap.

Davidson (1933) assumed that there is a wide range in the isoelectric points of the plant ampholytes allowing the occurrence of both electropositive and electronegative ampholytes within certain limits of hydrogen-ion concentration. This makes possible the simultaneous accumulation of cations and anions. A change in the reaction of the medium may modify to some extent the reaction within the plant cells and thus shift some of the electropositive ampholytes to the negative side, and vice versa. This might cause an increased absorption of cations and anions.

3. Donnan Equilibrium.—According to Briggs and Petrie (1928), the conception of a simple Donnan equilibrium operating between two homogeneous phases is inadequate to explain the phenomenon of the ionic intake by plants, since experimental evidence indicates that the product of the apparent internal concentration of cation and anion, estimated on the basis of the total volume of the tissue, may be greater than that of the concentrations of the two ions in the external medium instead of equal to it, as the above conception would require. According to these authors, the interior of the plant tissues comprises a number of phases, each of which may be in Donnan equilibrium with the external solution. The product of the apparent internal ionic concentrations resulting from the total effect of all these internal phases is shown under such circumstances to be greater than the external product. Even in such a system as this, however, if the ratio of the apparent internal to the external concentration is greater than unity, the same ratio of the corresponding anion would be less than unity. Experimental evidence, however, indicates that both may be greater than unity. Briggs and Petrie suggest that this behavior may be accounted for by assuming that one internal phase contains indiffusible cations and that another contains indiffusible anions.

Thomas (1930) believed that for plants in the presence of colloidal silica the increased absorption of phosphorus pentoxide may be explained better on the basis of the Gibbs-Donnan distribution law than by any other theory.

Osterhout and Stanley (1932) demonstrated that the accumulation of electrolytes in *Valonia* does not depend on Donnan equilibrium or the presence within the cell of ions and molecules that cannot pass out through the protoplasm.

4. Ionic Exchange.—It was considered by Casale (1921) that the ectoplasm yields hydrogen ions to the soil solution and thereby establishes a difference of potential between the plant cell and the soil particles. In the process of equalization of those charges, the more distant particles yield cations to the nearer ones, and these in turn yield them to the ectoplasm. Thus fertilizers act, in part, by changing the difference of potential between cells of the plant root and the soil particles.

McGeorge (1932) noted that air-dried tissues of alfalfa showed a definite property of absorbing bases and hydrogen and exchanging them for bases in chemically equivalent proportions. He suggested that living tissues may exhibit this same property and that it may be related to the permeability and cationic concentration within the plant system. It was suggested by Briggs (1932) that, from weak solutions, salt absorption by mature plant cells might consist essentially of an exchange of anions between the solution and the cell sap, and an exchange of cations between the solution and the cytoplasm.

5. Metabolic Activity.—The accumulation or "heaping up" of a given salt in the cell sap must necessarily involve the expenditure of energy to produce and maintain a condition not in conformity with physico-chemical equilibrium. Steward (1932, 1933), Steward, Wright, and Berry (1932), Asprey (1933), and Steward and Berry (1934, 1936) have studied the accumulation of salts in the tissues of a variety of storage organs including carrot, parsnip, beet, mangold, turnip, artichoke, dahlia, potato, and pear. Petrie (1933), Lunderårdh and Burström (1933), Potapov and Stankov (1934), Hoagland and Broyer (1936), and Prevot and Steward (1936) studied the accumulation of salts by the roots of plants. All these investigators considered that salt accumulation in cells is dependent upon metabolic phenomena that presumably supply the necessary energy. In storage tissues, the accumulation of salts is apparently a characteristic of the cells that are starting on a new cycle of intense metabolism which may culminate in cell division. The evidence indicates that the accumulation of salt resides in the surface cells whose respiration is markedly augmented. All the experiments indicate that the variables which affect respiration per unit weight also affect in a similar manner the salt absorption per unit volume, as represented by the internal sap concentration. Apparently the principal factor in the salt accumulation of storage tissues is their ability for renewed active metabolism and even growth. Consequently the amount of salt accumulated is closely related to the rate of aerobic respiration, which represents the major source of energy for these tissues.

Hoagland and Broyer (1936) made extensive studies on the accumulation of salts by the excised roots of barley. They obtained absorption ratios of 15 for potassium, 7 for the nitrate ion, and 10 for bromine when the solution in which the roots were placed contained over 7 milliequivalents of the potassium salt per liter. With sufficiently dilute solutions, absorption ratios in excess of 1,000 could be obtained. They found that the accumulation of salts in roots is dependent upon their age and the proportion of active metabolizing cells, a low initial salt content, and available carbohydrate. Apparently the same general factors that influence the accumulation of solutes by the storage tissues of plants also influence accumulation in the roots.

Prevot and Steward (1936) in barley roots found that the potential absorption surface extends back from the apex to the place where secondary roots appear. A pronounced longitudinal gradation in the capacity for salt accumulation was observed, the segments near the apex attaining higher concentrations than those more remote. This gradient is considered to be a consequence of the progressive development of cells from the root apex.

Since the penetration of ions into the cells that have just been mentioned apparently takes place from a solution of low to a solution of high concentration of these ions, the plant cell must do work in absorbing them. The processes involved would seem to be similar to those which occur in the secretion by the glands of animals, by which compounds are taken from a dilute concentration in the blood and excreted to the outside into a much higher concentration. Just how such a process is brought about is not known, so it is only expressing our ignorance when we state that the transference of an ion from a low concentration to a higher one is a process of secretion. The data that have been presented show that little is yet known concerning the means by which absorption of solutes occurs. The ratio of the internal concentration of a given solute in the cell to its external concentration has been termed the "absorption ratio." Some of the factors that influence its value will be discussed in the following topic.

II. FACTORS THAT AFFECT THE ABSORPTION RATIO

A. Light

Since light either directly or indirectly is the source of the energy of the plant cell, Hoagland and Davis (1923, 1929), Hoagland, Hibbard, and Davis (1926), and Hoagland (1930) undertook a study of the influence of light upon the removal of certain ions from the external solution by *Nitella clavata*. The solution and plants were exposed to light for different lengths of time, the amounts of certain elements remaining in the solution were observed, and the absorption ratio was determined. In one experiment the solution employed was 0.01*M* monopotassium phosphate plus potassium chloride in varying amounts. The general plan of the experiment together with certain of the data is shown in the following table:

Experiment number	Length of experiment, days	Concentration of KCl in external solution p.p.m.	Chlorine remaining in solution, p.p.m.					
			Full illumination	Dim light	Light 9 a.m. to 5 p.m.	Light 11 a.m. to 4 p.m.	Light 11 a.m. to 2 p.m.	Dark
1	4	17.8	0	1.5	5.0	13.5		
2	4	35.5	0	14	....	27	..	31
3	5	35.5	3	35	16	22	26	
4	4	35.5	2	27	10	15	25	34

In the case of 0.01*M* potassium nitrate, heavy tests for NO_3 were found in the sap for all cells in the light but only slight tests for those in the dark. In every experiment where the absorption of bromine was under consideration, the exposure to light strikingly increased the ability of the cells to accumulate this element. When a concentration of 5 milliequivalents of bromine was used, the concentration of bromine in the cell sap of *Nitella* kept in the dark did not exceed that of the outside solution, while the sap of those exposed to illumination for an equal period of time had a much higher concentration than the surrounding solution. The following figures will serve as an example of the results obtained. These results were noted at a temperature of 20°C.

Illumination	Time, hours	Concentration of bromine in cell sap, milliequivalents
Continuous light	25	6.4
Continuous darkness	25	2.6
Continuous light	50	12.3
Continuous darkness	50	3.3

It was observed by Benedict (1934) that ultraviolet rays between 2900 and 3100 Å caused an increase in the percentage of calcium in all the plants that he examined. He could not, however, formulate any conclusion from his data relative to the effect of these rays on the content of phosphorus in the plants. Ingold (1936) believed that permeability is increased in some cases by illumination, but it was difficult to determine how much absorption was influenced by this factor. Light apparently has a greater effect upon the absorption of cations than of anions.

It is evident, in the case of *Nitella* at least, that light greatly increases the absorption of certain ions. The cells contain only a very limited supply of organic material that might release energy by respiration, so that little kinetic energy is available except that received directly from the sun. It is suggested that if energy is needed for the absorption of solutes by the cells of the roots, it is supplied by the process of respiration. It is worthy of note here that in the absorption of bromine by *Nitella* the temperature coefficient between 10 to 20 and 14 to 24°C. ranged between 2 and 3. This is of such a high value that it would be considered characteristic of a chemical reaction rather than a physical process of diffusion.

B. Concentration of the Exterior Medium

The data on this subject have been obtained, for the most part, by immersing disks of plant tissue of known dimensions and volume in the solution and then either directly or indirectly determining to what degree the external solution had been depleted of the salt or ions in question. The tissues that have been used are carrot, beet, dahlia, and the tubers of *Helianthus* and Irish potato. The solution is either analyzed chemically, or its changes in concentration are determined by those in its conductivity. Since the volume of the tissue is known and since the decrease in the concentration of a given salt or ion in the solution can be determined after any period of absorption, it is possible to obtain values for the amount of salt or ion absorbed per unit of volume of tissue. The values thus obtained give an expression for the internal concentration of the salt. Thus the ratio of the concentration of the solute in the tissue to its concentration in the exterior medium can be obtained.

It has been observed by Nathansohn (1903, 1904) and Stiles and Kidd (1919) that the equilibrium reached in the absorption of a given solute by living tissue depends upon the external concentration of the solute. While the absolute amount of solute absorbed increased with the increasing concentration of the solute in the external solution, the amount absorbed relative to the external concentration nevertheless decreases rapidly from the lowest concentration upward. When the external concentration of the solute is low, the absorption ratio becomes more than unity, while if the external solution is relatively high, the absorption ratio is always less than unity. The following examples will illustrate these statements: Nathansohn (1903) found in the case of the marine alga *Codium* in twentieth-normal to normal solutions of sodium nitrate that the absorption ratio varied from 0.41 to 0.68, but it did not take up the nitrate ion in sufficient amount to produce equal concentrations inside and outside the tissue. He observed, however, that, in sea water where the concentration of the nitrate ion was low, its concentration was greater in the tissue than in the exterior medium. The effect of different concentrations and different salts upon the absorption ratio of carrot tissue as found by Stiles and Kidd (1919) is shown in the following table:

Salt	Initial external concentration	Length of experiment, hours	Absorption ratio
KCl	*N*/5,000	52	25
	N/500	52	17.6
	N/50	52	2.4
	N/10	52	0.8
NaCl	*N*/5,000	48	46.7
	N/500	48	27.0
	N/50	48	3.5
	N/10	48	0.8
$CaCl_2$	*N*/5,000	42.5	15.3
	N/500	42.5	2.8
	N/50	42.5	0.5
	N/10	42.5	0.24

These results show that the degree of absorption depends not only on the concentration of the external solution but also upon the nature of the ions absorbed. Less calcium ions are apparently absorbed than those of potassium and sodium. It is worthy of note that the absorption ratio for dead tissue never becomes greater than unity and that the concentration within the cells is practically identical with that of the external solution. Irwin (1925) noted when living cells of *Nitella* were placed in different concentrations of brilliant cresyl-blue solutions at pH 6.9 that the greater the concentration of the external dye solution the greater was the speed of accumulation of the dye in the cell sap and the higher the concentration of the dye in the sap at equilibrium. This behavior may be explained as a chemical combination of the dye with a constituent of the cell, since this explanation harmonizes with the fact that the temperature coefficient is about 5.0.

C. The Hydrogen-ion Concentration

There are wide differences of opinion in regard to the effect of a variation of the pH value of the cell or its surroundings upon the absorption of solutes. Any change in hydrogen-ion concentration of the cell or its surroundings alters the amphoteric substances in the cell, but it is not known how such a change alters the intake of ions. It was reported by Breazeale and McGeorge (1932) that plants are not able to absorb phosphate or nitrate ions from solutions of greater alkalinity than that represented by a pH of 7.6. Hydroxyl ions depress the absorption of phosphate and nitrate ions and the optimum pH for the absorption of nutrients is at or near neutrality. Osterhout and Kamerling (1934) concluded that any condition that checks the production of acid in the living cell will check the accumulation of nutrients and growth. Jacques and Osterhout (1934) found that increasing the pH value of the sea water, in which plants of *Valonia macrophysa* were growing, would increase the rate of entrance of potassium, and vice versa. It appeared that photosynthesis increases the rate of entrance of potassium by increasing the pH value of the protoplasm. In darkness there is little or no growth or absorption of electrolytes by this plant. It was observed by Jacques (1936) that the lower the pH value the greater is the tendency of potassium to enter the cell sap of *Nitella*. On the other hand, Pirschle and Mengdehl (1931) could find no definite relationship between absorption and the pH value.

III. THE UNEQUAL ABSORPTION OF IONS

The constituent ions of a salt are not necessarily absorbed in equivalent quantities by plant tissue or by the roots, as has been shown by the work of Nathansohn (1903, 1904), Meuer (1909), Ruhland (1909), Pantanelli (1915, 1918), Johnson (1915), Stiles and Kidd (1919), Redfern (1922), Hoagland (1923), Stiles (1924), Rudolfs (1925), Harris, Hoffman, and Lawrence (1925), Pirschle and Mengdehl (1931), and others. This fact has been determined directly by the chemical analysis of the solutions in which the living tissues had been placed or in which the roots of the plants were growing. The three following experiments will serve to illustrate the main facts observed and the chief points of interest in this regard. Redfern (1922) grew plants of corn and peas in a nutrient solution for 2 weeks and then transferred them to various solutions of calcium chloride. After varying periods of time, these solutions were analyzed and the amount of calcium and chloride remaining determined. A sample of the results where the plants were placed in a tenth-normal solution of calcium chloride is shown in the following table:

Corn			Peas		
Duration of experiment, hours	Percentage absorption of calcium	Percentage absorption of chloride	Duration of experiment, hours	Percentage absorption of calcium	Percentage absorption of chloride
23	17.5	6.5	36	17.7	3.6
48	25.3	6.5	60	12.7	4.0
57	27.0	5.9	84	14.6	3.8

The unequal absorption of ions by the barley plant from a variety of different salt solutions was observed by Hoagland (1923). His observations are summarized in the following table:

Absorption of Ions from Different Salts by the Barley Plant
Adapted from Hoagland (1923)

Salt used	Initial concentration, milli-equivalent	Ions absorbed, milliequivalents		Ions entering solution, milliequivalents			Length of absorption period, days
		Cation	Anion	Ca	Mg	K	
K_2SO_4	6.32	0.74	0.40	0.32	0.24	....	1⅙
KCl	6.32	1.28	1.30	0.24	0.18	....	
KNO_3	6.45	1.66	3.19	0.19	0.21	....	
KH_2PO_4	6.32	0.90	0.14	0.28	0.21	....	
$KHCO_3$	6.30	1.20	0.84	0.24	0.20	....	
$CaCl_2$	12.50	1.50	2.37	....	0.28	0.02	2
$Ca(NO_3)_2$	12.70	1.66	5.48	....	0.20	0.09	
$Ca(H_2PO_4)_2$	12.30	1.03	0.95	....	0.27	0.20	
$CaSO_4$	12.30	0.88	0.83	....	0.20	0.09	

Stiles (1924) studied the absorption of the constituent ions of single salts by the storage tissues of the carrot, parsnip, beet, and artichoke. Some observations of his work are shown in the following table:

Salt	Concentration, normality	Tissue	Duration of experiments, hours	Absorption, percentage of original amount	
				Cation	Anion
NH_4Cl	1/100	Carrot	48	57.2	32.2
$(NH_4)_3PO_4$	2/100	Carrot	23.5	28.8	6.0
$(NH_4)_3PO_4$	2/100	Parsnip	24.3	34.5	8.2
$(NH_4)_2SO_4$	1/100	Carrot	22.0	3.9	4.2
KCl	1/100	Carrot	22.0	4.6	6.3
NaCl	1/100	Carrot	47.1	18.5	8.5
Na_2SO_4	1/100	Carrot	22.0	15.0	8.0

Jacques (1934) noted that the inorganic constituents of the cell sap, from the cells of rhubarb, field sorrel, and wood sorrel, showed an average of 3.8:1 preponderance of cations over anions. True in general

A. Mechanism

Although it is an established fact that the ions (cations and anions) of a given salt are not absorbed equally by the plant, the mechanism of this unequal absorption is not very clear. According to Stiles (1924), there appear to be two possible consequences. In the first place, the excess of the more absorbed type of ion may be accompanied into the tissue by a quantity of one of the constituent ions of water, either hydrogen or hydroxyl, whichever is of the opposite sign to the more absorbed type of ion. If this theory is granted, then the external medium will be left either alkaline or acid depending on whether the anions or cations are absorbed in the greater amount. Let us consider a solution of sodium chloride as an example in which living tissue is immersed. If the sodium ions are absorbed in excess of the chlorine ions, the former will be accompanied into the tissue by an equivalent quantity of hydroxyl, leaving the same equivalent quantity of hydrogen ions to balance the excess of chlorine ions remaining in the external solution. The external medium would thus become acid. If, however, the chlorine ions are absorbed in excess, hydrogen ions accompany it into the tissue and the external solution becomes alkaline. The acidity and alkalinity of water-cultural solutions have been ascribed in some cases to the result of the unequal absorption of ions.

In the second place, it is possible that the excess of the more absorbed ion is replaced in the external solution by some other ion or ions of the same sign diffusing out from the tissue. Thus in the case of sodium chloride, the excess sodium absorbed might be compensated by the diffusion out from the tissue of magnesium, potassium, calcium, or other cations. If, however, chlorine is the more absorbed ion, its excess absorption could be compensated by the diffusion outward of nitrate, citrate, malate, or other anions.

The action of the protoplasm in the unequal absorption of ions has been considered by various authors. Höber and Höber (1928), in considering the cell sap of *Valonia macrophysa*, concluded that the protoplasmic membrane may consist of separate anion- and cation-permeable areas so that it might be impermeable to a given electrolyte although permeable to both its ions.

Brooks (1929) considered that the protoplasm consists of a mosaic of anion- and cation-permeable areas that are of the nature of charged porous films. These films may exaggerate differences between the diffusion velocities of the ions to which they are permeable. The diameter of these pores may vary according to the environment surrounding the cell. The penetrabilities of different ions are characteristic functions of their own diameters and of the diameter of the pores in the membrane. Briggs (1930) stated that the protoplasmic membrane exhibits alternating phases of ionic permeability. He considered that since the protoplasmic membrane is amphoteric in its nature it can be changed from an anionic to a cationic permeability by changes in the hydrogen-ion concentration. Collander (1930), and Collander and Barlund (1933), however, could not prove different degrees of permeability for different portions of the cell surface.

B. Reaction of the Exterior Solution

The factors that may influence the reaction of the solution in which a plant is growing are numerous and complex. Some of these are the type and concentration of the salts used, the rate of growth, and type of plant, in addition to the unequal absorption of ions. The intention is to state here some of the observations that have been made, in order to show the great variation of results and the complexity of the problem.

One of the earliest observations along this line was made by Skene (1915), who found that the cell walls of sphagnum moss caused the solutions of neutral salts in which they were placed to become acid. The amounts of acid hydrogen thus liberated varied with the salt. In the case of sodium chloride, potassium sulphate, calcium chloride, and sodium nitrate, it amounted to from 0.012 to 0.025 g.; in the case of sodium sulphate, sodium formate, and sodium butyrate, approximately 0.07 g. per 100 g. of dry sphagnum used. Skene found further that this property of sphagnum is exhibited generally by the cell walls of vascular plants but not to such a marked degree, since the number of grams of acid hydrogen liberated ranged only from 0.015 to 0.035 per 100 g. of dry matter. The cell walls of the sphagnums and other plant parts thus have the power to break up salts, removing the base and liberating the acid. It is supposed that this reaction is due to the negatively charged colloidal substances of the cell walls which adsorb the cations and set free the anions, thus causing an acid reaction in the exterior medium. Skene found that the different types of sphagnum with which he worked actually did utilize in growth the cations thus absorbed. Since plant cell walls in general exhibit this same property, although to a slighter degree, he suggested that the absorption of inorganic compounds by the root hairs and the epidermal cells of the root might take place in a similar manner (see also Dufrenoy, 1920).

Denny and Youden (1927) observed that when tissues of potato tubers, carrot roots, apple, whole seeds of corn, rye, and wheat came in contact with a tenth-normal solution of calcium chloride the solution soon became acidified and within an hour might increase as much as twentyfold. They showed that this increase in hydrogen ions was not due to the unequal role of absorption of anions and cations by the tissue, since they found that similar acidifications were produced by adding the salts to the water extract obtained by bringing the tissues in contact with water instead of the salt solution. In these cases, the salt came in contact not with the tissue itself but only with the soluble substances that leached out of the tissue. It was further found that the noncoagulable, noncolloidal, and nonprotein portions of the water extract underwent acidification upon the addition of salts, giving results analogous to those obtained by bringing the salts in contact with the tissues themselves. They further found that increases in hydrogen-ion concentration were obtained when salts were added to various organic acids such as malic, oxalic, succinic, aspartic, and others. The precipitation of insoluble salts of these acids with the resulting increase of the ionization of the remaining acid was shown to account in part for this development of acidity on adding solutions to organic acids, although acidifications were observed in cases where no precipitation occurred. They concluded, therefore, that the observed changes in hydrogen-ion concentration cannot be interpreted as indicating an isoelectric point for the tissue as a whole or furnishing proof that reaction has occurred between the ions of the salt solution and proteins with characteristic isoelectric points. The findings of Denny and Youden are in contrast to those of Rudolfs (1925), who found that when seeds were immersed in representative salt or acid solutions they were able to change the hydrogen-ion concentration of the solutions to definite points, and that a certain equilibrium was reached in all solutions after the seed had been immersed for a sufficient period of time. He considered that the chemical properties of the chief protein constituent of the seeds seemed responsible for the changes in the hydrogen-ion concentrations of the solutions.

The effect of the unequal absorption of ions upon the external medium has been especially studied in the case of cultural solutions. Redfern (1922) noted that the cultural medium in which corn and peas were grown remained neutral, although the cations were absorbed to a much greater degree than the anions. Hoagland (1923) found that single solutions of ammonium salts, potassium sulphate, sodium sulphate, and potassium bisulphate in which barley, peas, and cucumbers were grown generally increased in acidity. When barley, however, was grown in solutions of calcium nitrate, calcium chloride, magnesium nitrate, sodium nitrate, monocalcium phosphate, monomagnesium phosphate, or monosodium phosphate the actual acidity tended to decrease. In the case of any nitrate (except ammonium) the solution generally attained a reaction very close to neutrality. The changes of reaction of certain salt solutions are not always the same in different experiments. Thus, for example, Hoagland (1923) noticed in one experiment that even a slight shading of barley cultures markedly diminished the rate of increase of acidity in a potassium sulphate solution. After a short period of growth, the cultures nearest the window had a pH value of 4.0 and those at the opposite end of the row a pH value of 5.7. Haas and Reed (1926) considered that the changes in reaction of cultural solutions in which citrus seedlings were grown could be attributed directly to unequal absorption of ions together with the excretion of certain ions by the plants. In a complete nutrient solution, citrus seedlings in some cases brought about in a comparatively short period such a high concentration of hydrogen ions as to be injurious to the roots, but acidities greater than pH 3.0 to 3.2 were not observed to develop. In the case of walnut seedlings the acidity of the residual solution was entirely dependent upon the absorption rate of the two ions employed. In solutions of potassium phosphate with initial hydrogen-ion concentrations of pH 5.0 and lower, the preferential absorption of potassium by wheat seedlings resulted in increased acidity. In solutions with initial hydrogen-ion concentrations of 6.0 and 7.0, the increase in acidity was slight, owing to the buffer properties of the solutions (Davidson, 1927).

C. Interchange of Ions from the Plant to the Exterior Solution

When the unequal absorption of ions occurs, it frequently happens that the more absorbed ion is replaced by the diffusion outward from the plant of some other ion or ions of the same sign but of different chemical composition. This interchange of ions thus tends to keep the medium in which the plant is growing from becoming either acid or alkaline in its reaction to any degree. Thus Redfern (1922) found that when corn and peas were grown in a solution of calcium chloride it remained neutral in its reaction, although the calcium ions were absorbed greatly in excess of the chlorine ions. Potassium and magnesium, however, were found in appreciable quantities in the external medium, having diffused out from the roots, and it was considered that this interchange of bases caused the solution to remain neutral in spite of the unequal absorption of the calcium and chlorine. In the case of barley growing in single salt solutions (Hoagland, 1923), calcium and magnesium diffused outward as potassium was removed. When calcium salts were used, potassium and magnesium entered the solution from the plant. Thus in one case where the plant absorbed the potassium ions from a potassium sulphate solution to the extent of 1.44 milliequivalents, the calcium and magnesium ions entering the solution amounted to 0.50 and 0.53 milliequivalents, respectively. In another case where a calcium nitrate solution was used, the absorption of the calcium ions amounted to 1.66 milliequivalents, and the potassium and magnesion ions entering the solution amounted to only 0.09 and 0.20 milliequivalents, respectively. The anion HCO_3, according to Hoagland (1923), is the one which most abundantly replaces the anions that have been more abundantly absorbed. For example, in his experiments with barley growing in the solution of calcium chloride, the absorption of the cation amounted to 0.83 milliequivalent and the anion to 2.06 milliequivalents, while the amount of the HCO_3 ions diffusing into the external solution equaled 1.26 milliequivalents. In the case of a solution of calcium nitrate, the cations were absorbed to the extent of 1.66 milliequivalents and the nitrate ion to the extent of 5.48 milliequivalents. The quantity of HCO_3 ions that entered the exterior solution totaled 4.12 milliequivalents. Apparently, there has been no really satisfactory explanation advanced for the mechanism concerned in the interchange of ions between the plant and the external solution.

D. Factors That Influence the Unequal Absorption of Ions

The absorption and exchange of ions by a plant are influenced by a wide variety of conditions. Some of these are the concentration of the solution, the effect of one ion upon the absorption of another, the nature and concentration of the ions present, temperature, light, and numerous others.

1. Concentration of the External Solution.—The concentration of the solution has a marked influence upon the proportion of ions absorbed. For example, Redfern (1922) noted that peas growing in tenth-normal solution of calcium chloride absorbed 17.7 per cent of the calcium and 3.6 per cent of the chlorine in 36 hr., while in a thousandth-normal solution of the same salt during the same time they absorbed 23.1 per cent of the cation and 15.1 per cent of the anion. The relative absorption of ions by storage tissue is also influenced by the concentration of the solution, as is shown by the following experiments with carrot (Stiles, 1924):

Initial concentration of solution, normality	NH_4Cl, percentage of absorption		NaCl, percentage of absorption	
	48 hr.			
	Cation	Anion	Cation	Anion
1/10	12.9	5.9	25.2	8.3
1/100	57.2	32.2	36.1	15.8
1/1,000	97.1	80.7	38.1	22.8

These figures indicate that the two ions of the salts used tend to be absorbed more nearly equally as the concentration of the solution decreases.

2. The Influence of One Ion upon Another.—The topic here under discussion refers only to ions in dilute solutions that are not toxic to the plant and should not be confused with antagonism, which has to do with the influence of one ion on another in overcoming the toxic effect of solutions on plants.

Ions differ greatly in the rapidity with which they are absorbed. The initial rate of absorption of ions may be in a different order from that which prevails when equilibrium is reached. Thus Stiles and Kidd (1919) found that the cations with which they worked were absorbed initially by storage tissue after the following order: potassium, calcium, sodium, lithium, magnesium, but when equilibrium was approached the order of absorption was potassium, sodium, lithium, calcium, and magnesium. The absorption of anions under the same conditions was initially in the order sulphate, nitrate, chlorine, but this later gave way to the order nitrate, chlorine, sulphate.

Of the cations, potassium seems to be most rapidly absorbed in practically all cases observed (Haas and Reed, 1926), and calcium probably is the least rapidly absorbed. Of the anions, nitrate is the most rapidly and abundantly absorbed, while the sulphate ion has been observed by practically all workers to be the most slowly absorbed. In the majority of cases that have been observed, the cations have been absorbed in greater amount than the anions. The main exception has been the nitrate anion and sometimes the chlorine anion.

Pettinger (1932) found that the application of fertilizers containing chlorine increased the chlorine content of the cell sap of corn plants, the increase being partially proportional to the amount of chlorine supplied by the fertilizer. These plants absorbed the chlorine ion much more readily than the sulphate ion. As the chlorine content of the sap increased, the hydrogen-ion concentration also increased.

The effect of one ion or of several sets of ions upon the absorption of other ions has been noted by Hoagland (1923), Haas and Reed (1926), Lagatu and Maume (1927), Brown (1928), McCool and Weldon (1930), Cooper (1930), Larson (1931), Pettinger (1932), Bartholomew, Watts, and Janssen (1933), Colby (1933), Davidson (1933), and Jacobson and Swanback (1933). Apparently this effect occurs regardless of whether the ions have like or unlike electrical charges, and all three types of relations are involved, *i.e.*, cation to anion, cation to cation, and anion to anion. Haas and Hill (1926) noted that the presence of sodium decreased the rate of absorption of other cations, especially potassium and calcium in barley, citrus, and walnut plants, and potassium ions retarded the absorption of calcium in citrus seedlings. When the potassium in the cultural solution was low in amount, more calcium, magnesium, and phosphorus were absorbed from the solution than when potassium was abundant. Brown (1928) in growing wheat in sand cultures reported that high concentrations of potassium decreased the absorption of magnesium and that a high content of calcium inhibited the absorption of nitrates.

The work of Cooper (1930), on pasture grasses from various associations in New York, showed that there is a close correlation between the standard electrode potentials of elements and the amount of the various minerals in the ash of plants. The productive soils contained relatively large amounts of the elements that form strong ions, while unproductive soils were high in the elements that form weak ions. A negative correlation existed between potassium oxide and calcium oxide, and between calcium oxide and phosphorus pentoxide in the ash of these plants. The calcium seemed to be negatively correlated with the strong anionic materials, which suggested that plants growing on alkaline soils may absorb considerable quantities of the hydroxyl and bicarbonate ions.

It was considered by Bartholomew, Watts, and Janssen (1933) that a deficiency of potassium in a nutrient solution may result in an increased absorption of nitrogen and phosphorus by the leaves of the tomato, while an abundant supply of potassium may increase the absorption of phosphorus by the stems. The absorption of calcium and magnesium did not seem to be significantly influenced by the amount of nitrogen, phosphorus, or potassium absorbed by this plant.

Colby (1933) noted in young, French prune trees that a lack or scarcity of either potassium, magnesium, or phosphorus seriously depressed nitrate absorption and even caused a loss of this ion from the roots late in the season. The lack of calcium prevented root growth, and the trees absorbed very little nitrate. The total phosphorus absorbed during the season was depressed more by the lack of magnesium than by the lack of potassium and sulphur. The lack of calcium apparently prevented the absorption of a considerable quantity of any ion.

The work of Haas and Reed (1926) with citrus and walnut seedlings indicated that the amount of chlorine absorbed stood in the following order with respect to the accompanying cation: calcium, potassium, sodium, and magnesium. With barley plants Hoagland (1923) observed that the order of absorption for chlorine was potassium, sodium, magnesium, and calcium. The presence of chlorine seems to retard the absorption of nitrate, but other ions seem to have little effect upon its absorption. It has been observed that the nitrate ions penetrate more readily from an acid solution than from an alkaline one in experiments with *Nitella* (Hoagland and Davis, 1923). True and Bartlett (1915, 1916) found that the seedlings of *Lupinus albus* absorbed more salts from the mixtures of the nitrates of potassium, calcium, and magnesium than from equally concentrated solutions containing only one or two of these nitrates. The solutions, however, of the three nitrates that were most favorable to absorption were much inferior to corresponding solutions in which the three anions, dihydrogen phosphate, nitrate, and sulphate were present. It was observed by Hoagland, Davis, and Hibbard (1928) that the accumulation of bromium ions in the cell sap of *Nitella* was significantly retarded by chlorine or iodine ions also present in the cultural medium but was not retarded by sulphate, nitrate, or phosphate ions. The accumulation of bromium ions was definitely influenced by the nature of the cation, being most rapid when solutions of potassium bromide or rubidium bromide were used and least rapid with solutions of lithium bromide, calcium bromide, strontium bromide, and magnesium bromide.

Lagatu and Maume (1927) found in the grape that the assimilation of phosphorus checked the utilization of nitrogen, while Larson (1931) noted that in apple trees, with a higher percentage of nitrogen, the percentage of phosphorus was low, but if the percentage of nitrogen was low the percentage of phosphorus was high. McCool and Weldon (1930) grew plants of barley, oats, rye, and wheat in soils to which various amounts of sodium nitrate had been applied. The highest applications of sodium nitrate usually decreased the potassium content of the plants, while the effect of the smaller applications was variable. The calcium content decreased more with the greater application of nitrate, while the magnesium content was not greatly influenced by the amount of salt used. Davidson (1933), however, found that the potassium content of wheat plants grown in soils, to which sodium nitrate had been applied, was consistently higher than in the plants grown in the untreated soil. It was observed by Jacobson and Swanback (1933) in the tobacco plant that an increased proportion of nitrate nitrogen over ammoniacal nitrogen resulted in a large percentage of total calcium in the plant material. Bartholomew, Watts, and Janssen (1933) found that an abundant supply of nitrogen may increase the amount of potassium in the stems of tomatoes.

In the case of barley, Hoagland (1923) found that the slowly-absorbed sulphate ion decreased the rate of absorption of the associated cation. Thus potassium was absorbed much more rapidly from the nitrate and the chloride salt than from the sulphate salt of potassium. The absorption of cations was always greatest when they were accompanied by an anion like NO_3 that is rapidly absorbed (Haas and Reed, 1926). It was reported by Colby (1933), for French prune trees, that the lack of the sulphate ion had a less depressing effect on nitrate absorption than did the lack of any other of the six major elements of a complete nutrient solution.

Thomas (1933) grew Stayman winesap apple trees for six years in soil in large cylinders to which were added different combinations of nitrogen, phosphorus, and potassium nutrients. After this time, the amounts of these three elements that had been assimilated were determined. It was found that the percentage and absolute amount of each of the elements absorbed, and also the ratio in which it was absorbed, varied greatly with the nutrient treatments and to a lesser degree with the cultural system. Each of the respective treatments indicated that the omission of any one of the elements nitrogen, phosphorus, and potassium from the complete fertilizer is followed by a decreased absorption of the remaining elements. This decreased absorption resulted in a nutritional lack of balance as exhibited in the reduction of growth and of flowering. There appeared to exist a specific ratio in which nitrogen, phosphorus, and potassium are absorbed by the trees. The optimum ratio appeared to be 6:1:4, respectively.

The amount of any ion absorbed is not strictly related to the amount present in the plant, since in certain cases the phosphate and nitrate ions are rapidly absorbed, although the plant parts concerned are very rich in these ions. The rate of absorption of ions is not related to their velocities or any other of their physical chemical properties so far as they are known. Since the absorption of ions depends upon light; temperature; humidity; concentration and reaction of the solution; the amphoteric character, amount, and kind of protein present; and the kinds of ions present; it is a complex problem to handle experimentally, and no two experiments will give the same results unless every condition is identical.

IV. TRANSPIRATION AND THE INTAKE OF SALTS

Previous to 1914 the evidence was very confusing in regard to the relationship between the rate of transpiration of a plant and the amount of solutes taken up from the soil. It was considered by Schloesing (1869), Kohl (1886), Sachs (1887), Wollney (1898), Burgerstein (1897), Pfeffer (1900), Jost (1907), and others that the amount of solute taken up by the plant from the soil is proportional to the amount of water transpired. Their conclusions, however, were in most cases based on purely theoretical grounds, although in a few cases they were based on very meager and fragmentary experimental evidence. On the other hand, Ilienkoff (1865), Fittbogen (1873), and Haberlandt (1892) concluded also from a small amount of experimental evidence that there is no relation between transpiration and the quantity of mineral nutrients absorbed from the soil. The example of plants growing in the tropics has been frequently presented to substantiate the view that the transpiration rate under the conditions existing there is very low as compared to that of the temperate regions, yet the intake of salts is apparently abundant. There seems, however, to be insufficient evidence to prove that such a low rate of transpiration exists, as is evidenced by the contentions of Haberlandt (1897), who considered that the rate of transpiration in the tropics is lower than in central Europe, and Stahl (1894), Burgerstein (1897), and Giltay (1897), who claimed that the plants in the tropical rain forests transpire more than those in a temperate region such as Europe.

Beginning with Hasselbring (1914), well-controlled and rather extensive work has been conducted by Kiesselbach (1916), McLean (1919), Muenscher (1922), and Mendiola (1922) to determine the influence of the amount of transpiration upon the quantity of minerals in the plant. The method used, in general, has been to determine the ash content of plants whose rate of transpiration has been made to vary by changing the environmental conditions.

Hasselbring (1914) working in Cuba grew tobacco in large containers both under field conditions and under the shelter of cheesecloth after the method that is used in growing certain qualities of tobacco. Under the cheesecloth shelter the total light was reduced by one-third from that of the open. The temperature of the two stations showed practically no difference, but during the daylight hours the relative humidity under the shade was higher than that in the open. The shading used in this experiment did not diminish the production of total dry matter, but the leaves of the shade-grown plants had a much greater total area than those of the plants grown in the open. The following table gives a summary of the results obtained by the experiment, the figures representing the average for six plants:

Total dry weight, grams	Total water absorbed, cubic centimeters	Total ash in plants, grams	Water absorbed per gram of ash, cubic centimeters
	Plants grown in the open		
188.4	46,344	18.3	2,548
	Plants grown in the shade		
188.1	36,187	21.1	1,718

The plants in the open transpired nearly 30 per cent more water than those growing under the shelter, while the transpiration rate per unit of leaf surface was nearly twice as great for the plants in the sun as those growing in the shade. As indicated by the table, however, the plants that transpired the greater quantity of water contained the smaller amount of ash in percentage of the total dry matter and in absolute amount. Hasselbring, therefore, concluded that the absorption of salts is independent of the absorption of the water and that an accelerated transpiration rate does not exert an accelerating effect on the entrance of salts.

Kiesselbach (1916) studied the relation of transpiration to the ash content of corn grown under different climatic conditions in soils of

Content	Popcorn			Hogue Yellow Dent corn		
	Dry greenhouse	Humid greenhouse	Ratio dry to humid	Dry greenhouse	Humid greenhouse	Ratio dry to humid
Dry weight per plant, grams.....	67.9	71.9	......	100.9	90.5	
Ash content per cent, dry basis...	12.28	11.91	100:97	13.6	12.22	100:94
Total weight of ash, grams..........	8.34	8.56	100:102	13.18	11.06	100:84
Water transpired per plant, cc.....	24,279.0	14,304.0	100:59	33,640.0	19,067.0	100:57
Water transpired per gram of ash, cc.	2,910.0	1,670.0	100:57	2,550.0	1,720.0	100:67

different degrees of fertility and moisture content. The data obtained from growing plants in a dry and a humid greenhouse may be taken as an example to represent the effect of climate upon the relation of transpiration to the intake of salts. If the evaporation in the dry greenhouse is taken as 100 per cent, then that in the humid greenhouse would be only 54 per cent. Popcorn and Hogue Yellow Dent were the two varieties used.

These data indicate that there is no relationship between the amount of water transpired and the amount of ash in the plant. Kiesselbach noted that the greater the availability of the soil solutes the greater was the total amount of solute taken into the plant and the greater the amount taken in per unit of water transpired. Less water was transpired per gram of ash content in a relatively low soil saturation than where an abundance of water was present. Kiesselbach observed considerable variation in the ash and transpiration relationships of different corn varieties but considered that there is no absolute correlation among the percentage of ash, the quantity of water transpired per gram of ash, or the transpiration per gram of dry matter, a conclusion also reached by Mendiola (1922) in regard to tobacco plants grown in Pfeffer's nutrient solution under humid and dry conditions. Milo and Black Amber sorghums transpired a slightly smaller quantity of water per gram of ash than did corn, and sunflower plants transpired much more water per gram of ash content. McLean (1919) found that leaves taken from plants growing in the tropical rain forests of Brazil showed a higher ash content than sun plants. He assumed that the transpiration rate is lower for the leaves of the rain forest and considered, therefore, that the absorption of the mineral salts is independent of foliar transpiration.

Muenscher (1922) grew barley plants for 5 weeks in Knop's solution under conditions of high and low transpiration. He reduced the transpiration rate by increasing the humidity of the atmosphere, by reducing the light intensity by shading, and by increasing the concentration of the nutrient solution. It was found that the ash content expressed in percentage of total dry weight of the plants varied but slightly, regardless of whether the plants were grown under conditions of high or low transpiration and irrespective of how transpiration was reduced. The effect of reduced transpiration upon the total ash content, however, expressed in grams, depended upon how transpiration was reduced. With a uniform concentration of nutrient solution the ash content of the barley plants varied but slightly, even though the quantity of water transpired was reduced to less than one-half by increasing the humidity of the atmosphere. In the case of plants, however, in which the transpiration was reduced to less than one-half by shading and the photosynthetic activity was also reduced, the total ash content was correspondingly reduced.

When the total transpiration was reduced by increasing the concentration of the nutrient solution, the total ash content was only slightly reduced. Muenscher considered that his results indicated that there is little or no relation between transpiration and the absorption of salts in barley plants. He considered, however, that the amount of food available for growth, in which salts are used, is a very important factor in determining the quantities in which and the rapidity with which the essential salts enter the plant. It seems to be well established that the solutes enter a plant independently of the entrance of water.

Hoagland grew two sets of barley plants under such conditions that the amount of water absorbed by one set was practically twice that absorbed by the other. The results may be summarized by the following table:

Concentration	Milliequivalents						Water absorbed, cc.
	K	Ca	Mg	NO_3	PO_4	SO_4	
Initial	4.5	8.0	4.4	11.4	1.3	7.2	
Final	1.5	12.5	7.3	9.5	0.4	18.5	505
Final	3.2	9.6	5.2	9.7	1.4	13.1	265

The cultural solution thus became more concentrated with respect to certain ions and more dilute with respect to others as the absorption of water was increased. The ions that regularly increased in concentration were calcium, magnesium, and sulphate, while potassium and the nitrate ions decreased in concentration. The ions thus are not removed from the solution by the plant in the same proportion as water. Under some conditions at least, potassium, nitrate, and phosphate ions may be removed from the solution at a greater rate than water, while at the same time calcium, magnesium, and sulphate ions may be removed less rapidly than water. On this account, therefore, as the plant absorbs water and ions, the solution may become more dilute with respect to certain ions and more concentrated with respect to others.

A simple observation by Scofield (1927) will serve to illustrate this point further. He found that barley and wheat seedlings absorbed during a 24-hr. period a quantity of water equivalent to the original volume of the cultural solution, while at the end of that period the residual solution contained from 78 to 91 per cent of the salts originally present (see also Hoagland, 1923). Since plants do not absorb water and dissolved substances from the soil in the same proportion as these constituents occur in the solution, it is difficult to understand how transpiration may in any way influence the amount of salts in the plant. It is considered by many that the water current set up through the conducting elements in the stem is a factor in carrying the dissolved salts to the leaves. Presumably then, the more rapid the transpiration, the more rapid the current and the more salts thus translocated. However, as Muenscher (1922) stated, it must be borne in mind that although an upward current of water apparently exists, the existence of a transpiration stream in the sense of mass movement of a solution in normally growing plants has not yet been demonstrated.

Gračanin (1932) concluded from studies of the plants of barley, wheat, corn, and peas that no direct relation exists between the absorption of ions and the rate of transpiration.

Freeland (1935, 1936) studied the relationship between the amount of transpiration and the amount of various elements in the tomato, bean, *Coleus*, and sunflower. These plants were grown in sand, watered with Shive's nutrient solution. The rate of transpiration and absorption of water were varied by growing one set of plants under conditions of low relative humidity and the other set in air with high relative humidity. At the conclusion of the experiment, the plants were divided into leaves, stem, and roots. For the tomato the amounts of water absorbed from the two environments were 7,798 cc. in the low humidity and 3,087 cc. in the high humidity. The total quantity of ash, calcium, phosphorus, and potassium was greater in the plants that had the greater amount of transpiration. This increase in mineral content was mainly in the leaves and roots. In the plants of *Coleus*, the amounts of water absorbed were 19,430 cc. in low humidity and 4,679 cc. in high humidity. The ash, calcium, phosphorus, and potassium were greater in all three parts of the plants with higher rates of transpiration. Under the latter conditions, the potassium showed the greatest increase in the leaves, while the increase of calcium was greatest in the roots. Practically the same results were obtained for bean as for *Coleus*. However, the increase in the mineral elements and ash content was seldom directly proportional to the increase in dry weight or to the increase in water absorption. The data show, however, that low humidity, which increased the absorption of water from two to four times, resulted directly or indirectly in an increase in absorption of mineral salts. It is evident that available food and growth cannot account for this increase in mineral content. Freeland believed that the rate of movement of water through the plant may result in differences in the translocation of certain mineral ions. It is not possible, however, to ascribe the translocation results solely to the different amounts of water moving through the plants under high and low humidities, since there were some structural differences between the plants grown in these two environments.

V. NUTRIENT SOLUTIONS

It is impossible to study in any definite manner the nutrient requirements of a plant when it is growing in the soil, since by present known methods the soil solution cannot be controlled, analyzed, or varied with any degree of satisfaction. A water solution offers the best known means to study in a quantitative way the nutritional needs of the plant. Nearly all the cereals and beans, peas, buckwheat, and many other plants lend themselves favorably to growth in such solutions, in spite of the unusual conditions in which their roots are placed. The solutions that are thus used have been termed "nutrient" or "cultural solutions" or "water cultures." When plants are grown in pure sand to which the solution has been added, the medium is called a "sand culture." For quantitative work, however, the water culture has proved the most satisfactory. A nutrient or cultural solution is generally defined as one that contains all the essential elements in the form of salts that a plant absorbs from the soil. A solution that does not contain all the essential elements is sometimes termed a cultural solution, but it is generally inferred when the term is used that the solution contains all the known essential elements that the plant obtains from the soil.

A. Historical

The earliest recorded experiment with water cultures was that of Woodward (1699) who grew plants in spring, river, rain, and distilled water to determine whether it was the water or the solid particles of the soil that nourished the plants. Very little additional study, however, was made in this field until Sachs, Knop, and Nobbe from 1859 to 1865 developed the general procedure in water cultures that is in use at the present time. Sachs (1860) published the first standard formula for a culture solution for plants, and in 1865 Knop proposed a solution that has been one of the most widely used in the study of plant nutrition. Other nutrient solutions have been proposed by Tollens (1882), Schimper (1890), Pfeffer (1900), Crone (1902), Tottingham (1914), Shive (1915), Hoagland (1920), and many others. From 1859 to 1900 the preparation of nutrient solutions and the methods for their use were well established. By their use during that period 10 of the essential elements for plant growth were determined. It was found that the concentration of the standard solutions that were used could vary in concentration of total salts from 0.1 to 0.6 per cent and yet produce an optimum growth of the plants provided that the solutions were frequently renewed. The proportions of the different components of these solutions varied greatly, and it was considered that these had little apparent influence upon the growth and development of the plant, provided that a sufficient quantity of salt was present for the plant's needs.

From 1900 to the present time the investigations of cultural solutions have been characterized by a study of the effects of the physical as well as the chemical properties of the nutrient solutions upon the growth of the plant (Livingston, 1900). The osmotic properties of the solutions have been especially considered. The relationship to growth of the proportion of salts in the solution has been extensively studied, as has also the effect of light and temperature upon the relative value of different solutions for plant development. It has also been found that the unequal absorption of ions of a given salt may have a marked effect upon the reaction of the nutrient solution. Finally, a thorough reinvestigation is in progress to determine the number of elements that are essential to the life of the plant and the function they subserve in its metabolism.

A concise, reliable history of water cultures was published by Johnston (1932), and Sherman (1933) gives a résumé of the theories of plant nutrition from Aristotle to Liebig.

B. Kinds

It is the intention to present here only a few of the many formulas for nutrient solutions. Others may be obtained by consulting Tottingham (1914), Tollens (1882), and Schimper (1890).

Sach's solution (1860), grams		Knop's solution (1865), grams		Pfeffer's solution (1900), grams		Crone's solution (1902), grams	
KNO_3	1.00	$Ca(NO_3)_2$	0.8	$Ca(NO_3)_2$	0.8	KNO_3	1.00
$Ca_3(PO_4)_2$	0.50	KNO_3	0.2	KNO_3	0.2	$Ca_3(PO_4)_2$	0.25
$MgSO_4$	0.50	KH_2PO_4	0.2	$MgSO_4$	0.2	$MgSO_4$	0.25
$CaSO_4$	0.50	$MgSO_4$	0.2	KH_2PO_4	0.2	$CaSO_4$	0.25
$NaCl$	0.25	$FePO_4$	Trace	KCl	0.2	$FePO_4$	0.25
$FeSO_4$	Trace	H_2O	1,000 cc.	$FeCl_3$	Small amount	H_2O	1,000 cc.
H_2O	1,000 cc.			H_2O	1,000 cc.		

The nutrient solutions used prior to 1914 varied greatly in their total concentration and in the proportion of the ions present (Tottingham, 1914). The total concentration of salts ranged from 0.1 to 0.5 per cent, and the ratio of potassium to magnesium varied from 1:1 to 5:1, that of calcium to magnesium 2:1 to 3:1, that of nitrate to phosphate from 3:1 to 10:1, and that of nitrate to sulphate from 1.3:1 to 7:1.

In 1914, Tottingham, using Knop's solution as a basis, began an investigation to determine the influence of a wide range of variations in the proportions of the components of this solution. Three different total concentrations were used, having on a percentage basis a concentration of 0.01, 0.6, and 2 per cent and an osmotic pressure of 0.05, 2.5, and 8.15 atmospheres, respectively. In each of these concentrations the approximate osmotic pressure was divided into 10 equal parts, and the 10 parts were distributed among the four component salts of the solution. This arrangement gives for any given concentration of solution 84 solutions, each having the same osmotic pressure but no two of them containing the same relative proportion of salts. Tottingham found that the solution having the total concentration of 0.6 per cent or 2.5 atmospheres osmotic pressure gave the best growth for wheat in the seedling stage. The best combination of the 84 solutions in this series as judged by the dry weight of tops produced in the experiment was *T3R1S4*, in which the salts were distributed as follows:

Tottingham's best combination of Knop's solution

$$KH_2PO_4 \longrightarrow 0.0130M \qquad Ca(NO_3)_2 \longrightarrow 0.0144M$$
$$MgSO_4 \longrightarrow 0.0145M \qquad KNO_3 \longrightarrow 0.0049M$$

For iron, 2 drops of a suspension containing 0.0024 gr. of ferric phosphate per cubic centimeter was added to each liter of solution at the time of renewal. The nutrient solutions used prior to 1915 contained, excluding the iron, at least four different salts.

When the effect upon growth of the various proportions of these salts in solutions of this type is to be studied, the number of possible combinations becomes large and thus renders the experiment very tedious and cumbersome. Thus, as has been mentioned, if the osmotic pressure of a four-salt solution is distributed in tenths, there are 84 different combinations possible. If the number of salts in a solution could be reduced to three and the osmotic pressure distributed in tenths, the experiment would employ only 36 combinations. If this type of solution could be used, the experimental work would be greatly simplified and more rapid determinations could be made. If a two-salt solution could be used, only nine combinations would be possible on the basis that has just been mentioned. This, however, is impossible, since six elements at least are required in every nutrient solution and only two of these can be carried by any single salt.

In 1915 Shive carried on experiments with a three-salt solution. The potassium nitrate was omitted from the Knop's formula, the solution being composed of the three salts: monopotassium phosphate, calcium nitrate, and magnesium sulphate and a trace of ferric phosphate, thus containing all the elements of the four-salt solution. Shive experimented with three different total concentrations of this three-salt solution, the osmotic pressures of which amounted to 0.1, 1.75, and 4.0 atmospheres, respectively. For each of these three total concentrations, 36 different proportions of the three component salts were tested. It was found that the solution with an osmotic pressure of 1.75 atmospheres was the optimum one of the three for the growth of wheat and buckwheat plants. The production of dry matter by plants growing in this solution was equal to or greater than that produced by plants growing in any of the four-salt solutions. For the growth of young wheat plants, Shive found that the solutions *R5S2* and *R3S3* were the best as determined by the production of dry matter. The following are the formulas for these best three-salt solutions:

Best solution[1]		Next best solution[1]	
R5S2		*R3S3*	
KH_2PO_4	0.0180*M*	KH_2PO_4	0.0108*M*
$Ca(NO_3)_2$	0.0052*M*	$Ca(NO_3)_2$	0.0078*M*
$MgSO_4$	0.0150*M*	$MgSO_4$	0.0020*M*

[1] Ferric phosphate 0.0044 g. per liter of solution.

According to Loomis and Shull (1937), stock solutions for Shive's *R5S2* nutrient solution may be prepared as follows:

Salt	G. in 1 l.
KH_2PO_4	49.01
$Ca(NO_3)_2.4H_2O$	24.56
$MgSO_4.7H_2O$	73.95

Fifty cubic centimeters of each of the three stock solutions, plus 4 or 5 mg. of ferric phosphate, made to 1 l. form a full nutrient solution. When repurified salts are used, it is necessary to add some of the minor elements to obtain the proper development of plants. Ten cubic centimeters of a stock solution consisting of 0.10 g. manganese chloride, 0.05 g. zinc chloride, 0.05 g. boric acid, 0.01 g. copper chloride in 1 l. of water, are added at frequent intervals to the cultural solution.

Because of the simplicity of their preparation and the ease of manipulation in experiments, three-salt solutions of various types have practically superseded the four-salt solutions in the study of plant nutrition. A three-salt solution, however, has the disadvantage that if the proportion of one ion or radical of a salt is altered, the other ion of the salt is changed in a like proportion, so that the effects of an individual ion cannot be so well studied as in a four-salt solution. Six types of three-salt solutions are possible in order to supply potassium, phosphorus, calcium, magnesium, sulphur, and nitrogen to plants. These have been numbered and catalogued by Livingston and Tottingham (1918) and are now commonly called by their serial number when discussed in experimental work in plant nutrition. They are arranged as follows:

I	II	III	IV	V	VI
$Ca(NO_3)_2$	$Ca(NO_3)_2$	$Ca(H_2PO_4)_2$	$Ca(H_2PO_4)_2$	$CaSO_4$	$CaSO_4$
KH_2PO_4	K_2SO_4	KNO_3	K_2SO_4	KNO_3	KH_2PO_4
$MgSO_4$	$Mg(H_2PO_4)_2$	$MgSO_4$	$Mg(NO_3)_2$	$Mg(H_2PO_4)_2$	$Mg(NO_3)_2$

Solution III was tested by Livingston and Tottingham and found to be as good as solution I (Shive's) for the growth of wheat seedlings.

In the soil, many of the plant nutrients are present only in relatively small amounts in the soluble form, although there is generally a considerable reserve in the form of compounds that are only slightly soluble. In order to have a nutrient solution in which the available nutrients would more nearly approximate the concentration in the soil, Duggar (1920) used as nearly as possible salts that are quite insoluble. As a source for calcium he used calcium sulphate, calcium carbonate, tricalcium phosphate, monocalcium phosphate; as a source for magnesium, magnesium carbonate and trimagnesium phosphate were used, and iron was supplied in the form of iron phosphate and ferric citrate. The plants grew in the cultures with these salts just as vigorously as when the more soluble salts were used. Hoagland (1919) used various concentrations of nutrient solutions that were so constituted as to give approximately the same relation between the more important ions as that found in the water extract of soils at the time when the crop was actively absorbing. It was found that solutions with total concentrations ranging from 0.48 atmosphere to 1.45 atmospheres gave very similar results if renewal was frequent. The following solution was found by Hoagland (1920) to be as favorable for the growth of young barley plants as Shive's *R5S2* solution:

Ions	P.p.m.	Ions	P.p.m.
K	190	PO_4	117
Ca	172	NO_3	700
Mg	52	SO_4	202

C. Technique

It is the intention here to discuss only in a broad and general manner the methods of procedure in experimental work with nutrient solutions. The detailed directions can be found by consulting Livingston, Crocker, and Kellerman (1919) and the other references mentioned herein. Various methods can be used for obtaining the seedlings necessary for the experiment, the method of procedure depending to a great extent upon the kind of seeds that are to be germinated. For the cereals and some other seeds the following method has given very satisfactory results: The seeds are soaked in water for 5 or 6 hr. and then placed upon a germination net, which consists of ordinary mosquito netting that has been dipped quickly into melted paraffin of a high melting point so that the threadwork of the network is covered but the meshes remain open. The treated netting is tied as tightly as possible over the top of any wide-mouthed vessel or jar so that when the jar is filled with water or solution the surface practically coincides with the plane of the net. For ordinary demonstration experiments tap water may be used in the germinating vessel but for exact work a cultural solution should be used in order to avoid too abrupt a change when transferred from the germination container to the cultural vessel. The solution in the germination vessel should be changed frequently, preferably daily, by adding fresh solution through a glass tube that extends to the bottom of the container. The control of the temperature during germination is desirable and may be obtained by a water bath or any other means available to the experimenter. The germination jar should be placed in diffuse daylight until the seedlings have reached the desired stage of growth (see also Livingston, 1906). After the choice of the seedlings for the cultural experiment has been made, they are lifted from the net, preferably by paraffined forceps, and placed in a glass vessel containing a germination solution 2 cm. in depth so that the roots are in the solutions and the shoots above its surface. The seedlings are then taken from the vessel one by one as needed to be placed in position in the cultural jars, care being taken to avoid contamination.

The best cultural vessels are wide-mouthed glass jars, the ordinary Mason fruit jars of pint, quart, or 2-qt. capacity, being especially adapted for that purpose. Many different arrangements have been used for closing the cultural vessels and for supporting the plants growing therein. One of the most common methods is to use flat corks about ½ in. thick which have been thoroughly impregnated and lightly coated with paraffin. The desired number of holes about 3 cm. in diameter are made through the cork at convenient locations, as described by Tottingham (1914). The young plants are placed in position in these holes and held firmly just above the seed by means of ordinary cotton, which remains dry and becomes compressed as the stem enlarges. The cultural vessels should be filled with nutrient solution so that when the corks are tightly forced into position they will be within 1 to 2 cm. of the surface of the solution (see Livingston, 1906). If the growing plants require support, uprights of light glass or wooden rods anchored to the cork may be used. It is desirable that the uprights be light but substantial and easily attached, and they interfere as little as possible with the handling of the cultural vessels. McHargue (1923) has devised two methods of supports for the larger sand- or water-cultural vessels. In one case, four sockets made from galvanized iron into which wooden uprights may be fitted are placed equidistant around the top of the jar and held tightly in place by means of wire tightly twisted around it. In the other case, the earthenware cover is perforated in the central portion in numerous places in which the plants are held by cotton, and around the edge of the cover equidistant from each other are four holes in which light wooden supports can be tightly fitted. Perforated paraffin disks may also be used, according to Hammett (1928).

In order to make a comparative study of water cultures it is necessary that all plants be exposed to as nearly the same atmospheric conditions as possible. If the cultures are grouped in a given area, the plants shade each other so that they do not all receive the same amount of light. The transpiration loss of some plants is thus reduced and perhaps also their photosynthetic rate. The unequal atmospheric conditions may be overcome in part by changing the position of the plants at numerous intervals. This, however, requires close attention and a great amount of labor and is at the best unsatisfactory. One of the most convenient methods so far devised for placing the cultures under the same conditions is that used by Shive (1915). The plants are placed on the edge of a slowly revolving circular table approximately 4 ft. in diameter which makes a revolution every 4 min. and is continuously rotated during the time of the experiment except when stopped for short intervals for the renewal of solutions. By this means all the plants in the cultural solution are exposed to approximately the same light, heat, and cultural conditions. Trelease (1925) has devised a chamber with thermostatic control and a rotating table for plant cultures.

Gericke and Travernetti (1936) reported an experiment with tomatoes that is of marked importance in the practical application of cultural solutions. They used sheet-metal tanks, each of which was 2.5 ft. wide, 10 ft. long, and 8 in. deep; over them were wooden frames containing a wire screen that was held just above the top of the tanks. Two to three inches of excelsior and sawdust were placed on the screen to provide support for the plants. Of the five tanks, four were heated by means of electrical equipment and one was used as a control. In two of the four tanks the temperature was kept between 75 and 80°F. for the first 6 weeks and then raised to between 80 to 85°F.; the other two tanks were kept at 65 to 70°F., and, after 6 weeks, at 70 to 75°F. The control tank was kept at 55 to 65°F.

The tanks were filled with water to within 1 in. of the top, and each tank received 1 lb. of nutrient salts at the beginning of the experiment. During the 12-month period, six additional applications of salt were made per tank. The plants were started in soil and transplanted as 3-week-old seedlings. Two rows of 10 plants each were placed in each tank, thus giving to each plant an area of about 1.25 sq. ft. The plants in the heated tanks began to ripen fruit about 4 months after transplanting, while those in the unheated tank did not ripen fruit until 6 weeks later. In the heated tanks, the plants grew to a height of 25 ft. and produced fruit for at least 8 months.

The plants in the two tanks that were heated at 70 to 75°F. produced a total of 598.3 lb. of fruit, while those heated at 80 to 85°F. produced 625.6 lb. The control produced only 57.0 lb. of ripe fruit. The authors considered that on 100 sq. ft. of basin area 1 ton of tomatoes could be grown in 12 months, provided that ample room were provided for upward growth.

1. Renewal.—A common practice in water-culture experiments is to renew the solution at intervals of 3 days or, more generally, to remove the plants from the solution and place them in a fresh solution in clean containers. The vessels that have been used are emptied, cleaned, and then refilled at the next change. This method, however, is subject to criticism because it introduces several errors that could be entirely or at least partially overcome if the nutrient solution could be continually renewed. In the first place, if the nutrient solution is relatively dilute, the lack of an optimum amount of a given nutrient might be a limiting factor to growth if the solution were not renewed except at the intervals that have been mentioned. If a study is being made of the effect of a certain strength of solution upon plant growth, it is evident that a certain strength of concentration will not long continue unless a constant renewal is made, since the loss of water by transpiration continually increases the concentration of certain salts, at least in the solution, because they are not absorbed in the same relative amount as the water. In the third place, the influence of the organic compounds that arise from the cells that are lost from the root cap and the epidermis may have a detrimental influence upon the growing plants if allowed to accumulate for any period of time.

The depletion of oxygen and accumulation of carbon dioxide in the solution due to the respiration of the roots are factors to be considered as well as the disturbance of the physiological balance of the solution due to the unequal absorption of salts. Stiles

(1915) observed with rye plants growing in cultural solutions that if the solutions were changed frequently so as to maintain more nearly a constant composition, the solution could vary over a considerable range without producing much effect upon the rate of growth as measured by the amount of dry matter produced within a given time. In the case of barley and wheat seedlings (Brenchley, 1916) growing in nutrient solutions, the improvement in growth was more marked in concentrations up to fifth normal when the solutions were changed frequently. Trelease and Free (1917) investigated the effect of the renewal of solution upon the production of dry weight of the tops of wheat growing in Shive's best solution. The cultures ran for 41 days, and some of the results obtained were as follows:

Change	Dry weight, grams	Change	Dry weight, grams
Daily	1.243	After 2 weeks, then every 3 days	1.131
Every 3 days	1.012	After 1 month	0.654
After 1 week, then every 3 days	0.995	Not at all	0.621
Every 2 weeks	0.780		

In another experiment the culture solution was allowed to flow at the rate of 1 l. per day through the cultural jars, which had a capacity of 250 cc. The results obtained were as follows:

	Dry weight, grams
Continual renewal	1.678
Changed every 3 days	1.222
Not changed at all	0.666

Soybeans, grown in sand or water cultures with a continuous solution renewal of 1 l. of new solution per culture during each 24-hr. interval throughout the growing period, always produced plants that were superior in every respect to those grown in cultures with intermittent solution renewal (Allison and Shive, 1923).

It was shown by Marsh (1935) that plants grew better and more nearly uniform in cultural solutions if the excess solution was drained from the bottom of the container rather than siphoned from near the top of the cultural jar. This is apparently due to the elimination of precipitates and plant wastes that ordinarily collect near the bottom. Methods for supplying a continuous renewal of nutrient solution for water cultures have been described by Clark (1916), Trelease and Livington (1922), Johnston (1927), and Trelease and Thomson (1935), and methods for renewal of plant nutrients in sand cultures have been published by McCall (1916), Zurbicki (1933), and Mehrlich (1935).

2. Aeration.—Various methods have been devised for the aeration of water and sand cultures, some of which are described by Weatherwax (1916), Andrews (1920), and Allison (1921). One method that is frequently used is to pull a stream of air through the medium by use of a filter pump. Another is the drip method in which the cultural solution is replaced drop by drop, the fresh air supply being carried in, with, or by the drops. In the third case, partial aeration, at least, is secured by allowing a slow but steady flow of renewal solution into the cultural vessel. The benefits derived from aeration as measured by the amount of dry matter produced, height of plants or their general vigor have been rather varied, which, however, is to be expected, since plants differ greatly in their oxygen demands when growing under natural conditions. Hall, Brenchley, and Underwood (1914) greatly increased the root and top growth of barley and lupine in water cultures by pulling a stream of air through the medium. Free (1917), however, in experiments with buckwheat growing in Shive's nutrient solution found that the degree of aeration of the cultural solution had no important influence on the growth of these plants. His results were verified by Stiles and Jorgensen (1917), who found that aeration of the nutrient solution increased the rate of growth of barley growing in water cultures but that it had no effect on the growth of buckwheat under the same conditions. Pember (1917) noted that barley did not respond to aeration when grown in nutrient solutions that were renewed every two weeks. In experiments with corn plants, Beal (1918) and Andrews and Beal (1919) obtained much better growth in the aerated sand and water cultures than in the unaerated ones. In one case the plants growing in the aerated cultures were 65 cm. in height as compared to a height of 46 cm. for the plants in the unaerated vessels. Aeration showed its effect upon the anatomy of the plant, since the intercellular spaces were smaller and the tissues more compact in those plants growing in the aerated cultures. Oats, mustard, and peas growing in nutrient solutions respond very favorably to aeration, as is shown by the work of Andrews (1921), who found that the height of plants growing in aerated solutions was twice as much as the dry weight, in some cases five times greater than plants growing in solutions that were not aerated. In the case of soybeans, Allison (1921) and Allison and Shive (1923) stated that the best plants with respect to health, vigor, and yield were always secured in the cultures with continuous solution renewal and with constant and thorough aeration of the medium. They also found that the continuous renewal of the nutrient solution did not alone maintain the supply of dissolved oxygen necessary for the maximum development of the plants and that the roots showed a much greater response to aeration than the aerial portions of the plants.

Loehwing (1932) found that the ratio of the weight of tops to the weight of roots of sunflower plants, grown in aerated sand cultures, was much lower than that of the controls. The aerated plants not only showed accelerated growth, greater height, and larger tops and roots than did the controls, but they contained more calcium, phosphorus, sugars, hydrolyzable polysaccharides, and total nitrogen. The expressed sap of the aerated plants contained more total solids and sugars but had less ash and total nonnitrate nitrogen than the nonaerated ones.

Clark and Shive (1932) noted that the roots of the tomato plant grew throughout the container in the aerated nutrient solutions, while those of the nonaerated cultures grew only near the surface. They found that aeration had a more pronounced influence upon the growth of tops of this plant than upon the growth of the roots. The plants in the nonaerated cultures bloomed and fruited earlier than did those in the aerated cultures.

Bryant (1934), in studies of the roots of barley plants grown in aerated and nonaerated water cultures, found that after 60 days the nonaerated plants had 225 roots, while those in the aerated solution had 75 roots. The average length of the roots in the aerated cultures was 37.4 cm., as compared to 10.9 cm. for the nonaerated and the roots of the former were about 15 per cent greater in diameter than those of the latter. The cortex of the roots grown in the aerated cultures consisted of compact parenchyma with few and small intercellular spaces, while that of the roots grown in the nonaerated solution showed large air passages separated by narrow strands of parenchyma. The tissues of the nonaerated roots began to differentiate nearer the root tips than did those in the aerated cultures. In the mature regions of the aerated roots the cell walls were approximately twice as thick as the corresponding cell walls of the roots grown in the nonaerated solutions.

3. Preparation.—Since the method to be followed in the preparation of nutrient solutions will depend upon the type of solution to be used and the nature of the experiment, it will be impossible to give here anything but general methods of procedure. This perhaps can be best accomplished by giving in detail the methods to be followed in one or two specific cases. The first step in the preparation of nutrient solutions, especially if quantitative work is to be done, is to obtain salts of known purity either dry or containing a known amount of water. The next step is the preparation of stock solutions from which solutions of the desired composition and concentration may be made when desired. One of the greatest difficulties that has been experienced in the preparation of four-salt solutions and to some extent in three-salt solutions is to obtain a solution of the desired concentration and proportions and avoid any precipitate. In ordinary demonstration experiments with water or sand cultures, the formation of some precipitate is of small matter, but in a quantitative study it becomes an important factor and must be avoided. Tottingham (1914) in his work with Knop's solution found it best to employ the component salts in stock solutions, one part, *A*, containing only calcium nitrate and the other, *B*, containing the other three salts. The maximum possible concentration without precipitation of part *B* for a period of 14 days at temperatures of 20 to 23°C. was found to be 10 per cent when monopotassium phosphate was used and 9 per cent when dipotassium phosphate was used. This limits the concentration of solution *A* to 12.33 and 12 per cent, respectively. Tottingham found the maximum possible concentration of the complete Knop's solution without precipitation for a period of from 3 to 4 days at a temperature of 20 to 26°C. to be 2.8 and 2.2 per cent when monopotassium phosphate and dipotassium phosphate were used, respectively. It was found that the precipitate formed in unstable Knop's solution consisted almost entirely of calcium sulphate.

For a supply of iron a suspension of ferric phosphate of approximately 0.02 g. molecular concentration is generally recommended. A suspension of this type contains about 3 g. of ferric phosphate per liter or 3 mg. per cubic centimeter. Enough of this preparation may be made to last for several years. Enough of this suspension should be added to each liter jar at each filling to give 3 mg. of ferric phosphate.

Of interest in this regard is the work of Horner, Burk, and Hoover (1934) in devising methods for the preparation of synthetic and natural humates containing iron, aluminum, manganese, zinc, nickel, and copper. One or two applications of those humates will ordinarily supply sufficient iron, or similar metal, to water or sand sultures to last throughout the entire period of growth. The metals in the form of humates are stable in alkaline, neutral, and moderately acid media, and are not precipitated by phosphates. The general consideration of the iron supply in nutrient solutions is discussed under the heading of Iron in Chap. VI.

Five characteristics of a nutrient solution may be distinguished when its osmotic, ionic, and molecular relationships to plant response are considered: (1) the solution type or kinds of salt entering into the solution; (2) osmotic salt proportions, the division of the osmotic pressure of the solution into a definite number of parts, and their distribution among the salts in all the various proportions; (3) volume molecular salt proportions; when considered in its relationship to osmotic salt proportions it would mean that part of a volume molecular concentration of each salt which must be added to each solution to produce the osmotic proportion that is desired for that particular salt; (4) total volume molecular concentration or the total number of molecules of all kinds per liter; and (5) the total osmotic concentration, which is the resultant of the number of molecules and ions in the solution per liter. This calculated osmotic value or potential-osmotic pressure is commonly expressed in terms of atmospheres. Owing to ionization and in some cases also to hydration, the osmotic salt proportions and the volume molecular salt proportions of a solution will not vary proportionally from solution to solution. If a series of solutions are planned, therefore, so as to vary by definite increments of osmotic salt proportions, the volume molecular concentration will vary irregularly. The preparation of nutrient solutions for the study of their osmotic and salt relationships to plant growth, therefore, involves a rather tedious procedure and entails much care and labor if accurate and satisfactory results are to be obtained. In the case of a four-salt solution, if the osmotic pressure is divided into ten equal parts and distributed among the four salts in all the various proportions, 84 different solutions must be prepared. With a three-salt solution the same procedure requires the preparation of 36 different solutions. Let us here consider a yet more simple case in which the total osmotic pressure of 1 atmosphere of a three-salt solution of type I is distributed in eighths among the three component salts monopotassium phosphate, calcium nitrate, and magnesium sulphate in all the possible proportions. The molecular proportions of each of the salts and the partial molecular concentrations necessary to produce an osmotic pressure of 1.0 atmosphere for the 21 different solutions that are possible under this combination are given in the following table. This is taken from the *Report of the Committee of Biology and Agriculture of the National Research Council* (1919).

The method by which the figures presented in this table were obtained for each solution was somewhat as follows: For example, in the case of solution *R*1*S*1, a solution was prepared containing ⅛ g. molecule of monopotassium phosphate, ⅛ g. molecule of calcium nitrate, and

⅝ g. molecule of magnesium sulphate. This solution was then diluted until its freezing point indicated a total osmotic pressure of 1.0 atmosphere. In a solution that has thus been diluted, the molecular proportion of each salt can easily be calculated from the original volume, and since

PARTIAL MOLECULAR CONCENTRATIONS AND MOLECULAR PROPORTIONS (OSMOTIC PROPORTIONS) OF KH_2PO_4, $Ca(NO_3)_2$, AND $MgSO_4$ (TYPE I) IN 21 SOLUTIONS ALL HAVING A CALCULATED OSMOTIC VALUE OF APPROXIMATELY 1.00 ATMOSPHERE AT 25°C. BUT DIFFERING BY INCREMENTS OF ONE-EIGHTH IN SALT PROPORTIONS

Solution	Molecular proportions			Partial molecular concentrations necessary to produce the fractional pressures desired		
	KH_2PO_4	$Ca(NO_3)_2$	$MgSO_4$	KH_2PO_4	$Ca(NO_3)_2$	$MgSO_4$
R1S1	1	1	6	0.0027	0.0027	0.0161
S2	1	2	5	0.0025	0.0049	0.0123
S3	1	3	4	0.0024	0.0071	0.0094
S4	1	4	3	0.0022	0.0089	0.0067
S5	1	5	2	0.0022	0.0108	0.0043
S6	1	6	1	0.0020	0.0122	0.0120
R2S1	2	1	5	0.0053	0.0027	0.0132
S2	2	2	4	0.0049	0.0049	0.0099
S3	2	3	3	0.0047	0.0071	0.0071
S4	2	4	2	0.0045	0.0090	0.0045
S5	2	5	1	0.0041	0.0104	0.0021
R3S1	3	1	4	0.0076	0.0025	0.0101
S2	3	2	3	0.0072	0.0048	0.0072
S3	3	3	2	0.0068	0.0068	0.0045
S4	3	4	1	0.0065	0.0086	0.0021
R4S1	4	1	3	0.0099	0.0025	0.0074
S2	4	2	2	0.0094	0.0047	0.0047
S3	4	3	1	0.0090	0.0068	0.0022
R5S1	5	1	2	0.0123	0.0024	0.0049
S2	5	2	1	0.0118	0.0047	0.0023
R6S1	6	1	1	0.0145	0.0024	0.0024

the degree of ionization of each of the salts at that dilution can be obtained from prepared tables, the fractional molecular concentration of each salt necessary to produce the fractional osmotic pressure desired can thus be calculated. If, for example, at a given concentration the dissociation value is 50 per cent for a given salt, then the osmotic value of that salt is 1.5 times its molecular concentration, so that the molecular concentration divided by 1.5 would give the molecular concentration necessary to produce the fractional osmotic pressure that is desired (see table above).

In the next table are given the strength of stock solutions and the proportion of distilled water to make 1,000 cc. of each of the 21 nutrient solutions that have just been described.

NUMBER OF CUBIC CENTIMETERS OF DISTILLED WATER, AND OF THE SINGLE SALT-STOCK SOLUTIONS REQUIRED TO MAKE 1 L. OF EACH OF THE 21 NUTRIENT SOLUTIONS OF TYPE I WITH A TOTAL CONCENTRATION OF 1.00 ATMOSPHERE
Adapted from recommendations of Research Council, 1919

Solution	H_2O	KH_2PO_4, $\frac{M}{3.5}$	$Ca(NO_3)_2.4H_2O$, $\frac{M}{3.5}$	$MgSO_4.7H_2O$, $\frac{M}{3.5}$
*R*1*S*1	924.7	9.45	9.45	56.4
*S*2	931.1	8.75	17.15	43.0
*S*3	935.85	8.4	24.85	32.9
*S*4	937.7	7.7	31.15	23.45
*S*5	939.45	7.7	37.8	15.05
*S*6	943.3	7.0	42.7	7.0
*R*2*S*1	925.8	18.55	9.45	46.2
*S*2	931.05	17.15	17.15	34.65
*S*3	933.85	16.45	24.85	24.85
*S*4	937.0	15.75	31.50	15.75
*S*5	941.9	14.35	36.4	7.35
*R*3*S*1	929.3	26.6	8.75	35.35
*S*2	932.8	25.2	16.80	25.20
*S*3	936.55	23.8	23.80	15.75
*S*4	939.80	22.75	30.10	7.35
*R*4*S*1	930.7	34.65	8.75	25.9
*S*2	934.2	32.9	16.45	16.45
*S*3	937.0	31.50	23.80	7.7
*R*5*S*1	931.4	43.05	8.4	17.15
*S*2	934.2	41.3	16.45	8.05
*R*6*S*1	932.45	50.75	8.4	8.4

4. The Triangle System in the Studies of Nutrient Solutions, Sand Cultures, and Fertilizer Experiments.—In the study of nutrition experiments dealing with a large number of cultures of different salt combinations it is desirable to employ some diagrammatic scheme or arrangement to represent concretely the various salt proportions and to interpret more easily the results obtained. The triangle method, which was suggested by Schreinemocher (1893) and Bancroft (1902) for certain percentage experiments in physical chemistry, was first adapted by Schreiner and Skinner (1910) to nutrition experiments. Later the method was used by Tottingham (1914), Shive (1915), and McCall (1916) in both water- and sand-culture experiments and has now become the general method for interpreting the data obtained in that type of work. A summary of the method and its various applications has been given by

Schreiner and Skinner (1918), and a critical review of it has been made by Espino (1919).

As an example of the application of the triangle system let us here consider first the distribution of the 36 different combinations that are possible when the osmotic pressure of a three-salt solution is divided into tenths and proportioned in all the possible ways among the component salts. In Fig. 11, the distribution of the various proportions of type-I solution consisting of monopotassium phosphate, magnesium sulphate, and calcium nitrate is represented on the plant of an equilateral triangle according to the arrangement now in use for all three-salt solutions. The base of the triangle is considered as the potassium side, the left side as the calcium side, and the right as the magnesium side. Each

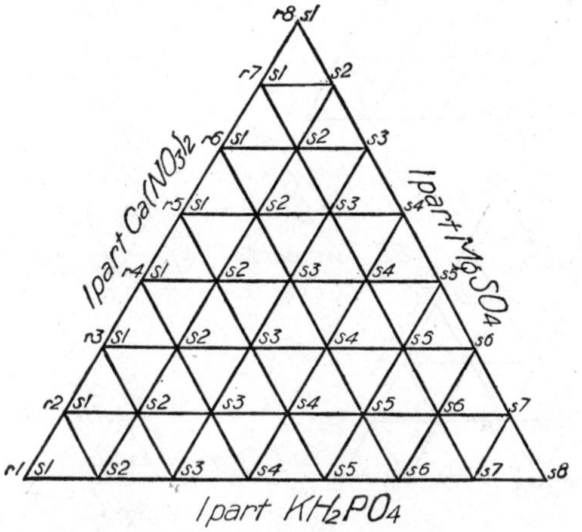

FIG. 11.—Diagram showing the 36 different combinations of a three-salt solution if the osmotic pressure is distributed in tenths among the three salts. (*Redrawn from Shive in "Physiological Researches."*) Description in the text.

of the sides of the triangle is divided into 7 equal parts and lines are drawn through each of these points parallel to each of the other two sides of the triangle. Each of the original points on the sides of the triangle and the points of intersection of the lines drawn represents one of the cultures under consideration. The base of the triangle and the lines parallel with it are numbered from the base, which is row 1 (*R*1), to the apex, which is row 8 (*R*8). The points of intersection are numbered as series (*S*) from the line on the left *S*1 to the extreme point on the right *S*8. All the cultures on the base line in the case of type-I solution have one-tenth of their osmotic pressure due to monopotassium phosphate, all the cultures on *R*2 have two-tenths of their osmotic pressure due to monopotassium phosphate, and so on until *R*8, which has eight-tenths of its osmotic pressure due to that salt. In the first line on the left of the triangle *S*1, all the cultures have one-tenth of their osmotic pressure due to calcium nitrate, while in *S*2, two-tenths of the osmotic pressure is due to that salt, and so on until *S*8, where eight-tenths of the pressure is due to the calcium nitrate. The proportion of the osmotic pressure due to magnesium sulphate at any points can be obtained by subtracting from ten the sum of the osmotic proportions of monopotas-

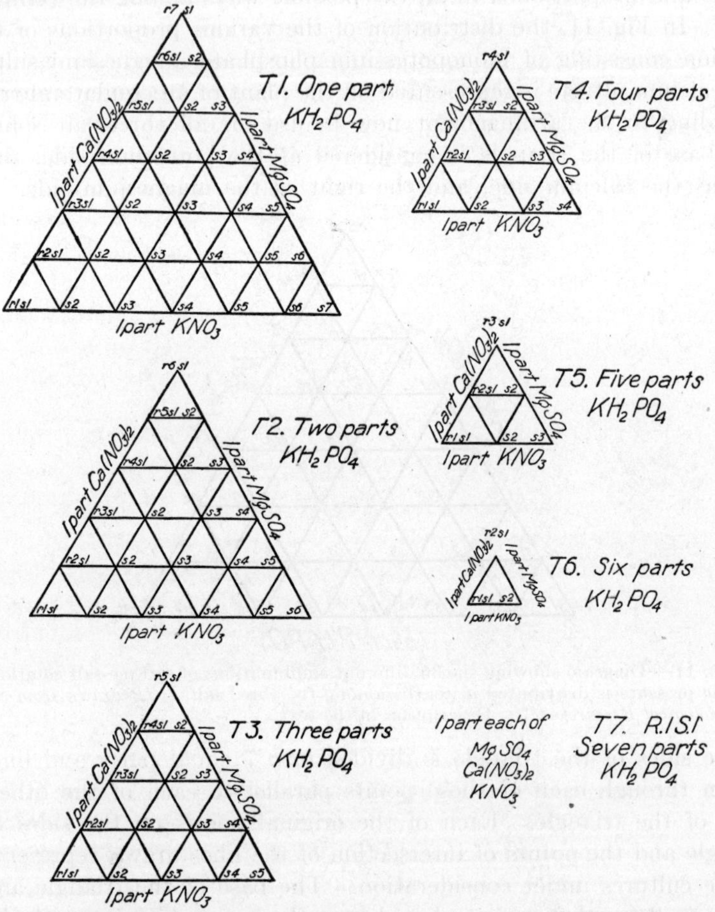

FIG. 12.—Diagrams showing the 84 different combinations of a four-salt solution if the osmotic pressure is distributed in tenths. (*Redrawn from Tottingham in "Physiological Researches."*) Description in the text.

sium phosphate and calcium nitrate for that point. Thus, for example, if a culture is at the point *R*3*S*4 it has three-tenths of its osmotic pressure due to monopotassium phosphate four-tenths to calcium nitrate, and three-tenths to magnesium sulphate. It is also customary to designate the type of solution being used by placing the type number before the symbol of concentration. Thus II*R*4*S*3 signifies that the solution in question is type II and that four-tenths of the osmotic pressure is due to potassium sulphate, three-tenths to calcium nitrate, and three-tenths to monomagnesium phosphate.

When a four-salt solution is under consideration and the osmotic pressure is distributed in tenths among the various salts, the representation by the triangle method becomes much more complicated, since 84 different combinations are possible. Figure 12 represents the schematic arrangement of the 84 combinations of Knop's solution as used by Tottingham (1914). For a procedure of this type seven triangles are necessary, one, however, forming only a single point. The triangles are designated as $T1$ to $T7$ corresponding to the tenths of the total osmotic pressure due to monopotassium phosphate. The base of each of these triangles is the potassium nitrate side, while the left and right sides are the calcium nitrate and magnesium sulphate sides, respectively. Taking any point as $T2R3S4$, for an example, we know that two-tenths of the total osmotic pressure of the nutrient solution at that location is due to monopotassium phosphate, three-tenths to potassium nitrate, four-tenths to calcium nitrate, and one-tenth to magnesium sulphate.

D. Factors That May Determine the Relative Value of Nutrient Solutions

One of the main aims in studying nutrient solutions has been to determine the best proportions, concentrations, and means of presenting the nutrient salts in order to secure the maximum response in the growth of the plants. It has frequently been stated that the aim is to determine or find the "best solution" with the idea of applying the proportions and concentrations of solutes thus found for a maximum response in agricultural practice. It is important, then, to consider some of the factors that may determine whether or not a nutrient solution will be the "best solution" for plant growth.

1. Temperature and Light.—It was reported by Gericke (1921) that the nutrient solutions which produced the best growth of Marquis wheat in the early growth phases varied with the temperatures under which the experiment was conducted. The plants were grown in the six types of three-salt nutrient solutions in which the total molecular concentration was distributed in eights among the component salts in all possible proportions. The temperatures used were 28 and 17°C., the former being considered about the optimum and the latter distinctly below the optimum for the early growth of wheat. The "good or best solutions" were considered to be those in which the growth values obtained lay within the upper one-fourth of the total range of values for the same temperature and the same solution type. The following table shows some of the best solutions for the two temperatures observed:

Temperature, degrees Centigrade	Type I KH_2PO_4 $Ca(NO_3)_2$ $MgSO_4$	Type II K_2SO_4 $Ca(NO_3)_2$ $Mg(H_2PO_4)_2$	Type III KNO_3 $Ca(H_2PO_4)_2$ $MgSO_4$	Type IV K_2SO_4 $Ca(H_2PO_4)_2$ $Mg(NO_3)_2$
28	R1S3	R1S3	R1S3	R1S4
	R1S4	R1S4	R1S4	R1S5
	R3S4	R2S1	R1S6	
	R4S1	....	R2S1	
17	R3S2	R3S3	R2S5	R4S1
	R3S3	R5S1	R3S4	R5S1
	R4S1	R5S2	R4S1	R5S2
	R5S1	....	R5S1	

These figures indicate that there is generally a marked difference between the proportions that give the best growth at higher temperatures and those that give the best growth at the lower temperatures. It appears, for example, that the proportion of potassium may be less for high than for low temperatures.

Trelease and Trelease (1926) grew wheat plants in nutrient solutions at temperatures of 14, 19, and 30°C. and studied the relation of temperature to the physiological values of these solutions as indicated by the growth of the roots. There was little or no difference at the first two temperatures, but at a temperature of 30°C. there were marked differences in the physiological values of the different solutions.

From a comparison of the relative value of a series of solutions for the growth of wheat plants under different degrees of illumination, Trelease and Livingston (1924) concluded that if a number of markedly different nutrient solutions are repeatedly compared by tests made under different climatic complexes, the relative physiological values assigned to the several solutions may be either the same or more or less different for the several tests according to the nature and magnitude of the climatic differences dealt with. They considered that the comparative physiological values of different salt solutions cannot be stated with any degree of precision without having described in an adequate manner the climatic condition under which the tests were made.

2. Concentration.—The experimental evidence in regard to the effect of the concentration of the nutrient solution on the growth of plants is confusing. Since the information is not very definite, it is the intention to present here only the results that have been reported and the criticisms that have been offered upon the question. Breazeale (1905) grew wheat seedlings in nutrient solutions with concentrations, respectively, of 15, 75, 155, 750, and 1,550 p.p.m. The seedlings were grown for 28 days and the solution changed daily. The plants increased in growth and vigor with the increasing concentration of the solutions up to 155 p.p.m., but above that point they were small and resembled those grown in the weaker solutions. Breazeale concluded that there is an optimum physical concentration of the nutrient solution in which plants thrive best aside from variations in the amounts present of the different nutrient materials. The work of Breazeale was substantiated by that of Hall, Brenchley, and Underwood (1914) who concluded from their experiments with barley, which was grown in both water and sand cultures with a standard nutrient solution of four concentrations, that the growth made by plants in the soil solution is, in the main, determined by the amount of nutrients that are available. They also considered that the concentration of the solution within certain wide limits, irrespective of the total amount of plant food available, is a factor in the rate of plant growth. Ayres (1917), Tottingham (1914), and Shive (1915) furnished evidence which indicates that the effect upon growth of any given set of salt proportions varies with the total concentration of the solution. Shive found that, in experiments with solutions of 0.01, 1.75, and 4.00 atmospheres osmotic pressure, the best growth of tops of young wheat was obtained in the solution with an osmotic pressure of 1.75 atmospheres, the poorest growth in the weakest solution, and the medium growth in the most concentrated solution. Trelease (1917) observed young wheat plants growing in a solution consisting of monopotassium phosphate, magnesium sulphate, potassium chloride, and calcium nitrate in the molecular proportions of 1.000, 0.587, 0.485 and 0.341, respectively, for these salts in which the concentration of the solution varied by six increments from 0.5 atmosphere to 7.0 atmospheres. The maximum yield of tops was obtained when the nutrient solution had a total osmotic concentration of 1.6 atmospheres. Between the concentrations of 1.6 and 7.0 atmospheres the dry weight of the tops was approximately a linear function of the concentration, decreasing as the concentration increased. When, however, the molecular proportion of the salts in the solution was changed to 1.000, 1.155, 7.282, and 0.699, the optimum concentration for growth was between 4.5 and 5.5 instead of at 1.6 atmospheres (see also Hoagland, 1919). Cameron (1911) and Stiles (1915) working with solutions similar to some of those used by the authors just mentioned were unable to obtain any marked variations in the amount of growth over a fairly wide range of concentrations. They considered that the slight differences observed might have been due to the depletion of some one element and that this, perhaps, could be overcome by the more frequent renewal of the solution.

Hoagland and Martin (1927) considered that all essential ions, except perhaps nitrate, are always present in the soil solution, and that no matter how rapidly the ion is absorbed, its concentration is not reduced to zero.

They believed that no fixed critical concentration exists for any single element. Thus they found that 1 p.p.m. of phosphate in a cultural solution was necessary to maintain a good growth of barley, while with other plants a much lower concentration would produce an adequate growth. They believed that the absorbing-power characteristic of different plants must be ascribed to their different abilities to draw indirectly on mineral supplies not initially present in the soil solution. Thus with buckwheat and barley, no difference between the two plants was manifested in their reduction of the amount of nitrate, calcium, magnesium and potassium in the soil solution. The content of those constituents, however, varied greatly, the buckwheat withdrawing from an equal volume of soil much more calcium, magnesium, phosphorus, potassium, and nitrogen than did the barley.

Nightingale and Farnham (1936) grew plants of sweet peas in two nutrient solutions with a concentration of 0.5 to 1.0 atmosphere and of 2 to 3 atmospheres. As compared with the more concentrated solution, the plants in the dilute solution grew more vigorously, were more succulent, and abscissed more buds. The leaf blades of the plants in the weaker solution were thin, somewhat circular in outline, and of a dark green color, while those of the plants in the concentrated solution were thick, oval or elliptical, and light green. The young, active cells were more abundant in the roots and tops of the plants grown in the more dilute solutions than they were in the plants of the more concentrated solution. In the former, the tissues differentiated and matured more slowly than in the latter. Gračanin (1935), in experiments with corn and wheat, found that the amount and rate of growth did not depend upon the amount of nutrient solution but upon its concentration.

The conflicting results obtained in regard to the effect of the concentration of the solution on growth may, in part at least, be due to factors other than the ones under consideration, which have been overlooked by those conducting the experiments.

3. Water vs. Sand Cultures.—A comparative study of sand and water cultures as media for plant growth has been made by Stolhane (1914), Hall, Brenchley, and Underwood (1914), McCall (1915), Shive and Martin (1918), Shive (1920), Bakke and Erdman (1923), Hoagland and Martin (1923), and others. A sand culture furnishes a medium that is apparently more nearly comparable to soil than is a water culture. It has been found by practically all investigators that the growth of roots as judged by the production of dry matter as well as by the profuse branching and general extent is much higher in sand cultures than in water cultures. This has generally been interpreted as due to better aeration of the sand cultures. In many cases, however, the best growth of tops as measured by both green and dry weight occurs in the water cultures. One of the first questions that arises is what effect the sand has upon the nutrient solution that is added to it. Shive (1920) with three-salt nutrient solutions showed that different grades of washed sand gave no evidence of the adsorption of salt or ions in sufficient amount to alter materially either the reaction or the total salt concentration of the solution in contact with the solid particles. The adsorptive capacity, however, of the unwashed sand was sufficient to reduce the total concentration of the solution from 1.76 to 1.61 atmospheres, an average reduction of 8.5 per cent, during the first 24-hr. period. The reaction of the nutrient solution was not markedly altered by its contact with the unwashed sand. By the renewal of the nutrient solution the initial adsorptive effect of the unwashed sand was soon eliminated, apparently by saturating its adsorptive capacity. As an example of the contrast between water and sand cultures the work of Bakke and Erdman (1923) may be mentioned. In the case of Marquis wheat grown to the heading stage in type III three-salt solution with a concentration of 1.00 atmosphere, they found, in general, a greater dry weight of tops produced in the water cultures and a greater dry weight of roots in the sand cultures. The solutions producing the maximum yield of tops for the water and sand cultures showed marked variations in salt proportions. The best water-culture solution was $R1S1$, while the best sand-culture solution was $R3S3$. The amount of water transpired was with but few exceptions greater from the water cultures than from the sand cultures. The sand cultures that transpired more water than the water cultures were characterized by a high proportion of monocalcium phosphate with only a small proportion of potassium nitrate and magnesium sulphate.

It was reported by Dunlap (1936) that, for the purpose of transplanting, seedlings can be grown to better advantage in sand cultures than in soil. The seedlings so grown withstand removal better than those grown in the soil, do not damp off, and can be kept for long periods of time without becoming too large for successful transplanting.

4. Physiological Balance.—It has already been noted that one of the outstanding lines of recent investigations in plant nutrition has been the attempt to determine whether certain combinations and ratios of salts or ions in nutrient solutions are more beneficial to plant growth than others.

The experimental data obtained have indicated that the question of physiological balance is an intricate and complex one, since it varies with the type and concentration of the solution, the type of plant under consideration, and the general conditions of the environment. Certain phases of this question are treated under other topics and it is the intention here to consider only a few of the experiments that have been conducted along this line.

The results obtained by Tottingham (1914) with wheat seedlings are very striking. He grew these plants in ordinary Knop's solution and in 84 different combinations of this solution in which the osmotic pressure was distributed in tenths among the component salts. He found, for example, that with a concentration of 0.01 per cent and an osmotic pressure of 0.05 atmosphere the solution producing the greatest dry weight of tops was 39 per cent superior to the ordinary Knop's solution of the same concentration. This solution contained 0.6, 3.1, 0.5, and 1.4 times as much of monopotassium phosphate, magnesium sulphate, calcium nitrate, and potassium nitrate, respectively, as occurs in the corresponding Knop's solution. In the case of roots, the best solution was 15.4 per cent superior to Knop's solution of the same total concentration. This solution contained 1.8, 0.8, 0.2, and 3.4 times as much of the respective salts as occurs in the corresponding Knop's solution. In the solutions in which the concentration was 0.6 per cent and the osmotic pressure approximately 2.5 atmospheres, the one that produced the greatest dry weight of tops was 11 per cent better than Knop's solution of the same concentration. The greatest dry weight of roots obtained in this series was 15.4 per cent better than that obtained in the same concentration of the ordinary Knop's solution.

Evidence that different plants require different physiological balances for their best development was furnished by Shive (1915, 1917). In a series of 36 combinations in a three-salt solution with an osmotic pressure of 1.75 atmospheres, it was found that the solution which gave the best growth for buckwheat was *IR4S2*, while the best solution for the growth of young plants of Marquis wheat was *IR5S2*. Johnston (1925) in a study of the relative value of the six types of three-salt solutions for the growth of Irish Cobbler potatoes found that types I and II were the best suited if the dry weight of the plants is used as a criterion.

Experimental evidence secured by Gericke (1922) showed that the pairing of ions or the type of salt in which they are made available to the plant exerts a marked influence upon growth. He grew wheat seedlings for 1 day in a solution of potassium nitrate with a concentration of 1.0 atmosphere and then removed the plants, rinsed the roots in distilled water, and placed them in a solution of monomagnesium phosphate of the same concentration. The next day the plants were transferred to a like solution of calcium sulphate and after 24 hr. again transferred to the potassium nitrate solution. It was found that the plants which were thus changed by this continued rotation at 24-hr. periods grew as large and produced approximately the same dry weight as those grown in the complete nutrient solutions of the same salts or in other three-salt solutions which supplied the same ions. It was further found that plants grown in these three single-salt solutions above mentioned were much superior to those grown in other types of single-salt solutions which were likewise changed every 24 hr.

In an experiment in which 10 different salts were selected to give six different types of nutrient solutions, it was found that the order of change from one salt to another was of only minor importance. It was also found that the differences in the pH value of the single-salt solutions were within the range of good plant growth and thus exerted little or no influence but that the kind of salt used in making the change was of great importance. Thus when potassium nitrate, monomagnesium phosphate, and calcium sulphate were used singly as above mentioned they gave much better yields than did the salts of any other type solution. Gericke considered that the cause for this might be the way in which the two essential ions are paired. Thus the best results were obtained when nitrate and potassium were present together, since the absence of the other essential elements for a period of 48 hr. was apparently of little consequence. Likewise, monomagnesium phosphate seemed to be the best salt with which to supply the magnesium and phosphorus to the plant, since if the plants, for example, were placed in magnesium sulphate or magnesium nitrate alone for 24 hr. they were decidedly injured. It seems apparent, then, that the pairing of nutrient elements or ions of opposite charge is an important factor in securing the maximum growth of plants in water cultures. This does not necessarily mean that a definite cation-anion ratio must exist, since ionic ratios can cover a considerable range of values without any apparent physiological effect. Results of a somewhat similar nature were obtained by Sewell (1924), who found that when wheat is growing in a soil with a sufficient supply of nitrogen, the plant will produce a maximum yield with less of the salts containing potassium and phosphorus than if the nitrogen supply is limited. On that account, if these elements are deficient in the soil a smaller application is required if there is a liberal supply of nitrogen present.

Trelease and Trelease (1925), from a study of the growth of wheat roots, concluded that the roots are slightly sensitive to small differences in salt proportions except when the partial concentration of calcium nitrate was below 5 per cent of the total concentration. In 1925, they grew wheat in 29 different cultural solutions with a wide range of proportions of the constituent salts. The plants grew vigorously in all these combinations, but, when inoculated with the spores of mildew (*Erysiphe graminis*), pronounced differences in susceptibility were found to exist. They considered from this that susceptibility can be influenced to a high degree by the composition of the solution supplied to the plants and that the tests gave no support to the view that low vigor of the host is an index to its susceptibility to invasion by a parasite.

Gericke (1929) found that a wide range of salt proportions was equally satisfactory for the early growth of Marquis wheat. Duncan (1931) found in growing corn in sand cultures that high nitrogen produced stocky, sturdy plants giving good yields, while low nitrogen gave tall, slender, late-maturing plants. High potassium produced large plants with a relatively extensive root growth, while low potassium produced a small top growth and a very limited formation of roots. It has been reported by Bartholomew, Watts, and Janssen (1933) for tomato, by Spencer and Shive (1933) for *Rhododendron ponticum*, and by Wynd (1933) for orchid seedlings that a wide range of salt proportions can be used equally well in the growth of those plants. Beckenbach, Wadleigh, and Shive (1936) studied statistically the effects on the growth of corn of varying the proportion of the different ions in the nutrient solutions. Within the limits of the concentrations employed, no correlation was found between the amounts of sulphate or phosphate in the solution and the growth of the plants. Increasingly high concentrations of nitrate, potassium and calcium gave the most marked correlation with increased growth, while increasingly high concentrations of magnesium within the same limits resulted in decreased growth. None of the three cation ratios—calcium/magnesium, potassium/magnesium, or potassium/calcium—showed any significance in the experiment.

The problem of physiological balance in the soil under agricultural practice is of vast importance if the application of fertilizers is to be placed upon a scientific basis. The most thorough work along this line has been done by Hibbard (1927). The plants that he especially considered were Marquis wheat and Worthy oats, and the salts used as fertilizers were monocalcium phosphate, sodium nitrate, and potassium sulphate. The wheat plants were grown in soil in the greenhouse, while the oat plants were grown in the field, but in each case the nitrogen, phosphorus pentoxide, and potassium oxide were varied by increments of 10 per cent so that there were 36 combinations or ratios. The plants were grown to maturity and the yields of grain taken as a measure of the results. Hibbard concluded that the data obtained indicate that the yields of cereals such as wheat, oats, and corn may be identical under somewhat wide variations of nutrient ratios. He considered that his work suggests that nothing can be gained by attempting to find a definite ratio or physiological balance in soil cultures through the triangular system of fertilizer tests. He considered further that many of the crop plants have become accustomed to wide variations in environmental conditions and for that reason are not suitable indicators for studies of this kind. Under field conditions the proportions of various salts available to the plant must vary from time to time because of the fluctuations of the salt content of the ordinary soils. Similar conclusions were drawn by Eisenmenger (1928) in studying the rate of elongation of wheat roots supplied with simple solutions of monopotassium phosphate, calcium nitrate, and magnesium sulphate and with solutions containing pairs of these salts. He considered that the wide range of salt proportions that allow nearly optimum growth is one of the conspicuous features of his results. Optimum salt proportions probably could be defined precisely if a sufficiently detailed study were made, but the growth rate in the optimum solution would be only slightly higher than in many solutions differing considerably with respect to the relative proportion of their constituent salts. Hibbard found, however, under the conditions of his experiment that the use of a single fertilizer ingredient did not give so good results in soil cultures as a combination of the three and suggested that only under conditions where the price for a mixture of the three fertilizer ingredients is prohibitive would the use of single fertilizers be advisable.

The relationship of the moisture content of the soil to the physiological balance has been studied by Shive (1919 to 1922). He found that good physiological balance and optimum total concentration of a nutrient solution are not alone sufficient to produce the best growth of which the solution is capable unless it is diffused as a film on the particles of a solid substratum. An optimum supply of moisture is thus necessary to impart to the soil its maximum physiological value so that the actual plants producing value of any fertilizer treatment is largely determined by the moisture conditions of the soil.

E. The Hydrogen-ion Relations in Nutrient Solutions

In discussing this subject a few facts should be noted concerning the effects of soil acidity on plant growth. The acidity of the soil has many general influences on fertility due to its effects on the physical, chemical, and biological conditions of this medium. According to Truog (1918) an acid condition of the soil usually lowers the availability of nearly all the essential elements, and favors the accumulation of toxic substances. Loehwing (1930) found, however, that the application of lime and potassium to sterile muck soils often proved harmful due largely to the low hydrogen-ion concentration of the cell sap which caused the precipitation of iron in the lower parts of the plant.

The data relating to the effect of the plant on the hydrogen-ion concentration of the culture solution and the effect of this reaction upon the plant are very confusing. This is to be expected, since the reactions under consideration are influenced by so many different factors. Among these are the type of plant, the rate of ion absorption, exosmosis from the roots, temperature, and the nature of the nutrient solution. It is the intention here to state only a few experimental results to show the general complexity of the question. Tarr and Noble (1922) grew seedlings of wheat, soybeans, and corn in nutrient solutions which were maintained at constant hydrogen-ion concentrations extending over a range from pH 3 to 8 at definite gradations of 1 pH They found that a reaction of pH 3 was prohibitive to growth in all cases. The maximum growth of wheat seedlings occurred at a reaction of approximately pH 4, while a lesser concentration exerted no harmful effect until a pH of 6 was reached, when chlorosis began to appear. The maximum growth of soybeans and corn was obtained at a reaction of pH 5. In both these cases chlorosis occurred at a reaction of pH 6 due to the insolubility of iron at all hydrogen-ion concentrations less (more neutral) than pH 6. According to Hoagland (1926), slight acidity pH 5 to 7 is not itself injurious to various types of plants, but alkaline solutions above pH 8 in general prove to be less favorable, and pH values above 9 are injurious to most agricultural plants. Bermuda grass, however, can grow vigorously when the medium in which it is growing has a pH value of 9. The effects of alkalinity on the plant are both direct and indirect, so that they are difficult to interpret, chiefly because it is difficult or impossible to maintain the desired concentration of the essential salts, since some of these precipitate out under alkaline conditions. Hoagland (1923) obtained some indications that barley, cucumbers, and peas grow more rapidly when a certain degree of acidity is maintained than when the reaction of the solution is allowed to change under the influence of the plant. In no case did he observe as good a growth at pH 7 as when the nutrient solution was slightly acid. It seems therefore to be uncertain whether the reaction to which a solution is brought by a plant growing in it is the most favorable for the growth of that plant. Duggar (1920) found in the case of wheat, corn, and Canada field peas growing in modified Shive and Livingston-Tottingham solutions among others that these plants under favorable conditions gave excellent growth even though the pH value of the solutions varied from 4.5 to 7.1. He concluded from these observations that there is a considerable range of pH values within which plants will grow well.

It was observed by Kachioni-Val'ter (1927) that the optimum growth and development of buckwheat in the presence of chlorine were at pH 5.0 to 6.5; that, at a pH of 7.0 to 9.0, growth and development were hindered; and that pH 4.5 had an intermediate effect. In the absence of chlorine, the optimum region of the pH value became narrower and approached a pH of 6.0.

It has been observed by numerous investigators, in the study of the relationship of the hydrogen-ion concentration to growth of plants, that there is frequently, between the extreme limits of acidity or alkalinity that are injurious and interfere with growth, a third region where the reaction affects the plant injuriously. Thus, if the growth of the plant or plant part is plotted against the hydrogen-ion concentration of the solution in which the plant is grown, the curve has a double maximum with a minimum located between the two maxima. That this phenomenon has not been observed in all cases should not be considered (Robbins, 1923) as necessarily indicating the nonexistence of such a curve, since there are many conditions that might conceal or obliterate the minimum located between the two maxima. The pH values of the solutions in which the plants were grown may have been so separated that the minimum point for growth could not have been observed. Then such factors as temperature, water supply, and salt content may have limited growth so that the maximum and minimum points could not be distinguished. It may be well to give a few examples of this. Salter and McIlvane (1917) noted a double maximum curve for the growth of wheat seedlings grown in nutrient solution with different pH values and found the minimum between these two maxima to lie at a pH value of 6, the same minimum point as that observed by Cole (1922) for corn seedlings. Hixon (1920) found a double maximum for growth in the case of peas, corn, wheat, and oats, a pH value of 5.0 representing the minimum point for peas, and pH 6.0 the minimum point for the other three plants. Arrhenius (1922) grew numerous crop plants in soils in which the hydrogen-ion concentration ranged from pH 3.0 to 10.0. The two hydrogen-ion concentrations giving the highest yield were as follows:

Dale oats	pH 7 and 5	Alfalfa	pH 8 and 7
Rubin wheat	pH 9 and 4	Flax	pH 6 and 4
Red clover	pH 6 and 5	Sugar beets	pH 8 and 6

The data indicate clearly the presence of a double maximum and also show that the same hydrogen-ion concentration may show wide variations in plant production.

A double-maximum curve also apparently results when the swelling, osmotic pressure, viscosity, and electrical conductivity of certain colloids are plotted against the pH value of the solution, and it has been found that the minimum between the two maxima is located at the isoelectric point of these substances. These facts led Robbins (1923) to consider that there may be an isoelectric point for plant tissue that may correspond to the pH value of the minimum point of growth. By means of experiments with the absorption of water and dyes by potato tuber tissue in contact with solutions of different hydrogen-ion concentrations, he obtained a double-maximum curve with a minimum between for water absorption. His results indicated that an ampholyte, possibly a protein, plays a chief role in this process, and its isoelectric point is approximately pH 6.0. He considered that this assumption of an isoelectric point for plant tissue and its influence on water absorption offers a satisfactory explanation for the double-maximum curves in the growth of plants and other physiological functions when the plants are grown in media of different hydrogen-ion concentrations.

The effect of the nature of the nutrient solution upon the rate of change of the pH value when plants are growing therein has been studied by Jones (1921), Jones and Shive (1922), McCall and Haag (1920), Loo (1927), Zinzadzé (1932, 1933), and Trelease and Trelease (1933, 1935). In the case of wheat growing in three-salt nutrient solutions, the rate of change of the hydrogen-ion concentration decreased with an increase in the total osmotic value of the solution and with its volume. The following table shows the change of the pH value of different types of nutrient solutions in which wheat had been grown for 52 hr. (Jones and Shive, 1922).

Kind of solution	pH value	
	Original	After 52 hr.
Crone's	6.6	6.6
Sach's	6.7	6.6
Knop's	4.6	5.7
Shive's *R5S2*	4.5	5.3
Tottingham's *T3R1S4*	4.6	5.5
Detmer's	4.7	5.9

They considered that the resistance offered to reaction change resulting from contact with the roots of growing plants was largely dependent upon the volume molecular proportions of the soluble phosphate salts contained in the solution. McCall and Haag (1920) stated that the hydrogen-ion concentration of various types of three-salt solutions is a function of the volume molecular proportion of the dihydrogen phosphate salt.

Loo (1927), and Trelease and Trelease (1933, 1935) considered that the type or proportion of the nitrogen salts used determined the pH value of the cultural solution. Wheat plants were grown in culture solutions with initial pH values adjusted to 4.3, 5.1, and 6.0 by suitable proportions of phosphoric acid, monopotassium phosphate, and dipotassium phosphate, respectively. A wide range of nitrate/ammonium ionic ratios was obtained by varying the proportions of potassium nitrate and ammonium sulphate, potassium nitrate and ammonium nitrate and calcium nitrate and ammonium nitrate. With low nitrate/ammonium ratios, the pH value of the solution decreased rapidly under the influence of the plants and approached in extreme cases a pH of 3.0. With high ratios, on the other hand, the pH values increased rapidly, tending to reach a limiting value of pH 6.5. When a nitrate/ammonium ratio of suitable value was used, a physiologically balanced solution was obtained in which the hydrogen-ion concentration tended to remain approximately constant during the 8-day period between the renewal of the solution.

Zinzadzé (1934) found that slightly soluble phosphate plus ammonium nitrate is efficient in stabilizing the reaction in nutrient solutions.

The majority of observations indicate that plants growing in nutrient solutions whose reactions are either above or below the neutral point tend to change the hydrogen-ion concentration toward neutrality, although this will depend in part upon the nature of the solution (Toole and Tottingham, 1918; Hoagland, 1917; and Meier and Halstead, 1921). Thus Bakke and Erdman (1923) noted that wheat plants at the heading stage changed the reaction of type-III salt solution in which they grew for 3½ days from pH 3.75 to 5.94, when water cultures were used, and to 6.66 when sand cultures were used. An example of the different reactions brought about by different plants is afforded by the work of Jacobson (1925) with wheat and rice plants. With Marquis wheat plants 97 days old placed in quart jars that contained solution type III *R*2*S*1 with the molecular proportions distributed in eighths, the pH value increased from 3.9 to 6.3 during the first 12 hr., after which the reaction of the solution remained fairly constant for the 72-hr. period of observation. The nitrogen concentration of the solution decreased from 85 to 24 p.p.m., a loss of practically 72 per cent during the first 12 hr., and by the end of the 36-hr. period only a trace of nitrogen remained. The results indicated that the potassium of the potassium nitrate was not taken up nearly so rapidly as the nitrate, and it combined with other elements after being liberated from its anion so that the combination thus formed was capable of partly neutralizing the acidity of the nutrient solution. In the case of rice plants, however, growing in four different types of nutrient solutions, the pH value decreased from 5.0 to about 3.5 during the first 24 hr., after which it remained practically constant for the 3 days that the experiment was continued. It was considered that the cations of the salts containing sulphate might have been removed from the solution more rapidly than the anion so that the acidity was increased. This has also been suggested by Hoagland (1918). It was observed in both the rice and wheat cultures that the excretion of carbon dioxide by the roots tended to decrease the pH value of the solution.

F. Criticism of the Data of Plant Nutrition

The basic purpose of the study of plant response in water and sand cultures is to determine the nature of the nutrients, their concentration, their proportions, and the method of their application that will give the best response when applied in agricultural practice. The question then arises as to how far the observations that have been made with sand and water cultures are comparable and applicable to plants growing under conditions in the field.

One of the chief objections to the use of water and sand cultures from the practical standpoint is that the plant is growing in a very unnatural medium. The differences between the plants grown in sand and in water cultures illustrate the fact that the medium in which the roots grow may affect the plant response.

In cultural solutions the nutrient ions are free to move, while in soils they are adsorbed on colloidal particles or fixed in crystals and thus exist under conditions of restraint (Jenny and Cowan, 1933). While many plants grow to maturity and produce seed in water and sand cultures, the production is, with but few exceptions, less than that of plants growing in the soil. The seed produced, however, has in the cases which have been tested shown little difference in its germinating power from that produced by plants growing in the soil.

Dunn (1921) found that the cortex of roots, grown in nutrient solutions, showed intercellular spaces, the time and place of their appearance varying with the kind of plant and solution used. These spaces were formed apparently by the disintegration of cells. The concentration of the solutions used in most cultural work is much higher than the soil solution upon which a plant normally has to depend (Crocker, 1919). Apparently, too, the solutes in cultural solutions are more readily available than they are in the soil, from both the standpoint of solubility and the rate of diffusion. Neither the frequent renewals nor continuous renewals of the solution in cultural work furnish a medium like that of the soil, although the exact status of the soil solution in regard to its variability of concentration, solubility, and composition is as yet quite uncertain. On that account, therefore, in the words of Stiles (1915), it is a very uncertain procedure to argue from a comparatively simple medium like a nutrient solution of mineral salts to a complex substance like the soil.

Jenny and Cowan (1933) grew plants of soybeans in different systems containing calcium as the only nutrient ion. In one system it was free in solution, in another it was adsorbed on colloids, while in a third it was fixed in the interior of crystals. They found that with a low content of calcium the plant yields were much higher for the free-calcium system than for the adsorbed one, indicating that the plant encounters considerable difficulty in obtaining the adsorbed ions. With higher concentrations of calcium, however, the plants grew best in the adsorbed system. They believed their work indicated that excreted hydrogen ions replaced the calcium ions on the clay, thus enabling calcium to enter the plant. They considered it probable that, inasmuch as the plant can feed successfully on adsorbed ions, the significance of the "soil solution" has been overestimated. According to them the solubility concept is not wholly adequate to account for plant growth in soils of humid regions, and they considered that ionic exchange must be given consideration in a discussion of the uptake of ions by plants. As pointed out by Davis (1921), a very great defect in most plant-nutrition work has been the comparatively small number of plants and cultures used in each experiment. The variation of plants growing even under field conditions is very great, so that unless a large number of individuals is considered in cultural work the results may be of no significance. Davis selected three combinations of Shive's three-salt solution (1.75 atmospheres) which had been reported, respectively, as the best, medium, and poor solution for the growth of wheat and grew in them Sonora wheat in 33 to 50 triplicate cultures for each of the three selected solutions. He found that when the probable error was obtained for the number of cultures used, there was no significant difference between the cultural solutions which had been reported as the best and medium, respectively, but that the solution reported as poor showed a decided inferiority.

The wide variation in cultures was illustrated by the chance selection of duplicate cultures. Thus where the mean of the dry weights of the series was 2.15 g., the mean of 16 pairs selected at random ranged from 1.82 to 2.55 g. Since in most of the experimental work that has been published on plant nutrition the cultures were grown only in duplicate or triplicate, considerable doubt is thrown upon the value of the data that have been reported.

One of the chief aims of nutrition work with plants has been to find the "best solution" for plant growth. The criterion for judging this has been the best response as measured by growth obtained from a series of solutions under conditions that too frequently have not been well defined. The standards by which the best response has been measured are the length and weight of tops, the amount of transpiration, the total amount of dry matter produced, and others. In outlining the standards by which the so-called "best growth" is to be judged, the purpose for which the plant is to be grown must obviously be taken into consideration. If a plant is to be grown for forage, a different standard must be considered from that which would have to be considered if only a grain crop were desired. Clements (1928) made chemical analyses of plants grown in the various positions in the triangle for starch, total sugars, hemicellulose, nitrate nitrogen, total organic nitrogen, and ash in the various parts, with the idea of determining the internal relations of the plant to the various salt combinations. It was found, however, that these criteria, for the most part, are unreliable for determining the best response of plants to nutrient solutions. Hoagland and Martin (1923) have pointed out that we are not justified in stating that a normal plant is the one making the best growth under a given set of environmental conditions. No solution has a fixed value so far as its ionic, atomic, or molecular properties are concerned for growth, since the relation of a plant to a cultural medium is dependent upon the environmental conditions. On that account, therefore, no statement of any value can be made in regard to the "best solution" unless a strict account of the various environmental conditions is presented—a requirement that is practically impossible. The stage of growth of the plant must also be taken into consideration when considering the value of a given nutrient solution, since the nutrient requirements of a plant vary with its stage of development, as will be discussed in the following chapter.

The consensus of opinion among workers in plant nutrition at the present time is that there is a wide range of solutions in regard to concentration, composition, and proportion of components in which plants will thrive equally well and that the response of plants to any solution will depend to a great extent upon the environmental conditions under which they are growing. In the words of Stiles (1916), we may state that the value of water- and sand-culture work is limited because of the difficulty of analyzing results due to complex factors that are not under control and because of the further difficulty of controlling in some cases even the factor whose action is being investigated.

Gericke (1930) considered that the actual salt requirement cannot be determined by growing plants to maturity in any so-called complete nutrient solution. Apparently the conditions that determine the relation between the magnitude of yield and the absorption of any given elements from a cultural solution involve not only the amounts of the elements absorbed but also the time required for absorption, as well as the time required for the various absorbed elements to function in growth.

On the other hand, it must be remembered that nutrient solutions furnish the best means that are available at present to study the nutrition of plants in a quantitative way, and much of our knowledge of plant nutrition has been gained by their use. Care should nevertheless be exercised in interpreting the results obtained with nutrient solutions in terms of the behavior of plants under natural conditions.

VI. ANTAGONISM AND BALANCED SOLUTIONS

It has long been known that solutions of certain single salts are toxic to living organisms placed therein but that in a mixed solution of these salts the organisms function normally and are not injured (True, 1915). The best example is the organisms living in sea water. They grow and develop in this mixture of salt solutions, but if placed in a solution of any one of the single salts that compose it they perish, in many cases in a relatively short time. The term "antagonism" is used to designate the hindrance that a given salt has upon the toxic action of another salt. Thus antagonism means that one toxic substance acts as an antidote to another.

As has been mentioned above, it is possible to mix two or more toxic substances in solution in such proportions that they lose their individual toxic properties. Such a solution has been termed a "physiologically balanced solution" (Loeb, 1906) and may be defined as a solution in which the toxic effects that each salt would have were it alone present in the solution are hindered by one or more antagonistic salts in the solution. It has also been defined as a solution containing salts in such proportions that they have none of the toxic properties of the individual salts. The term, however, has been used in a very broad sense and has referred in many cases to the proportional arrangement of the component parts of nutrient solutions regardless of their toxic properties. A physiologically balanced solution may contain only two salts, so that it is not necessarily a nutrient solution, but on the other hand a nutrient solution to be of the best advantage must be a balanced one.

The question of antagonism and balanced solutions in regard to plants has been extensively investigated by Osterhout and his pupils

(1906 to 1922). The intention here is not to discuss this problem in any detail but only to mention a few examples to illustrate the importance of this subject in relation to plant life. Osterhout (1914) has pointed out that a mixture of two equally toxic solutions must have precisely the same effect on growth as the pure solutions themselves, provided the effects of the salts are additive. By the additive effect is meant the effect produced when each salt acts independently of every other so that its toxicity is neither augmented nor diminished by the presence of other salts. If antagonism between the salts exists, there will be better growth in the mixed solutions than in the single ones. The amount of this increased growth expressed as percentage of the growth obtained in the pure solution is the most satisfactory measure of antagonism. Precautions must be taken in the determination of antagonism that only those parts of plants that come in contact with the solution should be used for measurement. Trelease and Trelease (1926) and Eisenmenger (1928) have demonstrated that the roots of young wheat seedlings can be used satisfactorily for demonstrating the growth-retarding effects of single-salt solutions and the antagonistic action of mixed solutions. The following table shows the relative effect of mixed- and single-salt solutions on the growth of wheat roots therein (Osterhout, 1907):

Medium	Growth of roots in 40 days, millimeters	Medium	Growth of roots in 40 days, millimeters
NaCl 0.12*M*	60	1,000 cc. NaCl 0.12*M* 78 cc. $MgCl_2$ 0.12*M* 10 cc. $CaCl_2$ 0.12*M*	327
$MgCl_2$ 0.12*M*	7		
KCl 0.12*M*	68		
$CaCl_2$ 0.12*M*	70	Dilute artificial sea water 0.12*M* NaCl 1,000 cc. 0.12*M* $MgCl_2$ 78 cc. 0.12*M* $MgSO_4$ 38 cc. 0.12*M* KCl 22 cc. 0.12*M* $CaCl_2$ 10 cc.	360
1,000 cc. NaCl 0.12*M* 10 cc. $CaCl_2$ 0.12*M*	254		
1,000 cc. NaCl 0.12*M* 22 cc. KCl 0.12*M* 10 cc. $CaCl_2$ 0.12*M*	324	Distilled water	740

Osterhout and others have found antagonism between the following cations:

NH_4 *vs.* Ca	NH_4 *vs.* Na	Na *vs.* Ba
K *vs.* Ca	NH_4 *vs.* K	K *vs.* Ba
Na *vs.* Ca	Na *vs.* K	Na *vs.* Sr
Mg *vs.* Ca	Mg *vs.* K	K *vs.* Sr
Mg *vs.* Sr	Fe *vs.* Mn	

Raber (1920) from experimental work with sodium acetate and sodium sulphate concluded that antagonism existed between anions, so any theory of the antagonistic action of salts must take into account the action of the anions as well as the cations.

Kearney and Harter (1907) noted the effect of calcium sulphate in overcoming the toxic effect of the magnesium and sodium salts that are found in alkali soils. The following table shows the critical concentrations for plants of lupine and corn grown in pure solutions of the various salts and in the same solutions after an excess of calcium sulphate had been added. The term "critical concentration" as here used was the concentration of solution that allowed the root tips of any one-half of the total number of seedlings tested to retain their capability of further elongation after being in the solution for 24 hr.

Salt	Critical concentrations			
	In pure solutions		After adding $CaSO_4$ in excess to solutions	
	Lupine	Corn	Lupine	Corn
$MgSO_4$	0.007*M*	0.25*M*	0.40*M*	0.60*M*
$MgCl_2$	0.007*M*	0.08*M*	0.20*M*	0.30*M*
Na_2CO_3	0.010*M*	0.02*M*	0.03*M*	0.05*M*
Na_2SO_4	0.040*M*	0.05*M*	0.30*M*	0.40*M*
$NaCl$	0.040*M*	0.04*M*	0.20*M*	0.25*M*
$HNaCO_3$	0.030*M*	0.05*M*	0.05*M*	0.10*M*

The neutralizing effect of calcium sulphate was much greater toward the magnesium salts than toward the sodium salts, while the order of toxicity of the different salts was also changed by its addition.

The causes of antagonism are little understood, and it is not the intention here to enter into any discussion of the various theories that have been proposed. The relationship of the changes in permeability produced by salts and their antagonistic relations as suggested by Osterhout (1915) may give some insight into the causes of antagonism. As has been mentioned in Chap. II, the substances that alter the permeability of the protoplasm may be divided into those which cause an increase and those which cause a decrease in its permeability. Osterhout considered that substances belonging to one class will antagonize those belonging to the other class, so that in order to predict which substances will antagonize each other it is necessary only to determine to which of these classes the substances belong. The amount of antagonism between salts may also be predicted, at least to a certain extent, since the greater their effect on the permeability the greater will be their antagonistic action. This theory of antagonism assumes that the protoplasmic membrane possesses a certain degree of permeability at which it functions normally, and when this permeability is increased or decreased beyond a certain limit the activities of the protoplasm are impaired. If an organism is placed, for example, in a solution of sodium chloride, the permeability of the protoplasm is increased to a point that is injurious to it; and if it is placed in a single solution of calcium chloride, the permeability will be decreased to the point of injury. If, however, these two salts are mixed in the proper proportion, their effects on the protoplasm will counteract each other, and the permeability of the protoplasm will be unaltered so that the cell or organism will function normally.

Bibliography

Allison, R. V., 1921: The relation of aeration to the development of the soybean plant in artificial culture, *N. J. Sta. Ann. Rept.*, **1921**: 338–345.

——— and J. W. Shive, 1923: Studies on the relation of aeration and continuous renewal of nutrient solution to the growth of soybeans in artificial culture, *Am. J. Bot.*, **10**: 554–567.

Andrews, F. M., and C. C. Beal, 1919: The effect of soaking in water and of aeration on the growth of *Zea mays*, *Bull. Torrey Bot. Club*, **46**: 91–100.

———, 1920: Apparatus for aerating plants, *Proc. Ind. Acad. Sci.*, **1919**: 165–167.

———, 1921: The effect of aeration on plants, *Proc. Ind. Acad. Sci.*, **1920**: 147–148.

Arrhenius, O., 1922: Hydrogen ion concentration, soil properties, and growth of higher plants, *Arch. Bot.*, **18**: 1–54.

———, 1923: Soil reaction and plant development, *Medd. Centralanst. Jordbruksomradet.* 245, 14 pp., *Expt. Sta. Rec.*, **50**: 532.

Asprey, G. F., 1933: Studies on antagonism. I, The effect of the presence of salts of monovalent, divalent, and trivalent kations on the intake of calcium and ammonium ions by potato tuber tissue, *Proc. Roy. Soc.*, B, **112**: 451–472.

———, 1933: Studies on antagonism. II, The effect of previous immersion of potato tuber tissue in solutions of monovalent, divalent, and trivalent cations on the subsequent absorption of the ammonium ion, *Proc. Roy. Soc.*, B, **113**: 71–82.

Ayres, A. H., 1917: Influence of the composition and concentration of the nutrient solution on plants grown in sand cultures, *Univ. Calif. Pub.*, **1**: 341–394.

Bakke, A. L. and L. W. Erdman, 1923: A comparative study of sand and solution cultures of Marquis wheat, *Am. J. Bot.*, **10**: 18–32.

Bancroft, W. D., 1902: Synthetic analysis of solid phase, *J. Phys. Chem.*, **6**: 178.

Bartholomew, R. P., V. M. Watts, and G. Janssen, 1933: The effect of variations in the nutrient media upon the nitrogen, phosphorus, and potassium content on plants, with special reference to the tomato, *Ark. Sta. Bull.* 288.

Beal, C. C., 1918: The effect of aeration on the roots of *Zea mays*, *Proc. Ind. Acad. Sci.*, **1917**: 177–180.

Benedict, H. M., 1934: Effect of ultraviolet radiation on growth and on the calcium and phosphorus contents of plants, *Bot. Gaz.*, **96**: 330–341.

Berry, W. E., and F. C. Steward, 1934: The absorption and accumulation of solutes by living plant cells. VI, The absorption of potassium bromide from dilute solution by tissue from various plant storage organs, *Ann. Bot.*, **48**: 395–410.

Blinks, L. R., and A. G. Jacques, 1930: The cell sap of *Halicystis*, *J. Gen. Physiol.*, **13**: 733–738.

Breazeale, J. F., 1905: Effect of the concentration of the nutrient solution upon wheat cultures, *Sci.*, **22**: 146–149.

———, 1923: Nutrition of plants considered as an electrical phenomenon, *J. Agr. Res.*, **24**: 41–55.

——— and W. T. McGeorge, 1932: Nutritional disorders in alkaline soils as caused by deficiency of carbon dioxide, *Ariz. Sta. Tech. Bull.* 41.

Brenchley, W. E., 1916: The effect of the concentration of the nutrient solution on the growth of barley and wheat in water cultures, *Ann. Bot.*, **30**: 77–91.

Briggs, G. E., and A. H. K. Petrie, 1928: On the application of the Donnan's equilibrium to the ionic relations of plant tissues, *Biochem. J.*, **22**: 1071–1083.

———, 1930: The accumulation of electrolytes in plant cells. A suggested mechanism, *Proc. Roy. Soc.*, B, **107**: 248–269.

———, 1932: The absorption of salts by plant tissues, considered as ionic exchange, *Ann. Bot.*, **46**: 301–322.

Brooks, S. C., 1929: The accumulation of ions in living cells. A non-equilibrium condition, *Protoplasma*, **8**: 389–412.

———, 1931: The accumulation of strong electrolytes in living cells, *Sci.*, **73**: 589–590.

Brown, D. D., 1928: The effect of different proportions of calcium nitrate, and potassium dihydrogen phosphate on the growth of wheat in sand cultures, *Soil Sci.*, **26**: 441–446.

Bryant, A. E., 1934: Comparison of anatomical and histological differences between roots of barley grown in aerated and in non-aerated culture solutions, *Plant Physiol.*, **9**: 389–391.

Burgerstein, A., 1897: Ueber die Transpirationgrösse von Pflanzen feuchter Tropengebiete, *Ber. bot. Ges.*, **15**: 154–165.

Cameron, F. K., 1911: "The Soil Solution," The Chemical Publishing Co., Easton, Pa.

Casale, L., 1921: A physical-chemical study of the absorptive power of soils and of the method whereby plants absorb nutritive materials from the soil, *Staz. Sper. Agrarie Ital.*, **54**: 65–113; *Bot. Abst.*, **10**: 115.

Clark, H. E., and J. W. Shive, 1932: Influence of continuous aeration upon the growth of tomato plants in solution cultures, *Soil Sci.*, **34**: 37–42.

Clark, O. L., 1916: A method for maintaining a constant volume of nutrient solutions, *Sci.*, **44**: 868–869.

Clements, H. F., 1928: Plant nutrition studies in relation to the triangular system of water cultures, *Plant Physiol.*, **3**: 441–459.

Colby, H. L., 1933: Seasonal absorption of nutrient salts by the French prune grown in solution culture, *Plant Physiol.*, **8**: 1–34.

Cole, A. M., 1922: Effect of variations in H-ion concentration on the growth of seedlings, *thesis*, *Univ. Mo.*

Collander, R., 1930: Permeabilitätsstudien an *Chara ceratophylla*. I, Die normale Zusammensetzung des Zellsaftes, *Acta Bot. Fennica*, No. 6, 1–20. Helsingforsiae.

——— and H. Barlund, 1933: Permeabilitätsstudien an *Chara ceratophylla*. II, Die Permeabilität für Nichtelektrolyte, *Acta Bot. Fennica*, No. 11, 1–114. Helsingforsiae.

Conrad, J. P., 1934: Physiological acidity and alkalinity of inorganic nitrogenous compounds in solution cultures, *J. Am. Soc. Agron.*, **26**: 364–372.

Cook, R. L., 1930: Effect of soil type and fertilizer on the nitrate content of the expressed sap and the total nitrogen content of the tissue of the small grains, *J. Am. Soc. Agron.*, **22**: 393–409.

Cooper, H. P., 1930: Ash constituents of pasture grasses, their standard electrode potential and ecological significance, *Plant Physiol.*, **5**: 193–215.

Cooper, W. C., M. J. Dorcas, and W. J. V. Osterhout, 1929: The penetration of strong electrolytes, *J. Gen. Physiol.*, **12**: 427–433.

——— and W. J. V. Osterhout, 1930: The accumulation of electrolytes. I, The entrance of ammonia into *Valonia macrophysa*, *J. Gen. Physiol.*, **14**: 117–127.

Crocker, W., 1919: Review of plant nutrition work, *Bot. Gaz.*, **67**: 174–178.

Crone, G., 1902: Ergebnisse von Untersuchungen über die Wirkung der Phosphorsäure auf die höhre Pflanzen und eine neue Nahrlösung, *Sitzungber. Niederrhein Ges. Nat. Heilkundee Bonn*, **1902**: 167–173.

Davidson, J., 1927: Effect of hydrogen-ion concentration on the absorption of phosphorus and potassium by wheat seedlings, *J. Agr. Res.*, **35**: 335–346.

———, 1933: The possible effect of hydrogen-ion concentration on the absorption of potassium and phosphorus by wheat plants under field conditions, *J. Agr. Res.*, **46**: 449–453.

Davis, A. R., 1921: The variability of plants grown in water cultures, *Soil Sci.*, **11**: 1–33.

Denny, F. E., and W. J. Youden, 1927: Acidification of unbuffered salt solutions by plant tissue in relation to the question of tissue isoelectric points, *Am. J. Bot.*, **14**: 395–414.

Dufrenoy, J., 1920: The biological significance of selective adsorption, *Am. Midland Nat.*, **6**: 159–165.

Duggar, B. M., 1920: The effect of conditions on the relation of seed plants to H-ion concentration of nutrient solutions, *Sci.*, **52**: 416.

———, 1920: Hydrogen ion concentration and the composition of nutrient solutions in relation to the growth of seed plants, *Ann. Mo. Bot. Gard.*, **7**: 1–50.

———, 1920: The use of insoluble salts in balanced solutions for seed plants, *Ann. Mo. Bot. Gard.*, **7**: 307–327.

Duncan, G. H., 1931: Influence of nutritional balance upon the development of corn plants, *Trans. Ill. State Acad. Sci.*, **23**: 143–148; *Exp. Sta. Rec.*, **74**: 32.

Dunlap, A. A., 1936: Sand culture of seedlings, *Conn. Sta. Bull.* 380.

Dunn, G. A., 1921: Note on the histology of grain roots, *Am. J. Bot.*, **8**: 207–211.

Eisenmenger, W. S., 1928: Toxicity, additive effects and antagonism of salt solutions as indicated by growth of wheat roots, *Bull. Torrey Bot. Club*, **55**: 261–305.

———, 1931: Factors modifying the toxicity of phenol, *Plant Physiol.*, **6**: 325–331.

Espino, R. B., 1919: Critical review of article by Schreiner and Skinner on the triangle system for fertilizer experiments, *Plant World*, **22**: 53–54.

Fittbogen, J., 1873: Untersuchungen über das für eine normale Production der Haferpflanze notwendige Minimum von Bodenfeuchtigkeit sowie über die Aufnahme von Bestandteilen des Bodens bei verschiedenem Wassergehalt desselben, *Landw. Jahrb.*, **2**: 252–371.

Free, E. E., 1917: The effect of aeration on the growth of buckwheat in water cultures. *Johns Hopkins Univ. Circ.*, **3**: 198–199.

Freeland, R. O., 1935: Effect of transpiration upon the absorption and distribution of mineral salts in plants, (Reprint) *Abstract, Doct. Diss. No.* 17, *Ohio State Univ. Press.*

———, 1936: Effect of transpiration upon the absorption and distribution of mineral salts in plants, *Am. J. Bot.*, **23**: 355–362.

Friedemann, W. G., 1929: A method for comparing the value of ammonia nitrogen and nitrate nitrogen, *Sci.*, **70**: 43–44.

Furnstal, A. H., and B. Johnson, 1936: Preparation of sintered pyrex glass aerators for use in water culture experiments with plants, *Plant Physiol.*, **11**: 189–194.

Gericke, W. F., 1921: Influence of temperature on the relations between nutrient salt proportions and the early growth of wheat, *Am. J. Bot.*, **8**: 59–63.

———, 1922: Water culture experimentation, *Sci.*, **56**: 421–422.

———, 1922: On the physiological balance in nutrient solutions for plant cultures, *Am. J. Bot.*, **9**: 180–182.

———, 1929: Some relations of maintained temperatures to germination and the early growth of wheat in nutrient solutions, *Philippine J. Sci.*, **38**: 215–240.

———, 1930: Plant food requirement of rice, *Soil Sci.*, **29**: 207–226.

——— and J. R. Travernetti, 1936: Heating of liquid culture media for tomato production, *Agr. Eng.*, **17**: 141–142.

Giltay, E., 1897: Vergleichende Studien über die Stärke der Transpiration in den Tropen und in mitteleuropäischem Klima, *Jahrb. wiss. Bot.*, **30**: 615–644.

Gračanin, M., 1932: La concentration des ions comme facteur de la résorption, *C. R. Acad. Sci. Paris*, **195**: 1311–1313.

———, 1932: The effect of light on the resorption of salts by plants, *Preslia Bull. Soc. Bot. Tchecoslovaque à Prague*, **11**: 1–5.

———, 1935: Die Konzentration der Nährlösung als Faktor des Wurzelwachstums, *Lab. Plant Nutrition Univ. Zagrebensis.*

Gregory, F. G., 1928: The differential effect of the ions of three-salt solutions on the growth of potato plants in sand culture, *Proc. Roy. Soc.*, B, **102**: 311–327.

Grunder, M. S., 1933: Yield and chemical composition of certain pasture crops, fertilized and unfertilized, *J. Agr. Res.*, **46**: 375–386.

Haas, A. R. C., and H. S. Reed, 1926: The absorption of ions by citrus and walnut seedlings, *Hilgardia*, **2**: 67–106.

Haberlandt, G., 1892: Anatomisch-physiologische Untersuchungen über das tropische Laubblatt, *Sitzungsber. Akad. Wiss. Wien.*, **101**: 785–816.

———, 1897: Ueber die Grösse der Transpiration im feuchten Tropenklima, *Jahrb. wiss. Bot.*, **31**: 273–288.

Hall, A. D., W. E. Brenchley, and L. E. Underwood, 1914: Soil solution and mineral constituents of the soil, *Phil. Trans. Roy. Soc.*, **204**: 179–200.

———, ———, and ———, 1914: The soil solution and the mineral constituents of the soil, *J. Agr. Sci.*, **6**: 278–301.

Hammett, F. S., 1928: A convenient device for plant solution culture work, *Sci.*, **67**: 106–107.

Harris, J. A., W. F. Hoffman, and J. V. Lawrence, 1925: Differential absorption of anions by varieties of cotton, *Proc. Soc. Exp. Biol. Med.*, **22**: 350–352.

Harvey, R. B., 1927: Ionization and adsorption isoelectric points, *Plant Physiol.*, **2**: 220.

Hasselbring, H., 1914: The relation between the transpiration stream and the absorption of salts, *Bot. Gaz.*, **57**: 72–73.

———, 1914: The effect of shading on the transpiration and assimilation of the tobacco plant in Cuba, *Bot. Gaz.*, **57**: 257–286.

Henricksen, H. C., 1933: Absorption of ash constituents by citrus trees, *Ann. Rept. Puerto Rico Sta.*, **1932**: 18–19.

Hibbard, R. P., 1917: Physiological balance in the soil solution, *Mich. Sta. Tech. Bull.* 40.

———, 1927: Negative results on physiological balance in soil cultures, *Plant Physiol.*, **2**: 1–38.

Hixon, R. M., 1920: The effect of the reaction of a nutrient solution on germination and the first stages of plant growth, *Medd. Vetenskapsak. Nobelinstitut*, **4**: 1–28.

Hoagland, D. R., 1917: The effect of hydrogen and hydroxyl ion concentration on the growth of barley seedlings, *Soil Sci.*, **3**: 547–560.

———, 1918: The relation of the plant to the reaction of the nutrient solution, *Sci.*, **48**: 422–425.

———, 1919: Relation of the concentration and reaction of the nutrient medium on the growth and absorption of the plant, *J. Agr. Res.*, **18**: 73–117.

———, 1919: Relation of nutrient solution to composition and reaction of cell sap to barley, *Bot. Gaz.*, **68**: 297–304.

———, 1920: Optimum nutrient solutions for plants, *Sci.*, **52**: 562–564.

———, 1923: The effect of the plant on the reaction of the culture solution, *Cal. Sta. Tech. Paper* 12.

———, 1923: The absorption of ions by plants, *Soil Sci.*, **16**: 225–246.

——— and A. R. Davis, 1923: The composition of the cell sap of the plant in relation to the absorption of ions, *J. Gen. Physiol.*, **5**: 629–647.

——— and ———, 1923: Further experiments on the absorption of ions by plants, including observations on the effect of light, *J. Gen. Physiol.*, **6**: 47–62.

——— and J. C. Martin, 1923: A comparison of sand and solution cultures with soils as media for plant growth, *Soil Sci.*, **16**: 367–388.

———, 1925: Physiological aspects of soil-solution investigations, *Hilgardia*, **1**: 227–257.

——— and A. R. Davis, 1925: Suggestions concerning the absorption of ions by plants, *New Phytol.*, **24**: 99–111.

———, 1926: Some phases of the inorganic nutrition of plants in relation to the soil solution, *Sci. Agr.*, **6**: 141–151, 177–189.

———, P. L. Hibbard, and A. R. Davis, 1926: The influence of light, temperature, and other conditions on the ability of *Nitella* cells to concentrate halogens in the cell sap, *J. Gen. Physiol.*, **10**: 121–146.

——— and J. C. Martin, 1927: Recent experiments concerning the adequacy of artificial culture solutions and of soil solutions for the growth of different types of plants, *Proc. First. Int. Cong. Soil Sci.*, Ithaca, 1926, **3**: 1–12.

———, A. R. Davis, and P. L. Hibbard, 1928: The influence of one ion on the accumulation of another by plant cells with special reference to experiments with *Nitella*, *Plant Physiol.*, **3**: 473–486.

——— and A. R. Davis, 1929: The intake and accumulation of electrolytes by plant cells, *Protoplasma*, **6**: 610–626; *Biol. Abst.*, **5**: 2655.

———, 1930: Observations on the nature of the process by which mineral elements are accumulated by plant cells, *Sci.*, **72**: 403–404.

———, 1931: Absorption of mineral elements by plants in relation to soil problems, *Plant Physiol.*, **6**: 373–388.

——— and T. C. Broyer, 1936: General nature of the process of salt accumulation by roots with description of experimental methods, *Plant Physiol.*, **11**: 471–508.

Höber, R., and J. Höber, 1928: Beobachtungen über die Zusammensetzung des Zellsaftes von *Valonia macrophysa*, *Pflügers Arch. Ges. Physiol.*, **219**: 260–272; *Biol. Abst.*, **5**: 2655.

Hollenberg, G. J., 1932: Some physical and chemical properties of the cell sap of *Halicystis ovalis*, *J. Gen. Physiol.*, **15**: 651–652.

Horner, C. K., D. Burk, and S. R. Hoover, 1934: Preparation of humate iron and other humate metals, *Plant Physiol.*, **9**: 663–670.

Ilienkoff, P. A., 1865: Einige Versuche zur Bestimmung des Einflusses, welchen die Bodenfeuchtigkeit auf die Vegetation ausübt, *Ann. Chem. Pharm.*, **136**: 160–165.

Iljin, W. S., 1924: The influence of salts on the alternation of concentration of cell sap in plants, *Studies in Plant Physiol. Lab. Charles Univ., Prague*, **2**: 5–25; *Bot. Abst.*, **15**: 799.

INGOLD, C. T., 1936: The effect of light on the absorption of salts by *Elodea canadensis*, *New Phytol.*, **35**: 132–141.

IRWIN, M., 1923: The behavior of chlorides in the cell sap of *Nitella*, *J. Gen. Physiol.*, **5**: 427–428.

———, 1925: On the accumulation of dye in *Nitella*, *J. Gen. Physiol.*, **8**: 147–182.

———, 1926: Mechanism of the accumulation of dye in *Nitella* on the basis of the entrance of the dye as undissociated molecules, *J. Gen. Physiol.*, **9**: 561–575.

JACOBSON, H. G. M., 1925: Changes in hydrogen-ion concentration in nutrient solutions. I, In a culture with wheat. II, In cultures with rice, *J. Am. Soc. Agron.*, **17**: 577–583; 583–586.

——— and T. R. SWANBACK, 1933: Relative influence of nitrate and ammoniacal nitrogen upon intake of calcium by tobacco plants, *Plant Physiol.*, **8**: 340–342.

JACQUES, A. G., and W. J. V. OSTERHOUT, 1930: The kinetics of penetration. II, The penetration of CO_2 into *Valonia*, *J. Gen. Physiol.*, **13**: 695–714.

——— and ———, 1932: The accumulation of electrolytes. IV, Internal versus external concentrations of potassium, *J. Gen. Physiol.*, **15**: 537–550.

——— and ———, 1934: The accumulation of electrolytes. VI, The effect of external pH, *J. Gen. Physiol.*, **17**: 727–750.

———, 1934, 1935: The accumulation of electrolytes. VII, Organic electrolytes, *J. Gen. Physiol.*, **18**: 235–242; 283–300.

———, 1936: The kinetics of penetration. XIII, Effect of pH on the entrance of potassium into *Nitella* at low concentrations, *Proc. Nat. Acad. Sci.*, **22**: 191–193.

JENNY, H., and E. W. COWAN, 1933: The utilization of adsorbed ions by plants, *Sci.*, **77**: 394–396.

JOHNSON, H. V., 1915: The absorption of ions by living and dead roots, *Am. J. Bot.*, **2**: 250–254.

JOHNSTON, E. S., 1925: Comparative study of the "six types" of nutrient solutions in relation to the growth of potato plants in sand cultures, *Soil Sci.*, **20**: 397–403.

———, 1927: An apparatus for controlling the flow of nutrient solutions in plant cultures, *Plant Physiol.*, **2**: 213–215.

———, 1932: Growing plants without soil, *Smithsonian Rept.*, **1931**: 381–387.

JONES, L. H., 1921: Some factors affecting the rate of change of hydrogen-ion concentration in nutrient solutions, *N. J. Sta. Rept.*, **1921**: 330–333.

——— and J. W. SHIVE, 1922: Influence of wheat seedlings upon the hydrogen ion concentration of nutrient solutions, *Bot. Gaz.*, **73**: 391–400.

JOST, L., 1907: "Lectures on Plant Physiology," Eng. ed., pp. 43, 79, Clarendon Press.

KACHIONI-VAL'TER, L. S., 1927: Influence of pH on development and growth of buckwheat, *Mém. Inst. Agron. à Lenigrade*, **4**: 151–174; *Biol. Abst.*, **4**: 351.

KEARNEY, T. H., and L. L. HARTER, 1907: The comparative tolerance of various plants for the salts common in alkali soils, *U. S. Bur. Plant Ind. Bull.* 113.

KIESSELBACH, T. A., 1916: Transpiration as a factor in crop production, *Nebr. Sta. Res. Bull.* **6**: 170–184.

KNOP, W., 1865: Quantitative Untersuchungen über den Ernährungsprocess der Pflanze, *Landw. Versuchs-Stat.*, **7**: 93–107.

KOHL, F. G., 1886: "Die Transpiration der Pflanzen," 124 pp., H. Bruhn.

LAGATU, H., and M. L. MAUME, 1927: Méthode de contrôle chimique du mode d'alimentation de la vigne, *C. R. Acad. Agr. France*, **13**: 439.

LAPIQUE, L., 1925: Sur l'absorption des sels par les cellules végétales épictèse et sélection, *Bull. Soc. Chim. Biol.*, **7**: 621–637; *Bot. Abst.*, **15**: 1161.

LARSON, C. A., 1931: The effect of continued application of irrigation water and commercial fertilizers on Wenatchee orchard soils, *unpub. thesis*, *Wash. State College.*

LIVINGSTON, B. E., 1900: On the nature of the stimulus which causes the change of form in polymorphic green algae, *Bot. Gaz.*, **30**: 289–317.

———, 1906: A simple method for experiments with water cultures, *Plant World*, **9**: 13–16.

——— and W. E. TOTTINGHAM, 1918: A new three-salt nutrient solution for plant cultures, *Am. J. Bot.*, **5**: 337–347.

———, 1919: Review of Gericke's work on the influence of temperature on the physiological balance of nutrient solutions, *Plant World*, **22**: 26–27.

———, K. F. KELLERMAN, and W. CROCKER, 1919: A plan for cooperative research on the salt requirements of representative agricultural plants, *Rept. Com. Div. Biol. Agr. Nat. Res. Council.*

LOEB, J., 1906: "Dynamics of Living Matter," Leincke & Buechner, Agents Columbia Press, New York.

LOEHWING, W. F., 1930: Sap hydrion concentration as a factor in plant metabolism, *Proc. Iowa Acad. Sci.*, **35**: 135–139.

———, 1932: Soil aeration in relation to plant metabolism, *Abst. Am. J. Bot.*, **19**: 844.

LOO, T. L., 1927: The influence of hydrogen ion concentration on the growth of the seedlings of some cultivated plants, *Bot. Mag.* (*Tokyo*), **41**: 33–41; *Exp. Sta. Rec.*, **61**: 627.

———, 1928: The effect of renewal of nutrient solutions upon the growth of culture plants and its relation to aeration, *Jap. J. Bot.*, **4**: 71–98.

LOOMIS, W. E., and C. A. SHULL, 1937: "Methods in Plant Physiology," xviii + 472 pp., McGraw-Hill Book Company, Inc., New York.

LUNDERÅRDH, H., and H. BURSTRÖM, 1933: Atmung und Ionenaufnahme, *Planta*, **18**: 683–699.

MACHT, D. I., and M. E. DAVIS, 1935: Toxicity of alpha- and beta-nicotines and nornicotines for *Lupinus albus*, *Am. J. Bot.*, **22**: 329–332.

MARSH, R. P., 1935: A modified cultural jar, *Sci.*, **82**: 256.

MCCALL, A. G., 1915: A new method for the study of plant nutrients in sand cultures, *J. Am. Soc. Agron.*, **7**: 249–252.

———, 1916: A method for the renewal of plant nutrients in sand cultures, *Ohio J Sci.*, **16**: 101–103.

———, 1916: Physiological balance of nutrient solutions for plants in sand cultures, *Soil Sci.*, **2**: 207–255.

——— and J. R. HAAG, 1920: The hydrogen-ion concentration of certain three-salt nutrient solutions for plants, *Soil Sci.*, **10**: 481–485.

MCCOOL, M. M., and M. D. WELDON, 1930: The effect of sodium nitrate on the composition of the expressed sap of the small grains, *J. Am. Soc. Agron.*, **22**: 434–437.

MCGEORGE, W. T., 1932: Chemical equivalent base exchange reactions in plants, *Plant Physiol.*, **7**: 119–124.

MCHARGUE, J. S., 1923: Convenient supports for plants in pot or water culture experiments, *Soil Sci.*, **16**: 359–363.

MCLEAN, R. C., 1919: Studies in the ecology of tropical rain forests, with special reference to the forests of Brazil, *J. Ecol.*, **7**: 5–54; 121–172.

MEHRLICH, F. P., 1935: A mechanism for controlled continuous flow of nutrient solutions, *Plant Physiol.*, **10**: 169–177.

MEIER, H. F. A., and C. E. HALSTEAD, 1921: Hydrogen-ion concentration relations in a three-salt solution, *Soil Sci.*, **11**: 325–353.

MENDIOLA, N. B., 1922: Effect of different rates of transpiration on the dry weight and ash content of the tobacco plant, *Philippine J. Sci.*, **20**: 639–655.

MEUER, R., 1909: Über die regulatorische Aufnahme anorganischen Stoffe durch die Wurzeln von *Beta vulgaris* und *Daucus Carota*, *Jahrb. wiss Bot.*, **46**: 503–567.

Muenscher, W. C., 1922: The effect of transpiration on the absorption of salts by plants, *Am. J. Bot.*, **9**: 311–330.

Nathansohn, A., 1903: Ueber Regulationserscheinungen im Stoffaustausch, *Jahrb. wiss. Bot.*, **38**: 241–290.

———, 1904: Über die Regulation der Aufnahme anorganischen Salze durch die Knollen von *Dahlia*, *Jahrb. wiss. Bot.*, **39**: 607–644.

———, 1904: Weitere Mitteilungen über die Regulation der Stoffaufnahme, *Jahrb. wiss. Bot.*, **40**: 403–442.

Nightingale, G. T., and R. B. Farnham, 1936: Effects of nutrient concentration on anatomy, metabolism and bud abscission of sweet pea, *Bot. Gaz.*, **97**: 477–517.

Osterhout, W. J. V., 1906: On the importance of physiologically balanced solutions for plants, *Bot. Gaz.*, **42**: 127–134.

———, 1907: On the importance of physiologically balanced solutions for plants, *Bot. Gaz.*, **44**: 259–272.

———, 1909: The nature of balanced solutions, *Bot. Gaz.*, **47**: 148–149.

———, 1909: The penetration of inorganic salts in living protoplasm, *Z. physikal. Chem.*, **70**: 408–413.

———, 1912: The permeability of protoplasm to ions and the theory of antagonism, *Sci.*, **35**: 112–115.

———, 1914: Quantitative criteria of antagonism, *Bot. Gaz.*, **58**: 178–186.

———, 1914: The measurement of antagonism, *Bot. Gaz.*, **58**: 272–276.

———, 1915: On the nature of antagonism, *Sci.*, **41**: 255–256.

———, 1915: The determination of additive effects, *Bot. Gaz.*, **60**: 228–234.

———, 1916: Antagonism and Weber's law, *Sci.*, **44**: 318–320.

———, 1916: The penetration of balanced solutions and the theory of antagonism, *Sci.*, **44**: 395–396.

———, 1917: Antagonism and permeability, *Sci.*, **45**: 97–103.

———, 1918: The basis of measurement of antagonism, *J. Biol. Chem.*, **34**: 363–368.

———, 1919: Decrease of permeability and antagonistic effect caused by bile salts, *J. Gen. Physiol.*, **1**: 405–409.

———, 1922: "Injury, Recovery and Death in Relation to Conductivity and Permeability," pp. 124–184, J. B. Lippincott Company, Philadelphia.

———, 1922: Some aspects of selective absorption, *J. Gen. Physiol.*, **5**: 225–231.

———, 1925: Is living protoplasm permeable to ions? *J. Gen. Physiol.*, **8**: 131–146.

——— and W. M. Stanley, 1932: The accumulation of electrolytes. V, Models showing accumulation and a steady state, *J. Gen. Physiol.*, **15**: 667–689.

———, S. E. Kamerling, and W. M. Stanley, 1934: The kinetics of penetration. VI, Some factors affecting penetration, *J. Gen. Physiol.*, **17**: 445–467.

———, ———, and ———, 1934: Kinetics of penetration. VII, Molecular *versus* ionic transport, *J. Gen. Physiol.*, **17**: 469–480.

——— and S. E. Kamerling, 1934: The kinetics of penetration. VIII, Temporary accumulation, *J. Gen. Physiol.*, **17**: 507–516.

Pantanelli, E., 1915: Assorbimento elettivo di ion nelle piante, *Bull. Orto Bot. R. Univ. Napoli*, **5**: 1–54.

———, 1915: Über Ionenaufnahme, *Jahrb. wiss. Bot.*, **56**: 689–733.

———, 1918: Decorso dell' assorbimento di ioni nelle piante, *Bull. Orto Bot. R. Univ. Napoli*, **6**: 1–37.

———, 1929: Über Ionenaufnahme, *Protoplasma*, **7**: 129–137.

Pearsall, W. H., and J. Ewing, 1924: The isoelectric point of some plant proteins, *Biochem. J.*, **18**: 329–339.

——— and ———, 1924: The diffusion of ions from living plant tissues in relation to protein isoelectric points, *New Phytol.*, **23**: 193–207.

Pember, F. R., 1917: Studies by means of both pot and solution culture of the phosphorus and potassium requirements of the barley plant during its different periods of growths, *R. I. Sta. Bull.* 169.

Petrie, A. H. K., 1933: The intake of ions by the plant and its relation to the respiration of the root, *Australian J. Exp. Biol. and Med. Sci.*, **11**: 25–34; *Biol. Abst.*, **8**: 45.

Pettinger, N. A., 1932: Effect of fertilizers on the chlorine content of the sap of corn plants, *J. Agr. Res.*, **44**: 919–931.

Pfeffer, W., 1900: "Physiology of Plants," Vol. I, 420 pp., Eng. ed., Ewart, Clarendon Press.

Pierre, W. H., and G. G. Pohlman, 1933: Preliminary studies on the exuded plant sap and the relation between the composition of the sap and the soil solution, *J. Am. Soc. Agron.*, **25**: 144–159.

Pirschle, K., and H. Mengdehl, 1931: Ionenaufnahme aus Salzlösung durch die höhere Pflanze, *Jahrb. wiss. Bot.*, **74**: 297–363.

Potapov, N. G., and N. T. Stankov, 1934: Periodicity of mineral nutrition within the twenty-four hours, *C. R. Acad. Sci. U. S. S. R.*, **2**: 43–45; *Chem. Abst.*, **28**: 5099.

Prevot, P., and F. C. Steward, 1936: Salient features of the root system relative to the problem of salt absorption, *Plant Physiol.*, **11**: 509–534.

Raber, O. L., 1920: The antagonistic action of anions, *J. Gen. Physiol.*, **2**: 541–544.

Redfern, G. M., 1922: On the absorption of ions by the roots of living plants, *Ann. Bot.*, **36**: 167–174.

Reed, H. S., and A. R. C. Haas, 1924: Some effects of certain calcium salts upon the growth and absorption of citrus seedlings, *Am. J. Bot.*, **11**: 15–18.

——— and ———, 1924: Some relations between the growth and composition of young orange trees and the concentration of the nutrient solution employed, *J. Agr. Res.*, **28**: 277–285.

Rippel, A., 1926: Quantitative Untersuchungen über Kationenaustausch in der Pflanze, *Jahrb. wiss. Bot.*, **65**: 819–885.

Robbins, W. J., 1923: An isoelectric point for plant tissue and its significance, *Am. J. Bot.*, **10**: 412–440.

Rosenfels, R. S., 1935: The absorption and accumulation of potassium bromide by *Elodea* as related to respiration, *Protoplasma*, **23**: 503–519.

Rudolfs, W., 1925: Effect of seeds upon hydrogen ion concentration equilibrium in solution, *J. Agr. Res.*, **30**: 1021–1026.

———, 1925: Selective absorption of ions by seeds, *Soil Sci.*, **20**: 249–252.

Ruhland, W., 1909: Zur Frage der Ionenpermeabilität, *Z. wiss. Bot.*, **1**: 747–763.

Sachs, J., 1860: Vegetationsversuche mit Ausschluss des Bodens über die Nährstoffe und sonstigen Ernährungsbedingungen von Mais, Bohnen, und anderen Pflanzen, *Landw. Versuchs-Stat.*, **2**: 219–268.

———, 1887: "Lectures on the Physiology of Plants," Eng. ed., Clarendon Press.

Salter, R. M., and T. C. McIlvaine, 1917: The effect of reaction of solution on germination of seeds and on growth of seedlings, *J. Agr. Res.*, **19**: 73–97.

Schimper, A. F. W., 1890: Zur Frage der Assimilation der Mineralsalze durch die grüne Pflanze, *Flora*, **73**: 207–261.

Schloesing, T., 1869: Tabac sous cloche et à l'air libre, *Ann. Sci. Nat. Bot.*, V, **10**: 366–369.

Schreinemocher, F. A. H., 1893: Konzentrierung oder Verdünnung einer Lösung bei konstanter Temperatur, *Z. physikal. Chem.*, **11**: 81–85.

Schreiner, O., and J. J. Skinner, 1910: Ratio of phosphate, nitrate and potassium on absorption and growth, *Bot. Gaz.*, **50**: 1–30.

——— and ———, 1910: Some effects of a harmful organic soil constituent, *U. S. Bur. Soils Bull.* 70.

——— and ———, 1918: The triangle system for fertilizer experiments, *J. Am. Soc. Agron.*, **10**: 225–246.

SCOFIELD, C. S., 1927: The effect of absorption by plants on the concentration of the soil solution, *J. Agr. Res.*, **35**: 745–757.

SEWELL, M. C., 1924: Relation of the molecular proportions in the nutrient solution to the growth of wheat, *J. Agr. Res.*, **28**: 387–393.

SHERMAN, C. C., 1933: Theories on the nutrition of plants, from Aristotle to Liebig, *Yale J. Biol. and Med.*, **6**: 43–60.

SHIVE, J. W., 1915: A study of physiological balance in nutrient media, *Physiol. Res.*, **1**: 327–399.

———, 1915: A three-salt nutrient solution for plants, *Am. J. Bot.*, **2**: 157–160.

———, 1917: A study of physiological balance for buckwheat grown in 3-salt solutions, *N. J. Sta. Bull.* 319.

——— and W. H. MARTIN, 1918: A comparison of salt requirements for young and for mature buckwheat plants in water cultures and sand cultures, *Am. J. Bot.*, **5**: 186–191.

———, 1919: The influence of the moisture content of sand cultures upon the physiological salt balance for plants, *N. J. Sta. Rept.*, **1919**: 358–363.

———, 1919: The influence of sand of different degrees of fineness upon the concentration and reaction of a nutrient solution, *N. J. Sta. Rept.*, **1919**: 363–366.

———, 1920: Relation of moisture in solid substrata to physiological salt balance for plants and to the relative plant-producing value of various salt proportions, *J. Agr. Res.*, **18**: 357–378.

———, 1920: The influence of sand upon the concentration and reaction of a nutrient solution for plants, *Soil Sci.*, **9**: 169–179.

———, 1920: The influence of varying degrees of soil moisture upon the physiological salt balance for plants, *N. J. Sta. Rept.*, **1920**: 387–395.

———, 1922: The relation of soil moisture to physiological salt balance for plants, *Soil Sci.*, **14**: 391–411.

——— and A. L. STAHL, 1927: Constant rates of continuous solution renewal for plants in water cultures, *Bot. Gaz.*, **84**: 317–323.

SIDERIS, C. P., B. H. KRAUSS, and E. MASUNAGA, 1928: The effect of salt concentration of the culture solution on the growth and composition of pineapple plants, *Am. J. Bot.*, **15**: 353–372.

SKENE, M., 1915: The acidity of sphagnum and its relation to chalk and mineral salts, *Ann. Bot.*, **29**: 65–87.

SORAUER, P., 1880: Studien über Verdünstung, *Forschungen auf dem Gebiete der Agr. Physik.*, **3**: 351–490.

SPENCER, E. L., and J. W. SHIVE, 1933: The growth of *Rhododendron ponticum* in sand cultures, *Bull. Torrey Bot. Club*, **60**: 423–440.

STAHL, E., 1894: Einige Versuche über Transpiration und Assimilation, *Bot. Ztg.*, **52**: 117–146.

STEWARD, F. C., 1932: The absorption and accumulation of solutes by living plant cells. I, Experimental conditions which determine salt absorption by storage tissue, *Protoplasma*, **15**: 29–58.

———, 1932: The absorption and accumulation of solutes by living plant cells. II, A technique for the study of respiration and salt absorption in storage tissue under controlled environmental conditions, *Protoplasma*, **15**: 497–516.

———, R. WRIGHT, and W. E. BERRY, 1932: The absorption and accumulation of solutes by living plant cells. III, The respiration of cut discs of potato tuber in air and immersed in water, with observations upon surface: volume effects and salt accumulation, *Protoplasma*, **16**: 576–611.

———, 1932: The absorption and accumulation of solutes by living cells. IV, Surface effects with storage tissue. A quantitative interpretation with respect to respiration and salt absorption, *Protoplasma*, **17**: 436–453.

———, 1933: Salt absorption and respiration of *Valonia*, *Carnegie Inst. Yearbook*, **32**: 281–283.

———, 1933: The absorption and accumulation of solutes by living plant cells. V, Observations upon the effects of time, oxygen and salt concentration upon absorption and respiration by storage tissue, *Protoplasma*, **18**: 208–242.

——— and W. E. Berry, 1934: The absorption and accumulation of solutes by living plant cells. VII, The time factor in the respiration and salt absorption of Jerusalem artichoke tissue (*Helianthus tuberosus*), with observations on ionic interchange, *J. Exp. Biol.*, **11**: 103–119.

——— and ———, 1936: The absorption and accumulation of solutes by living plant cells. VIII, The effect of oxygen upon respiration and salt accumulation, *Ann. Bot.*, **50**: 345–366.

Stiles, W., 1915: On the relation between the concentration of the nutrient solution and the rate of growth of plants in water culture, *Ann. Bot.*, **29**: 89–97.

———, 1916: On the interpretation of the results of water culture experiments, *Ann. Bot.*, **30**: 427–436.

——— and I. Jorgensen, 1917: Observations on the influence of aeration of the nutrient solution in water culture experiments with some remarks on the water culture methods, *New Phytol.*, **16**: 182–197.

——— and F. Kidd, 1919: The influence of external concentration on the position of the equilibrium attained in the intake of salts by plant cells, *Proc. Roy. Soc.*, B, **90**: 448–470.

——— and ———, 1919: The comparative rate of absorption of various salts by plant tissue, *Proc. Roy. Soc.*, B, **90**: 487–504.

———, 1924: The absorption of salts by storage tissue, *Ann. Bot.*, **38**: 617–635.

Stolhane, A. A., 1914: A comparative study of nutritive solutions in sand and water cultures, *Rec. Trav. Lab. Agron. Moskov. Selsk. Khoz. Inst.*, **10**: 289–336; *Exp. Sta. Rec.* **36**: 630.

Tarr, L. W., and S. C. Noble, 1922: The effect of hydrogen ion concentration upon the growth of seedlings, *Del. Sta. Bull.* 131.

Thomas, W., 1930: The effect of colloidal silica on the absorption of phosphoric acid by plants, *Sci.*, **71**: 422–423.

———, 1933: Absorption, utilization, and recovery of nitrogen, phosphorus and potassium by apple trees grown in cylinders and subjected to differential treatment with nutrient salts, *J. Agr. Res.*, **47**: 565–582.

Tollens, B., 1882: Ueber einige Erleichterungen bei der Cultur von Pflanzen in wasserigen Lösungen, *J. Landw.*, **30**: 537–540.

Toole, E. H., and W. E. Tottingham, 1918: The influence of certain added solids upon the composition and efficiency of Knop's nutrient solution, *Am. J. Bot.*, **5**: 452–461.

Tottingham, W. E., 1914: A quantitative chemical and physiological study of nutrient solutions for plant cultures, *Physiol. Res.*, **1**: 133–245.

Trelease, S. F., 1917: The relation of the concentration of the nutrient solution to the growth of young wheat plants in water cultures, *Johns Hopkins Univ. Circ.*, **3**: 225–227.

——— and E. E. Free, 1917: The effect of renewal of culture solutions on the growth of young wheat plants in water culture, *Johns Hopkins Univ. Circ.*, **3**: 227–228.

——— and B. E. Livingston, 1922: Continuous renewal of nutrient solution for plants in water culture, *Sci.*, **55**: 483–486.
——— and ———, 1924: The relation of climatic conditions to the salt-proportion requirements of plants in solution cultures, *Sci.*, **59**: 168–172.
———, 1925: Chamber with thermostatic control and rotating table for plant cultures, *Bull. Torrey Bot. Club*, **52**: 389–395.
——— and H. M. Trelease, 1925: Growth of wheat roots in salt solutions containing essential ions, *Bot. Gaz.*, **80**: 74–83.
——— and ———, 1926: Relation of temperature to the physiological values of salt solutions as indicated by growth of wheat roots, *Bull. Torrey Bot. Club*, **53**: 605–609.
——— and ———, 1926: Toxicity and antagonism in salt solutions as indicated by growth of wheat roots, *Bull. Torrey Bot. Club*, **53**: 137–156.
——— and ———, 1928: Susceptibility of wheat to mildew as influenced by salt nutrition, *Bull. Torrey Bot. Club*, **55**: 41–68.
——— and ———, 1933: Physiologically balanced culture solutions with stable hydrogen ion concentration, *Sci.*, **78**: 438–439.
——— and J. R. Thomson, 1935: Regulating the flow of solution for plant cultures, *Sci.*, **81**: 204.
——— and H. M. Trelease, 1935: Changes in hydrogen-ion concentration of culture solutions containing nitrate and ammonium nitrogen, *Am. J. Bot.*, **22**: 520–542.
True, R. H., and W. J. Gies, 1903: On the physiological action of some of the heavy metals in mixed solutions, *Bull. Torrey Bot. Club*, **30**: 390–402.
———, 1915: Antagonism and balanced solutions, *Sci.*, **41**: 653–656.
——— and H. H. Bartlett, 1915: The exchange of ions between the roots of *Lupinus albus* and cultural solutions containing one nutrient salt, *Am. J. Bot.*, **2**: 255–278.
——— and ———, 1915: The exchange of ions between the roots of *Lupinus albus* and cultural solutions containing two nutrient salts, *Am. J. Bot.*, **2**: 311–323.
——— and ———, 1916: The exchange of ions between the roots of *Lupinus albus* and culture solutions containing three nutrient salts, *Am. J. Bot.*, **3**: 47–57.
Truog, E., 1918: Soil acidity. I, Its relation to growth of plants, *Soil Sci.*, **5**: 169–195.
Wean, R. E., 1935: Automatic flowmeter for drip solutions in plant nutritional studies, *Sci.*, **82**: 336.
Weatherwax, P., 1916: An apparatus for aerating culture solutions, *Proc. Ind. Acad. Sci.*, **1914**: 157–160.
Wendt, G. V., 1930: Stimulationsversuche mit Jod, *Gartenflora*, **79**: 213–218; *Biol. Abst.*, **7**: 791.
Wodehouse, R. P., 1917: Direct determinations of permeability, *J. Biol. Chem.*, **29**: 453–458.
Wollney, W., 1898: Untersuchungen über den Einfluss der Luftfeuchtigkeit auf das Wachstum, *Bot. Cent.*, **76**: 249.
Woodward, J., 1699: Thoughts and experiments on vegetation, *Phil. Trans. Roy. Soc.*, **21**: 382–398.
Wynd, F. L., 1933: The sensitivity of orchid seedlings to nutritional ions, *Ann. Mo. Bot. Gard.*, **20**: 223–237.
Zinzadzé, C., 1932, 1933: Nutrition artificielle des plantes cultivées. I, Mélanges nutritifs à pH stable, *Ann. Agron.*, Nov.-Dec. **1932**: 809–853; Jan.-Feb. **1933**: 53–72.
———, 1934: Nutrient media with stable hydrogen ion concentration, *Sci.*, **79**: 480–481.
Zurbicki, Z., 1933: New method of arranging experiments with vegetables in sand cultures with flowing solution, *Plant Physiol.*, **8**: 553–558.

CHAPTER VI

THE ELEMENTS ABSORBED BY THE PLANT

I. NUMBER AND KIND

According to Palladin (1927), 35 elements have been detected in the analysis of plants. These are carbon, hydrogen, oxygen, nitrogen, sulphur, phosphorus, potassium, magnesium, calcium, iron, chlorine, bromine, iodine, fluorine, boron, silicon, sodium, lithium, rubidium, strontium, barium, zinc, mercury, aluminum, thallium, titanium, tin, lead, arsenic, selenium, manganese, cobalt, nickel, copper, and silver. Robinson, Steinkoenig, and Miller (1917) analyzed a large number of plants from different regions for the rarer elements. In addition to those just mentioned, they detected chromium, vanadium, and cesium. Vanadium was found, however, in only six cases and then only in traces. Cesium was detected in only three cases and in two of them cesium beryls were known to be present in that region. A general idea of the number and relative amounts of the elements that compose plants may be had by considering a given plant, although, as will be mentioned later, the elemental composition of a plant will depend upon the species and upon the conditions under which it is grown.

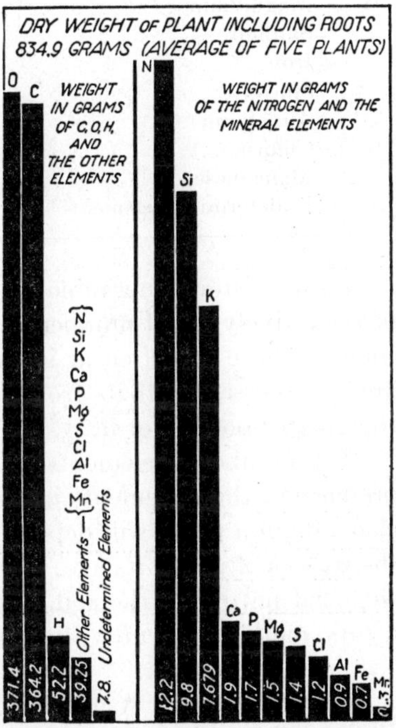

FIG. 13.—Diagram showing the elemental composition of a corn plant at the late dough stage.

An elemental analysis of the corn plant (Latshaw and Miller, 1924) may be taken as this example. These plants were analyzed at a stage when the grain was fully dented and in the late dough stage. The leaves were all intact, and the plants were evidently at a stage where absorption of solutes from the soil had practically ceased. The proportion of these different elements is shown in Fig. 13 and in the table on page 284.

Weight of the Elements That Composed the Stem, Leaves, Cob, Grain, and Roots of Pride of Saline Corn Plants Grown at Manhattan, Kans., in 1920
(Average of Five Plants)

Average dry weight of stems, leaves, cob, grain, and roots, 835.9 g.	Weight, grams	Percentage of the total dry weight
Oxygen	371.4	44.431
Carbon	364.2	43.569
Hydrogen	52.2	6.244
Nitrogen	12.2	1.459
Phosphorus	1.7	0.203
Potassium	7.7	0.921
Calcium	1.9	0.227
Magnesium	1.5	0.179
Sulphur	1.4	0.167
Iron	0.7	0.083
Silicon	9.8	1.172
Aluminum	0.9	0.107
Chlorine	1.2	0.143
Manganese	0.3	0.035
Undetermined elements	7.8	0.933

From the foregoing table it is seen that the bulk of a plant is composed of a relatively small number of elements, but as will be mentioned later some of those which occur in the plant in only relatively small amounts are just as essential to its growth and development as those which compose the greater portion of it.

The number of elements in a plant will depend principally upon the presence of these elements in the soil. The amount or proportion of the elements in a plant will depend upon many factors. Some of these are: the species of plant, its age, the distribution of its roots, the physical and chemical nature of the soil, the proportion and distribution of the elements, methods of cultivation, and the general climatic conditions.

A. Influence of the Plant

It is common knowledge that plants of different species grown under identical conditions differ in their elemental composition, especially in regard to the proportionate amount of the various elements that compose them.

Observations have been made by Parker and Truog (1920), Richardson (1920), Headden (1921), Skinner and Reid (1921), Greaves and Nelson (1925), Wherry (1926), von Wrangell (1927), Newton (1928), Gregory and Crowther (1928, 1931), Haley, Longenecker, and Olson (1931), Daniel

(1934), Harper, Daniel, and Murphy (1934), Davidson and Chambliss (1932), Maume and Dulac (1934), Archibald and Bennett (1935), Dadswell (1935), Hurd-Karrer (1935), Trelease and Martin (1936), and others in regard to the different proportions of the elements contained in plants of different species grown under the same conditions. Richardson (1920) determined the composition of the ash of certain dune plants and the sand in which they grew. Approximately 92 per cent of this sand was silica, while the remainder consisted of iron and aluminum oxides 4.3 per cent, calcium oxide 1.4 per cent, magnesium oxide 0.7 per cent, potassium oxide 1.0 per cent, and less than 1 per cent of organic matter. The composition of the ash of the sand cherry (*Prunus pumila*), artemisia (*Artemisia caudata*), *Andropogon* (*sp.*), and the scouring rush (*Equisetum hyemale*) growing in this sand will suffice to show the specific differences in the amount of certain elements absorbed by these plants.

ANALYSIS OF THE ASH OF PLANTS GROWING IN SAND DUNES

Constituent	Percentage of the ash			
	Prunus pumila	*Artemisia caudata*	*Andropogon*	*Equisetum hyemale*
Silica	1.5	12.1	65.4	49.4
Iron oxide	0.7	1.7	2.5	1.0
Aluminum oxide	0.02	0.4	2.6	2.3
Calcium oxide	44.1	35.5	10.2	13.4
Potassium oxide	10.9	11.6	6.7	6.0
Magnesium oxide	4.2	6.4	3.2	3.7

The most striking differences to be observed in the proportion of elements absorbed by these plants is in the amounts of silicon, calcium, and aluminum. *Andropogon* and *Equisetum* absorbed and accumulated large amounts of silicon, while *Prunus* and *Artemisia* accumulated only a relatively small amount. *Prunus* and *Artemisia*, on the other hand, accumulated from three to four times as much calcium as did *Andropogon* and *Equisetum*, while the amount of aluminum is from six to ten times greater in some plants than in others.

Headden (1921) noted that alfalfa growing in soils which contained both lithium and strontium had little lithium in the ash but strontium was present in considerable quantity. In tobacco, on the other hand, growing in the same soil, lithium was present in considerable quantity, but only a trace of strontium could be detected. Newton (1928) grew sunflowers, beans, wheat, barley, peas, and corn in the same nutrient solution in the same container. He harvested the wheat and barley at the heading stage, the corn at the time of tasseling, and the peas and beans at the blooming stage and determined the percentage of the various elements present in the plant. The similarities and differences in absorption which are characteristic of these plants are shown in the following data:

Plant	Percentage of the dry weight of				
	Ca	K	Mg	N	P
Sunflower	2.2	5.0	0.64	3.6	0.56
Beans	2.1	4.0	0.59	3.6	0.55
Wheat	0.8	6.7	0.41	4.5	0.49
Barley	1.9	6.9	0.54	4.7	0.52
Peas	1.6	5.3	0.50	4.5	0.19
Corn	0.5	3.9	0.40	2.9	0.39

The sunflower plant contained 2.7 times as much calcium as the wheat plant and 4.4 times as much as the corn. The sunflower also contained approximately 1.5 times as much magnesium as wheat and corn. Wheat and barley contained a percentage of potassium considerably higher than that for any of the other plants. Newton found that absorption from the cultural solution of the various elements was higher than by plants grown in a loam soil and that there was much less variability in absorption from the solution than from the soil. An observation of von Wrangell (1927) is worthy of note in considering the different rates of absorption by plants. He noted the differences in the amount of iodine contained in different types of plants growing in the same soil and found that lettuce and spinach contained 100 times more iodine than other agricultural plants under the same conditions.

Hurd-Karrer (1935) grew different species of plants in greenhouse plots in which the soil had been treated with 5 p.p.m. of selenium as sodium selenate. The amounts of selenium absorbed by the various plants are shown in the following table:

Plant	P.p.m. of dry weight	Plant	P.p.m. of dry weight	Plant	P.p.m. of dry weight
Mustard	1,240	Alfalfa	560	Proso millet	285
Broccoli	1,180	Peas	560	Corn	275
Sunflower	790	Oats	535	Crested wheat grass	255
Flax	685	Wheat	470		
Sweet clover	645	Barley	450	Brome grass	200
German millet	590	Spinach	315	Sorgo	130

Trelease and Martin (1936) reported that western wheat grass growing on various soils accumulated 1 to 60 p.p.m. of selenium, while *Astragalus bisulcatus*, growing on the same soil, accumulated 200 to 4,300 p.p.m. On soil containing 2.1 p.p.m. of selenium, *A. bisulcatus* accumulated 1,250 p.p.m., while *A. missouriensis* stored only 3.1 p.p.m. Daniel (1934) found that ordinarily legumes contained 3.9 times as much calcium, 1.7 times as much phosphorus, and 2.6 times as much nitrogen as the grasses. Plants that are low in calcium and phosphorus remained low in these elements when grown on a fertile soil. Those which are relatively high in these elements always contain relatively large amounts even though growing on a poor soil. The legumes have low nitrogen/phosphorus and high calcium/phosphorus and nitrogen/calcium, while the grasses have high nitrogen/phosphorus and low calcium/phosphorus and nitrogen/calcium.

Maume and Dulac (1934) showed that the percentages of nitrogen, phosphorus, and potassium, varied among the varieties of wheat at a like stage of development, when grown under the same conditions.

Since different plants are composed of different proportions of the various elements, the statement is frequently made that plants exercise "selective absorption" or "feeding power." Since these terms have been so frequently used and since they have been so differently interpreted, they should be mentioned here. It has been assumed by some that the differences in the composition of plants are due primarily to differences in the specific absorbing powers of the roots. They consider thus that the absorbing cells of the roots differ in their ability to exclude certain elements more than others. On the other hand, some consider that the difference in the amount of certain elements in plants is due to the specificity of the protoplasm, which requires different proportions of materials for its development in different plants. The protoplasm of the developing cells is thus considered to regulate the amount of any given element in the plant rather than the regulatory action of the absorbing cells of the root. The speed of absorption of a given compound in this case would thus be determined by the rate at which it was utilized by the cells in the interior of the plant. As has been mentioned in Chap. V, the absorption of solutes is apparently not so simple a process and the phenomenon of the heaping up of salts in the ionic form on the interior of the absorbing cells cannot be explained by either of the above-mentioned theories. Davis, Hoagland, and Lipman (1923) also point out that the differences in the extent of the root systems and the differences in the rate of production of carbon dioxide by these roots are factors that may determine to a considerable degree the differences in the amount of minerals absorbed by plants.

The relationship between the amount of nutrients absorbed by a plant and its development is a complex one and is very difficult of interpretation. In this regard the statement of Thomas (1934) is appropriate—"A comparison of the quantitative relations between the elements during the growth of a particular species, subjected to different growth conditions, cannot be made without a key of interpretation. Consequently comparisons between the composition of different species with respect to any of the dominant or accessory mineral elements cannot be made except by such a key."

B. Influence of Environment

The amount of any element that is found in the plant is influenced by the abundance and location of that element in the soil. Thus Scharrer and Strobel (1927) and Orr, Kelly, and Stuart (1928) found that they could greatly increase the iodine content of sugar beets, peas, oats, and alfalfa by the addition of this element to the soil in which these plants were growing. Parker and Pierre (1928) grew corn in various concentrations of phosphate (PO_4) from 0.05 to 50 p.p.m. During the younger stages of growth, increased concentration increased absorption; but in later periods, the absorption was not proportional to the concentration of the external solution in many cases. In case of potassium, the maximum growth of corn was secured with 2 p.p.m. The concentration necessary for growth, however, varies with the species. Thus soybeans and tomatoes give maximum growth at 3 to 5 p.p.m. of potassium, while maximum growth is secured for oats at 2 p.p.m. and alfalfa at only 1 p.p.m.

Godden (1926) reported that pastures in England responded to the application of artificial fertilizers by a considerable modification of the mineral contents of the plant. The constituents that showed the greatest variations were calcium and potassium, while phosphorus varied to a slightly lesser extent. Brown (1929) studied the mineral composition of apples that had been grown on 10 different soils. The amount of potassium in the apples showed a definite relationship with the amount in the soil, and there were indications of a similar relationship between the percentage of phosphorus, magnesium, and calcium. It was found by Vandecaveye and Baker (1932) in Washington that the application of lime, phosphate, or potassium, either alone or in combination, resulted in a higher content of calcium, phosphorus, and potassium in pasture grasses and hay. Thomas (1932) considered that the plant from the early stages of growth to blooming absorbs nutrients in approximately the same proportion as they are present in the medium in which it grows. Under some conditions, however, the amount of an element in a plant is not greatly influenced by the relative amount in the soil. Thus Brown (1932) found that the influence of fertilizers on the total ash content was not very evident in white clover, bluegrass, wild oats, and other plants growing in permanent pastures of Connecticut. Daniel and Harper (1934) studied the total calcium and phosphorus in the plant and soil and concluded that the amount of any single element in the soil will not give an accurate indication of the amount of that element which will be found in the plant.

The location of the various nutrients with respect to the roots may have a marked effect upon the amount of elements contained in the plant, as indicated by the work of Gile and Carrero (1917, 1921). They grew corn plants in such a way that the roots were divided among three solutions each of which lacked two elements so that the nitrogen was in one flask, the phosphorus in another, and the potassium in a third. It was found that the amount of these three elements in the plants thus grown was lower than under normal conditions, although the rate of absorption of nitrogen, for example, was greatly increased by the roots in that solution. The cause of the smaller amount is seemingly due to the slowness of distribution of these elements within the plant, since they have to be transferred over the plant by and from a comparatively small number of fibrovascular bundles.

The amount of a given element in a plant may be influenced by the proportion and kind of other elements contained in the soil or culture solution. Thus Ginsburg (1925) found that soybeans growing in a complete nutrient solution had an ash content of 13.3 per cent. When phosphorus was withheld from the solution, the ash content was 18.6 per cent and amounted to 18.4 per cent when magnesium was lacking. The plants growing in the incomplete cultural solutions, with the exception of those grown in solutions lacking sulphur, always absorbed higher percentages of calcium and lower percentages of both nitrogen and magnesium than did the checks. In the case of spinach, which was grown on soil that was heavily fertilized with different fertilizers or fertilizer mixtures, it was observed that the calcium, magnesium, and sulphur content of the plant soon reached an equilibrium regardless of the quantity or kind of nutrients applied but that the silicon, potassium, and sodium content showed wide fluctuation with the various applications of nutrients (True, 1919). Leaves of orange trees growing in cultures to which no calcium salts were added were found to be rich in potassium, while those of plants receiving no potassium salts in the nutrients were high in calcium (Reed and Haas, 1923).

Carolus (1933), in experiments with tomatoes, found that the use of 60 lb. per acre of ammonia produced a marked absorption of nitrogen, stimulated the absorption of magnesium, and slightly depressed the intake of potassium. The application of 120 lb. per acre of phosphorus pentoxide resulted in a large increase in the absorption of phosphorus and calcium, and in an increased intake of magnesium. A slight deficiency or excess of potassium in the plant caused a condition of physiological unbalance that resulted in both decreased growth and reduced yields, and impaired the quality of the fruit.

The amount of the various elements absorbed by plants is evidently influenced to a marked degree by the condition of the soil. Thus Brown (1929) considered in the case of the apple that the mechanical structure of the soil had a marked influence on the composition of the fruit. Greaves and Hirst (1929) reported that the ash content of growing wheat, oats, barley, and corn in Utah was increased 46, 31, 36, and 8 per cent, respectively, by the application of optimum quantities of irrigation water. Daniel and Harper (1934) also considered that the amount of moisture in the soil is an important factor in determining the amount of the absorption of various elements.

It was observed by Parker, Hester, and Carolus (1933) that the maximum amount and percentage of calcium in cabbage, peas, and beans occurred when the pH reaction of the soil was between pH 5.5 and 6.2. Magnesium was absorbed to a greater extent at the lower pH and was replaced by calcium as the neutral point was approached. Hester (1934) grew lima beans in soils with different pH values and in different calcium-ion concentrations, and found that at the pH values favorable to maximum crop production the soil carried a desirable nutrient balance for this crop.

Wynd and Fuller (1931) observed that tomato and cucumber plants which had been treated with ultraviolet radiation showed a decided increase in calcium and a decrease in phosphorus as compared to the controls. Steward and Arthur (1933, 1934) found that various species of plants, when grown under conditions of low light intensity, exhibited an increase in ash content upon irradiation. This increase in ash appeared 42 to 72 hr. after exposure to the rays. The effective wave lengths were within the range of 2900 to 3130 Å. The application of irradiated ergosterol in olive oil to the basal leaves brought about an increase in the ash of the plants.

Street (1934) found in the field pea that the crude ash was higher in plants grown in a 10-hr. than in a 13-hr. light period. Under the former condition, the plants were high in potassium and low in calcium and magnesium, while under the latter condition they were lower in potassium and higher in calcium and magnesium. Plants exposed to light for 17 hr. daily were markedly low in potassium and usually highest in calcium and magnesium.

If there is an abundance of the soluble inorganic salts present in the soil, most plants will absorb a greater quantity of them than is necessary for their maximum growth. This has been especially noted for the salts of phosphorus, potassium, and sulphur. Jordan (1913) in nutrition experiments with barley, peas, tomatoes, tobacco, buckwheat, and rape noted that both phosphorus and potassium compounds increased in amount in these plants after the maximum increase in growth had been reached. Similar observations have been made by Hoagland (1919 and 1926), who found that the concentration of the cell sap in regard to electrolytes is much higher than its concentration in the soil. A further indication that many of the elements are present in the plant cells in excess of their needs is the fact that a large proportion of them in many cases can readily be extracted with water. Thus Ames and Boltz (1912) found in the alfalfa plant that 90 per cent of the potassium, 85 per cent of the magnesium, 75 per cent of the nitrogen, and 50 per cent of the sulphur were soluble in water. A considerable portion of the nitrogen is known to be in simple soluble protein compounds, but apparently the other elements mentioned are only loosely combined and probably not at all.

Canals, Canaye, and Cabanes (1930) showed in the dialysis of the expressed saps of radish, carrot, turnip, beet, and other plants that most of the minerals occur in the dialyzable form. The loss of sodium and potassium was complete, while the loss of phosphorus and magnesium ranged from 95 to 100 per cent during dialysis. The loss of calcium was highly variable ranging from 30 to 94 per cent. The calcium of the leaves was predominantly in the dialyzable form, while in the roots a considerable portion was not in that condition. Lausberg (1935) found that leaves of *Ricinus* and *Vicia* lost large amounts of ash constituents when sprinkled or when placed in water for varying periods of time. Sprinkled leaves lost 9 to 32 per cent of their ash constituents, while the loss of potassium was 23 to 60 per cent, and that of calcium 6 to 10 per cent. When both sides of the leaves were sprinkled equally, twice the amount of minerals was lost from the upper as from the lower side. Halma and Haas (1931) noted that some of the inorganic constituents of citrus cuttings underwent marked changes in solubility during the rooting process.

It is a general observation that a plant will respond favorably to an increasing amount of a nutrient in the soil provided this nutrient was not originally present in an amount sufficient for maximum growth. Above a certain limit, however, the plant will cease to respond to increased application of the nutrient. It will generally continue, however, in such cases to absorb increasing amounts from the soil as indicated by an analysis of the tissues. The occurrence in a plant of a greater quantity of an element than is necessary for its needs is termed "luxury absorption" or "luxury consumption."

According to Thomas (1932) the optimum ratio of the constituents of a fertilizer may be defined as that ratio which on the addition of any one of its constituents—nitrogen, phosphorus, or potassium—results in no increased utilization by the plant of any of the other elements as determined by the time-absorption method.

The amount of a given element in a plant may be sufficient for its growth and development, but it may not be sufficient for the needs of the animal that consumes the plant. This condition prevails in regions where there is a scarcity of certain elements in the soil, or because they are in a form that is unavailable to the plant. The elements iodine, phosphorus, and copper are the ones that are the most generally lacking in the proper amount in the soil and consequently in the plant. The lack of iodine in the drink or diet of animals produces a type of goiter. The lack of sufficient phosphorus or copper in forage and grain causes the so-called "salt sickness" of animals. Certain areas of Florida, Minnesota, and Montana are especially lacking in available copper or phosphorus, and the herbivorous animals in those regions are not sufficiently supplied with these elements (Scott, 1929; Bryan and Becker, 1935; and Dove, 1934).

C. Indicators of the Need

Even after analyzing the soil for various constituents, it is practically impossible to interpret these data in relation to the influence of the constituents on the growth of the plant. A chemical analysis of the soil indicates more of its potential fertility than it does of the available nutrients that are actually concerned in crop production (Thornton, 1933). Many investigations have been made in recent years to diagnose the needs of plants by noting their general appearance, the characteristics of the cell sap, and by applying certain microchemical tests to the various tissues. A large number of workers, including Dastur and Cooper (1932); Emmert (1930–1935); Emmert and Ball (1933); Fraps (1931); Gilbert (1926); Gilbert and Hardin (1927); Gilbert, McLean, and Adams (1927); Gilbert and Smith (1929); Gilbert and Pember (1935); Hoffer (1926); Jones (1931); Kraybill (1930); Laurie and McElwee (1934); Lauritzen (1934); McMurtrey (1933); Mitchell (1934, 1935); Murneek and Gildehaus (1931); Pettinger (1931); Poehlman (1935); Pohlman and Pierre (1933); Smith, Crandall, and Frear (1932); Shaw (1934); Thornton (1932, 1933); Thornton, Conner, and Frasure (1934); Van Haarlem (1935); and others; have made the investigations. The work that has been done shows that with some plants, and under certain conditions, the three types of observations may be indicators of the fertilizer needs of the plant. Much more information, however, must yet be acquired before such observations can be considered as reliable and of practical application.

1. General Appearances.—In corn plants growing in the field, a scarcity or lack of potassium is evidenced by the premature death of the plants, chaffy ears borne on premature broken shanks, and badly decayed roots (Hoffer, 1926). In some plants, dead tissue around the margins and below the veins of the leaves is an indication of potassium deficiency. In tobacco, McMurtrey (1933) mentioned that typical potassium deficiency is distinguished by small, necrotic spots or specks at the tips or margins of the chlorotic leaves. The leaves curl upward at the tips and margins. In the tomato when the supply of phosphorus is restricted, growth is retarded, and the leaves acquire a typical brown or purplish tinge. In tobacco a shortage of phosphorus is indicated by yellowing or drying of the lower leaves; the roots are relatively long, little-branched, and reddish.

According to Hoffer (1926), the common symptoms of nitrogen starvation in corn are light yellow-green leaves and stalks. McMurtrey (1933) stated that in the absence of nitrogen, the plant became light-green with more or less yellowing, drying, or firing of the lower leaves. The roots were long, little-branched, and white.

Under this topic the so-called "deficiency diseases" of plants should be mentioned. These are characteristic abnormal conditions that result from the unavailability of Def certain elements to a plant. A comprehensive review of these disorders is given by Humphrey (1934). In this text these "diseases" are mentioned for the most part under the discussions of the various elements. Some of these diseases are: chlorosis, sand drown of tobacco, gray speck of oats, hollow heart of turnip, heart rot of beet, little leaf or rosette of fruit trees, white bud of corn, tea yellows, leaf scorch of apples, and frenching of the leaves of various plants. In some cases the lack of an element may not cause a definite disease, but its scarcity may make the plant more susceptible to the invasion of various fungi (Heck, 1934).

2. Characteristics of the Sap.—According to Pettinger (1931), the quantity of sap that can be extracted from the cornstalk tissues is closely related to the diameter of the stalk and to soil productivity. There is a general relationship between the color of the expressed sap and the productivity of the soil on which the plants are grown. Those saps that are colorless or only light brown after clarification indicate a fertile soil and a sufficiency of potassium. A clarified sap with a dark-brown color is associated with an infertile soil and a scarcity of potassium.

If the expressed sap contains less than 100 p.p.m. of nitrates, the soil is deficient in available nitrogen, but if it contains above 300 p.p.m. of nitrates, there is an ample supply in the soil for all the needs of the plant. If the sap contains less than 0.20 mg. of phosphorus pentoxide per cubic centimeter, the soil is deficient in available phosphorus, but if there is present more than 0.20 mg. of phosphorus pentoxide per cubic centimeter, the soil supply is abundant. When the concentration of potassium oxide in the sap is lower than 1.0 mg. per cubic centimeter, potassium is deficient in the soil. If potassium oxide is present in excess of 2.0 mg. per cubic centimeter of sap, the soil contains an abundance of potassium for the growth and maturation of the plant.

3. Microchemical Tests.—In corn the presence of iron in the tissues of the joints indicates a lack of potassium (Hoffer, 1926). The presence of iron is indicated by a red color when a few drops of a solution of potassium thiocyanate is applied to the fresh tissue and followed with a drop or two of dilute hydrochloric acid. The relative amount of potassium can be estimated by treating sections with a 10 per cent solution of platinic chloride and examining the crystals that are formed. The relative amount of potassium in the plant may be determined by treating a portion of macerated tissue with a special solution of cobalt nitrite. A heavy precipitate indicates abundant potassium, a medium precipitate indicates a doubtful supply, and no precipitate, or only the trace of one, indicates that the soil is deficient in that element (Thornton, 1933). The application of 2 or 3 drops of a special solution of diphenylamine to freshly cut tissue gives a characteristic blue color if nitrates are present. A dark-blue color indicates that there is available to the plant an abundant supply of nitrates, a pale-blue color indicates an adequate amount, and no color suggests that the supply in the soil is inadequate. If a portion of crushed plant tissue is treated with ammonium molybdate solution and a small portion of stannous chloride is added and the whole shaken, the inorganic phosphorus concentration is indicated by a series of color intensities and shades ranging from colorless or yellowish to a dark blue. A medium- to dark-blue color indicates a sufficiency to an abundance of this element in the soil, while it is doubtful if a sufficiency is present when the color is green to light blue.

D. Distribution of the Elements in the Plant

The distribution of the various elements in the different parts of the plant is shown in the following table. A knowledge of their distribution is of importance in interpreting the role that they serve in the functions of the plant.

PERCENTAGE DISTRIBUTION OF THE DIFFERENT ELEMENTS IN VARIOUS PARTS OF CORN PLANTS GROWN AT MANHATTAN, KANS., IN 1920

Organ	C	O	H	N	P	K	Ca	Mg	S	Fe	Si	Al	Cl	Mn
Leaves........	26.6	27.7	26.3	25.0	28.6	45.2	58.2	32.3	39.8	23.0	62.3	19.5	42.8	27.9
Stems........	24.5	23.7	22.7	13.8	10.5	32.2	18.0	21.0	22.7	14.6	8.6	2.9	36.9	12.6
Grain........	32.0	31.8	34.8	46.0	52.3	14.2	3.4	34.2	25.8	15.7	0.4	6.7	7.1	35.7
Roots........	7.0	7.1	9.5	6.3	4.2	3.6	19.5	6.8	10.7	44.1	27.6	66.3	5.4	14.9
Cobs.........	9.8	9.7	6.6	8.9	4.4	4.7	0.9	5.6	1.1	2.7	1.1	4.6	7.8	8.0

The distribution of the various elements in the plant parts varies to some extent with the stage of growth. The distribution of the elements in barley will suffice as an example (Pfeiffer, Rippel, and Pfotenhauer, 1919). The total ash in this plant increased up to the time of the milk stage. The total phosphorus increased up to harvest time, being highest in the leaves before heading and in the stems at heading time. The total calcium increased up to the milk stage and was highest in the leaves at heading time and in the stems during the milk stage. The total potassium increased up to the milk stage and was highest in the leaves before heading. The total nitrogen increased up to harvest time, being highest in the leaves before heading and in the stem at heading time.

II. FORM

A green plant obtains practically all its carbon and a part of its oxygen from the air and all the other elements from the soil. With the exception of some of the oxygen, all the elements that enter a plant must be presented to it in the form of soluble compounds. Oxygen is the only element that the plant can utilize when presented in elemental form, but it, too, enters the plant dissolved in water.

A. INORGANIC COMPOUNDS

It is a definitely established fact that the green plant is able to grow and develop to maturity if it is supplied with the necessary inorganic compounds in sufficient quantity for its needs. Under the proper conditions of light, climate, and soil fertility, the green plant is capable of manufacturing from the simple inorganic compounds that it obtains from the soil and the air all the numerous and varied organic substances that enter into its composition. The kind of inorganic compounds that best serve the needs of the plant will be discussed later in this chapter.

B. ORGANIC COMPOUNDS

The concentration of the soluble organic substances in the soil may be as high as that of the inorganic compounds, so it is of interest to know to what extent the green plant can absorb and utilize organic material.

Funchess (1916) found that large amounts of coumarin and vanillin applied to the soil depressed the yield of oats when they were planted immediately after the application of these compounds. Pyridine and quinoline were very beneficial to both oats and corn in all the soils tested. The addition of potassium and phosphorus greatly increased the beneficial effects of these two compounds. Asparagine and nucleic acid were beneficial to plants, and dihydroxystearic acid when applied alone had no effect on either oats or corn but, when added in combination with fertilizer, it produced a larger yield than when fertilizer alone was applied. Stokes, Leukel, and Barnette (1930) utilized the sewage from septic tanks for the irrigation of Napier grass and Japanese cane. The yield of green material from the irrigated plots was as much as 3.3 times that from the nonirrigated ones. Tanaka (1931) found that urea proved a good source of nitrogen for some plants. The sugars he used were mostly favorable to plant growth, while organic acids except in low concentrations gave mostly negative results. As a source of phosphorus, lecithin was more readily assimilated than phytin. Of the organic sulphur compounds, only cystine was assimilable to any extent. Malyschev (1932) found that excised roots of various plants could be grown in solutions of various salts to which sugar had been added. The most extensive experimental work has been done with the carbohydrates, organic acids, and nitrogenous compounds, although numerous other substances have been used.

1. Carbohydrates.—Acton (1889) supplied various carbohydrates and glycerin to *Acer pseudoplatanus*, *Phaseolus vulgaris*, *Quercus robur*, and *Euphorbia hirsuta* among other plants growing in water cultures. He found that glucose, saccharose, and glycerin were absorbed and utilized in the dark by these plants in the formation of starch but that glycogen, dextrin, and soluble starch were not so utilized. Laurent (1904) showed that corn, beans, and other plants could absorb glucose through their roots and utilize it, as indicated by the formation of starch in their leaves. The same results were also noted by Mazé and Perrier (1904) for corn grown in water cultures.

Knudson (1916) studied the influence of carbohydrates on plant growth by growing them under sterile conditions in an agar medium to which Pfeffer's nutrient solution and a sugar solution varying in concentration from 1 to 2 per cent had been added. The assimilability of the sugar was judged by noting the influence on the dry weight of the plants. Knudson found that corn thus grown was able to absorb by means of its roots and utilize in its metabolism glucose, fructose, saccharose, and maltose, the beneficial effects being in the order that the sugars are here named. Canada field peas responded in growth to the presence of saccharose, glucose, maltose, and lactose in the order stated. Timothy, vetch, and radish were also benefited by the presence of sugar. In the case of radish, the average dry weight of 71-day-old plants was as follows:

WEIGHT IN GRAMS

Case	Plants grown in Pfeffer's solution plus				
	Check	Sucrose 2 per cent	Maltose 2 per cent	Glucose 2 per cent	Lactose 2 per cent
I...........	0.791	1.237	1.572	1.713	1.128
II..........	1.208	....	1.950	2.461	1.900

The beneficial effects of sugar when supplied to the roots of the radish were also observed by Molliard (1907). Robbins (1918) grew the moss plant (*Ceratodon purpureus*) on sterile culture media and found that organic carbon in the form of levulose, glucose, galactose, lactose, sucrose, and maltose was absorbed and utilized, as was evidenced by the formation of starch in the plant in the dark. Mannite, glycerin, and starch, however, could not be utilized. He observed that the amount of growth with levulose as a source of carbon was from two to seven times greater than that with glucose as a source of carbon, and that with the presence of levulose the greater amount of growth occurred in the dark, while with glucose the greater amount occurred in the light. Light, however, seems to be necessary for the continued growth of this moss plant, even though a supply of available carbohydrates is furnished.

Brannon (1923), using the methods of Knudson, obtained data that indicate that different kinds of plants differ as to the sugar they are able to use most efficiently. Alfalfa grew best for him in solutions containing glucose, while timothy did better in solutions containing fructose. The radish used both sugars equally well. It seems that all the plants tested can use glucose and sucrose to some degree, while galactose, mannose, and fructose have been shown to be toxic to some plants under certain conditions. Roach (1926) studied the relation of soluble carbon compounds to the growth of certain soil algae. He found the relative average rates of growth in the different media to be as follows: glucose in the light, 100 per cent; maltose, 100 per cent; galactose, 94 per cent; sucrose, 84 per cent; fructose, 73 per cent; glycerin, 43 per cent; and mineral salts alone, 60 per cent. He also observed (1927) that the five species examined were capable of growing in complete darkness provided that a suitable organic compound was present in the medium.

Mayer (1929) obtained optimum root growth in a cane-sugar solution, while Wynd (1933) studied the growth of orchid seedlings on a series of different sugars over a period of 8 months. The order of excellence of growth on the different sugars was *d*-mannose > *d*-glucose > maltose >

d-fructose > sucrose > raffinose. No growth was obtained on *d*-galactose or the pentoses: 1-arabinose, 1-rhamnose, and 1-xylose.

The evidence that green plants can absorb and utilize carbohydrates is based in many cases upon the behavior of plants grown in the dark. The abnormal development and behavior of plants when grown in darkness, however, prevent one from drawing a general conclusion as to the response of green plants to organic nutrients under normal conditions of plant growth. The process of photosynthesis is a factor, too, that prevents a clear conception of the response of plants to organic nutrients, and this process cannot be eliminated in green plants when grown in the open in daylight even if the atmosphere is depleted of carbon dioxide. The process of respiration releases within the cells a considerable quantity of this gas, which is utilized at once by the chloroplasts in the process of photosynthesis. Excellent material for a study of organic nutrition would seem to be the albino plants, which frequently appear as sports in numerous species. These plants are apparently comparable in every regard to the green plants with the exception that they lack chlorophyll and functional plastids so that they cannot carry on photosynthesis.

Knudson and Lindstrom (1919) studied the effect of sugars upon the duration and growth of albino corn plants. Plants that were grown in Pfeffer's solution to which was added a two-tenth molecular solution of sucrose showed a very appreciable gain in weight during a 55-day experiment, while those grown in Pfeffer's solution alone showed a loss and perished 25 days sooner than those supplied with sugar. These authors attribute the failure of the albino plants to make a sustained growth to the slow diffusion of the sugar so that the plants were unable to absorb a sufficient amount of organic material to supply the needs of the entire plant. Their hypothesis is strengthened by the fact that after the leaves are dead, the roots continue to remain alive in the sugar solution several months after the check plants are dead. Robbins (1924) observed that excised root tips of peas, corn, and cotton made considerable growth in the dark in solution cultures containing mineral salts and glucose and fructose but that they made little growth under the same conditions when carbohydrates were lacking. The growth of these roots was markedly greater in the cultures containing glucose than in those containing fructose.

The roots under these conditions, however, do not grow to the extent that they do in the soil. Robbins (1929) considered that this lack of growth might be due to the toxicity of the nutrient solution, the failure to absorb sufficient glucose, lack of some essential element, lack of growth substances, or to a disturbance in the respiratory process.

2. Organic Acids.—Ravin (1913) investigated the influence of certain organic acids and their neutral and acid salts on the growth of the radish.

His results as summarized in the following table by Knudson (1916) indicate that the radish can absorb and make use of in growth such acids as malic, tartaric, succinic, and citric and their salts:

Culture medium	Dry weight grown in free atmosphere, milligrams	Dry weight grown in confined atmosphere, milligrams
Knop's solution	36.3	11.7
+ $KHSO_4$	41.0	16.9
+ K_2SO_4	42.0	17.0
+ Glucose	54.9	30.5
+ free organic acid	52.5	30.5
+ acid potassium salts of organic acids	53.9	33.0
+ neutral potassium, salts of organic acids	53.6	28.7

3. Organic Nitrogenous Compounds.—The utilization of organic nitrogenous compounds is discussed in detail under The Nitrogen Metabolism of the Green Plant (Chap. IX).

The absorbing system of the green plant is apparently adjusted for the absorption of the rapidly diffusing inorganic compounds and cannot readjust itself for the intake of sufficient quantities of the more complex organic compounds to supply all its needs for growth and repair. The experimental evidence indicates that green plants are capable of absorbing by their roots and utilizing in their metabolism a sufficient amount of organic matter to have a beneficial influence upon their growth. It seems evident, therefore, that the soluble organic matter of the soil may play a direct role in the nutrition of green plants.

III. ESSENTIAL VS. UNESSENTIAL ELEMENTS

From 1860 to 1900 much experimental work was done by Sachs, Knop, Nobbe, Pfeffer, and others to determine the elements that are essential to plant life. Their work indicated that 10 elements are absolutely necessary for all green plants. These are carbon, oxygen, hydrogen, nitrogen, phosphorus, potassium, calcium, magnesium, sulphur, and iron and have been considered until quite recently as the only ones necessary for the normal growth and development of plants. The other elements found in plants were considered of little or no physiological importance, and their presence was attributed to the fact that they entered the plant simply because of their occurrence in the soil in the soluble form; if comparatively large amounts were present, it was attributed to a specific power of the plant to accumulate them.

The experimental work by which the so-called "essential" elements were determined was considered to be above criticism until quite recently, when it was discovered that there are three important sources of error that had been overlooked: (1) many at least of the so-called "chemically pure" chemicals then used contained impurities that might supply elements other than the ten in sufficient amount to influence plant growth; (2) the various elements in the seed were not taken into consideration, and it is now known that the seed contains a sufficient amount of certain elements to supply the needs of the plant even up until the time of fruiting; (3) no precautions were taken to avoid the contamination of the cultures by certain elements that may dissolve in small amounts from the walls of the containers.

Haas (1930) noted that citrus plants grew normally in earthenware containers or in Swedish enamel mixing bowls without the addition of boron to the cultural solution. In sand cultures in galvanized iron cans, such solutions, however, gave poor results. Boron is present in the walls of many of the containers used in nutrition experiments, because borax is widely used in the preparation of pottery, enamel, agate ware, and certain kinds of glass. Silicon also is readily obtained by plants from the walls of the vessels commonly used in cultural studies. Consequently, Furnstal and Johnson (1936) devised sintered, pyrex glass aerators for water cultures to prevent any contamination by some of the rarer elements, which might invalidate results.

The elements needed by plants subserve two main functions: (*a*) Some of them are component parts of the cell structure. The amount of the different elements used for these purposes is relatively large, and any deficiency of them is soon noticed in the general growth of the plant. (*b*) Some of the elements apparently act as carriers of other ions, catalyzers of reactions, or antidoting agents. For these purposes the amount of some of the elements needed is very small and could easily be supplied in sufficient amount by impurities from the three sources that have just been mentioned. The observation that small amounts of certain of the so-called "unessential" elements improve the growth of certain plants and prevent certain deficiency diseases, and the discovery of the relation of vitamins in animal nutrition have perhaps given the impetus to renew the investigations concerning the role of many of the elements that have been considered unessential.

Mazé (1914–1919) has led the field in this work. He considered that zinc, silicon, chlorine, aluminum, boron, and perhaps other elements are essential to the growth and development of green plants. McHargue (1919–1935) in extended experiments, which will be discussed later in the chapter, has furnished very convincing data to prove that manganese is an essential element for plant growth. He considered that plants obtain such factors as are necessary for their growth from the mineral elements contained in a fertile soil, taking up small amounts of iron, manganese, copper, zinc, boron, nickel, and others with the aid of which are synthesized complex organic compounds in the plant. He considered that these organic compounds function as enzymes, catalysts, and vitamins and that when they are consumed by animals they are resynthesized into catalases, oxidases, hormones, and animal vitamins. Warington (1923, 1926), Brenchley (1914), Brenchley and Thornton (1925), Sommer and Lipman (1926), and many later investigators have shown that *Vicia faba* (Windsor bean) will not grow normally in water cultures unless small quantities of boron are added. A histological study showed that there is also a marked change in the anatomy of the root and nodules in the absence of this element (see Boron).

The increased study of the effects of the so-called unessential elements has given rise to the terms "rarer elements," "minor elements," and "less common elements." This does not mean that they are of infrequent occurrence in the soil, but that they are so named because apparently they are needed in only small amounts by the plants. These rarer elements, according to Thatcher (1932), and Young (1935), may affect the growth of plants in a number of ways. They may be necessary in very small amounts in certain physiological functions. Some may act as positive catalysts and others as negative catalysts. The latter type may depress certain reactions that are injurious to the plant. Some of these elements may inactivate toxins in the soil or may aid in oxidizing or reducing certain compounds in the soil to a form more readily available to the plant. The effects of manganese in maintaining the correct ratio of ferrous iron to ferric iron is given as an example of this. The rarer elements may have an effect upon colloids in overcoming the detrimental effects of certain ions. Thus the potassium ions in a pure aqueous solution are so strongly adsorbed that they injure the colloidal condition of the cell. When aluminum is added, neither the aluminum nor the potassium ion can monopolize the cell surface, and the effective action of both ions is decreased. These minor elements may have an action similar to vitamins in that the plant may grow to maturity and produce seed in their absence, but its general health will be greatly improved and its resistance to deficiency diseases greatly increased by their supplementary effects. Some of these elements may play an indirect role in the nutrition of the plant by balancing the nutrient solution in which it is growing. The concentration of these elements needed to produce the various effects mentioned above is low and variable, rarely below 1/500 p.p.m. and never greater than 400 p.p.m.

Thatcher (1934) proposed a classification of the chemical elements with respect to their functions in the metabolism of the green plant as follows: group I—hydrogen and oxygen—energy-exchange elements; group II—carbon, nitrogen, sulphur, and phosphorus—energy storers; group III—sodium, potassium, calcium, and magnesium—translocation regulators; group IV—manganese, iron (cobalt and nickel), copper and zinc—oxidation-reduction regulators. The other elements were placed into four other groups, but their function in the plant was not suggested. Haas and Reed (1927) found that young orange trees grown in sand cultures with the ordinary nutrient solutions showed marks of decline, while those receiving cultural solutions made with tap water did not show these symptoms. The leaves curled, became spotted and corky, and split along the veins. In other cases, the leaves were shed prematurely, and multiple buds were formed in the leaf axils or above the former leaf scars, but none of them was able to develop. They discovered that these plants could be rejuvenated by the addition to the sand culture solution of a suspension so as to give a concentration of 0.2 p.p.m. of aluminum, iodine, titanium, bromine, strontium, lithium, manganese, boron, and ammonia, respectively. The salts used were aluminum sulphate, potassium iodide, titanium sulphate, potassium bromide, strontium nitrate, lithium nitrate, manganese sulphate, ammonium nitrate, and boric acid. Within a week after this application, signs of rejuvenation were observed, new growth was in evidence, and none of the symptoms previously mentioned remained.

Cotton (1930) found that plants of buckwheat growing in a full nutrient solution containing minute quantities of manganese sulphate, zinc sulphate, boric acid, aluminum sulphate, sodium chloride, cobalt nitrate, copper sulphate, and lithium nitrate made decidedly better growth than those grown in a solution containing only the three main salts and iron. In growing oats in sand cultures, McHargue and Shedd (1930) noted that the yield of straw was increased 47 per cent by the addition of manganese to the culture, 53 per cent by the addition of manganese and copper, 65 per cent by the addition of manganese, copper, and zinc, 73 per cent by adding manganese, copper, zinc, and boron, and 63 per cent by the addition of manganese, copper, zinc, boron, and arsenic. The yields of grain increased in the same order 1,659, 2,134, 1,825, 1,637, and 1,157 per cent. Schropp and Scharrer (1933) showed that wheat, rye, maize, oats, peas, beans, clover, alfalfa, and tobacco growing in culture solutions were favorably influenced in their growth by the addition to the regular solutions of small amounts of a solution containing boron, manganese, copper, zinc, aluminum, nickel, cobalt, titanium, tin, iodine, and bromine. Meyer (1931) considered that the beneficial effects of the addition of kaolin to sand cultures was due to some of the rarer elements that it contained.

McHargue, Roy, and Pelphrey (1932) determined the amount of iron, copper, manganese, and zinc in various crop plants as shown below:

Plant	Ounces per ton of dry crop			
	Fe	Cu	Mn	Zn
Soybean	8.0	0.26	3.9	1.6
Sweet clover	4.2	0.29	1.6	1.6
Alfalfa	4.8	0.32	2.5	1.6
Bluegrass	10.2	0.45	2.6	0.5
Corn stover	6.4	0.22	5.1	1.9
Wheat straw	3.5	0.10	2.2	0.5

These data show that copper, manganese, and zinc are not absorbed in nearly so great an amount as the iron but yet in amounts great enough to be significant in the metabolism of the plant.

It was observed by Hoagland and Snyder (1933) that a supplementary solution of aluminum, iodine, bromine, titanium, tin, lithium, manganese, boron, zinc, copper, nickel, and cobalt caused strawberry plants to grow much better than in its absence. Pettinger, Henderson, and Wingard (1932) noted that the addition of magnesium, zinc, copper, boron, and arsenic to a basal nutrient solution increased the frost resistance of young corn plants.

Brenchley (1932), however, obtained only neutral or negative results by adding copper, lithium, titanium, and aluminum to cultures of barley, mustard, peas, or maize.

These results indicate that some elements other than those in the regular nutrient solutions exerted a beneficial influence upon the growth of these plants. As our knowledge now stands, we are not justified in saying that only 10 elements are necessary for plant life but must acknowledge that apparently others are essential. At least we are safe in saying that many of the so-called "nonessential" elements have a marked influence on the growth and development of plants.

IV. OCCURRENCE AND FUNCTION

A. Carbon, Hydrogen, Oxygen, and Nitrogen

Carbon, hydrogen, and oxygen enter into the composition of practically all organic compounds of the plant and make up approximately 95 per cent of its dry weight. The carbon is practically all obtained from the air in the form of carbon dioxide. The hydrogen and a considerable portion of the oxygen are obtained from the water absorbed by the plant. The remainder of the oxygen is obtained in the elemental state from the atmosphere. Nitrogen enters into the composition of the proteins, which in turn are apparently the active components of the protoplasm. The subject of nitrogen nutrition is discussed in detail in Chap. IX and will not be further mentioned here.

B. Phosphorus

1. Occurrence.—The amount of phosphorus in plants ranges from 0.2 to 0.8 per cent of the total dry substance. It is generally present in somewhat greater quantities than sulphur and magnesium and in smaller quantities than calcium. A considerable amount of the phosphorus in plants is soluble in water. In alfalfa (Ames and Boltz, 1912), 75 per cent; in immature cabbage (Peterson and Peterson, 1926), 61 per cent; and in the turnip, 80 per cent (Hartwell, 1917) of the total phosphorus present is soluble in water. The phosphorus in plants occurs in the greater amounts in the seeds and fruits. In the mature corn plant, 29 per cent of the total phosphorus is in the leaves, 11 per cent in the stem, 56 per cent in the grain and cobs, and 4 per cent in the roots (Latshaw and Miller, 1924). This is in accordance with the observation of MacGillivray (1926) that one-half the total phosphorus in the tomato plant is located in the fruit. The amount of phosphorus found in plant tissue will depend upon the amount present in the soil, and Jordan (1913) has shown that up to a certain point the production of plant substance increased with the increased phosphorus supply but that beyond this the phosphorus was not utilized in an increased plant growth. Hartwell (1917) found that young turnips which had been deprived of phosphorus and later placed in a nutrient solution well supplied with it showed crystals of ammonium-magnesium-phosphate within 24 hr. after the addition of the magnesium mixture. An examination for the relative abundance of the phosphorus showed it to be most abundant in the bundles and surrounding tissues near where the taproot merges into the thickened root. Next in amount is inside the cambium in the more active xylem elements and next outside the cambium in the cortex, and only traces are in the pith and leaf petioles. MacGillivray (1927) found that the percentage of phosphorus in tomato plants receiving an ample supply of phosphorus showed a steady increase in amount from the bottom to the top regions of the stem. Plants, however, grown in nutrient solutions lacking phosphorus did not show such an increase in phosphorus content but had approximately the same percentage phosphorus content in all regions of the stem from the bottom to the regions near the tip where it was higher. This indicates that high percentages of phosphorus are associated with regions of rapid growth. Phosphorus is apparently redistributed in plants to a greater extent than most of the other elements, and this property of redistribution may be one reason for plants having a smaller total amount of phosphorus than nitrogen, potassium, or other elements.

Webster (1928, 1929) found that the total phosphorus, in percentage of dry weight, was 0.54, 0.47, 0.42, 0.39, 0.37, and 0.36, for the seeds of soybean, cowpea, wheat, barley, kafir, and corn, respectively. The inorganic phosphorus represented only 2 to 5 per cent of the total phosphorus. The inorganic phosphorus in the roots of yellow corn, mung beans, alfalfa, and green onions only amounted to 0.012; 0.024, 0.136, and 0.189 per cent, respectively. According to Thoday and Evans (1932), the inorganic phosphorus in *Kleinia articulata* is localized mainly in the bundle zone. It was reported by Cockefair (1931) that phosphorus is distributed in the general meristematic tissues, in the synthetically active tissues, and in the fruiting organs. Storage organs are rich in phosphorus only in those parts where organic phosphorus has accumulated. The characteristics of plants suffering from a deficiency of phosphorus vary strikingly for different plants. Karraker and Bortner (1933) noticed in the plants of tobacco deprived of this element that the leaves had a dusky tinge, and the lower ones developed characteristic spots. The plants had a spindling appearance due to small petioles and stems in proportion to the size of leaves and the height of the plant. Fisher (1935) found that tomato plants lacking phosphorus appeared purplish in color, and the stems were covered with small deep-blue spots. The roots were long and brown with but few lateral branches.

of total P in seeds

2. Form.—Plants absorb and utilize phosphorus for the most part in the form of phosphates, although they can to a limited extent utilize some organic phosphorus, especially nucleic acid (Schreiner, 1923). Certain species of plants possess marked differences in their ability to acquire phosphorus from different sources (Baguley, 1912; Jordan, 1913; Truog, 1916; and Bauer, 1923). Cabbage, rape, sweet clover, alfalfa, and buckwheat were found to utilize to good advantage the phosphorus of phosphate rock, while barley, millet, corn, and oats were relatively less efficient in this regard.

Truog (1916) has proposed that according to the laws of mass action and chemical equilibrium, the reaction making the phosphorus of phosphate rock available to the plant may be represented as follows

$$Ca_3(PO_4)_2 + 2H_2CO_3 \rightleftharpoons Ca_2H_2(PO_4)_2 + CaH_2(CO_3)_2$$

According to this theory, if the soluble phosphate is removed, but only a portion of the soluble calcium bicarbonate, an equilibrium will finally be established in which little or no phosphorus will be brought in solution. Truog thus considered that plants containing a relatively high amount of calcium oxide have a relatively high feeding power and plants of a relatively low calcium oxide content have a relatively low feeding power for the phosphorus of phosphate rock (see also von Wrangell, 1922).

Hartwell and Damon (1927) found under conditions of an insufficient amount of phosphorus accompanied by liberal amounts of nitrogen and potassium that over long periods of time beets, cabbage, rape, and rutabagas are high phosphorus-response crops. In contrast to this, carrots, millet, tomatoes, and spring wheat are low phosphorus-response crops, and sorghums, soybeans, and oats occupy an intermediate position in this regard.

3. Amount.—It was noted by Tidmore (1930) that the quantity of phosphorus necessary for the maximum growth of corn, sorghum, and tomato depended upon the nature of the medium in which they grow. Thus, while these plants would not make a satisfactory growth in culture solutions with a phosphate concentration less than 0.1 p.p.m., they grew satisfactorily in a soil where the displaced solution contained less than 0.05 p.p.m. of phosphate. Teakle (1929) found that wheat plants showed vigorous growth and tillering when grown in nutrient solutions containing 1.0 p.p.m. to 50 p.p.m. of phosphate, and only small growth and greatly-reduced tillering when the phosphate content was between 0.05 and 0.1 p.p.m.

The amount of phosphorus in the sap of the plant generally varies directly with the amount in the soil. Thus by heavy application of phosphorus to the soil, Mather (1929) obtained an increase of this element in the plant sap of 34, 89, 100, 130, 135, 150, and 175 per cent for alfalfa, timothy, barley, rape, sweet clover, wheat, and alsike clover, respectively. In one instance, rape grown to maturity showed a 300 per cent increase. This effect has also been observed by Teakle (1929) for wheat; by Blair and Prince (1932) for corn, oats, rye, soybeans, hay, and potatoes; by Ames and Kitsuta (1932), Murphy (1933), and Neller (1935) for various crops; by Hartwell and Hammett (1911) for turnips; and by Jacob and others (1935) for millet, cabbage, and Sudan grass.

4. Physiological Role.—Phosphorus plays both a direct and indirect role in the metabolism of the plant, as indicated by the following topics.

a. Component of Compounds.—There are various organic compounds in plants in which phosphorus is one of the component elements.

1. *Phytin.*—This is an acid calcium-magnesium salt of inosite phosphoric acid. According to Posternak (1903) from 80 to 90 per cent of the phosphorus of seeds exists in this form. Averill and King (1926) reported that phytin makes up from 1 to 3.3 per cent of the air-dry material of seeds. Andrews and Bailey (1932) found that the amounts of phytin phosphorus in wheat bran, embryo, and endosperm were 1.4, 0.59, and 0.00 per cent, respectively. It was observed by McCance and Widdowson (1935) that the phytin phosphorus of corn, peanuts, peas, potatoes, and carrots amounted to 58.0, 57.5, 10.8, 23.0, and 15.8 per cent, respectively, of the total phosphorus of these parts, while apples, celery, and turnips were entirely lacking in this compound.

Knowles and Watkin (1932) found that three-fourths of the phytin in the mature wheat plant was in the spike. At maturity approximately

50 per cent of the total phosphorus of the grain was in the form of phytin, while in the straw 95 per cent of the total phosphorus was in this compound. The phytin phosphorus in the entire plant increased from 25 per cent of the total phosphorus in early spring to 60 per cent at harvest time. They observed in all parts of the wheat plant that the phytin phosphorus and the protein phosphorus rose and fell concurrently, which indicates a relationship between phytin and protein.

DeTurk, Holbert, and Howk (1933) noted that phytin is absent from the corn plant in all parts prior to pollination. Within two weeks after pollination, however, phytin appears in the developing seeds but is not present in any other part of the plant. They suggested that the activation of the zymogen of the enzyme responsible for the synthesis of phytin may be furnished by the pollen. Phytin disappears from corn seeds at a fairly rapid rate during germination.

2. *Phospholipides.*—These compounds are substituted fats containing phosphoric acid and nitrogen. Lecithin is the most abundant of the phospholipides in plants and is found in practically every living cell. Reed (1907) found that when phosphorus is withheld from plants the formation of lecithin does not occur, and that as a consequence there is an accumulation of fat in the cells. Webster and Dalblom (1930) stated that the lipoid phosphorus is only a very small fraction of the total phosphorus in the mung bean. Knowles and Watkin (1932) in wheat observed that the lipoid phosphorus makes up only 2.5 per cent of the total phosphorus of the spike, and from 6.5 to 4.0 per cent of the total phosphorus of the straw, decreasing in amount as maturity approaches.

3. *Hexose Phosphates.*—These are compounds in which the hexose molecule is combined with phosphoric acid. These compounds are apparently associated with respiration.

4. *Nucleic Acid.*—This compound is apparently associated with the nucleus. When phosphorus is withheld, no nuclear division occurs and meristematic activity is greatly diminished.

5. *Phosphoproteins.*—The evidence that such proteins are formed rests wholly on theory. The phosphorus evidently plays a role in protein formation since in its absence sugars increase in amount and coagulable proteins decrease. Under such conditions, plants have the general appearance of those that have been deprived of nitrogen.

b. Reduction.—It was found by Kraybill (1930) that nitrates are not reduced by tomato plants in the absence of phosphates. Eckerson (1929, 1931) noted in the depletion of phosphates by the tomato plants that the nitrate-reducing substances decrease rapidly in succession in the leaves, stem, and roots, and the nitrates accumulated in the same order. Jashnova (1930), as quoted by Nicol (1934), found that the nitrification of large amounts of ammonia was slowed up, if phosphates were present in insufficient quantity.

c. Respiration.—Lyon (1923, 1927) reported that phosphates exert an accelerating effect upon potato oxidase, so that carbon dioxide is produced by an oxidation of some component in a solution of glucose. Phosphate also catalyzes the slow oxidation of pyrogallol and of tannic acid by atmospheric oxygen. It also increases the rate of production of carbon dioxide by the anaerobic phase of respiration through its action as a catalyst toward oxidases. These accelerated actions are equally pronounced when enzymes are contained within the cells of *Elodea* or wheat seedlings.

According to Eckerson (1929), phosphorus, in all forms of life thus far examined, seems to stand on the threshold between potential and kinetic energy relations. If hexose phosphates are formed in plants it is supposed that they hold the same relation to respiration as in animals. It is considered that in the fermentation of sugars intermediary esters in a nascent state are formed. Sugars may be phosphorylated by the aid of coenzymes, through which action a readily decomposable form of sugar is formed. This type of sugar is susceptible to the action of destructive enzymes which transform its energy from the potential to the kinetic form.

Chatterjee (1933) noted that the acceleration of the outgo of carbon dioxide from the leaves of *Eugenia jambolana* and *Allium tuberosum* was greatest when the leaves were injected with a mixture of phosphate and glucose. The injection of phosphate alone, however, had little effect if the amount of glucose was small.

d. Root Growth.—In certain regions the application of phosphates is especially beneficial in promoting the maximum production of root crops. In the absence of phosphorus, root crops like turnips, mangolds, and rutabagas do not enlarge but remain permanently dwarfed. The general impression that the development of the roots of all plants is augmented by the application of phosphorus is probably due to the marked returns obtained from its use as a fertilizer (Russell, 1915; and Noll, 1923).

It was observed by Turner (1929) with barley, wheat, and cotton growing in water cultures, that the ratio of tops to roots decreased as the concentration of phosphate increased. This increase in phosphorus, however, retarded the development and growth in length of the lateral roots. He considered that the decreasing ratio of tops to roots under these conditions may be explained by assuming that the increase in volume and the storage of materials in the roots are stimulated by the presence of phosphorus. Sommer (1936) found in cultural solutions of tomato, wheat, cotton, and corn that high concentrations of phosphorus did not stimulate the development of roots. Apparently there is little or no experimental evidence to show that the proportion of roots to tops of ordinary crops is increased by the application of phosphorus.

e. Nitrogen Relations.—It has previously been mentioned that plants growing in a medium lacking sufficient phosphorus resemble nitrogen-starved plants. Such plants, however, contain more total nitrogen than those grown in complete nutrient solutions (MacGillivray, 1926). According to Kraybill (1930), the stems of phosphorus-deficient tomato plants contained three times as much water-soluble nitrogen and four times as much ammoniacal nitrogen, nitrate nitrogen, amide nitrogen, and α amino acid nitrogen as the stems from plants growing in complete nutrient solutions. McCool and Weldon (1930) noted that the phosphorus content of the plant sap was decreased by the application of sodium nitrate. In prune trees, Colby (1933) found in the leaf tissues that phosphorus starvation resulted in a relatively high content of nitrogen. Carolus (1935) found with collards, cowpeas, and tomatoes, growing under greenhouse conditions, that in general low availability of phosphorus was associated with a very high content of nitrogen in the plant, and a high content of phosphorus with a low content of nitrate nitrogen. Thus, in black cowpeas, the absence of adequate nitrogen increased the phosphate in the cell sap from 44 to 1,040 p.p.m. In collards a low content of phosphorus increased their nitrate nitrogen from 148 to 970 p.p.m. The same general behavior in the phosphorus and nitrogen relationship was also observed by Haas (1936) in experiments with citrus plants.

Emmert (1934) found that relatively large amounts of nitrogen decreased and small amounts of this element increased the phosphate content of the vascular bundles. He considered that this interrelation between nitrogen and phosphorus is probably caused by their utilization by the plant. The presence of an abundant supply of nitrogen as compared with phosphorus causes a utilization of this element and, consequently, a decrease in its amount in the conducting tissues. If nitrogen is deficient, then the phosphorus may accumulate because it is not used in the synthesis of nitrogenous, organic compounds.

f. Carbohydrate Relations.—It was observed by Reed (1907) that the transformation of starch into water-soluble carbohydrates was seriously impaired in the absence of phosphorus. Hartwell (1917) noted that, simultaneously with the increase in the absorption of phosphorus by the turnip root, which previously had been deprived of it, the leucoplasts containing the starch grains shrank in size as the grains were corroded and dissolved until finally all the starch had disappeared from the root tissue only to reappear when phosphorus was again withdrawn from the nutrient solution. MacGillivray (1926) noted in the tomato that the absence of phosphorus greatly increased both reducing and nonreducing sugars in the plant. This increase of sugars in the absence of phosphorus has also been noticed by Eckerson (1929) and Kraybill (1930). Eidelman (1930) stated that for average lengths of day and with relatively high temperatures photosynthetic activity tends to show a positive correlation with the content of phosphorus. The maxima, however, do not coincide under the conditions of a shortened day and lower temperatures.

g. Maturity.—It has long been observed that phosphate fertilizers induce earlier ripening of the grain crops growing on soil low in phosphorus. The same effect has also been noticed for cabbage and cotton. The effects seem to vary with the different types of phosphorus, the more soluble ones hastening maturity to a greater degree. Increasing the rates of application, however, above the needs of the crop as indicated by yields has not been accompanied by further increases in earliness (Noll, 1923). Phosphorus, however, is not the only element that promotes earliness, since moderate applications of nitrogen and potassium and sometimes calcium when they are present in insufficient amounts have similar effects on maturity to those of phosphorus.

h. General Effects.—In oats, McClelland (1931) found that the application of phosphorus increased the number of tillers, panicles, number of spikelets per panicle, yield, and size of seed. Eckerson (1929) noted that in the absence of phosphorus the protoplasm of young cells became less refractive, and that the chloroplasts disintegrated first in the stem and lower leaves and finally in the upper ones. Owen (1931) noted in tomatoes that if phosphates were omitted the potassium content of the foliage was depressed, and that if they were added to a medium deficient in this radical the content of potassium in the leaves was increased. Hurd-Karrer (1935) found that the presence of phosphorus greatly inhibited arsenic injury of plants.

C. Calcium

Calcium is essential to the growth and development of all green plants with the exception of some of the lower algae. Nonchlorophyllous flowering plants must also be supplied with this element, although such plants as a whole contain less calcium than green plants. Calcium, however, is not necessary for the development of the fungi.

1. Occurrence.—A considerable portion of the calcium found in plants is in forms that are soluble in water. In cabbage that is ready for storage, 60 per cent of the calcium present is soluble in cold water (Peterson and Peterson, 1926), but after the plants have been in storage for six weeks, the soluble calcium amounts to only 42 per cent of the total amount present. Ames and Boltz found that 40 per cent of the calcium in alfalfa can be extracted with water. In the tomato plant, Nightingale and others (1931) found that all the calcium of the fresh tissues of calcium-deficient plants was insoluble. The leaves contain the greatest amount of calcium, and the seed or grain contains the lowest amount. In the case of the corn plant (Latshaw and Miller, 1924), 58 per cent of the calcium was found in the leaves, 18 per cent in the stems, 20 per cent in the roots, and only 4 per cent of it in the grain and cob.

The amount of calcium in the plant varies greatly with the species and with environmental conditions. Thus, for example, the percentage of calcium based on dry weight has been found to vary from 3 to 8 per cent in the vines of the cucurbits, to be 2.7 per cent in mature tobacco, and to amount to only 0.4 per cent in timothy hay and wheat straw. As a rule, the legumes, lettuce, cabbage, and tobacco have a high percentage of calcium on a dry basis, while the members of the grass family have a relatively small amount. In the case of cucurbit vines, the amount of calcium increases from the early to the late stage of development, while the potassium, phosphorus, and nitrogen percentage decreases (Wilkins, 1917).

Cowell (1932) found the calcium content of the outermost leaves of cabbage may be 20 to 30 times as great as that of the inner leaves. In *Kleinia articulata*, Thoday and Evans (1932, 1933) observed crystals of calcium oxalate in the collenchymatous, hypodermal layers. This element was present in abundance in solution in the sap of cells of the pith and in patches of cells in the inner cortex, two-thirds of the total calcium being in the pith cells. Yanovsky, Nelson, and Kingsbury (1932) found that practically one-half of the ash of the hackberry was calcium oxide. It was found by Harper and Daniel (1934) that the calcium content of *Elodea* was as much as 8 to 10 per cent, while in *Typha latifolia* it occurred only in relatively small amounts. They considered that the absorption of calcium by aquatic plants may be an important factor in the clarification of the water of ponds and lakes. Larson (1928) found that the concentration of soluble calcium favorable for normal plant growth is apparently about 32 to 64 p.p.m., while 16 to 32 p.p.m. seems to be the minimum concentration. He considered that 32 p.p.m. is the most economical concentration, since above that point, although the yield is increased, the increases become successively smaller.

2. Physiological Role.—According to Bamford (1931), the necessity of calcium for plant growth was first demonstrated by Salm-Horstmar (1856). Numerous functions have been ascribed to calcium, although some of these are little understood. The roles may be classified under three main headings: (*a*) antidoting agent, (*b*) component part of the plant, and (*c*) general effects.

a. Antidoting Agent.—Calcium may function as an antidoting element in two ways: (1) it may play a part through the calcium/magnesium ratio, and (2) it may function in the neutralization of organic acids in the plant.

1. *The Calcium/Magnesium Ratio.*—Calcium reduces the toxic effects of single-salt solutions of sodium, potassium and magnesium, as was discussed under Antagonism and Balanced Solutions. Since calcium or magnesium salts in single solutions are toxic or injurious to plants and since either will neutralize, in part or in whole, the deleterious effects of the excess of the other, Loew (1903) proposed the hypothesis that one of the principal functions of calcium in plant metabolism is to neutralize the toxic action of magnesium and that a certain calcium-magnesium ratio varying with the type of plant is necessary for the proper growth and development of the plant. An enormous amount of experimental work was carried on by Loew and his pupils and by many others in an attempt to prove this hypothesis and to find the proper proportion of calcium and magnesium for various crop plants. Lipman (1916), however, after a thorough résumé of the literature on the subject, concluded that the experimental work presents no evidence in support of the hypothesis that a specific lime-magnesium ratio exists for any plant or group of plants. He considered that there is no more reason to assume that there should be a proper ratio in the soil between calcium and magnesium than that there should be a proper ratio between calcium and potassium or between calcium and any other essential element.

The conclusions of Lipman are further upheld by the work of Wyatt (1916) with wheat, soybeans, cowpeas, and alfalfa. The crop yields of these plants and the ratio of calcium to magnesium in them bear no direct relations to the ratio of the carbonates applied. Different ratios of calcium to magnesium within rather wide limits produced no marked differences in yield, although increasing the size of the applications increased the calcium and magnesium content of the plants. Prianischnikow (1923) found that weak solutions of the salts of magnesium, calcium, barium, and strontium in concentrations of 0.0005*N* to 0.005*N* increase the resistance of cells to free acid, and, of the cations used, calcium showed the strongest antagonism.

Moser (1933) could find no significant correlation between the calcium/magnesium ratio and crop yields. He believed that the significant factor in determining yields is the amount of active calcium in the soil. The addition of lime to the soil is beneficial not because it changes the calcium/magnesium ratio, but because it increased the replaceable calcium therein.

2. *Neutralization of Organic Acids.*—It has long been known that oxalic acid is one of the most abundant and the strongest of the acids in the plant and is supposed to be a by-product of carbohydrate and protein metabolism. Schimper (1888, 1890) suggested that one of the principal functions of calcium is to precipitate oxalic acid and soluble oxalates in the form of calcium oxalate, thus preventing the injurious effects that they might exert upon the plant. Calcium, for this purpose, however, is apparently not necessary in certain cases, for neither oxalic acid nor oxalates have been observed in some plants. The theory of Schimper was further weakened by the work of Bruch (1902), who showed that the grain plants tolerate relatively large amounts of oxalates, since wheat plants growing in nutrient solutions containing 18 to 278 p.p.m. reached the blooming stage. He further found that calcium oxalate could replace other calcium salts as a source of calcium in nutrient solutions, and Ames and Boltz (1912) found in the alfalfa plant that one-third of the calcium present was in an insoluble form other than oxalates, which indicates that calcium is essential for the formation of some of the organic compounds. Parker and Truog (1920) have more recently revived the theory that calcium plays an important role in the neutralization of acids within the plant. They observed that there is a close relationship between the calcium and nitrogen content in plants. In regard to the ratio of calcium to nitrogen, plants can be divided into two groups. The first group has a calcium nitrogen ratio of about 0.3 and contains almost entirely members of the grass family that have low calcium requirement and are more or less tolerant to soil acidity. The second group has an average calcium nitrogen ratio of about 0.55 and includes the legumes and those plants which require a large amount of calcium and are sensitive to soil acidity. These authors considered then that the larger the amount of nitrogen the more protein is formed and the more oxalic, acetic, succinic, and formic acids are produced in protein metabolism. Larger amounts of calcium are thus necessary to neutralize these acids than if a smaller amount of protein were manufactured. They assumed that, if calcium is lacking, the above acids accumulate in such quantity that they become injurious to the cells and inhibit further

protein production. Hobart (1924), however, considered that the neutralization of these acids can be performed by other bases and that the necessity of calcium can be explained by assuming that it acts as a stimulant to the protoplasm.

b. Component of Plant.—It is definitely known that calcium is one of the components of the middle lamella of the cell wall. It is considered by some that it may enter into the composition of the protoplasm and of certain types of proteins in the cell.

1. *Cell Wall.*—It was stated in Chap. I that calcium pectate is one of the compounds of the middle lamella. It was noted by Reed (1907) that when calcium was withheld from the plant, new cell walls were imperfectly formed if at all, although the nuclei of the cell divided mitotically. The most important observations on the function of calcium in the cell wall have been made by True and his workers (1918, 1921, 1922). They found that plants growing in various concentrations of a single salt of potassium or sodium were not able to carry on a sustained absorption but in the end gave up more ions to the medium than they were at any time able to take up. In solutions of calcium and magnesium salts, there was a concentration below which the roots were not able to absorb, and ions leached out into the medium, although, in solutions above this concentration, absorption took place in a greater or lesser degree. In mixtures containing calcium and other nutrient salts, the total quantity of ions absorbed by the plants exceeded by far the total number of calcium ions present. This indicates that in a solution of mixed salts, calcium in some way produces conditions that bring about the absorption of ions which without the presence of calcium would not be absorbed or would cause the leaching of ions from the plant. This behavior of the diffusion relations can, in part at least, be explained by the following microchemical observations:

It was found that when seedlings were grown in potassium solutions, ions readily entered the cells of the root, but that within 24 hr. calcium ions began to diffuse out of the calcium pectate middle lamella of the cells. Potassium pectate was formed in the cell walls instead of calcium pectate and soon dissolved out, since it is readily soluble in water. At this stage organic matter and salts began to leach rapidly from the cell vacuole. Approximately two-thirds of this material was organic and nonelectrolytes and consisted principally of sugars and amino acids. This action of the potassium salt upon the cell wall not only changed its character but also greatly changed the permeability of the protoplasmic membrane next to it. These changes are seen also by the inward passage of materials. Corn seedlings after 5 days in a potassium nitrate solution were placed in a one per cent solution of copper sulphate and within 1 hr. the copper ions had penetrated all the root tissue. Similar seedlings after 5 days in a calcium nitrate solution showed the penetration of copper ions only after 24 hr. in a similar copper solution. In solutions of magnesium salts, the calcium pectate in the middle lamella is replaced by magnesium, which is also an insoluble compound but more permeable than the calcium pectate. The fatal results from growing plants in magnesium solutions do not occur until the calcium of the middle lamellae has been wholly replaced by magnesium, after which sufficient magnesium ions are apparently free to penetrate the protoplasm and injure it. True considered that the integrity of the calcium pectate forming the middle lamella is maintained when a sufficient quantity of calcium ions are present in the external medium and with it the normal retention of the cell contents. When according to the laws of mass action this quantity of calcium ion falls below the equilibrium concentration, other cations replace the calcium in the middle lamella. No cation apparently can replace calcium in this regard, although magnesium comes more nearly to doing it than any other.

2. *Proteins.*—It is considered by some that calcium enters into the composition of proteins, and Loew (1903) believed that calcium protein compounds enter into the organized substance of the nucleus and plastids. He considered that these calcium compounds have a well-defined power of imbibition, which if disturbed by replacement substances like potassium-protein compounds leads to fatal results.

3. *Protoplasm.*—Sorokin and Sommer (1929) found that plants of *Pisum sativum* grown in culture solutions without calcium usually die within 2 or 3 weeks. The death of the plants is associated with the degeneration that takes place in the protoplasts of the meristematic cells of the root tips. The first visible effect of the absence of calcium is a decrease in the amount of cytoplasm and the appearance of large vacuoles in the meristematic cells, which thus resemble old cells. With the gradual depletion of the amount of calcium, the cells of the meristematic regions of the roots show transitional stages from interrupted and irregular mitoses to almost typical amitoses. Sorokin and Sommer considered that the absence of calcium does not have a direct effect on the disintegrating tissues due to the separation of the cells because of the absence of calcium pectate. According to these investigators, the effect of the absence of calcium is on the meristematic cells, either because it is a necessary chemical constituent of the protoplast or because its absence so affects the physical condition of the colloidal system that normal action of the protoplast is not possible.

Bamford (1931) studied the influence of a deficiency of calcium on the growth and cytological structure of the root tips of young wheat and corn seedlings growing in water cultures. Under such conditions, the cells of the root apex gradually lost all their stainable contents, but the cell walls remained intact. The protoplasm appeared to undergo a gradual disintegration, until it completely disappeared. At first the nucleus remained normal, but eventually it degenerated into a heterogeneous mass and disappeared. This process of disintegration at first affected only the epidermis and the peripheral layers of cells in the cortical region, but later progressed to the interior and finally upward in the roots. However, Lutman (1934) found that the lack of various essential elements, other than calcium, was reflected in the cytology of the cells by a premature vacuolization and by indications of maturity and senescence.

c. General Effects.—Under this heading are considered (1) the effects of calcium on the general appearance of the plant, (2) translocation of carbohydrates, and (3) the response of roots.

1. *General Appearance.*—In a deficiency of calcium in the soil or water culture, the lower leaves in the case of the cereals show an inrolling of the margins, which Loew (1903) and Tottingham (1914) ascribe to a high magnesium calcium ratio. Dickson (1918) observed this behavior and noticed that later brown spots appeared on these inrolled leaves, which finally became dry and twisted around the stem. Day (1929, 1935) working with *Pisum sativum* observed that, with plants deprived of calcium, the lower leaves were chlorotic, and the younger leaves were curled and tough; the plants under these conditions were shorter and less succulent. McMurtrey (1932, 1933) noted that the first indication of a deficiency of calcium in the tobacco plant was a peculiar hooking downward of the tip of the young leaves of the bud followed by the death of the tissues at their tips and margins. If later growth appears, the tips and margins show a cut-out appearance. At later stages of growth, calcium deficiency is manifested by the cessation of terminal growth and the development of chlorosis and necrotic spots in the older leaves (Davis, 1930). Mann (1924) found that young apple trees in sand cultures deficient in calcium produced leaves that were larger than those borne by trees supplied with a complete cultural solution and that in the case of the gooseberry a deficiency of calcium resulted in leaf water content higher than that shown in the complete nutrient. Resistance to evaporation in bright light was greatest in the case of the plants grown in a deficiency of calcium.

2. *Translocation of Carbohydrate.*—It was observed by Boehm (1875), that an abnormal accumulation of starch occurred in *Phaseolus vulgaris* grown in water cultures lacking that element—an observation that has since been verified for numerous plants by many investigators. The movement of starch from any tissue in a plant will depend primarily upon the diastase activity or the power of digestion and the depletion of the sugar in the surrounding tissues. It has been found (Groom, 1896) that potassium oxalate retards the action of diastase, and it has been considered that, with a sufficient supply of calcium, calcium oxalate is formed instead of the potassium compound, and this salt is not detrimental to diastatic action. Hartwell (1916) suggested that an accumulation of starch may accompany any general disturbance of the metabolism of the plant, so in the absence of calcium the accumulation of starch may be only an indirect result. It has also been observed (Grafe and Portheim, 1906) that the addition of levulose, dextrose, and saccharose to calcium-free solutions enables bean plants to live in them a longer period than where no sugar is added. The cause for the growth of plants not being inhibited at once by the absence of calcium is attributed to the fact that considerable carbohydrate is present in the form of sugar, and that this is sufficient for its immediate needs. In some cases, a deficiency of either magnesium or calcium causes an increased growth, but under such conditions the weight of the grain is less than where a sufficient supply of calcium is available, so that the total dry weight of the plant is less. This observation is held to indicate that a deficiency in calcium causes a disturbance in the translocation of carbohydrates and proteins and in their storage during seed formation rather than a hindrance to their synthesis. It was observed by Nightingale and others (1931) in the tomato that the calcium-deficient plants accumulated carbohydrates in large quantities apparently because absorption and accumulation of nitrates did not take place. When calcium-deficient plants were placed in the light and given a supply of calcium, absorption and assimilation of nitrates occurred within a few hours.

3. *Root Response.*—When calcium is withheld from water cultures, the size of the roots of the plants growing therein is greatly reduced. Reed and Haas (1923) showed

that the roots of citrus seedlings grown in the absence of calcium were injured, as indicated by the gelatinization of the superficial layers and the ultimate death of the roots. If injury has not progressed too far, the addition of calcium induces the production of lateral rootlets, the lowest of which mark the boundary between the dead and living portions of the root.

It was noted by Bamford (1931) that no lateral roots were developed by wheat and corn when calcium was absent from the culture solution. Albrecht and Davis (1929) considered that calcium plays some physiological role favoring nodule formation of soybeans.

D. Magnesium

Magnesium is usually present in the plant in somewhat smaller amounts than calcium. It is relatively more abundant in seeds and leaves than in the other parts of the plant. In the case of the corn plant, 34 per cent of the magnesium is in the grain, 32 per cent in the leaves, 21 per cent in the stems, 7 per cent in the roots, and 6 per cent in the cobs.

Daniel (1935) found that the average content of magnesium in 19 species of mature grasses was 0.156 per cent, and in 45 types of mature legumes it was 0.379 per cent. The magnesium content of the grasses ranged from 0.059 to 0.316 per cent, while the legumes varied from 0.329 to 1.024 per cent. Most soils are not deficient in magnesium. Wheeler and Hartwell (1904), however, showed that the application of magnesium salts to certain Rhode Island soils was beneficial to various plants and that a single application of Mg lime would cover the deficiency of Mg for many years to come.

Magnesium, as first shown by Willstätter (1906), is one of the constituents of chlorophyll, and when this element is lacking a disturbance in the formation of chlorophyll is soon noticed.

According to Beaumont and Snell (1935), plants differ markedly in their response to a deficiency of magnesium in the soil. Thus buckwheat and spinach are very sensitive, and turnips, mangels, corn, and tobacco considerably so. The small grains, grasses, clover, and potatoes, however, are only slightly affected. Dickson (1918) found that oat plants grown in a nutrient solution which had only approximately one-tenth of the requisite magnesium supply developed numerous broad leaves which at first were very bright green but which when older showed marked stripings between the veins because of the disappearance of the chlorophyll. Sand drown, a chlorosis of tobacco involving both the green and yellow pigments of the chloroplasts, has been shown (Garner and McMurtrey, 1923) to be due to a deficiency of magnesium in the soil.

According to Garner, McMurtrey, Bowling, and Moss (1930), the outstanding features of sand drown in tobacco in the field are:

1. A disintegration of both the green and yellow pigments of the chlorophyll, beginning at the tips of the lower leaves. The symptoms

of magnesium deficiency differ from those of potassium deficiency in that with magnesium deficiency the leaf surface ordinarily remains smooth, with no downward curvature of the tips and margins, and spotting, resulting from localized disintegration of the tissues, only rarely occurs.

2. An increase in the water content of the leaf, and a decided decrease in the amount of dry matter per unit of area.

General Symptoms of Mg Deficiency

According to Beaumont and Snell (1935), the symptoms of magnesium deficiency for plants in general are: (1) the development of chlorosis in the intervascular tissue of the older leaves, which produces a striped appearance in leaves with parallel veins and a mottled appearance in those with netted veins; and (2) (in severe cases) the browning of the margin of the leaves and the dropping of some of the leaves from the plant.

Reed and Haas (1924) observed that chlorosis of young citrus and walnut trees resulted from an insufficient application of magnesium. Brown, Houghland, Smith, and Carolus (1933) found in a number of potato-producing regions that a characteristic, chlorotic condition of the lower leaves developed, which apparently was due to a magnesium deficiency in the soil. Mameli (1918) found that within certain limits the development of chlorophyll is proportional to the magnesium supplied in the solution. Jones (1929) concluded that chlorosis of corn in certain regions of Massachusetts was due to the lack of magnesium, since the normally green plants contained a larger amount of this element than did the chlorotic ones.

Parbery (1935) noted that chlorotic orange leaves in Australia were characterized by marked deficiencies in magnesium. Lutman and Walbridge (1929) believed that lack of magnesium may lead to early chlorosis of the older leaves of the potato plant.

It was found by Garner, McMurtrey, Bowling, and Moss (1930) that a minimum content of 0.25 per cent of magnesium, or 0.4 per cent of magnesium oxide, in the leaf of the tobacco is required to prevent magnesium deficiency symptoms. In practically all cases the troubles arising from magnesium deficiency have been remedied by the application to the soil of various magnesium salts as fertilizers.

Magnesium appears to function in many cases as a carrier of phosphorus. The salts of magnesium undergo dissociation very easily and thus readily give up the anions that they carry. Phosphorus enters into the formation of nucleoproteins and lecithins, two compounds that enter into the composition of protoplasm. If magnesium functions as a carrier of phosphorus, then the content of magnesium should be relatively high in those regions where these compounds are being formed. This apparently is the case, since magnesium is more abundant and frequently increases in amount in the tissues—as the tips of stem and roots—where rapid development is taking place.

Magnesium is more abundant in oily seeds like those of flax and cotton than it is in starch-containing seeds like the cereals. The average proportion of magnesium in starchy seeds to that in oily seeds is about 2:5 (Loew, 1903). Since the formation of oil is preceded or accompanied by the formation of lecithin, it would appear that magnesium in the case of oil formation serves as a carrier of phosphorus, and this supposition is further supported by the fact that, when both lecithins and nucleoproteins are present in the cell, magnesium occurs in greater amounts than if only one of these compounds is present. In regard to oil formation, Reed (1907) observed that the filaments of *Vaucheria* growing in magnesium-free solutions were devoid of oil globules, although globules were present in those plants growing in the control solutions. He considered that there is an intimate relationship between the presence of magnesium and the formation of vegetable oils. Oat plants growing in a nutrient solution deficient in magnesium produced inferior grain, and the ratio of grain to straw was decreased (Dickson, 1918). Apparently only a small amount of magnesium is needed to satisfy the needs of the plant, and seemingly it can be used over and over again as a carrier. It has been observed that beans can reach their full vegetative growth with no supply of magnesium other than that stored in the seed.

When magnesium in the form of magnesium sulphate is present in relatively large amounts in nutrient solutions, the leaves of cereals growing therein show characteristic magnesium injury. The tips first wilt and then dry up and drop off. This injury occurs in a much greater degree under conditions conducive to high transpiration than to low. The addition of calcium salts is beneficial in overcoming this injury, and the results have been considered as proof that one of the functions of calcium is to overcome the toxic action of magnesium, but Gericke (1922) has shown that this injury can also be prevented by having a proper proportion of phosphorus present in the solution.

Trelease and Trelease (1931) reported that magnesium injury of seedling wheat is marked by a rolling and curling of the young leaf during its emergence from the sheath. This effect is first visible after the young wheat plants have grown in the cultural solution from 9 to 12 days and the third leaf is just appearing. Later, numerous tillers develop prematurely, growth of the seminal roots is inhibited, and a number of adventitious roots are produced. The plants of oats, rye, and barley exhibit the same general characteristics. Carmin (1931) noted that magnesium sulphate was more toxic to the roots than to the tops, and more toxic to the main roots than to the lateral ones.

E. Potassium

Potassium is absolutely essential for plants and cannot be entirely replaced by any other element. Sodium may be substituted for potassium to a certain degree, but by no means entirely (Colla, 1929; Tincker, 1932; Alten and Gottwick, 1933; McCalla, 1934; and Hartt, 1934). The plant can apparently absorb from the soil and utilize in its metabolism any soluble inorganic compound of potassium. The form in which it is presented to the plant apparently makes little difference in some cases, and in other cases the differences observed may be attributed to the nature of the soil, environmental conditions, or the nature of the plant.

1. Occurrence.—The range of distribution of potassium in the plant has been studied by McCallum (1905), Weevers (1911), Dowding (1925), Lukaszewicz (1926), Penston (1931), and Thoday and Evans (1933).

All meristematic tissues are rich in this element, as indicated by strong positive microchemical tests obtained in the growing tips and cambium. Potassium appears especially abundant in the growing parts in those cells in the region of secondary root outgrowths and the formation of new leaves. McCallum (1905) noticed that the cell in the *Equisetum* spore which will give rise to the primary root hair is rich in potassium, and as the tip elongates the potassium keeps in contact with it. The same thing holds true in the development of the pollen tube. The potassium in the pollen grain migrates into the tube and keeps in contact with the growing tip. Potassium is in the mesophyll cells of the leaf, the cells of the medullary rays, the phloem, and to some extent in other parenchymatous tissue, but little or none is found in the older xylem portions of the stem. In the spruce when the bud elongates, there is a translocation of the potassium within the embryonic cone to the next year's meristem and to the embryonic leaves (Dowding, 1925). This author noted also that in the wheat grain the cells of the aleurone layer contain a large amount of potassium but that the endosperm has only a trace. After 15 days of germination the potassium disappears from the aleurone layer, having apparently leached out into the surrounding medium, since there is no evidence that it is transferred into the endosperm. All parts of the embryo contain potassium, but the greatest amounts are contained in the root and stem apexes. Potassium is thus present in regions of meristematic division, photosynthesis, translocation, storage, and other special, physiological activities.

Chemical analysis of the corn plant at early maturity (Latshaw and Miller, 1924) shows that 45 per cent of the potassium in that plant is located in the leaves, 32 per cent in the stems, 14 per cent in the grain, approximately 5 per cent in the cob, and 4 per cent in the roots. Dickson (1918) found by analysis that potassium is localized first in the growing region of the young wheat seedling but that later it appears in large quantities in the region of grain formation. Potassium is well distributed over the plant, and André and Demoussy (1927) considered that it is the predominant soluble salt in plants because of its mobility.

The potassium present in potassium-deficient plants is apparently withdrawn from the older regions and transferred to the meristematic tissues (Janssen and Bartholomew, 1929; Nightingale, Schermerhorn, and Robbins, 1930; and James and Penston, 1933). It was observed by Hartt (1934) in sugar cane that potassium migrated from the dying leaves to the living ones and that under such conditions it was the only element that did so. Penston (1935) noted a pronounced variation in the potassium content of potato leaves throughout the day. It tended to reach a maximum about 3 to 4 P.M. and to fall at night. The greatest variation between the maximal and minimal content during the day amounted to 73.2 per cent. He concluded that potassium is being continually brought into the leaf, presumably in the transpiration stream, and removed through the phloem, since a loss of potassium becomes apparent in the evening when the rate of transpiration is decreased. In the expressed sap of perennial rye grass, taken from July to December, the absolute amount of potassium rose to a maximum at maturity or during late senescence, and then declined (Petrie, 1934).

In the individual cells, potassium seems to be absent from the nucleus. Stocklasa (1911) could find none in the chloroplasts, and Dowding (1925) stated that, in the spruce leaves during the summer, potassium occurs as a network of granules between the chloroplasts. He observed that in the epidermal cells of the hyacinth and onion potassium occurs in the corners of mature guard cells where oil and starch can be found, but that in the initial cells that have not yet divided there is no localization of this element. McCallum (1905) observed potassium in the cytoplasm, but Stocklasa could not positively recognize it there in many cases, but considered that the greater portion of it is in the vacuole of the cell.

James and Penston (1933) found potassium in the cytoplasm and vacuole but only on the surface of nuclei and plastids. In perennial rye grass, Petrie (1934) considered that potassium occurs entirely in the ionic form in the cell sap and cytoplasm. The amount of cell sap and cytoplasm probably rises to a maximum at maturity and then declines. The absolute amount of potassium is probably correlated with this drift.

The potassium present in the plant is for the most part in forms readily soluble in water (Janssen and Bartholomew, 1929). Kostytschew and Eliasberg (1920) reported that when plant tissue is extracted with cold water, the tissue yields potassium-free ash and lead acetate, and tannin precipitates were also found to contain no potassium. Weevers (1911) by water extraction removed 99 per cent of the total potassium from the pollen tubes of *Pinus sylvestris* and leaves of the horse-chestnut.

Mann and Wallace (1925) found that potassium leached readily from apple leaves when they were immersed in water. It was reported by Anderson, Swanback, and Street (1932) for corn and tobacco; Morris (1933) and Morris and Sayre (1935) for corn; and Tyner (1935) for rape, sweet clover, buckwheat, alfalfa, corn, and the sorghums that the potassium in plant tissue is in a form that is readily soluble in water. It would appear that if it is combined at all with the protoplasm, it is easily dissociated from it. Then apparently it is present in the cells chiefly in the form of inorganic compounds and as salts of the organic acids.

Practically all investigators including Bartholomew and Janssen (1929), Johnston and Hoagland (1929), Wallace (1931), Janssen and Bartholomew (1932), Wallace and Proebsting (1933), and Hartt (1934) have found that the amount of potassium in the plant is increased by its application to the soil. In many of these observations the luxury absorption of this element was also apparent.

It was observed by Hoagland and Martin (1933) that in many cases large amounts of potassium are fixed in a nonreplaceable form almost immediately after placing in the soil. They found in experiments with barley on widely different soils that if the amount of replaceable potassium is high, all or the major portion of the potassium absorbed by the plant may be balanced by the loss of replaceable potassium from the soil. High-replaceable potassium tends to induce luxury absorption of this element by the crop. As cropping continues, the proportion of potassium derived from nonreplaceable forms grows larger until a point is reached at which no further loss of replaceable potassium can occur. At that point the solubility of the nonreplaceable forms determines the supplying power of the soil. Proebsting (1933), however, with apple trees, buckwheat, and tomatoes concluded that neither the exchangeable nor the water-soluble potassium is an accurate means of measuring the ability of the soil to supply the plant with this element.

Plants vary widely with respect to the amount of potassium needed for their maximum growth. Janssen and Bartholomew (1930) noted with soybeans, cowpeas, oats, sweet clover, and other plants, that the greatest growth in water cultures was obtained when there were 2 to 3 p.p.m. of potassium in the solution. This was below the concentration at which the maximum absorption of potassium occurred, indicating that more potassium is absorbed by plants than is needed for their use. It was found by Bartholomew and Janssen (1929) that alfalfa and Hubam clover made optimum growth in a solution containing 0.5 p.p.m. of potassium, while oats, cowpeas, soybeans, and cotton made the best growth at a concentration of 2 p.p.m. Johnston and Hoagland (1929) found that the optimal growth of the tomato plant was obtained in a flowing solution with a concentration of approximately 5 p.p.m. of potassium.

Tyner (1935) observed that corn, rape, buckwheat, peas, and sorghums were poor feeders on feldspathic potassium, while alfalfa, alsike clover, red and sweet clovers could appropriate potassium to good advantage from that source. For buckwheat growing in flowing cultures, the minimum concentration of potassium necessary for good growth was approximately 3 p.p.m., while 1 p.p.m. was sufficient for good growth of red clover. Proebsting (1933) also observed that plants vary in their ability to withdraw potassium from the soil, and therefore he considered that the only reliable method for determining the ability of a plant to obtain adequate potassium from the soil was to grow the plant upon it.

Apparently plants can use to advantage the potassium that has been placed on the leaves. Domontovitsch (1930) found that plants which had been grown in a potassium-free medium responded favorably if the leaves were painted with a 2 to 3 per cent solution of some potassium salt.

Loring and Druce (1930) found that the potassium content of the ash of the potato plant had a greater proportion of the isotope with the higher mass than did the potassium in the soil. They considered that this indicated a highly selective ability of the plant. Heller and Wagner (1931), however, could find no accumulation of the isotopes of potassium in beet tops.

2. Physiological Role.—The exact role of potassium in the plant is not known. Its function is only surmised by noting the effects on the plant in its absence.

a. Formation of Carbohydrates.—It was first pointed out by Nobbe and his workers (1871) that carbohydrates are formed normally only when the potassium supply is sufficient. The production of sugar by sugar beets, mangolds, potatoes, and other crops that form a large amount of carbohydrate is reduced by a deficiency in potassium before any diminution in the vegetable growth of leaves and stems can be observed. In an experiment with mangolds, Russell (1915) reported that when potassium was deficient in the soil 7,255 lb. of leaves produced 14,684 lb. of root, while plants growing on the same soil with a sufficient supply of potassium and with very little more leaf weight (8,508 lb.) produced 40,128 lb. of root, or nearly three times as much as the plants growing under a deficiency of potassium. In all the green plants studied by Reed (1907), potassium was apparently essential for starch formation. The lack of this element gives rise so far as can be observed to no pathological condition of the chloroplasts, and under conditions of potassium deficiency they remained normal for 2 months and even increased in number. The loss of the power of synthesis by green plants in the absence of potassium has also been observed by Stocklasa (1908, 1912) and by Weiszmann (1923), but Smith and Butler (1921) could not sustain these general observations in their work with bluestem wheat growing in a solution deficient in potassium.

As has previously been mentioned, potassium is present in comparatively large amounts in the growing tips. It was first shown by Schimper (1890) that a potassium supply is necessary for the normal development of the growing apexes. Dickson (1918) grew oat plants in a nutrient solution containing only one-tenth of the optimum potassium supply. Plants grown after this manner produced only about one-half the amount of total dry matter of stem and leaves as did those growing in a solution with the proper amount of potassium. The total dry weight of the plants including grain, however, was two-thirds that of the normal plants. The decrease in grain was thus not proportional to the decrease in the dry matter of the vegetative parts of the plant. This result is evidently due to a lack of potassium for meristematic development, and as a consequence the vegetative growth of the plant was checked.

The data obtained from later experiments on the influence of potassium on the metabolism of carbohydrates are contradictory and confusing. Nightingale, Schermerhorn, and Robbins (1930) considered that potassium either directly or indirectly is essential for carbon dioxide assimilation. Carbohydrates, however, frequently accumulate in potassium-deficient plants because nitrogen assimilation is retarded. They considered that the translocation of sugars and digestion of starch can occur freely in plants that are extremely low in potassium. Janssen and Bartholomew (1930) found a high correlation between the percentage of potassium and the total weight of the starch and sugar in the plant. Hartt (1929) found in sugar cane that potassium-deficient plants had a greater percentage of total sugars, reducing sugars, and sucrose. Greater lignification occurred in the potassium-deficient plants, while greater cutinization was found in plants supplied with potassium. Tobler (1929) found that flax fertilized with potassium sulphate had more fiber than the control plants. The fibers were larger, more angular, had more wall laminations, and had intervening, intercellular spaces. Ramie, oats, and willow showed similar modifications, but hemp was only slightly affected. Tincker (1932) and Tincker and Darbishire (1933) found in dahlias, artichoke, and *Phaseolus multiflorus* that potassium accumulated in the roots and tubers as they increased in weight. In plants the evidence indicates that in general a potassium deficiency may diminish carbohydrate and protein synthesis. Richards (1932) found a positive correlation between the amount of potassium supplied and the rate of respiration in potassium-deficient plants. This is apparently due to the increase in the amount of carbohydrate resulting from an increased potassium supply. Hibbard and Grigsby (1934) found in plants of the pea, which had been deprived of potassium and calcium, that simple sugars, sucrose, starch, hemicellulose, and protein were produced but in less quantity than by plants growing in complete nutrient solutions. They believed that the deficiency of the two elements disturbed and retarded protoplasmic activity.

On the other hand, Lilleland (1930) could find no indication that the content of potassium in the soil had any influence on the sugar content of prunes. Janssen and Bartholomew (1932) found that cowpeas grown in a low-potassium nutrient solution generally were higher in reducing sugars, and total sugars, but lower in starch than plants grown in high-potassium nutrient solutions. Thick cell walls were associated with low potassium in the plant. The percentage of protein was usually greater for the potassium-deficient plants. Houghland and Schricker (1933) concluded that there is only a slight possibility of altering the starch content of the potato by varying the potassium supply. Phillips, Smith, and Dearborn (1934) noted that the potassium-deficient plants of tomato were high in solids and reducing sugars.

b. Relation to Nitrogen.—Potassium is considered to play a role in the formation of proteins in the plant. Stocklasa (1916) considered that in the presence of sunlight and with a sufficient supply of carbohydrate and other material, protein can be formed in either the presence or the absence of potassium. In the dark, however, the formation of protein occurs only in the presence of potassium, provided the necessary materials are at hand.

Loew (1934) considered that potassium is essential to the manufacture of the protein in the plant cell. The increase of protein reserves in seeds, in conjunction with an increase in potassium content, supports this view and explains the higher potassium content of leguminous seeds as compared with that in the seeds of grains and grasses. It has been observed by Gildehaus (1931) for apples, Janssen and Bartholomew (1932) for cowpeas and sugar beets, Colby (1933) for French prune trees, and Rippel, Behr, and Meyer (1933) for potatoes that the amount of nitrogen is markedly higher in those plants that have been grown in a potassium-deficient medium than in those that have grown in a sufficient supply of this element. Nightingale, Schermerhorn, and Robbins (1930) considered that potassium is apparently essential either directly or indirectly for the initial stages in nitrate reduction in the plant and probably for the synthesis of proteins of meristematic tissue. Janssen, Bartholomew, and Watts (1934) believed that they found evidence in tomato plants that a deficiency of potassium causes an accumulation of amino forms of nitrogen. The plants that were deficient in potassium had a lower percentage of nitrogen in a water-soluble form than those with a high potassium content.

c. Formation of Oils.—Potassium apparently plays a role in the formation of oils in plants. Soybean plants grown with a deficient potassium supply produce small and immature seeds which have an oil content lower than that of the normal seeds (Schuster, 1927). It is assumed that the carbohydrates have not under these conditions become fully transformed into oil.

d. Catalyst.—Most authors are agreed that potassium plays the role of a catalyst or condensing agent. Loew (1903) believed that the superiority of potassium over sodium in plant nutrition is due to its superiority as a condensing agent. He cites numerous cases in organic chemistry where condensation processes take place only in the presence of potassium or much more rapidly than with sodium or related elements. Stocklasa (1908, 1920) considered that potassium acts as a catalyst in the formation of organic compounds. He is of the opinion that potassium is concerned through its radioactivity in photosynthesis and that there is a relationship between this process and the beta and gamma rays that potassium emanates.

Bruno (1928) considered that potassium may influence plant growth and carbohydrate formation as a result of a photoelectric effect originating in the radioactivity of potassium. Loew (1934) believed that the specific action of potassium is due to the emission of beta rays by a potassium compound in the protoplasm. These rays promote the physiological oxidation processes in the protoplasm and transform the energy in the living cell. Further evidence that potassium acts as a condensing agent is furnished by the work of Reed (1907). In *Spirogyra*, which ordinarily can be stimulated to divide, mitotic nuclear division did not proceed in the absence of a potassium supply. The cells elongated but the nucleus did not divide. The assumption for this behavior is that nucleoprotein cannot be synthesized in the absence of potassium even if the necessary constituents are present. When such elongated cells are placed in a solution of potassium, normal mitotic division proceeded in some cases, but in a majority of them it did not. This behavior may be due to an insufficient supply of nucleoprotein, but potassium may also play some important function in the initial steps of mitotic division and, being absent at that time, could not function at a later stage.

Loew (1934) considered that, in the cell nuclei, potassium promotes synthetic chemical processes.

e. Action on Enzymes.—The effect of potassium on diastatic activity has been studied intensively. In most cases, potassium appears to accelerate the action of diastase. James (1930) believed that potassium aids in increasing the effectiveness of the catalytic surface, but the manner by which it does this is not known. It may increase the surface of the enzyme by induction of a greater dispersion, by increasing its absorbing and combining power, by a change in the degree of dissociation, or by the formation of new enzyme (Briggs, 1922). Cattle (1933) found that by whatever method the plants were compared, the lack of potassium depressed the diastatic activity of the leaves. The amount of this depression varied with the age of the leaf. James and Cattle (1935) considered that the effect of potassium is to increase the amount of diastase formed. It apparently does not enter into combination with the enzyme, since colloidal precipitates from extracts of plant materials usually contain diastase, but they may be entirely free of potassium. White (1936), studying the growth of colonies of *Lemna major* in solutions deficient in potassium, found that they were characterized by a high content of starch but by a low assimilation rate. The enzyme-containing extract of colonies grown in potassium-deficient mediums showed a diminished capacity for hydrolyzing starch.

Hartt (1929, 1934) found, however, in sugar cane that the potassium-deficient plants had the greatest diastatic activity in the tops, while in the roots it was the same for all conditions of potassium nutrition. She found further that potassium stimulated the action of invertase and catalase. Cattle (1933) noted that the catalase activity is decreased in potassium-starved plants.

3. General Response of Plants to Potassium.—Plants differ to a considerable degree in their response to potassium. Hartwell (1927) found that the yield on an area with insufficient potassium in comparison with the yield on an area with optimum fertilizer ranged from 16 per cent for onions to 88 per cent for oats. He grouped the plants in reference to their response to the applications of potassium as follows: low potassium-response crops—oats, rye, wheat, millet, and carrots; medium potassium-response crops—barley, rutabagas, parsnips, potatoes, and cabbage; high potassium-response crops—tomatoes, mangels, buckwheat, corn, and onions.

Potassium has been reported to modify the general appearance of the plant; to affect the quality of fruits, grains, and vegetables; to prevent certain diseases; and to influence the intake of other elements.

a. General Appearance.—It was reported by Hartt (1929, 1934) for the cuttings of sugar cane that the symptoms of potassium starvation are decreased growth, "dieback," and a deficiency of chlorophyll. Nightingale, Schermerhorn, and Robbins (1930) noted that potassium-deficient plants die prematurely if fruit is present, because most of the potassium is withdrawn from the vegetative parts and translocated to the fruit. Kraybill (1930) considered that dead tissue around the margins and between the veins of the leaves is an accurate indicator of potassium deficiency in the tomato plant. Johnston and Hoagland (1929) in the same plant noticed a characteristic spotting of the leaves that indicated a lack of potassium, while Fisher (1935) stated that symptoms of potassium deficiency appeared after 7 to 10 days as yellowish-brown spots on the leaves near the margin, and that the spots gradually spread until they coalesced and the entire leaf became yellow. The margins of the leaves turned under, and the lower leaves were pimpled between the veins. Wallace (1930) and Davis (1930) reported that the lack of potassium increased shoot growth, leaf scorch, and the defoliation of shoots.

b. Quality.—The quality of fruits and grains has been attributed to an optimum potassium supply. Schuster (1927) considered that the effect of the application of potassium on the quality of the crop will depend upon the amount of potassium in the soil and the property of the soil to correct the effect of the acid derived by the liberation of the applied potassium. He considered that the literature on potassium nutrition produces evidence which indicates that potassium aids in producing quality in potatoes, sugar crops, tobacco, soybeans, and wheat.

Lanham (1927) found that the application of potassium had no influence on the carrying quality of tomatoes grown in Texas. He considered that the carrying quality is mainly varietal and is influenced to some extent by the time of harvesting. Both the fruit and the plants from the fertilized plots contained more potassium than the checks, but the tomatoes containing the highest amount of potassium were not always those from the plots having the highest application of potassium. Robbins, Nightingale, Schermerhorn, and Blake (1929) stated that an adequate supply of potassium appears essential for the production of "chunky" sweet potatoes. It apparently functions in the formation of a type of protein necessary for the rapid development of the cambium, which is chiefly responsible for the thickness and chunkiness of the potato.

c. Relation to Other Elements.—Brown (1927) found that in the apple low values for the hydrogen-ion concentration are associated with total ash and potassium content so that potassium and acidity are inversely correlated. A correlation seems also to exist in the apple between potassium and phosphorus. Johnston and Hoagland (1929) found that a low supply of potassium in the soil cause an increased absorption of calcium, magnesium, and phosphorus. Fonder (1929) noted in alfalfa that there seemed to be an inverse correlation between the content of potassium and calcium. Colby (1933), with French prune trees noted under conditions of potassium starvation that the leaf tissues were higher in phosphorus, calcium, and magnesium but lower in ash than those trees that were supplied with sufficient potassium. Phillips Smith and Dearborn (1934) found that the effects of an early deficiency of potassium upon the tomato plant are a lower ash content and a higher calcium, magnesium and phosphorus content than are found in plants growing in a complete nutrient solution.

McCalla and Woodford (1935) found that the limiting of potassium to wheat plants resulted in a marked increase of calcium and magnesium in the dry matter.

d. Other Effects.—According to Brown (1929) good keeping quality of apples is associated with a high percentage of available potassium and phosphorus in the soil and with the amount of each in the fruit. Beaumont and Chandler (1933) found indications that a lack of potassium tended to make apples and peaches firmer at harvest but to make them softer during storage. Young, Janssen, and Ware (1932) and Ware and Young (1934) found that a plentiful application of potassium, alone or in combination with salts containing nitrogen and phosphorus, gave definite control of cotton wilt. Warne (1934) found that apples, gooseberries, and currants are especially susceptible to "leaf scorch" under conditions of potassium deficiency. The percentage of dry matter is higher in the scorch leaves, but the potassium content is almost always lower in the scorched than in the unscorched leaves.

Davis, Hill, and Johnson (1934) noted in strawberries that lack of hardiness was associated more markedly with deficient potassium than with any other factor. Wohack (1930) claimed that the frost-protecting effect of potassium is very evident even when it is applied late in the season. There seems to be some evidence that potassium tends to prevent lodging by stiffening the straw. Mann (1924) noted that young apple trees in sand cultures deficient in potassium had small leaves subject to leaf scorch, and that the gooseberry under similar conditions had leaves with lower water content which were less resistant to water loss than those growing in the controls. James (1931) found a significant, positive correlation between the amount of potassium and the water content of the plant throughout the growing season.

F. Sulphur

It has long been recognized that sulphur is necessary for plant development, but the amount needed by the plant was until recently underestimated. Much sulphur was lost by the ash-analysis method, which until comparatively recently was used in its determination. By more accurate methods of analysis it has been found that plants need from two to one hundred times as much sulphur as had formerly been supposed. Hart and Peterson (1911) stated that cereal crops remove from the soil about two-thirds as much sulphur as phosphorus, the legumes as much sulphur as phosphorus, and in the case of alfalfa a greater amount, while such crops as turnips and cabbage remove two to three times as much sulphur as phosphorus. In mustard, potato, onions, and chives, also, the sulphur content of the plant greatly exceeds that of the phosphorus (Crocker, 1923).

1. Occurrence.—Sulphur occurs in plant tissues chiefly in the form of proteins, volatile compounds, and sulphates. In the proteins it exists in the form of cystine. Some of the various volatile forms of sulphur are mustard oil, allyl sulphide, vinyl sulphide, and mercaptans. Those volatile compounds are most abundant in the mustard family, but some of them are to be found in the *Liliaceae* (onion and garlic), *Tropaeolaceae*, and other families. The mustard oils do not occur as such in the plant but are found in the form of glucosides, the best known of which are sinigrin found in black mustard and sinalbin found in white mustard. Sulphur occurs in the inorganic form in plants principally as sulphates, and as much as 65 per cent of the sulphur found in certain plants is in the form of soluble sulphates. Plants cannot utilize elemental sulphur; apparently it must be oxidized to the form of sulphates before it is of any benefit to them. According to Reimer and Tartar (1919), all the sulphates produce similar beneficial results when applied as fertilizers.

It was noted by Frazer (1935) that certain coniferous plants obtained several times as much elemental sulphur from localities where there was distinct atmospheric contamination from factories as when they grew in regions not subjected to this contamination.

It is also of interest to note that, when sulphur is applied as a fertilizer in the soil, a marked increase in the amount of sulphates in the plant is soon noted. In the case of rape, radishes, and clover that had been grown on a soil with only the normal amount of sulphur present, the plants grew vigorously but showed little or no sulphates in their tissues. This indicates that a considerable portion at least of the sulphur that is found in the plant in the form of sulphates is simply an excess supply of sulphur.

The following brief table adopted from Peterson (1914) shows the relative amount of the various types of sulphur compounds in certain plants:

Distribution of the Several Forms of Sulphur in Percentage of the Total Sulphur

Material	Soluble sulphate *A*	Volatile sulphur *B*	Soluble sulphur *C*	Insoluble sulphur *D*	Sum of fractions *B, C, D*	Loss on drying	Total unoxidized sulphur $(C - A) + D$
Rutabagas, average 4 samples	37	12	73	11	96	19	47
Cabbage	24	9	74	15	98	..	65
Sugar beets	34	..	62	38	100	11	66
Alfalfa hay	50	..	70	30	100	4	50
Rape, dried[1]	64	..	85	8	93	..	29

[1] Potassium, nitrogen and phosphorus, and sodium sulphate added to soil.

Sulphur is fairly evenly distributed throughout the plant. In a mature corn plant, 40 per cent of the total amount was in the leaves, 23 per cent in the stem, 26 per cent in the grain, and 11 per cent in the roots. Marsh (1922) observed that the amount of sulphur in the tomato plant increased from the tip toward the base, the oldest tissues having the highest sulphur content. In the case of the apple, the highest percentages of sulphur are found in the leaves and then in descending order in spurs, young bark, and old bark. The percentage of sulphur is at a minimum in the apple spur at the time of fruit-bud differentiation.

2. Physiological Role.—Sulphur functions as building material for the formation of protein and other constituents of the plant. The proteins of plants contain from 0.003 to 7.2 per cent, and it has been noted that plants with a high protein content or especially rich in organic sulphur compounds respond most markedly to the application of sulphur as a fertilizer. In Washington (Olson and St. John, 1921; Peterson, 1914) alfalfa grown on soils treated with sulphur was found to contain a higher percentage of protein than alfalfa on soils not so treated. This indicates that the sulphur under these conditions was used by the plant in the production of protein. This supposition is strengthened by the work of Tottingham (1918) on red clover, which indicated that a deficiency of sulphur restricted growth in that plant by limiting the synthesis of protein.

3. Formative Effects.—Aside from functioning as building material, sulphur produces certain formative effects in plants. These are manifested especially in three ways.

(*a*) An increased root or absorbing system. The effect of sulphur fertilizer on the root system of plants has been noted by Hart and Tottingham (1915), Pitz (1916), Miller (1919), Reimer and Tartar (1919), and others. It was found in the case of alfalfa that the root systems of plants which have been fertilized with various sulphur fertilizers were from two to three times as large as those of the unfertilized plants. This holds true also for clover and rape. It was suggested by Crocker (1923) that the increased root development which has been so frequently observed from the application of superphosphates and attributed to phosphorus might be due, in part at least, to the sulphates therein rather than entirely to the phosphates.

(*b*) The appearance of chlorophyll. It has been observed that plants grown on soils deficient in sulphur have a pale-green color, and this can be remedied by the application of sulphur as fertilizer. The plants thus treated show a dark, vigorous, green color (Eaton, 1922; Demolon, 1912). Powers (1930), in working with potatoes, corn, and tomatoes that had been dusted with land plaster or grown in soils that had been treated with sulphur, found that they developed a darker green color than the controls. He found that the amount of chlorophyll in alfalfa plants was increased as much as 18 per cent by the application of sulphur. The application of this element tended to increase the length of stems of this plant, to thicken the leaves, and to prevent their shattering in haying. Storey and Leach (1933) reported that the "tea-yellows" disease is produced by a deficiency of sulphur in relation to other elements. This disease may be prevented in the field by an application of either sulphates or elemental sulphur.

Just what role sulphur plays in this regard, however, is not known. This element is apparently necessary for the formation of chlorophyll, but it must act in an indirect manner, since it does not enter into the composition of the pigment.

(*c*) Effect on the nodules of the legumes. Duley (1916) was the first to note an increase in the number of nodules on the roots of sweet clover by the application of flowers of sulphur as a fertilizer. Pitz (1916) found that calcium sulphate when added to soil apparently had no marked effect on the total number of bacteria that grow on agar plates and did not increase to any extent ammonification or nitrification. The application of sulphur did, however, increase the growth of the legume bacteria and the number of nodules on red clover. Powers (1923) noticed in clover that the plants fertilized with sulphates had a higher nitrogen content than the controls. Since the number of nodules on the roots was increased, and since rape, which assimilated a large amount of sulphur, did not show any relationship between the content of nitrogen and sulphur, he attributed this increased nitrogen content of clover to the stimulating effect of the sulphates on the action of the legume bacteria.

4. General Effects.—It was noted by Nightingale, Schermerhorn, and Robbins (1932) for tomato, apple, narcissus, and asparagus, and by Eaton (1935) for soybeans, sunflower, rape, kale, and mustard that plants deficient in sulphur have the general appearance of those which have been completely deprived of nitrogen. These plants are extremely high in carbohydrates and contain much more nitrate than the plants receiving a sufficient sulphur supply. The rate of nitrate reduction and the oxidation of sugars is reduced but not totally inhibited. The digestion of starch and the translocation of sugars take place freely. Such plants have thick cell walls and a relatively high proportion of fibers and lignified tissue. The protoplasm appears much like that of plants lacking nitrogen in that it is limited in amount but not noticeably injured as it is when phosphorus, calcium, or potassium is deficient.

5. As a Fertilizer.—Sulphur is present in the soil in the form of either sulphates or organic matter. Most of the sulphates are quite soluble and are not readily adsorbed, so they are leached out rapidly, and only small amounts are found in the soil. In a large number of soils analyzed by Woodard (1922), the sulphur content was less than the phosphorus content. He estimated that the highest sulphur content found in these soils was sufficient for 39 years of alfalfa, 355 years of wheat, and 232 years of corn, while the lowest sulphur was sufficient for only 5 years of alfalfa, 46 years of wheat, and 30 years of corn. Good to excellent crops of alfalfa remove annually from 37 to 90 lb. of sulphur per acre from the soil, and, in general, hay from fields with the heaviest crops have the highest percentage of sulphur (Hall, 1922). The response of plants to the application of sulphur as a fertilizer will depend upon the type of plant and upon the soil. In many regions the sulphur supply is not a limiting factor in production, as shown by Bruce (1925) for Kansas soils and by Stewart (1920) for Illinois soils.

Hart and Tottingham (1915) working in Wisconsin found that plants most affected by the application of sulphates to the soil were members of the legume and mustard families. They observed, however, a noticeable stimulation to seed production in barley and oats but no increase in the quantity of straw. In the case of clover, they obtained an increase in dry matter due to calcium sulphate of about 23 per cent. With rape, the greatest increase occurred where the calcium sulphate was applied with a complete fertilizer, giving an increase of 17 per cent over the complete fertilizer. In general, the calcium sulphate was more effective than the more soluble sodium sulphate.

Shedd (1914, 1917) reported that wheat, corn, oats, timothy, alfalfa, rape, beets, potatoes, turnips, radishes, beans, tobacco, and grapes were benefited by the application of sulphur to certain Kentucky soils. He found that the sulphur content of 32 varieties of tobacco averaged 0.458 per cent of the total dry matter as compared to a phosphorus content of 0.302 per cent. Olson and St. John (1921) in Washington obtained increases of twofold or more in alfalfa yields by treating the soil with gypsum, acid phosphate, and sulphuric acid, while Powers (1923) from Oregon showed that marked results could be obtained by the application of sulphur for alfalfa, red clover, and alsike, but only a moderate response in yield was had with wheat and potatoes and little or no increase for corn, kale, or sunflowers.

G. Iron

1. Occurrence.—Iron is necessary for the growth and development of green plants and cannot be replaced by any other element. Although iron is universally distributed in plants, its detection microchemically is difficult, since it occurs in very small quantities. Molisch (1892) found that combined iron occurs in plants in the greatest amount and that very little of the iron is in a form that is readily soluble in water. He observed combined iron in the protoplasm, the vacuole, and the globoids of the aleurone bodies. Jones (1920) stated that, as a whole, plant tissues give the iron reaction more rapidly and intensely than animal tissues. The iron stain occurred in the chloroplasts, nuclei, and in large masses scattered throughout the cytoplasm. The leaves of some plants did not show the iron stain, while others did with a very definite reaction.

Plants vary widely in the amount of iron that they contain. Thus Stiebeling (1932) found that apples, cucumber, and watermelons had the least amount, containing less than 0.0004 per cent of this element. Bananas, celery, carrots, onions, and tomatoes had from 0.0004 to 0.00079 per cent, asparagus, beets, dock, and sorrel had from 0.0008 to 0.00159 per cent, while cowpeas, dandelion leaves, parsley, spinach and water cress contained over 0.0016 per cent. The iron in plants apparently occurs in two forms designated variously as "active and inactive," "free and combined," "soluble and insoluble," and "available and nonavailable." Sayre (1930) believed that the iron in corn stems is in the form of an oxide or hydroxide in a solid solution in some host substance.

2. Physiological Role.—If plants are deprived of iron, they do not normally produce chlorophyll, a fact that was first noted by Gris (1844). The new leaves and stems formed by plants deprived of iron are pale or yellow, a condition that has been termed "chlorosis." It has been definitely proved, however, that iron does not enter into the composition of chlorophyll (Willstätter and Stoll, 1913). The quantity needed by the plant at any time is very small. Gile (1916) estimated the available iron in the nutrient solution in which he grew normal healthy rice plants to be always less than 1 part per 10,000,000 parts of solution. The use to which iron is put in the plant is not known. Wolff (1913) and others considered that it acts in plants as a catalyzer, since such marked results are obtained with such small amounts, while Warburg (1925) has proposed the theory that iron is the oxygen-carrying component of the respiration ferment.

Hopkins (1930) believed that the concentration of the iron ions plays an important role in the cellular processes involving biological oxidation.

Oddo and Pollacci (1920) grew plants of *Zea mays*, *Solanum nigrum*, *Datura stramonium*, *Euphorbia sp.*, and *Aster sinensis* in a standard nutrient solution lacking iron but to which was added the magnesium salt of pyrrolecarbonic acid that has the formula

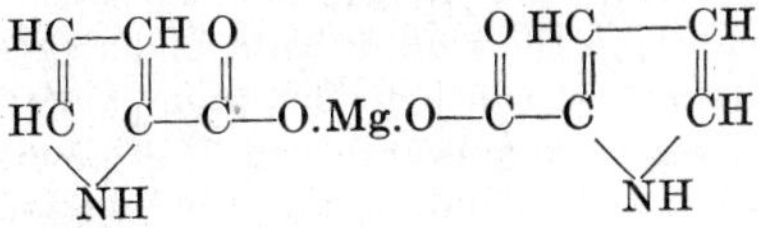

in a concentration equivalent to 0.0232 g. magnesium in 1 l. of water. These plants, although deprived of iron, when supplied with this organic compound formed chlorophyll. Since iron under ordinary conditions is indispensable for green plants, these authors suggested that the formation of chlorophyll in the presence of iron is due to the catalytic action of this element in the formation of the pyrrole nucleus, which in itself is the center of the chlorophyll complex. They believed that if this nucleus is already formed, the presence of iron is not necessary for the greening of the plant. This observation is in line with the statements of Willstätter (1906) and Mameli (1918) that the function of magnesium in the greening of the plastids is directly proportional to the presence of pyrrole. Deuber (1926), however, was unable to confirm the work of Pollacci and Oddo in this regard. He grew plants of corn, cowpea, soybean, and *Spirodela* in nutrient solutions containing a magnesium salt of pyrrolecarbonic acid substituted for iron. In no case did this compound prevent chlorosis of the leaves of the plant. He found further that the pyrrole salt in concentrations of 0.001 to 0.250 g. per liter was toxic to the plants used and that when the salt was applied as a paint to the leaves of cowpeas it had the same injurious effects as in solution.

Pollacci (1935), however, reported that *Chlorella vulgaris*, and other algae, thrived and remained green in cultural solutions devoid of iron but containing magnesium pyrrolecarboxylate instead. He considered that his results confirmed the previous conclusions of Oddo and Pollacci that the magnesium compound can replace iron in the nutrition of green plants. Oserkowsky (1933) noted that iron may be present in larger amounts in chlorotic leaves than in green ones. He believed this indicated that a specific kind of iron, which he designated as "active iron," is concerned in chlorophyll formation. He found that the chlorophyll content of leaves from chlorotic plants is proportional to the amount of active iron in the leaves. He considered that this active iron is in the ionic form or is in a compound that readily yields ionic iron.

3. Form and Amount Necessary for Plant Growth.—This is a subject that formerly was considered of little import in the nutrition of plants. All the directions for nutrient solutions stated that only a trace of iron was necessary and that it mattered little in what form it was presented to the plant, provided it was soluble in water. Quite recently, however, much work has been done on the question of iron nutrition, which shows that not only the quantity of iron available but the form in which it is presented to the plant can exert a marked influence upon plant growth and development. These two factors in turn are dependent upon the composition of the medium or nutrient solution—its hydrogen-ion concentration, the nature of the general environment—and to some extent upon the kind of plant under consideration. There is at the present time no criterion by which to determine whether a plant is receiving sufficient iron except by its general appearance of health and vigor.

Moore and Jones (1936), in England, obtained no striking differences in yield by the application of iron to potatoes and beets, but the plants on the treated plots had darker green leaves and a healthier appearance than those of the control plots. Gile (1916), however, remarks in this regard that although the color of the leaves will indicate a marked deficiency of iron, a slight deficiency may materially diminish the yield without appreciably affecting the appearance of the plant.

Jones and Shive (1921) used ferric phosphate and ferrous sulphate as a source of iron in Shive's nutrient solution (*R5C2*, 1.75 atmospheres) in which Marquis wheat was grown. The solution was renewed at intervals of about 4 days for a period of 90 days. The results obtained are shown in the following table:

Iron per liter of solution, milligrams	Series I $FePO_4$, dry weight, grams	Series II $FeSO_4$, dry weight, grams
0.01	4.55	10.95
0.10	5.75	14.65
0.25	8.20	15.95
0.75	12.05	27.40
1.00	14.90	18.75
1.50	16.25	19.15
2.00	20.50	23.50
3.00	18.00	20.90
5.00	16.80	18.45
Control no Fe..........8.40		

In series I the plants showed more or less chlorosis even when large amounts of iron were supplied, a condition which could easily be remedied in a day or two by the introduction of ferrous sulphate. There was no chlorosis in series II except when small amounts of iron were used. Ferrous sulphate, however, was apparently toxic in higher concentrations.

That plants may differ in regard to their utilization of iron was shown by Corson and Bakke (1917). They found a marked difference in the growth of wheat plants when the source of iron was ferric phosphate and ferrous sulphate, the former being the more efficient. Little or no difference, however, was observed in the behavior of Canada field peas to these two sources of iron. They considered that the iron stored in the seed might be a determining factor, a fact which was also suggested by Tottingham and Beck (1916) in their observations on the response of wheat plants to the salts of iron and manganese. Hartwell and Pember (1908) noted that ferrous sulphate was equally effective as a source of iron for barley and rye seedlings growing under the same conditions.

Marsh and Shive (1925) studied the relation of the growth of soybean to the amount and form of iron in several types of cultural solutions. The iron compounds used were ferric glycerophosphate, soluble ferric phosphate, ferric tartrate, and ferrous tartrate. Considering that the availability to the plant of iron in cultural solutions is closely correlated with the solubility of the iron compound in the solution, their data suggest that in a general way ferric glycerophosphate could be used to advantage in solutions in which the hydrogen-ion concentration is kept at a high level. Soluble ferric phosphate and ferrous sulphate are not very usable when the pH values approach the neutral point. In solutions of this type they suggest that ferric tartrate should prove the most effective source of iron for the plant. Marsh and Shive further found that if the iron supply is adjusted from day to day, large healthy plants are produced regardless of the type of cultural solution or the iron compound used. The iron in healthy plants was present in a smaller percentage, based on dry weight, than in the chlorotic plants or those suffering from iron toxicity. The iron in healthy plants, although low in concentration, seems to be uniformly distributed throughout the stem and leaves, while in the case of those suffering from toxicity of iron it is high in concentration. The iron content of chlorotic plants is low in the leaves but high in the stems. These authors suggest that in order to maintain the plant in healthy green condition, the supply of soluble iron in the cultural solution must be kept at as low a concentration as possible without inducing chlorosis from lack of available iron. A concentration of soluble iron slightly above this optimum may result in toxicity due to a too high content throughout the plant. Chlorosis, however, may occur due to the lodgment of iron in the roots and stems, thus preventing the translocation of this element to the leaves in sufficient amount (Hoffer, 1920). Marsh and Shive considered that the general appearance of the plant must serve as an index of the iron supply in each case, since the changing plant environment has pronounced influence upon the iron requirement.

Gines (1930) used six different types of iron salts in growing rice in nutrient solutions and found that each of these was capable of furnishing sufficient amounts of iron for the needs of the plant. A striking observation in this regard was noted by Gericke (1925) in the effects of two different light intensities on the growth of wheat in nutrient solutions devoid of iron. It appeared from the general behavior and appearance of the plants that the need for iron is much greater per unit of growth of plants exposed to high light intensities than to low ones. According to Ingalls and Shive (1931), the soluble-iron content of plants varies directly with the variation in the hydrogen-ion concentration brought about by changes in light intensity from day to night. Low hydrogen-ion concentration corresponds to high light intensity and high hydrogen-ion concentration to low light intensity. Plants in which the tissue fluids have low hydrogen-ion concentration values show high total and relatively

Plant	Leaf		
	Iron per gram of dry matter		
	Acidity, pH	Total	Soluble
Buckwheat	5.10	0.278	0.0477
Tobacco	5.74	0.325	0.0390
Tomato	5.87	0.297	0.0277
Asparagus	5.87	0.373	0.0256
Soybean	6.01	0.469	0.0246
Clover	6.10	0.571	0.0281

low soluble-iron content. Those plants in which the tissue fluids have high hydrogen-ion concentration show low total and relatively high soluble-iron content, as shown by the table on page 331.

The same conditions that in the presence of iron in the cultural medium are decidedly beneficial to growth become relatively harmful in the absence of iron. This observation is an important one, for it not only indicates that there can be no fixed minimum amount of iron for a unit yield of product, but it suggests that this may apply to other elements as well.

4. Relation to Hydrogen-ion Concentration.—Jones (1920) and Jones and Shive (1924 to 1923) replaced potassium nitrate with ammonium sulphate in Tottingham's solution and noted that the solution became more acid when plants were growing therein, changing from pH 4.8 to 4.2 during a 3½-day period, while the regular Tottingham solution became less acid during that time. One milligram of iron per liter in the form of ferric phosphate was not sufficient to supply the needs of plants in Tottingham's solution, but one-half milligram was ample when ammonium sulphate replaced the potassium nitrate. In Tottingham's solution 0.25 to 0.50 mg. of iron in the form of ferrous sulphate was sufficient, but in the modified solution this amount was toxic to the plants. McCall and Haag (1921) observed that wheat plants growing in solutions having pH values ranging from 4.02 to 7.0 suffered from a lack of available iron or from a change in metabolism that resulted in the immobility of iron within the plant. This observation is in line with that of Hoffer (1920), who considered that the rate of growth of young plants of corn and the degree of chlorosis may be determined by the relative ability of the plant to develop and maintain a sufficiently high hydrogen-ion concentration to keep the iron compounds mobile as occurs in normal green plants.

The solubility of ferric citrate, ferric phosphate, ferric sulphate, and ferrous sulphate in Tottingham's solution within a pH concentration of 4.2 and 6.0 was investigated by Tottingham and Rankin (1923). Using the growth measurements of young wheat plants as an indicator, the results showed that ferric citrate was the most favorable form of iron used. It remains soluble over a considerable range of pH values of the nutrient solution, and, when supplied at the rate of 10 mg. of iron per liter of nutrient solution, although it is not completely dissolved, it provides abundant iron for plant growth when the solution is renewed every 4 days. Reed and Haas (1924) found that the addition of certain organic compounds to an alkaline nutrient solution, in which ferric tartrate was the source of iron, increased the amount of soluble iron in the solution. The sodium and potassium salts of organic acids, *e.g.*, citrates and tartrates, effect marked solubility of iron without appreciably altering the pH of the solution. Starch and sugars added to the solution also dissolved appreciable amounts of iron. Deuber (1926) in using potassium ferrocyanide in concentrations equal to 0.033 and 0.066 p.p.m. of iron noted that the higher concentration produced a stoppage of growth in soybeans. Ferric ferrocyanide was a satisfactory source of iron when the solution had a reaction of pH 5.0, but, at a less acid reaction, growth of these plants and chlorophyll development were restricted.

A direct correlation was found by Barnette (1921) and Barnette and Shive (1923) between the decrease in hydrogen-ion concentration of the cultural solution and the appearance of chlorosis in plants when ferric phosphate and ferric sulphate were used as sources of iron. They could partially control the hydrogen-ion concentration of the solution in contact with the roots of the growing plants by the use of a large volume of solution renewed frequently. Hopkins and Wann (1926, 1927) also observed that iron was unavailable to *Chlorella* in nutrient solutions at the higher pH values. The addition of sufficient sodium citrate, however, kept the iron in solution even at alkaline reactions for an indefinite period. If too high an amount of sodium citrate is present, the ionization of the ferric citrate is decreased, so that although there may be a large total amount of iron present little or no growth will occur because of the low concentration of the ferric ion. The availability of iron is determined apparently in as large a measure by the composition, concentration, and general reactions of the cultural solutions, as it is influenced by the action of the plants growing therein.

It was noted by Rogers and Shive (1932) that the xylem always showed the lowest and the phloem the highest pH values of the plant tissues. The sclerenchymatous tissue had pH values that were the same as or slightly higher than the xylem. Small amounts of iron were always present in the xylem parenchyma, but in no instance did pronounced accumulations of iron occur there. Whenever marked accumulations of iron were present, they usually occurred in tissues with a high pH value, such as the phloem where it bordered on the xylem or other tissues with a low pH value. The iron in these accumulations can be regarded for the most part to be in the precipitated form and immobile, so that it does not function in the metabolism of the plant. It seems logical, therefore, to conclude that the xylem is the tissue in which the upward translocation of iron occurs. No accumulation of iron was found in plants with low pH tissues throughout. In such plants, the content of iron is low and uniformly distributed. The roots seemed more acid than the stem. They should thus readily absorb iron if it is available in the soil since most of them showed pH values at or above pH 6.0, the critical value for the precipitation of iron. The iron may be precipitated before it reaches the chlorophyll-bearing parts. Certain plants with a high pH of tissues may show little tendency toward chlorosis because apparently they contain certain organic solvents in the presence of which iron in the mobile form resists the precipitating action of high pH reactions. Oxalic, tartaric, and citric acids are some of the compounds that will prevent the precipitation of iron.

Oserkowsky (1932) in pear trees found no significant difference between the hydrogen-ion concentration and the iron concentration in the extracted tracheal sap of green and chlorotic branches from the same environment. Bennett and Oserkowsky (1933) found that the tracheal sap in several instances contained more copper than iron but no special relation of copper to the functioning of iron was indicated.

5. Absorption from the Soil.—Certain plants, especially rice and pineapple, on certain calcareous and manganese soils are affected with chlorosis and do not make a normal growth. This phenomenon has been encountered and studied most extensively on the calcareous soils of Porto Rico and the manganese soils of Hawaii. Gile (1911) applied a solution of ferrous sulphate to calcareous soils, but this showed no effect in reducing the chlorosis of the leaves of the pineapple. Crystals of the salt, however, applied in the soil in close proximity to the roots gave very evident results in reducing chlorosis. In another experiment (1916), rice plants were grown in acid, neutral, and alkaline solutions with 0.008 g. of iron per liter in various forms. The plants grown in the acid solution contained the highest percentages of iron. Plants in the neutral solution contained more iron than the alkaline solution when some forms of iron were used and an equal amount when others were used. The percentage of nitrogen-, phosphorus-, calcium-, magnesium-, and carbon-free ash remains the same in all plants regardless of the reaction of the solution. From these and other similar experiments, Gile (1911, 1916), Gile and Carrero (1920), Mazé (1912), Sidorin (1914), and Willis and Carrero (1921) agreed in considering that lime-induced chlorosis is due to a depression in the available iron. They are inclined to believe that the carbonate of lime in the soil reacts with the iron forming ferric carbonate and thus depressing the availability of the iron for the plants. In the manganese soils of Hawaii, Johnson (1924) has shown that it is present mainly in the dioxide form. He considered that the manganese dioxide either present as such or formed from manganese salts keeps the iron that is present in the soil oxidized to the ferric form, which is not readily available to the plants. Allyn (1927) believed that lime-induced chlorosis is not a result of iron becoming locked up in the soil, but rather that it is the result of a disturbance in the metabolism of iron after it has entered the plant.

The lack of iron available in the soil for pineapples has been overcome by spraying the plants with iron solutions. The most effective and economical spray that has been used in Hawaii is an 8 per cent solution of ferrous sulphate, although other forms of iron may be used. Plants that are already yellow are sprayed three or four times at intervals of a week and then as often thereafter as is necessary to keep them green and healthy. The cost of the iron spray above mentioned amounts to about 60 cts. per acre per spraying (Johnson, 1916). It is worthy to note that leaves nicked or injured recover from chlorosis more quickly than do the intact ones after being sprayed.

Burke (1932) reported that spraying trees with ferrous sulphate or applying it to the soil did not improve their chlorotic condition, while treatment internally with iron salts in the solid state or in solution gave beneficial results. Wallace (1935) obtained good results in overcoming chlorosis by using dry, powdered citrate according to the method of Burke. Finch, Albert, and Kinnison (1933, 1934) remedied chlorosis in certain plants in Arizona by the use of iron in various ways.

6. Immobility within the Plant.—It has been observed that when plants which have their needed iron supply are deprived of it, the new stem and leaves that develop are chlorotic or almost pure white so that there is a marked and abrupt contrast between the green leaves and the colorless ones. This phenomenon was observed by Molisch (1902) in squash, sunflower, corn, pineapple, and rice. The leaves of plants growing in an iron-free solution begin to die at the top in marked contrast to the behavior of plants that are deprived of or have a limited supply of nitrogen, potassium, or phosphorus, since in that case the lower leaves dry up while the uppermost ones are the last to perish. These observations point to the fact that iron is immobile in plants and is not readily withdrawn from where it occurs to newer organs. The immobility of iron was further demonstrated by Gile and Carrero (1916). The young leaves of rice plants growing in a solution lacking iron were brushed at the tips with a 0.2 to 0.4 per cent solution of ferrous sulphate. The tips of the leaves just emerging from the stalk were used, and, after they grew out, the part that had been brushed was a normal green, while the lower unbrushed part of the leaf was strongly chlorotic and remained so until treated with iron. Gile and Carrero found that spraying with iron compounds completely restored the green color of the pineapple leaves but that the new leaves formed after the treatment were chlorotic.

H. Silicon

1. Occurrence.—Silicon is an element that has an almost universal occurrence in plants, but the amount present varies greatly for different plants and for different parts of the same plant. Richardson (1920) considered that, for plants of all types and for all parts of plants, silicon is the greatest variable of all the elements that compose them. It is especially abundant in the aerial parts of *Equisetum*, the *Gramineae*, and in the bark of trees. Silicon has been found to compose 71 per cent of the ash of *Equisetum telemantia* and 51, 67, and 46 per cent, respectively, of the straw of barley, wheat, and oats. On the other hand, the silicon content of the stem and leaves of red clover and alfalfa equals only 2.7 and 9.5 per cent, respectively, of their ash content. The distribution of silicon in the different parts of the plant is shown in the analysis of the corn plant at the dent stage (Latshaw and Miller, 1924). The percentage of silicon present in the different parts was as follows:

Plant part	Per cent of silicon in dry matter	Plant part	Per cent of silicon in dry matter
Leaves	2.50	Grain	0.02
Stem	0.45	Cob	0.14

The deposits of silica are more frequent in tropical plants than in those of the temperate zone. As a result of the higher soil temperatures, silicic acid is dissolved and taken into the plant with the water of transpiration, but it is not used in plant metabolism (Frey-Wyssling, 1930). He considered that the deposits of silica and calcium in plant tissues are not excretions but are accumulations of materials that are of no immediate use to the plant.

2. Physiological Role.—The large amounts of silicon present in many plants led early to the belief that it might be an essential element in the nutrition of these plants. It has been shown, however, that those plants which normally contain such large amounts of silicon can be grown to maturity in nutrient solutions that have only a trace present. Sachs (1887) and Knop (1861 to 1862) grew corn and Kreuzhage and Wolff (1884) grew oats in nutrient solutions lacking silicon and the resulting plants contained only a fraction of the normal content of this element. Jodin (1883) grew corn through four generations without the application of silicon. Gile and his workers (1915 to 1916) found that rice that normally contains about 10 per cent of silicon in the dry matter contained only 0.2 per cent when grown in a silica-free nutrient solution. It is now known that these plants were not grown in the absence of silicon for the amount of silicon in the seed, accruing from the glass container, or occurring as impurities in the water or chemicals used, may be sufficient to supply the physiological demands for this element.

Although plants can be grown to maturity in the absence of silicon or at least when only a trace is present, certain benefits that we shall now discuss accrue to the plant by its presence. Palladin (1927) considered that silicon plays an important role in protecting plants from the attacks of various parasites. He stated that wheat and rye grown in nutrient solutions deficient in silicic acid often suffer very severely from rusts. He also considered that silicated cell walls are a protection against the attacks of plant lice and mentioned an experiment with *Lithospermum arvense* grown in a nutrient solution lacking silicic acid. Plants in this solution suffered much more than control plants grown in solutions in which an ample amount of silicon was present. Sommer (1926) reported that seeds of millet grown without silicon were badly infected by fungi and germinated very poorly with and without disinfection, while the seeds of plants grown with silicon were not infected by fungi and germinated well.

King and Davidson (1933) obtained an increased growth of diatoms by the addition of silicate to the cultural solution, but a high concentration had an unfavorable effect upon their growth. It was found by Germar (1934) that silica is deposited for the most part in the epidermis of the leaves of rye, wheat, and barley. If the transpiration is low, however, the silica is deposited for the most part in the stems. Strong illumination, a deficiency of nitrogen, and an excess of potassium favor the accumulation of silicon but the supply of phosphorus does not influence its intake. The leaves of cereals well supplied with silicon are more resistant to infection by mildew, *Erysiphe graminis*. This resistance is correlated with silicon deposition in the epidermis, which makes the layer more resistant to the enzymic penetration of fungi. Resistance to fungi that enter through the stomata is not increased by the infiltration of silicates.

The effect of silica upon the phosphorus nutrition of plants has been extensively studied. Kreuzhage and Wolff (1884) in cultural-solution experiments with oats observed that the total growth of these plants was not much increased by the presence of soluble silicates but that the proportion of grain formed was constantly raised. This effect is very similar to that brought about by the addition of phosphoric acid to cultural solutions which are deficient in that element, so they concluded that the action of silicon and phosphoric acid is in some way related.

Hall and Morison (1906) reported the results obtained from plots at the Rothamsted station which had been treated with soluble silicates during a period of more than 40 years. One series of plots received sodium nitrate with various combinations of fertilizers so as to provide plots receiving (1) nitrogen alone, (2) nitrogen and phosphoric acid, (3) nitrogen and potassium, and (4) a complete fertilizer. From 1864 to 1904, one-half of each of these plots was cross-dressed with sodium silicate. The effect of sodium silicate on the yield of barley for a period of 41 years is shown in the following table:

Plot	Fertilizer per acre					Yield per acre			
						Grain, bushel		Straw, hundredweight	
	$NaNO_3$, pounds	Acid phosphate, pounds	K_2SO_4, pounds	Na_2SO_4, pounds	$MgSO_4$, pounds	Without silicate	With silicate	Without silicate	With silicate
1	275	...	...	...	...	27.3	33.8	16.2	19.8
2	275	350	...	...	...	42.2	43.5	24.6	25.8
3	275	...	200	100	100	28.6	36.4	17.9	21.7
4	275	350	200	100	100	41.2	44.5	25.3	27.6

The table shows that the effect of the sodium silicate is chiefly on those plots which are low in phosphoric acid, and it appears to ripen the grain in a similar but not so striking manner as the application of phosphorus.

Hall and Morison concluded that the increased and earlier grain formation observed in the presence of silicates is due to an increased assimilation of phosphoric acid by the plant, but that there is no evidence that the function of phosphorus can be replaced by silicon. Jennings (1919) increased the dry weight of wheat seedlings from 18 to 29 per cent by the addition of 1 per cent silica to the culture solution in which they were growing. He also showed that the silicon content of these plants was increased as much as 10 to 15 per cent over those grown in the solutions containing no silica. Lemmerman and his workers (1922 to 1925) found that yields were increased by the application of silicates particularly where there was an absence of phosphorus but where potassium and nitrogen were applied in the requisite amounts. They found that colloidal silicic acid was the most favorable form in which to apply the silicon. These workers considered that silicon does not replace phosphoric acid in the nutrition of the plant but increases the amount of phosphoric acid that can be taken up from the soil and the efficiency with which it is utilized. Gile and Smith (1925) observed in experiments with millet grown in sand cultures that the addition of silica gels greatly increased the growth of plants supplied with acid phosphate. The growth of the plants was approximately proportional to the quantity of phosphoric acid in the plants but bore no relation apparently to the quantity of silicon taken up. These authors ascribe the beneficial action of the silica gels to those plants supplied with rock phosphate to the increased quantity of phosphoric acid. Schollenberger (1922), however, found that plots treated with various silicon compounds and soluble phosphates furnished no data on which to base conclusions as to the action of silica in phosphorus assimilation. Brenchley, Maskell, and Warington (1927) found that, under control conditions in water cultures in the absence of phosphorus, the addition of soluble silicates brought about a significant increase in the dry weight of barley. This addition of silicates caused an increase in the height of the main stem, hastened the development of leaves, and was the most marked in phosphate-free solutions. The soluble silicates were more active in causing the increase in dry weight than the glass silicates. In the heaviest application of silica the shoots contained ten times as much silica as they did in solutions that did not contain it. These authors considered that there is a variation in the response of mustard and barley to silicates on different types of soil. They suggested that silica might act within the plant by unlocking phosphate from relatively quiescent parts of the plant and enabling it to be transferred to regions where assimilation and growth are active.

Weihe (1931), although admitting that cultivated plants required silicic acid in small amounts, considered that the increases in yield obtained by its application are not to be explained entirely from the favorable effects of silicon inside the plant. He considered it probable that the finely divided silicic acid, as a result of the large surface exposed, causes adsorption phenomena that increase the water capacity of certain soils and thus improve the nutrition of the plant. Sreenivasan (1934) believed that the peptizing action of colloidal silica may facilitate the uptake of phosphorus from the more insoluble, mineral phosphates.

3. Lodging.—The term "lodging" refers especially to the plants of the small grains that have completely or partially fallen over. This condition is generally caused by a bend in the stem which usually occurs near one of the lower nodes. This bending occurs below the node if lodging takes place early in the season while the leaf sheaths are yet closely enveloping the stem; it occurs above the node if lodging develops after the sheaths have withered and become loose around the stems. The part immediately above the node is the smallest in diameter of any portion of the stem and during the period of development it is usually the highest in moisture content. The plant thus lodges because under certain environmental conditions some portion of the lower stem does not possess sufficient mechanical strength to support the parts above it. In the last analysis, lodging results from too low an amount of dry matter per unit length of culm.

Liebig considered that lodging is caused by the lack of sufficient silicon to properly strengthen the straw. Sach observed that lodging usually developed when the stand of plants was too thick, so that the culms partly shaded one another and took on some of the characteristics of etiolation. He believed that one of the causes of lodging is the lack of sufficient lignification of the stems. Pfeffer (1900) considered that the stiffness of straw is not due to silicon but that it depends upon the development of the internodes under the influence of illumination and exposure.

The causes of lodging are now known to be much more complex than formerly supposed, because lodging is the resultant of numerous and varied factors. Some of these are: fertility of the soil, rate of seeding, climate, heredity, and fertilization.

Lodging is frequently associated with hypernutrition, especially of a nitrogenous nature, and is very prevalent on fertile soils. Under the conditions of hypernutrition there is more tillering, a high proportion of straw growth, and thus more shading. This results in a relatively small amount of dry matter per unit of culm length and a low carbohydrate/nitrogen relation. Under such conditions there is a relatively low amount of polysaccharides per unit of stem. The amount of lignin present seems to vary with different plants or under different conditions. Welton (1928) considered that the stems which lodged were relatively low in lignin, while Davidson and Phillips (1930), and Phillips, Davidson, and Weihe (1931) found that the plants which lodged contained more lignin than those which did not lodge. Prutzkova (1932) and his workers, however, reported that the varieties of wheat which were subject to lodging contain a lower amount of lignin than those which did not lodge. There is much disagreement, however, as to the significance of lignin in contributing to the strength of tissues, so that its importance in relation to lodging may have been overestimated.

A high rate of seeding or the use of varieties that tiller prolifically results in a relatively thick stand that induces shading. Shading has to some extent the same general effects as hypernutrition. A season that is warm, cloudy, and rainy tends to induce a low carbohydrate/nitrogen relation in the plants and thus causes lodging. A relatively infertile soil, a low rate of seeding, and a cool, clear season are factors that tend to reduce the vegetative growth of plants and thus produce a relatively large amount of dry matter per unit length of the culms. Under such conditions a relatively high carbohydrate/nitrogen relation occurs, which results in thick cell walls and a relatively wide band of hypodermal tissue that stiffens the straw and makes the plants less susceptible to lodging. Gassner (1910) reported that if certain strains of oats were germinated at 25°C. they lodged very markedly in about 5 weeks after they had been transplanted in the field. Prutzkova (1932) found that varieties of wheat not subject to lodging contained higher amounts of ether-soluble substances, hemicelluloses and celluloses, and smaller amounts of sugar, than the lodging varieties. There was no difference, however, in the quantity of ash and nitrogen contained in the two types of plants.

Heredity plays some part in the lodging of grains. It is well known that the "hard wheats" have a soft straw, relatively easy to break, that they tiller profusely, and are subject to lodging. The "soft wheats," on the other hand, have hard, relatively stiff straw and only under exceptional conditions do they lodge. Salmon (1931) observed in general that the straw of soft wheats which showed resistance to lodging was more difficult to break than that of hard wheats which lodge easily. The correlations between lodging in various seasons were not statistically significant but those between the breaking strength of straw in different seasons were both high and highly significant. This is explained by the fact that the percentage of lodging is governed by weather conditions and the stage of growth to a much greater extent than is the breaking strength of the stem. Clark and Wilson (1933) could find no correlation between the tillering rates of wheat varieties and their lodging behavior. Brady (1934) observed that the number of vascular bundles, and width of lignified tissue and sclerenchyma were all closely associated with varietal differences in resisting lodging. All these characters, however, were so sensitive to soil variation that their use for the isolation of strains resistant to lodging could be employed only on a relative basis.

Hall (1934) with corn could find little or no relationship between the amount of lodging and the height of ear, cross section of stalk, size of brace roots, number of tillers, or weight of ears. However, in the selfed lines certain phases of the root development were to some extent associated with lodging.

Fertilization may have a marked effect upon lodging. Thus an excess of nitrogen produces an abundant vegetative growth with relatively weak culms. Tubbs (1930), however, noted that a large amount of nitrogen can be safely applied, provided the proper potassium nitrogen ratio is maintained. The plants that had a deficient supply of nitrogen and phosphorus showed lignification equal to that of fully manured plants. Anatomical sections from potassium-deficient plants, however, showed considerably less depth of staining, indicating less lignification. The thickness of the lignified parenchyma was quite narrow in these plants as compared to the fully fertilized ones. The role of potassium as a condensing agent in the formation of carbohydrates and possibly in lignification offers a possible explanation for the preceding observations. A deficiency of potassium had no influence upon the number of tillers produced, while a deficiency of phosphorus and nitrogen caused a marked reduction in that regard.

It was observed by Davidson and LeClerc (1923) and Phillips, Davidson, and Weihe (1931) that the application of sodium nitrate caused a distinct decrease in the silica content of the straw of wheat. Since this application was followed by lodging, they considered that the original view of Liebig for the necessity of silicon for strength of straw received new support. These observations also show clearly the superiority of potassium over sodium as a condensing agent in plant nutrition.

I. Boron

1. Occurrence.—Boron is very widespread in the plant kingdom, and Agulhon (1910) has determined its presence in the fungi, algae, ferns, and various members of the flowering plants. In the plants examined by him the amount of boron ranged from 1.2 per cent of the ash in the case of *Betula alba* to as low as 0.12 per cent of the ash in *Cannabis sativa*. Brenchley (1914) stated that, in general, annual plants seem to have the least boron in the composition of their ash and that the bark and wood are richer in this element than the leaves. The work of Cook (1916) showed that the absorption and distribution of boron vary greatly with different plants. In growing plants in soil to which manure treated with borax and calcined colemanite had been added, wheat and oats absorbed very little boron, while leguminous and succulent plants absorbed comparatively large amounts. The amount of boron in the entire plants of oats and peas grown in the same kind of soil varied in the case of oats from no detectable amount to 0.025 per cent of the total dry matter and in the case of the pea from 0.0005 to 0.04 per cent of the dry matter, the amounts depending on the amount and form of the boron applied. The fruit of the tomato plant contained only traces of boron, while the seed of the cowpea contained large amounts, from 0.0094 to 0.0222 per cent of boric acid on an ash-free basis. Wheat, beets, cowpeas, and potatoes contained boron principally in the tops and, with the exception of beets, comparatively little or none in the roots. String beans, soybeans, and cowpeas showed a more equal distribution of boron among the roots, tops, and fruits than any of the other plants tested.

McLean and Hughes (1936) determined the content of boron in *Vicia faba* and *Gossypium herbaceum* grown in solutions containing known quantities of boron. The highest percentage per unit of dry matter was in the leaves, and it increased with the age of the leaf. The roots had the smallest amount while, in the seeds, boron was confined to the cotyledons. In *Vicia faba* the percentage content in the cotyledons was 50 per cent above that of the stems.

2. Need.—The influence of boron on the growth and development of plants has been more extensively studied in recent years than that of any other element. The plants used have included the cereals, legumes, algae, cotton, flax, mustard, citrus, tomato, walnut, tobacco, grape, prune, peach, apricot, and others. The evidence indicates that boron is essential to the growth and development of the plant. The amount of this element that is necessary for the optimum functioning of plants is relatively low, while a very small excess becomes extremely toxic. Tolerance of this element varies widely with the kind of plant and the conditions under which it is grown. Of all the plants that have been investigated, cotton has the highest boron requirement and the cereals the lowest. It was reported by Haselhoff (1913) that beans were benefited by the application of $\frac{1}{10}$ p.p.m. of boron but corn was injured by that concentration. If boric acid was used as a source of boron, a concentration of 1 p.p.m. was toxic, but if borax was used the plants were not visibly affected. Warington (1923) found that in water cultures a continuous supply of boron was necessary for the healthy growth of *Vicia faba,* and the benefits of this element were evident in concentrations ranging from 1/12.5 p.p.m. to 40 p.p.m. Barley plants, however, were injured by concentrations as low as 1/2.5 p.p.m.

Geigel (1935) found that a concentration of boron above 1 p.p.m. was toxic to *Spirodela polyrhyza,* while with *Chlorella* a concentration of 30 p.p.m. was required to decrease growth, and it could survive a concentration of 140 p.p.m. The addition of dextrose or sucrose made *Chlorella* more tolerant to boron.

According to Haas (1930) and Haas and Klotz (1931), the addition of boric acid to a cultural solution in amounts sufficient to make a concentration of boron of $\frac{1}{10}$ to $\frac{1}{5}$ p.p.m. was sufficient to cause a rapid recovery of citrus plants suffering from a boron deficiency. Johnston and Fisher (1930) found that $\frac{1}{2}$ p.p.m. of boron was sufficient to cause a fourfold increase in fruit production in the tomato. Loomis and Wilson (1933) reported that $\frac{1}{10}$ and 5 p.p.m. of boron were inferior to $\frac{5}{10}$ p.p.m. in the response of this plant, while Fisher (1935) found 1.1 p.p.m. to be the optimum concentration. The maximum growth and fruitfulness of cotton occur only when boron is supplied in a concentration of 10 p.p.m. The boron requirement of cotton is the highest among some 60 plants that were compared. When the content of boron is ideal for cotton, the growth is depressed 25 per cent in beans, corn, and the smaller cereals (Eaton, 1932).

Johnston and Dore (1929) found 5.5 p.p.m. of boron in a nutrient solution to be toxic to tomato plants, while Newell (1930) under the conditions of his experiment noted that 2.75 p.p.m. injured these plants. Morris (1931) in working with wheat reported that a concentration of 50 p.p.m. of boron caused a retardation in growth of 40 per cent, while a concentration of 100 p.p.m. completely inhibited growth.

It was considered by Brenchley and Warington (1927) that the chemical combination in which boron is presented to the plant is immaterial, even the relatively insoluble borates being effective.

The accumulation of boron in the soil from the water used in irrigation has been especially detrimental to citrus plants in certain regions. In the irrigation waters of California the content of boron ranges from 0.15 to 0.30 p.p.m. in most cases, while in waters that have caused trouble, boron ranges from 0.5 to 2.5 p.p.m. In the mature leaves of orange trees that have been irrigated with water low in boron, there are present from 50 to 150 p.p.m. of boron, based on dry weight, while in those on plants irrigated with water high in boron, from 600 to 1,000 p.p.m. are present. There is a distinct relationship between the boron content of the leaves and that of the water supplied to the plant (Scofield and Wilcox, 1930, 1931).

Plants differ greatly in their rate of absorption of boron when grown under the same conditions. Thus Eaton and Blair (1935) observed that, on a dry basis, with plants grown in sand to which was added a solution containing 5 p.p.m. of boron, the amount of boron in p.p.m. absorbed was navy beans, 648; lima beans, 515; wheat, 453; pumpkin, 291; lettuce, 261; turnip, 245; Canadian field peas, 207; tomato, 159; alfalfa, 139; cotton, 123; asparagus, 120; and carrot, 69. This element tends to accumulate in some plant parts more than in others. Under the conditions mentioned above, the leaves of lemons contained 1,232 p.p.m., while the stem and roots contained only 54 p.p.m. The leaves of elm had 943 p.p.m., while there were only 22 p.p.m. in stems. The combined stems and roots of corn contained 60 p.p.m. but there were only 16 p.p.m. in the grain. The boron absorbed by plants is apparently mobile but upon reaching the leaves is changed to the immobile form.

Since plants differ in their rate of absorption of boron under identical conditions, Eaton and Blair (1935) considered that a specific rootstock might influence the amount of boron in the scion. They found that the concentration of boron accumulating in the leaves of the scion was directly influenced by the rootstock upon which it was grown. The concentration in the scion was reduced if grafted to varieties normally accumulating lesser concentrations and increased if grafted to varieties normally accumulating higher concentrations. The amount of boron absorbed, however, is not wholly dependent on the rootstock. Thus Eureka walnuts accumulated more boron in their leaves than the purple variety when both were grown on the same rootstock in the same nutrient solution. With other conditions equal, the rate of absorption of boron is apparently determined by the characteristic of the absorbing root cells and the equilibriums between the mobile and nonmobile forms of boron in the plant. McLean and Hughes (1936) considered that the role of boron in the plant is probably that of an activator or regulator of metabolic processes.

3. Deficiency Effects.—The effects of a deficiency of boron on the growth, development, and general appearance of plants vary with the type of plant, its age, and the general conditions under which it is grown. Some of the more general symptoms shown by various plants as a result of lack of a sufficiency of boron are given below.

Brenchley and Thornton (1925) observed that the absence of boron had a marked effect upon the development of nodules in the roots of *Vicia faba*. The number of nodules of macroscopic size was smaller in those plants that were not supplied with this element. In one case the number of nodules per plant was 353 and 199, respectively, for the plants receiving boron and those deprived of it. In the absence of boron the vascular supply of the nodule was defective, being either entirely absent or when present very weakly developed and running only a short distance into the nodule. The nodules that had no vascular strands remained minute and were usually buried in the cortical tissue. Very little nitrogen was fixed in these abnormal nodules, being in one case only one-tenth that fixed in plants where the nodules were normally developed.

It was found by Warington (1926) that the anatomical structure of *Vicia faba*, growing in a nutrient solution containing boric acid in the proportion of 1:2,500,000, is exactly similar to that of plants growing in the soil. If, however, boron is omitted, frequent disintegration of phloem and ground parenchyma, poor development of the xylem, and hypertrophy, discoloration, and disintegration of the cambial cells occur.

It was noted by Sommer and Sorokin (1928) that the meristematic tissues in the root tip of *Pisum sativum* are the first to show the effects of the absence of boron. In a relatively short time after the plants are transferred to boron-free solutions, the cells cease dividing and a general arrest of growth follows. The existing cells undergo premature development or pathological changes. Isolated xylem elements appear in regions occupied by the meristem or in the region of elongation in normal roots. The root apexes are enlarged owing to hyperplasia of the plerome and hypertrophy of the periblem regions. Fagundes (1934) obtained a marked increase in the growth of lima and broadbeans, and soybeans when boron was added to the cultural solution. More iron is necessary when the proper amount of boron is present, apparently because of the greater amount of material being synthesized in these plants.

McMurtrey (1929, 1935) found with tobacco plants grown in nutrient solutions deficient in boron that growth was greatly reduced, the most characteristic effect being the injury produced on the growing point. The first indication of this deficiency is the light-green color of the bud leaves, which is followed rapidly by a breakdown of the tissues. Frequently the leaves are distorted by growth around the injured areas. The stalk toward the top of the plant may manifest a one-sided or twisted growth. The final manifestation is a "dieback" of the terminal bud.

The effects of the absence of boron on the tomato have been studied by many workers. Johnston and Dore (1928, 1929) found that plants grown in a boron-deficient nutrient solution showed four distinct types of injury: (*a*) death of the terminal growing point of the stem, (*b*) breaking down of the conducting tissues of the stem, (*c*) a characteristic brittleness of the stem and petiole, and (*d*) roots of an extremely poor growth and of a brownish, unhealthy color. The total sugars and starches were more abundant in the leaves and stems of the boron-deficient plants, while a greater amount of benzene-soluble matter was found in the leaves of normal plants and in the stems of the boron-deficient plants. Johnston and Fisher (1930) reported that dark, dead areas appear on the fruit of the tomato plants grown in a solution deficient in boron during the fruiting phase. According to them, boron is fixed in the tissues and cannot be used repeatedly by the plant. It apparently functions as a simple nutrient element, which is in accord with the views of Warington (1923). She believed that it acts in a nutritive way rather than in a catalytic manner, since a continuous supply of it appears to be required throughout the life of the plant. According to Loomis and Wilson (1933), tomato plants did not develop properly when grown in a boron-deficient nutrient solution contained in soft-glass vessels.

It was found by Van Schreven (1935) that tomato plants suffering from a lack of boron were cured by the application of boric acid. The plants showing boron deficiency had up-curled leaf margins in sand cultures and down-curled margins in water cultures. This leaf curling was attributed to the thickening of the interveinal tissues due to the enlargement of individual cells, while the veins remained unchanged. The plastids did not decrease in number but the chlorotic symptom was due to their decrease in size and to the increase in the volume of the cell contents. Fisher (1935) stated that boron deficiency was apparent about 9 days after the plants were transferred to a solution lacking this element. The cotyledons and leaves turned a distinct purple color and within a few days the terminal shoot curled inward and died. The conductive tissue disintegrated and the petioles became extremely brittle. The fruits were covered with darkened or dried areas, which were apparently due to a breaking down of the cells making up this tissue.

In barley and wheat grown outdoors in large sand beds flooded daily with nutrient solution, Eaton (1930) found that shot blotch of barley, caused by *Helminthosporium sativum*, did not develop when boron was omitted from the cultural solution. In the summer plantings of barley, powdery mildew (*Erysiphe graminis*) was abundant on plants without boron and absent when it was applied. In the winter plantings it was abundant on all plants. There was no mildew on summer-grown wheat, but it was equally abundant on all winter wheats studied, regardless of the amount of boron applied.

Haas and Klotz (1931) found that a deficiency of boron in grapefruit and Valencia orange was marked by a gradual reduction in the size of the shoot produced, which in extreme cases resulted in the formation of "multiple buds." When boron was deficient, the cambium and parts of the phloem disintegrated, but the xylem tissues were affected to a lesser extent if at all. A copious amount of gum was formed and appeared in breaks in the cortex. The total amount of sugar and starch was doubled in the leaves of the plants deficient in boron. The addition of this element to the cultural medium caused a reduction in the sugar content of the leaves. The plants grown in the boron-deficient cultures showed a slightly greater diastatic activity than those receiving the boron.

It was reported by Gilbert and Pember (1931) that red clover, when grown in water cultures, is extremely sensitive to a deficiency of boron and manganese, and that the addition of minute amounts of these elements produced a marked increase in the growth of both roots and tops.

McHargue and Calfee (1932) found that the varieties of leaf and head lettuce which they observed required boron for normal growth. A deficiency of boron caused a burning of the leaves and an early death of the growing parts. The functions of boron could not be performed by magnesium, copper, zinc, nickel, cobalt, barium, strontium, iodine, or arsenic, which is in accordance with the report of Brenchley and Warington (1927), who tried to substitute 52 elements but found none that could replace boron.

Bobko and Belvoussov (1933) noted that the absence of boron caused a characteristic disease of the sugar beet known as "heart rot," which could be corrected by the application of small quantities of boron. Hill and Grant (1935) in the growth of turnips in pot sand cultures deficient in boron noted that there was a marginal yellowing of the foliage, followed by a purpling and scorching. The roots were small and shriveled or rotted at their juncture with the top. These symptoms could be remedied by the application of boron. An inverse relationship was found between the amount of boron supplied and the occurrence of hollow heart of the roots. There was also an inverse relationship between the amount of boron and the percentage of ash in the dry matter of the roots. It was found by Hoagland and Snyder (1933) that in the absence of boron the leaves of strawberry plants were deformed, dwarfed, cupped upward, puckered, and generally brown at the tips.

Schmucker (1934) in considering the effect of boron on pollen growth decided that boric acid can be designated as inorganic growth material because of its ability to form strongly acid compounds with certain organic substances. Warington (1933) found that reduced length of day rather than the lower temperature is the factor controlling the delay in appearance of the symptoms of a deficiency of boron during the spring and autumn as compared with the summer months. This behavior is noticeable in *Vicia faba*, *Phaseolus multiflorus*, and *Glycine hispida*. In the absence of boron, the influence of the length of day was less striking than when boron was present. The symptoms of boron deficiency were less pronounced under short-day than under full-day conditions. According to Warington (1934) the absorption of calcium is affected by the presence and quantity of boron. A relationship of some sort apparently exists between boron and calcium, but its exact nature is yet undetermined.

4. Excess Effects.—Boron is toxic to fruit trees when present in relatively small concentrations (Haas, 1929). Citrus and walnut leaves may become thin, mottled, chlorotic, and crinkled when boron is in excess. The addition of various amounts of ferrous sulphate to cultures of lemon seedlings tends to overcome the toxicity of boron. Unless badly defoliated, trees injured by boron may recover if the toxic agent is leached out with water. The leaves with an excess of boron are mottled and burned along the margins. Such leaves of citrus and walnut contain reduced amounts of calcium but increased amounts of potassium and do not mature. When tomato plants are supplied with an excess of boron, Newell (1930) reported a yellowing of the whole leaf, large brown spots developing thereon, and the occurrence of a mottled appearance. Brandenburg (1931) considered the dry rot of beet to be a boron-injury disease. With citrus plants, Webber (1935) found that leaf injury is generally the characteristic manifestation of boron toxicity, the older leaves showing the most pronounced symptoms. These are the yellowing of tips and margins, often followed by marginal or spotted browning. This condition may be followed by premature defoliation and the subsequent production of malformed leaves. Under conditions of excess boron, the trees of stone fruits seldom exhibit such symptoms of the leaf. The petioles and larger veins of the leaves of prunes and apricots may become brown and rough, and occasionally exude gum. Rodriquez (1933) considered that boric acid inhibits the solubility of iron. This suggests that chlorosis, which sometimes appears when boron is present in excess, may be due to a lack of availability of iron and not to the boron itself.

Histological studies of prune, peach, apricot, and grape showed that a toxic concentration of boron in the nutrient solution is not a stimulus to a specific reaction resulting in a specific histological change. The histological evidence of injury by boron is in general similar to that attributable to other causes. It seems probable that an excessive concentration of boron in the cell often injures the protoplasm to such an extent that it undergoes a progressive degeneration. If chloroplasts are present they are the first protoplasmic structures to be affected. In other cases, excess of boron may stimulate the cell to abnormal growth and division.

5. Fertilizer.—The effect of boron compounds in fertilizers and in manures treated with them has received considerable attention in the United States. Cook and Wilson (1917) applied manure, which had been treated with borax and calcined colemanite, to soil at Arlington, Va., and observed the effect of the boron compounds thus applied upon the growth and yield of wheat. The presence of 0.0022 per cent of boric acid added as borax to the first 6 in. of soil or 0.0029 per cent as colemanite was apparently nontoxic to this crop, although, in 2 years out of 3, borax applied at the above rate reduced the yield of grain 10 per cent, while colemanite had little if any effect. Only minute quantities of boron were absorbed by these plants, but more was absorbed from the borax than from the colemanite.

These workers (1918) observed that there is a decided difference in soils in rendering added boron nontoxic to plants. In some soils there is a tendency for the plant to absorb boron in proportion to the quantities added, while in others the same amount of boron is absorbed irrespective of the quantities added. The behavior of plants grown at Arlington, Va., and Bethesda, Md., will serve as an example in this regard. At the latter station, 0.0044 per cent of boric acid as borax and 0.0058 per cent as colemanite in the upper 6 in. of soil caused no injury to lettuce, spinach, or kale; but at the former station these concentrations caused a reduction in the crop of these plants.

Skinner, Brown, and Reid (1923), Neller and Morse (1921), and Blair and Brown (1921) found that the potato can tolerate a greater quantity of borax than corn and beans. The effect of borax on potatoes at Arlington, Va., when used in quantities of less than 5 lb. per acre applied in the furrows, was one of stimulation, but 20 lb. per acre produced injury and depressed the yield. In Maine, however, 5 lb. of borax applied in the drill at the time of planting produced definite injury. In the case of corn, 4 lb. of borax per acre in the drill caused a decrease in yield of both grain and stover. When sown broadcast, 20 lb. per acre was required to produce any decrease in yield, while practically no plant growth took place where the application exceeded 50 lb. of borax per acre. Skinner and Allison (1923) observed on the silt loam at Arlington, Va., that cotton was slightly injured by the broadcasting of 10 lb. of borax per acre and decidedly so by the application of 20 lb. Boric acid and to a lesser degree borax exerted a stimulating effect on plants in nutrient solution (Collins, 1927), the greatest stimulation appearing when 2.5 mg. of boron per liter of cultural solution was used. He, however, does not consider that boron is necessary to the production of mature plants in nutrient solutions. In field experiments, all compounds of borax tested reduced the dry weight of plants when applied in quantities equivalent to 1 to 7.5 lb. of boron acre, depending upon the type of soil.

The evidence indicates that boron is an essential element for plant growth but that only extremely small amounts are necessary to supply the needs of the plant. The application of the compounds of boron, however, as fertilizers must be made in relatively small amounts or they will have a toxic effect upon the plants. In most cases, applications as low as 5 lb. per acre cause detrimental effects on crop growth and production.

J. Manganese

1. Occurrence.—Manganese is universally distributed in plants (Bertrand and Rosenblatt, 1921 to 1922), and, according to Robinson (1917), it fluctuates more in amount than most of the other elements. Most investigators concur that it tends to accumulate in the leaves and that this element reaches its highest concentration in that part of the plant. In corn grown in Kansas, the manganese content was 0.017, 0.037, and 0.043 per cent of the dry matter of the stem, grain, and leaves, respectively (Latshaw and Miller, 1924). According to Kelley (1914), the leaves of the pineapple contain ten times as much manganese as the stalk. In the various plants grown in Kentucky, McHargue (1925) found manganese present in parts per million of air-dry material as follows: soybean leaves, 160; soybean seeds, 32; wheat bran, 125; wheat embryos, 150; corn embryos, 40; corn endosperm, 16; rice polishings, 100; and polished rice, 10. In Oregon, alsike clover utilized manganese in larger amounts than any other legume there grown (Jones and Bullis, 1921). The average manganese content for legumes considered in terms of milligrams per kilogram of air-dry material was alsike clover, 68; vetch, 42; white clover, 34; field peas and red clover, 33; sweet clover, 27; and alfalfa, 23. Gössl (1905) found that swamp plants and water plants, in general, contain more manganese than dry-land plants and that the evergreens have a greater quantity than the deciduous trees. It has been observed that the total amount of manganese in the plant will vary according to the nature of the soil, and McDonnell and Roark (1917) stated that the stems of *Chrysanthemum cinerariaefolium* if grown in Japan contain more manganese than those from other countries.

The amount of manganese in the seeds is an important point to consider, since it is assumed, by those who consider manganese an element essential to the normal growth of plants, that the seed, in some cases at least, contains a sufficient quantity to supply the plant during its growing period. Wester (1921) found that seeds from 13 plant families contained in every case manganese ranging in amount from 2 to 6 mg. for every 100 g. of dry material. McHargue (1923) investigated the iron and manganese content of a wide list of seeds, and the observations may be summed up in the table on page 347.

The average manganese content of the seeds of wheat, oats, and hemp, produced under natural conditions, is equal to or greater in amount than the average amount of

Plant	Per cent iron content on dry basis	Per cent manganese content on dry basis	Plant	Per cent iron content on dry basis	Per cent manganese content on dry basis
Wheat, 10 varieties....	0.0039	0.0047	Alsike clover ...	0.0517	0.0028
Oats, 6 varieties.......	0.005	0.005	Crimson clover.	0.0016	0.0029
Garden peas, 3 varieties	0.0096	0.0012	Tobacco........	0.024	0.007
Beans, 5 varieties	0.0103	0.0018	Hemp..........	0.0085	0.0165
Soybeans, 9 varieties ..	0.0074	0.0028	Sunflower......	0.0034	0.0023
Alfalfa................	0.010	0.0012			

iron found therein. This observation agrees with that of Headden (1915), who found that manganese is present in the grain of wheat in approximately the same proportion as iron, although the latter greatly predominates in the soil. Although the iron and manganese were present in approximately the same amounts in the grain, the former was present in the straw in amounts from two to six times greater than the latter. Headden examined samples of wheat seed that had been grown over wide areas and under a wide range of cultural conditions, climate, and soils and found that the manganese content ranged from 0.003 to 0.006 per cent of the dry seeds regardless of the conditions under which the crop was grown.

It was observed by Jacobson and Swanback (1932) in the tobacco plant that the highest percentage of manganese was found in the middle leaves, and the lowest in the upper ones. Bolin (1934) determined the amount of manganese in eight species of grass and found it ranged from 78.1 mg. per kilogram for Kentucky bluegrass to 207.5 mg. per kilogram for orchard grass. Alfalfa showed an average of only 46.6 mg. per kilogram.

2. Amounts Needed.—The amount of manganese needed by plants is small and apparently occurs in sufficient quantities as impurities in the various salts used in nutrient solutions. Thus McHargue (1922) found that the salts of calcium, magnesium, and iron generally used for plant nutrients may contain as impurities a supply of manganese sufficient for the plant's needs. This fact has apparently been overlooked by most investigators in plant nutrition. McHargue observed that crocks which were supposed to be perfectly glazed absorbed during the growing season from the soil with which they were filled a sufficient amount of manganese to supply the needs of the succeeding crop grown in them in a manganese-free medium. This experiment emphasizes the need for acid-proof glazed jars in experimental nutrition work. McHargue (1922) observed that when wheat seedlings were grown in Pfeffer's nutrient salt solution from which manganese was excluded, they showed no observable differences from those supplied with manganese during the first 6 or 8 weeks. Shortly after this period, the plants from which manganese was withheld began to give evidence of unequal growth. The first observable difference was the lack in the development of chlorophyll, the leaves becoming a yellowish green. The plants deprived of manganese made a stunted growth and produced no seed, whereas those to which manganese was available grew normally and produced seed. There was an increase of 135 per cent in the dry matter produced in the plants receiving manganese when compared with the dry weight of an equal number of other plants from which it was withheld. It was found that the seeds of radish, soybean, cowpea, field bean, and corn do not contain a sufficient amount of manganese for the growth of these plants to maturity. Samuel and Piper (1929) studied the relation of manganese to the growth of rye and oats. They found that these two plants differ in regard to their requirements for manganese. Thus rye developed to maturity in water cultures on an amount of manganese that did not allow the complete development of oats. A concentration of 1 part in 50,000,000 parts of nutrient solution allowed complete development of rye, while the minimum quantity of manganese that allowed healthy growth of Algerian oats was about 14 parts per 1,000,000 of dry matter at the flowering stage. Hopkins (1930) increased the growth of *Chlorella* 600-fold by the addition of manganese, in a concentration of $\frac{1}{5}$ p.p.m., to nutrient solutions that lacked this element. At a pH of 7.0 the increase in growth was 17-fold, while at a pH of 8.0 it was 170-fold. In 1934 he reported that certain unicellular algae and *Lemna minor* showed no growth in the absence of manganese from the nutrient solution. When, however, amounts of manganese ranging from $\frac{1}{100}$ to $\frac{1}{5}$ p.p.m. were added to this cultural solution, healthy growth occurred. If manganese was added to the solution before the process of disintegration of the plants had progressed too far, recovery occurred. Clark and Fly (1930) found that concentrations of manganese exceeding 1 p.p.m. became toxic for *Lemna minor*, and that the size, weight, reproduction, and ash content of this plant became very irregular. Clark (1933) obtained a fine growth of *Lemna major* in a nutrient solution containing manganese in a concentration as low as $\frac{1}{3000}$ p.p.m. Saeger (1933) showed that a manganese content of $\frac{1}{1000}$ p.p.m. was sufficient to maintain the vigorous growth of five species of *Lemnaceae*, while a concentration of 1 p.p.m. was excessive for optimum growth.

The reaction of the plant to manganese apparently depends upon the nature of the medium in which it is growing. Thus Jacobson and Swanback (1932) found that the tobacco plant showed toxic symptoms when growing in a nutrient solution containing as low a concentration of manganese as 1 p.p.m. In sand cultures, however, the application of 80 p.p.m. was necessary to produce toxicity, while in the field 486 lb. per acre did not cause injury. An increased acidity of the soil resulted in the presence of large amounts of active manganese and a greater intake by the plant. Leeper (1934) observed that the "manganese deficiency disease" is confined to soils of pH 6.7, or higher, and that it occurs especially on heavily limed, sandy soils. In soils with a sufficient supply of manganese for plant needs, the active manganese exceeded 100 p.p.m., while in the soils associated with manganese deficiency it was lower than 15 p.p.m. Gilbert (1934) considered that a manganese deficiency can be prevented in many soils by maintaining a suitable acidity that allows the manganese to remain available to the plant. Under the same soil conditions, different crops will respond differently to a deficiency of manganese (Odland and Crandall, 1932).

3. As a Fertilizer.—The influence of manganese as a fertilizer in soil and in water cultures has been studied by numerous investigators. The results indicate that the effect of the application of this element will depend upon the concentration of the various salts used and upon the type of plant under observation. Nagaoka (1903, 1904, 1906) showed that the application of 50 kilograms of manganese sulphate per hectare increased the rice harvest over one-third the first year, 16 per cent the second, only slightly the third year, and caused a depression in yield the fourth year, due probably to the increased acidity of the soil. Small quantities of manganese sulphate in soil cultures stimulated the growth of rice, peas, and cabbage according to the observations of Loew and Sawa (1902), while Loew and Honda (1904) report that *Cryptomeria japonica* growing in soil treated with manganese sulphate doubled in 1½ years its cubic contents over plants that had grown in soil not treated with this compound. Katayma (1906) applied both iron and manganese sulphate to soil at the rate of 0.01 per cent and observed an increase of 6.2 per cent straw and 7.2 per cent increase in grain yield of barley but a depression in the yield if larger amounts were applied. Brenchley (1910) stated that at the Rothamsted station concentrations of less than 1:100,000 of manganese sulphate stimulated the vegetative growth of barley but retarded the ripening of the grain. The application of 100 lb. of manganese sulphate per acre increased appreciably both ammonification and nitrification (Brown and Minges, 1916). The increase in activity was lower in applications above this amount, but a decrease occurred if amounts greater than 2,000 lb. per acre were applied.

Schreiner and Dawson (1927) observed that a highly calcareous soil occurring in certain regions of Florida failed to produce a crop even with a heavy application of inorganic fertilizer unless stable manure was applied. Striking results were obtained by the application of 25 to 50 p.p.m. of manganese sulphate in addition to the balanced inorganic fertilizer. In control greenhouse experiments with this soil, plants did not flower, developed chlorosis, and were stunted in growth. These effects rapidly disappeared on the application of manganese. Samuel and Piper (1929) found that on certain soils oats are unable to absorb a minimum quantity of manganese and that they suffer from the gray-speck disease, a manganese-deficiency disease that can be cured by the application of soluble manganese salt to the soil. On soils with abundant available manganese, considerably more manganese is absorbed by oats than is required for normal development.

Conner (1932) reported that applications of manganese to oats growing under neutral and alkaline conditions increased yields, and Skinner and Ruprecht (1930) found manganese sulphate to be beneficial to tomatoes growing in the calcareous soils of Florida which are deficient in manganese. Saeger (1933) stated that extracts of autolyzed yeast, raw carrots, and spinach will stimulate the growth of plants when added to a nutrient solution deficient in manganese.

McCool (1935) studied the effects of different light intensities on the injurious action of manganese on soybeans, buckwheat, snapdragons, and tobacco. The visible injury to the plants grown in the manganese-treated soil decreased as the light intensity became less. The percentage of manganese in the leaves of the soybean and buckwheat growing in the open decreased consistently as the light intensity increased.

4. Physiological Role.—It has been considered by Bertrand (1897, 1905, 1912), McHargue (1919–1932), Bishop (1928), Bryan (1929), Hopkins (1931), Conner (1932), Clark (1933), and others, that manganese is an essential element for the growth of plants. It has been observed that the absence or scarcity of manganese results in disturbed carbohydrate metabolism, chlorosis, retarded growth, a decrease in the content of ash, and the failure to reproduce.

Our understanding of the function performed by manganese in the metabolism of the plant rests almost entirely on theoretical grounds. Bertrand (1897, 1905, and 1912) considered that the addition of small amounts of soluble manganese increases the oxygen-carrying power of the oxidases. This enzyme is universally present in plants and fulfills a definite function in their metabolism. From their observations on the effect of manganese on the germination of seeds, Michiels and DeHeen (1906) considered that this element has a role in enzymic action. Kelley (1914) from his observations on the effect of manganese on the plants of pineapple believed that manganese may act in two general ways: (1) It may have an effect upon the inert bases of the soil by bringing about a greater degree of mobilization of calcium and magnesium therein, and (2) it may cause a stimulation of the necessary oxidation going on in the plant and in the soil. The first effect is substantiated by the work of Aso (1907), who found that a 41.8 per cent increase in the yield of rice was obtained from the first application of manganese sulphate, while similar applications to the same soil the following year produced only a little better than a 2 per cent increase. McHargue (1922) thinks that manganese is concerned in nitrogen assimilation and the synthesis of proteins and that it plays the role of a necessary catalyst in plant metabolism and also functions with iron in the synthesis of chlorophyll. He also considered that, since the lack of manganese affects the production of dry matter, the indications are that it has some very important function in carbon assimilation.

Bishop (1928) noted that manganese occurred in plants in the regions of most active chemical change, and he considered that it is concerned in chlorophyll formation and hence in carbon assimilation. Hopkins (1930) suggested that manganese functions in an indirect manner in plants by its action upon the oxidation of iron, because in vitro the reduction of iron from the ferric to the ferrous form by sodium acetate is hindered by the presence of manganese. Bertrand and Rosenblatt (1932) found in the leaves of dandelion, lettuce, cabbage, and celery, with different degrees of etiolation, that for a given plant the content of manganese varied directly with the amount of chlorophyll. McHargue and Calfee (1932) reported that chlorotic plants of *Lemna minor* could be restored to their original dark-green color by the addition to the nutrient solution of traces of manganese sulphate, but not by the addition of iron salts.

Miller (1933) found that wheat and tomato plants which were deficient in manganese were much lower in sugar than corresponding plants that had been supplied with a small amount of this element. Thus wheat tops with sufficient manganese contained on a dry basis 8.3 per cent of sucrose and only 1.6 per cent when manganese was lacking. The roots contained 3.3 per cent of sugar when manganese was present, but only a trace of this carbohydrate when it was absent. Pettinger, Henderson, and Wingard (1932) noted that a deficiency of manganese caused a peculiar type of chlorosis in corn growing in sand cultures. The fact that chlorosis is so often one of the criteria of a deficiency of manganese indicates that it is concerned in some manner in the formation of chlorophyll.

Chlorosis of plants seems to occur when they are deprived entirely of a manganese supply, and this may prevail also even when an abundance of this element is in the soil or the cultural solution (McCool, 1913). McGeorge (1923) considered that the chlorosis of pineapple leaves on plants grown in soils rich in manganese is due to a greater assimilation of lime indirectly caused by the presence of manganese in the soil. He believed that the principal physiological disturbance is the greater immobility of the iron in the plant, resulting from the excessive lime content of the leaves and stalks, since iron is present in an equally available form in both manganiferous and nonmanganiferous soil types of the same hydrogen-ion concentration. Rippel (1923) observed that manganese in the form of soluble salts produced chlorosis in barley plants grown in water cultures. This was overcome or cured by increasing the iron supply. Since the iron content of green and chlorotic plants was the same, it was concluded that manganese did not interfere with iron absorption but with the activity of iron within the plant.

Haas (1933) noted that walnut leaves affected with "yellows" contained a higher amount of manganese and iron than the healthy ones. He considered that the availability of these elements and not their total amount is the determining factor in their utilization by plants.

According to Gilbert (1934) the causes of manganese chlorosis have been definitely linked with soil alkalinity. The pH range within which chlorosis may occur appears to vary with the crop and climate and not with the soil type. Crops vary widely in their susceptibility to manganese chlorosis, spinach being very sensitive, while tomatoes are relatively immune. A relatively low temperature, by causing a slower growth rate, may enable the plant to avoid the chlorotic effect. Chapman (1931) noted that chlorotic plants always recovered their normal color uniformly over the whole of the treated leaves when sprayed with a 2 per cent solution of ferrous sulphate. When manganese was added with the iron or injected into the trees before spraying with iron, the leaves greened only in spots—suggesting a reduction in the solubility of iron by manganese in the leaf. Similar observations were made by Haas (1932), who found that excessive concentrations of manganese may produce chlorosis, although iron is present in the cultural solution. When manganese is deficient, apparently less iron accumulates in the leaves.

McLean and Gilbert (1925) were able to effect a complete cure of chlorosis of spinach, growing on heavily limed soils, by spraying the plants with a solution of 8 parts per 1,000,000 of manganese sulphate. When harvested, a 40 per cent increase in yield was obtained from the treated plants as compared with the untreated ones. The injection of solutions containing 5 to 50 p.p.m. of manganese into the leaves through the stomata brought about the recovery of the chlorotic spinach plants.

Bryan (1929), in experiments having to do with the effects of copper and manganese on cowpeas, beans, and sorghums in Florida, found that the responses were as striking when the chemicals were applied to the leaves as when they were applied to the soil. Gilbert (1934) also mentioned that chlorosis could readily be corrected by the application of a solution of manganese sulphate as a spray. Miller (1933) found with plants of wheat and tomato that the manganese necessary for the plant could be injected as a solution into the stems.

K. Chlorine: Its Occurrence, Function, and Value as a Fertilizer to the Plant

With the exception of the conifers, chlorine is present in practically all plants, according to Jung (1920), who studied microchemically the occurrence and distribution of this element in plant tissue. This author considered that chlorine occurs in the plant only in the form of chlorides which are generally dissolved in the cell sap. It usually increases in amount from the roots to the leaves, being most abundant in succulent parenchymatous tissue. It is especially abundant in fleshy roots and rhizomes and is very scant in the epidermis, fibrovascular bundles, pollen, and woody tissue. The mosses, ferns, and epiphytes contain little or no chlorine. All seeds that have been examined contain more or less chlorine. Most soils contain considerable amounts of it in the form of chlorides. Schmidt (1915) found that the entire chlorine content of the sugar-beet seed is soluble in water and can be precipitated directly by silver nitrate. This strongly indicates that the chlorine in this seed exists largely in the form of inorganic chlorides and that if it is involved in any organic compounds they must be very unstable.

In consequence of the results obtained by the earlier experimenters in plant nutrition, chlorine has generally been considered as one of the unessential elements for the growth of plants. Nobbe and Siegert (1862), and Leydhecker (1866), however, observed certain peculiarities in the development of buckwheat in a chlorine-free nutrient solution which led them to conclude that chlorine was necessary for the proper growth and development of this plant. As a result of this work, the buckwheat plant has been repeatedly mentioned in the literature as an example of a plant that requires chlorine for its development. Many other investigators, however, since that time, among whom are Shulov (1909), Tottingham (1919), Johnston (1917), and Trelease (1917, 1920), have concluded that, aside from the small amount of chlorine contained in the seed, the buckwheat plant does not need chlorine for its normal growth and development.

De Peralta (1927 found, however, that the addition of chlorine in the form of sodium chloride or potassium chloride to nutrient solutions was beneficial to the growth of young plants of rice.

A great amount of work has been done to determine the value of the chlorides as fertilizers. The results that have been published are very confusing. In many cases, chlorine is apparently beneficial to plant development, while in others it is deleterious in its influence on growth. The environmental conditions under which the experiments have been conducted have been very poorly defined, and this no doubt has been one of the main reasons for the contradictory data that have been reported. The student is referred to Tottingham (1919) for a thorough review of the literature on this subject. Tottingham (1915, 1917, 1919) carried on extensive and well-controlled experiments in chlorine nutrition. Wheat, buckwheat, radish, carrot, parsnip, sugar beet, and potato were grown in nutrient cultures, in soil cultures in the green-house, and in field plots. The chlorides of potassium and sodium were used chiefly as sources of chlorine, the sulphate or nitrate of potassium being replaced partly or wholly by the potassium chloride, while sodium chloride was applied either alone or supplementary to a complete nutrient mixture.

Skinner (1931) reported that potassium chloride could be used as a source of potassium for the fertilizing of cotton in the southeastern United States without apparent injury from the chlorine. Apparently cotton is more tolerant of this element than are many other plants.

In the case of wheat grown in water cultures, increasing amounts of potassium chloride depressed slightly the yield of dry matter in the tops and roots as well as the length of roots. The addition of sodium chloride, while it depressed the length of roots, led to a slightly increased yield of dry matter as compared with other solutions. Buckwheat grown in similar cultural solutions showed that the chlorine ion in all cases produced disturbances which were shown in a decreased root length, decreased production of dry matter of leaves and roots, and decreased water absorption per culture.

The application of chlorides to radishes growing in soil cultures in the greenhouse increased the dry weight of the roots. The production of dry matter in the roots was decreased in proportion to tops by the application of chlorine in the carrot but increased in the case of the sugar beet. The percentage of glucose in the radish and parsnip was depressed by both potassium chloride and sodium chloride, while in the case of the carrot and sugar beet they induced an increase of this sugar. Sodium chloride increased the amount of sucrose in the radish and carrot and depressed it in the case of the parsnip and sugar beet. The application of the chlorides led to the production of a more watery root of the radish, carrot, parsnip, and sugar beet, a result that has been observed by several other investigators (Bolin, 1916).

In greenhouse experiments with potatoes the percentage of starch in the Triumph variety was not altered by the application of potassium chloride, while the Rural New Yorker showed a decrease in starch under these conditions. Sodium chloride produced more watery tubers than the unfertilized controls. In field experiments no significant variations in the percentage of starch resulted from the varied treatments with the chlorides. Tottingham considered that the depressing effect of chlorine on the starch content of potatoes as reported by Pfeiffer (1898) and his workers does not occur under all conditions of culture. The application of sodium chloride alone in the field, however, produced tubers of an inferior quality that were possessed of a bitter alkaline flavor. Rudolfs (1919), from data secured in determining the effects of sodium chloride on oak, birch, and maple trees, considered that the application of small amounts of the salt exerted a stimulating effect, as shown by the luxuriant growth and the extremely dark color of the leaves. Lomanitz (1923) in growing alfalfa in cultural solutions found that the solutions containing sodium chloride gave higher yields than did the checks. The greatest increase was obtained with the solution in which one-tenth atmospheric osmotic concentration of sodium chloride was substituted for the one-tenth atmosphere of potassium nitrate of Tottingham's solution *TIR1S5*, the roots showing a greater relative increase than did the tops. In another experiment (1924) alfalfa was grown in nutrient solutions with varying amounts of sodium chloride. An increase in yield over the checks was observed in five out of six cultures. He considered that sodium chloride might function as a mineral nutrient.

It has been noted by later investigators that chlorine markedly increases the water content of tissue, affects the metabolism of carbohydrates, and the production of chlorophyll. Thus Garner, McMurtrey, Bowling, and Moss (1930) showed that the application of 20 to 30 lb. of chlorine per acre to certain types of soil stimulated the growth of tobacco plants so that the average increase in production amounted to 10 per cent. The immediate effect of muriate of potash is to reduce decidedly the symptoms of magnesia deficiency in the plants. Chlorine has a marked effect in increasing the water content of the tobacco plant during the growing season, and thus enables it to resist injurious desiccation. This behavior protects the leaves against the type of injury known as "drought spot." An excess of this element, however, may seriously interfere with the metabolism of carbohydrates, and this adverse nutritional effect may result from an application of chlorine in amounts as low as 40 lb. per acre. Wilson (1933) noted that an excess of chlorine interferes with normal carbohydrate metabolism in the tobacco plant so that the leaves have a high content of starch. The leaves become thick and brittle, curl upward at the margins, and have a high water content and a low amount of organic acids. The cell sap of plants grown in the field in the presence of excessive amounts of chlorides had a lower osmotic concentration and less bound water than the sap of plants to which a liberal application of complete fertilizer had been given. Watson (1936) found that the only significant result of the application of potassium chloride, on the carbohydrate metabolism of the potato was a marked reduction in the amount of sucrose in the leaves during midday. The amount of starch and reducing sugars was not noticeably affected. The rate of translocation of carbohydrates during periods of darkness was unaffected, but the rate of the removal of dry matter was increased. By augmenting the application of potassium chloride, the rate of photosynthesis was apparently increased in the afternoon and decreased in the morning. The percentage of water in the plant was also increased by the application of this fertilizer. These variations were evident when the application amounted to 25 to 100 lb. of potassium oxide per acre. Maiwald (1923) noted that the chlorophyll content of potato leaves was sharply reduced by the accumulation of the ions of potassium and chlorine, which apparently inactivated the iron in the plants. Dhein (1929) noticed that large applications of potassium chloride caused a reduction of the chlorophyll in the plants studied. Basslavskaya and Syroeshkina (1936) working with potatoes found that the action of chlorine was to reduce the amount of chlorophyll, and this action was more clearly shown on the plants in field plots than on potted plants.

Chlorine in many cases has a morphological effect upon the plants growing in the medium to which it is applied. Harter (1908) found that plants of wheat, oats, and barley grown from seeds in soil containing 0.7, 1.0, and 1.4 per cent, respectively, of sodium chloride developed a waxy bloom upon the leaf surface and a thickening of the cuticle, and this thickening increased with the concentration of the solution. The size of the epidermal cells decreased as the concentration of the salt in the soil increased, but when the amount of sodium chloride present was much below the minimum concentration that is injurious under field conditions no perceptible modification of the plant structure occurred. Hendry (1918) grew Windsor, kidney, lima, Red Mexican, Terpary, and other varieties of beans in sand to which were added 0.04, 0.16, 0.3, 0.5, and 1.0 per cent, respectively, of sodium chloride based on the dry weight of the sand. The visible effects of the sodium chloride upon the development of the plants were retardation of height, a reduction in the number and size of the leaves, a retardation of the blooming period, and a reduction in the number and size of the nodules formed. It was noted by Hendry and also by others that there is a distinct difference in the ability of different species of plants to withstand the toxic effects of chlorides when they are applied in excessive amounts to the medium in which the plant is growing, and Harrison and King (1925) have shown that the age of the plant is an important factor in determining the degree of toxicity of sodium chloride.

Asparagus is an example of a plant that can tolerate a large amount of sodium chloride and the salting of asparagus beds with sodium chloride has long been a traditional practice among gardeners. Rudolfs (1927) applied sodium chloride at the rate of 150, 300, and 500 lb. per acre to asparagus beds for 2 years. The application of 150 lb. did not increase the number of stalks over plants on unsalted plots. Doubling the amount of salt resulted in an increase in average number of stalks from six to seven, while an application of 500 lb. per acre brought the yield per plant up to seven and one-half stalks.

The manner in which the chlorides may bring about the responses that have been mentioned is not known. All the explanations that have been put forth are purely theoretical. Tottingham (1919) considered that the apparent relation of the chlorine content to the accumulation of starch in the potato tuber is a suggestion that enzyme activity might be the main controlling condition for this effect. The fact that chlorides have a stimulating effect upon the enzymes obtained from plant sources would seem to lend strength to this general theory. Since many of the vital activities of the plant are apparently controlled by enzyme action, the physiological responses might be brought about in this manner. Rudolfs (1919) from his observations on the effects of sodium chloride on trees considered that it is possible that the chlorine increases the acidity of the cell, which could accelerate the vital activities or enzymatic actions to a certain extent, but that by a too large increase of acidity the acceleration would change into a toxic action.

Our knowledge of the influence of chlorine upon the metabolism of the green plant is extremely meager. Tottingham (1919) is of the opinion that the most valuable general conclusion that can be drawn from the review of the literature is that the influence of chlorine upon plants seems to be impossible of any general statement. The facts observed indicate that the effect of chlorine will depend upon the nature of the plant under consideration and upon the nature of the soil or other medium in which the plant is growing, other than the chlorine content. Some consider that the influence of chlorine is in many cases indirect in that the reactions of the plant are due to changed conditions in the mineral content of the soil induced by the applied chlorides and not to their direct action upon the plant cells (Wheeler and Hartwell, 1902; Prianischnikow, 1905). Trelease (1917, 1920) reported extensive experiments which greatly illuminate this subject He found in growing wheat in water cultures that if the solutions were properly balanced, the same growth was obtained regardless of the addition of chlorine. If the cultural solutions, however, were not properly balanced, the addition of chlorine apparently diminished to some extent the retarding effect produced by the unbalanced solution, and a better growth was obtained. The addition of an unessential element may thus improve the properties of the medium that contains the essential ones in improper proportions.

The action of chlorides on plants is evidently influenced also by the general climatic conditions that prevail during growth. This fact is suggested by Tottingham (1917) and also by Lipman, Davis, and West (1926), who obtained evidence in the study of the tolerance of wheat, barley, and peas to sodium chloride that led them to state that the environmental conditions are a very important determinant of the kind of results that may be obtained in studying the influence of chlorine upon plant life.

L. Aluminum

1. Occurrence.—Aluminum has been found in the ash of all plants that have been examined. The amount of this element varies greatly for different plants, and Kratzmann (1914), who analyzed 130 species of plants, considered that they possess specific selective powers toward aluminum. As a rule, the percentage of aluminum in plants is very small, but, in a few cases, as certain species of club mosses, aluminum approximates 22 to 39 per cent of the ash. Latshaw and Miller (1924) found that aluminum made up 0.07, 0.013, 0.023, and 0.98 per cent of the dry weight of the leaves, stem, grain, and roots, respectively, of corn plants that had reached their full vegetative growth and development. Berthelot and André (1895) also stated that the amount of aluminum in the roots is much higher than that in the leaves. There is some question, however, in regard to the relatively high amounts of aluminum reported in the roots, since the adhering particles of soil even in exact preparation of material could exert a marked influence on the results. Meyers and Voegtlin (1914) analyzed several of the grain and vegetable foods for aluminum compounds with the results shown in the table on page 357.

At least 65 per cent of the aluminum in corn and carrots was soluble in water, while the total aluminum content was completely soluble in 0.3 per cent of hydrochloric acid. Only a trace of the aluminum present in wheat flour and in the sunflower seed was soluble in water. McLean and Gilbert (1927) found in the case of corn, cabbage, peppers, redtop, lettuce, and timothy that the greatest accumulation of aluminum

Food	Percentage of aluminum on dry basis	Food	Percentage of aluminum on dry basis
Corn	0.120	Sunflowers (hulled)	0.401
Corn oil cake	0.810	Cotton seed meal	0.996
Rye	0.273	Carrots	0.301
Millet	0.428	Irish potato	0.126
Wheat flour	0.045		

is in the cortex of the roots. In the cell they observed that it accumulates in the protoplasm rather than in the cell walls or vacuoles and that it is especially abundant in the nucleus. Little or nothing is known concerning the form in which aluminum occurs in plants.

Levy (1931) examined over 75 species of plants belonging to 37 families and found aluminum in all the phanerogams examined. The accumulation of this element was more rapid during early than during later growth. It was more abundant in leaves than elsewhere and more plentiful in green than in pale and etiolated leaves. As compared with iron and zinc, the amount of aluminum was low in seeds. In the vegetative parts of plants, aluminum was present in about the same quantity as iron.

2. Influence upon Plant Life.—The role of aluminum has been studied along three general lines: (*a*) the effect of this element upon growth and production, (*b*) the effect upon the soil relationships, and (*c*) the role of this element if any in the functions of the plant cell.

a. Growth and Production.—Yamano (1905) found that 0.2 per cent of ammonium alum injured wheat and rye plants in nutrient solutions after 3 weeks and that 0.8 per cent killed the plants in a few days. Prianischnikow (1911) grew wheat, oats, barley, peas and buckwheat in sand cultures fertilized with aluminum phosphate alone and with calcium carbonate alone. He concluded that aluminum phosphate is assimilated and that calcium carbonate has no appreciable depressing effect on the assimilability of aluminum phosphate. Baguley (1912) made comparative tests with the orthophosphates of iron, calcium, and aluminum on oats, peas, and Swedish turnips grown in sand and chalk. He obtained better results with the iron and aluminum phosphate than with calcium phosphate. Truog (1913) working at Wisconsin with 10 different kinds of plants concluded that precipitated ferric and aluminum phosphates produced, with but few exceptions, as good growth as, and in a few cases better growth than, the acid phosphate. Kratzmann (1914) found that salts of aluminum in a concentration of 0.005 per cent hindered the growth of corn, horse beans, lentils, and sunflowers, while a solution of 0.0001 per cent somewhat increased growth. Ruprecht (1915) found that a solution of aluminum sulphate greater than 40 p.p.m. had a very toxic action on clover seedlings. The effect of the aluminum salts was localized in the cells near the root hairs, and he attributed the death of the plants grown in aluminum salts to starvation incident upon the inability of the plant to obtain nutrient salts for normal metabolism. Miyake (1916) showed that aluminum chloride is toxic to rice seedlings in concentrations greater than 0.000133 N and that the toxicity of the salt is not due to the hydrogen formed by the hydrolysis of the salt solution. Mirasol (1920) reported that in the case of barley growing in aluminum sulphate the solution is toxic up to 0.0001 N but with a concentration 0.000033 N it seemed to have a stimulating effect. McLean and Gilbert (1927) found that lettuce, beets, timothy, and barley are the crops most sensitive to aluminum poisoning. The medium-sensitive crops are radishes, sorghum, cabbage, oats, and rye, while corn, turnips, and redtop are noticeably resistant to the toxic effects of aluminum, redtop being the most resistant of the three. Later (1928) they observed that very low concentrations of aluminum (3 to 13 p.p.m.) were stimulatory to plants, while higher concentrations were toxic. Nondialyzing forms of aluminum that could not pass through a collodion membrane in sufficient amount to cause toxicity were decidedly toxic when placed in contact with barley roots. High acidity is not necessarily an accompaniment of toxicity, since aluminum citrate is toxic at pH 6.5 or less acid.

Gilbert and Pember (1931) found that seedlings of Cos lettuce, growing in culture solutions with a pH range of 3.2 to 7.5, showed no appreciable differences in the production of dry weight, but if aluminum sulphate was added to the solutions, the yields in dry matter were considerably reduced. The yields in dry weight of barley plants, grown in soils of various and widely distributed types, were more closely correlated with the active aluminum content of the soil than with its acidity. The same authors (1935) showed that the various weed species commonly found in competition with lawn grasses vary greatly in their sensitivity to aluminum when grown in nutrient cultures. They suggested that the toxicity of aluminum in the soil solution might be one of the causes of the inhibition of the growth of weeds on acid soils. Eisenmenger (1935) noted that organic aluminum compounds exerted a distinct toxic effect on plants of corn, soybeans, and buckwheat and found that this toxicity could be most effectively overcome by the application of calcium hydroxide. He believed that the presence of phosphate lowers the amounts of ionic aluminum in the soil. He considered that low concentrations of aluminum served to increase slightly the rate of growth of these plants.

b. Relationship of the Soil to the Plant.—It is not the intention here to enter into any detailed discussion of the chemical reactions that concern aluminum in the soil but to mention only a few observations in which the beneficial results of lime or fertilizers are apparently due to their action upon the aluminum compounds of the soil. Hartwell and Pember (1918) noted that in treating acid soil with lime the growth of rye was very little benefited but that the yield of barley was increased from two- to threefold. In an acid nutrient solution, however, these two plants were affected equally, while aqueous extract of the acid soil like the soil itself affected the two kinds of plants very differently. These facts indicated that the toxicity of the so-called "acid" soils is not due only to the acid but to some ingredient not present in the ordinary nutrient solution. After a search for the cause of this toxicity they decided that aluminum is the causative agent under conditions such as have been described. The application of phosphoric acid or acid phosphate, while it increases the acidity of the soil, has a beneficial effect on the crop, since the solubility of the aluminum in the soil is reduced by the phosphate. Their results indicate that the practical advantage of phosphating and liming may often be due to the precipitation of active aluminum, which is toxic to many plants quite as much as supplying phosphorus as a nutrient or lime as a reducer of acids. Conner and Sears (1922) likewise considered that the toxicity of aluminum salts is due to the aluminum ion more than it is to the hydrogen ion on such plants as barley, and that this toxicity is reduced when much phosphate is used in the nutrient solution. They considered that acid soils are toxic to many plants largely because they obtain easily soluble aluminum salts. Magistad (1925) noted the difference in plants in the reaction to aluminum and considered that the benefits obtained in practice through the use of lime in the case of clover, alfalfa, oats, and rye result usually from a decrease in acidity and not from a decrease of aluminum present. In corn and barley, however, over the same range of acidity this benefit results from a decrease in both acidity and soluble aluminum. Hoffer and

Carr (1923) and Hoffer and Trost (1923) in investigations on corn diseases reported on the accumulation of iron and aluminum compounds in the nodal tissues of the plant, which causes a purplish brown discoloration of these tissues and finally disintegration. They found that the accumulation of aluminum in these tissues is affected by the genetic composition of the strain of corn and by conditions of the soil. When abundant aluminum injury occurs in the corn plant it is an indication that the soil is deficient in available phosphates. This injury from aluminum can be corrected by the application of lime and phosphates to the soil. Arndt (1922) likewise observed that the composition of the nutrient solution had a marked influence in determining the toxicity of aluminum salts. Coville (1923) obtained striking results in the growth of plants of *Rhododendron catawbiense* by the addition of aluminum sulphate to a rich garden soil in which the plants were growing. One set of plants was treated with distilled water, another with distilled water and magnesium sulphate, and a third with distilled water and aluminum sulphate. The increase in the diameter of the rosettes of leaves was 0, 30, and 250 per cent, respectively. Potted plants that for more than a year had made sickly growth in rich garden soil were revived in an 11-week period by two mild applications of aluminum sulphate. The leaves of the treated plants became a bright green color, while those of the controls remained sickly and dull. Coville believed that the aluminum sulphate acts to replace lime with aluminum, the released lime leaching away in the form of calcium sulphate, and as a result the soil reaction is changed from neutral or alkaline to acid, and this is a favorable medium for this plant, which thrives best on an acid soil.

c. Function in the Cell.—The physiological role of aluminum in plants is not definitely known, although marked reactions of plants due to this element have frequently been observed. Fluri (1909) found that aluminum salts cause the disappearance of starch from certain water plants. He used the salts: aluminum nitrate, sulphate, chlorate, and bichromate and observed that these salts caused a disappearance of starch even when the plants were well illuminated. The aluminum found in the cell was small and the results could not be attributed to chemical reaction. Fluri considered that the disappearance of starch under these conditions is probably due to three reactions induced by the aluminum salts: (1) increased permeability of the protoplasm causes a more rapid loss of sugar, (2) the diastatic action is increased, and (3) the action of photosynthesis is slowed down. Meurer (1909) considered that aluminum forms some chemical compound with the pectic bodies of the middle lamella, since he found that tissues killed with chloroform absorbed aluminum as readily as living tissue. Szücs (1913) believed that aluminum salts cause the protoplasm to set or harden, while Stocklasa (1911, 1918, 1922) considered that they combine with the cell wall to form colloidal salts, an action that renders the cell less permeable. He suggested that aluminum may act as a catalytic agent performing a function in the assimilation of carbon by increasing the rate of photosynthetic action. Mazé (1915) considered that it is an essential element in the growth of corn.

The influence of aluminum upon the color of the flowers of *Hydrangea hortensis* was discovered by Molisch (1897). It had long been known that this plant when grown in certain soils had reddish flowers and in other soils blue flowers. Cultures of soil treated with potassium sulphate produced plants that bore the ordinary reddish flowers, while the soil treated with aluminum sulphate produced plants that bore blue flowers.

Later work by various investigators indicates that the coloring of these flowers is dependent upon the reaction of the medium in which they grow as well as upon its composition. Thus Connors (1923) noted that plants of *Hydrangea*, growing in soils with a pH of 5.0, or lower, consistently produced blue flowers while those in a soil of pH 6.6, or higher, produced pink flowers. He was unable to determine whether the color of the flowers was influenced by the presence of iron or aluminum. Allen (1931) found in spraying the flowers of *Hydrangea* at various stages of development with dilute solutions of aluminum sulphate that a blue coloration of the flowers resulted. In general, the same results were obtained when the solutions were injected into the stem, the flowers being most affected on the side of the injection. He obtained no effects on the coloration of the flowers when they were sprayed with various salts of iron, copper, magnesium, or potassium. Later (1934) he attempted to determine the most efficient method of maintaining the aluminum in an available form in the soil, and to determine the effects of various materials upon the acidity. The best results were obtained from the use of aluminum sulphate, but it was difficult to determine the rate of application to a given soil in order to lower the pH sufficiently for the optimum solubility of the aluminum. Chouard (1933) reported that iron, chromium, and aluminum could produce the bluing agent in the flowers of *Hydrangea.* Poesch (1935) applied aluminum sulphate to the soil at the rate of 1 lb. of salt to 5 gal. of water, or in the dry form by using 0.5 teaspoon of salt per 6-in. pot. When the soil gave a reaction of pH 6.0 or lower, bluing of the flower resulted, the intensity of the color varying with the variety of plant used.

M. Sodium

Although sodium is not generally considered essential to the growth and development of the plant, its presence apparently has a marked effect upon the growth of some plants under certain conditions. Hellriegel (1898) found that sodium salts always gave increases in yield even when potassium salts were present in quantity. The same fact was also observed by Breazeale (1906) for wheat plants growing in culture solutions. Wheeler, Hartwell, Pember, and others (1904 to 1913), however, obtained no increase in growth by the addition of sodium when an optimum amount of potassium was present, but they observed, as did Breazeale (1906, 1916), that a larger amount of potassium was left in the solution by the growing plants when the potassium was supplemented by sodium. Sodium thus seems to retard the absorption of potassium or to conserve it in the metabolism of the plant. Wheeler, Hartwell, and others found that when the deficiency of potassium was great enough to bring about a 30 per cent depression in the green weight of plant produced, the addition of sodium gave an increase in growth amounting to 10 to 25 per cent within a period of from 2 to 3 weeks. Hartwell and Wessels (1913) observed no decrease in the percentage of sugar in the onion when the potassium deficiency was supplemented with sodium, although there was a decrease in the crop corresponding to the scarcity of potassium. In the case of mangel roots there was a tendency toward depression in both the total and the reducing sugars as a result of the addition of sodium to supplement a potassium deficiency. Pfeiffer, Rippel, and Pfotenhauer (1920) stated that sodium salts may to some extent replace potassium salts as a nutritive material for oats. In the production of dry matter sodium replaced potassium in the molecular ratio of 0.25:1. When relatively small quantities of potassium and relatively large quantities of sodium are present, this ratio is increased to 0.38:1. When such a replacement occurs, the sodium appears in the grain as well as in the leaves and stems.

According to Vincent and Herviaux (1929) the soluble chlorides in the leaves of barley at the blooming stage were 80 per cent potassium chloride and 20 per cent sodium chloride, while in the grain the ratio of potassium chloride to sodium chloride was 1.2:1. In young cabbage leaves, 60 per cent of the soluble chlorides was sodium chloride. They concluded that the quantity of sodium present in cultivated plants is largest in the organs of greater physiological activity. In beets the use of sylvanite, rich in sodium chloride, resulted in higher yields than the use of potassium chloride.

Apparently sodium is sometimes extremely toxic to certain plants. Thus Beaumont (1932) found in fertilizing "run out" pastures in Massachusetts that sodium nitrate was toxic to *Polytrichum commune*. He considered that toxicity was linked with anionic and cationic effects, since the nitrates were more toxic than the corresponding chlorides. The effects were found to be both immediate and cumulative since several smaller applications gave the same result as a larger application.

Sodium salts are apparently necessary for numerous marine algae, according to Osterhout (1912). His experiments show that the replacement in sea water of sodium by ammonium, calcium, magnesium, barium, and others is decidedly injurious. Sodium is apparently not needed for osmotic pressure in this case, since an equivalent concentration of other compounds does not satisfy the plants. Osterhout could not determine the value of sodium as a nutrient but considered that it might serve to antagonize the toxic action of other salts that are present in the sea water. From the experimental evidence at hand it seems that sodium may be able to perform one or more of the following roles depending, however, upon the environmental conditions and the type of plant under consideration: (1) It appears to be a conserver of potassium, since less is absorbed by the plant when sodium is added. (2) It may to a limited extent replace potassium as a plant nutrient (Schulze, 1913). It can apparently perform certain functions of potassium but by no means all of them. It may delay potassium starvation for the plant but cannot keep it off entirely. (3) When applied to a soil that is deficient in potassium it may liberate a certain amount of potassium which was adsorbed by the soil and make it available for the plant, thus bringing about an increase in growth. The experimental evidence with water cultures, however, makes it appear doubtful whether sodium functions in this manner, although it is doubtful whether water cultures would indicate anything in regard to base exchange that might occur in the soil. (4) In certain specific cases, it evidently acts as an antidoting agent against certain toxic salts present in the medium in which the plant is growing.

N. Copper

Brenchley (1914) concluded from a careful survey of the literature to that date that copper compounds under the most typical conditions act as poisons to the higher plants and that it is only under particular and peculiar conditions and in very great dilution that copper shows any stimulating action. It is, however, very widely distributed in plants and in considerable quantity. McHargue (1925) found that the copper content of various plants and plant parts ranged from a trace to 46 p.p.m. of dry material, while Maquenne and Demoussy (1920) found 0.25 mg. of it per liter of expressed sap of the potato and as high as 40 mg. per kilogram of dry leaf substance of lettuce.

Webster and Jansma (1929) examined wheat from 14 states and found that the amount of copper in the various samples of grain varied from 281 to 497 mg. per kilogram of ash. Pitcaithly and Worley (1933) reported that the amounts of copper in the leaves, bark, kernel, and fresh fruit of the Karaka tree (*Corynocarpus laevigata*) were 7.0, 2.7, 4.2, and 2.7 mg., respectively, per kilogram of fresh material. The response of plants to copper has been noted in its effects on increasing vigor and yield, on quality, on the control of chlorosis, exanthema and "frenching," and on the prevention of "salt sick" in live stock.

Marquenne and Demoussy (1920) found copper in the greatest abundance in cells that are active in growth and considered that its translocation is controlled by processes accompanying metabolism. They further found that the addition of other solutions of copper sulphate to the extent of 0.02 to 0.2 mg. per liter increased not only the length of life of various plants but also their dry weight (Forbes, 1917). A wide range of plants has been found by Allison, Bryan, and Hunter (1927) to respond in growth and production in a remarkable way to the application of 30 to 50 lb. per acre of copper sulphate to the raw peat soils of the Florida Everglades.

Sommer (1931) stated that amounts of copper as low as 0.06 mg. per liter of nutrient solution produced a marked effect on the dry weight of the plants growing therein as shown by the following table:

Plant	Weight of plants, in grams	
	Without copper	With copper
Sunflower	0.16	0.70
Tomato	0.30	2.60
Flax	1.40	4.50

Lipman and Mackinney (1931) reported that barley plants were unable to form seed without having copper in the root medium. Amounts as low as 1/16 to 1/8 p.p.m. were sufficient to produce seed.

Felix (1927), Knott (1932) in New York, and Harmer (1932) in Michigan have obtained marked results in increasing the yield and quality of onions and lettuce by the application of copper sulphate to muck soils. The amount of this salt necessary to produce the desired results varied from 100 to 300 lb. per acre, depending upon the kind of crop and the type of soil. Painting the leaves of lettuce with a dilute solution of copper sulphate or dusting them with 20–80 copper-lime dust at the rate of 75 lb. per acre had similar effects to the application of copper sulphate to the soil. Andressen (1932) reported that deciduous fruit trees affected with chlorosis, rosetting, and dieback were cured by the application of copper sulphate to the soil in amounts ranging from ¼ to 2 lb. per tree. The application of this salt was reflected in the increased copper in the leaves. It was also found that leaves that had been dipped in very dilute solutions of copper sulphate became green within 2 weeks. The treatment of the soil with potassium, magnesium, manganese, sulphur, iron, or manure had no effect on these maladies. This gave further evidence that the lack of copper was the cause of the disturbances. The ash content of the chlorotic parts of these plants was always much higher than that of the healthy material. Oserkowsky and Thomas (1933) found that Bartlett pear trees affected with xanthema could be cured or markedly improved by spraying with Bordeaux mixture or by the introduction of soluble copper-salts into the trunks. Similar treatments with iron citrate, manganese chloride, and zinc sulphate, respectively, were without effects. The amount of copper in the diseased leaves was lower than that of healthy leaves. It was found by Burge, Wickwire, and Orth (1933) and Orth, Wickwire, and Burge (1934) in Florida that the surface application of 4 lb. of copper sulphate around each of several 3-year-old orange trees, whose leaves were spotted, yellow, or frenched, greatly improved the condition of the trees. Chlorophyll determinations made 4 months after the application show that the leaves of the treated plants contained 4.6 times more of this pigment than did the leaves of the control trees. It was observed also that spraying with Bordeaux mixture greatly increased the vigor of the diseased plants. Willis and Piland (1934) noted that the application of copper sulphate to newly limed soils corrected the excessive absorption of iron by the corn plant. Haas and Quayle (1935) studied the copper content of the leaves and fruit of citrus and found that the amount of this element necessary for healthy growth varies with the soil and climate, although the solubility of the copper may be of considerable importance. Hopkins (1933) with cultures of *Lemna* and *Chlorella* could find no increase in growth that he could attribute to addition of copper.

Neal, Becker, and Shealy (1931) found that the forage plants from certain regions of Florida contained a small amount of copper due primarily to the scarcity of this element in the soils in which they grew. The cattle, sheep, or swine that consumed these plants developed salt sickness. This disease may be prevented or controlled by the application of copper to the soils in which the plants are growing or by the addition of copper to the ration of the animals. There is no definite knowledge concerning the role of copper in plants, and all ideas relative to its uses are based purely upon theory and not fact.

O. Zinc

Mazé (1915) considered that zinc is one of the elements essential for the best development of the corn plant, and his statement was verified to some extent by Lipman, Davis, and West (1926). Nakamura (1904), however, could observe no stimulatory effect of zinc upon the growth of rye, peas, onion, or mustard. Conner (1920) observed that the zinc salts, formed by the action of acid soils on galvanized iron containers, were very toxic to oats and buckwheat growing therein. Finch, Albert, and Kinnison (1933) reported that the pecan tree is susceptible to zinc chlorosis.

Sommer and Lipman (1926) furnished evidence that boron is indispensable to the development of Windsor bean, buckwheat, flax, mustard, castor bean, cotton, and barley and that zinc is essential to the dwarf sunflower and barley. Sommer (1928) noted in the case of buckwheat that the plants grown without zinc were much smaller than the controls. The average dry weight per culture grown without zinc was 1.94 g., while that with zinc was 8.45 g. In *Vicia faba* there was no visible difference between plants grown with and without zinc until the flowering stage was reached. Then most of the flower buds fell off, and if pods developed no seeds were formed when zinc was lacking, but where it was supplied, the plants bloomed freely and produced seed. The plants of buckwheat and sunflower that survived in the absence of zinc had the appearance of plants that had grown on poor soil.

Roberg (1932) observed in cultures of *Chlorella vulgaris* that 0.001 mg. of zinc per 100 cc. of nutrient solution stimulated growth but that 0.005 mg. per 100 cc. of solution retarded growth under all conditions. In an examination of the leaves of 11 different species of plants, Bertrand and Andreitcheva (1933) found that the zinc content ranged from 0.36 mg. per kilogram of fresh tissue in endive to 8.9 mg. in the same weight of cress. They considered that a high zinc content in leaves is correlated with a high chlorophyll content.

It was found by Chandler, Hoagland, and Hibbard (1933) that the spraying of peaches and apricots in the spring with zinc-lime when the symptoms of "little leaf" were most pronounced, improved the fruit but did not increase the amount of fruit set. The injection of zinc sulphate into holes 1.5 in. deep and 3 to 4 in. apart around the trunk or branch gave the most dependable results of any method of application used except spraying in the case of citrus and grape. These benefits seldom last longer than through the second summer. Little leaf is apparently a symptom of a deficiency of zinc. Hoagland, Chandler, and Hibbard (1935) reported that if zinc was withheld from nutrient solutions in which apricot, tobacco, squash, corn, mustard, tomato, and other plants were growing, these plants developed symptoms of little leaf.

Mowry and Camp (1934) found that the bronzing of tung trees can be overcome by a treatment with zinc sulphate applied either as a spray or directly to the soil. The fact that the spraying of the aerial parts of the tree corrects the trouble would indicate that the zinc itself is responsible for the recovery and that it is not a case of this element replacing or rendering soluble in the soil some other ion in which the plant is deficient.

It was observed by Reed and Dufrenoy (1935) that there was renewed activity in the cells of the mottled leaves of citrus and an accelerated growth of new shoots subsequent to the application of a spray of a very dilute solution of zinc sulphate. In the green leaves of new shoots whose growth had been promoted by this spray, neither calcium deficiency nor phloem necrosis was evident, and the chloroplasts developed to a fair size and formed starch. Barnette and Warner (1935) and Barnette, Camp, Warner, and Gall (1936) found in Florida that the chlorotic condition of corn plants, termed "white bud," could be prevented or overcome by the application of 20 lb. of zinc sulphate per acre mixed with an inorganic fertilizer and an alkaline peat. Following this application, the chlorotic plants regained their green color, renewed growth, and produced grain. They further found that peanuts, oats, velvet bean, cowpeas, millet, sugar cane, and Napier grass responded favorably to the application of zinc sulphate.

P. Iodine

According to Wynd (1934) the presence of iodine in the higher plants was first verified by Chatin (1850). It occurs in the plant apparently in both the inorganic and organic forms. Masuda and Nishida (1934) found in *Lamnaria japonica* that the amount of organic iodine varied from 13 to 85 per cent of the total iodine present. It has been shown by Stocklasa (1924, 1930), Scharrer and Strobel (1927), Scharrer and Schwaibold (1927), Hiltner (1928), Orr, Kelly, and Stuart (1928), McHargue, Roy, and Pelphrey (1930), Wendt (1930), Conner (1931), Mitchell (1931), Beaumont and Karns (1932), and McHargue, Young, and Calfee (1935) that plants absorb iodine more or less proportionately to the amount in the medium in which they grow, although because of reactions in the soil this fact at times is not very evident. Some of the plants that have been shown by these investigators to increase their content of iodine when this element was applied to the soil are beets, lettuce, peas, spinach, radishes, bean, turnip, red clover, oats, and various forage crops. The iodine content of these plants was increased from two to sixteen times over that of the controls. The amount of iodine in the soil has been increased by the application of potassium iodide or potassium iodate, or of raw phosphate rock, Chile nitrate, or limestone, which contain considerable quantities of iodine as impurities.

Frear (1934) analyzed 135 samples of potatoes for iodine from as many localities in Pennsylvania. The minimum iodine content on a dry basis was 10 p.p.b., the maximum 216 p.p.b., and the mean content 77.8 p.p.b. McClendon, Barrett, and Canniff (1934) found, from the analysis of 55 samples of potatoes in Minnesota, that the average iodine content in the eastern section was 0.085 mg. per kilogram and in the western section 0.226 mg. per kilogram. The cases of simple goiter per 1,000 drafted men during the World War was 19 for the eastern and 7.5 for the western portion of that state. This indicates an inverse relation between the amount of iodine in potatoes and the proportion of goiter in the drafted men.

It was found by Suzuki and Aso (1903) that an application of potassium iodide at the rate of 0.5 g. to 20 sq. m. of soil surface increased the production of oats, and a dosage of 0.05 gm. for the same surface had a beneficial influence on radishes. Brenchley (1924), however, could not find any significant response of mustard and barley to the application of iodine. She suggested that it might have some specific and effective action if means could be found to prevent it from being so rapidly converted from the elemental state to the form of compounds, which ordinarily occurs when it is applied to the soil. Stewart (1930) in Hawaii showed that iodine in the presence of copper was beneficial to the production of sugar cane. Wendt (1930) noted that root systems of various garden crops were larger, and that the flowering and fruiting of these plants were earlier when iodine was applied to the soil at the rate of 6 to 40 g. per hectare. It was observed by Cotton (1930) that buckwheat plants perished within 3 weeks when growing in nutrient solutions to which 40 to 127 p.p.m. of iodine had been added. The growth of these plants was markedly depressed when this element was added to the solution in quantities as low as 1.2 to 12 p.p.m. Wynd (1934) added potassium iodide to Shive's three-salt solution in portions of 1, 5, 10, and 20 p.p.m. of iodine, and found that all these concentrations had a depressing effect upon the growth of plants as indicated by a loss of green color and a progressive dropping of the lower leaves.

Stocklasa (1929) reported that certain halophytes growing in iodine-free and in iodine-containing mediums showed marked differences in growth. It appears that iodine and chlorophyll are closely associated, and that the presence of this element creates a condition favorable to the production of greener leaves. It was observed by Marine, Baumann, and Webster (1930) that the iodine-absorbing materials are present in the greatest concentration in the rapidly growing portions of plants but decrease during maturation. These iodine-absorbing materials decreased rapidly during drying and slowly during storage. It is likely that they belong to the oxidation system. At a concentration of 1 p.p.m. of potassium iodide in the nutrient solution, Wynd (1934) observed that the amount of peroxidase and invertase present in the plant and the rate of respiration decreased. These greatly increased, however, at higher concentrations, while oxidase activity was progressively lowered.

Q. Selenium

Selenium occurs in soils in varying amounts over wide areas, and in certain regions of Nebraska, South Dakota, Wyoming, and probably in other regions it is present in relatively large quantities. Some of the plants growing on these soils absorb a considerable amount of this element and are toxic or poisonous to the animals that consume them. According to Byers and Knight (1935), Byers (1935), Hurd-Karrer (1935), and Robinson (1936) the quantity of selenium absorbed by plants and their subsequent toxicity depend upon numerous factors. Chief among these are: the composition of the soil, the moisture relations of the soil, the concentration of this element in the soil, the species of plant, the stage of its development, the available sulphur, and the crops previously grown.

Hurd-Karrer (1933, 1934) found that wheat plants grown in soil to which sodium selenate had been added at the rate of 15 p.p.m. or less became characteristically chlorotic. The young leaves frequently became almost snow white, with green tips and green midveins. When the selenate was added to the pots containing older plants, white chlorosis appeared only on the leaves emerging subsequent to the addition of selenium. This investigator (Hurd-Karrer) noted that the toxicity of sodium selenate is determined by the amount of sulphur available to the plants. Thus in cultural solutions, concentrations of selenium as low as 0.1 p.p.m. produced distinct injury after a few weeks if no sulphate was present, whereas a concentration of 18 p.p.m. was required for this degree of injury if the solution contained 192 p.p.m. of sulphur. There was no visible injury to the plants when the proportion of selenium to sulphur was 1:12 or less. When the ratio was 1:8, or greater, the plants were chlorotic and stunted, while when the ratio was as high as 1:2, growth was almost completely inhibited. Hurd-Karrer (1935) grew 17 different crop plants in the greenhouse in Keyport clay loam to which had been added 5 p.p.m. of selenium in the form of sodium selenate. Under these conditions the mustard plants absorbed the greatest amount of the element, 1,240 p.p.m., while sorgo absorbed only 130 p.p.m.

It was found by Beath, Eppson, and Gilbert (1935) in Wyoming that plants of *Astragalus racemosus, A. bisulcatus, A. scobinatulus, A. pectinatus,* and *A. flaviflorus,* among others, were absorbers of selenium when growing in the proper soil. A group of plants, including *Atriplex*, *Solidago*, and several native grasses, sometimes gave negative and sometimes positive tests for selenium, while a large number of plants including white and purple loco, wheat grass, soapweed, four species of *Astragalus*, foxtail, wild flax, and Canada thistle did not absorb a sufficient quantity of selenium to be detected. Trelease and Martin (1936) in reviewing the literature on the absorption of selenium stated that western wheat grass growing on various soils accumulated 1 to 60 p.p.m. of selenium, while *Astragalus bisulcatus* growing on the same soils accumulated 200 to 4,300 p.p.m. The highest accumulation of this element, 8,840 p.p.m., was found in the latter plant. The difference in the ability of plants to absorb and store selenium is illustrated by *Astragalus bisulcatus* and *A. missouriensis* growing on a soil containing 2.1 p.p.m. of selenium. The former plant accumulated 1,250 p.p.m., while the latter under identical conditions stored only 3.1 p.p.m. On one type of soil containing 2.5 p.p.m. of selenium, wheat plants accumulated 45 p.p.m., while in a different soil containing 3.0 p.p.m. the plants absorbed only 0.5 p.p.m.

According to Levine (1925), selenium compounds in concentrations of 0.01 per cent and above exerted a detrimental influence on the growth of white lupine and timothy seedlings. However, he obtained increased growth of these plants in solutions of selenium dioxide and selenic acid with concentrations of 0.0001 and 0.001 per cent, respectively.

R. Other Elements

The influence upon plant growth of barium, strontium, titanium, thallium, vanadium, arsenic, fluorine, rubidium, cobalt, and nickel has been observed by numerous investigators. McHargue (1919) in his experiments with cowpeas, soybeans, oats, wheat, and corn grown in sand cultures found that neither barium nor strontium compounds can be considered important plant nutrients, but the presence of a small amount of the carbonate of each of these elements in the presence of calcium carbonate gives increased yields that are noteworthy. Both the barium and strontium carbonate accelerate especially the growth of the roots of plants under consideration. The compounds of both these elements are toxic to plants in the absence of calcium carbonate. He also found, as did Suzuki (1900), that neither can replace calcium in any degree in the metabolism of the plant. Osterhout (1916) noted that *Spirogyra* underwent a peculiar cytological reaction when placed in a 0.0001 *M* solution of barium chloride. The reaction observed was a characteristic contraction of the chloroplasts, while the protoplasm remained intact in its normal position. Numerous other salts including calcium chloride, magnesium chloride, manganese chloride, nickel chloride, cobalt chloride, sodium chloride, and potassium chloride were tried in the same manner but none produced this peculiar result. This reaction suggests that there is some specific constituent of the protoplasm of *Spirogyra* that reacts in this peculiar manner to barium.

The reaction of plants to titanium has been studied by Nemec and Kas (1923), who obtained increased crop yields by manuring *Sinapis alba*, *Pisum sativum*, and *Medicago sativa* with titanium as the insoluble titanate and the soluble titanium sodium citrate. The amount of phosphorus, silicon, and aluminum increased and decreased with the titanium content of the plants. Since the iron content decreased with increasing applications of titanium, it was considered that iron might be replaced by this element.

Sideris (1930) found that pineapple plants, growing in a nutrient solution lacking iron but containing 5 p.p.m. of titanium, formed chlorophyll as readily as when iron was added in the form of ferric chloride. Inman, Barclay, and Hubbard (1935), however, in using titanium chloride in amounts of 15 to 20 p.p.m. found in the plants with which they worked that titanium could not be substituted for iron in the formation of chlorophyll.

Arsenic in the form of disodium arsenate showed a beneficial influence in low concentrations upon beans, peas, wheat, and radish, according to the observations of Stewart (1922). The beneficial influence of arsenic was not manifest by any increase in dry weight but by the more vigorous and healthy appearance of the plant. He considered that the accumulation of arsenic in the soil from sprays might be beneficial rather than injurious to plants. Beans, peas, and wheat were stimulated by concentrations of disodium arsenate up to 75 p.p.m. Radishes were stimulated by concentrations up to and including 240 p.p.m. Beans were the most sensitive to this element. When the above plants were killed by the absorption of arsenic, they contained the following parts per million of dry weight of plant: wheat, 269; potato, 524; bean, 678; pea, 1,190; and radish, 940. Brenchley (1914) stated that the toxic effect of arsenic upon higher plants is much more marked with arsenious acid and its compounds than with arsenic acid and its derivatives.

Albert and Paden (1931) believed that the addition of calcium arsenate to light, sandy-loam soils may be expected eventually to interfere seriously with the growing of such arsenic-sensitive crops as cowpeas, oats, cotton, and various grasses. The effect of the amount of arsenic in the soil upon its content in the plant is shown in the following table:

Soil, arsenic content, p.p.m.	Plants, arsenate content, p.p.m.			
	Oats		Cowpeas	
	Roots	Tops	Roots	Tops
30	180	6	40	11
8	80	3	10	3

Gautier and Clausman (1919) reported that the application of 100 p.p.m. of fluorine, as calcium fluoride, to field plots definitely increased the yield in some cases. Mazé (1919) considered that fluorine is a necessary element for corn, but Wilson (1933) and Wöher (1920) found that it had a toxic effect upon the plants they observed. The former worker found that the tobacco plant responds to sodium fluoride by transforming free water into bound water in such quantities as to indicate that this compound induces a state of physiological drought. Bartholomew (1935) found that concentrations of fluorine as high as 50 p.p.m. did not significantly decrease the germination of Sudan grass, cowpeas, soybean, or red clover. The addition of soluble fluorides to the amount of 10 p.p.m. did not decrease the amount of dry matter produced by cowpeas growing in a nutrient solution. The fluorine was found mostly in the roots of these plants.

Scharrer and Schropp (1935) reported that the application of extremely small amounts of vanadium stimulated slightly the later growth of all the plants studied by them but that it exerted a detrimental influence when applied in larger amounts. The roots of corn were especially injured by this element even when applied in the smallest amounts.

McMurtrey (1932) found that a nutrient solution containing 1 p.p.m. of thallium was toxic to tobacco plants. This toxicity was manifested by a series of effects that are more or less the characteristic symptoms of "frenching," although these two effects are not identical. The first effect is a slowing-down of the growth rate and the development of a lighter green color along the veins of the upper leaves of the plant. As the younger leaves develop, they show at first a chlorosis which progresses along the smallest branches of the vascular system. The subsequent growth produces leaves which are decidedly distorted and may often consist essentially of only the midrib. This is followed by a proliferation of the lateral buds resulting in the so-called "witches-broom" effect. McCool (1933) noted that the presence of 2.1 p.p.m. of thallium sulphate in sandy loam slightly retarded the development of the roots and tops of soybean, wheat, buckwheat, alfalfa, corn, tobacco, and tomato plants, while an application of 8.5 p.p.m. was very injurious.

It was found by Loew (1878) that rubidium chloride permitted normal growth of buckwheat until flowering, after which time injury became progressively extended until the plants died. Arndt (1922) found that buckwheat, grown with rubidium salts instead of potassium salts showed premature dying of the roots, curling of the leaves, and deterioration of the chlorophyll, but that these symptoms could be allayed or removed by the addition of potassium salts. It was suggested by Blanck, Giesecke, and Henkeshoven (1933) that rubidium interferes with potassium assimilation although it may not have a specific toxic effect. Brenchley (1934) studied the action of rubidium sulphate and palladium chloride in nutrient solutions on the growth of barley, wheat, oats, peas, and beans and found that, over a wide range of concentrations, rubidium sulphate had no effect on the growth of these plants. A relatively low concentration of palladium chloride, however, was injurious, the tolerance varying with the species. Barley appeared to be the least and oats the most sensitive of the cereals.

Nakamura (1904) could observe no stimulatory effect of cobalt or nickel upon the growth of rye, peas, onion, or mustard. Bishop and Lawrenz (1932) found that the presence of cobalt in plants depends on the nature of the soil and the type of plant. Scharrer and Schropp (1933) reported that corn and barley were stimulated by a low concentration of cobalt but that wheat, oats, beans, and peas were injured. Cotton (1930) found that concentrations of 5.8 to 58.7 p.p.m. of nickel were very toxic to plants, while a concentration as low as 1.8 p.p.m. caused the spotting and chlorosis of leaves. According to Nakamura (1904), lithium in the form of lithium carbonate in amounts ranging from 10 to 100 mg. per kilogram of soil exerted a stimulatory action upon barley and peas. He also found that cesium chloride at the rate of 100 mg. per kilogram of soil increased the height of rice plants.

V. ELEMENTAL REQUIREMENTS AND THE AGE OF THE PLANT

The amount of nutrients necessary for the optimum growth of plants at the different stages of their growth is an important factor from the standpoint of crop production and is a subject that has received considerable experimental attention.

McCall and Richards (1918) studied the salt requirements of wheat plants at three stages of growth when grown in a three-salt solution composed of monopotassium phosphate, calcium nitrate, and magnesium sulphate. Of these salts, 36 different proportions were used, which necessitated the growing of 36 different cultures for each of the three growing periods. The three stages of development of the plant that were studied were: (*a*) the first 30 days from germination, (*b*) the second 30 days of growth, and (*c*) the period extending from the close of the second 30-day period to the maturity of the plant. For the early growth period, the solutions that gave the highest yield of tops were characterized by a high calcium nitrate content and a low proportion of magnesium sulphate, while the lowest yield of tops was associated with low calcium nitrate and a high proportion of magnesium sulphate. For this period the effect of monopotassium phosphate appeared to be overshadowed by the other two salts. The results for the second 30-day period showed results very similar to those of the first period. This indicates that the elemental requirements of the plants during the second 30-day period were substantially the same as for the first period. For the third and final growth period, the solutions that gave the highest yielding plants were characterized by a relatively high concentration of calcium nitrate and a low proportion of both magnesium sulphate and monopotassium phosphate, while the solutions producing low yields were characterized by a high proportion of monopotassium phosphate without regard to the ratio of the other two salts. Hibbard and Gershberg (1924) in the case of Marquis wheat using a cultural solution high in magnesium sulphate and relatively low in both calcium nitrate and monopotassium phosphate produced the best growth during the vegetative period between the time the plants were 3 weeks old and the time of blooming. Shive and Martin (1918) found in buckwheat that during the 4 weeks following germination the highest yields of tops and roots were produced by a solution of monopotassium phosphate 0.0144 M, calcium nitrate 0.0052 M, and magnesium sulphate 0.0200 M. During the second 4-week period, including seed production and ripening, the best solution was 0.0108 M

monopotassium phosphate, 0.0130 *M* calcium nitrate, and 0.0100 *M* magnesium sulphate.

In corn, Hornberger (1882) noted that when the tassels were beginning to form there was an abrupt slowing down of the rate of absorption of the various elements. This was followed by a period of rapid absorption, which was succeeded by the ripening period in which there was some loss of practically all the elements, due perhaps to diffusion outwardly of solutes and to the loss of leaves and roots. Jones and Huston (1914) found that for 8 weeks after the germination of corn a very rapid absorption of potassium occurred. This was followed by a period of relatively slow absorption and then, at the time of greatest starch formation in the grain, a period of rapid absorption. They observed no striking difference in the rate of phosphorus absorption during the growing season. The rate of nitrogen absorption reached its highest point about the end of the 8-week period. This was followed by a decreased rate, which again became high at the 16-week period. In corn grown in nutrient solutions, Duley and Miller (1921) found that an optimum nutrient during the last 30-day period largely determined ear production, but, when there was a copious supply of mineral elements at the end of the vegetative stage, the leaves and stems contained enough material to produce fair ears even if there was a small amount of nutrient supply during the last 30-day period of the plant. Burd (1919) studied the rate of absorption of certain elements from the soil by the barley plant at different stage of its growth under conditions that precluded leachings or gains to the soil or the removal of elements from the leaves by rain. During the first period of growth, ending 8 or 9 weeks from planting or about the time the heads begin to form, the increase of nitrogen and potassium conformed very closely to the gain in total weight and water content of the plant. The increase of calcium, magnesium, and phosphorus closely paralleled the formation of dry matter up to 8 to 9 weeks, respectively, after which they lagged behind. At this time the absolute amounts of potassium and nitrogen contained in the plant very closely approximated the amount present at maturity, while the potassium content may be even greater than at maturity. During the second period, which extended from the beginning of head formation to the beginning of the ripening of the grain, there was not only a decreased rate of absorption, as has been noted in corn, but there was a decided loss of potassium and nitrogen from the aerial parts of the plant. This loss took place concurrently with the migration of these same constituents into the developing head. The plant, however, had a tendency toward the close of the period to absorb again the soil constituents previously lost. The third period or stage occurring at the time of ripening of the grain was characterized by a practically complete cessation of absorption of all constituents and an actual loss of some of these. The striking thing in this experiment is the fact that the absorption of potassium and nitrogen during the first period of growth is approximately proportional to the growth attained and that in the succeeding periods the final dry-matter content of the crop more than doubles without any very substantial increase in nitrogen content and with an actual loss of potassium. Burd concluded from his work that it is generally desirable to have the large amounts of solutes in the soil solution at the commencement of the plant's growth cycle but that it is unnecessary and may be undesirable to maintain this condition during certain later stages of growth. Pember (1917) and Pember and McLean (1924) found that potassium was utilized most successfully by barley, oats, and wheat when supplied early in the growth period. Wheat was able to make a more effective use of a limited amount of potassium if supplied during a 4-week period from the fourth to seventh week after planting than if the same amount were supplied over a period of 12 weeks. The time of the application of phosphorus made little difference, but deficient amounts of nitrogen were most beneficial if supplied gradually over the entire growth period.

The tendency toward a delayed rate of absorption or the absolute loss of some constituents at a comparatively early age is probably characteristic of certain types of plants, as indicated by the work of Wilfarth, Römer, and Wimmer (1906) with wheat, barley, and potatoes. They found evidence of potassium and nitrogen losses from wheat and barley at comparatively early stages of development, but in the case of the potato they found no losses of potassium and nitrogen at any stages of development, nor did they observe any outstanding changes in the rate of absorption of these elements during its growing period.

Rice grown in Porto Rico (Gile and Carrero, 1915) showed a decrease in the percentage of potassium, phosphorus, and sulphur in the ash and of nitrogen in the dry matter with the age of the plant, which is a common observation for most plants. The amount of magnesium, phosphorus, iron, sulphur, and nitrogen in 103-day-old plants when the panicles were just out was the same as found in 123-day-old plants when the seed was ripe. The amount of potassium, however, was approximately twice as much in the mature as in the younger stage. The amount of iron in the ash of the green leaves and straw decreased regularly with the age of the plant, but the amount in the whole plant, including stem, leaves, and grain, remained constant. According to Chizhov (1926) all the necessary ash and nitrogenous substances are accumulated by winter and spring crops about the period of blooming or grain forming. In the sunflower this accumulation is complete at about the time of ripening, while in beans and potatoes it is completed at the end of the vegetative period.

Gericke (1924 to 1925) grew wheat plants in complete nutrient solutions for varying periods and then transferred them to cultural solutions lacking one of the following elements: potassium, calcium, iron, magnesium, phosphorus, nitrogen, and sulphur, and allowed them to complete their growth. The following is a summary of the results obtained:

After plants had grown for 4 weeks in a complete nutrient solution and had attained approximately one-seventh of the dry weight obtainable in such a medium and were transferred to solutions devoid of magnesium, phosphorus, or sulphur, they produced more grain and straw than did the plants grown to maturity in complete nutrient solutions which were three times renewed. The removal of plants grown for 4 weeks in a complete nutrient solution to a potassium-free medium resulted in a production of dry matter equal to that obtained from cultures grown in a complete nutrient solution. The removal of plants grown 4 to 10 weeks in complete nutrient solutions to solutions lacking nitrogen decreased the production of both grain and straw, but otherwise the plants appeared normal. The maximum development among all cultures was obtained from the plant grown 4 weeks in a complete nutrient solution and then transferred into a solution lacking phosphorus. A better growth of plants was obtained from the cultures grown to maturity in a complete nutrient solution which was not renewed than was obtained from cultures grown in a complete nutrient solution which was three times renewed. The elements required longest in available form in the growth mediums for the normal development of wheat were calcium and iron.

Brenchley (1929) tested the effect of depriving barley plants of phosphorus after varying initial periods during which it had been supplied and also supplying phosphorus after initial periods of deprivation. The providing of phosphorus for the first 6 weeks or longer permitted normal growth to be made as judged by appearance, increase in dry weight, and grain produced. If phosphorus was withheld for the first 4 weeks, tiller production was not affected, but no heads were produced. With longer periods of initial deprivation, growth was steadily depressed in all respects. The amount of phosphorus absorbed by the plant increased steadily in more or less direct proportion to the length of time phosphorus was given at the beginning of growth.

Sufficient phosphorus, however, was taken up in the first 6 weeks to enable the plant to make its maximum dry weight. The percentage of phosphorus in the dry matter rapidly increased from this time onward.

In connection with the elemental requirements of plants at different stages of their growth, it is important to consider the availability to the seedling of the elements stored in the seed and to note how the amounts thus supplied correspond to the total amounts found in the young seedling. Buckner (1915, 1919) studied the translocation of the mineral constituents of the bean and corn seed and of the potato tuber during germination. He found that approximately 47 per cent of the minerals stored in the cotyledons of garden beans was retained in them when they were no longer functional. The cotyledon of corn contained about 39 per cent of its original mineral content when it had ceased to supply any nutrients to the young seedling. In the case of the potato tuber, the young stem removed only about 50 per cent of the minerals stored in the fresh tuber. The percentage distribution of the various mineral constituents in the seedlings studied is found in the table below.

It was observed that there were no very striking differences in the quantities of the minerals translocated and no marked selective influences shown by the root, stem, or leaves of the seedlings for any particular mineral reserve contained in the seeds or tuber.

The intake of nitrogen, potassium, and phosphorus by wheat plants grown in nutrient solutions for fractional parts of a day was studied by Breazeale (1923). The absorption of these elements for 1 hr. or other fractional parts of a day was out of proportion to what might be expected if time alone governed absorption. Thus during the first hour the plants absorbed 0.0520 g. of potassium. During a 24-hr. period the same plants absorbed 0.2157 g. of this element. If the rate of absorption of the plants kept in the nutrient solution 1 hr. had been maintained for 24 hr., the plants would have absorbed 1.248 instead of 0.2157 g. of potassium. Thus over five times as much potassium was absorbed during the first-hour period as might be expected if time alone governed absorption.

PERCENTAGE DISTRIBUTION OF THE MINERAL CONSTITUENTS IN SEEDLINGS
Adapted from Buckner, 1915

Part	P_2O_5	CaO	MgO	K_2O	SiO_2
Beans:					
Exhausted cotyledons	47.2	54.5	45.7	45.1	40.8
Leaves	24.3	10.5	27.3	27.8	23.5
Stems	20.7	22.4	20.8	18.4	15.4
Roots	7.7	13.7	6.1	8.7	19.5
Corn:					
Exhausted cotyledons	43.2	43.9	47.0	20.0	58.6
Leaves	26.4	30.0	28.7	35.7	19.8
Stem	17.4	13.2	16.8	21.2	6.4
Root	12.9	12.9	7.4	23.1	15.1
Potato:					
Exhausted tuber	67.1	42.0	65.7	64.4	12.4
Sprouts	17.8	13.1	15.8	12.7	5.1

Davidson (1926) observed that 18-day corn seedlings had absorbed from the soil eight to ten times the amount of potassium contained in the seed. The amount absorbed during 1 week after the plants had been thinned in the hill was about twice that which had been absorbed during the previous 18 days. The phosphorus content of the seedlings was lower at both stages of growth in the seedling than in the seed. During the first 18 days the amount of nitrogen absorbed by the seedling was about equal to that in the original seed, but during the next 7 days it absorbed twice as much.

Gerdel (1931) working with corn noted that, with a continuous abundance of nitrogen, phosphorus, and potassium throughout the season, marked differentiation of the vegetative and reproductive cycles occurred associated with a narrow silking range, earliness of silking, and higher yield. With a low supply of nitrogen, phosphorus, and potassium throughout the season, he found that an overlapping and mergence of the two growth cycles occurred associated with a wide silking range, lateness of silking, and low yields.

Knowles and Watkin (1931) observed in wheat that the plant attained its maximum quantities of the nutrients in the following order: potassium, calcium, and nitrogen in 7, 5, and 3 weeks, respectively, and phosphorus, carbon, and silicon, 2 weeks before harvest. During the 6 weeks preceding harvest marked losses of calcium, chlorine, and potassium occurred. These losses could not have been caused by loss of parts or by leaching. It was considered that as maturity approaches there is a downward movement of these elements to the roots.

Échevin (1927, 1931) in studying the leaves of the hardwoods in autumn found that the various elements of the ash might increase or decrease prior to leaf fall. He concluded that no general conclusion can be drawn in regard to the movement of the various elements to and from the leaves previous to their fall from the tree. He found in young beech trees that the total amount of phosphorus in the entire plant remained constant throughout the year. In the spring, however, practically 50 per cent of the phosphorus of the stem and root migrated into the leaves. When the leaves had completed their growth this amount of phosphorus gradually returned to the stem and roots.

Davis and Moore (1934) noted in pear leaves that large amounts of calcium accumulated throughout the season so that at the time of leaf fall there was as much as 7.0 mg. per leaf. There was a seasonal increase of magnesium that reached its maximum 3 months prior to leaf fall. The amount of potassium and phosphorus remained more or less constant during the season until after September 20. Between that date and leaf fall the loss of potassium amounted to between 27 and 43 per cent. The amount of phosphorus remained constant until the last 2 or 3 weeks before leaf fall when its loss was approximately 25 per cent. Davis (1934) determined the amount of calcium, magnesium, and potassium in alternate-bearing prune trees. The production of a crop affects the amount of the various elements differently. A crop of fruit reduces the potassium and phosphorus in the wood, bark, and spurs, but in the leaves only the amount of potassium is diminished. There is no difference in the amount of phosphorus in the leaves of bearing and nonbearing trees. The calcium and magnesium content is higher in the leaves of bearing trees than in the leaves of nonbearing ones.

Bibliography

Acton, E. H., 1889: The assimilation of carbon by green plants from certain organic compounds, *Proc. Roy. Soc.*, **47**: 150–175.

Addoms, R. M., and F. C. Mounce, 1931: Notes on the nutrient requirements and the histology of the cranberry (*Vaccinium macrocarpon*) with special reference to mycorrhiza, *Plant Physiol.*, **6**: 653–666.

Agulhon, H., 1910: "Recherches sur la présence et le rôle du bore chez les végétaux," *thèse*, 163 pp., Paris.

Albert, W. B., and W. R. Paden, 1931: Calcium arsenate and unproductiveness in certain soils, *Sci.*, **73**: 622.

Albrecht, W. A., and F. L. Davis, 1929: Physiological importance of calcium in legume inoculation, *Bot. Gaz.*, **88**: 310–321.

Allen, R. C., 1931: Factors influencing the flower color of *Hydrangea, Proc. Am. Soc. Hort. Sci.*, **28**: 410–412.

———, 1934: Controlling the color of greenhouse hydrangeas (*Hydrangea macrophylla*) by soil treatments with aluminum sulfate and other materials, *Proc. Am. Soc. Hort. Sci.*, **32**: 632–634.

Allison, R. V., O. C. Bryan, and J. H. Hunter, 1927: The stimulation of plant response on the raw peat soils of the Florida Everglades through the use of copper sulphate and other chemicals, *Fla. Sta. Bull.* 190.

Allyn, W. P., 1927: The relation of lime to the absorption of iron by plants, *Proc. Ind. Acad. Sci.*, **43**: 405–409; *Exp. Sta. Rec.*, **66**: 121.

Alten, F., and R. Gottwick, 1933: Ein Beitrag zur Frage der Vertretbarkeit des Kaliums durch Rubidium und Caesium für die Pflanzenernährung, *Ernähr. der Pflanze*, **29**: 393–399.

Ames, J. W., and G. E. Boltz, 1912: Nitrogen and mineral constituents of the alfalfa plant, *Ohio Sta. Bull.* **247**: 755–773.

———, 1923: Sulfur experiments, *Ohio Sta. Mo. Bull.* **8**: 85–90.

——— and K. Kitsuta, 1932: Availability of rock phosphate as indicated by phosphorus assimilation of plants, *J. Am. Soc. Agron.*, **24**: 103–123.

——— and ———, 1933: Assimilation of phosphorus and potassium by barley plants grown according to Neubauer procedure and in undiluted soil, *Soil Sci.*, **35**: 197–207.

Anderson, P. J., T. R. Swanback, and O. E. Street, 1932: Potash requirements of the tobacco crop, *Conn. Sta. Bull.*, **334**: 133–217.

Anderssen, F. G., 1932: Chlorosis of deciduous fruit trees due to a copper deficiency, *J. Pom. and Hort. Sci.*, **10**: 130–146.

André, G., and E. Demoussy, 1927: Sur la répartition du potassium et du sodium chez les végétaux, *C. R. Acad. Sci. Paris*, **184**: 1501–1503.

Andrews, J. S., and C. H. Bailey, 1932: Distribution of organic phosphorus in wheat, *Ind. and Eng. Chem.*, **24**: 80–82.

Archibald, J. G., and E. Bennett, 1935: Yield and chemical composition of certain species of grass, *J. Agr. Res.*, **50**: 711–715.

Arendt, R., 1859: Untersuchungen über einige Vorgänge bei der Vegetation der Haferpflanze, *Landw. Versuchs-Stat.*, **1**: 31–66.

Arndt, C. H., 1922: The growth of field corn as affected by iron and aluminum salts, *Am. J. Bot.*, **9**: 47–72.

Arndt, J., 1922: New investigations of the problem of the substitution of potassium in the plant cell by rubidium, as well as the chlorine requirements of buckwheat and barley, *Ernähr. der Pflanze*, **18**: 177–180; 191–194.

Ashby, E., 1929: The interaction of factors in the growth of *Lemna*. IV, The influence of minute quantities of organic matter upon growth and reproduction, *Ann. Bot.*, **43**: 805–816.

Aso, K., 1907: On the continuous application of manganous chloride in rice culture. II, *Col. Agr. Imp. Univ. Tokyo Bull.*, **7**: 449–453.

Averill, H. P., and C. G. King, 1926: The phytin content of food stuffs, *J. Am. Chem. Soc.*, **48**: 724–728.

BAGULEY, A., 1912: The phosphate nutrition of plants, *J. Agr. Sci.*, **4**: 318–322.

BAMFORD, R., 1931: Changes in root tips of wheat and corn grown in nutrient solutions deficient in calcium, *Bull. Torrey Bot. Club*, **58**: 149–179.

BARNETTE, R. M., 1921: The influence of volume upon the rate of change in the hydrogen-ion concentration of nutrient solutions in contact with plant roots and the effect of this change upon iron availability, *N. J. Sta. Rept.*, **1921**: 345–347.

——— and J. W. SHIVE, 1923: The influence of solution volume upon plant growth in relation to reaction change and iron availability in culture solutions, *Soil Sci.*, **15**: 413–425.

——— and J. D. WARNER, 1935: A response of chlorotic corn plants to the application of zinc sulfate to the soil, *Soil Sci.*, **39**: 145–160.

———, J. P. CAMP, J. D. WARNER, and O. E. GALL, 1936: The use of zinc sulphate under corn and other field crops, *Fla. Sta. Bull.* 292.

BARRENSCHEEN, H. K., and W. ALBERS, 1928: Ueber die Rolle der Phosphorylierung im intermediären kohlenhydrat Stoffwechsel der Pflanzen, *Biochem. Z.*, **197**: 261–277.

BARTHOLOMEW, R. P., and G. JANSSEN, 1929: Luxury consumption of potassium by plants and its significance, *J. Am. Soc. Agron.*, **21**: 751–765.

——— and ———, 1929: The relation between concentrations of potassium in culture solutions and optimum plant growth, *Soil Sci.*, **27**: 189–203.

——— and ———, 1931: Rate of absorption of potassium by plants and its possible effect upon the amount of potassium remaining in soils from application of potassium fertilizers, *Ark. Sta. Bull.* 265.

———, 1935: Fluorine, its effect on plant growth and its relation to the availability to plants of phosphorus in phosphate rocks, *Soil Sci.*, **40**: 203–218.

BASSLAVSKAYA, S., and M. SYROESHKINA, 1936: Influence of the chloride ion on the content of chlorophyll in the leaves of potatoes, *Plant Physiol.*, **11**: 149–157.

BAUER, F. C., 1923: The foraging power of plants for phosphate rock, *J. Am. Soc. Agron.*, **15**: 99–110.

BEATH, O. A., H. F. EPPSON, and C. S. GILBERT, 1935: Selenium and other toxic minerals in soils and vegetation, *Wyo. Sta. Bull.* 206.

BEAUMONT, A. B., 1932: Toxicity of sodium nitrate for a species of moss, *Sci.*, **75**: 312–313.

——— and G. M. KARNS, 1932: Effect of an iodine fertilizer on iodine content of a food plant, *Sci.*, **78**: 567.

——— and R. F. CHANDLER, 1933: A statistical study of the effect of potassium fertilizers upon the firmness and keeping quality of fruits, *Proc. Am. Soc. Hort. Sci.*, **30**: 37–44.

——— and M. E. SNELL, 1935: The effect of magnesium deficiency on crop plants, *J. Agr. Res.*, **50**: 553–562.

BENNETT, J. P., and J. OSERKOWSKY, 1933: Copper and iron in the tracheal sap of deciduous trees, *Am. J. Bot.*, **20**: 632–637.

BERTHELOT, M., and G. ANDRÉ, 1895: Sur la présence de l'alumine dans les plantes et sur sa répartition, *C. R. Acad. Sci. Paris*, **120**: 288–290.

BERTRAND, G., 1897: Sur l'intervention du manganèse dans les oxydations provoquées par la laccase, *C. R. Acad. Sci. Paris*, **124**: 1032–1035.

———, 1897: Sur l'action oxydante des sels manganeux et sur la constitution chimique des oxydases, *C. R. Acad. Sci. Paris*, **124**: 1355–1358.

———, 1905: Sur l'emploi favorable du manganèse comme égrais, *C. R. Acad. Sci. Paris*, **141**: 1255–1257.

———, 1912: Sur le rôle des infiniment petits chimiques en agriculture, *Ann. Inst. Pasteur*, **26**: 852–867.

——— and M. Rosenblatt, 1921: Recherches sur la présence du manganèse dans le règne végétal, *Ann. Inst. Pasteur*, **35**: 815–819.

——— and ———, 1921: Sur la présence générale du manganèse dans le règne végétal, *C. R. Acad. Sci. Paris*, **173**: 333–336.

——— and ———, 1922: Sur la répartition du manganèse dans l'organisme des plantes supérieures, *Ann. Inst. Pasteur*, **36**: 230–232.

——— and ———, 1922: Recherches sur la variations de la teneur en manganèse des feuilles avec l'âge, *Ann. Inst. Pasteur*, **36**: 491–501.

——— and ———, 1932: The unequal manganese content in green leaves and etiolated leaves, *C. R. Acad. Sci. Paris*, **194**: 1405–1408.

——— and M. Andreitcheva, 1933: Sur la teneur comparée en zinc des feuilles vertes et des feuilles étiolées, *C. R. Acad. Sci. Paris*, **197**: 1374–1376.

Bishop, E. R., and M. Lawrenz, 1932: Cobalt in plant ash, *Sci.*, **75**: 264–265.

Bishop, W. B. S., 1928: The distribution of manganese in plants and its importance in plant metabolism, *Australian J. Exp. Biol. and Med. Sci.*, **5**: 125–141; *Exp. Sta. Rec.*, **67**: 118.

Blair, A. W., and B. E. Brown, 1921: The influence of fertilizers containing borax on the yield of potatoes and corn—season 1920, *Soil Sci.*, **11**: 369–385.

——— and A. L. Prince, 1932: The influence of phosphates on the phosphoric acid content of the plant, *J. Agr. Res.*, **44**: 579–590.

Blanck, E., F. Giesecke, and W. Heukeshoven, 1933: Vegetationsversuch mit Hafer zur Klärung der Frage nach der Vertretbarkeit des Kaliums durch Rubidium, *J. landw. Wiss.*, **81**: 205–222.

Bobko, E. V., and M. A. Belvoussov, 1933: The importance of boron for the sugar beet, *Ann. Agron.*, **3**: 493–504; *Chem. Abst.*, **27**: 5374.

Boehm, J., 1875: Ueber den vegetabilischen Nährwerth der Kalksalze, *Sitzungsber. Akad. Wiss. Wien*, **71**: 287–384.

Bolin, D. W., 1934: The manganese content of grasses and alfalfa from grazed plots, *J. Agr. Res.*, **48**: 657–663.

Bolin, P., 1916: Sodium chloride as a fertilizer, *Chem. Abst.*, **10**: 2268–2269.

Borrownicka-Ordzywolska, A., 1926: Effect of K and Na salts on metabolism in barley seedlings grown in the dark, *Bull. Int. Acad. Polonaise Sci. et Lett. B. Sci. Nat.*, **1925**: 80–846.

Brandenburg, E., 1931: Die Herz- und Trockenfäule der Rüben als Bormangel-Erscheinung, *Phytopath. Z.*, **3**: 499–517.

Brannon, J. M., 1923: Influence of certain sugars on higher plants, *Bot. Gaz.*, **75**: 370–390.

Breazeale, J. F., 1906: The relation of sodium to potassium in soil and solution cultures, *J. Am. Chem. Soc.*, **28**: 1013–1025.

———, 1916: Effect of sodium salts in water cultures on the absorption of plant food by wheat seedlings, *J. Agr. Res.*, **7**: 407–416.

———, 1923: Nutrition of plants considered as an electrical phenomenon, *J. Agr. Res.*, **24**: 41–54.

Brenchley, W. E., 1910: The influence of copper sulphate and manganese sulphate upon the growth of barley, *Ann. Bot.*, **24**: 571–583.

———, 1914: "Inorganic Plant Poisons and Stimulants," 110 pp., Cambridge University Press.

———, 1914: On the action of certain compounds of zinc, arsenic and boron on the growth of plants, *Ann. Bot.*, **28**: 283–302.

———, 1924: The effect of iodine on soils and plants, *Ann. App. Biol.*, **11**: 86–112.

——— and H. G. Thornton, 1925: The relation between the development, structure and functioning of the nodules on *Vicia faba* as influenced by the presence or absence of boron in the nutrient medium, *Proc. Roy. Soc.*, B, **98**: 373–400.

———, E. J. Maskell, and K. Warington, 1927: The inter-relation between silicon and other elements in plant nutrition, *Ann. App. Biol.*, **14**: 45–82.

——— and K. Warington, 1927: The rôle of boron in the growth of plants, *Ann. Bot.*, **41**: 167–187.

———, 1929: The phosphate requirement of barley at different periods of growth, *Ann. Bot.*, **43**: 89–110.

———, 1932: The action on the growth of crops of small percentages of certain metallic compounds when applied with ordinary artificial fertilizers, *J. Agr. Sci.*, **22**: 704–735.

———, 1934: The effect of rubidium sulphate and palladium chloride on the growth of plants, *Ann. App. Biol.*, **21**: 398–417.

Briggs, G. E., 1922: Experimental researches on vegetable assimilation and respiration. XVI, The characteristics of subnormal photosynthetic activity resulting from deficiency of nutrient salts, *Proc. Roy. Soc.*, B, **94**: 20–35.

Brown, B. A., 1932: The effects of fertilization on the chemical composition of vegetation in pastures, *J. Am. Soc. Agron.*, **24**: 129–145.

Brown, B. E., 1922: Effect of borax in fertilizer on the growth and yield of potatoes, *U. S. Dept. Agr. Bull.* 998.

———, V. C. Houghland, O. Smith, and R. L. Carolus, 1933: The influence of magnesium on different potato soil types, *Am. Potato J.*, **10**: 55–65.

Brown, J. W., 1927: Chemical studies in the physiology of apples. VI, Correlation in the individual apple between the mineral constituents and other properties, *Ann. Bot.*, **41**: 127–139.

———, 1929: Chemical studies in the physiology of apples. XI, The relation between the mineral constitution of apples and the soil on which they are grown, *Ann. Bot.*, **43**: 817–831.

Brown, P. E., and G. A. Minges, 1916: The effect of some manganese salts on ammonification and nitrification, *Soil Sci.*, **2**: 67–87.

Bruce, O. C., 1925: The relation of sulfur to alfalfa production, *J. Agr. Res.*, **30**: 937–949.

Bruch, P., 1902: Zur physiologischen Bedeutung des Kalciums in der Pflanze, *Landw. Jahrb.*, **30**: 127–144.

Bruno, A., 1928: Hypothèses nouvelles sur le mode d'action de potasse, *Rev. Viticulture*, **69**: 60–62; *Biol. Abst.*, **5**: 1623.

Bryan, O. C., 1929: The stimulating effect of external applications of copper and manganese on certain chlorotic plants of the Florida Everglades, *J. Am. Soc. Agron.*, **21**: 923–933.

——— and R. B. Becker, 1935: The mineral content of soil types as related to "salt sick" of cattle, *J. Am. Soc. Agron.*, **27**: 120–127.

Buckner, G. D., 1915: Translocation of mineral constituents of seeds and tubers of certain plants during growth, *J. Agr. Res.*, **5**: 449–459.

———, 1919: The translocation of the mineral constituents of the jack bean, *J. Am. Chem. Soc.*, **41**: 282–287.

———, 1919: The composition of the ash of crab grass (*Digitaria sanguinalis*) as affected by the soil in which it is grown, *J. Am. Chem. Soc.*, **41**: 1384–1385.

Burd, J. S., 1919: Rate of absorption of soil constituents at successive stages of plant growth, *J. Agr. Res.*, **18**: 51–72.

BURGE, W. E., G. C. WICKWIRE, and O. S. ORTH, 1933: The stimulating effect of copper on chlorophyll formation, *abst. Am. J. Bot.*, **20**: 669.

BURKE, E., 1932: Chlorosis of trees, *Plant Physiol.*, **7**: 329–334.

BYERS, H. G., 1935: Selenium occurrence in certain soils in the United States with a discussion of related topics, *U. S. Dept. Agr. Tech. Bull.* 482.

——— and H. G. KNIGHT, 1935: Selenium in soils, *Ind. and Eng. Chem.*, **27**: 902–904.

CANALS, E., J. CANAYE, and M. E. CABANES, 1930: Recherches physiochimiques sur les sucs végétaux, *Bull. Soc. Chim. Biol.*, **12**: 1022–1024; *Biol. Abst.*, **5**: 2654.

CARMIN, J., 1931: Growth and variability of wheat seedlings in magnesium sulphate solutions, *Bull. Torrey Bot. Club*, **58**: 179–190.

CAROLUS, R. L., 1933: Tomato fertilization. II, The effect of different fertilizer ratios on the chemical composition of tomatoes, *Va. Truck Exp. Sta. Bull.* 81.

———, 1935: Experiences with rapid chemical tests for the determination of nutrient deficiencies in vegetable crops, *Proc. Am. Soc. Hort. Sci.*, **33**: 579–583.

———, 1935: The relation of potassium, calcium, and sodium to magnesium deficiency, *Proc. Am. Soc. Hort. Sci.*, **33**: 595–599.

CATTLE, M., 1933: Studies of the physiological importance of the mineral elements in plants. V, The distribution of diastase, invertase and catalase in normal and potassium-starved bean plants, *New Phytol.*, **32**: 364–381.

CHANDLER, W. H., D. R. HOAGLAND, and P. L. HIBBARD, 1933: Little-leaf or rosette of fruit trees. III, *Proc. Am. Soc. Hort. Sci.*, **30**: 70–86.

CHAPMAN, G. W., 1931: The relation of iron and manganese to chlorosis in plants, *New Phytol.*, **30**: 266–283.

CHAPMAN, R. E., 1934: The composition of the salts of the leaves of some desert plants, *Ann. Bot.*, **48**: 777–780.

CHATIN, A., 1850: Existence de l'iode dans les plantes d'eau douce. Conséquences de ce fait pour la géognosie, la physiologie végétale, la thérapeutique et peut-être pour l'industrie, *C. R. Acad. Sci. Paris*, **30**: 352–354.

CHATTERJEE, U. N., 1933: Studies on the effect of phosphates on respiration of green leaves. 1, *Eugenia jambolana;* 2, *Allium tuberosum, Bull. Acad. Sci. United Prov. Agra and Oudh, Allahabad, India,* **2**: 179–192; *Biol. Abst.*, **9**: 1013.

CHIZHOV, B. A., 1926: The accumulation of dry matter, ash and nitrogenous substances in cultivated plants under field conditions, *J. Exp. Landw. Südosten Russ.*, **3**: 40–75.

CHOUARD, P., 1933: Preliminary studies of bluing and chlorosis of *Hydrangea opuloides*, *Bull. Mens. Natl. Hort. France*, **5**: 289–291. Quoted by Poesch, 1935.

CLARK, E. R., and H. K. WILSON, 1933: Lodging in small grains, *J. Am. Soc. Agron.*, **25**: 561–572.

CLARK, N. A., and C. L. FLY, 1930: The role of manganese in the nutrition of *Lemna*, *Plant Physiol.*, **5**: 241–248.

———, 1933: Manganese and the growth of *Lemna*, *Plant Physiol.*, **8**: 157–161.

COCKEFAIR, E. A., 1931: The role of phosphorus in the metabolism of plants, *Am. J. Bot.*, **18**: 582–598.

COLBY, H. L., 1933: Effect of starvation on distribution of mineral nutrients in French prune trees grown in solution cultures, *Plant Physiol.*, **8**: 357–394.

COLLA, S., 1929: Untersuchungen über Plasma und Plasmaströmung bei *Characeen*. II, Die Wirkung verschiedener Salze auf die Plasmaströmung, *Protoplasma*, **6**: 438–448.

COLLINGS, G. H., 1927: The influence of boron on the growth of the soybean plant, *Soil Sci.*, **23**: 83–107.

CONNER, S. D., 1920: The effect of zinc in soil tests with zinc and galvanized iron pots, *J. Am. Soc Agron.*, **12**: 61–65.

——— and O. H. Sears, 1922: Aluminum salts and acids at varying hydrogen-ion concentrations, in relation to plant growth in water cultures, *Soil Sci.*, **13**: 23–43.

———, 1932: Factors affecting manganese availability in soils, *J. Am. Soc. Agron.*, **24**: 726–733.

Conner, W. H., 1931: Study of the iodine content of Florida grown crops, *Fla. Sta. Ann. Rept.*, **45**: 65.

Connors, C. H., 1923: Color control in forcing *Hydrangeas*, *Proc. Am. Soc. Hort. Sci.*, **20**: 282–288.

Cook, F. C., 1916: Boron: its absorption and distribution in plants and its effect on growth, *J. Agr. Res.*, **5**: 877–891.

——— and J. B. Wilson, 1917: Effect of three annual applications of boron on wheat, *J. Agr. Res.*, **10**: 591–599.

——— and ———, 1918: Boron: its effect on crops and its distribution in plants and soil in different parts of the United States, *J. Agr. Res.*, **13**: 451–471.

Corson, G. E., and A. L. Bakke, 1917: The use of iron in nutrient solution for plants, *Proc. Iowa Acad. Sci.*, **24**: 477–482.

Cotton, M., 1930: Toxic effects of iodine and nickel on buckwheat grown in solution cultures, *Bull. Torrey Bot. Club*, **57**: 127–141.

Coville, F. V., 1923: The effect of aluminum sulphate on *Rhododendron* seedlings. *Am. Hort. Soc. Bull.* **1**: 6.

Cowell, S. J., 1932: A note on the calcium content of cabbage, *Biochem. J.*, **26**: 1422–1423.

Crocker, W., 1923: The necessity of sulfur carriers in artificial fertilizers, *J. Am. Soc. Agron.*, **15**: 129–141.

Dadswell, I. W., 1935: Some observations on the inorganic composition of wheat grown in eastern Australia, *Australian J. Exp. Biol. and Med. Sci.*, **13**: 33–42.

Damiens, A., and S. Blaignan, 1931: Sur le brome normal (règne végétal) graines comestible, blé, pain, *C. R. Acad. Sci. Paris*, **193**: 1460–1462.

——— and ———, 1932: Sur le brome normal (règne végétal) plantes et fruits comestible, *C. R. Acad. Sci. Paris*, **194**: 2077–2080.

Daniel, H. A., 1934: The calcium, phosphorus, and nitrogen content of grasses and legumes and the relation of these elements in the plant, *J. Am. Soc. Agron.*, **26**: 496–503.

——— and H. J. Harper, 1934: The relation between total calcium and phosphorus in mature prairie grass and available plant food in soil, *J. Am. Soc. Agron.*, **26**: 981–986.

———, 1935: The magnesium content of grasses and legumes and the ratios between this element and the total calcium, phosphorus and nitrogen in these plants, *J. Am. Soc. Agron.*, **27**: 922–927.

Dastur, R. H., and R. E. Cooper, 1932: The osmotic and suction pressures of the rice plant, *Oryza sativa*, when treated with different salts: a method of determining the salt requirements of plants, *Indian J. Agr. Sci.*, **2**: 99–137; *Chem. Abst.*, **26**: 5122.

Davidson, J., and J. A. LeClerc, 1923: Effect of various inorganic nitrogen compounds, applied at different stages of growth on the yield, composition, and quality of wheat, *J. Agr. Res.*, **23**: 55–69.

———, 1926: Removal of plant food in thinning corn, *J. Am. Soc. Agron.*, **18**: 962–967.

——— and C. E. Chambliss, 1932: Chemical composition of rice and its relation to soil fertility in China and Japan, *Sci.*, **75**: 294.

Davis, A. R., D. R. Hoagland, and C. B. Lipman, 1923: The feeding power of plants, *Sci.*, **57**: 299–301.

Davis, L. D., 1934: Some ash constituents of alternate-bearing sugar prune trees, *Proc. Am. Soc. Hort. Sci.*, **32**: 125–130.

——— and N. P. Moore, 1934: Seasonal changes in Bartlett pear leaves, *Proc. Am. Soc. Hort. Sci.*, **32**: 131–137.

Davis, M. B., 1930: Investigations on the nutrition of fruit trees. Some effects of deficiencies of nitrogen, potassium, calcium and magnesium with special reference to the behavior of certain varieties of apple trees, *J. Pom. and Hort. Sci*, **8**: 316–344.

———, H. Hill, and F. B. Johnson, 1934: Nutritional studies with *Fragaria*. II, *Sci. Agr.*, **14**: 411–432.

Day, D., 1928: Some effects on *Pisum sativum* of a lack of calcium in the nutrient solution, *Sci.*, **68**: 426.

———, 1929: Some effects of calcium deficiency on *Pisum sativum*, *Plant Physiol.*, **4**: 463–506.

———, 1935: Some chemical aspects of calcium deficiency effects on *Pisum sativum*, *Plant Physiol.*, **10**: 811–816.

Demolon, M. A., 1912: Sur l'action fertilisante du soufre, *C. R. Acad. Sci. Paris*, **154**: 524–526.

Deniges, G., 1920: Réaction de coloration extrêmement sensible des phosphates et des arséniates. Ses applications, *C. R. Acad. Sci. Paris*, **171**: 802–804.

De Peralta, F., 1927: Influence upon the development of young rice plants of sodium chloride added to a complete solution, *Philippine Agr.*, **15**: 471–479; *Exp. Sta. Rec.*, **61**: 719.

De Turk, E. E., J. R. Holbert, and B. W. Howk, 1933: Chemical transformation of phosphorus in the growing corn plant with results on two first generation crosses, *J. Agr. Res.*, **46**: 121–142.

Deuber, C. G., 1926: Can a pyrrole derivative be substituted for iron in the growth of plants? *Am. J. Bot.*, **13**: 276–285.

———, 1926: Potassium ferrocyanide and ferric ferrocyanide as sources of iron for plants, *Soil Sci.*, **21**: 23–27.

Dhein, A., 1929: Untersuchungen über Art und Wesen einiger Kalisalzwirkungen, *Landw. Jahrb.*, **70**: 255–309.

Dickson, J. G., 1918: The value of certain nutritive elements in the development of the oats plant, *Am. J. Bot.*, **5**: 301–325.

Domontovitsch, M. K., and P. A. Schelesnow, 1930: Feeding plants via the leaves with salts of potassium and of magnesium, *J. landw. Wiss.*, **7**: 134–143; *Exp. Sta. Rec.*, **67**: 22.

Dove, W. F., 1934: A study of the cause of nutritional deficiency diseases in the livestock and inhabitants of Maine with possible corrective methods, etc., *Maine Sta. Bull.* 375.

Dowding, E. S., 1925: The regional and seasonal distribution of potassium in plant tissues, *Ann. Bot.*, **39**: 459–474.

Duley, F. L., 1916: Relation of sulphur to soil productivity, *J. Am. Soc. Agron.*, **8**: 154–160.

——— and M. F. Miller, 1921: The effect of a varying supply of nutrients upon the character and composition of the maize plant at different periods of growth, *Mo. Sta. Tech. Bull.* 42.

Dymond, T. S., F. Hughes, and C. W. C. Jupe, 1905: The influence of sulphates as manure upon the yield and the feeding value of crops, *J. Agr. Sci.*, **1**: 217–229.

Eaton, F. M., 1930: The effect of boron on powdery mildew and spot blotch of barley, *Phytopath.*, **20**: 967–973.

———, 1932: Boron requirements of cotton, *Soil Sci.*, **34**: 301–306.

——— and G. Y. Blair, 1935: Accumulation of boron by reciprocally grafted plants, *Plant Physiol.*, **10**: 411–424.

Eaton, S. V., 1922: Sulphur content of soils and its relation to plant nutrition, *Bot. Gaz.*, **74**: 32–59.

———, 1935: Influence of sulphur deficiency on the metabolism of the soybean, *Bot. Gaz.*, **97**: 68–100.

Échevin, R., 1927: Variations de la matière minérale des feuilles au cours du jaunissement automnal, *Rev. Gén. Bot.*, **39**: 405–419; 488–514; *Biol. Abst.*, **5**: 2660.

———, 1931: Nitrogen, phosphorus and sulfur in woody plants with caducous leaves, *Rev. Gén. Bot.*, **43**: 517–677; *Chem. Abst.*, **26**: 1318.

Eckerson, S., 1929: Influence of phosphorus deficiency on the metabolism of tomato plants, *abst. Am. J. Bot.*, **16**: 852.

———, 1931: Influence of phosphorus deficiency on metabolism of the tomato (*Lycopersicum esculentum*), *Contr. Boyce Thompson Inst.*, **3**: 197–217.

Eidelman, Z. M., 1930: The influence of phosphorus and of day length on physiological functions in plants, *J. landw. Wiss.*, **7**: 387–402; *Exp. Sta. Rec.*, **67**: 116–117.

Eisenmenger, W. S., 1935: Toxicity of aluminum salts on tobacco plants, *J. Agr. Res.*, **51**: 917–924.

———, 1935: Toxicity of aluminum on seedlings and action of certain ions in the elimination of the toxic effects, *Plant Physiol.*, **10**: 1–25.

Emmert, E. M., 1930: A method for the rapid determination of phosphate in fresh plant tissue, *Plant Physiol.*, **5**: 411–417.

———, 1932: Field method for estimating nitrate, phosphate, and potassium in plants, *Plant Physiol.*, **7**: 315–321.

——— and F. K. Ball, 1933: The effect of soil moisture on the availability of nitrate, phosphate and potassium to the tomato plant, *Soil Sci.*, **35**: 295–307.

———, 1934: Tests for phosphate, nitrate and soluble nitrogen in conducting tissue of tomato and lettuce plants, as indicators of availability and yield, *Ky. Sta. Circ.* 43.

———, 1934: Tests for nutrients in conducting tissue as indicators of the nutritional status in horticultural crops, *Proc. Am. Soc. Hort. Sci.*, **32**: 604–609.

———, 1935: New methods for the determination of the availability of nitrogen and phosphorus to plants, *J. Am. Soc. Agron.*, **27**: 1–7.

Fagundes, A. B., 1934: O boro em nutriçâo vegetal, *Arq. Inst. Bio. Veg.* (*Rio de Janeiro*), **1**: 3–26; *Biol. Abst.*, **9**: 1234.

Felix, E. L., 1927: Correction of unproductive muck by the addition of copper, *Phytopath.*, **17**: 49–50.

Finch, A. H., D. W. Albert, and A. F. Kinnison, 1933: A chlorotic condition of plants in Arizona related to iron deficiency, *Proc. Am. Soc. Hort. Sci.*, **30**: 431–434.

———, ———, and ———, 1934: Progress of the control of citrus chlorosis or decline, *Proc. Am. Soc. Hort. Sci.*, **32**: 20–23.

Fisher, P. L., 1935: Responses of the tomato in solution cultures with deficiencies and excesses of certain essential elements, *Md. Sta. Bull.* 375.

Fluri, M., 1909: Der Einfluss von Aluminiumsalzen auf das Protoplasma, *Flora*, **99**: 81–126.

Fonder, J. F., 1929: Variations in potassium content of alfalfa due to stage of growth and soil type and the relationship of potassium and calcium in plants grown upon different soil types, *J. Am. Soc. Agron.*, **21**: 732–750.

Forbes, R. H., 1917: Certain effects under irrigation of copper compounds upon crops, *Univ. Cal. Agr. Sci.*, **1**: 395–494.

Fraps, G. S., 1931: How reliable are existing chemical methods for determining soil deficiencies in ash constituents of plants? *J. Am. Chem. Soc.*, **23**: 337–352.

Frazer, J., 1935: Sulphur content of plants, *Plant Physiol.*, **10**: 529–535.

Frear, D. E. H., 1934: A study of the iodine content of Pennsylvania potatoes, *J. Agr. Res.*, **48**: 171–181.

Frey-Wyssling, A., 1930: Über die Ausscheidung der Kieselsäure in der Pflanze, *Ber. bot. Ges.*, **48**: 179–183.

———, 1930: Vergleich zwischen der Ausscheidung von Kieselsäure und Kalziumsalzen in der Pflanze, *Ber. bot. Ges.*, **48**: 184–191.

Funchess, M. J., 1916: The effect of certain organic compounds on plant growth, *Ala. Sta. Bull.* 191.

Furnstal, A. H., and B. Johnson, 1936: Preparation of sintered pyrex glass aerators for use in water culture experiments with plants, *Plant Physiol.*, **11**: 189–194.

Garner, W. W., J. E. McMurtrey, C. W. Bacon, and E. G. Moss, 1923: Sand drown, a chlorosis of tobacco due to magnesium deficiency, and the relation of sulphates and chlorids of potassium to the disease, *J. Agr. Res.*, **23**: 27–41.

———, ———, J. D. Bowling, and E. G. Moss, 1930: Magnesium and calcium requirements of the tobacco crop, *J. Agr. Res.*, **40**: 145–169.

———, ———, ———, and ———, 1930: Role of chlorine in nutrition and growth of the tobacco plant and its effect on the quality of the cured leaf, *J. Agr. Res.*, **40**: 627–649.

———, C. W. Bacon, J. D. Bowling, and D. E. Brown, 1934: The nitrogen nutrition of tobacco, *U. S. Dept. Agr. Tech. Bull.* 414.

Gassner, G., 1910: Beobachtungen und Versuche über den Anbau und die Entwicklung von Getreidepflanzen in subtropischem Klima, *Jahres. Ver. Angew. Bot.*, **8**: 95–163. Quoted by Plitt, 1935.

——— and G. Goeze, 1932: The influence of potash nutrition on the rate of assimilation of wheat leaves, *Ber. bot. Ges.*, **50a**: 412–482.

Gautier, A., and P. Clausman, 1919: Action des fluorures sur la végétation, *C. R. Acad. Sci. Paris*, **169**: 115–122.

Geigel, A. R., 1935: Effect of boron on the growth of certain green plants, *J. Agr. Univ. Puerto Rico*, **19**: 5–28; *Biol. Abst.*, **10**: 312.

Gerdel, R. W., 1931: Effect of fertilizer and date of planting on the physiological development of the corn plant, *Plant Physiol.*, **6**: 695–714.

Gericke, W. F., 1922: "Magnesia injury" of plants grown in nutrient solutions, *Bot. Gaz.*, **74**: 110–113.

———, 1924: The beneficial effect to plant growth of the temporary depletion of some of the essential elements in the soil, *Sci.*, **59**: 321–324.

———, 1924: The beneficial effect to wheat growth due to depletion of available phosphorus in the cultural media, *Sci.*, **60**: 297–298.

———, 1925: Salt requirements of wheat at different growth phases, *Bot. Gaz.*, **80**: 410–425.

———, 1925: Effect of light on availability of iron to wheat plants in water cultures, *Bot. Gaz.*, **79**: 106–109.

Germar, B., 1934: Some functions of silicic acid in cereals with special reference to resistance to mildew, *Pflanzenernähr. Düng. u. Bodenk.* **A35**: 102–115; *Chem. Abst.*, **29**: 1461.

Gilbert, B. E., 1926: The adaptation of certain colorimetric methods to the estimation of nitrates, phosphates, and potassium in plant solutions, *Plant Physiol.*, **1**: 191–199.

——— and L. J. Hardin, 1927: The current mineral nutrient content of the plant solution as a possible means of chemical control of optimum fertilization, *J. Agr. Res.*, **35**: 185–192.

———, F. T. McLean, and W. L. Adams, 1927: The current mineral nutrient content of the plant solution as an index of metabolic limiting conditions, *Plant Physiol.*, **2**: 139–153.

——— and J. B. Smith, 1929: Nitrates in soil and plant as indexes of the nitrogen needs of a growing crop, *Soil Sci.*, **27**: 459–468.

——— and F. R. Pember, 1931: The sensitivity of red clover (*Trifolium pratense*) to small amounts of boron and manganese, *Plant Physiol.*, **6**: 727–729.

——— and ———, 1931: Further evidence concerning the toxic action of aluminum in connection with plant growth, *Soil Sci.*, **31**: 267–273.

———, 1934: Normal crops and the supply of available soil manganese, *R. I. Sta. Bull.* 246.

——— and F. R. Pember, 1935: Tolerance of certain weeds and grasses to toxic aluminum, *Soil Sci.*, **39**: 425–430.

——— and ———, 1935: The use of seedling plants to determine soil nutrient deficiencies, *R. I. 47th Ann. Sta. Rept.*, 80–83.

Gildehaus, E. J., 1931: The relation of nitrogen to potassium in the nutrition of fruit trees, *Bot. Gaz.*, **92**: 384–395.

Gile, P. L., 1911: Relation of calcareous soils to pineapple chlorosis, *Porto Rico Sta. Bull.* 11.

——— and J. O. Carrero, 1914: Assimilation of colloidal iron by rice, *J. Agr. Res.*, **3**: 205–210.

——— and ———, 1915: Ash composition of upland rice at various stages of growth, *J. Agr. Res.*, **5**: 357–364.

——— and ———, 1916: Assimilation of iron by rice from certain nutrient solutions, *J. Agr. Res.*, **7**: 503–528.

——— and ———, 1916: Immobility of iron in the plant, *J. Agr. Res.*, **7**: 83–87.

——— and ———, 1917: Absorption of nutrients as affected by the number of roots supplied with the nutrient, *J. Agr. Res.*, **9**: 73–95.

——— and ———, 1920: Cause of lime-induced chlorosis and availability of iron in the soil, *J. Agr. Res.*, **20**: 33–63.

——— and ———, 1921: Assimilation of nitrogen, phosphorus and potassium by corn when nutrient salts are confined to different roots, *J. Agr. Res.*, **21**: 545–575.

——— and J. G. Smith, 1925: Colloidal silica and the efficiency of phosphates, *J. Agr. Res.*, **31**: 247–260.

Gines, F. G., 1930: Relative effects of different iron salts upon growth and development of young rice plants, *Philippine Agr.*, **19**: 43–52; *Exp. Sta. Rec.*, **67**: 23.

Ginsburg, J. M., 1925: Composition and appearance of soybean plants grown in cultural solutions each lacking a different essential element, *Soil Sci.*, **20**: 1–15.

Godden, W., 1926: Report of the effect of fertilizers on the mineral content of pastures, *J. Agr. Sci.*, **16**: 98–104.

Gössl, J. V., 1905: Ueber das Vorkommen des Mangans in der Pflanze und über seinen Einfluss auf Schimmelpilze, *Beih. bot. Cent.*, I, **18**: 119–132.

Grafe, V., and L. Portheim, 1906: Untersuchungen über die Rolle des Kalkes in der Pflanze, *Sitzungsber. Akad. Wiss. Wien*, **18**: 324.

Greaves, J. E., and D. H. Nelson, 1925: The iron, chlorine and sulfur contents of grains and the influence of irrigation water upon them, *Soil Sci.*, **19**: 325–330.

——— and C. T. Hirst, 1929: The mineral content of grains, *Utah Sta. Bull.* 210.

Greenhill, A. W., 1930: The availability of phosphatic fertilizers as shown by an examination of the soil solution and of plant growth, *J. Agr. Sci.*, **20**: 559–572.

Gregory, F. G., and F. Crowther, 1928: A physiological study of varietal differences in plants. I, A study of the comparative yields of barley varieties with different manurings, *Ann. Bot.*, **42**: 757–770.

——— and ———, 1931: A physiological study of varietal differences in plants. II, Further evidence for the differential response in yield of barley varieties to manurial deficiencies, *Ann. Bot.*, **45**: 579–592.

Gris, E., 1844: Nouvelles expériences sur l'action des composés ferrugineux solubles appliqués à la végétation et spécialement en traitement de la chlorose et de la débilité des plantes, *C. R. Acad. Sci. Paris*, **19**: 1118–1119.

Grizzard, A. L., 1935: Effects of soil type and soil treatments on the chemical composition of alfalfa plants, *J. Am. Soc. Agron.*, **27**: 81–99.

Groom, P., 1896: Preliminary note on the relation between calcium and the conduction of carbohydrates in plants, *Ann. Bot.*, **10**: 91–96.

Haas, A. R. C., and H. S. Reed, 1927: Significance of traces of elements not ordinarily added to culture solutions for growth of young orange trees, *Bot. Gaz.*, **83**: 77–85.

———, 1928: Water solubility of dry matter in relation to calcium nutrition of normal orange and lemon leaves, *Bot. Gaz.*, **85**: 334–341.

———, 1929: Toxic effect of boron on fruit trees, *Bot. Gaz.*, **88**: 111–131.

———, 1930: Boron as an essential element for healthy growth of citrus, *Bot. Gaz.*, **89**: 410–413.

——— and L. J. Klotz, 1931: Some anatomical and physiological changes in citrus produced by boron deficiency, *Hilgardia*, **5**: 175–190.

——— and ———, 1931: Further evidence of the necessity of boron for health in citrus, *Bot. Gaz.*, **92**: 94–100.

———, 1932: Injurious effects of manganese and iron deficiencies on the growth of citrus, *Hilgardia*, **7**: 181–206.

———, 1933: Walnut-yellows in relation to ash composition, manganese, iron and other ash constituents, *Bot. Gaz.*, **94**: 495–511.

——— and H. J. Quayle, 1935: Copper content of citrus leaves and fruit in relation to exanthema and fumigation injury, *Hilgardia*, **9**: 143–177.

———, 1936: Phosphorus content of citrus and factors affecting it, *Soil Sci.*, **41**: 239–258.

———, 1936: Phosphorus relations of lemon cuttings grown in solution cultures, *Bot. Gaz.*, **97**: 794–807.

Haley, D. E., J. B. Longenecker, and O. Olson, 1931: Composition and quality of Pennsylvania cigar-leaf tobacco as related to fertilizer treatment, *Plant Physiol.*, **6**: 177–182.

Hall, A. D., and C. G. T. Morison, 1905: On the function of silica in the nutrition of cereals. I, *Proc. Roy. Soc.*, B, **77**: 455–477.

Hall, D. M., 1934: The relationship between certain morphological characters and lodging in corn, *Minn. Sta. Tech. Bull.* 103.

Hall, E. H., 1922: Sulfur and nitrogen content of alfalfa grown under various conditions, *Bot. Gaz.*, **73**: 401–412.

Halma, F. F., and A. R. C. Haas, 1931: Solubility changes of inorganic constituents in citrus cuttings, *Bot. Gaz.*, **91**: 213–218.

Harmer, P. M., 1932: Muck soil management for onion production, *Mich. Sta. Ext. Bull.*, **123**: 1–23.

Harper, H. J., and H. A. Daniel, 1934: Chemical composition of certain aquatic plants, *Bot. Gaz.*, **96**: 186–189.

———, ———, and H. F. Murphy, 1934: The total nitrogen, phosphorus, and calcium content of common weeds and native grasses in Oklahoma, *Proc. Okla. Acad. Sci.*, **14**: 36–44.

Harrison, G. T., and C. J. King, 1925: Age of seedlings as a factor in the resistance of maize to sodium chloride, *J. Agr. Res.*, **31**: 633–641.

HART, E. B., and W. H. PETERSON, 1911: Sulphur requirements of farm crops in relation to the soil and air supply, *Wis. Sta. Res. Bull.* 14.

——— and W. E. TOTTINGHAM, 1915: Relation of sulfur compounds to plant nutrition, *J. Agr. Res.*, **5**: 233–251.

HARTER, L. L., 1908: The influence of a mixture of soluble salts, principally sodium chloride, upon the leaf structure and transpiration of wheat, oats, and barley, *U. S. Plant. Ind. Bull.* 134.

HARTT, C. E., 1929: Potassium deficiency in sugar cane, *Bot. Gaz.*, **88**: 229–261.

———, 1934: Some effects of potassium upon the growth of sugar cane and upon the absorption and migration of ash constituents, *Plant Physiol.*, **9**: 399–451.

———, 1934: Some effects of potassium upon the amounts of protein and amino forms of nitrogen, sugars and enzyme activity of sugar cane, *Plant Physiol.*, **9**: 452–490.

HARTWELL, B. L., H. J. WHEELER, and F. R. PEMBER, 1907: The effect of the addition of sodium to deficient amounts of potassium upon the growth of plants in both water and sand cultures, *R. I. Sta. Rept.*, **20**: 299–357.

——— and F. R. PEMBER, 1908: The relative toxicity of ferrous sulfate to barley and rye seedlings, *R. I. Sta. 21st Rept.*, 286–294.

——— and F. S. HAMMETT, 1911: The effect of phosphorus manuring on the amount of inorganic phosphorus in flat turnip roots, *J. Ind. Eng. Chem.*, **3**: 831–832.

———, 1913: The percentage of total phosphorus in flat turnips as influenced by the amount available in soils, *R. I. Sta. Bull.* 154.

——— and P. H. WESSELS, 1913: Effect of sodium manuring on the composition of plants, *R. I. Sta. Bull.* 153.

———, 1916: Starch congestion accompanying certain factors which retard plant growth, *R. I. Sta. Bull.* 165.

———, F. S. HAMMETT, and P. H. WESSELS, 1917: Reactions of the phosphorus of the thickened root of the flat turnip, *J. Agr. Res.*, **11**: 359–371.

——— and F. R. PEMBER, 1918: The presence of aluminum as a reason for the difference in the effect of so-called acid soil on barley and rye, *Soil Sci.*, **6**: 259–281.

——— and ———, 1918: Aluminum as a factor influencing the effect of acid soils on different crops, *J. Am. Soc. Agron.*, **10**: 45–47.

———, 1927: Relative crop response to potash, *J. Am. Soc. Agron.*, **19**: 479–482.

——— and S. C. DAMON, 1927: The degree of response of different crops to various phosphorus carriers, *R. I. Sta. Bull.* 209.

HASELHOFF, E., 1913: Über die Einwirkung von Borverbindungen auf das Pflanzenwachstum, *Landw. Versuchs-Stat.*, **79–80**: 399–429.

HEADDEN, W. P., 1915: Occurrence of manganese in wheat, *J. Agr. Res.*, **5**: 349–355.

———, 1921: Titanium, barium, strontium, and lithium in certain plants, *Colo. Sta. Bull.* 267.

HEATH, H. C., 1930: Potash shale as a source of potassium for growing plants, *Bot. Gaz.*, **90**: 121–151.

HECK, A. F., 1934: Some indications of a relation of soil fertility and plant nutrition to cane diseases in Hawaii, *J. Am. Soc. Agron.*, **26**: 381–389.

HELLER, K., and C. L. WAGNER, 1931: Zur Anreicherung von Kaliumisotopen in Pflanzen, *Z. Anorgan. u. Allgem. Chem.*, **200**: 105–112; *Biol. Abst.*, **6**: 2206.

HELLRIEGEL, H., and H. WILFARTH, 1898: Vegetationsversuche über den Kalibedarf einiger Pflanzen, *Arb. landw. Ges.*, **34**: 1–101.

HENDRY, G. W., 1918: Relative effect of sodium chloride on the development of certain legumes, *J. Am. Soc. Agron.*, **10**: 246–250.

HESTER, J. B., 1934: The influence of soil acidity and soil type upon the growth and composition of the lima bean plant, *Proc. Am. Soc. Hort. Sci.*, **32**: 600–603.

HIBBARD, R. P., and S. GERSHBERG, 1924: The salt requirements of Marquis wheat in water culture for the vegetative phase of development, *Mich. Sta. Tech. Bull.* 64.

——— and B. H. GRIGSBY, 1934: Relation of light, potassium and calcium deficiencies to photosynthesis, protein synthesis and translocation, *Mich. Sta. Tech. Bull.* 141.

HILL, H., and E. P. GRANT, 1935: The growth of turnips in artificial cultures, *Sci. Agr.*, **15**: 652–659.

HILTNER, E., 1928: Ist eine wesentlich Jodanreicherung in Pflanzen möglich? *Fortschr. Landw.*, **3**: 1–4.

HOAGLAND, D. R., 1919: Relation of nutrient solution to composition and reaction of cell sap of barley, *Bot. Gaz.*, **68**: 297–304.

———, 1926: Some phases on the inorganic nutrition of plants in relation to the soil solution, *Sci. Agr.*, **6**(**5**): 141–151, 177–189.

——— and J. C. MARTIN, 1933: Absorption of potassium by plants in relation to replaceable, non-replaceable and soil solution potassium, *Soil Sci.*, **36**: 1–34.

——— and W. C. SNYDER, 1933: Nutrition of strawberry plant under controlled conditions: (*a*) effects of deficiencies of boron and certain other elements; (*b*) susceptibility to injury from sodium salts, *Proc. Am. Soc. Hort. Sci.*, **30**: 288–296.

———, W. H. CHANDLER, and P. L. HIBBARD, 1935: Little-leaf or rosette of fruit trees. V, Effect of zinc on the growth of plants of various types in controlled soil and water culture experiments, *Proc. Am. Soc. Hort. Sci.*, **33**: 131–141.

HOBART, F. G., 1924: Calcium in plant metabolism, *Pharm. J.*, **113**: 500; *Bot. Abst.*, **14**: 569.

HOFFER, G. N., and R. H. CARR, 1920: Iron accumulation and mobility in diseased corn stalks, *Phytopath.*, **10**: 56.

——— and ———, 1923: Accumulation of aluminum and iron compounds in corn plants and its probable relation to root rots, *J. Agr. Res.*, **23**: 801–825.

——— and J. F. TROST, 1923: The accumulation of iron and aluminum compounds in corn plants and its probable relation to root rots, *J. Am. Soc. Agron.*, **15**: 323–331.

———, 1926: Testing corn stalks chemically to aid in determining their plant food needs, *Ind. Sta. Bull.* 298.

HOLYNSKI, S., 1929: Assimilation of phosphorus and of potassium by sprouts and by young seedlings of rye, *Mém. Inst. Natl. Polon. Econ. Rurale Pulawy.*, **10**: 163–174; *Exp. Sta. Rec.*, **67**: 118.

HOPKINS, E. F., and F. B. WANN, 1926: Relation of hydrogen ion concentration to growth of *Chlorella* and of the availability of iron, *Bot. Gaz.*, **81**: 353–376.

——— and ———, 1927: Iron requirement for *Chlorella*, *Bot. Gaz.*, **84**: 407–427.

———, 1930: The necessity and function of manganese in the growth of *Chlorella sp.*, *Sci.*, **72**: 609–610.

———, 1930: Manganese as an essential element for a green alga, *abst. Am. J. Bot.*, **17**: 1047.

———, 1930: Iron-ion concentration in relation to growth and other biological processes, *Bot. Gaz.*, **89**: 209–240.

———, 1931: Manganese and the growth of *Lemna minor*, *Sci.*, **74**: 551–552.

———, 1933: The effect of small amounts of copper on the growth of *Chlorella* and *Lemna*, *abst. Am. J. Bot.*, **20**: 681.

———, 1934: Manganese an essential element for green plants, *N. Y. Sta.* (*Cornell*) *Mem.* 151.

HORNBERGER, R., 1882: Chemische Untersuchungen über das Wachsthum der Maispflanze, *Landw. Jahrb.*, **11**: 359–523.

HOUGHLAND, G. V. C., and J. A. SCHRICKER, 1933: The effect of potash on starch in potatoes, *J. Am. Soc. Agron.*, **25**: 334–340.

Humphrey, H. B., 1934: Rapport sur les recherches Américaines relatives aux maladies végétales dues aux carences minérales, *Deuxième Cong. int. Path. comparée*, 278–295.

Hurd-Karrer, A. M., 1933: Inhibition of selenium injury to wheat plants by sulfur, *Sci.*, **78**: 560.

———, 1934: Selenium injury to wheat plants and its inhibition by sulfur, *J. Agr. Res.*, **49**: 343–358.

———, 1935: Factors affecting the absorption of selenium from soil by plants, *J. Agr. Res.*, **50**: 413–428.

———, 1935: Plant physiology involved in the problem of the selenium diseases of live stock, *J. Wash. Acad. Sci.*, **25**: 335–336.

———, 1935: Selenium absorption by plants and their resulting toxicity to animals, *Smithsonian Pub.* 3361.

———, 1936: Inhibition of arsenic injury to plants by phosphorus, *J. Wash. Acad. Sci.*, **26**: 180–181.

——— and M. H. Kennedy, 1936: Inhibiting effect of sulphur in selenized soil on toxicity of wheat to rats, *J. Agr. Res.*, **52**: 933–942.

——— and F. W. Poos, 1936: Toxicity of selenium-containing plants to aphids, *Sci.*, **84**: 252.

Ingalls, R. A., and J. W. Shive, 1931: Relation of H-ion concentration of tissue fluids to the distribution of iron in plants, *Plant Physiol.*, **6**: 103–125.

Inman, O. L., G. Barclay, and M. Hubbard, 1935: Effect of titanous chloride on the formation of chlorophyll in *Zea mays*, *Plant Physiol.*, **10**: 821–822.

Inozemtzev, S. L., 1930: Nature of potassium compounds in plants, *Ergeb. Veg. Lab. arb. Pryanishnikov*, **15**: 85–101; *Chem. Abst.*, **27**: 5368.

Jacob, K. D., R. P. Bartholomew, B. E. Brown, W. H. Pierre, F. R. Reid, and J. W. Tidmore, 1935: Nutrient value of the phosphorus in calcined phosphate as determined by growth of plants in greenhouse experiments, *J. Agr. Res.*, **50**: 837–848.

Jacobson, H. G. M., and T. R. Swanback, 1929: Manganese toxicity in tobacco, *Sci.*, **70**: 283–284.

——— and ———, 1932: Manganese content of certain Connecticut soils and its relation to the growth of tobacco, *J. Am. Soc. Agron.*, **24**: 237–245.

James, W. O., 1930: Studies on the physiological importance of the mineral elements in plants. I, The relation of potassium to the properties and functions of the leaf, *Ann. Bot.*, **44**: 173–198.

———, 1931: Studies of the physiological importance of the mineral elements in plants. II, Potassium: its distribution, movement and relation to growth in the potato, *Ann. Bot.*, **45**: 425–442.

——— and N. L. Penston, 1933: Studies of the physiological importance of the mineral elements in plants. IV, The quantitative distribution of potassium in the potato plant, *Ann. Bot.*, **47**: 279–293.

——— and M. Cattle, 1935: Studies on the physiological importance of the mineral elements in plants. VII, The effects of potassium and chloride ions on the diastase of broad bean leaves, *New Phytol.*, **34**: 283–295.

Janssen, G., and R. P. Bartholomew, 1929: The translocation of potassium in tomato plants and its relation to their carbohydrate and nitrogen distribution, *J. Agr. Res.*, **38**: 447–465.

——— and ———, 1930: The influence of the potash concentration in the cultural medium on the production of carbohydrates in plants, *J. Agr. Res.*, **40**: 243–261.

——— and ———, 1932: The effect of potassium on the production of proteins, sugar and starch in cowpea and in sugar beet plants and the relation of potassium to plant growth, *J. Am. Soc. Agron.*, **24**: 667–679.

———, ———, and V. M. Watts, 1934: Some effects of nitrogen, phosphorus and potassium on the composition and growth of tomato plants, *Ark. Sta. Bull.* 310.

Jashnova, N. V., 1930: Nitrification in podzol soil, *Trans. Sci. Inst. Fertilizers*, 76, No. 415. Quoted by Nicol, 1934.

Javillier, M., and S. Imas, 1926: Le manganèse dans le grain de blé et dans les produits de mouture, *C. R. Acad. Agr. France*, **12**: 721–727.

——— and ———, 1926: Le zinc dans le grain de blé et dans les produits de mouture, *C. R. Acad. Agr. France*, **12**: 727–732.

Jennings, D. S., 1919: The effect of certain colloidal substances on the growth of wheat seedlings, *Soil Sci.*, **7**: 201–215.

Jodin, V., 1883: Du rôle de la silica dans la végétation du maïs, *Ann. Chim. et Phys.*, **30**: 485–494.

Johnson, M. O., 1916: The spraying of yellow pineapple plants on manganese soils with iron sulphate solutions, *Hawaii Sta. Press Bull.* 51.

———, 1917: Manganese as a cause of the depression of the assimilation of iron by pineapple plants, *J. Ind. Eng. Chem.*, **9**: 47–49.

———, 1924: Manganese chlorosis of pineapples: its cause and control, *Hawaii Sta. Press Bull.* 52.

———, 1928: Control of chlorosis of the pineapple and other plants, *Ind. and Eng. Chem.*, **20**: 724–725.

Johnston, E. S., 1917: Seasonal variations in the growth rates of buckwheat plants under greenhouse conditions, *Johns Hopkins Univ. Circ.*, n.s., **3**: 211–217.

——— and W. H. Dore, 1928: The relation of boron to the growth of the tomato plant, *Sci.*, **67**: 324–325.

——— and ———, 1929: The influence of boron on the chemical composition and growth of the tomato plant, *Plant Physiol.*, **4**: 31–62.

——— and D. R. Hoagland, 1929: Minimum potassium level required by tomato plants grown in water cultures, *Soil Sci.*, **27**: 89–109.

——— and P. L. Fisher, 1930: The essential nature of boron to the growth and fruiting of the tomato, *Plant Physiol.*, **5**: 387–392.

Jones, H. W., 1920: The distribution of inorganic iron in plants and animal tissues, *Biochem., J.*, **14**: 654–659.

Jones, J. P., 1929: Deficiency of magnesium the cause of a chlorosis in corn, *J. Agr. Res.*, **39**: 873–892.

———, 1931: The diagnostic value of plant symptoms in determining nutrient deficiencies of soils, *J. Am. Soc. Agron.*, **23**: 352–357.

Jones, J. S., and D. E. Bullis, 1921: Manganese in commonly grown legumes, *J. Ind. Eng. Chem.*, **13**: 524–525.

Jones, L. H., 1920: The effect of ammonium sulfate upon the availability of iron in nutrient solutions, *N. J. Sta. Rept.*, **1920**: 405–409.

——— and J. W. Shive, 1921: The influence of iron in the forms of ferric phosphate and ferrous sulphate upon the growth of wheat in a nutrient solution, *Soil Sci.*, **11**: 93–101.

——— and ———, 1921: Effect of ammonium sulfate upon plants in nutrient solutions supplied with ferric phosphate and ferrous sulphate as sources of iron, *J. Agr. Res.*, **21**: 701–729.

——— and ———, 1923: Influence of ammonium sulphate on plant growth in nutrient solutions and its effect on hydrogen-ion concentration and iron availability, *Ann. Bot.*, **37**: 355–377.

JONES, R. L., and F. R. PEMBER, 1925: The fertilizer nutrients required by barley, wheat, and oats as shown by both soil and water cultures, *Soil Sci.*, **19**: 169–201.

JONES, W. J., and H. A. HUSTON, 1914: Composition of maize at various stages of its growth, *Ind. Sta. Bull.* 175.

JORDAN, W. H., 1913: Studies in plant nutrition. I, *N. Y. Sta.* (*Geneva*) *Bull.* **358**: 11–30.

———, 1913: Studies in plant nutrition. II, *N. Y. Sta.* (*Geneva*) *Bull.* **360**: 53–77.

JUNG, J., 1920: Über den Nachweis und die Verbreitung des Chlors im Pflanzenreiche, *Sitzungsber. Akad. Wiss. Wien*, **129**: 297–340; *J. Chem. Soc.*, **121, 122**: 1098.

KARRAKER, P. E., and C. E. BORTNER, 1933: Symptoms of phosphorus deficiency in Turkish tobacco, *J. Am. Soc. Agron.*, **25**: 556–558.

——— and ———, 1934: Studies of frenching of tobacco, *Ky. Sta. Res. Bull.* 349.

KATAYMA, T., 1906: On the degree of stimulating action of manganese and iron salts on barley, *Bull. Col. Agr. Tokyo*, **7**: 91–93.

KELLEY, W. P., 1912: The function and distribution of manganese in plants and soils, *Hawaii Sta. Bull.* 26.

———, 1914: The function of manganese in plants, *Bot. Gaz.*, **57**: 213–227.

KIMBROUGH, W. D., 1931: The effect of fertilizer on the quality and keeping quality of watermelons, *Plant Physiol.*, **5**: 373–384.

KING, E. J., and V. DAVIDSON, 1933: The biochemistry of silicic acid. IV, Relation of silica to the growth of phytoplankton, *Biochem. J.*, **27**: 1015–1021.

KLEIN, G., 1930: Changes in the nitrogen of green plants, *Ergeb. Agr.-Chem.*, **2**: 143; *Chem. Abst.*, **26**: 2484.

———, K. TAUBÖCK, and H. LINSER, 1930, 1931: Urea and ureides in higher plants. I, Occurrence of urea in the vegetable kingdom and its transformation during the vegetative period. II, Appearance and disappearance of urea in plants, *Jahrb. Wiss. Bot.*, **73**: 193–225; **74**: 429–458.

KNOP, W., 1861–1862: Quantitativ-analytische Arbeiten über den Ernährungsprocess der Pflanzen, *Landw. Versuchs-Stat.*, **3**: 295–324; **4**: 173–185.

KNOTT, J. E., 1932: The effect of certain mineral elements on the color and thickness of onion scales, *N. Y. Sta.* (*Cornell*) *Bull.* 552.

KNOWLES, F., and J. E. WATKIN, 1931: The assimilation and translocation of plant nutrients in wheat during growth, *J. Agr. Sci.*, **21**: 612–637.

——— and ———, 1932: The amounts and distribution of some phosphorus and nitrogen compounds in wheat during growth, *J. Agr. Sci.*, **22**: 755–766.

KNUDSON, L., 1915: Toxicity of galactose for certain of the higher plants, *Ann. Mo. Bot. Garden*, **2**: 659–666.

———, 1916: Influence of certain carbohydrates on green plants, *N. Y. Sta.* (*Cornell*) *Mem.* 9.

———, 1917: The toxicity of galactose and mannose for green plants and the antagonistic action of other sugars toward these, *Am. J. Bot.*, **4**: 430–437.

——— and E. W. LINDSTROM, 1919: Influence of sugars on the growth of albino plants, *Am. J. Bot.*, **6**: 401–406.

KOSTYTSCHEW, S., and P. ELIASBERG, 1920: Über die Form der Kaliumverbindungen in lebenden Pflanzengeweben, *Z. physiol. Chem.*, **111**: 228–235.

KRATZMANN, E., 1913: Der microchemische Nachweis und die Verbreitung des Aluminiums im Pflanzenreich, *Sitzungsber. Akad. Wiss. Wien*, **122**: 311–336.

———, 1914: The physiological action of the salts of aluminum upon plants, *Sitzungsber. Akad. Wiss. Wien*, **123**: 211–233; *Exp. Sta. Rec.*, **34**: 525.

KRAYBILL, H. R., 1930: Plant metabolism studies as an aid in determining fertilizer requirements. I, *Ind. and Eng. Chem.*, **22**: 275–276.

Kreuzhage, C., and E. Wolff, 1884: Bedeutung der Kieselsäure für die Entwicklung der Haferpflanze, *Landw. Versuchs-Stat.*, **30**: 161–197.

Lanham, W. B., 1927: Effect of potash fertilizer on carrying quality of tomatoes, *Tex. Sta. Bull.* 357.

Larson, H. W. E., 1928: The relation of the concentration of calcium ion required by alfalfa to the amount present in the soil solution, *Soil Sci.*, **25**: 399–408.

Latshaw, W. L., and E. C. Miller, 1924: Elemental composition of the corn plant, *J. Agr. Res.*, **27**: 845–861.

Laurent, J., 1904: Recherches sur la nutrition carbonée des plantes vertes à l'aide de matières organiques, *Rev. Gén. Bot.*, **16**: 14–48, 66–80, 96–119, 120–128, 155–166, 188–202, 231–241.

Laurie, A., and E. W. McElwee, 1934: A method of studying nutrient deficiencies in greenhouse crops, *Proc. Am. Soc. Hort. Sci.*, **32**: 617–619.

Lauritzen, C. W., 1934: Displacement of soil solubles through plant roots by means of air pressure as a method of studying soil fertility problems, *J. Am. Soc. Agron.*, **26**: 807–819.

———, 1934: A method for studying soil-plant nutrient relationships, *Sci.*, **79**: 187.

Lausberg, T., 1935: Determination of the cuticular excretion of leaves, *Jahrb. wiss. Bot.*, **81**: 769–806; *Chem. Abst.*, **29**: 7387.

Leeper, G. W., 1934: Relationship of soils to manganese deficiency of plants, *Nature*, **134**: 972–973.

Lemmermann, O., 1922: Die ertragssteigernde Wirkung der Kieselsäure bei unzureichender Phosphorsäuredüngung, *Z. pflanz. Düng.*, 8vo., 72 pp.

———, H. Wiessmann, and K. Sammett, 1925: Untersuchungen über die Ursache der ertragssteigerenden Wirkung der Kiesselsäure, *Z. pflanz. Düng.*, A, **4**: 265–315.

Levine, V. E., 1925: The effect of selenium compounds upon growth and germination in plants, *Am. J. Bot.*, **12**: 82–91.

Levy, G., 1931: The presence, distribution and role of aluminium in plants, thesis, Paris; *Exp. Sta. Rec.*, **73**: 452.

Leydhecker, A., 1886: Ueber die physiologische Bedeutung des Chlor in der Buchweizenpflanze, *Landw. Versuchs-Stat.*, **8**: 177–187.

Ligon, W. S., and W. H. Pierre, 1932: Soluble aluminum studies. II, Minimum concentrations of aluminum found to be toxic to corn, sorghum and barley in culture solutions, *Soil Sci.*, **34**: 307–317.

Lilleland, O., 1930: Does potassium increase the sugar content of prunes? *Proc. Am. Soc. Hort. Sci.*, **27**: 15–18.

Lipman, C. B., 1916: A critique of the hypothesis of the lime-magnesium ratio, *Plant World*, **19**: 83–105, 119–135.

———, A. R. Davis, and E. S. West, 1926: The tolerance of plants for NaCl, *Soil Sci.*, **22**: 303–322.

———, 1928: The chemical elements indispensable to plants, *Sci. Month.*, **26**: 289–294.

——— and G. Mackinney, 1931: Proof of the essential nature of copper for higher green plants, *Plant Physiol.*, **6**: 593–599.

Lloyd, J. W., and E. P. Lewis, 1935: Substitution of commercial fertilizers for manure in vegetable production, *Ill. Sta. Bull.* 421.

Loew, O., 1878: Kann das Rubidium die physiologische Function des Kaliums in der Pflanzenzelle übernehmen? *Landw. Versuchs-Stat.*, **21**: 389–395.

———, 1891: Ueber die physiologische Function der Phosphorsäure, *Biol. Cent.*, **11**: 269–281.

——— and S. Sawa, 1902: On the action of manganese compounds on plants, *Bull. Col. Agr. Tokyo*, **5**: 161–172.

———, 1903: The physiological role of mineral nutrients in plants, *U. S. Dept. Agr. Bull.* 45.

——— and S. Honda, 1904: Über den Einfluss des Mangans auf Waldbäume, *Bull. Col. Agr. Tokyo,* **6**: 125–130.

———, 1934: The physiological importance of potassium and calcium, *Ernähr. der Pflanze,* **30**: 141–144; *Chem. Abst.,* **28**: 4448.

Lomanitz, S., 1923: A preliminary study of the effects of sodium chloride upon alfalfa grown in solution cultures, *Soil Sci.,* **16**: 183–195.

———, 1924: The influence of sodium chloride upon alfalfa grown in solution cultures, *Soil Sci.,* **18**: 353–371.

Loomis, W. E., and J. J. Wilson, 1933: Boron-iron relationship in the growth of tomato, *Proc. Iowa Acad. Sci.,* **40**: 53–56.

Loring, F. H., and J. G. F. Druce, 1930: Potassium: its radioactivity, its isotopes, and its association with the phenomena of life, *Chem. News,* **140**: 34–36.

Lukaszewicz, J., 1926: Über das Vorkommen von Kalium in manchen pflanzlichen Gebilden, *extr. du Bull. del'Académie Polonaise des Sci. et Lettres,* **1926**: 559–576.

Lutman, B. F., and N. L. Walbridge, 1929: The role of magnesium in the aging of plants, *Vt. Sta. Bull.* 296.

———, 1934: Cell size and structure in plants as affected by various inorganic elements, *Vt. Sta. Bull.* 383.

Lyon, C. J., 1923: The effect of phosphates on respiration, *J. Gen. Physiol.,* **6**: 299–306.

———, 1927: The role of phosphate in plant respiration, *Am. J. Bot.,* **14**: 274–284.

MacGillivray, J. H., 1926: The importance of phosphorus in the production of seed and non-seed portions of a tomato fruit, *Proc. Am. Soc. Hort. Sci.,* **22**: 374–379.

———, 1927: Effect of phosphorus on the composition of the tomato plant, *J. Agr. Res.,* **34**: 97–129.

Magistad, O. C., 1925: The aluminum content of the soil solution and its relation to soil reaction and plant growth, *Soil Sci.,* **20**: 181–227.

Magistris, H., 1932: Zur Biochemie und Physiologie organischer Phosphorverbindungen in Pflanze und Tier. III, Über den Austritt von anorganischem und organischem Phosphor aus Pflanzenzellen unter verschiedenen Bedingungen, *Biochem. Z.,* **253**: 64–80.

Maiwald, K., 1923: Wirkung hoher Nährstoffgaben auf den Assimilationsapparat, *Angew. Bot.,* **5**: 33–74.

———, 1930: Nährstoffaufnahme und Stofferzeugung beim Buchweizen (*Fagopyrum esculentum);* ein Beitrag zum t-Ertragsgesetz der höheren Pflanze, *Z. Pflanzenernähr. Düng. u. Bodenk.,* **17**: 12–42; *Biol. Abst.,* **5**: 2684.

Malyschev, N., 1932: Das Wachstum des isolierten Wurzelmeristems auf sterilem Nährboden, *Biol. Zent.,* **52**: 257–205.

Mameli, E., 1918: Sulla influenza del magnesis sopra la formazione della chlorofilla, *Univ. di Pavia. Atti Ist. Bot.,* **15**: 151–205.

Mann, C. E. T., 1924: The physiology of the nutrition of fruit trees. I, Some effects of calcium and potassium starvation, *Univ. Bristol Agr. Hort. Res. Sta. Ann. Rept.,* **1924**: 36–45.

——— and T. Wallace, 1925: The effects of leaching with cold water on the foliage of the apple, *J. Pom. and Hort. Sci.,* **4**: 146–161.

Maquenne, L., and E. Demoussy, 1920: Sur la distribution et la migration du cuivre dans les tissus des plantes vertes, *C. R. Acad. Sci. Paris,* **170**: 87–93.

——— and ———, 1920: Sur l'absorption du calcium par les racines des plantes et ses propriétés antitoxiques vis-à-vis du cuivre, *C. R. Acad. Sci. Paris,* **170**: 420–425.

——— and ———, 1920: Un cas d'action favorable du cuivre sur la végétation, *C. R. Acad. Sci. Paris*, **170**: 1542–1545.

Marine, D., E. J. Baumann, and B. Webster, 1930: The relative amounts of iodine-absorbing (reducing) material in various plants, *J. Biol. Chem.*, **89**: 213–219.

Marsh, R. P., and J. W. Shive, 1925: Adjustment of iron supply to requirements of soybean in solution culture, *Bot. Gaz.*, **79**: 1–28.

———, 1935: A modified culture jar, *Sci.*, **82**: 256.

Marsh, R. S., 1922: Preliminary studies on the sulfur content of the tomato, *Proc. Am. Soc. Hort. Sci.*, **19**: 83–84.

———, 1923: Seasonal variation of sulphur content in certain tissues of the apple tree, *Bot. Gaz.*, **75**: 400–413.

Martin, A. L., 1936: Toxicity of selenium to plants and animals, *Am. J. Bot.*, **23**: 471–483.

Masuda, E., and K. Nishida, 1934: Iodine of sea weeds, *J. Pharm. Soc. Japan*, **54**: 243–245.

Mather, T. H., 1929: The effect of fertilizers upon the forms of phosphorus and the amounts of phosphorus, nitrogen and silica in hays, *Sci. Agr.*, **10**: 35–63.

Maume, L., and J. Dulac, 1934: Varietal differences in the absorption of nitrogen, phosphoric acid and potash by wheat at the same physiological age and under the same environmental conditions, *C. R. Acad. Sci. Paris*, **198**: 199–202.

Mayer, G. G., 1929: Der Einfluss verschiedener Nährstoffzuführung auf das Längenwachstum isolierter Wurzeln, *diss. Hessischen Ludwigs-Univ.*

Mazé, P., and A. Perrier, 1904: Recherches sur l'assimilation de quelques substances ternaires par les végétaux à chlorophylle, *Ann. Inst. Pasteur*, **18**: 721–747.

——— and M. Lemoigne, 1912: Recherches sur la chlorose végétale provoquée par le carbonate de calcium, *C. R. Acad. Sci. Paris*, **155**: 435–437.

———, M. Ruot, and M. Lemoigne, 1913: Chlorose calcaire des plantes vertes. Rôle des excrétions des racines dans l'absorption du fer des sols calcaires, *C. R. Acad. Sci. Paris*, **157**: 495–498.

———, 1914: Influences respectives des éléments de la solution minérale sur le développement du maïs, *Ann. Inst. Pasteur*, **28**: 21–68.

———, 1915: Détermination des éléments minéraux rares nécessaires au développement du maïs, *C. R. Acad. Sci. Paris*, **160**: 211–214.

———, 1919: Recherche d'un solution purement minérale capable d'assurer l'évolution complète du maïs cultivé à l'abri des microbes, *Ann. Inst. Pasteur*, **33**: 139–173.

McCall, A. G., and P. E. Richards, 1918: Mineral food requirements of the wheat plant at different stages of its development, *J. Am. Soc. Agron.*, **10**: 127–134.

——— and J. R. Haag, 1921: The relation of the hydrogen-ion concentration of nutrient solutions to growth and chlorosis of wheat plants, *Soil Sci.*, **12**: 69–77.

McCalla, A. G., 1934: The effect of potassium supply on the composition and quality of wheat. I, *Can. J. Res.*, **11**: 687–700.

——— and E. K. Woodford, 1935: The effect of potassium supply on the composition and quality of wheat. II, *Can. J. Res.*, **13**: 339–354.

McCallum, A. B., 1905: On the distribution of potassium in animal and vegetable cells, *J. Physiol.*, **32**: 95–128.

McCance, R. A., and E. M. Widdowson, 1935: Phytin in human nutrition, *Biochem. J.*, **29**: 2694–2699.

McClelland, C. K., 1931: Effects of various plant foods on growth activities and development of oats, *J. Am. Soc. Agron.*, **23**: 304–311.

McClendon, J. F., E. Barrett, and T. Canniff, 1934: The iodine content of potatoes, *Biochem. J.*, **28**: 1209–1211.

McCool, M. M., 1913: The toxicity of manganese and the antidotal relations between this and various other cations with respect to green plants, *N. Y. Sta. (Cornell) Mem.* 2.

——— and M. D. Weldon, 1930: The effect of sodium nitrate on the composition of the expressed sap of the small grains, *J. Am. Soc. Agron.*, **22**: 434–438.

——— and R. L. Cook, 1930: Rate of intake, accumulation and transformation of nitrate nitrogen by small grains and Kentucky blue grass, *J. Am. Soc. Agron.*, **22**: 757–765.

———, 1933: Effect of thallium sulphate on the growth of several plants and on nitrification in soils, *Contr. Boyce Thompson Inst.*, **5**: 289–296.

———, 1935: Effect of light intensity on the manganese content of plants, *Contr. Boyce Thompson Inst.*, **7**: 427–438.

McDonnell, C. C., and R. C. Roark, 1917: Occurrence of manganese in insect flowers and insect flower stems, *J. Agr. Res.*, **11**: 77–83.

McGeorge, W. T., 1923: The chlorosis of pineapple plants grown on manganiferous soils, *Soil Sci.*, **16**: 269–275.

McHargue, J. S., 1919: The effect of manganese on the growth of wheat: a source of manganese for agricultural purposes, *J. Ind. Eng. Chem.*, **11**: 332–335.

———, 1919: Effect of certain compounds of barium and strontium on the growth of plants, *J. Agr. Res.*, **16**: 183–195.

———, 1921: The occurrence and distribution of manganese in plants, *Ky. Sta. Rept.*, **1921**: 36–37; **1922**: 35–36.

———, 1922: The role of manganese in plants, *J. Am. Chem. Soc.*, **44**: 1592–1598.

———, 1923: Effect of different concentrations of manganese sulphate on the growth of plants in acid and neutral soils and the necessity of manganese as a plant nutrient, *J. Agr. Res.*, **24**: 781–795.

———, 1923: Common earthenware jars a source of error in pot experiments, *J. Agr. Res.*, **26**: 231–233.

———, 1923: Iron and manganese content of certain species of seeds, *J. Agr. Res.*, **23**: 395–401.

———, 1925: The occurrence of copper, manganese, zinc, nickel and cobalt in soils, plants and their possible function as vital factors, *J. Agr. Res.*, **30**: 193–196.

———, 1925: The significance of the occurrence of copper, manganese and zinc in forage crops and foods, *J. Am. Soc. Agron.*, **17**: 368–372.

———, 1926: Mineral constituents of the cotton plant, *J. Am. Soc. Agron.*, **18**: 1076–1083.

———, 1927: Significance of the occurrence of manganese, copper, zinc, nickel and cobalt in Kentucky bluegrass, *Ind. and Eng. Chem.*, **19**: 274–276.

———, W. R. Roy, and J. G. Pelphrey, 1930: Iodine in some fertilizer materials, *Am. Fertilizer*, **73**: 40, 42, 63.

——— and O. M. Shedd, 1930: The effect of manganese, copper, zinc, boron and arsenic on the growth of oats, *J. Am. Soc. Agron.*, **22**: 739–747.

——— and R. K. Calfee, 1931: Effect of manganese, copper and zinc on the growth of yeast, *Plant Physiol.*, **6**: 559–566.

——— and ———, 1931: Effect of manganese, copper, and zinc on growth and metabolism of *Aspergillus flavus* and *Rhizopus nigricans*, *Bot. Gaz.*, **91**: 183–193.

——— and W. R. Roy, 1931: The mineral, nitrogen and fat content of some varieties of mature bean seed and of string beans, *J. Nutrition*, **3**: 479–481.

——— and R. K. Calfee, 1932: Effect of boron on the growth of lettuce, *Plant Physiol.*, **7**: 161–164.

——— and ———, 1932: Manganese essential for the growth of *Lemna minor*, *Plant Physiol.*, **7**: 697–703.

———, W. R. Roy, and J. G. Pelphrey, 1932: The iron, manganese, copper, zinc and iodine content of some important forage crops, *J. Am. Soc. Agron.*, **24**: 562–565.

——— and R. K. Calfee, 1933: Further evidence that boron is essential for the growth of lettuce, *Plant Physiol.*, **8**: 305–313.

———, D. W. Young, and R. K. Calfee, 1935: The effect of certain fertilizer materials on the iodine content of important foods, *J. Am. Soc. Agron.*, **27**: 559–565.

McKibben, R. R., 1928: The effect of sulphur on soils and on crop yields, *Md. Sta. Bull.* 296.

McLean, F. T., and B. E. Gilbert, 1925: Manganese as a cure for chlorosis of spinach, *Sci.*, **61**: 636.

———, 1927: Feeding plants manganese through the stomata, *Sci.*, **66**: 487–489.

——— and B. E. Gilbert, 1927: The relative aluminum tolerance of crop plants, *Soil Sci.*, **24**: 163–177.

——— and ———, 1928: Aluminum toxicity, *Plant Physiol.*, **3**: 293–302.

McLean, R. C., and W. L. Hughes, 1936: The quantitative distribution of boron in *Vicia faba* and *Gossypium herbaceum*, *Ann. App. Biol.*, **23**: 231–244.

McMurtrey, J. E., 1929: The effect of boron deficiency on the growth of tobacco plants in aerated and unaerated solutions, *J. Agr. Res.*, **38**: 371–380.

———, 1932: Effect of thallium on growth of tobacco plants, *Sci.*, **76**: 86.

———, 1932: Relation of calcium and magnesium to the growth and quality of tobacco, *J. Am. Soc. Agron.*, **24**: 707–716.

———, 1933: Distinctive effects of the deficiency of certain essential elements on the growth of tobacco plants in solution cultures, *U. S. Dept. Agr. Tech. Bull.* 340.

———, 1935: Boron deficiency in tobacco under field conditions, *J. Am. Soc. Agron.*, **27**: 271–273.

Meurer, R., 1909: Ueber die regulationische Aufnahme anorganischer Stoffe durch die Wurzeln von *Beta vulgaris* und *Daucus carota*, *Jahrb. wiss. Bot.*, **46**: 503–568.

Mevius, W., and I. Dikussar, 1930: Nitrites as sources of nitrogen for the higher plants, *Jahrb. wiss. Bot.*, **73**: 633–703; *Chem. Abst.*, **26**: 3005.

Meyer, A. H., 1931: Some neglected soil factors in plant growth, *J. Am. Soc. Agron.*, **23**: 606–625.

Meyers, C. N., and C. Voegtlin, 1914: Soluble aluminum compounds, *U. S. Health Rept.*, **29**: 1625–1629.

Michiels, H., and P. DeHeen, 1906: Note au sujet de l'action stimulante du manganèse sur la germination, *Bull. de l'Acad. Roy. Belg.*, **5**: 288–289.

Miller, H. G., 1919: The relation of sulfates to plant growth and composition, *J. Agr. Res.*, **17**: 87–103.

———, 1921: Further studies on relation of sulfates to plant growth and composition, *J. Agr. Res.*, **22**: 101–111.

Miller, L. P., 1933: Effect of manganese deficiency on the growth and sugar content of plants, *Am. J. Bot.*, **20**: 621–631.

Mirasol, J. J., 1920: Aluminum as a factor in soil acidity, *Soil Sci.*, **10**: 153–219.

Mitchell, H. L., 1934: Pot culture tests of forest soil fertility, *Black Rock Forest Bull.* 5.

———, 1935: A method for determining the nutrient needs of shade trees with special reference to phosphorus, *Black Rock Forest Papers*, Vol. 1, No. 1.

Mitchell, J. H., 1931: A study of relationships between the mineral content of the soil and plants grown on the soil, *S. C. Sta. Ann. Rept.*, **44**: 56.

Miyake, K., 1916: The toxic action of soluble aluminum salts upon the growth of the rice plant, *J. Biol. Chem.*, **25**: 23–28.

MOLISCH, H., 1892: "Die Pflanzen in ihren Beziehungen zum Eisen. Eine physiologische Studie," 8vo., 119 pp., G. Fischer, Jena; *Bot. Cent.*, **50**: 370.

———, 1897: Der Einfluss des Bodens auf die Blütenfarbe der *Hortensien*, *Bot. Ztg.*, **55**: 49–61.

———, 1902: Ueber vorübergehende Rothfärbung der Chlorophyllkörner in Laubblättern, *Ber. bot. Ges.*, **20**: 442–448.

MOLLIARD, M., 1907: Action morphogénique de quelques substances organiques sur les végétaux supérieurs, *Rev. Gén. Bot.*, **19**: 241–291, 329–349, 357–391.

MOORE, H. I., and A. P. JONES, 1936: Iron and plant growth, *Gard. Chron.*, **99**: 199–200.

MORRIS, H. S., 1931: Physiological effects of boron on wheat, *Bull. Torrey Bot. Club*, **58**: 1–30.

MORRIS, V. H., 1933: Form of potassium in corn tissues, *Ohio Sta. 51st Ann. Rept. Bull.*, **516**: 35–36.

——— and J. D. SAYRE, 1935: Solubility of potassium in corn tissue, *Plant Physiol.*, **10**: 565–568.

MORSE, F. W., 1927: Relation between water and potash in plant production, *J. Agr. Res.*, **35**: 939–946.

MOSER, F., 1933: The calcium magnesium ratio in soils and its relation to crop growth, *J. Am. Soc. Agron.*, **25**: 365–376.

MOTHES, K., 1932: Natural regulation of the vegetable protein metabolism, *Naturwiss.*, **20**: 102–103; *Chem. Abst.*, **26**: 3820.

MOWRY, H. and A. F. CAMP, 1934: A preliminary report on zinc sulfate as a corrective for bronzing of tung trees, *Fla. Sta. Bull.* 273.

MURNEEK, A. E., and E. J. GILDEHAUS, 1931: Leaf diagnosis and the interpretation of fertilizer requirements of plants, *Sci.*, **74**: 39–41.

MURPHY, H. F., 1933: Recovery of phosphorus from prairie grasses growing on central Oklahoma soils treated with superphosphate, *J. Agr. Res.*, **47**: 911–917.

———, 1936: The nitrogen, phosphorus and calcium content of the cotton plant at pre-blooming to early boll stages of growth, *J. Am. Soc. Agron.*, **28**: 52–57.

NAGAOKA, M., 1903, 1904: On the stimulating action of manganese upon rice, *Bull. Col. Agr. Tokyo*, **5**: 467–472; **6**: 135–136.

———, 1906: On the stimulating action of manganese upon rice. III, *Bull. Col. Agr. Tokyo*, **7**: 77–81.

NAKAMURA, M., 1904: Can lithium and cesium salts exert any stimulant action on phanerogams? *Bull. Col. Agr. Tokyo*, **6**: 153–160; *Bot. Cent.*, **95**: 669.

———, 1904: Can salts of zinc, cobalt and nickel in high dilution exert a stimulating action on agricultural plants? *Bull. Col. Agr. Tokyo*, **6**: 147–153; *Bot. Cent.*, **96**: 109.

NEAL, W. M., R. B. BECKER, and A. L. SHEALY, 1931: A natural copper deficiency in cattle rations, *Sci.*, **74**: 418–419.

NELLER, J. R., and W. J. MORSE, 1921: Effects upon the growth of potatoes, corn and beans resulting from the addition of borax to the fertilizer used, *Soil Sci.*, **12**: 79–132.

———, 1935: Phosphorus content and buffer capacity of plant sap as related to the physiological effect of phosphorus fertilizers in fibrous low-moor peat, *J. Agr. Res.*, **51**: 287–300.

NEMEC, A., and V. KAS, 1923: Studien über die physiologische Bedeutung des Titans im Pflanzenorganismus, *Biochem. Z.*, **140**: 583–590.

——— and M. GRAČANIN, 1930: Der Einfluss des Lichtes auf die Resorption von Kali und Phosphorsäure bei Neubaueruntersuchungen, *Z. Pflanzenernähr. Düng. u. Bodenk.*, **16**: 102–110.

Newell, H. J., 1930: Range of boron concentration in nutrient solutions for plants, *master's thesis, Univ. Md.*

Newton, J. D., 1928: The selective absorption of inorganic elements by various crop plants, *Soil Sci.*, **26**: 85–91.

Nicolas, G., 1921: Contribution à l'étude du mécanisme de l'action fertilisante du soufre, *C. R. Acad. Sci. Paris*, **172**: 85–87.

Nightingale, G. T., L. G. Schermerhorn, and W. R. Robbins, 1930: Some effects of potassium deficiency on the histological structure and nitrogenous and carbohydrate constituents of plants, *N. J. Sta. Bull.* 499.

———, R. M. Addoms, W. R. Robbins, and L. G. Schermerhorn, 1931: Effects of calcium deficiency on nitrate absorption and on metabolism in tomato, *Plant Physiol.*, **6**: 605–631.

———, L. G. Schermerhorn, and W. R. Robbins, 1932: Effect of sulphur deficiency on metabolism in tomato, *Plant Physiol.*, **7**: 565–596.

Nobbe, F., and T. Siegert, 1862, 1863: Über das Chlor als specifischer Nährstoff der Buchweizenpflanze, *Landw. Versuchs-Stat.*, **4**: 318–340; **5**: 116–136.

——— and ———, 1864: Ueber das Chlor als Pflanzennährstoff, *Landw. Versuchs-Stat.*, **6**: 108–120.

———, 1865: Ueber die physiologische Function des Chlor in der Pflanze, *Landw. Versuchs-Stat.*, **7**: 371–386.

———, J. Schröder, and R. Erdmann, 1871: Ueber die organische Leistung des Kalium in der Pflanze, *Landw. Versuchs-Stat.*, **13**: 321–423.

Noll, C. F., 1923: The effects of phosphate on early growth and maturity, *J. Am. Soc. Agron.*, **15**: 87–99.

Oddo, B., and G. Pollacci, 1920: Influenza del nucleo pirrolico sulla formazione della chlorofilla, *Gaz. Chim. Italiana*, **50**: 54–70.

Odland, T. E., and F. K. Crandall, 1932: The effect of the lack of available manganese in the soil on crop yields, *J. Am. Chem. Soc.*, **24**: 622–626.

Olson, G. A., and J. L. St. John, 1921: An investigation of sulfur as a plant food, *Wash. Sta. Bull.* 165.

Orr, J. B., F. C. Kelly, and G. L. Stuart, 1928: The effect of iodine manuring on the iodine content of plants, *J. Agr. Sci.*, **18**: 159–161.

Orth, O. S., G. C. Wickwire, and W. E. Burge, 1934: Copper in relation to chlorophyll and hemoglobin formation, *Sci.*, **79**: 33–34.

Oserkowsky, J., 1932: Hydrogen-ion concentration and iron content of tracheal sap from green and chlorotic pear trees, *Plant Physiol.*, **7**: 253–259.

———, 1933: Quantitative relation between chlorophyll and iron in green and chlorotic pear leaves, *Plant Physiol.*, **8**: 449–468.

——— and H. E. Thomas, 1933: Exanthema in pears and its relation to copper deficiency, *Sci.*, **78**: 315–316.

Osterhout, W. J. V., 1912: Plants which require sodium, *Bot. Gaz.*, **54**: 532–536.

———, 1916: Specific action of barium, *Am. J. Bot.*, **3**: 481–483.

Overholser, E. L., and F. L. Overley, 1933: Fertilizers as related to leaf area in apple production, *Proc. Am. Soc. Hort. Sci.*, **30**: 52–54.

Owen, O., 1931: The analysis of tomato plants. II, The effect of manurial treatment on the composition of tomato foliage, *J. Agr. Sci.*, **21**: 442–451.

Palladin, V. I., 1927: "Plant Physiology," Eng. trans. Livingston, 86 pp., P. Blakiston's Son & Company, Philadelphia.

Parbery, N. H., 1935: Mineral constituents in relation to chlorosis of orange leaves, *Soil Sci.*, **39**: 35–45.

Parker, F. W., and E. Truog, 1920: The relation between the calcium and the nitrogen content of plants and the function of calcium, *Soil Sci.*, **10**: 49–56.

——— and W. H. Pierre, 1928: The relation between the concentration of mineral elements in a culture medium and the absorption and utilization of these elements by plants, *Soil Sci.*, **25**: 337–343.

Parker, M. M., J. B. Hester, and R. L. Carolus, 1933: The effect of soil conditions on the growth and composition of certain vegetable crop plants as influenced by soil reaction, *Proc. Am. Soc. Hort. Sci.*, **30**: 452–457.

Pember, F. R., 1917: Studies by means of both pot and solution culture of the phosphorus and potassium requirements of the barley plant during its different periods of growth, *R. I. Sta. Bull.* 169.

——— and F. T. McLean, 1925: Economical use of nitrogen, phosphorus and potassium by barley, oats, and wheat in solution cultures, *R. I. Sta. Bull.* 199.

Penston, N. L., 1931: Studies on the physiological importance of the mineral elements in plants. III, A study by microchemical methods of the distribution of potassium in the potato plant, *Ann. Bot.*, **45**: 671–692.

———, 1935: Studies on the physiological importance of the mineral elements in plants. VIII, The variation in potassium content of potato leaves during the day, *New Phytol.*, **34**: 296–309.

Peterson, W. H., 1914: Forms of sulphur in plant materials and their variation with the soil supply, *J. Am. Chem. Soc.*, **36**: 1290–1300.

——— and C. B. Peterson, 1926: The water-soluble content of calcium and phosphorus in cabbage, *J. Agr. Res.*, **33**: 695–699.

——— and C. W. Lindow, 1928: Variations in the manganese content of certain vegetables, *Soil Sci.*, **26**: 149–153.

Petrie, A. H. K., 1934: The drift of the content of potassium and calcium with age on plants, *Australian J. Exp. Biol. Med. Sci.*, **12**: 99–110; *Chem. Abst.*, **29**: 2575.

Pettinger, N. A., 1931: The expressed sap of corn plants as an indicator of nutrient needs, *J. Agr. Res.*, **43**: 95–119.

———, R. G. Henderson, and S. A. Wingard, 1932: Some nutritional disorders in corn grown in sand cultures, *Phytopath.*, **22**: 33–51.

Pfeffer, W., 1900: "Physiology of plants," Eng. trans. by Ewarts, Vol. I, 416, 435 pp.; Vol. II, 33, 55 pp., Clarendon Press, London.

Pfeiffer, T., E. Frank, O. Lemmermann, and H. Schillbach, 1898: Über die Wirkung verschiedener Kalisalze auf die Zusammensetzung und den Ertrag der Kartoffeln, *Landw. Versuchs-Stat.*, **49**: 349–385.

———, A. Rippel, and C. Pfotenhauer, 1919: Über den Verlauf der Nährstoffaufnahme und Stofferzeugung bei der Gerstenpflanze, *Fühlung landw. Ztg.*, **68**: 81–101.

———, ———, and ———, 1920: The action of sodium in the presence of potassium as a plant nutrient, *J. Landw.*, **68**: 255–284; *Exp. Sta. Rec.*, **45**: 624.

Phillips, M., J. Davidson, and H. D. Weihe, 1931: Studies of lignin in wheat straw with reference to lodging, *J. Agr. Res.*, **43**: 619–627.

Phillips, T. G., T. O. Smith, and R. B. Dearborn, 1934: The effect of potassium deficiency on the composition of the tomato plant, *N. H. Sta. Tech. Bull.* 59.

Pierre, I., 1866: La silice et la verse des blés, *C. R. Acad. Sci. Paris*, **63**: 374–377.

Pitcaithly, N. P., and F. P. Worley, 1933: Distribution of copper in the Karaka tree (*Corynocarpus laevigata*), *J. Agr. Sci.*, **23**: 204–207.

Pitz, W., 1916: Effect of elemental sulfur and of calcium sulfate on certain of the higher and lower forms of plant life, *J. Agr. Res.*, **5**: 771–781.

Plagge, H. H., T. J. Maney, and B. S. Pickett, 1935: Functional diseases of the apple in storage, *Iowa Sta. Bull.* 329.

Plimmer, R. H. A., 1913: The hydrolysis of organic compounds by dilute acid and by dilute alkali, *Biochem. J.*, **7**: 72–80.

Poehlman, J. M., 1935: Some limitations on plant juice analyses as indicators of the nutrient needs of plants, *J. Am. Soc. Agron.*, **27**: 195–207.

Poesch, G. H., 1935: Coloring and fertilizing hydrangeas, *Ohio Sta. Mo. Bull.*, **22**: 92–94.

Pohlman, G. G., and W. H. Pierre, 1933: The phosphorus concentration of the exuded sap of corn as a measure of the available phosphorus of the soil, *J. Am. Soc. Agron.*, **25**: 160–171.

Pollacci, G., and B. Oddo, 1915: Influenza del nucleo pirrolico sulla formazione della chlorofilla. Nota preliminare, *Roy. Accad. Lincei*, **24**: 37–39.

———, 1935: Influence of the pyrrole nucleus on the formation of chlorophyll, *Ber. bot. Ges.*, **53**: 540–542; *Chem. Abst.*, **29**: 6275.

Posternak, S., 1903: Sur la matière phospho-organique de réserve des plantes à chlorophylle. Précédé de préparation, *C. R. Acad. Sci. Paris*, **137**: 202–206.

Powers, W. L., 1923: Progress of sulfur investigations with Oregon soils, *J. Am. Soc. Agron.*, **15**: 158–160.

———, 1930: The role of sulphur in plant nutrition, *J. Am. Soc. Agron.*, **22**: 371–374.

Prianischnikow, D. N., 1905: Results of vegetation experiments in the years 1901–1903, *Ann. Inst. Agron. Moscow*, **11**: 155–219; *Exp. Sta. Rec.* **18**: 320–321.

———, 1911: Über den Einfluss von kohlensaurem Kalk auf die Wirkung von verschiedenen Phosphaten, *Landw. Versuchs-Stat.*, **75**: 357–376.

———, 1923: Zur Frage über die Bedeutung des Calciums für die Pflanzen, *Ber. bot. Ges.*, **41**: 138–144.

Proebsting, E. L., 1933: Absorption of potassium by plants as affected by decreased exchangeable potassium in the soil, *J. Pom. and Hort. Sci.*, **11**: 199–204.

Prutzkova, M., M. Lebedeva, A. Melnikov, and S. Ostanin, 1932: Lodging of wheat, *Bull. App. Bot. Genetics and Plant Breeding*, Ser. A., No. 3: 67–87; *Exp. Sta. Rec.*, **72**: 325.

Quartaroli, A., 1929: Copper as an element necessary to plant life, *Ann. Chim. Applicata*, **19**: 467–519; *Chem. Abst.*, **24**: 2773.

Ravin, P., 1913: Nutrition carbonée des plantes à l'aide des acides organiques libres et combinés, *Ann. Sci. Nat. Bot.*, ser. IX, **18**: 289–446.

Reed, H. S., 1907: The value of certain nutritive elements to the plant cell, *Ann. Bot.*, **21**: 501–543.

——— and A. R. C. Haas, 1923: The pseudo-antagonism of sodium and calcium in dilute solutions, *J. Agr. Res.*, **24**: 753–759.

——— and ———, 1923: Studies on the effects of sodium, potassium, and calcium on young orange trees, *Calif. Sta. Tech. Paper* **11**: 1–23.

——— and ———, 1924: Nutrient and toxic effects of certain ions on citrus and walnut trees with especial reference to the concentration and pH of the medium, *Calif. Sta. Tech. Paper* **17**: 1–75.

——— and ———, 1924: Iron supply in a nutrient medium, *Bot. Gaz.*, **77**: 290–300.

——— and J. Dufrenoy, 1935: The effects of zinc and iron salts on the cell structure of mottled orange leaves, *Hilgardia*, **9**: 113–137.

——— and ———, 1935: The effects of zinc salts on the oxidation process in plant cells, *Sci.*, **82**: 249–250.

Reimer, F. C., and H. V. Tartar, 1919: Sulphur as a fertilizer for alfalfa in southern Oregon, *Ore. Sta. Bull.* 163.

Richards, F. J., 1932: Physiological studies in plant nutrition. III, Further studies on the effect of potash deficiency on the rate of respiration in leaves of barley, *Ann. Bot.*, **46**: 367–388.

Richardson, W. D., 1920: The ash of dune plants, *Sci.*, **51**: 546–551.

RIPPEL, A., 1923: Über die durch Mangan verursachte Eisenchlorose bei grünen Pflanzen, *Biochem. Z.*, **140**: 315–323.

———, G. BEHR, and R. MEYER, 1933: Zur Kenntnis der Wirkung des Kaliums auf höhere Pflanzen, *Z. Pflanzenernähr. Düng. u. Bodenk.*, **32**: 95–109; *Biol. Abst.*, **9**: 748.

ROACH, B. M. B., 1926: On the relation of certain soil algae to some soluble carbon compounds, *Ann. Bot.*, **40**: 149–203.

———, 1927: On the carbon nutrition of some algae isolated from soil, *Ann. Bot.*, **41**: 509–517.

ROBBINS, W. J., 1918: Direct assimilation of organic carbon by *Ceratodon purpureus*, *Bot. Gaz.*, **65**: 543–551.

———, 1922: Cultivation of excised root tips and stem tips under sterile conditions, *Bot. Gaz.*, **73**: 376–390.

——— and W. E. MANEVAL, 1924: Effect of light on growth of excised root tips under sterile conditions, *Bot. Gaz.*, **78**: 424–432.

———, 1929: A study of the metabolism of roots, *Mo. Sta. Bull.*, **272**: 41.

ROBBINS, W. R., 1927: Nutrient requirements of the sweet potato plant when grown in sand culture and effect of various nutrient salt proportions on the composition of the roots, *Ann. Rept. N. J. Sta.*, **48**: 178–179.

———, G. T. NIGHTINGALE, L. G. SCHERMERHORN, and M A. BLAKE, 1929: Potassium in relation to the shape of the sweet potato, *Sci.*, **70**: 558.

ROBERG, M., 1932: Ein Beitrag zur Stoffwechselphysiologie der Grünalgen. II, Über die Wirkung von Eisen-, Zink-, und Kupfersalzen, *Jahrb. wiss. Bot.*, **76**: 311–332; *Biol. Abst.*, **9**: 536–537.

ROBINSON, W. O., L. A. STEINKOENIG, and C. F. MILLER, 1917: The relation of some of the rarer elements in soils and plants, *U. S. Dept. Agr. Bull.* 600.

———, 1936: Selenium content of wheat from various parts of the world, *Ind. and Eng. Chem.*, **28**: 736–738.

RODRIGUEZ, A. G., 1933: Effect of boron on the availability of iron, *abst. Am. J. Bot.*, **20**: 679.

ROGERS, C. H., and J. W. SHIVE, 1932: Factors affecting the distribution of iron in plants, *Plant Physiol.*, **7**: 227–252.

RUDOLFS, W., 1919: Influence of sodium chloride upon the physiological changes of living trees, *Soil Sci.*, **8**: 397–427.

———, 1927: Influence of salt on the growth rate of asparagus, *Bot. Gaz.*, **83**: 94–98.

RUPRECHT, R. W., 1915: Toxic effect of iron and aluminum salts on clover seedlings, *Mass. Sta. Bull.*, **161**: 125–129.

RUSSELL, E. J., 1915: "Soil Conditions and Plant Growth," p. 35, Longmans, Green & Company, New York.

SAEGER, A., 1933: Manganese and the growth of *Lemnaceae*, *Am. J. Bot.*, **20**: 234–245.

SALM-HORSTMAR, F., 1856: Versuche und Resultate über die Nahrung der Pflanzen, *diss., Braunschweig.*

SALMON, S. C., 1931: An instrument for determining the breaking strength of straw and a preliminary report on the relation between breaking strength and lodging, *J. Agr. Res.*, **43**: 73–82.

SALTER, R. M., and J. W. AMES, 1928: Plant composition as a guide to the availability of soil nutrients, *J. Am. Soc. Agron.*, **20**: 808–837.

SAMUEL, G., and C. S. PIPER, 1929: Manganese as an essential element for plant growth, *Ann. App. Biol.*, **16**: 493–523.

SAYRE, J. D., 1930: Accumulated iron in the nodes of corn plants, *Plant Physiol.*, **5**: 393–399.

SCHARRER, K., and J. SCHWAIBOLD, 1927: Zur Kenntnis des Jods als biogenes Element. X, Mitteilung: Untersuchung einiger Kulturpflanzen auf ihren natürlichen Jodgehalt und dessen Steigerung durch Joddüngung, *Biochem. Z.*, **185**: 405–413.

——— and A. STROBEL, 1927: Die Jodanreicherung der Pflanzen durch Jodzufuhr, *Angew. Bot.*, **9**: 187–199.

——— and W. SCHROPP, 1933: Sand- und Wasserkulturversuche mit Nickel und Kobalt, *Z. Pflanzenernähr. Düng. u. Bodenk.*, **31**: 94–113.

——— and ———, 1935: The action of vanadium upon cultivated plants, *Z. Pflanzenernähr. Düng. u. Bodenk.*, **37**: 196–202; *Chem. Abst.*, **29**: 6624.

SCHIMPER, A. F. W., 1888: Ueber Kalkoxalatbildung in den Laubblättern, *Bot. Ztg.*, **46**: 65–69, 81–89, 97–107, 113–123, 129–139, 145–153.

———, 1890: Zur Frage der Assimilation der Mineralsalze durch die grüne Pflanze, *Flora*, **73**: 207–278.

SCHMIDT, E. W., 1915: The effects of chlorides on plant growth, *unpub. thesis, Univ. Wis.* Quoted by Tottingham, 1919.

SCHMUCKER, T., 1934: The influence of boric acid on plants, especially germinating pollen grains, *Planta*, **23**: 264–283; *Chem. Abst.*, **29**: 2201.

SCHOLLENBERGER, C. J., 1922: Silica and silicates in relation to plant growth and composition, *Soil Sci.*, **14**: 347–363.

SCHREINER, O., 1923: Organic phosphorus in soils, *J. Am. Soc. Agron.*, **15**: 117–125.

——— and P. R. DAWSON, 1927: Manganese deficiency in soils and fertilizers. *Ind. and Eng. Chem.*, **19**: 400–404.

SCHROPP, W., and K. SCHARRER, 1933: Water culture investigations with Hoagland's "A-Z" solution, *Jahrb. wiss. Bot.*, **78**: 544–563; *Exp. Sta. Rec.*, **71**: 171–172.

SCHULZE, B., 1913: Beitrag zur Frage der Düngung mit Natronsalzen, *Landw. Versuchs-Stat.*, **79-80**: 431–448.

SCHUSTER, G. L., 1927: Potash in relation to quality of crop, *J. Am. Soc. Agron.*, **19**: 506–517.

SCOFIELD, C. S., and L. V. WILCOX, 1930: The boron content of oranges, *Sci.*, **71**: 542.

——— and ———, 1931: Boron in irrigation waters, *U. S. Dept. Agr. Tech. Bull.* 264.

SCOTT, S. G., 1929: Phosphorus deficiency in forage feeds of range cattle, *J. Agr. Res.*, **38**: 113–130.

SHAW, J. K., 1934: Some experience with the Thornton test for potash in apple trees, *Proc. Am. Soc. Hort. Sci.*, **32**: 56–60.

SHEDD, O. M., 1913: The sulphur content of some typical Kentucky soils, *Ky. Sta. Bull.*, 174.

———, 1914: The relation of sulphur to soil fertility, *Ky. Sta. Bull.* 188.

———, 1917: Effect of sulphur on different crops and soils, *J. Agr. Res.*, **11**: 91–105.

SHIVE, J. W., and W. H. MARTIN, 1918: A comparative study of salt requirements for young and for mature buckwheat plants in solution cultures, *J. Agr. Res.*, **14**: 151–176.

———, 1918: A comparison of salt requirements for young and for mature buckwheat plants in water cultures and sand cultures, *Am. J. Bot.*, **5**: 186–191.

———, 1918: A comparative study of salt requirements for young and mature buckwheat plants in sand cultures, *Soil Sci.*, **6**: 1–32.

SHULL, C. A., 1920: Sulphur as a fertilizer, *Sci.*, **52**: 376–378.

SHULOV, I. S., 1909: Various smaller experiments with fertilizers and soils, *Ann. Inst. Agron. Moscow*, **15**: 116–125; *Exp. Sta. Rec.* **22**: 223–224.

SIDERIS, C. P., 1930: Titanium replacing the iron essential in plant growth, *Pineapple News*, **4**: 98–100.

SIDORIN, M. I., 1914: On the assimilation of iron by plants, *Rec. trav. lab. Agron. Moskow, Selsk. Khoz. Inst.*, **10**: 241–257; *Exp. Sta. Rec.* **36**: 633.

SKINNER, J. J., and F. R. REID, 1921: Nutrient requirements of clover and wheat in solution cultures, *Soil Sci.*, **12**: 287–301.

——— and F. E. ALLISON, 1923: Influence of fertilizers containing borax on the growth and fruiting of cotton, *J. Agr. Res.*, **23**: 433–445.

———, B. E. BROWN, and F. R. REID, 1923: The effect of boron on the growth and yield of crops, *U. S. Dept. Agr. Bull.* 1126.

——— and R. W. RUPRECHT, 1930: Fertilizer experiments with truck crops, *Fla. Sta. Bull.*, **218**: 1–65.

———, 1931: Influence of potash sources and chlorine content of fertilizers on yield of cotton, *J. Am. Soc. Agron.*, **23**: 13–21.

SMITH, J. B., F. K. CRANDALL, and D. E. FREAR, 1932: The relative effect of single and fractional applications of soluble nitrogen on nitrates in soil and plant and on the yield of certain vegetable crops, *J. Am. Soc. Agron.*, **24**: 203–221.

SMITH, T. O., and O. BUTLER, 1921: Relation of potassium to growth in plants, *Ann. Bot.*, **35**: 189–227.

SOMMER, A. L., 1926: Studies concerning the essential nature of aluminum and silicon for plant growth, *Univ. Calif. Pub.*, **5**: 57–81.

——— and C. B. LIPMAN, 1926: Evidence of the indispensable nature of zinc and boron for higher green plants, *Plant Physiol.*, **1**: 231–249.

———, 1927: The search for elements essential in only small amounts for plant growth, *Sci.*, **66**: 482–484.

———, 1928: Further evidence of the essential nature of zinc for the growth of higher green plants, *Plant Physiol.*, **3**: 217–223.

——— and H. SOROKIN, 1928: Effect of the absence of boron and of some other essential elements on the cell and tissue structure of the root tips of *Pisum sativum*, *Plant Physiol.*, **3**: 237–261.

———, 1931: Copper as an essential for plant growth, *Plant Physiol.*, **6**: 339–345.

———, 1936: The relationship of the phosphate concentration of solution cultures to the type and size of root systems and the time of maturity of certain plants, *J. Agr. Res.*, **52**: 133–148.

SOROKIN, H., and A. L. SOMMER, 1929: Changes in the cells and tissues of root tips induced by the absence of calcium, *Am. J. Bot.*, **16**: 23–40.

SREENIVASAN, A., 1934: The role of silicon in plant nutrition, *Current Sci.*, **3**: 193–197; *Chem. Abst.*, **29**: 2205.

STEWARD, F. C., 1935: Mineral nutrition of plants, *Ann. Rev. Biochem.*, **4**: 519–544.

STEWARD, W. D., and J. M. ARTHUR, 1933: Some effects of radiation from a quartz mercury arc upon the mineral composition of plants, *abst. Am. J. Bot.*, **20**: 673.

——— and ———, 1934: Some effects of radiation from a quartz mercury vapor lamp upon the mineral composition of plants, *Contr. Boyce Thompson Inst.*, **6**: 225–245.

STEWART, G. R., 1930: Report on chemistry, *Proc. Hawaiian Sugar Planters' Assoc.*, **49**: 328–351; *Chem. Abst.*, **24**: 5404.

STEWART, J., 1922: Some relations of arsenic to plant growth. I and II, *Soil Sci.*, **14**: 111–127.

STEWART, R., 1920: Sulphur in relation to soil fertility, *Ill. Sta. Bull.* 227.

STIEBELING, H. K., 1932: The iron content of vegetables and fruits, *U. S. Dept. Agr. Circ.* 205.

STOCKLASA, J., and J. PITRA, 1901: Ueber die Einwirkung der Kalisalze auf die Entwicklung der Gerste, *Z. landw. Versuch. im Oesterr.*, p. 567.

———, 1908: Beiträge zur Kenntniss der physiologischen Funktion des Kalis im Pflanzenorganismus, *Z. landw. Versuch.*, **11**: 52–61.

———, 1911: Catalytic fertilizers for sugar beets, *Bl. Zuckerrühenbau,* **18**: 193–197; *Exp. Sta. Rec.* **26**: 225.

——— and E. Senft, 1912: Ist das Kalium an dem Auf- und Abbau der Kohlenhydrate bei höheren Pflanzen Beteiligt? *Z. landw. Verwesen Öesterr,* **15**: 711–736.

———, 1916: Ist das Kaliumion an der Eiweisssynthese in der Pflanzenzelle beteiligt? *Biochem. Z.,* **73**: 107–160.

———, 1918: Über den Einfluss des Aluminiumions auf die Keimung des Samens und die Entwicklung der Pflanzen, *Biochem. Z.,* **91**: 137–223.

———, 1918: Über die Verbreitung des Aluminiumions in der Pflanzenwelt, *Biochem. Z.,* **88**: 292–322.

———, 1920: Der Mechanismus der physiologischen Wirkung der Radiumemanation und der Radioaktivität des Kaliums auf die biochemischen Vorgänge bei dem Wachstumsprocess der Pflanzen. II, *Biochem. Z.,* **108**: 140–172.

———, 1920: Bedeutung der Radioaktivität des Kaliums bei der Photosynthese, *Biochem. Z.,* **108**: 173–184.

———, 1920: Ueber die Radioaktivität des Kaliums und ihre Bedeutung in der chlorophyllosen und cholorophyllhaltigen Zelle. I, *Biochem. Z.,* **108**: 109–139.

———, 1922: Ueber die Resorption des Aluminiumions durch das Wurzelsystem der Pflanzen, *Biochem. Z.,* **128**: 35–47.

———, 1922: "Über die Verbreitung des Aluminiums in der Natur und seine Bedeutung beim Bau- und Betriebsstoffwechsel der Pflanzen," 8vo., x + 500 pp., Fischer, Jena.

———, 1924: De la fonction physiologique de l'iode dans organisme de la betterave à sucre, *C. R. Acad. Sci. Paris,* **178**: 120–122.

———, 1929: Über den Einfluss des Jodions auf das Wachstum und die Zellvermehrung der Halophyten, *Biochem. Z.,* **211**: 213–228; *Biol. Abst.,* **5**: 1616.

———, 1930: Die Resorption des Jods durch das Wurzelsystem der Pflanze, *Protoplasma,* **8**: 199–214.

Stokes, W. E., W. A. Leukel, and R. M. Barnette, 1930: Effect of irrigation with sewage effluent on the yields and establishment of Napier grass and Japanese cane, *Fla. Sta. Bull.* 215.

Storey, H. H., and R. Leach, 1933: A sulfur-deficiency disease of the tea bush, *Ann. App. Biol.,* **20**: 23–56.

Street, O. E., 1934: Carbohydrate-nitrogen and base element relationships of peas grown in water culture under various light exposures, *Plant Physiol.,* **9**: 301–322.

Suzuki, V., 1900: Can strontium and barium replace calcium in phanerogams? *Bull. Col. Agr. Tokyo,* **4**: 69–77.

——— and K. Aso, 1903: On the stimulating action of iodine and fluorine compounds on agricultural plants, *Bull. Col. Agr. Tokyo,* **5**: 4; *Bot. Cent.,* **93**: 143.

———, 1919: The translocation of the mineral constituents of the jack bean, *J. Am. Chem. Soc.,* **41**: 282–287.

Swain, R. E., and A. B. Johnson, 1936: Effect of sulfur dioxide on wheat development, *Ind. and Eng. Chem.,* **28**: 42–47.

Swanback, T. R., 1927: The effect of boric acid on the growth of tobacco plants in nutrient solutions, *Plant Physiol.,* **2**: 475–486.

Swanson, C. O., and R. W. Miller, 1917: The sulfur content of some typical Kansas soils and the loss of sulphur due to cultivation, *Soil Sci.,* **3**: 139–148.

Szücs, J., 1913: Über einige charakteristische Wirkungen des Aluminiumions auf das Protoplasma, *Jahrb. wiss. Bot.,* **52**: 269–332.

Tanaka, I., 1931: Nutrition of higher plants with organic compounds, *Japan J. Bot.,* **5**: 323–350; *Chem. Abst.,* **28**: 2388; *Exp. Sta. Rec.,* **68**: 458.

TEAKLE, L. J. H., 1929: The absorption of phosphate from soil and solution cultures, *Plant Physiol.*, **4**: 211–232.

THATCHER, R. W., and G. A. OLSON, 1912: Experiments in fertilizing alfalfa, *Wash. Sta. Pop. Bull.* 49.

———, 1932: Some new aspects of plant nutrition, *Sci.*, **76**: 281–285.

———, 1934: A proposed classification of the chemical elements with respect to their functions in plant nutrition, *Sci.*, **79**: 463–466.

THODAY, D., and H. EVANS, 1932: Studies in growth and differentiation. III, The distribution of calcium and phosphate in the tissues of *Kleinia articulata* and some other plants, *Ann. Bot.*, **46**: 781–807.

——— and ———, 1933: Studies in growth and differentiation. IV, The distribution of some solutes in the tissues of *Kleinia articulata, Ann. Bot.*, **47**: 1–20.

THOMAS, W., 1930: The conception of balance with respect to the absorption of nitrogen, phosphorus and potassium by plants and the influence of the balance of nutrition, *Sci.*, **72**: 425–427.

———, 1932: The reciprocal effects of the nitrogen, phosphorus, and potassium as related to the absorption of these elements by plants, *Soil Sci.*, **33**: 1–21.

———, 1934: Misconceptions relative to the mineral composition of plants, *Sci.*, **80**: 587.

THORNTON, S. F., 1932: A field and laboratory test on plant material for diagnosing phosphorus deficiencies, *Ind. Sta. Bull.* 355.

———, 1933: A simple and rapid chemical test on plant material as an aid in determining potassium needs, *J. Am. Soc. Agron.*, **25**: 473–481.

———, S. D. CONNER, and R. R. FRASURE, 1934: The use of rapid chemical tests on soil and plants as aids in determining fertilizer needs, *Ind. Sta. Circ.* 204.

TIDMORE, J. W., 1930: The phosphorus content of the soil solution and its relation to plant growth, *J. Am. Soc. Agron.*, **22**: 481–488.

———, 1930: Phosphate studies in solution cultures, *Soil Sci.*, **30**: 13–33.

TINCKER, M. A. H., 1932: The influence of the daily period of light and the supply of potassium on the rate of growth of tubers and other storage organs, *J. Roy. Hort. Soc.*, **57**: 326–331.

——— and F. V. DARBISHIRE, 1933: Studies on the formation of tubers and other storage organs. The influence upon translocation of the period of light and the supply of potassium, *Ann. Bot.*, **47**: 27–55.

TOBLER, F., 1929: Zur Kenntnis der Wirkung des Kaliums auf den Bau der Bastfaser, *Jahrb. wiss. Bot.*, **71**: 26–51.

TOTTINGHAM, W. E., 1915: The role of chlorine in plant nutrition, *Sci.*, **42**: 68.

——— and A. J. BECK, 1916: Antagonism between manganese and iron in the growth of wheat, *Plant World*, **19**: 359–370.

———, 1917: On the role of chlorine to plant growth, *Johns Hopkins Univ. Circ.*, **3**: 217–221.

———, 1918: The sulphur requirement of the red clover plant, *J. Biol. Chem.*, **36**: 429–438.

———, 1919: A preliminary study of the influence of chlorides on the growth of certain agricultural plants, *J. Am. Soc. Agron.*, **11**: 1–33.

——— and E. J. RANKIN, 1923: The availability of iron in nutrient solutions for wheat, *Am. J. Bot.*, **10**: 203–211.

TRELEASE, S. F., 1917: A study of salt proportions in a nutrient solution containing chloride as related to the growth of young wheat plants, *Johns Hopkins Univ. Circ.*, **3**: 222–225.

———, 1920: The relation of salt proportions and of concentrations to the growth of young wheat plants in nutrient solutions containing a chloride, *Philippine J. Sci.*, **17**: 527–603.

——— and H. M. Trelease, 1931: Magnesium injury of wheat, *Bull. Torrey Bot. Club*, **58**: 127–149.

——— and A. L. Martin, 1936: Plants made poisonous by selenium absorbed from the soil, *Bot. Rev.*, **2**: 373–396.

True, R. H., and R. B. Harvey, 1918: The absorption of calcium salts by squash seedlings, *Brooklyn Bot. Gard. Mem.*, **1**: 502–512.

———, O. F. Black, and J. W. Kelly, 1919: Ash absorption by spinach from concentrated soil solutions, *J. Agr. Res.*, **16**: 15–27.

———, 1921: The function of calcium in the nutrition of seedlings, *J. Am. Soc. Agron.*, **13**: 91–108.

———, 1922: The significance of calcium for higher green plants, *Sci.*, **55**: 1–6.

Truog, E., 1913: Availability of phosphates to various crops, *Wis. Sta. Bull.* 240.

———, 1916: The utilization of phosphates by agricultural crops, including a new theory regarding the feeding power of plants, *Wis. Sta. Res. Bull.* 41.

———, 1922: The feeding power of plants, *Sci.*, **56**: 294–298.

Tubbs, F. R., 1930: Physiological studies in plant nutrition. II, The effect of manurial deficiency upon the mechanical strength of barley straw, *Ann. Bot.*, **44**: 147–160.

Turner, T. W., 1929: Effect of mineral nutrients upon seed plants. II, Phosphates, *Bot. Gaz.*, **88**: 85–95.

Tyner, E. H., 1935: The feeding power of plants for the potassium in feldspar, exchangeable form, and dilute solution, *Soil Sci.*, **39**: 405–424.

Vandecaveye, S. C., and G. O. Baker, 1932: The effect of fertilizers on crop yields of different soils and on the composition of certain crops, *Wash. Sta. Bull.* 274.

Van Haarlem, J. R., 1935: Some results of mineral deficiency studies, *Ann. Rept. Mich. State Hort. Soc.*, **65**: 73–75.

Van Schreven, D. A., 1935: External and internal symptoms of boron deficiency in tomato, *Tijdschr. Plantenziekten*, **41**: 1–26; *Exp. Sta. Rec.*, **74**: 466–467.

Van Wijk, D. J. R., 1932: Leaching of phosphorus compounds and proteins from grasses by rain water, *J. South Africa Chem. Inst.*, **15**: 54–59.

Ville, G., 1889: Recherches sur les relations qui existent entre la couleur des plantes et la richesse des terres en agents de fertilité, *C. R. Acad. Sci. Paris*, **109**: 397–400.

Vincent, V., and J. Herviaux, 1929: Le sodium dans les végétaux cultivés, *Ann. Sci. Agron.*, **46**: 444–460; *Biol. Abst.*, **5**: 2662.

Voelcker, J. A., 1915: The influence of boron compounds on wheat and barley, *J. Roy. Agr. Soc. England*, **76**: 347–351.

Von Euler, H., and T. Philipson, 1932: Water-soluble growth factors, *Biochem. Z.*, **245**: 418–430; *Chem. Abst.*, **26**: 3281.

Von Sachs, J., 1887: "Lectures on the Physiology of Plants," 288 pp., Clarendon Press, London.

Von Wrangell, M., 1922: Laws governing the phosphoric acid nutrition of plants, *Landw. Jahrb.*, **57**: 1–78; *Exp. Sta. Rec.*, **48**: 426.

———, 1927: Das Jod als Pflanzennährstoff, *Natur. Wiss.*, **15**: 70–73.

Wallace, T., 1930: Experiments on the manuring of fruit trees. III, The effects of deficiencies of potassium, calcium and magnesium, respectively, on the contents of these elements and of phosphorus in the shoot and trunk regions of apple trees, *J. Pom. and Hort. Sci.*, **8**: 23–43.

———, 1931: Chemical investigations relating to potassium deficiency of fruit trees, *J. Pom. and Hort. Sci.*, **9**: 111–121.

——— and E. L. PROEBSTING, 1933: The potassium status of soils and fruit plants in some cases of potassium deficiency, *J. Pom. and Hort. Sci.*, **11**: 120–148.

———, 1935: Investigations on chlorosis of fruit trees. V, The control of lime-induced chlorosis by injection of iron salts, *J. Pom. and Hort. Sci.*, **13**: 54–67.

WARBURG, O., 1925: Iron, the oxygen-carrier of respiration-ferment, *Sci.*, **61**: 575–582.

WARE, J. O., and V. H. YOUNG, 1934: Control of cotton wilt and "rust," *Ark. Sta. Bull.* 308.

WARINGTON, K., 1923: The effect of boric acid and borax on the broad bean and certain other plants, *Ann. Bot.*, **37**: 629–672.

———, 1926: The changes induced in the anatomical structure of *Vicia faba* by the absence of boron from the nutrient solution, *Ann. Bot.*, **40**: 27–42.

———, 1933: The influence of length of day on the response of plants to boron, *Ann. Bot.*, **47**: 429–457.

———, 1934: Studies in the absorption of calcium from nutrient solutions with special reference to the presence or absence of boron, *Ann. Bot.*, **48**: 743–776.

WARNE, L. G. G., 1934: The distribution of potassium in normal and scorched foliage, *Ann. Bot.*, **48**: 57–67.

WATSON, D. J., 1936: The effect of potassium chloride on the diurnal changes of the carbohydrates of the potato leaf, *Ann. Bot.*, **50**: 59–77.

WEBBER, I. E., 1935: Histological characteristics of plants grown in toxic concentrations of boron, *J. Agr. Res.*, **50**: 189–194.

WEBSTER, J. E., 1928: Phosphorus distribution in grains, *J. Agr. Res.*, **37**: 123–125.

———, 1929: Phosphorus in plants, *Bull. Univ. Okla.* (*Proc. Okla. Acad. Sci.*), **9**: 47–49.

——— and F. JANSMA, 1929: A comparison of the copper content of Oklahoma wheat with those of other states, *Sci.*, **70**: 174.

——— and C. DALBLOM, 1930: Changes in the phosphorus content of growing mung beans, *J. Agr. Res.*, **41**: 819–824.

WEEVERS, T., 1911: Untersuchungen über die Lokalisation und Funktion des Kaliums in der Pflanze, *Rec. Trav. Bot. Néerland.*, **8**: 289–332; *Exp. Sta. Rec.* **26**: 823.

———, 1917: Die physiologische Bedeutung des Kaliums in den Pflanzen, *Biochem. Z.*, **78**: 354–357.

WEIHE, A., 1931: Ein Beitrag zur Kieselsäuredüngung, *diss., Univ. Kiel; Biol. Abst.*, **9**: 44.

WEISZMANN, H., 1923: Ueber den Einfluss des Kaliums auf die Entwicklung der Pflanzen und ihren morphologischen und anatomischen Bau, bei besonderer Berücksichtigung der landwirtschaftlichen Kulturpflanzen, *Z. Pflanzendüng.*, **2**: 1–79.

WELTON, F. A., 1928: Lodging in oats and wheat, *Bot. Gaz.*, **85**: 121–151.

——— and V. H. MORRIS, 1931: Lodging in oats and wheat, *Ohio Sta. Bull.* 471.

WENDT, G. V., 1930: Stimulationsversuche mit Jod, *Gartenflora*, **79**: 213–218; *Biol. Abst.*, **7**: 791–792.

WESTER, D. H., 1921: Über den Mangangehalt von (holländischen) Samen, *Biochem. Z.*, **118**: 158–163.

WHEELER, H. J., and B. L. HARTWELL, 1902: Conditions determining the poisonous action of chlorides, *R. I. Sta. 15th Ann. Rept.*, 287–304.

——— and ———, 1904: Magnesium as a manure, *R. I. Sta. 17th Ann. Rept.*, **1904**: 221–260.

———, 1905: Plant peculiarities as shown by the influence of sodium salts, *R. I. Sta. Bull.* 104.

——— and G. E. ADAMS, 1905: Concerning the agricultural value of sodium salts, *R. I. Sta. Bull.* 106.

WHERRY, E. T., and R. BUCHANAN, 1926: Composition of the ash of Spanish moss, *Ecology*, **7**: 303–307.

WHITE, H. L., 1936: The interaction of factors in the growth of *Lemna*. VII, The effect of potassium on growth and multiplication, *Ann. Bot.*, **50**: 175–196.

WILCOX, E. V., and W. P. KELLEY, 1912: The effect of manganese on pineapple plants and the ripening of the pineapple fruit, *Hawaii Sta. Bull.* 28.

WILFARTH, H., H. RÖMER, and G. WIMMER, 1906: Über die Nährstoffaufnahme der Pflanzen in verschiedenen Zeiten ihres Wachstums, *Landw. Versuchs-Stat.*, **63**: 1–70.

WILKINS, L. K., 1917: The high calcium content of some cucurbit vines, *N. J. Sta. Bull.* 310.

WILLIS, L. G., and J. O. CARRERO, 1921: Rice investigations, *Porto Rico Sta. Rept.*, **1921**: 8–9.

——— and J. R. PILAND, 1934: The influence of copper sulfate on iron absorption by corn plants, *Soil Sci.*, **37**: 79–84.

———, 1936: Bibliography of references on the literature on the minor elements and their relation to the science of plant nutrition, Chilean Nitrate Ed. Bur., New York, N. Y.

WILLSTÄTTER, R., 1906: Zur Kenntniss der Zusammensetzung des Chlorophylls, *Ann. Chem.*, **350**: 48–82.

——— and M. BENZ, 1908: Untersuchungen über Chlorophyll. VI, Ueber Krystallisirtes Chlorophyll, *Annalen*, **358**: 267–287.

——— and A. STOLL, 1913: "Untersuchung über Chlorophyllen, Methoden, und Ergebnisse," Julius Springer, Berlin.

WILSON, L. B., 1933: Effects of chlorine, bromine and fluorine on the tobacco plant, *J. Agr. Res.*, **46**: 889–899.

WINSTON, J. R., 1935: Reducing decay in citrus fruits with borax, *U. S. Dept. Agr. Tech. Bull.* 488.

WOHACK, F., 1930: Zur Frostschutzwirkung der Kalisalze, *Ernähr. der Pflanze*, **26**: 318–319; *Biol. Abst.*, **5**: 2692.

WÖHER, A., 1920: Uber die Giftwirkung von Arsen-, Antimon-, und Fluorverbindungen auf einige Kulturpflanzen, *Angew. Bot.*, **2**: 161–178.

WOLFF, J., 1913: De l'influence du fer dans le développement de l'orge et sur la spécificité de son action, *C. R. Acad. Sci. Paris*, **157**: 1022–1024.

WOODARD, J., 1922: Sulphur as a factor in soil fertility, *Bot. Gaz.*, **73**: 81–110.

WYATT, F. A., 1916: Influence of calcium and magnesium compounds on plant growth, *J. Agr. Res.*, **6**: 589–621.

WYND, F. L., and H. J. FULLER, 1931: Effects of ultra-violet radiation on the calcium and phosphorus of plants, *abst. Am. J. Bot.*, **18**: 895.

———, 1933: Sources of carbohydrate for germination and growth of orchid seedlings, *Ann. Mo. Bot. Gard.*, **20**: 569–581.

———, 1934: The effects of increasing the iodine content of the tomato plant on respiration and enzymatic activity, *Ann. Mo. Bot. Gard.*, **21**: 367–432.

YAMANO, Y., 1905: Can aluminum salts enhance plant growth? *Bull. Col. Agr. Tokyo*, **6**: 429–432.

YANOVSKY, E., E. K. NELSON, and R. M. KINGSBURY, 1932: Berries rich in calcium, *Sci.*, **75**: 564–566.

YOUNG, R. S., 1935: Certain rarer elements in soils and fertilizers and their role in plant growth, *N. Y. Sta. (Cornell) Mem.* 174.

YOUNG, V. H., G. JANSSEN, and J. O. WARE, 1932: Cotton wilt studies. IV, Effect of fertilizers on cotton wilt, *Ark. Sta. Bull.* 272.

ZAITSEVA, A. A., 1929: Influence of magnesium on chlorophyll accumulation in algae and higher plants, *Bull. Inst. Sci. Lesshaft.*, **15**: 137–175; *Biol. Abst.*, **5**: 2664.

CHAPTER VII

THE LOSS OF WATER FROM PLANTS

I. GENERAL NEED OF WATER BY PLANTS

It is common knowledge that water is absolutely essential to animal life, and that if it is withheld for a relatively short period from animal organisms death ensues. Water is just as essential to plant life, and if a plant is deprived of water it will perish, perhaps not so quickly as an animal but just as surely nevertheless. Let us now consider the relationship of water to the life of the plant in order to ascertain the need of a continuous water supply by a plant that is normally carrying on its various functions.

A. The Protoplasm

A large amount of water is necessary to the proper functioning of the protoplasm, so that it has been said that "All living matter lives in water." The younger, the more vigorous, the more impressionable, the more active, and the more rapidly growing protoplasm is the more water it contains. As an example of the large amount of water in the actively growing regions of plants, the tip of a young stalk of corn or the base of the growing leaves of the same plant may be cited. These regions of the corn plant are composed of 92 to 93 per cent of water, most of which is directly associated with the protoplasm, which completely fills the cells of these portions of the plant. When we consider that the flesh of a ripe watermelon contains no higher percentage of water than the growing regions just mentioned, we can realize the relatively large amount of water that is required in the protoplasm of plant cells to bring about normal growth and development.

If the normal amount of water in the protoplasm is removed, it becomes less and less active until a point is reached where death ensues. The case of seeds is a good example of what takes place when water is withdrawn from the protoplasm. The protoplasm in the cells that make up the seed is living, but so much water has been withdrawn from it in the process of curing that it is very inactive or dormant. If water, however, is supplied to the seed, the protoplasm obtains the amount necessary to carry on its functions actively, and the seed begins to grow.

B. Vacuoles of Cells

The relationship of the vacuole to the plant cell has been discussed in Chap. I. A certain amount of water in the vacuole is necessary to maintain the normal condition of turgor essential for the growth of the cells and for the erectness of the plant. The amount of water in the various parts of plants will depend upon the type of plant and the conditions under which it is examined. The following list shows the amount of water in the leaves of plants grown at Manhattan, Kans. These leaves were fully developed and in full vegetative vigor. The samples were collected at 10 A.M.

Kind of leaf	Percentage of water	Kind of leaf	Percentage of water
Cabbage	86	Pumpkin	80
Tomato	84	Corn	77
Cowpea	82	Apple	60

C. For Continual Translocation of Materials from One Part of Plant to Another

As far as is known, all foods and materials translocated in a plant move from cell to cell in a watery solution.

D. For Chemical Combination

Many of the various organic compounds found in plants are formed by the combination of water with certain materials that enter the plant from the soil and the air.

The water needed by any plant during its growing season for the purposes that have just been mentioned amounts to very little. Thus, for example, a normal corn plant that reached a height of 10 to 12 ft. and produced a good yield of grain had, including the stem, leaves, grain, cob, and roots, a moist weight of 2,707 g. at the height of its full vegetative growth when the vegetative parts were all intact and the grain was in the late dough stage and was well glazed and dented. The total weight of dry matter of this plant was 835 g. The water content of the plant was thus 1,872 g. This means that the water in this plant, which was needed for the vacuoles of the cells, for the protoplasm, and for the translocation of materials, did not exceed this amount. The amount of water necessary for chemical combination in this case would not amount to over 250 g. even if we consider that all the hydrogen and oxygen in the dry matter was obtained by the plant from water. According to these figures, a corn plant would thus need during its entire growing season only 2,122 g. (a little over 2 l. or approximately 2⅕ qt.) of water to supply it with a sufficient amount for the four uses that have just been mentioned. It is well known, however, that a single corn plant requires during its growing season a much greater quantity of water than that which has just been stated. A single corn plant during the growing season of 1924 at Manhattan, Kans., removed from the soil 54 gal. or 1⅘ bbl. of water. This is approximately ninety-eight times more than is needed for the cell vacuoles, translocation, protoplasm, and chemical combination. What,

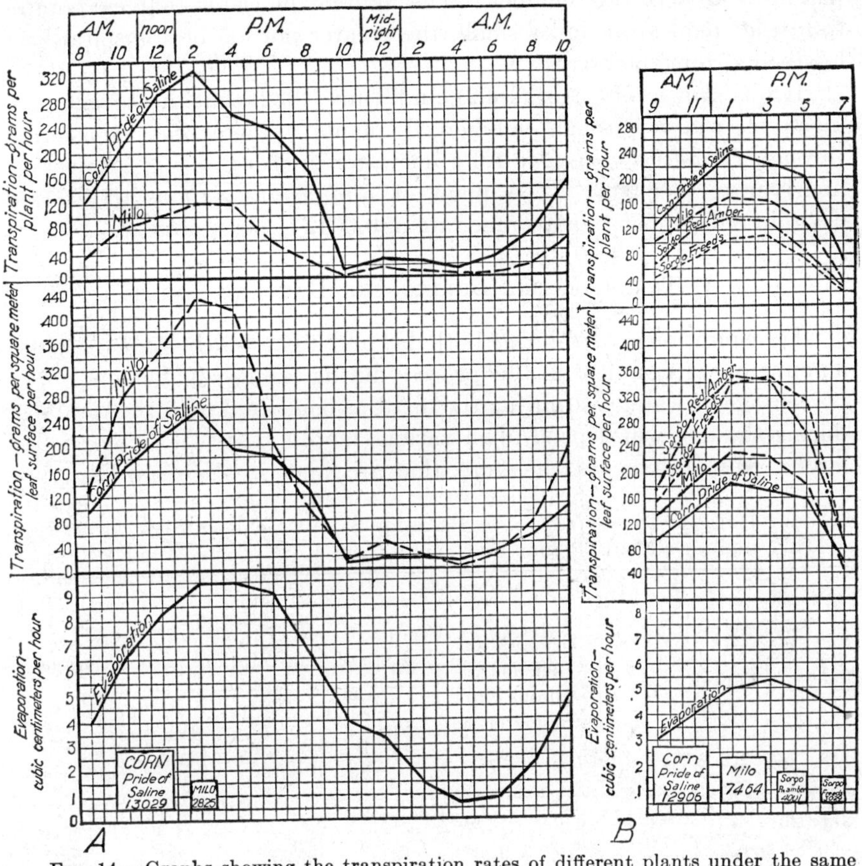

FIG. 14.—Graphs showing the transpiration rates of different plants under the same conditions and the distribution of their water loss during the day. The evaporation from a porous cup atmometer for the corresponding periods is also shown. The squares at the bottom of the figure represent the extent of the leaf surface of each plant expressed in square centimeters. *A*, the transpiration rates of Pride of Saline corn and Dwarf Yellow milo. *B*, the transpiration rates of Pride of Saline corn, Dwarf Yellow milo, Red Amber sorgo, and Freed's sorgo.

then, is the need of such a large supply of water for a plant during its growing season?

E. Replacement

A supply of water in the soil is needed by a plant at all times during its growing season to replace that which is being lost continuously from its stem and leaves by transpiration and by guttation.

II. THE EVAPORATION OF WATER FROM PLANTS. TRANSPIRATION

Transpiration is the loss of water in the form of vapor from the stem and leaves of the living plant. Practically all the water that is lost from plants escapes in this manner. The amount of water that evaporates from the stem, however, is small, the greater portion of transpirational loss being from the leaves.

It was shown by Unger (1862), and Thut (1932) that there is a movement of water through submerged plants. There was transferred through *Potamogeton crispus* 1.6 g. of water in 8 days, or at an average hourly rate of 0.0083 g. A plant of *Ranunculus fluitans* transferred 0.8 g. in 8 days, or at an average rate of 0.0042 g. per hour.

A. Transpiration during the Day and Night

Transpiration occurs during both day and night, but by far the greater amount of water is lost in this process during the daylight hours. The percentage of water lost during the night as compared with the total lost during the day will depend upon various factors that will be mentioned later. Any statements, therefore, that are made in this regard must be

Hourly Rate of Transpiration of Pride of Saline Corn and Dwarf Yellow Milo, during a 24-hr. Period at Garden City, Kans., July 26 and 27, 1916

Period	Evaporation per hour, porous cup atmometer, cc.	Corn, Pride of Saline		Milo, Dwarf Yellow	
		Rate of transpiration per hour, grams			
		Per plant	Per square meter of leaf surface	Per plant	Per square meter of leaf surface
July 26:					
6–8 a.m.	3.9	121	93	35	126
8–10 a.m.	6.7	212	163	79	281
10–12 noon	8.2	291	223	99	350
12–2 p.m.	9.5	329	253	122	432
2–4 p.m.	9.5	259	198	117	413
4–6 p.m.	9.1	236	181	58	206
6–8 p.m.	6.7	164	126	27	94
8–10 p.m.	3.9	11	8	3	12
10–12 mid.	3.4	25	19	13	44
July 27:					
12–2 a.m.	1.6	25	19	5	19
2–4 a.m.	0.8	18	14	2	6
4–6 a.m.	0.9	31	24	5	19
6–8 a.m.	2.7	71	55	21	75

general and relative only to the conditions prevailing at the time the observations were made.

The general distribution of the water loss during a 24-hr. period can be shown by an experiment selected from the work of Miller and Coffman (1918), as shown in Fig. 14 and in the table on page 410.

They found that from 94 to 96 per cent of the water transpired was lost during the daylight hours under the climatic conditions prevailing at Garden City, Kans. Briggs and Shantz (1916) observed at Akron, Colo., that the transpirational loss from various crop plants from sunset to sunrise amounted to only 3 to 6 per cent of the total transpiration for the 24-hr. period, according to the following table:

Plant	Percentage of total transpiration	
	Daylight	Darkness
Wheat	96	4
Oats	94	6
Rye	95	5
Sorghums	95	5
Alfalfa	97	3
Pigweed (*Amaranthus*)	97	3

Under the climatic conditions that prevail in the Great Plains area of the United States less water is transpired by plants during the forenoon than during the afternoon. The percentage of water transpired during the forenoon hours is, as a rule, from 40 to 50 per cent of the total water transpired during the daylight hours. As a general rule, the maximum rate of transpiration is reached somewhere between 11 A.M. and 3 P.M. Miller and Coffman (1918) found in 30 experiments with corn that the maximum rate of transpiration occurred eighteen times between 11 A.M. and 1 P.M. and twelve times between 1 P.M. and 3 P.M. In the case of 52 observations of various varieties of the sorghums, the maximum rate of transpiration occurred twenty-eight times between 11 A.M. and 1 P.M. and twenty-four times between 1 P.M. and 3 P.M.

Meyer (1932) noted that the peak of transpiration of young tulip trees (*Liriodendron tulipifera*) occurred shortly before noon. Of the 148 species of plants, in 16 plant associations in Ohio and Indiana, studied by Blaydes (1935), 93 per cent had their maximum rate of transpiration between 7 A.M. and 12 noon, over 89 per cent between 7 and 11 A.M., and 52 per cent between 9 and 11 A.M.

B. Amount of Water Lost

The total amount of water that is evaporated from plants during a growing season will depend primarily upon the prevailing climatic conditions, but the following data will serve as an example of the relatively large amount that escapes from the plant by transpiration. The data here presented were obtained by the author at Manhattan, Kans., in 1924 to 1925. The plants were grown in sealed containers and the soil was kept in good moisture tilth by the daily addition to it of the water that had been removed by the plants during that time (Fig. 21).

As shown in the table below, a single Kansas Sunflower corn plant during its growing season in 1924 lost 54 gal. of water. Let us consider how much water would thus be evaporated by an acre of corn plants during the growing season if the soil were supplied with sufficient moisture for normal growth and development.

Plant	Length of season	Total water lost during the season per plant, gallons
Cowpea	May 19 to Sept. 2, 1924	13
Potato, Irish cobbler, 1 hill, 2 to 3 plants	April 18 to July 30, 1924	25
Tomato, Louisiana Pink, pruned to 1 stem	May 19 to Sept. 2, 1924	34
Corn, Kansas Sunflower	May 5 to Sept. 8, 1924	54
Winter wheat, Kanred, 1 plant 15 stalks	Oct. 15 1924 to June 28, 1925	25
Sunflower, Russian	May 26 to Aug. 23, 1925	123
Sunflower, wild	May 24 to Aug. 23, 1925	130
Giant ragweed (*Ambrosia trifida*)	May 24 to Aug. 23, 1925	140

If the corn were in 44-in. rows, the plants 2 ft. apart in the row, with one plant per hill, we should have about 6,000 plants per acre if a perfect stand were obtained. The total amount of water that would be evaporated from the leaves and sheaths of an acre of corn during the growing season under these conditions would be 324,000 gal. or 1,296 tons of water. Since 1 in. of rainfall per acre weighs 113.25 tons, this is equivalent to approximately 11 acre in. of rainfall. A single corn plant that has reached its full leaf development has been observed to transpire as much as 32 qt. of water during a single week under the summer conditions of the Great Plains. This means an average of approximately 4½ qt. per plant per day. When the body weight of the plant is considered it is found that the plant during each day of the 7-day period that has just been mentioned evaporated a quantity of water equal to twice its body weight.

The question naturally arises as to what is the cause of such a large loss of water from plants. In order to understand this, we must consider the general structure of the leaves, and it is to that topic that we shall now devote our attention.

C. Factors That Influence the Rate of Transpiration

1. Leaf Structure.—It is not the intention here to enter into any detailed discussion of the anatomy of the leaf but only to mention in a general way those characteristics of its structure that are conducive to the loss of water by evaporation. The structure of leaves varies

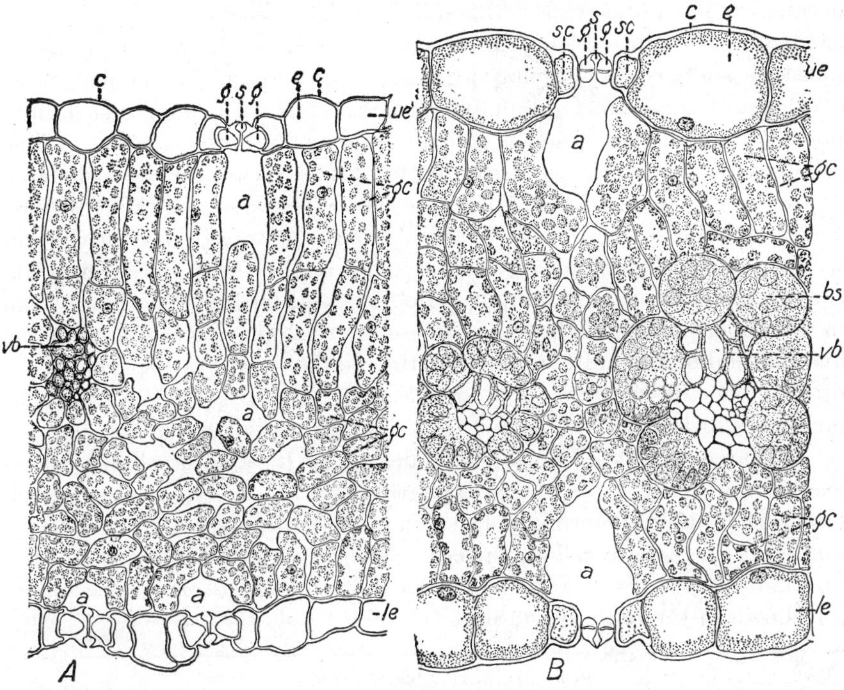

FIG. 15.—Cross section of leaves. *A*, alfalfa leaf. *B*, corn leaf. *a*, air space. *ue*, upper epidermis. *le*, lower epidermis. *gc*, green cells. *vb*, fibrovascular bundle. *bs*, bundle sheath. *e*, ordinary epidermal cell. *c*, cuticle. *sc*, subsidiary cells. *g*, guard cells. *s*, stoma.

greatly for different plants, as can be seen by comparing the cross sections of the leaves of corn and alfalfa as shown in Fig. 15 *A* and *B*. Certain general characteristics, however, are common to all leaves in relation to water loss, and it is these facts that we wish to consider here.

The leaf of most angiosperms is a relatively broad, flat, thin organ with a large amount of surface in proportion to its mass. A better conception of the large amount of surface exposed by leaves may be had by citing one or two examples. The leaf and sheath surface of an ordinary variety of corn plant grown under the conditions of the corn belt amounts to from 30 to 35 sq. ft. or is from four to six times as extensive as the surface of the soil in which the plants are growing. A wild sunflower plant growing under favorable conditions may have as much as 40 to 50 sq. ft. of leaf surface.

The ordinary leaf is composed, for the most part, of thin-walled parenchyma tissue through which numerous finely divided vascular bundles permeate and end free so that as many as 6,000 of these minute veins may end in a square centimeter of leaf. The entire leaf is covered by a layer of cells, generally only one cell in thickness, called the "epidermis." The outer wall of the epidermis is generally somewhat thickened and is cutinized and, as described in Chap. I, is impervious to moisture and gases in varying degrees. The epidermis of the leaf, however, is not completely intact but is perforated with numerous minute pores. Each of these pores is called a "stoma" (plural, stomata). The mesophyll or parenchyma tissue of the leaf is very loosely constructed so that there are many air spaces between the cells. In fact, in most leaves there is scarcely a mesophyll cell that does not come in contact with an intercellular space. In a typical dorsiventral leaf (Fig. 15*A*) the intercellular spaces are more numerous and larger in the lower than in the upper portion of the leaf. These intercellular spaces join with one another and finally unite with one of the relatively large air spaces that is directly beneath each pore or stoma (Fig. 15 *A* and *B*). These intercellular spaces form from 3 to 70 per cent of the total volume of the leaf.

The area of the cell surfaces bordering on these intercellular spaces exceeds by many times the area of the external leaf surface. Turrell (1933, 1934, 1936) determined the area of the exposed portion of the cells bordering on the intercellular spaces. The ratios between the area of the internally exposed cell walls and the external surfaces of various plants are shown in the following table:

Plant	Ratio of internal surface to external surface	Plant	Ratio of internal surface to external surface
Syringa vulgaris	13.2	Succulents	7.8
Vitis vulpina	11.6	Mesomorphic	11.6 to 16.3
Citrus limonia	22.2	Xeromorphic (in sun)	22.2 to 31.3
Berberis nervosa	9.8	Xeromorphic (in shade)	8.2 to 9.9

It is seen that the thin-walled cells of the mesophyll are in direct contact with the fibrovascular bundles, which, in turn, are in direct connection with the water supply of the roots and the soil. These cells, on the other hand, have a portion of their moist surface exposed to the intercellular air cavities which are connected either directly or indirectly with the atmosphere through the stomata of the leaf. The leaf structure thus is such that conditions are very favorable for the loss of water by evaporation. A moist porous cell-wall surface is exposed to the air on at least one side and is in contact on the other side under favorable conditions with a continuous water supply. The facts that have just been mentioned in regard to the leaves apply also in a more limited sense to the young stems of plants. Although the amount of water lost from the young stem parts is often considerable, the amount is negligible in comparison with the amount lost from the leaves. The large loss of water from plants by transpiration is thus due primarily to the structure of the leaf. The structure of the leaf is apparently an adaptation for the absorption of gases that it needs (Bessey, 1898) or for the excretion of those that arise in its metabolism, but it is poorly adapted for the retention of water, so that the continuous loss of water from it is a constant menace to the life of the plant.

It should be mentioned here that different plants exposed to the same environmental conditions do not have the same rate of transpiration. This may be illustrated, for example, by the experimental data obtained by Miller (1925) at the Kansas Agricultural Experiment Station in observations of corn, milo, pumpkin, cowpea, and soybean plants grown under field conditions, as shown in the table on page 416.

Blaydes (1928, 1935) collected data by means of the cobalt chloride-paper method on the transpiration rate of 148 species of plants growing under a wide range of conditions. The average standard water loss for all the species on which data were collected was 1.150 g. of water vapor per hour per 100 sq. cm. of leaf surface. On the same basis the greatest water loss was from *Plantago major*, *Polygonum amphibium*, and *Iris versicolor* with a rate of 3.802, 3.661, 3.452 g. per square decimeter per hour, respectively. The lowest transpiration rates were 0.074, 0.076, and 0.085 g. per 100 sq. cm. per hour, respectively, for *Juglans nigra*, *Bromus inermis*, and *Silphium perfoliatum*.

Weaver and Mogensen (1919) noted that the autumnal transpiration losses from conifers were as high per unit of surface or even higher than those from broad leaves. The winter losses from conifers were found to be only $1/251$ to $1/55$ as great as those in the autumn. The winter losses from conifers were scarcely greater than those from defoliated stems of broad-leaved trees.

Kelley (1930) in a study of the rate of transpiration of 21 species of deciduous fruit plants stated that *Ribes vulgare*, *Corylus maxima*, *Rubus occidentalis*, *Pyrus malus*, *Pyrus communis*, and *Cydonia maxima* showed a high rate of transpiration; *Fragaria chiloensis*, *Ribes grossularia*, and *Prunus domestica* a medium rate; and *Juglans nigra*, *Prunus amygdalus*, and *Prunus persica* the lowest rate of the plants examined. It was

Relative Transpiration of Corn, Milo, Pumpkin, Cowpeas and Soybeans under the Same Environmental Conditions in the Field at Manhattan, Kans.

Date	Corn, Kansas Sunflower		Milo, Dwarf Yellow		Pumpkin, Cheese		Cowpeas, New Era		Soybeans, Midwest		Evaporation
	Per plant	Per square meter of leaf surface	Per plant	Per square meter of leaf surface	Per plant	Per square meter of leaf surface	Per plant	Per square meter of leaf surface	Per plant	Per square meter of leaf surface	Porous cup atmometer, cc.
July 7–8:											
9 a.m.–4 p.m.	222	146	142	167	126	177	54	129	68	136	35.4
4 p.m.–9 a.m.	22	14	10	12	13	18	7	16	9	18	23.1
July 8–9:											
9 a.m.–4 p.m.	240	158	178	210	168	234	65	156	81	161	41.5
4 p.m.–9 a.m.	18	12	5	6	9	13	7	17	6	12	17.5
July 9–10:											
9 a.m.–4 p.m.	227	150	183	216	165	230	78	188	90	179	30.0
4 p.m.–9 a.m.	19	12	9	10	12	17	6	13	6	12	14.0

observed by Haas and Halma (1932) that the transpiration rate was highest for the lemon trees, less for grapefruit, and least for the orange. Ashby (1932) reported that transpiration from the wilting shoots of the creosote bush was greater than that from the privet. Compared with privet, the creosote stem offered twice the resistance to the passage of water under pressure. The root suction pressure of the latter was three times that of the former. Sperry (1936) noted in the Rocky Mountain National Park that the different species of pine varied markedly in their transpiration rates.

It was shown by Dowd (1931) that each variety of apple twigs has a distinct transpiration rate. These differences amounted to as much as 50 per cent in some cases. Markley and Sando (1931) found with but few exceptions that the skins of apples from New York were more permeable to water than those from Washington.

A striking example of the different effects of the changes in the environmental conditions upon the rate of transpiration of different plants is offered by the observations of Sayre (1919), who found that an increase in temperature of 23°F. and a decrease in humidity of 35 per cent increased the transpiration rate of the mullein plant only twenty-five times as compared to forty times for the tobacco plant. The two main factors that may contribute to these differences in response to the environment are the differences in anatomical structure and the specific ability of the plant to bring about changes in itself. Of the anatomical differences that apparently contribute to variation in the transpiration rate of different plants, the structure of the leaves is probably the most important. On account of their variation in structure, the leaves differ in their response to solar energy. Their absorption coefficients may not be the same, and the dissipation of the absorbed energy may be very different for different plants. We shall now discuss the more important structural differences in the anatomy of the leaves and the influence of the structural differences upon the rate of transpiration.

a. The Stomata.—The stomata or stomates (singular, stoma or stomate) are minute more or less elongated openings through the epidermis of plants. They occur in the epidermis of any part of the plant (Anderson, 1897; Jivanna Rao, 1923; and Barrett, 1929) except the root but are found in the greatest numbers in the leaves. A stoma or stomate is simply an intercellular opening between two specialized epidermal cells that are termed "guard cells." (Fig. 16*A*, *B*, and *C*.) In the most simple development, the stoma is formed by the division of a young epidermal cell, which has become ellipsoidal in outline, into two daughter cells. The dissolution of the middle lamella of the walls separating these two cells then occurs and the stoma or intercellular passage is formed by the splitting of two walls. The

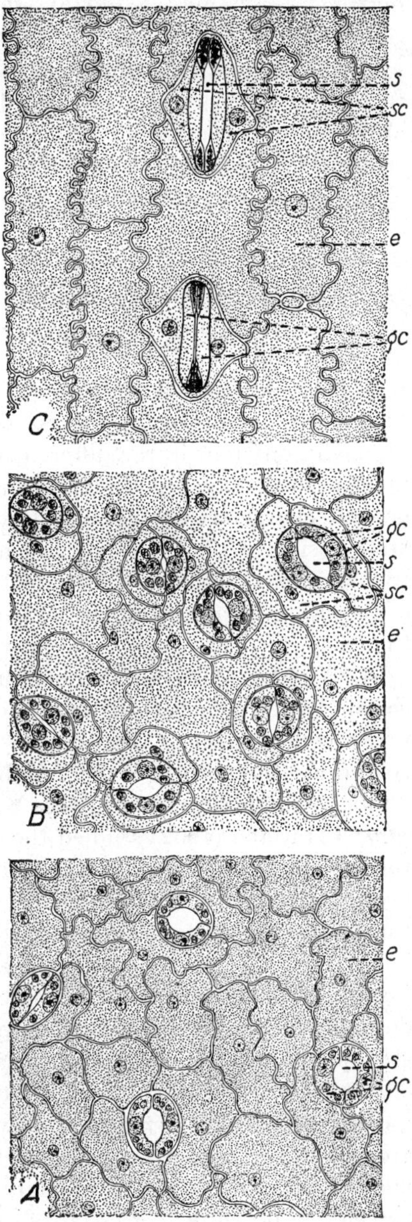

FIG. 16.—Stomata. Surface view of the epidermis of *A*, alfalfa leaf; *B*, leaf of cowpea; and *C*, leaf of corn. *s*, stoma. *gc*, guard cells. *sc*, subsidiary cells. *e*, ordinary epidermal cells.

stoma may be opened and closed in varying degrees by changes in the size and shape of the guard cells. The changes are caused by the variation of the turgor of the cells in connection with the structure of their cell walls. In some cases, the epidermal cells that border on the guard cells are also different from the ordinary epidermal cells and are called "subsidiary" or "accessory cells." The guard cells are richer in protoplasmic material, contain chloroplasts, are different in form, and have a higher concentration of the cell sap than the ordinary epidermal cells. The term "stoma" is also used to denote collectively the intercellular opening, the guard cells, and the subsidiary cells. In that case, the orifice is known as the "stomatal pore." In this text, however, the term stoma will be used to denote only the pore through the epidermis, while the guard cells and accessory or subsidiary cells will be collectively termed the "stomatal apparatus." The structure of the stomatal apparatus is markedly different in the different groups of plants. These peculiarities in structure may vary slightly under different conditions, but their general characteristics are constant. The peculiar type of guard cells and subsidiary cells of the grains and grasses is a good example of this (Schwendener, 1889). Rehfous (1917) considered that the structures of the stomatal apparatus are so constant within a given group of plants that they are of first importance in indicating phylogeny and relationships.

1. *Number.*—The number of stomata per unit of leaf surface is, within certain limits, a characteristic of the particular species or variety of plant under consideration. There seems to be evidence that environmental conditions do influence the number of stomata per unit of surface, although the information in that regard is rather meager and contradictory. Mariana (1902) studied the effect of the humidity of the air upon the development of the stomata on the cotyledonary leaves of buckwheat, beet, white lupine, crimson clover, and other plants and concluded that there were fewer stomata produced per unit of leaf area in a relatively moist than in a relatively dry atmosphere. He considered that humidity favors the development of the superficial areas of the leaves but that it does not increase the absolute number of the stomata. Duggar (1911) grew plants of wheat and corn in sand cultures containing, respectively, 38, 30, 20, 15, and 11 per cent of moisture and observed that the greatest number of stomata per unit area was found on the leaves of the plants growing in the sand with the highest moisture content, while the smallest number was on the leaves of plants grown in the sand with the lowest percentage of water. Kiesselbach (1916), however, in experiments with Hogue Yellow Dent and Pride of the North varieties of corn did not find any striking response in the relative number of stomata to variations in either soil moisture or soil fertility. Weiss (1866) also concluded from

extensive observations that the conditions under which plants are grown do not influence either the size or the number of their stomata.

It was noted by Reed and Hirano (1931) with *Citrus* that the density of the stomata varies in different portions of the leaf and between species. However, their density is decreased in all cases when the intensity of light is reduced. Tavčar (1926) believed that it is possible by crossing plants with larger and smaller numbers of stomata to change their density in the hybrid and thus alter to a certain degree the physiological processes that are concerned with them. Hirano (1931) noted that the species of *Citrus* in the tropics have more stomata per unit area than those growing in other regions. In studying barley, wheat, and corn, Van de Roovaart and Fuller (1935) found, with plants growing in a soil-moisture content ranging from 11 to 38 per cent, that there is a lower stomatal frequency under optimum conditions, because the ordinary epidermal cells are large, and hence there is a greater distance between the stomata. The ratio of the number of stomata to that of the ordinary cells, however, remains almost constant. The influence of a varying water content of the soil upon the number of stomata per unit of area is shown in the following table:

Soil moisture (wilting coefficient, 18 per cent), per cent	Stomata per unit area		
	Barley	Wheat	Corn
38	5.1	6.2	5.2
17	6.2	6.2	5.2
11	7.5	7.7	7.8

Niemann (1932) and Penfound (1932) also found that the number of stomata varied directly with the moisture content of the soil. Wood (1934) considered, however, that variations in stomatal frequency appear to be connected with generic and family characters to a greater extent than with environmental conditions.

Salisbury (1927) in an extensive study of the stomatal frequency of the woodland flora of England concluded that stomatal frequency is mainly dependent on the humidity of the environment, dry exposed conditions being associated with high frequencies and humid conditions with low frequencies. He noted also that the total pore space per unit area of leaf surface is greater in dry than in moist conditions, despite the larger size attained by the stomata in a humid environment. He makes the very important statement in this regard that there is a high positive correlation coefficient between the number of stomata and the number of epidermal cells per unit of area. The difference in frequency of stomata in wet and dry soil, small leaves as compared with large leaves, and different parts of the same leaf is due to differences in growth. Such differences in frequency are thus due to the spacing of the stomata and not to the differences in the number of stomata developed per number of epidermal cells in a unit area.

Salisbury proposed the stomatal index

$$I = \frac{S}{E + S} \times 100$$

where S equals the number of stomata per unit of area and E the number of ordinary epidermal cells in the same unit area. The stomatal index thus aims at expressing the percentage proportion of the ultimate divisions of the dermatogen of the leaf that have been converted into stomata. This infers that under a given set of conditions a species tends to form a definite proportion of stomatal initials. Salisbury considered that there is some evidence that high humidity tends to reduce the proportion of stomata formed, and aquatics would seem to have low stomatal indexes. Variations in the stomatal index appear to be due to internal factors, of which nutritional conditions are perhaps the most important. Hirano

Number of Stomata on the Leaves of Pride of Saline Corn, Dwarf Yellow Milo, and Blackhull Kafir during Nine Growing Seasons

Year	Number of stomata, thousands per square inch of leaf surface											
	Corn, Pride of Saline				Milo, Dwarf Yellow				Kafir, Blackhull			
	Upper	Lower	Total	Ratio lower to upper	Upper	Lower	Total	Ratio lower to upper	Upper	Lower	Total	Ratio lower to upper
Garden City, Kans.:												
1914	43	55	98	1.28	66	97	163	1.42	48	103	151	2.14
1915	40	50	90	1.25	64	76	140	1.17	44	89	133	2.02
1916	49	53	102	1.08	67	90	157	1.34				
1917	40	51	91	1.27	53	74	127	1.49	45	82	127	1.82
Manhattan, Kans.:												
1918	60	73	133	1.21	81	100	181	1.23	58	103	161	1.94
1919	39	51	90	1.51	59	85	144	1.27	55	91	146	1.65
1920	36	49	85	1.36	56	91	147	1.66	46	106	152	2.30
1922	43	56	99	1.30	61	87	148	1.42	51	98	149	1.92
1923	34	48	82	1.41	58	89	147	1.53	55	100	155	1.81
Average	42.6	54.0	96.6	1.27	62.8	87.6	150.4	1.40	50.2	96.5	146.7	1.92

(1931) noted in *Citrus*, that precipitation during the spring months, while the leaves were developing, showed a better correlation with the number of stomata than did the amount of annual rainfall.

Miller (1928) recorded the number of stomata per unit area for one variety each of corn, milo, and kafir during nine successive growing seasons. A summary of the data is given in the table on page 420.

The number was found to vary greatly from year to year. Thus for corn the number ranged from 82,000 in 1923 to 133,000 per square inch in 1918; for milo leaves from 140,000 in 1915 to 181,000 per square inch in 1918; and for kafir from 127,000 in 1917 to 161,000 per square inch in 1918. The number for each of the three plants was higher in 1918 than in any other of the growing seasons, but the variations in numbers during the other growing seasons were not consistent. Thus, for example, in 1923 the total number of stomata on corn leaves was only 82,000 per square inch, the lowest for corn of any of the nine seasons, while under the same conditions the leaves of kafir showed next to the highest number of stomata for that plant during the nine seasons.

The relative number of stomata on the lower and upper surfaces also showed marked fluctuations. For example, in corn the ratio of the number on the lower to that on the upper surface varied from 1.08 in 1916 to as much as 1.40 in 1923, and on kafir from 1.6 in 1919 to 2.3 in

Summary of the Observations on the Number of Stomata on the Leaves of Crop Plants

Plant	Number of years observed	Average number stomata, thousands per square inch of leaf surface			
		Upper	Lower	Total	Ratio lower to upper
Corn:					
Pride of Saline	9	43	54	97	1.255
Sherrod White Dent	4	48	60	108	1.250
Freed White Dent	6	44	57	101	1.295
Kansas Sunflower	5	49	59	108	1.204
Reid Yellow Dent	5	43	58	101	1.348
Sorghums:					
Sudan Grass	6	58	79	137	1.362
Feterita	6	71	98	169	1.380
Milo, Dwarf Yellow	9	63	88	151	1.396
Milo, White	4	65	86	151	1.322
Kafir, Blackhull	8	50	97	147	1.940
Kafir, Dawn	5	57	94	151	1.649
Kafir, Pink	4	53	100	153	1.886
Sorgo, Freed	5	65	98	163	1.507

Summary of the Observations on the Number of Stomata on the Leaves of Crop Plants.—(*Continued*)

Plant	Number of years observed	Average number stomata, thousands per square inch of leaf surface			
		Upper	Lower	Total	Ratio lower to upper
Sorghums: (*Continued*)					
Sorgo, Red Amber	5	51	86	137	1.686
Sorgo, Kansas Orange	4	53	86	139	1.622
Sorgo, Sumac	4	47	75	122	1.595
Broom Corn, Acme Dwarf	3	53	79	132	1.490
Wheat:					
Kanred	3	39	28	67	0.718
Blackhull	3	41	24	65	0.585
Triplet	2	40	28	68	0.700
Improved Turkey	2	45	28	73	0.622
Fulcaster	2	37	26	63	0.702
Harvest Queen	2	35	25	60	0.714
Rye:					
Abruzzi	3	33	30	63	0.909
Kansas	3	33	31	64	0.939
Rosen	3	32	28	60	0.875
Barley:					
Manchurian	3	37	39	76	1.054
Mariout	3	32	34	66	1.062
Kansas	3	36	35	71	0.972
Oats:					
Burt	3	25	27	52	1.080
Fulghum	3	27	28	55	1.036
Kherson	2	27	28	55	1.036
Texas	2	25	29	54	1.160
Alfalfa, common	3	109	89	198	0.816
Clover, white sweet	3	163	105	268	0.644
Clover, red	3	138	197	335	1.427
Cowpeas, New Era	3	74	157	231	2.121
Soybeans, Medium Yellow	2	95	107	202	1.126
Potato (several varieties)	3	33	104	137	3.151
Cabbage (several varieties)	3	91	146	237	1.604
Watermelon, Knobs Gem	3	127	234	361	1.842
Cucumber, Long Green	3	143	285	428	1.993
Squash, Hubbards	2	185	202	387	1.091
Tomato (several varieties)	3	62	131	193	2.112
Pumpkin, Cheese	2	135	182	317	1.348
Apple:					
Delicious	2	0	190	190	
Grimes Golden	2	0	266	266	
Jonathan	2	0	188	188	
Stayman Winesap	2	0	186	186	
Peach, Elberta	2	0	145	145	
Cherry, Early Richmond	2	0	161	161	

1920. The indications, from these observations, are that different plants may vary in their response to certain conditions as far as the number of stomata per unit of area is concerned.

Miller (1928) also observed the number of stomata of various other agricultural plants growing in the field during from two to nine growing seasons. The results of the observations, together with a summary of those just given, are shown in the tables on pages 421–422.

Of the plants observed, oats had the smallest number of stomata per unit of area—54,000 per square inch—while cucumber had the largest number—428,000 per square inch of leaf area. When stomata occur on both upper and lower surfaces, the majority of plants have more on the lower than on the upper surface. In these cases, the ratio of the number on the lower to that on the upper surface ranged from 1 in the case of barley and oats to 3 in the case of potato. The leaves of apple, peach, and cherry showed no stomata on the upper surface. Wheat, alfalfa, and white sweet clover consistently had more stomata on the upper surface of their leaves than on the lower, the ratio of the lower to the upper being 0.66, 0.82, and 0.64, respectively, while rye was found to have slightly more on the upper than on the lower surface.

Sawyer (1932) reported that the stomata on the leaves of the cranberry (*Vaccinium macrocarpon*) are confined entirely to the lower surface, and that in his observations their frequency was approximately 395,000 per square inch. The stomata of the strawberry are all on the lower surface, and, according to the observations of Darrow and Dewey (1934) ranged in number from 303 to 389 per square millimeter.

An interesting observation was made by Dodge (1923) in regard to the effect of orange rusts of blackberries and raspberries on the distribution of the stomata on the upper and lower surfaces of the leaves of these plants. In general, normal leaves of these plants had more stomata on the lower side of the leaves. Some few may be scattered singly or in groups of two or three on the upper surface along the large veins or near the edge of the leaf. Dodge found, however, that in blackberry plants which had been systemically infected with the orange rust for 2 or more years, from one-fourth to one-half of the total number of stomata were found on the upper surface.

It may be stated as a general rule (Eckerson, 1908) that there is an inverse relation between the number and size of stomata of a given type. Thus they are smallest where they occur in the largest numbers. For example, the stomata of oats are considerably larger than those of corn or the sorghums, while those of cucumber, squash, or watermelon are much smaller than the stomata of alfalfa, soybean, potato, or tomato. Reed and Hirano (1931), however, noted with *Citrus* that the size of the stomata seems to be influenced only slightly by the number per unit area.

The number of stomata vary on the different leaves of the same plant and on different parts of the individual leaf. Miller (1928) observed that the middle leaves of a mature corn plant have from 25 to 45 per cent more stomata per unit of surface than the lowest three or four leaves. The fact that the upper leaves have more stomata per unit area than the lower has been observed by Yapp (1912) for *Spiraea ulmaria*, L., by Rea (1921) for *Campanula rotundifolia*, and by Salisbury (1927) for a wide range of plants. The stomatal frequency in relation to the position of the leaf upon the plant is illustrated in the case of *Polygonatum multiflorum*, as shown by the data given by Salisbury in the table below.

The number of stomata per unit of area varies for different regions of the same leaf. Thus Salisbury (1927) found that the highest frequency was near the leaf tip and the lowest frequency toward the base, while the frequency for the middle region of the leaf was intermediate between the base and the apex. Salisbury also noted that, in some cases at least, the number of stomata per unit area increased from the midrib toward the margin. According to Maximow (1925), this is a part of the general law, first set forth by Zalenski, according to which the higher the leaf is attached to the stem the larger in comparison with the lower leaves are the total number and length of the vascular bundles per unit area, the smaller are the cells of the epidermis and mesophyll and also the cells of the stomata, but the more there are per unit of surface. Salisbury (1927) considered that the increase in the number of stomata on the leaves from the base of the plant toward the tip and from the base of an individual leaf toward its apex is determined by the water relations. Conditions favoring high rates of transpiration or a low rate of water intake favor high stomatal frequency. It was found that under constantly humid conditions the frequency gradient of stomata tends to disappear.

RELATION OF HEIGHT OF LEAF ON THE PLANT TO THE STOMATAL FREQUENCY IN *Polygonatum multiflorum*

Height of leaf above ground, cm.	Number of stomata per square millimeter	Height of leaf above ground, cm.	Number of stomata per square millimeter
21.5	53.6	51.8	77.0
27.0	59.0	53.8	83.6
30.2	76.5	55.2	88.3
50.2	73.0	56.5	91.0

The total number of stomata per plant is enormous. Thus the number on the entire leaf surface of a fully grown corn plant ranges from 140,000,000 to 240,000,000, depending upon the variety and size of the plant, while a single fully developed pumpkin leaf of average size has from 50,000,000 to 60,000,000 of stomata.

Although the stomata occur in very large numbers, they occupy only a relatively small portion of the epidermal surface of the leaves. Kiesselbach (1916) found that the dimensions of the stomatal aperture of the leaves of corn under the conditions of his experiment were 3.5 by 25.6μ and that the area of the pore was approximately 90 sq. μ. He estimated that the area of the stomata was about 1.5 per cent of the total leaf surface. Brown and Escombe (1900) considered that the area of the stomata of the sunflower plant approximated 1 per cent of its leaf surface. Eckerson (1908) found the average dimension of the stomata of over 30 different species of plants grown in the greenhouse to be 6.7 by 17.7μ, and when fully open they occupied about one-ninetieth of the total leaf surface. From the various observations that have been made it may be stated, in general, that the area of the stomata may equal from 1 to 3 per cent of the leaf surface of plants.

2. *Opening and Closing.*—Since the time of Von Mohl (1856), the opening or closing of the stomata has been attributed to the changes in turgor of the guard cells. The opening is apparently brought about by an increase in the turgor of the guard cells and the closing by a decrease in their turgor. The changes in turgor result in the opening or closing of the stoma because of the uneven thickenings of the cell walls of the guard cells, a fact that was first pointed out by Schwendener (1881). These uneven thickenings in the walls vary in their arrangement but are fairly constant for large groups of plants. The structure of the different types of guard cells and its relation to the opening and closing of the stomata have been extensively described by Copeland (1902). The intention here is to mention only one type of structure to illustrate the relationship of turgor and the structure of the cell walls of the guard cells to the opening and closing of the stomata.

In one of the most common types of guard cells, the cell wall is much thinner where it borders on the epidermal cell than on the opposite side next to the stoma. When the turgor of the cell is increased, this thinner wall is stretched or dragged out farther than the other and pulls the more rigid and thicker wall bordering the stoma with it, thus increasing the size of the pore. This action is illustrated by a simple experiment that is described in practically all elementary textbooks of botany. Two pieces of rubber tubing are connected by a Y tube to a water faucet or an air pump and are placed parallel to each other. The lower end of the tubes are stoppered, tied securely together, and the exterior side of each piece of tubing is pared thin by means of a sharp knife. When the pressure is turned into the tubing, the thin exterior walls bulge and draw the pieces of tubing apart, producing an ellipsoidal opening between them.

(*a*) *The Physiology of the Guard Cells.*—It has just been stated that the opening or closing of the stomata is due primarily to changes in the turgidity of the guard cells, but no mention was made of the manner by which these changes in turgor might be brought about. In order to understand the explanations that are commonly offered to account for these changes in turgor, a certain knowledge of the anatomy and physiology of the guard cells is necessary; it is the intention here to discuss it briefly.

The guard cells always contain green plastids in contrast to the ordinary epidermal cells. These plastids have until recently commonly been called "chloroplasts" and considered similar to those of the mesophyll, but there seems to be considerable evidence that such is not the case. Sayre (1926) noted in the case of patience dock (*Rumex patientia*) that the chloroplasts of the mesophyll appear as single structures and do not vary in size with their starch content. The green plastids of the guard cells, on the other hand, have the appearance of compound structures, vary in size with their starch content, and are somewhat smaller than the chloroplasts of the mesophyll cells. Chlorophyll develops only in the chloroplasts of the mesophyll when exposed to light, while the green color of the plastids of the guard cells develops in either light or darkness. If a green plant is placed in continued darkness, the chlorophyll of the chloroplasts of the mesophyll disappears, while the green coloring matter of the plastids of the guard cells does not. Sayre (1926) could not obtain the microchemical tests for chlorophyll from the plastids of the guard cells, but he did not consider the proof conclusive due to the difficulty of making the test on such small bodies. On account of these differences, Sayre considered that the plastids of the guard cells are different structurally, physiologically, and genetically from the chloroplasts of the mesophyll cells of the leaf. Sawyer (1932) noted that chloroplasts do not occur in the guard cells of the leaves of *Vaccinium macrocarpon* but do occur in the other epidermal cells on both surfaces of the leaf. The stomata of this plant are poorly adjusted for response to the changing environmental factors of light, temperature, and moisture.

(1) *Osmotic Relations of the Guard Cells.*—It has been established by Iljin (1914), Wiggins (1921), and Sayre (1923, 1926) that the osmotic value of the cell sap of the guard cells is greater than that of the ordinary epidermal cells. Sayre also reported that the osmotic value of the cell sap of the subsidiary cells of *Rumex patientia* is higher than that of the ordinary epidermal cells. The osmotic value of the cell sap of the guard cells is not constant but is highest when the stomata are open and lowest when they are closed, while that of the epidermal cells remains practically constant throughout the day. Iljin (1914) reported that the osmotic value of the cell sap of the guard cells of the numerous plants with which he worked was 70 to 80 atmospheres higher than that of the epidermal cells. These results are perhaps too high, since Iljin used potassium

nitrate as a plasmolytic agent and Sayre (1926) has pointed out that the protoplasmic membrane of the guard cells is readily permeable to this salt. Wiggins (1921) used a solution of calcium chloride to plasmolyze the cells and substantiated the observations of Iljin, although he was unable to obtain such great differences. The results of Wiggins expressed in the table below show that the osmotic value of the guard cells averaged about 13 atmospheres higher than that of the epidermal cells.

Plant	Osmotic value, atmospheres			
	Ithaca, N. Y.		Columbia, Mo.	
	Epidermis	Guard cells	Epidermis	Guard cells
Zebrina pendula	5.4	11.3		
Cyclamen	9.5	12.6	10.1	29.5
Iresine	11.3	18.1	11.6	39.5
Beets	....	....	12.6	31.5

The osmotic values of the cell sap of the guard cells and epidermal cells of *Rumex patientia* were determined by Sayre (1926) by using sucrose, glucose, and calcium chloride as plasmolyzing agents. His results with all three agents were practically the same and showed the osmotic value of the cell sap of the guard cells to be higher than that of the epidermal cells and the differences to be greater when the stomata were open than when they were closed. The table that follows gives the data obtained by using sucrose as a plasmolytic agent:

OSMOTIC VALUE OF THE CELL SAP OF THE GUARD, SUBSIDIARY, AND EPIDERMAL CELLS OF PATIENCE DOCK

Items considered	8:30 a.m.	10:30 a.m.	1:30 p.m.	3:30 p.m.	7:30 p.m.
Size of stomata, microns	3 by 30	6 by 31	12 by 34	9 by 31	0 by 0
Osmotic value of the cell sap, atmospheres of guard cells	16.6	16.6	21.0	20.2	13.2
Subsidiary cells	16.6	16.6	16.6	17.8	15.5
Epidermal cells	14.4	14.4	14.4	14.4	13.2

The experimental evidence seems to show conclusively that during the daylight hours the osmotic value of the cell sap of the guard cells is greater than that of the epidermal cells, this difference rising to a maximum during the day and falling off in the afternoon, paralleling somewhat the opening and closing of the stomata. The difference between the osmotic value of the cell sap of the guard cells and the epidermal cells varies, no doubt, with different plants and with the different conditions under which the observations are made.

(2) *Changes in the Cell Contents.*—Almost without exception, starch occurs in the plastids of the guard cells under any condition where these cells are fully formed. It is found in the guard cells where the plant has been grown from the seed in complete darkness, in the white area of variegated leaves, in albino plants, and in the guard cells of numerous plants that never produce starch in the chloroplasts of their mesophyll cells. It was formerly supposed that this starch was synthesized by the green plastids of the guard cells, but the evidence at present indicates that the carbohydrate content of the guard cells may be obtained entirely from the green cells of the mesophyll. It has been noted by various observers that the starch content of the plastids of the guard cells changes during the opening and closing of the stomata. Darwin (1898), Lloyd (1908), Iljin (1914), and Sayre (1926) found, in general, that the starch content in the guard cells is greatest in the early hours of the morning and gradually disappears until the stomata are at their maximum opening. It, however, is never wholly depleted. After the stomata begin to close, there is a gradual accumulation of starch in the guard cells until the maximum content is again reached early in the evening, after which no marked changes can be detected during the night. It has been observed that, when the starch content decreases, the osmotic value of the cell sap increases, and the reverse is the case when the starch content increases. Hagen (1918) observed that, when the stomata are open, sugar is present in large amounts, while Sayre (1926) showed that in the guard cells of *Rumex patientia* there was 2 to $2\frac{1}{2}$ times more sugar present when the stomata were open than when they were closed. Strugger and Weber (1926) in studying the stomata of *Galium mollugo* observed that as starch disappeared in the guard cells, it appeared in the companion cells, and vice versa. They concluded from this observation that the subsidiary cells cooperate in the movement of the stoma. This change of starch to sugar and sugar to starch seems to be a reversible reaction in which the total amount of carbohydrate material does not vary to any extent. This reaction has generally been assumed to be due to the enzyme diastase, but so far it has not been isolated from the guard cells, although Hagen (1918) by treating the guard cells with diastase converted the starch into sugar and thus caused the stomata to open.

The series of changes that result in the opening of the stomata of a turgid leaf are enumerated by Sayre (1926) and Loftfield (1921) as follows: In the morning, light initiates the action of diastase, probably by decreasing the acidity of the cell sap of the guard cells. The diastase changes the starch to sugar, which results in an increase of the osmotic value of their cell sap. This causes water to enter the guard cells, since the osmotic value of the cell sap of the epidermal cells remains constant. A swelling of the guard cells results, and this causes the stoma to open. The procedure during the closing of the stomata is perhaps the reverse of that during the opening.

There are marked changes in the pH values of the guard cells under various conditions. Thus Sayre (1926) found that they were slightly more alkaline when the stomata were open than when they were closed, but no quantitative data could be secured. By using the pressed juice from the leaves as a buffer solution and varying its acidity he could cause the stomata to open or close in light or in darkness. When placed in a solution with a pH value of 4.2 to 4.4 they opened after 2 hr., but when placed in a solution with a pH of 3.6 to 4.0 and 4.6 to 5.0 they remained closed.

Scarth (1932) found that there was a wide change in the reaction of the guard cells from pH 6.0 to 7.4 in the light to less than pH 5.0 in the dark. He concluded that in the plants studied all the stomatal movements whether due to changes of leaf turgor, of illumination, or of temperature, are accompanied by reversible transformation of carbohydrates and by a pH change in the guard cells. When the stomata are opening or are open, the guard cells have a higher pH value than when they are closing or are fully closed. He considered that the carbohydrate balance in the guard cells depends strictly on the pH value. The origin of the pH change, when governed by the light factor, depends mainly on the relative rate of photosynthesis and respiration in the leaf tissue. The absence of carbon dioxide favors, if it does not of itself induce, stomatal opening. A certain concentration of carbon dioxide causes the closure of the stomata, but with a higher concentration—15 per cent—there is a hydrolysis of starch causing an increased osmotic value in the guard cells and stomatal opening.

It was noted by Linsbauer (1926) that there are marked differences in the physical and chemical state of the protoplasm of the guard cells depending on whether they are open or closed. When open the cells are more resistant to the toxic action of "Rongelit" and silver nitrate than when they are closed. Leick (1927, 1928) considered that when the stomata are open there is a change in the electrostatic condition of the molecules of the proteins of the guard cells. This is indicated by the increased viscosity and permeability of the protoplasm of the guard cells and by their resistance to higher temperatures.

The rapidity with which the stomata close or open, however, was considered by Scarth (1926) and others to be too rapid to be due entirely to the hydrolysis of starch or its condensation. They considered that the action of diastase is so slow that long before the starch can undergo any great amount of change, the movement of the guard cells is completed.

Scarth (1926) found that illuminated and open guard cells are alkaline in reaction, probably from the utilization of carbon dioxide in photosynthesis, and certain colloidal constituents of these cells become much swollen as the pH value increases. He considered that high turgidity in the guard cells is mainly caused by colloidal imbibition. Closure is accompanied by increased hydrogen-ion concentration due to the accumulation of carbon dioxide of respiration when photosynthesis ceases on account of low light intensity. The colloids are dehydrated as they approach their isoelectric point, the bound water is freed, the cells quickly lose their surplus water to the surrounding cells, and the guard cells close. Scarth considered that if hyperacidity develops due to prolonged darkness, the colloids of the guard cells may become more acid than the isoelectric point, produce an acid swelling of the colloids, and thus bring about night opening of the stomata. According to this theory, we have a mechanism whose response is almost instantaneous to light changes. In such a case, the sugar and starch reactions that are commonly associated with stomatal opening and closing may be subsidiary to the main regulatory mechanism. It is worthy of note in this connection that the guard cells have been observed in the case of *Carnegiea gigantea* to function in the opening and closing of the stomata for as long as a century (Long, 1928).

(*b*) *Methods of Measuring the Changes of Stomatal Apertures.*—Darwin (1898) considered that the amount of water vapor escaping from the leaf could be taken as an index of the condition of the stomata. He used shavings of horn and other materials as hygroscopes. The degree of curvature of these hygroscopes, which was an indication of the rate of water loss, was read on a graduated quadrant. This method measured the change in the transpiration rate but not necessarily the condition of the stomata. Another method reported by him in 1904 in which the degree of stomatal opening was measured by the changes in the temperature of the leaves is open to the same objections. Various types of porometers have been devised and used by Darwin and Pertz (1912), Balls (1912), Jones (1914), Laidlow and Knight (1916), and Knight (1915, 1916) to measure changes in stomatal apertures. The general procedure has been to use a constant-pressure aspirator to draw air through a leaf either intact or detached from the plant. The speed of the air stream is measured by the rate at which water flows from the aspirator and is considered to indicate the relative size of the stomatal openings. The main objections to such porometers are the unnatural condition of drawing air through the intercellular spaces of the leaf and the uncertainty as to what the results indicate. Trelease and Livingston (1916) made a comparative study of the porometer and standardized cobalt paper in regard to the transpiration rate. They obtained results that show a general agreement of data between the two methods during the morning hours but a lack of accord later in the day. The evidence indicates that the porometer measures the diffusive capacity of the stomata but fails to take into account other factors influencing transpiration. Darwin (1912) considered the porometer method to have an advantage over any microscopic method, since it strikes an average of the degree of opening of many hundreds of stomata at each reading instead of the comparatively few that can be observed through a microscope. Lloyd (1908) devised a method of stomatal observation that is the most applicable and the most widely used of any of the methods yet presented for stomatal study. His method consists of stripping the epidermis from the leaf and quickly plunging it into absolute alcohol. The effect of the alcohol is to dehydrate the cell wall and to cause it to become very stiff and hard before removing the water from the cell. As a result, the cell walls retain their original shape, and the size of the stoma is unaltered as long as the material is kept dehydrated. The specimens can thus be collected, preserved, stained and mounted, and studied at leisure. The reliability of this method was verified by Loftfield (1921), who compared the dimensions of the stomata of leaves in position and of those of the stripped epidermis treated with absolute alcohol and found that the measurements were practically identical. Lloyd (1913) used a specially arranged microscope to observe the stomata in position in their natural state. The microscope was provided with a condenser, a 4-mm. objective with a long working distance, and a cooling cell in the form of a flask wired beneath the substage apparatus. By this method it is possible to observe accurately the size of the stomata without injury, but it has its objections in that it cannot be used at night or in poor light and is more or less cumbersome and too slow to be used under the conditions that frequently occur in the field.

Ashby (1931) compared the sensitivity of the method of Knight with that of Lloyd for measuring the area of the stomatal apertures. There was a satisfactory agreement of the two methods on the leaves of *Geranium* and *Verbena* when the stomata were wide open. At the time the apertures were nearly closed, however, the porometer method of Knight was approximately ten times more sensitive than the method devised by Lloyd. Gregory and Pearse (1934) described a porometer that measures the changes in stomatal aperture by the amount of resistance recorded. Jones (1932) described a method for observing stomata by the use of the ultraopaque microscope. Long and Clements (1934) found that solutions of collodion can be used to study the number and distribution of stomata. The solution is applied with a camel's hair brush to the selected areas. After the films have dried they are removed, and the stomatal impressions are studied.

(*c*) *Factors Affecting Opening and Closing.*—Any statements that may be made in regard to factors that affect the opening and closing of the stomata must be considered only as very general in their application, since in this regard, as in many others, different plants may not only respond differently to the same factors, but the nature of the response may be greatly influenced by other environmental factors.

(1) *Light.*—In general, it can be stated that light is the factor to which the stomata most universally respond. As a rule, the stomata of the majority of plants open in light and close in darkness, but the light requirements for opening and closing are not identical, even for closely related plants (Kuyper, 1915; Livingston and Estabrook, 1912; Gray and Peirce, 1919; Smith, 1922; Hart, 1929; Magness and Furr, 1930; Paetz, 1930; Bartholomew, 1931; Sitton, 1932; Deen and Bruner, 1933; Scarth, Whyte, and Brown, 1933; and Yocum, 1935).

Light appears to induce the opening of the stomata by initiating conditions favorable to the conversion of the starch in the guard cells to sugar, which increases the osmotic value of their cell sap. The closing of the stomata at night is, however, by no means universal. Of the 75 species of plants examined by Darwin (1898), 13 per cent did not close their stomata at night. Burgerstein (1920) examined 48 species of herbaceous and 27 species of woody plants from 9 to 10 P.M. and found that the stomata of 15 species were fully open, 13 species partially open, 14 species only slightly open, and 33 species fully closed. The stomata of nyctitropic plants are frequently open at night, as are also those of some of the marsh and aquatic plants, and Loftfield (1921) has noted the same condition for the stomata of the potato plant. Although the closure of the stomata in darkness is the rule for the majority of plants, it is not necessarily so complete as entirely to prevent the interchange of gases between the interior of the leaf and the external medium, although it does greatly diminish it. In many cases, too, some stomata remain open, although the great majority of them on the same leaf are completely or partially closed, a fact which was first reported by Leitbeg (1868).

It was observed by Loftfield (1921) that the reduction of light to less than one-half of normal is usually necessary to produce any effect upon the stomata of plants growing in the open, and when stomata are closing they respond to a decrease in light more rapidly than when opening. Leick (1927, 1928) reported that light increased the permeability of the guard cells. Sayre (1929) found that the stomata of *Rumex patientia* do not open in light of longer wave lengths than 690mμ. He suggested that red light beyond 690mμ does not affect plants except as it may be absorbed and changed to heat. Paetz (1930) reported that red light, 760 to 620mμ, and blue light, 500 to 410mμ, had the greatest effect upon stomatal movement, the red rays being twice as effective as blue rays of the same intensity. Since it is the red and blue rays that are concerned in photosynthesis, it was suggested that the movement of stomata may be correlated with photosynthetic activity. The stomata of the variegated leaves of *Caladium*, whose guard cells are entirely free from chlorophyll, opened in a saturated atmosphere but were unresponsive to light. Scarth (1932) stated that the portion of the spectrum that is absorbed by chlorophyll affects stomatal movement. It acts primarily by its role in photosynthetic activity. Deen and Bruner (1933) found during the eclipse of the sun in 1932 that the stomata of the gray birch began to close 25 min. before the beginning of the eclipse. After it had passed, the stomata opened slightly. Ordinarily at the hour of the eclipse, most of these stomata were fully open. Loftfield (1921) stated that the stomata open at night as a result of moonlight or a strong artificial light of much less intensity than 1 per cent of the sunlight maximum. He also observed that in response to light they opened more readily toward morning than toward midnight.

The question of the periodicity or rhythm of the movement of stomata has received considerable attention, but the experimental results that have been obtained are very conflicting. Darwin (1898) found that artificial darkness is more effective in producing closure in the afternoon than in the morning and that illumination opens closed stomata more readily in the morning than later in the day. Sayre (1926) observed that if the plants of *Rumex patientia* growing under normal conditions were placed in a dark chamber in the late afternoon, the stomata began to open at the usual time in the morning, had opened to 10 to 15 per cent of their maximum width by the middle of the forenoon, and had closed by noon. Thus rhythm occurred to some extent the following day but was not noticeable the third and fourth days. Kuyper (1915) found that in sugar cane there is some indication that after prolonged darkness the stomata tend to open between 8 and 10 A.M. Hart (1929) found that the daily rhythm of stomatal movements differed in varieties of wheat. The stomata of some varieties open very soon after sunrise and remain open during the greater portion of the day, while those of other varieties remain open only a relatively short period of time.

(2) *Temperature.*—The temperature of the air affects the speed with which the stomata open. Loftfield (1921) observed that within the limits of temperature at which protoplasm functions, the length of time for opening is reduced one half for every 10°C. rise in temperature. Zalenski (1921) noted that when the temperature rises to 35 to 50°C. starch in the guard cells is transformed to maltose and the turgor of these cells increases, in consequence of which the stomata open widely, a fact also observed in the wilting of plants. Of the 50 species observed, only millet and succulents showed no opening of the stomata at high temperatures.

It was considered by Scarth, Whyte, and Brown (1933) that the night opening of stomata occurs when the concentration of oxygen inside the leaf reaches a certain minimum. The effect of temperature is primarily to hasten the onset of this shortage of oxygen by the acceleration. They suggested that a low oxygen tension reduces the rate of evolution of carbon dioxide, so that its concentration in the leaf falls. Thus opening at night is really a response to a low content of carbon dioxide, just as is opening in the light. It was noted by Clements and Martin (1934) in the sunflower that the stomata did not begin to close until the temperature of the soil was lowered to about 40°F., but that they were nearly closed when the soil reached 37°F.

(3) *Humidity.*—It is generally considered that high humidity of the air permits the stomata to open wider and remain open longer than a low humidity. This factor is especially important when the water content of the leaf is near the minimum point for normal stomatal behavior.

Paetz (1930) found that in *Tradescantia fluminensis, Oxalis losiandra, Zea mays*, and certain other plants, the stomata opened at different rates upon exposure to light after the plants had been kept in the dark at varying degrees of humidity. The plants that had been kept in the most humid atmosphere opened the most rapidly, and vice versa. It was observed by Overley, Overholser, and Haut (1931) that, with a high temperature and a low relative humidity, an early closing of the stomata might result for the leaves of the apple even though the soil moisture was above the wilting point.

The effect of relative humidity and temperature upon stomatal behavior is reflected in the results obtained from various methods of drying hay. Thus Jones and Palmer (1932, 1933, and 1934) found that a double windrowing of alfalfa 2 hr. after cutting produced a hay with a more desirable color, a larger percentage of leaves remaining on the stem, and a lower moisture content than hay that was cured in the swath. They observed that the stomata on the leaves remained open longer when the hay was windrowed than when it was allowed to remain in the swath. At the end of 2 hr. in the swath, the stomata were nearly all closed, and plasmolysis of the cells had occurred. At the end of the same period, the stomata of the windrowed hay were partially closed and began to reopen. This reopening was followed by an increased rate of water loss. The temperature inside the windrow 3 hr. after cutting was 5° to 8°C. lower than it was in the swath, while the relative humidity of the former was 10 per cent higher than in the latter. In the subdued light of the windrow, the chlorophyll was better preserved and thus made a brighter hay.

(4) *Moisture.*—Next to light, the moisture supply of the leaves is the factor that most universally affects the opening or closing of the stomata. The effect of wilting on the movement of the stomata will be discussed under subsequent topics. It will suffice here to state that, if the supply of water in the leaf reaches a certain minimum which causes the guard cells to lose their rigidity, the stomata close regardless of the influence of light or of any of the other factors that have been mentioned.

It was noted by Veihmeyer and Hendrickson (1927), Magness and Furr (1930), Furr and Magness (1930), Furr and Degman (1931), and Sitton (1932) that the stomata of apple, apricot, peach, prune, and pecan open in the morning and remain open until late afternoon or evening if the moisture content in the root zone is above the wilting percentage. According to Veihmeyer and Hendrickson (1927), if the moisture content is above the wilting coefficient, no difference can be detected in regard to the opening of the stomata of trees growing in relatively dry or saturated soils.

(5) *Chemical Agents.*—The effects of the amount of carbon dioxide on the behavior of stomata have already been noted. Sayre (1926) found that the stomata of *Rumex patientia* open to 50 per cent of their maximum in ammonia vapor in darkness and close in an acid atmosphere in light. According to Haas-Poetzl (1930) the effectiveness of ions in prolonging the hydrolysis of starch in the guard cells, and thus in promoting the opening of the stomata, stand in the following order for *Zebrina pendula* and *Vicia faba:* Li > Na > K > Mg > Ca > Sr > Ba and Cl > NO_3 > SO_4 > Br > tartrate > citrate > CNS. Pleasant (1930) found for plants of pea, bean, and radish that the application of nitrates increased the speed of the opening and closing of the stomata. Weber (1931) noted that in the actively growing leaves of *Ranunculus ficaria* the guard cells were very permeable to urea, while in the embryonic stage they were permeable only to a slight degree. It was considered by Scarth (1932) that the influence of chemicals upon the movement of stomata is conditioned by their effect upon the pH value of the cells.

(*d*) *Daily Movement.*—The general behavior of the stomata during the 24-hr. period has been studied extensively by Loftfield (1921) for about 60 species of cultivated, native herbaceous plants, shrubs, and trees. He found that there is a great variation in the behavior of the stomatal aperture of the different leaves of the same plant and even on the same leaf (see also Leick, 1928). The stomata on the upper surface may behave differently from those on the lower. Thus in alfalfa those on the lower surface open more slowly and close earlier than those on the upper. On all leaves there are a certain number of stomata that are functionless, while a certain portion are more or less inactive. Loftfield observed that 80 per cent of the stomata of alfalfa acted alike as far as movement was concerned and that 18 per cent behaved approximately as the majority, while 2 per cent were completely inactive. It was also noticed by Leick (1927, 1928), and by Veihmeyer and Hendrickson (1927) that the behavior of the stomata is influenced by their position on the leaf, by the position of the leaves on the plant, and by the intensity of sunlight. The closure of some stomata is frequently prevented by particles of sand or dust wedging in the opening. The general behavior of the movements may be greatly modified by the prevailing physical conditions, so that as many reactions may be observed as there are combinations of physical factors. Loftfield considered that the stomatal behavior of the plants he examined provided a basis for classifying the stomata into three general types as typified by alfalfa, potato, and barley.

(1) *The Alfalfa Type.*—The stomata of this type under favorable conditions are open all day and closed all night. In alfalfa they open in from 2 to 6 hr. after daylight, remain open from 3 to 6 hr., and then gradually close during a period about twice as long as that required for opening. When conditions become less favorable for moisture, the stomata close partially for a time during the middle of the day, this period of midday closure increasing to complete closure as the conditions become less and less

favorable. With the appearance of midday closure, night opening develops and increases with the increase of day closure, until finally there is a partial opening of the stomata all night and a complete closure all day. This type of stomata behavior Loftfield considered to be characteristic of most thin-leaved mesophytes and includes sweet, red, and white clovers; peas; beans; turnip; radish; muskmelon; cucumber; watermelon; carrot; parsnip; apple; peach; pear; and many others.

(2) *The Potato Type.*—The daily stomatal movement in potato was found to be markedly different from that of alfalfa under the same favorable conditions. The stomata of potato are open continuously except for about 3 hr. following sundown. Midday closure is observable in alfalfa before wilting is evident, but midday closure does not occur in potato until the plants are visibly wilted, and only after very extreme lack of moisture do the stomata open at dusk. The stomata of the cabbage, onion, tulip, salsify, plantain, squash, pumpkin, and leek show the general or modified behavior of those of the potato. The stomata of *Scirpus validus*, *Equisetum hyemale*, and *Equisetum palustre* were found to be continuously wide open under normal conditions and even during wilting or death from water loss.

(3) *The Barley Type.*—The behavior of the stomata of the barley is typical of the cereals. They show no opening during the night no matter how slight the day opening. Also, there are many closed stomata during the day occurring more or less in groups. They open and close comparatively rapidly, but even under favorable conditions they are fully open for an hour or two only.

The optimum conditions for the full opening of the stomata vary widely for different groups of cereals. Thus in the case of corn and the sorghums, very warm to hot weather, bright sunshine, and a high water content of the leaves seem to be the optimum conditions; while in the case of wheat, oats, and barley, cool, humid conditions with relatively mild sunlight are the best.

The examples that have been mentioned refer to the stomata of the leaves. According to Loftfield, the stomata that are found on the stem differ considerably in structure and water relations and consequently in their behavior from those that occur on the leaves.

3. *Capacity for the Diffusion of Gases.* (*a*) *The Stomata as Paths of Gaseous Exchange.*—It has been proved through the work of Blackman (1895), Stahl (1894), and Buscalioni and Pollacci (1902) that practically all the interchange of gases between the interior of the leaf and the atmosphere takes place through the stomata. Blackman found in leaves which have stomata only on the lower surface that on an average only 3 per cent of the total carbon dioxide exchange occurred through the upper surface. When stomata occur on both upper and lower surface, the amount of carbon dioxide diffusing through each epidermis was approximately proportional to the number of stomata on each side of the leaf. Stahl (1894) observed that the loss of water by transpiration is confined, for the most part, to the leaf surface that contains the stomata. By using the cobalt chloride-paper method to determine the transpiration rate, he observed in leaves with stomata only on the lower surface that the paper on this surface colored pink in a few minutes, while that on the upper surface did not change color until after several hours. The relative loss of water through the stomata and cuticle was observed by Buscalioni and Pollacci (1902) by placing a collodion film over the leaf and noting that the film clouded very quickly only just above the stomata. The water that evaporates through the stomata is termed "stomatal transpiration," while that which evaporates from the surface of the cuticle is termed "cuticular transpiration." Sayre (1920) in the case of the mullein leaf estimated that the cuticular transpiration amounted to only from one-fortieth to one-twentieth that of the total water loss. Darwin (1916) found in the case of the ivy leaf that during the day

the cuticular transpiration was one-thirtieth of the total but that at night it amounted to as much as one-twelfth of the entire water loss.

Rudolph (1925) found that the cuticle on the leaves of many plants permits a considerable water loss, inasmuch as a third of the water loss in still air is cuticular. He showed also that sulphuric acid could diffuse readily through the cuticle of some weeds. Kamp (1930) observed that the thickness of the cuticle does not seem to govern the amount of water that escapes through it. Aside from their layer of cutin, apparently the epidermal cells themselves are a factor in controlling the water loss. When the stomata were all on the ventral surface, as in the case of *Coffea arabica*, the loss of water from the ventral surface was only twice that from the upper one, while in other plants the transpiration rate from the ventral side was forty-five times greater than that from the dorsal surface. Bartholomew (1931) found that during the day 85 to 95 per cent of the water lost by mature *Citrus* leaves came from the ventral surface, which contains the stomata. During the night the loss of water from the ventral surface of these leaves was only 58 to 66 per cent of the total amount of water lost. Since practically all gaseous exchange between the leaf and the air takes place through the stomata, the question arises as to the capacity of the stomata for carrying gases to and from the interior of the leaf and whether their capacity might be a limiting factor for the maximum activity of its various functions.

(*b*) *Diffusion of Gases through Small Openings.*—In 1900 Brown and Escombe made certain observations in regard to the intake of carbon dioxide by the leaves of *Catalpa bignonioides*. They found that these leaves which have stomata only on the lower surface could extract from the air under favorable conditions for photosynthesis about 0.07 cc. of carbon dioxide per square centimeter of leaf surface per hour (measured at 0°C. and 760 mm. Hg). Assuming that all this gas entered the plant through the stomata, whose area equaled only 0.009 of the lower leaf surface, the rate of diffusion would be at the rate of $0.07 \div 0.009$ or 7.77 cc. per square centimeter of the stomatal opening per hour. The surface of a strong solution of caustic soda when freely exposed to moderately still air, as is the surface of a leaf, absorbs carbon dioxide at ordinary temperatures at the rate of only 0.12 cc. per square centimeter per hour, and, even though the air current is increased, the maximum absorption is only about 0.177 cc. per square centimeter per hour. If all the carbon dioxide entered through the stomata of the lower surface of the leaf, the absorption per unit of area of these openings was from forty-three to sixty-four times or an average of more than fifty times as rapid as that of the surface of a constantly renewed concentrated solution of alkali. The observance of these facts led Brown and Escombe to take up the investigation of the free diffusion, especially of carbon dioxide and water vapor, through small apertures of known dimensions under conditions comparable to those existing in the leaf.

In the static diffusion of a gas, vapor, or solute, as the case may be, the amount of substances diffusing in a given time, all other conditions being the same, is directly proportional to the sectional area of the column. Brown and Escombe found, however, that by interposing at any point in the line of flow a thin septum pierced with a circular aperture, the rate of flow across unit area of the aperture is greater than it would be across an equal area of the unobstructed column at this point. When the size of the pore is diminished beyond a certain size relative to the cross section of the column, the amount of gas that passes becomes proportional to the linear dimensions of the aperture and not to the area. The following tables on page 438, giving data secured by Brown and Escombe (1900) and by Sayre (1926), will illustrate this point.

In order to understand more fully the diffusion of gases through small apertures let us consider experiments 3 and 5 in Table A. The diameters of the apertures are approximately 6 and 2 mm., respectively, and the ratio of their diameters is as 3:1, while the amount of carbon dioxide diffused in a unit of time stands in approximately the same ratio. The areas of the two openings, however, stand in the ratio of 9:1. Again let us consider experiments 3 and 5 in Table B. Here the diameters of the

TABLE A.—THE RELATION OF THE SIZE OF APERTURE TO DIFFUSION OF CARBON DIOXIDE
Brown and Escombe

No.	Diameter of aperture, mm.	CO_2 diffused per hour, cc.	Ratio of: Area of aperture	Diameter of aperture	CO_2 diffused in unit of time
1	22.70	0.238	1.00	1.00	1.00
2	5.86	0.055	0.068	0.25	0.23
3	6.03	0.062	0.07	0.26	0.26
4	3.23	0.039	0.023	0.14	0.14
5	2.00	0.023	0.0077	0.088	0.10

TABLE B.—DIFFUSION OF WATER VAPOR THROUGH SMALL OPENINGS UNDER UNIFORM CONDITIONS
After Sayre, 1926

No.	Diameter of aperture, mm.	Diffusion of H_2O vapor, grams	Ratio of: Area of aperture	Diameter of aperture	Diffusion of H_2O vapor
1	2.64	2.655	1.00	1.00	1.00
2	1.60	1.583	0.37	0.61	0.59
3	0.95	0.928	0.13	0.36	0.35
4	0.81	0.762	0.09	0.31	0.29
5	0.48	0.455	0.03	0.18	0.17
6	0.41	0.393	0.02	0.15	0.15

apertures are 0.95 and 0.48 mm., respectively, and their ratio is practically 2:1. The amount of water vapor diffusing through them in a unit of time is also in the ratio of 2:1. The areas of the two apertures, however, are in the ratio of 4:1. These data show another general law in regard to the diffusion of gases through small pores, *viz.*, that as the size of the openings decreases the velocity of flow through unit area increases, or, in other words, the velocity of flow varies inversely as the diameter of the pore.

Since the velocity of the diffusion flow of carbon dioxide and water vapor increases with the decrease in the diameter of the openings, it might reasonably be expected that a diaphragm could be so perforated with a series of very small holes that when interposed in the line of diffusive flow, diffusion could take place through such a partition approximately as rapidly as though no partition intervened at all, although the total area of the small holes might represent only a small fraction of the area of the partition.

Brown and Escombe investigated this question by using a thin celluloid kodak film 0.1 mm. in thickness, perforated at regular intervals with holes about 0.380 mm. in diameter. These films were placed over a solution of caustic potash 1 cm. from its surface, and the diffusion of carbon dioxide studied. The general results obtained are expressed in the following table:

No.	Diameter of holes, mm.	Distance of holes apart in terms of diameters	Number of holes per square centimeter	Percentage area of holes on unit area of septum	Observed flow expressed as percentage of septum diffusion on open-tube diffusion	Theoretical flow through porous septum expressed as percentage on open-tube diffusion
1	0.380	2.63	100.0	11.34	56.1	87.6
2	0.380	5.26	25.0	2.82	51.7	63.7
3	0.380	7.8	11.1	1.25	40.6	44.0
4	0.380	10.5	6.2	0.70	31.4	30.7
5	0.380	13.0	4.0	0.45	20.9	21.9

From the preceding table it is seen that the area of the pores in a diaphragm may be only a small percentage of the total area of the diaphragm, but that the amount of gas diffusing through them is much greater than the ratio of their area to that of the total area of the diaphragm. Thus in experiment 1 the total area of the pores is only 11.34 per cent of the area of the membrane, yet the flow of gas is equal to 56.1 per cent of the unobstructed tube. In experiment 4 the area of the perforations equals only 0.7 per cent of the area of the cross section, yet the flow is over 31 per cent of that of the unobstructed column. Brown and Escombe found in one case where the length of tube was 4 mm. that, with an area of pores totaling only 11.34 per cent of the total cross section, little obstructive influence on the flow of gas was produced, and the amount diffusing was practically equal to the open-tube diffusion. It may also be noted from the table that, when the pores are placed at least eight to ten times their diameter apart, the amount of diffusion approximates very closely the calculated rate, but that, when the holes are more closely spaced, the flow is less than that calculated. This is explained by assuming that in perfectly quiet air the lines of flux in a diffusive column are approximately parallel at some distance from the diaphragm but that as these stream lines approach the apertures they converge and gradually become parallel again on the other side of the diaphragm. This behavior is somewhat similar to the entrance and exit of water from a pipe. On entering the pipe it converges toward the opening and then again spreads out upon leaving it. In the case of gaseous diffusion, the rate of flow is slowed down if the pores are so close that the lines of flux of one pore interfere with those of another, but not if the apertures are so spaced that these lines of flow do not intercept one another, the maximum efficiency of flow being maintained. We shall now consider the application of these observations to the diffusion of carbon dioxide and water vapor through the stomata of the leaves.

(c) *The Stomata in Relation to Such Diffusion.*—Brown and Escombe (1900) considered that the ordinary green leaf has the structure for producing the phenomena of diffusion that is associated with a perforated diaphragm. They noted that the stomata on the lower side of the sunflower leaf are about 8 diameters apart, if they are regarded as circular openings, a distribution that allows each stoma to exercise its full efficiency for diffusion without any interference from the neighboring stomata. So far as is known to the author, however, it has never been determined how nearly the spacing of the stomata of all plants is within the limit for their maximum diffusion capacity. Jeffreys (1918) in a mathematical study of the laws of evaporation of water from circular surfaces calls into question the conclusions of Brown and Escombe in this regard. He concluded that the stomata in a leaf must close to one-fiftieth of that of their full aperture before they act independently of one another. The conclusions of Jeffreys have been adversely criticized by Lamour (1918), so that it would seem that a reinvestigation of the behavior of the stomata in this regard should be made.

The amount of gas that enters the leaf through an individual stoma will be conditioned by the gradient density of that gas between the exterior and interior of the leaf and by the linear dimensions of the stoma. In regard to the stoma of the leaf of the sunflower, Brown and Escombe found that, when fully opened, its depth was about 0.014 mm. and that it was elliptical in section. They determined that the area of this elliptical cross section was 0.0000908 sq. mm., which is about equal to that of a circle with a diameter of 0.0107 mm. They assumed from previous work in regard to diffusion through circular and elliptical openings (Stefan, 1873) that the elliptical stomatal tube would function the same as a cylindrical tube having the same area of cross section. Sayre (1926), however, considered that diffusion through an elliptical opening would be more nearly equal to that from a circular opening having the same perimeter rather than the same area.

Brown and Escombe considered that, if the stomata are wide open and the partial pressure of the carbon dioxide in the intercellular spaces is maintained practically at zero by the absorbent action of the parenchyma cells, the maximum amount of carbon dioxide that could pass into the sunflower leaf through the stomata would be 2.578 cc. per square centimeter per hour. The maximum rate of photosynthesis that has been recorded for the sunflower leaf by the weighing method of Sachs is approximately 1.8 g. of carbohydrate per square meter of leaf surface per hour. Such a rate of photosynthesis requires an intake of only 0.134 cc. of carbon dioxide per square centimeter per hour, or only about 5 per cent of that which could diffuse through the stomata of this leaf if they were wide open and the conditions for ideal diffusion were present. When the stomata of the sunflower are wide open, it would be necessary to have difference of pressure of carbon dioxide of only about 5 per cent between the inside of the leaf and the outer air in order to supply the gas needed by the leaf during maximum photosynthesis.

The theoretical carrying capacity of the stomata for water vapor was also determined by Brown and Escombe. At a temperature of 20°C. with the water vapor in the intercellular spaces maintained at 17.4 mm. of mercury—the maximum pressure corresponding to that temperature—with the pressure of the water vapor in the outer air at one-fourth this pressure, they considered that the maximum amount of water vapor which would diffuse through the open stomata of the sunflower leaf would amount to 1,730 cc. of water per square meter of leaf surface per hour. The maximum rates of transpiration that have been observed for the sunflower plant vary from 276 to 500 cc. per square meter of leaf per hour, according to the prevailing atmospheric conditions. Thus it appears that when the stomata of the sunflower are fully opened they are capable of carrying from three to six times as much water vapor as escapes from the plant at the maximum transpiration rate.

From the previous discussion it is seen that although the stomata occupy only from 1 to 3 per cent of the surface of the epidermis, they have, nevertheless, a carrying capacity for gases greatly in excess of that needed by the plant. Consequently they may close to a considerable degree and yet provide for all necessary gaseous exchange.

4. *Regulation or Control of Water Loss.*—After it had been observed by the earlier workers that the stomata are the main passageways for water vapor from plants, the general view came to be held that the stomata, by opening and closing, control or regulate the transpirational water loss from the plant. The opinions regarding the degree to which the stomatal apparatus may control transpiration have changed with increased knowledge of the stomata and the physiology of the guard cells. The earlier idea was that the stomata could close in anticipation of wilting and thus conserve the water supply of the leaf and prevent it from wilting. This has sometimes been called "active regulation." When it was found, however, that the stomata do not close until the leaf wilts to some extent, it was considered that stomatal regulation of water is passive, since under such conditions the stomatal closure is not due to any special activity of the guard cells but to a decrease in their turgor that is caused by the loss of water from the leaf cells being greater than the intake. It is now considered that passive regulation or hindrance of water loss is operative only after the stomata have closed to a considerable degree, since when they are fully open their carrying capacity for water vapor is from three to six times greater than that which escapes at the maximum rate of transpiration.

The question of stomatal regulation of water loss is complicated by several factors, the chief of which are the climatic conditions and the water supply of the soil. The actual rate of diffusion of water vapor through the stomata is not influenced alone by the size of the pore and the length of the tube but by the gradation of the density of the water vapor between the surface of the mesophyll cells and the outer air. This gradient is dependent upon the water-supplying power of the plant and upon the humidity of the outer air, which is, in turn, determined by the prevailing conditions. Thus the gradient of density of the water vapor from the inside of the leaf to the outside may vary independently of the size of the pores. The question of stomatal regulation of water loss has been considered along the three following general lines:

(*a*) *Stomatal Movement and the Water Content of the Leaf.*—The outgo of water from a leaf during the day is generally greater than the intake, while the reverse is generally true at night. The water content of the leaf thus may vary over a considerable range during a 24-hr. period even though no visible signs of wilting can be observed. This variation in water content may be demonstrated in various ways (Trelease, 1922) and will depend not only on the type and age of the plant but also to a large degree upon the prevailing atmospheric conditions. Livingston and Brown (1912) observed that the daily variation in the moisture content of a number of species of plants at Tucson, Ariz., was about 8 per cent on a moist basis. Lloyd (1913) noted for the leaves of the cotton plant grown under the conditions of Auburn, Ala., and Tucson, Ariz., that the loss of leaf water was from 7 to 15 per cent of the initial amount at sunrise. Shreve (1914) observed that the water content of the

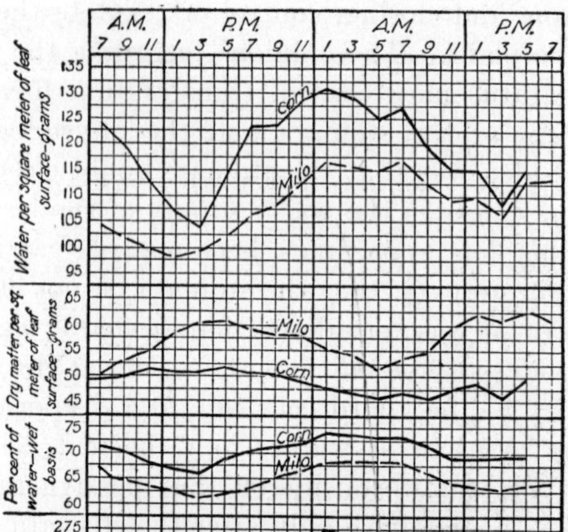

FIG. 17.—Graphs showing the variation in the amount of water and dry matter in the leaves of Pride of Saline corn and Dwarf Yellow milo during 24 hr. and the evaporation from a porous cup atmometer for the corresponding period.

leaves of *Parkinsonia microphylla* fluctuated as little as 2 per cent and as much as 9 per cent during the day at Tucson, Ariz. Knight (1922), however, found that the diurnal changes in leaf-water content in England amounted to not over 1 per cent. Miller (1917) studied extensively the daily variation of the water content of the leaves of corn and the two sorghums kafir and yellow milo growing under field conditions in the

Great Plains area of Kansas. The average difference between the maximum percentage of water in the leaves at night and the minimum percentage during the day was 5.4 for corn, 5.9 for kafir, and 6.0 for milo. The minimum percentage of water in the leaves during the day occurred from 11 A.M. to 5 P.M., but in the majority of cases it was reached between 1 and 3 P.M. The maximum percentage of water in the leaves occurred between 1 and 5 A.M., but in the majority of the 30 observations it occurred around 5 A.M. Hawkins (1927) reported that the lowest amount of water in the leaves of Acala and Pima cotton growing at Tucson, Ariz., occurred about 1:30 P.M. An example of the variation in the moisture content of the leaves of corn and milo during a 36-hr. period is shown in Fig. 17 and in the table on page 444.

Metzger, Holmes, and Bierman (1925) found in Maryland that the water content of the leaves of soybeans declined 7 per cent between 8 A.M. and noon. They recommended that on account of this lower moisture content these plants be cut for hay in the heat of the day so that they might cure more quickly. Willard (1931) from observations on alfalfa, and red, sweet, and alsike clovers concluded, however, that in Ohio there was not sufficient change in the water content of these plants to influence the time of day at which they should be cut for hay. Gilbert and Adams (1929) by withholding water from Swiss chard and barley caused a fall of 7.5 per cent in the water content of these plants. In the wheat plant, Hurd-Karrer and Taylor (1929) noted that the water content of the leaves gradually decreased from before flowering to maturity. For any one day Kiesselbach and Anderson (1931), in Nebraska, found a maximum difference of 4.4 per cent in the moisture content of alfalfa, and for all their observations an average difference of 2 per cent from 8 A.M. to 5 P.M. Yapp and Mason (1932) observed in the sunflower that the greatest quantity of water was in the lowest foliage and in the youngest leaves. Runyon (1936) noted that in the creosote bush the amount of moisture in the leaves differs with age, position on stem, habitat, and time of day. Isaac (1935) reported that between tides the decrease of water in *Laminaria digitata* was between 19.3 and 30.0 per cent, depending upon the prevailing atmospheric conditions.

Lloyd (1912) observed that the stomata continued to open when the water content of the leaves was decreasing and concluded, as did Knight (1917), that the stomata were ineffective in maintaining a constant water supply in the leaf and were thus of little or no value in controlling the water loss. Loftfield (1921) criticized these conclusions on the ground that there is a certain amount of the water present in a turgid leaf which may be regarded as a working margin and which may be lost without interfering with the opening of the stomata or the various other functions of the leaf. After the content of water has fallen below this working margin, the stomata begin to close owing to a fall in the turgor of the guard cells. Hendrickson (1926) noted in the case of peach trees that the decrease in moisture content of the leaves appeared at a period considerably in advance of the maximum opening of the stomata. Loftfield

Variation of the Water and Dry Matter in the Leaves of Pride of Saline Corn and Dwarf Milo at Garden City, Kans., Aug. 1 and 2, 1916

Time	Water per square meter of leaf, grams		Dry matter per square meter of leaf, grams		Percentage of water, wet basis		Percentage of water, dry basis	
	Corn	Milo	Corn	Milo	Corn	Milo	Corn	Milo
Aug. 1:								
7 a.m.	123	105	49	51	72	68	251	206
9 a.m.	119	102	50	53	70	66	238	191
11 a.m.	112	100	51	56	69	64	219	179
1 p.m.	107	98	51	58	68	63	208	168
3 p.m.	104	99	51	61	67	62	203	162
5 p.m.	115	102	52	62	69	62	223	166
7 p.m.	123	107	51	60	71	63	241	177
9 p.m.	124	109	50	58	71	65	247	187
11 p.m.	129	113	49	58	72	66	262	195
Aug. 2:								
1 a.m.	131	117	48	56	73	68	275	210
3 a.m.	129	116	47	55	73	68	273	210
5 a.m.	125	115	47	52	73	69	268	221
7 a.m.	127	117	47	54	73	68	270	217
9 a.m.	120	112	47	55	72	67	258	205
11 a.m.	116	110	50	60	70	65	230	183
1 p.m.	115	110	52	62	69	64	224	176
3 p.m.	108	107	47	62	70	63	230	173
5 p.m.	115	113	50	63	70	64	232	180
7 p.m.	...	114	..	62	..	65	...	184

observed that plants may wilt during the early forenoon to such an extent that the stomata completely close but may again regain their turgor, during that part of the day when the evaporation is the highest, to such an extent that the stomata again open. He attributed this recovery of turgor to the influence of the stomata in retarding water loss so that the cells of the leaf again accumulate a supply of water above the working margin.

(*b*) *Number of Stomata and Transpiration.*—Kuyper (1915) could establish no relationship between the transpiration rate and the number of stomata per unit area of different varieties of sugar cane. Muenscher

(1915) studied the relation of the number and size of stomata to the rate of transpiration of sunflower, bean, wheat, and corn among others. He found no constant relation between the amount of transpiration and the number of stomata per unit of leaf surface or between the amount of water lost and the number of linear units of stomatal pore, *i.e.*, the number of stomata per unit of leaf surface multiplied by the average length of the pore. Kelley (1930) also found no correlation between the stomatal number per unit area of leaf surface and the rate of transpiration. The species with the greatest density of stomata were in the group of plants having the lowest rate of transpiration.

In comparing different species of plants for transpiration rates, it should not be overlooked that they possess structural differences, other than the number of stomata which may greatly influence transpirational loss. The length of time the stomata are open is also a very important factor, and Loftfield (1921) pointed out that only when the stomata are open for comparable periods can we draw conclusions in regard to the number of stomata and the transpiration rate of different plants. Thus, for example, Muenscher (1915) found that the ratio of the linear units of the stomatal pore of the sunflower to that of corn was 4:3, while the transpiration rate for a 24-hr. period was 2:1. He concluded that there is no relationship between the number and size of the stomata and the transpiration rate. Loftfield (1921), however, found that the stomata of corn are open 8 hr., while those of sunflowers are open for 12 hr. Thus the water lost by the sunflower plant was during a 12-hr. period and by corn during an 8-hr. period, so that the hourly loss would stand in the ratio of 4:3, which is the same as the ratio of the linear units of the stomatal pores.

(*c*) *Changes in Stomatal Openings.*—It would be expected that the rate of transpiration could undergo sudden and relatively wide fluctuations with little or no alteration in the size of the stomata provided their full diffusion capacity for water vapor at the stage of observation is greater than the prevailing transpiration rate. This fact was observed by Lloyd (1908) for the cut stems of *Fouquieria splendens*, *Verbena ciliata*, and other plants placed in potometers for transpiration experiments. He noted that the rate of transpiration frequently continued to increase for a considerable period of time after the maximum stomatal opening had been reached and that in the later part of the day there were increases in transpiration without any increase in the size of the stomata. It was concluded from these observations that the stomata are nonregulatory in regard to the loss of water by transpiration. Loftfield (1921) considered that the evidence obtained by Lloyd in this regard is of doubtful value as relating to plants growing under natural conditions, since he found that the stomatal movements and rates of water loss from cut stems in potometers differed greatly from those of field or potted plants. Kuyper (1915) found that the maximum transpiration rate in sugar cane was frequently reached after the stomata had begun to close, so that the rate of transpiration under the conditions of his observations was independent of stomatal behavior. Knight (1917) found that the interrelationship of stomatal aperture and transpiration was not consistent. In some cases the graphs of transpiration and stomatal openings ran closely parallel, but in other cases there appeared to be no relationship at all between the two. Zhemchuznikov (1922, 1924) found in the case of sunflowers, corn, and sorghums growing in soil with 60 per cent of complete saturation that the curves of transpiration usually, but not always, paralleled those of the dimensions of the stomatal pore. He considered that this datum indicates a relationship between stomatal aperture and the rate of transpiration, although meteorological factors may modify it. The prevailing atmospheric conditions without doubt determine to a great extent the correlation between the degree of stomatal opening and the rate of transpiration. Thus Livingston (1909) and Darwin (1916) found in diffuse light and under relatively mild atmospheric conditions that the variation in the size of stomatal openings was parallel with the changes in the rate of transpiration. As the intensity of the sunlight increases, the stomata apparently respond more quickly in their opening than does the transpiration rate to the increasing intensity of the atmospheric conditions, so that they reach their maximum opening ahead of the maximum transpiration rate.

The general opinion concerning the relation of stomatal opening to the rate of transpiration can be summed up in the words of Loftfield (1921), and Walter (1925, 1926) to the effect that when the stomata are fully opened or nearly so, the transpiration rate is determined by the factors of evaporation alone, since the diffusion capacity of the stomata is so great as to interfere in no way with their action. As the stomata close, the influence of the evaporation factors is lessened, but even after the size of the apertures has been reduced over 50 per cent, stomatal regulation is still largely overshadowed by the influence of atmospheric conditions. When the stomata are almost closed, they begin to exert a controlling influence on the water loss regardless of the evaporating factors prevailing at the time. Jeffreys (1918) considered that nearly all the reduction due to the closure of the stomata occurs in the last 2 per cent reduction of the aperture.

The degree of closure of the stomata would appear to have an influence on the moisture content of the air of the intercellular spaces of the leaf. As the stomata close, the degree of saturation of the air in the intercellular spaces increases, so that there would be an increase in the gradient of density of water vapor between the interior of the leaf and the atmosphere. This would increase the rate of diffusion outwardly, so that it is conceivable that as a stoma closes, the rate of transpiration through it may decrease little or none at all over what it was before it began to close. When the closure of the stomata is complete, the air within the leaf must reach saturation. It is not known, however, how completely transpiration is prevented by the apparent complete closure of the stomata.

b. Hairy and Waxy Coverings.—In most plants certain of the young epidermal cells grow out either singly or in groups to form appendages of various shapes, which are termed "hairs" or "trichomes" (Haberlandt, 1914). These epidermal hairs are readily divided into two classes: those that contain protoplasm and sometimes chlorophyll, which are termed "glandular hairs," and those that have lost their protoplasmic contents and are filled with air. The glandular hairs evidently affect transpiration only by adding to the evaporating surface of the leaf, but the influence which the dead, air-containing hairs have on transpiration is considerably in doubt, although numerous textbooks state that the hairy covering of leaves reduces transpiration. Haberlandt (1914) stated that the transpiration rate of leaves of *Stachys lanata* was increased by 20 per cent in the shade and as much as 50 per cent in bright sunlight by shaving off the hairs from the upper surface. He considered that a dense woolly coat of hairs reduces transpiration considerably when the leaf is exposed to direct sunlight and to a much lesser extent in diffuse light.

Weigand (1910) studied the effect of artificial hairy coverings on the rate of evaporation, since he considered that such experiments would throw some light on the function that such coverings play in transpiration. He used as an evaporating surface a good quality of blotting paper saturated with water and then rolled with a pencil and placed tightly on a glass surface. The edges of the paper were dipped in paraffin and pressed down tightly upon the glass. The loss of water was determined by weighing. Cotton, made waterproof by paraffin in gasoline, was used for protective coverings of various degrees of thickness. The experiments were performed in still air and in air in motion, the wind being produced by an electric fan. The results showed that porous coverings like cotton, wool, or hair must be very thick to produce any appreciable effect in retarding evaporation if the air is still but become very efficient even in thin layers when the air is in motion.

Rhine (1924) noticed in the stomata of the juniper, fir, and Austrian pine a deposit of wax that apparently retarded the gaseous exchange between the leaf and its surroundings. Emerson and Hildreth (1933) applied various waxes, gums, resin oils, paraffin emulsions, and commercial nursery waxes to young evergreens to prevent excessive transpiration during transplanting. Heavy coatings of these various compounds reduced the transpiration almost to zero but caused severe injury to the plants. Since the temperature of the material applied is a limiting factor, pure oils and emulsions of oils and waxes were tried because they could be applied as sprays at ordinary temperatures. It was found that when applied in the pure form castor oil and corn oil were the only ones that did not injure the conifer needles. The emulsification of some of the injurious oils, however, greatly reduced or entirely removed their toxicity. In practice these sprays are best applied to the seed beds or nursery rows before transplanting. It is believed that this practice will also protect coniferous seedlings from excessive drying in winter.

The leaves of the mullein plant (*Verbascum thapsus*) have one of the most dense coverings of dead hairs of any of the more common plants, these hairs being many-celled structures much branched and having the branches both parallel and perpendicular to the leaf surface. Sayre (1919, 1920) studied the influence of these hairs upon transpiration. Three plants, which may be designated as M^{1+h}, M^{2+h}, and M^{3+h}, with the hairs of the leaves intact were exposed to the desired conditions and their rates of transpiration determined. The ratios of the daily transpiration rates of these three plants in the greenhouse in the light and in still air were as follows:

$$\frac{M^{1+h}}{M^{2+h}} = 0.98 \text{ and } \frac{M^{1+h}}{M^{3+h}} = 0.82.$$

The hairs were than removed from the upper surface of M^{2+h} and from the lower surface of M^{3+h}, after which the plants were designated as M^{2-h} and M^{3-h}, respectively. The plant M^{1+h} was left intact as a check. These three plants were placed in the greenhouse in the light and in still air, and their daily transpiration rates obtained. The ratios of the rates of the treated plants to the check were as follows:

$$\frac{M^{1+h}}{M^{2-h}} = 0.99 \text{ and } \frac{M^{1+h}}{M^{3-h}} = 0.83$$

These were practically the same ratios as were obtained when the hairs were intact.

In the light and in a gentle breeze produced by an electric fan the ratios of the water loss were

$$\frac{M^{1+h}}{M^{2-h}} = 0.91 \text{ and } \frac{M^{1+h}}{M^{3-h}} = 0.78$$

In a dark room in still air the ratios were

$$\frac{M^{1+h}}{M^{2-h}} = 0.56 \text{ and } \frac{M^{1+h}}{M^{3-h}} = 0.65$$

These data indicate that the removal of hairs from the mullein leaves does not appreciably alter their resistance to water loss in still air and light but does slightly decrease the resistance of the leaves to water loss in wind and light as compared with still air and light. The removal of the hairs, however, greatly decreases the resistance of the leaves to water loss in still air and darkness as compared with still air and light. The removal of the hairs affects the transpiration rate most noticeably when the plants are placed in the dark. The stomata are then closed and the transpiration is practically all cuticular. Sayre concluded that the effect of the hairs of the mullein leaf on transpiration in ordinary intensities of wind and light may be disregarded, since the removal of these hairs affects only the cuticular transpiration which is not over one-fortieth to one-twentieth of the stomatal transpiration upon which the removal of hairs apparently has no effect. Shapiro and de Forest (1932) found in southern California that *Salvia apiana*, which has a thick covering of hairs on each surface, lost the most water of any of the eight species studied. The effect of hairs upon transpiration of the leaf will no doubt vary according to their number and arrangement, and it must be admitted, in the words of Cowles (1910), that the known uses of the hairs of plants are small in comparison with their abundant development, and it is probable that most of them are of little or no advantage to the plant.

c. General Structure and Composition of the Leaf.—Little work is available to indicate the relation of the transpiring power of the leaf to its general structure or composition. It has always been assumed that the structural characteristics of the leaves of xerophytic plants are useful in reducing the loss of water by transpiration. The structural characters that are commonly associated with the xerophytic leaf are stated by Pool (1923) as follows: "The leaf is, as a rule, relatively thick and sometimes without chlorophyll at the center. The epidermis is frequently very thick and waxy, resinous, or heavily cutinized, and the stomata-bearing surface is often hairy. The stomata are sometimes protected by being enclosed and are frequently structurally rigid. Xerophytic leaves generally have prominent palisade cells often in very densely packed layers, extending from one epidermis to the other. Frequently the palisades in the lower part of the leaf are less prominently developed or less densely packed than in the upper portion of the leaf. When the sponge or lower chlorenchyma is present, it is, as a rule, comparatively densely packed with a relatively small amount of intercellular space. As a rule, the fibrovascular elements are more copiously developed in the xerophytic leaf than in the mesophytic one. Of all the structural differences between the xerophytic leaf and the mesophytic one, the most striking are the relative development of the palisade and sponge tissue and the relative proportion of the intercellular spaces. The entire chlorenchyma of the mesophyte is likely to be much more open and, as a whole, thinner than the xerophyte."

Pool (1923) studied by means of the cobalt chloride-paper method the transpiring power of some 40 representative species from the various plant communities from the short-grass mesas and plateaus east of Colorado Springs up through the foothills to an altitude of 12,000 ft. It was found that there was some correlation between the transpiring power and the leaf anatomy, leaves of the xerophytic types being found generally to have the lowest transpiring power. The indications were that the compactness of the chlorenchyma of the xerophytic leaves has a marked effect in preventing the evaporation of water in the interior of the leaf and thus reduces the ability to give off water in the form of vapor (see also Lesage, 1894). Pool, however, found many apparent exceptions to this rule. Thus, for example, the leaf of *Pentstemon unilateralis* showed a more xerophytic structure than that of *Populus angustifolia,* yet the index of transpiring power of the latter was 0.641 while that of the former was 0.908. Miller and Coffman (1918) made a somewhat similar observation in regard to the relationship of the leaf structure of corn and the sorghums and the rate of transpiration. The leaves of the sorghums show a marked xerophytic structure in comparison with the leaves of corn, yet the transpiration rate in numerous cases was over twice that of the corn.

Shapiro and de Forest (1932) studied the rate of transpiration of eight species of plants in the chaparral of southern California. The "relative transpiration indexes," or the ratios of the average rate of transpiration to that from an atmometer of moist blotting paper, were as follows:

Plant	Relative transpiration index	Plant	Relative transpiration index
Rhus integrifolia	0.066	*Quercus dumosa*	0.148
Rhus laurina	0.076	*Adenostema fasiculatum*	0.275
Salvia mellifera	0.102	*Pentstemon spectabilis*	0.427
Ceanothus cuneatus	0.119	*Salvia apiana*	0.483

With the possible exception of *Rhus integrifolia,* the general structural arrangement of the leaf furnished very little reliable information as to the relative loss of water.

Pool (1923) stated that he was unable to explain the discrepancies between leaf anatomy and the index of foliar transpiring power. He suggested that in his experiments it might be possible that the cobalt chloride-paper method was not sufficiently accurate to differentiate accurately between the more nearly similar forms. He believed also that the stomatal behavior should be considered in a study of this kind, since it is to be expected that such behavior would modify the transpiring power to a marked degree. Shreve (1924) considered that the differences in the anatomical structure of the mesophytic and xerophytic leaves do not account for the difference in resistance to water loss by plants in arid regions. Blaydes (1935) stated that according to the indications the standard water-vapor loss of a species is not necessarily correlated with the xerophytism as indicated by the usual habitat of that species. Many xerophytic species have relatively high rates of water loss, while many mesophytic species have low rates. A complete summary of the studies on the relation of leaf structure to transpiration is to be found in Burgerstein (1904, 1920).

It appears evident that internal factors other than the general structure of the leaf may influence the transpiration rate, although the extent to which they may do so is in most cases not definitely known. It has frequently been suggested that the osmotic value of the cell sap of the leaf cells might have some influence on the rate of water loss from the plant. The concentration of the cell sap, however, must vary over a wide range in order to have any appreciable effect upon the transpiration rate, as can be illustrated by the examples mentioned by Drabble and Drabble (1907). In these observations they found that osmotic concentration of the cell sap of plants growing in three different habitats was respectively equivalent to 3,728, 8,812, and 14,958 mm. of mercury.

They considered that the evaporation of water from the surfaces of these three solutions would be in the ratio of 1.011:1.005:1, respectively. The influence of the osmotic value upon the rate of water loss would thus be slight, since an increase in concentration of 400 per cent would reduce the rate of evaporation of water by only 1 per cent. Since the osmotic value of the cell sap varies with the accumulation or disappearance of osmotically active substances within the cells of the leaf, the metabolic changes that use or produce these substances must be taken into consideration for slight modifications of the transpiration rate.

Variations in the water-holding capacity of the internal tissues of the leaf may also influence the transpiring power of the plant. The emulsoid colloids of the protoplasm and the component compounds of the cell wall have an imbibitional attraction for water that varies with the changes in structure and composition of these two parts of the plant cell or of the structures with which they come in contact (Shreve, 1916). It is well known that acids, bases, and neutral salts affect the distribution of water between the dispersed and continuous phases of emulsoids. On that account, the alternation of the chemical constituents causes variations in the distribution of water among the cell wall, cell sap, and the protoplasmic membrane, so that the water-holding capacity of the tissues may be altered. The imbibitional forces of the cell walls are especially important factors in influencing the rate of transpiration and will be discussed under the following heading.

2. Incipient Drying or Incipient Wilting.—It has been observed in many cases when the hourly rate of transpiration is plotted that the graph shows a marked flattening in its slope as the evaporating power of the air increases. This flattening of the graph appears to be due to some change in the plant that causes a reduction in the transpiration rate below that which would be expected from the form of the curve during the earlier hours. This reduction in the increase of the transpiration rate has sometimes been interpreted to mean that the plant initiates certain changes, under conditions of increasing transpiration, which retard the loss of water and thus protect the plant against excessive water loss.

The causes of this retardation of the increase in the transpiration rate are apparently to be found in certain physical phenomena which were pointed out by Renner (1910) and by Livingston and Brown (1912). They suggested that the internal retardation in water loss which frequently occurs prior to the maximum rate of transpiration may be due largely to an increase in the rate of water loss itself. This fact can be illustrated, as shown by Livingston and Brown (1912), by mounting two filter-paper disks saturated with water over glass tubes in such a manner that one will be in continuous contact with water while the other will be deprived of a water supply. When these disks are exposed to conditions favoring evaporation, the one that is in contact with a water supply will maintain a moisture content that is practically constant, while the moisture content of the one mounted over the empty tube will become less and less. An experiment of this type was performed by Livingston and Brown and it was found that a fall in the moisture content of the paper of 6 per cent produced a corresponding decrease in the rate of evaporation of 5 per cent. A fall of 17 per cent in water content produced a rate of evaporation 8 per cent lower than when the paper was saturated, the effect becoming more and more pronounced as the paper dried out. The conditions that prevail in the leaf as the transpiration rate increases are apparently in some cases similar to those which prevail in the paper-disk experiment, since if at any time the rate of water loss from the leaf surpasses the rate of supply, the tissues of the leaf bordering on the intercellular spaces would become less moist after the manner of the filter paper above mentioned.

This drying out of the cell walls that border on the internal atmosphere of the leaf has been termed "incipient drying" or "incipient wilting" and is substantiated by experiments that have shown a decrease of the water content of the leaves as the evaporating power of the air increases. As incipient drying progresses, the water films gradually retreat into the pores or capillaries of the cell walls bordering on the intercellular spaces and thus increase the surface tension of these water surfaces. This increase of the surface tension of the water films decreases their vapor tension, which causes a fall in the rate of the vaporization of water from them. This means that the retardation of the increase of the transpiration rate under an increasing evaporating power of the air may, in some cases at least, be due to a rate of transpirational loss slightly greater than the rate of water intake by the leaves.

It is generally accepted that the water content of the leaf is one of the principal factors controlling the transpiration rate. According to Haines (1935) the content of water in the leaf depends partly upon the conditions of pressure or tension in the conducting tracts supplying the water. The amount of water in the leaf is thus only a sort of passive intermediary through which the transpiration rate might be affected indirectly by changes in the pressure or tension of the tracts. The operating factor is the "pressure deficit," which may be defined as the difference between the pressure in the conducting tracts and the pressure at the leaf surfaces. With *Acer pseudoplatanus*, Haines (1936) found that by increasing the pressure deficit the rate of transpiration was reduced. A decrease in this deficit caused a very large and immediate increase in the transpiration rate. These pressure deficits alone apparently cannot reduce transpiration by their effects upon the vapor pressure at the leaf cell surfaces but must also cause an increase in the resistance to the flow of water through the protoplasts of the cells of the leaf. When the transpiration of different plants is observed, it is seen that certain plants do not manifest this flattening of the transpiration graph. Thus Briggs and Shantz (1916) found that wheat, oats, and rye showed a flattening of their transpiration graph, while alfalfa and the sorghums under like conditions showed no such behavior. Huber (1923) noted that the transpiration of the lower twigs of *Sequoia gigantea* was six times that of the higher ones. If, however, they are allowed to stand in water for 24 hr., this transpiration difference disappears, which fact he considered proves that these observed differences are the result of unequal water saturation. Zalenski (1923), however, found in the plants that he studied that the intensity of transpiration per unit of surface increased from the lower to the upper leaves notwithstanding the more xeromorphous structure of the lower leaves.

We want now to inquire what internal factors may influence the imbibitional action on the transpiration rate. Two of the factors that apparently may greatly influence the degree of incipient drying in the leaf and thus influence its rate of transpiration are the leaf surface and the nature of the root system. We shall now consider the effect of these two factors.

a. Leaf Surface of the Plant.—In transpiration experiments with corn and sorghum plants Miller and Coffman (1918) observed that, in general, if two or more of these plants were placed under the same conditions, the total amount of water lost by each stood in the same relative order as the leaf surfaces. The results of an experiment conducted from 7 A.M. to 9 P.M., as shown in the following table, illustrate this relation:

LEAF SURFACE AND THE AMOUNT OF WATER TRANSPIRED BY CORN AND SORGHUMS FROM 7 A.M. TO 7 P.M.

Plant	Leaf surface, square centimeters	Water transpired, cubic centimeters
Corn, Pride of Saline	16,943	3,853
Sorgo, Red Amber	4,844	1,702
Sorgo, Freed	2,973	989

Although the amount of water transpired per plant stood in the same relative order as the extent of the leaf surface, the loss of water was not exactly proportional to it, so it follows that the rate of transpiration per unit of leaf surface was not the same for each plant (see table, page 454).

It is to be observed from the table (page 454) that the rate of transpiration of the sorghums per unit of leaf surface was approximately twice as

THE RATE OF TRANSPIRATION OF CORN AND SORGHUM EXPRESSED AS G.M.[2] HR. FOR EACH 2-HR. PERIOD FROM 7 A.M. TO 7 P.M.

Time	Corn, Pride of Saline	Sorgo	
		Red Amber	Freed
7 to 9 a.m.	122	205	190
9 to 11 a.m.	165	293	304
11 to 1 p.m.	216	374	376
1 to 3 p.m.	199	483	379
3 to 5 p.m.	161	383	272
5 to 7 p.m.	94	124	147

high as that of the corn, while the leaf surface of the Red Amber sorgo and Freed sorgo was only 28.5 and 16.5 per cent, respectively, of that of the corn plant. This same fact also holds when two varieties of plants of different extent of leaf surface are used. For example, Miller and Coffman (1918) observed in the case of a plant of Pride of Saline corn with a leaf surface of 14,568 sq. cm. and a plant of Sherrod White Dent corn with a leaf surface of 12,989 sq. cm. that the actual amount of water lost by the larger plant during a 6-hr. period was 918 cc. as compared to 784 cc. for the smaller plant. The rate of transpiration, however, for the larger plant was only 629 g. per square meter per hour as compared to 723 g. per square meter per hour for the smaller one. Cullinan (1920) made a similar observation on severely pruned and unpruned 2-year-old apple trees. The unpruned trees transpired more water per plant than did the pruned, but the latter transpired more water per unit of leaf surface. It has been further observed that the differences in the transpiration rate per unit of leaf surface were more marked when the differences in the leaf surfaces were the greatest and when the evaporating power of the air was the highest. Darrow and Sherwood (1931) found that the loss by transpiration from strawberry plants with a leaf area of 126.6 sq. cm. was 446 cc., while the loss of water from the same number of plants with a leaf area of 599 sq. cm. was only 206 cc. for the given period. Kelley (1932) found that the removal of approximately half the leaves from the shoots of various fruit trees increased the rate of transpiration per unit area of the remaining leaves by 22 to 93 per cent.

One of the chief causes for the higher transpiration rate of the plants with the smaller leaf surface is apparently their ability to maintain a more saturated condition of the external cell walls of the leaves than the plant with the larger surface. If the facilities for absorption and translocation are approximately the same in two plants with unequal leaf surface, then it is apparent that incipient drying would progress faster and farther in the leaves with the greatest extent of surface. As a consequence of this, the rate of transpiration per unit of area of the plant with the larger leaf surface would be less than that of the plant with the smaller surface.

b. Nature of the Root System.—The ability of the root system to absorb water must exert a marked influence upon the degree of incipient drying and therefore the rate of transpiration. This fact is illustrated by the root systems of corn and the sorghums. Miller (1916) observed that plants of corn, kafir, and milo growing under the same conditions in the field had practically the same number of roots of the first order and that these penetrated vertically to the same depth and showed no difference in their lateral spread. The roots of the second order, however, were twice as numerous in the sorghums as in corn, as is shown in the following table:

NUMBER OF ROOTS OF THE SECOND ORDER PER UNIT OF LENGTH OF ROOT OF THE FIRST ORDER OF CORN, KAFIR, AND MILO GROWING IN THE FIELD

Plant	Number of observations	Number of roots of second order per inch of root of first order
Corn, Pride of Saline	127	15
Milo, Dwarf Yellow	147	26
Kafir, Blackhull	100	28

If the number of fine roots can be taken as a criterion of the efficiency of absorption, it would appear that the sorghum plants would be able to supply the leaves with more water and prevent incipient drying to a greater degree than in the corn plant. The sorghums illustrate how in some plants both the root system and the extent of leaf surface may be important factors in keeping the water supply in the leaves sufficient to retard incipient drying. Maximow and Rybin (1925) in experiments with magnolia and cherry laurel also came to the conclusion that transpiration may be regulated by the conditions of absorption as well as by the action of the stomata.

3. Aerial and Meteorological Conditions.—The conditions that affect the evaporation of water act directly to influence the transpiration rate. Our knowledge, however, concerning the influence of physical environmental factors upon the rate of transpiration is as yet very limited. Efforts to find algebraic formulas or equations by which transpiration can be estimated from observed environmental conditions have produced none that is satisfactory. These efforts are handicapped primarily because the experimental data at the present time are not sufficient to enable a proper evaluation of radiation, wind, light, temperature, and humidity in relation to transpiration and because the effect of stomatal movement is very little understood. The two main environmental factors that affect the evaporation of water are the evaporating power of impinging solar radiation and the evaporating power of the air.

a. Evaporating Power of Solar Radiation.—By impinging solar radiation is meant the total energy received from the sun by a given surface in a unit of time. The intensity of solar radiation is frequently expressed as the number of calories falling per minute upon a square centimeter of surface normal to the sun's rays. The known methods for measuring solar radiation, however, are not very satisfactory (Livingston, 1911; Briggs and Shantz, 1916; and Pulling, 1919). Solar radiation may be direct from the sun or indirect as from the sky, clouds, or earth. With no clouds or other shade intervening, direct solar radiation is predominant, but when the sun is obscured by clouds indirect radiation may be the more influential of the two upon evaporation (Livingston, 1923).

The influence of solar radiation on evaporation is always positive and is confined to the daylight hours. Briggs and Shantz (1915) found that the radiation graph always rises in advance of the other cyclic environmental factors. This is to be expected, however, since the change in radiation is the primary cause of the cyclic changes of these factors. These investigators observed that radiation rose in advance of transpiration in the case of wheat, oats, and rye and approximately with it in the case of the sorghums, alfalfa, and *Amaranthus*. In the afternoon, radiation fell off somewhat in advance of the transpiration of wheat and oats but approximately with it in the case of rye, alfalfa, sorghum and *Amaranthus*. Martin (1935), working with *Helianthus annuus* in full sunlight, found that the amount of transpiration due to the effects of radiation was 38 to 81 per cent, depending upon the evaporating power of the air. They found that the correlation coefficient of transpiration with radiation ranged from 0.82 to 0.89 for the plants under consideration; the correlation coefficient of transpiration with temperature ranged from 0.77 to 0.86; and the correlation coefficient of transpiration with relative humidity ranged from 0.75 to 0.85. Manzoni and Puppo (1934) found that for wheat the correlation of transpiration with radiation was 0.882, with evaporation was 0.796, with temperature was 0.660, and with relative humidity was 0.538. These figures show that intercorrelations exist among the environmental factors, since the sum of the squares of the coefficients of the independent causative factors influencing transpiration cannot exceed unity.

Briggs and Shantz (1916) in transpiration experiments with wheat, oats, millet, and alfalfa under shaded and unshaded conditions concluded that the energy consumed in transpiration is only partially attributable to direct radiation. They considered that indirect radiation from the sky and surrounding objects and the heat energy received directly from the air contribute materially to the heat energy dissipated through transpiration. If, however, the total energy necessary to evaporate all the water lost from an acre of crop plants is compared to the total radiant energy falling upon that area, it is found that less than one-half of it is necessary to evaporate the water lost from the plants even at the time of the maximum transpiration rate.

The primary influence of solar radiation in determining transpiration is indicated by a consideration of the changes in the physical environmental conditions and of transpiration at night when solar radiation is zero. These conditions were determined during the hour from 8 to 9 P.M. and again from 3 to 4 A.M., together with the corresponding transpiration rate. It was found that a simultaneous diminution in the wet-bulb depression of one-fourth its maximum range and in temperature of one-third its maximum range resulted in only a 3 per cent drop in the transpiration rate. This would seem to indicate that the high correlation obtained between transpiration and air temperature and wet-bulb depression is largely due to the direct correlation between radiation and temperature or wet-bulb depression.

The study of the effect of impinging radiation upon transpiration has for the most part been confined to a consideration of the influence of light intensity upon this process, and it has been definitely shown that light has a direct accelerating action on transpiration aside from any secondary effect brought about by stomatal movement (Sampson and Allen, 1909). In order to eliminate any influence of stomatal action, Darwin (1914) covered the leaves with vaseline or cocoa fat and then slit them in various places for the escape of water. Under such conditions the acceleration in transpiration in diffuse daylight was found in the case of *Prunus laurocerasus* and *Hedera helix* to be as much as 50 per cent of the rate in darkness. The figures obtained by Darwin were perhaps too high, since he took data directly after making incisions in the leaves and did not take into consideration the increased water loss due to wound stimulation. Henderson (1926), who followed the same method but allowed the leaf to recover from the wound stimulation before collecting data, could not obtain the large increases in the transpiration rate obtained by Darwin. Electric light of the intensity of diffuse daylight was found to increase the mesophyll water loss in *Hedera helix* and *Eupatorium adenophorum* by an average amount of 5 per cent, of which 1 per cent was due to the heating effect. Thus the direct effect of the light on the transpiration rate amounted to 4 per cent. This amount varied from 1.5 to 9.4 per cent, indicating that the increase in water loss from the mesophyll due to light varies greatly for different plants and for different leaves on the same plant.

Chodat (1931) noted that the rate of transpiration of alpine plants decreased in high light intensities. He attributed this depression to changes in cellular permeability induced by the intense light. Because no depression in the rate was observed at low altitudes, it was considered that this regulatory action was a function of the ultraviolet region. Montemartini (1932) found with *Myosporum serratum* that light stimulated transpiration independently of the effects of temperature. He considered light to modify the action of the stomata or the attraction of the protoplasm for water. Bader (1930) considered that light and soil moisture are the most important external factors in modifying the transpiration of pine trees.

Light may increase transpiration in one or more of the following ways: (1) higher temperature of the leaves, although the impinging light may give some of the molecules of water sufficient kinetic energy to escape as vapor without generally raising the temperature of the leaves. (2) Greater permeability of the protoplasmic membrane as suggested by the work of Iwanoff and Thielmann (1923), who found a great increase in transpiration of the living, but not the dead, leaves of *Bromus inermis* and *Cyperus alternifolius* by changing the illumination from red-yellow to blue-violet rays. If light thus reduces the resistance to the passage of water through the protoplasm, the supply available to the cell wall will be increased, its imbibition will be higher, and the vapor tension at the evaporating surface of the cell wall will tend to rise. (3) Imbibitional changes in the cell-wall colloids in such a manner as to render the cell wall more permeable to water.

b. Evaporating Power of the Air.—The term "evaporating power of the air" is used to designate the combined influences of air temperature, air humidity, and air movement upon the evaporation of water. In contrast to impinging solar radiation, the evaporating power of the air acts on the evaporation of water, not only during the day but also during the night. It generally acts in a positive manner but sometimes in a negative way, as, for example, when the plant absorbs some moisture from the air or when dew is formed upon its surfaces. We wish now to consider individually the effect of temperature, relative humidity, and air movements upon the rate of transpiration.

1. *Temperature.*—The transpiration rate increases, in general, as the temperature rises. Briggs and Shantz (1915) observed that the behavior of transpiration in regard to temperature varies to a considerable degree with different plants. Thus they found in the case of wheat, oats, and alfalfa that the transpiration graph rose in advance of the temperature but rose with the temperature for rye and sorghums and after the tem-

perature in the case of *Amaranthus*. In the afternoon the transpiration for all plants always fell off more rapidly than the temperature, and when the transpiration had reached its night level, the temperature was still above the minimum by an amount equal to about one-third the daily range. In considering the relation to temperature of the average hourly transpiration rates of the above plants on clear days for periods of from 10 to 15 days, they found that the correlation coefficient of transpiration with temperature varied from 0.77 to 0.86 depending upon the plant. When, however, the correlation between the physical environmental factors and the transpiration of over 20 different crop plants during the growing season was determined as one population, the correlation coefficient between temperature and transpiration was found to be 0.64. This means that in the former case if temperature is considered as the primary causative factor, from 59 to 73 per cent of the transpiration was due to temperature, while in the latter observation only 41 per cent was due to this factor (Briggs and Shantz, 1916).

Whitfield (1932) found in the Pike's Peak region that the curve of transpiration showed a closer correlation with the curves of air temperature and relative humidity than with any of the other factors that were measured. It was observed by Kosaka (1933), where anthocyanin pigments occurred in the leaves, that the more highly colored leaves showed a higher transpiration rate than the less highly colored ones. He believed that in all probability this difference in transpiration rates is connected with heat absorption by the anthocyanin pigments. Arthur and Stewart (1933) studied the transpiration rates of tobacco plants under both visible and infrared radiation with controlled temperature and humidity. They found that between 73 and 78°F. an increase of the energy by 2.3 times doubled the water loss. This temperature relation appeared to be independent of the moisture content of the air within the range of 50 to 80 per cent relative humidity. Clements and Martin (1934) found with the sunflower that transpiration varied only slightly when the soil temperature was between 55 and 100°F., but that it dropped rapidly when the soil temperature reached 55°F., reached half at 38°F., and approached zero when the soil temperature neared 32°F.

2. *Relative Humidity*.—In general, it is known that a decrease in the relative humidity will induce a higher transpiration rate, but the amount of quantitative data on the effect of this factor on transpiration is very meager. Montgomery and Kiesselbach (1912) and Kiesselbach (1916) grew corn plants in two greenhouses in which the physical environmental conditions were practically the same with the exception of the relative humidity. During three growing seasons the humidity averaged 22 per cent higher for one greenhouse than for the other. The evaporation from a free water surface in the more humid greenhouse was 54 per cent,

and the transpiration per unit of leaf area was 62 per cent as much as in the drier greenhouse. Darwin (1914) found that the relation between relative humidity and the magnitude of transpiration in most cases is a linear one and that the straight line representing this relation is approximately in the position giving it an angle of 45 deg. with the horizontal. In some cases, however, the change in the rate of transpiration lagged behind the change in relative humidity so that the experimental graph became a curve instead of a straight line. Darwin also observed that transpiration is not zero in saturated air and apparently would not reach that point until the hypothetical humidity was 105 per cent. This means that the vapor tension of the leaf is higher than that of a saturated atmosphere and is about what it should be if the temperature of the leaf was 0.8°C. higher than that of the surrounding air. Darwin concluded that the temperature of the leaf above that of the atmosphere is produced by respiration. Dixon (1898, 1914) also obtained evidence that plants may transpire in a saturated atmosphere. Briggs and Shantz (1915) found in the case of the plants studied by them that transpiration rose in advance of the depression of the wet-bulb thermometer. They found that the correlation coefficient between transpiration and the wet-bulb thermometer depression varied from 0.75 to 0.85 for 10- to 15-day periods in clear weather.

3. *Wind.*—The effects of wind upon transpiration may be twofold. In the first place, moving air tends to remove the blanket of more or less saturated air that surrounds the leaf and thus tends to reduce the distance that the escaped molecular water must move in order to find free air conditions. The movement of air thus helps to increase the diffusion gradient and so increases the rate of transpiration. This influence may be understood by considering the opinion of Renner (1910). He considered that the rate of diffusion from leaves will be inversely modified by an increase in the distance between the region of minimum density outside the leaf and the maximum density within the leaf. This distance is increased in still air by the water-vapor cap that forms over the surface of the leaf, and the larger the leaf the greater the average thickness of the vapor cap, so that if the air were perfectly still, transpiration would vary inversely as the diameter of the leaves. On this account, Renner considered that wind increases the transpiration of large mature leaves by a much greater percentage than it does the small ones. In the second place, wind may influence transpiration if its velocity is sufficient to agitate the leaves and bring about a circulation of air in the intercellular spaces.

Copeland (1906) in experimenting with the coconut palm in the Philippine Islands observed that transpiration in one case in direct sunshine was four times as great in a wind with a velocity of 5 m.p.h. as in a calm, but the increase usually was not over 100 per cent. Brown (1910) found that, although gentle air currents stimulate transpiration, strong air currents may check it. In observations on *Silphium laciniatum*, Giddings (1914) noted that wind velocity increased the activity of the leaves until a velocity of about 8 m.p.h. was reached. If the velocity, however, rose above this, the activity of the leaves was impaired and the rate of transpiration became less. Knight (1917) found that a wind velocity of 0.26 m.p.h. caused an increase of 50 per cent in the rate of transpiration. Wilson (1924) noted that certain Australian plants had their transpiration checked when the wind velocity reached 20 m.p.h.

Seybold (1929, 1931, 1933) stated that wind influences only cuticular transpiration. Thus only hydrophytes that have a relatively high cuticular transpiration showed an appreciable increase in transpiration due to wind. Shapiro and de Forest (1932) noted that wind did not seem to affect the transpiration rates of xerophytic plants, although it did increase the evaporation from the atmometer. With a wind velocity of 10 m. per minute, Stalfelt (1932) with detached leaves obtained a very rapid rise in transpiration over that in still air. There was a slow rise from this velocity to 30 m. per minute, after which there was no further increase in transpiration with an increase in the velocity of the wind. The maximal increase of transpiration due to the wind was 140 per cent over that in still air. Wrenger (1935) studied the effects of wind on the rate of transpiration of 38 species of plants of various morphological types. In 1-hr. experiments, a wind velocity of 1.85 m. per second always brought about a considerable increase in the rate of transpiration. The rate of transpiration during the first 10 to 15 min. was always higher than that during the remainder of the period. The observed increase in the rate of transpiration varied from 12 to 150 per cent of the rate in still air, depending on the species, the effect generally being greater in mesophytes than in xerophytes.

Martin and Clements (1935) studied the effect of wind on the transpiration of *Helianthus annuus* for a short period of 8 days and for long periods of 6 to 8 weeks. During the short period, the rate of transpiration increased with an increased velocity of wind to 2 m.p.h. The rate of transpiration increased 20 to 30 per cent at the onset of the wind, and this value was maintained as long as the wind acted. For velocities higher than 2 m.p.h., there was a relatively large increase in transpiration directly after the application of the wind. This was followed by a fall in the rate, which in turn was followed by a gradual increase. This initial increase amounted to as much as 138 per cent at 16 m.p.h. and was followed by a slight wilting closure of the stomata and a reduction in the rate of transpiration. In about ½ hr. the leaves regained their turgidity, and the transpiration rate rose. A wind of 1 m.p.h. increased the rate of transpiration 30 per cent, while one of 16 m.p.h. increased it as much as 50 per cent. They could observe no difference in the effects of the wind at night as compared to the day.

In the long-time series the plants were grown under continuous wind of velocities of 5, 10, and 15 m.p.h., from the first leaf stage to the 6 to 8 weeks' stage. For the first week the transpiration rates of all three sets were practically the same as those of the controls. Thereafter, however, the rate began to increase with increasing wind velocity so that by the end of the series the three test sets were using water at rates about 20, 35, and 50 per cent higher per unit area than the controls. In this long-time series, the effect of wind on the transpiration rate was greater during the night than during the day. During the day the plants under a 15 m.p.h. wind velocity transpired about 1.5 times as rapidly per unit area as the controls, while at night they transpired four to five times as rapidly. If wind influences the rate of transpiration independently of the other physical environmental factors, it should show a higher correlation with transpiration than with these factors. This, however, is not the case, according to the work of Briggs and Shantz (1915) and Manzoni and Puppo (1934), who found that the wind velocity showed the lowest correlation with transpiration of any of the other physical environmental factors that were studied. When the correlation between the environmental factors and transpiration is considered for the growing season for a large variety of plants, Briggs and Shantz (1916) considered that only from 2 to 6 per cent of the transpirational loss can be attributed directly to the action of the wind.

Although the experimental evidence in regard to the influence of wind on transpiration is somewhat confusing, the indications are that the rate of water loss does not increase as the wind velocity increases above a gentle breeze. A method for controlling the rate of air movement in transpiration experiments has been devised by Blackman and Knight (1917).

4. Water Content of the Soil.—From what has been previously mentioned in regard to the relation of incipient drying to the rate of transpiration, it should be expected that the rate of transpiration would be higher if there were an abundant supply of water in the soil than if it were limited. A rapid movement of water to the roots would greatly aid in preventing incipient drying and thus aid in keeping up the transpiration rate (Shreve, 1924). Shreve (1905) and Hendrickson (1921) mentioned that the transpiration rate of plants is considerably increased by adding water to the relatively dry soil but submitted no quantitative data either in regard to soil moisture or in regard to the transpiration rates. Yuncker (1916) found in growing corn plants in a loam soil with moisture contents of 25, 45, and 65 per cent of saturation that the rate of transpiration for a 24-hr. period was 4.9, 5.8, and 7.2 g. per square meter per hour, respectively, for each of these moisture contents. Miller and Saunders (1923) determined the rate of transpiration of wilted and turgid plants. The soil with a wilting coefficient of 12 per cent had in the case of the turgid plants a moisture content of 20 to 22 per cent and was in good tilth, while that of the wilted plants was depleted to slightly above the wilting coefficient at the beginning of the experiment and slightly below it at its close. The transpiration rate for the turgid and wilted plants of cowpeas and soybeans is shown in the following table:

Transpiration Rate of Turgid and Wilted Plants

Date	Rate of transpiration, g.m.² hr.			
	Cowpeas		Soybeans	
	Turgid	Wilted	Turgid	Wilted
July 24:				
1 to 3 p.m.	69.8	32.0	82.6	26.2
3 to 5 p.m.	78.3	44.8	92.2	24.0
5 to 7 p.m.	57.9	25.6	37.6	13.0
July 25:				
7 to 9 a.m.	20.4	19.2	21.4	10.9
9 to 11 a.m.	71.5	12.8	64.3	15.3
July 26:				
11 a.m. to 1 p.m.	126.0	6.3	75.1	4.3
1 to 3 p.m.	165.2	9.6	92.9	4.3

It is a significant fact that when the evaporation rate is relatively low, as in the morning and evening, the differences in the transpiration rate of the turgid and wilted plants are the least. This is due evidently to the fact that incipient drying at these times has not progressed to such an extent as it does when the evaporating power of the air and solar radiation is high.

Frei (1924) grew plants of apple, sunflower, maize, and beans, among others, in a soil with an optimum water content of 60 to 70 per cent of the total moisture capacity and in the same type of soil with a minimum moisture content of 35 to 40 per cent of the water-holding capacity. The intensity of transpiration from the plants in the moist soil was always greater than in the dry soil. Also, an increase in the moisture content of the dry soil led to an increase in the intensity of transpiration until it equaled that of the plants growing in the soil of optimum moisture content. Plants, however, that were growing in the soil with the optimum water content showed a considerably lower transpiration rate when the soil moisture was reduced to the minimum (35 to 40 per cent) than was exhibited by the plant that had been grown in the soil with this water content. Dosdall (1919) grew *Equisetum fluviatile* in a soil with a moisture content of 35 per cent, in mud, and with the roots submerged in water and found that the plants growing in the mud lost the most water, while those in the soil of 35 per cent water content lost the least. Dole (1924) found, with young plants of *Pinus strobus* growing in a soil with a saturation point of 24 per cent, that the transpiration rate with a moisture content of 20 per cent was from two to six times greater than that from plants growing in a soil with a moisture content of 10 per cent.

Miller (1928) in 12 experiments, each of which extended over an 8-hr. period during the daylight hours, studied the rate of transpiration of corn plants in a water-saturated soil and in a soil in good tilth. The plants which had reached about one-half their vegetative growth were grown in soil in good tilth in sealed cans and, just previous to the experiment, water was added in an amount sufficient to saturate the soil. It was found that the average transpiration rate for the 12 plants growing in the soil in good tilth averaged 104 g. per square meter per hour, while the rate for the plants in the saturated soil averaged 107 g. per square meter per hour, a difference that is of little or no significance. It is apparent (Livingston, Hemmi, and Wilson, 1926) that after the water supply in the soil has increased to a certain degree, the effect of a deficiency in the oxygen-supplying power of the soil is manifest, so that factors other than the water-supplying power of the soil must be considered. Niemann (1932) found that sunflower plants transpired the maximal amount of water when the soil was 80 per cent saturated, while the plants of bean and wheat lost their maximal amount when the soil moisture was 60 per cent of saturation. Whitfield (1932) found that transpiration was the highest from plants growing in a soil with a medium water content, lower when growing in saturated soils, and least from those plants in the soils with a low water content. Penfound (1932) and Peterhänsel (1934) substantiated in general the observations of previous investigators along this line.

The experimental evidence in regard to the effect of the water content of the soil on transpiration thus seems to indicate that a plant will transpire the maximal amount of water when the soil has sufficient moisture for good tilth but that an increase above that amount has no effect. A decrease, however, below that for good tilth lowers the rate of transpiration so that when the water content has been reduced to the wilting coefficient or below and the plant is badly wilted, the rate of transpiration may be only one-twentieth to one-thirtieth of that which prevails when the optimum amount of moisture is present.

5. Chemicals.—The effect of chemicals on transpiration has been studied along two different lines: (*a*) their effect when applied to the soil or nutrient solution in which the plant is growing and (*b*) their effect when applied to the leaves in the form of sprays.

a. Chemicals Applied to the Soil.—The presence of any substance in the soil that will injure the absorbing parts, prevent their development, or decrease the permeability of the protoplasmic membrane to water will decrease the absorption of water and consequently lower the transpiration rate as compared with that of a control plant growing in a medium lacking this substance. The addition of chemicals to the soil may also diminish the transpiration rate, owing to the increased concentration of the soil solution, which lowers the diffusion gradient of water into the plant.

On the other hand, the presence in the soil of substances that increase the permeability of the cell membranes to water and stimulate the growth and development of absorbing organs, or act in an antagonistic way upon substances that would lower the transpiration, may greatly increase the rate of transpiration. Also, any substance that increases or hinders the growth of the plant in general increases or decreases the transpiration. It is thus seen that the relation of chemicals to transpiration is very complicated and is interwoven with changes in growth, nutrition, and permeability.

Burgerstein (1904) found, in general, that weak acids and extremely weak solutions of alkali accelerated transpiration, but that stronger solutions retarded it. He also found that corn plants growing in solutions of 0.1 and 0.25 per cent of calcium nitrate and magnesium sulphate increased their transpiration rates, while those in 0.5 and 1 per cent solutions decreased their rates as compared with the controls. Harter (1908) in the case of wheat plants found that, when sodium chloride was present in the soil in sufficient concentration to cause the modification of the structure of the plant, transpiration was considerably reduced, but that when sodium chloride was present in amounts too small to produce any measurable structural effect, the salt had a stimulating effect upon transpiration. Wheat plants grown in soils containing 0.06 to 0.08 per cent of sodium chloride showed an increased transpiration as compared with plants growing in a nonsaline soil, the lower concentration of salt inducing the higher rate of transpiration (Ricome, 1903). Sampson and Allen (1909) in experiments with *Zea mays* and *Helianthus annuus* studied the effect of the acids hydrochloric, nitric, and acetic and the alkalies potassium hydroxide, sodium carbonate, and ammonium hydroxide. For the plants growing in solutions, concentrations of 0.065 and 0.13 per cent were used, while concentrations of 1.0 and 1.5 per cent were added to the plants growing in the soil. In all cases the acids accelerated transpiration, while in most cases the alkalies retarded it. In some few cases the rate was increased. Reed (1910) noted the effect on transpiration of the wheat plant of the addition to 189 soils from different localities of 100 p.p.m. of sodium nitrate, potassium sulphate, and calcium hydrogen phosphate and 1,000 p.p.m. of calcium carbonate. Considering the transpiration from the control plants as 100, that for sodium nitrate was 90.4; for potassium sulphate, 94.8; for monocalcium phosphate, 100.8; and for calcium carbonate, 97.8. Similar experiments with solutions of various organic and inorganic acids ranging in concentration from 0.0004 *N* to 0.000025 *N*, however, showed very contradictory results. Bouyoucos (1911) in a very exhaustive study of the effect of salts on transpiration found that the transpiration rate was affected by different densities of nutrient and nonnutrient solutions, by different chemical salts, and by the ions of these salts. Reed made the observation that the addition of pyrogallol or tannic acid to soil extracts markedly increased the rate of transpiration. Large increases in transpiration were also obtained by treating soil extracts with absorbent compounds like carbon and ferric hydrate. The increased transpiration in this case is apparently due to the action of these substances upon the toxic or injurious compounds in the soil. Burgerstein (1885) found that solutions containing 1 part of camphor per 1,000 had an accelerating effect upon the transpiration of most plants. Newton (1925) performed experiments which show that the rate of root respiration is increased as related to transpiration when the concentration of the cultural solution is increased. The rate of transpiration is decreased when the concentration of the cultural solution is increased, but this is not accompanied by a corresponding decrease in the rates of root respiration. This indicates that as the concentration of the cultural solution is increased, the plant roots must expend more energy in absorbing a given volume of water. Schmucker (1928) noted that transpiration from leaves and branches was greatly depressed by narcotic vapors of relatively low tension. Heiling (1933) found that transpiration was increased by a low concentration of sulphur dioxide. Meyer (1931) applied various salts of sodium, potassium, and calcium to the soil in concentrations of 0.025, 0.05, 0.1, 0.2, and 0.4 per cent of its dry weight. In observations continued through 4 days, the addition of these salts resulted in nearly all cases in a decrease of transpiration. Williams (1935) observed with barley that the application of phosphorus to the soil had the effect of lowering the ratio of the amount of water transpired to the amount of dry matter produced. In some cases this was caused apparently by a decrease in transpiration and in others by its influence on growth.

b. Films from Sprays.—It has long been recognized that Bordeaux mixture applied as a spray exerts a marked influence upon the physiological functions of the plant aside from its action as an insecticide or fungicide. For example, spraying with Bordeaux mixture has been observed to prolong the life of the potato plant for 25 days and to increase the yield of tubers 100 bu. per acre over that of unsprayed plants during a season when no disease or insect pests were present. Considerable work has been done to determine the nature of the reactions of plants to Bordeaux mixture when applied as a spray, but it is the intention to consider here only its reaction upon the process of transpiration.

Prior to 1914, observations, for the most part, on the effect of Bordeaux mixture on the rate of transpiration were incidental to other experimental work, so that the reports on that subject previous to that date are very brief, with little or no information concerning the state of the plants or the conditions under which the observations were made. Rumm (1893) considered that Bordeaux mixture decreased the rate of transpiration, since he observed that the unsprayed abscised leaves of the grape wilted before the sprayed ones. Bayer (1902) and Schander (1904) also considered that decreased transpiration rates followed spraying with Bordeaux mixture. Schander suggested that the decreased rates might be due to the exclusion from the leaves of certain rays which in the unsprayed leaf tend to accelerate the process of transpiration. The increased vigor and yield of potato plants in the field when sprayed with Bordeaux mixture, even when no diseases and insects are a source of loss, were believed by Clinton (1909) to be largely due to the conservation of moisture in the leaves in dry seasons by the spray. This conservation of water he considered prevented tip burn. This theory was strengthened by the fact that the application of sprays with little sediment did not prolong the life of the plants or give increased yields as did Bordeaux mixture.

Frank and Krüger (1894) concluded that Bordeaux mixture applied to the leaves as a spray exerted an accelerating influence upon the rate of water loss by transpiration. These observations were later confirmed by Zucker (1896).

Since 1914 experimental work in considerable detail has been undertaken by various investigators to determine the effect of sprays and especially Bordeaux mixture upon the rate of transpiration. Duggar and Cooley (1914) instituted the method of procedure that has been generally followed by other workers along this line. The general plan has been to divide the plants under consideration into two groups, which may be designated test plants *T* and control plants *C*. These plants are then paired and the ratio $T:C$ of their units of transpiration during a given period is obtained. After this standardization period the plants *T* are sprayed and the ratio of the units of transpiration $T:C$ is again obtained for each pair of plants. If the ratio $T:C$ has increased over what it was during the standardization period, the transpiration rate has been increased by the spray. If the ratio has decreased, the transpiration rate has been lowered. Duggar and Cooley found that spraying abscised leaves of the castor bean with 4-6-50 Bordeaux mixture increased the rate of transpiration as much as 80 per cent. In the case of potted tomato plants 12 in. high growing in the greenhouse, the transpiration rate was increased as much as 40 and 25 per cent during a period from October 18 to November 4 by spraying with 4-6-50 and 2-3-50 Bordeaux mixtures, respectively. They also determined the effect upon the transpiration of tomato plants of calcium hydroxide, aluminum hydroxide, clay suspension, and lime sulphur as liquid sprays and powdered aluminum hydroxide, charcoal, and calcium carbonate as dust sprays. With the exception of calcium hydroxide during one period, all these applications produced during two 15-day periods an increase in the transpiration rate which varied from 8 to 15 per cent above that of the controls. In the case of potted potato plants, the effect of various sprays upon transpiration during two 5-day periods was as shown in the first table on page 467.

The lime-sulphur spray produced the least effect upon the transpiration rate. Lamp black added to Bordeaux mixture produced a higher

Spray used	Water transpired, cubic centimeters	
	May 6 to 11	May 11 to 15
Bordeaux mixture 4-6-50	5,266	4,633
Bordeaux mixture 2-3-50	6,424	5,740
Lime wash (60 g. CaO per liter)	5,845	6,136
Lime sulphur (1–25)	4,430	4,507
Bordeaux + lamp black	7,921	6,530
Lime wash + lamp black	5,966	5,856
Controls	4,138	4,333

rate of transpiration than the mixture alone. Martin (1916) at the New Jersey experiment station found that Bordeaux mixture (5-5-50) had a much more marked effect upon the rate of transpiration of abscised leaves than upon those of potted plants. Considering the ratio of $T:C$ before spraying as 1.00, the ratio of $T:C$ after spraying for abscised leaves was as follows:

Effect of Bordeaux Mixture upon the Rate of Transpiration of Abscised Leaves

Plant	Duration of experiment	$T:C$ after spraying
Ricinus communis	6 days	1.50
Datura meteloides	4 days	2.12
Phaseolus vulgaris	10 hr.	2.39
Beta cycla	10 hr.	1.57
Raphanus sativus	10 hr.	1.63
Hibiscus cardinalis	10 hr.	3.72

The average increase in transpiration due to this spray in these and other experiments was about 100 per cent over that of the standardization period. In the case of potted plants of cabbage, eggplant, pepper, and tomato, the increase in transpiration of the sprayed plants over that of the unsprayed plants was respectively 3, 11, 29, and 8 per cent. Martin also observed that a surface covering of dry powdered copper sulphate was less effective in the acceleration of transpiration than the film of Bordeaux mixture. Lutman (1916) found that the transpiration of potato plants was only slightly increased by spraying with Bordeaux mixture under the conditions of his experiments. The increase that he noted amounted to only 1 per cent, which is apparently insignificant. Shive and Martin (1917) determined the relative rate of transpiration of the sprayed and unsprayed leaves of tomato plants by means of the cobalt chloride method. They found that the rate of transpiration from the sprayed leaves was from 18 to 29 per cent higher than that of the unsprayed ones on the same plant.

Miller (1928) in a series of experiments during four growing seasons studied the effect on transpiration of Bordeaux mixture applied as a spray to tomato plants. The plants were grown in soil in large sealed containers under conditions of the field and were vegetatively normal and produced a normal crop of fruit. The data were collected during a period of 20 days when the plants were in full vegetative vigor. The method used was essentially that used by Duggar. The plants were arranged in pairs and the ratio of the transpiration rate of each pair was determined for a 10-day period. After this period one plant of each of the pairs was sprayed and their ratio obtained for another period of 10 days. The results for the 4 years are summarized in the following table:

EFFECT OF BORDEAUX MIXTURE UPON THE RATE OF TRANSPIRATION OF THE TOMATO PLANT

Year	Number of experiments performed	Spray used	Effect on transpiration, number		Transpiration of sprayed plants (unsprayed = 1)
			Increased	Decreased	
1924	6	4-8-50	4	2	0.990
1925	17	4-8-50	11	6	1.050
1926	47	4-8-50	20	27	0.994
1927	49	4-5-50	16	33	0.970

Miller thus was unable to obtain any of the striking results on transpiration from spraying as had been reported by other investigators. With the exception of Shive and Martin (1917), most of the experiments reported have been done under greenhouse conditions and under conditions, too, of relatively low transpiration as compared to the experiments at Manhattan, Kans. The results of Shive and Martin were obtained by the cobalt-paper method, while those of Miller were obtained by weighing. Whether the spray may in some way affect the reaction of the cobalt chloride paper is not known.

Anderson (1931) reported that the dusting of peanut plants with calcium sulphate influenced but little their rate of water loss. Childers (1935) found little or no alteration of the transpiration of tomato plants after spraying them with Bordeaux mixture. Cross (1927) found that butter fat and the emulsions of Wesson oil, of cotton oil, and of mineral oil reduced the rate of transpiration approximately 43 per cent. After spraying the twigs of the apple, peach, pear, plum, and sour cherry with saturated and unsaturated oils, Kelley (1930) noted the effects of the treatment upon the rate of transpiration. All the oils used retarded transpiration, this retardation ranging from 50 to 75 per cent. Spraying of the lower leaf surfaces retarded

Influence of Bordeaux Mixture on the Transpiration Rate of Various Plant Species (Wilson and Runnels, 1934)

Plant	Increase of transpiration due to application of 4-6-50 Bordeaux mixture		Portion of daily water loss that occurred during the night period (untreated plants), per cent
	Night period only (7 P.M. to 7 A.M.), per cent	24-hr. period, per cent	
Coleus	375	27	7
Bean	320	24	7
Tomato	250	24	8
Squash	186	29	11
Bean (lima)	183	19	7
Beet	171	9	4
Pepper	150	7	4
Castor bean	147	8	8
Chinese cabbage	141	32	17
Tobacco	136	12	6
Mustard	118	31	17
Muskmelon	114	16	14
Watermelon	100	23	17
Lettuce	82	16	19
Swiss chard	78	14	8
Cucumber	77	33	23
Parsnip	75	17	6
Chrysanthemum	75	13	10
Potato	67	11	11
Eggplant	64	15	5
Corn	60	12	6
Geranium	57	5	5
Cockscomb	56	6	5
Cauliflower	47	13	16
Carrot	43	18	13
Nasturtium	38	14	32
Onion	38	8	13
Sunflower	38	6	10
Cabbage	36	6	13
Hollyhock	35	10	11
Carnation	30	9	7
Spinach	28	11	22
Calendula	28	5	10
Soybean	27	10	18
Pea	25	20	8
Stock	25	−2	7
Snapdragon	18	6	20
Horseradish	17	8	24
Dill	11	5	17
Celery	8	3	37

transpiration, while spraying the upper leaf surfaces had no influence. The retarding effect of these sprays was noted within 30 min., and it continued for 3 days after their application.

Wilson and Runnels (1931 to 1935) studied extensively the action of sprays on the loss of water from plants. They applied 18 different spray solutions to *Coleus* and found that only a few of these caused an increase in the transpiration rate comparable to that caused by Bordeaux mixture. It was found that the sulphates of iron, nickel, and manganese, in combination with hydrated lime and water, caused nearly as great an increase in the rate of transpiration of *Coleus* as did Bordeaux mixture. They studied the effects of Bordeaux mixture upon 40 species of plants growing in the greenhouse. The most marked increases in transpiration due to this spray occurred from 7 P.M. to 7 A.M., and ranged from 8 to 375 per cent. For a 24-hr. period the increase ranged from −2 to 33 per cent. In the untreated plants the portion of the water loss which occurred during the night was 4 to 37 per cent. It appears, therefore, that this spray affects the cuticular rather than the stomatal transpiration, as shown in the table on page 469.

Tilford and May (1929) observed that the application of Bordeaux mixture and copper-lime dust to the leaves of potato lowered their temperature. A 4-4-50 Bordeaux mixture lowered the temperature of the leaves 1.45°C. below that of the controls, while a 4-6-50 mixture lowered their temperature 2.41°C. When the leaves were sprayed with a 4-4-50 Bordeaux mixture, plus lamp black, their temperature was 2.92°C. higher than that of the unsprayed ones. Rosen (1933) found in certain regions of Arkansas that spraying with Bordeaux mixture gave excellent control of tip burn of potato. He stated that around the sprayed plants there was in most instances a sudden drop in the air temperature ranging from 1.5 to 8.0°C.

6. Fungous Diseases.—Fungous diseases affect the rate of transpiration, in most cases, apparently in a manner that would be expected from the effect that the fungus has on the morphology of the host. In others the effect cannot be explained by any of the known reactions that the diseases have upon the plant. Müller-Thurgau (1894) seems to have been the first to note the effect of fungi upon the transpiration of the host. He noted by the use of hygrometric paper that, when pear leaves were infected with *Fusicladium pyrinum* (pear scab) and apple leaves with *Fusicladium dendriticum* (apple scab), the rate of transpiration was considerably increased, the transpiration of the diseased areas being higher than the healthy adjacent portions. Grape leaves infected with *Peronospora viticola* (downy mildew) showed a much lower rate of transpiration than the healthy leaves, and this was attributed to the fact that the conidiophores almost completely filled the stomata. Blodgett (1901) noted that leaves of *Rubus* which were infected heavily on the lower surface with *Gymnoconia interstitialis* transpired almost twice the amount of water as did the normal healthy leaves. Montemartini (1904) found in the case of 20 species of plants, each infected by a different fungus, that almost without exception the transpiration rate was higher for the diseased than for the healthy plants. Reed and Cooley (1913) by means of Freeman's method studied the effect of *Gymnosporangium* *juniperi-virginianae* upon the rate of transpiration of 8-year-old York Imperial and Ben Davis apple trees. The rate of transpiration per unit of surface of the diseased leaves was, in the majority of cases, less than that of the healthy ones. The average ratio of the transpirational loss of the diseased to the healthy leaves was 0.783 for the York Imperial and 0.742 for the Ben Davis. The ratios showed certain differences if they were grouped according to different stages in the development of the disease. The rate of transpiration was depressed to the greatest extent when the fungus had reached maturity and many of the cluster cups were open for the release of the aeciospores.

apples

This fungus brings about certain changes in the morphology of the leaf which would apparently lower the transpiration rate. The apple leaves bearing the cluster cups are thicker by three or four times than the normal leaf. The large intercellular spaces in the spongy tissue disappear owing to the formation of columnar cells, while the stomata are scarce or lacking and the substomatal cavities are absent altogether.

Weaver (1916) made a study of the influence of the rusts of wheat, rye, oats, barley, corn, sunflower, cocklebur, and carnation upon the transpiration rate. In the case of wheat infected with *Puccinia graminis tritici*, the transpiration rate of the diseased plants was increased 30 per cent above that of the healthy plants. This difference in rate was produced by the fungus when the rust pustules amounted to only nine-tenths of 1 per cent of the leaf surface. A striking increase in transpiration was noticeable when the pustule area was increased by only two-tenths of 1 per cent. Rye infected with *Puccinia rubigo-vera secalis* showed the same increase in the rate of transpiration per unit of leaf area over that of the healthy plants as did wheat. Thus an increase of 20 per cent in the transpiration rate was noted when the pustules represented only 3.8 per cent of the leaf surface.

In oats infected with *Puccinia graminis avenae*, in which the pustules equalled only 0.5 per cent of the leaf surface, a transpiration rate 37 per cent higher than that of the healthy leaves was obtained. Corn, sunflower, and carnation plants infected with rust behaved in a similar manner to the cereals that have just been discussed. Cocklebur plants that were inoculated with *Puccinia xanthii*, however, showed a transpiration rate which was only 50 to 60 per cent that of the healthy plants. This behavior of the cocklebur seems to be due to a change in the tissue of the leaves similar to that which occurs in the apple leaf when infected with *Gymnosporangium juniperi-virginianae.*

In the case of the rusts of the cereals, the increase in the transpiration rate occurs about the time that the pustules break through the epidermis. After the pustules are for the most part broken out, no increase in the transpiration rate is apparent. The cause of this increased transpiration rate is not known. Dufrenoy (1918) considered that a number of fungi accelerate transpiration owing to localized necrosis, rupture of the epidermis, or changes in the permeability of the cells.

In the case of the cereal rusts, however, Weaver considered that the cause of the accelerated transpiration is not wholly due to the torn epidermis. The whitish areas around the pustules indicate that some striking change has occurred in the cells composing these regions. The mesophyll cells that are penetrated by the haustoria certainly do not function normally. It is conceivable, as was suggested by Weaver, that the fungi may excrete into these cells substances which may cause a change in the permeability of the protoplasmic membrane that accelerates water loss. Weiss (1924) considered that the infection of wheat by either leaf or stem rust resulted in a lowered water economy of the host, as indicated by the higher water requirement of the rust-infected plants as compared to the healthy ones.

Kourssanow (1928) found that the loss of water from wheat plants infected by *Ustilago tritici* was slightly higher than that from uninfected ones. Nicolas (1930) studied 10 different species of plants parasitized by as many different types of fungi, and found that with but few exceptions the infected plants showed increased transpiration rates. The ratio of transpiration of the infected to that of the healthy plants varied from 1.01:1.0 to as high as 2.1:1.0. It was observed by Graf-Marin (1934) that barley plants infected with *Erysiphe graminis* transpired 67 per cent more water than did healthy plants. The proportion of water transpired at night was greater for mildewed plants than for healthy ones. This increased transpiration of the infected plants could not be attributed to the rupture of the cuticle or to an increased number of stomata. He considered that the increase of transpiration was due to an increase in the opening of the stomata and to the loss of water from the aerial mycelium of the fungus.

D. Temperature of the Leaves

It is commonly stated that transpiration is beneficial to plants because it cools the leaves and thus prevents their temperature from becoming so high in bright sunlight as to cause their death or at least to hinder their normal activities. The temperature of a leaf, however, depends upon numerous and varied factors, so that in order to understand the influence of transpiration upon it, a general discussion of the subject is necessary. The most important factors that influence the temperature of leaves are the temperature of the air, the supply of available moisture in the soil, the evaporating power of the air, currents of air, the thermal emissivity of the leaf, the intensity of light, and the angle at which it is incident to the leaf surface. Under identical conditions the temperature of one kind of leaf is different from that of another kind, while different regions of the same leaf have different temperatures. Any data, therefore, that are presented upon this subject must be regarded as relative only to the various conditions that prevailed when the temperature determinations were made.

1. Methods of Measuring.—The earlier determinations of leaf temperatures were made by pressing the thermometer against the leaves or by wrapping the leaves about the thermometer. This method is still used when only approximate results are desired. It is not accurate, however, due primarily to the relatively slow response of the ordinary thermometers to changes in temperature and to the inability to get a proper contact of leaf and thermometer with the leaf in its normal position.

More accurate determinations of leaf temperatures are now made by means of thermocouples, with which it is possible to measure not only accurately but very quickly either the internal or the surface temperature of the leaf. The thermocouple method has been used by Blackman and Matthaei (1905), Ehlers (1915), and Clum (1926) for measuring the internal temperature of leaves and by Shreve (1919), Miller and Saunders (1923), Eaton and Belden (1929), and Curtis (1936) for measuring the surface temperature.

In the determination of leaf temperatures by means of the thermocouple method, the general procedure of each investigator has been very similar, although the methods have varied somewhat in detail. The method used by Miller and Saunders (1923)—a modification of that used by Shreve (1919)—for determining the surface temperature of leaves will serve as an illustration of the general procedure for the determination of plant temperatures by the thermocouple method. A diagram of the apparatus used is shown in Fig. 18. It consists of two thermo junctions *TC* and *T′C′* approximately 5 mm. in length, which are formed by braiding two wires and then uniting them by an acid-free solder. The wires generally used are No. 36 copper and No. 36 constantan with insulated connecting lengths *A*, *B*, and *B′* totaling approximately 3 ft. One of the thermocouples *T′C′* is placed with a thermometer graduated to 0.1°C. in a stoppered Dewar flask surrounded by any suitable insulating material as glass wool (GW), and water, in any convenient container. The air in the flask remains practically constant, since under the most severe conditions it does not fluctuate more than 0.1°C. during a 15- to 20-min. period. The other thermo junction *TC* is attached to a clamp *C* in such a way that it can be conveniently placed upon the surface of the leaf. This clamp may consist of a pair of crucible tongs 44 cm. in length, modified by filing their distal ends to narrow dimensions. One of these filed ends is inserted in a small wedge shaped piece of cork *CK*, the edge of which measures 1 by 5 mm. and holds the thermocouple *TC*. On the other tip of the tongs is inserted a small cube of cork *CK* 5 by 5 mm. which serves as a support to the leaf when the thermocouple is placed upon it. By means of a clamp of this kind, the thermo junction can easily be placed as desired in direct contact with either the upper or the lower surface of the leaf and the temperature determined. Similarly, the temperature of the air can readily be determined by holding the open clamp in the air and taking precautions to shield the thermo junction from the direct rays of the sun. A galvanometer *G* and a damping switch *K* are placed in the circuit. Since the temperature of the thermo junction *T′C′* is known and can be kept approximately constant, the difference in temperature *T′C′* and *TC* can be calculated from the swing of the galvanometer, which has previously been calibrated. By such an instrument a difference in temperature of 0.1°C. can be accurately detected. This apparatus when in use in the field with leaves that are some distance above ground is placed upon a small table, which is shaded by a piece of heavy canvas placed about 2 ft. above its surface. When vines and low plants are being investigated, the apparatus is placed on a low box and shaded with an umbrella.

Curtis (1936) threaded the thermocouple wires through the leaf so that the thermocouple junctions as well as that part of the wires back to the insulation would touch the leaf surface. He found this method far superior to the one used by Miller and Saunders (1923) and Eaton and Belden (1929). The electrical equipment used by him consisted of a Leeds and Northrup Type K-2 potentiometer, a portable galvanom-

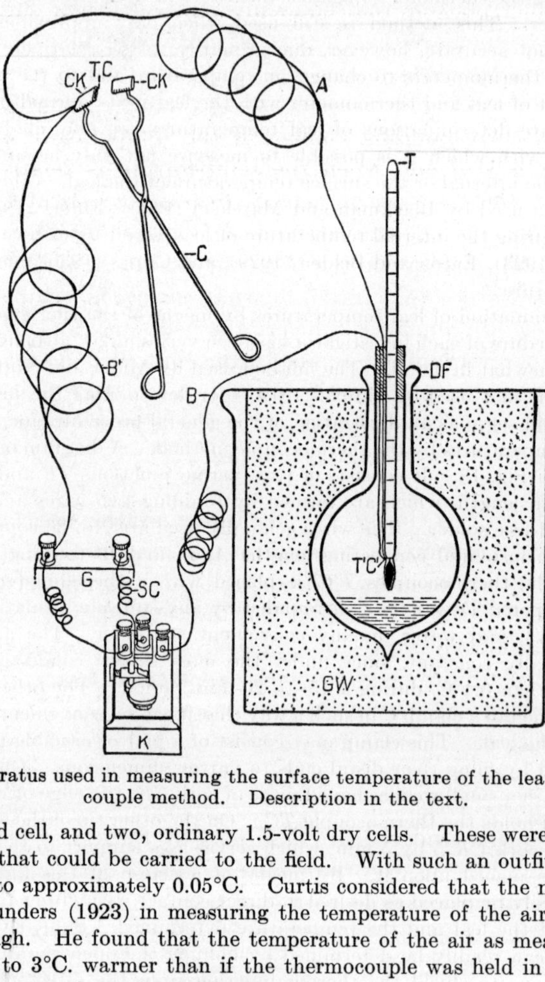

FIG. 18.—Apparatus used in measuring the surface temperature of the leaf by the thermocouple method. Description in the text.

eter, a standard cell, and two, ordinary 1.5-volt dry cells. These were all mounted in a shallow box that could be carried to the field. With such an outfit, temperatures could be read to approximately 0.05°C. Curtis considered that the method used by Miller and Saunders (1923) in measuring the temperature of the air gives readings that are too high. He found that the temperature of the air as measured by their method was 2 to 3°C. warmer than if the thermocouple was held in the shade at a distance of 1 to 2 cm. from the shading object.

2. Temperature. *a. Fluctuations in Temperature of the Air.*—In direct sunlight, when the temperature is relatively high and the air is in motion, the temperature of a leaf and the surrounding air is not constant but shows sudden and marked fluctuations even during so brief a period as a few seconds. These changes in temperature, which vary from a small fraction of a degree to as much as 4°C., are easily detected by the galvanometer through the thermo junctions but are of such short duration that as a rule they are not visibly recorded by a mercury-bulb thermometer, even when it is graduated to tenths of a degree. These rapid fluctuations in the temperature of the air are due, in all probability, to the fact that the air is not uniformly heated throughout but contains warmer or cooler pockets that suddenly replace the air surrounding the measuring instruments. The leaves of most crop plants respond very quickly to these changes in air temperature, since even a very slight increase or decrease in the temperature of the air is almost immediately followed by a corresponding change in the temperature of the leaves. When the air is still and when the temperature is relatively low, the fluctuations in the temperature are so few and so slight that they are seldom detected by the galvanometer. Curtis (1936) reported that leaves frequently change in temperature as much as 2.5°C. within a period of 10 to 15 sec. The rapid fluctuations of the temperature of the leaf and the surrounding air emphasize the necessity not only of having a rapid means of determining temperatures but also of making a large number of observations in order to obtain the relative temperature relations of a leaf and its surroundings. The thermocouple method for measuring temperatures is relatively rapid. The swing of the galvanometer for a temperature determination can be read in 2 or 3 sec. The time, however, that elapses during the recording of the deflection and the return of the galvanometer to equilibrium amounts to approximately 10 to 12 sec., so that an interval of about 15 sec. occurs between any two consecutive observations. Shull (1936), however, noted that it required from 1 to 2 min. for the leaves of *Pelargonium*, *Begonia*, and *Caladium* to reach thermal equilibrium after being transferred from diffuse light to direct illumination.

b. Temperature of Normal Turgid Leaves.—Miller and Saunders (1923) made numerous determinations of the temperature of the upper surface of the leaves of corn, sorghum, cowpeas, soybeans, watermelon, pumpkin, and alfalfa growing under field conditions in Kansas. The average of 1,000 determinations of the temperature of the air and of the upper surface of the turgid leaves of corn in direct sunlight between the hours of 9 A.M. and 4 P.M. and under a wide range of atmospheric conditions was 30.58°C. for the air and 30.64°C. for the leaves. The average of the same number of determinations made with the leaves of five varieties of sorghums under conditions of exposure similar to those of the corn leaves showed that the temperature of the leaves was 30.64°C., while the average of an equal number of determinations of the air temperature during the same period was 30.66°C. Five hundred observations on soybeans gave an average temperature of the air of 33.13°C. and an average of 33.66°C. for the temperature of the upper surface of the leaves. A similar number of determinations on the leaves of cowpeas showed an average temperature of 34.4°C., while the temperature of the air averaged 34.2°C. during the observations. These observations show that the temperature of the turgid leaves of corn, sorghum, pumpkin, watermelon, and soybean in direct sunlight under the general climatic conditions which prevail in Kansas fluctuates slightly above and below air temperatures but that their average temperature is essentially the same as that of the surrounding air. The temperature of the leaves of alfalfa under the same conditions as the plants above mentioned consistently showed less than 1°C. below that of the surrounding air. In the case of corn, sorghum, watermelon, pumpkin, soybean, and cowpeas, the heat absorbed by the leaves from the sun is quickly utilized in transpiration or rapidly disseminated into the surrounding air, so that the temperature of the leaves is always approximately that of the air. In the case of alfalfa the rate of transpiration is evidently rapid enough to reduce the temperature of the leaf slightly below that of the air. Eaton (1930) reported for cotton that the young leaves were cooler than the old ones. Thus the temperature of leaves 82 days old was 2.3°C. below that of the air, while those 20 to 26 days old had a temperature 4.1°C. below that of the atmosphere. Copeland (1932) reported from observations on chaparral in the Sierra Nevada that, in a breeze, leaves transpiring in the sun sometimes cooled below the shade temperature of the air. It was reported by Ezekiel and Taubenhaus (1932) that the leaves of cotton plants that had wilted from *Phymatotrichum* root rot were as much as 6.5°F. warmer than air, while the leaves of the control plants were usually cooler than the air.

However, Curtis (1936) in extensive experiments could find no case where the temperature of leaves in direct sunlight was below that of the surrounding air. He attributed the results obtained in this regard by Miller and Saunders (1923) to the fact that their method gave too high a reading for the temperature of the air.

The internal temperature of the leaves of *Fuchsia speciosa*, *Phaseolus vulgaris*, *Brassica oleracea*, and *Syringa vulgaris* was determined by Clum (1926) by means of thermocouples. The plants were in the direct sunlight in the open or in the greenhouse at Ithaca, N. Y. He found that the temperature of the leaves with which he worked was nearly always higher than the air during the day and that in direct sunlight it was 5 to 10°C. higher. The maximum difference observed between leaf and air temperature in the open was 13°C. and in the greenhouse, 16°C.

The type of leaf has an important bearing on its temperature in relation to the surrounding air. It seems reasonable to suppose that a thick, fleshy leaf will absorb and retain more heat than a thin one. This supposition is borne out generally in experimental observations. Thus Askenasy (1875) and Ursprung (1903) by placing thermometers in close contact with the surface of the fleshy leaves of *Sempervivum* found that the leaves attained a temperature in the sunlight of 18 to 25°C. above that of the surrounding air. Askenasy attributed the high temperature of these leaves primarily to the fact that the heat which they absorbed was not readily dissipated by air currents or radiation. As mentioned above, Clum (1926) obtained temperatures of leaves that were 5 to 10°C. higher than that of the surrounding air. The type of leaves with which he worked had, in general, a much greater mass in proportion to the surface exposed than did the leaves investigated by Miller and Saunders (1923), which in no case had temperatures appreciably different from that of the surrounding air. An example of the relation between the compactness and general mass of a plant organ and its temperature in relation to that of the air is afforded by the work of Brooks and Fisher (1926) with the apple. They found that the side of an apple exposed to the sun may have a temperature from 12 to 25°C. higher than the air temperature and from 10 to 16°C. higher than the shaded side of the same apple. In the case of the thin sunflower leaves, Brown and Escombe (1905) calculated their temperature to be only approximately 2° above or below that of the air.

Grainger and Allen (1936) by means of the thermocouple method found that apple buds were warmer than the air by day and usually cooler by night. When the buds were screened from the radiation of the sun, their temperature was cooler than that of the air by day as well as by night. The buds of black currants and raspberries were usually cooler than the air. The experiments that have just been discussed were conducted under summer conditions, but it is interesting to note that Ehlers (1915) found that evergreen conifer leaves under winter conditions maintain internal temperatures from 2 to 10°C. higher than that of the surrounding air.

In considering the temperature of a leaf, the air currents must be taken into consideration. A leaf in direct sunshine freely exposed to a breeze always has a lower temperature than one under like conditions of sunlight but in such a position as to be protected from air currents. Smith (1909) observed that breezes reduce the temperature of leaves in sunlight by amounts varying from 2 to 10°C. and that a thin leaf is much more noticeably affected than a thick one. Similar observations were also made by Miller and Saunders (1923) and by Ehlers (1915).

Miller and Saunders observed for the plants with which they worked that in diffuse light the temperature of attached turgid leaves is always consistently lower than the temperature of the surrounding air. The average temperature differences between the turgid leaves and the surrounding air under conditions of diffuse light varied from 0.1 to 3°C. In direct sunlight the temperature of a turgid leaf may be above the air temperature, but as soon as a cloud obscures the sun the temperature of the leaf almost immediately drops below the temperature of the air and remains there until the leaf is again exposed to direct sunlight (see the following table):

THE TEMPERATURE CHANGES OF THE AIR AND OF THE TURGID LEAVES OF SOYBEAN PLANTS IN DIRECT SUNSHINE AND IN DIFFUSE LIGHT CAUSED BY A HEAVY CLOUD OBSCURING THE SUN

Time, p.m.	Temperature, degrees centigrade, of	
	Air	Turgid leaf
	In direct sunlight	
1:10	32.0	34.0
1:11	32.1	33.9
1:12	32.7	33.3
1:13	30.7	30.7
1:14	31.0	32.3
1:15	31.3	31.8
1:16	32.8	32.6
1:18	32.0	34.2
	In diffuse light due to cloud passing over sun	
1:19	30.0	26.3
1:20	28.0	26.3
1:21	27.7	26.0
1:22	26.7	25.3
1:23	26.3	25.3
1:24	26.8	25.3
1:25	26.3	25.1
1:26	25.9	25.0

During the night the temperature of the leaves falls to the same level as that of the surrounding air or slightly below it.

c. Rate of Transpiration and the Temperature of the Leaf.—It has long been known that a wilted leaf has a higher temperature than a turgid one under the same conditions, but the actual differences in temperature and the correlation of the transpiration rate with this temperature difference have been but little studied. Darwin (1898) in the observation of withered detached leaves and normal attached leaves of *Tropaeolum majus* found in intermittent sunshine and relatively low humidity that the temperature of the withered leaves was from 1.2 to 3.9°C. higher than that of the attached leaf. Kiesselbach (1916) by inserting the bulb of a thermometer momentarily in the folds of corn leaves found that a transpiring leaf was uniformly cooler than a dry dead one, the difference in temperature amounting in one case to as much as 8.5°F. in direct sunshine and to 4.2°F. in the shade. Using a similar method, Loftfield (1921) made a few observations on the temperature of the leaves of alfalfa, potato, and sugar beet in relation to stomatal behavior. He found that usually the temperature of the leaves with stomata open was lower than that of the air, while the temperature of the leaves with closed stomata was higher than that of the air.

Rather extensive experiments in relation to transpiration and leaf temperature were conducted by Miller and Saunders (1923) and by Clum (1926). The former investigators measured the surface temperature of the leaves and the latter their internal temperature by means of the thermocouple method. The transpiration temperature experiments of Miller and Saunders were conducted with turgid and wilted plants of three varieties of corn, three varieties of sorghums, and one each of cowpeas and soybeans. The plants designated as turgid were those which showed no visible signs of wilting during the experiments. The soil in the containers in which the plants were growing was kept at a moisture content sufficient for good tilth by the frequent addition of water to replace that lost from the plants. In order to obtain wilted plants, certain containers were set aside to which no water was added to replace that lost by transpiration, and the plants were considered sufficiently wilted for the experiment when they did not regain their turgid condition during the night. The water content of the soil in which the wilted plants were rooted averaged about 2 per cent above the wilting coefficient. The rate of transpiration was determined at intervals of 2 hr., and the temperature determinations were made at times intermediate between these periods.

A typical response of the turgid and wilted leaves of plants in regard to temperature and the rate of transpiration is shown by the graphs in Fig. 19, representing the data obtained with corn leaves in direct sunlight.

The data obtained from several thousand observations on corn, sorghum, soybeans, and cowpeas showed that the temperature of a wilted leaf in direct sunlight is always higher than the temperature of a turgid leaf exposed to the same conditions. The average temperature of the wilted leaves of corn, sorghum, soybeans, and cowpeas during the hours of 9 A.M. to 4 P.M. was respectively 1.85, 1.55, 2.8, and 4.6°C. higher than that of the turgid leaves. The maximum temperature difference observed between the turgid and wilted leaves of corn in direct sunlight was 4.3°C. when the temperature of the air was 32.3°C., and the transpiration rate of the turgid leaves was approximately five times that from the wilted leaves. The maximum temperature difference between the wilted and turgid leaves of cowpeas was 6.7°C. when the temperature of the air was 37.6°C., and the transpiration of the wilted leaves was approximately only one-sixteenth that of the turgid plants.

The observations of Miller and Saunders agree, in general, with the results obtained by Clum (1926) with *Fuchsia speciosa*, *Phaseolus vulgaris*, *Brassica oleracea*, and *Syringa vulgaris*. He reduced transpiration by allowing the soil to dry and in some cases by vaselining the surface of the leaves. In general, the vaselined leaves and the leaves of the plants in the dry soil were 2 to 4°C. warmer than the controls. In no case, however,

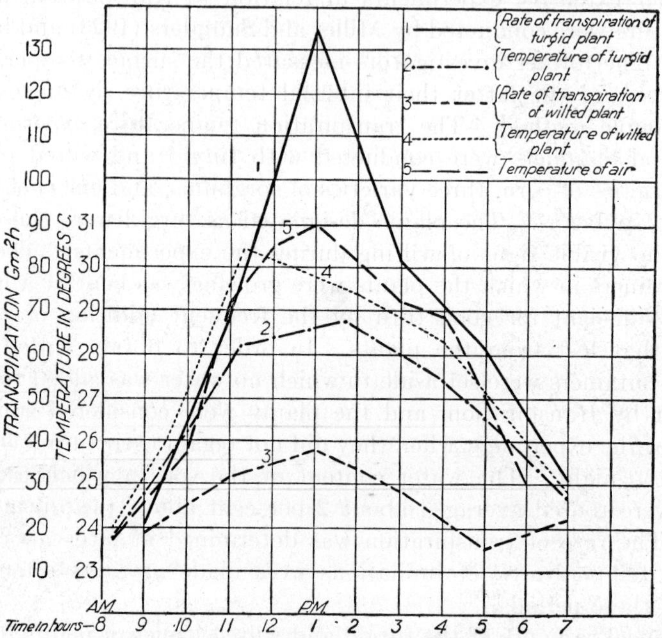

FIG. 19.—Graphs showing the transpiration rate of turgid and wilted corn plants and temperature of the leaves of these plants at different periods of the day.

did he find a definite correlation between the transpiration rate and the difference between the leaf and air temperatures or between differences in the transpiration rates of two leaves or plants and differences in their temperatures.

It was reported by Copeland (1932) that the transpiration of the chaparral in the Sierra Nevada cooled the leaves as much as 10°C., and he considered that a greater cooling effect would be recorded if more appropriate technique could be used.

Watson (1933) and Curtis (1936) considered that there are numerous factors other than transpiration which prevent the excessive heating of foliage leaves. The leaf may dissipate the absorbed energy by thermal emission, by conduction to internal tissues of a lower temperature, and by photochemical reactions. Curtis stated that the angle of exposure to incident radiation will greatly affect the absorption of radiant energy. Even when the light inception is normal to the leaf, 25 per cent of it may be reflected. When the leaf temperatures exceed that of the atmosphere, there may be considerable loss by conduction. There is also radiation to cooler objects near or far, and, when the sky is clear, to space, or to cold gases, especially carbon dioxide and water. Thus plant temperatures may be greatly influenced by water in the atmosphere, either as clouds or vapor, at a great distance from the plant, because of the influence of this water on the receipt or loss of infrared radiation. This effect may be totally independent of the effect on loss of water vapor close to the leaf and independent of the effects of clouds on the passage of visible radiation to the plant. The relative importance of these different methods of dissipating heat varies greatly with the environment of the plant as well as the conditions within the leaf, and its surface. Watson (1934) calculated the relative amounts of heat dissipated by thermal emission and transpiration for varying rates of transpiration and for varying temperature differences between the leaf and the air. The evidence thus obtained indicated that neither transpiration nor thermal emission can be called the more important factor.

d. The Thermal Death Point.—It has been observed by numerous investigators that the thermal death point of most plant cells lies between 45 and 55°C., although in extreme cases it may range between 62 and 105°C. for certain types of tissues (MacDougal, 1922; and Hopkins, 1936). In many cases at least, the fatal temperature must be prolonged some little time before death ensues. It has been shown that transpiration lowers the temperature of leaves even under severe conditions only to the extent of 2 to 5°C. with an average difference somewhat less than 3°C. This difference in temperature is not striking, and it would seem doubtful whether the increased temperature of the wilted leaves could to any marked degree injure the protoplasm or even retard its vital activities. Even if a leaf became heated to a critical temperature, the lowering of it from 2 to 3°C. by transpiration would not greatly prolong life. The temperature of a leaf does not long remain constant, due to rapid radiation, convection, and air currents, so that these factors may be more effective in protecting the leaves from overheating than is transpiration. The cause of the death of leaves under severe climatic conditions would thus seem to be due to the dehydration of the protoplasm rather than to a critical temperature induced by the lowering of the transpiration rate.

E. Evaporation

Since the days of Unger (1861), who regarded transpiration as a physical process modified by the nature or peculiarities of the plant, transpiration has frequently been defined as the evaporation of water from the plant altered to some extent by its structure and functions (see also Woods, 1893). Because of the similarity of evaporation and transpiration and because of the aim apparently to show that transpiration is a relatively simple physical process, the rate of evaporation has been frequently used as a standard to which transpiration is compared or referred. The rate of water loss from plant surfaces exposed to the air is always lower than the rate of evaporation from an equal area of water surface under the same environmental conditions (Livingston, 1906, 1913). Briggs and Shantz (1916) noted that the loss of water from wheat, oats, barley, sorghum, and alfalfa during a 10-day period of maximal transpiration at Akron, Colo., amounted to only 5 to 14 per cent of the loss during the same period from a free water surface. McLean (1919), however, stated that transpiration in the tropical rain forests was in his observations four-fifths of that from a free water surface. It is evident from these facts that certain internal factors exert a retarding influence upon the evaporation of water contained in the plant. This retardation or resistance to the evaporation of water will vary with the kind of plant, its age, the portion of the plant considered, and the changes of internal conditions.

Evaporation and transpiration are both determined largely by the same factors—solar radiation, temperature, saturation deficit, and wind—so that if transpiration is a purely physical process, then a high correlation between it and evaporation is to be expected. Briggs and Shantz (1916) found in comparing the rate of evaporation from a small shallow tank with the transpiration rate from wheat, oats, rye, alfalfa, and *Amaranthus* that the correlation coefficient ranged from 0.84 to 0.95. In the case of the highest correlation it would thus appear that nine-tenths of the transpiration was determined by the same factors that determine evaporation. They noted that the transpiration graph of these plants showed a tendency to rise somewhat earlier than the graph of evaporation from the small tank and to fall off somewhat later in the afternoon. Owing, however, to the greater slope of the transpiration graph in the late afternoon, they both reach the night level at the same time. These authors considered that the relation of transpiration to evaporation should be considered associative, both responding to the same environmental factors but not in the same way or to the same degree. The extent of this association, however, depends upon the manner in which evaporation is measured, so that a consideration of that subject is next in order.

1. Measurement. *a. Value of Data.*—A knowledge of evaporation in considering the water relations of plants is of importance in several ways. In the first place, the rate of evaporation from a free water surface or from a moist porous surface is considered the best single valued expression of the intensity of the weather factors influencing transpiration (Briggs and Shantz, 1917). Evaporation thus serves as a simple means of expressing the aerial conditions under which experimental work is performed. A study of the evaporating power of the environment prevailing in a given section during the growing season furnishes an approximate measure of the transpiration loss that would occur in that region. The intensity of evaporation thus gives a standard by which it is possible to correlate some of the general facts of plant distribution and to predict whether certain plants are adapted to the atmospheric conditions prevailing in the region under consideration. The rate of evaporation thus may be an indicator of the vegetative possibilities of a given region (Livingston, 1907, 1911).

b. Instruments for Measuring Evaporation.—Any instrument that is used for the measurement of evaporation is termed an "atmometer" or "evaporimeter." Numerous types of atmometers have been used, and one of the main handicaps in evaporation studies so far has been to obtain a standard instrument. It should be kept in mind that the water surface from which evaporation proceeds is as truly a control of the rate of water loss as are the atmospheric conditions. The loss of water from one instrument is thus different from the loss from another under the same conditions, and furthermore the evaporation from one instrument cannot be reduced to that of another. Thus suppose that under one set of conditions an instrument *A* loses twice the water per hour as an instrument *B*. If the environmental conditions are changed, this relation between the two rates of loss does not hold, and the instrument *A* does not lose twice the amount of water as does *B*. Hence the readings from any evaporating instrument cannot be reduced to the losses from another instrument of unlike form. It is of course impossible to produce two instruments that are exactly alike, but if their form and volume are approximately the same, they may be very accurately standardized, as will later be mentioned. The exposure of two instruments must be the same if the readings are to be at all comparable. If one of the two instruments that are alike in all details is placed 1 ft. from the level of the ground and the other 2 ft. from the ground, the evaporating rates will be different. The exposures thus need to be specific as well as the kind of instruments if the rates of evaporation are to be comparable (Livingston, 1915, 1917). These facts in regard to evaporation have in many cases been little appreciated or considered by those who have studied evaporation, so that the data are very confusing.

1. *Atmometers with a Free Water Surface.*—The two instruments of this type that have been the most extensively used have been designated the large tank and the small tank. The large open tank has been used extensively by the U. S. Department of Agriculture Office of Dry Land Agriculture. It is circular in form, made of galvanized iron, and is about 6 ft. in diameter and 2 ft. in depth, being sunk into the ground so that its surface and that of the ground are on approximately the same level. The loss in evaporation is measured in terms of depth by means of a vernier arrangement attached to the side of the tank, but a self-recording apparatus may also be attached. There are many objections to this means of measuring evaporation. The wind not only increases the surface exposed by waves or ripples but may frequently remove water by whipping it over the sides. There is no means of keeping the depth of the water constant, so that the evaporation alters the nature of the instrument. The removal of water by birds, insects, and toads and the addition of water by

rains and its loss by splashing also tend to make the data thus obtained less exact. Accurate readings for short periods have never been possible with the large tank, so that for hourly comparisons of evaporation with transpiration it is of little or no value.

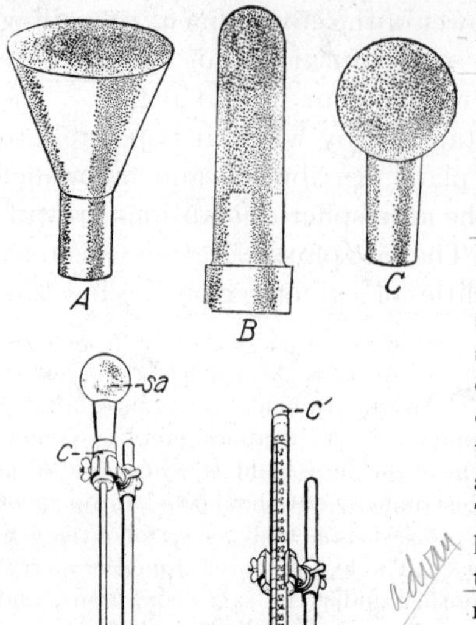

Fig. 20.—*A*, Bellani plate type of atmometer. *B*, cylindrical porous cup atmometer. *C*, spherical porous cup atmometer. *D*, apparatus for measuring the evaporation from an atmometer: *sa*, spherical porous cup atmometer; *c*, perforated rubber stopper; *c′*, notched cork; *b*, burette; *rt*, rubber tubing; *gt*, glass tubing.

For a comparison of transpiration and evaporation for the entire growing season, however, it has proved very satisfactory (Briggs and Shantz, 1916).

The small tank type of atmometer was devised and perfected by Briggs and Shantz (1916, 1917) and consists of a shallow copper tank 91 cm. in diameter and 2.5 cm. high and blackened on the inside. It is supported on a wooden disk 4 cm. in thickness and mounted about 1 m. above the ground on an automatic scale sufficiently sensitive to record the loss of a layer of water 10μ in thickness. The depth of the water is automatically maintained at about 1 cm. by means of a mariotte apparatus mounted on the scale platform. This instrument is automatic, accurate, and has few of the disadvantages of the large tank. It is the best yet devised for determining the loss of evaporation from a free water surface.

Stark and Whitfield (1930) described an evaporimeter that maintains a constant water level in the evaporating pan by utilizing the principle of the mariotte flask. A rain overflow is provided, which permits any increase of water in the evaporation pan to escape automatically and to be measured in a standard rain gauge. Gates and Black (1931) found the dragoyle a child's toy, very useful to measure ecological conditions.

2. *Atmometers with a Moist Porous Surface.* The most widely used porous surfaces for evaporation studies are those of porous clay, although filter and blotting papers have sometimes been used. The first known porous-clay atmometer was the Bellani porous plate devised by Bellani in 1820 (Fig. 20*A*). It consists of a horizontal porous-clay disk forming the top of a container completely filled with water so that the lower surface of the disk is in contact with the liquid while the upper surface is exposed to the air. The porous-clay cup is now the most universally used of any of the porous-surface evaporimeters. This instrument was independently devised three times, first by Babinet (1848), then by Mitscherlich (1904), and finally by Livingston (1906, 1910). These atmometers as now used are hollow cylinders or spheres of unglazed porcelain and have a capacity of about 50 cc. They are pro-

vided with a neck or collar which is either glazed or shellacked and are so arranged that they may be connected with a water supply by means of a rubber cork carrying a glass tube (Fig. 20*B* and *C*). The porous-surface atmometers are filled with water and the porous unglazed portions of the walls absorb water by capillarity or by imbibition, so that a moist porous surface is exposed to the air. These atmometers may be connected with a relatively large container whose supply of water need not be replenished for a number of days or weeks, the evaporation being determined by noting the amount of water necessary to fill the reservoir to its original level. The porous atmometers may be connected with a burette as shown in Fig. 20*D* when accurate readings are required at brief intervals of a few minutes to several hours. The porous cup atmometers present to the action of the air a water surface that is held in position by surface tension, adhesion, and the rigidity of the material that composes the instrument. The form of the surface is thus not disturbed by wind, insects, or splashing, as is so frequently the case with free water surfaces. The intake of water by rains can be almost prevented by various rain-excluding attachments that have been described by Livingston (1910), Livingston and Thone (1920), Shive (1915), Johnston (1918), Nichols (1923), and Wilson (1930).

Sayre (1928) describes a recording atmometer. The porous-clay cup atmometers generally used are white in color, but in order to obtain responses to the atmospheric complex that more nearly correspond with those of the plant or to study the effect of solar radiation on water loss, atmometers of darker colors have been employed. These are either made from colored clay or are painted with lampblack (Livingston, 1911, 1923, 1926).

Stanley (1932) described a method of making permanently black atmometer bulbs from white ones by dipping them into a 3*N* solution of silver nitrate. Bates (1919) devised an atmometer with an impervious upper surface and a perforated lower surface, between which surfaces a porous moist material was placed. He found that this type of an instrument followed very closely the transpiration losses from small conifers.

Wilson and Welton (1935) stated that a Livingston black atmometer could be used as an indicator for the proper time to water lawn grasses. The application of 1 in. of water, as soon as the atmometer had lost 320 cc. of water without a rain of ½ in. or more intervening, produced the best and most uniform growth of grass.

Before setting up a porous-clay cup atmometer it should be soaked in water for several hours at least, so as to replace the air in the interstices with water. The atmometer and connecting tubes must be filled with water and connected with the water supply in such a manner that no air bubbles enter the apparatus. If any bubbles enter they will collect at the top of the cup on the inside and thus hinder evaporation from that portion of its surface. The cup must always be placed higher than the water level in the reservoir, otherwise water will be forced out of the atmometer by hydrostatic pressure. It has been found, however, that the cup may be placed several feet above the level of the water in the burette or reservoir without perceptibly influencing the rate of water loss.

Similar forms of porous-clay cup atmometers may be standardized by operating them along with a cup that has been checked to a given standard. The atmometers are mounted upon a rotating table so that they will all be similarly exposed (Nichols, 1913). The evaporation is determined for a given period of time and the reading of the standardized cup is then divided by the reading from the cup that is to be standardized to determine the coefficient of correction. The calculation of this coefficient K may be stated algebraically as follows (Livingston, 1918): $K = CR/r$ where R is the reading of the standardized cup that is being used and C is its coefficient of correction, while r is the reading of the instrument that is being standardized. K is thus the coefficient by which the reading from the instrument under consideration must be multiplied to reduce it to the loss of the standard cup with a coefficient of unity. The readings from similar atmometers can thus all be reduced to the same basis, and the atmospheric conditions under which the observations are made can thus be more correctly interpreted than if no such reduction could be made.

Thomas and Ferguson (1917) considered that an attempt should be made to realize a standard which should be invariable under given conditions and capable of replacement at any place or time. They concluded from rather elaborate experiments that the free water surface contained in a cylindrical vessel and protected by a rim 3 cm. in depth fulfills these conditions, although many factors must yet be considered before definite statements in this regard can be made.

2. The Ratio of Transpiration to Evaporation.—The term "absolute transpiration" is used to denote the observed rate of water loss from the plant. In a general way, the curve of absolute transpiration usually follows very closely the curve of evaporation and may give no indication of any retardation or acceleration of transpiration that is due to causes other than the evaporating power of the surrounding air.

The ratio of the rate of transpiration to evaporation was termed by Livingston (1906, 1907) "relative transpiration" and by Briggs and Shantz (1916), the "transpiration coefficient." It has also been called the "transpiring power" of the plant, a term used to denote the reciprocal of the resistance or acceleration manifested by plants to water loss by transpiration as compared to evaporation from an atmometer. It represents the group of internal conditions influencing the transpiration rate. If the plant and an atmometer responded alike to changes in the environment, the rate of relative transpiration would give a straight line when plotted hour by hour.

The graph of relative transpiration, however, is periodic in form and exhibits a marked minimum during the night hours. This fluctuation in the rate of transpiration has been interpreted by Livingston (1906, 1913), Shreve (1914, 1916), Livingston and Hawkins (1915), Burns (1915), Trelease (1922), and others to represent departures in the transpiration rate of the plant from that which would prevail if the plant responded freely to its environment. When considered in this light the curve of relative transpiration should thus indicate the magnitude of the physiological retardation or acceleration of the transpiration rate due to closure of the stomata or to reversible changes taking place within the plant. In making this interpretation, it is assumed that an evaporimeter responds freely to its environment and that the rate of evaporation from it affords a perfect summation of the environmental condition determining evaporation from the plant. It should be kept in mind, however, that all atmometers do not respond to their environment in the same manner, so that the particular atmometer used should make a great difference in determining relative transpiration. Its value, therefore, may not be so definite an indicator of internal changes in the plant as was formerly supposed.

Briggs and Shantz (1917) compared the hourly transpiration rate of alfalfa with the hourly evaporation rate from various types of atmometers. The comparison rate was made by superimposing the hourly transpiration graph on each of the hourly evaporation graphs, choosing the scale of ordinates of the transpiration graph so that the total area under the transpiration graph was equal to the total area under the evaporation graph. The average hourly departure of each of the evaporation graphs from the superimposed transpiration graph was expressed in percentage of the mean hourly evaporation for the day. The summation of the data obtained is shown in the following table:

AVERAGE-HOURLY DEPARTURE OF EVAPORATION RATE FROM TRANSPIRATION RATE FOR ALFALFA, PERCENTAGE OF THE MEAN EVAPORATION RATE FOR VARIOUS TYPES OF ATMOMETERS

Type of atmometer	Day periods, 6 a.m. to 6 p.m., per cent	24-hr. period, per cent
White cylinder	38	49
Brown cylinder	29	41
White sphere	31	43
Bellani plate	30	41
Filter-paper atmometer	22	31
Shallow tank	12	17
Large deep tank	93	89

Since the greater part of the transpiration and evaporation occurs during the daylight hours, calculations are made for that period as well as for the 24-hr. period.

These figures indicate, for example, in the case of the white sphere atmometer, that the average error in predicting the amount of transpiration at each hour in the day from the evaporation, when the ratio of daily transpiration of the plant to the daily evaporation of the atmometer under consideration is known, would amount to 43 per cent of the mean hourly value of the transpiration in the case of the alfalfa plant. The departure from the hourly transpiration rate and the hourly evaporation rate was the least when evaporation was measured by the small shallow tank.

It is evident from the foregoing that an atmometer has not yet been secured that responds in the same manner to its environment as the plant. The atmometers depart widely from the plant during the night hours—the rate of evaporation being relatively much higher than transpiration in all the cases that have been observed. Knight (1917) found that the porous-clay cup atmometer and the plants with which he worked responded unequally to changes of wind velocity but that changes in temperature and relative humidity acted upon the rate of evaporation and transpiration equally. Freeman (1920) concluded that changes in relative transpiration do not necessarily represent changes in the intrinsic transpiring power of a plant unless conditions of air movement are maintained constant. The transpiration rate of different plants varies under the same environmental conditions, owing to internal factors, so that when a comparison of the responses of transpiration and evaporation to the environment is made, the kind of plant as well as the type of atmometer must be taken into consideration.

F. Methods of Measuring Transpiration

The rate of transpiration may be measured by determining the amount of water lost per plant during a given period of time or the rate of loss per unit of leaf surface per unit of time. One of the most common ways of expressing the loss is in grams per square meter of leaf surface per hour. Comparative transpiration between different plants may also be obtained in a qualitative way by means of color standards, as will later be described.

The determination of the rate of water loss per unit of leaf surface per unit of time makes the measurement of the leaf surface of the plants necessary. This may be accomplished in any of the following ways: (1) by tracing the outlines of the leaves on paper and then at a convenient time measuring the enclosed areas by means of a planimeter, (2) by cutting out the traced areas and comparing the total weight of paper thus obtained with that of a known area of paper, (3) by obtaining the form of the leaf on sensitized paper and then either measuring or weighing the enclosed areas as above mentioned, (4) by rough measurement of the leaves and determining their area by direct calculations, and (5) by measurement with the photoelectric cell. The essentials of the apparatus used in the last method comprise a light-tight cabinet or photometer within which is mounted a selenium cell having at its upper surface a fairly large plate-glass window illuminated from above by means of a balopticon. The leaves to be measured are placed upon the glass window, and the change in the electrical resistance of the selenium cell occasioned by the reduced amount of light passing into the cabinet is measured by a suitable electrical device. Gerdel and Salter (1928) stated that this device is accurate to within 3 per cent for an area of 500 sq. cm. and 0.8 per cent for areas of 2,000 sq. cm. Withrow (1935), Frear (1935), and Mitchell (1936) have reported for the measurement of leaf areas improved photoelectric instruments that are accurate to approximately 2 to 3 per cent. Bolas and Melville (1933) devised a light method for measuring leaves that is accurate to 2.5 per cent. Vyvyan and Evans (1932) determined leaf area by the use of a sheet of glass divided into areas of one square centimeter. This glass was placed over the leaf, the number of squares in contact with it were counted, and the amount of surface in the portions of the squares at the margins was estimated. Lott and LeMert (1932) reported a modified method for obtaining rapidly the impression of leaves on blueprint paper. Darrow (1932) described three methods for determining leaf area that had not as yet been reported:

1. *Leaf-product Method.*—This method is based on the correlation between the extent of leaf surface and the linear measurements of leaves. The product of the length by the breadth in correlation with the surface is the most satisfactory.

2. *Matching Method.*—A pasteboard form equal in area to the largest leaf to be encountered in the experiment is cut out, and its area determined and written on the form. On this form are drawn outlines of successively smaller leaves each of which is measured and the area recorded on the corresponding outline. The leaves of the plant are then matched on these forms and their areas are thus determined.

3. *The leaf may be measured intact* on the plant by inserting it under cellophane on an adjustable stand and using the planimeter.

Marshall (1933) devised an apparatus for the quick determination of the area of compound leaves by translating width and length measurements of end leaflets into terms of leaf area. The apparatus is so arranged that as two wires are extended for the width and length of the leaflet, water is discharged into a calibrated tube and the area of the leaflets is read from the height of the water column.

One of the main things to be desired in measuring the rate of transpiration is that the plant be placed during the experiment as nearly as possible under the same conditions of the soil and atmosphere as prevail in the field. The accuracy, rapidity, and convenience of the method are factors that also must be considered. No method, however, has been devised for the measurement of transpiration that is entirely satisfactory in all these regards. Numerous methods for measuring transpiration have been reported, detailed descriptions of which have been given by Burgerstein (1904, 1920), but it is the intention here to describe only the four methods that have been the most universally used, especially in experiments with agricultural plants. These are the potometer method, the weighing method, the cobalt chloride-paper method, and Freeman's method.

1. The Potometer.—This method consists of placing the severed plant or branch thereof directly in water and after proper sealing determining by weight or volume the amount of water lost. One of the most common methods of procedure is to connect the severed branch or plant with a burette by means of rubber tubing and to note the loss of water volumetrically. This method was formerly used very extensively, but at present it is very little used. It has the very serious objection that the plant is placed under abnormal conditions, and it has been definitely shown that the rate of transpiration under these conditions is in most cases not comparable at all to that occurring where the plant is normally rooted in the soil. Thus, as an example, Freeman (1908) found that the rate of transpiration for the daisy, coleus, portulaca, and geraniums growing in the soil was, respectively, 7.21, 2.77, 1.72, and 0.65 mg. per square centimeter per hour, but that, from like plants which were severed and placed in potometers under like conditions, the rate of transpiration was 1.44, 0.37, 0.47, and 0.65 mg. per square centimeter per hour, respectively. Not only was the rate of transpiration much less, but the order of the transpiration rate was entirely different from what it was with the plants growing in the soil. Loftfield (1921) has also pointed out the difference in the behavior of stomata of plants in potometers as compared to those growing in the soil.

2. The Weighing Method.—For quantitative data the weighing method of measuring transpiration is the most convenient and satisfactory. The amount of transpiration from the entire plant is obtained, while the conditions under which the plant is placed previous to and during the experiment can be made very comparable to those which occur in the field. The method consists, in general, of growing the plants in soil in containers, which can be sealed so that no water escapes except that which evaporates from the plant (Fig. 21), and determining the loss by weighing. The containers that have been most frequently used for experiments with crop plants were cylindrical galvanized-iron cans with a capacity of from 250 to 1,000 lb. of soil, which is tightly packed therein. The cans are covered with galvanized-iron lids supplied with perforations for the plant and for the addition of water (Fig. 21). This lid is either sealed with tape that is waterproof (Briggs and Shantz, 1913, 1914) or rendered so by shellac (Miller, 1916, 1923) or is covered with oilcloth or other waterproof material (Kiesselbach, 1916). The escape of water around the plant is prevented by using as a seal a mixture of beeswax and paraffin (Briggs and Shantz, 1913, 1914), beeswax and a small amount of Venetian turpentine (Miller, 1916, 1923), or plastic modeling clay (Kiesselbach, 1916). One of the greatest difficulties experienced in growing plants in soil in large containers is to replace evenly throughout the soil the water that has been removed by the plant. Kiesselbach (1916) placed in each container 16 ft. of ¾-in. brass tubing made into a spiral coil perforated every 8 in. and obtained a good distribution of moisture through the soil by adding water to a container connected with this coil. Briggs and Shantz (1913, 1914) placed 5-in. flower pots over a small bed of gravel placed near the upper surface of the soil and obtained in this manner a good distribution of water. A modification of this method was used by Miller (1916, 1923), as shown in Fig. 21. Two more or less cone-shaped masses of soil 6 in. in diameter and 15 in. in depth are removed from the upper portion of the container. From the bottom of this cavity a hole 1.5 in. in diameter is made to the bottom of the can by means of a soil tube. The entire cavity is then filled with coarse sand and gravel to within 5 or 6 in. of the top of the can. A 5-in. clay flower pot with the bottom removed is placed directly on top of this sand so that the rim of the pot is flush with the metal lid of the can. A 1-in. hole for the addition of water is in the lid of the can directly over each flower pot. By this device not only is water evenly distributed through the soil, but sufficient aeration is apparently also provided, since the roots do not collect between the soil and the surface of the container, as is so frequently the case in pot experiments, but are evenly distributed throughout the soil.

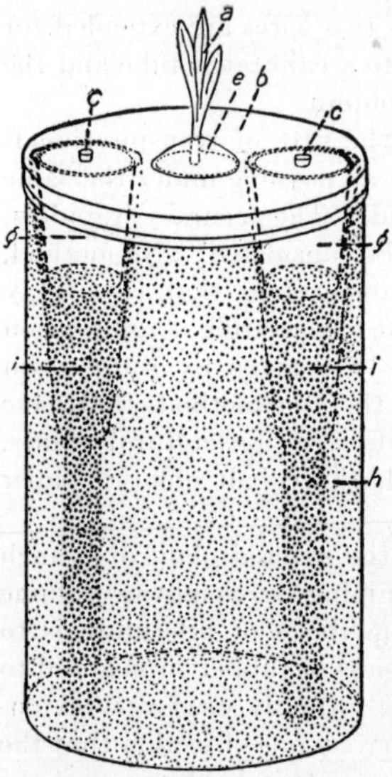

Fig. 21.—Diagram showing the method used in growing plants in soil in large metal containers for the determination of the loss of water by transpiration. *a*, plant. *b*, cover of galvanized iron. *c*, corks. *e*, wax seal. *g*, flower pots. *h*, soil. *i*, gravel beneath flower pots and extending to the bottom of the container.

When small plants and only a small quantity of soil are to be used, it has been found satisfactory to use unglazed porous-clay atmometers of the cylindrical, cone, or Bellani plate types as distributers of moisture. These are buried in the soil and connected with the surface by glass tubing so that water may be added directly or supplied automatically, as it is needed, from a reservoir (Livingston, 1908, 1918; Livingston and Hawkins, 1915). A small, inverted, porous flower pot buried in the soil and connected with the surface by means of glass tubes is also a very satisfactory watering device for small volumes of soil (Kiesselbach, 1916).

The plant containers are weighed at stated intervals by hand or on automatic scales connected with recording devices (Ganong, 1905; Briggs and Shantz, 1915; Transeau, 1911; and Blackman, 1914). By the latter method a continuous record of the water loss during the day and night is obtained. The use of automatic balances, however, is rather difficult in the field and is almost prohibited by price when a large number of plants must be considered. The most common method is to weigh the plants at certain stated intervals and in that manner determine their water loss for the period. The large containers may be lifted by means of a differential block and tackle and weighed with steelyard or spring balances (Kiesselbach, 1916; Briggs and Shantz, 1913, 1914), or they may be permanently mounted on small platforms provided with wheels or casters and pulled on and off the ordinary platform scales (Miller, 1916, 1923).

Since such a large mass of soil is required in many cases to grow normal plants, it is not possible to employ scales with a very high sensitivity, so that when the transpirational loss is desired for short intervals, the readings are not so accurate as are sometimes desired. The weighing method does not take into consideration the increased weight of the plant due to growth, but this increase in weight is very insignificant when compared to the large loss in weight due to transpiration.

3. The Cobalt Chloride Method.—Cobalt chloride ($CoCl_2.6H_2O$) occurs in the form of dark red crystals which may be dehydrated in different stages, forming successively red or violet $CoCl_2.4H_2O$, and $CoCl_2.2H_2O$, and finally the blue anhydrous $CoCl_2$. Taking advantage of this property of cobalt chloride, Stahl (1894) prepared a hygrometric paper by dipping filter paper in a solution of $CoCl_2.6H_2O$ and then drying it. Such a paper is blue when dry but changes its color to white or rose when moist. When this dry blue paper is placed upon the leaf and covered with glass or a thin piece of mica held firmly in place by a small spring clamp, it will in time become pink due to the escape of water vapor from the leaf surface. Stahl used this method to compare the rate of transpiration from the upper and lower leaf surfaces and from the leaves of different plants by comparing the time required for the paper to assume the pink color. He, however, did not attempt to standardize these readings to any standard physical evaporating surface. Livingston (1913) conceived the idea of comparing the power of a standard evaporating surface to give off water to the hygrometric cobalt chloride paper with that possessed by any leaf surface that might be considered. The method first devised by Livingston (1913) has been modified, improved, and studied by Bakke (1914), Bakke and Livingston (1916), Livingston and Shreve (1916), and Shreve (1917, 1919). The improved method of procedure is approximately as follows (Livingston and Shreve, 1916):

a. Preparation of the Hygrometric Paper.—Circular filter paper of a high grade of uniformity of texture is placed for 1 min. in an agitated solution of 3 g. of cobalt chloride in 100 cc. of water made distinctly acid by the addition of hydrochloric acid. It is then placed on a clean glass and the excess solution is squeezed out with a rubber roller. The paper is then partially dried in the oven between two filter papers, but while still pink it is removed from the oven and thoroughly pressed and dried between the two filter papers by means of a flatiron. The blue color of the paper must be more intense than that of the standard to be later mentioned, the intensity of color being varied by altering the time the paper remains in the solution. After the paper is dried, it is placed in a desiccator over a convenient dehydrating agent and left until needed.

b. Standard Colors.—Since the cobalt chloride paper takes up some moisture during the transfer from the desiccator to the leaf and since it is difficult to determine with any degree of accuracy the end point in color of the moist paper, Livingston and Shreve (1916) devised two permanent standard colors with which to mark the beginning and the end of the experiment. Both these colors are blue, one being just a little less intense than the color of the dried cobalt chloride paper while the other is much less intense but clearly blue. The permanent color standards are made by precipitating Prussian blue in the same type of filter paper as is used for the cobalt chloride paper. The paper is first dipped in a solution of ferric chloride, and after the excess solution is removed it is then dipped into a solution of potassium ferrocyanide, the excess solution removed, and the paper dried after the manner described for the preparation of the cobalt chloride paper. The two intensities of color are obtained by varying the concentration of the solutions and the time of treatment. The color-standard papers thus prepared do not alter their color when exposed to air and sunshine or with changes in the moisture content of the air. A composite strip of the cobalt paper and the two standard color papers is made by taping together pieces each 4 mm. square with the cobalt chloride square in the center.

These composite strips are taken from the desiccator and clamped upon the leaf as has previously been described. When the cobalt chloride paper matches in color that of the more intense standard, the experiment is considered started, and it is considered concluded when it reaches the intensity of the lighter color standard.

c. The Standard Evaporating Surface.—For a standard evaporating surface in this method, a free water surface (Livingston, 1913), a moist filter paper supplied from beneath with water (Bakke, 1914), and a porous-clay cup atmometer filed to a flat surface (Livingston and Shreve, 1916) have been used. In all cases, the cobalt chloride paper is held by suitable appliances at a distance of 1 mm. from the evaporating surface. The length of time that is required for the hygrometric paper to reach the color end point is determined after the same manner as when the paper is applied to the leaves.

d. The Index of Transpiring Power.—The time that is required for the hygrometric paper to reach its end point in color over the leaf is called the "leaf-test time" and over the evaporating surface the "water-test time." The ratio of the water-test time to the average leaf time is called the "index of transpiring power," a value that varies greatly at different times of the day; *e.g.*, Bakke (1916) found that within a single hour just after sunrise it increased 240 per cent in the case of sunflower. He also found (1914) that the younger leaves have a higher index of transpiring power than do the older ones. Let us consider a specific example to illustrate the application of the cobalt chloride method of determining the index of transpiring power. Let us assume that the water-test time is 15 sec. and the leaf-test time for the upper surface 300 sec. and for the lower surface 150 sec. The index of transpiring power for the upper surface would thus be $^{15}/_{300}$ or 0.05 and for the lower surface $^{15}/_{150}$ or 0.10. The average index of transpiring power for the whole leaf would thus be $\frac{0.05 + 0.10}{2}$ or 0.075. The evaporation from the leaf thus goes on at the rate of 0.075 times that from the standard evaporating surface under observation.

The cobalt chloride-paper method of determining transpiration is valuable since it makes possible the determination of the relative transpiring power of plants growing in the open soil under field conditions. The method thus is applicable and very convenient for determining the relation of the position of leaves on the plant to the rate of water loss, the xerophytism or mesophytism of plants, and the wilting point. Its chief disadvantage is that it is only a qualitative method and gives no quantitative data in regard to the amount of water actually lost from the plant.

Meyer (1927) used a modified method of procedure in the application of the cobalt chloride method so that the rate of water loss may be expressed in absolute units instead of only relative values. He determined the amount of water by weight that must be taken up by a given area of standardized cobalt chloride paper to reach the pink end point. From this datum it is thus possible to determine the number of units of water vapor being lost from a given area of leaf in a unit of time if the area of the paper used and the time required for it to reach the end point are known. In the method of Meyer the paper is so placed that the conditions of humidity, radiation, and air movement are practically constant beneath it, so that the only variable atmospheric factor is the temperature and thus all readings can be standardized to the same temperature. Meyer adopted the term "standard rate of water loss" for designating the amount of water vapor lost under the conditions of the experiment in terms of grams per hour per 100 sq. cm. of leaf surface at 20°C.

Blaydes (1932) determined the rates of water loss from *Helianthus tuberosus* simultaneously by the weighing and cobalt paper methods. He found that these methods do not correlate very closely because: (1) The weighing method measures the loss of water during an entire unit of time, while cobalt paper records the transpiration during a very short period of the unit of time interval; (2) the loss of water by the weighing method fluctuates with the environmental conditions, while with the cobalt chloride-paper method the direct effects of atmospheric environmental fluctuations are practically eliminated; (3) in the cobalt chloride method the diffusion gradient is much steeper at the beginning of the experiment than it is under the conditions of the weighing method.

Howes (1923) has devised a calcium chloride method for measuring the transpiring power of plants. Small tubes of calcium chloride are attached to the leaf by means of special clips. The calcium chloride retains its efficiency as a water absorber for a considerable period even when small amounts are used, and the results are not affected by the amounts of calcium chloride in the tubes.

Heinicke (1930) described a satisfactory means of studying the comparative transpiration rates, in which the water lost by the leaf is absorbed by calcium chloride contained in a small weighing bottle held to the leaf by a special clamp.

4. The Freeman Method.—This method was first used by Freeman in 1908 and is a modification and adaptation of the methods used by Ganeau de Lamarlière (1892) and E. and J. Verschaffelt (1890). The general details of the apparatus used in this method are shown in Fig. 22. The apparatus consists of a large glass cylinder of suitable size for a transpiration chamber for the twig or portion of the plant whose water loss is to be measured, two U tubes containing phosphorus pentoxide for absorption of moisture, and a water aspirator of known capacity. These are connected together by rubber tubing so that as water flows from the aspirator a known quantity of air may be drawn through the cylinder over the leaves under experiment and through the phosphorus pentoxide, which is loosely packed in order to allow the free passage of air. The glass cylinder containing the twig is closed at its distal end with a split cork, which fits tightly around the twig and carries a tube for the intake of air. The upper end of this cylinder is closed with a two-holed cork carrying a thermometer and a glass tube for connection with the rest of the apparatus.

In preparation for an experiment, the aspirator is filled with water, the U tubes containing phosphorus pentoxide are weighed and placed in the circuit with stopcocks closed, and lastly the stem of the plant is inserted in the cylinder. The stopcocks in the U tubes are then opened and the water started running from the aspirator, the speed with which the air is drawn over the twig being regulated according to its size and rate of transpiration. After all the water or a certain measured portion of it has flowed from the aspirator, the U tubes are closed, disconnected, and weighed, and the amount of moisture absorbed by them determined. This moisture, however, includes not only that which escapes from the plant but also that contained in the air that is drawn through the apparatus. The amount of moisture contained in the volume of air that has been drawn over the leaf is determined by running a control experiment at the same time with the transpiration experiment. The amount of water lost by transpiration is obtained by subtracting the increase in weight of the tubes used for the normal air from the increase in weight of those used with the plant. After the experiment, the leaf area may be determined by any convenient method and the rate of transpiration calculated. This method of measuring transpiration is adapted to cases where the plants are too large to be grown in containers or where the whole plant cannot be sacrificed for the experiment. It is adapted to give quantitative data on the comparative rate of transpiration of different plants. The method, however, has some serious objections. In the first place, the leaves are placed under very abnormal conditions of temperature in the transpiration cylinder—the temperature being much higher than that of the open air. In the second place, unless great care is exercised in the selection of a small amount of leaf surface and in the rate at which air is drawn through the apparatus, the moisture escaping from the leaves cannot be removed rapidly enough to prevent its condensation upon the walls of the cylinder. Such a collection of moisture of course renders the experiment void. The

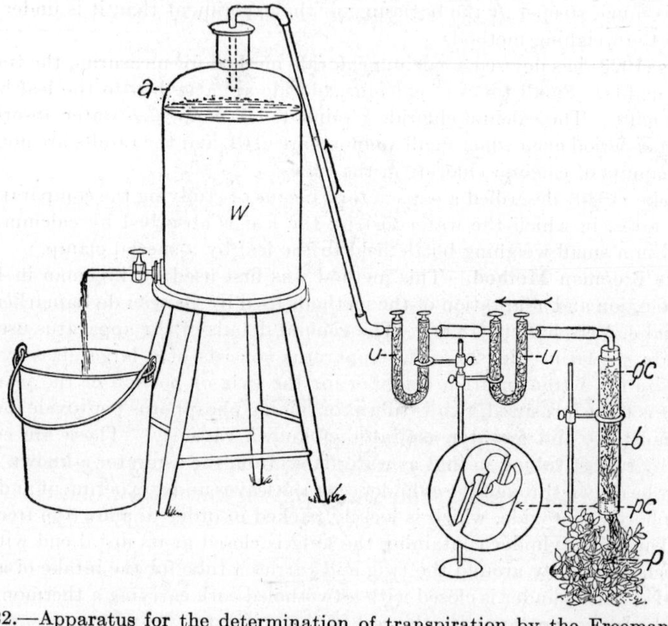

FIG. 22.—Apparatus for the determination of transpiration by the Freeman method. *a*, aspirator. *b*, branch of plant. *gc*, glass cylinder. *p*, plant. *pc*, perforated cork for the entrance of a portion of the plant and air. *u*, tubes containing phosphorus pentoxide. *w*, water.

period of time that must necessarily elapse in setting up the apparatus also tends to add considerable error.

Minckler (1936) used a modified Freeman method with an air pump and a gas meter to measure the flow of air. The maximum air flow in an ordinary Freeman experiment was 19.5 l. per hour, while with the modified method it was 5 to 10 l. per minute. This method permitted the use of a chamber sufficiently large to accommodate 20 to 30 or more relatively large leaves of trees. The duration of each experiment was only 6 min. As determined by this method, the daily transpiration of an American elm, 13 in. in diameter at a height of 4 to 5 feet from the ground, was 8.7 l., or 1,360 l. for a growing season of 150 days. The daily average for a red maple of similar size but with more foliage growing in a moist habitat was 51.7 l., or 7,770 l. for the growing season.

G. The Significance of Transpiration

Although transpiration has been studied, perhaps, more thoroughly than any other plant process, very little is yet known concerning the significance of this function in the life of the plant. The opinions that have been expressed in regard to the possible value of transpiration vary all the way from those which ascribe to this function benefits to the plant that are as important as photosynthesis and respiration to those which assume that transpiration brings nothing to the plant but harm (Curtis, 1926). When the present data in regard to transpiration are surveyed, it must be admitted that its obvious harmful effects outweigh any of the benefits that might be attributed to the process.

It has frequently been stated in texts that transpiration cools the leaves and prevents their death or injury by high temperatures. The change of water from the liquid to the vapor form within a leaf does absorb a considerable amount of heat and this vaporization would thus tend to lower the temperature of the leaf. As previously mentioned, however, the average leaves are cooled by transpiration rarely more than 2 to 5°C.—a difference which, so far as our knowledge of protoplasm goes, could be of no marked benefit in preventing injurious effects. The death of leaves during hot, dry weather is apparently due to the excessive loss of water from the protoplasm rather than to any injury caused by an excessive increase in temperature. Thus it would seem that the injury to leaves which frequently occurs on hot days is due chiefly to the injurious drying effects of transpiration itself. Transpiration thus apparently is more harmful than beneficial on hot days when the cooling effects are most needed (Curtis, 1926).

It has frequently been stated that transpiration increases the rate and amount of absorption of solutes from the soil, but, as stated in Chap. V, conclusive experimental evidence indicates that there is no relationship whatsoever between the rate of transpiration and the amount of nutrients absorbed.

It has also been claimed that transpiration plays an important part in transferring solutes from the roots to the leaves in the transpiration stream or current. As Curtis (1926), however, points out, if the rate of removal of nutrients from the roots were largely determined by transpiration, the removal of solutes from the absorbing organs should increase absorption. Increased transpiration should therefore increase absorption, but, as mentioned above, the experimental evidence indicates that it does not. The upward transpiration stream is in the xylem, so that if the transfer of solutes upward is in the phloem, as indicated by the work of Curtis, which will be discussed in Chap. XII, it seems apparent that transpiration can have little or no effect on their movement.

The loss of water by transpiration always results in a reduction of turgor of the cells, since the water lost cannot be instantaneously replaced. A slight reduction in turgor, however, has but little influence on the various processes of the plant, especially if there is an abundant supply of water in the soil to replace the loss from the plant. When, however, the moisture content of the soil is low, the decrease in turgor soon reaches a point where the reactions of the cells are greatly or completely retarded, so that death finally ensues.

It is estimated that more plants perish or are retarded in their development from a deficient water supply due primarily to its loss by transpiration than from all other factors combined, a fact which indicates that, if transpiration is of any benefit to a plant, it may be attended by very dangerous effects.

Barnes (1902) considered that the significance of transpiration is to be seen in its origin and in tracing its development. Thus it is considered that the most primitive plants lived in water and absorbed the carbon dioxide and oxygen needed in the process of photosynthesis and respiration directly from this medium. In the course of their development, plants have apparently not modified to any great extent the manner of obtaining these two gases, so that under the present organization of the plants, exposed wet cell walls cannot be waterproofed on that account. Thus, wherever moist cell surfaces are exposed to the atmosphere, evaporation of water must occur, so that transpirational loss must be considered unavoidable, although in itself it is a constant menace to the activity and life of the plant.

Clements (1934) believed that transpiration may play some important role. He stated that, although this process may cause the death of the plant under certain conditions, it is not due so much to the process itself as it is to the limiting factors that make the normal functioning of an essential process impossible.

H. The Water Requirement of Plants

The term "water requirement" in the most restricted sense is defined as the ratio of the number of units of water absorbed by the plant during the growing season to the number of units of dry matter produced by the plant during that time. In most cases, however, the dry matter considered in this relation is the amount harvested and in most instances includes all the plant except the roots. The water under consideration is in reality the amount lost through transpiration during the growing season, since of the total water absorbed the amount retained within the plant is very insignificant compared to the amount evaporated from it.

The term "water requirement" is a rather misleading one, since it seems to imply that it denotes the amount of water necessary for the growth of the plant. It merely, however, denotes that under certain conditions of climate, soil, and water supply, there was evaporated from the stem and leaves or retained within the plant and leaves water to the amount of so many times the number of units of dry matter produced. The water requirement of plants has been studied, for the most part, in those regions where water is a limiting factor in crop production, since it was formerly considered that there was a relationship between the ability of a plant to withstand drought and its water requirement. This relationship in the light of investigations, however, does not appear to exist.

1. Methods of Determining.—The methods of determining the water requirement of plants were, previous to 1910, not under very well-controlled conditions. A summary of the methods used previous to that time and the results thus obtained are discussed by Montgomery (1911), Briggs and Shantz (1913), and Kiesselbach (1916) in their review of the literature on the subject. Since 1910 the main work on water requirement has been reported by Briggs and Shantz (1913, 1914), Shantz and Piemeisel (1927) at Akron, Colo., Montgomery and Kiesselbach (1912), Kiesselbach (1916) in Nebraska, Miller (1916, 1923) in Kansas, and Khankoje (1914) and Thom and Holtz (1917) in Washington. The water requirement was determined by these investigators by growing the plants in large sealed containers after the manner described in the discussion of the weighing method for transpiration determinations. The containers are weighed at frequent intervals, and the amount of water that has been lost from the soil is replaced by adding a known amount of water to the container by means of the watering devices that have previously been described.

After the plant or plants in a given container have reached maturity, they are harvested and the amount of dry matter determined by drying to a constant weight at a temperature of 100 to 105°C. The amount of water taken from the soil is known from the weights of the container at the beginning of the experiment and at its close and from the amounts of water that have been added from time to time during the season. The water requirement is determined by dividing the number of units of water absorbed during the season by the number of units of dry matter produced. Thus, for example, if a plant absorbs from the soil 251,020 g. of water and produces 652 g. of dry matter during the season, its water requirement would be 251,020/652 or 385. The water requirement may be stated also in terms of the grain produced and is frequently determined in this manner when the plants under consideration are of importance in the production of grain. The measurement of the water requirement under field conditions has been attempted by Briggs and Shantz (1913),

Thom and Holtz (1917), and others. The general method is to determine the amount of water removed by the crop as shown by the difference between the initial and final water content combined with the rainfall entering the soil as determined by daily sampling. In some of the attempts, fallow plots are used to determine the loss of water from the soil by evaporation. Briggs and Shantz (1913) considered this method uncertain, owing to the difficulty of determining what proportion of the rainfall enters the soil and actually is used by the crop.

2. Factors Affecting the Water Requirement.—The water requirement of different plants varies greatly under the same conditions. Thus Briggs and Shantz (1913, 1914) and Shantz and Piemeisel (1927) found in their extensive work in Colorado that the water requirement of some 150 plants studied ranges from 216 for Kursk millet to 1,131 for *Franseria sp.*, a native weed. Thus *Franseria* evaporated five times as much water during the production of a unit of dry matter as did the millet. Shantz and Piemeisel (1927) summarized the variation in the water requirement in regard to the plants studied by them as follows: considering the water requirement of proso millet to be 1.00, the water requirement of the various crops would be: millets, 1.07; sorghums, 1.14; corn, 1.31; barley, 1.94; wheat, 2.09; oats, 2.18; rye, 2.37; legumes, 2.81; and grasses 3.10. In general, the millets, sorghums, and corns had the lowest water requirement, while the small grains, barley, wheat, oats, and rye lost almost twice as much water and the legumes almost three times as much per unit of dry matter produced as did the millets, sorghums, and corns.

Since the water requirement is the ratio between the amount of water taken up by the plant and the amount of dry matter produced, it is evident that its value will be affected by any factor that influences either transpiration or growth. Perhaps the most influential factor in this regard is the climatic complex that prevails during the growth of the plant.

a. Climatic Conditions.—The effect of air humidity on the water requirement is conclusively shown by the work of Montgomery and Kiesselbach (1912). Corn was grown in two greenhouses; in one the relative humidity was kept at 37 per cent during 12 hr. of the day, and in the other at 58 per cent for that time. The water requirement of the plants in the dry greenhouse was 340, while in the more humid greenhouse it was 191—the corn in the dry greenhouse having a water requirement 56 per cent higher than in the humid one. The range of the water requirement during different years due to prevailing climatic conditions is very great. Thus, for example, the lowest value for any one year at Akron, Colo., averaged only about 60 per cent of that of the highest value. Some idea of the variation in the water requirement due to climatic conditions may be obtained from the following table, which gives the water requirement of some of the more common crop plants and a summary of the weather conditions in terms of evaporation from a free water surface:

Water Requirement for the Years 1911 to 1917 at Akron, Colo., and the Evaporation in Inches from a Free Water Surface from Apr. 1 to Sept. 1

Plant	1911	1912	1913	1914	1915	1916	1917
Alfalfa	1,068	657	834	890	695	1,047	822
Oats, Burt	639	449	617	615	445	809	636
Barley, Hannchen	527	443	513	501	404	664	522
Wheat, Kubanka	468	394	496	518	405	639	471
Corn, N. W. Dent	368	280	399	368	253	495	346
Millet, Kursk	287	187	286	295	202	367	284
Sorghum, Red Amber	298	239	298	284	303	296	272
Evaporation, inches, April to September	48.8	37.7	43.0	41.8	33.4	47.1	42.7

Dillman (1931) reported that at Sewell, S. D., from 1912 to 1918, and at Mandan, N. D., from 1919 to 1922, the water requirement of alfalfa ranged from 602 to 1,036, of Kubanka wheat from 333 to 531, of Sudan grass from 272 to 347, of millet from 177 to 316, and of sorgo from 210 to 284. All the plants except sorgo and Sudan grass were highly responsive to seasonal changes.

Owing to changes in weather conditions, the loss of water from plants may vary as much as 600 per cent on successive days. Thus the weather conditions that prevail during a month or even so short a period as 10 days or a week may determine whether the water requirement is going to be high or low for that season. In this regard, Briggs and Shantz (1916) found that during a period of 10 days from July 7 to 16, 1914, many of the crop plants including wheat, barley, rye, corn, and Sudan grass lost 25 per cent of the total water transpired during the entire season. Richardson (1923) working in Australia found that in September to November, 80 per cent of the water lost from winter cereals was transpired, although the growing period was 7.5 months, while alfalfa in bloom transpired as much as 25 per cent of its total water in 3 days. The stage of development of the plant when certain climatic conditions occur also is an important factor in determining the value of the water requirement. If a high evaporating period occurs when the plant has as yet only a small leaf development, it may affect the water requirement but little, but if a high evaporation period of a week or 10 days should occur when the full leaf development has been attained, it will greatly increase the value of the water requirement for that season. An example from the work of Miller (1923) will illustrate that point. In 1920 the water requirement of Sudan grass at Manhattan, Kans., was only 81 per cent of what it was in the previous year. The evaporation from a free water surface for May, June, July, and August during 1920 was 98.9 per cent of that in 1919. It would seem that the difference in the water requirement for the two years should be small. The evaporation, however, during August when the Sudan grass had reached its full leaf development was in 1920 only 83 per cent of that of 1919—a difference which is very comparable to the difference in the water requirement for the two years.

b. Soil Fertility.—Many investigations have been made concerning the effects of the fertility of the soil on the water requirement. Almost without exception, the application of fertilizers to the soil has shown a reduction of the water requirement. In very productive soils this reduction is extremely slight, but in poor soils the water requirement may be reduced one-half and sometimes as much as two-thirds by the application of nutrients. When the supply of nutrients in the soil approaches exhaustion, the rate of growth of the plant is greatly reduced, but no corresponding change occurs in the transpiration rate. This fact is evidence that transpiration is not a measure of growth. An example to illustrate the effect of soil fertility on the water requirement may be taken from the work of Kiesselbach (1916). He grew corn in an infertile soil *a*, a fertile soil *c*, and an intermediate soil *b* composed of equal parts of *a* and *c*. The amount of dry matter produced and the water requirement obtained were as follows:

Soil	Dry matter produced, grams	Water requirement
a.........	113	550
b.........	184	479
c.........	270	392

By the application of sheep manure in equal amounts to soils *a*, *b*, and *c*, the dry matter was increased 194, 79, and 41 per cent and the total water transpired 106, 43, and 29 per cent in the order named, but the water requirement by this treatment was reduced 29, 17, and 8 per cent, respectively.

The effect of the fertility of the soil is shown indirectly when the relation of the volume of soil in which the plants are growing is considered in relation to the water requirement. Many of the experimental data that have been reported in regard to water requirement are in error because the plants were grown in such a small volume of soil that they did not mature normally, owing primarily to an insufficient nutrient supply for the plant. In such cases, as the volume of soil is increased the water requirement is reduced within certain limits. Thus Kiesselbach

(1916) found that when he used various amounts of soil he obtained the following results with corn:

Amount of soil used, pounds	Dry matter produced, grams	Water requirement
32	97.5	411
85	205.7	362
150	316.0	341
239	442.0	277
583	627.0	290
956	728.0	240

In these cases when 1¾ lb. of sheep manure was added to each can, the water requirement was 322, 311, 304, 272, 269, and 252, respectively. These results indicate that the nutrients contained in the volumes of soil used were in most cases insufficient for the plant's needs.

c. Soil Moisture.—The experiments that have been reported in regard to soil moisture and the water requirement show, as a rule, an increase in the water requirement when the soil-moisture content approaches either extreme (Briggs and Shantz, 1913). Kiesselbach (1916) obtained in corn a maximum production of dry matter when the water content of the soil was at approximately 70 per cent of saturation. The average for 3 years showed that the reduction of the moisture content below the optimum reduced the water requirement 7.9 per cent but that this was accompanied by a 30.7 per cent lower yield of dry matter. An increase in the soil-moisture content above the optimum for the three years increased the water requirement 8.2 per cent, but this was accompanied by a 16.7 per cent lower yield of dry matter.

Flory (1936) reported a 3-year comparative study of the rate of transpiration, development of leaf area, and production of dry matter in the climax prairie and a field of maize at Lincoln, Neb. At maturity the prairie vegetation had a leaf area of 950, 1,050, and 1,014 sq. in. per square foot of soil surface, respectively, for each of the successive years. By maturity the corn had developed 437, 546, and 467 sq. in. of leaf surface per square foot of soil surface. The average daily loss of transpiration per square foot of soil surface for the 3-year period was 0.59 lb. in the prairie and 0.39 lb. in the corn field. It was greater in the corn, however, when the period of tasseling was accompanied by a high water content of the soil. The total amount of water lost in the production of 1 g. of dry matter in the prairie was 1,376 g. of which 445 g. or 32 per cent was lost in runoff and surface evaporation. The total amount of water used or lost in the corn field for the production of 1 g. of dry matter was 786 g. About 65 per cent of this was lost in runoff and evaporation from the soil surface. Approximately 52 per cent more dry matter was produced per square foot of soil in the corn field than in the prairie.

d. Plant Diseases.—It was found by Johnston and Miller (1934) that leaf-rust infection increased the water requirement of susceptible varieties of wheat 31.7 to 104 per cent, depending on the length of the rust period. The water requirement of the resistant varieties was only very slightly increased by rust infection. Murphy (1935) found that when infected in the seedling stage a variety of oats susceptible to crown rust used 290.8 per cent more water per unit of dry matter than rust-free plants. The resistant varieties that had been inoculated used only 40 per cent more water than the controls.

I. Drought Resistance

The term "drought" may be defined in various ways depending upon the point of view from which it is considered. Drought may be defined as a condition of the soil or atmosphere, or of both, that prevents or hinders the plant in obtaining sufficient water for its functions. Drought as thus defined may be caused in two ways: (1) By the inability of the plant to absorb water from the soil, although an abundance is present. This condition is frequently termed "physiological drought." Plants that grow in marshes and bogs frequently have difficulty obtaining sufficient water to supply their needs although the soil in which they grow is saturated with it. Many of the plants growing under these conditions have xeromorphic anatomical adaptations that resemble those of plants that grow in regions of limited rainfall. The inability of plants to absorb water from a frozen soil when there is an abundance therein and when the aerial parts of the plant are in need of water is another example of physiological drought. This is exemplified frequently by the winter cereals growing in the Great Plains. During the winter they are sometimes injured by the loss of water from their leaves due to a high evaporating power of the air. The water thus lost cannot be replaced by the roots because they are in a cold or frozen soil. (2) Drought is more frequently caused by a lack of sufficient water in the soil or by a high evaporating power of the air, which depletes both the soil and plant of moisture. These two factors generally accompany each other, although this is sometimes not the case. The term "soil or edaphic drought" is used to define the condition of a lack of water in the soil. "Atmospheric drought" refers to the desiccation of the plant to the point of injury under conditions of relatively high soil moisture.

Maximow (1925) believed that the ability of a plant to resist drought is due to a multiplicity of morphological and physiological characteristics which investigators have so far failed to separate into their component parts. Some of the general characteristics of plants that may aid them in withstanding drought are a small leaf surface, an exceptional root development, a short growing season, and the ability to become dormant. Certain agricultural plants have several or all of these characteristics. On this account the three plants that have done the most for agriculture in the Great Plains are the hard winter wheats, alfalfa, and the sorghums. One of the first principles in dry-farming is to conserve the largest possible supply of moisture for the crop that is to be grown. In order to obtain the best returns under conditions of drought, the soil must be kept free of weeds and a thin stand of the desired crop must be grown. A weed evaporates as much water and frequently much more than a crop plant so that in a weedy field the loss of water from the weeds may be as great as or greater than it is from the crop plants. Under dry conditions, the plants must be well spaced so that each individual has a large mass of soil from which to draw water.

It is well known that plants exhibit different capacities for enduring drought at various stages of their development. Thus Tumanov (1929) stated that in wheats the "shooting" and "soft dough" periods were the most critical in regard to the effects of drought. He believed that varieties should be selected whose critical periods of growth do not coincide with the greatest combination of adverse meteorological factors. Ivanov (1923) cites the example of Turkestan wheat in this regard, which develops very rapidly during early stages of growth and is drought-escaping because it comes in head long before other types. Vassiliev (1932) believed that it is not possible to breed plants that are resistant to soil drought but that the breeding of plants resistant to atmospheric drought might be accomplished. Martin (1930) and Aamodt and Johnston (1936) give thorough discussions of the nature of drought and of drought resistance by plants.

III. LOSS OF WATER FROM PLANTS IN THE LIQUID FORM. GUTTATION

The loss of water in the form of liquid from the uninjured leaf or stem of the plant is called "guttation." Guttation occurs, for the most part, under conditions that are favorable for the absorption of water by the roots but conducive to reduced evaporation from the leaves. Cool nights following warm days provide excellent conditions for this phenomenon. The soil retains warmth and absorption is very active, while the air is cool and humid, and transpiration is reduced almost to zero. Although guttation occurs, for the most part, at night, it may occur in daylight if the proper conditions prevail and may be caused in many plants at any time by artificial means. The water of guttation usually collects in drops along the edges and at the tips of leaves, but in some cases it appears over the entire leaf surface when conditions are very favorable. The water exuded in the case of the sorghums and sometimes corn is so abundant that it runs down the leaves and collects in the leaf axils. Under the conditions of the Great Plains it is not uncommon for the sorghum leaves to be very badly wilted during the day yet recover their turgidity at night and form abundant water of guttation.

In the tropics, guttation occurs to a greater extent than under conditions of the temperate region, since at night the air is very humid, while the soil is warm, making conditions ideal for absorption. The water of guttation, especially on the leaves of cereals and grasses, is frequently mistaken for dew. The process of guttation has been observed for a large range of plants, and Burgerstein (1920) lists 333 genera and 115 families of plants for which guttation has been reported.

The quantity of water lost by guttation is often considerable, the largest recorded amount being 100 cc. for a single night from a young leaf of *Colocasia nymphaefolia*. In this case, the water is exuded near the tip of the leaf and is actually forced out or ejected in drops, which follow one another in rapid succession. It is stated by some observers that in the tropics in the early morning the bamboos in dense growths produce a fine mist due to the loss of water by guttation.

The water of guttation in most cases escapes from the plant through more or less specialized organs called "hydathodes." These organs are generally located on the leaves and their structure is subject to a considerable amount of variation. There are two main types of hydathodes in their relation to the vascular system. The one type has no direct connection with the water-conducting system. The hydathodes of this group are purely epidermal structures such as specialized hairs, conical protrusions, and other forms of modified epidermal cells.

The other type, called generally the "epithem hydathode," is characterized by the fact that it is in direct communication with the water-conducting system. One of the most common forms of hydathodes of this type opens to the outside through a water stoma beneath which is located a mass of thin-walled cells, loosely arranged and called the "epithem," which border directly upon the xylem elements of the fibrovascular bundle. The main distinction between the water stomata or water pores and the ordinary stomata consists in the fact that the guard cells of water stomata have lost their power of adjustment to a large extent. They are located at the end of a fibrovascular bundle and are generally much larger. Haberlandt (1914) stated that they usually become stationary at a very early stage of development, although in some cases from the first they are incapable of movement.

Lepeschkin (1923) noted that when the edge of the leaf bearing the epithem hydathode is severed and laid with its cut surface in water, guttation does not occur. Flood (1919) noted that when the tip of the leaf of *Colocasia antiquorum* is anaesthetized, the exudation of water continues at the normal rate. Both these investigators concluded that their experimental results indicate that guttation in the case of epithem hydathodes is simply a process of filtration, the force being supplied by sap pressure, which in some cases may also force the water through the ordinary stomata.

The water exuded from the purely epidermal hydathodes is apparently purely water of excretion, the cells of the hydathode themselves developing the force necessary for such an excretion process. It seems, however, that an exudation pressure or sap pressure must be set up in the conducting system before this type of hydathode will function. The role that the sap pressure here plays seems to be that of a stimulus which sets the action of these hydathodes in motion.

The water excreted in guttation is not pure but contains a small amount of total solids. One of the most recent and thorough investigations on this subject was conducted by Wilson (1923). He found that the total solids in the water of guttation from corn plants growing under nonsterile conditions amounted to 1,030 p.p.m. The total solids in water from timothy plants growing under sterile conditions was 573 p.p.m. in one case and 220 p.p.m. in another. In all cases, the total solids were more than one-half organic matter. In the guttation water from corn, oats, and timothy, reactions were obtained indicating the presence of nitrates, nitrites, catalases, and peroxidases, but the presence of reductases was uncertain. Materials were also present which were capable of reducing methylene blue. The pH value of the exudate from young plants was found neutral, but the acidity increased as the plants became older.

It was shown by Pavilinova (1926) with seedlings of maize growing in water cultures that the concentration of calcium in guttation water was directly proportional to its concentration in the culture solution. The quantity of calcium excreted was approximately 17 per cent of the quantity absorbed. The water of guttation had an acid reaction with an average pH of 5.3. Shardakov (1928) collected water of guttation from the hydathodes of various plants and examined it for calcium, potassium, chlorine, and phosphate. The salt concentration of water of guttation changes with time. These ions were more concentrated in the water of guttation than in the soil solution. He believed that these ions are absorbed as the solution passes through the hydathodes, because when these organs are removed, the ion concentration in the bleeding sap is greater than it was before their removal.

Hiltner (1930) believed that dew is absorbed by the aerial parts of plants and that this reduces the salt concentrations of the solutions supplied by the roots. Arens (1934) found that the dew on leaves is generally slightly alkaline, while the water of guttation is slightly acid.

Distilled water with a pH of 5.6 becomes alkaline in a few hours after being sprayed on the leaves. The alkalinity of this water adhering to the leaf surfaces is caused by exosmosis of potassium, calcium, magnesium, phosphorus, and organic substances.

The physiological significance of the hydathodes and of the loss of water by guttation is not definitely known. This escape of water from the plant would seem to be a means of regulating the turgor of the leaves and prevent it from becoming so high as to interfere with the metabolism of the cells when transpiration is almost or completely suppressed. This, however, is purely an assumption and is based upon no experimental evidence whatsoever.

Bibliography

AAMODT, O. S., and J. H. TORRIE, 1934: The relation between awns and yield in spring wheat, *Can. J. Res.*, **11**: 207–212.

——— and W. H. JOHNSTON, 1936: Studies on drought resistance in spring wheat, *Can. J. Res.*, **14**: 122–152.

ANDERSON, A. P., 1897: Stomata on the bud scales of *Abies pectinata*, *Bot. Gaz.*, **24**: 294–295.

ANDERSON, D. B., 1931: Influence of $CuSO_4$ dusts upon the transpiration rate of peanuts and other plants, *N. C. Sta. Ann. Rept.*, **53**: 110.

ARENDS, J., 1925: Ueber den Einfluss chemischer Agenzien auf Stärkegehalt und osmotischen Wert der Spaltöffnungschliesszellen, *Planta*, **1**: 84–115.

ARENS, K., 1934: Die kutikuläre Exkretion des Laubblattes, *Jahrb. wiss. Bot.*, **80**: 248–300; *Biol. Abst.*, **10**: 41.

ARTHUR, J. M., and W. D. STEWART, 1933: Transpiration of tobacco plants in relation to radiant energy in the visible and infrared, *Contr. Boyce Thompson Inst.*, **5**: 483–501.

ASHBY, E., 1931: Comparison of two methods of measuring stomatal aperture, *Plant Physiol.*, **6**: 713–719.

———, 1932: Transpiratory organs of *Larrea tridentata* and their ecological significance, *Ecology*, **13**: 182–188.

ASKENASY, E., 1875: Über die Temperatur, welche Pflanzen im Sonnenlicht annehmen, *Bot. Ztg.*, **33**: 441–444.

AVERY, G. S., 1933: Structure and development of the tobacco leaf, *Am. J. Bot.*, **20**: 565–592.

BABINET, M., 1848: Note sur un atmidoscope, *C. R. Acad. Sci. Paris*, **26**: 529–530.

BADER, A., 1930: Vergleichende Transpirationsuntersuchungen in der Gattung Pinus, *diss., Eberhard-Karl Universität (Tübingen); Biol. Abst.*, **8**: 1138.

BAKKE, A. L., 1914: Studies of the transpiring power of plants as indicated by the method of standardized hygrometric paper, *J. Ecol.*, **2**: 145–173.

——— and B. E. LIVINGSTON, 1916: Further studies on foliar transpiring power in plants, *Physiol. Res.*, **2**: 51–71.

BALLS, W. L., 1912: The stomatograph, *Proc. Roy. Soc.*, B, **85**: 33–44.

BARNES, C. R., 1902: The significance of transpiration, *Sci.*, **15**: 460.

BARRITT, N. W., 1929: The structure of the seed coat in *Gossypium* and its relation to the growth and nutrition of the lint hairs, *Ann. Bot.*, **43**: 483–489.

BARSS, A. F., 1932: Effect of moisture supply on development of *Pyrus malus*, *Bot. Gaz.*, **90**: 151–177.

Bartholomew, E. T., 1926: Internal decline of lemons. III, Water deficit in lemon fruits caused by excessive leaf evaporation, *Am. J. Bot.*, **13**: 102–117.

———, 1931: Certain phases of citrus leaf transpiration, *Am. J. Bot.*, **18**: 765–783.

Bates, C. G., 1919: A new evaporimeter for use in forest studies, *Month. Weather Rev.*, **47**: 283–294.

Bayer, L., 1902: Beitrag zur pflanzenphysiologischen Bedeutung des Kupfers in der Bordeaux-Brühe, *Inaug. Diss., Königsberg.*

Bellani, A., 1820: Descrizione di un mivo amidometro per servire di continuazione e fine alle reflessioni critiche intomo all' evaporazione, *Gior. Fis. Chim.*, **3**: 166–177.

Bessey, C. E., 1898: Some considerations upon the functions of the stomata, *Sci.*, **7**: 13–16.

Bialoglowski, J., 1935: Effect of humidity on transpiration of rooted lemon cuttings under controlled conditions, *Proc. Am. Soc. Hort. Sci.*, **33**: 166–169.

Blackman, F. F., 1895: Experimental researches on vegetable assimilation and respiration. I, On a new method of investigating the carbonic acid exchanges of plants, *Phil. Trans. Roy. Soc.*, **186**: 485–502.

———, 1895: Experimental researches on vegetable assimilation and respiration. II, On the paths of gaseous exchange between aerial leaves and the atmosphere, *Phil. Trans. Roy. Soc.*, **186**: 503–562.

——— and G. L. C. Matthaei, 1905: Experimental reseaches on vegetable assimilation and respiration. IV, A quantitative study of carbon-dioxide assimilation and leaf-temperature in natural illumination, *Proc. Roy. Soc.*, B, **76**: 402–460.

Blackman, V. H., and S. G. Paine, 1914: A recording transpirometer, *Ann. Bot.*, **28**: 109–113.

——— and R. C. Knight, 1917: A method of controlling the rate of air movement in transpiration experiments, *Ann. Bot.*, **31**: 217–220.

Blaikley, N. M., 1932: Absorption and conduction of water and transpiration in *Polytrichum commune*, *Ann. Bot.*, **46**: 289–300.

Blaydes, G. W., 1928: A survey of rate of water loss from leaves, *Ohio J. Sci.*, **28**: 99–118.

———, 1932: Water loss and stomatal investigations, *Abst. Doctor's Diss.* 8, *Ohio State Univ.*

———, 1935: Water vapor loss from plants growing in various habitats, *Ohio J. Sci.*, **35**: 112–130.

Blodgett, F. H., 1901: Transpiration of rust-infected *Rubus*, *Torreya*, **1**: 34–35.

Bolas, B. D., and R. Melville, 1933: The influence of environment on the growth and metabolism of the tomato plant. I, Methods, technique and preliminary results, *Ann. Bot.*, **47**: 673–688.

Bouyoucos, G. J., 1911: Transpiration of wheat seedlings as affected by soils, by solutions of different densities, and by various chemical compounds, *J. Am. Soc. Agron.*, **3**: 130–191.

Brierley, W. G., 1930: Transpiration rates in old and new canes of the Latham raspberry as measured by non-toxic dyes, *Proc. Am. Soc. Hort. Sci.*, **27**: 152–157.

———, 1931: Transpiration in new and old canes of the Latham raspberry, *Proc. Am. Soc. Hort. Sci.*, **28**: 188–193.

Briggs, L. J., and H. L. Shantz, 1913: The water requirement of plants. I, Investigations in the Great Plains area in 1910 and 1911, *U. S. Plant Ind. Bull.* 284.

——— and ———, 1913: The water requirement of plants. II, A review of the literature, U. S. *Plant Ind. Bull.* 285.

——— and ———, 1914: The relative water requirement of plants, *J. Agr. Res.*, **3**: 1–65.

——— and ———, 1915: Influence of hybridization and cross-pollination on the water requirement of plants, *J. Agr. Res.*, **4**: 391–403.

——— and ———, 1915: An automatic transpiration scale of large capacity for use with freely exposed plants, *J. Agr. Res.*, **5**: 117–133.

——— and ———, 1916: Daily transpiration during the normal growth period and its correlation with the weather, *J. Agr. Res.*, **7**: 155–213.

——— and ———, 1916: Hourly transpiration rate on clear days as determined by cyclic environmental factors, *J. Agr. Res.*, **5**: 583–651.

——— and ———, 1917: Comparison of the hourly evaporation rate of atmometers and free water surfaces with the transpiration rate of *Medicago sativa*, *J. Agr. Res.*, **9**: 277–293.

Brooks, C., and D. F. Fisher, 1926: Some high-temperature effects in apples: Contrasts in the two sides of an apple, *J. Agr. Res.*, **32**: 1–17.

Brown, H. T., and F. Escombe, 1900: Static diffusion of gases and liquids in relation to the assimilation of carbon and translocation in plants, *Phil. Trans. Roy. Soc.*, B, **193**: 223–291; *Ann. Bot.*, **14**: 537–542; *Proc. Roy. Soc.*, B, **67**: 124–128.

——— and ———, 1905: Researches on some of the physiological processes of green leaves, with special reference to the interchange of energy between the leaf and its surroundings, *Proc. Roy. Soc.*, B, **76**: 29–111.

Brown, M. A., 1910: The influence of air currents on transpiration, *Proc. Iowa Acad. Sci.*, **17**: 13–15.

Burgerstein, A., 1885: Ueber einige physiologische und pathologische Wirkungen des Kampfers auf die Pflanzen, inbesonders auf Laubesprosse, *Verh. Zool.-bot. Ges. Wien*, **34**: 543.

———, 1904: "Die Transpiration der Pflanzen," 283 pp., Jena, Gustav Fischer.

———, 1920: "Die Transpiration der Pflanzen," II, 262 pp., Jena, Gustav Fischer.

———, 1920: Änderungen der Spaltöffnungsweite unter dem Einfluss verschiedener Bedingungen, *Verh. Zool.-bot. Ges. Wien*, 113–131.

Burns, G. P., 1915: The relative transpiration of white pine seedlings, *Plant World*, **18**: 1–6.

Buscalioni, L., and G. Pollacci, 1902: Alteriovi ricerche sull' applicazione delle pellicole di collodio allo studio di alcuni processi fisiologici dell piante ed in particolare modo delle transpirazione vegetale, *Att. Inst. Bot. Univ. Pavia*, **7**.

Chapman, G. W., 1931: The cause of succulence in plants, *New Phytol.*, **30**: 119–128

Childers, N. F., 1935: Some effects of sprays on the growth and transpiration of tomatoes, *Proc. Am. Soc. Hort. Sci.*, **33**: 532–535.

Chodat, F., 1931: Influence de la lumière sur la transpiration végétale, *C. R. Séances Soc. Phys. et Hist. Nat. Genève*, **48**: 55–58.

Clapp, G. L., 1908: A quantitative study of transpiration, *Bot. Gaz.*, **45**: 254–267.

Clements, F. E., and E. V. Martin, 1934: Effect of soil temperature on transpiration in *Helianthus annuus*, *Plant Physiol.*, **9**: 619–630.

Clements, H. F., 1934: Significance of transpiration, *Plant Physiol.*, **9**: 165–172.

Clinton, G. P., 1909: Spraying potatoes in dry seasons, *Conn. Sta. Bien. Rept.*, 739–752.

Clum, H. H., 1926: The effect of transpiration and environmental factors on leaf temperatures. I, Transpiration, *Am. J. Bot.*, **13**: 194–216.

———, 1926: The effect of transpiration and environmental factors on leaf temperatures. II, Light intensity and the relation of transpiration to the thermal death point, *Am. J. Bot.*, **13**: 217–231.

Copeland, E. B., 1902: The mechanism of stomata, *Ann. Bot.*, **16**: 327–365.

———, 1906: On the water relations of the coconut palm, *Philippine J. Sci.*, **1**: 6–58.

———, 1932: Transpiration by chaparral and its effect upon the temperature of leaves, *Univ. Calif. Pub.* **17**: 1–21.

Cowles, H. C., 1910: "A Textbook of Botany," Vol. II, "Ecology," American Book Company, New York.

Cross, F. B., 1927: May water loss from plants be reduced by foliage sprays? *Proc. Okla. Akad. Sci. (Okla. Univ.)*, **7**: 129–131.

Cullinan, F. P., 1920: Transpiration studies with the apple, *Proc. Am. Soc. Hort. Sci.*, **17**: 232–240; *Bot. Abst.*, **9**: 226.

Curtis, O. F., 1926: What is the significance of transpiration? *Sci.*, **63**: 267–271.

———, 1936: Leaf temperatures and the cooling of leaves by radiation, *Plant Physiol.*, **11**: 343–364.

———, 1936: Transpiration and the cooling of leaves, *Am. J. Bot.*, **23**: 7–11.

Darrow, G. M., and H. Sherwood, 1931: Transpiration studies on strawberries, *Proc. Am. Soc. Hort. Sci.*, **28**: 225–230.

———, 1932: Methods of measuring strawberry leaf areas, *Plant Physiol.*, **7**: 745–747.

——— and G. W. Dewey, 1934: Studies on the stomata of strawberry varieties and species, *Proc. Am. Soc. Hort. Sci.*, **32**: 440–447.

Darwin, F., 1898: Observations on stomata, *Proc. Roy. Soc.*, B, **63**: 413–417.

———, 1898: Observations on stomata, *Phil. Trans. Roy. Soc.*, B, **190**: 531–621.

———, 1904: On a self-recording method applied to the movements of stomata, *Bot. Gaz.*, **37**: 81–105.

——— and D. F. M. Pertz, 1912: On a new method of estimating the aperture of stomata, *Proc. Roy. Soc.*, B, **84**: 136–154.

———, 1914: The effect of light on the transpiration of leaves, *Proc. Roy. Soc.*, B, **87**: 281–299.

———, 1916: On the relation between transpiration and stomatal aperture, *Phil. Trans. Roy. Soc.*, B, **207**: 413–437.

Deen, J. L., and M. H. Bruner, 1933: The effect of the 1932 eclipse upon the width of stomatal openings in gray birch, *Ecology*, **14**: 76–77.

DeVilliers, F. J., 1926: Factors influencing the transpiration rate of grapes, *S. African J. Sci.*, **23**: 486–487; *Biol. Abst.*, **3**: 1078.

Dillman, A. C., 1931: The water requirement of certain crop plants and weeds in the northern Great Plains, *J. Agr. Res.*, **42**: 187–238.

Dixon, H. H., 1898: Transpiration into a saturated atmosphere, *Proc. Roy. Irish Soc.*, III, **4**: 627–635; *Bot. Cent.*, **76**: 135.

———, 1914: "Transpiration and the Ascent of Sap in Plants," The Macmillan Company, New York.

Dodge, B. O., 1923: Effect of the orange-rusts of *Rubus* on the development and distribution of stomata, *J. Agr. Res.*, **25**: 495–501.

Dole, E. J., 1924: Studies on the effect of air, temperature and relative humidity on the transpiration of *Pinus Strobus*, *Vt. Sta. Bull.* 238.

Dosdall, L., 1919: Water requirement and adaptation in *Equisetum*, *Plant World*, **22**: 1–13; 29–44.

Dowd, O. J., 1931: Studies in the transpiration rate of apple varieties, *Proc. Am. Soc. Hort. Sci.*, **28**: 590–593.

Drabble, E., and H. Drabble, 1907: The relation between the osmotic strength of cell sap in plants and their physical environment, *Biochem. J.*, **2**: 117–132.

Dufrenoy, J., 1918: Les facteurs physiques de la transpiration chez les plantes et la transpiration des feuilles parasitées, *Rev. Gén. Sci. Ann.*, **29**: 565–566.

Duggar, B. M., 1911: "Plant Physiology," 92–93 pp., The Macmillan Company, New York.

——— and J. S. Cooley, 1914: The effect of surface films and dusts on the rate of transpiration, *Ann. Mo. Bot. Gard.*, **1**: 1–23.

——— and ———, 1914: The effect of surface films on the rate of transpiration. Experiments with potted potatoes, *Ann. Mo. Bot. Gard.*, **1**: 351–357.

——— and W. W. Bonns, 1918: The effect of Bordeaux mixture on the rate of transpiration, *Ann. Mo. Bot. Gard.*, **5**: 153–176.

Eaton, F. M., and G. O. Belden, 1929: Leaf temperatures of cotton and their relation to transpiration and varietal differences and yields, *U. S. Dept. Agr. Tech. Bull.* 91.

———, 1930: Cell-sap concentration and transpiration as related to age and development of cotton leaves, *J. Agr. Res.*, **40**: 791–805.

Eckerson, S. H., 1908: The number and size of stomata, *Bot. Gaz.*, **46**: 221–224.

Ehlers, J. H., 1915: The temperature of leaves of *Pinus* in winter, *Am. J. Bot.*, **2**: 32–70.

Eidelman, Z. M., 1933: Principal results of experiments on the artificial reduction of the leaf surface in various geographical locations, *Bull. Plant. Protection*, **3**: 65–72; *Biol. Abst.*, **9**: 747.

Emerson, J. L., and A. C. Hildreth, 1933: Preliminary report on reducing transpiration on transplanted evergreens, *Sci.*, **77**: 433–434.

Ezekiel, W. N., and J. J. Taubenhaus, 1932: Leaf temperatures of cotton plants with *Phymatotrichum* root rot, *Sci.*, **75**: 391–392.

Finch, A. H., and C. W. Van Horn, 1936: The moisture relations of pecan leaves, *Sci.*, **83**: 260.

Finnell, H. H., 1928: Effect of wind on plant growth, *J. Am. Soc. Agron.*, **20**: 1206–1210.

Firbas, F., 1931: Die Wirkung des Windes auf die Transpiration, *Ber. bot. Ges.*, **49**: 443–452.

Flood, M. G., 1919: Exudation of water by *Colocasia antiquorum*, *Sci. Proc. Roy. Dublin Soc.*, **15**: 505–512.

Flory, E. L., 1936: Comparison of the environment and some physiological responses of prairie vegetation and cultivated maize, *Ecology*, **17**: 67–103.

Frank, A. B., and F. Krüger, 1894: Ueber den Reiz, welchen die Behandlung mit Kupfer auf die Kartoffelpflanze hervorbringt, *Ber. bot. Ges.*, **12**: 8–11.

——— and ———, 1894: Ueber den direkten Einfluss der Kupfervitriolkalkbrühe auf die Kartoffelpflanze, *Arb. landw. Ges.*, **1894**: 1046.

Frear, D. E. H., 1935: Photoelectric apparatus for measuring leaf areas, *Plant Physiol.*, **10**: 569–574.

Freeman, G. F., 1908: A method for the quantitative determination of transpiration in plants, *Bot. Gaz.*, **46**: 118–129.

———, 1920: Studies in evaporation and transpiration, *Bot. Gaz.*, **70**: 190–211.

Frei, L., 1924: Influence of soil moisture on the transpiration capacity of plants, (Russian) *Bull. Soc. Nat. Leningrad*, **53**: 179–210; *Bot. Abst.*, **15**: 164–165.

Furr, J. R., and J. R. Magness, 1930: Preliminary report on relation of soil moisture to stomatal activity and fruit growth of apples, *Proc. Am. Soc. Hort. Sci.*, **27**: 212–218.

——— and E. S. Degman, 1931: Relation of moisture supply to stomatal behavior of the apple, *Proc. Am. Soc. Hort. Sci.*, **28**: 547–551.

Ganeau de Lamarlière, L., 1892: Recherches physiologiques sur les feuilles, développées à l'ombre et au soleil. VI, Transpiration, *Rev. Gén. Bot.*, **4**: 529.

Ganong, W. F., 1905: New precision appliances for use in Plant Physiology. II, *Bot. Gaz.*, **39**: 145–152.

Gates, F. C., and R. L. Black, 1931: The dragoyle as an ecological instrument, *Ecology*, **12**: 448–451.

GERDEL, R. W., and R. M. SALTER, 1928: Measurement of leaf area using the photoelectric cell, *J. Am. Soc. Agron.*, **20**: 635–643.

GIDDINGS, L. A., 1914: Transpiration of *Silphium laciniatum*, *Plant World*, **17**: 309–328.

GILBERT, B. E., and W. L. ADAMS, 1929: Moisture fluctuations in extracted plant solutions and in leaf tissue, *Plant Physiol.*, **4**: 529–536.

GIRTON, R. E., 1932: An inexpensive recording transpirometer, *Proc. Ind. Acad. Sci.*, **41**: 127–132.

GRAČANIN, M., 1932: Contribution à l'étude de la relation entre la transpiration et la résorption des ions, *C. R. Acad. Sci. Paris*, **195**: 899–901.

GRAF-MARIN, A., 1934: Studies on powdery mildew of cereals, *N. Y. Sta.* (*Cornell*) *Mem.* 157.

GRAINGER, J., and A. L. ALLEN, 1936: The internal temperatures of fruit tree buds, *Ann. App. Biol.*, **23**: 1–10.

GRAY, J., and G. J. PEIRCE, 1919: The influence of light upon the action of stomata and its relation to the transpiration of certain grains, *Am. J. Bot.*, **6**: 131–156.

GREGORY, F. G., and H. L. PEARSE, 1934: The resistance porometer and its application to the study of stomatal movement, *Proc. Roy. Soc.*, B, **114**: 477–493.

GUSTAFSON, F. G., 1934: The effect of a decrease in the amount of transpiration on the growth of certain plants, *Mich. Acad. Sci.*, *Arts and Letters*, **19**: 65–83.

HAAS, A. R. C., and F. F. HALMA, 1932: Relative transpiration rates in citrus leaves, *Bot. Gaz.*, **93**: 466–473.

HAAS-POETZL, I., 1930: Die Wirkung von Salzen auf die Spaltöffnungsbewegung, *Beih. bot. Cent.*, **47**: 255–270.

HABERLANDT, G., 1914: "Physiological Plant Anatomy," Eng. transl., pp. 493–498.

HAGEN, F., 1918: Zur Physiologie des Spaltöffnungsapparates, *Beitr. Allg. Bot.*, **1**: 260–291; *Bot. Abst.*, **13**: 63–64.

HAINES, F. M., 1932: A self-recording potometer with a note on transpiration under pressure, *Ann. Bot.*, **46**: 1051–1060.

———, 1935: Transpiration and pressure deficit. I, Apparatus and preliminary experiments, *Ann. Bot.*, **49**: 213–238.

———, 1936: Transpiration and pressure deficit. III, Observations by the thermopile method, *Ann. Bot.*, **50**: 1–22.

———, 1936: Transpiration and pressure deficit. IV, The effects of small deficits: Apparatus and preliminary experiments, *Ann. Bot.*, **50**: 281–291.

HART, H., 1929: Relation of stomatal behavior to stem rust resistance in wheat, *J. Agr. Res.*, **39**: 929–949.

HARTER, L. L., 1908: The influence of a mixture of soluble salts, principally sodium chloride, upon the leaf structure and transpiration of wheat, oats, and barley, *U. S. Plant Ind. Bull.* 134.

HAWKINS, R. S., 1927: Variations of water and dry matter in the leaves of Pina and Acala cotton, *Ariz. Sta. Tech. Bull.* **17**: 417–444.

HEILING, A., 1933: Influence of gaseous atmospheric impurities on plant transpiration, *Phytopath. Z.*, **5**: 435–492; *Biol. Abst.*, **27**: 3971.

HEINICKE, A. J., 1930: A method for studying the relative rates of transpiration of apple leaves and fruits, *Proc. Am. Soc. Hort. Sci.*, **26**: 312–314.

HENDERSON, F. Y., 1926: On the effect of light and other conditions upon the rate of water-loss from the mesophyll, *Ann. Bot.*, **40**: 507–535.

HENDRICKSON, A. H., 1921: Transpiration rate of deciduous fruit trees as influenced by irrigation and other factors, *Proc. Am. Soc. Hort. Sci.*, **18**: 145–147.

———, 1926: Certain water relations of the genus *Prunus*, *Hilgardia*, **1**: 479–524.

HILTNER, E., 1930: Dew and its importance in plant growth, *Pflanzenbau*, **3**: 1–70; *Exp. Sta. Rec.*, **64**: 211.

HIRANO, E., 1931: Relative abundance of stomata in *Citrus* and some related genera, *Bot. Gaz.*, **92**: 296–311.

HOPKINS, C. Y., 1936: Thermal death point of certain weed seeds, *Can. J. Res.*, **14**: 178–183.

HOWES, F., 1923: Researches on the vegetation of Natal. Ser. I. Section II. A new calcium chloride method of measuring the resistance to water loss in leaves, *Mem. Bot. Survey S. Africa*, **5**: 23–33; *Bot. Abst.*, **13**: 925.

HUBER, B., 1923: Transpiration in verschiedener Stammhöhe. I, *Sequoia gigantea*, *Z. Bot.*, **15**: 465–501; *Bot. Abst.*, **14**: 1218.

HURD-KARRER, A. M., and J. W. TAYLOR, 1929: The water content of wheat leaves at flowering time, *Plant Physiol.*, **4**: 393–397.

ILJIN, W. S., 1914: Die Regulierung der Spaltöffnungen in Zusammenhang mit Veränderung des osmotischen Druckes, *Beih. bot. Cent.*, **32**: 15–35.

———, 1932: Ueber Öffnen der Stomata bei starkem Welken der Pflanzen, *Jahrb. wiss. Bot.*, **77**: 220–251.

ISAAC, W. E., 1935: A preliminary study of the water loss of *Laminaria digitata* during intertidal exposure, *Ann. Bot.*, **49**: 109–117.

IVANOV, L. A., 1923: The present state of the question of drought resistance, *Bull. App. Bot. Plant Breeding*, **13**: 1–32.

——— and M. M. THIELMANN, 1923: Über den Einfluss des Lichtes verschiedener Wellenlänge auf die Transpiration der Pflanzen, *Flora*, **116**: 296–311.

JEFFREYS, H., 1918: Some problems of evaporation, *Phil. Mag. J. Sci.*, 6 ser., **35**: 270–280.

JIVANNA RAO, P. S., 1923: Stomata in cotton flower, *J. Madras Agr. Student Union*, **11**: 40–42; *Bot. Abst.*, **13**: 886.

JOHNSTON, C. O., and E. C. MILLER, 1934: Relation of leaf-rust infection to yield, growth, and water economy of two varieties of wheat, *J. Agr. Res.*, **49**: 955–981.

JOHNSTON, E. S., 1918: A simple non-absorbing atmometer mounting, *Plant World*, **21**: 257–261.

———, 1920: A method of studying the absorption-transpiration ratio in nutrient media, *Sci.*, **52**: 517–518.

JONES, I. D., 1932: Ultropaque microscope equipment as an aid in stomatal studies, *Proc. Am. Soc. Hort. Sci.*, **29**: 78–79.

JONES, T. N., and L. O. PALMER, 1932: Field curing of hay as influenced by plant physiological reactions, *Agr. Eng.*, **13**: 199–200.

——— and ———, 1933: Field curing of hay as influenced by plant physiological reactions. II, The role of leaves in the dehydration of hay plants, *Agr. Eng.*, **14**: 156–158.

——— and ———, 1934: Hay curing. III, Relation of engineering principles and physiological factors, *Agr. Eng.*, **15**: 198–201.

JONES, W. N., 1914: A self recording porometer and potometer, *New Phytol.*, **13**: 353–364.

KAMP, H., 1930: Untersuchungen über Kutikularbau und kutikuläre Transpiration von Blättern, *Jahrb. wiss. Bot.*, **72**: 403–465.

KELLEY, V. W., 1930: A comparison of the transpiration rates of twenty-one deciduous fruit species, *Ill. Sta. Bull.* 341.

———, 1930: Effect of certain hydrocarbon oils on the transpiration rate of some deciduous fruits, *Ill. Sta. Bull.* 353.

———, 1932: The effect of pruning of excised shoots on the transpiration rate of some deciduous fruit species, *Proc. Am. Soc. Hort. Sci.*, **29**: 71–73.

KHANKOJE, P., 1914: Some factors which influence the water requirement of plants, *J. Am. Soc. Agron.*, **6**: 1–23.

KIESSELBACH, T. A., and A. M. MONTGOMERY, 1911: The relation of climatic factors to the water used by the corn plant, *Neb. Sta. Rept.*, **1911**: 91–107.

———, 1916: Transpiration as a factor in crop production, *Neb. Sta. Res. Bull.* 6.

——— and A. ANDERSON, 1931: Quality of alfalfa hay in relation to curing practice, *U. S. Dept. Agr. Tech. Bull.* 235.

KNIGHT, R. C., 1915: A convenient modification of the porometer, *New Phytol.*, **14**: 212–216.

———, 1916: On the use of the porometer in stomatal investigation, *Ann. Bot.*, **30**: 57–76.

———, 1917: The interrelations of stomatal aperture, leaf water-content and transpiration rate, *Ann. Bot.*, **31**: 221–240.

———, 1917: Relative transpiration as a measure of the intrinsic transpiring power of the plant, *Ann. Bot.*, **31**: 351–359.

———, 1922: Further observations on the transpiration, stomata, leaf water-content, and wilting of plants, *Ann. Bot.*, **36**: 361–385.

KONDO, J. N., 1931: The influence of external factors as well as stages of development on the resistance of plants to dehydration, *Bull. App. Bot. and Plant Breeding*, **27**: 129–156.

KOSAKA, H., 1933: The relationship between different physiological aspects of plants and the appearance of pigments in the different parts of the plants. V, The relationship between the presence of anthocyanin pigments and transpiration in some cultivated plants, *J. Dept. Agr. Kyushu Imp. Univ.*, **4**: 95–126; *Chem. Abst.*, **28**: 2390.

KOURSSANOW, A. L., 1928: De l'influence de l'*Ustilago tritici* sur les fonctions physiologiques du froment, *Rev. Gén. Bot.*, **40**: 350–353.

KUYPER, J., 1915: Contributions to the physiology of stomata in *Saccharum officinarum*, *Arch. Siukerindus. Nederland. Indie.*, **23**: 1673–1700; *Exp. Sta. Rec.* **35**: 330.

LACHENMEIER, J., 1932: Transpiration und Wasserabsorption intakter Pflanzen nach vorausgegangener Verdunkelung bei Konstanz der Lichtintensität und der übrigen Aussenfaktoren, *Jahrb. wiss. Bot.*, **76**: 765–827; *Biol. Abst.*, **8**: 1659.

LAIDLOW, C. G. P., and R. C. KNIGHT, 1916: A description of a recording porometer and a note on stomatal behavior during wilting, *Ann. Bot.*, **30**: 47–57.

LAMOUR, J., 1918: On transpiration through leaf stomata, *Phil. Mag. J. Sci.*, 6 ser., **35**: 350–352, 433–434.

LEICK, E., 1927: Untersuchungen über den Einfluss des Lichtes auf die Öffnungsweite unterseitiger und oberseitiger Stomata desselben Blattes, *Jahrb. wiss. Bot.*, **67**: 771–848.

———, 1928: Über das verschiedenartige Verhalten der unterseitigen und oberseitigen Stomata desselben Blattes, *Ber. bot. Ges.*, **45**: 28–42.

LEITBEG, H., 1868: Beiträge zur Physiologie der Spaltöffnungsapparate, *Graz. Bot. Inst. Mitt.*, **1**: 123.

LEPESCHKIN, W. W., 1923: Über aktive und passive Wasserdrusen und Wasserpalten, *Ber. bot. Ges.*, **41**: 298–300; *Bot. Abst.*, **13**: 925.

LESAGE, P., 1894: Sur les rapports des palissades dans les feuilles avec la transpiration, *C. R. Acad. Sci. Paris*, **118**: 255–258.

LINSBAUER, K., 1926: Beobachtungen an Spaltöffnungen, *Planta*, **2**: 530–536; *Biol. Abst.*, **2**: 865.

———, 1927: Weitere Beobachtungen an Spaltöffnungen, *Planta*, **3**: 527–561; *Biol. Abst.*, **3**: 1079.

Livingston, B. E., 1906: The relation of desert plants to soil moisture and to evaporation, *Carnegie Inst. Wash. Pub.* 50.
———, 1907: Relative transpiration in cacti, *Plant World*, **10**: 110–114.
———, 1907: Evaporation and plant development, *Plant World*, **10**: 269–276.
———, 1908: Evaporation and plant habitats, *Plant World*, **11**: 1–9.
———, 1908: A method for controlling plant moisture, *Plant World*, **11**: 39–40.
———, 1909: Stomata and transpiration in *Tradescantia zebrina*, *Sci.*, **29**: 269–270.
———, 1910: A rain-correcting atmometer for ecological instrumentation, *Plant World*, **13**: 79–82.
———, 1910: Operation of the porous cup atmometer, *Plant World*, **13**: 111–119.
———, 1911: A radio-atmometer for comparing light intensities, *Plant World*, **14**: 96–99.
———, 1911: A study of the relation between summer evaporation intensity and centers of plant distribution in the United States, *Plant World*, **14**: 205–222.
———, 1911: Paper atmometers for studies in evaporation and plant transpiration, *Plant World*, **14**: 281–289.
———, 1911: Light intensity and transpiration, *Bot. Gaz.*, **52**: 417–438.
——— and A. H. Estabrook, 1912: Observations on the degree of stomatal movement in certain plants, *Bull. Torrey Bot. Club*, **39**: 15–22.
——— and W. H. Brown, 1912: Relation of the daily march of transpiration to variations in the water content of foliage leaves, *Bot. Gaz.*, **53**: 309–330.
———, 1913: The resistance offered by leaves to transpirational water loss, *Plant World*, **16**: 1–35.
———, 1915: Atmometry and the porous cup atmometer, *Plant World*, **18**: 21–30, 51–74, 95–111, 143–149.
——— and L. A. Hawkins, 1915: The water relation between plant and soil, *Carnegie Inst. Wash. Pub.* 204.
——— and E. B. Shreve, 1916: Improvements in the method for determining the transpiring power of plant surfaces by hygrometric paper, *Plant World*, **19**: 287–309.
———, 1917: Atmometric units, *Johns Hopkins Univ. Circ.*, **3**: 160–170.
———, 1918: Porous clay cones for the auto-irrigation of potted plants, *Plant World*, **21**: 202–208.
——— and F. Thone, 1920: A simplified non-absorbing mounting for porous porcelain atmometers, *Sci.*, **52**: 85–87.
———, 1923: Studies upon the influence of solar radiation on the rate of transpirational water-loss from plants, *Carnegie Inst. Wash. Yearbook*, **22**: 288–289.
———, 1923: Blackened spheres for atmometry, *Sci.*, **58**: 182–183.
———, 1926: A black collodion coating for atmometer spheres, *Sci.*, **63**: 362–363.
———, T. Hemmi, and J. D. Wilson, 1926: Growth of young wheat plants in auto-irrigated soils, as related to the water-supplying power of the soil and to the adjustment of the auto-irrigator, *Plant Physiol.*, **1**: 387–395.
———, 1927: Plant water relations, *Quart. Rev. Biol.*, **2**: 494–515.
Lloyd, F. E., 1908: The physiology of stomata, *Carnegie Inst. Wash. Pub.* 82.
———, 1912: The relation of transpiration and stomatal movements to the water content of the leaves in *Fouquieria splendens*, *Plant World*, **15**: 1–14.
———, 1913: Leaf-water, and stomatal movement in *Gossypium* and a method of direct visual observation of stomata *in situ*, *Bull. Torrey Bot. Club*, **40**: 1–26.
Loftfield, J. V. G., 1921: The behavior of stomata, *Carnegie Inst. Wash. Pub.* 314.
Long, F. L., 1928: Stomata which show functional movement for a century, *Sci.*, **69**: 218–219.

——— and F. E. Clements, 1934: The method of collodion films for stomata, *Am. J. Bot.*, **21**: 7–17.

Lott, R. V., and H. W. LeMert, 1932: A simple device for use in leaf area studies, *Proc. Am. Soc. Hort. Sci.*, **29**: 83–84.

Lutman, B. F., 1916: Some studies on Bordeaux mixture, *Vt. Sta. Bull.* 196.

MacDougal, D. T., 1922: How plants endure heat and cold, *Gard. Mag.*, **36**: 152–154.

Magness, J. R., and J. R. Furr, 1930: Stomatal activity in apple leaves, *Proc. Am. Soc. Hort. Sci.*, **27**: 207–211.

Manzoni, L., and A. Puppo, 1934: Sur la transpiration du blé en fonction das facteurs du climat, *C. R. Acad. Sci. Paris*, **198**: 1066–1068; *Biol. Abst.*, **9**: 1966–1967.

Mariana, G., 1902: The influence of humidity on the formation and development of stomata. *Separate from Att. Inst. Bot. Univ. Pavia*, 2 ser., **8**: 32; *Exp. Sta. Rec.* **14**: 943.

Markley, K. S., and C. E. Sando, 1931: Permeability of the skin of apples as measured by evaporation loss, *Plant Physiol.*, **6**: 541–547.

Marshall, R. E., 1933: An apparatus for the ready determination of areas of compound leaves, *J. Agr. Res.*, **47**: 437–439.

Martin, E. V., and F. E. Clements, 1935: Studies on the effect of artificial wind on growth and transpiration in *Helianthus annuus*, *Plant Physiol.*, **10**: 613–636.

Martin, J. H., 1930: The comparative drought resistance of sorghums and corn, *J. Am. Soc. Agron.*, **22**: 986–993.

Martin, W. H., 1916: Influence of Bordeaux mixture on the rates of transpiration from abscised leaves and from potted plants, *J. Agr. Res.*, **7**: 529–548.

Maximow, N., and V. Rybin, 1921: Non-stomatal regulation of transpiration, *J. Russ. Bot. Congress*, **1**: 25–26; *Bot. Abst.*, **14**: 287.

———, 1925: The physiological basis of drought resistance of plants, *Eng. Abst.*

———, 1929: "The Plant in Relation to Water," 451 pp., translated by R. H. Yapp, The Macmillan Company, New York.

McLean, R. C., 1919: Studies in the ecology of tropical rain forests. I, Humidity. *Ecology*, **7**: 5–54; II, Illumination, 121–172.

Metzger, J. E., M. G. Holmes, and H. Bierman, 1925: Soybeans, *Md. Sta. Bull.* 277.

Meyer, B. S., 1927: The measurement of the rate of water-vapor loss from leaves under standard conditions, *Am. J. Bot.*, **14**: 582–592.

———, 1931: Effects of mineral salts upon the transpiration and water requirement of the cotton plant, *Am. J. Bot.*, **18**: 79–93.

———, 1932: The daily periodicity of transpiration in the tulip poplar (*Liriodendron tulipifera*), *Ohio J. Sci.*, **32**: 104–114.

Miller, E. C., 1916: Comparative study of the root systems and leaf areas of the corn and the sorghums, *J. Agr. Res.*, **6**: 311–333.

———, 1916: Relative water requirement of corn and the sorghums, *J. Agr. Res.*, **6**: 473–485.

———, 1917: Daily variation of water and dry matter in the leaves of corn and the sorghums, *J. Agr. Res.*, **10**: 11–47.

——— and W. B. Coffman, 1918: Comparative transpiration of corn and the sorghums, *J. Agr. Res.*, **13**: 579–605.

———, 1923: Relative water requirement of corn and sorghums, *Kans. Sta. Tech. Bull.* 12.

——— and A. R. Saunders, 1923: Some observations on the temperature of the leaves of crop plants, *J. Agr. Res.*, **26**: 15–43.

———, 1925: Relative transpiration in crop plants, unpub. data, *Kans. Sta.*

———, 1928: Some observations on the number of stomata of crop plants, unpub. data, *Kans. Sta.*

———, 1928: The relation of soil moisture to transpiration, unpub. data, *Kans. Sta.*

———, 1928: The effect of Bordeaux mixture on the rate of transpiration, unpub. data, *Kans. Sta.*

MINCKLER, L. S., 1936: A new method of measuring transpiration, *J. Forestry*, **34**: 36–39.

MITCHELL, J. W., 1936: Measurement of the area of attached and detached leaves, *Sci.*, **83**: 334–336.

MITSCHERLICH, A., 1904: Ein Verdunstungsmesser, *Landw. Versuchs-Stat.*, **60**: 63–72.

MONTEMARTINI, L., 1904: Note di fisiopatologia vegetale, *Atti dell' Ist. Bot. di Pavia*, Ser. II, 63 pp., **9**: 1–66; *Bot. Cent.* **98**: 276–277.

———, 1932: Intorno all'azione della luce e di eccitanti chimici sopra la traspirazione, *Lavori R. Inst. Bot. Palermo*, **3**: 3–15; *Biol. Abst.*, **8**: 1138.

MONTGOMERY, E. G., 1911: Methods of determining the water requirements of crops, *Proc. Am. Soc. Agron.*, **3**: 261–283.

——— and T. A. KIESSELBACH, 1912: Studies in water requirements of corn, *Neb. Sta. Bull.* 128.

MUENSCHER, W. L. C., 1915: A study of the relation of transpiration to the size and number of the stomata, *Am. J. Bot.*, **2**: 487–504.

MÜLLER-THURGAU, H., 1894: Die Tätigkeit pilzkranker Blätter, 4 *Jahresb. der Vers. Stat. Obst. Wein- und Gartenbau Wadensweil*, **1894**: 54–57.

MURPHY, H. C., 1935: Effect of crown rust infection on yield and water requirement of oats, *J. Agr. Res.*, **50**: 387–412.

NEWTON, J. D., 1925: The relation of the salt concentration of the culture solution to transpiration and root respiration, *Sci. Agr.*, **5**: 318–320.

NEWTON, R., and W. M. MARTIN, 1930: Physicochemical studies on the nature of drought resistance in crop plants, *Can. J. Res.*, **3**: 336–383, 385–427.

NICHOLS, G. E., 1913: A simple revolving table for standardizing porous cup atmometers, *Bot. Gaz.*, **55**: 249–251.

———, 1923: The Musch rain-correcting mounting for porous porcelain atmometers. *Sci.*, **57**: 26–28.

NICOLAS, G., 1930: Sur la transpiration des plantes parasitées par des champignons, *Rev. Gén. Bot.*, **42**: 257–271.

NIEMANN, W., 1932: Über Beziehungen zwischen Blattgrösse und Spaltöffnungszahl in Abhängigkeit von der Bodenfeuchtigkeit, *Sonderabdruck Angew. Bot.*, **14**: Heft 1.

OVERLEY, F. L., E. L. OVERHOLSER, and I. A. HAUT, 1931: The relation of soil moisture and spray applications to stomatal behavior and growth of Jonathan apples, *Proc. Am. Soc. Hort. Sci.*, **28**: 543–546.

PAETZ, K. W., 1930: Untersuchungen über die Zusammenhänge zwischen stomatärer Öffnungsweite und bekannten Intensitäten bestimmter Spektralbezirke, *Planta*, **10**: 611–665; *Biol. Abst.*, **9**: 296–297.

PAVILINOVA, E., 1926: Physiological significance of guttation, *Bull. Inst. Rech. Biol. and Sta. Biol. Univ. Perm*, **4**: 471–478.

PENFOUND, W. T., 1932: The anatomy of the castor bean as conditioned by light intensity and soil moisture, *Am. J. Bot.*, **19**: 538–546.

PETERHÄNSEL, H., 1934: Experimental investigations on the relations between soil moisture and the use of water by plants, *Bot. Arch.*, **36**: 99–151; *Exp. Sta. Rec.*, **72**: 166.

PLEASANTS, A. L., 1930: The effect of nitrate fertilizer on stomatal behavior, *J. Elisha Mitchell Sci. Soc.*, **46**: 95–116; *Biol. Abst.*, **8**: 1659.

POOL, R. J., 1923: Xerophytism and comparative leaf anatomy in relation to transpiring power, *Bot. Gaz.*, **76**: 221–241.

PULLING, H. E., 1919: Sunlight and its measurement, *Plant World*, **22**: 151–172, 187–209.

REA, M. W., 1921: Stomata and hydathodes in *Campanula rotundifolia* L., *New Phytol.*, **20**: 56–72.

REED, H. S., 1910: The effect of certain chemical agents upon the transpiration and growth of wheat seedlings, *Bot. Gaz.*, **49**: 81–109.

——— and J. S. COOLEY, 1913: The transpiration of apple leaves infected with *Gymnosporangium*, *Bot. Gaz.*, **55**: 421–430.

——— and E. HIRANO, 1931: The density of stomata in citrus leaves, *J. Agr. Res.*, **43**: 209–222.

REHFOUS, L., 1917: Étude sur les stomates, *Univ. Genève Inst. Bot.* IX, No. 6, 110 pp.

RENNER, O., 1910: Beiträge zur Physik der Transpiration, *Flora*, **100**: 451–547.

RHINE, J. B., 1924: Clogging of stomata of conifers in relation to smoke injury and distribution, *Bot. Gaz.*, **78**: 226–232.

RICHARDSON, A. E. V., 1923: The water requirement of farm crops, *J. Dept. Agr. Victoria*, **21**: 193–212, 257–284, 321–339, 385–404, 449–481; *Exp. Sta. Rec.*, **50**: 733–735.

RICOME, H., 1903: Influence du chlorure de sodium sur la transpiration et l'absorption de l'eau chez les végétaux, *C. R. Acad. Sci. Paris*, **137**: 141–143.

RIPPEL, A., 1919: Der Einfluss der Bodentrockenheit auf den anatomischen Bau der Pflanzen insbesondere von *Sinapis alba*, etc., *Beih. bot. Cent.* I, **36**: 187–260.

ROSEN, H. R., 1933: Influence of spray applications on air temperatures surrounding sprayed potato plants, *Phytopath.*, **23**: 912–916.

RUDOLPH, K., 1925: Epidermis and epidermale Transpiration, *Bot. Arch.*, **9**: 49–94. Quoted by Crafts, 1933.

RUMM, C., 1893: Ueber die Wirkung der Kupferpräparate bei Bekämpfung der sogenannten Blattfallkrankheit der Weinrebe, *Ber. bot. Ges.*, **11**: 79–93.

RUNYON, E. H., 1936: Ratio of water content to dry weight in leaves of the creosote bush, *Bot. Gaz.*, **97**: 518–553.

SALISBURY, E. J., 1927: On the causes and ecological significance of stomatal frequency, with special reference to the woodland flora, *Phil. Trans. Roy. Soc.*, B, **216**: 1–65.

SAMPSON, A. W., and L. M. ALLEN, 1909–1916: Influence of physical factors on transpiration, *Minn. Bot. Studies*, **1–4**: 33–60.

SAWYER, W. H., 1932: Stomatal apparatus of the cultivated cranberry, *Vaccinium macrocarpon*, *Am. J. Bot.*, **19**: 508–513.

SAYRE, J. D., 1919: Comparative transpiration of tobacco and mullein, *Ohio J. Sci.*, **19**: 422–426.

———, 1919: Factors controlling variations in the rate of transpiration, *Ohio J. Sci.*, **19**: 491–509.

———, 1920: The relation of hairy leaf coverings to the resistance of leaves to transpiration, *Ohio J. Sci.*, **20**: 55–86.

———, 1923: Physiology of stomata of *Rumex patientia*, *Sci.*, **57**: 205–206.

———, 1926: Physiology of the stomata of *Rumex patientia*, *Ohio J. Sci.*, **26**: 233–267.

———, 1928: A recording atmometer, *Ecology*, **19**: 123–125.

———, 1929: Opening of stomata in different ranges of wave lengths of light, *Plant Physiol.*, **4**: 323–328.

SCARTH, G. W., 1926: The influence of the hydrogen-ion on the turgor and movement of the plant cells with especial reference to stomatal movement, *Proc. First Int. Cong. Plant Sci.*, Ithaca, N. Y., rev. by C. A. Shull, *Bot. Gaz.*, **82**: 453–454.

———, 1927: Stomatal movement: Its regulation and regulatory role—a review, *Protoplasma*, **2**: 498–511.

———, 1929: The influence of H-ion concentration on the turgor and movement of plant cells with special reference to stomatal behavior, *Proc. Int. Cong. Plant Sci.*, **2**: 1151–1162.

———, 1932: Mechanism of the action of light and other factors on stomatal movement, *Plant Physiol.*, **7**: 481–504.

———, J. Whyte, and A. Brown, 1933: On the cause of night opening of stomata, *Trans. Roy. Soc. Can.*, **Section 5**: 115–117.

Schander, R., 1904: Über die physiologische Wirkung der Kupfervitriol-kalkbrühe, *Landw. Jahrb.*, **33**: 517–584.

Schertz, F. M., 1929: The origin of chlorophyll and its relation to the blood pigment, *Sci.*, **70**: 258–259.

Schmucker, T., 1928: Über den Einfluss narkotischer Stoffe auf Transpiration und Wasserleitung, *Jahrb. wiss. Bot.*, **68**: 771–800.

Schwendener, S., 1881: Ueber Bau und Mechanik der Spaltöffnungen, *Monatsber. Akad. Wiss. Berlin*, **1881**: 833.

———, 1889: Die Spaltöffnungen der *Gramineen* und *Cyperaceen*, *Sitzungsber. Akad. Wiss. Berlin*, **1**: 65–77.

Seybold, A., 1929: Die pflanzliche Transpiration, *Ergeb. Biol.*, **5**: 29–165.

———, 1931, 1932: Weitere Beiträge zur Kenntnis der Transpirationsanalyse, *Planta*, **13**: 18–28; **14**: 77–93, 386–410; **16**: 518–525.

———, 1933: Zur Klärung des Begriffes Transpirationswiderstand, *Planta*, **21**: 353–367.

Shantz, H. L., and L. N. Piemeisel, 1927: The water requirement of plants at Akron, Colorado, *J. Agr. Res.*, **34**: 1093–1190.

Shapiro, A. A., and H. deForest, 1932: A comparison of transpiration rates in chaparral, *Ecology*, **13**: 290–295.

Shardakov, S., 1928: The physiological significance of guttation, *Bull. Inst. Rech. Biol. et Sta. Biol. Univ. Perm*, **6**: 193–208.

Shive, J. W., 1915: An improved non-absorbing porous cup atmometer, *Plant World*, **18**: 7–10.

——— and W. H. Martin, 1917: The effect of surface films of Bordeaux mixture on the foliar transpiring power in tomato plants, *Plant World*, **20**: 67–86.

Shreve, E. B., 1914: The daily march of transpiration in a desert perennial, *Carnegie Inst. Wash. Pub.* 194.

———, 1916: An analysis of the causes of variations in the transpiring power of cacti, *Physiol. Res.*, **2**: 73–127.

———, 1917: The transpiring power of plants, *Carnegie Inst. Wash. Year Book*, **16**: 66–68.

———, 1919: A thermo-electrical method for the determination of leaf temperature, *Plant World*, **22**: 100–104.

———, 1919: The role of temperature in the determination of the transpiring power of leaves by hygrometric paper, *Plant World*, **22**: 172–180.

———, 1924: Factors governing seasonal changes in transpiration of *Encelia farinosa*. *Bot. Gaz.*, **77**: 432–439.

Shull, C. A., 1919: Transpiration as energy dispersal, *School Sci. and Math.*, **19**: 1–6.

———, 1934: Lateral water transfer in leaves of *Ginkgo biloba*, *Plant Physiol.*, **9**: 387–389.

———, 1936: Rate of adjustment of leaf temperature to incident energy, *Plant Physiol.*, **11**: 181–188.

Sideris, C. P., and B. H. Krauss, 1928: Water relations of pineapple plants, *Soil Sci.*, **26**: 305–313.

Sitton, B. G., 1932: Some observations on stomatal movements in *Hicoria pecan*, *Proc. Am. Soc. Hort. Sci.*, **29**: 80–82.

Smith, A. M., 1909: On the internal temperature of leaves in tropical isolation with special reference to the effect of color on the temperature, *Ann. Roy. Bot. Gard. Peradniya*, **4**: 229–298.

Smith, G. E., 1935: On the orientation of stomata, *Ann. Bot.*, **49**: 451–477.

Smith, H. B., 1922: Stomatal behavior of plants in the greenhouse in winter, *Mich. Acad. Sci., Arts, and Letters*, pp. 109–117.

Smith, H. F., 1936: The effect of light on transpiration, *Ann. Bot.*, **50**: 155–159.

Smith, W. H., 1933: Evaporation of water from apples in relation to temperature and atmospheric humidity, *Ann. App. Biol.*, **20**: 220–235.

Spalding, V. M., 1905: Soil water in relation to transpiration, *Torreya*, **5**: 25–27.

Sperry, O. E., 1936: A study of the growth, transpiration, and distribution of the conifers of the Rocky Mountain National Park, *Bull. Torrey Bot. Club*, **63**: 75–103.

Sproat, B. B., and G. M. Darrow, 1935: Relation of leaf area to berry production in the strawberry, *Proc. Am. Soc. Hort. Sci.*, **33**: 389–392.

Stahl, E., 1894: Einige Versuche über Transpiration und Assimilation, *Bot. Ztg.*, **52**: 117–146.

Stålfelt, M. G., 1932: Der Einfluss des Windes auf die kutikuläre und stomatäre Transpiration, *Svensk. Bot. Tidsk.*, **26**: 45–69.

———, 1929: Die Abhängigkeit der Spaltöffnungsreaktionen von der Wasserbilanz, *Planta*, **8**: 287–340; *Biol. Abst.*, **6**: 681.

Stanley, J., 1932: A method of making permanent black atmometer bulbs, *Ecology*, **13**: 218–219.

Stark, O. K., and C. J. Whitfield, 1930: An improved evaporimeter, *Ecology*, **11**: 288–292.

Stefan, J., 1873: Versuche über die Verdunstung, *Sitzungsber. Akad. Wiss. Wien*, **68**: 385–428.

Steinberg, R. A., 1930: An automatic watering system with recorder for use in growing plants, *J. Agr. Res.*, **40**: 233–241.

Strugger, S., and F. Weber, 1926: Zur Physiologie der Stomata-Nebenzellen, *Ber. bot. Ges.*, **44**: 272–278.

Tavčar, A., 1926: Die Vererbung der Anzahl von Spaltöffnungen bei *Pisum sativum*, *Z. Pflanzenzücht.*, **11**: 241–259.

Thom, C. C., and H. F. Holtz, 1917: Factors influencing the water requirements of plants, *Wash. Sta. Bull.* 146.

Thomas, N., and A. Ferguson, 1917: On the reduction of transpiration observations, *Ann. Bot.*, **31**: 241–255.

Thut, H. F., 1932: The movement of water through some submerged plants, *Am. J. Bot.*, **19**: 693–709.

Tilford, P. E., and C. May, 1929: The effect of bordeaux mixture on the internal temperature of potato leaflets, *Phytopath.*, **19**: 943–949.

Transeau, E. N., 1911: Apparatus for the study of comparative transpiration, *Bot. Gaz.*, **52**: 54–60.

Trelease, S. F., and B. E. Livingston, 1916: The daily march of transpiring power as indicated by the porometer and by standardized hygrometric paper, *J. Ecol.*, **4**: 1–14.

———, 1922: Incipient drying and wilting as indicated by movements of cocoanut pinnae, *Am. J. Bot.*, **9**: 253–265.

———, 1922: Foliar transpiring power of the cocoanut, *Philippine J. Sci.*, **20**: 167–176.

Tumanov, J. J., 1927: Ungenügende Wasserversorgung und das Welken der Pflanzen als Mittel zur Erhöhung ihrer Dürreresistenz, *Planta*, **3**: 391–480.

———, 1929: Wilting and drought resistance, *Bull. App. Bot. and Plant Breeding*, **22**: 107–146.

Turrell, F. M., 1933: The internal exposed surface of foliage leaves, *Sci.*, **78**: 536–537.

———, 1934: A method for the measurement of the internal exposed surface of foliage leaves, *Univ. Iowa Studies Nat. Hist.*, **16**: 119–126.

———, 1936: The area of the internal exposed surface of dicotyledon leaves, *Am. J. Bot.*, **23**: 255–264.

Unger, F., 1861: Beiträge zur Anatomie und Physiologie der Pflanzen. XI, XII, Neue Untersuchungen über die Transpiration der Gewächse, *Sitzungsber. Akad. Wiss. Wien*, **44**: 181–227, 327–368.

———, 1862: Neue Untersuchungen über die Transpiration der Pflanzen, *Sitzungsber. Akad. Wiss. Wien*, **45**: 181–217.

Ursprung, A., 1903: Die physikalischen Eigenschaften der Laubblätter, *Bibl. Bot.*, **60**: 1–120.

Van de Roovaart, E., and G. D. Fuller, 1935: Stomatal frequency in cereals, *Ecology*, **16**: 278–279.

Van de Sande-Bakehuyzen, H. L., 1928: Studies upon wheat grown under constant conditions. I, II, *Plant Physiol.*, **3**: 1–31.

Vassiliev, I. M., 1932: Plant physiology in relation to drought control, (Russian), *Soc. Zern. Khoz.*, **1-2**: 10–12. Quoted by Aamodt and Johnston, 1936.

Veihmeyer, F. J., and A. H. Hendrickson, 1927: Soil-moisture conditions in relation to plant growth, *Plant Physiol.*, **2**: 71–82.

——— and ———, 1927: The relation of soil moisture to cultivation and plant growth, *Proc. Int. Cong. Soil Sci. Wash.*, **3**: 498–513.

Verschaffelt, E. and J., 1890: De transpiratie der planten in koolzuurorije lucht, *Bot. Jaarboek* (*uitgegeven door het kruidkundig genootschap Dodonaea te Gent*), **2**: 305.

Von Mohl, H., 1856: Welche Ursachen bewirken die Erweiterung und Verengung der Spaltöffnungen? *Bot. Ztg.*, **14**: 697–704, 713–720.

Vyvyan, M. C., and H. Evans, 1932: The leaf relations of fruit trees. I, A morphological analysis of the distribution of leaf surface on two nine-year-old apple trees (Laxton superb), *J. Pom. and Hort. Sci.*, **10**: 228–270.

Walter, H., 1925, 1926: Die Verdunstung von Wasser in bewegter Luft und ihre Abhängigkeit von der Grosse der Oberfläche, *Z. Bot.*, **18**: 1–47.

Watson, A. N., 1933: Preliminary study on the relation between thermal emissivity and plant temperatures, *Ohio J. Sci.*, **33**: 435–450.

———, 1934: Further studies on the relation between thermal emissivity and plant temperatures, *Am. J. Bot.*, **21**: 605–609.

Weaver, J. E., 1916: The effect of certain rusts upon the transpiration of their hosts, *Minn. Bot. Studies*, **4**: 397–406.

——— and A. Mogensen, 1919: Relative transpiration of coniferous and broad-leaved trees in autumn and winter, *Bot. Gaz.*, **68**: 393–424.

Weber, F., 1923: Enzymatische Regulation der Spaltöffnungsbewegung, *Natur. Wiss.*, **11**: 309–316.

———, 1931: Harnstoffpermeabilität ungleich alter Stomatazellen, *Protoplasma*, **14**: 75–82.

Weigand, K. M., 1910: The relation of hairy and cutinized coverings to transpiration, *Bot. Gaz.*, **49**: 430–444.

———, 1910: Some evaporation experiments in relation to excessive transpiration, *Sci.*, **31**: 434.

WEINBERGER, J. H., 1931: Growth of fruit and stomatal behavior of Elberta peaches as affected by weather conditions, *Proc. Am. Soc. Hort. Sci.*, **28**: 539–542.

WEISS, A., 1865–1866: Untersuchungen über die Zahlen- und Grössenverhältnisse der Spaltöffnungen, *Jahrb. wiss. Bot.*, **4**: 126–196.

WEISS, F., 1924: The effect of rust infection upon the water requirement of wheat, *J. Agr. Res.*, **27**: 107–118.

WHITFIELD, C. J., 1932: Ecological aspects of transpiration. I, Pike's Peak Region: Climatic aspects, *Bot. Gaz.*, **93**: 436–452.

———, 1932: Ecological aspects of transpiration. II, Pike's Peak and Santa Barbara regions: Endaphic and climatic aspects, *Bot. Gaz.*, **94**: 183–196.

WIGGINS, R. G., 1921: Variations in the osmotic concentrations of the guard cells during the opening and closing of stomata, *Am. J. Bot.*, **8**: 30–41.

WILLARD, C. J., 1931: The moisture content of forage at different times in the day, *J. Am. Soc. Agron.*, **23**: 853–859.

WILLIAMS, R. F., 1935: An analysis of the effect of phosphorus supply on the transpiration ratio in plants, *Australian J. Exp. Biol. and Med. Sci.*, **13**: 49–66.

WILSON, H. W., 1924: Studies on the transpiration of some Australian plants with notes on the structure of their leaves, *Proc. Roy. Soc. Victoria*, **36**: 175–237.

WILSON, J. D., 1930: A modified form of non-absorbing valve for porous-cup atmometers, *Sci.*, **71**: 101–102.

——— and H. A. RUNNELS, 1931: Bordeaux mixture as a factor increasing drought injury, *Phytopath.*, **21**: 729–738.

——— and ———, 1933: Some effects of Bordeaux mixture on transpiration, *Ohio Sta. Bimon. Bull.*, **18**: 147–151.

——— and ———, 1934: Influence of Bordeaux mixture and an oil emulsion on water requirement, *Ohio Sta. Bimon. Bull.*, **19**: 21–28.

——— and ———, 1934: The relative influence of calcium and magnesium in Bordeaux mixture on the transpiration rate. I, *Ohio Sta. Bimon. Bull.*, **19**: 158–163.

——— and ———, 1934: Influence of Bordeaux and oil sprays on the daily curve of transpiration, *Ohio Sta. Bimon. Bull.*, **19**: 179–186.

——— and ———, 1934: Transpirational response of various plants to Bordeaux mixture, *Ohio Sta. Bimon. Bull.*, **19**: 198–202.

——— and ———, 1935: Influence of certain sulfate-hydrated lime mixtures on transpiration, *Ohio Sta. Bimon. Bull.*, **175**: 143–146.

——— and ———, 1935: Influence of various sulfur-containing fungicides on transpiration, *Ohio Sta. Bimon. Bull.*, **175**: 146–148.

——— and F. A. WELTON, 1935: The use of an evaporation index in watering lawns, *Ohio Sta. Bimon. Bull.*, **20**: 112–119.

——— and H. A. RUNNELS, 1935: The relation of time to the effect of Bordeaux mixture on transpiration, *Ohio Sta. Bimon. Bull.*, **20**: 120–124.

WILSON, J. K., 1923: The nature and reaction of water from hydathodes, *N. Y. Sta. (Cornell) Mem.* 65.

WITHROW, R. W., 1935: A photoelectric device for the rapid measurement of leaf area, *J. Agr. Res.*, **50**: 637–643.

WOOD, J. G., 1934: The physiology of xerophytism in Australian plants. The stomatal frequencies, transpiration and osmotic pressures of sclerophyll and tomentose succulent-leaved plants, *J. Ecology*, **22**: 69–87.

WOODS, A. F., 1893: Some recent investigations of the evaporation of water from plants, *Bot. Gaz.*, **18**: 304–310.

WRENGER, M., 1935: Über den Einfluss des Windes auf die Transpiration der Pflanzen, *Z. Bot.*, **29**: 257–320.

YAPP, R. H., 1912: *Spirea Ulmaria*, L., and its bearing on the problem of xeromorphy in marsh plants, *Ann. Bot.*, **26**: 815–871.

——— and U. C. MASON, 1932: The distribution of water in the shoots of certain herbaceous plants, *Ann. Bot.*, **46**: 159–183.

YOCUM, L. E., 1935: The stomata and transpiration of oaks, *Plant Physiol.*, **10**: 795–801.

YUNCKER, T. G., 1916: A study of the relation of soil moisture to transpiration and photosynthesis in the corn plant, *Plant World*, **19**: 151–161.

ZALENSKI, V., 1921: Action of high temperature on the behavior of stomata, *J. Russ. Bot. Congress*, **1**: 62–63; *Bot. Abst.*, **14**: 287–288.

———, 1923: The intensity of transpiration in the upper and in the lower leaves, *Bull. Agr. Inst. Saratov*, **1**: 13–17; *Bot. Abst.*, **15**: 165–166.

ZHEMCHUZHNIKOV, E. A., 1922: To the question of stomatal regulation of plant's transpiration, *J. Agr. Res. Don and North Caucasus*, (**1922**): 29–39; *Exp. Sta. Rec.*, **52**: 517.

———, 1924: Transpiration by the stomata, (Russian) *Trans. Agr. Exp. Inst. Don and Northern Caucasus*, 1–34; *Bull.* 164; *Bot. Abst.*, **15**: 166.

ZUCKER, A., 1896: Beitrag zur direkten Beeinflussung der Pflanzen durch die Kupfervitriol-Kalkbrühe, *Inaug. Diss., Stuttgart.*

CHAPTER VIII

THE FORMATION OF CARBOHYDRATES BY THE GREEN PLANT

I. THE SYNTHETIC POWER OF THE GREEN PLANT

The materials that enter the green plant from its environment are for the most part inorganic compounds of the most simple character. Thus, for example, the plant obtains carbon dioxide from the air; water and the nitrates, sulphates, and phosphates of potassium, sodium, magnesium, iron, calcium, among others, from the soil. From these simple compounds or in their presence the plant is able to synthesize a large variety of substances of varying degrees of complexity, the most important and abundant of which are the carbohydrates, fats and oils, amino acids, proteins, glucosides, chlorophyll, and various other pigments, enzymes, vitamins, tannins, alkaloids, and numerous organic acids. Some of the substances named may be degeneration products under certain conditions, but, if they are, they are derived from more complex compounds which, in turn, have been synthesized by the plant.

The first evident synthetic product that is formed by the green plant in any appreciable quantity from the simple inorganic compounds that are obtained from the air and soil is one of the more simple forms of carbohydrate. The first carbohydrate thus formed is the nucleus or foundation of most of the other organic compounds in the plant. A general knowledge of the carbohydrates is thus necessary for a clear understanding not only of their formation but also of the formation of other plant substances.

II. THE CARBOHYDRATES OF THE GREEN PLANT

The carbohydrates are the most abundant compounds in the plant and make up the greater portion of its dry weight. The term "carbohydrates" signifies or implies that these compounds are composed of carbon and water, and the name originated from the fact that they contain carbon, hydrogen, and oxygen with the two latter elements generally present in the same proportion as in water. Some few carbohydrates, however, *e.g.*, rhamnose ($C_6H_{12}O_5$) do not show this simple relation between the hydrogen and oxygen atoms, while certain other compounds, *e.g.*, acetic acid ($C_2H_4O_2$) and lactic acid ($C_3H_6O_3$), which are not carbohydrates, contain hydrogen and oxygen atoms in the same proportion as in carbohydrates. From the present knowledge of the structure of carbohydrates they may be defined more specifically as aldehyde alcohols, ketone alcohols, which contain an "ose" or sugar group, or compounds that are converted by hydrolysis into one or both of these two types of alcohol.

The green plant alone possesses the power to synthesize carbohydrates from inorganic compounds. Although the production of carbohydrates has been accomplished artificially, this has been done in only an extremely limited way as far as the number of compounds and the quantity of material obtained are concerned. The members of the animal kingdom lack the power to synthesize carbohydrates from inorganic material and are dependent either directly or indirectly upon plants for their supply of these energy-producing foods.

In addition to being the basis for the formation of numerous plant products, the carbohydrates make up the greater part of the plant's structural framework and are also one of the chief foods of the protoplasm and the main source of energy in the plant. The carbohydrates are stored in relatively large quantities as reserve food in various parts of the plant, and when present in the cell sap in the soluble form they increase its osmotic value and thus play a part in maintaining the turgidity of the cells.

There are various classifications of the carbohydrates, but, in the main, their classification rests upon their solubility, taste, number of carbon atoms, the number of sugar groups that they contain, their reaction toward various reagents, their products of hydrolysis, and other physical and chemical reactions. The principal carbohydrates that occur in the green plant may be classified according to the following diagram:

- The principal carbohydrates that may occur in green plants
  - The sugars
    - Monosaccharoses
      - Pentoses $C_5H_{10}O_5$
        - Arabinose (R.)[1]
        - Xylose (R.)
        - Rhamnose (R.)
      - Hexoses $C_6H_{12}O_6$
        - Dextrose, glucose, or grape sugar (R.)
        - Levulose, fructose, or fruit sugar (R.)
        - Mannose (R.)
        - Galactose (R.)
    - Disaccharoses $C_{12}H_{22}O_{11}$
      - Sucrose, saccharose or cane sugar (N. R.)[2]
      - Maltose or malt sugar (R.)
      - Gentiobiose (R.)
      - Melibiose (R.)
    - Trisaccharoses $C_{18}H_{32}O_{16}$
      - Raffinose (N. R.)
      - Gentianose (N. R.)
      - Melicitose (N. R.)
  - The non-sugars
    - Polysaccharoses
      - Hexosans
        - Celluloses $(C_6H_{10}O_5)_n$ — See components of cell wall (Chap. I)
        - Amyloses or starches $(C_6H_{10}O_5)_n$
          - Starch (dextrosan)
          - Dextrins (dextrosans)
          - Inulin (levulosan)
        - Mannosans or Mannans $(C_6H_{10}O_5)_n$
        - Galactosans or Galactans $(C_6H_{10}O_5)_n$
      - Pentosans $(C_5H_8O_2)_n$
        - Araban
        - Xylan

[1] (R.) reducing sugars.
[2] (N. R.) nonreducing sugars.

From the preceding diagram it is observed that the carbohydrates may be divided into the sugars and the nonsugars and that, since each carbohydrate contains one or more sugar or saccharose groups, the term "saccharose" is used as a basis for the classification of the entire group of carbohydrates.

A. The Sugars

The sugars are the simplest carbohydrates. They are soluble in water and have a characteristic sweet taste that varies in degree according to the type of sugar.

1. Physical and Chemical Reactions.—The sugars give certain physical and chemical reactions that vary in number and degree according to the sugar under consideration. These reactions are in many cases specific and may be utilized as a means of identification of a given type of sugar. The most striking properties of sugars are their specific rotatory power, their reducing action, and their formation of osazones.

a. Specific Rotatory Power.—All the natural sugars are optically active in that they rotate the plane of polarized light when it passes through a solution of any one of these carbohydrates. The degree of rotation depends upon the kind of sugar, the kind of light, and the concentration and length of the column of solution through which the beam of light has to pass. The same amount of the same sugar dissolved in the same volume of water and placed in a tube of the same length, however, will always cause the same rotation of the plane of polarized light. The sugar is thus said to have a specific rotatory power. It is expressed as the number of degrees of angular deviation of the plane of polarized light caused by a column of the solution 100 mm. in length composed of 100 g. of the substance dissolved in 100 cc. of water at a temperature of 20°C. This deviation is expressed by the symbol $[\alpha]_D^{20}$, which indicates that the observation was made at 20°C. and with the D or sodium light. Sugars that rotate the beam of light to the right or in the direction in which the hands of the clock move are called "dextrorotatory," and the degree of rotation is preceded by the plus sign. Those sugars which rotate the beam of light to the left or in a direction counterclockwise are termed "levorotatory," and the nature of the rotation is indicated by the minus sign. Thus, for example, the specific rotation of d-glucose is $[\alpha]_D^{20} + 52.7°$ and d-fructose $[\alpha]_D^{20} - 92°$.

b. Reducing Action.—Any sugar that contains a potentially active "ose" or sugar group 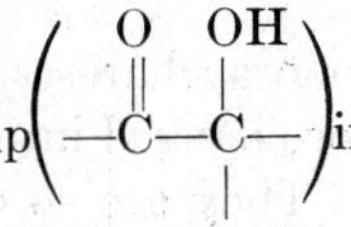 in its structure will behave as a reducing agent. Certain of the metallic compounds when in alkaline solutions are reduced when boiled with sugars of this type. Thus, for example, when treated in this manner, an ammoniacal silver solution gives a silver mirror, an alkaline solution of cupric hydroxide forms cuprous oxide, and alkaline mercury salts are reduced to metallic mercury. Any sugar that possesses this power of reduction is known as a "reducing sugar." The structure of certain of the di- and trisaccharose sugars is such, however, that the aldehyde and ketone groups are linked in such a manner that they lose their reducing power. Such sugars are termed "non-reducing sugars," and saccharose, sucrose, or cane sugar is the most common of this type.

The reagent most commonly used to determine the reducing qualities of sugars is Fehling's solution. This is prepared in the following manner: Part *A* is made by dissolving 69.278 g. of pulverized cupric sulphate in distilled water and making up to 1 l. Part *B* is made by dissolving 346 g. of sodium potassium tartrate (Rochelle salts) in distilled water and making up to 500 cc. and by dissolving 100 g. of sodium hydroxide in distilled water, again making up to 500 cc. Both parts of solution *B* are then mixed and kept until needed. Fehling's solution is then made by mixing equal parts of solutions *A* and *B* just before using.

The active substance in this solution, as far as reducing sugars are concerned, is cupric hydroxide, which is held in alkaline solution by the presence of the Rochelle salts. When this solution is heated with a solution of reducing sugar, oxygen is withdrawn from the cupric hydroxide, and it is converted into cuprous oxide. The cuprous oxide thus formed is insoluble in the solution and is separated from it in the form of a reddish-brown precipitate. Advantage is taken of this reaction of sugars with the alkaline solution of cupric hydroxide for their detection and quantitative determination, since the amount of the cuprous oxide produced bears a definite relation to the amount and type of sugar used.

c. Formation of Hydrazones and Osazones.—When sugars with aldehyde or ketone groups in their molecules are treated with phenylhydrazine ($C_6H_5.NH.NH_2$), they yield condensation products called "hydrazones," just as do other aldehydes and ketones. The phenylhydrazones are characteristic crystalline solids and are usually very easily soluble in water, so that they do not serve for the identification of individual sugars. When, however, these sugars are heated in aqueous solution with an excess of phenylhydrazine over that required to form the phenylhydrazone, a reaction takes place in which 1 molecule of the carbohydrate reacts with 3 molecules of the phenylhydrazine to form a substance callled a "phenylosazone." The osazones of sugars are compounds that dissolve with difficulty in cold water, crystallize well, and have a characteristic form and a definite melting point. The osazones are thus very characteristic for certain sugars and afford a means for their identification.

2. Kinds or Types.—The most important groups of sugars that occur in green plants are the monosaccharoses, the disaccharoses, and the trisaccharoses.

a. The Monosaccharoses.—The monosaccharoses are the simplest of the sugars and cannot be hydrolyzed or changed into simpler compounds still retaining the property of sugar. They are so named because they contain only one sugar group. There are two classes of the monosaccharoses in plants—the pentoses and the hexoses.

1. *The Pentoses.*—The pentoses are sugars that contain five carbon atoms and have the formula $C_5H_{10}O_5$. These sugars occur in the free state in plants in only very small amounts. Davis and Sawyer (1914) determined the free pentoses in the alcoholic extract of the leaves of the mangold, turnip, *Tropaeolum majus, Helianthus annuus,* carrot, and potato and found that they amounted to from 0.3 to 1 per cent of the dry weight. VerHulst, Peterson, and Fred (1923) found in the leaves of corn that the free pentoses amounted to 0.69, 0.52, 0.39, and 0.47 per cent of the dry weight, respectively, at the tasseling, silking, milk, and dent stages. Apparently, one of the components of the nucleic acids of plants is pentose. Paracaoutchouc is considered to be composed of a large number of groups of the formula C_5H_8 that have been polymerized to form an exceedingly large molecule. These C_5H_8 groups are considered reduction products of the pentoses (Spoehr, 1917, 1919). The pentoses occur in the plant for the most part in the combined state in the form of pentosans, natural gums, and hemicelluloses. As stated in Chap. I the pentoses may be obtained in various amounts by the hydrolysis of the hemicelluloses of the cell walls of seeds, pods, straw, and the xylem elements of the plant. Cherry and peach gum and certain mucilages of plants also yield pentoses on hydrolysis. The two most common pentoses that are obtained from the hydrolysis of the materials above mentioned are arabinose and xylose. Both of these are reducing sugars. Rhamnose ($CH_3C_5H_9O_5$), is a pentose in which 1 atom of hydrogen has been replaced by a methyl group. This compound is of interest in that it has the characteristics of the sugars yet does not have the proportion of hydrogen and oxygen that ordinarily occurs in the carbohydrates.

2. *The Hexoses.*—The hexoses are the most abundant of the monosaccharoses in the plant. They contain six carbon atoms and have the formula $C_6H_{12}O_6$. The most important hexoses that are found in the green plant in the free state or in the form of condensation products are glucose, fructose, mannose, and galactose. All these hexoses are reducing sugars, and all are optically active.

(*a*) *Glucose, Dextrose, or Grape Sugar.*—This sugar is found in considerable abundance in fruits—especially in the grape, on which account it received the name "grape sugar." It also occurs in the seeds, roots, and leaves of plants and is thus found in varying quantities in practically every living cell of the plant. This sugar occurs in the combined state in sucrose, maltose, starch, and various other di-, tri-, and polysaccharoses and may be obtained by hydrolysis from these products. *d*-Glucose is dextrorotatory, its specific rotatory power being $[\alpha]_D^{20} + 52.7°$, and for this property it has been termed "dextrose." It is approximately three-fifths as sweet as cane sugar. Glucose is an aldehyde alcohol, and the structure of *d*-glucose is represented as shown on page 528.

(*b*) *Fructose, Levulose, or Fruit Sugar.*—Fructose is a ketone alcohol and has the structure shown below. Owing to the ketose group in its molecule, it is a strong reducing agent and reduces Fehling's solution more rapidly than does glucose. Fructose is strongly levorotatory and for that reason has also been termed "levulose." Its specific rotatory power is $[\alpha]_D^{20} - 92.5°$. It is slightly sweeter than cane sugar and is the sweetest of the ordinary sugars (Archbold and Barter, 1935). Fructose is widely distributed in the plant and occurs in considerable quantities in fruits, generally in excess of glucose and sucrose. It occurs in the combined form in sucrose, in the trisaccharoses, raffinose, gentianose, and melicitose and in the polysaccharose inulin. It may be obtained from these substances by hydrolysis with acids or enzymes.

d-glucose

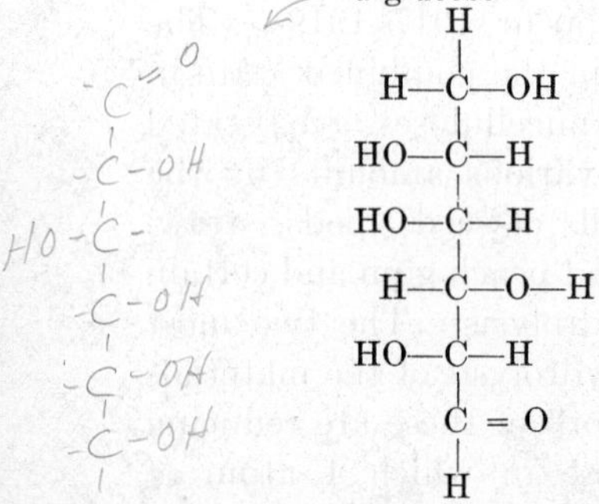

d-fructose

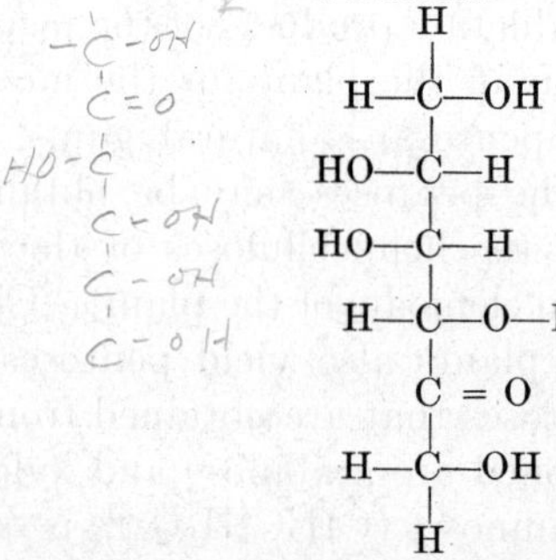

(*c*) *Mannose.*—Mannose is a sugar closely related to glucose and forms the same osazones as do that sugar and fructose. Mannose does not occur in the free state in plants but is readily obtained by hydrolysis from numerous compounds found in them. Thus the hemicelluloses of the cell wall of the date stone, ivory nuts, coffee, pea, and bean yield mannose among other sugars upon hydrolysis. The compounds in the plants which thus yield mannose are termed "mannans" or "mannosans." Mannose may also be obtained by the oxidation of mannitol, a hexatomic alcohol sometimes known as "mannite," which occurs in numerous plants. Mannose is dextrorotatory $[\alpha]_D^{20} + 14.3°$ and is a reducing sugar.

(*d*) *Galactose.*—Galactose very rarely occurs in the green plant in the free form and when so found is in only very small amounts. It exists in the plant almost entirely in the form of condensation products, which are sometimes collectively termed "galactans." This term includes the gums, mucilages, and hemicelluloses of the plant which yield galactose on hydrolysis. Galactose may thus be obtained from agar-agar, the pectins of carrot and pears, from gums occurring in peaches and plums, and from the cell walls of the seed coats of the seed of corn, lupine, bean, pumpkin, and many others. Galactose is also a constituent of the trisaccharose raffinose and may be obtained together with glucose and fructose from this source by hydrolysis. *d*-Galactose obtained from these sources just mentioned very closely resembles glucose in many of its properties, but it has one characteristic difference in that it forms mucic instead of saccharic acid when oxidized by concentrated nitric acid. Galactose is only very slightly sweet, reduces Fehling's solution more slowly than glucose, and is strongly dextrorotatory, $[\alpha]_D^{20} + 83.8°$.

Although galactose, mannose, and the pentose sugars, arabinose and xylose, are rarely found in the plant in the free form, they may nevertheless play an important part, in some cases at least, in plant nutrition. They are formed very slowly by hydrolysis and may thus be utilized by the cells almost as rapidly as they are produced, so no considerable amount ever accumulates.

b. The Disaccharoses.—A disaccharose sugar consists of two monosaccharoses or hexoses linked together with the elimination of 1 molecule of water. The molecular formula for disaccharoses is $C_{12}H_{22}O_{11}$. The manner in which the two monosaccharoses are joined together to form a sugar of this kind varies with the type of disaccharose. The linkage may be through the alcoholic groups of each monosaccharose, through the alcoholic group of one and the ketone or aldehyde group of another, or through the aldehyde or ketone groups of each of the two sugars. If the linkage is through the aldehyde or ketone group of both monosaccharose units, the reducing power of these groups is destroyed and the disaccharose thus resulting is a nonreducing sugar. The most important disaccharoses occurring in green plants are sucrose, maltose, gentiobiose, and melibiose.

1. *Sucrose, Saccharose, or Cane Sugar.*—Sucrose is the most abundant and the most widely distributed sugar in green plants, being found in varying amounts in practically all living plant cells. In the stalk of the sugar cane (*Saccharum officinarum*) it is present in quantities equal to from 12 to 20 per cent of the green weight and may compose as much as 90 per cent of the soluble material in the expressed juice of this plant. This sugar composes from 15 to 20 per cent or more of the fresh weight of sugar beets and is present in quantities ranging from 2 to 4 per cent in the sap of the sugar maple and in the stalks of the saccharine sorghums (*Sorghum saccharatum*), amounting to as much as 12 per cent of the green weight or 40 per cent on a dry basis. Sucrose is the sugar present in the greatest amount in most leaves, is the principal sugar in the nectar of flowers, and is found in most sweet fruits in amounts varying from 1 to 12 per cent.

Widdowson and McCance (1935) determined the glucose, fructose, sucrose, and starch in 41 different fruits. The results for some of the more common fruits are shown in the following table:

Fruit	Carbohydrates			
	Glucose, per cent	Fructose, per cent	Sucrose, per cent	Starch, per cent
Apples	1.7 to 2.2	5.0 to 7.2	2.4 to 3.6	0.02 to 0.35
Apricots	1.9	0.37	4.3	0
Bananas	5.8	3.8	6.6	3.0
Cherries	4.7 to 5.5	6.1 to 7.2	0	0
Cranberries	2.7	0.74	0.14	0
Currants	2.3 to 3.0	1.9 to 3.6	0.62	0
Dates	32.0	23.7	8.2	0
Grapes	8.2	7.2 to 8.0	0	0
Grape fruit	1.9	1.2	2.1	0
Lemon	1.4	1.4	0.4	0
Cantaloupe	1.2	0.8	3.3	0
Oranges	2.5	1.8	4.2	0
Peaches	1.4	0.93	6.6	0
Pears	2.2 to 3.5	6.0 to 7.0	1.0	0
Pineapple	2.3	1.4	7.9	0
Raspberry	2.3	2.4	1.0	0
Strawberry	2.6	2.3	1.3	0
Tomatoes	1.6	1.2	0	0.02

The structure of sucrose has not been definitely established, but, since it yields equal quantities of glucose and fructose by the addition of one molecule of water, it is considered an anhydride of these two sugars. Sucrose is dextrorotatory, $[\alpha]_D^{20} + 66.5°$, but when it is hydrolyzed with acid or the enzyme invertase to equal amounts of glucose and fructose the resulting mixture is levorotatory. For this reason the mixture resulting from such a hydrolysis has been termed "invert sugar." Sucrose is a nonreducing sugar, does not form a phenylhydrazone with phenylhydrazine, does not undergo direct alcoholic fermentation, and is, next to fructose, the sweetest of the common sugars.

2. *Maltose or Malt Sugar.*—Maltose is a reducing sugar, forms a characteristic osazone, is dextrorotatory, $[\alpha]_D^{20} + 137°$, and is broken down by hydrolysis into 2 molecules of glucose. It does not undergo alcoholic fermentation by yeast directly but must be broken down into glucose before such fermentation can occur. Maltose is widely distributed in plants, but generally it can be detected in only very small quantities. It is derived from starch by hydrolysis and should thus occur in plants wherever this process is taking place. It is formed in the greatest amount from this source during the germination of starchy seeds. In most cases, maltose is apparently hydrolyzed into glucose almost as rapidly as it is formed, so it is never detected in any considerable amount in the plant.

3. *Other Disaccharoses.*—Two of the less common disaccharoses of the plants that have been isolated are gentiobiose and melibiose. Gentiobiose is obtained by the partial hydrolysis of gentianose, a trisaccharose found in the roots of yellow gentian (*Gentiana lutea*). Gentiobiose is a reducing sugar and upon hydrolysis yields 2 molecules of glucose. Melibiose is obtained by the partial hydrolysis of the trisaccharose raffinose. It is a reducing sugar and when hydrolyzed breaks up into glucose and galactose.

c. The Trisaccharoses.—The trisaccharoses have the general formula $C_{18}H_{32}O_{16}$. When partially hydrolyzed they yield a monosaccharose and a disaccharose, but when completely hydrolyzed they yield three molecules of monosaccharoses. This indicates that the trisaccharoses are formed by the linking together of three monosaccharoses by the elimination of 2 molecules of water. There are five trisaccharoses known to occur in green plants, of which raffinose, melicitose, and gentianose are the most important. Raffinose occurs in cotton seeds, in the barley grain, and in the sugar beet. It is a nonreducing sugar and is strongly dextrorotatory, having a specific rotation of $[\alpha]_D^{20} + 104.5°$. According to Thatcher (1921), raffinose may break down by hydrolysis into three different compounds according to the catalyst used: (1) sucrose and galactose, (2) fructose and melibiose, or (3) fructose, glucose, and galactose.

Melicitose is a very sweet sugar that is known to occur in the sap of the European larch (*Larix europaea*), Douglas fir, and other conifers and frequently collects on the twigs in small white flakes that are called "manna." When this sugar is partially hydrolyzed it yields 1 molecule of glucose and 1 molecule of a disaccharose sugar called "turanose." When completely hydrolyzed it yields 2 molecules of glucose and 1 of fructose.

Gentianose occurs in the roots of the yellow gentian (*Gentiana lutea*). When partially hydrolyzed it yields fructose and gentiobiose and when completely hydrolyzed yields 1 molecule of fructose and 2 molecules of glucose.

B. The Nonsugars or Polysaccharoses

The polysaccharoses have the general formula $(C_6H_{10}O_5)_n$ or $(C_5H_8O_4)_n$ depending on whether they yield hexoses or pentoses on hydrolysis. They are on that account termed, respectively, "hexosans" or "pentosans." The hexosans may be further classified according to the type of hexose that they yield on hydrolysis into the dextrosans (glucosans), the levulosans (fructosans), the mannans, and the galactans. The polysaccharoses are substances with high molecular weights and are apparently composed of many molecules of a monosaccharose linked together through the elimination of water. The value of n has not been accurately determined for any of the polysaccharoses on account of their slight solubility in the solvents that are used for molecular-weight determinations. The value of n is estimated to be as low as 30 in some polysaccharoses and as high as 200 or more in others. The chief polysaccharoses that occur in the green plant are the celluloses, the amyloses or starches, the mannans, the galactans, and the two pentosans, araban and xylan.

1. The Celluloses.—Cellulose is the most abundant carbohydrate in the plant kingdom. It is one of the main components of the rigid cell wall that surrounds the protoplasm of practically every plant cell and, together with the various compounds that are united or closely associated with it, makes up in many cases the greater portion of the weight of the plant. The nature and properties of the celluloses are discussed fully under the subject of the constituents of the cell wall in Chap. I, so they will not be mentioned here.

2. The Amyloses or Starches.—The most important amyloses that occur in green plants are starch or amylum, dextrins, and inulin.

a. Starch or Amylum.—Next to cellulose, starch is the most abundant and widely distributed carbohydrate in the plant kingdom and is the most universal form of storage carbohydrates in plants. It may occur in the young cells near the growing points, in the medullary ray cells of the stems and roots, in the xylem and phloem parenchyma, in the mesophyll cells of the leaf, in the guard cells, in the cells of the cortex and pith, in the bundle sheaths, in seeds, fruits, tubers, and other places. There are few green plants in which starch does not appear in some of the tissues at one time or another in the life of the plant. Starch occurs in the greatest amount in certain seeds, roots, and tubers, which thus form one of the main sources of carbohydrates for the nourishment of the animal kingdom.

Starch is present in the seeds of about one-half of the families and genera of the monocotyledons and in the seeds of about one-sixth of the dicotyledons. When starch is present as a reserve product in seeds, other nonnitrogenous foods as fats and oils are usually absent or are present in only very small amounts. When fats and oils are the chief reserve foods, starch is generally absent. With but few exceptions, starch occurs in the plant in the insoluble form. In the soluble form it has been observed in the cell sap of the epidermal cells of *Saponaria officinalis* and *Arum italicum* and some other plants (Reichert, 1913).

1. *The Starch Grain.*—With the exception of the few cases of soluble starch that were mentioned above, starch always occurs in the green plant in the form of definite granules known as starch grains. These vary in size from macroscopic to almost ultramicroscopic dimensions (Fig. 23). Starch grains are exclusively a product of green plants and are always found within the plant cells. The number of grains that may occur in a single cell varies greatly. There may be as many as 100 or more of a small size and only a few or only a single grain of considerable size. As a general rule, the tissues above ground, with the exception of seeds, contain starch grains which are smaller and less fully developed than in the underground structures. The form or shape of starch grains varies in plants of different species, in the same plant in different parts, and even in the same part of a given plant. Lindet and Nottin (1923) observed in the case of the potato tuber that one-half the starch grains

measured from less than 1 to 10μ in diameter but that the starch grains of that size comprised only 1.5 per cent of the total weight of starch. Only 0.6 per cent of the grains measured from 60 to 80μ but made up about 20 per cent of the total weight. The maximum size observed was 100μ in diameter. The form of the grain may be affected by the age of the plant and by changes in the nutritive conditions, but the form is usually the same within certain limits for a given species. The general shape of the

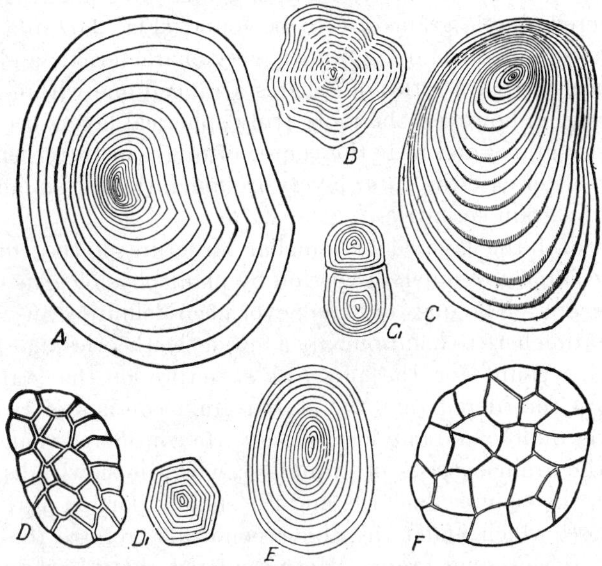

FIG. 23.—Starch grains. *A*, from lima bean. *B*, from corn grain. *C* from potato tuber. C_1, compound grain from potato tuber. *D*, compound grain from rice. D_1, one of the simple grains which compose the compound starch grain of rice. *E*, from wheat grain. *F*, compound grain from oats.

starch grains is so constant for a given species that the adulteration of flours and meals with starch flours from other sources can be detected and the starch used as an adulterant identified by a microscopic examination of the starch grains. The form of starch grains may be spherical, ellipsoidal, ovoid, lentiform, or, where closely packed or crowded together, polyhedral.

The structure of the starch grains is very characteristic in that each grain consists of numerous layers or lamellae which are laid down either concentrically or eccentrically around one or more morphological centers or points of origin. The point of origin in the starch grain has frequently been designated the hilum, nucleus, or locus of the grain. These first two terms, however, are now universally used in botanical literature in another sense, so that the "locus," "point," or "center of origin" are perhaps the best terms to use to designate the center around which the starch of the grain is deposited (Fig. 23). When the lamellae or layers of starch are deposited around a single center of origin the grain is termed a "simple" starch grain. When the deposition of starch is around two or more points of origin, the grain is said to be "compound." The number of centers around which starch may be deposited in a compound grain may vary from two to hundreds. When simple grains predominate there are usually a few scattered compound grains, *e.g.*, in the potato and wheat. In rice and oats the compound grains predominate, and only a few scattered simple grains are to be found (Fig. 23*D* and *F*). The layered structure of the starch grains is very characteristic and gives the general appearance of a series of spheres arranged one within the other. Strictly speaking, however, the more truly spherical layers occur only in the concentric grains and near the center of origin in the eccentric ones, since in the latter case the outer layers are much thicker on one side and thus make the grain eccentric.

The cause of the layered or lamellar structure of the starch grain has been the subject of much discussion by plant physiologists for almost a hundred years. The question has never been definitely answered, and it is the intention here to mention only a few of the theories that have been proposed to account for the physical structure of the starch grain. Nägeli (1858) considered that the starch grain consists of two distinct substances, granulose and amylocellulose. He considered that the larger portion of the grain consists of granulose, which is slowly dissolved by the action of the enzyme diastase, and the smaller portion of the grain of amylocellulose, which under the above conditions retains the form and structure of the original grain. He considered that the proportion of these two substances differs in the different layers of the grain, the dense layers containing relatively more amylocellulose than the soft or less dense layers, and the external parts of the grain more than the internal parts. The granulose part of the grain takes the blue color with iodine, while amylocellulose stains a dull or brownish red (Meyer, 1895). MacDougal (1895) considered that in the ordinary starch grain, amylose, consisting of two closely related forms α-amylose and β-amylose, and amylodextrin are present. He suggested that the dense layers of the grain are relatively rich in amylose and that the amylodextrins occur in larger quantities in the less dense porous layers. The amylose stains blue with iodine and the amylodextrin red. Kraemer (1902, 1905) suggested that the starch grain consists of colloidal and crystalloidal substances, these being arranged, for the most part, in distinct and separate lamellae. At the point of origin and in alternate lamellae the colloidal substance predominates but is associated with the crystalloid matter, while in the other layers the crystalloid substance occurs in the greater proportion. He held that differences in the starch grain show that starch instead of being a uniform substance is composed of several substances in varying proportions but more or less definitely arranged. Maquenne and Roux (1903, 1905) and Fernbach and Wolff (1904) considered that the starch granule consists of amylose and amylopectin. Amylose makes up the greater portion of the starch grain and is the substance that reacts with iodine. Amylopectin is not colored with iodine and is a mucilaginous substance that is thought to produce the gelatinization of starch in the form of starch paste.

Samec and Haerdtl (1920) in a study of the starches of potato, wheat, corn, rice, and horse-chestnut, among others, found that all contained an electrodialytically precipitable viscous portion that conducts electricity and a portion that is the opposite in all these respects. They considered the first-named portion to be the amylopectin described by Maquenne and the latter amylose. They found that the relative amount of these two substances differed in different starches. By special methods Walton (1929) separated from the starches of wheat and tapioca two types which were termed α- and β-amylose. The proportion of these two types was 24 to 76 for wheat and 17 to 83 for tapioca. Thurber (1933) found that the ratio of the two types of starches was approximately the same in the sweet potato and white potato. The viscosities of the two starches differed greatly, but Woodruff and Webber (1933) found no significant differences in the gels of the starches of hard and soft winter wheats.

Ling and Dinshaw (1923), Clayson and Schryver (1923), and Schryver and Thomas (1923) found in addition to amylose and amylopectin a substance resembling hemicellulose in many of the starches. This substance in wheat starch resists the action of taka-diastase and is soluble in hot water but separates from it in an amorphous form upon cooling. It is soluble in 0.1 N sodium hydroxide, from which solution it can be precipitated by acids.

Attempts have also been made to explain the lamellar structure of the starch grain upon the assumption that it is spherocrystalline in its composition. A spherocrystal is ordinarily defined as a more or less globular body composed of radially arranged crystals. Nägeli (1858) first suggested the crystalline structure of the starch grain and conceived it to be composed of minute crystalline structural units which he later termed "micellae." Schimper (1881) from studies with polarized light concluded that the grain is spherocrystalline in nature and consists of fibrous crystals arranged at right angles to the concentric layers. Meyer (1895) considered that the starch grain is composed of very thin, elongated needlelike or threadlike crystal units that are termed "trichites." These trichites are arranged in a radial manner at right angles to the main mass. The stratification of the grain he considered due to the varying thickness and number in successive layers of the more or less richly branched trichites. Sponsler (1922, 1923) studied the structure of the starch grain by means of X rays. He found that there is a regular and fairly uniform arrangement of atoms in the grain but that this regularity is destroyed when the grain is crushed, which shows that the regularity is not that of a crystalline structure. The evidence indicates that the starch grain is built up of concentric layers of units and that the structure is neither amorphous nor crystalline in the ordinary sense of the terms. Parts of the grain act as crystals in that for certain distances the layers or units are in planes, but taken as a whole the layers are curved. Van Sande-Bakhuyzen (1926) reported that the starch grains from seeds of plants which had been grown under constant conditions did not show lamellation. After heating these grains until they swelled, it was noted that they were formed of homogeneous radial needles. In the starch grains of the control seeds the needles were observed after heating, but they were in zones of different refractivity and solubility, which apparently cause the lamellation of the starch grains.

Schimper (1881) concluded from his observations on starch formation that starch grains never arise except in plastids. His observations were later confirmed by Meyer (1895) and this view of starch formation is now the commonly accepted one. The chloroplasts and leucoplasts are the plastids concerned in the formation of starch grains. The starch found in the chloroplasts may be termed "temporary," since it is stored there for only a very brief period until it can be translocated to other parts of the plant. Usually numerous grains are formed in a single chloroplast and sometimes project from the surface, giving the plastid an irregular appearance. These grains always remain small and are usually without a definite structure. According to Zirkle (1926), they are located within the central vacuole of the chloroplast.

According to Spoehr and Milner (1935) the yields of starch in percentage of dry weight of the leaf were 37.7 for *Nicotiana tabacum*, 13.4 for *Phaseolus coccineus*, 2.3 for *Linum usitatissimum*, 2.3 for the old leaves of *Hedera helix*, and 1.5 for the young leaves of this plant. Novello (1933) found that 52 of the 55 forms of palms examined by him elaborated starch in the chloroplasts. Three forms contained only lipides in their chloroplasts, and 23 forms contained both starch and lipides.

The most conspicuous and definite starch grains are those formed by the leucoplasts. These are the grains that make up the great bulk of the starch in the plant kingdom and compose thus the reserve starch in all parts of the plant. This type of starch is sometimes called "permanent starch," since it may remain intact for a considerable period of time. Apparently only a single starch grain originates in each leucoplast. The starch grain increases in size by the addition, accretion, or apposition of newly formed layers. It is not definitely known whether the grain in its development ruptures the leucoplast or whether the leucoplast stretches as the grain increases in size and thus remains intact and surrounds the exterior with a delicate film. Schimper (1881) showed that if the grain has its origin at the center of a spherical leucoplast, the strata of the grain will be laid down concentrically, because the conditions of growth are the same at every point of the surface. If the point of origin lies near the surface of the leucoplast, the stratification will be eccentric in character, since the deposition of starch will be more favored on one side. Lindet and Nottin (1923) noted in a study of the formation of potato starch that the sugar content of the cell sap during starch formation amounted to from 0.5 to 1.5 per cent. They considered that the leucoplasts contain during this process a mixture of sucrose, invert sugar, and maltose but could not detect any soluble starch.

2. *General Properties.*—The starch substance is not a unit body but exists in a number of stereoisomeric forms (Reichert, 1913, 1919). It not only differs in the grains of different plants and in the grains of the same plant, but differs in the starch substance of the outer layer and the enclosed part of the grain, in different lamellae, and in different parts of the same lamella. The lamellae vary in density and solubility, increasing in density and decreasing in solubility from within outward.

The outermost layer is the least soluble in both hot and cold water and is less digestible by weak acids and enzymes. While the starch grains are composed for the most part of amylose, they contain various polymers ranging in complexity from amylodextrin to amylocellulose (Mellanby, 1919). The grains of certain plants apparently do not contain ordinary starch but a modified form or mixture of ordinary starch and erythrodextrin, or erythrodextrin alone, since they stain a red violet or red with iodine. This reaction has been reported for the starch grains of the seeds of *Oryza sativa*, *Panicum miliaceum glutinosum*, and *Sorghum vulgare glutinosum*. The starch grains found in *Iris germanica*, *Chelidonium majus*, *Acer pseudoplatinus*, and *Gentiana lutea* among others also show this reaction (Meyer, 1886). From the preceding statements it would seem that it would be practically impossible to obtain a starch grain that could be considered typical in regard to its intimate structure (Kraemer, 1923).

Starch grains consist of from 80 to 85 per cent of starch substance, 15 to 20 per cent of water, and variable amounts of organic and inorganic substances including fats, protein, tannin, phosphates, and other matter. Samec and Haerdtl (1920) noted that all the starches examined by them contained phosphorus, the variation in amount being paralleled by the electric conductivity of the electrodialyzed solution. The major portion of the fatty material present in starch cannot be removed by solvents before hydrolysis, according to Taylor and Nelson (1920). When corn starch is hydrolyzed, fatty acids are liberated, palmitic acid being the principal one. The fat thus contained in starch is liberated when hydrolysis has reached the erythrodextrin stage. The table below adapted from Reichert (1913) shows the composition of some of the more common types of starch.

Composition of Starch

Kind of starch	Starch substance	Water	Ash	Fat
	Per cent			
Wheat	83.3	14	0.4	0.2
Corn	84.1	14	0.4	
Arrowroot	84.1	15.7	0.2	0.1
Tapioca	84.8	14.4	0.25	
Potato	79.6	19.2	0.3	0.04

It was concluded by Edwards and Ripperton (1933) that the electrolytes occurring naturally in starch are largely determined by the protoplasm of the cell rather than by a direct chemical equilibrium with the electrolytes of the cell sap. McNair (1932) identified the carbohydrates in 59 of the 295 families of gymnosperms and angiosperms. There is an indication that carbohydrates of a low molecular weight, with a low heat of combustion and a low potential energy, are more likely to be found in tropical families of plants than in those of temperate regions. Starch from temperate climates gelatinizes at a higher temperature than that from the tropics. Starch from plants of a temperate climate is less saturated toward iodine than that from tropical families.

Starch is insoluble in cold water but when it is boiled in water a colloidal suspension termed "starch paste" is obtained. The insolubility of uninjured raw starch grains in cold water seems to be due to some form of protective covering that is deposited by the leucoplast or by the cell sap. The idea, however, that the outer covering of the grain is a thin layer of cellulose or other nonstarch substance is no longer held. Reichert (1913), however, considered that some form of protective covering exists, since the rupture of the grain by grinding with sand or glass, the erosion by diastase, or the solution of the outer film by potassium hydroxide or other agents renders the grains soluble in cold water.

By boiling starch with dilute acids or by treating with acetone, a clear solution termed "soluble starch" is obtained. The formation of soluble starch with acetone was described by Fernbach (1912). His method as described by Haas and Hill (1913) consists in pouring a 1 to 2 per cent suspension of starch in a large excess of pure acetone and shaking vigorously. A flocculent precipitate is obtained which when filtered and ground up in a mortar with more acetone yields a light white powder that is completely soluble in cold water and gives the starch reactions. Raw starch, starch paste, and soluble starch give a blue color when treated with a solution of iodine in potassium iodide or a solution of iodine in alcohol. This color reaction is the most characteristic reaction of starch.

When starch is boiled for some time with dilute mineral acids it is converted into dextrins, maltose, and finally glucose as the end product. When starch paste is treated with a malt extract containing diastase, the starch is converted only to maltose as the end product (O'Sullivan, 1872). If, however, uninjured starch grains are subjected *in vitro* to the action of diastase under sterile conditions, absolutely no digestion occurs. If the grains, however, are broken or cracked or eroded, rapid erosion takes place similar to that observed when digestion takes place in the plant under natural conditions.

b. Dextrins.—The dextrins are transition products between starch and maltose. The carbohydrate called "soluble starch," which is found in the sap of the epidermal cells of *Arum italicum* and *Saponaria officinalis*, is perhaps a mixture of dextrins, while the starch grains that turn red with iodine, as has previously been described, are probably composed, in part at least, of dextrins. Weatherwax (1922) considered that the only carbohydrate in the endosperm of waxy maize is erythrodextrin. According to Parker (1935) the endosperm of sweet corn has a relatively high percentage of soluble polysaccharoses consisting of both dextrins and soluble starch. Although dextrins are probably being formed continuously in the hydrolysis of starch or in the formation of starch from sugar, they have not been observed to accumulate to any extent in the plant except in the few cases that have been mentioned.

The dextrins are readily soluble in water, from which they may be precipitated by alcohol. They are strongly dextrorotatory and are converted into glucose by hydrolysis with mineral acids. A large number of dextrins have been described, but only three fairly definite forms are recognized. These are (1) amylodextrin, which gives a blue color with iodine and is considered the chief constituent of soluble starch; (2) erythrodextrin, which gives a red color with iodine and has a neutral taste; and (3) achroodextrin, which has a sweetish taste and gives no color with iodine.

c. Inulin.—Inulin occurs as a reserve carbohydrate in the cell sap of numerous plants, especially the *Compositae*. It is found in the largest amounts in roots and tubers and occurs, if at all, in only very small amounts in the aerial parts. In the underground parts it may make up as much as 15 per cent of the dry weight (Shohl, 1923). Inulin is found in comparatively large amounts in the underground parts of *Inula helenium, Helianthus tuberosus, Dahlia variabilis, Lappa minor, Cichorium intybus, Solidago sp.*, and numerous others. Holm (1931) noted in the roots of numerous *Compositae* that inulin is never found in the epidermis, exodermis, phelloderm, sclerenchyma, or leptome. It is only occasionally present in the endodermis and pith. It is most abundant in the meristem near the primary hadrome and in the cortical parenchyma. In some of the monocotyledons it is found associated with starch, but as a general rule when inulin is present starch is absent. Some plants contain reserve carbohydrates of the same general type as inulin. Some of these are irisin in *Iris pseudacorus*, graminin in *Trisetum alpestre*, phleim in *Phleum pratense*, and triticin in *Triticum repens* (Dean, 1904).

Inulin is a white powder composed of finely divided granules. It is slightly soluble in cold water and very soluble in hot water, from which it may be precipitated by alcohol. An aqueous solution of inulin does not form a jelly or paste, is not colored by iodine, and is not fermentable. Inulin is slowly converted into fructose when boiled with water and converted very rapidly in the presence of an acid. It is also converted to fructose by the enzyme inulase. The methods for the extraction and preparation of inulin are described fully by Dean (1904), Willaman (1922), and Harding (1923).

3. Mannosans and Galactans.—The mannosans are present for the most part in the cell walls of plants and are generally classified as one of the groups of hemicelluloses. They are found especially in the cell walls of leaves, of wood in stems and roots, and in the cell walls of seeds and seed coats. The mannosans are strongly dextrorotatory, soluble in water with difficulty, and yield mannose upon hydrolysis. The galactans are polysaccharoses that yield galactose upon hydrolysis. They are slightly soluble in water and form gummy solutions. Like the mannosans, they are for the most part associated with the cell wall of various plant parts and are generally considered under the heading of "hemicelluloses."

4. Pentosans.—The pentosans have the general formula $(C_5H_8O_4)_n$ and occur, for the most part, in the cell walls of various plant parts. According to Thatcher (1921), they compose 22 to 25 per cent of the dry weight of wheat bran, 8 to 10 per cent of clover hay, 16 to 20 per cent of oats straw, 26 to 27 per cent of wheat straw, and as much as 60 to 92 per cent of various wood gums. The pentosan content of the leaves of corn amounted to 19.9, 17.6, 18.1, and 19.4 per cent, respectively, at the tasseling, silk forming, milk, and dent stages in analyses made by Ver Hulst, Peterson, and Fred (1923). The two chief pentosans are araban and xylan. Araban occurs in gum arabic and cherry gum. Xylan is found in the cell walls of the cells of straw, nut shells, seed coats, and corn cobs. The pentosans are soluble in cold water with difficulty but more soluble in hot water. They are very difficult to hydrolyze but finally yield the sugars arabinose and xylose.

III. PHOTOSYNTHESIS

A. Nature

It was mentioned at the beginning of this chapter that the first evident organic product that is synthesized by the green plant from the inorganic compounds that it absorbs is some form of carbohydrate. This carbohydrate is formed by the plant from carbon dioxide and water, and the term "carbon assimilation" or "photosynthesis" has been given to the process. Barnes (1893) first proposed the name "photosyntax," but later (1898) he changed to the term "photosynthesis," which is now universally used in America and to a considerable extent in England, but in Europe the term "carbon assimilation" is still in general use. Photosynthesis is commonly defined as the manufacture of some simple carbohydrate (probably sugar) from carbon dioxide and water by the chloroplasts in the presence of light. In this process, in addition to the formation of carbohydrate, oxygen always appears as a waste or end product. This liberation of oxygen by green plants was one of the earliest facts noted in plant physiology, and the conditions under which it is liberated were fairly well known long before the relationship of its appearance to the formation of carbohydrate was established.

The power of photosynthesis resides exclusively in the chloroplasts. These plastids thus have the power to transform the radiant energy of the sun into the potential energy of the carbohydrates. As will later be discussed, after the carbohydrates are once formed, the plant can apparently produce from them, regardless of light, most of the numerous complex compounds of which it is composed. The pigments of the chloroplasts are the agents in those plastids that absorb the radiant energy and aid at least in transforming it into the potential condition in the synthesized carbohydrates. A general knowledge of the nature of these pigments is therefore necessary for a proper understanding of the process of photosynthesis.

B. Pigments of the Chloroplasts

The pigments of the chloroplasts have received much consideration from the chemists and plant physiologists, and the literature concerning them is very voluminous. Due, however, to the lack of proper methods of extraction, purification, and analysis, the information in regard to these pigments was very confusing until the classical work of Willstätter and his workers from 1906 to 1926. In some of the most outstanding work of modern organic chemistry, they have succeeded in making the nature and chemistry of the leaf pigments as clear at least as that of many other plant substances. The composition and nature of the leaf pigments were found to be the same for all plants and constant for all conditions as is evidenced from the analysis of the pigments from various species of different families and from plants grown under a wide range of ecological conditions. The numerous works of Willstätter have been carefully translated and summarized by Schertz and Merz (1928), and a fine resumé is given by Jorgensen and Stiles (1917). The properties of chlorophyll, as well as methods for its determination, have also been clearly defined by Schertz (1928–1929) in numerous publications. Others who have more recently studied the properties and methods for the determination of the pigments of the chloroplasts are Sprague and Shive (1929), Guthrie (1929), Harriman (1930), Peterson (1930), Hilpert, Hofmeier, and Wolter (1931), Deleano and Dick (1934), Hicks and Panisset (1934), Zscheile (1934), and Kuhn (1935). According to Schertz (1929) the most important factors influencing the quantity of pigments are rainfall, soil moisture, nutrient elements in the soil, light, temperature, and humidity. Sprague and Shive (1929) showed in corn that the pigments in the chloroplasts did not increase proportionately with the growth rate. Thus chlorophyll content was more closely correlated with leaf area than with dry weight of leaves, while carotin and xanthophyll were more closely correlated in the opposite way.

According to the researches of Willstätter, the chloroplasts contain four pigments—two closely related chlorophyll pigments, chlorophyll component *a* and chlorophyll component *b*; and two yellow pigments, carotin and xanthophyll. A fifth pigment, fucoxanthin, is present in the brown algae. In the fresh green leaves the pigments occur in approximately the following proportions: chlorophyll *a*, 2 parts; chlorophyll *b*, three-fourths of 1 part; carotin, one-sixth of 1 part; and xanthophyll, one-third of 1 part per 1,000, respectively. The relative constant proportions of the various pigments of the leaf have been of much interest to investigators. Guthrie (1929) found that, on a dry basis, increasing the concentration of carbon dioxide decreased the total content of chlorophyll and carotinoids, but on the fresh weight basis a greater concentration effected little change. This treatment, however, did not change the ratios of the pigments. Increasing the length of day decreased the amount of all four pigments in the same proportion. Norris (1933) reported that in the development of chlorophyll and xanthophyll in etiolated plants there was a constant ratio of these pigments when the plants were exposed to light. The results indicated a close relationship between chlorophyll and xanthophyll, but not between these pigments and carotin. Oserkowsky (1932) found that chlorotic pear leaves were deficient in yellow pigments as well as in chlorophyll.

Various hypotheses have been advanced to explain the more or less constant ratio between the yellow and green pigments in leaves. Lubimenko (1927, 1928) suggested that these pigments are derived from a common substance. Because of the similarity between the constitution of the phytol of chlorophyll and of the phytol of the carotinoids, Smith (1930) concluded that this common substance either comes from the carotinoids or is produced from the same parent substance.

In the chloroplasts these pigments are mixed in a colloidal state with various colorless substances—fats, waxes, and salts of fatty acids among others. Thus in an alcoholic extract of dried leaves, the chlorophyll obtained is accompanied by six times its weight of other substances.

In the determination of chlorophyll, Harriman (1930) found that leaves which were dried at 78°C. lost approximately 70 per cent of their chlorophyll. At temperatures as low as 18° to 24°C., from 20 to 30 per cent of these pigments were destroyed. The leaves that were frozen with dry ice or desiccated at a reduced pressure over sulphuric acid at room temperatures lost little or no chlorophyll. Hilpert, Hofmeier, and Wolter (1931) believed that the chlorophyll in the leaf is chemically different from that which has been isolated. Zscheile (1934) considered that there is present in chlorophyll, a component "*c*" in addition to components *a* and *b*, which have commonly been reported.

1. The Chlorophyll Pigments. Chlorophyll Components ***a*** **and** ***b*****.**—The green pigment of the chloroplasts is called "chlorophyll." This pigment is the most abundant one in these plastids and consists of two components—chlorophyll component *a* of the composition $C_{55}H_{72}O_5N_4Mg$, blue black in the solid state and greenish blue in solution; and chlorophyll component *b* of the composition $C_{55}H_{70}O_6N_4Mg$, green black in the solid state and in solution pure green. The term "chlorophyll" as it is now commonly used refers to these two green pigments taken collectively and freed from all the others that are associated with it in the chloroplasts. Chlorophyll always contains 2.7 per cent magnesium, which is the only metal in the ash.

Chlorophyll component *a* is very easily soluble in ether and absolute alcohol, only moderately so in cold methyl alcohol, and rather easily in warm methyl alcohol. It is easily soluble in 95 per cent and with difficulty in 80 per cent ethyl alcohol. It is dissolved with difficulty in 90 per cent methyl alcohol even when warm and is almost insoluble in 80 per cent methyl alcohol. It is easily soluble in acetone, chloroform, carbon disulphide, and benzol. It dissolves in petroleum ether with difficulty even when warm.

The solubility of chlorophyll component *b* is in general somewhat less than that of component *a* but the difference is very marked in petroleum ether. Chlorophyll *b* is wholly insoluble in petroleum ether at room temperatures, and the solvent becomes scarcely colored even at boiling.

a. The Chemistry of Chlorophyll.—In addition to the idea of Willstätter and his workers with regard to the structure of the chlorophyll components, Conant, Dietz, Bailey, and Kamerling (1931), Fischer and Grassl (1935) and others have submitted formulas for the chlorophyll components, which have been based on various observations and determinations of these pigments. The general formulas herein reported are taken from the works of Willstätter and are apparently correct in regard to the proportion and general arrangement of the elements that compose the two chlorophylls. The actual arrangement of the elements in the molecule, however, is known only theoretically. Chlorophyll is a combination of the tricarboxylic acid chlorophyllin $(C_{31}H_{29}N_4Mg)(COOH)_3$ with methyl alcohol (CH_3OH) and phytol or phytyl alcohol ($C_{20}H_{39}OH$), which together compose about one-third of the chlorophyll molecule. It is thus an ester of the acid chlorophyllin.

The formulas for chlorophyll *a* and *b* are written:

Chlorophyll *a*: 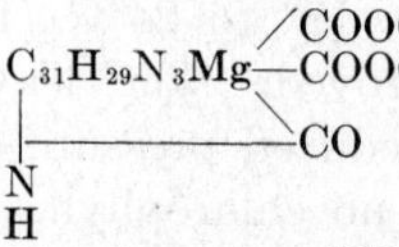

Chlorophyll *b*: 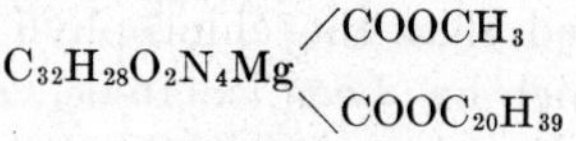

Although chlorophyllin is a tricarboxylic acid, only two of the acid (COOH) groups actually function in ester formation. The third acid group apparently does not exist as a free acid group. In chlorophyll *a* it is in what is known as the lactam ring

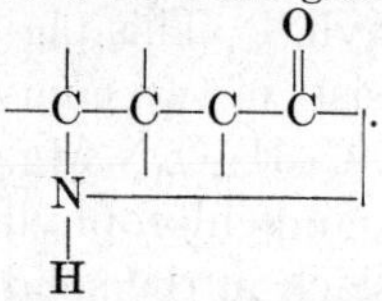

. In chlorophyll *b* it is considered by some to be in the lactone-arrangement, 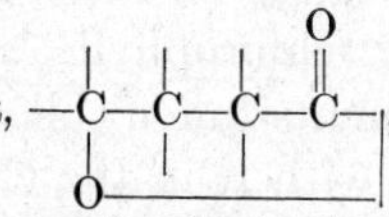 group, the two bonds in each case being attached to different structural units in the molecule (Thatcher, 1921).

When chlorophyll is treated with dilute alkalies, the alcoholic groups of the esters are split off by hydrolysis, producing the tribasic acids—chlorophyllin *a* and chlorophyllin *b*. When these chlorophyllins are heated with alkali, the carboxyl groups are driven off successively, giving products which are termed "phyllins" but which contain magnesium. The chlorophyllins are easily decomposed but were isolated by Willstätter and were found on analysis to contain magnesium bound to nitrogen in a complex way. The magnesium cannot be electrolytically dissociated as a magnesium salt. The magnesium-containing group is very easily affected by acids but is stable in the presence of alkalies.

When chlorophyll is treated with dilute acids, the magnesium is removed and replaced with two hydrogen ions without otherwise changing the molecule. A series of decomposition products result which are called "phytins" and which correspond to the alkali decomposition products of chlorophyll, except that the magnesium is lacking.

It should be noted here that chlorophyll in nature is accompanied by the enzyme chlorophyllase. By precipitation with acetone, Mayer (1930) obtained an active chlorophyllase from an aqueous extract of macerated leaves. This enzyme is active in an alcoholic medium and replaces the phytol group of the chlorophyll with the ethyl group from the alcohol. When an alcoholic extract of this type is evaporated nearly to dryness, dark-green crystals—crystalline chlorophyll—are obtained. When chlorophyll is extracted by solvents other than alcohol, it is obtained in the amorphous form.

There is a striking chemical similarity between chlorophyll and hemoglobin, the characteristic, red coloring matter of the blood of higher animals. On that account, much work has been done in an attempt to establish a direct relationship between these two pigments. Much of this work is beyond the scope of this text, and consequently only a few statements will be made here. Roddis (1936) believed that hemoglobin might be derived from chlorophyll by the substitution of iron for magnesium. He mentioned that the same color methods that are used in estimating hemoglobin may be used for estimating the amount of chlorophyll. Raber (1931) grew corn plants in the dark for 5 to 10 days in nutrient solutions to which liver extract had been added. These plants were distinctly greener than the controls. He considered that the physiological effect of the liver extract in checking the etiolation of green plants has a striking resemblance to the effect of this extract in checking pernicious anaemia in animals. Noack and Kiessling (1929), and Inman and Rothemund (1932) studied the relationship between the decomposition products of chlorophyll in the digestive tract of animals and the decomposition products artificially derived from this pigment. They believed that some of these products have a close relationship to the hemoglobin of the blood.

b. Optical Properties.—Chlorophyll in alcoholic solution exhibits the phenomenon of "fluorescence," a property that is possessed by a considerable number of substances, in which they exhibit one color by transmitted and another by reflected light. This phenomenon was so called because it was first observed in fluorspar and appears to be due to an alteration in the wave lengths of radiant energy brought about by the action of the molecules of the substance. Generally the action is to increase the wave lengths, but in some few cases it is the reverse. Usually the yellow, green, and blue rays undergo the change. In the alcoholic solution of chlorophyll, the much shorter wave lengths of green and blue are emitted as the longer wave lengths of the red. Thus an alcoholic solution of chlorophyll *a* is blue-green by transmitted light and blood-red by reflected light. An alcoholic solution of chlorophyll *b* is yellow-green by transmitted light and brown-red by reflected light.

One of the striking properties of chlorophyll is its ability to absorb certain rays of light. This absorption is so complete that dark bands appear in the spectrum. Since the absorption spectrum of every colored solution changes with its concentration, the spectrum of chlorophyll must be determined either throughout a range of concentrations or by using layers that vary in thickness. The spectrum of chlorophyll shown in Fig. 24 is that of an acetone solution of chlorophyll made by dilution of an acetone leaf extract of 40 cc. of acetone containing 0.02 g. of chlorophyll to five times its volume with 85 per cent acetone. This spectrum was obtained from a layer of this solution 1 cm. in thickness in a glass vessel with parallel walls. Figure 24 shows the absorption spectra

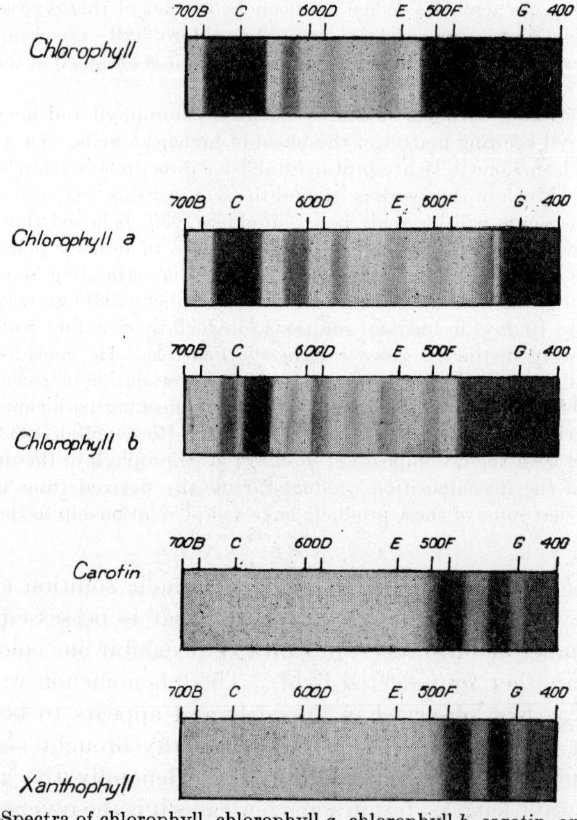

FIG. 24.—Spectra of chlorophyll, chlorophyll *a*, chlorophyll *b*, carotin, and xanthophyll. The nature of the solution in each case is described in the text. (*Redrawn from Schertz and Merz*, 1928.)

of chlorophyll *a* and *b* obtained from a layer of solution 2 cm. in thickness made by dissolving 0.04 g. of the pure components in 1,000 cc. of ether.

The spectrum of chlorophyll (Fig. 24) shows that there are five absorption bands. The most distinct band includes the red rays of 651 to 680mμ approximately on both sides of the Fraunhofer line *C*. This band appears in solutions of chlorophyll that are so dilute that none of the other absorption bands is evident. The next most distinct band and the greatest in width is one that includes most of the blue, indigo, and violet rays of 400 to 510mμ approximately. Three other bands of much less intensity are located between these two prominent ones. One of these is located in

the light orange and includes approximately the rays 600 to 620mμ. The second is in the orange-yellow region and includes the rays 550 to 590mμ, while the third is in the yellowish-green portion and includes the rays 525 to 540mμ. These bands are rather indistinct and not intense, the intensity decreasing from the red toward the blue end of the spectrum. The etherial solution of chlorophyll *a* gives a spectrum in which the absorption band in the red is not so distinct or wide as in the chlorophyll solution. The band in the blue end of the spectrum is only one-half the width of that of chlorophyll and includes only the more distant portion of that region. The solution of chlorophyll *b* shows a distinct absorption band in the red ultra to the line *C* and only a narrow rather indistinct one on the infra side. The band in the blue end of the spectrum is distinct and is approximately three-fourths the width of that of chlorophyll.

c. The State of Chlorophyll in the Chloroplasts.—The location of chlorophyll in the chloroplasts has been mentioned in Chap. I, but the condition or state in which it occurs has not been discussed. According to the work of Willstätter and others, the evidence seems to indicate that chlorophyll is present in the chloroplasts in a colloidal mixture. This assumption is substantiated by the behavior of chlorophyll to its solvents and from the evidence derived from spectroscopic examination.

Willstätter found that solvents which dissolve the pure extracted substance do not extract the chlorophyll from the dried leaf. Thus the pure pigment is readily soluble in acetone, but if the dried powder of nettle leaves is placed therein, it can remain in the acetone for ½ hr. without coloring the liquid. If a little water is added, however, the color becomes intensely green. This behavior indicates that chlorophyll occurs in the leaf in a different state of matter from extracted chlorophyll. From the above statements it would seem that chlorophyll in the chloroplasts is in the colloidal condition, that water added to the pure organic solvents dissolves the mineral substances in the leaf, and that the salt solutions so formed alter the colloidal condition of the chlorophyll in the chloroplasts and make it easily soluble. The pure solvents can, however, extract chlorophyll from fresh leaves, since abundant water is present in them. Further support for the theory of the above reactions is furnished by the fact that colloidal solutions of chlorophyll in water made from a pure extracted pigment behave in a way similar to dried leaf powder. If a solution of this type is mixed with ether, the ether remains colorless, but if a little salt solution of calcium chloride or calcium nitrate is added, the etherial layer becomes colored green upon shaking. The salt solution has changed some of the chlorophyll from the colloidal condition, and as a result it is soluble in the ether.

It has been observed by many (Tsweet, 1910) that the absorption bands of the spectrum of the living leaf are displaced toward the red end of the spectrum as compared with the bands in the spectrum of extracted chlorophyll. It is also known that the condition of chlorophyll in fresh leaves is altered if the leaves are plunged into boiling water and that the chlorophyll is then more easily extracted by its solvents. A leaf that has thus been placed in water gives absorption bands that occupy practically the same position as those in the chlorophyll extract. It is assumed that the chlorophyll has been changed by this treatment from a colloidal solution to a true solution and is now dissolved in the waxy substances with which it is associated and which have become liquid as a result of the temperature change. It has also been shown by Herlitzka (1912) that the spectrum of living leaves agrees with that of a colloidal solution of chlorophyll, while both differ in the same way from the spectrum of true solutions of chlorophyll.

d. Formation.—The formation of chlorophyll is a physiological process that occurs only in living cells and under conditions favorable to life. The substance or substances from which chlorophyll arises have never been isolated, and their existence is only inferred. It is considered by some that a pigment called "chlorophyllogen" is formed independently of light in the chloroplasts and that it is rapidly transformed into chlorophyll under the influence of light (Palladin, 1922). The absorption spectrum of the pigments obtained from etiolated seedlings has an absorption band between 640mμ and 620mμ, which is due to a red fluorescing pigment that has been termed "protochlorophyll." According to Eyster (1928), this is a pigment that develops without the aid of light and changes photochemically into chlorophyll upon exposure to light.

Noack and Kiessling (1929) stated that acid splits off magnesium from protochlorophyll, the same as it does from chlorophyll, to form a pigment that they have designated as "protopheophytin." According to Rothemund (1935), most of the protochlorophyll disappears when the plant is placed in the light and is thus apparently transformed into chlorophyll. Protochlorophyll, along with chlorophyll, is probably present in all green leaves. It forms in the dark, but it is not known whether it arises from chlorophyll, from the precursor of chlorophyll, or from both. It has never been demonstrated that the living green plant can change protochlorophyll into chlorophyll. It has only been shown that, after irradiation of the plant, the absorption spectrum of protochlorophyll becomes weaker and the chlorophyll spectrum appears, becoming more intense with time. Protochlorophyll, however, has never been prepared artificially.

Light is one of the requisites for the formation of chlorophyll. The stem and leaves of most plants grown in darkness are always yellow. Such plants are said to be etiolated. They quickly turn green when exposed to light. Some plants or plant parts, however, as the seedlings of some conifers, the fronds of young ferns, the plumule of *Nelumbo*, and certain unicellular algae, are exceptions to this, since they become green in darkness. According to Palladin (1922), the conifer seedlings, however, form much less chlorophyll in darkness than in light. Skinner (1932) showed that even in total darkness certain algae, placed in soil partially sterilized by heat, are able to increase as much as 100-fold when incubated at room temperature for long periods of time. Light of medium intensity is the most favorable for chlorophyll formation. It is considered that in more intense light the formation and decomposition of chlorophyll occur simultaneously, so, as a net result, the greening is less pronounced than it is in diffuse light.

It was demonstrated by Wiesner (1874) that chlorophyll is formed in the red, orange, and yellow rays produced by a potassium dichromate solution and in the green, blue, and violet region produced by copper sulphate. He also showed that plants do not become green under the nonluminous heat rays. Sayre (1928) grew plants under colored glass plates and studied the effect of different wave lengths of radiant energy on the formation of chlorophyll in the seedlings of corn, wheat, oats, barley, sunflower, and radish. The wave lengths longer than 680mμ are not effective in the formation of chlorophyll, but all other regions of the remaining visible and ultraviolet spectrum to 300mμ are effective, provided the energy value is sufficient. For equal energy values, the red rays are more effective than the green, and the green more than the blue. The effectiveness of radiant energy in chlorophyll formation appears to increase with wave length to about 680mμ and then ends abruptly. Inman (1935) found that the absorption band of chlorophyll in the region of 666mμ could easily be detected after only 10 sec. of exposure to the light. After the exposure of etiolated leaves for 150 min., a typical chlorophyll absorption spectrum was obtained. Ulvin (1934) found that the leaves of radish grown in continuous light contained more chlorophyll than those grown in a 10-hr. daily light period. Meier (1934) studied the development of chlorophyll in 18 kinds of unicellular algae and found that 11 of them developed chlorophyll while growing for 30 days in continuous light, in the intermittent light of day and night, or in continuous darkness. Three showed only poor development of this pigment under all these conditions. One grew best in continuous light, and one grew best in continuous darkness. It was reported by Colla (1930) that radiation of 3300 to 3900 Å. from a mercury-vapor lamp deepened the green color of the plant. This change in color was due to the increase in the amount of chlorophyll.

After the exposure of barley roots to diffuse daylight or electric light for 6 to 10 days, Gautheret (1932) detected chlorophyll in 70 per cent of them, and by the Molisch test the chloroplasts gave the reaction for sugar.

The rate of production of chlorophyll depends upon temperature. Greening occurs most quickly in etiolated plants between 18 to 30°C. Wiesner (1877) found within that range of temperature that greening occurs in about 1.6 hr., while at 10°C. 3.5 hr. was required. Lubimenko and Hubbenet (1932) noted that the greening of etiolated wheat seedlings takes place within definite limits of temperature, beginning between 2 to 4°C., attains a maximum between 26 and 30°C., and ceases at or near 48°C.

A supply of iron is necessary for the formation of chlorophyll, and the probable function of that element in this regard is discussed in detail in Chap. VI. Many other elements also influence the development of chlorophyll. The effect of potassium, phosphorus, calcium, and magnesium on the color of plants is considered under the discussion in Chap. VI of the functions of these elements in the metabolism of the plant. Ville (1889) found in the case of hemp that the greatest decrease in chlorophyll and carotin was brought about by a deficiency in nitrates, while Schertz (1921) associated the mottling of *Coleus* plants with the same cause. Deuber (1926) in the soybean noted that the chloroplast pigments increased with the increasing concentrations of iron and sulphur in the cultural solution. The lack of iron or potassium in the nutrient solution resulted in a more marked depression of the chloroplast pigments than a lack of sulphur. When the sulphur content reached a concentration of 46 p.p.m., and potassium reached a concentration of 28 p.p.m., a marked decrease in the pigment content of the leaves occurred. Deuber, however, could not establish an exact proportional relation between the concentration or amount of any of the elements studied and the amount of pigment formed. Schertz (1929) noted that phosphorus, potassium and nitrogen may each be correlated with an effect on the formation of the chloroplast pigments. For the potato, high phosphorus produces more pigment than does potassium, but less than nitrogen. High yields of potatoes are correlated with heavy potassium fertilization, and this in turn with a low chlorophyll content of the leaves. Low yields are correlated with high nitrogen fertilization and a high chlorophyll content of the leaves. It was thought by Echevin (1934) that the parallelism in the disappearance of phospholipins and chlorophyll in the autumn indicated a relationship between these phosphorus compounds and the activity of the green pigment. In the leaves of pineapple, Tam and Magistad (1935) found indications that the amount of available nitrogen determines to a large degree the amount of chlorophyll formed, provided that all of the other necessary materials and conditions are not limiting factors. Ulvin (1934) reported that with sugar cane more chlorophyll was produced by plants receiving nitrogen in the form of nitrate than by those receiving this element in the form of ammonia.

Since magnesium and nitrogen enter into the molecular structure of chlorophyll, it is to be expected that a variation in the amount of these elements in the soil or in the cultural solution would influence the production of this pigment. However, the cause of chlorosis in plants that is due to variations of the calcium, sulphur, potassium, magnesium, phosphorus, or iron supply is not so easily explained, since these elements do not enter into the composition of chlorophyll. This question is generally answered by saying that the production of chlorophyll depends upon the general vigor and tone of the plant, which is known to vary with certain combinations and with certain concentrations of these elements.

Palladin (1891) has shown that carbohydrates in the leaves are essential to the formation of chlorophyll. Etiolated leaves that contain a considerable supply of carbohydrates green rapidly when placed in light, but those which contain little or no soluble carbohydrate remain yellow. The leaves that are deficient in carbohydrate, however, green rapidly when floated in the light upon a solution of cane or grape sugar. The presence of oxygen is also essential for the production of chlorophyll. Etiolated leaves in an atmosphere devoid of oxygen remain yellow in light. It has been assumed by some that oxygen, mineral nutrients, carbohydrates, and a favorable temperature are necessary for the formation of chlorophyllogen, which arises independently of light. The transformation of chlorophyllogen to chlorophyll in the living cells is considered to be dependent upon light.

Ruth (1922) noted that the chlorophyll content per unit area of the primordial leaves of the bean was increased slightly by spraying with Bordeaux mixture. Henrici (1926) found in South Africa that there was a pronounced difference between the amount of chlorophyll present in the morning and in the afternoon. Following a heavy rain there was an increase in the chlorophyll content, while drought caused a decrease.

e. Inheritance.—According to Demerec (1935) chlorophyll shows more peculiarities in its inheritance than any other known plant characteristic. In maize there are approximately 100 genetically different characters for chlorophyll. Some of these manifest themselves only in the seedling stage, others only in the mature plants, while still others appear in both the seedling and mature stages. Four types of plants are distinguished in regard to their chlorophyll content. They are: the albino, virescent, pale-green, and variegated forms. In the albino types, the chlorophyll is absent, and because of this the plants die at an early seedling stage. The virescent types are those which are albinos in the early stage but later develop chlorophyll. The pale greens are those plants in which the quantity of pigment is reduced at certain stages of development. The plants included in this type vary from plants that contain little chlorophyll to those having almost the normal amount. The variegated type includes those plants in which the content of chlorophyll varies in different regions of the leaf. These differences may show either by a mottling or by a striping of the leaf.

Miles (1915) believed that plastids are entirely absent from albino plants, but Randolph (1922) concluded that the failure of a plant to become green is not due to the absence of plastids or plastid primordia. The albino plants thus lack some physiological mechanism essential for the development of plastid color. This condition is brought about by the genetic constitution of the plant. A comparison of the size of plastids of various chlorophyll types of maize indicates that the size of the plastids is correlated with the amount of pigment present. Thus Trajkovich (1924) determined the chlorophyll content and measured the plastids of four chlorophyll types. He found a correlation between these two factors as shown in the following table:

Type	Relative amount of pigment		Average size of plastids, microns
	Green	Yellow	
Green	100	100	6.19
Xantha	19.7	100	4.62
Yellow	6.1	100	3.19
White	0	0	Primordial

At this point it is well to mention the effects of light rays on the coloring of apples, although this process is apparently in no manner concerned with the formation of chlorophyll. Arthur (1931) reported that the ideal light source for coloring apples is one that has little infrared but considerable ultraviolet extending to a wave length of 290mμ. Pearce and Streeter (1931) stated that the region in the solar spectrum from 3600 to 4500 Å. is the most influential in the coloring of McIntosh apples.

2. The Carotinoids.—Various yellowish pigments, which have been termed collectively the "carotinoids," are always associated with the chlorophylls in the chloroplasts. This group generally includes two types of pigments that are very similar in chemical composition. These are the carotins, or carotenes, and the xanthophylls. Fucoxanthin is a brownish-red pigment that is associated with the chlorophylls, carotins, and xanthophylls in the brown algae, and it imparts to them their characteristic color. The carotinoids may also occur in fruits, roots, and other plant parts where chlorophyll is absent. According to Howard (1925), the carotinoids of the fruits examined by her are granular or crystalline in form and are frequently contained in the chromoplasts. The amount of the carotinoids present in plant tissues apparently depends upon the environmental conditions and upon the type of plant. Thus Smith and Smith (1931) found that bagged Elberta peaches developed a higher content of carotinoids than the unbagged ones, while the reverse was true of nectarines and apricots. Whiteside, Edgar, and Goulden (1934) considered that although the carotinoid pigmentation of wheat is an inherent characteristic, the environmental conditions have an appreciable effect upon the amount present.

Mackinney (1935) observed that the content of chlorophyll and carotinoids fluctuated together, but that the ratio of the carotins to xanthophylls was markedly lower in the chlorotic leaves of the barley plant than in the normal leaves.

Murneek (1934) noted that the concentration of the carotinoids appeared to reach a maximum in the leaves of *Cosmos*, *Salvia*, and *Soja* at the time of flowering, after which they decreased. He stated there are strong indications that more yellow pigments are present in the female than in male plants of dioecious species. Roberts and Livingston (1935), however, could find no relationship between the content of carotinoids and the fruitfulness of the apple or tobacco. Schertz (1921 to 1929), Palmer (1934), Miller (1935), and Murneek (1936) discussed methods for the estimation of carotinoids and made suggestions for their nomenclature.

a. Carotins or Carotenes.—The pigments of this group are found not only in the chloroplasts but are widely distributed in both the amorphous and crystalline forms in seeds, fruits, and roots. According to Howard (1925) the carotins of carrot occur in both the crystalline and amorphous forms in the protoplasm. The carotins occur in great abundance in carrot roots, are in the grain of yellow corn and in ripe tomatoes, and are the pigments that give these parts their characteristic colors. Strain (1934) and Spoehr, Smith, Strain, and Milner (1935) stated that it is now definitely established that there are a large number of carotin pigments in the leaves and other plant parts. These are mostly mixtures of isomers of very similar structure and differ only slightly in most of their physical and chemical properties. Three of the more prevalent types of carotin are α-, β-, and γ-carotin; the second named generally occurs in the greatest amount. According to Mackinney (1935), dodder is one of the richest sources for γ-carotin. Mackinney and Milner (1933) found that the carotin which they isolated from carrot leaves contained approximately 10 per cent α-carotin, and traces of γ-carotin, thus closely resembling the nature of the pigments in the carrot root. Mackinney (1935) showed that both α- and β-carotins were present in the carrot root, and that there was no significant change in the ratio of the amounts of these two pigments with the development of the plant. The amount of carotin pigments varies widely in the roots of carrots. Thus Bills and McDonald (1932) found that some varieties of carrots had over eighty times as much carotin as did others. According to Matlack and Sando (1934) and Miller, Mackinney, and Zscheile (1935), the carotins from different localities and from different plant sources yield identical products. The carotins have the same general formula, $C_{40}H_{56}$. They are insoluble in water, but are very soluble in ether, chloroform, and carbon disulphide. They become bleached in air, and in dry air increase in weight 35 per cent by the addition of oxygen. The crystals are rhombohedra, and are orange-red by transmitted, and greenish-blue by reflected light. Smith (1932) reported that the carotins from the leaves of alfalfa, cauliflower, spinach, sunflower, chard, and sugar beet are like those of carrot in that they have 10 double bonds in the molecule. Lycopene or lycopersicin, an isomer of carotin, occurs in red tomatoes, and has 13 double bonds. β-Carotin has been termed "provitamin A." If a molecule of this pigment would separate in the middle and a molecule of water be added to each portion, the resulting product would be two molecules of vitamin A (Zechmeister, 1934). Deleano and Dick (1932, 1933), Strain (1934), and Russell, Taylor, and Chichester (1935) have reported methods for the preparation and determination of the carotins.

b. Xanthophylls.—These pigments have the general formula $C_{40}H_{56}O_2$. According to Spoehr, Smith, Strain, and Milner (1935), and Strain (1936) the leaves contain not less than 12, and probably more, xanthophylls. The heterogeneity of leaf xanthophylls and the variation of the properties of their constituents probably account for the divergent values which have been reported for the xanthophyll content of leaves.

The crystals of these pigments are pleochromatic, often with a steel luster. They are yellow in transmitted light, and are red only when two or more cross each other. The xanthophylls may absorb 36.5 per cent of their weight of oxygen. They are very easily soluble in chloroform and alcohol, soluble in carbon disulphide with difficulty, and insoluble in petroleum ether. The carotins and xanthophylls are very different in their color intensities, but there is no simple ratio of their color strengths, since this varies with the solvent and with the concentration. The carotins are stronger in color and are more red in carbon disulphide and ether, while the xanthophylls are rather greenish tinted in these solvents.

The carotins and xanthophylls show spectra that consist of two absorption bands in the blue and indigo blue in addition to the end absorption that starts at the beginning of the violet. In Fig. 24, the spectra of these two pigments were obtained by using a solution, 1 cm. thick, of 5 mg. of pigment per liter of ether. The absorption bands of xanthophyll are displaced slightly toward the violet, as compared to those of the carotin.

c. Fucoxanthin.—This brownish-red pigment is associated with carotin, xanthophyll, and chlorophyll in the brown algae. It has the formula $C_{40}H_{54}O_6$ and undergoes a 7 per cent increase in weight in air, owing to the absorption of oxygen. This pigment does not exhibit fluorescence.

3. Autumnal Coloration of Leaves.—A thorough discussion of the factors involved in the coloring of leaves in autumn presented by Stout (1936) and Roberts (1937). Only a few general facts will be mentioned.

The autumnal coloration of leaves is confined to the deciduous trees and shrubs, which shed their leaves at the approach of winter. According to Stout (1936) the areas of the globe in which this coloration is manifested are very limited. In the southern hemisphere there is only one small region, which is located in the southern portion of South America. In the northern hemisphere there are three relatively large but separated areas in which autumnal coloration appears. One is in eastern Asia, one is in southwestern Europe with extensions eastward and northward, and one in North America extending from the Gulf of St. Lawrence south to Florida and westward through the Mississippi Valley to the Great Plains area. The variety, intensity and brilliance of coloration vary greatly over this area in North America, depending primarily upon the kinds of plants native to a given region, upon the general climate, and upon the autumnal weather that prevails during a given season. The most gorgeous autumnal colorations that occur in the United States are in New England, New York, Pennsylvania, Michigan and northern Ohio. This is due primarily to the kind of trees and shrubs that are dominant in that general region and to the ideal weather conditions for this coloration that generally prevail prior to leaf fall.

A wide variation of colors is very prevalent and very lasting in many plant parts during their natural growth and development. Thus the bright colorings of apples, plums, peaches, cherries, and many berries are familiar to all. In roots there are the scarlet radish, the purple-topped turnip, the yellow carrot, and the red garden beet. The leaves of many plants are brilliantly colored during the entire season as shown in Coleus, Cockscomb, Joseph's Coat, striped grasses, and many others. The various colors are due to internal regulations in the plant which result from a combination of hereditary factors and environmental conditions. These colorations in plants are very similar to those seen in autumn but differ in that they are relatively permanent during the life of the plant, while the autumnal colors are of only relatively short duration.

The autumnal colorations may be roughly divided into the yellows and the reds. The former is due primarily to the carotinoids and the latter to anthocyanin. During the growing season, the yellow pigments in the chloroplasts are masked by the chlorophylls, which are the most abundant coloring matters in the leaves. When cell activity begins to cease as autumn approaches, the chlorophylls are not replaced as they break down, and the carotins and xanthophylls become evident. These pigments or their oxidation products are responsible for the yellow and brown autumnal coloration of leaves. The yellow colors are especially prominent in the leaves of willow, poplar and tulip trees.

The autumnal colors of red and purple as seen in the leaves of sumac, red oak, sweet gum, red maple and sugar maple are due primarily to the anthocyanin pigments of the cell sap. The particular type of pigment or pigments found in the leaves is to a great extent hereditary. The colors, however, vary in intensity according to the region in which the plants grow and according to the type of season that prevails previous to or during their formation.

C. Historical

In 1772 Priestley discovered that green plants confined in an atmosphere rich in fixed air (carbon dioxide) produced after a period of time a considerable amount of dephlogisticated air (oxygen). He, however, had no understanding of the factors concerned in the process. At the same time, Scheele working in Sweden reported just the opposite results, since the plants with which he worked produced fixed air (carbon dioxide). Upon learning of the results of Scheele, Priestley repeated his investigations, but his results were confusing, since sometimes the plants increased the oxygen in the air and sometimes they depleted it, and Priestley was unable to explain these variations. He, however, is given the credit for being the first to observe that oxygen is given off by green plants.

Jean Ingen-Housz, physician to the emperor of Austria, was interested in the effect of pure air upon the health of man and was led to investigate plants in this regard by the first report of Priestley. Ingen-Housz showed that plants were able to purify foul air in a few hours when placed in sunlight and that sunlight had no effect in increasing the amount of oxygen without the action of the plant. He also noted that the absorption of air and the exhaling of oxygen were more active the brighter the sunlight, and that at night, or under extremely shaded conditions, the plants acted upon the air just as animals. He observed that only the green parts of plants gave off oxygen and that roots, flowers, and fruits rendered the air impure. These results were published in 1774 but were not presented in the terms of modern chemistry until 17 years later.

In 1800 Jean Senebier, a minister of Geneva, published his researches on the influence of light upon vegetation. He considered that the oxygen given off came from the carbon dioxide absorbed by the plant, and by means of colored glass he was able to ascribe to the red rays of the spectrum the chief action of this process. In 1804, Nicolas de Saussure published his classical work, "Chemical Researches on Plants." Although a contemporary of the men just mentioned, his experiments were so well carried out and his data so well presented in the light of modern chemistry that Spoehr (1919) says that one might well believe that a century had elapsed between the time of De Saussure and that of his contemporaries. His experiments, which were entirely quantitative, showed that the elements of water were fixed in the plant at the same time as the carbon and that there was an increase in weight as a result of the process. He also noted that there was no normal nutrition of the plant without the absorption of nitrates and mineral substances.

The discovery of these four men, however, found no interest among the botanists of their time. They were absorbed in the description and classification of plants and had no time for the consideration of facts about plants discovered by men foreign to the field of botany—for Priestley, Ingen-Housz, Senebier, and De Saussure were chemists and physicists. The discoveries of these men were soon forgotten or misinterpreted and little more was accomplished in this line, with the exception of the proof by Boussingault that a plant may obtain all its carbon from the air, until Sachs took up the work in 1860 to 1865.

Sachs and his pupils established, for the most part, the foundation upon which our modern views of photosynthesis rest. He was the first to discover that the chlorophyll corpuscles (chloroplasts) are the organs concerned in the appropriation of carbon dioxide and the separation of oxygen. Sachs also showed that chlorophyll, except in a few cases, is formed only in the presence of light, and he further distinguished between the effect of light in this process and in the release of oxygen by the leaf. Starch grains in the chloroplasts had been observed by Von Mohl as early as 1837 and by several subsequent investigators, but the significance of these grains was in no way suspected. It remained for Sachs, however, to connect the appearance of starch with the fixation of carbon and water and the elimination of oxygen in the chloroplasts in the presence of light. He was thus the first to observe and prove that a carbohydrate as well as oxygen is a product of the fixation of carbon and water in the green plant. He also grasped the fact that the formation of carbohydrates by this process is the starting point for the production of all other organic compounds of the plant, as shown by his statement (1882): "In the nutrition of plants it is only necessary in the first place to decompose carbon dioxide under the influence of light in the cells containing chlorophyll, with the cooperation of certain mineral matter absorbed by the roots and to produce at the cost of its carbon an organic substance—starch—(carbohydrate) which then represents the starting point, so to speak, from which all the organic substances of the plant proceed by progressive chemical changes."

D. Products

Oxygen and carbohydrates are the two end products that are definitely known to be produced in the process of photosynthesis.

1. The Evolution of Oxygen.—It has already been stated that the evolution of oxygen by green plants in sunlight was the first fact observed in photosynthesis. This emission of oxygen is most easily observed in submerged aquatic plants, since in sunlight bubbles of gas escape from these plants and rise to the surface of the water. The gas which is thus emitted, however, is not pure oxygen but consists of a mixture of oxygen, nitrogen, and a small amount of carbon dioxide. The two latter gases are not emitted from the interior of the cells but are swept out of the intercellular spaces by the stream of oxygen gas. The oxygen content of the emitted gas varies from 25 to 85 per cent, increasing with the increased rate of bubble formation. Spoehr (1926) considered that when the rate of photosynthesis is high and the escaping gas stream correspondingly high, the oxygen sweeps out the nitrogen and carbon dioxide from the intercellular spaces faster than it can be replaced from the surrounding water. When the rate of photosynthesis is low, the escaping gas stream is slow due to slow oxygen formation, so that the nitrogen and carbon dioxide which are being swept out are replaced in the intercellular spaces from the surrounding water. Under these conditions the gas that escapes contains a higher percentage of nitrogen and carbon dioxide than when the photosynthetic rate is high. The emission of oxygen from aquatic plants may also be observed after the following manner: The water in which the plants are submerged is covered with a film of oil, and the container is placed in the dark until the plants have exhausted the oxygen supply in the water. A leuco compound of methylene blue or indigo carmine is then added, and shortly after the plants are placed in sunlight the liberation of oxygen is made manifest by the formation of the bright-colored dye close to the plant. Oxygen is the only gas that has definitely been proved to be emitted by the plant during photosynthesis. It was found by Rudolfs and Heukelekian (1931) that the dissolved oxygen, in waters containing large quantities of blue-green and green algae, could be decreased from supersaturation to 17 per cent by darkness, and could be increased to 282 per cent saturation by diffuse light.

The ratio of the volume of carbon dioxide absorbed to the volume of oxygen produced is termed the "photosynthetic" or "assimilation" quotient. It is now commonly written as CO_2/O_2 but in the earlier literature it was expressed as O/CO_2 or O_2/CO_2. It was observed by De Saussure (1804) and Boussingault (1864) that the photosynthetic ratio approximated unity. Their experiments, however, did not take into account the respiration of the plants. The determination of the photosynthetic ratio is difficult, because the process of respiration is proceeding at the same time and in the opposite direction in that oxygen is being absorbed and carbon dioxide given off.

Separation of these two reactions has been found very difficult. The first work on this problem was undertaken by Bonnier and Mangin (1886), who employed four methods for the separation of the gaseous exchanges due to photosynthesis and respiration, of which the following are the most important: (*a*) comparison of the gas exchange of the plant in the light and in the dark. The amount of carbon dioxide absorbed in the light includes not only that obtained from the surrounding air but also that which has been released by the cells in the process of respiration. The oxygen that is released by the plant in the process of photosynthesis includes not only that which escapes from the plant but also that used by the cells in the process of respiration. By placing the plants in the dark for a given period, the volume of carbon dioxide C' released and the volume of oxygen O' absorbed by the cells in respiration can be determined. By placing the plants in the light for the same period, the volume of oxygen O'' released into the air and the volume of carbon dioxide C'' absorbed from the air under these conditions is measured. The total volume of oxygen released in the photosynthetic process would thus be $O' + O'' = O$ and the total volume of carbon dioxide absorbed would be $C' + C'' = C$. The true photosynthetic quotient would thus be $\frac{C' + C''}{O' + O''} = \frac{\text{C}}{\text{O}} = \frac{\text{CO}^2}{\text{O}_2}$. This method is open to objection, since it is not definitely known that respiration proceeds at the same rate in the light as in the dark. (*b*) The second method used by Bonnier and Mangin is based on an observation of Bernard (1878) that the process of photosynthesis may be suppressed by an anesthetic without the suppression of respiration. Two similar quantities of leaves, one anesthetized, the other not, are placed in light and the gaseous exchange due to photosynthesis estimated according to the measure just mentioned. This method is not exact, since the amount of the anesthetic to apply in order to just suppress photosynthesis and not retard respiration is not known. (*c*) The third method for determining the gaseous exchange in photosynthesis consists in the removal by means of baryta water of all the carbon dioxide from the vessel containing one set of leaves. This also removes any escaping carbon dioxide evolved in respiration. The other set of leaves is placed under normal conditions of carbon dioxide supply. Thus, as in the second method, the difference between the oxygen content and the carbon dioxide content of the two vessels at the end of the experiment gives the true values of the oxygen and carbon dioxide evolved. Spoehr (1916) considered that this method is inexact, as the rate of carbon dioxide emission in the two vessels is not the same, since Spoehr and McGee (1924) showed that when the carbon dioxide content of the air surrounding a leaf is changed from a higher to a lower concentration the leaf showed a primary increased rate of carbon dioxide emission. When the carbon dioxide content of the air changed from a lower to a higher concentration, the leaf showed a reduced rate of carbon dioxide emission.

By the three methods, Bonnier and Mangin (1886) concluded that the volume of oxygen emitted in photosynthesis was somewhat greater than that of the carbon dioxide. Maquenne and Demoussy (1913) using the first-named method of Bonnier and Mangin investigated the gaseous exchange of some 30 different species of plants and concluded that the photosynthetic quotient is equal to unity. The same conclusion was also reached by Willstätter and Stoll (1918), who determined the photosynthetic quotient under conditions in which the photosynthetic activity was from twenty to thirty times higher than that of respiration. This was done since the chief errors in the determinations of the photosynthetic quotient are occasioned in determining the rate of respiration. If the rate of photosynthesis is high as compared to respiration, the errors occasioned in determining the respiration rate will have an insignificant effect on the determination of the photosynthetic quotient. It was found by Kostytschew (1921) that the photosynthetic quotient varies with time. When leaves were first exposed to sunlight the value of CO_2/O_2 was considerably greater than unity, about one-third of the carbon dioxide absorbed being fixed without oxygen evolution. After a short time, more oxygen is given off than carbon dioxide is absorbed (Spoehr, 1926). After a time, however, the CO_2/O_2 ratio will become practically unity.

The value of the photosynthetic quotient is of importance, since it yields some information in regard to the first products formed in photosynthesis, because in the determination of the nature of a reaction it is of the first importance to know the quantitative relation between the initial substances and the products of the reactions (Jorgensen and Stiles, 1917). When the photosynthetic quotient is exactly unity, it indicates that the carbon dioxide absorbed has been reduced to $C + O_2$, or in the hydrate form to a compound with the general formula $C_nH_{2n}O_n$. This, however, does not give any information as to the exact compound formed, since it might be formaldehyde or any other compound of that formula (page 93 of Spoehr, 1926). It is interesting to note the value of the calculated photosynthetic quotient for different compounds. For formic acid it would be 2.0; for oxalic acid, 4.0; for glyoxalic acid, 2.0; for glycollic acid, 1.3; for malic acid, 1.3; and for formaldehyde, 1.0.

Moldavan (1934) suggested that the carbon dioxide absorbed during the photosynthetic process and the oxygen absorbed during respiration differ isotopically from the oxygen eliminated in photosynthesis and the carbon dioxide that escapes during respiration.

2. The Production of Carbohydrates in the Leaf.—Before discussing the production of carbohydrates in the photosynthetic process, it would be best to enumerate those that have been found in the leaf. Exclusive of cellulose and the pectic substances that occur in the cell walls of the leaf cells, the carbohydrates that have been detected in the leaves or other photosynthetic organs of the plant are starch, sucrose, *d*-glucose, *d*-fructose, pentosans, dextrins, pentoses, and inulin. Some of the earlier workers (Brown and Morris, 1893) considered that maltose also is present, but later investigators (Davis, Daish, and Sawyer, 1916; and Dastur and Samant, 1933) were not able to detect it. Maltose is undoubtedly formed in the hydrolysis of the starch in the leaves, but it is evidently almost immediately transformed into other sugars, so that it is not detected by the ordinary chemical methods. Starch, sucrose, *d*-glucose, and *d*-fructose are the most abundant carbohydrates in leaves, the others mentioned above being found in only very small amounts.

Although carbohydrates are known to be the end products of photosynthesis, it is not definitely known which one is the first to be formed in the chloroplasts. The determination of the first carbohydrate formed in the leaf is attended with many difficulties. In the first place, it is impossible by any of the present methods of analysis to separate quantitatively the various groups of carbohydrates in the leaf, owing to the similarity of their properties and the relatively small amount of materials present. It would seem thus that to make any progress in this regard more exact methods for separation must be developed. The detection or isolation of the first carbohydrate formed in photosynthesis is also difficult, because as many as four functions involving carbohydrates may be proceeding simultaneously in the leaf. The first carbohydrate is being formed, and from it almost immediately other carbohydrates may be formed by condensation for temporary storage as starch and the more complex sugars. These temporary storage products are, in turn, being hydrolyzed for translocation. An appreciable amount of carbohydrates is also broken down in the process of respiration which is proceeding at all times in the cells of the leaf. It is thus impossible by present-day methods to separate the carbohydrates first formed in photosynthesis from those which are formed as intermediate products in the condensation of the temporary reserves and in the hydrolysis of these reverses for translocation. On this account, therefore, it is no wonder that so many divergent views are held as to the kind of carbohydrate first formed in the photosynthetic process.

a. Early Observations.—The basis of our knowledge of the carbohydrates of the leaves was laid by Sachs (1862, 1864, 1884) when he proved that the appearance of starch in the chloroplasts is a direct outcome of the fixation of carbon under the influence of light and chlorophyll. He stated that starch is the first visible product of photosynthesis and that it is translocated from the leaf in the form of sugar. He seems to have considered that all the carbohydrates synthesized in the leaf pass at one time of their history through the starch stage in the chloroplasts. Sachs, however, worked almost exclusively with the leaves of sunflower, squash, bean, and other plants that are abundant starch formers.

As early as 1856 Boehm observed that the chloroplasts of *Asphodelus luteus, Allium fistulosum,* and others do not normally form starch. Later (1883) he placed the ends of etiolated leaves of *Phaseolus vulgaris* in sugar solutions and noted that they formed starch, while leaves of galanthus, hyacinth, and iris laid in 10 per cent sugar also showed an abundance of starch. He reasoned from these results that the same sequence of events occurred in the photosynthetic process. Schimper (1885) not only held that starch is converted into sugar in the plant but, from his observations on the increase of starch in leaves which were artificially supplied with sugar solutions, he concluded that glucose precedes starch in the process of photosynthesis. He considered that starch is formed from the glucose when the concentration of the latter exceeds a certain maximum which differs in different plants. Finally, in 1885 to 1886, Meyer emphasized the fact that the formation of starch in plants was by no means universal, since he observed that many of the monocotyledonous plants and especially the *Liliaceae* form very little starch in their leaves and some none at all. A very important contribution to the knowledge of the carbohydrate content of the leaves was the observation of Kayser (1883) that both sucrose and reducing sugars occur in the leaves of grape, beet, potatoes, and onion. He was the first to isolate sucrose in the crystalline form from the leaves of the grape.

The work of Boehm, Meyer, and Schimper indicated that the formation of starch in the leaves is preceded by the formation of the more simple carbohydrates. The chloroplasts like the leucoplasts form starch only when the supply of formative carbohydrates is in excess of the metabolic needs and translocating power of the cell in which they are contained or when the solution of the formation product has reached a certain stage of concentration.

The work of Kraus (1869) is also frequently quoted in order to show that starch is probably not the first carbohydrate formed in photosynthesis. He observed that the formation of starch in the cells of *Spirogyra* did not occur until 5 min. after the commencement of illumination, while the evolution of oxygen commenced at once. It is now generally considered by botanists very unlikely that such a complex and insoluble compound as starch should be the first product formed in the process of photosynthesis. The general opinion is that the first carbohydrate produced in the process is some form of sugar. The evidence as to which sugar it is will be presented in the following topics:

b. Daily Variations of the Carbohydrates in the Leaves.—The first quantitative determinations of the sugars in the leaf were made by Girard (1883, 1884). He observed that the proportion of sucrose in the leaves of the sugar beet was frequently twice as great at the close of the day as the following morning, while the amount of hexose sugars remained about the same. In 1893 Brown and Morris published their classical work on the daily variation of the carbohydrates in the leaves of the nasturtium. The evidence they obtained was in accord with that of Meyer and Schimper that the starch in the chloroplasts is a secondary product and not the first one of photosynthesis. They also showed that the dissolution of the starch is brought about by the enzyme diastase. Brown and Morris noted that sucrose showed the greatest increase during the day of any of the carbohydrates of the leaf and that during the night it was the one most diminished.

Parkin (1912) found in the case of the leaves of snowdrop (*Galanthus nivalis*) that during any single day the percentage of hexoses remained fairly constant no matter in what hour of the 24 the leaves were examined. The sucrose content, however, fluctuated greatly, increasing during the day and diminishing during the night, as may be observed by the following example:

Content	Per cent, dry basis, leaves picked at	
	9 a.m.	4:30 p.m.
Sucrose	9.8	14.7
Hexoses	12.4	11.6
Total	22.2	26.3

Parkin found also that the proportion of sucrose to hexoses decreased as the season advanced. In the early season the quantity of cane sugar was more than double that of the reducing sugars, whereas seven weeks later the two types were present in approximately the same quantities.

Knowledge of the daily variations in the carbohydrates of the leaves was greatly increased by the work of Davis and Daish (1913, 1914), Davis and Sawyer (1916), Davis, Daish, and Sawyer (1916), Davis (1916), and Davis and Sawyer (1916), who developed new methods and perfected old ones for the determination of the carbohydrates of leaf material, and knowledge has been further increased by the more recent work of Langner (1927), Clements (1930), Barton-Wright and Pratt (1930), Kishi and Monobe (1931), Belval (1932), Cole (1932), Denny (1932), Puhr and Hume (1932), Heinicke (1933), Sampson and Samisch (1935), Hartt (1935), and Stanescu (1936). In the leaves of the mangold, Davis, Daish, and Sawyer observed the following facts: (1) During the earlier stages of growth when leaf formation is the principal function, sucrose was present in the leaf tissue in excess of the hexoses, but later in the season the latter sugars largely predominate. (2) All the sugars in the leaf increase in quantity from the first to the final stages of growth. (3) Of the total sugars, the hexoses form a progressively increasing proportion as the season advances. (4) In the first stages of growth practically all the hexoses and half the sucrose disappear during the night, but as the season proceeds a smaller proportion of the sugars disappear each night. (5) The pentosans form a larger and larger proportion of the matter insoluble in alcohol as the season advances.

In the leaves of the potato at the stage of growth when the tubers are beginning to develop, these investigators observed that the principal sugar present is sucrose, which increases in amount from sunrise to 2 P.M. and then falls off during the rest of the day and night. The hexoses, on the other hand, are present in the leaves in very small amounts, amounting to less than 1 per cent of the dry weight of the leaf. After the sucrose had reached its maximum at 2 P.M., the hexoses began to increase in the leaf, owing apparently to the hydrolysis of the sucrose. At the same time, soluble starch or dextrin is first detected in the leaf and its amount increased regularly up to 6 P.M., when the true starch reached its maximum for the day. After this time, both the soluble and the true starch decreased in amount, being converted, apparently, directly into hexoses. In none of the leaves of potato or other starch-bearing leaves were they able to detect maltose. The enzyme maltase evidently acts at once upon the maltose formed in the digestion of starch, so that there is no accumulation whatsoever of this sugar in the leaf.

Kishi and Monobe (1931) found in mulberry leaves that the total carbohydrates were least in amount at sunrise and greatest at sunset. Denny (1932) noted that herbaceous plants show marked changes in the content of carbohydrate from night until morning, while woody plants showed only slight changes during the same period. Puhr and Hume (1932) found that the total sugar in the leaves of corn increases during the day and decreases at night. The maximum amount of sugar occurred from 1 to 4 P.M., and the minimum amount from 1 to 4 A.M. The insoluble carbohydrates reached a maximum between 7 P.M. and 1 A.M., and a minimum between 4 and 7 A.M. Hartt (1935) reported that sucrose increased in the leaves of sugar cane during the day and decreased during the night. This sugar did not completely disappear from the leaves at any time during the day. It was found by Tschesnokov and Bazyrina (1930) that translocation in the potato did not begin until 4 to 6 P.M., reaching a maximum about 8 P.M. but continuing all night. In peas the maximum translocation occurred at the same time as maximum photosynthesis. They believed that this difference is due to the fact that the temporary storage product of potatoes is insoluble, while that of the peas is soluble. Heinicke (1934) believed that the foliage of apple leaves, which functions efficiently late in the season, increases the reserve carbohydrates in the stem tissues, and thus promotes a large leaf area the following spring. McCarty (1935) observed in the grasses, *Elymus ambiguus* and *Muhlenbergia gracilis*, that the trend of starch and sucrose during the season was from a low concentration early in the growth cycle to a high concentration at the close of the growing season. Sampson and Samisch (1935) found in the leaves of *Quercus gambelli* and *Q. kelloggii* in California that the carbohydrates increased very rapidly and remained constant after the initial rise in July. In the Utah species of oaks, however, the carbohydrates continued to accumulate until September. Stanescu (1936) noticed that toward the end of the vegetative period the accumulation of starch occurred later in the morning than during the summer.

In the leaves of *Tropaeolum majus*, *Cucurbita ficifolia*, *Vitis vinifera*, and others, Gast (1917) noted that sucrose was generally in excess of the other sugars, being less at night than during the daytime, and the proportion of sucrose to hexoses decreased during the night. Miller (1924) determined the changes of the carbohydrates in the leaves of corn and the sorghums at 2-hr. intervals during a 24-hr. period. The total sugars in the leaves of these plants began to increase between 4 and 6 A.M., reached a maximum that varied from 12 M. to 5 P.M., and decreased gradually from that time until daylight the following morning. The insoluble carbohydrates were estimated as starch, according to the official method of acid hydrolysis, and thus included besides starch the pentosans and other insoluble carbohydrates that are converted into reducing sugars by boiling with hydrochloric acid. The insoluble carbohydrates thus estimated generally reached a maximum later in the day than the sugars and, after they had reached a maximum, showed little decrease until about midnight, after which they decreased rapidly until daylight. The nonreducing sugars in the leaves of the plants studied were, in nine experiments out of ten, always in excess of the reducing sugars. The nonreducing sugars increased markedly during the day and decreased during the night, while the reducing sugars, as a rule, showed very little increase and the amount present at the different periods of the day was very irregular. The maximum increase in the reducing sugars in the leaves during the day amounted only to from one-tenth to one-third of the increase in the nonreducing sugars. An example of the daily variation of the carbohydrates in the leaves of corn is shown in the following table and Fig. 25. These data were collected from the uppermost fully developed leaves of plants 4 to 5 ft. in height.

Stanescu (1927) noted that in the leaves of *Acer negundo* and *Medicago sativa* the monosaccharoses remained practically constant during the day and were much smaller in amounts than the disaccharoses, which showed a marked increase during the day and a diminution at night. In the leaves of *Urtica dioica* and *Polyanthes tuberosa* the two sugars remained constant during the day, but the disaccharoses were always present in the greater amount. Kishi and Monobe (1931), Cole (1932), and Puhr and Hume (1932) observed that sucrose predominates in amount in the leaves and fluctuates during the day and night the most markedly of all the sugars.

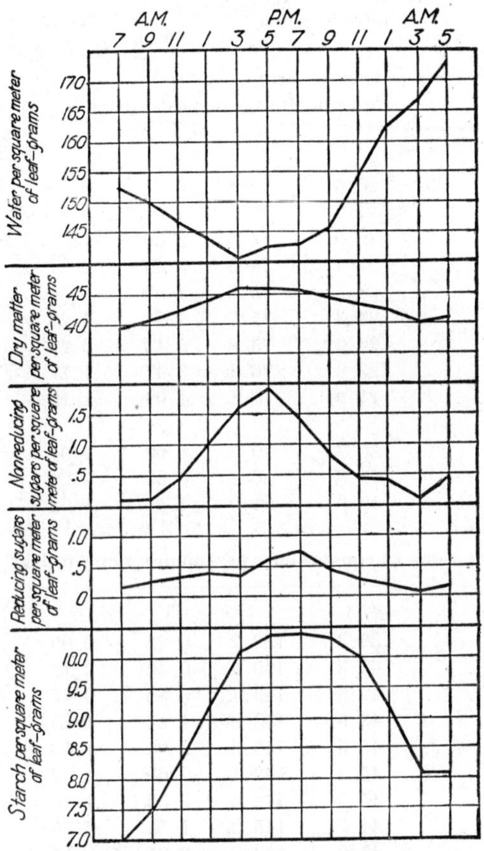

FIG. 25.—Graphs showing the variation in the water, dry matter, and carbohydrates of corn leaves during a 24-hr. period.

c. Variations in Different Portions of the Leaf.—It was noted by Parkin (1912) that the sugar content of the leaf of the snowdrop increased from above downward while the ratio of the sucrose to the hexose diminished. In the mangold leaf, Davis, Daish, and Sawyer (1916) found that the proportion of sugars and matter soluble in alcohol was always far greater in the petioles and midribs than in the blades of the leaves. In the midribs and petioles the hexoses always predominated greatly over the cane sugar. The proportion of hexoses to cane sugar was always very small in the blade of the leaf as compared with the value in the midrib. In passing from blades to midribs, from midribs to the tops of petioles,

DAILY VARIATION OF WATER, DRY MATTER, AND CARBOHYDRATES IN THE LEAVES OF PRIDE OF SALINE CORN, AT MANHATTAN, KANS., JULY 3 AND 4, 1919

Period ending	Dry matter	Water	Sugars			Starch[1]
			Total	Non-reducing	Reducing	
Percentage of the constituents						
July 3:						
7 a.m.	20.6	79.4	0.71	0.25	0.42	17.84
9 a.m.	21.2	78.8	1.01	0.31	0.65	18.72
11 a.m.	22.5	77.5	2.00	1.11	0.79	19.75
1 p.m.	23.4	76.6	3.44	2.34	0.90	21.12
3 p.m.	24.5	75.5	4.59	3.54	0.81	22.34
5 p.m.	24.3	75.7	5.92	4.23	1.41	22.85
7 p.m.	24.2	75.8	5.14	3.22	1.68	22.80
9 p.m.	23.6	76.4	3.47	1.77	1.05	23.13
11 p.m.	21.9	78.1	1.99	1.19	0.70	23.29
July 4:						
1 a.m.	20.7	79.3	1.60	1.04	0.48	21.61
3 a.m.	19.8	80.2	0.60	0.28	0.29	20.18
5 a.m.	19.0	81.0	1.51	1.02	0.40	19.88
Grams of constituents per square meter of leaf						
July 3:						
7 a.m.	39.3	152.2	0.28	0.09	0.16	7.01
9 a.m.	40.3	149.9	0.40	0.12	0.26	7.54
11 a.m.	42.3	146.1	0.85	0.47	0.33	8.35
1 p.m.	43.9	143.9	1.51	1.03	0.39	9.27
3 p.m.	45.3	140.3	2.08	1.60	0.37	10.12
5 p.m.	45.5	142.4	2.69	1.92	0.64	10.14
7 p.m.	45.3	142.6	2.33	1.46	0.76	10.33
9 p.m.	44.7	145.2	1.55	0.79	0.46	10.33
11 p.m.	43.0	153.8	0.85	0.51	0.30	10.01
July 4:						
1 a.m.	42.2	162.5	0.67	0.44	0.20	9.12
3 a.m.	40.0	166.0	0.24	0.11	0.12	8.07
5 a.m.	40.6	173.3	0.61	0.41	0.16	8.07

[1] The term "starch" as here used includes, besides the starch, the pentosans and other insoluble carbohydrates that undergo conversion into reducing sugars on boiling with hydrochloric acid according to the official method of acid hydrolysis for the estimation of starch.

and from the tops of petioles to the bottoms, the ratio of hexoses to cane sugar steadily and rapidly increased. Thus in the blade the ratio of hexoses to cane sugar was 0.307, at the top of petioles 1.42, and at the bottom of petioles 2.48. The excess of reducing sugars over nonreducing sugars in the leaf stems of the sugar beet and in the stalks of wheat has been noted by Colin (1914) and Colin and Belval (1922). Weevers (1923) determined the sucrose and hexose content of the green and yellow parts of variegated leaves. He examined the leaves of *Pelargonium zonale, Acer negundo,* and *Humulus lupulus* among others and found that both cane sugar and hexoses were in the green parts whereas, with few exceptions, sucrose was found only in the yellow parts.

Belval (1932) reported that the sugar in the blade of the leaf is sucrose and that it is progressively transformed into invert sugar in the veins and petioles.

d. The First Sugar Formed.—Owing to observations that (1) cane sugar is generally present in the leaves in considerable excess of the hexoses and shows a marked increase during the day and a decrease during the night while the hexoses remain practically constant and (2) the cane sugar of the leaves is changed into hexoses to be translocated from the leaf, as is evidenced by the excess of hexoses over cane sugar in the midribs and stems, Perrey (1882), Brown and Morris (1893), Davis, Daish, and Sawyer (1916), Parkin (1912, 1925), Gast (1917); and Colin (1916, 1917) considered that cane sugar is the first product of photosynthesis. They considered that the hexoses present in the leaf are more easily accounted for as the products of the hydrolysis of cane sugar than as its precursors. Parkin (1912, 1925) considered that the carbohydrates resulting from the photosynthetic process may be in no case truly up-grade products. The earliest sugar to appear in the free state may be split off from some complex molecule, and this sugar might easily be sucrose in the higher plants. The quantitative variations of cane sugar in the leaves during the day and night and its excess over the hexoses together with the relationship of these two sugars in the midribs and veins do not necessarily indicate that cane sugar is the first sugar formed in photosynthesis. It cannot be determined definitely from chemical analysis whether the marked increase in the amount of nonreducing sugars in the leaves during the day is due to the fact that they are the primary sugars of photosynthesis or to the fact that they are formed from the more simple sugars and accumulate in the leaves during the day as temporary storage products (Miller, 1924). It is just as reasonable to suppose that hexoses are formed and converted into cane sugar as a temporary storage product as it is to reason that cane sugar is the first one formed. Dixon and Mason (1916) by microchemical tests of the photosynthetic cells of several plants found that there was a considerable concentration of hexoses in the chloroplasts or in the protoplasm immediately surrounding them. Other experiments suggest that while sucrose is concentrated in the large vacuoles, invertase is held apart from it in the protoplasm. They considered that probably the hexoses are first formed and that, when the concentration reaches a certain limit, condensation to sucrose due to invertase or some saccharogenic enzyme takes place, and the sucrose thus formed is stored in the vacuole. Since the available space of the protoplasm is small, in sunlight the amount of sucrose will be greatly in excess of that in the protoplasm. On that account, the rise of sucrose on illumination shown by the analysis of leaves is not a cogent argument for regarding it as the primary sugar. The observations of Weevers (1923) also suggested that the hexoses are formed in the plants before the sucrose is formed. Thus the leaves of *Pelargonium zonale*, which were kept in the dark until they were depleted of sugar, produced when exposed to light first hexoses, then cane sugar, and finally starch. The fact that in the variegated leaves in most cases hexoses were found in the green portions while sucrose was confined to the yellow portion also indicated that the hexoses are the first sugars formed. Cane sugar frequently appears in considerable quantities in leaves that contain a large amount of starch. It is contended by those who consider sucrose to be the first products of photosynthesis that this sugar is not a transition product, since so far as any one knows cane sugar does not arise directly from starch. Priestley (1924) suggested that the hexoses could be the primary photosynthetic sugars and that the starch in the leaves could be formed from them. The starch or other carbohydrates could thus be used in the synthetic metabolism of the protoplasm, while cane sugar might arise as a secondary product of the metabolism of the cell and thus have no direct relationship with either the hexoses or the starch. In regard to the question concerning the first sugar formed in photosynthesis, an observation of Miller (1924) is worthy of note. He found that the total increase in the dry matter of the leaves during the day could not be accounted for by the increase in the sugars and insoluble carbohydrates during the same period. The increase in the total sugars and insoluble carbohydrates in the leaves during the day approximated only from 46 to 92 per cent of the total increase in the dry weight of the leaves for the same period. Since the cellulose content of the leaves did not increase during that period, it would appear that certain compounds were being formed which are not detected by the methods of carbohydrate determination as now used. Barton-Wright and Pratt (1930), working with the leaves of daffodil, narcissus, and pseudo narcissus, Clements (1930), with sunflower, soybean, and potato, and Virtanen and Nordlund (1934), with red clover and wheat, believed that sucrose is not the first product of photosynthesis. They considered that the hexoses are manufactured first, and that sucrose is later formed as a temporary storage product from these simple sugars.

From a chemist's viewpoint it would seem that the plant would first form the more simple carbohydrates from carbon dioxide and water and then by condensation form the more complex ones like starch and cane sugar. Before making this assumption, however, one should consider the words of Pfeffer: "It is a very confusing error to presume that an organism in its metabolic economy must follow a course which seems to man, under the influence of existing chemical and physical knowledge, to be the most plausible course" (Spoehr, 1926).

E. Chemistry

The chemical equation for photosynthesis is generally written as follows: $6CO_2 + 6H_2O \rightarrow C_6H_{12}O_6 + 6O_2$. This equation is based upon the fact that the volume of carbon dioxide absorbed is equal to the volume of oxygen emitted and that carbohydrates accumulate as products of the process. The equation thus represents the proportion of carbon dioxide and water used and the amount of oxygen released in the formation of hexose sugar. This equation is simple and shows little. It shows, so to speak, the materials that enter the hopper of the machine and the finished product that comes out at the spout but shows nothing concerning the intermediate stages in the process. Before considering the chemistry of photosynthesis, some statements should be made in regard to the general nature of the process, so that the difficulties involved in the experimental study of this subject may perhaps be better appreciated.

The outstanding fact of photosynthesis is the building up, from carbon dioxide and water, of compounds of higher energy content, the energy being supplied by the radiant energy of the light from the sun. The process of photosynthesis is thus an endothermal reaction, and there are no known analogous reactions in which there is a storing of energy approaching in amount that taking place in this process (Spoehr, 1926). It is through this process, as will be mentioned later, that the energy of the sun becomes available for use on the earth (see Peirce, 1934).

The conditions under which the synthesis of carbohydrates occurs are also baffling, since this process proceeds rapidly in the living plant under ordinary conditions of illumination and at ordinary temperatures. Outside the living plant, however, such synthesis can be brought about only very slowly and under conditions much different from those of the living plant. The photosynthetic process is thus closely associated with or dependent upon the life activities of the plant, but the relationship between this process and the protoplasm or living matter is in no manner definitely defined.

1. The Function of Chlorophyll.—Although it is known that chlorophyll and probably the other pigments of the chloroplasts are necessary for photosynthesis, the manner in which these pigments play a part is not definitely known. Chlorophyll has very characteristic chemical and physical properties and it is possible that it may play both a physical and chemical role in the photosynthetic process. Two of the physical properties of chlorophyll are the characteristic absorption spectrum and its power of fluorescence. It is now generally agreed that the chlorophyll absorbs the light energy necessary for photosynthesis, but the relation of the chlorophyll to the carbon dioxide that is ultimately reduced as a result of the absorbed energy is not known. Beyond the fact that chlorophyll absorbs certain rays of light which furnish the energy for photosynthesis, its functions are purely speculative. It is the intention here to mention only briefly some of the theories relative to photosynthesis. The student can readily discern from the discussion the almost total lack of experimental evidence on the subject, as is indicated by the wide diversity of views on the role that chlorophyll plays.

It has been considered by some that chlorophyll absorbs the energy of light and transmits it to another substance. Thus Usher and Priestley (1906, 1911) considered that the absorbed light energy is transferred to the carbon dioxide, which is thereupon reduced to formaldehyde. Tsweet (1911) was of the opinion that chlorophyll, since it exhibits fluorescence, functions in changing polychromatic light into monochromatic red light which is specifically absorbed by the carbon dioxide. As pointed out by Stiles (1925), however, carbon dioxide possesses no power for the absorption of light within the range of the visible spectrum. The carbon dioxide in the plant cell is present in an aqueous solution, and little is known concerning the absorption of the spectrum by the ions H^+ and $\overline{HCO_3}$ as they occur under this condition.

Baly, Heilbron, and Barker (1921) considered that light is absorbed by chlorophyll and radiated out again at infrared frequencies, which are utilized in the formation of formaldehyde and carbohydrates from carbon dioxide and water. Warburg and Negelein (1923), however, could observe no photosynthesis in *Chlorella* when exposed to infrared light of wave lengths of 800 to 900mμ and only a slight amount in wave lengths of 700 to 780mμ.

Kautsky, Hirsch, and Davidshöfer (1932) concluded that chlorophyll absorbs the light energy necessary for the process of photosynthesis, and passes it on to the system concerned in the assimilation of carbon dioxide. They stated that this energy transfer is associated with a reduction in the fluorescence of the chlorophyll. Dhar (1935) considered that chlorophyll and carotinoids act as photosensitizers and as reducing agents in the photoreduction of carbon dioxide.

Chlorophyll is considered by some to play a chemical part in photosynthesis as well as a physical role. It was suggested by Willstätter and Stoll (1918) that in the process of photosynthesis a compound is first formed between the chlorophyll and carbon dioxide and that light acts on this compound converting it into a substance of peroxide constitution which is then acted upon by an enzyme (Stiles, 1925). The presence of catalase in the photosynthetic regions has recently been determined. Von Euler, Hertzsch, Forssberg, and Hellström (1931) found in certain variegated plants that catalase activity is greater in the green than in the white or yellow portions of the leaves. Inman (1935) believed his observations indicated that whenever green plants are capable of evolving oxygen there is generally present within the cells active catalase that is available for acting on hydrogen peroxide. Emerson (1929) considered that the evidence indicates that photosynthesis involves an autocatalytic reaction, with chlorophyll playing some part in photosynthesis in addition to the absorption of light.

By some it is held that chlorophyll is acted upon by light to form a photochemical primary product that reacts with a derivative of carbon dioxide. Under this view it is considered that the actual reduction of the carbon dioxide or its derivatives is purely chemical, the light producing, by means of the chlorophyll, the reducing substance.

Dixon and Poole (1920) and Dixon and Ball (1922) performed experiments to determine whether the electrons were actually ejected from the chlorophyll by the incident radiation or were merely displaced within groups of atoms or from molecule to molecule of chlorophyll. They considered that photosynthesis is apparently caused by light the frequency of which is too low to effect the expulsion of the electrons from the chlorophyll molecule. The photosynthetic reactions must thus concern the chlorophyll molecule itself. The wave lengths that are effective in displacing the electrons of chlorophyll as indicated by its sensitizing action on the photographic plate are the same as those that are effective in photosynthesis. The theory that the chlorophyll molecule combines with carbon dioxide in the photosynthetic process has also been suggested by Gordon (1929), James (1934), and Baly (1935).

It would seem thus that the atomic groups of the leaf pigment enter into the reactions of photosynthesis and participate in the combinations and decompositions that ultimately lead to the formation of carbohydrates. From these observations, theories of photosynthesis which assume that the chlorophyll itself enters into the reactions are to be preferred to those suggestions which suppose that the reaction is accomplished externally to the chlorophyll by means of energy absorbed and transformed by it.

The role of the yellow pigments in the chloroplasts in photosynthesis is not known. The general indications are that they at least may act in a supplementary way in the process even if they may not be absolutely necessary (Ewart, 1918).

2. The Absorption of Carbon Dioxide by the Leaf.—The entrance of carbon dioxide into the leaf may rightly be considered the first stage in the photosynthetic process. It has been shown in Chap. VII that the capacity of the stomata for the diffusion of carbon dioxide into the leaf is far beyond that which is actually needed for the photosynthetic process provided that the partial pressure of this gas in the leaf is kept at approximately zero. The diffusion of carbonic acid in water, however, is a slow process, and this together with its low solubility in water and its low partial pressure in the atmosphere has long been recognized as conditions which to all appearances would be unfavorable to photosynthesis. According to Spoehr and McGee (1924), 1,000 cc. of water will dissolve 0.033 cc. of carbon dioxide at 25°C. from a concentration of 3 parts per 10,000 by volume. On the other hand, 1,000 g. of the leaves of sunflower with an area of 3.3 sq. m. and with a water content of 850 cc. would dissolve under normal conditions only 0.027 cc. of carbon dioxide if the water were considered as the absorption agency. This amount of leaf material, however, during the process of photosynthesis absorbs as much as 1,500 cc. of carbon dioxide per hour. The leaves of the nettle (*Urtica dioica*) have also been observed to have a high absorption capacity for carbon dioxide. Spoehr (1926), however, was unable to detect any absorption factor in the case of the leaves of spinach, turnip, lettuce, rhubarb, grass, and others, since the amount of gas that was absorbed was equivalent only to that which could be dissolved in the quantity of water present.

Since some leaves absorb a quantity of carbon dioxide greater than can be accounted for by the water present, it has been assumed that there is a substance in the leaf for absorbing this gas so that its absorption is accelerated and its concentration in the photosynthetic cells raised. Willstätter and Stoll (1918) considered that this substance is not related to the pigments of the chloroplasts but that it is an organic compound. Siegfried (1905) found that carbon dioxide combines with amino acids and concluded that the major portion of the carbon dioxide in the leaf forms definite compounds with the amino groups of the proteins and thus increases the concentration of carbon dioxide in the cells. He further considered that these compounds would dissociate to release the carbon dioxide as needed in the cells. Spoehr (1926), however, considered that the amount of nitrogen present is too small to account for all the carbon dioxide absorbed and believed that the indications are that the absorption of carbon dioxide is due to a carbonate-to-bicarbonate reaction. It is not known, however, whether the absorption capacity of a leaf for carbon dioxide has any relationship to the process of photosynthesis.

3. Theories Concerning Photosynthesis.—The great difference between the chemical constitution of a carbohydrate and the carbon dioxide and water out of which it is formed suggests that these compounds are not transformed into sugar at a single step but that a series of compounds increasing more and more in complexity are produced until sugar finally appears as the finished product. Numerous theories have been proposed as to the probable course of the reactions and in regard to the intermediate products that might be formed. It is interesting to note that most of the experimental work relative to the chemistry of photosynthesis has been done in an attempt to verify these theories.

a. Organic Acid Theory.—The first theory to explain the reactions that occur between the intake of carbon dioxide and water and the production of carbohydrates was made by Liebig (1843). He considered that the various organic acids as oxalic, tartaric, and malic were formed as intermediate products in photosynthesis. This theory was based on no experimental evidence and has found little or no support.

b. Formaldehyde Theory.—This theory in regard to the intermediate products in photosynthesis has received by far the most consideration of any that have been proposed. In fact, with but a few exceptions all the theories of photosynthesis that have been proposed assume that formaldehyde is the final stage of reduction in the process. The main differences of opinion in regard to the various theories of photosynthesis have centered largely around the question as to whether the reduction of carbon dioxide to formaldehyde is accomplished in one step or whether there are intermediate products (Spoehr, 1926). The formaldehyde theory was first proposed by Baeyer (1870), and, since this hypothesis has so greatly influenced the experimental work on photosynthesis, it seems desirable to review in considerable detail some of the facts regarding its origin. In 1861 Butlerow discovered that formaldehyde in aqueous alkaline solution condensed to an optically inactive syrup possessing some of the properties of hexose sugars. Baeyer (1870) in discussing Butlerow's results considered formaldehyde in solution to be $CH_2(OH)_2 \rightarrow H_2CO + H_2O$ and suggested that the results obtained by him might be due to the condensation of 6 molecules of $CH_2(OH)_2$ into a hexose sugar with the loss of 6 molecules of water. Baeyer also suggested that this might be the manner in which grape sugar is formed in the plant. He considered that the sunlight might split the carbon dioxide into carbon monoxide and oxygen, the oxygen escaping and the carbon monoxide being held by the chlorophyll. The monoxide is reduced to formaldehyde ($CO + H_2$) by the decomposition of water, and the formaldehyde is then condensed to sugar. Baeyer's hypothesis was formulated without any experimental evidence and was published only as a suggestion, yet it has directed the greater portion of the investigational work on photosynthesis and has become so commonplace in its mention that it is considered by many to be an established fact. Many modifications of Baeyer's theory have been proposed, but these differ but little from the original, and a detailed discussion of these theories is outside the province of this work. The experimental work that has been conducted in an attempt to prove the formaldehyde theory of photosynthesis may be summarized under four general headings as follows:

1. *The Reduction of Carbon Dioxide.*—From the chemical standpoint, the reduction of carbon dioxide is the most important as well as the most difficult process in photosynthesis, and many attempts with many methods have been tried to bring it about. Although these attempts have been made under varied conditions, many of which are not comparable to those of the plant, and have given very contradictory results, they at least show the behavior of carbon dioxide under different conditions and may thus give some little light either directly or indirectly upon the mode of reduction in the photosynthetic process.

Previous to 1907 all the attempts to reduce carbon dioxide had resulted in the production of formic acid. In that year Fenton by passing a stream of carbon dioxide for 18 to 20 hr. through pure water in contact with several rods of amalgamated magnesium obtained a solution that indicated the presence of formaldehyde as indicated by the color tests. Weak alkalies seemed to aid materially in this reaction. In 1906 Loeb by means of the silent electrode and by using something to take up the oxygen, *e.g.*, chlorophyll, obtained formaldehyde and formic acid with ease and in quantity from carbon dioxide and water. He also noted the formation of glycolaldehyde from hydrogen and carbon dioxide under these conditions. The synthesis of a hexose sugar from this compound is apparently a rather simple process. Gibson (1908) formulated a photoelectric theory of photosynthesis in which it was considered that formaldehyde is synthesized from carbon dioxide in the presence of water by feeble electric changes, which he considered occurred in the photosynthetic tissues when adequately illuminated, the light energy being transformed by the chlorophyll into electric energy. He considered that Loeb's experiment strengthened a part of this theory, and it has later (Dixon and Poole, 1920) been established that a slight photoelectric effect occurs in chlorophyll under illumination. Since the work of Loeb and Fenton, the experiments on the reduction of carbon dioxide have centered mainly around the influence of ultraviolet light. Work on the effect of ultraviolet light as obtained especially from the quartz mercury-vapor lamp has been reported by Berthelot and Gaudechon (1910), Usher and Priestley (1911), Stocklasa and Zdobnicky (1911, 1912), Spoehr (1916, 1923, 1926), and Baly, Heilbron, and Barker (1921, 1923). The results that have been reported are very contradictory. Some workers have obtained formaldehyde only in the presence of nascent hydrogen, while others have not noted the need of such. Some have obtained formaldehyde in the polymerized form, some both formaldehyde and sugar, while yet others could obtain no formaldehyde at all. Thus Baly, Heilbron, and Barker (1921) reported that an aqueous solution of carbon dioxide gives formaldehyde when exposed to light of wave length of 200mμ. Spoehr (1916, 1923) in numerous well-controlled experiments was, however, unable to obtain any definite evidence of the formation of formaldehyde from carbon dioxide and water under the action of ultraviolet light. Under the conditions of his experiments, however, he was able to obtain formic acid in small amounts. The reasons for such contradictory results are not known, but Spoehr (1923) suggested that where a direct reduction of carbon dioxide to formaldehyde by means of ultraviolet light has been reported there existed certain essential conditions or factors not known by the investigator, or there were some misinterpretations of the experimental observations.

It is doubtful if the conditions under which carbon dioxide was reduced in the various experiments just reported are comparable to the conditions that exist in the plant. Apparently the ultraviolet or any rays that lie outside the visible spectrum play little or no part in photosynthesis. Photosynthesis, for example, proceeds normally when ultraviolet rays are excluded from the plant, as in the case of their growing under ordinary glass. It would thus, in the words of Spoehr (1916), seem that we have no clear record of a reduction of carbon dioxide or its salts to formaldehyde which could be applied to the process going on in the leaf under the conditions that therein exist.

Considerable work has been done to determine the behavior of isolated chlorophyll and the other pigments in regard to the reduction of carbon dioxide and the formation of formaldehyde. It was reported by Usher and Priestley (1906, 1911) that films of extracted chlorophyll in the presence of carbon dioxide and moist air produced formaldehyde and hydrogen peroxide under the influence of light. The production of formaldehyde under similar conditions was observed by Schryver (1910), who considered that it was in combination with the chlorophyll.

Since then it has been observed by Wager (1914), Warner (1914), Jorgensen and Kidd (1917), Osterhout (1915, 1918), and Ewart (1920) that formaldehyde is formed when chlorophyll is exposed to the light but only in the presence of oxygen in a moist atmosphere. The formation of formaldehyde occurs regardless of the presence of carbon dioxide. When the extracted chlorophyll is thus exposed to the light, the first change observed is its yellowing and ultimate bleaching. In the first stage while bleaching is complete, the quantity of formaldehyde reaches a maximum and then diminishes. It is considered by most of the investigators just mentioned that the aldehyde which appears under these conditions is not produced by the reduction of carbon dioxide but rather by the oxidation or decomposition of chlorophyll under the influence of light and oxygen. It is on that account, therefore, suggested by Ewart (1907) and Wager (1914) and others that the production of sugar may be initiated by the photooxidation of chlorophyll and the subsequent polymerization of the aldehyde thus formed rather than by the direct photosynthesis of carbon dioxide and water. That chlorophyll or some component may decompose during photosynthesis was suggested by the work of Osterhout and Haas (1918). They found, in experiments with *Ulva*, *Potamogeton*, *Spirogyra*, and other plants that had been kept in the dark, that they began photosynthesis as soon as they were exposed to sunlight and steadily increased their rate until a constant speed was attained. Since no limiting factors as light and carbon dioxide were involved, these authors considered that the behavior indicated that sunlight decomposes a substance whose products either catalyze photosynthesis or enter directly into the reaction.

Osterhout (1918) and Burk (1927) employed aniline dyes to ascertain if aldehyde could be produced by them as in the case of chlorophyll. A considerable number gave positive results. Osterhout found methyl green and iodine green to be the most reliable. Burk used malachite green. Osterhout sprayed aqueous solutions of the dyes upon filter paper until the paper was a deep green color when dried. These were then exposed to sunlight under bell jars for several days until the color disappeared. A positive test for aldehyde was obtained in the majority of cases. Baly, Heilbron, and Barker (1921) found also that the photosynthesis of formaldehyde from carbon dioxide and water could be photocatalyzed by certain colored basic substances as colloidal uranium, colloidal ferric hydroxides, malachite green, and methyl orange. Under these conditions the formation of formaldehyde took place in visible light.

When carbon dioxide was exposed to ultraviolet light, Mezzadroli (1928) obtained the maximum yield of reducing substance with calcium bicarbonate in the presence of very finely powdered metallic magnesium. Rajvansi and Dhar (1932) obtained formaldehyde from carbon dioxide and water in the presence of sunlight and chromium oxide, cobalt carbonate, nickel carbonate, manganous chloride, copper carbonate, or nickel sulphate. Qureshi and Mohammad (1932) obtained formaldehyde when a colloidal solution of chlorophyll *a* and solutions of pure malachite green, methyl orange, and copper acetate were exposed to sunlight in sealed bulbs from which carbon dioxide was excluded.

2. *The Condensation of Formaldehyde to Sugar.*—Since it was assumed in Baeyer's hypothesis that formaldehyde is condensed to sugar in the presence of alkalies in the photosynthetic process, numerous experiments have been conducted in an attempt to bring this about in the laboratory. Positive results have been obtained but in many cases under conditions of high temperatures and in concentrations of alkalies that are in no way comparable to the conditions that prevail in the plant (Nef, 1913; Ewart, 1919). It has also been determined that a great complexity of reactions occurs when a solution of formaldehyde is treated with an alkali, so that not only some sugar but also a large number of other compounds are formed. The condensation of formaldehyde to sugar is thus not such a simple process as had formerly been supposed. The attempts to form sugar from formaldehyde under conditions comparable to those existing in the green plant have given conflicting results. Spoehr (1916) placed 3 per cent solutions of formaldehyde in tenth-normal solutions of weak alkalies in glass flasks and exposed them to sunlight for as long as 4 months. The solutions of formaldehyde in calcium carbonate, potassium bicarbonate, zinc carbonate, and basic magnesium carbonate showed no trace of sugar formation. The solutions of formaldehyde in lead hydroxide, calcium hydroxide, and potassium carbonate formed sugar in the dark as well as in the sunlight.

The formation of sugars on the exposure of aqueous solutions of formaldehyde to ultraviolet light was demonstrated by Moore and Webster (1918). In the same year it was reported by Baly, Heilbron, and Barker (1921), Baly (1922, 1924), and Heilbron (1923) that an aqueous solution of formaldehyde is polymerized to reducing sugars in light of wave length of 290mμ. In the presence of paraldehyde, sodium phenoxide, and certain metallic salts that absorb light of wave lengths 290mμ the yield of formaldehyde was materially increased, since these substances protected it from polymerization when it was formed.

In contrast to these investigators, Ewart (1920) concluded that the production of formaldehyde does not form a stage in the synthesis of sugar in the plant and that alkalies do not polymerize formaldehyde to sugar except to a very slight extent. Pronounced sugar formation occurs only when the alkali acts on a polymer such as *p*-formaldehyde, the polyhydrate, or *m*-formaldehyde, and in such cases the endothermal reaction may be very violent and not applicable to the conditions in the plant. Baly (1922) and Heilbron (1923) state that ordinary formaldehyde is not condensed to sugar under ordinary conditions, but they consider that the formaldehyde molecule produced in the process of photosynthesis is in a highly reactive state and has been designated by them as "activated formaldehyde." The formaldehyde of this type is quickly condensed to sugar and cannot be detected in the plant. The formaldehyde that is detected in the leaf they consider to be a secondary product from reactions other than photosynthesis. Baly (1928) by a suspension of pure aluminum hydroxide maintained by a stream of carbon dioxide exposed to ultraviolet light obtained a much greater yield of carbohydrates than if no suspensions were used. The best yield of carbohydrates, however, was obtained by passing a stream of carbon dioxide in water through a suspension of nickel carbonate or cobalt carbonate, exposed to the light from an ordinary tungsten-filament electric lamp. It was found that these particles used for the surfaces must be absolutely free from alkalies or the formation of carbohydrates will not occur. At times, even when freed from all alkali, the carbonates of nickel and cobalt were entirely ineffective in promoting photosynthesis. These suspensions could be activated, however, by heating to 120°C. or by exposure in thin layers to ultraviolet light. These experiments of Baly are important in that they indicate that surface phenomena are of primary importance in the reduction of carbon dioxide to carbohydrates and that sugar is formed directly from carbon dioxide by a process which is physically similar to that of the living plant. The amount of carbohydrate obtained, however, is very small, so it has not been possible to carry on a complete analysis of the different materials obtained in the process. It has, however, been fairly definitely determined that glucose and fructose are present. Emerson (1929), and Zscheile (1932), however, could not duplicate these results of Baly.

3. *The Detection of Intermediate Products in the Leaves.*—Most of the experiments for the determination of intermediate products of photosynthesis have been confined to attempts to identify formaldehyde in the green parts of plants. Many of the reports are of no value, since it has later been found that the methods used were at fault and were in many cases not specific for the substance being determined. The identification of formaldehyde in green leaves, however, has been reported by numerous and accurate investigators among whom are Grafe and Vieser (1906), Kimpflin (1907), Gibson (1908), Fincke (1913), Barton-Wright and Pratt (1930), Pollacci and Bergamaschi (1931), and Sommer, Bishop, and Otto (1933). There is thus little doubt but that formaldehyde occurs in the green parts of plants under certain conditions. It was noted by Von Euler and others (1931) that the green portion of certain etiolated leaves contained from two to five times more aldehyde than the yellow or white portions. There is no evidence, however, that it is a product of photosynthesis, since, as stated by Spoehr and others, there are a number of components of plants which may yield formaldehyde when exposed to light, or it may arise in the general metabolic reactions of the plant.

4. *Nutrition Experiments with Organic Compounds.*—It has been stated in Chap. VI that plants are capable of utilizing in their metabolism certain organic compounds if presented to their roots. This utilization is manifested by increased growth and by the formation of secondary or condensation products as evidenced, for example, by the appearance of starch in the tissue after the plant has been given sugar as a nutrient. Various organic compounds, principally formaldehyde, which are considered intermediate products of photosynthesis have been presented to plants to determine whether the plant could utilize them in the formation of carbohydrates. Formaldehyde has been presented to plants in nutrient solutions, by its introduction into the water in which aquatic plants are growing and in the form of vapor in the atmosphere bathing the aerial parts of the plant.

Bokorny (1891 to 1911) was the first to try the effect of formaldehyde vapor on plants, and his work has been followed by that of Grafe and Vieser (1909, 1911) and Baker (1913). These authors agree that formaldehyde can be used to some extent in the light for the synthesis of carbohydrates and that starch in some cases appears as a secondary product, although in some cases there were chances for errors in the experiments. They also noted that formaldehyde was not utilized in the dark and that the poisonous effects of formaldehyde were always more marked in the dark than in the light. Spoehr (1916) found that formaldehyde mixed with air quickly oxidized to formic acid in sunlight and suggests that formic acid might be the source of carbon in the formaldehyde experiments that have been mentioned. Jacoby (1919, 1922) placed abscised leaves of the nasturtium in the dark with the petioles immersed in water and with the laminae of the leaves surrounded by a humid, carbon dioxide-free air, containing the vapor of formaldehyde. The leaves subjected to the formaldehyde for 24.5 to 32 hr. increased in dry weight from 12 to 15 per cent as compared with an increase of 1.7 to 5.1 per cent in the controls. He considered that the formaldehyde under these experiments was fixed by the plant. Results similar to Jacoby's have been obtained by Sabalitschka and Riesenberg (1924) with *Phaseolus multiflorus* and *Pelargonium.*

The utilization of formaldehyde from nutrient solutions has been studied by Bokorny (1911), Nicholas and Nicolas (1922), and Sabalitschka and Weidling (1926). They found that plants can polymerize formaldehyde thus presented to carbohydrates equally well in the light or dark. The latter investigators found the optimum concentration of formaldehyde in the nutrient solution to be 0.024 per cent. Higher concentrations damaged the enzymes and retarded or inhibited the condensation. They considered that this offers strong confirmation of the hypothesis that formaldehyde is a stage in the production of starch under normal conditions.

It has generally been considered that if formaldehyde or other organic substances could be utilized in plant growth or in the formation of carbohydrates within the plants, this was evidence that they were intermediate products in photosynthesis. There seems, however, to be no grounds for such a conclusion. The organic compounds that plants may absorb and utilize in the formation of sugar and then starch are very numerous, and the experiments that have been performed, especially with formaldehyde, only make that fact the more evident but prove in no way that formaldehyde or any of the other substances so utilized are intermediate products in the process of photosynthesis.

c. Recent Theories.—A summary of the more recent ideas concerning the process of photosynthesis is presented here. James (1934) considered that a diffusion reaction, a photochemical reaction, and at least one thermal "dark" reaction are involved in the process. He proposed the following steps in photosynthesis, which are based largely on the statements of Willstätter and Stoll (1918), and Warburg (1919):

1. The carbon dioxide arriving at the surface of the chloroplasts reacts as carbonic acid, or some simple derivative, with the magnesium of the chlorophyll to form a dissociable addition compound.
2. Under the influence of light this compound undergoes internal rearrangement of the molecule giving an unstable compound, chlorophyll-formaldehyde-peroxide, in which the oxygen is only loosely held.
3. This compound in the presence of a peroxidase readily breaks down by two stages, each releasing ½ molecule of oxygen.
4. The gaseous oxygen escapes from the system, and the formaldehyde eventually polymerizes to sugar.

Baly (1935) suggested that in the process of photosynthesis a complex of chlorophyll *a* and hydrated carbon dioxide is converted by the action of light into chlorophyll *b* and activated formaldehyde, the latter being at once polymerized into hexoses. The chlorophyll *b* thus produced undergoes the dark reaction and is reduced to chlorophyll *a*. The findings of Burr (1936) that there is not a catalyst in the plant for carbon dioxide hydration throws doubts upon the view that carbon dioxide must become hydrated to carbonic acid before reacting with chlorophyll. The rate of the consumption of carbon dioxide in photosynthesis is far greater than the rate of uncatalyzed hydration of carbon dioxide.

Dhar (1935) submitted the following procedure in photosynthesis:

1. Partial activation of carbon dioxide and water, due to their adsorption by chlorophyll and other plant pigments.

2. Further activation of the adsorbed carbon dioxide and water by the absorption of part of the energy available from respiration and from the oxidation of carotin.

3. Absorption of light by chlorophyll, and other pigments, and the dissociation of activated water molecules into H and OH. The activated carbon dioxide molecules are then reduced to formaldehyde by the atomic-H produced from the sensitized photolysis of water.

4. The polymerization of formaldehyde to reducing sugar.

5. The formation of hydrogen peroxide from OH, and its rapid decomposition into water and hydrogen.

The student can easily judge from this discussion of the chemistry of photosynthesis that but little is known concerning the mechanism of the process. The facts that are known are that the process takes place under the influence of the visible portion of the spectrum, that chlorophyll, water, oxygen, and carbon dioxide are necessary, that carbohydrates are formed and oxygen eliminated, and that the process proceeds under the general conditions which are suitable to the activities of living organisms. These facts are not new and have been known in a general way for more than 70 years. The enormous amount of research upon the process of photosynthesis during the past half century has thrown little or no additional light upon the subject. The problem is evidently too complex for specialists in any one field of science to solve, and, in the words of Spoehr (1916): "The subject of photosynthesis requires for its investigation the masterly application of many branches of science—physics, chemistry, plant physiology and the frank cooperation of the most skilled workers in these various fields."

F. Factors That Affect Photosynthesis

The physiological processes that occur in plants are multiconditioned, so that in an investigation of the influence of one factor or condition, the interrelation of this factor with the others concerned in any process must be taken into consideration. It is, therefore, impossible to speak of a single condition as being the cause of an observed effect in a plant (Livingston, 1917), but one must consider that a given result depends upon numerous other external and internal conditions.

The conception of multiconditioned processes was first recognized by Liebig (1843) and was expressed in his law of the minimum in regard to the yield of field crops. This law as commonly stated says that "The yield of any crop always depends on that nutritive constituent which is present in minimum amount." This means, as stated by Hooker (1917), that if the available amounts of the essential plant nutrients are divided by their respective growth values, the smallest quantity obtained gives the maximum amount of growth possible. Liebig, however, apparently did not consider water supply, oxygen supply, temperature, and other factors which influence the yield of a crop as well as the supply of nutrients, so it is not nearly so easy to express the law of the minimum as might be inferred from its original rather simple presentation.

Thomas (1929) reviewed the work of Lagatu and Maume (1924, 1925) who showed that the absorption of nitrogen, phosphorus, and potassium by the grape is irreconcilable with Liebig's law. They found that:

1. The application of an incomplete fertilizer deficient in one of the elements, nitrogen, phosphorus, or potassium, results in an increased absorption of the other two.

2. The lack of balance in a fertilizer containing only two of the principal fertilizer constituents—nitrogen, phosphorus, and potassium—may be such that it will retard the absorption of the omitted one. Under such conditions the plant may take up less of the element in which the soil is deficient than if there had been no addition of the other two principal fertilizer constituents.

3. The lack of nutritional balance, resulting from the addition of an unbalanced fertilizer (except in a nitrogen deficiency), is indicated by decreased yields.

Bartholomew, Watts, and Janssen (1933) stated that Liebig's law of the minimum is not applicable to potassium salts, because the elimination of potassium from a nutrient solution results in an increased absorption of nitrogen and phosphorus by the plants.

Macy (1936) in discussing the limitations of nutrients on the development of plants proposed that the percentage of the limiting nutrient in the plant is directly proportional to the response of the plant to an increase in the supply of that nutrient.

The process of photosynthesis is one that is conditioned by numerous separate factors. It was realized by Pfeffer (1900) and Pantanelli (1904) that the optimum value in photosynthesis was not a fixed value in the case of any condition but might depend upon other conditions. It remained, however, for Blackman (1905) to enunciate this fact clearly and to crystallize the facts in regard to the interaction of factors not only in the process of photosynthesis but in regard to other physiological processes as well. In a consideration of the interaction of factors in the process of photosynthesis he stated his law of limiting factors: "When a process is conditioned as to its rapidity by a number of separate factors, the rate of the process is limited by the pace of the 'slowest' factor." The classical example used by Blackman was the interaction of carbon dioxide and light. Suppose that a leaf has enough light falling upon it to supply sufficient energy to decompose 5 cc. of carbon dioxide per hour. If the leaf is supplied, for example, with only 1 cc. of carbon dioxide per hour and then this amount is increased up to 5 cc. per hour, the rate of photosynthesis will increase from one- to fivefold. After the rate of carbon dioxide is increased beyond 5 cc. per hour, the rate of photosynthesis will remain the same as it was when the rate of carbon dioxide application was 5 cc. per hour. Light in this case is thus the limiting factor in the photosynthetic process. If the intensity of the light is increased until it is sufficient to decompose 10 cc. of carbon dioxide per hour, an increase in the pressure of the carbon dioxide will increase the rate of photosynthesis until the leaf is supplied with 10 cc. of carbon dioxide per hour. If a larger amount of carbon dioxide is supplied, the

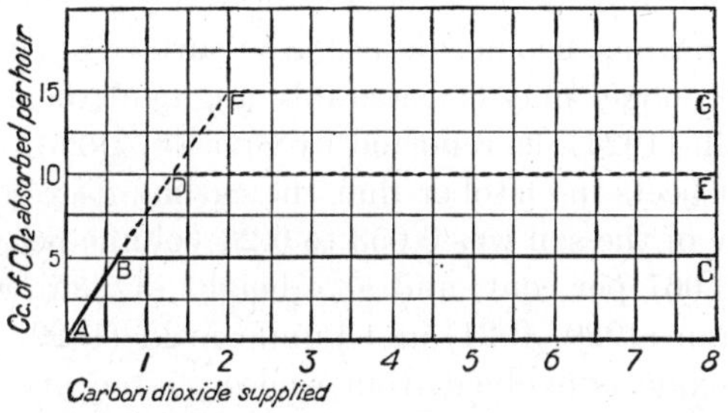

Fig. 26.—Diagram to illustrate the action of a limiting factor. Description in the text. (*After F. F. Blackman, "Annals of Botany,"* 1905.)

rate of photosynthesis will remain the same, since light will at that point again become the limiting factor (Fig. 26). Thus it is impossible to investigate the relation between carbon dioxide supplied and the rate of photosynthesis without considering the factor of light.

The law of limiting factors is an elaboration of the law of the minimum and is of general application where a process depends upon a number of factors. There is considerable difference of opinion regarding the general applicability of the law of limiting factors. Hooker (1917) considered that, although the individual processes obey the law, it does not apply to the questions of general development. Crocker (1918) considered that it is a question whether the law of limiting factors applies to plant activities as generally as or with anything like the rigidity assumed by some. He suggested that the question should not be so much what external factor is the limiting one as what internal condition or inhibitor must this factor act upon to initiate the reaction that is under consideration. Harder (1921), and Singh and Lal (1935) believed that, when the limiting factor comes into operation, the curve is smooth and regular with no sharp break as reported by Blackman.

The factors that influence the rate of photosynthesis may, in common with other factors that influence the general process of plants, be divided or grouped into external factors and internal factors, depending upon whether they are conditions that exist in the external medium or those that exist within the plant owing to its structure or composition. The external factors that influence photosynthesis are the carbon dioxide supply, light, temperature, water supply, and numerous other factors essential to the general growth and vigor of the plant. The known internal factors that play a part in the process are the chlorophyll content and the protoplasmic factors.

1. The Carbon Dioxide Supply. *a. Amount in the Air.*—The amount of carbon dioxide in the air is approximately 3 parts per 10,000 by volume, and this proportion is very constant whether the samples are taken over the sea or over the land. In industrial centers or in cities where large amounts of carbon dioxide are being liberated, the percentage may be increased somewhat, as also occurs near the surface of the soil where the decay of organic matter may considerably augment the average amount. Thus Lundegårdh (1924), as reported by Spoehr (1926), found in a well-fertilized field of beets in October that the carbon dioxide content of the air at the surface of the soil was 0.053 to 0.28 volume per cent; above the leaves, 0.04 to 0.067 per cent; and at a height of 1 m., 0.03 to 0.07 per cent. Bornemann (1920, 1921) and Lundegårdh (1922) considered that more carbon dioxide is evolved from well-manured and well-cultivated land than from that which is not manured or cultivated and considered that one of the advantages of good cultivation is the release of large amounts of carbon dioxide at the soil surface. Keuhl (1926) believed that mineral fertilizers are beneficial to a certain extent in that they cause a greater production of carbon dioxide by hastening bacterial action in the soil. Reinau (1927) believed that the carbon dioxide arising from the soil is of more significance to plants than that in the atmosphere above the plants. He considered that there should be sufficient organic material in the soil to provide for a sufficient flow of this gas around the plant. Appleman (1927) found that the volumetric concentration of carbon dioxide, in the soils of Maryland, was as high as 5 per cent in some cases. Gerlach (1921), however, could obtain no positive evidence that stable or green manures could act as producers of carbon dioxide in amounts that might be beneficial to crops. The fact that the concentration of carbon dioxide is considerably higher near the soil surface than in the upper atmosphere may be of great importance in the growth of plants, since, as will be shown later, carbon dioxide can be used by plants to advantage in much stronger concentration than that which prevails in the general atmosphere.

The carbon dioxide content of the soil atmosphere is always higher than that of the atmosphere above ground. Lundegårdh (1924) stated that it varies from 0.12 to 2.5 per cent by volume, while Banal (1926) stated that the amount is from 5 to 10 per cent higher than that of the general atmosphere.

The carbon dioxide of the air that is combined by the plant is replaced by the decay of organic matter, by the respiration of animals and plants, by the combustion of wood, coal, gasoline, and other products, by the disintegration of rocks, especially limestone, and by the gases escaping from volcanoes. Owing to these replacements the carbon dioxide content of the air remains practically constant. The carbon dioxide content of water should also be noted here. Water can absorb from the air, at 760 mm., 0.04 per cent of its own volume of carbon dioxide at 10°C., 0.03 per cent at 15°C., and 0.02 per cent at 20°C. Sea water contains in both the free and combined state about fifty times as much carbon dioxide as does the atmosphere, but the greater portion of it is not available to the plant.

b. Amount Used by the Plant.—The relatively small amount of carbon dioxide in the air was for a long time a stumbling block in the determination of the source of the plant's carbon supply. It was difficult to conceive that an element which constitutes from 40 to 45 per cent of the dry weight of the plant kingdom could be obtained from the small quantity that exists in the atmosphere. An enormous volume of air must be depleted of carbon dioxide to supply the carbon for a crop, as may be illustrated in the case of corn by the following example:

In 10,000 l. of air there are approximately 3.1 l. of carbon dioxide. Since 1 l. of carbon dioxide weighs approximately 2.0 g. and since carbon constitutes three-elevenths of the weight of carbon dioxide, 3 l. of carbon dioxide contain approximately 1.7 g. of carbon. Thus in 10,000 l. of air there is present in the form of carbon dioxide 1.7 g. of carbon. A single mature corn plant including stem, leaves, grain, cob, and roots contains 364 g. of carbon (Latshaw and Miller, 1924). Thus a single corn plant must deplete $364 \times 10{,}000 \div 1.7$ or 2,141,176 l. of air of its carbon dioxide to obtain the carbon that it has combined. Since 1 l. of air weighs approximately 1.3 g., the weight of the air that would be depleted by such a plant would be $2{,}141{,}176 \times 1.3$ or 2,783,528 g., which is equivalent to 2,783.5 kilos or 3.1 tons. A single corn plant thus to obtain its supply of carbon must deplete 3.1 tons of air of its carbon. Considering 6,000 corn plants to the acre, an acre of such plants would thus deplete 18,600 tons of air of its carbon dioxide in order to obtain the total amount of carbon that these plants contain.

Heinicke and Hoffman (1932) found that the amount of carbon dioxide in 100 l. of normal air varied from 38 to 70 mg. in a warm, unventilated greenhouse. In some cases as much as 25 per cent of the carbon dioxide was removed by passing the ordinary atmosphere over a leaf, but as a rule the removal was less than 10 per cent. The amount of carbon dioxide absorbed during a 1-hr. period per 100 sq. cm. of leaf surface varied from less than 1 to as much as 25 mg.

The evidence in regard to the ability of the plant to use carbon monoxide is confusing. Under some conditions it apparently can be used by the plant to advantage, while under others it has a toxic effect. Thus Bottomley and Jackson (1903) found that young plants of nasturtium grown in sterilized sand and in air in which the carbon dioxide had been replaced by an equal quantity of carbon monoxide did not grow, but that when the amount of carbon monoxide was increased to twenty times the carbon dioxide, the plants grew well and were normal and healthy, provided that oxygen was added in amount sufficient to keep its pressure the same as air. Plants placed in air free from carbon dioxide but containing 10 per cent of carbon monoxide produced abundant starch when placed in the sunlight. They noted also that when carbon monoxide was so used, only one-half the normal amount of oxygen was given off. On the other hand, however, Richards and MacDougal (1904) found that carbon monoxide was very toxic to seedlings of corn, sunflower, wheat, and rice that were grown in an atmosphere in which the nitrogen had been replaced with carbon monoxide but with a normal oxygen pressure.

It has long been known that a plant can obtain all the carbon from the air that it needs for its growth and development, but since the soil atmosphere contains such a relatively high concentration of carbon dioxide the question has frequently arisen whether the plant is able to draw on this supply through the roots and utilize it in photosynthesis. Two early investigators, Godlewski (1873) and Moll (1878), were unable to obtain any starch in the leaves of plants that were grown in air free of carbon dioxide but with the roots growing in soil rich in organic matter. Later investigators, Cailletet (1911), Moillard (1912), Maquenne (1911), Pollacci (1917, 1920), Breazeale (1923), Banal (1926), Stocklasa (1929), Bergamaschi (1932), and Livingston and Beall (1934), however, have obtained evidence that plants may absorb carbon dioxide from the soil to advantage in the process of photosynthesis. Thus, for example, Pollacci (1920) grew plants in an air free from carbon dioxide but with the roots in soil rich in humus or in nutrient solutions containing carbon dioxide and observed that plants of *Zea mays*, *Acer pseudoplatanus*, *Quercus aesculus*, and others lived under such conditions and stored starch. Livingston and Beall (1934) believed that as much as 5 per cent or more of the carbon dioxide used by the plant comes from the soil. They thought that the most probable means of transport of this gas is the transpiration stream. Zijlstra (1909) and Moll (1909) considered that carbon dioxide was only translocated for short distances in plants, and on that account the amount of this gas absorbed by the roots would be of little significance in the process of photosynthesis. There thus appears to be evidence that a green plant may absorb some carbon dioxide from the soil and thus supplement its supply from the air.

c. Concentration and the Rate of Photosynthesis.—The experimental evidence in regard to the increase in the photosynthetic rate due to increasing the normal carbon dioxide supply of the atmosphere is rather confusing. This is due to the fact that many workers did not consider other factors that are known to play a part and to the fact that the interaction of numerous factors in this regard is not understood by the investigators today. A rather detailed discussion of the experimental work in regard to the effect of the concentration of carbon dioxide on the photosynthetic rate will be given, since this subject is of interest not only from a purely scientific standpoint but from the practical viewpoint as well. In 1873, Godlewski reported that the optimum value of the carbon dioxide content for photosynthesis for clear days was between 5 and 10 per cent for the plants with which he worked, while Kreusler (1885 to 1890) noted that the photosynthetic rate increased with the increase of carbon dioxide concentration and reached its maximum when the carbon dioxide content was thirty-five times that of the normal, or approximately 1 per cent of the atmosphere. Under more exact control methods Brown and Escombe (1902) found with concentrations of carbon dioxide lying between 0.6 part and 6 parts per 10,000 of air, when the air was passed over leaves attached to the plant and exposed to various intensities of sunlight, that the rate of absorption of carbon dioxide was strictly proportional to its partial pressure. Thus for example, in the case of the leaves of the sunflower where the mean concentrations of carbon dioxide over the leaf were 2.22 parts per 10,000 and 14.82 parts per 10,000, respectively, the ratio of the partial pressures was as 1:6.6, while the carbon dioxide absorbed by the leaves under these conditions was, respectively, 248.2 cc. and 1,802.8 cc. or in the ratio of 1:7.2. In another case, the ratio of the partial pressures of the carbon dioxide was 1:4.4 and the ratio of the carbon dioxide absorbed by the leaves, 1:5.3. Blackman and Smith (1911) studied the effect of different concentrations of carbon dioxide upon the rate of photosynthesis of *Elodea* and *Fontinalis*, submerged water plants, under like conditions of temperature and light, the carbon dioxide supply being varied through a range of 0.0025 to 0.054 g. per 100 cc. of water. It was found that in the weaker solutions of carbon dioxide the rate of photosynthesis increased directly with the carbon dioxide supply, provided light was not a limiting factor, but that with higher concentrations the rate of photosynthesis became depressed. Boysen-Jensen (1918) and Lundegårdh (1921) working with leaves of land plants found results similar to those obtained by Blackman and Smith. Warburg (1919) and Harder (1921) working on water plants failed to find a direct proportionality between carbon dioxide concentration and the rate of photosynthesis. Thus Warburg (1919) in working with *Chlorella* found that photosynthesis is proportional to the carbon dioxide concentration between 0.05 and 10 where the unit concentration is that of carbon dioxide in water in equilibrium with the atmosphere. Above this limit further increases in carbon dioxide concentration resulted in increased photosynthesis until the process became constant.

The effects of the concentration of carbon dioxide on the photosynthetic rate have been studied more recently by Christopher (1933), Heinicke and Hoffman (1933), Miller and Burr (1935), and Burgess (1935). It was shown by Heinicke and Hoffman (1933), and Christopher (1933) that the rate of photosynthesis falls off rapidly if the air supply to the leaves is less than 2 l. per hour per square centimeter. Miller and Burr (1935) found that many types of potted plants reduce the content of carbon dioxide to 0.01 per cent by volume, and that they maintain this concentration in light of about 2,000 foot-candles intensity. Deneke (1931) stated that the absorption of carbon dioxide by *Ficus elastica* and *Tradescantia pendula*, among others, reached a maximum at a wind velocity of 100 m. per minute, and that it could not be increased appreciably by higher velocities. In this connection, too, it was observed by Blackman and Smith (1911) that plants vary in their ability to assimilate carbon dioxide from the same concentration. Thus under a given set of conditions *Fontinalis* assimilated only one-half as much carbon dioxide as did *Elodea*. They suggested that this difference in utilization might be due to the difference in the absorption ability of the two plants. The behavior of plants toward different concentrations of carbon dioxide and light intensity in relation to photosynthesis has been recently again investigated by Maskell (1928) and James (1928) in regard to both land plants and submerged plants with the thought of accounting for some of the differences recorded by various workers on this problem. In experiments of this type it should be remembered that only the concentration of carbon dioxide in the external medium is open to the direct control of the experimenter, and that the internal rates of diffusion are due to many and varied factors which in many cases are beyond his control. The rate of diffusion from the external surface of the plant to the surface of the chloroplasts is apparently the important determining factor. Thus Maskell (1928) considered that the rate of apparent photosynthesis when carbon dioxide is a limiting factor may be regarded as being determined by a potential, the external carbon dioxide concentration, and a series of resistances, namely the stomatal resistance to diffusion, that in the intercellular space system, the liquid diffusion path up to the surface of the chloroplast, and, furthermore, "resistances" in the photochemical and chemical phases of photosynthesis. The form of the relation between the carbon dioxide supply and photosynthesis may be greatly modified by the alteration in the relative magnitude of the resistance in the diffusion phase, which may be altered by numerous and varied conditions. When the relation of the carbon dioxide concentration to the rate of photosynthesis is considered in this light it is seen why inconsistent results have been obtained. The whole curve expressing the relation between the rate of photosynthesis and the concentration of carbon dioxide when external factors are kept constant may be regarded as consisting of three parts (Stiles, 1925). The first ascending part of the curve, corresponding to the condition where, with increasing carbon dioxide concentration, photosynthesis increases proportionately, is approximately a straight line. This merges with or without a break into a part parallel or approximately so to the axis of the carbon dioxide concentration where light, stomata, protoplasm, rate of diffusion, etc., are limiting the process. This region of the curve then passes over into a descending part, where with increasing carbon dioxide concentration, the rate of photosynthesis is lowered on account of toxic action due to the high concentration of the gas.

The discovery that green plants can increase their photosynthetic rate with an increase of the carbon dioxide content of the atmosphere has stimulated investigations in regard to the utilization of this principle in agricultural practice. Such experiments have been confined mostly to plants growing in the greenhouse, although considerable work has been done under field conditions.

d. Application in Agricultural Practice.—Brown and Escombe (1902) and Demoussy (1903 to 1904) did the first work in applying increased amounts of carbon dioxide to plants upon a relatively large scale. Demoussy obtained increases in dry weight of 158 per cent for those plants growing under conditions of carbon dioxide fertilization. Brown and Escombe, on the other hand, obtained only negative results. They grew the plants in closed roomy compartments in which the carbon dioxide supply varied from three to one hundred times that of normal. They supplied the plants for 12 hr. with this concentration, after which they were exposed to normal conditions for the next 12 hr. They found that a concentration of carbon dioxide as high as 0.1 per cent was decidedly injurious to plants under these conditions and their control plants were always the more vigorous. The treated plants developed abnormally in that their leaf area was reduced; the internodes developed more slowly, while the flower buds did not open, and no fruit was produced. Negative results were also obtained by Cummings and Jones (1918) when plants were grown in closed containers with an increase of the normal carbon dioxide concentration.

It has thus developed, if any reliable results are to be obtained by carbon dioxide fertilization, that the plants must be grown in open compartments under as nearly normal conditions as possible with the exception of the concentration of the carbon dioxide (Bolas and Henderson, 1928). The increased humidity of the air in a closed container is apparently one of the detrimental factors when experiments are conducted in that manner. The purity of the carbon dioxide is a very important factor in this type of experiment (Owen and Williams, 1923; Cerighelli, 1921). The distribution of the carbon dioxide must also be considered. Cummings and Jones (1918, 1920) allowed the gas to diffuse slowly but continuously around the plants during the application period, the only precautions being that the escaping gas did not come in direct contact with the leaves but was liberated a few inches from the plant. In most cases, however, the amounts of carbon dioxide that have been used have not been definitely measured. The gas has simply been liberated alongside or among the plants growing in open containers or in the field. Under such conditions it has been found that there is little or no danger of increasing the carbon dioxide content to toxic amounts. Cummings and Jones, using open boxes 26 in. high, 26 in. long, and 18 in. wide, added during an 8-hr. period as much as 348 l. of gas, the dry weight of the plants increasing with the increased quantities added up to that amount.

In practically all cases where the carbon dioxide supply to field and greenhouse crops has been increased, beneficial results have been obtained as measured by increased yield of grain, fruit, or amount of dry matter produced. In Europe, H. Fischer (1912 to 1927) has been one of the leading investigators in using carbon dioxide as an aerial fertilizer. In some cases he has purified the carbon dioxide from the gases that emanate from smelter furnaces and has conveyed it in pipes to greenhouses and field plots. By this means increased yield and more vigorous growth were obtained with potatoes, beets, tomatoes, and bush beans. Favorable results from the application of carbon dioxide have also been obtained by Riedel (1919, 1921), Cummings and Jones (1918, 1920), Jess (1921), Bornemann (1920), Owen, Small, and Williams (1926), and Gradenwitz (1920). Thus Riedel (1919, 1921) reported that the yields from tomato plants in a greenhouse into which carbon dioxide was introduced through perforated pipes was 175 per cent of the weight of the yield in a greenhouse with ordinary air. Cucumbers under like conditions increased 235 per cent over controls, while spinach, potatoes, and barley increased 150, 190, and 100 per cent, respectively, over the controls. Field plots gave yields varying from one and one-half to three times those of the ungassed plots. Jess (1921) in outdoor experiments with Irish potatoes increased the average weight of the tubers from 140 to 330 g. by the application of carbon dioxide. In the greenhouse the yield of tomatoes was increased from 29.5 kilos for an untreated house to 81.3 kilos for the treated one, while the yield of cucumbers was increased from 138 to 235 kilos. In America the most extensive work on the fertilization of plants with carbon dioxide has been reported by Cummings and Jones (1918, 1920). These experiments were all under greenhouse conditions and the plants were grown in large open containers in the manner that has previously been described. The plants used in these experiments included peas, beans, radishes, potatoes, lettuce, endive, Swiss chard, nasturtium, and strawberries. In the case of beans, the dry weight of the seeds produced by the treated plants was from 32 to 204 per cent higher than the checks, while the dry weight of the pods was from 75 to 143 per cent higher than the controls. The percentage of proteins was practically the same for both treated and untreated plants, while the carbohydrates showed an increase. In this connection it is worthy of note that Kostytschew observed that under conditions of increased carbon dioxide content the rate of photosynthesis is decidedly higher in legumes than in nonleguminous plants. Cummings and Jones (1918) found also, in the case of peas, that the total dry matter of the plant was from 18 to 117 per cent greater than the checks, while the dry weight of the seed increased from 167 to 291 per cent over the control. The results for the potatoes showed that the total number of tubers was 338 for the treated and 231 for the untreated, while the gain in weight of the tubers ranged from 7 to 325 per cent of the controls. Endive, lettuce, and Swiss chard all showed a marked response to treatment with carbon dioxide. In the case of lettuce and Swiss chard, the number of leaves was increased approximately 10 per cent. The leaves of lettuce, however, were darker green and tougher in the treated than in the untreated. The average gain for strawberries in weight was 80 per cent, while the increase in the number of plants was 55 per cent.

Nasturtium plants showed marked and definite response to treatment with carbon dioxide. The gross weight was greater, more flowers bloomed, and blooming began earlier than with the untreated ones. Small and White (1930), and White (1930) tried different methods for supplying carbon dioxide to greenhouse plants. The application of this gas in excess of the amount contained in ordinary air increased the yield of tomatoes but was not profitable. Some of the effects on tomato plants were: a shortening of the period of time between flower opening and the ripening of the fruit, a high percentage of the bloom developing into fruit, and a lower percentage of fruit retarded in ripening. Reid (1930) found that for squash a content of carbon dioxide of 1 per cent, as compared to 0.02 per cent, increased both the fresh and dry weight of the plant. A much greater root system was produced, and a much greater amount of strengthening tissue was developed. A smaller leaf area was formed, and there was a larger amount of starch and reducing substances in the leaves. More of the total nitrogen was found in the roots and stems, and less in the leaves of those plants which were grown in the higher concentration of carbon dioxide. Wilson, Fred, and Salmon (1933) found for clover that as the carbon dioxide in the atmosphere was augmented from 0.03 to 0.1 per cent, the leaf area, excess of carbohydrates, and the number of nodules were increased. Further increments in the amount of carbon dioxide, however, did not induce correspondingly greater development. Georgi, Orcutt, and Wilson (1933) noted that if red-clover plants were supplied with the same quantity of carbon dioxide, but at different partial pressures, the higher the partial pressure of the gas the greater the increase in dry weight, in the amount of nitrogen fixed, and in the number of nodules. They considered that the most important factor is the pressure at which the gas is applied and not the total quantity available. Johnston (1935) found that wheat growing in air enriched with carbon dioxide showed an increase in the number of tillers, weight of straw, number of heads, and number of grains. Bolas and Melville (1935) found that when fertilized with carbon dioxide the tomato plants growing in a small greenhouse increased their yield of fruit 23.9 per cent for the first half of the season, and 13.9 per cent for the whole season.

Lemmerman (1920, 1926), Gerlach (1919, 1926), and Spirgartis (1923) are the only investigators who do not report an increase in plant production from application of carbon dioxide. A general review of the experiments in regard to the application of increased amounts of carbon dioxide to plants is given by Kuiper (1920). While the application of carbon dioxide apparently has possibilities for increasing crop production, the experimental evidence cannot yet be welded into a scientific system or into a working basis (Rippel, 1926). Johnston (1935) stated that the practicability of carbon dioxide fertilization in field experiments is yet an unsolved problem. The mere escape of the gas is in itself not sufficient, but a circulating system must be provided in order to secure a more uniform distribution. The methods of distribution thus far devised, however, are very unsatisfactory. The results obtained will no doubt be varied due to the interrelations of carbon dioxide with light, temperature, moisture conditions, time and duration of treatments, and the specific effect on different plants.

2. Supply of Light. *a. Intensity.*—According to Withrow (1936) the intensity of sunlight at midday in summer in the temperate zone is approximately 10,000 to 12,000 foot-candles. According to experiments in the greenhouse, this is greatly in excess of what plants actually need for their optimum development. Experimental work indicates that an intensity of 2,000 to 3,000 foot-candles is sufficient for an optimum growth, although plants vary widely in that regard. Most plants will grow fairly well and produce seed in an intensity of light approximating 500 foot-candles. They may produce seed in a light intensity as low as 250 foot-candles, although they may be dwarfed in vegetative growth. Artificial light is as yet so costly that it is practical for greenhouse culture only in cases where it might aid in the maturing of a crop that would otherwise be totally lost. As will be mentioned in Chap. XIV, increasing the length of day by the use of artificial light is very practical because light intensities of one-third to 10 foot-candles are sufficient for that purpose. Arthur and Porter (1935) constructed an insulated greenhouse that allowed the rays of the sun to penetrate only through the roof. The energy of the sun was supplemented by artificial light which, together with the rays of the sun, heated the house in addition to furnishing the energy necessary for photosynthesis. In the region of New York City the best growth was obtained during the winter months by supplementing the daylight with artificial light for 5 to 6 hours per day.

The intensity of light at which photosynthesis will occur in various water plants has been given considerable attention. Gail (1922) and Shelford and Gail (1922) found that photosynthesis in the red algae begins at depths at which the red and orange light is reduced to approximately 1 per cent, and it continues to a depth at which the intensity of light is approximately only 0.0032 per cent of the incident rays. Schomer (1934) found in the lakes of Wisconsin that the depth for optimum photosynthesis varied with the plant and with the weather conditions. On bright days the depth for optimum photosynthesis was at 5 m. for *Ceratophyllum* and *Elodea,* while on dark, cloudy days it was at the surface. The maximum rate of photosynthesis of algae occurred at a depth of 7 m. on bright days, while on cloudy days it was at 0.5 m. Tschudy (1934) found that most of the photosynthesis in red and brown algae occurs at a depth of less than 10 m. The red algae can utilize the light that penetrates to a depth of 25 m., while the brown algae seem unable to synthesize carbohydrates below 15 m. of water.

The experimental work on the influence of the intensity of light upon the rate of photosynthesis previous to 1905 is of doubtful value, since the factors other than the intensity of light were not given due consideration. For a review of the early literature on this topic the student is referred to Stiles (1925) and to Brown and Heise (1917). The latter reviewers concluded that this early experimental work does not indicate that carbon dioxide assimilation in plants is proportional to the light intensity. They considered that these experiments indicate a progressively smaller increase of the rate of photosynthesis for each increase in light intensity and that this progressive decrease in the rate of augmentation continues until a point is reached at which further increase in light produces no measurable increase in assimilation.

In 1905 Blackman and Matthaei, however, investigated the influence of light intensity upon the photosynthetic rate with a full realization of the importance of other factors and under well-controlled conditions. They concluded that if the temperature and carbon dioxide are in excess, the rate of photosynthesis is proportional to the intensity of the incident light. For every temperature there is a maximum light intensity which will produce a maximum rate of photosynthesis at that temperature. It has been observed by Osterhout and Haas (1918), Warburg (1919), and Spoehr and McGee (1923) that when plants have been kept in darkness for a period and then exposed to sunlight, the rate of photosynthesis is low at first and then steadily increases until a maximum is reached. The term "photochemical induction" has been given by Stiles (1925) to this phenomenon. Harder (1930) noted in *Frontenalis* and *Cladophora* that carbon assimilation under constant conditions of light, temperature, water, and carbon dioxide reached a maximum slowly in plants that had been darkened for several hours. Li (1929) noted that when certain plants are changed from light of high available energy for photosynthesis to light of low available energy there is an inhibition of this process, and vice versa.

The intermittent application of light has been observed to have a marked influence on the photosynthetic rate. Thus Warburg (1919), as reported by Stiles (1925) and Spoehr (1926), found in the case of *Chlorella* that with alternations of 8,000 per minute the rate of photosynthesis was 100 per cent above that in continuous illumination. With an alternation of four light and four dark periods per minute, the rate of photosynthesis was only 10 per cent above that of continuous illumination. Emerson and Arnold (1932) made measurements of photosynthesis in continuous and flashing light of high intensity. For each flash of light, there were present about 4,800 molecules of chlorophyll for each molecule of carbon dioxide reduced. The length of time required for one unit in the photosynthetic mechanism to complete the cycle of the photochemical and "Blackman" reactions was found to be about 0.02 sec. at 25°C. Spoehr (1926) considered that this behavior might be explained by assuming that during the dark period carbon dioxide is synthesized by the chloroplasts and moved away so that when they are again exposed to light there are available higher concentrations of the necessary material for photosynthesis, which would tend to increase its rate. This phenomenon occurs only under conditions of high light intensity and is not noticeable under conditions of low illumination.

In the process of photosynthesis, carbon dioxide and water are used and oxygen is given off as a by-product, while in the process of respiration, which is proceeding at the same time, oxygen is utilized and carbon dioxide and water are given off as waste products. Since the rate of photosynthesis increases with the intensity of light, while the process of respiration is independent of this factor, there will be a light intensity for every plant at every temperature at which the building-up process of photosynthesis will just equal the tearing-down process of respiration. Under such conditions carbon dioxide and oxygen will be neither absorbed nor set free. This point in the intensity of the incident light has been called the "compensation point" (Plaetzer, 1917). The value of the compensation point varies widely for different species and for different temperatures.

It was observed by Blackman and Matthaei (1905) that if light is a limiting factor, equal areas of leaves of different plants equally illuminated give equal photosynthetic rates. They also observed that plants differ in their photosynthetic rates. Thus at 29.5°C. cherry laurel leaves reached their maximum rate of photosynthesis at 0.36 of full sunlight, while *Helianthus tuberosus* required 0.69 of full sunshine to reach its maximum photosynthetic rate. Dastur and Asana (1932) found with *Allium cepa* and *Helianthus annuus* that the amount of carbohydrate formed in leaves in polarized light was not statistically different from that formed in ordinary light. Later Dastur and Gunjikar (1934) noted that the carbohydrate content of leaves exposed to elliptically polarized light was less than that of leaves exposed to ordinary light.

b. Wave Lengths of Light.—There are numerous factors entering into a solution of this problem which are difficult of estimation and control but which greatly influence any results that may be obtained. Many of these factors were not considered by the early investigators, while the more recent investigators, who realize these difficulties, have not been able satisfactorily to measure or control them. Some of the difficulties that enter into a solution of this problem are: (1) the measurement of the energy in the different portions of the spectrum, (2) the methods of obtaining the different wave lengths used, (3) methods for measuring the rate of photosynthesis, (4) the relative absorption of the different wave lengths by different plants, (5) and variation in the light supply.

The relative distribution of energy in various spectra varies widely, so, unless the sources of energy used are the same, widely different results will be obtained. Sunlight also varies in intensity and composition at the surface of the earth, from time to time, so different portions of the spectrum may vary independently of each other. It should also be noted that, if equal intensities of different lights are presented to the plant, the plant exhibits selective absorption so the greatest rate of photosynthesis will be brought about by the rays that are most absorbed regardless of what would occur if the plant absorbed them equally. On account of the difficulties that have just been enumerated, the investigations in regard to photosynthesis and wave length have been productive of but few data that are conclusive.

The range of wave lengths of light in which photosynthesis may occur was investigated by Timiriazeff (1890, 1904) and Ursprung (1917) by means of the iodine-starch method. By this method the spectrum is thrown upon the leaf for a period, after which the chlorophyll is removed and the leaf is tested for starch with iodine after the usual manner, the depth of color being taken as an index to the rate of starch formation in the various regions of the spectrum. In *Phaseolus multiflorus* (Ursprung, 1917) exposed to various sources of light, starch formation was found to occur throughout the whole visible spectrum if sufficient time were allowed. Starch formation was thus found to occur between the wave lengths 330 and 760mμ. The maximum activity apparently was in the region 687 to 656mμ. In the normal solar spectrum, starch formation in the blue requires a longer time than in the red. Ursprung (1918) reported that he was able to observe slight starch formation in the infrared portion of the spectrum in *Phaseolus vulgaris* which had been exposed for 40 hr. The starch-iodine method for determining the rate of photosynthesis for an experiment of this type is, however, open to serious objections. The formation of starch is, without doubt, secondary to photosynthesis and can occur in the dark as well as in the light, provided a supply of sugars is present. The formation of starch might be influenced by light in an entirely different manner from that influencing the process of photosynthesis. The role of the ultraviolet rays in photosynthesis apparently is of little or no consequence, as has been mentioned in the discussion of the theories of photosynthesis. Plants can carry on normal photosynthesis under conditions that exclude ultraviolet rays, while the experimental attempts to produce photosynthesis in plants in ultraviolet light are not convincing.

The relative value of the different rays of light in the process of photosynthesis was studied by Kniep and Minder (1909) for *Elodea canadensis*. They measured the energy value of red, blue, and green rays incident upon the plant and determined the rate of photosynthesis by the oxygen-bubble method. Their observations indicated that the red and the blue lights of the same intensity gave the same rate of photosynthesis, but they could observe no photosynthesis in green light. This work, however, although a great step in the direction of properly controlled experiments of this type, gave results that were unreliable, since the screens used allowed rays other than those desired to pass, unless successive screens were used, which diminished the intensity of the light to a point where the intensity became a limiting factor. Lubimenko (1923) in experiments with 8 species of plants used monochromatic light with red rays from 760 to 600mμ and blue rays from 480 to 400mμ and assumed the intensity of the blue rays to be 85 per cent of the red. The experiments carried on at 20°C. for 6 hr. in a clear sky with a carbon dioxide content ranging from 9 to 11 per cent showed that the rate of photosynthesis was greater in red than in blue light, being double in the former as compared with the latter. The rate of photosynthesis was retarded with time in both red and blue light. In species adapted to diffuse light of low intensity, as *Hedera helix* and *Aspidistra elatior*, the activity of the blue rays equalled or exceeded that of the red rays. Wurmser (1925) in experiments with *Ulva* and other sea algae reported that the liberation of oxygen in green light of 490 to 590mμ was higher than that in the red rays of 590 to 700mμ, the rate of green to red being as 1.15:1.00.

Stiles (1925) considered that the experimental evidence that has been presented indicates that, with equal intensity of incident light, photosynthesis is influenced by wave lengths, being greatest in the red between the B and C lines and least in the blue violet. Warburg and Negelein (1923) agreed with this in that they consider that the rate of photosynthesis decreases with decreasing wave length. They further considered that there is no apparent relation between the efficiency of photosynthesis and the absorption bands of chlorophyll. Thus the yield of photosynthesis in the red, a region of high absorption, is greater than in the green, a region of low absorption, and the yield in the green is higher than that in the blue, the region of highest absorption (Spoehr, 1926). Since there are different degrees of absorption of the different wave lengths by the leaf cells due to specific differences of the cells themselves or to the thickness of the leaves, it would appear to be impossible to find values for the relative rates of photosynthesis, in light of different wave lengths, which will hold in general for all plants and all conditions.

Dangeard (1927) noted in *Chlorella vulgaris* and *Scenedesmus acutus* that the amount of growth in the various regions of the spectrum was closely correlated with the intensity of absorption by the chlorophyll. Tottingham and Moore (1931) found that the most consistent compositional response of plants grown under Vitaglass was an increased percentage of the lipides in the dry matter. Meier (1932) reported that the rays of ultraviolet light from 2536 to 3022 Å. killed cultures of some of the green algae. The wave lengths longer than 3022 Å. had no appreciable lethal effect. The ozone in the upper atmosphere serves as a filter for the injurious rays of the sun and thus protects organisms against them. Arnold (1934) reported that an unidentified unit in the mechanism of the photosynthesis of *Chlorella pyrenoidosa* is rendered inactive by the absorption of 1 quantum of ultraviolet light of a wave length of 2537 Å. Burns (1933, 1934) found that the infrared rays longer than 1,100mμ are detrimental to photosynthesis, and that white pine and Norway spruce were able to utilize all the visible spectrum except a portion of the blue and violet. It was reported by Dastur and Samant (1933) that the production of starch, total carbohydrates, and total sugars is much greater in diffuse daylight than in artificial light. He did not attribute these differences to any variations in the intensity of the light but to the difference in its quality. Dastur and Mehta (1935) concluded from the use of artificial light that photosynthetic activity is highest in white light, intermediate in red light, and very feeble in the blue-violet region of the spectrum. Thus the rate of photosynthesis does not depend alone on the total energy content delivered to the plant, but also on the distribution of radiation of different wave lengths. Arthur (1929) found that plants do not develop normally unless exposed to both the red and blue rays of the spectrum. The ultraviolet region of sunlight, which is not transmitted by greenhouse glass, is of no definite benefit to growing plants, and the ultraviolet region beyond a wave length of 290mμ is very injurious to green plants.

3. Temperature. *a. Coefficient.*—The temperature coefficient, as has been stated in Chap. II, is denoted by Q_{10} and is the ratio of the rate of a reaction or process at a given temperature to the rate of the process at a temperature 10° lower. The temperature coefficient for purely physical processes is relatively low, being from 1.2 to 1.4 at the temperatures that ordinarily prevail in the plant. The temperature coefficient for chemical reactions are, on the other hand, relatively high, ranging from 2 to 3, while relatively low temperature coefficients ranging from 1 to 1.4 are characteristic of photochemical reactions.

It was observed by Blackman and Matthaei (1905) that between 5 and 25°C. the temperature coefficient of photosynthesis in cherry laurel was 2.1 and for artichoke 2.5, which suggests that it may be different for different species. Osterhout and Haas (1918) noted that the temperature coefficient of photosynthesis in the case of *Ulva rigida* between 17 and 27° was 1.81. Since light reactions ordinarily have low temperature coefficients, they assumed that the process of photosynthesis involves a light reaction with a low coefficient followed by an ordinary chemical reaction with a high temperature coefficient, so that as a result the temperature coefficient of the process as a whole is low. This is in line with the idea that photosynthesis is a complex of reactions falling under a diffusion phase, a photochemical or light phase, and a chemical or dark phase (James, 1928). In a critical review of the literature in regard to the influence of temperature on the rate of photosynthesis, Brown and Heise (1917) considered that the work, for the most part, indicates temperature coefficients which are of the same order of magnitude as photochemical coefficients. The temperature coefficient of photosynthesis is apparently influenced by the chlorophyll content of the leaves. Thus Willstätter and Stoll (1918) found that the leaves of *Ulmus* with low chlorophyll content showed a temperature coefficient of 1.34 as compared to 1.5 between 15 to 25°C. for leaves with a high content of the pigment.

When neither light nor the concentration of carbon dioxide is a limiting factor, the rate of photosynthesis is increased by an increase in temperature just as are many chemical reactions. However, when the light intensity is low, temperature has little effect—a characteristic of photochemical reactions. It is thus considered that when the light intensity is low a photochemical reaction determines the rate of photosynthesis, whereas when the light intensity and the supply of carbon dioxide are both high, a "dark," chemical reaction, the rate of which is determined by temperature, limits the whole process. The existence of the "dark" reaction was first established by workers in the laboratory of F. F. Blackman, and it is now frequently termed the "Blackman" reaction (Gordon, 1929; and Kohn, 1935).

Time, successive periods, hours	Rate of photosynthesis, grams of CO_2 combined per 50 cm.^{2} of leaf per hour, at	
	8.8°C.	37.5°C.
2	0.0039	0.0574
2	0.0039	0.0352
2	0.0038	0.0278
2	0.00385	0.0218

The temperature coefficient of photosynthesis has been investigated with the idea that it might throw some light on the nature of the reactions involved; however, the temperature coefficient in photosynthesis is not so indicative of the nature of the process as it is in a simple reaction. The process of photosynthesis consists of a series of reactions that are influenced by numerous factors, any one of which, other than temperature, might be the limiting one.

b. Range.—The threshold temperature of photosynthesis varies greatly for different plants. Matthaei (1905) found that photosynthesis took place in the leaves of cherry laurel at −6°C. Henrici observed photosynthesis in lichens at −20°C. and in alpine plants at −16°C. Ewart (1896) stated that the evolution of oxygen ceases between 0 and 2°C. in the case of warm, temperate, and subtropical plants and between 4 and 8°C. for tropical plants. Provided that no other factor is a limiting one, the rate of photosynthesis increases with an increase in temperature until about 25°C., as observed by Matthaei (1905) and Blackman and Matthaei for cherry laurel. Above a temperature of 25°C., however, complications arise. Below 25°C. the rate of photosynthesis remains constant at a given temperature hour after hour, but above 25°C. it decreases with time so that the initial rate is not maintained. The falling off of the initial rate is irregular; it is more rapid the higher the temperature; and at any temperature above 25°C., it is greater at first and subsequently becomes less rapid. The data in the table on page 594, taken from Matthaei (1905), illustrate the behavior of photosynthesis in the leaf of cherry laurel at a relatively low and a relatively high temperature.

Owing to this falling off in the rate of photosynthesis at the higher temperatures, the previous history of the leaf as regards temperature would determine the value of the photosynthetic rate. Thus the temperature at which the highest rate is observed will depend upon the time that has elapsed between the commencement of the experiment and the measurement of the photosynthetic rate. On this account, some regard the optimum temperature as the highest temperature that can be maintained continuously without a resulting depression of the process under consideration.

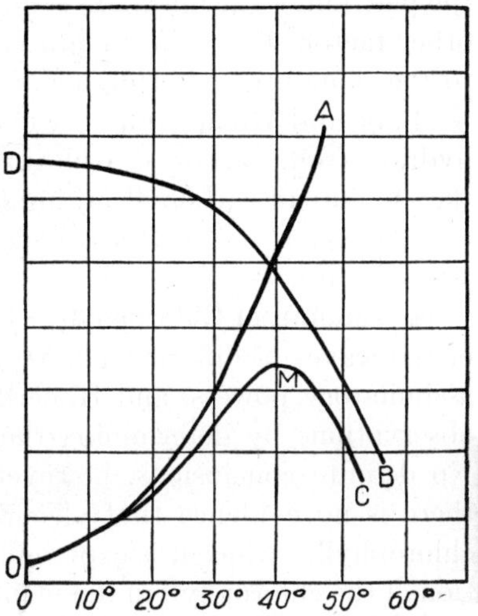

Fig. 27.—Graphs showing the relation between temperature and enzyme action. (*Redrawn from Jorgensen and Stiles*, 1917, *after Duclaux*, 1899.) The graph *OA* shows the relation between temperature and enzyme action if the enzyme activity remains unimpaired. The graph *DB* represents the relation between temperature and quantity of enzyme. The graph *OMC* represents the actual graph between temperature and enzyme action.

The behavior of photosynthesis at higher temperatures appears similar to that of hydrolytic enzymes (Blackman, 1905). Thus Kjeldahl (1879) noted that malt diastase hydrolyzed increasing quantities of starch up to 63°C., after which the reaction falls off rapidly, becoming nothing at 86°C. This decrease in the reaction is apparently due to the destruction of the enzyme by heat. This characteristic behavior of enzymes was further investigated by Tammann (1892, 1895) and by Duclaux (1899) and has been termed "Tammann's principle." This principle is shown by the following curves (Fig. 27): The curve *OA* shows the relation between temperature and the enzyme action, provided the reaction is in no way hindered by the temperature. The curve *DB* represents the relation between the temperature and the quantity of enzymes, since it is being destroyed by heat. The curve *OMC* represents the actual curve between temperature and the enzyme action. The apparent production of an optimum is thus due to two opposed processes, the increased hydrolytic action of the enzyme and the destruction of the enzyme due to heat. Evidence was obtained by Molisch (1925) which indicated that an enzyme might be concerned in the photosynthetic process. He observed that green leaves from a wide range of plants dried at 30 to 40°C. and pulverized liberated oxygen when illuminated. Leaf material, however, that was heated to 96°C. did not thus behave, while material prepared from leaves that were killed by freezing could liberate oxygen in illumination. If Tammann's principle or some similar one is involved in photosynthesis, the position of the optimum is not a definite fixed point but depends upon other factors that will remain constant only if the plants are subjected to the same previous treatments (Jorgensen and Stiles, 1917).

Baly and Hood (1929) claimed in the artificial formation of carbohydrate from carbon dioxide in the presence of nickel carbonate that the amount of carbohydrate produced was a linear function of temperature between 5 and 31°C., beyond which the yield fell off with rising temperature.

4. Amount of Chlorophyll.—In 1879 Weber observed that equal areas of the leaves of different plants under the same conditions had different assimilatory powers, and Haberlandt (1882) attempted to explain these observations by determining the number of chloroplasts per unit area. No definite conclusions, however, can be drawn by this method, since there is no evidence that all chloroplasts contain the same amount of chlorophyll. Since it is experimentally impossible to vary the concentration of the chlorophyll in the plant, the relation of the chlorophyll content to rate of photosynthesis has been studied by using plants from different habitats, young and old leaves, green and yellow leaves of different varieties, and etiolated plants.

a. Plants of Different Habitats and of Different Varieties.—It was observed by Lubimenko (1905, 1908) that shade plants could accomplish the same amount of photosynthesis with a lower illumination than the sun plants and that the chlorophyll content of the former was higher than that of the latter. The sun plants with the low chlorophyll content showed the maximum rate of photosynthesis at the highest light intensities, while the shade plants showed a decreased rate at the same intensities. He considered that leaves with a high chlorophyll content have a high absorption coefficient and that the optimum temperature and light intensity for photosynthesis decrease with chlorophyll content. Henrici (1918) made observations somewhat in accord with those of Lubimenko, since she noted that lowland plants may have as much as 2.3 times more chlorophyll than the alpine plants of the same species, and in strong light the rate of photosynthesis in the alpine plants was higher than in the lowland plants. She stated further that alpine plants assimilate more rapidly than lowland plants when red rays predominate around sunset but less rapidly than lowland plants when blue rays predominate.

Willstätter and Stoll (1918) in their discussion of the relation of the chlorophyll content to photosynthesis introduced the term "assimilation number" or "photosynthetic number." The photosynthetic number may be defined as the number of units of carbon dioxide absorbed per hour per unit of chlorophyll. It is the ratio of the number of grams of carbon dioxide absorbed per hour by a unit of leaf to the number of grams of chlorophyll contained in that unit. If the rate of photosynthesis depends only on the chlorophyll content, then the photosynthetic number should be constant, but if other factors than chlorophyll content influence the rate, the photosynthetic numbers will show variations. These authors determined the photosynthetic number for a large number of plants at a temperature of 25°C., with a 5 per cent carbon dioxide supply and a supply of light equal to 48,000 lux. The following examples taken at random show the variation of the photosynthetic number for different plants: *Cucurbita pepo,* 12.1; *Helianthus annuus,* 10.9; *Hydrangea opuloides,* 6.5; and *Acer pseudoplatanus,* 5.2. In experiments with light-green or yellow and normal green varieties of plants, Plester (1912), Willstätter and Stoll (1918), and Fleischer (1935), observed that the rate of photosynthesis increased with the chlorophyll content, but they were not able to establish any relation between these two factors.

Briggs (1935) stated that when illumination is weak the assimilation of young leaves increases with age even when the chlorophyll carbon content is maintained at a constant value. He suggested that chlorophyll combines with carbon dioxide, is then activated by light energy, and subsequently is broken down into free chlorophyll and other products. Apparently when the concentration of carbon dioxide and the intensity of illumination are very great, all the chlorophyll will be in the form of the activated compound, so that the rate of reaction will be proportionate to the amount of chlorophyll present.

Ireland and Yeats (1933) found a positive correlation between the content of chlorophyll and the yield of kafir. Sprague and Curtis (1933) believed that the mean values for chlorophyll concentration and total chlorophyll for selfed lines of corn are reasonably reliable indexes of the total yield that will be produced by their hybrids. Ulvin (1934) reported that the percentage of the dry matter of corn leaves increased in direct proportion to the chlorophyll content.

b. Etiolated Plants.—In experiments with etiolated plants of barley, *Vicia faba,* and *Phaseolus,* Irving (1910) and Briggs (1920) concluded that etiolated leaves do not possess any appreciable power of photosynthesis. They observed further that the development of photosynthetic activity in the young leaves lagged behind greening and did not appear until after the leaves had attained a full green color, when it developed very rapidly. They considered that photosynthetic activity is dependent upon some other factor than chlorophyll and is not developed so quickly as the pigment that remains idle until the other factor becomes active. Willstätter and Stoll (1918) were not able to duplicate the results of Irving (1910), but according to Briggs (1920) the difference in the results is due to the difference in the age of the material used. In the case of seedlings of plants such as *Helianthus annuus*, *Acer*, and *Cucurbita*, where the first photosynthetic organ is one that also serves as a storage organ, Briggs (1922) observed that the photosynthetic activity is fully developed at germination. With this type, photosynthesis under natural conditions of growth in the field showed no lag behind greening. In such plants as *Phaseolus*, *Ricinus*, and *Zea* where the seedling develops a specialized photosynthetic organ different from the storage organ, the photosynthetic activity was not developed until some time after germination, and photosynthesis under natural conditions shows a corresponding lag. Inman (1935) noted that, after etiolated plants of corn, wheat, and oats were transferred from the dark to the light, the evolution of oxygen began about the time that the eye could detect the appearance of a green color in the leaves.

At the present time it is definitely known that only those cells which contain chlorophyll are capable of photosynthesis, but the quantitative role of the pigment in this process has not been established.

5. Water Supply.—Water plays both a direct and an indirect role in the process of photosynthesis. It plays a direct role in that it enters into chemical combination with carbon dioxide to form carbohydrates in the process. It plays an indirect role in that it maintains the turgidity of the leaf cells and thus keeps the stomata open and allows an unobstructed intake of carbon dioxide. Since the amount of water used in chemical combination is relatively small, the greatest hindrance to photosynthesis due to a limited water supply comes from the secondary effect of the closure of the stomata. Thoday (1910) found that the turgid leaves of *Helianthus annuus* carried on photosynthesis approximately ten times more rapidly than did leaves that were wilted to drooping during the experiment. In withering leaves of *Tropaeolum majus*, Schroeder and Horn (1922) and Horn (1923) noted that starch was broken down into sugar and that the cane sugar steadily increased with the decreasing water content. Iljin (1923) found that when the water content of the leaves of *Bidens tripartita* was reduced 43 to 44 per cent, a reduction of 53 to 78 per cent of photosynthesis occurred. In *Phlomis pungens* a water loss of 34 per cent caused a reduction of 13 per cent in photosynthesis. In these experiments the full activity of photosynthesis was not regained at once upon the recovery of normal turgidity and in some cases showed a reduction of 29 per cent 16 hr. after complete recovery of turgor. Iljin (1915) considered that plants that are adapted to a dry habitat lose less water per amount of carbon dioxide decomposed than do mesophytes. Owing to water loss the stomata of the mesophytes close, and as a result the intake of carbon dioxide is reduced and photosynthesis is checked, while in the xerophytic plants it is going on at the normal rate. It was observed by Brilliant (1924) that, with a loss of 41 to 63 per cent of the water content of the leaves of *Hedera helix* and *Impatiens parviflora*, photosynthesis was almost stopped. The process was at its maximum when the water content was reduced only 5 to 15 per cent. Young plants are not so much affected by the loss of water as are more mature ones. Dastur (1924, 1925) also found that there is a direct correlation between the decrease in the rate of transpiration and the fall in the water content, per unit of leaf area. Those plant tissues, placed most unfavorably in regard to their water supply, are the first to lose their power of photosynthesis, this inhibition appearing first in the marginal and intravascular regions and then gradually proceeding inwardly toward the central and main fibrovascular bundles. Yuncker (1916), however, found that the greater amount of dry matter per unit of leaf area was produced in those corn plants that showed the lower amount of transpiration and that grew in the drier soil. He considered that this was due to the more active respiration and growth in the leaves with the higher water content. The leaves with the lower water content were carrying on respiration less rapidly, and as a consequence dry matter accumulated. Skvortzov (1931), Dastur and Desai (1933), Smyth (1934), Childers and Cowart (1935), and Heinicke and Childers (1935) have shown that the water content of the leaf is closely allied with the process of photosynthesis. Any fluctuation in the water content is almost immediately followed by a similar fluctuation in the rate of photosynthesis. Iljin (1929) found more sugar in leaves in dry than in wet habitats. The leaves of *Rumex crispus* and *R. conglomeratus* contained almost twice as much sugar when withered as when turgid. The lack of translocation may have been a factor in this case. Molisch (1921) noted that wilting hastens the disappearance of starch from starch-filled leaves, and Schroeder and Herrman (1931) found in wilted nasturtium leaves that the sucrose content increased at the expense of starch. Hartt (1934) found evidence that a greater synthesis of sucrose occurred in the leaves of the sugar cane when the plant was supplied with water than in those from which it was withheld, although some photosynthesis occurred at the wilting point. McKay (1935) found that photosynthesis began in 6 to 10 min. after water was added to drought-treated plants of the xerophytic moss, *Grimmia montana*. Bernbeck (1924) showed that a deficiency of water retards the rate of photosynthesis independently of the closing of the stomata.

6. Other External Factors.—It has been shown by Hoffman (1932, 1933) and Christopher (1935) that lime-sulphur, flotation, and oil sprays decrease the rate of photosynthesis, the decreases ranging from 17 to 40 per cent.

Fisher (1934) found in various apple varieties that the size of the fruit was positively correlated with the leaf extent per apple. He considered that this relationship is due to the greater amount of food synthesized per fruit. The weight of fruit produced per unit of leaf surface was least in the McIntosh variety and greatest in Rome Beauty. He suggested that these results may be due to differences in the photosynthetic efficiency of the leaves of these two varieties. Gustafson and Stoldt (1936) found with the tomato that the efficiency of the plant in setting fruit is highest when the leaf area per fruit is small. After setting, however, the size of the fruit is increased by a larger leaf area.

It was found by Eidelman and Bankul (1933) that the removal of some of the leaf blades of wheat and barley intensified the photosynthetic rate in the remaining ones. Lubimenko and Stscheglova (1932) removed 20 to 30 per cent of the leaf surface of barley, bean, oats, and buckwheat by perforating the leaves. The amount of carbon dioxide absorbed was diminished for the first 2 days after this treatment. A considerable increase in the absorption of carbon dioxide then occurred, and the process reached a maximum in 6 to 8 days. In some cases the perforated leaves were absorbing more carbon dioxide than the controls after a period of 18 days. By perforating the leaves of wheat and barley to a degree not to exceed 10 per cent of the surface, Shcheglova and Chernysheva (1933) obtained increases in the rate of photosynthesis that persisted for several days. Eldredge (1935) found with corn in Iowa that total leaf removal, by stripping at weekly intervals from the four-leaf stage to the tasseling stage, caused reductions in yield in almost direct proportion to the percentage of leaves that were unrolled when the injury was inflicted. Thus the reduction in yield varied from 9 per cent in early June to 100 per cent on July 20, at the pre-tasseling stage. The severe shredding that removed about 50 per cent of the open leaves caused losses in yield ranging from 2 per cent, when this mutilation occurred in early June, to 50 per cent when it was performed at the pre-tasseling stage.

Curry and Trelease (1935) showed that with 99.9 per cent of deuterium oxide the rate of photosynthesis in *Chlorella* was only about 0.4 per cent of that with ordinary water. Wallace (1932) noted that sublethal doses of ether increased the rate of photosynthesis. This was due possibly to the increase in the opening of the stomata.

7. Internal Factors.—The chlorophyll and the protoplasmic structure of the chloroplast are mutually dependent upon one another. Chlorophyll cannot carry on photosynthesis when separated from the plastid, and the plastid devoid of chlorophyll is totally unable to carry on this process. The nature of the mutual dependence of the chloroplast upon the general cytoplasm of the cell, however, is not definitely known. It is considered by some (Ewart, 1898) that chloroplasts isolated from the general cytoplasm can still function in photosynthesis, while others (Kny, 1897, 1898) believed that they are unable to do so under such conditions. In general, however, it may be stated that the process of photosynthesis is intimately associated with the vital activities of the plant, so any factors that affect the protoplasm will have their effect upon photosynthesis.

Pickett (1933) determined the average cross-sectional area of the intercellular spaces of various varieties of apples. In a portion of leaf magnified 900 times the areas of these intercellular spaces were as follows:

Variety	Average area, sq. in.	Variety	Average area, sq. in.
Delicious	8.8	York	12.7
Jonathan	9.5	Wealthy	12.7
Gano	10.4	Liveland	17.7
Winesap	12.4		

Pickett (1934) found that the leaves of Liveland apples have a greater area of intercellular spaces than those of the Delicious variety. The rate of photosynthesis is the higher in the former variety, and he believed that the more extensive intercellular spaces are conducive to greater photosynthetic activity.

The data in regard to the effect of the oxygen supply upon photosynthesis are conflicting. Harvey (1928) found that 14 species of marine algae including greens, reds, and browns are able to produce oxygen from carbon dioxide when illuminated in the complete absence of oxygen. The oxygen appeared in the external medium within a second after illumination. Willstätter and Stoll (1918) found that plants exhibit marked differences in regard to the amount of oxygen necessary for photosynthesis. Thus *Pelargonium* is quite sensitive, while *Cyclamen* is very tolerant to the lack of this gas. The influence of the lack of oxygen on photosynthesis is apparently thus largely one of degree, some plants being easily inhibited in photosynthesis, while others withstand the absence of oxygen for longer periods without any change in the photosynthetic rate. Willstätter and Stoll considered that some oxygen, however, is absolutely essential for photosynthesis. It seems doubtful in any of the experiments that have been reported whether the oxygen supply was completely exhausted, since it is very difficult to free the plant entirely of the gas. An amount sufficient for the plant under consideration might have been available. Singh and Lal (1935) studied the rate of photosynthesis of the young, mature, and old leaves of wheat, flax, and sugar cane. Their data showed a low value for the young leaves, the highest value when the maximum vegetative growth occurred, and a slowing down with age until photosynthesis ceased at senescence.

Work by Ewart (1897, 1898), Pantanelli (1903), Ursprung (1917), and Yap (1920) indicates that the rate of photosynthesis may be checked by continuous exposure to light or by exposure to intense light. This inactivation might be due to the destruction of the chlorophyll or to injury of the protoplasm of the chloroplast or to both these factors, as suggested by Pantanelli (1903). It was noted by Yap (1920) that the rate of photosynthesis of the sugar cane decreased from 10 A.M. to 4 P.M., the process being the most active from 8 to 10 A.M. He considered in this case that the intensity of light decreased the rate of photosynthesis. Ursprung (1917) found that the inactivation of photosynthesis by continuous exposure to light in the case of *Phaseolus* began first in the orange-red portion of the spectrum where the highest photosynthetic rate was proceeding and then spread as the light intensity was increased to the region of shorter wave lengths. The term "solarization" was given by Ursprung to this effect of light upon the photosynthetic activity.

Holman (1930) stated that the term "solarization," as used in reference to plants, refers to a decrease or complete disappearance of starch following the long exposure of leaves to light of an intensity that initially was favorable to abundant starch formation. It is assumed that the effect of solarization is upon the process of photosynthesis, and it is only noticeable through the disappearance of the starch. Solarization apparently has no permanent effects upon the activities of the leaf, since after a period of darkness it again regains its original photosynthetic power. Stanescu (1924) observed that solarization is, in general, more pronounced in leaves with a medium rate of starch production. He was not able to find a definite relation between the appearance of solarization and the intensity of light, although those leaves which grew with their surface turned toward the south were solarized more frequently and more clearly. Emerson (1935) found in *Chlorella* that a light intensity fourteen times greater than that to which the cells are accustomed causes no injurious effects provided that the carbon dioxide supply remains adequate. If the supply of carbon dioxide is lowered, however, injury to the carbon-assimilatory mechanism results, but it is not accompanied by any injury to the chlorophyll. He believed that solarization can be greatly delayed or prevented by increasing the concentration of carbon dioxide.

The rate of photosynthesis is decreased by the accumulation of the products of the process. This behavior has been observed particularly with leaves that have been removed from the plant, since no means of translocation exists under these conditions. In many plants, the sugar which is formed in excess of that which can be translocated at once is converted into temporary starch in the chloroplasts. In these cases it is considered that this accumulation of starch in the chloroplasts impairs their activity by interfering with the photosynthetic process, which is considered by some to occur at the surface of these plastids. The retardation of photosynthesis by the accumulation of sugar in those leaves which do not form starch is apparently associated with the osmotic relations of the cells, and when a certain concentration of the cell sap is reached photosynthesis ceases.

The effect of the carbohydrate content of the leaf upon the rate of photosynthesis was studied by Saposchinkoff (1890, 1893). In the case of the detached leaves of *Vitis labrusca* and *Vitis vinifera* it was found that photosynthesis ceased when the carbohydrate content reached 17 to 25 per cent of the dry weight in the former and 23 to 29 per cent in the latter. These observations were confirmed by Ewart (1896), who also observed that leaves cease photosynthesis after they have been fed sugar for a time. Certain plants store up sugar, which has been produced in excess of that which is translocated from the leaf, as temporary starch in the chloroplasts. This starch appears after from 6 min. to 2 hr. or more, depending upon the plant and the conditions prevailing (Andrews, 1925). Plants that thus form starch have been termed "starch" plants, while those which form little or no starch are termed "sugar" plants. It has been stated by Müller (1904) that starch leaves exceed the sugar leaves in the total production of carbohydrates during the day, the limit of accumulation being higher in the starch leaves. He found that the maximum rate of photosynthesis remained almost stationary during the remainder of the day. Winkler (1898) showed that in ordinary starch-forming leaves 0.2 to 0.5 per cent of sugar is the critical concentration at which they will form starch. For sugar leaves the concentration attains a concentration of sugar as high as 18 per cent in the case of the sugar cane before starch will appear. The onion was the only plant in his experiments that resisted all feeding attempts to form starch. Leaves not detached from the stems but emptied of their starch by a sufficient period in darkness carry on photosynthesis and accumulate starch at a rate very much greater than normal leaves when they are exposed anew to the light.

Lubimenko (1927) noted in leaves low in starch that its accumulation occurs only in cells surrounding the vascular bundles. After starch plants are depleted of this carbohydrate and again exposed to favorable conditions for its production the first grains appear nearest the vascular bundles. According to Langner (1927), the amount of starch present in the leaves at any given time is not necessarily a direct result of photosynthesis. The mode of accumulation of starch is not the same for all leaves or even for those of the same species. It is a process primarily dependent upon internal factors that are not yet known (Stanescu, Aronescu, and Mihăilescu, 1929). Chapman and Camp (1932) found in the variegated leaves of *Pelargonium hortorium* that starch does not occur in the nongreen portion unless the leaf is placed in a solution of glucose, the optimal concentration of which is approximately 0.5 molecular. It was noted by Heinicke (1932) that the ringing of apple twigs reduced the photosynthetic activity of the leaves above the girdle as much as 33 to 50 per cent. This reduction might be due either to the harmful influence of accumulated products or to a lack of water and nutrients.

Suit and Hibbert (1934) introduced bacterial cultures of *Bacillus subtilis* containing the levan enzyme into young potato plants at stated intervals during the growing season. From four of the 30 plants so treated, they claim to have obtained tubers that were free from starch except in a peripheral ring of tissue a fraction of an inch in width. They believed this indicated that the type of carbohydrate formed in plants is influenced by internal conditions. It was observed by Gruzit and Hibbard (1916) that the dry weight per unit area of leaves of seedlings grown in a complete culture was less than that of leaves of seedlings grown in a solution lacking an element as potassium, calcium, or phosphorus. The explanation for this lies in a reduced translocation of the carbohydrate that accumulates in the leaf and results in a retarded photosynthesis, since the actual amount of material formed is greater in those plants growing in a complete nutrient solution.

In experiments with *Phaseolus vulgaris* grown in cultural solutions devoid of potassium, magnesium, iron, or phosphorus, Briggs (1922) found that the photosynthetic activity was lowered. This result may be attributed to two causes. The lack of certain elements might cause an accumulation of carbohydrates, since the growing parts of the plant in their absence cannot utilize the carbohydrates as rapidly as the leaves can synthesize them. On the other hand, the deficiency of salts may act directly upon the photosynthetic mechanism by reducing the actual leaf surface or by decreasing the rate per unit of leaf. Briggs considered that the chief factor involved in the latter case is a decrease in the "reactive surface" of the chloroplast. This is not necessarily the same as the actual visible chloroplast surface. The internal surface of the plastid is just as important as the visible surface in the process and may be greatly influenced by nutritive conditions. In conclusion, it should be stated in regard to protoplasmic factors that any factor or set of factors that affects the protoplasm in general will influence, to some extent at least, the photosynthetic process. In this connection the influence of fungi upon the photosynthetic activity should be mentioned. Long (1919) observed that rusts and mildews interfered with photosynthesis in all cases that she observed. Thus in *Avena* affected by *Puccinia coronata*, when it was well developed but not eruptive, the photosynthetic activity was reduced to 72 per cent and in the erupted stage to 48 per cent. *Puccinia graminis* on wheat reduced the rate of photosynthesis to 50 per cent in the stage before the pustules became brown and in the eruptive stage to 39 per cent.

Trelease and Trelease (1929) found that the leaves of wheat, with an abundant store of carbohydrates supplied by photosynthesis or artificially, were much more susceptible to mildew than those which were low in carbohydrates.

G. The Energy Relations of Photosynthesis

The process of photosynthesis is one in which the radiant energy of the sun is transformed into potential energy in the form of the organic compounds that are synthesized. In order to study the energy relations in the process it is necessary to know the amount of the compounds synthesized and the amount of energy available to the plant, so it is those topics that will now be considered.

1. Methods of Determining the Rate of Photosynthesis.—Three general methods have been used to measure the photosynthetic activity of the leaves and stems. These methods are based on (*a*) the liberation of oxygen, (*b*) the intake of carbon dioxide, and (*c*) the amount of organic substance or dry matter produced. It should be mentioned at the beginning that none of the methods that have been devised for the measurement of photosynthesis is satisfactory, since none takes into consideration all the factors concerned and since frequently the three methods do not yield the same results. The conflicting results obtained by the different methods are no doubt due to the fact that, although photosynthesis is a series or chain of reactions more or less dependent upon each other, each of the three methods is based upon a different chemical reaction in the process. Thus the intake of carbon dioxide occurs at the beginning of photosynthesis, and the increase in dry matter is the last stage in the process, and it is not known at which step in the reaction oxygen is liberated.

In the measurement of photosynthesis, the process of respiration is a difficulty regardless of the method used. Respiration goes on simultaneously in the same cells with photosynthesis, so there is a reduction and an oxidation process occurring at the same time. In photosynthesis, carbon dioxide is absorbed and carbohydrates are formed with the release of oxygen; while, in respiration, oxygen is absorbed, carbohydrates are broken down, and carbon dioxide released. The process of photosynthesis generally proceeds at a more rapid rate than does respiration. The quantity of carbon dioxide, therefore, absorbed from the exterior will be less than that actually used in the process of photosynthesis by an amount equal to the carbon dioxide evolved in respiration and utilized in photosynthesis before it can leave the cells. If the evolution of oxygen is used to measure photosynthesis, the amount of this gas that escapes is not all that has been liberated in the process, since some is utilized in respiration before it can escape from the cell. If photosynthesis is measured by the increase in dry weight, the loss in weight due to respiration must also be considered. The method usually used to correct for respiration is to determine its rate in the same plant parts or upon checks in the dark and then correct for this factor. The principal objection to this method of procedure is the uncertainty as to whether respiration goes on at the same rate in the light as in the dark. The value obtained for the rate of photosynthesis without correcting for the process of respiration is termed the "apparent rate of photosynthesis" as compared to the "true rate of photosynthesis," which has been corrected for the process of respiration. Hicks (1934) in *Lemna minor* found that the rate of photosynthesis was six times that of the respiratory rate under certain conditions.

a. Liberation of Oxygen.—The methods that have been devised to demonstrate the liberation of oxygen in the process of photosynthesis may be classified under the following headings: (1) The use of bacteria, (2) the use of oxygen-absorbing compounds, (3) the bubble-counting method, and (4) gas-analytical methods.

1. *The Use of Bacteria.*—Two types of bacteria have been used to demonstrate the liberation of oxygen: (*a*) luminous bacteria, which emit light only in the presence of oxygen, and (*b*) motile bacteria, which are active only in the presence of this gas. The plant material, however, that may be used in experiments of this type is limited to a size that can be observed under a relatively high power of the microscope.

2. *The Use of Oxygen-absorbing Compounds.*—Certain dyestuffs that are reduced to colorless substances react with oxygen, reforming the original colored compound. These reduced compounds are termed "leucobases" and are used to demonstrate the liberation of oxygen by plants. One of the most frequently used dyes for this purpose is indigo or indigo carmine. When an aqueous solution of this dye is reduced by sodium hyposulphite ($NaHSO_2$), the blue color disappears. When a water plant is illuminated in a solution of this reduced dye, the blue color reappears in the vicinity of the leaves due to the oxidation brought about by the oxygen evolved in photosynthesis.

If a leaf is immersed in defibrated venous blood and exposed to the light, the evolution of oxygen produces the red color of oxyhemoglobin in the vicinity of the leaf. The methods that have so far been mentioned are purely qualitative and have no value for quantitative work.

3. *Bubble Counting.*—When cut parts of an aquatic plant are placed in water with a supply of carbon dioxide and illuminated, bubbles of gas escape from the cut ends or surfaces. The major portion of this gas is oxygen, and it was shown by Sachs (1864) that the rate at which these bubbles are liberated may be taken as a relative measure of the rate of photosynthesis. This method is a comparatively simple one and has been used extensively in botanical laboratories, for both qualitative and quantitative work. A piece of water plant is securely submerged in water in which carbon dioxide is dissolved, and the gas as it escapes is collected in an inverted funnel, or test tube if it is so desired. The rate of photosynthesis is estimated by the number of bubbles that escape in a unit of time. For accurate quantitative work, however, this method is open to criticism. In the first place, the bubbles are not pure oxygen. Some of the oxygen that they contain passes out into the water, and some nitrogen from the water enters the bubbles. The bubbles are not always of the same size, being smaller when photosynthesis is rapid than when it is slow. Lastly, in some cases oxygen does not always escape from the plant in the form of bubbles at the same rate as it is formed.

Wilmot (1921) devised a "glass bubbler" that when fitted to the cut stem of a submerged water plant caused the liberated bubbles to be of a definite and constant size. This apparatus also delivered the bubbles directly into a cup so that they did not come into direct contact with the solutions that were being used.

4. *Gas-analytical Methods.*—These methods involve the analysis of the air in which a plant has been confined or the analysis of samples of air that has passed over the plant. The latter procedure is the more simple and satisfactory. The analysis of the air in experiments of this type is similar in procedure to ordinary gas analysis with the exception of certain modifications in apparatus that are necessary to expose the leaf or leaves to the air and to collect the air after it has been in contact with them. A discussion of the methods of gas analysis, however, is beyond the scope of this work. A concise discussion of the forms of apparatus and of absorbents that have been used is given by Spoehr (1926).

b. Absorption of Carbon Dioxide.—The determination of the amount of carbon dioxide used is the best known method for the measurement of photosynthesis in land plants. One of the objections to the method is that generally only a single leaf can be used and the behavior of detached leaves in photosynthetic activity is known to differ greatly from leaves that are attached to the plant. The carbon dioxide content of the air is first determined, and then the content of this gas in the air is determined after it has been pulled in a stream over the leaf in the dark. By this means the rate of respiration of the leaf is determined. The apparatus is then placed in the light and the carbon dioxide content of the air that has passed over the leaf is determined as before. The correction is made for respiration and the amount of carbon dioxide used in photosynthesis is calculated. This method was first used by Kreusler (1885) and has been used and modified since that time by numerous workers including Matthaei (1905), Blackman and Matthaei (1905), Brown and Escombe (1905), Willstätter and Stoll (1918), and Spoehr and McGee (1923, 1924). This method, although relatively simple in its procedure, involves many details that must be very carefully manipulated or the results will be of no value. Among these details are: (1) the regulation of the air stream, (2) the constant temperature of this stream, (3) the carbon dioxide content, (4) the regularity of illumination, and (5) the construction of forms of apparatus suitable for the collection of the gas and at the same time placing the plant under favorable conditions for photosynthesis.

The amount of carbon dioxide used in photosynthesis may be demonstrated in a rather simple manner by the eudiometric method first employed by Pfeffer (1871) and modified by Ganong (1908). The leaf or plant part is placed in a closed vessel of known volume to which is attached a graduated burette. A known amount of carbon dioxide is introduced into the apparatus, and after a period of time the amount remaining in the vessel is determined by using a concentrated solution of alkali as an absorbent, the carbon dioxide content being determined by the height to which the liquid rises in the graduated burette. Similar methods for the demonstration and measurement of photosynthesis have also been described by Osterhout (1918, 1919), by Osterhout and Haas (1918), and by Heinicke and Hoffman (1933).

c. Increase in Organic Matter. 1. *Qualitative Methods.*—The increase of the carbohydrates in photosynthesis is easily demonstrated in a qualitative way in those leaves that form starch. The presence of starch or its increase in amount may be easily demonstrated by the iodine method, which was first reported by Sachs (1884) and which has become the most used qualitative method for demonstrating starch in plant tissue. The plant is placed in the dark for a period before exposure to light, or portions of the leaf are shaded from light while others are exposed. When determinations are to be made, the leaf is detached, dipped in hot water, and rendered colorless by the extraction of the chlorophyll with ethyl alcohol. After this the leaves are placed in a solution of iodine in potassium iodide in water. The leaves or portions thereof that have been kept in the dark or shielded from the light will remain unstained, while the portions that have been exposed to light are colored from violet to black according to the amount of starch present. This method, however, is applicable only to those leaves which form starch.

The increase in carbohydrates by photosynthesis in sugar leaves can be demonstrated in a qualitative way by noting the increased reducing power of the cell sap with Fehling's solution or a similar one. This increase in reducing power may be determined in a microchemical way or by working directly with the expressed sap of the leaves.

2. *Quantitative Methods.* (*a*) *Weighing.*—This method was first used by Sachs in 1884 and is sometimes called the "half-leaf method." As originally devised, it consisted in removing one-half of an attached leaf at the beginning of the experiment by cutting along the midrib and then determining the dry weight of this severed portion. After exposure to light for a given period, the other half of the leaf was removed after the same manner and its dry weight determined. The method was later modified by Sachs and others in that only small portions of the leaf were removed from each side, thus enabling more uniform samples to be taken and eliminating errors due to irregular veins and the lack of symmetry in the two halves of the leaf.

Ganong (1908) devised a leaf punch that removes a disk of leaf with an area of 1 sq. cm. by means of which, if proper precautions are used, samples of very uniform weight and structure may be obtained. By beginning toward the tips of the leaves with this instrument, a series of samples may be taken at different periods of the day without seriously disturbing the conducting system of the leaf. Miller (1917) in determining the changes in the dry matter in the leaves of corn and the sorghums during the day and night obtained samples with a Ganong leaf punch after the following manner:

Thirty representative plants were selected and a leaf chosen on each plant for furnishing all the samples for an experiment extending over the desired length of time. Figure 28 shows the manner in which the leaf samples were obtained. The figure shows the appearance of a leaf at the end of a 40-hr. experiment. At 7 A.M. a single sample was taken from each of the 30 selected leaves at *a* toward the tip of the leaf. At 9 A.M. a sample was taken from each of the 30 leaves at *b* directly opposite *a*. At 11 A.M. the samples were collected from the leaves at *c* directly below *a*, and then at the next 2-hr. period at *d*, and so on. The samples for a 40-hr. experiment were thus obtained from a portion of the leaves representing less than 6 in. of their respective lengths. The samples were collected in weighed vials which were then quickly stoppered, weighed, placed in an oven at 100 to 108°C., and dried to constant weight.

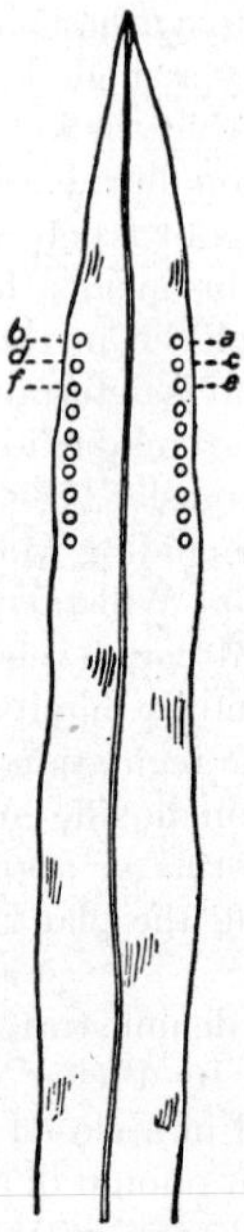

FIG. 28.—Diagram to illustrate the "leaf-punch" method for determining the variations in dry matter, water, carbohydrates, or nitrogen during a 24-hr. period. Description in the text.

The increase in dry weight that is obtained for a given area of leaf in an experiment of this kind does not represent the total weight of the product formed by photosynthesis during that period. Translocation of products from the leaf takes place during the period of photosynthesis, while a certain amount of the product of photosynthesis is utilized in respiration, which is proceeding simultaneously with photosynthesis. The increase in dry weight thus obtained represents the difference between the total weight of the products formed and the weight of the material that has been moved from the leaf or utilized in respiration. Sachs attempted to correct for the translocation by placing a detached leaf in water as a control, since translocation from the leaf does not occur under such conditions. Brown and Escombe (1905), however, showed that a detached leaf does not utilize carbon dioxide in the same proportion as an attached leaf, so such a correction is unreliable.

Sachs also considered that the total weight of photosynthate made during the day in a given area of leaf could be determined by adding to the gain during the day the loss in weight of the same area during the night. This should give the total increase in weight due to photosynthesis, provided that respiration and translocation proceed at the same rate during the daylight hours as they do during the night—which, however, is not definitely known to be a fact. The following data taken from the work of Miller (1917) show how the dry matter varies during the day and night in the leaves of Dwarf Yellow milo.

VARIATION IN DRY MATTER IN THE LEAVES OF DWARF YELLOW MILO DURING A 24-HR. PERIOD AT GARDEN CITY, KANS.

Time	Dry matter per square meter of leaf, grams	Increase or decrease in dry matter per square meter of leaf, grams
5 a.m.	50.3	
7 a.m.	50.8	+0.5
9 a.m.	53.4	+2.6
11 a.m.	55.7	+2.3
1 p.m.	58.3	+2.6
3 p.m.	61.0	+2.7
5 p.m.	61.5	+0.5
7 p.m.	60.2	−1.3
9 p.m.	58.0	−2.2
11 p.m.	57.8	−0.2
1 a.m.	55.6	−2.2
3 a.m.	54.9	−0.7
5 a.m.	52.1	−2.8

These data show that during the 24-hr. period there was manufactured in the leaf from 5 A.M. to 5 P.M. 11.2 g. of dry matter per square meter of leaf in excess of that which was translocated and used in respiration. From 5 P.M. to 5 A.M., 9.4 g. of dry matter disappeared per square meter of leaf. If the rate of translocation and respiration was the same during the day as during the night, the total amount of photosynthate during the day from 5 A.M. to 5 P.M. was 20.6 g. per square meter of leaf.

As pointed out by Thoday (1910), there are numerous objections to the dry-weight method of determining the rate of photosynthesis other than those above mentioned. The dry weight obtained includes the increase not only of carbohydrates but of proteins and oils as well. Changes in dry weight may consist in part of ash to as much as 5 per cent. The changes in ash content should be determined and a deduction made for the same. Apparent increases in dry weight would result from any shrinking in area of the leaf due to water loss during the day. Thoday (1910) considered that the dry-weight method gives useful results for rates of increase greater than 2 mg. per square decimeter per hour.

(*b*) *The Twin-leaf Method.*—Denny (1930) believed that the twin-leaf method, or opposite-leaf method, can be used to advantage in many cases. This method is applicable only for those plants that have opposite leaves. Although two opposite leaves may vary more than two half leaves, they are under more nearly normal conditions than two half leaves. The percentage difference in the fresh weight of opposite leaves may be as much as 4.2 or as little as 0.3 per cent.

(c) *Saccharification.*—This method consists of hydrolyzing the entire carbohydrate content of the leaf to hexoses and estimating the products thus formed as glucose. The method is essentially the same as that recommended by the Association of Official Agricultural Chemists but must be modified to meet the conditions that are peculiar to the material under consideration. In this method a correction must also be made for respiration and translocation, if the actual amount of photosynthate is to be determined.

Pickett (1935) determined the rate of photosynthesis in apple leaves by the punch method of Sachs, by the carbon dioxide-absorption method of Heinicke and Hoffman, and by the saccharification method. He reported that all these methods have disadvantages, it being necessary for one to choose the method that is best suited for the conditions under which one has to work.

2. The Rate of Photosynthesis.—It is of interest to know the rate at which carbohydrates are manufactured by the plant in the process of photosynthesis. The average rate of photosynthesis was determined by Miller (1917) for corn, kafir, and milo under the conditions that prevail in the field in Kansas during the growing season. The data for a large number of experiments for three growing seasons are averaged in the table below. These data have been corrected for translocation and respiration and thus approximate the actual amount of photosynthate manufactured by the leaves during the periods mentioned.

AVERAGE RATE OF PHOTOSYNTHESIS FOR EACH SQUARE METER OF LEAF FOR CORN, KAFIR, AND MILO DURING EACH 2-HR. PERIOD OF THE DAY

Plant	a.m.		p.m.		
	7 to 9	9 to 11	11 to 1	1 to 3	3 to 5
Corn	3.8	1.9	1.4	1.2	1.4
Kafir	3.0	2.1	1.2	2.1	1.2
Milo	2.3	2.7	3.8	3.5	1.4

The high rate of photosynthesis in corn and kafir from 7 to 9 A.M. as compared with other periods of the day may be due to water relations. During the growing season in the Great Plains, the water content of the leaves, due to a high rate of transpiration, is apparently below the optimum for some of the physiological processes for the majority of plants.

It also has been observed by Miller that the average rate of photosynthesis during a 10-hr. period for pumpkin, cowpeas, and soybeans was respectively 1.8, 0.85, and 0.8 g. per square meter per hour under field conditions. It is of interest to note that the rate for pumpkin is approximately that reported by Sachs (1884) for *Helianthus annuus*. The rates of photosynthesis just mentioned are much lower than those obtained by workers who have determined the rate by the amount of carbon dioxide taken up. They, however, used concentrations of carbon dioxide that were many times higher than normal. Thus Blackman and Matthaei (1905) using 6.3 per cent concentration of carbon dioxide obtained a rate of 3.9 g. per square meter per hour for *Helianthus annuus* at 30°C. in strong sunlight. Willstätter and Stoll (1918) in a 5 per cent concentration of carbon dioxide at 25°C. with a light intensity of 48,000 lux obtained a rate of 5.5 g. per square meter per hour with *Helianthus annuus* and under the same conditions, but with an atmosphere of 4 per cent carbon dioxide a rate of 4.3 g. per square meter per hour for *Cucurbita pepo*. In the ordinary atmosphere, McLean (1920) obtained a maximum absorption of carbon dioxide by the leaves of sugar cane that was equivalent to only 0.35 g. per square meter per hour of carbohydrate. Heinicke (1932) found that the highest rate of carbon assimilation in apple leaves was 35 mg. of carbon dioxide per hour per square centimeter, the most common rate being 5 to 10 mg. Under identical external conditions, the same plants may vary as much as 500 per cent in their photosynthetic rate.

3. The Total Incident Radiation.—In order to understand the energy relations of the green plant, a general idea must be had of the total incident radiation of the sun. The thermopile, pyroheliometer, radiometer, radiomicrometer, and bolometer are some of the instruments that have been used to determine the intensity of solar radiation, but a description of these instruments does not belong here. Shirley (1931) discussed the available sources of light, and the characteristics of the instruments that are available for light measurements.

The intensity of solar radiation has been measured by means of the heat produced when the radiation is absorbed on a black surface at right angles to the rays. The data thus obtained are generally converted into calories per square centimeter per minute (Spoehr, 1926). The intensity of solar radiation is not a constant but will depend upon the height above sea level, the condition of the atmosphere, and the position of the sun. The maximum intensities of solar radiation that have been reported vary from 1.15 to 1.75 cal. per square centimeter per minute depending upon the position and general condition when the observations were made. Spoehr (1926) considered that 1.35 cal. per square centimeter per minute is a conservative estimate of the energy that reaches the earth's surface during the 6 hr. of the most intense solar radiation of the day. Thus there would be received 1.35 by 60 by 6 or 486 cal. per square centimeter during that period. Taking 90 days as a growing season, the total radiation falling upon that surface during that time would be

43,740 cal. Since there are approximately 41,000,000 sq. cm. per acre, the total energy received in 90 days by that surface would be 1,793,340,000,000 cal. According to Spoehr (1926), the heat of combustion of 1 kg. of anthracite coal is equal to 8,000 kg. cal., so that the total incident radiation on an acre during a 90-day period during the growing season would be equivalent to the energy derived from the combustion of 224,167 kg. or 246 tons of anthracite coal.

This represents sufficient energy to heat a seven-room house for 50 years during the months heat is required in the North Temperate Zone. Another way of expressing this is that if only 1 per cent of all the energy falling for 6 months on the land surface of the globe was fixed, it would equal the energy contained in the world's consumption of coal for 200 years—assuming that 1,000,000 tons were used annually. The energy falling annually on the total surface of the United States is equivalent to 7,000 trillion horsepower, which is approximately fourteen times all the energy used in industry in this country from all known sources including that furnished by man and beasts (Egloff, 1936). Thus, the amount of energy reaching the surface of the earth from the sun is enormous.

4. The Coefficient of Absorption of Radiant Energy for Leaves.—As stated by Brown and Escombe (1905), the coefficient of absorption of radiant energy by a leaf is the difference between the solar radiant energy falling upon the leaf in full sunshine and the amount transmitted through it. These investigators measured the total radiant energy by means of a radiometer. The leaf was then interposed so that the light passed through it before falling upon the radiometer. The amount of energy absorbed by the leaf was thus determined, but no account was taken of the reflection or radiation from the leaf surface. This error would tend to increase unduly the estimate of the coefficient of absorption. Brown and Escombe found that the coefficient of absorption was fairly constant for mature and healthy leaves of the same species of plant, but there was considerable variation in the leaves of different species. Thus the coefficient of absorption of the leaves of *Helianthus annuus* was 0.686; for *Polygonum Weyrichii*, 0.647; for *Verbascum olympicum*, 0.758; and for *Archium majus*, 0.728. They observed in most cases no significant differences in the absorption coefficient of young and old leaves. Brown and Escombe measured the absorption and transmission of the radiant energy by the white and green portions of the leaf of *Acer negundo*. They found that the coefficient of absorption of the green portion of the leaf was 0.787, while that of the white portion was 0.745. If the transmission through the albino leaf is taken as 100, that of the green leaf would be 83.5, a difference of 16 per cent, which may be regarded as expressing the increase in the absorptive power due to the coloring matter of the leaf. In this regard, Timiriazeff (1904) noted that the chlorophyll of a certain area of leaf absorbed only 27, 29, and 23.5 per cent of the radiant energy in the case of maple, lime, and oak, respectively. Puriewitsch (1914) by means of the bolometer found that the leaves of different plants absorbed only 18 to 27 per cent of the incident solar radiation. Long (1919) by means of a photometer noted the total incident light absorbed by the leaves to be 27.3, 43.4, and 29.2 per cent of the total incident light for dandelion, bean, and sunflower, respectively. These values reported by Puriewitsch and Long are much lower than those reported by Brown and Escombe.

Shull (1929) found that the amount of reflection from leaves varies with the wave length, the maximum reflection being of rays in the region of 540 to 560mμ. The reflection in this region of the spectrum varies from 6 to 8 per cent in the darkest green leaves to 20 to 25 per cent in the lightest green ones. The amount of reflection is not dependent upon hairiness or the smoothness of the cuticle. Thus the leaves of *Verbascum thapsus* and *Arbutilon theophrasti* show but little more reflection than the leaves of nonhairy plants. The amount of reflection decreases with the age of the leaf, which apparently is due to the greater amount of chlorophyll. The presence of fungi, however, on the surface increases reflection. The presence of anthocyanin shifts the position of maximal reflection. Thus in the autumn-colored leaves of *Psedera* the maximal reflection occurred at 640mμ instead of 540 to 560mμ—the usual region for maximal reflection in green leaves.

Briggs (1929) stated that photosynthetic efficiency is the ratio of the energy fixed in the leaf to the energy absorbed by it. Neither factor can be measured directly. In a leaf the system involved in the reflection of light is very complicated. Light undergoes some reflection every time it strikes an interface. The fraction of light reflected depends upon the refractive indexes, which themselves depend upon the wave length of the light and upon the angle of incidence. In land plants, reflection will occur at the upper surface of the leaves, on the inside where the cell walls abut on the intercellular spaces, at the interfaces of the cell sap, cytoplasm, and cell wall. For those leaves of *Liriodendron* which had become yellow, the reflection was as high as 50 per cent. Shull (1930) criticized the data on energy relations obtained by Brown and Escombe (1905) and stated that there should be thorough reinvestigations of the energy exchange between the leaf and its environment, and of the internal transformation of energy in the process of photosynthesis with reference to the variability of the reflection and absorption of energy in all regions of the spectrum.

Kalitin (1931) using an Angstrom pyranometer found the following reaction of various leaves to incident light.

Plant	Percentage of incident light		
	Reflected	Absorbed	Transmitted
Acer	29	39	32
Populus	42	24	34
Tilia	33	27	40
Ulmus	35	16	49
Cornus	35	25	40

Schanderl and Kaempfert (1933) with an improved actinometer found that ferns allowed 37 per cent of the incident radiation to be transmitted. Desert plants were found to be the most efficient absorbers of light in that only 15 to 30 per cent of the direct rays was transmitted, while in diffuse light the absorption was complete.

Dastur and Gunjikar (1934) studied the coefficient of absorption of the leaves of 12 plants to ordinary light rays and to light that was 89 per cent polarized. The average absorption coefficient for the normal light was 0.895, and for the polarized light, 0.917.

The cause for these differences in the absorption coefficients is difficult to determine and only emphasizes the fact that our knowledge concerning even the absorption of energy by the leaves is very limited.

In considering the absorption coefficient of the leaves, the quality of the rays transmitted or absorbed must be taken into consideration. The green leaf is selective in its absorption of the various wave lengths of solar radiation incident upon it, a behavior that would be suspected, to a degree at least, owing to the selective property of chlorophyll. This selective power of the leaf is shown by the quantity of light which is transmitted when leaves are superimposed upon each other. If the leaf possessed no power of selection, it would be expected that the transmitted portion of the radiant energy would diminish in geometric proportion as the number of leaves was increased in arithmetic proportion (Brown and Escombe, 1905). Thus, for example, if the leaf of *Helianthus annuus* having a transmission coefficient of 0.313 is superimposed upon another, the coefficient of transmission should be $(0.313)^2 = 0.098$, with three such leaves $(0.313)^3 = 0.030$, and so on, if the leaf had no power of selective absorption. This, however, is not the case, since Brown and Escombe found that the coefficient of transmission through two leaves of sunflower was 0.174 and through three leaves 0.116, while the radiation that should have been transmitted in the absence of selective absorption would have been 0.098 and 0.030, respectively. A similar observation was made by Timiriazeff with a solution of chlorophyll. He found, for example, that although the chlorophyll which had been dissolved out of a given area of maple leaf arrested 27 per cent of the energy of direct sunlight, this absorption was increased to only 31 per cent by a triple concentration of the solution.

5. The Utilization of Solar Energy in the Leaves.—A portion of the energy that is absorbed by the leaf is utilized in photosynthesis and in transpiration, and a certain portion is lost by thermal emission into the surrounding air, provided the temperature of the surrounding medium is lower than that of the leaf. The question naturally arises as to what proportions of the absorbed energy are used in photosynthesis and transpiration. Knowledge in this regard is, however, very limited and fragmentary, because experimental evidence on this subject is difficult to obtain. This difficulty is due to the fact that so many factors must be controlled and measured by methods that in many cases are doubtful in their reliability and accuracy. Some of the factors that must be controlled, measured, or taken into account are the total radiation, the coefficient of absorption, the amount of carbon dioxide used, the amount of water vapor given off, the intensity of respiration, the effect of the accumulation of products on photosynthesis, and the influence of the general factors upon the action of the stomata and the interchange of gases.

Briggs (1929) stated in this regard "that the amount of energy fixed in the immediate products of photosynthesis is obscured partly by the fact that the molecular mechanism of the reaction is not established and partly by the complex setting of the reactions in the living cells."

a. Determination of the Energy Used in Photosynthesis.—The energy used in photosynthesis has been estimated from the amount of carbon dioxide absorbed, from the amount of oxygen given off, and from the increase in dry weight. In all cases the determinations are reduced to the actual carbohydrate formed, correction always being made for respiration. In calculating the energy utilized by these methods, however, the difficulty that the proportions of the various carbohydrates are unknown is encountered, and, since each differs in its energy value, the energy utilized in producing the total carbohydrate products can be only approximated. Brown and Escombe (1905) determined the weight of carbon dioxide absorbed and multiplied this by the factor 0.640 to determine the weight of the carbohydrate produced, the factor being based on the carbohydrate content of the leaf as determined by previous investigators. Puriewitsch (1914) measured the increase in dry weight per unit area by the half-leaf method and the heat of combustion per unit area and thus determined the heat of combustion of the products of assimilation. The values obtained by him were higher than the value for glucose, starch, sucrose, or cellulose and suggest the production of fat or protein in the leaf during the period of photosynthesis.

b. The Energy Used in Photosynthesis.—The most extensive work that has been done to determine the energy relations of the leaf is that of Brown and Escombe (1905).

The work of Brown and Escombe shows that as little as 0.5 per cent of the total incident radiation in full sunshine may be utilized in the process of photosynthesis, but that as much as 4.5 per cent may be utilized when the sunshine approximates only 13 per cent of full intensity. Puriewitsch (1914) by measuring the incident energy with the bolometer and determining the energy used in photosynthesis by the increase in the heat of combustion per unit of leaf area found the following utilization of the total incident energy: *Acer platanoides*, 0.6 to 2.7 per cent; *Polygonum sacchalinese*, 1.1 to 7.7 per cent; *Helianthus annuus*, 4.5 per cent; and *Saxifraga cordifolia*, 5.0 per cent. Some of these percentages are considerably higher than those of Brown and Escombe. This may be due to the fact that the products of photosynthesis had not yet accumulated and thus did not exert a hindering influence on the rate in certain experiments. The figures of both Brown and Escombe and Puriewitsch make it evident that with a high light intensity only a small part of the incident radiation is utilized in photosynthesis. As the light intensity decreases, however, the proportion of the total incident radiation used in photosynthesis increases. This is what would be expected if light were present in excess and not a limiting factor.

The energy relations of photosynthesis in crop plants growing in the field may be approximated and are of some interest. In the case of the pumpkin plant the rate of photosynthesis is approximately 1.8 g. per square meter per hour. If this increase in photosynthate is considered as starch and its heat of combustion taken as 4,200 g.-cal., the energy used in photosynthesis would be somewhat less than 1 per cent of the total incident radiation if such radiation were considered as 1.35 cal. per square centimeter per minute. In the case of cowpeas and soybeans where the rate of photosynthesis is only about 0.8 g. per square meter per hour the energy used in photosynthesis would be only about 0.4 per cent of the total incident radiation.

Miller (1928) observed that the maximum daily increase in the dry weight of corn plants that had reached their full leaf development was 18.3 g. under field conditions in Kansas. Assuming this total increase in weight as starch and its heat of combustion as 4,200 g.-cal. and considering 6,000 corn plants per acre, the total energy fixed during the day would be 451,000,000 cal. The total incident radiation on an acre for a 6-hr. day would amount to 19,926,000,000 cal. considering 41,000,000 sq. cm. to the acre. The energy utilized in photosynthesis would thus be 451 ÷ 19,926 or 2.2 per cent of the total incident radiation on the acre of surface. Transeau (1926) estimated that the corn plant utilized in photosynthesis only about 1.6 per cent of the total energy available to it. This, however, does not take into consideration the material utilized in respiration. Pütter, as quoted by Spoehr (1926), estimated that the utilization of the total incident solar energy by the crop plants in the production of organic matter was as follows, allowance being made for the material used in respiration: summer wheat, 3.26 per cent; summer rye, 2.60 per cent; oats, 3.31 per cent; potatoes, 3.02 per cent; and beets, 2.12 per cent.

6. The Economic Importance of Photosynthesis.—The immediate source of the energy supply that is needed continuously by all living organisms, both plant and animal, is food. This food is manufactured from simple inorganic substances by the green plant with the aid of energy derived from the sun. Food thus represents a supply of energy that has been stored by the plant from the rays of the sun. All life on the earth thus depends upon the energy that comes from the sun through the intermediary of the green plant. The green plant is thus the sole agent that has the power to transform the kinetic energy of the sun into the potential form. This is accomplished through the process of photosynthesis; and sugar, starch, cellulose, lignocellulose, fats and oils, proteins, and the multitude of plant organic compounds are stores of potential energy that has thus been transformed. This storage of the energy of the sun has been going on for millions of years and has always been in excess of that needed for the maintenance of plant and animal life. This fact is evidenced at the present time by the forests that until comparatively recently furnished man with fuel for his bodily warmth and for industrial power. The energy stored up in plant materials in the past ages became "fossil energy" and is now stored in the form of coal and oil.

The present industrial civilization is a machine-using age dependent upon a continuous and abundant supply of stored energy that may be readily released to the kinetic form for the operation of these machines. At present, coal and oil are the two main sources of this stored energy which, as before stated, has been made possible through the process of photosynthesis through millions of years.

The question that is now of considerable concern, to those who are looking to the future, is, what sources can be drawn upon for a supply of energy when the supply of stored energy, oil especially, is greatly depleted? Alcohol derived from grains, cellulose, and other plant products has been the source that is most commonly suggested. Such a substitution, however, is much more involved and difficult than is commonly supposed. The production of alcohol from plant products as a source of power for industry has serious objections. The question of disturbing the equilibrium of the present food supply is not one of the least of these objections.

Thus Spoehr (1922) stated that the total average annual yield of corn in the United States (2,740,000,000 bu.) if converted into alcohol would produce only 7,500,000,000 gal. of alcohol, which would be equivalent to 5,000,000,000 gal. of gasoline. This is about one-fourth of the present yearly consumption. Thus the present production of grain from all sources in this country would scarcely more than supply the present needs for this fuel, without leaving any for a food supply. The industrial age has been made possible because an enormous supply of stored energy is close at hand or immediately available, and climatic changes, diseases, insect pests, and numerous other factors have no influence upon the supply. Such a certain and undisturbed source of energy would not thus be available if the supply of energy were furnished by the crop of the current year.

Although the green plant is the sole converter of solar energy through the process of photosynthesis, it is, as has previously been mentioned, a very inefficient machine, utilizing only from 0.5 to 3 per cent of the total energy falling upon it. It would appear that the solution of the future fuel supply lies in the discovery of the physics and chemistry of photosynthesis and by their application artificially transforming the kinetic energy of the sun rather than by depending upon this transformation by the slow-acting plant. If man can solve the problem of this transformation of energy by the plant and improve it, he will have at hand an almost unlimited and inexhaustible source of energy.

Bibliography

ANDREWS, F. M., 1925: Starch formation, *Proc. Ind. Acad. Sci.*, **35**: 182–184.

APPLEMAN, C. O., 1927: Percentage of carbon dioxide in soil air, *Soil Sci.*, **24**: 241–245.

ARCHBOLD, H. K. and A. M. BARTER, 1935: A fructose anhydride from the leaves of the barley plant, *Biochem. J.*, **29**: 2689–2693.

ARNOLD, W., 1934: The effect of ultra-violet light on photosynthesis, *J. Gen. Physiol.*, **17**: 135–143.

——, 1933: The order of the Blackman reaction in photosynthesis, *J. Gen. Physiol.*, **17**: 145–149.

——— and H. I. KOHN, 1934: The chlorophyll unit in photosynthesis, *J. Gen. Physiol.*, **18**: 109–112.

ARTHUR, J. M., 1929: Some effects of radiant energy on plants, *J. Optic Soc. Am. and Rev. Sci. Inst.*, **18**: 253–263.

——, 1931: Color development in apples after picking, *Am. J. Bot.*, **18**: 896–897.

——— and L. C. PORTER, 1935: A new type of insulated greenhouse heated and lighted by Mazda lamps, *Contr. Boyce Thompson Inst.*, **7**: 131–146.

BAEYER, A., 1870: Ueber die Wasserentziehung und ihre Bedeutung für das Pflanzenleben und die Gährung, *Ber. chem. Ges.*, **3**: 63–78.

BAKER, S. M., 1913: Quantitative experiments on the effects of formaldehyde on living plants, *Ann. Bot.*, **27**: 411–443.

BALY, E. C. C., and J. B. DAVIS, 1927: Photosynthesis of naturally occurring compounds. III, Photosynthesis in vivo and in vitro, *Proc. Roy. Soc.*, A, **116**: 219–226.

——— and N. R. Hood, 1929: The photosynthesis of naturally occurring compounds. IV, The temperature coefficient of the photosynthesis of carbohydrates from carbonic acid, *Proc. Roy. Soc.*, A, **122**: 393–398.

———, I. M. Heilbron, and W. F. Barker, 1921: Photocatalysis. I, The synthesis of formaldehyde and carbohydrates from carbon dioxide and water, *J. Chem. Soc. Trans.*, **119**: 1025–1035.

———, 1922: Photosynthesis, *Nature*, **109**: 344–346.

———, I. M. Heilbron, and W. F. Barker, 1923: Photochemical production of formaldehyde, *Nature*, **112**: 323.

———, 1924: Photosynthesis, *Ind. and Eng. Chem.*, **16**: 1016–1018.

———, 1928: Photosynthesis, *Sci.*, **68**: 364–367.

———, 1935: The kinetics of photosynthesis, *Proc. Roy. Soc.*, B, **117**: 218–239.

Banal, R., 1926: Un fertilizzante fondamentale: l'anidride carbonica. Problemi ed esperienze sull' assimilazione del carbonio, *L'Italia agricola*, **63**: 118–121.

Barnes, C. R., 1893: On the food of green plants, *Bot. Gaz.*, **18**: 403–411.

———, 1898: So-called "assimilation," *Bot. Cent.*, **76**: 255–259.

Bartholomew, R. P., V. M. Watts, and G. Janssen, 1933: The effect of variations in the nutrient media upon the nitrogen, phosphorus, and potassium content in plants with special reference to the tomato, *Ark. Sta. Bull.* 288.

Barton-Wright, E. C., and M. C. Pratt, 1930: Studies in photosynthesis. I, The formaldehyde hypothesis, *Biochem. J.*, **24**: 1210–1216.

——— and ———, 1930: Studies of photosynthesis. II, The first sugar of carbon assimilation and the nature of the carbohydrates in the narcissus leaf, *Biochem. J.*, **24**: 1217–1234.

Bazyrina, K. N., and V. Tschesnokov, 1930: Der Einfluss der Luftdüngung auf die Pflanzen, *Planta*, **11**: 463–472; *Biol. Abst.*, **8**: 873.

Belval, H., 1932: Transformations of sugars in the banana, the formation and disappearance of starch, *Rev. Gén. Bot.*, **44**: 513–525.

Bergamaschi, M., 1932: Sull' assorbimento del biossido di carbonio per opera delle radici e sulla sua utilizzazione nella fotosintesi clorofilliana, *Biol. Abst.*, **6**: 1891.

Bernard, C., 1878: "Leçons sur les phénomènes de la vie commune aux animaux et aux végétaux," Paris.

Bernbeck, O., 1924: Wind und Pflanze, *Flora*, **117**: 293–300.

Berthelot, D., and H. Gaudechon, 1910: Synthèse photochimique des hydrates de carbone aux dépens des éléments de l'anhydride carbonique et de la vapeur d'eau, en l'absence de chlorophyll, *C. R. Acad. Sci. Paris*, **150**: 1690–1693.

Bills, C. E., and F. G. McDonald, 1932: The carotene content of ten varieties of carrots, *Sci.*, **76**: 108.

Blackman, F. F., and G. L. C. Matthaei, 1905: Experimental researches in vegetable assimilation and respiration. IV, A quantitative study of carbon-dioxide assimilation and leaf-temperature in natural illumination, *Proc. Roy. Soc.*, B, **76**: 402–460.

———, 1905: Optima and limiting factors, *Ann. Bot.*, **19**: 281–295.

———, 1910: Experimental researches on vegetable assimilation and respiration. VIII, A new method for estimating the gaseous exchanges of submerged plants, *Proc. Roy. Soc.*, B, **83**: 374–388.

——— and A. M. Smith, 1911: Experimental researches on vegetable assimilation and respiration. IX, On assimilation in submerged water plants and its relation to the concentration of carbon dioxide and other factors, *Proc. Roy. Soc.*, B, **83**: 389–412.

Boehm, J. A., 1856: Beiträge zur näheren Kenntniss des Chlorophylls, *Sitzungsber. Akad. Wiss. Wien*, **22**: 479.

———, 1883: Ueber Stärkebildung aus Zucker, *Bot. Ztg.*, **41**: 33–38, 49–54.

Bokorny, T., 1891: Ueber Stärkebildung aus Formaldehyd, *Ber. bot. Ges.*, **9**: 103–106.

———, 1892: Ernährung von grünen Pflanzenzellen mit Formaldehyd, *Landw. Jahrb.*, **21**: 445–465.

———, 1897: Ueber die organische Ernährung von grünen Pflanzen und ihre Bedeutung in der Natur, *Biol. Cent.*, **17**: 1–20, 33–48.

———, 1908: Ueber die Assimilation des Formaldehyds und die Versuche dieses Zwischenproduct bei Kohlensäure-Assimilation nachzuweisen. Assimilation von Glycerin und Zucker, *Arch. ges. Physiol.*, **125**: 467–490.

———, 1909: Weitere Mitteilung über CO_2-Assimilation und Ernährung von Pflanzen mit Formaldehyd, *Arch. ges. Physiol.*, **128**: 565–586.

———, 1910: Die bisherigen Versuche das Formaldehyd direkt in Blättern nachzuweisen, *Chem. Z.*, **33**: 1141–1143, 1150–1151.

———, 1911: Ernährung von grünen Pflanzen mit Formaldehyd und formaldehydabspaltenden Substanzen, *Biochem. Z.*, **36**: 83–97.

Bolas, B. D., and F. Y. Henderson, 1928: The effect of increased atmospheric carbon dioxide on the growth of plants. I, *Ann. Bot.*, **42**: 509–523.

——— and R. Melville, 1935: The effect on the tomato plant of carbon dioxide produced by combustion, *Ann. App. Biol.*, **22**: 1–15.

Bonnier, G., and L. Mangin, 1886: Recherches sur l'action chlorophyllienne séparée de la respiration, *Ann. Sci. Nat. Bot.*, 7e, **3**: 1–44.

Bornemann, O., 1920: Kohlensäure und Pflanzenwachstum, *Mitteil. landw. Ges.*, **35**: 693–695; *Bot. Abst.*, **8**: 2.

———, G. Lemmerman, and F. Riedel, 1921: Zur Kohlenstoffernährung der Kulturpflanzen, *Mitteil. landw. Ges.*, **36**: 481–485, 496–498.

Bose, J. C., 1924: "Physiology of Photosynthesis," 286 pp., Longmans, Green and Company, New York.

Bottomley, W. B., and H. Jackson, 1903: Some preliminary observations on the assimilation of carbon monoxide by green plants, *Proc. Roy. Soc.*, **72**: 130–131.

Boussingault, J. B., 1864: De la végétation dans l'obscurité, *Ann. Sci. Nat. Bot.*, 5e, **1**: 314–324.

Boysen-Jensen, P., 1918: Studies on the production of matter in light and shadow plants, *Bot. Tidssk.*, **36**: 219–262.

Breazeale, J. F., 1923: The absorption of carbon by the roots of plants, *J. Agr. Res.*, **26**: 303–313.

Briggs, G. E., 1920: Experimental researches on vegetable assimilation and respiration. XIII, The development of photosynthetic activity during germination, *Proc. Roy. Soc.*, B, **91**: 249–268.

———, 1922: Experimental researches on vegetable assimilation and respiration. XV, The development of photosynthetic activity during germination of different types of seeds, *Proc. Roy. Soc.*, B, **94**: 12–19.

———, 1922: Experimental researches on vegetable assimilation and respiration. XVI, The characteristics of subnormal photosynthetic activity resulting from deficiency of nutrient salts, *Proc. Roy. Soc.*, B, **94**: 20–35.

———, 1929: Experimental researches on vegetable assimilation and respiration. XX, The energetic efficiency of photosynthesis in green plants. Some new data and a discussion of the problems, *Proc. Roy Soc.*, B, **105**: 1–35.

———, 1933: Experimental researches on vegetable assimilation and respiration. XXI, Induction phases in photosynthesis and their bearing on the mechanism of the process, *Proc. Roy. Soc.*, B, **113**: 1–42.

———, 1935: Photosynthesis in intermittent light in relation to current formulations of the photosynthetic mechanism, *Biol. Rev.*, **10**: 460–482.

Brilliant, B., 1924: Le teneur en eau dans les feuilles et l'énergie assimilatrice, *C. R. Acad. Sci. Paris*, **178**: 2122–2125.

Brown, H. T., and G. H. Morris, 1893: A contribution to the chemistry and physiology of foliage leaves, *J. Chem. Soc.*, **63**: 604–677.

——— and F. Escombe, 1902: The influence of varying amounts of carbon dioxide in the air on the photosynthetic process of leaves and on the mode of growth of plants, *Proc. Roy. Soc.*, **70**: 397–413.

——— and ———, 1905: Researches on some of the physiological processes of green leaves, with special reference to the interchange of energy between the leaf and its surroundings, *Proc. Roy. Soc.*, B, **76**: 29–112.

——— and W. E. Wilson, 1905: On the thermal emissivity of a green leaf in still and moving air, *Proc. Roy. Soc.*, B, **76**: 122–137.

Brown, W. H., and G. W. Heise, 1917: The application of photochemical temperature coefficients to the velocity of carbon dioxide assimilation, *Philippine J. Sci.*, C, **12**: 1–24.

Buchanan, J. H., and G. G. Naudain, 1923: Influence of starch on strength of wheat flour, *Ind. and Eng. Chem.*, **15**: 1050–1051.

Burgess, A. H., 1935: The effect of the rate of flow of air upon assimilation and of fluids upon other natural processes, *Ann. Bot.*, **49**: 567–578.

Burk, D., 1927: Photosynthesis with malachite green, *J. Am. Chem. Soc.*, **49**: 1819–1825.

——— and H. Lineweaver, 1935: The kinetic mechanism of photosynthesis, *Symposia Quant. Biol.*, **3**: 165–183.

Burns, G. R., 1933: Photosynthesis in various portions of the spectrum, *Plant Physiol.*, **8**: 247–262.

———, 1934: Long and short wave-length limits of photosynthesis, *Plant Physiol.*, **9**: 645–652.

Burr, G. O., 1936: Carbonic anhydrase and photosynthesis, *Proc. Roy. Soc.*, B., **120**: 42–46.

Butlerow, A., 1861: Bildung einer zuckerartigen Substanz durch Synthese, *Ann. Chem. Pharm.*, **120**: 295–298.

Cailletet, M. L., 1911: Sur l'origine du carbone assimilé par les plantes, *C. R. Acad. Sci. Paris*, **152**: 1215–1217.

Cerighelli, R., 1921: Influence de l'anhydride carbonique sur le développement des plantes—emploi de CO_2 comme en grais atmosphérique, *Ann. Sci. Agron.*, **38**: 68–75.

Chapman, A. G., and W. H. Camp, 1932: Starch synthesis in the variegated leaves of *Pelargonium*, *Ohio J. Sci.*, **32**: 197–217.

Childers, N. F., and F. F. Cowart, 1935: The photosynthesis, transpiration and stomata of apple leaves as affected by certain nutrient deficiencies, *Proc. Am. Soc. Hort. Sci.*, **33**: 160–163.

Chouchak, D., 1929: L'assimilation chlorophyllienne de l'acide carbonique par les feuilles vertes dans un champ électrique, *Rev. Gén. Bot.*, **41**: 465–468.

Christopher, E. P., 1933: Influence of the amount of air supplied to apple leaves upon their CO_2 assimilation, *Proc. Am. Soc. Hort. Sci.*, **30**: 435–439.

———, 1934: The intensity of light striking leaves of apple trees at different times of day, *Proc. Am. Soc. Hort. Sci.*, **32**: 86–92.

———, 1935: The effect of flotation sulphur spray on the CO_2 assimilation of apple leaves, *Proc. Am. Soc. Hort. Sci.*, **33**: 149–151.

CLAYSON, D. H. F., and S. B. SCHRYVER, 1923: The hemicelluloses. I, The hemicellulose of wheat flour, *Biochem. J.*, **17**: 493–496.

CLEMENTS, H. F., 1930: Hourly variations in carbohydrate content of leaves and petioles, *Bot. Gaz.*, **89**: 241–272.

———, 1932: Mannose and the first sugar of photosynthesis, *Plant Physiol.*, **7**: 547–550.

COLE, R. C., 1932: The diurnal and seasonal changes in the sugar content of the sap and tissue of potato plants as affected by soil fertilization, *Soil Sci.*, **33**: 347–362.

COLIN, M. H., 1914: Sur la saccharogénie dans la betterave, *C. R. Acad. Sci. Paris*, **159**: 687–689.

———, 1916–1917: Le saccharose dans le betterave. Formation et disparition, *Rev. Gén. Bot.*, **28**: 289–299, 321–328, 368–380; **29**: 21–32, 56–64, 89–96, 113–127. Quoted by Stiles, 1925.

———, 1918: The origin of inulin in plants, *C. R. Acad. Sci. Paris*, **166**: 224–227; *Exp. Sta. Rec.*, **39**: 524.

——— and H. BELVAL, 1922: La genèse des hydrates de carbone dans le blé. Présence de lèvulosanes dans la tige, *C. R. Acad. Sci. Paris*, **175**: 1441–1443.

COLLA, S., 1930: Formation of chlorophyll in plants exposed to ultra-violet rays, *Ann. Bot.* (*Rome*), **18**: 329–349; *Exp. Sta. Rec.*, **67**: 20.

COMBES, R., 1928: Étude critique de la méthode de Sachs appliquée à la mesure des migrations des substances, *C. R. Acad. Sci. Paris.*, **187**: 666–668.

———, 1928: Influence des traumatisme sur les migrations des substances chez les végétaux, *C. R. Acad. Sci. Paris*, **187**: 993–995.

CONANT, J. B., E. M. DIETZ, C. F. BAILEY, and S. E. KAMERLING, 1931: Studies in the chlorophyll series. V, The structure of chlorophyll A, *J. Am. Chem. Soc.*, **53**: 2382–2393.

———, ———, and S. E. KAMERLING, 1931: The dehydrogenation of chlorophyll and the mechanism of photosynthesis, *Sci.*, **73**: 268.

COWART, F. F., 1935: Apple leaf structure as related to position of the leaf upon the shoot and to type of growth, *Proc. Am. Soc. Hort. Sci.*, **33**: 145–148.

CROCKER, W., 1918: Law of the minimum—a review of Hooker's work, *Bot. Gaz.*, **65**: 287–288.

CUMMINGS, M. B., and C. H. JONES, 1918: The aerial fertilization of plants with carbon dioxide, *Vt. Sta. Bull.* 211.

——— and ———, 1920: The aerial fertilization of plants with carbon dioxide, *J. Chem. Soc.*, **118**: 267.

CURRY, J., and S. F. TRELEASE, 1935: Influence of deuterium oxide on the rate of photosynthesis, Sci., 82: 18.

DANGEARD, P. A., 1927: Recherches sur l'assimilation chlorophyllienne et les questions qui s'y rattachent, *Botaniste*, **19**: 1–397; *Biol. Abst.*, **3**: 474.

DASTUR, R. H., 1924: Water content, a factor in photosynthesis, *Ann. Bot.*, **38**: 779–788.

———, 1925: The relation between water content and photosynthesis, *Ann. Bot.*, **39**: 769–786.

——— and R. D. ASANA, 1932: Effect of plane-polarized light on the formation of carbohydrates in leaves, *Ann. Bot.*, **46**: 879–891.

——— and J. J. CHINOY, 1932: Carbon dioxide assimilation of the leaves of the rice plant, *Oryza sativa*, *Indian J. Agr. Sci.*, **2**: 431–454.

——— and B. L. DESAI, 1933: The relation between water-content, chlorophyll content, and the rate of photosynthesis in some tropical plants at different temperatures, *Ann. Bot.*, **47**: 69–88.

——— and K. M. Samant, 1933: A study of the products of photosynthesis in leaves in artificial and in natural light, *Ann. Bot.*, **47**: 295–304.

——— and ———, 1933: A method for the determination of carbohydrates in leaves, *Indian J. Agr. Sci.*, **3**: 460–477; *Exp. Sta. Rec.*, **70**: 580.

——— and L. K. Gunjikar, 1934: Effect of elliptically polarized light on the formation of carbohydrates in leaves, *Ann. Bot.*, **48**: 1003–1012.

——— and ———, 1935: Energy absorption by leaves in normal and plane polarized light, *Ann. Bot.*, **49**: 273–281.

——— and R. J. Mehta, 1935: The study of the effect of blue-violet rays on photosynthesis, *Ann. Bot.*, **49**: 809–821.

Davidson, G. F., and W. A. Richardson, 1936: The molecular structure of cellulose and starch, *Sci. Prog.*, **31**: 68–77.

Davis, W. A., and A. J. Daish, 1913: A study of the methods of estimation of carbohydrates, especially in plant extracts. A new method for the estimation of maltose in presence of other sugars, *J. Agr. Sci.*, **5**: 437–468.

——— and ———, 1914: Methods of estimating carbohydrates. II, The estimation of starch in plant material. The use of taka-diastase, *J. Agr. Sci.*, **6**: 152–168.

——— and G. C. Sawyer, 1914: The estimation of carbohydrates. IV, The presence of free pentoses in plant extracts and the influence of other sugars on their estimation, *J. Agr. Sci.*, **6**: 406–412.

———, 1916: Studies of the formation and translocation of carbohydrates in plants. II, The Dextrose-laevulose ratio in the mangold, *J. Agr. Sci.*, **7**: 327–351.

———, A. J. Daish, and G. C. Sawyer, 1916: Studies of the formation and translocation of carbohydrates in plants. I, The carbohydrates of the mangold leaf, *J. Agr. Sci.*, **7**: 255–326.

——— and G. C. Sawyer, 1916: Studies of the formation and translocation of carbohydrates in plants. III, The carbohydrates of the leaf and leaf stalks of the potato. The mechanism of the degradation of starch in the leaf, *J. Agr. Sci.*, **7**: 352–384.

Dean, A. L., 1904: On inulin, *Am. Chem. J.*, **32**: 69–84.

Deleano, N. T., and J. Dick, 1932: Beiträge zur Kenntnis des Carotins. Neue Methoden zur Darstellung, Nachweis und Bestimmung, *Bull. Soc. Chem. Roumania*, **14**: 117–170.

——— and ———, 1933: Beiträge zur Kenntnis des Carotins. I, *Biochem. Z.*, **259**: 110–133.

——— and ———, 1934: Détermination de la chlorophylle dans les feuilles de *Salix fragilis* et de *Populus alba*, *Bull. Soc. Chem. Biol.*, **16**: 749–756.

Demerec, M., 1935: Behavior of chlorophyll in inheritance, *Symposia Quant. Biol.*, **3**: 80–86.

Demoussy, E., 1903–1904: Sur la végétation dans des atmosphères riches en acide carbonique. *C. R. Acad. Sci. Paris*, **136**: 325–328; **138**: 291–293 (1904); **139**: 883–885 (1904).

Deneke, H., 1931: Über den Einfluss bewegter Luft auf die Kohlensäureassimilation, *Jahr. wiss. Bot.*, **74**: 1. Quoted by Burgess, 1935.

Denny, F. E., 1930: The twin leaf method of studying changes in leaves, *Am. J. Bot.*, **17**: 818–842.

———, 1932: Changes in leaves during the night, *Contr. Boyce Thompson Inst.*, **4**: 65–83.

———, 1933: Bases for calculations in measuring changes in leaves during the night, *Contr. Boyce Thompson Inst.*, **5**: 181–194.

———, 1934: Starch determination methods involving solubility in acids, *Contr. Boyce Thompson Inst.*, **6**: 381–394.

De Saussure, N. T., 1804: "Recherches chimiques sur la végétation," Paris.

Deuber, C. G., 1926: Influence of mineral elements upon the development of chloroplast pigments of soybeans, *Bot. Gaz.*, **82**: 132–152.

Dhar, N. R., 1935: Chemistry of photosynthesis, *Symposia Quant. Biol.*, **3**: 151–164.

Dixon, H. H., and T. G. Mason, 1916: The primary sugar of photosynthesis, *Nature*, **97**: 160.

——— and H. H. Poole, 1920: Photosynthesis and the electronic theory, *Sci. Proc. Roy. Dublin Soc.*, **16**: 63–77.

——— and ———, 1920: Photosynthesis and the electronic theory, *Notes Bot. School Trinity Col. Dublin*, **3**: 90–104.

——— and N. G. Ball, 1922: Photosynthesis and the electronic theory, *Notes Bot. School Trinity Col. Dublin*, **3**: 199–205.

Duclaux, E., 1899: "Traité de microbiologie." 2, "Diastases, Toxines et Venins," Masson et Cie, Paris, 768 pp.

Dunlap, A. A., 1931: Carbohydrate variations accompanying the mosaic disease of tobacco, *Am. J. Bot.*, **18**: 328–336.

Echevin, R., 1934: L'évolution des phospholipides des feuilles au cours du jaunissement automnal, *C. R. Acad. Sci. Paris*, **198**: 1254–1256.

Edwards, D. W., and J. C. Ripperton, 1933: Some factors affecting the electrolytes of the starch granules, *J. Agr. Res.*, **47**: 179–191.

Egloff, G., 1936: Modern energy supplies, *Sci. Month.*, **42**: 291–300.

Eidelman, Z. M., and E. A. Bankul, 1933: Effect of the mechanical reduction of the leaf surface and different nutritive conditions on the accumulation of dry matter in cereals, *Lenin. Acad. Agr. Sci., U.S.S.R., Inst. Plant Protect.*, **1933**: 113–130; *Exp. Sta. Rec.*, **70**: 596.

Eldredge, J. C., 1935: The effect of injury in imitation of hail damage on the development of the corn plant, *Iowa Res. Bull.* 185.

Emerson, R., 1929: The relation between maximum rate of photosynthesis and concentration of chlorophyll, *J. Gen. Physiol.*, **12**: 609–622.

———, 1929: Photosynthesis as a function of light intensity and of temperature with different concentrations of chlorophyll, *J. Gen. Physiol.*, **12**: 623–639.

———, 1929: On the behavior of nickel carbonate in relation to photosynthesis, *J. Gen. Physiol.*, **13**: 163–169.

——— and W. Arnold, 1932: The photochemical reaction in photosynthesis, *J. Gen. Physiol.*, **16**: 191–205.

——— and L. Green, 1934: Kinetics of photosynthesis, *Nature*, **134**: 289–290.

———, 1935: The effect of intense light on the assimilatory mechanism of green plants, and its bearing on the carbon dioxide factor, *Symposia Quant. Biol.*, **3**: 128–137.

Ewart, A. J., 1896: On assimilatory inhibition in plants, *J. Linn. Soc. Bot.*, **31**: 364–461.

———, 1897: The effect of tropical insolation, *Ann. Bot.*, **11**: 439–480.

———, 1898: The action of cold and of sunlight upon aquatic plants, *Ann. Bot.*, **12**: 363–397.

———, 1898: Can isolated chloroplastids continue to assimilate? *Bot. Cent.*, **75**: 33–36.

———, 1907: On the supposed extracellular photosynthesis of carbon dioxide by chlorophyll, *Proc. Roy. Soc.*, B, **80**: 30–36.

———, 1918: On chlorophyll, carotin, and xanthophyll and on the production of sugar from formaldehyde, *Proc. Roy. Soc. Victoria*, **30**: 178–209.

———, 1919: On the synthesis of sugars from formaldehyde, *Proc. Roy. Soc. Victoria*, **31**: 378–387.

———, 1920: On the synthesis of sugar from formaldehyde and its polymers, its quantitative relations and its exothermic character, *Proc. Roy. Soc. Victoria,* **32**: 168–188.

Eyster, W. H., 1928: Photochlorophyll, *Sci.*, **68**: 569–570.

Fenton, H. J. H., 1907: The reduction of carbon dioxide to formaldehyde in aqueous solution, *J. Chem. Soc. Trans.*, **91**: 687–694.

Fernbach, A., and J. Wolff, 1904: Nouvelles observations sur la formation diastasique de l'amyloicellulose, *C. R. Acad. Sci. Paris,* **138**: 819–821.

———, 1912: Sur une nouvelle forme d'amidon soluble, *C. R. Acad. Sci. Paris,* **155**: 617–618.

Fincke, H., 1913: Über den Nachweis von Formaldehyd in Pflanzen, *Biochem. Z.*, **52**: 214–225.

Fischer, H., 1912: Carbonic-acid gas as a plant nutrient, *Gartenflora,* **61**: 298–307; *Exp. Sta. Rec.*, **28**: 837.

———, 1913: The effect of an increased amount of carbonic acid in the air on green plants, *Jahresber. Angew. Bot.*, **11**: 1–8; *Exp. Sta. Rec.*, **31**: 140.

———, 1919: The carbon nutrition of cultivated plants, *Gartenflora,* **68**: 165–168; *Exp. Sta. Rec.*, **42**: 137.

———, 1920: Plant growth and carbon dioxide, *Naturwiss.*, **8**: 413–417; *Exp. Sta. Rec.*, **45**: 218.

———, 1921: Carbon dioxide manuring, *Angew. Bot.*, **3**: 129–144; *Exp. Sta. Rec.*, **49**: 522.

———, 1927: Die Kohlensäure-Ernährung der Pflanzen, *Ber. bot. Ges.*, **45**: 331–339.

——— and J. Grassl, 1935: Weiterer Beitrag zur Feinstruktur von Chlorophyll *b*, *Liebigs Ann.*, **517**: 1–21.

Fisher, D. V., 1934: Leaf area in relation to fruit size and tree growth, *Sci. Agr.*, **14**: 512–518.

Fleischer, W. E., 1935: The relation between chlorophyll content and rate of photosynthesis, *J. Gen. Physiol.*, **18**: 573–597.

Gail, F. W., 1922: Photosynthesis in some of the red and brown algae as related to depth and light, *Pub. Puget Sound Biol. Sta.*, **3**: 141–176.

Galwialo, M. J., 1925: Zur Frage nach der Photosynthese der Kohlenhydrate, *Biochem. Z.*, **158**: 65–75.

Ganong, W. F., 1906: Normal apparatus, *Bot. Gaz.*, **41**: 209–213.

———, 1908: "Plant Physiology," Henry Holt & Company, pp. 107–109.

Gast, W., 1917: Quantitative Untersuchungen über den Kohlenhydratstoffwechsel im Laubblatt, *Z. physiol. Chem.*, **99**: 1–54.

Gautheret, R. J., 1932: The production of chlorophyll in roots exposed to light, particularly the root of barley, *C. R. Acad. Sci. Paris,* **194**: 1510–1513.

Georgi, C. E., F. S. Orcutt, and P. W. Wilson, 1933: Further studies on the relation between the carbon assimilation and nitrogen fixation in leguminous plants, *Soil Sci.*, **36**: 375–382.

Gerlach, P., 1919: Kohlensäuredüngung, *Mitt. deut. landw. Ges.*, **34**: 77–82.

———, 1921: Carbon dioxide nutrition of plants and stable manure, *Mitt. deut. landw. Ges.*, **36**: 147–150; *Exp. Sta. Rec.*, **46**: 424.

———, 1926: Ueber Kohlensäuredüngung, *Z. Pflanzenernähr. und Düng.*, **5**: 65–69.

Gibson, R. J. H., 1908: A photoelectric theory of photosynthesis, *Ann. Bot.*, **22**: 117–120.

Girard, A., 1883: Recherches sur la saccharogénie dans la betterave, *C. R. Acad. Sci. Paris,* **97**: 1305–1308.

———, 1884: Recherches sur la saccharogénie dans la betterave, *C. R. Acad. Sci. Paris,* **99**: 808–811.

GIROUD, A., R. RATSIMAMANGA, and C. P. LEBLOND, 1934: Parallelism between the occurrence of vitamin C and chlorophyll, *C. R. Soc. Biol.*, **117**: 612–614; *Chem. Abst.*, **29**: 1131.

———, ———, and ———, 1935: Relation between vitamin C and the carotenoids. Ripening of fruits, *C. R. Soc. Biol.*, **118**: 874–876; *Chem. Abst.*, **29**: 3702.

GODLEWSKI, E., 1873: Abhängigkeit der Sauerstoffausscheidung der Blätter von dem Kohlensäuregehalt der Luft, *Arb. bot. Inst. Würzburg*, **1**: 343–370.

GORDON, R. B., 1929: Suggested equations for the photosynthetic reaction, *Ohio J. Sci.*, **29**: 131–132.

GRADENWITZ, A., 1920: Carbonic acid gas to fertilize the air, *Sci. Am.*, **123**: 549.

GRAFE, V., and E. VIESER, 1909–1911: Untersuchungen über das Verhalten grüner Pflanzen zum gasförmigen Formaldehyd. I, II, *Ber. bot. Ges.*, **27**: 431–446; **29**: 19–26 (1911).

GRIFFIN, A., 1935: The effect of interrupted translocation upon loss of chlorophyll in leaves during autumn coloration, *Butler Univ. Bot. Studies*, **3**: 129–137.

GRUZIT, O. M., and R. P. HIBBARD, 1916: The influence of an incomplete culture solution on photosynthesis, *Rept. Mich. Acad. Sci.*, **18**: 50–52.

GUSTAFSON, F. G., and E. STOLDT, 1936: Some relations between leaf area and fruit size in tomatoes, *Plant Physiol.*, **11**: 445–451.

GUTHRIE, J. D., 1929: Effect of environmental conditions on the chloroplast pigments, *Am. J. Bot.*, **16**: 716–747.

HAAS, P., and T. G. HILL, 1913: "The Chemistry of Plant Products," Longmans, Green and Company, 400 pp.

HABERLANDT, G., 1882: Vergleichende Anatomie des assimilatorischen Gewebesystems der Pflanzen, *Jahrb. wiss. Bot.*, **13**: 74–188.

HANES, C. S., 1936: The determination of starch in plant tissue with particular reference to the apple fruit, *Biochem. J.*, **30**: 168–175.

HARDER, R., 1921: Kritische Versuche zu Blackmans Theorie der "begrenzenden Factoren" bei der Kohlensäureassimilation, *Jahrb. wiss. Bot.*, **60**: 531–571.

———, 1930: Über die Assimilation der Kohlensäure bei konstanten Aussenbedingungen. I, *Planta*, **11**: 263–293.

HARDING, T. S., 1923: History of inulin, its discovery and methods of preparation, *Sugar*, **25**: 636–638; *Exp. Sta. Rec.*, **51**: 409.

HARRIMAN, P. A., 1930: Effect of various methods of storage on the chlorophyll content of leaves, *Plant Physiol.*, **5**: 599–606.

HARTT, C. E., 1934: Water and cane ripening, *Hawaiian Planters' Rec.*, **38**: 193–206.

———, 1935: The fluctuations of sugars in the leaf blades of the cane sugar plant during the day and night, *Hawaiian Planters' Rec.*, **39**: 298–326.

HARVEY, E. N., 1928: Photosynthesis in absence of oxygen, *Plant Physiol.*, **3**: 85–89.

HAWORTH, W. N., and E. L. HIRST, 1936: The chemistry of the carbohydrates and glucosides, *Am. Rev. Biochem.*, **5**: 81–100.

HEILBRON, I. M., 1923: The photosynthesis of plant products, *Nature*, **111**: 502–504; *Bot. Abst.*, **12**: 981.

HEINICKE, A. J., 1932: Assimilation of carbon dioxide by apple leaves under natural conditions, abst., *Am. J. Bot.*, **19**: 845.

———, 1932: The assimilation of carbon dioxide by apple leaves as affected by ringing the stem, *Proc. Am. Soc. Hort. Sci.*, **29**: 225–229.

——— and M. B. HOFFMAN, 1932: An apparatus for determining the absorption of carbon dioxide by leaves under natural conditions, *Sci.*, **77**: 55–58.

——— and ———, 1933: The rate of photosynthesis of apple leaves under natural conditions. I, *N. Y. Sta. (Cornell) Bull.* 577.

——— and N. F. CHILDERS, 1935: The influence of water deficiency in photosynthesis and transpiration of apple leaves, *Proc. Am. Soc. Hort. Sci.*, **33**: 155–159.

HENRICI, M., 1918: Chlorophyllgehalt und Kohlensäureassimilation bei Alpen- und Ebenenpflanzen, *Verhand. Naturforsch. Ges. Basel*, **30**: 43–136.

———, 1926: The chlorophyll content of grasses in Bechuanaland, *Union So. Africa Dept. Agr. Vet. Ed. and Res.*, **12**: 259–270.

HERLITZKA, A., 1912: Über den Zustand des Chlorophylls in der Pflanze und über kolloidales Chlorophyll, *Biochem. Z.*, **38**: 321–330.

HIBBERT, H., 1933: The plant synthesis of carbohydrates and polysaccharides, *Can. J. Res.*, **8**: 103–104.

HICKS, P. A., 1934: Interaction of factors in the growth of *Lemna*. V, Some preliminary observations upon the interaction of temperature and light on the growth of *Lemna*, *Ann. Bot.*, **48**: 515–523.

——— and T. E. PANISSET, 1934: The quantitative determination of minute amounts of chlorophyll, *New Phytol.*, **33**: 199–210.

HILPERT, S., H. HOFMEIER, and A. WOLTER, 1931: Über den Zustand des Chlorophylls in der Pflanze, *Ber. chem. Ges.*, **64**: 2570–2577.

HOFFMAN, M. B., 1932: The effect of certain spray materials on the carbon dioxide assimilation by McIntosh apple leaves, *Proc. Am. Soc. Hort. Sci.*, **29**: 389–393.

———, 1933: Carbon dioxide assimilation by apple leaves as affected by lime sulfur sprays. II, Field experiments, *Proc. Am. Soc. Hort. Sci.*, **30**: 169–175.

HOLM, T., 1931: The apparent influence of inulin on the meristem in roots of *Compositae*, *Beih. bot. Cent.*, **47**: 359–377; *Exp. Sta. Rec.*, **66**: 520.

HOLMAN, R., 1930: On solarization of leaves, *Univ. Calif. Pub.* (*Bot.*), **16**: 139–151.

HOOKER, H. D., 1917: Liebig's law of the minimum in relation to general biological problems, *Sci.*, **46**: 197–204.

HOOVER, W. H., E. S. JOHNSTON, and F. S. BRACKETT, 1933: Carbon dioxide assimilation in a higher plant, *Smithsonian Misc. Coll.*, **87**: No. 16: 1–19.

HORN, T., 1923: Das gegenseitige Mengenverhältnis der Kohlenhydrate im Laubblatt in seiner Abhängigkeit von Wassergehalt, *Bot. Archiv.*, **3**: 137–172; *Bot. Abst.*, **13**: 1019.

HOWARD, G., 1925: Pigment studies with special reference to carotinoids in fruits, *Ann. Mo. Bot. Gard.*, **12**: 145–212.

ILJIN, W. S., 1915: Transpiration and assimilation in steppe plants, *Bull. Acad. Imp. Sci. Petrograd.*, 6 ser., **9**: 346–367; *Exp. Sta. Rec.*, **36**: 734.

———, 1923: Der Einfluss des Wassermangels auf die Kohlenstoffassimilation durch die Pflanzen, *Flora*, **116**: 360–378.

———, 1929: Standortsfeuchtigkeit und der Zuckergehalt in den Pflanzen, *Planta*, **7**: 59–71.

INGEN-HOUSZ, J., 1779: "Experiment upon Vegetables, Discovering Their Great Power of Purifying Common Air in the Sunshine and of Injuring in the Shade and Night," London.

INMAN, O. L., and P. ROTHEMUND, 1932: The decomposition of chlorophyll by the animal body, abst., *Am. J. Bot.*, **19**: 839.

———, 1935: Formation of chlorophyll and the beginning of photosynthesis, *Plant Physiol.*, **10**: 401–403.

———, 1935: The evolution of oxygen in the process of photosynthesis, *Symposia Quant. Biol.*, **3**: 184–190.

IRELAND, J. C., and P. A. YEATS, 1933: Chlorophyll content of grain sorghums, *Bot. Gaz.*, **95**: 300–315.

IRVING, A. A., 1910: The beginning of photosynthesis and the development of chlorophyll, *Ann. Bot.*, **24**: 805–818.

JACOBY, M., 1919, 1922: Über das Formaldehyd als Übergangsstufe zwischen der eigentlichen Assimilation und der Kohlenhydratbildung in der Pflanze. I, II, *Biochem. Z.*, **101**: 1–6; **128**: 119–121.

JAMES, W. O., 1928: Experimental researches on vegetable assimilation and respiration. XIX, The effect of variations of carbon dioxide supply upon the rate of assimilation of submerged water plants, *Proc. Roy. Soc.*, B, **103**: 1–42.

———, 1934: The dynamics of photosynthesis, *New Phytol.*, **33**: 8–40.

———, 1935: The rate of photosynthesis, *Nature*, **136**: 220–221.

JESS, M., 1921: Carbonic acid gas as a fertilizer, *J. Agr. Prat.*, **35**: 250–252; *Exp. Sta. Rec.*, **45**: 344.

JOHNSTON, E. S., 1935: Aerial fertilization of wheat plants with carbon-dioxide gas, *Smithsonian Misc. Coll.*, **94**: No. 15: 1–9.

JORGENSEN, I., and F. KIDD, 1917: Some photochemical experiments with pure chlorophyll and their bearing on theories of carbon assimilation, *Proc. Roy. Soc.*, B, **89**: 342–361.

——— and W. STILES, 1917: Carbon assimilation. A review of recent work on the pigments of the green leaf and the processes connected with them, *New Phytol.*, repr., **10**: 1–180.

KALITIN, N. N., 1931: Determination of reflection and absorption of the solar radiation by the plant leaves, *J. Soc. Bot. Russ.*, **16**: 101–110.

KAUTSKY, H., A. HIRSCH, and F. DAVIDSHÖFER, 1932: Energy transfers at surfaces. VI, Carbon dioxide assimilation. I, *Ber. chem. Ges.*, **65**B: 1762–1770.

KAYSER, R., 1883: Ueber das Vorkommen von Rohrzucker und einigen seiner Unwandlungsproducte im Organismus der Pflanzen, *Landw. Versuchs-Stat.*, **29**: 461–473.

KEUHL, H. J., 1926: The concentration of carbon dioxide in the air above cultivated areas, *Z. Pflanzenernähr. u. Düng.*, **6**: 318–378.

KIMPFLIN, G., 1907: Sur la présence du menthanal dans les végétaux vertes, *C. R. Acad. Sci. Paris*, **144**: 148–150.

KISHI, Y., and S. MONOBE, 1931: On the quantitative changes of the chemical constituents of mulberry leaves during 24 hours, (In Japanese), *Bull. Kjusu. Imp. Univ. Fukuoka*, **4**: 531–599; *Biol. Abst.*, **8**: 335.

KJELDAHL, J., 1879: Recherches sur le ferment production du sucre, *C. R. trav. lab. Carlsberg.*

KLEIN, G., and O. WERNER, 1926: Formaldehyd als Zwischenprodukt bei der Kohlensäureassimilation, *Biochem. Z.*, **168**: 361–386.

KNIEP, H., and F. MINDER, 1909: Ueber den Einfluss verschiedenfarbigen Lichtes auf die Kohlensäureassimilation, *Z. Bot.*, **1**: 619–650.

KNY, L., 1897: Die Abhängigkeit der Chlorophyllfunction von den Chromatophoren und von Cytoplasma, *Ber. bot. Ges.*, **15**: 388–403.

———, 1898: Vermögen isolirter Chlorophyllkörper im Lichte Sauerstoff auszuscheiden? *Bot. Cent.*, **73**: 426–439.

KOHN, H. I., 1935: Inhibition of photosynthesis in *Chlorella pyrenoidosa* by the iodoacetyl radical, *J. Gen. Physiol.*, **19**: 23–34.

KOSTYTSCHEW, S., 1921: Studien über Photosynthese. I, Das Verhältnis CO_2/O_2 bei der Kohlensäureassimilation, *Ber. bot. Ges.*, **39**: 319–328; *Bot. Abst.*, **12**: 744.

———, 1922: Studien über Photosynthese. IV, Die CO_2 Assimilation der Leguminosen, *Ber. bot. Ges.*, **40**: 112–119.

KRAEMER, H., 1902: The structure of the starch grain, *Bot. Gaz.*, **34**: 341–355.

———, 1905: Further observations on the structure of the starch grain, *Bot. Gaz.*, **40**: 305–310.

———, 1923: The starch grains, *Sci.*, **57**: 175.

Kraus, G., 1869: Einige Beobachtungen über den Einfluss des Lichts und der Wärme auf die Stärkeerzeugung im Chlorophyll, *Jahrb. wiss. Bot.*, **7**: 511–531.

Kreusler, U., 1885: Ueber eine Methode zur Beobachtung der Assimilation und Athmung der Pflanzen und über einige diese Vorgänge beeinflussenden Momente, *Landw. Jahrb.*, **14**: 913–965.

———, 1887: Beobachtungen über die Kohlensäure-Aufnahme und -Ausgabe (Assimilation und Athmung) der Pflanzen. II, Mittheilung. Abhängigkeit von Entwicklungszustand—Einfluss der Temperatur, *Landw. Jahrb.*, **16**: 711–755.

———, 1888: Beobachtung über die Kohlensäure-Aufnahme und -Ausgabe (Assimilation und Athmung) der Pflanzen. III, Mittheilung. Einfluss der Temperatur, *Landw. Jahrb.*, **17**: 161–175.

———, 1890: Beobachtungen über Assimilation und Athmung der Kohlensäureausscheidung seitens getödteter Exempläre; Kohlensäureverbrauch, wenn Ober- und Unterseite der Blätter dem Licht zugewendet, *Landw. Jahrb.*, **19**: 649–668.

Kuhn, R., 1935: Plant pigments, *Ann. Rev. Biochem.*, **4**: 479–496.

Kuiper, K. J., 1920: Koolsuurbemesting nachtverlichting en andere groeibevorderende factoren in de plantenteelt. (Effects of CO_2 manuring, night illumination and other growth-stimulating factors upon plant culture), *Cultura*, **32**: 332–344, 355–368; *Bot. Abst.*, **8**: 96.

Kursanov, A. L., 1934: The photosynthesis of green fruits and its dependence on the normal activity of the leaves, *Planta*, **22**: 240–250; *Chem. Abst.*, **28**: 5498–5499.

Kylin, H., 1918: Zur Kenntnis der wasserlöslichen Kohlenhydrate der Laubblätter, *Z. physiol. Chem.*, **150**: 77–88.

Lagatu, H., and L. Maume, 1924: Évolution remarquablement regulière de certains rapports physiologiques (chaux, magnésie, potasse) dans les feuilles de la vigne bien alimentée, *C. R. Acad. Sci. Paris*, **179**: 782–785.

——— and ———, 1924: Étude, par l'analyse périodique des feuilles, de l'influence des engrais de chaux, de magnésie et de potasse sur la vigne, *C. R. Acad. Sci. Paris*, **179**: 932–935.

——— and ———, 1925: Relation linéaire entre les quantités successives d'acide phosphorique et d'azote dans la feuille de la vigne bien alimentée, *C. R. Acad. Sci. Paris*, **180**: 1179–1181.

Langner, W., 1927: Über die Verteilung der Stärke in Laubblättern zu verschiedenen Tageszeiten, *Jahrb. wiss. Bot.*, **67**: 291–333.

Lemmerman, O., 1920: Ueber die Kohlensäureernährung der Pflanzen, *Mitteil. landw. Ges.*, **35**: 696–699; *Bot. Abst.*, **8**: 5.

———, 1926: Untersuchungen über die Bedeutung der Bodenkohlensäure für die Ernährung der Pflanzen und über die Wirkung einiger Humus-bezw. Kohlensäuredünger, *Z. Pflanzenernähr. u. Düng.*, **5**: 70–85.

Li, T., 1929: The immediate effect of change of light on the rate of photosynthesis, *Ann. Bot.*, **43**: 587–601.

Liebig, J., 1843: "Chemistry in Its Application to Agriculture and Physiology," 3d ed., Peterson, Philadelphia.

———, 1843: Die Wechselwirthschaft, *Ann. Chem. Pharm.*, **46**: 58–97.

Lindet, L., and P. Nottin, 1923: Évolution des grains de fécule dans le tubercule de la pomme de terre, *C. R. Acad. Sci. Paris.*, **176**: 149–155.

Ling, A. R., and R. N. Dinshaw, 1923: Studies on starch. I, The nature of polymerized amylose and amylopectin, *J. Chem. Soc.*, **123**: 2666–2688.

Livingston, B. E., 1917: A quarter-century of growth in plant physiology, *Plant World*, **20**: 1–15.

——— and R. Beall, 1934: The soil as direct source of carbon dioxide for ordinary plants, *Plant Physiol.*, **9**: 237–259.

Loeb, W., 1905–1906: Zur Kenntnis der Assimilation der Kohlensäure, *Z. Elektrochem.*, **11**: 745–752; **12**: 282–312.

Long, F. L., 1919: The quantitative determination of photosynthetic activity in plants, *Physiol. Res.*, **2**: 277–300.

Loomis, W. E., and K. H. Burnett, 1931: Photosynthesis in corn, *Proc. Iowa Acad. Sci.*, **38**: 150.

Lubimenko, W., 1905: Sur la sensibilité de l'appareil chlorophyllien des plantes ombrophiles et ombrophobes, *Rev. Gén. Bot.*, **17**: 381–415.

———, 1908: La concentration du pigment vert et l'assimilation chlorophyllienne, *Rev. Gén. Bot.*, **20**: 162–177, 217–238, 253–267, 285–297.

———, 1923: Action spécifique des rayons lumineux de diverses couleurs dans la photosynthése, *C. R. Acad. Sci. Paris*, **177**: 606–608; *Bot. Abst.*, **13**: 568.

———, 1927, 1928: Les pigments de plastes et leur transformation dans les tissus vivants de la plante, *Rev. Gén. Bot.*, **39**: 547–559, 619–637, 698–710 (1927); **40**: 23–29, 88–94, 146–155, 226–243, 303–318, 372–381.

——— and E. P. Gubbenet, 1930: Effect of temperature on rate of accumulation of chlorophyll in etiolated germinated plants, *Bull. Inst. Sci. Lesshaft.* (*Russian*), **16**: 165–188; *Biol. Abst.*, **6**: 1892.

——— and E. R. Hubbenet, 1932: The influence of temperature on the rate of accumulation of chlorophyll in etiolated seedlings, *New Phytol.*, **31**: 26–57.

——— and O. A. Stscheglova, 1932: Effect of protoplasmic irritation on photosynthesis, *Planta*, **18**: 383–404; *Chem. Abst.*, **27**: 4830–4831.

Lundegårdh, H., 1921: Ecological studies in the assimilation of certain forest plants and short plants, *Svensk. Bot. Tidssk.*, **15**: 46–95.

———, 1922: Theoretical and practical basis for carbon dioxid fertilization, *Angew. Bot.*, **4**: 120–151; *Exp. Sta. Rec.*, **49**: 421.

———, 1924: "Der Kreislauf der Kohlensäure in der Natur," Gustav Fischer, Jena, 308 pp., 47 figures.

MacDougal, D. T., 1895: Nature and life history of starch grains, *Bot. Gaz.*, **20**: 458–459.

MacKinney, G., and H. W. Milner, 1933: Carrot leaf carotene, *J. Am. Chem. Soc.*, **55**: 4728–4729.

———, 1935: Development of the chlorophyll and carotenoid pigments in barley seedlings, *Plant Physiol.*, **10**: 365–373.

———, 1935: Properties of carotenes from certain roots and leaves at various stages of development, *J. Biol. Chem.*, **108**: 45–49.

———, 1935: On the plastid pigments of marsh dodder, *J. Biol. Chem.*, **112**: 421–424.

Macy, P., 1936: The quantitative mineral nutrient requirements of plants, *Plant Physiol.*, **11**: 749–764.

Magistad, O. C., 1935: Carotene and xanthophyll in pineapples, *Plant Physiol.*, **10**: 187–191.

Maquenne, L., 1903: Sur la rétrogradation de l'empois d'amidon, *C. R. Acad. Sci. Paris*, **137**: 88–90.

——— and E. Roux, 1905: Sur la constitution la saccharification et la rétrogradation des empois de fécule, *C. R. Acad. Sci. Paris*, **140**: 1303–1308.

———, 1911: À propos d'une communication récente de M. L. Cailletet, *C. R. Acad. Sci. Paris*, **152**: 1818–1819.

——— and E. Demoussy, 1913: Sur la valeur des coefficients chlorophylliens et leurs rapports avec les quotients respiratoires réels, *C. R. Acad. Sci. Paris*, **156**: 506–512.

Martin, E. V., 1934: Effect of solar radiation on transpiration of *Helianthus annuus*, *Plant Physiol.*, **10**: 341–354.

Martin, W. E., 1936: Distribution of certain sugars in Bosc pears, *Plant Physiol.*, **11**: 139–147.

Marx, D., 1929: The effect of small electric currents on the assimilation of *Elodea canadensis*, *Ann. Bot.*, **43**: 163–172.

Maskell, E. J., 1927: Field observations on starch production in the leaves of the potato, *Ann. Bot.*, **41**: 327–344.

———, 1928: Experimental researches on vegetable assimilation and respiration. XVII, The diurnal rhythm of assimilation in leaves of cherry laurel as limiting concentrations of carbon dioxide, *Proc. Roy. Soc.*, B, **102**: 467–488.

———, 1928: Experimental researches on vegetable assimilation and respiration. XVIII, The relation between stomatal opening and assimilation—a critical study of assimilation rates and porometer rates in leaves of cherry laurel, *Proc. Roy. Soc.*, B, **102**: 488–533.

Matlack, M. B., and C. E. Sando, 1934: A contribution to the chemistry of tomato pigments. The coloring matter in American red and purple tomatoes (*Lycopersicum esculentum*), *J. Biol. Chem.*, **104**: 407–414.

Matthaei, G. L. C., 1902: The effect of temperature on carbon dioxide assimilation, *Ann. Bot.*, **16**: 591–593.

———, 1905: Experimental researches on vegetable assimilation and respiration. III, On the effect of temperature on carbon-dioxide assimilation, *Phil. Trans. Roy. Soc.*, **197**: 47–105.

Mayer, H., 1930: Untersuchungen über die Chlorophyllase, *Planta*, **11**: 294–330.

McCarty, E. C., 1935: Seasonal march of carbohydrates in *Elymus ambiguus* and *Muhlenbergia gracilis* and their reaction under moderate grazing use, *Plant Physiol.*, **10**: 727–738.

McKay, E., 1935: Photosynthesis in *Grimmia montana*, *Plant Physiol.*, **10**: 803–809.

McLean, F. T., 1920: Field studies of the carbon dioxide absorption of coconut leaves, *Ann. Bot.*, **34**: 367–389.

McNair, J. B., 1932: Some properties of plant substances in relation to climate of habitat—volatile oils, saponins, cyanogenetic glucosides, and carbohydrates, *Am. J. Bot.*, **19**: 168–192.

Meier, F. E., 1932: Lethal action of ultra-violet light on a unicellular green alga, *Smithsonian Misc. Coll.*, **87**: No. 10: 1–11.

———, 1934: Colonial formation of unicellular green algae under various light conditions, *Smithsonian Misc. Coll.*, **92**: No. 5: 1–14.

———, 1934: Effects of intensities and wave lengths of light on unicellular green algae, *Smithsonian Misc. Coll.*, **92**: No. 6: 1–27.

———, 1936: Growth of a green alga in isolated wave-length regions, *Smithsonian Misc. Coll.*, **94**: No. 17: 1–12.

Mellanby, J., 1919: The composition of starch. I, Precipitation by colloidal iron. II, Precipitation by iodine and electrolytes, *Biochem. J.*, **13**: 28–36.

Meyer, A., 1881: Ueber die Structur der Stärkekörner, *Bot. Ztg.*, **39**: 841–852, 856–864.

———, 1885: Ueber die Assimilationsproducte der Laubblätter angiospermer Pflanzen, *Bot. Ztg.*, **43**: 417–423, 433–440, 449–458, 465–472, 481–491, 497–507.

———, 1886: Bildung der Stärkekörner in den Laubblättern aux Zuckerarten, Mannit und Glycerin, *Bot. Ztg.*, **44**: 81–88, 105–113, 129–137, 145–151.

———, 1886: Ueber Stärkekörner, welche sich mit Jod roth färben, *Ber. bot. Ges.*, **4**: 337–362.

———, 1895: "Untersuchungen über die Stärkekörnerwesen und Lebensgeschichte der Stärkekörner der hoheren Pflanzen," Gustav Fischer, Jena, 318 pp.

Meyer, B. S., and D. S. Rader, 1936: Simple apparatus for the quantitative determination of photosynthetic and respiratory ratios, *Plant Physiol.*, **11**: 437–443.

Mezzadroli, G., 1928: Action of ultra-violet rays on the formation of sugars and aldehydes by breaking down solutions of bicarbonates and CO_2 in the presence of colloidal, reducing catalysts, *Rev. Indus. y. Agr. Tucumán* (*Italian*), 1928; *Biol. Abst.*, **6**: 703.

Miles, F. C., 1915: A genetic and cytological study of certain types of albinism in maize, *J. Genetics*, **4**: 193–214.

Miller, E. C., 1917: Daily variation of water and dry matter in the leaves of corn and the sorghums, *J. Agr. Res.*, **10**: 11–47.

———, 1924: Daily variation of the carbohydrates in the leaves of corn and the sorghums, *J. Agr. Res.*, **27**: 785–808.

———, 1928: Unpub. data, *Kans. Sta.*

Miller, E. S., 1935: Improved methods for the purification of the common carotenoids and the quantitative measurement of their absorption spectra, *Bot. Gaz.*, **96**: 447–467.

——— and G. O. Burr, 1935: Carbon dioxide balance at high light intensities, *Plant Physiol.*, **10**: 93–114.

———, G. Mackinney, and F. P. Zscheile, Jr., 1935: Absorption spectra of alpha and beta carotenes and lycopene, *Plant Physiol.*, **10**: 375–381.

Moillard, M., 1912: L'humus est-il une source directe de carbone pour les plantes vertes supérieures? *C. R. Acad. Sci. Paris*, **154**: 291–294.

Moldavan, A., 1934: Isotopic CO_2 and O_2 in plants, *Sci.*, **79**: 455.

Molisch, H., 1921: Über den Einfluss der Transpiration auf das Verschwinden der Stärke in den Blättern, *Ber. bot. Ges.*, **39**: 339–344.

———, 1925: Carbon dioxide assimilation in dead leaves, *Z. Bot.*, **17**: 577–593; *Exp. Sta. Rec.*, **57**: 118.

Moll, J. W., 1878: Ueber die Herkunft des Kohlenstoffs der Pflanzen, *Arb. bot. Inst. Würzburg*, **2**: 105–113.

———, 1909: Communication on presenting the dissertation of Mr. K. Zijlstra entitled "Kohlensäuretransport in Blättern," *Proc. Akad. Amsterdam*, **2**: 649–668. Quoted by Livingston and Beall, 1934.

Moore, B., and T. A. Webster, 1918: Action of light rays on organic compounds and the photosynthesis of organic from inorganic compounds in presence of inorganic colloids, *Proc. Roy. Soc.*, B, **90**: 168–186.

Müller, A., 1904: Die Assimilationsgrösse bei Zucker und Stärkeblättern, *Jahrb. wiss. Bot.*, **40**: 443–498.

Murneek, A. E., 1934: Relation of carotinoid pigments to sexual reproduction in plants, *Sci.*, **79**: 528.

———, 1936: Determination of the chloroplast pigments of plants, *Sci.*, **83**: 327.

Nägeli, C., 1858: "Die Stärkekörner. Morphologische, physiologische, chemisch-physikalische und systematisch-botanische," Monographie, Zurich, 625 pp.

Nef, J. U., 1913: Dissociationsvorgänge in der Zuckergruppe, *Ann. Chem.*, **403**: 204–383.

Nelson, J. M., and M. Auchincloss, 1933: The effects of glucose and fructose on the sucrose content in potato slices, *J. Am. Chem. Soc.*, **55**: 3769–3772.

Nicholas, E., and G., Nicolas, 1922: Influence du formol sur les végétaux supérieurs, *C. R. Acad. Sci. Paris*, **175**: 1437–1439.

Noack, K., and W. Kiessling, 1929: Formation of chlorophyll and its relation to the blood pigment, *Z. physiol. Chem.*, **182**: 13–49.

———, 1929: Formation of chlorophyll and its relation to the blood pigment, *Naturwiss.*, **17**: 104.

NORRIS, R. J., 1933: Observations on the development of chlorophyll and carotinoid pigments in etiolated plants, *Bull. Basic Sci. Res. Univ. Sci.*, **5**: 23–32; *Biol. Abst.*, **9**: 1481.

NOVELLO, R., 1933: Alcune osservazioni sui cloroplasti delle palme, *Atti R. Accad. Naz. Lincei Rend.*, **18**: 245–248; *Biol. Abst.*, **10**: 33.

OSERKOWSKY, J., 1932: Relation between the green and the yellow pigments in chlorotic leaves, *Plant Physiol.*, **7**: 711–716.

OSTERHOUT, W. J. V., 1915: Artificial photosynthesis by means of chlorophyll, *Sci.*, **42**: 68.

———, 1918: A demonstration of photosynthesis, *Am. J. Bot.*, **5**: 105–111.

———, 1918: A simple method of demonstrating the production of aldehyde by chlorophyll and by aniline dyes in the presence of sunlight, *Am. J. Bot.*, **5**: 511–513.

——— and A. R. C. HAAS, 1918: On the dynamics of photosynthesis, *J. Gen. Physiol.*, **1**: 1–16.

——— and ———, 1918: Dynamical aspects of photosynthesis, *Proc. Nat. Acad. Sci.*, **4**: 85–91.

——— and ———, 1918: A simple method of measuring photosynthesis, *Sci.*, **47**: 420–422.

———, 1919: Apparatus for the study of photosynthesis and respiration, *Bot. Gaz.*, **68**: 60–62.

O'SULLIVAN, C., 1872: On the transformation-products of starch, *J. Chem. Soc.*, **25**: 579–588.

OWEN, O., 1923: Carbon dioxide in relation to glasshouse crops, *Ann. App. Biol.*, **10**: 312–317.

——— and P. H. WILLIAMS, 1923: The preparation of an atmosphere rich in carbon dioxide, *Ann. App. Biol.*, **10**: 318–325.

———, T. SMALL, and P. H. WILLIAMS, 1926: Carbon dioxide in relation to glasshouse crops. III, The effect of enriched atmospheres on tomatoes and cucumbers, *Ann. App. Biol.*, **13**: 560–577.

PALLADIN, V. I., 1922: "Plant Physiology." English translation by B. E. Livingston. P. Blakiston's Son & Company, Philadelphia.

PALLADIN, W., 1891: Ergrünen und Wachsthum der etiolirten Blätter, *Ber. bot. Ges.*, **9**: 229–232.

PALMER, L. S., 1934: The biological and chemical nomenclature for the carotenoids, *Sci.*, **79**: 488–490.

PANTANELLI, E., 1904: Abhängigkeit der Sauerstoffausscheidung belichteter Pflanzen von äusseren Bedingungen, *Jahrb. wiss. Bot.*, **39**: 167–228.

PARKER, M. W., 1935: Physical and chemical properties of the soluble polysaccharides in sweet corn, *Plant Physiol.*, **10**: 713–725.

PARKIN, J., 1912: The carbohydrates of the foliage leaf of the snowdrop (*Galanthus nivalis*) and their bearing on the first sugar of photosynthesis, *Biochem. J.*, **6**: 1–47.

———, 1925: The first sugar of photosynthesis and the role of cane sugar in the plant—a reply to Professor J. H. Priestley, *New Phytol.*, **24**: 57–64.

PEARCE, G. W., and L. R. STREETER, 1931: A report on the effect of light on pigment formation in apples, *J. Biol. Chem.*, **92**: 743–749.

PEIRCE, G. J., 1934: The greatest factory in the world, *Sci. Month.*, **39**: 464–467.

PERREY, A., 1882: Sur l'origin des matières sucrées dans la plante, *C. R. Acad. Sci. Paris*, **94**: 1124–1125.

PETERSON, P. D., 1930: Methods for the quantitative extraction and separation of the plastid pigments of tobacco, *Plant Physiol.*, **5**: 257–261.

Pfeffer, W., 1871: Die Wirkung farbigen Lichtes auf die Zersetzung der Kohlensäure in Pflanzen, *Arb. bot. Inst. Würzburg*, **1**: 1–76.

———, 1900: "The Physiology of Plants." English translation, Vol. I, Clarendon Press.

Pickett, W. F., 1933: A comparative study of the intercellular spaces of apple leaves, *Proc. Am. Soc. Hort. Sci.*, **30**: 156–161.

———, 1934: Photosynthetic activity and internal structure of apple leaves are correlated, *Proc. Am. Soc. Hort. Sci.*, **32**: 81–85.

———, 1935: A comparison of three methods of measuring photosynthetic activity of apple leaves, *Proc. Am. Soc. Hort. Sci.*, **33**: 152–154.

Plaetzer, H., 1917: Untersuchungen über die Assimilation und Atmung von Wasserpflanzen, *Verhandl. Phys. med. Ges. Würzburg*, **45**: 31–102.

Plester, W., 1912: Kohlensäureassimilation und Atmung bei varietäten derselben Art, die sich durch ihre Blattfärbung unterscheiden, *Beitr. Biol. Pflanzen*, **11**: 249–304.

Pollacci, G., 1917: Carbon assimilation in green plants, *Atti Ist. Bot. Univ. Pavia*, Ser. II, **17**: 29–51.

———, 1920: Sul carbonio delle piante verdi (the carbon of green plants), *Atti Ist. Bot. Univ. Pavia*, Ser. II, **17**: 29–51; *Bot. Abst.*, **13**: 1212.

——— and M. Bergamaschi, 1931: Sulla formazione di aldeide formica in piante vive durante la fotosintesi clorofilliana, *Atti Ist. Bot. Univ. Pavia*, **2**: 1–64; *Biol. Abst.*, **8**: 874.

Porowski, G. I., 1925: Über die Lichtabsorption von Blättern einiger Bäume, *Biochem. Z.*, **165**: 420–426.

Presley, J. T., 1936: Rubber content of golden-rod leaves affected by light, *Sci.*, **83**: 436.

Priestley, J., 1774: "Experiments and Observations on Different Kinds of Air," 342 pp.

Priestley, J. H., 1924: The first sugar of photosynthesis and the role of cane sugar in the plant, *New Phytol.*, **23**: 255–264.

Puhr, L. F., and A. N. Hume, 1932: Variations in amounts of carbohydrates in the leaves of corn, *S. Dak. Sta. Bull.* 270.

Puriewitsch, K., 1914: Untersuchungen über Photosynthese, *Jahrb. wiss. Bot.*, **53**: 210–254.

Qureshi, M., and S. S. Mohammad, 1932: Photochemical reduction of carbon dioxide in aqueous solution, *J. Phys. Chem.*, **36**: 2205–2216; *Biol. Abst.*, **8**: 1914.

Raber, O., 1931: A possible relationship between hemoglobin and chlorophyll as shown by the use of liver extract, *Sci.*, **73**: 457–458.

Rajvansi, A. R., and N. R. Dhar, 1932: Photosynthesis in tropical sunlight. III, Synthesis of formaldehyde, *J. Phys. Chem.*, **36**: 568–574.

Reichert, E. T., 1913: The differentiation and specificity of starches in relation to genera, species, etc., *Carnegie Inst. Wash. Pub.* 173.

———, 1919: A biochemic basis for the study of problems of taxonomy, heredity, evolution, etc., with special reference to the starches and tissues of parent-stocks and hybrid-stocks and the starches and hemoglobins of varieties, species and genera, *Carnegie Inst. Wash. Pub.* 270.

Reid, M. E., 1930: Growth and nitrogen metabolism of squash seedlings. III, With respect to high and low carbohydrate synthesis, *Am. J. Bot.*, **17**: 579–601.

Reinau, E. H., 1927: Forschungsergebnisse zur Kohlensäurefrage, *Angew. Bot.*, **9**: 12–20.

Richards, H. M., and D. T. MacDougal, 1904: The influence of carbon monoxide and other gases upon plants, *Bull. Torrey Bot. Club*, **31**: 57–66.

Riedel, F., 1919: Die Ausnützung der Hochofenabgase zur Kohlensäuredüngung, *Stahl u. Eisen*, **39**: 1497–1506.

———, 1921: Importance of carbon dioxide as a fertilizer for vegetables, *Gartenwelt*, **25**: 302–304; **34**: 336–338 (1921); *Exp. Sta. Rec.*, **45**: 834.

Rippel, A., 1926: Kohlensäure und Pflanzenerzeugung, *Z. Pflanzennähr. u. Düng.*, **5**: 49–64.

Roberts, H. F., 1937: The causes of autumn coloration, *Sci. Mo.*, **45**: 427–435.

Roberts, R. H., and N. Livingston, 1935: Plant pigments and reproduction, *Sci.*, **82**: 596.

Roddis, L. H., 1936: Chlorophyllometry, *Plant Physiol.*, **11**: 211–212.

Rosecrans, C. Z., 1934: The determination of CO_2 in the atmosphere of a closed system, *Sci.*, **80**: 483–484.

Rosen, H. R., and W. M. Roberds, 1933: A device for measuring intensity of illumination, *Sci.*, **78**: 241–242.

Rothemund, P., 1935: Protochlorophyll, *Symposia Quant. Biol.*, **3**: 71–79.

Rudolfs, W., and H. Heukelekian, 1931: Effect of sunlight and green organisms on reaeration of streams, *Ind. and Eng. Chem.*, **23**: 75–78.

Russell, W. C., M. W. Taylor, and D. F. Chichester, 1935: Colorimetric determination of carotene in plant tissue, *Plant Physiol.*, **10**: 325–340.

Ruth, W. A., 1922: The effect of Bordeaux mixture on the chlorophyll content of bean leaves, *Am. J. Bot.*, **9**: 535–551.

Sabalitschka, T., and H. Riesenberg, 1924: Über die Ernährung von Pflanzen mit Aldehyden. II, III, IV, *Biochem. Z.*, **144**: 545–550, 551–555; **145**: 373–378.

——— and H. Weidling, 1926: Polymerisation des Formaldehyds durch *Elodea canadensis* zu höheren Kohlenhydraten, *Biochem. Z.*, **172**: 45–57.

Sachs, J., 1862: Über den Einfluss des Lichtes auf die Bildung des Amylums in den Chlorophyllkörnern, *Bot. Ztg.*, **20**: 365–373.

———, 1864: Über die Auflösung und Wiederbildung des Amylums in den Chlorophyllkörnern bei wechselnder Beleuchtung, *Bot. Ztg.*, **22**: 289–294.

———, 1884: Ein Beitrag zur Kenntniss der Ernährungsthätigkeit der Blätter, *Arbeiten bot. Inst. Würzburg*, **3**: 1–33.

Samec, M., and H. Haerdtl, 1920: Studien über Pflanzenkolloide. IX, Zur Kenntniss verschiedener Stärkearten, *Beih. Kolloid Chem.*, **12**: 281–300.

Sampson, A. W., and R. Samisch, 1935: Growth and seasonal changes in composition of oak leaves, *Plant Physiol.*, **10**: 739–751.

Saposchinkoff, W., 1890: Bildung und Wanderung der Kohlenhydrate in den Laubblättern, *Ber. bot. Ges.*, **8**: 233–242.

———, 1891: Ueber die Grenzen der Anhäufung der Kohlenhydrate in den Blättern der Weinrebe und anderer Pflanzen, *Ber. bot. Ges.*, **9**: 293–300.

———, 1893: Beitrag zur Kenntniss der Grenzen der Anhäufung von Kohlenhydraten in den Blättern, *Ber. bot. Ges.*, **11**: 391–393.

Sauer, M. E., 1936: The chlorolytic action of serum, *Plant Physiol.*, **11**: 159–166.

Sayre, J. D., 1928: The development of chlorophyll in seedlings in different ranges of wave lengths of light, *Plant Physiol.*, **3**: 71–77.

Saywell, L. G., and D. P. Robertson, 1932: Carbohydrate content of tomato fruit, *Plant Physiol.*, **7**: 705–710.

Schanderl, H., and W. Kaempfert, 1933: Über die Strahlungsdurchlässigkeit von Blättern und Blattgeweben, *Planta*, **18**: 700–750.

Schertz, F. M., 1921: A chemical and physiological study of mottling of leaves, *Bot. Gaz.*, **71**: 81–130.

———, 1925: Some physical and chemical properties of carotin and the preparation of the pure pigment, *J. Agr. Res.*, **30**: 469–475.

———, 1925: Some physical and chemical properties of xanthophyll and the preparation of the pure pigment, *J. Agr. Res.*, **30**: 575–587.

———, 1928: The extraction and separation of chlorophyll (a and b), carotin and xanthophyll in fresh green leaves, preliminary to their quantitative determination, *Plant Physiol.*, **3**: 211–216.

———, 1928: The quantitative determination of chlorophyll, *Plant Physiol.*, **3**: 323–334.

———, 1928: The preparation of chlorophyll, *Plant Physiol.*, **3**: 487–499.

———, 1928: The chloroplast pigments, their functions and the probable relation of chlorophyll to the vitamines, *Quart. Rev. Biol.*, **3**: 459–485.

———, 1929: Seasonal variation of the chloroplast pigments in several plants on the mall at Washington, D. C., *Plant Physiol.*, **4**: 133–139.

———, 1929: The effect of potassium, nitrogen and phosphorus fertilizing upon the chloroplast pigments, upon the mineral content of the leaves and upon production in crop plants, *Plant Physiol.*, **4**: 269–278.

———, 1929: The pure pigments, carotin and xanthophyll and the Tsweet adsorption method, *Plant Physiol.*, **4**: 337–348.

Schimper, A. F. W., 1880: Untersuchungen über die Entstehung der Stärkekörner, *Bot. Ztg.*, **38**: 881–902.

———, 1881: Untersuchungen über das Wachsthum der Stärkekörner, *Bot. Ztg.*, **39**: 185–194, 201–211, 217–227.

———, 1885: Über die Bildung und Wanderung der Kohlenhydrate in den Laubblättern, *Bot. Ztg.*, **43**: 737–743, 753–763, 769–787.

Schomer, H. A., 1934: Photosynthesis of water plants at various depths in the lakes of northeastern Wisconsin, *Ecology*, **15**: 217–218.

Schroeder, H., and T. Horn, 1922: Das gegenseitige Mengenverhältnis der Kohlenhydrate im Laubblatt im seiner Abhängigkeit vom Wassergehalt, *Biochem. Z.*, **130**: 165–198.

——— and F. Herrman, 1931: Über die Kohlenhydrate und den Kohlenhydratstoffwechsel der Laubblätter. I, Die Zunahme des Saccharosegehaltes beim Welken, *Biochem. Z.*, **235**: 407–425.

Schroeder, R. A., 1935: The effect of some summer oil sprays upon the carbon dioxide absorption of apple leaves, *Proc. Am. Soc. Hort. Sci.*, **33**: 170–172.

Schryver, L. B., 1910: The photochemical formation of formaldehyde in green plants, *Proc. Roy. Soc.*, B, **82**: 226–232.

Schryver, S. B., and E. M. Thomas, 1923: The hemicelluloses. II, The hemicellulose content of starches, *Biochem. J.*, **17**: 497–500.

Sen, P. K., and V. H. Blackman, 1933: On conditions leading to the injection of leaves submerged in water, *Ann. Bot.*, **47**: 663–671.

Senebier, J., 1800: "Physiologie végétale," Geneve.

Shcheglova, O. A., and E. V. Chernysheva, 1933: The influence of mechanical reduction of the leaf surface on the development, accumulation of dry matter and yield of spring wheat and barley, *Bull. Plant Protection*, III, ser., **3**: 73–111; *Biol. Abst.*, **9**: 748.

Shelford, V. E., and F. W. Gail, 1922: A study of the light penetration into sea water made with the Kunz photoelectric cell with particular reference to the distribution of plants, *Pub. Puget Sound Biol. Sta.*, **3**: 141–176.

Shirley, H. L., 1931: Light sources and light measurements, *Plant Physiol.*, **6**: 447–466.

Shohl, A. T., 1923: Analysis of the Jerusalem artichoke, *J. Am. Chem. Soc.*, **45**: 2754–2756.

SHULL, C. A., 1929: A spectrophotometric study of reflection of light from leaf surfaces, *Bot. Gaz.*, **87**: 583–607.

———, 1930: The mass factor in the energy relations of leaves, *Plant Physiol.*, **5**: 279–282.

SIEGFRIED, M., 1905: Über die Bindung von Kohlensäure durch amphotere Amidokörper, *Z. physiol. Chem.*, **44**: 85–96.

SINGH, B. N., and K. N. LAL, 1935: Limitations of Blackman's law of limiting factors and Harder's concept of relative minimum as applied to photosynthesis, *Plant Physiol.*, **10**: 245–268.

——— and ———, 1935: Investigations of the effect of age on assimilation of leaves, *Ann. Bot.*, **49**: 291–307.

SJOSTROM, O. A., 1936: Microscopy of starches and their modifications, *Ind. and Eng. Chem.*, **28**: 63–74.

SKINNER, C. E., 1932: The soil as a habitat for growth of green algae, *Soil Sci.*, **34**: 25–29.

SKVORTZOV, S. S., 1931: Influence of desiccating winds on photosynthesis, *Bull. App. Bot., Genetics, and Plant Breeding (Leningrad)*, **25**: 45–66; *Chem. Abst.*, **26**: 5605.

SMALL, T., and H. L. WHITE, 1930: Carbon dioxide in relation to greenhouse crops. IV, The effect on tomatoes of an enriched atmosphere maintained by means of a stove, *Ann. App. Biol.*, **17**: 81–90.

SMITH, A. M., 1919: The temperature coefficient of photosynthesis—a reply to a criticism, *Ann. Bot.*, **33**: 517–536.

SMITH, J., 1932: Carotene. IV, The hydrogenation of carotenes obtained from different sources of dihydrocarotene and of lycopin, *J. Biol. Chem.*, **96**: 35–51.

SMITH, L. E. W., and O. SMITH, 1931: Light and carotinoid content of certain fruits and vegetables, *Plant Physiol.*, **6**: 265–275.

SMITH, S. H. C., 1930: The yellow pigments of green leaves: Their chemical constitution and possible function in photosynthesis, *Contr. Marine Biol., Stanford Univ. Press*, 145–160.

SMYTH, E. M., 1934: The determination of CO_2 content of an atmosphere in a closed system, *Sci.*, **80**: 294.

SOMMER, A. L., E. R. BISHOP, and I. G. OTTO, 1933: Detection and estimation of formaldehyde within the cell of a green plant by the Allison apparatus, *Plant Physiol.*, **8**: 564–567.

SPIRGATIS, P., 1923: The growth factor carbon dioxide, *Bot. Arch.*, **4**: 381–403; *Exp. Sta. Rec.*, **52**: 424.

SPOEHR, H. A., 1913: Photochemische Vorgänge bei der diurnalen Entsäurung der Succulenten, *Biochem. Z.*, **57**: 95–111.

———, 1916: Theories of photosynthesis in the light of some new facts, *Plant World*, **19**: 1–16.

———, 1917: The pentose sugars in plant metabolism, *Plant World*, **20**: 365–379.

———, 1919: The carbohydrate economy of cacti, *Carnegie Inst. Wash. Pub.* 287.

———, 1919: The development of conceptions of photosynthesis since Ingen-Housz, *Sci. Mo.*, **9**: 32–47.

———, 1922: Photosynthesis and the possible use of solar energy, *J. Ind. Eng. Chem.*, **14**: 1142–1145.

———, 1923: The reduction of carbon dioxide by ultraviolet light, *J. Am. Chem. Soc.*, **45**: 1184–1187.

——— and J. M. MCGEE, 1923: Studies in plant respiration and photosynthesis, *Carnegie Inst. Wash. Pub.* 325.

——— and J. M. McGee, 1924: Investigations in photosynthesis. An electrometric method of determining CO_2, *Ind. and Eng. Chem.*, **16**: 128–130.

——— and ———, 1924: The effect of fluctuations in the CO_2 content of the atmosphere and the rate of respiration of leaves, *Am. J. Bot.*, **11**: 493–502.

——— and ———, 1924: Absorption of carbon dioxide the first step in photosynthesis, *Sci.*, 59: 513–514.

———, 1926: "Photosynthesis," Chemical Catalog Company, Inc., New York, 393 pp.

——— and H. W. Milner, 1935: Leaf starch: its isolation and some of its properties, *J. Biol. Chem.*, **111**: 679–687.

———, J. H. C. Smith, H. H. Strain, and H. W. Milner, 1935: Photosynthesis—plant pigments, *Carnegie Inst. Year Book*, **34**: 192–198.

Sponsler, O. L., 1922: The structure of the starch grain, *Am. J. Bot.*, **9**: 471–492.

———, 1923: Structural units of starch determined by X-ray crystal structure method, *J. Gen. Physiol.*, **5**: 757–777.

Sprague, H. B., and J. W. Shive, 1929: A study of the relations between chloroplast pigments and dry weights of tops in dent corn, *Plant Physiol.*, **4**: 165–192.

——— and N. Curtis, 1933: Chlorophyll content as an index of the productive capacity of self lines of corn and their hybrids, *J. Am. Soc. Agron.*, **25**: 709–723.

Stanescu, P. P., 1924: Les variations quantitatives de l'amidon dans les feuilles des plantes vertes pendant une journée (24 heures), *C. R. Acad. Sci. Paris*, **178**: 117–119.

———, 1927: Contributions à l'étude de la photosynthèse, *Ann. Scien. de l'Université de Jassy*, **15**: 161–212.

———, 1927: "Les variations quantitatives des products de la photosynthèse dans les feuilles des plantes vertes pendant un jour (24 heures)," thèse, Bucarest.

———, A. Aronescu, and I. G. Mihăilescu, 1929: Sur un mode particulier d'accumulation transitoire de l'amidon dans les feuilles des amylophiles, *Bull. Acad. Sci. Roumaine*, **12**, No. 6: 1–3.

———, 1929: Ueber das Welken und Austrocknen der Pflanzen, *Bull. Acad. Sci. Roumaine*, **12**: 171–175.

———, ———, and ———, 1932: Observations sur l'évolution diurne de l'accumulation transitoire de l'amidon dans les feuilles des plantes vertes, *Bull. Acad. Sci. Roumaine*, **15**: 80–83.

———, 1936: Daily variations in products of photosynthesis, water content, and acidity of leaves toward the end of the vegetative period, *Am. J. Bot.*, **23**: 374–379.

Stiles, W., 1925: "Photosynthesis—The Assimilation of Carbon by Green Plants," Longmans, Green & Company, New York, 268 pp.

Stocklasa, J., and W. Zdobnicky, 1911: Ueber photochemische Synthese der Kohlenhydrate aus CO_2 und H_2 in Anwesenheit von KOH und in Abwesenheit von Chlorophyll, *Monatsh. Chem.*, **32**: 53–75.

——— and ———, 1912: Ueber die photochemische Synthese der Kohlenhydrate unter Einwirkung der Ultravioletten Strahlen, *Biochem. Z.*, **41**: 333–372.

———, 1929: Dynamik und Energetik der Kohlensäure-assimilation. Die Ernährung der Pflanze, *Deutsch. Kalisyndikat, Berlin*, **25**: 97–105. Quoted by Livingston and Beall, 1934.

Stout, A. B., 1936: Autumn coloration, *J. N. Y. Bot. Gard.*, **37**: 7–11.

Strain, H. H., 1934: The separation of carotenes by adsorption on magnesium oxide, *Sci.*, **79**: 325–326.

———, 1934: Carotene. VIII, Separation of carotenes by absorption, *J. Biol. Chem.*, **105**: 523–535.

———, 1936: Leaf xanthophylls, Sci., **83**: 241–242.

Suit, R. F., and H. Hibbert, 1934: A starchless potato induced by the introduction of foreign enzymes, *Sci.*, **79**: 78–79.

Tageeva, S. V., 1931: Study of photosynthesis in connection with photoperiodism, *Bull. App. Bot., Genetics, and Plant Breeding (Leningrad)*, **27**: 197–245; *Chem. Abst.*, **26**: 5605.

Tam, R. K., and O. C. Magistad, 1935: Relationship between nitrogen fertilization and chlorophyll content in pineapple plants, *Plant Physiol.*, **10**: 159–168.

Tammann, G., 1892: Die Reactionen der ungeformten Fermente, *Z. physiol. Chem.*, **16**: 271–328.

———, 1895: Zur Wirkung ungeformter Fermente, *Z. physikal. Chem.*, **18**: 426–442.

Taylor, T. C., and J. M. Nelson, 1920: Fat associated with starch, *J. Am. Chem. Soc.*, **42**: 1726–1738.

Thatcher, R. W., 1921: "The Chemistry of Plant Life," McGraw-Hill Book Company, Inc., New York, 268 pp.

Thoday, D., 1909: Experimental researches on vegetable assimilation and respiration. V, A critical examination of Sachs-method for using increase of dry weight as a measure of carbon dioxide assimilation in leaves, *Proc. Roy. Soc.*, B, **82**: 1–56.

———, 1910: Experimental researches on vegetable assimilation and respiration. VI, Some experiments on assimilation in the open air, *Proc. Roy. Soc.*, B, **82**: 421–450.

Thomas, A. W., 1915: The chemical constitution of starch—a review, *Biochem. Bull.*, **4**: 379–397.

Thomas, W., 1929: Balanced fertilizers and Liebig's law of the minimum, *Sci.*, **70**: 383–384.

Thurber, F. H., 1933: Chemical and physical properties of sweet potato starch, *Ind. and Eng. Chem.*, **25**: 565–568.

Timiriazeff, C., 1890: Enregistrement photographique de la function chlorophyllienne par la plante vivante, *C. R. Acad. Sci. Paris*, **110**: 1346–1347.

———, 1904: The cosmical function of the green plant, *Proc. Roy. Soc.*, **72**: 424–461.

Tottingham, W. E., and J. G. Moore, 1931: Some phases of plant development under vita glass, *J. Agr. Res.*, **43**: 133–163.

Trajkovich, H., 1924: Inheritance of xantha seedlings in maize, *N. Y. Sta. (Cornell) Mem.*, **82**: 1–13.

Transeau, E. N., 1926: The accumulation of energy by plants, *Ohio J. Sci.*, **26**: 1–11.

Trelease, S. F., and H. M. Trelease, 1929: Susceptibility of wheat to mildew as influenced by carbohydrate supply, *Bull. Torrey Bot. Club*, **56**: 65–92.

Tschesnokov, V., and K. Bazyrina, 1930: Die Ableitung der Assimilate aus dem Blatt, *Planta*, **11**: 473–484.

Tschudy, R. H., 1934: Depth studies on photosynthesis of the red algae, *Am. J. Bot.*, **21**: 546–556.

Tsweet, M. S., 1910: "Chromophylls in the Plant and Animal Worlds" (Russian), Warsaw, 379 pp.; *Exp. Sta. Rec.*, **26**: 710.

———, 1911: Eine Hypothese über den Mechanismus der photosynthetischen Energie-übertragung, *Z. physiol. Chem.*, **76**: 413–419.

Ulvin, G. B., 1934: Chlorophyll production under various environmental conditions, *Plant Physiol.*, **9**: 59–81.

Ursprung, A., 1917: Über die Stärkebildung im Spektrum, *Ber. bot. Ges.*, **35**: 44–69.

———, 1918: Über die Absorptionskurve des grünen Farbstoffes lebender Blätter, *Ber. bot. Ges.*, **36**: 73–85.

———, 1918: Ueber die Bedeutung der Wellenlänge für die Stärkebildung, *Ber. bot. Ges.*, **36**: 86–100.

USHER, F. L., and J. H. PRIESTLEY, 1906: The mechanism of carbon assimilation in green plants: the photolytic decomposition of carbon dioxide *in vitro*, *Proc. Roy. Soc.*, B, 78: 318–327.

——— and ———, 1906: A study of the mechanism of carbon assimilation in green plants, *Proc. Roy. Soc.*, B, **77**: 369–376.

——— and ———, 1911: The mechanism of carbon assimilation. III, *Proc. Roy. Soc.*, B, **84**: 101–112.

VAN DER PLANK, J. E., 1936: The estimation of sugars in the leaf of the mangold (*Beta vulgaris*), *Biochem. J.*, **30**: 457–462.

VAN DE SANDE-BAKHUYZEN, H. L., 1926: The structure of starch grains from wheat grown under constant conditions, *Proc. Soc. Exp. Biol. and Med.*, **23**: 302–305.

VAN PAAUW, F., 1932: The indirect action of external factors on photosynthesis, *Rec. trav. bot. Néerl.*, **29**: 497–620.

VER HULST, J. H., W. H. PETERSON, and E. B. FRED, 1923: Distribution of pentosans in the corn plant at different stages of growth, *J. Agr. Res.*, **23**: 655–665.

VILLE, G., 1889: Recherches sur les relations qui existent entre la couleur des plantes et la richesse des terres en agents de fertilité, *C. R. Acad. Sci. Paris*, **109**: 397–400.

VINES, S. H., 1886: "Lectures on the Physiology of Plants," University Press (John Wilson & Son, Inc.), Cambridge, Mass., x + 710 pp.

VIRTANEN, A. I., and M. NORDLUND, 1934: Synthesis of sucrose in plant tissue, *Biochem. J.*, **28**: 1729–1732.

VON EULER, H., and H. HELLSTRÖM, 1929: Über die Bildung von Xanthophyll, Carotin und Chlorophyll in belichteten und unbelichteten Gerstenkeimlingen, *Z. physiol. Chem.*, **183**: 177–183.

———, W. HERTZSCH, A. FORSSBERG, and H. HELLSTRÖM, 1931: Vergleichende Versuche über verschiedene Arten von Chlorophylldefekten, *Z. Indukt. Abstamm. u. Vererb.*, **60**: 1–16; *Biol. Abst.*, **8**: 602.

WAGER, H., 1914: The action of light on chlorophyll, *Proc. Roy. Soc.*, B, **87**: 386–407.

WALLACE, R. H., 1932: The effect of ether and chloroform upon the rate of photosynthesis in detached leaves of *Acer negundo*, abst., *Am. J. Bot.*, **19**: 847.

WALTER, H., 1926: Die Verdunstung von Wasser in bewegter Luft und ihre Abhängigkeit von der Grösse der Oberfläche, *Z. Bot.*, **18**: 1–47.

WALTON, R. P., 1929: Separation and characterization of starch amyloses, diss., *Columbia Univ.*, New York, 12 pp.

WARBURG, O., 1919: Über die Geschwindigkeit der photochemischen Köhlensäurezersetzung in lebenden Zellen. I, *Biochem. Z.*, **100**: 230–270.

——— and E. NEGELEIN, 1923: Über den Einfluss der Wellenlänge auf den Energieumsatz bei der Kohlensäureassimilation, *Z. physiol. Chem.*, **106**: 191–218.

———, 1925: Versuche über die Assimilation der Kohlensäure, *Biochem. Z.*, **166**: 386–406.

WARNER, C. H., 1914: Formaldehyde as an oxidation product of chlorophyll extracts, *Proc. Roy. Soc.*, B, **87**: 378–386.

WATSON, D. J., 1936: The effect of potassium chloride on the diurnal changes of the carbohydrates of the potato leaf, *Ann. Bot.*, **50**: 59–83.

WEATHERWAX, P., 1922: A rare carbohydrate in waxy maize, *Genetics*, **7**: 568–572.

WEBER, C., 1879: Ueber specifische Assimilationsenergie, *Arb. bot. Inst. Würzburg*, **2**: 346–352.

WEEVERS, T., 1923: The first carbohydrates that originate during the assimilatory process—a physiological study with variegated leaves. *Koninklijke Akad. von Wetensch.*, 6, *Amsterdam Proc.*, **27**: 1–11, Eng. summary.

WEIER, T. E., 1933: Plastid structure and its possible effect on photosynthesis, abst., *Am. J. Bot.*, **20**: 677.

WHITE, H. L., 1930: Carbon dioxide in relation to glasshouse crops. V, An analysis of the response of the tomato crop to an atmosphere enriched with carbon dioxide, *Ann. App. Biol.*, **17**: 755–766.

WHITESIDE, A. G. O., J. EDGAR, and C. H. GOULDEN, 1934: The influence of environment on the carotenoid content of hard red spring wheat, *Cereal Chem.*, **11**: 615–625.

WIDDOWSON, E. M., and R. A. MCCANCE, 1935: The available carbohydrate of fruits, *Biochem. J.*, **29**: 151–156.

WIESNER, J., 1874: Untersuchungen über die Beziehungen des Lichtes zum Chlorophyll, *Sitzungsber. Akad. Wiss. Wien*, **69**: 327–385.

———, 1877: "Die Entstehung des Chlorophylls in der Pflanze," Wien.

WILLAMAN, J. J., 1922: The preparation of inulin, with special reference to artichoke tubers as a source, *J. Biol. Chem.*, **51**: 275–285.

WILLSTÄTTER, R., and A. STOLL, 1918: "Untersuchungen über die Assimilation der Kohlensäure," Julius Springer, Berlin.

——— and ———, 1928: "Investigations on Chlorophyll. Methods and Results," English translation by F. M. Schertz and A. R. Merz. Science Press, Lancaster, Pa., 380 pp.

WILMOT, A. J., 1921: Experimental researches on vegetable assimilation and respiration. XIV, Assimilation by submerged plants in dilute solutions of bicarbonates and of acids: an improved bubble-counting technique, *Proc. Roy. Soc.*, B, **92**: 304–327.

WILSON, P. W., E. B. FRED, and M. R. SALMON, 1933: Relation between carbon dioxide and elemental nitrogen assimilation in leguminous plants, *Soil Sci.*, **35**: 145–166.

WINKLER, H., 1898: Über die Stärkebildung in den verschiedenartigen Chromatophoren, *Jahrb. wiss. Bot.*, **32**: 525.

WITHROW, R., 1936: Private communication.

WOODRUFF, S., and L. R. WEBBER, 1933: A photomicrographic study of gelatinized wheat starch, *J. Agr. Res.*, **46**: 1099–1108.

WURMSER, R., 1925: Sur l'activité des diverses radiations dans la photosynthèses, *C. R. Acad. Sci. Paris*, **181**: 374–376.

YANOVSKY, E., and R. M. KINGSBURY, 1931: New sources of inulin, *J. Am. Chem. Soc.*, **53**: 1597–1601.

YAP, G. G., 1920: A study of the photosynthesis of sugar cane, *Philippine J. Agr.*, **8**: 269–276.

YUNCKER, T. G., 1916: A study of the relation of soil moisture to transpiration and photosynthesis in the corn plant, *Plant World*, **19**: 151–161.

ZECHMEISTER, L., 1934: "Carotenoide," 8 vol., Julius Springer XII + 338 pp., 85 Figs., Berlin; Review, *Bot. Gaz.*, **96**: 378–380.

ZIJLSTRA, K., 1909: Kohlensäuretransport in Blättern, *Rec. trav. bot. Néerl.*, **6**: 99–210. Quoted by Livingston and Beall, 1934.

ZIRKLE, C., 1926: The structure of the chloroplast in certain higher plants. I, II, *Am. J. Bot.*, **13**: 301–341.

ZSCHEILE, F. P., 1932: Attempted photosynthesis of carbohydrates *in vitro* with visible light, *J. Am. Chem. Soc.*, **54**: 973–976.

———, 1934: An improved method for the purification of chlorophylls a and b; Quantitative measurement of their absorption spectra. Evidence for the existence of a third component of chlorophyll, *Bot. Gaz.*, **95**: 529–562.

CHAPTER IX

THE NITROGEN METABOLISM OF THE GREEN PLANT

I. GENERAL RELATION OF NITROGEN TO THE PLANT

Nitrogen enters into the structure of chlorophyll, the amino acids, amides, alkaloids, protein, and the protoplasm of the plant. It occurs for the most part in organic combination, and, as will be noted in a subsequent heading, ammonia, nitrates, and nitrites are found, as a rule, only in traces if at all. The percentage of nitrogen in the various plant parts will vary with the age, type of tissue, and the kind of plant as well as with the period of the day when the examination is made. In the corn plant at the period of the late milk stage when the grain was glazed and dented (Latshaw and Miller, 1924), the leaves contained 25 per cent of the total nitrogen in the plant, the stem 14 per cent, the grain 46 per cent, the cobs 9 per cent, and the roots 6 per cent. The nitrogen content of the dry weight of the leaves varied from 1.2 to 1.4 per cent, that of the stem from 0.5 to 1.0 per cent, and that of the grain from 2 to 2.2 per cent. Miller (1929) observed the nitrogen content of various leaves when the plants were in a vigorous vegetative condition previous to blooming. The average nitrogen content for different periods of the day was for soybeans 5.8 per cent, for cowpeas 5.6 per cent, for garden beans 4.5 per cent, for pumpkin 4.5 per cent, for feterita 3.6 to 4.1 per cent, for kafir 3.6 to 4.2 per cent, and for corn 3.2 to 3.8 per cent of the dry weight of these leaves.

Armstrong and Albert (1931) found in cotton that shortly before bud formation 80 to 90 per cent of the nitrogen in the plant was in the leaves, while at the end of fruit setting 40 to 60 per cent of the total nitrogen of the plant was in the bolls. In some seasons, from 40 to 50 per cent of the nitrogen of the crop was absorbed in about 2 weeks. Cameron and Appleman (1933) noted that the average crop of oranges removes about 35 lb. of nitrogen per acre. At no time do the roots of the orange tree contain more than 21 per cent of the total nitrogen of the tree. Excluding the leaves, the aerial portions of the plant contain more nitrogen than the roots. In Connecticut, it was observed by Morgan and Street (1935) that the leaves of tobacco attained a maximum nitrogen content approximately 40 days after setting, and these leaves lost nitrogen after the plants were topped. By the time of cutting, the tobacco plants had extracted 114 lb. of nitrogen per acre from the soil. Of this amount 39.3 per cent was in the crop leaves, 21 per cent in the sucker leaves, 26 per cent in the stalks, and 13.2 per cent in the roots.

A large amount of nitrogen in the soil produces a rank growth of foliage, giving an inordinate growth of stem and leaves. Thus Gračinin (1932) found an inverse relation between the growth in length of the roots of corn, barley, wheat and rye and the amount of nitrogen in the nutrient medium, and the curve of growth for the aerial organs rose with this increase in nitrogen. The roots were more sensitive to a change in the supply of nitrogen than the tops. According to Verner (1933), an increased leaf area is the most striking result obtained by fertilizing apple trees with nitrates. Crowther (1935) stated that one of the main functions of nitrogen is the initiation of meristematic activity. He noted in cotton that the total growth of the plant depends primarily on the rate of development of the leaf surface. Variations in the growth rate of leaves are determined for the most part by variations in the supply of available nitrogen. The size of the plant is thus largely a measure of the rate of nitrogen metabolism. The rate of uptake of nitrogen is determined primarily by the concentration of nitrogen in the soil solution, while the speed of development of the plant is proportional to the inflow of nitrogen. In the tomato, Fisher (1935) observed that an excess of nitrogen stimulated vegetative growth at the expense of flowers and fruit development. It was found by Das (1936) in Hawaii that increasing the application of nitrogen to sugar cane increased the leaf area, rate of leaf and joint formation, rate of stem elongation, lodging, and tillering, and decreased the concentration of sucrose in the expressed sap.

If plants can be grown to maturity, the excess foliage resulting from an abundant supply of nitrogen has no apparent effect on the yield. Frequently, however, under such conditions ripening occurs prematurely, due to various causes, so that considerable amounts of food materials are not transferred from the vegetative parts to the seed or grain, and as a result the grain is shriveled.

It was observed by Thatcher (1921) for cabbage and lettuce, by Pearsall and Ewing (1929) for radish, turnip, and dandelion, by Gourley (1930) for the fruit of the apple, by Armstrong and Albert (1931) in the leaves, stalk, and fruit of cotton, and by Garner and others (1934) for tobacco leaves that an abundant supply of nitrogen increases the water content of the tissues of these various plants. Pearsall and Ewing (1929) considered that this increased succulence results from a lower rate of transpiration of the plants with an abundant supply of nitrogen as compared to those with a limited supply. This reduction in transpiration may be due to an accumulation of amino acids, which in some manner lower the hydrogen-ion concentration and thus allow a greater swelling of the protoplasmic colloids that would retard the loss of water by transpiration.

The amount of nitrogen in the plant is determined in general by the concentration of this element in the soil. Thus Gainey and Sewell (1932) in a study of the so-called "nitrogen spots," commonly observed in the wheat fields of Kansas and other midwestern States, reported that these spots are characterized by a higher nitrogen content than that of the adjoining soil. The plants growing in these spots have a higher nitrogen content than those growing in the surrounding soil. Richardson and Gurney (1933, 1935) found in Australia that nitrate accumulated in fallow soil, and this was a major factor in accounting for the heavy increase in crop yield from fallow as compared to stubble land. Nicol (1934) stated that in agricultural soils the nitrogen as nitrates may amount to as much as 300 p.p.m. of moist soil.

Hamner (1936) reported that there is a marked increase in the respiration rate upon the application of nitrate to the plant provided there is an initial reserve of carbohydrate.

In the absence of a sufficient supply of available nitrogen, the leaves are stunted, and a yellowing occurs. Thus Fisher (1935) noted that tomato plants grown in a cultural solution deficient in nitrogen showed a yellowing of the entire plant in 6 to 8 days. The lower leaves dried slowly and were dropped. A decreased amount of protoplasm was formed, and there was a general reduction in the growth of leaves and stem. According to Valleau and Johnson (1927), and Haas (1929) the frenching of the leaves of tobacco and other plants occurs when there is a deficiency of nitrates but disappears upon the application of sufficient available nitrogen to the soil. Under greenhouse conditions, Shear (1933) could control this disease by the application of nitrogenous fertilizers, but that method failed in the field.

Jones (1936) applied an acidulated (H_2SO_4) solution of diphenylamine to plant tissues to determine the need for nitrogen. When deposited on a section of tissue, a drop of this solution imparts a blue to blue-black coloration, depending upon whether there is a trace or a relatively large amount of nitrate present. A lack of blue color indicates an absence of nitrates, signifying a deficiency of nitrogen in the plant and its root environment. These tests should be made upon older tissues, because the younger portions may show a low amount of nitrates due to their usage in the synthesis of organic nitrogenous compounds.

II. SOURCES OF NITROGEN SUPPLY OF PLANT

The possible sources of the nitrogen supply of the plant are the various nitrogenous compounds in the soil and the combined and elemental nitrogen of the atmosphere. The absorption of the nitrogenous compounds of the soil will first be considered.

A. Absorption from the Soil

The combined nitrogen in the soil exists in the form of inorganic and organic compounds. The former are principally the nitrates, nitrites, and ammonium salts, while the latter include a long list of the decomposition products of the proteins, protoplasm, and other complex nitrogenous substances of plants and animals. A partial list of the various organic nitrogenous compounds that have been found in the soil is given in Chap. III.

1. Inorganic Nitrogen.—Early in the history of plant nutrition it became generally accepted that plants, in general, absorb the combined inorganic nitrogen preferably in the form of nitrates. In some cases, however, plants apparently did equally well when nitrogen was applied to the soil in the form of ammonium salts or as nitrates. Since these experiments were not made under sterile conditions, it is not definitely known whether these salts were absorbed as such or not. Since 1889, however, a considerable amount of work has been done, apparently under sterile conditions, to determine the relative value of nitrates, nitrites and ammonium salts in plant nutrition.

a. Nitrates and Ammonium Salts.—The investigations from 1889 until about 1930 established the fact that some plants grow equally well with either nitrates or ammonium salts as a source of nitrogen, and that others, although assimilating ammonium salts in the absence of nitrates, seem to grow better when nitrates are applied. From 1930 until the present time, the factors that influence the relative value of nitrates and ammonium salts in the nutrition of plants have been studied.

1. *Relative Value.*—Müntz (1889), Pitsch (1887 to 1896), Griffiths (1891), and Gerlach and Vogel (1905) worked with soils under apparently sterile conditions and showed that beans, barley, hemp, and corn were able to grow in the absence of nitrates, but they did not prove conclusively that ammonium salts were the sole source of the nitrogen obtained by the plants. According to Treboux (1904), *Lemna minor*, mosses, diatoms, and certain green algae used by him utilized effectively ammonium

nitrogen. Krüger (1905) concluded that ammonium salts and nitrates are equally suitable for mustard, oats, and barley, and that, if anything, ammonium salts are better than nitrates for potatoes, while for beets nitrates are superior to ammonium salts. It has been shown by Mazé (1898, 1900) and Prianischnikow (1908, 1923) that corn can utilize ammonium salts as readily as nitrates, while Brigham (1917) concluded that ammonium sulphate is a much better source of nitrogen than sodium nitrate. In the case of sugar cane, McGeorge (1923) concluded that this plant will not grow without some nitrate nitrogen. The cane in an ammoniacal nutrient solution sent out strong, vigorous shoots, but practically no root development took place, while an excellent growth of both roots and tops occurred in a solution containing nitrates. Stewart (1925) and others found that the pineapple plant is capable of assimilating all its nitrogen in the form of ammonium salts, although the best growth of plants was made by those growing in solutions in which the nitrogen was present in the form of nitrates. The indications were that both nitrates and ammoniacal nitrogen can be utilized by these plants. The relative value of ammonium salts and nitrates in the nutrition of wheat and peas was observed by Hutchinson and Miller (1909), who grew these plants in sand and water cultures and found that ammonium sulphate was assimilated by them directly and that in the case of peas there was no difference between the plants supplied with ammonium and those that had sodium nitrate. Wheat plants showed a decided preference for nitrogen in the form of nitrates.

The nitrogen nutrition of rice has been studied by numerous investigators, and they all report that, under some conditions at least, the ammonium salts are superior to the nitrates in the growth of this plant. Kellner (1884) noted that swamp rice, in pots, during its early development grew better with ammonium salts than with nitrates but that during later growth nitrates proved the more effective. He also showed that the formation of nitrates in paddy soils took place slowly, while ammonia was formed in large quantities. Kelley (1911) applied ammonium sulphate to some rice plants and sodium nitrate to others. The application of the former was followed by an increase in the yield of straw and grain, but sodium nitrate was ineffective. In sand cultures unhealthy and stunted plants resulted if nitrates were used as the only source of combined nitrogen, while the use of ammonium salts resulted in vigorous apparently normal plants. The same beneficial returns from ammonium salts in contrast to nitrates have also been obtained by Nagaoka (1904), Daikuhara (1905), Espino (1920), and Trelease and Paulino (1920).

Zahnley and Duley (1934) found that ammonium sulphate, sodium nitrate, and urea were equally effective for growth when applied to Kentucky bluegrass, Washington bent grass, and other grasses.

The relative value of nitrates and ammonium salts in the nutrition of plants depends upon numerous factors. Among these are: reaction of the medium in which the roots are growing; age of the plants; kind of plant; concentration of the medium; light; and the presence of other ions (Beaumont and Moore, 1933). The two most important factors are the following:

(*a*) *Reaction of the Medium.*—The pH value of the medium in which the roots are growing has a marked influence upon the rate of absorption and utilization of nitrates and ammonium salts. Pirschle (1929) was one of the first to show that in a pronounced acid medium, plants cannot obtain sufficient nitrogen from ammonium salts. Later (1931) he noted that plants tolerate a much wider pH range for the utilization of nitrates than for ammonium salts, and that with neither compound is there a single pH value at which all plants make their optimum growth. Tiedjens and Robbins (1931) found that tomato plants grew well when nitrogen was supplied as ammonium sulphate in a medium with a pH of 8.0, and that the best growth with calcium nitrate occurred at a pH 5.0 to 6.0, while with ammonium nitrate the best growth was obtained at a pH of 4.0 to 5.0. They observed that, although seedling trees of peach and apple could absorb and assimilate ammonium salts over a wide pH range, they did not assimilate a sufficient amount of nitrogen for maximum growth unless the culture solution had a pH value of 7.0 to 8.0. Naftel (1931) noticed that the absorption of NH_4 by the cotton plant increased as the acidity of the culture solution decreased, while the absorption of NO_3 was only slightly affected by the reaction of the medium. The highest absorption of total nitrogen usually occurred at pH 6.0 and was greatest when both forms of nitrogen were present. Arrington and Shive (1935) found with tomato plants that the rate of absorption of cation nitrogen from a medium with a pH of 7.0 was more than three times as high as the rate at pH 4.0. For anion nitrogen, the rate of absorption was somewhat greater at pH 4.0 than at pH 7.0.

The observation that plants absorb nitrogen in the form of ammonium salts more rapidly from the nutrient medium at a relatively high pH, while the nitrogen in the form of nitrates is absorbed more rapidly at a lower pH value has been reported also by Addoms and Mounce (1932) for young cranberry plants, by Tiedjens and Blake (1932) for apple trees, by Clark (1933) for strawberry plants, by Davidson and Shive (1934) for peach trees, and by Tiedjens (1934) for tomato, cotton, and the seedling plants of several varieties of apple.

Clark and Shive (1934) grew tomato plants in solutions containing approximately equal portions of nitrogen in the NH_4 and NO_3 forms. Their observations relative to the influence of the reaction of the medium on the rate of absorption agree with those of other investigators. They observed further, however, in 41-day-old plants that the maximum rate of absorption of nitrogen as NH_4 was higher than that of nitrogen as NO_3. The rate of absorption of NO_3 by the older plants was not so greatly influenced by the pH value of the solution as it was in the younger plants.

Conrad (1934) considered that a plant absorbs the NO_3 ion more rapidly from a more acid medium because it can secure more easily therefrom the requisite hydrogen-ions to accompany the NO_3 ions into the plant. The same reasoning is applied for the absorption of the NH_4 ions from the more alkaline solutions. When both ions are present in approximately equal concentrations in neutral solutions, more NH_4 ions will be absorbed because less energy is required to incorporate the nitrogen of the NH_4 ions into protein than in the case of the NO_3 ions.

(*b*) *The Age and Kind of Plant.*—It was reported by Pardo (1930) that young rice plants cannot use nitrates but thrive on ammonium sulphate, while the older plants do well on either form. Young sugar cane uses nitrates to better advantage than it does ammonium salts. Holley and others (1931) noted that young cotton plants grew more rapidly during the first 2 or 3 weeks in a nitrate solution than in one containing ammonium salts. Although the production of dry matter was always favored more when the plants were in nitrate solutions, it was favored least at the age of 5 to 6 weeks. Pirschle and Mengdehl (1931) observed that young plants used equal amounts of nitrogen, as the anion and cation of ammonium nitrate, but that older plants absorbed the NH_4 ion to a greater extent than the NO_3 ion. According to Thelin and Beaumont (1934) young plants of rice and wheat accumulated nitrogen better from ammonium sulphate than from nitrates. In a perennial rye grass, Lewis (1936) noted that except at an early stage the percentage of nitrogen in the herbage was higher when nitrates were applied to the soil than when ammonium nitrogen was applied.

Stahl and Shive (1933), and Sessions and Shive (1933) found with oats that the absorption of nitrogen as NH_4 is highest during the earliest periods, then gradually declines, and reaches its minimum about maturity. The absorption of nitrogen as NO_3 is at a minimum at the earlier period, rapidly increases to reach a maximum at or about blooming, and then declines toward maturity. The total nitrogen in the plants varied with the total nitrogen of the solution in which they were growing although not in a direct proportion. In buckwheat the absorption of nitrogen as NH_4 is at the maximum at the beginning of the blooming period, declines rapidly after blooming, and is low during the later stage of growth. The quantity of nitrogen absorbed as NO_3 is very low at the beginning of the blooming period, increases rapidly during blooming and then declines rapidly. The maximum absorption of nitrogen occurs at the time of maximum absorption of nitrogen as NH_4, which is at the beginning of the flowering period. The maximum rate of absorption of NH_4 is approximately six times the maximum for NO_3. During the greater part of the life of buckwheat the absorption of NH_4 is the predominating factor in determining the rate of nitrogen assimilation.

Dastur and Malkani (1933) found that rice seedlings in contact with different concentrations of ammonium salts for varying periods of time gave the following order for the quantity of ammonium ions absorbed: $SO_4 < PO_4 < NO_3 < Cl$ for early stages, and $SO_4 < NO_3 < PO_4 < Cl$ for the later stage. The order of the quantity of NO_3 ion absorbed was: $NH_4 < Mg < Ca < K < Na$ for the early stage, and $NH_4 < K < Mg < Ca < Na$ for the later stage. The absorption of NH_4 was greater in the early stage of growth, but in later growth the absorption of NO_3 predominated.

2. *Availability.*—It was observed by Prianischnikow (1908, 1923) and Shulov (1912) that plants which take up nitrogen exclusively as ammonium salts generally contain very distinctly higher percentages of nitrogen than other plants supplied with nitrates. These observations have been substantiated by Pardo (1930, 1933) for sugar cane, by Loo (1931) for corn, and by Tiedjens and Robbins (1931) for tomato.

Kinoshita (1897) with seedlings of various kinds, and Pantanelli and Severini (1910, 1911) for wheat, rice, flax, and corn found that the nitrogen in ammonium compounds is in a form that can be utilized more readily by the plant in the formation of organic nitrogenous compounds than the nitrogen of the nitrates. Klein (1930) for numerous plants; Addoms and Mounce (1931) for cranberries; Tiedjens and Robbins (1931) for tomato, peach, and apple seedlings; Tiedjens and Blake (1932) for apple trees; and Tiedjens (1934) for cotton, tomato, and seedling apple trees reported that in all cases the nitrogen in the form of ammonium salts was utilized directly by these plants in the synthesis of nitrogenous compounds. Lewis (1936) for perennial rye grass, and Clark (1936) for tomato observed that more nitrogen accumulated in the tissues when nitrates were used than when ammonium salts were applied. They considered that this indicated that the nitrogen in the form of ammonium salts was quickly utilized and thus did not accumulate. It has been noticed by several investigators that many plants require a much lower concentration of ammonium than nitrate nitrogen in the nutrient solution to produce an equal volume of growth. This also is believed to indicate that the nitrogen in the form of ammonium is much more readily utilized than when it is in the form of nitrates. Nightingale (1934, 1936) and others have observed that the intake of nitrogen in the form of NH_4 is followed by a more rapid synthesis of amides, amino acids, and other soluble organic compounds than when nitrates are supplied to the plants.

Under certain conditions, however, some of the above-mentioned plants may grow better and produce more dry matter with nitrates than with ammonium. This was observed by Beaumont and others (1931) for tobacco; by Holley, Pickett, and Dulin (1931) for cotton; by Pardo (1933) for sugar cane; and by Arrington and Shive (1935) and Clark (1936) for the tomato plant. Beaumont, Eisenmenger, and Moore (1933) grew plants of timothy, bluegrass, and other grasses, and alsike, red, and white clovers in nutrient solutions containing sodium nitrate, ammonium sulphate, and urea. As judged by the appearance of plants, by the weight of dry matter produced, and by the chemical composition, the order of assimilation of nitrogen by the grasses was sodium nitrate, urea, and ammonium sulphate; by the clovers: urea, sodium nitrate, and ammonium sulphate.

3. *General Effects.*—Frear (1931) observed a negative correlation between the nitrate nitrogen in the sap of the leaves of the beet and the weight of the leaves. It was found by Hopkins, Wilson, and Peterson (1932) that the addition of nitrogen in the form of nitrates to clover plants growing in agar, generally resulted in a decrease in the number of nodules when the concentration was 2 to 3 parts per 100,000. The amount of nitrogen fixed by the nodules was also reduced, and if the concentration of nitrate was sufficiently increased, fixation ceased. Harrison (1934) noted in bluegrass supplied with calcium nitrate that the leaf blades were shorter and more upright and that the rhizomes were more numerous, stockier, and more branched than when supplied with ammonium sulphate. Lewis (1936) found that the per-

centage of phosphorus pentoxide in perennial rye grass was higher when the nitrogen was applied in the form of ammonium than when it was applied as nitrates.

4. *Toxicity.*—Numerous explanations have been proposed to explain the fact that many plants respond better to nitrates than to ammonium salts. By some it is considered that the basic ions combined with the nitrates may exert a stimulating effect while the acid ion of the ammonium salt may be injurious to certain plants. It is also considered that nitrates may be more available than the ammonium salts, since the former are very soluble and not fixed, while the latter salts are fixed and held more firmly and do not circulate so freely in the soil. On that account the ammonium salts do not come in contact with the root hair to the same extent that the nitrates do and hence are not so readily absorbed by the plant. It has been suggested by Willis and Carrero (1923) that the observation that nitrates are less suitable than ammonium salts for the fertilization of young rice plants may be based on the influence on the plants of the residues of the fertilizer used rather than on the inferiority of nitric nitrogen as a nutrient. They consider that under the conditions under which they worked in Puerto Rico the injury noted, which has generally been ascribed to the toxic effects of the nitrites derived from nitrates by reduction, was a manifestation of chlorosis caused by the action of the basic residues of the nitrate salts used as nutrients. Furthermore, the nitrogen of calcium nitrate may be equally suitable with that of ammonium sulphate in the physiological processes of the rice plant when it is used under cultural conditions in which the reaction of the unassimilated residue of the nitrate does not interfere with the absorption and utilization of iron.

It was stated by Willis and Rankin (1932) that ammonium hydroxide as a fertilizer is toxic to plants except under well-prescribed conditions, and Tidmore (1933) reported that the concentration of acid phosphate as a buffer must be very high in order to prevent injury by ammonium hydroxide. However, if the ammonium was supplied in small amounts under such conditions, excellent response in plant growth was obtained. Beaumont and Moore (1933), and Blackman (1934) believed that the toxicity from NH_4 salts of strong acids is due primarily to the accumulation of the ammonium ions and to increased acidity. Thelin and Beaumont (1934) found that the toxicity of ammonium sulphate was more pronounced with wheat than with rice, and that toxicity increased with the age of the plants.

b. Nitrites.—It was observed by Schultz (1903) that, in dilute solutions, nitrites seemed to be more reasily absorbed by plants than the nitrates. In some cases (Stutzer, 1906) a dilute solution of nitrites seems to be beneficial to some plants and injurious to others. Fairly concentrated solutions of nitrites are always more toxic to plants than the same concentration of nitrates. It was observed by Perciabosco and Rosso (1909) that plants which are supplied with nitrites contain distinctly higher percentages of nitrogen than when given nitrates. According to Nightingale (1936), nitrites generally do not accumulate in plants, and thus can be detected in appreciable amounts only after the application of nitrates to plants deficient in nitrogen.

Mevius and Dikussar (1930) and Dikussar (1935) reported that sweet corn can utilize nitrogen of nitrites in either neutral or alkaline solutions. The optimum concentration is approximately 50 mg. of nitrite nitrogen per liter, but at a pH of 7.0 as much as 200 mg. per liter is not injurious.

With nitrates, an increase of two- to three-fold in the exterior medium causes only a slight increase in the amount of nitrogen in a plant. An increase in the amount of nitrite in a medium, however, is followed by a rapid increase in assimilation and in the total amount of nitrogen in the plant.

Eggleton (1935) reported that nitrites occur naturally in the spring growth of grass, and that the amount is increased when the grass is fertilized with ammonium sulphate or sodium nitrate. He suggested that nitrites play a significant part in the grass "tetany" of cattle, which is usually associated with herbage of high inorganic content. Both nitrites and amino compounds increase upon the application of ammonium salts. It appears that the leaves can transform ammonium salts to nitrites.

2. Organic Nitrogen.—The effect of organic nitrogen on the growth of plants has been studied in two general ways. The first is concerned with the direct assimilation of organic nitrogen by the plant, and the second has to do with the stimulating effects of minute quantities of organic nitrogenous matter upon the rate of growth.

a. Assimilation.—Previous to 1911 considerable experimental work had been done with the nutrition of plants with organic nitrogenous compounds, but the value of much of the data presented is uncertain, since in many cases the plants were not grown under conditions that would prevent bacterial action. More or less satisfactory evidence of the assimilation of organic nitrogenous compounds by green plants had been obtained for the following compounds: methyl, amyl and allyl amines, acetamide, leucine, urea, aspartic acid, asparagine, glutamine, uric acid, hippuric acid, and tyrosine. The gains of nitrogen, however, were very small, and in many cases negative results were obtained by other investigators. The reader is referred to Hutchinson and Miller (1911) for a thorough and detailed review of the literature on the subject previous to this date. They grew wheat and peas in cultural solutions under sterile conditions. The ordinary nutrient solutions were used with organic substances as a source of nitrogen. They found that these plants could assimilate acetamide, urea, barbituric acid, alloxan, formamide, glycine, amino propionic acid, oxamide, sodium aspartate, and peptone.

Under sterile conditions where no reduction of nitrogenous compounds to nitrites or nitrates took place, Schreiner and Skinner (1912) reported that the following compounds entered the plant and reacted with the protoplasm in building up plant protein: nucleic acid, hypoxanthine, xanthine, guanine, creatinine, creatine, histidine, arginine, alloxan, asparagine, leucine, glycocoll, and others. Brigham (1917) found that corn could directly assimilate and use some of the nitrogen in asparagine, casein, cottonseed meal, hemoglobin, linseed meal, uric acid, peptone, guanine, alanine, urea, creatine, malt, and glycocoll. Eight of the substances that were directly available to the plant produced better growth when acted upon by *Bacillus subtilis*, probably because of ammonification. These were peptone, guanine, alanine, linseed meal, cottonseed meal, casein, hemoglobin, and urea. Ciamician and Ravenna (1919 to 1920) have reported the effect of a large number of the more uncommon organic nitrogenous compounds on plants. Crowther (1925) found that wheat and white mustard could obtain nitrogen from various organic compounds among which were dried blood, peptone, amino acetic acid, oxamide, and pyridine. Coville (1926) in preliminary experiments noted that buttermilk applied to young blueberry plants increased their height 32 to 37 per cent besides increasing the general appearance of their vigor. Skimmed milk when applied to the soil gave better results than buttermilk, owing perhaps to the fact that the sweet milk penetrated the soil more thoroughly. Azaleas also responded to treatment in much the same manner as the blueberry plants. The indications are that casein is the most active fertilizer of any of the derivatives of the buttermilk and skimmed milk used in the experiments.

Pirschle (1929), and Yamaguchi (1930) believed that plants can utilize urea directly. Klein (1930) reported that amino acids may be absorbed directly but are not equally utilizable, aspartic and glutamic acids being superior to glycine and alanine. Macht (1934) considered that the levo varieties of the amino acids exhibit a greater physiological activity than the dextro varieties. Addoms and Mounce (1932) found that glycine itself was absorbed by cranberry plants. Other investigators have reported that certain organic substances are not used by the plant directly. Thus Willis and Rankin (1930), and Beaumont and others (1931) found that the nitrogenous material in cottonseed meal is changed to nitrate and ammonium before being absorbed and utilized by the plant. Knudson (1933) obtained negative results in the use of glycine, lucine, and aspartic acid as nutrients in the culture of orchids.

The studies of the effect of organic nitrogenous compounds on the nutrition of green plants seem to indicate that they can commonly supplement but in only a few cases replace the inorganic nitrogenous salts. The experimental work would seem to show at least that green plants growing in a heavily manured soil in the field or in a rich compost as frequently used in greenhouses can absorb directly and utilize in their metabolism larger or smaller amounts of complex organic nitrogenous compounds.

b. Stimulation.—It has been observed in numerous cases that small amounts of certain organic nitrogenous matter exert a stimulating effect upon the rate of growth and the reproduction of certain plants. These organic substances have been termed "auximones."

1. *Auximones.*—Such substances may be similar to or identical with the growth substance discussed in Chap. XIV. It should be remembered that, however essential vitamins may be for animal growth, they may be waste products in plants. The vitamins needed by the animal may not be identical with the accessory substances required for plant growth. The terms "phytamins" and "nutrilites" have been suggested for accessory plant-growth substances.

(*a*) *Effects.*—It was observed by Bottomley (1917 to 1920) that the growth of the water plants *Lemna minor, Salvinia natans, Azolla filiculoides,* and *Limnobium stoloniferum* in Detmer's solution was greatly accelerated by the addition of various amounts of extract from partially decayed peat and other vegetable matter. In one case where the addition of this organic material amounted to 368 p.p.m. after 6 weeks, the treated plants of *Lemna minor* showed a multiplication to twenty times and an increase in weight to sixty-two times that of the plants growing in the water containing mineral salts only. An examination of the pond water in which the plants grew vigorously showed 1,577 p.p.m. of dissolved solids, of which 488 p.p.m. were organic material. The inference was that in their natural habitat these plants utilize a certain portion of these organic nitrogenous compounds in their metabolism. The extracts from the partially decayed peat were proved to be derivatives of nucleic acid. Bottomley found that the nitrogen-fixing bacteria have the power of elaborating products from carbohydrate and elementary nitrogen, which also greatly increase the growth of *Lemna* plants.

Mockeridge (1917, 1924) investigated the effect of varying quantities of nucleic acid derivatives on the growth of plants of *Lemna minor* and found that, although the effect obtained was approximately proportional to the quantity of material supplied, it was not directly proportional to it. The growth of these plants was also accelerated by the addition of sterilized cultures of *Azotobacter* to the medium in which they were growing. Since yeast was found to have plant-growth promoting power on *Lemna minor* and since it contains nucleic acid radicals, the *Azotobacter* were examined for traces of these constituents. Purine and pyrimidine bases, phosphoric acid, and a carbohydrate were all shown to be present, and since all the necessary radicals for the formation of nucleic acid were thus isolated, it was concluded that the beneficial action on the growth of *Lemna minor* by sterilized cultures of *Azotobacter* was due to these organic substances. Ashby (1929) found that small quantities of organic matter added to inorganic solutions greatly increased the rate of growth of *Lemna minor.* Hartley and Greenwood (1933) and Tyagny-Ryadno (1933) reported that barnyard manure in small amounts has a greater effect on farm crops than can be explained by its contents of total nitrogen, phosphorus, and gross organic matter. Popoff (1933) found that extracts of the roots, stems, and growing tips of seedlings of maize greatly stimulated the growth of *Euglena gracilis.*

Clark and Roller (1924) and Clark (1924), however, concluded that the growth of *Lemna minor* in mineral solutions depends upon the suitable concentration of salts and that small amounts of organic matter are not essential for growth and reproduction and cannot be classed with vitamins, which are necessary for animal growth.

Clark (1930) and Clark and Roller (1931) found that the rate of reproduction of *Lemna minor* and *Lemna major* was greatly increased by the addition of sterile organic extracts to a nonsterile medium, but that these extracts had no influence when they were added to a sterile medium. Saeger (1925) and Wolfe (1926) in experiments with plants closely related to those used by Bottomley reported that such plants will grow and reproduce in both Detmer's and Knop's solutions which are properly diluted or balanced. In some cases the addition of organic compounds slightly accelerated the growth and in others depressed it. They considered that the necessity of organic accessories (auximones) in the nutrition of green plants can thus not be accepted as an established fact.

(*b*) *Origin.*—According to Nicol (1934) the origin of accessory substances, whether required for the proper growth of plants or for that of the animals which ingest them, may be as follows:

(1) Derived from inorganic compounds of nitrogen.
- *a.* Directly built up within the plant.
- *b.* Synthesized by microorganisms in the soil.

(2) Derived from organic compounds of nitrogen.
- *a.* Synthesized by nodule bacteria.
  1. Used wholly within the host plant.
  2. Excreted and then absorbed by associated plants.
- *b.* Organic manures of plant or animal origin.

2. *Legumes upon Nonlegumes.*—According to Nicol (1934) it has been observed by Lyon and Bizzell (1911), Evans (1916), and Stallings (1926) that timothy and redtop grown with clover and alfalfa, oats grown with peas, and wheat grown with soybeans showed either an increased content of nitrogen or a better growth than when grown alone.

Lipman (1910, 1912) found that oats could secure an adequate supply of nitrogen in a soil devoid of nitrogen if its roots were separated by only a porous membrane from the roots of field peas growing in the adjoining soil. There was little or no growth of the oats when a glazed impermeable membrane separated the roots of the two plants. These effects occurred before there was any disintegration of the roots of the pea so that no compounds beneficial to the oats could originate after that manner. Lipman believed that compounds of nitrogen pass from the roots of some legumes and may be available to the roots of nonleguminous plants growing in the adjoining medium. Virtanen and Von Hausen (1930) noted that red clover growing in sand excreted nitrogenous substances therein. Virtanen, von Hausen, and Karström (1933) found that the roots of inoculated peas excreted various forms of nitrogenous organic compounds into the surrounding medium in the following proportions.

Type of Nitrogen	Percentage of Total Nitrogen
Amino	77.4
Amide	3.3
Volatile basic	2.7

Nicol (1934) stated that, "It can be considered an established fact that the production of nitrogenous compounds produced by one plant is available to another plant, although the mode of excretion is obscure."

B. Absorption from the Air

About 80 per cent of the atmosphere is composed of nitrogen in the elemental form. There are also present small quantities of nitrogen in the form of ammonia, nitrates, and nitrites. Rainfall and snow thus contain nitrogen in this form taken up in their passage through the atmosphere. Russell and Richards (1919) over a period of 28 years at Rothamsted found that with an average rainfall of 28.82 in. the ammoniacal content amounted annually to 2.64 lb. per acre, while the nitric nitrogen was about one-half as much, or 1.33 lb. per acre. It was observed by

Wilson (1921) at Ithaca, N. Y., that with an average yearly rainfall of 29.31 in. from May, 1915, to May, 1920, 12.51 lb. of nitrogen per acre were brought down annually. Of this amount 11.5 lb. were in the form of ammoniacal nitrogen, while 1.01 lb. were in the form of nitrate nitrogen. There is, however, considerable doubt as to what extent the green plant can absorb this free and combined nitrogen of the atmosphere. A historical résumé of the early work in this field is essential in order to appreciate the general opinion of the subject that prevails at this time.

1. Historical.—De Saussure (1804) considered that the plants with which he worked did not take up appreciable quantities of the nitrogen supplied to them in the free and gaseous form. He believed that the source of the nitrogen of plants is more probably the nitrogenous compounds of the soil and the small amount of ammonia that he demonstrated to exist in the atmosphere. In 1837 Boussingault took up the question of the source of the nitrogen of plants where De Saussure had left it more than 30 years before and for 20 years carried on experiments with a large variety of plants to determine if any of these absorbed free nitrogen. His conclusions were that neither leguminous nor nonleguminous plants were able, either when their supplies of combined nitrogen were limited to that contained in the seed sown or when their vigor of growth was stimulated by artificial supplies of combined nitrogen, to assimilate the free or uncombined nitrogen of the atmosphere. Ville, a French contemporary of Boussingault, however, in his works published from 1849 to 1857, reported results that are exactly the opposite of those obtained by the latter. He found that the plants in his experiments showed a very considerable gain of nitrogen during growth whether they were subjected to a current of unwashed air or to one of ammonia-free air and also when the plants were grown in free air and their activity of development increased by the use of nitrates or other nitrogenous matters as manure. This gain of nitrogen he considered to be due to the assimilation of free or uncombined nitrogen of the air.

Lawes, Gilbert, and Pugh in 1860 published the results of their experiments at Rothamsted (1857 to 1859) in which they grew grain plants both with and without a supply of combined nitrogen other than that contained in the seed sown. From these experiments, which were conducted over a wide range of conditions, character, and amount of growth and in the amount of combined nitrogen involved, they concluded that in no case was there any evidence of the assimilation of uncombined nitrogen. Their results of similar experiments with leguminous plants indicated that no assimilation of free nitrogen by the plants occurred. They concluded, however, that, since in practice leguminous plants assimilate from some source so much more nitrogen than grain plants under equal circumstances of combined nitrogen, it would be desirable if the evidence of further experiments with these plants under conditions of more healthy growth could be obtained. Their inference in this regard was correct, since it was shown later that they had sacrificed in their attempt for accuracy the conditions that are necessary for the symbiotic relationship of bacteria and the roots of the legumes. Finally, in 1888 Hellriegel and Wilfarth published the results of their classical work on the nitrogenous nutrition of legumes. They observed that legumes differ fundamentally from the grains in their nutrition with respect to nitrogen. The latter can satisfy their nitrogen need only by means of assimilable combinations existing in the soil, while the former in addition to the nitrogen of the soil have at their service a second source upon which they draw for nitrogen and this source is the elementary nitrogen of the atmosphere. They stated further that the legumes do not possess by themselves the faculty of assimilating the free nitrogen from the air but that it is necessary for certain microorganisms of the soil to enter into a symbiotic relationship with the plants in order to secure this result. Beyerinck (1888) and Prazmowski (1890) were able to secure pure cultures of the tubercle bacterium and gave to it the name of *Bacillus radicicola.* Thus by the work of these four investigators the relationship of the legumes in regard to their nitrogen supply was placed upon a basis that is now universally accepted.

2. Nonleguminous Plants.—Although it is commonly considered that the nonleguminous plants cannot utilize the free nitrogen of the atmosphere, considerable work has recently been done to investigate the validity of this assumption. Recently Lipman and Taylor (1922, 1924) have called attention to the fact that the assumption that nonleguminous plants cannot assimilate free nitrogen is based almost exclusively upon the work of Boussingault (1854 to 1855) in which he claimed that neither the leguminous nor nonleguminous plants could assimilate free nitrogen. They further pointed out that the work of Ville (1853 to 1857), which indicated that all plants could utilize free nitrogen, has been entirely overlooked.

a. The Algae.—Since numerous blue-green and green algae inhabit the soil and since some of these resemble the bacteria in their morphology or are closely related to the fungi in their structure, it has been thought that they might fix free nitrogen and thus be an important factor in the nitrogen relations of the soil and the higher plants. Accordingly a considerable amount of experimental work has been done to determine whether these plants possess the power to fix free nitrogen.

Frank (1888 to 1889) obtained positive results for the blue-green algae *Nostoc* and *Gloeocapsa,* but his experiments were not performed under sterile conditions. Kossowitch (1894) was the first to use pure cultures of algae in this type of work. He used *Cystococcus* and showed clearly that under sterile conditions this alga could not fix free nitrogen either in the absence or in the presence of glucose. Schramm (1914) worked with seven species of the green algae and found that in the absence of combined nitrogen no growth took place, so he concluded that these algae could not under these conditions assimilate free nitrogen. Similar results were also obtained by Muenscher (1923) for *Chlorella*. Wann (1920 to 1921) obtained experimental results that led him to conclude that seven species of the green algae (*Chlorophyceae*) exhibited the ability to fix free nitrogen when grown in pure cultures on mineral nutrient agar media containing either ammonium nitrate or calcium nitrate as a source of nitrogen and glucose. The work of Wann was repeated by Bristol and Page (1923), who showed that his data were not correct owing to the faults of the methods of nitrogen determination that he used.

Moore and others (1919 to 1921) concluded from their observations that in the absence of all other sources of nitrogen save the elemental nitrogen of the atmosphere, but with an abundance of carbon dioxide, fresh-water unicellular algae can fix nitrogen, grow, and form proteins. The rate of growth is greatly accelerated, however, if nitrates or oxides of nitrogen are available. They considered that the green seaweed (*Enteromorpha compressus*) can fix elemental nitrogen from sea water in the presence of sunlight but not in darkness, and thus indirectly from the air. Their results are open to criticism, since they did not work with sterile cultures and depended only on the turbidity of the water as an indicator of the presence of bacteria. The results of all the experimental work thus indicate that the algae examined cannot fix free nitrogen, and we must conclude, in the words of Bristol and Page (1923): "It is conceivable that green algae might assimilate atmospheric nitrogen under certain conditions, as yet unknown, but at present there is no trustworthy evidence that they can do so."

However, Drewes (1928) and Allison and Morris (1930) found convincing evidence that the blue-green algae, *Anabaena variabilis* and *Phormidium molle*, are important nonsymbiotic nitrogen-fixing organisms. In 75 days, in the absence of sugar, they fixed on the average 5 mg. of nitrogen per 100 cc. of culture medium. When 1 g. of sucrose was added, 8.5 mg. of nitrogen was fixed. The fixation occurred only in sunlight and was negligible in darkness. Copeland (1932) found that the fixation of nitrogen by the blue-green algae was shown only by certain strains, and he considered that this power is a physiological specialization that occurs only under conditions of nitrogen deficiency.

b. The Higher Nonleguminous Plants.—The ability of the higher green nonleguminous plants to fix atmospheric nitrogen has been investigated recently by several investigators.

Mameli de Calvino and Pollacci (1911 to 1923) worked with a wide range of plants in relation to their source of nitrogen. The plants used included lichens, mosses, and aquatic and terrestrial phanerogams and were grown under the best known conditions to overcome incomplete sterilization of culture, error due to presence of nitrogen compounds in the air, and incomplete development of the plants. Their data showed that all the plants tested had taken up nitrogen in greater or lesser quantities from the air in which respiration and assimilation took place and that more free nitrogen was taken from the air when the culture medium was supplied with small quantities of nitrogen compounds. Some of the species of the higher plants which they claim can assimilate some free nitrogen directly from the air without the aid of symbiotic microorganisms are *Raphanus sativus, Acer negundo, Cucurbita pepo*, and *Polygonum fagopyrum.*

Lipman and Taylor (1922, 1924) claimed to have secured conclusive evidence from experiments carried on with wheat and barley in culture solutions with and without nitrate nitrogen that the cultures gained nitrogen at the expense of the elementary nitrogen of the atmosphere. They reported that in wheat plants there was a gain of nitrogen from the air which varied from 13 to 21 per cent of the total amount of nitrogen found in the plant. The cultures used by these investigators were not sterile, but they concluded that the small number of bacteria present and the low amount of organic matter in the cultures available for them could not account for the gains in nitrogen observed. Lipman and Taylor believed that the fixation of nitrogen occurs in the leaves of the plants. With the dwarf variety of *Pisum sativum*, however, under as nearly sterile conditions as could be obtained in nutrient solutions, Burk (1927) was unable to observe any fixation of nitrogen.

The results reported by Lipman and Taylor have been skeptically received. It would seem that if any nitrogen is fixed by the nonleguminous green plants it is very small in amount and is inconsequential when compared with that absorbed by the roots in the combined form.

Brown (1933) believed from his experiments that plants of *Lolium* met some of their nitrogen requirements by the use of atmospheric nitrogen, but Ludwig (1934) in experiments with willow cuttings found that no significant nitrogen fixation occurred under the conditions involved. It has not been proved to the satisfaction of all that the fixation of nitrogen by the nonleguminous green plants so far reported has not been due to the action of bacteria, and according to Fisher (1917) it should be considered that where claims for nitrogen fixation are made a need exists for the proof that the technique was not in error.

The absorption by the aerial parts of the plant of the combined nitrogen in the air is considered by some (Moore, 1921) to be significant in the metabolism of the plant. He considered that nitrites may be absorbed in considerable quantities from rain and dew, since his experiments indicated that nitrite was present in these waters in amounts of 1 part per 10,000,000 to 2 p.p.m. These concentrations seem at first glance to be so extremely low that the nitrite would be of no nutritional value to the plant, but Moore points out that the demand of a plant for nitrogen is relatively small and that a concentration of nitrites or nitrates in the atmosphere or cell-sap equivalent to 1 part nitrogen to 120,000,000 parts water could supply a sufficient quantity for assimilative purposes.

III. THE PROTEINS OF THE GREEN PLANT

The proteins are complex organic substances composed of carbon, hydrogen, oxygen, and nitrogen, usually sulphur, and sometimes phosphorus and are compounds which, for the most part, yield α-amino acids upon hydrolysis. The proteins occur in some form or another in every living plant or animal cell and are essential to all life processes, since they are the main constituents of the protoplasm. The proteins occur in plants in the dissolved state in the cell sap, in a semidissolved state in the protoplasm, and are stored in the insoluble form as reserves in the cells of seeds, tubers, bulbs, buds, roots, and other plant parts. The reserve proteins make up the major portion of the proteins found in plants and occur in the cells with nonnitrogenous reserves of carbohydrates and oils. The undissolved protein in plant cells may be in the form of crystals, in the form of aleurone grains, and in the form of aleurone. Crystalline proteins are found in the seeds of flax, squash, cotton, sunflower, rape, mustard, peanuts, and in the potato tuber. The term "aleurone grain" is used in a broad sense to denote all noncrystalline protein bodies of a more or less definite shape, and Osborne (1924) stated that the aleurone grain may be irregular, semicrystalline with faces and angles on a part of the surface or distorted spheres. The aleurone grains may be rather complicated in their structure, as in the case of those in the castor bean. These grains consist of an amorphous protein matrix in which are embedded a protein crystalloid and a globoid consisting of a double phosphate of calcium and magnesium. The matrix and its inclusions are surrounded by an outer layer of protein that is not so readily soluble as the matrix (Fig. 29). The protein that occurs in

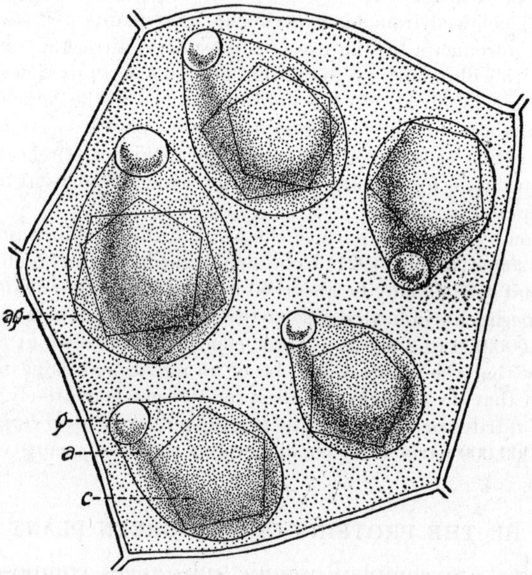

FIG. 29.—Section of a cell from the endosperm of the castor bean (*Ricinus communis*) showing the aleurone grains. *ag*, aleurone grain. *c*, crystal of protein. *a*, amorphous protein. *g*, globoid.

plant cells in an amorphous, finely granular form is generally designated as "aleurone."

In the seeds of the grains and grasses the outermost layer of cells of the endosperm is termed the "aleurone layer." This layer is composed of short, prismatic, rather thin-walled cells and receives its name from the fact that these cells contain, for the most part, protein either as aleurone grains or in the amorphous form regardless of storage materials in the rest of the endosperm (Fig. 33).

The amount of protein in a plant or plant part may vary greatly with the strain of plant used and the conditions under which the plant is grown. Thus Alway and Nesom (1930) found that the protein content of various strains of Canary grass (*Phalaris arundinaceae*) varied from

6.6 to 25.2 per cent. It ranged from 2.8 to 11.9 per cent in the culms, from 8.5 to 23.5 per cent in the leaves, and from 9.4 to 30.5 per cent in the panicles. Enlow and Coleman (1929) noted that the grasses of Florida which were mowed frequently averaged much higher in protein than those which were cut at the end of the season. The protein content of most plants could be maintained at a higher level by frequent light applications of a nitrogenous fertilizer.

A. Physical Properties

The proteins have no definite melting points or boiling points but carbonize on heating and give off gases. They are colloidal in character and thus when in solution do not pass through parchment, collodion, or other membranes. All of the solutions of plant proteins are optically active, rotating the plane of polarized light to the left. The proteins differ widely as to their solubility in water and in salt solutions, and these differences in solubility are used as a basis for their classification. The proteins are insoluble in strong salt solutions and are precipitated out of their solutions without change in their composition by saturating the solution with various neutral salts as sodium chloride, ammonium sulphate, and magnesium sulphate. Dilution of such saturated solutions will cause the precipitated protein to redissolve. All the natural proteins when in a colloidal solution may be coagulated by heat, alcohol, or enzymes, forming a semisolid gel, which cannot again be rendered soluble except by decomposition.

B. Chemical Properties

The average percentage of the elements that compose plant proteins is approximately as follows: carbon, 52 per cent; oxygen, 22 per cent; hydrogen, 7 per cent; nitrogen, 17 per cent; and sulphur, about 0.6 per cent. Phosphorus is also present in certain proteins in small amounts. The percentages of carbon, hydrogen, and oxygen vary within comparatively narrow limits, but the sulphur content varies to a greater extent in the different proteins than any other of their constituent elements. Thus in legumin it amounts to only 0.385 per cent of the dry weight of the protein. Since it has been shown that many plant proteins yield cystine on hydrolysis, it is considered that in proteins much of the sulphur is in that form. By the present methods of determining cystine sulphur, however, the amount of sulphur in the proteins that cannot be accounted for by the cystine determinations ranges from 24 to 80 per cent.

The amount of nitrogen is very similar in many proteins. Thus from Osborne (1924) it is noted that the globulin from squash seed contains 18.5 per cent nitrogen, edestin from the hemp seed 18.7, excelsin from the Brazil nut 18.3, globulin from the cottonseed 18.6, and legumin from peas 18.0 per cent. Of the more than 20 plant proteins mentioned by Osborne (1924) the nitrogen content ranges from 15.9 per cent for the legumin of the horse bean to 19.0 per cent for corylin of the hazelnut. In general, it may be stated that the proteins from plants contain from 2 to 3 per cent more nitrogen than the animal proteins.

The molecule of even the most simple protein is very large. The molecular weights of some of the proteins have been estimated by various methods, but exact data are still lacking in this regard. The complexity of the protein molecule is shown by the calculated formula of zein and gliadin as given by Thatcher (1921):

Zein from corn $C_{736}H_{1,161}N_{184}O_{208}S_3$

Gliadin from wheat $C_{685}H_{1,068}N_{196}O_{211}S_5$

1. The Amino Acids.—The proteins are apparently compounds formed by the union of amino acids, since their products of hydrolysis are all acids of this type. Since these acids are the units out of which the protein molecule is built and since the characteristics of the proteins are due in part to the characteristics of the amino acids, a few general statements in regard to their nature should be made. An amino acid may be regarded as an organic acid, in which one or more hydrogens of the noncarboxyl group is replaced by the amino group NH_2, or, viewed from another standpoint, an amino acid may be considered as a substituted ammonia, one hydrogen of ammonia being replaced by an organic acid.

Thus amino-acetic acid or glycine, $\begin{array}{c} NH_2 \\ | \\ H—C—COOH \\ | \\ H \end{array}$, is derived from acetic acid, $\begin{array}{c} H \\ | \\ H—C—COOH \\ | \\ H \end{array}$. From propionic acid, $\begin{array}{c} H\ \ H \\ |\ \ \ | \\ H—C—C—COOH \\ |\ \ \ | \\ H\ \ H \end{array}$, two amino acids may be derived: α-amino propionic acid or alanine,

$$\begin{array}{c} H\ \ NH_2 \\ |\ \ \ | \\ H—C—C—COOH \\ |\ \ \ | \\ H\ \ H \end{array}$$

or β-amino propionic acid,

$$\begin{array}{c} NH_2\ \ H \\ |\ \ \ \ \ | \\ H—C——C—COOH. \\ |\ \ \ \ \ | \\ H\ \ \ \ H \end{array}$$

The amino acids that result from the hydrolysis of protein and out of which proteins are supposed to be built are all α-amino acids, which means that the NH_2 group, or one of the NH_2 groups, is attached to the carbon atom nearest to the COOH group. Since the amino acids contain both the basic NH_2 group and the acid COOH group, they easily unite together to form larger molecules, the linkage taking place between the NH_2 group of one amino acid and the COOH group of the other with the elimination of water. A compound thus formed will contain a free NH_2 group and a free COOH group, so that another amino acid may be attached, further to increase the size and complexity of the molecule, and so on. In the laboratory as many as 18 of these acids have been caused to unite in this manner, forming compounds that exhibit some of the characteristics of the natural proteins (Fischer, 1899 to 1906, quoted by Thatcher, 1921). The following amino acids have been obtained by the hydrolysis of plant proteins or have been found in the free form in different plant tissues especially in germinating seeds and in growing buds:

Amino acids	Derivative name	Formula
Glycine.........	α-aminoacetic acid	$CH_2.NH_2COOH$
Alanine.........	α-aminopropionic acid	CH_3CHNH_2COOH
Valine	α-aminoisovaleric acid	
Leucine.........	α-aminoisocaproic acid	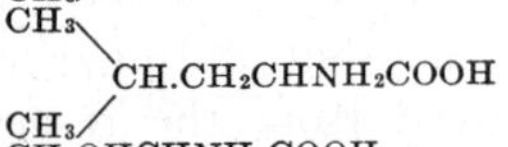
Serine...........	α-amino β-hydroxypropionic acid	$CH_2OHCHNH_2COOH$
Aspartic acid....	α-aminosuccinic acid	$COOH.CH_2.CHNH_2COOH$
Glutamic acid ...	α-aminoglutaric acid	$COOH.CH_2.CH_2.CHNH_2COOH$
Arginine........	α-amino δ-guanidine valeric acid	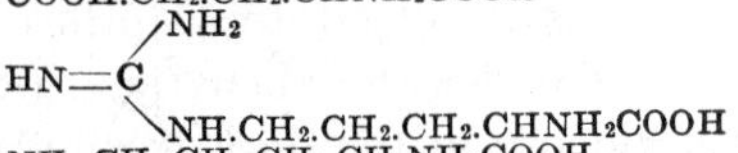
Lysine..........	α-ε-di-amino caproic acid	$NH_2.CH_2.CH_2.CH_2.CH.NH_2COOH$
Cystine.........	di-(α-amino β-thio propionic acid)	$HOOC.CHNH_2.CH_2SSH_2C.CHNH_2COOH$
Phenyl alanine ..	α-amino β-phenyl propionic acid	$C_6H_5.CH_2CHNH_2COOH$
Tyrosine........	α-amino β-parahydroxy phenyl propionic acid	$OH.C_6H_4CH_2CHNH_2COOH$
Proline..........	α-pyrrolidine carboxylic acid	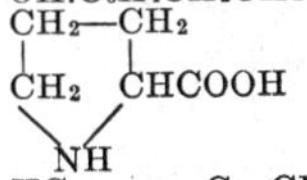
Histidine........	α-amino β-imidazol propionic acid	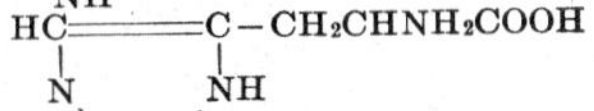
Tryptophane....	α-amino β-indole propionic acid	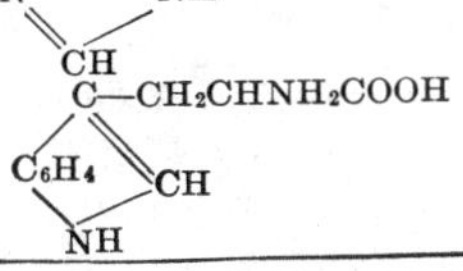

Pollard and Chibnall (1934) found that the proteins from grass contain cystine in amounts varying from 0.3 to 0.95 per cent, while the proteins of alfalfa contain as much as 1.2 per cent of this amino acid. Csonka and Jones (1934) noted that the amount of cystine, tryptophane, and tyrosine of the soybean varied with the variety. The percentage of these amino acids on a dry basis of the oil-free meal was 0.35 to 0.48 for cystine, 0.93 to 1.2 for tryptophane, and 2.3 to 2.6 for tyrosine. According to Maranon (1932) the old leaves of camphor showing signs of chlorophyll degradation have a larger amount of nitrogen in the form of amino acids than either the young or the full-grown leaves. Dhar and

Mukherjee (1934) synthesized amino acids by exposing solutions of glycerol and nitrates, or glucose and nitrates to sunlight in the presence of titanium oxide as a catalyst.

2. The Amides.—An amide is a compound resulting from the replacement of 1 or more atoms of hydrogen in ammonia by a univalent acid radical. Thus 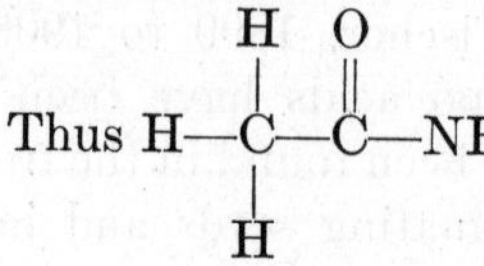 is acetamide.

The amides are highly reactive substances, and, according to Vickery, Pucher and Clark (1936), the amount present in plants at any time is largely dependent on the circumstances under which the plants have been grown. It was noted by Barbera (1933) that the apical parts of alfalfa are very rich in amino acids and amides. These two types of compounds collectively represent approximately 50 per cent of the nitrogen substances in the fresh sap of these parts. McCalla (1934) found with the seeds of cereals that during germination the amide groups are more readily removed from the reserve protein than are any other of its structural units.

It was reported by Mothes (1926, 1931) that no amides can be detected in leaves when they are exposed to the light or are supplied with glucose in the dark. Schulze (1911) assumed that the amides are used in the synthesis of protein. When the carbohydrates of the leaves are reduced to a low point, amides are produced and ammonia appears. In the absence of oxygen *neither* amides *nor* ammonia appear, but there is an increase of the amino, basic, and other nitrogen fractions. Oxidation is thus apparently essential for the production of amides.

Bennett-Clark (1933) believed that the disappearance of amino acids and proteins, either in normal metabolism or under special experimental conditions, is associated with the formation of either amides or ammonium salts. Kultzscher (1931, 1932), and Vickery and Pucher (1931) considered that the ammonia which accumulates in the plant as the end product of physiological changes may become toxic unless it is rendered inactive by transformation into amides or ammonium salts. They believed that the formation of amides is a defensive mechanism for the detoxication of ammonia, the latter resulting from metabolic processes or from an excessive reduction of nitrates.

a. Asparagine.—This amide is the most abundant one in plants and has been more thoroughly studied than any other. It is the amide of aspartic acid and has the formula $CONH_2CH_2CHNH_2COOH$. It was first noted in asparagus and is named after that plant. It occurs in asparagus, onion bulbs, and in the seedlings of beets, peas, beans, and numerous others. It is especially abundant in seedlings when they are supplied with an abundance of nitrates (Zaleski, 1898; Nightingale and Schermerhorn, 1928; Rahn, 1932). The abundance of asparagine in the seedlings of legumes is shown by the following data of Schulze as presented by Murneek (1935):

Age of seedlings, days	Protein nitrogen, per cent	Asparagine nitrogen, per cent	Other nitrogen, mostly amino acids, per cent
6	5.5	1.2	1.7
12	1.7	4.0	2.4
24	1.8	5.1	1.4

Prianischnikow (1924) advanced the hypothesis that asparagine is synthesized in the plant in response to an accumulation of ammonia. This amide can then be translocated and stored for subsequent synthesis in which nitrogen is required. Chibnall (1922, 1924) showed that asparagine disappears at night from the leaves of the bean and that it comes largely from the hydrolysis of protein. According to Wassilieff (1908), Schulze and Winterstein (1910), and Schulze (1911) nitrogen may be translocated in the form of asparagine to the developing seeds and pods of legumes, and it then remains in the mature seed or fruit.

It was observed by Mothes (1926, 1931) that in the presence of some carbohydrates any excess of ammonia is stored in asparagine. If sufficient carbohydrates are available the asparagine is used in the synthesis of proteins. When ammonium salts are fed to leaves high in carbohydrates, proteins are formed quickly. If carbohydrates are present in limited quantities, asparagine will be formed, but when carbohydrates are absent, ammonia will accumulate in the cells until poisoning of the leaves occurs. Thus asparagine apparently has a dual function in the removal of excess ammonia and in the storage of nitrogen. McKie and Barnett (1936) considered that asparagine is formed from amino acids. A union of aspartic acid with ammonia forms ammonium aspartate, from which, by dehydration, asparagine is produced. According to the work of Smirnow (1923), malic and succinic acids seem to be the two acids that in the presence of ammonia are concerned in the synthesis of asparagine.

b. Glutamine.—This amide is apparently widely distributed in plants. Most seeds that contain oil as a reserve food produce glutamine instead of asparagine upon germination. Glutamine, the amide of glutamic acid, has the formula

$$CONH_2CH_2CH_2CHNH_2COOH.$$

Vickery, Pucher, and Clark (1934) grew tomato plants in sand cultures at a pH of 6.7 using calcium nitrate and ammonium sulphate as sources of nitrogen. It was found that the glutamine nitrogen composed over 15 per cent of the total nitrogen in the stems when ammonium sulphate was used as s source of nitrogen, and less than 2 per cent when the plants were grown in nitrate. Greenhill and Chibnall (1934) noticed under certain conditions that perennial rye grass, when supplied with ammonium sulfate as a source of nitrogen, produced a white exudation upon the upper portion of the blades which consisted almost entirely of glutamine. It appears that even in dilute concentrations glutamine exercises some selective influence on the permeability of the leaf cells of this plant.

Vickery, Pucher, and Clark (1936) found in beets supplied with ammonium sulphate that the maximum content of glutamine was 5.4 per cent of the dry tissue of the roots. They believed that glutamic acid is formed after the manner common to that of other amino acids, *i.e.*, the ammonium salt of glutamic acid is formed, and that from this compound glutamine is formed by subsequent dehydration.

3. Qualitative Reactions of the Proteins.—Most of the proteins show similar reactions with a large number of reagents. This is due to the fact that they are complex compounds built up in the same general manner from a large number of relatively simple molecules. None of the various chemical reactions of the proteins is characteristic of them alone, but if a substance gives several of these reactions it may be considered to belong to the protein group. Two of the more common reactions of the proteins are the precipitation reactions and the color reactions.

a. Precipitation.—The proteins are precipitated out of solution by the various mineral acids. In this regard, if nitric acid is allowed to flow gently into a protein solution a white ring forms at the junction of the acid and the solution. This is one of the most delicate protein reactions and is known as "Heller's test." The proteins are precipitated from solutions by the reagents that precipitate alkaloids from their solution. Among these reagents are phosphotungstic acid, tannic acid, picric acid, and phosphomolybdic acid.

The proteins are also precipitated from solution by the addition of small amounts of the salts of various heavy metals such as the chlorides, sulphates, and acetates of iron, copper, mercury, and lead. The precipitation reactions that have just been mentioned are due to the fact that a chemical reaction occurs between the protein and the reagents used. The protein is profoundly altered by these reactions and cannot again be dissolved except by treatment with some reagent that will remove the metal or acid radicals that have combined with the protein. With the exception of some of the mineral acids, the protein cannot be recovered in its original form after such precipitation. This type of precipitation is thus entirely different from the precipitation caused by the saturation of a protein solution with the salts of the alkali metals in which the precipitate proteins can again be readily redissolved and recovered in their original form.

b. Color.—The proteins give certain characteristic color reactions with certain reagents. This is due to the fact that certain radicals or groups are present in the protein molecule; accordingly, some proteins may respond with a color reaction to a certain reagent, while others will not. Certain of the decomposition products of proteins also give the color reactions, provided that disintegration has not progressed to a point where the reacting groups are destroyed. The following are the five most important color reactions of the proteins:

1. *Biuret Reaction.*—When a protein is treated with a solution of sodium hydroxide, and a dilute solution of copper sulphate is added drop by drop, a reddish-violet or violet-blue color is produced. When the protein has been hydrolyzed to peptone, a red coloration results with this test. This reaction is known as the "biuret test," since biuret, $NH_2CONHCONH_2$, gives this characteristic color with the reagents named. This reaction is also given by groups containing two—CONH—radicals connected by a carbon, nitrogen, or sulphur atom. Since most of the proteins contain several such groups, the biuret test is a very general one for proteins.

2. *Xanthoproteic Reaction.*—When proteins are treated with strong nitric acid, a yellow coloration is produced, which intensifies upon heating. If an excess of sodium hydroxide or ammonia is added, the color is changed to orange or orange red. The

production of this color is apparently associated with the presence of the benzene ring and in protein tests is evidently the effect of nitric acid on tyrosine.

3. *Millon's Reaction.*—When Millon's reagent is added to a solution or suspension of protein in water, there is obtained either in the cold or after boiling a pink coloration of the solution or a pink to brownish-red coloration of the precipitated protein. Millon's reagent is prepared by dissolving 1 part mercury in 2 parts concentrated nitric acid (sp. gr. 1.42), first cold and then warm, and after the complete solution of the mercury, one volume of the solution is diluted with two volumes of water. The Millon reaction is given by all benzene derivatives in which one of the hydrogen atoms has been replaced by a hydroxyl group, and thus the reaction is a test for proteins containing tyrosine.

4. *Adamkiewicz's or Hopkins-Cole's Reaction.*—If a protein solution is treated with one volume of concentrated sulphuric acid and two volumes of glacial acetic acid, a reddish-violet color is produced. It has been shown that this color is due to the reaction of the glyoxylic acid, which is present in the glacial acetic acid, upon the tryptophane group in the protein. A more intense reaction is obtained if glyoxylic acid is used instead of acetic acid.

5. *Sulphur Reaction.*—If a drop of a solution of lead acetate is added to a solution containing a protein, and sufficient sodium hydroxide is then added to dissolve the precipitate that is formed, a black or brown coloration will be produced upon heating to boiling, provided that the protein under consideration contains cystine, the amino acid containing sulphur. The reaction depends upon the splitting off of sulphur from the cystine and the subsequent formation of lead sulphide.

4. Amphoteric Nature of Proteins.—The proteins are amphoteric substances, *i.e.*, they are substances that may behave as either acids or bases (Loeb, 1924). The amphoteric nature of the proteins is due to the carboxyl (—COOH) and NH_2—linkages of their constituent amino acids. The behavior of a protein as an acid or as a base will depend upon the reaction of the medium in which it is dispersed, and there is one reaction of this medium at which the protein will behave neither as an acid nor as a base. This reaction is termed the "isoelectric point" of the protein. It is the pH value of the medium in which the protein is dispersed, at which the protein behaves as neither a cation nor an anion, or where the sum of the number of protein anions is equal to the number of protein cations present (Robbins, 1923). Since proteins as well as certain other colloids bear an electric charge when dispersed in water, the isoelectric point may be defined as that pH value of the dispersal medium at which the dispersed protein will behave as if uncharged (Pearsall and Priestley, 1923). The isoelectric point may be stated in yet another manner by saying that, as amphoteric electrolytes, proteins combine with either acids or bases, but at a particular hydrogen-ion concentration exist most nearly uncombined. Whether proteins are ionized at all at their isoelectric point or whether they are ionized equally as an acid or a base is a disputed point. The value of this singular point in ionization and its behavior is characteristic of each protein and is its isoelectric point (Cohn, Gross, and Johnson, 1919). If a pure protein is dispersed in a medium which contains electrolytes and which is more acid than its isoelectric point, the protein bears a positive charge, behaves as a base, combines with anions only, and thus forms a protein-acid salt. If the medium is more alkaline than the isoelectric point, the protein is negatively charged, behaves as an acid radical, combines with cations only, and forms salts of the base-protein type. The isoelectric points of plant proteins have been determined by Cohn, Gross, and Johnson (1919), Pearsall and Priestley (1923), Robbins (1923), Pearsall and Ewing (1924), and Csonka, Murphy, and Jones (1926). The following values are taken from Pearsall and Ewing (1924) and Csonka, Murphy, and Jones (1926):

Protein	Isoelectric point, pH	Protein	Isoelectric point, pH
Vicilin	3.4	Leucosin	4.5
Legumin	4.4 to 4.6	Tuberin	4.4
Legumelin	4.2	Edestin	5.6
Glutenin	4.4 to 4.5	Globulins	
Gliadin	3.5 to 5.5	Carrot	4.1
Globulins		Tomato seed	4.9
Cottonseed	5.4	Wheat bran	5.5
Navy bean	4.5	Pea	5.4
Squash	5.4	Georgia velvet bean	5.4
Flax seed	5.4	Locust bark	5.5

Since the cell sap of the parenchymatous tissues in plants is usually only slightly acid (pH 5.5 to 6.5), Pearsall and Ewing pointed out that the plant proteins are normally on the alkaline side of their isoelectric points and hence would behave as anions.

The position of the isoelectric point is one of the most significant properties of the proteins. At the isoelectric point the proteins are least soluble and exhibit a minimum tendency to swell in water. Thus at its isoelectric point a protein will most readily part with its water, a fact which is of great significance in the synthesis of substances in plants. In solutions of greater acidity or alkalinity than the isoelectric point, the swelling of proteins increases, as do their viscosity, osmotic pressure, and conductivity. Little or nothing is as yet known concerning the significance of the amphoteric nature of the proteins in the life activities of the plant.

C. Classification of Plant Proteins

The classification of the proteins is based on the little knowledge possessed concerning their chemical nature as derived from their products of hydrolysis and on their solubilities, precipitation by salts, and coagulation by heat. It has been made primarily upon the characteristics of those common to the animal kingdom, since they were the first examined and have been the most thoroughly investigated. Although many of the plant proteins show certain reactions not exactly characteristic of animal proteins, for the most part no separate classification has been made and they have been placed in those general groups to which they most nearly correspond in behavior. The following outline shows the groups of plant proteins that have so far been studied.

Classification of the proteins that have been found in plants

- The simple proteins
  - The albumins
    - Leucosin in seeds of wheat, rye, and barley
    - Legumelin in seeds of vetch, soybean, lentil, cowpea, and horse bean
    - Phaselin, in seeds of kidney bean
    - Ricin in seeds of castor bean
  - The globulins
    - Legumin in seeds of pea, bean, vetch, and lentil
    - Vignin in seeds of cowpea
    - Glycinin in seeds of soybean
    - Phaseolin (crystalline) in the seed of kidney bean (*Phaseolus vulgaris*)
    - Excelsin (crystalline) in the seed of Brazil nut
    - Edestin (crystalline) in the seed of hemp
    - Avenalin (crystalline) in the seed of oats
    - Cucurbitin (crystalline) in the seed of squash
    - Vicilin in seed of pea, lentil, horse bean
    - Arachin in seed of peanut
    - Corylin in seed of hazelnut
    - Amandin in seed of peach and almond
    - Juglansin in seed of walnut and butternut
    - Maysin in seed of corn
    - Tuberin in the potato tuber
    - Also globulins in various types from coconut and the seeds of flax, castor bean, cotton, sunflower, radish, rape, mustard, buckwheat, tomato, lima bean, and others
  - The glutelins
    - Glutenin in seed of wheat
    - Oryzenin in seed of rice
    - Maize glutelin in seed of corn
  - The prolamins
    - Gliadin in seed of wheat
    - Hordein in seed of rye
    - Zein in seed of corn
    - Kafirin in seed of sorghum (*Andropogon sorghum*)
- The conjugated proteins have not been isolated as yet in plants but considered to occur (see discussion in text)
- The derived proteins occur in plants in relatively small amounts as temporary intermediate products in the digestion or in the synthesis of proteins

1. Simple Proteins.—The simple proteins are those which yield only amino acids when they are hydrolyzed. They are practically the only plant proteins that have thus far been isolated and studied. There are four main groups of the simple plant proteins: (*a*) the albumins, (*b*) the globulins, (*c*) the glutelins, and (*d*) the prolamins.

a. Albumins.—The albumins are those proteins which are soluble in pure water and are coagulated by heat. The albumins of animal origin are not precipitated by saturating their neutral solutions with sodium chloride or with magnesium sulphate. The plant proteins that are classified as albumins, however, are often precipitated by saturation with one or the other of these salts but are included in the albumin group on the basis of their solubility in water at the neutral or slightly acid point and their coagulability by heat. The albumins occur in plants in small quantities, and on that account their nature has not been very thoroughly determined. According to Osborne (1924), most seeds and probably the cell sap in most cases contain proteins that belong in the albumin group. The plant albumins that have been isolated and studied are leucosin from the seeds of wheat, rye, and barley; legumelin from the seeds of soybeans, vetch, horse bean, lentil, and cowpea; phaselin from the seeds of kidney bean; and ricin from the castor bean.

b. Globulins.—A typical globulin is a protein that is insoluble in water but soluble in saline solutions (5 to 10 per cent) and is coagulated in solution by heat. The globulins of animal origin are precipitated from solution by saturation with magnesium sulphate or by half saturation with ammonium sulphate. The plant proteins do not always conform to these conditions of solubility. According to Osborne (1924), the plant globulins are precipitated by partial saturation with ammonium sulphate at very different degrees of saturation. Some are precipitated with half saturation, while others are not precipitated until their solutions are nearly saturated with this salt. Most of the seed globulins are only imperfectly coagulated by heating their solutions even to boiling, and some can be heated in this manner for a considerable time without showing any apparent change. Some of the plant globulins including edestin from hemp seed, excelsin from the Brazil nut, and globulins from the squash seed, flaxseed, oat kernel, and castor bean can be obtained in the crystalline form. Others are obtained in the form of spheroids that are sometimes mixed with crystals. The globulins include most of the proteins that thus far have been isolated and studied in plants. The names and sources of some of the most common globulins that have been studied are listed in the classification diagram.

c. Glutelins.—The glutelins represent a small group of proteins that are exclusively of plant origin. They are proteins that are not dissolved by neutral aqueous solution, by salt solutions, or by alcohol but are dissolved in dilute acid or alkali. The most characteristic protein of this type is the glutenin of wheat. Oryzenin and maize glutelin are two other glutelins that have been isolated from rice and corn, respectively. It is considered by Osborne that other glutelins may be present in the seeds of other cereals, but owing to the difficulty in extracting them no definite products have been obtained.

d. Prolamins.—The prolamins are proteins that are exclusively of plant origin and are known to occur only in the seeds of cereals. The proteins of this group are freely soluble in ethyl alcohol of 70 to 80 per cent, some dissolving in alcohol of 90 to 92 per cent, and the solutions are unaltered by boiling. They are insoluble in water or salt solutions but soluble in dilute acids or alkalies. On hydrolysis they yield a relatively large amount of proline and amide nitrogen and received their name on that account. The principal prolamins thus far isolated and studied are gliadin from the seeds of wheat and rye, hordein in the seed of barley, zein in the seed of corn, and kafirin in the seeds of sorghum (*Andropogon sorghum*). The glutelins and prolamins are collectively termed "glutens."

2. Conjugated Proteins.—The conjugated proteins are compounds of proteins with some other nonprotein group. They are nucleoproteins, glycoproteins, phosphoproteins, hemoglobins, and lecithoproteins. They have not been as yet definitely proved to occur in plants. According to Osborne (1924), some of the proteins in spinach and alfalfa leaves show behaviors in extraction which indicate that they may be a type of conjugated protein. It is frequently stated that nucleoproteins occur in abundance in plants, especially in the nuclei. The compounds, however, that occur in the nuclei of plant cells and in the embryos of many seeds are not true nucleoproteins but are salts or esters of protein and nucleic acid. A true nucleoprotein is one which, when subjected to peptic digestion or treated with dilute acid, gives a protein and a nuclein. This nuclein after treatment with caustic alkali breaks down into a second protein and nucleic acid (Haas and Hill, 1913). No nucleoproteins of this type have been isolated from plants.

3. Derived Proteins.—The term "derived proteins" is given to the decomposition or hydrolytic products of the simple proteins. Three of the most mentioned derived proteins are the proteoses, the peptones, and the peptides. The proteoses are the protein products of hydrolysis that are soluble in water, not coagulated by heat, but precipitated by the saturation of solution with ammonium sulphate. The peptones are products of further hydrolysis that are soluble in water, not coagulated by heat, not precipitated by ammonium sulphate, but give the biuret reaction. The peptides are the simplest products of the hydrolysis of the proteins and include the polypeptides and the individual amino acids. These compounds may or may not give the biuret reaction, depending upon the degree of hydrolysis, since amino acids do not give this test. The above-mentioned derived proteins have been observed in plants yet are not considered as permanent compounds but more as temporary intermediate products of the hydrolysis of proteins or as compounds from which proteins are being synthesized. Extensive work by Jodidi and Markley (1923), Jodidi (1924, 1925), and Jodidi and Wangler (1925) has shown that amino acids and polypeptides are present in considerable amounts in the ungerminated seeds of wheat, oats, corn, and rye. Thus in the case of Kanred, Fultz, Marquis, and Kubanka wheats the polypeptide nitrogen amounted, respectively, to 3.89, 4.67, 4.98, and 5.13 per cent of the total nitrogen of the seed. The ungerminated kernels of these varieties contained free amino acids, the nitrogen contained in these acids amounting to 1.4, 1.8, 1.8, and 2.3 per cent, respectively, of the total nitrogen content of the kernels of Kubanka, Fultz, Marquis, and Kanred.

In considering the kinds of proteins in plants it should be remembered that the investigations in this regard have been relatively few and that they have been confined, for the most part, to the reserve proteins of seeds and other storage organs. The proteins of the physiologically active tissue of the plant as the leaves and growing regions have been but little studied. It is highly probable that when such studies are made, the present ideas concerning the occurrence of conjugated proteins and derived proteins in plants will be greatly changed.

D. Synthesis of Proteins by the Green Plant

The general opinion is that proteins are formed by the linkage of amino acids. The evidence in this regard is that polypeptids have been formed artificially in the laboratory by linking together amino acids and that proteins yield amino acids upon hydrolysis. It has also been observed that in the maturing of seeds the proteins are formed at the expense of the amino acids. The origin of the amino acids, however, is not definitely known. It is known fairly definitely that proteins are not formed in the plant unless a supply of carbohydrates is present, but the manner in which the carbohydrates are changed to amino acids is not known. The formation of amino acids from the fatty acids is apparently a rather simple reaction, but the transformation of carbohydrates into fatty acids is accomplished, artificially at least, only with considerable difficulty. It was observed by Chibnall (1922, 1924) in the leaves of the bean that the nitric n trogen and monoamino nitrogen varied directly with the protein nitrogen, which indicates that they may be connected with protein synthesis. Baly and Heilbron (1922, 1923) suggested as a theory of the formation of proteins that "activated" formaldehyde H—C—O—H, which they consider is photosynthetically formed in the leaves from carbon dioxide and water, reacts readily with potassium nitrite to give formhydroxamic acid, and the formhydroxamic acid at once reacts with more molecules of activated formaldehyde to produce a great variety of the complex substances that are found in the living plant. This reaction, they believe, takes precedence over the polymerization of the activated formaldehyde to reducing sugars, which they consider is produced only when the activated formaldehyde is produced at a rate greater than that at which it can react with the nitrite and the formhydroxamic acid, , thus formed.

1. Reduction of Nitrates in the Plant.—Since the nitrogen in the protein molecule is generally in the form of NH_2 groups, the nitrogen that has been absorbed by the plant in the form of nitrates must be reduced in order to convert the grouping NO_3 into the necessary NH_2 form. The first step in this reduction is the conversion of the nitrates to nitrites. The transition from nitrate to nitrite is a strongly endothermic reaction and can occur only by the transformation of other forms of energy such as that of light into chemical energy or by a linked chemical reaction with oxidation of previously formed reduced chemical substances, *e.g.*, the oxidation of carbohydrates. It is the intention here to mention the occurrence of nitrates in the plant and to discuss the conditions under which they are reduced both outside and inside the plant.

a. Occurrence.—As has previously been mentioned, a considerable portion of the nitrogen absorbed by plants is in the form of nitrates. The occurrence of nitrates in the plant, however, has for the most part been observed only at random, and the quantitative data in this regard are limited. Woo (1919) observed with *Amaranthus retroflexus* that the rate of nitrate absorption increased with the age of the plants until they were 20 in. high and near the blooming stage. At this stage the nitrate nitrogen amounted to 1.01, 1.94, 2.82, and 0.06 per cent of the dry weight of the roots, stem, branches, and leaves, respectively, as compared with 0.42, 0.22, and 0.3 per cent in the roots, stem, and leaves when the plants were only 1 to 4 in. high. Campbell (1924) determined the nitrate content of *A. retroflexus* and 25 other weeds just before the blooming stage and at maturity. He found that, although the nitrate nitrogen was prominent in some of these plants during the young and immature stages, it had completely disappeared at full maturity. In all species grown under normal conditions the highest percentage of nitrate was found in that stage of development designated as "just before blooming." Anderson (1924) found nitrates in the expressed sap of the shoots of a large number of plants including oats, lupine, wheat, and grasses. He observed that the amount of nitrate found in the plant would depend upon the time of day the examination was made and upon the soil in which the plants were grown. *Solanum dulcamara* showed considerably less nitrate in the morning than later in the day, and *Mercurialis perennis* gave positive nitrate reactions in October and negative in June. The amount of nitrate in the etiolated leaves of wheat, oats, and cabbage was higher than that found in the normal green leaves. Doneen (1934) mentioned that forest trees show different periods for the maximum absorption of nitrates ranging from early seasonal absorption, as in the pine, to late absorption in the autumn, as in the horse-chestnut. The most rapid absorption of nitrates does not always occur at the time of maximal terminal growth of the top. The French prune trees showed a primary peak of nitrate absorption near the end of June with a final secondary peak late in October followed by a decline to winter dormancy.

Woo (1919) observed in the root, stem, and leaves of *Amaranthus retroflexus* that there was a considerable accumulation of nitrates, especially in the stem and branches. In the soybean, Webster (1928) observed that nitrates, as a rule, are much higher in the stem and roots than in the leaves, while the various parts of the plant are lowest in nitrates at about the time of seed maturity. It was observed by Shmuk and Poliakov (1927) in *Sorghum saccharatum* that the nitrogen in the plant reached a maximum at flowering and thereafter rapidly decreased. It was observed by Frear (1930) in the beet that the sap of the midrib contained more nitrate than any other portion of the leaf. Dittrich (1930) noticed that some plants such as *Chenopodiaceae*, *Solanaceae*, and *Utricaceae* are high in nitrates, while others such as *Boraginaceae* and *Gramineae* are low in these salts. In the former group the stem contains more nitrates than the roots and leaves, while in the latter the nitrate content decreases from the roots to the top of the plant. Chapman (1933) found in the tulip tree that the nitrates were more abundant in the leaf, and decreased in order in the roots, bark, and wood.

Campbell (1924) reported that *A. retroflexus* and *Atriplex patula* grown upon manure piles and heaps of decaying hog hair at early maturity showed more than a trace of nitrate that at full maturity was still present. Webster (1928) observed that the addition of nitrates to the soil resulted in a temporary increase in the nitrate storage in all organs of the plant, but this was soon reduced to normal. The percentage of ammonia was very small, but there seem to be reasons for considering that it is one of the intermediate steps in both the synthesis and decomposition of proteins.

According to Strowd (1920) the nitrates in the cell sap of soybean plants increase to some extent with the increase of the nitrates in the soil, but this increase is not proportional to the increased nitrate supply. Thomas (1927) found that the application of sodium nitrate as a fertilizer to apple trees markedly increased the total nitrogen content of the leaves and growing twigs. Cook (1930) for small grains; McCool and Cook (1930) for barley, oats, and bluegrass; Williams (1932) for peach shoots; and Doneen (1934) for wheat found that the application of nitrates to the soil increased temporarily at least the concentration of nitrates in the cell sap. The rate of transformation of the nitrate nitrogen in the plant is very rapid as indicated by a decrease therein within 3 hr. after the source of nitrogen was removed. On the other hand Shmuk and Poliakov (1927) for *Sorghum saccharatum*, Chapman (1933) for the tulip tree, and Varadachar (1933) for the sunflower have reported that a variation of the concentration of nitrates in the soil had no effect on their concentration in the cell sap.

Nitrites have been found in small amounts only if at all in the plant tissues that have been examined (Aso and Sekine, 1914). They were detected by Webster (1928) in the leaves of the soybean in the earlier stages, but their presence in other parts was doubtful. Boncquet (1916) could detect no nitrites in any plant tissue that was normal. In the mosaic tissue of tobacco, however, considerable nitrites were found.

b. Conditions under Which Nitrates Are Reduced.—It was observed by Laurent (1890) that, when a sterilized solution of potassium nitrate was exposed to sunlight, oxygen and potassium nitrite were formed.

In investigations concerning the reduction of nitrates, Baudisch (1921, 1923) found that glucose solutions do not reduce nitrates or nitrites even when heated under pressure, but the smallest trace of iron salt is sufficient to reduce a large amount of nitrite on warming with glucose in weakly alkaline solution. Under these conditions, nitrates remain entirely unattacked. He found, however, that oxygen in the presence of ferrous salts reduces nitrates instantaneously even in the cold with the formation of nitrites and that there is a direct relation between the amount of oxygen dissolved in the water and the amount of nitrite formed. The reduction of an alkaline solution of nitrate to nitrite can be brought about in sunlight in the summer if the oxygen that is split off is taken care of chemically, a process that may be brought about by the addition of complex ferrous salts such as potassium ferrocyanide. Baudisch considered that the nitrite thus formed from nitrates by light energy or with ferrous hydroxide and oxygen is, in the presence of these same factors reduced to nitrosyl (R—NO), which will combine readily with aldehydes and produce formhydroxamic acid,

$$\begin{array}{c} \text{N—OH} \\ \| \\ \text{H—C—OH} \end{array}.$$

The conditions under which nitrates are reduced in the green plant are not very well defined. Apparently the process goes on in both light and darkness and in any part of the plant, but there is some little evidence that certain conditions are more favorable than others for the reaction. Thus Schimper (1888) found that nitrates were depleted in the green leaves exposed to daylight and were not destroyed in the leaves kept in the dark. He also observed that shade leaves were richer in nitrates than leaves in the sun, and that the disappearance of nitrates in the leaves in the light occurred only in green leaves and not in etiolated ones. Moore (1921) also considered that the reduction of nitrates to nitrites occurs in the leaves in the presence of sunlight. Warburg and Negelein (1920) transferred *Chlorella pyrenoidea* from Knop's solution to 0.1 N sodium nitrate and observed that reduction of nitrates took place in the dark but was increased by illumination. The rate of reduction, however, was independent of carbon assimilation. Dittrich (1930), Eisenmenger (1933), and Sommer (1936) could obtain little or no reduction of nitrates in the dark. Dittrich (1930) and Lease and Tottingham (1935) considered that the reduction of nitrates to nitrites is closely associated with ultraviolet radiation. Gautreret (1934) reported that red light is more effective in reducing nitrates in plants than any of the other visible rays of the same intensity. Tottingham and Lease (1934) reported that the blue to violet light increased the percentage of protein in the young wheat plant and caused a limited replacement of nitrate by nitrite.

Acqua (1899) and Thomas (1927) obtained evidence that some of the nitrates absorbed by the young roots are utilized directly in or near the root tips. It has been reported by Nightingale and Robbins (1928) for *Narcissus*, Davidson and Shive (1934) for peach trees, and Nightingale (1936) for apple and peach trees that the reduction of nitrates and the initial synthesis of amino acids and amides occurred for the most part in the roots. In the tomato, reduction of nitrates occurred in both the stem and the roots (Nightingale, 1936). Dittrich (1930) noted in numerous plants that the sap of the leaves and underground storage organs had a greater reducing activity than the sap of the stems. In tobacco, Eisenmenger (1933) believed that the midrib of the leaf is the region in which considerable reduction of nitrates occurs. Eckerson (1924) found further that the reduction of nitrates to nitrites and ammonia took place in tomato-plant extract of slightly alkaline reaction, in the presence of fructose or glucose and some free oxygen, in darkness as well as in light and that the reduction was most active in solutions having a pH value of approximately 7.6. Loehwing (1927) found in the tomato that an alkaline reaction and the presence of sugars are indispensable for nitrate reduction and for the formation of amino acids, while Dittrich (1930) found that the optimum pH value reported by Eckerson held in nitrate reduction over a wide range of plants. The most complete information that we have concerning the transformation of the nitrates after they enter the plant is afforded by the careful work of Eckerson (1924). This investigator grew tomato plants of the Bonny Best variety in rich soil until they were 8 in. high and then transferred them to quartz sand which was watered with a dilute nutrient solution lacking nitrogen. When the plants thus grown were ready for experimentation, they contained an abundance of glucose, some fructose, and a little sucrose and an abundance of starch in parenchyma cells of the stem and leaves. The tissues gave no reaction for nitrates, nitrites, or ammonia, and no amino acids could be detected. Calcium nitrate was then applied to the sand, and the subsequent chemical changes followed by microchemical means. Within 24 hr. after the nitrate was applied, nitrates were present in all parts of the plants. The tops of a few plants gave a slight reaction for nitrites but no reaction for ammonia. At the end of 36 hr. all the plants had considerable nitrite localized in the cortical cells at the tips of the stems, in the cortical cells near the phloem, and in the phloem parenchyma at the nodes of the stem. In all regions there was a trace of ammonia. After 48 hr. there was slightly less nitrite but more ammonia, and a slight decrease in the amount of starch could be detected in the cortical cells at the tips of the plant and in the youngest leaves. Within 3 to 5 days there was very much less nitrite, a little ammonia, but a great increase in amino acids. These amino acids were aspartic acid, alanine, leucine, cystine, and histidine and appeared at the nodes and in petioles and blades of young leaves and just behind the stem tips. Succinic acid, malic acid, and asparagin were also present.

c. Nitrate-reducing Substances.—Considerable experimental work has been done to determine the presence of nitrate-reducing enzymes in the plant, but the results are somewhat conflicting, especially in regard to the behavior of the extract to heat. Pozzi-Escot (1903) found that the aqueous extracts of the stem of the burdock have the power to reduce nitrates to nitrites and ammonia and concluded that the enzymes which reduce nitrates are very abundant in actively growing plants. Kastle and Elvove (1904) also found that the potato sprouts and eggplant contain a reducing substance or substances capable of effecting the reduction of nitrates to nitrites. It was found that the reduction takes place most rapidly at 40 to 50°C. and that it is augmented by an increase in the amount of nitrate present and also by an increase in the quantity of the reducing extract. The extracts capable of effecting this reduction lose this property when heated to boiling.

Irving and Haukinson (1908) claimed to have found a nitrate-reducing enzyme in *Elodea, Potamogeton, Iris, Vicia,* and numerous members of the grain family. They considered that the only conditions necessary for nitrate reduction seem to be the presence of the enzyme found in the roots, stem, and leaves and a suitable carbohydrate. Anderson (1924) detected nitrites in the shoots of 25 of the 105 species examined, and in the leafy parts of 23 of these, including sunflower, artichoke, pea, radish, tomato, clover, and wheat, the presence of a nitrate-reducing substance was confirmed. This nitrate-reducing substance was found to be thermolabile and an oxidizable substance rather than an enzyme. Eckerson (1924) reported that tomato-plant juice boiled and brought to a pH value of 7.6 reduced nitrates as rapidly as the unheated juice.

Loehwing (1927) considered that the reduction of nitrates is not caused by enzymatic action because the boiled, expressed sap reduces as effectively as the unboiled. Eckerson (1930, 1931) studied the weekly variations of the nitrate reductase in the various parts of the apple tree during the year. The ability of the tree to reduce nitrate was highest in early spring when the buds were swelling, and at zero during the 5 weeks following flowering. It was low during the summer but increased in the early autumn. It was high from late autumn to late winter but dropped extremely low during February, after which there was a rapid rise to the maximum in early spring. The high reductase content during the autumn and winter was localized in the fine roots. The maximal amount of reductase in early spring was in both the fine roots and buds. There was very little reductase in the leaves at any time.

In 1932 Eckerson found that the quantity of reductase in soybeans amounted to only 31 to 36 per cent of that in the unshaded plants when grown under conditions of only 34 per cent of total solar energy. When grown under an 8-hr. day and a 16-hr. night, the amount of reductase was only 1 to 5 per cent of that of the plants in full daylight.

Both potassium and phosphorus appear to be necessary for the synthesis of reductase, because when either is deficient there is little or no reductase formed. A deficiency of calcium and sulphate also decreases the production of reductase.

2. Protein Metabolism of the Green Leaf.—The observation of protein changes in the leaf is a difficult task. In the first place, many metabolic processes are continually taking place in the leaf, so that numerous nitrogenous substances are present at the same time. It is impossible to determine which of these substances are synthetic and which are deterioration products. In the second place, the methods for the extraction of proteins and protein products, especially from parenchymatous tissue, are not well developed. Then, too, the amount of protein in leaves is relatively small, and it is difficult to differentiate the extracted proteins into those which are uncombined in the cell and those which are constituent parts of the protoplasm.

a. Extraction of Proteins.—Several methods for the general extraction of proteins from leaves have lately been published. Thus Chibnall and Schryver (1920, 1921) treated the macerated leaves of spinach, cabbage, and scarlet runner bean with water saturated with ether and obtained, in such an extract, two-thirds of the total nitrogen of the leaves. Osborne and Wakeman (1920) by grinding the fresh leaves of spinach and centrifuging at high speed obtained solutions containing a certain portion of their protein. These same workers and Leavenworth (1921) found that by grinding and pressing fresh young alfalfa plants it was possible to obtain large quantities of extract suitable for protein determinations. In the case of the leaves of the sugar beet, Tottingham, Schulz, and Lepkovsky (1924) obtained results that indicated the desirability of extracting the macerated tissue with water as a general procedure in separating the nitrogenous constituents of plant cells. Lincoln and Mulay (1929) gave methods of extracting the nitrogenous material from the leaves, bark, and twigs of the pear. Chibnall, Miller, Hall, and Westall (1929) by the use of ether water obtained good yields of protein from the leaves of grass and forage crops. The above statements show that the methods for the extraction of proteins and nitrogenous products from leaves vary greatly, and the indications are that for the best results special methods of extraction must be applied to different types of leaves. Thus no one method is equally applicable in any considerable number of cases.

1. *Previous History of Tissue and Nature of Nitrogenous Extract.*—Much of the study that has been made on the nitrogenous changes in the leaves is open to criticism because of the methods used in the preparation of the material for analysis. For example, many of the results that have been obtained are based on the analysis of materials that were air dried or desiccated at low temperatures. The errors that are likely to occur from such a procedure have been stated by Chibnall (1922), who showed that proteolytic enzymes are present in the leaves of runner bean that were dried at 40 to 50°C., for 100 to 120 hr. He also observed evidences that indicated the presence of an asparaginase. A comparison of the distribution of the total nitrogen in the aqueous extract of fresh leaves and of leaves that have been dried at 40°C. for 120 hr. will indicate the changes that go on in the nitrogenous compounds of the leaves during the drying process.

COMPARISON OF THE DISTRIBUTION OF THE NITROGEN IN THE AQUEOUS EXTRACT FROM THE FRESH AND DRIED LEAVES OF THE RUNNER BEAN, PERCENTAGE OF TOTAL NITROGEN

Material	Total water-soluble N	Protein-free and proteose-free water-soluble N	Ammonia	Amide N of asparagine	Nitric N	Humin N	Basic N	Amino N	Other N
Fresh ...	21.57	17.22	0.44	0.53	2.73	0.99	2.00	7.76	2.77
Dried, 40°C..	33.68	29.55	5.32	1.51	1.62	0.82	2.84	13.22	4.22

Chibnall's work shows that, when leaves are dried at a low temperature, protein autolysis takes place, which brings about an increase in the simple water-soluble nitrogenous products consisting, for the most part, of ammonium salts, asparagin, and amino acids. There is a diminution in the amount of leaf proteins during the drying, but they are not appreciably changed in character.

The effect of the method of preparation of material upon the nitrogenous products obtained in extraction was shown by Tottingham, Schulz, and Lepkovsky (1924) in the case of the leaves of the sugar beet as follows:

Substance	Treatment of tissue		
	Extracted fresh, per cent	Extracted after drying at 40°C. without aeration, per cent	Extracted after freezing by ice salt mixture, per cent
Total soluble nitrogen......	4.0	2.2	3.3
Soluble protein nitrogen....	3.4	0.5	2.3

This has been especially studied in leaves affected with the mosaic disease. In this regard Jodidi (1918 to 1920) reported on spinach and cabbage, Dunlap (1930) on numerous plants, and Cordingley, Grainger, Pearsall and Wright (1934) on tobacco. Coons and Klotz (1925) noted that the nitrogenous constituents of celery plants affected by *Cercospora apii* and *Septoria apii* were different from those of the healthy plants.

2. *Separation of the Proteins of the Cell.*—The proteins in the cells of the leaf and of other parenchymatous tissue are those in solution in the vacuole, those held uncombined by the protoplasmic structure, and those combined in the structural material of protoplasm. By the ordinary methods of protein extraction, the proteins from these three sources are obtained in a general mixture. It is evident that before any knowledge can be gained in regard to the nitrogen metabolism of the leaf, methods must be devised to separate the uncombined proteins of the cell from the protoplasm. Chibnall (1923; see also Thomas, 1927) devised a method for extracting protein from leaves which he considered separates the protein of the vacuole of the cell from the protein of the protoplasm. This method consists of plasmolyzing the cells by means of certain organic agents, as ether or butyl alcohol, and then pressing out the major part of the vacuole content in a press. The remainder of the vacuole content can be washed out by repeatedly allowing the pressed residues to imbibe very dilute acid (0.002*N*) and subsequent pressing. This operation, he considered, does not rupture the leaf cells, and the protoplasm from which some of the water-soluble constituents may have been washed out is retained in them. After the removal of the vacuole contents the protoplasmic material remaining in the residue can be extracted by grinding with water. This material passes into colloidal solution and can be flocculated by the addition of acid. In the case of spinach leaves, the protein found in the vacuole amounted to from 1 to 2 per cent of the total protein of the leaves and had properties distinct from those proteins of the protoplasm.

Chibnall and Grover (1926) undertook to obtain the proteins of the protoplasm of the cells of the leaf as follows: The leaves were treated with ether, enclosed in filter cloth, and placed under pressure in a hand press, so as to remove the soluble proteins of the vacuole. The leaf residues were then washed free from the remainder of the vacuole material by alternate soaking in water and pressing. The protoplasm from which some water-soluble substances may have been removed was obtained in colloidal solution by grinding with water in a meat chopper, and the ground material filtered or strained through fine gauze. The colloidal solution thus contains proteins from the protoplasm of the leaf cells. The proteins thus obtained could be separated into (*a*) "combined" proteins, which are in loose combination with substances soluble in alcohol and (*b*) "soluble" proteins, which are uncombined and pass fairly readily into solution when the protoplasm is ground with water. These investigators examined these soluble proteins of the protoplasm of the leaves of *Spinacia oleracea, Vicia faba, Phaseolus multiflorus, Medicago sativa, Brassica oleracea,* and *Zea mays* and numerous others. It was found that these proteins are glutelins with very similar chemical properties. They have an isoelectric point from pH 4.0 to 5.0 at which their solubility is at a minimum. The hydrogen-ion concentration of the leaf cell sap was found in all cases to be alkaline with respect to the isoelectric point of the proteins so that they were probably present in the living cells as anions. Chibnall and Grover were, however, unable to obtain any "soluble" protein from the leaves of *Populus tremula, Wistaria chinensis, Rheum rhaponticum, Polygonum cuspidatum, Rumex acetosella,* and *Vitis vinifera* among others. In this regard it should be noted that Thomas (1927) found, in the case of apple leaves which had been collected at early bud formation, at the stage of active maximum growth, and at the period of chlorophyll degeneration, that the insoluble cytoplasmic proteins showed little or no variation in their nitrogen composition throughout the season.

b. Diurnal Variation of the Nitrogen in the Leaves.—In a consideration of the nitrogen metabolism of the leaves, a knowledge of the variation of their nitrogen content as to both quantity and quality during a 24-hr. period is of primary importance, since the changes in these two factors are indicators, at least, of the processes that are occurring.

1. *Methods of Determining.*—The determination of the changes of the nitrogen content of the leaves during a 24-hr. period is attended with considerable difficulty. This is due to the relatively small amount of nitrogen present, to the difficulty of separating the various nitrogenous compounds, and to the wide variation of the other more abundant constituents of the leaf during the period in question. Four methods, each of which has certain disadvantages, have been used to determine the variation of the nitrogen content of the leaves during the day and night: (*a*) Estimation of the nitrogen in percentage of the dry weight of the leaves. Most of the data on the diurnal variation of the nitrogen have been obtained by this method. This method is unreliable, since the total dry weight of the leaves changes markedly during the day, so that when the diurnal changes in the nitrogen content of the leaves are estimated in this manner, it is not the absolute amounts of nitrogen in the leaves that are being compared but only the concentrations of nitrogen in terms of total solids. (*b*) Estimation of the nitrogen by weight in terms of a certain number of leaves. This method is inaccurate, as shown by Chibnall (1924), for the reason that the error in sampling is high. Leaves differ considerably in texture, weight, and general composition, so that when the errors of the determination are taken into consideration, the changes observed in the nitrogen content cannot be considered reliable. These inaccuracies are partially but not entirely overcome by the twin-leaf method of Denny (1933). (*c*) Estimation of the nitrogen in percentage of the fresh weight of the leaves. Chibnall (1923) considered that this method gives the most reliable index to the daily fluctuations of the nitrogen in the leaves. It is evident, however, that this method

would be of value only under conditions that cause little or no fluctuation in the water content of the leaves. Under the climatic conditions under which Chibnall worked in England, the water content of the leaves remained almost constant. Such a constancy of the water content of the leaves is generally not the case, especially in regions of a relatively high evaporation. Miller (1917) observed in Kansas that the minimum amount of water in the leaves of corn and sorghums during the day may be from 88 to 97 per cent of the maximum amount. He also observed that the amount of water in the leaves of pumpkin, soybeans, and cowpeas at 1 P.M. may be only 66, 76, and 86 per cent, respectively, of the water content of the leaves at 8 A.M. It is evident, therefore, that the determination of the percentage of nitrogen on the fresh weight of leaves would in such cases not give a very accurate index to the actual changes of the nitrogen content. (*d*) Estimation of the nitrogen by weight in terms of a definite leaf area. This method has been used by Miller (1917, 1924) in the determination of the diurnal changes of carbohydrates, dry matter, and water in leaves. Its procedure is as follows: The nitrogen is determined in percentage of the dry weight of the leaves and this value is multiplied by the dry weight of a certain number of leaf areas that have been obtained by means of a calibrated punch (see Chap. VIII, Fig. 28). The two main objections to this method are the shrinkage and expansion of the leaves due to changes in the moisture content and to the errors in sampling due to lack of uniformity in the structure of the leaves. Experience has shown, however, that the errors from these two sources are comparatively small, so that it appears that this method is one of the most accurate of the four that have been mentioned.

2. *Total Nitrogen of the Leaves.*—The data of Suzuki (1897) showed that the total nitrogen content of the leaves of *Wistaria brachybotrys*, *Phaseolus mungo*, *P. vulgaris*, *Solanum tuberosum*, and *Polygonum fagopyrum* decreased during the night 16.6, 9.9, 10.9, 5.1, and 7.5 per cent, respectively, of the amount found in the leaves at the close of the day. The changes in the total nitrogen of the leaves of *Acer negundo* was determined by Schulze and Schütz (1909). The data showed that at all times of the growing season the absolute amount of nitrogen in the leaves was higher in the evening than in the morning, as shown in the following table:

VARIATION IN THE TOTAL NITROGEN OF THE LEAVES OF *Acer negundo*
After Schulze and Schütz

Date	Per cent of dry substance		Absolute amount in grams, 200 leaves	
	Morning	Evening	Morning	Evening
May 7	4.30	4.50	0.734	0.817
June 6	3.90	3.98	0.973	1.231
July 5	4.06	4.14	1.211	1.306
Aug. 2	3.74	3.75	0.864	0.898
Sept. 3 to 6	3.16	3.18	0.791	0.971

The total nitrogen of the leaves as expressed in percentage of the dry weight was reported by Kosutany (1897) for the grape and by Pigorini

(1914) for the mulberry to be higher in the morning than in the evening. By the same method Frank and Otto (1891) in the case of red clover and alfalfa and Otto and Kooper (1910) in the case of *Aesculus hippocastanum*, *Philadelphus coronarius*, *Syringa vulgaris*, and *Sambucus nigra* found that the total nitrogen was higher in the evening than in the morning. Chibnall (1924) in his investigations of the nitrogen metabolism of the leaves in the scarlet runner bean noted that the fall in the total nitrogen of the leaves expressed in percentage of fresh weight at night amounted to 2.5 per cent with a probable error of only ± 0.2 per cent.

Miller (1926) studied the daily variation of the total nitrogen in the leaves of various crop plants. The nitrogen was determined in percentage of the dry weight, and the results were reduced to the amount of nitrogen per unit of leaf area. A small portion of the data obtained in 1925 for the leaves of soybeans, cowpeas, and garden beans is shown in the table on page 682.

In 18 observations for soybeans in 1925, the total nitrogen content of the leaves expressed as grams per square meter of leaf at 8 P.M. showed fifteen times an increase and four times a decrease over the amount present at 8 A.M. The average nitrogen content of the leaves for the 18 cases at 8 P.M. showed an increase of 5.9 ± 0.9 per cent of the nitrogen content at 8 A.M. For the same year in 17 observations, the nitrogen content of the leaves of cowpeas, expressed in the same manner as above, showed at 8 P.M. an increase in 13 cases and a decrease in 4 cases over the amount of nitrogen present at 8 A.M. The average nitrogen content of the leaves of cowpeas at 8 P.M. for the 17 observations showed an increase of 6.7 ± 0.8 per cent over the amount of nitrogen present at 8 A.M. The leaves of garden beans in 16 observations showed eleven times an increase and five times a decrease of their nitrogen content at 8 P.M. as compared to the amount present at 8 A.M. The average nitrogen content at 8 P.M. showed an increase of only 2.9 ± 1.2 per cent over that present in the leaves at 8 A.M. The results in the case of garden beans do not show conclusively that the nitrogen content of the leaves increases during the day, at least under the conditions under which these observations were made. In 6 observations of the nitrogen changes in the leaves of Dwarf Milo in 4 different years, the amount of nitrogen in the evening showed an increase four times and a decrease two times, over the amount present in the morning. The results seem to indicate that in this sorghum there is little change in the total nitrogen content of the leaves during the day. The same statement can also be made for the nitrogen changes in the leaves of the corn. Denny (1933) by the use of the twin-leaf method could find no change in the amount of total nitrogen during the day and night in the leaves of beans and peanuts. The observed facts would seem to indicate that the total nitrogen variation in the leaves during the day may be marked in some plants and very little in others. The experimental work that has so far been done indicates that an actual increase in the total nitrogen content of the leaves occurs during the day, and that this increase disappears during the night.

3. *Nitrogenous Compounds of the Leaves.*—The information in this regard is meager and more or less contradictory. Suzuki (1897) reported

DAILY VARIATION OF THE TOTAL NITROGEN OF THE LEAVES OF SOYBEANS, COWPEAS, AND GARDEN BEANS AT MANHATTAN, KANS., 1925

Plant	Time	Water per square meter of leaf, grams	Dry matter per square meter of leaf, grams	Total nitrogen per square meter of leaf, grams	Percentage of total nitrogen	
					Dry basis	Wet basis
July 10–11:						
Soybeans......	8 a.m.	96.9	32.8	1.863	5.68	1.43
	8 p.m.	108.7	38.6	2.138	5.54	1.45
	8 a.m.	107.5	31.8	1.847	5.81	1.32
Cowpeas......	8 a.m.	215.0	50.7	2.808	5.54	1.05
	8 p.m.	230.5	55.7	3.002	5.39	1.04
	8 a.m.	228.9	50.1	2.790	5.57	1.00
Garden beans .	8 a.m.	163.9	38.3	1.758	4.59	0.86
	8 p.m.	162.0	43.6	1.979	4.54	0.96
	8 a.m.	167.5	38.7	1.880	4.86	0.91
July 13:						
Soybeans......	8 a.m.	107.2	32.4	1.918	5.92	1.37
	8 p.m.	108.8	39.0	2.215	5.68	1.49
Cowpeas......	8 a.m.	207.1	43.4	2.439	5.62	0.97
	8 p.m.	219.7	49.7	2.788	5.61	1.03
Garden beans .	8 a.m.	166.3	42.3	1.941	4.59	0.93
	8 p.m.	175.5	47.7	2.027	4.25	0.90
July 17–18:						
Soybeans......	8 a.m.	97.9	32.5	1.755	5.40	1.34
	8 p.m.	99.5	36.8	2.013	5.47	1.47
	8 a.m.	90.6	31.7	1.962	6.19	1.60
Cowpeas......	8 a.m.	209.4	51.5	2.837	5.51	1.08
	8 p.m.	213.5	55.2	3.097	5.61	1.15
	8 a.m.	224.5	49.3	2.746	5.57	1.00
Garden beans .	8 a.m.	172.9	52.2	2.176	4.17	0.96
	8 p.m.	174.2	53.0	2.178	4.11	0.95
	8 a.m.	186.3	50.4	2.192	4.35	0.92
July 24:						
Soybeans......	8 a.m.	110.7	38.8	2.312	5.96	1.54
	8 p.m.	111.2	39.0	2.215	5.68	1.47
Cowpeas......	8 a.m.	239.2	54.5	2.992	5.49	1.01
	8 p.m.	239.5	55.3	2.925	5.29	0.99
Garden beans .	8 a.m.	199.3	52.5	2.136	4.07	0.84
	8 p.m.	209.8	52.5	2.168	4.13	0.82

that in the leaves of *Phaseolus* the asparagine nitrogen decreased during the night to 57 per cent and the amino nitrogen to 91 per cent of what it was during the day. He considered that the reserve proteins in the leaves are decomposed into amino compounds during the night and transferred to other parts of the plant. Schulze and Schütz (1909) in the case of the leaves of *Acer negundo* could observe little change in the amino acid content during the day. They observed, however, a greater amount of protein nitrogen in the evening than in the morning, and the protein nitrogen varied with the total nitrogen.

In *Phaseolus vulgaris* var. *multiflorus*, Chibnall (1924) reported that on the basis of the fresh weight of the leaves there was a fall in the protein nitrogen during the night of 1.8 per cent with a probable error of only ±0.2 per cent. The nonprotein nitrogen showed a definite decrease of 9 per cent with a probable error of only ±0.6 per cent. Both the ammonia nitrogen and amide nitrogen of asparagine remain unchanged at 0.001 and 0.005 per cent of the fresh weight, respectively. The experimental work all strongly indicates that protein translocation takes place in the leaf cells at night either as unchanged protein or as its decomposition products. This loss of protein is from the cytoplasm and not from the small amounts present in the vacuole. Chibnall (1924), in order to obtain data regarding the process of protein degradation in the leaves, cut leaves with long petioles from *P. vulgaris* var. *multiflorus* and allowed them to remain with their petioles in water for 4 or 5 days. At the same time that the leaves were removed, others were analyzed directly after cutting. It was found that the leaves with their petioles in water showed, in terms of the total leaf nitrogen, a large increase in nonprotein nitrogen with a corresponding decrease in the protein nitrogen, and that this increase in nonprotein nitrogen is due, in large part, to increases in asparagine and free amino nitrogen. Thus the amide nitrogen in the freshly plucked leaves constituted only 0.64 per cent of the total leaf nitrogen, while the detached leaves with their petioles in water contained 3.78 per cent of amide nitrogen on the same basis. Likewise the free amino nitrogen amounted to only 6 per cent of the total leaf nitrogen in the case of the freshly plucked leaves but to 14.1 per cent in the detached leaves after 4 days. Chibnall believed that there is a continuous production of asparagine from the reserve proteins of the leaves, that its role in the metabolism of the mature plant is that of a translocatory substance, and that it is the chief medium whereby nitrogen in a form suitable for subsequent metabolic processes can be transferred from one part of the plant to another.

It was found by Kishi and Monobe (1931) in the leaves of the mulberry that the total nitrogen, the albuminoid nitrogen, and the nonalbuminoid nitrogen in the fresh leaves showed practically no change during a 24-hr. period. On the basis of the dry matter of the leaves, these were greatest at sunrise, least at noon and 3 P.M., and less at sunset than at sunrise.

c. Formation of Protein in the Leaves.—The formation of protein in the leaves is indicated by the fact that (1) the amount of nitrate in the leaves is relatively small as compared with that in the stem and roots, as reported for *Vicia faba* by Emmerling (1880, 1887, 1900); (2) incised leaves with their petioles in nutrient solutions form protein, while those with their petioles in distilled water or in solutions lacking nitrogen do not (Chrapowicki, 1889; Sapoznikow, 1895), and (3) leaves can increase their protein content in the dark, provided that the petioles are placed in a nutrient solution containing the essential salts and sugar (Zaleski, 1897 to 1907). It has also been observed by Chibnall (1924), as previously mentioned, that the protein nitrogen of the leaves decreases during the night, a fact which indicates that the protein increases during the day and is translocated from the leaves during the night.

Work has been done by Molisch (1916), Lakon (1916), Meyer (1918), and Ullrich (1924) to determine what portion of the green cell is concerned in the formation of protein. All these workers concluded that the main portion of the protein formed in the leaves is synthesized in the chloroplasts. The general procedure of these workers may be illustrated by the work of Lakon (1916), who studied the variegated leaves of *Acer negundo*, *Acer pseudoplatanus*, *Sambucus nigra*, *Vinca major*, *Tradescantia zebrina*, *Abutilon vexillarium*, and *Aegopodium podagraria* among others to determine the distribution of protein in the differently colored parts. The leaves of these plants contain contrasts in color ranging from stripes of dark green and light green to stripes of dark green and pure white. The leaves were first placed in scalding water and then transferred to warm alcohol until they became colorless, after which they were tested for protein by the xanthoproteic, biuret, or Millon's reaction. In the leaves of *Acer negundo*, all three protein tests gave only a very slight reaction in the white portion of the leaf, while the green portion showed a very intense reaction, the contrast between the two portions being so great that it could be observed at a considerable distance. In the leaves of *Acer pseudoplatanus*, which have intense green, bright-green, and pure white stripes, the white stripes showed only a trace of coloration, the light green an intense coloration, and the dark-green regions a very intense coloration with both the biuret and Millon's reagents. In the pure white part of leaves there is generally no chlorophyll apparatus, since the plastids do not develop or else remain rudimentary. Since the intensity of the protein reaction increases from the white portions of the leaf to the dark-green parts, it is assumed that the intensity of the protein reaction depends upon the relative number of the chloroplasts and that the faint protein test observed in the colorless regions of the leaves is due to the reaction of the cytoplasm. Lakon thus concluded that the main part of the protein in the leaves is formed in the chloroplasts. When leaves of *Phaseolus, Lactuca, Brassica, Cucurbita, Zea,* and *Tropaeolum* are detached from plants (grown without nitrogen) and such leaves are placed under favorable conditions of light and furnished with a supply of nitrate, they show the formation of nitrogenous material that is proved by tests to be accumulated in the chloroplasts (Ullrich, 1924).

3. Light and Protein Synthesis.—Evidence has been submitted above to show that protein may be synthesized in the chloroplasts of the cells of the leaves. The question then arises as to whether the leaves are the only portion of the plant in which proteins are manufactured and as to what conditions are necessary for their formation. Suzuki (1897, 1898) grew young barley plants in the dark after the following manner: When the plants had reached a height of 15 cm. they were transferred to a complete nutrient solution that contained the nitrogen in the form of nitrates and allowed to remain for 7 days. One set was then allowed to remain in the nutrient solution while the other was transferred to a 10 per cent sugar solution that was daily renewed. After 7 days the two sets of seedlings were analyzed. It was found that the plants supplied with sugar had been completely exhausted of their nitrates and had greatly increased in their amount of protein and in their dry weight as compared to the plants grown in the solution lacking sugar. Similar results were also obtained in like experiments with beans and potatoes. It was concluded by Suzuki that his results indicated that plants can assimilate nitrates in darkness and form protein provided there is a sufficient supply of sugar present in the cells. This same conclusion was reached by Hansteen (1896) in nutrition experiments with *Lemna minor,* by Prianischnikow (1899) in the study of the regeneration of protein in the onion, and by Muenscher (1923) in a study of protein synthesis in *Chlorella.* Gouwentak (1929) found that the formation of protein may occur in leaves in the dark when the intake of nitrogen-containing substances is possible. Newby and Pearsall (1930) noted in the leaves of *Vitus* and *Rheum* that the ratio of protein to soluble nitrogen changes with the age of the leaf and increases when the content of water falls. Fujita (1931) observed that the lower the degree of illumination the less protein there was in the leaves.

There is considerable experimental evidence indicating that protein synthesis may occur in any part of the plant regardless of light. Thus Müller-Thurgau (1894) grew plants under such a condition that part of the roots of an individual plant extended into a complete nutrient solution containing assimilable nitrogen and the other part extended into a nutrient solution lacking the nitrogen. The roots in the nitrogen solution were profusely branched and showed a thrifty growth, while those growing in the solution lacking nitrogen were very poorly branched and showed lack of vigor. Müller-Thurgau believed this indicated that nitrogen can be assimilated in the plant without going to the leaves.

Iwanoff (1901) worked with the roots of *Brassica napus*, *Daucus carota*, and *Solanum tuberosum* and concluded that an increase of protein can take place in the roots of these plants in the absence of light only when certain conditions prevail. The conditions necessary are a small amount of protein, a large supply of amides, and a considerable quantity of usable carbohydrates. Bambacioni (1923) made chemical analyses of plants of *Vicia faba*, *Ricinus communis*, and *Cucurbita pepo*, which were grown in a full nutrient solution and in the same solution lacking nitrogen, and considered that his data indicated that all parts of the plant may manufacture organic nitrogenous substances. He found that the increase of these was greater in the roots than in the aerial parts and believed it probable that protein substances migrate from the roots to the aerial parts, since roots cut off and immersed in complete solutions contain more nitrogen than the roots of the intact plants immersed in such a solution.

It has been reported by Björksten (1930), Loose and Pearsall (1933), and Gile (1935) that light has no influence on the synthesis of proteins provided the necessary components are present in the plant. Platenius (1932) found an inverse relation betweem amino and protein nitrogen which indicated that protein is the result of a polymerization of amino acids. According to Loew (1933) this formation must be very rapid since intermediate products cannot be demonstrated when asparagine and like compounds are converted into protein.

The experimental evidence in nearly all cases indicates that protein formation may take place in any of the living cells of the plant, provided that a supply of carbohydrates and of the proper nitrogenous compounds and other constituents is present and that light is not directly essential to the process. Protein synthesis may take place to a greater extent in the leaves than in other parts of the plant, due primarily to the abundant supply of carbohydrate in them as a result of photosynthesis. Recent work by Tottingham and Lowsma (1928), however, indicates that the shorter visible light rays may greatly enhance the absorption of nitrate and significantly increase the synthesis of protein as evidenced by the early growth of wheat. Tottingham, Stephens, and Lease (1934) found with the young plants of Progress wheat that the absorption of nitrates was promoted by the longer ultraviolet radiations. Mothes (1932) found substances in the plants of tobacco and bean that act as regulators of the metabolism of proteins in the plant. One type accelerates protease activity during the blooming stage, while another type inhibits proteolysis during ripening.

4. The Seasonal Variation of Nitrogen.—In the apple tree, according to Murneek (1930), growth and development during the early spring are largely at the expense of the nitrogen stored during the previous year. Consequently all the woody structures of the tree including the roots show a progressive decrease in nitrogen from early spring until active growth has ceased. The nonbearing spurs have a maximum content of nitrogen at the time of bud swelling and a minimum when vegetative elongation has ceased. The bearing spurs draw heavily on the nitrogen reserves of the older parts of the branches and possibly on those of the roots.

An unusually large proportion of the supply of nitrogen in the tree is utilized for the growth of leaves. There is a seasonal percentage decrease of nitrogen in leaves from 4 to a fraction of 1 per cent. Murneek found that 35 to 45 per cent of the total nitrogen of the leaves is reabsorbed by the tree prior to their abscission. The soluble nitrogen of the leaves usually increases at this period, while the insoluble fraction decreases.

In a 6-year-old Bartlett pear tree (Mulay, 1931, 1932), the total nitrogen in the shoots began to increase at the end of October, reached a peak in December, remained constant until February, and showed a sharp fall in March when new growth began. In May it had reached a new low. When absolute amounts are considered, the nitrogen changes in the wood about equaled those in the bark. The soluble nitrogen in the bark constituted about 20 to 30 per cent of the total nitrogen, and both types of nitrogen showed the same general trend. In December, insoluble nitrogen increased at the expense of the soluble nitrogen. Between February and March, the rate of proteolysis was much faster than the rate of transport, and there was a temporary accumulation of soluble nitrogen. The soluble nitrogen in the wood followed the same trend as the total nitrogen.

In beet leaves Pearsall (1931) found that the basic nitrogen fraction was relatively abundant in their youngest stages of growth but was sparse in the older stages. The amide nitrogen increased very rapidly at the beginning of the photosynthetic period. According to Richards and Templeman (1936), nitrogen content in the leaves of barley decreased with age in the earlier leaves but increased in the later leaves for some time after their expansion.

5. Formation of Protein in Seeds.—Since the greater amount of the protein that occurs in plants is stored as a reserve in seeds, it is of interest to consider the question of its origin. Wassilieff (1908) observed that the disappearance of asparagine at the maturity of seeds is accompanied by an increase in protein material. The production of this amide nitrogen is accompanied by a decrease in amino nitrogen. It was noted by Schulze and Winterstein (1910) and Schulze (1911) in the case of legumes that in the unripe seed there were found along with the proteins small quantities of asparagine, monoamino acids, arginine, and histidine. They also observed that in the hulls of the unripe seeds tryptophane occurred in appreciable quantities, while in the developing seed but little occurred. In the leaves and stems, asparagine occurred in considerable amounts, while in the maturing seed it occurred in only very small quantities. These investigators considered these data to indicate that the above-mentioned nitrogenous compounds are the materials from which the protein of the seed is synthesized. Zaleski (1911) observed in the maturing seeds of the pea and corn which had been taken from the plant and kept in moist air for varying periods that the protein increased in amount while the other nitrogenous compounds decreased as illustrated in the following data in the case of ripening peas (Raber, 1928):

Time	Protein, per cent	Amino acids, per cent	Nitrogen bases, per cent	Other nitrogen compounds, per cent
Beginning..........	79.2	8.7	10.8	1.4
After 5 days........	89.2	4.6	5.6	0.8

Teller (1935) harvested wheat in Arkansas daily during the development of the grain and for several days after ripening he found that the larger proportion of the nonprotein, nitrogenous compounds, which are predominant at the earliest stage of kernel formation, are quickly changed into the different types of proteins characteristic of the developed grain. Zeleny (1935) examined the grain of corn at four stages of development. He found that during these stages of development the nonprotein nitrogen decreased from 41 to 4.6 per cent, and he considered that the simpler amino compounds undergo rapid condensation as the grain approaches maturity. All the available evidence indicates that the simple organic compounds furnish the materials for the formation of the proteins.

The development of the wheat grain has been studied more in detail than the development of any other seed, and considerable information has been obtained concerning the formation of its protein. It will be necessary to discuss here in detail the development of the wheat grain, since the reactions involving protein formation are closely involved with and cannot be separated from the reactions of other processes. In the formation of the wheat grain, three stages may be distinguished (Brenchley and Hall, 1908 to 1910): (*a*) a period during which the pericarp is the most prominent feature, (*b*) a main period during which the endosperm is filled, and (*c*) the ripening period characterized by the desiccation of the grain.

According to Eckerson (1917), at the time of spike formation the largest amount of potassium nitrate is found in the tissues of the root and stem. A continuous stream of sugars and amino acids moves from the leaves of the culms into the young spike. Fructose, glucose, some arginine, histidine, considerable asparagine, and a large amount of potassium nitrate are present in the vascular elements of the culms. The cells of the primordia of the flower contain much fructose and asparagine. The primordia of the stamens, pistils, and glumes are alike chemically, but when the sporogenous tissue is developing, phosphates begin to enter the stamens and pistils while the magnesium oxalate crystals in the spike disappear and magnesium appears in the sporogenous cells. Phosphates continue to move into the sporogenous tissue until the pollen grain and egg are formed. It appears that the greatest absorption of phosphates is during the shooting period. It seems thus that the greatest absorption of phosphates is at the time that they are being assimilated most actively, while nitrates are absorbed in large quantities by the young plant and are utilized much later in the development.

After the fertilization of the egg there is a continuous stream of nutrient materials moving into the endosperm from the leaves and glumes. These materials include fructose and glucose, asparagine, arginine, histidine, and leucine. As the outer membrane of the nucellus becomes suberized, the current of sugar and amino acids passes from the base of the grain up through the raphe, enters the embryo sac at the chalazal region, and passes down through the endosperm cells to the embryo. Starch is formed in the endosperm cells soon after wall formation. It continues to increase in amount until desiccation begins or as long as the leaves and glumes are green and form an excess of sugar. The nitrogen compounds present, aside from aleurone and protoplasm in the endosperm, just before the ripening of the grain are a large amount of asparagine, considerable arginine, histidine, and some leucine but no glutamine. Thatcher (1913, 1915) found that there was an increase in the actual quantity of all the materials in the kernel with the exception of sugar during each successive period of growth. The decrease in the amount of sugar is due to its condensation into starch. The carbohydrate/protein ratio is at first greater and then becomes less in the developing kernel than in the pericarp material at the beginning of the experiment.

Eckerson (1917) stated that in the full-grown, still green wheat kernel containing about 90 per cent moisture there is no storage protein in the endosperm. The aleurone layer and the layer of cells immediately below it contain more protoplasm than the other endosperm cells and give a protein reaction, but this is not storage protein. No storage protein is formed in the endosperm until desiccation begins. The main feature of the ripening process is thus desiccation, during which the condensation of nitrogenous compounds to proteins occurs. The percentage of water in barley was found by Harlan (1920) to be the highest at flowering time, when 80 per cent of the caryopsis was water. From flowering time until maturity the water content gradually decreased until at maturity it amounted to 40 per cent. This percentage decrease is very uniform, and Harlan found in his experiments that it amounted to 2 per cent per day, while Brenchley (1912) found that it averaged about 1 per cent per day. The maximum dry weight of grain is attained a day or two before the grain should be regarded as ripe in practice. This is in agreement with the observations of Spitzer, Carr, and Epple (1919), who found that the total nitrogen of soft corn is only slightly lower than that of mature corn but that the amide nitrogen is much higher in the former than in the latter. It is also in agreement with the observation of Olson (1917), who stated that the nitrogenous compounds of the wheat kernel are at first very simple but change to the more complex forms as the period of desiccation advances.

6. Factors That Influence the Percentage of Protein in Grains.—The amount of protein in seeds and grains varies with different species and varieties of plants and with the conditions under which the plants are grown. The factors that influence the percentage of proteins in the cereals have received considerable attention, since the amount of protein in the grain, especially in wheat, determines to a certain degree its market value. A certain amount of protein is necessary in the wheat grain in order to produce a high quality of flour. In certain regions the wheats are especially low in protein so that high-protein wheats must be mixed with them in order to obtain a flour of good baking qualities.

The factors or causes that influence the percentage of protein in grains are apparently numerous and include the general weather conditions prevailing during the growing season, the time of nitrogen application, the available nitrogen, length of the growing season, available water, and certain unknown genetic factors. These factors are all more or less interrelated, so that in any discussion concerning them it is impossible to separate one from the other.

a. Time of Application of Nitrogen.—It is generally agreed that the available nitrogen is a very important and in many cases the greatest factor in determining the yield and quality of wheat (Burke, 1925; Ames, 1910; Whitson, Wells, and Vivian, 1902). As stated by Gericke

(1927), however, the peculiarities of the varieties may determine the effect of fertilizer upon the protein content of the grain. The properties of wheat varieties can and do markedly affect the efficiency of any fertilizer treatment.

Davidson and LeClerc (1917, 1918, 1923) applied nitrogen to wheat plants when they were 2 in. high, at heading, and at the milk stage. They found that the application of nitrogen in any of the inorganic forms used at the early stages of growth was instrumental in producing the highest yields of wheat, while the application of nitrogen in any form at the time of heading produced the highest protein content. When nitrogen was applied at the milk stage, it had no effect on the yield or upon the composition or quality of the grain. The period between the resumption of growth of wheat in the spring and at the time of heading was divided into three parts (Davidson, 1922), each of three corresponding sets of plots receiving nitrates at one of these subperiods. It was found that the effectiveness of increasing yields decreased consistently as the time of the application of nitrates approached the heading stage, while the reverse was true as regards the effectiveness of the nitrates in increasing the protein content of the grain.

Gericke (1920, 1922) applied nitrogenous fertilizers to the soil at planting time of wheat, oats, and rye in some cases and at different periods thereafter in others. In experiments with spring wheat, nitrogen was applied at planting time, at 17, 33, 48, 72, and 110 days after planting. The protein content of the grain produced under each of these applications taken in the same order as above was 8.6, 9.3, 10.4, 11.8, 13.2, and 15.2 per cent of the dry weight. The data thus show an increase of 77 per cent in the protein content of the grain obtained from plants that received their nitrogen 110 days after planting over those which were treated with nitrogen at the time of planting. Practically the same results were obtained with oats and rye. Later, as mentioned above, it was found that the response in the protein content of the grain was different for different varieties of plants. The observations of Watson (1936) regarding the time of application of nitrogen and its percentage in the grain agreed with those of Gericke above mentioned.

Gericke (1930) sowed wheat in a soil deficient in nitrogen and thus restricted the stalk formation of the early growth to one culm per plant. After 90 days from this planting a heavy application of nitrogen as sodium nitrate was added to the soil. Soon thereafter new culms arose from these plants and eventually produced grain. The grain of the original culm ripened before that of the newly formed ones. The highest percentage of protein was in the grain from the newly produced culms. The variation in the percentage of protein in the grains from older and from younger culms amounted to 30, 39, 57, 59, and 119 per cent respectively for the varieties Bunyip, Fulcaster, Cedar, Hard Federation, and Sonora. McCalla (1933) found that the organic nitrogen produced in the vegetative parts of the wheat plant after heading was not synthesized to protein but accumulated in the vegetative parts. It was noted by Doneen (1934) that if sufficient quantities of nitrogen for large yields of grain were present in the soil, the addition of sodium nitrate only retarded growth and did not increase the yield or the content of nitrogen in the grain. When, however, the nitrogen in the soil was limited, the addition of this element stimulated growth and increased the yield and the amount of nitrogen in the grain. The composition of the soil, however, was found by LeClerc (1910), Shaw and Walters (1911), Thatcher (1913), and LeClerc and Yoder (1914) to have little effect on the protein content of the grain under the conditions of their experiments and with the varieties with which they worked. In this connection it might also be mentioned that Lipman and Blair (1916) found that the liming of the soil increased the percentage of protein in soybeans. Stark (1924) in the case of soybeans noted that the application of limestone and organic matter increased the percentage of protein, while potassium in addition to these compounds gave a decrease in the percentage of protein.

b. Soil Moisture.—In general, it may be said that the protein content of grains varies inversely with the total rainfall (Hopkins, 1935). Neidig and Snyder (1922, 1924, 1926) found in the case of Marquis and Palouse bluestem wheats that under field conditions available nitrogen shows its greatest effect on protein content and yield of wheat when climatic factors are most favorable, and that both the climatic factors and available nitrogen play an important part in the quality of the wheat. It is considered that the importance of climate is chiefly due to its effect upon the availability of the nutrients in the soil, which, in turn, affect the nutrition of the plant. Neidig and Snyder (1924) noted that under field conditions a high moisture content properly distributed during the growing season in an average soil will produce a high-yielding wheat with a low protein content. Thatcher (1913) found that the increase in protein content of wheat in eastern Washington decreased with an increase in rainfall. Greaves and Carter (1923) observed a decrease in the nitrogen content of wheat, oats, and barley as the irrigation water used in their growth was increased. The decrease in the nitrogen content of the grain of wheat, oats, and barley amounted to 21, 40, and 19 per cent, respectively. Stewart and Greaves (1909) also observed that wheat grown on arid nonirrigated land contained more protein than was found in wheat on adjoining irrigated land. Greaves and Nelson (1925) showed that corn grown with and without irrigation water and manure gave a decrease in the nitrogen content of the kernel due to irrigation water and an increase due to the manure, the nitrogen content of the grain being a function of the available nitrogen content of the soil.

As a rule, however, the composition of corn is modified little if at all by variations of nutrient treatments or of climatic factors, as is indicated by the fact that the nitrogen content of corn seed from widely separated sections is approximately the same.

c. *Kind of Plant.*—The effect of changes in soil and climate upon the amount of protein in grain or seed depends to a great extent upon the kind of plant. Thus Tottingham (1924) and Tottingham and Kerr (1926) found in the case of Leaming Yellow Dent corn grown in soil in the greenhouse that variations in the amounts and time of application of sodium nitrate did not modify the nitrogenous content of the cured seed.

Dewiche and Tottingham (1930) grew corn, barley, and clover at two stations in Wisconsin with a difference of 3.5° in latitude. The crude protein varied but little for the grain of corn and barley under these two conditions, while that in the seed of the clover was significantly different.

Ivanov (1927) reported that peas, corn, lentils, and vetches grown under different conditions of soil moisture and climate showed no variation in the content of protein of their seeds, and that there was maintained a definite relationship between proteins and carbohydrates. Schuster and Graham (1927) found indications that the composition of the soybean cannot be changed by soil treatment.

Tottingham (1923) found that the composition of red clover (*Trifolium pratense*) and buckwheat was greatly altered by changes in the environmental conditions, while Waldron (1933) in North Dakota reported that the later and lower yielding wheat had the highest content of protein.

Tottingham and his workers found that the leaves of corn are characterized by low proportions of total soluble nitrogen and soluble protein as contrasted with high values for these factors in wheat and sugar beets, plants whose composition responds readily to changes of climatic conditions. Corn leaves are characterized by low proportions of reducing sugars, while the sugar-beet leaf contains much more reducing sugar than sucrose in the tests that have been made. They believed that sucrose and less soluble forms of protein are quite likely to be less altered by the effects of temperature and other climatic factors than are such reactive compounds as glucose, soluble proteins, and other soluble nitrogenous compounds.

7. Yellow Berry in Hard Winter Wheats.—In connection with the protein content of wheat, the nature of yellow berry and the factors that influence it should be taken into consideration. The term "yellow berry" has been defined by Roberts and Freeman (1908) as the appearance in hard, flinty wheats of grains of a light-yellow color, opaque, soft, and starchy. These opaque yellow grains may have this character throughout, or sometimes from a small fraction to half of a grain will be yellow and starchy while the remainder of the kernel will be hard, flinty, and translucent. The difference in color between the flinty grains and the "yellow berries" is due to differences in the structure and content of the cells of the endosperm. Considerable attention has been given to the causes of yellow berry. Lyon and Keyser (1905) concluded that the amount of yellow berry increases with the lateness of ripening and that crops of large yield and low nitrogen content contain more yellow berries than do crops of opposite kinds. Roberts and Freeman (1908) found that there was a diminution in the amount of yellow berry corresponding to the shortening of the fall growing period on account of late planting. They found, in general, that late ripening increased it and that higher mean temperatures 3 weeks before ripening were found to be correlated with low percentage of yellow berry. Headden (1915) reported that the application of nitrogen in the form of sodium nitrate greatly reduced the amount of yellow berry and in some cases prevented it altogether under the conditions prevailing in Colorado. He found also that yellow berry could be greatly increased by the application of available potassium. He considered that the occurrence of yellow berry indicates that potassium is present in excess of that necessary to form a potassium/nitrogen ratio favorable to the formation of hard, flinty kernels. Roberts (1919) concluded that the operation of common causes for the production of yellow berry overshadows any differences that may be due to hereditary tendencies and precludes a definite statement concerning the influence of hereditary factors in this regard. Some isolated forms of wheat are, however, apparently freer from the condition of yellow berry than others growing in the same field. Jones and Mitchell (1926) found that thorough tillage of summer fallow in contrast to poor tillage resulted in larger yields and correspondingly less yellow berry. They considered that yellow berry is the manifestation of nutritional disturbances resulting from an insufficiency of nitrogen and other elements for adequately meeting the requirements of a normally developing crop. Since no means appear feasible as yet for protecting dry-farmed lands against the gradual depletion of their native supply of nitrogen, high efficiency in the control of yellow berry cannot be maintained indefinitely by tillage.

In conclusion it should be stated that yellow berry is apparently different from ordinary soft wheat, since it is produced under apparently the same conditions as the flinty kernels, not merely in the same field, but on the same plant or in the same head.

IV. NITROGEN BASES IN PLANTS

The nitrogen-containing compounds in plants may, for the most part, be regarded as derived from ammonia by the replacement of one or more of the hydrogen atoms by organic radicals of various types. The nature of these replacements in the formation of amino acids and amides has already been discussed, and it is the intention here to make a statement concerning the general nature of the amines. The replacement of one or more hydrogen ions of ammonia by alkyl groups gives rise to compounds strongly basic in nature and known as "amines." There are three different classes of amines designated as primary, secondary, and tertiary according as one, two, or three of the hydrogen atoms of ammonia have been replaced by the organic radicals mentioned above. The nitrogen bases in plants are amines of these three classes and may be divided into three main groups, (*a*) the plant amines, (*b*) the alkaloids, and (*c*) the purine bases.

A. Plant Amines or Natural Bases

These are simple, open, chain amines. Some of the plant amines are trimethyl amine, a volatile compound that occurs in the seeds of *Mercurialis annua* and in the flowers of *Crataegus oxyacantha* and in the leaves of *Chenopodium vulvaria;* and betaine and choline, which occur together frequently in the embryo of *Hordeum sativum, Triticum sativum, Vicia sativa*, and other plants. Betaine occurs also in the juice of the beet and in the tubers of the Jerusalem artichoke, while choline occurs in the root of *Acorus calamus*, in the endosperm of *Cocos nucifera*, and in many other plants (Haas and Hill, 1913). There is no known physiological use for these simple amines in the plant. They are regarded by some as intermediate products in the synthesis or decomposition of proteins, but there are no data to warrant such a statement.

B. Alkaloids

The alkaloids are compounds that for the most part are composed of carbon, hydrogen, oxygen, and nitrogen, but a few, as coniine and nicotine, contain no oxygen. The alkaloids differ widely in structure but are related, for the most part, to heterocyclic compounds, as pyridine, quinoline, and isoquinoline. The term "alkaloid" literally means an alkali-resembling substance, since the compounds of this type have properties similar to those of the alkalies in being able to form salts with acids that are usually soluble in water.

The alkaloids are precipitated from their solutions by the so-called "alkaloid reagents" among which are tannins, phosphomolybdic acid, phosphotungstic acid, picric acid, and potassium mercuric iodide. The majority of the alkaloids are optically active, rotating the plane of polarized light to the left, although a few, such as coniine, are dextrorotatory. The alkaloids, in general, are colorless, odorless, crystalline, and have a bitter taste, although there are a number of exceptions in one or more of these regards. They are generally insoluble in water but are easily soluble in organic solvents as chloroform, carbon tetrachloride, and ether. The alkaloids generally do not occur in the plant as such but are combined with acids in the form of a salt. Some of the acids that are combined in this manner are tannic, malic, citric, succinic, and oxalic. The alkaloids are generally set free in extraction by the addition of an alkali as lime or barium, since only the free bases and not their salts are soluble in the solvents mentioned above.

The alkaloids are very limited in their distribution in the plant kingdom, occurring, for the most part, in the *Leguminosae*, *Papaveraceae*, *Ranunculaceae*, *Rubiaceae*, and

Solanaceae. They may occur in solution in the cell sap in the young parenchyma or may be stored in older tissue in the solid state. They are generally found in the seeds and fruits, but some occur in the leaves, stem, and roots (Haas and Hill, 1913).

Some of the alkaloids are: coniine in *Conium maculatum;* nicotine in tobacco; atropine, which is the principal alkaloid in the deadly nightshade; cocaine in cocoa leaves; quinine, which is the most important alkaloid of the 24 found in cinchona bark; strychnine in the seeds of *Strychnos nux vomica;* and morphine, which is the principal alkaloid of the 20 that are found in the juice of the poppy. According to McNair (1931) the ratio of the number of tropical plant families containing alkaloids to the total number of tropical families is 0.16 as compared to the ratio of 0.12 between such families in the temperate zone. For the tropical alkaloids the melting point is between 200 and 250°C., while in the temperate group it is between 100 and 150°C. The number of carbon atoms on the alkaloid molecule increases from the tropics to the temperate region. The average and minimum number of nitrogen atoms increases from the tropics to the temperate regions but the maximum number decreases. The maximum number of oxygen atoms increases and the minimum number decreases as one proceeds from the tropics. The alkaloids of tropical origin are generally less toxic to man than those of the temperate region.

The origin of the alkaloids in the plant is not definitely known. Some consider that they arise in the formation of proteins, while others consider that they are formed in the disintegration of the proteins. In general, they are considered as nitrogenous waste products that the plant, although it is incapable of eliminating them, renders noninjurious by storing in this condition (Weevers, 1930). It would seem, however, that the connection of alkaloids with protein metabolism is not a common relationship, since it is manifested by such a limited number of species. Emde (1929) considered that the formation of alkaloids is dependent upon the activity of photosynthesis, because the amount of photosynthesis decreases progressively from the equator to the poles. The maximum of plant alkaloids is produced in the tropics, while the plants of the polar regions produce no alkaloids.

C. The Purine Bases

The purine bases are complex compounds containing a nucleus with four carbon atoms and four nitrogen atoms arranged alternately to form a double-ring group. They are a group of compounds, all of which are derivatives of the compound purine ($C_5H_4N_4$). Under this list are included caffeine or theine in coffee and tea; theobromine in the fruit of *Theobroma cacao;* xanthine; hypoxanthine; guanine in sprouting seeds; and adenine, which occurs in the juice of beets. Uric acid, a purine derivative which does not occur in plants but which is a waste product of the metabolism of animals, is of interest in regard to the nature of the purine bases found in plants. The close relationship between this substance and the purine bases found in plants suggests that the latter might be considered as waste products in the protein metabolism of the plant. Weevers (1930) believed that the xanthine derivatives have their origin in the decomposition of proteins, accumulate for a time, and are again used for new proteins.

V. LOSS OF NITROGEN FROM PLANTS

One of the most striking differences between the metabolism of animals and plants is that the former continually excrete nitrogen compounds to the outside world, while plants rarely if ever excrete or eliminate nitrogen in any form. The nitrogenous compounds eliminated by animals are chiefly urea, uric acid, ammonia, and others, all of which result from the breaking down of proteins and the protoplasm. The same compounds are apparently formed in the life processes of the plant cell but are again directly utilized in the formation of new protein compounds and not eliminated as in animals.

Fosse (1913) proved that urea is present in the leaves of carrot, spinach, in potato tubers, root of the turnip, and in the seedlings of pumpkin, wheat, rye, clover, beet, pea, and corn. In the resting seed of wheat, corn, and pea it was present in amount equal to 1 cg. per kilogram of weight. Fosse has proved that the urea detected was formed within the plant and was not absorbed directly from the soil where it is known to be present. Chibnall (1922), in working with the changes of protein in the leaves of the runner bean, found that their products of protein disintegration pass into urea derivatives that have a striking analogy to what is found in the animal kingdom.

Klein and Tauböck (1927, 1931, 1932) and Klein, Tauböck, and Linser (1930) found an abundance of urea in seedlings but none in the mature seeds. Young leaves, petioles, and leaf buds usually gave positive tests. In herbaceous plants urea is present in the largest amount in the roots. It occurs usually in the greatest amount in the more actively growing parts of the plant. These investigators concluded that urea is formed by the decomposition of arginine or of argininelike bodies. When arginine is taken up by plants in sterile cultures it is broken down and an increase in the amount of urea occurs. The etiolated seedlings of soybeans have 30 per cent more urea than the green ones of the same age.

In the germination of seeds where the processes of metabolism are very intense, the question of the loss of nitrogen in the gaseous form is somewhat in doubt. Schulz (1862) and Atwater and Rockwood (1886) reported appreciable losses of nitrogen during germination, but Boussingault (1838) and Hellriegel (1855) and more recently Davidson (1923) found only negative evidence of nitrogen loss. Davidson (1923) grew cowpeas and two varieties of soft winter wheat under sterile conditions in Kjeldahl flasks so that no transference of seedlings was necessary in the analyses. The seedlings were grown for 15 days, and the nitrogen content of the growing seedlings and that of the sterilized seeds were found to be identical. One is thus justified in concluding that no nitrogen in the gaseous form is lost during the early stages of these seedlings. Davidson (1926) found that wheat seedlings floating on distilled water differ in composition from the stock seed from which they were obtained. In 7 days from the time the seeds were set to germinate, or 3 to 4 days after germination, the nitrogen content of the seedlings was approximately 10 per cent higher than the nitrogen content of the stock seeds. This fact seems to be due to the leaching of the nitrogen from the ungerminated seeds and its reabsorption by the growing seedlings. The reserve in these seeds is mostly carbohydrate and it would seem advisable that the loss of nitrogen should be investigated in the germination of seeds in which the major portion of the reserve material is of an oily nature.

Miller (1910) in studying the chemical changes in the germination of the sunflower seed found evidence indicating that there might be a loss of nitrogen from these seeds during the first 3 days of their germination. The total nitrogen in 100 seeds before germination amounted to 346 mg. and at the end of the 3-day period, when the hypocotyls were 2.5 to 3.5 cm. in length, the total nitrogen of 100 seedlings amounted to 292 mg.—a loss amounting to approximately 15 per cent of the total nitrogen present in the seed. The total nitrogen in the seedlings at four other more advanced stages showed no loss of nitrogen whatsoever. Newcomb and Miller (1927) determined the nitrogen content of the seeds and seedlings of the sunflower and squash. They obtained data indicating that a loss of nitrogen occurred during the germination of these seeds, the loss amounting to from 5 to 6 per cent of the nitrogen content of the seed. This loss occurred while the hypocotyl and radicle were attaining a length of 4 to 5 cm. Several sources of error were overlooked, however, and a more careful study of this subject is needed before any definite statements can be made. In the squash plant Reid (1930) suggested that a relatively small amount of nitrogen might be lost as a result of the respiratory process in the leaves. Although the question of the loss of nitrogen from the germinating seed seems as yet not to be definitely settled, the loss of nitrogen from plants growing in the field seems to be confined wholly to the diffusion outward of nitrogenous salts into the soil, to the leaching of the stems and leaves by rains, and to the falling off of certain parts of the plants.

VI. CARBON/NITROGEN RELATIONS

Reproduction in plants, as well as their vegetative growth, may be controlled, to some extent at least, by varying their external and internal conditions. Recently it has been found that rather definite carbon and nitrogen relations in the plant influence its response in regard to vegetative growth and reproduction. The relationship of the carbon and nitrogen content of the plant has come to be expressed as the "carbon/nitrogen ratio," a term, however, that has been rather loosely used. To some it is the carbohydrate/nitrogen ratio; to others the carbohydrate/insoluble-nitrogen ratio is the important one; others consider the effective ratio to be the starch/nitrogen; while yet others express the relationship as the ratio of the total carbon to the total nitrogen.

In 1916 Fisher reported that the vegetative condition was characterized by a carbohydrate content relatively low in proportion to the nitrogen or, in other words, by a low carbohydrate/nitrogen ratio. He found that the reproductive condition was characterized by a carbohydrate content relatively high in proportion to the nitrogen content, a condition that is designated as a high carbohydrate/nitrogen ratio. Kraus and Kraybill (1918) in their nutrition studies of the tomato plant brought out clearly that the carbohydrate/nitrogen relation influences its growth and reproduction. They recognized four conditions that had a decided effect upon the behavior of this plant. These may be summarized as follows:

(1) A very high carbohydrate/nitrogen ratio accompanying a weakly vegetative condition. Here nitrogen appears to be a limiting factor of growth and the high ratio is apparently due to the small amount of nitrogen present. (2) A high carbohydrate/nitrogen ratio accompanying abundant fruit production. In this type, nitrogen compounds are available, but the high ratio is due to an excess of carbohydrates. (3) A low carbohydrate/nitrogen ratio accompanying a vigorous vegetative condition. In this type there appears to be an available supply of both carbohydrate and nitrogen and the balance between them is such as to produce the best vegetative conditions without leaving a residue of carbohydrate. (4) An exceedingly low carbohydrate/nitrogen ratio accompanying a weakly vegetative condition. In this type carbohydrates appear to be the limiting factor of growth, and the low ratio is due to the small amount of carbohydrate present, while the amount of nitrogen appears to be an indifferent factor.

Nightingale (1922) in experiments with salvia, buckwheat, soybeans, and radish concluded that the carbohydrate/insoluble-nitrogen ratio is the effective one in relation to vegetative growth and reproduction in these plants. He considered that nitrates may be stored in the plant until the proper conditions arise for synthesis to other forms of nitrogen and that the presence of nitrates as such in the plant does not appear to affect materially the type of growth of the plant. Gurjar (1920) noted in the case of the tomato that the carbon/nitrogen ratio may be as high as 19 and as low as 2 but that the fruiting took place between the ratios 4 and 6. Work (1924) considered that a nitrogen content of 0.3 per cent on a green basis in this plant is conducive to a vigorous vegetative growth.

Kraus and Kraybill (1918) also found that withholding moisture from plants grown under conditions of relative abundance of available nitrogen results in much the same condition of fruitfulness and carbohydrate storage as the limiting of the supply of available nitrogen. They further concluded that fertilizers containing available nitrogen or that which may be readily made available are mainly effective in producing vegetative response and that they may either increase or decrease fruitfulness according to the relative available carbohydrate supply. They also pointed out that pruning and girdling are largely effective in promoting or retarding fruitfulness by their effects in balancing the carbohydrate supply within the plant or the means for its manufacture with the available moisture and nitrogen supply. These observations of Kraus and Kraybill have greatly stimulated investigations in regard to the carbohydrate/nitrogen relation, since it offers a means of outlining methods of practice in regard to fertilizing, pruning, and ringing based upon physiological relations. It is not the intention here, however, to consider any of the methods of agricultural practice but to mention only some of the carbohydrate/nitrogen relationships that so far have been observed.

A. Fruit-bud Formation

The effects of the amount of nitrogen, and the time and manner of its application upon fruit-bud production have been extensively studied. Only a brief résumé of some of the work will be given here because a detailed consideration of this topic belongs to the field of practical horticulture rather than to the field of plant physiology.

1. The Carbohydrate/Nitrogen Relation.—The relationship of carbohydrate and nitrogen in regard to fruit-bud formation has been extensively studied in the apple by Hooker (1920 to 1924), Roberts (1920), Harvey and Murneek (1921), Harvey (1923), and Kraybill (1923).

Hooker (1920 to 1924) in his investigations of the seasonal changes in the chemical composition of apple spurs found that conditions leading to high starch and low nitrogen content at the time of fruit-bud differentiation appear to be essential for productivity. Fruit-bearing spurs that develop leaf buds have a low starch and a high nitrogen content, and barren spurs, which neither blossom nor produce fruit buds, have a low starch and a low nitrogen content. The starch/nitrogen ratio was more indicative than the total carbohydrate/nitrogen ratio. Starting with an apple spur in the off year, it contains slightly less ash including potassium and phosphorus, less nitrogen, and less starch than a spur at the beginning of the bearing year, but it is higher in dry weight, reducing sugars, and polysaccharoses. During the spring, water is absorbed, the stored starch is hydrolyzed to sugar, and the sugar is utilized along with phosphorus and nitrogen in the production of new growth. During this period of the utilization of reserves and production of new vegetative tissue the acidity, polysaccharoses, and, to some extent, the potassium content increase. Eventually the supply of available carbohydrates is reduced to a minimum, whereupon terminal-bud formation takes place on spurs. After this, growth practically stops and is accompanied by a falling off of acidity and polysaccharose content. By the photosynthetic activity of the newly formed leaves a fresh supply of reducing sugar is formed, some of which is stored in the form of starch and other polysaccharoses, resulting in an increase of dry weight. At the same time, the potassium and phosphorus content increase somewhat while the nitrogen content continues to fall off. As a result of this condition, leading to a high starch and a low nitrogen content late in June, fruit-bud differentiation occurs.

It was observed by Roberts (1921), under conditions of low nitrogen nutrition, that the type and appearance of foliage and growth of the apple tree are more influenced by the composition of the wood as a result of previous treatment than they are by the current seasonal conditions. Thus, for example, trees that were previously poor vegetatively made little growth when maintained in a low nitrogen nutrient, while trees that were previously very vegetative made an excellent growth after being transferred to a low nitrogen nutrient.

Harvey and Murneek (1921) and Harvey (1923) investigated the effects of defoliation and ringing of apple twigs and spurs with especial reference to the carbohydrate/nitrogen ratio. In the case of defoliation, it was found that the defoliated spurs, as compared to the checks, contained more nitrate nitrogen, total nitrogen, and insoluble nitrogen, more reducing sugars, and less total carbohydrates and a smaller value for the carbohydrate/nitrogen ratio. In general, the bearing and sterile spurs showed extreme values, while the nonbearing spurs assumed an intermediate position between them. Howlett (1923) considered that the lack of an adequate supply of nitrogen, rather than a lack of carbohydrates, is more often the limiting factor to normal flower development in the case of the apple.

It was observed by Potter and Phillips (1930) for the Baldwin apple that the insoluble nitrogen was the constituent which was most consistently associated with the formation of fruit buds. Its accumulation accompanied their formation, while the accumulation of carbohydrates depressed their development.

Finch (1935) noted in apple trees that the degree of vegetativeness is closely related to the chemical composition of the tree especially as regards the amount of carbohydrates and nitrogen. The starch/nitrogen ratio and the total carbohydrate/nitrogen ratio were highest in the fruitful terminal shoots of trees of biennially bearing varieties. These ratios were the lowest in the unfruitful, terminal shoots of the trees of biennially bearing varieties.

2. The Application of Nitrogen.—Hooker also found that the chief effects of spring applications of nitrogenous fertilizers to healthy apple trees are, on bearing trees, an increased set of fruit associated with a greater nitrogen content in the spurs during the period of fruit setting and, in nonbearing trees, an increased rate of growth. Spring applications of nitrogenous fertilizers do not favor starch accumulation at the period of fruit-bud differentiation, and consequently they could not be expected to favor the process. The later in the season nitrogenous fertilizers are applied the greater is the nitrogen content of the spurs the following spring immediately before growth begins.

Hofman (1930) believed that the application of nitrogen in the autumn to orchards in Virginia produced the best growth and yields. Harley, Masure, and Magness (1932) believed that the principal factor in the initiation of fruit-bud formation in apple is the ratio of the amount of foliage to the amount of fruit. In Delicious apples the time of bud differentiation extends over a relatively long period during the summer. Thus apparently some buds are differentiated as early as June 17, while the formation of others could be influenced as late as July 29 by some treatments. Harlan and Collison (1933) noted that under the conditions of low fertility the use of nitrogen fertilizer produced increased yields of apples. Applications of nitrogen in the early spring produced larger yields than did the application in summer. The application of sodium nitrate after blooming did not produce any greater yield than did the application in the spring.

Lagassé (1935) reported that the application of nitrogen over a 5-year period to Yellow Transparent apple trees increased the diameter and the terminal growth of the tree but had no effect on the total yield of fruit.

It was noticed by Plagge (1930) that single or successive, annual applications of sodium nitrate to Jonathan and Grimes apple trees increased the susceptibility of the fruit to soggy breakdown. Potter and others (1930) found that at the time of the maximum increase in growth of the apple in July, each fruit absorbed daily approximately 1 mg. of nitrogen and about 300 mg. of carbohydrate. When nitrogen is applied to apple trees, Marth (1934) obtained the best results when the fertilizer was spread just outside the branches where the greater proportion of the fibrous roots were located in the upper soil. Gardner (1923), in the case of the strawberry, reported that when moisture and temperature are not limiting factors, the number of flower clusters, number of flowers, and size of berries are dependent on nutritive conditions within the plant during the preceding fall and winter and are practically independent of soil-fertility conditions during the spring and at the time of fruiting. Treatments that would increase production through modifying fertility, therefore, should be given during the summer and fall months. He further stated that the maximum production of flower clusters, flowers, and berries was associated with those summer and fall treatments that led to the greatest accumulation of starch and total carbohydrates at the time of fruit-bud differentiation. Taylor (1932) found in Alabama that the application of nitrogen in the autumn to strawberry plants, and then again about 90 days prior to the first harvest gave the best production of a large number of flower clusters effective in producing a crop. Schilletter (1932) considered that low temperature or lack of moisture taken separately or in combination are associated with the time of fruit formation of the strawberry. Gardner also noted that low carbohydrate and low starch content at the time of fruit-bud differentiation led to the production of female flowers in a variety that is normally hermaphroditic. In the strawberry, a low carbohydrate content is apparently associated with the female condition, high carbohydrate content with the male condition, and an intermediate carbohydrate content with hermaphroditism. Talley (1934) analyzed the aerial portions of the staminate and pistillate plants of hemp at flowering. The staminate plants were higher in total carbohydrates, polysaccharides, and sugars than the pistillate plants. Nitrogen was more abundant in the pistillate than in the staminate regions of the respective plants. Platenius (1932) considered that the formation of the reproductive organs of celery is not determined by nutritional factors that involve the entire plant. He thought it possible that the ultimate cause of the formation of the seed stalk is localized accumulation of certain chemical compounds in the meristematic tissue possibly involving only a few cells.

B. Length of Day

It was observed by Garner and Allard (1920) in their investigations on the effect of the length of day and night upon growth and reproduction, that there are some plants which bloom and produce seed under light conditions of relatively short duration, but which when the length of day is increased become strongly vegetative and unfruitful. Other plants, however, require a long day to be reproductive and become weakly vegetative when the length of day is shortened. These observations led to the assumption by some that the light requirements might be associated in some manner with the quantity of carbohydrates manufactured during the light period and its relationship to the nitrogen supply. A survey of the light responses, however, indicates that the problem is a complicated one and is difficult to interpret on the basis of photosynthetic activity. Thus the effects produced are due not to the total quantity of radiation received but rather to the actual length of the exposure period. Plants exposed for long daily periods to sunlight of only one-fourth full intensity behaved similarly to plants exposed to full intensity for the same period and not in the same manner as plants exposed to full intensity for one-fourth the length of time. It is not known, however, whether the difference in total radiation was accompanied by differences in the photosynthetic activity, and other factors might have regulated or influenced the rate of photosynthesis.

The effect of the light period on the fertilizer requirements of plants was studied by Nightingale and Kraus (1924). They found in the case of tomatoes that light, within the limits of a 6-hr. day, did not appear markedly to limit the building up of nitrates to insoluble forms of nitrogen, provided there was present an available supply of carbohydrates. Buckwheat, soybeans, radish, and salvia of the varieties used were limited, however, in the building of nitrates to insoluble forms of nitrogen by a 7-hr. day, as it occurred in the greenhouse, even though there was present an available supply of carbohydrates.

Werner (1934) reported for the potato plant that high temperatures, long days, and an abundant supply of nitrogen favored vegetative growth in all plant parts except the tubers. Early tuberization was induced with low temperature, short days, and a deficiency of nitrogen. Maximum tuberization occurred with days of intermediate length, low temperature, and an abundant supply of nitrogen.

C. Different Stages of Growth

The value of the total-carbon/total-nitrogen ratio was determined by Hicks (1928) for different stages of growth of Starling winter wheat and of Nevin and Marquis spring wheats. The stages considered were (1) the seedling stage up to the active photosynthetic action of the first leaf and the exhaustion of the seed's store of food, (2) the vegetative cycle up to the heading period, and (3) the fruiting stage.

The seedling stage is characterized by a rapid drop in the carbon and nitrogen percentage. Carbon is lost by respiration, but most of the reserve in the endosperm is taken into the embryo. No nitrogen is lost, but the actual amount is increased by the withdrawal of the reserves from the endosperm, although the percentage is decreased owing to the increased dry weight of the young plant. The carbon/nitrogen ratio in the seedling stage remains low, is constant, and is of practically the same value as that of the ungerminated embryo. The vegetative cycle opens with a low carbon and fairly high nitrogen and with a low carbon/nitrogen balance. After the emergence of the first leaf and the beginning of active photosynthesis, there is a rise in the carbon curve. The carbon/nitrogen ratio shows a marked increase throughout the whole vegetative period. At the time of the highest carbon/nitrogen ratio, flower initials appear, and a general movement in the plant of the carbohydrates and the soluble nitrogen constituents through the stem and into the reproductive parts occurs. After fertilization, the carbon/nitrogen ratio decreases in the developing organs until a very low value is shown by the mature embryo. Hicks further observed that the three strains of wheat showed differences in the rate of carbon accumulation. In winter wheat it accumulates very slowly, the plant maintaining an intermediate carbon/nitrogen ratio. A much higher carbon/nitrogen ratio is required to produce flowers in the winter wheat than in the spring types. Thus a carbon/nitrogen ratio of 14 to 17 covers the range of conditions favorable to flowering in the spring wheats, while a ratio of 31 was required by the winter wheat. Hicks concluded that although each strain examined had its own initial carbon/nitrogen ratio value at which flowers are initiated, it represents in every case the maximum of the ascending carbon/nitrogen ratio curve.

D. Variations of Ratio in Different Organs of the Plant

The carbon/nitrogen ratio for different parts of the wheat plant was determined by Hicks (1928) at different stages of growth. In the case of a plant in the vegetative phase at the tenth leaf stage, the carbon/nitrogen ratio at the base of the stem was 13.7, at the middle of the stem 12.4, and at the upper portion 11.9. In the case of the tenth leaf, the lower half had a carbon/nitrogen ratio of 8.1 while the upper half had a ratio of 6.5. The carbon content was practically constant throughout the length of the stem, but the nitrogen increased from the base upward. In the case of the roots of plants in the fruiting stage, the carbon/nitrogen ratio of the secondary roots was 19, while that of the root hairs was 15. Hicks concluded from these observations that as a general rule in developing annual plants "the younger the tissue the lower the carbon/nitrogen ratio."

E. Effect on Root and Shoot Growth

Much study has been made on the effect of the relation of nitrogen and carbohydrates to root and shoot growth. Chandler (1919), in peach trees, found that root development is increased less than top growth by the addition of nitrates, and Gericke (1922) stated that an abnormally large root development of wheat seedlings, in proportion to that of tops, is primarily associated with a deficiency of nitrogen in the medium in which the plants are rooted. As mentioned in Chap. III, Turner (1922) found that barley and corn showed significant increases in ratios of tops to roots as the nitrate concentration of the solution was increased, but that flax plants did not show such an increased ratio. When nitrogen is lacking in the nutrient medium, the plant is able to utilize in new growth the nitrogen that is stored within the plant. It has been shown by Roberts (1921) that apple trees can absorb nitrogen, store it as a reserve, and later utilize it in growth. An abundant supply of carbohydrates stimulates root development. Thus it has been observed by Dachnowski (1914), Knudson (1916), Curtis (1918), and others that when twigs, cuttings, or plants are grown in sugar solutions an increased root development results.

It was observed by Davies (1931) that willow shoots develop in the area of highest total nitrogen per gram of dry weight of the cutting, while the roots regenerate in the area of lowest total nitrogen per gram of dry weight of cutting. Harrison (1934) found that it was possible to influence the amounts and relative proportions of the various parts of bluegrass by limiting the supply of nitrogen. As the supply of nitrogen was decreased, the relative amount of top growth diminished, while that of the roots and rhizomes increased.

Starring (1923) studied the effect of the carbohydrate/nitrogen content upon the production of roots on the cuttings of tomato and *Tradescantia* and found that the best production of roots is obtained from those cuttings which have a high content of carbohydrate. When carbohydrates are absent or low in amount, the cuttings develop no roots; and if there is sufficient carbohydrate merely to initiate the growth of roots, they make little or no growth. No wide differences in root development could be secured by varying the nitrogen content.

Reid (1924, 1926) investigated the behavior of cuttings taken from tomato plants that had been grown under conditions to produce the extremes in the carbohydrate/nitrogen ratio. Her results indicated that a high-nitrogen supply plus a readily available supply of carbohydrates appears to furnish favorable conditions for shoot growth. A somewhat limited nitrogen supply plus a readily available supply of carbohydrates appears to furnish conditions for root growth, and, in general, the cutting has a tendency to do what the plant, from which it is taken, has been doing. She observed also that shoots are produced in greater number in light than in darkness. A greater quantity of roots is produced in darkness than in light by high-carbohydrate cuttings having a moderately high internal nitrogen reserve and growing in solutions lacking nitrates. If, however, high-carbohydrate cuttings with a lower nitrogen reserve are grown under the same conditions, a greater quantity of roots is produced in light than in darkness. Hicks (1928) obtained similar results in the case of normal cuttings of *Salix viminalis* and in those injected with potassium nitrate and various sugars. She considered that the initiation of growth is due to stimulated respiration giving energy for the withdrawal of nitrogen into the bast and its consequent upward translocation, particularly to the buds. The injected sugars hasten this considerably, while potassium nitrate in excess prevents the commencement of growth. Hicks concluded that shoots grow at the area of the lowest carbon/nitrogen ratio and roots at the highest. Reid (1926) also found that different kinds of seedlings show very different growth responses, which are often related to the chemical composition of the seed. Thus the higher the nitrogen in proportion to the carbohydrate content of the seed the higher the shoot/root ratios of the seedlings. The shoot/root ratios in seedlings ranged from 2.01 for low-protein wheat to 8.37 for sunflower. In the case of high-protein corn and high-protein wheat, the shoot/root ratios are higher than the corresponding types that are low in protein. It was also found that with the exception of cotton seedlings, the shoot/root ratios of all types were increased by the addition of nitrates, but that the seedlings of *Leguminosae* were the least responsive in this respect. Reid makes the significant statement that it is apparently inadvisable to draw conclusions as to the effect of mineral nutrients upon the subsequent growth of plants when the observations are based on growth responses during the first 3 weeks of growth only. Before studying the effect of an element or compound on growth it would therefore seem desirable to conduct experiments in which the storage supply of the elements or compound in question has been at least partially exhausted by the seedling.

F. Significance in Agricultural Practice

The investigations of the carbon/nitrogen ratio indicate its importance upon the vegetative growth and reproduction of plants. When, however, an attempt is made to formulate methods of agricultural practice based upon the carbon/nitrogen ratio, the complexity of the problem and the lack of definite information on this subject become evident. Little or nothing is known concerning the question of cumulative and reciprocal influences or the effect of the length of the growing season as influenced by weather upon the relations of carbon to nitrogen. In this regard it is necessary to consider not only the accumulation of reserves but also the relative rates of nitrogen absorption and of carbohydrate synthesis at the critical periods of growth (Knight, 1924). Hooker (1925) enumerated various ways in which the carbohydrate/nitrogen relations may be varied. Thus the carbohydrate content may be increased by girdling, by thinning, and by the application of nitrogen at certain times. It can be decreased by shading, by shortening the daily period of illumination, by defoliation, by heading, and by manuring with nitrogen. The nitrogen content, on the other hand, can be increased by manuring, by shading, by defoliation, and by certain types of pruning. It may be decreased by girdling or defoliation early in the season or by local suppression of growth through pruning, and in some cases the insoluble nitrogen may be increased by shortening the day. The nitrogen content may also be decreased by sod or intercrops.

In conclusion it may be stated that the facts that have been observed concerning the carbon/nitrogen ratio have thrown light only on many established practices, as, for example, manuring, pruning, and girdling, and have thus shown that certain fairly definite physiological processes occur as a result of these practices. The knowledge concerning the carbon/nitrogen ratio is thus at present not sufficient to formulate any methods of agricultural practice other than those which have long been established. The information that has been obtained in regard to the carbon/nitrogen ratio has helped to a certain extent to account for some of the conflicting results that have been obtained from certain agricultural practices. It is now known that the results obtained from agricultural practices will depend upon both the internal and external conditions that prevail both before and after the plant has been treated.

Bibliography

Acqua, C., 1910: The place in which nitrates are utilized by plants, *Exp. Sta. Rec.*, **24**: 531.

Addoms, R. M., and F. C. Mounce, 1931: Notes on the nutrient requirements and the histology of the cranberry (*Vaccinium macrocarpon*) with special reference to mycorrhiza, *Plant Physiol.*, **6**: 653–666.

——— and ———, 1932: Further notes on the nutrient requirements and the histology of the cranberry, with special reference to the sources of nitrogen, *Plant Physiol.*, **7**: 643–656.

Allison, F. E., and H. J. Morris, 1930: Nitrogen fixation by blue-green algae, *Sci.*, **71**: 221–223.

Alway, F. J., and G. H. Nesom, 1930: Protein content of reed canary grass on peat soils, *J. Agr. Res.*, **40**: 297–321.

Ames, J. W., 1910: The composition of wheat. Influence of various factors on the phosphorus, potassium, and nitrogen content of the wheat plant, *Ohio Sta. Bull.* 221.

Anderson, V. L., 1924: Some observations on the nitrate-reducing properties of plants, *Ann. Bot.*, **38**: 699–706.

Armstrong, G. M., and W. B. Albert, 1931: A study of the cotton plant with especial reference to its nitrogen content, *J. Agr. Res.*, **42**: 689–703.

Arrington, L. B., and J. W. Shive, 1935: Rates of absorption of ammonium and nitrate nitrogen from culture solutions by ten-day-old tomato seedlings at two pH levels, *Soil Sci.*, **39**: 431–435.

Ashby, E., 1929: The interaction of factors in the growth of *Lemna*. IV, The influence of minute quantities of organic matter upon growth and reproduction, *Ann. Bot.*, **43**: 805–816.

Aso, K., and T. Sekine, 1914: Nitrites in plants. *Beih. bot. Centrbl.*, **32**: 146–147; *Exp. Sta. Rec.*, **33**: 627.

Atwater, W. O., and E. W. Rockwood, 1886: On the loss of nitrogen by plants during germination and growth, *Am. Chem. J.*, **8**: 327–343.

Baly, E. C. C., I. M. Heilbron, and D. P. Hudson, 1922: Photocatalysis. II, The photosynthesis of nitrogen compounds from nitrates and carbon dioxide, *J. Chem. Soc.* (London), **121**: 1078–1088.

———, ———, and H. J. Stern, 1923: Photocatalysis. III, The photosynthesis of naturally occurring nitrogen compounds from carbon dioxide and ammonia, *J. Chem. Soc.* (London), **123**: 185–197.

Bambacioni, V., 1923: Contribuzione alla conoscenza del luogo di formazione del sostanza organiche azotate nei vegetali, *Atti R. Accad. Lincei Roma Rendiconte Cl. Sci. Mat. e. Nat.*, **32**: 108–110.

———, 1926: Ulteriori osservazioni sul luogo in cui si compie l'organizzazione dell'-azoto nei vegetali superiori, *Ann. Botanica*, **17**: 4–23; *Biol. Abst.*, **2**: 873.

Barbera, G., 1933: Proteine e sostanze azotate non proteiche contenute nel succo di erba medica, *Ann. Chim. Applicata*, **23**: 462–470; *Biol. Abst.*, **10**: 27.

Baudisch, O., 1921: The mechanism of reduction of nitrates and nitrites in process of assimilation, *J. Biol. Chem.*, **48**: 489–502.

———, 1923: On the formation of organic compounds from inorganic by the influence of light, *Sci.*, **57**: 451–457.

Beaumont, A. B., C. J. Larsinos, P. Piekenbrock, and P. R. Nelson, 1931: The assimilation of nitrogen by tobacco, *J. Agr. Res.*, **43**: 559–567.

———, W. S. Eisenmenger, and W. J. Moore, 1933: Assimilation of fixed nitrogen by grasses and clovers, *J. Agr. Res.*, **47**: 495–503.

——— and W. J. Moore, 1933: The absorption and assimilation of ammonic and nitric nitrogen by plants, *Am. Fertilizer*, **79**: 7–9, 24–27.

Bennet-Clark, T. A., 1933: The role of the organic acids in plant metabolism, *New Phytol.*, **32**: 197–230.

Beyerinck, M. W., 1888: Die Bacterien der Papilionaceen-Knöllchen, *Bot. Ztg.*, **46**: 725–735, 741–750, 757–771, 781–790, 797–802.

Björksten, J., 1930: Zur Kenntnis der Synthese von Eiweissstoffen und ihrer Bausteine bei höheren Pflanzen, *Biochem. Z.*, **225**: 1–80.

Blackman, G. E., 1934: The ecological and physiological action of ammonium salts on the clover content of turf, *Ann. Bot.*, **48**: 975–1001.

Boncquet, P. A., 1916: Presence of nitrites and ammonia in diseased plants, *J. Am. Chem. Soc.*, **38**: 2572–2576.

Bottomley, W. B., 1917: Some effects of organic growth promoting substances (auximones) on the growth of *Lemna minor* in mineral culture solutions, *Proc. Roy. Soc.*, B, **89**: 481–508.

———, 1919: The isolation from peat of certain nucleic acid derivatives, *Proc. Roy. Soc.*, B, **90**: 39–44.

———, 1920: The effect of nitrogen-fixing organisms and nucleic acid derivatives on plant growth, *Proc. Roy. Soc.*, B, **91**: 83–95.

———, 1920: The growth of *Lemna* plants in mineral solutions and in their natural medium, *Ann. Bot.*, **34**: 345–352.

———, 1920: The effect of organic matter on the growth of various water plants in cultural solutions, *Ann. Bot.*, **34**: 353–365.

Boussingault, J., 1854–1855: *Ann. Chim. et Phys.*, 3 ser., tomes 41, 43.

———, 1838: Recherches chimiques sur la végétation entreprises dans le but d'examiner si les plantes prennent l'azote à l'atmosphère, *Ann. Chim. et Phys.*, **67**: 5–54.

BOUSSINGAULT, M., 1864: De la végétation dans l'obscurité, *C. R. Acad. Sci. Paris*, **58**: 917–922.

BRENCHLEY, W. E., and A. D. HALL, 1908–1910: The development of the grain of wheat, *J. Agr. Sci.*, **3**: 193–217.

———, 1909: On the strength and development of the grain of wheat (*Triticum vulgare*), *Ann. Bot.*, **23**: 117–140.

———, 1912: The development of the grain of barley, *Ann. Bot.*, **26**: 903–928.

BRIGHAM, R. O., 1917: Assimilation of organic nitrogen by *Zea mays* and the influence of *Bacillus subtilis* on such assimilation, *Soil Sci.*, **3**: 155–197.

BRISTOL, B. M., and H. J. PAGE, 1923: A critical enquiry into the alleged fixation of nitrogen by green algae, *Ann. App. Biol.*, **10**: 378–409.

BROWN, R., 1933: Nitrogen fixation in the genus *Lolium*, *Nature*, **131**: 169–170.

BURK, D., 1927: Does the pea plant fix atmospheric nitrogen? *Plant Physiol.*, **2**: 83–90.

BURKE, E., 1925: The influence of nitrate nitrogen upon the protein content and yield of wheat, *J. Agr. Res.*, **31**: 1189–1199.

BURKHART, L., 1934: Metabolism of etiolated seedlings as affected by ammonium nutrition, *Plant Physiol.*, **9**: 351–358.

BURRELL, R. C., 1926: Effect of certain deficiencies on nitrogen metabolism of plants, *Bot. Gaz.*, **82**: 320–328.

CAMERON, S. H., and D. APPLEMAN, 1933: The distribution of total nitrogen in the orange trees, *Proc. Am. Soc. Hort. Sci.*, **30**: 341–348.

CAMPBELL, E. G., 1924: Nitrogen content of weeds, *Bot. Gaz.*, **78**: 103–115.

CHANDLER, W. H., 1919: Some results as to the response of fruit trees to pruning, *Proc. Am. Soc. Hort. Sci.*, **16**: 88–101.

CHANNON, H. J., and C. A. M. FOSTER, 1934: The phosphatides of the wheat germ, *Biochem. J.*, **28**: 853–864.

CHAPMAN, A. G., 1933: Some effects of varying amounts of nitrogen on the growth of tulip poplar seedlings, *Ohio J. Sci.*, **33**: 164–181.

CHIBNALL, A. C., and S. B. SCHRYVER, 1920: The isolation of proteins from leaves, *Proc. Physiol. Soc.*, *J. Physiol.*, **54**: 1–2.

——— and ———, 1921: Isolation of proteins from leaves, *Biochem. J.*, **15**: 60–75.

———, 1922: The distribution of nitrogen in the leaves of the runner bean, *Biochem. J.*, **16**: 344–362.

———, 1922: The effect of low-temperature drying on the distribution of nitrogen in the leaves of the runner bean, *Biochem. J.*, **16**: 599–607.

———, 1922: Distribution of nitrogen in the dead leaves of the runner bean, *Biochem. J.*, **16**: 608–610.

———, 1923: A new method for the separate extraction of vacuole and protoplasmic material from the leaf cells, *J. Biol. Chem.*, **55**: 333–342.

———, 1923: Diurnal variations in the total nitrogen content of foliage leaves, *Ann. Bot.*, **37**: 511–518.

———, 1924: Investigation on the nitrogenous metabolism of the higher plants. V, Diurnal variations in the protein nitrogen of runner bean leaves, *Biochem. J.*, **18**: 387–394.

———, 1924: The role of asparagine in the metabolism of the mature plant, *Biochem. J.*, **18**: 395–404.

———, 1924: Leaf protein metabolism in normal and abnormal runner bean plants, *Biochem. J.*, **18**: 405–407.

——— and C. E. GROVER, 1926: A chemical study of leaf cell cytoplasm. I, The soluble proteins, *Biochem. J.*, **20**: 108–118.

———, 1926: Leaf cytoplasmic proteins, *J. Am. Chem. Soc.*, **48**: 728–732.

———, E. J. MILLER, D. H. HALL, and R. G. WESTALL, 1933: The proteins of grasses. II, A new method of preparation, *Biochem. J.*, **27**: 1879–1884.

CHRAPOWICKI, W., 1889: Beobachtungen über die Eiweissbildung in den chlorophyllführenden Pflanzen, *Arbeiten St. Petersburger Natur. Ges.*, **18**: 1–27.

CIAMICIAN, G., and C. RAVENNA, 1919 and 1920: Sulla influenza di alcune sostanze organiche sullo sviluppo della piante, *Atti R. Accad. Lincei Rend*, **28**: 13–20; **29**: 7–13; *Bot. Abst.*, **6**: 194; **7**: 309.

——— and ———, 1919 and 1920: Sul contegno di alcune sostanze organiche nei vegetali, *Gaz. Chim. Italiana*, **49**: 83–126; **50**: 13–46; *Bot. Abst.*, **5**: 292; **7**: 308.

CLARK, H. E., and J. W. SHIVE, 1934: The influence of the pH of a culture solution on the rates of absorption of ammonium and nitrate nitrogen by the tomato plant, *Soil Sci.*, **37**: 203–226.

——— and ———, 1934: The influence of the pH of a culture solution on the assimilation of ammonium and nitrate nitrogen by the tomato plant, *Soil Sci.*, **37**: 459–476.

———, 1936: Effect of ammonium and of nitrate nitrogen on the composition of the tomato plant, *Plant Physiol.*, **11**: 5–24.

CLARK, J. H., 1933: Reaction of the nutrient medium as affecting growth of strawberry plants, *Proc. Am. Soc. Hort. Sci.*, **30**: 283–287.

CLARK, N. A., 1924: The soil organic matter and growth-promoting accessory substances, *Ind. and Eng. Chem.*, **16**: 249–250.

——— and E. M. ROLLER, 1924: "Auximones" and the growth of the green plant, *Soil Sci.*, **17**: 193–199.

———, 1930: "Auximones" and the stimulation of *Lemna* by organic matter, *Sci.*, **71**: 268–269.

——— and E. M. ROLLER, 1931: The stimulation of *Lemna major* by organic matter under sterile and non sterile conditions, *Soil Sci.*, **31**: 299–310.

COHN, E. J., J. GROSS, and O. C. JOHNSON, 1919: The isoelectric points of the proteins in certain vegetable juices, *J. Gen. Physiol.*, **2**: 145–160.

COLBY, H. L., 1933: Seasonal absorption of nutrient salts by the French prune grown in solution culture, *Plant Physiol.*, **8**: 1–34.

CONRAD, J. P., 1934: Physiological acidity and alkalinity of inorganic nitrogenous compounds in solution cultures, *J. Am. Soc. Agron.*, **26**: 364–372.

COOK, R. L., 1930: Effect of soil type on the nitrate content of the expressed sap and the total nitrogen content of the tissue of the small grains, *J. Am. Soc. Agron.*, **22**: 393–409.

COONS, G. H., and L. J. KLOTZ, 1925: The nitrogen constituents of celery plants in health and disease, *J. Agr. Res.*, **31**: 287–300.

COPELAND, J. J., 1932: Nitrogen fixation by *Myxophyceae*, abst., *Am. J. Bot.*, **19**: 844.

CORDINGLEY, H., J. GRAINGER, W. H. PEARSALL, and A. WRIGHT, 1934: The effect of mosaic disease upon certain metabolic products in the tobacco plant, *Ann. App. Biol.*, **21**: 78–89.

COVILLE, F. V., 1926: Buttermilk as a fertilizer for blueberries, *Sci.*, **64**: 94–96.

CROWTHER, E. M., 1925: A note on the availability of organic nitrogen compounds in pot experiments, *J. Agr. Sci.*, **15**: 300–302.

———, 1935: Comparative trials of calcium cyanamide and other nitrogenous fertilizers on arable crops, *Empire J. Exp. Agr.*, **3**: 129–143.

CROWTHER, F., 1914: Studies in growth analysis of the cotton plant under irrigation in the Sudan. I, The effects of different combinations of nitrogen applications and water supply, *Ann. Bot.*, **48**: 877–913.

Csonka, F. A., J. S. Murphy, and D. B. Jones, 1926: The isoelectric points of various proteins, *J. Am. Chem. Soc.*, **48**: 763–768.

——— and D. B. Jones, 1934: The cystine, tryptophane and tyrosine content of the soybean, *J. Agr. Res.*, **49**: 279–282.

Curtis, O. F., 1918: Stimulation of root growth in cuttings by treatment with chemical compounds, *N. Y. Sta. (Cornell) Mem.* 14.

Dachnowski, A., 1914: The effects of acid and alkaline solutions upon the water relation and the metabolism of plants, *Am. J. Bot.*, **1**: 412–439.

Daikuhara, G., 1905: On the application of magnesium in the form of magnesium sulphate for the needs of the rice plant, *Bull. Imp. Cent. Agr. Expt. Sta. Japan*, **1**: 24–29.

Das, U. K., 1936: Nitrogen nutrition of sugar cane, *Plant Physiol.*, **11**: 251–317.

Dastur, R. H., and T. J. Malkani, 1933: The intake of nitrogen by the rice plant (*Oryza sativa*), *Indian J. Agr. Sci.*, **3**: 157–206; *Biol. Abst.*, **9**: 43; *Exp. Sta. Rec.*, **69**: 796.

Davidson, J., and J. A. LeClerc, 1917, 1918: The effect of sodium nitrate applied at different stages of growth on the yield, composition, and quality of wheat. I, II, *J. Am. Soc. Agron.*, **9**: 145–154; **10**: 193–198.

———, 1922: The effect of nitrates applied at different stages of growth on the yield, composition, and quality of wheat, *J. Am. Soc. Agron.*, **14**: 118–122.

———, 1923: Is gaseous nitrogen a product of seedling metabolism? *Bot. Gaz.*, **76**: 95–102.

——— and J. A. LeClerc, 1923: Effect of various inorganic nitrogen compounds, applied at different stages of growth on the yield, composition, and quality of wheat, *J. Agr. Res.*, **23**: 55–69.

———, 1926: Changes in nitrogen, potassium, and phosphorus content of wheat seedlings during germination and early stages, *Bot. Gaz.*, **81**: 87–95.

Davidson, O. W., and J. W. Shive, 1934: The influence of the hydrogen ion concentration of the culture solution upon the absorption and assimilation of nitrate and ammonium nitrogen by peach trees grown in sand cultures, *Soil Sci.*, **37**: 357–386.

Davies, P. A., 1931: Distribution of total nitrogen in regeneration of the willow, *Bot. Gaz.*, **91**: 320–326.

Denny, F. E., 1933: Bases for calculations in measuring changes in leaves during the night, *Contr. Boyce Thompson Inst.*, **5**: 181–194.

Dewiche, E. J., and W. E. Tottingham, 1930: Effect of climate on nitrogen content of maize, barley and red clover, *J. Am. Soc. Agron.*, **22**: 681–689.

Dhar, N. R., A. K. Bhattacharya, and N. N. Biswas, 1933: Photonitrification of soil, *Soil Sci.*, **35**: 281–284.

——— and S. K. Mukherjee, 1934: Photosynthesis of amino acids *in vitro*, *Nature*, **134**: 499.

Dittrich, W., 1930: Zur Physiologie des Nitratumsatzes in höheren Pflanzen (unter besonderer Berücksichtigung der Nitratspeicherung), *Planta*, **12**: 69–119.

Domontovich, Shestakov, and Polossin, 1933: Ueber die Änderung der Wasserstoffionenkonzentration der Nährlösung durch die Tätigkeit der Lupinenwurzeln, *Abh. Sowiet Sekt., Int. Bodenk. Gesells. K.*, **4**: 214. Quoted by Nicol, 1934.

Doneen, L. D., 1934: Nitrogen in relation to composition, growth and yield of wheat, *Wash. Sta. Bull.* 296.

Drewes, K., 1928: Über die Assimilation des Luftstickstoffs durch Blaualgen, *Cent. Bakt. Par., Abt.* 2, **76**: 88–101.

Dunlap, A. A., 1930: The total nitrogen and carbohydrates, and the relative rates of respiration in virus infected plants, *Am. J. Bot.*, **17**: 348–357.

Eckerson, S., 1917: Microchemical studies in the progressive development of the wheat plant, *Wash. Sta. Bull.* 139.

———, 1924: Protein synthesis by Plants. I, Nitrate reduction, *Bot. Gaz.*, **77**: 377–390.

———, 1930: Seasonal distribution of reductase in the various organs of an apple tree, *Am. J. Bot.*, **17**: 1045–1046.

———, 1931: Seasonal distribution of reductase in the various organs of an apple tree, *Contr. Boyce Thompson Inst.*, **3**: 405–412.

———, 1932: Conditions affecting nitrate reduction by plants, *Contr. Boyce Thompson Inst.*, **4**: 119–130.

Eggleton, W. G. E., 1935: The assimilation of inorganic nitrogenous salts, including sodium nitrite by the grass plant, *Biochem. J.*, **29**: 1389–1397.

Eisenmenger, W. S., 1933: The distribution of nitrogen in tobacco when the supplies of nitrogen and of light are varied during the growing period, *J. Agr. Res.*, **46**: 255–266.

Ellett, W. B., H. H. Hill, and W. G. Harris, 1915: The effect of association of legumes and non legumes, *Va. Sta. Tech. Bull.* 1.

Emde, H., 1929: Alkaloid und Kohlensäureassimilation, *Naturwiss.*, **17**: 699–703; *Biol. Abst.*, **5**: 2666.

Emmerling, A., 1880: Studien über die Eiweissbildung in der Pflanze. I, *Landw. Versuchs-Stat.*, **24**: 113–160.

———, 1887: Studien über die Eiweissbildung in der Pflanze. VII, *Landw. Versuchs-Stat.*, **34**: 1–91.

———, 1900: Studien über die Eiweissbildung in der Pflanze, *Landw. Versuchs-Stat.*, **54**: 215–281.

Emmert, E. M., 1929: The determination of nitrate in green tomato and lettuce tissues, *Plant Physiol.*, **4**: 519–528.

———, 1936: Effect of drouth on the nutrient levels in the tomato plant, *Soil Sci.*, **41**: 67–70.

Engel, H., 1929: Beiträge zur Kenntnis des Stickstoffumsatzes grüner Pflanzen, *Planta*, **7**: 133–163; *Biol. Abst.*, **5**: 1616.

Enlow, C. R., and J. M. Coleman, 1929: Increasing the protein content of pasture grasses by frequent light applications of nitrogen, *J. Am. Soc. Agron.*, **21**: 845–853.

Espino, R. B., 1920: Some aspects of the salt requirements of young rice plants, *Philippine J. Sci.*, **16**: 455–525.

Evans, M. V., 1916: Some effects of legumes on associated non legumes, *J. Am. Soc. Agron.*, **8**: 348–357.

Fagan, T. W., 1925: The variation in the moisture and nitrogen content of the potato during growth and storage, *Welsh J. Agr.*, **1**: 110–115; *Bot. Abst.*, **14**: 1441.

Feher, D., and I. Vagi, 1924, 1926: Untersuchungen über die Einwirkung von Nitriten auf das Wachstum der Pflanzen, *Biochem. Z.*, **153**: 156–158; **174**: 262–270.

Finch, A. H., 1935: Physiology of apple varieties, *Plant Physiol.*, **10**: 49–72.

Fischer, E., 1899–1906: "Untersuchungen über Aminosäuren, Polypeptide, und Proteine," Julius Springer, Berlin, 770 pp.

Fisher, H., 1916: Zur Frage der Kohlensäureernährung der Pflanze, *Gartenflora*, **65**: 232–237.

———, 1917: Das Problem der Stickstoffbindung, Festlegung des Luftstickstoffs bei niederen Pflanzen, *Ber. bot. Ges.*, **35**: 423–454.

FISHER, P. L., 1935: Response of the tomato in solution cultures with deficiencies and excesses of certain essential elements, *Md. Sta. Bull.* 375.

FOSSE, M. R., 1912: Recherches sur l'urée, *C. R. Acad. Sci. Paris*, **155**: 851–852.

———, 1913: Formation de l'urée par deux moisissures, *C. R. Acad. Sci. Paris*, **156**: 263–265.

———, 1913: Formation de l'urée par les végétaux supérieurs, *C. R. Acad. Sci. Paris*, **156**: 567–568.

———, 1913: Recherche de l'urée dans les végétaux, *C. R. Acad. Sci. Paris*, **156**: 1938–1941.

———, 1914: Présence simultanée de l'urée et de l'uréase dans le même végétal, *C. R. Acad. Sci. Paris*, **158**: 1374–1376.

FRANK, B., 1888: Untersuchungen über die Ernährung der Pflanze mit Stickstoff und über den Kreislauf desselben in der Landwirthschaft, *Landw. Jahrb.*, **17**: 421–553.

———, 1889: Ueber den experimentellen Nachweis der Assimilation freien Stickstoffs durch erdbodenbewohnende Algen, *Ber. bot. Ges.*, **7**: 34–42.

——— and R. OTTO, 1891: New experiments concerning the assimilation of nitrogen by plants, *Exp. Sta. Rec.*, **3**: 64.

FREAR, D. E., 1930: The clarification of plant juices: Nitrate concentration in large and small leaves, *Sci.*, **72**: 634–635.

———, 1930: Estimation of the nitrate nitrogen in plant juice: A study of the expression and clarification of the juice, *Plant Physiol.*, **5**: 359–371.

———, 1931: A negative correlation observed between the nitrate nitrogen in the juice of beet leaves and the weight of the leaves, *J. Agr. Res.*, **42**: 53–56.

FUJITA, T., 1931: Über die Beziehungen zwischen dem Eiweissgehalt der Blätter und dem Grad der Sonnenbeleuchtung, *Bull. Kjusu. Imp. Univ. Fukuoka*, **4**: 358–368; *Biol. Abst.*, **6**: 2515.

GAINEY, P. L., and M. C. SEWELL, 1932: The role of nitrogen in the production of wheat fields, *J. Agr. Res.*, **45**: 129–148.

GARDNER, V. R., 1923: Studies in the nutrition of the strawberry, *Mo. Sta. Res. Bull.* 57.

GARNER, W. W., and H. A. ALLARD, 1920: Effect of the relative length of day and night and other factors of the environment on growth and reproduction in plants, *J. Agr. Res.*, **18**: 553–607.

GAUTRERET, R., 1934: Recherches sur la réduction du nitrate d'argent par les chloroplastes, *C. R. Acad. Sci. Paris*, **198**: 1252–1254; *Biol. Abst.*, **9**: 1012.

GEORGI, C. E., 1935: Influence of the carbohydrate-nitrogen relation on nodule production by red clover, *J. Agr. Res.*, **51**: 597–612.

GERICKE, W. F., 1920: On the protein content of wheat, *Sci.*, **52**: 446–447.

———, 1922: Certain relations between the protein content of wheat and the length of the growing period of the head-bearing stalks, *Soil Sci.*, **13**: 135–139.

———, 1922: Differences effected in the protein content of grain by applications of nitrogen made at different growing periods of the plants, *Soil Sci.*, **14**: 103–109.

———, 1922: Certain relations between root development and tillering in wheat; significance in the production of high-protein wheat, *Am. J. Bot.*, **9**: 366–369.

———, 1927: Why applications of nitrogen to land may cause either increase or decrease in the protein content of wheat, *J. Agr. Res.*, **35**: 133–141.

———, 1930: Variation in the percentage of protein in the grain of a single wheat plant, *Sci.*, **71**: 73–74.

GERLACH, M., and J. VOGEL, 1905: Ammoniakstickstoff als Pflanzennahrstoff, *Cent. Bakt. und Parasit.* II, **14**: 124–138.

GILE, P. L., 1935: Absorption of nitrates by corn in the dark, *Sci.*, **81**: 520–521.

GOURLEY, J. H., 1930: The relation of nitrogen to the storage quality of fruit, *Sci. Agr.*, **11**: 130–136.

GOUWENTAK, C. A., 1929: Untersuchungen über den N-Stoffwechsel bei *Helianthus annuus*, *Rec. trav. bot. Néerl.*, **26**: 19–96.

GRĂCANIN, M., 1932: Forschungen über die Beziehung zwischen Langenwachstum der Wurzeln und dem Stickstoffgehalt im Nährsubstrat, *Arbeiten der südslav. Akad. Wiss. u. Kunst.*, Bd. 244, Zagreb.

GREAVES, J. E., and E. G. CARTER, 1923: The influence of irrigation water on the composition of grains and the relationship to nutrition, *J. Biol. Chem.*, **58**: 531–541.

——— and D. H. NELSON, 1925: The influence of irrigation water and manure on the composition of the corn kernel, *J. Agr. Res.*, **31**: 183–191.

GREENHILL, A. W., and A. C. CHIBNALL, 1934: The exudation of glutamine from perennial rye grass, *Biochem. J.*, **28**: 1422–1427.

GRIFFITHS, A., 1891: On the direct absorption of ammoniacal salts by certain plants, *Chem. News*, **64**: 147.

GURJAR, A. M., 1920: Carbon nitrogen ratio in relation to plant metabolism, *Sci.*, **51**: 351–352.

HAAS, A. R. C., 1929: Effect of nitrate salt upon the growth and composition of tobacco leaves, *Bot. Gaz.*, **88**: 96–102.

HAAS, P., and T. G. HILL, 1913: "An Introduction to the Chemistry of Plant Products," Longmans, Green and Company, New York, 401 pp.

HAMNER, K. C., 1936: Effects of nitrogen supply on rates of photosynthesis and respiration in plants, *Bot. Gaz.*, **97**: 744–764.

HANSTEEN, B., 1896: Beiträge zur Kenntniss der Eiweissbildung und der Bedingungen der Realisirung dieses Processes in phanerogamen Pflanzenkörper, *Ber. bot. Ges.*, **14**: 362–371.

HARLAN, H. V., 1920: Daily development of kernels of Hannchen barley from flowering to maturity at Aberdeen, Idaho, *J. Agr. Res.*, **19**: 393–429.

HARLAN, J. D., and R. C. COLLISON, 1933: Experiments with commercial nitrogenous fertilizers on apple orchards, *N. Y. Sta. (Geneva) Bull.* 623.

HARLEY, C. P., M. P. MASURE, and J. R. MAGNESS, 1932: Effects of leaf area, nitrate of soda, and soil moisture on fruit bud formation in the Delicious apple, *Proc. Am. Soc. Hort. Sci.*, **29**: 193–198.

HARRISON, C. M., 1934: Responses of Kentucky blue grass to variations in temperature, light, cutting and fertilizing, *Plant Physiol.*, **9**: 83–106.

HARTLEY, K. T., and M. GREENWOOD, 1933: The effect of small applications of farm yard manure on the yield of cereals in Nigeria, *Empire J. Exp. Agr.*, **1**: 113–121.

HARVEY, E. M., and A. E. MURNEEK, 1921: The relation of carbohydrates and nitrogen to the behavior of apple spurs, *Ore. Sta. Bull.* 176.

———, 1923: A study of growth in summer shoots of the apple with special consideration of the rôle of carbohydrates and nitrogen, *Ore. Sta. Bull.* 200.

HEADDEN, W. P., 1915: Yellowberry in wheat, its cause and prevention, *Colo. Sta. Bull.* 205.

HELLRIEGEL, H., 1855: Beiträge zur Keimungsgeschichte der ölgebenden Samen, *J. Prakt. Chem.*, **64**: 94–107.

——— and H. WILFARTH, 1888: Untersuchungen über die Stickstoffnahrung der *Gramineen* und *Leguminosen*, *Beilage Z. Rüben zucker-Ind. Reich*, 234 pp.

HICKS, P. A., 1928: The carbohydrate/nitrogen ratio in the wheat plant, *New Phytol.*, **27**: 1–46.

———, 1928: Distribution of carbon/nitrogen ratio in the various organs of the wheat plant at different periods of its life history, *New Phytol.*, **27**: 108–116.

———, 1928: Chemistry of growth as represented by carbon/nitrogen ratio, *Bot. Gaz.*, **86**: 193–209.

Hill, H. H., 1930: Plant juice clarification for nitrate nitrogen determinations, *Sci.*, **71**: 540.

Hofman, F. W., 1930: Time of applying nitrogen to apple trees, *Proc. Am. Soc. Hort. Sci.*, **27**: 19–22.

Holley, K. T., T. A. Pickett, and T. G. Dulin, 1931: A study of ammonia and nitrate nitrogen for cotton. I, Influence on absorption of other elements, *Ga. Sta. Bull.* 169.

Hooker, H. D., 1920: Seasonal changes in the chemical composition of apple spurs, *Mo. Sta. Res. Bull.* 40.

———, 1922: Certain responses of apple trees to nitrogen applications of different kinds and at different seasons, *Mo. Sta. Res. Bull.* 50.

———, 1922: Some effects of fall application of nitrogen to apple trees, *Proc. Am. Soc. Hort. Sci.*, **19**: 241–243.

———, 1924: Changes produced in apple trees by various types of pruning, *Mo. Sta. Res. Bull.* 72.

———, 1925: A survey of investigations by American horticulturists on carbohydrate-nitrogen relations, *J. Pom. and Hort. Sci.*, **5**: 34–42.

Hopkins, E. F., and J. H. Gourley, 1930: The effect of nitrate applications on the soluble carbohydrate in apples, *Proc. Am. Soc. Hort. Sci.*, **27**: 32–36.

——— and E. W. Greve, 1931: The effect of nitrate applications on the soluble carbohydrates in apples, *Proc. Am. Soc. Hort. Sci.*, **28**: 501–506.

Hopkins, E. W., P. W. Wilson, and W. H. Peterson, 1932: Influence of potassium nitrate on nodule formation and nitrogen fixation by clover, *Plant Physiol.*, **7**: 597–611.

Hopkins, J. W., 1935: Influence of weather conditions on the nitrogen content of wheat, *Can. J. Res.*, **12**: 228–237.

Howlett, F. S., 1923: Nitrogen and carbohydrate composition of the developing flowers and young fruits of the apple, *Proc. Am. Soc. Hort. Sci.*, **20**: 31–37; *N. Y. Sta. (Cornell) Mem.* 99.

Hulme, A. C., 1936: Biochemical studies in the nitrogen metabolism of the apple fruit. II, The course followed by certain nitrogen fractions during the development of the fruit on the tree, *Biochem. J.*, **30**: 258–268.

Hutchinson, H. B., and N. H. J. Miller, 1909: Direct assimilation of ammonium salts by plants, *J. Agr. Sci.*, **3**: 179–195.

——— and ———, 1911: The direct assimilation of inorganic and organic forms of nitrogen by higher plants, *Cent. Bakt.* II, *Abt.*, **30**: 513–548.

Irving, A. A., and R. Haukinson, 1908: The presence of a nitrate reducing enzyme in green plants, *Biochem. J.*, **3**: 87–96.

Ivanov, N. N., 1927: On the stability of the chemical composition of leguminous plants and corn, *Bull. App. Bot. and Plant Breeding*, **17**: 225–259; *Biol. Abst.*, **4**: 694.

Iwanoff, M., 1901: Versuche über die Frage, ob in den Pflanzen bei Lichtabschluss Eiweissstoffe sich bilden, *Landw. Versuchs-Stat.*, **55**: 78–94.

Jodidi, S. L., E. H. Kellogg, and R. H. True, 1918: Nitrogen metabolism in normal and in blighted spinach, *J. Agr. Res.*, **15**: 385–404.

———, S. C. Moulton, and K. S. Markley, 1920: The mosaic disease of spinach as characterized by its nitrogen constituents, *J. Am. Chem. Soc.*, **42**: 1061–1070.

———, 1920: A mosaic disease of cabbage as revealed by its nitrogen constituents, *J. Am. Chem. Soc.*, **42**: 1883–1893.

——— and K. S. Markley, 1923: The occurrence of polypeptides and free amino acids in the ungerminated wheat kernel, *J. Am. Chem. Soc.*, **45**: 2137–2144.

———, 1924: Physiological studies on cereals. II, The occurrence of amino acids and polypeptides in the ungerminated oat kernel, *J. Franklin Inst.*, **198**: 201–211.

———, 1925: Physiological studies on cereals. III, The occurrence of polypeptides and amino acids in the ungerminated maize kernel, *J. Agr. Res.*, **30**: 587–592.

——— and J. G. Wangler, 1925: Physiological and biochemical studies on cereals. IV, On the presence of amino acids and polypeptides in the ungerminated rye kernel, *J. Agr. Res.*, **30**: 989–994.

———, 1930: Isolation and purification of the alcohol-soluble protein (prolamin) occurring in English rye grass (*Lolium perenne*), *J. Agr. Res.*, **40**: 361–370.

———, 1931: The non protein nitrogen of the Alaska pea with special reference to the chemical nature of humin nitrogen, *J. Agr. Res.*, **43**: 811–825.

Jones, J. S., and G. A. Mitchell, 1926: The cause and control of yellowberry in Turkey wheat grown under dry farming conditions, *J. Agr. Res.*, **33**: 281–293.

Jones, L. H., 1936: Diagnosis of plant troubles with diphenylamine, *Plant Physiol.*, **11**: 207–209.

Karmarkar, D. V., 1934: The seasonal cycles of nitrogenous and carbohydrate materials in fruit trees, *J. Pom. and Hort. Sci.*, **12**: 177–221.

Kastle, J. H., and E. Elvove, 1904: On the reduction of nitrates by certain plant extracts and metals, and the accelerating effect of certain substances on the progress of the reduction, *Am. Chem. J.*, **31**: 606–641.

Kelley, W. P., 1911: The assimilation of nitrogen by rice, *Hawaii Sta. Bull.* 24.

Kellner, O., 1884: Agriculturchemische Studien über die Reiscultur, *Landw. Versuchs-Stat.*, **30**: 18–41.

Kezer, A., and D. W. Robertson, 1927: The critical period of applying irrigation water to wheat, *J. Am. Soc. Agron.*, **19**: 80–116.

Kinoshita, Y., 1897: On the assimilation of nitrogen from nitrates and ammonium salts by phanerogams, *Bull. Coll. Agr. Tokyo*, **2**: 200–202.

Kishi, Y., and S. Monobe, 1931: On the quantitative changes of the chemical constituents of mulberry leaves during 24 hours. (In Japanese.) *Kjusu. Imp. Univ. Fukuoka Bull.*, **4**: 531–599; *Biol. Abst.*, **8**: 335.

Klein, G., and K. Tauböck, 1927: Physiologie des Harnstoffs in der höheren Pflanze. II, *Österr. Bot. Z.*, **76**: 195–221; *Biol. Abst.*, **6**: 693.

———, ———, and H. Linser, 1930: Harnstoff und Ureide bei den höheren Pflanzen. I, Das Vorkommen von Harnstoff im Pflanzenreich und sein Wandel im Laufe der Vegetationsperiode, *Jahrb. wiss. Bot.*, **73**: 193–225.

———, 1930: Changes in the nitrogen of green plants, *Ergeb. Agr.-Chem.*, **2**: 143; *Chem. Abst.*, **26**: 2484.

——— and K. Tauböck, 1931: Harnstoff und Ureide bei den höheren Pflanzen. II, Untersuchungen über das Auftreten und Verschwinden von Harnstoff bei der Pflanze, *Jahrb. wiss. Bot.*, **74**: 429–458.

——— and ———, 1931: Harnstoff und Ureide bei den höheren Pflanzen. III, Das Vorkommen von Ureiden. Quantitative Bestimmung von freiem und gebundenem Harnstoff, *Biochem. Z.*, **241**: 413–459; *Biol. Abst.*, **8**: 1919.

——— and ———, 1932: Argininstoffwechsel und Harnstoffgenese bei höheren Pflanzen, *Biochem. Z.*, **251**: 10–50.

——— and ———, 1932: Argininstoffwechsel und Harnstoffgenese bei höheren Pflanzen. II, *Biochem. Z.*, **255**: 278–286.

KNIGHT, R. C., 1924: The carbohydrate/nitrogen ratio, *Sci. Prog.*, **19**: 34–42.

KNOWLTON, H. E., and M. B. HOFFMAN, 1930: Nitrogen fertilization and keeping quality of apples, *Proc. Am. Soc. Hort. Sci.*, **27**: 28–31.

KNUDSON, L., 1916: Influence of certain carbohydrates on green plants, *N. Y. Sta. (Cornell) Mem.* 9.

———, 1933: Investigation on the use of certain amino acids by green plants, abst., *Am. J. Bot.*, **20**: 680.

KOSSOWITSCH, P., 1894: Untersuchungen über die Frage ob die Algen freien Stickstoff fixiren, *Bot. Ztg.*, **52**: 97–116.

KOSUTANY, T., 1897: Untersuchungen über die Entstehung des Pflanzeneiweisses, *Landw. Versuchs-Stat.*, **48**: 13–32.

KRAUS, E. J., and H. R. KRAYBILL, 1918: Vegetation and reproduction with special reference to the tomato, *Ore. Sta. Bull.* 149.

KRAYBILL, H. R., 1923: Effect of shading and ringing upon the chemical composition of apple and peach trees, *N. H. Sta. Tech. Bull.* 23.

———, 1932: Effect of plant nutrition on the composition of wheat, *Cereal Chem.*, **9**: 71–80.

KRÜGER, W., 1905: Ueber die Bedeutung der Nitrifikation für die Kulturpflanzen, *Landw. Jahrb.*, **34**: 761.

KULTZSCHER, M., 1931, 1932: Die biologische NH_3-Entgiftung in höheren Pflanzen in ihrer Abhängigkeit von der Wasserstoffionenkonzentration des Zellsaftes, *Planta*, **14**: 782–783; **17**: 699–757.

LAGASSÉ, F. S., 1935: Some responses of Yellow Transparent apple trees in Delaware to various nitrogen treatments, *Del. Sta. Bull.* 195.

LAKON, G., 1917: Der Eiweissgehalt panachierter Blätter geprüft mittels des makroskopischen Verfahrens von Molisch, *Biochem. Z.*, **78**: 145–154.

LATSHAW, W. L., and E. C. MILLER, 1924: Elemental composition of the corn plant, *J. Agr. Res.*, **27**: 845–861.

LAURENT, E., 1890: Expériences sur la réduction des nitrates par les végétaux, *Ann. Inst. Pasteur*, **4**: 722–744.

LAWES, J. B., J. H. GILBERT, and E. PUGH, 1860: On the sources of nitrogen of vegetation; with special reference to the question whether plants assimilate free or uncombined nitrogen, *Phil. Trans. Roy. Soc.*, **151**: 431–578.

LEASE, E. J., and W. E. TOTTINGHAM, 1935: Photochemical responses of the wheat plant to spectral regions, *J. Am. Chem. Soc.*, **57**: 2613–2616.

LEATHER, J. W., 1934: Final report of the agricultural chemist to the government of India. 99 pp. Quoted by Nicol, 1934.

LECLERC, J. A., 1910: Tri-local experiments on the influence of environment on the composition of wheat, *U. S. Dept. Agr. Bull.*, **128**: 7–18.

———, and P. A. YODER, 1914: Environmental influences on the physical and chemical characteristics of wheat, *J. Agr. Res.*, **1**: 275–291.

LEWIS, A. H., 1936: The effects of ammonium and nitrate nitrogen on the growth of perennial rye grass, *J. Agr. Sci.*, **26**: 247–257.

LI, H. W., and T. N. LIU, 1935: Defoliation experiments with kaoliang, *J. Am. Soc. Agron.*, **27**: 486–491.

LINCOLN, F. B., and A. S. MULAY, 1929: The extraction of nitrogenous materials from pear tissues, *Plant Physiol.*, **4**: 233–250.

LIPMAN, C. B., and J. K. TAYLOR, 1922: Proof of the power of the wheat plant to fix atmospheric nitrogen, *Sci.*, **56**: 605–606.

——— and ———, 1924: Do green plants have the power of fixing elementary nitrogen from the atmosphere? *J. Franklin Inst.*, October, **1924**: 475–506.

LIPMAN, J. G., 1910: A method for the study of soil fertility problems, *J. Agr. Sci.*, **3**: 297–300.

———, 1912: The associative growth of legumes and non-legumes, *N. J. Sta. Bull.* 253.

———, 1913: A further discussion of certain methods used in the study of "The associative growth of legumes and non legumes," *J. Am. Soc. Agron.*, **5**: 72–79.

——— and A. W. BLAIR, 1916: Factors influencing the protein content of soybeans, *Soil Sci.*, **1**: 171–178.

LOEB, J., 1924: "Proteins and the Theory of Colloidal Behavior," McGraw-Hill Book Company, Inc., New York, 380 pp.

LOEHWING, W. F., 1927: Preliminary report on amino acid synthesis in plants, *Proc. Iowa Acad. Sci.*, **34**: 115–118.

LOEW, O., 1933: Über die Bildung von Eiweiss in den Pflanzen, *Angew. Bot.*, **15**: 518–539; *Biol. Abst.*, **10**: 31.

LOO, T. L., 1931: Studies on the absorption of ammonia and nitrate by the roots of *Zea mais* seedlings in relation to the concentration and the actual acidity of the culture solution, *J. Faculty Agr. Hokkaido Imp. Univ.*, **30**: 1–118; *Exp. Sta. Rec.*, **68**: 458.

LOOSE, L., and W. H. PEARSALL, 1933: Synthesis of protein by green plants, *Nature*, **131**: 362–363.

LUDWIG, C. A., 1934: Do germinating willow cuttings fix atmospheric nitrogen? *Am. J. Bot.*, **21**: 557–561.

LYON, T. L., and A. KEYSER, 1905: Nature and causes of "yellow-berry" in hard winter wheat, *Neb. Sta. Bull.*, **89**: 23–36.

——— and J. A. BIZZELL, 1911: A heretofore unnoted benefit from the growth of legumes, *N. Y. Sta.* (*Cornell*) *Bull.* 294.

——— and ———, 1913: A discussion of certain methods used in the study of "The associative growth of legumes and non legumes," *J. Am. Soc. Agron.*, **5**: 65–72.

MACHT, D. I., 1934: Growth of *Lupinus albus* seedlings in solutions of some amino acids, *Am. J. Bot.*, **21**: 72–76.

MAMELI DE CALVINO, E., and G. POLLACCI, 1911: Sull' assimilazione diretta dell' azoto atmosferico libero nei vegetali, *Atti dell' Instituto Bot. di Pavia*, Ser. II, **14**: 159–257.

——— and ———, 1911: Ancora sull' assimilazione diretta dell' azoto atmosferico libero nei vegetali, *Atti dell' Instituto Bot. di Pavia*, Ser. II, **16**: 197–203.

———, 1923: The direct assimilation of free nitrogen by plants. A review of recent work on the subject, *Torreya*, **23**: 99–104.

———, 1923: Sobre la assimilacion directa del nitrogeno libre por las plantas, *Estoc. Exp. Agron.* (Cuba) *Circ.* **59**: 3–7.

MANGEL, C. E., 1927: Pre-harvest factors which affect wheat quality, *Cereal Chem.*, **4**: 376–388.

MARAÑON, J., 1932: Nitrogen distribution in the leaves of Philippine camphor trees, *Philippine J. Sci.*, **49**: 461–468.

MARTH, P. C., 1934: Response of nitrogen fertilizers applied in different areas under Rome apple trees, *Proc. Am. Soc. Hort. Sci.*, **32**: 73–75.

MAZÉ, P., 1898: L'assimilation de l'azote nitrique et de l'azote ammoniacal par les végétaux supérieurs, *C. R. Acad. Sci. Paris*, **127**: 1031–1033.

———, 1900: Recherches sur l'influence de l'azote nitrique et de l'azote ammoniacal sur le développement du maïs, *Ann. Inst. Pasteur*, **14**: 26–45.

McCALLA, A. G., 1933: The effect of nitrogen nutrition on the protein and non protein nitrogen of wheat, *Can. J. Res.*, **9**: 542–570.

———, 1934: Amide nitrogen in germinating seeds, *Can. J. Res.*, **10**: 430–434.

McCool, M. M., and R. L. Cook, 1930: Rate of intake, accumulation and transformation of nitrate nitrogen by small grains and Kentucky blue grass, *J. Am. Soc. Agron.*, **22**: 757–765.

———, 1932: Use of peats in composts to increase nitrification and plant growth, *Contr. Boyce Thompson Inst.*, **4**: 257–271.

McGeorge, W. T., 1923: The assimilation of nitrogen by sugar cane, nitrates *vs.* ammonium salts, *Planter's Rec.*, **27**: 347–352.

McKie, P., and J. Barnett, 1936: The nitrogen metabolism of the pea seedling, *Biochem. J.*, **30**: 1010–1013.

McNair, J. B., 1931: Some properties of alkaloids in relation to climate or habitat, *Am. J. Bot.*, **18**: 416–424.

Metzger, W. H., 1935: The residual effect of alfalfa cropping periods of various lengths upon the yield and protein content of succeeding wheat crops, *J. Am. Soc. Agron.*, **27**: 653–659.

Mevius, W., and I. Dikussar, 1930: Nitrite als Stickstoffquellen für höhere Pflanzen. Ein Beitrag zur Frage nach der Assimilation des Nitratstickstoffs, *Jahrb. wiss. Bot.*, **73**: 633–703.

Meyer, A., 1918: Eiweissstoffwechsel und Vergilben der Laubblätter von *Tropaeolum majus*, *Flora*, **11-12**: 85–127.

Miller, E. C., 1910: A physiological study of the germination of *Helianthus annuus*, *Ann. Bot.*, **24**: 693–726.

———, 1917: Daily variation of water and dry matter in the leaves of corn and the sorghums, *J. Agr. Res.*, **10**: 11–47.

———, 1924: Daily variation of the carbohydrates in the leaves of corn and the sorghums, *J. Agr. Res.*, **27**: 785–808.

———, 1926: Daily variation of the nitrogen in the leaves, unpub. data, *Kans. Sta.*

———, 1929: Nitrogen in the leaves of crop plants, unpub. data, *Kans. Sta.*

Miller, J. E., and A. C. Chibnall, 1932: The proteins of grasses. I, Preliminary communication, *Biochem. J.*, **26**: 392–402.

Mockeridge, F. A., 1917: Some effects of organic growth-promoting substances (auximones) on the soil organisms concerned in the nitrogen cycle, *Proc. Roy. Soc.*, B, **89**: 508–534.

———, 1924: The formation of plant growth-promoting substances by microorganisms, *Ann. Bot.*, **38**: 723–734.

Molisch, H., 1916: Die Eiweissproben makroskopisch angewendet auf Pflanzen, *Z. Bot.*, **8**: 124–131.

Moore, B. and T. A. Webster, 1919: The formation of nitrites from nitrates in aqueous solution by the action of sunlight and the assimilation of the nitrites by green leaves in sunlight, *Proc. Roy. Soc.*, B, **90**: 158–167.

———, 1920: Studies of photosynthesis in fresh water algae. I, The fixation of both carbon and nitrogen from the atmosphere to form organic tissue by the green plant cell. II, III, *Proc. Roy. Soc.*, B, **91**: 201–215.

———, E. Whitley, and T. A. Webster, 1921: Studies of photosynthesis in marine algae. I, Fixation of carbon and nitrogen from inorganic sources in sea water. II, *Proc. Roy. Soc.*, B, **92**: 51–60.

Morgan, M. F., and O. E. Street, 1935: Rates of growth and nitrogen assimilation of Havana seed tobacco, *J. Agr. Res.*, **51**: 163–172.

Mothes, K., 1926: Ein Beitrag zur Kenntnis des N-Stoffwechsels höherer Pflanzen, *Planta*, **1**: 472–552.

———, 1931: Zur Kenntnis des N-Stoffwechsels höherer Pflanzen. III, Beitrag, *Planta*, **12**: 686–731.

———, 1932: Natural regulation of the vegetable protein metabolism, *Naturwiss.*, **20**: 102–103; *Chem. Abst.*, **26**: 3820.

Muenscher, W. C., 1923: Protein synthesis in *Chlorella*, *Bot. Gaz.*, **75**: 249–268.

Mulay, A. S., 1931: Seasonal changes in total, soluble, soluble protein, non protein, and insoluble nitrogen in current year's shoot of Bartlett pear, *Plant Physiol.*, **6**: 519–529.

———, 1932: Seasonal changes in the composition of the non protein nitrogen in the current year's shoots of Bartlett pear, *Plant Physiol.*, **7**: 107–118.

———, 1932: Seasonal changes in the composition of the insoluble nitrogen fraction in the current year's shoots of Bartlett pear, *Plant Physiol.*, **7**: 323–327.

Müller-Thurgau, H. 1894: Einfluss des Stickstoffs auf die Wurzelbildung, *Jahresbericht Versuchs-Stat. Wädensweil*, **4**: 48–52.

Müntz, A., 1889: Sur le rôle de l'ammoniaque dans la nutrition des végétaux supérieurs, *C. R. Acad. Sci. Paris*, **109**: 646–648.

Murneek, A. E., 1930: Quantitative distribution and seasonal fluctuation of nitrogen in apple trees, *Proc. Am. Soc. Hort. Sci.*, **27**: 228–231.

———, 1935: Physiological rôle of asparagine and related substances in nitrogen metabolism of plants, *Plant Physiol.*, **10**: 447–464.

Naftel, J. A., 1931: The absorption of ammonium and nitrate nitrogen by various plants at different stages of growth, *J. Am. Soc. Agron.*, **23**: 142–158.

Nagaoka, M., 1904: On the behavior of the rice plant to nitrates and ammonium salts, *Bull. Col. Agr. Tokyo Imper. Univ.*, **6**: 285–334; *Exp. Sta. Rec.*, **16**: 560.

Neidig, R. E., and R. S. Snyder, 1922: The effect of available nitrogen on the protein content and yield of wheat, *Idaho Exp. Sta. Bull.* 1.

——— and ———, 1924: The relation of moisture and available nitrogen to the yield and protein content of wheat, *Soil Sci.*, **18**: 173–181.

——— and ———, 1926: The relation of the yield and protein content of wheat to the nitrogen content of the soil under ten years of different systems of cropping, *Idaho Sta. Res. Bull.* 5.

Newby, H. L., and W. H. Pearsall, 1930: Observations on nitrogen metabolism in the leaves of *Vitis* and *Rheum*, *Proc. Leeds Phil. and Lit. Soc., Sci. Sec.*, **2**: 81–85.

Newcomb, M. A., 1927: The loss of nitrogen from oily seeds, unpub. data, *Master's thesis, Kans. State Col.*

Nicol, H., 1934: The derivation of the nitrogen of crop plants with special reference to associated growth, *Biol. Rev.*, **9**: 383–410.

Nightingale, G. T., 1922: Light in relation to the growth and chemical composition of some horticultural plants, *Proc. Am. Soc. Hort. Sci.*, **19**: 18–29.

——— and E. J. Kraus, 1924: Does light affect the fertilizer requirements of plants? *Wis. Sta. Bull.* **362**: 52–54.

———, L. G. Schermerhorn, and W. R. Robbins, 1928: The growth status of the tomato as correlated with organic nitrogen and carbohydrates in roots, stem and leaves, *N. J. Sta. Bull.* 461.

———, and W. R. Robbins, 1928: Some phases of nitrogen metabolism in polyanthus narcissus (*Narcissus tazetta*), *N. J. Sta. Bull.* 472.

——— and L. G. Schermerhorn, 1928: Nitrate assimilation by asparagus in the absence of light, *N. J. Sta. Bull.* 476.

———, 1934: Ammonium and nitrate nutrition of dormant Delicious apple trees at 48°F., *Bot. Gaz.*, **95**: 437–452.

———, 1936: The biochemistry of the nitrogenous constituents of the green plants, *Ann. Rev. Biochem.*, **5**: 513–524.

Olson, G. A., 1917: Gluten formation in the wheat kernel, *Wash. Sta. Bull.* 142.

———, 1923: A study of factors affecting the nitrogen content of wheat and of the changes that occur during the development of wheat, *J. Agr. Res.*, **24**: 939–955.

ORCUTT, F. S., and E. B. FRED, 1935: Light intensity as an inhibiting factor in the fixation of atmospheric nitrogen by Manchu soybeans, *J. Am. Soc. Agron.*, **27**: 550–558.

OSBORNE, T. B., and A. J. WAKEMAN, 1920: The proteins of green leaves. I, Spinach leaves, *J. Biol. Chem.*, **42**: 1–26.

———, ———, and C. S. LEAVENWORTH, 1921: The proteins of the alfalfa plant, *J. Biol. Chem.*, **49**: 63–91.

———, 1924: "The Vegetable Proteins," mon., Longmans, Green and Company, New York, 151 pp.

OTTO, R., and W. D. KOOPER, 1910: Beitrag zur Abnahme bezw. Rückwanderung der Stickstoffverbindungen aus den Blättern während der Nacht sowie zur herbstlichen Rückwanderung von Stickstoffverbindungen aus den Blättern, *Landw. Jahrb.*, **39**: 167–172.

PANTANELLI, E., and G. SEVERINI, 1910, 1911: Some experiments on the utilization of ammonium salts by green plants, *Staz. Sper. Agr. Ital.*, **43**: 449–544; **44**: 873–908; *Exp. Sta. Rec.*, **25**: 223; **27**: 634.

PARDO, J. H., 1930: Utilization of certain nitrogen compounds by the sugar cane, *Int. Sugar J.*, **32**: 11–19.

———, 1933: Ammonia vs. nitrate as nitrogen sources for sugar cane, *Proc. Fourth Cong. Int. Soc. Sugar Cane Tech. Bull.* 13; *Chem. Abst.*, **27**: 2712.

———, 1933: A note on the utilization of ammonium and nitrate by higher plants, *New Phytol.*, **32**: 308–310.

———, 1935: Ammonium in the nutrition of higher green plants, *Quart. Rev. Biol.*, **10**: 1–31.

PEARSALL, W. H., and J. H. PRIESTLEY, 1923: Meristematic tissues and protein isoelectric points, *New Phytol.*, **22**: 185–191.

——— and J. EWING, 1924: The isoelectric points of some plant proteins, *Biochem. J.*, **18**: 329–339.

——— and ———, 1929: The relation of nitrogen metabolism to plant succulence, *Ann. Bot.*, **43**: 27–33.

———, 1931: The distribution of the insoluble nitrogen in *Beta* leaves of different ages, *J. Exp. Biol.*, **8**: 279–285.

PERCIABOSCO, F., and V. ROSSO, 1909: L'assorbimento diretto dei nitrite nelle piante, *Staz. Sper. Agrar. Ital.*, **42**: 5–36; *Exp. Sta. Rec.*, **23**: 724.

PIGORINI, L., 1914: Studi sulla foglia di gelso; sulla composizione chimica della foglia al mattino e alla sera, *Atti dei. Lincei.*, **23**: 433–437.

PIRSCHLE, K., 1929: Zur Assimilation des Harnstoffes durch die höhere Pflanze, *Biochem. Z.*, **212**: 466–474.

———, 1929: Nitrate und Ammonsalze als Stickstoffquellen für höhere Pflanzen bei konstanter Wasserstoffionenkonzentration, *Ber. bot. Ges.*, **47**: 86–92.

———, 1931: Nitrate und Ammonsalze als Stickstoffquellen für höhere Pflanzen bei konstanter Wasserstoffionkonzentration. III, *Planta*, **14**: 583–676; *Biol. Abst.*, **9**: 289.

——— and H. MENGDEHL, 1931: Ionenaufnahme aus Salzlösungen durch die höhere Pflanze, *Jahrb. wiss. Bot.*, **74**: 297–363.

PITSCH, O., 1887, 1893, 1896: Versuche zur Entscheidung der Frage, ob salpetersäure Salze für die Entwickelung unserer landwirtschaftlichen Kulturgewächse unentbehrlich sind oder nicht, *Landw. Versuchs-Stat.*, **34**: 217–258; **42**: 1–95; **46**: 357–370.

PLAGGE, H. H., 1930: Do successive applications of nitrogenous fertilizers influence the development of physiological disorders of apples in cold storage, *Proc. Am. Soc. Hort. Sci.*, **27**: 23–27.

PLATENIUS, H., 1932: Carbohydrate and nitrogen metabolism in the celery plant as related to premature seeding, *N. Y. Sta. (Cornell) Mem.* 140.

POLLARD, A., and A. C. CHIBNALL, 1934: The proteins of grasses. III, The cystine content of certain grasses and other pasture plant proteins, *Biochem. J.*, **28**: 326–336.

POPOFF, M., 1933: Ueber die pflanzlichen Auxine und ihre Wirkung auf Einzellige, *Biol. Zent.*, **53**: 661–668.

POTTER, G. F., H. R. KRAYBILL, S. W. WENTWORTH, J. T. SULLIVAN, and P. T. BLOOD, 1930: Some effects of defloration on fruit spur composition and fruit bud formation in the Oldenburg apple, *N. H. Sta. Tech. Bull.* 41.

——— and T. G. PHILLIPS, 1930: Composition and fruit bud formation in non bearing spurs of the Baldwin apple, *N. H. Sta. Tech. Bull.* 42.

POZZI-ESCOT, M-EMM, 1903: The reducing enzymes, *Am. Chem. J.*, **29**: 517–563.

PRAZMOWSKI, A., 1890: Die Wurzelknöllchen der Erbse. I, Die Aetiologie und Entwickelungsgeschichte der Knöllchen, *Landw. Versuchs-Stat.*, **37**: 161–238.

PRIANISCHNIKOW, D. N., 1899: Die Rückbildung der Eiweissstoffe aus deren Zerfallsprodukten, *Landw. Versuchs-Stat.*, **52**: 347–382.

———, 1908: Zur physiologischen Characteristik der Ammoniumsalze, *Ber. bot. Ges.*, **26**: 716–724.

———, 1922: Ueber den Aufbau und Abbau des Asparagins in den Pflanzen, *Ber. bot. Ges.*, **40**: 242–248; *Bot. Abst.*, **12**: 751.

———, 1923: Sur l'assimilation de l'ammoniaque par les plantes supérieures, *C. R. Acad. Sci. Paris*, **177**: 603–606.

———, 1924: Asparagin und Harnstoff (Physiologische Parallele), *Biochem. Z.*, **150**: 407–423.

PRIESTLEY, J. H., 1922: Some recent work on proteins in its relation to physiology, *Sci. Prog.*, **16**: 201–207.

RABER, O., 1928: "Principles of Plant Physiology," The Macmillan Company, New York, 362 pp.

RAHN, H., 1932: Untersuchungen über den N-Stoffwechsel pflanzlicher vegetativer Speicherorgane, *Planta*, **18**: 1–51.

REID, M. E., 1924: Relation of kind of food reserves to regeneration in tomato plants, *Bot. Gaz.*, **77**: 103–111.

———, 1924: Quantitative relations of carbohydrates to nitrogen in determining growth responses in tomato cuttings, *Bot. Gaz.*, **77**: 404–418.

———, 1926: Growth of tomato cuttings in relation to stored carbohydrate and nitrogenous compounds, *Am. J. Bot.*, **13**: 548–575.

———, 1926: Growth of seedlings in relation to composition of seed, *Bot. Gaz.*, **81**: 196–203.

———, 1929: Growth of seedlings in light and in darkness in relation to available nitrogen and carbon, *Bot. Gaz.*, **87**: 81–117.

———, 1929: Effect of variations in the amounts of available carbon and nitrogen on the growth of wheat seedlings, *Am. J. Bot.*, **16**: 770–779.

———, 1930: Growth and nitrogen metabolism of squash seedlings. II, With respect to stages of development and the influence of light, *Am. J. Bot.*, **17**: 396–416.

RICHARDS, F. J., and W. G. TEMPLEMAN, 1936: Physiological studies in plant nutrition. IV, Nitrogen metabolism in relation to nutrient deficiency and age in barley leaves, *Ann. Bot.*, **50**: 367–402.

Richardson, A. E. V., and H. C. Gurney, 1933: The effect of nitrogenous fertilizers on the growth and yield of wheat and barley in South Australia. I, Wheat grown after fallow and after stubble. II, The effect of a previous crop in the response of wheat to nitrogen fertilizers and the effect of increasing amounts of nitrogen on barley, *Empire J. Exp. Agr.*, **1**: 193–205, 325–332.

——— and ———, 1935: Nitrogen in relation to cereal culture, *J. Dept. Agr. So. Australia*, **28**: 954–971.

Richardson, G. M., 1934: Critique on the biological estimation of amino nitrogen, *Proc. Roy. Soc.*, B, **115**: 142–169.

Robbins, W. J., 1923: An isoelectric point for plant tissue and its significance, *Am. J. Bot.*, **10**: 412–440.

Roberts, H. F., and G. F. Freeman, 1908: The yellow berry problem in Kansas hard winter wheats, *Kans. Sta. Bull.* 156.

———, 1919: Yellow berry in hard winter wheat, *J. Agr. Res.*, **18**: 155–171.

Roberts, R. H., 1920: Off year apple bearing and apple spur growth, *Wis. Sta. Bull.* 317.

———, 1921: Nitrogen reserves in apple trees, *Proc. Am. Soc. Hort. Sci.*, **18**: 143–145.

———, 1923: Effect of defoliation upon blossom bud formation, *Wis. Sta. Res. Bull.* 56.

Robinson, M. E., 1929: The protein metabolism of the green plant, *New Phytol.*, **28**: 117–150.

Russell, E. J., and E. H. Richards, 1919: The amount and composition of rain falling at Rothamsted, *J. Agr. Sci.*, **9**: 309–337.

Saeger, A., 1925: The growth of duckweeds in mineral nutrient solutions with and without organic extracts, *J. Gen. Physiol.*, **7**: 517–526.

Sampson, A. W., 1935: Growth and seasonal changes in composition of oak leaves, *Plant Physiol.*, **10**: 739–751.

Sapoznikow, W., 1895: "Eiweisstoffe und Kohlenhydrate der grünen Blätter als Assimilationsproducte," 61 pp., Tomsk (1894); (Russian) *Bot. Cent.*, **63**: 246–251.

Schiletter, J. C., 1932: Time of bud differentiation in the Dunlap strawberry, *Proc. Am. Soc. Hort. Sci.*, **29**: 325–329.

Schramm, J. R., 1914: The relation of certain grass-green algae to elementary nitrogen, *Ann. Mo. Bot. Gard.*, **1**: 157–184.

Schreiner, O., and J. J. Skinner, 1912: Nitrogenous soil constituents and their bearing on soil fertility, *U. S. Bur. Soils Bull.* 87.

Schultz, M., 1903: "Studien über den Einfluss von Nitriten auf die Keimung von Samen und auf das Wachstum von Pflanzen," *Inaug-Diss., Königsberg.*

Schulz, M., 1862: Chemische Beiträge zur Kenntniss des Keimprocesses bei einigen Phanerogamen, *J. Prakt. Chemie*, **71**: 129–173.

Schulze, B., and J. Schütz, 1909: Die Stoffwandlungen in den Laubblättern des Baumes, insbesondere in ihren Beziehungen zum herbstlichen Blattfall, *Landw. Versuchs-Stat.*, **71**: 299–352.

Schulze, E., 1897: Über die Verbreitung des Glutamins in den Pflanzen, *Landw. Versuchs-Stat.*, **48**: 33–55.

———, and E. Winterstein, 1910: Studien über die Proteinbildung in reifenden Pflanzensamen, *Z. physiol. Chem.*, **66**: 431–476; *Bot. Cent.*, **116**: 63.

———, 1911: Studien über die Proteinbildung in reifenden Pflanzensamen, *Z. physiol. Chem.*, **71**: 31–48.

Schuster, G. L., and J. M. Graham, 1927: Effect of various fertilizers and lime on composition of soybeans, *J. Am. Soc. Agron.*, **19**: 574–577.

Sessions, A. C., and J. W. Shive, 1933: The effect of culture solutions on growth and nitrogen fractions of oat plants at different stages of their development, *Soil Sci.*, **35**: 355–374.

Shaw, G. W., and E. H. Walters, 1911: A progress report upon soil and climatic factors influencing the composition of wheat, *Calif. Sta. Bull.* 216.

Shear, G. M., 1933: Field and laboratory studies on frenching of tobacco, *Va. Sta. Tech. Bull.* 49.

Shmuk, A., and M. Poliakov, 1927: The accumulation of nitrates in plants, *Arb. Kubansch. Landw. Inst.*, **5**: 161–173; *Biol. Abst.*, **3**: 478.

Shull, C. A., and J. W. Mitchell, 1933: Some physico-chemical properties of seed extracts, *Bot. Gaz.*, **95**: 258–278.

Shulov, I., 1912: Sterile cultures of higher plants. Assimilation of nitrogen as ammonia and nitrates, *J. Exp. Landw.*, **13**: 200–206.

Smirnow, A. I., 1923: Über die Synthese der Säureamide in der Pflanzen bei Ernährung mit Ammoniaksalzen, *Biochem. Z.*, **137**: 1–34; *Bot. Abst.*, **13**: 67.

Sommer, A. L., 1936: Reduction of nitrites by the expressed juice of higher green plants, *Plant Physiol.*, **11**: 429–436.

Spitzer, G., R. H. Carr, and W. F. Epple, 1919: Soft corn, its chemical composition and nitrogen distribution, *J. Am. Chem. Soc.*, **41**: 1212–1221.

Spoehr, H. A., 1911: The relation between photosynthesis of CO_2 and nitrate reduction, *Sci.*, **34**: 63–64.

Stahl, A. L., and J. W. Shive, 1933: Studies on nitrogen absorption from culture solutions: I, Oats, *Soil Sci.*, **35**: 373–400.

——— and ———, 1933: Further studies on nitrogen from culture solutions. II, Buckwheat, *Soil Sci.*, **35**: 469–483.

Stallings, J. H., 1926: The form of legume nitrogen assimilated by non legumes when grown in association, *Soil Sci.*, **21**: 253–276.

Stark, O. K., 1924: Environmental factors affecting the protein and oil content of soybeans and the iodine number of soybean oil, *J. Am. Soc. Agron.*, **16**: 636–646.

Starring, C. C., 1923: Influence of the carbohydrate-nitrate content of cuttings upon the production of roots, *Proc. Am. Soc. Hort. Sci.*, **20**: 288–292.

Steiner, M., 1929: Weitere Untersuchungen über flüchtige Stickstoffbasen bei höheren Pflanzen, *Beitr. Biol. Pflanzen*, **17**: 247–258; *Biol. Abst.*, **5**: 2675.

Stewart, C. R., E. C. Thomas, and J. Horner, 1925: The comparative growth of pineapple plants with ammonia and nitrate nitrogen, *Soil Sci.*, **20**: 227–243.

Stewart, R., and J. E. Greaves, 1909: A study of the production and movement of nitric nitrogen in an irrigated soil, *Utah Sta. Bull.* 106

Strowd, W. H., 1920: The relation of nitrates to nodule production, *Soil Sci.*, **10**: 343–356.

Stuart, N. W., 1932: Nitrogen and carbohydrate metabolism of young apple trees as affected by excessive applications of sodium nitrate, *N. H. Sta. Bull.* 50.

——— and C. O. Appleman, 1935: Nitrogenous metabolism in Irish potatoes during storage, *Md. Sta. Bull.* 372.

Stutzer, A., 1906: Die Wirkung von Nitrit auf Pflanzen, *Bot. Cent.*, **105**: 569–570.

Suzuki, U., 1897: The formation of asparagin in plants under different conditions, *Imp. Univ. Col. Agr. Tokyo Bull.*, **2**: 409–457.

———, 1897: An important function of leaves, *Imp. Univ. Col. Agr. Tokyo Bull.*, **3**: 241–252.

———, 1898: On the formation of proteids and the assimilation of nitrates by Phanerogams in the absence of light, *Imp. Univ. Col. Agr. Tokyo Bull.*, **4**: 488–587.

———, 1898: Ueber die Assimilation der Nitrate in Dunkelheit durch Phanerogamen, *Bot. Cent.*, **75**: 289–292.

Talley, P. J., 1934: Carbohydrate-nitrogen ratios with respect to the sexual expression of hemp, *Plant Physiol.*, **9**: 731–748.

Taylor, R. W., 1932: Influence of nitrogenous fertilizers applied at different dates on the number of flower clusters, flowers, and fruits of the strawberry, *Proc. Am. Soc. Hort. Sci.*, **29**: 311–317.

Teller, G. L., 1935: Changes in nitrogen compounds in the wheat grain at different stages of development, *Plant Physiol.*, **10**: 499–509.

Thatcher, R. W., 1913: A report on the investigations concerning the chemical composition of wheat, 1906 to 1912 inclusive, *Wash. Sta. Bull.* 111.

———, 1913: The progressive development of the wheat kernel, *J. Am. Soc. Agron.*, **5**: 203–213.

———, 1915: The progressive development of the wheat kernel. II, *J. Am. Soc. Agron.*, **7**: 273–283.

———, 1921: "The Chemistry of Plant Life," McGraw-Hill Book Company, Inc., New York, 268 pp.

Thelin, G., and A. B. Beaumont, 1934: The effect of some forms of nitrogen on the growth and nitrogen content of wheat and rice plants, *J. Am. Soc. Agron.*, **26**: 1012–1017.

Thomas, W., 1927: Nitrogenous metabolism of *Pyrus malus*. I, Influence of temperature of desiccation on water-soluble nitrogenous constituents and separation of water-soluble protein from non-protein constituents, *Plant Physiol.*, **2**: 55–67.

———, 1927: Nitrogenous metabolism of *Pyrus malus*. II, The distribution of nitrogen in the insoluble cytoplasmic proteins, *Plant Physiol.*, **2**: 67–71.

———, 1927: Nitrogenous metabolism of *Pyrus malus*. III, The partition of nitrogen in the leaves, one and two year branch growth and non-bearing spurs throughout a year's cycle, *Plant Physiol.*, **2**: 109–137.

———, 1927: Nitrogenous metabolism of *Pyrus malus*. IV, The effect of sodium nitrate applications on the total nitrogen and its partition products in the leaves, new and one year branch growth throughout a year's cycle, *Plant Physiol.*, **2**: 245–271.

———, 1927: The seat of formation of amino acids in *Pyrus malus* L., *Sci.*, **66**: 115–116.

Tidmore, J. W., 1933: Ammonium hydroxide versus calcium nitrate for cotton seedlings, *J. Am. Soc. Agron.*, **25**: 617–622.

Tiedjens, V. A., and W. R. Robbins, 1931: The use of ammonia and nitrate nitrogen by certain crop plants, *N. J. Sta. Bull.* 526.

———, 1932: Growing cotton and other crop plants with ammonium nitrogen, *Sci.*, **75**: 648.

——— and M. A. Blake, 1932: Factors affecting the use of nitrate and ammonium nitrogen by apple trees, *N. J. Sta. Bull.* 547.

———, 1934: Factors affecting assimilation of ammonium and nitrate nitrogen, particularly in tomato and apple, *Plant Physiol.*, **9**: 31–57.

Tottingham, W. E., 1923: Temperature effects in plant metabolism, *J. Agr. Res.*, **25**: 13–31.

———, E. R. Schulz, and S. Lepkovsky, 1924: The extraction of nitrogenous constituents from plant cells, *J. Am. Chem. Soc.*, **46**: 203–208.

———, 1924: Physiological stability in maize, *Sci.*, **59**: 69–70.

——— and H. W. Kerr, 1926: Climatic effects in the metabolism of maize, *Plant Physiol.*, **1**: 415–416.

——— and H. Lowsma, 1928: Effects of light upon nitrate assimilation in wheat, *J. Am. Chem. Soc.*, **50**: 2436–2445.

——— and E. J. Lease, 1934: A photochemical aspect of nitrate assimilation in plants, *Sci.*, **80**: 615–616.

———, H. L. Stephens, and E. J. Lease, 1934: Influence of shorter light rays upon absorption of nitrate by the young wheat plant, *Plant Physiol.*, **9**: 127–142.

Treboux, O., 1904: Zur Stickstoffernährung der grünen Pflanze, *Ber. bot. Ges.*, **22**: 570–572.

Trelease, S. F., and P. Paulino, 1920: The effect on the growth of rice of the addition of ammonium and nitrate salts to soil cultures, *Philippine J. Agr.*, **8**: 293–313.

Tyagny-Ryadno, M., 1933: The relations of *Bacillus mycoides* with ammonification, nitrification and soil fertility, *J. Agr. Sci.*, **23**: 335–358.

Ullrich, H., 1924: The role of chloroplasts in protein formation in green plants, *Z. Bot.*, **16**: 513–562; *Exp. Sta. Rec.*, **56**: 814.

Valleau, W. D., and E. M. Johnson, 1927: Tobacco frenching—a nitrogen deficiency disease, *Ky. Sta. Bull.* 281.

Varadachar, K. S., 1933: Synthesis of proteins in plants. I, Conversion of nitrates into protein in *Helianthus annuus*, *J. Indian Inst. Sci.*, 16A, **12**: 129–138.

Verner, L., 1933: Effect of nitrate fertilization on apple fruits, *Proc. Am. Soc. Hort. Sci.*, **30**: 32–36.

Vickery, H. B., and G. W. Pucher, 1931: The chemical changes that occur during the curing of Connecticut shade grown tobacco, *Conn. Sta. Bull.* 324.

———, ——— and H. E. Clark, 1934: Glutamine in the tomato plant, *Sci.*, **80**: 459–461.

———, ——— and ———, 1936: Glutamine metabolism of the beet, *Plant Physiol.*, **11**: 413–420.

Ville, M. G., 1853–1857: "Recherches expérimentales sur la végétation," Paris.

Virtanen, A. I., and S. Von Hausen, 1930: The capability of grass plants to take advantage of the nitrogen fixed by the nodule bacteria of leguminous plants, *Contr. Lab. Valio, Helsinki.* (Finnish with English summary.) Quoted by Nicol, 1934.

———, ——— and H. Karström, 1933: Untersuchungen über die Leguminose-Bakterien und -Pflanzen. XII, Die Ausnutzung der aus den Wurzelknöllchen der Leguminosen herausdiffundierten Stickstoffverbindungen durch Nichtleguminosen, *Biochem. Z.*, **258**: 106–117.

——— and S. Saastamoinen, 1933: Ueber die Stickstoffbindung bei Erlen (*Alnus*), *Acta Chem. Fenn. B.*, 6. Quoted by Nicol, 1934.

Waldron, L. R., 1933: Yield and protein content of hard red spring wheat under conditions of high temperature and low moisture, *J. Agr. Res.*, **47**: 129–148.

Wann, F. B., 1920: The fixation of free nitrogen by green plants, *Sci.*, **51**: 247–248.

———, 1921: The fixation of free nitrogen by green plants, *Am. J. Bot.*, **8**: 1–30.

Warburg, O., and E. Negelein, 1920: Über die Reduktion der Salpetersäure in grünen Zellen, *Biochem. Z.*, **110**: 66–115.

Wassilieff, N., 1908: Eiweissbildung im reifenden Samen, *Ber. bot. Ges.*, **26**: 454–467.

Watson, D. J., 1936: The effect of applying a nitrogenous fertilizer to wheat at different stages of growth, *J. Agr. Sci.*, **26**: 391–414.

Webster, J. E., 1928: Nitrogen metabolism in the soybean, *Plant Physiol.*, **3**: 31–43.

Weevers, T., 1930: Die Funktion der Xanthinderivate im Pflanzenstoffwechsel, *Arch. Néerl. Sci. Exactes et Nat. Sér.*, 3B, **5**: 111–195; *Biol. Abst.*, **5**: 2676.

Weinberger, J. H., and F. P. Cullinan, 1934: Nitrogen intake and growth response in peach trees following fall and spring fertilizer applications, *Proc. Am. Soc. Hort. Sci.*, **32**: 65–69.

Welton, F. A., and V. H. Morris, 1930: Effect of fertility in the carbohydrate-nitrogen relation in the soybean, *Plant Physiol.*, **5**: 607–612.

Werner, H. O., 1934: The effect of a controlled nitrogen supply with different temperatures and photoperiods upon the development of the potato plant, *Neb. Sta. Res. Bull.* 75.

Westgate, J. M., and R. A. Oakley, 1914: Percentage of protein in non legumes and legumes when grown alone and in association with field mixtures, *J. Am. Soc. Agron.*, **6**: 211–215.

White, H. L., 1936: The interaction of factors in the growth of *Lemna.* VIII, The effect of nitrogen on growth and multiplication, *Ann. Bot.*, **50**: 403–417.

Whitson, A. R., F. J. Wells, and A. Vivian, 1902: Influence of the soil on the protein content of crops, *Wis. Sta. 19th Ann. Rept.*, **1902**: 192–209.

Williams, C. F., 1932: Seasonal variations in nitrogen concentration in twigs of peach trees under sand hill conditions, *Proc. Am. Soc. Hort. Sci.*, **28**: 39–44.

Williams, R. J., and E. Rohrman, 1935: Pantothenic acid as a nutrilite for green plants, *Plant Physiol.*, **10**: 559–563.

Willis, L. G., and J. O. Carrero, 1923: Influence of some nitrogenous fertilizers on the development of chlorosis in rice, *J. Agr. Res.*, **24**: 621–640.

——— and W. H. Rankin, 1930: Free-ammonia injury with concentrated fertilizers, *Ind. and Eng. Chem.*, **22**: 1405–1407.

——— and ———, 1932: Is ammonium hydroxide toxic to cotton plants? *Sci.*, **76**: 214–215.

Wilson, B. D., 1921: Nitrogen in the rainwater at Ithaca, N. Y., *Soil Sci.*, **11**: 101–110.

Wilson, R. W., E. B. Fred, and M. R. Salmon, 1933: Relation between carbon dioxide and elemental nitrogen assimilation in leguminous plants, *Soil Sci.*, **35**: 145–166.

Wolfe, H. S., 1926: The auximone question, *Bot. Gaz.*, **81**: 228–231.

Woo, M. L., 1919: Chemical constituents of *Amaranthus retroflexus*, *Bot. Gaz.*, **68**: 313–344.

Work, P., 1924: Nitrate of soda in the nutrition of the tomato, *N. Y. Sta.* (*Cornell*) *Mem.* 75.

Yamaguchi, S., 1930: Studies on the resorption of urea by the root of *Zea mays* seedlings in sterile culture, *J. Fac. Sci. Hokkaido Imp. Univ.*, Ser. 5, Bot. 1, 37–55; *Biol. Abst.*, **6**: 2516.

Zahnley, J. W., and F. L. Duley, 1934: The effect of nitrogenous fertilizers on the growth of lawn grasses, *J. Am. Soc. Agron.*, **26**: 231–234.

Zaleski, W., 1897: Zur Kenntniss der Eiweissbildung in den Pflanzen, *Ber. bot. Ges.*, **15**: 536–542.

———, 1898: Zur Keimung der Zwiebel von *Allium cepa* und Eiweissbildung, *Ber. bot. Ges.*, **16**: 146–151.

———, 1901: Die Bedingungen der Eiweissbildung in den Pflanzen, *Bot. Cent.*, **87**: 277–282.

———, 1907: Über die autolytische Ammoniakbildung in den Pflanzen, *Ber. bot. Ges.*, **25**: 357–360.

———, 1907: Über den Aufbau der Eiweissstoffe in den Pflanzen, *Ber. bot. Ges.*, **25**: 360–368.

———, 1909: Über die Rolle des Lichtes bei der Eiweissbildung in den Pflanzen, *Ber. bot. Ges.*, **27**: 56–62.

———, 1911: Zur Kenntnis der Stoffwechselprozesse in reifenden Samen, *Beih. Bot. Cent.*, **27**: 63–82; *Bot. Cent.*, **117**: 57.

——— and W. Shatkin, 1913: Untersuchungen über den Eiweissaufbau in den Pflanzen. I, Über den Eiweissaufbau in den Zwiebeln von *Allium cepa*, *Biochem. Z.*, **55**: 72–78.

Zeleny, L., 1935: The distribution of nitrogen in the seed of *Zea mays* at different stages of maturity, *Cereal Chem.*, **12**: 536–542.

CHAPTER X

THE FAT METABOLISM OF THE GREEN PLANT

I. FATTY SUBSTANCES IN PLANTS. CRUDE LIPIDES

When dry plant tissues are extracted with ethyl ether, chloroform, petroleum ether, carbon disulphide, benzene, acetone, or any of the other "fat solvents," a large number of substances are obtained which are designated collectively as the "ether extract," the "chloroform extract," etc., according to the particular solvent that has been used in the extraction. These substances are also frequently termed the "fatty substances" of the plant. Recently Sando (1928) in order to simplify the classification of the fatty substances suggested the name "crude lipides" for the tissue constituents obtained by extraction with the fat solvents, and that term will be used in the following discussion.

The crude lipides that have thus been extracted may be divided into two main groups: (*a*) Those substances which contain in their chemical constitution the fatty acids and which yield these acids upon hydrolysis. This group of substances includes the fats, fatty oils, waxes, sterol esters, phospholipides, and galactolipides. Until 1925, however, there was no comprehensive name to include all such substances. In that year Bloor proposed the name "lipides" to designate these types of compounds, and this term will be used in the following classification of the fatty substances of plants. (*b*) Those substances which are not compounds of fatty acids. These include the essential oils, hydrocarbons, resins, and coloring matters. Although some of these substances are similar to the lipides in possessing a greasy feel, they have little or no resemblance to them from a chemical standpoint.

The classification of the crude lipides that may occur in plant tissue is given in the outline on page 727.

A. Lipides

The lipides are the most abundant, the most widely distributed throughout the plant, and apparently are the most important, from a biological standpoint, of the fatty substances that occur in plants. As defined by Bloor (1925), they include those substances in plants and animals which are insoluble in water but soluble in ether, chloroform, benzene, or other fat solvents and which have a relationship to the fatty acids as esters either actual or potential. The compounds included under the term "lipides" form a group which is as strictly defined as either the proteins or carbohydrates and which are practically of the same physiological importance.

- The crude lipides: The fatty substances which may be extracted from plant tissues by ether, chloroform, benzene, and other fat solvents
  - The lipides: Esters of fatty acids and alcohols and their hydrolytic products which are soluble in fat solvents
    - The simple lipides
      - The fats or fatty oils: Esters of the fatty acids with glycerol
      - The waxes: Esters of fatty acids with monohydric alcohols
    - The compound lipides: Esters of the fatty acids containing groups in addition to alcohols and fatty acids
      - The phospholipides
        - Lecithin
        - Cephalin
      - Galactolipides
        - Phreosin
        - Kerasin
    - The derived lipides: Substances derived from the simple and compound lipides by hydrolysis
      - The fatty acids
        - The saturated fatty acids
        - The unsaturated fatty acids
      - Sterol alcohols
      - Wax alcohols
  - Substances not compounds of fatty acids
    - Resins
    - Hydrocarbons
    - Essential oils
    - Coloring matters
    - Sterols: Not derived from hydrolysis of esters

1. Composition.—The fatty acids and various types of alcohols are always component parts of the lipides and should be discussed here in some detail, since the reactions of the lipides are due in a large part to the character of these two types of compounds that enter into their composition.

a. Fatty Acids.—According to their position in the scheme of classification, the fatty acids should be considered under the heading of the derived lipides. In order to give the student a clear understanding of the nature of the lipides, however, and the part that the fatty acids play in determining their characteristics, a discussion of the fatty acids is given here. The fatty acids are monobasic and are so named because they occur in natural fats and because in the free state they resemble the fats in physical properties. The lower fatty acids as formic, acetic, and propionic are widely distributed in the various cells throughout the plant, either in the free or in the combined form. The higher fatty acids with which we are here more particularly concerned occur in the free state to a considerable extent in germinating seeds and perhaps to a limited extent in other plant parts. By far the greater portion that occurs in plants, however, is combined with the alcohols to form the various types of esters that are designated as lipides. The classification and identification of a large proportion of the lipides are based upon the fatty acids that they contain. The fatty acids may be divided into two main series: (1) the saturated and (2) the unsaturated.

1. *Saturated.*—The saturated fatty acids have the general formula $C_nH_{2n}O_2$ and contain the maximal hydrogen possible. A saturated acid is thus saturated with respect to its hydrogen. The more important saturated acids are shown in the following table:

THE SATURATED FATTY ACIDS

Name of the acid	Formula
Formic	$HCOOH$
Acetic	CH_3COOH
Propionic	C_2H_5COOH
n Butyric	C_3H_7COOH
Isobutyric	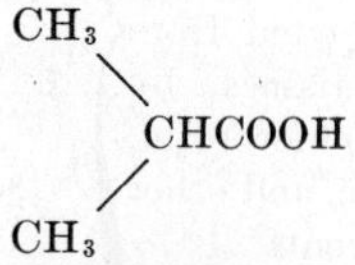
n Valeric	C_4H_9COOH
n Caproic	$C_5H_{11}COOH$
Caprylic	$C_7H_{15}COOH$
Capric	$C_9H_{19}COOH$
Lauric	$C_{11}H_{23}COOH$
Myristic	$C_{13}H_{27}COOH$
Palmitic	$C_{15}H_{31}COOH$
Stearic	$C_{17}H_{35}COOH$
Arachidic	$C_{19}H_{39}COOH$
Lignoceric	$C_{23}H_{47}COOH$
Cerotic	$C_{25}H_{51}COOH$

The acids of this series up to capric acid are liquid at ordinary temperatures, while the members containing 10 or more carbon atoms are solids. The lowest members of this saturated series containing four or less carbon atoms are miscible with water in all proportions. As the molecular weight increases beyond this, however, the solubility of the acids in water rapidly diminishes to zero. The acids of this series from capric acid down are easily distilled with steam and are termed the "volatile fatty acids." The acids next above capric pass over in a current of steam only in traces, while the acids higher distill scarcely at all. Practically all these saturated acids above normal valeric occur, for the most part, in the form of glycerides in various plants. Thus caproic, caprylic, and capric acids occur as glycerides in coconut and palm-nut oils. Lauric acid occurs as a glyceride in laurel oil, coconut oil, palm oil, and myrtle wax. Myristic acid occurs as a glyceride in palm-kernel oil. Palmitic acid is found as glycerides in barberry wax and in palm oil. Stearic acid occurs in the form of glyceride in shea butter and cacao butter while arachidic acid is found in relatively large amounts as glycerides in peanut oil.

2. *Unsaturated.*—The organic chemist designates these acids as the unsaturated acids, but since the term "unsaturated fatty acids" is used extensively by biologists, it was thought best to designate them by this name in this text. When the fatty acid molecule does not contain the maximum amount of hydrogen possible, the acid is said to be unsaturated. This means that at some point in the chain one or more pairs of carbon atoms are united by a double union. This double union may occur in a number of different positions, and the properties of the acid differ with the different positions of the double unions. The unsaturated fatty acids may be made to combine with hydrogen, halogens, or oxygen by chemical means. The greater the degree of unsaturation the more unstable is the acid and the more readily does it combine with the elements mentioned above. The unsaturated acids are liquids at ordinary temperatures. They fall into different series according to their degree of unsaturation. The more important ones are given in the following table:

THE UNSATURATED FATTY ACIDS

Name of the acid	Formula
Oleic series:	$C_nH_{2n-2}O_2$
	or $C_nH_{2n-1}COOH$
Tiglic	C_4H_7COOH
Hypogeic	$C_{15}H_{29}COOH$
Oleic	$C_{17}H_{33}COOH$
Isomer of oleic:	
Rapic	$C_{17}H_{33}COOH$
Erucic	$C_{21}H_{41}COOH$
Linolic series:	$C_nH_{2n-4}O_2$
	or $C_nH_{2n-3}COOH$
Linolic	$C_{17}H_{31}COOH$
Linolenic series:	$C_nH_{2n-6}O_2$
	or $C_nH_{2n-5}COOH$
Linolenic	$C_{17}H_{29}COOH$
Hydroxy series:	$C_nH_{2n-2}O_3$
	or $C_nH_{2n-1}OCOOH$
Ricinoleic acid	$C_{17}H_{33}OCOOH$

The oleic series takes its name from oleic acid, which occurs as glyceride in most fats and oils. Tiglic acid occurs as a glyceride in croton oil, while hypogeic acid is found in the form of glyceride in peanut and maize oil. Oleic is the most widespread and most abundant of any of the fatty acids that occur in the plant. The glyceride of this acid is found in the fats and oils of the coconut, palm kernel, olive, peanut, cotton seed, maize, flax seed, soybean, and many other seeds. Linolic acid occurs in the combined form in linseed oil, cotton-seed oil, olive oil, peanut oil, soybean oil, and maize oil, while linolenic acid and its isomers are found in linseed oil, hemp oil, sunflower oil, soybean oil, and other plant oils.

The hydroxy-fatty acids have one or more hydroxyl groups that replace hydrogen in the chain. One of the common hydroxy-fatty acids is ricinoleic, which is unsaturated and forms the glyceride that is the principal constituent of castor oil.

b. Glycerol and Other Alcohols.—Alcohols of various kinds are combined with the fatty acids to form numerous types of lipides. Of these alcohols, the trihydroxyalcohol, glycerol, or glycerin ($C_3H_5(OH)_3$), is by far the most abundant and forms esters or glycerides with many different fatty acids. At ordinary temperatures, it is a viscid fluid and mixes with water in all proportions. It is miscible with ethyl alcohol, but only slightly soluble in ether and is insoluble in chloroform, petroleum ether, and carbon disulphide. It is highly hygroscopic, absorbing more than one-half its weight of water from the air and cannot be completely freed from water *in vacuo* over sulphuric acid. Glycerol is formed in the alcoholic fermentation of sugar. It was discovered during the World War that glycerol can be prepared from sugar by the action of yeast in the presence of large amounts of sodium sulphite and large quantities were obtained in this manner.

The alcohols that are combined with the fatty acids to form waxes or sterol esters are monohydric and are insoluble in water. Some of these which are found in plants are ceryl alcohol ($C_{26}H_{53}OH$), which occurs in the combined state as ceryl palmitate in poppy wax; myricyl alcohol ($C_{30}H_{61}OH$), which occurs free and as the ester of cerotic acid in carnauba wax, which is obtained from the leaves of the Brazil palm; and cholesterol ($C_{27}H_{45}OH$), which is a cyclic, monohydric alcohol. Such compounds are sometimes termed "sterols." Cholesterol is found in animal tissue, but so far has not been detected in plants. It is of interest here, however, since phytosterol, which is apparently an isomer of cholesterol, is found in practically all the fatty materials of plants, constituting from 0.13 to 0.30 per cent of such materials. The term "phytosterol," however, is now used generically, since phytosterol is considered to be composed of a mixture of two substances: sitosterol ($C_{27}H_{45}OH$), which has been obtained from the oil of wheat, rye, and maize; and stigmasterol ($C_{30}H_{47}OH$), which has been found in the phytosterol of rape oil and cacao butter.

Anderson and Moore (1923) found that corn oil contains 2 per cent of unsaponifiable matter, largely of phytosterol, which is identical with sitosterol. Both cottonseed oil and linseed oil contain at least two phytosterols differing in optical properties. The corn pollen of white flint corn contains phytosterol palmitate (Anderson, 1923). The fat extracted from this pollen with absolute alcohol contains 25 per cent unsaponifiable matter, primarily a mixture of optically inactive phytosterols, which are free from water of crystallization. Anderson (1924) also found that the unsaponifiable matter derived from the endosperm of corn contained considerable quantities of ordinary sitosterol associated with dihydrositosterol (see Anderson and others, 1926; Terrone and others, 1927; Matlack, 1935; and Moyer, 1935).

2. Classification.—According to Bloor (1925), the lipides may be divided into three main groups: (*a*) the simple, (*b*) the compound, and (*c*) the derived.

a. Simple.—The simple lipides include those compounds which are the esters of the fatty acids with various alcohols. The fats and fatty oils and the waxes are the main groups of the simple lipides.

1. *Fats and Fatty Oils.*—This group of compounds is composed of esters of the fatty acids with glycerol, termed the "triglycerides" of fatty acids. These compounds are commonly called "oils" if they remain liquid at ordinary temperatures and "fats" when they are solid at such temperatures. The oils of this type are frequently designated as "fatty oils" to distinguish them from the volatile oils and others that have no chemical relation to the triglycerides. In this discussion the term "fat" will be used in the general sense as including all the triglycerides. The fats as they occur in nature are practically always mixtures in varying proportions of triglycerides. In some fats one triglyceride predominates, in others another, and in still others several are present in considerable amounts, so that probably no natural fat consists solely of a single triglyceride. Furthermore, the individual fatty-acid radicals of any single glyceride may be different (Collin and Hilditch, 1929; and Bengis and Anderson, 1934). Thus each of the three fatty-acid radicals may be different, two may be of one type and the third another, or all three may be of the same kind. The type of glyceride that predominates in a given fat gives that fat its more pronounced characteristics. Although a large number of triglycerides occur in nature, those of myristic acid, lauric acid, palmitic acid, stearic acid, oleic acid, linolic acid, and linolenic acid make up the great bulk of the natural fats and oils. This fact, as well as the mixed nature of the fats and oils from plants, is shown in the table on page 732, adapted from Alsberg and Taylor (1928), Baughman and Jamieson (1920, 1921, 1922), and Jamieson, Baughman, and Brauns (1921, 1922).

(*a*) *Physical Properties.*—The triglycerides of the higher fatty acids are insoluble in water, while those of the lower are slightly soluble. The triglycerides above those of caprylic acid are lighter than water and all float thereon. They are readily soluble in ether, chloroform, benzene, and other organic solvents in the cold but are much more soluble in these solvents when heated. They are only slightly soluble in ethyl and

PERCENTAGE OF THE MORE IMPORTANT FATTY ACIDS COMBINED IN FATS AND OILS

Fat or oil	Lauric	Myristic	Palmitic	Stearic	Oleic	Linolic	Linolenic
Coconut	45	20	5	3	6		
Palm kernel	55	12	6	4	10		
Olive	..	....	14.6	...	75.4	10	
Peanut	..	....	8.5	6	51.6	26	
Cotton seed	..	....	23.4	...	31.6	45	
Maize	..	....	6.0	2	44.0	48	
Linseed	..	3	....	...	....	77	17.0
Soybean	..	....	11.0	2	20.0	64	3.0
Cantaloup seed oil	..	0.3	10.2	4.5	27.2	56.6	
Sunflower oil	..	....	3.5	2.9	33.4	57.5	
Hubbard squash seed oil	..	....	13.0	6.0	37.0	44.0	

methyl alcohol and acetone, at ordinary temperatures, but readily soluble when heated. Castor oil is an exception, however, and is readily soluble in ethyl alcohol in the cold.

The pure triglycerides are colorless, odorless, and tasteless, the properties of color, odor, and taste, when present, being due entirely to foreign substances mixed with or dissolved in them. The pure triglycerides are for the most part optically inactive. Castor oil is an exception, since it contains an optically active fatty acid. Some of the unpurified triglycerides show optical properties, but these properties are due to such nonfat impurities as resins and sterols.

(*b*) *Chemical Properties.* (1) *Hydrolysis.*—All the fats and fatty oils, under proper conditions, may be hydrolyzed to glycerin and their component fatty acids. Thus, for example, in the case of tripalmitin, hydrolysis proceeds after the following manner:

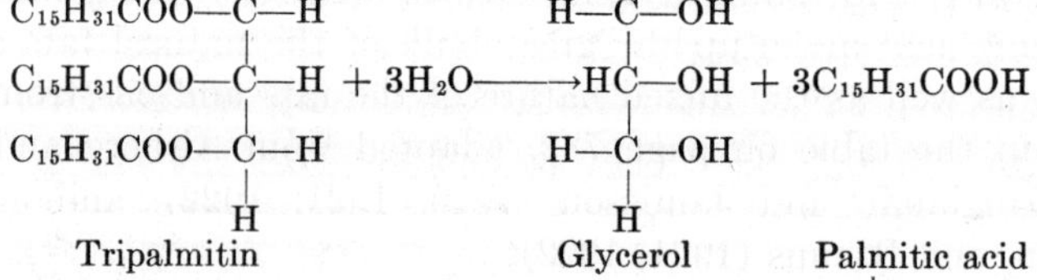

Tripalmitin Glycerol Palmitic acid

The fats may be hydrolyzed: (*a*) By treating them with superheated steam in the presence of a suitable catalytic agent. Four compounds that may be used as catalyzers are sulphuric acid, hydrochloric acid, lime, and magnesia. Sulphuric acid is the most efficient agent since it apparently aids in bringing the fat into a finer emulsion. (*b*) By the action of the enzyme lipase. This enzyme may be isolated from plant tissue and used in the laboratory for hydrolyzing fats. Lipase hydrolyzes the fats in plants under natural conditions as will be described later. The hydrolysis of fats is apparently the first step in their breaking up for utilization in the metabolism of the plant.

(2) *Saponification.*—When fats are boiled with caustic alkali they are hydrolyzed into glycerol and fatty acids, the alkali combining with the fatty acid to form soap, a process termed "saponification." The term saponification literally means soapmaking. The process may be illustrated by using the fat tripalmitin as an example:

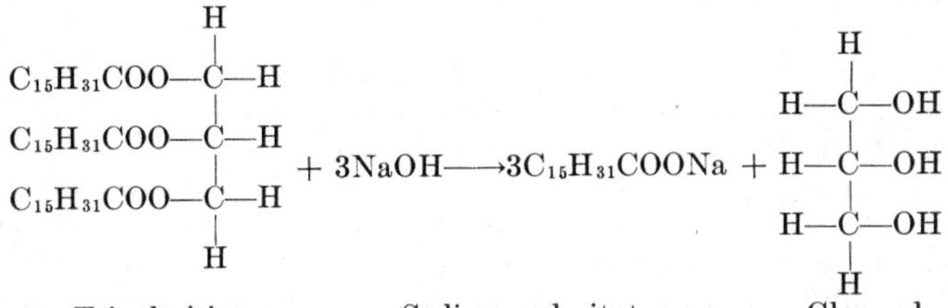

Tripalmitin — Sodium palmitate (a soap) — Glycerol

In the laboratory the saponification of fats is usually carried out by alcoholic alkali solutions, since the reaction is much more rapid with this reagent than with alkali in water. The process is of interest in plant physiology because it is considered to play some part in the translocation and utilization of fats, as will be discussed later.

(3) *Drying of Oils.*—When most of the fatty oils are exposed to the air they tend to thicken or dry. This is due in some cases to polymerization and in others to oxidation. Some oils only thicken, while others become dry and form a more or less hard mass or, if spread out sufficiently, a thin elastic film. The drying process is due primarily to the oxidation of the oils and is characteristic of those oils which contain fatty acids with unsaturated carbon unions; the larger the number of these unsaturated unions the more readily do the oils take up oxygen. According to the degree to which they dry, oils are grouped into three classes (Haas and Hill, 1913):

a) *Nondrying.*—The oils in this group only thicken but do not become hard or form films. They are composed, for the most part, of the triglyceride of oleic acid. To this group belong olive oil, peanut oil, almond oil, rice oil, and the oil from the kernels of cherry, plum, peach, and apricot seeds.

b) *Drying.*—The oils of this group are composed principally of the triglycerides of the unsaturated fatty acids of the linoleic and linolenic series and contain a relatively small amount of oleic acid. The drying oils absorb oxygen from the air and form an elastic skin or film, a characteristic that is utilized in the manufacture and application of paints. Some of the oils that belong to this group are linseed, hemp seed, sunflower, cedar nut, walnut, and poppy seed.

c) *Semidrying.*—According to the classification of Haas and Hill (1913), the oils of this group have drying properties lying midway between those of the drying and nondrying oils. They differ from the true drying oils in containing triglycerides of acids of the linoleic series. Some of the oils that fall in this group are cottonseed, pumpkin seed, soybean, sesame and croton, watermelon seed, barley and rye seed, rape, mustard seed, and radish seed.

(4) *Values for Determining the Character of Fats and Oils.*—There are certain chemical values that are used as measures of certain characteristics of the fats. These values are determined in the technical chemical analyses of fats but are also of importance in determining certain changes that may occur in the metabolism of these compounds in the plant. It is not the intention here to enter into any detailed discussion of the procedure or methods to be used in determining these values, but to discuss them only in so far as they will be an aid in understanding the physiological changes that occur in the fat metabolism of the plant. A detailed discussion may be obtained by consulting Haas and Hill (1913), Leathes and Raper (1925), and "The Methods of Analysis of Official Agricultural Chemists" (1925). The chemical values of the fats that will be mentioned here are (*a*) the acid, (*b*) the Hehner, (*c*) the saponification, (*d*) the iodine, and (*e*) the acetyl.

a) *Acid Value.*—By the acid value of a fat is meant the number of milligrams of potassium hydroxide necessary to neutralize 1 g. of the fat. It is thus a measure of the amount of free fatty acid present in the fat and indicates the degree of hydrolysis it has undergone. The acid value, however, indicates nothing as to the nature of the free acids present in the fat. The acid value of a fat may be obtained by dissolving a measured portion in neutral 95 per cent alcohol, heating to boiling, and titrating with 0.1 *N* alkali, using phenolphthalein as an indicator.

In connection with the acid value of a fat, the total insoluble acids and the total soluble acids should be considered.

b) *Hehner Value.*—By the Hehner value of a fat is meant the percentage of fatty acids insoluble in water that is yielded on the saponification of the fat. The determination of the insoluble fatty acids of a fat gives some information as to its composition. Thus if the fat contained in a certain plant part shows, during the course of the development of the plant, a decrease in the percentage of insoluble fatty acids, it is an indication that the higher fatty acids are being replaced by those of the lower groups, which are soluble in water. The total soluble acids of a fat represent both the free and the combined acids of the oil which are soluble in water and represent the fatty acids of lower molecular weight. The determination of the soluble acids supplements the results obtained in the estimation of the total insoluble fatty acids.

c) *Saponification Value.*—The saponification equivalent or value of a fat is the amount of potassium hydroxide in milligrams that is neutralized during the saponification of 1 g. of the fat by the combined and free fatty acids that it contains. It is in reality an indication of the mean molecular weight of the fatty acids that enter into the composition of the fat. Thus fats that are the glycerides of the higher fatty acids give relatively low saponification values, while those that are glycerides of the lower fatty acids give relatively high saponification values. The saponification values of some of the more common plant fats are as follows:

Fat or oil	Saponification value
Coconut oil	246 to 260
Peanut oil	190 to 196
Cottonseed oil	193 to 195
Linseed oil	192 to 195
Soybean oil	193
Olive oil	185 to 196
Corn oil	188 to 193
Rape-seed oil	170 to 179

d) *Iodine Value.*—By the iodine value of a fat is meant the amount of iodine, expressed in percentage of the weight of the oil, that the combined and free unsaturated fatty acids of the fat will take up. Thus if the iodine value of the fat content of a plant decreases during any given period, it indicates that the unsaturated fatty acids of the oil are becoming saturated, probably by the addition of oxygen. The iodine value of some of the more common plant fats is shown in the following table:

IODINE VALUES OF PLANT FATS
(After Alsberg and Taylor, 1928)

Fat or oil	Iodine value	Fat or oil	Iodine value
Linseed oil	173 to 201	Peanut oil	83 to 100
Soybean oil	137 to 143	Palm oil	51 to 57
Sunflower oil	119 to 135	Cacao butter	32 to 41
Corn oil	111 to 130	Palm kernel oil	13 to 17
Cottonseed oil	108 to 110	Coconut oil	8 to 10
Rape-seed oil	94 to 102		

e) *Acetyl Value.*—The acetyl value of a fat gives the number of milligrams of potassium hydroxide required to neutralize the acetic acid that is set free when 1 g. of the acetylated fat is saponified. The acetyl value is really a measure of the amount of the hydroxyl groups that a fat contains. The acetyl values of some of the fats are as follows (after Haas and Hill, 1913):

Linseed oil	3.98	Palm oil	18.0
Olive oil	10.64	Grape-seed oil	144.0
Rape-seed oil	14.7	Castor oil	153.0

2. *Waxes.*—The waxes are generally defined as fatty-acid esters of some alcohol other than glycerol, usually a monohydric alcohol such as ceryl alcohol, cetyl alcohol,

myricyl alcohol, and phytosterol. The waxes may also contain acids other than those which have been mentioned in considering the fats and oils. According to McNair (1931), plant waxes include as components higher fatty acids, mono- or dihydroxy saturated alcohols of high molecular weight, esters formed from these, ketones, and hydrocarbons. The waxes from the tropics have lower melting points and higher molecular weights than those from the temperate zone. Most of the vegetable waxes are produced by the integuments of the organism and not internally as are the fats. The term "waxes" is used only in the chemical sense and has reference to their composition regardless of their physical condition, so that waxes occur as both solids and liquids. The waxes of the plant kingdom are all solids. Some of the waxes of plant origin are ceryl palmitate in poppy wax and myricyl cerotate and ceryl cerotate in carnauba wax. Some of the waxes of animal origin are beeswax, wool wax, and spermaceti.

Waxes are somewhat less soluble in the ordinary fat solvents than are fats. The waxes, as a general rule, have to be boiled with alcoholic potash much longer than the glycerides in order to be saponified. This is due in part, apparently, to their low solubility even in hot alcohol. Their saponification value is low because the molecular weights of the alcohols as well as the fatty acids that compose them are high. Waxes are commonly found on fruits and leaves and are generally spread over the cuticle as a wax covering. In fruits this covering is termed the "bloom" when it is noticeable. These wax coverings may consist of grains, small rods, or crusts. The greatest thickness of wax deposit that has been observed is on the leaves of some of the palms where in some cases the coverings are more than 5 mm. in thickness. In the waxlike coating of the apple, Sando (1923) obtained from the ether extract triacontane ($C_{30}H_{62}$), melting at 63.5 to 64°C.; heptacosanol ($C_{27}H_{56}O$), melting at 81 to 81.5°; and malol, a crystalline alcohol.

Pollard, Chibnall, and Piper (1931) and Smith and Chibnall (1932) found that *n*-hexacosanol is the chief constituent of the wax of the perennial forage grasses, cocksfoot, and rye grass.

Sahai and Chibnall (1932) identified *n*-nonacosane, cerotic acid, and ceryl alcohol among the constituents of the leaf wax of Brussels sprouts. Chibnall and others (1933) found that a long-chain primary alcohol, *n*-triacontanol, is the principal component of the wax from the leaves of alfalfa, and Pollard, Chibnall, and Piper (1933) reported that *n*-octacosanol, a primary alcohol, is the principal component of the wax from the young blades of wheat. The waxes of the three last mentioned plants contain paraffin of a complex nature, and the wax from the leaves of alfalfa and wheat contains mixed fatty acids of an unknown composition.

Markley and Sando (1931, 1933) noted in the formation of the waxlike coating of the apple that there is an increase in the ursolic acid, the oily fraction, and the total ether extract throughout the growing period and during storage. They considered that the difficulty in removing spray residues cannot be attributed directly to an accumulation of oil of the sprays but rather to an increase of all the ether-soluble, waxlike constituents due to physiological stimulation. The application of a mineral oil may exert a semisolvent action on the natural waxy coating. This would cause the particles of poison to be more easily and firmly embedded. After the disappearance of the dissolving oil, the natural waxy constituent would form again, and this more compact coating would protect the arsenate particles against the cleaning fluid.

b. Compound Lipides.—The term "compound lipides" as defined by Bloor (1925) includes the esters of the fatty-acid-containing groups in addition to an alcohol and a fatty acid. This group of substances has been designated by various authors as "lipins," "lipoids," and "fatlike bodies." MacLean and MacLean (1927) in defining the term "lipins" stated that they are substances of a fatlike nature, yielding upon hydrolysis fatty acids or derivatives thereof and containing either nitrogen or nitrogen and phosphorus. The compound lipides are divided into (1) the phospholipides and (2) the galactolipides.

1. *Phospholipides.*—This group has also been called the "phospholipins," "phosphatides," and sometimes "the lecithins." The phospholipides are substituted fats containing phosphoric acid and nitrogen. They thus contain phosphorus, nitrogen, and fatty acids and on hydrolysis yield glycerophosphoric acid, fatty acids, and basic bodies, such as choline and amino ethyl alcohol. The phospholipides are soluble in the majority of the ordinary fat solvents with the exception of acetone. The phospholipides are precipitated from the crude ether or fat solvent extract by the addition of an excess of acetone (MacLean and MacLean, 1927). The two most abundant phospholipides are lecithin and cephalin.

(*a*) *Lecithin.*—Lecithin is considered the most important plant phospholipide. It may be regarded as a fat in which one of the three fatty acids is replaced by phosphoric acid combined with choline. The type of fatty acids combined in lecithins varies widely. In a lecithin preparation from the soybean, Levene and Rolf (1925) found stearic, palmitic, oleic, linoleic, and linolenic acids. Lecithin is very labile and undergoes changes as the result of oxidation. It combines with acids and bases and acts readily as an adsorbent. The occurrence of lecithin in the plant will be mentioned in a subsequent topic.

(*b*) *Cephalin.*—This compound closely resembles lecithin and is always found closely associated with it. It is distinguished from lecithin by its insolubility in alcohol and is separated in the laboratory from lecithin by the utilization of this property. Cephalin is considered to contain two fatty acids—one of the common fatty-acid series and another peculiar to cephalin, termed "cephalic acid." It also contains three basic bodies, two of choline and one of amino ethyl alcohol. The methods for the separation of cephalin have never been very well defined, and it is probable that pure cephalin has never been prepared.

2. *Galactolipides.*—The galactolipides are compounds of the fatty acids with a carbohydrate and contain nitrogen but no phosphoric acid. They have also been designated as "glycolipins," "cerebrosides," and "galactosides." The galactolipides are bodies of a glucoside nature and upon hydrolysis yield galactose, a fatty acid, and a base.

c. Derived Lipides.—The term "derived lipides" is given those substances which are derived from the simple and compound lipides by hydrolysis and which are soluble in the fat solvents. They include the fatty acids of the various series and the large molecular alcohols which are soluble in the fat solvents and are found in nature combined with the fatty acids. These alcohols include cholesterol, myricyl alcohol, and cetyl alcohol. The derived lipides have been discussed in relation to the compounds of which they are a part and will not be mentioned further here.

3. Distribution in the Plant Cell.—Fatty substances occur apparently in every living plant cell in exceedingly variable quantities. They are present in the cells of the apical meristem and were determined by Priestley (1924) to be for the most part phospholipides. Fats, fatty acids, and soaps can be detected in the walls of plant cells at a very early stage and appear to be present as early as the cellulose and pectic substances. Hansteen-Cranner (1924 to 1926) noted the presence of fats and phospholipides in the cell walls of parenchymatous tissues. He considered these of importance in regard to the permeability of these tissues. Priestley (1924) stated that the cell walls of the phloem tissues are free from fat, although fat is included in the cell content. The external protective coverings of the plant owe their imperviousness to water to various oxidation and condensation products of fatty acids, especially hydroxy-fatty acids, as mentioned in Chap. I. In the photosynthetic tissues, fats and phospholipides are always present, and the pigments of the green leaf are probably dissolved in colloidal solutions of fatty substances. By far the major portion of the fatty substances in plants is found in the protoplasm, but the physical relation between these substances and the protoplasm is, as yet, not clearly understood. Wakker (1886) considered that the oil substances are distributed evenly in very minute drops in the protoplasm, while Tschirch and Kritzler (1900) believed that the fatty substances of oily seeds are not in drops but are homogeneously mixed with the protoplasm. Czapek (1913) stated that fatty substances in seeds are ultramicroscopically divided in the protoplasm and that oil drops cannot be demonstrated. Haberlandt (1914) stated that in oily seeds the greater proportion of the oil is contained in the interstices between the delicate meshwork of the protoplasm, but that when the oil is present in the tissues in small amounts it is suspended in the protoplasm in small drops. Dangeard (1923) by means of intravital staining of living cells of the seeds of *Ricinus* and conifers found the cytoplasm to be alveolar and in the alveoli the aleurone grains and the oil are located. The oil in these alveoli is not a continuous phase but is in the form of spherules, which are thus inclusions in the cytoplasm. The dispersion of fatty substances or lipoids in the cells of the embryo and endosperm was studied by Pack (1925) with especial reference to juniper seeds. As afterripening and germination proceeded at 5°C. the lipoid drops gradually became smaller and increased in number. This dispersion was found to proceed and accompany the appearance of sugar and starch in the hypocotyl region. As germination progresses, the lipoid drops in the root and shoot axil reach a degree of dispersion equal to the limit of the microscope. During germination there is a decrease in the number of these drops because the lipoid material is used in the manufacture of other compounds. The oil reserve observed in the seeds of hawthorn, castor bean, peach, and juniper shows that it is often stored in the form of large drops. At times these drops are collected into masses that fill the meshes of the protoplasm. Resting seeds, with little fatty content, deposit it as fine drops. The lipoids of active cells and tissues are in a very fine state of division. This dispersion of the oily substances increases the oil surface many hundreds of times and is favorable to hydrolysis by enzymes, so that translocation, absorption, and transformation of the oil are favored.

Pack stated that in the deposition of fatty materials in storage organs, one first notices fine oil droplets or bodies in the younger cells. These droplets increase in size with the age of the cells, so that in the old cells large masses or drops of oil can be seen. These changes are the reverse of the dispersion of the lipoids during their digestion in the active tissues. Condensation attends the storage of fats while dispersion accompanies their digestion.

Under certain conditions bodies termed "elaioplasts" are formed in certain cells and have been considered by some to be oil formers. According to Beer (1909), they are formed by the aggregation of plastids and their decomposition products, and, although they frequently contain oil, they are in no sense special oil-forming organs. The oil vacuoles sometimes observed in certain cells originate by the coalescence of numerous minute oil droplets secreted by the protoplasm in a manner similar to that in which the ordinary sap vacuole arises. Although these oil vacuoles may be large and project into the sap vacuole, they continue to be surrounded by a thin film of cytoplasm. The oil vacuoles are in no sense plastids, nor are they formed by elaioplasts but are simply accumulations of fatty substances (Rivett, 1918).

4. Occurrence of Fats and Fatty Oils in the Plant.—The fats and fatty oils make up the greater portion of the fatty substances that are found in plants. Although they are widely distributed in the plant world, the commercial supply of these products is obtained for the most part from a comparatively small number of species (approximately 200). The fats and fatty oils are reserve foods and are utilized by the protoplasm in respiration, growth, and repair after the same manner as carbohydrates. When fats and oils are stored in tissues in relatively large amounts, carbohydrates, as a rule, are present in relatively small amounts. When carbohydrates are present in relatively large amounts, fats and oils generally occur only in limited quantity.

a. In Fruits and Seeds.—The fats and oils are found in the greatest abundance in fruits and seeds. While they occur in the root, stem, branches, and leaves of plants, they are rarely present in these organs in quantities large enough to be of much significance. In some seeds, as the cereals, the fats and oils are confined almost entirely to the embryo. In the oil palm, both the embryo and endosperm contain a large amount of fat. In the olive, a large amount of oil is stored in the pulp surrounding the kernel and a smaller amount is in the kernel itself. The oil or fat in these different parts of the seed may have characteristics that are quite different from each other. The table that follows gives the percentages of oil that may occur in various seeds and plant parts. The percentage of oil varies in the same species under different conditions, so that any data in this regard should be considered as more or less relative. The figures that are commonly given for the fat and oil content mean simply the fat-solvent extract and include, thus, numerous substances other than fats and oils.

THE PERCENTAGES OF FATS AND OILS IN THE DRY MATERIAL OF SEEDS AND OTHER PLANT PARTS

Plant part	Percentage of fat or oil	Plant part	Percentage of fat or oil
Coconut	65	*Madia sativa* seed	32
Brazil nut	70	Olive pulp	50
Castor bean	60	Kafir seed	2.5
Sunflower seed	45–50	Feterita seed	2.4
Flax seed	30–35	Milo seed	2.3
Cotton seed	15–20	Corn seed	5.0
Peanut	40–50	Wheat seed	2.1
Hemp	30–35	Rye seed	1.8
Walnut	50–65	Rice seed	1.9
Cacao bean	40–50	Buckwheat seed	2.5
Poppy seed	40–50	Field pea	1.1
Pumpkin seed	41	Bean	1.5
Almond	40–50	Bluegrass leaves	2.8
Soybean	15–20	Buffalo grass leaves	1.8
Cantaloup seed	30	Orchard grass stem and leaves	2.9
Spurge seed	35–45	Alfalfa stem and leaves	2.3
Rape seed	33–43	Cowpeas stem and leaves	2.6
Sesame seed	50–55	Soybeans stem and leaves	2.5
Colza seed	43–53	Cabbage leaf	1.7

The starch and oil content of the seeds of 216 of the 280 families of angiosperms and gymnosperms was tabulated by McNair (1930). He found that the seeds of 64.3 per cent of these families contain oil but no starch, 24.1 per cent contain both oil and starch, and 7 per cent contain starch but not oil. Sullivan and Near (1933) found that wheat and its products under different storage conditions showed marked variations in the amount of ether extract. The changes apparently correlated with increase of acidity and of content in the sample.

b. In Stem and Leaves.—The fatty components of the leaves had been studied very little until Chibnall and Channon (1927, 1929) and Channon and Chibnall (1927, 1929) examined the leaves of cabbage. They found that the crude lipides of these leaves amounted to about 3.5 per cent of the total solids. Chibnall and Sahai (1931) reported that in the leaves of cocksfoot (*Dactylis glomerata*) the crude lipides amounted to as much as 7 per cent of the dry weight, while the total reducing sugar present amounted to only 3 per cent on the same basis. The relative amounts of some of the various constituents of the crude lipides of the leaves of cabbage have been reported by Chibnall and Sahai (1931) as follows:

Component	Percentage of ether extract
Chlorophyll, carotene, and xanthophyll	10.6
Substances containing phosphorus	23.4
Iron compounds	3.0
Glycerides, waxes, and free acids	17.5
Glycerol	1.3
Unsaponifiable material:	
Saturated fraction	12.3
Unsaturated fraction:	
Sterols	4.5
Unidentified material	13.3

Heyl, Wise, and Speer (1929) and Speer, Wise, and Hart (1929) examined the fatty components of spinach. They found that 53 per cent of the fatty acids were present in the free state and 47 per cent as glycerides. From 550 g. of fatty acids, 26.5 g. of solid acids was obtained. These solid acids consisted chiefly of palmitic, stearic, and cerotic acids. About 145 g. of liquid acids was recovered, and of this amount 12.7 per cent was linolic, 34.7 per cent linoleic, and 26.3 per cent oleic. The volatile acids were present only in traces, if at all.

From the unsaponifiable fraction of the ether extract, they obtained a hydrocarbon of the formula $C_{20}H_{42}$, a common phytosterol with the formula $C_{27}H_{46}O_2$, two alcohols of the formulas $C_{22}H_{46}O$ and $C_{24}H_{50}O_2$, and an unsaturated, oily compound having the formula $C_{27}H_{54}O$.

The crude lipides of the leaves that have been examined contain relatively large amounts of phosphatides, unsaponifiable materials, and unsaturated compounds. The role of lipides in the metabolism of the leaf is impossible of evaluation at the present time because of the occurrence of lipides in relatively small amounts and because of their close association with the pigments of the leaf and other impurities.

c. In Trees and Shrubs.—Fats and oils form a considerable portion of the food reserve in trees and shrubs. Under certain conditions, carbohydrates appear to predominate in amount; and under others, fats and oils. In some plants these two reserves are about equal in amount (Sinnott, 1918). It has been observed for a long time that under certain conditions the carbohydrate reserves are transformed into fats and oils and vice versa. The conditions under which such changes are brought about are still very obscure. In trees growing at Edmonton, Can., a progressive decrease of the sugars from the winter maximum toward the summer was found (Lewis and Tuttle, 1920; Tuttle, 1919, 1921). Oils and fats were present in nearly all cells of the phloem, cortex, and the medullary-ray cells. In a majority of cases, starch disappeared during October and early November, while oils and fats were abundant. In leaves at about the time of leaf fall, 15 species were characterized by the absence of starch, but all showed a relatively high oil and fat content. Sinnott (1918) considered that starch is more common, in the woody plants, in regions remote from the centers of conduction and in cells with well-lignified and small-pitted walls, while fats and oils were more abundant in and near the phloem, close to vessels, and in cells with thin, unlignified walls with large pits. Where movement of liquids apparently is slow and difficult, the reserve food persists as starch, while where such movement is easy, starch disappears at the beginning of winter and fats and oils are produced. The character of the food reserves thus depends primarily upon the ease with which water or substances carried by it have access to the cells.

5. Occurrence of Compound Lipides in the Plant.—Our knowledge of the compound lipides in plants is very limited, and it is doubtful if any of them have yet been obtained in the pure form. The methods used for their isolation from animal tissues are apparently not applicable for their extraction from plant tissues. The chief obstacle in investigating the compound lipides in plants is due, according to MacLean and MacLean (1927), to the presence in the plant of a large amount of carbohydrate material which tends to form physical complexes with the compound lipides. It is not known whether the compound lipides of plants are identical or not with those of animal origin. The occurrence of compound lipides has been reported for a wide range of plants and has been especially studied in seeds and seedlings. These compounds have been variously reported as phospholipins, phosphatides, lecithin, lipoids, and lipins. The compound lipides that have been reported in plants are, for the most part, phospholipides, especially lecithin, but it is probable that cephalin is also present in plants (Levene and Rolf, 1925). According to MacLean and MacLean (1927), the presence of carbohydrates in the compound lipides of plants is almost universal. Although some of this carbohydrate is only in physical association with these substances, there exist bodies of wide occurrence in plants with which the carbohydrates are chemically combined and thus identical with or similar to the galactolipides that are found in animals. The only compound lipides that have been studied to any extent in plants are the phospholipides, and our attention will be devoted entirely to them. The phospholipides are widely distributed in plants and, although occurring in very small amounts, are present in practically every living plant cell (Kossel, 1891).

a. Distribution.—Knop (1859) isolated from plant tissue a fatty extract that contained phosphorus, and thus he was probably the first to observe the presence of phospholipides in plants. Jacobson (1889) showed that one of the decomposition products of the phospholipides occurring in plants is choline, while Schulze and Likiernik (1891) hydrolyzed phospholipides from seeds and obtained fatty acids, glycerophosphoric acid, and choline, thus establishing the similarity of the plant phospholipides to those of animal origin. Winterstein and Hiestand (1908) found phospholipides in the potato and seeds of cereals, vetch, and lupine. Euler and Nordenson (1908) isolated a lecithinlike substance from the carrot, and Winterstein and Stegmann (1909) were able to obtain a phospholipide from the leaves of *Ricinus*. It was determined by Winterstein and Smolenski (1909) that wheat flour contains a mixture of phospholipides that differ in their physical and chemical properties, while Smolenski (1909) obtained a phospholipide from the wheat embryo. Njegovan (1911) obtained three phosphatides from the seeds of *Lupinus albus* differing in their content of phosphorus, nitrogen, and fatty acids. Shorey (1898) found that the raw sap of the sugar cane contained a mixture of several lecithins, as indicated by the different acids and bases obtained upon hydrolysis.

Hansteen-Cranner (1914–1926) made extensive studies of the distribution of the phospholipides in the protoplasmic membrane and in the cell walls. He considered that the cell wall is permeated with phospholipides which are in intimate connection and relation with those of the plasma membrane. His observations were substantiated by Grafe (1925), Grafe and Horvat (1925), and Grafe and Magistris (1925, 1926), who prepared water-soluble phospholipides from beets and carrots among other plant tissues. The methods used by these investigators were considered to extract the phospholipides from the cell wall and the plasma membrane without altering the permeability of the general protoplasmic membrane. Steward (1928), however, by the careful extraction of the tissues of potato and beet concluded that leaching experiments do not yield evidence that phospholipides are present in the surface layer of the protoplasm and diffuse from living tissue into distilled water. In a study of the ether-soluble substances of cabbage leaves, Chibnall and Channon (1927) could not detect any true phospholipides. All the phosphorus present was apparently in combination with calcium, glycerol, and fatty acids, while nitrogen was virtually absent. The substance that was isolated was thus apparently the insoluble calcium salts of glycerophosphoric acids in which the choline was replaced by calcium. The phospholipides that have been extracted from plants are no doubt mixtures of similar compounds and vary widely from different sources in regard to their nitrogen and phosphorus content and in the types of their fatty acids. This is illustrated by the work of Webster (1929), who determined the percentage of nitrogen, phosphorus, and iodine value of the phospholipides secured from the seeds of wheat, corn, sorghums, and oats. The results obtained were as follows:

Source of phospholipides	Percentage of nitrogen	Percentage of phosphorus	Iodine value of the fatty acids
Wheat	0.99	1.26	81.5
Corn	1.14	1.65	65.3
Soybean	0.70	0.50	92.5
Oats	1.61	0.54	88.8

The amount of the phospholipides that occurs in plant tissue apparently varies with the type of tissue, the kind of plant, and the environmental conditions.

Thus Chibnall and Sahai (1931) observed in the leaves of Brussels sprouts that lecithin and cephalin were absent but that calcium phosphatidate was relatively abundant. It appears that in this leaf the synthesis of phospholipides ceases under these conditions and that the formation of an ester with choline to produce lecithin or with amino ethyl alcohol to give cephalin does not occur. In the leaves of cocksfoot (*Dactylis glomerata*), however, Smith and Chibnall (1932) found lecithin and cephalin as well as the magnesium or calcium salt of phosphatidic acid. Jordan and Chibnall (1933) noted that all portions of the seedling of the runner bean contain lecithin and cephalin in progressively smaller amounts from the cotyledons to the pinnate leaves. In the resting seed, a small amount of magnesium phosphatidate is present, and this increases in amount as germination progresses. At the onset of chlorophyll formation, the magnesium in this compound is replaced by calcium, and the chief phospholipide of the mature pinnate leaves is calcium phosphatidate. According to Channon and Foster (1934) lecithin, cephalin, and the calcium, magnesium, and potassium salts of phosphatidic acid are present in the wheat embryo.

b. Relation of the Amount of Phospholipides to the Protein and Oil Content.—It was considered by Stocklasa (1895 to 1898) that lecithins accompany the proteins in plants, while Schulze and Likiernik (1891), and Schulze, Steiger, and Maxwell (1891) considered that the phospholipides vary directly with the protein content. According to Guerrant (1927), it was considered by Parrozzani (1909) that the ratio between the phospholipides and the amides is more definite than that of proteins and phospholipides. The amount of oils, proteins, and phospholipides in various seeds was determined by Guerrant (1927). Some of his results are shown in the table on page 745.

Guerrant (1927) considered that his results indicated a slight correlation between the phospholipide content and the protein, fat, and phosphorus content of the various classes of seeds. In general, the highest percentages of phospholipides are found in the seeds having the highest protein and the highest fat content, but this is not always the case.

c. Pho pholipides in Developing and Germinating Seeds.—It was considered by Schulze and Frankfurt (1894) that unripe seeds contain less phospholipides than ripe ones, but M'Clenahan (1909) observed just the opposite in the development of the black walnut. Guerrant (1927) determined the phospholipide content of seven varieties of sorghum seeds at the milk stage, dough stage, and maturity. He obtained an increase in these compounds as maturity was approached, while the percentage of protein, fats, and total phosphorus also increased. Sufficient data, however, were not obtained to show any definite correlation between the rate of formation of any two of these substances.

It was observed by Schulze and Steiger (1889) that phospholipides decreased when seeds were germinated in the dark. Maxwell (1891), Zalenski (1902), and Iwanoff (1902), in the case of the seeds of corn, bean, and cotton, found an increase of these substances when germinated in the light. Guerrant (1927) germinated two samples of grain sorghum

Seed	Number of		Percentage on a dry basis of			
	Samples	Varieties	Protein	Oils	Total phosphorus	Phospholipide phosphorus
Corn	6	6	11.5	4.3	0.196	0.0144
Wheat	5	4	12.5	2.2	0.289	0.0306
Oats	2	2	13.3	4.1	0.262	0.0368
Sorghums	26	12	12.1	2.9	0.245	0.0378
Barley	1	1	13.1	1.8	0.275	0.0442
Clover	1	1	35.2	5.4	0.271	0.0506
Peas	5	5	27.0	1.7	0.292	0.0760
Peanut	1	1	33.9	48.6	0.332	0.0880
Beans	10	8	30.4	12.6	0.387	0.0926
Cotton	1	1	23.5	25.22	0.476	0.1456

seed in both light and darkness and found that the lipoid phosphorus showed a greater increase when germination occurred in the light. Green and Jackson (1905) noted in the germination of the castor bean that the amount of lecithin diminished during the earlier stages of germination, the reserve supply becoming almost exhausted. After the young seedling had begun to develop, there was a gradual increase in the amount, which was maintained during the later stages and remained fairly constant until the endosperm was depleted. They considered that the endosperm of this seed contains the several groups necessary for the formation of lecithin. The decomposition of the oil furnishes the fatty-acid component and the glycerin for the glycerophosphoric acid. The phosphorus is present in the seed, and the nitrogenous body, choline, may arise from the decomposition of the protein.

d. Function of the Compound Lipides.—Since the compound lipides are so widespread in plants and are found especially in those parts of the plant where life activities are very pronounced, it seems evident that they must play an important part in the activities of the protoplasm. In this regard, however, nothing definite is known, and only theories have been proposed. The compound lipides have been considered by some to exercise an influence on the permeability of the protoplasmic membrane, but their influence in this regard is apparently not of the importance that was once thought. It has been considered by some that the compound lipides play a role in the fat metabolism of the plant, but nothing is known concerning their function in this regard. Palladin and Stanewitsch (1910) considered that there is a relationship between the lecithins and respiration, since they found that the respiration intensity of seeds was lower after treatment with those solvents which extracted the most lipides.

Korsakow (1910) considered that the lipides influence the activity of proteolytic enzymes, and Gallagher (1923) believed that the production of peroxidase of potato is intimately associated with the lecithin of the tuber, and that the so-called "oxygenase" is in reality an autooxidizable, lecithinlike substance. It has been pointed out by Bang (1911) that the compound lipides are the most labile of the constituents of the protoplasm and that many of the striking properties of protoplasm may be due to these substances, especially the phospholipides, rather than to the proteins. The compound lipides evidently play an important role in the activities of the living cell, but, in the words of MacLean and MacLean (1927), it is not yet possible to dissociate these compounds from the other cell constituents and attribute to them alone any specific properties associated with the cell. Steward (1929), in a review of the literature concerning the phospholipides of plants, stated that the experimental evidence for the theories concerning the role of phospholipides in plant cells is either lacking or is of a controversial nature.

6. Synthesis of the Lipides.—The study of the synthesis of the lipides in the plant has been confined almost entirely to the changes occurring in the development of oily seeds. The term "lipides" as here used refers almost without exception to the crude lipides, since the quantity of oily matter has been estimated by determining the ether extract.

a. Relation of the Carbohydrates to the Formation of Fats and Oils.—It was first observed by De Luca (1861, 1862) that as the sugar content of the developing olive diminished progressively, the amount of oil correspondingly increased, so that when the oil content had reached its maximum the sugar had disappeared. The same fact was observed by Leclerc du Sablon (1896) in studying the development of the kernels of the walnut and almond. Some of the data obtained by him are shown on page 747.

It was observed by Schwartze and Alsberg (1923), Gallup (1927, 1928), and Caskey and Gallup (1931) that the gossypol content of the cotton seed varied directly with the amount of oil present. They considered that gossypol is associated in some manner with the formation of oil.

It was found by Rushkovskii (1930) that soluble carbohydrates decreased from 37.5 to 4.5 per cent during the ripening of the sunflower seed. Caskey and Gallup (1931) noted that reducing sugars decreased from 2.8 to 0.15 per cent, and total sugars from 4.6 to 0.32 per cent during the period of the formation of oil in the seed of cotton. Sahasrabuddhe and Kale (1933) thought that the oil in the niger seed (*Guizotia abyssinica*) is formed from the carbohydrates that are present in the young seed. Thor and Smith (1935) observed in the formation of the pecan nut that the decrease in the sugar content of the whole fruit during the period of oil synthesis accounted for not more than 5 per cent of the final amount of oil. It appeared that practically all the oil in the kernel of the pecan was formed from materials brought into the fruit from other parts of the tree at the time of oil formation. It was not observed whether the source of this material was reserve carbohydrates in the twigs, trunk, or roots, or whether it was a product of photosynthetic activity during the period of the formation of the oil.

Although the disappearance of carbohydrates and the appearance of fats and oils in a developing organ have been considered as evidence that carbohydrates are transformed into fats and oils, this fact in itself is not conclusive evidence that such is the case. In order to obtain more information on this subject, Gerber (1897) studied the respiratory quotient of the developing olive and the seeds of the almond, castor bean, rape, peach, and others. He found that during the period when

PERCENTAGE OF OIL, GLUCOSE, SUCROSE, STARCHES, AND DEXTRINS IN THE DEVELOPING KERNEL OF THE WALNUT AND ALMOND

Date	Oil	Glucose	Sucrose	Starches and dextrins
		Walnut		
July 6	3	7.6	0.0	21.8
Aug. 1	16	2.4	0.5	14.5
Aug. 15	42	0.0	0.6	3.2
Sept. 1	59	0.0	0.8	2.6
Oct. 4	62	0.0	1.6	2.6
		Almond		
June 9	2	6.0	6.7	21.6
July 4	10	4.2	4.9	14.1
Aug. 1	37	0.0	2.8	6.2
Sept. 1	44	0.0	2.6	5.4
Oct. 4	46	0.0	2.5	5.3

the sugar content had reached its maximum and before any fats and oils had yet been formed, the respiratory quotient was slightly less than unity. During the period when the sugar content was diminishing and the oil content was increasing, the respiratory quotient was greater than unity, as, for example, in the olive where it ranged from 1 to 1.46. After the sugar had disappeared from the developing seeds and the oil content had reached its maximum, the respiratory quotient again fell to less than unity. Since citric, tartaric, or malic acid, which give rise to a respiratory quotient greater than unity, were not found in these developing seeds, and since there was no production of alcohol, which is also accompanied by a coefficient greater than unity, Gerber concluded that there is a relationship between the disappearance of carbohydrate and the respiratory quotient. He thus considered that the respiratory quotient of these developing organs indicates that the oil is being produced at the expense of the carbohydrate.

b. Changes in the Fats and Oils during Their Formation. 1. *Amount.*—As an oily seed develops, the percentage of oil gradually increases. Except for the period immediately following blooming and directly preceding maturity, there is throughout the development of the seed a gradual and rather uniform gain in the oil content. There is a very sharp increase in the percentage of oil during the first few weeks after blooming and a slow gain until near the end of the ripening period in the case of the soybean (Garner, Allard, and Foubert, 1914). At this time a decrease in the size of the seed and in the oil content occurs, probably due to an intense respiration at that period. The changes in the oil content during the development of the seed may be illustrated by the data obtained by Dillman (1928) for the developing flax seed:

Time of sampling, number days after flowering	Percentage of oil in the developing seed	Time of sampling, number days after flowering	Percentage of oil in the developing seed
8	1.8	24	37.5
9	2.3	27	40.2
12	8.5	30	40.7
15	26.0	33	40.7
18	34.6	36	40.6
21	35.5	39	40.7

Caskey and Gallup (1931) found in the cotton seed that the most rapid increase in oil occurred during the period of 21 to 30 days after flowering. Eyre (1931) noted in flax that there was a rapid formation of oil over a period of 15 days during which time most of the oil was deposited. Thor and Smith (1935) observed in the formation of the pecan nut that 60 to 70 per cent of the final weight of oil was deposited during a period of 4 to 6 weeks.

The factors influencing the percentage of oil in oily seeds have been studied by Garner, Allard, and Foubert (1914) and Stark (1924). By partially defoliating the plants, the number and size of beans produced were decreased, while the percentage of oil was somewhat increased. The general climatic conditions during the growing season seem to have a more marked effect on the oil content than does the length of the season. The difference in the amount of oil per seed may be as much as 100 per cent, while the percentage of oil may vary as much as 7 or 8 per cent. These authors considered that the maximum production of oil in soybeans requires conditions of nutrition favorable to the accumulation of carbohydrates during the vegetative period. Kayser (1925) showed that the application of commercial fertilizers to flax may influence the quantity and quality of the oil of the seeds. Johnson (1932) considered that the date of planting would influence the oil content of the flaxseed in some seasons but not in others. Stark (1924) found that the application of limestone and inorganic matter to the soil decreased the percentage of oil in the soybean, and the addition of rock phosphate to the above fertilizer further decreased the oil content of the seed. He observed that, by varying the soil conditions, greater differences could be obtained within the variety than generally exist between most varieties.

Tottingham (1932) noted that a consistent increase in the lipide content of various plants occurred when they were grown under Vitaglass.

2. *Nature.*—Ivanow (1911) studied the changes that occur in the carbohydrates and oils in the seeds of flax, hemp, rape, and sunflower during their ripening. His observations indicate that there is a connection between the carbohydrates and the higher fatty acids. The first fatty acids that are developed in these seeds from carbohydrates are the saturated ones and evidently belong to the higher series of the fatty acids, since they are not volatile as indicated from their Reichert-Meissl values. It was observed that the acid values of the ether extract decreases sharply with the ripening of the seed. Thus in flax, in one experiment, the acid value decreased from 15.4 to 5.6, in hemp from 5.8 to 2.7, in rape from 74.3 to 9.4, and in sunflower from 46 to 8 during their first ripening period. The evidence seemed to indicate that the unsaturated

Date	Seed and stage of development	Per cent of oil	Acid value	Iodine value
	Flax:			
July 5	Flowering	2.3	15.4	120.6
July 18	Seed soft and green	11.0	3.6	151.0
Aug. 3	Seed unripe	32.5	4.0	168.0
Aug. 25	Seed ripe	35.0	5.6	175.0
	Rape:			
July 7	Seed clear and soft	10.0	74.3	
July 25	Seed green and opaque	37.5	16.0	99.8
Aug. 8	Seed unripe	48.3	13.8	95.3
Aug. 23	Seed ripe	49.9	9.4	97.5
	Hemp:			
Aug. 17	Two weeks after bloom	15.3	5.8	152.1
Aug. 28	Seed unripe	27.0	2.5	155.0
Sept. 13	Seed unripe	33.5	2.7	154.0

acids are derived from the saturated ones, and Ivanow suggested the following scheme to illustrate the procedure in the formation of oil from carbohydrates:

Carbohydrate ⟨ glycerin ——————————————————————— ⟩ oil
Carbohydrate ⟨ saturated fatty acids ⟶ unsaturated fatty acids ⟩ oil

The data on page 749 taken from his work show the changes that occur in the oil of ripening seeds.

This theory has been further substantiated by Eyre (1931) and Johnson (1932) in the study of the development of oil in flaxseed, and by the work of Sahasrabuddhe and Kale (1933) on the development of the oil in the niger seed. The changes in the nature of the oil during the development of the flaxseed were as follows:

Eyre (1931)		Johnson (1932)	
Days after flowering	Acid value	Days after flowering	Iodine value
14	8.6	5	106
16	8.4	9	116
19	4.8	11	124
25	1.0	13	131
32	0.6	14	148
66	0.5	15	153

Priestley (1924) considered that when the fatty acids are finally deposited in the endodermis, cuticle, or cork layers they undergo oxidation and condensation and in so doing lose, for the most part, their power of dissolving in fat solvents. They are thus changed in the same general way as are certain of the vegetable oils in the drying process that they undergo on exposure to the air.

c. General Observations.—Although the fatty oils are quite different chemically from the carbohydrates, there is an intimate and significant physiological relationship between these two classes of compounds in the plant. There is considerable evidence to indicate that fats and oils are formed from carbohydrates and that in germination of oily seeds, as will be mentioned later, the fats give rise to carbohydrates. The fact that the disappearance of the carbohydrate and the increase in the oil content proceed in unripe seeds detached from the plant is considered as further supporting the evidence that oil is derived from carbohydrate in the plant. Outside the living cell, however, the transformation of carbohydrates into fats and oils or the reverse has been accomplished only in a slight way (Smedley and Lubrzynska, 1913). Smedley (1912) prepared fatty acids from carbohydrates by breaking the hexoses into three carbon compounds and then oxidizing these into pyruvic acid, $CH_3COCOOH$, which was then condensed into acids with longer carbon chains. These reactions are accomplished only with great difficulty in the laboratory, and there is no evidence whatsoever that the formation of oil in the plant follows this procedure.

The particular carbohydrate which is transferred directly into the fats and oils, however, is not known (Vallee, 1903). According to Terroine (1920), saccharose is the last sugar to disappear in the maturation of oily seeds and the first to appear upon their germination. This, however, is not sufficient evidence to state that it is the carbohydrate which is transformed directly into the oil.

Some evidence has been obtained which indicates that fats and oils may be formed in plants from other materials than carbohydrates. Thus M'Clenahan (1909) considered that the tannins might afford the material for the formation of the oil in the developing kernel of the walnut. Scurti and Tommasi (1910-1911) observed that the ether-soluble substance of the green olive fruit consists almost entirely of oleanol, a higher alcohol ($C_{31}H_{50}O_3$), and that later in the process of development it is still present but that, in addition, higher fatty acids both saturated and unsaturated are found. When the ripening of the fruit is progressing, the ether extract consists almost entirely of neutral oil. Oleanol occurs in the leaves of the olive, and it was considered by these investigators that this alcohol is produced in the leaf and is transported to the fruit when it is transformed into neutral oil. They examined other fat-containing seeds and fruits, and in all cases an alcohol closely related to oleanol was found in the leaves and fruits, and, as it diminished in amount, neutral oil increased. If a procedure like that above occurs, it is not in accord with the high respiratory quotients reported by Gerber (1897) during the ripening process, since the conversion of oleanol into fatty acids would be accompanied by a respiratory quotient less than unity.

Although oily seeds are generally rich in protein and the accumulation of oil proceeds simultaneously with protein, there is no evidence of any relationship between the two processes. In the case of the soybean, however, Stark (1924) noted that the conditions which produce an increase in the percentage of protein result in a decrease in the oil content, and vice versa.

It is not known whether the formation of oil is brought about by the protoplasm itself or results from a set of reactions catalyzed by the plant enzymes, but which could also be performed by the extracted catalysts. The synthesis of fats from fatty acids and glycerol in the presence of oil-free material from oily seeds was undertaken by Dunlap and Gilbert (1911). They found that flaxseed is a good emulsifying agent, so that, when glycerol, oleic acid, and oil-free castor-bean powder are mixed with it and thoroughly ground together, an emulsion is formed which persists for many days, and as a result of such a mixture they obtained evidence of the synthesis of oil. The method used for detecting the synthesis of oil in such a mixture was as follows: Equal portions of the mixture were placed in a series of flasks, with the exception that in the control flask the oil-free castor-bean material was replaced by an equal amount of flaxseed. The acid content of the flasks was determined at successive periods with the following results:

Period	Milligrams of oleic acid in one gram of the mixture	
	Control, without castor-bean powder	Experiment, with castor-bean powder
At once	315.5	337.3
After 2 days	319.0	322.0
After 6 days	314.8	311.0
After 11 days	314.9	248.6

The decrease in the amount of acid present could be accounted for only by the esterification of the oleic acid. The results, however, do not show whether the combination of the oleic acid and the glycerol produces a mon- or diglyceride before the triglyceride is produced.

An interesting observation was made by Abegg (1929) on the effects of the waxy gene in maize on fat metabolism. He noted that the average acid value of the ether-soluble crude oil from nonwaxy maize was 13, while that of the waxy was 34. The average saponification number of the crude fats from the nonwaxy endosperm was 177 and that from the waxy 200. These marked differences between the nonwaxy and waxy types of grain are restricted to the endosperm tissue. The evidence indicates that the waxy gene in maize produces marked effects upon the fat metabolism of the endosperm tissue. It is considered that the activities of this gene are connected with the synthetic and hydrolytic powers of maize endosperm lipase. It appears that the waxy-type lipase does not affect a condensation of fatty acids with glycerol to the same degree as does the nonwaxy, hence the waxy endosperm contains a crude fat the acid value of which is higher than that from the nonwaxy type.

7. Changes in the Fats and Fatty Oils during Germination. *a. Changes in the Amount of Oil.*—In 1842 it was observed by De Saussure that as the germination of rape and hemp seeds progressed, the amount of oil in the seeds diminished. He undertook this work in order to test the suggestion that the oil in seeds might subserve the same purpose during germination as the starch in the cereal grains. It is now known that the fats and oils are reserve foods, and all the numerous investigators who have studied oily seeds, since the time of De Saussure, agree that the oils in seed disappear during the course of germination. This behavior of the oily matter in seeds may be illustrated by the data obtained by Miller (1910, 1912) in the case of sunflower seeds and seedlings, as shown in the following table:

Parts of seeds or seedlings	Seeds	Seedlings				
		4 days	5 days	7 days	10 days	14 days
Percentage of ether extract on dry basis						
Cotyledons.	55.6	55.9	51.2	33.9	16.7	13.5
Hypocotyls and roots.	47.4	24.2	9.3	9.7	9.8	11.4
Grams of ether extract per 100 seeds and seedlings						
Cotyledons.	3.786	3.00	2.53	1.298	0.501	0.317
Hypocotyls and roots.	0.219	0.188	0.123	0.214	0.273	0.238

It is noted from the above data that after 4 days the percentage of oil in the cotyledons is the same as or somewhat higher than that in the cotyledons of the seed, while the actual amount of oil has markedly diminished. This fact has been frequently observed in the germination of oily seeds and may be attributed to one or two causes. In the first place, it may be due to the depletion of the protein and carbohydrate reserve in a relatively greater amount than the oily reserve during that period. On the other hand, it may be due to the combining of oxygen with the unsaturated oils, thus increasing their weight to the same extent as or a greater extent than it has been diminished by depletion during the period under consideration. In most oily seeds, the oily reserve begins to disappear at a very early stage. In the case of the sunflower seed 4 days after planting, when the hypocotyls and roots have a length of only 2.5 to 3.5 cm., over one-fifth of the oil has already disappeared.

It was observed by Von Ohlen (1930, 1931) and MacLachlan (1936) that the total oil content of the cotyledons of soybeans diminished markedly as germination proceeded. According to the former investigator the depletion of oil began at the base of the cotyledon and progressed toward the apex. The palisade tissue of the cotyledons was depleted of oil much more slowly than were the other tissues. According to Peters (1861), Schmidt (1891), and Miller (1910, 1912) the actual amount of ether extract in the roots and hypocotyls remains practically constant during germination.

b. Chemical Changes in the Oily Reserve.—The changes in the chemical nature of the oily reserve in seeds during germination have been studied for many seeds by numerous investigators. Miller (1912) studied in considerable detail the changes in the oily reserve of the sunflower seed during its germination, and his work is summarized in tabular form as an example of the general changes that the oily reserve undergoes during germination.

CHEMICAL NATURE OF THE ETHER EXTRACT OF THE COTYLEDONS, HYPOCOTYLS, AND ROOTS OF THE SUNFLOWER SEEDS AND SEEDLINGS

Items considered	Seeds	Stage I 3½ days	Stage II 5 days	Stage III 7 days	Stage IV 10 days	Stage V 14 days
			Cotyledons			
Dry weight of the cotyledons of 100 seedlings	4.884	3.153	2.785	2.319	2.021	1.621
Ether extract (percentage of dry material)	54.1	52.6	49.2	36.0	16.8	8.7
Grams of ether extract per 100 cotyledons	2.642	1.658	1.370	0.834	0.339	0.141
Percentage of the original oil present	....	62.7	51.8	31.8	12.8	5.3
Acid value of the ether extract	1.6	1.7	1.5	4.6	27.5	66.8
Percentage of free fatty acid in ether extract[1]	0.008	0.009	0.009	2.3	13.8	33.5
Saponification value of ether extract	189.3	190.9	192.1	184.3	186.1	204.2
Saponification value of the neutral oil	188.4	189.1	190.6	188.2	184.4	226.1
Total insoluble acids of ether extract	95.4	96.0	95.7	93.6	87.6	84.8
Total insoluble acids of the neutral oil	94.6	95.2	94.7	95.9	88.9	
Total soluble acids of ether extract[2]	0.005	0.005	0.007	0.005	0.008	4.5
Total soluble acids of the neutral oil[2]	0.005	0.003	0.006	0.003	0.002	3.0
Iodine number of the ether extract	125.5	126.4	124.5	125.8	120.1	111.7
Iodine number of the neutral oil	124.9	126.0	124.1	126.0	119.9	111.9
Acetyl saponification value of ether extract	222.7	225.0	225.1	215.0	236.3	263.6
Acetyl value of ether extract	40.1	39.0	40.6	38.2	36.5	37.5
			Hypocotyls and roots			
Dry weight of the hypocotyls of 100 seedlings	....	0.605	0.885	1.21	1.52	2.17
Ether extract (percentage of dry material)	......	9.9	7.4	3.8	3.1	1.8
Grams of ether extract per 100 hypocotyls	......	0.060	0.065	0.046	0.047	0.040
Acid value of ether extract	....	18.2	24.0	27.1	58.3	97.8
Percentage of free fatty acid in ether extract[1]	......	9.1	12.0	13.4	26.4	49.1
Saponification value of ether extract	....	190.3	189.2	198.4	236.9	238.3
Total insoluble acids in ether extracts	....	89.6	85.6	86.5	67.1	57.2
Total soluble acids in ether extract[2]	....	1.0	1.6	1.8	16.1	18.5
Iodine number of ether extract	....	117.7	118.8	96.1	72.6	48.3
Acetyl saponification value of ether extract	......	251.4	300.0	274.8		
Acetyl value of ether extract	....	71.2	114.0	95.1		

[1] Estimated as oleic acid.
[2] Estimated as butyric acid.

1. *Fatty Acids of the Oil.*—It was first observed by Fleury (1865) that nonvolatile acids appeared during the germination of oily seeds. Müntz

(1871) in the study of the seeds of the rape, poppy, and radish observed for the first time that during germination the oil in the seeds gives rise to free fatty acids. In the case of these particular seeds, the amount of free fatty acids in the oil increased from 10 per cent in the resting seed to as much as 98 per cent in the 10-day seedlings. The increase in the amount of free fatty acid in the germination of oily seeds has since that time been observed by Leclerc du Sablon (1895, 1897), Miller (1910, 1912), Ivanow (1912) and others.

It was suggested by Müntz (1871) that the oil during germination is split up into glycerin and fatty acids. In 1890 Green discovered the enzyme lipase in the castor-bean seedling and showed that the fats and oils are hydrolyzed into fatty acids and glycerin by its action. Although glycerin is easily obtained by the hydrolysis of fats and oils under artificial conditions, no one has as yet been able to detect it in the germinating seeds and seedlings. The changes that occur in the fatty acids of the oil of seeds and seedlings have for the most part been observed in the ether extract of the entire seedling. In order, however, to understand the significance of the changes in the oily reserve in the metabolism of the seedling, the changes in the oil in the cotyledons or in the endosperm must be considered separately from the oily material of the roots and hypocotyls.

(*a*) *Fatty Acids of the Cotyledons.*—The changes in the fatty acids of the oil of the sunflower seed during its germination were studied in considerable detail by Miller (1912). In this seed the oily reserve is stored in the cotyledons and amounts to approximately 50 per cent of their dry weight. Miller observed that the ether extract of the cotyledons has an acid value of 1.6 in the resting seed and that this remains constant until one-half of the oil has disappeared. In the 7-day seedlings when the cotyledons are above ground and are spread out perpendicular to the hypocotyl and when the oily reserve has decreased to one-third its original amount, the acid value rises to 4.6. The acid value continues to increase so that when only 5.3 per cent of the original amount of oil remains the acid value amounts to 66.8. The free fatty acids, estimated as oleic, constitute less than 1 per cent of the ether extract of the seed. This remains constant until over two-thirds of the oil has disappeared, after which it increases to 2.3 per cent of the ether extract. This percentage continues to increase until at the end of 2 weeks the free fatty acids constitute 33.5 per cent of the oily material in the seed. Some of the free fatty acids present in the cotyledons are those of low molecular weight and are soluble in water, but most of them are of higher molecular weight and are insoluble in water.

The iodine value of the oils of seeds at different stages of germination has been studied by Ivanow (1912), Jegorow (1904), and Miller (1912).

It has been observed that as the germination of flax, hemp, and rape progressed, the iodine value of the oil decreased. In the sunflower seed, the iodine value of the ether extract is 125. This value remains constant for a considerable period of time and drops to only 120 when the seedlings are 7 days old. After 14 days, when only 5.3 per cent of the original oil remains in the cotyledons, the iodine value falls to 111.8. Miller (1912) considered that this decrease in the iodine value of the oil of the seed during germination indicates that unsaturated fatty acids are becoming saturated. Ivanow (1912), however, considered that this conclusion is erroneous and that the decrease in the iodine value indicates that the unsaturated fatty acids are being utilized. He noted that the unsaturated fatty acids are the first to disappear during germination. Thus in the flax and hemp seedlings the linolenic acids first disappeared, then the linoleic, and finally the oleic series. In the germination of soybeans, MacLachlan (1936) observed that so far as the degree of saturation was concerned the fatty acids of the cotyledons were no different from those of the ungerminated seeds.

The saponification value of the oil in the cotyledons of the sunflower seed does not change until the oily reserve has been depleted to 5.3 per cent of the original amount. There is a marked increase in the saponification value at this period, which indicates the presence of glycerides or free fatty acids of lower molecular value than those originally contained in the cotyledons of the seed.

The acetyl value of the ether extract of the sunflower seed is 40.1, a value that remains practically constant during all stages of the germination. There is thus apparently no increase in the number of the hydroxyl groups of the oil during any stage of germination. These observations are in accord with those of Von Fürth (1904) for the sunflower and castor bean.

(*b*) *Fatty Acids of the Roots and Hypocotyls.*—Miller (1912) observed in the roots and hypocotyls of the sunflower seedlings that the saponification value of the ether extract remained practically constant at 190 until the seventh day when it increased to 198. After that it increased rapidly until at the end of 2 weeks it had risen to 238.

The amount of insoluble fatty acids remains constant during the 7-day period and is about 10 per cent less than the amount of insoluble acid present in the ether extract of the cotyledons at the same stages. After the first week the amount of insoluble acid falls rapidly and amounts to only 57.2 per cent of the oily material present at the end of 14 days, as compared to 90 per cent at the beginning of germination. The amount of soluble fatty acids during the first 7 days of germination changes but little, being about 2 per cent, the amount being somewhat higher than that contained in the oily matter of the cotyledons at these stages. During the second week of germination the amount of soluble fatty acids increases rapidly and amounts to 18.5 per cent of the oily matter present at the end of 14 days. The acid value increases from 18.2 at the first stage of the seedling to 97.8 at the last stage examined. During the first week of germination there is evidence that a considerable portion of the free acid is composed of acids of higher molecular weight, which are insoluble in water. During the last week of germination, however, the free acids present seem to be composed entirely of those of low molecular weight, which are soluble in water.

The iodine value of the ether extract of the cotyledons and roots remains at 118 until the end of 7 days when it falls to 96 and finally to 48.3 at the end of 2 weeks. The iodine value is thus considerably lower at the earlier stages of germination than it is in the oily material of the cotyledons.

The acetyl value of the ether extract of the hypocotyls and roots is approximately twice as great as that of the ether extract of the cotyledons and increases in value as germination progresses.

The properties of the ether extract of the hypocotyls and roots during the first 7 days of germination bear a striking resemblance to those of the ether extract of the cotyledons during that time. Observations, however, indicate that the changes in the oily material of the hypocotyls and roots during the earlier stages of germination consist in a gradual but evident breaking down of the higher free fatty acids and glycerides into those of lower molecular weight, the saturation of the fatty acids, or the utilization of the unsaturated ones and an increase in the amount of hydroxyl groups present. During the later stages of germination, however, these changes are rapid and very marked, as indicated by the different values determined.

MacLachlan (1936) found during the germination of the soybean that the newly formed fatty acids of the roots, stem, and leaves were considerably more saturated than those of the cotyledons. A marked increase in the amount of sterols occurred in the roots, stem, and leaves of the young plants. The results indicate that a close relationship exists between the metabolism of the sterol and the utilization of oil in the cotyledon, and that the sterol is a vital constituent of the plant and not a waste product. It was also noted by Beumer (1933) in peas and beans that a marked synthesis of sterol occurred during germination in both light and darkness. Kao (1936) in the germinating mung bean noted that there was an increase in the sterol fraction and a decrease in the crude phospholipides. The fatty-acid content greatly increased in the growing portion of the plant while that of the cotyledons remained practically constant.

2. *Relation of Carbohydrates to the Disappearance of the Fats and Oils.*—It was first observed by De Saussure (1842) that as the oily reserve disappeared in the germination of oily seeds, the carbohydrate content

of the seeds and seedlings increased. Hellriegel (1855) examined the seedlings of rape at five different stages of growth. He found that the oil decreased from 47 per cent in the seed to 36 per cent in the oldest seedlings, and that the sum of the sugar, organic acids, and tannin increased at the same time from 7.7 to 15.4 per cent. He also observed that the seeds contained 3.4 per cent cane sugar and that during the early stages of germination it disappeared. This fact has been noted by other investigators since that time, and it is inferred that the sugar is utilized directly at the beginning of germination before the oily reserve has been transformed or before it has become available.

Sachs (1859, 1863) concluded that the oil in seeds is transformed during germination into starch or sugar, but he did not consider that starch is a transition between oil and sugar or sugar between oil and starch. The fact that sugar increases in the germination of oily seeds as the oily reserve decreases has also been observed by Peters (1861), Laskovsky (1874), Detmer (1875), Green (1890), Leclerc du Sablon (1893 to 1897), Maquenne (1898), Mazé (1900), Kirkwood and Gies (1902), Miller (1910), Von Ohlen (1930, 1931), and others. The changes in the amounts of oil, protein, and carbohydrate in the seedlings of the sunflower as found by Miller (1910) are shown in the following table.

Constituents	Seed		4 days		5 days		7 days		10 days		14 days	
	Cot.	Hyp. and roots	Cot.	Hyp. and roots	Cot.	Hyp. and roots	Cot.	Hyp. and roots	Cot.	Hyp. and roots	Cot.	Hyp. and roots
Average percentage of constituents												
Ether extract..	55.6	47.4	55.9	24.2	51.2	9.3	33.9	9.7	16.7	9.8	13.5	11.4
Total sugars ..	4.1	4.1	1.1	9.2	1.8	22.5	3.1	18.6	3.1	12.6	2.1	7.7
Reducing sugar ...	Tr.	Tr.	None	8.7	None	20.5	1.4	17.3	3.1	12.6	1.1	7.7
Protein ..	24.4	25.3	22.0	14.2	21.0	9.4	19.8	9.3	21.0	7,0	20.4	6.5
Grams of constituents per 100 seeds and seedlings												
Ether extract..	3.786	0.219	3.00	0.188	2.53	0.123	1.298	0.214	0.501	0.273	0.317	0.238
Total sugars ..	0.279	0.019	0.06	0.071	0.09	0.298	0.12	0.431	0.09	0.351	0.05	0.161
Reducing sugar ...	Tr.	Tr.	None	0.068	None	0.271	0.054	0.382	0.09	0.351	0.025	0.161
Protein ..	1.66	0.117	1.18	0.110	1.03	0.125	0.76	0.205	0.63	0.195	0.48	0.135

Thus in the case of the sunflower the amount of sugar in the resting seed amounts to 4.1 per cent. Of this amount all but a trace is nonreducing sugar and has been identified as cane sugar. The total weight of sugar in the cotyledons of the seedlings at all stages is comparatively small. At the beginning of germination the nonreducing sugar falls rapidly until the 4-day stage, after which it gradually increases until the cotyledons begin to unfold. Up until that time the only sugar present in the cotyledons is of the nonreducing type, but, when the cotyledons assume the foliage functions, reducing sugars begin to make their appearance, and in the 10-day seedlings it is the only sugar present.

The percentage of total sugar rapidly rises in the hypocotyls and roots, and when they have reached a length of 3 to 4½ in. it amounts to 20 per cent of their dry weight. After that period it increases, and in the 10-day seedling it amounts to 12.6 per cent. The actual amount of sugar, however, reaches its maximum at the end of the 7-day period when the cotyledons have just spread out, and from that time on it gradually decreases. A small amount of starch appears in the starch sheath of the hypocotyls and roots and in the parenchyma cells of the vascular bundles of the cotyledons during the process of germination.

Pierce, Sheldon, and Murlin (1933) found in germinating castor beans that there was a continual decrease in the amount of oil present in the entire seedling and that there was a continual increase in the amount of sugar to 40 per cent of the dry weight when the hypocotyls had attained a length of 80 to 140 mm. The carbon balance between fat loss and carbohydrate gain, including the crude fiber, was not close except at the beginning of germination. An undetermined residue, which increased steadily with the increase in total carbohydrate, accounted for more and more of the carbon as time progressed.

Murlin (1933) studied the respiration of single castor beans at various stages of germination. He found that the respiratory quotients had a value of 0.30 to 0.58, indicating the conversion of oil to carbohydrates. Daggs and Harco-Wardlaw (1933) by the use of a modified oxycalorimeter found that the respiratory quotient of the germinated seed increased as the stage of germination increased. This indicates a change from an oxygen-poor to an oxygen-rich substance, which change is probably the transformation of oil to sugar.

The marked increase in the amount of carbohydrate in the seedlings of oily seeds raises the question as to its origin. Miller (1910) has shown that in the sunflower the changes in the protein reserve could not account for the increase in carbohydrate. The only other possible source of the sugar, then, is the oil, but how sugar is formed from this is not known. Detmer (1875) considered that when oil is hydrolyzed in the seedling the glycerin formed is changed into unknown bodies, and the free fatty acids might be the source of the carbohydrates. Green (1890) considered that the glycerin formed gives rise directly to sugar and that the free fatty acids give rise to vegetable acids. Leclerc du Sablon (1895) thought that the glycerin with the acids still combined with it goes directly to form carbohydrate, since he found that the amount of free fatty acids, as well as sugar, increased during the process of germination, and since glycerin could not be detected. He thus considered that fatty acids were liberated in such a way that glycerin is not set free. Maquenne (1898) and Ivanow (1912) considered that the unsaturated fatty acids contribute to the formation of sugar. Pirschle (1926) found considerable amounts of acetaldehyde in germinating fatty seeds of several different genera and families and considered that it might possibly be the compound through which the fatty acids pass in being transformed to carbohydrates. It was observed by Matthes (1927) that, in the germinating seeds of sunflower, the oil was saponified and the fatty acids were broken into shorter and shorter chains, and eventually became oxy-acids and oxaldehydes, which were later on built up into sugars.

A study of the fatty-acid-to-sugar process was made by Rhine (1926). He extracted enzymes from germinating fatty seeds by the methods used in extracting lipase. One part of fatty acid to 10 parts of enzyme extract was placed in small flasks and slightly acidified with HCl and kept at 40°C. with frequent agitation for varying periods of time. The results of this simple experiment indicated that a slight decomposition of fatty acids occurred, and that a small amount of a water-soluble reducing substance was produced.

In later and more elaborate experiments, higher temperatures, increased acidity, and an abundance of oxygen were substituted for enzyme action upon the mixture of fatty acids and water. In an 800-cc. Kjeldahl flask fitted with a reflux condenser and heated at temperatures varying from 75 to 95°C. were placed 100 cc. of fatty acid and 200 cc. of water. Air from a pressure tank was allowed to bubble through the mixture and served to keep a good emulsion and to furnish an oxygen supply. After varying periods of such treatment, it was found that water-soluble, ether-soluble, and nonvolatile aldehydes were produced. These aldehydes had more oxygen than that of the ordinary aldehyde or carbonyl group. The actual structure of these substances, however, is not known, but Rhine considered that they approached more nearly the sugars in structure than any products yet reported in the literature. Also produced were short-chain acids which were ether-, water-, and alcohol-soluble and which were fermentable by bread yeast, yielding carbon dioxide and fermenting in about the same time as glucose with the same yeast.

B. Fatty Substances Not Compounds of Fatty Acids

Two of the more important groups of substances, not compounds of fatty acids that may be found in the crude lipides, are the essential oils and resins. These substances, according to Thatcher (1921), include those compounds to which the characteristic odors of plants are due, along with others similar in structure and possessing characteristic uniformity in composition, and belong to several widely different chemical groups. The essential oils may be divided into two major groups: (1) the hydrocarbon oils or terpenes, some forms of which constitute the main proportions of the oils of turpentine, bergamot, lemon, eucalyptus, fennel, and pennyroyal; (2) the oxygenated and sulphuretted oils. The oxygenated essential oils include alcohols, aldehydes, ketones, acids, and phenols derived from either five-membered or six-membered closed-ring hydrocarbons. These compounds include, among others, menthol, oil of bitter almonds, oil of cinnamon, camphor, and thymol. The two most common sulphuretted oils are oil of mustard and oil of garlic. The latter is present in onions, garlic, radishes, and many other plants. The difference in flavor of these plants is due to the fact that the allyl sulphide is united with other different groups in the glucoside arrangement.

The resins are the oxidation products of the terpenes, but their exact structure is little understood. They are divided into two classes, the balsams and the solid or hard resins. Canada balsam and crude turpentine are examples of the first class and consist of resinous substances dissolved with fluid terpenes. Asafoetida and myrrh are resinous substances that are mixtures of gums and true resins.

McNair (1932) reported that 29 per cent of the families of higher plants contain volatile oils. Approximately 28 per cent of the families that contain essential oils have resin in appreciable quantities, while 100 per cent of the resin families possess essential oils. Both these substances are located in the same anatomical structures. Apparently, essential oils contain substances from which resins may be formed by either condensation or polymerization, or by a combination of these two processes.

The use of the essential oils and resins to the plant is not known. They are formed sometimes normally in healthy tissue and at other times they are the result of injury or disease. They are frequently classed as excretions, since aside from possible protective influences to the plant no biological use of these substances is known.

Bibliography

Abegg, F. A., 1929: Some effects of the waxy gene in maize on fat metabolism, *J. Agr. Res.*, **38**: 183–193.

Alsberg, C. L., and A. E. Taylor, 1928: The fats and oils, *Leland Stanford Junior Univ., Food Res. Inst.*, 1.

Anderson, R. J., and M. G. Moore, 1923: A study of the phytosterols of corn oil, cotton seed oil and linseed oil, *J. Am. Chem. Soc.*, **45**: 1944–1953.

———, 1923: Composition of corn pollen. II, Concerning certain lipoids, a hydrocarbon and photosterol occurring in the pollen of white flint corn, *J. Biol. Chem.*, **55**: 611–628.

———, 1924: The phytosterols of the endosperm of corn, *J. Am. Chem. Soc.*, **46**: 1450–1460.

——— and R. L. Shriner, 1926: The phytosterols of corn oil, *N. Y. Sta. (Geneva) Tech. Bull.*, **124**: 10–25.

———, ———, and G. O. Burr, 1926: The phytosterols of wheat germ oil, *N. Y. Sta. (Geneva) Tech. Bull.*, **124**: 26–41.

———, F. P. NABENHAUER, and R. L. SHRINER, 1926: The distribution of dihydrositosterol in plant fats, *N. Y. Sta. (Geneva) Tech. Bull.*, **124**: 42–51.

———, 1926: Properties of cholesterol obtained from different sources, *N. Y. Sta. (Geneva) Tech. Bull.*, **124**: 57–67.

BANG, I., 1911: "Chemie und Biochemie der Lipoide," J. F. Bergman, Wiesbaden. 187 pp.

BAUGHMAN, W. F., and G. S. JAMIESON, 1920: The composition of Hubbard squash seed oil, *J. Am. Chem. Soc.*, **42**: 152–157.

———, D. BRAUNS and G. S. JAMIESON, 1920: Cantaloup seed oil, *J. Am. Chem. Soc.*, **42**: 2398–2402.

——— and G. S. JAMIESON, 1921: The chemical composition of corn oil, *J. Am. Chem. Soc.*, **43**: 2696–2702.

——— and ———, 1922: The chemical composition of soy-bean oil, *J. Am. Chem. Soc.*, **44**: 2947–2952.

BEER, R., 1909: On elaioplasts, *Ann. Bot.*, **23**: 63–72.

BENGIS, R. O., and R. J. ANDERSON, 1934: The chemistry of the coffee bean. II, The composition of the glycerides of the coffee bean oil, *J. Biol. Chem.*, **105**: 139–151.

BEUMER, H., 1933: Über die Sterinbildung bei der Keimung, *Biochem. Z.*, **259**: 469–470.

BLOOR, W. R., 1925: Biochemistry of the fats, *Chem. Rev.*, **2**: 243–300.

BULL, H. B., 1936: "The Biochemistry of the Lipids," Burgess Pub. Co., Minneapolis, Minn., 135 pp.

CASKEY, C., and W. D. GALLUP, 1931: Changes in the sugar, oil, and gossypol content of the developing cotton boll, *J. Agr. Res.*, **42**: 671–673.

CHANNON, H. J., and A. C. CHIBNALL, 1927: The ether-soluble substances of cabbage leaf cytoplasm. IV, Further observations on diglyceridephosphoric acid, *Biochem. J.*, **21**: 1112–1120.

——— and ———, 1929: The ether-soluble substances of cabbage leaf cytoplasm. V, The isolation of n-nonacosane and di-n-tetradecyl ketone, *Biochem. J.*, **23**: 168–175.

——— and C. A. M. FOSTER, 1934: The phosphatides of the wheat germ, *Biochem. J.*, **28**: 853–864.

CHIBNALL, A. C., and H. J. CHANNON, 1927: The ether-soluble substances of cabbage leaf cytoplasm. I, Preparation and general characters, *Biochem. J.*, **21**: 225–232.

——— and ———, 1927: The ether-soluble substances of cabbage leaf cytoplasm. II, Calcium salts of glyceridephosphoric acids, *Biochem. J.* **21**: 233–246.

———, E. F. WILLIAMS, A. L. LATNER, and S. H. PIPER, 1933: The isolation of n-triacontanol from lucerne wax, *Biochem. J.*, **27**: 1885–1888.

——— and H. J. CHANNON, 1929: The ether-soluble substances of cabbage leaf cytoplasm. VI, Summary and general conclusions, *Biochem. J.*, **23**: 176–184.

——— and P. N. SAHAI, 1931: Observations on the fat metabolism of leaves. I, Detached and starved mature leaves of Brussels sprout (*Brassica oleracea*), *Ann. Bot.*, **45**: 489–502.

COLLIN, G., and T. P. HILDITCH, 1929: Regularities in the glyceride structure of vegetable seed fats, *Biochem. J.*, **23**: 1273–1289.

———, 1933: The kernel-fats of some members of the *Palmae*, *Biochem. J.*, **27**: 1366–1372.

CRUZ, A. O., and A. P. WEST, 1932: Composition of Philippine rice oil (Ramai variety), *Philippine J. Sci.*, **48**: 5–12.

——— and ———, 1932: Composition of Philippine soybeans and soybean oil, *Philippine J. Sci.*, **48**: 77–89.

CZAPEK, F., 1913: "Biochemie der Pflanzen," **1**: 710–733, Gustav Fischer.

DAGGS, R. G., and H. S. HALCRO-WARDLAW, 1933: The conversion of fat to carbohydrate in the germinating castor bean, *J. Gen. Physiol.*, **17**: 303–309.

DANGEARD, P., 1923: Remarques sur l'état de l'huile à l'intérieur des graines oléagineuses, *C. R. Acad. Sci. Paris*, **177**: 67–69; *Bot. Abst.*, **13**: 487.

DELEANO, N. T., 1909, 1910: Recherche chimique sur la germination, *Cent. Bakt.* II, **24**: 130–146; *Arch. Sci. Biol.* (St. Petersburg), **15**: 1–24.

DELUCA, S., 1861, 1862: Recherches sur la formation de la matière grasse dans les olives, *C. R. Acad. Sci. Paris*, **53**: 380–384; **55**: 470–473, 506–510.

DE SAUSSURE, T., 1842: Ueber das Keimen der Oelsamereien, *Froriep's Notizen*, **24**: 24.

DETMER, W., 1875: "Phys.-chem. Untersuchung über die Keimung ölhaltiger Samen," Gustav Fischer, Leipzig.

———, 1880: "Vergleichende Physiologie des Keimungsprocesses," pp. 336–338, Jena.

DILLMAN, A. C., 1928: Daily growth and oil content of flaxseeds, *J. Agr. Res.*, **37**: 357–377.

DUNLAP, F. L., and L. O. GILBERT, 1911: The synthesis of fats by the action of enzymes, *J. Am. Chem. Soc.*, **33**: 1787–1791.

EULER, H., and E. NORDENSON, 1908: Zur Kenntnis des Möhrencarotens und seiner Begleitsubstanzen, *Z. physiol. Chem.*, **56**: 223–235.

EVANS, E. I., 1935: Antioxidant properties of vegetable lecithin, *Ind. and Eng. Chem.*, **27**: 329–331.

EYRE, J. V., 1931: Notes on oil development in the seed of a growing plant, *Biochem. J.*, **25**: 1902–1908.

FLEURY, M., 1865: Recherches chimiques sur la germination, *Ann. Chim. et Phys.*, 4e série, **4**: 38.

FRANCIS, C. K., and W. G. FRIEDEMANN, 1917: The fats and fatty acids of the grain sorghums, *Okla. Sta. Bull.*, **117**: 3–14.

GALLAGHER, P. H., 1923: Mechanism of oxidation in the plant. I, The oxygenase of Bach and Chodat: function of lecithins in respiration. *Biochem. J.*, **17**: 515–529.

GALLUP, W. D., 1927: The gossypol content and chemical composition of cotton seeds during certain periods of development, *J. Agr. Res.*, **34**: 987–992.

———, 1928: A chemical study of the development of cotton bolls and the rate of formation of gossypol in the cotton seed, *J. Agr. Res.*, **36**: 471–480.

GARNER, W. W., H. A. ALLARD and C. L. FOUBERT, 1914: Oil content of seeds as affected by nutrition of the plant, *J. Agr. Res.*, **3**: 227–249.

GERBER, C., 1897: Étude de la transformation des matières sucrées en huile dans les olives, *C. R. Acad. Sci. Paris*, **125**: 658–661.

———, 1897: Recherches sur la formation des réserves oléagineuses des graines et des fruits, *C. R. Acad. Sci. Paris*, **125**: 732–735.

———, 1901: Recherches sur la respiration des olives et sur les relations existant entres les valeurs du quotient respiratoire observé et la formation de l'huile, *J. Bot. Année*, **15**: 9, 88, 121; *Bot. Cent.*, **89**: 622.

GRAFE, V., 1925: Zur Physiologie und Chemie der Pflanzenphosphatide, *Biochem. Z.*, **159**: 444–448.

——— and V. HORVAT, 1925: Die wasserlöslichen Phosphatide aus der Wurzel der Zuckerrübe. I, *Biochem. Z.*, **159**: 449–467.

——— and H. MAGISTRIS, 1925: Zur Chemie und Physiologie der Pflanzenphosphatide. II, Die wasserlöslichen Phosphatide aus *Aspergillus oryzae*, *Biochem. Z.*, **162**: 366–398.

——— and ———, 1926: Über die Phosphatide aus *Daucus carota*, *Z. wiss. Biol. Abt. E. Planta*, **2**: 429–437.

———, 1933: Die Phosphatide der Pflanzenzelle, *Protoplasma*, **17**: 602–609.

Green, J. R., 1890: On the germination of the seed of the castor oil plant (*Ricinus communis*), *Proc. Roy. Soc.*, **48**: 370–392.

——— and H. Jackson, 1905: Further observations on the germination of the seeds of the castor oil plant, *Proc. Roy. Soc.*, B, **77**: 69–85.

Guerrant, N. B., 1927: Some relations of the phospholipins in seeds to other constituents, *J. Agr. Res.*, **35**: 1001–1021.

Haberlandt, G., 1914: "Physiological Plant Anatomy," English translation. The Macmillan Company, New York.

Hanriot, M., 1898: Sur la transformation de la graisse par oxydation directe, *C. R. Acad. Sci. Paris.*, **127**: 561–563.

Hansteen-Cranner, B., 1914: Über das Verhalten der Kulturpflanzen zu den Bodensalzen. II, Beiträge zur Biochemie und Physiologie der Zellwand lebender Zellen, *Jahrb. wiss. Bot.*, **53**: 536–598.

———, 1919: Beiträge zur Biochemie und Physiologie der Zellwand und der plasmatischen Grenzschichten, *Ber. bot. Ges.*, **37**: 380–391.

———, 1926: Untersuchungen über die Frage ob in der Zellwand lösliche Phosphatide vorkommen, *Planta Archiv. wiss. Bot.*, **2**: 438–445.

———, 1926: Untersuchungen über die Frage, ob in der Zellwand lösliche Phosphatide vorkommen, *Planta*, **2**: 438–445; *Bot. Abst.*, **1**: 610, 1927.

Hellriegel, H., 1855: Beitrag zur Keimungsgeschichte der ölgebenden Samen, *J. prakt. Chemie*, **64**: 94.

Heyl, F. W., E. C. Wise, and J. H. Speer, 1929: The unsaponifiable fraction from spinach fat, *J. Biol. Chem.*, **82**: 111–116.

Hicks, G., 1895: Oil producing seeds, *U. S. Dept. Agr. Year Book*, pp. 185–204.

Hilditch, T. P., 1928: The chemical composition of vegetable seed fats in relation to the natural orders of plants, *Proc. Roy. Soc.*, B, **103**: 111–117.

———, 1935: The fats: new lines in an old chapter of organic chemistry, *J. Soc. Chem. Ind.*, **54**: 139–145.

Ivanow, S., 1911: Ueber Oelsynthese unter Vermittlung der pflanzlichen Lipase, *Ber. bot. Ges.*, **29**: 595–602.

———, 1912: Über die Verwandlung des Öls in der Pflanze, *Jahrb. wiss. Bot.*, **50**: 375–386.

———, 1912: Über den Stoffwechsel beim Reifen ölhaltiger Samen mit besonderer Berücksichtigung der Ölbildungsprozesse, *Beih. bot. Cent.*, I, **28**: 159–191.

Iwanoff, L., 1902: Ueber die Umwandlungen des Phosphors beim Keimen der Wicke, *Ber. bot. Ges.*, **20**: 366–372.

Jacobson, H., 1889: Ueber einige Pflanzenfett, *Z. physiol. Chem.*, **13**: 32–65.

Jamieson, G. S., W. F. Baughman, and D. H. Brauns, 1921: The chemical composition of peanut oil, *J. Am. Chem. Soc.*, **43**: 1372–1382.

——— and ———, 1922: The chemical composition of sunflower-seed oil, *J. Am. Chem. Soc.*, **44**: 2952–2957.

Jegorow, M., 1904: Ueber Stoffmetamorphose bei der Samenkeimung von *Cucurbita maxima*, *Bot. Cent.*, **101**: 597–599.

Johnson, I. J., 1932: The relation of agronomic practice to the quantity and quality of the oil in flax seed, *J. Agr. Res.*, **45**: 239–255.

Jordan, R. C., and A. C. Chibnall, 1933: Observations on the fat metabolism of leaves. II, Fats and phosphatides of the runner bean (*Phaseolus multiflorus*), *Ann. Bot.*, **47**: 163–186.

Kao, H. C., 1936: Lipoid metabolism in the mung bean during germination, *Biochem, J.*, **30**: 202–207.

KAYSER, R., 1925: Hat eine Mineraldüngung Einfluss auf die Eigenschaften von Oelpflanzen und ändert sich durch die Düngung das Oel in seiner Zusammensetzung? *Bot. Arch.*, **10**: 349–386.

KIRKWOOD, J. E., and W. J. GIES, 1902: Chemical studies of the cocoanut with some notes on the changes during germination, *Bull. Torrey Bot. Club.*, **29**: 321–359.

KNOP, W., 1859: Ueber ein phosphorhaltiges Oel der Erbsen, *Landw. Versuchs-Stat.*, **1**: 26–30.

KORSAKOW, M., 1910: Ueber den Einfluss der Zellipoide auf die Autolyse der Weizenkeime, *Biochem. Z.*, **28**: 121–126.

KOSSEL, A., 1891: Chemische Zusammensetzung der Zelle, *Arch. Physiol.*, **1891**: 181–186.

LASKOVSKY, N., 1874: Ueber einige chemische Vorgänge bei der Keimung der Kürbissamen, *Landw. Versuchs-Stat.*, **17**: 219–244.

LEATHES, J. B., and H. S. RAPER, 1925: "The Fats," a mon., Longmans, Green and Company, New York, 243 pp.

LECLERC DU SABLON, M. 1893: Sur la germination du *Ricin*, *C. R. Acad. Sci. Paris*, **117**: 524–527.

———, 1894: Sur la germination des graines oléagineuses, *C. R. Acad. Sci. Paris*, **119**: 610–612.

———, 1895: Recherches sur la germination des graines oléagineuses, *Rev. Gén. Bot.*, **7**: 145–165, 205–215, 258–269.

———, 1896: Sur la formation des réserves non azotées de la noix et de l'amande, *C. R. Acad. Sci. Paris*, **123**: 1084–1086.

———, 1897: Sur la germination des amandes, *Rev. Gén. Bot.*, **9**: 5–16.

———, 1897: Sur les réserves oléagineuses de la noix, *Rev. Gén. Bot.*, **9**: 313–317.

LEVENE, P. A., and I. P. ROLF, 1925: Plant phosphatides. I, Lecithin and cephalin of the soybean, *J. Biol. Chem.*, **62**: 759–766.

———, 1926: Plant phosphatides. II, Lecithin, cephalin and so-called cuorin of the soybean, *J. Biol. Chem.*, **68**: 285–293.

LEWIS, F. J., and G. M. TUTTLE, 1920: Osmotic properties of some plant cells at low temperatures, *Ann. Bot.*, **34**: 405–416.

MACDOUGAL, D. T., 1922: The probable action of lipoids in growth, *Proc. Am. Phil. Soc.*, **61**: 33–53; *Sci.*, **55**: 653–654.

MACLEAN, H., and I. MACLEAN, 1927: "Lecithin and Allied Substances," mon., Longmans, Green, and Company, New York, 220 pp.

MAQUENNE, L., 1898: Sur les changements de composition qu' éprouvent les graines oléagineuses au cours de la germination, *C. R. Acad. Sci. Paris*, **127**: 625–628.

MARKLEY, K. S., and C. E. SANDO, 1931: Progressive changes in the waxlike coating on the surface of the apple during growth and storage, *J. Agr. Res.*, **42**: 705–723.

———, S. B. HENDRICKS, and C. E. SANDO, 1932: Further studies on the waxlike coating of apples, *J. Biol. Chem.*, **98**: 103–107.

——— and C. E. SANDO, 1933: Possible changes in the waxlike coating of apples caused by certain spray and other treatments, *Plant Physiol.*, **8**: 475–478.

——— and ———, 1933: Progressive changes in the cuticle of apples during growth and storage, *J. Agr. Res.*, **46**: 403–412.

MATLACK, M. B., 1935: A phytosterol and phytosterolin from the sweet potato, *Sci.*, **81**: 536.

MATTHES, E., 1927: Physiologische Untersuchungen über Umwandlungen des Oeles im keimenden Sonnenblumensamen, *Bot. Arch.*, **19**: 79–135; *Biol. Abst.*, **2**: 872.

MAXWELL, W., 1891: On the behavior of the fatty bodies and the role of the lecithins during normal germination, *Am. Chem. J.*, **13**: 16–24.

Mazé, M., 1900: Recherches sur la digestion des réserves dans les graines, en voie de germination et leur assimilation par les plantules, *C. R. Acad. Sci. Paris.*, **130**: 424–427.

M'Clenahan, F. M., 1909–1913: The development of fat in the black walnut. I, II, *J. Am. Chem. Soc.*, **31**: 1093–1098; **35**: 485–493.

McNair, J. B., 1930: The taxonomic and climatic distribution of oil and starch in seeds in relation to the physical and chemical properties of both substances, *Am. J. Bot.*, **17**: 662–669.

———, 1931: Some properties of plant waxes in relation to climate of habitat, *Am. J. Bot.*, **18**: 518–526.

———, 1932: The interrelation between substances in plants: essential oils and resins, cyanogen and oxalate, *Am. J. Bot.*, **19**: 255–272.

Miller, E. C., 1910: A physiological study of the germination of *Helianthus annuus*, *Ann. Bot.*, **24**: 693–726.

———, 1912: A physiological study of the germination of *Helianthus annuus*. II, The oily reserve, *Ann. Bot.*, **26**: 889–901.

Moyer, L. S., 1935: Electrophoresis of sterols. II, Ergosterol, *J. Gen. Physiol.*, **18**: 749–753.

Müntz, M., 1871: Sur la germination des graînes oléagineuses, *Ann. Chim. et Phys.*, 4e série, **22**: 472.

Murlin, J. R., 1933: The conversion of fat to carbohydrate in the germinating castor bean. I, The respiratory metabolism, *J. Gen. Physiol.*, **17**: 283–301.

Njegovan, V., 1911: Beiträge zur Kenntnis der pflanzlichen Phosphatide, *Z. physiol. Chem.*, **76**: 1–26.

Noller, C. R., I. J. Millner, and J. J. Gordon, 1933: Capric acid from the seed fat of the California bay tree, *J. Am. Chem. Soc.*, **55**: 1227–1228.

Pack, D. A., 1925: Dispersion of lipoids, *Bot. Gaz.*, **79**: 334–338.

Palladin, W., and E. Stanewitsch, 1910: Die Abhängigkeit der Pflanzenatmung von den Lipoiden, *Biochem. Z.*, **22**: 351–369.

Parrozzani, A., 1909: Il fosforo organico ed il rapporto fra l'azoto amidico e quello delle altre combinazioni azotate non proteiche nei semi maturi, *Staz. Sper. Agr. Ital.*, **42**: 890–901.

Peters, E., 1861: Zur Keimungsgeschichte des Kurbissamens, *Landw. Versuchs-Stat.*, **3**: 1–18.

Pierce, H. B., D. E. Sheldon, and J. R. Murlin, 1933: The conversion of fat to carbohydrate in the germinating castor bean. III, The chemical analysis and correlation with respiratory exchange, *J. Gen. Physiol.*, **17**: 311–325.

Pirschle, K., 1926: Acetaldehyd als Zwischenprodukt bei der Keimung fetthaltiger Samen, *Biochem. Z.*, **169**: 482–489; *Biol. Abst.*, **2**: 463.

Policard, A., and G. Mangenot, 1923: Recherches cytologiques sur l'état de l'huile dans les graines oléagineuses. La graine mure, *C. R. Acad. Sci. Paris*, **176**: 1841–1844.

——— and ———, 1923. Sur l'état de l'huile dans la cellule de réserve des graines oléagineuses. La graine en germination, *C. R. Acad. Sci. Paris*, **177**: 346–348.

Pollard, A., A. C. Chibnall, and S. H. Piper, 1931: The wax constituents of forage grasses. I, Cocksfoot and perennial rye grass, *Biochem. J.*, **25**: 2111–2122.

———, ———, and ———, 1933: The isolation of *n*-octacosanol from wheat wax, *Biochem. J.*, **27**: 1889–1893.

Priestley, J. H., 1924: The fundamental fat metabolism of the plant, *New Phytol.*, **23**: 1–20.

Rhine, J. B., 1926: Studies on the oxidation of certain fatty acids, *Plant Physiol.*, **1**: 349–361.

Riebsomer, J. L., and G. A. Nesty, 1934: An examination of the fatty oil from pumpkin seed. The constitution of linoleic acid, *J. Am. Chem. Soc.*, **56**: 1784–1785.

Rivett, M. F., 1918: The structure of the cytoplasm in the cells of *Alicularia scolaris*, *Ann. Bot.*, **32**: 207–214.

Rohrabaugh, P. W., 1934: Penetration and accumulation of petroleum spray oils in the leaves, twigs and fruit of citrus trees, *Plant Physiol.*, **9**: 699–730.

Rushkovskii, S., 1930: Changes in the chemical composition of sunflower seeds with the sowing time, *Masloboino Delo*, **1930**: 43–56; *Chem. Abst.*, **26**: 4356.

Sachs, J., 1859: Ueber das Auftreten der Stärke bei der Keimung ölhaltiger Samen, *Bot. Ztg.*, **17**: 177–183, 185–188.

———, 1863: Ueber die Keimung des Samens von *Allium Cepa*, *Bot. Ztg.*, **21**: 57–65.

Sahai, P. N., and A. C. Chibnall, 1932: Wax metabolism in the leaves of Brussels sprout, *Biochem. J.* **26**: 403–412.

Sahasrabuddhe, D. L., and N. P. Kale, 1933: A biochemical study of the formation of oil in niger seed (*Guizotia abyssinica*), *Indian J. Agr. Sci.*, **3**: 57–88; *Biol. Abst.*, **9**: 750.

Sando, C. E., 1923: Constituents of the wax-like coating on the surface of the apple, *J. Biol. Chem.*, **56**: 457–468.

———, 1928: Lipides and their estimation in vegetable tissues, *Plant Physiol.*, **3**: 155–184.

Schmidt, R. H., 1891: Ueber Aufnahme und Verarbeitung von fetten Oelenpflanzen, *Flora*, **74**: 300–371.

Schulze, E., and E. Steiger, 1889: Ueber den Lecithingehalt der Pflanzensamen, *Z. physiol. Chem.*, **13**: 365–384.

———, ———, and W. Maxwell, 1891: Untersuchungen über die chemische Zusammensetzung einiger Leguminosensamen, *Landw. Versuchs-Stat.*, **39**: 269–326.

——— and A. Likiernik, 1891: Ueber das Lecithin der Pflanzensamen, *Z. physiol. Chem.*, **15**: 405–414.

——— and S. Frankfurt, 1894: Über den Lecithingehalt einiger vegetabilischer Substanzen, *Landw. Versuchs-Stat.*, **43**: 307–318.

Schwartze, E. W., and C. L. Alsberg, 1923: Quantitative variation of gossypol and its relation to the oil content of cotton seed, *J. Agr. Res.*, **25**: 285–295.

Scurti, F., and G. Tommasi, 1909–1910: Sulla formazione del grasso nei frutti oleaginosi; 1910–1911: *Ann. della R. Staz. Chim. Agrar. Sperim. di Roma*, **4**: 253–286; **5**: 103.

——— and ———, 1911: Sulla formazione del grasso nei frutti oleaginosi, *Ann. della R. Staz. Chim. Agrar. Sperim. di Roma*, **5**: 223.

——— and ———, 1912: In regard to the formation of fatty acid in oleaginous fruits. *R. Soc. Chim. Ital.*, **3**: 358–363; *Exp. Sta. Rec.*, **26**: 801.

Shorey, E. C., 1898: The lecithins of sugar cane, *J. Am. Chem. Soc.*, **20**: 113–118.

Sinnott, E. W., 1918: Factors determining character and distribution of food reserve in woody plants, *Bot. Gaz.*, **66**: 162–175.

Smedley, I., 1912: The biochemical synthesis of fatty acids from carbohydrates, *J. Physiol.*, **45**: 25–27.

——— and E. Lubrzynska, 1913: The biochemical synthesis of the fatty acids, *Biochem. J.*, **7**: 364–374.

Smith, J. A. B., and A. C. Chibnall, 1932: The glyceride fatty acids of forage grasses. I, Cocksfoot and perennial rye grass, *Biochem. J.*, **26**: 218–234.

——— and ———, 1932: The phosphatides of forage grasses. I, Cocksfoot, *Biochem. J.*, **26**: 1345–1359.

Smolenski, K., 1909: Zur Kenntniss der aus Weizenkeimen darstellbaren Phosphatide, *Z. physiol. Chem.*, **58**: 522–526.

SPEER, J. H., E. C. WISE, and M. C. HART, 1929: The composition of spinach fat, *J. Biol. Chem.*, **28**: 105–110.

STANLEY, O. B., 1931: Fat deposits in certain *Ericaceae*, *Butler Univ. Bot. Studies*, **2**: 33–41.

STARK, R. W., 1924: Environmental factors affecting the protein and oil content of soybeans and the iodine number of soybean oil, *J. Am. Soc. Agron.*, **16**: 636–646.

STEWARD, F. C., 1928: An experimental examination of the evidence for the presence of phosphatides in the limiting surface of the living protoplast, *British J. Exper. Biol.*, **6**: 32–41.

———, 1928: On the evidence for phosphatides in the external surface of the plant protoplast, *Biochem. J.*, **22**: 268–275.

———, 1929: Phosphatides in the limiting protoplasmic surface. I, A review with special reference to the plant protoplast, *Protoplasma*, **7**: 602–621.

STOCKLASA, J., 1895: Die Assimilation des Lecithins durch die Pflanze, *Sitzungsber. (Math. Naturw., Kl.) K. Akad. Wiss. Wien.*, **104**: 712–722.

———, 1898: Ueber die Entstehung und Unwandlung des Lecithins in der Pflanze, *Z. physiol Chem.*, **25**: 398–405.

STRASBURGER, E., F. NOLL, H. SCHENCK, and G. KARSTEN, 1908: "Textbook of Botany," English translation. The Macmillan Company, New York.

SULLIVAN, B., and C. NEAR, 1933: Change of crude lipides of wheat on storage, *Ind. and Eng. Chem.*, **25**: 100–103.

TERROINE, E. F., 1920: État actuelle de nos connaissances sur la formation des graisses au cours de la maturation des graines et fruits oléagineux et sur l'utilisation des graisses au cours de la germination, *Ann. Sci. Nat.* 10, **26**: 1–63; *Bot. Gaz.*, **75**: 224.

———, R. BONNET, G. KOPP, and J. VECHOT, 1927: La formation des stérols est-elle liée au métabolisme des matières graisses? *Bull. Soc. Chim. Biol.*, **9**: 678–691: *Biol. Abst.*, **2**: 1506.

THATCHER, R. W., 1921: "The Chemistry of Plant Life," McGraw-Hill Book Company, Inc., New York.

THOR, C. J. B., and C. L. SMITH, 1935: A physiological study of seasonal changes in the composition of the pecan during fruit development, *J. Agr. Res.*, **50**: 97–122.

TOTTINGHAM, W. E., 1932: Are leaf lipides responsive to solar radiation? *Sci.*, **75**: 223–224.

TSCHIRCH, A., and H. KRITZLER, 1900: Microchemische Untersuchungen über die Aleuronekörner, *Ber. Pharm.*, **10**: 214–223.

TUTTLE, G. M., 1919: Induced changes in reserve materials in evergreen herbaceous leaves, *Ann. Bot.*, **33**: 201–210.

———, 1921: Reserve food materials in vegetative tissue, *Bot. Gaz.*, **71**: 146–151.

VALLEE, C., 1903: Sur la présence du saccharose dans les amandes et sur son rôle dans la formation de l'huile, *C. R. Acad. Sci. Paris.*, **136**: 114–117.

VON FÜRTH, O., 1904: Ueber das Verhalten des Fettes bei der Keimung ölhaltiger Samen, *Hofmeisters Beiträge zur chem. Physiol. u. Path.*, **4**: 430.

VON OHLEN, F. W., 1930, 1931: A microchemical study of soybeans during germination, *Am. J. Bot.*, **17**: 1046; **18**: 30–49.

WAKKER, J. H., 1886: Studien über die Inhaltskörper der Pflanzenzelle, *Jahrb. wiss. Bot.*, **19**: 423–496.

WEBSTER, J. E., 1929: Iodine value of fatty acids from plant phosphatides, *Ohio J. Sci.*, **29**: 39–42.

WINTERSTEIN, E., and O. HIESTAND, 1908: Beiträge zur Kenntniss der pflanzlichen Phosphatide. II, *Z. physiol. Chem.*, **54**: 288–330.

——— and K. Smolenski, 1909: Beiträge zur Kenntnis der aus Cerealien darstellbaren Phosphatide, *Z. physiol. Chem.*, **58**: 506–521.

——— and L. Stegmann, 1909: Ueber einen eigenartigen phosphorhaltigen Bestandteil der Blätter von *Ricinus*, *Z. physiol. Chem.*, **58**: 527–528.

Zalenski, W., 1902: Beiträge zur Verwandlung des Eiweissphosphors in den Pflanzen, *Ber. bot. Ges.*, **20**: 426–433.

CHAPTER XI

THE PROCESS OF DIGESTION IN THE GREEN PLANT

I. FOODS OF THE GREEN PLANT

Before considering the process of digestion it will be necessary to discuss the foods of plants in order to understand the significance of this process in their life functions. A food may be defined as any substance that may be used directly, or after hydrolysis, by the protoplasm for growth and repair or as a source of energy. Since this definition applies to animals as well as to plants, it raises the question whether plants and animals use the same kind of foods in their life functions. It is common knowledge that animals are nourished with carbohydrates, fats, and proteins, but the fact that the protoplasm of plants requires for its nourishment and energy the same foods as are used by animals is not so well known. Such, however, is the case, as will be proved in the subsequent discussion.

A. Plant Sources of Animal Foods

It is well known that animals obtain their food supply either directly or indirectly from plants, since they live entirely upon plants or prey upon other animals that are dependent upon plants for their food supply. Animals obtain this food supply from the seeds and grains, fruit, stems, roots, and leaves of plants. Seeds and grains are one of the main sources of food for animals, including man. Some of the more important seeds and grains that are a source of food for animals are wheat, corn, rye, oats, barley, rice, peas, beans, sorghum grains, nuts, cotton seed, sunflower seed, buckwheat, millet, and peanuts. Some of the stems that are a source of food to animals are the stems of the grains, grasses, and legumes, Irish potatoes, artichokes, asparagus, sugar cane, and young bamboo, while some of the roots that furnish food are sweet potato, yams, beets, carrots, parsnips, salsify, turnips, rutabagas, and cassava. Some of the leaves that furnish food to animals are those of the grains, grasses and legumes, celery, rhubarb, onion, spinach, lettuce, cabbage, Swiss chard, kale, and rape. The foods that are stored in these plant parts are the proteins, carbohydrates, and fats together with certain minerals and vitamins. These can satisfy all the needs of the animals that feed upon them, in so far as their food supply is concerned. The question then naturally arises concerning the benefits of the foods stored in plants to the plants themselves. Are these foods stored in the organs just mentioned of use to the plant or are they beneficial only to the animal kingdom?

B. Value to the Plant of Its Stored Foods

Let us first consider the function of the foods stored in seeds. A normal seed contains a young plant which will, if the proper conditions are supplied, begin to grow and develop. The young seedling uses the foods stored in the seed to form new protoplasm and new tissues and from them it also derives the energy to push the stem out of the soil and the roots downward. As the young plant increases in size, the amount of food in the seed continues to decrease until it all disappears. This fact may be illustrated by considering the germination of the sunflower seed. This seed contains protein and oil as stored foods, the former making up 26 per cent and the latter about 50 per cent or more of the dry weight of this seed. This seed does not have an endosperm and these foods are stored in the cotyledons of the embryo. The depletion of these foods as germination progresses is shown in the following table:

Grams of Oil and Protein in the Cotyledons of 100 Seeds and Seedlings of the Sunflower Seed at 2, 4, 7, and 10 days, Respectively

Substance	Stage				
	Seed	2 days	4 days	7 days	10 days
Protein	1.7	1.2	1.0	0.8	0.6
Oil	4.0	3.0	2.5	1.3	0.5

A plant can thus grow to a considerable size without any food except that which is contained in the seed. These foods are transferred to the young plant parts and built into new tissue in much the same manner as this process occurs in animals.

In the aerial stems and branches of plants there is found in the region of the buds a large amount of protein, oil, or carbohydrates stored in the cells. As the buds develop into the new branches and leaves, this stored food is rapidly depleted and used in the formation of these new plant organs. In modified stems like the Irish potato and artichoke, the young sprouts that grow from them when they are planted are nourished until they are above ground and sometime thereafter from the food supply in the old tuber beneath the soil.

The three types of foods are stored in all roots of plants but in especially large quantities in roots of the fleshy type. The use to which the stored foods are put can be best illustrated by considering beets, carrots, parsnips, and other biennial plants. These plants during their first season's growth store up food in the roots, and during the second season develop a new stem that produces a crop of seed. The foods stored in these roots during the first year furnish the food necessary for this development of stem and seeds so that when the seed has matured the roots have been depleted of their stored food and during that time little or no material has been absorbed by the roots or manufactured by the aerial parts. In the majority of cases, the foods that are found in leaves have been manufactured there and are only awaiting transportation to the roots, stem, or seeds of the plant. Certain fleshy leaves, however, as the cabbage and onion, are real storage organs where the foods are stored permanently.

From what has been mentioned above, it is seen that the proteins and carbohydrates, or fats, are stored in all parts of the plant where growth is taking place or is going to take place. When the new organs or parts of the plant are formed, they draw upon these foods and form their tissues from them. We know this to be the case because these parts have no access to any food except that which is stored and also because the translocation of this food from its place of storage to the new parts can readily be detected. Plants thus use for their growth and development the same three types of foods as are used by animals in their growth and development. They differ in regard to the food proposition in that plants can manufacture the proteins, fats, and carbohydrates that they use as foods from the simple materials which they obtain from the air and the soil, while animals have no power to do this but have to depend entirely upon plants to furnish these foods already prepared for them. Thus it may be stated that plants are independent, while animals are dependent in regard to their food supply.

The simple inorganic compounds that plants obtain from the air and soil and from which they manufacture the three types of foods are frequently termed "plant foods." These compounds, however, are not foods in the strict sense of the word, since they cannot be used directly by the protoplasm. They are the crude materials from which foods are formed, in the same fashion as finished walls are formed from bricks, sand, and cement or finished cloth is made from the crude cotton or woolen fibers. The question now arises as to what changes if any the foods found in plants must undergo before they can be utilized by the protoplasm of the plant.

II. NATURE OF DIGESTION

The greater portion of the carbohydrates, proteins, fats, and oils of plants is stored in plants in forms that are insoluble in water. The storage of foods in an insoluble form in the plant cells has several advantages. In the first place, more food can be stored in a given space in the insoluble than in the soluble form. In the second place, foods are more stable in the insoluble form than in the soluble form and are less liable to be destroyed by decay. In the third place, the conversion of soluble foods to the insoluble form keeps the osmotic value of the cell sap at a level that is not injurious to the cell.

The foods in the insoluble form, however, cannot leave the cells in which they are stored until they are changed to a form that is soluble in water. They must therefore be changed to a soluble form before they can be translocated. Furthermore, all foods must be in a soluble form before they can be utilized for growth or as a source of energy. Even if a food is soluble in water it must in some cases be reduced to a more simple form before the protoplasm can utilize it in its metabolism. With these general facts concerning the foods of plants, we can now understand better what is meant by digestion.

Digestion may be defined as the processes whereby the foods of plants and animals are changed from an insoluble to a soluble form or from a complex to a more simple form. In nearly every case the action on foods in digestion is one of hydrolysis. The food is made to take up water and is then broken down, a fact that will be considered in the discussion of the action of digestive enzymes. As a result of this change, insoluble foods are rendered translocatable and changed to a form that is usable by the protoplasm in growth and development. An example of digestion in which the food is changed from an insoluble to a soluble form is the transformation of starch to sugar. Starch is an insoluble compound and cannot be transferred through the walls of the cells in which it is stored or be directly utilized by the protoplasm. When starch by the process of digestion is changed to maltose or eventually to glucose, we have a carbohydrate that is soluble in water, that can readily move from cell to cell, and that can be utilized by the protoplasm. Cane sugar is an example of a food that is soluble in water and easily translocated, but which must be digested into the more simple sugars glucose and fructose before it can be assimilated by the protoplasm of plants and animals.

III. ENZYMES

One of the most striking characteristics of living organisms is the ease with which they are able to split up compounds of a highly stable nature. The oxidation and hydrolysis of compounds in plants and animals occur at ordinary temperatures very rapidly, while in the laboratory these changes are effected in many cases only at higher temperatures and by the use of strong reagents. Little is known, however, concerning the manner in which these reactions are carried out by living organisms. It is well established, however, that the process of digestion in both plants and animals is not brought about directly by the protoplasm itself but by substances secreted by it. These substances behave as catalysts and are called "enzymes." This leads us to a discussion of catalysts in general and of the nature of enzymes in particular.

A. General Nature of Catalysts

In regard to the speed with which they are completed, chemical reactions may be placed in two main groups: (1) those reactions that are completed practically instantaneously, *e.g.*, the neutralization of a base by an acid or the precipitation of copper hydroxide by the addition of a solution of potassium hydroxide to a solution of copper sulphate; (2) those which require a measurable time for their completion, *e.g.*, the hydrolysis of cane sugar to glucose and fructose by a mineral acid or the saponification of a fat or oil by sodium hydroxide. It is with this latter type of reaction that catalysts are primarily concerned. A catalyst may be defined as a body which alters the speed of a chemical reaction that requires a measurable time to arrive at its final state. This alteration in the speed of a reaction may be either an acceleration or a retardation. Catalysts, however, are generally considered in regard to their acceleration of reactions, and the process of catalysis has been defined as "the acceleration of a chemical change by the presence of some foreign substance" (Bayliss, 1925). It is almost a universal opinion that catalysts cannot in themselves initiate reactions but that they only accelerate or retard reactions that are proceeding in their absence. Some of the general properties of catalysts are: (1) They are not chemically combined with the final products of the reaction but may be recovered at the end of the reaction measurably unchanged in quantity and quality, except where they may be destroyed by subsidiary reactions. (2) A relatively small amount of catalysts produces very large results in the reacting mixture. (3) The effect of a small amount of catalyst is in the end the same as that of a larger quantity if sufficient time is allowed, the only difference being the rate at which the new products appear.

B. General Nature of Enzymes

In beginning a discussion of this topic, the first step should be to give a definition of the term "enzyme." A definition, however, is difficult of formulation, since so little is known concerning the nature of enzymes. In the light of our present knowledge, perhaps as exact and inclusive a definition as can be formulated is that of Bayliss (1925) who defined an enzyme as "a catalyst produced by living organisms."

These biochemical catalysts, however, differ widely in their reaction towards the environment. According to Tauber (1936) some of these, as pepsin, trypsin, maltase, lipase, and others, can act independently of the living cell and are destroyed by heat. Others, however, as for example the catalyst affecting the synthesis of urea in the liver act only in the unimpaired cell and are destroyed by heat. There are certain other nonspecific biochemical catalysts, as glutathione and ascorbic acid, which although elaborated by the living cell can act independently of the living cell and are not destroyed by heat.

Enzymes are colloidal substances and do not readily diffuse through artificial membranes. The rate at which they diffuse through membranes is largely dependent upon the nature of the membrane, but as a rule they will not pass through parchment membranes. This is one important characteristic that distinguishes enzymes from other organic catalysts. Most enzymes, however, will pass through a porcelain filter, and this characteristic is utilized for their separation from the materials with which they are associated. Since enzymes are colloids, they have the property of carrying down with them by adsorption some of the constituents of the solutions from which they are precipitated. The methods used for the precipitation of enzymes also precipitate most proteins, so that proteins are generally present in an ordinary enzyme preparation. On that account it is difficult to determine the chemical composition of enzymes with any degree of certainty, and so far all attempts to isolate an enzyme as a pure substance of definite chemical composition have been without satisfactory results. In general, the more the enzymes are purified the fewer characteristic reactions do they show, although in some cases their activity, weight for weight, can be greatly increased. On the other hand, in many cases the attempts to purify enzyme preparations by the means at present at our disposal greatly decrease the activity of the enzyme. Since, as will be mentioned later, enzymes are specific in their action, it would seem probable that they differ in their chemical structure. On that account the chemical reactions of the different enzymes will be considered under their individual headings. According to Waldschmidt-Leitz (1933), it was considered by Willstätter (1922) that enzymes consist of a carrier that is colloidal in nature, and of one or several chemically active groups. This hypothesis accounts for the physico-chemical and chemical behavior of enzymes and has been widely accepted. The colloidal carrier is considered to determine the stability and the magnitude of catalytic activity of the active groups, while the nature of the active groups is responsible for the specificity of the enzymes. Willstätter, Graser, and Kuhn (1922) believed that the colloidal carrier of an enzyme can be varied by changing the procedure of purification, and that a small change in the nature of the active group, such as the introduction of two hydrogen atoms, produces great changes in the specificity of enzymes.

Hogounenq and Loiseleur (1925) considered that an enzyme consists of a colloidal organic complex and a mineral element. The organic complex constitutes a colloidal support upon which the electrolyte is held by adsorption, retaining its crystalloid properties and specific chemical activity. The electrolyte is considered specific for a given enzyme, *e.g.*, chlorine for pepsin and manganese for oxidases. The adsorption of the electrolyte is determined by the electrical sign of the protein micelles, and the sign is determined by the functional grouping in the molecule. Thus pepsin is supposed to consist of a colloidal protein support having predominantly acid groups that determine its negative sign. It absorbs such cations as sodium, setting free chlorine, which attacks the NH_2 groups of the proteins digested. Trypsin is a similar colloidal complex in which amidogenic groups predominate, giving the micelles a positive charge. Zeile and Hellström (1930) and Zeile (1931) found by photospectrometric measurements that an iron-porphyrin complex is the active group of the catalase of the pumpkin. Kuhn, Hand, and Florkin (1931) considered that this complex is also the active component in certain peroxidases. According to Balls (1931), Sumner (1933), and Falk (1935), it is generally agreed that practically all enzymes are of a protein nature, and the fact that many enzymes are digested and inactivated by proteases lends support to that general belief (Glick and King, 1933; Falk, 1935; and Tauber, 1936). The active substance of an enzyme cannot exist alone in solution but apparently must always be accompanied by some protective colloid with which it is in some sort of combination. The active principle of an enzyme, however, has never been isolated. There are, however, some observations which indicate that the enzyme is not of a colloidal nature. Thus, Miller (1929) reported that a very active soluble tyrosinase obtained from a mushroom was not colloidal in nature since it dialyzed through a cellophane tube. He considered that this observation invalidates the general application of the theory of Willstätter that enzymes exist in a particular, colloidal state.

Enzymes are sensitive to heat. The optimum temperature for most plant enzymes is between 37 and 55°C. In most cases, enzymes are injured at a temperature above 60°C. and all are completely destroyed at a temperature of 100°C. Their resistance to heat differs with different types of enzymes, with the kind of material associated with them, and with the amount of water present, as will be mentioned when the different types of enzymes are considered. The destruction of enzymes by heat is attributed to the alteration of their colloidal nature probably by coagulation. In general, the activity of enzymes is doubled for every rise of 10°C., but this effect is diminished at higher temperatures because of the destruction of the enzymes by heat. The relation of the increased activity of an enzyme to temperature, the effects of the increased temperature on the destruction of the enzyme, and the resultant activity of increases in temperature are shown in Fig. 27. Enzymes are inactivated at lower temperatures but again regain their activity when the temperature is again raised.

It has generally been considered that enzymes are sensitive to light, especially ultraviolet and polarized light. On that account experiments with the activity of enzymes have generally been conducted in the dark. Recently, however, Fuller (1931, 1932) found that when tomato and bean plants were severely injured by ultraviolet rays the activity of their amylase, invertase, peptase, and catalase was significantly increased. When solutions of taka-diastase and commercial invertase were irradiated *in vitro* by a quartz mercury lamp, they underwent a partial inactivation, but this was partially due to the infrared radiations.

In the study of the action of enzymes, an antiseptic must be employed to prevent the hydrolytic action of bacteria. The most commonly used antiseptics are toluene, chloroform, sodium fluoride, phenol, thymol, and formaldehyde. Practically all these compounds exert more or less of an inhibiting influence on enzymic action, depending upon the enzyme under observation. Of these, toluene is the most suitable and is the most commonly employed antiseptic in the study of enzymes.

C. General Nature of Enzyme Action

Enzymes have two properties that are common to all catalysts in that they are not capable of initiating a new reaction, but merely of changing the rate of one already in progress, and in that they do not appear combined with the final products of the reactions that they influence. Enzymes differ from catalysts in that they are more specific in their reactions. Thus a catalyst, as the hydrogen ions of mineral acids, will hydrolyze proteins, carbohydrates, fats, and other esters, while a given enzyme will act on only one of these compounds. They also differ from catalysts in that they do not always carry the process to the same state of completeness as does an inorganic catalyst that influences the same reaction. Thus the enzyme diastase hydrolyzes starch to malt sugar, while a mineral acid hydrolyzes the starch to glucose. Certain of the proteolytic enzymes also do not carry the digestive process so far as do acids. Thus an acid hydrolyzes the protein to amino acids, while the enzymes frequently leave complex polypeptides that are unattacked. Thus an enzyme catalyzes only a single reaction as the hydrolysis of starch to maltose or of maltose to glucose, while the acid catalyzes both reactions. This difference between enzymes and the inorganic catalysts may be more apparent than real, since the reactions that occur under the influence of enzymes might proceed to complete hydrolysis if the appropriate conditions were present. Thus it has been observed, for example, that if malt extract is allowed to act for a long time on starch, the hydrolysis proceeds beyond malt sugar, and glucose is formed. Enzyme action in many cases apparently comes to an end or equilibrium point, owing to the accumulation of the products of the reaction, so that by the dilution or removal of products the reaction can proceed farther.

The mechanism by which an enzyme accomplishes its catalytic effects is not definitely known. Two general theories concerning the mode of this catalytic action have been advanced and will be briefly mentioned here, since a full discussion of the nature of enzyme action belongs to the province of advanced biochemistry rather than to the field of plant physiology. The few statements in regard to enzyme action that are given herein are made in order that the student may be familiar with the complexity of the problem and with the meager knowledge available on the subject. The following presentation is based primarily on the discussion by Thatcher (1921).

The earliest and simplest theory of enzyme action that was proposed was that the enzyme simply creates an environment favorable for the particular chemical reaction to take place and does not enter into the combination with the substrate at all. The favorable environment was considered to be brought about by the exposure of large surfaces of the substance under consideration to the action of the active agent of the enzyme by the means of surface adsorption of this substrate on the colloidal material of the enzyme. According to Alexander (1934), high efficiency for an enzyme demands an intermediate degree of dispersion, which involves a large exposure of active interfacial surface coupled with a high kinetic activity.

The theory that is now most generally accepted is that there is an actual combination between the substrate and the enzyme. This combination is only temporary and breaks down with a resultant change in the substrate and the freeing of the enzyme for recombination with additional substrate. The theory of this mode of enzyme reaction is based on the following: (1) the fact that if an enzyme and its substrate, each of which is filterable through a given filter, are mixed, the active material in the combined solution will not pass through the filter; (2) the interpretation of the curves representing the reaction velocities of typical reactions that are proceeding under the influence of an enzyme; and (3) the fact that the limitation of a given enzyme to a specific substrate is definitely related to the molecular configuration of the molecules of the substrate.

There are two general theories in regard to the nature of the supposed combination of the enzyme and substrate. One is that the combination is molecular, and that a definite chemical compound is formed. The other is that the result of the union is purely a physical or colloidal complex. The latter theory is substantiated by the effects of electrolytes and heat on enzyme action and by the appearance of the reacting substances under the microscope.

According to Falk (1935) the only conclusion that appears justified at present is that addition compounds of enzyme and substrate are formed, which then break down to form the products of the enzymatic action. Such addition compounds, however, have not been isolated, and the evidence for their presence is wholly indirect.

In many cases the chemical action brought about by enzymes never goes to completion, so that there is some of the substrate remaining intact. This is explained by saying that the action of enzymes is reversible and that the same enzyme will accelerate the velocity of the reaction in either direction, depending upon the environmental conditions of the reacting materials. The synthesis of fats and oils from fatty acids and glycerin in the presence of lipase, as mentioned in Chap. X, is an example of the synthetic action of an enzyme. The synthesis of isomaltose from glucose by maltase is another example of the reversible action of enzymes. It is considered that the synthesis of the various plant products is brought about by enzyme action, although the experimental evidence in this regard is very meager. Enzyme action under artificial conditions is distinguished from direct protoplasmic action in that it occurs in the presence of an amount of antiseptic that is destructive to the life of protoplasm. Some enzymes, however, are very sensitive to certain antiseptics, as will later be mentioned.

D. Nomenclature

The substance whose chemical change is catalytically affected by an enzyme is termed the "substrate." The substrate is generally defined as the substance upon which the enzyme acts.

The activity of enzymes is greatly influenced by the presence in the solution of other bodies. Any substance that increases the catalytic activity of an enzyme is termed an "accelerator" or "activator," while one that inhibits or destroys its action is called an "inhibitor" or "paralyzer." The substances that thus influence the rate of enzyme action are generally electrolytes, such as mineral acids, alkalies, and salts, although this is not always the case. The accelerating action of the amino acids on the activity of certain enzymes is an example of these exceptions. All enzymes, however, do not respond alike to the same substance. Thus the activity of some enzymes is greatly increased by the presence of a small amount of mineral acid, while the action of another may be partially or completely inhibited by the same acid in the same concentration. Some of the substances that almost always exercise an inhibitive or toxic action on enzyme activity are hydrogen sulphide, formaldehyde, hydrocyanic acid, chloroform, and the salts of mercury, silver, and copper.

It has been found that many enzymes are inactive unless accompanied by some other substance which is normally present in the protoplasm that produces them. Such substances have been named "coenzymes." This accompanying substance can usually be separated from the enzyme by dialysis, the coenzyme passing through the parchment membrane. After the coenzyme is thus separated, it is not injured by boiling, and the activity of the enzyme may be restored by adding the boiled substance to the liquid that remains behind in the dialyzer. Coenzymes have been studied in connection with the enzymes of animals and in connection with the activity of zymase of yeast cells, but few observations have been made in this regard with the enzyme studies of green plants.

The term "antienzyme" has been applied to a substance occurring naturally in a living organism, or which is produced after the injection of an enzyme, that prevents the action of the specific enzyme upon the tissues. Antienzymes have been considered more particularly in the physiology of animals, but in many cases the evidence of the existence of such substances and their action rests upon theory rather than experimental evidence. The evidence of antienzymes in the green plants is extremely meager. The only work in this regard that is known to the author is that of Czapek (1905), who considered that he obtained evidence of antioxidases in his studies of the reactions occurring during the tropistic movements of plants.

Since enzymes are produced by the protoplasm, it is apparent that, at a given stage in their synthesis, they have not yet acquired their catalytic powers, although possessing, in general, the other properties of enzymes. An enzyme that has reached this stage of its formation is termed a "proenzyme" or "zymogen." These terms are used, however, only when the substance can be obtained free from the cells in which it is formed and can be converted after its extraction into the active enzyme by purely chemical means (Bayliss, 1925). When such a substance has once been activated, it cannot, according to present knowledge, be converted again to the proenzyme or zymogen state. Calcium salts and dilute acids are considered to be the most energetic activators of proenzymes. The activator of a proenzyme has also been termed a "kinase." According to Thatcher (1921), proenzymes are believed to be present in plant cells in the form of definite characteristic granules which may be observed under the microscope but which disappear when the enzyme becomes active. According to him, proinulase has been reported as occurring in artichoke tubers, prolipase in castor beans (Tanka, 1910), and prooxidase in tobacco leaves.

When organic catalysts were first isolated from plant and animal tissues, they were termed "ferments" because of the resemblance of the changes produced to alcoholic fermentation. After it was found that fermentation was due to the presence of living organisms as yeast, these organic catalysts were termed "soluble" or "unorganized ferments" to distinguish them from living organisms producing similar changes, which were termed "insoluble" or "organized ferments." The double use of the word ferment caused much confusion and induced Kühne (1878) to suggest that the name "enzymes" be given to those substances which were then designated as unorganized ferments. The name enzyme comes from the Greek words *en zymos* and literally means "in yeast." Kühne stated that the term was meant to include those unorganized ferments which behave in a manner similar to those substances which occur in yeast. When it was shown in 1897 by Büchner that if yeast cells are crushed and filtered the filtrate is just as active in fermentation as the intact yeast cells, the difference between organized and unorganized ferments disappeared and the term enzyme came into general use.

E. Classification

Enzymes are specific in their action, each one activating a particular chemical change upon one substance, so that there are as many enzymes as there are substances upon which they act. On this account, a system of nomenclature has been adopted which assigns to each enzyme the name

- Enzymes that function in the process of digestion in plants
  - Hydrolases
    - Carbohydrases
      - Invertins, invertases, saccharases, or sucrases
      - Diastases or amylases
      - Dextrinases
      - Inulases
      - Glucases or maltases
      - Cellulases or cytases
      - Pectinases
      - Melibiase
      - Gentianase
    - Glucosidases
      - Emulsin
      - (Maltase)
      - Phytase
      - Rhamninase
      - Myrosin
    - Esterases and Lipases
    - Proteases
      - Pepsins or peptases
      - Trypsins or tryptases
        - Papaïn
        - Bromelin
        - Cradina
        - Other trypsins
      - Erepsins or ereptases
  - Coagulases
    - Rennins or rennets
    - Pectases

of the material upon which it acts, followed by the suffix *-ase*. Thus sucrase is the enzyme acting upon sucrose, amylase the enzyme acting upon amylum (starch), and cellulase the enzyme acting upon cellulose. This system of nomenclature is now generally followed, although in some cases, where the original name of the enzyme has long been in use, the older terminology is used, *e.g.*, pepsin, trypsin, diastase, invertase, and so on. The enzymes that have been the most studied and the ones with which we are primarily concerned in the study of digestion accelerate hydrolytic reactions and may be termed "hydrolases." Certain enzymes that cause coagulation are termed "coagulases" and in some cases apparently play a role in the digestive process. The enzymes that are concerned in digestion may be classified after the manner as shown at the bottom of page 781.

F. Occurrence

Digestive enzymes of one type or another are found in all living plant tissue. A few examples will suffice to illustrate this point. Robertson, Irvine, and Dobson (1909) found diastase, invertase, maltase, inulase, and emulsin in the stem and leaves of the beet. Bailey (1912) found invertase, diastase, proteases, and lipase in the banana, and in the stem, leaves, roots, and seeds of alfalfa, Jacobson and Holmes (1914) detected lipase, diastase, coagulases, emulsin, invertase, pectinase, and certain proteases. The fungi are especially rich in digestive enzymes and are the sources of many of the commercial enzyme preparations. Thus in *Echinodontium tinctorium*, Schmitz (1920) demonstrated the presence of esterase, maltase, lactase, invertase, diastase, inulase, cellulase, rennet, and others. Thatcher (1915) found oxidases, esterases, and proteases in ripening apples but could not detect any of the carbohydrate-splitting enzymes. Paton (1921) found catalase, diastase, invertase, reductase, and pectinase in the pollen of 18 species of plants. Pepsin, trypsin, erepsin, and lipase were in the pollen from some plants but not in that of others.

The abundance of enzymes in a given part of a plant depends apparently on the conditions under which it is placed. There is some evidence that fungi in some cases may secrete an enzyme corresponding to the kind of substrate that is supplied, but no observations in that regard have been made upon green plants (Kertesz, 1931). After enzymes are formed in the plant they may remain active for a long period of time. Thus White (1909) found that the seeds of wheat, oats, corn, barley, and rye which had been stored for 20 years and which had lost their germinating power, contained diastase and erepsins that had retained their activity apparently unimpaired. Miehe (1923) examined rye seeds that were at least 112 and probably 280 years old. Although the embryos were disintegrated, the endosperms were intact and the starch contained in them reacted to the iodine test. Enzymes were obtained from these seeds by water extractions and changed starch to sugar and differed from those of fresh rye-seed extracts only in acting more slowly. Maestrini (1923) studied acidulated aqueous extracts of germinating barley that had been kept for 3 or 4 years. Diastase did not appear to be perceptibly modified in its enzymic action by this treatment. The proteases, however, showed a marked reduced hydrolytic capacity, and lipase and emulsin were destroyed in less than 2 years (Longenecker and Haley, 1935).

We shall now discuss the distribution, general properties, reactions, and methods of preparation of the various digestive enzymes found in the green plant.

1. The Carbohydrases of Plants. *a. The Invertases, Invertins, Saccharases, or Sucrases.*—The enzyme that hydrolyzes sucrose to glucose and levulose has been variously named "invertase," "invertin," "saccharase," or "sucrase." This hydrolysis is one of the most simple of the enzymic reactions and proceeds after the following manner:

$$\underset{\text{Sucrose}}{C_{12}H_{22}O_{11}} + H_2O \xrightarrow[\text{of invertase}]{\text{In presence}} \underset{\text{Glucose}}{C_6H_{12}O_6} + \underset{\text{Levulose}}{C_6H_{12}O_6}$$

The occurrence of invertase in green plants has been reported by Brown and Morris (1893); Kastle and Clark (1903); Martinand (1907); Vinson (1907); Robertson, Irvine, and Dobson (1909); Colin (1915); McGuire and Falk (1919, 1923); and others. Invertase has been found in the roots and leaves of the radish; in the leaves and tubers of the artichoke; in the leaves, tubers, and sprouts of the potato; in the leaves of the dahlia; in the leaves and roots of the dandelion; in the entire plant of corn and *Sorghum saccharatum;* in the leaf lamina, petioles, and collar of the sugar beet; in all parts of the leaves, stem, and fruit of the grape; in the leaves, stem, and pods of the soybean; in the leaves and roots of the sweet potato; in the fig; in the banana fruit; and in many other plants and plant parts. Invertase is thus apparently of almost universal occurrence in green plants. It has not, however, been definitely established whether the invertases from different sources are identical substances (Doby, 1915). Invertase occurs in great abundance in yeast, and it is from this source that the invertase of commerce is mostly obtained. It may also be prepared from cultures of *Aspergillus niger* (Stoward, 1912). Most of the studies on the nature and action of invertase have been made on the invertase of yeast. Invertase is now used rather extensively in the industries. In the hydrolysis of a number of commercial products in which sucrose is the principal saccharine constituent, this enzyme has been found more suitable than an acid as a catalyst. It is used in products which exhibit a marked buffer effect toward added acid and which possess a flavor adversely influenced by the concentration of acid required to effect hydrolysis (Paine, Walton, and Badollet, 1925). According to Paine and Balch (1925), an analytical procedure involving the use of the enzymes invertase and melibiase has been adapted to practical use for the precise determination of the sugars sucrose and raffinose in the chemical control of beet-sugar factories.

1. *Preparation and Study of Invertase.* (*a*) *Preparation from Yeast.*—Although this text deals only with the physiology of the green plant, the preparation of invertase from yeast is here discussed, since much of our knowledge of the action of invertase has been learned from enzyme preparations obtained from yeast. Methods for the preparation of invertase from yeast have been given by O'Sullivan and Tompson (1890), Hudson (1910, 1914), and Nelson and Born (1914). The modified method of Nelson and Born as used by Hudson (1914) is as follows: Equal amounts of water and yeast are autolyzed at room temperature for 5 days with toluene as an antiseptic. The material is then filtered, and 95 per cent alcohol is added in equal volume. After the precipitate settles, the alcohol is decanted and the residue dissolved in water. Alcohol is again added as before and again decanted. After the precipitate is again dissolved in water, a concentrated solution of lead acetate is added until no more precipitate is formed. After filtering off the precipitate, potassium oxalate is added to throw down the excess lead, and the precipitate is filtered off. The filtrate is then treated with kaolin and stirred constantly for 1 hr., the treatment being repeated several times. The resulting liquid is then placed in collodion bags with toluene and dialyzed in running water from 3 to 20 days. After this dialysis, a precipitate cannot be obtained by the addition of alcohol until some sodium chloride, potassium acetate, or acetic acid is added to alter the colloidal nature of the solution. If a precipitate is thus obtained, it is washed with absolute alcohol and dried over sulphuric acid. The purest preparation of invertase obtained after this manner will give a faint biuret and strong Millon and xanthoproteic reactions as long as the enzyme remains active, and there is a rough parallelism between the nitrogen content and activity when the preparation is fairly pure. An active preparation of invertase has a nitrogen content of 2.2 per cent, but when the nitrogen content falls below 1 per cent the enzyme is no longer active (Mathew and Glenn, 1911). The invertase of yeast as ordinarily prepared is considered to be a union of protein with a carbohydrate gum, the protein constituent being the active part of the preparation. Nelson and Born (1914) noted that a 10 per cent solution of invertase prepared from yeast after the above method was not precipitated by any of the ordinary protein precipitants. After the hydrolysis of this solution with 2.5 per cent sulphuric acid and neutralization with barium carbonate, copious precipitates were obtained with mercuric chloride, mercuric acetate, phosphomolybdic acid, and phosphotungstic acids. This indicates that in the original invertase preparation either the protein part is combined with the carbohydrate or the latter acts as a protective colloid.

(*b*) *Preparation from Green Plants.*—Falk and McGuire (1921) used the following method for the preparation of invertase from bananas: The pulp is ground finely in a food chopper and then mixed in a porcelain mortar with normal sodium chloride, 100 cc. to each 400 g. of pulp, toluene is added, and the mixture is filtered through paper. The filtrate is then dialyzed in collodion bags for 18 to 24 hr. against running water to remove salts, carbohydrates, and other products. The resulting solution will keep for a considerable period of time under toluene and is used for invertase experiments. Invertase in leaves has been studied, for the most part, by drying the leaves at air temperature, pulverizing the dried material to a fine powder, and taking a weighed portion for the study of enzyme action. It should always be remembered that the manner in which an enzyme is obtained may affect not only its activity but its physical properties as well. Thus Falk and McGuire in studying invertase in the banana found a water-soluble, a sodium chloride-soluble, and another invertase in the extracted residue of the fruit pulp. All three preparations were identical, as far as could be observed in their action on cane sugar, but were entirely different in their physical behavior.

The general method for the study of the action of invertase is to add 10 to 15 cc. of an invertase solution to 10 to 25 cc. of a 1 to 20 per cent sucrose solution together with sufficient toluene to act as an antiseptic. This mixture is incubated at 35 to 38°C. for varying periods of time, depending upon the activity of the invertase preparation. The periods of incubation vary from 2 to 24 hr. As a rule, the invertase preparations from leaves and fruits are relatively slow in their action as compared to the preparations from yeast. After the incubation, a portion of the mixture is treated with Fehling's solution, heated, and the degree of invertase action determined in terms of the number of milligrams of copper oxide produced by the action of 1 cc. of the original enzyme solution, correction always being made for the controls. The rate of invertase action may be obtained qualitatively by estimating the amount of reduction of Fehling's solution by the color of the solution or by the relative amount of the precipitate produced.

2. *Nature of the Action of Invertase.*—Invertase has been the best studied of any of the plant enzymes. Since the action of this enzyme is relatively simple, starting with a well-known substance and producing well-known end products, the hydrolysis of cane sugar by it has been chosen by many investigators as the best process with which to study the velocity of enzyme action in relation to different factors.

Invertase is not destroyed or injured by its action on sucrose. O'Sullivan and Tompson (1890) found that a sample of invertase which had induced inversion of 100,000 times its own weight of cane sugar was still active, so there is apparently no limit to the amount of sugar that may be hydrolyzed by a given amount of this enzyme, provided that the proper conditions are maintained for its action.

It should be mentioned here that there is apparently a difference between the action of acids and invertase on cane sugar, although both substances are catalysts and both produce the same end products in the hydrolysis of this sugar. Thus in different concentrations of cane sugar, the same amount of acid gives the same velocity constant. For example, a 5, 10, or 15 per cent sucrose solution will be inverted by a 1 per cent acid solution in the same time to the same extent, say halfway. In the case of invertase, however, the same quantity of enzyme in the different concentrations of sucrose invert in the same time, not the same fractions of each solution, but the same absolute amounts.

The velocity of the inversion of cane sugar is directly proportional to the concentration of invertase. The velocity of reaction is nearly independent of the concentration of the sucrose in the more concentrated solutions, while in very dilute sugar solutions, the velocity increases with the increase of concentration of the substrate and finally reaches a maximum (Nelson and Vosburgh, 1917). It was shown by Nelson and Bodansky (1925) that the invert sugar formed in hydrolysis exerts a retarding influence on the velocity of the reaction of invertase. This is due to the fact that invert sugar has a different retarding effect upon hydrolysis according to whether it is in a freshly liberated or in a final mutarotated form. If the invert sugar were mutarotated immediately upon its liberation, the course of reaction would be different from the ordinary observed course of the reaction.

According to Nelson and Griffin (1916), invertase is of a colloidal nature and the reaction between the enzyme and a cane-sugar solution depends on the contact of two phases. They found that the activity of invertase is not affected when it is adsorbed to a solid like charcoal or to a colloid like saponin, serum, or egg albumin distributed uniformly throughout the substrate (Nelson and Hitchcock, 1921). Nelson and Vosburgh (1917) considered that adsorption is one of the controlling factors in the kinetics of invertase action, since the velocity of the inversion curve, where the concentration of the cane sugar is used as the abscissas, has the same general shape as adsorption curves. There are yet considerable contradiction and uncertainty, however, concerning the nature of the reaction involved in the hydrolysis of cane sugar by invertase. The results obtained by Nelson and Vosburgh (1917) and others indicate that it is a heterogeneous reaction and not a unimolecular homogeneous one.

It was observed by Nelson and Hitchcock (1921) that not all the preparations of yeast invertase are alike in their action, but that some are abnormal in allowing the hydrolysis of cane sugar to slow up more than others after the first 20 per cent of the inversion. This abnormality could be rendered normal by the addition of boiled normal invertase or 0.1 *N* sodium chloride in some cases, while in others it was not affected by either of these additions. The cause of this abnormal behavior is not known.

It is generally considered that invertase exerts a reversible reaction, and Kohl (1908) stated that inversion gives place to the opposite process sooner in darkness than in light. Hudson and Paine (1914), however, concluded that invertase from yeast accomplishes a complete hydrolysis of sucrose to yield invert sugar and that the reaction does not establish mobile equilibrium and is not a reversible or balanced reaction within the limits of detection.

3. *Factors Influencing the Action.*—The method by which invertase is prepared, the age of the preparation, the dilution of the preparation, and the general treatment of the preparation must all be taken into consideration in discussing the influence of various factors on the activity of

invertase. Thus Hudson and Paine (1910) noted that alcohol reduces the activity of invertase. This reduction is not noticeable below 20 per cent alcohol at 30°C., is almost instantaneous at 50 per cent, and decreases to zero at 80 per cent. Cane sugar in alcohol has a protective value on the enzyme. Invertase may be precipitated with alcohol without much destruction, provided the strength of the alcohol is approximately 90 per cent. A solid preparation obtained in this manner was 78 per cent as active as the original solution. If cane sugar is present, the invertase may be precipitated with 70 per cent alcohol, and the resulting precipitate give 94 to 96 per cent of the original reaction. Vosburgh (1921) found that invertase solutions are subject to loss in activity when diluted, the loss varying with the substances present in the water with which the invertase is diluted. Dilution with distilled water is attended by less loss than is dilution with weak acids. When an invertase solution is added to a solution containing sucrose, the losses in activity due to dilution are less than when the invertase is diluted by similar solutions not containing sucrose. Different preparations of invertase differ in this regard. In studying the action of invertase of banana extracts, McGuire and Falk (1923) showed that under certain conditions this enzyme exhibited apparently spontaneous increases of activity of considerable magnitude. In some cases the original activity was almost doubled, the activity increasing with the aging of the extract. After reaching a maximum value in a certain number of days, the invertase activity decreased. This increase of invertase activity was independent of the composition of the solution used for extraction and was not due to bacterial growth or cell decomposition accompanied by enzyme liberation. The behavior observed might be due to the fact that the extract contained materials that in the absence of life processes form new enzyme. The invertase in the extract may lose its activity steadily and the formation of new enzyme may be at first greater than the destruction and then become less. This theory would account for the increase and subsequent decrease in the activity of the enzyme.

In their classical paper on invertase, O'Sullivan and Tompson (1890) stated that small amounts of acid are favorable to the action of this enzyme but that alkalies even in small amounts are very destructive to its action. These observations, in general, have been substantiated by the work of later investigators. Hudson and Paine (1910) found that acids and alkalies in small concentrations influence the activity of the enzyme but do not permanently destroy it. The destruction by acids at 30°C. reaches a barely noticeable rate at a concentration of one-hundredth normal and increases with the acidity until it becomes almost instantaneous at a one-twentieth normal concentration. The alkaline destruction of invertase is almost instantaneous at a concentration of 0.045 *N*.

The activity of invertase in weak alkaline solutions is zero, rises to a maximum in very weakly acid ones, and decreases with stronger acid solutions. The activity of the enzyme in weak solutions of hydrochloric, nitric, sulphuric, phosphoric, boric, oxalic, tartaric, citric, and acetic acids depends almost entirely on the concentration of the hydrogen ions, and the acids thus show typical differences that correspond to their degree of dissociation. The optimum activity of invertase, according to Nelson and Bloomfield (1924), at 25, 30, and 35°C. is between pH 4.5 and 5.0. They found that the effect of temperature and hydrogen-ion concentration is independent of the concentration of sucrose, and the sucrose concentration, at which the hydrolysis in the presence of invertase attains a maximum velocity, is independent of temperature and hydrogen-ion concentration. They considered that the hydrolysis of cane sugar in the presence of invertase involves at least two distinct stages. One of these stages is influenced by the concentration of the sucrose but is independent of the temperature and hydrogen-ion concentration, while the other is influenced by both these factors (Compton, 1921).

O'Sullivan and Tompson (1890) stated that the speed of inversion by invertase increases with the temperature until 55 to 60°C. is reached, and that at 65°C. the enzyme is slowly destroyed. The optimum temperature noted by them was about 54°C. Hudson and Paine (1910) found that at 60 to 65°C. the destruction of invertase with hot water was rapid. A dry sample of the enzyme, however, may be heated for some time at 100°C. without any appreciable deterioration. As the temperature is raised, acids and alkalies react more readily on invertase. Hudson and Paine considered that the destruction of this enzyme by acids, alkalies, and hot water is caused by the hydrolysis of the enzyme molecule. The presence of fructose and sucrose protects the destruction of invertase, and O'Sullivan and Tompson stated that a solution of invertase in the presence of cane sugar will withstand a temperature 25°C. higher than in the absence of the sugar. It is assumed that the enzyme forms a combination with the sugar which is more resistant to the destructive action of acids, alkalies, hot water, and alcohol than invertase itself.

It was observed by Joslyn and Sherrill (1933) that sucrose was inverted by invertase at temperatures as low as −12 to −16°C., but that this enzyme was completely inactivated at −40°C.

The effect of tannin on the solubility of the invertase of the date was noted by Vinson (1908). The invertase of the date remains insoluble in all ordinary solvents throughout its green stages but becomes readily soluble on ripening. This change of the invertase toward solvents coincides very closely in point of time with the passage of the tannin of the date into the insoluble form. Vinson showed experimentally that, although soluble tannin hinders the solubility of date invertase in water, it may be extracted with glycerol, provided the glycerol is added at the same time as the tannin.

b. The Diastases or Amylases.—The enzyme or aggregate of enzymes that hydrolyze starch to malt sugar is termed "diastase" or "amylase." The general reaction of this hydrolysis may be represented by the following equation:

$$\underset{\text{Starch}}{(C_6H_{10}O_5)_n} + \underset{\substack{\text{In presence}\\ \text{of diastase}}}{\tfrac{1}{2}nH_2O} \longrightarrow \underset{\text{Maltose}}{\tfrac{1}{2}nC_{12}H_{22}O_{11}}$$

This process of digestion proceeds from starch to amylodextrin, to erythrodextrin, to achrodextrin, to maltose. The substance designated as diastase is thus not composed of a single enzyme but of a series of amylases and dextrinases, none of which, however, has been isolated in a pure condition.

According to Bayliss (1925), Dubrunfaut (1830) prepared an extract of malt which converted starch into sugar just as strong acids were known to do. This, according to Bayliss, was the first account of an enzyme in solution. Three years later Payen and Persoz precipitated from malt extract by alcohol a substance which could be dried and preserved and which had a hydrolytic action on starch. To this preparation they gave the name "diastase." Diastase thus was the first plant enzyme to be reported, and due primarily to its importance in the brewing industry, it has been the most extensively studied enzyme in the plant kingdom. Its presence in leaves was first proved by Kosmann (1877), and since then (Baranetzky, 1878; Krauch, 1879; and Brasse, 1884) it has been found in all green plants or parts thereof that have been examined, so that apparently it is one of the most widespread as well as abundant enzymes of plants. Diastase occurs in many fungi, and the commercial taka-diastase, which is used in medicine and in digestion experiments in the laboratory, is derived from the fungus *Aspergillus oryzae.* The diastase produced by this fungus has been known to science since about 1875 but has been in use in Japan empirically for centuries. Commercial taka-diastase was introduced by Takamine in 1898 and is prepared by growing the fungus on wheat bran, extracting it with water, and precipitating the enzyme and other substances by the addition of alcohol in such quantity as to give a concentration of 70 per cent (Sherman and Tanberg, 1916).

Otis (1930) reported that a digestive ferment with the trade name of "Pangestion" is capable of converting 80 parts of potato starch into water-soluble substances in 5 min., and that the treated starch solution produces positive tests with Fehling's solution after 10 min. at room temperature. It is not a pure preparation of diastase, for it also contains active proteolytic and lipolytic enzymes.

Lothrop and Paine (1931) found that all American honeys contain diastase. Those darker in color usually show a high diastatic value, while honey from alfalfa and orange bloom shows low values. Moderate heating lowers the diastatic value of honey. Europeans object to American honey because its naturally low diastatic activity leads them to believe that it has been heated.

McGinty (1919) noted that the diastatic activity of the potato increased as the tuber developed. Tysdal (1934) studied the relation of the protected diastatic activity to winter hardiness in alfalfa. The protected diastatic activity is the diastatic activity observed after the extract has been subjected to a temperature of 70°C. for 10 min. He found that the hardier a variety of alfalfa, the greater was its protected diastatic activity. The concentration of sugar and the concentration of amino acids influenced the protected diastatic activity. Since an increase in some of these substances occurs during hardening, and since these substances increase the resistance of diastatic enzymes to heat, Tysdal suggested a relationship between winter hardiness and protected diastatic activity.

Bradley and Kellersberger (1912) could detect no starch in the radish, although it showed the highest diastatic activity of any of the tissues studied. The potato tuber that was rich in starch was low in diastase. This enzyme appeared the most abundant in the leaves where starch is never stored permanently. From these observations, these investigators concluded that the diastase in plants exercises little or no synthetic activity.

1. *Types.*—The diastatic enzymes of plants are classified according to their origin, general action, and the nature of the products formed.

(*a*) *Origin and Action.*—There are apparently two well-defined types of diastase that have been termed by Brown and Morris (1890) "diastase of secretion" and "diastase of translocation." The former receives its name because it is secreted by specialized cells, as will be described in a subsequent paragraph. Its distribution is rather limited, being confined for the most part to the germinating seeds of the grains and grasses. Diastase of translocation is so named because it is the type of this enzyme that digests starch for translocation to various parts of the plant. It is widely distributed and is found in practically all living parenchymatous tissues of green plants and is probably the product of the protoplasm of the cells in which it occurs. In this regard, Neidig (1914) found in alfalfa hay, clover hay, and corn stover that the activity of the diastase therein was roughly proportional to the number of living cells in the parts at the time of cutting. Diastase of translocation is always found in the greatest quantity in the actively growing regions of the plant

(Prunet, 1892; Green, 1894; and Sjöberg, 1922, 1923). Diastase of secretion has the power actively to liquefy and saccharify starch paste, while diastase of translocation does not have the power to liquefy starch paste but has the power to saccharify solutions of soluble starch. Diastase of secretion etches and corrodes the starch grains, giving them the notched and irregular markings that are easily observed in the germinating seeds of grains and grasses (Fig. 30). Diastase of translocation dissolves the starch grains gradually over the whole surface so that no markings can be observed. Diastase of secretion is much more rapid in its action on starch than is diastase of translocation. According to Brown and Morris (1893), diastase of secretion is from two to three times more active than the diastase of translocation obtained from the leaves of various plants. The amount of diastase in leaves, however, is ample to transform all the starch in them to maltose, and Brown and Morris estimated that the quantity of diastase in the leaves of *Pisum sativum* examined by them was sufficient to hydrolyze twenty-four times their own dry weight of starch in 24 hr.

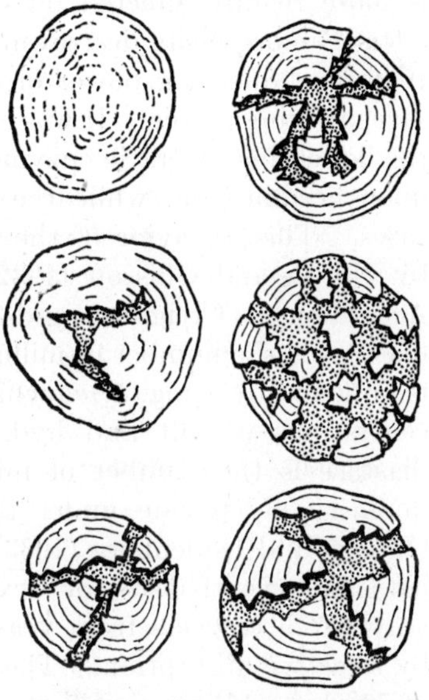

FIG. 30.—Starch grains of the endosperm of the corn kernel showing different stages of digestion by diastase of secretion.

The diastases obtained from different plants are similar in many respects but are not identical (Effront, 1922). They all digest starch, but the conditions under which their optimum activity occurs vary widely, as will be discussed in detail under another heading. Sherman, Walker, and Caldwell (1919) noted that when similarly purified by washing with very dilute alkali, wheat, corn, and rice starches showed the same digestibility in the sense that under the action of the same amount and kind of diastase they were all transformed into sugar at the same rate. Their observations were made with malt diastase and taka-diastase. Stone (1896) found that the starches of potato, sweet potato, maize, rice, and wheat varied greatly in their susceptibility to the action of diastase, so that under precisely the same conditions certain of the starches required eighty times as long as others for saccharification. It was found that the order in which the starches were the most easily digested, beginning with the one that was the most readily changed was as follows: for malt extract, the starch of sweet potato, potato, wheat, and corn; for saliva, the starch of potato, sweet potato, corn, rice, and wheat; for pancreatic diastase, the starch of potato, sweet potato, and corn; while for taka-diastase, the potato was more quickly changed than any other.

(*b*) *Products.*—The plant diastases are apparently capable of two types of activity. They possess liquefying power, which is the ability to destroy the viscosity of starch paste, and saccharifying power, which is the ability to produce maltose from liquid starch. It is therefore considered that amylase consists of an amyloclastic enzyme and a saccharogenic enzyme. Ohlsson (1922, 1926, 1930) believed that the former is more readily affected by the hydrogen-ion concentration, while the latter is more easily destroyed by heat. Gore and Józsa (1932) reported that the liquefying power of diastase is much more prominent than its saccharifying power. It was claimed by Chrzaszcz and Joscht (1917) that the liquefying diastatic enzyme is lacking in resting seeds but is developed upon germination, while the saccharogenic enzyme is present in both cases. This behavior of the diastatic enzymes has been substantiated by Nordh and Ohlsson (1932) and by Freeman and Hopkins (1936). According to Creighton and Naylor (1933) the amyloclastic power of diastase is the number of milligrams of a 1 per cent starch dispersion that is liquefied by 1 mg. of enzyme in 30 min. at 40°C. under optimum concentrations of salt and hydrogen ions. The saccharogenic power of diastase is the number of milligrams of maltose formed by the same quantity of enzyme under the same conditions as above mentioned. Ohlsson and Uddenberg (1933) and Ohlsson and Edfeldt (1933) found for oats and rye that α-, or dextrinogenic, and the β-, or saccharogenic, amylase of extracts from yeasts may be obtained free from each other by selective absorption. The former is obtained by heating the extract at 70°C. for 15 min., while β-amylase is obtained by treating the mixture with 0.1 *N* hydrochloric acid. According to Freeman and Hopkins (1936), α-malt diastase splits starch initially into dextrin. A mixture of the α- and β- amylases in low concentration functions additively in hydrolyzing starch.

2. *Comparison with Animals.*—The diastases from animal sources are, as a rule, more active than those of vegetable origin. Sherman and Schlesinger (1915) made a comparative study of the composition and activity of pancreatic and malt diastase preparations. They found that both preparations were amorphous nitrogenous compounds soluble in water or 50 per cent alcohol but insoluble in concentrated alcohol or acetone. Both showed typical protein reactions with Millon's, xanthoproteic, tryptophane, and biuret tests. Both contained 15 to 16 per cent nitrogen with the amino acids in proportion and within the range of variation shown by casein, edestin, and hemoglobin. When allowed to

act on soluble starch for 30 min. at 40°C. under optimum conditions of reaction and salt concentration, the pancreatic diastase showed a much higher activity than any yet obtained from malt. Under such conditions the pancreatic diastase formed 10,000 times its weight of maltose, while malt diastase formed about 4,000 times its weight of maltose. For long digestion experiments at 40°C. the malt diastase was observed to form a total of 9,300 times its weight of maltose, whereas the pancreatic diastase had formed 1,200,000 times its weight of maltose. Sherman, Caldwell, and Doebbeling (1934) obtained a malt amylase, by repeated precipitations with ammonium sulphate and subsequent dialyses, which formed approximately 10,000 times its weight of maltose in a 2 per cent starch solution in 30 min. at 40°C. in a dilution of $\frac{1}{9}$ p.p.m. of the enzyme. This preparation shows as high a sugar-forming activity as the most active preparation of pancreatic amylases that has so far been obtained. Malt diastase is always more stable than pancreatic diastase on standing in water, but both deteriorate on dialysis. Koga (1923) was able to show by means of serological reactions the difference between animal and plant diastases.

3. *Preparation of Plant Diastase.*—Plant diastase has been prepared, for the most part, from the germinating seeds of the cereals, especially barley (Osborne, 1895; Thatcher and Koch, 1914; Sherman and Schlesinger, 1913, 1915; Glimm and Sommer, 1927). Numerous methods for its preparation have been reported, but those used by Sherman and Schlesinger (1915) and Naylor, Spencer, and House (1925) represent the most improved technique for the preparation of diastase from germinating grains. The general procedure is as follows:

A definite weight of the air-dry germinated grain is ground and soaked for 3 hr. in 2.5 times its weight of cold distilled water. The entire mass is then placed in collodion bags and dialyzed against running tap water for 24 hr. at as nearly icebox temperature as possible and then filtered. The filtrate is then treated with solid ammonium sulphate (42 g. to 100 cc. of filtrate) and the precipitate that contains the active diastase is separated by means of a centrifuge, dissolved in a small quantity of water, and dialyzed in collodion bags against running tap water and finally against distilled water until free from sulphates. At the end of this dialysis the sacks may contain a deposit of material which has precipitated with the removal of the dialyzable substances from the solution. This insoluble material, which is without diastatic power, is removed from the solution by centrifuging. The solution is placed in fresh collodion bags and concentrated by evaporation through these bags until the volume is reduced one-half. (This evaporation in the case of rye lowers the activity of diastase and consequently is omitted.)

The solution is then treated with an equal volume of cold alcohol, keeping the temperature below 15°C. The material precipitated by this amount of alcohol is separated by means of the centrifuge and rejected. To the liquid, sufficient cold alcohol is added to make the alcohol content 70 to 80 per cent by volume. The precipitate thus obtained is separated by means of the centrifuge and dried in a partial vacuum over sulphuric acid in the dark, at a temperature not exceeding 15°C. The end product that is obtained from the above procedure is considered to be as nearly pure diastase as any preparation that has so far been made.

In the preparation of diastase after the above manner the greatest destruction of enzyme usually occurs in the first precipitation with alcohol. The losses in the final precipitation and in drying are less serious. In general, considerable losses of diastatic power also occur during dialysis, but in order to eliminate impurities it has not been possible to omit this part of the preparation. The loss of activity is not due in any large measure to the passage of the enzyme as such through the dialyzing membrane, as shown by the tests of the dialysate. Sherman and Schlesinger (1915) considered that this behavior indicates that the diastase is a compound composed of an albumin with a proteose or peptone, as suggested by Osborne (1895). Since this compound is destroyed apparently through hydrolysis on heating in water, it seems probable that a fraction of the enzyme would be hydrolyzed in water solution even at low temperatures. The removal of the products of hydrolysis by dialysis would result in a further hydrolysis of the enzyme and loss of diastatic power without any accumulation of enzyme in the dialysate. If carried out at icebox temperature, the loss of diastatic power is much diminished, and the removal of impurities is little retarded.

Hanes (1932) used the following method for the preparation of diastase from barley. Two hundred grams of barley are washed thoroughly and allowed to soak in distilled water 10 hr. The grain is then spread on filter paper to germinate in a dark chamber. When the plumules have grown to 0.5 to 0.75 of the length of the seed, the grains are placed in a beaker and killed by stirring with a little toluene. They are then ground to a paste. The paste is extracted with 0.1 per cent acetic acid for 6 to 24 hr. with occasional stirring. The coarser material is removed either by squeezing through a fine cloth in a press or by centrifuging. The turbid liquid is filtered until it is almost clear, and the enzyme is precipitated by adding 95 per cent alcohol to make a final concentration of 70 per cent by volume. The liquid is set aside in the icebox for 12 to 24 hr. during which time the precipitate settles. The bulk of the liquid is decanted and the residue filtered on a Büchner funnel through filter paper. The paper bearing the precipitate is ground in a mortar with distilled water, and the paper is removed by filtration. The filtrate containing the enzyme is then made up to a definite volume and toluene is added as a preservative.

Work by McGinty (1919), Baker and Hulton (1929), Schultz and Landis (1932), and Oparin and Risskina (1932) gives special information concerning the preparation of diastase.

Diastase may be prepared from leaves or any other plant parts that contain a relatively small amount of diastase after the following method, which was used by Bartholomew (1914) in his study of the diastase of red algae:

The fresh material to be studied is immersed in 95 per cent alcohol for 20 min. and after freezing from alcohol is plunged into acetone for 10 min. It is then partially dried *in vacuo* and placed in an oven at 35 to 40°C. until dry. The material is then pulverized and placed in three times its volume of 20 per cent alcohol and left standing 18 to 20 hr. and then filtered through cheesecloth and filter paper if possible. To the filtrate is added two and one-half times its volume of 95 per cent alcohol. The precipitate is then filtered off and washed with equal parts ether and alcohol, dried in a desiccator, and pulverized for use. The amount of diastase material obtained in this manner from ordinary vegetative tissue is very small. Thus, for example, Bartholomew from 51 g. of dried algal material obtained only 0.15 g. of diastase preparation.

The presence of diastase in leaves or other plant parts may be demonstrated by using a water extract of the finely macerated tissues. This water extract should be filtered through cheesecloth or other porous cloth but not through filter paper, since it has been found that if the extract is filtered clear it loses much of its diastatic power

(Vines, 1891). Apparently the diastase of the leaves is closely held in the protoplasm (Palladin and Popoff, 1922; Palladin and Popova, 1924) and when freed is adsorbed to a considerable extent by the fine particles of tissue that do not pass through filter paper. The activity of such adsorbed diastase, however, is not impaired by being adsorbed and exerts its full power when in contact with the substrate. In regard to the adsorption of diastase, the work of Ambard (1920, 1921) is worthy of note. He found that it is possible to adsorb, quantitatively, the amylase present in dilute solutions upon raw commercial rice starch and to remove foreign substances without the loss of enzyme. The adsorption by the raw starch is specific, since it has no power to adsorb invertase. When diastase has been adsorbed by starch after this manner, repeated washings with various solutions do not remove the enzyme. Upon the addition of a neutral salt as sodium chloride the enzyme is released by the starch and again goes into solution without any loss of activity.

4. *Study of Diastatic Activity.*—Two methods have been used to determine the activity of diastase: (*a*) Liquefaction. This method measures the power of the enzyme to convert completely a known amount of starch paste into products that no longer give the color reaction with iodine. (*b*) Saccharification. By this method is measured by means of Fehling's solution the amount of reducing sugar produced by the enzyme in a given period of time. These two steps in the action of diastase upon starch are more or less independent of each other, since the conditions for their optimum activity are not the same. For quantitative work the saccharification method is now considered the more reliable.

Various methods have been used for preparing starch for the substrate in the study of diastatic activity, but for the most part they may be grouped under two headings: (*a*) The dispersion of starch in water under different conditions of time, temperature, and pressure but without chemical treatment. A substrate prepared in this manner is commonly called "starch paste." The concentrations that have been used vary from 0.25 to 2 per cent. A common method for the preparation of such a paste is to heat the desired amount of water to boiling and then add the desired amount of starch, thoroughly pulverized and suspended in as small an amount of cold water as possible, to the boiling water and continue to boil for 2 to 5 min. with constant stirring. After cooling, the paste is ready for use. Some investigators discard the starch sediment, which settles out when such a paste is allowed to stand, and use for experimental purposes the more or less clear supernatant liquid. (*b*) Starch that has been subjected to acids or other chemical treatment to render it soluble so that when mixed with water a more or less clear suspension results. Starch treated in this manner may be obtained commercially and is commonly called "soluble starch." The degree of treatment with acids varies greatly, so that while some of the soluble starches react readily with iodine some no longer give that test.

The methods used in determining the activity of diastase are based primarily upon a method devised by Lintner (1885). These methods in general consist in using a series of test tubes each containing from 2 to 10 cc. of a given concentration of raw or soluble starch. To each of the test tubes is added a different amount of diastase extract, generally in a series of 0.1, 0.2, 0.3, 0.4 cc., and so on. After standing under toluene for periods of time ranging from 2 to 48 hr. at 37 to 40°C., 5 cc. of Fehling's solution is added to each tube. The tubes are then placed in boiling water for 10 min. and examined at the end of that time to determine the lowest concentration of diastase in which the copper is all reduced. This may be determined qualitatively by noting the absence of blue color or quantitatively by the usual weighing method. The sample of diastase that produces this degree of conversion into maltose under a given set of conditions is taken as the standard and rated as having an activity of

100. To this standard the activity of diastase under different conditions or from different sources is compared.

The method for determining the degree of diastatic activity by means of iodine may be illustrated by the procedure used by Appleman (1911) in his work with the potato. One cubic centimeter of a 1 per cent soluble starch solution was placed in each of 10 test tubes surrounded by ice. Increasing amounts of the diastase extract were added to the tubes beginning with 1 cc. and increasing the amount 0.1 cc. for each succeeding tube, using toluene as a preservative. The tubes were then incubated at 40°C. for 48 hr., placed again in ice water, and filled nearly full with water, and 3 drops of iodine solution added to each tube. The first tube in the descending series which showed a blue or violet color was considered the index for comparing diastatic activity.

5. *Factors Influencing the Action.*—The literature in regard to the action of the medium upon the activity of diastase is very confusing, and there are almost as many contradictory results as there are investigators. This confusion is doubtless due to two main causes. In the first place, the method by which the diastase is obtained greatly influences its response to its environment (Osborne and Campbell, 1896); and, in the second place, the methods for the determination of the diastatic activity also influence the response of the enzyme to its environment. The nature of the antiseptic used also has an influence upon the activity of the enzyme. Thus Sherman and Wayman (1921) found in low concentrations that chloroform did not affect commercial malt extract, but, when the enzymes were purified, chloroform diminished their action. Toluene had no appreciable effect on either pure or impure diastase preparation, while formaldehyde with a concentration of 0.0000116 *M* gave a 3 per cent reduction of enzymic activity. All diastases studied were retarded in their action by copper sulphate. It is the intention to discuss here only briefly some of the observations that have been made. A detailed account of the work pertaining to this topic may be found in the literature cited at the end of this chapter.

(*a*) *Chemicals.*—Much study has been devoted to the effect of chemicals upon the behavior of plant diastase. In the following discussion, mention is made only of a sufficient number of cases to indicate the general enzymic behavior under various chemical treatments.

(1) *Inorganic Salts.*—Sherman and Thomas (1915) studied extensively the effect of the chlorides, nitrates and sulphates, and primary phosphates of sodium and potassium with reference to the concentrations favoring optimum activity of malt diastase. All the electrolytes mentioned above increased the activity of this enzyme. The activities observed at the optimum concentrations of these activating compounds varied, depending upon the purity of the enzyme preparation and of the starch used, from less than two- to more than thirty-fold the corresponding activities in the absence of the electrolyte. Berczeller and Freud (1922) found that malt diastase was permanently inactivated by iodine, chlorine and bromine, and that starch exercised a protective action against this inactivation, although the enzyme eventually lost its digestive power. According to Haehn and Schweigart (1923), the chlorides of potassium, calcium, sodium, barium, and magnesium activated potato diastase under the conditions of their experiment, while diastatic hydrolysis was delayed by salts of zinc, cadmium, lead, and copper. Salt-free diastase prepared by dialysis and ultrafiltration was entirely inactive, no change occurring during 24 hr. The preparation could be rendered active by the addition of sodium chloride. Ambard (1923) considered that neutral salts in some way act upon the substrate. The most widely accepted hypothesis is that they accelerate the enzymic activity by increasing the colloidal dispersion of the enzyme.

Englis and Lunt (1925) found that diastatic activity in the leaves of nasturtium decreased with increased rates of application of potassium to sand cultures, while an intermediate rate of application of potassium to peat cultures gave the highest activity of diastase. Apparently there was a correlation of diastatic activity with favorable growing conditions rather than with the presence or absence of any element. Doby and Hibbard (1927) found that amylase is strongly activated by the chloride ion and slightly activated by the potassium, nitrate, and fluoride ions. Englis and Gerber (1929) obtained only slight response of diastatic action to the application of acid phosphates to soybeans in pot cultures. Józsa and Gore (1932) noted that sodium chloride and other salts, when added to flour suspensions, greatly increased the solubility of the diastatic, liquefying, and saccharifying enzymes of flour. James and Cattle (1933) observed that the breakdown of starch by potato diastase was accelerated by the application of potassium chloride, while the formation of sugar was not affected.

(2) *Acidity.*—In studying the effects of various acids upon the action of diastase, Sherman and Thomas (1915) found that the acids showed a greater activating influence than the neutral salts, while the acid phosphates of sodium and potassium gave as high activation as free acids. The acids all showed optimum activation in those concentrations which had essentially the same actual acidity. This optimum hydrogen-ion concentration was in each case between pH 4.2 and 4.6. Acetic and propionic acids, in quantity ten times the optimum, decreased the activity about one-half, while hydrochloric, nitric, and sulphuric acids reduced the activity more than one-half when present in concentrations 2½ times the optimum.

Harley, Fisher, and Masure (1931) showed that the juice of the apple contains an effective inhibitor of diastase. This inhibiting substance is soluble in water and alcohol, is acid in reaction, and can be separated from the enzyme by dialysis or by precipitation of the diastase by neutral alcohol.

A large amount of work has been done to determine the optimum hydrogen-ion concentration for diastatic activity. It should be remembered that the apparent optimum reaction may vary to a considerable degree depending upon the temperature (Chrzaszcz, Bidzinski, and Krause, 1925), the buffer used, the accumulation of products formed, and the source of the enzyme. Sherman and Thomas (1915) found that the optimum hydrogen-ion concentration for malt diastase was pH 4.2 to 4.6. Sjöberg (1922) found the optimum acidity for a series of plant diastases to be between pH 5.0 and 5.4. Sherman, Thomas, and Baldwin (1919) found that the optimum pH value for malt diastase was 4.4 to 4.5, taka-diastase 4.8, and pancreatic diastase 7.0. Sherman, Thomas, and Caldwell (1924) found that the isoelectric point of a malt-diastase preparation was pH 4.3 to 4.5, which coincides with that of its optimum enzymic action upon starch. Maslow and Davidson (1926) found that the optimum reaction for the dextrin-liquefying activity of the dextrinase of taka-diastase was pH 4.0 at 34°C. According to Polak and Tychowski (1928), there are present in malt extract two diastases whose optimum pH values are different. α-Diastase works best at pH 5.1, while β-diastase reacts best at pH 4.3. Sjöberg (1922) noted that the optimum pH for sugar formation by diastase was 5.0 to 5.5, while that for the disappearance of starch was 4.0 to 6.0.

Doby and Burger (1932) found in the potato tuber that the optimum pH value for β-amylase was approximately 7.0 under varied conditions. Amylase activity increased during the resting period of 2 months to approximately five times the initial activity. This was partially due to the alteration of the pH value of the tuber and partially because of the formation of new enzyme. The optimum reaction of the medium for the saccharogenic amylase of the leaves of the sugar beet was pH 6.0.

Hopkins, Cole, and Green (1933) found that the optimum pH for the saccharifying action of barley diastase was 4.5 to 4.7. Ohlsson and Uddenberg (1933), and Ohlsson and Edfeldt (1933) found that the β-amylase of oats and rye had an optimum activity at pH 4.3, while the α-amylase had its optimum action at pH 5.5. Tysdal (1934) noted that the hydrogen-ion concentration influenced the protected activity of diastase in alfalfa. This activity increased from 4 mg. of maltase, produced at pH 5.2, to 11 mg. at a pH of 6.7.

(3) *Organic Materials.*—It was found by Brown and Morris (1893) that tannin exercised an inhibiting action on the diastase of leaves and that when it was present in considerable amounts it was difficult to obtain an extract of the leaf with any marked degree of hydrolyzing power. They demonstrated that tannin prevented the diastase from going into solution.

It was discovered by Sherman and his workers (1919 to 1922) that numerous amino acids including glycine, aspartic acid, arginine, tyrosine, cystine, and lysine increased the activity of certain animal diastases. These amino acids also had the power to restore to full activity diastase that had been partially inactivated by copper sulphate and to prevent the deterioration of this enzyme in water. Haehn and Schweigart (1923) noted that glycine, alanine, and leucine activated the diastase of the potato.

Filipowicz (1931) found that ovalbumin, gelatin, peptone, glycine, and aspartic acid retarded the action of malt diastase upon starch at pH values below 4.5 and accelerated it at values above that figure. He attributed these effects to the amino acid units.

Denny (1930) reported that potassium cyanide added to a mixture of potato juice and soluble starch markedly increased the activity of amylase as measured by the gain in reducing sugar.

The mechanism whereby ethylene, thiocyanates, and ethylene chlorhydrin produce artificial ripening or break the dormancy of plants has been sought in the action of these compounds on diastatic activity. Englis and Zannis (1930) found that ethylene had no effect upon the action of corn-meal diastase upon corn starch. They concluded that the effects of ethylene in ripening fruits and vegetables is primarily concerned with color changes rather than with a true ripening process involving marked changes in certain of the food reserves. Denny (1931, 1932, 1933) could find no simple relation between the effects of potassium thiocyanate and ethylene chlorhydrin upon the amylase activity of taka-diastase, malt diastase, pancreatin, pangestion, and of an extract from germinating barley seeds. Since these two chemicals produce almost identical results in the initiation of sprouting of dormant potatoes, the indications are that the changes which occur are not due to their direct effect upon diastase. Clark, Fowler, and Black (1931), however, reported that the hydrolysis of soluble starch by diastase was increased in the presence of ethylene chlorhydrin, potassium thiocyanate, or thiourea.

(*b*) *Temperature.*—The maximal and optimum temperatures for diastatic action will depend upon the source of the enzyme and the nature of and amount of the substrate as well as the presence of the end products. Chrzaszcz (1923) by the use of watery extracts of rye, wheat, oats, corn, buckwheat, barley, broom corn, and millet found that diastases from different sources did not respond to temperature in the same manner at the supposedly optimum hydrogen-ion concentration, *i.e.*, pH 4.9. The optimum temperatures for barley diastase were 49 to 54°C.; for wheat and rye, 49 to 55°; for oats, 51 to 53°; for buckwheat, 50 to 55°; for corn, 56 to 57°; and for millet, 58 to 59°. Barley diastase was completely destroyed at 61 to 62°C., wheat at 63 to 64°, and others at 65 to 66°. Sjöberg (1922) found that the temperature coefficient for sugar formation by diastase decreases between 20 and 40°C., while the temperature coefficient for starch disappearance is constant between those two points.

Fisher, Harley, and Brooks (1930) considered that "water core" is a result of the nonuniform conversion of starch induced by relatively high atmospheric temperatures.

(c) *Light.*—It was observed by Brown and Morris (1893) that the amount of diastase in leaves increased when they were placed in the dark. They considered that the secretion of diastase is stimulated by the absence of soluble carbohydrates. Sjöberg (1923), however, in experiments with the leaves and blossoms of the tulip, concluded that the formation of diastase is independent of light. Pincussen (1923) found that the reaction of light from a quartz lamp on malt diastase depended on the concentration of the enzymes, on accompanying compounds, and on the reaction of the medium. The greatest injury from light occurred at the optimum reaction for enzyme activity. Green (1894) exposed portions of a preparation of malt diastase to light for periods varying from 2 to 10 days and tested the action of these exposed preparations upon starch paste. He concluded that light, both solar and electric, exercises a destructive influence upon diastase, the deleterious influence being confined to the rays of the violet end of the spectrum, the other rays being slightly favorable instead of destructive. He considered also that the coloring matter of the barley husk acts as a screen in preserving the diastase in that grain from the destructive effects of light. The destruction by light of diastase in the leaves is only 15 to 20 per cent as compared with 68 per cent for malt diastase under the same conditions. Green (1897) considered that either the diastase is more resistant in the leaves or there is some constituent that has the power of screening the enzyme from the deleterious rays. Chlorophyll might serve this purpose, and protein is known to protect the enzyme in this regard. Green considered that when a leaf is illuminated by the whole spectrum, the beneficial rays convert the zymogen into diastase and the latter is speedily destroyed by the deleterious rays. The latter action is the more rapid so that the effect of the whole spectrum is to diminish the amount of diastase in solution.

Baly and Semmens (1924) in experiments with the action of diastase upon the starch of potato, wheat, and corn obtained results which indicate that plane-polarized light exercises a powerful acceleration on the hydrolysis of starch by weak solutions of diastase. Later (1925, 1926) they claimed that starch grains in pure distilled water may be hydrolyzed by a beam of polarized light and that after 2 to 4 days' exposure the etching on the starch grains may be observed. Macht (1928) substantiated their results when he found that polarized light hastened the conversion of starch into sugar in the presence of diastase. Macht and Hill (1925) noted also that when a suspension of *Saccharomyces cerevisiae* was added to solutions of sucrose the fermentation of sugar proceeded much more rapidly in polarized than in nonpolarized light. The work of Baly and Semmens (1924), however, has been adversely criticized by Jones (1925), who considered that the results reported for the action of polarized light on starch grains are doubtful.

The effects of polarized light and ordinary light upon the action of diastase have been studied by Navez and Rubenstein (1928, 1932), and Navez (1930). They reported that there is present in a mixture of starch and diastase a substance that partially inhibits hydrolysis. This substance is equally sensitive to ordinary and to plane-polarized light. The addition of light during hydrolysis increases the rate of liberation of reducing substances over that of the same reaction in the dark. This inhibitor is linked to the diastatic complex, as is evidenced by the fact that irradiation of starch alone does not produce the same effects as irradiation of the total mixture.

Bunker and Anderson (1928) studied the effects of polarized light on the hydrolysis of starch by taka-diastase, using intact starch grains, swollen starch grains, broken grains, and soluble starch from numerous sources. In all these cases, they obtained negative results. Semmens (1932) found in a vine leaf that the starch disappeared from that portion which was exposed to moonlight, while it remained intact in the protected part. He considered that the depletion of the starch in the exposed portion was due to the influence of polarized light of the moon on diastatic activity.

Hutchinson and Ashton (1933) found that full irradiation from a mercury-arc light retarded the dextrinogenic and saccharogenic activity of diastase directly with the intensity. The monochromatic effects on malt diastase were generally inhibitory for the dextrinogenic phase and stimulatory for the saccharogenic phase of diastatic activity.

c. The Inulases.—Inulase is an enzyme that hydrolyzes inulin to levulose or fruit sugar. It was named by Green (1887), who first observed it in the germinating tubers of the Jerusalem artichoke (*Helianthus tuberosus*) and later (1899) found it in the bulbs of *Leucocrium* and *Scilla.* Green could not find inulase in the resting stage of the tubers and bulbs, but if the macerated tissue was kept at 35°C. for a time, the enzyme was found to be present. It thus appears to occur in the resting tissues in the zymogen form. Inulase may occur with diastase but has no action whatsoever upon starch. Green found that inulase works most advantageously in a neutral or very faintly acid medium, 0.001 per cent hydrochloric acid being very favorable, but higher concentrations than this were very destructive. An alkali medium was extremely deleterious. These observations in regard to the medium have been confirmed by Kizel (1915). The optimum activity of inulase occurs at 50 to 55°C. but loses its activity when heated to 70 to 75°. Wolff (1916) found in the roots of *Dahlia variabilis* and *Cichorium intybus* a substance that coagulates very actively the juices extracted from these plants and precipitates inulin from the solution.

In 1893 Bourquelot found inulase in the fungi *Aspergillus niger* and *Penicillium glaucum*. It has since been studied in these plants by Dean (1903) and Young (1918). It occurs in abundance in these fungi and most of our knowledge concerning inulase has been derived from these sources. Dean (1903) verified the work of Bourquelot (1893) and found that the enzyme from these fungi reacted in much the same manner as the inulase of the Jerusalem artichoke described by Green (1887).

d. The Glucases or Maltases.—The enzyme that hydrolyzes maltose to glucose has been termed "glucase" or "maltase." The reaction occurs after the following manner:

$$\underset{\text{Maltose}}{C_{12}H_{22}O_{11}} + H_2O \xrightarrow[\text{of maltase}]{\text{In presence}} \underset{\text{Glucose}}{2C_6H_{12}O_6}$$

The occurrence of a maltose-splitting enzyme in plants was first suggested by the work of Brasse (1885). He noted that, when a preparation of malt diastase was allowed to act on starch at 32 to 42°C., glucose was formed, but that, when the action occurred at 50 to 57°, only maltose was formed. Brasse, however, did not realize that he was working with a special enzyme capable of digesting maltose. According to Daish (1916), maltase was first observed in corn seeds and seedlings by Cuisinier (1885), who termed it "glucase" and patented a process of converting starch into glucose based on its use. He stated that this enzyme is present in the organs and seeds of a large number of plants, a statement which is now known to be correct but which could not be substantiated by many subsequent workers for reasons that will be mentioned later in this discussion. Observations by Beyerinck (1895), Huerre (1909), Wierzchowski (1913), and Daish (1916) have substantiated the statement of Cuisinier that maltase is widely distributed in plants. It has been demonstrated in the leaves of a large number of plants and in the resting seeds and seedlings of corn, rice, millet, oats, wheat, barley, rye, buckwheat, and others. Maltase is also present in many of the fungi, and the commercial taka-diastase contains this enzyme in larger amounts than does germinating barley.

Maltase has three or four characteristics that are considerably different from other carbohydrate-splitting enzymes, and it is due to these characteristics that many investigators failed to detect maltase in the plant material with which they worked. The optimum temperature for activity is from 38 to 40°C., but above that temperature the maltase is rapidly destroyed. Much of the plant material that has been examined for maltase was dried at a temperature of from 40 to 50°C. so that the enzyme was completely inactivated in the preparation. This also explains the reason for the observation of Brasse (1885) that a malt-diastase preparation at 32 to 42°C. formed glucose when acting on starch, while at 50 to 57°C. only maltose was formed. The maltase obtained from different sources varies widely in its temperature range, which indicates that there are several varieties of the enzyme (Huerre, 1909). Maltase is practically all destroyed by precipitation by alcohol, a treatment that injures only to a limited degree most of the enzymes. It is very difficult to obtain this enzyme from plant tissue by extraction with water. The enzyme is an endocellular one, and unless the material is extremely well pulverized very little of the enzyme is released. The best action of maltase has been obtained by using the pulverized powder of the seeds or the crushed pulp of the vegetative portions of the plant. It is very sensitive to chloroform and shows only a slight action in the presence of that antiseptic. Toluene has been found to be very satisfactory as a preservative in studying this enzyme.

e. The Cellulases or Cytases.—The enzymes that are capable of dissolving the cellulose walls of plants have been termed "cytases," "cellulases," or "cytohydrolytic enzymes." The existence of cytase was first noted by DeBary in 1886 when he observed that the hyphae of the fungus *Peziza sclerotiorum* were able to penetrate the cell walls of plants. In 1888 a similar enzyme was reported by Ward for a fungus belonging to the genus *Botrytis.* Since then similar enzymes have been found in numerous green plants and in other fungi. The action of cytase is relatively slow and is confined entirely to normal cellulose and to hemicelluloses. In all the cases that have been observed the cell wall at first becomes hyaline, then gradually becomes more and more transparent, and finally fades away into solution. The cytases are destroyed at a temperature of 60°C. or above.

Brown and Morris (1890, 1898) noted in the early stages of the germination of barley that the parenchymatous cell walls of the endosperm directly under the epithelium cells of the scutellum and the cells of the aleurone layer are disintegrated and ultimately dissolved. This dissolution of the cell walls always precedes any visible action upon the starch grains contained within these cells. This dissolving action they attributed to cytase secreted by the epithelial cells of the scutellum and aleurone layer. They were, however, unable to obtain any action of this enzyme upon the cellulose of the date, asparagus, coffee, or onion seeds, but in the case of the Jerusalem artichoke, potato, carrot, and turnip this enzyme preparation from malt dissolved cell walls. The most thorough work that so far has been done on cytase is that of Newcombe (1899). He prepared a water extract of the seedlings of barley and *Lupinus albus* and the embryos and endosperm of the date seed and prepared cytase by precipitation from the extract after the manner used in the preparation of diastase. He used the following method to determine the action of cytase: Sections were made of the tissues to be examined and were freed from starch if present by treating for 24 hr. with saliva or a pancreatic preparation. The sections were then placed in a solution containing 150 mg. of the enzyme preparation to 10 cc. of water with chloroform as an antiseptic and allowed to stand at 32 to 34°C. The action of the cytase was determined by microscopical examination. In the cell walls of the endosperm of barley, the cytase from various sources dissolved them down to the middle lamella in from 5 to 10 hr., but the cell walls of the aleurone layer were more resistant. In the cell walls of the white lupine, 9 days were required to reduce the cell walls to a hyaline condition, while complete dissolution was not effected even in thin sections until 3 or 4 weeks and then only at the edges of the sections. The enzyme prepared from the embryo and endosperm of the date was capable of digesting the heavy cell walls of the endosperm cells of that seed. After 30 days the whole surface of the section was corroded and the exposed walls frayed out. All the walls became hyaline and the whole section became smaller owing to the gradual dissolving action. Newcombe found that the cytase preparations obtained from the seedlings of the date, white lupine, barley, and *Aspergillus oryzae* were apparently not specific so far as could be observed.

f. Other Carbohydrases.—Some of the carbohydrases that have been found in plants, though in most cases playing an unimportant role in the life of the green plant, are (1) melibiase, which hydrolyzes melibiose to glucose and galactose; (2) gentianase, which hydrolyzes gentiobiose to glucose; and (3) pectinase, an enzyme capable of hydrolyzing pectose to arabinose.

2. The Glucosidases of Plants.—Before discussing the action of the enzymes of this group, a few statements should be made concerning the compounds upon which they act. The glucosides for the most part are compounds that contain glucose as the characteristic basic group. Upon hydrolysis the glucosides yield glucose, together with one or more other substances, usually of an aromatic nature. Some of the organic constituents that may be combined with sugar to form glucosides are phenols, alcohols, aldehydes, acids, and mustard oils. Most of the common glucosides are derived from *d*-glucose. Some, however, are derivatives of galactose or rhamnose, but in others the exact sugar that is combined in the glucoside is not known. Some of the glucosides are amygdalin found in the almond and other plants of the *Rosaceae;* sinigrin in the *Cruciferae;* salicin in the willow; arbutin in the leaves of bearberry; phlorizin in the bark of apple, pear, cherry, and plum; populin in the bark of the poplar tree; esculin in the bark of horse-chestnut; phytin in the coats of seeds; and saponins, which occur in more than 400 species of plants.

In this connection some mention should be made of the tannins, although they are not at present classified as glucosides. Tannins are a group of substances that are almost universally distributed in plants. They occur in leaves, wood, bark, unripe fruits, walnut hulls, and many other plant parts. They are especially abundant in the pathological plant growths known as "galls" and their dry weight may be composed of from 40 to 75 per cent of this group of substances. Tannins make up about 15 per cent of the dry weight of tea leaves and about 40 per cent of the dry weight of oak bark.

The tannins are amorphous, have an astringent taste, give ink colors with ferric salts, precipitate gelatine from solution, and form insoluble compounds with it and are precipitated from solution by copper or lead acetate, stannous chloride, and other metallic salts and by a strong aqueous solution of potassium bichromate. They are readily soluble in hot water and form solutions that are in reality colloidal gels.

Tannins are usually hydrolyzed by acids into a variety of products, one being nearly always *d*-glucose and the other a hydroxy derivative of the aromatic series. Although they usually contain sugar, they do not show the characteristic properties of the true glucosides and are considered as a separate class of substances.

The tannins are generally considered as waste products, rather than as substances that can be utilized in the nutrition of the plant, for the following reasons: (*a*) The known uses of tannin in the life of the plant are very few and limited as compared with its abundance and almost universal occurrence. (*b*) It is present in great abundance in the bark, which will be eventually lost from the plant. (*c*) It is produced in large amounts in unfolding leaves, in germinating seeds, in the growth of galls, and in fruit development where intense metabolism is going on. It seems certain that in these cases tannins are the result and not the causative agents of the very rapid growth and changes that take place in these regions.

Tannins are of frequent occurrence in the green fruits and give to them their characteristic puckery taste. As the fruit ripens, the tannins may disappear or, as in the persimmon, remain in the ripe fruit encased in special large cells or sacks with impermeable membranes so that when the fruit is eaten by animals the tannins do not affect the organs of taste and thus cause no disagreeable effects.

Michel-Durand (1928 to 1932) concluded that the free tannins may be utilized by the plant, but that the great variation in their structure and composition precludes the possibility that all members of the group serve the same purpose. In the young etiolated plants of *Castanea*, the tannins arise from carbohydrates stored in the seed, while the tannins in the leaves and branches of *Aesculus* come from the disintegration of proteins. Hence, no single explanation of the origin of tannins can be given. Light in general favors their formation and may influence not only the quantity but the nature of the tannins that are produced. Baens, Yenko, West, and Curran (1934) noted that the tannin content of the bark of trees varied from 5 to 45 per cent, and that there were marked variations in the amount of tannin in the same species grown under different conditions.

a. Emulsin.—The name "emulsin" was first given to the enzyme that hydrolyzes amygdalin to benzoic aldehyde, hydrocyanic acid, and glucose. This reaction proceeds after the following manner:

$$\underset{\text{Amygdalin}}{C_{20}H_{27}NO_{11}} + 2H_2O \xrightarrow[\text{of emulsin}]{\text{In presence}} \underset{\substack{\text{Benzoic}\\\text{aldehyde}}}{C_7H_6O} + \underset{\substack{\text{Hydrocyanic}\\\text{acid}}}{HCN} + \underset{\text{Glucose}}{2(C_6H_{12}O_6)}$$

This enzyme has been the most thoroughly studied of any of the glucosidases. It has been found that it will also hydrolyze phlorizin, salicin, arbutin, or any of the β-glucosides. Emulsin has been found in the almond, cherry laurel, in certain of the *Euphorbiaceae*, in certain lichens, and in the fungi *Aspergillus niger* and *Penicillium glaucum*. According to Green (1899), the enzyme is most active at a temperature of 30 to 40°C., but its power gradually declines above that and is entirely destroyed at 80°C.

According to Robinson (1930) hydrocyanic acid is known to occur in about 50 natural orders of plants. McNair (1932) stated that cyanogenetic glucosides are found in 13 per cent of the plant families of the gymnosperms and angiosperms. It is generally considered to exist exclusively in glucosidal combination in the living plant, although some workers have postulated the existence of free cyanide in plant tissues. Ten cyanophoric glucosides have been isolated in crystalline form from plants and of these seven are derivatives of benzaldehyde cyanhydrin.

In plants, the cyanophoric glucosides are, with few exceptions, accompanied by active hydrolytic enzymes capable of liberating cyanide. In general, the concentration of these glucosides is highest in the young growing organs, but in some instances it is highest in the bark.

In the *Rosaceae*, the cyanophoric glucosides seem to be constant constituents of the plant throughout its life cycle, while in the grass family they disappear as maturity approaches and are not found in the seed. In *Amygdalus communis*, *Vicia angustifolia*, and *Phaseolus lunatus* the concentration of hydrocyanic acid in the plant has been diminished by cultivation. Climatic conditions, especially drought, cause variations in the content of *d*-hurrin in *Sorghum vulgare* and of the cyanophoric glucoside of *Lotus corniculatus*. Temperatures below freezing cause a rapid increase in the amount of cyanophoric glucosides in *Sorghum vulgare* and *Prunus laurocerasus*. The function of hydrocyanic acid in the plant is unknown. Its rapid disappearance under conditions of starvation in some plants lends support to the view that such nitrogen is readily utilizable by the plant. In some cases, the cyanophoric glucosides are apparently excretory products and may serve as protective substances.

It is common knowledge that the ordinary sorghums and other species of *Andropogon* form hydrocyanic acid in their tissues following the breakdown of glucosides that apparently occurs under unfavorable conditions of growth. *d*-Hurrin ($C_{14}H_{17}O_7N$) is the cyanophoric glucoside in sorghums and on hydrolysis yields hydrocyanic acid and *p*-oxybenzaldehyde (Harris and Goss, 1934). The stem and leaves contain this glucoside, but none is present in the roots. Slade (1903) found most of it in the stems, while Acharya (1933) in India found most of it in the leaves. Apparently the conditions under which the plants are grown determine the location of the major portion of this glucoside. The hydrocyanic acid content decreases from the early stages of growth to maturity. In sorghums, the seedlings up to 40 days of age, plants stunted by drought, second-growth tissues, and plants directly after being frozen (Swanson, 1921) have the highest content of hydrocyanic acid. The largest amounts may be from 0.1 per cent to 0.2 per cent of the dry matter of the plants. The content of hydrocyanic acid appears to be the lowest in the morning and to increase to a maximum in midafternoon. The hydrocyanic acid rapidly disappears from the sorghum plants after they are harvested and cured.

b. Maltase, Phytase, and Rhamninase. 1. *Maltase.*—According to Thatcher (1921), maltase acts on all α-glucosides but does not produce the slightest effect upon β-glucosides.

2. *Phytase.*—The glucoside phytin is decomposed into phosphoric acid and inosite ($C_6H_{12}O_6$), by the action of the enzyme phytase.

3. *Rhamninase.*—According to Green (1899), the glucoside xanthorhamnin, which is found in the seeds of *Rhamnus infectorius* (Persian berry), is hydrolyzed into rhamnin and glucose by an enzyme that has been termed "rhamninase."

c. Myrosin.—This name has been given to the enzyme that acts upon the glucoside, sinigrin, or myronate of potassium, decomposing it into sulphocyanate of allyl or mustard oil, glucose, and potassium hydrogen phosphate. The reaction occurs after the following manner:

$$\underset{\text{Sinigrin}}{C_{10}H_{18}NKS_2O_{10}} \longrightarrow \underset{\text{Sulphocyanate}}{C_3H_5CNS} + \underset{\text{Glucose}}{C_6H_{12}O_6} + \underset{\text{Potassium hydrogen sulphate}}{KHSO_4}$$

Myrosin not only acts on sinigrin but has the power to decompose practically all sulphur-containing glucosides that have been examined.

According to Sandberg and Holly (1932), myrosin consists of two components, a glucosidase and thiosulfase.

3. The Esterases or Lipases of Plants.—The enzymes of this group accelerate the cleavage of all the simple esters of the trihydroxy alcohols and fatty acids as well as many synthetic esters. These enzymes are frequently called "fat splitting" because they hydrolyze the fats and oils to glycerin and fatty acids. The term "lipase" has been almost universally used in a collective sense in speaking of the enzymes of this group. Recent work, however, has shown that the ordinary preparations of lipase contain at least two fat-splitting enzymes, which differ in their solubilities and rate of action. It has also been observed that, although lipases from different sources show striking similarities in their behavior (Barton, 1920), they at the same time show minor differences which indicate that they are not identical enzymes. Thus, Longenecker and Haley (1935) noted that the lipase of the castor bean exhibited a differential rate of digestion of various oils, the decreasing rates of hydrolysis being in the following order: peanut, castor bean, corn, cotton seed, soybean, rape, olive, and linseed oils.

The presence of a fat-splitting enzyme in plants was suggested by the observations of Pelouze (1855) on the seeds of flax, mustard, and rape. Lipase, however, was first proved to exist in plants by Green (1890) in working with the germinating castor bean. Sometime later Sigmund (1890, 1892) found it in both the resting and germinating seeds of the rape, opium, poppy, hemp, flax, and corn. These observations were later confirmed and extended by Connstein, Hoyer, and Wartenberg (1902) and Hoyer (1904) who found that some oily seeds, at least, contain the enzyme whether germinating or not, and that an enzyme obtained from a given seed is capable of hydrolyzing fats and oils from other sources. Theis, Long, and Brown (1929) found lipase in the developing seed of flax and noted that its activity decreased as the seed matured. Sullivan and Howe (1933) obtained this enzyme from the wheat grain and its various products.

a. Preparation and Study of Lipase.—Numerous studies have been made concerning the preparation and action of this enzyme, but they have been confined almost entirely to the lipase of the castor bean (*Ricinus communis*). This is probably due to the fact that the enzyme from this source is easily prepared, very stable, and relatively very active. To understand the methods used in the preparation of lipase, a few facts concerning the nature of the enzyme should be considered. Lipase in the castor bean occurs apparently in the zymogen stage. This zymogen appears to be somewhat soluble in fats or in a mixture of fat and ethyl ether but is insoluble in ether alone (Taylor, 1906). The lipase zymogen is activated by acids. According to Haley and Lyman (1921), the active enzyme is unstable and is rapidly destroyed in an acid medium in the absence of fats. In the presence of fats, however, the enzyme shows much greater stability. The lipase found in the seeds of *Chelidonium majus* does not exist in the zymogen state, so that it is not necessary to treat with an acid in order to activate it. The method used by Haley and Lyman (1921) in preparation of the lipase of the castor bean may be taken as an example of the procedure generally followed in the preparation of the enzymes of this group. Fifty grams of hulled castor-bean kernels are ground as finely as possible, mixed with 150 cc. of 1 per cent acetic acid, and allowed to stand for 3 days at room temperature (the acid treatment varying from 15 min. to several days with other investigators, depending on the concentration and types of acid used). The mixture is then filtered, washed with water, and dried at room temperature, after which it is extracted in a Soxhlet apparatus with petroleum ether of a low boiling point. The extracted material is passed through a 40-mesh sieve and again extracted if necessary to completely free it from oil. The extracted material may then be dried in vacuum or over any drying agent in a desiccator, and the light fluffy powder thus obtained may be kept indefinitely in the dry state.

Longenecker and Haley (1935) prepared lipase from the castor bean in the following manner: The beans were hulled, macerated, and the oil removed by extraction with petroleum ether having a boiling point between 20 and 40°C. The residue that remained after the extraction was pulverized sufficiently to pass through a 60-mesh sieve.

It has been observed by Falk (1913, 1917) and Falk and Sugiura (1915) that two lipases are present in a lipase preparation that has been obtained after the manner mentioned above. One of these is soluble in water and the other insoluble. The former has, comparatively, a greater action toward ethyl butyrate than toward triacetin, while the latter has a greater action toward triacetin than toward ethyl butyrate. The water-soluble portion of such a lipase preparation has been termed "esterase" by Falk, and the other portion has been termed "lipase." He obtained the esterase from a dialyzed and filtered water extract of oil-free castor-bean powder, 0.5 g. to 60 cc. of water. The lipase preparation was obtained by extracting the water-extracted castor-bean powder with a 1.5 *N* sodium chloride solution, 1.0 g. of material to 100 cc. of solution, and dialyzing until salt free. After such dialysis the lipase is in a suspension in the water. Both the "esterase" and "lipase" preparations act on the higher glycerides, but the latter is somewhat more active than the former. Falk (1915) found that soybeans contain apparently the same lipases as castor beans, the only difference being that the reaction of the castor-bean lipase was somewhat more intense than that obtained from the soybean.

The methods of studying the action of lipases have been numerous, and the differences in results obtained by different investigators may be attributed, in part at least, to this cause. Since the action of lipase on fats and oils is relatively slow, many investigators have used ethyl butyrate, triacetin, and other simple esters as a substrate in their study of the action of this enzyme. From the standpoint of plant physiology, however, the study of the hydrolyzing action of the lipases on the fats and oils is more desirable, since these compounds are the ones normally acted upon by these enzymes in the digestive process in plants.

The method used by Haley and Lyman (1921) and Longenecker and Haley (1935) in studying the action of lipase is as follows: Place 0.1 g. of lipase powder, 1.0 g. olive oil, and 0.6 cc. of hydrochloric acid, varying in concentration from 0.02 to 0.2 per cent, in a test tube. The contents are thoroughly mixed and placed in the incubator at 38°C. for 24 hr., using toluene as an antiseptic. The order in which the ingredients are added in making up the digestive mixture is of much importance. Thus Haley and Lyman found that if the acid is added to the enzyme material before the oil, complete loss of the lipase is apt to occur. After the period of incubation the mixture is washed from the test tube with 50 per cent alcohol into wide-mouth titration bottles. The resulting mixture is then titrated with 0.1 *N* sodium hydroxide in 50 per cent alcohol using phenolphthalein as an indicator. The degree of hydrolyzation of the oil is determined by the amount of sodium hydroxide required for the neutralization of the fatty acids formed in excess of that required to neutralize the controls.

Nicloux (1904 to 1906) made extensive studies concerning the lipolytic activities of the castor bean in a manner quite different from that pursued by other investigators. By mechanical means he separated the cytoplasm of the ground material from the aleurone grains and other cell contents. The cytoplasmic material extracted in this manner had the property of hydrolyzing fats. It acted on the fats in the same manner as enzymes and followed all the laws of enzyme action. He termed the active matter in the cytoplasmic material "lipaseidine" and claimed that it was destroyed by water as soon as it was no longer protected by fats. It seems very probable that Nicloux was working with the lipase of the castor bean, which had been obtained in a different manner from that which is generally followed.

Kirsh (1935) found that the molds, *Penicillium oxalicum* and *Aspergillus flavus*, when grown on a bran-soybean medium, produced a water-soluble enzyme capable of hydrolyzing olive oil. The production of this lipase was greatest when the fungus was incubated at 28°C. for 3 to 4 days during sporulation.

b. Factors Influencing the Action.—The influence of the acidity of the medium upon the action of lipase has been extensively studied. The early investigators on this subject noted in the germination of oily seeds that the hydrolysis of oils took place slowly at first and then increased rapidly. This led to the conclusion that for rapid hydrolysis by lipase a certain amount of free acid is necessary. This conclusion was verified in many cases, since when a small amount of free acid was added at the beginning of an experiment the hydrolysis of the fat was rapid from the first and was soon completed. Considerable contradiction, however, existed in the literature in regard to the effect of acids upon the hydrolysis of fats and oils by lipase. These contradictions are due primarily to the failure of the investigators to take into account the two following facts: (1) that lipase in the seeds in many cases exists in the zymogen condition and (2) that after the zymogen is once activated the enzyme is rapidly inactivated by acids unless oil is present. Armstrong (1905) and Armstrong and Gosney (1913) found that any acid is effective in activating lipase, but the enzyme prepared with a weak acid such as acetic is distinctly superior to that obtained by the use of a stronger acid. They found that aspartic, glutamic, and succinic acids were very efficient in activating lipase preparations. In this regard it is also of interest that Falk and Nelson (1912) found that glycine, phenylalanine, and other amino acids as well as polypeptides aided the hydrolysis of methyl acetate, ethyl butyrate, and olive oil and that this action was to a certain extent selective in character. The above observations are significant, since the compounds mentioned occur in relatively large amounts in germinating seeds at a very early stage.

The effect of various concentrations of hydrochloric acid on the hydrolysis of olive oil by castor-bean lipase that had been activated in its preparation by 1 per cent acetic acid is shown by the following data, which were taken from the work of Haley and Lyman (1921). In this experiment 0.1 g. of lipase powder, 1.0 g. of olive oil, and 0.6 cc. of hydrochloric acid were mixed and incubated for 24 hr. at 38°C.

HCl 0.6 cc., per cent	Extent of hydrolysis NaOH used in titration, cubic centimeters	Fat hydrolyzed, per cent
0.00	1.0	3
0.02	2.2	6
0.04	16.5	49
0.06	18.2	54
0.08	24.6	73
0.10	24.6	73
0.12	27.2	79
0.14	30.6	90
0.16	29.7	88
0.18	28.3	84
0.20	27.6	81

They found that the optimum pH for castor-bean lipase activity was 5.0 and that lipolytic activity ceased entirely at pH 3.0. Tanka (1910) considered that the lipase of the castor-oil bean is contained in the resting seed as a zymogen and that the acid, found by some to be necessary for the activity of the extracts, is of use only for the conversion of the zymogen into enzyme, which in reality is most active in a neutral medium. In this regard the observations of Falk and Hamlin (1913) in respect to the effect of manganous sulphate on the acceleration of the activity of castor-bean lipase preparation should be noted. They considered that the acceleration of lipase activity is due to the fact that the inactive zymogen of lipase in the castor bean is converted into the active enzyme by an oxidation reaction for which the presence of an "oxygen carrier" or catalytic agent is necessary. Thus manganous sulphate does not accelerate directly the action of the enzyme present, but, by aiding the conversion of the inactive zymogen into active enzyme, it increases the total amount of enzyme present or replaces the enzyme which may become inactive on account of hydrolysis. It was shown by Falk (1913) that solutions of methyl and ethyl alcohol and acetone exerted an inhibitory action on the hydrolysis of ethyl butyrate by castor-bean lipase preparation, the amount of inhibition increasing with the concentration. Solutions of glucose and glycerol showed no inhibiting effects except in concentrated solutions. As compared with aqueous solutions, decreased activities of lipase were shown in the presence of all the univalent salts of the chlorides and nitrates of barium, calcium, and magnesium, of sodium oxalate, and of dilute solutions of sodium sulphate. Increased activities were shown by lipase in the presence of dilute solutions of the chlorides of barium and calcium, of more concentrated solutions of sodium sulphate and magnesium sulphate, and of the chlorides and sulphates of manganese. Theis, Long, and Brown (1929) noted that the activity of the lipase of the flax seed was retarded by the presence of potassium iodide, potassium chloride, potassium nitrate, sodium chloride, lithium chloride, and magnesium sulphate.

Armstrong and Gosney (1913) considered that the fatty acids and glycerol formed in the hydrolysis of oil by lipase inhibit the interaction of enzyme and oil, but there are few data in this regard.

In this connection the reverse action of lipase on its substrate—the formation of glycerides from the glycerol and acids—should be considered. Taylor (1906) using triacetin as a substrate considered that this reverse action of castor-bean lipase is one of very low velocity and may be disregarded in ordinary tests. The hydrolytic reaction of lipase, however, is never complete and there is a definite point of equilibrium in the reaction. The synthetic action of lipase has been discussed in Chap. X.

The enzyme lipase when prepared in the powdered form is very resistant to heat and may be heated when dry to 100°C. without injury. Falk (1917) found that when lipase extractions in water were heated to 100°C. for a few minutes they were completely inactivated. The same preparations when heated dry at 100 to 110°C. lost from 50 to 80 per cent of their activity. The temperature range of lipase activity is about 23 to 42°C. with the optimum temperature in the vicinity of 38°C.

The degree of the emulsification of the fats and fatlike materials is an important one in determining the lipolytic action of lipase. The lipase powder of castor bean and soybean is apparently effective in bringing about emulsification. The addition of ether or petroleum ether to certain fatty mixtures has been found greatly to facilitate the hydrolytic action of lipase, due, evidently, to the solution of the fats. The amount of water present in the digestive mixture is also an important factor. The rate at which interaction takes place is dependent presumably on the conditions at the colloid surfaces, and Armstrong and Gosney (1913) considered that in the case of lipase this interaction must be supposed to take place at or between the surfaces separated, at the most, by a thin film of water. Falk (1917) considered that the inactivation of esterase and lipase preparations by acids, bases, neutral salts, alcohols, acetone, esters, and heat leads to the hypothesis that the active enzyme grouping in these substances possesses the enol-lactim structure —C(OH) = N— which becomes inactive by tautomerization to the keto-lactam structure, —CO—NH—.

4. The Proteases of Plants.—The proteases of one type or another are apparently universally distributed in all plants and plant parts. As a general rule, they are more difficult to prepare for study than the carbohydrate enzymes, and their action proceeds much more slowly. Even the most active of the proteolytic enzymes require from several hours to several days to produce an appreciable amount of digestion products under laboratory conditions as contrasted with only a few minutes required by diastase or invertase to transform starch into sugar or cane sugar into the invert form. The stage to which the proteolytic action has proceeded as well as the amount of digestion products formed is also difficult to determine, so that much of the work that has been done with plant proteases has been only qualitative in character.

The common qualitative methods that have been the most generally used are to test the enzyme mixture for tryptophane, to examine it for leucine or tyrosine crystals, to test for peptone by the biuret color reaction, or to note the general dissolving action of the enzyme. Some of the quantitative methods that have been used are (*a*) the determination by weight of the amount of coagulable protein that remains in the solution, (*b*) the determination of the total nitrogen of the digestion products, (*c*) the measurement of the increased amino nitrogen by the Van Slyke method, (*d*) the determination of the increase in the electrical conductivity, (*e*) the determination of the changes in optical activity, (*f*) the determination of the extent of protein hydrolysis from the number of free carboxyl groups formed (Sörensen's method, 1908), and (*g*) the use of the biuret and tryptophane color reactions in a quantitative way by noting the differences in color intensities. Sherman and Neun (1916) considered that the determination of the total nitrogen or the amino nitrogen of the digestion products is the most accurate and delicate means of measuring proteolysis. Some of the methods for detecting the degree of proteolytic action will be discussed in detail when some of the enzymes are considered.

a. Historical.—Vines (1909) considered that the delivery of Sir Joseph Hooker's presidential address at the British Association in 1874 and the publication of Darwin's book on "Insectivorous Plants" in the following year may be taken as the starting point of the scientific investigations of proteolysis in plants. Hooker found that if cubes of the white of eggs were immersed in the pitchers of *Nepenthes* for 24 hr. their surfaces were more or less dissolved or gelatinized and pieces of fibrin totally dissolved. He was not certain, however, whether this action was due to the juices secreted by the pitchers or to the action of bacteria. Tait (1875), however, considered that digestive enzymes of a peptic nature occurred in these plants and prepared the enzyme in crude form.

Von Gorup-Besanez (1874, 1875, 1876) appears to have been the first to examine seeds for proteolytic enzymes. Using the glycerin method of extraction, he obtained protease preparations from the resting seeds of hemp, flax, and vetch and from germinating barley. He failed to find enzymes of this type in the ungerminated seeds of barley, lupine, and corn. In 1887 Green found a proteolytic enzyme in the seeds of lupine. This enzyme worked in an acid medium and converted fibrin into peptone, leucine, and tyrosine. He considered that it existed in the seed in the form of a zymogen and was transformed into the active state by acid. In 1890 the same author discovered that a tryptic enzyme was developed in the castor bean during germination. This enzyme was capable of splitting fibrin with the formation of peptone and crystal bodies including tyrosine. He considered that this enzyme also existed in zymogen form in the seed. Neumeister (1894) found positive tests for proteases in barley, beet, poppy, corn, and wheat seedlings but negative tests in many others. Fernback and Hubert (1900) and Weis (1900) showed that the germinating barley of malt contained a proteolytic enzyme. Finally, the work of Fermi and Buscalioni (1899), who were the first to make a systematic examination of a large number of plants, showed that proteases occur in a great diversity of plants and plant parts and give strong evidence to the present conception that proteases are of almost universal occurrence in plants.

b. Kinds.—The proteolytic enzymes are generally grouped into three classes: (1) the pepsins, (2) the trypsins, and (3) the erepsins.

1. *The Pepsins or Peptases.*—A pepsin or peptase is a proteolytic enzyme that hydrolyzes protein no further than to the form of proteoses or peptones. It does not act on proteoses or peptones whether they are produced by its own digestion or supplied to it in the artificial form. Enzymes of this type occur in the gastric juice of all normal animals. It was considered by the early investigators that many of the proteolytic enzymes of plants were of the nature of pepsin, but later investigations have proved that they were mistaken and that enzymes of the pepsin type occur only rarely in plants.

The only case in which pepsin has apparently been proved to occur in plants is that of *Drosera,* which was studied by White (1910). She studied the leaves of the four species of this plant, and, with the best methods so far known for enzyme extraction, the only proteolytic enzyme that could be obtained was one which had the characteristics of a pepsin, since the end product under well-controlled experiments of digestion was always peptone. Her method of procedure is given here because it may be used as a model for the extraction of other plant proteases and the study of their reaction.

The leaves were cut from the plant and all traces of foreign matter were removed from them. They were then washed in cold boiled water and again in a strong solution of chloroform, which acts as an antiseptic. The leaves were then chopped into minute pieces with a sterilized knife and the fragments weighed and put into a bottle containing about 100 cc. of lukewarm boiled water and 30 drops of chloroform. The bottle was shaken vigorously for 2 hr., when its contents were filtered. To the filtrate was added half its volume of a saturated solution of ammonium sulphate, which was found to be more efficient in the precipitation of this enzyme than alcohol. The precipitate thus obtained was filtered off on a filter paper that had been previously sterilized with boiling water and was dried in a sulphuric acid desiccator. When almost dry the precipitate was scraped off the filter paper with a sterilized knife and dissolved in cold boiled water. A measured quantity of this solution was then placed in test tubes, and the material to be tested for digestion added. One of the test tubes was boiled and another left unboiled. About 20 drops of chloroform was then added, and the contents were well shaken. The ends of the test tubes were plugged with cotton and placed in an oven at 35°C. After 30 to 50 hr. the tubes were removed from the incubator and tested for the activity of the enzyme.

Quintanilha (1926) found that *Drosophyllum lusitanicum* digests proteins by an enzyme of the pepsin type which is secreted for the most part by the sessile glands. The stalked glands serve largely in capturing the prey and indirectly stimulate the secretion of enzymes by the sessile glands. Hopkins (1929) reported that a protease and a peptidase could be obtained by a water extraction of green malt. Stern and Stern (1932) demonstrated a peptic and a tryptic enzyme in the secretions of *Nepenthes hibberdii* and *N. mixta*.

2. *The Trypsins or Tryptases.*—The term "trypsin" is given to an enzyme that not only hydrolyzes proteins to proteoses and peptones but also splits these up further into amino acids. The protein may be completely hydrolyzed into amino acids by an enzyme of this type or only partly so, since in some cases the hydrolysis is not complete, some polypeptides remaining as a residue. The first enzyme of the trypsin type to be identified in plants was papain, first studied by Wurtz and Bouchut (1879), whose tryptic behavior was first definitely determined by Martin (1884 to 1885). The discovery by Green (1887, 1890) of a tryptic enzyme in the germinating seeds of lupine and the castor bean, the investigations by Chittenden (1894) of bromelin in the pineapple, and the study of the pitcher liquid of *Nepenthes* by Vines (1897, 1898) definitely established that enzymes of the trypsin type are widespread in plants. We shall now consider in detail some of the plant trypsins.

(*a*) *Papain.*—This enzyme has been studied more than any other protease of the plant kingdom. It is elaborated by the papaw (*Carica papaya*), a plant that is native to the Caribbean, Gulf of Mexico, and South American regions, whence it has spread to many parts of the world. This enzyme is secreted into the milky latex that is characteristic of this plant and is distributed in its roots, stem, leaves, and fruit. The enzyme has been designated variously "papain," "papayotin," "papaytin," or "papayacin." It is obtained by collecting the latex from incisions made in the plant parts and drying it in the sun. This dried preparation is reduced to a powder and may be obtained on the drug market.

The proteolytic properties of the latex of this plant have long been known, and Vines (1909) noted that statements concerning its digestive action are recorded as early as 1750. It was observed that if meat was steeped in this liquid for a short time it became very tender. Its digestive action on the mucous membrane of the intestines of swine that had eaten the fruit was also observed. The latex was first scientifically investigated by Wurtz and Bouchut (1879) and Wurtz (1880). They extracted the latex with distilled water and added to this extract ten times its volume of alcohol. The precipitate that was formed by this treatment was collected on a filter and dried. To the white powder thus formed they gave the name "papain" and regarded it as the digestive enzyme of the latex. Sen (1931) found that the yield of papain increases as the fruit develops until it reaches a maximum when the fruit is full grown but not yet ripe. A plucked fruit exudes no papain. The first milky juice exuded after lancing gives the largest yield and best quality of papain. Martin (1884, 1885) proved the formation of leucine and tyrosine in the course of the digestion of fibrin and albumin by papain and thus considered that it behaved as a trypsin. Later Chittenden (1892) reported the formation of large proportions of peptone together with small amounts of leucine and tyrosine from the digestion of coagulated egg white, fibrin, and raw and cooked meats by the action of this enzyme. These observations were substantiated by Chittenden, Mendel, and McDermott (1898), who showed that the chief products of digestion in the case of papain were albumose and peptone, and later these observations were again confirmed by Mendel and Underhill (1901). In 1902 Vines found by means of the tryptophane test that papain actively digests peptone in neutral solution, more actively in the presence of acid (0.5 per cent citric), and less actively in the presence of alkali (0.5 per cent sodium carbonate).

Ringer and Grutterink (1927) observed that the maximal digestion by papain occurred at pH 2.5 and 11.3, the optimum varying with the type of protein used. Small quantities of electrolytes had a restraining influence on the activity of this enzyme in both an acid and an alkaline medium. In later experiments, Vines (1903, 1905) suggested that papain might be a mixture of two proteases, basing his opinion on the behavior of the enzyme toward different antiseptics. He found that chloroform, sodium fluoride, thymol, salicylic acid, toluene, and formalin did not inhibit the peptonizing action of papain, but that they impeded further proteolysis in the formation of tryptophane. He found that hydrocyanic acid was the only antiseptic in the presence of which papain split off any considerable amount of tryptophane. It is of interest to note that several of the earlier workers who observed leucine and tyrosine as digestive products of this enzyme used hydrocyanic acid as an antiseptic. Mendel and Blood (1910) considered that the extreme differences in the rate at which papain accomplishes the earlier and later steps in digestion furnish strong evidence that it is a mixture of at least two enzymes, although conclusive proof that this is the case has not been furnished.

The properties of papain were studied in considerable detail by Mendel and Blood (1910). They observed that the digestion of peptone by papain, in presence of most of the common antiseptics used in digestion experiments, is very slow judged by the production of tryptophane. In the presence of 0.15 per cent hydrocyanic acid, however, hydrolysis is rapid. This accelerated effect is not limited to the hydrolysis of peptone but is also shown in the digestion of raw and coagulated egg white, fibrin, and edestin, whether the measure of digestion is the appearance of tryptophane, leucine, and tyrosine or the rate of solution of insoluble protein. It was also found that hydrogen sulphide in saturation also produces an acceleration of digestion comparable to that effected by hydrocyanic acid. Brill and Brown (1922) found that dilute sodium chloride showed a slightly activating effect followed by an inhibitory effect in more concentrated solutions. Sodium carbonate, sodium bicarbonate, calcium chloride, and magnesium sulphate had no marked influence, potassium chloride and sodium citrate showed marked activating influence, while acetic acid and lactic acid showed strong inhibitory effects.

The behavior of papain at relatively high temperatures has received considerable attention. Chittenden (1892) noted that in acid solution, papain digested more meat protein at 70°C. than at any lower temperature. According to Delezenne and Mouton (1906) and Pozerski (1909), papain digests proteins of egg white and serum so rapidly at 80 to 90°C. that if a suitable mixture of enzyme and uncoagulated protein faintly acidified is heated to boiling over a flame, scarcely any coagulum is obtained. This rapid digestion by papain at high temperatures was verified by Mendel and Blood (1910). Since at 40°C., a temperature which is generally considered favorable for enzyme action, the proteins are not measurably digested by papain, during short periods of time, the question arises as to the cause of the rapid rate of digestion at the higher temperatures. The rapid rate of digestion by papain at these relatively high temperatures is apparently due to the fact that it is not so easily destroyed by heat as other enzymes, and that the acceleration of the rate of action by heat more than compensates for the rate of destruction of the enzyme, so that for short periods under favorable conditions, the rate of digestion may be very high. This assumption is strengthened by the observations of Pratt (1915), who stated that at 70°C. the activity of the enzyme in the presence of a large amount of the enzyme is not greatly weakened but that with decreasing percentages of the enzyme, the loss in activity becomes more marked. Finally, Pratt found that, if a papain solution is heated rapidly to 100°C., allowed to boil 5 sec., and immediately cooled, it loses all its proteolytic activities.

Several methods for the determination of the activity of papain are given here in brief in order to give the student a general idea of the procedure that has been followed in studying this enzyme. Details, however, in this regard may be obtained by consulting the literature that has been mentioned in the preceding discussion. Mendel and Blood (1910) in the study of the digestion of peptone by papain used the following method: 3 cc. of 5 per cent peptone in 1 per cent sodium chloride were mixed with 3 cc. of 1 per cent papain in 1 per cent sodium chloride solution. Enough hydrocyanic acid was added to make a 0.15 per cent solution. Chloroform, 0.5 per cent; thymol, 0.5 per cent; sodium fluoride, 1 per cent; or several drops of toluene were also used as antiseptics. The mixture was digested in glass-stoppered bottles at 40°C. for 17 hr. It was tested at the end of the experiment for tryptophane by adding saturated bromine water drop by drop to 2-cc. portions of the digest.

The digestion of uncoagulated egg white by this enzyme was studied after the following manner: To 15 cc. of an egg-white solution prepared by diluting slightly beaten white of egg with three volumes of 1 per cent salt solution and filtering, 1 cc. of 1 per cent papain solution was added. This was made up to 30 cc., antiseptics were added after the manner mentioned above, and the digestion was allowed to proceed at 40°C. for 17 hr. The undigested protein was precipitated by adding 30 cc. of a 10 per cent solution of trichloracetic acid and heating for 5 min. in a water bath at 100°C. to coagulate the proteins completely and dissolve any albumoses. The mixture was then filtered while hot through a dried weighed filter paper, washed free from acid with water, and dried to constant weight. The amount of protein in the original mixture of egg white and papain was determined in a blank by the trichloracetic acid method.

Pratt (1915) prepared a papain solution by dissolving 0.75 g. of powdered papain in 150 cc. of distilled water. Although papain is not completely soluble in water, the active principle is dissolved upon being kept for 30 min. at 40°C., so that upon filtering a clear solution is obtained. Pratt used milk as nearly free from butter fat as possible. Into 150-cc. flasks were measured 25 cc. of milk, and to this was added water, varying from 5 to 25 cc., and sufficient papain solution in each case to make the total addition of water and papain solution 25 cc. After the enzyme solution was added, the contents of the flasks were thoroughly mixed by shaking, after which the flask was placed at 40°C. for 30 min. The flasks were then cooled by ice water until their contents could be examined. The contents of each flask were washed into 500-cc.

beakers, sufficient water being used to make the volume up to 75 cc. The undigested protein was precipitated by slowly adding 0.5 cc. of copper sulphate solution followed by 5 cc. of glacial acetic acid, the solution being vigorously stirred during the precipitation. After the precipitate had settled it was collected and washed upon weighed ashless paper and dried to constant weight at 100°C. The weight of protein digested by the papain was calculated by subtracting from similar determinations made on blanks to which no enzyme was added. Duplicate analyses in this manner agreed within 2 per cent. No antiseptic was used for such short periods of incubation.

(*b*) *Bromelin.*—This tryptic enzyme occurs in large quantities in the juice of the pineapple (*Ananas sativa*) and in other plants of the family *Bromeliaceae*. This enzyme was discovered by V. Marcano of Venezuela, according to a brief note published by Russell (1891) in the *Bulletin of Pharmacy*. Russell suggested the name "bromelin" for the enzyme after the natural order to which the plants containing it belong. This enzyme has been studied almost exclusively in the pineapple and was first scientifically investigated by Chittenden and his pupils (1891, 1894). The juice of the pineapple is acid, the acidity being equal to that of a solution of hydrochloric acid of about 0.45 per cent. The fresh juice is strongly proteolytic and its protein digesting power is manifested in an acid, neutral, or slightly alkaline medium, but in general it acts most energetically in a neutral medium. Ambros and Harteneck (1929) found that the activity of bromelin of the pineapple decreased as the fruit ripened. This enzyme from the ripe pineapple may be activated by hydrocyanic acid or by juice from the unripe fruit. An activator of bromelin is apparently present in the unripe pineapple.

When blood fibrin is warmed at 40°C., or thereabouts, with filtered pineapple juice, the fibrin swells up and then quickly disintegrates and in part dissolves. For studying the proteolytic power of pineapple juice under different conditions, Chittenden used the following procedure: About 100 cc. of filtered pineapple juice was warmed at 40°C. for a given period of time with 10 g. of moist, freshly coagulated egg albumin that had been completely freed from all soluble matter by washing with water. When the period of digestion was completed, the undissolved matter was collected in a weighed filter, washed with water, and dried at 110°C. to a constant weight. By subtracting the weight of the insoluble residue so obtained from the weight of dry albumin, equivalent to the moist albumin used in the experiment, the amount of protein matter converted into the soluble form was ascertained. In this manner, it was found, for example, that after 3½ hr. in one experiment 34.5 per cent of the protein was rendered soluble in the juice of natural acidity and 40 per cent in the juice that had been neutralized. This enzyme produces in digestion proteoses, peptones, tyrosine, and leucine, which may be detected by proper means.

The temperature relations of bromelin are interesting. Chittenden (1894) observed that neutral pineapple juice was capable of digesting more protein at 60°C. than under like conditions at 40°C., but that warming the enzyme solution alone at 60°C. for 15 min. diminished the proteolytic power of the enzyme 50 per cent. This suggested that the products of digestion protect the enzyme from the destructive action of high temperatures, since, when the neutralized juice is heated to 70 to 80°C. for 15 min. in the absence of digestive products, it is rapidly destroyed. It was observed by Pozerski (1909) that at high temperatures the action of bromelin was very rapid, behaving in this regard similarly to papain, the results being due apparently to the same causes as are mentioned under that topic.

Chittenden (1894) precipitated the enzyme from the juice by saturation with sodium chloride, magnesium sulphate, and ammonium sulphate and he considered the product obtained from sodium chloride to be superior to that obtained with the other salts. Effront (1917) stated that bromelin may be prepared by adding two volumes of alcohol to the pineapple juice and filtering off the precipitate, which contains little or no enzyme. To the filtrate thus obtained, five volumes of 95 per cent ethyl alcohol are added, and the precipitate formed, which contains the enzyme, is collected, dried, and pulverized for use.

Impure preparations of bromelin are strongly autodigestive in either an acid or an alkaline medium, such digestion, according to Caldwell (1905), beginning when the breaking up of protein impurities has been completed and proceeding to the total destruction of the enzyme. Bromelin, however, when prepared in a relatively pure condition is not at all autodigestive, so that the presence of some protein in the juice is apparently a prerequisite for such action. Caldwell (1905) obtained by the following method a preparation of this enzyme which contained only very slight traces of protein: The dialyzed sodium chloride precipitate was dissolved in a small amount of water and reprecipitated by the addition of 95 per cent alcohol. This precipitate was again dissolved in water and again precipitated by adding crystals of ammonium sulphate. A little of the associated protein was left behind at every precipitation, and five repetitions of the process gave a preparation containing only slight traces of protein. Precipitation was much hastened by placing the solutions at 4°C. After being dried upon the water bath at 40°C., the preparation was completely soluble in water and was not autodigestive in any medium even after prolonged standing.

(c) *Other Plant Trypsins.*—In 1880 Bouchut discovered a tryptic enzyme in the juice of the stem, leaves, and fruit of the common fig (*Ficus carica*) to which Mussi (1890) gave the name "cradina." Green (1887, 1890, 1892) found enzymes of the trypsin type in the germinating seeds of the lupine and castor-bean seedlings and in the

Kachree gourd (*Cucumis utilissimus*), while Butkewitsch (1900, 1901) observed tryptic enzymes in the germinating seeds of *Lupinus angustifolius* and *Lupinus luteus*. Vines (1902) found the liquid of the pitchers of *Nepenthes* contained a tryptic enzyme and later (1903) found similar enzymes in the juice of the cucumber and melon and in the bulbs of the tulip and hyacinth. The early observations of vegetable trypsins were, with but few exceptions, in seedlings and storage organs. In 1919 Fisher examined the leaves of numerous plants for the presence of trypsins, using finely powdered legumin as a substrate after the following manner: 10 g. of ground fresh leaves were placed with 3 g. of legumin in 250 cc. of water in a flask with a layer of toluene as an antiseptic. The flasks were tightly corked and placed in an incubator at 37°C. for 3 days. The material was filtered and the residue thoroughly washed with water, until the volume was 400 cc. A 40-cc. portion was then transferred to a flask, decolorized by shaking with alumina cream, filtered, and the residue thoroughly washed. The filtrate was then tested by Sorensen's method to determine the extent of protein hydrolysis as follows: To the filtrate was added 15 cc. of a fresh solution consisting of 50 cc. of commercial formaldehyde, 25 cc. of absolute alcohol, and 10 cc. of a thymol phthalein solution. (The thymol phthalein solution consisted of 0.5 g. of thymol phthalein dissolved in 1,000 cc. of 93 per cent alcohol.) Fifth-normal barium hydroxide was then run in until a distinctly green or blue color was obtained which matched that obtained by similar treatment of the control flask. The results show that the leaves of barley, oats, corn, rye, red clover, alfalfa, vetch, beans, peas, buckwheat, white mustard, and others contain enzymes of the trypsin type that are capable of hydrolyzing legumin. The reaction, however, is relatively slow as compared with the hydrolysis of peptone by the same leaf extract. In 3 days the amount of legumin hydrolyzed by the various leaf materials only amounted to from 1.5 to 9.6 per cent. Enzymes of a tryptic nature thus seem to be widely distributed in plants and plant parts. Their action is apparently relatively slow, but this may be due to the fact that the conditions under which they are placed in the laboratory may not be comparable to the conditions that prevail in the cells in which they normally act. Mounfield (1936) found that the aqueous extract of sprouted wheat contained a protease that showed an optimum activity at a pH of 4.1 for the decomposition of edestin at 40°C.

3. *The Erepsins or Ereptases.*—In 1901 Cohnheim described and gave the name "erepsin" to the enzyme formed in the mucous membrane of the small intestine which actively digests peptone, prolamins, polypeptides, and casein to the amino acids but does not act upon the higher proteins. Vines (1903) obtained evidence that various parts of widely different species of plants contain enzymes that resemble erepsin in their reaction. He based his conclusion on his observation that many plant extracts split off tryptophanes rapidly from peptone but entirely fail to hydrolyze fibrin as indicated by any apparent solution or by the formation of tryptophane. Independently of Vines, Javillier (1903) found that *Lolium perenne* contained among other enzymes one that hydrolyzes peptone into products that no longer give the biuret reaction. It has since been shown that erepsins are almost universally distributed in plants and plant parts. They have been found in the embryo of wheat, in green peas, in the expressed juice of the melon, cucumber, and tomato, in the rind of the apple and orange, in the juice of grape, in the leaves

of lettuce, cabbage, spinach, barley, oats, corn, rye, clover, alfalfa, vetch, bean, peas, buckwheat, mustard, in the potato tuber and artichoke, in the onion and bulbs of the hyacinth and tulip, and in the roots of turnip, tomato, carrot, and beet, in the leaves of various seedlings, and in many other plants and plant parts.

The erepsin of cabbage was studied in considerable detail by Blood (1910). She prepared the enzyme after the following manner: The leaves were ground and then pressed and the juice so obtained was dialyzed for 40 hr. against running water, with toluene as an antiseptic. The clear juice obtained by filtering the dialysate was saturated with ammonium sulphate and left in the cold room over night. The precipitate was then washed with saturated ammonium sulphate and dissolved in water. This solution was filtered and the filtrate dialyzed against running water, with toluene for an antiseptic, until free from ammonium sulphate, which required 6 days. A preparation of this type can be kept for a considerable period of time without loss of enzymic action.

In the digestive experiments, Blood followed the following procedure: To 5 cc. of 5 per cent peptone in 1 per cent sodium chloride solution 5 cc. of the enzyme solution was added and made to a volume of 12 cc. This was then digested in the incubator at 40°C. for 17 hr., using toluene as an antiseptic. To test for digestion, 2-cc. portions were removed, acidified with acetic acid when necessary, and treated with bromine water drop by drop until a maximum color was obtained.

The erepsin of cabbage obtained and tested in this manner split tryptophane and tyrosine from commercial peptones, clotted milk, and liquefied gelatin. It did not digest fibrin, edestin, or coagulated egg white in neutral, acid, or alkaline solution. It was active over a considerable range of acidity and alkalinity.

5. The Coagulases of Plants. *a. The Rennets or Rennins.*—Rennet is an enzyme which when introduced into milk causes coagulation, the milk becoming gelatinous and compact. The formation of this milk clot is due to the alteration of the casein of the milk. Rennet is secreted by the gastric glands of animals and is associated in relatively large amounts with animal pepsin.

The enzyme rennet or rennin is apparently widespread in the plant kingdom and is similar to if not identical with that found in the animal body. Experimental evidence indicates that, in the majority of cases at least, it is always closely associated with proteolytic enzymes. The occurrence of rennet in the plant kingdom was known as early as the sixteenth century. It was at that time known by certain tribes of Lapland and by the peasants of the Alps that the leaves of the bitterwort (*Pinguicula vulgaris*) had the power to curdle milk and the leaves were used by them for that purpose. It was also known in England, about that time, that the stem and leaves of the yellow bedstraw (*Galium verum*) coagulated milk, and until recently, at least, this plant was used in cheese making in certain parts of England.

When the proteases of plants were discovered, it was found that in many cases they possessed the power of curdling milk. Thus papain, bromelin, the trypsin of the castor bean and the ereptase of cabbage all have this power, which is evidently due to rennins accompanying them. The most extensive studies of plant rennins have been made by Gerber (1907 to 1912), and the student is referred to his works for a detailed study of this enzyme.

Rennin has been found in the juice of the stem, leaves, and fruit of the fig, in the juice of the papaw, in the brown algae, and in the seeds of *Datura stramonium*, *Pisum sativum*, *Lupinus hirsutus*, and others. In some of the seeds, it occurs apparently in the zymogen state. This enzyme was found by Gerber to be present in the *Cruciferae*, in the *Ranunculaceae*, in the *Rubiaceae*, in the *Compositae*, in the *Euphorbiaceae*, and others.

As an example of the methods of isolation and study of this enzyme the work of Lea (1883), who studied the rennin of *Withania coagulans*, may be cited. He found that a 5 per cent sodium chloride solution at 38°C. gave the best extraction of the enzyme from the seeds. The active principle is soluble in glycerin and can be extracted from the seeds by this means. This enzyme loses its activity if boiled in water for a minute or two. Alcohol precipitates the enzyme from its solution, and the precipitate after being washed with alcohol may again be dissolved without losing its coagulating power. When a normal clot is formed by this rennin, the reaction of the clot remains neutral or faintly alkaline. The clot is thus a true one and is not due to the action of acid and resembles in appearance the clot formed by animal rennin.

b. The Pectases.—The enzymes of this group coagulate pectic bodies with the formation generally of gelatinous calcium pectate. The presence of calcium salts is apparently necessary for the action of the pectases and it has been found that pectic gels can be formed in the absence of the enzyme, provided the necessary conditions are present.

G. The Formation of Enzymes in Plants

Two types of enzymes are distinguished in plants, the classification of these two types being based on the relation of their place of origin to the place of their activity. Most of the enzymes of plants function in the cells in which they are formed and are called "intracellular" or "endocellular" enzymes. Some plant parts produce enzymes that function outside the cells that produce them, and such enzymes are commonly termed "extracellular enzymes." Enzymes of this type are secreted, for the most part, by specialized cells that have a more or less glandular appearance. The study of the formation of extracellular enzymes in plants has been confined almost entirely to the secretion of extracellular enzymes by insectivorous plants and to the germinating seeds of barley, wheat, corn, and other cereal grains.

The secretion of enzymes by certain cells of insectivorous plants received considerable attention from early investigators who demonstrated that these cells excreted enzymes into the liquid in the "pitchers" of *Nepenthes* and similar plants or to the exterior as in the case of *Drosera* or sundew. They found that these enzymes were secreted and digestion of food brought about only when the secreting cells were stimulated in some way or another.

A discussion of the observations of Hepburn (1918) in his studies of the liquid of *Nepenthes* will suffice to show the general status of our present knowledge concerning the secretion of enzymes by insectivorous plants. He examined the liquid in the pitchers of 14 species and hybrids of *Nepenthes,* always studying it prior to the opening of the pitcher. Through an opening he stimulated the glands of the inner wall of the pitcher by stroking with a camel's-hair brush or by placing several round, solid, glass beads therein and shaking the pitcher at intervals during one or more days. After stimulation by the brush or the insertion of the beads, cotton plugs were inserted in the openings of the pitchers and the liquid removed for study the following day. Hepburn studied especially the action of the proteolytic enzymes secreted by the pitchers and found that the liquid from the stimulated pitchers produced a more rapid digestion than did that from the non-stimulated pitchers. The latter exerted proteolytic action only in the presence of acid and failed to produce this action in its absence. The liquid from the stimulated pitchers exerted proteolytic action in either the presence or the absence of an acid. Hepburn concluded that increased proteolytic activity due to stimulation might be caused by one or more of the following results: (1) by changing the hydrogen-ion concentration, thus rendering conditions more favorable for the action of proteases already present; (2) by causing the activation of a zymogen already present; and (3) by producing an increased secretion of proteases by the glands of the pitcher. Hepburn showed conclusively that the liquid taken from the young unopened pitchers was sterile and that the proteolytic action of the liquid is due to a protease and not to bacterial action. Holter and Linderstrom-Lang (1933) found two proteases in *Drosera rotundifolia.* One of these was secreted by the glandular hairs and showed a maximum activity at pH 3.2; the other was an endoenzyme in the leaf tissues and showed its optimum activity at pH 4.3.

In order to discuss the secretion of enzymes by the germinating seeds of the *Gramineae,* it is necessary to have a knowledge of the arrangement and structure of the various seed parts. The structure of the barley grain has been described in detail by Brown and Morris (1890), and that of the corn to some extent by Torrey (1902) and Sargant and Robertson (1905). The general structure of the seeds of the grains and grasses is approximately the same, so that the structure of the corn grain to be herein considered may be taken as representative of this group of seeds.

In Fig. 31 it is seen that a grain of corn consists of an endosperm and embryo surrounded by several coats and the pericarp. The endosperm occupies the greater portion of the seed and partly surrounds the embryo. The endosperm consists for the most part of large cells filled with starch grains, part of which are embedded in a matrix of protein matter. Just beneath the coats of the grain and completely encircling the outer part of the endosperm is a layer of columnar cells filled with protein and termed the "aleurone layer." The embryo is on the furrowed, flattened side of the grain. It is thickest at the end attached to the cob and gradually slopes to a point reached at about three-fourths to four-fifths the length of the grain. The embryos consist of the young plant and the scutellum. The young plant or plantlet consists of the plumule and radicle with the connecting hypocotyl. The scutellum is a shieldlike expansion of the hypocotyl and makes up the greater part of the bulk of the embryo. It is considered by some to be the homologue of the cotyledon and serves as a special organ for absorbing the food from the endosperm and transferring it to the developing plantlet.

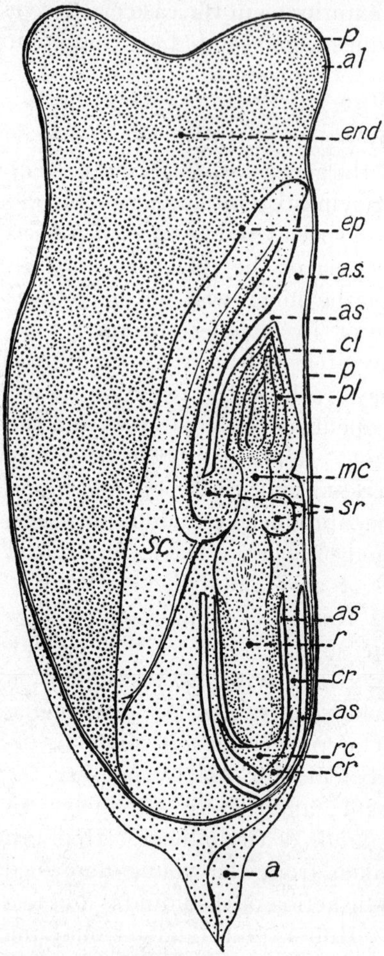

FIG. 31—Longitudinal section of a corn grain. *end*, endosperm. *sc*, scutellum. *p*, pericarp. *al*, aleurone layer. *ep*, epithelial layer. *as*, air space. *cl*, coleoptile. *pl*, plumule. *mc*, mesocotyl. *sr*, initials of secondary roots. *r*, radicle. *rc*, root cap. *cr*, coleorhiza. *a*, point of attachment to cob.

The surface of the scutellum that is in contact with the endosperm consists of a layer of columnar cells which is termed the "epidermal" or "epithelial" layer of the scutellum. In the case of corn, this epithelial layer dips down at frequent intervals into the endosperm, thus producing a greater surface for secretion. These convolutions of the epithelial layer, however, are not present in most of the other members of the grain family (Sargant and Robertson, 1905). These epithelial cells are about three times as long as broad and have one end resting on the scutellar parenchyma and the other in close contact with the endosperm. The protoplasmic contents of the epithelial cells stand in marked contrast to those of the other cells of the scutellum in being finely granulated, semitransparent, and possessing prominent nuclei (Fig. 32).

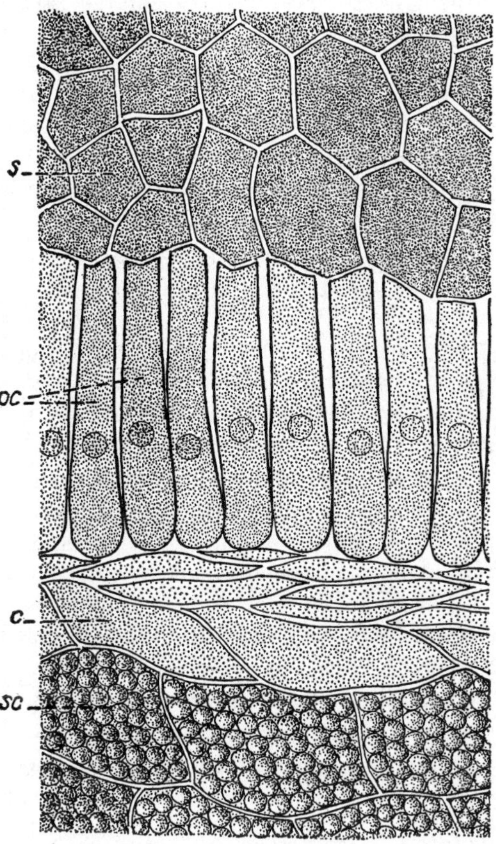

FIG. 32.—Section through the scutellum, epithelial layer, and endosperm of the corn grain. *s*, scutellum. *epc*, epithelial cells. *c*, flattened cells of the endosperm. *sc*, starch cells of the endosperm.

Avery (1930) discussed the analogy of the different parts of the seeds of corn, oats, and wheat to the various organs and portions of organs in the adult plants.

Within a few hours after the seeds of barley, corn, or other grains are moistened and placed under favorable conditions for germination, the protoplasm of the epithelial cells undergoes marked changes, becoming much coarser in structure, the granulations increasing in size and number and clouding the cell contents so that the nucleus which is conspicuous in the resting stage of the seed cannot readily be observed. This cloudiness reaches its maximum in from 24 to 36 hr. and shows little change in most cases until the endosperm is depleted, after which the cells again become hyaline. In the case of the date seed, however, the epithelial cells clear long before the endosperm is digested (Reed, 1904). The epithelial cells also greatly enlarge at the beginning of germination, so that between the first and third day in the germination of corn they enlarge to three to four times their original size. Soon after the above-mentioned activities appear in the epithelial layer, the cell walls of the cells of the endosperm nearest to the scutellum begin to dissolve and the starch grains therein show evidence of pitting, while temporary starch appears as grains in the parenchymatous tissue of the scutellum bordering upon the epithelial cells.

All investigators are agreed that the epithelial cells are capable of forming cytase, diastase, and probably proteases, which are excreted into the adjoining endosperm. Pozen (1934) reported that the enzymes cytase, amylase, protease, peptidase, oxidase, and phytase are secreted by the embryo of barley during germination.

The cytological changes that occur in the cells of the epithelial layer during the formation of enzymes have been studied by Torrey (1902) and Reed (1904). Torrey considered that the nucleus is the seat of this enzyme formation. He noted in the case of corn that at the beginning of germination the nuclei contained dark, staining granules that exuded into the cytoplasm and finally collected at the wall next to the endosperm where they disappeared. Immediately after this dissolution, the first activity of enzymes is noted in the cell walls and starch grains of the endosperm. Torrey thus considered these accumulations at the base of the epithelial cells to be deposits of actual enzyme substance. Reed (1904), however, was unable to confirm the observations of Torrey relative to the extrusion of solid matter from the nucleus but noted that the chromatin of the nuclei increased as germination progressed while the nucleoli diminished in size. In this connection the statements of Horning and Petrie (1927) in regard to the function of mitochondria as mentioned in Chap. I should again be mentioned. They considered that the mitochondria of the epithelial cells migrate into the endosperm, carrying on their surfaces the enzymes that act upon the food reserves contained therein.

The function of the aleurone layer has also received considerable attention (Fig. 33). Beyerinck (1895) considered that this layer secreted maltase; Green (1899), Haberlandt (1890), and Brown and Escombe (1898) concluded that it secreted cytase in the case of barley; while Puriewitsch (1897) and Linz (1896) came to the opposite conclusion. There is much evidence, however, that the aleurone layer has no enzymic secretory powers. The cells that compose it are relatively thick walled and undergo apparently little change in their contents until the endosperm is entirely depleted. No erosion of the starch grains or dissolution of the cell walls occurs next to the aleurone layer except as these changes move out from the scutellum. From thousands of observations on barley grains, Mann and Harlan (1915) concluded that the aleurone layer of that seed exercises no secretory power at all during germination. They considered that all the enzymes formed in the seed during germination are secreted by the epithelial layer. They considered that the aleurone layer may act as a protective layer to the endosperm and embryo and that the nitrogenous material in these cells might serve as a supply of nitrogen for the young plant after it had full power to manufacture its own carbohydrates. The origin of the aleurone layer has been investigated by Jodidi and Peklo (1929) and should be mentioned here in a discussion of its function. From certain cytological and biochemical observations they concluded that symbiotic fungi produce the aleurone layer in the case of English rye grass, wheat, and barley. This is an entirely new line of thought and should be investigated further. Schander (1934) found that the germination of rice and other cereals was greatly impaired if the aleurone layer was removed by the polishing of the grains.

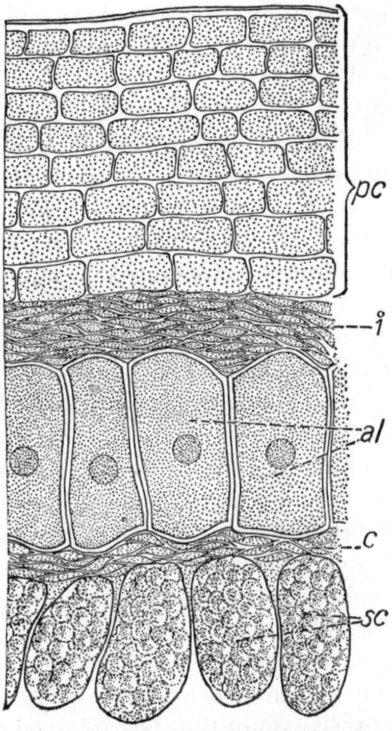

FIG. 33.—Section through the outer portion of a grain of corn. *pc*, pericarp. *i*, integuments. *al*, aleurone layer. *c*, flattened cells containing some protein. *sc*, starch cells containing mostly starch but also some protein.

Although all investigators have agreed that the epithelial cells of the scutellum secrete enzymes, there has been much dispute as to the extent to which these cells supply the enzymes concerned in the digestion of the reserves in the endosperm.

It has been found that embryos of many of the *Gramineae* can be grafted onto other endosperms or placed in contact with starch preparations and still develop and function normally by digesting and absorbing their food from the substratum upon which they are placed. On this account and from certain observations on the behavior of endosperms, Brown and Morris (1890), Brown and Escombe (1898), and Pond (1906) concluded that the endosperm has no power at all of self-digestion. On the other hand, Krabbe (1890), Grüss (1893 to 1897), Hansteen (1894), Puriewitsch (1897), Paton, Nanji, and Ling (1924), and others have reported observations that contradict these conclusions. The work of Bruschi (1908) showed the cause for the contradictory results that had been obtained in this regard. He found that the starchy endosperm of corn, barley, wheat, and rye can digest itself in the absence of the scutellum though to a very different degree. The vitality of the endosperm cells of various species and varieties differs greatly, so that the enzymic action that would be developed will vary greatly. If an investigator worked with a set of seeds in which the vitality was very low, he would naturally observe little or no digestion of the endosperm. If he happened to study a variety in which the vitality of the endosperm cells remained high, he would observe considerable self-digestion. Bruschi found that some digestion may occur in endosperms that have lost all their vitality. This is apparently due to the fact that the cells contain proenzymes or zymogens that exist after the death of the cells and can form active enzymes after soaking in water at ordinary temperatures.

Bach, Oparin, and Wähner (1927) reported that the dormant seeds of wheat contain active enzymes although in minute quantities. During germination these enzymes increase rapidly, attain a maximum, and then decrease in amount during the last phases of germination. The presence of free oxygen is necessary for their formation. A similar increase in the amount of enzymes is observed during the ripening of the seed, but a rapid decrease occurs toward the end of ripening. Tascher and Dungan (1928) found that the diastatic activity of the corn seed and seedling increased during germination. The stage at which corn was harvested did not influence this activity to any appreciable extent. The vigor of seed cannot thus be explained as due to differences in diastatic activity but is due apparently to the greater quantity of reserves.

The diastatic activity or power of wheat flour is an important factor in its use in baking. A considerable amount of work has been done to find the cause of the marked fluctuations of this power in various flours. Swanson (1935) stated that the diastatic activity varies in different classes of wheat. It was found to be highest in durum wheats. Soft winter wheats as a group have a lower diastatic activity than hard winter wheats. Those wheats grown under very dry conditions are lower in this activity than those grown under more moist conditions. The diastatic activity is not increased by wetting until the moisture is sufficient to start germination. He considered that the diastatic enzymes in the endosperm, and consequently in the flour, probably exist in the form of proenzymes which may be activated by water. The amount of such enzymes present probably depends on conditions prevailing during kernel formation and ripening. The amount of diastatic enzymes produced by the epithelial layer is increased during germination so that flour from malted wheat is rich in these enzymes. The water added in tempering wheat, however, does not affect this diastatic activity. Swanson showed that the greatest amount of diastase was produced in the germ end of the grain during germination. His observations were:

Part analyzed	White wheat		Red wheat	
	Sugar content, mg.	Diastatic activity, mg.	Sugar content, mg.	Diastatic activity, mg.
Check ungerminated seed	68	240	50	277
Whole germinated seed	415	1050	565	1750
Germ end of germinated seed	720	1800	775	1225
Brush end of germinated seed	130	750	340	850

Bailey and Sherwood (1935) in a discussion of the biochemistry of breadmaking stated that during the dough mixing the diastatic enzymes therein become active and hydrolyze the starch into fermentable sugars. The amylase of wheat flour is of the beta type. Malted wheat flour has its α-amylase greatly increased in malting, while the β-, or saccharogenic, amylase remains constant. The addition of 1 per cent malted flour to high or low diastatic flours effects an increase in sugar production of 100 per cent in some cases. Several hypotheses have been proposed to account for this accelerated saccharogenesis. The first is that the dextrinogenic action of the α-amylase may result in a new substrate which the β-amylase can more readily convert into maltose. The second is that the malted flour may contain active proteases that function in liberating amylases from protein complexes with which they are associated. The third is that the malted wheat flour may contribute a substrate that is more readily hydrolyzed than the native starch of the normal flour. The diastatic activity of flours is influenced by the variety of wheat, the fineness of the starch particles, and the degree of hydration of the starch.

The effect of the water content and temperature upon the secretion of diastase in the barley grain was studied by Pickler (1919). The grains were placed under certain conditions of temperature and humidity until the absorption of water had reached an equilibrium. A certain portion of each grain sample was then taken, dried and pulverized to a powder, and the action of each sample of powder on a 0.2 per cent solution of starch paste noted. It was found that the temperature changes made little difference in the rate of secretion of the enzyme, but that the changes in water content exerted a marked effect. This may be illustrated by the following concrete example: At 10°C. the water content of two samples of seed was 11 and 29.9 per cent, respectively. The former had a diastatic activity of 36.25, while the latter had an activity of 24.25. That is, it required 36.25 hr. for a diastase sample from the seed with a water content of 11 per cent to digest a given amount of starch paste, while it required only 24.25 hr. for a diastase sample from the seed with a water content of 29.9 per cent to digest the same amount of starch paste. At 20°C. with water contents of 11.5 and 30.9 per cent, respectively, the diastatic activities were 36.5 and 23.1. These results show that the water content is a very important factor in determining the rate of formation of diastase by the scutellum of the barley grain.

IV. A COMPARISON OF DIGESTION IN PLANTS AND ANIMALS

The process of digestion in plants and animals is similar in that, for the most part, plants digest the same foods as animals. These foods are digested by similar enzymes and the products of digestion are the same. As a general rule, the process of digestion in animals is more rapid than in plants.

The process of digestion in plants is different from that in animals primarily in regard to its location. In the higher animals digestion occurs in certain definite specialized organs, while in plants, with but few exceptions, it does not occur in any special organs but may take place in any part of the plant body.

Digestion in plants is, for the most part, intracellular, occurring within the cells which produce the enzymes that act upon the food contained therein. The only exceptions are the digestion of some food by insectivorous plants and of the reserve foods in the endosperms of the *Gramineae* and date seed by the enzymes secreted by the epithelial cells of the scutellum and the absorbing portions of the date embryo.

Digestion in animals occurs in practically all cases outside the cells and is termed "extracellular digestion." Thus in the higher animals digestion occurs in the alimentary tract and is brought about by enzymes excreted therein by cells bordering upon it or by special glands connected with it. Also, in the case of unicellular animals as the amoebae, which simply wrap themselves around food particles and digest them, the digestive process is really an extracellular one.

Bibliography

ACHARYA, C. N., 1933: Investigations on the development of prussic acid in cholam (*Sorghum vulgare*), *Indian J. Agr. Sci.*, **3**: 851–869; *Exp. Sta. Rec.*, **71**: 312–313.

ALEXANDER, J., 1934: Enzymes, vitamins and the zone of maximum colloidality, *Sci.*, **80**: 79–80.

———, 1935: The nature of enzymes, *Sci.*, **81**: 44–45.

AMBARD, L., E. PELBOIS, and M. BRICKA, 1920: Similitude de l'hydrolyse du sucre par les acides et de l'hydrolyse de l'amidon par l'amylase. Étude de certaines réactions mono-moléculaire, *Bull. Soc. Chim. Biol.*, **2**: 42–63.

———, 1920, 1921: Fixation de l'amylase par l'amidon cru et l'empois d'amidon, *C. R. Soc. Biol.*, **83**: 1458–1460; **84**: 230–232.

———, 1921: Sur l'amylase. Son dosage, mécanisme de la digestion amylolytique, *Bull. Soc. Chim. Biol.*, **3**: 51–65.

———, 1923: De l'amylase. Liaison du ferment et des substances qu'il digère, *Bull. Soc. Chim. Biol.*, **5**: 693–716.

AMBROS, O., and A. HARTENECK, 1929: Über natürliche Aktivierung von Proteasen pflanzlicher Milchsäfte, *Z. physiol. Chem.*, **181**: 24–41.

APPLEMAN, C. O., 1911: Physiological behavior of enzymes and carbohydrate transformations in after-ripening of the potato tuber, *Bot. Gaz.*, **52**: 306–315.

ARMSTRONG, H. E., 1905: Studies on enzyme action. Lipase, *Proc. Roy. Soc.*, B, **76**: 606–608.

——— and E. ORMEROD, 1906: Studies on enzyme action. Lipase II, *Proc. Roy. Soc.*, B, **78**: 376–385.

———, E. F. ARMSTRONG, and E. HORTON, 1908: Studies on enzyme action. XII, The enzymes of emulsin, *Proc. Roy. Soc.*, B, **80**: 321–331.

——— and H. W. GOSNEY, 1913: Studies on enzyme action. Lipase III, *Proc. Roy. Soc.*, B, **86**: 586–600.

——— and ———, 1915: Studies on enzyme action. Lipase IV, The correlation of synthetic and hydrolytic activity, *Proc. Roy. Soc.*, B, **88**: 176–189.

AVERY, G. S., 1930: Comparative anatomy and morphology of embryos and seedlings of maize, oats and wheat, *Bot. Gaz.*, **89**: 1–40.

BACH, A., A. OPARIN, and R. WÄHNER, 1927: Untersuchungen über den Fermentgehalt von reifenden, ruhenden und keimenden Weizensamen, *Biochem. Z.*, **180**: 363–370.

BAENS, L., F. M. YENKO, A. P. WEST, and H. M. CURRAN, 1934: Tannin content of Philippine barks and woods, *Philippine J. Sci.*, **55**: 177–190.

BAILEY, C. H., and R. C. SHERWOOD, 1935: Biochemistry of bread making, *Ind. and Eng. Chem.*, **27**: 1426–1430.

BAILEY, E. M., 1912: Biochemical and bacteriological studies on the banana, *J. Am. Chem. Soc.*, **34**: 1706–1730.

BAKER, J. L., and H. F. E. HULTON, 1921: Amylases of the cereal grains. Rye, *J. Chem. Soc., London*, **119**: 805–809.

——— and ——— 1929: Amylases of the cereal grains—oats, *J. Chem. Soc., London*, **1929**: 1655–1660.

BALLS, A. K., 1931: Recent advances in enzyme chemistry: the mechanism of peptide hydrolysis, *Sigma Xi Quart.*, **19**: 98–112.

BALY, E. C. C., and E. S. SEMMENS, 1924: The selective photochemical action of polarised light, *Proc. Roy. Soc.*, B, **97**: 250–253.

——— and ———, 1925: Selective action of polarized light upon starch grains, *Nature*, **116**: 817.

BARANETZKY, J., 1878: "Die stärkebildenden Fermente in den Pflanzen," Felix, Leipzig.

BARENDRECHT, H. P., 1913: Enzyme-action, facts and theory, *Biochem. J.*, **7**: 549–561.

BARTHOLOMEW, E. T., 1914: Concerning the presence of diastase in certain red algae, *Bot. Gaz.*, **57**: 136–147.

BARTON, A. W., 1920: The lipolytic activity of the castor and soybean, *J. Am. Chem. Soc.*, **42**: 620–632.

BAYLISS, W. M., 1925: "The Nature of Enzyme Action," mon., Longmans, Green and Company, New York, 200 pp.

BERCZELLER, L., and J. FREUD, 1922: Über die Wirkung der Halogene auf Diastasen, *Biochem. Z.*, **139**: 476–481.

BERGMANN, M., and W. F. ROSS, 1935: On proteolytic enzymes. VIII, The proteolytic systems of papain, *J. Biol. Chem.*, **111**: 659–666.

BEYERINCK, M. W., 1895: Ueber Nachweis und Verbreitung der Glukase, das Enzym der Maltose, *Cent. Bakt. u. Parasitenkunde*, II, **1**: 221–229, 265–271, 329–342.

BLAGOWESTSCHENSKI, A., 1926: Über die synthetische Wirkung der pflanzlichen Proteasen, *Biochem. Z.*, **168**: 1–6.

BLOOD, A. F., 1910: The erepsin of the cabbage, *J. Biol. Chem.*, **8**: 215–225.

BOIS, E., and A. NADEAU, 1935: Contribution à l'étude d'*Acer saccharum*, *Nat. Canadien*, **62**: 106–112.

BOUCHUT, E., 1880: Sur un ferment digestif contenu dans le suc de figuier, *C. R. Acad. Sci. Paris*, **91**: 67–68.

BOURQUELOT, E., 1893: Inulose et fermentation alcoolique indirect de l'inuline, *C. R. Soc. Biol.*, **5**: 481–483.

———, 1893: Inulase et fermentation alcoolique indirecte de l'inuline, *C. R. Acad. Sci. Paris*, **116**: 1143–1145.

———, 1893: Remarques sur les ferments solubles sécrétés par l'*Aspergillus niger* et le *Penicillium glaucum*, *C. R. Soc. Biol.*, **5**: 653–654.

———, 1903: Généralities sur les ferments solubles qui déterminent l'hydrolyse des polysaccharides, *C. R. Acad. Sci. Paris*, **136**: 762–764.

BRADLEY, H. C., and E. KELLERSBERGER, 1912: The problem of enzyme synthesis. III, Diastase and starch of plant tissue, *J. Biol. Chem.*, **13**: 425–429.

BRASSE, L., 1884: Sur la présence de l'amylase dans les feuilles, *C. R. Acad. Sci. Paris*, **99**: 878–879.

———, 1885: Action de la diastase du malt sur l'amidon cru, *C. R. Acad. Sci. Paris*, **100**: 454–456.

BRAUN, K., and E. BEHRENDT, 1903: Beitrag zur fermentative Spaltung der Fette, *Ber. chem. Ges.*, **36**: 1142–1145, 1900–1911.

BRILL, H. C., and R. E. BROWN, 1922: The digestive properties of Philippine papaïn, *Philippine J. Sci.*, **20**: 185–191.

BROWN, H. T., and J. HERON, 1879: Contributions to the history of starch and its transformations, *J. Chem. Soc., London*, **35**: 596–654.

——— and G. H. MORRIS, 1890: Researches on the germination of some of the *Gramineae*, *J. Chem. Soc., London*, **57**: 458–528.

——— and ——— 1893: A contribution to the chemistry and physiology of foliage leaves, *J. Chem. Soc., London*, **63**: 604–677.

——— and F. ESCOMBE, 1898: On the depletion of the endosperm of *Hordeum vulgare* during germination, *Proc. Roy Soc.*, **63**: 3–25.

BRUSCHI, D., 1908: Researches on the vitality and self digestion of the endosperm of some *Graminaceae*, *Ann. Bot.*, **22**: 449–463.

BUNKER, J. W. M., and E. G. E. ANDERSON, 1928: Polarized light and starch hydrolysis, *J. Biol. Chem.*, **77**: 473–488.

BUTKEWITSCH, W., 1900: Ueber das Vorkommen proteolytischer Enzyme in gekeimten Samen und über ihre Wirkung, *Ber. bot. Ges.*, **18**: 185–189.

———, 1901: Über das Vorkommen eines proteolytischen Enzyms in gekeimten Samen und über seine Wirkung, *Z. physiol. Chem.*, **32**: 1–53.

———, 1908: Zur Frage über die Umwandlung der Stärke in den Pflanzen und über den Nachweis der amylolytischen Enzyme, *Biochem. Z.*, **10**: 314–344.

Caldwell, J. S., 1905: The effect of toxic agents upon the action of bromelin, *Bot. Gaz.*, **39**: 409–419.

Chittenden, R. H., E. P. Joslin, and F. S. Meara, 1891: On the ferments contained in the juice of the pineapple together with some observations on the composition and proteolytic action of the juice, *Trans. Conn. Acad. Arts and Sci.*, **8**: 283–308.

———, 1892: Papoid-digestion, *Trans. Conn. Acad. Arts and Sci.*, **9**: 297–332.

———, 1894: The proteolytic action of bromelin—the ferment of pineapple juice, *J. Physiol.*, **15**: 249–310.

———, L. B. Mendel, and H. E. McDermott, 1898: Papaïn proteolysis with some observations on the physiological action of the products formed, *Am. J. Physiol.*, **1**: 256–276.

Chrzaszcz, T., and A. Joscht, 1917: Über die Verschiebung einzelner amylolytischer Kräfte der Malzamylase und deren Verhalten beim Aufbewahren in Gegenwart verschiedener Reagentien, *Biochem. Z.*, **80**: 211–241.

———, 1923: Einfluss der Temperatur auf die Wirkung der Amylase. Die Wirkung der verzuckernden Kraft der Amylase, *Biochem. Z.*, **142**: 417–439.

———, Z. Bidzinski, and A. Krause, 1925: Über den Einfluss der Wasserstoffionkonzentration auf die Dextrinierung der Stärke durch gereinigte Malzamylase, *Biochem. Z.*, **160**: 155–171.

Clark, R. H., F. L. Fowler, and P. T. Black, 1931: The activation of amylase, *Trans. Roy. Soc. Canada*, Sec. 3, **25**: 99–105.

Cohnheim, O., 1901: Die Umwandlung des Eiweiss durch die Darmwand, *Z. physiol. Chem.*, **33**: 451–465.

Colin, M. H., 1915: Sur la distribution de l'invertine dans les tissus de la betterave, aux différentes époques de la végétation, *C. R. Acad. Sci. Paris*, **160**: 777–779.

Compton, A., 1921: Studies in the mechanism of enzyme action. I, Role of the reaction of the medium in fixing the optimum temperature of a ferment, *Proc. Roy. Soc.*, B, **92**: 1–6.

Connstein, W., E. Hoyer, and H. Wartenberg, 1902: Ueber fermentative Fettspaltung, *Ber. chem. Ges.*, **35**: 3988–4007.

Creighton, M., and N. M. Naylor, 1933: A study of the starch-digesting and the sugar forming enzymes of wheat, *Iowa State Col. J. Sci.*, **7**: 253–260.

Cuisinier, L., 1885: Une nouvelle matière sucrée diastasque la Ceréalose et pour sa fabrication, *French Patent* 171958, Oct. 30.

Czapek, F., 1903: Antifermente im Pflanzenorganismus, *Ber. bot. Ges.*, **21**: 229–242.

———, 1905: The anti-ferment reaction in tropistic movements of plants, *Ann. Bot.*, **19**: 75–98.

Daish, A. J., 1916: The distribution of maltase in plants. II, The presence of maltase in foliage leaves, *Biochem. J.*, **10**: 49–55.

———, 1916: The distribution of maltase in plants. III, The presence of maltase in germinated barley, *Biochem. J.*, **10**: 56–76.

Davis, A. R., 1915: Enzyme action in the marine algae, *Ann. Mo. Bot. Gard.*, **2**: 771–836.

Davis, W. A., 1916: The distribution of maltase in plants. I, The function of maltase in starch degradation and its influence on the amyloclastic activity of plant materials, *Biochem. J.*, **30**: 31–48.

Dean, A. L., 1903: Experimental studies on inulase, *Bot. Gaz.*, **35**: 24–35.

———, 1905: On proteolytic enzymes. I, *Bot. Gaz.*, **39**: 321–339.

———, 1905: On proteolytic enzymes. II, *Bot. Gaz.*, **40**: 121–134.

Delezenne, C., and H. Mouton, 1906: Sur l'allure anomale de quelques protéolyses produites par la papaïne, *C. R. Soc. Biol.*, **60**: 68–70.

———, ———, and E. Pozerski, 1906: Sur la digestion brusque de l'ova-albumine et du sérum sanguin par la papaïne, *C. R. Soc. Biol.*, **60**: 309–312.

Denny, F. E., 1930: Increasing amylase activity of potato juice by the addition of potassium cyanide, *abst.*, *Am. J. Bot.*, **17**: 1050.

———, 1931: The effect of thiocyanates upon amylase activity. I, Potato amylase, *Contr. Boyce Thompson Inst.*, **3**: 277–286.

———, 1931: The effect of potassium cyanide upon the amylase activity of potato juice, *Contr. Boyce Thompson Inst.*, **3**: 297–308.

———, 1932: Direct versus indirect effects upon potato amylase by chemicals which induce sprouting of dormant tubers, *Contr. Boyce Thompson Inst.*, **4**: 53–63.

———, 1933: Effect of potassium thiocyanate and ethylene chlorhydrin upon amylase activity, *Contr. Boyce Thompson Inst.*, **5**: 441–450.

Doby, G., and R. P. Hibbard, 1927: Nutrient ions of plants and the ion activation of plant enzymes, *J. Biol. Chem.*, **73**: 405–416.

——— and J. Burger, 1932: Potato amylase, *Fermentforschung*, **13**: 201–211; *Chem. Abst.*, **26**: 4078.

——— and L. Brázay, 1932: The amylase of sugar beet leaves, *Fermentforschung*, **13**: 212–227; *Chem. Abst.*, **26**: 4078.

——— and E. Szladits, 1932: Enzymes and salt ions. III, Saccharogenic amylase of leaves of different fertilized potatoes, *Z. physiol. Chem.*, **206**: 177–186; *Exp. Sta. Rec.*, **70**: 754.

Doby, P., 1915: Über Pflanzenenzyme. IV, Die Invertase der Kartoffelblätter, *Biochem. Z.*, **71**: 495–500.

Duggar, B. M., and A. R. Davis, 1914: Enzyme action in *Fucus vesiculosis*, *Ann. Mo. Bot. Gard.*, **1**: 419–426.

Effront, J., 1905: Sur le développement de l'amylose pendant la germination des graines, *C. R. Acad. Sci. Paris*, **141**: 626–628.

———, 1917: "Biochemical Catalysts in Life and Industry. Proteolytic Enzymes." English translation by S. C. Prescott. John Wiley & Sons, Inc., New York.

———, 1922: Sur les propriétés distinctives des amylases de différentes provenances, *C. R. Acad. Sci. Paris*, **174**: 18–21.

Emmerling, O., 1902: Ueber die Eiweissspaltung durch Papayotin, *Ber. chem. Ges.*, **35**: 695–700.

Englis, D. T., and H. A. Lunt, 1925: Effect of the concentration of potassium salts in soil media upon the carbohydrate metabolism of plants. The diastatic activity of the nasturtium, *Soil Sci.*, **20**: 459–463.

——— and L. Gerber, 1929: A study of diastase activity in plants: The effect of phosphates in the soil media, *Soil Sci.*, **28**: 221–223.

——— and C. D. Zannis, 1930: The effect of ethylene upon the activity of diastase and invertase, *J. Am. Chem. Soc.*, **52**: 797–802.

Fales, H. A., and J. M. Nelson, 1915: The effect of sodium chloride upon the action of invertase, *J. Am. Chem. Soc.*, **37**: 2769–2786.

Falk, K. G., and J. M. Nelson, 1912: Studies on enzyme action. I, Some experiments with the castor bean lipase, *J. Am. Chem. Soc.*, **34**: 735–745.

——— and ———, 1912: Studies on enzyme action. II, The hydrolytic action of some amino acids and polypeptides of certain esters, *J. Am. Chem. Soc.*, **34**: 828–845.

——— and M. L. Hamlin, 1913: Studies on enzyme action. III, The action of manganous sulfate on castor bean lipase, *J. Am. Chem. Soc.*, **35**: 210–219.

———, 1913: Studies on enzyme action. V, The action of neutral salts on the activity of castor bean lipase, *J. Am. Chem. Soc.*, **35**: 601–616.

———, 1913: Studies on enzyme action. VI, The specificity of lipase action, *J. Am. Chem. Soc.*, **35**: 616–624.

———, 1913: Studies on enzyme action. IX, Extraction experiments with the castor bean lipase, *J. Am. Chem. Soc.*, **35**: 1904–1915.

——— and K. Sugiura, 1915: Studies on enzyme action. XII, The esterase and lipase of castor beans, *J. Am. Chem. Soc.*, **37**: 217–230.

———, 1915: Studies on enzyme action. XIII, The lipase of soybeans, *J. Am. Chem. Soc.*, **37**: 649–653.

———, 1917: Studies on enzyme action. XIV, Further experiments on lipolytic actions, *J. Biol. Chem.*, **31**: 97–123.

———, G. McGuire, and E. Blout, 1919: Studies on enzyme action. XVII, The oxidase, peroxidase, catalase, and amylase of fresh and dehydrated vegetables, *J. Biol. Chem.*, **38**: 229–244.

——— and ———, 1921: Studies on enzyme action. XIX, The sucrolytic actions of bananas, *J. Gen. Physiol.*, **3**: 595–611.

——— and ———, 1922: Studies on enzyme action. XXI, Banana gel and banana sucrose, *J. Biol. Chem.*, **54**: 655–671.

———, 1935: The mechanism of enzyme actions, *Sci.*, **81**: 471–475.

Feist, K., 1908: Die Spaltung des Amygdalin unter dem Einfluss von Emulsin, *Archiv. Pharm.*, **246**: 206, 509.

Fermi, C., and Dr. Buscalioni, 1899: Die proteolytischen Enzyme im Pflanzenreiche, *Cent. Bakt.*, II, **5**: 24–27, 63–67, 91–95, 125–135, 146–158.

Fernbach, A., and L. Hubert, 1900: Sur la diastase proteolytique du malt, *C. R. Acad. Sci. Paris*, **130**: 1783–1785.

Filipowicz, B., 1931: Influence of proteins and protein derivatives on the enzymic hydrolysis of starch by malt diastase, *Biochem. J.*, **25**: 1874–1884.

Fisher, D. F., C. P. Harley, and C. Brooks, 1930: The influence of temperature on the development of water core, *Proc. Am. Soc. Hort. Sci.*, **27**: 276–280.

Fisher, E. A., 1919: Contributions to the study of the vegetable proteases. I, Introductory, *Biochem. J.*, **13**: 124–134.

Frankel, E. M., 1917: Studies on enzyme action. XV, Factors influencing the proteolytic activity of papain, *J. Biol. Chem.*, **31**: 201–215.

Freeman, G. G., and R. H. Hopkins, 1936: The mechanism of degradation of starch by amylases. I, Nature of the early fission products, *Biochem. J.*, **30**: 442–445.

——— and ———, 1936: The mechanism of degradation of starch by amylases. II, Kinetics of action of α- and β- amylases on starch, its components and partial degradation products, *Biochem. J.*, **30**: 446–450.

——— and ———, 1936: The mechanism of degradation of starch by amylases. III, Mutarotation of fission products, *Biochem. J.*, **30**: 451–456.

Fuller, H. J., 1931: Some effects of ultra-violet radiation upon plant enzymes, abst., *Am. J. Bot.*, **18**: 895.

———, 1932: Some effects of radiations from a mercury vapor arc in quartz upon enzymes, *Ann. Mo. Bot. Gard.*, **19**: 505–531.

Gardiner, W., 1897: The histology of the cell wall with special reference to the mode of connection of cells, *Proc. Roy. Soc.*, **62**: 100–112.

Gerber, C., 1907: La présure des *Crucifères*, *C. R. Soc. Biol.*, **62**: 1223–1225.

———, 1907: Action du phosphate neutre de sodium sur la coagulation du lait de vache par les présures végétales, *C. R. Soc. Biol.*, **63**: 640–642.

———, 1907: Action du phosphate neutre de potassium sur la coagulation du lait de vache par les présures végétales, *C. R. Soc. Biol.*, **63**: 642–644.

———, 1907: La présure des *Crucifères*, *C. R. Acad. Sci. Paris*, **145**: 92–94.

———, 1907: La présure des *Rubiacées*, *C. R. Acad. Sci. Paris*, **145**: 284–286.

———, 1907: Les agents de la coagulation du lait contenus dans le suc du Mûrier de Chine (*Broussonetia papyrifera*), *C. R. Acad. Sci. Paris*, **145**: 530–532.

———, 1908: Action des acides sur la coagulation du lait par les présures végétales, *C. R. Acad. Sci. Paris*, **146**: 1111–1114.

———, 1908: Suc présurants des *Renonculacées*, *C. R. Soc. Biol.*, **64**: 522–523.

———, 1909: La présure du papayer.—Son action sur le lait bouilli aux diverses températures, *C. R. Soc. Biol.*, **66**: 227–229.

———, 1909: Relations entre les ferments présurants et les ferments proteolytiques végétaux. Leur rôle dans la plante, *C. R. Soc. Biol.*, **67**: 332–334.

———, 1909: Méthode générale de préparation des présures végétales, *C. R. Soc. Biol.*, **66**: 890–892.

———, 1909: Action comparée des présures végétales sur la peptone et la caséine. I, Type *Composées*. II, Algues brunes, *C. R. Soc. Biol.*, **66**: 1122–1125, 1125–1127.

———, 1909: Activité présurante des divers organes des *Composées*, *C. R. Congr. Soc. Sav. Rennes*, **1909**: 152–155.

———, 1909: Coagulation du lait cru par la présure du Papayer (*Carica papaya*), *C. R. Acad. Sci. Paris*, **148**: 497–500.

———, 1909: Répartition de la présure dans les membres et tissus végétaux, *C. R. Acad. Sci. Paris*, **148**: 992–994.

———, 1912: Relations entre l'activité présurante du latex du *Euphorbes* et l'espèce considérée ou la partie du végétal considérée. Résistance de cette présure à la chaleur, *C. R. Soc. Biol.*, **73**: 578–580.

———, 1912: Action des sels neutres sur la caséification du lait par la présure du latex des *Euphorbes*, *C. R. Soc. Biol.*, **73**: 582–584.

Glick, D., and C. G. King, 1933: The protein nature of enzymes. An investigation of pancreatic lipase, *J. Am. Chem. Soc.*, **55**: 2445–2449.

Glimm, E., and W. Sommer, 1927: Beitrag zur Reinigung der Malzamylase, *Biochem. Z.*, **188**: 290–325.

Gore, H. C., and S. Józsa, 1932: Sugar formation by diastatic enzymes of flour, *Ind. and Eng. Chem.*, **24**: 99–102.

——— and ———, 1932: Diastatic enzymes in certain foods, *Ind. and Eng. Chem.*, **24**: 102.

———, 1936: Action of amylases on starch, *Ind. and Eng. Chem.*, **28**: 86–88.

Green, J. R., 1887: On the changes in the proteids in the seed which accompany germination, *Phil. Trans. Roy. Soc.*, B, **178**: 39–61.

———, 1887: On the germination of the tuber of the Jerusalem artichoke (*Helianthus tuberosus*), *Ann. Bot.*, **1**: 223–236.

———, 1888: Vegetable rennet, *Nature*, **38**: 274–276.

———, 1890: On the germination of the seed of the castor-oil plant, *Proc. Roy. Soc.*, **48**: 370–392.

———, 1892: On the occurrence of vegetable trypsin in the fruit of *Cucumis utilissimus*, *Ann. Bot.*, **6**: 195–202.

———, 1893: On vegetable ferments, *Ann. Bot.*, **7**: 83–137.

———, 1894: Researches on the germination of the pollen grain and the nutrition of the pollen tube, *Phil. Trans. Roy. Soc.*, B, **185**: 385–409.

———, 1894: The influence of light on diastase, *Ann. Bot.*, **8**: 370–373.

———, 1897: On the action of light on diastase and its biological significance, *Phil. Trans. Roy. Soc.*, B, **188**: 167–190.

———, 1899: "The Soluble Ferments and Fermentation," Cambridge University Press, 480 pp.

GRIFFIN, E. G., and J. M. NELSON, 1916: The influence of certain substances on the activity of invertase, *J. Am. Chem. Soc.*, **38**: 722–730.

GRÜSS, J., 1893: Ueber den Eintritt von Diastase in das Endosperm, *Ber. bot. Ges.*, **11**: 286–292.

———, 1894: Ueber das Verhalten des diastatischen Enzyms in der Keimpflanze, *Jahrb. wiss. Bot.*, **26**: 379–437.

———, 1895: Die Diastase im Pflanzenkörper, *Ber. bot. Ges.*, **13**: 2–13.

———, 1896: Beiträge zur Physiologie der Keimung, *Landw. Jahrb.*, **25**: 385–452.

———, 1897: Ueber die Secretion des Schildchens, *Jahrb. wiss. Bot.*, **30**: 645–664.

HABERLANDT, G., 1890: Die Kleberschicht des Gras-Endosperms als Diastase ausscheidendes Drüsengewebe, *Ber. bot. Ges.*, **8**: 40–48.

HAEHN, H., and H. SCHWEIGART, 1923: Zur Kenntnis der Kartoffelamylase, *Biochem. Z.*, **143**: 516–526.

HALEY, D. E., and J. F. LYMAN, 1921: Castor bean lipase, its preparation and some of its properties, *J. Am. Chem. Soc.*, **43**: 2664–2670.

HANES, C. S., 1932: Studies on plant amylases. I, The effect of starch concentration upon the velocity of hydrolysis by the amylase of germinated barley, *Biochem. J.*, **26**: 1406–1421.

HANSTEEN, B., 1894: Ueber die Ursachen der Entleerung der Reservestoffe aus Samen, *Flora*, **79**: 419–429.

HARADA, T., 1931: Preparation of *Aspergillus oryzae* enzymes, *Ind. and Eng. Chem.*, **23**: 1424–1427.

HARLEY, C. P., D. F. FISHER, and M. P. MASURE, 1931: A starch splitting enzyme in apple tissues, *Proc. Am. Soc. Hort. Sci.*, **28**: 561–565.

HARRIS, M. R., and W. L. GOSS, 1934: A seedling disease of sorghum and Sudan grass, *Calif. Sta. Month. Bull.*, **23**: 109–118.

HENRI, V., 1902: Théorie générale de l'action de quelques diastases, *C. R. Acad. Sci. Paris*, **135**: 916–919.

——— and A. LALOU, 1903: Action de l'émulsine sur la salicine et l'amygdaline. Théorie de l'action de l'émulsine, *C. R. Acad. Sci. Paris*, **136**: 1693–1694.

——— and M. NICLOUX, 1904: Influence des proportions d'huile et d'acide sur la vitesse de saponification par la lipaséidine, *C. R. Soc. Biol.*, **57**: 175–177.

HEPBURN, J. S., 1918: Biochemical studies of the pitcher liquor of *Nepenthes*, *Proc. Am. Phil. Soc.*, **57**: 112–130.

———, E. Q. ST. JOHN, F. M. JONES, and W. F. BAKER, 1922: Studies of the North American *Sarraceniaceae*, *J. Biol. Chem.*, **50**: *Proc.* XLVI.

HILTNER, R. S., 1900: The fatal effect of green sorghum, *Neb. Sta. Bull.* 63.

HOGOUNENQ, L., and J. LOISELEUR, 1925: Sur la constitution des diastases proteolytique et le mécanisme de leur actions, *Bull. Soc. Chim. Biol.*, **7**: 955–973.

HOLTER, H., and K. LINDERSTROM-LANG, 1933: Beiträge zur enzymatischen Histochemie. III, Ueber die Proteinasen von *Drosera rotundifolia*, *Z. physiol. Chem.*, **214**: 222–240.

HOPKINS, R. H., 1929: The proteolytic enzymes of green malt. I, Adsorption and elution, *Biochem. J.*, **23**: 1022–1029.

———, G. F. COLE, and J. W. GREEN, 1933: Some observations on barley diastase, *J. Inst. Brewing*, **39**: 487–493; *Biol. Abst.*, **8**: 1409.

HORNING, E. S., and A. H. K. PETRIE, 1927: The enzymatic function of mitochondria in the germination of cereals, *Proc. Roy. Soc.*, B, **102**: 188–206.

HORTON, E., 1921: On the use of taka-diastase in estimating starch, *Jour. Agr. Sci.*, **11**: 240–257.

HOYER, E., 1904: Ueber fermentative Fettspaltung, *Ber. chem. Ges.*, **37**: 1436–1447.

———, 1907: Ueber fermentative Fettspaltung, *Z. physiol. Chem.*, **50**: 414–435.

HUBER, H., 1929: Über den Zustand und die Rolle der Gerbstoffe in der Pflanze, *Jahrb. wiss. Bot.*, **70**: 278–327; *Biol. Abst.*, **5**: 2669.

HUDSON, C. S., 1908: The inversion of cane sugar by invertase. I, II, *J. Am. Chem. Soc.*, **30**: 1160–1166, 1564–1583.

———, 1909: The inversion of cane sugar by invertase. III, *J. Am. Chem. Soc.*, **31**: 655–664.

——— and H. S. PAINE, 1910: The inversion of cane sugar by invertase. IV, The influence of acids and alkalies on the activity of invertase, *J. Am. Chem. Soc.*, **32**: 774–779.

——— and———, 1910: The inversion of cane sugar by invertase. V, The destruction of invertase by acids, alkalis and hot water, *J. Am. Chem. Soc.*, **32**: 985–989.

———, 1910: The inversion of cane sugar by invertase. VI, A theory of the influence of acids and alkalis on the activity of invertase, *J. Am. Chem. Soc.*, **32**: 1220–1222.

——— and H. S. PAINE, 1910: The inversion of cane sugar by invertase. VII, The effect of alcohol on invertase, *J. Am. Chem. Soc.*, **32**: 1350–1357.

———, 1910: The quantitative determination of cane sugar by the use of invertase, *J. Ind. Eng. Chem.*, **2**: 143–146.

———, 1914: The inversion of sucrose by invertase. VIII, An improved method for preparing strong invertase solutions from top or bottom yeast, *J. Am. Chem. Soc.*, **36**: 1566–1571.

——— and H. S. PAINE, 1914: The inversion of sucrose by invertase. IX, Is the reaction reversible? *J. Am. Chem. Soc.*, **36**: 1571–1581.

HUERRE, R., 1909: Sur la maltase du maïs, *C. R. Acad. Sci. Paris*, **148**: 300–302, 505–507.

———, 1909: Influence de la réaction du milieu sur la activité des maltases du maïs, *C. R. Acad. Sci. Paris*, **148**: 1121–1123.

———, 1909: Sur la maltase du sarrasin, *C. R. Acad. Sci. Paris*, **148**: 1526–1528.

HUTCHINSON, A. H., and M. R. ASHTON, 1933: The effect of radiant energy on diastase activity, *Can. J. Res.*, **9**: 49–64.

JACOBSON, C. A., and A. HOLMES, 1914: Alfalfa investigations. V, Enzymes present in alfalfa, *J. Am. Chem. Soc.*, **36**: 2170–2182.

JAMES, W. O., and M. CATTLE, 1933: Studies on the physiological importance of the mineral elements in plants. VI, The influence of potassium chloride on the rate of diastatic hydrolysis of starch, *Biochem. J.*, **27**: 1805–1809.

JAVILLIER, M., 1903: Sur quelques ferments proteolytiques associés à la présure chez les végétaux, *C. R. Acad. Sci. Paris*, **136**: 1013–1015.

JODIDI, S. L., and J. PEKLO, 1929: Symbiotic fungi of cereal seeds and their relation to cereal proteins, *J. Agr. Res.*, **38**: 69–93.

JONES, W. N., 1925: Polarized light and starch grains, *Ann. Bot.*, **39**: 651–653.

JOSLYN, M. A., and M. SHERRILL, 1933: Inversion of sucrose by invertase at low temperatures, *Ind. and Eng. Chem.*, **25**: 416–417.

JÓZSA, S., and H. C. GORE, 1932: Development of diastatic enzymes of malt and flour. Effects of certain salts, carbohydrates, and papain, *Ind. and Eng. Chem.*, **24**: 95–98.

KASTLE, J. H., and M. CLARK, 1903: On the occurrence of invertase in plants, *Am. Chem. J.*, **30**: 422–427.

KAYSER, E., 1891: Note sur les ferments de l'ananas, *Ann. Inst. Pasteur*, **5**: 456–463.

KELLERMAN, K. F., 1903: The effects of various chemical agents upon the starch converting power of taka diastase, *Bull. Torrey Bot. Club*, **30**: 56–70.

KERSTAN, G., 1934: The physiological significance of the glucosides in *Aesculus* and *Salix*, *Planta*, **21**: 677–698; *Chem. Abst.*, **28**: 6772.

KERTESZ, Z. I., 1931: Discharge of saccharase from mycelium of *Penicillium glaucum*, *Plant Physiol.*, **6**: 249–264.

KIRSH, D., 1935: Lipase production by *Penicillium oxalicum* and *Aspergillus flavus*, *Bot. Gaz.*, **97**: 321–333.

KIZEL, A., 1915: The effect of the reaction of the medium on the activity of inulase of *Aspergillus niger*, *Isv. Imp. Akad. Nauk.* (*Bull. Acad. Imp. Sci.*) *Petrograd*, **6**: 1077–1092; *Exp. Sta. Rec.*, **37**: 203.

KNUDSON, L., 1920: The secretion of invertase by plant roots, *Am. J. Bot.*, **7**: 371–380.

KOGA, T., 1923: Zur Frage der Differenzierung tierischer und pflanzlicher Diastasen, *Biochem. Z.*, **142**: 159–164.

KOHL, F. G., 1908: Über die Reversibilität der Enzymwirkungen und den Einfluss äusserer Faktoren auf die Enzyme (Invertase, Maltase), *Beih. bot. Cent.*, **23**: 64*b*–64*o*.

KOSMANN, M. C., 1877: Recherches chimiques sur les ferments contenus dans les végétaux et sur les effets produits par l'oxydation du fer sur les matières organiques, *Bull. Soc. Chim.*, **27**: 251–256.

KRABBE, G., 1890: Untersuchungen über das Diastaseferment unter specieller Berücksichtigung seiner Wirkung auf Stärkekörner innerhalb der Pflanzen, *Jahrb. wiss. Bot.*, **21**: 521–561.

KRAUCH, C., 1879: Beiträge zur Kenntniss der ungeformten Fermente im Pflanzenreich, *Landw. Versuchs-Stat.*, **23**: 77–104.

KUHN, R., D. B. HAND, and M. FLORKIN, 1931: Über die Natur der Peroxydase, *Z. physiol. Chem.*, **201**: 255–266.

LEA, S., 1883: On a "Rennet" ferment contained in the seeds of *Withania coagulans*, *Proc. Roy. Soc.*, **36**: 55–58.

LING, A. R., and D. R. NANJI, 1923: On the presence of maltase in germinated and ungerminated barley, *Biochem. J.*, **17**: 593–596.

LINTNER, C. J., 1885: Studien über Diastase, *J. Prak. Chem.*, **34**(**2**): 378–394.

LINZ, F., 1896: Beiträge zur Physiologie der Keimung von *Zea mais*, *Jahrb. wiss. Bot.*, **29**: 267–319.

LONGENECKER, H. E., and D. E. HALEY, 1935: Ricinus lipase, its nature and specificity, *J. Am. Chem. Soc.*, **57**: 2019–2021.

LOTHROP, R. E., and H. S. PAINE, 1931: Diastatic activity of some American honeys, *Ind. and Eng. Chem.*, **23**: 71–74.

MACHT, D. I., and J. H. HILL, 1925: The influence of polarized light on yeast and bacteria, *Proc. Soc. Exper. Biol. Med.*, **22**: 474–475.

———, 1928: Effect of polarized light on plants, *abst.*, *Am. J. Bot.*, **15**: 621.

MAESTRINI, D., 1923: Contributo alla conoscenza degli enzimi. IX, Sulla resistenza all' invecchiamento di alcun enzimi dell'orzo germoglato, *Atti R. Accad. Lincei Roma Rendiconti Cl. Sci. Fis. Mat. e Nat.*, **32**: 397–398; *Bot. Abst.*, **13**: 178.

MANN, A., and H. V. HARLAN, 1915: Morphology of the barley grain with reference to its enzyme secreting areas, *U. S. Dept. Agr. Bull.*, **183**: 1–32.

MAQUENNE, L., and E. ROUX, 1906: Recherches sur l'amidon et sa saccharification disastasique, *Ann. Chem. et Phys.*, 8^e, **9**: 179–220.

MARTIN, S. H. C., 1884: Papaïn-digestion, *J. Physiol.*, **5**: 213–230.

———, 1885: The nature of papaïn and its action on vegetable proteid, *J. Physiol.*, **6**: 336–360.

MARTINAND, V., 1907: Recherche de l'invertine ou sucrase et du saccharose dans les divers organes de la Vigne et dans quelque fruits, *C. R. Acad. Sci. Paris*, **144**: 1376–1379.

Maslow, H. L., and W. C. Davidson, 1926: The effect of hydrogen ion concentration upon the starch liquefying activity of the amylase of *Aspergillus oryzae*, *J. Biol. Chem.*, **68**: 83–93.

Mathew, A. P., and T. H. Glenn, 1911: The composition of invertase, *J. Biol. Chem.*, **9**: 29–56.

McGinty, R. A., 1919: Diastase activity in relation to stage of development and carbohydrate content of the tuber *Solanum tuberosum*, *Ann. Mo. Bot. Gard.*, **6**: 223–251.

McGuire, G., and K. G. Falk, 1919: The saccharogenic actions of potato juice, *J. Gen. Physiol.*, **2**: 215–227.

——— and ———, 1923: Studies on enzyme action. XXIII, The spontaneous increase in sucrase activity of banana extracts, *J. Am. Chem. Soc.*, **45**: 1539–1551.

McNair, J. B., 1932: Some properties of plant substances in relation to climate of habitat-volatile oils, saponins, cyanogenetic glucosides, and carbohydrates, *Am. J. Bot.*, **19**: 168–193.

Mendel, L. B., and F. P. Underhill, 1901: Observations on the digestion of proteids with papaïn, *Trans. Conn. Acad. Arts and Sci.*, **11**: 1–14.

———, 1902: Observations on vegetable proteolytic enzymes, with special reference to papaïn, *Am. J. Med. Sci.*, **124**: 310–318.

——— and A. F. Blood, 1910: Some peculiarities of the proteolytic activity of papaïn, *J. Biol. Chem.*, **8**: 177–213.

Michaelis, L., and M. Rothstein, 1920: Zur Theorie der Invertasewirkung, *Biochem. Z.*, **110**: 217–233.

Michel-Durand, E., 1928–1930: Physiological investigations on tannins, *Rev. Gén. Bot.*, **40**: 705–734 (1928); **41**: 35–48, 106–118, 142–153, 237–251, 307–336, 388–399, 441–464, 513–528, 582–592, 637–652, 700–720, 772–783 (1929); **42**: 46–63, 115–128, 164–184, 233–253, (1930); *Chem. Abst.*, **27**: 3238.

———, 1929: "Recherches sur les composés tanniques," Librairie générale de l'enseignement, Paris, 293 pp., 11 Figs.

———, 1932: Origin of tannins from proteins and carbohydrates, *Rev. Gén. Bot.*, **44**: 161–214; *Chem. Abst.*, **27**: 3238.

Miehe, H., 1923: Über die Lebensdauer der Diastase, *Ber. bot. Ges.*, **41**: 263–268.

Miller, E. R., 1929: The chemical nature of enzymes, *Sci.*, **70**: 356.

Mounfield, J. D., 1936: The proteolytic enzymes of sprouted wheat, *Biochem. J.*, **30**: 549–557.

Murry, D. R. P., 1933: The activation of papain by cyanide and other substances. I, *Biochem. J.*, **27**: 541–556.

Mussi, Dr., 1890–1891: Cradina, a proteolytic enzyme in the fig (*Ficus carica*), abst., *Pharm. J.*, 3 ser., **21**: 560–561.

Navez, A. E., and B. B. Rubenstein, 1928: Starch hydrolysis as affected by polarized light, *J. Biol. Chem.*, **80**: 503–513.

———, 1930: On the alleged effect of polarized light on films of starch, *Sci.*, **72**: 13.

——— and B. B. Rubenstein, 1932: Starch hydrolysis as affected by light, *J. Biol. Chem.*, **95**: 645–660.

Naylor, N. M., M. Spencer, and M. House, 1925: The preparation and properties of amylase from germinated wheat and rye, *J. Am. Chem. Soc.*, **47**: 3037–3039.

Neidig, R. E., 1914: The survival of amylase in dried fodders, *J. Am. Chem. Soc.*, **36**: 1312–1314.

Nelson, J. M., and S. Born, 1914: A study of the chemical constitution of invertase, *J. Am. Chem. Soc.*, **36**: 393–403.

——— and E. G. Griffin, 1916: Adsorption of invertase, *J. Am. Chem. Soc.*, **38**: 1109–1115.

——— and W. C. Vosburgh, 1917: Kinetics of invertase action, *J. Am. Chem. Soc.*, **39**: 790–811.

——— and D. I. Hitchcock, 1921: The activity of adsorbed invertase, *J. Am. Chem. Soc.*, **43**: 1956–1961.

——— and ———, 1921: Uniformity in invertase action, *J. Am. Chem. Soc.*, **43**: 2632–2655.

——— and F. Hollander, 1923: Uniformity in invertase action, *J. Biol. Chem.*, **58**: 291–305.

——— and G. Bloomfield, 1924: Some characteristics of invertase action, *J. Am. Chem. Soc.*, **46**: 1025–1043.

——— and O. Bodansky, 1925: Mutarotation as a factor in the kinetics of invertase action, *J. Am. Chem. Soc.*, **47**: 1624–1638.

——— and C. Hemperly, 1933: Co-ferments and banana respiration, *J. Am. Chem. Soc.*, **55**: 1102–1106.

Neumeister, R., 1894: Ueber das Vorkommen und die Bedeutung eines eiweisslösenden Enzyms in jugendlichen Pflanzen, *Z. Biol.*, **30**: 447–463.

Newcombe, F. C., 1898: Cellulose-enzyme, *Bot. Cent.*, **73**: 105–108.

———, 1899: Cellulose-enzymes, *Ann. Bot.*, **13**: 49–81.

Nicloux, M., 1904: Étude de l'action lipolytique du cytoplasma de la graine de ricin, *C. R. Acad. Sci. Paris*, **138**: 1288–1290.

———, 1904: Sur le pouvoir saponifiant de la graine de ricin, *C. R. Soc. Biol.*, **56**: 702–704; *C. R. Acad. Sci. Paris*, **138**: 1175–1177.

———, 1904: Étude de l'action lipolytique du cytoplasma de la graine de ricin, *C. R. Soc. Biol.*, **56**: 839–843.

———, 1904: La propriété lipolytique du cytoplasma de la graine de ricin n'est pas due à un ferment soluble, *C. R. Acad. Sci. Paris*, **138**: 1352–1354; *C. R. Soc. Biol.*, **56**: 868–870.

———, 1904: Mécanisme d'action du cytoplasma (lipaseïdine) dans la graine en voie de germination: Réalisation synthétique in vitro de ce mécanisme, *C. R. Acad. Sci. Paris*, **139**: 143–145.

———, 1904: Mécanisme d'action du cytoplasma "lipaseïdine" dans la graine en voie de germination réalisation synthétique en vitro de ce mécanisme, *C. R. Soc. Biol.*, **57**: 84–86.

———, 1905: La saponification des corps gras, *Rev. Gén. des Sciences*, 16ᵉ année, No. 23, pp. 1029–1037.

———, 1906: Studies on enzyme action—Lipase, *Proc. Roy. Soc.*, B, **77**: 454–455.

Nordh, G., and E. Ohlsson, 1932: Amylases in resting and germinating seeds. I, Barley, *Z. physiol. Chem.*, **204**: 89–100; *Chem. Abst.*, **26**: 3003.

Ohlsson, E., 1922: Sur l'existence de deux ferments amylolytique dans la diastase du malt, *C. R. Soc. Biol.*, **87**: 1183–1184.

———, 1926: On the two components of malt diastase, *C. R. trav. lab. Carlsberg*, **16**: 1.

———, 1930: Über die beiden Komponenten der Malzdiastase, besonders mit Rücksicht auf die Mutarotation der bei der Hydrolyse der Stärke gebildeten Produkte, *Z. physiol. Chem.*, **189**: 17–63.

——— and C. E. Uddenberg, 1933: Amylases in resting and germinating seeds. II, Rye, *Z. physiol. Chem.*, **221**: 165–173; *Chem. Abst.*, **28**: 504–505.

——— and O. Edfeldt, 1933: Amylases in resting and germinating seeds. III, Oats, *Z. physiol. Chem.*, **221**: 174–179; *Chem. Abst.*, **28**: 505.

Olsson, U., 1921: Über Vergiftungserscheinungen an Amylasen, *Z. physiol. Chem.*, **117**: 91–145.

Oparin, A., and S. Risskina, 1932: Über die Aktivität der Amylase in den Blättern der Zukerrübe, *Biochem. Z.*, **252**: 8–15.

OSBORNE, T. B., 1895: The chemical nature of diastase, *J. Am. Chem. Soc.*, **17**: 587–604.

——— and G. F. CAMPBELL, 1896: The chemical nature of diastase. II, *J. Am. Chem. Soc.*, **18**: 536–542.

OSBORNE, W. A., 1898: Beiträge zur Kenntniss des Invertins, *Z. physiol. Chem.*, **28**: 399–425.

O'SULLIVAN, C., and F. W. TOMPSON, 1890: Invertase: A contribution to the history of an enzyme or unorganized ferment, *J. Chem. Soc.*, **57**: 834–931.

OTIS, C. H., 1930: A source of diastase, *Sci.*, **71**: 220.

PAINE, H. S., and R. T. BALCH, 1925: Application of enzymes to beet sugar factory control, *Ind. and Eng. Chem.*, **17**: 240–246.

———, C. F. WALTON, and M. S. BADOLLET, 1925: Industrial applications of invertase, *Ind. and Eng. Chem.*, **17**: 445–450.

PALLADIN, V. I., and E. M. POPOVA, 1924: On the origin of diastase in plants, *Bull. App. Bot. Plant-breeding,* **13**: 355–369 (Russian); *Bot. Abst.*, **14**: 1448.

PALLADIN, W., and H. POPOFF, 1922: Über die Entstehung der Amylase und Maltase in den Pflanzen, *Biochem. Z.*, **128**: 487–494.

PATON, F. J., D. R. NANJI, and A. R. LING, 1924: On the hydrolysis of the endosperm of *Phytelephas macrocarpa* by its own enzymes, *Biochem. J.*, **18**: 451–454.

PATON, J. B., 1921: Pollen and pollen enzymes, *Am. J. Bot.*, **8**: 471–501.

PELOUZE, J., 1855: Mémoire sur la saponification des huiles sous l'influence des matières qui les accompagnent dans les graines, *C. R. Acad. Sci. Paris*, **40**: 605–611.

PICKLER, W. E., 1919: Water content and temperature as factors influencing diastase formation in the barley grain, *Plant World,* **22**: 221–238.

PINCUSSEN, L., 1923: Fermente und Licht. I, Diastase. II, Urease, *Biochem. Z.*, **134**: 459–469, 470–475.

POLAK, F., and A. TYCHOWSKI, 1928: Über die Umwandlung der α-Diastase in β-Diastase, *Biochem. Z.*, **192**: 463–478.

POND, R. H., 1906: The incapacity of the date endosperm for self-digestion, *Ann. Bot.*, **20**: 62–78.

POZEN, M. A., 1934: Enzymes in brewing, *Ind. and Eng. Chem.*, **26**: 1127–1133.

POZERSKI, E., 1909: Contribution à l'étude physiologique de la papaïne, *Ann. Inst. Pasteur*, **23**: 321–359.

———, 1909: Digestion brusque produite par la bromeline. Comparaison avec la papaïne, *Ann. Inst. Pasteur*, **23**: 341–346.

PRATT, D. S., 1915: Papaïn. Its commercial preparation and digestive properties, *Philippine J. Sci.*, A, **10**: 1–37.

PRUNET, A., 1892: Sur le mécanisme de la dissolution de l'amidon dans la plante, *C. R. Acad. Sci. Paris,* **115**: 751–754.

PURIEWITSCH, K., 1897: Physiologische Untersuchungen über die Entleerung der Reservestoffbehälter, *Jahrb. wiss. Bot.*, **31**: 1–76.

QUINTANILHA, A., 1926: O proglema das plantas carnivoras. Estudo cito-fisiológico da digestão no *Drosophyllum lusitanicum*, *Bol. Soc. Broteriana* (Portuguese), **4**: 44–129; *Biol. Abst.*, **2**: 874.

REED, H. S., 1904: A study of the enzyme-secreting cells in the seedlings of *Zea mais* and *Phoenix dactylifera*, *Ann. Bot.*, **18**: 267–287.

RICHAND, A., 1900: "Recherches physiologique sur l'inulase et sur l'inuline," thèse 8°, Carre et Naud, Paris, 95 pp.

RINGER, W. E., and B. W. GRUTTERINK, 1927: Einfluss der Reaktion auf die Eiweissverdauende Kraft des Papains. II, *Z. physiol. Chem.*, **164**: 112–142.

ROBERTSON, R. A., J. C. IRVINE, and M. E. DOBSON, 1909: A polarimetric study of the sucroclastic enzymes in *Beta vulgaris*, *Biochem. J.*, **4**: 258–273.

Robinson, M. E., 1930: Cyanogenesis in plants, *Biol. Rev. and Biol. Proc. Cambridge Phil. Soc.*, **5**: 126–141; *Biol. Abst.*, **5**: 2675.

Russell, J. B., 1891: Bromelin, *Bull. Pharm.*, **5**: 77.

Sandberg, M., and O. M. Holly, 1932: Note on myrosin, *J. Biol. Chem.*, **96**: 443–447.

Sargant, E., and A. Robertson, 1905: The anatomy of the scutellum in *Zea mais*, *Ann. Bot.*, **19**: 115–124.

Schander, H., 1934: The significance of the aleurone layer in *Oryza* and other *Gramineae*, *Z. Bot.*, **27**: 433–515; *Chem. Abst.*, **29**: 489.

Schenker, R., 1921: Zur Kenntnis der Lipase von *Aspergillus niger*, *Biochem. Z.*, **120**: 164–196.

Schmitz, H., 1920: Enzyme action in *Echinodontium tinctorium*, *J. Gen. Physiol.*, **2**: 613–616.

Schultz, A. S., and Q. Landis, 1932: Vegetable amylases. Study of diastase action in the absence of maltose, *J. Am. Chem. Soc.*, **54**: 211–220.

Semmens, E. S., 1926: Hydrolysis of starch grains by light polarized by small particles, *Nature*, **117**: 821–822.

———, 1926: Some further notes on the hydrolysis of starch grains under polarized light, *Plant Physiol.*, **1**: 201–202.

———, 1932: Hydrolysis in green plants by moonlight, *Nature*, **130**: 243.

Sen, H. D., 1931: A study on the extraction of papain, the active digestive principle from *Papaya*, *J. Agr. Sci.*, **21**: 209–219.

Sherman, H. C., E. C. Kendall, and E. D. Clark, 1910: Studies on amylases. I, An examination of methods for the determination of diastatic power, *J. Am. Chem. Soc.*, **32**: 1073–1087.

——— and M. D. Schlesinger, 1913: Studies on amylases. V, Experiments upon the purification of the amylase of malt, *J. Am. Chem. Soc.*, **35**: 1617–1624.

——— and A. O. Gettler, 1913: Studies on amylases. VII, The forms of nitrogen in amylase preparations from the pancreas and from malt, as shown by the Van Slyke method, *J. Am. Chem. Soc.*, **35**: 1790–1794.

——— and A. W. Thomas, 1915: Studies on amylases. VIII, The influence of certain acids and salts upon the activity of malt amylase, *J. Am. Chem. Soc.*, **37**: 623–643.

——— and M. D. Schlesinger, 1915: Studies on amylases. IX, Further experiments upon the purification of malt amylase, *J. Am. Chem. Soc.*, **37**: 643–648.

——— and ———, 1915: Studies on amylases. X, Comparison of certain properties of pancreatic and malt amylase preparations, *J. Am. Chem. Soc.*, **37**: 1305–1319.

——— and A. P. Tanberg, 1916: Experiments upon the amylase of *Aspergillus oryzae*, *J. Am. Chem. Soc.*, **38**: 1638–1644.

——— and J. C. Baker, 1916: Experiments upon starch as substrate for enzyme action, *Am. J. Chem. Soc.*, **38**: 1885–1904.

——— and D. E. Neun, 1916: An examination of certain methods for the study of proteolytic action, *J. Am. Chem. Soc.*, **38**: 2199–2216.

——— and J. A. Walker, 1917: Influence of certain electrolytes upon the course of the hydrolysis of starch by malt amylase, *J. Am. Chem. Soc.*, **39**: 1476–1493.

———, A. W. Thomas, and M. E. Baldwin, 1919: Influence of hydrogen-ion concentration upon enzymic activity of three typical amylases, *J. Am. Chem. Soc.*, **41**: 231–235.

———, F. Walker, and M. L. Caldwell, 1919: Action of enzymes upon starches of different origin, *J. Am. Chem. Soc.*, **41**: 1123–1129.

——— and ———, 1919: Influence of aspartic acid and asparagin upon the enzymic hydrolysis of starch, *J. Am. Chem. Soc.*, **41**: 1866–1873.

——— and M. Wayman, 1921: Effect of certain antiseptics upon the activity of amylases, *J. Am. Chem. Soc.*, **43**: 2454–2461.

——— and F. WALKER, 1921: The influence of certain amino acids upon the enzyme hydrolysis of starch, *J. Am. Chem. Soc.*, **43**: 2461–2469.

——— and M. L. CALDWELL, 1921: A study of the influence of arginine, histidine, tryptophane and cystine upon the hydrolysis of starch by purified pancreatic amylase, *J. Am. Chem. Soc.*, **43**: 2469–2476.

——— and ———, 1922: Influence of amino acid in protecting amylase from inactivation by mercury, *J. Am. Chem. Soc.*, **44**: 2923–2926.

——— and———, 1922: Influence of lysine upon the hydrolysis of starch by purified pancreatic amylase, *J. Am. Chem. Soc.*, **44**: 2926–2930.

——— and N. M. NAYLOR, 1922: Influence of some organic compounds upon the hydrolysis of starch by salivary and pancreatic amylases, *J. Am. Chem. Soc.*, **44**: 2957–2966.

———, A. W. THOMAS, and M. L. CALDWELL, 1924: The isoelectric point of malt amylase, *J. Am. Chem. Soc.*, **46**: 1711–1717.

———, M. L. CALDWELL, and M. ADAMS, 1927: Establishment of the optimal hydrogen-ion activities for the enzymic hydrolysis of starch by pancreatic and malt amylases under varied conditions of time and temperature, *J. Am. Chem. Soc.*, **49**: 2000–2005.

———, ———, and H. H. BOYNTON, 1930: A quantitative study of the influence of acetate and of phosphate on the activity of malt amylase, *J. Am. Chem. Soc.*, **52**: 1669–1672.

———, ———, and S. E. DOEBBELING, 1934: Further studies upon the purification and properties of malt amylase, *J. Biol. Chem.*, **104**: 501–509.

SIGMUND, W., 1890: Ueber fettspaltende Fermente im Pflanzenreiche, *Monats. Chem. Wien*, **11**: 272–276.

———, 1892: Beziehungen zwischen fettspaltenden und glycosidspaltenden Fermente, *Sitzungsber. Akad. Wiss. Wien*, **101**: 549–559.

SJÖBERG, K., 1922: Beiträge zur Kenntnis der Amylase in Pflanzen. I, Über die Bildung und das Verhalten der Amylase in lebenden Pflanzen, *Biochem. Z.*, **133**: 218–294.

———, 1922: Beiträge zur Kenntnis der Amylase in Pflanzen. II, Die Temperaturempfindlichkeit der Amylase von *Phaseolus vulgaris*, *Biochem. Z.*, **133**: 294–330.

———, 1923: Beiträge zur Kenntnis der Amylase in Pflanzen. III, *Biochem. Z.*, **142**: 274–279.

SLADE, H. B., 1903: Prussic acid in sorghum, *J. Am. Chem. Soc.*, **25**: 55–59.

SÖRENSEN, S. P. L., 1908: Enzymstudien, *Biochem. Z.*, **7**: 45–101.

STERN, K. G., and E. STERN, 1932: Über die Proteinasen insectivorer Pflanzen, *Biochem. Z.*, **252**: 81–96.

STONE, W. E., 1896: The carbohydrates of wheat, maize, flour and bread and the action of enzymic ferments upon starches of different origin, *U. S. Dept. Agr. Off. Exp. Sta. Bull.* 34.

STOWARD, F., 1912: On the influence exercised by certain acids on the inversion of saccharose by sucrase, *Biochem. J.*, **6**: 131–140.

SULLIVAN, B., and M. A. HOWE, 1933: Lipases of wheat. I, *J. Am. Chem. Soc.*, **55**: 320–324.

SUMNER, J. B., 1926: The isolation and crystallization of the enzyme urease, *J. Biol. Chem.*, **69**: 435–441.

———, 1933: The chemical nature of enzymes, *Sci.*, **78**: 335.

SWANSON, C. O., and J. W. CALVIN, 1913: A preliminary study of the conditions which affect the activity of the amylolytic enzymes in wheat flour, *J. Am. Chem. Soc.*, **35**: 1635–1640.

———, 1921: Hydrocyanic acid in Sudan grass and its effect on cattle. *J. Am. Soc. Agron.*, **13**: 33–36.
———, 1935: Some factors which affect the diastatic activity in wheat, *Cereal Chem.*, **12**: 89–107.
TAIT, L., 1875: Insectivorous plants, *Nature*, **12**: 251–252.
TAKAMINE, J., and O. KOKICHI, 1920: The properties of a specially prepared enzymic extract, polyzime, comparing its starch liquefying power with malt diastase, *J. Am. Chem. Soc.*, **42**: 1261–1265.
TANKA, Y., 1910: The action of acids in the enzymic decomposition of oil by castor seeds, *J. Col. Eng. Tokyo*, **5**: 25–42. Quoted by Bayliss, 1925.
———, 1912: The preparation of lipase-powder acting in neutral medium, *J. Col. Eng. Tokyo*, **5**: 125–136.
TASCHER, W. R., and G. H. DUNGAN, 1928: Seedling vigor and diastatic activity of dent corn as related to composition of endosperm and stage of maturity, *J. Am. Soc. Agron.*, **20**: 133–141.
TAUBER, H., 1936: "Experimental enzyme chemistry," Burgess Pub. Co., Minneapolis, Minn., 118 pp.
TAYLOR, A. E., 1906: On the action of lipase, *J. Biol. Chem.*, **2**: 87–105.
TEODORESCO, E. C., 1914: Température mortelle pour quelques diastases d'origine animale et végétale, *Rev. Gén. Bot.*, **25**: 599–630.
THATCHER, R. W., and G. P. KOCH, 1914: The quantitative extraction of diastases from plant tissues, *J. Am. Chem. Soc.*, **36**: 759–770.
———, 1915: Enzymes of apples and their relation to the ripening process, *J. Agr. Res.*, **5**: 103–116.
———, 1921: "The Chemistry of Plant Life," McGraw-Hill Book Company, Inc., New York, 268 pp.
THEIS, E. R., J. S. LONG, and C. E. BROWN, 1929: Studies in the drying oils. XII, Changes in linseed oil, lipase and other constituents of the flaxseed as it matures, *Ind. and Eng. Chem.*, **21**: 1244–1248.
THOMAS, A. W., 1917: A noteworthy effect of bromides upon the action of malt amylase, *J. Am. Chem. Soc.*, **39**: 1501–1503.
TORREY, J. C., 1902: Cytological changes accompanying the secretion of diastase, *Bull. Torrey Bot. Club.*, **29**: 421–435.
TYSDAL, H. M., 1934: Determination of hardiness in alfalfa varieties by their enzymatic responses, *J. Agr. Res.*, **48**: 219–240.
VINES, S. H., 1891: On the presence of a diastatic ferment in green leaves, *Ann. Bot.*, **5**: 409–412.
———, 1897, 1898, 1901: The proteolytic enzyme of *Nepenthes*. I, II, III, *Ann. Bot.*, **11**, 563–584; **12**: 545–555; **15**: 563–573.
———, 1902: Tryptophane in proteolysis, *Ann. Bot.*, **16**: 1–22.
———, 1903: Proteolytic enzymes in plants. I, II, *Ann. Bot.*, **17**: 237–264, 597–616.
———, 1904–1906, 1908–1910: The proteases of plants. I, II, III, IV, V, VI, VII, *Ann. Bot.*, **18**: 289–317; **19**: 149–162, 171–187; **20**: 113–122; **22**: 103–113; **23**: 1–18; **24**: 213–222.
VINSON, A. E., 1907: The function or invertase in the formation of cane and invert sugar dates, *Bot. Gaz.*, **43**: 393–407.
———, 1908: The endo- and ektoinvertase of the date, *J. Am. Chem. Soc.*, **30**: 1005–1020.
VON GORUP-BESANEZ, E., 1874: Ueber das Vorkommen eines diastatischen und peptonisirenden Fermentes in den Wickensamen, *Ber. chem. Ges.*, **7**: 1478–1480.
———, 1875: Weitere Beobachtungen über diastatische und peptonbildende Fermente im Pflanzenreiche, *Ber. chem. Ges.*, **8**: 1510–1516.

——— and H. Will, 1876: Fortgesetzte Beobachtungen über peptonbildende Fermente im Pflanzenreiche, *Ber. chem. Ges.*, **9**: 673–678.

Vosburgh, W. C., 1921: Some errors in the study of invertase action, *J. Am. Chem. Soc.*, **43**: 1693–1705.

Waksman, S. A., 1920: A method of testing the amylolytic action of the diastase of *Aspergillus oryzae*, *J. Am. Chem. Soc.*, **42**: 293–299.

Waldschmidt-Leitz, E., 1933: The chemical nature of enzymes, *Sci.*, **78**: 189–190.

Ward, M., 1888: A lily disease, *Ann. Bot.*, **2**: 319–383.

Weis, F., 1900: Ueber das proteolytische und ein eiweisscoagulirendes Enzym in keimender Gerste (Malz), *Z. physiol. Chem.*, **31**: 79–97.

White, J., 1909: The ferments and latent life of resting seeds, *Proc. Roy. Soc.*, B, **81**: 417–442.

———, 1910: The proteolytic enzyme of *Drosera*, *Proc. Roy. Soc.*, B, **83**: 134–139.

Wierzchowski, Z., 1913: Studien über die Einwirkung von Maltase auf Stärke, *Biochem. Z.*, **56**: 209–219.

———, 1913: Über das Auftreten der Maltase in Getreidearten, *Biochem. Z.*, **57**: 125–131.

Willstätter, R., 1922: Über Isolierung von Enzymen, *Ber. chem. Ges.*, **55**: 3601–3623.

———, J. Graser, and R. Kuhn, 1922: Zur Kenntnis des Invertins, *Z. physiol. Chem.*, **123**: 1–78.

Wolff, J., 1916: Sur une substance coagulant l'inuline et l'accompagnant dans les tissus végétaux, *C. R. Acad. Sci. Paris.*, **162**: 514–516.

Wurtz, A., 1879: "Recherches cliniques et chimiques sur la papaïne," Paris.

——— and E. Bouchut, 1879: Sur le ferment digestif du *Carica papaya*, *C. R. Acad. Sci. Paris*, **89**: 425–429.

———, 1880: Sur la papaïne. Nouvelle contribution à l'histoire des ferments solubles, *C. R. Acad. Sci. Paris*, **91**: 787–791.

Young, V. H., 1918: Some factors affecting inulase formation in *Aspergillus niger*, *Plant World*, **21**: 75–87, 114–133.

Zeile, K., and H. Hellström, 1930: Über die aktive Gruppe der Leberkatalase, *Z. physiol. Chem.*, **192**: 171–192.

———, 1931: Über die aktive Gruppe der Katalase. II, *Z. physiol. Chem.*, **195**: 39–48.

CHAPTER XII

THE TRANSLOCATION OF MATERIALS IN PLANTS

I. NEED OF TRANSLOCATION

Most of the green plants are highly specialized organisms. They are thus composed of numerous interdependent parts or organs that perform special functions. Each organ must be supplied by other organs with the materials necessary for its functions, while it in turn furnishes other organs with certain materials needed by them. Thus the epidermal cells of the roots absorb the compounds that enter the plant from the soil. These substances are for the most part not used entirely or directly by these cells that absorb them but are moved to other parts of the plant where they are utilized in the various life functions. For example, only a very small portion of the water that is absorbed by the epidermal cells of the roots is utilized by these cells. The greater portion of it is moved through the roots and stem to supply the needs of their cells and then on to the leaves to replace the water lost by transpiration. The movement of sugar is another striking example of translocation. This carbohydrate is manufactured only in the green cells of the leaf but is needed by every living cell of the plant for the manufacture of other foods, for respiration, for osmotic effects, and for other purposes. Sugar thus must be moved from the leaves to every living cell in the plant. The reserve foods of plants are practically all stored in places that are some distance removed from the regions in which they will eventually be utilized. Thus when the demand for them arises, they must be translocated to those parts of the plant in which they are needed.

II. TISSUES CONCERNED

In the broadest sense, all the living cells and certain of the nonliving cells play a role in the movement of materials in plants. In the lower plants where materials need to be moved for only short distances, no specialization of cells for translocation occurs and all cells are equally concerned in the process. Also, if, in the higher plants, materials are moved for relatively short distances only, the cells show no special adaptations for translocation. Thus the materials taken from the soil by the epidermal cells of the root are moved to the endodermis across the parenchyma cells of the cortex. Likewise, the raw materials needed by the chlorenchyma cells of the leaf, as well as their manufactured products, are moved across them to and from the fibrovascular bundles. These parenchyma cells, as well as those of the pith and cortex of the stem, show no special adaptations for translocation other than that they are possessed of thin cell walls that are readily permeable to solvent and solutes.

The tissues of the green plant that are specialized for the translocation of materials are the xylem, the phloem and possibly the latex system.

A. The Xylem

The cells of the xylem especially concerned in translocation are the tracheids, the vessels or tracheae, the wood parenchyma, or xylem parenchyma, and the xylem medullary rays. The tracheids are elongated cells with tapering ends. They may have relatively thin walls that are strengthened by spiral or annular thickenings, or the walls may be thick, lignified, and possessed of numerous pits, generally of the bordered type. The tracheids are living cells for varying periods after their formation but eventually lose their protoplasmic contents so that they consist only of their cell walls. The vessels or tracheae, according to Eames and MacDaniels (1925), include those xylem elements which have a large lumen and marked perforations, especially in the end cell walls. The development of these perforations in the end walls brings a series of cells into a definite tubelike system. The conduction in the vessels is thus direct and is in contrast with the indirect lines of conduction in a group of tracheids. The vessels frequently have thin walls that are strengthened by annular or spiral thickenings. In other cases, the walls are thickened but generally not to the degree of the walls of the tracheids, and the vessels then are possessed of pits in their lateral walls. These pits are usually more numerous and smaller than those of the tracheids. The vessels also lose their protoplasmic contents at an early period after their development. Malhotra (1931) found the average length in centimeters of tracheae was 50 for grape; 49 for peach; 46 for apricot; 42 for plum; 34 for prune, lemon, and Delicious apple; 24 for cherry; 20 for walnut; and 17 for Bartlett pear. The "wood" or "xylem parenchyma" is the name given to those living cells which are associated with the tracheids and vessels. These cells may be thin walled or thick walled and supplied with simple pits. They retain their protoplasmic contents and remain alive for a long period of time.

The xylem rays, or wood rays, are ribbonlike strips of tissue that extend radially into the xylem. These strips are composed of parenchyma cells that have their long axes at right angles to the long axes of the tracheids, vessels, and wood parenchyma. These cells are living cells and contain their protoplasmic contents for a long period of time, and, as mentioned in Chap. I, they remain alive longer than any other known plant cells. The ray cells have cell walls of varying degrees of thickness and have numerous simple pits on all their walls. According to Eames and MacDaniels (1925), the xylem rays may be one, two, or several cells in width, and in height they may range from one cell to many and thus may vary in height from a fraction of a millimeter to 8 to 10 mm. in extreme cases. In a given stem some of the xylem rays may extend from the cambium to the pith, while others extend from the cambium to varying distances in the xylem.

In this general description of the conducting elements of the xylem, it should be noted that those tracheids and vessels which have lost their protoplasmic contents and are yet concerned in conduction are in contact with some parts of their wall surface with one or more living cells. Abundant pits are also present in these contact areas. The living cells with which these nonliving elements are in contact may be either the xylem parenchyma or xylem rays, or both these elements.

B. The Phloem

The phloem is composed of the sieve tubes, companion cells, phloem parenchyma, and phloem rays. A sieve tube is an elongated living cell with a thin cellulose wall. It has a large vacuole, and the cytoplasm forms a thin peripheral layer. No nucleus is present in a sieve tube. The sieve tubes are joined end to end and are arranged more or less definitely in longitudinal rows. Their protoplasmic contents are connected through pores or perforations in their end and lateral walls. These pores are located in more or less definite areas which are termed "sieve plates." Although these sieve plates may occur over the end and side walls, they are limited to a considerable extent to the end walls (Fig. 34). The cell sap of many of the sieve tubes contains a relatively large amount of protein compounds of a slimy nature (Hill, 1901, 1908). The outstanding characteristics of the sieve tubes are thus the pores in their walls, which are perforated with conspicuous cytoplasmic strands, the absence of a nucleus, and the relatively large amount of protein compounds in their vacuoles.

Crafts (1931, 1933) found that young sieve tubes are nucleated and show the characteristics of ordinary living cells. As they mature the nuclei disintegrate, and the protoplasm apparently changes in its organization and becomes more permeable.

The companion cells are elongated, slender, parenchymatous cells that have a small vacuole and dense protoplasmic contents with a prominent nucleus. One companion cell is associated with each sieve tube, but it is often difficult to determine the sieve tube with which a given companion cell is associated (Fig. 34). A companion cell has numerous simple pits connecting with the sieve tube with which it is associated.

The phloem parenchyma cells have dense protoplasmic contents and possess simple pits in both their longitudinal and end walls. The relative amount of phloem parenchyma varies greatly with different types of phloem and with different plants. In some cases it makes up the greater portion of the phloem, while in others it is entirely absent. The phloem rays have the same general structure and arrangement as the xylem rays. It should be noted in this connection that the elements which compose the phloem are all living cells, which is in marked contrast with the elements of the xylem tissue.

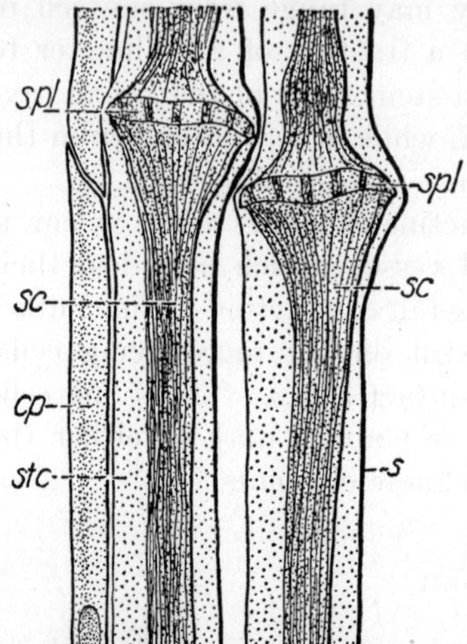

FIG. 34.—Sieve tubes and a companion cell from the pumpkin vine. *spl*, sectional view of sieve plates. *s'pl'*, surface view of sieve plates. *sc*, protoplasmic contents of sieve tube, here contracted to the center of the cell. *stc*, cell cavity caused by the contraction of the protoplasmic contents. *s*, cell wall of sieve tube. *cp*, companion cell.

C. The Latex System

Certain families of the Angiosperms including the *Papaveraceae, Lobeliaceae, Campanulaceae, Asclepiadaceae, Apocynaceae, Euphorbiaceae, Moraceae, Araceae,* and *Musaceae,* are characterized by the possession of an apparent additional conducting system in the form of laticiferous elements or latex tubes, which owe their name to the milky appearance of their contents. There are two distinct types of laticiferous organs: the articulated latex tubes or vessels and the nonarticulated latex tubes or latex cells. According to Eames and MacDaniels (1925), the former originate from rows of meristematic cells by either the complete or partial absorption of the separating walls early in the development of the vessel. In the mature plant a much branched anastomosing system is formed by the joining of more or less parallel ducts through connecting living cells and by the penetration of lateral outgrowths into the surrounding tissue. The latex cell on the other hand is structurally a single cell which as the plant grows develops into a branching system ramifying throughout the entire plant body. In this method of growth the tips of the growing cell penetrate between the cells of the tissue after the manner of a parasitic fungus. In this type of laticiferous duct, there is no anastomosing as in the latex vessel.

The laticiferous structures are living and are coenocytic in nature. The liquid that fills these structures is called "latex," is under pressure, and is apparently the cell sap of these cells. In color it is white, yellow, orange, or blood red depending on the plant in which it occurs. The latex is composed of water in which are dissolved or suspended mineral salts, sugars, proteins, tannins, gums, resins, alkaloids, starch, caoutchouc and proteolytic enzymes. It is commercially important as the source of opium, papain, India rubber, gutta-percha, and chicle (Haberlandt, 1914).

According to Barnes (1930) the function of conduction has been ascribed to the laticiferous system because of its relation to the nutritive cells of the leaves and the abundance of food in the latex. The carbohydrate and nitrogenous foods sometimes compose 30 per cent of the dry matter therein; they are most abundant at the beginning of active growth and development and are least in amount when growth is checked and a resting period is at hand. It may be, however, that the latex is concerned in the storage of materials in the plant or serves as a protection against predatory insects or animals. A thorough review of the physiology of latex is given by Moyer (1937).

III. MOVEMENT OF MATERIALS

A. Translocation of Water

The movement of water from the roots to the leaves, especially in the case of trees, has been of much general interest for several centuries. One of the most common questions asked by the general public of those individuals whom they know to be interested in the functions of plants concerns the manner in which the sap is moved through the stems of plants. It should be stated here that at present no answer that is entirely satisfactory can be given to this question. Although a large amount of experimental work has been undertaken to demonstrate the manner of water movement in plants, the evidence that has been obtained is in no way conclusive in many respects. In the following topics it is the intention to relate not only those facts which have been definitely established in regard to the translocation of water through the plant but also some of the theories concerning certain phases of this process that as yet have not been satisfactorily explained by the experimental evidence at hand.

1. Pathway of Water Movement.—The pathway through which most of the water travels from its entrance into the root until it is lost from the leaves into the atmosphere has been definitely established. The water enters the epidermal cells of the root, crosses the cells of the cortex, and enters the endodermis whence it moves into the lumina of the tracheids and vessels of the xylem of the root. It moves upward in these elements in the roots and thence through the same unit structures in the stem to the fibrovascular bundles of the leaf. From the xylem elements of the fibrovascular bundles of the leaf, it moves through the various chlorenchyma cells to those bordering on the intercellular air spaces or on the substomatal air cavities. Here it is lost from the cell walls in the form of vapor and escapes through the stomata to the external atmosphere. Some water may travel upward through the pith and cortex of the root and stem of certain plants, but apparently the amount moved in these regions is relatively small, except in young plants, compared with that moved through the channels that have just been mentioned. The lateral movement of water apparently takes place for the most part through the xylem rays and phloem rays in the case of young dicotyledons and through the general parenchymatous tissue of the stem in the case of the monocotyledons.

The pathway of water in the xylem has been determined directly by Peirce (1929), Harvey (1930, 1931), and Baker and James (1933) with the use of colored liquids and has been proved in an indirect manner by noting the lack of sufficient water to supply the leaves when the xylem elements are blocked or severed. The portion of the xylem concerned in water conduction has, however, in the case of woody plants been in considerable dispute. The work of Overton (1925), Overton and Smith (1926), and MacDougal, Overton, and Smith (1929) has thrown much light upon this controversy. Their work indicates that the portion of the xylem concerned in water translocation varies markedly in different plants. Thus by using acid fuchsin it was found for the willow (*Salix lasiolepis*) that the colored liquid rose in every annual layer in stems that were 11 or more years of age. It was also found that the liquid passed only through the summerwood of each layer. The colored zones were less than 1 mm. in thickness, while the unstained zones were from 2.5 to 4.5 mm. wide. In the alder (*Alnus oregona*) the same concentric zonation of water conduction is shown, but the movement of water in this stem does not take place through the late summerwood but is confined exclusively to the early formed springwood. In the walnut where heart- and sapwood are present all the conduction is in the sapwood through the late summer- and early springwood. In the pine (*Pinus radiata*) there is no localization of conduction in the xylem and the water moves through practically all parts of each annual ring.

Bowen (1931, 1933) found that the amount of water conducted over the external surface of moss plants, growing in either damp or dry areas, exceeded that conducted internally. The water that was conducted externally ascended in the form of capillary films between the leaves and the stem.

2. Water-conducting Cells.—The absorbing power of the epidermal cells of the root for water is always greater than that of the soil solution under the general conditions of plant growth. There is an increasing gradient in the absorbing power of the cortical cells for water from the epidermal cells of the root toward the endodermis. The absorbing power of the endodermis, however, in all the cases that have been investigated, drops considerably below that of the cortical cells bordering on it.

The cells concerned in the absorption and translocation of water across the root to the xylem elements of the central cylinder are all living cells. The tracheids and vessels of the root and stem with the exception of those which have been recently formed are, however, devoid of protoplasmic contents.

a. Gases in Conducting Cells.—The amount of water contained in these nonliving water-conducting elements has long been a matter of dispute. The observation of the contents of the tracheids and vessels under natural conditions is a difficult task, and in many of the observations the tissues examined were placed under abnormal conditions. It is thus impossible to determine whether the conditions of the cell contents observed were those normally prevailing or the result of the conditions imposed by the experiment. It has long been known that certain of the elements of the xylem contain gas bubbles with the liquid therein contained. At one time it was supposed that the presence of these bubbles lessened the total weight of a theoretical column of water and thus facilitated the ascent of sap in stems containing these so-called "Jaminian chains." It is now, however, almost universally recognized that the presence of a gas bubble in a tracheid or vessel at once eliminates that element from the conducting system. Those xylem elements which no longer function in the translocation of water are filled with gases, or the gases are intermingled as bubbles with the liquids yet remaining in them. The condition and composition of the gaseous mixture which fills these woody elements have been studied in detail by MacDougal (1925, 1926, 1927). He found that the gases of the central cylinder in the walnut may include carbon dioxide at a partial pressure 600 times as great as in the atmosphere, but with an oxygen pressure of only one-half that of the atmosphere. The gases in the central cylinder of the Monterey pine included carbon dioxide at a pressure 200 times as great as the atmosphere with the oxygen pressure reduced to one-fourth of that normally present in the air. The amounts of these gases vary during the season of year. Thus the carbon dioxide content of the oak was 3 per cent in October and 9 per cent in June. In the case of the willow the carbon dioxide content was 13.2 per cent in June and 8.4 per cent in August. The oxygen content of the walnut stem varied from 8.8 per cent in June to 12 to 13 per cent in July.

Gibbs (1930) found in jack pine that the heartwood contained approximately three times the amount of gas found in the sapwood. The heartwood, however, contained only 12 per cent of water as compared to 52 per cent in the sapwood. Chase (1934) in a study of the composition of gas drawn from the trunks of elm, oak, cottonwood, and white pine observed the following facts: (1) The percentage of carbon dioxide is highest during the growing season and is lowest in winter; (2) the percentage of oxygen varies inversely with the percentage of carbon dioxide at all times; (3) the volume or percentage of these two gases is less than their total in the atmosphere; and (4) the gas from the outer rings of sapwood is lower in carbon dioxide and higher in oxygen than that from the heartwood.

According to MacDougal, Overton, and Smith (1929), pressures and suctions on the gaseous system within the trunk of a tree are readily transmitted vertically for distances many times the length of the vessels. Tangential transmission, however, of suctions and pressures is at a very slow rate and even slower in a radial direction. The internal gaseous system of the tree is, however, in communication with the atmosphere radially through the cambium and vertically through the ends of the branches. The tensions in the air system have been shown to vary from less than ½ atmosphere to not to exceed 2 atmospheres. The relative volumes of the water and air systems within a tree are subject to variations during the season, so that specific conducting elements may at one time be partially or wholly filled with gas and at another time entirely filled with water.

Gibbs (1935) considered that the actual amount of gas in trees examined by him varied but little during diurnal changes but that it did vary with the seasonal fluctuation in water content. Priestley (1935) believed that under a deficit of water during the summer a large proportion of the water-conducting elements of trees become filled with gases. He did not propose, however, the means by which these vessels might be again refilled with water.

Using a specially devised method, Haines (1935) found that gases were sometimes present and sometimes absent in the conducting xylem of the twigs of elder, lilac, hawthorn, and ash. The presence or absence of gases depended largely on conditions external to the plant. He believed that a mechanism must exist in these plants for causing gas in bubble form to go again into solution.

b. Water in Conducting Cells.—According to MacDougal, Overton, and Smith (1929), it has been observed by Holle (1915) that when leaves were completely wilted the water columns remained intact in the conducting elements. These authors also stated that Bode (1923) in observations of the wilted plants of *Tradescantia*, *Cucurbita*, *Impatiens*, *Helianthus*, *Sambucus*, and *Phaseolus* noted that gas bubbles entered the conducting elements of these parts only when living cells in contact with them were touched or wounded. Bode thus concluded that bubbles may appear in conducting elements containing water only by being drawn in or by wounding. MacDougal, Overton, and Smith (1929) stated that in the stems examined by them the elements that are operative in conduction contain exclusively water, while those that are inoperative in conduction may contain gas exclusively or water and gas bubbles in the form of chains.

Gibbs (1935) considered his data indicated that there is a continuous water column in the conducting elements of the birch, larch, balsam, and spruce. The hardwoods examined by him showed a maximum water content in spring and a sharp drop in the summer. The degree of this variation differed from year to year. In some cases this loss of water from the woods during the summer was not replaced until the following spring. The softwoods showed no marked seasonal changes in their content of water.

It is thus apparently well established that the elements of the xylem concerned in conduction are completely filled with water and that on that account there is a continuous column of water through them from the endodermis to the chlorenchyma cells of the leaf. As mentioned in Chap. II, it has been shown that there is an increasing gradient of absorbing power for water in the chlorenchyma cells of the leaf from the fibrovascular bundles of the leaf toward its margins.

The evidence thus indicates that water columns are continuous in the plant from the soil particles through the epidermal cells, cortex, and endodermis of the root;

through the conducting xylem of the root and stem to the veins of the leaves; and thence through the cell walls and vacuoles of the chlorenchyma cells to the surfaces of the menisci formed in the submicroscopical cavities of the cell walls bordering on the intercellular spaces in the leaf. These water connections extending throughout the plant have been termed by MacDougal (1925 to 1926) the "hydrostatic system" of the plant. Woodhouse (1933) proposed the collective term "sap hydraulics" to designate the entrance of water into the plant, its passage through the stem, and its eventual loss from the stem and leaves.

3. Ascent of Water.—In attempting to determine the forces that are concerned in the ascent of water in plants, the investigator is handicapped because of the relatively small number of the external manifestations of the action of these forces as well as the difficulty encountered in measuring and correctly interpreting these manifestations. The external manifestations by which the action of the hydrostatic system of plants in elevating water may be estimated are, according to MacDougal (1925), (*a*) the amount of the intake of water, (*b*) the amount of transpiration, (*c*) the amount of water forced from the plant in the form of liquid, and (*d*) the registration of the daily reversible variations in stems as a result of the altered cohesive tensions in the ascending column of water. Some of these manifestations are more evident under conditions of low evaporation, while others are more evident under conditions of high evaporation. The forces that are concerned in the ascent of water in plants are apparently active all the time but are more manifest under one of these conditions than under the other. On that account, it is convenient in discussing the ascent of water to consider (*a*) the ascent of water in plants under conditions of low evaporation and (*b*) the ascent of water in plants under the conditions of high evaporation.

a. Low Evaporation.—Under conditions that are favorable for both low transpiration and the absorption of water from the soil, water is forced into and up through the xylem elements under considerable pressure. This fact is made manifest by the exudation of water from the stem and leaves of intact plants, the water so exuded being designated as "water of guttation." This action is also shown when bore holes are made into the stems and branches or when they are severed, since under these conditions water is forced out or exuded from the hole or stump apparently by some pressure within the stem or root. This exudation of water may also occur from perforations in the stem or from the stump of stems or branches under conditions of high transpiration, although a considerable period of time must elapse before the exudation develops. The force or pressure with which water is exuded from holes made in the stem or from the fresh stumps of stems or branches has been termed "root pressure," "sap pressure," "endodermal pressure," "bleeding pressure," or "exudation pressure." These terms have also been applied to the forces causing the exudation of water of guttation. The term "root pressure" has been the one most commonly used to designate not only the force with which water is being pushed up through the stem of the intact plant but also the force with which water is being exuded from some injured part. The term has been used because it has generally been considered that the force under consideration is generated in certain of the root cells. In some cases, however, as will be mentioned later, the exudation of water from cut or injured stems is brought about by forces generated in the living cells contiguous to the exuding region. In such cases "root pressure" is not the appropriate term to use. When water is exuded in considerable quantities from the stumps of cut stems or branches or from bore holes in the stem, as in the case of the grape vine, sugar maple, *Begonia*, or *Fuchsia*, the process has been termed "bleeding."

Lowry and Tabor (1931) reported that the stub of a single corn plant at the silking stage will exude in 3 days, under favorable conditions, more than 500 cc. of sap. By renewing the cut surface at frequent intervals, as much as 1,700 cc. of sap may be collected during a 15-day period. Some cut stems exude only a relatively small quantity of water, but there is no correlation between the amount of water exuded and the pressure with which it is being forced out of the stem.

The exudation pressures have been measured by manometers fitted to a bore hole in the side of large stems or to the stump, as in the case of the smaller stems or branches. Although many studies have been made of exudation pressures, the data concerning their value are very meager. Boehm (1892) found exudation pressures as high as 9 atmospheres in *Aesculus hippocastanum*, and Molisch (1902) reported pressures of over 6 atmospheres in the native trees of Austria. Clark (1874) reported root pressures of 2 to 2.5 atmospheres for *Betula lenta*, about 1 atmosphere for the sugar maple and 1.25 atmospheres for the grape vine. Eckerson (1908) studied the exudation pressures of the stumps of over 20 herbaceous plants growing under greenhouse conditions and found that they ranged from 0.43 to 1.4 atmospheres. MacDougal (1925) found that pressures of from 2 to 4 atmospheres developed in bore holes in the side of the trunk of Monterey pine. The pressures observed will depend to a considerable degree, however, upon the general conditions prevailing at the time the observations are made. White (1938) devised methods for measuring the root pressure of the excised roots of the tomato plant and found that pressures of 90 lb. per sq. in. were developed. This pressure is sufficient to raise water to a height of 200 ft. Root pressure thus may be a more important factor in sap movement than has generally been assumed.

It has generally been assumed that the exudation of water from cut stems and branches is due to the action of the cells of the root. There is considerable evidence, however, that exudation pressures are due, in many cases at least, to the action of cell masses contiguous to the cut surfaces (Molisch, 1902). It was considered by MacDougal (1925, 1926) that his work on the exudation pressures of the Monterey pine and the cactus *Carnegiea gigantea* indicated that the pressures developed were due to the local action of the living cells. Thus in the pine no root pressure could be observed, but exudation pressures as high as 4 atmospheres were observed in the stem. In the case of the cactus, exudation pressures as high as 2 atmospheres could be developed when no roots were present, and root pressure was thereby entirely absent. In the cases of guttation from epithem hydathodes, or in the case of continuous loss of water from a cut stem for a considerable period, as in the grape vine, the exudation is apparently caused by the action of certain cells of the root or stem other than those contiguous to the regions of water loss.

Heyl (1933) decided from experiments with the decapitated stems of *Ricinus* and *Brassica* that the intensity of bleeding closely paralleled soil temperatures. Experiments with narcotics showed definitely that the mechanism of bleeding is located in the roots and that it involves active secretion of water by living cells. The manner of exudation may not be the same for all plants, but this mechanism evidently depends upon complicated processes that are not purely osmotic in nature. Thus, James and Baker (1933) stated that there is no adequate evidence that the liquid escaping from the cut surfaces of plants issues from the vessels. They believed that any pressure on the sap that exudes from cut surfaces must be transmitted through the living cells and not through the vessels. On the other hand Ingold (1935) found indications that bleeding from birch trees occurs only when the vessels are injured.

In this connection the sap flow of the sugar maple should be mentioned. As is well known, large quantities of sap will escape from this tree during late winter and early spring through tap holes made in the lower part of the trunk. The causes for this escape of sap have long been a subject of discussion and are as yet little understood. Jones, Edson, and Morse (1903) found that the root pressure was very infrequent and that when it was observable it did not exceed 1.5 pounds. On the other hand, negative pressures were frequent during the time of flow and reached values as high as 6 pounds. They thus considered that the flow of sap cannot be attributed to root pressure, since the small positive pressure thus developed is too infrequent and slight to be the main causative factor. Gas expansion in the conducting elements occurs during the time of flow, but the force thus generated is considered to be inadequate to account for the results obtained. Jones, Edson, and Morse considered that the activities of the protoplasm of the living cells of the xylem are the dominant factors in causing this sap flow. They considered that the changes in temperature and the changes in gas pressure within the trunk are the stimuli that cause the protoplasm to become active.

Jones and Bradlee (1933) held that the periodic bleeding occurring at the close of winter, when the enclosing bark and the wood layer are tapped, represents an outpouring at the point of least resistance. In an untapped tree, the internal pressures aid in forcing the water to the extreme upper sections of the tree before transpiration begins.

Since the gradient of absorbing power for water increases from the epidermal cells of the root through the cortex to the endodermis, the migration of water from the epidermis to the endodermis can be attributed to osmotic and imbibitional forces, provided that a saturation deficit is maintained at the endodermis. In the leaves in the cases that have been examined, there is an increasing gradient of the absorbing power of the cells for water from the xylem elements of the veins across the chlorenchyma cells to the intercellular spaces. Here the movement of water from the veins toward the intercellular spaces may be explained on the basis of osmotic and imbibitional forces, provided a saturation deficit is maintained at the cell surfaces bordering upon the intercellular spaces.

The manner in which the water is transferred from the cortical cells of the root to the xylem elements and then forced upward is not known. The theories that have been proposed to account for root pressure are purely speculative or at the best are based upon very meager evidence. Little or nothing is known concerning the osmotic conditions of the parenchyma cells and xylem elements within the central cylinder of the root and in the conducting tissues of the stem. In the few studies that have been made, the suction force of the parenchyma cells of the central cylinder as well as the surrounding endodermal cells was found to be lower than that of the cortical cells. Whether this osmotic relationship is of common occurrence in plants, however, is not known. If it were, the entrance of water into the central cylinder by the ordinary methods of diffusion would not occur.

The structure of the endodermis suggests that it plays an important part in the entrance of water. The term "endodermal pressure," which is sometimes applied to the force that causes the exudation of water from cut stems, was so named because it was assumed that the endodermal cells force the water into the central cylinder of the root. If the endodermis is designated as the organ that forces water into the central cylinder and thence up through the stem, the nature of its action, according to our present knowledge of osmotic relationships, can be expressed only by saying that it is a process of "secretion" or "excretion." The use of these terms, however, is only an indirect way of stating that we know nothing concerning the manner in which the endodermis functions in the entrance of water into the central cylinder of the root.

Priestley (1920) and Priestley and Armstead (1922) formulated a theory of the mechanism by which exudation pressures might be developed. They considered that the solute concentration of the xylem, in certain regions at least, is maintained at a higher level than that of the adjoining cells, and that this supply of solutes is maintained by the endodermis, which prevents their outward leakage. They obtained evidence which indicated that the parenchyma cells around the xylem but within the central cylinder have specially permeable protoplasmic membranes that permit the exchange of organic substances between their cell sap and the solution in the xylem tubes. This exchange of solutes appears to involve both physical adsorption and physiological absorption. If the concentration of solutes in the xylem elements is greater than that in the cells adjoining them, it is easy to explain the entrance of water into them and the development of an exudation pressure.

Priestley and Armstead were able under suitable conditions to produce exudation pressures in stem and leaves as well as in the roots. In some instances, however, these pressures in the stem and leaves were only temporary, since the absence of a functional endodermis permits too rapid a leakage of the necessary solutes from the xylem for these pressures to be long maintained. In order to account for the fact that the water of guttation contains only a relatively small amount of solutes, these authors assumed that the solutes are absorbed from the xylem by the bordering cells during the upward passage of the sap. They thus assumed that in one region the sap in the xylem vessels may be receiving solutes from the surrounding cells, while in another region the same solute may be diffusing away from the xylem into the neighboring tissue.

b. High Evaporation.—As is well known, the greater portion of the water lost from plants is lost under conditions of relatively high evaporation. Under such conditions the water in the conducting elements of the xylem is under tension due to negative pressure, so that if a branch or stem is severed, water will be drawn into the cut ends instead of being forced outward. On that account, it has long been recognized that under conditions of high evaporation, root pressure or endodermal pressure is not the main force in causing the movement of water through the stem to the leaves. It has been almost universally considered that the forces which are operative in causing the upward movement of water reside in the stem and leaves, but the nature of these forces and even the cells concerned in the process have not, as a rule, been established with any degree of certainty.

The ascent of water in trees has especially attracted the attention of investigators, since the large quantity of water that is lost from them, the great distances that it has to be moved, and the great amount of energy that must be expended to bring this about are striking. On that account, most of the studies on the ascent of water have been confined to woody plants, and most of the theories proposed to account for this phenomenon have been based upon the conditions that are known to prevail or considered to prevail in woody stems. It is beyond the scope of this work to discuss in detail all the theories that have been proposed to account for the ascent of water in plants. Many of these theories are at present only of passing interest, since they were based upon conditions that are now known not to exist in the plants. A thorough and critical review of the literature on this subject may be obtained by consulting Copeland (1902) and Dixon (1910, 1914), and the student who is interested in this regard is referred to these sources for detailed information. It is the intention here to mention only briefly the earlier theories that have been proposed to explain the ascent of water in stems and to confine the discussion principally to the ideas that dominate the field in this regard at the present time.

The theories that have been proposed to explain the ascent of water in plants under conditions of high evaporation may be divided into two main classes: (1) vital and (2) physical. The authors of the vital theories assume that the forces that are active in raising water in plants are manifested, for the most part, by the living cells of the root, stem, and leaves, although most of them consider that purely physical factors may assume a definite, although a minor, part. Those who propose the physical theories contend that the major forces concerned in the elevation of water are purely physical in their nature and consider that their manifestation by the plant is identical with or similar to that which would occur in a purely physical apparatus under like conditions. Most of those who favor the physical explanation of the ascent of sap agree that the living cells of the plant play their part in the process, although their action is classed as of only minor importance. The matter of contention between the advocates of these two theories is thus apparently a question of whether vital or purely physical forces are the dominant ones in the process. It would seem that the forces which are dominant might depend upon the conditions surrounding the plant. Thus under some conditions, *e.g.*, low evaporation, vital forces might be the dominant ones, and physical forces might be exerting only a slight influence. Under another set of conditions, *e.g.*, under conditions of high evaporation, physical factors might dominate the process and thus overshadow the activity of the living cells.

1. *Vital Theories.*—Westermaier (1883 to 1884) formulated one of the first vital theories in regard to the ascent of water. He considered that

the upward passage of water in the stem was effected in the wood parenchyma, and that the tracheids and vessels acted as water reservoirs rather than conducting elements. Godlewski (1884) assumed a periodic change in the permeability of the protoplasmic membranes of the cells of the medullary rays of the xylem, on which account water was taken up on one side from the bordering xylem elements and then released into the xylem elements bordering upon the upper part of the opposite side. This was considered possible since it was assumed that the air pressure in the conducting xylem at the upper portion of the medullary ray was lower than the pressure inside those xylem elements in contact with the lower part of this ray. Janse (1887) supported the idea of Godlewski and was one of the first to note that if the lower portion of a branch is killed, the leaves above are affected within a few days. These theories, however, were proposed upon little or no experimental evidence and upon the assumption of certain conditions in the stem that are now known not to exist.

The later advocates of the vital theory of water ascent have been Ursprung (1905 to 1907), Ewart (1905, 1906, 1908), Schwendener (1892, 1909), Leclerc du Sablon (1910), Bose (1923) and Molisch (1928 to 1929). These advocates, however, base their assumptions upon a considerable amount of experimental evidence which, although convincing them of the nature of the forces involved in the ascent of water in plants, has not so appealed to many other workers in this field. Ursprung and Ewart, although agreeing that certain physical forces play a prominent part in the ascent of water, considered that the process is dependent to a certain extent upon the living cells of the stem. They considered that their experimental evidence indicates that the continuous ascent of water in the xylem is possible only in the presence of its living cells. The exact functions, however, that these living cells perform are not clearly set forth. Bose and Molisch, however, attribute the ascent of water practically entirely to the physiological activity of the living cells of the plant. The work of Bose (1923) has attracted much attention and on that account will be mentioned here in considerable detail, together with some of the adverse criticisms of the methods used and of the conclusions that he has deduced from his experimental evidence.

Bose observed by means of special apparatus that the cells of the lower and upper halves of the pulvinules of the lateral leaflets of the telegraph plant (*Desmodium gyrans*) execute alternate contractions, the period of a complete pulsation varying according to the circumstances from 1 to 4 min. Bose also claimed that by means of proper apparatus pulsating activities are noticeable in growing regions of the plant. These growth pulsations consist of a series of alternate expansions and contractions. It was ascertained that these pulsating activities can be modified in very definite directions by changing the physiological conditions. Thus pulsatory activity is depressed or arrested under diminished internal pressure. A rise of temperature stimulates it while a fall in temperature depresses it. A small amount of an anesthetic increases this rhythmic activity, while it is completely stopped by poisons.

Bose observed that the ascent of sap is modified by the same conditions that so strikingly affect the pulsatory activity. Thus, diminished internal hydrostatic pressure causes a depression in the ascent of sap, a rise of temperature increases the rate, and a fall in temperature depresses it. Small doses of an anesthetic increase the ascent of sap, while larger doses arrest it. Poisons also arrest the ascent of sap. From these observations Bose concluded that he had "proved conclusively that it is the pulsatory activity of living cells which maintains the ascent of sap in plants." He considered also that the response of transpiration to various factors indicates that it is "not a physical process but a physiological one carried on by a rhythmic tissue forming part of a rhythmic system continuous throughout the plant for the absorption and distribution of water." Thus he considered that transpiration is a physiological excretion of water by the leaf cells, the excreted water being removed by the physical process of evaporation. He also considered that the excretion of water can be shown when the leaves are covered with vaseline and considered that this is a further proof that transpiration is an active physiological process.

By electrical methods Bose considered that he demonstrated that a definite layer of tissue, the innermost layer of the cortex in dicotyledonous plants, is in a state of active pulsation consisting of alternate contraction and expansion. He thus concluded that these are the cells that carry the water upward. Since there is no such pulsatory activity to be observed in the wood, Bose considered that the xylem is by no means essential for the conduction of water and that it acts only as a reservoir, the water being pumped into or withdrawn from it according to different circumstances.

Molisch (1928, 1929) reported that he successfully repeated some of the experiments of Bose. In one experiment he employed a cut piece of stem of *Antirrhinum* which was held at the cut end by a piece of sponge. When the water supply was low the leaves on this stem became badly wilted. A few drops of a dilute solution of camphor applied on the sponge, however, caused the leaves to revive quickly and become fully erect in the course of 2 or 3 min. The results of this experiment are thought to be indicative of the fact that the movement of water in a plant is a physiological rather than a physical process. By means of one of the instruments of Bose, Molisch reported that the flow of sap along the stem was observed to consist of a series of pulsations. These pulsatory activities were greatly increased by drugs that enhance cardiac activity in the animal. Molisch stated: "There is an inherent activity in the stem itself, independent of those in the leaves and roots which cause the movement of sap. The sap can be made to flow either upward or downward according to different stimulation. The law of directing movement of sap is that it moves from the stimulated to unstimulated or depressed regions. A local depression or stimulation starts the machinery into action and makes the sap proceed toward the depressed or away from the overstimulated region."

Dixon and his workers (1924) repeated certain phases of Bose's work but were unable to verify his results. The work and conclusions of Bose have been adversely criticized by Shull (1923), MacDougal and Overton (1927, 1929), and Benedict (1927). One of the chief criticisms is that the relationship of the pulsations or rhythmic activity reported by Bose and Molisch to the ascent of sap has not been definitely established. It does not follow that because pulsation activities and the ascent of sap respond in a like manner to the same factors, one is the cause of the other. The method of determining the rate of water ascent by noting the time it took a wilted leaf to show signs of recovery when the stem is supplied with water is considered to be untenable. The fact that the leaf shows signs of recovery does not indicate that the water entering the stem has ascended to the leaf by that time any more than that water coming from the nozzle of a full garden hose is the identical water which at that instant is entering the distal end of the hose (MacDougal and Overton, 1927). It has been estimated that the pulsatory movement of sap claimed by Bose (1923) and Molisch (1928 to 1929) would imply a transfer through 200 to 400 living cells per second. Benedict (1927) measured the cross-sectional area of the living cells in the tree trunks of 10 different species and computed the maximum pumping capacity of the most rapid pulsating rates found by Bose. The actual rate of sap flow under maximal conditions of transpiration proved to be from 8,000 to 30,000 times as rapid as would be possible under the theory proposed by Bose.

Peirce (1934) believed that the movement of water from the roots through the stem and branches to the leaves is dependent upon the living cells between the root and the leaves. He stated that where living cells are present water ascends in stems, but where only dead cells exist there is no ascent of water to replace that lost by transpiration. He held, however, that the physical conditions of the soil and atmosphere influence to a considerable degree the amount of water translocated. Since each vessel, tracheid, and living cell is a constituent unit of the whole system, and since the proportion of these various parts may vary, the relative amount and speed of movement of water in one plant part may be different from those in another. Peirce (1935) showed that the rate of water movement in stems varied with temperature, thus indicating further that living cells are concerned in the process. Priestley (1935) noted that when the leaves are removed from woody stems, water continues to enter with as much force as if they were present. He thus concluded that the imbibitional forces in the leaves are not necessary for the ascent of water, but that living cells in the stem are the effective agents. Peirce (1936) from later experiments involving the relation of temperature to the ascent of sap concluded that living cells condition or affect but do not accomplish or effect the movement of water through vascular plants.

It appears evident under the conditions of high evaporation that all the forces which function in the ascent of water in plants cannot be attributed to the direct action of living cells. The living cells, however, evidently play an important part, especially those of the roots and leaves. Whether those in the conducting tissue play any part is evidently not definitely known. Dixon (1924) stated that the fact that water rises in stems of plants even when surrounded by a saturated atmosphere and the fact that dead leaves inadequately draw up water indicate that the living cells play some part in the ascent of sap. Smith, Dustman, and Shull (1931) believed that the rise of water under the conditions stated by Dixon can be explained solely on a basis of the saturation deficit that exists in leaf and stem tissues when the branches are cut. Although submerged in water, the amount of internal water continues to determine the intake of water until the saturation deficit is satisfied. They showed by experiments involving a number of branches of various plants that if the saturation deficit was satisfied there was no ascent of water.

It would be safe to state, however, that the living cells of the roots and leaves at least maintain good contact and thus keep intact the passageways across them for the movement of water even if they do not in themselves exert any vital force in the ascent of water in the stem.

2. *Physical Theories.*—Under this heading are grouped those theories which assume that the ascent of water is for the most part due to purely physical forces and that the living cells of the stem play only a minor part

in the process. Before discussing the physical theories of the ascent of sap, however, some of the experimental evidence which has been presented to show that the living cells are unessential in the ascent of water should be mentioned.

(*a*) *Absence of Living Cells.*—The general method of procedure that has been followed to demonstrate that the living cells of the stem are not concerned in the ascent of sap has been to kill a portion of the stem by exposing it to hot water, steam, or chemicals or to place the cut end of a branch into a solution of a chemical that is known to kill protoplasm when in contact with it. These methods, however, are subject to criticism, as will later be mentioned.

Boucherie (1840, 1841), while injecting timber with various compounds in order to test their effects upon its preservation, found that if a tree were cut across at its base and supplied with a poisonous solution, it drew this liquid up to its highest leaves. It would also draw up a second solution if supplied. The significance of this experiment in the ascent of sap, however, was apparently not recognized by this investigator or by contemporary botanists. Boehm (1889) noted that if a portion of a stem were killed, water would still ascend to the leaves, but the classical work of Strasburger (1891 to 1893) gave the first conclusive evidence that the living cells of the stem are at most of only minor importance in the ascent of water. He placed the base of cut stems in poisonous solutions, among which were copper sulphate and picric acid. In the case of branches set in a 5 to 15 per cent solution of copper sulphate, the liquid rose to a height of from 6 to 20 m. in from 16 to 20 days. In another experiment he killed the branches for a distance of 10 to 12 m. at the basal end with boiling water and observed that eosin rose in them from the cut end to a height of almost 17 m. The leaves at the uninjured end of the stem, however, remained alive for only a few days when they dried up and fell off. In another experiment Strasburger used a young oak tree 22 m. in height. This was sawed off near the ground and kept upright with the lower end in a solution of picric acid. After 8.5 days it was found that the highest branch was completely impregnated with picric acid, while fuchsin which had been added 3 days later than the picric acid was also found in the same regions. Roshardt (1910) performed experiments on 131 species of plants, including shrubs and herbaceous plants of 59 families, by killing portions of the stem, branches, or petioles by steam, ether, low temperatures, or other means and then determining the effect of this treatment upon the sap flow. A considerable number of his experiments were made upon grasses, among which were *Poa pratensis*, *Bromus sterilis*, *Bromus hordeaceus*, *Agropyrum repens*, and *Secale cereale*. He noted that water ascended the stem for a considerable time after certain portions of the lower part had been killed but that the leaves dried up after from 8 to 17 days. There was no distinct wilting, but the leaves began to dry at the tips.

Overton (1911) in experiments with the umbrella plant (*Cyperus alternifolius*) found that when 5 to 10 cm. of the stem was killed by treatment with picric acid, 95 per cent alcohol, or copper sulphate for 36 to 48 hr., sufficient water ascended through the poisoned portions to supply the transpiration needs for 90 days and to allow the development of new branches.

MacDougal, Overton, and Smith (1929) sawed off a willow tree 11 cm. in diameter and about 8 m. high and placed it upright with the basal end in a concentrated solution of picric acid. It was allowed to transpire for 4 days, after which it was transferred to acid fuchsin. In 4 days the tree absorbed 10 l. of the picric acid solution and about the same amount of acid fuchsin in the succeeding 4 days, both solutions being drawn up the tree for a distance of 6 m. This experiment was repeated with a walnut tree 10 cm. in diameter with results very similar to those observed for the willow.

The results obtained from the experiments that have just been mentioned indicate that the living cells of the stem play no important role in the ascent of sap. There has, however, been some question of the conclusions drawn from these experiments. It is pointed out that the killing of the cells in a given length of the stem does not show that those living cells above or below the injured portion are not concerned in the process. Some suggest that even though poisonous solutions are drawn up the stem, some of the living cells may not be injured. In the experiments in which the stems have been immersed in poisonous liquids, the leaves remain alive for only a relatively short period of time after the treatment. It has been contended that the leaves should remain alive for a long period of time after such treatments if the death of the living cells of the stem has no influence on the ascent of water. In the cases where a portion of the stem has been killed by the various means, Dixon (1905, 1909, 1915) and Overton (1911) considered that the leaves are not injured because of a lack of water being moved to them through the dead stem but by the transportation to them in the water stream of poisonous or plasmolytic substances that have been formed by the decomposition of the cells in the treated region. Both these investigators noted that the leaves showed injury in proportion to the length of the stem that was killed. This opinion was further substantiated by Dixon, who forced water through a steamed portion of the stem and found that it was tinged with a brown substance, while the water that was forced through an untreated stem emerged a clear liquid. Some of the investigators are of the opinion that the treatments of the stem cause the formation of substances which clog the lumina of the xylem elements and thus prevent the normal ascent of water. It has also been suggested that these treatments of stems might rupture the water columns and thus bring about a decrease in water movement. Roshardt (1910) considered that the decrease in water content under such treatment is due to the lack of energy which is normally furnished by the living cells and which the dead cells are unable to supply. It appears evident that the methods that have been used to render the living cells of the stems inactive in the study of the ascent of water bring about other changes in the stem which cannot be ignored.

(*b*) *Imbibition and Capillary Theories.*—The imbibition theory was proposed by Unger (1868) and championed by Sachs (1878, 1879), who became its chief exponent. This theory assumed that the water moved upward in the stem entirely in the walls of the xylem elements by the process of imbibition. It was considered that the creation of a saturation deficit in the cell walls at the top of a plant would cause a flow of water toward the point of loss from the adjoining portions of the cell wall and that this demand for water would thus be transmitted downward through the walls to the roots where it would be satisfied. The forces of imbibition are known to be sufficient to elevate the water in this manner much beyond the height of the tallest trees. The imbibition theory of the ascent of sap after the manner above mentioned, however, became obsolete when it was demonstrated beyond doubt that practically all the water moves upward in the lumina of the xylem elements. This fact, proved by numerous investigators among whom are Elfving (1882), Vesque (1884), Darwin and Phillips (1885), Errer (1886), Strasburger (1891), Dixon and Joly (1895), and Smith (1896), shows that if the lumina are blocked by gelatin, paraffin, ice, water vapor, or by compressing the stems until the lumina are closed, the supply of water to the upper portion of the plant is checked to the extent that severe wilting results. The ascent of water in the lumina is also demonstrated by means of dyes. There is no doubt that some water ascends in the walls of the xylem elements, but the amount is insignificant in comparison with that which moves upward through the lumina of the tracheids and vessels (Vines, 1896).

The elevation of water has been ascribed by some writers to the force of capillarity in the lumina of the xylem elements. The lifting power of a capillary is determined by the form and size of the meniscus of the liquid therein contained. The form of the meniscus is, in turn, influenced by the nature of the walls of the capillary. It was early recognized that the lumina of the xylem elements are of such a size that their lifting power by capillarity is not very large and cannot account for a rise of water, according to Strasburger (1891), to distances exceeding 3 m.

Stamm (1929) studied the capillary structure of the softwoods by determining the rate of electroendosmotic flow, the hydrostatic flow, and the force of surface tension in these woody tissues. The values he obtained in this manner agreed with the values obtained by microscopical measurements. Reed and Bartholomew (1927), and Reed (1930) believed that the transport of water in the fruits of oranges and lemons occurs through the hydrophylic layers of the cell walls rather than across the protoplasm and cell sap. Peirce (1934) believed that water moves along the inner surfaces of tracheids and vessels, and that the maintenance of a movable film of water on the inner surface of a vessel or tracheid depends upon the quantity of water in reserve in the cell wall. The size and relative proportion of vessels and tracheids, among other things, determine the heights that plants attain because these two factors determine the volume and speed of water movement. The terms "imbibition" and "capillarity" have been used more or less interchangeably by some writers. As here interpreted, however, capillarity is confined to small but well-defined lumina or tubes, while imbibition is confined to the submicroscopical pores or interstices of the cell wall.

(*c*) *Cohesion Theory.*—In 1894 Dixon and Joly published their cohesion theory of the ascent of water in plants. Askenasy (1895, 1896) independently developed a similar theory which was published shortly after the abstract of Dixon and Joly had appeared. This theory has been further elaborated by Dixon, and much work has been undertaken to substantiate it. It is so much in accord with the general conditions known to prevail in the plant and has been so well substantiated by experimental evidence that it has been almost universally accepted as the best explanation of the ascent of water in stems under conditions of high evaporation.

The cohesion theory of the ascent of water is based essentially upon the two following facts: (1) the cohesive force of water or the quality that water possesses of resisting tensile stress and (2) the imbibitional force of the evaporating cell walls of the leaf or the tractional force exerted by the water menisci of the submicroscopical cavities of these cell walls.

(1) *Cohesive Force of Water.*—The cohesive force of water has been known since about 1850, and, according to Copeland (1902), Nägeli (1866) first suspected its importance in the ascent of water. The publication of the cohesion theory of the ascent of water greatly stimulated interest in the cohesive properties of water, and studies in that regard have been reported by Dixon (1910, 1915), Ursprung (1913, 1915, 1916), and Renner (1915). These investigators all found that water has great cohesive power and demonstrated that the tensile strength of water containing dissolved air is at least 300 atmospheres. It has been shown that the xylem sap has a greater tensile strength than pure water and that when it is in position in the conducting xylem it may withstand tensions greater than 300 atmospheres without separating from the walls of these conducting elements.

(2) *Imbibitional and Osmotic Forces.*—The imbibitional forces developed by drying organic matter range from zero to 1,000 atmospheres, depending upon the degree of drying. The force of imbibition in the cell walls of the leaves is thus enormous and is much more than ample to lift a column of water to the height of the tallest tree. It has been estimated that the total force necessary to lift water to a height of 350 ft. through the xylem elements of a tree would not exceed 20 atmospheres. Only a small fraction of the possible imbibitional power of the cell walls of the leaves would be necessary to overcome this force. It is considered

that a saturation deficit of 2 per cent would be more than ample to lift water to the top of the tallest living tree, while a deficit of 0.1 per cent would exert a pull sufficient to raise water to the height of the ordinary annual plants. Ursprung (1925) proved that the suction power of the vacuoles of the leaf cells would furnish ample force to move the column of water in stems at an adequate rate to replace the water evaporated from the leaves, provided osmotic forces play a determining part in the lifting of the water in the stem.

It was calculated by Dixon (1897) that it would require a pressure of 100 atmospheres to tax the cell walls of the mesophyll cells of the leaf to their breaking point. Thus it is apparent that the osmotic pressures of the leaves never seriously tax the tensile strength of these cell walls.

There is some dispute as to what part the osmotic forces of the cells of the mesophyll of the leaves and the cortex of the roots play in the ascent of water. It is considered that these forces at least keep these cells turgid and thus prevent them from collapsing under the tension that pulls the water through them. The osmotic pressure of these cells, however, must equal or exceed the tension in the water that passes through them (Dixon, 1896).

(3) *Description of Theory.*—The cohesion theory of the ascent of water assumes that the water in the conducting tracts of the xylem exists in the form of unbroken columns and that these columns are continuous with each other both vertically and laterally through the cell walls. A continuous meshwork of water thus exists throughout the whole length of the stem, roots, and leaves. The terminals of this water meshwork are the menisci of the water in the submicroscopical cavities of the cell walls of the epidermal cells of the root and of the cells bordering upon the intercellular spaces of the leaf or the epidermal cells. This theory also assumes that the water in the conducting tracts is brought into a state of tension by the imbibitional forces in the evaporating cell walls of the leaf and that these water columns are able to withstand the tensile pull that is thus exerted upon them.

Thus in the process of transpiration, water particles are set free from the menisci of the submicroscopical cavities of the cell walls of the evaporating cells of the leaf. This loss of water particles produces incipient drying or a saturation deficit, the menisci become more concave, and their surface tension is increased. As a result of this, water is abstracted from the protoplasm of the cells which the evaporating walls enclose. The protoplasm in turn abstracts water from the vacuole, so, on account of this disturbance in water equilibrium, osmotic and imbibitional forces that abstract water from adjoining cells or cell parts are set in motion. This disturbance in equilibrium is transmitted eventually to the water of the xylem elements of the leaf. A pull is thus set up on the water in this conducting element, which, in turn, is carried down the petiole to the conducting xylem elements of the stem and root and then to the soil. Owing to the weight of the water columns and to their adhesion to the walls of the conducting tracts, the water is, under the conditions above mentioned, thrown into a state of tension which may in some cases be as great as 200 atmospheres (MacDougal, Overton, and Smith, 1929). As a result of this pull exerted by the water menisci of the evaporating cell walls of the leaf, water is moved upward through the root and stem to the leaves.

Under conditions of high transpiration the tensile stress of the water in the stem and roots must be transmitted across the cortex of the root to the submicroscopical cavities of the outer cell walls of the epidermal cells of the root. On this account, as the menisci of the evaporating cell walls of the leaf become more concave, the menisci of the exterior cell walls of the epidermal cells behave in a similar manner. The rendering of the menisci of the evaporating cell walls of the leaf more concave tends to bring them in equilibrium with a lower vapor pressure and consequently reduces evaporation, while the increased concavity of the menisci in the epidermal cells of the root renders them more ready to condense water vapor. The menisci of the cell walls of the epidermal cells of the root entrap the more active molecules of water, while the menisci of the evaporating cell walls of the leaf retain the less active ones. Dixon (1910) stated:

> Thus we may regard the flow of water up the highest tree as due to the evaporation and condensation produced by the difference between the vapor pressure at the surface of the epidermal cells of the root and that which obtains in the cell walls of the evaporating cells of the leaf.

Dixon (1914) considered that the loss of water by a plant under conditions of very low evaporation might be a process of secretion by the mesophyll cells of the leaves. If this is true, the energy that is necessary for the lifting of the water is considered to be supplied by respiration. Dixon estimated that the energy necessary to raise 1 cc. of water 100 m. in a tree and to overcome the resistance would amount to only 0.5 cal. To evaporate 1 cc. of water at 20°C. requires 592.5 cal.; hence, in order to obtain the energy needed to raise the water in a 100-m. tree the amount evaporated would be diminished by only 0.001 part. Even taking the highest and most excessive estimate of the resistance in the highest trees, the amount of water evaporated would be reduced by only one-hundredth to one-thirtieth if, in addition to evaporation, the energy absorbed by the leaf had to do the work of transporting the water from the roots. If under conditions of extremely low evaporation water is excreted from the leaf cells, ample energy can be derived from respiration to lift the water to the tops of the tallest trees.

Smith, Dustman, and Shull (1931) considered that respiration is not necessary for the ascent of water. They believed that the ascent of water is due entirely to physical and chemical forces, if osmotic action can be considered as purely physical and chemical in nature. Whether evaporation is exceedingly rapid or extremely slow, the cause of sap movement is considered to be due to surface tension and imbibitional and osmotic forces that are in no manner related to excretory and secretory activity of the living cells. Homan, Young, and Shull (1934) stated that the maximal tension or negative pressure in the tracheae depends on the relative humidity of the atmosphere. Since the relative humidities of natural atmospheres are frequently as low as 20 to 30 per cent, it is evident that the structure of plants enables them to prevent the evaporation of water to a marked degree. The epidermal cells with their cutinized surfaces and regulated stomata operate efficiently in maintaining the humidity of the atmosphere surrounding the mesophyll cells in an approximate state of saturation. Hungate (1934) believed that the cohesion theory offers the best explanation to date of the ascent of water in plants.

(4) *Evidence for Theory.*—Two general methods have been used to demonstrate the plausibility of the cohesion theory of the rise of water in plants. These methods have to do both with artificial systems and with living plant materials.

Renner (1911 to 1912) measured the suction force of transpiring twigs and found under the conditions of his experiment that it amounted to 10 to 20 atmospheres. He considered that the saturation deficit in the leaves was the cause of the water movement in the stem. By applying artificial suction to the leaves at a time when no saturation deficit existed, he determined the rate of flow produced by a given suction force. The rate of flow through the stem was then determined when rapid transpiration was proceeding, and the suction that was being exerted by the leaves was calculated from these data, since it was considered that the rate of flow through the stem was proportional to the suction forces exerted by the leaves. Nordhausen (1916, 1919, 1921) showed that suction forces exist in the leaves, although by his methods he did not obtain so high values as did Renner. In this connection it should be mentioned that Thut (1928) devised an apparatus that shows definitely the lifting power of evaporation under ordinary laboratory conditions. In 1932 he reported a special method for demonstrating the suction force of the twigs of *Acer negundo*, *Thuja occidentalis*, and *Platanus occidentalis*, among others.

If the water in the xylem elements is thrown into a state of tension by the suction forces of the leaves, these elements should show a contraction when transpiration is proceeding and should again expand when transpiration decreases or ceases entirely. MacDougal (1921, 1925 to 1926) and MacDougal and Shreve (1924) have observed that the trunks and branches of trees show daily alterations in diameter. This swelling and shrinking were attributed to changes in the wood due to changes in the transpiration rate. It was found that variations in diameter depended directly upon the depletion and repletion of the water supply of the wood, and the effect of increasing tension in the water upon the diameter of the conducting elements was noted. Bode (1923) made a direct microscopical examination of the vessels of wilted plants and found that they would shrink with increased transpirational pull and that they would expand again when the transpiring surfaces were removed from the plant. Haasis (1932) found that the fall and winter swelling of the trunks of Monterey pine and redwood was preceded by a progressive contraction during the dry season of the spring and summer. The relative humidity and the amount of available moisture in the soil most appreciably affected this shrinkage. He stated that a tree might be shrinking in one part and swelling in another. Gibbs (1935) found indications that water is continuous in the xylem of the birch, larch, balsam, and spruce. It has been shown by Dixon (1914) that the resistance of a current of water moving through wood at the velocity of the transpiration stream is equivalent to a head of water equal in length to the wood traversed. In a tree 100 m. in height, the tension of the ascending water would be approximately 20 atmospheres, a force that would in no way tax its cohesive powers.

Huber (1928) found that trees and nonsucculent xerophytes have the least resistance to the conduction of water per unit length of their stem. Their total resistance, however, may be high owing to the long stems of trees or to the difficulty of obtaining water under arid conditions. Groom (1910), Farmer (1918), Holmes (1918), Rivett and Rivett (1920), Coster (1931), Malhotra (1931, 1932), and Huber (1932) investigated rates of water flow in the conductive tissues of different plants. Coster (1931) studied the speed of the transpiration stream in herbs, shrubs, lianas, and trees in Germany and in the rain forests of West Java. Under optimum transpiration conditions the maximal speeds were 50 to 80 cm. per minute for herbs and shrubs, 100 to 200 cm. per minute for shrubs and trees, and 100 to 250 cm. per minute for lianas. The experimental values are lower than those which actually occur in nature since the severing of the plant slows the transpiration stream. In general, the speed is greater the larger the cross-sectional area of the elements and the taller the plants. Haines (1935) found that the greater the water deficit, the greater the resistance to flow by the walls and the protoplasm of the leaf cells. Pond (1903) concluded that the transpiration stream of the aquatic plants studied by him appeared to be caused primarily by forces in the basal part of the plant. This force seemed to be root pressure.

(5) *Objections to the Theory.*—The development of high tensions in the water columns of the plant introduces the greatest difficulty associated with the cohesion theory of the ascent of water, *viz.*, the permeability of the walls of the wood cells to gases. Why do gases not pass into the conducting elements when tensions are developed? Bailey (1916) found that the surface tension of the sap in the pit membranes of different conifers could be overcome by pressures that would be inadequate for the cohesion theory. Wright (1928) believed that the torus of these bordered pits could not exert a valve action and that pressures of 3 atmospheres would cause the passage of gases through them. Apparently many of the water-conducting elements fill with gas during the summer or during conditions of water deficiency and then are again filled with water during the winter or during conditions of an available water supply. Woodhouse (1933) believed that air enters the xylem cells through their walls before a sufficient tension is created to raise the water to the height of trees. He considered that the cohesion theory of sap ascent in itself is inadequate, but that combined with the imbibition theory a satisfactory hypothesis for the ascent of water is obtained. By this hypothesis the walls of the xylem imbibe and support the weight of water moving through it by tensile strength. He considered that the flow of water occurs through the largest vessels in which an imbibed tensile fluid can be maintained under the conditions of the moment. He believed that this explains the zonation of con-

ducting areas in the xylem, since it has vessels of varying diameters in definite seasonal zones.

Münch (1926, 1927, 1930) claimed that there is a downward stream of water containing osmotically active materials in the bark. When these dissolved materials are removed at the base of the tree, the water is forced into the xylem and refills the vessels emptied during the summer. He stressed the rapid streaming in the sieve tubes as significant.

(*d*) *Other Theories.*—A theory suggesting the manner of translocation of water in plants was advanced by Münch (1930). The basis of this theory lies in the fact that when two osmotic pressures operate against one another through a liquid connection, a movement of solution occurs from the direction of the greater pressure toward that of the lesser pressure until this movement has equalized the osmotic activities at the two ends of the path. Simultaneously, water will be taken in under the influence of the greater pressure and thrust out against that of the lesser.

In applying this principle, each living cell is not regarded as an osmotic unit, but the whole aggregation of living cells, the "symplast," is considered as being linked together by plasmatic fibrils into a single osmotic system. When a point in the system develops relatively high osmotic and hydrostatic pressures and external water is available, liquid movements will be set up from this position toward any other where these pressures are lower. If the root is intact there is no such easy escape; consequently the back pressure prevents such extensive movements, and the only movement that takes place is associated with the filling of the higher regions. This filling seems to consist mainly of passing fluid into the partially empty elements, distention of the living cells, and reduction to a limited extent of the tension in the vessels. When high hydrostatic tensions exist in the tracheae, positive hydrostatic pressures exist in the living cells surrounding them. The osmotic value of the cell sap of living cells is always higher than that of the tracheal sap, hence there will be a tendency for water to be withdrawn from the vessels across the semipermeable membranes of the surrounding cells. An easier line of escape for the moving fluid will be through the sieve tubes, and perhaps through the protoplasmic connections of living cells, since such movements do not involve the penetration of surface membranes.

Lund (1931, 1932) stated that an electric current flows downward in the outer cortex and upward in the wood axis. There is a possibility that one of the functions of this continuously upward electric current in the wood is to supply electrical energy for the electroendosmotic flow of sap in an upward direction in the conductive elements of the wood. This suggestion may apply to a downward flow in the cortex as well as transport across the stem.

4. Downward Movement of Water.—It has been known since the time of Hales (1727) that under certain conditions there may be a downward as well as an upward movement of water through the xylem. A few examples of this downward movement selected from the literature are cited herewith. Hales (1727) among other observations noted that a forked branch cut from a tree will remain fresh and continue to transpire for many days provided that one limb of the fork with its leaves is immersed in water. Dixon (1924) stated that if the tip of an upper leaf of a potato plant is cut under an eosin solution, the colored liquid is quickly drawn back into the conducting elements of the xylem of the leaf and thence into the same tissue in the petiole. From there it not only makes its way into the upper branches and leaves but also passes down the stem supporting these parts and into the xylem tissue of the tuber.

Mason and Maskell (1928) found in the case of the cotton plant that an eosin solution taken in through cuts in the leaf blades was in 6 hr. found to have penetrated both upward and downward in the xylem. The stain passed in with the foliar trace and occupied a median position in the wood, the peripheral region of the wood being unstained.

Arndt (1929) studied the movement of solutions through the xylem of the coffee tree. In one case the terminal internode was cut and an eosin solution applied through a funnel and rubber tube. In 15 hr., 50 cc. of the solution was absorbed and the corollas of flowers in lateral branches 80 cm. from the cut internode were stained. In 23 hr. after the beginning of the experiment all the leaves of the trees were stained and the total distance of eosin movement exceeded 200 cm. In one the stain descended 125 cm., passed horizontally 60 cm., and then ascended 100 cm. In none of the experiments did the eosin solution in moving either up or down the stem pass through the cambium into the phloem, and the cut surface of the phloem in contact with the dye did not conduct it. MacDougal (1925) noted that the outermost wood of the Monterey pine is not the path of an upward-moving current in the normal tree, but he could not, however, demonstrate a downward moving current in that part of the xylem. The path of reverse currents has been traced by Birch-Hirschfeld (1920) by means of lithium nitrate and eosin.

Harvey (1931) found that when suction was applied to a cut apple branch and a dye solution was applied simultaneously to a similarly decapitated twig located at a considerable distance from the former, the dye entered the main stem and streamed not only toward the source of suction but also rapidly away from it. Baker and James (1933) injected dyes into the wood of *Acer pseudoplatanus* and noted that they invariably moved both upward and downward. This movement never occurred in the bark or pith. The rate of upward movement exceeded the rate of downward movement only when transpiration was high.

The examples that have been cited above leave no doubt in regard to a downward current of water in the xylem. The cause of this movement, however, is not definitely known. Dixon (1924) considered that these reversals in the current of water are readily explained by the cohesion theory of the ascent of water. He argued that since transpiration causes the water in the conducting xylem to pass into a state of tension, this tension determines a flow from any source wherever situated and the continuous transpiration from the leaves draws the supply through the plant along the channels of least resistance. This path of least resistance, however, is not always the shortest path in the wood. Transverse motion across several xylem-conducting elements seldom occurs and the separate linear series of conducting elements are practically isolated from each other laterally. On that account, Dixon considered that this enables the tension developed by the transpiring cells of the leaves, although it raises a water column in one series of the conducting xylem, to draw a solution in a neighboring filament of xylem elements terminating above in some local supply. Smith, Dustman, and Shull (1931) believed that the downward translocation in the xylem, which occurs when leaves on wilted plants are truncated and supplied at the cut end with eosin, is a consequence of a root-saturation deficit that will be relieved from any direction by a free water supply. They claimed that this behavior is not connected with the ordinary transpiration or translocation currents in the plant.

Arndt (1929) noted that during the daytime when the leaves are actively transpiring, the eosin solution moved upward more rapidly than downward but that at late evening, at night, and in early morning the downward movement was not greatly reduced, while the upward movement was so greatly reduced that the downward movement exceeded it. Arndt considered, since only slightly reduced downward movement occurs at night and since topping a plant affects this movement but little, that it is independent of transpiration. If the postulation of Dixon (1924) that there are separate conduits for transportation in each direction is correct, Arndt considered that in the case of the coffee tree these conduits are so close together that a microscopical examination after short-interval experiments does not show stained and unstained areas as would be expected. He considered that the observations indicate that living cells other than those of the leaf must play an important role in sap conduction, since it does not seem probable that there could be simultaneously ascending and descending streams unless there are conduits that are hydrostatically isolated from each other by an impermeable partition of living cells.

5. Lateral Movement of Water.—The data concerning the lateral transfer of water in stems are very limited, and the observations in that regard have been confined almost entirely to woody stems. It is known that the lateral transfer is very slow and that translocation in a radial direction usually takes place more slowly than it does in a tangential direction. In many cases, the location of pits in the conducting xylem elements is such as to facilitate translocation in a tangential direction, while the medullary rays offer passageways for radial transfer. MacDougal, Overton, and Smith (1929) considered, however, that the movements which take place through the medullary rays are largely or wholly osmotic and that mass movements in a radial direction do not occur except possibly in some cases in the pines where the medullary-ray cells are of a tracheid nature.

Experimenting with a 13-year-old willow, they found that there was no radial transfer from one annual layer to another. Thus when dye was injected into the oldest 10 annual layers of this tree in such a manner as to exclude the dye from the outermost three layers, they found that to the height of the 10-year level there was no dye in the outer three annual layers. There was, however, a tangential movement of the dye in those annual layers to which the dye had access, and at a considerable distance above the bore hole the tangential movement in the summerwood was completely around the annual layer. The number of annual layers in which there was no dye remained constant from the bore hole to the 10-year level. The number of annual layers, toward the center of the tree, which contained the dye became progressively smaller until at the 10-year-old level only the innermost of the terminal portions of the wood of the tenth year showed the dye. These investigators considered that this experiment demonstrates that there is not a radial mass movement nor a radial transfer of the forces concerned in sap flow from one annual layer to another through the living cells of the medullary rays or through abutting xylem elements. They considered, however, that the possibility exists for an osmotic transfer of water through these structures.

It was noted by MacDougal, Overton, and Smith (1929) that, although the injection experiments showed no mass movement in a radial direction from the inner to the outer annual layers, the leaves at the tips of all branches above the bore hole showed the presence of the dye. Since these leaves are connected only with the wood of the current year and since there was no dye in this layer at the level of the bore hole, the question arises as to what path is followed by the transpiration stream in moving from the innermost layer of the xylem to the outmost one if there is no radial mass movement.

These authors explain this point by using as an example a 2-year-old unbranched stem. A stem of this type consists of a central column of xylem of the first year's growth whose terminal end is bluntly conical. This central column is surrounded by a thimble-shaped mass of wood whose upper half, which is the longitudinal growth of the present season, is solid. The lower portion, which is the lateral formation to the previous year's wood, is a hollow cylinder enclosing the last year's xylem. The leaves attached to the upper portion of the current year's growth exert a pull on the water in the xylem elements, and this is transmitted through the dome-shaped portion of the preceding year's growth to its conducting xylem. The transpiration stream thus moves from the xylem of the first year to that of the second year in a more or less vertical direction. The transpiration pull on the water columns of the first year may thus be transmitted to the columns in the xylem of the second year and thence to the xylem of the third year and so on to all the columns present in any annual layer.

It was noted by Arndt (1929), using a coffee tree, that, when a single lateral root was cut under an eosin solution, a one-sided staining was evident to the top of the tree after a period of 20 hr. In another case, this one-sided staining was still evident after 72 hr., the staining of the leaves of the laterals corresponding to the stained sector of the xylem. As the top of the tree was approached, the dye gradually diffused tangentially so that all the branches tended to become stained. In some experiments it was noted that as the dye passed upward it tended to diffuse laterally, but in all cases the radial and tangential diffusion was slow. Baker and James (1933) observed in *Acer pseudoplatanus* that radial transfer of water occurred from early to late wood and from one annual ring to another. They claimed to have demonstrated the radial and tangential anastomosing of vessels.

Auchter (1923) obtained evidence which suggested that water may move laterally in the peach tree. When the roots were removed from one side of the tree, the total water content of the whole plant was reduced and the growth checked to some extent. The moisture contents on the two sides of the tree, however, were identical and the leaves on the cut side did not wilt on bright, clear days. Apparently in these trees water moved through and around the stem without much difficulty. Furr and Taylor (1933) observed that the cross transfer of water was accomplished very readily in both old and young lemon trees, when irrigation water was applied on alternate sides of the tree on successive irrigation dates.

B. Translocation of Inorganic Compounds

1. Upward Transfer.—It has been commonly stated in the literature since the time of Sachs that the water and inorganic materials are carried upward for the most part through the xylem. Although it is apparently established beyond a doubt that the xylem is the pathway of the upward movement of water, the statements concerning the movement of solutes are apparently based upon inconclusive evidence.

The xylem has been considered as the pathway of the movement of the inorganic solutes on account of the following evidence: (1) It has been assumed that since the inorganic salts are dissolved in the water of the soil and since the pathway of the water is in the xylem, these inorganic materials would be absorbed and carried with it. (2) The inorganic salts have actually been found in the water-conducting elements and in the sap that has been obtained therefrom. (3) When cut stems are placed in solutions of dyes or salts, the solutes move for considerable distances in the xylem tissues. (4) Experiments in which the movement of dyes and salts has been studied in rooted and uninjured plants give indications that translocation of the solutes occurs in the xylem.

This evidence has been critically considered by Curtis (1923). He pointed out, as has been discussed in Chap. V, that it has apparently been definitely established that there is no direct relation between the absorption of water and the absorption of inorganic salts. He also considered that the fact that inorganic salts have been found in the sap of the xylem is not conclusive evidence that they are translocated in this tissue. He stated that unless it is known that the water moves by mass flow through open tubes and not by diffusion or through occasional membranes, or unless it is known that neighboring cells do not remove solutes from a passing stream perhaps transferring them and reintroducing them at a lower level, the mere presence of these solutes in the xylem sap is no proof that they are being moved therein.

The methods by which the samples of xylem sap have been obtained may be open to question. Curtis considered that sap obtained from a cut or bleeding stem may be different from that in a normal uninjured stem, since the movement of water through a series of cell membranes may be different from the flow that would occur if the membranes were cut open. Some of the methods for obtaining sap may, however, be free from this objection (Bennett, Anderssen, and Milad, 1927). Objections may also be made to the method of placing cut stems in dyes and salt solutions, since normally the water-conducting system may be considered as a closed system with no actual openings. Curtis considered that numerous sources of error have been overlooked in the experiments in which the translocation of salts and dyes has been studied in intact plants so that the results obtained have given no clear indications of the tissues concerned in the movement of the inorganic salts.

a. Through Xylem.—The observations of Anderssen (1929) in regard to the tracheal sap of the pear tree, however, indicate that the xylem functions in part in the translocation of the inorganic solutes. Thus he found that the individual inorganic constituents of the tracheal sap showed a large increase in concentration from early winter until early spring. From Nov. 10 to May 10 the calcium increased from 16.6 to 84.7 p.p.m., magnesium from 0.8 to 23.5 p.p.m., potassium from 23.6 to 59.6 p.p.m., and PO_4 from 10.6 to 25.2 p.p.m. On May 10 the calcium, magnesium, and PO_4 ions far exceeded their concentration in the soil extract. The total electrolytes were almost twice as concentrated in the outer annual ring as in the inner rings.

Bodenberg (1927) studied the tissues concerned in the translocation of inorganic salts in willow and *Acer negundo* by the use of lithium and cesium nitrates. These salts are not normally present in these plants and their presence can be detected spectroscopically. The removal of a section of xylem completely obstructed the upward movement of both salts, while the removal of a section of the phloem retarded only the movement into the parts above the ring. He believed that the upward and downward movement of mineral salts occurs principally through the xylem in the transpiration stream. He considered, however, that this movement may be aided by diffusion, by protoplasmic streaming, and by transfer along the threads of protoplasm.

Clements (1930) removed a ring of phloem from the stems of red and black raspberries, grape, apple, peach, plum, and cherry, and noted that the portions of the plant above the girdle continued to obtain mineral materials and nitrates from below. Since a ring of phloem was removed, it seemed apparent that the xylem served as a channel of transport for these materials. Clements considered that the distribution of the tracheae in the seasonal development of xylem is to a large extent the factor that determines the ease with which plants can tolerate girdling.

Crafts and Kennedy (1930), Morgan (1931), and Crafts (1933) studied the movement of toxic salts to the roots when the salts were applied as sprays to the leaves. After these salts entered the leaves they eventually entered the xylem and were transported downward to the roots where they slowly diffused from the xylem into the living cells, which were later killed. The rate of penetration of a toxic substance into plants and its translocation are influenced by temperature, incident radiation, humidity, and the rate of evaporation. Loomis, Smith, Bissey, and Arnold (1931, 1933) found that the movement of sodium chlorate within the plant is principally in the xylem and is most rapid in the direction of the transpiration stream. The removal of the phloem did not appreciably hinder the translocation of the sodium chlorate, but the checking of transpiration by vaselining the leaves did reduce this movement. Maskell and Mason (1930) and Mason and Maskell (1931) believed that in the cotton plant the evidence indicates that the bulk of the phosphorus, potassium, calcium, and inorganic nitrogen absorbed by the roots is carried in the transpiration stream through the xylem to the leaves. When a ring of bark, including the phloem, is removed between the foliage region and the roots, phosphorus, potassium, and inorganic nitrogen accumulated in the tissues of the stem and leaves above the ring and diminished in these parts below. They found that nitrogen, phosphorus, potassium, and other constituents of the ash ascend the stem mainly in the wood, are reexported from the foliage, and then move downward through the phloem toward the root. The calcium also ascends in the xylem, but there is no evidence that it moves downward from the leaves through the phloem. The ratios of nitrogen, phosphorus, and potassium in downward movement to the carbohydrate moving in the same direction appear to be in excess of the ratios required for the growth of the lower part of the plant. It is suggested that the excess of these elements may be liberated into the tracheal sap and thus again ascend the plant. They stated that this return toward the roots of many of the mineral nutrients that have been moved upward with the transpiration stream must influence the uptake of more nutrients by the roots. This may explain in part at least the reason why an increase in the transpiration rate does not affect the rate of the intake of salts.

The most serious objections that have been offered to ringing experiments is that the xylem may be altered by the ringing. Numerous authors have reported that tyloses or gums may plug the xylem elements in the region of a ring or other wound. However, in many of the cases that have thus been reported, no precautions were taken to protect the exposed xylem. Even when the ringed portion is protected by wax, vaseline, or paraffin, the xylem is in some way affected by the ringing, since it has been noted that under these conditions there is an increased resistance to the flow of water through it. The xylem, however, at the most, is only partially plugged in well-managed experiments, and this partial plugging cannot fully account for the hindrance in solute movement when the phloem is severed.

b. Through Phloem.—Curtis (1923 to 1925) conducted experiments to determine if the phloem is concerned in the translocation of the inorganic salts. The general method used by him was to ring twigs or branches by a complete or partial removal of the bark or to remove a portion of the xylem and to study the effect of this ringing upon the upward movement of nitrogen, ash, and carbohydrates. In these experiments all the ring wounds were covered with warm paraffin to prevent the entrance of bacteria and to prevent the tissue from drying out.

In the experiments with privet, peach, and lilac, the branches were selected in pairs, so the ringed and control branches were similar in size and position. Immediately after ringing, nitrogen as sodium nitrate was added to the soil. In some cases, calcium chloride and sodium chloride were also added. Observations were made upon the leaves above the ringed portions for area, dry weight, ash, and nitrogen content, and these quantities compared with those of the leaves from similar regions of the control branches. In the case of the peach, the total amount of nitrogen moving up the unringed stem was evidently much greater than that moving up the ringed stem, since there were from ten to twenty times as many leaves on the control stem and they were from 30 to 70 per cent larger than those upon the ringed twig. With the lilac the control leaves contained 1.8 to 2.2 times as much nitrogen per unit of dry weight as those from the ringed stem. The ash analysis also indicated that ringing hinders the movement of inorganic salts, since the check leaves contained from 1.5 to 2.0 times more ash than did those from the ringed stems.

In one series of experiments, Curtis (1925), by a technique that need not be described here, completely severed the xylem in one portion of a split stem, leaving the phloem intact. In another series, the xylem on one portion of the stem was left intact and the phloem severed by ringing. By the use of such divided stems where water was supplied by one set of roots and nitrogen by others, data were obtained which indicated that if the roots supplied with nitrogen were connected with the top by the xylem only, there was little transfer of nitrogen to the tops, while if they were connected by a strip of phloem only, considerable transfer of that element occurred.

Curtis also found that if a narrow band or strip of phloem containing from one-tenth to one-fourth of the normal amount was left intact in ringing, the transfer of solutes was normal. It has been shown many times that a small portion of the xylem cylinder will carry sufficient water to keep the leaves apparently normal. Curtis (1925) considered that, since a small amount of the phloem is sufficient to allow for an approximately normal solute transfer, while its complete severance causes a marked reduction, and since a small part of the xylem is sufficient to allow for approximately normal water conduction, while its complete severance practically stops water movement, it would appear that water moves chiefly in the xylem and solute movement occurs chiefly in the phloem. In later work MacDaniels and Curtis (1930) found that the phloem of the apple tree is the more important tissue concerned in the translocation of solutes since the severing of the phloem had the same effect as the severance of both the xylem and phloem.

There seems to be strong evidence which indicates that the inorganic solutes move upward in considerable amounts through the phloem. The cases of translocation, however, that have been studied are relatively few and have been confined to only a few species of plants. On that account, a much greater number and variety of plants must be studied before any general conclusion can be drawn.

2. Lateral Transfer.—The lateral transfer of inorganic solutes has received some attention. The most thorough work in this regard was done by Auchter (1923), who studied the lateral transfer of nitrates in 10-year-old peach trees, 5-year-old apple trees, 3-year-old privet bushes, and 5-year-old oak trees. In these experiments some plants received nitrates about all their roots, some had nitrates applied to only one-half of their roots, while others had all their roots removed from one side and nitrates applied evenly under both sides. In practically all cases the leaves increased in nitrogen content per unit of dry weight, per unit of green weight, and per unit of area on the nitrated sides. The leaves on those halves which had not received nitrate actually lost nitrogen in the cases where analyses were made before and after treatments. The check halves of the trees acted much the same as full-check trees, while the nitrated halves responded approximately the same as fully nitrated trees. In one experiment with Grimes Golden apple trees, the soil was cultivated on one side for 2 years, while the other side was not. The average terminal growth of the branches on the side that had been cultivated was 12 to 14 in., while that on the uncultivated side was only 3 to 4 in. It was considered that the additional nitrates in the cultivated soil were responsible for this difference in growth and that it indicated that they were used by the halves of the trees which were directly above the cultivated areas. Similar results in regard to fertilization and cultivation have been obtained by Blake (1921) for peach trees and by Knowlton (1921) for Rome Beauty apple trees.

In the case of dormant privet cuttings transplanted so that one-half of their roots were in quartz sand and the other half in fertile soil plus nitrate, Auchter (1923) found that the two halves of these stems broke their rest period unevenly. The sides above the fertile soil started to grow 3 weeks sooner and had made twice the amount of growth after a period of 5 weeks. Since nitrates are known to break the rest period of plants, it was considered that they had moved upward in the stem only on the side to which they were presented to the roots and that there was little or no lateral transfer.

Bodenberg (1929) believed that inorganic solutes cannot pass laterally through the stems of woody plants. He was unable to detect lithium nitrate spectroscopically in regions remote from the vertical line of flow in a willow stem after a period of 295 hr. MacDaniels and Curtis (1930) from their observations on the effect of the spiral ringing of apple trees reported that the lateral transfer of solutes in the stem of this tree is relatively slow, although it does occur when forced by spiral ringing. They stated that nutrient salts from below tend to move in straight lines parallel to the long axis of the conducting elements. Zimmerman, Hitchcock, and Crocker (1931) noted that ethylene gas enters a plant and moves rapidly in all directions from its point of entrance. It is not known, however, whether this gas diffuses through the intercellular spaces or whether it moves in the transpiration stream.

From these studies of woody plants it appears that the mineral nutrients absorbed by the roots on one side of a plant are, in a large measure, translocated to and used by the trunk, limbs, and leaves directly above them and that there is very little cross transfer of such nutrients in the plant. Since there are some indications of the cross transfer of water in stems, the mineral nutrients evidently travel independently of the water movement.

In the case of the corn plant, Gile and Carrero (1921) distributed the roots in three different flasks containing different nutrients. In one flask was a solution of potassium salts, in another the phosphorus salts, and in a third the nitrogen salts. There was a diminished assimilation of these nutrients under these conditions as compared with the plants growing in a solution that contained all these salts. It was considered that this diminished assimilation was not due to the inability of the roots to absorb the ions with sufficient rapidity but to the slowness with which the three ions were translocated to the cells where they were needed. These three ions thus each accumulated in different parts of the plant as a result of having been absorbed by different roots.

The same general results were obtained in a similar experiment performed by Carpenter (1936) with the roots of sugar cane. In this case all the roots of the plant except ten were removed. One of the intact roots was placed in distilled water and each of the other nine in individual solutions containing one of the elements necessary for the optimum growth of the plant. Henricksen (1933) found that the ash constituents in citrus trees move only through those conducting tissues which are connected to a definite portion of the root system.

White (1927) found that the entire vascular cylinder in the crown of the strawberry is composed of a network of short anastomosing bundles. These provide an efficient means of rapid transfer of water and solutes across the stem. Thus, in striking contrast to most plants, water and nutrients supplied to one side of a row of plants can apparently supply uniformly both sides of these plants.

3. Downward Transfer.—The downward movement of inorganic salts has received special study in leaves prior to their abscission. Before considering the migration of materials from the leaf in the autumn, a brief summary of the nature of leaf abscission will be given. According to Sampson (1918), Von Mohl (1860) was the first to state that previous to the fall of the leaf there is formed near the base of the petiole, a definite separation layer in which abscission always occurs by the separation of the cells from each other. The xylem elements are not included in this layer for they are eventually ruptured mechanically, following which the leaf falls. Von Mohl also stated that abscission and the formation of a protective tissue are two very distinct processes, and that the latter might either precede or follow the former.

The nature of the processes involved in the formation of the abscission layer and the fall of plant parts has been studied by Wiesner (1871, 1904, 1905), Molisch (1886), Tison (1900), Lee (1911), Fitting (1911), Hannig (1913), Lloyd (1914, 1916), Sampson (1918), and Heinicke (1919), and their observations may be summarized as follows: The formation of the abscission layer begins a short time before the maturity of the leaf and continues for a considerable period. Cell division begins to occur in two to four layers of cells across the base of the petiole. This begins in the epidermal and cortical regions and gradually extends inwardly to the phloem. Generally there results an abscission layer eight to twelve cells in thickness. These cells remain smaller and they have thinner walls than the adjacent cells of the petiole. After the formation of this abscission layer there is a disintegration of the middle lamella and of the cellulose of the secondary wall of the cells that compose this layer. The cells are thus no longer held together, and leaf fall results.

The external factors that accelerate leaf fall are extremely diversified. According to Lloyd (1914), they include high and low light intensities, high and low water supplies, high and low temperatures, low concentration of anesthetics, toxic condition of acids and salts, and wounding of the blade. Low concentrations of oxygen and high concentrations of anesthetics retard leaf fall.

The migration of the inorganic compounds from the leaves of deciduous plants previous to their fall has been studied by Tucker and Tollens (1900), Otto and Kooper 1910), Michel-Durand (1913, 1932), Swart (1914), Serex (1917), Rippel (1921), Combes and Kohler (1922), Seiden (1926), Deleano (1932), Deleano and Bordeianu (1932, 1934), Echevin (1932), Komatsu and Ozawa (1932), McHargue and Roy (1932), Deleano and Andreesco (1932), Murneek and Logan (1932), Thomas (1933), Denny

(1933), Polovrăgeanu (1933), Trandafirescu (1934), Sampson and Samisch (1935), and others. The leaves that have been studied included those from forest trees, fruit trees, and shrubs. With the exception of the findings of McHargue and Roy (1932) for certain forest trees, all the investigators give data that indicate the migration of nitrogen, phosphorus, potassium, magnesium, and iron from the leaves previous to their fall in autumn. The percentage of migration of a given element based on the maximal amount present during the life of the leaf has been termed the "reabsorption coefficient." The value of the reabsorption coefficient reported by various investigators has ranged from 1 to 90 per cent for the different leaves and elements considered.

It has been observed by Hornberger (1882) for corn, Snyder (1893) for wheat, Wilfarth, Römer, and Wimmer (1905) for a number of crop plants, Jones and Huston (1914) for corn, Burd (1919) for barley, and Miller (1937) for wheat that there is a migration of certain elements, especially potassium, from the stem and leaves to the roots and then to the soil. Penston (1935) has reviewed the work of several workers in regard to the loss of minerals from the plant.

4. Upward Migration.—There is evidence that during the development of new growth or of seeds that the vegetative parts are depleted of inorganic compounds to supply the raw materials for the synthesis of organic compounds in these developing parts. Thus Jones and Huston (1914) observed in the corn plant that the nitrogen in the stalks after the ears began to form decreased from 53.5 lb. per acre in August to 31.8 lb. in October, although the nitrogen content of the whole plant was increasing. The phosphorus decreased in the stalk from 15 to 6.3 lb. per acre during this same period, while it increased in the ear from 6.0 to 35.5 lb. per acre. The migration of potassium apparently was not so marked, since in the stalk it decreased only from 73.5 to 70 lb. per acre, while it increased in the ear from 10 to 25 lb. per acre.

Burke and Morris (1933) reported that the leaves and developing parts of an apple tree receive a considerable portion of their nutrients from the materials stored in the various regions of the plant, while the remainder is absorbed from the soil.

Fagan and Watkin (1931) noted in the developing oat plant that there was a continuous migration of ash, phosphorus, potassium, iron, and nitrogen out of the straw into the spikelet from the time of its formation to the ripening of the grain. Miller (1937) noted that a portion of the minerals and nitrogen in the heads of wheat was obtained from the soil, while the remainder was drawn from storage forms in the stem and leaves of the plant.

C. Injection Experiments

A considerable number of injection experiments have been undertaken in an attempt to supplement the supply of water and nutrients to plants and to control fungous diseases and insect pests. These experiments should be mentioned here, as the factors involved are primarily those of translocation.

The injection of iron salts into the trunks of trees in order to overcome chlorosis has recently received considerable attention. Lipman and Gordon (1925) cured chlorosis, in lemon trees that had been chlorotic for several years, by the injection of a solution of ferrous sulphate. Bennett (1927) introduced iron salts directly into the trunk of pear trees that were affected with chlorosis. Over 6,000 trees were treated and a cure was effected in 90 per cent of the cases at a cost of 1.5 to 15 cents per tree, depending upon the size of the trees. Beneficial results were observed on those trees which did not fully recover from the chlorotic condition. Wann (1929) found that dry iron salts placed in an ordinary uncapped medicine capsule and inserted in holes bored in the stem of the plant proved effective in overcoming chlorosis of grapevines and of peach and apple trees. The iron salts used were ferric citrate, ferric phosphate, and ferric ammonium citrate, but the best response was obtained with the ferric phosphate. The leaves above the point of injection were the first to recover, after which the recovery spread gradually over the whole plant. Thomas and Haas (1928) were able to improve the chlorotic orange trees temporarily by the injection of iron salts. Subsequent injections, however, were necessary, and in a new place, since the wood was injured for some distance around the point of injection. Special methods for the injection of nutrient solutions into woody plants have been reported by Collison, Harlan, and Sweeney (1932); Roach (1934); and Thomas and Roach (1934) for apple trees. Collison and his workers believed that the injection method is a poor substitute for the natural intake of nutrients from the soil, but Thomas and Roach obtained increases in growth that were comparable to those obtained when the nutrients were applied to the soil.

Goff (1897) undertook to supply water in an artificial way to the roots of young transplanted trees. He placed a supply of water at a height equal to or greater than that of the tree and connected this with the cut end of a root. He reported that the buds on the treated plants developed more quickly than on the control plants and that the effects of this treatment were noticeable within 48 hr.

The injection of chemicals into plants for the control of fungous diseases has been undertaken by a considerable number of investigators. Bolley (1903 to 1907) found that trees would take up water solutions in large quantities when applied through a bore hole with an attached feeder. Chemical substances were carried to the most distal parts of the trees in a few hours. Thus in 10 hr. substances could be detected microscopically in the tops of cottonwood trees 30 to 45 ft. high. He was unable, however, to prevent fungous diseases by any injection experiments, but he considered that in some cases he held them in check. Mangin (1898), Morkrjetsky (1903), and Simon (1906) also considered that they obtained some control over fungous diseases by the injection of trees, potatoes, cauliflower, cabbage, and other plants. Injection experiments by Rankin 1917 and Rumbold (1915, 1920) using 80 or more chemicals with chestnut trees all gave negative results in the control of the chestnut-blight fungus. Scherer (1927) in injection experiments with white birch, gray birch, American elm, American chestnut, and apple trees obtained negative results in the control of fungous diseases and insect pests with the exception of apple trees that were injected with thymol. These seemed to show definite resistance to the progress of *Bacillus amylovorus*.

Injection experiments with potassium ferrocyanide have been tried for the control of insect pests, especially those in woody plants. The results that have been reported are extremely conflicting. Sanford (1914, 1915) and Shattuck (1915) have reported favorable results, while Surface (1914) and Flint (1915) could obtain no control whatsoever.

The conduction and behavior of the potassium cyanide when introduced into plants have been studied by Moore and Ruggles (1915) and Elliot (1917). Elliott found in smaller plants that the injury due to this chemical was very marked. On a hot, dry day the wilting of the leaves of herbaceous plants and small woody plants was noticeable under conditions of high evaporation within 10 min., and the fumes due to the potassium cyanide were readily detected coming from the leaves. Plants treated on cool, damp days showed little injury within 24 hr. The injury to these plants was also less extensive than to those treated on hot, dry days. The diffusion of potassium cyanide laterally is greatest when conduction is the slowest, and in woody stems Moore and Ruggles (1915) found that hydrocyanic acid passed through a particular area which had its point of departure on the upper side of the injection hole and that the chemical diffused but little in a lateral direction. Craighead and

St. George (1930) by an appropriate method allowed wood alcohol, copper sulphate, carbon disulphide, formaldehyde, mercuric chloride, potassium cyanide, and other compounds to penetrate the wood of pine trees for the control of the bark beetle. In a few hours, from 2 qt. to several gallons of liquid could be injected and distributed throughout all the outer annual rings to the uppermost branches and leaves. With some of the chemicals, complete brood mortality was obtained in some cases, provided that the application was made before the sapwood became stained blue by fungi and the ascending stream thus disturbed. The efficiency of the control of fungous diseases and injurious insects is limited apparently by the fact that a solution of any chemical that will injure the protoplasm of the cells of the fungus or insect will also injure the protoplasm of the host cells. The control of insect pests in woody stems by injections is further handicapped because the lateral conduction of solutions is very limited and the injected materials will come in contact for the most part only with those insects which are directly above the point of injection.

D. Translocation of Organic Compounds

1. Storage of Reserves.—Before discussing the translocation of organic compounds, a few facts should be stated in regard to their occurrence, storage, and depletion in the plant. The principal organic materials stored in plants are the proteins, carbohydrates, fats, and oils. The cells that are concerned in their storage are the medullary-ray cells, the parenchyma cells of the xylem and phloem, the parenchyma cells of the cortex, and the cells of seeds, bulbs, tubers, and rhizomes.

a. Formation of Reserves.—The place of the origin of these organic materials is an important point to consider in a discussion of their translocation. Are they manufactured at places other than that of their occurrence and then translocated to where they are found or are they manufactured in the cells in which they are stored?

Sugar is apparently the only organic compound for whose formation light is directly essential, although, as mentioned in Chap. VIII, the kind of sugar that is first formed is not known. From this primary sugar, all the other carbohydrates may be formed in any living cells regardless of light. The indications are that the proteins, fats, oils, and other organic compounds may also be formed in any living cell independently of light, provided that sugar, of one type or another, and the necessary inorganic materials are present. If such is the case, it is conceivable in the formation and storage of organic compounds that sugar of one kind or another is the only organic compound that needs to be translocated for long distances in the plant. Since it is formed, for the most part, only in the leaves and since it is found in either the free or combined form in every living cell, there is no doubt that it is moved farther and in greater abundance in the plant than any other organic compound.

There is evidence, however, in numerous cases that there is a movement of organic materials, other than the freshly manufactured sugars, from the vegetative part of the plant to the seeds and fruits in their formative stage. Thus Trowbridge, Haigh, and Moulton (1915) noted that the dry matter of the roots and stubble of wheat increased up to the milk stage, after which it decreased in amount, being moved to the parts of the plant above ground. The stalks and leaves contained their maximal amount of dry matter at blooming time, after which it decreased until the ripening of the grain. The protein in the stems and leaves decreased from 240 lb. per acre at blooming to 120 lb. at the maturity of the grain. The most rapid depletion during that period was from the milk stage to ripening. During that time the total carbohydrates decreased from 1,700 to 1,400 lb. per acre in the leaves and stems. Data quoted by Jones and Huston (1914) from German experiments showed during a 22-day period of grain formation that the nitrogen in the straw decreased from 83.5 to 25.0 lb. per acre, while the nitrogen content of the head increased from 23.4 to 60.8 lb. per acre. The dry matter in the straw decreased from 77.9 to 60.1 lb. per acre, while it increased in the head from 11.9 to 32.6 lb. per acre. The work of Jones and Huston (1914) with maize indicates that there is little depletion of the organic materials in the vegetative parts of this plant during the formation of the grain. Thus they found that the dry matter, fats, and carbohydrates in the stalk and blades remained practically constant, although they rapidly increased in the ear. Although the carbohydrates remained constant in amount in the vegetative parts, they increased from 4 to 40 lb. per acre during a 40-day period of grain formation.

Murneek (1928, 1932) found in the apple after flowering that the direction of movement of most organic compounds seems to be primarily if not exclusively toward the fruit that had set. The relative amount of foliage on a branch determines partially at least the size and quality of the apple fruit, but it has little or no effect on the weight and composition of the seed. The apple fruit draws heavily upon the products of photosynthesis during the latter part of its development. It was observed by Weinberger and Cullinan (1932) in peach trees that the average size of fruit on branches that had been ringed was greater by 20 per cent than those grown on branches not ringed. The organic materials that normally would have been translocated down the stream were prevented from moving and were thus available to the developing fruit.

Jones and Bisson (1932), and Bisson and Jones (1932) observed in the development of the garden pea that in the pod there was an early maximum of the absolute weight of nitrogen, total sugars, starch, and ash, after which there was a decline. In the peas there was an increase in the absolute weight of nitrogen, starch, crude fiber, and ash throughout the entire growth of 48 days. The absolute weight of sucrose reached its maximum in 32 days, after which it declined. Rewald and Riede (1933) noted that the seeds of the soybean during their development grew at the expense of the other parts of the plant. Fagan and Watkin (1931), and Miller (1937) found that a portion of the carbohydrates and nitrogenous materials in the seeds of wheat was obtained by depleting the stem and leaves of these substances.

The distribution of sugars and dry matter in the maturing corn stem from tasseling to maturity was studied by Welton, Morris, and Hartzler (1930). They found in the varieties Clarage and Burr Leaming that the percentages of dry matter and sucrose were larger in the upper than in the lower part of the stem. The percentage of dry matter increased and the percentage of reducing sugars decreased progressively from the tasseling stage to maturity in both varieties. Sayre, Morris, and Richey (1931), and Brunson and Latshaw (1934) found in corn that protein and carbohydrates tended to accumulate in other portions of the plant when the set of grain was prevented. The composition of the cobs was influenced more than any other part. They were higher in protein, fat, and ash, and lower in fiber in poorly filled plants than in those which had normal grain development. A sweet-corn plant usually does not produce so great a weight of grain as a dent-corn plant of equal size. By pollination studies Henson (1932) showed that this behavior results in part from the fact that the elaborated carbohydrates are more completely transferred to the grain from the vegetative parts of dent corn than from those of sweet corn.

In the development of the endosperm of corn, Lampe (1931) stated that the reducing sugars decreased as the cells approached maturity. The sucrose content of the endosperm in the several varieties rose gradually to its maximum at about the fifteenth day after pollination and then decreased but not so rapidly as the reducing sugars. In the nonsweet corns the disappearance of sugars from the endosperm was essentially complete at maturity, while in the sweet types a small amount of sugar, mostly sucrose, remained in the mature endosperm. This was mostly confined to the central region where the dextrin was located. Kretovitch (1933) stated that, in the wheat grain, sucrose, to the extent of 3 per cent, is present in the embryo. The embryo makes up 19 per cent of the weight of the grain. There is no sugar in the aleurone layer, and glutenin and albumin compose the major portion of the protein in this part of the seed.

b. Time of Storage.—The time of the storage of organic compounds as well as the tissues in which they are laid down should be noted, since considerable work has been done in order to determine the relation of the amount of these reserves to fruit formation, to the amount and quality of vegetative growth, to the conservation of plants, or to their eradication.

In timothy (Trowbridge, Haigh, and Moulton, 1915) the dry matter in the bulbs increased throughout the growing period, but the amount became constant before the ripening of the seed.

Working with alfalfa, Graber and his coworkers (1927) did not find any exact period of growth when the storage of organic food reserves began in the roots but noted that such storage was greatly accelerated during the periods of blooming and seed formation, as shown in the following table:

DISTRIBUTION OF ORGANIC MATERIALS IN THE TOPS AND ROOTS OF ALFALFA AT DIFFERENT PERIODS DURING THE GROWING SEASON

Adapted from Graber, Nelson, Leukel, and Albert, 1927

Stage of growth	Per cent of moist weight		Per cent of dry weight									
	Dry matter		Total sugars		Soluble starches, dextrins		True starch		Hemi-cellulose		Total nitrogen	
	Roots	Tops	Roots	Tops	Roots	Tops	Roots	Tops	Roots	Tops	Roots	Tops
Early-spring stage, 4⁄13	26.8	20.5	11.2	6.1	2.8	1.9	6.0	1.9	11.3	10.9	2.1	3.8
Flower-bud stage, 6⁄21	28.0	20.1	5.3	3.9	5.5	1.5	17.8	2.1	12.1	12.0	2.3	2.5
Full-bloom stage, 7⁄10	36.9	24.6	5.1	3.1	14.4	2.2	17.6	1.9	12.5	14.5	2.2	2.2
Seed-pod stage, 8⁄7	56.0	43.0	5.0	1.6	6.7	1.4	20.0	2.0	9.6	12.6	2.7	2.6
Fall dormancy, 10⁄22	44.8	26.8	4.6	6.1	4.7	1.6	24.9	3.1	8.2	11.9	2.5	2.9

Aldous (1929) found that in the stems of the sumac (*Rhus glabra*) and buckbrush (*Symphoricarpos vulgaris*) the largest amount of reserve materials was reached about Aug. 15 in both cases. In the former the minimum amount of stored materials was present about June 10, while in the latter the minimum was reached as early as May 10. The location of the food reserves in the two plants is of interest. In the sumac the major portion is stored in the parenchyma cells of the bark, while in the buckbrush practically all of it is stored in the medullary rays and parenchyma cells of the xylem.

In the 2- to 3-year-old portions of apple twigs, Traub (1927) found that the total reserve carbohydrate of the xylem tissues was higher than that of the cortex phloem at all times of the year and that this reserve reached its maximum after leaf fall. The pentosan content formed a relatively large proportion of the total reserve carbohydrates. They made up from 45 to 68 per cent of the total carbohydrates of the cortex phloem, from 65 to 93 per cent of the carbohydrates of the outer xylem, and from 58 to 83 per cent of those of the inner xylem. The acid hydrolyzable hexosans made up from 2.3 to 26.5 per cent of the carbohydrates of the cortex phloem and from 10.1 to 31.8 per cent of those of the outer xylem.

During the summer this hexosan content increased, but with the approach of cold weather there was a change of hexosans into sugar beginning about October. The total sugars in these parts also increased at this time, owing to translocation from the leaves. During the winter the hexosans disappeared almost entirely from the cortex phloem and were reduced more than 50 per cent in the xylem tissues. There was a general depletion of hexosans in May in all the tissues, owing to the translocation of sugar used in growth. The total sugar content of the phloem cortex was always markedly higher than that of the xylem tissues.

The nitrogen maximum in all parts was reached in March and April preceding growth in length. After a decline during the growth in length, the nitrogen content was relatively constant during the summer. From the fall of leaves through the winter it remained practically constant. The amino nitrogen was highest during the growing season, decreased as the rest period approached, and reached a minimum during the dormant period. The protein nitrogen was highest during the dormant season, decreased as the active growing season approached, and reached a minimum in June. The protein nitrogen was consistently higher in the cortex phloem than in the xylem tissues. Gardner (1925) observed that starch deposition in the shoots of the Bartlett pear begins very shortly after the cessation of length growth. Starch is deposited first at the tip of the shoot and then progressively downward toward the base. In the various tissues of the current season's growth, starch appears in the following order: medullary rays, wood parenchyma, and bark. Microchemical tests for sugar showed that the bark is much higher in sugar content than is the wood. The cortical parenchyma contains the largest amount of sugar.

Gäumann (1927) noted that the carbohydrates in the sapwood of *Picea excelsa* were at the maximum in October and April, and at the minimum in February and July. In *Abies pectinata* they were at the maximum in November and April, and at the minimum in February and June.

Kraybill, Sullivan, and Miller (1930) noted that in the Stayman apple tree there was a maximal content of starch in all parts of the tree after leaf fall in the autumn. This was utilized in the new growth in the spring, and starch storage began anew after petal fall in May. Grandfield (1930) found that the organic food reserves in *Solidago rigida*, *Verbena stricta*, and *Veronia baldwinii* were at their low point just before these plants began to show buds. Aldous (1930) noted that the organic reserves in the roots of little bluestem, side oats, Indian grass, hairy grama, buckbrush, sumac, ironweed, and vervain decreased in organic reserves to about the time of flowering after which these plants began to restore these reserves. Arny (1932) reported that there was a decline in the reserves of carbohydrates and nitrogenous compounds in the underground storage organs of leafy spurge, Australian field cress, and sow and Canada thistle from spring until blooming. Storage of these compounds then began in these parts and continued at a moderate rate until the close of the season. It was observed by Martin (1934) that in *Melilotus alba* the growth of tops predominated during the first of the growing season, while root growth was more prominent during the latter part. Grandfield (1935) stated that the total nitrogen and carbohydrate contents reached their maximal accumulation in the roots and crowns of alfalfa at or near the time of blooming. Long (1935) believed that the strawberry plant behaves in general like a biennial plant in that it stores large quantities of carbohydrates and nitrogen during the summer and autumn, and uses them the following spring for new growth. Culpepper, Caldwell, and Moon (1935) found that as the strawberry ripened it showed a marked increase of sugar, which makes up about 70 to 80 per cent of the soluble solids and about 50 per cent of the total solids of the ripened fruit.

Werner (1935) found in the potato that whenever carbohydrates were manufactured in excess of the capacity of the plant to build new tissues they accumulated in the potato tuber after a transitory period of storage in the leaves, stems, and stolons. Such accumulations occurred when the day length was shortened, temperature lowered, and the nitrogen supply reduced.

c. Factors Influencing.—Studies have been made by numerous investigators concerning the frequency and time of cutting of the aerial parts of plants on the amount of stored reserves in the underground parts. In some cases, the purpose of these studies has been to determine the methods of agricultural practice that will be conducive to the storage of a reserve in these parts to produce an optimum growth and development of the plants the following season. In other cases the purpose has been to determine the best practice to follow in order to obtain a maximal yield of hay or forage.

It was noted by Winkler (1929) that the nonpruned vines of *Vitis vinifera* contained from three to four times as much total carbohydrate at the beginning of the growing season as the normally pruned vines. In the case of kudzu (Pierre and Bertram, 1929), the fewer the number of cuttings the greater is the production of food reserves. The roots of plants receiving six cuttings per season decreased in weight during a period of 2 years. Those receiving four cuttings increased in weight 150 per cent, those receiving two cuttings increased about 400 per cent, while those from plants receiving only one cutting increased in weight about 1,250 per cent. The percentage of reserve starch and nitrogen was found to be less than one-half as much in the roots from plants receiving six cuttings as in the roots of those receiving four or a fewer number of cuttings.

It has been found by Aldous (1930) for certain prairie grasses, Shutt, Hamilton and Selwyn (1930) for meadow foxtail, Graber (1931) for blue grass, Parker and Sampson (1931) for *Bromus hordeaceus* and *Stipa pulchra*, Robertson (1933), and Biswell and Weaver (1933) for various grasses that the yield of hay or forage from these plants was least from those which had been cut or clipped frequently during the growing season and highest from those which were cut only at their maturity. Janssen, McClelland, and Metzger (1930) reported that the amount of sugar in the stalks of sorghum was slightly depressed by stripping them of the leaves 3 or 4 days previous to cutting.

The effects of defoliation or injury to the leaves upon the production of grain has been extensively studied in corn. The dominant idea in such studies has been to approximate "hail injury" and to determine the loss in yield that such injury might cause. It was reported by Culpepper and Magoon (1930) that defoliation reduced the yield of grain to the greatest extent when it was performed 10 to 15 days after silking, and that beyond this time the undesirable consequences of defoliation were lessened progressively as the treatment was delayed. Hume and Franzke (1929) and Dungan (1930, 1934) found that the removal of blades or portions of blades most severely affected the yield of corn when the treatment was applied during the early silk stage. Leaf removal before or after this stage is progressively less harmful to yields. Tearing the sides of the blades from the midribs but leaving them intact at the base resulted in a yield that was 70 per cent higher than the one obtained after the complete removal of the blades. The reduction in yields corresponded roughly at all stages with the percentage of the leaf area removed.

It has been reported by Sturkie (1930) for *Sorghum halapense*, Graber and Ream (1931) for bluegrass, Robertson (1933), Biswell and Weaver (1933) and Sprague (1933) for various grasses, Thornton and Nicol (1934) and Grandfield (1935) for alfalfa, Leukel and Barnette (1935) for Bahia grass, and Virtanen and Nurmia (1936) for sweet clover that frequent cutting or clipping of the tops during the growing season resulted in a marked decrease in the storage reserves of the root. Thornton and Nicol (1934) stated that the clipping of alfalfa did not significantly alter the number of nodules on the roots.

According to Thomas (1932) apple trees supplied in the spring with different combinations of sodium nitrate, potassium sulphate, and monohydrogen calcium phosphate in low concentrations showed marked differences in their storage reserves. Thus the concentrations of simple sugar in the trees that did not receive any sodium nitrate was much lower than in nitrogen-treated trees.

d. General Facts.—It was observed by Jacobson (1934) that the stems of Leoti Red sorgo had 39.9, 40.1, 38.6, 45.3, and 44.3 per cent sugar on a dry basis at the milk, soft dough, early hard dough, late hard dough, and ripe stages, respectively, while for the same stages, Atlas sorgo had 30.2, 40.2, 38.6, 35.4, 38.6, and 36.5 per cent sugar.

Gerry and Hall (1935) stated that the gum or oleoresin exudes from cuts on the living pine tree from its earliest seedling stage. This exudate does not come from the ordinary sap-conducting tissues but from the areas of parenchyma cells in the outer sapwood. These cells contain actively functioning protoplasm for a longer period of time than the associated cells that conduct water. Resin ducts are spaces that arise in the midst of clusters of parenchyma cells by a separation or splitting apart of these cells. In the slash pine there are more than 450 horizontal resin passages per inch of tangential surface, while the vertical ones number about 200 per square inch of cross section. The oleoresins are products of the protoplasm, but as yet not one fact is known with certainty concerning the process by which they are elaborated. In the oleoresinous exudate there are present glucosidal complexes of resin acids, volatile oils, and reactive sugars.

2. Utilization of Reserves.—These reserve organic compounds are used by the plant in respiration and in the formation of new plant parts, after a rest period as in the case of renewed growth in the spring or for the production of new organs and parts when these have been removed. Thus it has been found by Salmon, Swanson, and McCampbell (1925), Nelson (1925), Graber (1927), Albert (1927), and Leukel (1927) that in alfalfa new top growth is initiated in a large measure at the expense of previously deposited root reserves and that unless such reserves are sufficiently replenished during the period of successive cuttings, a reduction in the reserve contents of the root occurs, which diminishes the amount of new top and root growth, following each cutting. The effects of too frequent cutting at immature stages are to prevent the plants from forming organic food reserves in sufficient quantities for storage in the roots for their future development, as well as that of the tops. When alfalfa is cut too frequently, the vigor of succeeding growths decreases rapidly, and many plants may die during the growing season. It has also been observed that the amount of winter injury to these plants may be correlated with the time and number of the cuttings during the summer. The vigor of bluegrass, redtop, and timothy is also influenced by the number of cuttings during their growing season.

Gerhardt (1929) noted that in the common milkweed the leaf, stem, and root yielded carbohydrates and nitrogenous compounds to the developing seed. The rubber in this plant amounts to as much as 3.5 per cent, and it is stored for the most part in the leaves. Pentosans and hemicelluloses are stored in the stem, while starch and nitrogen are the main reserves in the root. Crafts and Kennedy (1930) believed that the large quantities of starch stored in the roots of *Convolvulus arvensis* makes possible the persistent vegetative activity of this plant. The rapidity with which this reserve can be depleted or replenished indicates the presence of a very efficient conducting system. Davis (1931) found that in the nonbearing sugar prune trees, the amount of starch was consistently higher than in the bearing trees. The roots of bearing trees contained a relatively small amount of starch, whereas those of the nonbearing trees were high in this carbohydrate. During the greater part of the summer the amount of total nitrogen in the bearing trees was lower than that in the nonbearing ones. Stuart and Appleman (1935) reported that potato sprouts were richer in total nitrogen than the tubers from which they grew. The nonprotein nitrogen was much higher in the medulla of the potato than in the cortex. It was noted by Vickery, Pucher, Leavenworth, and Wakeman (1935) in the tobacco plant that during the development of the seeds there was evidence that organic and inorganic substances were being transferred from various parts of the plant to them.

Bartel, Martin, and Hawkins (1935) found that tillers increased the development of the main stalks of dwarf hegari when the leaves of the main stalk were removed previous to heading. The longer the tillers were left on the plant, the greater was the weight of the main stalk. Although each tiller develops a root system of its own, a vascular connection exists between the main stalk and the stalk of the tiller.

In practice it has long been contended that if obnoxious brush and briars are cut at certain seasons of the year, few if any sprouts will arise from the roots and stump and that thus the plants would be eradicated. If, however, these were cut at other seasons of the year, numerous and vigorous sprouts would be developed, so that within a year or two the growth would be as heavy as or heavier than before. It seems very probable that this growth of sprouts is correlated with the amount of organic reserves stored in the stump and roots. If the cutting is done when the reserves are almost depleted, few sprouts develop; while if an abundant supply is present, new growth is very vigorous. Brown (1930) found under the conditions of Connecticut that July was the best time to mow pasture land to kill the undesirable plants.

3. Distance Moved.—Although it is definitely known that the reserve organic materials are utilized by the plant, the question arises as to how far these materials are moved from their place of storage to their regions of utilization. This question is of special interest in the case of woody plants where the distances between the various plant organs are relatively great. Since foods are stored in large amounts in stems and roots in late summer and autumn and since these disappear when growth is renewed in the spring, the general conclusion was early formed that foods are translocated from the roots and lower parts of the stem to the developing shoots and leaves.

It has been stated by Hartig (1858), Leclerc du Sablon (1906), and Butler, Smith, and Carey (1917) that the percentage of carbohydrates in the roots of various woody plants may be higher than that of the stem. This has been considered to indicate that the roots might act as storage organs for foods to be supplied to the developing parts of the stem. Curtis (1920), however, pointed out that the actual amount of carbohydrate and other reserves present in the roots may be less than that in the stem, since the total mass of stem is much greater than that of the roots, as shown by Chandler (1919), who calculated from the percentage concentrations given by Butler the relative amounts of foods available in the roots and tops of the apple tree. Aldrich (1932) considered that the roots of the Imperial apple obtained nitrogen from the soil during the winter months. This was also the opinion of Sullivan and Kraybill (1930) for Stayman apple trees in Indiana, and of Weinberger and Cullinan (1934) who noted that nitrogen might be absorbed by the roots of peach trees in autumn and early winter. This nitrogen was held in the roots until growth started in the spring, when it was rapidly translocated to twigs and buds. Murneek (1933) found that although the roots of an 18-year-old apple tree weighed only a third as much as the tops, the total quantity of starch present was about equal to that in the aerial parts.

Curtis (1920) ringed stems of *Acer saccharum* in early spring at different distances from the tips. In one series the rings were in the first-year wood, in another in the second- and third-year wood, and in another in wood ranging from 5 to 15 years of age. Of the 15 twigs ringed in the 1-year-old wood, the average growth in length during the following month was 0.84 cm., while that of the check stems was 2.22 cm. The terminal growth of those stems which were ringed in the second and third year's growth was 2.03 cm. as compared with 2.25 cm. for the corresponding checks. Those twigs ringed in the 5- to 15-year-old wood showed no difference in growth from the controls. Similar results were also obtained for the pear tree.

In apple twigs it was found that shoot growth was fairly vigorous when no food farther back than that obtained from a branch about

1 cm. in diameter was available. When the ring was no farther back than the 5- to 10-year-old wood, the growth of the shoots above the ring approached that of the unringed twigs. Soon after the shoots have started, much or all the food necessary for continuing growth is apparently produced by the leaves of the same shoot, so that in some cases the growth of ringed twigs with leaves was fully as great as that of twigs not ringed.

Haller (1930) studied the relation of the distance and direction of the fruit from leaves to the size and composition of apples. Branches were ringed, leaving above each ring one fruit and 30 leaves. These were left intact at varying distances from the fruit. It was found that apples can draw upon elaborated foods that are synthesized at a considerable distance from them. There was no decrease in the size or sugar content of the fruit when no leaves were nearer than 4.5 ft. in Grimes Golden, 6 ft. in Ben Davis and York Imperial, and 10 ft. in Baldwin, compared with the fruit produced when the same number of leaves were immediately adjacent.

Winkler (1932) stated that the average cluster of grapes requires all the food that has been synthesized by at least 16 leaves. This indicates that the fruit must use the foods that have been manufactured at a distance considerably removed from the region of its utilization. He considered that food moved upward and downward, in this case, for a distance of at least 3 ft.

Zimmerman and Connard (1934) by the arch method grafted the branches of tomato, coleus, salvia, and tobacco, and showed that both mineral salts and elaborated foods moved upward and downward through the stem.

Using the Bartlett pear, Gardner (1925) found that when the growth of twigs began in the spring, the stored starch disappears first from the tip and then from the lower regions of the shoot successively. In the various tissues of the shoot, the stored starch disappears in the following order as growth begins: bark, medullary rays, wood parenchyma, and pith. In some cases, at least, root growth occurs before shoot growth begins in the spring, as shown by Goff (1898) for a large number of woody plants. The observations of Jones, Edson, and Morse (1903) on the amount of water in the trunk of the sugar maple would also indicate that root growth precedes stem growth. They found that the amount of water in the stem increased from 36.5 per cent on Mar. 1 to 47 per cent on Apr. 28, at which time the buds opened and the water content of the stems fell off. Crider (1928) and Rogers (1929) have noted that roots of weeds grow under conditions that are unfavorable for the development of the aerial portions. Since apparently the roots of woody plants begin to grow earlier in the spring and continue later in the autumn than the twigs, it would seem probable that they would utilize a considerable portion of their stored foods. Since a twig ringed in that part of a stem from 5 to 15 or more years old produces a growth above the ring which approximates that of a normal stem, the indications are that the upward movement of food from points below the ring is not essential for the development of the twig. This statement is made on the assumption that the major portion of the organic materials is translocated through the phloem. Curtis considered that the distance from which food is withdrawn back of the tip to be used in shoot growth may depend upon the different species, individual plants, and the conditions of the present and previous seasons. The indications are that there is normally no movement upward to the tips of the branches from the roots and little or none from the main trunk.

4. Tissues Concerned.—No uniformity of opinion regarding the tissues that are concerned in the translocation of organic materials in plants has been reached by investigators. Some claim that the major movement is through the xylem and consider that the phloem is totally inadequate for that purpose. Others consider that the phloem carries the major portion of these compounds. Both sides present considerable experimental evidence that is worthy of brief consideration here.

a. The Xylem.—Those who consider the xylem as the main pathway for the translocation of organic materials base their opinion upon two general observations: (1) the physical unsuitability of the phloem for their translocation; (2) the composition and amount of the organic substances in the tracheal sap.

1. *Physical Unsuitability of the Phloem.*—Dixon and Ball (1922) and Dixon (1922 to 1924) presented theoretical evidence to show that the phloem is unsuitable for the translocation of organic substances. They mentioned that the phloem tissue is always relatively small in cross sections; that it is formed of short cells and comparatively short, narrow tubes with many cross partitions that must be traversed by the stream carrying the organic substances if this tissue is used as a pathway for their translocation. They considered that the resistance to a translocation stream must be high owing to the fact that a large proportion of the cross section of the phloem is occupied by viscous contents of protoplasm and proteins. They stated that the velocities of transport comparable with diffusion alone could be maintained only under such conditions, and that such velocities could not account for the quantity of material that is moved through the stem and roots.

To illustrate these points they cited as an example the translocation of carbohydrates from the leaves of the potato to its tubers. They selected a branch connecting the stem with a tuber that contained 50 g. of carbohydrate that had been deposited during a period of 100 days and determined that the area of the cross section of the phloem tissue of this branch amounted to 0.422 sq. mm., not allowing for the cell walls and nonfunctional elements. Assuming that all the carbohydrates in the tuber passed through the phloem tissue and that they were carried in a solution whose concentration was 10 per cent, 500 cc. of this solution would have passed in 100 days. The velocity of the solution in its passage would thus have been $\frac{500}{0.0042 \times 100 \times 24}$ or approximately 50 cm. per hour. The concentration of the solution, however, probably never exceeds 5 per cent; hence the rate of translocation must have approximated 100 cm. per hour. In the yam, Mason and Lewin (1924) believed that higher values than those for the rate of diffusion would have to be assumed. Dixon and Ball assumed that if a 10 per cent sugar solution were supplied by the leaves to the phloem elements and this were converted into an insoluble carbohydrate in the tuber 50 cm. distant as soon as it reached this destination, then a rate of transport by diffusion alone of only 0.2 mm. per day would be expected. They therefore considered that the observed rate of translocation of carbohydrates is so high that it cannot be accounted for by diffusion alone through the phloem elements, even though protoplasmic streaming is taken into account. They considered that these facts force one to conclude that the phloem elements are not adequate to transmit the amounts of carbohydrates actually known to travel downward in the stems and petioles of plants. Dixon and his students maintained that the major portion of the organic compounds are moved upward and downward through the conducting elements of the xylem in the xylem sap. Dixon considered that his observations on the seedlings of *Vicia* strengthens this view. He noted (1923) that the tracheal elements in these seedlings are differentiated in the petioles of the cotyledons much earlier than are the sieve tubes, so that before the latter are formed, considerable transport of organic substances to the embryo takes place. He believed that this is evidence that the tracheae and not the sieve tubes carry the organic substances.

Dixon and his pupils believed that the phloem is a means for the transmission of organic substance in a radial direction and considered that the medullary rays receive their organic materials from it. The phloem elements with the medullary rays and wood parenchyma present a large surface to the conducting xylem, which thus facilitates the interchange of the organic substances between the living cells and these nonliving conducting elements.

Curtis (1925), however, calculated the cross-sectional areas of the conducting tissues of potato rhizomes and found that the xylem in this case has an even smaller cross-sectional area than the phloem. The small area of the cross section of the phloem cannot thus be considered as evidence that the sugar travels in the xylem. Furthermore, according to Curtis, sugars have never been demonstrated to be present in the conducting elements of the potato xylem. It should be noted in this regard that the relative areas of the conducting xylem elements and the living cells of the stem differ widely for different plants, as shown by Dixon and Marshall (1915).

2. *Composition and Amount of Organic Compounds in Tracheal Sap.*—Hartig (1858) ringed a number of trees and noted after a time that the starch below the ring had disappeared. He concluded from this that

the reserve food products must have moved up through the xylem. As pointed out by Curtis (1920), however, he overlooked the fact that the foods stored below the ring might have been used locally for growth or have been moved downward through the phloem and used in root growth. Hartig considered that the materials formed in the leaves are passed down in the bark and stored in the wood parenchyma and medullary rays and in the spring are brought into solution and passed upward through the conducting elements of the xylem.

Schroeder (1871) made some observations on the sap of *Acer platanoides* and *Betula*. He noted that the sugar content of the sap from the birch decreased from early to late spring. The sugar concentration of the maple sap was greater in the upper part of the stem than in the lower, while the opposite condition prevailed in the birch.

Fischer (1888, 1890) by microchemical means demonstrated the presence of reducing substances in the conducting xylem elements of numerous trees and found that they were less in amount in the summer and autumn than in the spring. Dixon and Atkins (1915, 1916) examined the tracheal sap of a number of deciduous woody plants and found that the concentration of the soluble carbohydrates in the transpiration stream was greater than that of the electrolytes. The sugars were both the reducing and nonreducing types, and, estimated as glucose, their concentration ranged from 0.3 to 2 per cent depending upon the date of examination. The greatest concentration of sugar occurred in the early spring. This was followed by a rapid dilution in spring and early summer with a minimum occurring in the summer or autumn. A rise in concentration took place during the winter, culminating in a spring maximum just prior to the opening of the buds. These authors concluded that the conveyance upward of sugars is a continual and primary function of the xylem. The sheath of wood parenchyma around the vessels functions as a storage place from which sugars pass into the rising transpiration stream. They believed that the medullary rays convey carbohydrates from the bark to these sheaths.

The sap of the sugar maple has a sugar concentration of between 2 and 3 per cent. This flow of sap is considered good evidence that the conducting elements of the wood are able to carry this sugar both upward and downward at a comparatively rapid rate. No sap in this case comes from the bark (Adams, 1923). MacDougal (1926) found considerable amounts of reducing substances in the sap of the xylem of Monterey pine, the more concentrated solutions being in the outer rings.

Anderssen (1929), in the case of the tracheal sap of the pear, found that the free reducing substances and sucrose showed a high concentration in late winter and early spring, a rapid decrease in late spring, and a very low amount during the summer. These substances were found to be limited to the outer annual ring in 3-year-old pear branches during the late winter and spring. He thought this showed that the starch in the inner rings when hydrolyzed does not pass into the adjacent tracheae but is transferred along the medullary rays toward the outer tracheae into which it passes.

The nitrogen in the tracheal sap of the pear was found to consist almost entirely of organic nitrogen, and no nitrates could be detected until after the sap had been concentrated. On Nov. 10 the sap contained 60 p.p.m. of total nitrogen and on May 10, 163 p.p.m., of which 110 p.p.m. were present as amino nitrogen and 49 p.p.m. as amide nitrogen.

Some evidence has been presented that indicates that the living cells of stems and petioles are not necessary for the translocation of organic materials. This has been taken as proof that the organic matter moves through the xylem rather than through the phloem. The work of Deleano (1911) has been the most frequently quoted in proof of this. He chloroformed the petioles of grape leaves and found that the translocation of organic materials from the leaves continued, although at a reduced rate. Some movement of organic materials from the leaves also took place after the petioles had been killed by heat.

b. The Phloem.—It is the intention to state here in a general way the type of experiments performed to determine the relation of the phloem to the translocation of organic materials and the theories that have been proposed to account for their movement.

1. *The Kind of Experiments.*—Most of the experiments may be classified under the following headings:

(*a*) *Ringing.*—The view that the phloem elements are the channels for the movement upward and downward of organic compounds in the plant rests mainly upon the results obtained from ringing experiments in which the phloem portion of the stems has been severed by removing a ring of bark extending to the cambium. Experiments of this type were some of the earliest performed in plant physiology, and in general they have indicated that the downward passage of organic material is prevented or greatly hindered by the ringing. The death of a tree due to girdling is attributed to the fact that the downward movement of organic materials manufactured in the leaves is prevented to the extent that the roots die from starvation.

It has generally been assumed that the removal of the phloem in no way injured or affected the functioning of the xylem, but recent observations indicate that it does so unless special precautions are taken. On that account, the evidence obtained in the earlier ringing experiments cannot be taken as conclusive proof that the results obtained were due alone to the severance of the phloem tissue. It would be well now to review some of the more recent ringing experiments and consider the evidence thus obtained in regard to the part played by the phloem in the movement of organic material.

Curtis (1920) by ringing the twigs of *Philadelphus pubescens, Acer saccharum, Crataegus sp., Pyrus malus,* and others found that the defoliated stems did not continue to grow after the starch supply above a ring was depleted. If the stem above a ring, however, was not defoliated, the leaves were able to furnish a sufficient supply of material to allow for considerable growth. The carbohydrates stored in the xylem below the ring can apparently not be removed through the xylem but may be transferred radially to the phloem, where they may be carried downward if there is no second ring below.

Mason (1922) in the case of the bitter cassava (*Manihot utilissima*) found that the final weight of the tuberous roots of the ringed plants was only about one-fourth that of the unringed stems. Gardner (1925) in a study of the conductive tissues in the shoots of Bartlett pear concluded that the phloem is the tissue largely concerned in the longitudinal movement of foods. By using potassium hydroxide a ring was made which did not injure the xylem, yet it did not conduct the food stored between two rings past them, while the starch above and below the rings was removed. Both old and new tracheal tubes of excised pear branches allowed water and solutions of glucose and asparagine to pass through under pressure. In his opinion there is little reason to believe that the outer and inner xylem act differentially in the conduction of foods and water, because only the outer xylem conducts foods, while the whole cross section is used for water transport. He found that ringing did not induce the formation of tyloses within the tracheae of the pear shoots but that it did affect the xylem by actual mechanical injury by clogging and by drying the outer vessels. The inner xylem, comprising 75 to 80 per cent of the ringed shoots, was uninjured and was available for carbohydrate conduction but was evidently not used for that function. In *Robinia pseudoacacia,* tyloses effectively blocked the passage of water and solutes of glucose and asparagine through the innermost xylem. The conduction of foods in this species is thus apparently limited to the phloem or to the outermost xylem.

Loomis (1934, 1935) reported that in woody plants ringing stopped the upward movement of the organic nitrogenous compounds formed in the roots and the downward movement of the carbohydrates formed in the leaf. Under ordinary conditions, the protoplasm of the phloem cells is impermeable to these organic compounds so that they do not pass into the xylem. Their movement is thus confined to the plasmodesmal connections between the living cells of the phloem. Loomis noted that, after 4 to 5 weeks, ringing lost its effectiveness in checking the upward and downward movement of organic compounds. He suggested that the cells of the ringed segments became increasingly permeable with time so that these compounds diffused into the xylem, and were thus moved past the interruption in the phloem. Phillis and Mason (1936) in ringing experiments upon the cotton plant observed that in intact stems, nitrogen traveled upward in considerable amounts from the basal to the apical region, while little or none entered the apical region of the ringed group. They inferred that some of the nitrogen at least which entered the apical region of normal plants traveled in the phloem, and they concluded that nitrogen and carbohydrate could travel simultaneously in opposite directions in this tissue.

(b) *Insects.*—The position taken by certain aphids on the vascular bundles of leaves has been considered an indicator of the tissue in which the organic materials travel. Thus Davidson (1921) by fixing the aphids *in situ* found that they seek the phloem part of the vascular bundle when tapping the plant for nutrients. The observation of Schneider-Orelli (1909) on the influence of a leaf miner upon translocation is worthy of note. He found that when it had bored through veins the storage of starch in the tissues was affected. It was observed that the destruction of the xylem and the greater portion of the bundle sheath could be brought about without causing an accumulation of starch in the distal part of the leaf, but that an injury of phloem resulted in an accumulation of starch proportional to the extent of the injury.

(c) *Protoplasm.*—It has previously been mentioned in this chapter that the sieve tubes, companion cells, and phloem parenchyma are living cells and have a high colloidal content compared with the vessels and tracheids of the xylem. It has also been mentioned that the presence of the contents of these cells has been considered by some to prevent the translocation of organic matter in sufficient amount, by simple diffusion, to account for the total quantity that is actually moved. Some, on the other hand, consider that the protoplasm plays some role in this translocation, and considerable experimental work has been done to determine if it is actually concerned in this process.

DeVries (1885) pointed out that diffusion alone is apparently too slow to account for the movement of organic compounds that actually occurs. He suggested that the rotation and circulation of the protoplasm might aid in the translocation process, since he and others had noted protoplasmic streaming in the phloem tissue. Curtis (1929) observed streaming in the companion cells and phloem parenchyma in numerous plants but in no sieve tubes except those of *Elodea.* He considered that injury in preparing sections might account for the failure to observe streaming in the mature sieve tubes and believed that it is very likely that rotation and streaming of some sort do actually occur in them under normal conditions.

Czapek (1897) and Deleano (1911) found that chloroforming the petioles without killing the cells delayed starch removal from the leaves. Czapek did not consider the streaming of the protoplasm as essential but seemed to favor some secretive action by the protoplasm. Steward (1928), however, believed that streaming of the protoplasm might be the cause that facilitates the diffusion of the organic compounds through protoplasmic membranes.

Curtis (1928 to 1929) studied the effects of chilling upon the translocation of carbohydrates. In the case of leaves, he encased the petioles with rubber tubing through which water of a desired temperature could be run. He found that the total carbohydrate content in the leaves with the chilled petioles was always higher than that in the controls, the ratio of the total carbohydrate of the chilled leaves to the checks being 1.37. The ratio of the sugar content of the chilled leaves to the checks was 1.49. The chilling of the petioles to 2 to 3°C. clearly interfered with the removal of sugar from the leaf, but chilling to 7 to 9° did not cause a significant retardation. Using plants grown in sand and watered with known chemicals and treating the stems as mentioned above, Curtis found that chilling interfered with the upward movement of inorganic nutrients. Thus the gain in total ash of the unchilled plants was 58.3 per cent of that at the beginning, while that of the chilled stems was only 35 per cent.

The temperatures at which chilling interferes with the translocation of materials is approximately that at which streaming in protoplasm ceases. The evidence indicates that living cells take an active part in both the upward and downward movement of materials and that this movement is probably facilitated by the streaming of the protoplasm. It should be noted that chilling apparently has little or no effect upon water movement, since plants bled as freely from the cut ends of chilled stems as they did from the unchilled.

Schumacher (1930, 1932) concluded that the sieve tubes are the channels of transport for organic materials. He observed the location of the potassium salt of fluorescein by means of ultraviolet light. He found that this compound is taken up by leaves and is transported by the protoplasm into the sieve tubes and then downward.

2. *Theories of Movement.*—It is apparently well established that the translocation of the major portion of the organic compounds in plants is through the phloem. The means, however, by which this movement of these compounds is accomplished are as yet not known (Kruseman, 1931).

A rather elaborate theory of transport in the phloem was proposed by Münch (1926, 1927, 1930). According to Curtis (1935) this theory suggests that there is a unidirectional flow of phloem contents that is caused by a pressure gradient. This pressure gradient is established and maintained by differences in osmotic concentrations of supplying and receiving tissues. In applying this theory it is assumed that a semipermeable membrane separates the living cells of the phloem from the xylem tissues. The plasmodesms are supposed to allow for the flow of solutions from one living cell to another, while the sieve pores allow for similar flow through the specialized sieve tubes. Thus, according to this theory, if the supplying cells or tissues have a high osmotic concentration, the receiving cells a low concentration, and the plasmodesms and sieve pores allow for mass flow, there should be a transport of solution from the cells with high concentration and high turgor to those with low.

The theory of Münch has been adversely criticized by Weevers and Westenberg (1931), Curtis (1935), and Mason, Maskell, and Phillis (1936). Three of the numerous objections as stated by Curtis are as follows: (*a*) The theory does not account for the observed simultaneous movement in both directions through the phloem, (*b*) the pressure gradients available are insufficient to cause a flow of solution through the conducting tissues into receiving cells and exudation of water from the receiving cells, and (*c*) existing osmotic and turgor gradients have been observed in many cases to lead in a reverse direction from that required by the proposed hypothesis. Curtis (1920), Fernald (1925), and Curtis and Scofield (1933) found in certain cases that the osmotic relationships of plant tissues were directly opposite to those demanded by the hypothesis of Münch.

Crafts (1931, 1932, 1933, 1936) considered that the movement of organic materials in plants occurs through the entire cross section of the phloem including the cell walls. The phloem exudation in cucurbits indicates that the movement is not of local origin but is a direct manifestation of a process occurring throughout the plant. As a result of this process, organic foods are moved over long distances in a relatively short time. It appears that this conduction is of the nature of an elongated, somewhat elastic osmometer working under a positive pressure and activated by a concentration gradient of its osmotically active contents. Crafts stated that the mechanism of xylem exudation appears to be of a different nature for the ordinary osmotic action cannot account for the rapid flow obtained from plants. When root pressure is manifested, the xylem vessels have their total contents replaced many times during the day. Any theory of root pressure that fails to explain the simultaneous absorption and secretion of water and inorganic ions into nonliving vessels is inadequate. Although the movements of organic and inorganic materials may not be directly correlated, they are both in progress at the same time and involve the same general tissues.

Phloem sap may be coagulated in the stem by heat, and if a heated portion of a stem is cut no phloem exudation occurs, while heating has no visible effects upon xylem exudation. Exudation from the cut xylem depends upon conditions affecting water loss, since the water deficit must be satisfied before an actual exudation occurs. Phloem exudation will occur from severely wilted plants (Mothes, 1931), and, although the volume of the exudate is reduced in proportion to the water deficit, the dry weight composition tends to increase.

Steward and Priestley (1932) adversely criticized the theory propounded by Crafts that there is a movement of organic materials through the cell walls. Mason and Maskell (1934) reported that their experimentation strengthened the hypothesis that movement of materials along the phloem is determined independently for each material by the concentration gradient of its mobile form in the channel of transport. Mason, Maskell, and Phillis (1936) considered that neither the Münch theory of mass flow, nor the theory of protoplasmic streaming gives a satisfactory explanation of the movement of organic materials in the cotton plant. The student is referred to the monograph by Curtis (1935) for a detailed discussion of the movement of solutes in the plant.

5. Lateral Translocation.—The lateral transfer of organic materials in plants is for the most part in a radial direction. This movement takes place principally through the medullary rays but is never so marked as the longitudinal movement. Translocation around the stem is apparently almost negligible. It has long been observed that the annual rings of trees on the side directly under the larger limbs or on the sunny side of trees bordering on a forest are much larger than those on the opposite side. Czapek (1897) in ringing experiments found that ringing could be nullified only by vertical bridges connecting the upper and lower portions of the bark. Oblique and zigzag bridges were ineffective in the transference of organic materials across the ring. This indicates that there is little lateral translocation. Auchter (1923) obtained evidence in the case of fruit trees which indicates that the foods manufactured on one side of these are used and stored mainly in that side or are translocated for the most part, to the roots directly beneath. Caldwell (1930) noted that the removal of the leaves from one side of the crown of the Swede turnip during the early stages of growth caused the development of a root that was flattened on the defoliated side. The removal of the leaves from one side of a sunflower plant was followed by an asymmetrical development of it. When differentiated treatment of the roots of 2-year-old privet cuttings was performed, there resulted a considerable difference in the foliage on each side.

6. Translocation of Carbohydrates.—It has previously been mentioned that sugars are translocated farther and in greater quantities than any other organic compound and are almost without exception the form in which carbohydrates are moved. Mason and Maskell (1928) in their thorough and detailed study of the transport of carbohydrates in the cotton plant have given us the most extensive information concerning carbohydrate translocation that is at present available. On that account their results will be reviewed here in considerable detail.

They found that the concentration of the total sugars in the cell sap of the bark was greatly in excess of that of the wood or leaf. Thus in one 24-hr. period the sugar content of the xylem varied from 0.05 to 0.07 per cent, the sugar content of the bark from 3.0 to 3.5 per cent, and the sugar content of the leaf from 1.5 to 1.8 per cent. In another observation it was found that the total sugar concentration in the boll was approximately 0.5 per cent higher than that of the bark, the concentration ranging from 5 to 6 per cent in both. Under the same conditions the sugar concentration of the leaves varied from 1.5 to 2.5 per cent. The concentration curves of these three plant parts, however, ran parallel, for the most part, but those for the bark and boll lagged behind that of the leaf. The sugar concentration rose in all three of these parts during the day and fell during the night, but the concentration in the wood was not appreciably affected by the change from day to night. The diurnal variations in the concentration of total sugars in the leaf were more highly correlated with variations of the concentrations in the bark than in the wood. In two experiments in which samples of the bark were taken from different levels on the stem it was shown that the concentration of the total sugars in the upper part of the stem was higher than that in the lower portion.

Reducing sugars were present in the leaves greatly in excess of sucrose. In the boll, the excess of reducing sugars over sucrose was especially marked. In the stem, however, sucrose was in excess of the reducing sugar, the concentration of sucrose in the bark increasing during the day and decreasing during the night, while the reducing sugars remained relatively constant. Phillis and Mason (1933) noted that sucrose is the only sugar in the cotton plant that shows well-marked and consistent diurnal changes in the lamina and petiole of the leaf accompanied by corresponding changes in the rate of transport.

Mason and Maskell (1928) believed that their chemical analyses suggest that in the cotton plant the translocation of sugar occurs in the bark. As the major part of the fluctuations in the total sugars of the bark was found to be due to changes in the sucrose, they believed that the bulk of the carbohydrate translocated through this stem travels in that form. They considered that sugar moves from the leaf to the bark by a process similar to physical diffusion in that concentration changes in the leaf are followed by similar changes in the bark. They assumed that even though the exchange of the solute may take place against a concentration gradient, the possibility of a definite causal relationship between concentration changes in the leaf and bark is not excluded. Since a concentration gradient exists in the bark down the stem, the movement of sugars down the stem toward the roots resembles movement by simple diffusion, in that the direction of movement is from a region of high to one of low concentration.

Since the sucrose concentration of the bark was found to be on an average 2 per cent higher than that of the boll and since the rate of transport of sugar into the boll tended to increase as the sucrose gradient from bark to boll increased, it was concluded that the sucrose gradient between bark and boll is probably an important factor in determining variations in the rate of carbohydrate transport through the pedicel to the boll. It was also shown that the rate of transport of sugar into the boll was approximately four times greater by day than by night, a fact that is in harmony with the preceding conclusions, since the variation in the sucrose concentration in the bark follows closely in the changes in sugar concentration in the assimilating organs due to alternation of day and night.

By well-controlled ringing experiments, Mason and Maskell obtained further evidence that the transport of carbohydrates in the cotton plant occurs in the bark. (1) Ringing caused an accumulation of carbohydrates, not only in the bark and wood just above the ring but also in the leaves 2 ft. distant from it. (2) Ringing interrupted the downward transport of carbohydrate in the stem but did not interrupt the downward movement of a dye in the wood. It was thus concluded that the movements in these two parts are fundamentally different. (3) When the wood and bark were completely separated from each other, the movement of sugar took place through the bark at nearly a normal rate, but no movement of sugar through the wood could be detected. Baldwin (1934) found that from 3 to 4 years were required for girdled hardwoods to die. From 20 to 300 times as much sugar accumulated above the cut as below, and from two to three times as much as in normal trees. Two years after girdling, the sugar reserves above the cut were lower in amount than in the trunk of normal trees. In order to determine what portion of the bark is concerned in the translocation of sugar, Mason and Maskell subdivided the bark into three zones—outer, middle, and inner of approximately equal cross-sectional area—and analyzed each of these portions. They found that sucrose increased in concentration from outside to inside, while reducing sugars increased in the reverse direction. The radial gradient of reducing sugars was relatively stable, but the gradient in sucrose fluctuated markedly, increasing or decreasing as the sugar supply in the bark increased or decreased. There was a high positive correlation between the percentage of sieve tubes in any zone of bark and the sucrose concentration in that zone, and a high negative correlation between the percentage of sieve tubes and the concentration of reducing sugars.

The sugars, in particular the sucrose, in the sieve tubes fluctuated markedly with changes in the sugar supply to the bark. The changes in the sugars of the parenchyma were, however, very small. The gradient in sugar concentration from sieve tubes to bark parenchyma was found in some cases to be 500 times as steep as the vertical gradient in the sieve tubes. There was a high correlation between the concentration of total sugars in the whole bark and the concentration in the sieve tubes, so that the gradient in the whole bark could be used as an estimate of the gradient in the sieve tubes. These observations appear to offer conclusive proof that the longitudinal translocation of sugars for the most part takes place through the sieve tubes.

It should be kept in mind, however, that all the individual carbohydrates in plant material are capable of conversion into each other, and that in being transported from the leaves to the roots, a given carbohydrate may have been transformed numerous times (Virtanen and Nordlund, 1934).

It has been noted by Jost (1907), Mangham (1910 to 1922), Dixon and Ball (1922), and others that physical diffusion alone seems much too slow to account for the large quantities of carbohydrates that are moved longitudinally in plants. Mangham (1917) proposed a theory to account for the translocation of carbohydrates through the phloem elements that does not involve the physical diffusion of sugars through the protoplasmic membranes. He assumed that in the plant protoplasm there are present constituents capable of adsorbing sugars from solution; that for any given concentration of sugar present in the liquid phase of the protoplasm and the cell sap continuous with it there would be a definite concentration of sugar present at the adsorbing surface; and that any alteration of concentration in any region would lead to a readjustment of concentration equilibrium which would be transmitted or propagated as a wave through the system of these adsorbing particles and the solution immediately in contact with them. The connecting threads of the sieve tubes are assumed to provide a continuous protoplasmic pathway so the sieve plates would cause little if any obstruction to the progress of the wave of readjustment of concentration equilibrium. He considered that the sieve tubes would thus permit the passage of the waves with more effect than would be possible in any other type of cell except laticiferous cells.

The work of Mason and Maskell (1928), however, indicates that sugars move through the sieve tubes by a process similar to diffusion. Thus if sugars move in this tissue by diffusion, it would seem possible that the rate of movement of sugar at any point would be proportional to the concentration gradient in the channels of transport and to the cross-sectional area of the channels. Mason and Maskell investigated this point by examining the changes in rate produced by reducing the area of the channels of transport. It was found that a reduction by ringing in a cross-sectional area of the bark of approximately 45 per cent caused a reduction in the rate of sugar translocation of only 22.4 per cent. Thus the rate of transport across unit cross-sectional area of the bark in the constricted area was increased by reducing the area available for transport. This increase in rate was accompanied by and could be explained in terms of an increase in the gradient of sugar concentration across the constricted region.

The rate of sugar movement through the sieve tubes is, however, much higher than that which ordinarily occurs in purely physical diffusion. Mason and Maskell (1928) stated in the case of the cotton plant that the observed diffusion constant of sugar in the sieve tubes is about 40,000 times as great as the diffusion constant for sugar in a 2 per cent solution of sucrose in water and that it is almost identical with the diffusion constant for molecules of the size of the sucrose molecule diffusing in air. Although there seems to be good evidence that the carbohydrates travel as sugars in the sieve tubes, no satisfactory explanation has been given of the manner in which these products are moved with adequate speed in the required amount (MacDougal, 1926). The rate of movement of sugar from bark to wood appears, on the other hand, not very much greater than would be accounted for by purely physical diffusion.

The translocation of carbohydrates in germinating seeds and in seedling plants has been extensively studied by Sachs (1862), Detmer (1880), LeClerc and Breazeale (1911), Boysen-Jensen (1915), Choate (1921), Toole (1924), Yocum (1925), and others. Only a few of these observations will be discussed here, as detailed information may be obtained by consulting the references mentioned above. It should be kept in mind that the transformation of carbohydrates from one form to another is very frequent and rapid in plants, especially in seedlings, so that any given molecule of sugar passes no considerable distance without changing into other sugars or into insoluble carbohydrates for temporary storage.

In the wheat seedling, Choate (1921) detected reducing sugars in the coleorhiza, root, endosperm, coleoptile, plumule, and scutellum after 12, 18, 24, 36, 48 and 96 hr., respectively, while starch appeared in the root cap in 12 hr. after the beginning of germination. In the endosperm, reducing sugar is first found near the basal end of the embryo but is eventually found throughout the whole tissue. At the end of 7 days, starch was still present in the endosperm, although all the grains were corroded. Yocum (1925) noted that the greatest amount of reducing sugar in the plumules and roots was present at the ninth day. There was no evidence of sucrose formation taking place in the seed. When the starch showed a rapid loss the dextrins increased and then dropped off at about the same rate as the total sugars, but at no time were there sufficient dextrins in the roots and plumules to indicate that translocation of dextrins occurs. The starch content of the seed and seedling dropped off rapidly the first 7 days, the decline being extreme from the third to the sixth day. The amount of sugar was highest at 6 days. The proportional increase of plumule and roots in dry weight was greatest at this period, so the greatest starch hydrolysis and the greatest sugar content accompanied the greatest growth. Starch was found in the plumules and roots at all stages. It appears that if the seedling has good growing conditions, it will store temporary starch in all its tissues. The seedlings of wheat depend entirely upon the endosperm for their carbohydrate supply for 6 days. From the sixth to the twelfth day they depend only partially on this source, after which they depend entirely upon photosynthesis for this supply. Toole (1924) in the case of germinating corn could find no reducing sugars in the cells of the scutellum, although sucrose was present in about the same amount as in the dry embryo. He thought that sucrose might represent the translocation form of carbohydrates in the corn seedling.

7. Translocation of Fats and Oils.—Our knowledge concerning the translocation of fats and oils is very meager and is confined almost entirely to observations of the germination of fatty seeds. The known facts may perhaps be best presented by reviewing the results that have been reported.

Sachs, as early as 1859, observed considerable quantities of oil in the hypocotyls of seeds and concluded that it had moved out from the seed in that form. He believed, however, that the greater portion of the reserve oil in seeds is transformed to carbohydrates for transport and use.

Schmidt (1891) reported extensive experiments on the translocation of oil in seedlings. He believed that in many cases oils are transported as such to the different regions of the seedling, where it is broken up into the necessary products for growth. His conclusions were based on the following experiments: Pea seedlings were grown in the dark until the stored foods of the cotyledons were exhausted. A longitudinal incision was made just above the ground and a strip of filter paper saturated with oil was inserted in the incision. By using colored oils, their presence in the intercellular spaces and in the cytoplasm could be detected. When almond oil mixed with oleic acid was used, he observed that the rise of the oil was increased as the amount of fatty acid was increased. He reported that the neutral oil with which he worked did not enter the cell unless there was present at least 10 per cent free acid. When pure fatty acid was used, its presence could be detected in the cytoplasm within a few hours. Schmidt explained these results by assuming that soaplike compounds were formed with the free acid and that the emulsification of the neutral fat by these soapy substances occurred. He thus assumed that the neutral oil passed through the cell walls from cell to cell as a fine emulsion.

Rhine (1926) repeated the work of Schmidt. He found that the movement of oil in plants treated after the manner of Schmidt was through the intercellular spaces of the outer region of the cortex. Contrary to the observations of Schmidt, however, he found that a fatty acid in the pure state did not rise so rapidly as did the neutral fats of these acids. Thus linseed oil rose at the rate of 2 cm. per hour, while the fatty acids of this oil rose at the rate of only 1 cm. per hour. The fatty acids entered the cytoplasm of the cells from the intercellular spaces and appeared in the cytoplasm in the form of small droplets. Linseed oil, however, with a free acid content of less than 2 per cent penetrated the cell walls as well as the free acid.

Rhine believed that since the oil in the cells during translocation is in the emulsified state, while the oil on the outside of the cells in these experiments was pure and homogeneous, the cases could not be considered comparable. He believed that the movement of oils through the cell walls must be demonstrated from the emulsion state with water if the continuous phase of the experimental results is to be taken as an indication of what actually occurs in the plant. When pea seedlings were placed in an emulsion of linseed fatty acids, no intake whatsoever of the oils took place.

Since the seedlings that had been used in the foregoing experiments were starved and etiolated, it was considered that an abnormal condition might prevail in the walls bordering on the intercellular spaces. In order to investigate this point, two sets of seedlings were treated. One set was well supplied with water and the other partially desiccated by drying with a fan. In the seedlings that had an abundance of water, the rise of the oil in most cases was zero and no penetration of the cell walls occurred. The increased drying effects of air currents increased the rate of the ascent of the oil and facilitated its penetration. It was thus definitely shown that the observed intake of oil by plants was due to an unnatural drying out of the cell walls—a condition that does not prevail in the cells of germinating fatty seed through which oil has been supposed to pass.

It was suggested by Schmidt (1891) that soaps formed from the fatty acids might facilitate the movement of oils through the cells. If soaps are found to function in that manner, the tissues must be alkaline in order that they may not decompose. Rhine (1926) determined the pH value of the hypocotyls of a number of seedlings of fatty seeds with the following results:

Hypocotyls of	Range of pH
Peanut	5.4 to 5.8
Sunflower	5.2 to 5.6
Castor bean	5.2 to 5.6
Hemp	6.2 to 6.6

The phloem of these tissues was examined with special care, since it has been reported to be alkaline, but its acidity was found to be the same as that of the adjoining tissues. Under the conditions of acidity that were found, it would be difficult to conceive of the existence and movement of soaps through the tissues.

The respiratory quotient of the hypocotyls or other growing parts of the seedling should serve as an index to the nature of the food that is being translocated through them. If oil is being translocated as such and moved into the hypocotyls, it would be expected that oil would be used in respiration, and in that case the respiratory quotient would be relatively low, approximately 0.6. If the hypocotyl is receiving sugar, it would be used in respiration, and the respiratory quotient of the hypocotyl would be approximately 1.00. Rhine found that the respiratory quotient of the hypocotyls of the seedlings of sunflower and cotton averaged 0.770, while that of the pea, wheat, barley, and buckwheat averaged 0.775. The respiratory quotient of the hypocotyl of these germinating oily seeds was thus almost the same as that of starchy seeds. This indicates that the food supplied to the hypocotyls by the storage region of fatty seeds reaches them in a form that yields the same value for the respiratory quotient as the food supplied by the typical carbohydrate storage seeds. This food is commonly supposed to be sugar.

Rhine studied the oil gradient from the source of supply to the growing tip in the case of the seedlings of cotton, sunflower, squash, cocklebur, castor bean, peanut, and flax. In no case did he find a gradient from the storage region to the growing tip, but in every case there was a distinct gradient in the other direction; *i.e.*, the amount of fat increased with the approach toward the tip of the hypocotyl, as shown in the following table:

Seedling	Percentage of ether extract on wet basis		
	Tips	Middles	Bases
Cotton	5.75	2.24	1.33
Sunflower	5.31	1.34	0.66
Squash	2.59	0.60	0.22

If oils are moved as such from cell to cell, some would be expected to be present in the cell walls. Rhine, however, was unable to observe in either cross or longitudinal sections any evidence of its passing through the walls.

Rhine concluded that from a physical standpoint the movement of oils as such through the cells is improbable. In moving downward into the hypocotyls, the oil must pass through the cytoplasm whose specific gravity is 1.00 or greater. The various extracts of the hypocotyls made by Rhine showed in all cases a specific gravity lower than that of water. Oil movement downward if it occurs would thus have to take place against the pressure of a denser medium, a procedure that would seem rather improbable.

The presence of oil in the hypocotyls of seedlings is no proof that it has been moved there as such. Starch is present in most cases in the growing regions of plants, yet no one considers that it has been moved from its place of storage in that form.

The evidence thus indicates that oils, especially in seedlings, are converted into sugars at their place of storage and translocated to the different parts of the plant in that form. After the arrival of sugar at these regions, a portion of it at least may be converted into oil as a temporary storage product. It seems to be proved that oil as such is rarely if ever moved from cell to cell or from one region of the plant to another.

Knight, Chamberlin, and Samuels (1929), Ginsburg (1931), Rohrabaugh (1934), and Young (1934, 1935) have studied the penetration of petroleum oils applied as sprays.

Ginsburg (1931) noted that emulsified oils penetrated the leaves of apple, peach, and tomato much more slowly than did pure oils. Only those oils of very low viscosity penetrated the upper surface of these leaves. Young (1934, 1935) found that petroleum oil sprayed on apple leaves entered the parenchyma cells and tracheae of the veins and petioles. The oil in the twigs, however, was outside the parenchyma cells of the cortex, rays, pith, and tracheae. The petroleum oil placed upon or injected into the leaves of rutabaga, cucumber, potato, turnip, and onion passed downward into the stems, roots, bulbs, and other plant parts.

Some of the oil was in the tracheae but most of it was between the parenchyma cells of the cortex, where apparently under the conditions of the observations most of it was conducted.

Knight, Chamberlin, and Samuels (1929) noted in *Citrus* that petroleum oils applied as sprays began to appear in the palisade layer of cells in about 4 hr. These oils eventually accumulated in the cells along the larger veins and passed into the vascular system. They next appeared in the phloem, rays, pith, and in the old wood fibers of the xylem, but none was observed in the wood of the current season's growth. They considered that most of these oils were deposited in the smaller twigs and limbs. Rohrabaugh (1934), however, found in his experiments with *Citrus* that oil sprays entered by capillarity, and only penetrated under the most favorable circumstances to the depth of five or six cells beneath the epidermis. The less volatile oils remained in the citrus leaves throughout their life, which was more than 2 years.

Van den Honert (1932) studied the rate of transport of potassium oleate along the interface between water and ether by measuring the resulting pH change at the distal end of the interface. The precipitation of oleic acid on the interface sets free potassium hydroxide, which is allowed to neutralize hydrochloric acid in a bulb into which the interface extends. The potassium hydroxide under these conditions is transported at a rate 68,000 times that at which it could have diffused through the water layer. He suggested that in the sieve tubes the protoplasm-vacuole interface is the one in which transport occurs, and that this fact is probably the explanation for the very rapid transport of organic materials in plants.

8. Translocation of Proteins.—Maskell and Mason (1930), and Mason and Phillis (1934) stated that in the cotton plant the movement of nitrogen through the bark from the leaves to the roots occurs against a negative gradient of organic, crystalloidal nitrogen. The indications are that there is no well-defined positive gradient in either amino acids or residual nitrogen in the sieve tubes. It should be noted, however, that a positive gradient in the sieve tubes may be masked by a negative one in the companion cells. The gradient in the sieve tubes may consist of two components, a dynamic gradient in translocatory nitrogen, superimposed upon a steeper reverse gradient of relatively immobile, storage nitrogen. There is apparently little or no movement of proteins as such through cell membranes or through the special conducting elements of the plant. Only a few proteins are soluble in water, and, since they are slowly diffusing substances, their movement must necessarily be very slow. The contents of the sieve tubes give the protein reactions and have been considered on that account to be the elements in which proteins are moved longitudinally in the plant. These elements are rich in protoplasm and colloidal material, but how much of this if any is protein in transit cannot be determined.

The proteins can apparently be formed in any region of the plant, provided that a supply of sugar and the necessary inorganic materials are present. It seems probable, therefore, that proteins are not, as a rule, manufactured in superabundance in any special portion of the plant and then translocated for considerable distances for utilization or storage, but that they are formed in the immediate region in which they are found. The frequently observed fact that, when a nitrogen supply is abundant in the tops of plants the carbohydrates are used for top growth in larger proportions than when the nitrogen is more limited, is indicative that protein formation proceeds wherever the necessary materials are present and that the proteins are utilized more or less locally (Chandler, 1919). It was mentioned in Chap. IX that proteins are manufactured in the leaves in considerable amounts during the day and moved from them especially at night, but to what distances they are moved is not known. Amino acids and amides have been detected in both the xylem and phloem elements, but whether they are of local origin or are in transit from regions more remote has not been definitely determined.

The movement of protein from their place of storage to their point of utilization has been studied for the most part in seedlings. The richest stores of proteins occur in seeds and are rapidly broken down and moved to the growing portions of the seedling during germination. The general procedure has been to study the kind and quantity of the hydrolytic products in the original storage parts and in the newly formed organs of the seedlings. A few of the observations that have thus been made will now be reviewed.

Schulze (1900) and Schulze and Castoro (1903) found in the seedlings of *Lupinus albus* and other leguminous seedlings that amino acids appeared as leucine, tyrosine, and hexone bases. In the seedlings of the Alaska pea, Thompson (1915) noted that the amount of the total nitrogen in the amino form increased from 6.58 per cent of the total nitrogen in the 1-day-old seedlings to 28.27 per cent in those 7 days old. The distribution of the nitrogen in the 7-day seedlings was as follows:

Part of seedling	Percentage on a dry basis of		Percentage of total nitrogen in the amino form
	Total N	Amino N	
Plumule	7.02	0.254	36.18
Cotyledons	3.36	0.421	12.53
Root	5.33	1.705	32.00

Stark (1927) found in the seedlings of Manchu soybeans grown at 20°C. that the α-amino nitrogen increased from 1.86 per cent of the total nitrogen at the age of 1 day to 23.2 per cent at the end of 39 days. In the case of the Midwest variety, the increase was from 2.1 to 17.8 per cent.

These data indicate that the proteins in the seed are broken down into amino acids, which are translocated as rapidly as they are formed to the plumules and roots. In corn seedlings (Pettibone and Kennedy, 1916) the total nitrogen in the seed decreased from 75.0 mg. per 20 seedlings when the plumules were ½ in. in length to 23.1 mg. when they had reached a length of 10 in. The total nitrogen in the plumules during the period increased from 12.3 to 57.5 mg., while in the roots it had increased from 8.2 to 307 mg. Amino acids were present in considerable amounts in all three parts. An analysis of the sap exuded from the stumps of several seedlings showed that amino acids and other protein decomposition products resembling proteoses and peptones were present. Observations similar to these have been observed by Choate (1921) working with wheat seedlings.

Jodidi (1925 to 1927) observed also in etiolated corn seedlings that 48 per cent of the proteins present in the seed are converted into water-soluble diffusing nitrogen compounds within 8 days. During this period there is a rise in the amino nitrogen and a diminution in the peptide nitrogen. This is taken to mean that amino acids increase at the expense of polypeptides, which, along with proteoses, are among the first degradation products of proteins.

Since nitrogen increases in the plumules and roots as it disappears from the seed, and since amino acids are present in these parts and in the flowing sap, it seems evident that the reserve proteins of the seeds are transported to their points of utilization in the seedling in the form of amino acids.

Bibliography

Adams, J., 1923: The translocation of carbohydrates in the sugar maple, *Nature*, **112**: 207.

Albert, W. B., 1927: Studies on the growth of alfalfa and some perennial grasses, *J. Am. Soc. Agron.*, **19**: 624–655.

Aldous, A. E., 1929: The eradication of brush and weeds from pasture lands, *J. Am. Soc. Agron.*, **21**: 660–666.

———, 1930: Effect of different clipping treatments on the yield and vigor of prairie grass vegetation, *Ecology*, **11**: 752–759.

———, 1930: Relation of organic food reserves to the growth of some Kansas pasture plants, *J. Am. Soc. Agron.*, **22**: 385–393.

Aldrich, W. W., 1932: Nitrogen intake and translocation in apple trees following fall, winter, and spring sodium nitrate applications, *Proc. Am. Soc. Hort. Sci.*, **28**: 532–538.

ANDERSSEN, F. G., 1929: Some seasonal changes in the tracheal sap of pear and apricot trees, *Plant Physiol.*, **4**: 459–477.

ARNDT, C. H., 1929: The movement of sap in *Coffea arabica*, *Am. J. Bot.*, **16**: 179–190.

ARNY, A. C., 1932: Variations in the organic reserves in underground parts of five perennial weeds from late April to November, *Minn. Sta. Tech. Bull.* 84.

ASKENASY, E., 1895: Ueber das Saftsteigen, *Verh. naturh.-med. Ver. Heidelberg*, **5**: 325–345; *Bot. Cent.*, **62**: 237–238.

———, 1896: Beiträge zur Erklärung des Saftsteigens, *Verh. naturh.-med. Ver. Heidelberg.*, **5**: 429–448.

ATKINS, W. R. G., 1916: "Some Recent Researches in Plant Physiology," Chap. XI, The Function of the Wood, pp. 186–215.

AUCHTER, E. C., 1923: Is there normally a cross transfer of foods, water, and mineral nutrients in woody plants? *Md. Sta. Bull.* 257.

BAILEY, I. W., 1916: The structure of the bordered pits of conifers and its bearing upon the tension hypothesis of the ascent of sap in plants, *Bot. Gaz.*, **62**: 133–142.

BAKER, H., and W. O. JAMES, 1933: The behavior of dyes in the transpiration stream of sycamores (*Acer pseudoplatanus*), *New Phytol.*, **32**: 245–260.

BALDWIN, H. I., 1934: Some physiological effects of girdling northern hardwoods, *Bull. Torrey Bot. Club*, **61**: 249–257.

BARNES, C. R., 1930: "A Text Book of Botany. Vol. II. Physiology." Revised and enlarged by C. A. Shull. American Book Co., N. Y. and Cincinnati.

BARTEL, A. T., J. H. MARTIN, and R. S. HAWKINS, 1935: Effect of tillers on the development of grain sorghums, *J. Am. Soc. Agron.*, **27**: 707–714.

BARTHOLOMEW, E. T., 1926: Internal decline of lemons. III, Water deficit in lemon fruits caused by excessive leaf evaporation, *Am. J. Bot.*, **13**: 102–117.

BENEDICT, H. M., 1927: Application of Bose's theory of sap rise to ten species of trees, *Am. J. Bot.*, **14**: 623.

BENNETT, J. P., 1927: Chlorosis studies, *Calif. Sta. Rept.*, **1927**: 85–87.

———, F. G. ANDERSSEN, and Y. MILAD, 1927: Methods of obtaining tracheal sap from woody plants, *New Phytol.*, **26**: 316–323.

BIRCH-HIRSCHFELD, L., 1920: Untersuchungen über die Ausbreitungsgeschwindigkeit gelöster Stoffe in der Pflanze, *Jahrb. wiss. Bot.*, **59**: 171–262.

BISSON, C. S., and H. A. JONES, 1932: Changes accompanying fruit development in the garden pea, *Plant Physiol.*, **7**: 91–106.

BISWELL, H. H., and J. E. WEAVER, 1933: Effect of frequent clipping on the development of roots and tops of grasses in prairie sod, *Ecology*, **14**: 368–390.

BLAKE, M. A., 1921: Effect of an early application of nitrogen on peach trees deficient in vigor, *Proc. Amer. Soc. Hort. Sci.*, **18**: 139–145.

BODE, H. R., 1923: Beiträge zur Dynamik der Wasserbewegung in den Gefässpflanzen, *Jahrb. wiss. Bot.*, **62**: 92–127.

BODENBERG, E. T., 1927: Tissues involved in the transfer of mineral salts in plants, *Pub. Puget Sound Biol. Sta.*, **5**: 231–244; *Exp. Sta. Rec.*, **62**: 122.

———, 1929: Lateral transfer of lithium nitrate in *Salix*, *Am. J. Bot.*, **16**: 229–237.

BOEHM, J. A., 1878: Les causes de l'ascension de la sève, *Ann. Sci. Nat. Bot.*, VI, **6**: 223–236.

———, 1889: Ursache des Saftsteigens, *Ber. bot. Ges.*, II, **7**: 46–56.

———, 1890: Ueber Ursache der Wasserbewegung in transpirirenden Pflanzen, *Verh. Zool.-Bot. Ges. Wien*, **40**: 149–159.

———, 1892: Ueber einen eigenthümlichen Stammdruck, *Ber. bot. Ges.*, **10**: 539–544.

———, 1893: Capillarität und Saftsteigen, *Ber. bot. Ges.*, **11**: 203–212.

Bolley, H. L., 1904, 1905, 1907: Tree feeding and tree medication, *N. Dak. Sta. Ann. Rept.*, **14**: 55–58 (1904); **15**: 56 (1905); **17**: 35 (1907).

Bose, J. C., 1923: "Physiology of the Ascent of Sap," Longmans, Green and Company, New York and London, 277 pp.

Boucherie, A., 1840: Rapport sur un Mémoire relatif à la conservation des bois, *C. R. Acad. Sci. Paris*, **11**: 894–898.

———, 1841: Nouvelles recherches sur la conservation des bois, *C. R. Acad. Sci. Paris*, **12**: 337–339.

Bourdouil, C., 1931: The formation and transformation of the starch in bananas, *Rev. bot. app. Agr. trop.*, **11**: 656–660; *Chem. Abst.*, **26**: 753.

Bowen, E. J., 1931: Water conduction in *Polytrichum commune*, *Ann. Bot.*, **45**: 175–200.

———, 1933: The mechanism of water conduction in the *Musci* considered in relation to habitat. I, Mosses growing in wet environments, *Ann. Bot.*, **47**: 401–422.

———, 1933: The mechanism of water conduction in the *Musci* considered in relation to habitat. II, Mosses growing in damp situations, *Ann. Bot.*, **47**: 635–661.

———, 1933: The mechanism of water conduction in the *Musci* considered in relation to habitat. III, Mosses growing in dry environment, *Ann. Bot.*, **47**: 889–912.

Boysen-Jensen, P., 1915: Über synthetische Vorgänge im pflanzlichen Organismus. II, Vorkommen, Bedeutung und Bildung des Rohrzuckers bei der Keimung von *Pisum sativum*, *Jahrb. wiss. Bot.*, **56**: 431–446.

Brown, B. A., 1930: Effect of time of cutting on the elimination of bushes in pasture, *J. Am. Soc. Agron.*, **22**: 603–605.

Brunson, A. M., and W. L. Latshaw, 1934: Effect of failure of pollination on composition of corn plants, *J. Agr. Res.*, **49**: 45–54.

Burd, J. S., 1919: Rate of absorption of soil constituents at successive stages of plant growth, *J. Agr. Res.*, **18**: 51–72.

Burke, E., and H. E. Morris, 1933: Nutrient elements used by leaves and growth of apple trees, *Plant Physiol.*, **8**: 537–544.

Butler, O. R., T. O. Smith, and B. E. Carey, 1917: Physiology of the apple. Distribution of food materials in the tree at different periods of vegetation, *N. H. Sta. Tech. Bull.* 13.

Caldwell, J., 1930: Studies in translocation. II, The movement of food materials in plants, *New Phytol.*, **29**: 27–44.

———, 1930: Studies in translocation. I, Movement of food-materials in the Swedish turnip, *Proc. Roy. Soc. Edinburgh*, **50**: 130–141; *Biol. Abst.*, **5**: 2657.

Capen, R. G., and J. A. LeClerc, 1933: Chemical composition of native Alaskan hays harvested at different periods of growth, *J. Agr. Res.*, **46**: 665–668.

Carpenter, C. W., 1936: Notes on Pythium root rot. VIII, Absorption of essential elements by segregated roots of sugar cane, *Hawaiian Planters' Rec.*, **40**: 137–142.

Chandler, W. H., 1914: Sap studies with horticultural plants, *Mo. Sta. Res. Bull.* 14.

———, 1919: Some results as to the response of fruit trees to pruning, *Proc. Amer. Soc. Hort. Sci.*, **16**: 88–101.

Chase, W. W., 1934: The composition, quantity and physiological significance of gases in tree stems, *Minn. Sta. Tech. Bull.* 99.

Choate, H. A., 1921: Chemical changes in wheat during germination, *Bot. Gaz.*, **71**: 409–426.

Clark, W. S., 1874: The circulation of sap in plants, *Mass. Bd. Agr. Ann. Rept. Sec.*, **21**: 159–204.

Clements, H. F., 1930: The upward movement of inorganic solutes in plants, *Res. Studies State Col. Wash.*, **2**: 91–106.

COLLISON, R. C., J. D. HARLAN, and M. P. SWEENEY, 1932: Direct tree injection in the study of tree nutrition problems, *N. Y. Sta. (Geneva) Tech. Bull.* 192.

COMBES, R., and D. KOHLER, 1922: Ce que déviement les hydrates de carbone quand meurent les feuilles des arbres, *C. R. Acad. Sci. Paris*, **175**: 590–592.

———, 1926: Émigration des substances azotées des feuilles vers les tiges et les racines des arbres au cours du jaunissement automnal, *Rev. Gén. Bot.*, **38**: 430–448, 510–517, 565–579, 632–645, 673–686.

COPELAND, E. B., 1902: The rise of the transpiration stream: an historical and critical discussion, *Bot. Gaz.*, **34**: 161–193, 260–283.

COSTER, C., 1931: Die Geschwindigkeit des Transpirationsstromes, *Planta*, **15**: 540–566; *Chem. Abst.*, **8**: 1659.

CRAFTS, A. S., and P. B. KENNEDY, 1930: The physiology of *Convolvulus arvensis* (Morning glory or bindweed) in relation to its control by chemical sprays, *Plant Physiol.*, **5**: 329–344.

———, 1931: Movement of organic materials in plants, *Plant Physiol.*, **6**: 1–43.

———, 1931: The function of the sieve tube in conduction, abst., *Am. J. Bot.*, **18**: 893.

———, 1932: Phloem anatomy, exudation and transport of organic nutrients in cucurbits, *Plant Physiol.*, **7**: 183–226.

———, 1933: The use of arsenical compounds in the control of deep-rooted perennial weeds, *Hilgardia*, **7**: 361–372.

———, 1933: Sieve-tube structure and translocation in the potato, *Plant Physiol.*, **8**: 81–104.

———, 1936: Further studies on exudation in cucurbits, *Plant Physiol.*, **11**: 63–79.

CRAIGHEAD, F. C., and R. A. ST. GEORGE, 1930: A new technique in tree medication for the control of bark beetles, *Sci.*, **72**: 433–435.

CRIDER, F. J., 1928: Winter root growth of plants, *Sci.*, **68**: 403–404.

CULPEPPER, C. W., and C. A. MAGOON, 1930: Effect of defoliation and root pruning on the chemical composition of sweet corn kernels, *J. Agr. Res.*, **40**: 575–583.

——— and J. S. CALDWELL, 1932: Relation of age and of seasonal conditions to composition of root, petiole and leaf blade in rhubarb, *Plant Physiol.*, **7**: 447–479.

———, ——— and H. H. MOON, 1935: A physiological study of development and ripening in the strawberry, *J. Agr. Res.*, **50**: 645–696.

CURTIS, O. F., 1920: The upward translocation of foods in woody plants. I, Tissues concerned in translocation, *Am. J. Bot.*, **7**: 101–124.

———, 1920: The upward translocation of foods in woody plants. II, Is there normally an upward transfer of storage foods from the roots or trunk to the growing shoots? *Am. J. Bot.*, **7**: 286–296.

———, 1923: The effect of ringing a stem on the upward transfer of nitrogen and ash constituents, *Am. J. Bot.*, **10**: 361–383.

———, 1925: Studies on the tissues concerned in the transfer of solutes in plants. The effect on the upward transfer of solutes of cutting the xylem as compared with that of cutting the phloem, *Ann. Bot.*, **39**: 573–585.

———, 1928: Studies indicating that translocation is dependent on the activity of living cells and the possible bearing of this on solute distribution, *Am. J. Bot.*, **15**: 630–631.

———, 1929: Studies on solute translocation in plants. Experiments indicating that translocation is dependent on the activity of living cells, *Am. J. Bot.*, **16**: 154–168.

——— and H. T. SCOFIELD, 1933: A comparison of osmotic concentrations of supplying and receiving tissue and its bearing on the Münch hypothesis of the translocation mechanism, *Am. J. Bot.*, **20**: 502–512.

——— 1935: "The Translocation of Solutes in Plants," McGraw-Hill Book Company, Inc., New York and London, 273 pp.

CZAPEK, F., 1897: Über die Leitungswege der organischen Baustoffe im Pflanzenkörper, *Sitzungsber. Akad. Wiss. Wien*, **106**: 117–170.

DARWIN, F., and R. W. PHILLIPS, 1885: On the transpiration stream in cut branches, *Proc. Cambridge Phil. Soc.*, **5**: 330–367.

———, S. H., VINES, and J. JOLY, 1896: A discussion on the ascent of water in trees, *Ann. Bot.*, **10**: 630–661.

DAVIDSON, J., 1921: Biological studies of *Aphis rumicis*, *Ann. App. Biol.*, **8**: 51–65.

DAVIS, L. D., 1931: Some carbohydrate and nitrogen constituents of alternate-bearing sugar prunes associated with fruit bud formation, *Hilgardia*, **5**: 119–154.

DELEANO, N. T., 1911: Über die Ableitung der Assimilate durch die intaken, die chloroformierten und die plasmolysierten Blattstiele der Laubblätter, *Jahrb. wiss. Bot.*, **49**: 129–186.

——— and M. ANDREESCO, 1931: L'absorption et la migration negative de l'azote dans les feuilles de *Salix fragilis*, *Bull. Soc. Sci. Cluj Roumanie*, **6**: 209–220.

———, 1932: Die Anwendung der biometrischen Methode zum Forschen der Rolle von Mineralstoffen im Leben der Pflanze, *Bull. Cult. si Fermentarii Tutunului*, **21**: no. 2.

———, 1932: Die Rückwanderung der Stickstoffsubstanz aus den Laubblättern und ihre Speicherung im Stamme und Wurzel, *Bull. Soc. Sci. Cluj Roumanie*, **7**: 45–49.

——— and M. I. ANDREESCO, 1932: Beiträge zum Studium der Rolle und Wirkungsweise der Mineral- und organischen Stoffe im Pflanzenleben. I, Der quantitative Stoffwechsel der Mineral -und organischen Substanzen in den *Salix fragilis*-Blättern während ihrer Entwicklung, *Beitr. Biol. Pflanzen*, **19**: 249–286; *Biol. Abst.*, **8**: 2170.

——— and C. BORDEIANU, 1932: Beiträge zum Studium der Rolle und Wirkungsweise der Mineral- und organischen Stoffe im Pflanzenleben. II, Der quantitative Stoffwechsel der Mineral- und organischen Substanzen in den Blättern und geschälten Samen von *Aesculus hippocastanum* während Entwicklung, *Beitr. Biol. Pflanzen*, **20**: 179–197.

——— and ———, 1934: Beiträge zum Studium der Rolle und Wirkungsweise der Mineral- und organischen Stoffe im Pflanzenleben. II. *Acad. Romana Mem. Sec. Stiint Mem.* 10, Bucuresti.

DENNY, F. E., 1933: Changes in leaves during the period preceding frost, *Contr. Boyce Thompson Inst.*, **5**: 297–312.

DETMER, W., 1880: "Vergleichende Physiologie des Keimungsprocesses der Samen," Fischer, Jena.

DEVRIES, H., 1885: Ueber die Bedeutung der Circulation und der Rotation des Protoplasma für den Stofftransport in der Pflanze, *Bot. Ztg.*, **43**: 2–6, 17–26.

DIXON, H. H., and J. JOLY, 1894: On the ascent of sap, *Proc. Roy. Soc.*, **57**: 3–5.

——— and ———, 1894: On the ascent of sap, *Ann. Bot.*, **8**: 468–470.

——— and ———, 1895: On the ascent of sap, *Phil. Trans. Roy. Soc.*, 186 B, I, 563–576.

——— and ———, 1895: The path of the transpiration-current, *Ann. Bot.*, **9**: 403–420.

—, 1896: Note on the role of osmosis in transpiration, *Proc. Roy. Irish Acad.*, **3**: 767–775.

———, 1897: The tensile strength of cell walls, *Ann. Bot.*, **11**: 585–588.

———, 1903: The cohesion theory of the ascent of sap, *Sci. Proc. Roy. Dublin Soc.*, 10, No. 4.

———, 1905: Note on the supply of water to leaves on a dead branch, *Sci. Proc. Roy. Dublin Soc.*, **11**: 7–12.

———, 1906: On the transpiration current in plants, *Proc. Roy. Soc.*, B, **79**: 41–57.

———, 1909: Vitality and the transmission of water through the stems of plants, *Notes Bot. School Trinity Col. Dublin*, **2**: 5–18; *Sci. Proc. Roy. Dublin Soc.*, **12**: 21–34.

———, 1910: Transpiration and the ascent of sap, *Progressus rei botanicae*, **3**: 1–66.

———, 1910: Transpiration and the ascent of sap, *Ann. Rept. Smithsonian Insi.*, **1910**: 407–425.

———, 1914: "Transpiration and the Ascent of Sap in Plants," Macmillan & Company, Ltd., London, 216 pp.

——— and W. R. G. Atkins, 1915: On the constituents and concentration of the sap in the conducting tracts and on the circulation of carbohydrates in plants, *Sci. Proc. Roy. Dublin Soc.*, **14**: 374–392.

——— and E. S. Marshall, 1915: A quantitative examination of the elements of the wood of trees in relation to the supposed function of the cells in the ascent of sap, *Proc. Roy. Dub. Soc.*, **16**: 358; *Notes Bot. School Trinity Col. Dublin*, **2**: 264.

———, 1915: Notes on the spread of morbid changes through plants from branches killed by heat, *Notes Bot. School Trinity Col. Dublin*, **2**: 221–224.

———, 1915: Changes produced in the sap by the heating of branches, *Notes Bot. School Trinity Col. Dublin*, **2**: 225–229.

———, 1915: On the tensile strength of sap, *Notes Bot. School Trinity Col. Dublin*, **2**: 230–235.

——— and W. R. G. Atkins, 1916: Osmotic pressure in plants. IV, On the constituents and concentration of the sap in the conducting tracts, and on the circulation of carbohydrates in plants, *Notes Bot. School Trinity Col. Dublin*, **2**: 275–293.

——— and J. Joly, 1916: On the composition of the sap in the conducting tracts of trees at different levels and at different seasons of the year, *Sci. Proc. Roy. Dublin Soc.*, **15**: 51.

——— and N. G. Ball, 1922: Transport of organic substances in plants, *Nature*, **109**: 236–237.

———, 1922: Transport of organic substances in plants, *Nature*, **110**: 547–551.

——— and ———, 1923: On the channels of transport from the storage organs of the seedlings of *Lodoicea*, *Phoenix* and *Vicia*, *Sci. Proc. Roy. Dublin Soc.*, **17**: 185–196.

———, 1923: Transport of organic substances in plants, *Notes Bot. School Trinity Col. Dublin*, **3**: 207–215.

———, 1924: "The Transpiration Stream," University of London Press, 80 pp.

———, 1933: Bast sap, *Sci. Proc. Roy. Dub. Soc.*, **20**: 487–494.

Doyle, J., and P. Clinch, 1927: Seasonal changes in conifer leaves with special reference to enzymes and starch formation, *Proc. Roy. Irish Acad.*, B, **37**: 373–414.

Dungan, G. H., 1930: Relation of blade injury to the yielding ability of corn plants, *J. Am. Soc. Agron.*, **22**: 164–171.

———, 1934: Losses to the corn crop caused by leaf injury, *Plant Physiol.*, **9**: 749–766.

Dustman, R. B., and L. C. Shriver, 1931: The chemical composition of *Ambrosia trifida* at successive growth stages, *J. Am. Soc. Agron.*, **23**: 190–194.

Eames, A. J., and L. H. MacDaniels, 1925: "An Introduction to Plant Anatomy," McGraw-Hill Book Company, Inc., New York, 363 pp.

Echevin, R., 1932: Autumn migration of phosphorus from tree leaves, *C. R. Acad. Sci. Paris*, **194**: 2160–2162.

Elfving, F., 1882: Ueber die Wasserleitung im Holz, *Bot. Ztg.*, **40**: 707–723.

Elliot, J. A., 1917: The conduction of potassium cyanide in plants, *Phytopath.*, **7**: 443–448.

Errera, L., 1886: Ein Transpirationsversuch, *Ber. bot. Ges.*, **4**: 16–18.

Ewart, A. J., 1904: The ascent of water in trees, *Proc. Roy. Soc.*, **74**: 554–556.

———, 1906, 1908: The ascent of water in trees, *Phil. Trans. Roy. Soc.*, B, **198**: 41–85; **199**: 341–392.

Fagan, T. W., and J. E. Watkin, 1931: The distribution of the nitrogenous and mineral constituents in the oat plant at different stages of growth, *Welsh J. Agr.*, **7**: 229–246.

Farmer, J. B., 1918: On the quantitative differences in the water conductivity of the wood in trees and shrubs. I, The evergreens. II, Deciduous plants, *Proc. Roy. Soc.*, B, **90**: 218–250.

Fernald, E. I., 1925: The imbibition of bud-development as correlated with the osmotic concentration of cell sap, *Am. J. Bot.*, **12**: 287–305.

Fischer, A., 1888: Glycose als Reservestoff der Laubhölzer, *Bot. Ztg.*, **46**: 405–417.

———, 1890: Beiträge zur Physiologie der Holzgewächse, *Jahrb. wiss. Bot.*, **22**: 73–160.

Fitting, H., 1911: Untersuchungen über die vorzeitge Entblätterung von Blüten, *Jahrb. wiss. Bot.*, **49**: 187–263.

Flint, W. P., 1915: The effect of cyanide on the locust-borer and the locust tree, *Sci.*, **42**: 726–727.

Frear, D. E., 1931: A negative correlation observed between the nitrate nitrogen in the juice of beet leaves and the weight of the leaves, *J. Agr. Res.*, **42**: 53–56.

Furr, J. R., and C. A. Taylor, 1933: The cross-transfer of water in mature lemon trees, *Proc. Am. Soc. Hort. Sci.*, **30**: 45–51.

Gardner, F. E., 1925: A study of the conductive tissues in shoots of the Bartlett pear and the relationship of food movement to dominance of the apical buds, *Calif. Sta. Tech. Paper* 20.

Gäumann, E., 1927: Der jahreszeitliche Verlauf des Kohlehydratgehaltes in Tannen- und Fichtenstamm, *Ber. bot. Ges.*, **45**: 591–597.

Gerhardt, F., 1929: Propagation and food translocation in the common milkweed, *J. Agr. Res.*, **39**: 837–853.

Gerry, E., and J. A. Hall, 1935: Biochemical phases of oleoresin production, *Plant Physiol.*, **10**: 537–543.

Gibbs, R. D., 1930: Sinkage studies. II, The seasonal distribution of water and gas in trees, *Can. J. Res.*, **2**: 425–439; *Biol. Abst.*, **5**: 2658.

———, 1935: Studies of wood. II, On the water content of certain Canadian trees and on changes in the water-gas system during seasoning and flotation, *Can. J. Res.*, **12**: 727–760.

———, 1935: Studies of wood. III, On the physiology of the tree with special reference to the ascent of sap and the movement of water before and after death, *Can. J. Res.*, **12**: 761–787.

Gile, P. L., and J. O. Carrero, 1921: Assimilation of nitrogen, phosphorus and potassium by corn when nutrient salts are confined to different roots, *J. Agr. Res.*, **21**: 545–573.

Ginsburg, J. M., 1931: Penetration of petroleum oils into plant tissue, *J. Agr. Res.*, **43**: 469–474.

Godlewski, E., 1884: Zur Theorie der Wasserbewegung in den Pflanzen, *Jahrb. wiss. Bot.*, **15**: 569–630.

Goff, E. S., 1897: The application of artificial root pressure to recently transplanted trees, *Wis. Sta. Ann. Rept.*, **14**: 272–282.

———, 1898: The resumption of root growth in the spring, *Wis. Sta. Ann. Rept.*, **15**: 220–228.

Graber, L. F., N. T. Nelson, W. A. Leukel, and W. B. Albert, 1927: Organic food reserves in relation to the growth of alfalfa and other perennial herbaceous plants, *Wis. Sta. Res. Bull.* 80.

———, 1931: Food reserves in relation to other factors limiting the growth of grasses, *Plant Physiol.*, **6**: 43–72.

——— and H. W. Ream, 1931: Growth of blue grass with various defoliations and abundant nitrogen supply, *J. Am. Soc. Agron.*, **23**: 938–944.

Grandfield, C. O., 1930: The relation of organic food reserves to the effect of cutting pasture weeds at different stages of growth, *J. Am. Soc. Agron.*, **22**: 709–713.

———, 1935: The trend of organic food reserves in alfalfa roots as affected by cutting practices, *J. Agr. Res.*, **50**: 697–710.

Groom, P., 1910: Remarks on the ecology of *Coniferae*, *Ann. Bot.*, **24**: 241–269.

Haas, A. R. C., and F. F. Halma, 1929: Chemical relationship between scion and stock in *Citrus*, *Plant Physiol.*, **4**: 113–121.

——— and L. J. Klotz, 1935: Physiological gradients in citrus fruits, *Hilgardia*, **9**: 181–217.

Haasis, F. W., 1932: Seasonal shrinkage of Monterey pine and redwood trees, *Plant Physiol.*, **7**: 285–295.

———, 1933: Shrinkage and expansion in woody cylinders of living trees, *Am. J. Bot.*, **20**: 85–91.

———, 1934: Diametral changes in tree trunks, *Carnegie Inst. Wash. Pub.* 450.

Haberlandt, G., 1914: "Physiological Plant Anatomy." Translation by M. Drummond. Macmillan & Company, Ltd., London, XV + 777 pp.

Haines, F. M., 1935: Observations on the occurrence of air in conducting tracts, *Ann. Bot.*, **49**: 367–379.

———, 1935: Transpiration and pressure deficit. II, *Ann. Bot.*, **49**: 521–565.

Haller, M. H., 1930: The relation of the distance and direction of the fruit from the leaves to the size and composition of apples, *Proc. Am. Soc. Hort. Sci.*, **27**: 63–68.

Hannig, E., 1913: Untersuchungen über das Abstossen von Blüten u. s. w., *Z. Bot.*, **5**: 417–469.

Hartig, T., 1858: Über die Bewegung des Saftes in den Holzpflanzen, *Bot. Ztg.*, **16**: 329–335, 337–342.

Harvey, E. M., 1931: Movement of water in plants as affected by a mutual relation between the hydrostatic and pneumatic systems, *Plant Physiol.*, **6**: 495–506.

Harvey, T. B., 1930: Tracing the transpiration stream with dyes, *Am. J. Bot.*, **17**: 657–662.

Hawkins, R. S., R. L. Matlock, and C. A. Hobart, 1933: Physiological factors affecting the fruiting of cotton with special reference to boll shedding, *Ariz. Sta. Tech. Bull.* 46.

———, S. P. Clark, G. H. Serviss, and C. A. Hobart, 1934: Varietal differences in cotton boll shedding as correlated with osmotic pressure of expressed tissue fluids, *J. Agr. Res.*, **48**: 149–157.

Heinicke, A. J., 1919: Concerning the shedding of flowers and fruits and other abscission phenomena in apples and pears, *Proc. Am. Soc. Hort. Sci.*, **16**: 76–83.

Henricksen, H. C., 1932: Absorption and movement of plant nutrients, *Puerto Rico Sta. Agr. Notes*, **60**: 1–3; *Exp. Sta. Rec.*, **70**: 167.

———, 1933: Absorption of ash constituents by citrus trees, *Ann. Rept. Puerto Rico Exp. Sta.*, **1932**: 18–19.

Henson, P. R., 1932: The effect of the su factor on translocation and storage of carbohydrates in the corn plant, abst., 25th *Ann. Meeting Am. Soc. Agron.*

Heyl, J. G., 1933: The influence of external factors on sap of plants, *Planta*, **20**: 294–353; *Exp. Sta. Rec.*, **71**: 23.

HILL, A. W., 1901: The histology of the sieve-tubes of *Pinus*, *Ann. Bot.*, **15**: 575–613.

———, 1908: The histology of the sieve-tubes of angiosperms, *Ann. Bot.*, **22**: 245–291.

HOLLE, H., 1915: Untersuchungen über Welken, Vertrocknen und Wiederstraffwerden, *Flora*, **108**: 73–126.

HOLMES, M. C., 1918: A study in the anatomy of Hazel-wood with reference to conductivity of water, *Ann. Bot.*, **32**: 553–567.

HOMAN, C., T. F. YOUNG, and C. A. SHULL, 1934: Evaporation in its thermodynamic relation to the cohesion theory and to imbibitional and osmotic pressures of cell constituents, *Plant Physiol.*, **9**: 653–661.

HORNBERGER, R., 1882: Chemische Untersuchungen über das Wachsthum der Maispflanze, *Landw. Jahrb.*, **11**: 359–523.

HUBER, B., 1923: Theoretische Betrachtungen zur Kohäsionstheorie der Wasserbewegung in der Pflanze, *Biol. Zent.*, **43**: 30–49.

———, 1924: Beiträge zur Kenntnis der Wasserbewegung in der Pflanze. II, Die Strömungsgeschwindigkeit und die Grösse der Widerstände in den Leitbahnen, *Ber. bot. Ges.*, **42**: 27–32.

———, 1928: Weitere quantitative Untersuchungen über das Wasserleitungsystem der Pflanzen, *Jahrb. wiss. Bot.*, **67**: 859–877.

———, 1932: Beobachtung und Messung pflanzlicher Saftströme, *Ber. bot. Ges.*, **50**: 89–109.

HUME, A. N., and C. FRANZKE, 1929: The effect of certain injuries to leaves of corn plants upon weights of grain produced, *J. Am. Soc. Agron.*, **21**: 1156–1164.

HUNGATE, R. E., 1934: The cohesion theory of transpiration, *Plant Physiol.*, **9**: 783–794.

INAMDAR, R. S., and K. V. VARADPANDE, 1930: The problem of carbohydrate transport in plants. A glucose effect on the permeability of cell membranes to sugar molecules as studied by the intensity of respiration when leaves of *Artocarpus integrifolia* were injected with varying concentrations of glucose solution, *Proc. Indian Sci. Cong. (Calcutta)*, **17**: 300–301; *Exp. Sta. Rec.*, **66**: 621.

INGOLD, C. T., 1935: Note on exudation and exudation pressures in birch, *New Phytol.*, **34**: 437–441.

JACOBSON, L. A., 1934: Determination of total sugars in stalks of sorghum varieties, *unpub. data, Hays branch, Kans. Sta.*

JAMES, W. O., and H. BAKER, 1933: Sap pressure and the movements of sap, *New Phytol.*, **32**: 317–343.

JANSE, J. M., 1887: Die Mitwirkung der Markstrahlen bei der Wasserbewegung im Holz, *Jahrb. wiss. Bot.*, **18**: 1–69.

JANSSEN, G., 1929: The relationship of organic root reserves and other factors to the permanency of alfalfa stands, *J. Am. Soc. Agron.*, **21**: 895–911.

———, C. K. MCCLELLAND, and W. H. METZGER, 1930: Sap extraction of sorghum and the localization of juice and sugars in internodes of the plant, *J. Am. Soc. Agron.*, **22**: 627–639.

JODIDI, S. L., 1925: Nitrogen metabolism in etiolated corn seedlings, *J. Agr. Res.*, **31**: 1149–1164.

———, 1927: Isolation and identification of some organic nitrogenous compounds occurring in etiolated corn seedlings, *J. Agr. Res.*, **34**: 649–656.

JONES, C. H., A. W. EDSON, and W. J. MORSE, 1903: The maple sap flow, *Vt. Sta. Bull.* 103.

——— and J. L. BRADLEE, 1933: The carbohydrate contents of the maple tree, *Vt. Sta. Bull.* 358.

JONES, H. A., and C. S. BISSON, 1932: Changes in the composition of the garden pea after harvest, *Plant Physiol.*, **7**: 273–282.

JONES, W. J., and H. A. HUSTON, 1914: Composition of maize at various stages of its growth, *Ind. Sta. Bull.* 175.

JOST, L., 1907: "Lectures on Plant Physiology," English translation, Clarendon Press, London, pp. 167–172.

KENNEDY, P. B., and A. S. CRAFTS, 1927: The application of physiological methods to weed control, *Plant Physiol.*, **2**: 503–506.

KNIGHT, H., J. C. CHAMBERLIN, and C. D. SAMUELS, 1929: On some limiting factors in the use of saturated petroleum oils as insecticides, *Plant Physiol.*, **4**: 299–321.

KNIGHT, R. C., 1920: Storage and translocation, *Sci. Prog.*, **15**: 200–206.

KNOWLTON, H. E., 1921: A preliminary experiment on half tree fertilization, *Proc. Am. Soc. Hort. Sci.*, **18**: 148–149.

KOMATSU, S., and S. OZAWA, 1932: Biochemical study of the autumn fall of the leaf, *Mem. Col. Sci. Kyoto Imp. Univ.*, **A15**: 57–65; *Chem. Abst.*, **26**: 4629.

KRAYBILL, H. R., J. T. SULLIVAN, and L. P. MILLER, 1930: Seasonal changes in the composition of Stayman apple trees. I, Carbohydrates, *Proc. Am. Soc. Hort. Sci.*, **27**: 206.

KRETOVITCH, W., 1933: Distribution of sugar and nitrogenous substances in wheat grain, *Biochem. J.*, **27**: 1687–1690.

KRUSEMAN, W. M., 1931: De Invloed van Temperatuur en Narcose op Het Transport der Assimilaten. Purmerend.

LAMPE, L., 1931: A microchemical and morphological study of the developing endosperm of maize, *Bot. Gaz.*, **91**: 337–377.

LECLERC DU SABLON, M., 1904, 1906: Recherches physiologiques sur les matières de réserves des arbres, *Rev. Gén. Bot.*, **16**: 341–368; **18**: 5–25, 82–96.

———, 1910: Sur le mécanisme de la circulation de l'eau dans les plantes, *Rev. Gén. Bot.*, **22**: 125–136; *Bot. Cent.*, **116**: 87.

LECLERC, J. A., and J. F. BREAZEALE, 1911: Translocation of plant food and elaboration of organic plant material in wheat seedlings, *U. S. Bur. Chem. Bull.* 138.

LEE, E., 1911: The morphology of leaf-fall, *Ann. Bot.*, **25**: 51–106.

LEONARD, O. A., 1936: Seasonal study of tissue function and organic solute movement in the sunflower, *Plant Physiol.*, **11**: 25–61.

LEUKEL, W. A., 1927: Deposition and utilization of reserve foods in alfalfa plants, *J. Am. Soc. Agron.*, **19**: 596–624.

——— and J. M. COLEMAN, 1930: Growth behavior and maintenance of organic foods in Bahia grass, *Fla. Sta. Bull.* 219.

———, J. P. CAMP, and J. M. COLEMAN, 1934: Effect of frequent cutting and nitrate fertilization on the growth behavior and relative composition of pasture grasses, *Fla. Sta. Bull.* 269.

——— and R. M. BARNETTE, 1935: Cutting experiments with Bahia grass grown in lysimeters, *Fla. Sta. Tech. Bull.* 286.

LI, H. W., and T. N. LIU, 1935: Defoliation experiments with kaoliang, *J. Am. Soc. Agron.*, **27**: 486–491.

LIPMAN, C. B., and A. GORDON, 1925: Further studies on new methods in the physiology and pathology of plants, *J. Gen. Physiol.*, **7**: 615–623.

LLOYD, F. E., 1914: Abscission, *The Ottawa Naturalist*, **28**: 41–75.

———, 1916: The abscission of flower buds and fruits in *Gossypium* and its relation to environmental changes, *Trans. Roy. Soc. Canada*, **10**: 55–61.

———, 1916: Abscission in *Mirabilis Jalapa*, *Bot. Gaz.*, **61**: 213–230.

LONG, J. H., 1935: Seasonal changes in nitrogen and carbohydrate content of the strawberry plant, *Proc. Am. Soc. Hort. Sci.*, **33**: 386–388.

LOOMIS, W. E., R. BISSEY, and E. V. SMITH, 1931: Chlorates as herbicides, *Sci.*, **74**: 485.

———, E. V. Smith, R. Bissey, and L. E. Arnold, 1933: The absorption and movement of sodium chlorate when used as an herbicide, *J. Am. Soc. Agron.*, **25**: 724–739.

———, 1934: The translocation of nitrogen in woody plants, *Proc. Am. Soc. Hort. Sci.*, **32**: 61–64.

———, 1935: Translocation and growth balance in woody plants, *Ann. Bot.*, **49**: 247–272.

Lowry, M. W., and P. Tabor, 1931: Sap for analysis by bleeding corn plants, *Sci.*, **73**: 453.

Lund, E. J., 1931: Electric correlation between living cells in cortex and wood in the Douglas fir, *Plant Physiol.*, **6**: 631–652.

———, 1932: Control of the flux equilibrium of electro-chemical processes and electric polarity in the Douglas fir by temperature, *Plant Physiol.*, **7**: 297–307.

———, 1932: Comparison of the effects of temperature on the radial and longitudinal electric polarities in wood and cortex of the Douglas fir, *Plant Physiol.*, **7**: 505–516.

MacDaniels, L. H., and O. F. Curtis, 1930: The effect of spiral ringing on solute translocation and the structure of the regenerated tissue of the apple, *N. Y. Sta. (Cornell) Mem.* 133.

MacDougal, D. T., 1921: Growth in trees, *Carnegie Inst. Wash. Pub.* 307.

——— and F. Shreve, 1924: Growth in trees and massive organs of plants, *Carnegie Inst. Wash. Pub.* 350.

———, 1925: Absorption and exudation pressures in plants, *Proc. Am. Phil. Soc.*, **64**: 102–130.

———, 1925: Reversible variations in volume, pressure and movements of sap in trees, *Carnegie Inst. Wash. Pub.* 365.

———, 1926: The hydrostatic system of trees, *Carnegie Inst. Wash. Pub.* 373.

———, 1925–1927: Composition of gases in tree trunks, *Carnegie Inst. Wash. Year Book* **25**: 160–162; **26**: 162–163.

——— and J. B. Overton, 1927: Sap flow and pressure in trees, *Sci.*, **65**: 189–190.

———, ———, and G. M. Smith, 1929: The hydrostatic-pneumatic system of certain trees: movements of liquids and gases, *Carnegie Inst. Wash. Pub.* 397.

Malhotra, R. C., 1931: A contribution to the physiology and anatomy of tracheae, with special reference to fruit trees. I, Influence of tracheae and leaves on the water conductivity, *Ann. Bot.*, **45**: 593–620.

———, 1932: A contribution to the physiology and anatomy of tracheae with special reference to fruit trees. II, Water conductivity in higher plants and its relation to tracheae, *Ann. Bot.*, **46**: 11–28.

———, 1932: The distribution of some reserve substances in hard wheat at successive growth stages and their possible utilization, *J. Agr. Sci.*, **22**: 485–496.

Mangham, S., 1910–1911: The translocation of carbohydrates in plants, *Sci. Prog.*, **5**: 256–285, 457–479.

———, 1915: Observations on the osazone method of locating sugars in plant tissues, *Ann. Bot.*, **29**: 369–391.

———, 1917: On the mechanism of translocation in plant tissues. An hypothesis, with special reference to sugar conduction in sieve-tubes, *Ann. Bot.*, **31**: 293–311.

———, 1922: Transport of organic substances in plants, *Nature*, **109**: 476–477.

Mangin, L., 1898: Sur la nutrition et la défense de la vigne par injection, *J. Prak. Agr.*, **1898**: 918–920.

Martin, J. N., 1934: The relative growth rates and interdependence of tops and roots of the biennial white sweet clover, *Melilotus alba*, *Am. J. Bot.*, **21**: 140–159.

Maskell, E. J., and T. G. Mason, 1930: Studies on the transport of nitrogenous substances in the cotton plant. I, II, III, *Mem. Cotton Res. Sta. Trinidad*,

2; *Ann. Bot.*, **43**: 205–231, 615–652 (1929); **44**: 1–29; 233–267, 657–688 (1930).

Mason, T. G., 1922: A note on the growth and the transport of organic substances in bitter cassava (*Manihot utilissima*), *Sci. Proc. Roy. Dublin Soc.*, **17**: 105–112; *Bot. Abst.*, **12**: 759.

——— and C. J. Lewin, 1924: On the rate of carbohydrate transport in the greater yam, *Dioscorea alata*, *Sci. Proc. Roy. Dublin Soc.*, **18**: 203–205.

——— and E. J. Maskell, 1928: Studies on the transport of carbohydrates in the cotton plant. I, A study of diurnal variation in the carbohydrates of leaf, bark and wood, and of the effects of ringing, *Ann. Bot.*, **42**: 189–253.

——— and ———, 1928: Studies on the transport of carbohydrates in the cotton plant. II, The factors determining the rate and the direction of movement of sugars, *Ann. Bot.*, **42**: 571–636.

——— and ———, 1928: Studies on the transport of carbohydrates in the cotton plant. I, II, *Mem. Cotton Res. Sta. Trinidad*, Ser. B, *Physiology*.

——— and ———, 1931: Further studies on transport in the cotton plant. I, Preliminary observations on the transport of phosphorus, potassium and calcium, *Ann. Bot.*, **45**: 125–173.

——— and ———, 1934: Further studies on transport in the cotton plant, *Ann. Bot.*, **48**: 119–141.

——— and E. Phillis, 1934: Studies on the transport of nitrogenous substances in the cotton plant. VI, Concerning storage in the bark, *Ann. Bot.*, **48**: 315–333.

———, E. J. Maskell, and E. Phillis, 1936: Further studies on transport in the cotton plant. III, Concerning the independence of solute movement in the phloem, *Ann. Bot.*, **50**: 23–58.

Maximow, N. A., and M. A. Krotkina, 1930: Investigations on the after effect of low temperature on the length of vegetative period, *Trudy pro Prikladnoi Botanike Genetike i Selekcii*, **23**: 427–478.

McCarty, E. C., 1932: Some relations between the carbohydrates and the growth rate in the wild oat, *Avena fatua*, *Occasional Papers Riverside Jr. Col. Calif.*, **6**: 1–32; *Chem. Abst.*, **26**: 4081.

McHargue, J. S., and W. R. Roy, 1932: Mineral and nitrogen content of the leaves of some forest trees at different times in the growing season, *Bot. Gaz.*, **94**: 381–393.

McLean, D. M., 1933: The effect of harvesting at different stages of maturity upon the yield and chemical composition of barley, *Sci. Agr.*, **13**: 698–727.

McLean, R. C., and L. R. Hutchings, 1935: Streamline flow and the movement of solutes in the transpiration stream, *Plant Physiol.*, **10**: 773–780.

Michel-Durand, E., 1913: Variations des substances hydrocarbonée des feuilles au cours du développement, *C. R. Acad. Sci. Paris*, **156**: 1926–1929.

———, 1932: Sur l'azote, le soufre, le phosphore, le potassium des feuilles de *Prunus laurocerasus* au moment de lur chute, *C. R. Acad. Sci. Paris*, **195**: 1419–1420.

Miller, E. C., 1937: A physiological study of hard winter wheat at different stages of growth, Unpub. data, *Kansas Exp. Sta.*

Molisch, H., 1886: Untersuchungen über Laubfall, *Bot. Cent.*, **25**: 393–394.

———, 1902: Ueber localen Blütungsdruck und seine Ursachen, *Bot. Ztg.*, **60**: 45–63.

———, 1928, 1929: The movement of sap in plants, *Nature*, **122**: 168–169; *Sci.*, **69**: 217–218.

Moore, W., and A. G. Ruggles, 1915: The action of potassium cyanide when introduced into the tissues of a plant, *Sci.*, **42**: 33–36.

Morgan, A., 1931: Experiments on the control of hoary cress, *Victoria J. Agr.*, **29**: 504–515.

Morkrjetsky, A., 1903: Über die innere Therapie der Pflanzen, *Z. Pflanzenkr.*, **13**: 251–265. Quoted by Rumbold, 1920.

Mothes, K., 1931: Zur Kenntnis des N-Stoffwechsels höher Pflanzen, *Planta*, **12**: 686–731.

Moyer, L. S., 1937: Recent advances on the physiology of latex, *Bot. Rev.*, **3**: 522–544.

Münch, E., 1926: Dynamik der Saftströmungen, *Ber. bot. Ges.*, **44**: 68–71.

———, 1927: Versuche über den Saftkreislauf, *Ber. bot. Ges.*, **45**: 340–356.

———, 1930: "Die Stoffbewegungen in der Pflanze," Jena, 234 pp.

Murneek, A. E., 1928: Nitrogen and carbohydrate distribution in organs of bearing apple spurs, *Mo. Sta. Res. Bull.* 119.

———, 1932: Relation of leaf area to fruit size and food reserves in apple seeds and branches, *Proc. Am. Soc. Hort. Sci.*, **29**: 230–234.

——— and J. C. Logan, 1932: Autumnal migration of nitrogen and carbohydrates in the apple tree, *Mo. Sta. Res. Bull.* 171.

———, 1933: Carbohydrate storage in apple trees, *Proc. Am. Soc. Hort. Sci.*, **30**: 319–321.

Nägeli, C., 1866: Die Theorie der Capillarität, *Bot. Mitth.*, **2**: 471–501. Quoted by Copeland, 1902.

Nelson, N. T., 1925: The effect of frequent cutting on the production, root reserves and behavior of alfalfa, *J. Am. Soc. Agron.*, **17**: 100–113.

Nichols, P. F., 1930: Variations in content of sugars and related substances in olives, *J. Agr. Res.*, **41**: 89–96.

Nicol, H., 1934: Yield, duration and drought-resistance of lucerne as influenced by frequency and time of cutting, *Empire J. Exp. Agr.*, **2**: 380–390.

Nightingale, G. T., 1933: Effect of temperature on metabolism in tomato, *Bot. Gaz.*, **95**: 35–58.

Nordhausen, M., 1916: Ueber die Saugkraft transpirierender Sprosse, *Ber. bot. Ges.*, **34**: 619–639.

———, 1919: Zur Kenntniss der Saugkraft und der Wasserversorgung transpirierender Sprosse, *Jahrb. wiss. Bot.*, **58**: 295–335.

———, 1921: Weitere Beiträge zum Saftsteigeproblem, *Jahrb. wiss. Bot.*, **60**: 307–353.

Norman, A. G., 1933: A preliminary investigation of the development of structural constituents in the barley plant, *J. Agr. Sci.*, **23**: 216–227.

Otto, R., and W. D. Kooper, 1910: Beitrag zur Abnahme bezw. Rückwanderung der Stickstoffverbindungen aus den Blättern während der Nacht sowie zur herbstlichen Rückwanderung von Stickstoffverbindungen aus den Blättern, *Landw. Jahrb.*, **39**: 167–172.

——— and ———, 1910: Untersuchungen über Stickstoffassimilation in den Laubblättern, *Landw. Jahrb.*, **39**: 999–1004.

Overton, J. B., 1911: Studies on the relation of the living cells to transpiration and sap flow in Cyperus. I, II, *Bot. Gaz.*, **51**: 28–63, 102–120.

———, 1925–1926: Stem anatomy and sap conduction, *Carnegie Inst. Wash. Year Book*, **25**: 155–158.

——— and J. B. Smith, 1926–1927: Additional observations on stem anatomy and sap conduction, *Carnegie Inst. Wash. Year Book*, **26**: 159–160.

Parker, K. W., and A. W. Sampson, 1931: Growth and yield of certain *Gramineae* as influenced by reduction of photosynthetic tissue, *Hilgardia*, **5**: 361–381.

Peirce, G. J., 1929: To demonstrate the course of sap ascent in plants, *Sci.*, **69**: 186–187.

———, 1934: Observation on sap hydraulics, *Am. J. Bot.*, **21**: 211–227.

———, 1935: The ascent of sap, *Madrono*, **3**: 1–3.

———, 1936: Are living cells involved in the ascent of sap? *Am. J. Bot.*, **23**: 159–162.

PENSTON, N. L., 1935: Return of mineral elements to the soil by plants, *Nature*, **136**: 268–269.

PETTIBONE, C. J. V., and C. KENNEDY, 1916: Translocation of seed protein reserves in the growing corn seedling, *J. Biol. Chem.*, **26**: 519–525.

PHILLIS, E., and T. G. MASON, 1933: Studies on the transport of carbohydrates in the cotton plant. III, The polar distribution of sugar in the foliage leaf, *Ann. Bot.*, **47**: 585–634.

——— and ———, 1936: Further studies on transport in the cotton plant. IV, On the simultaneous movement of solutes in opposite directions through the phloem, *Ann. Bot.*, **50**: 161–174.

PIERRE, W. H., and F. E. BERTRAM, 1929: Kudzu production with special reference to influence of frequency of cutting on yields and formation of root reserves, *J. Am. Soc. Agron.*, **21**: 1079–1101.

PLATENIUS, H., 1934: Chemical changes in carrots during growth, *Plant Physiol.*, **9**: 671–680.

———, 1934: Physiological and chemical changes in carrots during growth and storage, *N. Y. Sta. (Cornell) Mem.* 161.

POLOVRĂGEANU, I., 1933: Sur la variation quantitative des substances minérales au cours du développement des feuilles de *Populus pyramidalis* et *Robinia pseudo-acacia*, *Inst. Art. Grafice E. Mavin, Bucuresti VI, Univ. Bucuresti.*

POND, R. H., 1903: The biological relation of aquatic plants to the substratum, *Contrib. Biol. Great Lakes, Rept. U. S. Fish Comm.*, 483–526.

PRIESTLEY, J. H., 1920: The mechanism of root pressure, *New Phytol.*, **19**: 189–200.

——— and D. ARMSTEAD, 1922: Physiological studies in plant anatomy. II, The physiological relation of the surrounding tissue to the xylem and its contents, *New Phytol.*, **21**: 62–80.

———, 1935: Sap ascent in the tree, *Sci. Prog.*, **30**: 42–56.

RANKIN, W. H., 1917: The penetration of foreign substances introduced into trees, *Phytopath.*, **7**: 5–14.

REED, H. S., and E. T. BARTHOLOMEW, 1927: Water translocation in young fruits, *Sci.*, **66**: 382–385.

———, 1930: The swelling of citrus fruits, *Am. J. Bot.*, **17**: 971–983.

RENNER, O., 1911: Experimentelle Beiträge zur Kenntnis der Wasserbewegung, *Flora*, **103**: 175–247.

———, 1911: Zur Physik der Transpiration. I, *Ber. bot. Ges.*, **29**: 125–132.

———, 1912: Zur Physik der Transpiration. II, *Ber. bot. Ges.*, **30**: 572–575.

———, 1912: Versuche zur Mechanik der Wasserversorgung. I, Der Druck in den Leitungsbahnen von Freilandpflanzen. II, Über Wurzeltätigkeit, *Ber. bot. Ges.*, **30**: 576–580, 642–648.

———, 1915: Theoretisches und Experimentelles zur Kohäsionstheorie der Wasserbewegung, *Jahrb. wiss. Bot.*, **56**: 617–667.

———, 1918: Versuche zur Mechanik der Wasserversorgung, *Ber. bot. Ges.*, **36**: 172–179.

REWALD, B., and W. RIEDE, 1933: Das Verhalten von Fett, Phosphatiden und Eiweiss während der Samenreifen, *Biochem. Z.*, **260**: 147–152; *Biol. Abst.*, **8**: 876.

RHINE, J. B., 1926: Translocation of fats as such in germinating fatty seeds, *Bot. Gaz.*, **82**: 154–170.

RIPPEL, A., 1921: Die Frage der Eiweisswanderung beim herbstlichen Vergilben der Laubblätter, *Biol. Zent.*, **41**: 508–523.

Rivett, M., and F. Rivett, 1920: The anatomy of *Rhododendron ponticum* and of *Ilex aquifolium* in reference to specific conductivity, *Ann. Bot.*, **34**: 525–550.

Roach, W. A., 1934: III, Injection for the diagnosis and cure of physiological disorders of fruit trees, *Ann. App. Biol.*, **21**: 333–343.

Robertson, J. H., 1933: Effect of frequent clipping on the development of certain grass seedlings, *Plant Physiol.*, **8**: 425–447.

Rogers, C. F., 1929: Winter activity of the roots of perennial weeds, *Sci.*, **69**: 299–300.

Rohrabaugh, P. W., 1934: Penetration and accumulation of petroleum spray oils in the leaves, twigs, and fruit of citrus trees, *Plant Physiol.*, **9**: 699–730.

Roshardt, P. A., 1910: Ueber die Beteiligung lebender Zellen am Saftsteigen bei Pflanzen von niedrigem Wuchs, *Beih. bot. Cent.*, I, **25**: 243–357.

Ruhland, W., 1911: Untersuchungen über den Kohlenhydratstoffwechsel von *Beta vulgaris, Jahrb. wiss. Bot.*, **50**: 200–257.

Rumbold, C., 1915: Methods of injecting trees, *Phytopath.*, **5**: 225–229.

———, 1920: The injection of chemicals into chestnut trees, *Am. J. Bot.*, **7**: 1–21.

———, 1920: Effect on chestnuts of substances injected into their trunks, *Am. J. Bot.*, **7**: 45–57.

Sachs, J., 1862: Zur Keimungsgeschichte der Gräser, *Bot. Ztg.*, **20**: 145–150.

———, 1862: Microchemische Untersuchungen, *Flora*, **45**: 289–301, 313–320, 325–336.

———, 1878: Ein Beitrag zur Kenntniss des aufsteigenden Saftstroms in transpirirenden Pflanzen, *Arb. Bot. Inst. Würzburg*, **2**: 148–184.

———, 1879: Ueber die Porosität des Holzes, *Arb. Bot. Inst. Würzburg*, **2**: 291–332.

Salmon, S. C., C. O. Swanson, and C. W. McCampbell, 1925: Experiments relating to the time of cutting of alfalfa, *Kans. Sta. Tech. Bull.* 15.

Sampson, A. W., and R. Samisch, 1935: Growth and seasonal changes in composition of oak leaves, *Plant Physiol.*, **10**: 739–751.

Sampson, H. C., 1918: Chemical changes accompanying abscission in *Coleus blumei*, *Bot. Gaz.*, **66**: 32–53.

Sanford, F., 1914: An experiment on killing tree scale by poisoning the sap of the tree, *Sci.*, **40**: 519–520.

———, 1915: In regard to the poisoning of trees by potassic cyanide, *Sci.*, **41**: 213–214.

Sayre, J. D., V. H. Morris, and F. D. Richey, 1931: The effect of preventing fruiting and of reducing the leaf area on the accumulation of sugars in the corn stem, *J. Am. Soc. Agron.*, **23**: 751–753.

Scherer, C. M., 1927: Tree injection for control of fungous diseases and insect pests, abst., *Phytopath.*, **17**: 51.

Schmidt, R. H., 1891: Ueber Aufnahme und Verarbeitung von fetten Ölen durch Pflanzen, *Flora*, **74**: 300–370.

Schneider-Orelli, O., 1909: Die Miniergänge von *Lyonetia clerkella* und die Stoffwanderung in Apfelblättern, *Cent. Bact.*, II, **24**: 158–181.

Schroeder, J., 1871: Die Frühjahrsperiode der Birke (*Betula alba*) und des Ahorns (*Acer platanoides*), *Landw. Versuchs-Stat.*, **14**: 118–146.

Schulze, E., 1900: Über den Umsatz der Eiweissstoffe in der lebenden Pflanze. II, *Z. physiol. Chem.*, **30**: 241–312.

——— and N. Castoro, 1903: Beiträge zur Kenntnis der Zusammensetzung und des Stoffwechsels der Keimpflanzen, *Z. physiol. Chem.*, **38**: 199–258.

Schumacher, W., 1930: Untersuchungen über die Lokalisation der Stoffwanderung in den Leitbündeln höherer Pflanzen, *Jahrb. wiss. Bot.*, **73**: 770–823.

———, 1932: Direct detection of material transport in sieve tubes of plants, *Naturwiss.*, **20**: 364–365; *Chem. Abst.*, **26**: 4628.

———, 1933: Untersuchungen über die Wanderung des Fluoreszenz in den Siebröhren, *Jahrb. wiss. Bot.*, **77**: 685–732.

SCHWENDENER, S., 1892, 1893: Zur Kritik der neuesten Untersuchungen über das Saftsteigen, *Sitzungsber. Akad. Wiss. Wien*, **44**: 911–946; *Bot. Cent.*, **53**: 290.

———, 1909: "Vorlesungen über mechanische Probleme der Botanik," W. Engelmann, Leipzig.

SEIDEN, R., 1926: Vergleichende Untersuchungen über den Einfluss verschiedener äusserer Faktoren insbesondere auf den Aschengehalt in den Pflanzen, *Landw. Versuchs-Stat.*, **104**: 1–50.

SEREX, P., 1917: Plant food materials in the leaves of forest trees, *J. Am. Chem. Soc.*, **39**: 1286–1296.

SHATTUCK, C. H., 1915: Effect of cyanide of potassium on trees, *Sci.*, **41**: 324.

SHAW, R. H., and P. A. WRIGHT, 1921: A comparative study of the composition of the sunflower and corn plant at different stages of growth, *J. Agr. Res.*, **20**: 787–793.

SHULL, C. A., 1923: Review of physiology of the ascent of sap by Bose, *Bot. Gaz.*, **76**: 316–317.

SHUTT, F. T., S. N. HAMILTON, and H. H. SELWYN, 1930: The protein content of grass, chiefly meadow foxtail (*Alopecurus pratensis*), *J. Agr. Sci.*, **20**: 126–135.

SIMON, W. J. M., 1906: Recherches sur l'alimentation artificielle des plantes, *J. Soc. Nat. Hort. France*, III, **7**: 678–686; *Gard. Chron.*, **3**: ser., **41**: 8 (1907).

SMITH, E. F., 1896: The path of the water current in cucumber plants, *Am. Nat.*, **30**: 372–378, 451–457, 554–562.

SMITH, F., R. B. DUSTMAN, and C. A. SHULL, 1931: Ascent of sap in plants, *Bot. Gaz.*, **91**: 395–410.

SMITH, H. F., 1933: The physiological relations between tillers of a wheat plant, *Australia Council Sci. and Ind. Res. J.*, **6**: 1; *Biol. Abst.*, **9**: 1716. Quoted by Bartel, Martin and Hawkins.

SMITH, O., 1931: A study of growth and development of the potato plant, *Proc. Am. Soc. Hort. Sci.*, **28**: 279–284.

SNOW, R., 1925: The transport of organic foodstuffs in plants, *Nature*, **116**: 360.

SNYDER, H., 1893: The draft of the wheat plant upon the soil in the different stages of its growth, *Minn. Sta. Bull.*, **29**: 152–160.

SPRAGUE, H. B., 1933: Root development of perennial grasses and its relation to soil conditions, *Soil Sci.*, **36**: 189–209.

STAMM, A. J., 1929: The capillary structure of softwoods, *J. Agr. Res.*, **38**: 23–67.

STARK, O. K., 1927: The protein metabolism of the soybean, *Am. J. Bot.*, **14**: 532–548.

STEWARD, F. C., 1928: A preliminary study of the diffusion of sugars through membranes of living plant tissue, abst., *Am. J. Bot.*, **15**: 629–630.

——— and J. H. PRIESTLEY, 1932: Movement of organic materials in plants: A note on a recently suggested mechanism, *Plant Physiol.*, **7**: 165–171.

STRASBURGER, E., 1891: Ueber den Bau und die Verrichtungen der Leitungsbahnen in den Pflanzen, *Histol. Beitr. Jena*, **3**: 609.

———, 1893: Ueber das Saftsteigen, *Histol. Beitr. Jena*, **5**: 1–94; abst. *Bot. Cent.*, **60**: 270–272.

STREETS, R. B., 1934: The treatment of deciduous fruit trees and nut trees infected by *Phymatotrichum omnivorum* with ammonium compounds, *Sci.*, **79**: 417–418.

STUART, N. W., and C. O. APPLEMAN, 1935: Nitrogenous metabolism in Irish potatoes during storage, *Md. Sta. Bull.* 372.

STURKIE, D. G., 1930: The influence of various top-cutting treatments on rootstocks of Johnson grass (*Sorghum halepense*), *J. Am. Soc. Agron.*, **22**: 82–93.

SULLIVAN, J. T., and H. R. KRAYBILL, 1930: Seasonal changes in the composition of Stayman apple trees. II, Forms of nitrogen, *Proc. Am. Soc. Hort. Sci.*, **27**: 220.

SURFACE, H. A., 1914: Cyanide of potassium in trees, *Sci.*, **40**: 852–853.

SUROZ, J. I., 1891: Oel als Reservestoff der Bäume, *Beih. bot. Cent.*, **1**: 342–343.

Swarbrick, T., 1927: Studies in the physiology of fruit trees. I, The seasonal starch content and cambial activity in one- to five-year-old apple branches, *J. Pom. and Hort. Sci.*, **6**: 137–156.

Swart, N., 1914: "Die Stoffwanderung in ablebenden Blättern," Jena, 118 pp.

Thomas, E. E., and A. R. C. Haas, 1928: Injection method as a means of improving chlorotic orange trees, *Bot. Gaz.*, **86**: 355–362.

Thomas, L. A., and W. A. Roach, 1934: Injection of fruit trees: Preliminary experiments with artificial manures, *J. Pom. and Hort. Sci.*, **12**: 151–166.

Thomas, M., 1933: The natural abscission of twigs in Canadian black poplar (*Populus serotina* Hartig), *Naturalist*, **1933**: 79–86.

Thomas, W., 1919: Soil investigations. Unpublished data.

———, 1932: Composition of current and previous season's branch growth in relation to vegetative and reproductive responses in *Pyrus malus*, *Plant Physiol.*, **7**: 391–445.

Thompson, T. G., 1915: The total amino nitrogen in the seedlings of the Alaska pea, *J. Am. Chem. Soc.*, **37**: 230–235.

Thornton, H. G., and H. Nicol, 1934: Some effects of clipping the tops upon the root development of lucerne, *Medicago sativa*, *J. Agr. Sci.*, **24**: 532–539.

Thut, H. F., 1928: Demonstration of the lifting power of evaporation, *Ohio J. Sci.*, **28**: 292–298.

———, 1932: Demonstrating the lifting power of transpiration, *Am. J. Bot.*, **19**: 358–364.

Tison, A., 1900: Recherches sur la chute des feuilles chez les Dicotyledonées, *Mém. Soc. Linn. Normandie*, **20**: 121–327. Reviewed by Sampson, 1918.

Toole, E. H., 1924: The transformations and course of development of germinating maize, *Am. J. Bot.*, **11**: 325–350.

Trandafirescu, E., 1934: La variation quantitative des matières minérales et organiques au cours du développement des feuilles de *Aesculus hippocastanum*, *Quercus robur*, *Juglans regia* et *Zea mays*, *Contr. Univ. Bucuresti Lab. Anyl. Chem.*

Traub, H. P., 1927: Regional and seasonal distribution of moisture, carbohydrates, nitrogen, and ash in 2–3 year portions of apple twigs, *Minn. Sta. Tech. Bull.* 53.

Trowbridge, P. F., L. D. Haigh, and C. R. Moulton, 1915: Studies of the timothy plant. II, The changes in the chemical composition of the timothy plant during growth and ripening with a comparative study of the wheat plant, *Mo. Sta. Tech. Bull.* 20.

Tucker, G. M., and B. Tollens, 1900: Über den Gehalt der Platenenblätter an Nahrstoffe und die Wanderung dieser Nahrstoffe beim Wachsen und Absterben der Blätter, *J. Landw.*, **48**: 39–63.

Tukey, H. B., and E. L. Green, 1934: Gradient composition of rose shoots from tip to base, *Plant Physiol.*, **9**: 157–163.

Unger, F., 1868: Weitere Untersuchungen über die Bewegung des Pflanzensaftes, *Sitzungsber. Akad. Wiss. Wien*, **58**: 392–418.

Ursprung, A., 1905: Untersuchungen über die Beteiligung lebender Zellen am Saftsteigen, *Beih. bot. Cent.*, I, **18**: 147–158.

———, 1906: Die Beteiligung lebender Zellen am Saftsteigen, *Jahrb. wiss. Bot.*, **42**: 503–544.

———, 1907: Ueber die Ursache des Welkens, *Beih. bot. Cent.*, I, **21**: 67–75.

———, 1907: Studien über die Wasserversorgung der Pflanzen, *Biol. Cent.*, **27**: 33–60.

———, 1907: Abtötungs- und Ringelversuche an einigen Holzpflanzen, *Jahrb. wiss. Bot.*, **44**: 287–349.

———, 1913: Zur Demonstration der Flüssigkeitskohäsion, *Ber. bot. Ges.*, **31**: 388–400.

———, 1913: Ueber die Bedeutung der Kohäsion für das Saftsteigen, *Ber. bot. Ges.* **31**: 401–412.

———, 1915: Ueber die Kohäsion des Wassers im Farnannulus, *Ber. bot. Ges.*, **33**: 153–163.

———, 1915: Zweiter Beitrag zur Demonstration der Flüssigkeitskohäsion, *Ber. bot. Ges.*, **33**: 253–265.

———, 1916: Dritter Beitrag zur Demonstration der Flüssigkeitskohäsion, *Ber. bot. Ges.*, **34**: 475–487.

———, 1925: Einige Resultate der neuesten Saugkraftstudien, *Flora*, **18-19**: 566–599.

VAN DEN HONERT, T. H., 1932: Mechanism of the transport of organic materials in plants, *Proc. Acad. Sci. Amsterdam*, **35**: 1104–1112; *Chem. Abst.*, **27**: 752.

VESQUE, J., 1884: Recherche sur le mouvement de la sève ascendante, *Ann. Sci. Nat. Bot.*, VI, **19**: 159–199.

VICKERY, H. B., G. W. PUCHER, C. S. LEAVENWORTH, and A. J. WAKEMAN, 1935: Chemical investigations of the tobacco plant, *Conn. Sta. Bull.* 374.

VINES, S. H., 1896: The suction force of transpiring branches, *Ann. Bot.*, **10**: 429–444.

VIRTANEN, A. I., and M. NORDLUND, 1934: Synthesis of sucrose in plant tissue, *Biochem. J.*, **28**: 1729–1732.

——— and M. NURMIA, 1936: Studies on the winter hardiness of clover. I, Effect of cutting on the carbohydrate reserves in red clover roots, *J. Agr. Sci.*, **26**: 288–295.

VLADESCU, I. D., 1934: La variation quantitative des substances minérales et organiques pendant le développement de la *Nicotiana tabacum, Contr. Univ. Bucaresti Lab. Anyl. Chem.*

VON MOHL, H., 1860: Über den Ablösungsprozess saftiger Pflanzenorgane, *Bot. Ztg.*, **18**: 273–274.

WALTER, H., 1931: "Die Hydratur der Pflanzen und ihre physiologischökologische Bedeutung," Jena. Quoted by James and Baker, 1934.

WANN, F. B., 1929: Experiments on the treatment of chlorosis in Utah, abst, *Am. J. Bot.*, **16**: 844.

WEEVERS, T., and J. WESTENBERG, 1931: Versuche zur Prüfung der Münchschen Theorie der Stoffbewegungen in der Pflanze, *Rec. trav. bot. Néerl.*, **36**: 1173.

WEINBERGER, J. H., and F. P. CULLINAN, 1932: Further studies on the relation between leaf area and size of fruit, chemical composition and fruit bud formation in Elberta peaches, *Proc. Am. Soc. Hort. Sci.*, **29**: 23–27.

——— and ———, 1934: Nitrogen intake and growth response in peach trees following fall and spring fertilizer applications, *Proc. Am. Soc. Hort. Sci.*, **32**: 63–69.

WELTON, F. A., V. H. MORRIS, and A. J. HARTZLER, 1930: Distribution of moisture, dry matter, and sugars in the maturing corn stem, *Plant Physiol.*, **5**: 555–564.

WERNER, H. O., 1935: The effect of temperature, photoperiod and nitrogen level upon tuberization in the potato, *Am. Potato J.*, **12**: 274–280.

WESTERMAIER, M., 1883: Zur Kenntniss der osmotischen Leistungen des lebenden Parenchyms, *Ber. bot. Ges.*, **1**: 371–383.

———, 1884: Bedeutung todter Röhren und lebender Zellen für die Wasserbewegung, *Sitzb. preuss. Akad. Wiss.*, **48**: 1105–1117.

WHITE, P. R., 1927: Studies of the physiological anatomy of the strawberry, *J. Agr. Res.*, **35**: 481–492.

———, 1938: "Root Pressure"—an unappreciated force in sap movement, *Am. J. Bot.*, **25**: 223–227.

WIESNER, J., 1871: Untersuchungen über die herbstliche Entlaubung der Holzgewächse, *Litz. Akad. Wiss. Wien*, **64**: 465–509.

———, 1904: Über Sommerlaubfall, Trieblaubfall, und Hitzelaubfall, *Ber. bot. Ges.*, **22**: 64–72, 316–323, 501–505.

———, 1905: Über Frostlaubfall nebst Bemerkungen über die Mechanik der Blattablösung, *Ber. bot. Ges.*, **23**: 49–60.

Wilfarth, H., H. Römer, and G. Wimmer, 1905: Über die Nährstoffaufnahme der Pflanzen in verschiedenen Zeiten ihres Wachstums, *Land. Versuchs-Stat.*, **63**: 1–70.

Willard, C. J., 1927: An experimental study of sweet clover, *Ohio Sta. Bull.*, **405**: 3–84.

———, 1930: Root reserves of alfalfa with special reference to the time of cutting and yield, *J. Am. Soc. Agron.*, **22**: 595–602.

Winkler, A. J., 1929: The effect of dormant pruning on the carbohydrate metabolism of *Vitis vinifera*, *Hilgardia*, **4**: 153–173.

———, 1932: The lateral movement of elaborated foods in the grape vine, *Proc. Am. Soc. Hort. Sci.*, **29**: 335–338.

Woodhouse, E. D., 1933: Sap hydraulics, *Plant Physiol.*, **8**: 177–202.

Wright, J. G., 1928: The pit closing membrane in the wood of the lower Gymnosperms, *Trans. Roy. Soc. Canada*, **22**: 63–94.

Yocum, L. E., 1925: The translocation of the food materials of the wheat seedling, *J. Agr. Res.*, **31**: 727–744.

Young, P. A., 1934: Penetration, distribution, and effect of petroleum oils in apple, *J. Agr. Res.*, **49**: 559–571.

———, 1935: Oil mass theory of petroleum-oil penetration into protoplasm, *Am. J. Bot.*, **22**: 1–8.

———, 1935: Distribution and effect of petroleum oils and kerosenes in potato, cucumber, turnip, barley, and onion, *J. Agr. Res.*, **51**: 925–934.

Zimmerman, P. W., A. E. Hitchcock, and W. Crocker, 1931: The movement of gases into and through plants, *Contr. Boyce Thompson Inst.*, **3**: 313–321.

——— and M. H. Connard, 1934: Reversal of direction of translocation of solutes in stems, *Contr. Boyce Thompson Inst.*, **6**: 297–302.

CHAPTER XIII

THE PROCESS OF RESPIRATION IN PLANTS

1. Energy Relations of the Plant.—The various processes that go on in a green plant involve the execution of work. Some of these processes are the formation of carbohydrates, fats, proteins, and other organic compounds from the simple materials obtained from the soil and the air; the evaporation of water from the stem and leaves; the penetration of the roots through the soil; the elevation of the stem into the air; the accumulation of solutes; and the movements and activities of the protoplasm not included in the processes mentioned above. To accomplish the various kinds of work, a supply of kinetic energy must be furnished continually to the different plant parts concerned in the performance of the various functions. Let us now inquire into the source of the energy of the plant and its distribution, transformance, and general availability in the various activities of plant life.

The sun is the source of the energy that is necessary for all life, and this energy is transmitted in the form of heat, light, and chemical rays. About 70 per cent of this energy which falls upon the plant is absorbed by it, while the remainder is either reflected or transmitted. It has been mentioned in Chap. VIII that slightly over 70 per cent of this absorbed energy is utilized in the evaporation of water in transpiration. From 1.5 to 4.5 per cent of it is utilized in the synthesis of carbohydrates, fats, proteins, and other organic compounds and is thus transformed into potential or stored energy. Another portion of the absorbed energy may be used directly in the various processes of the parts by which it is absorbed and in raising the temperature of the plant, while the remainder is lost by radiation into the surrounding air.

Although the rays of the sun are the original source of the energy supply of the plant, the structure of the plant and the conditions under which it grows prevent it from utilizing this energy directly in all its parts. The structure of the plant is such that its interior portions are not exposed to the rays of the sun, so that they cannot utilize directly the energy of the sunshine. Also the living cells must be furnished with a continuous supply of energy during darkness as well as light. It is evident, therefore, that some supply of energy must be available for immediate use in all the cells of the plant under conditions of both light and darkness. In every living plant cell there is present a supply of proteins, carbohydrates, fats, and other organic compounds in varying amounts. Every living plant cell thus has in the form of these compounds, a supply of potential energy, which, if changed into the kinetic form, would be available for the life processes of the cell. The living cell has the power so to transform this stored energy whenever the proper conditions exist. The processes associated with the transformation of potential energy in the living cell into the kinetic form are primarily chemical decompositions in which the stored foods and the protoplasm of the cell are involved. The most important process concerned in the liberation of stored energy in the cells is termed "respiration." Respiration may be defined as a process that goes on in every living cell whereby energy is released in a form that is available to the protoplasm. The term "energesis" has been suggested (Barnes, 1905) as a name for the processes involved in energy release, but this has never come into general use.

According to Stiles and Leach (1932) respiration has sometimes been called "dissimilation," in contrast to the process of assimilation. Malpighi (1679) reported that seeds require a supply of free oxygen to germinate. Jean Ingen-Housz in 1774 and De Saussure (1804) noted that plants in the light apparently produced oxygen, and, in the dark, carbon dioxide. The latter investigator noted the difference between the green and colorless parts of plants in this regard, but it remained for Sachs (1882) to explain the difference between the two basic processes concerned in these gaseous exchanges.

2. Importance of Respiration.—The energy necessary for the performance of any function by a living organism or part of an organism is obtained by the process of respiration in the protoplasm. Two conditions, at least, are necessary for cells to perform the functions of life: (*a*) there must be substances present that can readily be oxidized, and from the oxidation of which the available energy can be obtained; and (*b*) there must be oxygen present with which to oxidize these materials. Most organisms can perform some work in the absence of free oxygen, but usually it is only a limited amount and for only a limited time (Meldrum, 1933).

3. Nature of Respiration.—In the process of respiration under ordinary conditions, oxygen is absorbed, organic compounds, especially carbohydrates and fats, disappear, and carbon dioxide and water are formed as end products. The materials used and the waste products formed in the process are well defined, but the changes that occur in the cell whereby the organic compounds are utilized and the waste products produced are not clearly understood. We can thus observe the beginning and end of the process but know little or nothing of the intermediate changes that occur.

a. Direct Oxidation of Foods.—Respiration is generally considered to be an oxidation process because the final products—carbon dioxide, water, and kinetic energy—are the same as those derived from the combustion of organic substances in the air. The oxidation of organic substances in respiration, however, is striking in that it occurs at ordinary temperatures generally below 35°C., while oxidation of the same substances outside the living organism is accomplished only by the aid of very high temperatures or of powerful reagents. The oxidation of organic substances in respiration apparently occurs in the protoplasm itself or in close proximity to it, and it is considered that the oxidation is accomplished by the action of oxidizing enzymes which are secreted by the protoplasm. Herzfeld and Klinger (1919) stressed the possibility that certain molecular structures may weaken the union of the oxygen atoms in the oxygen molecule so that "activation" (increased oxidizing power) results, whereby loose chemical combinations are made with H_2O, with OH ions, and with O_2 absorbing surfaces. Such organic compounds as the lower fatty acids or their salts might thus be oxidized to carbon dioxide and water, fulfilling the conditions for oxidation processes in the living organism, namely active O_2 and easily oxidizable simpler metabolic products, thereby rendering the assumption of certain oxidases unnecessary.

According to Stiles and Leach (1932), Blackman (1928) presented a scheme of the respiratory process involving a chain of four consecutive processes as follows:

1. Hydrolysis of starch and disaccharoses to hexoses.
2. Activation: This includes the steps leading to the formation of heterohexoses with the less stable type of internal ring structure. This involves an isomeric change of the hexose molecule.
3. Glycolysis: This stage, involving the action of the zymase complex, includes the degradation of activated hexose to methyl glyoxal, pyruvic acid, acetaldehyde, and probably lactic acid.
4. Aerobic Respiration: This is concerned with the production of carbon dioxide and water from the intermediate substances produced in glycolysis.

The carbon dioxide formed in the last process may account for only a fraction of the carbon that appears to be involved in glycolysis. The conclusion is thus drawn that a part of the products of glycolysis are in some manner reconstructed into the system. To this process the name "oxidative anabolism" is given. In this scheme oxygen plays a double role in respiration. It not only produces an oxidation of the products of glycolysis, but it also affects the preglycolytic stages in that it brings about an alteration in the rate of production of the direct substrate for glycolysis.

In 1897 Büchner obtained zymase from yeast. It is now apparent that the substance originally termed zymase is a mixture of enzymes, and it is now usually called the "zymase complex." The following are some of the enzymes present or accompanying the substance termed "zymase": (1) hexosephosphatase, concerned in the formation of hexosephosphate; (2) phosphatase, which hydrolyzes hexosephosphate into hexose and phosphate; (3) glycolase, which acts on hexose to form pyruvic aldehyde or methylglyoxal; (4) carboxylase, which acts on α-ketonic acids, *e.g.*, pyruvic acid, to give acetaldehyde and carbon dioxide.

Allison, Hoover, and Burk (1933) claimed that they obtained, by extracting commercial sucrose with alcohol, a factor that was specific and essential for respiration. This factor is soluble in water and in absolute alcohol, but insoluble in the ordinary fat solvents. These characteristics classify it as a coenzyme.

According to Stiles and Leach (1932) it is misleading to speak of respiration as combustion. There are only two points of resemblance between them. The substrate and the end products may be alike; and, when this is true, the amount of energy released will be the same in respiration and in combustion. It should, however, be remembered that in no case has it been definitely proved that the carbohydrates, fats, or proteins are directly attacked outside the protoplasm, and it will also be shown in the discussion of the oxidizing enzymes that they are incapable, as prepared by present methods, of oxidizing under laboratory conditions any of the carbohydrates or fats that are apparently used in respiration.

b. Decomposition of Protoplasm.—It is believed by some that the facts of respiration may be explained more satisfactorily by assuming that the process is concerned in the decomposition of the living matter itself. Barnes (1905) assumed that in the labile protein components of the protoplasm there occurs a shifting of atomic groups within the molecule as a result of the addition of hydroxyl groups from the water present. Dissociation of these components follows this procedure, the rate depending upon the prevailing conditions. The groups or radicals of the protoplasm that have thus been dissociated may contain structural components that have been derived from sugars, fats, or proteins. The portion of the protoplasm that has been dissociated is then rebuilt at the expense of the sugars, fats, or proteins present in the cell, and as a consequence they disappear during the process of respiration. An attempt was made by Palladin (1896) to determine the relationship of the amount of protoplasm to the rate of carbon dioxide production. He determined the amount of carbon dioxide given off by wheat seedlings in the dark, and in parallel experiments he determined the amount of protein present which was indigestible in gastric juice. It was assumed that such proteins represented those that were components of the protoplasm. Palladin found that with an adequate supply of carbohydrate, the elimination of carbon dioxide was proportional to the amounts of indigestible proteins present in the seedlings. Although this method is far from exact, it nevertheless gives some indication of the relationship between the protoplasm and the process of respiration. The part of the protoplasm, however, that is the most actively concerned in respiration has received but little attention. The leaf cells of the Indian pipe (*Monotropa uniflora*) contain a colorless chromogen that oxidizes very easily if the proper conditions are provided. Osterhout (1917) determined that the oxidation of this chromogen in the cell is much more rapid in the nucleus than in the cytoplasm. From that he inferred that respiration is much more rapid in the nucleus than in the other parts of the protoplasm. It is not known, however, whether the nuclei of other plant cells show a similar behavior.

4. Organic Materials Used in Respiration.—One of the striking results of respiration is the loss in weight of the organism, which, however, is not ordinarily noticed in plants on account of the manufactured food equaling or exceeding the amount lost. If food is withheld from an animal, or if a plant is prevented from manufacturing food, the loss in weight due to respiration is very apparent. The most striking examples of the loss in weight from respiration are shown in the case of germinating seeds. The seedling plant always weighs less in dry matter than the seed from which it originated. Thus corn seedlings, including the residues of the seeds 2 weeks after planting, contain from 5 to 10 per cent less dry matter than the original seeds. Miller (1910) noted in sunflower seedlings 10 days old that the dry weight of 100 seedlings was 5.7 g. as compared with 7.3 g., which was the dry weight of an equal number of seeds. This was a loss in weight of 1.6 g. or approximately 20 per cent of the original dry weight of the seeds. Thus, one-fifth of the foods stored in the seeds was consumed in respiration in order to generate sufficient energy to push the plant roots into the soil, to raise the stem or cotyledons above ground, and to carry on the various life activities of the seedling.

Haller, Harding, Lutz, and Rose (1931) noted in strawberries that at 80°F., the loss per day of citric acid amounted to 25 per cent of the amount present at the time of packing. Bolas and Melville (1933) found in the tomato plant that the loss in weight due to respiration during a 17-hr. period of darkness amounted to 4.6 per cent of the dry weight of the plant.

a. Fats, Oils, and Carbohydrates.—The materials which disappear in the process of respiration are, for the most part, the carbohydrates, fats and oils. It is well established that the carbohydrates furnish the greater portion of the materials that are used in respiration. Even when fats and oils are the only source of oxidizable material, it is very probable that they are transformed into sugars before they are ultimately utilized in respiration.

The relationship of the supply of carbohydrates, especially of the sugars, to the intensity of respiration has received considerable consideration. Palladin (1896) and Palladin and Komleff (1902) found that etiolated bean leaves floated on a solution of sucrose or with their petioles immersed therein respired much more rapidly when transferred to distilled water than did the controls. In one case, 100 g. of leaves low in carbohydrate gave off at room temperature 89.6 mg. of carbon dioxide per hour for a given period. After the same leaves were floated upon a solution of cane sugar in darkness for 2 days they gave off carbon dioxide at the rate of 147.8 mg. per hour. Maige and Nicolas (1910) showed that, in general, respiration increased as the concentration of the sugar in the medium increased until the concentrations were sufficient to produce plasmolysis when the intensity of respiration declined. In ripening bananas, Gore (1914) found that the graph of the rate of respiration paralleled that of the rate of starch hydrolysis in the fruit, while there were only slight changes in the amounts of fats and proteins. His results were substantiated by Olney (1926), who found that the value of the respiratory quotient during the ripening of this fruit was approximately unity, which indicated that the oxidation in this case consisted solely of the combustion of carbohydrates. It was noted by Knudson (1916) that there was an increased respiration in plants when sugars were added to the nutrient solutions in which they were growing. According to Hopkins (1927), the sugar content of wounded potatoes increased from 53 to 68 per cent of the original sugar present. The rate of respiration was also increased after the wounding, and he assumed that the basis of this augmentation of the intensity of respiration was the increase in the sugar content of the cells. The curves of the sugar content and of the rate of respiration, however, did not exactly parallel.

The hexoses are apparently the sugars utilized in respiration. Thus Hasselbring and Hawkins (1915) could observe no correlation between the total sugars in the sweet potato and respiratory activity. A simultaneous decrease, however, in the reducing sugars and the respiratory intensity was observed, which indicated that they were the immediate source of the materials used in respiration. In plants *d*-glucose is apparently the most easily oxidizable sugar. The amount of *d*-levulose has been found to be in excess of *d*-glucose in the leaves, which is taken to indicate that the former is the more stable and is thus used to a lesser degree in respiration. Spoehr and McGee (1922) found that the intensity of respiration in the excised leaves of *Helianthus annuus* was not increased when levulose was fed to them, while the presentation of *d*-glucose, sucrose, and mannose stimulated respiration. As will be mentioned later, the addition of amino compounds with these sugars stimulated the rate of respiration in all cases except with levulose.

When the carbohydrate content of plants is low, the rate of the emission of carbon dioxide increases with the increasing supply of sugars. This may be conceived of as a matter of mass action, in that as the concentration of the available material to be oxidized becomes greater, the rate of reaction is increased. This is true, however, only within the limits of the capacity of the protoplasm present to carry on respiration. Thus, the intensity of respiration in etiolated leaves of the bean was greatly accelerated by floating them for a 2-day period upon a sugar solution. A further increase in respiration, however, could not be obtained by extending the period of sugar feeding to 40 days, although the concentration of the sugar in the leaf cells was greatly increased. During a 2-day period upon the sugar solution, sufficient sugar was absorbed for the maximum rate of respiration for the quantity of protoplasm present, so that the further addition of sugar was without influence upon the process. The total quantity of carbohydrate in the cell will thus not influence the rate of respiration except to determine the length of time that the process may continue therein.

b. Proteins.—It was formerly considered that proteins were also directly utilized in respiration, but experiments have shown that under normal conditions they are not utilized to any appreciable extent in this way. In the absence of an adequate supply of carbohydrates, however, the proteins are apparently utilized in the process. Thus, Palladin (1888) found that if wheat plants were left in the dark until they were deprived of their carbohydrate reserves, they lost a portion of their protein. Deleano (1912) found that excised leaves of the grape during the first 100 hr. used only carbohydrates in respiration, while the protein remained intact. After the carbohydrate supply was depleted, however, the process of respiration was strikingly altered and the protein was split up into soluble products. Meyer (1918) made the same observation with leaves and noted that the protein supplied only the carbon chains for respiration and that no loss of nitrogen from the leaves could be observed.

Yemm (1935) considered that during the later stages of starvation it is highly probable that the breakdown of protein is one of the important reactions concerned with carbon dioxide production. According to Stiles and Leach (1932) the minimum constant rate of respiration of starved leaves has been termed "protoplasmic respiration," in contrast to "floating respiration" in which carbohydrates and oily reserves are utilized. This protoplasmic respiration may result from the utilization of protein as a respiratory substance after the other reserve materials have been exhausted.

It is evident that the amount of energy which could be obtained by the decomposition of proteins under normal conditions would be very small, for this decomposition is only a cleavage into the amino acids. Furthermore, the plant is capable of re-forming these amino acids again into proteins without the direct utilization of energy save that which is derived from the oxidation of carbohydrates. Spoehr and McGee (1923) believed, therefore, that the total gain in energy for the plant from protein decomposition would be nothing and that unless the proteins yield energy of a certain form or at a particular rate, which is necessary for the proper functioning of certain processes, it is difficult to see what advantage would result or what the mode would be of deriving energy by continual decomposition and subsequent resynthesis of the proteins. It may safely be assumed that the rate of energy release from protein decomposition is much slower than that from carbohydrates, but of the possible different forms of energy liberated in the decomposition of these two types of compounds little or nothing is known.

Since it is known that amino acids have a marked influence upon the action of certain enzymes, Spoehr and McGee suggested that the decomposition of proteins may provide materials essential to the proper functioning of these agents. As the amino acids are amphoteric electrolytes, they may also function in maintaining the hydrogen-ion concentration of the cell contents within definite limits. Spoehr and McGee investigated the effects of glycocoll and asparagine upon the intensity of respiration in excised leaves of *Helianthus annuus* and found that the leaves which were fed glycocoll respired more actively than those which had not thus been treated. The rate of respiration was increased by feeding the leaves with *d*-glucose, but this rate was further increased by feeding in addition glycocoll or asparagine. When the leaves were given sucrose there was an initial drop in the rate of carbon dioxide emission, which probably represented the time required for the sugar to migrate into the leaves, after which there was an irregular rise that became prominent after 35 hr. When glycocoll was added in addition to the sucrose, there was a marked rise in the rate of respiration immediately after the initial drop. Hafenrichter (1928), however, in his study of the germination of the soybean concluded that there was no evidence under the conditions of his experiments to support the theory that amino acids stimulate respiration.

c. Effect of Food Type.—Hafenrichter (1928) pointed out that since the stored food materials of seeds undergo a series of changes during the germination and development of the seedlings, there would be qualitative and quantitative differences in the materials available for respiration. The manner in which these qualitative and quantitative differences may affect respiration under various conditions has not, however, been investigated to any extent, and it is not known whether or not the required amount of energy under any set of conditions is always supplied by a definite metabolic process. It has been demonstrated for animals that it is possible to maintain the same energy relations by substituting foods that are qualitatively different provided that the caloric values remain constant. It has been observed, as has been previously mentioned, that plants apparently prefer one of two closely related carbohydrates for respiration and it has also been found that some plants may utilize different amounts of one food in the presence of different quantities of another. This indicates that external as well as internal conditions determine the kind of materials that may be utilized in respiration. From a study of the respiration of the seedlings of two varieties of soybean, Hafenrichter concluded that the difference in the respiration of different plants cannot be explained on the basis of quantitative differences in the reserve food materials. He obtained evidence, however, which suggested that plants exhibit a selection of organic compounds in their metabolism and that this is reflected in their respiration.

5. Aerobic and Anaerobic Respiration.—Under the conditions of a normal supply of free oxygen, the process of respiration is marked by an absorption of oxygen, by a consumption of organic compounds, especially of carbohydrates, and by the appearance of water and carbon dioxide as end products. When respiration proceeds after this manner, it is termed "normal," "aerobic," or "oxygen" respiration. If carbohydrates are the compounds concerned, the chemical procedure in the process is represented by the equation

$$\underset{\text{Glucose}}{C_6H_{12}O_6} + 6O_2 \longrightarrow 6CO_2 + 6H_2O$$

If ethyl alcohol is the compound concerned, the equation is written:

$$\underset{\text{Ethyl alcohol}}{C_2H_5OH} + 3O_2 \longrightarrow 2CO_2 + 3H_2O$$

In some cases, carbon monoxide appears to be a product of respiration. Thus Langdon and Gailey (1920) reported from 1 to 12 per cent of this gas but no carbon dioxide in the hollow pneumatocysts of *Nereocystis luetkeana.*

When plants are placed in an atmosphere devoid of free oxygen, the process of normal respiration is greatly modified so that the end products are carbon dioxide and a number of more or less complex compounds including ethyl alcohol, acetic, formic, oxalic, propionic, other organic acids, and probably molecular hydrogen instead of simply carbon dioxide and water. Under this condition, however, the evolution of carbon dioxide, as a rule, is at a much lower rate than during aerobic respiration. When respiration proceeds after this fashion it is termed "intramolecular" or "anaerobic" respiration, or "fermentation." Under the conditions of anaerobic respiration, the decomposition of the organic compounds may be accomplished in some instances by the rearrangement of the atoms within the molecule without the presence of any oxygen supply. Alcoholic fermentation is one of the best examples of this type of procedure. The process is represented by the equation

$$\underset{\text{Glucose}}{C_6H_{12}O_6} \longrightarrow \underset{\text{Ethyl alcohol}}{2C_2H_5OH} + \underset{\text{Carbon dioxide}}{2CO_2}$$

Fermentation and anaerobic respiration are either the same process, or are very similar. According to Stiles and Leach (1932) this assumption is substantiated by the following facts: (*a*) Hexose sugars are apparently utilized in both processes. (*b*) In both processes carbon dioxide is produced. (*c*) The presence of ethyl alcohol has been demonstrated in many cases of anaerobic respiration. Generally, it is found in smaller amounts than in fermentation, but some of it may be utilized in secondary reactions. (*d*) Zymase, which is concerned in alcoholic fermentation, has been found in various parts of the higher plants. (*e*) Acetaldehyde is probably formed as an intermediate product in both cases. (*f*) Phosphates increase the rate of both reactions.

The action of zymase on sugars, in the production of ethyl alcohol and carbon dioxide, is frequently termed "zymasis."

The partial oxidation of organic compounds under anaerobic conditions may also occur under the influence of oxygen that has been derived from the breaking down of other organic compounds. It is therefore conceivable that the oxygen thus derived from one part of a complex molecule may oxidize even another part of the same molecule. In no case, however, in anaerobic respiration is the oxidation of complex compounds entirely completed to carbon dioxide and water.

6. Relationship of Anaerobic and Aerobic Respiration.—According to Kidd (1917) Rollo as early as 1798 noted that plant tissues, as a vital process, produced carbon dioxide in the absence of free oxygen, but his statement was not accepted for many years afterward. Pfeffer (1881 to 1885) termed this anaerobic production of carbon dioxide "intramolecular respiration," as he believed that it was a normal process in respiration and genetically connected with a subsequent oxidation by the oxygen of the atmosphere under ordinary conditions. An impetus was given to this manner of considering anaerobic respiration when Stocklasa and Czerny (1903) reported the extraction of an enzyme from beet roots, potatoes, peas, and other plant tissues and also from animal tissues, which appeared very similar in its action to the enzyme zymase from yeast, which has the power of converting sugar into ethyl alcohol and carbon dioxide. It was through the extensive studies of Palladin and Kostytschew, however, from 1888 to 1920, and through the suggestions of Barnes (1905) and Peirce (1909, 1912) that the theory of the direct relationship between anaerobic and aerobic respiration came to be generally accepted. The student is referred to Kostytschew (1927) for a detailed review of the literature pertaining to this subject.

It is now generally believed that anaerobic and aerobic respiration are intimately connected in the vital activities of the plant. The present theories of respiration generally assume that anaerobic respiration constitutes a preliminary step in the process. This initial decomposition of the organic materials in the process of respiration is believed to be independent of free oxygen. This may be brought about by a rearrangement of the atoms within the molecule or by the intervention of oxygen derived from other compounds. The end products derived from this initial decomposition are believed to be ethyl alcohol (Palladin and Kostytschew, 1906), organic acids, and carbon dioxide. It is believed by some that ethyl alcohol and carbon dioxide are the principal compounds formed in this initial step (Genevois, 1927; Blackman, 1928). Under normal conditions, however, plants and animals utilize free oxygen to oxidize to carbon dioxide and water, the products produced in anaerobic respiration. If this is true, then anaerobic respiration is proceeding under the conditions of aerobic respiration, but it is not discernible, because its end products are almost immediately oxidized.

Most organisms, however, can live for only a short time by the process of anaerobic respiration and must be supplied with free oxygen if they continue to function normally. The inability to survive may be due to one or both of the following causes: In the first place, the products of anaerobic respiration may be toxic to the protoplasm, so that unless they are oxidized by the free oxygen their accumulation will render the protoplasm completely inactive. The production of blackheart in Irish potatoes is a good example of the detrimental effects that may be initiated by anaerobic conditions (Bennett and Bartholomew, 1924).

Thomas (1929) found that Newton Wonder apples, under anaerobic conditions, sooner or later suffered from "invasive alcohol poisoning," which was associated with, although not necessarily caused by, an accumulation of ethyl alcohol produced in zymasis. He reported later (1931) that apples stored in air at low temperatures above 0°C. are likely to show "low temperature, internal breakdown," and also to develop "soft and deep scald." After these pathologic conditions develop, ethyl alcohol and acetaldehyde progressively accumulate, but there is no evidence that their formation precedes the incidence of these diseases. Zymasis apparently follows many types of tissue injury. Thomas and Fidler (1933) noted that increasing concentrations of free oxygen progressively retarded zymasis. Fidler (1933) stated that senescence in apples was accompanied by a change in the respiratory metabolism of the cells of the fruits. This change caused an accumulation in them of considerable amounts of ethyl alcohol and acetaldehyde. In 1934, he described methods for estimating these two compounds in apples.

In the second place, if sufficient free oxygen is not present, the decomposition processes in the anaerobic stage may reach an equilibrium due to the accumulation of the products thus formed. There may be needed in the process some substance to disturb constantly in one direction or another the equilibrium that tends to be reached, and oxygen appears to be the agent in this regard.

The initial decomposition of the compounds that occur in anaerobic respiration is considered by some to be the process that liberates the energy necessary for the vital activities of the cell. The energy derived from the oxidation of the end products of anaerobic respiration is assumed to be greatly in excess of what is actually needed and is lost by radiation, conduction, and convection into the surrounding medium (Peirce, 1909 to 1912). The amount of energy liberated, however, in anaerobic respiration is small compared with that released in aerobic respiration. Thus in the transformation of a gram molecule of glucose to alcohol and carbon dioxide, 57 kg.-cal. of heat would be liberated. The oxidation of the alcohol thus produced to water and carbon dioxide would liberate 652 kg.-cal. of heat. The complete oxidation of a gram molecule of glucose would consequently give 709 kg.-cal. In order to liberate a given amount of energy, more than twelve times as much glucose must be used in fermentation or anaerobic respiration as would be required in aerobic respiration (see also McGinnis and Taylor, 1923).

The following observations are generally presented as evidence to substantiate the assumption that anaerobic respiration is a normal process of the organism under conditions of a free oxygen supply, and that it is intimately connected with oxygen respiration:

a. All the plants that have thus far been examined are able to produce carbon dioxide in the absence of free oxygen. This behavior is believed to be so general that it cannot be considered simply a biological adaptation to abnormal conditions but must be classed as a vital process of all cells.

b. When plants are transferred from an anaerobic condition to one that has a normal supply of free oxygen, the production of carbon dioxide is markedly increased over that which normally prevails under aerobic conditions. This excess carbon dioxide production is assumed to be due to the oxidation of the accumulated end products of anaerobic respiration.

This phenomenon has been observed by Gustafson (1930) for tomato fruits of all ages, and by Leach and Kent (1934), and Leach (1936) for various seeds. The latter investigators noted that when the seeds were changed from the anaerobic to the aerobic condition, the absorption of oxygen was very rapid. They suggested that under anaerobic conditions some product is formed which has a strong affinity for oxygen, and that it thus becomes oxidized rapidly as soon as free oxygen is supplied.

c. Any changes that are made in the external conditions, excepting the oxygen supply, do not modify the value of the ratio of the carbon dioxide output of anaerobic respiration to that of aerobic respiration. This indicates that the intensity of both types of respiration are influenced by the same factors and to the same extent. Thus Palladin (1894) showed that etiolated bean leaves which were artificially supplied with carbohydrates increased the formation of carbon dioxide in the absence of oxygen in the same proportion as did leaves that were similarly treated and placed under conditions of a normal supply of oxygen. Karlson (1925) found that the effects of the vapors of ether, benzene, and alcohol upon aerobic and anaerobic production of carbon dioxide by wheat seedlings were closely similar.

d. The oxidizing enzymes of plants are unable to attack sugar directly, but according to Kostytschew (1927) they are able to oxidize to carbon dioxide and water the substances contained in fermented solutions.

e. It has been demonstrated by Zaleski (1915), Kostytschew (1908, 1927), and others that green plants can oxidize ethyl alcohol. This is assumed to account for the failure to detect alcohol in the tissues under normal conditions, a fact which has been used as an argument against the theory that anaerobic respiration occurs normally.

It was also found by Kostytschew (1908) that sugar solutions fermented by yeast stimulated the aerobic respiration in plants.

According to Thomas (1925) it is now generally accepted that in the usual respiration of glucose by higher plants, a zymase system cooperates with an oxidation enzyme system to produce carbon dioxide. As long as the oxidation system maintains its normal activity, the zymase activity of cells is not demonstrable. When, however, the oxidation system is inhibited under anaerobic conditions, zymase activity becomes evident by the accumulation of ethyl alcohol. It is considered that acetaldehyde is the precursor of ethyl alcohol in alcoholic fermentation. Gustafson (1934) found acetaldehyde in tomato fruits under all conditions of treatment, but the amount did not increase as the time of anaerobosis lengthened. Ethyl alcohol, however, was found in all tomato fruits under all conditions of treatment, and the amount increased with the length of time the anaerobic conditions continued.

Some investigators, however, including Maquenne and Demoussy (1921), believed that anaerobic and aerobic respiration do not have a common origin but that they proceed from different causes and should both be regarded as autonomous functions as truly in their intimate mechanism as in the influence they exert upon the life of the green plant. A direct proof that anaerobic respiration is the initial step in normal respiration and is thus intimately associated with aerobic respiration is difficult and could be supplied only if it were possible to demonstrate which products of anaerobic metabolism are oxidized in the presence of oxygen. In the oxidation of organic compounds in respiration there are a great many steps between the original compound and the formation of carbon dioxide and water, and the most important products and reactions in the process cannot be measured or detected by the chemical methods at present available.

A few facts should be mentioned here in regard to the relative intensity of anaerobic and aerobic respiration as measured by the production of carbon dioxide. As a general rule, the production of carbon dioxide in anaerobic respiration is less than that under aerobic conditions ranging from 18 to 85 per cent of the latter. The production of carbon dioxide in anaerobic respiration is greater, however, than that which would be produced by the formation of ethyl alcohol and carbon dioxide by fermentation alone, so that some of the carbon dioxide evolved must have its origin from other sources than from the activity of zymase. Some cases have been reported in which the intensity of anaerobic respiration as measured by the evolution of carbon dioxide equaled or exceeded its production in aerobic respiration. Thus Hill (1913) found that the production of carbon dioxide by ripe cherries, blackberries, and grapes for a considerable period of time was as rapid as the aerobic production. The fruits that showed this, however, had apparently finished their growth and were ripe. He found, however, that green peaches and germinating wheat respired more than twice as intensely under aerobic as under anaerobic conditions.

Gustafson (1932) observed that certain of the cacti often produce nearly as much carbon dioxide in the absence of free oxygen as they do in the air.

Edwards (1933) reported that seeds of rice and arum germinate in the almost complete absence of oxygen. The coleoptile may elongate two or three times its original length, owing apparently to the lengthening of cells already formed. The seedlings of rice and arum possess either a type of respiratory metabolism that produces nontoxic intermediary products when complete oxidation is prevented or a greater tolerance to toxic products than other plants.

Emerson (1927) demonstrated the presence of two types of respiration in *Chlorella* and other green algae. The first is the typical respiration during the course of which a photosynthetic product is consumed. The rate of this type of respiration is slightly increased by cyanide. The second type is apparent when glucose is added to the suspending medium and the rate of respiration doubles. The mechanism responsible for this increase differs from that concerned with typical respiration in that it may be inhibited by cyanide.

7. Aerobic Respiration and Photosynthesis Contrasted.—Jean Ingen-Housz (1774) noted that green plants in darkness rendered the air impure just as did animals and that in daylight the fruits, flowers, and roots acted in the same manner. He was thus apparently the first to define in this general manner the conditions under which plants carry on respiration, although he did not recognize the significance of his observations. It remained for De Saussure (1804) to show that this absorption of oxygen and the formation of carbon dioxide were stages in the respiration of plants, and that the respiration of plants was identical with the respiration of animals. He also recognized the formation of water in the process and was one of the first to explain the relation of respiration in growth and the other vital activities of the plant.

For a long time, however, the processes of respiration and photosynthesis were not clearly defined, and it was considered that they did not occur concurrently in the same cells. It was argued that it was impossible that two processes which were direct opposite could occur in the cells at the same time. This erroneous idea still persists, and it is frequently stated that one of the marked differences between green plants and animals is that the plants take up carbon dioxide and give off oxygen while animals absorb oxygen and eliminate carbon dioxide. This statement is only partially correct and leaves the general impression that green plants do not carry on respiration. The best way in which to differentiate clearly between photosynthesis and respiration is to contrast them in regard to the time and place of their occurrence, the compounds used, the end products formed, and the relation of these two processes to the energy supply of the plant. These points may be contrasted after the following manner:

Photosynthesis	Respiration
1. Photosynthesis occurs only in the chlorophyllous cells of plants.	1. Respiration occurs in every living cell regardless of its color and regardless of whether it is of plant or animal nature.
2. Photosynthesis occurs only in the sunlight or under artificial illumination.	2. Respiration occurs all the time. It occurs in darkness as well as in light.
3. In the process of photosynthesis water and carbon dioxide are used.	3. In the process of respiration water and carbon dioxide are the end or waste products.
4. In the process of photosynthesis oxygen is given off as a waste product.	4. In the process of respiration oxygen is used.
5. Food is built up in the process of photosynthesis.	5. Food is destroyed in the process of respiration.

Photosynthesis	Respiration
6. Photosynthesis increases the weight of a plant.	6. Respiration decreases the weight of a plant.
7. Energy is stored in the process of photosynthesis.	7. Energy is released in the process of respiration.
8. The chemical equation of photosynthesis is written:	8. The chemical equation of respiration is written:
$6CO_2 + 6H_2O \rightarrow C_6H_{12}O_6 + 6O_2$	$C_6H_{12}O_6 + 6O_2 \rightarrow 6CO_2 + 6H_2O$

8. Respiration and Breathing.—In the preceding topics, the process of respiration in plants has been discussed in regard to the place of its occurrence, the materials used, and the waste products produced. It is the intention here to describe the means by which the oxygen is supplied to the cells for use in respiration and the manner in which the waste products are eliminated from the plant body. In discussing these processes, it is necessary to distinguish between the terms "respiration" and "breathing." These two terms are commonly used synonymously, but from a biological standpoint each has a distinct meaning and the difference between them should be clearly fixed. Breathing is the forcible inhaling and exhaling of air by the animal body, while respiration is an energy-releasing process that occurs in every living cell. In the higher animals there are special organs—the nose, windpipe, lungs, and certain muscles—that are employed to draw the air into the body and to force it out again. The air entering the body contains the oxygen supply needed in respiration, while the air leaving the body contains the carbon dioxide and water that are the waste products of the process. After the air has entered the lungs, a portion of its oxygen is taken up by the blood stream and carried to the cells of the body. This oxygen is taken up by the cells while the carbon dioxide and water formed in the cells as a result of respiration diffuse into the blood stream and are carried primarily to the lungs, where they are eliminated with the exhaled air.

Plants, in contrast with animals, do not have any special organs for drawing air into the body and forcing it out again. The air simply diffuses in and out through the stomata, lenticels, and cuticle of the younger plant parts. After entering the plant, it circulates through the intercellular spaces in the plant and possibly to a considerable degree in solution in the water movements of the plant. As the air thus comes in contact with the cells of the plant, they absorb the oxygen therefrom and give up the carbon dioxide thereto that has been formed in respiration. The entrance and exit of these gases from the cells are due primarily to the difference in gradient that exists between the respective gases in the cells and in the surrounding air. The oxygen exterior to the plant diffuses inwardly through the lenticels, stomata, and intercellular spaces to restore the oxygen equilibrium that has been disturbed by the absorption of oxygen by the cells. Likewise, the carbon dioxide diffuses outwardly through these openings, since its concentration within the intercellular spaces is higher than that of the surrounding air. Plants thus differ markedly from animals in regard to the manner in which they obtain their air supply and by which they eliminate the end products of respiration into the external atmosphere. The general statement can thus be made that animals breathe but plants do not. The student should keep clearly in mind that respiration is a chemical process which goes on continuously in every living cell whether of plant or animal nature, while breathing is a mechanical process that supplies air to and eliminates it from the animal body but does not occur in plants at all.

9. Methods of Measuring the Intensity of Aerobic Respiration.—The intensity of aerobic respiration in plants may be measured by the loss in weight, the amount of oxygen absorbed, or the amount of carbon dioxide produced by the plant or plant part concerned. A detailed description of the various methods is beyond the scope of this work, and the student is referred to Harrington and Crocker (1923) and Kostytschew (1927) for references and more complete information until 1927. More recently, methods for studying respiration have been reported by Hottes and Hafenrichter (1928); Harding, Maney, and Plagge (1929); Maney, Harding, and Plagge (1929); Emerson (1930); Stiles and Leach (1931); Bennet-Clark (1932); Tang (1932); Wynd (1932, 1935); Leach (1932); Balch (1933); Jeffrey and Cruess (1933), and Mitchell (1935).

It is the intention here to describe in detail only two of these methods and to describe only the general procedure to be followed in others. Most of the methods that have been devised for determining the rate of respiration are based on the measurement of the carbon dioxide that is produced. These methods may be grouped as follows according to the principles involved in their manipulation:

a. One of the oldest and most generally used types of procedure is to allow the carbon dioxide that is respired to diffuse into a stream of carbon dioxide free air, which is constantly being drawn through the apparatus containing the respiring material. This air then passes through a solution of alkali which depletes it of the carbon dioxide that it thus acquires. The amount of carbon dioxide may then be determined by titration, by the increased weight of the absorbent, or by the weight of the precipitate, provided an insoluble carbonate is formed. Apparatus of this type have been devised or modified by Pfeffer (1885), Kostytschew (1908), Gurjar (1917), Ginsburg (1925), and Beaumont, Willaman, and DeLong (1927). For relatively large amounts of actively respiring materials, methods of this type are excellent for measuring the carbon dioxide produced. They have the advantage in that the air is constantly renewed over the respiring material so the disturbing effects of the changes in the partial pressures of carbon dioxide and oxygen that occur in closed containers are avoided. The methods of this type, however, neglect entirely the question of the absorption of oxygen, which is not always proportional to the amount of carbon dioxide given off. They are not adapted for use with limited quantities of small or inactive material, as the experimental error is relatively large, so the amount of carbon dioxide given off in a reasonable time may fall within the limit of error.

One of the most common methods of the above type which can easily be used by students in the laboratory is conducted after the following manner: The apparatus, which consists of a connected series of closed containers, U-shaped calcium chloride tubes, and potash tubes, is attached at one end to a filter pump, by means of which a continuous stream of air is drawn through the apparatus. As the air enters, it is drawn through a concentrated solution of sodium or potassium hydroxide and thence through a solution of barium hydroxide, which will indicate whether or not the entering air has been deprived of its carbon dioxide. The carbon dioxide-free air now passes over the respiring material, which is in a closed container, and thence through calcium chloride tubes, which deplete it of moisture. The moisture-free air is thence drawn into weighed potash bulbs containing a concentrated solution of sodium or potassium hydroxide where the carbon dioxide is absorbed. The air that has absorbed moisture from the alkali solution next passes through weighed calcium chloride tubes where this moisture is again absorbed. After the experiment has proceeded the required time, the alkali tubes and the last two calcium chloride tubes are weighed. The initial weight of the alkali tubes is subtracted from their final weight plus the increased weight of the two calcium chloride tubes. This difference in weight represents the weight of the carbon dioxide that has been eliminated by the respiring material during the period of the experiment.

b. Numerous types of microrespirometers have been devised for the detection of very slight changes in the composition of the gases under consideration. Instruments of this type have been devised and used by Tashiro (1913), Osterhout and Haas (1917), Lund (1919), and others. Some of these methods are too complicated for general use and are limited in regard to the quantity of material that may be used. None of them provides for the simultaneous determination of both sides of the respiratory exchange in the same material.

c. An indicator method has been perfected by Haas (1916) and Osterhout (1918) for the detection of small quantities of carbon dioxide respired. The method has the advantage of sufficient sensitiveness so that the experimental period can be very short. The method is also useful for the rapid and simultaneous determination in several different experiments. By a special arrangement the apparatus can be used to determine approximately the amounts of oxygen absorbed. Indicator methods for the determination of the oxygen absorbed by respiring material have been described by Osterhout (1918). The use of the indicator method requires care to avoid errors due to buffer effects or to the production of acid-reacting substances other than carbon dioxide or of alkaline-reacting substances (Harrington and Crocker, 1923).

d. A method that has been used to a considerable extent is to place the respiring material in a closed container and occasionally sample the air for analysis in some form of microapparatus. This method is described in detail by Thoday (1913) by whom it was used with success.

e. The absorption by a solution of alkali of the carbon dioxide respired in a closed container with manometric measurement of the oxygen absorbed has been used by Stich (1891) and others. This type of apparatus gives both the oxygen consumption and the carbon dioxide produced. The method, however, has a disadvantage, since the carbon dioxide is absorbed as rapidly as it is produced and thus does not compensate for the oxygen absorbed by the respiring material. This fact causes relatively large changes in pressure within the apparatus, which might affect the behavior of the plant in the experiment.

An apparatus for the measurement of the gaseous exchange in the respiration of seeds and small parts of plants, devised by Harrington and Crocker (1923) and later modified by Davis (1925), is shown in Fig. 35. This instrument, based upon the principle of the respirometer of Ganong (1908), is one of the most accurate and easy to operate of the many respirometers that have been devised. With this apparatus, the oxygen consumption and the carbon dioxide evolved are determined in the same apparatus and for the same period of time, using the whole volume of air instead of a sample. This apparatus also has an advantage in that the carbon dioxde given off by the respiring material accumulates and tends to compensate for the oxygen absorbed so no large changes in pressure occur during the experimental period.

This apparatus consists of a cylindrical glass tube *c* in the form of a bottle with an extension *c′*, which serves as a receptacle for the seeds or material to be studied. Into the mouth of this cylinder is fitted a ground-glass stopper *s*, upon which are mounted a short open manometer *m* with a 2-mm. bore and an inlet tube provided with a stop cock *sc* and a small chamber *ch*. The stopper fits deep into the neck of the cylinder so that a mercury seal *mc* can be made. The apparatus has been used in three sizes of about 20, 40, and 80 cc. capacity. Before conducting an experiment, the combined capacity of the corked cylinder *c* including the seed chamber *c′* and the pores of the manometer to the level of the mercury and of the intake tube to the stop cock *sc* must be determined. This is found by measuring the volume of water necessary to fill these parts of the apparatus.

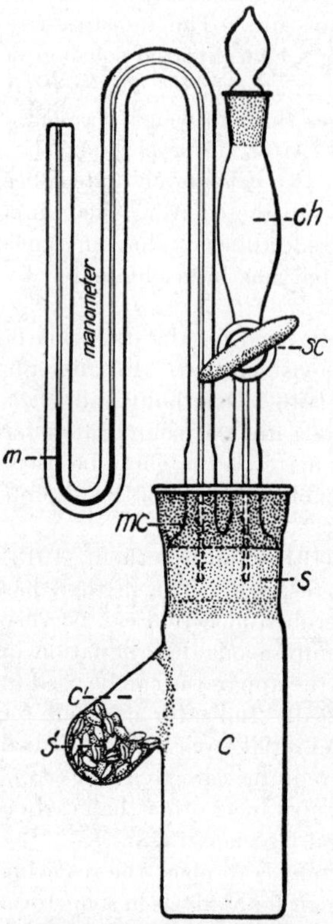

FIG. 35.—Diagram of respirometer. *c*, chamber of respirometer. *c′*, seed chamber. *s′*, seeds. *s*, ground-glass stopper with attached manometer and supply tube provided with a small chamber *ch* and a stop cock, *sc*. *mc*, mercury seal. *m*, mercury. (*After Davis*, 1925.)

After the volume of the instrument has been determined, the soaked seeds *s′* whose volume and weight have been obtained are placed in the seed chamber *c′*. The glass stopper is then inserted in the cylinder, the stop cock *sc* and cork of the intake tube being left open to equalize the pressure, and the mercury seal made. The apparatus is then placed in any convenient thermostat, preferably a water bath, and as soon as the instrument has come to the desired temperature the stop cock *sc* in the intake tube is closed and the chamber *ch* stoppered. Care must be taken that the respirometer stands vertically and that the mercury in the manometer *m* is at the same level in the two arms. The experiment is now ready to proceed, and the temperature of the medium surrounding the respirometer is taken, and the barometric pressure is recorded.

After the experiment has proceeded for the desired period, which is generally from 24 to 48 hr., from 1 to 3 cc., as required, of 20 per cent sodium hydroxide is measured into the chamber *ch* of the intake tube, which is then left unstoppered. The manometer reading is then taken by holding a millimeter scale behind the arm of the manometer. The stop cock *sc* is now opened and the alkali carefully admitted into the respiratory cylinder *c*, care being taken that no air escapes from the cylinder. The temperature of the medium surrounding the respirometer and the reading of the barometer are recorded. After absorption is completed, the manometer reading is again determined.

From the manometer and barometric readings, the temperature, and the volume of the apparatus corrected for the volumes of the respiratory material and of the alkali admitted, the three significant volumes of air under standard conditions (0°C., 760 mm.) can be determined. From these three volumes, the volume of oxygen absorbed and the volume of carbon dioxide given off may be calculated after the following manner (Sherman, 1921; Harrington and Crocker, 1923):

V_r = volume of the respirometer
V_s = volume of the imbibed seeds
V_a = volume of the absorbent alkali
T_1 = initial absolute temperature
T_2 = final absolute temperature
B_1 = initial barometric pressure
B_2 = final barometric pressure
m_1 = initial manometer reading
m_2 = final manometer reading

$$A.\ (V_r - V_s)\frac{273}{T_1} \cdot \frac{B_1}{760}$$

$$B.\ (V_r - V_s)\frac{273}{T_2} \cdot \frac{B_2 \pm m_1}{760}$$

$$C.\ (V_r - V_s - V_a)\frac{273}{T_2} \cdot \frac{B_2 \pm m_2}{760}$$

$B - C$ = the volume of carbon dioxide eliminated
$A - C$ = the volume of oxygen absorbed

From this the respiratory quotient CO_2/O_2 is easily calculated. The number of milligrams of carbon dioxide eliminated and of oxygen absorbed per gram of imbibed weight for any given period can also be determined; 1 cc. of carbon dioxide weighs 1.96 mg. and 1 cc. of oxygen weighs 1.428 mg. at 0°C. and 760 mm. barometric pressure.

Stiles and Leach (1931), and Leach (1932) described an instrument known as a "katharometer," which they developed for the measurement of small amounts of respiration. The principles involved in the instrument are: A wire becomes heated when an electric current is passed through it. The temperature of this wire is determined by the thermal conductivity of the medium surrounding it. The thermal conductivities of different gases are different, so that if in the respiration chambers oxygen is replaced by carbon dioxide there is a change in the thermal conductivity of the gases in the vessel, and hence in the temperature and electrical resistance of the wire. The output of carbon dioxide is thus followed by determining the changes in electrical resistance of a heated wire in the respiratory chamber. This method is sufficiently sensitive to detect a change of 0.001 per cent in volume of carbon dioxide. However, it is only applicable when the concentrations of oxygen and carbon dioxide are changing.

10. The Respiratory Quotient.—The ratio of the volume of carbon dioxide given off to the volume of oxygen absorbed during respiration is termed the "respiratory quotient" and is written CO_2/O_2. Great diversity of opinion, however, exists as to the importance to be attached to the value of this ratio as an index to the reactions occurring in respiration.

a. Factors Influencing.—A discussion of the factors that may cause the fluctuation of this ratio will serve to show some of the conditions prevailing in the tissues that may be indicated by its changes in value.

1. *Kind of Material.*—The types of materials consumed in respiration will influence the value of the respiratory quotient. Thus under the ordinary aerobic conditions of respiration, if carbohydrates are the

compounds utilized in the process and if all the oxygen absorbed by the plant is consumed in their combustion, the volume of oxygen absorbed would be equal to the volume of carbon dioxide given off according to the equation:

$$\underset{\text{Glucose}}{C_6H_{12}O_6} + 6O_2 \longrightarrow 6CO_2 + 6H_2O$$

Thus the respiratory quotient would be unity. If organic acids that are relatively rich in oxygen were the materials utilized, the respiratory quotient would be greater than unity. Thus the respiratory quotient for the complete oxidation of tartaric acid would be 1.6, while that of oxalic acid would be 4.0. If fats are the materials utilized in respiration, the respiratory quotient is less than unity, since most of these compounds are relatively low in oxygen. The fats are relatively low in oxygen and generally absorb oxygen to satisfy their unsaturated condition before their disintegration is initiated in the process of respiration. If the fats are converted into sugar before they are ultimately used in respiration, this transformation requires a quantity of oxygen which is not indicated by the quantity of carbon dioxide produced when the sugar thus formed is oxidized. This accounts for the low respiratory quotient of plant parts which contain fats as reserve foods. During the ripening of oily seeds the respiratory quotient is also greater than 1. As mentioned in Chap. X, the fats in oily seeds are formed at the expense of carbohydrates. In the transformation of carbohydrates to fats a certain amount of oxygen is set free. The oxygen that has thus been freed is utilized directly in respiration, and consequently a smaller amount is absorbed from the exterior than would otherwise occur. Consequently the volume of carbon dioxide produced is greater than the volume of oxygen absorbed, and the respiratory quotient is greater than unity.

2. *Temperature.*—The respiratory quotient may also vary according to the temperature at which the observations are made. This has been explained by the fact that both the absorption of oxygen and the exhalation of carbon dioxide vary with differences of temperature, but they vary in different degrees. Harrington (1923) noted in respiratory studies of dormant apple seeds that the respiratory quotient increased with an increase in temperature. He believed that this was due to an impoverishment in easily oxidizable substances and might also be indicative of an oxygen deficiency in the respiring tissues. The respiratory quotient decreased with a decreasing temperature. This indicated a storage of oxygen that became very considerable at 10 and 0°C. In the respiration of these seeds the temperature coefficient was affected by the previous treatment. It was higher after treatments that tended toward after-ripening and lower after treatments that induced deeper dormancy. The influence of external and internal conditions upon the respiratory quotient

is, however, but little understood, and no definite statements can be made in that regard.

Tang (1931, 1932) reported that the temperature characteristics for the production of carbon dioxide by *Lupinus albus* are different from those for the consumption of oxygen, so that the respiratory quotient changes accordingly with temperature.

3. *Organic Acids.*—A portion of the oxygen that is absorbed by the plant may be utilized for purposes other than that of respiration. This extra oxygen may be used in the formation of organic acids. The depression of the value of CO_2/O_2 due to this factor is noticed in the storage of organic acids in ripening fruits and in succulents (Richards, 1915).

According to Bennet-Clark (1933) it has long been known that some members of the *Crassulaceae* possess the peculiarity of producing a large quantity of an acid, now known to be malic acid, when they are placed in darkness. Since the quantities of carbon dioxide that they produce are very small in comparison with the quantity of malic acid formed, the view became established that this formation of acid was an abnormal type of respiration—a respiration in which a product is formed that is less highly oxidized than carbon dioxide, the normal end product of complete respiration. Bennet-Clark considered that no more than 1 molecule of malic acid is formed from each molecule of sugar, so that 2 or 3 atoms of carbon are converted into other substances that appear to be polysaccharoses. He believed that the carbon dioxide which is liberated must originate from some intermediate product that arises in the conversion of sugar into malic acid.

When succulent plants are placed in darkness, the amount of oxygen absorbed in the process of respiration is in excess of the carbon dioxide evolved, and in extreme cases the respiratory quotient may approach zero (Aubert, 1892). After the plants have been in darkness for a time, the accumulation of organic acids decreases, and the rate of evolution of carbon dioxide gradually increases with a corresponding increase in the value of the respiratory quotient, which, however, does not reach unity. When these succulent plants are exposed to sunlight the acids are decomposed and carbon dioxide is freed. Some of this carbon dioxide is used in photosynthesis. This procedure is considered to be an advantage to fleshy plants because they are handicapped by the relatively slow diffusion of gases through their tissues. The respiratory quotient under these conditions greatly exceeds unity, and, according to Bennet-Clark (1932), maximal values of 1.39 and 2.05 have been observed in *Sedum praealtum.* He believed that these high values are inconsistent with the view that the disappearance of malic acid is due to oxidation, and that this procedure strengthens the view that malic acid forms a link in the carbon cycle in succulent plants other than that of the oxidation of carbon dioxide.

4. *Other Factors.*—Carbon dioxide may be formed without the absorption of oxygen. This occurs under anaerobic conditions, as has been mentioned previously. The same effects may occur, but to a lesser degree if the supply of oxygen is rather limited. Under these conditions the respiratory quotient is greater than unity.

Plants with leaves colored red by anthocyanin behave similarly to succulents in respiration. The respiratory quotient is lower in red than in green leaves, and this is apparently due to a more rapid absorption of oxygen by the red leaves. This difference in the value of the quotients can be related to a greater accumulation of organic acids in the leaves containing anthocyanin than in those from which this pigment is absent.

b. Germination of Seeds.—The respiratory quotient of a large number of different types of germinating seeds has been studied by Godlewski (1882), Bonnier and Mangin (1884), Gerber (1900), Sherman (1921), Harrington (1923), Ermakoff and Iwanoff (1931), Stiles and Leach (1932, 1933), and Dastur and Desai (1935). If carbohydrates are the only reserves present and the supply of free oxygen is unlimited, the respiratory quotient will approximate unity. If fats and oils are the only foods present in the seed and a supply of free oxygen is readily available, the respiratory quotient will be considerably less than unity until these foods have been converted into sugars. In practically all seeds that have been examined, however, the respiratory quotient during the first few hours of germination is unity or greater than unity. As germination proceeds, the value of the quotient falls. In seeds containing fats and oils the fall is very marked and continues for a considerable time, after which it rises and approaches unity.

These fluctuations in the value of the respiratory quotient may be explained for the most part upon the basis of the nature of the food available for respiration and upon the structure of the seed coats of the seed. Let us consider first the influence of the types of food. Practically all seeds, whether of a starchy or an oily nature, contain a small reserve of sugars that serve the seedling in the initial stage of germination, so that while this is being utilized the respiratory quotient would be approximately unity. Even starchy seeds contain more or less oil, which is generally concentrated in the embryo. If, as would be expected, this oil is drawn upon during very early germination, the large use of oxygen in its conversion to sugar would give a temporarily low value to the respiratory quotient of such seeds. In the germination of oily seeds, the value of the respiratory quotient will be near that for complete oxidation (approximately 0.70), provided that the stored fats are easily oxidized. If these fats, however, are difficult of oxidation, the respiratory quotient will be low and near to that for the conversion of fats to sugar.

After the fats and oil in seed have been converted into carbohydrates, the value of the respiratory quotient approaches unity.

The structure of the seed coats may influence the respiratory quotient after the following manner: In the initial phase of germination, the presence of the unbroken testa around the seed prevents the free diffusion of gases, so that a relatively small amount of oxygen enters, and anaerobic respiration occurs. The rupture of the seed coats causes a quick rise in the absorption of free oxygen and a less rapid rise in the output of carbon dioxide, so that the value of the respiratory quotient falls.

Some observations have been made on the value of the respiratory quotient of other plant parts. Thus Cerighelli (1924) reported that the respiratory quotient of the roots of wheat, sunflower, corn, peas, buckwheat, rhubarb, and others at the beginning of development was always less than unity. In annuals low in stored foods, it remained less than unity during growth, while in those which accumulated reserve foods it became greater than unity.

According to Maquenne and Demoussy (1913), the respiratory quotients of the leaves examined by them were begonia, 1.11; lilac, lily, maize, and pea, 1.07; pear, 1.10; tobacco and wheat, 1.03; and rhubarb, 1.02.

11. Respiration and the Production of Heat.—It is a familiar fact that respiration produces heat in the animal body, but that this occurs in plants is not known so generally. The heat produced by respiration in plants is quickly dissipated into the surrounding air, primarily because they have no means of controlling its loss and because they have a large surface in proportion to their mass. Fleshy fruits, bulbs, and tubers may, however, retain sufficient heat to raise their temperature slightly above their surroundings. Thus ripening bananas maintain themselves at temperatures distinctly above that of their surroundings, while Richards (1897) stated that the temperature of the Irish potato was 0.17°C. and that of the onion 0.5°C. higher than the surrounding air. Wright and Whiteman (1931) noted that the internal temperature of cases of holly stored at 50°F. was from 2 to 6°F. higher than the surrounding air. The production of heat by respiration in plants, however, cannot readily be observed unless the proper precautions are taken to prevent its loss.

The method that is now the most commonly used for demonstrations in laboratory studies was developed and perfected independently by Ganong (1908) and Peirce (1908). The general procedure is to place the plant parts to be studied in Dewar flasks or thermos bottles, insert a thermometer, carefully plug with cotton, insulate the containers as thoroughly as possible, and note the changes in temperature as indicated by the thermometer. This experiment, however, is not so easily performed as would at first appear, since, to obtain reliable results, certain conditions, of which some are difficult to obtain, must be satisfied. According to Potter (1917), some of these conditions are (*a*) a sufficiently large amount of material, (*b*) a supply of oxygen, (*c*) a means for the disposal of the carbon dioxide produced, and (*d*) efficient control for the prevention of putrefaction and fungal growth. This last condition is the most difficult to obtain in this experiment. The heat equivalent of the material under consideration, of the water used, and of the container must also be known and taken into consideration if accurate results are to be obtained.

The results obtained by Peirce (1912) with peas are commonly cited as illustrations of the liberation of heat by plants in respiration. He placed 75 g. of air-dry peas that had been rinsed with a solution of mercuric chloride in a Dewar flask, added 100 cc. of sterilized distilled water, and noted the changes in temperature. The temperature of the flask rose from 18°C. at the beginning of the experiment to 38°C. at the end of 5 days, after which it remained constant. In the 7 days this quantity of peas liberated 2,593 cal. or an average of 370 cal. per day. In similar experiments Darsie, Elliott, and Peirce (1914) determined the average daily increase in temperature of several kinds of seeds due to respiration, when infection from bacteria and fungi was prevented. They found that the increase in temperature to be expected from 10 g. of viable seeds was 1.82°C. for hemp, 0.80° for barley, 0.75° for clover, 0.73° for wheat, 0.55° for oats, and 0.49° for corn. Their experiments indicated that the quantity of heat liberated in respiration when they were placed under conditions favorable for germination gave some indication of the quality of the seeds, of their viability, and of the vigor of their growth immediately following germination.

The question arises as to the value of the heat liberated in respiration to the plant. Peirce (1909, 1912) was of the opinion that although the essential product of respiration is energy, it is the energy that is immediately converted into work which is utilized by the plant in its vital activities. If the aerobic phase of respiration is merely considered a process of purification whereby useless or injurious substances are converted into forms which can be eliminated, then the heat thus liberated may be considered as of no special benefit to the plant. It may be considered as a waste product of respiration just as carbon dioxide and water.

The heating of moist grain when placed in storage is due primarily to the heat generated in the respiration of the cells of the seeds. A high temperature and moisture content are two factors that especially stimulate respiration, as will be discussed in subsequent topics. In this self-heating of grains, respiration may raise the temperature to a point where other chemical reactions may occur (Gore, 1911). Bacterial action may also play a part under these conditions. When grain with a high moisture content is stored, the heat generated from respiration cannot readily be dissipated into the atmosphere, so that it accumulates and raises the temperature to a point where it is injurious to the seed and where it may even cause the disintegration of the tissues. Hot weather accelerates the heating of stored moist grain in two ways. It affects it biochemically by increasing the rate of enzyme action and physically by reducing the rate of conduction of heat into the atmosphere. Bailey (1918) found that all the normally plump spring wheat that heated contained over 14.3 per cent of moisture. He observed that sound, plump, hard wheat containing less than 14.5 per cent of moisture would keep without heating in storage in a temperate climate. A lower moisture content must be employed, however, in storing shriveled or frosted wheat and possibly also with plump wheat in a mild climate.

It was observed by Gilman and Barron (1930) that the rise in temperature due to the germination of oats, wheat, and barley was 6.8, 2.6, and 3.6°C., respectively. The presence of *Aspergillus flavus* on these grains, however, raised their temperature (at 18 per cent moisture) to 26, 7.4, and 16°C., respectively, while *Aspergillus niger* had approximately the same effect. It was found by Larmour, Clayton, and Wrenshall (1935) with red spring wheat that when the action of fungi is prevented by the vapor of toluene or carbon tetrachloride, the production of carbon dioxide occurs at a slow rate, and no heating occurs even when the moisture content is raised to 25 per cent.

Bakke and Noecker (1933) found with oats that there is a general rise in the consumption of oxygen as the moisture content increased. The optimum moisture content for heat production is between 24 and 40 per cent, depending on aeration. Molds and microorganisms are almost always associated with grain in storage. When the water content is below 15 per cent, under conditions of farm storage, respiration is so low that the little amount of heat produced is rapidly radiated. The optimum temperature for heat production by moist seeds is between 30 and 35°C. Swanson (1934) observed that wheat which was stored at 60°F. produced very little growth of mold when the moisture amounted to 20 per cent and none when it was 18 per cent or lower. However, if the grain was placed at 95°F. some growth of mold was observed when the moisture content was as low as 14 per cent. Mold growth may be inhibited by the exclusion of air.

The question of the spontaneous heating and ignition of hay has long been discussed and many conflicting hypotheses have been proposed to account for these reactions. Hoffman (1935) stated it is commonly accepted that the production of heat in a mass of uncured or wet hay at temperatures up to 70°C., or higher, is due mainly to the process of respiration of the living cells of the plant and of the microorganisms present. Chemical reactions are considered to account for increases in temperature above which microorganisms are inactive. He found in the heating haymow that along with the operation of biological agencies there occurs a purely chemical oxidation as evidenced by a loss of oxygen considerably in excess of the carbon dioxide formed. This chemical oxidation is more marked beyond the temperature range usually ascribed to the activity of microorganisms.

12. Factors Affecting Respiration. *a. Temperature.*—It is apparently well established that the rate of respiration increases with the temperature, until the latter begins to injure the tissues. Clausen (1890) found the temperature coefficient for the respiration of germinating wheat to be 2.86 between 0 and 10°C., and 1.09 between 30 and 40°C., with an average value of 2.71 between 0 and 40°C.

Gerhart (1930) stated that the temperature coefficient for respiration in the fruit of the strawberry was 2.5 for temperatures below 25°C. Bělehrádek and Bělehrádková (1929) found that the temperature coefficient of respiration first increased and later decreased with the age of the tissues. They attributed this to the change with age of the colloidal nature of the protoplasm. This modifies the rate of diffusion in the reacting protoplasmic phase. Matthaei (1905) observed that at 5.8°C. 2 g. of leaves respired 0.1 mg. of carbon dioxide per hour, while at 33.1°C. the rate was 1.35 mg. per hour. Gore (1911) obtained an average temperature coefficient of 2.38 in 49 respiration experiments with 40 different fruits. Olney (1926) noted that the rate of respiration in ripening bananas at 12°C. was only one-third of that at 20°C. Bigelow, Gore, and Howard (1905) noted that the respiration of apples was twice as high in cellar storage as in cold storage. Morse (1908) studied the intensity of respiration of apples at different temperatures by determining the amount of carbon dioxide evolved. Some of his results are shown in the following table:

MILLIGRAMS OF CARBON DIOXIDE EXHALED PER KILO OF APPLES PER HOUR

Experiment	0°C.	5° to 10°C.	18 to 25°C.	Experiment	0°C.	5 to 10°C.	18 to 25°C.
1	2.3	8.0	16.4	4	2.8	8.8	18.2
2	2.6	7.3	18.7	5	2.2	7.9	18.0
3	3.8	8.7	12.6	6	5.2	13.2	21.9

The effect of the temperature upon the rate of respiration of Wagener apples directly after picking was determined by Burroughs (1922). At

0°C. the carbon dioxide eliminated amounted to only 3.1 mg. per kilogram of apples per hour as compared with 7.5, 12, and 24.4 mg. per kilogram per hour, respectively, for temperatures of 9, 20, and 30°C. Hopkins (1924) noted that the carbon dioxide eliminated by Cobbler potatoes at 0, 4.4, 7.2, and 10.0°C. was, respectively, 5.0, 5.1, 6.0, and 8.0 mg. per kilogram of weight per hour. It was observed by Drain (1926) that apples of different varieties showed marked differences in their response in the respiration rate to temperature changes. Thus 6 days after picking, the carbon dioxide produced by Maiden Blush apples was 280 mg. per kilogram per hour at 25°C. as compared with 46.8 mg. produced by those at 0°C. Apples of the Oldenburg variety, however, produced 55.8 mg. of carbon dioxide per kilogram of weight at 25°C. as compared with 39.3 mg. at 0°C. The localization of the oxidase activity in the different varieties is of interest in regard to the intensity of respiration. The Maiden Blush variety shows this localization in the periphery, which suggests that the localization may be influenced by gas exchange. Many apple varieties show greater oxidase activity near the core, which suggests that they have a gas exchange through the calyx tube and core.

Bushnell (1925) found that the rate of respiration in the aerial parts of the potato increased with temperature during the night. The reduction in the amount of carbohydrate that resulted from increased respiration at higher temperatures roughly corresponded to the reduction in tuber growth under these conditions. Kertesz (1933) showed in green peas that the inactivation of respiration with water at 80, 90, and 100°C. was slow after 90 per cent inactivation had been reached. A complete inactivation was obtained by a treatment at 100°C. for 2 min., while Mack (1930) found in germinating wheat that the rate of carbon dioxide production increased with a rise in temperature. Harding (1935) noted that temperature was the outstanding factor influencing the intensity of respiration in stored Grimes Golden apples. Werner (1935) gives a complete review of the literature on the physiological reactions of potatoes in storage.

The composition of the air contained in the intercellular spaces of apples, potatoes, and carrots stored at different temperatures was determined by Magness (1920). Some of his results are shown in the table on page 957.

These figures show that marked variations may occur in the composition of the gaseous content in tissues under varying conditions of temperature. It is to be noted that at the lower temperatures the sum of the percentages of oxygen and carbon dioxide closely approximates their total percentage in normal air. As the temperature increases, however, the sum of these gases increases, which indicates that at higher temperatures more than 1 molecule of carbon dioxide is liberated for each molecule of oxygen absorbed. At the higher temperatures it is probable that some anaerobic respiration occurred, owing to the relatively small amount of oxygen present. In the tissues of ripening Bartlett pears, Magness and Ballard (1926) also found that the

Storage temperature, degrees Centigrade	Percentage of		
	Carbon dioxide	Oxygen	Nitrogen
Apples			
2	6.7	14.2	79.1
6	8.4	12.9	78.7
11	12.2	10.7	77.1
20	17.2	5.5	77.3
30	21.4	3.2	75.4
Potatoes			
11	19.6	10.9	
22	34.4	5.7	
Carrots			
11	12.2	13.1	
24	28.6	5.2	

concentration of carbon dioxide was higher and that of oxygen lower in the fruit that was held at higher temperatures than in that held in cold storage. The high concentration of carbon dioxide and the correspondingly low oxygen content in this fruit ripening at a temperature of about 15°C. appear to be associated with the development of high aroma and flavor. The main factors determining the amount of carbon dioxide and oxygen in the intercellular spaces are (1) the rate of oxidation or the rate at which oxygen is taken up from and carbon dioxide given off into the intercellular spaces, (2) the permeability of the skin or epidermal coverings to carbon dioxide and oxygen, and (3) the difference in the pressure of carbon dioxide and oxygen within the tissue, which determines the rate of gaseous exchange when the permeability factor is a constant.

Rigg and Henry (1935) stated that there are three possible sources of the gases contained in the internal cavities of marine algae: (1) atmospheric air, (2) gases dissolved in sea water, and (3) the respiratory metabolism of the plant.

It was stated by Willaman and Brown (1930) that the amount of carbon dioxide in apple twigs varied directly with temperature. This amount may be as high as 360 mg. per kilogram of twigs at −22°C., and from 150 to 250 mg. per the same weight at 0°C. Dowd (1933) noted that the percentage of carbon dioxide in the internal atmosphere of apples tended to decrease during the growing period. Kertesz (1934) observed that the content of carbon dioxide of peas remained at approximately 1.6 per cent under general conditions. However, when the pods were sealed with paraffin and stored at room temperature, the content of this gas increased to 43 to 50 per cent. This indicates that ordinarily there is a very vigorous exchange of gases through the intact pods.

Besides the direct effect of temperature upon respiration, there is evidence to indicate that there is a stimulating effect upon the process caused by temperature changes. Thus Müller-Thurgau (1882) noted that respiration was much more intense in potatoes which had been kept at 0°C. for some time and then returned to the original temperature than in those tubers which had not been so treated. Similar results were observed by Palladin (1899) with bean seedlings and by Burroughs (1922) for freshly picked apples. The latter investigator noted that this stimulation in respiration was more marked for immaturely than for maturely picked fruit.

Kimbrough (1924) determined the respiration rate of potato tubers that had been stored at 32, 40, and 72°F. and found that the lower the temperature of storage the higher was the initial intensity of respiration when they were exposed to warmer temperatures. It was also observed that the initial respiration rate was higher the longer the tubers had been in storage up to 3 weeks, after which longer storage did not further increase the rate. It was noted by Hopkins (1924) in studying the respiration of potatoes at low temperatures that there was a marked acceleration in the rate at 0°C., so that for a considerable period the intensity was greater than at 4.5°C. The curve of respiration for temperatures from 0.8 to 11.5°C. showed a minimum point at 3.0°C., and as the temperature increased it showed an increase at 0°C. and then a decline. Carrick (1928) found that Winesap apples frozen from 3 to 6 hr. at temperatures of from −7.5 to −8.5°C. increased their respiration at 0°C. as much as 85 per cent above their previously normal performance at that temperature. This acceleration lasted for several days, gradually declining so that slight effects were to be noted 1 month after the treatment. When the respiration rate was measured at 20°C., the treated fruit showed an increase in the rate, although generally with a smaller gain than when measured at 0°C. When the fruit was frozen sufficiently to kill 80 per cent of the cells, the excretion of carbon dioxide at 0°C. was only one-third that which was normally produced at that temperature. In 1930, he found that the respiratory rate of Emperor grapes decreased 26 per cent between 1.75 and 0°C., and 40 per cent between 0 and −3°C. DeLong, Beaumont, and Willaman (1930) reported that if apple twigs were transferred from a lower to a higher temperature, a peak of evolution of carbon dioxide occurred for several hours, following which a constant level was attained. The lower the previous temperature the higher was the peak of the evolution of carbon dioxide. They suggested that this behavior indicated that the carbon dioxide is less soluble at the higher temperature and thus diffuses outward more readily. Dexter (1934) found that the rate of respiration of cabbage and winter wheat continuously decreased during storage at 2°C. regardless of the simultaneous increase in the percentage of sugar.

These changes in the rate of respiration can apparently be explained, in some cases at least, by the changes in the sugar content that accompany the changes in temperature. Thus it was noted by Müller-Thurgau (1882), Appleman (1911), and Hopkins (1924) that sugar accumulated in the tubers of the potato at 0°C. Appleman also found that after 2 to 3 weeks at this temperature the diastatic activity had greatly increased over that of the tubers stored at room temperatures.

Wright (1932) stored potatoes at 32, 36, 40, 50, 60, and 70°F. from Nov. 5 to Mar. 24, and found that at 32°F. the sugar in the tubers increased steadily during the experiment. This accumulation was less at 36°F. than at 32°F., while at 40°F. there was only a slight increase. At 50, 60, and 70°F. the sugar content decreased during the period of storage in direct proportion to the increase in temperature. The causes of this accumulation of sugar are not definitely known. It has been suggested that the lower temperatures are unfavorable for respiration, while the activity of diastase is not affected to so great a degree, so that sugar is thus formed in excess of its consumption by respiration and as a consequence accumulates. It is thus assumed that the increased respiration of tubers and fruits following their exposure to low temperatures is caused by the increased sugar supply available for the process (Butler, 1913). This assumption has been strengthened by the observations of Hopkins (1924), who noted, as previously mentioned, that the respiration of potatoes at lower temperatures showed a marked acceleration at 0°C. over that occurring at

4.5°C. It was found that sugar accumulated in these tubers at 0°C. but began to disappear rapidly when stored at 4.5°C. The respiration curve and that of the change in sugar content thus approximately paralleled each other.

Barker (1936) considered there is strong evidence that sucrose is closely related to the supply of substrate for respiration, and that neither glucose nor fructose is directly associated with this function.

Barker (1933) found that the sugar-respiration relation was influenced by a temperature effect that consisted of an enduring depression of the rate of respiration by exposure to low temperature. This lag persisted for several weeks after a return to higher temperatures. The factors causing this depression are apparently the result of the accumulation under conditions of low temperature of an inhibiting substance and its subsequent activation by higher temperatures. The accumulation of inhibiting substance is greater the lower the temperature. It is greatest at −1°C., and declines to zero at 8°C. The developmental effect of warmth is very rapid at 10°C., and falls to a negligible rate at −1°C. Newton and Anderson (1931) reported that the rate of respiration for different wheat varieties at −7°C. was in the inverse order of hardiness. This probably accounts for the observed fact that hardy varieties maintain their sugar reserves better than nonhardy plants during the winter.

b. Moisture.—The amount of moisture in the tissues, within certain limits, is one of the determining factors in the intensity of respiration. This fact has been especially observed in the case of seeds and grains. In air-dry seeds respiration is proceeding at a relatively low rate, but directly following an increase in their moisture content the intensity of respiration increases. Thus Kolkwitz (1901) found that barley kernels containing 10 to 11 per cent of moisture liberated only 0.33 to 1.5 mg. of carbon dioxide per kilogram per hour, while after they had absorbed 33 per cent of moisture, 2,000 mg. of this gas were liberated in a similar period. White (1909) reported that the respiration rate of air-dry seeds of wheat with a moisture content of 12 per cent was 0.03 part of what it was in active seedlings, while the intensity in air-dry seeds of *Acacia melanoxylon* amounted to only 0.0001 of that from the germinating seeds. The coefficient of increase in the respiration rate due to changes in the moisture content was observed in wheat by Bailey and Gurjar (1918) to be as follows:

Increase in moisture, per cent	Acceleration in respiration, per cent
12 to 13	0.16
13 to 14	0.17
15 to 16	1.41
16 to 17	3.02

In corn seed, Bailey (1921) observed that respiration became marked when the moisture content exceeded 13 per cent, and that when the moisture content was raised to 15 to 16 per cent an increase in the respiration rate of 400 per cent was obtained. Coleman, Rothgeb, and Fellows (1928) in a study of the respiration rate of the seeds of the sorghums noted the following responses in yellow milo and feterita seeds to changes in moisture content at a temperature of 37.8°C.

Percentage of moisture in the grain	Milligrams of carbon dioxide evolved per 24 hr. per 100 g. of dry matter	
	Yellow milo	Feterita
11	0.53	1.00
12	0.70	1.10
13	1.00	1.58
14	1.56	3.00
15	3.27	5.80
16	7.58	11.80

Observations similar to those which have been mentioned in regard to the relation of the moisture content and respiration have been made by Duvel (1904), Babcock (1912), and others.

The length of time the excess moisture is present in seeds bears a relation to the rate of respiration, as may be observed from data mentioned by Bailey and Gurjar (1918) for wheat. The quantity of carbon dioxide respired by a sample of wheat which contained 15.2 per cent moisture and which was stored 55 days after it was dampened was approximately four times as great as that respired by freshly dampened wheat of the same moisture content. The carbon dioxide evolved from a sample which contained 15.7 per cent of moisture and which was stored for 108 days was about eight times as great as the wheat of the same moisture content that had just been moistened. It is considered that the dampened grain is favorable for the formation of sugar, which tends to accumulate as time goes on so that there is a supply sufficient for a relatively high rate of respiration.

It is worthy of note here that the humidity of the atmosphere may affect the rate of respiration in stored grain by increasing its water content. Thus Bailey (1921) noted that at 35 per cent humidity the moisture content of corn was 8.5 per cent, while at 85 per cent humidity it was 17.8 per cent.

The effect of moisture upon the rate of respiration varies to a considerable degree for different plants or plant parts. Thus Jacquot and Mayer (1926) in the case of the seeds of corn and the peanut found that the rate of carbon dioxide production per unit of moist weight increased with increased water content up to 30 to 35 per cent, when the rate began to fall off, while in the bean the respiration rate increased up to 60 per cent before it began to decline. When the carbon dioxide production, however, was determined per unit of dry matter, it was found to increase up to the complete imbibition of the bean seeds (117 per cent) and up to 58 to 60 per cent in peanuts. The respiratory quotient was greater than unity in all the seeds at low water content and as complete saturation was approached. Fraymouth (1928) studied the relationship of respiration to the moisture content of certain terrestrial algae and other lower plants. In the algae the respiration rate increased rapidly on the addition of water to the air-dry material until the amount of water had reached 100 per cent. There was then a gradual rise in rate until a water content of 600 per cent was reached, when gaseous exchange was apparently seriously impaired. After prolonged drought, respiration in these algae was practically zero, but immediately after the addition of water, it was found to be almost as intense as for fresh material of a similar water content. In the lichen (*Parmelia*) the respiration rate showed no great increase until a water content of 60 per cent was reached, after which it rapidly increased until a water content of 200 per cent was attained. In *Hypnum sp.* there was a gradual rise in the respiratory rate with increasing amounts of associated water until 250 per cent was reached, after which the tissues were saturated with water and gaseous exchange was apparently impaired. The effects of desiccation upon respiration and certain organic constituents of the cacti have been reported by MacDougal, Long, and Brown (1915) and Long (1917).

It was reported by Singh and Varadpande (1930) that a deficiency of water in leaves lowered their rate of respiration, and Smyth (1934) found that the respiration rate of certain lichens increased with a rise in water content to saturation. Livingston (1934) estimated that approximately 8 per cent of the water incorporated in growth was derived from respiration. The quantitative importance of this water may vary according to environmental conditions.

Several experiments have been reported, however, that show some striking variations from the majority of the observations in regard to the relationship between the moisture content of the tissues and respiration. Smith (1915) by means of a vacuum desiccator diminished the water content of the leaves of the snowdrop and the stem tips of nasturtium and young asparagus and found that the respiratory rate increased with a diminishing water content until about 30 per cent of the water had been withdrawn. The rate of respiration then remained about the same until the water was diminished to about 50 per cent of that originally present. Below this limit to complete dryness, respiration decreased proportionately to the amount of water lost. Palladin and Sheloumova (1918) used potato tubers and exposed a part of the material for drying in the air at room temperature, with the result that during the first few days, in spite of a great loss of water, the intensity of respiration was gradually increased, but after further loss of water there was a distinct lowering of the respiratory rate. When the same tubers were immersed in water, the respiratory intensity was again stimulated for a few days, after which it fell to the normal rate. They thus could not establish any definite correlation between the amount of water lost and the intensity of respiration. Luthra (1924) in experiments with pears kept in moist and in dry air noted that both transpiration and respiration of some varieties were increased in the dry air.

The question naturally arises as to why respiration is accelerated by an increased moisture content. Bailey and Gurjar (1918) assumed in the case of grains that the colloids of the cells imbibe water and form elastic gels and that if sufficient water is supplied the gels become more and more dilute. Diffusion through the stronger gels is relatively slow, while through the dilute gels it is relatively rapid. Increasing the moisture content thus results in progressively less viscous gels and a corresponding increase in the rate of diffusion. It is assumed that the rate of respiration depends upon the rate of diffusion between the various cells of the tissues, so that the less viscous the gels of these cells the more rapid the rate of respiration.

c. Injury.—It has been known for a considerable time that wounded plant parts respire more intensely than those which are intact. One of the first to observe this fact was Boehm (1887), who called attention to the abnormal production of carbon dioxide by potatoes that had been injured by cuttings. Richards (1896) by using potatoes and carrots cut in four pieces, slit leaves, split hypocotyls, and decapitated roots of seedlings demonstrated increases of as much as 100 per cent in the intensity of respiration when measured by the output of carbon dioxide. In the case of the potatoes and carrots he noted a very marked and sudden increase during the first 2 hr., which again fell rapidly, although to a point still considerably above the normal respiration, and then rose more slowly later to the maximum production of carbon dioxide. The respiratory activity reached its maximum 24 to 30 hr. after the injury had been inflicted. Lutman (1926) found an increased respiration in potatoes during the second and third days after cutting, which was followed by a gradual fall until the normal rate of the uninjured tuber was attained. If the cutting was repeated in 5 to 7 days, the respiration curve did not rise so high as after the first cutting, but a similar intensity could be attained by increasing the cut surface. Repeated cuttings lowered the respiration curve until finally the tubers would not respond. By lengthening the intervals between cuttings to 11 to 12 days, the respiration curve more nearly approximated the one obtained after the first cutting. Johnstone (1925) by means of a cork borer removed cylindrical plugs from sweet potatoes and noted that from one lot of roots with an injured area of 7,385 sq. mm. the evolution of carbon dioxide increased from 64.3 mg. per kilogram per hour before wounding to 126.8 mg. during the same time after wounding. It was noted by Coleman, Rothgeb, and Fellows (1928) that cracked grains of the sorghums showed a higher intensity of respiration than did the intact grains.

It was noted by Richards (1897) that the temperature of injured parts of plants was materially higher than that of the surrounding tissues. In the potato no increase in temperature was observed until 24 hr. after injury or at the time of maximum respiratory activity due to injury. This increase in temperature was only 0.4 per cent and was very local, since 2 cm. back of the wound no perceptible differences in temperature could be noted. Similar reactions were also observed in the radish and carrot. Onion bulbs showed the most marked changes in temperature due to injury of any of the material used. The rise in temperature due to injury was as much as 3.3°C. and in contrast with the potato the temperature effect extended a considerable distance into the surrounding tissue.

Harris (1929, 1930) reported that pruning or any injury to the roots of fruit trees resulted in a temporary increase in their rate of respiration. Lall (1934) noted that the cutting of apples produced an increase in the rate of respiration which disappeared at a temperature of 12°C. in about 7 days. Audus (1935) and Godwin (1935) showed that handling starved laurel leaves greatly increased their rate of respiration. Barker (1935) could find no increase in respiration due to handling firm potatoes but obtained a marked increase by handling those that had softened from desiccation. The stimulus that causes this increased respiratory rate arises apparently from a deformation of the cells by bending or rubbing.

The probable causes of this increased respiratory activity of tissues following their injury will now be considered. One cause is apparently the facilitation of the gaseous exchange. Thus when Stich (1891) sealed together the cut ends of a potato with neutral gelatin, no carbon dioxide was evolved. Johnstone (1925) noted in injured sweet potatoes that the rate of respiration increased 97.3 per cent when the injured area was exposed, while it increased only 17.1 per cent when this area was not directly exposed to the air. The sudden and marked increase in the output of carbon dioxide directly following injury is probably due to the escape of the gas that has accumulated in the intercellular spaces of the tissues. This assumption is strengthened by the fact that only bulky tissues show this marked increase in the evolution of carbon dioxide directly following injury. The absorption of oxygen is also rendered more favorable by the freer access of air across the tissues due to injury. This facilitation of gaseous exchange is indicated by the composition of the air taken from the intercellular spaces of uninjured and injured apples, as shown in the following table (Magness, 1920):

State of apple	Storage, temperature degrees centigrade	Percentage of	
		Carbon dioxide	Oxygen
Whole apple.........	1	6.6	14.6
Ends peeled.........	1	1.7	15.8
Whole apple.........	20	17.8	7.0
Ends peeled.........	20	7.4	9.9
Whole apple.........	30	23.9	1.8
Ends peeled.........	30	12.6	8.9

It has been noted by Drain (1926) for apples and Lutman (1926) for potatoes that an injury which is not accompanied by the rupture of the skin does not increase the respiratory rate. This indicates also that the increase in respiration rate due to injury is due primarily to an acceleration of the gaseous exchange between the tissues and the surrounding air.

Lutman (1926) believed that the increased respiration due to injury in the potato may be regarded, in the main, as an irritable response of the protoplasm to the stimulus of the contact with the oxygen of the air of the tuber cells, which under ordinary conditions are protected from such a contact by the external layer of cork cells. He was able to remove a large part of the effect of injury on respiration by washing the cut surfaces in running water for 1 to 3 hr. The activators of respiration or the substances upon which they act are apparently soluble in water.

The chemical changes that occur in the injured regions of tissues have also been observed to a limited degree. Thus Grüss (1907) found in the cells around wounded potatoes an accumulation of sugar in the subphellogen and an increase in the oxidizing enzymes and in the diastatic activity. Friedrich (1908) in potatoes noted in the cells bordering the cut surfaces an increase in the total nitrogen and acidity and a decrease in total carbohydrates but an increase in reducing sugars. He considered that the sugars increased the intensity of respiration, while the acids were waste products of this process. Lutman (1926), however, did not consider this increase in sugar content to have any effect upon the respiratory rate. It was observed by Hopkins (1927) that the sugar content of wounded potatoes increased from 53 to 68 per cent of the original amount present. This sugar content began to increase after injury, reached a maximum after several days, and then fell off. He believed that this increase in sugar was brought about by activities leading to callus formation.

d. Kind and Age of Tissues.—The intensity of respiration varies markedly with the kind and age of the tissues. This variation of the rate of respiration of different tissues may be illustrated by the observations of Benoy (1929), who determined the rate of respiration of a number of different vegetables at 30°C. for a 30-hr. period following their harvesting. With respect to the amount of carbon dioxide evolved for equal weights of dry matter, they stood in the following order: asparagus, lettuce, green bean, okra, green onion, carrot, tomato, beet, green mango, and red pimento. It should be noted here that the amount of glucose which accumulated per 100 g. of material was the highest (13.68 g.) in asparagus and the lowest (1.29 g.) in the red pimento. Beaumont and Willaman (1924) found that the output of carbon dioxide from the buds and twigs of a hardy variety of apple, as Charlamoff, at temperatures of 5.3 and 12.4°C. was greater than that from the Delicious variety, which is considered nonhardy. This difference in carbon dioxide evolution held on the basis of the number of milligrams per unit of length of twig, weight, volume, or number of buds. This difference in the intensity of respiration was not due to any differences in the extent of the rest period or dormancy.

The difference in the rate of respiration of different tissues has been especially studied in seeds and seedlings by determining the relative respiratory intensity of the embryo and endosperm. In this regard, Burlakow (1898) stated that the respiratory activity in the embryo of wheat was twenty times that of the endosperm. Kolkwitz (1901) found that when wheat kernels were divided transversely into equal portions, the embryo end respired three times as much carbon dioxide as the opposite end when equal weights of material were compared, while Karchevski (1903) reported the aerobic respiration of the embryo of wheat to be twelve times as great as that of the endosperm. The embryo of the rice seed, according to Bailey and Gurjar (1920), also shows a higher rate of respiration than does the endosperm. The same authors also noted (1918) that shriveled wheat respired at a rate two to three times greater than did the plump grains. They attributed this to the higher ratio of embryo weight to endosperm in the shriveled grain as compared with that of the plump wheat. In many of the studies of the relative rate of respiration in the embryo and endosperm, the embryos have been removed. As the tissues are more or less injured by this operation, this method is open to criticism. The production of embryoless seeds of wheat in certain instances furnished Lyon (1928) an opportunity to compare the respiration rate of these seeds with the normal ones without inflicting any injury upon their tissues. It was found that when these two types of seeds were placed under conditions favorable to germination, the seeds devoid of embryos respired 22.1 mg. of carbon dioxide per gram of seed in 6 days, while the normal seeds gave off 26.5 mg. for the same unit weight. As the embryo made up only 3 per cent of the entire seed by weight, its rate of respiration was decidedly greater than that of the endosperm.

The age of the tissues has a marked influence upon the intensity of respiration. Thus Nicolas (1918) noted that the respiratory intensity of young leaves and stems was from three to seven times greater than that of corresponding fully developed organs taken from the older parts of the plant. Bezagu (1919) observed that respiration was relatively weak in young leaves and increased to a maximum at the time the leaf reached its full development, after which it decreased as the leaf became older. Bergman (1925) and Wakabayashi (1925) found the rate of respiration of the flowers of cranberries much less than that of buds. The shoots of the current year's growth respired two to three times as rapidly as those of the previous year and at about the same rate as the flowers. The young fruits for a period of 1 to 3 days after the petals fell respired very rapidly, the rate exceeding that of the buds, but decreased rapidly with the development of the fruit. Hover and Gustafson (1926) in observations on the respiratory activity of the leaves of corn, oats, sorghum, and wheat noted a decreased activity with increasing age until middle age and thereafter a gradual increase as determined per unit of dry weight. They assumed that this increase was due to the decrease of the dry matter in the leaves after middle age. Maige and Nicolas (1910) considered that both the respiratory intensity and the respiratory quotient increased with the increased turgidity of the cells. An increase in turgidity, however, may be only an indicator of favorable conditions for growth and vital activity and thus not a direct causative factor in increasing respiration.

Kidd, West, and Briggs (1921) determined the respiratory index for *Helianthus annuus* at different stages of growth. They defined the respiratory index as the milligrams of carbon dioxide liberated per gram of dry weight per hour at 10°C. when the amount of respiratory material is not limited and when the external concentration of the oxygen is that of the atmosphere. This index is consequently a measure of the effective amount of respiring cell matter per gram of dry weight. The observed respiratory indexes for the sunflower plant and its parts at different ages are shown in the following table:

RESPIRATORY INDEXES OF SUNFLOWER PLANTS AND PLANT PARTS AT DIFFERENT STAGES OF GROWTH

Days from germination	Number of plants used	Dry weight of a single plant, grams	Respiratory index			
			Entire plant	Stem	Leaves	Stem apex
1	30	0.0225	2.90			
2	25	0.0223	3.00			
4	25	0.0242	3.00			
13	10	0.1009	2.80			
22	8	0.630	3.00			
29	2	4.065	2.30			
36	1	12.850	1.21	0.81	1.56	
43	1	22.050	1.03	0.69	1.38	
50	1	45.150	0.94	0.46	1.52	2.56
59	1	93.200	0.66	0.33	1.32	1.72
89	1	294.700	0.48	0.31	0.90	1.13
99	1	377.400	0.37	0.25	0.45	0.89
112	1	818.300	0.26	0.098	0.38	0.75

It is to be observed from this table that the respiratory index of the entire plant falls off continuously from 3 to about one-tenth of this value at the end of the life cycle. The respiratory index of the stem, of individual leaves, and of the flowers, respectively, decreased with the age of the organ. The initial respiratory index of successive leaves decreases with the age of the plant, indicating that the respiratory index of the meristematic tissue decreases with age. The fall in the respiratory index of the meristematic tissue and young leaves indicated that the fall in the respiratory index of the whole plant was not due entirely to the increase in proportion of mechanical tissue. The fall in the value of the respiratory index with age followed closely that of the value of the relative growth rate, indicating a close connection between the "internal" factor for respiration and the "internal" factor for growth.

Gustafson (1929) found a decrease in the production of carbon dioxide during growth in the John Baer variety of tomato fruits until a minimum point was reached at about the time increase in size stopped. This was followed by an increase in carbon dioxide production, which reached its maximum when the fruits were orange red in color, after which there was a final decrease in respiration. Gustafson attributed the increase in respiration to a lowering of the hydrogen-ion concentration in the cell sap, while the final decrease was thought to be due to a slowing down of all life activities as the fruit approached its maturity. Hafenrichter (1928) found, however, in the germination of the soybean that if the growth rate of the entire plant was considered instead of its parts, the maximum rate of respiration was reached earlier in the development of these plants than was the maximum rate of growth.

It was observed by Kidd and West (1929) that the rate of respiration during the senescence of the apple rises at first and then falls. Bergman (1929) noted that the rate of respiration in developing blueberries was highest when the fruits had reached the pink-red stage, and that it then decreased until the berries had reached the blue-ripe stage. Harley (1929) found that the accumulation of acetaldehyde was more rapid in the late-picked pears than in those which had been picked earlier. Overholser, Hardy, and Locklin (1931) reported that the greater firmness of the flesh of strawberries is apparently not directly correlated with a low intensity of respiration in either immature or mature fruit.

Smith (1929) believed that the reason more mature potato tubers respire less than younger ones is that an increase in the development of the periderm prevents the exchange of gases.

It was observed by Harding (1929) in Grimes Golden apples, Bonazzi (1931) in the sugar cane, and Luthra and Chima (1931) in grape that the rate of respiration was more intense in the younger than in the older tissues. Johansson (1933) noted that the maximal respiration values in the bark occurred at the time of greatest growth.

e. Anesthetics.—It was observed by Irving (1911) that very small doses of chloroform applied as vapor to leaves increased their respiration. This increased rate could be maintained if the chloroform was applied continuously, but when it was withdrawn the respiration rate dropped to normal. When medium doses were applied, an initial outburst of carbon dioxide resulted, which was followed by a decline much below normal, the rate and extent of this decline increasing with the amount of chloroform used. When strong doses were applied, there was no initial increase of carbon dioxide, and its production rapidly dropped to zero. Thoday (1913) substantiated the observations of Irving in regard to the production of carbon dioxide and noted further that, in the stimulation of respiration due to small doses of chloroform, the absorption of oxygen and carbon dioxide increased in like proportion, but that, when the doses of chloroform were large, the absorption of oxygen was no longer closely correlated with the production of carbon dioxide. The behavior of different leaves in this regard is of interest. Thus in the leaves of *Tropaeolum*, which contained no tannin, the absorption of oxygen was depressed more than the production of carbon dioxide. In leaves of cherry laurel and *Helianthus*, however, which contained tannins and the oxidation of which imparted a brown or black color to the disorganized leaves, the absorption of oxygen was very rapid for a short time and, though quickly falling, remained at a much higher level than the production of carbon dioxide.

With wheat seeds soaked in 7.3 and 3.6 per cent ether solutions, Thomas (1918) obtained an increase in the rate of respiration, which was followed by a decrease. Smith (1921) treated the seedlings of wheat when the roots were 1 to 1.5 in. in length with 1, 3.65, and 7.3 per cent of ether in water and found that the first effect was an increase in the rate of respiration, which in turn was followed by a decrease. With all these concentrations the rate of respiration was ultimately reduced to approximately the same level, and the stronger the concentration the more rapid the fall. When respiration was reduced below normal, recovery was possible on the removal of the ether and appeared to be complete if sufficient time was allowed. If respiration was too far depressed, however, no recovery was possible. Haas (1919) exposed *Laminaria* to 0.3 per cent chloroform, 0.8 per cent formaldehyde, and 10 per cent ethyl alcohol and observed that the rate of respiration was respectively 1.4 to 2.8, 2.7, and 6.2 times that of the controls. Upon the exposure of the alga to these compounds, the initial effect was an increase in respiration, which was followed by a decrease if they were sufficiently toxic. No decrease in respiration, however, was observed when the concentration was too low to be toxic.

The probable action of anesthetics upon the oxidative mechanism of the cells is indicated by the observations of Ray (1923). He found that when *Ulva* was killed in such a manner as to destroy the oxidizing enzymes no carbon dioxide was produced unless hydrogen peroxide and ferric sulphate, which acts like a peroxidase, were present. If the concentration of the iron was low, there was an increase of the carbon dioxide production when chloroform was added. If the concentration of iron was high, the rate of carbon dioxide production appeared to decrease from the start. It was also observed that if oleic, tannic, succinic, or malic acid were treated with hydrogen peroxide and ferric sulphate, carbon dioxide was produced at a rate that could be measured by the indicator method. In the case of acids with a double bond, the rate of production of carbon dioxide could be varied by the addition of an anesthetic. The changes in the rate of production of carbon dioxide under the influence of an anesthetic, such as chloroform, showed a striking resemblance to the reaction of an organism. Ray considered that there is a definite relation between the rate of production of carbon dioxide and the ability of the compounds to absorb iodine. An analysis of the effects of chloroform on the production of carbon dioxide by the living and killed cells of *Ulva* and unsaturated organic acids indicates that the same process is taking place in all three cases. The conclusion of Ray was that the action of chloroform on the oxidative mechanism of the cell was chemical in nature, and that it acted either by catalysis or by the formation of a loose compound with some portion of the system.

f. Afterripening Agents.—The effects of ethylene, which is now extensively used in the ripening of fruits, upon respiration might well be noted under this topic. Denny (1924) found that concentrations of this gas of 1 part in 1,000, 1 part in 10,000, 1 part in 100,000, and even 1 p.p.m. of air increased the respiration of green lemons. The increase in carbon dioxide output ranged from 100 to 250 per cent and appeared to be greatest at the intermediate concentrations. After the discontinuance of the application the rate of respiration decreased. Davis and Church (1931) also have reported that the treatment greatly stimulated the rate of respiration in Japanese persimmons. Regeimbal, Vacha, and Harvey (1927) treated ripening bananas with 1 part ethylene to 1,000 parts air for 15 to 20 min. and then determined their rate of respiration. In all cases the rate of respiration expressed in milligrams of carbon dioxide per hour was doubled or trebled within a few minutes, after which the rate fell off to a value lower than normal. It was assumed that the high initial increase might have been due either to the increase of oxidation or to the increase in the permeability of the membranes, allowing the diffusion outward of the carbon dioxide already in the cells. The rise in the respiratory rate after a second dose of the ethylene seemed to indicate that the oxidation rate rather than the permeability rate was increased. This stimulation wears off in less than an hour. It should be noted in this regard that the treated bananas had one-fifth to one-fourth more sugar in them than the untreated ones, while the starch content was proportionately decreased. The activity of diastatic enzymes as well as respiratory enzymes was apparently increased by the treatment with ethylene. It is not known, however, whether permeability changes indicated by the gas allow the enzyme and substrate to come together, activate the enzymes, or actually increase the amount of enzymes present.

It was reported by Guthrie (1931) that the expressed juice of potatoes that had been treated with ethylene chlorhydrin, ethyl alcohol, and acetaldehyde showed an increase in pH value and the power to reduce iodine in solution. The treatment with thiourea and potassium thiocyanate produced only a small change in the hydrogen-ion concentration and reducing properties. Hanes and Barker (1931) reported that the rate of respiration of potato tubers at 15°C. was increased when exposed to an atmosphere containing a low concentration of hydrogen cyanide. Shill (1931) stated that citrus plants treated with hydrocyanic acid increased their rate of respiration activity about 75 per cent, and that after 35 hr. the rate of respiration gradually decreased to normal. Guthrie, Denny, and Miller (1932) found that treatment with ethylene chlorhydrin produced an increase in the peroxidase, catalase, and sulphydryl content and in the pH value in both the dormant and nondormant corms of gladiolus.

Miller (1932, 1933, 1934, 1935) reported large increases in the rate of respiration when the whole tubers of the potato are subjected to the vapors of such chemicals as ethylene chlorhydrin, ethylene mercaptan, ethylene bromohydrin, hydrogen sulphide, ethyl disulphide, acetaldehyde, and halogenated derivatives. These increases occur very soon after the beginning of the 24-hr. period of treatment and reach their maximum the second or third day. The rate of respiration then declines approximately to that of the untreated tubers. There is neither a close correlation between the effect of these chemicals on the rate of respiration and their efficiency in breaking dormancy, nor between the increase in sucrose and the breaking of dormancy. Miller, Guthrie, and Denny (1934) stated that the decrease in citric acid and the increase in pH value begin approximately at the same time as the increase in the rate of respiration. They believed that the acid is used in respiration, and that the pH value increases as a consequence of the utilization of the acid. The large increase of glutathione that follows the treatment by ethylene chlorhydrin is not the cause of the increased respiratory rate, since the increment of this compound occurs 40 to 60 hr. after the beginning of the treatment, which is much later than the increase in the rate of respiration. Tubers treated with ethyl alcohol always show a decrease in the rate of respiration.

It should be noted here also that Haas (1919) found that when *Laminaria* was killed by alcohol, acetone, formaldehyde, or ethyl bromide, as well as by drying, the rate of respiration was even greater than under normal conditions. This rate, however, was not long maintained but gradually fell off and eventually became negligible.

g. Concentration of Oxygen and Carbon Dioxide.—The effects of the concentration of oxygen upon the rate of respiration will depend not only upon the plant or plant part under consideration but also upon its condition when the observation is made. When the supply of oxygen is sufficient for the normal respiration of the plant, the presence of a surplus produces in most cases no perceptible effects. The intensity of respiration in many plants is not markedly modified when the oxygen supply is reduced to one-half that normally present in the air or when it is increased to five to ten times that which normally occurs. According to Stich (1891), the absorption of oxygen becomes insufficient for most plants when the supply in the air falls below 5 to 8 per cent. If a plant is confined in a container, it can generally absorb the last trace of oxygen from the medium. When the oxygen supply is increased, the plant may continue to respire normally for a period of time even when the partial pressure of oxygen has been increased to twenty or thirty times the normal amount. In a few hours or days under such conditions, respiration declines and death ultimately ensues.

The variation of plants in their oxygen needs under different conditions of their development is illustrated by the embryo of *Alisma*, as reported by Crocker and Davis (1914). They found that the hypocotyl was capable of elongating 1.2 times its original length in the total absence of free oxygen. For greening and branching, however, it required an air pressure of at least 5 mm. to supply the minimum amount of oxygen, while for the development of the primary root an even greater amount was required. The amount of oxygen that must be supplied to produce a certain intensity of respiration or rate of growth will depend in many cases upon the permeability of the membranes that it must penetrate in reaching the seat of its utilization. This is illustrated by the coats of seeds through which oxygen diffuses with difficulty. When such coats are present, a higher supply of oxygen is required to produce a certain rate of respiration or growth than when they are removed. Thus Shull (1914) found with *Xanthium* seeds that when the oxygen supply was increased, it brought about, in some manner, an immediate and rapid increase in the rate of oxygen absorption, resulting in an immediate germination of the seeds. Shull considered that it was difficult to say what function or functions were affected by this treatment. It is certain that oxygen acted as a limiting factor on some function, but whether it was by limiting the process of respiration, by limiting enzyme formation, or by limiting the action of oxygen carriers could not be stated. Becker (1912) suggested that in some instances oxygen may act as a stimulus for certain functions without being directly concerned in respiration.

Blackman (1930) stated that the respiration rate of apples decreased with diminishing oxygen supply until a minimum concentration of 5 to 10 per cent of this gas was reached. As the concentration of oxygen was increased above this point, the rate of respiration continued to increase until this gas composed approximately 100 per cent of the gaseous medium. Zimmerman (1930) reported that, with cuttings of *Salvia* in 2 in. of water, roots were formed at the base of the stem; in 5 in. they had a tendency to root along the water surface; while, in 7 in. of water, roots were only sparsely produced, if at all. When the water was aerated, roots were produced equally at all three depths. Willow cuttings, however, produced roots when the oxygen supply in the water was as low as 1 p.p.m.

Loehwing (1934) noted that the aerated plants of sunflower and soybeans grown in sand and loam soil differed from the unaerated controls by being taller, heavier, and possessing a more fibrous and a more highly branched root system. The aerated plants had a higher absolute amount of ash, calcium, potassium, phosphorus, crude fiber, starch, sugars, and total nitrogen. The expressed sap of the aerated plants was more alkaline.

Jones (1933) considered that a deficiency of oxygen is probably the principal factor which causes a lack of germination in rice seeds continuously submerged in water. Earlier seeding is preferable because the temperatures of the air and water are lower early in the spring. Since more oxygen is dissolved in water at low temperatures, there is more of this gas available to the germinating seeds earlier in the season.

Soil preparation	Average yield in bushels per acre	
	1933	1934
Check (no tile)	311.6	330.6
Over plain tile	362.2	363.5
Over perforated tile	381.2	400.9

Bushnell (1935) in Ohio laid 4-in. ordinary drain tile and tile perforated with 36 holes per section at a depth of 10 in. in plowed soil. Potatoes were planted over these tile, and the yields were obtained as shown on page 970.

These data indicate that the potato plant is sensitive to soil aeration, and that in certain soils the lack of sufficient oxygen may be a limiting factor in crop yields.

The quantity of oxygen in the intercellular spaces of fruits and other plant parts due primarily to poor gaseous exchange is frequently considerably below that of the air and may thus in some cases be a limiting factor in the respiration of these parts. Injury to plants due to an inadequate supply of oxygen is shown by the effects upon cranberry plants that have been submerged by flooding (Bergman, 1921, 1925). This injury was more pronounced on cloudy than on clear days, owing to the difference in the amount of oxygen released in photosynthesis. Shading the flooded plants had the same effect as clouds in this regard. Low temperatures operated to reduce the amount of injury by lowering the respiration rate and thus decreasing the need for oxygen while increasing the capacity of the water to absorb this gas. The flowers and tips of the shoots were the parts of the plants most seriously affected by flooding, due to their high respiration rate (Bergman, 1930).

The amount of carbon dioxide in the atmosphere surrounding the plant may have some effect upon the respiration rate. Kidd (1917) noted that the rate of respiration was lowered by the accumulation of carbon dioxide, and that this effect was more marked upon the process of aerobic respiration. These depressing effects were evidently not due to permanent disorganization, as they always passed away as soon as the depressing concentrations of carbon dioxide were removed. In their study of the respiration of apple twigs at 0°C., potato tubers at 22°C., and wheat at 40°C., Willaman and Beaumont (1928) found that if carbon dioxide was allowed to accumulate in the respiratory chamber, the rate of carbon dioxide production continuously decreased with time. When aspiration of the atmosphere surrounding the tissues was commenced after 30 or 40 hr. of accumulating carbon dioxide, the respiration rate as measured by the output of carbon dioxide immediately assumed a much higher value than it had during the accumulation period. The work of Spoehr and McGee (1924) indicates that under such conditions, the establishment of an equilibrium between the carbon dioxide in the air surrounding the tissues and that which is dissolved in them is the result that is being observed. Thus they found that when the carbon dioxide content of the air surrounding a leaf was changed from a lower to a higher concentration, the leaf showed a reduced rate of carbon dioxide emission for a period following the change and then finally again attained the same rate as before the change in carbon dioxide content was made. When the carbon dioxide of the air surrounding the leaf was changed from a higher to a lower concentration, the leaf showed a primary increased rate of carbon dioxide emission and a subsequent decrease to the original rate.

The accumulation of carbon dioxide in some cases, however, apparently has no marked effect upon certain activities. Thus Newcombe and Bowerman (1918) in observations upon 2,000 seedlings of 12 different species reported that ventilation had no effect in producing better seedlings in small or large chambers in the dark and that ventilation had no visible effect on the sensitive reactions of these plants to gravity and light.

Schulz (1926) and Michaels (1932) with potato, and Clements (1935) with apple, studied the relationship between the number of lenticels per unit of surface and the rate of outward diffusion of carbon dioxide. They concluded that the carbon dioxide within these plant parts escapes with equal speed regardless of the number of lenticels. Gerhardt and Ezell (1934) found that carbon dioxide was absorbed to a much greater extent by Bosc pears than by Jonathan apples, when these fruits were stored in this gas. Thornton (1931, 1932, 1933, 1935) found that, in the presence of 20 to 70 per cent of carbon dioxide and 20 per cent of oxygen, the rate of uptake of the latter gas by different tissues differs greatly, so that they may show an increase, decrease, or no change in the rate of absorption. Thus under the preceding conditions the absorption of oxygen was increased 83 per cent for dormant potatoes, 239 per cent for nondormant ones, 117 per cent for those undergoing sprouting, 89 per cent for onion bulbs, 11 per cent for tulip buds, and 95 per cent for beet roots. On the other hand, the amount of absorption was decreased 33 per cent for asparagus, 17 per cent for bananas and strawberry fruit, and 34 per cent for shelled lima beans, while there was no variation for the roots of carrots.

h. Various Salt Solutions.—The effects of sodium chloride and calcium chloride upon respiration have been studied by Lyon (1921) and Inman (1921). It was observed in the case of *Elodea* that solutions of sodium chloride caused an increase in respiration that was followed by a decrease, while solutions of calcium chloride caused only a decrease. In a mixture containing 99.65 mols of sodium chloride and 0.35 mol of calcium chloride, the rate remained normal, while a mixture of 98.6 mols of sodium chloride and 1.38 mols of calcium chloride caused a great increase in respiration. In highly hypertonic solutions of sea water, the rate of respiration of *Laminaria agardhii* was rapidly reduced, while in hypotonic solutions it was reduced somewhat less rapidly. Hypertonic solutions of sodium chloride, calcium chloride, and mixtures of sodium chloride and calcium chloride in the proportion of 50 to 1 all caused a decrease in the respiration of wheat seedlings.

The stimulating action of phosphates upon oxidation reactions has been reported by Spoehr (1924) and Lyon (1924, 1927). The former investigator noted that glucose and other hexoses were oxidized by air, in solutions containing disodium phosphate and methylene blue, with the formation of carbon dioxide. This reaction was greatly accelerated by small quantities of iron salt. Sodium ferropyrophosphate was a more active catalyst than methylene blue in causing the oxidation of carbohydrates with air. Lyon (1924, 1927) reported that phosphates exerted a promoter action upon potato oxidase so that carbon dioxide was produced by an oxidation of some component in a solution of glucose. He also stated that phosphate catalyzes the slow oxidation of pyrogallol by atmospheric oxygen. He assumed that phosphate increases the rate of production of carbon dioxide by anaerobic processes because of its role in the early stages of alcoholic fermentation, and that it effects an increase in the production of carbon dioxide by the aerobic phase of respiration through its action as a catalyst toward oxidases. In *Elodea canadensis* and wheat seedlings Lyon was able by the application of phosphate to obtain an increase in the production of carbon dioxide amounting to 55 per cent for the former and 35 per cent for the latter plant. Lyon also observed that arsenates exerted a catalytic effect on the oxidation of pyrogallol by atmospheric oxygen and by metallic iron and on the production of carbon dioxide by *Elodea.* Warburg (1921, 1925, 1928) believed that iron is the oxygen-carrying component of the respiration ferment. He stated that if respiration is considered as an oxidation, the respiration ferment is the sum of all the catalytically active iron compounds that occur in the cell. The respiration ferment is thus considered a substance that takes up the atmospheric oxygen and transfers it to the organic substance and in its chemical constitution is related to the hemoglobin or red pigment of the blood.

It was observed by Smith (1921) that the respiratory rate of wheat roots which had attained a length of about 2 in. was greatly accelerated and then depressed after they were placed in a 0.0025 *M* solution of lactic acid. As the concentration of the acid was increased, this preliminary rise in respiratory intensity became less marked until a concentration was reached where the rate began to fall at once. Recovery was complete even when the respiratory rate was reduced to 25 per cent of normal by a 2 *M* solution of the acid. These results were apparently not due to acidity or osmotic pressure but to some specific action of the lactic acid. They also indicate that, in the case of wheat, lactic acid is not an important intermediate product of the respiratory process.

Hutchinson (1922) reported that, when radish seedlings were placed in 0.003 and 0.002 per cent solutions of adrenalin, they showed a falling off of 20 per cent in carbon dioxide production at the end of 15 min. and then a gradual rise. More dilute solutions of 0.0005 and 0.0002 per cent gave a much smaller decrease, with a final rise to normal in about 1 hr.

Thomas (1931) found that the rate of respiration in the tissue of carrots in buffered solutions increased with increasing hydrogen-ion concentration. The relation of the respiratory rate to the absorption of salts should be considered here. Briggs and Petrie (1931) established the view that a general parallelism exists between the rate of respiration in the tissues and the rate of conductivity of the external solution. They considered that the respiratory rate is an important factor in determining the concentration of electrolytes in the water in which pieces of plant tissue have been placed. Steward, Wright, and Berry (1932) believed that the accumulation of salt in a tissue is dependent upon respiratory phenomena that presumably supply the energy necessary for it. Becker (1936) considered, however, that the total energy available from respiration is far in excess of the presumed energy expenditure for accumulation alone, and that if the fraction of the total energy utilized in accumulation were to vary, the relation between the processes would doubtless differ from the one here described.

i. Light and Electricity.—Spoehr (1915) found respiration in the sunlight higher than in the dark. He believed that the air may possess higher oxidative power during the hours of illumination than during darkness, owing to the ionizing power of the sun's rays. He assumed that deionization of the air has its influence on oxidation. In wheat seedlings the ratio of the day rate of respiration to the rate at night in ordinary air during two experiments was 1.042 and 1.091, while in deionized air these ratios were 1.010 and 1.014, respectively. The rate of respiration in deionized air thus was approximately the same during the day and night. For the autoxidative process the dissociation or liberating of free valences in the oxygen molecule is necessary, and by this is explained the accelerating influence of light and heat upon oxidative processes.

Middleton (1927) found that when the air passing over barley seedlings was artificially ionized by means of polonium, the rate of respiration was increased by as much as 29.1 $\pm$ 5.6 per cent. In the case of the leaves of *Pelargonium zonale* with air ionized to a degree varying from 7.28×10^2 to 3.64×10^2 that of normal air, Whimster (1927) observed a percentage increase of 85.7 $\pm$ 7.1 in the respiration rate. After the removal of the polonium there was a definite aftereffect, the succeeding 2 hr. showing a percentage increase of 28.0 $\pm$ 7.3 over the control period. This effect of ionized air on respiration is apparently due to the ions themselves and not to the gaseous products produced by the action of the ionizing agent.

Portsmouth (1934) found that the respiration rate in barley seedlings in ionized air increased about 2 per cent over the control. Ionized air apparently had little or no effect on the rate of respiration in the leaves of the geranium. Sapozhnikova (1928) reported that ionization of the air decreased the rate of respiration in the plants that he observed. DeBoer (1930) grew *Phycomyces blakesleeanus* and *Polyporus destructor* in ionized air in which the degree of ionization varied from a few hundred to several million times that of normal air. He, however, could observe no effects of ionization on the rate of respiration.

Tang (1932) reported that the consumption of oxygen by the germinating seeds of *Lupinus albus* can be reversibly inhibited by carbon monoxide to a maximum of 36 per cent with a mixture of 24 per cent oxygen and 76 per cent carbon monoxide at 18°C. in darkness. This inhibition, however, is completely removed when the seeds are illuminated. Parija and Saran (1934) found that red light had no effect on the rate of respiration of starved leaves, while blue and violet light increased it after the manner of white light. They considered that light might cause this increase by the activation of the enzyme system or by changing the permeability of the cells. Groner (1936) found that if albino corn were suddenly exposed to light after a period in darkness, an increase in the output of carbon dioxide resulted. Wynd and Reynolds (1935) give a thorough review of the effects of ultraviolet light on rate of respiration and respiratory enzymes.

In 1914 Knight and Priestley reported some observations on the effects of electricity on respiration. Direct currents of a density 10^{-6} to 10^{-4} amp. had no effect on the respiration of peas other than that due to accompanying changes of temperature. The proportion of the currents actually traversing the peas, however, was small, the greater portion being taken by the water films of the seeds. Overhead discharges producing a current of density less than 3×10^{-6} amp. had no effect upon respiration. When higher currents were employed, a definite increase of the carbon dioxide output was observed. This increase, however, could be wholly attributed to the rise of temperature caused by the discharge. In the field where the currents are too small to produce any appreciable rise in temperature, electrification will have no effect on respiration, and explanations of the acceleration of growth due to electricity must be sought in other functions of the plant.

Marsh (1935) derived a quantitative relation between the velocity of respiratory rate and the inherent e.m.f. of the cells considered. Rosene and Lund (1935), and Rosene (1935) believed that a quantitative relation exists between cell oxidation and the continuously maintained e.m.f. in the roots of the onion.

j. Other Factors.—Other factors that have been observed to influence the rate of respiration are as follows:

1. *Sprays.*—The effects of the application of oil sprays on the rate of respiration have been reported by several investigators. Kelley (1930) noted that all oil sprays applied to the dormant twigs of apple increased their rate of respiration but retarded it after the unfolding of the first leaves. Green and Johnson (1931) found that the respiratory rate of bean leaves was increased 7.5 per cent when sprayed with dark petroleum oils containing more than 16 per cent sulphonatable residues. The rate, however, was decreased 5.0 per cent when the leaves were sprayed with light-colored oils containing less than 16 per cent of sulphonatable residue. Green (1936) stated that the rate of respiration was increased more markedly by oil sprays that were poorly refined than by those of a better quality. The average gain in the respiratory rate of various plants due to the application of oil sprays was 4.2 per cent. According to Hoffman (1935) the application of lime-sulphur sprays to leaves slightly increased their rate of respiration.

2. *Materials.*—It was observed by Gregory and Richards (1929) that the respiratory rate of barley plants was subnormal when nitrogen and potassium were deficient but normal when there was a phosphorus deficiency. Benoy and Webster (1930) in a study on fresh vegetables could find no relationship between the rate of respiration and the composition of the parts concerned. When different sugars were injected into starved leaves, Singh (1930) found that the intensity of the respiratory rate at first increased and then decreased. Pal and Chatterji (1936) found that the injection of a mixture of 40 units of insulin in 100 cc. of water into the leaves of *Hispage*

madablota and *Allium tuberosum* increased the rate of respiration as compared to those which had been injected with water. LeVan (1930) found that a stimulation of the production of carbon dioxide occurred in seedlings of *Lupinus albus* when placed in solutions of aluminum chloride, sodium chloride, magnesium sulphate, and ferrous nitrate. These increases ranged from 15 to 50 per cent. Overholser and Claypool (1931) reported that the average intensity of the respiratory rate in strawberries from plots that had received nitrogen applications was practically 10 per cent greater than that in berries from plots to which no nitrogen had been applied. Kertesz (1933) found that the sugar which disappears in green peas after shelling is utilized largely in respiration and is not transformed into starch as has commonly been assumed.

3. *Diseases.*—It was reported by Richter, Dvoretskaia, and Grechushnikov (1929) that there was a decreased respiratory rate in badly diseased plants. They attributed this decrease to a prolonged deficiency of foodstuffs. Whitehead (1934) stated that, except for the short period extending from the end of tuber dormancy to the unfolding of the first leaves, potato plants infected with potato leaf roll respire at a much higher rate than do healthy plants.

13. Substances or Enzymes Concerned in or Associated with Oxidation in Plants.—It was demonstrated by Schönbein as early as 1856 that there are present in plants and animal juices substances that act as oxidizing agents for certain compounds. Traube (1877) termed these substances "oxidizing ferments," and Bertrand (1896) proposed the term "oxydases" as a group or class name. This term has come into general use, although in England and America it is now commonly written "oxidases." As the behavior of these substances was more thoroughly studied it was found that they varied widely in their composition and in the nature of their reactions. As a result, numerous terms have been introduced into the literature, so that the nomenclature in regard to substances concerned in or accompanying oxidation by plant and animal tissues has become rather confusing. Before beginning a detailed discussion of the nature and action of the oxidizing substances of the plant it would perhaps be best, for clearness, to define four terms that have been the most generally used. These are "oxidases," "oxygenases," "peroxidases," and "catalases."

An *oxidase* is a complex of substances which has the power at ordinary temperatures to bring about the oxidation of certain compounds in the presence of gaseous or dissolved free oxygen.

An *oxygenase* is a substance that is capable of causing the formation of organic peroxides. By some authors an oxygenase is defined as a preformed organic peroxide resulting from auto-oxidation. Some consider that the oxygenase itself takes up oxygen and thus becomes a peroxide, while others believe that it acts only as a catalyzer in the production of peroxides. An oxygenase is sometimes called a "peroxide-forming enzyme."

A *peroxidase* is a substance that is capable of rendering active the oxygen of peroxides. The oxygen is freed from the peroxides in the atomic or active form and is thus capable of causing the oxidation of certain compounds.

A *catalase* is a substance that is capable of decomposing hydrogen peroxide into water and molecular oxygen without being able apparently to activate the oxygen thus liberated toward oxidizable substances.

a. The Oxidases.—When the juice or watery extract of a plant tissue causes the oxidation of compounds, which cannot be brought about by oxygen alone to any appreciable extent at ordinary temperatures, it is said to contain a direct-oxidizing enzyme or oxidase. Some of these oxidation reactions are manifested by the bluing of guaiacum, the conversion of hydroquinone into quinone, the liberation of iodine from potassium iodide, and the formation of phenolphthalein from phenolphthalin by the liberation of 2 atoms of hydrogen. The action of oxidases is manifested by the blackening of cut potatoes, the staining of lumber (Bailey, 1910), and the blackening of tea leaves upon curing. Since the oxidases are in most cases destroyed by heat, mineral acids, and poisons, and since they apparently act in a catalytic manner, they have been termed "oxidizing enzymes," although they possess certain characteristics that are different from those of enzymes. Thus in most cases they do not behave as true catalysts, as they are unable to accomplish the oxidation of unlimited amounts of oxidizable materials.

It was stated by Bach (1897), Kastle and Lovenhart (1901), and Bach and Chodat (1903) that the action of the oxidases indicated that they have the power to produce peroxides, and they believed that an oxidase is composed of two substances—an oxygenase and a peroxidase. They suggested that the oxygenase has the power of fixing the atmospheric oxygen in such a manner as to produce a peroxide, but the manner in which this occurs is not clear. Some suggest that the oxygenase catalyzes the formation of organic peroxides from auto-oxidizable substances in the cells, while others believe that the oxygenase itself combines with the atmospheric oxygen and thus becomes a substituted organic peroxide of a very unstable nature (Moore and Whitley, 1909).

It was stated by Kertesz (1933, 1934) that oxygenase catalyzes the oxidation of certain compounds with the *o*-dihydroxy-grouping characteristics of the catechol tannins. The oxidation is accompanied by the formation of the dark-colored *o*-quinone, the presence of which causes the darkening of peaches.

Gallagher (1923) found evidence that the production of peroxide in the potato is intimately associated with the lecithin of the tuber. He believed that the substance from which the peroxide is derived may either be the lecithin itself or a compound intimately associated with it. Gallagher isolated from fresh potatoes a substance that bore a close relation to the lipins. In contact with air or oxygen a solution of this substance acquired the property of causing the immediate oxidation of guaiacum in the presence of peroxidase. He thought this indicated that the so-called "oxygenase" of the potato is really an auto-oxidizable lecithin-like substance. He believed this evidence substantiates the view that there is a relationship between lecithins and plant respiration. Onslow (1919) believed that the peroxides are formed under the influence of oxygenase from aromatic substances having the dihydroxy grouping of catechol. The peroxidase is considered to catalyze the decomposition of the peroxide that is formed by the oxygenase, with the liberation of active oxygen. This active oxygen then combines with the oxidizable substances with which it comes in contact. This theory of the nature and reaction of oxidases has been generally accepted and has largely dominated the field of thought in that regard.

The two constituents of the oxidases have been separated by Bach and Chodat (1903) and Reed (1916) in a fairly satisfactory manner. Reed found that the filtrate from finely ground potato peel in 55 per cent alcohol contained a peroxidase, while the residue after washing with alcohol gave a water extraction containing oxygenase. When these two preparations were mixed, the oxidation increased in the case of pyrogallol about six times over that where oxygenase alone was used.

1. *Classification of Oxidases.*—The oxidases of plants that have been studied have been grouped into two main classes according to the substances upon which they act. These are the laccases or phenolases and the tyrosinases.

(*a*) *The Laccases or Phenolases.*—It has long been known that a blue coloring matter is produced when a tincture of gum guaiacum is treated with certain oxidizing agents, and the first knowledge of oxidases is closely associated with this fact (Kastle, 1910). The laccases or phenolases include those oxidases which are capable of reacting directly, in the presence of free oxygen, with a fresh alcoholic solution of gum guaiacum with the production of its characteristic blue oxidation product (Gallagher, 1923). Each of the oxidases in this group is capable of oxidizing one or more of the phenol compounds. Some of the compounds that can be oxidized under the influence of the phenolases are pyrogallol, benzidine, α-naphthol, phloroglucin, pyrocatechol, hydroquinone, phlorizin, resorcin, guaiacol, *o*-cresol, *m*-cresol, *p*-cresol, *o*-toluidine, *m*-toluidine, *p*-toluidine, phenolphthalin, tannins, guaiac tincture, and others.

In 1883, Yoshida reported that the hardening of the milk sap of the lac tree (*Rhus vermicifera*) was apparently an oxidation process that was catalyzed under the influence of a nitrogenous substance having the characteristics of an enzyme. In 1894 and 1895 Bertrand confirmed these results and called the oxidizable substances present in the sap "laccol" and the compound that catalyzed the oxidation "laccase." On account of the resemblance of "laccol" to the phenols, Bertrand studied the effects of laccase upon a number of them and found that they were oxidized under its influence. He found guaiacum to be a useful reagent for laccase and detected its presence by this means in many plants and plant parts. Laccase has been found to be not a specific enzyme in the narrow sense, because it will bring about the oxidation of many phenols and phenol derivatives. It is not able, however, to oxidize tyrosine or any of its derivatives. It thus apparently acts only upon substances containing a certain grouping in their structure.

Substances that give the same or similar reactions as laccase toward the phenol compounds are widely distributed in the plant kingdom and have been named collectively the "laccases," the "phenol oxidases," or the "phenolases." Most of the oxidases that have been studied in plants belong to this group, but practically no progress has been made toward their further classification. Some idea of the widespread distribution of the laccases may be had from the work of Onslow (1920, 1921), who examined 320 species of angiosperms representing more than 180 natural orders and 309 genera and found that 62 per cent of the orders examined contained oxidases of the laccase type. She found the laccases or phenolases in the fruit of the apple, pear, quince, plum, cherry, peach, apricot, strawberry, grape, fig, and mulberry. They have also been found in apple bark and leaves, in alfalfa, clover, turnips, potatoes, peas, cabbage, silks of green corn, asparagus, and in many other plants and plant parts. Bunzel (1913, 1914) noted that the seeds of the beet had the highest phenolase activity of any of the plant parts, while the leaves stood next in order. In the normally developing leaves of the potato plant the phenolase activity was greatest in the early stages of development. It fell off with the growth of the plants and again rose when growth was completed. Bunzel (1916) noted in the case of the onion, potato plant, and sugar beet that the relative activity of the phenolases, as manifested toward the reagents which he used, varied only in intensity for the different parts of the same plant but found that the enzyme from each plant was markedly different from those of the other plants.

Drain (1923) in tests made at various periods during the time of fruit development, at harvest, and late in the storage season of apples, found that the majority of them fell in that class where the greatest phenolase activity was noted in the vicinity of the core and core line. He believed that oxygen may enter this fruit largely through the fibrovascular bundles that connect the calyx tube and the core cavity.

(*b*) *The Tyrosinases.*—It has long been noted that when certain fungi were exposed to the air they turned pink or red and finally black, while other species became blue due, as we now know, to the activities of the laccases. Bertrand (1896) proved that the former colorations were due to the oxidation of tyrosine by the absorption of oxygen under the influence of a specific oxidizing enzyme which he named "tyrosinase." He proved that this enzyme also occurs in the roots of the beet and in the tubers of the potato and dahlia, and it has since been demonstrated to be rather widely distributed in the plant and animal kingdoms. It is especially abundant in wheat bran. Miller (1929) considered that the insoluble tyrosinase of the seed coat of the velvet bean might be a peroxide.

Tyrosinase exhibits two general reactions. It oxidizes tyrosine, with the production finally of the black pigment melanin and the oxidation of *p*-cresol with the formation of a deep orange color. The former is the most common reaction shown by this enzyme. When tyrosinase is added to a solution containing tyrosine, the solution passes through a series of color changes ranging from pink to rose, to red, to violet, to blue black, and finally to the formation of a black precipitate. The tyrosinases are nondialyzable and are destroyed by heat. They are apparently more or less specific, since they act upon a group of compounds that are closely associated in structure.

2. *Methods of Determining Activity.*—The presence and activity of oxidases in plant extracts and tissues have been studied by color reactions and by the determination of the quantity of oxygen absorbed.

(*a*) *Color Methods.*—Much of the knowledge of the oxidases has been obtained from the study of their action upon compounds that undergo a change of color upon oxidation. The compounds generally employed for this purpose are those which do not oxidize spontaneously, or do so only very slowly, when exposed to the air in very dilute solutions, but which when brought into contact with living tissues or their extracts show a change in color.

The color reactions have been especially useful in determining the distribution of oxidases, and their behavior toward various indicators has also shown that these enzymes vary in the type of reactions that they catalyze. The color methods, however, give no quantitative indication of the amount of oxidation that has taken place, although it has been so assumed in many experiments. The amount of oxidation necessary to produce the colored appearance is, in most cases, very small, so that after the color appears, the oxidation may continue without necessarily changing the intensity of the color. There is also no definite means of determining the amount of oxidation that a particular color reaction represents. It should also be noted in this connection that the amount of oxidation necessary to produce a color in the various reagents varies over a wide range (Reed, 1916). Some of the most common preparations used in the color determinations of oxidase activity and their method of application are given below.

(1) *Guaiacum Reaction.*—Two grams of gum guaiacum are dissolved in 100 cc. of 95 per cent ethyl alcohol. To 5 cc. of plant extract is added 10 drops of this solution. If oxidases of the laccase type are present, a blue color develops.

(2) *Benzidine Reaction.*—One gram of benzidine is dissolved in 100 cc. of 50 per cent ethyl alcohol. To 5 cc. of plant extract is added 10 drops of this solution. This is allowed to stand for ½ hr., and if oxidases are present, a blue color develops.

(3) *α-Naphthol Reaction.*—A 1 per cent solution of α-naphthol in 50 per cent ethyl alcohol is prepared. When a few drops of this reagent is added to plant extracts, a lilac or lavender color results if oxidation occurs.

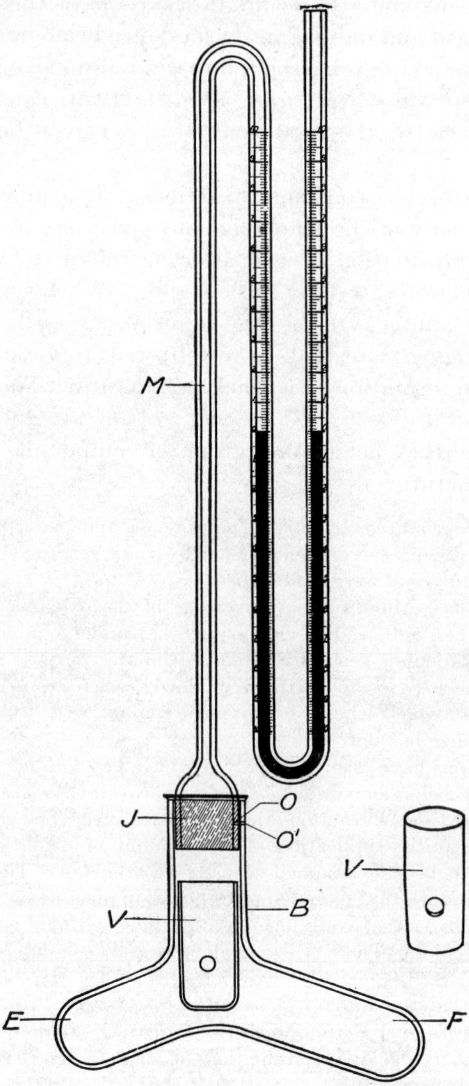

FIG. 36.—Bunzel apparatus for determining oxidase activity. Description in the text. (*Courtesy of Emil-Greiner Co.*)

(4) *Phenolphthalin Reaction.*—The phenolphthalin solution is prepared after the following manner: To 1 cc. of 0.1 *N* sodium hydroxide solution is added 0.1 g. of phenolphthalin. This is dissolved as thoroughly as possible, after which it is diluted with 25 cc. of water, filtered, and made up to 100 cc.

To 10 cc. of plant extract is added 5 cc. of the phenolphthalin solution and the mixture allowed to stand for 15 min., after which it is made alkaline with 0.05 *N* sodium hydroxide. If oxidase activity has occurred, the solution acquires a pink or red color due to the formation of phenolphthalein by the oxidation of the colorless phenolphthalin.

(5) *Röhmann-Spitzer Reaction.*—Equal parts of the α-naphthol solution and of a 1 per cent solution of *p*-phenylenediamine hydrochloride in distilled water are mixed just before the oxidase test is made. Five drops of this mixture is added to 2 cc. of plant extract. If oxidases are present, the colors developed within 15 min. vary from pink to deep reddish purple. This is one of the most delicate of the oxidation tests.

Guthrie (1930) reported an iodimetric method for determining the action of oxidase, while Fong and Cruess (1929) studied various indicators as a means of determining the extent of the activity of oxidases occurring in fruits.

(*b*) *Measurement of Oxygen Absorbed.*—The measurement of the oxygen used in oxidation offers a means of determining quantitatively the intensity of oxidase activity. Apparatus for this purpose has been devised by Bunzel (1912, 1914) and Harvey (1920). The simplified apparatus of Bunzel is shown in Fig. 36. This consists of two parts—a manometer *M* and a basal part *B*, which has two compartments *E* and *F*. The two parts of the apparatus are connected by a ground-glass junction *J*. The total volume of the apparatus minus the vial *V* up to the zero point on the manometer is 76.1 cc. In preparing for an experiment the manometer *M* is detached at the ground joint *J*, and a measured quantity of the plant juice is run into the compartment *E*, and the solution containing the oxidizable substance is run into compartment *F*. Enough water is then added to make the total volume of the liquid in the apparatus 18 cc. Rose (1915) used the materials in the following proportions: 3 cc. of plant extract, 3 cc. of water, and 12 cc. of a 1 per cent solution of pyrogallol. The vial *V* with 1 cc. of concentrated alkali is next inserted in the basal portion of the apparatus. The volume of this vial and contents is 3.5 cc. The manometer is replaced in such a manner that the two openings *O* and *O′* in the ground joint coincide, and the apparatus is brought to the temperature at which the experiment is to be carried out. The apparatus is then closed by turning the manometer in the ground joint through an angle of 90 deg. The plant extract and the oxidizable material are mixed by shaking the apparatus. The basal portion of the apparatus may be immersed in a water bath of the desired temperature, and continuous shaking by mechanical means may be carried on for the duration of the experiment. After 18 cc. of liquid and the vial and contents are added, the volume of air in the apparatus is 54.1 cc., so that every change in pressure of 1 cm. of mercury corresponds to the absorption of 0.71 cc. of oxygen.

3. *Factors Influencing Activity.*—The oxidases, as a rule, increase in oxidizing power with an increase in temperature up to 42 to 45°C., but above that they become less and less active and are completely inactivated at 100°C. They are very sensitive to acids and apparently are most active in a neutral or slightly alkaline medium. Reed (1916) found that potato oxidase was completely inactivated by 0.0005 to 0.007 *N* acids. Bunzel (1916) found that the inhibiting range of hydrogen-ion concentration for oxidase activity in the case of the potato tuber was pH 3.55 to 3.70. In the buds and leaves of the tulip it was pH 2.30 to 2.80, and for the buds and leaves of magnolia pH 2.45 to 3.05.

Cruess, Jeffrey, and Pancoast (1932) noted that the oxidase activity of the juice expressed from peaches, apricots, olives, avocados, asparagus, and spinach, was greatly affected by pH values. The action of these oxidases almost ceased at pH values below 2.5. Cruess and Fong (1929) stated that in any comparison of the activity of fruit oxidase the pH value of the surrounding medium must be carefully controlled. According to Haber (1928) the soil reaction had little or no effect on the oxidase activity of the tomato. Rose, Kraybill, and Rose (1920) studied the effects of various salts upon the rate of oxidation of pyrogallol by apple bark. They found that tenth-normal solutions of the chlorides of potassium, sodium, lithium, cesium, ammonium, calcium, manganese and ferric iron decreased oxidation, while it was slightly increased by solutions of the same concentration of the sulphates of these cations. The nitrates of potassium, sodium, and magnesium in tenth-normal concentration had no effect on oxidation, while the nitrates of carbon, barium, manganese, and ferric iron decreased it. Oxidation was increased in the presence of the tartrates, oxalates, citrates, acetates, and carbonates.

In their experiments the increased oxidation seemed to be due, in part at least, to the low acidity of the mixture of bark, pyrogallol, and salt, but marked decreases in oxidase activity were not necessarily accompanied by high acidity of the mixture. They believed that ions other than hydrogen and hydroxyl may be important in influencing oxidase activity so that in neutralizing these ions in the study of oxidases the possible effects of the salts formed should be considered.

The relative oxidase activity of healthy and diseased tissues has received considerable attention. Woods (1902) found a greater oxidase activity in the spotted leaves of tobacco affected with mosaic than in the normal ones. Bunzel (1913) noted that the juice from the leaves of beets affected with curly top had an oxidase content two to three times greater than the healthy and normally developing ones. An abnormally high oxidase activity was also shown in those plants whose growth had been retarded. Bunzel (1914) also found that the juice of the tubers and foliage of potato plants affected with curly dwarf showed a greater oxidase activity than the juice from these parts of healthy plants. The oxidase activity of tumor tissue in the beet caused either by freezing or by *Bacterium tumefaciens* was observed by Harvey (1920) to be greater than in the adjacent healthy tissue. Rose (1915) found that the extracts of apple-tree bark affected with Illinois canker caused greater and more rapid oxidation of pyrogallol than did the extract of healthy bark. The diseased bark extract was less acid than the healthy bark extract, hence he concluded that oxidase activity was in approximately inverse ratio to the acidity of the extract in the range of concentrations used by him. His conclusions were substantiated by the fact that the addition of acid to the solution in the apparatus decreased oxidation, while the addition of alkali increased it. Rose (1919) found in later work that apple bark attacked by the blister canker caused about twice as much oxidation with pyrocatechol, guaiacol, and benzidine as did the healthy bark. He considered that the lower tannin content of the diseased bark might have some influence on the oxidase activity.

Guthrie (1931) noted that the treatment of dormant potato tubers or gladiolus corms with ethylene chlorohydrin caused an increase in the capacity of their extracted juices to reduce sulphur to hydrogen sulphide. This action was increased by boiling the juice and was decreased by the addition of unboiled to boiled juice. This action is explained by assuming the presence of oxidase in unboiled juice. Christensen and Samisch (1934) stated that the exposure of plant material to high-frequency sound waves decreased considerably the oxidase activity, but no complete inactivation was obtained after an exposure of 12 hr.

4. *The Relation to Physiological Functions.*—The phenomena of oxidation in plants are good examples of the ability of living organisms to carry out changes at ordinary temperatures without the presence of chemical agents. Thus the living plant is continually oxidizing materials in the process of respiration under conditions under which oxidation cannot be accomplished in the laboratory. In an attempt to explain the mechanism of respiration, the oxidases have generally been assigned a role in the process, although their relationship to it is almost entirely hypothetical. The activity of oxidases has been studied, for the most part, by the action of plant juices upon phenol compounds under the conditions of the laboratory. In one instance, however, Kastle and Buckner (1917) injected phenolphthalin into the stalks of corn and obtained evidence that the phenolphthalin had been oxidized to phenolphthalein in the cells into which it had penetrated. They thus considered that oxidation similar to that obtained by oxidases under laboratory conditions prevails in the plant.

It is, however, difficult to assign a function to the oxidases in respiration, since, by the present known methods of measuring their activity, they have no direct action at all upon the ordinary substances consumed in respiration. The decomposition products of respiration are also incapable of being oxidized directly by the oxidases under the laboratory conditions that have so far been tried. Appleman (1916) found no correlation between oxidase activity and the rate of respiration in potato tubers. He stated that although oxidases may play some role in the process they are certainly not the controlling factor in regulating the rate of respiration.

Navez (1929) believed that some respiratory oxidative process controlled the onset of geotropic curvature. Bonazzi (1931) reported that there was more oxidase in the younger than in the older tissues of the sugar cane. Dexter (1934) found that oxidase activity was unchanged in the plants of cabbage and winter wheat when stored at 2°C. He considered, therefore, that the amount or the activity of oxidase is not correlated with the rate of respiration, with an increase in sugar content, or with a change in cold hardiness. Tyson (1930) noted in the sugar beet that oxidase activity was greatest in those tissues where growth was inhibited. Samisch (1935) found that the oxidase activity from the juice of ripe apricots was greatest in the epidermis and in the cells surrounding the fibrovascular bundles.

Crocker and Harrington (1918) observed in the case of the seeds of Johnson grass and *Amaranthus* that the intensity of oxidase activity decreased markedly with age. One-year-old seeds of Johnson grass showed only two-thirds the oxidase activity of fresh seeds, while 3-year-old seed had only one-third the activity of 1-year-old seeds. Two-year-old *Amaranthus* seeds were only about one-half as active as fresh seeds. The oxidase activity of the dry seeds of Johnson grass showed no difference in intensity from that of germinating seeds. It was also observed that the oxidase activity of the nonliving bracts that enclose the mature caryopses of Sudan and Johnson grass was as high as or higher than that of the caryopses.

Davis (1931) reported that phenolase activity was greater in young seeds with a high percentage of germination than in old seeds with a low percentage of germination. In wheat and cucumber, the phenolase and catalase activities parallel.

It was reported by True and Stockberger (1916) that the intensity of oxidase activity roughly paralleled the distribution of the latex in the poppy (*Papaver somniferum*). With the exception of the roots, the oxidase activity ran roughly parallel with the alkaloid content. The alkaloids appear not to exist as such in the poppy plant but appear as products of the action of the oxidases on constituents present in the latex in the presence of oxygen. Annett (1922) noted that the latex of the Indian poppy has a powerful oxidizing action on guaiacum tincture, pyrogallol, benzidine, and tyrosine in the absence of hydrogen peroxide. He found that opium stored for 3 years possessed an oxidizing enzyme that acted on benzidine and suggested that the loss of morphine from opium on storage might be due to the action of oxidizing enzymes.

There appears to be some connection between the occurrence of oxidases and certain sap pigments in the stems, leaves, and flowers. A vast amount of work has been reported in this regard by Wheldale (1910); Keeble, Armstrong, and Jones (1912); Keeble and Jones (1913); Combes (1913); Everest (1914); Atkins (1915); Robinson (1924); and many others. A fine review of the literature in this field is given by Atkins (1916). The discussion of this relationship of oxidases, however, is beyond the scope of this work.

b. The Peroxidases.—When the juices or watery extracts of tissues produce oxidation only after the addition of hydrogen peroxide, they are said to contain an indirect oxidizing enzyme, which was called "peroxidase" by Linossier (1898). The usual test for a peroxidase is to note if the juice or extract gives a blue color with a solution of guaiacum after the addition of hydrogen peroxide. Guthrie (1930) described a method for estimating peroxidase activity. It is based on the formation of indophenol from a mixture of *p*-phenylenediamine and α-naphthol in a citrate-buffered solution with a pH value of 4.5. A peroxidase is a peroxide-activating enzyme activating both hydrogen peroxide and organic peroxides (Bach and Chodat, 1903). A peroxidase apparently combines with the oxygen from the peroxides to form an intermediate compound which is a more energetic oxidizing agent than the original source of the oxygen. The final stage of oxidation is then effected by the intermediate compound.

The peroxidases are probably more widely distributed in plants than any other enzyme with the exception of catalase. Of the 320 species examined by Onslow (1921), 95 per cent gave the peroxidase reaction with hydrogen peroxide and guaiacum. Clark (1911) reported that 75 per cent of the plants examined by him contained peroxidase. It is especially abundant in the root of the horse-radish, in the tuber of the potato, and in the beet. Onslow (1920) found peroxidase in the blackberry, red and black currants, gooseberry, grape fruit, pineapple, melon, and tomato. McHargue (1920) and Coupin (1925) found that peroxidase predominates in the seeds of most species of plants.

Peroxidase is apparently the most stable of the enzymes associated with oxidation in plant tissues.

Apparently peroxidases from different sources behave differently. Thus Elliot (1932) found that the peroxidase from milk would oxidize both nitrate and tryptophane in the presence of H_2O_2, but that from horse-radish would not.

1. *Preparation.*—Methods for the preparation of peroxidase have been reported by Onslow (1919), Getchell and Walton (1931), and Elliot (1932). According to Onslow it is obtained from the pear or potato as follows:

Thin slices of the material are rapidly pounded in a mortar after sufficient 95 per cent alcohol has been added to prevent exposure to the air. After filtering off the alcohol with a filter pump, the residue is again ground with more alcohol and again filtered. This process should be repeated two or three times, and the grinding should be very thorough. The insoluble residue or white powder that is finally obtained upon drying contains the peroxidase. A cold-water extract is then made of this residue and filtered. This filtered extract does not darken on exposure to air, and in the presence of hydrogen peroxide it oxidizes guaiacum, guaiacol, benzidine, and other phenol compounds. In the absence of hydrogen peroxide it is without action on these substances.

Elliot (1932) gives in detail the following method for the preparation of peroxidase from horse-radish: Wash, chop, and pass through a meat mincer about 1,500 g. of horse-radish. Extract this for 2 hr. with 1 l. of water and squeeze the liquid portion through a cloth. Repeat this process until three or four extractions are made. The combined extracts are then centrifuged or filtered on fluted filter paper, and the translucent liquid that is thus obtained is saturated with ammonium sulphate. This completely precipitates the enzyme with certain impurities. The precipitate is then obtained by centrifuging or by filtering on a Büchner funnel, and lixiviated with 250 cc. of water. The turbid solution is then dialyzed for 3 days until salt free, and the liquid, which doubles in volume during the dialyses, is filtered until clear; twice its volume of denatured alcohol is added. The resulting precipitate is collected by centrifuging and dried in a vacuum desiccator. The product obtained is not readily soluble and should be ground to a paste with a little water before diluting. The entire time necessary for this preparation of peroxidase approximates 7 days.

2. *Factors Influencing Activity.*—It has been observed that SO_2 prevents the darkening of apple tissue. Overholser and Cruess (1923) considered that this inhibitory action is due to the effect of this radical upon the organic peroxide in the tissue and not to the influence upon either the oxidases or the chromogen from which the dark compound is formed by oxidation. Balls and Hale (1934) found that glutathione or cysteine salts applied to sliced apples prevent their discoloration upon drying. They considered this effect to be due to the sulphydryl component upon the peroxidase. Denny (1935) found that treating apple tissue with a solution of thiourea of a concentration as low as 0.1 per cent prevented the browning of the cut surface upon drying in the air. Favorable results were also obtained with pear, banana, and egg plant, but the darkening of potatoes was only slightly inhibited. Balls and Hale (1935) stated that the darkening of the freshly cut surfaces of apple is a reaction that is catalyzed by peroxidase, and that the formation of hydrogen peroxide by a respiratory enzyme which uses molecular oxygen is a necessary preliminary step.

It was noted by Fong and Cruess (1929) and Fong (1931) that the temperature required for the inactivation of the peroxidase of apricots, pears, prunes, figs, tomatoes, bananas, and dates varied with the pH value. The resistance to heat was greatest between pH 5 to 7, decreasing with both increased and decreased values beyond this range until at pH 2.0 and 12.0 the peroxidase was inactivated at room temperatures.

3. *Relation to Physiological Functions.*—It was noted by Flemion (1931) that the activity of peroxidase increased during the afterripening of *Sorbus aucuparia* until germination, and then decreased. Garner, Bacon, Bowling, and Brown (1934), in a study of the nitrogen nutrition of tobacco, noted that the peroxidase activity stood in direct relation to the vegetative activity in the leaf. Josyln and Marsh (1933) considered that peroxidase is not the chief causative agent involved in the deterioration of plant tissue following freezing.

c. The Catalases.—It was noted by Schoenbein (1863) that all the animal and plant tissues with which he worked had the power to decompose hydrogen peroxide with the evolution of molecular oxygen. A further study of this phenomenon by him and others led to the general conclusion that it was a property possessed by all the various enzymes in plant and animal tissues. In 1901, however, Loew isolated from tobacco leaves a body capable of decomposing hydrogen peroxide into molecular oxygen but possessing no other enzymic properties. This body he called "catalase," and with but few exceptions (Reed, 1916; Harvey, 1924) it has been found in the cells of all living tissues that have been examined and is thus apparently the most widely distributed of any of the enzymes. Catalase is classified as an enzyme upon its sensitiveness toward heat, acids, and various poisons, but, as will be discussed later, it exhibits certain behaviors that are not characteristic of enzymes.

1. *Preparation.*—Any description of a method for the preparation and study of catalase must necessarily be general, since the details will depend upon the type of material used, upon the amount of catalase present, and upon other conditions under which the experiment is conducted. For details concerning the preparation and study of various plant tissues, the experiments of Appleman (1910 to 1916), Heinicke (1923, 1924, 1933) Rhine (1924), Davis (1925), Lantz (1927), Knott (1927), Overholser (1928), Bunzel (1930), and Dexter (1934) should be consulted.

The material to be studied is generally ground or macerated in a small quantity of water with an excess of calcium carbonate, since catalase is very sensitive to acids. Quartz sand is sometimes used in the grinding in order to pulverize the tissue more thoroughly. The pulverizing process is usually limited to 2 to 5 min., since in some cases it has been found that the activity of catalase is sensitive to agitation. In some cases the pulverized material is forced through bolting cloth or a sieve to obtain a uniform sample. The fineness to which the material must be reduced to give the best results must be determined by actual experiment, as it has been observed in some cases that the activity of the enzyme may be decreased by pulverizing the tissue too finely. In the case of the seeds of Johnson grass, Crocker and Harrington (1918) found that the maximal activity was obtained when the material was forced through a 70- to 80-mesh screen. They considered that the best results were obtained when the average diameter of the sieve mesh was several times the diameter of the cells.

As soon as maceration is completed, water is added to the mixture—the amount added being determined by the activity of the catalase in the material under consideration. In the case of apple bark and leaves, Heinicke (1923, 1924) added sufficient water to give a dilution of 1 to 50. In working with seeds, Crocker and Harrington (1918) used from 0.5 to 0.025 g. of material for each determination but found 0.14 g. the most desirable. This preparation should not stand for longer than an hour or two, since in most cases the activity of catalase deteriorates very rapidly on aging. This enzyme, as a rule, also loses its activity very rapidly when powdered material is stored in the desiccator. On that account, it is best to grind the material and prepare the mixture just previous to making the catalase determination.

The methods that have been employed in making catalase determinations have been, for the most part, modifications of the method described by Appleman (1910). A description of the apparatus and procedure used by Davis (1926) will serve to illustrate, in general, the methods that are followed (Fig. 37). The apparatus consists of a burette A with a two-way stopcock G at the upper end and a connection with a leveling bulb B at the lower end. An ordinary pinchcock burette may be used with a T-tube inserted in the upper end and connected directly at the bottom with the leveling tube (Heinicke, 1924). A wide-mouthed bottle C of approximately 100 cc. capacity is fitted with a three-holed stopper H. Through one of these openings a small separatory funnel D with a stopcock t is inserted, the top of the funnel being connected with the bottle by the rubber tube E and a glass tube inserted through a hole in the stopper. The bottle is connected through its third hole with the opening R at the top of the burette by means of the rubber tube F.

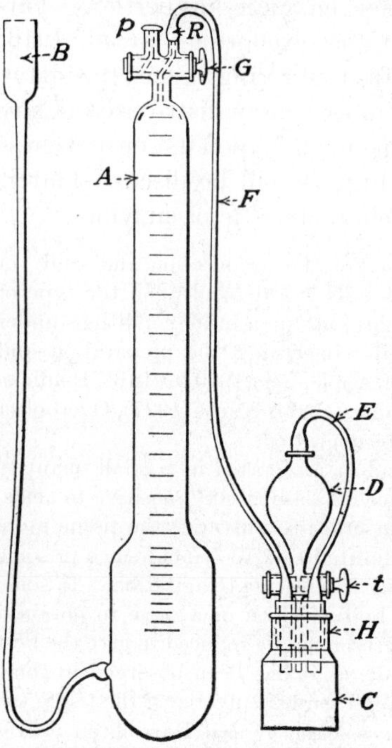

FIG. 37.—Apparatus for the determination of catalase. Description in the text. (*Redrawn from Davis, 1925.*)

After the apparatus is thus set up, a given portion of the mixture whose catalase activity is to be determined is transferred into the bottle C, and the stopper tightly inserted. The stopcock t being closed, 5 cc. of peroxide (dihydrogen) neutralized with 0.1 N sodium hydroxide is added to the separatory funnel D. The bottle is next placed in a water bath at a temperature of 20 to 25°C., and after the contents have reached this temperature the experiment is ready to proceed. The stopcock G is so adjusted that the burette A is connected with the outside air through P. The water level B of the leveling bulb is so placed that it corresponds to the water level in A, after which the exit P is closed and R is opened. The peroxide is then introduced into C through the stopcock t. The bottle C is then gently and uniformly shaken either by hand or by some mechanical apparatus. As the oxygen is liberated, it passes over through the tube F and causes the water level in A to fall. The water level of the leveling tube B is adjusted and the volume of oxygen liberated determined at any desired intervals.

2. *General Nature and Behavior.* Catalase is evidently different in its nature from the ordinary enzymes, since it is not unlimited in its power to decompose hydrogen peroxide, and the indications are that it is consumed in the reaction, and that a given amount is capable of decomposing a definite amount of hydrogen peroxide. According to Loew (1901), catalase exists in two forms, one insoluble in water and the other soluble, which he designated as α-catalase and β-catalase, respectively. Appleman (1910) found that the insoluble catalase of the potato could be separated from the soluble form by ordinary filter paper. Approximately 50 per cent of the total enzyme passed through the filter, indicating that the two types were present in approximately equal proportions. None of this catalase, however, would pass through a Chamberland-Pasteur

filter. It is believed by some (Becking and Hampton, 1920) that the two different catalases are different degrees of peptization of the same substance.

(*a*) *Temperature.*—Catalases from different sources show considerable variation in temperature relations. In the potato, Appleman (1910) found that the catalase was completely destroyed at 50°C. The velocity coefficient was 1.5 from 0 to 10°C., but at higher temperatures there was an apparent decrease in this coefficient, which was probably due to the actual destruction of the catalase as the temperature increased. Lantz (1927) found that there was a gradual accumulation of catalase in corn germinating at 10°C., so that ultimately the catalase content was nearly equal to that at 20 to 30°C. At 42°C. the catalase content was markedly reduced. The point of total destruction of catalase for most of the cases reported, however, ranges from 65 to 80°C.

Crocker and Harrington (1918) found that the air-dry seeds of Johnson grass were all killed and their catalase completely destroyed by heating at 100°C. for 5 hr. Heating at 81°C. for ½ hr. reduced the catalase activity by 30 per cent but improved the germination of the seeds. Heating at 81°C. for 17 hr. reduced the catalase activity approximately 90 per cent of its original value and killed all the seeds. The heating of the seeds of *Amaranthus retroflexus* at 81°C. from 4 to 48 hr. reduced the catalase activity but little, while the power of germination was destroyed. The catalase in the air-dry seeds of *Amaranthus* is thus comparatively heat stable, while the substances essential to vitality are comparatively thermolabile.

It was found by Green, M'Endarfer, Orth, and Burge (1929) that cold weather decreased and hot weather increased the catalase content in pine needles. Carrick (1929) found that catalase activity in the vascular tissue of McIntosh apples was increased after an exposure to a temperature of −7.5°C. for 3.6- and 9-hr. periods, but was decreased after a like exposure for 20 hr. It was noted by Leggatt (1929) in spring wheat that all the thermolabile catalase was not destroyed until heated at 100°C. for 90 min. Lantz (1930) reported that the catalase content of germinating corn diminished rapidly at 42°C. Gonzalez (1933) found that the rise in temperature due to smudge heating increased the activity of catalase in the leaves, buds, bark, and wood of mango trees. Arighi, Joslyn, and Marsh (1933) stated that the catalase content of peas and spinach varies with the temperature used in blanching. It is higher in those treated at 40°C. than in those treated at 20 or 50°C.

(*b*) *Water Content.*—The effect of drying the material upon catalase activity has been observed in numerous cases. Crocker and Harrington (1918) noted that drying the seeds of peach and Johnson grass after they had been in the germinator reduced the catalase activity. Lantz (1927), however, found that the drying of corn seedlings at room temperature *in vacuo* over sulphuric acid did not affect their catalase content. Heinicke (1924) reported that the maceration of dry tissue of apple bark resulted in a marked retardation of its catalase activity, but that maceration of wet tissue had little influence on its activity. The longer the tissue is ground, however, the greater is the reduction in its catalase activity. The addition of water in grinding may tend to neutralize the medium, or it may reduce the heat generated in grinding. Knott (1925) found that catalase preparations deteriorate rapidly at 10 to 20°C., and that placing the bottles on ice immediately after maceration and dilution appears to be the best method of keeping preparations of spinach and celery. Tomato catalase, however, lost its activity in 24 hr. even when on ice.

(c) *Acidity.*—Loew (1901) and Appleman (1910) pointed out that catalase is very sensitive to acids and that the maximal activity and minimal destruction occur in a neutral or slightly alkaline medium. The ground plant tissues either contain or develop sufficient acid to injure greatly the catalase activity. This injury to the enzyme is avoided by grinding the tissue with calcium carbonate. The effect of the hydrogen-ion concentration upon the activity of catalase apparently varies with different tissues. Thus Harvey (1920) found that the juice expressed from overgrowths on *Ricinus*, *Bryophyllum*, and beets had a hydrogen-ion concentration consistently lower than that obtained from the healthy tissues regardless of whether these overgrowths were caused by freezing or by *Bacillus tumefaciens.* The tumor juice in all cases showed a catalase activity much greater than that of the healthy tissue. In the case of the mosaic leaves of tobacco, however, the catalase activity was greatly increased over that of the normal leaves, while the pH values of the diseased and healthy leaves were approximately the same. Weiss and Harvey (1921) found that the hydrogen-ion concentration of potato tubers affected with the wart disease was consistently higher than that of healthy tubers from the same plant, the values being represented by pH 6.00 and pH 6.49, respectively. The catalase activity, however, was much greater in the wart tissue, the values being represented by 17.9 cc. of oxygen for diseased and 7.8 cc. for the healthy tissue. Overholser (1928) noted that when the pH value of the pulp of pears was reduced to 3.55 by the addition of 0.1 *N* hydrochloric acid the catalase was almost completely checked. As the pH value was raised by the addition of 0.1 *N* sodium hydroxide, the catalase activity increased to a maximum, with a pH of from 6.00 to 6.50.

(d) *Light and Darkness.*—It was observed by Euler, Myrbäck, and Myrbäck (1930) that the catalase content for various plants grown in the dark was lower from the fourth to the eighth day than that of those grown in the light. After the tenth day, however, the amount of enzyme in the two types of plants tended to become equal. Schoppel (1933) noted in the germination in the dark of the light-sensitive seeds of *Nicotiana tabacum*, that catalase activity slowly increased, reached a maximum in about 30 hr., and then decreased. In the light there was a marked increase in catalase activity after 37 hr.

3. *Relation to Physiological Activity.*—Since catalase occurs in practically all living cells, the question naturally arises as to what function it performs. Loew (1901) thought that hydrogen peroxide might be produced in the cell as a result of the respiratory process and that its accumulation would be harmful. He believed that its destruction is brought about by the catalase; this would not only prevent injury, but the oxygen thus liberated could be again used in respiration. The formation of hydrogen peroxide in respiration, however, has not been proved, and even if it were formed it is probable that it would not be injurious to the plant, as indicated by the work of Bach and Chodat (1903), who were able to cultivate plants in a medium containing 0.68 per cent hydrogen peroxide. Catalase, however, acts only on hydrogen peroxide and not on organic peroxides. These compounds, rather than hydrogen peroxide, are concerned in respiration, and Reed (1916) considered that the function of catalase in protecting the organism against hydrogen peroxide is very limited if it exists at all. The relation of catalase activity to respiration and to the general metabolic activity or vitality of plants has been extensively studied and will be considered in the discussion of those phenomena.

(*a*) *Respiration.*—The great majority of workers have found some connection between catalase activity and respiration, although the data show some contradictions, but the exact nature of this connection is not known.

(1) *Correlation.*—The catalase activity and the intensity of respiration have been found to parallel each other in many cases, and on that account it has been concluded that the one is the cause of the other.

Appleman (1910, 1916, 1918) found that the catalase activity in the potato tuber and in fresh sweet corn showed a striking correlation with respiratory activity. Thus respiration was very high when the corn was first pulled, but this rate fell off rapidly with storage. The catalase activity in a collateral set of ears showed a decline with storage which was almost directly proportional to the decline in respiratory intensity. Crocker and Harrington (1918) found that the catalase activity of the embryo of Stoner wheat was approximately twenty-eight times that of the endosperm. Since it has been found by Burlakow (1898) that the respiratory intensity of the wheat embryo is about twenty times that of the endosperm, it would appear that here also catalase activity parallels high respiratory intensity. In *Spirogyra*, Burge and Burge (1924) also found a parallelism between respiration and catalase activity. Rosenberg (1910) and Zaleski and Rosenberg (1911) likewise considered that catalase stands in a direct relationship with the respiratory process.

Ranjan and Mallik (1931) stated that catalase activity was correlated with the amount of monosaccharoses and that it was influenced more by the formation of hexoses than by the actual amount present. They considered that respiratory and catalase activity are affected indirectly through hexose formation by either hydrolysis or synthesis. Gustafson, Clark, Shaw, and Warweg (1932) noted in the John Baer tomato that high catalase activity accompanied a high rate of respiration and growth. Ransom (1935) found in both the seeds and fruit of *Polygonum scandens* that the catalase activity and the rate of respiration followed curves similar to those of afterripening and of germination.

(2) *Divergence.*—Sherman (1921) found in the germination of *Crataegus* seeds that the catalase activity increased continuously up to the twelfth day, while the respiratory activity increased only to the sixth day. In the germination of *Amaranthus* the fluctuations of catalase activity and respiratory activity were frequently in opposite directions.

It was found by Morinaga (1925) that the catalase of germinating rice increased under aerobic conditions with a reduced quantity of oxygen but did not increase in the course of anaerobic germination. He concluded that in the case of rice the ratio of increase of catalase activity is a function of the free oxygen in the medium. Magness and Ballard (1926) found that there was an increase in catalase activity in Bartlett pears as they were held in storage after picking, followed by a decrease as the fruit approached prime eating conditions. The decrease in catalase, however, came much earlier in the life of the fruit than did the decrease in carbon dioxide output.

Rhine (1924) noted in the germination of the seeds of wheat, feterita, clover, mustard, radish, and buckwheat that there was always a drop in catalase at the onset of germination, the low point being reached in from ½ hr. to several days, depending upon the type of seeds. This decrease in catalase activity was nevertheless followed by a sharp rise. The intensity of respiration, however, increased rapidly from the first hour of germination. The early curve of respiration thus diverged widely from the curve of catalase. Similar observations were made by Lantz (1927) in germinating corn. He found some correlation between catalase activity and respiration at 20°C., but at 10 and 30°C. this correlation was not apparent, and he concluded that there is, in this case at least, no evidence to assume that catalase is the enzyme chiefly concerned in physiological oxidation.

Bunzel and Kenyon (1933) could find no parallelism between oxidase activity and catalase activity in various parts of the potato plant, and there was some evidence that these stood in some sort of a reciprocal relationship. The oxidase activity was much greater in the tuber than in the foliage, while the catalase activity was higher in the latter and lower in the former. Oxidase activity was greatest in early stages of growth of the foliage, while catalase activity was the lowest at this stage. Marks (1935) found that the catalases from certain species of marine plants were inactivated either directly or indirectly by free oxygen. In barley, Pope (1932, 1933) found a negative correlation between respiratory rate and catalase activity. To explain his results on the basis that respiratory rate and total catalase production are correlated, it would be necessary to assume that in barley a high respiratory rate is associated with proportionately greater consumption of the catalase-activating agency. Thus the surplus of catalase would be actually smaller in amount than when the rate of respiration is low. Miller, Guthrie, and Denny (1934) found in treating potato tubers to overcome dormancy that the catalase and peroxidase activities were not directly connected with the variation in the carbon dioxide output. The enzymes were activated markedly but the changes tended to follow, rather than to precede or to coincide with, the changes in the rate of carbon dioxide production.

Since the catalase activity drops markedly at the beginning of germination and then increases, while respiration increases from the first, Rhine (1924) believed that the most plausible theory of catalase formation in the plant seems to be that it is formed as an enzyme according to the theory of need—the presence of the substance which it attacks acting as a stimulus for the production of the enzyme. The decrease of catalase with germination could thus be explained by supposing the catalase reserve to be used up in attacking the respiratory products faster than it is produced. The high stimulation of the suddenly increased quantity of by-products would thus lead in a short time to the increased production of catalase observed. If this is the case, the catalase curve would follow but never precede the respiration curve.

(*b*) *Vitality.*—It has been shown in some cases that catalase activity parallels the general metabolic activity of the organism, and numerous experiments have been conducted to determine if the catalase activity can be used as an indicator of the vitality and vigor of the plant or plant parts. The relation of catalase activity to the maturity, dormancy, and vitality of seeds has been especially studied. Crocker and Harrington (1918) found in the seeds of Johnson grass and Sudan grass that the fall in catalase began with harvest and continued for an indefinite period. It was found in these seeds that a considerable fall in catalase occurred with little or no fall in vitality. Thus in 9-year-old seeds of Sudan grass the percentage of germination was 98 per cent, although the catalase activity had fallen to less than one-half that of the 1-year-old seeds. In the case of the seeds of Johnson grass and Sudan grass, the catalase activity seemed to be an indicator of age rather than of vitality. In the seeds of *Amaranthus retroflexus*, however, there was no regular fall of catalase activity with the age or vitality of the seed. In maple seed, Jones (1920) noted that after 8 months of storage under laboratory conditions the catalase activity had declined to only one-half that of the fresh seeds. In the seeds of various varieties of lettuce 20 years old, Davis (1925) found that the relation of catalase to vitality was not such as to give reliable data on the viability of seeds, as the catalase content of dead seeds frequently was sufficiently high to indicate fair germination. He found, however, that, if seeds were soaked for a time in warm water at a temperature not sufficiently high to affect their viability, the catalase of dead seeds disorganized very rapidly, while that of the viable seeds was not affected. These observations are contrary to those reported by Nemec and Duchon (1921, 1922, 1923) in their work with seeds of oats and peas. They considered that the viability of seeds could be determined by means of the catalase content of these seeds, since they found that the ratio of catalase of poor seeds to that of viable seeds also expressed very closely the percentage germination of the seeds. The disappearance of catalase from seeds during and after death no doubt bears a close relation to such factors as temperature, moisture, and aeration to which the seeds have been subjected during that time (Davis, 1925).

The catalase activity as a rule increases markedly during germination, and in the seeds of Sudan grass and Johnson grass it doubled during that time (Crocker and Harrington, 1918). As is well known, the lower seeds of *Xanthium* under natural conditions germinate much more readily than the upper ones. It was found by Shull and Davis (1923) that, in the field during the germination season, the lower seeds increased in catalase activity while the upper lagged behind, showing practically no increase. They concluded that the catalase differences in *Xanthium* seeds are in harmony with the physiological differences that operate to bring about delayed germination of upper seeds with intact coats. Davis (1930) noted in the seeds of *Xanthium* and *Ambrosia* that during the period in which the embryo became dormant both the respiration and catalase activity were reduced. In the case of *Xanthium* during the period of afterripening or the removal of dormancy at low temperature, there was a rise in the catalase activity and also in the respiratory activity of the seed. Sherman (1921) noted in the seeds of *Acer saccharum* and *Juniperus virginiana* that catalase activity increased as dormancy ended and germination began.

Tyson (1930) found in sugar beets that catalase activity was correlated with the vigor, growth, and size of the plant. Neller (1931) reported that the catalase activity of the roots of the bindweed (*Convolvulus arvensis*) was greatly decreased after the aerial parts of the plant had been sprayed with chlorates. Landon (1934) believed that the measurement of the degree of variation in catalase activity is useful in studying the herbicidal properties of chemicals, provided that the toxic action peculiar to that specific chemical is considered. Leggatt (1933) found that the vitality of wheat seed may be estimated rather accurately from the determination of total and of thermostable catalase. A determination of this type may be completed within a few hours, while a germination test requires 12 days. It was reported by Baldwin (1935) that there was a close relationship between the viability of certain coniferous seeds and the "catalase quotient." He defined the catalase quotient as the ratio of the volume of oxygen evolved from seeds that had been in the germinator for a few days to the oxygen evolved from resting seeds.

The relation of catalase activity to the general vigor and response of the plant has been studied in the apple tree by Heinicke (1922, 1923, 1924). He found that the catalase activity in apple-leaf tissue was influenced by the factors that affected the nutritive or physiological conditions of it. Thus the wilting of leaves reduced the catalase activity. Leaves from trees growing in sod showed less catalase activity than did those from trees under cultivation, while nitrogenous fertilizers increased the catalase activity in leaves from trees in sod. Ringing reduced the activity of catalase. The catalase activity was less in those leaves grown in the shade as compared with those having good exposure to light. Heinicke found that the catalase activity of the phloem was greater than that of the xylem. The activity of young phloem and xylem was greater than the old phloem and xylem, respectively. Catalase in dormant vegetative buds was greater in amount than in dormant flower buds. Ezell and Crist (1927), however, in studying the effect of nutritive conditions upon the oxidase and catalase activity found in the case of radish, lettuce, and spinach plants that the correlation between the activity of catalase and growth or size of the plants was better than with oxidase and was significantly negative. Heinicke (1923) believed that the activity of catalase is a more sensitive measure of the metabolic status of the tissues than the usual chemical analysis, and that it may serve, along with other measures, as an indicator of the physiological responses of plants to various cultural conditions or treatments. If it could so be used, it would be a convenient indicator. The determinations are quickly and easily made, and numerous samples can be taken at frequent intervals from the same source without serious injury to the experimental subject. The effect of disturbances in physiological functions upon the action of catalase was observed by Moore and Willaman (1917) in tomato plants that had been fumigated with hydrocyanic acid. There was a reduction in the activity of catalase and oxidase, but within a few hours after the fumigation the oxidase activity had returned to normal, while the catalase and respiratory activities exceeded the normal.

(c) *Growth.*—It was reported by Lopriore (1928) that catalase is especially abundant in sexual organs, and that it increases as the time for fertilization approaches. Ripe pollen is richer in this enzyme than the unripe pollen. Camp (1929) noted in 12 species of plants that the tissues related to the male structures whether vegetative or floral showed a distinctly greater catalase activity than the tissues related to the female structures.

The relation of catalase to growth is uncertain in that some investigators have found a positive and others a negative correlation. Thus Harding (1930) found that catalase activity was correlated with the fertilization of Grimes Golden apple trees in that the fruit from those which had been treated with nitrate was consistently higher in catalase. The degree of catalase activity indicated early in the season whether or not tissue breakdown in the first was to occur, since immediately prior to this condition a high catalase activity was registered. Similar observations were made by Neller (1931) with regard to breakdown in Jonathan apples. Garner, Bacon, Bowling, and Brown (1934) noted in the tobacco plant that the catalase activity stood in direct relation to the vegetative activity in the leaf and reached its maximum in the later stages of leaf growth.

Haber (1928) found in the tomato plant that catalase activity in various plant parts was negatively correlated with growth and yield. Chance (1931) could find no parallelism between the catalase activity of inbred strains of corn seedlings and that of the corresponding crosses. Pope (1932, 1933) observed in barley that catalase activity was roughly proportional to the reciprocal growth rate. There were, however, three very definite elevations of catalase activity in the general curve. These were during early germination, during the development of crown roots, and during early jointing. Each of these peaks thus occurred at the time of the inception of a new and definite stage in the functional activity of the plant.

Euler, Gard, and Rislund (1931) found that low catalase and low chlorophyll amounts occur together. Since they considered that catalase is a porphyrin derivative it must have certain groups in common with chlorophyll. Newton and Brown (1931) reported that the catalase activity of the expressed juice of the leaves of winter wheat during the later summer and autumn was directly proportional to the winter hardiness of the varieties studied.

(*d*) *Afterripening.*—Denny, Miller, and Guthrie (1930) studied the effects of ethylene chlorohydrin, sodium thiocyanate, and thiourea upon the action of the enzymes of potatoes that had been treated with these chemicals to break the rest period. All these chemicals appeared to increase the activity of the catalase, peroxidase, and reducing agents in the juice. The increases in catalase and peroxidase began in 24 hr. after treatment with ethylene chlorohydrin, but the response to the other two chemicals was less marked and occurred more slowly. The effect was apparently on the potato tissues and not on the enzyme, since the addition of these chemicals to the expressed juice produced no effect. Flemion (1931) found that the stratified seeds of *Sorbus aucuparia* during afterripening increased in catalase content until at the termination of the process it was ten times that of the dry controls. The catalase content of the seedlings, however, was 20 times that of the seeds. No condition was found under which germination took place without an increase in catalase activity. Haut (1932) noted that as the afterripening of seeds of the apple, peach, and cherry progressed at low temperature, there was a distinct increase in catalase activity.

Bibliography

ALLISON, F. E., S. R. HOOVER, and D. BURK, 1933: A respiration coenzyme, *Sci.*, **78**: 217–218.

ANNETT, H. E., 1922: The enzymes of the latex of the Indian poppy (*Papaver somniferum*), *Biochem. J.*, **16**: 765–770.

APPLEMAN, C. O., 1910: Some observations on catalase, *Bot. Gaz.*, **50**: 182–192.

———, 1911: Physiological behavior of enzymes and carbohydrate transformation in after-ripening of the potato tuber, *Bot. Gaz.*, **52**: 306–315.

———, 1915: Relation of catalase and oxidases to respiration in plants, *Md. Sta. Bull.* 191.

———, 1916: Relation of oxidases and catalase to respiration in plants, *Am. J. Bot.*, **3**: 223–234.

———, 1918: Respiration and catalase activity in sweet corn, *Am. J. Bot.*, **5**: 207–209.

ARCHBOLD, H. K., and A. M. BARTER, 1934: Chemical studies in the physiology of apples. XV, The relation of carbon dioxide output to the loss of sugar and acid in Bramley's seedling apples during storage, *Ann. Bot.*, **48**: 957–966.

ARIGHI, A. L., M. A. JOSLYN, and G. L. MARSH, 1936: Enzyme activity in frozen vegetables, *Ind. and Eng. Chem.*, **28**: 595–598.

ASO, K., 1903: On the chemical nature of the oxydases, *Bull. Col. Agr. Tokyo*, **5**: 481–491; *Bot. Cent.*, **96**: 106 (1904).

———, 1905: On the nature of oxidases, *Beih. bot. Cent.*, I, **18**: 319–326.

ATKINS, W. R. G., 1915: The oxidases and their inhibitors in plant tissues. II, The flowers and leaves of the Iris, *Notes Bot. School Trinity Col. Dublin*, **2**: 210–212.

———, 1915: Oxidases and their inhibitors in plant tissues. IV, The flowers of Iris, *Notes Bot. School Trinity Col. Dublin*, **2**: 236–246.

———, 1916: "Some Recent Researches in Plant Physiology," Chap. 13, The Oxidases in Relation to Pigmentation and the Anthocyanin Pigments, Whittaker and Co., London, pp. 250–292.

AUBERT, E., 1892: Recherches sur la respiration et l'assimilation des plantes grasses, *Rev. Gén. Bot.*, **4**: 203–282, 321–331, 337–353, 421–441, 497–502, 558–568.

AUDUS, L. J., 1935: Mechanical stimulation and respiration rate in the cherry laurel, *New Phytol.*, **34**: 386–402.

BABCOCK, S. M., 1912: Metabolic water: Its production and role in vital phenomena, *Wis. Sta. Tech. Bull.* 22.

BACH, A., 1897: Du rôle des peroxydes dans les phénomènes d'oxydation lente, *C. R. Acad. Sci. Paris*, **124**: 951–954.

——— and R. CHODAT, 1903: Ueber den gegenwärtigen Stand der Lehre pflanzlicher Oxydationsfermente, *Biochem. Cent.*, **1**: 417–421.

——— and ———, 1903: Untersuchungen über die Rolle der Peroxyde in der Chemie der lebenden Zelle. IV, Ueber Peroxydase, *Ber. chem. Ges.*, **36**: 600–605.

——— and ———, 1903: Untersuchungen über die Rolle der Peroxyde in der Chemie der lebenden Zelle. V, Zerlegung der sogenannten Oxydasen in Oxygenasen und Peroxydasen, *Ber. chem. Ges.*, **36**: 606–609.

BAILEY, C. H., 1917: The catalase activity of American wheat flours, *J. Biol. Chem.*, **32**: 539–545.

———, 1917: The moisture content of heating wheat, *J. Am. Soc. Agron.*, **9**: 248–252.

——— and A. M. GURJAR, 1918: Respiration of stored wheat, *J. Agr. Res.*, **12**: 685–713.

———, 1919, 1920: The storage of wheat, *Operative Miller*, **24**: 352, 381–382; **25**: 5–6.

——— and A. M. GURJAR, 1920: Respiration of cereal plants and grains. II, Respiration of sprouted wheat. III, Respiration of rice paddy and milled rice. IV, The respiration of frosted wheat plants, *J. Biol. Chem.*, **44**: 5–7, 9–12, 13–15.

———, 1921: Respiration of shelled corn, *Minn. Sta. Tech. Bull.* 3.

BAILEY, E. M., 1912: Biochemical and bacteriological studies on the banana, *J. Am. Chem. Soc.*, **34**: 1706–1730.

BAILEY, I. W., 1910: Oxidizing enzymes and their relation to sap stain in lumber, *Bot. Gaz.*, **50**: 142–147.

BAKKE, A. L., and N. L. NOECKER, 1933: The relation of moisture to respiration and heating in stored oats, *Iowa Sta. Res. Bull.* 165.

BALCH, W. B., 1933: A large respirometer, *Sci.*, **77**: 116–117.

BALDWIN, H. I., 1935: Catalase activity as a measure of viability of tree seeds, *Am. J. Bot.*, **22**: 635–645.

BALLS, A. K., and W. S. HALE, 1934: On peroxidase: Role of peroxidase in the darkening of apples, *Am. Chem. Soc., Division Agr. and Food Chem. and Division Biol. Chem.*, 3 pp. Quoted by Denny, 1935.

——— and ———, 1935: Peroxidase in the darkening of apples, *Ind. and Eng. Chem.*, **27**: 335–337.

BARKER, J., 1933: Analytical studies in plant respiration. IV-V, The relation of the respiration of potatoes to the concentration of sugars and to the accumulation of a depressant at low temperatures. I, The effect of temperature history on the respiration/sugar relation. II, The form of the normal respiration/sugar relation and the mechanism of depression, *Proc. Roy. Soc.*, B, **112**: 316–358.

———, 1935: A note on the effect of handling on the respiration of potatoes, *New Phytol.*, **34**: 407–408.

———, 1936: Analytic studies in plant respiration. VI, The relation of the respiration of potatoes to the concentration of sugars and to the accumulation of a depressant at low temperatures. III, The relation of the respiration to the concentration of sucrose, *Proc. Roy. Soc.*, B, **119**: 453–473.

BARNES, C. R., 1905: The theory of respiration, *Bot. Gaz.*, **39**: 81–98.

BARTLETT, A. W., 1920: Note on an improved method for demonstrating the absorption of oxygen in respiration, *New Phytol.*, **19**: 151–152.

Beaumont, J. H., and J. J. Willaman, 1924: Preliminary report on the respiration of apple twigs during the winter, *Proc. Am. Soc. Hort. Sci.*, **21**: 99–104.

———, ———, and W. A. DeLong, 1927: Notes on apparatus for low temperature respiration studies, *Plant Physiol.*, **2**: 487–495.

Becker, H., 1912: Über die Keimung verschiedenartiger Früchte und Samen bei derselben Spezies, *Beih. bot. Cent.*, I, **29**: 21–143.

Becker, Z. E., 1936: A comparison between the action of carbonic acid and other acids upon the living cell, *Protoplasma*, **25**: 161–175.

Becking, L. G. M. B., and H. C. Hampton, 1920: Measurement of the catalytic power of catalase, *Am. J. Bot.*, **7**: 261–275.

Bělehrádek, J., and M. Bělehrádková, 1929: Influence of age on the temperature coefficient of the respiration rate in leaves of *Scolopendrium scolopendrium*, *New Phytol.*, **28**: 313–317.

Bennet-Clark, T. A., 1932: A method of automatically recording the oxygen intake of living tissues, *Sci. Proc. Roy. Dub. Soc.*, **20**: 281–291.

———, 1932: The respiratory quotients of succulent plants, *Sci. Proc. Roy. Dub. Soc.*, **20**, 293.

———, 1933: The role of organic acids in plant metabolism. II, *New Phytol.*, **32**: 128–159.

Bennett, J. P., and E. T. Bartholomew, 1924: The respiration of potato tubers in relation to the occurrence of black heart, *Calif. Sta. Tech. Paper* 14.

Benoy, M. P., 1929: The respiration factor in the deterioration of fresh vegetables at room temperature, *J. Agr. Res.*, **39**: 75–80.

——— and J. E. Webster, 1930: Note on the relation of rate of respiration to chemical composition in fresh vegetables, *Plant Physiol.*, **5**: 181–182.

Bergman, H. F., 1921: The effect of cloudiness on the oxygen content of water and its significance in cranberry culture, *Am. J. Bot.*, **8**: 50–58.

———, 1925: The respiratory activity of various parts of the cranberry plant in relation to flooding injury, *Am. J. Bot.*, **12**: 641–660.

———, 1929: Changes in the rate of respiration of the fruits of the cultivated blueberry during ripening, *Sci.*, **70**: 15.

———, 1930: Respiration of cranberry plants in relation to water injury, *Proc. Ann. Meeting Am. Cranberry Growers' Assoc.*, **60**: 21–25; *Exp. Sta. Rec.*, **69**: 343.

———, 1930: The oxygen content of water in winter flooded bogs under ice, *Wis. State Cranberry Growers' Assoc.*, **44**: 32–40.

Bertrand, G., 1894: Sur la latex de l'arbre à laque, *C. R. Acad. Sci. Paris*, **118**: 1215–1218.

———, 1895: Sur la laccase et le pouvoir oxydant de cette diastase, *C. R. Acad. Sci. Paris*, **120**: 266–269.

———, 1896: Sur les rapports qui existent entre la constitution chimique des composés organiques et leur oxydabilité sous l'influence de la laccase, *C. R. Acad. Sci. Paris*, **122**: 1132–1134.

———, 1896: Sur les rapports qui existent entre la constitution chimique des composés organiques et leur oxydabilité sous l'influence de la laccase, *Bull. Soc. Chim. Paris*, **15**: 793. Quoted by Appleman, 1916.

———, 1896: Sur une nouvelle oxydase, ou ferment soluble oxydant, d'origine végétale, *C. R. Acad. Sci. Paris*, **122**: 1215–1217.

Bezagu, M., 1919: Variations de la respiration des cellules de feuille avec l'âge, *C. R. Acad. Sci. Paris*, **169**: 701–702.

Bigelow, W. D., H. C. Gore, and B. J. Howard, 1905: Studies on apples. I, Storage, respiration and growth, *U. S. Bur. Chem. Bull.* 94.

BLACKMAN, F. F., and P. PARIJA, 1928: Analytic studies in plant respiration. I, Respiration of a population of senescent ripening apples, *Proc. Roy. Soc.*, B, **103**: 412–446.

———, 1928: Analytic studies in plant respiration. III, Formation of a catalytic system for the respiration of apples and its relation to oxygen, *Proc. Roy. Soc.*, B, **103**: 491–523.

———, 1930, 1931: Respiration and oxygen concentration, *Fifth Int. Bot. Congr. Cambridge* (1930); *Rept. Proc.* Pub. Cambridge (1931).

BOAS, F., 1936: Ionenwirkung, Katalasetätigkeit und Leistung der Zelle, *Angew. Bot.*, **18**: 13–16.

BOEHM, J. A., 1887: Ueber die Respiration der Kartoffel, *Bot. Ztg.*, **45**: 671–675, 680–691.

BOLAS, B. D., and R. MELVILLE, 1933: The influence of environment on the growth and metabolism of the tomato plant. I, Methods, technique and preliminary results, *Ann. Bot.*, **47**: 673–688.

BONAZZI, A., 1931: Studies in sugar cane physiology. II, Sugar cane gases, *Bot. Gaz.*, **92**: 442–445.

BONNIER, G., and L. MANGIN, 1884: Sur les variations de la respiration des graines germantes avec le développement, *Bull. Soc. Bot. France*, **31**: 306–309.

BRIGGS, G. E., and A. H. K. PETRIE, 1931: Respiration as a factor in the ionic equilibria between plant tissues and external solutions, *Proc. Roy. Soc.*, B, **108**: 317–326.

BÜCHNER, E., 1897: Alkoholische Gärung ohne Hefezellen, *Ber. chem. Ges.*, **30**: 117–124, 1110–1113.

BUNZEL, H. H., 1912: The measurement of the oxidase content of plant juices, *U. S. Bur. Plant Ind. Bull.* 238.

———, 1913: A biochemical study of the curly-top of sugar beets, *U. S. Bur. Plant Ind. Bull.* 277.

———, 1914: A simplified and inexpensive oxidase apparatus, *J. Biol. Chem.*, **17**: 409–411.

———, 1914: Oxidases in healthy and in curly-dwarf potatoes, *J. Agr. Res.*, **2**: 373–404.

———, 1915: On alfalfa laccase, *J. Biol. Chem.*, **20**: 697–706.

———, 1916: The mode of action of the oxidases, *J. Biol. Chem.*, **24**: 91–102.

———, 1916: The relative oxidase activity of different organs of the same plant, *J. Biol. Chem.*, **24**: 103–110.

———, 1917: The relationship existing between the oxidase activity of plant juices and their hydrogen ion concentrations with a note on the cause of oxidase activity in plant tissues, *J. Biol. Chem.*, **28**: 315–333.

——— and H. HASSELBRING, 1917: The supposed action of potassium permanganate with plant peroxidases, *Bot. Gaz.*, **63**: 225–228.

———, 1930: A simple apparatus for measuring catalase activity in plant and animal tissues, *Sci.*, **72**: 505–506.

——— and M. B. KENYON, 1933: On potato catalase, *Bull. Torrey Bot. Club*, **60**: 469–474.

BURGE, W. E., and E. L. BURGE, 1924: Effect of temperature and light on catalase content of *Spirogyra*, *Bot. Gaz.*, **77**: 220–224.

BURLAKOW, G., 1898: Ueber Athmung des Keimes des Weizens, *Triticum vulgare*, abst., *Bot. Cent.*, **74**: 323–324.

BURROUGHS, A. M., 1922: Changes in the respiration rate of ripening apples, *Proc. Am. Soc. Hort. Sci.*, **19**: 225–235.

BUSHNELL, J., 1925: The relation of temperature to growth and respiration in the potato plant, *Minn. Sta. Tech. Bull.* 34.

———, 1935: Sensitivity of the potato plant to soil aeration, *J. Am. Soc. Agron.*, **27**: 251–253.

Butler, O., 1913: A note on the significance of sugar in the tubers of *Solanum tuberosum*, *Bull. Torrey Bot. Club.*, **40**: 110–118.

Camp, W. H., 1929: Catalase activity and sex in plants, *Am. J. Bot.*, **16**: 221–224.

Cannon, W. A., 1932: On the variation of the oxygen content of cultural solutions, *Sci.*, **75**: 108–109.

——— and E. A. Purer, 1935: On the removal of oxygen from water by cut branches, *Sci.*, **81**: 100.

Carrick, D. B., 1928: The effect of freezing on the respiration of the apple, *N. Y. Sta.* (*Cornell*) *Mem.* 110.

———, 1929: The effect of freezing on the catalase activity of apple fruits, *N. Y. Sta.* (*Cornell*) *Mem.* 122.

———, 1930: Some cold-storage and freezing studies on the fruit of the vinifera grape, *N. Y. Sta.* (*Cornell*) *Mem.* 131.

Cerighelli, R., 1924: Sur le quotient respiratoire de la racine et ses variations au cours du développement de la plante, *C. R. Acad. Sci. Paris*, **178**: 645–647.

Chance, H. L., 1931: The relation between catalase activity and vigor in inbred strains and crosses of corn seedlings, *Am. J. Bot.*, **18**: 696–703.

Christensen, R. J., and R. Samisch, 1934: Effect of high frequency sound waves on oxidase activity, *Plant Physiol.*, **9**: 385–386.

Clark, E. D., 1910: "The Plant Oxidases," *thesis*, *Columbia Univ.*, New York.

———, 1911: The nature and function of the plant oxidases, *Torreya*, **11**: 23–31, 55–61, 84–92, 101–110.

Clausen, H., 1890: Beiträge zur Kenntniss der Athmung der Gewächse, *Landw. Jahrb.*, **19**: 893–930.

Clements, H. F., 1935: Morphology and physiology of the pome lenticels of *Pyrus malus*, *Bot. Gaz.*, **97**: 101–117.

Coleman, D. A., B. E. Rothgeb, and C. H. Fellows, 1928: Respiration of sorghum grains, *U. S. Dept. Agr. Tech. Bull.* 100.

Combes, R., 1913: Production expérimentale d'une anthocyane identique à celle qui se forme dans les feuilles rouges et automne, en partant d'un composé extrait des feuilles vertes, *C. R. Acad. Sci. Paris*, **157**: 1002–1005.

Coupin, H., 1925: Sur les peroxydases dans les graines sèches, *C. R. Acad. Sci. Paris*, **180**: 685–687.

Crocker, W., and W. E. Davis, 1914: Delayed germination in seeds of *Alisma plantago*, *Bot. Gaz.*, **57**: 285–322.

——— and G. T. Harrington, 1918: Catalase and oxidase content of seeds in relation to their dormancy, age, vitality and respiration, *J. Agr. Res.*, **15**: 137–174.

Cruess, W. V., and W. Y. Fong, 1929: The effect of pH value and hydrogen peroxide concentration of fruit oxidase activity, *Plant Physiol.*, **4**: 363–366.

———, R. N. Jeffrey, and H. M. Pancoast, 1932: Observations on the effect of pH value on oxidase activity, abst., *Am. J. Bot.*, **19**: 841.

Darsie, M. L., C. Elliott, and G. J. Peirce, 1914: A study of the germinating power of seeds, *Bot. Gaz.*, **58**: 101–137.

Dastur, R. H., and R. M. Desai, 1935: The carbon dioxide carbohydrate ratio in the aerobic and the anaerobic respiration of rice, *Ann. Bot.*, **49**: 53–69.

Davis, W. B., and C. G. Church, 1931: The effect of ethylene on the chemical composition and the respiration of the ripening Japanese persimmons, *J. Agr. Res.*, **42**: 165–183.

Davis, W. C., 1931: Phenolase activity in relation to seed viability, *Plant Physiol.*, **6**: 127–137.

DAVIS, W. E., and R. C. ROSE, 1912: The effect of external conditions upon the after-ripening of the seeds of *Crataegus mollis*, *Bot. Gaz.*, **54**: 49–62.

———, 1925: The use of catalase as a means of determining the viability of seeds, *Proc. 18th Ann. Meeting Assoc. Off. Seed Analysts N.A.*, 6–12.

———, 1925: Unpub. data, *Kans. State Col.*

———, 1930: Primary dormancy, after-ripening and the development of secondary dormancy in embryos of *Ambrosia trifida*, *Am. J. Bot.*, **17**: 58–76.

———, 1930: The development of dormancy in seeds of cocklebur (*Xanthium*), *Am. J. Bot.*, **17**: 77–87.

DE BOER, S. R. A., 1930: The effect of ionized air on the rate of respiration of fungi, *Ann. Bot.*, **44**: 989–1001.

DELEANO, N. T., 1912: Studien über den Atmungsstoffwechsel abgeschnittener Laubblätter, *Jahrb. wiss. Bot.*, **51**: 541–592.

———, 1912: Untersuchungen über die in Weinblättern enthaltenen Kohlenhydrate und stickstoffhaltigen Körper, *Z. physiol. Chem.*, **80**: 79–94.

DE LONG, W. A., J. H. BEAUMONT, and J. J. WILLAMAN, 1930: Respiration of apple twigs in relation to winter hardiness, *Plant Physiol.*, **5**: 509–534.

DENNY, F. E., 1924: Effect of ethylene upon respiration of lemons, *Bot. Gaz.*, **77**: 322–330.

———, L. P. MILLER, and J. D. GUTHRIE, 1930: Enzyme activities of juices from potatoes treated with chemicals that break the rest period, *Am. J. Bot.*, **17**: 483–509.

———, 1935: Thiourea prevents browning of plant tissues and juices, *Contr. Boyce Thompson Inst.*, **7**: 55–61.

DE SAUSSURE, T., 1804: "Recherches chimiques sur la végétation," Paris.

DEXTER, S. T., 1934: Respiratory rate and enzyme activity as related to the hardened condition of plants, *Plant Physiol.*, **9**: 831–837.

DOWD, O. J., 1933: Preliminary studies on the internal atmosphere of apples, *Proc. Am. Soc. Hort. Sci.*, **30**: 162–163.

DOYLE, J., and P. O'CONNOR, 1930: Seasonal changes in the catalase content of conifer leaves, *Ann. Bot.*, **44**: 907–917.

DRAIN, B. D., 1923: Oxidase activity in varieties of apples, *Proc. Am. Soc. Hort. Sci.*, **20**: 25–30.

———, 1926: Temperature and respiratory enzymes of apples, *Bot. Gaz.*, **82**: 183–195.

DUVEL, J. W. T., 1904: The vitality and germination of seeds, *U. S. Bur. Plant Ind. Bull.* 58.

EAVES, C. A., 1935: Preliminary study of the effect of a series of temperature changes upon respiratory activity of apples during the post-climacteric in senescent decline, *Sci. Agr.*, **16**: 28–39.

EDWARDS, T. I., 1933: The germination and growth of *Peltandria virginica* in the absence of oxygen, *Bull. Torrey Bot. Club*, **60**: 573–582.

ELLIOT, K. A. C., 1932: Oxidations catalyzed by horse radish and milk-peroxidases, *Biochem. J.*, **26**: 1281–1290.

EMERSON, P., 1930: A method for measuring time-rate carbon dioxide production, *J. Am. Soc. Agron.*, **22**: 819–820.

EMERSON, R., 1927: The effect of certain respiratory inhibitors on the respiration of *Chlorella*, *J. Gen. Physiol.*, **10**: 469–477.

ERMAKOFF, A. I., and N. N. IWANOFF, 1931: Über die Atmung der Samen von Ölpflanzen, *Biochem. Z.*, **231**: 79–91.

EULER, H., and I. BOLIN, 1909: Zur Kenntniss biologisch wichtiger Oxydation, *Z. physiol. Chem.*, **61**: 79–92; *Bot. Gaz.*, **49**: 395 (1910).

———, K. Myrbäck, and S. Myrbäck, 1930: Zum Bestimmen der Katalase in Pflanzenmaterial, *Z. physiol. Chem.*, **186**: 212–222.

———, U. Gard, and G. Rislund, 1931: Catalase and sugar determinations in chlorophyll-deficient plants, *Z. physiol. Chem.*, **203**: 165–177; *Chem. Abst.*, **26**: 1317.

Everest, A. E., 1914: The production of anthocyanins and anthocyanidins. I, II, *Proc. Roy. Soc.*, B, **88**: 326–332; **87**: 444–452.

Ewart, A. J., 1915: A comparative study of oxidation by catalysts of organic and inorganic origin, *Proc. Roy. Soc.*, B, **88**: 284–320.

Ezell, B. D., and J. W. Crist, 1927: Effect of certain nutrient conditions on activity of oxidase and catalase, *Mich. Sta. Tech. Bull.* 78.

Falk, K. G., G. McGuire, and E. Blout, 1919: Studies on enzyme action. XVII, The oxidase, peroxidase, catalase and amylase of fresh and dehydrated vegetables, *J. Biol. Chem.*, **38**: 229–244.

Fidler, J. C., 1933: Studies in zymasis. IV, The accumulation of zymastic products in apples during senescence, *Biochem. J.*, **27**: 1614–1621.

———, 1933: Studies in zymasis. V, Seasonal fluctuations in zymasis and in carbon dioxide/alcohol number ratios in apples in the absence of oxygen, *Biochem. J.*, **27**: 1622–1628.

———, 1934: Studies in zymasis. VII, The estimation of ethyl alcohol and acetaldehyde content of apples, *Biochem. J.*, **28**: 1107–1120.

Flemion, F., 1931: After-ripening germination and vitality of seeds of *Sorbus aucuparia*, *Contr. Boyce Thompson Inst.*, **3**: 413–439.

Fong, W. Y., and M. V. Cruess, 1929: The effect of pH value on the inactivation temperature of fruit oxidase, *Plant Physiol.*, **4**: 537–543.

——— and ———, 1929: Comparison of several indicators for fruit oxidase, *Am. J. Bot.*, **16**: 799–802.

———, 1931: The effect of pH value on the inactivation temperature of fruit oxidase, *Trans. Sci. Soc. China*, **7**: 88–94; *Chem. Abst.*, **26**: 496.

Fraymouth, J., 1928: The moisture relations of terrestrial algae. III, The respiration of certain lower plants, including terrestrial algae with special reference to the influence of drought, *Ann. Bot.*, **42**: 75–100.

French, C. S., H. I. Kohn, and P. S. Tang, 1934: Temperature characteristics for the metabolism of *Chlorella*. II, The rate of respiration of cultures of *Chlorella pyrenoidosa* as a function of time and of temperature, *J. Gen. Physiol.*, **18**: 193–207.

———, 1934: Temperature characteristics for the metabolism of *Chlorella*. III, The catalytic decomposition of hydrogen peroxide by *Chlorella pyrenoidosa*, *J. Gen. Physiol.*, **18**: 209–213.

Friedrich, R., 1908: Ueber die Stoffwechselvorgänge infolge der Verletzung von Pflanzen, *Cent. Bakt.*, II, **21**: 330–349.

Gallagher, P. H., 1923: Mechanism of oxidation in the plant. I, The oxygenase of Bach and Chodat: Function of lecithins in respiration, *Biochem. J.*, **17**: 515–529.

———, 1924: Mechanism of oxidation in the plant. II, An investigation of substances which are capable of behaving as peroxydases, *Biochem, J.*, **18**: 29–38.

———, 1924: Mechanism of oxidation in the plant. III, Peroxydase. Observations on the thermostability of the peroxydase of the mangold, *Biochem. J.*, **18**: 39–46.

Ganong, W. F., 1908: "A Laboratory Course in Plant Physiology," Henry Holt and Company, New York, 129 pp.

Garner, W. W., C. W. Bacon, J. D. Bowling, and D. E. Brown, 1934: The nitrogen nutrition of tobacco, *U. S. Dept. Agr. Tech. Bull.* 414.

GENEVOIS, L., 1927: Über Atmung und Gärung in grünen Pflanzen, *Biochem. Z.*, **186**: 461–473.

———, 1928: Über Atmung und Gärung in grünen Pflanzen. II, Der Stoffwechsel der Phanerogamen, *Biochem. Z.*, **191**: 147–158.

GERBER, C., 1900: Étude comparée de la respiration des graines oléagineuses pendant leur développement et pendant leur germination, *Actes 1st Cong. Internat. Bot. Paris*, **1900**: 59–101.

GERHARDT, F., and B. D. EZELL, 1934: Retention of carbon dioxide gas in the intercellular atmosphere of pears and apples, *Sci.*, **80**: 253–254.

GERHART, A. R., 1930: Respiration in strawberry fruits, *Bot. Gaz.*, **89**: 40–64.

GETCHELL, R. W., and J. H. WALTON, 1931: Some factors influencing the activity of peroxidase, *J. Biol. Chem.*, **91**: 419–433.

GILMAN, J. C., and D. H. BARRON, 1930: Effect of molds on temperature of stored grain, *Plant Physiol.*, **5**: 565–573.

GINSBURG, J. M., 1925: A modified respiration apparatus for plant and soil studies, *Soil Sci.*, **19**: 411–417.

GODLEWSKI, E., 1882: Beiträge zur Kenntniss der Pflanzenathmung, *Jahrb. wiss. Bot.*, **13**: 491–543.

GODWIN, H., 1935: The effect of handling on the respiration of cherry laurel leaves, *New Phytol.*, **34**: 403–406.

GONZALEZ, L. G., 1933: Influence of smudging on the respiration and catalase activity of the mango, *Mangifera indica*, *Philippine Agr.*, **21**: 533–540; *Exp. Sta. Rec.*, **70**: 482.

GORE, H. C., 1911: Studies on fruit respiration, *U. S. Bur. Chem. Bull.* 142.

———, 1914: Changes in composition of peel and pulp of ripening bananas, *J. Agr. Res.*, **3**: 187–205.

GORTNER, R. A., 1910: A contribution to the study of oxidases, *J. Chem. Soc.* (*London*), **97**: 110–120.

GRĂCANIN, M., 1926: Ein Beitrag zur Kenntnis der Katalasewirkung bei autotrophen Pflanzen, *Biochem. Z.*, **168**: 429–442.

GREEN, F. C., M. E. M'ENDARFER, O. S. ORTH, and W. E. BURGE, 1929: Further study on the effect of summer and winter temperatures on the catalase of pine needles: a reply to criticism, *Proc. Roy. Irish Acad.*, **39B**: 156–159; *Biol. Abst.*, **5**: 2677.

GREEN, J. R., and A. H. JOHNSON, 1931: Effect of petroleum oils on the respiration of bean leaves, *Plant Physiol.*, **6**: 149–159.

———, 1936: Effect of petroleum oils on the respiration of bean plants, apple twigs and leaves, and barley seedlings, *Plant Physiol.*, **11**: 101–113.

GREGORY, F. G., and F. J. RICHARDS, 1929: Physiological studies in plant nutrition. I, The effect of manurial deficiency on the respiration, assimilation and assimilation rate in barley, *Ann. Bot.*, **43**: 119–161.

GRONER, M. G., 1936: Respiration of green and chlorophyll-deficient types in maize, *Am. J. Bot.*, **23**: 381–385.

GRÜSS, J., 1907: Enzymewirkungen aus Wundrand der Kartoffelknolle, *Z. Pflanzenkrank.*, **17**: 69–70.

GURJAR, A. M., 1917: The adaptation of Truog's method for the determination of carbon dioxid to plant respiration studies, *Plant World*, **20**: 288–293.

GUSTAFSON, F. G., 1929: Respiration of tomato fruits, *Plant Physiol.*, **4**: 349–357.

———, 1930: Intramolecular respiration of tomato fruits, *Am. J. Bot.*, **17**: 1011–1028.

———, 1932: Anaerobic respiration of cacti, *Am. J. Bot.*, **19**: 823–834.

———, I. CLARK, D. A. SHAW, and E. WARWEG, 1932: Catalase activity in tomato fruits at different stages of their development, *Plant Physiol.*, **7**: 155–164.

———, 1934: Production of alcohol and acetaldehyde by tomatoes, *Plant Physiol.*, **9**: 359–367.

———, 1934: Äthylalkohol und Acetaldehyd in gewissen Arten von Kakteen, *Biochem. Z.*, **272**: 172–179.

———, 1936: Influence of oxygen and carbon dioxide concentrations on the respiration of tomato fruits, *Am. J. Bot.*, **23**: 441–445.

GUTHRIE, J. D., 1930: An iodimetric method for determining oxidase activity, *J. Am. Chem. Soc.*, **52**: 3614–3618.

———, 1931: A method for the determination of peroxidase activity, *J. Am. Chem. Soc.*, **53**: 242–244.

———, 1931: The inhibiting effect of oxidase on the reduction of sulphur by potato and gladiolus juice, *Contr. Boyce Thompson Inst.*, **3**: 125–130.

———, 1931: The effect of various chemical treatments of dormant potato tubers on the peroxidase, catalase, pH and reducing properties of the expressed juice, *Contr. Boyce Thompson Inst.*, **3**: 499–508.

———, F. E. DENNY, and L. P. MILLER, 1932: Effect of ethylene chlorhydrin treatments on the catalase, peroxidase, pH and sulphydryl content of Gladiolus corms, *Contr. Boyce Thompson Inst.*, **4**: 131–140.

HAAS, A. R. C., 1916: A simple and rapid method of studying respiration by the detection of exceedingly minute quantities of carbon dioxide, *Sci.*, **44**: 105–108.

———, 1917: Rapid respiration after death, *Proc. Nat. Acad. Sci.*, **3**: 688–691.

———, 1917: Anesthesia and respiration, *Sci.*, **46**: 462–464.

———, 1919: Respiration after death, *Bot. Gaz.*, **67**: 347–366.

———, 1919: Effect of anesthetics upon respiration, *Bot. Gaz.*, **67**: 377–405.

———, 1930: Oxidation-reduction indicators as a means of determining overheating in walnuts during dehydration, *Bot. Gaz.*, **89**: 200–204.

HABER, E. S., 1928: Catalase and oxidase of the tomato as influenced by the soil reaction, *Iowa State Col. J. Sci.*, **3**: 29–39; *Bot. Abst.*, **4**: 49.

HAFENRICHTER, A. L., 1928: Respiration of the soybean, *Bot. Gaz.*, **85**: 271–299.

HALLER, M. H., P. L. HARDING, J. M. LUTZ, and D. H. ROSE, 1931: The respiration of some fruits in relation to temperature, *Proc. Am. Soc. Hort. Sci.*, **28**: 583–589.

———, ———, and D. H. ROSE, 1932: The interrelation of firmness, dry weight and respiration in strawberries, *Proc. Am. Soc. Hort. Sci.*, **29**: 330–334.

HANES, C. S., and J. BARKER, 1931: The physiological action of cyanide. I, The effects of cyanide on the respiration and sugar content of the potato at 15°C., *Proc. Roy. Soc.*, B, **108**: 95–119.

HARDING, P. L., 1929: Respiration studies of Grimes apples under various controlled temperatures, *Proc. Am. Soc. Hort. Sci.*, **26**: 319–324.

———, T. J. MANEY, and H. H. PLAGGE, 1929: Apparatus for the determination of carbon dioxide in the respiration of apples, *Sci.*, **70**: 125–126.

———, 1930: Relation of catalase activity to temperature, respiration and nitrogen fertilization of Grime's Golden apples, *Proc. Am. Soc. Hort. Sci.*, **27**: 37–42.

———, 1935: Physiological behavior of Grimes Golden apples in storage, *Iowa Sta. Tech. Bull.* 182.

HARLEY, C. P., 1929: Relation of picking time to acetaldehyde content and core breakdown of Bartlett pears, *J. Agr. Res.*, **39**: 483–493.

HARRINGTON, G. T., 1923: Respiration of apple seeds, *J. Agr. Res.*, **23**: 117–130.

——— and W. CROCKER, 1923: A new and efficient respirometer for seeds and other small objects: Directions for its use, *J. Agr. Res.*, **23**: 101–115.

HARRIS, G. H., 1929: Studies on tree root activities. I, An apparatus for studying root respiration and factors which influenced it, *Sci. Agr.*, **9**: 553–565.

———, 1930: The influence of top on root as determined by root respiration of young fruit trees, *Proc. Am. Soc. Hort. Sci.*, **26**: 329–334.

———, 1930: Studies on tree root activities. II, Some factors which influence tree root respiration, *Sci. Agr.*, **10**: 564–585.

———, 1930: Studies on tree root activities. III, *Sci. Agr.*, **11**: 191–199.

Harvey, R. B., 1919: Apparatus for the measurement of oxidase and catalase activity, *J. Gen. Physiol.*, **2**: 253–254.

———, 1920: Hydrogen ion changes in the mosaic disease of tobacco plants and their relation to catalase, *J. Biol. Chem.*, **42**: 397–400.

———, 1920: Relation of catalase, oxidase and H concentrations to the formation of overgrowths, *Am. J. Bot.*, **7**: 211–223.

———, 1924: Enzymes of thermal algae, *Sci.*, **60**: 481–482.

Hasselbring, H., and L. Alsberg, 1910: Studies upon oxidases, *Sci.*, **31**: 637.

——— and L. A. Hawkins, 1915: Respiration experiments with sweet potatoes, *J. Agr. Res.*, **5**: 509–519.

Haut, I. C., 1932: Catalase activity in relation to the after ripening of fruit tree seeds, *Proc. Am. Soc. Hort. Sci.*, **29**: 375–379.

Heinicke, A. J., 1922: Catalase activity as an indicator of nutritive condition of fruit tree tissues, *Proc. Am. Soc. Hort. Sci.*, **19**: 209–214.

———, 1923: Factors influencing catalase activity in apple leaf tissue, *N. Y. Sta. (Cornell) Mem.* 62.

———, 1924: Catalase, activity in dormant apple twigs: its relation to the condition of the tissue, respiration and other factors, *N. Y. Sta. (Cornell) Mem.* 74.

———, 1933: Apparatus for making autographic records of catalase activity of plant tissues, and the procedure involved, *J. Agr. Res.*, **46**: 1137–1143.

Herzfeld, E., and R. Klinger, 1919: Chemische Studien zur Physiologie und Pathologie. VI, Zur Biochemie der Oxydationen. (Zellatmung; Oxydationsfermente; zur Theorie der Narkose), *Biochem. Z.*, **93**: 324–352; *Bot. Abst.*, **8**: 95 (1921).

Hill, G. R., 1913: Respiration of fruits and growing plant tissues in certain gases, with reference to ventilation and fruit storage. *N. Y. Sta. (Cornell) Bull.* 330.

Hoffman, E. J., 1935: Oxidation and gas formation in the spontaneous heating of hay, *J. Agr. Res.*, **51**: 527–546.

Hoffman, M. B., 1935: The effect of lime sulphur spray on the respiration rate of apple leaves, *Proc. Am. Soc. Hort. Sci.*, **33**: 173–176.

Hopkins, E. F., 1924: Relation of low temperatures to respiration and carbohydrate changes in potato tubers, *Bot. Gaz.*, **78**: 311–326.

———, 1927: Variations in sugar content in potato tubers caused by wounding and its possible relation to respiration, *Bot. Gaz.*, **84**: 75–88.

Hottes, C. F., and A. L. Hafenrichter, 1928: A constant rate aspirator, *Sci.*, **67**: 320–321.

Hover, J. M., and F. G. Gustafson, 1926: Rate of respiration as related to age, *J. Gen. Physiol.*, **10**: 33–39.

Hutchinson, D. M., 1922: Comparative studies on respiration. XXIII, The effect of adrenalin on the production of CO_2 by animals and by plants, *Am. J. Physiol.*, **62**: 192–196.

Iljin, W. S., 1923: Einfluss des Welkens auf die Atmung der Pflanzen, *Flora*, **116**: 379–403.

Inamdar, R. S., and B. N. Singh, 1927: Studies in the respiration of tropical plants. I, Seasonal variations in aerobic and anaerobic respiration in the leaves of *Artocarpus integrifolia*, *J. Indian Bot. Soc.*, **6**: 133–212; *Exp. Sta. Rec.*, **61**: 816–817.

INMAN, O. L., 1921: Comparative studies on respiration. XVI, Effects of hypotonic and hypertonic solutions upon respiration, *J. Gen. Physiol.*, **3**: 533–539.

———, 1921: Comparative studies on respiration. XVII, Decreased respiration and recovery, *J. Gen. Physiol.*, **3**: 663–667.

IRVING, A. A., 1911: The effect of chloroform upon respiration and assimilation, *Ann. Bot.*, **25**: 1077–1099.

IVANOW, S. L., 1913: Der Einfluss von Oelvorräten der Samen und der Temperatur auf den Respirationscoefficient CO_2/O_2, *Russ. J. Exp. Landw.*, **14**: 75–87.

JACQUOT, R., and A. MAYER, 1926: Hydration et respiration des graines, *Ann. Physiol. Physiochem. Biol.*, **2**: 408–425; *Biol. Abst.*, **1**: 611–612.

JEFFREY, R. N., and M. V. CRUESS, 1933: Gasometric method of estimating oxidase activity, *Plant Physiol.*, **8**: 327–331.

JOHANSSON, N., 1933: The relation between the respiration of the tree stem and its growth, *Biol. Abst.*, **8**: 50.

JOHNSTONE, G. R., 1925: Effect of wounding on respiration and the exchange of gases, *Bot. Gaz.*, **79**: 339–340.

———, 1925: Physiological studies of two varieties of *Ipomoea batatas*, *Bot. Gaz.*, **80**: 145–167.

JONES, H. A., 1920: Physiological study of maple seeds, *Bot. Gaz.*, **69**: 127–152.

JONES, J. W., 1933: Effect of reduced oxygen pressure on rice germination, *J. Am. Soc. Agron.*, **25**: 69–81.

JOSLYN, M. A., and G. L. MARSH, 1933: The role of peroxidase in the deterioration of frozen fruits and vegetables, *Sci.*, **78**: 174–175.

KARCHEVSKI, M., 1903: Influence of the fluctuation of temperature on the respiration of seeds and embryos of wheat, *Exp. Sta. Rec.*, **14**: 839.

KARLSON, A., 1925: Comparative studies on respiration. XXVIII, The effects of anesthetics on the production of carbon dioxide by wheat under aerobic and anaerobic conditions, *Am. J. Bot.*, **12**: 619–624.

KASTLE, J. H., and A. S. LOVENHART, 1901: On the nature of certain of the oxidizing ferments, *Am. Chem. J.*, **26**: 539–566.

———, 1910: The oxidases and other oxygen-catalysts concerned in biological oxidations, *Public Health and Marine Hospital Service Hygienic Lab. Bull.* 59.

——— and G. D. BUCKNER, 1917: Evidence of the action of oxidases within the growing plant, *J. Am. Chem. Soc.*, **39**: 478–482.

KEEBLE, F., and E. F. ARMSTRONG, 1912: The distribution of oxidases in plants and their role in the formation of pigments, *Proc. Roy. Soc.*, B, **85**: 214–218.

———, ———, and W. N. JONES, 1913: The formation of anthocyan pigments of plants. IV, The chromogens, *Proc. Roy. Soc.*, B, **86**: 308–317.

———, ———, and ———, 1914: The formation of anthocyan pigments of plants. VI, *Proc. Roy. Soc.*, B, **87**: 113–131.

KELLEY, V. W., 1930: Effect of certain hydrocarbon oils on respiration of foliage and dormant twigs of the apple, *Ill. Sta. Bull.* 348.

KERTESZ, Z. I., 1933: The oxidase system of a non-bearing yellow peach, *N. Y. Sta. (Geneva) Tech. Bull.* 219.

———, 1933: Some carbohydrate changes in shelled green peas, *N. Y. Sta. (Geneva) Bull.* 622.

———, 1933: Inactivating the respiration of cannery peas by heat, *The Canner*, Feb. 25, 1933.

———, 1934: The browning of yellow peaches. Enzyme studies reveal why yellow peach varieties show different extent of darkening, *Fruit Products J.*, **13**: 304–306.

———, 1934: Carbon dioxide content of the gas from pea pods, *Plant Physiol.*, **9**: 339–350.

KIDD, F., 1917: The controlling influence of carbon dioxide. III, The retarding effect of carbon dioxide on respiration, *Proc. Roy. Soc.*, B, **89**: 136–157.

———, C. WEST, and G. E. BRIGGS, 1921: A quantitative analysis of the growth of *Helianthus annuus*. I, The respiration of the plant and of its parts throughout the life cycle, *Proc. Roy. Soc.*, B, **92**: 368–384.

——— and ———, 1929: Physiology of fruit. I, Changes in the respiratory activity of apples during their senescence at different temperatures, *Proc. Roy. Soc.*, B, **106**: 93–109.

KIMBROUGH, W. D., 1924: The relation of respiration to storage and transportation of potatoes, *Proc. Ann. Meeting Potato Assoc. Am.*, **11**: 53–54.

———, 1925: A study of respiration in potatoes with special reference to storage and transportation, *Md. Sta. Bull.* 276.

KNIGHT, R. C., and J. H. PRIESTLEY, 1914: The respiration of plants under various electrical conditions, *Ann. Bot.*, **28**: 135–163.

KNOTT, J. E., 1925: Some conditions influencing the determination of catalase activity in plant tissue, *Proc. Am. Soc. Hort. Sci.*, **22**: 398–407.

———, 1927: Catalase in relation to growth and to other changes in plant tissue, *N. Y. Sta. (Cornell) Mem.* 106.

KNUDSON, L., 1916: Influence of certain carbohydrates on green plants, *N. Y. Sta. (Cornell) Mem.* 9.

KOLKWITZ, R., 1901: Ueber die Athmung ruhender Samen, *Ber. bot. Ges.*, **19**: 285–287.

KOSTYTSCHEW, S., 1908: Über die Anteilnahme der Zymase am Atmungsprozesse der Samenpflanzen, *Biochem. Z.*, **15**: 164–195.

———, 1910: Über den Vorgang der Zuckeroxydation bei der Pflanzenatmung, *Z. physiol. Chem.*, **67**: 116–137.

———, 1927: "Plant Respiration." English translation by C. J. Lyons. P. Blakiston's Son & Company, Philadelphia.

LALL, G., 1934: Chemical studies in the physiology of apples. XIV, A method of estimating chemical change and rate of respiration in stored apples, *Ann. Bot.*, **48**: 273–292.

LANDON, R. H., 1934: The effect of certain chemicals on the catalase activity in plants, *Am. J. Bot.*, **21**: 583–591.

LANGDON, S. C., and W. R. GAILEY, 1920: Carbon monoxide a respiration product of *Nereocystis leutkeana*, *Bot. Gaz.*, **70**: 230–239.

LANTZ, C. W., 1927: Respiration in corn with special reference to catalase, *Am. J. Bot.*, **14**: 85–105.

———, 1930: Temperature and catalase activity in germinating corn, *Proc. Iowa Acad. Sci.*, **35**: 147–149.

LARMOUR, R. K., J. S. CLAYTON, and C. L. WRENSHALL, 1935: A study of respiration and heating of damp wheat, *Can. J. Res.*, **12**: 627–645.

LEACH, W., 1932: Further experimental methods in connexion with the use of the katharometer for the measurement of respiration, *Ann. Bot.*, **46**: 583–596.

——— and K. W. KENT, 1934: Researches on plant respiration. III, The relationship between the respiration in air and in nitrogen of certain seeds during germination. (a) Seeds in which fats constitute the chief food reserve, *Proc. Roy. Soc.*, B, **116**: 150–169.

———, 1936: Researches on plant respiration. IV, The relation between the respiration in air and in nitrogen of certain seeds during germination. (b) Seeds in which carbohydrates constitute the chief food reserve, *Proc. Roy. Soc.*, B, **119**: 507–521.

LEGGATT, C. W., 1929: Catalase activity as a measure of seed viability, *Sci. Agr.*, **10**: 73–110.

———, 1933: A further note on catalase activity as a measure of seed vitality, *Can. J. Res.*, **9**: 571–573.

LeVan, W. C., 1930: The effect of metals on the respiration of *Lupinus albus*, *Am. J. Bot.*, **17**: 381–395.

Linossier, M. G., 1898: Contribution à l'étude des ferments oxydants. Sur la peroxydase du pus, *C. R. Soc. Biol.*, **50**: 373–375.

Lipman, C. B., 1936: Normal viability of seeds and bacterial spores after exposure to temperatures near the absolute zero, *Plant Physiol.*, **11**: 201–205.

Livingston, B. E., 1933: A computation concerning the importance of respiration water to young wheat seedlings, abst., *Am. J. Bot.*, **20**: 671.

———, 1934: Possible importance of respiration water to young wheat seedlings, *Bull. Torrey Bot. Club*, **61**: 217–223.

Loehwing, W. F., 1934: Physiological aspects of the effect of continuous soil aeration on plant growth, *Plant Physiol.*, **9**: 567–583.

Loew, O., 1901: Catalase, a new enzyme of general occurrence, with special reference to the tobacco plant, *U. S. Dept. Agr. Rept.* 68.

Long, E. R., 1917: Desiccation and respiration in succulent plants, *Carnegie Inst. Wash. Year Book*, **16**: 80–82.

———, 1918: Further results in desiccation and respiration of *Echinocactus*, *Bot. Gaz.*, **65**: 354–358.

Lopriore, G., 1928: Die Katalase-Reaktion und die Biologie des Pollens, *Ber. bot. Ges.*, **46**: 413–423.

Lund, E. J., 1919: A simple method for measuring carbon dioxide produced by small organisms, *Biol. Bull.*, **36**: 105–114.

Luthra, J. C., 1924: Some experiments on the effects of dry and moist air on the rate of respiration and breakdown of ripe pears, *New Phytol.*, **23**: 131–142.

——— and I. S. Chima, 1931: Some studies in respiration and other metabolic activities in berries of the grape vine, *Indian J. Agr. Sci.*, **1**: 695–714; *Exp. Sta. Rec.*, **67**: 260.

Lutman, B. F., 1926: Respiration of potato tubers after injury, *Bull. Torrey Bot. Club*, **53**: 429–457.

Lyon, C. J., 1921: Comparative studies on respiration. XVIII, Respiration and antagonism in *Elodea*, *Am. J. Bot.*, **8**: 458–464.

———, 1924: The effect of phosphates on respiration, *J. Gen. Physiol.*, **6**: 299–306.

———, 1927: Phosphate ion as a promoter catalyst of respiration, *J. Gen. Physiol.*, **10**: 599–615.

———, 1927: Arsenate as a catalyst of oxidation, *J. Gen. Physiol.*, **10**: 617–622.

———, 1927: The role of phosphate in plant respiration, *Am. J. Bot.*, **14**: 274–284.

Lyon, M. E., 1928: The occurrence and behavior of embryoless wheat seeds, *J. Agr. Res.*, **36**: 631–637.

MacDougal, D. T., E. R. Long, and J. G. Brown, 1915: End results of desiccation and respiration in succulent plants, *Physiol. Res.*, **1**: 289–327.

Mack, W. B., 1930: The relation of temperature and partial pressure of oxygen to respiration and growth in germinating wheat, *Plant Physiol.*, **5**: 1–69.

Magness, J. R., 1920: Composition of gases in intercellular spaces of apples and potatoes, *Bot. Gaz.*, **70**: 308–316.

——— and H. C. Diehl, 1924: Physiological studies on apples in storage, *J. Agr. Res.*, **27**: 1–38.

——— and W. S. Ballard, 1926: The respiration of Bartlett pears, *J. Agr. Res.*, **32**: 801–833.

MAIGE, A., and G. NICOLAS, 1910: Recherches sur l'influence des variations de la turgescence sur la respiration de la cellule, *Rev. Gén. Bot.*, **22**: 409–422.

——— and ———, 1910: Recherches sur l'influence des solutions sucrées de divers degrés de concentration sur la respiration, la turgescence et al croissance de la cellule, *Ann. Sci. Nat. Bot.*, 9e ser., **12**: 315–368.

MALPIGHI, M., 1679: "Anatomes plantarum pars Altera," Londini.

MANEY, T. J., P. L. HARDING, and H. H. PLAGGE, 1929: A new type of respiration chamber, *Sci.*, **70**: 44.

MANGENOT, G., 1928: Sur la localisation cytologique des peroxydases et des oxydases, *C. R. Acad. Sci. Paris*, **186**: 710–712.

MAQUENNE, L., and E. DEMOUSSY, 1913: Sur la valeur des coefficients chlorophylliens et leur rapports avec les quotients respiratoires réels, *C. R. Acad. Sci. Paris*, **156**: 506–512.

——— and ———, 1921: Sur la respiration des feuilles dans le vide ou des atmosphères pauvres en oxygène, *C. R. Acad. Sci. Paris*, **173**: 373–377.

MARKS, G. M., 1935: The inactivation of catalases from certain marine plants by oxygen, *Biochem. J.*, **29**: 509–512.

MARSH, G., 1935: Kinetics of an intracellular system for respiration and bioelectric potential at flux equilibrium, *Plant Physiol.*, **10**: 681–698.

MATTHAEI, G. L. C., 1905: Experimental researches on vegetable assimilation and respiration. III, The effect of temperature on carbon dioxide assimilation, *Phil. Trans. Roy. Soc.*, B, **197**: 47–86.

MCGINNIS, F. W., and G. S. TAYLOR, 1923: The effect of respiration upon the protein percentage of wheat, oats and barley, *J. Agr. Res.*, **24**: 1041–1048.

MCHARGUE, J. S., 1920: The significance of the peroxidase reaction with reference to the viability of seeds, *J. Am. Chem. Soc.*, **42**: 612–615.

MCKENZIE, K. A., 1932: Respiration studies with lettuce, *Proc. Am. Soc. Hort. Sci.*, **28**: 244–248.

MELDRUM, N. U., 1933: "Cellular respiration," Methuen and Company, Ltd., London, 116 pp.

MEYER, A., 1918: Eiweissstoffwechsel und Vergilben der Laubblätter von *Tropaeolum majus*, *Flora*, **11–12**: 85–127.

MICHAELS, W. H., 1931: Respiration of the shoot as affected by temperature changes of the root, *Bot. Gaz.*, **91**: 167–182.

———, 1932: Relation of lenticels and surface area to respiration in the potato tuber, *Bot. Gaz.*, **94**: 416–418.

MIDDLETON, N. I., 1927: The effect of ionized air on the rate of respiration of barley seedlings, *Ann. Bot.*, **41**: 345–356.

MILLER, E. C., 1910: A physiological study of the germination of *Helianthus annuus*, *Ann. Bot.*, **24**: 693–726.

MILLER, E. R., 1929: The insoluble tyrosinase of the velvet bean seed coat, *Plant Physiol.*, **4**: 507–517.

MILLER, L. P., 1932: Effect of various chemicals on the respiratory rate and sugar content of potato tubers, abst., *Am. J. Bot.*, **19**: 840.

———, 1933: Effect of various chemicals on the sugar content, respiratory rate and dormancy of potato tubers, *Contr. Boyce Thompson Inst.*, **5**: 213–234.

———, 1934: Time relations in effect of ethylene chlorohydrin in increasing and of ethyl alcohol in decreasing the respiration of potato tubers, *Contr. Boyce Thompson Inst.*, **6**: 123–128.

———, 1934: The effect of the alkyl halides on the respiration of potato tubers, *Contr. Boyce Thompson Inst.*, **6**: 279–296.

———, J. D. Guthrie, and F. E. Denny, 1934: Induced changes in respiration rates and time relations in the changes in internal factors, *Contr. Boyce Thompson Inst.*, **8**: 41–62.

———, 1935: Further experiments on the effect of halogenated alipathic compounds on the respiration of potato tubers, *Contr. Boyce Thompson Inst.*, **7**: 1–18.

Mitchell, J. W., 1935: A method of measuring respiration and carbon fixation of plants under controlled environmental conditions, *Bot. Gaz.*, **97**: 376–387.

Moore, B., and E. Whitley, 1909: The properties and classification of the oxidizing enzymes and analogies between enzymic activity and the effects of immune bodies and complements, *Biochem. J.*, **4**: 136–168.

Moore, W., and J. J. Willaman, 1917: Studies in greenhouse fumigation with hydrocyanic acid: physiological effects on the plant, *J. Agr. Res.*, **11**: 319–338.

Morgulis, S., M. Beber, and I. Rabkin, 1926: Studies on the effect of temperature on the catalase reaction. I, Effect of different hydrogen peroxide concentrations. II, Loss of catalase activity. III, Temperature effects at different hydrogen ion concentrations. IV, A theory of catalase reaction, *J. Biol. Chem.*, **68**: 521–563.

Morinaga, T., 1925: Catalase activity and the aerobic and anaerobic germination of rice, *Bot. Gaz.*, **79**: 73–85.

Morse, F. W., 1908: The effect of temperature on the respiration of apples, *J. Am. Chem. Soc.*, **30**: 876–881.

Müller-Thurgau, H., 1882: Ueber Zuckeranhäufung in Pflanzentheilen in Folge niederer Temperatur, *Landw. Jahrb.*, **11**: 751–828.

Navez, A. E., 1929: Respiration and geotropism in *Vicia faba*. I, *J. Gen. Physiol.*, **12**: 641–667.

Neller, J. R., 1931: Effect of chlorates upon the catalase activity of the roots of bindweed, *J. Agr. Res.*, **43**: 183–189.

———, 1931: Relation of catalase activity to physiological breakdown in Jonathan apples, *Plant Physiol.*, **6**: 347–354.

Nemec, A., and F. Duchon, 1921: Peut-on déterminer la valeur des semences par voie biochimique? *C. R. Acad. Sci. Paris*, **173**: 933–935.

——— and ———, 1922: Sur une méthode indicatrice permettant d'évaluer la vitalité des semences par voie biochemique, *C. R. Acad. Sci. Paris*, **174**: 632–634.

——— and ———, 1923: Sur une nouvelle méthode biochimique pour le détermination de la force vitale des semences, *Ann. Sci. Agron., Franc., et Etrangère*, **40**: 121–150.

Newcombe, F. C., and E. A. Bowerman, 1918: Behavior of plants in unventilated chambers, *Am. J. Bot.*, **5**: 284–295.

Newton, R., and W. R. Brown, 1931: Catalase activity of wheat leaf juice in relation to frost resistance, *Can. J. Res.*, **5**: 333–336.

——— and J. A. Anderson, 1931: Respiration of winter wheat plants at low temperatures, *Can. J. Res.*, **5**: 337–354.

Nicolas, G., 1918: Contribution à l'étude des variations de la respiration des végétaux avec l'âge, *Rev. Gén. Bot.*, **30**: 214–225.

Olney, A. J., 1926: Temperature and respiration of ripening of bananas, *Bot. Gaz.*, **82**: 415–462.

Onslow, M. W., 1919: Oxidizing enzymes. I, The nature of the "peroxid" naturally associated with certain direct oxidizing systems in plants, *Biochem. J.*, **13**: 1–9.

———, 1920: Oxidising enzymes. II, The nature of the enzymes associated with certain direct oxidising systems in plants, *Biochem. J.*, **14**: 535–540.

———, 1920: Oxidising enzymes. III, The oxidising enzymes of some common fruits, *Biochem. J.*, **14**: 541–547.

———, 1921: Oxidising enzymes. IV, The distribution of oxidising enzymes among the higher plants, *Biochem. J.*, **15**: 107–113.

———, 1921: Oxidising enzymes. V, Further observations on the oxidising enzymes of fruits, *Biochem. J.*, **15**: 113–118.

———, 1923: Oxidising enzymes. VI, A note on tyrosinase, *Biochem. J.*, **17**: 216–220.

———, 1924: The oxidising enzymes. VII, The oxygenase of the higher plants, *Biochem. J.*, **18**: 549.

——— and M. E. Robinson, 1925: Oxidising enzymes. VIII, The oxidation of certain parahydroxy-compounds by plant enzymes and its connection with "tyrosinase," *Biochem. J.*, **19**: 420–423.

——— and ———, 1928: Oxidizing enzymes. X, The relationship of oxygenase to tyrosinase, *Biochem. J.*, **22**: 1327–1332.

Oparin, A., and N. Djatschkow, 1928: Über die Fermentbildung in reifenden Samen, *Biochem. Z.*, **196**: 289–293.

Oppenheimer, C., 1913: *Die Fermente*, **11**: 754.

Osterhout, W. J. V., and A. R. C. Haas, 1917: An adaptation of Winkler's method to biological work, *J. Biol. Chem.*, **32**: 141–146.

———, 1917: The rôle of the nucleus in oxidation, *Sci.*, **46**: 367–369.

———, 1918: A method of studying respiration, *J. Gen. Physiol.*, **1**: 17–22.

———, 1918: An indicator method of measuring the consumption of oxygen, *J. Gen. Physiol.*, **1**: 167–171.

———, 1918: Comparative studies on respiration. I, Introduction, *J. Gen. Physiol.*, **1**: 171–181.

Ota, J., 1925: Continuous respiration studies of dormant seeds of *Xanthium*, *Bot. Gaz.*, **80**: 288–299.

Overholser, E. L., and W. V. Cruess, 1923: A study of the darkening of apple tissue, *Calif. Sta. Tech. Paper* 7.

———, 1928: A study of the catalase of the fruits of pear varieties, *Am. J. Bot.*, **15**: 285–307.

——— and L. L. Claypool, 1931: The relation of fertilizers to respiration and certain physical properties of strawberries, *Proc. Am. Soc. Hort. Sci.*, **28**: 220–224.

———, M. B. Hardy, and H. D. Locklin, 1931: Respiration studies of strawberries, *Plant Physiol.*, **6**: 549–558.

Pal, N. L., 1934: Studies on the respiration of conjugating *Spirogyra* with special reference to fat metabolism, *New Phytol.*, **33**: 241–273.

——— and U. N. Chatterji, 1936: Effect of insulin on plant respiration, *Nature*, **137**: 535–536.

Palladin, W., 1888: Ueber Eiweisszersetzung in den Pflanzen bei Abwesenheit von freiem Sauerstoff, *Ber. bot. Ges.*, **6**: 205–212.

———, 1888: Ueber Zersetzungsproducte der Eiweissstoffe in den Pflanzen bei Abwesenheit von freiem Sauerstoff, *Ber. bot. Ges.*, **6**: 296–304.

———, 1893: Recherches sur la respiration des feuilles vertes et des feuilles étiolées, *Rev. Gén. Bot.*, **5**: 449–473.

———, 1894: Sur le rôle des hydrates de carbone dans la résistance à l'asphyxie chez les plantes supérieures, *Rev. Gén. Bot.*, **6**: 201–209.

———, 1896: Recherches sur la corrélation entre la respiration des plantes et les substances azotées actives, *Rev. Gén. Bot.*, **8**: 225–248.

———, 1899: Influence de la lumière sur la formation des matières protéiques actives et sur l'énergie de la respiration des parties vertes des végétaux, *Rev. Gén. Bot.*, **11**: 81–105.

———, 1899: L'influence des changements de température sur la respiration des plantes, *Rev. Gén. Bot.*, **11**: 241–247.

——— and A. KOMLEFF, 1902: Influence de la concentration des solutions sur l'énergie respiratoire et sur la transformation des substances dans les plantes, *Rev. Gén. Bot.*, **14**: 497–516.

——— and S. KOSTYTSCHEW, 1906: Anaerobe Athmung, Alkoholgärung und Acetonbildung bei den Samenpflanzen, *Z. physiol. Chem.*, **48**: 214–239.

———, 1909: The theory of respiration of plants, *Biochem. Z.*, **18**: 151–206; *Exp. Sta. Rec.*, **21**: 532.

——— and A. M. SHELOUMOVA, 1918: Effect of loss of water on respiration of plants, *Bull. Acad. Sci. Russ.*, 6 ser., No. 8, 801–808 pp.; *Exp. Sta. Rec.*, **39**: 731.

———, 1922: "Plant Physiology." English translation by Livingston. P. Blakiston's Son & Company, Philadelphia.

PARIJA, P., 1928: Analytic studies in plant respiration. II, The respiration of apples in nitrogen and its relation to respiration in air, *Proc. Roy. Soc.*, B, **103**: 446–491.

——— and A. B. SARAN, 1934: The effect of light on the respiration of starved leaves, *Ann. Bot.*, **48**: 347–354.

PEIRCE, G. J., 1908: A new respiration calorimeter, *Bot. Gaz.*, **46**: 193–202.

———, 1909: What is the use of respiration? *Plant World*, **12**: 193–198.

———, 1912: The liberation of heat in respiration, *Bot. Gaz.*, **53**: 89–112.

PFEFFER, W., 1881–1885: Ueber intramoleculare Athmung, *Untersuch. Bot. Inst. Tübingen*, **1**: 636–685.

POPE, M. N., 1932: Catalase activity in relation to the growth curve in barley, *J. Agr. Res.*, **44**: 343–355.

———, 1933: Catalase activity and respiration in the leaves of growing barley, *J. Agr. Res.*, **46**: 35–41.

PORTSMOUTH, G. B., 1934: The effect of ionized air on the respiration of green plants, *Ann. Bot.*, **48**: 1013–1027.

POTTER, M. C., 1917: Note on a method of demonstrating the heat of respiration, *Ann. Bot.*, **31**: 435–438.

RANJAN, S., and A. K. MALLIK, 1931: A study of the catalase reaction, with special reference to respiration in plants, *New Phytol.*, **30**: 355–381.

RANSOM, E. R., 1935: The inter-relations of catalase, respiration, after-ripening and germination in some dormant seeds of the *Polygonaceae*, *Am. J. Bot.*, **22**: 815–825.

RAY, G. B., 1923: Comparative studies on respiration. XXIV, The effects of chloroform on the respiration of dead and living tissue, *J. Gen. Physiol.*, **5**: 469–477.

———, 1923: Comparative studies on respiration. XXV, The action of chloroform on the oxidation of some organic acids, *J. Gen. Physiol.*, **5**: 611–623.

———, 1923: Comparative studies on respiration. XXVI, The production of CO_2 from organic acids in relation to their iodine absorption, *J. Gen. Physiol.*, **5**: 623–629.

———, 1923: Comparative studies on respiration. XXVII, The mechanism of oxidation in relation to chloroform anesthesia, *J. Gen. Physiol.*, **5**: 741–749.

REED, G. B., 1914: The oxidases of acid tissue, *Bot. Gaz.*, **57**: 528–530.

———, 1915: Evidence for the general distribution of oxidases in plants, *Bot. Gaz.*, **59**: 407–409.

———, 1916: The significance of color changes in oxidase reagents, *Bot. Gaz.*, **61**: 430–432.

———, 1916: The measurement of oxidation potential and its significance in the study of oxidases, *Bot. Gaz.*, **61**: 523–527.

———, 1916: On the mechanism of oxidase action, *Bot. Gaz.*, **62**: 53–64.

———, 1916: The mode of action of plant peroxidases, *Bot. Gaz.*, **62**: 233–238.

———, 1916: The separation of oxidase reactions from the catalase reaction, *Bot. Gaz.*, **62**: 303–310.

———, 1916: The relation between oxidase and catalase in plant tissues, *Bot. Gaz.*, **62**: 409–412.

———, 1916: The relation of oxidase reactions to changes in hydrogen ion concentration, *J. Biol. Chem.*, **27**: 299–302.

Regeimbal, L. O., G. A. Vacha, and R. B. Harvey, 1927: The effect of ethylene on the respiration of bananas during ripening, *Plant Physiol.*, **2**: 357–359.

Rhine, L. E., 1924: Divergence of catalase and respiration in germination, *Bot. Gaz.*, **78**: 46–67.

Richards, H. M., 1896: The respiration of wounded plants, *Ann. Bot.*, **10**: 531–582.

———, 1897: The evolution of heat by wounded plants, *Ann. Bot.*, **11**: 29–63.

———, 1915: Acidity and gas interchange in cacti, *Carnegie Inst. Wash. Pub.* 209.

Richter, A. A., E. I. Dvoretskaia, and A. I. Grechushnikov, 1929: Perseverance of cultivated plants. I, Transpiration and katabolism of organisms infected with fungous diseases, *J. Exp. Landw. Südosten Eur. Russlands* (*Russian*), **7**: 99–114; *Biol. Abst.*, **6**: 966.

Rigg, G. B., and B. S. Henry, 1935: On the origin of the gases in the float of bladder kelp, *Am. J. Bot.*, **22**: 362–365.

Rischawi, L., 1876: Einige Versuche über die Athmung der Pflanzen, *Landw. Versuchs-Stat.*, **19**: 321–340.

Robbins, W. J., 1929: A study of the metabolism of roots, *Mo. Sta. Bull.* 272.

Roberts, R. H., and J. E. Kraus, 1934: Respiratory types and photoperiodism, *Sci.*, **80**: 122–123.

Robinson, M. E., 1924: A comparison of certain oxidizing enzymes of the higher and lower plants, *Biochem. J.*, **18**: 543–548.

Rohmann, E., and W. Spitzer, 1895: Ueber Oxydationswirkungen thierischer Gewebe, *Ber. chem. Ges.*, **28**: 567–572.

Rose, D. H., 1915: Oxidation in healthy and diseased apple bark, *Bot. Gaz.*, **60**: 55–65.

———, 1919: Blister canker of apple trees: a physiological and chemical study, *Bot. Gaz.*, **67**: 105–146.

———, H. R. Kraybill, and R. C. Rose, 1920: Effect of salts upon oxidase activity of apple bark, *Bot. Gaz.*, **69**: 218–236.

Rosenberg, A., 1910: Über die Rolle der Katalase in den Pflanzen, *Ber. bot. Ges.*, **28**: 280–288.

Rosene, H. F., and E. J. Lund, 1935: Linkage between output of electric energy by polar tissue and cell oxidation, *Plant Physiol.*, **10**: 27–47.

———, 1935: Proof of the principle of summation of cells E. M. F., *Plant Physiol.*, **10**: 209–224.

Sachs, J., 1882: "Vorlesungen über Pflanzenphysiologie," Leipzig.

Samisch, R., 1935: The location of oxidase in the apricot, *Am. J. Bot.*, **22**: 291–293.

Sapozhnikova, K. V., 1928: Respiration of wheat seed in ionized air, *C. R. Acad. Sci., U.R.S.S.*, **A8**: 119–124; *Biol. Abst.*, **5**: 1618.

Schöenbein, C. F., 1863: Ueber die katalytische Wirksamkeit organischer Materien und deren Verbreitung in der Pflanzen- und Thierwelt, *J. Prakt. Chem.*, **89**: 323–344.

Schröppel, F., 1933: Catalase, peroxidase and respiration in the germination of the light-sensitive seeds of *Nicotiana tabacum*, *Beih. bot. Cent.*, **51**: 377–407; *Chem. Abst.*, **28**: 503.

Schulz, K. G., 1926: Vergleichende Untersuchungen über die Atmungsvorgänge bei verschiedenen Kartoffelsorten, *Landw. Versuchs-Stat.*, **105**: 23–73.

Sherman, H., 1921: Respiration of dormant seeds, *Bot. Gaz.*, **72**: 1–30.

Shill, A. C., 1931: The respiration of *Citrus* as affected by hydrocyanic acid gas fumigation, *Univ. Calif. Pub.*, **5**: 167–180.

Shull, C. A., 1911: The oxygen minimum and the germination of *Xanthium* seeds, *Bot. Gaz.*, **52**: 453–477.

———, 1914: The role of oxygen in germination, *Bot. Gaz.*, **57**: 64–69.

——— and W. B. Davis, 1923: Delayed germination and catalase activity in *Xanthium*, *Bot. Gaz.*, **75**: 268–281.

Singh, B. N., 1930: Studies in the respiration of tropical plants, *Proc. Indian Sci. Cong.* (*Calcutta*), **17**: 306–310; *Exp. Sta. Rec.*, **66**: 619–620.

——— and K. V. Varadpande, 1930: Studies in the respiration of tropical plants. Effect of injecting water, glucose, and phosphates on a failing system of respiration in the leaves of *Artocarpus integrifolia* and its significance on the mechanism of respiration, *Proc. Indian Sci. Cong.*, **17**: 311–312; *Exp. Sta. Rec.*, **66**: 722–723.

Smith, A. M., 1915: The respiration of partly dried plant organs, abst. in *Rept. Brit. Assoc. Adv. Sci.*, **85**: 725; *Exp. Sta. Rec.*, **36**: 824.

Smith, E. P., 1921: Comparative studies on respiration. XIX, A preliminary stage in the process of ether anesthesia, *J. Gen. Physiol.*, **4**: 157–163.

———, 1922: Comparative studies on respiration. XXII, The effect of lactic acid on the respiration of wheat, *Am. J. Bot.*, **9**: 307–311.

Smith, O., 1929: Effects of various treatments on the carbon dioxide and oxygen in dormant potato tubers, *Hilgardia*, **4**: 273–306.

Smyth, E. S., 1934: A contribution to the physiology and ecology of *Peltigera canina* and *P. polydactyla*, *Ann. Bot.*, **48**: 781–818.

Spoehr, H. A., 1915: Variations in respiratory activity in relation to sunlight, *Bot. Gaz.*, **59**: 366–387.

——— and J. M. McGee, 1923: Studies in plant respiration and photosynthesis, *Carnegie Inst. Wash. Pub.* 325.

——— and ———, 1924: The effect of fluctuations in the CO_2 content on the atmosphere on the rate of respiration of leaves, *Am. J. Bot.*, **11**: 493–502.

———, 1924: The oxidation of carbohydrates with air, *J. Am. Chem. Soc.*, **46**: 1494–1503.

Steward, F. C., R. Wright, and W. E. Berry, 1932: The absorption and accumulation of solutes by living plant cells. III, The respiration of cut discs of potato tuber in air and immersed in water with observations upon surface: volume effects and salt accumulation, *Protoplasma*, **16**: 576–611.

Stich, C., 1891: Die Athmung der Pflanzen bei verminderter Sauerstoffspannung und bei Verletzungen, *Flora*, **74**: 1–57.

Stiles, W., and W. Leach, 1931: On the use of the katharometer for the measurement of respiration, *Ann. Bot.*, **45**: 461–488.

——— and ———, 1932: "Respiration in Plants," Dial Press (Lincoln MacVeagh), Inc., New York and Oxford.

——— and ———, 1932: Researches on plant respiration. I, The course of respiration on *Lathyrus odoratus* during germination of the seed and the early development of the seedling, *Proc. Roy. Soc.*, B, **111**: 338–355.

——— and ———, 1933: Researches on plant respiration. II, Variations in the respiratory quotient during germination of seeds with different food reserves, *Proc. Roy. Soc.*, B, **113**: 405–429.

———, 1935: Respiration, *Bot. Rev.*, **1**: 249–268.

Stocklasa, J., and F. Czerny, 1903: Isolirung des die anaerobe Athmung der Zelle der höher organisirten Pflanzen und Thiere bewirkenden Enzyms, *Ber. chem. Ges.*, **36**: 622–634.

Steward, F., 1908: On endospermic respiration in certain seeds, *Ann. Bot.*, **22**: 415–448.

SWANSON, C. O., 1934: Some factors involved in damage to wheat quality, *Cereal Chem.*, **11**: 173–199.

TANG, P., 1931: Temperature characteristics for the oxygen consumption of germinating seeds of *Lupinus albus* and *Zea mays*, *J. Gen. Physiol.*, **14**: 631–641.

———, 1931: Temperature characteristics for the production of CO_2 by germinating seeds of *Lupinus albus* and *Zea mays*, *J. Gen. Physiol.*, **15**: 87–95.

———, 1932: On the respiratory quotient of *Lupinus albus* as a function of temperature, *J. Gen. Physiol.*, **15**: 561–569.

———, 1932: A respirometer vessel for the study of metabolism of seeds, *J. Gen. Physiol.*, **15**: 571–574.

———, 1932: The effects of CO and light on the oxygen consumption and on the production of CO_2 by germinating seeds of *Lupinus albus*, *J. Gen. Physiol.*, **15**: 655–663.

———, 1932: Temperature characteristics for the anaerobic production of CO_2 by germinating seeds of *Lupinus albus*, *J. Gen. Physiol.*, **16**: 65–73.

TASHIRO, S., 1913: A new method and apparatus for the estimation of exceedingly minute quantities of carbon dioxide, *Am. J. Physiol.*, **32**: 137–145.

TERROINE, E. F., R. BONNET, and P. H. JOËSSEL, 1923: Influence de la température sur le rendement énergétique dans la germination, *C. R. Acad. Sci. Paris*, **177**: 212–214.

THODAY, D., 1913: On the capillary eudiometric apparatus of Bonnier and Mangin for the analysis of air in investigating the gaseous exchanges of plants, *Ann. Bot.*, **27**: 565–573.

———, 1913: The effect of chloroform on the respiratory exchanges of leaves, *Ann. Bot.*, **27**: 697–717.

THOMAS, H. S., 1918: Comparative studies on respiration. IV, The effect of ether on the respiration of wheat, *J. Gen. Physiol.*, **1**: 203–209.

THOMAS, L. A., 1931: The respiration of carrot tissue in buffer solutions, *Australian J. Exp. Biol. and Med. Sci.*, **8**: 225–236.

THOMAS, M., 1925: The controlling influence of carbon dioxide. V, A quantitative study of the production of ethyl alcohol and acetaldehyde by cells of the higher plants in relation to concentration of oxygen and CO_2, *Biochem. J.*, **19**: 927–947.

———, 1929, 1931: The production of ethyl alcohol and acetaldehyde by apples in relation to the injuries occurring in storage. I, Injuries to apples occurring in the absence of oxygen, and in certain mixtures of carbon dioxide and oxygen. II, Injuries to apples and pears occurring in the presence of oxygen and in the absence of accumulation of CO_2 in the storage atmosphere, *Ann. App. Biol.*, **16**: 444–457 (1929); **18**: 60–74 (1931).

——— and J. C. FIDLER, 1933: Studies in zymasis. VI, Zymasis by apples in relation to oxygen concentration, *Biochem. J.*, **27**: 1629–1642.

THORNTON, N. C., 1931: The effect of carbon dioxide on fruits and vegetables in storage, *Contr. Boyce Thompson Inst.*, **3**: 219–244.

———, 1932: Carbon dioxide storage. III, The influence of carbon dioxide on the oxygen uptake by fruits and vegetables, abst., *Am. J. Bot.*, **19**: 843.

———, 1932: Carbon dioxide storage. IV, The influence of carbon dioxide on the acidity of fruits and vegetables, abst., *Am. J. Bot.*, **19**: 843.

———, 1933: Carbon dioxide storage. III, The influence of carbon dioxide on the oxygen uptake by fruits and vegetables, *Contr. Boyce Thompson Inst.*, **5**: 371–402.

———, 1933: Carbon dioxide storage. IV, The influence of carbon dioxide on the acidity of plant tissue, *Contr. Boyce Thompson Inst.*, **5**: 403–418.

———, 1935: Carbon dioxide storage. VIII, Chemical changes in potato tubers resulting from exposure to carbon dioxide, *Contr. Boyce Thompson Inst.*, **7**: 113–118.

Traube, M., 1877: Die chemische Theorie der Fermentwirkungen und des Chemismus der Respiration, *Ber. chem. Ges.*, **10**: 1984–1992.

True, R. H., and W. W. Stockberger, 1916: Physiological observations on alkaloids, latex and oxidases in *Papaver somniferum*, *Am. J. Bot.*, **3**: 1–11.

Tyson, J., 1930: Influence of soil conditions, fertilizer treatments and light intensity on growth, chemical composition and enzymic activities of sugar beets, *Mich. Sta. Tech. Bull.* 108.

Wakabayashi, S., 1925: The injurious effect of submergence on the cranberry plant, *N. J. Sta. Bull.* 420.

Warburg, O., 1921: Physikalische Chemie der Zellatmung, *Biochem. Z.*, **119**: 134–166.

———, 1925: Iron, the oxygen-carrier of respiration ferment, *Sci.*, **61**: 575–582.

———, 1928: The chemical constitution of respiration ferment, *Sci.*, **68**: 437–443.

Weiss, F., and R. B. Harvey, 1921: Catalase, hydrogen-ion concentration, and growth in the potato wart disease, *J. Agr. Res.*, **21**: 589–592.

Werner, H. O., 1935: A review of the literature on the physiological aspects of the storage of potatoes, *Ann. Rept. Neb. Sta. Bd. Agr.*, **1935**: 680–703; 16th *Ann. Rept. Neb. Potato Improv. Assoc.*, **1934-1935**: 36–59.

Wheldale, M., 1910: Plant oxidases and the chemical interrelationships of color varieties, *Prog. rei. Bot.*, **3**: 457.

Whimster, K., 1927: The effect of ionized air on the assimilation and respiration of green leaves, *Ann. Bot.*, **41**: 357–374.

White, J., 1909: The ferments and latent life of resting seeds, *Proc. Roy. Soc.* B, **81**: 417–442.

Whitehead, T., 1934: The physiology of potato leaf roll. I, On the respiration of healthy and leaf-roll infected potatoes, *Ann. App. Biol.*, **31**: 48–77.

Willaman, J. J., and J. H. Beaumont, 1928: The effect of accumulated carbon dioxide on plant respiration, *Plant Physiol.*, **3**: 45–61.

——— and W. R. Brown, 1930: Carbon dioxide dissolved in plant sap and its effect on respiration measurements, *Plant Physiol.*, **5**: 535–542.

Willstätter, R., and H. Weber, 1926: Zur quantitativen Bestimmung der Peroxydase, *Liebigs Ann. Chem.*, **449**: 156–174.

Woods, A. F., 1902: Observations on the mosaic disease of tobacco, *U. S. Bur. Plant Ind. Bull.* 18.

Wright, R. C., 1932: Some physiological studies of potatoes in storage, *J. Agr. Res.*, **45**: 543–555.

Wynd, F. L., 1932: An apparatus for the determination of carbon dioxide production in physiological plant studies, *Ann. Mo. Bot. Gard.*, **19**: 499–502.

———, 1935: An improved apparatus for the determination of carbon dioxide production in physiological plant studies, *Ann. Mo. Bot. Gard.*, **22**: 361–363.

——— and E. S. Reynolds, 1935: Studies in ultra-violet and respiratory phenomena. I, Review of work published before June 1935, *Ann. Mo. Bot. Gard.*, **22**: 771–835.

———, H. J. Fuller, and E. S. Reynolds, 1935: Studies in ultra-violet and respiratory phenomena. II, The effects of ultra-violet on respiration and respiratory enzymes of higher plants, *Ann. Mo. Bot. Gard.*, **22**: 837–852.

Yemm, E. W., 1935: The respiration of barley plants. I, Methods for the determination of carbohydrates in leaves, *Proc. Roy. Soc.*, B, **117**: 483–504.

———, 1935: The respiration of barley plants. II, Carbohydrate concentration and carbon dioxide production in starving leaves, *Proc. Roy. Soc.*, B, **117**: 504–525.

Yoshida, H., 1883: Chemistry of lacquer, *J. Chem. Soc. (London)*, **43**: 472–486.

Yoshisuke, R., 1932: Davis' modified catalase method prophesying the germinating power of seeds, *Bull. Sci. Fakult. Terkeltura, Kjusu Imp. Univ.*, **5**: 103–114 (115–116 Eng. Summary); *Chem. Abst.*, **29**: 4797.

Zaleski, W., and A. Rosenberg, 1911: Zur Kenntnis der Rolle der Katalase in den Pflanzen, *Biochem. Z.*, **33**: 1–15; *Bot. Cent.*, **119**: 431 (1912).

———, 1915: Über die Alkoholoxydation durch die Samenpflanzen, *Biochem. Z.*, **69**: 289–293.

Zerban, F. W., 1918: The role of oxidases and of iron in the color changes of sugar cane juice, *J. Ind. Eng. Chem.*, **10**: 814–817.

Zimmerman, P. W., 1930: Oxygen requirements for root growth of cuttings in water, *Am. J. Bot.*, **17**: 842–861.

CHAPTER XIV

THE PROCESS OF GROWTH IN PLANTS

I. NATURE OF GROWTH

Growth is one of the outstanding characteristics of living matter, but at the same time it is a process which is very difficult to define. In discussing the nature of growth, it should be remembered that it may be considered from the standpoint of the entire organism or from that of its individual parts. Thus growth may refer to the changes that occur in the protoplasm, in the cell wall, in the cell considered as a unit, in the organs or parts of organs, or in the entire plant considered as a whole. Any statements concerning growth, therefore, will be modified somewhat by the viewpoint from which they are considered.

Regardless of the standpoint from which growth is considered, the first step in the process is, with but few exceptions, the formation of new protoplasm. This step has been termed the "fundamental stage of growth." This formation of new living matter is always accompanied or immediately followed by a permanent increase in the amount of the nonliving materials that are always closely associated with the protoplasm. This increase in the amount of nonliving matter is plainly evident through the enlargement of the plant part and by an increase in its weight. These changes are due chiefly to an increased water content and to the deposition of materials in the cell walls. Since this stage in growth is readily observed, it has been termed the "evident stage of growth." This increase in the size and weight of the plant part under consideration is permanent as long as that part remains alive. In the light of these facts, growth has been defined by some authors as a permanent increase in weight, attended by a permanent change in form, these changes having been induced primarily by an increase in the quantity of protoplasm.

While there are instances where this definition of growth does not strictly describe the changes that occur, it is applicable in most cases. Two of the examples that have frequently been cited as cases where this definition does not apply are the sprouting potato tuber and germinating seeds. Thus the total dry weight of the tuber and sprout will be less than that of the original tuber, owing to loss in weight through respiration. The sprouts, however, have greatly increased in weight and changed markedly in form from the buds that produced them. The same relation exists in germinating seeds where the seedling and seed combined for several weeks contain less dry matter than the original seed. The plumule and radicle of the embryo, however, have increased their weight and have changed in form, although the seed as a whole has decreased in dry matter.

To some, however, growth implies more than simply an increase in weight and size. They believe that differentiation and development should also be included in the term. The development of a plant from the fertilized egg is not simply an increase in the weight and size of the fertilized egg; otherwise a mature plant would be only an enlarged egg and not a highly specialized individual. It is evident that if the differentiation of a plant or plant part is not considered as a phase of growth, it must at least be regarded as intimately associated with it.

The growth of an organism consists primarily of the formation of new cells, of their enlargement and specialization. In order, therefore, to understand growth in plants, we must consider in detail the manner in which a plant cell grows.

II. STAGES IN THE GROWTH OF A PLANT CELL

In the formation and development of a plant cell, three more or less distinct stages or types of activity are to be observed. The transition from one stage to another is not sudden and complete, but the different stages can readily be distinguished and can be seen to be differently affected by the immediate environment of the cells (Priestley, 1929). The different stages of growth which may be distinguished are (*a*) the embryonic or formative stage, (*b*) the stage of elongation, and (*c*) the stage of differentiation.

A. Embryonic or Formative

This stage includes that period in which the cell is formed from a preexisting cell. In the vascular plants, the formation of new cells is localized in more or less definite regions known as "meristems." These regions are fairly definitely confined to the tips of branches, near to the tips of roots, to the cambium, to the bases of the internodes, especially in the monocotyledons, and to the bases of the growing sheaths and leaves of monocotyledons.

The cells of the meristem which thus give rise to new cells are small, with dense protoplasmic contents and prominent nuclei. The nuclei are of the normal size of other plant cells, but in such small cells they bulk large in proportion to the rest of the protoplasm. The protoplasm is especially free from vacuoles and characteristic inclusions, no granular food reserves being evident, while the plastids are very minute or invisible. With the exception of the cambium, the cells may be 12 to 14 sided but frequently are cubical and are closely packed together with no intercellular spaces. The cell walls are the thinnest of any in the plant body, and, according to Priestley (1929), they frequently take stains, which indicate that they are impregnated with protoplasm. He considered that the walls of these cells are the channels of transport for the food supplies of this tissue.

In the meristematic cells, protein synthesis is dominant, while carbohydrate formation is at a minimum. In this formative stage, therefore, the protoplasm is primarily engaged in protoplasmic synthesis. According to Priestley (1929), this is maintained only when the liquid surrounding the cell has a hydrogen-ion concentration near the isoelectric point of the main constituent protein of the cell. When the mass of a meristematic cell has increased to a certain point, cell division follows, and the narrow range of cell size in these regions indicates that the division follows the attainment of a certain upward limit of the cell mass. MacDougal (1916 to 1921) came to the conclusion from his extensive experiments with growth that the first step in the process is an increase in the volume of the protoplasm by hydration. This increase in volume is usually but not always accompanied by the incorporation of materials other than water in the colloids of the protoplasm with the entailed additional capacity for absorbing water. The actual manner in which this incorporation occurs, however, is not known. MacDougal believed that the protoplasm of plants consists of a comparatively inert base of a polysaccharose character in colloidal combination with proteins, amino acids, lipins, and salts. The enlargement of protoplasmic mass due to hydration is therefore determined by the character of this colloidal mixture as well as by the quantity and type of salts and by the acidity or alkalinity of the medium. These factors might all be altered by the action of the metabolic products of the meristematic cells.

McDougal (1925) called this procedure the accretion stage of the cell. During this stage, proteins and carbohydrates are formed by synthesis, dehydration, and condensation.

Hottes (1929) found in the roots of *Vicia faba*, the tips of which were mechanically inhibited from further elongation, that nuclear and cell division ceased when the cell reached a minimum size. He considered that turgor is perhaps the principal inhibitor of nuclear division. Turner (1929) and Farr (1931, 1933) noted that the hair or fiber of cotton is formed by the outward extension of a single epidermal cell of the seed coat. These epidermal cells, which develop into fibers, are formed during a period of at least 10 days following the beginning of the opening of the flower.

Priestley (1930) and Priestley, Scott, and Malins (1933) noted that the cambium changes in the spring from a "granular gel" condition to a swollen "semifluid sol" condition. Some time later, cell division begins. It commences in the buds, proceeds downward in the cambium of the old wood, and finally reaches the younger roots, in which a slow cambial activity may persist for most of the season. Thus the initiation of cell division may be weeks later in the roots than in the trunk. The dormant cambial layer is usually about one cell in thickness, and, with the exception of their granular nature at this stage, there is little to distinguish these cells from the adjoining ones; thus separation of the bark and wood is difficult. The first sign of growth activity is a swelling of the cambial cells. The radial walls rapidly become thinner as they extend, and the contents of the cell appear to change from a solid to a liquid consistency. After this stage is attained, the bark readily slips over the wood, and separation occurs in the plane of the cambium. The new cells are produced almost entirely on the inner side of the cambium. The student is referred to Priestley (1930) for a thorough and critical discussion of the concept of sliding growth in the cambium.

According to Wright (1933) cambial growth in the trunk of *Pinus sylvestris* begins in May, before the opening of the buds, while the most vigorous growth occurs from May to early July. After this, slow but continuous growth occurs until October.

The retardation of growth of an organ may be directly connected with the lessening of the water capacity of the colloidal masses. This theory of the role of colloidal hydration in growth has been substantiated by the work of MacDougal (1920, 1921) and MacDougal and Spoehr (1917) with artificially prepared mixtures consisting largely of amorphous, condensed carbohydrates such as agar to which had been added a small proportion of albumen or amino acid. These mixtures were found to respond to the action of acids, alkalies, and salts in a manner similar to that of the plant.

Instances of growth are known in which water only has been added to the colloids of the protoplasm, but in these cases in all probability the solid particles are variously rearranged. MacDougal (1919) found no increase in dry matter in the leaves of *Crassulaceae*, in the joints of cacti, and in fruits, although growth as measured by the change in form was taking place. The proportion of water and solid matter thus underwent but little change, their incorporation being at a rate that kept them near the initial proportion.

B. Elongating

1. Nature.—After a new cell has been formed, the protoplasm tends toward carbohydrate synthesis rather than protein synthesis. The increased carbohydrate synthesis is indicated by the increase in the thickness of the cell walls and by the inclusion of starch grains in the protoplasm (Priestley, 1928). The amount of protoplasm increases little if at all during this stage of cell development.

The protoplasm of the new cell now exhibits syneresis similar to that shown by many colloids in which cavities or canals are formed throughout the colloidal mass. These syneretic cavities enlarge by the absorption of water to form the "vacuoles," by which name they are generally called. These vacuoles continue to expand by the absorption of water and eventually coalesce to form the vacuole of the cell. The expansion of these vacuoles and the consequent increase in volume of the cell take place primarily by the force of osmotic action. The vacuole continues to hold some of the colloidal material and in addition the various solutes that enter the cell or are formed therein (MacDougal, 1919).

MacDougal (1925) stated that the greatest distension of a cell does not occur in the isoelectric zone of its proteins but at the hydrogen-ion concentration at which the protoplasmic layer is the least permeable. Freeland (1933) noted in *Bryophyllum* that the hydrogen-ion concentration and total acidity of the margins of proliferating leaves increased for 4 days after the inception of this process and then decreased. The osmotic value in the margins of the proliferating leaves was in general higher than in the margins of the paired, inactive leaves. During proliferation, starch and reducing sugars accumulated in the growing foliar bud, and the amounts of total sugars, sucrose, amino nitrogen, amide nitrogen, and nitrate nitrogen increased in the leaf. In addition, there was an increase in the amounts of catalase, diastase, and oxidase in the foliar bud. Loomis (1932, 1934) reported that the growth of maize depends upon a liberal supply of water at the growing point. Such a supply is reduced, and hence growth is checked by the following factors, listed in order of effectiveness: direct sunlight, deficient soil moisture, and low relative humidity. Growth decreases rapidly as the temperature approaches 10°C. This may be due in part to a slowing of the chemical processes of cell division and to a decreased rate of translocation of food material.

During the elongating period, the newly formed cell increases in size, especially in length (Fig. 3). When this elongation takes place, the protoplasm, which has increased but little in amount over that of the newly formed cell, spreads out in a thin film along the cell wall. The volume of a plant cell may thus increase several hundred times and yet contain very little more protoplasm than before expansion began. In this respect, the plant cell differs from the animal cell, for in the latter no such increase in volume takes place without a corresponding increase in the amount of protoplasm. By this expansion of cells due to the absorption of water very rapid growth occurs without the expenditure of the energy necessary to manufacture relatively large quantities of protoplasm. This ability to grow rapidly is a distinct advantage to an organism where there is a time limit to its period of growth. The total period during which the elongation of the plant cell occurs is termed the "grand period of growth."

2. Rate.—Except in rare cases, growth in plants proceeds at such a rate that it cannot be detected by the naked eye. The most rapid growth in length is that of the filaments of stamens in the wheat flower. According to Andrews (1932) this growth may be 2 to 3.5 mm. per minute. If such a growth rate were maintained for 24 hr., the stamen would attain a length of 6 to 8 ft. Under favorable conditions the pollen tube of *Zea mays* may grow 6 in. in 24 hr. When an appreciable period of time is considered, the bamboo grows the most rapidly of any plant, and a total increase in the length of its stem of 2 ft. in 24 hr. is not infrequent. Under the conditions of summer, the squash vine grows approximately 6 in. during a 24-hr. day. Florell and Faulkner (1934) studied the daily increase in height of the peduncle of wheat at heading. The maximum daily elongation during this period ranged from 4.3 to 9.1 cm. Lichens, however, may grow as little as 2 to 5 mm. per year under ordinary, favorable conditions.

Based on the data obtained from various species of plants growing under many conditions, it is estimated that their average growth in length during the growing season does not exceed 0.3 in. for a 24-hr. period. Booth (1929) noted in oats that the growth in length of the grain was most rapid during the first 6 days after fertilization. Following this the rate of elongation gradually diminished until the maximum length was attained on the fifteenth day. The grain increased rapidly in width and thickness during the first 10 days and reached its limit in these dimensions before growth in length ceased. Dastur and Kapadia (1931) stated that the maximum axial growth of the tendril of *Momordica charantia* is 8.0 to 9.4 cm. in 24 hr., while that of *Lagenaria vulgaris* is 17.5 cm. during a like period.

C. Differentiating

About the time a cell reaches its maximal enlargement, it begins to undergo changes that will transform it into a specialized cell. It may become a sclerenchyma cell, a sieve tube, a tracheid, or any other of the numerous types of cells that compose a plant, depending upon its location and the resultant of internal and external factors. When a cell has once become specialized, it seldom if ever reverts to another form but remains the same type of cell as long as it exists. After a cell is differentiated, it functions solely in aiding other cells to grow and develop.

It was stated by Loomis (1932) that growth is an increase in size due to cell division and cell enlargement, while "differentiation" is the sum of the morphological changes that start during cell enlargement and end with the death of the cell.

Inulin occurs in *Helianthus tuberosus*, but not in *H. annuus*. According to Thoday (1933), grafts, made between the two species by Daniel (1921) and Colin (1922), showed that the transport of this carbohydrate occurred freely from one to the other, but each part retained its normal characteristic. In a double graft in which a piece of the sunflower was interpolated between the basal and apical parts of an artichoke, each of the three portions functioned in their usual manner. There must, therefore, be a specific difference in the metabolism of the parenchymatous cells of the two stems. Metabolic differences are shown in the various parts of the same plant. Thus in rhubarb, the rhizome contains very little acid and the nitrogen is present largely in the form of amino acids or amides and a small amount of ammonia. In the petioles, however, two-thirds of the total nitrogen is in the form of ammonia, and acids are abundant.

McNair (1934) believed that as a general rule the more highly the plant is organized, the more complex are likely to be its chemical products. Thus he believed that the volatile oils of the families highest in evolutionary development have constituents with a large number of double bonds, more aromatic compounds, and more compounds with small amounts of substances of low molecular weight.

III. GROWTH OF PLANT ORGANS

In the great majority of plants the regions of cell expansion are strictly localized. Thus the following are the regions in the higher plants where increase in the size of the cells occurs: (*a*) Near the root tip. The elongating region is restricted to a portion of the root not over 10 mm. in extent. The region of maximal elongation generally occurs only about 3 mm. back of the root tip. (*b*) At the tips of stems. The stem, as a rule, has a longer elongating region than the root. (*c*) At the cambium. (*d*) At the base of the internodes in the monocotyledons. (*e*) At the base of the sheaths and leaves of monocotyledons. In the leaves of dicotyledons, growth is rather evenly distributed throughout the entire surface but with slightly more activity at the basal region than at the apical one. In general, growth is more localized in a mature plant than in a younger one and more so in a perennial than in an annual plant.

Since each cell of a plant passes through the three more or less distinct stages in its growth and development, each plant part or organ in its development passes through the same stages. Thus each organ or plant part has its formative stage, its elongative stage or grand period of growth, and its stage of differentiation or maturation. The enlargement of any plant part begins at a slow rate that gradually increases to a maximum, after which the rate progressively decreases until enlargement ceases. This behavior of growth is illustrated by the rate of elongation of the growing zone of a root of *Vicia faba*, as quoted by Raber (1928).

Day	Growth, millimeter	Day	Growth, millimeter
1	2	6	14
2	4	7	7
3	18	8	4
4	18	9	0
5	17		

From these data it is seen that elongation during the first 2 days is relatively slow but reaches a maximum during the next 3 days, after which it rapidly declines to zero.

The higher plants, as a rule, continue to develop new organs as long as they live, forming stems, leaves, roots, flowers, and seeds in a somewhat indefinite repetition. Even in an annual plant no definite number of organs is developed. This is in marked contrast to most animals, which develop a single set of organs that serve them throughout their life. Although growth in an annual plant is limited to one season, this process in perennial plants continues as long as the plant lives. Thus a tree that is several thousand years old will continue to produce organs and increase in height and thickness of stem.

IV. NATURE OF GROWTH CURVES

The rate of growth of plants or plant parts is commonly measured by their increase in extent or by their increase in weight. As has been previously mentioned, the rate of growth of a plant or any portion of it is relatively slow at first, increases rapidly to a maximum, and finally decreases until it ceases. This behavior is especially noticeable when the elongation of plant parts is considered. When the rate of growth is plotted against time, a typical S curve is obtained.

Martini, Harlan, and Pope (1930) did not believe that the S curve of growth is wholly applicable to the development of barley kernels. They considered that factors other than growth must be considered.

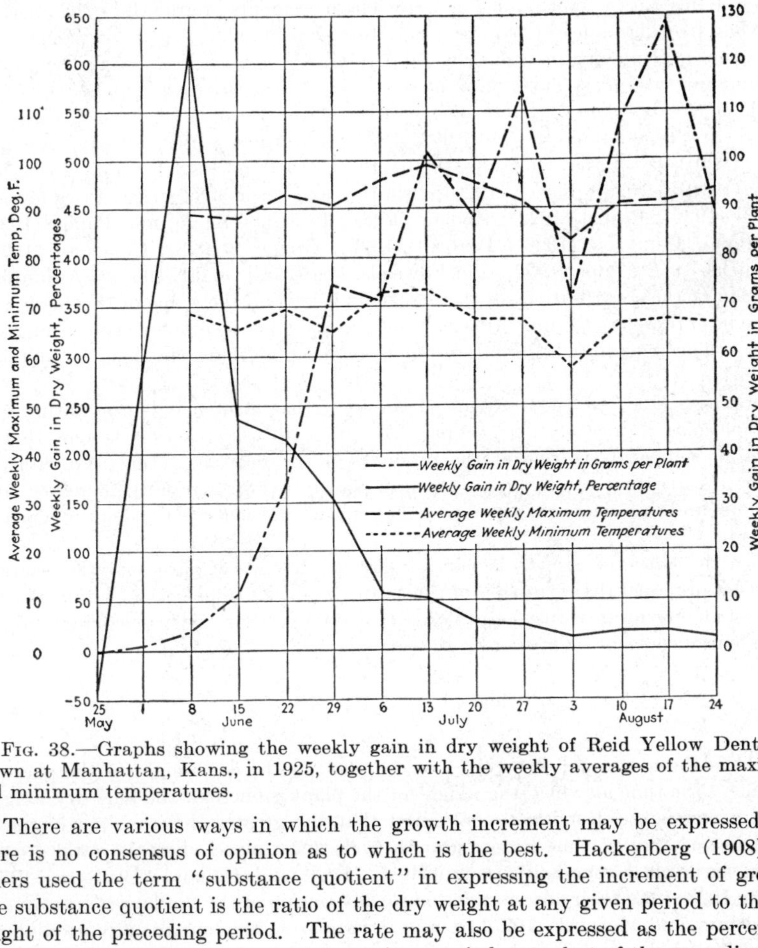

FIG. 38.—Graphs showing the weekly gain in dry weight of Reid Yellow Dent corn grown at Manhattan, Kans., in 1925, together with the weekly averages of the maximum and minimum temperatures.

There are various ways in which the growth increment may be expressed, but there is no consensus of opinion as to which is the best. Hackenberg (1908) and others used the term "substance quotient" in expressing the increment of growth. The substance quotient is the ratio of the dry weight at any given period to the dry weight of the preceding period. The rate may also be expressed as the percentage increase of dry weight or of extent for a given period over that of the preceding one. The increase in leaf surface or in leaf weight is also considered by some to be a good indicator of growth increment.

Pearl, Winsor, and Miner (1928) stated that the growth of an etiolated seedling from a sterilized seed in the dark on a nonfood, sterile medium is an expression of the "inherent vitality" of that individual. Inherent vitality is defined as the total potential capacity of an organism to perform vital actions in the complete absence of exogenous derivation of food or energy. The "total vitality" is the capacity of an organism to perform vital activities in the presence of its endogenous sources and of sufficient exogenous supplies.

In the table on page 1027 are given data obtained by Miller (1925) in the growth of the corn plant at Manhattan, Kans. The weekly increments of growth are therein expressed by the increase in dry weight, the percentage increase in dry weight during each period over that of the succeeding period, the increase in the leaf surface, and the increase in the height of the plant for each period. The rate of transpiration per plant and the weather conditions in regard to temperature are also included. A portion of this data is plotted in Fig. 38. The increments in growth in this case are somewhat irregular due, in part at least, to a considerable degree to rather wide fluctuations in temperature during several periods.

Numerous attempts have been made to give mathematical expression to the growth curves of plants. Some of the investigators who have worked with this problem are Robertson (1907 to 1908, 1923); Blackman (1919, 1920); Brenchley (1919, 1920); Reed and Holland (1919); Reed (1919 to 1923); Kidd, West, and Briggs (1919 to 1928); Fisher (1920); Priestley and Evershed (1922); Priestley and Pearsall (1922); Rippel (1922); Hooker (1925); Inamdar, Singh, and Pande (1925); van Sande-Bakhuyzen (1926); Pearsall (1927); Gregory (1928); Copeman (1928); Porterfield (1928); Pope (1932); and Edwards, Pearl, and Gould (1934). A complete discussion of the contributions of these authors is beyond the scope of this book, and it is the intention here to summarize as briefly as possible the opinions concerning this subject. The student who is interested is referred to these citations for detailed information.

Robertson (1907 to 1908) advanced the view that growth is of the nature of a monomolecular autocatalytic reaction. An autocatalytic reaction is one that is capable of self-catalysis, one of the products of the reaction acting as a catalytic reagent. A reaction of this type proceeds slowly at first, rapidly increases to a maximum, and then declines in its rate. The curve for such a reaction is thus typically S shaped.

Since the curve of growth is very similar to that of an autocatalytic reaction, Robertson proposed that growth is of this nature and that some special catalyst of the nature of an enzyme governs the growth rate of an organism. He considered that the growth curve could be expressed by the formula

$$\log \frac{x}{A - x} = K(t - t_1)$$

where A is the maximum dry weight of the plant, x the dry weight of the plant at any time, t the time at which the weight of the plant is one-half the final dry weight, and K a constant. Robertson considered that the greatest increase in volume or weight in any unit of time in any growth cycle takes place when the total growth due to the cycle is one-half accomplished. Reed and Holland (1919) and Reed (1919 to 1923) from their observations on the elongation of the sunflower, apricot, and other plants, and Murneek (1925) from observations on the tomato and cotton plants concluded that the growth rate approximated the course of an autocatalytic reaction. A close similarity between the observed and calculated growth values led them to the conclusion that the growth rate is governed by constant internal factors rather than by external factors. They obtained no strong correlation of growth either with temperature or with transpiration summation. In the case of apricot, Reed (1919)

WEEKLY GROWTH CHANGES OF REID YELLOW DENT CORN AT MANHATTAN, KANS. DURING GROWING SEASON OF 1925
Averages of Five Plants

Date	Average dry weight per plant, grams	Weekly gain in dry weight per plant		Total leaf surface per plant, grams	Weekly increase in leaf surface per plant		Rate of growth in height, inches	Transpiration per plant, cc.	Average weekly maximum temperature, °F.	Average weekly minimum temperature, °F.
		Grams	Per cent		Grams	Per cent				
May 11	Seed planted. Average dry weight of seed 0.3 g.									
11–25	0.16	0.14[1]	47.0[1]	70.5	....	.....	3.5	59		
25–June 1	0.65	0.49	306.2	348.1	277.6	393.7	2.5	102		
June 1–8	4.66	4.01	617.0	1,554.8	1,206.7	346.9	5.0	1,079	88.8	68.7
8–15	15.74	11.08	237.7	4,260.7	2,705.9	174.0	6.0	2,874	88.4	65.1
15–22	49.30	33.56	213.2	9,891.5	5,620.8	131.9	14.0	4,110	92.7	69.6
22–29	123.70	74.40	151.0	17,069.2	7,277.7	73.5	17.0	8,290	90.4	64.8
29–July 6	195.00	71.30	57.6	21,585.8	4,516.6	26.4	17.0	22,610	95.8	72.9
July 6–13	297.00	102.02	52.3	26,699.2	5,113.4	23.6	18.0	27,220	99.4	73.6
13–20	385.50	88.48	29.8	28,629.4	1,930.2	7.2	33.0[2]	30,420	95.3	67.9
20–27	499.70	114.20	29.6	Full leaf development				22,190	90.1	67.6
27–Aug. 3	570.80	71.10	14.2					15,530	83.1	58.7
Aug. 3–10	679.00	108.20	18.9					21,610	90.7	66.3
10–17	807.50	128.50	18.9					15,890	91.0	67.4
17–24	895.90	88.40	10.9					17,130	93.9	66.3

[1] Loss.
[2] Tasseling.

considered that an organism grows at a definite rate which is at any moment proportional to the amount of growth yet to be made.

Blackman (1919, 1920) suggested that the growth of an annual plant can be treated as a process following the compound-interest law expressed by the formula

$$W_1 = W_0 e^{rt}$$

or, written in another manner,

$$\log_e \frac{W_1}{W_0} = rt$$

In this formula W_1 is the dry weight of the plant at the end of the time t, W_0 the initial seedling or seed rate, e the base of the natural logarithms, and r the rate of interest or increase. Blackman termed this rate of increase in dry matter the "efficiency index of production," since it represents the efficiency of the plant as a producer of new material and gives a measurement of the plant's economy. It may also be termed the "economy constant" of the plant and is comparable to the velocity constant of a chemical reaction. The rate of increase of dry material increases during daylight according to external conditions and during the night becomes a negative quantity. On that account, Blackman (1920) stated that the efficiency index when applied to plants growing under natural conditions is only an average measure of the activity of the plant in the production of new material. He also stated that caution must be exercised in making deductions as to a plant's general economy from the efficiency index calculated over periods. The efficiency index nevertheless gives a measure of the plant's average efficiency during any particular period.

Kidd, West, and Briggs (1920) and Briggs, Kidd, and West (1920) calculated the growth rate by means of the formula

$$\frac{R}{100} = \frac{W_2 - W_1}{W_1}$$

where W_1 and W_2 are the dry weights, respectively, of the plants at the beginning and end of the period under consideration. This formula gives the rate of increase in dry weight for the period as a percentage of the dry weight at the beginning of the period. It was adversely criticized by Fisher (1920), who believed that the growth rate should be calculated as a percentage of the mean value for the period.

In a consideration of growth formulas, it should be noted that the rate of change at any time is conditioned by the state of the system at that time; that the rate of change of the system is conditioned by its previous history as well as by the state of the general environment (Gregory, 1928).

V. PERIODIC ELONGATION

According to Friesner (1920), it was first observed by Sachs (1874) that plants exhibit a rhythm of elongation. He found that plants exposed to the alternation of light and darkness showed a maximal elongation shortly after sunrise and the minimum shortly after sunset. Prantl (1873) found a similar behavior in the increase in width of leaves and under normal conditions the maximum was reached in the morning from six to nine and the minimum in the evening from six to nine. He observed that by shifting the hours of illumination and darkness he could alter the time of maximal and minimal expansion, since for each change in the time of illumination and darkness there was a corresponding change in the times of maximal and minimal expansion. These investigators could find no evidence of periodicity when the plants were grown in the dark from the beginning, and Baranetzky (1878) observed that plants which exhibit this regular periodicity lose this when placed in darkness. The time for its complete loss varied from 2 to 3 days and in one case to 14 days. He also observed that plants grown from the beginning in darkness did not exhibit this periodicity except in the case of the shoots of *Brassica rapa*, where it continued to exist.

MacMillan (1891), however, noted that the increase in the diameter of the potato tuber developing in continuous darkness is not regular but rhythmic. The maxima of growth may occur from one to four times during a 24-hr. period. These are not of long duration and are followed by periods of slower growth or entire cessation of growth. The maxima of some days were greater absolute maxima than those of other days, which indicate a grand period of growth for the tuber. MacMillan believed that there is some connection between the periodic growth of the tuber and the periodic growth of the aerial stem, but that there is also an independent periodicity in the growth of the tuber, which is obscured and modified by the secondary induced periodicity, which is related to the aerial stem and its mode of growth.

It was observed by Lewis (1901) that the roots of *Allium cepa* showed two waves in their rate of cell division, the maxima coming at midnight and noon and the minima at 4 A.M. and 4 P.M. His work was substantiated by Kellicott (1904), who also observed that the periods of rapid cell division coincided with a low rate of elongation and that during rapid elongation the rate of cell division was the lowest.

Friesner (1920) in the study of the rate of elongation of the radicles of *Cucurbita pepo*, *Lupinus albus*, *Pisum sativum*, *Vicia faba*, and *Allium cepa* showed that elongation in all these plants proceeded in a wavelike fashion, two to four waves being exhibited in a 24-hr. period. He also found that the curve of cell division in these plants exhibited a number of oscillations and that the maxima of elongation were near the minima of cell division, and *vice versa*. He made the important discovery that the exact time of the occurrence of the maximal and minimal cell division is dependent upon the time of the initiation of activity and not upon the time of day.

The question now arises as to what are the causes of the rhythm found in both the elongation and the cell division of plants. It apparently is not due to external influences of changes in illumination and temperature, since this occurs in darkness under conditions of constant temperature. It is apparently not due to heredity, since it is generally lost when the plants are placed in continuous darkness and uniform temperatures. Friesner (1920) considered that when a large number of cells are undergoing mitosis, the total energy available within the tip is directed more to mitosis than toward elongation, so that the one process will be near to its maximum when the other is near to its minimum. These processes alternate with each other in the individual cell and in the root as a whole. The fact that these rhythms have a definite interval and that the maxima and minima in the different curves depend for the time of their appearance upon the time when activity begins indicates that this alternation between mitosis and elongation is entirely an internal cause and is not related to external conditions. Friesner suggested that a distinction should be made between the terms "periodicity" and "rhythm." He considered that the term "periodicity" as used by the earlier writers means a regular oscillation which is caused by the alternation of day and night or by other external changes and which is lost when the environmental conditions are rendered constant. The term "rhythm" he would restrict to mean any oscillation in activity which is definite and regular and not related to any external influence.

VI. GROWTH-PROMOTING AND GROWTH-INHIBITING SUBSTANCES

It was assumed by Sachs that specific organ-forming substances are needed for growth and that the exhaustion of these substances in the growing regions is the cause of the inhibition of the growth of dormant buds. Pfeffer (1900) considered that regulatory influences exerted by enzymes and other chemical substances play a role in growth. Fitting (1905–1906) studied the conduction of the photostimulus in the coleoptile of oats, and Boysen-Jensen (1910) concluded that this stimulus is conducted by the downward movement of a growth-promoting substance on the shaded side of this organ. Armstrong and Armstrong (1911) stressed the function of substances resembling hormones in the metabolism of plants, and Gericke (1924) concluded from nutritional studies of wheat that plants have a physiology which is subject to growth-inhibiting and growth-stimulating agencies.

It is now generally considered that the growth of plants and plant parts is promoted or inhibited by certain substances other than the inorganic and organic compounds that have been shown to be necessary for plant life. Most of these substances are apparently manufactured by the plant, although some seem to be obtained directly from the soil. These substances have been variously termed "vitamins," "hormones," "growth-promoting substances," or "growth-inhibiting substances."

A. Growth-promoting Substances

The fact that small amounts of organic extracts appear to stimulate plant growth to a marked degree has been taken to mean that substances other than those used directly in food building are necessary for the proper functioning of the plant. Thus the application of animal manure to the soil greatly increases the development of the plant. It has generally been assumed that this stimulation was due to the compounds of nitrogen, phosphorus, and potassium that it contained. Breazeale (1927), however, found that this stimulating property rests largely in the water-soluble organic matter and not in the inorganic compounds that it contains. He believed that the beneficial effects of manure and crop rotation are due in large part to the fact that their vegetative matter decomposing develops or sets free certain organic compounds that are essential to the growth of plants.

The cause of the dominance of the apical bud on stems and tubers and of the polarity of stems has always been a question of great interest, and much of the literature on growth-promoting and growth-inhibiting substances has been devoted to theories that attempt to explain such behavior. In extensive studies concerning the development of roots, buds, and shoots in *Bryophyllum calycinum*, Loeb (1915 to 1924) proposed several theories to account for the growth of certain plant organs and for the inhibition of others. One of these theories assumes that the growth of the bud depends upon the flow of certain substances from the leaf to the bud. The bud that receives these substances first will grow out first and thereby prevent the flow to the other buds whose growth is thereby inhibited. The apparent inhibition of growth in certain regions is thus due simply to the fact that under certain conditions the substances required for growth flow to some other place and are retained there. The removal of this inhibition consists in creating conditions that force the substances to flow to other places at which growth will then occur.

Appleman (1918) obtained evidence from sprouting potato tubers which indicates the presence of growth-promoting substances therein. Thus the vigor of potato sprouts bears a direct relation to the size of the seed piece or, in other words, to the amount of tissue surrounding the eye. When a certain minimum is reached, the vigor of the sprout decreases as the size of the seed piece is reduced, the plant resulting therefrom remaining weak during its vegetative period and yielding a small crop of tubers. The weak sprouts are not due to the lack of ordinary food materials, since a sufficient quantity of these for future growth remains in the tuber piece. It has also been shown that the ordinary inorganic nutrients are not the materials that are lacking under these conditions. Appleman concluded that the potato tuber contains a limited amount of a special growth-promoting substance, and that if the amount of tissue surrounding the growing bud is too small, there is not enough of this substance available for normal growth.

1. Metaxenia.—Any effect produced on the endosperm of an angiosperm by pollen from a plant having a different kind of endosperm is called "xenia." Swingle (1928) suggested the term "metaxenia" for the direct effect of the pollen on the parts of the seed and fruit lying outside the embryo and endosperm. This effect of the pollen has been especially noted and studied in the date by Swingle (1928, 1931) and Nixon (1928, 1934, 1935). They reported that the pollen affects the shape of the seed, and the size, shape, character and time of ripening of the fruit. The time of ripening can be varied as much as 10 days, depending upon the source of the pollen that is used for fertilization. Swingle believed that the simplest, and the most probable, explanation of this behavior is that the embryo or endosperm, or both, secrete hormones, or soluble substances analogous to them. These diffuse into the tissues that constitute the fruit and there exert a specific effect varying according to the particular male parent used to fecundate the embryo and endosperm.

Nebel (1930, 1934) reported that apples grown on the same tree but originating from different kinds of pollen were unlike. These differences though small were indicated in the length, color, weight of fruit, and length of seed. He noted also that the chemical composition of the fruit may be influenced by metaxenia, as indicated by differences in pH value, and in the amount of sugar and titratable acid. Hibbard (1933) and Degman and Auchter (1934) with the apple, and Tufts and Hansen 1933) with the pear, however, could detect no effects due to metaxenia.

Harrison (1931) stated that he had observed metaxenia in cotton as indicated by the length of time required for the fruit to mature, by the length of lint, and by the quality of the fiber on the seed.

2. Vitamins.—The term "vitamin" is used by animal physiologists to denote certain substances that are necessary in small amounts for the proper functioning of the animal organism. Williams (1928) suggested that the word "nutrilite" be used to designate all those vitaminlike substances which function in minute amounts in the general nutrition of organisms.

The vitamins present in animal tissues and in products of animal activity have for the most part not originated there but have been obtained from a vegetative source. Clark (1929) found that diatoms and fresh-water algae can form vitamins from purely inorganic constituents. The synthesis of vitamins by bacteria is in doubt. Yeasts contain a large amount of the water-soluble vitamin B, but it is not known whether it is necessary for the metabolism and reproduction of this plant. In India it has been observed that rice from different localities varied in nutritive value, and this has been correlated with soil differences. In millet the vitamin B content of the grain and in wheat the vitamin A content of the grain were highest when grown on manured land. There are apparently substances in manure that enable the plant to manufacture vitamins A and B (Rowlands and Wilkinson, 1930; Richardson and Mayfield, 1932). Mineral fertilizers, however, influence the production of vitamins little or not at all.

Haber, Nelson and Swanson (1931) found that the content of vitamin A in the leaves of *Coleus* was much higher when the plants were grown in fertile soil than when grown in sand. Clark and Thomas (1934) reported that if conditions are favorable to the growth of green plants, the formation of vitamin A is not affected by the absence of microorganisms. Within certain limits the variation in the amount of light has little or no effect on the production of this vitamin. Gunderson and Skinner (1934) stated that the unicellular alga, *Chlorococcum sp.*, synthesized vitamin A, or its provitamin, in large quantities, and vitamins B and G in lesser amounts but not vitamin C when grown in complete darkness on a medium of inorganic salts and dextrose. Vitamins A, B and G thus must have been elaborated by the cells without light, bacterial symbiosis, or complex nutrients. Richardson and Mayfield (1932), and Virtanen (1936) stated that the quantity, potency, or available amount of a certain vitamin in a plant of a given species is not uniform. This production appears to vary with the variety of plant, its maturity, and the climatic and soil conditions of its habitat.

It was observed by MacLeod, Armstrong, Heap, and Tolbert (1935) in five varieties of sweet potatoes that the content of vitamin A increased as much as 3 and 4 per cent after a storage period of 2 months. This increase in vitamin A after harvest indicates either (1) that the carotene in the sweet potato is not fully developed when the roots are harvested and thus has not yet developed into the precursor of vitamin A, or (2) that the carotene responsible for the formation of vitamin A is present in a form less available to the animal body when the roots are first harvested than after a period of storage. Dove and Murphy (1936) stated that there is a wide difference in the vitamin C content of apples, and that the ascorbic acid content of the leaves has a direct relation to the vitamin C content of the fruit.

Although it is evident that plants manufacture the vitamins that are so essential to the growth and development of animals, the evidence as to their role in the plant body is very limited. The fact that they are frequently present in the largest amounts in those parts of the plant where the metabolic processes are the most active has been considered to indicate that they play a role in the nutrition of the plant (Tschirch, 1921). Havas (1935) added 3- to 5-cc. portions of ascorbic acid in varying concentrations (1 to 10,000, 5 to 10,000, 2.5 to 1,000, and 5 to 1,000 parts of water) to seeds germinating under sterile conditions. The two lowest concentrations proved to be the stimulatory ones. Their addition caused no stimulation of germination but produced an acceleration of growth indicated by an increase of 25 to 30 per cent in the length and the weight of shoots, and an increase of 50 per cent in the weight of the roots. The seedlings of oats were much less sensitive to this treatment than those of wheat. Hausen (1935) added ascorbic acid to nutrient solutions and obtained 35 to 75 per cent increases in the dry weight of the plants. The treated plants also showed a much higher content of vitamin C than the controls. Virtanen (1936) found that the addition of crystalline vitamin C to the medium in sterile cultures of peas caused an increase of 40 to 100 per cent in the dry weight. It is thus considered that vitamins are essential to the best growth and development of green plants.

3. Growth Hormones.—According to Snow (1932) the best known of the "growth substances" that normally occur in plants is at present the substance formed by the tip of the coleoptile in seedlings of members of the grass family. This growth substance, "auxin," "growth regulator," "growth hormone," or "phytohormone" accelerates the elongation of the coleoptile.

a. Historical.—The first indication that the tip of the coleoptile normally produces such a substance was obtained by Paal (1914, 1918) while investigating the transmission of a phototropic stimulus across a discontinuity. After excising the tips of the coleoptiles of *Coix lacryma*, he replaced them eccentrically on the stump and found that the coleoptiles curved strongly away from the side covered by this tip. He assumed this indicated that the tip produced a growth-accelerating substance which passed across the moist discontinuity and down that side of the coleoptile. The result was that the treated side grew faster than the other and caused the coleoptile to curve. Söding (1926, 1929) substantiated the presence and behavior of a similar substance in the coleoptiles of the seedlings of the oat. Went (1928), however, fully established the presence of this growth substance in the tips of coleoptiles of various seedlings and devised means for its quantitative determination. From 1928 to the present, an enormous amount of work has been done to determine the occurrence, nature, transport and the stimulatory action of growth substances. The facts obtained must of necessity be discussed here only briefly, and the student is referred for a detailed review of the literature to Snow (1932); Went (1935); Stiles (1935); Thimann (1935); Boysen-Jensen, Avery, and Burkholder (1936); and Went and Thimann (1937).

b. Occurrence and Preparation.—Growth substances have been obtained from many plant and animal sources. Thus they occur in the coleoptiles of the seedlings of various grains, in seedlings of various other species of plants, in buds, root tips, leaves, green algae, mammalian urine, and saliva. It was found by Nielsen (1930, 1931), Boysen-Jensen (1931, 1932), and Nielsen and Hartelius (1932) that relatively large quantities of growth substances which accelerate the growth of coleoptiles but retard or completely arrest the growth of roots are produced by various fungi and bacteria. Two genera, *Rhizopus* and *Aspergillus*, produce relatively large amounts of these growth substances to which Nielsen gave the general name, "rhizopin." Seubert (1925) found that growth substances are associated with malt extract, pepsin, and diastase. Thimann and Dolk (1933) found that aeration greatly increased the production of growth substance by *Rhizopus*.

Some cytological work has been done to determine which cells of a certain region produce the growth substances. Tetley and Priestley (1927) could find no definite region in the coleoptile of barley for the production of growth substances, since all the cells appeared vacuolated and inactive. They concluded that the response of the coleoptile to light is not the result of any chemical substance but rather of differential permeability of tissue on the two sides of an unequally light-stimulated organ. Perry (1932) found that the structure of the epidermal layer of the coleoptile suggests a secretory function. Over the tip of the coleoptile the cells making up the dermatogen are numerous with large, deep-staining nuclei, dense cytoplasm, small vacuoles, and many granules. Continuous with these cells and extending down the sides of the coleoptile are the epidermal cells, which become longer as their distance from the tip increases. He suggested that the growth substance is produced in the tip cells, and that it is transported to the region of bending by the rapid streaming of cytoplasm in the relatively long epidermal cells. They could find no evidence of the secretion or translocation of growth substance in unequally illuminated coleoptiles. Avery and Burkholder (1936) studied cytologically the development of the coleoptile of *Avena* and concluded that cell division is not involved in coleoptile growth at the time that it is used in tests for growth substance.

The growth substances are obtained by the diffusion method and by the extraction method. In the diffusion method, the growth substance diffuses into water or into an aqueous gel. The latter type of medium was first used by Went (1928) and is now the most universally used. The parts that contain the growth substances are decapitated, and the cut end is placed on 3 per cent agar, into which the substances diffuse. After this diffusion has progressed a sufficient time, the agar is cut into cubical blocks of a standard size and used for experimentation.

The extraction method may be illustrated by describing the procedure used by Thimann (1934). The fresh material is killed by immersion in chloroform, after which about one-fifth its volume of $\dot{N}$ hydrochloric acid is added, and the mixture thoroughly ground. The chloroform layer is removed, fresh chloroform is added, and the acidified tissue is again ground. The chloroform is evaporated and the lipoidal material is extracted in a small volume of water. This extracted material, which contains the growth substance, may then be mixed with agar and its action determined by the usual methods.

According to Went and Thimann (1937) the extraction method of Thimann (1934) demonstrated that growth substance is distributed throughout the seedling plant, although the greatest amount occurs at the tips. It also showed that the growth substances are present in the plant in the free-moving form and in the bound form. The chloroform method determines the total amount of growth substance present at a given time, while the diffusion method is successful only if there is continual production of the material.

c. Determination of Action.—Two general methods have been used for determining the action of growth substances. The one that has been the most universally used was first reported by Went (1928) and may be called the "coleoptile method." This method involves a determination of the degree of curvature of a decapitated coleoptile on which has been placed the growth substance being tested. A block of agar containing the growth substance is applied unilaterally to the decapitated coleoptile sometime during a 40-min. period following its decapitation. The time allowed for action varies from 110 to 120 min., after which the degree of curvature of the coleoptile is determined. It was established by Went (1928) that there is a direct proportionality between the growth substance and the curvature produced. Thimann and Bonner (1932) reported that the curvature is proportional to the concentration of the growth substance and not to the total amount present.

Various units have been proposed for determining the degree of curvature due to the growth substance. Kögl and Haagen-Smit (1931) introduced the term "Avena unit" or "A. E." (Avena Einheit). This unit is defined as the amount of growth substance that, when contained in a block of 3 per cent agar measuring 2 by 2 by 0.5 mm., imparts a curvature of 10 deg. in 2 hr. at 22 to 23°C. to one decapitated coleoptile of *Avena* upon which the agar block is placed eccentrically. They found that from the tips of the coleoptiles of barley, there diffused outward a solution containing 300 A. E. per milligram. Urine contained 400 A. E. per milligram. From urine these workers isolated a pure crystalline substance that had an efficiency of 30,000,000 A. E. per milligram. One milligram of this substance thus is sufficient to impart a curvature of 10 deg. to 30,000,000 *Avena* coleoptiles.

Bonner (1932) described a unit of growth substance as that amount of "rhizopin" which when dissolved in 1 cc. of water and mixed with 1 cc. of 3 per cent agar in the form of agar blocks of standard technique and applied to the side of a decapitated *Avena* coleoptile will produce in 110 min. a deflection of 1 deg.

In about 3 hours after the coleoptile is decapitated, the uppermost part of the stump begins to act physiologically like a tip; in that it begins to produce growth substance (Söding, 1929). This behavior is known as the "physiological regeneration of the tip. To avoid this complication in experimental work the coleoptiles are decapitated a second time after 2 hr. Physiological regeneration of the stump is much retarded by replacing the original tip.

The "pea method" for the quantitative determination of growth substances was devised by Went (1934). The seedlings of *Pisum sativum* were grown in sand cultures in the dark until the epicotyls were 5 to 20 cm. in length. Pieces 2 to 20 cm. in length were cut from the stem about 5 mm. below the growing point and the top of each piece of stem was split longitudinally for 1 to 3 cm. If such a piece of stem is immersed in an aqueous solution, the two halves bend outward. If a growth substance is present in the solution, the free ends of the two halves begin to curve inward after approximately 1 hr. at 25°C. The inward curvature is greater the more concentrated the growth substance, and equilibrium is generally attained in about 6 hr. By using a range of concentrations of the substance to be tested and finding the lowest concentration in which a recognizable reaction occurs, the test can be made quantitative.

d. Mode of Action.—Proof has been furnished by Heyn (1931, 1934), Heyn and van Overbeek (1931), Söding (1931), Thimann and Bonner (1933), Bonner (1933, 1934), and Went (1935) that the application of growth substances to coleoptiles increases the plasticity of their cell walls. Opinion differs as to whether the initiation of growth is caused by an increase in turgor or by an increase in elasticity or plasticity of the walls; and also as to whether or not the laying-down of the new particles is itself a necessary accompaniment of growth. Growth substances seemingly induce only the elongation of the cells that have been formed already, for in the zone of elongation of the coleoptile the cells have ceased to divide. Almoslechner (1934) believed that certain "hormones" are concerned in initiating cell division. Elasticity, however, may or may not increase. The changes in elasticity are correlated with growth but not with growth substances. Increased elasticity is a result of growth, while increased plasticity is the cause of growth. It was observed by duBuy (1936) that with increasing age, the response of *Avena* coleoptiles to growth substances decreased. Plants are only sensitive to these materials at a certain stage of their development.

The amount of growth substance that enters a given region is too small to bear any simple stoichiometrical relation to the substances that are formed in the cell wall during the resultant growth. The growth substance is thus a true hormone in that it is active in minute amounts and in an indirect manner. It has been found that if growth substances are applied to plasmolyzed cells no increase in plasticity occurs, and hence no growth occurs. Since the increase in plasticity occurs only when the protoplasm is in contact with the cell wall, it is inferred that growth substance acts on the cell wall through the protoplasm. This belief is further substantiated by the fact that growth substances may increase the rate of respiration as much as 27 per cent. It was observed by Dolk and Thimann (1932) that if growth substance is supplied to the *Avena* coleoptile in a solution buffered to an alkaline pH, its effect in causing cell elongation is abnormally small. It is common experience that solutions of growth substance must be acid to give the maximal effect of cell elongation. Bonner (1934) found that acid solutions increase the growth rate by increasing the plasticity of the cell wall. There is no increase in cell acidity, however, under the influence of growth substance. It is considered that the increase of growth rate in acid buffers is due to the conversion of growth substances, already in the plant, from an inactive to an active form.

e. Chemistry.—The chemistry of the growth substances has been studied by many workers, including Kögl and his helpers (1931, 1933), Dolk and Thimann (1932), Went (1934), and Thimann and Went (1934). It was suggested by Kögl that the term "auxin" be used for any phytohormone that causes cell enlargement. He distinguished three auxins—auxin *a*, auxin *b* and heteroauxin. Auxin *a* is a monocyclic, trihydroxycarboxylic acid with one double bond and with the formula $C_{18}H_{32}O_5$. It is stable in acid but not in alkali, has a molecular weight of 328, and is heat- and light-stable. Auxin *b* is a monocyclic, hydroxyketocarboxylic acid with one double bond and with the formula $C_{18}H_{30}O_4$. It is stable in light and heat, has a molecular weight of 310, and is destroyed by both acids and alkalies. Heteroauxin is β-indole acetic acid, or 3-indole acetic acid, and has the formula $C_{10}H_9O_2N$. It is destroyed by acids but is stable in alkali and has a molecular weight of 175. This is the auxin that is present in yeast and other fungi.

Hitchcock (1935) and Zimmerman and Wilcoxon (1935) showed that there are numerous organic compounds that may be classed as growth substances. These compounds were dissolved in solutions in which cut stems were placed, were injected into plant parts, or were absorbed by lanolin which was rubbed on the stem or leaf. The principal plant responses initiated by these substances were the local initiation of adventitious roots or stems, proliferations, swelling and bending of stems, acceleration of growth and epinasty of leaves. These compounds moved both upward and downward in the stem. Some of the substances used were indole-3-*n*-propionic acid, phenylacrylic acid, phenolpropionic acid, carbon monoxide, ethylene, propylene, acetylene, α-naphthaleneacetic acid, β-naphthaleneacetic acid, fluoreneacetic acid, anthraceneacetic acid, and α-naphthylacetonitrile. They considered that there are probably many compounds other than those now known which would be equally effective on growth. If plants manufacture their own growth substances, it is not probable that any one plant can produce all those compounds that are now known to be effective. Likewise it is not logical to assume that all plants naturally make and use one and the same growth substance. The growth of a given plant may not be regulated always by the same substance but this may vary with the environmental conditions of the plant.

Wallace (1927) found that $1/100$ p.p.m. of ethylene in air induced intumescences in the twigs of the Transparent apple. Williams, Lyman, Goodyear, Truesdail, and Holaday (1933) reported that a single acid substance apparently occurs in all tissues, and that it is capable of stimulating the growth of yeast. They called this substance "pantothenic acid," a term meaning "from everywhere." Tincker (1935) failed to obtain accelerated growth or flowering by presenting small quantities of ketohydroxyoestrin and theelol to flowering plants in different ways. Havas and Caldwell (1935) reported that glandular extracts had little or no effect on the growth and development of plants. Levine (1934) reported that different portions of plants which had been painted or injected with the carcinogenic agents of animals showed a proliferation of tissue around the treated parts giving rise to the swellings of the stem accompanied with small irregular masses of new tissues. Davis (1934) found that plants of *Vicia faba*, *Allium*, *Narcissus*, and *Pisum sativum* which were injected with a solution of thyroid material in water showed a stimulation in the time of flowering and in the height and number of flowering stalks.

The classification of ethylene as a growth substance has been questioned by Thimann (1934) and Michener (1935). Since, in observed cases, ethylene affects the production of growth substances but not its transport or utilization, they believed that ethylene influences only the enzyme concerned in the production of growth substance.

f. Transport and Polarity.—According to Snow (1932) the transport of the growth substance in coleoptiles is polar since it travels only in a morphologically downward direction. If the lower end of a portion of the coleoptile is placed on an agar block and a cube of agar containing the growth substance is placed on the upper end, the growth substance diffuses downward into the agar block below. If the cylinder is inverted, however, no translocation of growth substance occurs. If the cube of agar containing the growth substance is placed at the bottom of the inverted coleoptile it will diffuse upward into the pure agar on the other end. The growth substance thus must be transported in a manner different from that by which the common organic substances are moved for they travel in either direction according to the circumstances. It has been demonstrated that the growth substance does not travel in the two small vascular bundles to the coleoptile. Avery (1935) noted in leaves of *Nicotiana* that the transport of auxin is polar, in that it moves from the leaf tip to base through the midrib, to which it is conveyed by the lateral veins. Snow (1933) found that the cambial stimulus from the leaves can pass across a protoplasmic discontinuity, which indicates that the stimulus is caused by a hormone.

van Weij (1932) distinguished between the "velocity of transport" and the "intensity of transport" of growth substances. The velocity of transport is measured by the time that elapses before the first appreciable trace of growth substance reaches the basal end after traveling through a coleoptile of known length. The intensity of transport is measured by the amount of growth substance that subsequently reaches the lower end per unit of time. The intensity of transport through paths of different lengths is approximately the same. This behavior differs from simple diffusion, since in diffusion the intensity varies inversely with the length of the path. The intensity of transport varies with the temperature, reaching a maximum at about 30°C. and falling off above or below this point. The velocity of transport, however, is nearly independent of temperature. The concentration of growth substance in the blocks of agar on the lower ends of the coleoptiles continues to rise until it is much higher than in the upper blocks in which the growth substance is contained at the start. The increase in amount does not appreciably retard the diffusion into the lower part. Koch (1934) stated that the transverse transport of growth substance in the *Avena* coleoptile was slower than the longitudinal transport.

g. Tropisms.—Numerous investigators have believed that the geotropic reactions of stems and roots are controlled in part at least by growth substances. Thus it was shown by Hawker (1932, 1933) that if the root tips of *Vicia faba* were placed on blocks of gelatin, and these blocks subsequently placed eccentrically on the cut ends of decapitated roots, the latter curved to the sides covered by the blocks. This showed that these root tips excreted a growth-retarding substance, since blocks of pure gelatin similarily placed caused no curvatures. These results supported the theory of Cholodny (1924, 1926, 1927, 1928) and Went (1928) that the opposite curvatures of root and coleoptile in response to gravity are caused by the growth substance formed by the tips of these organs which ordinarily travel in straight lines to the elongating zones and arrive there in equal concentrations on all sides. Growth is thus equal on all sides of the stem and root and they extend in a straight line. As a result, however, of the stimulus of gravity, these substances are somewhat diverted so that they travel obliquely and accumulate in greater concentration on the lower side of the elongating zone. This procedure accelerates growth on the lower side of the stem as compared to the upper and the stem grows away from the stimulus of gravity. In the root, growth is retarded on the lower side and this organ responds in a positive manner to the stimulus. This general theory was substantiated by Keeble, Nelson, and Snow (1929); Navez and Robinson (1932); and Boysen-Jensen (1933), who found a greater quantity of growth substance on the lower than on the upper side of a root that had been geotropically induced for 2 to 4 hr.

It has generally been assumed that a growth substance must be responsible for the response of plant organs to light. Sierp and Seybold (1926) found that in the coleoptile of the oats the uppermost portion of the tip back to a distance of ¼ mm. is sensitive to light, and that this sensitivity decreases rapidly towards the base. Navez (1933) found that the apical portions of seedlings of *Lupinus albus*, grown under illumination, released into agar blocks twice as much growth substance as those grown in the dark. According to Boysen-Jensen (1936), it has been proved that the concentration of growth hormone is decreased on the illuminated side and increased on the shaded side of a bending organ. From studies of the *Avena* coleoptile it is assumed that unilateral illumination does not affect the formation of growth substance but brings about its displacement toward the shaded side during the course of its downward movement. The subsequent rate of growth on each side is proportional, within limits, to the concentration of the growth substance present. The cause of light exercising an influence upon the distribution of the growth substance is not understood. Keeble and Nelson (1935) believed that negative and positive traumatic curvatures may be interpreted in terms of growth substances and their concentration on opposite sides of a wounded root.

h. Root Formation.—According to Snow (1932) it was first reported by Cholodny (1924, 1926, 1928) that the coleoptile tips of maize retard the growth of decapitated maize roots when placed upon them. These observations were confirmed by Keeble, Nelson, and Snow (1931). The same workers in 1930 observed that if the decapitated roots of the seedlings of *Zea mays* and *Pisum sativum* were washed from 15 to 20 min. they grew much more rapidly than those not so treated. They considered that the wound substances which are produced at the cut surface and which retard growth are largely removed by washing. Hammett (1929) showed that the portion of the root in which mitotic activity occurred contained an acid-stable, alkali-labile substance that stimulated root growth.

Went (1929) reported that he extracted a heat-stable substance from germinating seeds and barley leaves which promoted the formation of new roots when it was applied to cuttings. In 1934 he gave the name "rhizocalin" to this substance. In 1934 he described a method for the quantitative determination of this substance. Rhizocalin mixed with lanolin is an excellent material for the induction of root formation. Navez (1933) found that the decapitation of a root of the *Lupinus* seedling stopped its elongation for a varying period of time. After this cessation there was a period of renewed elongation at a rate lower than the original one. The tipping of the decapitated root with a tip of the coleoptile of *Zea mays* or *Avena* induced a decrease in the rate of elongation of the root. The diffusate of root tips has a similar action on the rate of root elongation. The evidence indicates that the growth substances of the coleoptile tip or root tip inhibit the rate of elongation of the decapitated root. Boysen-Jensen (1933) obtained growth substance from the root tips of *Zea curagua* and *Vicia faba*. It appeared to be concentrated in the first 6 mm. of the root tip. It was noted by Kögl, Haagen-Smit, and Erxleben (1934) that the immersion of roots of the oats seedlings into a dilute solution of auxin *a* greatly inhibited their longitudinal growth. It was stated by Cholodny (1934) that the decapitated *Avena* coleoptiles gave significant curvatures in 2 hr. after the application of the root tips. The application of the root tips of *Lupinus*, *Cucurbita*, and *Helianthus*, however, gave no response. Loo and Loo (1935) found that the extracts of willow roots and of mulberry leaves stimulated the elongation of the primary root of *Zea mays*.

Zimmerman, Crocker, and Hitchcock (1932, 1933) exposed a large number of species of plants to carbon monoxide. They found that under the conditions of the experiment, this gas induced the initiation of roots in numerous young stems. This gas also stimulated the growth of preexisting root primordia in the older part of stems. The root hairs also were more abundant on the roots exposed to this gas than on the controls. Butyric acid, acetylene, propylene and ethylene gave similar results in inducing the formation of root primordia. Zimmerman and Hitchcock (1933) called these chemicals "root-forming substances," since they are specific for the formation of adventitious roots. When only the root tip was treated, new roots formed directly back of this tip. If the root was treated along the region of elongation, root formation was limited to this region. If a zone 8 in. back of the tip was treated, elongation was retarded and new roots were formed all the way down to the tip region. Thus the indications are that the root-forming substance has a direct action on the tissues with which it comes in contact. Hitchcock and Zimmerman (1936) noted that when the cuttings or shoots of *Ilex*, *Taxus*, *Hibiscus*, *Acer*, and *Chrysanthemum* were treated with preparations of indoleacetic, indolepropionic, indolebutyric, or naphthaleneacetic acids, earlier rooting was induced, more roots were produced, and the roots emerged from a greater area of the stem than was observed in the controls.

Cooper (1935) mixed 1 part of pure, synthetic β-indylacetic acid with 2,000 parts lanolin and applied 10 mg. of this paste unilaterally on small areas of the cuttings of lemon, lantana, fig and *Tradescantia*. In all cases the cuttings showed a much greater root development than the untreated ones.

Zimmerman and Hitchcock (1935) applied lanolin preparation of α-naphthaleneacetic acid, indolebutyric acid, indoleacetic acid, indolepropionic acid, Δ-3-indolevaleric acid, and phenylacetic acid to growing roots. When these substances were thus applied along the region of elongation, new branch roots appeared through the epidermis in 3 to 5 days. They also caused swelling and retardation in the elongation of these roots. Application of the growth substance back of the region of elongation was relatively ineffective. During the same year these workers reported that the application of heteroauxin, indolepropionic, indolebutyric, naphthaleneacetic, phenylacetic, and phenylpropionic acids to the soil induced responses on the plants (tomato and tobacco) that were similar to those produced by the application of these substances in lanolin or oil to the aerial parts of the plants. Davies, Atkins, and Hudson (1936) reported that when willow twigs were set in Pfeffer's solution, containing various growth substances in concentrations of 1 part per 100,000, the order in which both roots and shoots appeared was: ascorbic acid, β-indylacetic acid, β-indylpropionic acid, and control. Zimmerman, Hitchcock, and Wilcoxon (1936) added the nine esters—methyl α-naphthalene acetate, ethyl α-naphthalene acetate, methyl phenyl acetate, ethyl phenyl acetate, *n*-butyl acetate, isobutyl acetate, methyl β-indole acetate, methyl β-indole propionate, and methyl β-indole butyrate—to the list of growth substances. These esters when absorbed by the roots from water solution and moved upward cause formative responses in the aerial parts.

i. General.—It was noted by Wallace (1931) that the concentration of ether necessary to prevent the evening closure of the leaves of *Mimosa pudica*, *Oxalis stricta*, and *Marsilia macropus* was different from that necessary to prevent their opening in the morning. The optimum temperature for the sensitivity of *Mimosa* was 40°C.; the minimum, 14°C.; and the maximum, 60°C. Sensitivity appeared independent of relative humidity. The period of greatest sensitivity was about 5 A.M., while the period of least sensitivity was from 1 to 7 P.M. The leaves were not sensitive after an exposure to darkness of 4 to 5 days. von Euler and Philipson (1932) reported that the coleoptile of *Avena* yielded an extract containing a substance which stimulated by 20-fold the growth of the yeast, *Saccharomyces cerevisiae*.

Snow (1935) found that the application of 1 to 2 p.p.m. of pure synthetic heteroauxin and of pure auxin *a* in gelatin to the decapitated stems or hypocotyls of young sunflowers greatly activated cambial growth. He concluded that cambial growth is activated by the same growth substance that is formed by young leaves and promotes extension in the stems. Leonian (1936) noted that a growth substance extracted from garden peas greatly stimulated the growth of certain of the unicellular green algae.

Levine (1936) reported that scars and swellings are produced when certain injured parts of the stems of tobacco, tomato, sunflower, *Ricinus*, and *Datura* are painted with numerous substances including glycine, glutamic acid, and cystine. LaRue (1936) reported that certain growth substances in agar or in lanolin inhibited the development of the abscission layer in the petioles of *Coleus* when applied to these parts.

Some investigators, however, do not consider that it is necessary to assume the existence of special growth-controlling substances to explain growth behavior. Gustafson (1927) believed that growth, especially in fruits, is influenced mainly by nutrition. As the cells begin rapidly to divide, rapid growth ensues, provided that food is supplied; and when the supply of food becomes limited, cell formation ceases and the fruit begins to mature. He considered that the growth of shoots and roots can be explained in the same manner. Thus, in the tomato when the nutrients were not diverted to fruit formation, the rate of growth of the shoot continued at approximately a uniform rate in contrast with an increasingly lower rate when fruiting was taking place. Pearsall (1923) stated that, although there appears to be some evidence in plants of the effect of specific substances, it is not known whether we are dealing with specific substances that are products of cell metabolism or simply with combinations of direct effects.

B. Growth-inhibiting Substances

The absorption by the plant of compounds from the soil which inhibit plant growth has been discussed in detail in Chap. III. The discussion here will deal especially with the inhibitors that retard the development of stems and roots.

1. Buds and Leaves.—Loeb (1915 to 1924) proposed a theory to account for the dominance of the apical stem upon the basis of the formation of growth-inhibiting substances by the plant. He assumed that the growing apex of a stem might form a definite substance which is antagonistic to bud growth and which, migrating toward the base, would impede or inhibit the growth of the lower buds. The apical buds are the first to be freed from this substance and begin to grow out, producing at the same time fresh supplies of the inhibitor, the concentration of which increases gradually toward the base of the stem. The subapical buds are therefore retarded in growth increasingly in this direction, until finally the point is reached where the concentration is great enough to inhibit lateral shoot growth completely.

Appleman (1918) obtained evidence that the terminal buds of the potato tuber inhibit the development of the lateral ones. Thus when all the eyes of a tuber are subjected to the same external conditions, the buds on the terminal or seed end will grow out first, the number of such buds developing depending upon the variety, size of the tuber, and the vigor of the terminal sprouts. The number of eyes on the tuber, however, bears no direct relation to the number of sprouts that will grow out from the terminal end. If the terminal buds grow normally, they inhibit the growth of the more basal ones. If, however, the terminal sprouts are destroyed or otherwise retarded, sprouts will appear on the stem half of the tuber, which otherwise would have remained dormant. If the tubers are cut transversely into halves, the suppressing influence of the terminal eyes is thus removed and the eyes on the stem end show an equal if not greater capacity for the production of vigorous sprouts than the terminal bud (Fig. 39).

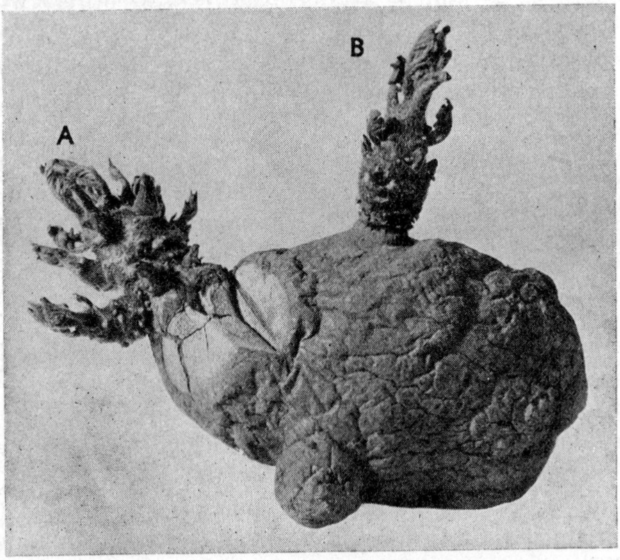

FIG. 39.—A potato tuber bearing an axillary sprout *B* and a terminal sprout *A*. The axillary branch developed because the tuber was notched and thus the growth-inhibiting substance secreted by the terminal branch prevented from reaching the axillary region and inhibiting growth there.

Reed and Halma (1919) believed that the manner of growth of the young shoots of pear and Chinese lemon observed by them afforded evidence for the hypothesis that a growth-inhibiting substance is produced in the apical portion of the shoot, and that it travels downward toward the base, causing a condition of dormancy in the subapical buds. It was observed by Priestley and Evershed (1922) and Priestley and Pearsall (1922) that the growth curve of the roots formed from cuttings of tomato and *Tradescantia* showed a progressive series of S curves. In all cases the first curve ended just before the secondary root formation and the second reached its low level at the appearance of the tertiary roots. They attributed these phases of retardation to the operation of inhibiting factors such as accumulated end products that were left over from the previous growth.

Snow (1925, 1929, 1931, 1932) concluded that the axillary buds in *Phaseolus* are inhibited in development by a substance that originates at the growing tip. The terminal bud of *Vicia faba*, the youngest leaf, the next leaf, and the third leaf possessed respectively 120, 22, 15, and 3 units of growth substance. In *Pisum sativum* the inhibiting effect exerted by the shoot upon the axillary buds comes from three or four of the developing leaves. A leaf of this plant begins to inhibit when it has reached a length of 2.0 to 2.5 mm.; the inhibitory action begins to fall when a length of 20 mm. is attained and ceases entirely when it has reached a length of 40 mm.

It was observed by Keeble, Nelson, and Snow (1931) that the growth of the decapitated roots of maize was retarded by the tips of coleoptiles or the tips of roots. When these tips were placed eccentrically upon decapitated roots, they curved toward the side covered by the tip. This indicated that the growth substance excreted by these tips inhibits the elongation of the cells of the root. Thimann and Skoog (1933, 1934) found that the terminal bud of *Vicia faba* produced a growth substance similar to that obtained from *Rhizopus suinus*, but none was observed in the undeveloped lateral buds. When the terminal bud was decapitated, the lateral buds developed rapidly and each produced approximately one-half as much growth substance as the terminal bud of the intact plant. This behavior may account for the well-known fact that the axillary buds of the stem grow rapidly when the growing tip of the main stem is removed. Tukey and Brase (1933) noted that in "double-worked" fruit trees, the intermediate stem piece influenced the entire plant by producing a dwarfing effect. Van Overbeek (1935) considered that the difference in growth of normal corn and *nana*, a dwarf type, is due to the fact that dwarf corn produces a smaller amount of growth substance than the normal plant.

2. Fruits, Seeds, and Flowers.—The effect of flowering and fruiting upon the vegetative growth of plants has been studied by Murneek (1925, 1926, 1932, 1933) who also summarized the observations of Egorov (1915), Mason (1922) and Taranovsky (1923). It appears that in all types of plants an antagonism exists between vegetative growth and reproduction. In plants with determinate growth, as in the grains and grasses, the vegetative structures undergo rapid senile changes and die promptly upon fertilization. Those plants with an indeterminate type of growth, such as tomato, cotton, pumpkin, some legumes, and most perennial woody plants, do not show this clear-cut antagonism between vegetation and reproduction. Vegetative growth is curtailed or ceases when a maximum set of fruit or seed is being formed, but there is a resumption of vegetative development after the maturing of the fruit. Many apple trees that bear heavily during a given season frequently do not yield fruit the following year and may require several years for complete recovery.

Murneek suggested that some definite and powerful mechanism must be at the disposal of the reproductive organs to accomplish the transfer of the food supply to the newly formed embryo. There is nothing known regarding the nature of this mechanism beyond the conjecture that it may be either of a general physiological or of a specific enzymatic or hormonic character.

According to Murneek the cultivated apple exhibits four waves of abscission of nonsetting flowers and immature fruits, which occur at intervals of approximately 2 weeks and appear to be governed primarily by internal, hereditary factors. The causes for these abscissions may be (*a*) structurally and functionally abnormal flowers, (*b*) unpollinated or unfertilized flowers, (*c*) fertilized flowers in which the embryos abort early, and (*d*) the abscission of some fruits owing to the competition for a local food supply.

Mason (1922) observed in cotton that there was a marked retardation in the growth of both the central stem and lateral branches during flowering and fruiting. This retardation was proportional to the number of flowers that were formed. Kearney and Harrison (1932) found that when two species of cotton were pollinated with a mixture of pollen from both species, selective fertilization in favor of like pollen occurred. These results were explained on the assumption that the presence of like pollen induces a reaction in the stigmatic tissues which renders them less suitable for the development of unlike pollen. The development of the pollen tube appears necessary for the reaction.

McCollum (1934) noted in the cucumber that growing fruits exerted an inhibiting influence upon plant development until the seed coats of the developing seeds began to harden and mature. Parthenocarpic fruits did not produce a striking inhibitive effect on the plant. It is assumed that this inhibitory effect is due to the production of a growth-regulating substance by the developing, fertilized ovaries. Austin (1935) stated that the removal of the flowers from the soybean does not affect the growth of this plant the same as it does many others, but that the exfoliated plants cease growing at the same time as the controls. The changes that accompany the development of fruits in the normal soybean plant are: a diminution of water, a marked decrease in the potassium content in all parts, and a low amount of phosphorus in the stem tips.

3. Volatile Substances.—It was observed by Elmer (1932, 1936) that a volatile substance emanating from ripe apples inhibits the normal sprout development of germinating potatoes. The apical growth of the sprouts is practically stopped, and small stem tubers instead of normal sprouts may develop. Immature apples, however, did not have this effect. Pears and the fruit of the hawthorn also produced the same effect as the apples. Kidd and West (1932) reported that ripe apples, pears, peaches, tomatoes, and bananas produce a vapor that has the effect of stimulating at once the beginning of ripening. Smith and Gane (1933) and Kidd and West (1934) emphasized the need of having ripe fruit to obtain these effects. Nelson and Harvey (1935) believed that a gaseous, unsaturated hydrocarbon, or similar compound, is present in self-blanching celery during the natural blanching of the leaves.

Denny and Miller (1935), Denny (1935, 1936) and Gane (1935) obtained epinastic response of leaves from the emanations of fruits of the apple, pear, tomato, banana, cantaloupe, squash, and eggplant; of petals of geranium, verbena, hollyhock, and petunia; of anthers of the lily; of pistils of the unopened flowers of rose, squash, and hollyhock; of leaves of the rose, spinach, onion, Virginia creeper, potatoes, and tomato; of roots of the radish, turnip, and beet and of the stem of the tomato. Epinasty was noticed after 4 to 6 hours of exposure to these emanations. Some of the plants that are sensitive to the volatile products of ripe apples are *Mimosa pudica, Ricinus communis, Helianthus, Salvia,* and *Solanum lycopersicum.* Some of the seedling plants that are affected by these emanations are pea, radish, turnip, white mustard, cress, peanut, flax, and runner bean.

Elmer (1932, 1936), Botjes (1933), and Denny and Miller (1935) believed that this gaseous emanation is ethylene. Gane (1934) reported that ethylene was identified among the gaseous products produced by ripe apples.

VII. DORMANT OR RESTING PERIODS

The rest or dormant period in plants is a period when the plant or plant parts to all appearances do not grow. Woody plants especially exhibit this characteristic, and it is very common in seeds, bulbs, tubers, buds, and spores. Plants and plant parts vary greatly in the extent of this dormant period. In some instances it may last for only a few days, while in some seeds it may last to 100 years or more. Woody plants go into the dormant condition in the temperate zone at the close of the summer, while crocus, hyacinth, spring beauty, and tulip die to the ground in the spring, and their bulbs become dormant by early summer and almost never begin to grow again before fall, no matter what sort of weather prevails. The majority of seeds will not germinate at once when placed under the conditions that ordinarily produce germination. Howard (1915) found that, if 2 weeks is sufficient time to allow seeds for making immediate germination, fully 75 per cent of the species of Missouri exhibit dormancy. Of the 76 species of seed examined by him, 23.6 per cent grew in less than 3 weeks, 39.4 per cent in the fall or following spring, while 37 per cent did not germinate at all during the time of his experiments.

Duvel (1905) and Goss (1924) found that of the 107 species of seeds buried by them, 50 species grew after a period of 20 years, the germination ranging from 1 to almost 100 per cent. Ohga (1923) found that seeds of *Nelumbo nucifera* taken from strata of soil in a railroad cut and estimated to be from 120 to 400 years old still retained their vitality and readily germinated after their coats were filed to allow the intake of water. The rest or dormant period in the seeds of wild plants is more general and more persistent than in the cultivated ones.

Eastham (1914) found no loss in the vitality of oats after storage for 10 years. Sifton (1920) reported that wheat seed retained its vitality for 5 years, but that 75 per cent of its vitality was lost between the eleventh and fifteenth years. The longevity of oats was found to be greater, and 14 per cent of the kernels were alive after 19 years. The vitality of timothy seed began to decrease at once, and after 12 years it had decreased 88.5 per cent. Darlington (1931) tested the viability of seeds that had been kept for 50 years in sterile, dry soil in sealed pint bottles. He obtained 52, 38, 62, and 8 per cent germination, respectively, for *Rumex crispus*, *Oenothera biennis*, *Verbascum blattaria*, and *Brassica nigra*. Robertson and Lute (1933) in Colorado tested the viability of seeds that had been stored under ordinary conditions for 10 years, and found that wheat, barley, and oats had decreased in vitality during that time 7, 14, and 13 per cent, respectively. The vitality of Rosen rye and of soybeans had decreased only 10 per cent in 5 years, but after that time it declined rapidly. Black Amber sorgo decreased only 2 per cent in viability in 6 years. Yellow Dent corn showed a high rate of germination for the first four years then decreased in vitality 13 per cent in the fifth year, and 20 per cent after 8 years. Stevens (1935) found that the viability of good seed of sweet clover and alfalfa decreased steadily from an original 95 per cent to 60 per cent in 20 years. Red clover dropped to approximately 10 per cent. Soybeans declined steadily, yet after 9 years they had an average vitality of 30 per cent. Blake (1935) decided after testing the germination of the seeds of 42 species of grasses from the tall-grass prairie that there were wide fluctuations in their germinating power. As a general rule, it was low at harvest, highest in spring and autumn, and lowest in winter and midsummer. Most of these seeds lost their vitality in 5 to 6 years.

A. Causes of Dormancy in Plants and Plant Parts

According to Crocker (1916), dormancy is brought about by factors inhibitory to general processes preceding or accompanying growth. The rest period is considered by some to be a direct response to the changing external conditions, while others consider it to be the result of fixed heredity. It seems reasonable to consider it a result of a combination of these two factors. Thus, for example, the structure of a seed which possesses impermeable seed coats or an immature embryo which causes it to be dormant is apparently due primarily to hereditary factors. The environmental conditions under which the seed develops, however, will serve to alter to a considerable degree the nature of the seed structures. The conditions under which such a seed is placed after its separation from the plant will also determine to a marked degree the extent of its resting period.

The work of Howard (1910) indicates that in many cases dormancy is induced primarily by changes in the environment. Thus, of 283 species of woody plants brought into the greenhouse in November and January more than one-half grew easily inside 2 weeks without any treatment. In some species the winter rest is a habit induced by unfavorable outward circumstances, and if these conditions are removed some species can be induced to grow readily. The peach tree apparently belongs to this type of plants. It has a comparatively short dormant period and frequently becomes active during the warmer days of winter, with the result that the fruit buds are killed.

The causes of dormancy in seeds have been studied extensively. According to Crocker (1916), their dormancy is due to one or a combination of the following causes:

1. The possession of thick or hard coats that prevent the intake of water and probably also of oxygen. Some of the seeds that possess such coats are those of the legumes except peanut, several species of *Nelumbo*, velvet leaf, Kentucky coffee bean, asparagus, morning glory, and okra.

Some other seeds in which dormancy is either entirely or partly due to the impermeability of the seed coats to water are vetch (Jones, 1928), Kentucky coffee bean (Raleigh, 1930), *Tilia* (Spaeth, 1932, 1934; Barton, 1934), and cotoneaster (Giersbach, 1934). Johnson (1935) stated that delayed germination in the seeds of *Avena fatua* is determined by a condition of the seed coat that develops after fertilization. There is apparently a correlation between the germinability of these seeds and their position in the panicle.

The possession of coats that are hard or impermeable to water is one of the most common causes of dormancy in seeds. In some species all the seeds are hard, but more frequently only a portion of the seeds are hard, and these are of varying degrees of hardness. This in nature insures the distribution of germination of a single crop over a number of years. Most of the long-lived seeds have hard, impermeable coats.

In some cases, the seed coats are readily permeable to water but are so firm that they prevent the expansion of the embryo. The seeds of *Alisma plantago* and *Amaranthus retroflexus* are apparently inhibited in their germination on this account. Crocker and Davis (1914) reported that when the seeds of *Alisma plantago* lie in water, the embryo itself does not much more than one-fourth consummate its possible imbibitional and osmotic swelling. The embryo thus lives in water for years restrained in its swelling by the seed coat against which it must exert a pressure of approximately 100 atmospheres. When the carpel wall is removed, the seed swells rapidly and increases 40 per cent of its air-dry weight in the course of 2 hr. The elongation of the embryo when the seed coat is removed was 30 per cent of its length in 5 hr.

2. The possession of coats that interfere with the absorption of oxygen. The classical example of this type is the seed coats of the cocklebur (*Xanthium*). This character appears in both the upper and lower seeds of the bur but is more marked in the former. It is considered that the germination of many other seeds is greatly influenced by the rate of the diffusion of oxygen through the seed coats. Spaeth (1932) stated that the dormancy of the seeds of *Tilia* is due in part to the impermeability of the nucellar membrane to oxygen.

3. The possession of immature embryos. When some seeds are separated from the parent plant, the young plant in the seed has not yet reached its full embryonic development. This development is reached after the seed has fallen off, and several weeks or months must elapse before it becomes sufficiently developed to germinate. The seeds of

Ginkgo biloba, *Gnetum gnemon*, *Erythronium denscanis*, *Ranunculus ficaria*, *Tilia*, and *Cotoneaster* are some of those whose dormancy is due to an undeveloped embryo.

4. The possession of an embryo which, although to all appearances mature, must undergo certain changes before it will germinate. Most of the seeds of the *Rosaceae*, including peach, apple, cherry, plum, and hawthorn, are of this type. Before these seeds will germinate they must "afterripen," as will subsequently be discussed.

B. Overcoming Dormancy

1. Stems.—The principal methods that have been used in breaking the dormant period of these plant parts are (*a*) raising the temperature, (*b*) lowering the temperature, (*c*) treatment with gases and vapors, (*d*) immersion in chemical solutions, and (*e*) desiccation. Howard (1915) found that the most effective treatments for overcoming the dormancy of 65 species of herbaceous perennials with which he worked were freezing, desiccation, and ether vapors. With woody stems, freezing for 8 days followed by treatment with ether caused the quickest growth. Coville (1920) showed that the blueberry and other woody plants which have become dormant will not resume normal growth in the spring unless they have been subjected to chilling. Dormant plants that have not been chilled do not resume growth until long after those which have been exposed to winter weather. It was found by Denny and Stanton (1928) that the vapors of ethylene dichloride and ethylene chlorohydrin were very effective in breaking the rest period of the lilac, crab apple, flowering almond, *Azalea*, and others. A treatment of 24 to 48 hr. was usually sufficient. The gain in time varied from 2 weeks in the flowering almond to 2 months in the crab apple. Other chemicals which gave favorable results were propylene chlorohydrin, furfurole, and acetylene tetrachloride.

The effect of various treatments is apparently limited to that portion of the plant which has been treated. Thus, when a branch of blueberry was placed through a hole in the window and exposed to winter weather, it leafed out in the spring, while the portion that remained inside and was unchilled did not do so (Coville, 1920). By applying the various chemical treatments to individual buds of the lilac, Denny and Stanton (1928) showed that dormancy in this plant is not systemic but is localized in the buds. Thus of the pair of buds at the tip of a twig, one could be made to start growth by treating without breaking the dormancy of the other member of the pair.

Gardner (1929) noted that dormant pear trees which had been stored in the greenhouse at ordinary temperatures remained in that condition, while those which had been stored at 2°C. or outdoors during the winter grew after such treatment. Kramer (1934) reported that the treatment of the dormant seedlings of yellow poplar, red oak, and white oak with ethylene chlorohydrin broke the dormancy in 85 to 100 per cent of the cases. Mack and Livingston (1933) noted that the effects of ethylene on plants are influenced by oxygen and temperature relations, light, water, and the nature and physiological state of the tissues.

2. Bulbs and Tubers.—Denny (1930) and Denny and Miller (1934) stated that the treatment of the corms of gladiolus with 0.12 to 0.20 per cent ethylene chlorohydrin per liter in a closed container for 3 to 5 days hastened sprouting. The advance in time of germination ranged from 60 to 180 days, varying with the variety and stage of dormancy.

The treatment of potato tubers for the overcoming of dormancy has received considerable attention. According to Appleman (1918), treatment of the buds with ether or chloroform forces them to open 3 to 6 weeks ahead of normal. Treatments that furnish a better oxygen supply to the tissues also shorten the dormant period. Thus wounding of the tuber and the removal of the skin shorten the dormant period. If the tubers are wrapped in cotton saturated with hydrogen peroxide, the rest period is shortened. The hydrogen peroxide is absorbed by the tuber, and oxygen is released through the catalase that it contains.

Denny (1926, 1928) tested the value of a large number of chemicals for the hastening of germination in potatoes. The vapors of ethylene chlorohydrin and solutions of sodium and potassium thiocyanate gave excellent results. The treated tubers showed 100 per cent germination, and in 2 months the plants were 2 ft. high and tubers were forming, while the check tubers had not yet appeared above the ground. Favorable results were also obtained from trichloroethylene, dichloroethylene, carbon bisulphide, ethylene dichloride, and ethyl bromide. According to Vacha and Harvey (1927), ethylene also breaks dormancy in the potato.

Rosa (1928) reported that the more mature the potato tubers are at harvest, the shorter is their dormant period. Storage at 28 to 30°C. has a marked accelerating effect upon subsequent sprouting, as compared to storage at lower temperatures. The dormancy of fully matured potatoes was overcome in 6 to 15 days, depending on the variety, by a treatment with ethylene in concentrations of $\frac{1}{400}$ to 1/2,200 of air. Wright and Peacock (1934) found that the rest period of immature potatoes was longer than that of mature ones. Miller (1933) showed that ammonium dithiocarbonate, thiosemicarbazide, hydrogen sulphide, and ethyl mercaptan are effective in breaking the dormancy of freshly harvested potatoes. Denny and Miller (1935) found that a concentration of ethylene chlorohydrin can be obtained which, when applied continuously for 7 days, will break the dormancy of potato tubers so that, when they are planted 2 weeks after treatment, satisfactory germination will be secured.

3. Seeds.—The treatments that have been used to hasten the germination of seeds are scarification, treatment with chemicals, alternation of temperatures, freezing and thawing, storage, and placing under conditions favorable to afterripening. Dormancy that is due to hard or thickened coats is overcome by scratching or filing the coats or by dissolving them with strong sulphuric acid so that water may be absorbed. The dormancy of hard-coated seeds is also overcome by placing them in the soil to be subjected to freezing and thawing during the winter season. Rose (1915) found that the hard-coated seeds of legumes, lettuce, mustard, okra, and snapdragon could be made to germinate more quickly by being blown against needle points. It was found by Jones (1928) that the seed coats of *Nelumbo lutea* do not absorb water after soaking in water for 18 months at room temperatures, but that when the seed coats were broken without injury to the embryo, the seeds quickly germinated. These seeds may be prepared for germination by treatment for 5 hr. with concentrated sulphuric acid, followed by washing in tap water and drying on a screen to eliminate immediate germination. The threshing machine acts as an effective abrasive agent on hard seeds and hastens their germination. Thus certain legume seeds when hulled by hand have as high as 90 per cent of hard seeds that will not immediately germinate, while there are only 20 per cent of such hard seeds when hulled by machine. The longevity of some seeds, however, is lessened by being scarified (Wolfe and Kipps, 1926).

Gracanin (1928) found that the speed of germination of the seeds of beet, rye, sunflower, *Festuca*, and others was accelerated by treatment with *o*-phosphoric acid. Deuber (1931) and Bramble (1932) showed that the dormant seeds of sugar maple, Norway maple, and certain varieties of oats could be stimulated into germination with solutions of thiourea and ethylene chlorohydrin and by the vapors of ethylene chlorohydrin. Baldwin (1932) improved the germination of the seeds of *Picea rubra* by immersion in absolute alcohol. Flemion (1934) stated that the germination of freshly harvested seeds of peach, apple, and hawthorn could be obtained by placing the embryos under germinating conditions after the removal of the outer and inner seed coats. Brown (1933) reported that the delinting of cotton seeds with sulphuric acid gave an increased rate of germination and a 20 per cent increase in yield over the controls.

a. Temperature.—It was found by Harrington (1923) that seeds of redtop, parsnip, celery, orchard grass, bluegrass, Bermuda grass, and Johnson grass germinate much better with favorable alternations of temperature than at constant temperatures, the exact alternation depending on the kind of seed. In alternating the temperatures, the upper temperature should be maintained only a small part of the day, never more than 8 and usually not more than 6 hr., and the change to the lower temperature should be fairly rapid. The best germination of the seeds of Johnson grass was obtained after being subjected to a temperature of 30°C. for 18 to 22 hr. and at 45°C. for 2 to 6 hr. each day. The best results were obtained for parsnip, celery, redtop, and orchard-grass seeds when placed at 20°C. for 16 to 18 hr. and at 30°C. for 6 to 8 hr. Morinaga (1926) reported that alternating temperatures were necessary to germinate the intact seeds of Bermuda grass and *Typha latifolia*, but that breaking the seed coats or treating with sulphuric acid allowed germination at constant temperatures.

Fivaz (1931) found that the seeds of *Ribes rotundifolia* and *Ribes cynosbati* germinated more promptly and more abundantly when subjected to daily alternations of temperature from 10 to 25°C. than when the temperature remained constant. Crocker and Barton (1931) reported that apple seeds germinate when the temperature fluctuates between 5 and 10°C. Flemion (1931) found that the dormant seeds of *Sorbus aucuparia* are rendered active by alternating the temperatures daily or weekly from 1 to 5°C.

Busse and Burnham (1930) found that the cotyledons and hypocotyls of the seedlings of cotton and flax showed abnormalities when the seeds were treated with liquid air. Busse (1930) stated that dormancy of seeds due to impermeable seed coats may frequently be broken by cooling these seeds to −190°C. with liquid air. Lipman and Lewis (1934) reported that seeds of various kinds which had been dried over calcium chloride and subjected to temperatures of −189 to −193°C. for 60 days, germinated when placed under suitable conditions. Tang (1931) found that with wheat seeds, higher temperatures in general gave higher percentages of germination until an optimum was reached, after which still higher temperatures gave lower percentages of germination.

According to Kondo and Okamura (1931), it is difficult to maintain the vitality of rice seeds containing 14 per cent moisture. By lowering the moisture content of rice over calcium chloride, however, vitality may be preserved readily at temperatures of 30 to 40°C. Robbins and Petsch (1932) noted that it required a temperature of 80°C. for 2 hr. to kill 75 per cent of the embryos when the moisture content of the grain was 5.2 per cent. However, a temperature of only 57°C. was necessary to kill the same percentage of embryos when the seed contained 30.5 per cent of moisture. According to Ludwig (1932) the minimum temperature for the germination of cotton seed is approximately 12°C. Barton (1932) found that the sealed storage of *Delphinium* seeds at any temperature is more effective than open storage. Under open storage conditions, temperatures of 8 and −15°C. are superior to room temperature for preserving the vitality of these seeds. Livingston and Haais (1933) obtained complete germination of "French" rice seed within 14 days at all low temperatures excepting 8 and 12°C., and excepting 41 to 55°C. for high temperatures. Edwards

(1934) stated that the optimum temperature for the germination of "Black Eyebrow" soybeans was between 33 and 36.5°C. The literature on the temperature relations of germination was reviewed critically by Edwards (1932).

b. Oxygen.—The supply of oxygen is an important factor in germination, and it is generally considered that the increased germination due to various treatments is in many cases caused by the favorable effect the treatment has upon the diffusion of oxygen through the membranes of the seed. The exact changes in the oxygen relations are, however, not understood. Morinaga (1926) noted that the seeds of cattail (*Typha latifolia*), which germinated poorly or not at all in air, germinated promptly when the oxygen supply was decreased by diluting the air with 48 to 80 per cent hydrogen or nitrogen. When 20 per cent oxygen was added to the air, germination was only 1.3 per cent. Some oxygen, however, is necessary for germination, since when all of it is removed the seeds will not germinate. Morinaga also found that Bermuda grass germinates better with a reduced oxygen supply, but the effects are not so marked as in the case of the seeds of *Typha latifolia*. The breaking of the seed coats of *Typha latifolia* removes the necessity of reducing the oxygen supply, and they germinate equally well with from 1 to 90 per cent of full atmospheric oxygen. Thornton (1935) found that as the temperature was increased from 21 to 31°C. the upper and lower seeds of the cocklebur required a lower partial pressure of oxygen for germination.

c. Light.—Light has an effect upon the germination of certain seeds. Thus Gardner (1921) reported that the seeds of *Rumex crispus*, *Datura stramonium*, and *Phoradendron flavescens* are light sensitive. The germination of the first and last were promoted by light, while that of *Datura stramonium* was hindered by light. It was noted by Morinaga (1926) that light, nitrate, and nitrite with alternating temperatures increased the germination of the seeds of Bermuda grass. The highest germination of the seeds of bluegrass was obtained with these three factors in combination. Light was found to be effective in the germination of celery seeds only when unfavorably high temperatures were used.

Andersen (1931) placed seeds of Canada bluegrass in a 0.02 *N* potassium nitrate solution in light at 30°C. for 6 hr. and then in darkness for 18 hr. for 28 days, and obtained a germination rate of 60 to 70 per cent. Seeds that were placed in the dark in a dilute solution of nitric acid gave a 60 to 70 per cent germination, while those in water in the dark gave a germination of only 20 to 30 per cent. Hutchings (1932) found that freshly collected seeds of the monkey flower (*Mimulus ringens*) are light sensitive. At low light intensities, the germination of this seed is roughly proportional to the intensity of light. Thus with intensities of 6.0, 3.5, and 1.5 foot-candles, there were germinations of 75, 34.5, and 11.5 per cent, respectively.

Axentjev (1930) found that the light-inhibiting effect on germination was entirely dependent on the presence and integrity of the seed coats in *Amaranthus retroflexus*, *Phacelia tanacetifolia*, *Androsace maxima*, and *Bromus squarrosus*, but not entirely due to the coats in *Cucumis melo*, *C. sativus*, *Cucurbita pepo* and *Nigella arvensis*. In light-stimulated seeds the light effect was entirely due to the integrity of the seed coats in *Rumex crispus* and *Epilobium hirsutum*, but the seed-coat condition played little or no part in the light stimulation of *Oenothera biennis* or *Silene densiflora*. The coats evidently act by restricting the oxygen available to the embryos. Apparently, in some cases, light interferes with oxidation processes within the seed, while in others it favors these processes.

Because lettuce seeds are frequently light sensitive, the germination of these seeds has attracted much attention. According to Shuck (1934, 1935), American lettuce seed is in a physiologically unstable condition that makes it particularly sensitive during germination to the effects of light, moisture, and temperature. In the laboratory, germination is promoted by the exposure of the seeds to light, by the use of a very moist substratum, and by starting the germination at a low temperature. The light requirement may be satisfied by continuous exposure to light under germinating conditions or by exposing the moist seeds to light before placing them in a dark chamber. He considered that light causes certain little-understood, photochemical changes within the seed. Shuck believed that there is a growth-inhibiting substance in lettuce seed, since after a sufficient number of seeds had been germinated in the dark on a cotton medium for 10 days, the germination of similar seeds was completely inhibited on this medium; but when this cotton was washed in water and again used as a substratum, 98 per cent of a similar lot of seeds germinated. Thompson (1935) found indications that immaturity at the time of harvest is an important factor contributing to the dormancy and light-sensitivity in lettuce seed. Much of this difficulty may be eliminated by using only fully matured seeds. Perhaps Borthwick and Robbins (1928) worked with fully matured seeds for they stated that the only requirements for the germination of lettuce seeds are an adequate supply of moisture, a temperature below 25°C., and good aeration. Thornton (1936) obtained good germination of lettuce seed at 35°C. in the presence of carbon dioxide in either light or darkness.

Flint (1934) found that the red, orange, and yellow rays of light were the most effective in promoting the germination of lettuce, while the green, blue, and violet rays inhibited germination. He found further that non-light-sensitive seeds could be made sensitive by subjecting them for a period, while moist, to blue light and then drying. Seeds thus treated would not germinate in the dark but would do so under red light. Flint and McAlister (1935) discovered a band in the region of 7600 Å which inhibits the germination of light-sensitive lettuce seed far more effectively than do similar inhibitory influences previously noted in the regions 4200 to 5200 Å. The relative effectiveness of radiation in the violet, blue, and green regions of the spectrum (4200 to 5200 Å) in inhibiting germination in light-sensitive lettuce seed is found to be the same as its relative effectiveness in causing phototropic response in the etiolated coleoptiles of oats.

C. Afterripening

The term "afterripening" refers to the series of chemical or physical changes, occurring within the plant or plant parts, which bring to a close the dormant period and make growth again possible (Jones, 1920). It is evident that these changes will vary for different plants and for different treatments, so that no prediction can be made with certainty concerning the changes that will occur.

1. Changes Accompanying.—Howard (1915) believed that the specific effect of all rest-period-breaking agents on dormant woody tissue is due to the stimulation of enzymes. He thought that the diastatic, proteolytic, fat-splitting, and oxidizing enzymes are stimulated into greater activity by the treatment with the various reagents. He considered that the rest period sets in on account of the inhibition of enzyme activity due to the overaccumulation of the products of their action.

In the potato tuber, Appleman (1914, 1916) could observe no changes in the amount of diastase and invertase under normal growing conditions until sprouting had occurred. The protein, lipoid and organic extractives, and the inorganic phosphorus also remained constant until that time. The abrupt changes in regard to these substances was concurrent with sprouting. The juices from the tubers at the end of the rest period caused a greater acceleration of the oxidation of pyrogallol than the juice from new immature tubers of the same varieties. It was observed by Appleman and Miller (1926) that the ripening and maturing process in immature potatoes may continue during storage, so that by the end of the rest period, immature potatoes large enough for seed have practically the same percentage composition and respiratory response as potatoes allowed to mature on the vine, provided that they are stored under the same conditions.

The stimulating effect produced by chilling in overcoming dormancy in woody plants is believed by Coville (1920) to be intimately associated with the transformation of starch into sugar, since tests for starch and sugar showed a decrease of starch and an increase of sugar as winter progressed. The theory offered in explanation of this is that the starch grains are at first separated by living cell membranes from the diastatic enzymes, but that, when the plant is chilled, the vital activity of these membranes is weakened, the enzymes are allowed to pass through, and the starch is changed to sugar. This change is regarded as the mechanism for creating high osmotic pressures to start the plants into growth.

The changes accompanying afterripening have been especially studied in seeds. The period of afterripening may be greatly shortened by proper treatment; thus the seeds of the hawthorn (*Crataegus mollis*) have a latent period of 1 year or more. Davis and Rose (1912) showed that if these seeds are removed from the carpels and kept very moist at a temperature of 5 to 6°C., the latent period may be shortened to 2.5 to 3 months. If the testas are removed and the embryo treated as mentioned, the dormant period may be reduced to 30 days. Eckerson (1913) studied microchemically the changes accompanying the afterripening of these seeds and observed that the first initial change in the embryo was an increase in acidity, which was correlated with an increased water-holding capacity and an increase in the activity of catalase and peroxidase. Near the end of the afterripening period there was a sudden increase in acidity and in the water content, and oxidase made its first appearance. The enzymes, acidity, and water absorption all increased in amount until the hypocotyl was 3 to 5 cm. in length. At that time the fats decreased and sugar appeared. Eckerson found that afterripening in this case could be greatly shortened by treating the embryos with dilute hydrochloric, acetic, or butyric acid. The water-holding capacity, acidity, and amount of peroxidase increased much more rapidly and oxidase much earlier in the embryos thus treated than in the untreated ones. This increase in acidity, water-holding capacity, and catalase and oxidase activity during afterripening was also observed by Rose (1919) in the seeds of *Tilia* and by Jones (1920) in the seeds of the sugar maple.

Pack (1921) in the study of the embryo of *Juniperus* has reported the most extensive data in regard to the changes accompanying afterripening. The following changes were observed: (1) rapid or complete imbibition followed by a steady slow decrease in water content until germination, (2) increased hydrogen-ion concentration in the embryo and an increment of titratable acid, (3) a steady and enormous increase in the degree of dispersion of stored fats, (4) a decrease in the amount of stored fat and protein with an increase of sugar, (5) the translocation of food in the form of fat or fatty acids from the endosperm to the embryo, (6) a sevenfold increase in the amino acid content, (7) an increase of the soluble proteins with a marked hydrolysis of the stored proteins, (8) a slight increase in the respiratory rate, (9) an increase in the respiratory quotient, (10) a decrease in intramolecular respiration, (11) a twofold increase in catalase activity, (12) a slight growth of the embryo, and (13) a rise in the vigor of the seeds as indicated by their resistance to fungous attacks. The changes that accompany afterripening in the seeds of *Juniperus* are thus represented by the accumulation of cell-building materials and enzymes.

The changes accompanying afterripening have been especially studied in recent years. A brief summary of the various changes that have been observed is given in the table on page 1056.

2. Ripening of Fruits and Vegetables.—In the ripening process certain chemical changes occur. Many of these changes are the same regardless of the manner in which the ripening occurs. According to the conditions under which it occurs, ripening of fruits and vegetables may be designated as natural or artificial.

a. Natural Ripening.—It was found by Rosa (1928) in the ripening of cantaloupes that there was a progressive increase in the percentage of dry matter, in total sugars, in soluble solids, and in the specific gravity of the juice. The sucrose increased more rapidly than the reducing sugars decreased, indicating that sugars continued to move into the fruit. The pectic material remained the same, while the protopectin was high in amount in the unripe melons but changed into pectin and pectic acid as the fruit ripened. Kertesz (1930) noted that, as soon as peas were shelled, their content of sucrose decreased, while the percentage of alcohol-insoluble residue increased.

Allen (1932) found that apples, peaches, pears, and apricots require sunlight for the development of red color. The softening of the flesh is one of the most important changes that occur during ripening. All the fruits analyzed gained significantly in sugar as they colored on the tree.

Changes Occurring during Afterripening

Investigator	Plant part	Treatment used	Results
Gardner (1929)	Bartlett pear trees	Cold	Increase in hexose sugars, sucrose, and organic acids and depression of freezing point of cell sap and in starch.
Denny (1930)	Potato tubers	Ethylene chlorohydrin Sodium thiocyanate	Increase in sucrose, decrease in starch. No change in reducing sugars.
	Lilac plants	Thiourea Ethylene chlorohydrin	Increase in catalase, invertase, moisture in buds, soluble nitrogen, and respiratory rate. Decrease in total sugars.
Denny and Miller (1932)	Potato tubers	Organic sulphur compounds	Increase in catalase and peroxidase, pH from 0.1 to 0.4 over controls, sucrose 2 to 4 times, respiration 4 to 5 times.
Hasselbring (1932)	Narcissus bulbs	Immersion in water at 43.5°C.	More available carbohydrate and acceleration of metabolic activity.
Denny (1933)	Gladiolus corms	Ethylene chlorohydrin	Increase in sucrose and soluble nitrogen. Decrease in reducing sugars and in insoluble nitrogen.
Evans (1933)	Seeds of *Magnolia grandiflora*	Chemical changes occurring between time of the rupturing of coat and protrusion of hypocotyl.	Embryo increased in water content by 50 to 60 per cent, epidermal cells increased in acidity. Starch appears in the cotyledons. Absorbs food from endosperm. Oils decrease in endosperm. Starch disappears in the endosperm.
Flemion (1933, 1934)	Seeds of:		
	Rhodotypos kerrioides	At 50°C.	Increase in catalase, peroxidase, lipase, titratable acid, sucrose, and water. Decrease in ether-soluble substances.
	Symphoricarpos racemosus		Increase several fold in catalase and peroxidase.
Guthrie (1933, 1934)	Potato tubers	Ethylene chlorohydrin	Slight increase in conductivity of tissue. Increased leaching of electrolytes. The expressed juice increased in pH from 6.0 to 7.0, and decreased in citric, sulphuric, and nitric acids. Isolated glutathione from juice.
Thornton (1933)	Potato tubers	40 to 50 per cent CO_2 with 20 per cent O_2 at 25°C.	Increased respiration, reducing action of expressed juice. Decreased glutathione during treatment but increased over controls when planted.

After picking they showed a decrease in starch and a gain in sugar. Caldwell (1934) stated that the young fruits of oranges, grapefruit, apples, cherries, strawberries, and others had a hydrogen-ion concentration approximately the same as the other tissues of the plant. As these fruits mature, however, the acid content increases as much as 80-fold in citrus fruits. This great increase in the hydrogen-ion concentration markedly increases the imbibitional capacity of the protoplasmic colloids and of certain cell-wall constituents.

b. Artificial Ripening.—The term "artificial ripening" has been used to designate the hastening of natural ripening by the application of some artificial stimulus. The artificial ripening and coloring of fruits have apparently been practiced for a long time, but during the past 20 years renewed interest has been taken in this problem and methods have been devised for artificial ripening on a large scale of citrus fruits, bananas, dates, persimmons, pineapples, cantaloupes, tomatoes, celery, and others. The first method consisted in placing kerosene stoves in the car or bin, and it was thought for some time that heat was the effective agent in the process. Later, however, it was found that the gases generated by the stoves were effective when distributed by pipes among the fruits and vegetables to be treated. Denny (1924) determined that the change in color in lemons was due to some unsaturated hydrocarbon compound. He tried ethylene and found that the coloring was very effectively and quickly done. Ethylene is now extensively employed in the ripening process. According to Chace and Denny (1924) and Harvey (1928), it has distinct advantages over the older methods of ripening. The time is shorter; there is less chance for spoiling, since a lower temperature is maintained; and a better color and flavor result. The best temperature for treatment is 65 to 70°F., and the best concentration 1 cu. ft. of gas to 1,000 cu. ft. of air. The time of treatment varies, but 48 hr. is generally sufficient to ripen the greenest fruits.

Harvey considered that the effect of the ethylene is to activate the enzymes which digest starch to sugar and oxidize acids and tannins. He found that the treated stalks of celery increased their sugar content 20 to 30 per cent over the checks. Increases in the sugar content were also observed in treated tomatoes and bananas. Regeimbal and Harvey (1927) noted that treated pineapples were sweeter and the activity of the proteolytic enzymes more marked than in the controls. It was demonstrated by Babb (1928) that the blanching of celery with ethylene does not affect its vitamin-B content. Other investigators, including Chace and Church (1927) and Hibbard (1930), however, could not observe an increased carbohydrate content in the fruits and vegetables treated with ethylene. Harvey (1928) suggested that the strength of the gas used in the treatment might have its influence upon the changes in the sugar content.

The effect of ethylene in artificial ripening has also been studied by Harvey (1925) on celery, Rosa (1925) on tomatoes, and Harvey (1926) on bananas. Wolfe (1931) stated that bananas in an atmosphere containing 1/100 to 1/10,000 parts ethylene turn yellow, increase slightly in sugar, and decrease in starch. According to Davis and Church (1931), ethylene seems to act on Japanese persimmons by its effect on the general metabolism. The treatment stimulated softening, color development, and respiratory activity. Hartshorn (1931) reported that "the carbide treatment" hastened the ripening processes of green bananas, as shown by the rate of softening, respiration, starch hydrolysis, flavor, and color changes. Apparently the effect is mainly a shortening of the period of low activity that normally occurs at the beginning of the ripening processes. Dustman (1934) found that stored apples were not affected chemically when treated with ethylene, ethylene chlorohydrin and ultraviolet irradiation. Treatment with ethylene, however, hastened the color changes from green to yellow, and softened the tissues.

D. Regeneration

According to White (1931), the successful culturing of excised root tips was reported independently by Kotte (1922) and Robbins (1922). Chambers (1923) found that individual cells would separate and migrate from root fragments, provided the latter were of a definite size and were cut from a definite portion of the root. Robbins and Maneval (1923) found that the excised root tips of beans, morning glory, alfalfa, mustard, wheat, sunflower, and flax were capable of making considerable growth in sterilized Pfeffer's solution containing 2 per cent of dextrose, and that they were capable of growing in sterile Pfeffer's solution alone for only 12 days. White (1932) studied the effect of hydrogen-ion concentration, temperature, aeration, and illumination upon the growth of excised wheat roots. In 1933 he reported the best compounds of iron and sulphur to use in the culture of root and stem tips. He grew over 400 excised stem tips of *Stellaria media* in hanging-drop cultures for 3 weeks without changing the medium. During this time there occurred division and differentiation of the meristematic tissue into leaves, stems, and floral organs. In 1934, White grew excised root tips (10 mm. long) of the tomato for over a year, making 52 transfers in the following medium:

Salt[1]	Millimols	Salt	Millimols
$Ca(NO_3)_2$	0.60	KCl	0.87
$MgSO_4$	0.30	KH_2PO_4	0.09
KNO_3	0.80	Fe_2SO_4	0.006

[1] Plus 2 per cent of sucrose and the filtered extract of 0.01 per cent by weight of dried, brewer's yeast.

This solution was placed in 50-cc. flasks and immediately autoclaved at 15 lb. pressure for 20 min. One isolation of a root tip in vitro produced during the year 35,000 growing points and more than 400,000 mm. or approximately 1,300 ft. of tissue. Robbins, Bartley, and White (1936) found that if the root tips of corn and pea, 1 to 2 mm. in length, were divided longitudinally into equal parts, both pieces developed into complete roots in the nutrient medium used. Rea and Karper (1932) showed that the sorghums can readily be propagated from cuttings.

LaRue (1935) reported that the seeds of corn, wheat, and barley which were split lengthwise produced roots and shoots. The embryos of wheat, Sudan grass, and sorghum that were deprived of endosperm were capable of growing into complete, although weak plants. Wheat seedlings from which the shoots were removed produced new shoots, Sudan grass seedlings produced roots but no shoots, while barley and oat seedlings died. Later (1936) he found that pieces 0.5 mm. in length cut from the immature embryos of dandelion, wild lettuce, oxeye daisy, and tomato could be grown into complete plants in culture solutions. The cotyledons of the tomato excised from immature embryos grew into complete plants, but, in the other species, the cotyledons failed to regenerate. He used White's solution, omitting the yeast extract, but added synthetic heteroauxin. No tissues regenerated without either yeast extract or heteroauxin in the medium. All the plants grown in culture were found to be capable of growth in the soil.

VIII. FACTORS AFFECTING GROWTH OF PLANTS

According to Reed (1919), growth may be considered a function of two variables. The first is the genetic constitution of the individual, and the second is the resultant of all those factors which make up what is commonly called the "environment" of the organism. The factors of the first group are essentially internal, while those of the second are essentially external.

A. Internal Factors

In the fertilized egg there are certain genetic tendencies which when expressed later during the life of the growing plant tend to produce under any set of external conditions a certain definite form of growth (Hanna, 1925). Thus a plant of a dwarf variety, although it may be made to vary slightly in its size and composition by its environment, cannot be made to attain any considerable size by any known cultural methods. The effect of inherent factors on growth is also illustrated by the similar growth rates of green and albino corn seedlings, which were observed by Kempton (1924). That sister plants differing so greatly in a major physiological characteristic should have such similar growth rates is a very striking fact. This behavior of growth of these two types of plants is believed to be controlled by inherent factors that are not associated with chlorophyll and photosynthesis.

Some of the conditions that are classified as internal factors other than the inherent nature of the protoplasm are the concentration of the cell sap, the number and size of the growing points, and the accessibility of manufactured food. Although these factors are classified as internal, they may be altered to a marked degree by the external environment. The growth rate of a plant is thus a function of the internal factors, which at present are indefinitely defined, and the various conditions of the external environment. The external conditions merely accelerate or retard the manner in which the internal factors tend to express themselves (Briggs, 1928; Gregory, 1928). Changes of environment thus need not always have the same effect on the rate of growth. Even though two organisms have the same rate of growth under a given set of conditions, the internal conditions in the two cases may be different and may be differently affected by the changes in the environment.

In a discussion of the influence of internal factors upon growth, the effect of the size of the seed upon plant production could well be considered. The size of the seed, although governed by genetic tendencies, is influenced to some degree by its position on the parent plant and by the general environmental conditions. The reports of the experiments on the effect of the size of the seed upon the vigor and productivity of the plants derived from them are conflicting. This is apparently due to the fact that the comparative results vary greatly with the manner of testing (Kiesselbach, 1924). Kidd and West (1918) and Rotunno (1924) in reviewing the literature on the subject found that the evidence was in favor of the conclusion that larger seeds give rise to more vigorous plants and a better yield. Rotunno in his experiments with small, medium, and large seeds of the radish found that the percentage germination of all three sizes of seeds average the same within the limits of error. The medium seeds averaged the heaviest roots and produced with but one exception the highest percentage of salable roots.

Rudolfs (1923) working with bean seedlings from large and small seeds in the dark found that those from large seeds showed the better growth and that temperature acting as an accelerating factor increased this advantage. Thus at 10°C. large and small seeds required 9.8 days to attain a growth of 140 mm. At 15°C. the large seeds required 7.8 days and small seeds 8.8 days. At 20°C. the large seeds required only 3.1 days, while the small seeds require 8.5 days. In the large seeds a normal temperature coefficient for growth was shown, while in the small seeds the temperature coefficient was abnormally small. This seems to suggest that the amount of food stored in the seeds is of major importance for early growth.

Kiesselbach (1924) in reviewing the literature concerning the yields from seed grades of cereal crops summarized the published data as follows: (1) When the space planting was such as to permit maximum development, the individual plant grain yield was 19 per cent less for the small than for the large seed. This difference is apparently due to the immediate advantage of a greater reserve food supply in the larger seed, which results in a more vigorous initial growth. (2) When planted in equal number at a rate optimum for the large seeds, the small seeds yielded 11 per cent less grain per acre than the large. (3) When planted in equal weights at a rate optimum for the large seed, the small seed yielded 3 per cent less and the unselected seed 2 per cent more per acre than the large seed. The relatively low yield of plants from small seeds was largely overcome by planting a greater number of seeds. The results of experiments by Kiesselbach with winter and spring wheats and oats are very comparable to these reported results, and he concluded that there is no material or practical gain in the grain yield, under farm conditions, from the practice of grading small grain seed, provided it is free from inert matter.

Schmidt (1924) found with seeds of soybeans, buckwheat, lima beans, and corn that germination in the soil occurred more rapidly in light seeds than in heavy ones. Seeds of medium weight grade, or slightly heavier, were superior in germinating power by approximately 8 per cent over seeds of very heavy or very light weight. In Golden Bantam corn the number of ears, green weight of ears, and air-dry weight of stalks and husks varied in the same order as did the weight of the seeds that were planted. The superiority of the plants from heavier seeds over those grown from lighter ones decreased notably as the plants approached maturity and sometimes the difference had disappeared entirely by that time. Kotowski (1929) noted that the differences in plants of peas, broad beans, and cabbage due to seed size may be distinguished during the early growth of the plant but disappear before blooming and apparently have no influence on the yield. Mitchell (1934) found, however, that seed weight is an important factor in determining the size attained by 1-year-old seedlings of Scotch and White pines. According to Paton (1929) this effect is noticeable for 6 years. McComb (1934) also emphasized the effect of the size of seed upon the nature of the young seedling.

B. External Factors

Some of the external factors that may influence growth are temperature, light, moisture, electricity, and the amount and composition of the materials in the soil.

1. Temperature.—It has long been known that the growth of plants is closely correlated with temperature, since the rising temperature of the environment is attended within certain limits by an increased rate of growth.

a. Growth Curve in Relation to Temperature.—In general, the growth curve in relation to temperature shows three cardinal points: the minimum, the lowest temperature at which growth is exhibited; the optimum, the temperature at which growth is the greatest; and the maximum, the highest temperature at which growth will occur. The position of these three points, especially the optimum temperature, has varied rather widely in the reports of different investigators. Some of these differences may be attributed to the fact that the growth rate varies for different plants, for the different organs, and for different ages of the same plant or plant part (Newcombe, 1923). Thus for a 12-hr. period the optimum temperature for the growth of the coleoptile for oats was 30, for bluegrass 27, and for *Sorghum vulgare* 31°C. It has also been shown that the length of the time of exposure is an important factor in determining the position of the optimum, minimal, and maximal temperatures for growth.

Lehenbauer (1914) in corn seedlings found that the optimum temperature for growth for a 12-hr. period was 32°C., but that this was not the optimum when the length of the period of exposure was altered. At the temperatures above 31°C. the initial rate of growth was not maintained and there was a marked falling off of this rate during prolonged periods of exposure. Similar observations were made by Leitch (1916) in the growth of *Pisum sativum.* The minimum temperature for growth in this plant was −2°, the maximum 44.5°, and the optimum 28 to 30°C. In this instance, the relation of growth to temperature could be expressed as a uniform curve from −2 to 29°C., the curve showing a very close resemblance to that of respiration. Above 20°C., however, for each higher temperature a different curve had to be constructed to express the rate of growth in successive time intervals. This falling off of the growth rate with time is similar to the decrease in rate of enzyme action and photosynthesis under the influence of high temperatures during prolonged periods.

No definition of optimum temperature in regard to growth is thus of value unless it involves a consideration of time. The optimum temperature for growth has been defined by some as the highest temperature that will allow growth for some specified time period without decrease in rate. Lehenbauer believed it should be defined as the temperature which for a specified time period of exposure produces the greatest growth. He observed in corn seedlings that when the experiments were made over a 3-hr. period, the optimum temperature was 29°C. If the observations were over a 6-hr. period, the optimum temperature was 30°; and if the periods were from 9 to 27 hr., the optimum was 32°C. Above 27 hr. the optimum shifted to 31°C.

A definition of the maximal temperature for growth must also involve a consideration of time. The maximal temperature might thus be defined as the highest temperature at which growth ceases after a specific time exposure. With corn, Lehenbauer noted that a temperature of 40°C. allowed growth in all the plants tested for a period of at least 21 hr. With a temperature of 41°C., five of the six plants exhibited no growth after 15 hr., while the other plant ceased to grow after 18 hr. At 42°C., three plants ceased to grow after 15 hr. and three after 18 hr. At 43°C. three grew no more after 9 hr., two ceased to grow after 12 hr., and the sixth stopped growth after 15 hr. It may thus be stated that 43°C. is the maximal temperature for growth of these seedlings for an exposure period of 15 hr., and 42°C. for an 18-hr. period. It was observed that plants kept at 41 to 43°C. for 6 hr. after growth had ceased were not killed, since when the temperature was gradually lowered growth was again resumed.

Plitt (1935) noted that oat seedlings grown from seeds which were germinated at 25°C. exhibited a higher percentage of dry matter than those germinated at 5°C. The seedlings were more rapidly developed at the higher temperatures, and photosynthetic activity started earlier. Smith and Cochran (1935) noted that the maximum rate of growth of pollen tubes occurred at 70°F.

MacDougal (1917) noted that a retardation in the growth of wheat and corn seedlings occurred at more than one place in the temperature scale and at different times of the day. In some instances this was attributed to direct temperature effect, while in others this explanation was not deemed adequate, and the cause was attributed to varying imbibition capacity coincident with alternations of acidity or alkalinity.

Since growth is intimately associated with hydration and imbibition, it is of interest to note the effect of temperature upon these processes under different conditions of the medium. Sections of petioles of *Solanum* were placed in distilled water and in 4.2 and 2.6 per cent citric acid at temperatures ranging from 18 to 38°C. The swelling in distilled water was nearly three times as great at the higher temperature, while in the acid solution a retardation took place that limited the total at the higher temperature to something over a half that possible at the lower temperature. The total swelling in the acid at the lower temperature occupied an hour, while at the higher temperature it was a matter of 10 to 15 min. The total imbibitional capacity of the lower temperature was not reached for 8 to 10 hr., while at the higher temperature it was reached under 2 hr. The effect of temperature on growth is thus apparently determined to a marked degree by the chemical reactions of the tissues under observation.

Berkley and Berkley (1933) stated that cotton plants were least resistant to high temperatures at the higher relative humidities. They defined the "thermal death point" as that temperature at a given relative humidity that will kill the protoplasm immediately. A thermal death point is not valid without stating the age of the plant and the relative humidity at the time of its consideration.

It has been assumed by many in studying the effect of temperature on growth that the process is a unified one or a series of successive reactions. MacDougal (1920) emphasized the statement that growth is a group or "constellation" of activities and that the rate of one of these dependent on temperature may be the determining one when the particular process forms either the retarding or leading agency. On that account, he asserted that the relation of growth to temperature between 10 to 50°C. cannot be expressed by any simple formula. At times, however, the rates of metabolism, respiration, hydration, diffusion, and other processes may coincide in such a manner as to make possible the application of a simple formula for the effects of temperature within a limited range.

Kondo and Okamura (1930) reported that the best temperature for growth in length of the rice plant was 30 to 32°C.; for growth in weight, 34°C.; and for grain production, 30°C. The optimum temperature range for the growth of rice was thus 30 to 34°C. Shippy (1930) found that the range of temperature permitting the formation of callus from apple cuttings and grafts includes 0 to 40°C. For general callusing, temperatures below 20°C., rather than higher ones, have been the more satisfactory. Edwards, Pearl, and Gould (1934) and Pearl, Edwards, and Miner (1934) stated that the processes of growth in a relatively simple structure as a seedling are not homogeneous. Not all parts of the developing organism are growing at the same time or at the same rate. It appears that the processes are so interrelated that any excessive development of one function is counterbalanced by the reduced activity of another function. Nightingale (1935) grew peach and apple trees in sand at temperatures of 45, 50, 55, 60, and 95°F., and found that during the current season the maximal yield of roots and tops occurred at 65°F.

b. Resistance of Plants to Low Temperatures.—The large economic losses caused by the injury or death of plants during the autumn, winter, or spring by low temperatures have stimulated a considerable amount of research to determine the cause of this injury, the characteristics that retard it, and methods for its prevention, and to develop or discover plants that are inherently resistant to the effects of such temperatures. A thorough review of the literature pertaining to the effects of low temperatures on plants may be found in Abbe (1895), Salmon (1917), Rosa (1921), and West (1924), and the student is referred to these references for detailed information on the subject.

1. *Causes of Injury.*—The cause of injury by freezing has been assumed to be due primarily to the withdrawal of water from the cell. Upon freezing, water is usually withdrawn from the cell and freezes in the intercellular spaces. As the temperature falls, more water is frozen on the outer walls and more is drawn from the cell by imbibition. Since the force with which the remaining water is held increases rapidly with the progressive loss of water by virtue of the increasing concentration of the cell sap, the amount of water frozen at each successive degree is smaller and smaller.

According to Müller-Thurgau (1880, 1886) death from low temperatures is due to a denaturing of the protoplasm that is caused by the exosmosis of water to the intercellular spaces where it forms ice masses. Molisch (1897) considered it due to an injurious concentration of the cell sap caused by the removal of water in ice formation. Gorke (1906) considered that death results from this increased concentration together with the loss of some specific substances at their eutectic points. He considered that, under the ordinary conditions of life, these substances protect the protoplasm against the injurious effects of other components which remain in solution at freezing. Dexter

(1934) concluded from work with wheat crowns that the withdrawal of water by ice formation was not a fully reversible process, but that in hardier plants it was more nearly so. Thus when the plants are frozen beyond recovery, the water that has been removed by ice formation is not reabsorbed sufficiently to give the previous condition of turgidity.

Maximow (1912, 1914) believed that death from low temperatures is due to the rupturing of the protoplasm by masses of ice crystals that are formed in it. In experiments with pollen grains, Chandler and Hildreth (1935) believed that the theory of Maximow is correct. According to Stiles (1930), the death of a plant by freezing is caused by the formation of ice crystals in the protoplasm, which produces a disturbance in the relations between the dispersed and continuous phases, with an aggregation of the former. Such changes are usually irreversible, so that, upon thawing, the original colloidal system of the living protoplasm is not re-formed and is thus no longer living. Evidence in support of this view is derived from actual observations on the freezing of living and nonliving colloidal systems, and on the structural effect of varying the rate of freezing of such systems.

With very rapid freezing, it is possible to obtain the original system again upon thawing. By such treatment, cases are on record in which living plant and animal cells have remained alive through freezing and subsequent thawing. In frost-resistant plants, however, it is probable that water is bound to hydrophile colloids of the protoplasm and is nonfreezable, so that the formation of ice crystals does not occur. Chambers and Hale (1932) studied strips of the epidermal cells taken from onion bulbs that had been frozen at between -7 and $-10°C$. Ice readily appeared between the cell wall and the protoplasm, causing plasmolysis of the latter. The ice sometimes broke the outer membrane of the protoplasm, but the tonoplast and vacuole remained intact. Ice was not definitely seen to form within the vacuole but the general impression was gained that it did form. Schaffnit and Lüdtke (1932) considered that death of a plant from frost or heat is a problem of metabolism. The entire process of metabolism need not be suspended. The cessation of activity of only one enzyme may be sufficient to cause death. Joslyn and Marsh (1933) stated that the physical changes which occur during freezing depend chiefly upon ice formation and osmotic relations; these involve changes in volume and texture. The chemical changes are those concerned with the hydrolysis of pectins and sucrose, and other changes are in color and flavor. Lutz (1935) believed that the recorded evidence indicated that at least in many cases the rapidity of thawing had no measurable influence on the injury resulting from exposure to a freezing temperature. In his experiments with onions and apples, the influence of temperature in thawing did not become pronounced until a temperature of 95°F. was applied. The killing by cold depends apparently in some cases on ice formation rather than on the direct effect of low temperature. The withdrawal of water from the cell by freezing may bring about death by the following causes: (1) the desiccation of the protoplasm, (2) the mechanical injury to the protoplasmic membrane, and (3) the precipitation of proteins by salting out due to the greater concentration of salts in the sap or to its increase in acidity as water is withdrawn. It should be noted here that chilling above the freezing point has an injurious effect upon some plants. Thus Sellschop and Salmon (1928) found that an exposure of 24 to 36 hr. at 0.5 to 5°C. was fatal to rice, velvet beans, and cotton, while with the same exposure cowpeas were completely defoliated. Peanuts and Sudan grass exposed to these temperatures for 48 hr. were apparently uninjured at first but died in about 2 weeks. Soybeans, potato, buckwheat, tomatoes, and flax were exceedingly hardy, showing no injury when exposed at these temperatures for 84 to 96 hr.

2. *Characteristics of Hardiness.*—Since certain varieties of plants are known to be hardy under conditions of low temperature, the question naturally arises as to what

characteristics enable them to survive while others perish. Strausbaugh (1921) stated that the dormant condition reached by hardy forms of plums appeared to involve fundamental changes in the colloidal condition of the protoplasm, whereby there was a marked retention of water against the force of dehydration. According to Newton (1924), the imbibition pressure of fresh leaves in the winter-hardened condition is in most cases directly related to hardiness. The imbibitional pressure of hardened leaves appears to depend on the physical state of the cell colloids of the living tissue, since this property was lost when the tissues were killed. The quantity of hydrophilic colloids contained in the pressed juice was found proportional to winter hardiness. Ackerman (1917) observed a high correlation between the sugar content of the hardened leaves and hardiness. Although the concentration of the cell sap may play some role in winter hardiness, it does not always show a relationship in that regard (Salmon and Fleming, 1918). According to Martin (1927), the hardy wheats are characterized by a low moisture content of the tissues, a high percentage of total solids in the cell sap, a high osmotic concentration of the cell sap, a high percentage of bound water, and a low rate of respiration at low temperature. Zacharowa (1925) in the freezing of roots found a positive correlation between cold resistance and alkaline reaction and a negative correlation between acid reaction and cold resistance. The water content of the soil may exert a marked influence on injury of plants by freezing. Thus Hill and Salmon (1927) noted that plants grown in a dry soil were injured more severely by freezing than similar plants in a wet soil. The specific heat of the water in the wet soil prevents a rapid change in temperature.

Recently much study has been given to the characteristics of plants that enable them to withstand low temperatures and other severe environmental conditions. Some of the observations in this regard are summarized in the table on pages 1067 and 1068.

3. *Hardening Plants.*—Plants in many instances may be made to endure low temperatures, without serious injury, by hardening. They may be hardened by being exposed to gradually decreasing temperatures, scantily supplied with water, severely pruned, or watered with various salt solutions. Apparently any treatment that checks the growth of the plant increases its resistance to cold. Harvey (1918) reported that cabbage plants became resistant to ½ hr. of freezing at −3°C. by exposure to 3°C. for 5 days. He considered that the principal hardening process in cabbage was to change the proteins to forms less easily precipitated, which was indicated by an increase in the amino acid content. Hardened plants contain a larger amount of bound water than the unhardened ones. The percentage of moisture frozen in hardened cabbage leaves at −3 and −4°C. was only two-thirds of that frozen in tender cabbage. That is, the actual amount of water remaining unfrozen at a given temperature was greater in hardened than in tender leaves, although the total moisture content was less.

Martin (1927) stated that during the hardening of wheats there is a decrease in the moisture content, an increase in the total solids in the sap, and an increased osmotic pressure and imbibition pressure of colloids as measured by the ability of the tissues to hold the sap against the forces of freezing and pressure. Hardened plants possess a larger amount of hydrophilic colloids, probably pentosans (Hooker, 1920), than the untreated ones. They lose less moisture by transpiration per unit of leaf surface than do tender plants (Rosa, 1921). Weimer (1929) found a correlation between the amount of bound water and cold resistance of hardened and unhardened alfalfa plants but could not show a similar correlation between varieties shown to differ in hardiness.

The term "bound water," as herein used, refers to that water which is associated with the colloidal material of the cell in such a manner that it will not freeze or at least it will not form crystals as does the free water under freezing conditions. Sayre

SUMMARY OF SOME RECENT OBSERVATIONS ON HARDINESS

Investigator	Plant or plant part	Characteristics of hardiness
DeLong (1929)	Wheat	Frost injury to standing wheat interrupts and prevents the translocation of materials from the stems and leaves to the grain.
Janssen (1929)	Wheat	The percentage of winterkilling varies with the date of seeding. Winterkilling was greatest in the seeding of Oct. 19, followed by Oct. 6, and Aug. 31. Temperature at which plant grows previous to freezing is important. Plants grown at 33°C. were killed when subjected to a temperature of −10°C., whereas plants grown at 5°C. were not injured when subjected to the same temperature.
Walter (1929)	Evergreens	Concentration of cell sap increased as winter progressed. Resistance to frost increased with an increase of osmotic value. The more slowly the latter increases the more resistant the plant is to cold. Cold injury is due primarily to desiccation.
Hottes and Wilson (1930)	Wheat	Resistance to high temperatures varies inversely with water content of seed.
Chandler (1931)	Pollen	Unwetted pollen remained viable after being subjected to temperatures of −15 to −18°C. for many hours. Moistened pollen was killed under similar conditions. Freezing kills cells by the physical effect of ice masses.
Newton, Brown, and Anderson (1931).	Winter wheat	Exposure of the expressed sap of unhardened plants to frost decreased the coagulable protein and increased amino acids. This splitting of proteins in hardened plants may be a result of frost instead of an adaptation against it and the value of sugar as a protection to winter plants may lie partly in its ability to delay this action.
Newton and Brown (1931)	Winter wheat	Maximal precipitation of proteins in expressed sap of leaves of unhardened plants when they were subjected for 5 hours to a temperature of −7°C. The maximum precipitation occurred at pH 5.1 and 7.3. Dialysis of electrolytes increased subsequent to precipitation of proteins by frost. This precipitation is irreversible.
Peltier and Tysdal (1931)	Alfalfa	Hardy alfalfa varieties become dormant earlier and harden more rapidly. They retain their hardiness longer in the spring.
Friedberg (1932)	Wheat	Cold-resistant varieties tiller deeper than sensitive wheats. The capacity to develop a tiller from the coleoptile is a varietal characteristic by which the wheat renews growth when the growing tip is killed by cold.
Martin (1932)	Spring wheat	Late heading of wheats of the Pacific coast indicates resistance to cold.
Timmons and Salmon (1932).	Alfalfa	The resistance to artificial low temperatures was correlated with the severity of the winters of the regions to which the various strains had been adapted.
Waldron (1932)	Wheat	Plants injured by frosts in April and May showed decreased yields.
Dunn (1933)	Cabbage, Brussels sprouts, and alfalfa	Pectic substances may be partially responsible for cold resistance. Proteins are apparently converted into other compounds in the hardening process, and the protoplasmic constituents are probably of primary importance in the hardening of the plants to cold.

SUMMARY OF SOME RECENT OBSERVATIONS ON HARDINESS.—(*Continued*)

Investigator	Plant or plant part	Characteristics of hardiness
Salmon (1933)	Wheat	Plants frozen artificially at night were injured less than similar plants frozen during the day. Plants in dry soil injured more than those in wet soil. If soil frozen previous to the test, no difference observed in injury to plants. The less injury in wet soil due to lag in temperature.
Schaffnit and Wilhelm (1933).	Potatoes and tomatoes	Application of potassium fertilizer lessens freezing injury by lowering the freezing point.
Collison and Harlan (1934).	Baldwin apple trees	Trees given nutritional treatment conducive to growth and yield showed less winter injury than those not receiving this favorable treatment.
Dexter (1934, 1935)	Wheat	Hardiness associated with a higher percentage of dry matter and with low concentration of soluble salts. At low temperatures there is an increase in soluble organic nitrogen, and a decrease in respiratory rate that is not correlated with a change in sugar content or enzyme activity. Usually an increase in sugar that may or may not be accompanied by increased hardiness.
Greathouse and Stuart (1934).	Red clover	Winter hardiness associated with greater concentration of carbohydrate and nitrogen, lower content of moisture, a large amount of unfreezable water, a lower specific conductivity, and a slightly higher pH value.
Megee (1935)	Alfalfa roots	The higher the conductivity of expressed juice the less the hardiness. No direct relationship existed between winter hardiness and heat of swelling, moisture equivalent, freezing point, chemical composition, respiration, and amount of moisture. Heredity is the most plausible explanation of winter hardiness.
Worzella (1935)	Winter wheat	Cold resistance inherited in the same manner as other quantitative characters.
Woods (1935)	Raspberries	Increased nutrients produce more vigorous growth and such plants are more susceptible to winter injury than the less vigorous canes.
Laude (1937)	Winter wheat, rye, barley, and oats	Water content and amount of expressed sap increased as active growth began after dormancy. The total solids in the sap decreased. The loss of cold resistance was usually more rapid in the varieties possessing greater midwinter hardiness.

(1932) considered that no satisfactory definition of bound water can be made. He stated that all the water which does not show some of the common properties of liquid water may be considered as bound water. He described three methods for determining bound water.

Harvey (1930) found that the threshold for producing hardiness in cabbage was about 5°C. Alternate exposures for 12 hr. at 0°C. and 12 hr. at 10°C. produced greater hardiness than continuous exposure at the average of these two temperatures.

Walton and Fort (1931) stated that mannitol and dextrin were found in frozen sugar cane in Louisiana. Acetic acid fermentation also appeared under such conditions. Bourne (1935), however, did not notice these facts in frozen cane in Florida. Schaffnit and Lüdtke (1932) reported that with winter wheat, winter vetch, and cabbage the lowering of the temperature altered the entire relationship of the nitrogen compounds. Meyer (1932) found that the threshold temperature for the hardening of evergreen leaves was about 6°C. A few hours' exposure daily to this or a lower temperature suffices to keep them in a hardened condition. The basis for cold resistance lies in some physicochemical properties of the protoplasm which as yet are not well understood and cannot be discerned by the gross measurements that at present are generally employed.

Dexter (1933) stated that the hardening of plants is favored by conditions which tend toward the accumulation or conservation of carbohydrates and other reserve foods, which favor photosynthesis, and which reduce the respiratory rate and the extension of vegetative parts. Tysdal (1933) found that the hardier varieties of alfalfa hardened more rapidly under short days than under normal days than did the less hardy ones. The short day reduced growth, a process that is conducive to hardening. Plants hardened under white light could withstand cold better than those hardened under red and blue light. Peltier and Kiesselbach (1934) and Suneson and Peltier (1934) noted that the seedlings of winter wheat which were yet dependent on the endosperm excelled in hardiness those which had emerged 1 or 2 weeks previously and were independent of it.

Calvert (1935) found in wheat that the total water and the free water per 100 g. of expressed sap were significantly higher in the morning than in the afternoon, while the content of bound water was the reverse of this. The total water, the free water, and the bound water per gram of dry matter in the expressed sap were all significantly greater in the forenoon than in the afternoon. The free water exceeded the bound water in the morning, while the bound water exceeded the free water in the afternoon. Dexter (1935) found indications that there is no loss of electrolytes in any of the varieties of alfalfa during the process of hardening.

4. *Measurement.*—It was stated by Dexter, Tottingham, and Graber (1930, 1932) that the degree of resistance of plants to injury by cold weather may be measured by the diffusion of electrolytes and other substances from chilled or frozen tissues into water, when such tissues are thawed. Tottingham, Shands, and Delwiche (1931) considered that Chibnall's method of extraction for the separation of the vacuolar sap and cytoplasmic fluid in plant cells can be used for investigating the winter hardiness of wheat. Peltier and Tysdal (1932) reported a method for growing and hardening alfalfa in the greenhouse.

Dexter (1932) described a method by which leaves may be tested for hardiness by absorbing the expressed juice on ashless filter pads and, after the extraction of the pad with distilled water, determining the quality of sap by measurements of the electrical conductivity. In every case, extracts of tender varieties have higher electrical conductivities than those from hardier plants. Extractions by freezing or by pressure served to differentiate the varieties on a hardiness basis, while the extraction of plants injured by heat, ether, or grinding did not. A considerable part of freezing injury may be due to the mechanical effect of the pressure of the ice that is formed.

5. *Hardiness and Drought Resistance.*—According to Maximow (1929) an increase in the accumulation of sugars, other soluble carbohydrates, and hydrophilic colloids gave a greater resistance to frost. The same factors also aid a plant in enduring drought. Waldron (1931) in a study involving five varieties of wheat under semiarid conditions found that there was a positive relationship between frost resistance and drought resistance. Tysdal (1933) found that plants which were kept severely wilted for 10 to 14 days were much more resistant to cold than those which had been watered regularly. As reported by Schröder (1909), Savage and Jacobson (1935), and Vassiliev and Vassiliev (1936), plants resistant to drought have the same general characteristics as those resistant to cold.

2. Light.—In addition to its influence on the formation of chlorophyll and the process of photosynthesis, light exerts a formative effect upon plants.

a. Growth in Darkness.—When a plant is grown in darkness, striking structural and morphological changes result. It is greatly elongated, the tip of the stem has a characteristic hooked form, and the leaves remain very narrow and scalelike. A plant grown under such conditions and with such an appearance is said to be "etiolated." According to Priestley (1926), the main morphological and structural features of etiolation are determined by a redistribution of meristematic growth at the shoot apex, which results from the greater difficulty experienced by the meristem in drawing nourishment from the vascular supply, because when grown in the dark the walls between the vascular strands and meristem are rendered relatively impermeable by the retention in them of the protein and fatty substances that form the surface of the protoplast. The preponderance of growth in length over superficial growth in the shoot in etiolated plants suggests that the superficial layers of meristem are making less growth, while new tissues are being added to the stem as the result of the activity of the more deeply situated meristematic cells. Priestley obtained evidence from micro- and macrochemical data which indicated that the fatty substances released from the differentiating cells are only slowly removed from the walls between these protoplasts under conditions of etiolation. This slow removal of fatty substances from the walls is obviously in close correlation with the deep-seated meristematic activities. Trumpf (1924) found that exposing etiolated plants to light for only 1 to 2 min. and then returning them to darkness produced marked changes in the appearance of the plant. The most striking results were the marked expansion of the leaf lamina and the disappearance of the plumular hook. Priestley (1925) by the exposure to light of the etiolated plants of *Vicia faba* and *Pisum sativum* produced the changes observed by Trumpf and also greatly shortened the stems, so that the etiolated plants had much the same appearance as normal ones, except that they lacked the green color. It appears that the light action in these cases is photocatalytic.

b. Rate of Growth in Light and Darkness.—If stem elongation is taken as the measure of growth, it will be found that in most instances the rate of growth during the night will exceed that during the day, provided that the temperature does not fall too low during the night. Thus Osmaston (1918) noted that the increase in height of the bamboo was 9.65 ft. in 14 days of which 3.7 ft. occurred by day and 5.95 ft. by night. Porterfield (1928) reported that the ratio of the night growth of bamboo to that of daylight was 1.8 to 2.8 as reported by investigators in Java, Ceylon, and India. In Algiers, however, the growth of this plant was greater during the day than during the night. Porterfield found the ratio of day growth to night growth to be 1.6. Brown and Trelease (1918) found that the shoots of *Cestrum nocturnum*, which is one of the most rapidly growing plants around Manila, P. I., wilted during the day and decreased in length, but that later in the afternoon they returned to their original length and increased it during the night. Prescott (1921), according to Hanna (1925), found that the growth rate of corn in Egypt as measured by increase in height showed two maxima during the 24 hr., one a short time after sunrise and the other in the late afternoon about sunset. The total growth, however, during the 12 hr. of night from 8 P.M. to 8 A.M. was generally greater than that during the remaining 12 hr. of the daytime. Miller (1925) observed at Manhattan, Kans., that corn plants which were about 4 ft. high increased in height 6 cm. from 7 A.M. to 7 P.M. and 8 cm. from 7 P.M. to 7 A.M.

Mason (1925) reported that in the date palm, normal growth, which is manifested by the pushing up of the leaves from the growth center, is made chiefly in the time between sunset and sunrise but also at a reduced rate in the daytime when direct sunlight is cut off by the clouds. In full sunlight, elongation of the date palm ceases entirely. Leaf growth in this plant, however, may be induced in darkness obtained by enclosing the plant in a dark chamber at any hour of the day. Partial growth may also be obtained by screening the plant from direct sunlight, exposing it to reflected light, or placing it with a north exposure.

Coster (1927) in a study of 19 species of plants found that the majority of them grew more slowly on hot days than during the nights, but these diurnal differences were decreased or absent on rainy days. Some species showed equal growth rates during the day and night. Darrow (1929) noted in the brambles that there was a periodic, diurnal march of elongation with great regularity of two maxima and two minima. Thus the growth in length was most rapid from 1 to 2 P.M. until 11 P.M., and from daylight until 10 or 11 A.M. The slowest rate of growth in length was from 11 P.M. until daylight and from 10 to 11 A.M. to 1 to 2 P.M. Elongation was slightly greater during the day than at night.

Porterfield (1930) noted in observations on the growth rate of *Phyllostachys nigra* that 87.8 per cent of the cases exhibited greater day growth, 9.6 per cent greater night growth, and 2.5 per cent the same rate of growth during each period. Stone (1933) reported that, of the total growth of potato leaves in the greenhouse, 54.1 per cent occurred during the night and 45.9 per cent during the day. Karling (1934) found that the trunk of *Achras zapota* reached its maximum diameter between 6 and 7 A.M., and gradually decreased to a minimum at approximately 5 P.M. However, rain, humidity, and changes in temperature altered the rhythm to a considerable degree.

The question arises as to what is the cause of the greater elongation of stems during the night. In many cases it is apparently due to a better water supply for growth. The demands for water by transpiration are so great during the day that there is not a sufficient amount available in the growing regions for the maximal elongation rate of the cells. Brown and Trelease (1918) believed that the absence of growth and even shrinkage of *Cestrum nocturnum* during the day is connected with excessive transpiration. This plant grows better when exposed to the sun than it does in the shade. During the daytime a sufficient amount of water is apparently available in the leaf for photosynthesis, and the plant accumulates food to be used in elongation at night. Mason (1925) believed that in the desert it is the action of the rays of a wave length of 570mμ in the yellow to 405mμ in the violet and probably, also, the ultraviolet rays that inhibit the growth of the palm leaf. In the absence of direct sunlight, the growth of these leaves was apparently synchronous with the closing of the stomata, the checking of transpiration, and increased turgescence in the meristematic tissue. In some succulents the rate of growth varies during the day and night according to the acidity of the tissue. Thus, according to Long (1915), MacDougal noted in *Opuntia blakeana* that the growth increase in diffuse light was 6.9 per cent, while in bright light it was 12.4 per cent. In the daylight in this instance the acidity decreased, while at night acids were accumulated.

Increased rate of growth is not always due to an increase in the osmotic pressure, since Borowikow (1912) has found in the sunflower that the most rapidly growing seedling is not always the most turgescent one. Vogt (1915) in observations on the growth of the coleoptile of barley obtained evidence that the greater growth in darkness is due largely to the stimulating effect of the previous illumination. He believed that a knowledge of the effects of light on the hydration of colloids will be necessary to explain the behavior of growth in light and darkness.

c. Different Rays.—Under this heading is reviewed some of the work that has been done on the effects of the rays of the ordinary spectrum, X rays, and polarized light on the growth of plants.

1. *Rays of the Spectrum.*—Schanz (1919) mentioned the fact that in Holland many growers prefer crude glass to regular window glass for forcing. The former cuts out more of all rays than the latter and is especially effective in screening out the ultraviolet. On this account, Schanz carried on some experiments to determine the effect of the different rays of the spectrum upon plant growth. He stated that cucumber, *Fuchsia, Begonia,* beet, and potato became taller the more the short rays of light were eliminated. The maximal height was obtained under red light and the minimal under the blue-violet light. He concluded, in general, that light of short wave lengths, particularly the ultraviolet, was detrimental to the growth of plants and recommended the use of glass for greenhouses which eliminates these rays. He believed

also that the ultraviolet rays have an important relation to the development of anthocyanin in the epidermal layer, since this pigment failed to develop in red-leaved lettuce and other plants when these rays were excluded.

When plants were grown in daylight from which all wave lengths shorter than 529mμ were eliminated, Popp (1926) found that they developed the following characteristics as compared with plants grown in the entire spectrum of daylight: a greater final height in soybeans, tomatoes, and *Coleus* but a decided decrease in height in sunflowers, petunia, buckwheat, and Sudan grass. There was a considerable decrease in the thickness of stems and a reduction in the number of branches and side shoots. There was a good development of chlorophyll but a reduction in the amount of anthocyanin of the leaves and flowers. The number of flowers was reduced and there was a delay in their time of appearance. There was also a very weak development of seeds, fruits, and general storage organs. The decrease in starch and total carbohydrates was considerable, but usually there was an increase in total nitrogen. When all wave lengths shorter than 472mμ were removed, the same effects were produced but to a somewhat lesser degree. When only ultraviolet rays were eliminated, none of the foregoing results was obtained with any of the plants used, although there was a small increase in the length of stems in all the plants except buckwheat. In general, there was very little difference between plants that received all the rays of the spectrum of daylight and those from which all ultraviolet rays were eliminated. The results of the work of Popp indicate that, although the ultraviolet rays are not indispensable, the blue-violet end of the spectrum is necessary for the normal vigorous growth of plants. This conclusion is substantiated by the work of Shirley (1928), who grew plants under five different light qualities with three intensities and found that the plants produced more dry matter per unit of light intensity under the complete solar spectrum than under any portion of it.

Eltinge (1928) found that raying with an unscreened quartz mercury light caused injury to all the plants used. This ultraviolet radiation had little if any effect upon the decomposition of the chlorophyll or upon the hydrogen-ion concentration of the plants. Raying with a lamp screened by Vitaglass, which allowed the passage of the rays 578 to 289mμ, was beneficial to some plants, while it produced little visible effects in others. Raying with a lamp screened by quartzlite glass (578 to 313mμ) injured none and benefited many of the plants. Eltinge believed that her results emphasize the fact that each plant has its own ultraviolet requirement for its best growth, which can be determined only by experiment. Tincker (1930), in using various types of commercial glass which allow the ultraviolet rays to pass, found that a heavier crop of radishes and carrots was produced under them than under ordinary glass but that the rate of growth of lettuce was not affected.

Colla (1927) subjected young wheat seedlings in the dark to the action of light rays of a length of 3300 to 3900 Å. After 8 hr. of continuous illumination, these plants curved toward the light and the new leaves developed sufficient chlorophyll to give a pallid-green color to the plants. Fuller (1930) showed that plants of tomato and cucumber that had been treated with ultraviolet irradiation increased in growth approximately 34 per cent over the controls. The ash content in the treated plants was greater by 13 per cent than that of the controls. It was observed by Sheard and Johnson (1930) that marked changes in the difference of potential between the base and tip of intact leaves can be produced by ultra- and infraviolet radiations. In darkness or subdued light the base of the leaf is positive with respect to the tip. Under intense artificial illumination, behind ordinary glass, the difference in potential decreases so that the tip of the leaf may become positive relative to the base.

Shull and Lemon (1931) stated that in general two physiologically different regions of ultraviolet radiation have been recognized—the abiotic, or destructive, region, and the biotic, or biologically valuable, region. The abiotic are the shorter and the biotic the longer rays (Fuller, 1931). The ultraviolet rays of sunlight are not harmful to ordinary plants, nor do they seem to be useful, since plants can be grown very successfully without them. Seed coats allow only the longer ultraviolet radiations to penetrate. The lowest limit of rays transmitted by any of the seed coats tested was at 3020 Å, and the penetration of those shorter than 3630 Å was always very slight. In the corn seed, different portions of the same seed coat are not alike in permeability to radiations. Bucholtz (1931) reported that the expanded leaves of *Mnium* and the stamen hairs of *Tradescantia* were not injured by strong intensities of the rays, 2378 to 3654 Å. Wynd and Fuller (1931) found that tomato and cucumber plants which had been stimulated by ultraviolet light showed a definite increase in calcium and a decrease in phosphorus, in percentage on a dry basis.

Masure (1932) subjected the dry seeds of *Pisum sativum* to a radiation of 3650 Å, and found that this wave length exerted a stimulative action on the subsequent rate of growth of the hypocotyls. Singh, Kapoor, and Choudhri (1936) found that the treatment of seeds with ultraviolet radiations for 10 min., or less, accelerated germination and induced luxuriant growth, while an exposure for periods of 20 min. retarded both germination and development.

Fuller (1932) found that radiations from an unscreened quartz mercury lamp produced decided injury in the tissues of plants of tomato and bean. The major part of this injury was due to the infrared radiations. Meier (1934) found that the wave lengths of ultraviolet ranging from 2652 to 3022 Å killed *Chlorella vulgaris*. It is stated by Popp and Brown (1933) in their comprehensive review of the literature on the effect of ultraviolet radiation upon plants that the influence of these rays upon seed plants, other than their destructive action, is yet to be ascertained.

Arthur and Stewart (1935) grew plants of buckwheat under mazda, neon, mercury-vapor, and sodium-vapor lamps, and found that if the dry weight produced under the mazda lamp were taken as one, the values for the other lamps were neon, 1.10; sodium vapor, 0.90; and the mercury vapor, 0.66. Calculations of the dry weights that might have been produced, if equal amounts of energy in the visible region had been used, were mazda, 1.00; sodium vapor, 1.41; neon, 1.21; and mercury vapor, 0.62. Johnston (1932) and Brackett and Johnston (1932) found that the infrared region of the spectrum was of little or no benefit in chlorophyll formation by the tomato. The near infrared region of the spectrum is of considerable biological importance. Haig (1934, 1935) noted in the response of the seedlings of oats to light that if the illuminations were confined to the extreme tip of the coleoptiles only the low-energy portion of the response curve was obtained, but when it was confined to the basal portion the high-energy part appeared. The tip had a distinct maximum at 4800 Å, while the base curve exhibited no maximum in the visible range. The high portions of the base curve may represent a broad maximum somewhere between 4000 and 4700 Å. The reactive time for positive phototropic responses for short exposures to white light decreased with increasing intensity to 1,000 meter-candles.

Johnston (1934, 1935) found that the sensitivity of the oat plant increased rapidly from 4100 to 4400 Å, fell to 4575 Å, then again rose to a second maximum at 4750 Å. From this point the sensitivity decreased rapidly to 5000 Å, from which point it gradually decreased to 5461 Å—the threshold of sensitivity on the long-wave-length side. Growth was retarded to the greatest extent in blue light.

Burkholder and Pratt (1936) noted that the leaflets of *Mimosa pudica* responded readily in blue to long ultraviolet and in long-red radiations, but showed little or no response in the orange, yellow-green, or infrared rays. Ball (1936) found that the shorter red rays were as effective as white light in producing inhibition of anthesis in the flowers of *Turnea ulmifolia*. There was a decrease in efficiency of the red rays beyond 700mμ. The yellow and violet regions were much less effective than the red in this regard. The infrared rays produced little or no effect.

Atkins (1936) found that the maximal phototropic response of *Avena* to the rays of the spectrum was in the region of 4480 Å with diminution in the ultraviolet and in the red. With *Lepidium sativum*, the initial maximum was about 4640 Å in the blue with a second maximum in the red. Plants containing red anthocyanin pigments in the epidermis gave a sluggish response, the pigments absorbing light without inducing curvatures.

2. *X Rays.*—Johnson (1926) found that the catalase activity was depressed in the young seedlings of *Helianthus annuus* whose seeds had been exposed to medium doses of X rays, but that in older seedlings little change could be noted. When stunted growth accompanied heavy irradiation, the rate of respiration was decreased. Cells from the root tip of seedlings whose growth had been inhibited by irradiation showed elongation, large vacuoles, and the absence of a nucleus. Potatoes irradiated before sprouting with a slight dose of X rays produced 27 per cent more tubers per hill, but the average weight was 18 per cent less than the controls, thus making the average total weight per hill practically the same (Johnson, 1928). When the seeds of sunflower were irradiated with medium doses of X rays, an inhibiting effect proportional to the dosage was noted in the growth of the seedlings for about 3 weeks. After that, however, the plants grew rapidly until at maturity, when there was little difference in height or time of blossoming between the controls and the plants from treated seeds. Williams (1923, 1925) noted that the cells of *Saxifraga umbrosa* and *Elodea* after a short exposure to the β and γ rays from radium showed an increase in the rate of circulation of the protoplasm and an increase in the permeability of the protoplasmic membrane as indicated by the exosmosis of solutes. Large doses produced shrinkage and vacuolation effects that were irreversible. Extensive experiments have been conducted by Stadler (1930) on the genetic effect of X rays on barley and corn. A discussion of these effects, however, belongs primarily to the province of genetics and will not be considered here.

Some of the more recent work on the effect of X rays on plant activities is shown in the table on page 1076.

Johnson (1936) reported that of the 70 species of plants from 35 families that have been studied in regard to reaction to X rays, 15 species were apparently unaffected, 15 were slightly affected, especially during their early growth, while the remaining 40 species were noticeably affected. Members of the *Chenopodiaceae*, *Umbelliferae*, and *Brassicaceae* frequently gave indications of injury soon after treatment, but, by the time maturity was reached, showed little effect. Those plants noticeably injured by radiation were marked by decrease in total height, increased branching in *Caryophyllaceae*, and frequent occurrence of dichotomous branching particularly in the *Carduaceae*. Other effects were irregularities in shape, margins, and chlorophyll development of the leaves and delayed and reduced flowering. The members of the *Solanaceae*, *Scrophulariaceae*, and *Carduaceae* were noticeably X-ray susceptible, as were the genera *Ambrosia*, *Ricinus*, *Lavatera*, *Cobaea*, and *Ipomoea*.

3. *Polarized Light.*—The effect of polarized light on growth was noted by Macht (1926, 1928). Plants of *Lupinus albus* exposed to polarized light grew much better than did the controls. The growth of wheat and sunflowers was also hastened, although not to such a marked degree. It was also found, as mentioned in Chap. XI, that polarized light hastened the conversion of starch to sugar.

RESULTS OF TREATMENTS WITH X RAYS

Investigator	Parts treated	Effects of treatment
Reinhard and Tucker (1928)	Seedlings (*Vicia faba*)	Showed their greatest sensitivity to irradiation at 10 and 11 A.M., and 1 and 5 P.M.
Sprague and Lenz (1929)...	Potatoes	Definite lesions on leaves, may reduce number of, but increase size of, tubers.
Wigoder and Patten (1929).	Dry and soaked seeds of *Vicia faba*, barley, white mustard, timothy, and peas	Failure to develop roots and shoots. The maximal effect occurred with actively dividing cells in the first 4 mm. of the root tip. Mitotic rate greatly decreased 3 hr. after irradiation and after 3 days was absent entirely. Cell division resumed after 5 to 8 days, when many abnormal, multinucleate cells were observed.
McKay and Goodspeed (1930).	Pollen of cotton	Seeds obtained from crossing normal pistils with X-ray treated pollen were planted. Many of the seeds were infertile.
Cattell (1931)............	Wheat seedlings	Growth of lateral roots reduced one-half in length. The effects detected in a few hours after treatment.
Johnson (1931)...........	Tomato and various other seedlings, tulip bulbs, and wild potato tubers	Tulip bulbs and wild potato tubers grown from irradiated material showed slight increase over controls. Marked leaf anomalies in tomato leaves. Leaflets often absent, badly twisted, or joined together. Older leaves spotted or variegated. More branches than control and abnormalities of floral parts.
Moore and Haskins (1932).	Seeds of grapefruit	Plants from these seeds flowered prematurely.
Johnson (1933)...........	Plants of *Atriplex hortensis*	Red forms showed greatest decrease in height. Green varieties showed the greatest number of leaf abnormalities.
Shull and Mitchell (1933)..	Seeds of corn, wheat, oats, and sunflower	Treatment of seeds produced stimulative effects upon seedlings. The conditions for the best results are metallic screens, high voltage, low amperage, and short exposures.
Benedict and Kersten (1934).	Seeds of wheat	Increase in diastatic activity and reducing sugars with treatment for five seconds. Both decreased if irradiated longer. Irradiated seeds absorb water less rapidly than controls.
Francis (1934)...........	Seeds of wheat	Retarded growth of seedlings. Depressed respiratory rate. Linear growth retarded. Lateral roots most sensitive and coleoptiles least sensitive.
Skoog (1934)............	*Avena* coleoptiles	Great decrease in amount of growth substance diffusing from them. Does not decrease transport of growth substances.
Haskins and Moore (1935).	Seeds of *Citrus*	High percentage of seedlings showed a deficiency of chlorophyll. Many grades of albinism noted. Many monstrosities.
Havas (1935)............	Various seeds and seedlings	Radiations from a radioactive mud obtained from Hungary showed stimulating effects on some plants and depressing effects upon others at an early stage of growth.
Nogüchi (1935)..........	Seeds of sunflower	Seedlings from treated seeds showed abnormal development of vegetative parts. Abnormality of leaves the most common symptom. The minimum time for inducing leaf abnormality is irradiation for 3 to 4 min.
Reynolds (1935)..........	Tomato plants	Injury by irradiation from an unscreened mercury-vapor arc is more difficult to show on those plants which have developed during the summer months than those which have grown during the winter months. In summer, 1 per cent, and, in winter, 0.2 per cent of solar radiation is ultraviolet.
Collins and Maxwell (1936).	Dry seeds of corn	Found a range of treatment that caused all the plants to die in the seedling stage without reducing the rate of germination.
Meier (1936)............	*Chlorella vulgaris*	The maximal lethal sensitivity was at 2600 Å. The wave length of 3130 A, which is but slightly longer than the shortest wave length of solar radiation reaching the surface of the earth, had no lethal effect.

Harvey (1922) grew a large number o
artificial illumination with electric lights.
with ordinary enameled reflectors at a dista
and gave about 457 lumens per square foo
wheat, oats, barley, rye, flax, buckwheat, wh
a number of weeds grew from seed to maturit
toes, tomatoes, red and alsike clover, and
seed. Adams (1925) grew the castor bea
artificial light, while Tjebbes and Upholf (19
obtained increased yields from greenhouse c
light. Harrington (1926) grew cereals in
winter through the use of artificial light to s
light. By this means he was able to grow
generations of wheat in one year.

Redington (1929) reported that *Zea m*
Cucurbita pepo, *Pisum sativum*, and *Linum*
growth with 16 hr. of light per day than in c
however, generally occurred earlier in con
found that the exposure of the plants of cc
melon, and sunflower to the normal lengt
produced an inhibitory effect on the growt
hypocotyl but a stimulatory effect on the gr
cotyledons. Arthur, Guthrie, and Newell (
is sensitive to continuous light. In cabbage
was doubled in a 19-hr. day as compared t
found in the soybean that the amount of gro
was in direct proportion to the length of
that numerous garden vegetables, grains
ornamentals grown in the greenhouse in
elongated than the controls. Flowering wa
inhibited in others. In continuous light,
extensive and the stems showed a thinner co
more pith than the controls.

Darrow (1933) stated that in Alaska t
and tomatoes occurs under continuous light
Wickwire, and Fuller (1936) found that
irritability in *Mimosa pudica* did not fall du
high day level. In continuous darkness it

2. *Short and Long Days.*—The work of C
and of Tincker (1925, 1928, 1929) showed
toward the length of light exposure, arrang
groups according to the effect of the treatm
(*a*) Those species and varieties which ar

Semmens (1930) found that the leaves of alfalfa, nasturtium, and grape grown in polarized light with alternating periods of light and darkness showed temporary negative photoperiodism, leaf fall, disappearance of starch, and signs of starvation. He believed that plants require both polarized and unpolarized light for the best growth.

d. Intensity of Light.—According to Withrow (1936), the processes involving visible light are of two types—those which require light intensities above 100 foot-candles and those which require lower light intensities. Photosynthesis falls within the first class and, aside from the length of day factor, is apparently the limiting factor that prevents the normal growth of plants in greenhouses during cloudy weather. In other respects, plants in greenhouses appear to grow quite normally in light intensities of 100 foot-candles or less. Light intensities of 20 foot-candles, or less, are apparently sufficient for the synthesis of chlorophyll, for preventing etiolated characteristics, for proper permeability, and for phototropic response. To sustain vigorous growth of plants in windowless greenhouses without the aid of sunlight would probably require intensities of 1,000 foot-candles for a period of 12 hr. per day. Low light intensities prevailing during the winter months frequently involve intensities of as low as 100 to 300 foot-candles during late morning and early afternoon, and plants grow very slowly under such conditions. It would probably be necessary to double at least these lower values with artificial radiation to satisfactorily accelerate growth.

Under artificial light, Harvey (1922) found that some plants grew and bloomed over a wide range of light intensities, while others bloomed only in a very limited range. Thus, with the light intensity measured in foot-candles, the range of corn was 500 to 3,000; tobacco, 800 to 10,000; *Amaranthus retroflexus*, 500 to 10,000; squash, 500 to 750; and wheat, 650 to 800. Popp (1926) grew four varieties of soybeans for 7 days at a temperature of 19 to 23°C. under light intensities of 4,285, 1,536, 560, 390, 250, and 26 foot-candles, respectively. He observed that the lower the light intensity the more rapid was the rate of stem elongation during the period of initial growth. The greatest general height was attained by plants under a light intensity averaging 560 foot-candles and the lowest under 26 foot-candles. The thickness of the stem was directly proportional to the light intensity, and there was a gradual decrease in vigor with decreasing light intensity. Adams (1924) believed that light of lower intensity but more prolonged will produce the same result as light of higher intensity but of shorter duration. Observations as regards height, weight, and time of flowering on numerous plants grown in the greenhouse and in the open showed that the results were as satisfactory under 2 hr. exposure to light at midday as under 3 hr. exposure during the morning and afternoon.

their time of blooming by the duration of the exposure to light. (*b*) Those plants which are caused to flower by the action of short days. Such plants are commonly called "short-day" plants. (*c*) Those plants which are forced into flowering through the action of long days. These have been designated "long-day" plants.

(*a*) *Flowering.*—Some of the plants that have been found to require short days to produce flowering are Maryland Mammoth tobacco, Biloxi soybeans, ragweed, *Aster linariifolius*, *Phaseolus vulgaris*, *P. multiflorus*, *Cosmos*, and *Chrysanthemum*. Some of those which require a long day for the production of flowers are Mandarin soybeans, *Hibiscus*, goldenrod, climbing hempweed, timothy, radish, and perennial rye grass. The plants that are placed in the short-day group are those which can bloom only under a daily light period of 12 hr. or less, while those which are placed in the long-day group require a length of daylight of more than 12 hr. (Kellerman, 1926).

The influence of the relative length of day upon the date of flowering can be illustrated by considering certain varieties of soybeans. If the Biloxi variety is planted about the middle of April in the latitude of Washington, D. C., approximately 125 days elapse between germination and the time of blooming. If, however, these plants are exposed to daylight for only 7 to 12 hr. the time elapsing between germination and blooming is only 28 days. The fact that certain varieties of plants are early or late maturing is apparently connected with the different responses of these plants to the relative length of day and night. Thus for plantings in the field through the month of May, the average number of days from germination to blossoming for the Mandarin, Peking, Tokyo, and Biloxi varieties of soybeans was approximately 27, 56, 70, and 105 days, respectively. When these plants are placed in the daylight for only 7 to 12 hr. daily, they all bloom in approximately 28 days and thus all become early maturing. On the other hand, the late-maturing varieties may be maintained in the active vegetative condition for a prolonged period of time by continued exposure to a relatively long daily period of illumination.

It is impossible to review here all the work that has been done on photoperiodism. Some of the more important studies have been by Darrow and Waldo (1930, 1933) on the strawberry; Garner and Allard (1930) and Austin (1935) on soybeans; Forster, Tincker, Vasey, and Wadham (1932), Tincker (1932) and Hurd-Karrer (1933) on wheat; Kondo, Okamura, Isshiki, and Kasahara (1932) and Pan (1936) on rice; Starkey (1932), Laurie and Poesch (1932), and Post (1934) on chrysanthemums; Greene, Withrow, and Richman (1932) and Withrow and Richman (1933) on greenhouse flowers; Evans and Allard (1934) on timothy; Ramaley (1934) and Withrow and Benedict (1936) on numerous greenhouse plants; Robinson (1933) on flax; Allard (1935) and Kramer (1936) on various woody plants. Johnston (1936) reviews some of the work in this regard.

currents as low as 1×10^{-10} amp. per plant. The optimum intensity of current appeared to be below 1×10^{-8} amp.

Briggs, Campbell, Heald, and Flint (1926) in electrocultural experiments extending over 8 years at the Arlington Experimental Farms found no well-defined increase in yield due to electrical treatment, as the results obtained were well within the limit of experimental error. Collins, Flint, and McLane (1929) found that treatment with a current intensity of 10^{-9} amp. was slightly positive for corn seedlings under laboratory conditions.

Shibusawa and Shibata (1927) found that an electric current of very low intensity was induced in plants in the greenhouse by applying high voltage to a thin, wire network suspended at a distance of 15 to 20 cm. above the plants. Under such treatment the yield of buckwheat was increased 12.6 per cent and that of tobacco 21.7 per cent. Singh (1932) subjected seedlings in the soil to an electric current by means of electrodes. After 13 days this treatment increased the root growth as much as 400 per cent but did not visibly affect the growth of the tops. Musso (1933) found that oats headed earlier, formed a greater number of stalks, and yielded 20 per cent more dry weight than the controls when a strong, positive potential was applied to the soil. Fraser and Pidgeon (1933), however, could find no increase in the yield of wheat by treating the seeds electrically.

The knowledge concerning the proper conditions under which electrical treatments should be given is very meager, and much more experimental work is needed before any definite statements can be made concerning the effects of electricity on growth and crop production.

The mode of the action of an electrical current in those cases where increased yields have been reported is not known. Blackman (1924) noted that in some cases the treated crops had a deeper green than those of the controls. The effect of the discharge is apparently of the nature of a stimulus, since the additional energy in most of the experiments was less than one-thousandth of the energy that the plant was receiving from the sun.

It should be mentioned here that the atmosphere becomes heavily charged with electricity during the dust storms that occur in the Great Plains area, and it is believed by some that this causes the marked injury to cereals which has been observed after such storms.

b. Electrical Phenomena in Tissues.—Waller (1900) and Waller (1924, 1925) reported that, when light falls upon a plant tissue, electricity is produced by the tissue under certain conditions. This photoelectric current is of interest on account of the distinct significance that it appears to possess in reference to plant metabolism. It was believed that the nature of the photochemical reactions in the plant may be indicated by this electrical response. This photoelectric response is shown mainly by the green parts of plants and is apparently dependent upon the metabolic activity of the chlorophyll, although it may occur in the absence of chlorophyll, as shown in the cases of *Hydrangea* petals and in blanched celery. The direction of the current is apparently determined by the kind of leaf and by the conditions to which it is subjected. Dixon and Bennet-Clark (1927) found that a moderate electrical current in a tissue caused a fall in electrical resistance which became less and less rapid. After a few minutes the resistance began to rise and reached the normal in about an hour. They suggested that the ability of a current to stimulate a cell is determined by the potential difference across the membrane of the cell.

The electrical phenomena in plant tissues have been studied more recently by Marsh (1928), Vysotskii (1928), Umrath (1928), Lund (1929, 1931, 1932), Lund and Bush (1930), Glass (1933), Rehm (1936), and Seifriz (1936).

It was observed by Marsh (1928) that the e.m.f. of a given length of onion root was the algebraic sum of the e.m.fs. of the cells of such length. He considered that the electric polarity of the root or a part of it is a fluctuating quantity which may be increased, diminished, or reversed by changes in the environment or the internal processes of cells. Lund (1929) reported the existence of an easily measurable, inherent electric polarity in the main stems of the Douglas and white firs. This e.m.f. varies from 30 to 200 mv. in different trees. The apical growing point is generally positive with respect to the more basal parts of the stem. This e.m.f. is equal to the algebraic sum of the e.m.fs. of the individual segments of the stem. He reported that the cells of the onion root, of the leaf of *Bryophyllum*, and of Douglas fir generate electric currents continuously. The orientation of the electric polarities is such that they act in series and parallel and sometimes may be observed to oppose one another's electromotive forces like electric batteries. In the Douglas fir the unstimulated stem and each lateral branch exhibit an electric polarity. The apex is electropositive to the basal parts of the stem, while the unstimulated apex of the main stem is electropositive to the apex of each lateral branch. This positive electric dominance of the main apex corresponds to its dominance in growth.

Lund and Bush (1930) reported that the distal part of the petiole of *Bryophyllum* was positive to the growing points and selected points in the lamina. Lund (1931) found that the cortex of the Douglas fir is the origin of a characteristic e.m.f. The orientation of this radial e.m.f. in the cortex is opposite that in the wood. This e.m.f. increases toward the apex, while that in the wood decreases. In 1932 he reported that the longitudinal and radial e.m.fs. in the intact stem below the apex are both greatly decreased by lowering the temperature and increased by raising it. The radial e.m.f. of the isolated wood axis is affected in a similar manner. The e.m.fs. of dead stems is not affected by temperature changes. Lund believed that electric currents continually flow along certain circuits in the tree, thus correlating the living parts of the whole tree and making it a definite and continuous electrified system.

Rehm (1936) reported that in the scarlet runner bean (*Phaseolus multiflora*) certain electrical polarities in the region of the axillary buds possessed characteristics that suggest the possibility of their playing a role in bud inhibition. They possessed a relatively constant orientation in both young and old tissues which was maintained under varied conditions. He stated that these potential differences are due to the activities of living cells.

Seifriz (1936) in studies of the nature of protoplasm reported that the tissues of apple, pear, banana, onion, carrot, and potato, among others, are transformed into a soft, usually dark-colored mass when they are the cathode of a 110-volt, direct electric current. When they were used as the anode, little or no change was produced. The primary cause of the degradation is thought to be proteolysis catalyzed by intracellular proteolytic enzymes that are rendered active by the reducing conditions existing at the cathode.

4. Hydrogen-ion Concentration.—The effects of the hydrogen-ion concentration on the various activities of the plant have been stated in the discussion of these activities in the text, so that little remains to be said in regard to its direct effect upon growth. McCall (1923) emphasized the fact that the intensity of the acidity of the soil in many instances is of greater significance in biochemical processes than is the quantity of acid present. Kurz (1923) reported that the hydrogen-ion concentration *per se* was not the main factor in determining the distribution of the species which he considered. Reed and Haas (1924) noted that the injury to walnut seedlings in the absence of calcium was due to the lack of this element rather than to the high concentration of hydroxyl ions. Knudson (1925) stated that the hydrogen-ion concentration may affect many factors in plants other than the alteration of their growth.

Numerous workers have studied the effects of the reaction of the medium upon plant development. A summary of these observations is tabulated in the following table:

Investigator	Plants	Optimum pH	Investigator	Plants	Optimum pH
Wherry (1924)...	Timothy	9.0	Katchioni-Walther (1930).	Buckwheat	5.0 to 6.5
	Alfalfa, barley, and orchard grass	8.0			
	Alsike clover and beet	7.5	Meyer (1930)......	*Nelumbo lutea*	4.5 to 9.0
	Radish, rutabaga, and wheat	7.0	Wilson (1930)......	Onion	6.0 to 7.0
	Lupine, peas, and red clover	6.0	Emmert (1931)....	Tomatoes	8.4
				Lettuce	7.5
	Oats	5.0			
	Flax and turnip	4.0	Waltman (1931)...	Everbearing strawberry	5.3 to 5.5
Sideris (1926)....	Pineapple	5.0	Burkholder (1932).	Bean	Could not determine
Powers (1927)....	Alfalfa and alsike	5.6 to 6.0			
	Hungarian vetch	5.3			
	Spearmint	6.0	Gourley (1932).....	Iris	7.0 to 7.4
Wann (1927).....	*Chlorella*	3.4	Hicks (1932).......	*Lemna trisulca*	4.9 to 7.3
Arrhenius (1929).	Rye, oats, potato, and timothy	5.0	Wessels (1932).....	Cauliflower	5.5 to 6.6
				Potato	4.8 to 5.4
	Beets, barley, and alfalfa	7.0 to 7.2			

Aslander (1929) believed that plants in a fertile soil can tolerate a higher acidity than those growing in an infertile soil. Tait and Knott (1933) concluded from a study of the muck soils of New York that it is impossible to designate any particular soil reaction as the optimum for a given crop. The effect of the soil reaction in plant growth is to some extent an expression of the effects of such soil reaction on the availability of materials. Cooper (1932) suggested that a well-balanced supply of available cationic and anionic nutrient materials, with relatively high oxidation-reduction potentials, is much more effective than the hydrogen-ion concentration of the soil in determining the dominance of various pasture plants. Howell (1932) stated that the seedlings of western yellow pine grow best in an acid medium, but under field conditions other factors may have more influence on growth than the soil reaction.

The relation of the hydrogen-ion concentration to the growth of the cell was discussed by Pearsall (1925). He suggested that one of the conditions for the development of dividing cells in normal nongrowing tissue is the development of a gradient of hydrogen-ion concentration such that the protoplasm or its principal proteins can be reduced to an isoelectric condition. It is at this point that the proteins show minimum swelling, minimum osmotic pressure, maximum precipitation, and least viscosity. If the protoplasm and principal proteins were made approximately isoelectric by an alteration in the hydrogen-ion concentration, they would tend to lose water to adjacent media of higher osmotic potential, and as a result the condition in the protoplasmic or protein aggregate would favor synthesis. In stems there is a superficial layer of growing cells next to the cuticle that is usually markedly acid (pH 3.5 to 4.5), while the ground tissue usually has a pH value between 5 and 6. When lateral roots develop, they do so outside the xylem with its markedly acid reaction and not opposite the phloem, which is relatively alkaline. The surroundings of the dividing cells usually have a reaction of about pH 4.5 to 5.0 as against the normal tissue reaction of pH 5.5 to 6.5. Small (1929) found that the tissue of the xylem is more acid than that of the phloem. The various tissues of the broad bean, potato, and sunflower never showed a reaction less than pH 7.0.

5. Gases.—In 1908 Crocker and Knight found that the flowers of carnation are extremely sensitive to traces of illuminating gas. They observed that concentrations of 1 part in 40,000 parts of air killed the young buds and prevented the opening of those already showing the petals. A concentration of 1 part of gas to 80,000 parts of air caused the closing of the open flowers after 12 hr. exposure. Doubt (1917) reported that the following plants were found to be well adapted for use in detecting illuminating gas in greenhouses: *Lycopersicum esculentum*, *Salvia splendens*, *Mimosa pudica*, *Ricinus communis*, and *Datura stramonium*. Traces of gas 50 p.p.m. caused epinastic growth of the petals of all of these plants. This amount of gas is far below the limit of odor. The following plants were not injured by gas unless there was enough present to be detected by odor: *Caladium esculentum*, *Lupinus perennis*, *Eriobotrya japonica*, *Phoenix canariensis*, and others.

The effects of illuminating gas on plants have been reported by Hitchcock, Crocker, and Zimmerman (1930, 1931, 1932, 1934), Deuber (1932, 1933, 1934), and Zinkernagel (1932). This gas hastens the abscission of leaves, fruits, and flowers of numerous plants. Apparently the abscission layer of these parts is stimulated to produce many new cells, which results in the separation of the leaf from the stem. The dormancy of the buds of young oak and catalpa trees may be overcome by treatment with illuminating gas. The epinasty of leaves is one of the universal effects of this gas upon leafy plants.

Hitchcock, Crocker, and Zimmerman (1934) reported that the toxicity of gassed soils is due primarily to the constituents of illuminating gas that combine with the soil particles. After this combination, the gas constituents are not readily removed by aeration or leaching. According to these investigators (1932), the injury by gas to plants was lessened when it was first passed through water, sodium hydroxide, soil or moss peat before reaching the plant. Zinkernagel (1932) found that the treatment of onion roots for a few hours with illuminating gas caused the spindle fibers of cells in the early stages of division to disappear. This interfered with the movement of the chromosomes to the poles. Later in the absence of this gas the spindle reformed and cell division was completed. When in gas, the roots did not respond to gravity but grew upward.

Crocker, Zimmerman, and Hitchcock (1932) tested 202 species and varieties of plants with 38 different gases. Approximately 44 per cent of these plants showed induced epinasty but only in the presence of five of these gases: ethylene, acetylene, propylene, carbon monoxide, and butylene. If the concentration of ethylene were taken as one, then the concentration required for a response was 500 for acetylene and propylene, 5,000 for carbon monoxide, and 500,000 for butylene. All the effective gases were carbon gases with unsaturated bonds. Their effectiveness is not correlated with their solubilities in water, but may be related to their ability to combine with one or more constituents of the protoplasm.

According to Rodriguez (1932) the use of smoke on pineapple plants in the field in Puerto Rico resulted in the general flowering of all plants and in early fruit production. The quantity of the smoke was apparently unimportant. One or more constituents of the smoke, and not the increased temperature, was responsible for the flowering.

Deuber (1934) placed the roots of tomato plants that had been exposed to illuminating and ethylene gases into nutrient solutions containing methylene blue in concentrations of 1 to 10 p.p.m. The results indicated that this pigment thus presented to the roots helped these plants to recover from the effects of various gases. Zimmerman and Crocker (1934) found that the plants of 65 different genera were susceptible to injury from the vapors emanating from soil or tankage moistened with a solution of mercuric chloride.

The influence of sulphur dioxide on plants has been studied by various workers. Zimmerman and Crocker (1930, 1934) found that the leaves of tomato, salvia, *Coleus*, geranium, castor bean, buckwheat, roses, and several other plants were injured after treatment for 1 hr. in 3 to 4 p.p.m. of this gas. The age of the tissues was a factor in the susceptibility of leaves to injury. The middle-aged leaves were the least resistant and the younger ones the most resistant. Interveinal portions of the leaves were more readily injured than the tissues along the veins. The wilted plants were more resistant than the turgid ones. Plants treated during the night were more resistant than those treated during the day. Buckwheat was the most susceptible, being injured by a fumigation of 0.46 p.p.m. for 7 hr., while orchids were the most resistant, withstanding 60 p.p.m. for several hours.

Zimmerman, Crocker, and Hitchcock (1933) treated plants from 108 species with carbon monoxide, and, of these, 45 showed epinastic growth of leaves, while several showed hyponasty.

The effects of ethylene on plants have been more extensively studied than those of any other gas. Some of the workers who have reported in this regard are Harvey (1913), Knight and Crocker (1913), Harvey and Rose (1915), Harvey (1915), Zimmerman, Hitchcock, and Crocker (1931, 1932), and Deuber (1936).

Harvey (1913) found that the castor bean is a very delicate indicator of the presence of ethylene, since it responded by a nastic drooping of 15 to 30 deg. when exposed to as little as 1 part ethylene to 10,000,000 parts of air. Knight and Crocker (1913) showed that tobacco smoke of a concentration of 1,000 p.p.m. caused a reduction in the rate of elongation of the sweet pea, while 5,000 p.p.m. completely stopped elongation. They determined that ethylene was the component that thus affected growth, and that the sweet pea was a more delicate indicator of the presence of this gas than any chemical test.

Harvey and Rose (1915) believed that ethylene is the most harmful constituent of illuminating gas except in extremely high concentrations of gas when other substances may play a part. Harvey (1915) investigated the changes in the metabolism of the sweet pea which were caused by the action of ethylene. In the treated plants the simple soluble substances increased at the expense of the higher soluble and insoluble forms. Thus the sugars, amino acids, amides, polypeptids, and lipoids increased by 8 per cent, while the insoluble substances as proteins, starch, and cellulose correspondingly diminished. The water content of the treated and untreated plants, however, remained the same. The osmotic pressure of the treated plants increased as did also the permeability of the protoplasmic membrane.

Ethylene apparently always causes epinasty in leaves, and may also cause abscision of leaves and flowers, depending upon the environmental conditions and the kind of plant. The effects of illuminating gas on plants are due in a large degree to the ethylene in it. According to Crocker, Zimmerman and Hitchcock (1932) this gas does not act directly by inducing epinasty of petioles but probably acts indirectly by modifying the equilibrium position of the petiole with gravity, as shown by Neljubow (1901, 1911).

Priestley (1922) suggested that the effect of the injurious components of illuminating gas on the stems and roots of etiolated plants is due to their inhibition of the formation of an unbroken primary endodermis normally present in these structures. The unsaturated hydrocarbons prevent the formation of a functional endodermis by preventing the normal accumulation of unsaturated acids in the region of the future Casparian strip. Woffenden and Priestley (1924) traced the effect of coal gas upon the stem of *Sambucus* to its action upon cork and lenticel formation. When the cork cells are cut off from the phellogen, they are very susceptible to the presence of coal gas. The fatty acids diffusing from the protoplasm fail to remain in and on the walls, so that no suberin lamella can be formed. This is due apparently to the greater mobility of unsaturated fatty acids in the presence of the gaseous unsaturated hydrocarbons present in the coal gas. They believed that coal gas will not be toxic to cork-enveloped tissue except in the case of leakage from buried gas mains around the roots.

6. Other Factors.—It was reported by Thornton (1930) that treatment with carbon dioxide in concentrations of 5 to 30 per cent was effective in prolonging the life of cut roses when stored at 38 to 50°F. He stated (1931) that the percentage of carbon dioxide necessary to cause injury to fruits and vegetables in storage is related directly to the firmness of the tissue and inversely to the amount of moisture upon their surfaces. Miller and Brooks (1932) could not find any change in the percentage of the various carbohydrates of peaches and cherries when they were treated in storage with carbon dioxide. Pearl, Edwards, Winsor, and Winsor (1934) stated that the seedlings of cucumber showed a diminished growth rate and less efficient translocation of food materials as the ventilation was progressively lessened.

It was shown by Sampson and McCarty (1930) in California that the rate of growth of the needle grass (*Stipa pulchra*) in winter is controlled by atmospheric temperatures; in spring and early summer it is partly related to the internal factors, among which the food supply and growth habits are predominant. Root growth occurred in the autumn after the top growth had practically ceased. Bailey (1933) found that soaking the seeds of dwarf bean in distilled water for varying periods of time resulted in a progressive decrease in the growth rate of the plants produced and in the time necessary for them to reach maturity. Livingston (1934), in a comprehensive paper, emphasized that the general organization of the organism and its environment must be studied together if we are to understand the former. Brain (1935) found that plants which were rotated on a horizontal clinostat showed an increase in root growth and a decrease in shoot growth. He believed that this fact indicated that a different growth mechanism is involved in the opposite tropisms of shoot and root.

Brink (1924) stated that the growth of the tubes of cultured pollen grains required sugar. There was an increase of growth in length of 60 to 142 per cent by the addition of sterile yeast to the sugar medium. The growth-promoting substance contained in the yeast was water soluble, heat stable, and active in small amounts. The addition of extracts of the potato and of pistils of flowers to the artificial medium regularly increased the growth of pollen tubes. Reid (1930) believed that the seedlings of the Hubbard squash utilized their nitrogenous reserves in a different manner at opposite seasons of the year. Thus the leaves of plants grown in December contained much more of the total nitrogen than those grown during May. The plants grown in May contained much more of their nitrogen in the roots. Gustafson and Laing (1931) considered that the decrease in the productivity of the later developed flower clusters in tomatoes under ordinary conditions is due to a decrease in available nutrient material, which has been utilized by the first-formed fruit. Hence a lack of nutrient material and not the presence of inhibiting substances is the cause of the smallness of the later formed fruit. Harrison (1931) noted that the shorter the grass was cut and the greater its leaf area was reduced, the smaller was the quantity of roots produced, and that the application of mineral fertilizers did not compensate for the lack of top growth. Gould, Pearl, and Edwards (1934) noted that when a considerable portion of the cotyledonary tissues of the canteloupe was removed before planting, the growth in length of the hypocotyl was in excess of that expected from the available food present. They stated that this illustrates "the physiological factor of safety" in that the seedling does not use all the available stored material.

Walter (1923) noted that the rootlets of germinating seeds of water cress, nasturtium, and peas would grow only in an atmosphere with a relative humidity of not less than 97 to 98 per cent. Nightingale and Mitchell (1934) grew two sets of tomato plants under identical conditions for growth, except that one set was grown under a relative humidity of 35 per cent and the other under a relative humidity of 95 per cent. Those grown under the latter condition were far better plants when judged by every criterion of development and appearance.

Lloyd and Barnes (1932) considered that the chemical activity of water appears to be associated with the colloidal form, trihydrol, which is present in large quantities even at relatively high temperatures. The growth and photosynthesis of *Spirogyra* were accelerated to a greater degree in water cultures rich in trihydrol than in those rich in monohydrol and dihydrol.

The effects of the tillers of corn upon the yield of grain have been studied by Lyon (1905), Montgomery (1909), Williams (1912), Williams and Etheridge (1912), Thompson (1926), McClelland (1928), and Dungan (1931). Their observations indicated that the removal of tillers reduced the yield of grain to a greater or lesser degree in practically all cases.

C. Vernalization

1. General Consideration.—The terms "Iarovization," "Jarovization," "Yarovization," and "Iarovtsii" have been used by the Russians to indicate a certain treatment of seeds. Each means literally "making springlike." The English have used "Springification," "Springization," and "Vernalization" to describe this treatment. The last named has become widely used and will be used in this discussion.

According to Thomson (1936) vernalization is the name applied to a treatment given to seeds before sowing to hasten the time of flowering of the plants that will develop from them. Martin (1933) and Laude (1933) stated that it is a method of treating the seeds of winter cereals so that they may mature from spring sowing, and of treating the seeds of spring cereals and other spring crops to advance the time of reproduction and maturity. Thomson (1936) stated that the essence of the vernalization treatment of seeds is that the stages incident to the developing plant are passed before the seeds are sown and yet the seeds are in a condition to be sown by the ordinary methods. Briefly this consists of starting germination and then holding it in check by the proper moisture, temperature, and light relations, as experience has indicated. The seeds are germinated until the radicles are just beginning to break the seed coats and then restraining treatments are applied. These treatments cannot be successfully applied to dormant seeds, but the embryo must have started development and growth. After the seeds have been treated in this manner, they may be planted immediately or dried and kept for a later planting.

According to McKinney and Sando (1933), Martin (1933), Laude (1933), and McKee (1935), in winter cereals the vernalization process consists of adding water to the seeds in an amount that will barely produce visible germination. This requires a period of 1 to 2 days in the processing chamber with the temperature at 10 to 12°C. The seeds are then transferred to a temperature of 3 to 5°C., stirred frequently, and their moisture content kept constant by the additions of water. The time required in the cool room will vary, depending upon the temperature and variety of the seeds, but the average time is from 35 to 45 days. Lysenko (1932) stated that vernalization, at least for some plants, should be completed in darkness. He also stated that with plants requiring high temperatures for germination, such as corn, foxtail, millet, soybeans, Sudan grass, and other sorghums, the treatment during the germination stage is much the same as with the cereal grains. After germination has begun, however, the seeds are kept at 20 to 30°C. for 5 to 10 days.

Vernalization is really an old and well-known seed treatment under a new name. The fact that the low temperature requirement of winter cereals can be satisfied in the early stages of germination has been known since 1837. A process very similar to what is now called "vernalization" was well described in 1849 in the "New American Farm Book," a standard reference book on agriculture at that time. In 1857 Klippart wrote in the annual report of the Ohio State Board of Agriculture, "To convert winter wheat into spring wheat, nothing more is necessary than that the winter wheat should be allowed to germinate slightly in the fall or winter but kept from vegetating by a low temperature of freezing until it can be sown in the spring. This is usually done by soaking and sprouting the seed and freezing it while in this state and keeping it frozen until the season for spring sowing has arrived. Only two things seem requisite, germination and freezing. The experiment of converting winter wheat into a spring wheat has met with great success. It retains many of its primitive winter wheat qualities and is inferior in no respect to the best varieties of spring wheat and produces at the rate of 28 bu. per acre." Klebs (1913), Gassner (1918), Maximow and Pojarkova (1925), and Maximow and Krotkina (1930) have contributed to this work.

2. Economic Value.—Vernalization has been used in field practice to a considerable extent in Russia during the past 6 years and has attracted attention in western Europe and in North America. McKinney and Sando (1930, 1933); Gfeller, Derrick, and Fraser (1933); Sprague (1934); Peltier and Kiesselbach (1934); McKinney, Sando, and others (1934); Burr and Turner (1935); McKee (1935); and Bell (1936) have studied the practicability of vernalization. The available evidence indicates that the process does not offer great possibilities in commercial wheat production in the United States. There are available in this country so many spring varieties well adapted to the spring-wheat regions that there appears to be no need for the vernalization of winter varieties. It also seems doubtful if vernalization of spring varieties will return yields proportional to the cost of treating. It has been suggested that it might be used for spring seeding in winter-wheat areas where the fall sowing has perished. In many cases, however, the need for spring sowing cannot be determined in sufficient time to permit vernalization. In England, winter-sown wheat always yields best, and its time for planting relieves the congestion of work. Vernalized winter wheat yields better there than spring wheat but produces less than winter-sown wheat. In Russia, however, there is a great need for wheat that will head before the dry season occurs, and it appears that vernalization will supply the remedy. The drought and long days of Russia are against growing such crops as maize and millet, but if the vernalization of short-day plants can be accomplished there seems to be no doubt but that they will grow.

Sprague (1934) reported that the vernalization of certain corn hybrids resulted in statistically significant hastening of sexual maturity, but the difference was so slight as to be of no agronomic importance, and a general reduction in germination and vigor was associated with the treatment.

According to Laude (1933), the obvious difficulties, such as the necessity for accurate control of temperature, moldy seed, low germination, poor stands, and those difficulties inherent in drying the seeds or in planting moist, partly germinated seeds, are such as to indicate that vernalization offers little of immediate value to the practical farmer. Vernalization, however, has its value in experimental culture in the greenhouse as shown by McKinney and Sando (1930). It advances the time of growth and reproduction in winter cereals, so that two or three generations may be grown in the same time that is required for one generation under ordinary treatment.

3. Theory of the Process.—According to Maximow (1934, 1935), Purvis (1934), and Thomson (1936), Lysenko, the Russian authority on vernalization, stated that the results obtained from this process are based upon the following concepts of plant behavior: (*a*) growth and development are not identical phenomena but are independent of each other; (*b*) the entire process of the development of an annual seed plant consists of a series of individual steps and stages; (*c*) these stages always proceed in a strict sequence, and a subsequent stage cannot occur until the preceding stage has been completed; and (*d*) in the same plant different stages of development require, for their completion, different external conditions.

According to Lysenko there are at least two outstanding stages in the development of a plant:

a. The Thermo-stage.—The indications are that this stage must be completed before the initials of the reproductive organs can be formed. The conditions required for the completion of this stage are low temperatures of from 0 to 20°C., suitable moisture, and adequate aeration. The time required for the completion of this stage varies with the type of plant and the prevailing environmental conditions. The effect on the thermo-stage is incurred only when the dormant period is broken and the embryo is induced to start growth without being allowed to penetrate the seed coat. This indicates that seeds thus treated have ceased to be seeds in the physiological sense and have become equivalent to growing plants. To all external appearances, however, these may differ in no respect from resting seeds. Many varieties of winter wheat fail to head when sown in the spring because the temperatures are too high for the accomplishment of this thermo-stage. Once this stage has been completed, the other stages in the cycle of development can normally occur. Thomson (1936) stated that it has long been known that if the seedlings of beet or members of the genus *Brassica* get slightly frosted the plants will bolt, although normally they do not flower until after winter. Full-grown cabbages were transplanted to a warm greenhouse in October and 2 years later were yet growing. Those that were transplanted at the same time into a cool greenhouse flowered in 22 weeks, while cabbages left in the open until December and then transplanted into a warm greenhouse flowered in 6 weeks. It has been noted in American agricultural practice that if winter wheat is sown late and germination starts, but is not completed, before cold weather begins, so that the seedling plants do not appear until spring, the plants develop and head in a normal manner. However, if the seeds remain dormant until spring, and then germinate, the resulting plants will not head. This suggests that the winter cold influences germinating seeds but not dormant ones.

Draghetti (1933) used the term "crypto-vegetation" to signify the state of vegetation maintained by certain plants under the low-temperature conditions of winter. As a result of climate this state is characterized by the cessation or intermittent functioning of photosynthesis. He found that the wheat plants continued to absorb nutrient salts, especially the nitrogenous ones, to perform metabolic activities, and to increase in weight during the winter. These facts show that the plants do not have a definite winter rest period.

b. The Photo-stage.—The changes involved in the thermo-stage are not capable of initiating reproduction. A photo-stage is necessary, and it can only become effective after the thermo-stage is completed. Light and darkness play no part in the thermo-stage. The photo-stage requires high temperature and can be effected only under conditions of a long day or under continuous illumination. This long-day requirement does not hold for the entire cycle of development of plants but only during this one particular stage through which they must pass immediately after the thermo-stage. The photo-stage can be most rapidly effected in wheat in continuous illumination and less rapidly under long days, but it is delayed indefinitely under short-day conditions.

McKinney and Sando (1935) believed that sexual reproduction in spring and winter wheat is not dependent upon a critical temperature or a photoperiod since this process occurs over a wide range of such factors. The time, however, when sexual reproduction occurs is greatly influenced by the temperature and the photoperiod. Most spring wheats complete their life cycle quickly when given a long day and temperatures at 70°F. or above, throughout the life cycle. On the other hand, winter wheats complete their life cycle most rapidly when given a short day and low temperatures during the earlier stages of growth and a long day and high temperatures during later stages of development.

Bibliography

Abbe, C., 1895: Influence of cold on plants,—a résumé, *Exp. Sta. Rec.*, **6**: 777–781.

Abbott, C. E., 1935: Blossom-bud differentiation in citrus trees, *Am. J. Bot.*, **22**: 476–485.

Abbott, O., 1923: Chemical changes at beginning and ending of rest period in apple and peach, *Bot. Gaz.*, **76**: 167–184.

Ackerman, A., 1917: Beiträge zur Kenntniss der Kältresistanz der Winterweizen, *Z. Pflanzenzucht.*, **5**: 349–359.

Adams, J., 1920: Relation of flax to varying amounts of light, *Bot. Gaz.*, **70**: 153–156.

———, 1923: The effect on certain plants of altering the daily period of light, *Ann. Bot.*, **37**: 72–94.

———, 1924: Does light determine the date of heading out in winter wheat and winter rye? *Am. J. Bot.*, **11**: 535–539.

———, 1924: Duration of light and growth, *Ann. Bot.*, **38**: 510–523.

———, 1924: The effect on tomato, soybean, and other plants of altering the daily period of light, *Am. J. Bot.*, **11**: 229–233.

———, 1925: Some further experiments on the relation of light to growth, *Am. J. Bot.*, **12**: 398–413.

ALLARD, H. A., 1932: Length of day in relation to the natural and artificial distribution of plants, *Ecology*, **13**: 221–234.

———, 1932: Flowering behavior of the hog peanut in response to length of day, *J. Agr. Res.*, **44**: 127–137.

———, 1935: Response of the woody plants *Hibiscus syriacus*, *Malvaviscus conzattii* and *Buginvillea glabra* to length of day, *J. Agr. Res.*, **51**: 27–34.

ALLEN, F. W., 1932: Physical and chemical changes in the ripening of deciduous fruits, *Hilgardia*, **6**: 381–441.

ALLEN, R. L., 1849: "American Farm Book," C. M. Saxton, New York, XII + 325 pp.

ALMOSLECHNER, E., 1934: Die Hefe als Indikator für Wuchsstoffe, *Planta*, **22**: 515–542. Quoted by Avery, 1935.

ANDERSEN, A. M., 1931: The use of dilute nitric acid on the germination of seeds of *Poa compressa*, abst., *Am. J. Bot.*, **18**: 889.

ANDREWS, F. M., 1932: A measurement of growth, *Proc. Ind. Acad. Sci.*, **41**: 77–80.

ANTHONY, R. D., and W. S. CLARKE, JR., 1932: Growth record on fertilized apple trees grown in metal cylinders, *J. Agr. Res.*, **44**: 245–266.

APPLEMAN, C. O., 1911: Physiological behavior of enzymes and carbohydrate transformations in after-ripening of the potato tuber, *Bot. Gaz.*, **52**: 306–315.

———, 1912: Changes in Irish potatoes during storage, *Md. Sta. Bull.* 167.

———, 1914: Biochemical and physiological study of the rest period in the tubers of *Solanum tuberosum*, *Md. Sta. Bull.* 183.

———, 1916: Biochemical and physiological study of the rest period in the tubers of *Solanum tuberosum*, *Bot. Gaz.*, **61**: 265–294.

———, 1918: Physiological basis for the preparation of potatoes for seed, *Md. Sta. Bull.* 212.

———, 1918: Special growth-promoting substances and correlation, *Sci.*, **48**: 319–320.

——— and E. V. MILLER, 1926: A chemical and physiological study of maturity in potatoes, *J. Agr. Res.*, **33**: 569–579.

ARMSTRONG, H. E., and E. F. ARMSTRONG, 1911: The function of hormones in regulating metabolism, *Ann. Bot.*, **25**: 507–519.

ARMSTRONG, J. M., and W. J. WHITE, 1935: Factors influencing seed setting in alfalfa, *J. Agr. Sci.*, **25**: 161–179.

ARRHENIUS, O., 1929: Soil acidity. Plant growth and its practical application, *Proc. Int. Cong. Plant Sci., Ithaca*, **1**: 53–54.

ARTHUR, J. M., J. D. GUTHRIE, and J. M. NEWELL, 1930: Some effects of artificial climates on the growth and chemical composition of plants, *Am. J. Bot.*, **17**: 416–482.

——— and W. D. STEWART, 1931: Plant growth under shading cloth, abst., *Am. J. Bot.*, **18**: 897.

———, 1932: Artificial light and plant growth, *Agr. Eng.*, **13**: 288–291.

——— and W. D. STEWART, 1935: Relative growth and dry weight production of plant tissue under mazda, neon, sodium and mercury vapor lamps, *Contr. Boyce Thompson Inst.*, **7**: 119–130.

ASHBY, E., and T. A. OXLEY, 1935: The interaction of factors in the growth of *Lemna*. VI, An analysis of the influence of light intensity and temperature on the assimilation rate and the rate of frond multiplication, *Ann. Bot.*, **49**: 309–336.

ASLANDER, A., 1929: Concentration of the nutrient medium *versus* its hydrogen-ion concentration as manifest by plant growth, *Svensk Bot. Tidskr.*, **23**: 96–140; *Exp. Sta. Rec.*, **67**: 22.

Atkins, G. A., 1936: The effect of pigment on phototropic response: A comparative study of reactions to monochromatic light, *Ann. Bot.*, **50**: 197–218.

Auchter, E. C., and C. P. Harley, 1924: Effect of various lengths of day on development and chemical composition of some horticultural plants, *Proc. Am. Soc. Hort. Sci.*, **21**: 199–214.

Austin, J. P., 1935: Some effects of the photoperiod on development of *Impatiens balsamea*, *Plant Physiol.*, **10**: 545–552.

Austin, S., 1933: Vegetation and reproduction in the soy bean, *Sci.*, **78**: 363–364.

———, 1935: Effects of exfloration on plant metabolism, *Plant Physiol.*, **10**: 225–243.

Avery, G. S., 1935: Differential distribution of a phytohormone in the developing leaf of *Nicotiana*, and its relation to polarized growth, *Bull. Torrey Bot. Club*, **62**: 313–330.

——— and P. R. Burkholder, 1936: Polarized growth and cell studies on the *Avena* coleoptile, phytohormone test object, *Bull. Torrey Bot. Club*, **63**: 1–16.

Axentjev, B. N., 1930: Über die Rolle der Schalen von Samen und Früchten, die bei der Keimung auf Licht reagieren, *Beih. bot. Cent.*, **46**: 119–202; *Biol. Abst.*, **8**: 1911.

Babb, M. F., 1928: The effect of ethylene upon the vitamin B content of celery, *Sci.*, **68**: 231.

Bachmann, F., and F. Bergann, 1930: Über die Werkigkeit von Strahlen verschiedener Wellenlänge für die phototropische Reizung von *Avena sativa*, *Planta*, **10**: 744–755.

Bailey, L. H., 1891: Some preliminary studies of the influence of the electric arc lamp upon greenhouse plants, *N. Y. Sta.* (*Cornell*) *Bull.* 30.

Bailey, W. M., 1933: Structural and metabolic after-effects on soaking seeds of *Phaseolus*, *Bot. Gaz.*, **94**: 688–713.

Bakke, A. L., 1913: The effect of smoke and gases on vegetation, *Contr. Bot. Dept. Iowa State Col.*, No. 54; *Proc. Iowa Acad. Sci.*, **20**: 169–188.

Baldwin, H. I., 1932: Alcohol separation of empty seed, and its effect on the germination of red spruce, *Am. J. Bot.*, **19**: 1–12.

———, 1934: Effect of after-ripening treatment on germination of white pine seeds of different ages, *Bot. Gaz.*, **96**: 372–376.

Ball, N. G., 1936: The effect of nocturnal illumination by different regions of the spectrum on the subsequent opening of flower buds, *New Phytol.*, **35**: 101–116.

Barnes, T. C., 1932: The physiological effect of trihydrol water, *Proc. Nat. Acad. Sci.*, **18**: 136–137.

Barton, L. V., 1932: Effect of storage on the vitality of *Delphinium* seeds, *Contr. Boyce Thompson Inst.*, **4**: 141–153.

———, 1934: Dormancy in *Tilia* seeds, *Contr. Boyce Thompson Inst.*, **6**: 69–90.

———, 1935: Storage of vegetable seeds, *Contr. Boyce Thompson Inst.*, **7**: 323–331.

Bell, G. D. H., 1936: Experiments on vernalization, *J. Agr. Sci.*, **26**: 155–171.

———, 1936: Vernalization: its meaning and practical application, *Agr. Prog.*, **13**: 76–82.

Benedict, H. M., 1915: Senile changes in leaves of *Vitis vulpina* and certain other plants, *N. Y. Sta.* (*Cornell*) *Mem.* 7.

——— and H. Kersten, 1934: Effect of soft X-rays on germination of wheat seeds, *Plant Physiol.*, **9**: 173–178.

Berkley, D. M., and E. E. Berkley, 1933: Super optimal and thermal death temperatures of the cotton plant as affected by variations in relative humidity, *Ann. Mo. Bot. Gard.*, **20**: 583–603.

BERKLEY, E. E., 1931: Studies of the effects of different wave lengths of day, with variations in temperature on vegetative growth and reproduction in cotton, *Ann. Mo. Bot. Gard.*, **18**: 573–601.

BIDDULPH, O., 1935: Histological variations in *Cosmos* in relation to photoperiodism, *Bot. Gaz.*, **97**: 139–155.

BLACKMAN, V. H., 1919: The compound interest law and plant growth, *Ann. Bot.*, **33**: 353–360.

———, 1920: The significance of the efficiency index of plant growth, *New Phytol.*, **19**: 97–100.

———, A. T. LEGG, and F. G. GREGORY, 1923: The effect of a direct electric current of very low intensity on the rate of growth of the coleoptile of barley, *Proc. Roy. Soc.*, B, **95**: 214–228.

———, 1924: Field experiments in electroculture, *J. Agr. Sci.*, **14**: 240–267.

——— and A. T. LEGG, 1924: Pot culture experiments with an electric charge, *J. Agr. Sci.*, **14**: 268–287.

BLAKE, A. K., 1935: Viability and germination of seeds and early life history of prairie plants, *Ecol. Monographs*, **5**: 405–460.

BONNER, J. F., 1932: The production of growth substance by *Rhizopus suinus*, *Biol. Zent.*, **52**: 565–582.

———, 1933: Studies on the growth hormones of plants. IV, On the mechanism of the action, *Proc. Nat. Acad. Sci.*, **19**: 717–719.

———, 1933: The action of the plant growth hormone, *J. Gen. Physiol.*, **17**: 63–76.

———, 1934: The relation of hydrogen ions on the growth rate of the *Avena* coleoptile, *Protoplasma*, **21**: 406–423.

———, 1934: Studies on the growth hormone of plants. V, The relation of cell elongation to cell wall formation, *Proc. Nat. Acad. Sci.*, **20**: 393–397.

——— and K. V. THIMANN, 1935: Studies on the growth hormones of plants. VII, The fate of growth substance in the plant and the nature of the growth process, *J. Gen. Physiol.*, **18**: 649–658.

BOOTH, E. G., 1929: Daily growth of the oat kernel and effect on germination of immaturity and controlled low temperatures, *Minn. Sta. Tech. Bull.* 62.

BORODIN, D. N., 1933: Yarovization formula for winter oats and barleys, abst., *Am. J. Bot.*, **20**: 677.

BOROWIKOW, G. A., 1912: Über die Ursachen des Wachstums der Pflanzen, *Biochem. Z.*, **48**: 230–246; **50**: 119–128; *Plant World*, **16**: 347–349 (1913).

BORTHWICK, H. A., and W. W. ROBBINS, 1928: Lettuce seed and its germination, *Hilgardia*, **3**: 275–304.

BOTJES, J. O., 1933: Aethyleen als vermoedelijke oorzaak van de groeiremmende werking van rype appels, *Tydschr. Plantenziek.*, **39**: 207–211.

BOURNE, B. A., 1935: Effects of freezing temperatures on sugar cane in the Florida everglades, *Fla. Sta. Tech. Bull.* 278.

BOYSEN-JENSEN, P., 1910: Über die Leitung des phototropischen Reizes in Avenakeimpflanzen, *Ber. bot. Ges.*, **28**: 118–120.

———, 1931: Über Bildung eines Wachstumsregulators durch *Aspergillus niger*, *Biochem. Z.*, **239**: 243–249.

———, 1932: Über die Bildung und biologische Bedeutung des Wachstumsregulators bei *Aspergillus niger*, *Biochem. Z.*, **250**: 270–280.

———, 1933: Über den Nachweis von Wuchsstoff in Wurzeln, *Planta*, **19**: 345–350; *Biol. Abst.*, **9**: 746.

———, 1933: Significance of growth substance for the growth and the geotropic curvature of the roots of *Vicia faba*, *Planta*, **20**: 688–698; *Chem. Abst.*, **28**: 1380.

———, 1936: "Growth Hormones in Plants." Translated by G. S. Avery and P. R. Burkholder. McGraw-Hill Book Company, Inc., New York and London, XIV + 268 pp.

Brackett, F. S., and E. S. Johnston, 1932: The functions of radiation in the physiology of plants. I, General methods and apparatus, *Smithsonian Misc. Coll.*, **87**, No. 13: 1–10.

Brain, E. D., 1935: Studies in the effects of prolonged rotation of plants on a horizontal klinostat. I, Growth rate, *New Phytol.*, **34**: 97–112.

Bramble, W. C., 1932: Breaking the dormancy of tree seedlings by chemical treatment, *Sci.*, **75**: 193–194.

Breazeale, J. F., 1927: Vitamin-like substances in plant nutrition, *Ariz. Sta. Tech. Bull.* 16.

Brenchley, W. E., 1919: Some factors in plant competition, *Ann. App. Biol.*, **6**: 142–170.

———, 1920: On the relations between growth and the environmental conditions of temperature and bright sunshine, *Ann. App. Biol.*, **6**: 211–244.

Briggs, G. E., F. Kidd, and C. West, 1920: A quantitative analysis of plant growth. I, II, *Ann. App. Biol.*, **7**: 103–123, 202–223.

———, ———, and ———, 1920: Methods in the quantitative analysis of plant growth—a reply to criticism, *Ann. App. Biol.*, **7**: 403–406.

———, 1925: Plant yield and the intensity of external factors.—Mitscherlich's Wirkungsgesetz, *Ann. Bot.*, **39**: 475–502.

———, 1928: A consideration of some attempts to analyse growth curves, *Proc. Roy. Soc.*, B, **102**: 280–285.

Briggs, L. J., A. B. Campbell, R. H. Heald, and L. H. Flint, 1926: Electroculture, *U. S. Dept. Agr. Dept. Bull.* 1379.

Brink, R. A., 1924: The physiology of pollen. I, The requirements for growth, *Am. J. Bot.*, **11**: 218–229.

———, 1924: The physiology of pollen. II, Further consideration regarding the requirements for growth, *Am. J. Bot.*, **11**: 283–295.

———, 1924: The physiology of pollen. III, Growth *in vitro* and *in vivo*, *Am. J. Bot.*, **11**: 351–365.

———, 1924: The physiology of pollen. IV, Chemotropism: effects on growth of grouping grains, formation and function of callose plugs; Summary and conclusions, *Am. J. Bot.*, **11**: 417–436.

Brown, A. B., 1935: Cambial activity, root habit and sucker shoot development in two species of poplar, *New Phytol.*, **34**: 163–179.

Brown, A. H., 1933: Effects of sulphuric-acid delinting on cotton seeds, *Bot. Gaz.*, **94**: 755–770.

Brown, W. H., and S. F. Trelease, 1918: Alternate shrinkage and elongation of growing stems of *Cestrum nocturnum*, *Philippine J. Sci.*, C, **13**: 353–360.

Bruman, A. J., 1932: "Yarovization" a new factor in plant breeding technique, *J. Heredity*, **23**: 465–466.

Bucholtz, A. F., 1931: The effect of monochromatic ultra-violet light of measured intensities on behavior of plant cells, *Ann. Mo. Bot. Gard.*, **18**: 489–508.

Burge, W. E., G. C. Wickwire, and H. J. Fuller, 1936: Daily variation in the sensitivity of *Mimosa* with special reference to the action of light, *Bot. Gaz.*, **97**: 674–677.

Burkholder, P. R., and R. Pratt, 1936: Leaf-movements of *Mimosa pudica* in relation to light, *Am. J. Bot.*, **23**: 46–52.

——— and ———, 1936: Leaf-movements of *Mimosa pudica* in relation to the intensity and wave length of the incident radiation, *Am. J. Bot.*, **23**: 212–219.

———, 1936: The rôle of light in the life of plants. I, Light and physiological processes. II, The influence of light upon growth and differentiation, *Bot. Rev.* **2**: 1–52, 97–168.

Burkholder, W. H., 1932: Effect of the hydrogen-ion concentration of the soil on the growth of the bean and its susceptibility to dry root rot, *J. Agr. Res.*, **44**: 175–181.

Burnett, L. C., and A. L. Bakke, 1930: The effect of delayed harvest upon the yield of grain, *Iowa Sta. Res. Bull.* 130.

Burr, S., and D. M. Turner, 1935: Vernalisation of tomatoes, *Gard. Chron.* III, **98**: 288.

Busse, W. F., 1930: Effect of low temperatures on germination of impermeable seeds, *Bot. Gaz.*, **89**: 169–179.

——— and C. R. Burnham, 1930: Some effects of low temperature on seeds, *Bot. Gaz.*, **90**: 399–412.

Caldwell, J. S., 1934: Hydrion concentration changes in relation to growth and ripening in fruits, *U. S. Dept. Agr. Tech. Bull.* 403.

Calvert, J., 1935: Drought resistance in wheat: the bound and free water of expressed sap from wheat leaves in relation to time and soil moisture, *Protoplasma*, **24**: 505–524.

———, 1935: Diurnal variation in "bound and free water" and other factors in the sap expressed from the leaves of *Phalaris tuberosa*, *Protoplasma*, **24**: 525–530.

Cattell, W., 1931: The effects of X-rays on the growth of wheat seedlings, *Sci.*, **73**: 531–532.

Chace, E. M., and F. E. Denny, 1924: Use of ethylene in the coloring of citrus fruit, *Ind. and Eng. Chem.*, **16**: 339–340.

——— and E. G. Church, 1927: Effect of ethylene on the composition and color of fruits, *Ind. and Eng. Chem.*, **19**: 1135–1139.

Chambers, R., and H. P. Hale, 1932: The formation of ice in protoplasm, *Proc. Roy. Soc.*, B, **110**: 336–352.

Chambers, W. H., 1923: Cultures of plant cells, *Proc. Soc. Exp. Biol. and Med.*, **21**: 71–72.

Chance, H. L., 1934: The influence of various types of defoliation and leaf wounding upon the growth and yield of beans, *Am. J. Bot.*, **21**: 85–108.

Chandler, W. H., 1931: Freezing of pollen: Evidence as to how freezing kills plant cells, abst., *Am. J. Bot.*, **18**: 892.

——— and A. C. Hildreth, 1935: Evidence as to how freezing kills plant tissue, *Proc. Am. Soc. Hort. Sci.*, **33**: 27–35.

Child, C. M., and A. W. Bellamy, 1920: Physiological isolation by low temperature in *Bryophyllum*, *Bot. Gaz.*, **70**: 249–267.

Cholodny, N., 1924: Ueber die hormonale Wirkung der Organspitze bei der geotropischen Krümmung, *Ber. bot. Ges.*, **42**: 356–362.

———, 1926: Beiträge zur Analyse der geotropischen Reaktion, *Jahrb. wiss. Bot.*, **65**: 447–459.

———, 1927: Wuchshormone und Tropismen bei den Pflanzen, *Biol. Zent.*, **47**: 604–626.

———, 1928: Beiträge zur hormonalen Theorie von Tropismen, *Planta*, **6**: 118–134.

———, 1934: Über die Bildung und Leitung des Wuchshormons bei den Wurzeln, *Planta*, **21**: 517–530; *Biol. Abst.*, **10**: 37.

Clark, N. A., 1926: Plant growth promoting substances, hydrogen-ion concentration and the reproduction of *Lemna*, *Plant Physiol.*, **1**: 273–279.

———, 1929: Vitamins and the world of plants, *Sci. Mo.*, **28**: 343–348.

——— and B. H. Thomas, 1934: Microorganisms and vitamin production in green plants, *Sci.*, **79**: 571–572.

Clark, W. G., 1935: Note on the effect of light on the bioelectric potentials in the *Avena* coleoptile, *Proc. Nat. Acad. Sci.*, **21**: 681–684.

Coggeshall, M., 1931: Influence of acetic, propionic, normal butyric and sulphuric acids and potassium acetate on elongation of primary roots of seedlings of white lupine, *Plant Physiol.*, **6**: 389–447.

Colin, M. H., 1922: L'inuline dans les plantes greffées. La greffe Soleil annuel-Topinambour, *Rev. Gén. Bot.*, **34**: 145–155, 202–213.

Colla, S., 1927: L'azione dei raggi ultra violetti sulle piante etiolate, *Boll. Soc. Ital. Biol. Sperim.*, **2**: 724–726; *Biol. Abst.*, **4**: 354.

Collins, G. N., L. H. Flint, and J. W. McLane, 1929: Electrical stimulation of plant growth, *J. Agr. Res.*, **38**: 585–600.

——— and L. R. Maxwell, 1936: Delayed killing of maize seedlings with X-rays, *Sci.*, **83**: 375–376.

Collison, R. C., and J. D. Harlan, 1934: Winter injury of Baldwin apple trees and its relation to previous tree performance and nutritional treatment, *N. Y. Sta. (Geneva) Bull.* 647.

Cooper, H. P., 1932: Relation of hydrogen-ion concentration of soils to the growth of certain pasture plants, *Plant Physiol.*, **7**: 527–532.

Cooper, W. C., 1935: Hormones in relation to root formation on stem cuttings, *Plant Physiol.*, **10**: 787–794.

Copeman, P. R., 1928: Autocatalysis and growth, *Ann. App. Biol.*, **15**: 613–622.

Coster, C. H., 1927: Die täglichen Schwankungen des Längenzuwachses in den Tropen, *Rec. trav. bot. Néerl.*, **24**: 257–306; *Biol. Abst.*, **2**: 1512; *Exp. Sta. Rec.*, **61**: 817.

Coville, F. V., 1920: The influence of cold in stimulating the growth of plants, *J. Agr. Res.*, **20**: 151–160.

Crocker, W., and L. I. Knight, 1908: Effect of illuminating gas and ethylene upon flowering carnations, *Bot. Gaz.*, **46**: 259–276.

——— and W. E. Davis, 1914: Delayed germination in seeds of *Alisma plantago*, *Bot. Gaz.*, **57**: 285–322.

———, 1916: Mechanics of dormancy in seeds, *Am. J. Bot.*, **3**: 99–121.

——— and L. V. Barton, 1931: After-ripening, germination, and storage of certain Rosaceous seeds, *Contr. Boyce Thompson Inst.*, **3**: 385–404.

———, P. W. Zimmerman, and A. E. Hitchcock, 1932: Ethylene-induced epinasty of leaves and the relation of gravity to it, *Contr. Boyce Thompson Inst.*, **4**: 177–218.

———, A. E. Hitchcock, and P. W. Zimmerman, 1935: Similarities in the effects of ethylene and the plant auxins, *Contr. Boyce Thompson Inst.*, **7**: 231–248.

Curtis, O. F., and H. I. Chang, 1930: The relative effectiveness of the temperature of the crown as contrasted with that of the rest of the plant upon the flowering of celery, *Am. J. Bot.*, **17**: 1047–1048.

Daniel, L., 1921: À propos des greffes de Soleil sur Topinambour, *C. R. Acad. Sci. Paris*, **172**: 610–612.

Darlington, H. T., 1931: The 50-year period for Dr. Beal's viability experiment, *Am. J. Bot.*, **18**: 262–266.

Darrow, G. M., 1929: Effect of light, temperature, and transpiration on elongation of canes of raspberry and other brambles, *Proc. Am. Soc. Hort. Sci.*, **26**: 308–311.

———, 1930: Experimental studies on the growth and development of strawberry plants, *J. Agr. Res.*, **41**: 307–325.

——— and G. F. Waldo, 1930: Fruit-bud formation in the strawberry in spring in southeastern states, *Sci.*, **72**: 349–350.

——— and ———, 1933: Photoperiodism as a cause of the rest period in strawberries, *Sci.*, **77**: 353–354.

———, 1933: Tomatoes, berries and other crops under continuous light in Alaska, *Sci.*, **78**: 370.

Dastur, R. H., and G. A. Kapadia, 1931: Mechanism of curvature in the tendrils of *Cucurbitaceae*, *Ann. Bot.*, **45**: 279–301.

Davies, P. A., 1930: The relation of total nitrogen to regeneration in the willow, *Sci.*, **72**: 275–276.

Davies, W., G. A. Atkins, and P. C. B. Hudson, 1936: Effect of ascorbic acid and indolyl acetic acid on regeneration of willow branches and germination, *Nature*, **137**: 618.

Davis, E. E., 1934: Influence of thyroxin on the growth of plants, *Plant Physiol.*, **9**: 377–384.

Davis, W. B., and C. G. Church, 1931: The effect of ethylene on the chemical composition and the respiration of the ripening Japanese persimmon, *J. Agr. Res.*, **42**: 165–182.

———, 1935: Detection and measurement of freezing injury in Valencia oranges, *Am. J. Bot.*, **22**: 557–566.

Deats, M. E., 1925: The effect on plants of the increase and decrease of the period of illumination over that of the normal day period, *Am. J. Bot.*, **12**: 384–393.

DeBeer, G. R., 1924: "Growth," Edward Arnold & Company, London, 120 pp.

Degman, E. S., and E. C. Auchter, 1934: Metaxenia studies in apples, *Proc. Am. Soc. Hort. Sci.*, **32**: 213–220.

DeLong, G. E., 1929: The effect of cutting garnet wheat at different stages of maturity and on consecutive dates after the occurrence of frost, *Sci. Agr.*, **9**: 566–574.

Denny, F. E., 1924: Hastening the coloration of lemons, *J. Agr. Res.*, **27**: 757–771.

———, 1926: Hastening the sprouting of dormant potato tubers, *Am. J. Bot.*, **13**: 118–126.

———, 1926: Second report on the use of chemicals for hastening the sprouting of dormant potato tubers, *Am. J. Bot.*, **13**: 386–396.

———, 1928: Chemical treatments for controlling the growth of buds of plants, *Ind. and Eng. Chem.*, **20**: 578–581.

——— and E. N. Stanton, 1928: Chemical treatments for shortening the rest period of pot-grown woody plants, *Am. J. Bot.*, **15**: 327–337.

——— and ———, 1928: Localization of response of woody tissues to chemical treatments that break the rest period, *Am. J. Bot.*, **15**: 337–344.

———, 1928: The importance of temperature in the use of chemicals for hastening the sprouting of dormant potato tubers, *Am. J. Bot.*, **15**: 395–404.

———, 1929: Role of mother tuber in growth of potato plant, *Bot. Gaz.*, **87**: 157–194.

———, 1930: Shortening the rest period of *Gladiolus* by treatment with chemicals, *Am. J. Bot.*, **17**: 602–613.

———, 1930: Sucrose and starch changes in potatoes treated with chemicals that break the rest period, *Am. J. Bot.*, **17**: 806–818.

———, 1930: Direct *vs.* indirect effects in the treatment of potato tubers with chemicals, abst., *Am. J. Bot.*, **17**: 1049–1050.

——— and L. P. Miller, 1932: Effect of ethylene chlorhydrin vapors upon dormant lilac tissues, *Contr. Boyce Thompson Inst.*, **4**: 513–528.

———, 1933: Effect of ethylene chlorohydrin vapors upon the chemical composition of *Gladiolus* corms, *Contr. Boyce Thompson Inst.*, **5**: 435–440.

——— and L. P. Miller, 1934: Hastening the germination of dormant *Gladiolus* cormels with vapors of ethylene chlorohydrin, *Contr. Boyce Thompson Inst.*, **6**: 31–38.

——— and———, 1935: Production of ethylene by plant tissue as indicated by the epinastic response of leaves, *Contr. Boyce Thompson Inst.*, **7**: 97–102.

——— and ———, 1935: Further experiments on shortening the rest period of potato tubers, *Contr. Boyce Thompson Inst.*, **7**: 157–181.

———, 1935: Testing plant tissue for emanations causing leaf epinasty, *Contr. Boyce Thompson Inst.*, **7**: 341–347.

———, 1936: Gravity-position of tomato stems and their production of the emanation causing leaf epinasty, *Contr. Boyce Thompson Inst.*, **8**: 99–104.

DEUBER, C. G., 1931: Chemical treatments to shorten the rest period of tree seeds, *Sci.*, **73**: 320–321.

———, 1932: Effects of illuminating gas on young oak trees, abst., *Am. J. Bot.*, **19**: 845.

———, 1932: Stimulative effects of illuminating gas on trees, *Sci.*, **75**: 496–497.

———, 1933: Accelerating seed germination with gas, *Am. Gas Assoc. Month.*, **15**: 313–315.

———, 1933: Stimulative effects of gas on small trees, *Am. Gas. Assoc. Month.*, **15**: 380–383.

———, 1933: Observations of shade trees in the vicinity of gas leaks, *Am. Gas. Assoc. Month.*, **15**: 465–466.

———, 1934: Methylene blue and gassed plants, *Sci.*, **79**: 389–390.

———, 1936: Effects on trees of an illuminating gas in the soil, *Plant Physiol.*, **11**: 401–412.

DEXTER, S. T., 1930: A method for determining hardiness in plants, *Sci.*, **71**: 220–221.

———, W. E. TOTTINGHAM, and L. F. GRABER, 1930: Preliminary results in measuring the hardiness of plants, *Plant Physiol.*, **5**: 215–223.

———, ——— and ———, 1932: Investigations of the hardiness of plants by measurement of electrical conductivity, *Plant Physiol.*, **7**: 63–78.

———, 1932: Studies of the hardiness of plants: A modification of the Newton pressure method for small samples, *Plant Physiol.*, **7**: 721–726.

———, 1933: Effect of several environmental factors on the hardening of plants, *Plant Physiol.*, **8**: 123–139.

———, 1933: Decreasing hardiness of winter wheat in relation to photosynthesis, defoliation and winter injury, *Plant Physiol.*, **8**: 297–304.

———, 1934: Salt concentration and reversibility of ice-formation as related to the hardiness of winter wheat, *Plant Physiol.*, **9**: 601–618.

———, 1935: Growth, organic nitrogen fractions, and buffer capacity in relation to hardiness of plants, *Plant Physiol.*, **10**: 149–158.

———, 1935: Salt concentration and reversibility of ice formation as related to the hardiness of alfalfa, *Plant Physiol.*, **10**: 403–406.

DICKSON, H., 1932: Polarity and the production of adventitious growing points in *Marchantia polymorpha*, *Ann. Bot.*, **46**: 683–702.

DIXON, H. H., and T. A. BENNET-CLARK, 1927: Response of plant tissue to electric currents, *Notes, Bot. School Trinity Col. Dublin*, **4**: 33–54.

DOLK, H. E., and K. V. THIMANN, 1932: Studies on the growth hormone of plants. I, *Proc. Nat. Acad. Sci.*, **18**: 30–46.

DOUBT, S. L., 1917: The response of plants to illuminating gas, *Bot. Gaz.*, **63**: 209–224.

DOVE, W. F., and E. MURPHY, 1936: The vitamin C content of apples and its relation to human welfare, *Sci.*, **83**: 325–327.

DRAGHETTI, A., 1933: Importance of winter cryptovegetation in the forced culture of wheat, *Ann. Tech. Agr.*, **6**: 484–501; *Exp. Sta. Rec.*, **72**: 596.

DUBUY, H. G., 1936: The change in the response of *Avena* coleoptiles to growth regulators produced by aging, *Proc. Nat. Acad. Sci.*, **22**: 272–275.

DUGGAR, B. M., 1937: "Biological Effects of Radiation," McGraw-Hill Book Company, Inc., New York and London, 2 vol., 1342 pp.

DUNGAN, G. H., 1931: An indication that corn tillers may nourish the main stalk under some conditions, *J. Am. Soc. Agron.*, **23**: 662–670.

DUNN, S., 1933: Relation of hydrophilic colloids to hardiness in cabbage, brussels sprouts, and alfalfa plants as shown by the dye adsorption test, *Plant Physiol.*, **8**: 275–286.

DUSTMAN, R. B., 1934: Effect of ethylene, ethylene chlorohydrin, and ultra-violet light on carbohydrate content of stored apples, *Plant Physiol.*, **9**: 637–643.

DUVEL, J., 1905: The vitality of buried seeds, *U. S. Bur. Plant. Ind. Bull.* 83.

EASTHAM, A., 1914: The longevity of some common seeds, *Agr. Gaz. Canada*, **1**: 544–546.

EATON, F. M., 1924. Assimilation-respiration balance as related to length of day reactions of soybeans, *Bot. Gaz.*, **77**: 311–321.

EATON, S. V., 1931: Effects of variation in dry-length and clipping of plants on nodule development and growth of soy bean, *Bot. Gaz.*, **91**: 113–144.

ECKERSON, S. H., 1913: A physiological and chemical study of after-ripening, *Bot. Gaz.*, **55**: 286–299.

EDWARDS, T. I., 1932: Temperature relations of seed germination, *Quart. Rev. Biol.*, **7**: 428–443.

———, 1934: Relations of germinating soy beans to temperature and length of incubation time, *Plant Physiol.*, **9**: 1–30.

———, R. PEARL, and S. A. GOULD, 1934: The growth and duration of life of *Celosia cristata* seedlings at different temperatures, *J. Gen. Physiol.* **17**: 763–781.

———, ——— and ———, 1934: Influence of temperature and nutrition on the growth and duration of life of *Cucumis melo* seedlings, *Bot. Gaz.*, **96**: 118–135.

EGOROV, M. A., 1915: À propos du rôle des éléments de cendre dans la vie des plantes, *Zhur. Opytn. Agron.*, **16**: 270–280. Quoted by Murneek, 1932.

ELMER, O. H., 1932: Growth inhibition of potato sprouts by the volatile products of apples, *Sci.*, **75**: 194.

———, 1936: Growth inhibition in the potato caused by a gas emanating from apples, *J. Agr. Res.*, **52**: 609–626.

ELTINGE, E. T., 1928: The effect of ultra-violet radiation upon higher plants, *Ann. Mo. Bot. Gard.*, **15**: 169–240.

EMMERT, E. M., 1931: The effect of soil reaction on the growth of tomatoes and lettuce and on the nitrogen, phosphorus and manganese content of the soil and plant, *Ky. Sta. Res. Bull.* 314.

EVANS, C. R., 1933: Germination behavior of *Magnolia grandiflora*, *Bot. Gaz.*, **94**: 729–754.

EVANS, M. W., and H. A. ALLARD, 1934: Relation of length of day to growth of timothy, *J. Agr. Res.*, **48**: 571–586.

EWART, A. J., 1908: On the longevity of seeds, *Proc. Roy. Soc. Victoria*, **21**: 1–210.

FARR, W. K., 1931: Cotton fibers. I, Origin and early stages of elongation, *Contr. Boyce Thompson Inst.*, **3**: 441–458.

———, 1933: Cotton fibers. III, Cell divisions in the epidermal layer of the ovule subsequent to fertilization, *Contr. Boyce Thompson Inst.*, **5**: 167–172.

FERNALD, E. I., 1925: The inhibition of bud development as correlated with the osmotic concentration of sap, *Am. J. Bot.*, **12**: 287–305.

———, 1931: Freezing point depressions of asparagus shoots determined by a thermoelectric method, abst., *Am. J. Bot.*, **18**: 892.

FISHER, R. A., 1920: Some remarks on the methods formulated in a recent article on the quantitative analysis of plant growth, *Ann. App. Biol.*, **7**: 367–372.

FITTING, H., 1905, 1906: Die Reizleitungsvorgänge bei den Pflanzen, *Ergebn. Physiol.*, **4**: 684–763; **5**: 155–249.

———, 1909: Die Beeinflussung der Orchideenblüten durch die Bestäubung und durch andere Umstände, *Z. Bot.*, **1**: 1–86.

Fivaz, A. E., 1931: Longevity and germination of seeds of *Ribes*, particularly *R. rotundifolium*, under laboratory and natural conditions, *U. S. Dept. Agr. Tech. Bull.* 261.

Flemion, F., 1931: After-ripening, germination and vitality of seeds of *Sorbus aucuparia*, *Contr. Boyce Thompson Inst.*, **3**: 413–439.

———, 1933: Physiological and chemical studies of after-ripening of *Rhodotypos kerrioides* seeds, *Contr. Boyce Thompson Inst.*, **5**: 143–159.

———, 1934: Physiological and chemical changes preceding and during the after-ripening of *Symphoricarpos racemosus* seeds, *Contr. Boyce Thompson Inst.*, **6**: 91–102.

———, 1934: Dwarf seedlings from non after-ripened embryos of peach, apple and hawthorn, *Contr. Boyce Thompson Inst.*, **6**: 205–209.

Flint, L. H., 1934: Light in relation to dormancy and germination in lettuce seed, *Sci.*, **80**: 38–40.

——— and E. D. McAlister, 1935: Wave lengths of radiation in the visible spectrum inhibiting the germination of light sensitive lettuce seed, *Smithsonian Misc. Coll.*, **94**: No. 5.

Florell, V. H., and B. Faulkner, 1934: Growth of wheat during the heading period, *J. Am. Soc. Agron.*, **26**: 954–965.

Forster, H. C., M. A. H. Tincker, A. J. Vasey, and S. M. Wadham, 1932: Experiments in England, Wales and Australia on the effect of the length of day on various cultivated varieties of wheat, *Ann. App. Biol.*, **19**: 378–412.

Francis, D. S., 1934: The effect of X-rays on growth and respiration of wheat seedlings, *Bull. Torrey Bot. Club*, **61**: 119–153.

Fraser, J. G. C., and L. M. Pidgeon, 1933: Electrolysis of seed of cereals, *Sci. Agr.*, **14**: 141–148.

Freeland, R. O., 1933: Some morphological and physico-chemical changes accompanying proliferation of *Bryophyllum* leaves, *Am. J. Bot.*, **20**: 467–480.

Friedberg, R., 1932: The tillering of wheat: Relation between the depth of the tillering node and cold resistance, *Ann. Agron. (Paris)*, **2**: 215–228; *Exp. Sta. Rec.*, **72**: 44.

Friesner, R. C., 1920: Daily rhythms of elongation and cell division in certain roots, *Am. J. Bot.*, **7**: 380–406.

Fuller, H. J., 1930: Stimulatory effects of ultra-violet radiation upon higher plants, *Sci.*, **72**: 535–536.

———, 1931: Stimulatory effects of radiation from a quartz mercury vapor arc upon higher plants, *Ann. Mo. Bot. Gard.*, **18**: 17–39.

———, 1932: The injurious effects of ultra-violet and infra red radiations on plants, *Ann. Mo. Bot. Gard.*, **19**: 79–84.

Gane, R., 1934: Production of ethylene by some ripening fruits, *Nature*, **134**: 1008.

———, 1934: Identification of ethylene among the volatile products of ripe apples, *Dept. Sci. and Ind. Res., Food Invest. Bd. Rept. (Gr. Brit.)*, **1934**: 122–123; *Exp. Sta. Rec.*, **74**: 40.

———, 1935: The formation of ethylene by plant tissues, and its significance in the ripening of fruits, *J. Pom. and Hort. Sci.*, **13**: 351–358.

Gardner, F. E., 1929: Composition and growth initiation of dormant pear shoots as influenced by temperatures, *Plant Physiol.*, **4**: 405–434.

Gardner, W. A., 1921: Effect of light on germination of light-sensitive seeds, *Bot. Gaz.*, **71**: 249–288.

GARNER, W. W., and H. A. ALLARD, 1920: Effect of the relative length of day and night and other factors of the environment on growth and reproduction in plants, *J. Agr. Res.*, **18**: 553–607.

——— and ———, 1923: Further studies in photoperiodism, the response of the plant to relative length of day and night, *J. Agr. Res.*, **23**: 871–921.

———, C. W. BACON, and H. A. ALLARD, 1924: Photoperiodism in relation to hydrogen-ion concentration of the cell sap and the carbohydrate content of the plant, *J. Agr. Res.*, **27**: 119–157.

——— and H. A. ALLARD, 1925: Localization of the response in plants to relative length of day and night, *J. Agr. Res.*, **31**: 555–567.

——— and ———, 1930: Photoperiodic response of soy beans in relation to temperature and other environmental factors, *J. Agr. Res.*, **41**: 719–737.

——— and ———, 1931: Effect of abnormally long and short alternations of light and darkness on growth and development of plants, *J. Agr. Res.*, **42**: 629–639.

——— and ———, 1931: Duration of the flowerless condition of some plants in response to unfavorable lengths of day, *J. Agr. Res.*, **43**: 439–445.

———, 1933: Comparative responses of long-day and short-day plants to relative lengths of day and night, *Plant Physiol.*, **8**: 347–356.

GASSNER, G., 1918: Beiträge zur physiologischer Characteristik Sommer-und Winter-annueller Gewächse, insbesondere der Getriedepflanzen, *Z. Bot.*, **10**: 417–480.

GAUTHERET, R., 1933: Cultures de méristèmes de racines de *Zea mays*, *C. R. Acad. Sci. Paris*, **197**: 85–87.

GERICKE, W. F., 1918: Effects of rest and no-rest periods upon growth of *Solanum*, *Bot. Gaz.*, **65**: 344–353.

———, 1924: Growth-inhibiting and growth-stimulating substances, *Bot. Gaz.*, **78**: 440–455.

GFELLER, F., R. A. DERRICK, and J. G. C. FRASER, 1933: Yarovization or vernalization of cereal crops, *Sci. Agr.*, **13**: 728–729.

GIERSBACH, J., 1934: After-ripening and germination of *Cotoneaster* seeds, *Contr. Boyce Thompson Inst.*, **6**: 323–338.

GILBERT, B. E., 1926: Interrelation of relative day length and temperature, *Bot. Gaz.*, **81**: 1–25.

———, 1926: The response of certain photoperiodic plants to differing temperature and humidity conditions, *Ann. Bot.*, **40**: 315–321.

GILL, N., 1933: The relation of flowering and cambial activity, *New Phytol.*, **32**: 1–12.

GLASS, H. B., 1933: Effect of light on the bioelectric potentials of isolated *Elodea* leaves, *Plant Physiol.*, **8**: 263–274.

GOODSPEED, T. H., 1929: The effect of X-rays and radium on species of the genus *Nicotiana*, *J. Heredity*, **20**: 243–259.

GORKE, H., 1906: Über chemische Vorgänge beim Erfrieren der Pflanzen, *Landw. Versuchs-Stat.*, **65**: 149–160.

GOSS, W. L., 1924: The vitality of buried seeds, *J. Agr. Res.*, **29**: 349–362.

GOULD, S. A., R. PEARL, T. I. EDWARDS, and J. R. MINER, 1933: Available food, relative growth and duration of life in seedlings of *Cucumis melo*, *Proc. Nat. Acad. Sci.*, **19**: 228–233.

———, ——— and ———, 1934: On the effects of partial removal of the cotyledons upon the growth and duration of life of canteloup seedlings without exogenous food, *Ann. Bot.*, **48**: 575–599.

GOURLEY, J. H., 1932: Response of iris to soil reaction, *Plant Physiol.*, **7**: 739–742.

GRAČANIN, M., 1928: Orthophosphorsäure als Stimulator der Keimungsenergie und Aktivator der Keimfähigkeit der Samen, *Biochem. Z.*, **195**: 457–468.

GREATHOUSE, G. A., and N. W. STUART, 1934: A study of physical and chemical properties of red clover roots in the cold hardened and unhardened condition, *Md. Sta. Bull.* 370.

GREENE, L., R. B. WITHROW, and M. F. RICHMAN, 1932: The response of greenhouse crops to electric light supplementing daylight, *Ind. Sta. Bull.* 366.

GREGORY, F. G., 1926: The effect of climatic conditions on the growth of barley, *Ann. Bot.*, **40**: 1–27.

——— and L. BATTEN, 1926: A critical statistical study of experimental data on the effect of minute electric currents on the growth rate of the coleoptile of barley, *Proc. Roy. Soc.*, B, **99**: 122–130.

———, 1928: Studies in the energy relations of plants. I, The effect of temperature on the increase in area of leaf surface. II, The effect of temperature on increase in area of leaf surface and in dry weight of *Cucumis sativus*, *Ann. Bot.*, **42**: 469–507.

———, 1928: The analysis of growth curves.—a reply to criticism, *Ann. Bot.*, **42**: 531–541.

GROVES, J. F., 1917: Temperature and life duration of seeds, *Bot. Gaz.*, **63**: 169–189.

GUNDERSON, M. F., and C. E. SKINNER, 1934: Production of vitamins by a pure culture of *Chlorococcum* grown in darkness on a synthetic medium, *Plant Physiol.*, **9**: 807–815.

GUSTAFSON, F. G., 1926: Growth studies on fruits, *Plant Physiol.*, **1**: 265–272.

———, 1927: Growth studies on fruits: an explanation of the shape of the growth curve, *Plant Physiol.*, **2**: 153–161.

——— and H. E. LAING, 1931: Growth studies of fruits. The effect of the removal of alternate flower clusters on the development of fruits in the tomato, *Mich. Acad. Sci., Arts, and Letters*, **13**: 95–99.

GUTHRIE, J. D., 1933: Effect of chemical treatments of dormant potato tubers on the conductivity of the tissue and on the leaching of electrolytes from the tissue, *Contr. Boyce Thompson Inst.*, **5**: 83–94.

———, 1933: Changes in the glutathione content of potato tubers treated with chemicals that break the rest period, *Contr. Boyce Thompson Inst.*, **5**: 331–350.

———, 1934: Metabolism of citric, sulphuric and nitric acid in the potato tuber. An explanation for the high pH of the juice of tubers treated with ethylene chlorhydrin, *Contr. Boyce Thompson Inst.*, **6**: 247–268.

HABER, E. S., P. M. NELSON, and P. P. SWANSON, 1931: Association of vitamins with nutritional conditions in plants, *Iowa Sta. Ann. Rept.*, **1931**: 105.

HABERLANDT, G., 1922: Cell-division hormones and their relation to wound repair, fertilization, parthenogenesis, and adventive embryogeny, *Biol. Zent.*, **42**: 145–172; *Exp. Sta. Rec.*, **49**: 219.

HACKENBERG, H., 1909: Ueber die Substanzquotienten von *Cannabis sativa* und *Cannabis gigantea*, *Beih. bot. Cent.*, I, **24**: 45–67.

HAIG, C., 1934: The spectral sensibility of *Avena*, *Proc. Nat. Acad. Sci.*, **20**: 476–479.

———, 1935: The phototropic responses of *Avena* in relation to intensity and wave length, *Biol. Bull.*, **69**: 305–324.

HAMMETT, F. S., 1929: The chemical stimulus essential for growth by increase in cell number, *Proc. Am. Phil. Soc.*, **68**: 151–161.

HANNA, W. F., 1924: Growth of corn and sunflowers in relation to climatic conditions, *Bot. Gaz.*, **78**: 200–215.

———, 1925: The nature of the growth rate in plants, *Sci. Agr.*, **5**: 133–138.

HARRINGTON, G. T., 1916: Agricultural value of impermeable seeds, *J. Agr. Res.*, **6**: 761–796.

———, 1923: Use of alternating temperatures in the germination of seeds, *J. Agr. Res.*, **23**: 295–333.

HARRINGTON, J. B., 1926: Growing wheat and barley hybrids in winter by means of artificial light, *Sci. Agr.*, **7**: 125–130.

HARRISON, C. M., 1931: Effect of cutting and fertilizer applications on grass development, *Plant Physiol.*, **6**: 669–684.

HARRISON, G. J., 1931: Metaxenia in cotton, *J. Agr. Res.*, **42**: 521–544.

HARTSHORN, R., 1931: Some effects of acetylene on the ripening processes of bananas, *Plant Physiol.*, **6**: 467–484.

HARVEY, E. M., 1913: The castor bean plant and laboratory air, *Bot. Gaz.*, **56**: 439–442.

———, 1915: Some effects of ethylene on the metabolism of plants, *Bot. Gaz.*, **60**: 193–214.

——— and R. C. ROSE, 1915: The effects of illuminating gas on root systems, *Bot. Gaz.*, **60**: 27–45.

———, 1931: A method for comparing growth rates by means of a protractor, *Sci.*, **74**: 46–47.

HARVEY, R. B., 1918: Hardening process in plants and developments from frost injury, *J. Agr. Res.*, **15**: 83–113.

———, 1922: Growth of plants in artificial light, *Bot. Gaz.*, **74**: 447–451.

———, 1922: Growth of plants in artificial light from seed to seed, *Sci.*, **56**: 366–367.

———, 1925: A new method of blanching celery, *Minn. Hort.*, **53**: 41.

———, 1926: The ripening of fruits by ethylene gas, *Minn. Hort.*, **54**: 140.

———, 1928: Artificial ripening of fruits and vegetables, *Minn. Sta. Bull.* 247.

———, 1928: Ethylene as a ripener of fruits and vegetables, *Sci.*, **67**: 421–422.

———, 1930: Time and temperature factors in hardening plants, *Am. J. Bot.*, **17**: 212–217.

———, 1931: The action of toxic agents used in the eradication of noxious plants, *J. Am. Soc. Agron.*, **23**: 481–489.

HASKINS, C. P., and C. N. MOORE, 1935: Growth modification in citrus seedlings grown from X-rayed seeds, *Plant Physiol.*, **10**: 179–185.

HASSELBRING, H., 1932: Effect of hot water treatment on the carbohydrate changes in narcissus bulbs during storage, *Plant Physiol.*, **7**: 145–152.

HAUSEN, S., 1935: Effect of vitamin C (ascorbic acid) on the growth of plants, *Nature*, **136**: 516.

HAVAS, L., 1935: Ascorbic acid (vitamin C) and the germination and growth of seedlings, *Nature*, **136**: 435.

———, 1935: Follicular (Oestrus) hormone and plant tumours, *Nature*, **136**: 516.

——— and J. CALDWELL, 1935: Some experiments on the effects of animal hormones on plants, *Ann. Bot.*, **49**: 727–748.

———, 1935: Some effects of radioactive mud upon germination of seeds and growth of seedlings, *J. Agr. Sci.*, **25**: 198–216.

HAWKER, L. E., 1932: Experiments on the perception of gravity by roots, *New Phytol.*, **31**: 321–328.

———, 1933: The effect of temperature on the geotropism of seedlings of *Lathyrus odoratus*, *Ann. Bot.*, **47**: 501–515.

HAYNES, D., and H. K. ARCHBOLD, 1928: Chemical studies in the physiology of apples. X, A quantitative study of chemical changes in stored apples, *Ann. Bot.*, **42**: 965–1017.

HEDRICKS, E., and R. B. HARVEY, 1924: Growth of plants in artificial light. II, Intensities of continuous light required for blooming, *Bot. Gaz.*, **77**: 330–334.

HEYN, A. N. J., 1931: Der Mechanismus der Zellstreckung, *Rec. trav. bot. Néerl.*, **28**: 113–244.

——— and J. van Overbeek, 1931: Weiteres Versuchsmaterial zur plastischen und elastischen Dehnbarkeit der Zellmembran, *Proc. Kon. Akad. Wet. Amsterdam*, **34**: 1190–1195.

———, 1934: Weitere Untersuchungen über den Mechanismus der Zellstreckung und die Eigenschaften der Zellmembran. III, Die Änderungen der Plastizität der Zellwand bei verschiedenen Organen, *Jahrb. wiss. Bot.*, **79**: 753–789.

Hibbard, A. D., 1933: Metaxenia in apple and squash, *Proc. Am. Soc. Hort. Sci.*, **30**: 140–142.

Hibbard, R. P., 1930: The physiological effect of ethylene gas upon celery, tomatoes and certain fruits, *Mich. Sta. Tech. Bull.* 104.

Hicks, L. E., 1932: Ranges of pH tolerance of the *Lemnaceae*, *Ohio J. Sci.*, **32**: 237–244.

Hill, D. D., and S. C. Salmon, 1927: The resistance of certain varieties of winter wheat to artificially produced low temperatures, *J. Agr. Res.*, **35**: 933–939.

Hitchcock, A. E., W. Crocker, and P. W. Zimmerman, 1932: Effect of illuminating gas on the lily, narcissus, tulip and hyacinth, *Contr. Boyce Thompson Inst.*, **4**: 155–176.

———, ———, and ———, 1932: The effect on plants of passing illuminating gas through the soil, abst., *Am. J. Bot.*, **19**: 846.

———, ———, and ———, 1934: Toxic action in soil of illuminating gas containing hydrocyanic acid, *Contr. Boyce Thompson Inst.*, **6**: 1–30.

———, 1935: Indole-3-*n* propionic acid as a growth hormone and the quantitative measurement of plant response, *Contr. Boyce Thompson Inst.*, **7**: 87–95.

———, 1935: Tobacco as a test plant for comparing the effectiveness of preparations containing growth substances, *Contr. Boyce Thompson Inst.*, **7**: 349–364.

——— and P. W. Zimmerman, 1935: Absorption and movement of synthetic growth substances from soil as indicated by the response of aerial parts, *Contr. Boyce Thompson Inst.*, **7**: 447–476.

——— and ———, 1936: Effect of growth substances on the rooting response of cuttings, *Contr. Boyce Thompson Inst.*, **8**: 63–80.

Hooker, H. D., 1920: Pentosan content in relation to winter hardiness, *Proc. Am. Soc. Hort. Sci.*, **17**: 204–207.

———, 1925: Plant growth, *Proc. Nat. Acad. Sci.*, **11**: 710–713.

Hostermann, G., 1922: Kulturversuche mit elektrischem Licht, *Gartenwelt*, **26**: 74–75; *Bot. Abst.*, **11**: 383.

Hottes, C. F., 1929: Studies in experimental cytology, *Plant Physiol.*, **4**: 1–30.

——— and H. K. Wilson, 1930: Resistance of wheat to high temperatures under different methods of heating, *J. Am. Soc. Agron.*, **22**: 108–113.

Howard, W. L., 1910: An experimental study of the rest period in plants. The winter rest, *Mo. Sta. Res. Bull.* 1.

———, 1915: An experimental study of the rest period in plants. The summer rest of bulbs and herbaceous perennials, *Mo. Sta. Res. Bull.* 15.

———, 1915: An experimental study of the rest period in plants. Pot grown woody plants, *Mo. Sta. Res. Bull.* 16.

———, 1915: An experimental study of the rest period in plants. Seeds, *Mo. Sta. Res. Bull.* 17.

———, 1915: An experimental study of the rest period in plants. Physiological changes accompanying breaking of the rest period, *Mo. Sta. Res. Bull.* 21.

Howell, J., 1932: Relation of western yellow pine seedlings to the reaction of the culture solution, *Plant Physiol.*, **7**: 657–671.

Huelin, F. E., 1933: Effects of ethylene and of apple vapors on the sprouting of potatoes, *Dept. Sci. and Ind. Res., Food Invest. Bd. Rept.* (*Gr. Brit.*), **1932**: 51–53.

Hurd-Karrer, A. M., and A. D. Dickson, 1934: Carbohydrates and nitrogen relations in wheat plants with reference to type of growth under different environmental conditions, *Plant Physiol.*, **9**: 533–565.

———, 1933: Comparative responses of a spring and winter wheat to day length and temperature, *J. Agr. Res.*, **46**: 867–888.

Hutchings, S. S., 1932: Light in relation to seed germination of *Mimulus ringens*, *Am. J. Bot.*, **19**: 632–643.

Inamdar, R. S., S. B. Singh, and T. D. Pande, 1925: The growth of the cotton plant in India. I, The relative growth rates during successive periods of growth and the relation between growth rate and respiratory index throughout the life cycle, *Ann. Bot.*, **39**: 281–311.

Isbell, C. L., 1931: Regenerative capacities of leaf and leaflet cuttings of tomato and of leaf and shoot cuttings of potato, *Bot. Gaz.*, **92**: 192–201.

Janssen, G., 1929: Physical measurements of the winter wheat plant at various stages in its development, *Plant Physiol.*, **4**: 477–491.

Jensen, I. J., 1925: Winter wheat studies in Montana with special reference to winter killing, *J. Am. Soc. Agron.*, **17**: 630–631.

Johnson, E. L., 1926: Effect of X-rays upon growth, development, and oxidizing enzymes of *Helianthus annuus*, *Bot. Gaz.*, **82**: 373–403.

———, 1928: Growth and germination of sunflowers as influenced by X-rays, *Am. J. Bot.*, **15**: 65–77.

———, 1928: Tuberization of potatoes increased by X-rays, *Sci.* **68**: 231.

———, 1931: On the alleged stimulating action of X-rays upon plants, *Am. J. Bot.*, **18**: 603–614.

———, 1931: Effect of X-irradiation upon growth and reproduction of tomato, *Plant Physiol.*, **6**: 685–694.

———, 1933: The influence of X-radiation on *Atriplex hortensis*, *New Phytol.*, **32**: 297–307.

———, 1936: Susceptibility of seventy species of flowering plants to X-radiation, *Plant Physiol.*, **11**: 319–342.

Johnson, L. P. V., 1935: General preliminary studies on the physiology of delayed germination in *Avena fatua*, *Can. J. Res.*, **13**: 283–300.

Johnston, E. S., F. S. Brackett, and W. H. Hoover, 1931: Relation of phototropism to the wave length of light, *Plant Physiol.*, **6**: 307–313.

———, 1932: The functions of radiation in the physiology of plants. II, Some effects of near infra red radiation on plants, *Smithsonian Misc. Coll.*, **87**, No. 14.

———, 1934: Phototropic sensitivity in relation to wave length, *Smithsonian Misc. Coll.*, **92**, No. 11.

———, 1935: Photoperiodism: a specific growth response to light, *Smithsonian Rept.*, **1934**: 313–323.

———, 1936: Sun rays and plant life, *Smithsonian Rept.* **1936**: 353–371.

Jones, J. A., 1920: Physiological study of maple seeds, *Bot. Gaz.*, **69**: 127–152.

———, 1928: Overcoming delayed germination of *Nelumbo lutea*, *Bot. Gaz.*, **85**: 341–344.

Jones, J. P., 1928: A physiological study of dormancy in vetch seed, *N. Y. Sta. (Cornell) Mem.* 120.

Jones, W. N., 1925: Polarity phenomena in sea kale roots, *Ann. Bot.*, **39**: 359–372.

Jorgensen, I., and J. H. Priestley, 1914: The distribution of the overhead electrical discharge employed in recent agricultural experiments, *J. Agr. Sci.*, **6**: 337–349.

——— and W. Stiles, 1917: Atmospheric electricity as an environmental factor, *J. Ecology*, **5**: 203–209.

JOSLYN, M. A., and G. L. MARSH, 1933: Changes occurring during freezing and storage and thawing of fruits and vegetables, *Calif. Sta. Bull.* 551.

JOZEFOWICZ, M., 1930: The effect of certain treatments on the germination of tomato seeds, *Ann. App. Biol.*, **17**: 504–513.

———, 1930: Some observations on tomato plants from seed submitted to high temperatures, *Ann. App. Biol.*, **17**: 514–521.

KARLING, J. S., 1924: A preliminary account of the influence of light and temperature on growth and reproduction in *Chara fragilis*, *Bull. Torrey Bot. Club*, **51**: 469–488.

———, 1934: Dendrograph studies on *Achras zapota*, *Am. J. Bot.*, **21**: 161–193.

KATCHIONI-WALTHER, L. S., 1930: Influence of H-ion concentration on the development and growth of buckwheat, *Exp. Sta. Rec.*, **62**: 318–319.

KATER, J. M., 1927: A cytological study of dormancy in the seed of *Phaseolus vulgaris*, *Ann. Bot.*, **41**: 629–643.

KEARNEY, T. H., and G. J. HARRISON, 1932: Pollen antagonism in cotton, *J. Agr. Res.*, **44**: 191–226.

KEEBLE, F., M. G. NELSON, and R. SNOW, 1929: Integration of plant behavior. I, Separate geotropic stimulations of tip and stump in roots, *Proc. Roy. Soc.*, B, **105**: 493–498.

———, ———, and ———, 1930: The integration of plant behavior. II, The influence of the shoot on the growth of roots in seedlings, *Proc. Roy. Soc.*, B, **106**: 182–188.

———, ———, and ———, 1930: A wound substance retarding growth in roots, *New Phytol.*, **29**: 289–293.

———, ———, and ———, 1931: The integration of plant behavior. III, The effect of gravity on the growth of roots, *Proc. Roy. Soc.*, B, **108**: 360–365.

———, ———, and ———, 1931: The integration of plant behavior. IV, Geotropism and growth substance, *Proc. Roy. Soc.*, B, **108**: 537–545.

——— and ———, 1935: The integration of plant behavior. V, Growth substance and traumatic curvature of the root, *Proc. Roy. Soc.*, B, **117**: 92–119.

KEINHOLTZ, R., 1924: The effect of high temperatures on the germination and subsequent growth of corn, *Philippine J. Sci.*, **25**: 311–347.

KELLERMAN, K. F., 1926: A review of the discovery of photoperiodism: the influence of the length of daily light periods upon the growth of plants, *Quart. Rev. Biol.*, **1**: 85–94.

KELLICOTT, W. E., 1904: The daily periodicity of cell division and of elongation in the root of *Allium*, *Bull. Torrey Bot. Club*, **31**: 529–550.

KEMPTON, J. H., 1924: The rate of growth of green and albino maize seedlings, *J. Agr. Res.*, **29**: 311–312.

KERTESZ, Z. I., 1930: The chemical changes in peas after picking, *Plant Physiol.*, **5**: 399–412.

KIDD, F., and C. WEST, 1918: Physiological pre-determination: The influence of the physiological condition of the seed upon the course of subsequent growth and upon the yield. II, Rev. Lit., Chap. I, "Parental conditions affecting the potentiality" of the seed, *Ann. App. Biol.*, **5**: 112–142.

———, ———, and G. E. BRIGGS, 1920: What is the significance of the efficiency index of plant growth? *New Phytol.*, **19**: 88–96.

——— and ———, 1932: Effects of ethylene and of apple vapors on the ripening of fruits, *Dept. Sci. and Ind. Res., Food Invest. Bd. Rept.* (*Gr. Brit.*) **1932**: 55.

——— and ———, 1934: The influence of the composition of the atmosphere upon the incidence of the climacteric in apples, *Dept. Sci. and Ind. Res., Food Invest. Bd. Rept.* (*Gr. Brit.*), **1933**: 51–57.

KIESSELBACH, T. A., 1924: Relation of seed size to yield of small grain crops, *J. Am. Soc. Agron.*, **16**: 670–682.

KLEBS, G., 1913: Physiologie der Fortpflanzung. Handwörterb., *Naturwiss.*, **4**: 276–296.

KLIPPART, J. H., 1858: An essay on the origin, growth, diseases, varieties, etc., of the wheat plant, *Ann. Rept. Ohio State Bd. Agr.*, **12**: 562–816.

KNIGHT, L. I., and W. CROCKER, 1913: Toxicity of smoke, *Bot. Gaz.*, **55**: 337–371.

KNOTT, J. E., 1926: Further localization of the response in plant tissues to relative length of day and night, *Proc. Am. Soc. Hort. Sci.*, **23**: 67–70.

———, 1927: Catalase in relation to growth and to other changes in plant tissue, *N. Y. Sta. (Cornell) Mem.* 106.

———, 1932: Rapidity of response of spinach to change in photoperiod, *Plant Physiol.*, **7**: 123–130.

KNUDSON, L., 1925: Hydrogen-ion concentration and plant growth, *J. Am. Soc. Agron.*, **17**: 711–716.

KOCH, K., 1934: Investigations on the transverse and longitudinal transport of the growth substance in plant organs, *Planta*, **22**: 190–220.

KÖGL, F., and A. J. HAAGEN-SMIT, 1931: I, Mitteilung über pflanzliche Wachstumsstoffe. Über die Chemie des Wuchsstoffs, *Proc. K. Akad. Wetensch. Amsterdam*, **34**: 1411–1416.

———, ———, and H. ERXLEBEN, 1933: Über ein Phytohormon der Zellstreckung. Reindarstellung des Auxins aus menschlichem Harn, *Z. physiol. Chem.*, **214**: 241–262.

———, H. ERXLEBEN, and A. J. HAAGEN-SMIT, 1934: Über die Isolierung der Auxine *a* und *b* aus pflanzlichen Materialien, *Z. physiol. Chem.*, **225**: 215–229.

——— and ———, 1934: Über die Konstitution der Auxine *a* und *b*, *Z. physiol. Chem.*, **227**: 51–73.

———, A. J. HAAGEN-SMIT, and H. ERXLEBEN, 1934: Über den Einfluss der Auxine auf das Wurzelwachstum und über die chemische Natur des Auxins der Graskoleoptilen, *Z. physiol. Chem.*, **228**: 104–112; *Chem. Abst.*, **29**: 195.

———, 1935: Über Wuchsstoffe der Auxin- und der Bios-Gruppe, *Ber. chem. Ges.*, **68**: 16–28.

KONDŌ, M., and T. OKAMURA, 1930: On the influence of various temperatures in duration of storage and various moistures of rice upon the preservation of germination power of hulled rice, *Ber. Ōhara-Inst. Landw. Forsch., Kurischiki, Japan*, **4**: 315–341.

——— and ———, 1930: Beziehung zwischen der Wassertemperatur und dem Wachstum der Reispflanzen, *Ber. Ōhara-Inst. Landw. Forsch., Kurischiki, Japan*, **4**: 395–411.

——— and ———, 1931: Beziehung zwischen Wassertemperatur und Wachstum der Reispflanzen, *Ber. Ōhara-Inst. Landw. Forsch., Kurischiki, Japan*, **5**: 67–85.

——— and ———, 1931: Storage of rice. V, On the influence of a desiccating material upon the preservation of the germinating power of hulled rice having different moisture contents and stored at different temperatures, *Ber. Ōhara-Inst. Landw. Forsch., Kurischiki, Japan*, **5**: 221–242.

———, ———, S. ISSHIKI, and Y. KASAHARA, 1932: Untersuchungen über "Photoperiodismus" der Reispflanzen, *Ber. Ōhara-Inst. Landw. Forsch., Kurischiki, Japan*, **5**: 243–280.

KORSTIAN, C. F., 1927: Factors controlling germination and early survival in oaks, *Yale Forestry Bull.* 19.

KOTOWSKI, F., 1929: Effect of size of seed on plant production, *Proc. Int. Cong. Plant Sci.*, **2**: 974–987.

KOTTE, W., 1922: Kulturversuche mit isolierten Wurzelspitzen, *Beitr. allgem. Bot.*, **2**: 413–434.

———, 1922: Wurzelmeristem in Gewebekultur, *Ber. bot. Ges.*, **40**: 269–272.

KRAMER, P. J., 1934: Methods of breaking dormancy in certain forest trees, *J. Forestry*, **32**: No. 7.

———, 1936: Effect of variation in length of day on growth and dormancy of trees, *Plant Physiol.*, **11**: 127–137.

KRASSINSKY, N., A. A. KONDRASHOVA, and N. I. VINOGRADOVA, 1936: On photoperiodism and changes in the enzymatic system, *Ann. Bot.*, **50**: 293–304.

KURZ, H., 1923: Hydrogen ion concentration in relation to ecological factors, *Bot. Gaz.*, **76**: 1–30.

LA RUE, C. D., 1933: Intumescences on poplar leaves. II, Physiological considerations, *Am. J. Bot.*, **20**: 159–175.

———, 1935: Regeneration in monocotyledonous seedlings, *Am. J. Bot.*, **22**: 486–492.

———, 1936: Tissue cultures of spermatophytes, *Proc. Nat. Acad. Sci.* **22**: 201–209.

———, 1936: The effect of auxin on the abscission of petioles, *Proc. Nat. Acad. Sci.*, **22**: 254–259.

LAUDE, H. H., 1933: Review of the literature regarding vernalization, *Rept. Agr. Seminar, Kans. State Col.*, Manhattan, Kans.

———, 1937: Cold resistance of winter wheat, rye, barley and oats in transition from dormancy to active growth, *J. Agr. Res.*, **54**: 899–817.

———, 1937: Comparison of the cold resistance of several varieties of winter wheat in transition from dormancy to active growth, *J. Agr. Res.*, **54**: 919–926.

LAURIE, A., and G. H. POESCH, 1932: Photoperiodism. The value of supplementary illumination and reduction of light on flowering plants in the greenhouse, *Ohio Sta. Bull.* 512.

LEGGATT, C. W., 1929: The comparative value of scarified and unscarified sweet clover seed, *Sci. Agr.*, **9**: 611–614.

———, 1929: The agricultural value of hard seeds of sweet clover in Alberta, *Sci. Agr.*, **9**: 683–688.

LEHENBAUER, P. A., 1914: The growth of maize seedlings in relation to temperature, *Physiol. Res.*, **1**: 247–288.

LEITCH, I., 1916: Some experiments on the influence of temperature on the rate of growth in *Pisum sativum*, *Ann. Bot.*, **30**: 25–46.

LEONIAN, L. H., 1936: Effect of auxins from some green algae upon *Phytophthora cactorum*, *Bot. Gaz.*, **97**: 854–859.

LEVINE, M., 1934: A preliminary report on plants treated with the carcinogenic agents of animals, *Bull. Torrey Bot. Club*, **61**: 103–118.

———, 1936: The response of plants to localized application of various chemical agents, *Bull. Torrey Bot. Club*, **63**: 177–200.

LEWITH, S., 1890: Über die Ursache der Widerstandsfähigkeit der Sporen gegen hohe Temperaturen. Ein Beitrag zur Theorie der Disinfection, *Arch. Path. Pharm.*, **26**: 341–354.

LIPMAN, C. B., and G. N. LEWIS, 1934: Tolerance of liquid-air temperatures by seeds of higher plants for sixty days, *Plant Physiol.*, **9**: 392–394.

LIVINGSTON, B. E., and W. B. MACK, 1931: Temperature relations of CO_2 production by young wheat seedlings, as influenced by oxygen pressure, time and presence of ethylene, abst., *Am. J. Bot.*, **18**: 889–890.

——— and F. W. HAASIS, 1933: Relation of time and maintained temperature to germination percentage for a lot of rice seed, *Am. J. Bot.*, **20**: 596–615.

———, 1934: Environments, *Sci.*, **80**: 569–576.

Lloyd, F. E., and T. C. Barnes, 1932: Changes in the cells of *Spirogyra* associated with the presence of water polymers, *Proc. Nat. Acad. Sci.*, **18**: 422–427.

Lodewick, J. E., 1930: Effect of certain climatic factors on the diameter growth of longleaf pine in western Florida, *J. Agr. Res.*, **41**: 349–363.

Loeb, J., 1915: Rules and mechanism of inhibition and correlation in the regeneration of *Bryophyllum calycinum*, *Bot. Gaz.*, **60**: 249–277.

———, 1916: Further experiments on correlation of growth in *Bryophyllum calycinum*, *Bot. Gaz.*, **62**: 293–302.

———, 1916: On the association and possible identity of root forming and geotropic substances or hormones in *Bryophyllum calycinum*, *Sci.*, **44**: 210–211.

———, 1916: "The Organism as a Whole," G. P. Putnam's Sons, New York, 379 pp.

———, 1917: A quantitative method of ascertaining the mechanism of growth and of inhibition of growth of dormant buds, *Sci.*, **45**: 436–439.

———, 1917: Influence of the leaf upon root formation and geotropic curvature in the stem of *Bryophyllum calycinum* and the possibility of a hormone theory of these processes, *Bot. Gaz.*, **63**: 25–51.

———, 1917: The chemical basis of axial polarity in regeneration, *Sci.*, **46**: 547–551.

———, 1918: Chemical basis of correlation. I, Production of equal masses of shoots by equal masses of sister leaves in *Bryophyllum calycinum*, *Bot. Gaz.*, **65**: 150–174.

———, 1924: "Regeneration from a physico-chemical viewpoint," McGraw-Hill Book Company, Inc., New York, 143 pp.

Long, E. R., 1915: Growth and colloid hydration in cacti, *Bot. Gaz.*, **59**: 491–497.

Loo, T., and S. Loo, 1935: Studies on the culture of isolated root-tips under sterile conditions. I, The effect of leaf extract on the growth of root-tips, *Science Repts., Contr. Lab. Plant Physiol. Nat. Cent. Univ. Nanking, China*, Ser. B, Biology, **2**: 51–79.

Loomis, W. E., 1923: Some relations of hardening to transplanting, *Proc. Am. Soc. Hort. Sci.*, **20**: 206–215.

———, 1927: Temperature and other factors affecting the rest period of potato tubers, *Plant Physiol.*, **2**: 287–302.

———, 1932: Growth-differentiation balance *vs.* carbohydrate-nitrogen ratio, *Proc. Am. Soc. Hort. Sci.*, **29**: 240–244.

———, 1932: The daily growth of *Zea mays*, abst., *Am. J. Bot.*, **19**: 838.

———, 1934: Daily growth of maize, *Am. J. Bot.*, **21**: 1–6.

Ludwig, C. A., 1932: The germination of cottonseed at low temperatures, *J. Agr. Res.*, **44**: 367–380.

Lund, E. J., 1929: Continuous bioelectric currents; a mechanism for cell correlation, abst., *Am. J. Bot.*, **16**: 846–847.

———, 1929: The electric polarity of the Douglas fir and the internal distribution of its electric correlation currents, abst., *Am. J. Bot.*, **16**: 858.

———, 1929: Electric polarity in the Douglas fir, *Pub. Puget Sound Biol. Sta.*, **7**: 1–28.

———, 1929: The relative dominance of growing points in the Douglas fir, *Pub. Puget Sound Biol. Sta.*, **7**: 29–37.

——— and M. Bush, 1930: Electric correlation potentials in the leaf of *Bryophyllum*, *Plant Physiol.*, **5**: 491–508.

———, 1931: The unequal effect of O_2 concentration on the velocity of oxidation in loci of different electric potential, and glutathione content, *Protoplasma*, **13**: 236–258.

———, 1931: External polarity potentials in the apex of the Douglas fir before and after mechanical stimulation, *Plant Physiol.*, **6**: 507–517.

———, 1931: Electric correlation between living cells in cortex and wood in the Douglas fir, *Plant Physiol.*, **6**: 631–652.

———, 1932: Control of the flux equilibrium of electrochemical processes and electric polarity in the Douglas fir by temperature, *Plant Physiol.*, **7**: 297–307.

———, 1932: Comparison of the effects of temperature on the radial and longitudinal electric polarities in wood and cortex of the Douglas fir, *Plant Physiol.*, **7**: 505–516.

Lutz, J. M., 1935: The influence of rate of thawing on freezing injury of apples, potatoes and onions, *Proc. Am. Soc. Hort. Sci.*, **33**: 227–233.

Lyon, T. L., 1905: Tillering of corn, *Neb. Sta. Bull.*, **91**: 16–24.

Lysenko, T. D., 1932: On the problem of Iarovization of corn, millet, Sudan grass, sorghum, and soy bean, *Bull. Jarovizacii*, No. 2 and 3.

———, 1932: Fundamental results of research on vernalization of agricultural plants, *Bull. Jarovizacii*, No. 4, 1–57. Quoted by Maximow, 1934.

MacDougal, D. T., 1903: The influence of light and darkness upon growth and development, *Mem. N. Y. Bot. Gard.*, **2**: 1–319.

———, 1916: The mechanism and conditions of growth, *Mem. N. Y. Bot. Gard.*, **6**: 5–26.

——— and H. A. Spoehr, 1917: Growth and imbibition, *Proc. Am. Phil. Soc.*, **56**: 289–352.

———, 1917: Rate and course of growth of succulents, *Carnegie Inst. Wash. Year Book*, **16**: 83–85.

———, 1917: Growth of wheat (*Triticum*) and corn (*Zea*), *Carnegie Inst., Wash. Year Book*, **16**: 85–87.

———, 1918: Development of conceptions included in growth, *Carnegie Inst. Wash. Year Book*, **17**: 55–57.

———, 1918: The growth mechanism of protoplasm, *Carnegie Inst. Wash. Year Book*, **17**: 57–58.

———, 1918: The nature and course of growth in higher plants, *Carnegie Inst. Wash. Year Book*, **17**: 58–59.

———, 1918: Temperature and the hydration and growth of colloids and of cell masses, *Carnegie Inst. Wash. Year Book*, **17**: 68–70.

———, 1919: Growth in organisms, *Sci.*, **49**: 599–605.

———, 1919: Hydration and growth, *Proc. Am. Phil. Soc.*, **58**: 346–372.

———, 1920: Hydration and growth, *Carnegie Inst. Wash. Pub.* 297.

———, 1920: The physical factors in the growth of the tomato, *Bull. Torrey Bot. Club*, **47**: 261–269.

———, 1921: Growth of trees, *Sci.*, **54**: 338.

———, 1921: Growth in trees, *Carnegie Inst. Wash. Pub.* 307.

———, 1921: The distinctive agencies in the growth of the cell, *Proc. Soc. Exp. Biol. Med.*, **19**: 103–110.

——— and F. Shreve, 1924: Growth in trees and massive organs of plants, *Carnegie Inst. Wash. Pub.* 350.

———, 1925: Accretion and distention in plant cells, *Am. Nat.*, **59**: 336–345.

Macht, D. I., 1926: Concerning the influence of polarized light on the growth of seedlings, *J. Gen. Physiol.*, **10**: 41–52.

———, 1928: Effect of polarized light on plants, *Am. J. Bot.*, **15**: 621.

———, 1930: A study of twenty-three octyl alcohols on growth of *Lupinus* seedlings, *Am. J. Bot.*, **17**: 572–578.

———, 1931: Effect of some inorganic and organic mercurials on growth of *Lupinus albus*, *Am. J. Bot.*, **18**: 598–602.

——— and J. D. Meyer, 1933: Effect of eighteen normal aliphatic alcohols on growth of *Lupinus albus*, *Am. J. Bot.*, **20**: 145–149.

Mack, W. B., and B. E. Livingston, 1933: Relation of oxygen pressure and temperature to the influence of ethylene on carbon dioxide production and on shoot elongation in very young wheat seedlings, *Bot. Gaz.*, **94**: 625–687.

MacLeod, F. L., M. R. Armstrong, M. E. Heap, and L. A. Tolbert, 1935: The vitamin A content of five varieties of sweet potatoes, *J. Agr. Res.*, **50**: 181–187.

MacMillan, C., 1891: On the growth-periodicity of the potato tuber, *Am. Nat.*, **25**: 462–469.

Marsh, G., 1928: Relation between continuous bioelectric currents and cell respiration. IV, The origin of electric polarity in the onion root, *J. Exp. Zool.*, **51**: 309–325.

Martin, J. F., 1932: The cold resistance of Pacific Coast spring wheats at various stages of growth as determined by artificial refrigeration, *J. Am. Soc. Agron.*, **24**: 871–880.

Martin, J. H., 1927: Comparative studies of winter hardiness in wheat, *J. Agr. Res.*, **35**: 493–537.

———, 1933: The practical application of iarovization, *J. Am. Soc. Agron.*, **26**: 251.

———, 1933: Iarovization in field practice, *U. S. Plant Ind., Div. Cereal Crops Mimeographed Sheets.*

Martini, M. L., H. V. Harlan, and M. N. Pope, 1930: Some growth curves of barley kernels, *Plant Physiol.*, **5**: 263–272.

Mason, S. C., 1925: The minimum temperature for growth of the date palm and the absence of a resting period, *J. Agr. Res.*, **31**: 401–415.

———, 1925: Partial thermostasy of the growth center of the date palm, *J. Agr. Res.*, **31**: 415–455.

———, 1925: The inhibitive effect of direct sunlight on the growth of the date palm, *J. Agr. Res.*, **31**: 455–469.

Mason, T. G., 1922: Growth and correlation in sea island cotton, *West Indian Bull.*, **19**: 214–238.

———, 1922: Growth and abscission in sea island cotton, *Ann. Bot.*, **36**: 457–484.

Masure, M. P., 1932: Effect of ultra-violet radiation on growth and respiration of pea seeds with notes on statistics, *Bot. Gaz.*, **93**: 21–41.

Matzke, E. B., 1936: The effect of street lights in delaying leaf-fall in certain trees, *Am. J. Bot.*, **23**: 446–452.

Maximow, N. A., 1912: Chemische Schutzmittel der Pflanzen gegen Erfrieren, *Ber. bot. Ges.*, **30**: 52–65, 293–305, 504–516.

———, 1914: Experimentelle und kritische Untersuchungen über das Gefrieren und Erfrieren der Pflanzen, *Jahrb. wiss. Bot.*, **53**: 327–420.

——— and A. Pojarkova, 1925: Über die physiologische Natur der Unterschiede zwischen Sommer- und Wintergetreide, *Jahrb. wiss. Bot.*, **64**: 702–730.

———, 1929: Internal factors of frost and drought resistance in plants, *Protoplasma*, **7**: 259–291.

——— and M. A. Krotkina, 1930: Investigation on the after effect of low temperature on the length of vegetative period. Quoted by Maximow, 1934.

———, 1934: The theoretical significance of vernalization, *Imp. Bur. Gen. Aberystwyth* (Wales) *Herbage Pub. Ser. Bull.* 16.

———, 1935: "Vernalization and phasic development of plants," *Imp. Bur. Plant Genetics, School of Agr.*, Cambridge, Eng., 151 pp.

McCall, A. G., 1923: The influence of acidity itself on plant growth without regard to other factors, *J. Am. Soc. Agron.*, **15**: 290–297.

McClelland, C. K., 1928: Study of suckers in corn, *Miss. Sta. Bull.*, **231**: 11.

McClelland, T. B., 1928: Studies of the photoperiodism of some economic plants, *J. Agr. Res.*, **37**: 603–628.

McCollum, J. P., 1934: Vegetative and reproductive responses associated with fruit development in the cucumber, *N. Y. Sta.* (*Cornell*) *Mem.* 163.

McComb, A. L., 1934: The relation between acorn weight and the development of one year chestnut oak seedlings, *J. Forestry*, **32**: 479–484.

McKay, J. W., and T. H. Goodspeed, 1930: The effects of X-radiation on cotton, *Sci.*, **71**: 644.

McKee, R., 1935: Vernalization experiments with forage crops, *U. S. Dept. Agr. Circ.*, 377.

McKinney, H. H., and W. J. Sando, 1930: The behavior of winter wheat in artificial environments, *Sci.*, **71**: 668–670.

——— and ———, 1933: Russian methods for accelerating sexual reproduction in wheat. Further information regarding "Iarovization," *J. Heredity*, **24**: 165–166.

——— and ———, 1933: Earliness and seasonal growth habit in wheat as influenced by temperature and photoperiodism, *J. Heredity*, **24**: 169–179.

———, *et al.*, 1934: Field experiments with vernalized wheat, *U. S. Dept. Agr. Circ.* 325.

——— and W. J. Sando, 1935: Earliness of sexual reproduction in wheat as influenced by temperature and light in relation to certain phases, *J. Agr. Res.*, **51**: 621–641.

McNair, J. B., 1934: The evolutionary status of plant families in relation to some chemical properties, *Am. J. Bot.*, **21**: 427–452.

McRostie, G. P., 1924: Longevity of alfalfa and clover seeds, *Sci. Agr.*, **4**: 236–238.

Megee, C. R., 1935: A search for factors determining winter hardiness in alfalfa, *J. Am. Soc. Agron.*, **27**: 685–698.

Mehrlich, F. P., 1931: Factors affecting growth from the foliar meristems of *Bryophyllum calycinum*, *Bot. Gaz.*, **92**: 113–141.

Meier, F. E., 1934: Lethal response of the alga *Chlorella vulgaris* to ultra-violet rays, *Smithsonian Misc. Coll.*, **92**: No. 3.

———, 1936: Growth of a green alga in isolated wave length regions, *Smithsonian Misc. Coll.* **94**: No. 17.

———, 1936: Lethal effect of short wave lengths of the ultraviolet on the alga *Chlorella vulgaris*, *Smithsonian Misc. Coll.* **95**: No. 2.

Meyer, B. S., 1932: Further studies on cold resistance in evergreens with special reference to the possible rôle of bound water, *Bot. Gaz.*, **94**: 297–321.

Meyer, W. C., 1930: Dormancy and growth studies of the American lotus, *Nelumbo lutea*, *Plant Physiol.*, **5**: 225–234.

Michener, H. D., 1935: Effects of ethylene on plant growth hormone, *Sci.*, **82**: 551–552.

Miller, E. C., 1925: Unpub. data, *Kans. Sta.*

Miller, E. V., and C. Brooks, 1932: Effect of carbon dioxide content of storage atmosphere on carbohydrate transformation in certain fruits and vegetables, *J. Agr. Res.*, **45**: 449–460.

Miller, L. P., 1933: Effect of sulphur compounds in breaking the dormancy of potato tubers and in inducing changes in the enzyme activities of the treated tubers, *Contr. Boyce Thompson Inst.*, **5**: 29–82.

Mitchell, H. L., 1934: Pot culture tests of forest soil fertility, *Black Rock Forest Bull.* 5.

Molisch, H., 1897: "Über das Gefrieren und Erfrieren der Pflanzen," Jena.

Montgomery, E. G., 1909: The economic value of tillers, *Neb. Sta. Bull.*, **112**: 31–34.

Moore, C. N., and C. P. Haskins, 1932: Note on premature flowering in grapefruit from X-rayed seeds, *Sci.*, **76**: 167–168.

——— and ———, 1933: Physiological variations in certain crop plants following seed exposure to high voltage X-rays, *Bot. Gaz.*, **94**: 801–807.

MORINAGA, T., 1926: Effect of alternating temperatures upon the germination of seed, *Am. J. Bot.*, **13**: 141–159.

———, 1926: The favorable effect of reduced oxygen supply upon the germination of certain seeds, *Am. J. Bot.*, **13**: 159–166.

MORROW, K. S., R. B. DUSTMAN, and H. O. HENDERSON, 1931: Changes in the chemical composition of mangels and rutabagas during storage, *J. Agr. Res.*, **43**: 919–930.

MÜLLER-THURGAU, H., 1880, 1886: Über das Gefrieren und Erfrieren der Pflanzen, *Landw. Jahrb.*, **9**: 133–189 (1880); **15**: 453–610 (1886).

MURNEEK, A. E., 1925: Correlation and cyclic growth in plants, *Bot. Gaz.*, **79**: 329–333.

———, 1925: The effect of fruit on vegetative growth in plants, *Proc. Am. Soc. Hort. Sci.*, **21**: 274–276.

———, 1926: Is fruiting of the apple an exhaustive process? *Proc. Am. Soc. Hort. Sci.*, **22**: 196–200.

———, 1926: Physiology of reproduction in horticultural plants. I, Reproduction and metabolic efficiency in the tomato, *Mo. Sta. Res. Bull.* 90.

———, 1926: Effects of correlation between vegetative and reproductive functions in the tomato (*Lycopersicum esculentum*), *Plant Physiol.*, **1**: 3–56.

———, 1932: Growth and development as influenced by fruit and seed formation, *Plant Physiol.*, **7**: 79–90.

———, 1933: The nature of shedding of immature apples, *Mo. Res. Bull.* 201.

MUSSO, J. O., 1933: A new method of using electricity in growing plants, *Fortschr. Landw.*, **8**: 272–274; *Chem. Abst.*, **27**: 5776.

NAVEZ, A. E., and T. W. ROBINSON, 1932: Geotropic curvature of *Avena* coleoptiles, *J. Gen. Physiol.*, **16**: 133–145.

———, 1933: Growth promoting substance and illumination, *Proc. Nat. Acad. Sci.*, **19**: 636–637.

———, 1933: "Growth promoting substance" and elongation of roots, *J. Gen. Physiol.*, **16**: 733–739.

NEBEL, B. R., 1930: Xenia and metaxenia in apples, *N. Y. Sta. (Geneva) Tech. Bull.* 170.

———, 1934: Metaxenia and xenia in apples. IV, *Gartenbauwiss.*, **9**: 45–64; *Chem. Abst.*, **29**: 7389.

NELJUBOW, D., 1901: Ueber die horizontale Nutation der Stengel von *Pisum sativum* und einigen anderen Pflanzen, *Beih. bot. Cent.*, **10**: 128–138.

———, 1911: Geotropismus in der Laboratoriumsluft, *Ber. bot. Ges.*, **29**: 97–112.

NELSON, R. C., and R. B. HARVEY, 1935: The presence in self-blanching celery of unsaturated compounds with physiological action similar to ethylene, *Sci.*, **82**: 133–134.

NEWCOMBE, F. C., 1923: Optimum temperatures for growth of some grass coleoptiles, *Papers, Mich. Acad. Sci., Arts, and Letters*, **3**: 203–210.

NEWTON, R., 1924: Colloidal properties of winter wheat plants in relation to frost resistance, *J. Agr. Sci.*, **14**: 178–191.

———, 1924: The nature and practical measurement of frost resistance in winter wheat, *Univ. Alberta Col. Agr. Res. Bull.* 1.

——— and W. R. BROWN, 1931: Frost precipitation of proteins of plant juice, *Can. J. Res.*, **5**: 87–110.

———, ———, and J. A. ANDERSON, 1931: Chemical changes in nitrogen fractions of plant juice on exposure to frost, *Can. J. Res.*, **5**: 327–332.

——— and A. G. MCCALLA, 1934: Effect of frost on wheat at progressive stages of maturity. I, Physical characteristics of the kernels, *Can. J. Res.* **10**: 414–429.

NICHOLS, G. E., 1934: The influence of exposure to winter temperatures upon seed germination in various native American plants, *Ecology*, **15**: 364–373.

NIELSEN, N., 1930: Untersuchungen über einen neuen wachstumsregulierenden Stoff: Rhizopin, *Jahrb. wiss. Bot.*, **73**: 125–191.

———, 1931: Über Wuchsstoffe der Hefe, *Biochem. Z.*, **237**: 244–246.

——— and V. HARTELIUS, 1932: The separation of growth promoting substances, *C. R. trav. lab. Carlsberg*, **19**: 1–17; *Chem. Abst.*, **26**: 4358.

NIGHTINGALE, G. T., 1922: Light in relation to the growth and chemical composition of some horticultural plants, *Proc. Am. Soc. Hort. Sci.*, **19**: 18–29.

——— and J. W. MITCHELL, 1934: Effects of humidity on metabolism in tomato and apple, *Plant Physiol.*, **9**: 217–236.

———, 1935: Effect of temperature on growth, anatomy, and metabolism of apple and peach roots, *Bot. Gaz.*, **96**: 581–639.

NIXON, R. W., 1928: The direct effect of pollen on the fruit of the date palm, *J. Agr. Res.*, **36**: 97–128.

———, 1928: Immediate influence of pollen in determining the size and time of ripening of the fruit of the date palm, *J. Heredity*, **19**: 240–254.

———, 1934: Metaxenia in dates, *Proc. Am. Soc. Hort. Sci.*, **32**: 221–226.

———, 1935: Metaxenia and interspecific pollination in *Phoenix*, *Proc. Am. Soc. Hort. Sci.*, **33**: 21–26.

NOGÜCHI, Y., 1934: Metaxenia in the Japanese persimmon. Shape and sweetness, *Japan. J. Bot.*, **7**: 61–71.

———, 1935: Modification of leaf structure by X-rays, *Plant Physiol.*, **10**: 753–762.

NOVIKOV, V. A., 1928: Cold resistance of plants. II, *J. Exp. Landw. Südosten Eur.-Russlands*, **6**: 71–100.

OAKLEY, R. A., and H. L. WESTOVER, 1921: Effect of length of day on seedlings of alfalfa varieties and the possibility of utilizing this as a practical means of identification, *J. Agr. Res.*, **21**: 599–607.

OHGA, I., 1923: Longevity of seed of *Nelumbo nucifera* (Japanese), *Bot. Mag. Tokyo*, **27**: 87–95; *Bot. Abst.*, **14**: 899.

OSMASTON, B. B., 1918: Rate of growth of bamboos, *Indian Forester*, **44**: 52–57; *Exp. Sta. Rec.*, **44**: 518–519.

PAAL, A., 1914: Über phototropische Reizleitungen, *Ber. bot. Ges.*, **32**: 499–502.

———, 1918: Über phototropische Reizleitung, *Jahrb. wiss. Bot.*, **58**: 406–458.

PACK, D. A., 1921: After-ripening and germination of *Juniperus communis*, *Bot. Gaz.*, **71**: 32–60.

———, 1921: Chemistry of after ripening, germination and seedling development of Juniper seeds, *Bot. Gaz.*, **72**: 139–151.

PALLADIN, W., 1923: "Plant Physiology," English translation, P. Blakiston's Son & Company, Philadelphia.

PAN, C. L., 1936: Length of exposure to light in relation to plant growth in rice, *J. Am. Soc. Agron.*, **28**: 58–63.

PANCHAUD, J., 1934: Action du milieu extérieur sur le métabolisme végétal, *Rev. Gén. Bot.*, **46**: 586–603; *Bot. Gaz.*, **96**: 579.

PATON, R. R., 1929: The relation of size of seedling trees to their vigor, *Ohio Sta. Bimon. Bull.*, **141**: 191–194.

PEARL, R., A. A. WINSOR, and J. R. MINER, 1928: The growth of seedlings of the canteloup *Cucumis melo* in the absence of exogenous food and light, *Proc. Nat. Acad. Sci.*, **14**: 1–4.

———, ———, and F. B. WHITE, 1928: The form of the growth curve of the canteloup (*Cucumis melo*) under field conditions, *Proc. Nat. Acad. Sci.*, **14**: 895–901.

———, T. I. Edwards, and J. R. Miner, 1934: The growth of *Cucumis melo* seedlings at different temperatures, *J. Gen. Physiol.*, **17**: 687–700.

———, ———, A. A. Winsor, and C. P. Winsor, 1934: Aeration and growth of canteloup seedlings (*Cucumis melo*), *Am. J. Bot.*, **21**: 242–250.

Pearsall, W. H., 1923: Studies in growth. IV, Correlations in development, *Ann. Bot.*, **37**: 261–275.

———, 1925: Hydrogen-ion concentration and growth, *Sci. Prog.*, **20**: 58–67.

——— and A. M. Hanby, 1926: Growth studies. V, Factors affecting the development and form of leaves, *Ann. Bot.*, **40**: 85–103.

———, 1927: Growth studies. VI, On the relative sizes of growing plant organs, *Ann. Bot.*, **41**: 549–557.

Peltier, G. L., and H. M. Tysdal, 1931: Hardiness studies with 2-year-old alfalfa plants, *J. Agr. Res.*, **43**: 931–957.

——— and ———, 1932: A method for the determination of comparative hardiness in seedling alfalfas by controlled hardening and artificial freezing, *J. Agr. Res.*, **44**: 427–444.

——— and T. A. Kiesselbach, 1934: Inducing early fruiting in winter wheat by seed activation, *J. Am. Soc. Agron.*, **26**: 503–506.

——— and ———, 1934: The comparative cold resistance of spring small grains, *J. Am. Soc. Agron.*, **26**: 681–687.

Penrose, L. S., 1925: A note on the relation of rate of growth to structure in plants, *New Phytol.*, **24**: 294–299.

Perry, J. T., 1932: A possible hormone-secreting region in the grass coleoptile, *Sci.*, **76**: 215–216.

Pfeiffer, N. E., 1926: Microchemical and morphological studies of effect of light on plants, *Bot. Gaz.*, **81**: 173–195.

———, 1928: Anatomical study of plants grown under glasses transmitting light of various ranges of light, *Bot. Gaz.*, **85**: 427–436.

Pieters, A. J., 1925: Difference in internode lengths between, and effect of variation of light duration upon, seedlings of annual and biennial white sweet clover, *J. Agr. Res.*, **31**: 585–596.

Plitt, T. M., 1932: Some photoperiodic and temperature responses of the radish, *Plant Physiol.*, **7**: 337–339.

———, 1935: Temperature as a predetermining factor in the development of *Avena sativa*, *Plant Physiol.*, **10**: 269–289.

Poesch, G. H., 1931: Studies on photoperiodism of the *Chrysanthemum*, *Proc. Am. Soc. Hort. Sci.*, **28**: 389–392.

Pope, M. N., 1932: The growth curve in barley, *J. Agr. Res.*, **44**: 323–341.

Popp, H. W., 1926: A physiological study of the effect of light of various ranges of wave length on the growth of plants, *Am. J. Bot.*, **13**: 706–737.

———, 1926: Effect of light intensity on growth of soybeans and its relation to the auto-catalyst theory of growth, *Bot. Gaz.*, **82**: 306–320.

——— and F. Brown, 1933: A review of recent work on the effect of ultra-violet radiation upon seed plants, *Bull. Torrey Bot. Club.*, **60**: 161–210.

Porterfield, W. M., 1928: A study of the grand period of growth in bamboo, *Bull. Torrey Bot. Club*, **55**: 327–406.

———, 1930: Daily periodicity of growth in the bamboo, *Phyllostachys nigra*, with special reference to atmospheric conditions of temperature and moisture, *Bull. Torrey Bot. Club*, **57**: 533–557.

———, 1935: The relation of shoot roots to shoot elongation in the bamboo *Phyllostachys nigra*, *Am. J. Bot.*, **22**: 878–888.

Post, K., 1934: Production of early blooms of Chrysanthemums by the use of black cloth to reduce the length of day, *N. Y. Sta.* (*Cornell*) *Bull.* 594.

Powers, W. L., 1927: The effect of hydrogen-ion concentration on the growth of certain plants, *Soil Sci.*, **24**: 1–4.

Prankerd, T. L., 1936: Studies in the geotropism of *Pteridophyta*. VI, On rhythm in gravi-perception and reaction to gravity, *Proc. Roy. Soc.*, B, **120**: 126–141.

Prantl, K., 1873: Ueber den Einfluss des Lichts auf das Wachsthum der Blätter, *Arbeiten Bot. Inst. Würzburg*, **1**: 371–384.

Prescott, J. A., 1921: Some observations on the growth of maize in Egypt, *Sultanic Agr. Soc. Bull.* 7. Quoted by Hanna, 1925.

Priestley, J. H., and A. F. C. H. Evershed, 1922: Growth studies. I, A quantitative study of the growth of roots, *Ann. Bot.*, **36**: 225–237.

——— and W. H. Pearsall, 1922: Growth studies. II, An interpretation of some growth curves, *Ann. Bot.*, **36**: 239–249.

——— and ———, 1922: Growth studies. III, A volumeter method of measuring the growth of roots, *Ann. Bot.*, **36**: 485–488.

———, 1922: The toxic action of traces of coal gas upon plants, *Ann. App. Biol.*, **9**: 146–155.

———, 1925: Light and growth. I, The effect of brief light exposure upon etiolated plants, *New Phytol.*, **24**: 271–283.

———, 1926: Light and growth. II, On the anatomy of etiolated plants, *New Phytol.*, **25**: 145–170.

———, 1928: The meristemic tissues of the plant, *Biol. Rev.*, **3**: 1–20.

———, 1929: Cell growth and cell division in the shoot of the flowering plant, *New Phytol.*, **28**: 54–81.

———, 1930: Studies in the physiology of cambial activity. I, Contrasted types of cambial activity. II, The concept of sliding growth, *New Phytol.*, **29**: 56–73, 96–140.

———, 1930: Studies in the physiology of cambial activity. III, The seasonal activity of the cambium, *New Phytol.*, **29**: 316–354.

———, L. I. Scott, and M. E. Malins, 1933: A new method of studying cambial activity, *Proc. Leeds Phil. Soc.*, 2, **VIII**: 365–374.

Purvis, O. N., 1934: An analysis of the influence of temperature during germination on the subsequent development of certain winter cereals and its relation to the effect of length of day, *Ann. Bot.*, **48**: 919–956.

Raber, O. L., 1928: "Principles of Plant Physiology," The Macmillan Company, New York.

Raleigh, G. J., 1930: Chemical conditions in maturation, dormancy, and germination of seeds of *Gymnocladus dioica*, *Bot. Gaz.*, **89**: 273–294.

Ramaley, F., 1931: Growth of plants under continuous light, *Sci.*, **73**: 566–567.

———, 1931: Some Caryophyllaceous plants influenced in growth and structure by artificial illumination supplemental to daylight, *Bot. Gaz.*, **92**: 311–320.

———, 1934: Influence of supplemental light on blooming, *Bot. Gaz.*, **96**: 165–174.

Rea, H. E., and R. E. Karper, 1932: Propagating sorghum by cuttings, *Am. J. Bot.*, **19**: 464–476.

Redington, G., 1929: The effect of the duration of light upon the growth and development of the plant, *Biol. Rev. and Biol. Proc. Cambridge Phil. Soc.*, **4**: 180–208.

Reed, E., 1923: Hypothesis of formative stuffs as applied to *Bryophyllum calycinum*, *Bot. Gaz.*, **75**: 113–142.

Reed, H. S., 1919: Growth and variability in *Helianthus*, *Am. J. Bot.*, **6**: 252–271.

——— and F. F. Halma, 1919: On the existence of a growth inhibiting substance in the Chinese lemon, *Univ. Calif. Pub.*, **4**: 99–112.

——— and ———, 1919: The evidence for a growth-inhibiting substance in the pear tree, *Plant World*, **22**: 239–247.

——— and R. H. Holland, 1919: The growth rate of an annual plant *Helianthus*, *Proc. Nat. Acad. Sci.*, **5**: 135–144.

———, 1920: Slow and rapid growth, *Am. J. Bot.*, **7**: 327–333.

———, 1920: The dynamics of a fluctuating growth rate, *Proc. Nat. Acad. Sci.*, **6**: 397–410.

———, 1920: The nature of the growth rate, *J. Gen. Physiol.*, **2**: 545–561.

———, 1921: A method for obtaining constants for formulas of organic growth, *Proc. Nat. Acad. Sci.*, **7**: 311–316.

———, 1921: Correlation and growth in the branches of young pear trees, *J. Agr. Res.*, **21**: 849–875.

———, 1923: A note on the statistics of cyclic growth, *Proc. Nat. Acad. Sci.*, **9**: 65–67.

——— and A. R. C. Haas, 1924: The effect of hydroxyl-ion concentration on the growth of walnut roots, *Am. J. Bot.*, **11**: 78–85.

———, 1927: Growth and differentiation in plants, *Quart. Rev. Biol.*, **2**: 79–101.

———, 1929: Quantitative aspects of the problem of growth and differentiation, *Proc. Int. Cong. Plant Sci.*, **2**: 1095–1106.

———, 1932: Criteria of plant growth and their reliability, *Philippine J. Agr.*, **3**: 155–161.

Reeves, R. G., and C. C. Valle, 1932: Anatomy and microchemistry of the cotton seed, *Bot. Gaz.*, **93**: 259–277.

Regeimbal, L. O., and R. B. Harvey, 1927: The effect of ethylene on the enzymes of pineapple, *J. Am. Chem. Soc.*, **49**: 1117–1118.

Rehm, W. S., 1936: Maintained electrical polarities in region of the axillary buds in *Phaseolus multiflorus*, *Plant Physiol.*, **11**: 365–382.

Reid, M. E., 1929: Relation of composition of seed and the effects of light on growth of seedlings, *Am. J. Bot.*, **16**: 747–769.

———, 1930: Growth and nitrogen metabolism of squash seedlings. I, Variations at different seasons of the year, *Am. J. Bot.*, **17**: 272–289.

Reinhard, M. C., and K. L. Tucker, 1928: Hourly variations in effect of equal Roentgen-ray doses on growth of *Vicia faba* seedlings, *Am. J. Roentgenol.*, **19**: 71–73; *Biol. Abst.*, **4**: 356.

Reynolds, E. S., 1935: The reactions of plants to ultra-violet, *Ann. Mo. Bot. Gard.*, **22**: 759–769.

Richardson, J. E., and H. L. Mayfield, 1932: Availability of vitamins in plant tissues, *Sci.*, **76**: 498–500.

Richey, F. D., and G. E. Sprague, 1932: Some factors affecting the reversal of sex expression in the tassel of maize, *Am. Nat.*, **66**: 433–443.

Rippel, A., 1922: Die gesetzmässige Erforschung von Reaktionsgleichgewicht und Reactionsgeschwindigkeit bei den höheren Pflanzen, *J. Landw.*, **70**: 9–44; *Bot. Abst.*, **12**: 472–473.

Robbins, W. J., 1922: Effect of autolyzed yeast and peptone on growth of excised root tips in the dark, *Bot. Gaz.*, **74**: 59–79.

——— and W. E. Maneval, 1923: Further experiments on the growth of excised root tips under sterile conditions, *Bot. Gaz.*, **76**: 274–287.

———, 1926: Relation of hydrogen-ion concentration to the growth of plants, *Mo. Sta. Bull.*, **244**: 30.

———, S. Brody, A. G. Hogan, C. M. Jackson, and C. W. Greene, 1928: "Growth," Yale University Press, 182 pp.

——— and K. F. Petsch, 1932: Moisture content and high temperature in relation to the germination of corn and wheat grains, *Bot. Gaz.*, **93**: 85–92.

———, M. Bartley, and V. B. White, 1936: Growth of fragments of excised root tips, *Bot. Gaz.*, **97**: 554–579.

Robertson, D. W., and A. M. Lute, 1933: Germination of the seed of farm crops in Colorado after storage for various periods of years, *J. Agr. Res.*, **46**: 455–462.

Robertson, T. B., 1907, 1908: On the normal rate of growth of an individual and its biochemical significance, *Arch. Entwicklungsmech. Organismen*, **25**: 580–614.

———, 1923: "The Chemical Basis of Growth and Senescence," J. B. Lippincott Company, Philadelphia, 389 pp.

Robinson, B. B., 1933: Some physiological factors influencing the production of flax fiber cells, *J. Am. Soc. Agron.*, **25**: 312–327.

Rodriguez, A. G., 1932: Influence of smoke and ethylene on the fruiting of the pineapple, *J. Puerto Rico Dept. Agr.*, **16**: 5–18.

Rosa, J. T., 1921: Investigation on the hardening process in vegetable plants, *Mo. Sta. Res. Bull.* 48.

———, 1925: Ripening of tomatoes with ethylene, *Proc. Am. Soc. Hort. Sci.*, **22**: 315–322.

———, 1928: Relation of tuber maturity and of storage factors to potato dormancy, *Hilgardia*, **3**: 99–124.

———, 1928: Effects of chemical treatments on dormant potato tubers, *Hilgardia*, **3**: 125–142.

———, 1928: Changes in composition during ripening and storage of melons, *Hilgardia*, **3**: 421–443.

Rose, D. H., 1915: A study of delayed germination in economic seeds, *Bot. Gaz.*, **59**: 425–444.

Rose, R. C., 1919: After-ripening and germination of seeds of *Tilia*, *Sambucus*, and *Rubus*, *Bot. Gaz.*, **67**: 281–309.

Rotunno, N. A., 1924: Effect of size of seed on plant production with special reference to radish, *Bot. Gaz.*, **78**: 397–413.

Rowlands, M. J., and B. Wilkinson, 1930: The vitamin B content of grass seeds in relationship to manures, *Biochem. J.*, **24**: 199–204.

Rudolfs, W., 1923: Influence of temperature and initial weight of seeds upon the growth-rate of *Phaseolus vulgaris* seedlings, *J. Agr. Res.*, **26**: 537–541.

Sachs, J., 1874: Ueber den Einfluss der Lufttemperatur und des Tageslichts auf die stündlichen und täglichen Aenderungen des Längenwachsthums der Internodien, *Arbeit. Bot. Inst. Würzburg*, **1**: 99–192.

Saeger, A., 1933: Gas injury to pure cultures of *Spirodela*, *Plant Physiol.*, **8**: 479–480.

Salmon, S. C., 1917: Why cereals winterkill, *J. Am. Soc. Agron.*, **9**: 353–380.

——— and F. L. Fleming, 1918: Relation of the density of cell sap to winter hardiness in small grains, *J. Agr. Res.*, **13**: 497–506.

———, 1933: Resistance of varieties of winter wheat and rye to low temperature in relation to winter hardiness and adaptation, *Kans. Sta. Tech. Bull.* 35.

Sampson, A. W., and E. C. McCarty, 1930: The carbohydrate metabolism of *Stipa pulchra*, *Hilgardia*, **5**: 61–100.

Savage, D. A., and L. A. Jacobson, 1935: The killing effect of heat and drought on buffalo grass and blue grama grass at Hays, Kansas, *J. Am. Soc. Agron.*, **27**: 566–583.

Sayre, J. D., 1932: Methods of determining bound water in plant tissue, *J. Agr. Res.*, **44**: 669–688.

SCHAFFNER, J. H., 1923: The influence of relative length of daylight on the reversal of sex in hemp, *Ecology*, **4**: 323–334.

———, 1927: Control of sex reversal in the tassel of Indian Corn, *Bot. Gaz.*, **84**: 440–449.

SCHAFFNIT, E., and M. LÜDTKE, 1932: The effect of cold on plant cells. II, Metabolism of cultivated plants at different temperatures and with changing nutrition, *Phytopath. Z.*, **4**: 329: *Chem. Abst.*, **27**: 2473.

——— and ———, 1932: Beiträge zur Kenntnis von Kältewirkungen auf die pflanzliche Zelle. II, Über den Stoffwechsel landwirtschaftlicher Kulturpflanzen bei verschiedenen Temperaturen und wechselnder Ernährung, *Phytopath. Z.*, **4**: 329–386.

——— and A. F. WILHELM, 1933: Beiträge zur Kenntnis von Kältewirkungen auf die pflanzliche Zelle. III, Kühlversuche mit verschieden ernährten Pflanzen und Untersuchungen über deren Stoffwechselphysiologie, *Phytopath. Z.*, **5**: 505–566.

SCHANZ, F., 1919: Wirkungen des Lichts verschiedener Wellenlänge auf die Pflanzen, *Ber. bot. Ges.*, **37**: 430–442.

SCHEIBE, A., 1932: Die Keimung des Hafers in ihrer Abhängigkeit von der physiologischen Konstitution des Saatgutes, *Pflanzenbau*, **8**: 579–649.

SCHLEY, E. O., 1913: Chemical and physical changes in geotropic stimulation and response, *Bot. Gaz.*, **56**: 480–490.

SCHMIDT, D., 1924: The effect of the weight of the seed on the growth of the plant, *N. J. Sta. Bull.* 404.

SCHRÖDER, D., 1909: Ueber den Verlauf des Welkens und die Lebenszähigkeit der Laubblätter, *diss., Göttingen*, 110 pp.

SEIFRIZ, W., 1936: Breakdown of fruit and vegetable tissue due to an electric current, *Plant Physiol.*, **11**: 195–200.

SELLSCHOP, J. P. F., and S. C. SALMON, 1928: The influence of chilling above the freezing point on certain crop plants, *J. Agr. Res.*, **37**: 315–339.

SEMMENS, E. S., 1930: Hydrolysis in the living plant by polarized light, *Bot. Gaz.*, **90**: 412–427.

SEUBERT, E., 1925: Über Wachstumsregulatoren in der Koleoptile von *Avena*, *Z. Bot.*, **17**: 49–88.

SHEAR, G. M., 1931: Studies on inanition in *Arachis* and *Phaseolus*, *Plant Physiol.*, **6**: 277–294.

SHEARD, C., and A. F. JOHNSON, 1930: The effects of infra-red, visible and ultra-violet irradiation on changes in electrical potentials and currents in plants, *Sci.*, **71**: 246–248.

SHIBUSAWA, M., and K. SHIBATA, 1927: The effect of electric discharges on the rate of growth of plants, *J. Inst. Elec. Eng. of Japan*, **473**: 1–42; *Biol. Abst.*, **4**: 2257–2258.

SHIPPY, W. B., 1930: Influence of environment on the callusing of apple cuttings and grafts, *Am. J. Bot.*, **17**: 29–327.

SHIRLEY, H. L., 1928: The influence of light intensity and quality upon the growth of plants, *Am. J. Bot.*, **15**: 621–622.

———, 1929: The influence of light intensity and light quality upon the growth of plants, *Am. J. Bot.*, **16**: 354–390.

———, 1931: The influence of light and temperature upon the utilization by young seedlings of organic reserves in the seed, *Am. J. Bot.*, **18**: 717–728.

———, 1932: Light intensity in relation to plant growth in a virgin Norway pine forest, *J. Agr. Res.*, **44**: 227–244.

SHUCK, A. L., 1934: Some factors influencing the germination of lettuce seed in laboratory practice, *N. Y. Sta. (Geneva) Tech. Bull.* 222.

———, 1935: Light as a factor influencing the dormancy of lettuce seeds, *Plant Physiol.*, **10**: 193–196.

———, 1935: A growth inhibiting substance in lettuce seeds, *Sci.*, **81**: 236.

SHULL, C. A., and H. B. LEMON, 1931: Penetration of seed coats by ultra-violet radiation, *Bot. Gaz.*, **92**: 420–429.

——— and J. W. MITCHELL, 1933: Stimulative effects of X-rays on plant growth, *Plant Physiol.*, **8**: 287–296.

SIDERIS, C. P., 1926: Behavior of pineapple plants in water cultures of varying hydrogen ion concentrations, *Exp. Sta. Assoc. Hawaiian Pineapple Canners, Univ. Hawaii Bull.* 4.

———, 1926: Review of physiological and pathological studies on the pineapple plant, *Hawaiian Pineapple Canners' Sta. Bull.* 8.

SIERP, H., and A. SEYBOLD, 1926: Untersuchungen über die Lichtempfindlichkeit der Spitze und des Stumpfes in der Koleoptile von *Avena sativa*, *Jahrb. wiss. Bot.*, **65**: 592–610.

SIFTON, H. B., 1920: Longevity of the seeds of cereals, clovers and timothy, *Am. J. Bot.*, **7**: 243–251.

SINGH, B. N., G. P. KAPOOR, and R. S. CHOUDHRI, 1936: Growth studies in relation of ultra-violet radiation, *Bot. Gaz,*, **97**: 649–665.

SINGH, T. C. N., 1932: A note on the response of grain (*Cicer arietinum*) seedlings to electricity, *New Phytol.*, **31**: 64–65.

SKOOG, F., 1934: The effect of X-rays on growth substance and plant growth, *Sci.*, **79**: 256.

——— and K. V. THIMANN, 1934: Further experiments on the inhibition of the development of lateral buds by growth hormone, *Proc. Nat. Acad. Sci.*, **20**: 480–485.

SMALL, J., 1929: Hydrogen-ion concentration in plant cell and tissues, *Protoplasma, Monographien*, II, 421 pp.

SMITH, A. J. M., and R. GANE, 1933: Influence of a gaseous product from apples on the germination of seeds, *Dept. Sci. and Ind. Res., Food Invest. Bd. Rept. (Gr. Brit.)*, **1932**: 156–158.

SMITH, O., and H. L. COCHRAN, 1935: Effect of temperature on pollen germination and tube growth in the tomato, *N. Y. Sta. (Cornell) Mem.* 175.

SNOW, R., 1925: The correlative inhibition of the growth of axillary buds, *Ann. Bot.*, **39**: 841–859.

———, 1929: The transmission of inhibition through dead stretches of stem, *Ann. Bot.*, **43**: 261–267.

———, 1929: The young leaf as the inhibiting organ, *New Phytol.*, **28**: 345–358.

———, 1931: Experiments on growth and inhibition. II, New phenomena of inhibition, *Proc. Roy. Soc.*, B, **108**: 305–317.

———, 1932: Experiments of growth and inhibition. III, Inhibition and growth promotion, *Proc. Roy. Soc.*, B, **111**: 86–105.

———, 1932: Growth-regulators in plants, *New Phytol.*, **31**: 336–354.

———, 1933: The nature of the cambial stimulus, *New Phytol.*, **32**: 288–296.

———, 1935: Activation of cambial growth by pure hormones, *New Phytol.*, **34**: 347–360.

SÖDING, H., 1926: Über den Einfluss der jungen Infloreszenz auf das Wachstum ihres Schaftes, *Jahrb. wiss. Bot.*, **65**: 611–635.

———, 1929: Weitere Untersuchungen über die Wuchshormone der Haferkoleoptile, *Jahrb. wiss. Bot.*, **71**: 184–213; *Biol. Abst.*, **5**: 1621.

———, 1931: Wachstum und Wanddehnbarkeit bei Haferkoleoptile, *Jahrb. wiss. Bot.*, **74**: 127–151.

SONAVNE, K. M., 1928: Longevity of crop seeds, *Agr. J. India*, **23**: 271–276.
SPAETH, J. N., 1932: Dormancy in *Tilia* seeds, *Sci.*, **76**: 143–144.
———, 1934: A physiological study of dormancy in *Tilia* seed, *N. Y. Sta.* (*Cornell*) *Mem.* 169.
SPRAGUE, G. F., 1934: Experiments on iarovizing corn, *J. Agr. Res.*, **48**: 1113–1120.
SPRAGUE, H. B., and M. LENZ, 1929: The effect of X-ray on potato tubers for "seeds," *Sci.*, **69**: 606.
STADLER, L. J., 1930: Some genetic effects of X-rays in plants, *J. Heredity*, **21**: 3–19.
STARKEY, E. J., 1932: Photoperiodism and chrysanthemum production, *Sci.*, **76**: 594–595.
STEINBAUER, G. P., 1932: Growth of tree seedlings in relation to light intensity and concentration of nutrient solution, *Plant Physiol.*, **7**: 742–744.
STEINBERG, R. A., 1931: An apparatus for growing plants under controlled environmental conditions, *J. Agr. Res.*, **43**: 1071–1084.
STEVENS, O. A., 1935: Germination studies of aged and injured seeds, *J. Agr. Res.*, **51**: 1093–1106.
STEWART, F. C., 1936: The relative vigor and productivity of potato plants from basal and apical sets cut from tubers in different stages of sprouting, *N. Y. Sta.* (*Geneva*) *Bull.* 658.
STILES, W., 1930: On the cause of cold death of plants, *Protoplasma*, **9**: 459–468.
———, 1935: Growth hormones, *Sci. Prog.*, **30**: 313–317.
STONE, W. E., 1933: Normal growth of potato leaves in greenhouse and field, *J. Agr. Res.*, **46**: 565–578.
STRAUSBAUGH, P. D., 1921: Dormancy and hardiness in the plum, *Bot. Gaz.*, **71**: 337–358.
SUMMERS, F., 1924: Factors governing bud formation: a chapter of plant physiology, *New Phytol.*, **23**: 20–50, 78–102, 113–131.
SUNESON, C. A., and G. L. PELTIER, 1934: Effect of stage of seedling development upon the cold resistance of winter wheats, *J. Am. Soc. Agron.*, **26**: 687–692.
SWINGLE, W. T., 1928: Metaxenia in the date palm, *J. Heredity*, **19**: 257–268.
———, 1931: Metaxenia or the direct effect of the pollen parent on the maternal fruit tissues and its application in date culture, abst., *Am. J. Bot.*, **18**: 894.
TAIT, G. M., and J. E. KNOTT, 1933: Correcting the unproductiveness of acid and alkaline muck soils for the growing of vegetable crops, *N. Y. Sta.* (*Cornell*) *Bull.* 572.
TANG, P., 1931: An experimental study of the germination of wheat seed under water as related to temperature and aeration, *Plant Physiol.*, **6**: 203–248.
TARANOVSKY, W. G., 1923: Über Veränderungen welche in Pflanzen durch das Kastrationverfahren hervorgerufen werden, *Zhur. Opytn. Agron.*, **23**: 127–164.
TETLEY, M., and J. H. PRIESTLEY, 1927: The histology of the coleoptile in relation to its phototropic response, *New Phytol.*, **26**: 171–186.
THATCHER, R. W., 1913: Enzymes of apples and their relation to the ripening process, *J. Agr. Res.*, **5**: 103–116.
THIMANN, K. V., and J. BONNER, 1932: Studies of the growth hormone of plants. II, The entry of growth substance into the plant, *Proc. Nat. Acad. Sci.*, **18**: 692–701.
——— and———, 1933: The mechanism of the action of the growth substance of plants, *Proc. Roy. Soc.*, B, **113**: 126–149.
——— and H. E. DOLK, 1933: Conditions governing the production of the plant growth hormone by *Rhizopus* cultures, *Biol. Zent.*, **53**: 49–66.
——— and F. SKOOG, 1933: Studies on the growth hormone of plants. III, The inhibiting action of the growth substance on bud development, *Proc. Nat. Acad. Sci.*, **19**: 714–716.

———, 1934: Studies on the growth hormone of plants. VI, The distribution of the growth substance in plant tissues, *J. Gen. Physiol.*, **18**: 23–34.

——— and F. Skoog, 1934: On the inhibition of bud development and other functions of growth substance in *Vicia faba*, *Proc. Roy. Soc.*, B, **114**: 317–339.

——— and F. W. Went, 1934: On the chemical nature of the root-forming hormone, *Proc. K. Akad. Wetensch. Amsterdam*, **37**, No. 7.

———, 1935: Growth substances in plants, *Ann. Rev. Biochem.*, **4**: 545–568.

——— and J. B. Koepfli, 1935: Identity of the growth-promoting and root-forming substances of plants, *Nature*, **135**: 101–102.

Thoday, D., 1933: Some physiological aspects of differentiation, *New Phytol.*, **32**: 274–287.

Thompson, H. C., 1926: Results of sweet corn suckering experiments, *N. Y. Sta. (Cornell) Bull.*, **450**: 1–13.

———, H. S. Mills, and P. H. Wessels, 1930: Results of sweet corn suckering experiments on Long Island, *N. Y. Sta. (Cornell) Bull.*, **509**: 3–11.

Thompson, R. C., 1935: Some factors associated with dormancy of lettuce seed, *Proc. Am. Soc. Hort. Sci.*, **33**: 610–616.

Thomson, J. R., 1936: Vernalisation, *Sci. Prog.*, **30**: 644–651.

Thornton, N. C., 1930: The use of carbon dioxide for prolonging the life of cut flowers with especial reference to roses, *Am. J. Bot.*, **17**: 614–626.

———, 1931: The effect of carbon dioxide on fruits and vegetables in storage, *Contr. Boyce Thompson Inst.*, **3**: 219–244.

———, 1933: Carbon dioxide storage. V, Breaking the dormancy of potato tubers, *Contr. Boyce Thompson Inst.*, **5**: 471–481.

———, 1935: Factors influencing germination and development of dormancy in cocklebur seeds, *Contr. Boyce Thompson Inst.*, **7**: 477–496.

———, 1936: Carbon dioxide storage. IX, Germination of lettuce seeds at high temperatures in both light and darkness, *Contr. Boyce Thompson Inst.*, **8**: 25–40.

Timmons, F. L., and S. C. Salmon, 1932: The resistance of certain varieties and regional strains of alfalfa to controlled low temperatures, *J. Am. Soc. Agron.*, **24**: 642–655.

Tincker, M. A. H., 1925: The effect of length of day upon the growth and reproduction of some economic plants, *Ann. Bot.*, **39**: 721–754.

———, 1928: The effect of length of day upon the growth and chemical composition of the tissues of certain economic plants, *Ann. Bot.*, **42**: 101–140.

———, 1929: The effect of length of day on plants, *J. Roy. Hort. Soc.*, **54**: 354–378.

———, 1930: Some experiments with ultra-violet ray glasses, *J. Roy. Hort. Soc.*, **55**: 79–87.

———, 1932: Response of some common garden plants to the daily period of light, *J. Roy. Hort. Soc.*, **57**: 321–325.

———, 1935: Experiments with follicular and other hormones and plant growth, *Ann. App. Biol.*, **22**: 619–629.

———, 1935: Rose seeds: their after-ripening and germination, *J. Roy. Hort. Soc.*, **60**: 399–417.

Tjebbes, K., and J. C. Uphof, 1921: Der Einfluss des elektrischen Lichtes auf das Pflanzenwachstum, *Landw. Jahrb.*, **56**: 315–328; *Bot. Abst.*, **10**: 206.

Tottingham, W. E., R. G. Shands, and E. D. Delwiche, 1931: Test of Chibnall's method of extraction for investigating winter hardiness of plants, *Plant Physiol.*, **6**: 167–176.

Trumpf, C., 1924: Ueber den Einfluss intermittierender Belichtung auf das Etiolement der Pflanzen, *Bot. Arch.*, **5**: 381–410. Quoted by Priestley, 1926.

Tschirch, A., 1921: Besitzt die Pflanze Hormone? *Viertelijahrsschr. Naturforsch. Ges. Zurich*, **66**: 201–211; *Bot. Abst.*, **12**: 465–466.

Tufts, W. P., and C. J. Hansen, 1933: Xenia and metaxenia in Bartlett pear, *Proc. Am. Soc. Hort. Sci.*, **30**: 134–139.

Tukey, H. B., and K. D. Brase, 1933: Influence of the scion and of an intermediate stem piece upon the character and development of roots of young apple trees, *N. Y. Sta. (Geneva) Bull.* 218.

Turner, M. A., 1929: Ginning percentage and lint index of cotton in relation to the number of cotton fibers per seed, *J. Text. Ind.*, **20**: 233–273.

Tysdal, H. M., 1933: Influence of light, temperature and soil moisture on the hardening process in alfalfa, *J. Agr. Res.*, **46**: 483–515.

Umrath, K., 1928: Elektrische Potentiale pflanzlicher Gewebe, *Protoplasma*, **4**: 539–546.

Ursprung, A., and G. Blum, 1924: Eine Methode zur Messung des Wand- und Turgordruckes der Zelle, nebst Anwendungen, *Jahrb. wiss. Bot.*, **63**: 1–110.

Vacha, G. A., and R. B. Harvey, 1927: The use of ethylene, propylene, and similar compounds in breaking the rest period of tubers, bulbs, cuttings, and seeds, *Plant Physiol.*, **2**: 187–192.

Van de Sande-Bakhuyzen, H. L., 1926: Growth and growth formulas in plants., *Sci.*, **64**: 653–654.

———, 1926: Physiological phenomena at the time of flowering, *Proc. Soc. Exp. Biol. and Med.*, **24**: 143–145.

——— and C. L. Alsberg, 1927: The growth curve in annual plants, *Physiol. Rev.*, **7**: 151–187.

———, 1928: Studies upon wheat grown under constant conditions. I, II, *Plant Physiol.*, **3**: 1–31.

———, E. P. Griffing, and C. L. Alsberg, 1937: "Studies on Wheat Growth under Constant Conditions," *Food Res. Inst., Stanford Univ.*, Calif., 400 pp.

Van Overbeek, J., 1935: The growth hormone and the dwarf type of growth in corn, *Proc. Nat. Acad. Sci.*, **21**: 292–299.

———, 1936: Different action of auxin-*a* of hetero-auxin (preliminary note), *Proc. Nat. Acad. Sci.*, **22**: 187–190.

Van Weij, H. G., 1932: Der Mechanismus des Wuchsstofftransporte, *Rec. trav. bot. Néerl.*, **29**: 379–496.

Vassiliev, I. M., and M. G. Vassiliev, 1936: Changes in carbohydrate content of wheat plants during the process of hardening for drought resistance, *Plant Physiol.*, **11**: 115–125.

Vickery, H. B., and G. W. Pucher, 1931: The chemical changes that occur during the curing of tobacco leaves, *Sci.*, **73**: 397–399.

Virtanen, A. I., 1936: Vitamins and plants, *Nature*, **137**: 779–780.

Vogt, E., 1915: Über den Einfluss des Lichts auf das Wachstum der Koleoptile von *Avena sativa*, *Z. Bot.*, **7**: 193–270; *Bot. Gaz.*, **60**: 423.

Von Euler, H., and T. Philipson, 1932: Water soluble growth factors, *Biochem. Z.*, **245**: 418–439; *Chem Abst.*, **26**: 3281.

Vysotskii, K. A., 1928: The electro-ionization of the cotton plant, *Proc. Turkestan Plant Breeding Sta.*, **13**: 109–110; *Biol. Abst.*, **8**: 603.

Waldo, G. F., 1930: Fruit-bud development in strawberry varieties and species, *J. Agr. Res.*, **40**: 393–409.

———, 1930: Fruit-bud formation in everbearing strawberries, *J. Agr. Res.*, **40**: 409–417.

Waldron, L. R., 1931: Frost injury to spring wheat with a consideration of drought resistance, *J. Am. Soc. Agron.*, **23**: 625–638.

———, 1932: Differences of injury by frost to wheat plants grown comparably, *J. Am. Soc. Agron.*, **24**: 494–500.

WALLACE, R. H., 1927: The production of intumescences in transparent apple by ethylene gas as affected by external and internal conditions, *Bull. Torrey Bot. Club*, **54**: 499–542.

———, 1931: Studies on the sensitivity of *Mimosa pudica.* I, The effect of certain animal anesthetics upon sleep movements, *Am. J. Bot.*, **18**: 102–111.

———, 1931: Studies on the sensitivity of *Mimosa pudica.* II, The effect of animal anesthetics and certain other compounds upon seismonic sensitivity, *Am. J. Bot.*, **18**: 215–235.

———, 1931: Studies on the sensitivity of *Mimosa pudica.* III, The effect of temperature, humidity, and certain other factors upon seismonic sensitivity, *Am. J. Bot.*, **18**: 288–307.

WALLER, A. D., 1900: The electrical effects of light upon green leaves, *Proc. Roy. Soc.*, **67**: 129–137.

WALLER, J. C., 1924: On the fluctuations of potential in green leaves under the influence of light, *J. Physiol.*, **58**, *Proc.* XXXVIII–XXXIX.

———, 1925: Plant electricity. I, Photoelectric currents associated with the activity of chlorophyll in plants, *Ann. Bot.*, **39**: 515–538.

———, 1929: Plant electricity. II, Towards an interpretation of the photoelectric currents of leaves, *New Phytol.*, **28**: 291–302.

WALSTER, H. L., 1920: Formative effect of high and low temperatures upon growth of barley: A chemical correlation, *Bot. Gaz.*, **69**: 97–126.

WALTER, H., 1923: Protoplasma- und Membranquellung bei Plasmolyse, *Jahrb. wiss. Bot.*, **62**: 145–213.

———, 1929: Die osmotischen Werte und die Kälteschäden unserer wintergrünen Pflanzen während der Winterperiode 1929, *Ber. bot. Ges.*, **47**: 338–348.

WALTMAN, C. S., 1931: Effect of hydrogen-ion concentration on the growth of strawberries in sand and in soil, *Ky. Sta. Bull.*, **321**: 331–352.

WALTON, C. F., and C. A. FORT, 1931: Mannite and dextran in the jelly of molasses from juice of frozen and deteriorated cane, *Ind. and Eng. Chem.*, **23**: 1295–1297.

WANN, F. B., and E. F. HOPKINS, 1927: Further studies on growth of *Chlorella* as affected by hydrogen-ion concentration, *Bot. Gaz.*, **83**: 194–201.

WANSER, H. M., 1922: Photoperiodism of wheat: a determining factor in acclimatization, *Sci.*, **56**: 313–315.

WEAVER, J. E., and W. J. HIMMEL, 1929: Relation between the development of root systems and shoot under long- and short-day illumination, *Plant Physiol.*, **4**: 435–459.

WEIMER, J. L., 1929: Some factors involved in the winter killing of alfalfa, *J. Agr. Res.*, **39**: 263–283.

WELTON, F. A., 1921: Longevity of seeds, *Ohio Sta. Mo. Bull.*, **6**: 18–25.

WENT, F. A. F. C., 1933: Growth substance (auxin) in plants, *Nature*, **132**: 452.

WENT, F. W., 1926: On growth accelerating substances in the coleoptile of *Avena sativa*, *Proc. K. Akad. Wetensch. Amsterdam Verslag Gewone Vergader. Afd. Natuurk*, **35**: 723–732; *Biol. Abst.*, **7**: 1293.

———, 1928: Wuchsstoff und Wachstum, *Rec. trav. bot. Néerl.*, **25**: 1–116.

———, 1929: On a substance causing root formation, *Proc. K. Akad. Wetensch. Amsterdam*, **32**: 35–39; *Exp. Sta. Rec.*, **67**: 661.

———, 1934: A test method for Rhizocaline, the root forming substance, *Proc. K. Akad. Wetensch. Amsterdam*, **37**, No. 7.

———, 1934: On the pea test method for auxin, the plant growth hormone, *Proc. K. Akad. Wetensch, Amsterdam*, **37**, 547–555; *Exp. Sta. Rec.*, **72**: 457.

———, 1935: Auxin, the plant growth-hormone, *Bot. Rev.*, **1**: 162–182.

——— and K. V. Thimann, 1937: "Phytohormones," The Macmillan Company, New York and London, XI + 294 pp.

Wessels, P. H., 1932: Soil-acidity studies with potatoes, cauliflower and other vegetables on Long Island, *N. Y. Sta.* (*Cornell*) *Bull.* 536.

West, C., G. E. Briggs, and F. Kidd, 1920: Methods and significant relations in a quantitative analysis of plant growth, *New Phytol.*, **19**: 200–207.

———, 1924: Effects of freezing and low temperatures upon plants, *Sci. Prog.*, **18**: 386–392.

White, P. R., 1931: Plant tissue cultures. The history and present status of the problem, *Arch. Exp. Zellforsch.*, **10**: 501–518.

———, 1932: Influence of some environmental conditions on the growth of excised root tips of wheat seedlings in liquid media, *Plant Physiol.*, **7**: 613–628.

———, 1933: Plant tissue cultures. Results of preliminary experiments on the culturing of isolated stem-tips of *Stellaria media*, *Protoplasma*, **19**: 97–116.

———, 1933: The SH radical and some other sources of sulfur as affecting growth of isolated root tips of wheat seedlings, *Protoplasma*, **19**: 132–135.

———, 1933: Liquid media as substrata for the culturing of isolated root tips, *Biol. Zent.*, **53**: 359–364.

———, 1933: Concentrations of inorganic ions as related to growth of excised root tips of wheat seedlings, *Plant Physiol.*, **8**: 489–508.

———, 1934: Potentially unlimited growth of excised tomato tips in a liquid medium, *Plant Physiol.*, **9**: 585–600.

Whyte, R. O., and P. S. Hudson, 1933: Vernalization, *Brit. Imp. Bur. Plant Genetics*, **9**: 1–27.

Wickwire, G. C., L. Kneer, and W. E. Burge, 1931: The relation of quantity of stimulating effect of insulin, glycerol, urea and dipotassium phosphate on sugar utilization by *Spirogyra*, abst., *Am. J. Bot.*, **18**: 894–895.

Wigoder, S. B., and R. Patten, 1929: Some effects of Röntgen rays on seedlings, *Proc. Roy. Irish Acad.*, **39B**: 146–155; *Biol. Abst.*, **6**: 704.

Williams, C. B., 1912: Corn sucker investigations, *N. C. Sta. Rept.*, **35**: 16.

——— and W. C. Etheridge, 1912: Economic value of corn suckers, *N. C. Sta. Rept.*, **35**: 37–40.

Williams, M., 1923: Observations on the action of X-rays on plant cells, *Ann. Bot.*, **37**: 217–225.

———, 1925: Some observations on the action of radium on certain plant cells, *Ann. Bot.*, **39**: 547–562.

Williams, R. J., 1928: Nutrilites, *Sci.*, **67**: 607–608.

———, C. M. Lyman, G. H. Goodyear, J. H. Truesdail, and D. Holaday, 1933: "Pantothenic acid"; a growth determination of universal biological occurrence, *J. Am. Chem. Soc.*, **55**: 2912–2927.

Wilson, A. L., 1930: Hydrogen-ion concentration in relation to the growth of onions, *Proc. Am. Soc. Hort. Sci.*, **27**: 524–528.

Withrow, R. B., and M. W. Richman, 1933: Artificial radiation as a means of forcing greenhouse crops, *Ind. Sta. Bull.* 380.

———, 1934: Plant-forcing with electric lights, *Ind. Sta. Circ.* 206.

———, 1936: Light and its effects on plant growth, *Agr. Eng.*, **17**: 150–152.

——— and H. M. Benedict, 1936: Photoperiodic responses of certain green annuals as influenced by intensity and wave length of artificial light used to lengthen the daylight period, *Plant Physiol.*, **11**: 225–250.

WOFFENDEN, L. M., and J. H. PRIESTLEY, 1924: The toxic action of coal gas upon plants. II, The effect of coal gas upon cork and lenticel formation, *Ann. App. Biol.*, **11**: 42–54.

WOLFE, H. S., 1931: Effect of ethylene on the ripening of bananas, *Bot. Gaz.*, **92**: 337–367.

WOLFE, T. K., and M. S. KIPPS, 1926: Comparative value of scarified and of unhulled seed of biennial white sweet clover for hay production, *J. Am. Soc. Agron.*, **18**: 1127–1129.

WOODS, J. J., 1935: Availability of nutrients in raspberry plots in relation to winter injury, *Sci. Agr.*, **16**: 1–7.

WORZELLA, W. W., 1935: Inheritance of cold resistance in winter wheat with preliminary studies on the technic of artificial freezing tests, *J. Agr. Res.*, **50**: 625–635.

WRIGHT, R. C., W. T. PENTZER, T. M. WHITEMAN, and D. H. ROSE, 1931: Effect of various temperatures on the storage and ripening of tomatoes, *U. S. Dept. Agr. Tech. Bull.* 268.

——— and W. M. PEACOCK, 1934: Influence of storage temperatures on the rest period and dormancy of potatoes, *U. S. Dept. Agr. Tech. Bull.* 424.

WRIGHT, W., 1933: Radial growth of the xylem and the starch reserves of *Pinus sylvestris:* a preliminary survey, *New Phytol.*, **32**: 77–96.

WYND, F. L., and H. J. FULLER, 1931: Some effects of ultra-violet radiation upon the calcium and phosphorus content of higher plants, *Ann. Mo. Bot. Gard.*, **18**: 565–572.

ZACHAROWA, T. M., 1925: Über den Einfluss niedriger Temperaturen auf die Pflanzen, *Jahrb. wiss. Bot.*, **65**: 61–87; *Exp. Sta. Rec.*, **57**: 419.

ZIMMERMAN, P. W., and A. E. HITCHCOCK, 1929: Root formation and flowering of dahlia cuttings when subjected to different day lengths, *Bot. Gaz.*, **87**: 1–14.

——— and W. CROCKER, 1930: The effect of sulphur dioxide fumes on plants, abst., *Am. J. Bot.*, **17**: 1042.

———, ———, and A. E. HITCHCOCK, 1930: The response of plants to illuminating gas, *Proc. Am. Soc. Hort. Sci.*, **27**: 53–56.

———, A. E. HITCHCOCK, and W. CROCKER, 1931: The effect of ethylene and illuminating gas on roses, *Contr. Boyce Thompson Inst.*, **3**: 459–481.

———, W. CROCKER, and A. E. HITCHCOCK, 1932: Initiation and stimulation of roots from exposure to carbon monoxide, abst., *Am. J. Bot.*, **19**: 846.

———, ———, and ———, 1932: The effect of carbon monoxide on plants, abst., *Am. J. Bot.*, **19**: 847.

———, ———, and ———, 1933: Initiation and stimulation of roots from exposure of plants to carbon monoxide gas, *Contr. Boyce Thompson Inst.*, **5**: 1–18.

———, ———, and ———, 1933: The effect of carbon monoxide on plants, *Contr. Boyce Thompson Inst.*, **5**: 195–211.

——— and A. E. HITCHCOCK, 1933: Initiation and stimulation of adventitious roots caused by unsaturated hydrocarbon gases, *Contr. Boyce Thompson Inst.*, **5**: 351–369.

——— and W. CROCKER, 1934: Plant injury caused by vapors of mercury and compounds of mercury, *Contr. Boyce Thompson Inst.*, **6**: 167–187.

——— and ———, 1934: Toxicity of air containing sulfur dioxide gas, *Contr. Boyce Thompson Inst.*, **6**: 471–478.

——— and F. WILCOXON, 1935: Several chemical growth substances which cause initiation of roots and other responses in plants, *Contr. Boyce Thompson Inst.*, **7**: 209–229.

——— and A. E. HITCHCOCK, 1935: The response of roots to "root forming" substances, *Contr. Boyce Thompson Inst.*, **7**: 439–445.

———, ———, and F. WILCOXON, 1936: Several esters as plant hormone, *Contr. Boyce Thompson Inst.*, **8**: 103–112.

ZINKERNAGEL, H., 1932: Über den Einfluss des Leuchtgases auf Wurzeln von *Allium cepa*, *Ber. bot. Ges.*, **50**: 134–153; *Biol. Abst.*, **8**: 1660.

AUTHOR INDEX

(**Boldface** type indicates bibliographic reference)

A

Aamodt, 503, **506**
Abbe, 1064, **1095**
Abbott, C. E., **1095**
Abbott, O., **1095**
Abegg, 752, **761**
Aberson, 161, **174**
Acharya, 806, **830**
Ackerman, 1066, **1095**
Acqua, 674, **705**
Acton, **373**
Adams, G. E., **405**
Adams, J., 896, **912,** 1077–1079, 1081, **1095, 1096**
Adams, M., **844**
Adams, W. L., 291, **383,** 443, **511**
Addoms, 153, **174, 373, 396,** 646, 648, 651, **705**
Agulhon, 340, **374**
Ahmann, 20, **51**
Aiyer, **178**
Albers, **375**
Albert, D. W., 334, 363, **381**
Albert, W. B., 131, **174,** 367, **374,** 642, 643, **705,** 887, 890, **912, 919**
Albrecht, 313, **374**
Aldaba, 2, **51**
Aldous, 887–889, **912**
Aldrich, 147, **174,** 205, **213,** 892, **912**
Alexander, 778, **830**
Allan, 138, 140, **174**
Allard, 701, **711,** 748, **763,** 1078–1083, **1096, 1104, 1106**
Allen, A. L., 477, **511**
Allen, C. E., 3, **51**
Allen, F. W., 1055, **1096**
Allen, L. M., 457, 464, **517**
Allen, R. C., 360, **374**
Allen, R. L., **1096**
Allison, F. E., 346, **401,** 656, **705,** 934, **996**
Allison, R. V., 134, **174,** 247, 248, **271,** 362, **374**
Allyn, 334, **374**
Almoslechner, 1036, **1096**
Alsberg, 731, 735, 746, **761, 767, 1005, 1128**
Alten, 316, **374**
Alvarado, **51**
Alway, 195, 197, **212,** 658, **705**
Ambard, 795, 796, **831**
Ambros, 818, **831**
Ames, 290, 301, 303, 307, 309, **374, 399,** 689, **705**
Andersen, 1052, **1096**
Anderson, A., 141, 144, 148, **180,** 443, **513**
Anderson, A. P., 417, **506**
Anderson, D. B., 1, 5, 10, 17, 19, **51,** 468, **506**
Anderson, E., 13, **51**
Anderson, E. G. E., 800, **832**
Anderson, J. A., 959, **1010,** 1067, **1118**
Anderson, P. J., 317, **374**
Anderson, R. J., 731, **761, 762**
Anderson, V. L., 671, 675, **705**
Anderssen, 363, **374,** 877, 896, **913**
André, 316, 356, **374, 375**
Andreesco, M., 882, **916**
Andreesco, M. I., **916**
Andreitcheva, 364, **376**
Andrews, F. M., 247, 248, **271,** 603, **618,** 1023, **1096**
Andrews, J. S., 303, **374**
Annett, 984, **996**
Anthony, **1096**
Appel, **52**
Appleman, C. O., 580, **618, 712,** 796, **831,** 891, **927,** 958, 983, 987, 989–991, **996,** 1031, 1042, 1049, 1053, 1054, **1096**
Appleman, D., 642, **707**
Appleyard, 160, **184**
Archbold, 528, **618, 996, 1108**
Archibald, 285, **374**
Arends, **506**
Arendt, **374**
Arens, 505, **506**
Arighi, **996**
Arker, 131, **174**
Armstead, 859, **925**
Armstrong, E. F., **831,** 984, **1006,** 1030, **1096**
Armstrong, G. M., 131, **174,** 642, 643, **705**
Armstrong, H. E., 809, 811, 812, **831,** 1030, **1096**
Armstrong, J. I., 46, **52**
Armstrong, J. M., **1096**
Armstrong, M. R., 1033, **1116**
Arndt, C. H., 359, **374,** 873–875, **913**
Arndt, J., 368, **374**
Arnold, L. E., 878, **922**
Arnold, W., 589, 592, **618, 624**
Arny, 138, 166, **174,** 888, **913**
Aronescu, 603, **638**
Arrhenius, 263, **271,** 1087, **1096**
Arrington, 646, 648, **705**
Arthur, 289, **401,** 459, **506,** 551, 588, 593, **618,** 1074, 1078, 1079, **1096**
Artschwager, 16, 17, **52**
Asana, 590, **622**
Ashby, **374,** 416, 431, **506,** 652, **705,** 1078, **1096**
Ashton, 800, **838**
Askenasy, 477, **506,** 867, **913**
Aslander, 1087, **1096**
Aso, 365, **374, 402,** 673, **706, 996**

Asprey, 106, **112**, 223, **271**
Astbury, 6, **52**
Astruc, **52**
Atkins, G. A., 1041, 1075, **1097, 1102**
Atkins, W. R. G., 42, 44, 45, 47, **52, 54**, 896, **913, 917**, 984, **996**
Atwater, 696, **706**
Aubert, 951, **996**
Auchincloss, **632**
Auchter, 876, 880, 881, 901, **913**, 1032, **1097, 1102**
Audus, 963, **997**
Austin, J. P., 1080, **1097**
Austin, S., 1045, **1097**
Averill, 303, **374**
Avery, **506**, 825, **831**, 1034, 1038, **1097, 1099**
Axentjev, 1052, **1097**
Ayres, 257, **271**

B

Babb, 1057, **1097**
Babcock, 960, **997**
Babinet, 484, **506**
Bach, 828, **831**, 976, 977, 985, 991, **997**
Bachmann, **1097**
Bacon, **382**, 986, 995, **1002**, 1082, **1106**
Bader, 458, **506**
Badollet, 784, **842**
Baens, 805, **831**
Baeyer, 571, **618**
Baguley, 302, 357, **375**
Bailey, C. F., 544, **622**
Bailey, C. H., 303, **374**, 829, **831**, 954, 959, 960, 962, 965, **997**
Bailey, E. M., 782, **831, 997**
Bailey, I. W., 4, 7, 11, **52, 60**, 871, **913, 997**
Bailey, K., 24, **52**
Bailey, L. H., **1097**
Bailey, W. M., 1090, **1097**
Baker, G. L., **52**
Baker, G. O., 288, **404**
Baker, H., 852, 857, 873, 875, **913, 920**
Baker, J. C., **843**
Baker, J. L., 794, **831**
Baker, S. M., 575, **618**
Baker, W. F., **837**
Bakke, 193, **212**, 258, 259, 264, **271**, 330, **379**, 491, 492, **506**, 954, **997, 1097, 1100**
Balch, R. T., 784, **842**
Balch, W. B., 946, **997**
Baldwin, H. I., 202, **212**, 903, **913**, 994, **997**, 1051, **1097**
Baldwin, M. E., 797, **843**
Ball, F. K., 291, **381**
Ball, N. G., 569, **624**, 894, 895, 904, **917**, 1074, **1097**
Ballard, 956, 992, **1009**
Balls, A. K., 776, **831**, 986, **997**
Balls, W. L., 6, **52**, 430, **506**
Baly, 568, 569, 572–574, 576, 596, **618, 619**, 670, **706**, 799, 800, **831**
Bambacioni, 686, **706**
Bamford, K. F., **52**
Bamford, R., 308, 311, 313, **375**
Banal, 580, 582, **619**
Bancroft, 252, **271**
Bang, 746, **762**
Bank, **112**
Bankul, 600, **624**
Baptiste, 101, 106, **112**
Barakov, 160, **174**
Baranetzky, 789, **831**, 1028
Barbera, 662, **706**
Barclay, 367, **387**
Bardell, 151, 153, **174**
Barendrecht, **831**
Barker, J., 959, 963, 969, **997, 1004**
Barker, W. F., 568, 572–574, **619**
Barlund, 230, **272**
Barnes, C. R., 496, **506**, 541, **619**, 851, **913**, 932, 934, 940, **997**
Barnes, T. C., 1091, **1097, 1114**
Barnett, 663, **717**
Barnette, 294, 332, 364, **375, 402**, 890, **921**
Barrenscheen, **375**
Barrett, 365, **392**
Barritt, 417, **506**
Barron, 954, **1003**
Barss, **506**
Bartel, 891, **913**
Barter, 528, **618, 996**
Bartholomew, E. T., 211, **214**, 431, 437, **507**, 794, **831**, 866, **913, 925**, 941, **998**
Bartholomew, R. P., 234, 235, 261, **271**, 316–321, 368, **375, 387, 388**, 578, **619**
Bartlett, A. W., **997**
Bartlett, H. H., 13, **53**, 107, 234, **282**
Bartley, 1058, **1123**
Barton, A. W., 807, **831**
Barton, L. V., 1048, 1051, **1097, 1101**
Barton-Wright, **55**, 560, 566, 575, **619**
Bartoo, 151, **174**
Basslavskaya, 355, **375**
Bates, 485, **507**
Batjer, 12, **54**
Batten, 1084, **1107**
Baudisch, 673, **706**
Bauer, 302, **375**
Baughman, 731, **762, 764**
Baumann, 365, **392**
Bayer, 466, **507**
Bayliss, 774, 780, 789, **832**
Bazyrina, 561, **619, 639**
Beal, 248, **271**
Beall, 582, **630**
Beams, 25, **52**
Beath, 366, **375**
Beaumont, A. B., 313, 314, 324, 361, 365, **375**, 646–649, 651, **706, 723**
Beaumont, J. H., 946, 958, 965, 971, **998, 1001, 1016**
Beber, **1010**
Beck, A. J., 330, **403**
Beck, W. A., 40, **52**, 78, 94, **112**
Beckenbach, J. R., 140, 141, **175**, 261
Becker, H., 970, **998**
Becker, R. B., 290, 363, **377, 395**
Becker, Z. E., 973, **998**
Becking, 989, **998**
Beckwith, 138, 139, **175**

Becquerel, 83, **112**
Bedford, 167, **183**
Beer, I., 125, **180**
Beer, R., **52**, 739, **762**
Behr, 320, **399**
Behrendt, **832**
Belden, 473, 474, **510**
Bělehrádek, 955, **998**
Bělehrádková, 955, **998**
Bell, 1093, **1097**
Bellamy, **1100**
Bellani, 484, **507**
Belval, 560, 565, **619, 622**
Belvoussov, 344, **376**
Belzung, 33, **52**
Benedict, 255, **271**, 862, 863, **913**, 1076, 1080, **1097, 1130**
Bengis, 731, **762**
Bennet-Clark, 25, 46, 48, 49, **52, 55**, 662, **706**, 946, 951, **998**, 1086, **1103**
Bennett, E., 285, **374**
Bennett, J. P., 333, **375**, 877, 883, **913**, 941, **998**
Bennett, M. E., 135, **175**
Benoy, 965, 974, **998**
Benz, **406**
Berczeller, 796, **832**
Bergamaschi, 575, 582, **619, 634**
Bergann, **1097**
Bergman, 130, 133, **175**, 965, 967, 971, **998**
Bergmann, **832**
Berkley, D. M., 1063, **1097**
Berkley, E. E., 1063, 1081, **1097, 1098**
Bernal, 6, **52**
Bernard, 29, **52**, 557, **619**
Bernbeck, 599, **619**
Berry, 223, **271, 280, 281**, 973, **1014**
Berthelot, D., 572, **619**
Berthelot, M., 356, **375**
Bertram, 889, **925**
Bertrand, 346, 350, 351, 364, **375, 376**, 975, 977–979, **998**
Berzin, **67**
Bessey, 415, **507**
Beumer, 757, **762**
Beyerinck, 655, **706**, 801, 826, **832**
Bezagu, 965, **998**
Bhattacharya, **709**
Bialoglowski, **507**
Biddulph, 1083, **1098**
Bidzinski, 797, **833**
Bierman, 443, **515**
Bigelow, 955, **998**
Bills, 552, **619**
Birch-Hirschfeld, 873, **913**
Bishop, E. R., 369, **376**, 575, **637**
Bishop, W. B. S., 350, 351, **376**
Bissey, 878, **921, 922**
Bisson, 886, **913, 920**
Biswell, 129, **175**, 889, 890, **913**
Biwas, **709**
Bizzell, 160, 165, **175, 181**, 653, **716**
Björksten, 686, **706**
Black, O. F., **404**
Black, P. T., 798, **833**
Black, R. L., 484, **510**
Blackman, F. F., 436, 473, **507**, 578, 579, 583, 584, 589, 590, 593–595, 607, 611, **619**, 933, 941, 970, **999**
Blackman, G. E., 649, **706**
Blackman, V. H., 211, **212**, 490, **507, 636**, 1026, 1028, 1083–1085, **1098**
Blagowestschenski, **832**
Blaignan, **379**
Blaikley, **507**
Blair, A. W., 303, 345, **376**, 691, **716**
Blair, G. Y., 342, **381**
Blake, A. K., 1047, **1098**
Blake, M. A., 323, **399**, 646, 648, **723**, 881, **913**
Blanck, 161, **183**, 368, **376**
Blaydes, 411, 415, 450, 493, **507**
Blinks, 88, **112**, 219, **272**
Bloch, 17, **53**
Blodgett, 470, **507**
Blood, A. F., 816, 817, 821, **832, 840**
Blood, P. T., **720**
Bloomfield, 788, **841**
Bloor, 726, 731, 736, **762**
Blout, **835, 1002**
Blum, 94, **113, 119, 120, 1128**
Boas, **999**
Bobko, 344, **376**
Bodansky, 786, **841**
Bode, 854, 871, **913**
Bodenberg, 877, 881, **913**
Boehm, J. (A), 312, **376**, 559, **619, 620**, 856, 864, **913**, 962, **999**
Bois, **832**
Bokorny, 575, 576, **620**
Bolas, 488, **507**, 585, 587, **620**, 935, **999**
Bolin, D. W., 347, **376**
Bolin, I., **1001**
Bolin, P., 354, **376**
Bolley, 884, **914**
Boltz, 290, 301, 307, 309, **374**
Bonazzi, 967, 983, **999**
Boncquet, 673, **706**
Böning, 50, 51, **53**
Böning-Seubert, 50, 51, **53**
Bonner, J., 3, **53**, 1035, 1036, **1126**
Bonner, T. F., **1098**
Bonnet, **768, 1015**
Bonnier, 556, 557, **620**, 952, **999**
Bonns, **510**
Booth, 1023, **1098**
Bordeianu, 882, **916**
Born, 784, **840**
Bornemann, 580, 586, **620**
Borodin, **1098**
Borowikow, 1072, **1098**
Borowski, 163, **175**
Borrownicka-Ordzywolska, **376**
Borthwick, 1053, **1098**
Bortner, 302, **389**
Bose, **620**, 861–863, **914**
Botjes, 1046, **1098**
Bottomley, 582, **620**, 652, **706**
Boucherie, 864, **914**
Bouchut, 814, 815, 819, **832, 846**
Bourdouil, **914**
Bourne, 1068, **1098**

Bourquelot, 801, **832**
Boussingault, J. (B.), 188, **212,** 556, **620,** 654, 655, 696, **706**
Boussingault, M., **707**
Bouygues, 132, **175**
Bouyoucos, 465, **507**
Bowen, E. J., **212,** 852, **914**
Bowen, J. L., 72, **113**
Bowen, R. H., **53**
Bowerman, 971, **1010**
Bowling, 313, 314, 354, **382,** 986, 995, **1002**
Boynton, **844**
Boysen-Jensen, 583, **620,** 905, **914,** 1030, 1034, 1039, 1040, **1098, 1099**
Brackett, **627,** 1074, **1099, 1110**
Bradlee, 858, **920**
Bradley, 790, **832**
Brady, **53,** 339
Bragg, **53**
Brain, 1090, **1099**
Bramble, 1050, **1099**
Brandenburg, 345, **376**
Brannon, J. M., 294, **376**
Brannon, M. A., **53**
Brase, 1044, **1128**
Brasse, 789, 801, 802, **832**
Braun, **832**
Brauner, 82, **113**
Brauns, 731, **761, 764**
Brazay, **834**
Breazeale, 123, 170, 172, **175, 181,** 191, 197, 203, **212, 214,** 220, 227, 256, 257, **272,** 360, 372, **376,** 582, **620,** 905, **921,** 1030, **1099**
Bredow, 33, **53**
Brenchley, 131, **178,** 247, 248, 257, 258, **272, 274,** 299, 338, 340–342, 349, 362, 365, 367, 369, 371, **376, 377,** 688, 689, **707,** 1026, **1099**
Brewig, 141, **185**
Bricka, **831**
Brierley, 189, **212, 507**
Briggs, G. E., 222, 230, **272,** 322, **377,** 597, 598, 604, 613, 615, **620, 621,** 966, 973, **999, 1007,** 1026, 1028, 1059, **1099, 1111, 1130**
Briggs, L. J., 190–194, 196–198, **212, 272,** 411, 453, 456, 458–461, 482–484, 486, 487, 490, 491, 497–499, 501, **507, 508,** 1085, **1099**
Brigham, 645, 650, **707**
Brill, 816, **832**
Brilliant, 599, **621**
Brink, 1090, **1099**
Bristol, 26, **53,** 656, **707**
Britton, 170, **180**
Brody, **1122**
Brooks, C., 477, **508,** 799, **835,** 1090, **1117**
Brooks, S. C., 85, 88, 97, 99, 108, **113,** 217, 229, **272**
Brown, A., 431, 433, **518**
Brown, A. B., **1099**
Brown, A. H., 1051, **1099**
Brown, A. J., 82, 84, 110, **113**
Brown, B. A., 288, **377,** 891, **914**
Brown, B. E., 314, 345, **376, 377, 387, 401**
Brown, C. E., 807, 811, **845**
Brown, D. D., 234, **272**
Brown, D. E., **177, 382,** 986, 995, **1002**
Brown, F., 1074, **1120**
Brown, H. T., 425, 437–440, 477, **508,** 558, 560, 565, 583, 585, 607, 609, 612–616, **621,** 783, 790, 791, 798, 799, 802, 823, 826, 827, **832**
Brown, J. G., 26, **61,** 961, **1008**
Brown, J. W., 288, 289, 323, 324, **377**
Brown, M. A., 460, **508**
Brown, P. E., 160, **185,** 349, **377**
Brown, R., 85, **113,** 657, **707**
Brown, R. E., 816, **832**
Brown, W. H., 198, 208, **212, 213,** 441, **451,** 452, **514,** 588, 593, **621,** 1071, 1072, **1099**
Brown, W. R., 957, 996, **1010, 1016,** 1067, **1118**
Broyer, 223, **275**
Bruce, 327, **377**
Bruch, 309, **377**
Bruman, **1099**
Bruner, M. H., 431, 432, **509**
Bruner, W. E., **186**
Bruno, 321, **377**
Brunson, 886, **914**
Brunstetter, **53**
Bruschi, 828, **832**
Bryan, 291, 350, 352, 362, **374, 377**
Bryant, **175,** 207, **212,** 248, **272**
Buchanan, J. H., **621**
Buchanan, R., **406**
Buchinger, 95, **113**
Büchner, 933, **999**
Bucholtz, 1074, **1099**
Buckner, 372, **377,** 983, **1006**
Bugarsky, 99, **113**
Bull, **762**
Bullis, 346, **388**
Bunker, 800, **832**
Bunzel, 978, 981, 982, 987, 993, **999**
Burd, 370, **377,** 883, **914**
Burge, E. L., 992, **999**
Burge, W. E., 363, **378, 396,** 989, 992, **999, 1003,** 1079, **1099, 1130**
Burger, 797, **834**
Burgerstein, 236, **272,** 432, 450, 464, 465, 489, 504, **508**
Burgess, 584, **621**
Burk, 249, **275,** 573, **621,** 657, **707,** 934, **996**
Burke, 334, **378,** 689, **707,** 883, **914**
Burkhart, B. A., 12, 13, **53**
Burkhart, L., **707**
Burkholder, P. R., 1034, 1074, **1097, 1099, 1100**
Burkholder, W. H., 1087, **1100**
Burlakow, 965, 991, **999**
Burnett, K. H., **630**
Burnett, L. C., **1100**
Burnham, 1051, **1100**
Burns, G. P., 486, **508**
Burns, G. R., 592, **621**
Burr, G. O., 576, 584, **621, 632, 761**
Burr, S., 1093, **1100**
Burrell, **707**
Burroughs, 955, 957, **999**
Burstrom, 223, **277**
Burt, 188, **212**
Buscalioni, Dr., 813, **835**
Buscalioni, L., 436, **508**
Büsgen, **175**

Bush, 1086, **1114**
Bushnell, 956, 971, **999, 1000**
Busse, 1051, **1100**
Buston, 15, **53**
Butkewitsch, 820, **832**
Butler, O., 319, **401,** 958, **1000**
Butler, O. R., 892, **914**
Butlerow, 571, **621**
Byers, 366, **378**

C

Cabanes, 290, **378**
Cable, **62**
Cailletet, 582, **621**
Cake, 13, **53**
Caldwell, J., 901, **914,** 1037, **1108**
Caldwell, J. S., 198, 199, **212,** 819, **833,** 889, **915,** 1056, **1100**
Caldwell, M. L., 791, 793, 797, **843, 844**
Calfee, 344, 351, 365, **393, 394**
Calvert, 1069, **1100**
Calvin, **844**
Cameron, F. K., 9, **65, 187,** 257, **272**
Cameron, S. H., 642, **707**
Camp, A. F., 364, **395**
Camp, J. P., 364, **375,** 603, **921**
Camp, W. H., 621, 995, **1000**
Campbell, A. B., 1085, **1099**
Campbell, E. G., 671, 672, **707**
Campbell, G. F., 796, **842**
Campbell, W. G., **52**
Canals, 290, **378**
Canaye, 290, **378**
Candlin, 10, 13, **53**
Canniff, 365, **392**
Cannon, 133, 134, 141, **175, 176, 1000**
Capalungan, 194, **213**
Capen, **914**
Carey, 892, **914**
Carlson, 136, **176**
Carmin, 315, **378**
Carolus, 289, 306, 314, **377, 378, 397**
Carpenter, 882, **914**
Carr, 359, **386,** 689, **722**
Carre, 22, **53**
Carrero, 288, 334, 335, 371, **383, 406,** 649, **725,** 881, **918**
Carrick, 958, 989, **1000**
Carter, 691, **712**
Casale, 222, **272**
Caskey, 746, 748, **762**
Castoro, 13, **53, 66,** 911, **926**
Cattell, 1076, **1100**
Cattle, 322, **378, 387,** 797, **838**
Cavers, 36, **53**
Cerighelli, 585, **621,** 952, **1000**
Chace, 1057, **1100**
Chamberlin, 909, 910, **921**
Chambers, R., 1065, **1100**
Chambers, V. H., **53**
Chambers, W. H., 1058, **1100**
Chambliss, C. E., 285, **379**
Chance, 995, **1000, 1100**
Chandler, R. F., 324, **375**
Chandler, W. H., 43, 45, **53,** 364, **378, 386,** 703, **707,** 892, 911, **914,** 1065, 1067, **1100**
Chang, **1101**
Channon, **707,** 740, 743, **762**
Chapman, A. G., 603, **621,** 672, **707**
Chapman, G. W., 352, **378, 508**
Chapman, R. E., **378**
Chase, 853, **914**
Chatin, 364, **378**
Chatterjee, 305, **378**
Chatterji, 974, **1011**
Chernysheva, 600, **636**
Chibnall, 18, **53,** 661, 663, 670, 676–681, 683, 684, 696, **707, 708, 712, 717, 720,** 736, 740, 741, 743, 744, **762, 764, 766, 767**
Chichester, 552, **635**
Child, **1100**
Childers, 468, **508,** 599, 621, **627**
Chima, 967, **1008**
Chinoy, **622**
Chittenden, 814–816, 818, 819, **833**
Chizhov, 371, **378**
Choate, 905, 912, **914**
Chodat, F., 458, **508**
Chodat, R., 976, 977, 985, 991, **997**
Cholodny, 1039, 1040, **1100**
Chouard, 360, **378**
Chouchak, **621**
Choudhri, 1074, **1125**
Chrapowicki, 684, **708**
Christensen, 983, **1000**
Christopher, 584, 600, **621**
Chrysler, **54**
Chrzaszcz, 792, 797, 798, **833**
Church, C. G., 968, **1000,** 1058, **1102**
Church, E. G., 1057, **1100**
Ciamician, 651, **708**
Clague, 46, **54**
Clapp, **508**
Clark, E. D., **843,** 985, **1000**
Clark, E. R., 339, **378**
Clark, G. L., 6, 7, **54, 55**
Clark, H. E., 248, **272,** 646, 648, 662, 663, **708, 724**
Clark, I., 992, **1003**
Clark, J. H., 646, **708**
Clark, M., 783, **838**
Clark, N. A., 348, 350, **378,** 652, **708,** 1032, **1100**
Clark, O. L., 247, **272**
Clark, R. H., 798, **833**
Clark, S. P., **919**
Clark, W. G., **1101**
Clark, W. S., 137, **176,** 856, **914**
Clarke, **1096**
Clausen, 955, **1000**
Clausman, 368, **382**
Claypool, 975, **1011**
Clayson, 21, **54,** 535, **622**
Clayton, 954, **1007**
Clements, F. E., 135, 142, **176,** 431, 433, 459, 461, 496, **508, 515**
Clements, H. F., 267, **272,** 456, **508,** 560, 566, **622,** 877, 878, **914,** 971, **1000**
Clevenger, 48, 49, **54**
Clinch, **917**
Clinton, 466, **508**

Clowes, 111, **113**
Clum, 473, 476, 477, 479, 480, **508**
Coburn, 144
Cochran, 1063, **1125**
Cockefair, 302, **378**
Coffman, 411, 449, 453, 454, **516**
Coggeshall, **1101**
Cohn, 665, 666, **708**
Cohnheim, 820, **833**
Colby, 234, 235, **272,** 306, 320, 323, **378, 708**
Cole, A. M., 263, **272**
Cole, G. F., 798, **837**
Cole, R. C., 560, 563, **622**
Coleman, D. A., 959, 963, **1000**
Coleman, J. M., 659, **710, 921**
Colin, 565, **622,** 783, **833,** 1023, **1101**
Colla, 316, **378,** 549, **622,** 1073, **1101**
Collander, 230, **272**
Collings, 138, **176**
Collin, 731, **762**
Collins, E. J., 84, **113**
Collins, G. H., 346, **378**
Collins, G. N., 1076, 1085, **1101**
Collison, 134, 172, 173, **176,** 700, **712,** 884, **915,** 1068, **1101**
Combes, **622,** 882, **915,** 984, **1000**
Compton, 788, **833**
Conant, 544, **622**
Cone, 44, **56**
Connard, 893, **930**
Conner, S. D., 291, 350, 358, 363, **378, 379, 403**
Conner, W. H., 365, **379**
Connors, 359, **379**
Connstein, 807, **833**
Conrad, C. M., 21, 23, **54**
Conrad, J. P., 148, 173, **176, 272,** 646, **708**
Cook, F. C., 340, 345, **379**
Cook, M. T., 168, **176**
Cook, R. L., **272, 393,** 672, **708, 717**
Cooley, 466, 470, **510, 517**
Coons, 678, **708**
Cooper, H. P., 219, 234, **273,** 1087, **1101**
Cooper, R. E., 291, **379**
Cooper, W. C., 216, **273,** 1041, **1101**
Copeland, E. B., 425, 460, 476, 480, **508, 509,** 860, 867, **915**
Copeland, J. J., 656, **708**
Copeman, 1026, **1101**
Cordingley, 678, **708**
Cormack, 151, 152, 154, 158, **176**
Corson, 330, **379**
Coster, C., 871, **915**
Coster, C. H., 1071, **1101**
Cotton, 300, 365, 369, **379**
Cottrell, 138, 139, 144, **176**
Coupin, 159, 985, **1000**
Coville, 359, **379,** 651, **708,** 1049, 1054, **1101**
Cowan, 265, 266, **276**
Cowart, 599, **621, 622**
Cowdry, 36, **54**
Cowell, 307, **379**
Cowles, 448, **509**
Crafts, 17, **54,** 849, 878, 891, 900, 901, **915, 921**
Craighead, 884, **915**
Crandall, 291, 349, **401**
Creighton, 792, **833**
Crider, **176,** 893, **915**
Crist, 12, **54,** 127, 136, 147, **176, 186,** 995, **1002**
Crocker, 244, 266, **273, 277,** 324, 326, **379,** 579, **622,** 881, **930,** 946, 947, 949, 970, 984, 987, 989, 991, 993, 994, **1000, 1004,** 1040, 1047, 1048, 1051, 1088–1090, **1101, 1109, 1112, 1131**
Crone, 241, 242, **273**
Cross, C. F., **54**
Cross, F. B., 468, **509**
Crowther, E. M., 642, 651, **708**
Crowther, F., 284, **383, 384, 708**
Crozier, 49, **54, 113**
Cruess, 946, 981, 982, 986, **1000, 1002, 1006, 1011**
Crump, 195, **213**
Cruz, **762**
Csonka, 661, 666, **709**
Cuisinier, 801, **833**
Cullinan, 454, **509, 724,** 886, 892, **929**
Culpepper, **54,** 889, **915**
Cummings, 585, 586, **622**
Curran, 805, **831**
Curry, **176,** 600, **622**
Curtis, N., 597, **638**
Curtis, O. F., 473–476, 480, 481, 495, **509,** 703, **709,** 876, 877, 879–881, 892, 894–896, 898–901, **915, 916, 922, 1101**
Czaja, 6, **54**
Czapek, 159, **176,** 738, **763,** 780, **833,** 899, 901, **916**
Czerny, 940, **1014**

D

Dachnowski, 703, **709**
Dadswell, H. E., 11, **54**
Dadswell, I. W., 285, **379**
Daggs, 759, **763**
Daikuhara, 645, **709**
Daish, 558, 560, 563, 565, **623,** 801, **833**
Dalblom, 304, **405**
Damiens, **379**
Damm, 18, **54**
Damon, E. B., 88, **113, 117**
Damon, S. C., **178,** 303, **385**
Dandeno, 165, 168, **176,** 188, **213**
Dangeard, 592, **622,** 738, **763**
Daniel, H. A., 284, 285, 287–289, 307, 313, **379, 384**
Daniel, L., 1023, **1101**
Darbishire, 319, **403**
Darlington, 1047, **1101**
Darrow, 189, **213,** 423, 454, 488, **509, 519,** 1071, 1079, 1080, **1101, 1102**
Darsie, 953, **1000**
Darwin, 428, 430, 432, 433, 436, 446, 457, 460, 478, **509,** 866, **916**
Das, 192, 197, **215,** 643, **709**
Dastur, 291, **379,** 558, 590, 592, 593, 599, 614, **622, 623,** 647, **709,** 952, **1000,** 1023, **1102**
Davidshöfer, 568, **628**
Davidson, G. F., 6, 10, **54, 623**
Davidson, J., 11, **54,** 221, 231, 234, 235, **273,** 285, 339, 340, 372, **379, 397,** 690, 696, **709,** 899, **916**
Davidson, O. W., 646, 674, **709**
Davidson, V., 336, **389**

Davidson, W. C., 797, **839**
Davies, P. A., 703, **709, 1102**
Davies, W., 1041, **1102**
Davis, A. R., 48, **59,** 100, 104, **114,** 218, 219, 224, 234, 235, 266, **273, 275,** 287, 356, 363, **379, 390, 833, 834**
Davis, E. E., 1037, **1102**
Davis, E. F., 173, 174, **176**
Davis, F. L., 313, **374**
Davis, J. B., **618**
Davis, L. D., 373, **380,** 891, **916**
Davis, M. B., 312, 323, 324, **380**
Davis, M. E., **277**
Davis, W. A., 527, 558, 560, 563, 565, **623, 833**
Davis, W. B., 968, 994, **1000, 1014,** 1058, **1102**
Davis, W. C., 984, **1000**
Davis, W. E., 947, 970, 987, 994, **1000, 1001,** 1048, 1054, **1101**
Dawson, 349, **400**
Day, 312, **380**
Dean, A. L., 540, **623,** 801, **833**
Dean, B. E., 132, **176**
Dearborn, 320, 323, **397**
Deats, 1083, **1102**
De Beer, **1102**
De Boer, 973, **1001**
De Candolle, 163, **176**
Deen, 431, 432, **509**
De Forest, 448, 461, **518**
Degman, 434, **510,** 1032, **1102**
De Haan, 106, **113**
De Heen, 350, **394**
Deleano, 542, 552, **623, 763,** 882, 896, 899, **916,** 937, **1001**
Delezenne, 816, **833, 834**
Delf, 100–102, 111, **113**
De Long, G. E., 1067, **1102**
De Long, W. A., 946, 958, **998, 1001**
De Luca, 746, **763**
Delwiche, 1069, **1127**
Demaree, 134, **176**
Demerec, 550, **623**
Demolon, 326, **380**
Demoussy, 316, 362, **374, 391, 392,** 557, 585, **623, 630,** 943, 952, **1009**
Deneke, 584, **623**
Deniges, **380**
Denny, 80, 82, 111, **113,** 230, 231, **273,** 560, 561, 609, **623,** 679, 681, **709,** 798, **834,** 882, **916,** 968, 969, 986, 993, 996, **1001, 1004, 1010,** 1046, 1049, 1050, 1056, 1057, **1100, 1102, 1103**
DePeralta, 166, 168, **182,** 353, **380**
Derrick, 1093, **1106**
Desai, B. L., 599, **622**
Desai, R. M., 952, **1000**
De Saussure, N. T., 554–556, **624,** 654
De Saussure, T., 752, 753, 757, **763,** 932, 944, **1001**
Deshpande, **182**
Detmer, 758, 759, **763,** 905, **916**
De Turk, 304, **380**
Deuber, 141, 329, 332, **380,** 549, **624,** 1050, 1088, 1089, **1103**
De Villiers, **509**
De Vries, 87 **114,** 899, **916**
Dewey, 423, **509**
Dewiche, 692, **709**
Dexter, 958, 983, 987, **1001,** 1064, 1068, 1069, **1103**
Dhar, 568, 573, 577, **624, 634,** 661, **709**
Dhein, 355, **380**
Dick, 542, 552, **623**
Dickson, A. D., 1083, **1110**
Dickson, H., **1103**
Dickson, H. H., 569, **624**
Dickson, J. G., 85, **118,** 312, 313, 315, 316, **380**
Diehl, **1009**
Diehm, 12, **68**
Dietz, 544, **622**
Dikussar, **394,** 649, **717**
Dillman, 166, **174,** 499, **509,** 748, **763**
Dinshaw, 535, **629, 842**
Dittmer, 138, **176**
Dittrich, 672, 674, **709**
Dixon, 25, 42–44, **54, 55,** 94, 103, **114,** 209, **213,** 319, 460, **509,** 565, 572, 860, 862, 863, 865–869, 871–874, 894–896, 904, **916, 917,** 1086, **1103**
Djatschkow, **1011**
Dobson, 782, 783, **842**
Doby, G., 797, **834**
Doby, P., 783, **834**
Dodd, 170, **176**
Dodge, 423, **509**
Doebbeling, 793, **844**
Dole, 463, **509**
Dolk, 1034, 1036, **1103, 1126**
Domontovitsch, 318, **380, 709**
Doneen, 671, 672, 691, **709**
Dorcas, 216, **273**
Dore, W. H., 7, **67,** 341, 343, **388**
Dore, W. S., **55**
Dorée, **54, 55**
Dosdall, 463, **509**
Doubt, 1088, **1103**
Dove, 291, **380,** 1033, **1103**
Dowd, 416, **509,** 957, **1001**
Dowding, 316, 317, **380**
Doyle, **917, 1001**
Drabble, E., 40, **55,** 450, **509**
Drabble, H., 40, **55,** 450, **509**
Draghetti, 1095, **1103**
Drain, 956, 964, 978, **1001**
Drewes, 656, **709**
Druce, 318, **391**
DuBuy, 1036, **1103**
Duchon, 994, **1010**
Duclaux, 595, **624**
Dufrénoy, **273,** 364, **398,** 472, **509**
Duggar, 244, 262, **273,** 418, 466, 468, **509, 510, 834, 1103**
Dulac, 285, 287, **392**
Duley, 124, 144, **176, 177, 181,** 326, 370, **380,** 645, **725**
Dulin, 648, **713**
Dumont, 94, **114**
Duncan, G. H., 261, **273**
Duncan, W. H., 140, 141, **177**
Dungan, 828, **845,** 889, **917,** 1091, **1104**
Dunlap, A. A., 259, **273, 624,** 678, **710**
Dunlap, F. L., 751, **763**

Dunn, G. A., 265, **273**
Dunn, S., 1067, **1104**
Dunne, 45, 47, **55**
d'Urbal, 104, **116**
Durr, **56**
Dustman, 162, **177,** 863, 870, 874, **917, 927,** 1058, **1104, 1118**
Duvel, J., 1046, **1104**
Duvel, J. W. T., 960, **1001**
Dvoretskaia, 955, **1013**
Dymond, **380**

E

Eames, **213,** 848–850, **917**
Eastham, 1046, **1104**
Eaton, F. M., 40, **55,** 149, **177,** 341–343, **380, 381,** 473, 474, 476, **510, 1104**
Eaton, S. V., 326, 327, **381,** 1079, **1104**
Eaves, **1001**
Échevin, 373, **381,** 550, **624,** 882, **917**
Eckerson, 2, 7, **55,** 103, **114,** 135, 136, **177,** 304–307, **381,** 423, 425, **510,** 674–676, 688, 689, **710,** 856, 1054, **1104**
Edfeldt, 792, 798, **841**
Edgar, 551, **641**
Edson, 857, 893, **920**
Edwards, D. W., 538, **624**
Edwards, T. I., 943, **1001,** 1026, 1051, 1064 1090, 1091, **1104, 1106, 1120**
Effront, 791, 819, **834**
Eggleton, 650, **710**
Egloff, 612, **624**
Egorov, 1044, **1104**
Ehlers, 473, 477, **510**
Eidelman, 306, **381, 510,** 600, **624**
Eisenmenger, **114,** 261, 269, **273,** 358, **381,** 648, 674, **706, 710**
Eldredge, 600, **624**
Elfving, 866, **917**
Eliasberg, 317, **389**
Ellett, **710**
Elliot, J. A., 884, **918**
Elliot, K. A. C., 985, 986, **1001**
Elliott, 953, **1000**
Elmer, 1045, 1046, **1104**
Eltinge, 1073, **1104**
Elvove, 675, **714**
Emde, 695, **710**
Emerson, J. L., 447, **510**
Emerson, P., 946, **1001**
Emerson, R., 569, 575, 589, 602, **624,** 943, **1001**
Emmerling, A., 684, **710**
Emmerling, O., **834**
Emmert, 291, 306, **381, 710,** 1087, **1104**
Emmett, 23, **55**
Engel, **710**
Englis, 797, 798, **834**
Enlow, 659, **710**
Epple, 689, **722**
Eppson, 366, **375**
Erdmann, L. W., 258, 259, 264, **271**
Erdmann, R., **396**
Ermakoff, 952, **1001**
Ernest, A., 159, 160, **185**
Ernest, E. C. M., 94, 96, **114**
Errera, 866, **918**
Erxleben, 1040, **1112**
Escombe, 425, 437–440, 477, **508,** 583, 585, 607, 609, 612–616, **621,** 826, 827, **832**
Espino, 253, **273,** 645, **710**
Esselen, **55**
Estabrook, 431, **514**
Estioko, 166, 168, **182**
Etheridge, 1091, **1130**
Euler, 743, **763,** 990, 995, **1001, 1002**
Euler, von, **404,** 569, 575, **640,** 1041, **1128**
Evans, C. R., **1104**
Evans, E. I., **763**
Evans, H., 302, 307, 316, **403,** 488, **520**
Evans, M. V., 653, **710**
Evans, M. W., 1080, **1104**
Everest, 984, **1002**
Evershed, 1026, 1043, **1121**
Ewan, 122, **181**
Ewart, 134, **177,** 569, 573, 574, 594, 600, 601, 603, **624, 625,** 861, **918, 1002, 1104**
Ewing, **278,** 643, 666, **719**
Eyre, 748, 750, **763**
Eyster, 31, 35, **55,** 548, **625**
Ezekiel, 137, **185,** 476, **510**
Ezell, 971, 995, **1002, 1003**

F

Faber, **55**
Fagan, **710,** 883, 886, **918**
Fagundes, 343, **381**
Fales, **834**
Falk, 776, 779, 783–785, 787, 808, 810–812, **834, 835, 840, 1002**
Famintzin, 34, **55**
Farmer, 871, **918**
Farnham, 258, **278**
Farr, C. H., 154–156, **177**
Farr, W. K., 2, 7, **55, 56,** 1020, **1104**
Farris, 138, 139, **177**
Faul, 26, **56**
Faulkner, 1023, **1105**
Faurel, 44, **60**
Feher, **710**
Feist, **835**
Felix, 362, **381**
Fellenberg, von, 21, 22, **68**
Fellers, 46, **54**
Fellows, 959, 963, **1000**
Fenton, 572, **625**
Ferguson, 486, **519**
Fermi, 813, **835**
Fernald, 900, **918, 1104**
Fernbach, 535, 538, **625,** 813, **835**
Fidler, 941, **1002, 1015**
Filipowicz, 798, **835**
Finch, 334, 363, **381, 510,** 700, **710**
Fincke, 575, **625**
Findlay, 78, **114**
Fink, **56**
Finnell, **510**
Firbas, **510**
Fischer, A., 896, **918**

Fischer, E., 661, **710**
Fischer, F., 10, **56**
Fischer, H., 544, 586, **625**
Fischer, M. H., 87, **114**
Fisher, D. F., 477, **508,** 797, 799, **835, 837**
Fisher, D. V., 600, **625**
Fisher, E. A., 820, **835**
Fisher, H., 657, 698, **710**
Fisher, P. L., 302, 323, 341, 343, **381, 388,** 643, **711**
Fisher, R. A., 1028, **1104**
Fittbogen, 236, **273**
Fitting, 41, **56,** 882, **918,** 1030, **1104, 1105**
Fivaz, 1051, **1105**
Fleck, **65**
Fleischer, 597, **625**
Fleming, 1066, **1123**
Flemion, 986, 996, **1002,** 1051, 1056, **1105**
Fletcher, 164, **177**
Fleury, 754, **763**
Flint, L. H., 1053, 1085, **1099, 1101, 1105**
Flint, W. P., 884, **918**
Flood, 504, **510**
Florell, 1023, **1105**
Florkin, 776, **839**
Flory, 501, **510**
Fluri, 359, **381**
Fly, 348, **378**
Fonder, 323, **381**
Fong, 981, 982, 986, **1000, 1002**
Forbes, **381**
Forssberg, 569, **640**
Forster, 1080, **1105**
Fort, 1068, **1129**
Fosse, 696, **711**
Foster, **707,** 744, **762**
Foubert, 748, 763
Fowler, 798, **833**
Francis, C. K., **763**
Francis, D. S., 1076, **1105**
Frank, A. B., 466, **510**
Frank, B., 127, **177,** 655, 681, **711**
Frank, E., **397**
Frankel, **835**
Frankfurt, 744, **767**
Franzen, **56**
Franzke, 889, **920**
Fraps, 291, **381**
Fraser, 1085, 1093, **1105, 1106**
Frasure, 291, **403**
Fraymouth, 961, **1002**
Frazer, 325, **382**
Frear, 291, 365, **382, 401,** 488, **510,** 648, 672, **711,** **918**
Fred, 161, **177,** 527, 540, 587, **640, 641, 719, 725**
Free, 111, **114,** 133, **175, 177,** 247, 248, **273, 281**
Freeland, 240, **273,** 1022, **1105**
Freeman, G. F., 488, 489, 493, **510,** 693, **721**
Freeman, G. G., 792, **835**
Frei, 463, **510**
Fremy, 20, **56**
French, **1002**
Freud, 796, **832**
Freudenberg, 6, 10, **56**
Frey, 6, **56**
Frey-Wysslıng, 6, **56,** 335, **382**
Friedberg, 1067, **1105**
Friedemann, **273, 763**
Friedrich, 964, **1002**
Friesner, 1029, **1105**
Fujita, A., **114, 116**
Fujita, T., 685, **711**
Fuller, G. D., 419, **520**
Fuller, H. J., 289, **406,** 777, **835, 1016,** 1073, 1074, 1079, **1099, 1105, 1131**
Funchness, 294, **382**
Furnstal, **273,** 298, **382**
Furr, 431, 434, **510, 515,** 876, **918**
Furth, von, 756, **768**

G

Gaddum, 23, **56**
Gail, 44, **56,** 588, **625, 636**
Gailey, 939, **1007**
Gainey, 643, **711**
Gall, 364, **375**
Gallagher, 746, **763,** 976, 977, **1002**
Gallup, 746, 748, **762, 763**
Galwialo, **625**
Gane, 1046, **1105, 1125**
Ganeau de Lamarlière, 493, **510**
Ganong, 188, **213,** 490, **510,** 607, 608, **625,** 947, 953, **1002**
Gard, 995, **1002**
Gardiner, 29, **56, 835**
Gardner, F. E., 888, 893, 898, **918,** 1049, 1056, **1105**
Gardner, V. R., 701, **711**
Gardner, W. A., 1052, **1105**
Garner, 169, **177,** 313, 314, 354, **382,** 643, 701, **711,** 748, **763,** 986, 995, **1002,** 1078–1083, **1106**
Gary, 15, **65**
Gassner, 339, **382,** 1093, **1106**
Gast, 562, 565, **625**
Gates, 189, **213,** 484, **510**
Gaudechon, 572, **619**
Gäumann, 888, **918**
Gautheret, **177,** 549, **625**
Gautier, 368, **382**
Gautreret, 674, **711, 1106**
Geigel, 341, **382**
Gelfan, 88, **113**
Gemmer, 138, 140, **177**
Genevois, 941, **1003**
Georgeson, 138, 139, 145, **177**
Georgi, 587, **625, 711**
Gerber, C., 747, 748, 751, **763,** 822, **835, 836,** 952, **1003**
Gerber, L., 797, **834**
Gerdel, 373, **382,** 488, **511**
Gerhardt, **56,** 891, **918,** 971, **1003**
Gerhart, 955, **1003**
Gericke, 141, **177,** 246, 255, 260, 261, 268, **274,** 315, 331, 371, **382,** 689, 690, 703, **711,** 1030, **1106**
Gerlach, M., 644, **711**
Gerlach, P., 580, 587, **625**
Germar, 336, **382**
Gerritz, 23, **56**

Gerry, 890, **918**
Gershberg, 369, **386**
Getchell, 985, **1003**
Gettler, **843**
Gfeller, 1093, **1106**
Gibbs, 5, 6, 9, **56, 65,** 853, 854, 871, **918**
Gibson, R. G. H., 137, **177**
Gibson, R. J. H., 572, 575, **625**
Giddings, 460, **511**
Giersbach, 1048, **1106**
Gies, **282,** 758, **765**
Giesecke, 368, **376**
Giglioli, 84, **114**
Gilbert, B. E., 291, 344, 349, 352, 356–358, **382, 383, 394,** 443, **511,** 1081, **1106**
Gilbert, C. S., 366, **375**
Gilbert, J. H., 654, **715**
Gilbert, L. O., 752, **763**
Gildehaus, 291, 320, **383, 395**
Gile, 288, 328, 330, 333–337, 371, **383,** 686, **711,** 881, **918**
Gill, **1106**
Gilman, 954, **1003**
Gilson, 15, **57**
Giltay, 236, **274**
Gines, 331, **383**
Ginesburg, 289, **383,** 909, **918,** 946, **1003**
Girard, 559, **625**
Giroud, **626**
Girton, **511**
Glass, 1086, **1106**
Glenn, 784, **840**
Glick, 776, **836**
Glimm, 793, **836**
Godden, 288, **383**
Godlewski, 582, 583, **626,** 861, **918,** 952, **1003**
Godwin, 963, **1003**
Goeze, **382**
Goff, 138, 139, **178,** 884, 893, **918**
Goldsmith, **114**
Gonzalez, 989, **1003**
Goodspeed, 24, **68,** 1076, **1106, 1117**
Goodyear, 1037, **1130**
Gordienko, 124, **178**
Gordon, A., 883, **921**
Gordon, J. J., **766**
Gordon, R. B., 569, 594, **626**
Gore, 792, 797, **836, 838,** 936, 954, 955, **998, 1003**
Gorke, 1064, **1106**
Gortner, 41, 43, **58, 1003**
Gorup-Besanez, von, 813, **845, 846**
Gosney, 809 811, 812, **831**
Goss, M. J., 11, 12, **64**
Goss, W. L., 806, **837,** 1046, **1106**
Gössl, 346, **383**
Gottwick, 316, **374**
Gould, 1026, 1064, 1091, **1104, 1106**
Goulden, 551, **641**
Gourley, 23, **59,** 140, 141, **175,** 643, **712, 713,** 1087, **1106**
Gouwentak, 685, **712**
Graber, 887, 889, 890, **919,** 1069, **1103**
Gračanin, 218, 240, 258, **274, 395, 511,** 642, **712, 1003,** 1050, **1106**
Gradenwitz, 586, **626**
Grafe, 312, **383,** 575, **626,** 743, **763, 764**
Graf-Marin, 472, **511**
Graham, 692, **721**
Grainger, 477, **511,** 678, **708**
Grandfield, 148, **178,** 888, 890, **919**
Grant, 344, **386**
Graser, 775, **846**
Grassl, 544, **625**
Gray, G. P., 49, **57**
Gray, J., 431, **511**
Greathouse, 44, **57,** 1068, **1107**
Greaves, 284, 289, **383,** 691, **712, 722**
Grechushnikov, 975, **1013**
Green, E. L., **928**
Green, F. C., 989, **1003**
Green, J. R., 745, 755, 758, 760, **764,** 791, 799–801, 805–807, 813, 814, 819, 826, **836,** 974, **1003**
Green, J. W., 798, **837**
Green, L., **624**
Greene, C. W., **1122**
Greene, L., 1080, **1107**
Greenhill, **383,** 663, **712**
Greenwood, 652, **712**
Gregory, **274,** 284, **383, 384,** 431, **511,** 974, **1003,** 1026, 1028, 1059, 1083, 1084, **1098, 1107**
Greve, **713**
Griffin, A., **626**
Griffin, E. G., 786, **837, 840**
Griffing, **1128**
Griffiths, 644, **712**
Grigsby, 320, **386**
Gris, A., 33, **57**
Gris, E., 328, **384**
Grizzard, **384**
Groner, 974, **1003**
Groom, 312, **384,** 871, **919**
Gross, 665, 666, **708**
Grover, 678, 679, **707**
Groves, **1107**
Grunder, **274**
Grundmann, 189, **213**
Grüss, 13, **57,** 828, **837,** 964, **1003**
Grutterink, 815, **842**
Gruzit, 604, **626**
Gubbinet, **630**
Guerrant, 744, **764**
Guilliermond, 36, 37, **57**
Gunderson, 1032, **1107**
Gunjikar, 590, 614, **623**
Gurewitsch, 112, **114**
Gurjar, 698, **712,** 946, 959, 960, 962, 965, **997, 1003**
Gurney, 643, **721**
Gursky, 138, **178**
Gustafson, 50, 51, 57, **511,** 600, **626,** 942, 943, 965–967, 992, **1003–1005,** 1042, 1091, **1107**
Guthrie, 542, **626,** 969, 981, 983, 985, 993, 996, **1001, 1004, 1010,** 1056, 1078, 1079, **1096, 1107**

H

Haag, 264, **277,** 332, **392**
Haagen-Smit, 1035, 1040, **1112**
Haarlem, van, 291, **404**

Haas, A. R. C., 44, 47, 48, **57,** 159, 161, **177, 178,** 217, 231, 233–235, **274, 279,** 289, 290, 298, 299, 306, 312, 314, 332, 341, 344, 352, 363, **384, 398,** 416, **511,** 573, 589, 593, 607, **633,** 643, **712,** 884, **919, 928,** 947, 968, 969, **1004, 1011,** 1087, **1122**
Haas, P., 234, 539, **626,** 669, 694, 695, **712,** 733–735
Haasis, 871, **919,** 1051, **1113**
Haas-Poetzl, 435, **511**
Haber, **57,** 982, 995, **1004,** 1032, **1107**
Haberlandt, **57,** 154, **178,** 236, **274,** 447, 504, **511,** 596, **626,** 738, **764,** 826, **837,** 851, **919, 1107**
Hackenberg, **1107**
Haehn, 796, 798, **837**
Haerdtl, 535, 537, **635**
Hafenrichter, 938, 939, 946, 967, **1004, 1005**
Hagen, 428, **511**
Haig, 1074, **1107**
Haigh, 885, 887, **928**
Haines, **213,** 452, **511,** 854, 871, **919**
Halcro-Wardlaw, 759, **763**
Hale, H. P., 1065, **1100**
Hale, W. S., 986, **997**
Hales, 872
Haley, 284, **384,** 783, 807–810, **837, 839**
Halket, **213**
Hall, A. D., 131, 161, **178,** 248, 257, 258, **274,** 337, **384,** 688, **707**
Hall, D. H., 676, **708**
Hall, D. M., 340, **384**
Hall, E. H., 327, **384**
Hall, J. A., 890, **918**
Haller, 23, **57,** 893, **919,** 935, **1004**
Halma, 44, 45, 49, **57, 65,** 290, **384,** 416, **511, 919,** 1043, **1122**
Halstead, 264, **277**
Hamilton, 889, **927**
Hamlin, 810, **834**
Hamly, 85, **114**
Hammett, 245, **274,** 303, **385,** 1040, **1107**
Hamner, 643, **712**
Hampton, 989, **998**
Hanby, **1120**
Hancock, 6, **52**
Hand, 776, **839**
Hanes, **626,** 794, **837,** 969, **1004**
Hanna, 1059, 1071, **1107**
Hannig, 42, 882, **919**
Hanriot, **764**
Hansen, 1032, **1128**
Hansteen, 685, **712,** 828, **837**
Hansteen-Cranner, 23, **57,** 154, **178,** 738, 743, **764**
Harada, **837**
Harder, 579, 583, 589, **626**
Hardin, 291, **382**
Harding, P. L., 935, 946, 956, 967, 995, **1004, 1009**
Harding, T. S., 540, **626**
Hardy, F., 202, **213**
Hardy, M. B., 967, **1011**
Harlan, H. V., 689, **712,** 827, **839,** 1025, **1116**
Harlan, J. D., 700. **712,** 884, **915,** 1068, **1101**
Harley, 700, **712,** 797, 799, **835, 837,** 967, **1004, 1097**
Harlow, 5, 10, 11, **58**
Harmer, 362, **384**
Harmon, 140, 141, **178**
Harper, H. J., 285, 288, 289, 307, **379, 384**
Harper, R. A., 31, **58**
Harrar, **183**
Harriman, 542, 543, **626**
Harrington, G. T., 946, 947, 949, 950, 952, 984, 987, 989, 991, 993, 994, **1000, 1004,** 1051, **1107, 1108**
Harrington, J. B., 1079, **1108**
Harrington, O. E., 41, **59**
Harris, E. S., 88, **117**
Harris, F. S., 124, 128, **178**
Harris, G. H., 963, **1004, 1005**
Harris, J. A., 40–42, 44, **58, 59, 62,** 227, **274**
Harris, M. R., 806, **837**
Harris, W. G., **710**
Harrison, C. M., 648, **712,** 1091, **1108**
Harrison, G. J., 42, **59,** 1032, 1045, **1108, 1111**
Harrison, G. T., 355, **384**
Harrison, W. H., 131, **178**
Hart, E. B., 324, 326, 327, **385**
Hart, H., 431, 433, **511**
Hart, M. C., 741, **768**
Hartelius, 1034, **1119**
Harteneck, 818, **831**
Harter, 17, **68,** 270, **276,** 355, **385,** 464, **511**
Hartig, 892, 895, 896, **919**
Hartley, 652, **712**
Hartshorn, 1058, **1108**
Hartt, 316, 317, 319, 322, **385,** 560, 561, 599, **626**
Hartwell, 169, **178,** 301, 303, 306, 312, 313, 322, 330, 356, 358, 360, **385, 405**
Hartzler, 886, **929**
Harvey, E. M., 699, 700, **712,** 852, 873, **919,** 1089, **1108**
Harvey, E. N., 601, **626**
Harvey, R. B., **274, 404,** 968, 981, 982, 987, 990, **1005, 1013, 1016,** 1046, 1050, 1057, 1066, 1068, 1077, 1079, **1108, 1118, 1122, 1128**
Harvey, T. B., 852, **919**
Haselhoff, 341, **385**
Haskins, C. P., 1076, **1108, 1117**
Haskins, H. D., 124, **179**
Hasselbring, 46, **63,** 236, 237, **274,** 936, **999, 1005,** 1056, **1108**
Haukinson, 675, **713**
Hausen, 1033, **1108**
Hausen, von, 653, **724**
Haut, I. A., 434, **516**
Haut, I. C., 996, **1005**
Havas, 1037, 1076, **1108**
Häveler, 126, **178**
Hawker, 1039, **1108**
Hawkins, L. A., 205, **213,** 486, 490, **514,** 936, **1005**
Hawkins, R. S., 172, **178,** 443, **511,** 891, **913, 919**
Hawley, 11, **54, 59**
Haworth, 9, **59, 626**
Hawthorne, **114**
Haynes, 22, **53, 59, 66, 1108**
Hays, 138, 139, **178**
Headden, 138, 139, **178,** 284, 285, 347, **385,** 693, **712**
Heald, 1085, **1099**
Heap, 1033, **1116**

Heath, **385**
Heck, 292, **385**
Hedrick, 168, **178**
Hedricks, **1108**
Heilbron, 568, 572–574, **619, 626,** 670, **706**
Heiling, 465, **511**
Hein, 35, **59**
Heinicke, 132, 148, **178,** 493, **511,** 560, 561, 581, 584, 599, 603, 607, 611, **626, 627,** 882, **919,** 987, 988, 990, 994, 995, **1005**
Heise, 588, 593, **621**
Heller, 318, **385**
Hellriegel, 137, **178,** 360, **385,** 655, 696, **712,** 758, **764**
Hellström, 569, **640,** 776, **846**
Helwert, **56**
Hemmi, **214,** 463, **514**
Hemperly, **841**
Henderson, F. Y., 457, **511,** 585, **620**
Henderson, H. O., **1118**
Henderson, L., 211, **213**
Henderson, R. G., 300, 351, **397**
Hendricks, **765**
Hendrickson, 123, **179,** 192, 197, 205, **213, 215,** 434, 435, 444, 462, **511, 520**
Hendry, 355, **385**
Henri, **837**
Henrici, 550, 596, **627**
Henricksen, **274,** 882, **919**
Henry, 957, **1013**
Henslow, 189, **213**
Henson, 886, **919**
Hepburn, 823, **837**
Herlitzka, 548, **627**
Heron, **832**
Herrick, 94, 95, **114**
Herrman, 599, **636**
Hertzsch, 569, **640**
Herviaux, 361, **404**
Herzfeld, 933, **1005**
Herzog, 6, **59**
Hess, 7, **59**
Hesse, 157, **179**
Hester, 289, **385, 397**
Heukelekian, 556, **635**
Heukeshoven, 368, **376**
Heyl, F. W., 741, **764**
Heyl, J. G., 857, **919**
Heyn, 3, **53,** 1036, **1108, 1109**
Heyward, 140, 141, **179**
Hibbard, A. D., 1031, **1109**
Hibbard, P. L., 224, 235, **275,** 364, **378, 386**
Hibbard, R. P., 41, **59,** 261, **274, 275,** 320, 369, **386,** 604, **626, 627,** 797, **834,** 1057, **1109**
Hibbert, 604, **639**
Hicks, G., **764**
Hicks, L. E., 1087, **1109**
Hicks, P. A., 542, 606, **627,** 702–704, **712, 713**
Hiestand, 743, **768**
Hilditch, 731, **762, 764**
Hildreth, 447, **510,** 1065 **1100**
Hill, A. W., 29, **56,** 849, **920**
Hill, D. D., 1066, **1109**
Hill, E. S., **117**
Hill, G. R., 943, **1005**
Hill, H., 324, 344, **380, 386**
Hill, H. H., **710, 713**
Hill, J. H., 800, **839**
Hill, T. G., 39, **59,** 234, 539, **626,** 669, 694, 695, **712,** 733–735
Hilpert, 542, 543, **627**
Hiltner, E., 365, **386,** 505, **512**
Hiltner, R. S., **837**
Himmel, 132, **187,** 1082, **1129**
Hirano, 419, 420, 423, **512, 517**
Hirsch, 568, **628**
Hirst, C. T., 289, **383**
Hirst, E. L., **59, 626**
Hitchcock, A. E., 881, **930,** 1037, 1040, 1041, 1082, 1088–1090, **1101, 1109, 1131, 1132**
Hitchcock, D. I., 786, **841**
Hixon, 263, **274**
Hoagland, 48, **59,** 100, 104, **114,** 218–220, 223, 224, 227, 228, 231, 232, 234, 235, 239, 241, 244, 257, 258, 262, 264, 265, 267, **274, 275,** 287, 290, 300, 317, 318, 323, 344, 364, **378, 379, 386, 388**
Hobart, C., **919**
Hobart, C. A., **919**
Hobart, F. G., 310, **386**
Höber, J., 229, **275**
Höber, R., 229, **275**
Hoffer, 291, 292, 331, 358, **386**
Hoffman, C. T., **59**
Hoffman, E. J., **1005**
Hoffman, M. B., 581, 584, 600, 607, **626, 627, 715,** 955, 974, **1005**
Hoffman, W. F., 44, **58, 59,** 227, **274**
Höfler, 101, **115**
Hofman, 700, **713**
Hofmeier, 542, 543, **627**
Hogan, **1122**
Hogounenq, 776, **837**
Höhn, 151, **179**
Holaday, 1037, **1130**
Holbert, 304, **380**
Holch, 140, 141, **179**
Hole, **179**
Holland, 1026, **1122**
Hollander, **841**
Holle, 208, **213,** 854, **920**
Hollenberg, **275**
Holley, 647, 648, **713**
Holly, 807, **843**
Holm, 540, **627**
Holman, 602, **627**
Holmes, A., 782, **838**
Holmes, M. C., 871, **920**
Holmes, M. G., 443, **515**
Holter, 823, **837**
Holtz, 497, 498, **519**
Holyński, **386**
Homan, 870, **920**
Honda, 349, **391**
Honert, van den, 910, **929**
Hood, 596, **619**
Hooker, 20, **51,** 121, **179,** 577, 579, **627,** 699, 700, 704, **713,** 1026, 1066, **1109**
Hooper, **59**
Hoover, S. R., 249, **275,** 934, **996**

Hoover, W. H., **627, 1110**
Hopkins, C. Y., 481, **512**
Hopkins, E. F., 23, **59,** 328, 332, 348, 350, 351, 363, **386, 713,** 936, 956, 958, 964, **1005, 1129**
Hopkins, E. W., 648, **713**
Hopkins, J. W., 691, **713**
Hopkins, R. H., 792, 798, 814, **835, 837**
Hopmann, 158, **179**
Horn, 598, **627, 636**
Horn, van, **510**
Hornberger, 370, **386,** 883, **920**
Horner, C. K., 249, **275**
Horner, J., **722**
Horning, 36, 37, **59,** 826, **837**
Horton, **831, 837**
Horvat, 743, **763**
Hostermann, 1079, **1109**
Hottes, **114,** 946, **1005,** 1020, 1067, **1109**
Hougen, 137, **187**
Houghland, 314, 320, **377, 386**
House, 793, **840**
Hover, 965, **1005**
Howard, A., 130, **179**
Howard, B. J., 955, **998**
Howard, G., 551, 552, **627**
Howard, G. L. C., 130, **179**
Howard, W. L., 1046, 1047, 1049, 1053, **1109**
Howe, C. G., 158, **179**
Howe, M. A., 807, **844**
Howell, 1088, **1109**
Howes, 493, **512**
Howk, 304, **380**
Howlett, 700, **713**
Hoyer, 807, **833, 837**
Hubbard, 367, **387**
Hubbenet, 549, **630**
Huber, B., 101, **115,** 453, **512,** 871, **920**
Huber, H., **838**
Hubert, 813, **835**
Hudson, C. S., 784, 786–788, **838**
Hudson, D. P., **706**
Hudson, P. C. B., 1041, **1102**
Hudson, P. S., **1130**
Huelin, **1109**
Huelsen, **114**
Huerre, 801, 802, **838**
Hughes, F., **380**
Hughes, H. D., 170, **187**
Hughes, W. L., 341, 342, **394**
Hulme, **713**
Hulton, 794, **831**
Hume, 560, 561, 563, **634,** 889, **920**
Humphrey, 292, **387**
Hungate, 870, **920**
Hunter, C., 132, **179**
Hunter, J. H., 362, **374**
Hurd, 50, 51, **59**
Hurd-Karrer, 43, 47, **59, 60,** 285, 286, 307, 366, **387,** 443, **512,** 1080, 1083, **1110**
Huston, 370, **389,** 883, 885, 886, **921**
Hutchings, L. R., **923**
Hutchings, S. S., **1110**
Hutchins, 130, **179**
Hutchinson, A. H., 800, **838**
Hutchinson, D. M., 973, **1005**
Hutchinson, H. B., 645, 650, **713**

I

Ilienkoff, 236, **275**
Ilin, 41, **60**
Iljin, 41, **60, 275,** 426, **512,** 598, 599, **627, 1005**
Imai, **63**
Imas, **388**
Inamdar, **115, 920, 1005,** 1026, **1110**
Ingalls, 331, **387**
Ingen-Housz, 554, 555, **627**
Ingold, 106, **115,** 225, **276,** 857, **920**
Inman, 367, **387,** 545, 549, 569, 598, **627,** 972, **1006**
Inozemtzev, **387**
Ireland, 597, **627**
Irvine, 782, 783, **842**
Irving, A., 31, **65**
Irving, A. A., 597, 598, **627,** 675, **713,** 967, **1006**
Irwin, 100, **115,** 216, 227, **276**
Isaac, 443, **512**
Isbell, **1110**
Isshiki, 1080, **1112**
Ivanov, L. A., 458, 503, **512**
Ivanov, N. N., 692, **713**
Ivanow, S., 749, 750, 755, 756, 760, **764**
Ivanow, S. L., **1006**
Iwanoff, L., 744, **764**
Iwanoff, M., 686, **713**
Iwanoff, N. N., 952, **1001**

J

Jackson, C. M., **1122**
Jackson, H., 582, **620, 764**
Jackson, L. C., 72, **116**
Jacob, 303, **387**
Jacobson, C. A., 782, **838**
Jacobson, H., 743, 745, **764**
Jacobson, H. G. M., 234, 235, 265, **276,** 347, 348, **387**
Jacobson, L. A., 890, **920,** 1069, **1123**
Jacoby, 575, **628**
Jacques, 88, 104, **115, 117,** 219, 227, 229, **272,** 276
Jacquot, 960, **1006**
James, 316, 317, 322, 324, **387,** 569, 576, 584, 593, **628,** 797, **838,** 852, 857, 873, 875, **913, 920**
Jamieson, 731, **762, 764**
Janse, 861, **920**
Jansma, 362, **405**
Janssen, 234, 235, 261, **271,** 316–321, 324, **375, 387, 388, 406,** 578, **619,** 889, **920,** 1067, **1110**
Jashnova, 304, **388**
Javillier, **388,** 820, **838**
Jean, 127, 147, **179**
Jeffrey, 946, 981, **1000, 1006**
Jeffreys, 440, 446, **512**
Jeffs, 155, **179**
Jegorow, 755, **764**
Jenkins, J. M., 131, **186**
Jenkins, S. H., **64**
Jennings, 337, **388**
Jenny, 265, 266, **276**
Jensen, C. A., **181**

Jensen, I. J., **1110**
Jensen, J. H., 137, **179**
Jess, 586, **628**
Jivanna, 417, **512**
Jodidi, 46, **60**, 669, 678, **713, 714**, 827, **838**, 912, **920**
Jodin, 335, **388**
Joëssel, **1015**
Johann, 85, **118**
Johansson, 967, **1006**
Johnson, A. B., **402**
Johnson, A. F., 1073, **1124**
Johnson, A. H., 974, **1003**
Johnson, B., **273**, 298, **382**
Johnson, E. L., 1075, 1076, 1080, **1110**
Johnson, E. M., 643, **724**
Johnson, F. B., 324, **380**
Johnson, H. V., 227, **276**
Johnson, I. J., 138, **174**, 749, 750, **764**
Johnson, L. P. V., 1048, **1110**
Johnson, M. O., 334, **388**
Johnson, O. C., 665, 666, **708**
Johnston, C. O., 129, **179**, 502, **512**
Johnston, E. S., 242, 247, 260, **276**, 317, 318, 323, 341, 343, 353, **388**, 485, **512**, 587, **627, 628**, 1074, **1099, 1110**
Johnston, W. H., 503, **506**
Johnstone, 962, 963, **1006**
Joly, 866, 867, **916, 917**
Jones, A. P., 330, **395**
Jones, C. H., 585, 586, **622**, 857, 858, 893, **920**
Jones, D. B., 661, 666, **709**
Jones, F. M., **837**
Jones, H. A., 886, **913, 920**, 994, **1006**
Jones, H. W., 328, **388**
Jones, I. D., 42, 44, **59, 62**, 431, **512**
Jones, J. A., 1050, 1053, 1055, **1110**
Jones, J. P., 291, 314, **388**, 1048, **1110**
Jones, J. S., 346, **388**, 693, **714**
Jones, J. W., 970, **1006**
Jones, L. H., 124, **179**, 264, **276**, 330, 332, **388**, 643, **714**
Jones, L. R., 168, **179**
Jones, R. L., **389**
Jones, T. N., 434, **512**
Jones, W. J., 370, **389**, 883, 885, 886, **921**
Jones, W. N., 430, **512**, 800, **838**, 984, **1006, 1110**
Jordan, R. C., 744, **764**
Jordan, W. H., 290, 301, 302, **389**
Jorgensen, 99, 101, 102, 107, 110, **119**, 248, **281**, 542, 557, 573, 596, **628**, 1084, **1110**
Joscht, 792, **833**
Joslin, **833**
Joslyn, 788, **838**, 986, 989, **996, 1006**, 1065, **1111**
Jost, 236, **276**, 904, **921**
Jozefowicz, **1111**
Józsa, 792, 797, **836, 838**
Jung, 352, **389**
Jungers, **60**
Jupe, **380**

K

Kachioni-Val'ter, 263, **276**
Kaempfert, 614, **635**
Kaho, 111, **115, 764**
Kale, 746, 750, **767**
Kalitin, 613, **628**
Kamerling, S. E., 216, 227, **278, 544, 622**
Kamerling, Z., **60**
Kamp, 437, **512**
Kao, 757, **764**
Kapadia, 1023, **1102**
Kapoor, 1074, **1125**
Karchevski, 965, **1006**
Karling, 1071, **1111**
Karlson, 942, **1006**
Karmarkar, **714**
Karns, 365, **375**
Karper, 1058, **1121**
Karraker, 302, **389**
Karsten, **768**
Karström, 653, **724**
Kas, 367, **395**
Kasahara, 1080, **1112**
Kaserer, 166, **179**
Kastle, 675, **714**, 783, **838**, 976, 977, 983, **1006**
Katayma, 349, **389**
Katchioni-Walther, 1087, **1111**
Kater, **1111**
Kautsky, 568, **628**
Kayser, E., **838**
Kayser, R., 559, **628**, 749, **765**
Kearney, 270, **276**, 1045, **1111**
Keeble, 984, **1006**, 1039, 1040, 1044, **1111**
Keinholtz, **1111**
Keller, 41, **60**
Kellerman, 244, **277, 838**, 1080, 1082, **1111**
Kellersberger, 790, **832**
Kelley, V. W., 415, 445, 454, 468, **512**, 974, **1006**
Kelley, W. P., 346, 350, **389, 406**, 645, **714**
Kellicott, 1029, **1111**
Kellner, 645, **714**
Kellogg, **713**
Kelly, F. C., 288, 365, **396**
Kelly, J. W., **404**
Kempton, 1059, **1111**
Kendall, **843**
Kennedy, C., 912, **925**
Kennedy, M. H., **387**
Kennedy, P. B., 878, 891, **915, 921**
Kent, 942, **1007**
Kenyon, 993, **999**
Kerr, H. W., 692, **723**
Kerr, T., 4, 7, 11, **52, 60**, 106, **115, 180**
Kerstan, **838**
Kersten, 1076, **1097**
Kertesz, 782, **839**, 956, 957, 975, 976, **1006**, 1055, **1111**
Kessler, 189, **213**
Keuhl, 580, **628**
Keyser, 693, **716**
Keyssner, **56**
Kezer, 123, **184, 714**
Khankoje, 497, **513**
Kidd, 226, 227, 233, **281**, 573, **628**, 940, 966, 967, 971, **1007**, 1026, 1028, 1046, 1060, **1099, 1111, 1130**

Kiesselbach, 121, 123, 141, 144, 148, **180**, 236–238, **276**, 418, 425, 443, 459, 478, 490, 491, 497, 498, 500, 501, **513**, **516**, 1060, 1069, 1093, **1112**, **1120**
Kiessling, 545, 548, **632**
Killian, 44, **60**
Kimbrough, **389**, 958, **1007**
Kimpflin, 575, **628**
King, C. G., 303, **374**, 776, **836**
King, C. J., 355, **384**
King, E. J., 336, **389**
King, F. H., 138, 139, **180**
King, R., 93, **115**
King, R. L., 25, **52**
Kingsbury, 307, **406**, **641**
Kinman, 148, **180**
Kinnison, 334, 363, **381**
Kinoshita, 647, **714**
Kinsman, 13, **51**
Kipps, 1050, **1131**
Kirkwood, 758, **765**
Kirsh, 809, **839**
Kishi, 560, 561, 563, **628**, 683, **714**
Kisser, 125, 154, **180**
Kite, 87, **115**
Kitsuta, 303, **374**
Kizel, 801, **839**
Kjeldahl, 595, **628**
Klason, 11, **60**
Klebs, 1093, **1112**
Klein, **389**, **628**, 648, 651, 696, **714**
Klercker, 136, **180**
Klinger, 933, **1005**
Klippart, 1092, **1112**
Klotz, 341, 344, **384**, 678, **708**, **919**
Kneeland, **52**
Kneer, **1130**
Kniep, 591, **628**
Knight, H., 909, 910, **921**
Knight, H. G., 366, **378**
Knight, L. I., 1088, 1089, **1101**, **1112**
Knight, R. C., **180**, 430, 431, 442, 443, 446, 460, 487, **507**, **513**, 704, **715**, **921**, 974, **1007**
Knop, 241, 242, **276**, 297, 335, **389**, 743, **765**
Knott, 362, **389**, 987, 990, **1007**, 1083, 1087, **1112**, **1126**
Knowles, 303, 304, 373, **389**
Knowlton, **714**, 881, **921**
Knudson, 163, **180**, 294, 296, 389, 651, 703, **715**, **839**, 936, **1007**, 1087, **1112**
Kny, 600, **628**
Koch, G. P., 793, **845**
Koch, K., 1038, **1112**
Koepfli, **1127**
Koga, 793, **839**
Kögl, 1035, 1036, 1040, **1112**
Kohl, 236, **276**, 786, **839**
Kohler, 882, **915**
Kohn, 594, **618**, **628**, **1002**
Koketsu, 191, 202-204, **213**, **214**
Kokichi, **845**
Kolkwitz, 959, 965, **1007**
Komatsu, 882, **921**
Komleff, 936, **1012**
Kondo, J. N., **513**
Kondō, M., 1051, 1064, 1080, **1112**
Kondrashova, 1083, **1113**
Koonce, 123, **184**
Kooper, 681, **719**, 882, **924**
Kopaczewski, 23, **60**
Kopp, **768**
Korsakow, 746, **765**
Korstian, 40, 42–44, **60**, **1112**
Kosaka, 459, **513**
Kosmann, 789, **839**
Kossel, 742, **765**
Kossowitch, 160, **180**, **715**
Kostytschew, 317, **389**, 557, **628**, 940–943, 946, **1007**, **1012**
Kosutany, 680, **715**
Kotowski, **115**, 1061, **1112**
Kotte, 1058, **1113**
Kourssanow, 472, **513**
Kozlowski, **60**
Krabbe, 109, **115**, 828, **839**
Kraemer, 534, 537, **628**, **629**
Kramer, J., 140, 141, **186**, **187**
Kramer, P. J., 203, 211, **213**, 1049, 1080, **1113**
Krassinsky, 1083, **1113**
Krassovsky, 138, **180**
Kratzmann, 356, 357, **389**
Krauch, 789, **839**
Kraus, E. J., 698, 699, 702, **715**, **718**
Kraus, G., 49, **60**, 559, **629**
Kraus, J. E., **1013**
Krause, 797, **833**
Krauss, **280**, **518**
Kraybill, 291, 304, 306, 322, **389**, 698, 699, **715**, **720**, 888, 892, **921**, **927**, 982, **1013**
Kretovitch, 887, **921**
Kreusler, 583, 607, **629**
Kreuzhage, 335, 336, **390**
Kritzler, 738, **768**
Krogh, 166, **180**
Krotkina, **923**, 1093, **1116**
Krüger, F., 466, **510**
Krüger, W., 645, **715**
Kruseman, 900, **921**
Kružilin, 138, 139, **181**
Kuhn, 542, **629**, 775, 776, **839**, **846**
Kuiper, 587, **629**
Kultzscher, 662, **715**
Kunkel, 208, **213**
Kunze, 159, **180**
Kursanov, **629**
Kurz, 1087, **1113**
Küster, 35, **60**, 108, **115**
Kuyper, 431, 433, 444, 446, **513**
Kylin, **629**

L

Lachenmeier, **513**
Lagassé, 700, **715**
Lagatu, 234, 235, **276**, 578, **629**
Laidlow, 430, **513**
Laing, 1091, **1107**
Laird, 138, 139, **180**
Laitakari, 138, 140, **180**
Lake, **55**

Lakon, 684, **715**
Lal, 579, 601, **637**
Lall, 963, **1007**
Lalou, **837**
Lambrecht, **60**
Lamour, 440, **513**
Larmour, 954, **1007**
Lampe, 886, **921**
Landis, 794, **843**
Landon, 994, **1007**
Langdon, 939, **1007**
Lange, 166, **180**
Langner, 560, 603, **629**
Lanham, 323, **390**
Lantz, 987, 989, 992, **1007**
Lapique, 220, 221, **276**
Larsinos, **706**
Larson, C. A., 234, 235, **276**
Larson, H. W. E., 308, **390**
La Rue, 1042, 1058, 1059, **1113**
Laskovsky, 758, **765**
Latner, **762**
Latshaw, 283, 301, 307, 316, 335, 346, 356, **390,** 581, 642, **715,** 886, **914**
Lau, 160, **180**
Laude, 1068, 1091, 1092, **1113**
Laurent, E., 673, **715**
Laurent, J., 294, **390**
Laurie, 291, **390,** 1080, **1113**
Lauritzen, 291, **390**
Lausberg, 290, **390**
Lavison, 85, **115**
Lawes, 654, **715**
Lawrence, J. V., 41–44, **58,** 227, **274**
Lawrence, Z. W., **58**
Lawrenz, 369, **376**
Lea, 822, **839**
Leach, R., 326, **402**
Leach, W., 932–934, 937, 940, 942, 946, 949, 952, **1007, 1014**
Lease, 674, 686, **715, 723**
Leather, **715**
Leathes, 734, **765**
Leavenworth, 676, **719,** 891, **929**
Leavitt, 151, **180**
Lebedeva, **398**
Leblond, **626**
LeClerc, 340, **379,** 690, 691, **709, 715,** 905, **914, 921**
LeClerc du Sablon, 746, 755, 758, 760, **765,** 861, 892, **921**
Lee, B., 17, 18, **60**
Lee, E., 882, **921**
Lee, H. A., 138, 144, **180**
Leeper, 348, **390**
Legg, A. T., 1083, 1084, **1098**
Legg, V. H., 18, **60**
Leggatt, 989, 994, **1007, 1008, 1113**
Lehenbauer, 1062, **1113**
Leick, 429, 432, 435, **513**
Leitberg, 432, **513**
Leitch, 1062, **1113**
LeMert, 488, **515**
Lemmerman, G., **620**
Lemmerman, O., 337, **390, 397,** 587, **629**
Lemoigne, **392**
Lemon, 1073, **1125**
Lenz, 1076, **1126**
Leonard, **921**
Leonian, 1042, **1113**
Leontjew, **60**
Lepeschkin, 25, **61,** 98, 101, 103, 104, **115,** 504, **513**
Lepkovsky, 13, 14, **67,** 676, 677, **723**
Lesage, 449, **513**
Leukel, 294, 402, 887, 890, **919, 921**
Le Van, 975, **1008**
Levene, 737, 742, **765**
Levine, M., 1037, 1042, **1113**
Levine, V. E., 366, **390**
Levy, 357, **390**
Lewin, 895, **923**
Lewis, A. C., 1029
Lewis, A. H., 647, 648, **715**
Lewis, E. P., **390**
Lewis, F. J., 742, **765**
Lewis, G. N., 1051, **1113**
Lewis, M. R., 147, **174, 187,** 191, 205, **213, 215**
Lewith, **1113**
Lewitsky, 36, **61**
Leydhecker, 353, **390**
Li, H. W., **715, 921**
Li, T., 589, **629**
Liebig, 159, **180,** 571, 577, **629**
Lieske, 10, **56**
Ligon, **390**
Likiernik, 743, 744, **767**
Lilienfeld, 124, 125, **180**
Lilleland, 320, **390**
Lincoln, 676, **715**
Linderstrom-Lang, 823, **837**
Lindet, 532, 537, **629**
Lindow, **397**
Lindstrom, 296, **389**
Lineweaver, **621**
Ling, 535, **629,** 828, **839, 842**
Link, 13, **61**
Linossier, 985, **1008**
Linsbauer, 429, **513**
Linser, **389,** 696, **714**
Lintner, 795, **839**
Linz, 826, **839**
Lipman, C. B., 26, **61,** 287, 299, 308, 356, 362, 363, **379, 390, 401,** 655–657, **715,** 883, **921, 1008,** 1051, **1113**
Lipman, J. G., 653, 691, **716**
Litvinov, **61**
Liu, **715, 921**
Livingston, B. E., **116,** 133, 168, 170, **177, 179–181,** 199, 202–205, 208, 209, **213–215,** 241, 244, 245, 247, 256, **277, 282,** 430, 431, 441, 446, 451, 452, 456, 463, 482–486, 490–492, **506, 514, 519,** 577, 582, **629, 630,** 961, **1008,** 1049, 1051, 1090, **1113, 1116**
Livingston, L. G., 29, **61**
Livingston, N., 552, **635**
Lloyd, F. E., 111, **116,** 428, 431, 442, 443, 445, **514,** 882, **921,** 1091, **1114**
Lloyd, J. W., **390**
Locklin, 967, **1011**
Lodewick, **1114**
Loeb, J., 268, **277,** 665, **716,** 1030, 1042, **1114**

Loeb, W., 572, **630**
Loehwing, 48, 49, **61,** 132, **181,** 248, 262, **277,** 674, 675, **716,** 970, **1008**
Loew, 308, 311, 312, 315, 320–322, 349, 368, **390, 391,** 686, **716,** 987, 988, 990, 991, **1008**
Loftfield, 428, 431–433, 435, 443–446, 479, 489, **514**
Logan, 882, **924**
Loiseleur, 776, **837**
Lomanitz, 354, **391**
Lombardi, 44, **61**
Long, E. R., 961, **1008,** 1072, **1114**
Long, F. L., 26, **61,** 430, 431, **514, 515,** 604, 613, **630**
Long, J. H., 888, **921**
Long, J. S., 807, 811, **845**
Longenecker, H. E., 783, 808, 809, **839**
Longenecker, J. B., 284, **384,** 807, **839**
Loo, S., 1040, **1114**
Loo, T., 1040, **1114**
Loo, T. L., 264, **277,** 647, **716**
Loomis, **116,** 122, **181,** 243, **277,** 341, 343, **391, 630,** 878, 898, **921, 922,** 1022, 1023, **1114**
Loose, 686, **716**
Lopriore, 995, **1008**
Loring, 318, **391**
Lothrop, 790, **839**
Lott, 488, **515**
Lovenhart, 976, **1006**
Lowry, 856, **922**
Lowsma, 686, **723**
Lubimenko, V. 35, **61**
Lubimenko, W., 543, 549, 591, 596, 600, 603, **630**
Lubrzynska, 750, **767**
Lüdtke, 5, 7, **61,** 1065, 1068, **1124**
Ludwig, 657, **716,** 1051, **1114**
Lukaszewicz, 316, **391**
Lund, 872, **922,** 947, 974, **1008, 1013,** 1086, **1114, 1115**
Lunderardh, 223, **277,** 580, 583, **630**
Lunn, **177**
Lunt, 797, **834**
Lute, 1047, **1123**
Luthra, 962, 967, **1008**
Lutman, 43, 44, **61,** 311, 314, **391,** 467, **515,** 962, 964, **1008**
Lutz, 935, **1004,** 1065, **1115**
Lyman, C. M., 1037, **1130**
Lyman, J. F., 808–810, **837**
Lyness, 141, **180**
Lyon, C. J., 94, 95, **116,** 305, **391,** 972, **1008**
Lyon, M. E., 965, **1008**
Lyon, T. L., 160, 163, 165, **175, 181,** 653, 693, **716,** 1091, **1115**
Lysenko, 1092, 1094, **1115**

M

Ma, **61**
McAlister, 1053, **1105**
McBain, **62**
McCall, 247, 252, 258, 264, **277,** 332, 369, **392** 1087, **1116**
McCalla, 316, 324, **392,** 662, 691, **716, 1118**
McCallum, A. B., 316, 317, **392**
McCallum, G. A., **112**
McCampbell, 890, **926**
McCance, 303, **392,** 529, **641**
McCarty, 562, **631, 923,** 1090, **1123**
McClelland, C. K., 307, **392,** 889, **920,** 1091, **1116**
McClelland, T. B., 1082, **1116**
M'Clenahan, 744, 751, **766**
McClendon, 365, **392**
McClure, 129, 149, **187**
McCollum, 1045, **1117**
McComb, 1061, **1117**
McCool, 39, 40, 44, **62,** 136, 234, 235, **277,** 306, 350, 351, 368, **393,** 672, **717**
MacDaniels, **213,** 848–850, 880, 881, **917, 922**
McDermott, 815, **833**
McDonald, A. H. E., 170, **184**
McDonald, F. G., 552, **619**
McDonnell, 346, **393**
MacDougal, 26, **61, 62,** 109, **116, 214,** 481, **515,** 534, 582, **630, 634, 765,** 852–857, 862–864, 869, 870, 873–875, 896, 905, **922,** 961, **1008,** 1020–1022, 1063, 1072, **1115**
McDougall, 156, **181**
McElivee, 291, **390**
M'Endarfer, 989, **1003**
McGee, 557, 570, 589, 607, **637, 638,** 936, 938, 971, **1014**
McGeorge, 222, 227, **272, 277,** 351, **393,** 645, **717**
MacGillivray, 301, 306, **391,**
McGinnis, 942, **1009**
McGinty, 790, 794, **840**
McGuire, 783–785, 787, **835, 840, 1002**
McHargue, 245, **277,** 298, 300, 344, 346, 347, 350, 351, 362, 365, 367, **393, 394,** 882, 883, **923,** 985, **1009**
Macht, **277,** 651, **716,** 800, **839,** 1075, **1115**
McIlvaine, 263, **279**
Mack, 956, **1008,** 1049, **1113, 1116**
McKay, E., 41, **62,** 599, **631**
McKay, J. W., 1076, **1117**
McKee, C., 166, **174**
McKee, R., 1092, 1093, **1117**
McKenzie, **1009**
McKibben, **394**
McKie, 663, **717**
McKinley, 173, **181**
Mackinney, 362, **390,** 552, **630, 632**
McKinney, 1092, 1093, 1095, **1117**
MacLachlan, 753, 756, 757
McLane, 1085, **1101**
McLean, D. M., **923**
McLean, F. T., 291, 352, 356, 357, 370, **383, 394, 397,** 611, **631**
MacLean, H., 736, 737, 742, 746, **765**
MacLean, I., 736, 737, 742, 746, **765**
McLean, R. C., 236, 238, **277,** 341, 342, **394, 515,** 923
MacLeod, 1033, **1116**
MacMillan, 1029, **1116**
McMurtrey, 291, 312–314, 343, 354, 368, **382, 394**
McNair, 538, **631,** 695, **717,** 736, 740, 761, **766,** 805, **840,** 1024, **1117**
McQuilkin, 140, **181**
McRostie, **1117**
Macy, 578, **630**
Maestrini, 783, **839**

Magistad, 40, **62,** 123, **181,** 197, 203, **214,** 358, **391,** 550, **630, 639**
Magiştris, **391,** 743, **763**
Magness, **62,** 431, 434, **510, 515,** 700, **712,** 956, 963, 992, **1008, 1009**
Magnus, **65**
Magoon, **53,** 889, **915**
Magowan, 154, **181**
Mahood, **62**
Maige, 936, 966, **1008**
Maiwald, 355, **391**
Malhotra, 15, **62,** 848, 871, **922**
Malins, 1020, **1121**
Malkani, 647, **709**
Mallik, 992, **1012**
Malpighi, 932, **1009**
Malyschev, 294, **391**
Mameli, 314, 329, **391,** 656, **716**
Maneval, **399,** 1058, **1122**
Maney, **397,** 946, **1004, 1009**
Mangel, **716**
Mangenot, **766, 1009**
Mangham, 904, **922**
Mangin, L., 556, 557, **620,** 884, **922,** 952, **999**
Mangin, M. L., 4, **61, 62**
Mann, A., 827, **839**
Mann, C. E. T., 312, 317, 324, **391**
Manzoni, **515**
Maquenne, 362, **391, 392,** 535, 557, 582, **630,** 758, 760, **765, 839,** 943, 952, **1009**
Marañon, 661, **716**
Mariana, 418, **515**
Marine, 365, **392**
Mark, 6, **63**
Markley, 416, **515,** 669, **713, 714,** 736, **765**
Marks, 993, **1009**
Marsh, G., 974, **1009,** 1086, **1116**
Marsh, G. L., 986, 989, **996, 1006,** 1065, **1111**
Marsh, R. P., 247, **277,** 330, 331, **392**
Marsh, R. S., 325, **392**
Marshall, E. S., 895, **917**
Marshall, R. E., 489, **515**
Marston, 36, **62**
Marth, 140, 141, **181, 716**
Martin, A. L., 285, 286, 366, **392, 404**
Martin, E. V., 433, 456, 459, 461, **508, 515, 631**
Martin J. C., **185,** 257, 258, 267, **275,** 317, **386**
Martin J. F., 1067, **1116**
Martin J. H., 44, **62,** 503, **515,** 891, **913,** 1066, 1091, 1092, **1116**
Martin, J. N., 888, **922**
Martin, S. H., 47, 50, **62**
Martin, S. H. C., 814, 815, **839**
Martin, T. L., **181**
Martin, W. E., **631**
Martin, W. H., 258, **280,** 369, **400,** 467, 468, **515, 518**
Martin, W. M., **516**
Martinand, 783, **840**
Martini, 1025, **1116**
Marwick, 6, **52**
Marx, **631**
Masashi, 48, **63**
Maskell, 338, **377,** 584, **631,** 873, 878, 900–903, 905, 910, **922, 923**
Maslow, 797, **839**
Mason, S. C., 1071, 1072, **1116**
Mason, T. G., **62,** 202, **214,** 565, **624,** 873, 878, 895, 898–903, 905, 910, **922, 923, 925,** 1044, 1045, **1116**
Mason, U. C., 443, **522**
Masoni, 49, **62**
Massey, 168, **181**
Masuda, 364, **392**
Masunaga, **280**
Masure, 700, **712,** 797, **837,** 1074, **1116**
Mathaei, **1009**
Mather, 303, **392**
Mathew, 784, **840**
Matlack, 552, **631,** 731, **765**
Matlock, **919**
Mattern, 46, 50, **63**
Matthaei, 473, **507,** 589, 590, 593–595, 607, 611, **619, 631,** 955
Matthes, 760, **765**
Matzke, **1116**
Maume, L., 285, 287, **392,** 578, **629**
Maume, M. L., 234, 235, **276**
Maximow, 138, 139, **181,** 424, 455, 502, **515, 923,** 1065, 1069, 1093, 1094, **1116**
Maxwell, L. R., 1076, **1101**
Maxwell, W., 744, **765, 767**
May, 470, **519**
Mayer, A., **1006**
Mayer, G. G., **392**
Mayer, H., 545, **631**
Mayer, M., 960, **1006**
Mayfield, 1032, **1122**
Mazé, M., 758, **766**
Mazé, P., 294, 298, 334, 359, 363, 368, **392,** 645, **716**
Mazoni, 456, 461, **515**
Meacham, 48, 49, **68**
Meara, **833**
Megee, 1068, **1117**
Mehrlich, 247, **277, 1117**
Mehta, M. M., 9–11, **62**
Mehta, R. J., 593, **623**
Meier, F. E., 549, 592, **631,** 1076, **1117**
Meier, H. F. A., 264, **277**
Meldrum, 932, **1009**
Mellanby, 537, **631**
Melnikov, **398**
Melville, 488, **507,** 587, **620,** 935, **999**
Mendel, 815–817, **833, 840**
Mendiola, 236, 237, **277**
Mengdehl, 227, **279,** 647, **719**
Menke, **63**
Menon, 12, **63**
Mer, 153, **181**
Merkalz, von, **65**
Merkle, **178**
Merz, 542, **641**
Metzger, J. E., 443, **515**
Metzger, W. H., 148, 162, **178, 181, 717,** 889, **920**
Meuer, 227, **277,** 359, **394**
Mevius, 154, **181, 394,** 649, **717**
Meyer, A., 534–537, 559, **631, 632,** 684, **717,** 937, **1009**
Meyer, A. H., 300, **394**

Meyer, B. S., 40 42, **63**, 411, 465, 492, **515**, **632**, 1069, **1117**
Meyer, J. D., **1115**
Meyer, K. H., 6, **63**
Meyer, R., 320, **399**
Meyer, W. C., 1087, **1117**
Meyers, A., 33, **63**
Meyers, C. N., 356, **394**
Mezzadroli, 573, **632**
Michaelis, **116**, **840**
Michaels, 971, **1009**
Michel-Durand, 805, **840**, 882, **923**
Michener, 1038, **1117**
Michiels, 350, **394**
Middleton, 973, **1009**
Miehe, 782, **840**
Mihailescu, 603, **638**
Mikosch, 33, **63**
Milad, 877, **913**
Miles, 551, **632**
Millar, 39, 44, **62**, 126, 147, **181**
Miller, C. F., 283, **399**
Miller, E. C., 34, **63**, 129, 134, 138, 139, 145, 146, **179**, **181**, 196, **214**, 283, 301, 307, 316, 335, 346, 356, **390**, 411, 415, 421, 423, 424, 442, 449, 453–455, 462, 463, 468, 473–477, 479, 480, 490, 491, 497, 499, 502, **512**, **515**, **516**, 562. 565, 566, 581, 608–610, 616, **632**, 642, 680, 681, 697, **715**, **717**, 753–756, 758, **766**, 883, 886, **923**, 935, **1009**, 1026, 1071, **1117**
Miller, E. J., 676, **708**
Miller, E. R., 776, **840**, 979, **1009**
Miller, E. S., 552, 584, **632**
Miller, E. V., 1054, 1090, **1096**, **1117**
Miller, H. G., 326, **394**
Miller, J. E., **717**
Miller, L. P., 50, **63**, 351, 352, **394**, 888, **921**, 969, 993, 996, **1001**, **1004**, **1009**, **1010**, 1046, 1049, 1050, 1056, **1102**, **1117**
Miller, M. F., 124, **181**, 370, **380**
Miller, N. H. J., 645, 650, **713**
Miller, R. W., **402**
Millner, **766**
Mills, **1127**
Milner, 536, 552, 553, **630**, **638**
Minckler, 494, **516**
Minder, 591, **628**
Miner, 1025, 1064, **1106**, **1119**, **1120**
Minges, 349, **377**
Minina, **182**
Mirasol, 357, **394**
Mitchell, E. M., 14, **63**
Mitchell, G. A., 693, **714**
Mitchell, H. L., 291, **394**, 1061, **1117**
Mitchell, J. H., 365, **394**
Mitchell, J. W., 488, **516**, **721**, 946, **1010**, 1076, 1091, **1119**, **1125**
Mitscherlich, 484, **516**
Miyake, 48, **63**, 357, **394**
Mockeridge, 652, **717**
Mogensen, 415, **520**
Mohammad, A., **182**
Mohammad, S. S., 573, **634**
Mohl, von, 425, **520**, 882, **929**
Moillard, 582, **632**
Moinat, 191, **214**
Moldavan, 558, **632**
Molisch, 162, **182**, 328, 334, 359, **395**, 595, 599, **632**, 684, **717**, 856, 857, 861–863, 882, **923**, 1064, **1117**
Moll, 582, **632**
Molliard, **182**, 295, **395**
Molz, 94, 95, **116**, 210
Monobe, 560, 561, 563, **628**, 683, **714**
Montemartini, 458, 470, **516**
Montgomery, A. M., **513**
Montgomery, E. G., 459, 497, 498, **516**, 1091, **1117**
Moon, **54**, 889, **915**
Moore, B., 574, **632**, 656, 657, 673, **717**, 976, **1010**
Moore, C. N., 1076, **1108**, **1117**
Moore, H. I., 330, **395**
Moore, J. G., 592, **639**
Moore, M. G., 731, **761**
Moore, N. P., 373, **380**
Moore, W., 884, **923**, 995, **1010**
Moore, W. J., 646, 648, 649, **706**
Moravek, **116**
Morgan, A., 878, **923**
Morgan, M. F., 642, **717**
Morgulis, **1010**
Morinaga, 992, **1010**, 1051, 1052, **1118**
Morison, 337, **384**
Morkrjetsky, 884, **924**
Morris, G. H., 558, 560, 565, **621**, 783, 790, 791, 798, 799, 802, 823, 827, **832**
Morris, H. E., 883, **914**
Morris, H. J., 656, **705**
Morris, H. S., 341, **395**
Morris, V. H., **65**, 317, **395**, **405**, **724**, 886, **926**, **929**
Morrish, 166, **182**
Morrow, **1118**
Morse, F. W., **395**, **1010**
Morse, W. J., 168, **179**, 345, **395**, 857, 893, **920**
Moser, 308, **395**
Moss, 313, 314, 354, **382**
Mothes, **395**, 662, 663, 686, **717**, **718**, 901, **924**
Mottier, 32, 34, 36, **63**
Moulton, C. R., 885, 887, **928**
Moulton, S. C., **713**
Mounce, **373**, 646, 648, 651, **705**
Mounfield, 820, **840**
Mouton, 816, **833**, **834**
Mowry, 364, **395**
Moyer, 731, **766**, 851, **924**
Muenscher, 236, 238–240, **278**, 444, 445, **516**, 656, 685, **718**
Mukerji, 166, **182**
Mukherjee, 662, **709**
Mulay, 676 687, **715**, **718**
Müller, A., 603, **632**
Müller-Thurgau, 127, **182**, 470, **516**, 685, 686, **718**, 957, 958, **1010**, 1064, **1118**
Münch, **175**, 872, 900, 901, **924**
Müntz, A., 644, **718**
Müntz, M., 754, 755, **766**
Murlin, 759, **766**
Murneek, 14, **63**, 291, **395**, 552, **632**, 663, 687, 699, 700, **712**, **718**, 882, 886, 892, **924**, 1026, 1044, 1045, **1118**
Murphy, E., 1033, **1103**

Murphy, H. C., 129, **182**, 502, **516**
Murphy, H. F., 194, **213**, 285, 303, **384, 395**
Murphy, J. S., 666, **709**
Murry, **840**
Mussi, 819, **840**
Musso, 1085, **1118**
Myrbäck, K., 990, **1002**
Myrbäck, S., 990, **1002**

N

Nabenauer, **762**
Nadeau, **832**
Naftel, 646, **718**
Nagaoka, 349, **395**, 645, **718**
Nägeli, 6, **63**, 534, 535, **632**, 867, **924**
Nakajima, 189, **214**
Nakamura, 363, 369, **395**
Nanji, **839, 842**
Nathansohn, 226, 227, **278**
Naudain, **621**
Navez, 800, **840**, 983, **1010**, 1039, 1040, **1118**
Naylor, G. L., **63**
Naylor, N. M., 792, 793, **833, 840, 844**
Nazarova, 41, **60**
Neal, 363, **395**
Near, 740, **768**
Nebel, 1031, **1118**
Nedrow, 141, **182**
Nef, 574, **632**
Negelein, 568, 592, **640**, 673, **724**
Neidig, 691, **718**, 790, **840**
Neljubow, 1090, **1118**
Neller, 163, **182**, 303, 345, **395**, 994, 995, **1010**
Nelson, D. H., 284, **383**, 691, **712**
Nelson, E. K., 46, 50, **63**, 307, **406**
Nelson, J. M., 538, **632, 639**, 784, 786, 788, 810, **834, 837, 840, 841**
Nelson, M. G., 1039, 1040, 1044, **1111**
Nelson, N. T., 887, 890, **919, 924**
Nelson, P. M., 1032, **1107**
Nelson, P. R., **706**
Nelson, R. C., 1046, **1118**
Nemec, 367, **395**, 994, **1010**
Nesom, 658, **705**
Nesty, **767**
Neumeister, 813, **841**
Neun, **843**
Newby, 685, **718**
Newcomb, 697, **718**
Newcombe, 124, 125, **182**, 803, **841**, 971, **1010**, 1061, **1118**
Newell, H. J., 341, 345, **396**
Newell, J. M., 1078, 1079, **1096**
Newton, J. D., 161, **182**, 284–286, **396**, 465, **516**
Newton, R., **516**, 959, 996, **1010**, 1066, 1067, **1118**
Nicholas, 576, **632**
Nichols, G. E., 485, **516, 1119**
Nichols, P. F., **924**
Nicloux, 809, **837, 841**
Nicol, 166, **182, 186**, 304, 643, 653, **718**, 890, **924, 928**
Nicolas, **396**, 472, **516**, 576, **632**, 936, 965, 966, **1008, 1010**
Nielsen, 1034, **1119**
Niemann, C., **56**
Niemann, W., 419, 463, **516**
Nightingale, 258, **278**, 307, 312, 316, 319, 320, 322, 323, 327, **396, 399**, 648, 649, 662, 674, 698, 702, **718**, 924, 1064, 1091, **1119**
Nishida, 364, **392**
Nixon, 1031, **1119**
Njegovan, 743, **766**
Noack, 545, 548, **632, 633**
Nobbe, 126, 137, **182**, 297, 318, 353, **396**
Noble, 262, **281**
Noecker, 954, **997**
Nogüchi, 1076, **1119**
Noll, C. F., 305, 306, **396**
Noll, F., **768**
Noller, **766**
Nordenson, 743, **763**
Nordh, 792, **841**
Nordhausen, 870, **924**
Nordlund, 566, **640**, 904, **929**
Norman, 11–15, 21, 22, 24, **59, 63, 64, 924**
Norris, F. W., 13, 15, 21, 24, **52, 54, 63, 64**
Norris, R. J., 542, **633**
North, 206, **214**
Nottin, 532, 537, **629**
Novello, 536, **633**
Novez, **182**
Novikov, **1119**
Nurmia, 890, **929**
Nutman, 140, 141, **182**, 204, **214**
Nutter, 14, **65**

O

Oakley, **725, 1119**
O'Connor, **1001**
Oddo, 328, **396, 398**
Odland, 169, **182**, 349, **396**
O'Dwyer, 13, **64**
Offord, 104, **116**
Ohga, 39, **64**, 202, **214**, 1046, **1119**
Ohlen, von, 753, 758, **768**
Ohlsson, 792, 798, **841**
Okamura, 1051, 1064, 1080, **1112**
Olney, 936, 955, **1010**
Olson, G. A., 325, 327, **396, 403**, 689, **718, 719**
Olson, O., 284, **384**
Olsson, **841**
Onslow, 977, 978, 985, **1010, 1011**
Oparin, 794, 828, **831, 841, 1011**
Oppenheimer, C., **1011**
Oppenheimer, H. R., **116**
Orcutt, 587, **625, 719**
Ormerod, **831**
Orr, 288, 365, **396**
Orth, 363, **378, 396**, 989, **1003**
Osborne, T. B., 658–660, 668, 669, 676, **719**, 793, 794, 796, **842**
Osborne, W. A., 72, **116, 842**
Oserkowsky, 192, 329, 333, 363, **375, 396**, 543, **633**
Oskamp, 140, 141, **182**
Osmaston, 1070, **1119**
Ostanin, **398**

Osterhout, **64,** 88, 99, 100, 104–107, 110, **113, 115–117,** 216, 218, 219, 222, 227, 268–270, **273, 276, 278,** 361, 367, **396,** 573, 589, 593, 607, **633,** 935, 947, **1011**
Ostertag, **56**
O'Sullivan, 539, **633,** 784, 785, 787, 788, **842**
Oswald, 72, **117**
Ota, **1011**
Otis, 789, **842**
Otto, I. G., 575, **637**
Otto, R., 681, **711, 719,** 882, **924**
Overbeek, van, 1036, 1044, **1109, 1128**
Overholser, **396,** 434, **516,** 967, 975, 986, 987, 990, **1011**
Overley, **396,** 434, **516**
Overton, E., 108, **117**
Overton, J. B., 212, **214,** 852, 854, 862–865, 869, 874, 875, **922, 924**
Owen, 307, **396,** 585, 586, **633**
Oxley, 1078, **1096**
Ozawa, 882, **921**

P

Paal, 1033, **1119**
Paauw, van, **640**
Pack, 738, 739, **766,** 1055, **1119**
Paden, 367, **374**
Paetz, 431, 432, 434, **516**
Page, 656, **707**
Paine, H. S., 784, 786–788, 790, **838, 839, 842**
Paine, S. G., **507**
Pal, 974, **1011**
Palladin, V. I., 283, 336, **396,** 548, **633,** 795, **842**
Palladin, W., 550, **633,** 745, **766,** 795, **842,** 934, 936, 937, 940–942, 957, 961, **1011, 1012, 1119**
Palmer, L. O., 434, **512**
Palmer, L. S., 552, **633**
Pan, 1080, **1119**
Panchaud, 1078, **1119**
Pancoast, 981, **1000**
Pande, 1026, **1110**
Panisset, 542, **627**
Pantanelli, 227, **278,** 578, 601, 602, **633,** 647, **719**
Parbery, 314, **396**
Pardo, 647, 648, **719**
Parija, 974, **999, 1012**
Parker, F. W., 161, **182,** 284, 288, 309, **396, 397**
Parker, K. W., 154, **182,** 889, **924**
Parker, M. M., 289, **397**
Parker, M. W., 539, **633**
Parkin, 560, 563, 565, **633**
Parrozzani, 744, **766**
Pascoe, 40, 42, **59**
Paton, F. J., 828, **842**
Paton, J. B., 782, **842**
Paton, R. R., 1061, **1119**
Patten, 1076, **1130**
Paulino, 645, **724**
Pavilinova, 505, **516**
Pavlychenko, 138, **182**
Payne, 138, 139, 145, **177**
Peacock, 16, **64,** 1050, **1131**
Pearce, 551, **633**
Pearl, 1025, 1026, 1064, 1090, 1091, **1104, 1106, 1119, 1120**
Pearsall, 154, **182, 278,** 643, 665, 666, 678, 685–687, **708, 716, 718, 719,** 1026, 1042, 1043, 1088, **1120, 1121**
Pearse, 431, **511**
Peblois, **831**
Peirce, 431, **511,** 567, **633,** 852, 863, 866, **924, 925,** 940, 942, 953, 954, **1000, 1012**
Peklo, 827, **838**
Pelouze, 807, **842**
Pelphrey, 300, 365, **393, 394**
Peltier, 1067, 1069, 1093, **1120, 1126**
Pember, **178, 181,** 248, **279,** 291, 330, 344, 358, 360, 370, **383, 385, 389, 397**
Penfound, 419, 464, **516**
Penrose, **1120**
Penston, 316, 317, **387, 397,** 883, **925**
Pentzer, **1131**
Perciabosco, 649, **719**
Peren, 138, 140, **183**
Perrey, 565, **633**
Perrier, 294, **392**
Perry, 1034, **1120**
Persecke, **183**
Pertz, 430, **509**
Peterhänsel, 464, **516**
Peters, 753, 758, **766**
Peterson, C. B., 301, 307, **397**
Peterson, P. D., 542, **633**
Peterson, W. H., 301, 307, 324, 325, **385, 397,** 527, 540, **640,** 648, **713**
Petrie, 36, 37, **59,** 222, 223, **272, 279,** 317, **397,** 826, **837,** 973, **999**
Petsch, 1051, **1123**
Pettibone, 912, **925**
Pettinger, 127, **183,** 233, 234, **279,** 291, 292, 300, 351, **397**
Pfeffer, 87, 109, **117, 118,** 236, 241, 242, **279,** 297, 339, **397,** 578, 607, **634,** 940, 946, **1012,** 1030
Pfeiffer, F., 161, **183**
Pfeiffer, M., **64**
Pfeiffer, N. E., 1083, **1120**
Pfeiffer, T., 293, 354, 360, **397**
Pfenniger, 13, **66**
Pfotenhauer, 293, 360, **397**
Philipson, **404,** 1041, **1128**
Phillips, M., 10–12, **54, 64,** 339, 340, **397**
Phillips, R. W., 866, **916**
Phillips, T. G., 320, 323, **397,** 700, **720**
Phillis, 899–902, 910, **923, 925**
Pickering, 167, 172, **183**
Pickett, B. S., **397**
Pickett, T. A., 648, **713**
Pickett, W. F., 601, 610, **634**
Pickler, 829, **842**
Pidgeon, 1085, **1105**
Piekenbrock, **706**
Piemeisel, 497, 498, **518**
Pierce, 758, **766**
Pierre, I., **397**
Pierre, W. H., 220, **279,** 288, 291, **387, 390, 397, 398,** 889, **925**
Pieters, **1120**

Pigorini, 680, **719**
Piland, 363, **406**
Pincussen, 799, **842**
Piper, C. S., 348, 349, **399**
Piper, S. H., **53**, 736, **762, 766**
Pirschle, 227, **279**, 646, 647, 651, **719**, 760, **766**
Pitcaithly, 362, **397**
Pitra, **401**
Pitsch, 644, **719**
Pitz, 326, **397**
Plaetzer, 590, **634**
Plagge, **397**, 701, **720**, 946, **1004, 1009**
Plank, van der, **640**
Platenius, 686, 701, **720, 925**
Platt, **176**
Pleasants, **516**
Plester, 597, **634**
Plimmer, **397**
Plitt, 1063, 1081, **1120**
Plowe, 87, 88, 106, **118**
Poehlman, 291, **398**
Poesch, 360, **398**, 1080, **1113, 1120**
Pohlman, 170, **176**, 220, **279**, 291, **398**
Poirault, 29, **64**
Pojarkova, 1093, **1116**
Polak, 797, **842**
Poliakov, 672, **722**
Policard, **766**
Politis, 36, **64**
Pollacci, 328, 329, **396, 398**, 436, **508**, 575, 582, **643**, 656, **716**
Pollard, **53**, 661, **720**, 736, **766**
Polle, 123, **183**
Polossin, **709**
Polovrăgeanu, 883, **925**
Pond, 827, **842**, 871, **925**
Pool, 449, 450, **516**
Poole, 569, 572, **624**
Poos, **387**
Pope, 993, 995, **1012**, 1025, 1026, **1116, 1120**
Popenoe, **58**
Popesco, 151, **183**
Popoff, H., 795, **842**
Popoff, M., 652, **720**
Popova, 795, **842**
Popp, 1073, 1074, **1120**
Porodko, 124, 125, **183**
Porowski, **634**
Porter, 588, **618**
Porterfield, 1026, 1070, 1071, **1120**
Portheim, 312, **383**
Portsmouth, 973, **1012**
Post, 1080, **1121**
Posternak, 303, **398**
Potapov, 223, **279**
Potter, G. F., 700, 701, **720**
Potter, M. C., 953, **1012**
Powell, **64**
Powers, 195, **214**, 326–328, **398**, 1087, **1121**
Pozen, 826, **842**
Pozerski, 816, **834, 842**
Pozzi-Escot, 675, **720**
Prankerd, **1121**
Prantl, 1028, **1121**
Pratt, D. S., 817, **842**
Pratt, I. W., **183**
Pratt, M. C., 560, 566, 575, **619**
Pratt, R., 1074, **1099**
Prazmowski, 655, **720**
Preece, 13, **64**
Prescott, J. A., 1071, **1121**
Prescott, S. C., **834**
Presley, **634**
Preston, 7, **65**
Prevot, 223, **279**
Prianischnikov, 168, **183**
Prianischnikow, 308, 356, 357, **398**, 645, 647, 663, 685, **720**
Priestley, J., 554, 555, **634**
Priestley, J. H., 4, 15–18, 31, 35, **60, 65, 68**, 206, 207, 211, **214**, 566, 568, 572, 573, **634, 640**, 665, 666, **719, 720**, 737, 738, 750, **766**, 854, 859, 863, 901, **925, 927**, 974, **1007**, 1019–1022, 1026, 1034, 1043, 1070, 1084, 1090, **1110, 1121, 1126, 1131**
Prince, 303, **376**
Pringsheim, **65**
Proebsting, 317, 318, **398, 405**
Prunet, 791, **842**
Prutzkova, 339, **398**
Pucher, 662, 663, **724**, 891, **929, 1128**
Pugh, E., 654, **715**
Pugh, G. W., 85, **118**
Puhr, 560, 561, 563, **634**
Pulling, 202, 209, **214**, 456, **517**
Pumphrey, **118**
Puppo, 456, 461, **515**
Purer, 134, **176, 1000**
Puriewitsch, 613, 615, 616, **634**, 826, 828, **842**
Purvis, 1094, **1121**

Q

Quartaroli, **398**
Quayle, 363, **384**
Quintanilha, 814, **842**
Quirk, 47, **67**
Qureshi, 573, **634**

R

Raber, 106, **118**, 270, **279**, 545, **634**, 687, **720**, 1024, **1121**
Rabkin, **1010**
Raciborski, 162, **183**
Rader, **632**
Rahn, 663, **720**
Rajvansi, 573, **634**
Raleigh, 1048, **1121**
Ramaley, 1079, 1080, **1121**
Randolph, 35, **65**, 551
Ranjan, 992, **1012**
Rankin, E. J., 332, **403**
Rankin, W. H., 649, 651, **725**, 884, **925**
Ransom, 992, **1012**
Raper, 734, **765**
Rastek, **183**
Ratsimamanga, **626**
Ravenna, 651, **708**
Ravin, 296, **398**

Ray, 968, **1012**
Rea, H. E., 137, **185**, 1058, **1121**
Rea, M. W., 47, 49, 50, **62**, **65**, 424, **517**
Read, **186**
Ream, 890, **919**
Redfern, 227, 231, 232, **279**
Redington, 1079, **1121**
Reed, E., **1121**
Reed, G. B., 977, 979, 981, 987, 991, **1012**, **1013**
Reed, H. S., 45, 49, **65**, 162, 163, 168, 171, **184**, 211, **214**, 217, 231, 233–235, **274**, **279**, 289, 299, 304, 306, 310, 312, 314, 315, 319, 321, 332, 364, **384**, **398**, 419, 423, 465, 470, **517**, 826, **842**, 866, **925**, 1026, 1043, 1059, 1087, **1121**, **1122**
Reeves, **1122**
Regeimbal, 968, **1013**, 1057, **1122**
Rehfous, 418, **517**
Rehm, 1086, **1122**
Reichert, 532, 537, 538, **634**
Reid, F. R., 170, **180**, 284, 345, **387**, **401**
Reid, M., 129, **183**
Reid, M. E., 587, **634**, 697, 703, 704, **720**, 1079, 1091, **1122**
Reimer, 325, 326, **398**
Reinau, 580, **634**
Reinhard, H., **65**
Reinhard, M. C., 1076, **1122**
Renner, 208, 209, **214**, 451, 460, **517**, 867, 870, **925**
Resch, **64**
Rewald, 886, **925**
Reynolds, 974, **1016**, 1076, **1122**
Rhine, J. B., 447, **517**, 760, **766**, 907–909, **925**
Rhine, L. E., 987, 992, 993, **1013**
Rhodes, A., 124, 125, **182**
Rhodes, E., 16, **65**, 206, 207, **214**
Rhodes, R. D., **112**
Rich, 132, **179**
Richand, **842**
Richard, **60**
Richards, E. H., 653, **721**
Richards, F. J., 320, **398**, 687, **720**, 974, **1003**
Richards, H. M., 582, **634**, 951, 953, 962, 963, **1013**
Richards, P. E., 369, **392**
Richardson, A. E. V., 499, **517**, 643, **721**
Richardson, G. M., **721**
Richardson, J. E., 1032, **1122**
Richardson, W. A., 6, 10, **54**, **623**
Richardson, W. D., 284, 285, 335, **398**
Richey, 886, **926**, 1082, **1122**
Richman, M. F., 1080, **1107**
Richman, M. W., **1130**
Richter, 975, **1013**
Ricome, 464, **517**
Ridgway, 157, **183**
Riebsomer, **767**
Riede, 886, **925**
Riedel, 586, **620**, **635**
Riesenberg, 576, **635**
Rigg, 153, **183**, 956, **1013**
Ringer, 815, **842**
Rippel, **279**, 293, 320, 351, 360, **397**, **399**, **517**, 587, **635**, 882, **925**, 1026, **1122**
Ripperton, 538, **624**
Rischawi, **1013**
Rislund, 995, **1002**
Risskina, 794, **841**
Ritter, 5, 11, **65**
Rivett, F., 871, **926**
Rivett, M., 871, **926**
Rivett, M. F., 739, **767**
Roach, B. M. B., 295, **399**
Roach, W. A., 884, **926**, **928**
Roark, 346, **393**
Robbins, W. J., 221, 263, **279**, 295, 296, **399**, 665, 666, **721**, **1013**, 1051, 1058, **1122**, **1123**
Robbins, W. R., 316, 319, 320, 322, 323, 327, **396**, **399**, 646–648, 674, **718**, **723**
Robbins, W. W., 1053, **1098**
Roberds, **635**
Roberg, 364, **399**
Roberts, E. A., 39, 157, **184**, 202
Roberts, H. F., 553, **635**, 693, **721**
Roberts, R. H., 13, 14, **67**, 552, **635**, 699, 700, 703, **721**, **1013**
Robertson, A., 823, 824, **843**
Robertson, D. P., **635**
Robertson, D. W., 123, **184**, **714**, 1047, **1123**
Robertson, I. M., 49, 50, **65**
Robertson, J. H., 889, 890, **925**
Robertson, R. A., 782, 783, **842**
Robertson, T. B., 1026, **1123**
Robinson, B. B., 1080, **1123**
Robinson, M. E., **721**, 805, **843**, 984, **1011**, **1013**
Robinson, T. W., 1039, **1118**
Robinson, W. O., 283, 346, 366, **399**
Rockwood, 696, **706**
Roddis, 545, **635**
Rodriguez, 345, **399**, 1089, **1123**
Rogers, C. F., 333, 893, **925**
Rogers, C. H., **398**
Rogers, W. S., 127, 138, 140, 141, **184**
Rohmann, **1013**
Rohrabaugh, **767**, 909, 910, **926**
Rohram, **725**
Rolf, 737, 742, **765**
Roller, 652, **708**
Römer, 371, **406**, 883, **930**
Roovaart, van de, 419, **520**
Rosa, 1050, 1055, 1057, 1064, 1066, **1123**
Rose, D. H., 935, 982, **1004**, **1013**, 1050, **1123**, **1131**
Rose, R. C., 982, **1001**, **1013**, 1054, 1055, 1089, **1108**, **1123**
Rosecrans, **635**
Rosen, 470, **517**, **635**
Rosenberg, 992, **1013**, **1017**
Rosenblatt, 346, 351, **376**
Rosene, 974, **1013**
Rosenfels, **279**
Roshardt, 864, 865, **926**
Ross, **832**
Rosso, 649, **719**
Rothemund, 545, 548, **627**, **635**
Rothgeb, 959, 963, **1000**
Rothstein, **840**
Rotmistrov, 138, 139, **184**
Rotunno, 1060, **1123**
Roux, 535, **630**, **839**
Rowlands, 1032, **1123**
Roxas, 138, 139, **184**

Roy, 300, 365, **393, 394,** 882, 883, **923**
Rubenstein, 800, **840**
Rubinstein, **118**
Rudolfs, 227, 231, **279,** 354–356, **399,** 556, **635,** 1060, **1123**
Rudolph, 36, **65,** 437, **517**
Ruggles, 884, **923**
Ruhland, 50, **65,** 104, 108, **118,** 227, **279, 926**
Rumbold, 884, **926**
Rumm, 465, **517**
Runnels, 469, 470, **521**
Runyon, 443, **517**
Ruot, **392**
Ruprecht, 350, 357, **399, 401**
Rushkovskii, 746, **767**
Russel, 144, 148, **180**
Russell, E. J., 146, 160, **184,** 305, 318, **399,** 653, **721**
Russell, J., 170, **184**
Russell, J. B., 818, **843**
Russell, W. C., 552, **635**
Russell-Wells, **63**
Ruth, 550, **635**
Ryan, 49, **57**
Rybin, 455, **515**
Rysselberghe, 101–103, 110, **118**

S

Saastamoinen, **724**
Sabalitschka, 576, **635**
Sachs, 33, 159, **184,** 193, **214,** 236, 241, 242, **279,** 297, 335, **404,** 555, 559, 606–608, 611, **635,** 758, **767,** 865, 905, 906, **926,** 932, **1013,** 1028, 1029, **1123**
Saeger, 348, 350, **399,** 652, **721, 1123**
Sahai, 736, 740, 741, 744, **762, 767**
Sahasrabuddhe, 746, 750, **767**
St. George, 885, **915**
St. John, E. Q., **837**
St. John, J. L., 325, 327, **396**
Salisbury, 419, 420, 424, **517**
Salm-Horstmar, 308, **399**
Salmon, M. R., 587, **641, 725**
Salmon, S. C., 339, **399,** 890, **926,** 1064–1068, **1109, 1123, 1124, 1127**
Salter, 263, **279, 399,** 488, **511**
Samant, 558, 592, **623**
Samec, 535, 537, **635**
Samisch, 560, 562, **635,** 883, **926,** 983, 984, **1000, 1013**
Sammett, **390**
Sampson, A. W., 154, **182,** 457, 464, **517,** 560, 562, **635, 721,** 883, 889, **924, 926,** 1090, **1123**
Sampson, H. C., 882, **926**
Samuel, 348, 349, **399**
Samuels, 909, 910, **921**
Sandberg, 807, **843**
Sande-Bakhuyzen, van de, **520,** 536, **640,** 1026, **1128**
Sanders, 9, **65**
Sando, C. E., 416, **515,** 552, **631,** 726, 736, **765, 767**
Sando, W. J., 1092, 1093, 1095, **1117**
Sands, 14, 15, **65**
Sanford, 884, **926**
Sani, 163
Saposchinkoff, 603, **635**
Sapozhnikova, 973, **1013**
Sapoznikow, 684, **721**
Saran, 974, **1012**
Sargant, 823, 824, **843**
Sauer, **635**
Saunders, 164, **184,** 462, 473–477, 479, 480, **515**
Savage, 1069, **1123**
Sawa, 349, **390**
Sawyer, G. C., 527, 558, 560, 563, 565, **623**
Sawyer, W. H., 423, 426, **517**
Sayre, C. B., 126, 127, **184**
Sayre, J. D., **65,** 317, 328, **395, 399,** 416, 426–429, 432–434, 436–438, 440, 448, 485, **517,** 549, **635,** 886, **926,** 1068, **1123**
Saywell, **635**
Scarth, 5, 25, **65,** 429–433, 435, **517, 518**
Schafer, **65**
Schaffner, 1081, **1124**
Schaffnit, 1065, 1068, **1124**
Schander, H., 827, **843**
Schander, R., 466, **518**
Schanderl, 614, **635**
Schanz, 1072, **1124**
Scharrer, 288, 300, 365, 368, 369, **400**
Scheele, 554
Scheibe, **1124**
Schelesnow, **380**
Schellenberg, 11, **66**
Schenck, **768**
Schenker, **843**
Scherer, 884, **926**
Schermerhorn, 316, 319, 320, 322, 323, 327, **396, 399,** 662, **718**
Schertz, **518,** 542, 549, 552, **635, 636, 641**
Schiletter, 701, **721**
Schillbach, **397**
Schimper, 30, 33, **66,** 241, 242, **279,** 309, 319, **400,** 535–537, 559, **636,** 673
Schleisinger, 792–794, **843**
Schley, **1124**
Schloesing, 236, **279**
Schmidt, D., 1061, **1124**
Schmidt, E. W., 36, **66,** 353, **400**
Schmidt, R. H., 753, **767,** 906–908, **926**
Schmidt, von, 15, **68**
Schmitz, 782, **843**
Schmucker, 344, **400,** 465, **518**
Schneiderhan, 168, **184**
Schneider-Orelli, 899, **926**
Schoenbein, 987, **1013**
Schollenberger, 338, **400**
Schomer, 588, **636**
Schorger, **66**
Schramm, 655, **721**
Schreinemocher, 252, **279**
Schreiner, 162, 163, 168, 171, 172, **184, 185,** 252, 253, **279, 280,** 302, 349, **400,** 650, **721**
Schreven, van, 343, **404**
Schricker, 320, **386**
Schröder, D., 1069, **1124**
Schröder, H., 83, 84, **118**
Schroeder, H., 598, 599, **636**
Schroeder, J., 896, **926**
Schroeder, R. A., **636**

Schropp, 300, 368, 369, **400**
Schröppel, 991, **1013**
Schryver, L. B., 573, **636**
Schryver, S. B., 10, 13, 21, 22, **53,** 54, **64, 66,** 535, **622, 636,** 676, **707**
Schultz, A. S., 794, **843**
Schultz, M., 649, **721**
Schulz, E. R., 676, 677, **723**
Schulz, K. G., 971, **1013**
Schulz, M., 696, **721**
Schulze, B., 138, 139, **185,** 361, **400,** 680, 683, **721**
Schulze, E., 12, 13, **66,** 662, 663, 687, **721,** 743, 744, **767,** 911, **926**
Schumacher, 900, **926**
Schürhoff, **66**
Schuster, 321, 323, **400,** 692, **721**
Schütz, 680, 683, **721**
Schwaibold, 365, **400**
Schwartze, 746, **767**
Schwarz, 151, 157, **185**
Schweigart, 796, 798, **837**
Schwendener, 418, 425, **518,** 861, **927**
Scofield, C. S., 239, **280,** 342, **400**
Scofield, H. T., 900, **915**
Scott, D. A., **62**
Scott, L. I., 207, **214,** 1020, **1121**
Scott, S. G., 291, **400**
Scurti, 751, **767**
Sears, 358, **379**
Seelhorst, von, 123, 127, **186**
Seiden, 882, **927**
Seifriz, 6, 24–26, 28, **66,** 87–89, 106, 108, **118,** 1086, **1124**
Sekine, 673, **706**
Sellschop, 1065, **1124**
Selwyn, 889, **927**
Semmens, 799, 800, **831, 843,** 1077, **1124**
Sen, B., 103, **118**
Sen, H. D., 815, **843**
Sen, P. K., **636**
Senebier, 554, 555, **636**
Senft, **402**
Serex, 882, **927**
Serviss, **919**
Sessions, 647, **721**
Seubert, 1034, **1124**
Severini, 647, **719**
Sewell, 168, 172, **185,** 260, **280,** 643, **711**
Seybold, 461, **518,** 1039, **1125**
Shands, 1069, **1127**
Shantz, 123, **185,** 190–194, 196–198, **212,** 411, 453, 456, 458–461, 482–484, 486, 487, 490, 491, 497–499, 501, **507, 508, 518**
Shapiro, 448, 461, **518**
Shardakov, S., 505, **518**
Sharp, **66**
Shatkin, **725**
Shattuck, 884, **927**
Shaw, D. A., 992, **1003**
Shaw, G. W., 691, **722**
Shaw, J. K., 291, **400**
Shaw, R. H., **927**
Shcheglova, 600, **636**
Shealy, 363, **395**
Shear, 643, **722, 1124**
Sheard, 1073, **1124**
Shedd, 44, **66,** 300, 327, **393, 400**
Sheldon, 759, **766**
Shelford, 588, **636**
Sheloumova, 961, **1012**
Shepperd, 138, 139, **185**
Sherman, C. C., 242, **280**
Sherman, H., 949, 952, 992, 994, **1013**
Sherman, H. C., 789, 791–794, 796–798, **843, 844**
Sherrill, 788, **838**
Sherwood, H., 189, **213,** 454, **509**
Sherwood, R. C., 829, **831**
Shestakov, **709**
Shibata, 1085, **1124**
Shibusawa, 1085, **1124**
Shill, 969, **1013**
Shippy, 1064, **1124**
Shirley, 611, **636,** 1073, 1078, **1124**
Shive, 134, **174,** 199, **214,** 241, 243, 246–248, 252, 257, 258, 260–262, 264, **270, 272, 276, 280,** 330–333, 369, **375, 387, 388, 392, 399, 400,** 467, 468, 485, **518,** 542, **638,** 646–648, 674, **705, 708, 709, 721, 722**
Shmuk, 672, **722**
Shohl, 540, **636**
Shorey, 171, **184,** 743, **767**
Showalter, **118**
Shreve, E. B., 442, 450, 451, 462, 473, 486, 491, 492, **514, 518**
Shreve, F., 870, **922, 1115**
Shriner, **761, 762**
Shrinkhande, 14, **64**
Shriver, **917**
Shuck, 1052, 1053, **1124, 1125**
Shull, C. A., 75, 83, 84, 93, 110, **119,** 164, **185,** 200, 202, 208, 210, **215,** 243, **277, 400,** 475, **518,** 613, **637, 722,** 862, 863, 870, 874, **920, 927,** 970, 994, **1014,** 1073, 1076, **1125**
Shull, S. P., 75, 110, **119**
Shulov, 353, **400,** 647, **722**
Shutt, 889, **927**
Sideris, **280,** 367, **400, 518,** 1087, **1125**
Sidorin, 334, **400**
Siegert, 353, **396**
Siegfried, 570, **637**
Siemens, **185**
Sierp, 151, **185,** 1039, **1125**
Sifton, 1046, **1125**
Sigmund, 807, **844**
Simon, 884, **927**
Simmonds, **65**
Singh, B. N., 579, 601, **637,** 961, 974, **1005, 1014,** 1074, **1125**
Singh, K., 137, **185**
Singh, L., **66**
Singh, S. B., 1026, **1110**
Singh, T. C. N., 1085, **1125**
Sinnott, 741, 742, **767**
Sisson, 2, 7, **56**
Sitton, 431, 434, **519**
Sjöberg, 791, 797, 799, **844**
Sjogren, 123, **184**
Sjostrom, **637**
Skeen, 88, **119**
Skene, 230, **280**

Skinner, C. E., 548, **637**, 1032, **1107**
Skinner, H. J., 24, **66**
Skinner, J. J., 168, 171, **181, 184, 185**, 252, 253, **279, 280**, 284, 345, 346, 350, 353, **401**, 650, **721**
Skoog, 1044, 1076, **1125–1127**
Skvortzov, 599, **637**
Slade, 806, **844**
Small, J., 47, 49, 50, **62, 65, 66**, 1088, **1125**
Small, T., 586, 587, **633, 637**
Smedley, 750, **767**
Smirnow, 663, **722**
Smith, A. J. M., 1046, **1125**
Smith, A. M., 49, 50, **65**, 477, **519**, 583, 584, **619, 637**, 961, **1014**
Smith C. L., 746, 748, **768**
Smith, E. F., 47, **67**, 866, **927**
Smith, E. P., 967, 972, **1014**
Smith, E. V., 878, **921, 922**
Smith, F., 863, 870, 874, **927**
Smith, F. B., 160, **185**
Smith, G. E., **519**
Smith, G. M., 26, **62**, 852, 854, 864, 869, 874, 875, **922**
Smith, H. B., 431, **519**
Smith, H. F., **519, 927**
Smith, J., 552, **637**
Smith, J. A. B., **53**, 736, 744, **767**
Smith, J. B., 169, **178, 182**, 291, **383, 401**, 852, **924**
Smith, J. G., 337, **383**
Smith, J. H. C., **114**, 552, 553, **638**
Smith, L. E. W., 551, **637**
Smith, O., 17, **67**, 314, **377**, 551, **637, 927**, 967, **1014**, 1063, **1125**
Smith, R. S., 163, **180**
Smith, S. H. C., 543, **637**
Smith T. O., 319, 320, 323, **397, 401**, 892, **914**
Smith, W. H., **519**
Smith, W. W., 12, **67**
Smolenski, 743, **767, 769**
Smyth, E. M., 599, **637**
Smyth, E. S., 189, **215**, 961, **1014**
Snell, 313, 314, **375**
Snow, L. M., 152, 153, **185**
Snow, R., **927**, 1033, 1038–1041, 1044, **1111, 1125**
Snyder, E., 140, 141, **178**
Snyder, H., 883, **927**
Snyder, R. S., 691, **718**
Snyder, W. C., 300, 344, **386**
Söding, 1034–1036, **1125**
Sohns, **56**
Sommer, A. L., 154, **185**, 299, 305, 311, 336, 343, 362, 363, **401**, 575, **637**, 674, **722**
Sommer, W., 793, **836**
Sonavne, **1126**
Sorauer, **280**
Sörensen, 812, **844**
Sorokin, 154, **185**, 311, 343, **401**
Spaeth, J. N., 1048, **1126**
Spaeth, R. A., 111, **119**
Spalding, 152, **185**, 188, 189, **215, 519**
Speer, 741, **764, 768**
Spencer, E. L., 261, **280**
Spencer, M., 793, **840**
Sperry, O. E., 416, **519**
Sperry, T. M., 138, 139, **185**
Spier, 5, **65**
Spirgatis, 587, **637**
Spitzer, G., 689, **722**
Spitzer, W., **1013**
Spoehr, **61**, 527, 536, 552–555, 557, 567, 570–576, 580, 589, 592, 606, 607, 611, 612, 617, 618, **637, 638, 722**, 936, 938, 971–973, **1014**, 1021, **1115**
Sponsler, 3, 6, 7, **67**, 536, **638**
Sprague, G. E., **927**, 1082, **1122**
Sprague, G. F., 1093, **1126**
Sprague, H. B., 542, 597, **638**, 890, 1076, **1126**
Sproat, **519**
Sreenivasan, 338, **401**
Stadler, 1075, **1126**
Staehelin, **67**
Stahl, A. L., **280**, 647, **722**
Stahl, E., 236, **280**, 436, 491, **519**
Stalfelt, 461, **519**
Stalling, 653, **722**
Stamm, 866, **927**
Stanescu, 560, 562, 602, 603, **638**
Stanewitsch, 745, **766**
Stanfield, 41, **67**
Stange, 39, **67**
Stankov, 223, **279**
Stanley, J., 485, **519**
Stanley, O. B., **768**
Stanley, W. M., 216, 222, **278**
Stansel, 140, 141, **186**
Stanton, 1049, **1102**
Stark, O. K., 77, 79, **119**, 484, **519**, 691, **722**, 912, **927**
Stark, R. W., 748, 749, 751, **768**
Starkey, 1080, **1126**
Starrett, 17, **52**
Starring, 703, **722**
Stefan, 440, **519**
Stegmann, 743, **769**
Steiger, 744, **767**
Steinbauer, 1078, **1126**
Steinberg, **519, 1126**
Steiner, **722**
Steinkoenig, 283, **399**
Stephens, 686, **723**
Stern, A. L., **67**
Stern, E., 814, **844**
Stern, H. J., **706**
Stern, K. G., 814, **844**
Stevens, C. L., 148, **185**
Stevens, N. E., 17, **67**
Stevens, O. A., 1047, **1126**
Steward, F. C., 97, **119**, 223, **279–281, 401**, 743, 746, **768**, 899, 901, **927**, 973, **1014**
Steward, W. D., **271**, 289, **401**
Stewart, C. R., 645, **722**
Stewart, F. C., **1126**
Stewart, G. N., 99, **119**
Stewart, G. R., **185**, 365, **401**
Stewart, J., 367, **401**
Stewart, R., 327, **401**, 691, **722**
Stewart, W. D., 459, **506**, 1074, 1078, **1096**
Stich, 947, 963, 969, **1014**

Stiebeling, 328, **401**
Stieler, **185**
Stiles, 87, 88, 99, 101, 102, 107, 110, **119**, 220, 226–229, 233, 246, 248, 257, 266, 267, **281**, 542, 557, 568, 569, 585, 588, 589, 592, 596, **628, 638**, 932–934, 937, 940, 946, 949, 952, **1014**, 1034, 1065, 1084, **1110, 1126**
Stoa, 166, **174**
Stockberger, 984, **1016**
Stocker, 202, **215**
Stocklasa, 159, 160, 162, **185**, 317, 319–321, 359, 365, **401, 402**, 572, 582, **638**, **744**, **768**, 940, **1014**
Stoddart, 40, **67**, 156, **185**
Stokes, 294, **402**
Stoldt, 600, **626**
Stolhane, 258, **281**
Stoll, 328, **406**, 557, 568, 570, 576, 593, 597, 598, 601, 607, 611, **641**
Stone, 33, **67, 844**, 1071, **1126**
Storey, 326, **402**
Stout, A. B., 35, **67**, 553, **638**
Stout, G. J., **176**
Stoward, 783, **844, 1014**
Strain, 552, 553, **638**
Strasburger, **768**, 864, 866, **927**
Stratton, **67**
Strausbaugh, 1066, **1126**
Street, 290, 317, **374, 402**, 642, **717**
Streeter, 551, **633**
Streets, **927**
Strobel, 288, 365, **400**
Strowd, 672, **722**
Strugger, 428, **519**
Stscheglova, 600, **630**
Stuart, G. L., 365, **396**
Stuart, N. W., 288, **722**, 891, **927**, 1068, **1107**
Sturkie, 890, **927**
Stutzer, 649, **722**
Suguira, 808, **835**
Suit, 604, **639**
Sullivan, B., 740, **768**, 807, **844**
Sullivan, J. T., **720**, 888, 892, **921, 927**
Sullivan, M. X., 163, **185**
Summers, **1126**
Sumner, 776, **844**
Suneson, 1069, **1126**
Surface, 884, **927**
Suroz, **927**
Suzuki, U., 680, 682, 685, **722**
Suzuki, V., 365, 367, **402**
Swain, **402**
Swanback, 234, 235, **276**, 317, 347, 348, **374, 337, 402**
Swanson, C. O., **402**, 806, 828, 829, **844, 845**, 890, **926**, 954, **1015**
Swanson, P. P., 1032, **1107**
Swarbrick, **928**
Swart, 882, **928**
Sweeney, 884, **915**
Swingle, 1031, **1126**
Syroeshkina, 355, **375**
Szladits, **834**
Szücs, 359, **402**

T

Tabor, 856, **922**
Tageeva, **639**
Tait, G. M., 1087, **1126**
Tait, L., 813, **845**
Takamine, **845**
Talley, 701, **722**
Tam, 550, **639**
Tammann, 595, **639**
Tanaka, 294, **402**
Tanberg, 789, **843**
Tang, P., 946, 950, 974, **1015**, 1051, **1126**
Tang, P. S., **1002**
Tangl, 99, **113**
Tanka, 780, **845**
Taranovsky, 1044, **1126**
Tarr, 262, **281**
Tartar, 325, 326, **398**
Tascher, 828, **845**
Tashiro, 947, **1015**
Taubenhaus, 137, **185**, 476, **510**
Tauber, 774, 776, **845**
Tauböck, **389**, 696, **714**
Tavcar, 419, **519**
Taylor, A. E., 731, 735, **761**, 808, 811, **845**
Taylor, C. A., 876, **918**
Taylor, G. S., 942, **1009**
Taylor, J. K., 655–657, **715**
Taylor, J. W., 443, **512**
Taylor, M. W., 552, **635**
Taylor, R. W., 701, **723**
Taylor, T. C., 538, **639**
Teakle, 303, **403**
Teller, 688, **723**
Templeman, 687, **720**
Ten Eyck, 138, 139, **186**
Tenney, 171, **186**
Teodoresco, **845**
Terroine, 731, 751, **768, 1015**
Tester, 192
Tetley, 1034, **1126**
Thatcher, **67**, 174, **186**, 299, **403**, 531, 540, 544, **639**, 643, 660, 661, 689, 691, **723**, 761, **768**, 778, 780, 782, 793, 806, **845, 1126**
Theis, 807, 811, **845**
Thelin, 647, 649, **723**
Thielman, **67**
Thielmann, 458, **512**
Thiessen, 6, **67**
Thimann, 1034–1036, 1038, 1044, **1098, 1103, 1125–1127, 1130**
Thoday, 18, **67**, 210, **215**, 302, 307, 316, **403**, 598, 609, **639**, 947, 967, **1015**, 1023, **1127**
Thom, 497, 498, **519**
Thomas, A. W., **639**, 796, **843–845**
Thomas, B. H., 1032, **1100**
Thomas, E. C., **722**
Thomas, E. E., 884, **928**
Thomas, E. M., 13, **66**, 535, **636**
Thomas, H. E., 363, **396**
Thomas, H. S., 967, **1015**
Thomas, L. A., 884, **928, 1015**
Thomas, M., 882, **928**, 941, 943, **1015**
Thomas, N., 486, **519**

Thomas, W., **186,** 222, 235, **281,** 287, 288, 290, **403,** 578, **639,** 672, 674, 678, 679, **723,** 890, **928**
Thompson, H. C., 1091, **1127**
Thompson, R. C., 1053, **1127**
Thompson, T. G., 911, **928**
Thomson, 247, **282,** 1091, 1092, 1094, **1127**
Thone, 485, **514**
Thor, 746, 748, **768**
Thornton, H. G., 166, **186,** 299, 342, **377,** 890, **928**
Thornton, N. C., 972, **1015, 1016,** 1052, 1053, 1056, 1090, **1127**
Thornton, S. F., 291, **403**
Thurber, 535, **639**
Thut, **410, 519,** 870, **928**
Tidmore, 303, **387, 403,** 649, **723**
Tiedjens, 646–648, **723**
Tilford, 470, **519**
Timiriazeff, 591, 613, 614, **639**
Timmons, 1067, **1127**
Tincker, 132, **186,** 316, 319, **403,** 1037, 1073, 1079, 1080, **1105,** 1127
Tinker, **113**
Tisdale, 131, **186**
Tison, 882, **928**
Tjebbes, 83, **119,** 1079, **1127**
Tobler, 319, **403**
Tolbert, 1033, **1116**
Tollens, **55,** 241, 242, **281,** 882, **928**
Tommasi, 751, **767**
Tompson, 784, 785, 787, 788, **842**
Toole, 264, **281,** 905, 906, **928**
Torrey, 823, 826, **845**
Torrie, **506**
Tottingham, 13, 14, **67,** 241, 242, 244, 245, 249, 252, 255, 257, 259, 264, **277, 281,** 312, 326, 327, 330, 332, 353, 355, 356, **385, 403,** 592, **639,** 674, 676, 677, 686, 692, **709, 715, 723,** 749, **768,** 1069, **1103, 1127**
Trajkovich, 551, **639**
Trandafirescu, 883, **928**
Transeau, 490, **519, 639**
Traub, 140, 141, **186,** 887, **928**
Traube, 108, 975, **1016**
Travernetti, 246, **274**
Treboux, 644, **724**
Treitel, 135, **186**
Trelease, H. M., 256, 260, 264, 269, **282,** 315, **404,** 604, **639**
Trelease, S. F., **176,** 247, 256, 257, 260, 264, 269, **281, 282,** 285, 286, 315, 353, 356, 366, **403, 404,** 430, 441, 486, **519,** 600, 604, **622, 639,** 645, **724,** 1071, 1072, **1099**
Tröndle, 104, **119**
Trost, 359, **386**
Trowbridge, 885, 887, **928**
True, 107, 234, 268, **282,** 289, 310, **404, 713,** 984, **1016**
Truesdail, 1037, **1130**
Trumpf, 1070, **1127**
Truog, 40, 48, 49, **62, 68,** 262, **282,** 284, 302, 309, 357, **396, 404**
Tschesnokov, 561, **619, 639**
Tschirch, 738, **768,** 1033, **1128**
Tschudy, 588, **639**
Tsweet, 547, 568, **639**
Tubbs, **404**
Tucker, G. M., 882, **928**
Tucker, K. L., 1076, **1122**
Tucker, L. R., **68**
Tucker, M., 123, **186**
Tufts, 1032, **1128**
Tukey, **928,** 1044, **1128**
Tumanov, **520**
Tupper-Carey, 4, **68,** 207
Turner, D. M., 1093, **1100**
Turner, M. A., 1020, **1128**
Turner, T. W., 128, **186,** 305, **404,** 703
Turpin, 160, 161, **186**
Turrell, 414, **520**
Tutin, 22, **68**
Tuttle, 742, **765, 768**
Twiss, 34, **68**
Tyagny-Ryadno, 652, **724**
Tychowski, 797, **842**
Tyner, 317, 318, **404**
Tysdal, 790, 798, **845,** 1067, 1069, **1120, 1128**
Tyson, 984, 994, **1016**

U

Uber, 24, **68**
Uddenberg, 792, 798, **841**
Ullrich, 684, 685, **724**
Ulvin, 549, 550, 597, **639**
Umrath, 1086, **1128**
Underhill, 815, **840**
Underwood, L. E., 248, 257, 258, **274**
Underwood, L. M., 131, **178**
Unger, 410, 481, **520,** 865, **928**
Uphof, 1079, **1127**
Ursprung, 94, **119, 120,** 477, **520,** 591, 601, 602, **639,** 861, 867, 868, **928, 929, 1128**
Usher, 568, 572, 573, **640**
Uspenskaja, **118**

V

Vacha, 968, **1013,** 1050, **1128**
Vagi, **710**
Valentine, **58**
Valle, **1122**
Valleau, 643, **724**
Vallee, 751, **768**
Vandecaveye, 288, **404**
Varadachar, 672, **724**
Varadpande, **115, 920,** 961, **1014**
Vasey, 1080, **1105**
Vassiliev, I. M., 503, **520,** 1069, **1128**
Vassiliev, M. G., 1069, **1128**
Vechot, **768**
Veihmeyer, 123, 148, **176, 179,** 192, 197, 205, **213, 215,** 434, 435, **520**
Venkatraman, **186**
VerHulst, 527, 540, **640**
Verner, 642, **724**
Verschaffelt, 493, **520**
Vesque, 866, **929**
Vickery, 662, 663, **724,** 891, **929, 1128**
Vieser, 575, **626**
Villano, 138, 139, **184**

Ville, G., **404,** 549, **640**
Ville, M. G., 654, 655, **724**
Vincent, 361, **404**
Vines, **640,** 795, 812, 814–816, 820, **845,** 866, **916, 929**
Vinogradova, 1083, **1113**
Vinson, 783, 788, **845**
Virtanen, 166, **186,** 566, **640,** 653, **724,** 890, **904, 929,** 1032, 1033, **1128**
Vittal, **186**
Vivian, 689, **725**
Vladescu, **929**
Voegtlin, 356, **394**
Voelcker, **404**
Vogel, 644, **711**
Vogt, 1072, **1128**
Vosburgh, 786, 787, **841, 846**
Vysotskii, 1086, **1128**
Vyvyan, 138, 140, 141, **184,** 488, **520**

W

Wadham, 1080, **1105**
Wadleigh, 261
Wadsworth, 192, 197, **215**
Wager, 573, **640**
Wagner, C. L., 318, **385**
Wähner, 828, **831**
Wakabayashi, 965, **1016**
Wakeman, 676, **719,** 891, **929**
Wakker, 738, **768**
Waksman, 12, **68,** 171, **186, 846**
Walbridge, 314, **391**
Waldo, 1080, **1101, 1128**
Waldron, 692, **724,** 1067, 1069, **1128, 1129**
Waldschmidt-Leitz, 775, **846**
Walker, F., 791, **843, 844**
Walker, J. A., **843**
Wallace, R. H., 600, **640,** 1037, 1041, **1129**
Wallace, T., 317, 323, 334, **391, 404, 405**
Waller, A. D., 1085, **1129**
Waller, J. C., 1085, **1129**
Walster, **1129**
Walter, **120,** 446, **520, 640, 929,** 1067, 1091, **1129**
Walters, 691, **722**
Waltman, 1087, **1129**
Walton, C. F., 784, **842,** 1068, **1129**
Walton, J. H., 985, **1003**
Walton, R. P., 535, **640**
Wangler, 669, **714**
Wann, 332, **386,** 656, **724,** 883, **929,** 1087, **1129**
Wanser, **1129**
Warburg, **68,** 328, **405,** 568, 576, 583, 584, 589, 592, **640,** 673, **724,** 972, **1016**
Ward, **846**
Ware, 324, **405, 406**
Warington, **299,** 338, 341–344, **377, 405**
Warne, 324, **405**
Warner, C. H., 573, **640**
Warner, J. D., 138, **176,** 364, **375**
Wartenberg, 807, **833**
Warweg, 992, **1003**
Wassilieff, 663, 687, **724**
Watenpaugh, 125
Watkin, 303, 304, 373, **389,** 883, 886, **918**
Watson, A. N., 480, 481, **520**
Watson, D. J., 354, **405, 640,** 690, **724**
Watts, 234, 235, 261, **271,** 321, **388,** 578, **619**
Wayman, 796, **843**
Wean, **282**
Weatherwax, 247, **282,** 539, **640**
Weaver, 121, 124, 127, 132, 134, 136–143, 147, 156, **176, 179, 186, 187,** 415, 471, 472, **520,** 889, 890, **913, 929,** 1082, **1129**
Webber, H. A., **68**
Webber, I. E., **405**
Webber, L. R., 535, **641**
Weber, C., 596, **640**
Weber, F., **120,** 428, 435, **519, 520**
Weber, H., **1016**
Webster, B., 365, **392**
Webster, J. E., 302, 304, 362, **405,** 672, 673, **724,** 743, **768,** 974, **998**
Webster, T. A., 574, **632, 717**
Weevers, 316, 317, **405,** 565, 566, **640,** 695, **724,** 900, **929**
Weidling, 576, **635**
Weier, 32, 33, **68, 640**
Weigand, 447, **520**
Weihe, A., 338, **405**
Weihe, H. D., 339, 340, **397**
Weihing, 149, **180, 187**
Weij, van, 1038, **1128**
Weimer, 17, **68,** 1066, **1129**
Weinberger, **521, 724,** 886, 892, **929**
Weis, 813, **846**
Weiss, A., 418, **521**
Weiss, F., 472, **521,** 990, **1016**
Weiszmann, 319, **405**
Weldon, 40, **62,** 137, **187,** 234, 235, **277,** 306, **393**
Wells, F. J., 689, **725**
Wells, S. D., **68**
Welton, 204, **215,** 339, **405,** 485, **521, 724,** 886, **929, 1129**
Wendt, **282,** 365, **405**
Went, F. A. F. C., **1129**
Went, F. W., 1034–1036, 1039, 1040, **1127, 1129, 1130**
Wentworth, **720**
Werner, H. O., 702, **724,** 889, **929,** 956, **1016**
Werner, O., **628**
Wessels, 360, **385,** 1087, **1127, 1130**
West, A. P., **762,** 805, **831**
West, C., 966, 967, **1007,** 1026, 1028, 1046, 1060, 1064, **1099, 1111, 1130**
West, E. S., 138, 139, **187,** 356, 363, **390**
Westall, 676, **708**
Westenberg, 900, **929**
Wester, 346, **405**
Westermaier, 860, **929**
Westgate, **725**
Westover, **1119**
Wetzel, 50, **65,** 189, **215**
Wheeler, H. J., 313, 356, 360, **385, 405**
Wheeler, R. V., 18, **60**
Wheldale, 984, **1016**
Wherry, 284, **406**
Whimster, 973, **1016**
Whitaker, 156, **187**
White, F. B., **1119**

White, H. L., 322, **406**, 587, **637, 641, 725**
White, J., 782, 813, **846**, 959, **1016**
White, P. R., 856, 882, **929**, 1058, **1130**
White, V. B., 1058, **1123**
White, W. J., **1096**
Whitehead, 975, **1016**
Whiteman, **1131**
Whiteside, 551, **641**
Whitfield, 41, **68**, 459, 464, 484, **519, 521**
Whitley, **717**, 976, **1010**
Whitney, **187**
Whitson, 689, **725**
Whitten, 84, **120**
Whyte, J., 431, 433, **518**
Whyte, R. O., **1130**
Wickwire, 363, **378, 396**, 1079, **1099, 1130**
Widdowson, 13, 15, **68**, 303, **392**, 529, **641**
Wierzchowski, 801, **846**
Wiesner, **69**, 549, **641**, 882, **929, 930**
Wiessmann, **390**
Wiggins, 426, 427, **521**
Wigoder, 1076, **1130**
Wijk, van, **404**
Wilcox, E. V., **406**
Wilcox, L. V., 342, **400**
Wilcoxon, 1037, 1041, **1131, 1132**
Wilfarth, 371, **385, 406**, 655, **712**, 883, **930**
Wilhelm, 1068, **1124**
Wilkins, F. S., 170, **187**
Wilkins, L. K., 307, **406**
Wilkinson, 1032, **1123**
Will, **846**
Willaman, 540, **641**, 946, 957, 958, 965, 971, 995, **998, 1001, 1010, 1116**
Willard, 129, 149, **187**, 443, **521, 930**
Williams, C. B., 1091, **1130**
Williams, C. F., 672, **725**
Williams, E. F., **53, 762**
Williams, H. F., 189, **215**
Williams, M., 1075, **1130**
Williams, P. H., 585, 586, **633**
Williams, R. F., 465, **521**
Williams, R. J., **725**, 1032, 1037, **1130**
Willis, 334, 363, **406**, 649, 651, **725**
Willstätter, 313, 328, 329, **406**, 542, 544, 547, 557, 568, 570, 576, 593, 597, 598, 601, 607, 611, **641**, 775, 776, **846, 1016**
Wilmot, 606, **641**
Wilson, A. L., 1087, **1130**
Wilson, B. D., 173, **187**, 654, **725**
Wilson, H. K., 339, **378**, 1067, **1109**
Wilson, H. W., 461, **521**
Wilson, J. B., 345, **379**
Wilson, J. D., 202, 204, 205, **214, 215**, 463, 469, 470, 485, **514, 521**
Wilson, J. J., 341, 343, **391**
Wilson, J. K., 163, 173, **181, 187**, 505, **521**
Wilson, K., 151, 152, **187**
Wilson, L. B., 354, 368, **406**
Wilson, P. W., 587, **625, 641**, 648, **713**
Wilson, R. W., **725**
Wilson, W. E., **621**
Wimmer, 371, **406**, 883, **930**
Wingard, 300, 351, **397**
Winkler, A. J., 889, 893, **930**
Winkler, H., 603, **641**
Winsor, A. A., 1025, 1090, **1119, 1120**
Winsor, C. P., 1090, **1120**
Winston, **406**
Winterstein, 663, 687, **721**, 743, **768, 769**
Wise, 741, **764, 768**
Wisselingh, van, 15, **68**
Withrow, R. R., 587, **641**, 1077, 1080, 1081, **1107, 1130**
Withrow, R. W., 488, **521**
Wodehouse, 100, **120**, 218, **282**
Woffenden, 16, **65**, 1090, **1131**
Wohack, 324, **406**
Wöher, 368, **406**
Wolf, 49
Wolfe, H. S., 196, **215**, 652, **725**, 1057, **1131**
Wolfe, T. K., 1050, **1131**
Wolff, E., 335, 336, **390**
Wolff, J., 328, **406**, 535, **625**, 801, **846**
Wollney, 236, **282**
Wolter, 542, 543, **627**
Woo, 671, 672, **725**
Wood, F. M., 5, 9, **69**
Wood, J. G., 189, **215**, 419, **521**
Woodard, 327, **406**
Woodford, 324, **392**
Woodhead, 17, **69**
Woodhouse, 855, 871, **930**
Woodroof, J. G., **187**
Woodroof, N. C., **187**
Woodruff, S., 535, **641**
Woodruff, W. M., 46, **52**
Woods, A. F., 482, **521**, 982, **1016**
Woods, J. J., 1068, **1131**
Woodward, 241, **282**
Work, P., 698, **725**
Work, R. A., 147, **174, 187**, 191, 205, **213, 215**
Worley, 110, **113**, 362, **397**
Wormall, **65, 69**
Wortmann, 136, **187**
Worzella, **187**, 1068, **1131**
Wrangell, von, 284, 286, 303, **404**
Wray, 154, **182**
Wrenger, 461, **521**
Wrenshall, 954, **1007**
Wright, A., 678, **708**
Wright, J. G., 871, **930**
Wright, P. A., **927**
Wright, R., 223, **280**, 973, **1014**
Wright, R. C., 958, **1016**, 1050, **1131**
Wright, W., 1021, **1131**
Wurdack, **69**
Wurmser, 592, **641**
Wurtz, 814, 815, **846**
Wyatt, 308, **406**
Wynd, 261, **282**, 289, 295, 364, 365, **406**, 946, 974, **1016**, 1074, **1131**

Y

Yamaguchi, 651, **725**
Yamano, 357, **406**
Yanovsky, 307, **406, 641**
Yap, 601, 602, **641**
Yapp, 424, 443, **522**

Yeager, 140, **187**
Yeats, 597, **627**
Yemm, 937, **1016**
Yenko, 805, **831**
Yocum, 431, **522,** 905, 906, **930**
Yoder, 691, **715**
Yoshida, 977, **1017**
Yoshisuke, **1017**
Youden, 230, 231, **273**
Young, D. W., 365, **394**
Young, P. A., 909, **930**
Young, R. S., 299, **406**
Young, T. F., 870, **920**
Young, V. H., 324, **405, 406,** 801, **846**
Yuncker, 462, **522,** 599, **641**

Z

Zacharowa, 1066, **1131**
Zahnley, 645, **725**
Zaitseva, **406**
Zalenski, V., 424, 433, 453, **522**
Zalenski, W., 744, **769**
Zaleski, 662, 684, 687, **725,** 942, 992, **1017**
Zamis, 798, **834**
Zavitz, 165, **187**
Zdobnicky, 572, **638**
Zechmeister, 552, **641**
Zeile, 776, **846**
Zeleny, 688, **725**
Zerban, **1017**
Zhemchuzhnikov, 446, **522**
Ziegenspek, 158, **187**
Zijlstra, 582, **641**
Zimmerman, 881, 893, **930,** 970, **1017,** 1037, 1040, 1041, 1082, 1088–1090, **1101, 1109, 1131, 1132**
Zinkernagel, 1088, **1132**
Zinzadzé, 264, **282**
Zirkle, 31, 34, 35, **69,** 536, **641**
Zscheile, F. P., 542, 543, 575, **641**
Zscheile, F. P., Jr., 552, **632**
Zucker, 466, **522**
Zurbicki, 247, **282**
Zycha, 104, **120**

SUBJECT INDEX

A

Abscission layer, 882
Absolute transpiration, 486
Absorbing compounds, oxygen, 606
Absorbing power, 93–96
 methods of determining, 94, 95
 observations of, 95–96
Absorption, active, 218–224
 of CO_2 by leaf, 569, 570
 coefficient of radiant energy, 612–615
 contrasted with permeability, 96
 depth of, by roots, 145–148
 of inorganic compounds, 293
 luxury, 290
 method of, photosynthesis, 607
 of organic compounds, 292–297
  carbohydrates, 294–296
  nitrogenous compounds, 297
  organic acids, 296–297
 passive, 217, 218
 of solutes, 216–271
 of water, 188–212
Absorption ratio, definition of, 224
 factors affecting, 224–227
  concentration of exterior medium, 225–227
  hydrogen-ion concentration, 227
  light, 224, 225
Accelerator, 779
Accessory cells, 418
Acetaldehyde, 934, 940, 969
Acetamide, utilization, 650
Acetic acid, 728
 excreted by roots, 159
 formula for, 728
 relation of, to diastase, 797
  to invertase, 788
  to lipase, 809
  to protein formation, 309
  to respiration, 939
Acetyl value, 735
Acetylene, 1041
Achroodextrin, 539
Acid (*see* various kinds)
Acid value of fat, 734
Acidity of cell sap, 45–51
 amount of, 45–47
 causes of, 46–47
 factors influencing, 47–51
  age and vigor, 50–51
  kind of tissue, 50
  previous treatment, 49–51
  reaction of medium, 47–49
  time of sampling, 49
 influence of, on catalase, 990
  on diastase, 797
Acidity of cell sap, quantities considered, 46
 in water cultures, 231
Acids, influence on transpiration, 464
Activated formaldehyde, 574, 670
Activation in respiration, 933
Activator, of enzymes, 779
Active absorption of solutes (*see* Physiological absorption)
Active iron, 328, 329
Active regulation of stomata, 441
Adamkiewicz's reaction, 665
Additive effects, 269
Adenine, in plants, 695
 in soil, 171
Adrenalin, 973
Aeration, effect of, on plant growth, 130, 131
 influence of, on decay, 171, 172
 relation of, to temperature, 133
 response to, of roots, 131–133
 of soils, 129–135
  methods of 129–130
 of water cultures, 131, 134, 247, 248
 by water plants, 131
Aerial conditions, evaporating power, of air, 458–462
  of relative humidity, 459, 460
  of temperature, 458, 459
  of wind, 460, 462
 relating to transpiration, 455–462
 of solar radiation, 456, 457, 458
Aerobic respiration, 933, 939–940
 contrasted with photosynthesis, 944, 945
 defined, 939
 relation of, to anaerobic, 940–944
Afterripening, agents, 968, 969
 changes accompanying, 1053–1056
 defined, 1053
 effect of, on catalase, 996
  on respiration, 968, 969
 ripening (*see* Ripening)
Age, relation of, to absorption of ammonium and nitrate, 647
  to acidity, 50–51
  to concentration of cell sap, 44
  to elemental requirements, 369–373
Air temperature, 474
Alanine, formula for, 661
 in soil, 171
 in tomato, 675
 utilization of, 651
Albino plants, 296, 428, 550, 551
Albumins, 667, 668
Alcohol vs. petroleum, 617
Aleurone, 658
Aleurone grains, 657, 658

Aleurone layer, 658
of corn, 824, 826, 827
Alfalfa, absorption, of elements, 290
of phosphorus, 302
aciditiy of sap, 48, 49
amino acid and amides, 662
boron absorbed, 342
calcium content, 307, 309
calcium-magnesium ratio, 308
clipping, and organic reserves, 890
copper content, 300
depth of absorption, 148
enzymes, 782
erepsin, 821
hemicelluloses, 13
hydrogen-ion concentration and growth, 1087
influence on other plants, 653
iodine content, 288
iron content, 300, 347
leaf structure, 413
temperature, 475
lithium content, 285
manganese content, 300, 347
oil of stem and leaves, 740
organic compounds, 887
organic reserves, 887, 888
oxidases of, utilization, 890
phosphorus, absorption, 302
content, 302
solubility, 301
potassium, amount needed, 318
water soluble, 317
proteins, conjugated, 669
of leaf, 676
radiation and transpiration, 456
response, to chlorine, 354
to potassium, 288
to thallium, 368
root hairs, 157
roots, 136, 139, 144, 149
depth of penetration, 139, 144
effect of soil texture, 136
weight, 149
seed coats, 84
selenium content, 286
silicon content, 335
solubility of inorganic salts, 290
stomata, 417, 422, 423
influence of humidity and temperature, 434
type, 435, 436
strontium, content, 285
sulphur content, 325, 326, 327
response to, 326, 327, 328
temperature of leaves, 475, 476
titanium, response to, 458, 459
transpiration, atmometers, 487
day and night, 411
water content, 443
and hay, 443
water culture, 354
acidity, 263
water requirement, 499
zinc content, 300
Alfalfa type of stomata, 435, 436
Algae, aeration by, 131
Algae, fixation of nitrogen, 655, 656
Alkalies, influence on transpiration, 464
Alkaloids, 694, 695
Alloxan, 650
Almond oil, 740, 761
and carbohydrates, 746, 747
formation of, 747
Almond seed coats, 80, 81, 82
Aluminum, 284, 356
influence of, on function in the cell, 359–360
on growth and production, 357, 358
on plant life, 357–360
on relationship of the soil, 358, 359
occurrence of, 356, 357
Amandin, 667
Amaranthus, 411, 456, 459, 482, 671, 672, 984, 994, 1048, 1077
Ambrosia, 95, 412
Amides, 648, 653
defined, 662
kinds of, 662–664
asparagine, 662, 663
glutamine, 663, 664
in protein synthesis, 662
Amines, 650, 694
Amino acids, classification of, 661
described, 660
influence of, on diastase, 798
in leaves, 683
nitrates vs. ammonium, 648
in protein formation, 660, 674
in protoplasm, 1020
in respiration, 938
in seedlings, 911
in seeds, 669
utilization of, 650, 651
Ammoniacal nitrogen, in air, 653, 654, 656, 657
in stomatal opening, 435
utilization of, 644–645
Ammonium salts, availability of, 647, 648
general effects of, 648, 649
influence of, hydrogen-ion concentration, 646
in plant nutrition, 644–649
relative value of, 644–647
toxicity of, 649
Amount of chlorophyll and photosynthesis, 596–598
in etiolated plants, 597, 598
in plants of different habitats and varieties, 596, 597
Ampholyte, defined, 221
relation of, to solute intake, 221
Ampotheric nature of proteins, 665, 666
Amygdalin, 804, 805
Amylase (*see* Diastase)
Amylo-cellulose, 534, 537
Amylodextrin, in starch grain, 534, 537, 539
Amyloid in root-hair wall, 158
Amylopectin, 535
Amyloses (*see* Starches)
Amylum (*see* Starch)
Anaerobic respiration, 939–940
defined, 939
relation of, to aerobic, 940–944

Anesthetics, relation of, to guttation, 504
to permeability, 106
to photosynthesis, 557
to respiration, 967, 968
to transpiration, 862
Anions, absorption of, 227–229, 232–236
influence of, on permeability, 106
Antagonism, 268
causes, 270, 271
of chlorine, 356
of vegetative and reproductive parts, 1044, 1045
Antagonism and balanced solutions, 268–271
Anthocyanin, 38
and respiration, 951
Antienzyme, 780
Apparent rate of photosynthesis, 606
Apple, absorption, of ions, 235
of nitrogen in winter, 892
acidifying power, 330
acidity, 46
affected by walnut, 168
calcium, evidence of need, 312
CO_2 exhaled, 955–957
in intercellular spaces, 964
in respiration, 971
catalase, 994, 995
changes in pectic materials, 23
color of, 551
downward movement of water, 873
erepsin, 820
fruit-bud formation, 699–700
fungi and transpiration, 470–471
hemicelluloses, 13
influence on other plants, 167
inhibition of diastase, 797
intercellular spaces and photosynthesis, 601
keeping quality and potassium, 324
metaxenia, 1031
mineral composition, 288
movement of organic compounds, 886, 892, 893
of nitrogen, 892
of solutes, 880
nitrate reducing substances, 675, 676
nitrate vs. ammonium, 648
organic reserves, 887, 888
oxidases, 978, 982
oxygen, in intercellular spaces, 964
in respiration, 970
photosynthesis, 610
potassium, water soluble, 318
respiration, 955–958, 965, 967
respiratory quotient, 950
seed coats, 83
soil water and transpiration, 463
stomata, 422, 423, 434
sulphur content, 325
response to, 327
temperature, 477
tracheae, 848
translocation, lateral, 880
of oil, 909, 910
upward, solutes, 877
transpiration, relative, 416, 454
vegetative vs. reproductive tissues, 1045
water content of leaves, 408
Appleman's diastase method, 796
Application of CO_2 in practice, 585, 586, 587
Araban, 540, 541
Arabinose, 13, 529
Arachidic acid, 728, 729
Arachin, 667
Arachys hypogaea (*see* Peanut)
Arbutin, 804
Arginine, action on diastase, 798
formula for, 661
in seeds, 687, 689
in soil, 171
utilization of, 650
Arrow root, 538
Arsenic, in nutrition, 367, 368
Artichoke, 228
Artificial ripening, 1057–1058
Asafoetida, 761
Ascent of water, 855–872
under low evaporation, 855–859
under high evaporation, 859–872
Asparagine, defined, 662
in leaves, 663, 683
occurrence of, 662, 663
in respiration, 938
in seeds, 47, 687
utilization of, 650
Asparagus, boron absorbed, 342
cytase, 803
hemicelluloses, 13
iron content, 331
oxidases, 978
relation to chlorine, 355
respiration, 961, 965
response to sulphur, 327
seeds, 4
Aspartic acid, action of, on diastase, 798
formula for, 661
in tomato, 675
utilization of, 650, 651
Assimilation of organic nitrogen, 650–651
Assimilation number, 597
Assimilation quotient, 556
Atmometer, defined, 483
free water surface of, 483, 484
moist porous surface of, 484, 485, 486
standardization of, 485, 486
Atmospheric drought, 502
Atropine, 695
Autocatalytic reaction, 1026
Autumnal coloration of leaves, 553, 554
Auximones, 651, 652
defined, 651, 652
effects of, 652
origin of, 653
Auxin, 1033
chemistry of, 1036
kinds of, 1036–1038
Available iron, 328
Avena unit, 1035
Avenalin, 667
Azaleas, 651
Azelaic acid, 18

B

Back pull of soil water, 200–202
Bacteria method in photosynthesis, 606
Baeyer hypothesis, 571
Balanced solutions, 268, 269, 270
Balsam, 761
Bamboo, 1070–1071
Banana, acidity in, 47
  pectic substances in, 23
Barium, 367
Bark, translocation in, 903–905
Barley, absorbing power, 95
  absorption, depth of, 147
    of elements, 290
    of ions, 228, 235
    of phosphorus, 302
  acidifying power, 231
  acidity, 46, 48
  age, elemental requirements, 370, 371
  ash content, 289
  CO_2 fertilization, 586
    result of, 586
  chemicals in soil and transpiration, 465
  composition, 285, 286
  cytase, 802
  depth of absorption, 147
  diastase, 793, 794
  distribution of elements, 293
  erepsin, 821
  etiolation, 597
  excretion of acids, 159, 160
  fixation of nitrogen, 656, 657
  formation of grain, 689
  heat of respiration, 953, 954
  influence, of electricity, 973, 974
    of other plants, 165
    on other plants, 165
  interchange of ions, 232
  maltase, 801
  manganese, amount needed, 349
  mitochondria, 37
  moisture and respiration, 959
  nitrogen content, 692
  phospholipides, 745
  phosphorus content, 302
  proteases, 813
  proteins, 667, 691
  raffinose, 531
  relation to light, 1079
  respiration, 959, 973
  response of, to aluminum, 357, 358
    to ammonium salts, 644, 645
    to boron, 343, 344
    to chlorine, 355
    to iodine, 365
    to iron, 330
    to lithium, 369
    to manganese, 349
    to nitrates, 644, 645
    to palladium, 369
    to potassium, 322
    to rubidium, 369
    to silicon, 338
  root hairs, 157
Barley, roots, 124, 131, 139
    number and length, 137
    penetration in dry soil, 123
    response, to chemicals, 127, 128, 129
      to moisture, 124
  rusts and transpiration, 471
  salt requirement, 290
  seed coats, 83, 84
  selenium content, 286
  silicon content, 335
  stomata, 419, 422
    type, 436
  transpiration and solutes, 238, 239
  unequal absorption of ions, 228
  water content, 443
    in cultures, 247, 248, 257, 258
      acidity in, 262
  water requirement, 498, 499
  wilting coefficient, 194
Barley type of stomata, 436
Bean, absorption, carbohydrates, 294
  acidity, 48, 50
  asparagine, 663, 683
  boron, absorbed, 341, 342
    location, 340
  calcium, content, 289
    effect of deficiency, 312
  CO_2 fertilization, 586
  composition, 285, 286
  conduction of water, 854
  erepsin, 821
  etiolation and photosynthesis, 598
  growth, inhibiting substances, 1044
  iron content, 347
  length of ray and photosynthesis, 591
  manganese content, 347
  minerals in seedlings, 372
  nitrogen, content, 642
    in leaves, 642, 681, 682
  nutrition, 604
  oil of seed, 740
  permeability, 101, 103, 104
  phospholipides, 744, 745
  phosphorus content, 302
  photosynthesis, 598, 604
  potassium, accumulation, 320
  protein, 668, 679
  proteolytic enzymes, 677
  relation to light, 591, 1078
  respiration, 942
  response of, to ammonium, 644
    to arsenic, 367
    to boron, 341
    to calcium, 312
    to hydrogen-ion concentration, 289
    to nitrates, 644
    to palladium, 369
    to rubidium, 369
    to sulphur, 327
  root hairs, 153
  roots, 101, 103, 163
    depth of penetration, 139
    number, 126, 127
    oxidation, 163
  salt requirement, 372

Bean, seed coats, 84, 85
soil water and transpiration, 463
solarization, 602
solvent action, 162
starch grains, 533
temperature of leaves, 476, 480
transpiration and solutes, 240
water in conducting cells, 854
wilting coefficient, 195, 199
(*See also* Lupine, *Vicia*)
Beet, absorption, 225, 228
boron, location, 340
CO_2 fertilization, 586
result of, 586
composition of cell sap, 290
concentration of sap, 43
content of boron, 340
of iodine, 288
of sulphur, 325
depletion of soil, 169
diffusion of solutes, 106
energy, utilized, 617
enzymes, 782
excretion of acid, 159, 160
glutamine, 663, 664
guard cells, 427
oxidase, 982
phospholipides, 743
protease, 813
protein, 692
raffinose, 531
respiration, 965
response of, to aluminum, 357
to ammonium salts, 645
to chlorine, 353, 354
to light rays, 1072
to nitrates, 645
to phosphorus, 303
to potassium, 318
to sulphur, 327
roots, 127, 139
sulphur content, 325
tyrosinases, 979
unequal absorption of ions, 228
water cultures, 263
acidity, 263
Begonia, acidity of, 47, 50
Bellani plates, 484
Beneficial influence, plant to plant, 164, 165
Benzidine reaction, 979
Benzoic aldehyde, 805
Bergamot, oil, 761
Best growth, 267
Best solution, 243, 255, 267
Bindweed, 891, 994
Biuret reaction, 664
Blackberry, rusts and stomata, 423
Blackheart, of potatoes, 941
Blackman reaction, 589, 594
Bleeding, 856
Bleeding pressure, 855
Bloom on fruits, 736
Blueberry, 651
Bluegrass, 204, 300, 740, 891
Borax fertilizer, 345, 346
Bordeaux mixture, effect on transpiration, 465–470
Boron, 340–346
contamination with, 298
deficiency of, effects, 342–344
excess of, effects, 344, 345
in fertilizer, 345, 346
injury from, 345
need of, 341, 342
occurrence of, 340, 341
Bose's theory, water ascent, 861, 862
Bound water, 1066, 1068
Breathing vs. respiration, 945, 946
Broccoli, 286
Bromelin, 781, 818, 819
Bronzing, 364
Bubble method of photosynthesis, 606
Buckbrush, organic reserves, 887, 888
Buckwheat, 39, 48, 159, 162, 258, 290, 300, 302, 313, 317, 318, 331, 353, 357, 365, 368, 740, 1074
Buds and leaves, growth-inhibiting substances, 1042–1044
Buffer systems in cell sap, 47
Buttermilk, utilization, 651
Butyric acid, 728, 1041

C

Cabbage, 127, 157, 158, 289, 301, 302, 303, 307, 325, 358, 408, 422, 671, 740, 743, 821
Caffeine, 695
Calcium, 307–313
influence of, on root-hair formation, 154
occurrence of, 307, 308
physiological role of, 308–313
as antidoting agent, 308, 309, 310
calcium magnesium ratio, 308, 309
neutralization of acids, 309, 310
as component of plants, 310, 311
cell wall, 310, 311
proteins, 311
protoplasm, 311
in general appearance, 312
in general effects, 312, 313
in root response, 312, 313
in translocation of carbohydrates, 312
Calcium magnesium ratio, 308, 309
Calcium nitrogen ratio, 309
Calcium oxalate, relation to nutrition, 309
Calculated wilting coefficient, 191–193, 199
Callose, 23–24, 157
Camphor, 761
influence on transpiration, 465
Cane sugar (*see* Sucrose)
Capillarity, water ascent, 865, 866
Capric acid, 728, 729
Caproic acid, 728, 729
Caprylic acid, 728, 729
Carbohydrases, 781–803
cellulase, 802–803
diastase, 789–800
glucase, 801–802
inulase, 800–801
invertase, 783–789

Carbohydrates, absorbed by roots, 294–296
in cell sap, 38
classification of, 524
polysaccharoses, 531–541
sugars, 525–531
definition, 523, 524
first formed, 565, 566
formation of, relation to potassium, 318, 319, 320
in leaf, 558–567
in plant, 523–541
in respiration, 935–937
relation of, to chlorophyll, 550
to fats and oils, 746–752
to phosphorus, 306
to photosynthesis, 603
stored foods, 772
translocation of, 902–906
effect of calcium, 312
Carbon, amount of, in plant, 284
function of, 294
Carbon assimilation, 541
Carbon bisulfide, 1050
Carbon dioxide, absorption by leaf, 569, 570
in air and soil, 580, 581
application of, in agriculture, 585–587
in concentration and photosynthesis, 583, 584, 585
diffusion of, 438
excretion of, by roots, 159–161
relative amounts, 159, 160, 161
solvent action, 161, 162
method of studying photosynthesis, 607
respiration, 946–949
origin of, 580, 581
reduction of, 572, 573, 574
in relation to respiration, 969, 970
solvent action of, 161–162
supply available, 580–585
used by plant, 581, 582, 583
Carbon monoxide, 582
Carbon nitrogen relations, 697–705
defined, 697
effect of, on fruit buds, 699, 700
on growth stages, 702, 703
influence of length of day, 701, 702
influence of light on, 1082
response of organs of plant, 703
response of root and shoot, 703, 704
Carboxylase, 934
Carnauba wax, 730, 736
Carotene (*see* Carotin)
Carotin, 552
Carotinoids, 551–553
Carrot, 102, 106, 228, 230, 233, 290, 303, 342, 354, 552, 956, 957, 962, 973
Casein, utilization, 650, 651
Casparian strip, 206, 207, 211
Castor bean, 467, 658, 667, 686, 738, 743, 780, 819, 990, 1079
Catalase, 299
and acid relations, 990
defined, 976
drying, effect on, 989, 990
Catalase, effect of, on afterripening, 996
on photoperiodism, 1083
and growth, 995, 996
nature of, 987
preparation and study of, 987, 988
relation of, to light, 990, 991
to photoperiodism, 1083
to physiological activity, 991–996
to respiration, 991–993
to vitality, 993–995
temperature relations, 989
x ray, effect on, 1075
Catalase quotient, 994
Catalysts, 774
potassium, 321
Cations, absorption of, 227–236
influence on permeability, 105–107
Celery, 770, 990, 1051, 1057
Cell (*see* Plant cell)
Cell sap, 37–51
acidity in, 45–51
buffer system of, 47
composition, 38
concentration of, factors influencing, 39–45
age of tissue, 44
growth of, 44–45
habitat, 39–41
kind of plant, 41–42
portion of plant, 42–44
time of sampling, 44
general law, 40
need for water in, 407, 408
origin of, 37–38
osmotic value, 90
relation of, to transpiration, 450
sources of materials in, 38–39
Cell wall, chemical composition of, 7–24
commercial uses, 24
composition and properties of, 1–24
and infiltration constituents, 23–24
membrane of, 79–86
methods of studying, 5–6
ash analysis, 6
differential solubilities, 6
differential staining, 5
direct observation, 5
hydration, 6
measurement of refraction indexes, 6
polarized light, 6
use of fluorescent light, 6
of spierer lens, 6
x-ray analysis, 6
microincineration of, 24
origin of, 1–5
physical and chemical properties of, 5–7
primary, 4
of root hair, 157–158
secondary, 4
theories of structure of, 6
Cellobiose, 9
Cellulase, 802, 803
Cellulose, as carbohydrate, 531, 532
kinds of, 8–12
in root hair, 158
Center of origin, starch grains, 533

Cephalin, 727, 737, 744
Cerebrosides, 737
Cerotic acid, 728, 736, 741
Ceryl alcohol, 730, 735
Cetyl alcohol, 735, 737
Changes in stomatal opening, 445, 446, 447
Characteristics of the sap indicating elemental deficiency, 292
Chemical analysis for permeability, 97
Chemical combination and water, 408, 409
Chemical properties, proteins, 659–666
Chemical theory of membrane action, 109–111
Chemicals, applied to the soil, 464, 465
diastatic activity of, 796
films from sprays of, 465–470
influence of, on transpiration, 464–470
Chemistry, of chlorophyll, 544, 545
of photosynthesis, 567–577
Chemotropism of roots, 124–126, 135–136
Cherry, 422
gum, 527, 541
oxidases, 978
respiration, 943
stomata, 422, 423
translocation of solute, 877
Chlorella, 348, 363, 592, 602, 685, 943, 1074
Chlorides, in cell sap, 42
as fertilizers, 352–356
in protoplasm, 25
Chlorine, antagonistic ion, 356
as fertilizer, 352–356
influence of, on asparagus, 355
on roots and tubers, 354
occurrence of, in plants, 283, 284
and relation to water content of plants, 354
Chloroform, influence of, on maltase, 802
on permeability, 106
on respiration, 967, 968
Chlorophyll, *a* and *b*, 543, 544
absorption of light, 546
amount of, and photosynthesis, 596–598
chemistry of, 544, 545
fluorescence of, 545
formation of, 548–550
function of, 567, 568, 569, 576
inheritance of, 550, 551
optical properties of, 545–547
relation of, to calcium, 549
to iodine, 365
to iron, 328
to magnesium, 313, 549
to manganese, 352
to phosphorus, 549
to sulphur, 326
state of, in chloroplasts, 547, 548
Chlorophyllase, 545
Chlorophyllin, 544
Chlorophyllogen, 548
Chloroplasts, described, 30–32
of guard cells, 426
inheritance of, 35–36
origin of, 35
pigments of, 541–554
reactive surface of, 604
Chloroplasts, relation of, to photosynthesis, 600
to starch grains, 537
Chlorosis, defined, 328
deficiency disease, 292
due to lack, of iron, 328, 330, 331
of magnesium, 313, 314
of manganese, 351, 352
injections for, 883, 884
Cholesterol, 730, 737
Choline, 171, 743
Chondriosomes, 36–37
Chromoplasts, 32
Chunky sweetpotatoes, 323
Cinnamon, 761
Citric acid, 747
relation of, to invertase, 788
and respiratory quotient, 747
Citrus plants and fruits, absorption of ions in, 234
boron in, excess effect, 344, 345
changes on ripening, 23
effect, of afterripening, 1057
of magnesium chlorosis, 314
of petroleum sprays, 910
number of stomata in, 419, 423
Classification, of amides, 662, 663
of amino acids, 661
of carbohydrates, 524
of cell parts, 1
of cell-wall compounds, 8
of enzymes, 781
of lipides, 727, 731–737
of proteins, 666–670
Climate, relation of, to protein formation, 689–692
to transpiration, 455–462
to water requirement, 498, 499, 500
to wilting, 198–200
Clipping and organic reserves, 889
Clover, absorption of phosphorus, 302
acidity of sap, 48
CO_2 fertilization, 587
production, 160
catalase, 992
copper content, 300
depletion of soil, 169
erepsin, 821
excretion of acids, 160
heat of respiration, 953
hydrogen-ion concentration and growth, 1087
influence on, of electricity, 1084
of light, 1079
on other plants, 167
iron content, 300, 331, 347
manganese content, 300, 347
nitrogen content, 692
nitrogen vs. ammonium, 648
oxidases, 978
phospholipides, 745
potassium, amount needed, 318
water soluble, 318
response of, to aluminum, 357
to boron, 344
to fluoride, 368
to magnesium, 313
to sulphur, 326, 327

Clover, roots, 136, 139
depth of root penetration, 139
effect of soil texture, 136
selenium content, 286
silicon content, 335
stomata, 422, 423
water content and hay, 443
water cultures, 263
acidity in, 263
zinc content, 300
Coagulases, 781, 821, 822
pectases, 822
rennets or rennins, 821, 822
Cobalt chloride method, index of transpiring power, 492, 493
preparation of hygrometric paper, 491, 492
standard colors in, 491
for standard evaporating surface, 492
transpiration determined by, 415, 491, 492, 493
Cocaine, 695
Cocklebur, absorption of water, 200–201
catalase, 994
factors in photoperiodism, 1081
relation to light, 1081
respiration, 970
rusts and transpiration, 471
seed coats, 80, 81, 83
temperature coefficient, 110
transpiration, 471
water movement, 200, 201
Coefficient, energy absorption, 612–615
temperature, 593
transpiration, 486
wilting, 190
Coenzyme, 780, 934
Coffee tree, 873, 874, 875
Cohesion theory, ascent of water, 867–872
of cohesive force of water, 867
description of, 868–870
evidence of, 870, 871
of imbitional and osmotic forces, 867, 868
objection to, 871, 872
Cold resistance, 1064–1070
Coleoptile of corn, 824
Coleoptile method, 1035
Coleorhiza, 824
Collard, 154, 155, 156, 158
Collodion film, 436
Collodion membrane, 217
Colloidal solution, of chlorophyll, 547, 548
defined, 70
Colloidal theory of membrane action, 111–112
Colloids, attraction of, for ions, 220, 221
relation of, to growth, 1020, 1021
to ion intake, 220, 221
to stomatal behavior, 429, 430
to water loss, 451
(*See also* Absorption; Cell wall; Membranes; Polysaccharoses; Proteins; Protoplasm)
Coloring matter, 726
in cell walls, 23
Combined iron, 328
Combined proteins, 678
Companion cells, 849
Compensation point, 590
Composition of plant, 283–292
distribution of elements in, 292–293
influence on, of environment, 288–291
of plant, 284–288
Compound celluloses, 10
Compound-interest law and growth, 1028
Compound lipides, 736, 737
defined, 727, 730
distribution of, 743, 744
function of, 745, 746
occurrence of, 742–746
relation of, to oil and protein, 744
in seeds, 744–745
Concentration of CO_2 and photosynthesis, 583, 584, 585
Concentration of cell sap, 39–45, 450
critical, 270
methods of expressing, 71, 72
relation of, to absorption ratio, 225–227
to nutrient solutions, 256–258
to starch formation, 603
to transpiration, 450, 451
Condensation of formaldehyde, 574, 575
Condition of nitrate reduction, 673, 674, 675
Conductivity of cell sap, influenced, by age, 44
by kind of plant, 42
by portion of plant, 44
Coniine, 694, 695
Conjugated proteins, 669
Continuous light, growth, 1078–1079
Copper, 362, 363
effect of, on chlorosis, 363
on muck soils, 362, 363
relation of, to cell wall, 310, 311
to chlorosis, 362
to exanthema, 362
to frenching, 362
to salt sickness, 291, 362
Cork formation, 15–17, 1090
Corn, absorption, of elements, 288, 289
of ions, 227
of organic compounds, 294
of phosphorus, 302
of sugar, 296
acidifying power, 230, 231
acidity, 46, 48, 50, 51
aeration of roots, 133
age, elemental requirements, 370, 373
albino plants, 296
alcohol source, 618
aluminum content, 356, 357
amino acids, 669, 670
antagonism, 270
ash content, 289
calcium content, 307
carbohydrates in leaves, 563, 564, 566
CO_2 production, 161, 162
utilization of, 581, 582
carotin, 552
cell sap, 39
quantity of, 292
cell walls and calcium, 310, 311
cells of root, 27
chemicals in soil and transpiration, 464
coloring matter, 552

Corn, composition, 284
concentration of cell sap, 39–44
copper content, 300
depletion of organic reserves, 885, 886
depth of absorption, 146, 147
diastase, 798
distribution in, of elements, 285, 286, 293
of sulphur, 325
embryo, 824
endosperm, 824
energy utilized, 616, 617
enzymes of roots, 163
erepsin, 821
etiolation and photosynthesis, 598
excretion by, of acids, 159
of digestive enzymes, 163
of organic substances, 164
formation of grain, 688, 689
formation of protein, 685
frost resistance and minor elements, 300
fungi and transpiration, 471, 472
growth, 1027
heat of respiration, 953
hemicellulose, 13
hydrogen-ion concentration, 50
humidity and transpiration, 459
influence on, of electricity, 1085
of other plants, 165–170
on other plants, 165
inhibiting substances, 1044
interchange of ions, 232
iron content, 300
iron supply, 334
kerosene on seeds, 84
leaf structure, 413, 449
leaf surface, 453
lipase, 807
magnesium content, 313
maltase, 801
manganese content, 300, 346
maximal temperature, 1062
minerals, in seedlings, 372, 373
mitochondria, 37
moisture and respiration, 959
nitrate vs. ammonium, 644, 645, 647
nitrogen content, 642, 692
evidence of need of, 291
in leaves, 642, 681
number of roots, 126, 127, 455
oil, 732, 740
optimum temperature, growth, 1062
organic acids, 46
oxidase, 978
pentosans, 540
pentose, 527
phospholipides, 743, 744, 745
phosphorus content, 301, 302
phytosterol, 730
polypeptides, 669, 670
potassium, content, 316
amount needed, 318
evidence of need, 291
water soluble, 317
photosynthesis, 608, 610
primordia of leucoplasts, 34, 35
Corn, proteases, 813
proteins, 668, 679
translocation of, 912
quantity of phosphorus, 303
relative weight of roots and tops, 128, 129
respiration, 960
respiratory quotient, 952
response of, to aluminum, 358, 359
to ammonium salts, 644, 645
to barium, 367
to fluorine, 368
to light, 1077
to magnesium, 313, 314
to manganese, 348
to moisture, 196, 463
to nitrate, 644, 645
to nitrite, 649
to potassium, 288
to pyrrole-carbonic acid, 329
to salt distribution, 288, 289
to thallium, 368
to vanadium, 368
reversal of sex, 1081
root formation, 1040
root hairs, 157, 158
roots, anchorage, 127, 128
depth of penetration, 139, 141
lateral extent, 145
response of, to chemicals, 126–128
to cultivation, 141
to moisture, 123, 124
to oxygen, 133, 134, 135
solvent action, 163
weight of, 149
salt requirement, 370, 371
seed, 824–826
selenium content, 286
silicon, content, 335
need of, 335, 336
soil water and transpiration, 463
solutes, lateral transfer, 881
starch, 535, 538
starch grain, 533, 791
stomata, 417–424, 445
temperature of leaves, 475–481
titratable acids, 50
transpiration and solutes, 237, 238, 240
day and night, 410, 411
relative, 416, 454
unequal absorption of ions, 227, 228
utilization, organic nitrogen, 650, 651
variation of dry matter, 444, 609
water, of leaf, 442, 444
lost, 412
used, 408
water content of leaves, 408
water cultures, 258, 261
acidity, 262, 263
water requirement, 498, 499, 500, 501
wilting coefficient, 194, 195, 196, 199
zinc content, 300
Corylin, 660, 667
Cotton, aluminum content, 357
boron, absorbed, 342
requirement, 341

Cotton, cell-sap concentration, 40
of chlorine, 42
of sulphate, 42
composition, 9
downward movement of water, 872
formation of fiber wall, 2
gossypol, 746
hydrogen-ion, effect on absorption, of ammonium, 646
of nitrate, 646
magnesium in seeds, 315
metaxenia, 1032
movement of water, 873
nitrate vs. ammonium, 648
nitrogen in plant, 642
oil, 732, 740, 748, 908, 909
phospholipides, 744, 745
phytosterol, 730
potassium, amount needed, 318
raffinose, 531
relation to light, 1081
response of, to boron, 341
to phosphorus, 306
roots, 137, 139
weight, 149
temperature, growth, 1063
translocation, of carbohydrates, 902–905
of proteins, 910
of solutes, 878
Cottonseed meal, 650, 651
Cowpeas, acidity, 48
boron, content of, 340
location of, 340
calcium magnesium ratio, 308
nitrogen content, 642
in leaves, 642, 681, 682
oil of stem and leaves, 740
phosphorus content, 302
photosynthesis, rate, 610
potassium, amount needed, 318
effect of deficiency, 320
protein, 667
response of, to barium, 367
to fluoride, 368
to pyrrole-carbonic acid, 329
to strontium, 367
stomata, 417, 422
temperature of leaves, 475, 476, 478
transpiration, relative, 416
turgidity and transpiration, 462
water content of leaves, 408
water lost, 412
wilting and transpiration, 462
Cradina, 781
Cranberry, acidity of sap, 46
absorption in, ammonium vs. nitrates, 646, 648
organic nitrogen, 651
age and respiration, 965
cuticle, 17
oxygen and respiration, 971
stomata, 423, 426
Creatine, 171, 650, 651
Creatinine, 650
Critical concentration, 270
Critical water-supplying power of soil, 203
Criticism of nutrient solutions, 265–268
Crone's solution, 242
Crude lipides, 726, 741
Cryptoplasm, 28
Cucumber, 158, 289, 422, 423, 586
Cucurbita (*see* Pumpkin; Squash)
Cucurbitin, 667
Cultivated plants, extent of root systems, 139
Cultivation, benefits of, 129
influence of, on root development, 141
Cultural solutions, 241
Cultures, sand, 241
water, 241
Cumarin, 171
Cuticle, 17–18
Cuticular transpiration, 436, 448, 461
Cutin, 15, 17–18
Cutinization, effect of potassium, 319
Cystine, 661
action of, on diastase, 798
formula for, 661
in tomato, 675
Cytase (*see* Cellulase)
Cytohydrolytic enzymes, 802
Cytoplasm, 29–30
Cytoplasmic membrane, 86–89
formation of, 86
nature of, 86–89
permeability, 89
measuring permeability, 97–101

D

Dahlia, 225, 540, 832
Daily variation of carbohydrates in leaves, 559–563
Dandelion, 102, 111
Dark reaction, 594
Darkness vs. light, growth, 1070–1072
Date, 4, 13, 14, 803, 1031, 1032
Date palm, growth in light and darkness, 1071
Death point, temperature, 481
Decomposition products of roots, influence, 170–174
Deficiency diseases, 291
Deionized air, 973
Deleterious influence of one plant on another, 167, 174
Derived lipides, 737
Derived proteins, 669, 670
Desert legumes, 195
Detection of intermediate products in leaves, 575
Deutreium oxide, relation to photosynthesis, 600
Dewberry, changes in pectic substances, 23
Dextrinase, 781
Dextrins, 539
Dextro-rotatory sugars, 525
Dextrosans, 531
Dextrose (*see* Glucose)
Diastase, activity of, in tubers, 958
in afterripening, 1053
compared with diastase from animals, 792–793
from different plants, 791

Diastase, effect on, of aluminum salts, 359
of oxalate, 312
of potassium, 322
factors influencing, 796–800
chemicals, 796–798
light, 799, 800
temperature, 798, 799
in guard cells, 428
history of, 789
in honey, 790
occurrence of, 789–793
preparation of, 793–795
relation of, to winter hardiness, 790
secretion of, 790
study of activity, 795–796
taka-, 789
translocation of, 790
types of, 790–792
origin and action, 790–792
products, 792
Diastatic activity, of flour and grain, 828–830
Dichlorethylene, 1050
Dieback, 322, 343
Differentiating stage of growth, 1023, 1024
Differentiation, 1023
Diffusion, through membranes, 76–79
through small openings, 437, 439
in solutions, 72–74
through stomata, 439, 440, 441
Digestion in plants and animals, 770–830
bibliography of, 830–846
comparison of, in plants and animals, 830
extracellular, 830
intracellular, 830
nature of, 772, 773
Direct method, of measuring permeability, 100
of measuring wilting coefficient, 191–193
Direction of root growth to chemicals, 124–126
Disaccharoses, 529–531
Diseases, respiration, 975
Dissimilation, 932
Distance organic reserves moved, 892–894
Distribution of elements, 292, 293
Dodder, source of carotin, 552
Donnan equilibrium, relation of, to solute intake, 222
Dormancy, causes of, 1047–1049
defined, 1046
localized, 1049
overcoming, 1049–1053
of bulbs and tubers, 1049, 1050
of seeds, 1046, 1047, 1050–1053
by light, 1052, 1053
by oxygen, 1052
by temperature, 1051
of stems, 1049
Double membranes, 81, 82
Downward movement of water, 872–874
Downward transfer of solutes, 882, 883
Drought, 502
Drought resistance, 502, 503
Drought spot, 354
Drying oils, 734
Duggar's solution, 244

E

Eclipse, stomatal movement, 432
Economic importance of photosynthesis, 617, 618
Economy constant, 1028
Ectoplast, 29, 87
Ectosarc, 87
Edestin, 659, 666, 667
Efficiency index of production, 1028
Elaioplasts, 739
Electrical conductivity, measuring permeability, 99–100
Electrical phenomena, solute absorption, 220
in tissues, 1085–1086
Electrical theory of photosynthesis, 572
Electricity, effect of, on crop production, 1083–1085
influence of, on growth, 1083–1086
on respiration, 973, 974
Elemental requirement and age, 369–373
Elements, absorbed by plants, 283–373
bibliography of, 373–406
distributed in plants, 292, 293
essential vs. unessential, 297–301
forms in which absorbed, 293–294
inorganic compounds, 293
organic compounds, 293, 294
function of, 298, 301–369
influence of, on photosynthesis, 604
number and kind of, 283, 284
indicator of needs, 291–292
characteristics of sap, 292
general appearance, 291
microchemical tests, 292
influence, of environment, 288–291
of plant, 284–288
occurrence of, 301–369
Elodea, 104, 151, 153, 591, 675, 972, 1075
Elongating stage of growth, 1021–1023
nature of, 1021–1023
rate of, 1023
Elongation, periodic, 1021–1023
Embryonic stage of growth, 1019–1021
Emulsin, 781, 805
Endodermal pressure, 855, 858
Endodermis, function of, 207, 858
passage cells in, 207
permeability of, 85
structure of, 205–207
Endoplasm, 29
Endosmosis, 79
Endosperm, 824–830
respiration, 965
Energesis, 932
Energy, absorbed by leaves, 612–615
for absorption of solutes, 225
released in respiration, 942
utilized, in photosynthesis, 615–617
in transpiration, 456–458
for water ascent, 869
Energy relations, in photosynthesis, 605–618
of plant, 931, 932
Enollactim, 812

Enzymes, in cell sap, 38
  classification of, 781, 782
  defined, 774
  endocellular, 822
  formation of, in plant, 822, 830
  intracellular, 822
  longevity of, 782
  nature of, 774–777
  nature of action of, 777–779
  nomenclature for, 779–781
  occurrence of, 782, 783
  origin of name, 781
  relation of, to antiseptics, 777
    to heat, 776
    to light, 777
    to photosynthesis, 595
  secreted by roots, 162, 163
Epictesis, 220
Epidermal cells, 413
  osmotic value, 426, 427
Epidermis, 414
Epithelial layer of scutellum, 824–830
Epithem hydathodes, 504
Equivalent-weight solution, 72
Erepsin, 781, 820, 821
Ereptase (*see* Erepsin)
Erucic acid, 729
Erythrodextrin, 537, 539
Esculin, 804
Essential elements, 298
Essential oils, 726, 761
Esterases (*see* Lipases)
Ether extract, 726
Etherial oils, 23
Ethyl alcohol, in respiration, 940, 941, 942, 943, 969
Ethyl bromid, 1050
Ethylene, 968, 1057
  influence of, on diastase, 798
    on intumescences, 1037
    on respiration, 968
    on root growth, 1041
  production of, by plant parts, 1046
Ethylene chlorohydrin, 1049–1050
  influence of, on diastase, 798
    on dormancy, 1049, 1050
    on respiration, 969
Ethylene dichlorid, 1050
Etiolated plants, characteristics of, 1070
  defined, 1070
  photosynthesis of, 597, 598
Eucalyptus, 761
Evaporating power of air, 458–462
  defined, 458
  effect of, on transpiration, 456–462
    on wilting, 200
  of solar radiation, 456–458
Evaporation, measurement of, 482–486
  ratio of, to transpiration, 486–488
  relation of, to transpiration, 481, 482
    to water ascent, 855–872
    to wilting, 198–200
  value of, 482, 483
Evaporimeter (*see* Atmometer)
Evident plasmolysis, 90
Evident stage of growth, 1018
Exanthema, 362
Excelsin, 659, 667
Excretions of roots, 158–164
  acids, 159–162
  carbon dioxide, 159–162
  enzymes, 162–163
  organic substances, 163, 164
Exosmosis, 79
Extraction of protein, 659, 676
Exudation pressure, 855

F

Factors, affecting water requirement, 498–502
  influencing digestion, 786–789
    lipase, 809–812
    organic storage, 889, 890
    protein in grains, 689–692
      kind of plant, 692
      soil moisture, 691, 692
      time of application, 689–692
    transpiration, 413–472
  limiting, 577–580
Fat metabolism in plant, 726–761
  bibliography of, 761–769
Fat solvents, 726
Fats, artificial synthesis, 752, 753
  changes in, during formation, 548–550
    during germination, 752–760
      amount of oil, 752, 753
      carbohydrates and fats, 757–760
      fatty acids, of the oil, 754–757
        of cotyledons, 755, 756
        of hypocotyls and roots, 756, 757
  chemical properties of, 732–735
  described, 727, 731
  distribution of, in cell, 737–739
  drying of oils, 733, 734
  fatty oils, amount in plant, 740
  hydrolysis of, 732, 733
  occurrence of, with carbohydrates, 741, 742
    in fruit and seeds, 739, 740
    in plants, 739–746
    in stem and leaves, 740, 741
    in trees and shrubs, 741, 742
  physical properties of, 732
  relation of, to carbohydrates, 746, 747
  in respiration, 935–937
  saponification of, 733
  stored food in, 772
  synthesis of, 746–752
  translocation of, 906–910
  values for determining characteristics of, 734, 735
    (*See also* Lipides)
Fatty acids, 727–730
  from carbohydrates, 670
  percentage of, in fats and oils, 732
  relation of, to fat formation, 749
    to translocation, 906–908
Fatty oils (*see* Fats)
Fatty substances, 726, 727
  in cell walls, 81, 738
  compounds not of fatty acids, 761

Fatty substances, in protoplasm, 738
Feeding power of plants, 287
Fehling's solution, preparation, 526
Fennel, 761
Fermentation, 939
Ferments, 781
  soluble, 781
  unorganized, 781
Fertilizers, boron, 345
  effect of, on toxic substances, 172
  manganese, 349
  as oxidizers, 171
  sulphur, 327
Feterita (*see* Sorghum)
Films, influence of, on transpiration, 465–470
Flax, acidity, 48
  cell wall of fiber, 1–2
  copper influence, 362
  formation of oil, 748, 749, 750, 752
  influence of potassium, 319
  magnesium in seeds, 315
  nitrates vs. ammonium, 647
  oils, 732, 740, 908
  protease, 813
  relation to light, 1078
  response to temperature, 1065
  roots, 127
    response of, to aeration, 132
      to fertilizers, 127
      to nitrogen, 128, 129
  selenium content, 286
  water cultures, 263
    acidity in, 263
Floating respiration, 937
Flowering, photoperiodism, 1080–1082
Fluorescence, 545
Foods of plant, 770–772
  oxidation in respiration, 932–934
  sources to animals, 770
  type of, effect on respiration, 938, 939
  value of, to plant, 771, 772
Formaldehyde, activated, 574
  condensation of, in plants, 574, 575
  detection of, in leaves, 575
  formation of, by plants, 571–577
  photosynthetic quotient, 558
  utilization of, 575, 576
Formaldehyde theory of photosynthesis, 571, 576
Formation, of carbohydrates, 523–618
    bibliography of, 618–641
    and relation to potassium, 318, 319, 320
  of chlorophyll, 548–550
  of protein in seeds, 687–689
Formative stage of growth, 1019–1021
Formhydroxamic acid, 670, 673
Formic acid, 728
  excreted by roots, 159
  in protein formation, 309
  in respiration, 939
Fossil energy, 617
Four-salt solution, 242, 249, 254
Free iron, 328
Freed sorgo (*see* Sorghum)
Freeman's method, transpiration, 493, 494
Freezing, causes of injury, 1064–1065
  characteristics of hardiness, 1065, 1066
  effect on plants, 1064–1070
  hardening plants, 1066–1069
  hardiness and drought resistance, 1069
  measurement, 1069
Frenching of leaves, 292, 362, 368
Fructose, description of, 528
  in fruits, 530
  in leaves, 558
  in nutrition, 295
Fruit sugar (*see* Fructose)
Fruit-bud formation, 699–701
  application of nitrogen to, 700, 701
  carbohydrate nitrogen relation in, 699, 700
Fucoxanthin, 553
Function of chlorophyll, 567, 568, 569
Fundamental stage of growth, 1018
Fungi, relation of, to transpiration, 470–472
    to photosynthesis, 604
Furfurole, 1049

G

Galactans (*see* Galactosans)
Galactolipides, 727, 737
Galactosans, 12, 528, 531, 540
Galactose, 13, 295, 528, 529
Galactoside, 737
Galacturonic acid, 13
Ganong leaf punch, 608
Ganong respirometer, 947
Gas analytical method of photosynthesis, 607
Gases, in conducting cells, 853, 854, 871
  effect of, on cork formation, 1090
    on growth, 1088–1090
Gentianase, 781, 803
Gentianose, 531
Gentiobiose, 531
Geotropism and oxidase, 983
Germination, amino acids in, 911
  catalase activity in, 987–995
  changes of, in carbohydrates, 905, 906
    in fats and oils, 752–761
    in proteins, 910–912
  enzyme formation during, 822–830
  respiration during, 946–955
  respiratory quotient during, 952
  of seeds, 1050–1053
  translocation of oil during, 906–910
Glandular hairs, 447
Glass bubbler, 606
Gliadin, 660, 666, 667, 669
Globulin, 652, 666, 667, 668
Glucase, 801, 802, 806
Glucosans, 531
Glucose, described, 527
  in fruits, 530
  in hemicellulose, 13
  in leaves, 558
  utilization of, by plants, 294–296
Glucosidases, 781, 803–807
  emulsin, 805, 806
  myrosin, 806, 807
  phytase, 806

Glucosidases, rhamninase, 806
Glucosides, 523
in cell sap, 38
as form of mustard oils, 324
occurrence of, 523
Glucuronic acid, 13
Glutamic acid, 651, 661
Glutamine, defined, 663
occurrence of, 663, 664
in seeds, 689
utilization of, 650
Glutelins, 669
Glutenin, 666, 667, 668
Glutens, 669
Glycerine (*see* Glycerol)
Glycerol, 730, 752
Glycerophosphoric acid, 743, 745
Glycine, 650, 651, 661
action of, on diastase, 798
Glycinin, 667
Glycocol, in respiration, 938
in soil, 171
utilization of, 650, 651
Glycolipins, 737
Glycollic acid, 558
Glycolysis, 933
Glycoproteins, 669
Glycuronic acid, 13
Glyoxalic acid, 558
Gossypol, 746
Gram-molecular solution, 71
Graminin, 540
Grand period of growth, 1024
Granulose, in starch grain, 534
Grape, carbohydrate storage, 889
erepsin, 820
fungi and transpiration, 470
oil of seed, 735
oxidases, 978
pectic materials, 23
respiration, 967
response to sulphur, 327
sugars in leaf, 562
tracheae, 848
Grape sugar (*see* Glucose)
Grapefruit, 46
Grasses, distribution of protein, 658, 659
glutamine, 663
nitrates vs. ammonium, 645, 648, 649
nitrites, 650
response of, to arsenic, 368
to magnesium, 313
water requirement, 498
wilting coefficient, 194
Grey-speck disease, 292
Growth, afterripening, 1053–1058
bibliography of, 1095–1132
compound interest law of, 1028
curves of, 1025–1028
dormant or resting period in, 1046–1059
effect of cell sap on, 44–45
evident stage of, 1018
factors affecting, 1059–1095
external, 1061–1095
electricity, 1083
Growth, factors affecting, external, gases, 1088–1090
hydrogen-ion concentration, 1087–1088
light, 1070–1083
growth in darkness, 1070
intensity of, 1077, 1078
length of exposure, 1078–1083
continuous light, 1078–1079
short and long day, 1079–1083
polarized light, 1075–1077
rate of growth in light and darkness, 1070–1072
rays of spectrum, 1072–1075
x rays, 1075–1076
temperatures, 1061–1070
internal, 1059–1061
fundamental stage of, 1018
grand period of, 1024
nature of, 1018, 1019
of organs, 1024
periodicity of, 1029
periodic elongation during, 1028, 1029
of plant, 1018–1095
rate of cell-wall, 3
regions of, 1024
regeneration in, 1058–1059
relation of, to catalase, 995, 996
to hydrogen-ion concentration, 1087–1088
rhythm of, 1029
stages of, in plant cell, 1019–1024
differentiation, 1023–1024
elongation, 1021–1023
embryonic, 1019–1021
substances inhibiting, 1042–1046
substances promoting, 1030–1042
Growth curves, 1025–1028
relation of, to temperature, 1061–1064
Growth hormones, 1033–1042
Growth-inhibiting substances of stems and roots, 1042–1046
buds and leaves, 1042–1044
fruit, seeds, and flowers, 1044, 1045
volatile substances, 1045, 1046
Growth-promoting substances, 1030–1042
defined, 1030
growth hormones, 1033–1042
chemistry of, 1036–1038
determination of action of, 1035, 1036
general, 1041, 1042
historical, 1033, 1034
mode of action of, 1036
occurrence and preparation of, 1034, 1035
in root formation, 1040, 1041
transport and polarity of, 1038, 1039
tropisms, 1039, 1040
metaxenia, 1031, 1032
vitamins, 1032–1033
Guaiacum reaction, 979
relation to oxidase, 976, 977
Guanine, 650, 651, 695
Guard cells, 418
changes in content of, 428–430
length of life for, 430
osmotic relations of, 426, 427, 428
osmotic value of, 427

Guard cells, physiology of, 425, 426
Gums, 23–24, 527, 528
Guttation, 503–506
defined, 503
water of, 855

H

Habitat, effect on cell sap, 39–41
Hail injury, 889
Hair root, 150
Hairs, kinds of, 447
relation of, to transpiration, 447, 448
Half-leaf method, photosynthesis, 608
Hard wheat, 339
Hardening of plants, 1066, 1068–1069
Hardiness, characteristics of, 1065–1066
drought and, 1069
influence of potassium on, 324
Hay, heat of respiration, 955
Heart rot of beet, 292, 344
Heat, in combustion, photosynthesis, 615
in respiration, 952, 955
Hegari (*see* Sorghums)
Hehner value, 734
Helianthus annuus (*see* Sunflower)
Heller's test, 664
Hemicelluloses, defined, 12
hydrolytic products, 12–13
occurrence of, 13–14
origin of, 15
properties of, 12
in starch, 535
utilization of, 14–15
Hemoglobin, conjugated proteins, 669
relation of, to chlorophyll, 545
utilization of, 650, 651
Hemp, 667, 734, 740, 749, 908, 953
Heptacosanol, 735
Heteroauxin, 1036, 1037
Hexacosanol, 736
Hexosans, 531
Hexosephosphatase, 934
Hexosephophates, 304, 934
Hexoses, 527–529
defined, 527
kinds of, 527–529
in leaves, 560, 561, 563, 565
used in respiration, 936
Hilum of starch grains, 533
Hippuric acid, 650
Histidine, 650
in developing seeds, 687, 688
formula for, 661
in tomato, 675
History, of growth-promoting substances, 1033–1034
of nitrogen nutrition, 654–655
of nutrient solutions, 241–242
of photosynthesis, 554–555
of proteases, 812–813
Hoagland's solution, 244
Hollow heart, 292
Hopkins-Cole's reaction, 665
Hordein, 667, 669
Hormones, 1030, 1036
growth, 1033–1042
chemistry of, 1036–1038
determination of action in, 1035, 1036
general, 1041, 1042
historical, 1033, 1034
mode of action of, 1036
occurrence and preparation of, 1034, 1035
in root formation, 1040, 1041
transport and polarity of, 1038, 1039
tropisms, 1039, 1040
Horsechestnut, 535
Humidity, effect of, on stomatal movement, 443, 434
influence of, on transpiration, 459, 460
correlation with transpiration, 460
Hurrin, 806
Hyaloplasm, 87
Hydathodes, 504, 505, 506
Hydrangia, influence of aluminum on color, 359, 360
Hydration and growth, 1020
Hydrazones, 526
Hydrocelluloses, 9–10
Hydrocyanic acid, 805, 806, 816, 884
Hydrogen, amount of in plant, 284
function of, 294
Hydrogen-ion concentration, of cell sap, 45–51
influence of, on absorption ratio, 227
on catalase, 990
on diastase, 797, 798
on invertase, 788
on lipase, 810
on nitrogen absorption, 646
on oxidases, 981, 982
on root hairs, 154, 155
of oily seeds, 908
optimum, for growth, 263–264
relations of, to growth, 1087–1088
to iron supply, 332, 333
to nutrient solutions, 262–265
to stomatal opening, 429
Hydrolases, 781
Hydrostatic pressure, in cell, 93
in osmosis, 77, 78
Hydrostatic system, 855
Hydrotropism of roots, 121–123
Hydroxy-fatty acids, 16–18, 171, 730
Hygrometric paper, preparation, 491
Hygroscope, 430
Hygroscopic coefficient of soil, 192, 195, 196
relation to wilting coefficient, 192
Hygroscopic water, 192
Hypertonic solution, 90
Hypogaeic acid, 729, 730
Hypotonic solution, 90
Hypoxanthine, 650, 695

I

Iarovization, 1091
Iarovtsii, 1091
Imbibition, defined, 75
force of, 208
relation of, to diffusion, 76

Imbibition, relation of, to entrance of water, 208–209
to hardened plants, 1066
to water ascent, 865, 866, 867, 868
to water loss, 452
Immobility, of iron, 334, 335
Impermeable membrane, 75
Inactive iron, 328
Incipient drying, 208, 452
relation of, to leaf surface, 453, 454, 455
to root system, 455
to transpiration, 451–455
Incipient plasmolysis, 90
Incipient wilting, 452
relation of, to leaf surface, 453, 454, 455
to root system, 455
to transpiration, 451–455
Index of oxygen-supplying power, 130
Index, root hair, 152
soil moisture, 204
stomatal, 419, 420
of transpiring power, 449, 450, 492, 493
indicator of permanent wilting, 193
Indicators of elemental needs, 291–292
characteristics of cell sap, 292
general appearances, 291, 292
microchemical tests, 292
Indirect method, of measuring permeability, 97–100
of measuring wilting coefficient, 191–193
Influence, of environment on element requirements, 288–291
amount of elements, 289, 290
fertilizer, 288
light, 289, 290
location of elements, 288
proportions, 289
of plant on element absorption, 284–288
Inherent vitality, 1026
Inheritance to chlorophyll formation, 550, 551
Inhibitor, 779
Injection experiments, 883–885
Injury and respiration, 962–964
Inorganic compounds (*see* Solutes)
Inorganic nitrogen in plant nutrition, 644–650
nitrates and ammonium salts, 644–649
availability of, 647, 648
general effects of, 648, 649
relative value of, 644–647
age and kind of plant, 647
reaction of medium, 646
toxicity of, 649
nitrites, 649, 650
Inorganic salts (*see* Solutes)
Insects, translocation organic foods, 899
Insoluble iron, 328
Intake, of solutes, 216–271
bibliography of, 271–282
factors affecting absorption ratio of, 225–227
concentration of exterior medium, 225–227
hydrogen-ion concentration, 227
light, 224, 225
mode of entrance for, 216–224
active or physiological absorption, 218–224
passive or physical absorption, 217, 218
Intake, of solutes, relation of, to transpiration, 236–240
unequal absorption of ions in, 227–236
factors influencing, 232–236
concentration of external solution, 232, 233
different ions, 233–236
mechanism, 229, 230
reaction of exterior solution, 230, 231
of water, 188–212
bibliography of, 212–215
Intensity, of light and photosynthesis, 587–590
affect on growth, 1077–1078
of transport, 1038
Intercellular spaces, area, 414, 601
Intermittent light and photosynthesis, 589
Internal breakdown, 941
Internal factors, affecting transpiration, 450, 451
of growth, 1059–1061
of photosynthesis, 600, 604
of respiration, 966
Internal oxygen, 134
Intramolecular respiration, 939, 940
Intumescences, 1037
Inulase, 800, 801
Inulin, 539, 540
in leaves, 558
Invasive alcohol poisoning, 941
Invert sugar, 530
Invertase, 781
factors influencing, 786–789
in industry, 783
nature of action of, 783, 785, 786
occurrence of, 783
preparation of, 784, 785
from green plants, 784, 785
from yeast, 784
secretion of, by roots, 163
Invertin (*see* Invertase)
Iodine, 364–366
method of measuring, diastase action, 795
in photosynthesis, 607
in plant nutrition, 291
value of, 735
Iodine content of plants, 288
Ionic exchange, solute intake, 222
Ions, antagonistic, 268
unequal absorption of, 227–236
factors influencing, 232–236
concentration of external solution, 232, 233
influence of one ion on another, 233–236
in interchange from plant to exterior, 231, 232
mechanism of, 229, 230
in reaction of exterior solution, 230, 231
Iris, 96
Irisin, 540
Iron, absorption of, from soil, 333, 334
form and amount necessary, 329–332
immobility of, in plant, 334, 335
influence of, on photosynthesis, 604
location of, in cell, 328
in nutrient solutions, 242, 249
occurrence of, in plants, 328
physiological role, 328, 329
relation to chlorophyll formation, 549

Iron, relation to hydrogen-ion concentration, 332, 333
in respiration, 972
spraying with, 333, 334
Isoelectric point, of cell sap, 221
defined, 221
of proteins, 665, 666
relation of, to guard cells, 430
Isotonic solution, 90

J

Jaminian chains, 853
Jarovization, 1091
Johnson grass (*see* Sorghum)
Juglansin, 667
Juglone, 174

K

Kafir (*see* Sorghum)
Kafirin, 667, 669
Kale, 327, 345
Karyolymph, 30
Kerasin, 727
Kerosene on corn seed, 84
Ketolactam, 812
Kinase, 780
Knop's solution, 242
Kudzu, 889

L

Laccases, 977, 978
Laccol, 978
Lactam ring, in chlorophyll, 544
in lipase, 812
Lactic acid, 523
influence on respiration, 972
Lactim, 812
Lactone ring, 544
Lamella of starch grain, 534–536
Laminaria, 99, 105, 106, 107, 111, 364, 968
Latent osmotic pressure, 90
Lateral extent of roots, 145
Lateral movement, of organic compounds, 901, 902
of solutes, 880–882
of water, 874–876
Latex, 851
Latex system, 850, 851
Lauric acid, 728, 729
Law, of limiting factors, 578, 579, 580
of minimum, 577, 578
Leaf scorch of apples, 292
relation to potassium, 324
Leaf surface, 413, 453
methods of measuring, 488, 489
relation of, to incipient drying, 453, 454, 455
to incipient wilting, 453, 454, 455
to photosynthesis, 600
to wilting coefficient, 197
Leaf-product method, measuring leaf area, 488
Leaf-punch method, 608
Leaf-test time, 492
Leaves, absorption by, of CO_2, 569, 570
of energy, 612–615
of water, 188, 189, 190
carbohydrates in, 558–567
daily variations in, 559–563
first formed, 565–567
variations in different portions, 563–565
formation of protein in, 684, 685
general structure of, 413, 414, 448
intercellular spaces in, area of, 414
nitrogen in, 679–684
compounds of, 682–684
diurnal variation in, 679–684
total, 680–682
number of stomata in, 418
paths of gaseous exchange in, 436, 437
structure and transpiration in, 413–451
temperature of, 475–481
water content of, 441, 442, 443
influence on water loss, 441–444
Lecithins, described, 727, 737, 744
relation of, to magnesium, 315
to oxygenase, 976, 977
to peroxidase, 976, 977
to phosphorus, 304
Lecithoproteins, 669
Legumelin, 666, 667
Legumes, effect on nonlegumes, 653
Legumin, 659, 660, 667, 668
isoelectric point, 666
Lemna, 348, 351, 363, 652, 685
Lemon, 49
oil of, 761
Length of day, relation to carbon-nitrogen relation, 701, 702
Lettuce, 286, 307, 342, 344, 345, 357, 358, 362, 586
Leucine, in developing seeds, 689
in digestion, 816
formula for, 661
in seedlings, 911
in tomato, 675
utilization of, 650
Leuco compounds, 556, 606
Leucoplasts, 32–33, 536, 537
Leucosin, 666, 667, 668
Levorotatory sugars, 525
Levulosans, 531
Levulose (*see* Fructose)
Liebig's law of minimum, 577, 578
Light, chemical changes by, 1082
influence of, on absorption ratio, 224, 225
on chlorophyll formation, 548, 549
on diastase, 799, 800
on dormancy, 1052–1053
on enzyme action, 777
on germination, 1052–1053
on growth, 1070–1083
on permeability, 104, 432
on photosynthesis, 587–593
on protein synthesis, 685, 686
on respiration, 973, 974
on reproduction, 1081
on root hairs, 155
on roots, 129, 137
on stomatal movement, 431, 432, 433

Light, influence of, on transpiration, 457, 458
on value of nutrient solution, 255, 256
Lignification, influence of potassium on, 319
Lignin, 11–12
Lignocellulose, 10–12
Lignoceric acid, 728
Limiting factors, 577–580
Linolenic acid, 729, 730, 741
Linolic acid, 729, 741
Linseed meal, 650, 651
Lintner's method, diastase activity, 795, 796
Lipaseidine, 809
Lipases, 781, 807–812
factors influencing action of, 809–812
preparation and study of, 807–809
relation of, to sugar formation, 760
Lipides, changes in, during germination, 752–760
classification of, 727
compound, 736, 737
defined, 727, 736
distribution of, 743, 744
function of, 745, 746
relation of, to oil and protein, 744
in seeds, 744–745
defined, 726, 727
derived, 727, 737
simple, 727, 731–736
composition of, 727–731
fatty acids, 727–730
glycerol, 730
defined, 727
distribution of, in cell, 737–739
occurrence of, in plant, 739–746
synthesis of, 746–752
changes in, during formation, 748–750
general observations of, 750–752
influence of magnesium on, 315
relation of, to carbohydrate, 746–748
to potassium, 321
Lipins (*see* Lipides, classification of, compound)
Lipoid theory of permeability, 108–109
Lipoids (*see* Lipides, classification of, compound)
Liquefication method, diastase action, 795
Lithium, in alfalfa, 285
in nutrition, 369
in plants, 283
in tobacco, 285
Little leaf, 292, 364
Living membrane (*see* Cytoplasmic membrane)
Locus of starch grain, 533
Lodging, amount of lignin in, 339, 340
defined, 338
fertilization in, 340
heredity in, 339
hypernutrition in, 339
rate of seeding and, 339
silicon content and, 338, 339, 340
Long-day plants, 1080
Loss, of nitrogen from plants, 695–697
of water from plants, 407–506
bibliography of, 506–522
Lupine, absorption of ions, 234, 235
acidity, 48
aeration of roots, 131, 135
amino acids, 911
Lupine, antagonism, 270
chemotropism of roots, 124, 125
excretion of acids, 160
exosmosis, 107
growth substances, 1039
hemicellulose in seeds, 13
hydrotropism of roots, 121, 122
phospholipides, 743
protein, 667
rate of growth, 1029
response of, to aluminum, 357
to electricity, 974
to insulin, 794, 795
to silenium, 366
roots, response to oxygen, 135
structure of root, 206
translocation of proteins, 911
water culture, 248
Lupinus (*see* Lupine)
Luxury absorption, 290
Luxury consumption, 290
Lysine, action on diastase, 798
formula for, 661

M

Magnesium, carrier of phosphorus, 314
constituent of chlorophyll, 313
influence of, on photosynthesis, 604
injury from, 315
occurrence of, in plants, 313
in oily seeds, 315
relation of, to chlorosis, 314
Malic acid, in cell sap, 38
photosynthetic quotient for, 558
in respiration, 747, 951
Malol, 736
Malt, 651
Malt diastase, 792, 793, 799
Malt sugar (*see* Maltose)
Maltase (*see* Glucase)
Maltose, 294–295
described, 530
from starch, 539
Manganese, amount needed, 347–349
essential element in, 350
as fertilizer, 349, 350
occurrence of, in plants, 346, 347
physiological role of, 350–352
catalysts, 351
chlorophyll formation, 351
sprays of, 352
Manganese deficiency disease, 348, 349
Mangels, 313
Mangold, 318, 319, 560
Manna, 531
Mannans (*see* Mannosans)
Mannite, 528
in sugar cane, 1068
Mannitol (*see* Mannite)
Mannosans, 12, 528, 531, 540
Mannose, 13, 295
described, 528, 529
Mannuronic acid, 13
Matching method, measuring leaf area, 489

Maturity, influence of phosphorus, 306, 307
Maximal temperature, growth, 1062
Maysin, 667
Melibiase, 781, 803
Melibiose, 531
Melicitose, 531
Membranes, artificial, 74
bibliography of, 112–120
cell-wall, 79–86
collodion, 217
cytoplasmic, 86–112
defined, 74, 75
diffusion through, 76–79
kinds of, 74, 75
nature of, 74–75
of plant cell, 79, 86
of seed coats, 80–86
theories of action in, 107–112
Mesophytes, leaf structure of, 449
transpiration in, 449, 450
Mesophyll of leaf, 414
Metabolic activity, solute intake, 223, 224
Metaxenia, 1031, 1032
Meteorological conditions, effect of, on transpiration, 455–462
evaporating power of air, 458–462
relative humidity, 459, 460
temperature, 458, 459
wind, 460, 462
evaporating power of solar radiation, 456–458
Methods, of aerating nutrient solutions, 247, 248
soils, 129, 130
of determining action of growth substances, 1035–1036
catalase, 987–988
energy in photosynthesis, 615
injury from freezing, 1069
need of elements, 291–292, 643, 644
oxidase activity, 979–981
relation of solute intake, 236–241
total nitrogen of leaves, 679, 680
unequal absorption of ions, 227
utilization of organic compounds, 293–297
water supplying power, 202, 203
wilting coefficient, 190–193
of extraction of proteins, 659, 676, 677, 678, 679
of measuring antagonism, 269
evaporation, 482–486
evolution of oxygen, 556, 557
influence of sprays on transpiration, 466
leaf surface, 488, 489
leaf temperature, 473, 474
oxygen supplying power, 130
permeability of protoplasm, 97–101
photosynthesis, 605–610
photosynthetic ratio, 556–558
respiration, 946–949
stomatal aperature, 430, 431
transpiration, 488–494
water requirement, 497, 498
of overcoming dormancy, 1049, 1053
of preparation, of diastase, 793–795
of growth substances, 1034–1035
of invertase, 784–785
of lipase, 807–809
Methods, of preparation, of nutrient solutions, 249–252
of peroxidase, 985, 986
of proteases, 813–821
of renewing nutrient solutions, 246, 247
of studying, diastase, 795–796
lipase, 807–809
proteases, 812
Micellae, of cell wall, 6
of starch grain, 535
Microchemical tests for elemental deficiencies, 292
Microincineration, 24
Middle lamella, composition of, 4–6
defined, 3–4
origin of, 3
relation of, to calcium nutrition, 310, 311
Mildew, effect of, on photosynthesis, 604
relation of, to carbohydrates, 604
Millet, 160, 286, 302, 303, 357, 456, 498, 499
Millon's reaction, 665
Milo (*see* Sorghum)
Minerals, translocation in seedlings, 372–374
Minimum, law of, 577, 578
Minor elements, 299
function of, 299, 300, 301
Mitochondria (*see* Chondriosomes)
Mode of water entrance into roots, 207–212
Moisture, effect of, on nitrogen content, 691, 692
on stomatal opening, 434
Moisture equivalent of soil, 192
relation of, to wilting coefficient, 192
Moisture-holding capacity of soil, 193
relation to wilting coefficient, 193
Monosaccharoses, 526–529
hexoses, 527–529
pentoses, 527
Moonlight, stomatal response, 432
Morphine, 695
Moss, organic nutrition, 295
Mulberry, 561
Mullein, hairs, 447, 448
transpiration, 416, 436, 448
Münch's theory, ascent of water, 872
of solute intake, 900
Mustard, 48, 104, 155, 157, 163, 248, 286, 327, 367, 369, 645, 651
Mustard oils, 803
Mutarotation, 786
Myricyl alcohol, 730, 736, 737
Myristic acid, 728, 729
Myrosin, 781, 806
Myrrh, 761

N

Naphthol reaction, 980
Narcissus, 327
Natural ripening, 1055–1057
Nature of respiration, 932–935
decomposition of protoplasm, 934, 935
direct oxidation of foods, 932–934
Need of water by plants, 407–409
for chemical combination, 408
for protoplasm, 407
for replacement, 409

Need of water by plants, for translocation, 408
for vacuole, 407, 408
Neutralization of organic acids by calcium, 309, 310
Nicotine, 695
Nitella, 100, 104, 218, 219, 224, 225, 227, 234, 235
Nitrate, reducing substances, 675, 676
Nitrates, availability of, 647, 648
conditions for reduction of, 673–675
general effects of, 648, 649
influence, of deficiency, 643
hydrogen-ion concentration, 646
on respiration, 643
on root development, 128, 129
on stomatal opening, 435
occurrence of, 671–673
in plant nutrition, 644–649
in rain, 653, 654
reducing substances for, 675, 676
reduction of, in plants, 671–676
relation to phosphorus, 304, 305
relative value of, 644–647
in soil, 643
toxicity of, 649
Nitrites, in air, 657
in leaves, 673
in nutrition, 649, 650
occurrence of, in plant, 673, 674, 675
in rainwater, 653
Nitrogen, absorption of, from air, 653–657
from soil, 644–653
inorganic, 644–650
organic, 650–653
amino, 653
amount of, in air, 653–657
in plant, 284, 642
in protein, 659
asparagine, 663
bases, 694, 695
bibliography of, 705–725
carbon relations, 697–705
in compounds, 683
deficiency effect of, 643
effect of, on growth, 642
fixation of, by leguminous plants, 654, 655
by nonleguminous plants, 655–657
function of, 301
influence of phosphorus on, 306
on root development, 128, 129, 643
on water content, 643
in leaves, 642, 679–684
compounds, 682–684
diurnal variation in, 679–684
methods of determining, 679, 680
total, 680–682
loss of, from plants, 695–697
in metabolism, 642–705
method of determining need for, 643, 644
movement of, in stems, 892
occurrence of, in plant, 642, 643
in relation to the plant, 642–644
seasonal variation in, 687
sources of, 644–657
time of application and protein, 689–691
in twigs, 887
Nitrogen spots, 643
Nitrogenous compounds, in cell sap, 38
inorganic, absorbed, 644–650
organic, absorbed, 297, 650–653
Nodules, relation of, to boron, 342
to sulphur supply, 326, 327
Nomenclature of enzymes, 779–781
Nonacosane, 736
Nonavailable iron, 328
Nondrying oils, 733
Nonessential element, 301
Nonleguminous plant, fixation of nitrogen in, 656, 657
Nonreducing sugars, defined, 526
in leaf, 559–563
Nonsugars (*see* Polysaccharoses)
Normal cellulose, 8–9
Normal respiration, 939
Normal solution, 72
Nucleic acid, 171, 304, 650, 652, 669
Nucleoproteins, 30, 314, 669
Nucleus, 30
in respiration, 935
of starch grain, 533
Number and kind of elements, 283, 284
Nutrient solutions, 241–271
best, 243, 255
characteristic, and response, 250–252
criticism of data on, 265–268
factors determining value of, 255–262
concentration, 256–258
physiological balance, 259–262
temperature and light, 255, 256
water vs. sand cultures, 258, 259
historical, 241, 242
hydrogen-ion relations, 262–265
kinds of, 242–244
technique for, 244–255
aeration, 247–248
preparation, 249–252
renewal, 246, 247
triangle system, 252–255
Nutrients, depletion in soil, influence on plants, 168–170
Nutrilites, 652

O

Oats, absorption of elements, 288
of ions, 235
of organic compounds, 294
of phosphorus, 302
acidity of cell sap, 48
amino acids, 669, 670
ash content, 289
boron content, 340
CO_2 production, 160
catalase, 994
coleoptile, 1034–1036, 1039, 1040
diastase, 798
energy utilized, 617
erepsin, 821
excretion of acids, 159, 160
growth hormones, 1034–1036, 1039, 1040
heat of respiration, 953, 954

Oats, influence, by other plants, 653
of electricity, 1083–1084
on other plants, 165
iodine content, 288
iron content, 347
light rays and growth, 1074, 1075
magnesium needs, 313
maltase, 801
manganese content, 347
amount needed, 349, 350
nitrate content, 671
nitrates vs. ammonium, 645
pentosans, 540
phospholipides, 743, 745
photosynthesis, 604
polypeptides, 669, 670
potassium, amount needed, 318, 319
response to, 319
protein, 667, 668, 691
radiation and transpiration, 456, 457
relation to light, 1079
response of, to aluminum, 357, 358
to ammonium, 645
to arsenic, 300
to barium, 367
to boron, 300
to chlorine, 355
to cobalt, 369
to copper, 300
to iodine, 365
to magnesium, 315
to manganese, 300
to nickel, 369
to nitrates, 645
to palladium, 369
to phosphorus, 303
to potassium, 288
to rubidium, 369
to strontium, 367
to sulphur, 327
to zinc, 300
root formation, 1040
roots, 123, 137, 138, 139, 143, 144, 149
depth of penetration, 139, 143, 144
number and length, 137, 138
response to moisture, 123
weight, 149
working depth, 143, 144
rusts, and photosynthesis, 604
and transpiration, 471
salt requirements, 370
silicon, content, 335
use of, 336
starch grain, 533
stomata, 422, 423
temperature and transpiration, 458, 459
transpiration, 411, 458, 471, 482
day and night, 411
water culture, 248, 261
acidity, 263
requirement, 498, 499
wilting coefficient, 194
Observed wilting coefficient, 190, 191, 199
Octacosanol, 736
Oil (*see* Lipides)
Oleanol, 751
Oleic acid, 729, 730, 741, 752
Oleorsin, 890
Olive, oil formation in, 746, 747, 751
Onion, acidity, 47
bulb scale, 80, 82, 83, 85
carbohydrate and polarized light, 590
cell division, 1029
cytase, 803
depletion of nutrients, 169
permeability of scape, 100, 101, 102, 111
phosphorus content, 302
respiration, 953
response of, to cobalt, 369
to copper, 362, 363
to light, 1082
to nickel, 369
to potassium, 322
to sulphur, 324
to zinc, 363
starch formation, 603
Optical properties, chlorophyll, 545, 546, 547
protein, 659
sugars, 525
triglycerides, 732
Optimum temperature, growth, 1061, 1062, 1064
Orange, 46, 49, 299, 300, 314, 344, 642
Organic acids, absorbed by roots, 296, 297
in cell sap, 38
relation of, to respiratory quotient, 951
theory of, in photosynthesis, 571
Organic compounds, absorbed, 293–297
carbohydrates, 295–296
organic acids, 296–297
organic nitrogen, 297
distance moved, 892–894
influence of, on diastase, 798
nutrition experiments with, 575, 576
in respiration, 935–939
of soil, 171
storage of, 885–890
time, 887–889
in tracheal sap, 895–897
translocation of, 885–912
carbohydrates, 902–906
fats and oils, 906–912
lateral, 901
proteins, 910–912
tissues concerned, 894–901
utilization, 890, 891
Organic matter, rate of photosynthesis for, 607–610
qualitative method for, 607
quantitative methods for, 608–610
saccharification, 610
twin-leaf method, 609, 610
weighing method, 608, 609
Organic nitrogen, 650–653
assimilation of, 650, 651
stimulation of, 651–653
by auximones, 652, 653
effects, 652
origin, 653
of legumes upon nonlegumes, 653

Organic substances, decomposition of roots, 171, 172, 173
excreted by roots, 163, 164
Organs, growth of, 1024
Origin, of materials, in cell sap, 38, 39
of auximones, 653
of plastids, 33–36
of vacuole, 37, 38
Oryzenin, 667
Osazones, 526
Osmosis, defined, 77, 78, 79
examples of, 76–79
Osmotic concentration, of cell sap, 90
of external solution, 91–93
relation of, to exterior solution, 91–93
Osmotic forces, relation of, to water ascent, 867, 868
to water intake, 209–212
Osmotic pressure, of cell sap, 39–45
defined, 72, 77, 78
distribution of, in nutrient solutions, 250–252
examples of, 76–79
latent, 90
potential, 90
Osmotic relationships of plant cell, 89–96
Osmotic value, of cell sap, 90
of guard cells, 427
in relation to transpiration, 450, 451
Oxalic acid, in cell sap, 38
excreted by roots, 159
in plants, 309
in protein formation, 309
photosynthetic quotient, 558
relation of, to invertase, 788
in respiration, 939, 950
in soil, 171
Oxidases, 299
action of, 975, 976
in apples, 956
classification of, 977–979
laccases or phenolases, 977, 978
tyrosinases, 978, 979
defined, 975
effect of phosphorus of, 305
factors influencing, 981–983
healthy vs. diseased tissues, 982, 983
methods of studying, 979–981
occurrence of, 978
in potatoes, 956, 976
relation of, to physiological functions, 983–984
to respiration, 942
to sap pigments, 984
secreted by roots, 162, 163
Oxidation, of enzymes concerned, 975
of foods, 932–934
Oxidative anabolism, 933
Oxidizing enzymes, 976
Oxidizing ferments, 975
Oxycelluloses, 9–10
Oxydase (*see* Oxidase)
Oxygen, absorbed by oxidases, 981
amount of, in plant, 284
determined in respiration, 947–949
evolution of, in photosynthesis, 555–558
function of, 294
Oxygen, influence of, on stomatal movement, 433
on transpiration, 132
internal, 134
method of studying photosynthesis, 606, 607
relation of, to chlorophyll formation, 550
to dormancy, 1048, 1052
to germination, 1052
to needs of plant, 133, 134
in respiration, 969
response of roots to, 131–135
in soil, 129–135
sources of, 134
supply of, for photosynthesis, 601
and root development, 129–135
in water cultures, 131
Oxygen respiration, 939
Oxygenase, 746
defined, 975
relation of, to oxidase, 976, 977
Oxygenated oils, 761
Oxygen-supplying power of soil, 130, 463

P

Palmitic acid, 728, 729, 741
Pangestion, 789
Pantothenic acid, 1037
Papaïn, 781, 815–818
Papayacin, 815
Papayotin, 815
Papaytin, 815
Paracaoutchouc, 527
Paralyzer, 779
Parenchyma, 848, 850
Parsnip, 228, 354
Passage cells, 207
Passive absorption, of solutes, 217, 218
of water, 209, 212
Passive regulation of stomata, 441
Paths of gaseous exchange, stomata, 436, 437
Pea, absorption in, of elements, 290
of ions, 227, 228
acidifying power, 231
acidity, 49
boron, absorbed, 342
content of, 340
calcium content, 289
catalase, 994
cell-sap concentration, 39
changes in pectic substances, 23
composition, 285, 286
CO_2 fertilization, 586
diastase in leaves, 791
effect of light, 290
erepsin, 821
etiolated plants, 1061
excretion of digestive enzymes, 163
fixation of nitrogen, 657
formation of protein, 687
growth-inhibiting substances, 1044
heat of respiration, 953
influence on, of electricity, 974
influence of, on other plants, 653
interchange of ions, 232
iodine content, 288

Pea, iron content, 347
light and growth, 1070
manganese content, 347
movement of organic compounds, 886
nitrogen content, 692
oil of seed, 740
optimum temperature, growth, 1062
oxidase, 978
permeability, of roots, 104
of seeds, 110
phospholipides, 745
potassium, amount needed, 318
effect of deficiency, 320
protein, 667
rate of growth, 1029
relation of, to calcium, 312
to growth, 1029
respiration, 956
respiratory quotient, 952
response of, to aluminum, 357
to arsenic, 367, 368
to boron, 343
to calcium, 312
to cobalt, 369
to lithium, 369
to manganese, 349
to nickel, 369
to nitrate vs. ammonium, 645
to organic compounds, 650
to palladium, 369
to rubidium, 369
to titanium, 367
to zinc, 363
root formation, 1040
root hairs, 157, 158
roots, 104, 127, 135, 163, 296
response of, to fertilizers, 127
to nitrogen, 127
to oxygen, 135
salt requirement, 290
selenium content, 286
solvent action of CO_2, 161, 163
translocation of fats, 906
transpiration and solutes, 240
unequal absorption of ions, 227, 228
water cultures, 248, 258
acidity, 262, 263
wilting coefficient, 194
Pea method, 1035
Peach, 168, 422
carotinoids, 551
dormancy, 1047
gum, 527
movement of organic compounds, 886
nitrates vs. ammonium, 648
oil reserves, 738
oxidases, 978
pectic substances, 23
protein, 667
respiration, 943
seed coats, 83
stomata, 422, 423, 434
tracheae, 848
translocation in, of oil, 909, 910
of solutes, lateral, 880
Peach, translocation in, of solutes, upward, 877, 879
water movement, 876
Peanut, moisture and respiration, 960
oils of, 732, 740, 908
phospholipides, 745
protein, 667
respiration, 960
response to low temperature, 1065
seed coat, 80, 81, 111
Pear, 373, 848, 888, 893, 896, 897, 1043
Pecan, oil formation, 746, 747
Pectases, 22, 23, 781, 822
Pectates, description of, 21
relation of, to middle lamella, 310, 311
of root hair wall, 157
Pectic acid, 21
Pectic substances, 19–23
characteristics of, 19
classification of, 20
pectic acid, 21
pectin, 22
pectates, 21
pectose, 21
defined, 19
in leaf, 558
in ripening fruits, 23
Pectin, 22, 23
occurrence in fruits, 23
in root hairs, 157–158
Pectinase, 781, 803
Pectinogen, 21
Pectose, 21, 22
changes in fruit, 23
Pennyroyal, 162, 761
Pentosans, 531, 540, 541
in hardening, 1066
in hemicelluloses, 12, 14
in leaf, 561
in soil, 171
in twigs, 888
Pentoses, defined, 527
occurrence of, 13, 527, 529
in soil, 171
Pepper, 199
Pepsin, 781
occurrence and preparation of, 813, 814
Peptase (*see* Pepsin)
Peptides, 669
Peptone, defined, 669
in digestion, 813, 815
in seedlings, 912
utilization of, 650, 651
Periodic elongation, 1028, 1029
Periodicity, of growth, 1029
of stomatal movement, 433
Permanent starch, 536, 537
Permanent wilting, defined, 190
Permeability, cell-wall membrane, 79–86
contrasted with absorption, 96
of cytoplasmic membrane, 89
defined, 74, 75
effect of direction on, 81, 82
of endodermis, 85, 207
extreme changes in, 107

Permeability, factors influencing, light, 104, 432
solutes, 103–107
water, 101–103
methods of measuring, 97–101
in plasma membrane, 88
in seed coats, 80–86
Permeable membrane, 74, 75
Peroxidases, 985, 986
defined, 975
factors influencing activity of, 986
occurrence of, 985
preparation of, 985, 986
production of, 746
in relation to physiological functions, 986
tests for, 985
Peroxide, 976, 977
Pfeffer's solution, 242
Phaneroplasm, 28
Phaselin, 667
Phaseolin, 667
Phaseolus (*see* Bean)
Phellonic acid, 15, 18
Phenolases, 977, 978
Phenol oxidases, 977–979
Phenolphthalin reaction, 980, 981, 983
Phenyl alanine, 661
Phleim, 540
Phloem, 849, 850
in conduction, of inorganic salts, 779, 780
of organic materials, 897–901
unsuitability for conduction, 894–895
Phloem rays, 850
Phloionic acid, 15
Phlorizdin, 804
Phosphates, absorbed, 301, 302
in cell, 25
influence of, on respiration, 972
Phosphatides (*see* Phospholipides)
Phospholipides, described, 727, 737
distribution of, 23, 304, 743, 744
function of, 745, 746
in relation to oil and protein, 744, 745
in seeds, 743
Phospholipins (*see* Phospholipides)
Phosphomolybdic acid, 664
Phosphoproteins, 304, 669
Phosphorus, amount of, 301, 302, 303
in cell sap, 303
form absorbed, 302, 303
for growth, 303
influence of, on photosynthesis, 604
on root growth, 305, 306
occurrence of, 301–302
physiological role of, 303–307
carbohydrate relations, 306
components of compounds, 303, 304
hexosephosphates, 304
nucleic acid, 304
phospholipides, 304
phosphoproteins, 304
phytin, 303
general effects, 307
maturity, 306
nitrogen relations, 306
reduction, 304
Phosphorus, physiological role of, respiration, 305
root growth, 305
relation of, to salt sickness, 291
to silicon, 336–338
solubility, 301
Phosphotungstic acid, 664
Photochemical induction, 589
Photoperiod, 1078
Photoperiodism, defined, 1078
discussed, 1079–1083
Photosyntax, 541
Photosynthesis, 541–618
absorption of CO_2 by, 569, 570
bibliography of, 618–641
chemistry of, 567–577
coefficient of energy absorption of, 612–615
contrasted with respiration, 944, 945
defined, 541
economic importance of, 617, 618
experimental evidence on, 570–577
condensation of formaldehyde, 574, 575
intermediate products, 575
organic nutrition, 575–576
reduction of CO_2, 572, 573, 574
factors influencing, 577–605
amount of chlorophyll, 596–598
CO_2 supply, 580–587
internal factors, 600–605
light, 587–593
other factors, 600
protoplasmic factors, 600–605
temperature, 593–596
water supply, 598–600
function of chlorophyll, 567–569
history of, 554, 555
incident radiation, 611, 612
methods of measuring, 605–610
absorption of CO_2, 607
increase in organic matter, 607
liberation of oxygen, 556, 557, 606
nature of, 541
products of, 555–567
rate of, 610–611
apparent, 606
true, 606
relation of energy to, 605–618
theories of, 570–577
utilization of energy in, 615–617
Photosynthetic number, 597
Photosynthetic quotient, 556, 557, 558
Phreosin, 727
Phyllins, 544
Physical absorption of solutes, 217
Physical properties, of fats and oils, 732
of proteins, 659
Physical theories of water ascent, 863–872
absence of living cells, 864, 865
cohesion theory, 867–872
imbibition or capillary theory, 865–867
Physiological absorption, 218–224
forces concerned in, 220–224
amphoteric nature of protoplasm, 221
attraction of colloids for ions, 220, 221
Donnan equilibrium, 222
ionic exchange, 222

Physiological absorption, forces concerned in, metabolic activity, 223, 224
Physiological balance of nutrient solutions, 259–262
Physiological drought, 502
Physiological regeneration of tip, 1035
Physiological balanced solution, 268
Phytamines, 652
Phytase, 781, 806
Phytins, 545, 804
  occurrence of, 303, 304
Phytohormone, 1033, 1036
Phytosterol, 730, 736, 741
Phytyl alcohol, 544
Picric acid, 664
Pigments, 523, 984
  of chloroplasts, 541–554
Pigweed (*see* Amaranthus)
Pineapple, acidity, 46, 47
  bromelin, 818
  manganese content, 346
  nitrate vs. ammonium, 645
  relation of, to manganese, 351
    to titanium, 367
  ripening, 1057
  spraying with iron, 333, 334, 335
Pisum sativum (*see* Pea)
Pits, 1, 848, 849
Plant, elemental composition of, 284
  influence on, of one on another, 164–174
  influence of, on elemental composition, 284–288
  kind of, ammonium vs. nitrate absorption, 647
  relation of, to wilting coefficient, 194–198
  synthetic power of, 523
  water requirement of, 496–502
Plant cell, 1–51
  absorbing power of, 93, 94, 96
  bibliography of, 51–69
  parts of, 1
  stages in growth of, 1019–1024
  turgor, 93
  turgor pressure, 93
  wall pressure, 93
Plant diseases and water requirement, 502
Plant foods, 770
Plasma membrane, 87
  difference in two surfaces of, 88
    effects of chemicals, 88
    potential difference, 88
  methods of measuring, 88
  permeability of, 88
Plasmodesms, 29, 30
Plasmolysis, 89–90
Plasmolytic method, measuring permeability, 98–99
Plastids, 30–36
  defined, 30
  inheritance of, 35–36
  origin of, 33–36
  relation of, to starch grains, 536, 537
Polarity, growth substance, 1038
Polarized light, effect of, on diastase, 799, 800
  on growth, 1075, 1077
Polypeptides, 669
Polysaccharoses, 531–541
  amyloses, 532–540
  celluloses, 531–532
  galactosans, 540
  mannosans, 540
  pentosans, 540, 541
  starches, 532–540
Polyuronides, 14
Poppy and oxidase, 984
Populin, 804
Porometer, 430
Porous-cup atmometers, 484, 485, 486
Potassium, 316–324
  deficiency of, 318, 319, 320, 324
  general response of, 322–324
    general appearance, 322, 323
    other effects, 324
    quality, 323
    relation to other elements, 323, 324
  high-response plants, 322
  influence of, on photosynthesis, 604
  low-response plants, 322
  luxury absorption of, 317, 318
  medium-response plants, 322
  occurrence of, 316–318
  physiological role of, 318–322
    action on enzymes, 322
    catalysts, 321, 322
    in formation of carbohydrates, 318, 319, 320
    in formation of oils, 321
    in relation to nitrogen, 320, 321
  vs. sodium, 316, 361
  water soluble, 316, 317, 318
Potato, absorption in tissue, 225
  acidifying power, 230
  acidity, 50
  action of potassium, 318
  afterripening, 1054
    influence on respiration, 969
  aluminum content, 357
  boron, location, 340
  carbohydrate, of leaf, 561
  CO_2, exhaled, 956–959
    produced, 160
    in respiration, 971
  catalase, 988, 991
  cell-sap concentration, 43, 44
  chloroplast, 33
  chlorosis and magnesium, 313, 314
  copper content, 362
  cytase, 803
  depletion of soil, 169
  depth of absorption, 147
  diffusion of solutes, 106
  downward movement of water, 872
  energy utilized, 617
  erepsin, 821
  fertilized with CO_2, 586
  formation of suberin, 16, 17
  growth, light, 1071
  growth-inhibiting substances, 1042, 1043
  growth-promoting substances, 1031
  influence on plants, 168
  iodine content, 365
  levan enzyme, 604

Potato, minerals in tuber, 372
  moisture and respiration, 961
  nitrate reducing substance, 675
  nitrate vs. ammonium salts, 645
  overcoming dormancy, 1049–1050
  oxidase, 976, 977, 978, 983
  oxygen and respiration, 970, 971
  pentoses, 527
  permeability to water, 101, 102
  peroxidase, 985
  phospholipides, 743
  potassium, deficiency, 320
    type, 318
  protein formation, 686
  proteins, 667
  relation to light, 1079, 1082
  respiration, 953, 962–964
  response of, to ammonium, 645
    to borax, 345
    to chlorine, 353, 355
    to low temperature, 1065
    to nitrates, 645
    to potassium, 322
    to sulphur, 326
  roots, 127, 139
  salt requirement, 372
  sprays and transpiration, 466, 470
  starch grains, 534
  starch of, 533, 535, 538
  stomata, 422, 423
  stomatal type, 436
  sulphur content, 324
  temperature of, 953
  tyrosinase, 979
  utilization, organic reserves, 890
  water cultures, 260
  water lost, 412
Potato type of stomata, 436
Potential difference, protoplasmic membrane, 88
Potential osmotic pressure, 90
Potometer method, measuring transpiration, 489
Powdery mildew and boron, 344
Primary wall, 4
Production of carbohydrates in leaf, 558–567
  daily variation, 559–563
  early observation, 559
  first sugar formed, 565–567
  variation in different portions, 563–565
Proenzyme, 780
Proinulase, 780
Prolamins, 667, 668, 669
Proline, 661
Prolipase, 780
Prooxidase, 780
Propionic acid, 728
  influence of, on diastase, 797
  in respiration, 936
Propylene, 1041
Propylene chlorhydrin, 1049
Proteases, 781, 812–821
  erepsins or ereptases, 820, 821
  historical, 812, 813
  kinds of, 813–821
  methods of preparing, 813–820
  methods of studying, 812
Proteases, pepsins or peptases, 813, 814
  trypsins or tryptases, 814–820
Protein, amides, 662–664
  amino acids of, 660–662
  amphoteric nature of, 665, 666
  in cell walls, 4, 23
  chemical properties of, 659–666
  classification of, 666–670
  combined, 678
  composition of, 657, 658, 659, 660
  conjugated, 669
  defined, 657
  derived, 669, 670
  diurnal variations of nitrogen, 679–684
    methods of determining, 679, 680
    nitrogen compounds of leaf, 682–684
    total nitrogen in leaves, 680–682
  extraction of, from leaves, 676–679
    effect of previous treatment, 677, 678
    separation from cells, 678, 679
  formation of, and light, 685, 686
    in leaves, 684, 685
    in seeds, 687–689
      factors influencing, 689–692
  isoelectric point of, 221, 665, 666
  metabolism in, in leaf, 676–685
  occurrence of, 657
  physical properties of, 659
  of plant, 657–693
  in protoplasm, 1020
  qualitative reactions of, 664, 665
  relation of, to calcium, 311
    to oil formation, 751
    to solute intake, 221
  in respiration, 937–938
  role of manganese in formation of, 350
  role of potassium in formation of, 320, 321
  simple, 667–669
  solubility, 659
  soluble, 679
  stored food, 772
  sulphur content, 325
    relation to formation, 325, 326
  synthesis, 670–694, 1020
  translocation, 910–912
Proteoses, defined, 669
  in seedlings, 912
Protochlorophyll, 548
Protopectin, 21
Protoplasm, 24–37
  amphoteric nature of, 221
  composition of, 25–26
  defined, 24
  longevity of, 26, 28
  need of, for water, 407
  parts of, 1, 2, 24, 28
  properties of, 24–26
  in relation to calcium, 311
  in relation to solute intake, 221
  in relation to translocation of organic matter, 899, 900
  in respiration, 934, 935
Protoplasmic membrane (*see* Cytoplasmic membrane)
Protoplasmic respiration, 937

Provitamin A, 552
Pseudocellulose (*see* Hemicellulose)
Pumpkin, acidity of sap, 50
boron absorbed, 342
chemotropism of roots, 125
conduction of water, 854
fixation of nitrogen, 656
growth, 1029
hemicelluloses, 13
nitrogen in leaves, 642
oils, 740
photosynthesis, 610, 611
protein formation, 686
root hairs, 154
seed coats, 80
stomata, 422
temperature of leaves, 475, 476
transpiration, relative, 416
water content, of leaves, 408
Purine bases, 695
Pyridine, 171, 651
Pyrogallol, 972
Pyrrole-carbonic acid, 329

Q

Qualitative reaction of proteins, 664, 665
by color, 664, 665
Adamkiewicz's or Hopkins-Cole's reaction, 665
biruet reaction, 664
Millon's reaction, 665
sulphur reaction, 665
xanthoproetic reaction, 664, 665
by precipitation, 664
Quinine, 695
Quinone, 171
Quotient, assimilation, 556
photosynthetic, 556
respiratory, 949-952

R

Radiation (*see* Solar radiation)
Radioactivity of potassium, 321
Radish, absorption of organic acids, 297
aeration of roots, 135
carbohydrates, 295
catalase, 992, 995
cell-sap concentration, 39
composition, 290
fertilized with CO_2, 586
fixation of nitrogen, 656
manganese in seeds, 348
permeability of roots, 103, 104
relation to light, 1080, 1082
response of, to aluminum, 358
to arsenic, 367
to chlorine, 353, 354
to iodine, 365
to sulphur, 327
root hairs, 155
roots, response of, to oxygen, 135
and temperature, 135, 136
transpiration, 467
Raffinose, 531
Ragweed (*see* Ambrosia)
Ralig, 12
Ramie, 319
Rape, 48, 290, 302, 303, 317, 318, 325, 327, 740, 749
Raphanus sativus (*see* Radish)
Rapic acid, 729
Rarer elements, 299
Raspberry, rusts and stomata, 423
temperature of buds, 477
Rate of water movement in soil, 202–205
factors influencing, 203–205
Ratio, absorption, 224
calcium magnesium, 308, 309
calcium nitrogen, 309
carbon nitrogen, 697–705
photosynthetic, 556
respiratory, 747
Rays, affecting growth, 1072–1075
utilized, in chlorophyll formation, 549
in condensation, 568, 572
in photosynthesis, 546, 547, 568, 590, 593
Reabsorption coefficient, 883
Reactive surface of chloroplasts, 604
Recent theories of photosynthesis, 576, 577
Red amber (*see* Sorghum)
Redtop, 169, 358, 653, 891
Reducing sugars, defined, 526
in leaf, 560–567
Reductases, 162, 163
excreted by roots, 163
Reduction, of CO_2, 572, 573, 574
of nitrates, 304, 305
Regeneration, 1058–1059
Relative transpiration, 486, 487, 488
Relative weight of roots and tops, 123, 124, 128, 129
Renewal of nutrient solutions, 246, 247
Rennets (*see* Rennins)
Rennins or rennets, 781, 821, 822
Replacement of water, 409
Resins, 23, 171, 726, 761, 890
Resistances, photosynthesis, 584, 585
Respiration, aerobic and anaerobic, 939–943
bibliography of, 996–1017
vs. breathing, 945, 946
contrasted with photosynthesis, 944, 945
defined, 932
diseases of, 975
energy relations in, 931, 932
energy released in, 942
enzymes of, 975–996
factors affecting, 955–975
afterripening agents, 968, 969
anesthetics, 967–968
CO_2 content, 969–972
electricity, 973–974
food type, 938–939
injury, 962–964
light, 973–974
materials, 974
moisture, 959–962
oxygen supply, 969–972
phosphorus content, 305

Respiration, factors affecting, salts, 972–973
sprays, 974
temperature, 955–959
tissues, 964–967
importance of, 932
in leaves, 557
loss in weight during, 935
methods of measuring, 946–949
nature of, 932–935
organic material used in, 935–939
in plants, 931–966
production of heat during, 952–955
relation of, to catalase, 991–993
respiratory quotient of, 949–952
Respiratory index, 966
Respiratory quotient, 949–952
defined, 949
factors influencing, 949–952
kind of material, 949, 950
organic acids, 951
other factors, 951
temperature, 950, 951
germination of seeds, 952
of seeds, 747, 748
Respirometers, 946–949
Rhamninase, 781, 806
Rhamnose, 523, 527
Rhizopin, 1034, 1035
Rhubarb, 46, 229
Rhythm of growth, 1029
Rhythm of stomatal movement, 433
Rice, aeration, 130, 131
ammonium salts vs. nitrates, 645, 647, 649
iron need, 331
iron supply, 334
maltose, 801
manganese content, 346
oil of, 740
oxygen supply, 943
proteins, 667
respiration, 943
response of, to aluminum, 357
to caesium, 369
to manganese, 349
roots, depth, 139
salt requirement, 371
starch, 535
starch grain, 533
temperature, growth, 1064
walter cultures, acidity, 265
wilting coefficient, 194
Ricin, 667
Ricinoleic acid, 729, 730
Ricinus (*see* Castor bean)
Ringing experiments, 877–890, 891, 892, 897, 898, 903
Ripening, after-, 1053–1058
artificial, 1057, 1058
natural, 1055–1057
Röhmann-Spitzer reaction, 981
Root hairs, 149–158
advantage of, 151
cell wall in, 157–158
concentration of cell sap in, 39
Root hairs, conditions affecting formation of, 152–154
elements, 154
hydrogen-ion, 153, 154
obstruction, 153
temperature, 153
toxic substances, 153
water, 152, 153
defined, 150
growth of, 154–156
concentration of medium, 155–156
hydrogen-ion, 156
manner of, 154
rate of, 155
relation of, to light and temperature, 155
location of, 150
longevity in, 156, 157
origin of, 151, 152
Root pressure, 210, 855, 856
Root systems, 137–145
depth of absorption of, 145–148
extent of, 138–145
relation of, to incipient drying, 455
incipient wilting, 455
wilting coefficient, 196
Root-hair index, 152
Roots, 121–174
beneficial and toxic effects of, 164–174
bibliography of, 174–187
chemotropism in, 124–126
decomposition products of, 170–174
depth of absorption by, 145–148
development and cultivation of, 141
excretions of, 158–164
acids, 159–162
enzymes, 162–163
organic substances, 163–164
extent of, 138–145
factors affecting development of, 121–137
moisture, 121–124
oxygen, 129–135
solutes, 124–129
hairs on, 149–158
influence on, of calcium, 312, 313
of phosphorus, 305, 306
of solutes, 126–128
of sulphur, 326
of wilting coefficient, 196
length of, 137–138
mode of water entrance in, 207–212
nitrate reduction in, 674
number of, 137–138
penetration of, in dry soil, 123
rate of growth of, 148
structure of, 205–207
weight of, 123, 124, 128, 129, 148–149
working depth of, 142
Rosette, 292
Rotatory power, sugars, 525
Rusts, influence of, on photosynthesis, 604
on stomata, 423
on transpiration, 470, 471, 472
Rutabagas, 303, 325
Rye, absorption of ions, 235
acidifying power, 230

Rye, acidity, 230
aluminum content, 357
amino acids, 669, 670
CO_2 production, 162
cell-sap acidity, 46
diastase, 798
energy utilized, 617
erepsin, 821
maltase, 801
oil of seed, 740
polypeptides, 669, 670
protein, 667
relation to light, 1078
response of, to aluminum, 357, 358
to cobalt, 369
to iron, 330
to manganese, 348
to nickel, 369
to potassium, 322
to zinc, 363
roots, depth of penetration of, 139, 143, 144
number and length of, 137, 138
working depth of, 143, 144
rusts and transpiration, 471
sitosterol, 730
solvent action, 162
stomata, 422
temperature and transpiration, 458, 459
transpiration, day and night, 411
water culture, 247
water requirement, 498, 499

S

Saccharase (*see* Invertase)
Saccharification methods, diastatic activity, 795
photosynthesis, 610
Saccharose (*see* Sucrose)
Sach's method, photosynthesis, 608
Sach's solution, 242
Salicin, 804
Salt sickness, 291, 362, 363
Salts (*see* Solutes)
Sand cultures, 241
Sand drown, 292, 313, 314
Sap flow, 857
Sap pressure, 855
Saponification, of fats, 733
Saponification value, 735
Saponin, 804
Saturated fatty acids, 727, 728, 729
Saturation deficit, 208, 211, 863
Schweitzer's reagent, 8
Scutellum of corn, 824, 825
Seasonal variation of nitrogen, 687
Sebacic acid, 18
Secondary wall, 4
Seed coats, membranes of, 80–86
permeability of, to solutes, 82–86
to water, 80–82
Seeds, afterripening of, 1053
dormancy of, 1046, 1047
imbibition of water in, 200
longevity of, 1046–1047
Seeds, means for measuring water-supplying power of soil to, 200–203
overcoming dormancy in, 1050–1053
respiratory quotient in, 952
size and growth of, 1060–1061
Selective absorption of elements, 287
Selenium, 366, 367
Semidrying oils, 734
Semipermeable membrane, cell wall, 82–86
cytoplasm, 75
defined, 75
seed coats, 80–86
Separation of protein of cell, 678, 679
Serine, 661
Sesamum, root excretions, 164
Shive's solution, 243
Short-day plants, 1080
Shot blotch and boron, 343
Sieve plates, 849
Sieve theory of membrane action, 107–108
Sieve tubes, 849, 850
conduction of sugar in, 903–905
Significance of transpiration, 495, 496
Silicates, in nutrition, 337–338
occurrence of, 335
Silicon, effect of, on phosphorus nutrition, 336, 337, 338
function of, in plants, 336, 337, 338
lodging, 338–340
occurrence of, 335
physiological role of, 335–338
Simple lipides, 727, 731–737
fats and fatty oils, 727, 731–735
waxes, 727, 735, 736
Simple proteins, 667, 668, 669
Sinapis (*see* Mustard)
Siningrin, 804
Sitosterol, 730, 731
Sodium, in nutrition, 360–362
as substitute for potassium, 361
Sodium chlorate, translocation, 878
Soft wheat, 339
Soil, aeration in, 129–135
chemicals and transpiration in, 464, 465
cones for water-supplying power of, 203
fertility and water requirement of, 500, 501
ionic exchange and solute intake in, 222
iron content of, 333, 334
moisture of, and nitrogen content, 691
and root development, 121–124
stomatal number, 419
and water requirement, 501, 502
moisture index for, 204
nutrients and root development in, 124–129
soil water and transpiration in, 462–464
texture and root development in, 136–137
type and wilting coefficient of, 193, 194
water-supplying power of, 200–205
Soil fertility and water requirement, 500, 501
Soil moisture, effect of, on water requirement, 501, 502
index of, 204
Solar radiation, 456
and coefficient of absorption, 612–615
determination of energy used in, 616, 617

Solar radiation, and energy used in photosynthesis, 615–617
expressed, 456
influence of, on cyclic factors, 456
on transpiration, 456–458
measurement of, 611
total incident of, 611, 612
utilization of, by leaves, 615–617
Solarization, 602
Soluble iron, 328
Soluble starch, 538, 539, 795
Soluble proteins, 679
Solutes, in cell walls, 23
defined, 71
influence of, on diastatic activity, 796, 797
on oxidase activity, 982
on permeability, 105–107
on respiration, 972, 973
on root development, 124–129
intake of, by plants, 216–271
bibliography of, 271–282
mode of entrance of, into root, 216–224
permeability to, 103–107
relation of, to transpiration, 236–241
translocation of, 876–883
Solution theory of membrane action, 108–109
Solutions, bibliography of, 112–120
colloidal, 70, 71
culture, 241
defined, 70, 71
diffusion, 72–74, 76–79
expressing concentration, 71, 72
hypertonic, 90
hypotonic, 90
isotonic, 90
nutrient, 241–268
true, 70
Solvent, 71
Sörensen method, 812
Sorghum, acidity, 48
catalase, 987, 989
concentration of cell sap, 42, 44
depth of absorption, 146
excretion from roots, 164
hydrocyanic acid, 806
influence on plants, 172, 173, 174
leaf structure, 449
leaf surface, 453
light rays, effect, 1073
moisture and respiration, 960
nitrogen content, 642
in leaves, 642, 681
oil, 740
oxidase, 984
phospholipides, 745
phosphorus content, 302
photosynthesis, 608, 609, 610
potassium, water soluble, 317
proteins, 667
relation to soil moisture, 196
respiration, 960
response of, to aluminum, 358
to fluoride, 368
to phosphorous, 303
Sorghum, roots, depth of penetration, 139
lateral extent, 145
number, 455
response to cultivation, 141
to oxygen, 134
weight, 149
selenium content, 286
solvent action of CO_2, 161
stomata, 421, 422
structure of leaf, 449
sugar content, 890
temperature and growth, 1061
of leaf, 475, 476, 479
and transpiration, 458, 459
tillers, 891
transpiration, day and night, 410, 411
relative, 416, 454
water of leaf, 442, 444
water requirement, 498, 499
wilting coefficient, 194, 195, 196
Sources of water supply, to plant, 188–190
Soybeans, absorption of elements, 289
acidity, 48
amino acids, 912
ash content, 289
boron, location, 340
calcium-magnesium ratio, 308
copper content, 300
iron content, 300, 331, 347
length of day, 1080
light rays affecting, 1073
low temperature, effect, 1065
manganese content, 300, 347
nitrate content, 642
nitrate reducing substances, 676
nitrogen content, 642
in leaves, 642, 681, 682
oil formation, 749
oils of, 732, 740
phospholipides, 743
phosphorus content, 302
photosynthesis rate, 610
potassium, amount needed, 318
proteins, 667
relation to light, 1080
response of, to aluminum, 358
to barium, 367
to fluoride, 368
to iron, 330, 331
to phosphorus, 303
to potassium, 288
to pyrrole-carbonic acid, 329
to strontium, 367
to sulphur, 327
to thallium, 368
root oxidation, 163
roots, response to aeration, 132
soil vs. plant, 672
solvent action, 161
stomata of, 422, 423
temperature of leaves, 475, 476, 478, 479
translocation of proteins, 912
transpiration, relative, 416
water content and hay, 443

Soybeans, water culture, 247, 266
acidity of, 262
zinc content, 300
Spectrum, of carotin, 546
of chlorophyll, 546, 547
of xanthophyll, 546
Sphagnum moss, acidity, 230
Spherocrystals, starch grains, 535, 536
Spinach, 286, 313, 345, 352, 586, 741
Spirogyra, 101
Spraying, iron salts, 334
manganese salts, 352
Sprays, influence on respiration, 974
relation of, to conduction of oil, 909, 910
to photosynthesis, 600
to transpiration, 447, 465–470
Springification, 1091
Springization, 1091
Squash, acidity, 50, 51
aeration of roots, 135
CO_2 fertilization, 587
globulin, 659
iron supply, 334
oil of, 732, 908, 909
proteins, 667
relation to light, 1077
root hairs, 158
roots, 137
response to oxygen, 135
seed coats, 80
stomata, 422
water-conducting cells, 854
Standard colors, cobalt chloride, 491
Standard evaporating surface, 492
Standard rate of water loss, 492
Starch, composition of, 538
in fruits, 530
general properties of, 537–539
grain of (*see* Starch grain)
in guard cells, 428–429
in leaves, 558, 602, 603
occurrence of, 532
paste of, 538
permanent, 536, 537
relation of, to potassium, 319
soluble, 538
temporary, 536
Starch grain, center of origin of, 533
composition of, 537–539
dimensions of, 532, 533
form of, 533
kinds of, 534
occurrence of, 532–533
relation of, to plastids, 536, 537
structure of, 533–536
Starch paste, 538, 795
Starch plants, 603
Starches, 532–540
dextrins, 539
in guard cells, 428, 435
inulin, 539, 540
State of chlorophyll in the chloroplasts, 547, 548
Stearic acid, 728, 729, 741
Sterols, 730
Stigmasterol, 730
Stoma, 414, 418
Stomata, capacity of, 436–441
daily movement in, 435, 436
defined, 414, 417
diffusion through, 439, 440, 441
dimensions of, 425
factors influencing number of, 418–425
formation of, 417
location of, 417
movement and transpiration in, 445, 446, 447
number, 418–425
and transpiration, 444–447
occurrence of, 417
opening and closing in, 425–436
cause, 425, 428, 429, 430
factors influencing, 431–435
methods of measuring, 430, 431
regulation of water loss, 441–447
Stomatal apparatus, 418
Stomatal frequency, 424
Stomatal index, 419, 420
Stomatal initials, 420
Stomatal pore, 418
Stomatal regulation, of water loss, 441
Stomatal transpiration, 436, 448
Stomatal types, 435–436
Storage of organic reserves, 885–890
effect of photoperiod, 1082
factors influencing, 889, 890
formation of reserves, 885–887
general facts, 890
time of storage, 887–889
Straight head of rice, 130, 131
Strawberry, CO_2 fertilization, 586
changes in pectic substances, 23
influence on, of boron, 344
of minor elements, 300
leaf area, transpiration, 454
respiration, 955
stomata, 423
translocation of solutes, 882
Stripping and organic reserves, 889
Strontium, in alfalfa and tobacco, 285
in nutrition, 367
in plants, 285
Strychnine, 695
Suberic acid, 15, 18
Suberin, 15–17
Submicroscopical cavities, 868, 869
Subsidiary cells, 418
Substance quotient, 1025
Substrate, 779
Succinic acid, 171
Succulents and respiration, 951
Sucrase (*see* Invertase)
Sucrose, described, 529, 530
first formed, 565, 566
in leaves, 558, 560, 561, 562
occurrence of, 529, 530
utilization of, by plants, 294–296
Suction force, of cell, 93
of twigs, 870, 871
Suction pressure of cell, 93
Suction tension of cells, 93
Sudan grass (*see* Sorghum)

Sugar cane, migration of potassium, 316, 318
nitrates vs. ammonium, 645, 647, 648
number of stomata, 444
respiration, 967
sugar, 529
Sugar group, 524, 525
Sugar maple, 857, 892, 896, 898
Sugar plants, 603
Sugars, artificial synthesis of, 760
defined, 525
first formed, 565–567
in guard cells, 428
kinds or types of, 526–531
disaccharoses, 529–531
monosaccharoses, 526–529
trisaccharoses, 531
physical and chemical reactions of, 525, 526
formation of osazones, 526
reducing action, 525, 526
specific rotatory power, 525
relation of, to respiration, 958, 959
in respiration, 936–937
translocation of, 902–906
Sulphates, occurrence of, 42, 324, 325
Sulphocynate, 807
Sulphur, amount used by plants, 324
distribution of, in the plant, 325
as fertilizer, 327, 328
form, in plants, 324
in which utilized, 325
formative effects of, 326, 327
general effects of, 327
luxury absorption of, 325
occurrence of, in plant, 324, 325
physiological role, 325, 326
in protein, 659
Sulphur reaction, protein, 665
Sulphuretted oils, 761
Sumac, organic reserves, 887
Sunflower, absorption of energy, 614
acidity, 50, 51
aluminum content, 357
area of stomata, 440
changes in oil in germination, 753–760
chemicals in soil and transpiration, 464
chemotropism of roots, 125
composition, 285, 286
conduction of water, 854
copper influence, 362
energy absorption, 612
energy of photosynthesis, 616
energy transmitted, 614
etiolation and photosynthesis, 598
formation of oil, 749
inhibitor of diastase, 797
iron content, 347
iron supply, 334
light rays, effect, 1073
manganese content, 347
need of zinc, 364
oils, 732, 740, 908, 909
pentoses, 527
permeability of cells, 103, 104, 110
photosynthesis, 611
plastids, 34
Sunflower, polarized light and carbohydrate, 590
relation to light, 1078
respiration, 936, 937, 967
respiratory index, 966
respiratory quotient, 952
response of, to aluminum, 357
to sulphur, 327
root formation, 1040
root hair, 154
roots, 9
penetration in dry soil, 123
response of, to aeration, 132
to oxygen, 135
rusts and transpiration, 471
seed coats, 84
selenium content, 286
soil water and transpiration, 463
solar radiation, 456
stomata, 445
transpiration, and soil temperature, 459
and solutes, 240
water in conducting cells, 854
water cultures, 248
water lost, 412
water supply and photosynthesis, 598, 599
wind and transpiration, 461
x rays, 1075
Surface film of cytoplasm, 87
Surface layer of cytoplasm, 87
Surface tension of cytoplasm, 86
Sweet potato, 17
formation of suberin, 17
influence of potassium, 323
injury and respiration, 962, 963
Symplast, 872
Synthesis of proteins, 670–694
Synthetic power of plant, 523

T

Taka-diastase, 535, 789
Tammann's principle, 595
Tanks, evaporation, 483–484
Tannic acid, influence of, on transpiration, 465
precipitation of proteins by, 664
Tannin, 804, 805
absorption of oxygen by, 967
action of, on diastase, 798
on invertase, 788
in cell wall, 81
influence of, on oxidase, 983
nature of, 804
occurrence of, 23
relation of, to oil, 751
Tartaric acid, 747, 788, 950
Tartrate, ferric and ferrous, 331
Tautomerization, 812
Tea yellows, 292, 326
Temperature, action of, on invertase, 788
of air, 474, 475
correlation of, with transpiration, 459
death point, 481
influence of, on bromelin, 818, 819
on chlorophyll formation, 549
on diastase, 798, 799

Temperature, influence of, on dormancy, 1049–1051
on glucase, 802
on lipase, 811
on nutrient solutions, 255, 256
on oxidase, 981
on papaïn, 816, 817
on permeability, 101–104
on photosynthesis, 593–596
on respiratory quotient, 950, 951
on root hairs, 155
on secretion, 829, 830
on stomatal movement, 433
on transpiration, 458, 459
of leaves, 472–481
factors influencing, 472, 473
methods of measuring, 473, 474
range of, in photosynthesis, 594, 595
relation of, to growth, 1061–1070
to internal breakdown, 941
to oxygen needs, 133, 134
to respiration, 955–959
to secretion of enzymes, 830
response of roots to, 135–136
Temperature coefficient, 109, 593
of permeability, 109–111
of photosynthesis, 593
Temporary starch, 536, 602
Tensile stress, water, 869
Terpenes, 761
Tetany of cattle, 650
Tetrachloride, 1049
Theine, 695
Theobromine, 695
Theories of membrane action, 107–112
Theory, of membrane action, 107–112
chemical, 109, 110, 111
colloidal, 111, 112
sieve, 107, 108
solution, 108, 109
of organic translocation, 900–901
of photosynthesis, 570–577
formaldehyde, 571–576
condensation of, 574, 575
detection of products, 575
nutrient experiments, 575, 576
reduction of CO_2, 572, 573, 574
organic acid, 571
recent, 576, 577
Thermal-death point, 481, 1063
Thermocouple, 473
Thermotropism of roots, 136
Thiocyanates, influence on diastase, 798
Thiourea, influence on diastase, 798
on respiration, 969
Three-salt solution, 243, 244, 251, 252
Thymal, 761
Tiglic acid, 729
Tillers of corn, 1091
Timothy, acidity, 48
aluminum content, 356
benefitted by legumes, 653
calcium content, 307
depletion of soil, 169
organic materials, storage, 887
Timothy, relation to light, 1080
response of, to aluminum, 357
to selenium, 366
to sulphur, 327
toxic extracts, 173
utilization of organic reserves, 891
Tissues, concerned in translocation of organic materials, 894–901
phloem, 897–901
kind of experiments, 897–900
insects, 899
protoplasm, 899–900
ringing, 897–899
theories of movement, 900, 901
unsuitability of, 894–895
xylem, 894–897
kind and age of, respiration, 964–967
Titanium, 367
Tobacco, absorption by, of elements, 290
of ions, 235
acidity, 50, 51
calcium content, 307
evidence of need, 312
indicator of potassium deficiency, 291
influence upon plants, 167
iron content, 331, 347
lithium content, 285
manganese content, 347
nitrates vs. ammonium, 648
nitrogen relations, 642
relation to light, 1080
response of, to boron, 343
to fluoride, 368
to potassium, 323
to sulphur, 327
to thallium, 368
root hair, 157
salt requirement, 290
sand drown and magnesium, 313, 314
strontium content, 285
transpiration, and light, 459
and solutes, 236, 237
utilization, organic reserves, 891
Toluene, action of, on diastase, 796
on invertase, 785
on maltase, 802
Tomato, absorption of elements, 289, 290
acidity, 48
affected by walnut, 168
amino acids, 675
boron, absorbed, 342
content, 340
needs, 341
calcium deficient plants, 312
carotins, 552
catalase, 990
copper, influence, 362
effect of fertilizers, 289
erepsin, 821
ethyl alcohol, 943
fertilization with CO_2, 586, 587
glutamine, 663
growth-inhibiting substances, 1043
growth-promoting substances, 1042

Tomato, hydrogen-ion, absorption, 646
ammonium, 646
nitrates, 646
influence of low temperature, 1065
influence on other plants, 167
iron content, 331
nitrate reduction, 648, 674, 675
nitrates vs. ammonium, 647
nutrient solution, 246
peroxidase, 985
phosphorus content, 301, 302
indications of need, 302
plastids, 33
potassium, amount needed, 318
indication of deficiency, 322, 323
water soluble, 318
quality, 323
regeneration of roots, 1058
relation of, to light, 289, 290
to potassium, 322
respiration, 942
response of, to boron, 343
excess effect, 345
to fertilizers, 127
to phosphorus, 303
to potassium, 288
to sulphur, 326, 327
to thallium, 368
ripening, 1057
root hairs, 158
salt requirement, 290
sprays and transpiration, 465–470
stomata, 422, 423
sulphur content, 325
synthesis of protein, 674, 675
translocation of oil, 909, 910
transpiration and solutes, 240
water content, of leaves, 408
water cultures, 248, 261
water lost, 412
wilting coefficient, 194
Tonoplast, 29
Total incident radiation, 611, 612
Total vitality, 1026
Tottingham's solutions, 242
Toxic substances, of roots, 171
Tracheae, 848
Tracheids, 848
Tradescantia, 101, 103, 110, 153, 854, 1041, 1043, 1074
Translocation, 847–912
bibliography of, 912–930
influence of calcium on, 312
of minerals in seedlings, 372
and movement of materials, 851–912
carbohydrates, 902–906
fats and oils, 906–910
inorganic compounds, 876–885
organic compounds, 885–912
proteins, 910–912
water, 851–876
need of, 847
tissues concerned, 847–851
Transpiration, coefficient of, 486
cooling effect of, 478–481
Transpiration, correlation of, with temperature, 456, 459
cuticular, 436
during day and night, 410, 411
defined, 410
effect on, of humidity, 459, 460
of temperature, 458, 459
of wind, 460–462
factors influencing rate of, 413–472
aeration of soil, 132
aerial and meteorlogical conditions, 455–462
chemicals, 464–470
concentration of sap, 450
day and night, 410, 411
fungi, 470, 472
hairs, 447, 448
incipient drying, 451, 452, 453
incipient wilting, 451, 452, 453
leaf structure, 448–451
oil sprays, 447, 468, 470
water in soil, 462–464
waxy covering, 447
methods of measuring, 488–494
cobalt chloride, 491, 492, 493
Freeman's method, 493, 494
potometer, 489
weighing, 489, 490, 491
ratio of, to evaporation, 486, 487, 488
relation of, to intake of solutes, 236–240
to leaf temperature, 478–481
to number of stomata, 444, 445
to stomatal opening, 445, 446, 447
relative, 415, 416, 417, 486, 487, 488
significance of, 495, 496
stomatal, 436
Transpiring power, 486
Transport, growth substance, 1038
Trees and shrubs, extent of root system, 140
Triacontane, 736
Triacontanol, 736
Triangle system in nutrition, 252–255
Trichites, 535
Trichlorethylene, 1050
Trichoblasts, 151
Trichomes, 447
Triglycerides, 731, 732
Trisaccharoses, 531
Triticin, 540
Triticum (*see* Wheat)
Tropisms, growth substances, 1039
True rate of photosynthesis, 606
True solution, 70
Trypsin, 781
occurrence of, in plants, 814–820
preparation of, 814–820
Tryptases (*see* Trypsin)
Tryptophane, in digestion, 812, 813, 815, 816
formula for, 661
in seeds, 687
Tuberin, 666, 667
Turanose, 531
Turgidity, of cell, 93
of leaves, 475
relation of, to temperature, 478, 479, 480
to transpiration, 462

Turgor pressure, 93
Turnip, 290, 301, 313, 327, 342, 344, 357
Turpentine, 761
Twin-leaf method, photosynthesis, 609, 610
Tyloses, 878
Tyrosinases, 978, 979
Tyrosine, action of, on diastase, 798
  in digestion, 815, 816
  formula for, 661
  in seedlings, 911
  in soil, 171
  utilization of, 650

U

Ultrafiltration theory of membrane action (*see* Sieve theory)
Ultraviolet light, in carbohydrate synthesis, 572, 574
  on coloring of apples, 551
  in condensation, 574
  effect of, on growth, 1072–1075
  in protein formation, 686
Unequal absorption of ions (*see* Ions)
Unessential elements, 298
  vs. essential elements, 297–301
Unsaturated fatty acids, 727, 729, 730
Upward migration of solutes, 883
Upward transfer of inorganic compounds, 876–880
Urea, 650, 651
  occurrence of, 696
  and stomatal opening, 435
  utilization of, 650, 651
Uric acid, 650, 695
Uronic acids, 13–14
Utilization of organic reserves, 890, 891
Utilization of solar energy, in photosynthesis, 615–617

V

Vacuole, origin (*see* Cell sap)
Vacuome, 37–38
Valeric acid, 728
Valine, 661
Valonia, 100, 110, 218, 219, 222, 227, 229
Vanillin, 171
Van Slyke method, 812
Velocity of transport, 1038
Vernalization, 1091–1095
  defined, 1091
  described, 1091, 1092
  economic value of, 1093
  theory of, 1094–1095
    the photo stage, 1095
    the thermo stage, 1094–1095
Vertical penetration of roots, 141–145
Vessels (*see* Tracheae)
Vicia faba, absorbing power of, 95, 210
  amount of boron needed, 342
  cell walls of root, 9
  etiolated plants, 1070
  growth, 1020, 1024, 1029, 1037
  inhibiting substances, 1044
  legumin, 659
Vicia faba, need in, of boron, 299
    of zinc, 363
  nitrate-reducing substance, 675
  osmotic value of sap, 95
  permeability of pith cells, 110
  permeability of root cells, 110
  photosynthesis, 597
  protein formation, 686
  proteins, 667
  response of, to aluminum, 357
    to boron, 342, 343
    to light, 1070
  tropisms, 1039
  turgor pressure, 95
Vicilin, 666, 667
Vignin, 667
Vigor, relation to acidity, 50–51
Virescent plants, 550
Visible changes within the cell for permeability, 98
Vita glass, 1073
Vital theories of water ascent, 860–863
Vitality and catalase, 993–995
  inherent, 1026
  total, 1026
Vitamins, 1030, 1032–1033
  defined, 1032
  function of, 1033
  synthesis of, 1032, 1033
Volatile fatty acids, 728
Volatile substances and growth, 1045, 1046

W

Wall pressure, 93
Walnut, absorption in, of boron, 342
      excess, 344, 345
    of ions, 234
  formation of oil, 751
  need for magnesium, 314
  oil, 740
  proteins, 667
  relation, carbohydrates to oil, 746, 747
  toxic effect, 168, 173
  yellows, 352
Water, amount in soil and transpiration, 462–464
  amount lost from plant, 411–413
  amount needed, 408
  cohesive force, 867
  content of leaf, 408
    effect of potassium, 324
  control of loss of, 441–447
  cultures in, vs. sand cultures, 258, 259
  diffusion of, through cell wall, 79–82
  effect of, on root hair, 152
  entrance of, into root, 205–212
  evaporation of, from plant, 410–503
  factors influencing loss of, 413–417
  of guttation, 503–506
  imbibition of, 75, 76
  influence of, on root development, 121–124
    on photosynthesis, 598, 599
    on root hairs, 152
    on secretion of enzymes, 830
  intake of, by leaves, 189–190
    by plants, 188–212

Water, loss of, from plants, 407–506
need of, by plants, 407–409
percentage of, in cell sap, 38
in tissue, 409
permeability of protoplasm, 100–103
in relation to respiration, 959–962
requirement of, by plants, 496–502
standard rate of loss of, 492
sources of, to plant, 188–190
tensile stress of, 869
test time, 492
translocation of, 851–876
ascent of water, 855–872
in absence of living cells, 864–865
under high evaporation, 859–872
under low evaporation, 855–859
other theories, 872
physical theories, 863–872
vital theories, 860–863
downward movement of, 872–874
lateral movement of, 874–876
pathway of movement of, 851–853
in water-conducting cells, 833–855
gases in, 853, 854
water in, 854, 855
vapor diffusion of, 438
working margin for, 443, 444
Water cultures, 241
vs. sand cultures, 258, 259
Water loss, regulation or control of, 441–447
stomatal movement and water content of leaf, 441–444
Water requirement, 496–502
defined, 496, 497
factors affecting, 498–502
climatic conditions, 498, 499, 500
plant diseases, 502
soil fertility, 500, 501
soil moisture, 501, 502
Watermelon, 158, 422, 423, 475, 476
Water-supplying power of soil, critical, 203
at wilting, 200–205
Water-test time, 492
Wave lengths of light, in photosynthesis, 590–593
Waxes, 727, 735, 736
Waxy coverings, transpiration, 447
Waxy gene, 752
Weighing method, of measuring photosynthesis, 608, 609
of measuring transpiration, 489, 490, 491
Weight of roots, 123, 124, 128, 129, 148–149
Wheat, absorption of ions, 235
acidifying power, 230
acidity, 46, 49, 50, 51
age, elemental requirement, 369, 370, 371, 373
aluminum content, 357
amino acids, 669, 670
ash content, 289
balanced solution, 269
boron, absorbed, 342
boron content, 340
bound water, 1069
buffers in sap, 47
calcium content, 307
calcium-magnesium ratio, 308
Wheat, CO_2 fertilization, 587
CO_2 production, 162
carbon nitrogen ratio, 702–703
catalase, 991
cell-sap concentration, 39, 40
composition, 285, 286
copper content, 300
diastase, 798
energy utilized, 617
erepsin, 820
fixation of nitrogen, 656, 657
formation of grain, 688, 689
heat of respiration, 953, 954
hemicelluloses, 13
influence on, of electricity, 1084
of other plants, 168
on other plants, 165, 166, 172
iron content, 300, 347
magnesium injury, 315
maltase, 801
manganese content, 300, 347
mitochondria, 37
moisture and respiration, 959, 960
movement of organic materials, 885, 886
nitrate content, 671
nitrate vs. ammonium salts, 645, 647, 649
oil of seeds, 740
oxidation by roots, 162
pentosans, 540
permeability of seed coat, 83, 84, 85
phospholipides, 743, 745
phosphorus content, 302
polypeptides, 669, 670
potassium content, 316
indicator of deficiency of, 324
proteases, 813
protein, 667, 668, 689–692
relation to light, 1078
respiration, 955, 959, 965, 967
of endosperm, 965
of embryo, 965
respiratory quotient, 952
response of, to aluminum, 357
to ammonium, 645
to arsenic, 367
to barium, 367
to borax, 345
to boron, 343, 344
to chlorine, 353, 355
to iron, 330
to manganese, 347, 348
to nitrates, 645
to palladium, 369
to phosphorus, 303
to potassium, 319
to rubidium, 369
to strontium, 367
to sulphur, 327
to thallium, 368
root hairs, 153
roots, 123, 127, 128, 137, 139, 143, 144, 149, 162
depth of penetration, 139, 143, 144
number, 137, 138
penetration in dry soil, 123

Wheat, roots, response of, to aeration, 132
to chemicals, 127, 128
to moisture, 123
to rust, 129
weight, 149
rusts and transpiration, 471
salt requirement, 369, 371
seed coats, 83
selenium content, 286
silicon content, 335
sitosterol, 730
starch, 533, 535, 538
stomata, 419, 422, 423
temperature and growth, 1063
temperature, and transpiration, 458, 459
translocation of carbohydrates, 905, 906
transpiration, 411, 456, 471, 482
correlation with radiation, 456, 457
day and night, 411
and solutes, 240
utilization of organic compounds, 651
water content, 443
water cultures, 247, 256, 257, 259, 260, 261, 266
acidity in, 262, 263, 264, 265
water lost, 412
water requirement, 498, 499
water-supplying power, 203
wilting coefficient, 194, 195, 196
yellow berry, 692, 693
zinc content, 300
White bud, 292, 364
Wilted plants, photosynthesis in, 598–599
temperature of, 478–479
transpiration in, 479–480
Wilting, incipient, 452
permanent, 190
relation of, to leaf surface, 197
to root system, 196
to transpiration, 462
Wilting coefficient, calculated, 191–193
defined, 190
factors influencing, 193–198
kind of plant, 194–198
type of soil, 193, 194
methods of determining, 190–193
direct, 190, 191
indirect, 191–193
observed, 190, 191
relation of, to hygroscopic coefficient, 192, 193
Wilting coefficient, relation of, index of transpiring power, 193
moisture equivalent, 192
moisture-holding capacity, 193
Wind, influence on transpiration, 460–462
Witches-broom, 368
Working depth of roots, 142, 143–144
Working margin, of water, 443, 444

X

Xanthine, 171, 650, 651, 695
Xanthium (*see* Cocklebur)
Xanthophyll, 553
Xanthoproteic reaction, 664, 665
Xenia, 1031
Xerophytism, leaf structure and, 449
transpiration and, 449, 450
X rays and growth, 1075–1076
cell-wall structure, 6, 7
starch-grain structure, 536
Xylan, 541
Xylem, 848, 849
conduction in, inorganic salts, 877–879
organic materials, 894–897
organic compounds in tracheal sap, 895–897
rays in, 848
Xylose, occurrence, 13, 529

Y

Yarovization, 1091
Yeast, 781, 784, 940
Yellow berry, 692, 693
Yellows, walnut leaves, 352

Z

Zalenski's law, 424
Zea mays (*see* Corn)
Zein, 660, 667, 669
Zinc, 363–364
effect of, on bronzing, 364
on little leaf, 364
on whitebud, 364
Zymase, 780, 933, 934, 940, 943
Zymasis, 941
Zymogen, 780
of lipase, 810
of proteolytic enzymes, 823